Instructor Resource Center on CD (0-13-100569-3)

Thousands of digital resources, plus electronic versions of all the instructor's print resources located on CD, containing:

- *Figures and Tables* — more than 1000 labeled and unlabeled illustrations, tables, and photos from the text, formatted for large lecture hall presentations.
- *Media Activities* – interactive activities with animations from OneKey.
- *Animations* — animations from OneKey.
- *PowerPoint* — presentations for each chapter available in two formats: with media activity animations embedded and with book figures and tables embedded. Labels for figures and tables are all editable in PowerPoint.
- *Lecture Resources* — suggestions for what resources to incorporate into your lecture.
- *Lecture Activities* — presentation and assessment ideas for use during class that promote active learning, including discussion questions, demonstrations, and group and collaborative activities.
- *Review Questions* — all review questions, including Self Test, Essay Challenge questions from OneKey, end of chapter questions, and figure caption questions from the text.
- *Test Questions* — electronic version of the Test Item File available in editable Word Files and formatted for import to WebCT or Blackboard courses. We also provide the TestGen test generation software.
- OneKey *OneKey* — conveniently organized by textbook chapter, this tool combines all resources. OneKey is available in CourseCompass, BlackBoard, or WebCT platforms.
- *Personal Response System (PRS) Questions* — a wireless polling system that enables instructors to pose questions, record results, and display those results instantly in the classroom. The PRS allows each student to respond privately and provides instructors with valuable insight into student learning.

OneKey OneKey (www.prenhall.com/audesirk7)

OneKey offers the best teaching and learning resources all in one place. It is all you need to plan and administer your course. Conveniently organized by textbook chapter, these compiled resources help save you time. OneKey is available in CourseCompass, WebCT, and BlackBoard platforms.

- *OneKey in CourseCompass (0-13-100573-1)* — request an instructor Access Code online or from your sales rep to get started.
- *OneKey in BlackBoard Premium (0-13-100571-5)* — available as a downloadable course cartridge with resources ready to incorporate into an existing course or to be used to construct a new one.
- *OneKey in WebCT Premium (0-13-100572-3)*— available as a downloadable course cartridge with resources ready to incorporate into an existing course or to be used to construct a new one.

Instructor Resource Guide (0-13-100563-4)

The print alternative to the digital instructor resources provides:

- *Resource Directory* — a list of all resources available to the instructor, including figures and tables, animations, lecture activities, test questions, and more.
- *Lecture Resources* — suggestions for what resources to incorporate into your lecture.
- *Lecture Activities* — presentation and assessment ideas for use during class that promote active learning, including discussion questions, demonstrations, and group and collaborative activities.

Instructor Toolbox (0-13-146510-4)

A lecture organizer that integrates the entire Audesirk package by chapter, containing:

- *Resource Directory* — a list of all the resources available to instructors.
- *Transparency Acetates* — more than 500 four-color figures and tables from the text.
- *Lecture Resources*
- *Lecture Activities*
- *Review Questions and Answers* — solutions to end of chapter questions and figure caption questions in the text.

Test Item File (0-13-1005642)

The print version of the electronic test bank offers:

- More than 3000 questions compiled by a carefully selected team of educators.
- Three question types—factual, conceptual, and applied—to test your students on all facets of biology in their lives.
- Questions with a variety of difficulty levels.

STUDENT RESOURCES

Resources	Self Test Questions	Essay Challenge Questions	Hints and Feedback for All Answers	Media Activities	Web Investigations	Issues in Biology	Bizarre Facts	Recent Articles	Links	Figures and Tables
OneKey	•	•	•	•	•	•	•	•	•	
Student Study Companion	•	•			•	•	•			
Student Lecture Notebook										•
Research Navigator with Guide to Evaluating Online Resources								•		

*Content listed in multiple resources is identical.

Laboratory Manuals

Explorations in Basic Biology, Tenth Edition (0-13-145312-2)

by Stanley Gunstream

This best-selling laboratory manual can be used with *Biology: Life on Earth*. It includes 41 self-contained, easy-to-understand experiments that blend traditional experiments with investigative exercises. The format provides flexibility that allows instructors to adapt each exercise to their particular lab needs.

Instructor's Manual to Explorations in Basic Biology, Tenth Edition (0-13-145313-0)

Biological Explorations: A Human Approach, Fifth Edition (0-13-145314-9)

by Stanley Gunstream

Specifically designed for courses in general biology where the human organization is emphasized—and for a growing number of courses in human biology—this lab

OneKey (www.prenhall.com/audesirk7)

OneKey for *Biology: Life on Earth, 7e* is all students need for anywhere-anytime access to course materials conveniently organized by textbook chapter to reinforce and apply what they have learned in class. It contains:

- *Self Test* — multiple choice, labeling, and fill-in-the-blank questions that test students' mastery of the major chapter concepts, as well as their understanding of the connections these biological concepts have to their lives.
- *Essay Challenge* — essay questions that challenge students to think conceptually about biology and apply it to their own lives.
- *Hints and Feedback* — all questions include helpful tips and immediate feedback to all responses.
- *Media Activities* — activities that offer students a visual view of concepts and test their knowledge using animations and simulations.
- *Web Investigations* — activities related to the chapter Case Study that lead students to the Web site to answer a series of investigative questions.
- *Issues in Biology* — short articles and investigative questions that help students explore significant issues in biology today.
- *Bizarre Facts* — fun, short articles about the oddities of the biological world, including links that enable students to further explore the unusual phenomena.
- *Links* — links to Web sites on topics addressed in the chapter.

Student Accelerator CD

Included in every student copy of the text, the Accelerator CD is intended for use with OneKey. This CD contains high bandwidth files such as movies and animations. When pages that require these files load on OneKey, the Web site will first attempt to find them on the CD. If the Accelerator CD is loaded in the CD drive of the computer, the file will load much more quickly.

Student Study Companion (0-13-145755-1)

The print alternative to OneKey provides:

- *Self Test*
- *Issues in Biology*
- *Essay Challenge*
- *Bizarre Facts*
- *Web Investigations*

Student Lecture Notebook (0-13-146537-6)

This portable lecture companion provides figures and tables from the textbook with space for taking notes.

Research Navigator (Available with OneKey)

This tool equips students with the means to start a research assignment or paper or to access full-text articles. It is complete with extensive help on the research process and three exclusive databases of credible and reliable source material, including the EBSCO Academic Journal and Abstract Database, *The New York Times* Search by Subject Archive, and "Best of the Web" Link Library. It enables students to efficiently and effectively make the most of their research time and stay up-to-date on the issues.

The Prentice Hall Guide to Evaluating On-Line Resources with Research Navigator (0-13-113674-7)

Complete with access codes necessary for entering Research Navigator, this guide includes valuable information on how to find and determine the validity of information on-line, how to most effectively use Research Navigator, and how to determine the ethical use of information.

manual contains 32 outstanding exercises by the author of *Explorations in Basic Biology, Tenth Edition.*

Instructor's Manual to Biological Explorations: A Human Approach, Fifth Edition (0-13-147096-5)

Thinking About Biology: An Introductory Biology Manual, Second Edition (0-13-145820-5)

by Mimi Bres and Arnold Weisshaar

This lab manual is designed for a one-semester, non-majors introductory biology laboratory course with a human focus. It features a deceptively simple approach to solving complex problems as well as instructional flexibility. Each exercise can be easily adapted to accommodate your laboratory session.

Instructor's Manual to Thinking About Biology, Second Edition (0-13-145821-3)

BIOLOGY
Life on Earth

SEVENTH EDITION

Teresa Audesirk
Gerald Audesirk
University of Colorado at Denver

Bruce E. Byers
University of Massachusetts, Amherst

Upper Saddle River, New Jersey 07458

Library of Congress Cataloging-in-Publication Data

Audesirk, Teresa.
Biology : life on Earth / Teresa Audesirk, Gerald Audesirk, Bruce E. Byers.-- 7th ed.
p. cm.
Includes bibliographical references and index.
ISBN 0-13-100506-5
1. Biology. I. Audesirk, Gerald. II. Byers, Bruce E. III. Title.

QH308.2.A93 2005
570--dc22

2003070689

Executive Editor: Teresa Ryu Chung
Project Manager: Ann Heath
Editor in Chief, Science: John Challice
Associate Development Editor: Anne Madura
Production Editor: Tim Flem/PublishWare
Media Editor: Travis Moses-Westphal
Assistant Editor: Colleen Lee
Assistant Vice President of Production and Manufacturing: David W. Riccardi
Executive Managing Editor: Kathleen Schiaparelli
Editor in Chief of Development: Carol Trueheart
Assistant Managing Editor, Science Media: Nicole Bush
Media Production Editors: Karen Bosch, Elizabeth Wright, and PublishWare
Marketing Manager: Andrew Gilfillan
Director of Creative Services: Paul Belfanti
Creative Director: Carole Anson
Art Director: Maureen Eide
Page Composition: PublishWare
Manufacturing Manager: Trudy Pisciotti
Buyer: Alan Fischer
Managing Editor, AV Production & Management: Patricia Burns
AV Project Managers: Adam Velthaus and Connie Long
Art Studio: Imagineering
Illustrators: Imagineering; Rolando Corujo; Hudson River Studios; Howard S. Friedman; David Mascaro; Edmund Alexander; Roberto Osti
Assistant Managing Editor, Science Supplements: Becca Richter
Supplements Production Editor: Dana Dunn
Editorial Assistant: Mary Burket
Production Assistant: Nancy Bauer
Photo Research: Yvonne Gerin
Photo Coordinator: Carolyn Gauntt
Photo Research Administrator: Beth Boyd
Cover Photograph: Frans Lanting / Minden Pictures

Pearson Prentice Hall
Pearson Education, Inc.
Upper Saddle River, NJ 07458

Printed in the United States of America

Printed on Recycled Paper

10 9 8 7 6 5 4 3 2 1

ISBN 0-13-100506-5 (student edition)

ISBN 0-13-146760-3 (instructor's edition)

ISBN 0-13-192010-3 (school edition)

Pearson Education Ltd., *London*
Pearson Education Australia Pty., Limited, *Sydney*
Pearson Education Singapore, Pte. Ltd.
Pearson Education North Asia Ltd., *Hong Kong*
Pearson Education Canada, Ltd., *Toronto*
Pearson Educación de Mexico, S.A. de C.V.
Pearson Education—Japan, *Tokyo*
Pearson Education Malaysia, Pte. Ltd.

Brief Contents

About the Authors

Terry and Gerry Audesirk grew up in New Jersey, where they met as undergraduates. After marrying in 1970, they moved to California, where Terry earned her doctorate in marine ecology at the University of Southern California and Gerry earned his doctorate in neurobiology at the California Institute of Technology. As postdoctoral students at the University of Washington's marine laboratories, they worked together on the neural bases of behavior, using a marine mollusk as a model system.

Terry and Gerry are now professors of biology at the University of Colorado at Denver, where they have taught introductory biology and neurobiology since 1982. In their research lab, funded by the National Institutes of Health, they investigate the mechanisms by which neurons are harmed by low levels of environmental pollutants.

Terry and Gerry share a deep appreciation of nature and of the outdoors. They enjoy hiking in the Rockies, running near their home in the foothills west of Denver, and attempting to garden at 7000 feet in the presence of hungry deer and elk. They are long-time members of many conservation organizations. Their daughter, Heather, has added another focus to their lives.

Bruce E. Byers, a midwesterner transplanted to the hills of western Massachusetts, is a professor in the biology department at the University of Massachusetts, Amherst. He's been a member of the faculty at UMass (where he also completed his doctoral degree) since 1993. Bruce teaches introductory biology courses for both nonmajors and majors; he also teaches courses in ornithology and animal behavior.

A lifelong fascination with birds ultimately led Bruce to scientific exploration of avian biology. His current research focuses on the behavioral ecology of birds, especially on the function and evolution of the vocal signals that birds use to communicate. The pursuit of vocalizations often takes Bruce outdoors, where he can be found before dawn, tape recorder in hand, awaiting the first songs of a new day.

To Heather, Jack, and Lori and in memory of Eve and Joe

T. A. & G. A.

To Bob and Ruth, with gratitude

B. E. B.

Contributors: Instructor and Student Resources

James A. Hewlett
Finger Lakes Community College

My true love is teaching a mixed-majors undergraduate biology class. The mention of recombinant DNA brings out clamors for detailed protocols from majors and concerns regarding the ethical and social aspects of such a practice from liberal arts students. Art students become scientific illustrators as part of a class project, while business majors are enthused over the marriage of biology and the NASDAQ. These combinations make teaching undergraduates enjoyable.

Stephen Kilpatrick
University of Pittsburgh, Johnstown

Teaching nonmajors is a fun challenge. The natural world provides a lot of examples of amazing organisms and processes, so I can first get students' attention and then sneak in some basic biology and help the students understand how it applies to their lives.

Kelli Prior
Finger Lakes Community College

I enjoy generating interest in biology's applications outside the classroom, from the food we eat to what we hear about on the news. I see to it that students realize that they need not become biologists to appreciate this relevance to their everyday lives.

Joanne Russell
Manchester Community College

Nonmajors bring a different perspective and different areas of expertise to the class and give me the opportunity to take a fresh look at biology. Many nonmajors have avoided the biology class as long as possible! It is very gratifying to see these same students develop an understanding of biological issues and the relevance of biology in their daily lives.

Linda Smith-Staton
Pellissippi State Technical Community College

I love teaching nonmajors because it's so much fun to share the "a-ha, that's how it works" and "wow, that's so cool" moments with them. We work hard, but laugh a lot in my classes. The most rewarding thing about teaching nonmajors is when they come back to let me know about changes they've made in their lives because of my class. Some have quit smoking, started recycling programs where they work, obsessed about their calcium intake, or started feeding their plants!

Teresa Snyder-Leiby
State University of New York, New Paltz

I enjoy these classes because I get to show people how alive and vital plants are to people and the environment, and how so many of our everyday needs are met by plants or chemicals produced by plants.

Mark Sugalski
New England College

I especially enjoy teaching students with little science experience because it is gratifying to show them that, not only can biology be fun and interesting, but it is relevant to their everyday lives.

David Tapley
Salem State College

I enjoy the diversity of backgrounds and worldviews that nonmajors bring to the classroom. While the diverse nature of the students can present a challenge, it also enriches the classroom experience for both the students and the instructor. But the most rewarding aspect of teaching nonmajors is seeing a student truly grasp a concept that he or she may have at first found unfathomable.

Michelle Withers
Louisiana State University

As our society becomes more dependent on science and technology, our responsibility to educate the nonscientific community becomes greater. Teaching biology to nonmajors is a valuable opportunity to reach this community. My goal is to have students leave my class with the tools necessary to make educated decisions about issues that affect them and the rest of society.

David Zimmer
Cayuga Community College

There is probably no greater challenge than teaching the broad and complex field of biology to nonmajors, but our strongest ally is the subject matter itself. As we explore the weird and wondrous diversity of solutions organisms have evolved to overcome life's hurdles, students gain a new perspective on their lives and their place in the universe. And watching that transformation is something to behold!

Essays

*Asterisk denotes "ETHICS" essay

**See our Web site at http://www.prenhall.com/audesirk7 for additional "A Closer Look" essays.:

Contents

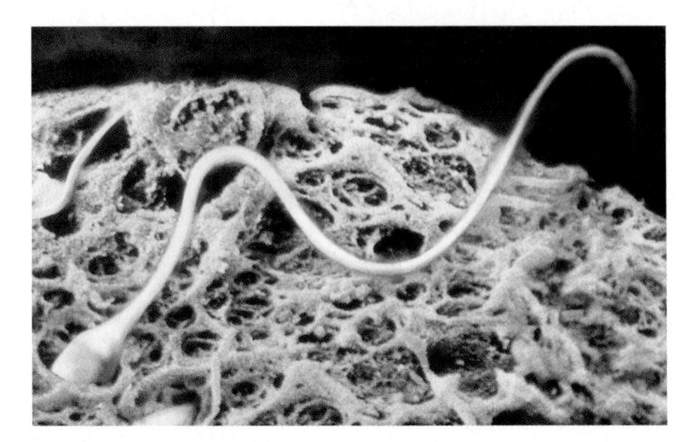

465
239

11 The Continuity of Life: Cellular Reproduction 184

12 Patterns of Inheritance 208

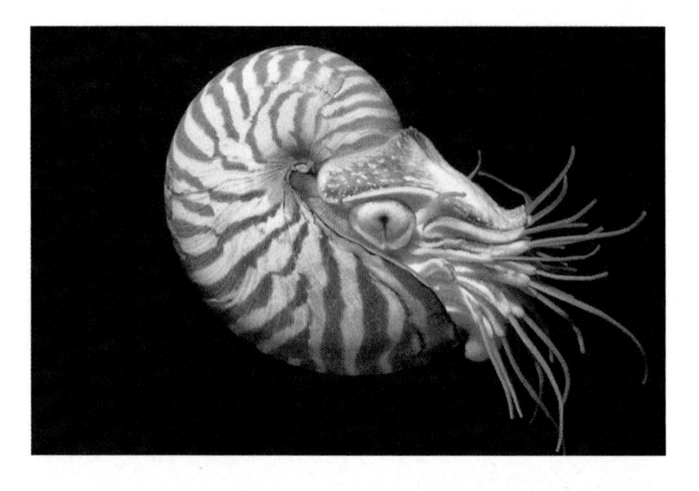

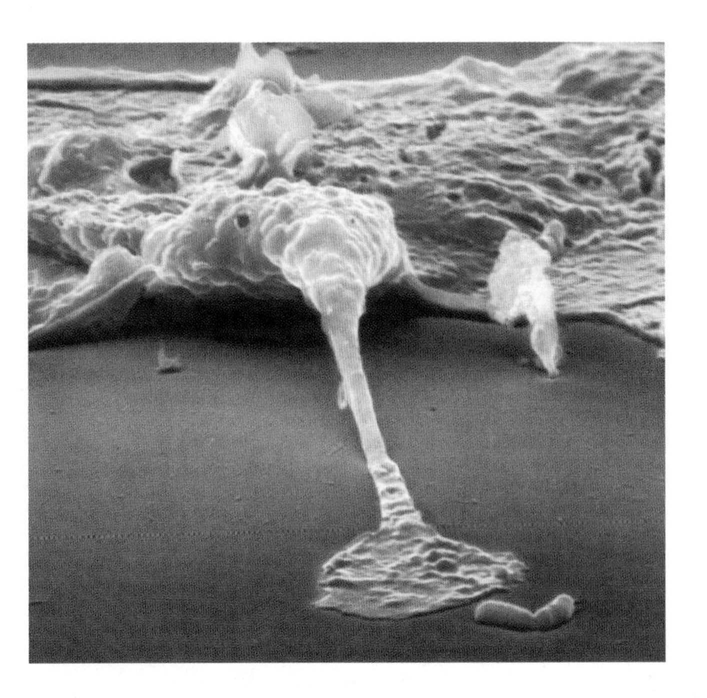

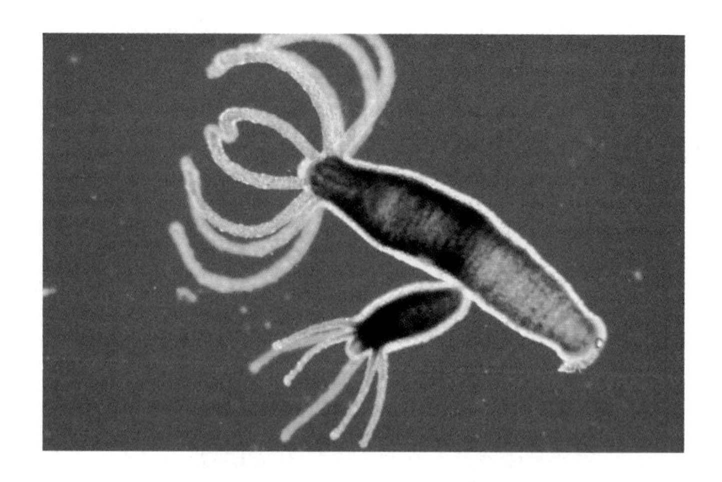

Preface

Animal behavior, cloning, seed germination, stem cell research, biodiversity, population growth, artificial kidneys, global warming, pollination strategies, sequencing genomes—the staggering number of important and fascinating biological topics continues to grow. In fact, one can make a compelling argument that "biology" is no longer a single discipline. In the midst of this wealth of information, how does an instructor of an introductory course choose what to cover, and at what level to cover it? What basic concepts and principles form the foundation of biology? What topics are the most compelling to students? Which types of biological knowledge will best prepare them to make informed choices relating to their own lives and to the greater world that their personal choices impact? In talking to educators faced with these tough decisions, a single overriding consensus emerged: "We need to help our students become scientifically literate."

Scientific literacy endows a student with the mental tools to cope with expanding knowledge. It requires a foundation of basic factual knowledge—a mental framework into which new information can be integrated. But it also encompasses the ability to grasp and evaluate information from the news and popular press. A scientifically literate individual does not think in isolated "soundbites," but recognizes the interrelatedness of concepts and the need to bring information from many areas to bear on an issue. A scientifically literate individual also needs to be able to use today's amazing resources, most significantly the Internet, to find accurate information; he or she must be able to distinguish true scientific information from pseudoscience.

Biology: Life on Earth Effectively Manages a Wealth of Scientific Information

Our seventh edition of *Biology: Life on Earth*—which is more than a textbook, but rather a complete package of teaching aids for the instructor and learning aids for students—has been revised with three specific goals:

- To help instructors manage the presentation of biological information with the goal of producing scientifically literate students
- To help each student to acquire information according to his or her own learning style
- To help students relate this information to their own lives so as to understand its importance and relevance

Biology: Life on Earth

. . . Is Organized Clearly and Uniformly

Throughout each chapter, students will find aids that help them navigate through the information.

- "At a Glance" at the start of each chapter brings together the chapter's major subheadings and now includes the titles of Essays as well. Instructors can easily assign—and students can easily locate—key topics within the chapter.
- Major sections are introduced as questions to which the student will find answers in the section, while minor subheadings are presented as summary statements that reflect content. A crucial outgrowth of this organizational scheme is that it imparts an understanding of biology as a hierarchy of interrelated concepts, rather than a set of isolated, independent subjects.
- The "Summary of Key Concepts" pulls together important concepts using the major subheadings as an organizing feature. Both "At a Glance" and the "Summary of Key Concepts" use the heading-based numbering system featured within the chapters themselves, allowing instructors and students to move efficiently among the different components within a chapter.
- Information is integrated and easily managed within the total package. Media Activity tabs within each chapter direct the student to OneKey, which contains relevant activities, animations, and practice tests; the Media Activity numbers in the book correspond with OneKey for easy navigation. Detailed descriptions of each Media Activity are found at the end of each chapter.

. . . Contains Revitalized Illustrations for Greater Clarity, Consistency, and Reader Interest

Benefiting from the advice of reviewers, a talented biological illustrator, and careful scrutiny by the authors and development editor, we have extensively revised the illustration program. For the seventh edition, we have:

- ***Expanded the consistent use of color***. We have been vigilant in tracking the use of color to provide consistency in illustrating specific atoms, structures, or processes. We have also made the colors more vibrant to better distinguish individual parts of a figure, to help engage the readers' interest, and to focus attention on the most important aspects of the illustration.
- ***Improved overall quality***. We have redrawn the more diagrammatic figures for greater interest and accuracy.
- ***Enhanced label clarity***. We have revised the size, placement, and font of figure labels for more consistency and readability.
- ***Organized content more efficiently***. We have modified the placement of parts of multipart figures for easier navigation through the figure.
- ***Explained figure content more clearly***. Through the judicious use of "talking boxes," we have placed more explanatory statements within figures for greater clarity.
- ***Modified figure captions to enhance function***. Our figure titles summarize the content; we have made the captions more concise, and have added thought questions to several captions within each chapter.

. . . Has Been Updated and Reorganized

We incorporate new discoveries that students might encounter in news articles, but place them within an accurate biological context to help foster scientific literacy. Further, organizational changes within chapters and units now make the basic information even more accessible to students. Although each chapter has been carefully revised, highlights of this revision include the following:

- ***Unit Two: Inheritance.*** We have extensively revised and reorganized these chapters to help students assimilate and appreciate this potentially daunting subject matter. Chapter 13 incorporates and interprets important new advances in the exploding area of biotechnology in a student-friendly manner.
- ***Unit Three: Evolution.*** We now have two chapters covering animal diversity, one on vertebrates and the other on invertebrates. This presents the information in chapters whose length is more manageable for both students and instructors. New advances in discovering the human lineage are also incorporated in Chapter 17.
- ***Unit Five: Animal Physiology***. Updated Essays, Case Studies, and the Links to Life highlight new advances, such as those in the treatment of diabetes and spinal cord injuries, and in artificial hearts, stem cell research, and contraception. Animal behavior has been moved to the Ecology unit to better emphasize behavior as an adaptive response to the environment, and to create a clearer physiological focus for Unit Five.
- ***Unit Six: Ecology***. While basic ecological principles remain unchanged, human impact is constantly growing. Students will find updated information on acid rain, global warming, tropical deforestation, human population growth, and the collapse of fisheries, as well as some good news about the ozone hole and the success of global cooperation in addressing this problem.

. . . Engages and Motivates Students

Scientific literacy cannot be imposed on students; they must actively participate in acquiring both the necessary information and skills. Thus, it is crucial for students to recognize that biology is about their personal lives as well as the life all around them. To help engage and motivate students, this new edition incorporates the following:

- ***Links to Life***. A new feature in the seventh edition, Links to Life ends each chapter on a relevant note. These short, informally written segments relate to subjects that are both very familiar to the student and relevant to the chapter. For example, "Health Food?" (Chapter 2) examines the antioxidant properties of chocolate; "Dehydrating Drinks" (Chapter 31) explains how alcoholic beverages affect the kidneys; and "Treading Lightly—How Big Is Your 'Footprint'?" (Chapter 39) discusses the relationship between a student's lifestyle and humanity's imprint on the biosphere.
- ***Case Studies***. In the seventh edition, we retain and update our popular Case Studies and introduce many new ones, while making the sixth edition versions available in OneKey. Case Studies are based on recent news items, situations in which students might find themselves, or particularly fascinating biological topics. Each Case Study is revisited at the end of the chapter, allowing students to explore the topic further in light of what they have learned and, often, to find answers to questions raised in the initial study. Additionally, students will find an in-depth investigation of each Case Study in OneKey.
- ***Bioethics***. Many topics explored in the text have ethical implications for human life. These include genetic engineering and cloning, the use of animals in research, and human impact on other species. They are now identified with a bioethics icon that alerts both students and instructors to the possibility for further discussion and exploration.
- ***Essays***. We retain our full suite of essays, including "Earth Watch," environmental essays that explore pressing issues such as the loss of biodiversity, endocrine-disrupting chemicals, the ozone hole, and Earth's carrying capacity for people; and our medically related "Health Watch" essays, which investigate topics such as sexually transmitted diseases, the dangers of artificial steroids, and how smoking damages the lungs.
- ***Caption queries***. New in the seventh edition, we introduce questions at the end of several selected figure captions in each chapter. These questions encourage the student to apply his or her new knowledge of the structure or process being illustrated to a larger biological question.

. . . Provides Print and Media Resources That Offer User-Friendly Ways to Manage, Access, and Explore Content

- ***OneKey* (www.prenhall.com/audesirk7).** OneKey provides instructors and students with the best teaching and learning resources all in one place. OneKey for *Biology: Life on Earth 7e* is all students need for anywhere/anytime access to course materials, conveniently organized by textbook chapter to reinforce and apply what they have learned in class. OneKey is everything instructors need to plan and administer their courses. All instructor resources are in one place to maximize effectiveness and minimize time and effort. OneKey for convenience, simplicity, and success ... for instructors and students.
- ***Instructor's Resource Center on CD.*** No need to search for the right CD, disk, or Web site; the Instructor Resource Center on CD contains all digital resources in one place. It can be browsed by chapter and resource type, or you can search the entire contents of the CD using a powerful search tool. The Resource Manager makes it easy to post images, questions, animations, and more to your Web site.

- ***Instructor Toolbox.*** All print supplements are packaged together for convenience and organized by chapter. These include the Resource Directory, Chapter at-a-Glance, Lecture Activities, Lecture Resources, Key Terms, End-of-Chapter and Caption Questions and Answers, and Overhead Transparencies.
- ***Instructor Resource Guide.*** This print companion to the Instructor Resource Center on CD includes a complete Resource Directory, Chapter at-a-Glance, Lecture Activities, Lecture Resources, Key Terms, End-of-Chapter and Caption Questions, and Answers.
- ***Accelerator CD-ROM.*** Every copy of the book is shipped with an Accelerator CD-ROM. Insert the CD before accessing OneKey to have a faster Internet experience. Accelerator CDs minimize the time students spend waiting for content, so that even dial-up connections run at CD-ROM speed.
- ***Student Study Companion.*** For students without an Internet connection, all questions and review materials from OneKey are included in the printed Student Study Companion.

Acknowledgments

Biology: Life on Earth is truly a team effort. Ann Heath, our Project Manager, has handled the tremendous task of pulling it all together with consummate skill and patience. Our Development Editor Anne Madura scrutinized every word and figure, looking for ways to make the text more clear, consistent, and student-friendly. While Art Director Maureen Eide developed and executed a fresh design for this new edition, Art Editor Adam Velthaus deftly coordinated the art program; the numerous modifications were artfully rendered by Imagineering Studios under the direction of Jack Haley. Kim Quillin, a biological illustrator, carefully reviewed the art with an eye for accuracy, color consistency, and layout. Photo Researcher Yvonne Gerin tirelessly tracked down excellent photos. Sybil Sosin tackled the job of copyediting with meticulous attention to detail. Tim Flem, our Production Editor, brought the art, photos, and manuscript together into a seamless whole, and dealt remarkably cheerfully with last-minute changes. Media Editor Travis Moses-Westphal and Assistant Editor Colleen Lee coordinated production of all the media and study aids that contribute so much to the total package that is *Biology: Life on Earth*. Our marketing manager, Andrew Gilfillan, helped create a marketing strategy that will effectively communicate our message to our audience. The entire project was overseen with energy and imagination by Executive Editor Teresa Ryu Chung. We thank her for her faith in this project, for the fantastic team she assembled, and for arranging a schedule that allowed us to continue to be what we are—not only authors, but parents, teachers, and researchers.

Terry and Gerry Audesirk
Bruce E. Byers

SEVENTH EDITION REVIEWERS

Jerry L. Cook, *Sam Houston State University*
Thomas Emmel, *University of Florida*
Dennis Forsythe, *The Citadel*
Harvey Friedman, *University of Missouri—St. Louis*
Teresa L. Fulcher, *Pellissippi State Technical Community College*
John Geiser, *Western Michigan University*
David Grise, *Southwest Texas State University*
Richard Hanke, *Rose State College*
James Hewlett, *Finger Lakes Community College*
Tom Langen, *Clarkson University*
Richard Manning, *Southwest Texas State University*
Ken Marr, *Green River Community College*
Kathleen A. Marrs, *Indiana University—Purdue University Indianapolis*
Hugh Miller, *East Tennessee State University*
Jeanne Mitchell, *Truman State University*
Gary B. Peterson, *South Dakota State University*
Jennifer Roberts, *Lewis University*
Connie Russell, *Angelo State University*
Doug Schelhaas, *University of Mary*
Brian Schmaefsky, *Kingwood College*
Marilyn Shopper, *Johnson County Community College*
John Sollinger, *Southern Oregon University*
Sally Sommers Smith, *Boston University*
Bruce Stallsmith, *University of Alabama-Huntsville*
Barbara Stebbins-Boaz, *Willamette University*
Jyoti Wagle, *Houston Community College*
Lisa Weasel, *Portland State University*
Emily Willingham, *University of Texas-Austin*
Brenda Young, *Daemen College*

SEVENTH EDITION REVIEWERS: INSTRUCTOR AND STUDENT RESOURCES

Deborah Dardis, *Southeastern Louisiana University*
Carolyn Peters, *Spoon River College*
Connie Russell, *Angelo State University*
Marilyn Shopper, *Johnson County Community College*
John Sollinger, *Southern Oregon University*

SEVENTH EDITION STUDENT FOCUS PARTICIPANTS AND REVIEWERS

Anthony Age, *Louisiana State University*
Ryan F. Berni, *Louisiana State University*
Lacey Elizabeth Brandt, *Louisiana State University*
Man Kong Chan, *Pasadena City College*
Erin Colton, *California State University Northridge*
Elyse Cottone, *Pasadena City College*
Paul Fortenberry, *Louisiana State University*
Marissa Ishida, *Pasadena City College*
Armig Matosian, *Pasadena City College*
Danielle Reynolds, *California State University Northridge*
Rebekah Ruswick, *Pasadena City College*
Leah Sharp, *Louisiana State University*
Tiffany Thompson, *California State University Northridge*
Christina Thomsen, *California State University Northridge*
Kirsten Wagner, *California State University Northridge*
Joey Wang, *Pasadena City College*
Jenna Wilcox, *California State University Northridge*
Shan Xhe, *Pasadena City College*

PREVIOUS EDITION REVIEWERS

W. Sylvester Allred, *Northern Arizona University*
Judith Keller Amand, *Delaware County Community College*
William Anderson, *Abraham Baldwin Agriculture College*
Steve Arch, *Reed College*
Kerri Lynn Armstrong, *Community College of Philadelphia*
G. D. Aumann, *University of Houston*
Vernon Avila, *San Diego State University*
J. Wesley Bahorik, *Kutztown University of Pennsylvania*
Bill Barstow, *University of Georgia-Athens*
Colleen Belk, *University of Minnesota, Duluth*
Michael C. Bell, *Richland College*
Gerald Bergtrom, *University of Wisconsin*
Arlene Billock, *University of Southwestern Louisiana*
Brenda C. Blackwelder, *Central Piedmont Community College*
Raymond Bower, *University of Arkansas*
Marilyn Brady, *Centennial College of Applied Arts and Technology*
Virginia Buckner, *Johnson County Community College*
Arthur L. Buikema, Jr., *Virginia Polytechnic Institute*
J. Gregory Burg, *University of Kansas*
William F. Burke, *University of Hawaii*
Robert Burkholter, *Louisiana State University*
Kathleen Burt-Utley, *University of New Orleans*
Linda Butler, *University of Texas-Austin*
W. Barkley Butler, *Indiana University of Pennsylvania*
Jerry Button, *Portland Community College*
Bruce E. Byers, *University of Massachusetts-Amherst*
Sara Chambers, *Long Island University*
Nora L. Chee, *Chaminade University*
Joseph P. Chinnici, *Virginia Commonwealth University*
Dan Chiras, *University of Colorado-Denver*
Bob Coburn, *Middlesex Community College*
Joseph Coelho, *Culver Stockton College*
Martin Cohen, *University of Hartford*
Walter J. Conley, *State University of New York at Potsdam*
Mary U. Connell, *Appalachian State University*
Jerry Cook, *Sam Houston State University*
Joyce Corban, *Wright State University*
Ethel Cornforth, *San Jacinto College-South*
David J. Cotter, *Georgia College*
Lee Couch, *Albuquerque Technical Vocational Institute*
Donald C. Cox, *Miami University of Ohio*
Patricia B. Cox, *University of Tennessee*
Peter Crowcroft, *University of Texas-Austin*
Carol Crowder, *North Harris Montgomery College*
Donald E. Culwell, *University of Central Arkansas*
Robert A. Cunningham, *Erie Community College, North*
Karen Dalton, *Community College of Baltimore County—Catonsville Campus*
Lydia Daniels, *University of Pittsburgh*
David H. Davis, *Asheville-Buncombe Technical Community College*
Jerry Davis, *University of Wisconsin, LaCrosse*
Douglas M. Deardon, *University of Minnesota*
Lewis Deaton, *University of Southwestern Louisiana*
Fred Delcomyn, *University of Illinois-Urbana*
David M. Demers, *University of Hartford*
Lorren Denney, *Southwest Missouri State University*
Katherine J. Denniston, *Towson State University*
Charles F. Denny, *University of South Carolina-Sumter*
Jean DeSaix, *University of North Carolina-Chapel Hill*
Ed DeWalt, *Louisiana State University*
Daniel F. Doak, *University of California-Santa Cruz*
Matthew M. Douglas, *University of Kansas*
Ronald J. Downey, *Ohio University*
Ernest Dubrul, *University of Toledo*
Michael Dufresne, *University of Windsor*
Susan A. Dunford, *University of Cincinnati*
Mary Durant, *North Harris College*
Ronald Edwards, *University of Florida*
Rosemarie Elizondo, *Reedley College*
George Ellmore, *Tufts University*
Joanne T. Ellzey, *University of Texas-El Paso*
Wayne Elmore, *Marshall University*
Carl Estrella, *Merced College*
Nancy Eyster-Smith, *Bentley College*
Gerald Farr, *Southwest Texas State University*
Rita Farrar, *Louisiana State University*
Marianne Feaver, *North Carolina State University*
Susannah Feldman, *Towson University*
Linnea Fletcher, *Austin Community College-Northridge*
Charles V. Foltz, *Rhode Island College*
Douglas Fratianne, *Ohio State University*
Scott Freeman, *University of Washington*
Donald P. French, *Oklahoma State University*
Don Fritsch, *Virginia Commonwealth University*
Teresa Lane Fulcher, *Pellissippi State Technical Community College*
Michael Gaines, *University of Kansas*
Irja Galvan, *Western Oregon University*
Gail E. Gasparich, *Towson University*
Farooka Gauhari, *University of Nebraska-Omaha*
George W. Gilchrist, *University of Washington*
David Glenn-Lewin, *Iowa State University*
Elmer Gless, *Montana College of Mineral Sciences*
Charles W. Good, *Ohio State University-Lima*
Margaret Green, *Broward Community College*
Lonnie J. Guralnick, *Western Oregon University*
Martin E. Hahn, *William Paterson College*
Madeline Hall, *Cleveland State University*
Georgia Ann Hammond, *Radford University*

Blanche C. Haning, *North Carolina State University*
Helen B. Hanten, *University of Minnesota*
John P. Harley, *Eastern Kentucky University*
William Hayes, *Delta State University*
Stephen Hedman, *University of Minnesota*
Jean Helgeson, *Collins County Community College*
Alexander Henderson, *Millersville University*
Timothy L. Henry, *University of Texas-Arlington*
James Hewlett, *Finger Lakes Community College*
Alison G. Hoffman, *University of Tennessee-Chattanooga*
Leland N. Holland, *Paso-Hernando Community College*
Laura Mays Hoopes, *Occidental College*
Michael D. Hudgins, *Alabama State University*
David Huffman, *Southwest Texas State University*
Donald A. Ingold, *East Texas State University*
Jon W. Jacklet, *State University of New York-Albany*
Rebecca M. Jessen, *Bowling Green State University*
J. Kelly Johnson, *University of Kansas*
Florence Juillerat, *Indiana University—Purdue University at Indianapolis*
Thomas W. Jurik, *Iowa State University*
Arnold Karpoff, *University of Louisville*
L. Kavaljian, *California State University*
Jeff Kenton, *Iowa State University*
Hendrick J. Ketellapper, *University of California, Davis*
Jeffrey Kiggins, *Blue Ridge Community College*
Harry Kurtz, *Sam Houston State University*
Kate Lajtha, *Oregon State University*
Patricia Lee-Robinson, *Chaminade University of Honolulu*
William H. Leonard, *Clemson University*
Edward Levri, *Indiana University of Pennsylvania*
Graeme Lindbeck, *University of Central Florida*
Jerri K. Lindsey, *Tarrant County Junior College-Northeast*
John Logue, *University of South Carolina-Sumter*
William Lowen, *Suffolk Community College*
Ann S. Lumsden, *Florida State University*
Steele R. Lunt, *University of Nebraska-Omaha*
Daniel D. Magoulick, *The University of Central Arkansas*
Paul Mangum, *Midland College*
Michael Martin, *University of Michigan*
Linda Martin-Morris, *University of Washington*
Kenneth A. Mason, *University of Kansas*
Margaret May, *Virginia Commonwealth University*
D. J. McWhinnie, *De Paul University*
Gary L. Meeker, *California State University, Sacramento*
Thoyd Melton, *North Carolina State University*
Joseph R. Mendelson III, *Utah State University*
Karen E. Messley, *Rockvalley College*
Timothy Metz, *Campbell University*
Glendon R. Miller, *Wichita State University*
Neil Miller, *Memphis State University*
Jack E. Mobley, *University of Central Arkansas*
John W. Moon, *Harding University*
Richard Mortenson, *Albion College*
Gisele Muller-Parker, *Western Washington University*
Kathleen Murray, *University of Maine*
Robert Neill, *University of Texas*
Harry Nickla, *Creighton University*
Daniel Nickrent, *Southern Illinois University*
Jane Noble-Harvey, *University of Delaware*
David J. O'Neill, *Community College of Baltimore County-Dundalk Campus*
James T. Oris, *Miami University of Ohio*
Marcy Osgood, *University of Michigan*
C. O. Patterson, *Texas A&M University*
Fred Peabody, *University of South Dakota*
Harry Peery, *Tompkins–Cortland Community College*
Rhoda E. Perozzi, *Virginia Commonwealth University*
Bill Pfitsch, *Hamilton College*
Ronald Pfohl, *Miami University of Ohio*
Bernard Possident, *Skidmore College*
Ina Pour-el, *DMACC—Boone Campus*
Elsa C. Price, *Wallace State Community College*
Marvin Price, *Cedar Valley College*
James A. Raines, *North Harris College*
Paul Ramp, *Pellissippi State Technical College*
Mark Richter, *University of Kansas*
Robert Robbins, *Michigan State University*
Chris Romero, *Front Range Community College*
Paul Rosenbloom, *Southwest Texas State University*
K. Ross, *University of Delaware*
Mary Lou Rottman, *University of Colorado-Denver*
Albert Ruesink, *Indiana University*
Christopher F. Sacchi, *Kutztown University*
Alan Schoenherr, *Fullerton College*
Edna Seaman, *University of Massachusetts, Boston*
Patricia Shields, *George Mason University*
Anu Singh-Cundy, *Western Washington University*
Linda Simpson, *University of North Carolina-Charlotte*
Russel V. Skavaril, *Ohio State University*
John Smarelli, *Loyola University*
Shari Snitovsky, *Skyline College*
Jim Sorenson, *Radford University*
Mary Spratt, *University of Missouri, Kansas City*
Benjamin Stark, *Illinois Institute of Technology*
William Stark, *Saint Louis University*
Kathleen M. Steinert, *Bellevue Community College*
Barbara Stotler, *Southern Illinois University*
Gerald Summers, *University of Missouri-Columbia*
Marshall Sundberg, *Louisiana State University*
Bill Surver, *Clemson University*
Eldon Sutton, *University of Texas-Austin*
Dan Tallman, *Northern State University*
David Thorndill, *Essex Community College*
William Thwaites, *San Diego State University*
Professor Tobiessen, *Union College*
Richard Tolman, *Brigham Young University*
Dennis Trelka, *Washington and Jefferson College*
Sharon Tucker, *University of Delaware*
Gail Turner, *Virginia Commonwealth University*
Glyn Turnipseed, *Arkansas Technical University*
Lloyd W. Turtinen, *University of Wisconsin-Eau Claire*
Robert Tyser, *University of Wisconsin-La Crosse*
Robin W. Tyser, *University of Wisconsin-LaCrosse*
Kristin Uthus, *Virginia Commonwealth University*
F. Daniel Vogt, *State University of New York-Plattsburgh*
Nancy Wade, *Old Dominion University*
Susan M. Wadkowski, *Lakeland Community College*
Jyoti R. Wagle, *Houston Community College-Central*
Michael Weis, *University of Windsor*
DeLoris Wenzel, *University of Georgia*
Jerry Wermuth, *Purdue University-Calumet*
Jacob Wiebers, *Purdue University*
Carolyn Wilczynski, *Binghamton University*
P. Kelly Williams, *University of Dayton*
Roberta Williams, *University of Nevada-Las Vegas*
Sandra Winicur, *Indiana University-South Bend*
Bill Wischusen, *Louisiana State University*
Chris Wolfe, *North Virginia Community College*
Stacy Wolfe, *Art Institutes International*
Colleen Wong, *Wilbur Wright College*
Wade Worthen, *Furman University*
Robin Wright, *University of Washington*
Brenda L. Young, *Daemen College*
Cal Young, *Fullerton College*
Tim Young, *Mercer University*

CHAPTER

1 An Introduction to Life on Earth

Life on Earth is confined to a thin film encompassing Earth's surface: the biosphere. Earth, seen from the moon, is an oasis of life in our solar system.

AT A GLANCE

CASE STUDY Life on Earth—and Elsewhere?

"Viewed from the distance of the moon, the astonishing thing abut the earth, catching the breath, is that it is alive. The photographs show the dry, pounded surface of the moon in the foreground, dead as an old bone. Aloft, floating free beneath the moist gleaming surface of bright blue sky, is the rising earth, the only exuberant thing in this part of the cosmos."

[Lewis Thomas in "The Lives of a Cell" (1974)]

When natural philosopher Lewis Thomas viewed the early photographs of Earth taken by astronauts from the surface of the moon (see the photo on the opposite page), he—like most of humanity—felt a sense of awe. The dry and barren surface of the moon in the foreground reminds us of how truly special Earth is—blanketed with green plants, blue oceans, and white clouds. But is Earth itself "alive"? There is no question that life has invaded nearly every nook and cranny of Earth. The toughest life-forms are also the simplest—single-celled organisms collectively described as *extremophiles*. These "survivalist microbes" inhabit the most inhospitable environments on Earth. Some thrive in vents in the deep ocean floor, with pressure 30 times that on Earth's surface and which spew water heated to temperatures over 100 °C. Others have been discovered in ice cores 1200 feet below the surface of an Antarctic lake frozen for hundreds of thousands of years. Extremophiles inhabit the highly acid environments produced by mining wastes and hot springs, and have been discovered in rock samples taken from 4 miles beneath Earth's surface. These life-forms are as foreign to us as alien life from another solar system. Indeed, their existence on Earth fuels guarded optimism that life may exist or may have once existed in the seemingly hostile conditions found on other planets. What is life? How did it evolve? Could life survive on the barren surface of the moon, or in the harsh environments of other planets?

1.1 What Are the Characteristics of Living Things?

What is life? If you look up *life* in a dictionary, you will find definitions such as "the quality that distinguishes a vital and functioning being from a dead body," but you won't find out what that quality is. All of us have an intuitive understanding of what it means to be alive. Nevertheless, defining "life" is difficult, partly because living things are so diverse and nonliving matter, in some cases, seems so lifelike. A more fundamental difficulty in defining life is that living things cannot be described as simply the sum of their parts. The quality of life emerges as a result of the incredibly complex, ordered interactions among these parts. Because it is based on these *emergent properties*, life is a fundamentally intangible quality that defies simple definition. We can, however, describe some of the characteristics of living things that, taken together, are not shared by nonliving objects. These characteristics are as follows:

Media Activity
1.1 Defining Life

- Living things have a complex, organized structure that consists largely of organic molecules.
- Living things respond to stimuli from their environment.
- Living things actively maintain their complex structure and their internal environment, a process called *homeostasis*.
- Living things acquire and use materials and energy from their environment and convert them into different forms.
- Living things grow.
- Living things reproduce themselves, using a molecular blueprint called *DNA*.
- Living things, as a whole, have the capacity to evolve.

Let's explore these characteristics in more detail.

Living Things Are Both Complex and Organized

Compared with nonliving matter of similar size, living things are highly complex and organized. A crystal of table salt (Fig. 1-1a), for example, consists of just two chemical elements—sodium and chlorine—arranged in a precise cubical arrangement; salt crystals are organized but simple. The oceans (Fig. 1-1b) contain some atoms of all the naturally occurring elements, but these atoms are randomly distributed; the oceans are complex but not organized. In contrast, even the tiny waterflea (Fig. 1-1c) contains dozens of different elements linked together in thousands of specific combinations that are further organized into ever larger and more complex assemblies to form structures such as eyes, legs, a digestive tract, and even a small brain.

Life on Earth consists of a hierarchy of structures. Each level is based on the one below it and provides the foundation for the one above it (Fig. 1-2). All of life is built on a chemical foundation of substances called **elements**, each of which is a unique type of *matter*. An **atom** is the smallest particle of an element that retains the properties of that element. For example, a diamond is made of the element carbon. The smallest possible unit of the diamond is an individual carbon atom; any further

(a) Organized

(b) Complex

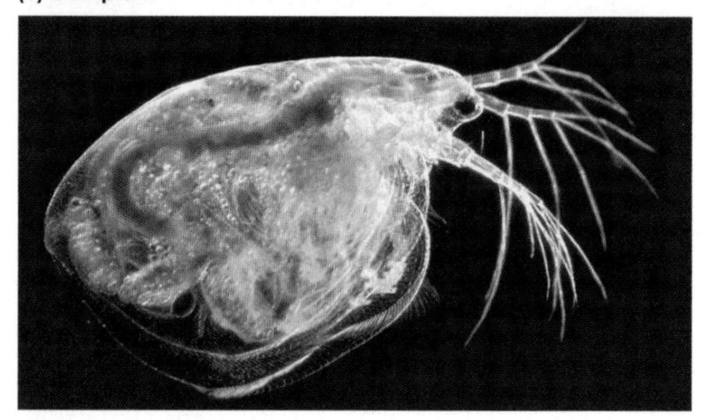
(c) Organized and complex

FIGURE 1-1 Life is both complex and organized
(a) Each crystal of table salt, sodium chloride, is a cube, showing great organization but minimal complexity. **(b)** The water and dissolved materials in the ocean represent complexity but very little organization. **(c)** Living things have both complexity and organization. The waterflea, *Daphnia pulex*, is only 1 millimeter long (1/1000 meter; smaller than the letter "i"), yet it has legs, a mouth, a digestive tract, reproductive organs, light-sensing eyes, and even a rather impressive brain in relation to its size.

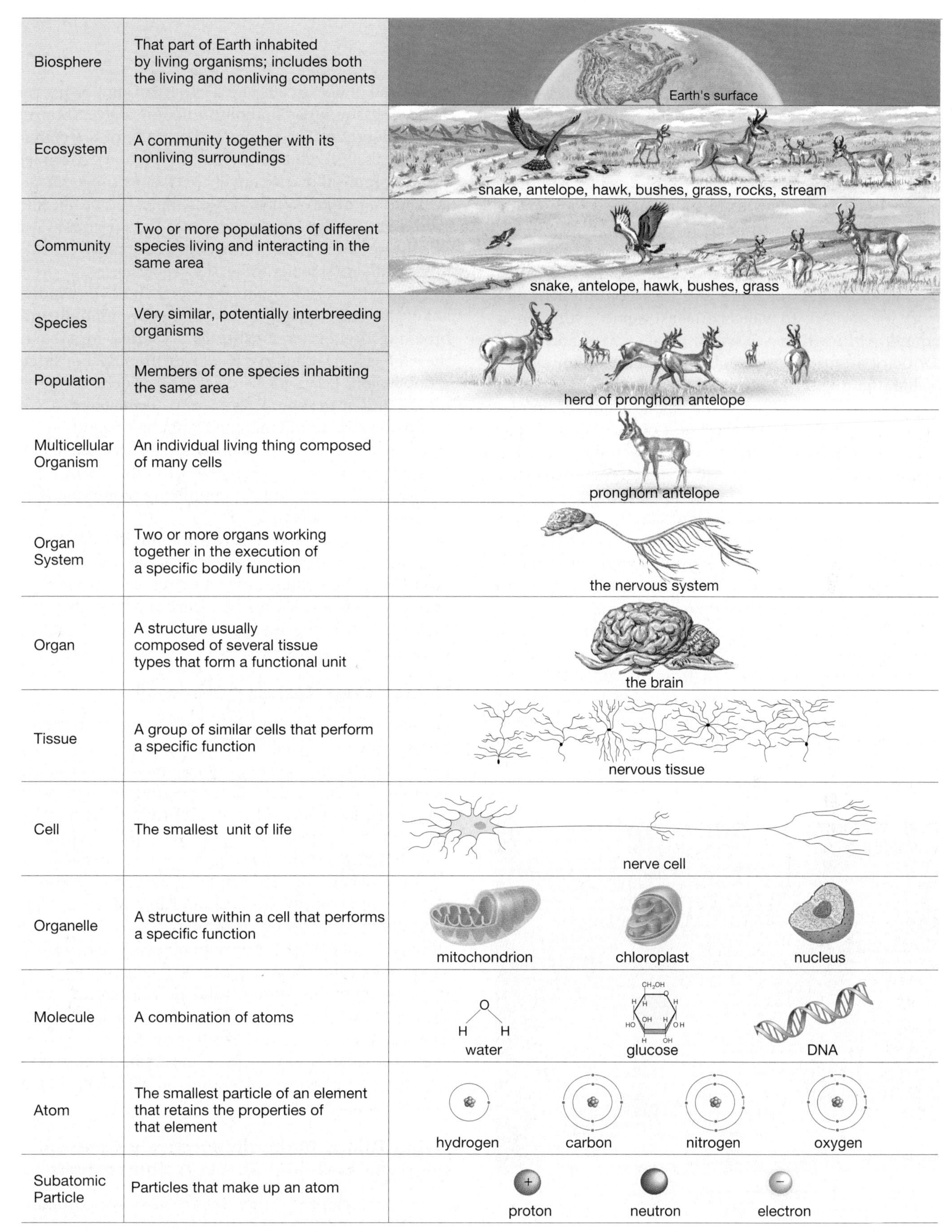

Level	Definition	Example
Biosphere	That part of Earth inhabited by living organisms; includes both the living and nonliving components	Earth's surface
Ecosystem	A community together with its nonliving surroundings	snake, antelope, hawk, bushes, grass, rocks, stream
Community	Two or more populations of different species living and interacting in the same area	snake, antelope, hawk, bushes, grass
Species	Very similar, potentially interbreeding organisms	herd of pronghorn antelope
Population	Members of one species inhabiting the same area	
Multicellular Organism	An individual living thing composed of many cells	pronghorn antelope
Organ System	Two or more organs working together in the execution of a specific bodily function	the nervous system
Organ	A structure usually composed of several tissue types that form a functional unit	the brain
Tissue	A group of similar cells that perform a specific function	nervous tissue
Cell	The smallest unit of life	nerve cell
Organelle	A structure within a cell that performs a specific function	mitochondrion; chloroplast; nucleus
Molecule	A combination of atoms	water; glucose; DNA
Atom	The smallest particle of an element that retains the properties of that element	hydrogen; carbon; nitrogen; oxygen
Subatomic Particle	Particles that make up an atom	proton; neutron; electron

FIGURE 1-2 Levels of organization of matter
All life has a chemical basis, but the quality of life itself emerges on the cellular level. Interactions among the components of each level and the levels below it allow the development of the next-higher level of organization.
EXERCISE Think of a scientific question that can be answered by investigating at the cell level, but that would be impossible to answer at the tissue level. Then think of a question answerable at the tissue level but not the cell level. Repeat the process for two other pairs of adjacent levels of organization.

division would produce isolated **subatomic particles** that would no longer be carbon. Atoms may combine in specific ways to form assemblies called **molecules**; for example, one carbon atom can combine with two oxygen atoms to form a molecule of carbon dioxide. Although many simple molecules form spontaneously, only living things manufacture extremely large and complex molecules. The bodies of living things are composed primarily of complex molecules. The molecules of life are called **organic molecules**, meaning that they contain a framework of carbon, to which at least some hydrogen is bound. Although the chemical arrangement and interaction of atoms and molecules form the building blocks of life, the quality of life itself emerges on the level of the cell. Just as an atom is the smallest unit of an element, so the **cell** is the smallest unit of life (Fig. 1-3). The differences between a living cell and a conglomeration of chemicals illustrate some of the emergent properties of life.

All cells contain **genes**, units of heredity that provide the information needed to control the life of the cell; subcellular structures called **organelles**, miniature chemical factories that use the information in the genes and keep the cell alive; and a **plasma membrane**, a thin sheet surrounding the cell that both encloses the **cytoplasm** (the organelles and the watery medium surrounding them) and separates the cell from the outside world. Some life-forms, mostly microscopic, consist of just one cell, but larger life-forms are composed of many cells, each with a specialized function. In multicellular life-forms, related cells combine to form **tissues**, which perform a particular function. For example, nervous tissue is composed of nerve cells and a variety of supporting cells. Various tissue types combine to form a structural unit called an **organ** (for example, the brain, which contains nervous tissue, connective tissue, and blood). A group of several organs that collectively perform a single function is called an **organ system**; for example, the brain, spinal cord, sense organs, and nerves form the nervous system. All the organ systems functioning cooperatively make up an individual living thing, the **organism**.

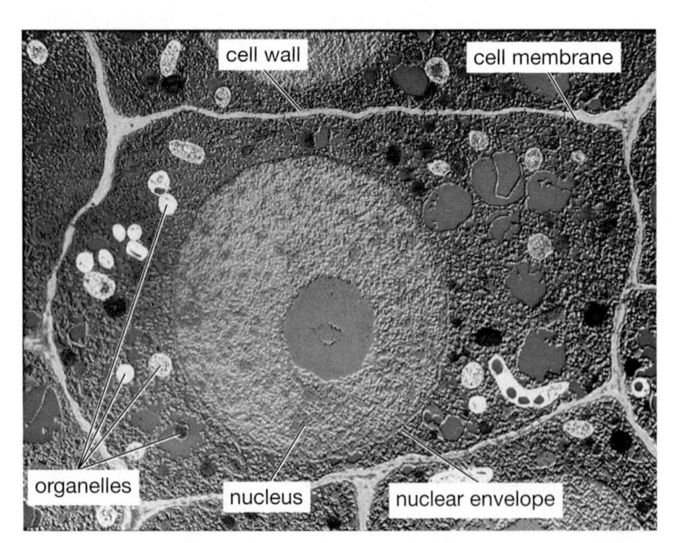

FIGURE 1-3 The cell is the smallest unit of life
This micrograph of a plant cell clearly shows the supporting cell wall that surrounds and supports plant (but not animal) cells. Just inside the cell wall, a thin plasma membrane (found in all cells) controls what substances enter and leave the cell. The nucleus, surrounded by a membrane called the *nuclear envelope*, contains the cell's DNA. The cell also contains several types of specialized organelles. Some store food; some break down food to provide usable energy; and, in plants, some organelles capture light energy.

Broader levels of organization reach beyond individual organisms. A group of very similar, potentially interbreeding organisms constitutes a **species**. Members of the same species that live in a given area are considered a **population**. Populations of several species living and interacting in the same area form a **community**. A community and its nonliving environment—including land, water, and atmosphere—constitute an **ecosystem**. Finally, the entire surface region of Earth inhabited by living things (and including its nonliving components) is called the **biosphere**.

The organization of this text roughly follows the pattern of organization of life on Earth. We will begin with atoms and molecules, move to cells and principles of heredity, continue with organisms and how they function, and conclude with ecology, the study of the interactions among organisms.

Living Things Respond to Stimuli

Organisms perceive and respond to stimuli in their internal and external environments. Animals have evolved elaborate sensory organs and muscular systems that allow them to detect and respond to light, sound, chemicals, and many other stimuli from their surroundings. Internal stimuli are perceived by receptors for stretch, temperature, pain, and various chemicals. For example, when you feel hungry, you perceive contractions of your empty stomach and low levels of sugars and fats in your blood. You then respond to external stimuli by choosing appropriate objects to eat, such as a sandwich rather than a plate. Yet animals, with their elaborate nervous systems and motile bodies, are not the only organisms that perceive and respond to stimuli. The plants on your windowsill grow toward light, and even the bacteria in your intestines manufacture different digestive enzymes depending on whether you drink milk, eat candy, or both.

Living Things Maintain Relatively Constant Internal Conditions Through Homeostasis

Complex, organized structures are not easy to maintain. Whether we consider the molecules of your body or the books and papers on your desk, organization tends to disintegrate into chaos unless energy is used to sustain it. (We will explore this tendency more fully in Chapter 6.) To stay alive and function effectively, organisms must keep the conditions within their bodies fairly constant; in other words, they must maintain **homeostasis**

FIGURE 1-4 Living things maintain homeostasis
Sweating helps cool former Olympic and World Heavyweight boxing champion George Foreman. **QUESTION** In addition to reducing body temperature, how else does sweating affect homeostasis?

FIGURE 1-5 Living things acquire energy and nutrients from the environment
The green plants seen here capture energy from the sun and nutrients from the air, water, and soil. The insect gains both energy and nutrients from the plants, while the toad will extract energy and nutrients from its insect prey. Without a continuous supply of solar energy, nearly all life would cease.

(derived from Greek words meaning "to stay the same"). One of the many conditions that organisms regulate is body temperature. Among warm-blooded animals, for example, vital organs such as the brain and heart are kept at a warm, constant temperature despite wide fluctuations in environmental temperature.

Homeostasis is maintained by a variety of automatic mechanisms. In the case of temperature regulation, these mechanisms include sweating during hot weather (Fig. 1-4), metabolizing more food in cold weather, and behaviors such as basking in the sun or even adjusting the thermostat in the room. Of course, not everything stays the same throughout an organism's life. Major changes, such as growth and reproduction, occur, but these are not failures of homeostasis. Rather, they are specific, genetically programmed parts of the organism's life cycle.

Living Things Acquire and Use Materials and Energy

Organisms need materials and energy to maintain their high level of complexity and organization, to grow, and to reproduce (Fig. 1-5). Organisms acquire the atoms and molecules they need from air, water, or soil or from other living things. These materials, called **nutrients**, are extracted from the environment and incorporated into the molecules of the organisms' bodies. The sum total of all of the chemical reactions needed to sustain an organism's life is called its **metabolism**.

Organisms obtain **energy**—the ability to do work, such as carrying out chemical reactions, growing leaves in the spring, or contracting a muscle—in one of two basic ways. Plants and some single-celled organisms capture the energy of sunlight and store it in energy-rich sugar molecules, a process called **photosynthesis**. In contrast, neither fungi nor animals can photosynthesize, nor can most bacteria; these organisms must consume the energy-rich molecules contained in the bodies of other organisms. In either case, acquired energy is converted into a form that the organism can use or store for future use.

Ultimately, the energy that sustains nearly all life comes from sunlight, captured by photosynthetic organisms and incorporated into energy-rich molecules. Organisms that cannot photosynthesize depend on photosynthetic life-forms for food, either directly or indirectly. Thus, energy flows from the sun through nearly all forms of life. That energy is eventually released again as heat.

Living Things Grow

At some time in its life cycle, every organism becomes larger—that is, it *grows*. This characteristic is obvious for plants, birds, and mammals, all of which start out very small and undergo tremendous growth during their lives. Even single-celled bacteria, however, grow to about double their original size before they divide. In all cases,

FIGURE 1-6 Living things reproduce
As they grow, these polar bear cubs will resemble, but not be identical to, their parents. The similarity and variability of offspring are crucial to the process of evolution.

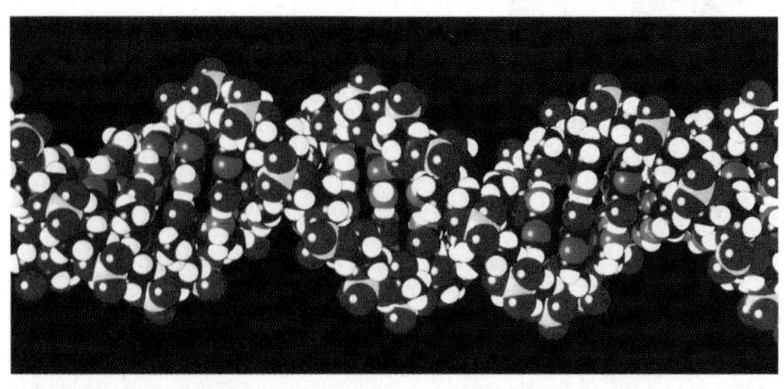

FIGURE 1-7 DNA
A computer-generated model of DNA, the molecule of heredity. As James Watson, its co-discoverer, put it: "A structure this pretty just had to exist."

growth involves the conversion of materials acquired from the environment into the specific molecules of the organism's body.

Living Things Reproduce Themselves

Organisms reproduce, giving rise to offspring of the same type (Fig. 1-6) and creating *continuity of life*. The processes by which this occurs vary, but the result—the perpetuation of the parents' genetic material—is the same. The *diversity of life* occurs in part because offspring, though arising from the genetic material provided by their parents, are usually somewhat different from their parents, as explained briefly below and in more detail in Units Two and Three. The specific mechanisms by which traits are passed from one generation to the next, using genetic information that is recombined in various ways, produce these variable offspring.

DNA Is the Molecule of Heredity

All known forms of life contain their hereditary information within a molecule called **deoxyribonucleic acid**, or **DNA** (Fig. 1-7). Much of Unit Two will be devoted to exploring the structure and function of this remarkable molecule, segments of which are called *genes*. An organism's DNA is its genetic blueprint or molecular instruction manual, a guide to the construction—and, at least in part, the operation—of its body. When an organism reproduces, it passes a copy of its DNA to its offspring. The accuracy of the DNA copying process is astonishingly high: only about one mistake occurs for every billion bits of information contained in the DNA molecule. These occasional errors, called **mutations**, produce variety. Without mutations, all life-forms might be identical. Indeed, there is reason to hypothesize that, without mutations, there would be no life. Variations, caused by mutations and superimposed on a background of overall genetic fidelity, make possible the final property of life, the capacity to evolve.

Living Things As a Whole Have the Capacity to Evolve

Although the genetic makeup of a single organism remains essentially the same over its lifetime, the genetic composition of a species as a whole changes over many generations. Over time, mutations and variable offspring create diversity in the genetic material of a species. In other words, the species *evolves*. The scientific theory of **evolution** states that modern organisms descended—with modification—from preexisting life-forms and that, ultimately, all forms of life on Earth share a common ancestor. The most important force in evolution is **natural selection**, the process by which organisms with *adaptations* (characteristics that help an organism cope with the rigors of its environment) survive and reproduce more successfully than do others that lack those traits. Adaptive traits arising from genetic mutation are passed on to the next generation.

1.2 How Do Scientists Categorize the Diversity of Life?

Although all living things share the general characteristics discussed earlier, evolution has produced an amazing variety of life-forms. The classification and structures of Earth's diverse organisms are discussed in detail in Chapters 18 through 23.

Organisms can be grouped into three major categories, called **domains**: Bacteria, Archaea, and Eukarya. This classification reflects fundamental differences among the cell types that compose these organisms. Members of both the Bacteria and the Archaea usually consist of single, simple cells. Members of the Eukarya have bodies composed of one or more highly complex cells. This domain includes three major subdivisions or **kingdoms**: the Fungi, Plantae, and Animalia, as well as a diverse collection of mostly single-celled organisms collectively known as "protists" (Fig. 1-8). There are exceptions to any simple set of criteria used to characterize

(a) The domain Bacteria

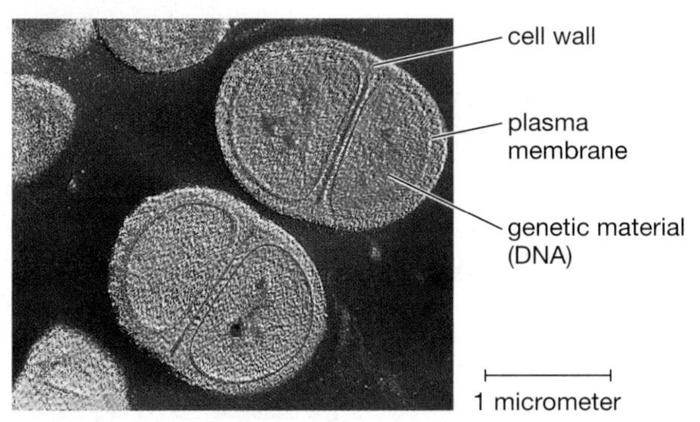

A color-enhanced electron micrograph of a dividing bacterium. Bacteria are unicellular and prokaryotic; most are surrounded by a thick cell wall. Some bacteria photosynthesize, but most absorb food from their surroundings.

(b) The domain Archaea

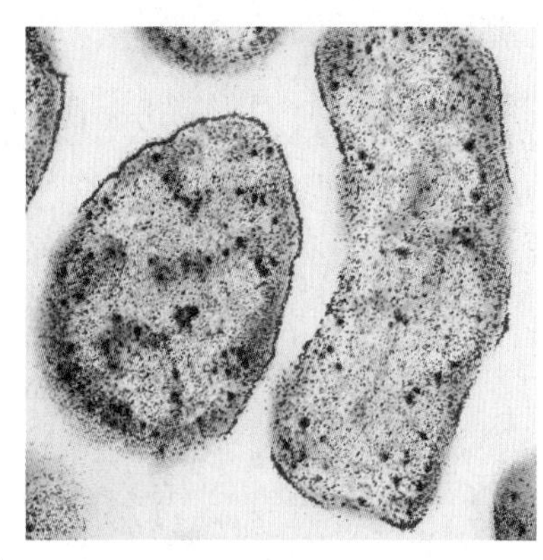

A color-enhanced electron micrograph of an archaean. The cell wall appears red, and DNA is scattered inside. Many archaeans can survive extreme conditions. This Antarctic species lives at temperatures as low as –2.5°C.

(c) A protist (domain Eukarya)

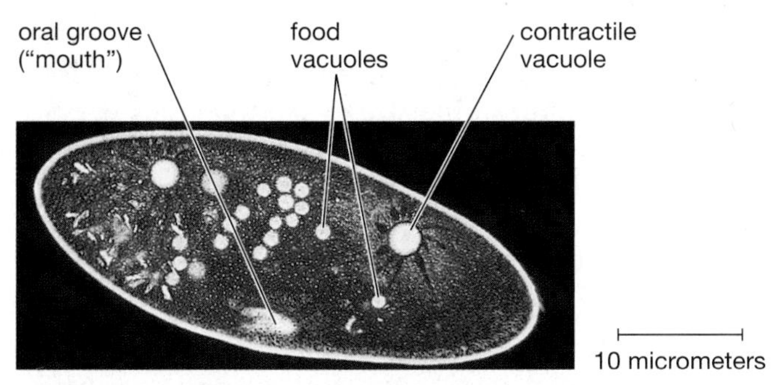

This light micrograph of a *Paramecium* illustrates the complexity of these large, normally single, eukaryotic cells. Some protists photosynthesize, but others ingest or absorb their food. Many, including *Paramecium*, are mobile, moving with cilia or flagella.

(d) The kingdom Fungi (domain Eukarya)

An exotic mushroom found in Peru. Most fungi are multicellular. Fungi generally absorb their food, which is usually the dead bodies or wastes of plants and animals. The food is digested by enzymes secreted outside the fungal body. Most fungi cannot move.

(e) The kingdom Plantae (domain Eukarya)

This butterfly weed represents the flowering plants, the dominant members of the kingdom Plantae. Flowering plants owe much of their success to mutually beneficial relationships with animals, such as these pearl crescent butterflies, in which the flower provides food and the insect carries pollen from flower to flower, fertilizing them. Plants are multicellular, nonmotile eukaryotes that acquire nutrients by photosynthesis.

(f) The kingdom Animalia (domain Eukarya)

A wrasse rests on a soft coral. Animals are multicellular; animal bodies consist of a wide assortment of tissues and organs composed of specialized cell types. Most animals can move and respond rapidly to stimuli. The coral is a member of the largest group of animals: the invertebrates, which lack a backbone. This group also includes insects and mollusks. The wrasse is a vertebrate; like humans, it has a backbone.

FIGURE 1-8 The domains and kingdoms of life

Table 1-1 Some Characteristics Used in Classification of Organisms

Domain	Kingdom	Cell Type	Cell Number	Energy Acquisition
Bacteria	(Under discussion)	Prokaryotic	Unicellular	Absorption, photosynthesis
Archaea	(Under discussion)	Prokaryotic	Unicellular	Absorption
Eukarya*	Fungi	Eukaryotic	Multicellular	Absorption
	Plantae	Eukaryotic	Multicellular	Photosynthesis
	Animalia	Eukaryotic	Multicellular	Ingestion

*The Eukarya also includes several groups of mostly unicellular organisms collectively called "protists."

the domains and kingdoms, but three characteristics are particularly useful: cell type, the number of cells in each organism, and how it acquires energy (Table 1-1).

The Domains Bacteria and Archaea Consist of Prokaryotic Cells; the Domain Eukarya Is Composed of Eukaryotic Cells

There are two fundamentally different types of cells: **prokaryotic** and **eukaryotic**. *Karyotic* refers to the **nucleus** of a cell, a membrane-enclosed sac containing the cell's genetic material (see Fig. 1-3). *Eu* means "true" in Greek; eukaryotic cells possess a "true," membrane-enclosed nucleus. Eukaryotic cells are generally larger than prokaryotic cells and contain a variety of other organelles, many surrounded by membranes. *Pro* means "before" in Greek; prokaryotic cells almost certainly evolved before eukaryotic cells (and, as we will see in Chapter 17, eukaryotic cells almost certainly evolved from prokaryotic cells). Prokaryotic cells do not have a nucleus; their genetic material resides in their cytoplasm. They are usually small—only 1 or 2 micrometers in diameter—and lack membrane-bound organelles. The domains Bacteria and Archaea consist of prokaryotic cells; as its name implies, the cells of Eukarya are eukaryotic.

Media Activity
1.2 Experimental Design

Bacteria, Archaea, and the Protists Are Mostly Unicellular; Members of the Kingdoms Fungi, Plantae, and Animalia Are Primarily Multicellular

Most members of the domains Bacteria and Archaea and the protists from the domain Eukarya are single-celled, or **unicellular**, although a few live in strands or mats of cells with little communication, cooperation, or organization among them. Most members of the kingdoms Fungi, Plantae, and Animalia are many-celled, or **multicellular**; their lives depend on intimate communication and cooperation among many specialized cells.

Members of the Different Kingdoms Have Different Ways of Acquiring Energy

All organisms need energy to live. Photosynthetic organisms capture energy from sunlight and store it in molecules such as sugars and fats. These organisms, including plants, some bacteria, and some protists, are therefore called **autotrophs**, meaning "self-feeders." Organisms that cannot photosynthesize must acquire energy prepackaged in the molecules of the bodies of other organisms; hence, these organisms are called **heterotrophs**, meaning "other-feeders." Many archaea, bacteria, protists, and all fungi and animals are heterotrophs. Heterotrophs differ in the size of the food they eat. Some, such as bacteria and fungi, absorb individual food molecules; others, including most animals, eat whole chunks of food (*ingestion*) and break them down to molecules in their digestive tracts (*digestion*).

1.3 What Is the Science of Biology?

Biology utilizes the same principles and methods as other sciences. In fact, a basic tenet of modern biology is that living things obey the same laws of physics and chemistry that govern nonliving matter.

Scientific Principles Underlie All Scientific Inquiry

All scientific inquiry, including biology, is based on a small set of assumptions. Although these assumptions can never be proven absolutely, they have been so thoroughly tested and validated that we might call them scientific principles. These are the principles of *natural causality, uniformity in space and time*, and *common perception*.

Natural Causality Is the Principle That All Events Can Be Traced to Natural Causes

Over the course of human history, two approaches have been taken to the study of life and other natural phenomena. The first assumes that some events happen through the intervention of supernatural forces beyond our understanding. For example, the ancient Greeks believed that the god Zeus hurled thunderbolts from the sky and that the god Poseidon caused earthquakes and storms at sea. In contrast, science adheres to the principle of **natural causality**: all events can be traced to natural causes that are potentially within our ability to comprehend. For example, epilepsy was once thought to be the result of a visitation from the gods. Today, we realize that epilepsy is a disease of the brain in which groups of nerve cells are uncontrollably activated.

The principle of natural causality has an important corollary: the natural evidence we gather has not been deliberately distorted to fool us. This corollary may seem obvious, yet not so very long ago some people argued that fossils are not evidence of evolution; rather, they were placed on Earth by God to test our faith. If we cannot trust the evidence provided by nature, then the entire enterprise of science is futile.

The Natural Laws That Govern Events Apply Everywhere and for All Time

A second fundamental principle of science is that natural laws—laws derived from the study of nature—are uniform in space and time and do not change with distance or time. The laws of gravity, the behavior of light, and the interactions of atoms, for example, are the same today as they were a billion years ago, and hold true in Moscow as well as in New York, or even on Mars. Uniformity in space and time is especially vital to biology, because many important biological events, such as the evolution of today's diversity of living things, happened before humans were around to observe them. Some people believe that each of the different types of organisms was individually created at one time in the past by the direct intervention of God, a philosophy called *creationism.* Scientists freely admit that this idea cannot be disproved, but creationism is contrary to both natural causality and uniformity in time. The overwhelming success of science in explaining natural events through natural causes has led scientists to reject creationism.

Scientific Inquiry Is Based on the Assumption That People Perceive Natural Events in Similar Ways

A third basic assumption of science is that, generally, all human beings perceive natural events in fundamentally the same way, and that these perceptions provide us with reliable information about the natural world. Common perception is, to some extent, a principle peculiar to science. Value systems, such as those involved in the appreciation of art, poetry, and music, do not assume common perception. We may perceive the colors in a painting in a similar way (the scientific aspect of art), but we do not perceive the aesthetic value of the painting identically (the humanistic aspect of art). Values also differ radically among individuals, commonly owing to their culture or religious beliefs. Because value systems are subjective, not objective, science cannot solve certain types of philosophical or moral problems, such as the morality of abortion.

The Scientific Method Is the Basis for Scientific Inquiry

Given these assumptions, how do biologists study the workings of life? Scientific inquiry is a rigorous method for making observations of specific phenomena and searching for the order underlying those phenomena. Biology and other sciences commonly use the **scientific method**, which consists of four interrelated operations: *observation*; *hypothesis*; *experiment*; and *conclusion*. All scientific inquiry begins with an **observation** of a specific phenomenon. The observation, in turn, leads to questions, such as "How did this come about?" Then, in a flash of insight—or more typically after long, hard thought—a hypothesis is formulated. A **hypothesis** is a supposition, based on previous observations, that is offered as an explanation for the observed phenomenon. To be useful, the hypothesis must lead to predictions that can be tested by additional controlled observations, or **experiments**. These experiments produce results that either support or refute the hypothesis, and a **conclusion** is drawn about its validity. A single experiment is never an adequate basis for a conclusion; the results must be repeatable not only by the original researcher but also by others.

Simple experiments test the assertion that a single factor, or **variable**, is the cause of a single observation. To be scientifically valid, the experiment must rule out a variety of other possible variables as the cause of the observation. For this reason, scientists design **controls** into their experiments. Controls, in which all the variables not being tested remain constant, are then compared with the experimental situation, in which only the variable being tested is changed. In the 1600s, Francesco Redi used the scientific method to test the hypothesis that flies do not arise spontaneously from rotting meat (see "Scientific Inquiry: Does Spoiled Meat Produce Maggots?").

The scientific method can be used not only to generate new knowledge, but also to solve everyday problems. Let's consider an everyday situation in which you can apply the scientific method. Late for an appointment, you rush to your car and make the *observation* that it won't start. Immediately, you form a *hypothesis*: the battery is dead. Quickly, you design an *experiment*: you replace your battery with the battery from your roommate's new car and try to start your car again. The result seems to confirm your hypothesis, because your car starts immediately.

But wait! You haven't provided controls for several variables. Perhaps your battery was fine all along, and you just needed to try to start the car again. Or perhaps the battery cable was loose and simply needed to be tightened. Realizing the need for a good *control*, you replace your old battery, making sure the cables are secured tightly, and attempt to restart the car. If your car repeatedly refuses to start with the old battery but then starts immediately when you put in your roommate's new battery, you have isolated a single *variable*, the battery. Although you may have missed your appointment, you can now safely draw the *conclusion* that your old battery is dead.

The scientific method is powerful, but it is important to recognize its limitations. In particular, scientists can seldom be sure that they have controlled *all* the variables other than the one they are trying to study. Therefore, scientific conclusions must always remain tentative

SCIENTIFIC INQUIRY Does Spoiled Meat Produce Maggots?

The critical experiments of Italian physician Francesco Redi (1621–1697) beautifully demonstrate the scientific method and also help illustrate the principle of natural causality, on which modern science is based. Redi investigated why maggots appear on spoiled meat. Before Redi, the appearance of maggots was considered to be evidence of *spontaneous generation*, the production of living things from nonliving matter.

Redi *observed* that flies swarm around fresh meat and that maggots appear on meat left out for a few days. He formed a testable *hypothesis*: the flies produce the maggots. In his *experiment*, Redi wanted to test just one variable, the flies' access to the meat. He took two clean jars and filled them with similar pieces of meat. He left one jar open (the *control* jar) and covered the other with gauze to keep out flies (the *experimental* jar). He did his best to keep all the other variables the same (for example, the type of jar, the type of meat, and the temperature). After a few days, he observed that maggots swarmed over the meat in the open jar, but no maggots appeared on the meat in the covered jar. Redi *concluded* that his hypothesis was correct and that maggots are produced by flies, not by the meat itself (Fig. E1-1). Only through controlled experiments could the age-old hypothesis of spontaneous generation be laid to rest.

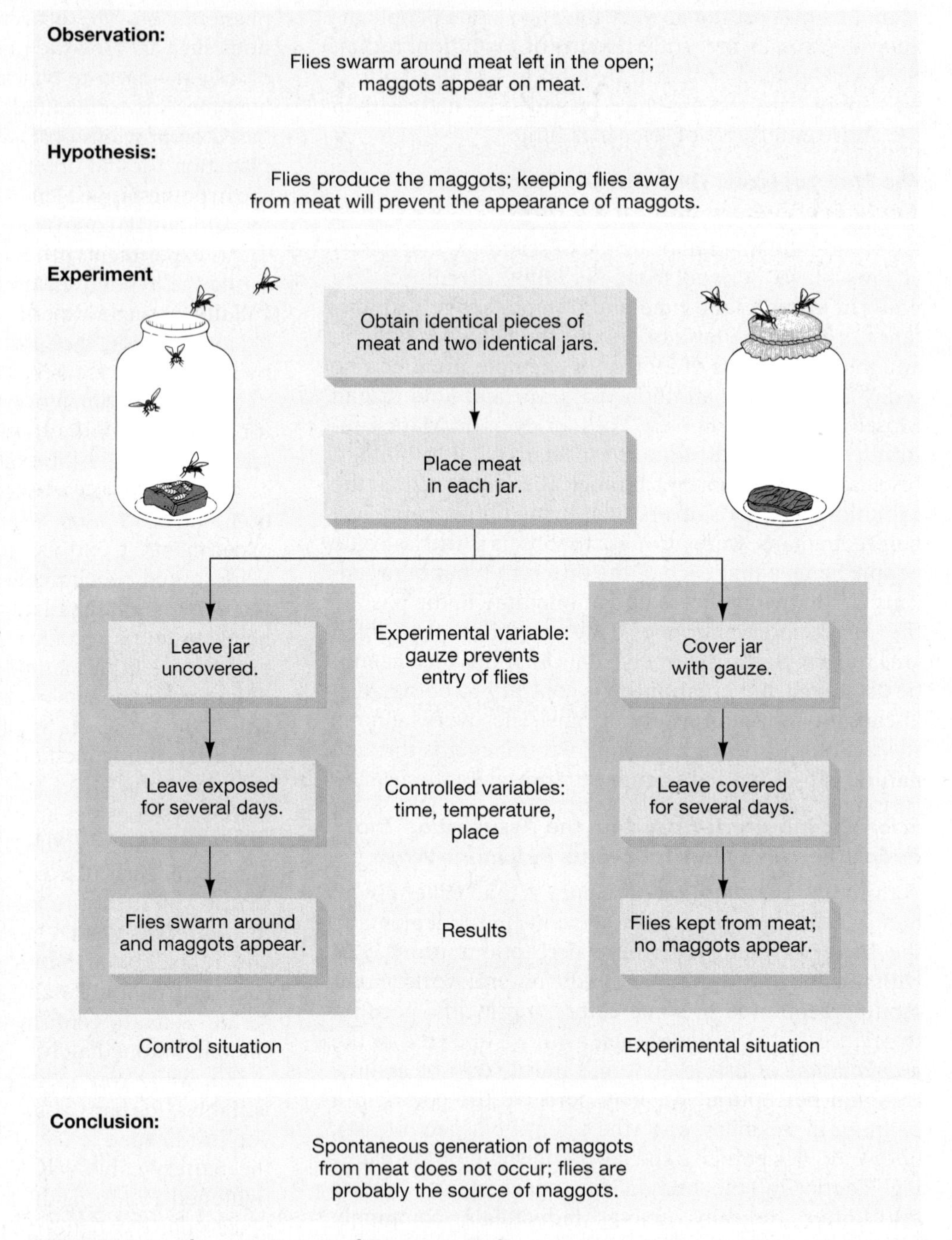
Observation:
Flies swarm around meat left in the open; maggots appear on meat.
Hypothesis:
Flies produce the maggots; keeping flies away from meat will prevent the appearance of maggots.
Experiment
Obtain identical pieces of meat and two identical jars.
Place meat in each jar.
Leave jar uncovered.
Experimental variable: gauze prevents entry of flies
Cover jar with gauze.
Leave exposed for several days.
Controlled variables: time, temperature, place
Leave covered for several days.
Flies swarm around and maggots appear.
Results
Flies kept from meat; no maggots appear.
Control situation
Experimental situation
Conclusion:
Spontaneous generation of maggots from meat does not occur; flies are probably the source of maggots.

FIGURE E1-1 The experiments of Francesco Redi
QUESTION Redi falsified spontaneous generation, but did his experiment conclusively demonstrate that flies cause maggots? What kind of follow-up experiment would be necessary to really determine the source of maggots?

and are subject to revision if new observations or experiments demand it.

A final important element of science is *communication*. No matter how well designed an experiment is, it is useless if it is not communicated thoroughly and accurately. Redi's experimental design and conclusions survive today because he carefully recorded his methods and observations. If experiments are not communicated to other scientists in enough detail to be repeated, their conclusions cannot be verified. Without verification, scientific findings cannot be safely used as the basis for new hypotheses and further experiments.

A wonderful aspect of scientific inquiry is that whenever a scientist reaches a conclusion, the conclusion immediately raises further questions that lead to further hypotheses and more experiments (why did your battery die?). Science is a never-ending quest for knowledge.

Science Is a Human Endeavor

Scientists are real people. They are driven by the same ambitions, pride, and fears as other people, and they sometimes make mistakes. As you will read in Chapter 9, ambition played an important role in the discovery of the structure of DNA by James Watson and Francis Crick. Accidents, lucky guesses, controversies with competing scientists, and, of course, the intellectual powers of individual scientists contribute greatly to scientific advances. To illustrate what we might call "real science," let's consider an actual case.

When they study bacteria, microbiologists must use pure *cultures*—that is, plates of bacteria that are free from contamination by other bacteria, molds, and so on. Only by studying a single type at a time can they learn about the properties of that particular bacterium. Consequently, at the first sign of contamination, a culture is usually thrown out, often with mutterings about sloppy technique. On one such occasion, however, in the late 1920s, Scottish bacteriologist Alexander Fleming turned a ruined culture into one of the greatest medical advances in history.

One of Fleming's bacterial cultures became contaminated with a patch of a mold called *Penicillium*. Before throwing out the culture dish, Fleming observed that *no bacteria were growing near the mold* (Fig. 1-9). Why not? Fleming hypothesized that perhaps *Penicillium* releases a substance that kills off bacteria growing nearby. To test

Penicillin diffuses outward from a penicillin-coated disk of paper, killing the nearby bacteria that coat this petri dish.

FIGURE 1-9 Penicillin kills bacteria
QUESTION Why do some molds produce substances that are toxic to bacteria?

this hypothesis, Fleming grew some pure *Penicillium* in a liquid nutrient broth. He then filtered out the *Penicillium* mold and applied the liquid in which the mold had grown to an uncontaminated bacterial culture. Sure enough, something in the liquid killed the bacteria. Further research into these mold extracts resulted in the production of the first *antibiotic*: penicillin, a bacteria-killing substance that has since saved millions of lives. Fleming's experiments are a classic example of the use of scientific methodology. They began with an observation and proceeded to a hypothesis, followed by experimental tests of the hypothesis that led to a conclusion. But the scientific method alone would have been useless without the lucky combination of accident and a brilliant scientific mind. Had Fleming been a "perfect" microbiologist, he would not have had any contaminated cultures. Had he been less observant, the contamination would have been just another spoiled culture dish. Instead, it was the beginning of antibiotic therapy for bacterial diseases. As French microbiologist Louis Pasteur said, "Chance favors the prepared mind."

Scientific Theories Have Been Thoroughly Tested

Scientists use the word *theory* in a way that is different from its everyday usage. If Dr. Watson were to ask Sherlock Holmes, "Do you have a theory as to the perpetrator of this foul deed?" in scientific terms, he would be asking Holmes for a hypothesis—an "educated guess" based on observable evidence, or clues. A **scientific theory** is far more general and more reliable than a hypothesis. Far from being an educated guess, a scientific theory is a general explanation of important natural phenomena, developed through extensive and reproducible observations. In common English, it is more like a *principle* or a *natural law*. For example, scientific theories such as the atomic theory (that all matter is composed of atoms) and the theory of gravitation (that objects exert attraction for one another) are fundamental to the science of physics. Likewise, the *cell theory* (that all living things are composed of cells) and the theory of evolution are fundamental to the study of biology. Scientists describe fundamental principles as "theories" because a basic premise of scientific inquiry is that it must be performed with an open mind. If compelling evidence arises, a theory will be modified.

A modern example of the need to keep an open mind in the light of new scientific evidence is the discovery of *prions*, which are infectious proteins (see Chapter 19). Before the early 1980s, all known infectious disease agents possessed genetic material—either DNA or the related molecule, RNA. When neurologist Stanley Prusiner from the University of California at San Francisco published evidence in 1982 that scrapie (an infectious disease that causes the brains of sheep to degenerate) is actually caused and transmitted by a protein with no genetic material, his results were met with widespread disbelief. Prions have since been found to cause "mad cow disease," which has killed not only cattle but also over 130 people who ate beef from infected cattle. Prior to the discovery of prions, the concept of an infectious protein was unknown to science. But by being willing to modify accepted beliefs to accommodate new data, scientists maintained the integrity of the scientific process while expanding their understanding of disease. For his pioneering work, Stanley Prusiner was awarded the Nobel Prize for Physiology or Medicine in 1997.

Scientific theories arise through *inductive reasoning*. **Inductive reasoning** is the process of creating a generalization as a result of making many observations that support it, and none that contradict it. Simplistically, the theory that Earth exerts gravitational forces on objects arose from repeated observations of objects falling down toward Earth and from a complete lack of observations of objects "falling up." Likewise, the cell theory arises from the observation that all organisms that have the attributes of life are composed of one or more cells, and that nothing that is not composed of cells shares all these attributes. Once a scientific theory has been formulated, it can be used to support *deductive reasoning*. In science, **deductive reasoning** is the process of generating hypotheses about how a specific experiment or observation will turn out, based on a well-supported generalization such as a scientific theory. For example, based on the cell theory, if a new organism is found that shares all the attributes of life, scientists can confidently deduce or hypothesize that it will be composed of cells. Of course, the new organism should be carefully scrutinized under the microscope to determine its cellular structure; if compelling new evidence arises, a theory can be modified.

1.4 Evolution: The Unifying Theory of Biology

One of the most important theories in biology is evolution. Ever since its formulation in the mid-1800s by two English naturalists, Charles Darwin and Alfred Russel Wallace, the theory of evolution has been supported by fossil finds, geological studies, radioactive dating of rocks, genetics, molecular biology, biochemistry, and breeding experiments. People who refer to evolution as "just a theory" profoundly misunderstand what scientists mean by the word *theory*.

Evolution is the unifying theory that explains the origin of diverse forms of life as a result of changes in their genetic makeup. As noted earlier, the theory of evolution states that modern organisms descended, with modification, from preexisting life-forms. In the words of biologist Theodosius Dobzhansky, "Nothing in biology makes sense, except in the light of evolution." Why don't snakes have legs? Why are there dinosaur fossils but no living dinosaurs? Why are monkeys so like us, not only in appearance, but also in the structure of their genes and proteins? The answers to those questions, and thousands more, lie in the processes of evolution (which we'll examine in detail in Chapters 14 through 17). Evolution is so vital to our understanding and appreciation of biology that we must review its important principles before going farther.

Three Natural Processes Underlie Evolution

In the mid-1800s, Darwin and Wallace formulated the theory of evolution that is still the basis of our modern understanding. Evolution arises as a consequence of three natural processes: *genetic variation* among members of a population, *inheritance* of those variations by offspring of parents who carry the variation, and *natural selection*, the survival and enhanced reproduction of organisms with favorable variations.

Much of the Variability Among Organisms Is Inherited

Look around at your classmates and notice how different they are. Although some of this variation is due to differences in environment and lifestyles, it is mainly influenced by our genes. For example, most of us could pump iron for the rest of our lives and never develop a body like that of "Mr. Universe." From where does genetic variation arise? The genetic instructions—the genes—of all organisms are segments of molecules of deoxyribonucleic acid, or DNA. Occasionally, the DNA suffers an accident; perhaps radiation strikes the DNA molecule, causing a mutation (a change in a gene), thereby altering its information content. Mutations occasionally occur as a result of mistakes in copying DNA during cellular reproduction. Many mutations have no effect or are harmless, while others make the organism less able to function. But in rare cases, mutations may improve an organism's ability to survive and reproduce. As a result of mutations, many of which occurred millions of years ago and have been passed from parent to offspring through countless generations, members of the same species tend to be slightly different from one another.

Natural Selection Tends to Preserve Genes That Help an Organism Survive and Reproduce

On average, organisms that best meet the challenges of their environment will leave the most offspring; these offspring will inherit the genes that made their parents successful. Thus, natural selection preserves genes that help organisms flourish in their environment. To create a hypothetical example, a mutated gene that causes beavers to grow larger teeth will allow the beavers with this mutation to chew down trees more efficiently, build bigger dams and lodges, and eat more bark than can "ordinary" beavers. Because these big-toothed beavers will obtain more food and better shelter, they will probably raise more offspring who will inherit their parents' genes for larger teeth. Over time, less-successful, smaller-toothed beavers will become increasingly scarce; after many generations, all beavers will have larger teeth.

Structures, physiological processes, or behaviors that aid in survival and reproduction in a particular environment are called **adaptations**. Most of the features that we admire so much in our fellow life-forms, such as the long limbs of deer, the wings of eagles, and the mighty trunks of redwood trees, are adaptations molded by millions of years of natural selection acting on random mutations.

In the long run, however, what helps an organism survive today can become a liability tomorrow. If environments change—for example, as ice ages come and go—then the genetic makeup that best adapts organisms to their environment will also change over time. When new mutations happen to occur that increase the fitness of an organism in the altered environment, these mutations in their turn will spread throughout the population.

Over millennia, the interplay of environment, genetic variation, and natural selection results in evolution: a change in the genetic makeup of species. In environments that are reasonably constant through time, such as the oceans, some well-adapted forms persist relatively unchanged and are sometimes called "living fossils." For example, sharks (Fig. 1-10) have retained essentially the same body form for tens of millions of years, because their sleek shape, powerful tail, acute sense of smell, and formidable teeth have made them superbly successful predators.

In changing environments, some species do not experience the chance genetic changes that allow them to adapt. The rate of environmental change outstrips the rate of genetic changes, and those species become *extinct*—that is, no members of the species remain. The dinosaurs (Fig. 1-11) were mighty reptiles that could not adapt to changing conditions 65 million years ago. Other species experience chance mutations that adapt them to meet new challenges. The mutation that produced the first stout, fleshy fins in what came to be known as lobefin fishes, for instance, enabled those fish to crawl on the bottom of shallow waters; over evolutionary time, this adaptation provided the basis for limbs that could crawl, run, jump, and even fly over dry land. The result of evolution is a tremendous variety of species. Within particular habitats, these species have evolved complex interrelationships with one another and with their nonliving surroundings. The diversity of species and the complex interrelationships that sustain them are encompassed by the term **biodiversity**. In recent decades, the rate of environmental change has been drastically accelerated by a single species, *Homo sapiens* (modern humans). Few wild species are able to adapt to this rapid change. In habitats most affected by humans, many species are being driven to extinction. This concept is explored further in "Earth Watch: Why Preserve Biodiversity?"

1.5 How Does Knowledge of Biology Illuminate Everyday Life?

Some people regard science as a "dehumanizing" activity, feeling that too deep an understanding of the world robs us of vision and awe. Nothing could be farther from the truth, as we repeatedly discover anew in our own lives. Years ago, we (Teresa and Gerald Audesirk, two of the authors of this text) watched a bee foraging at a spike of lupine flowers. Members of the pea family, lupines have a complicated structure, with two petals on the lower half of the flower enclosing the pollen-laden male reproductive structures (*stamens*) and sticky pollen-capturing female

FIGURE 1-10 Ancient adaptations
This tiger shark possesses features that have characterized sharks for tens of millions of years: streamlined shape; a long, powerful tail; an acute sense of smell; and rows of sharp teeth.

FIGURE 1-11 A fossil of *Triceratops*
This *Triceratops* died in what is now Montana about 70 million years ago. No one is certain what caused the extinction of the dinosaurs, but we do know that they were unable to evolve new adaptations to keep up with changes in their habitat.

EARTH WATCH Why Preserve Biodiversity?

"The loss of species is the folly our descendants are least likely to forgive us."

E. O. Wilson, Professor, Harvard University

What is biodiversity, and why should we be concerned with preserving it? *Biodiversity* refers to the total number of species within an ecosystem and to the resulting complexity of interactions among them; in short, it defines the "richness" of an ecological community.

Over the 3.5-billion-year history of life on Earth, evolution has produced an estimated 8 to 10 million unique and irreplaceable species. Of these, scientists have named only about 1.4 million, and only a tiny fraction of this number have been studied. Evolution has not, however, merely churned out millions of independent species. Over thousands of years, organisms in a given area have been molded by forces of natural selection exerted by other living species as well as by the nonliving environment in which they live. The outcome is the *community*, a highly complex web of interdependent life-forms whose interactions sustain one another. By participating in the natural cycling of water, oxygen, and other nutrients, and by producing rich soil and purifying wastes, these communities contribute to the sustenance of human life as well. The concept of biodiversity has emerged as a result of our increasing concern over the loss of countless forms of life and the habitats that sustain them.

The Tropics are home to the vast majority of all the species on Earth, perhaps 7 or 8 million of them, living in complex communities. The rapid destruction of habitats in the Tropics—from rain forests to coral reefs—as a result of human activities is producing high rates of extinction of many species (Fig. E1-2). Most of these species have never been named, and others never even discovered. Aside from ethical concerns over eradicating irreplaceable forms of life, as we drive unknown organisms to extinction, we lose potential sources of medicine, food, and raw materials for industry.

For example, a wild relative of corn that is not only very disease-resistant but also *perennial* (that is, lasts more than one growing season) was found growing only on a 25-acre plot of land in Mexico that was scheduled to be cut and burned within a week of the discovery. The genes of this plant might one day enhance the disease-resistance of corn or create a perennial corn plant. The rosy periwinkle, a flowering plant found in the tropical forest of the island of Madagascar (off the eastern coast of Africa) produces two substances that are now widely marketed for the treatment of leukemia and Hodgkin's disease, a cancer of the lymphatic organs. Only about 3% of the world's flowering plants have been examined for substances that might fight cancer or other diseases. Closer to home, loggers of the Pacific Northwest frequently cut and burned the Pacific yew tree as a "nuisance species" until the active ingredient that has since gone into making the anticancer drug Taxol® was discovered in its bark. Animals, too, have proven useful in fighting cancer: in 1997, researchers isolated a potent anticancer compound from a species of coral that dwells in the Indian Ocean.

FIGURE E1-2 Biodiversity threatened
Destruction of tropical rain forests by indiscriminate logging threatens Earth's greatest storehouse of biological diversity. Interrelationships such as those that have evolved between this *Heliconia* flower and its hummingbird pollinator, and this frog and the bromeliad on which it lives sustain these diverse communities and are threatened by human activities.

Many conservationists are also concerned that as species are eliminated, either locally or through total extinction, the communities of which they were a part might change, becoming less stable and more vulnerable to damage by diseases or adverse environmental conditions. Some experimental evidence supports this viewpoint, but the interactions within communities are so complex that these hypotheses are difficult to test. Clearly, some species have a much larger role than others in preserving the stability of a given ecosystem. Which species are most crucial in each ecosystem? No one knows. Human activities have increased the natural rate of extinction by a factor of at least 100 and possibly by as much as 1000 times the prehuman rate. By reducing biodiversity to support increasing numbers of humans and wasteful standards of living, we have ignorantly embarked on an uncontrolled global experiment, using planet Earth as our laboratory. In their book *Extinction* (1981), Stanford ecologists Paul and Anne Ehrlich compare the loss of biodiversity to the removal of rivets from the wing of an airplane. The rivet-removers continue to assume that there are far more rivets than needed, until one day, upon takeoff, they are proven tragically wrong. As human activities drive species to extinction while we have little knowledge of the role each plays in the complex web of life, we run the risk of removing "one rivet too many."

---

reproductive structures (*stigma*) within a tube-like structure. We had recently learned that in young lupine flowers (Fig. 1-12a), the weight of a bee pushing on these petals compresses the stamens, pushing pollen out of the tube and onto the bee's abdomen (Fig. 1-12b). In flowers that are ready to be fertilized, the stigma protrudes through the lower petals; when a pollen-dusted bee visits, it usually leaves behind a few grains of pollen.

Did our newfound insights into the functioning of lupine flowers detract from our appreciation of them?

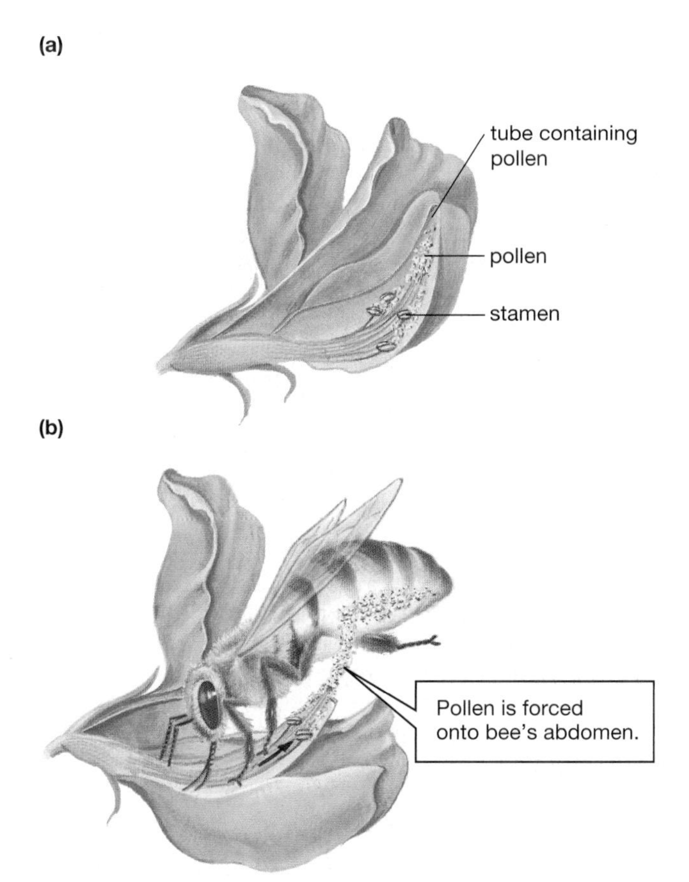

FIGURE 1-12 Complex adaptations help ensure pollination **(a)** In young lupine flowers, the lower petals form a tube enclosing the reproductive structures, including the male stamens which shed pollen within the tube. **(b)** The weight of a foraging bee compresses the tube, thrusting the reproductive structures forward, and forcing pollen out of the tube onto the bee's abdomen. Some pollen adheres to the abdomen and may come off on the sticky female stigma of the next flower that the bee visits, thus pollinating the flower.

FIGURE 1-13 Wild lupines and subalpine fir trees Thousands of people visit Hurricane Ridge in Washington State's Olympic National Park each summer to gaze in awe at Mt. Olympus, but few bother to investigate the wonders at their feet.

Far from it. Rather, we now looked on lupines with new delight, understanding something of the interplay of form and function, bee and flower, that shaped the evolution of the lupine. A few months later we ventured atop Hurricane Ridge in Olympic National Park in Washington State, where the alpine meadows burst with color in August (Fig. 1-13). As we crouched beside a wild lupine, an elderly man stopped to ask what we were looking at so intently. He listened with interest as we explained the structure to him; he then went off to another patch of lupines to watch the bees foraging. He too felt the increased sense of wonder that comes with understanding.

We try to convey to you that dual sense of understanding and wonder throughout this text. We also emphasize that biology is not a completed work but an exploration that has really just begun. As Lewis Thomas, a physician and natural philosopher, eloquently stated: "The only solid piece of scientific truth about which I feel totally confident is that we are profoundly ignorant about nature. Indeed, I regard this as the major discovery of the past hundred years of biology . . . but we are making a beginning."

We cannot urge you strongly enough, even if you are not contemplating a career in biology, to join in the journey of biological discoveries throughout your life. Don't think of biology as just another course to take, just another set of facts to memorize. Biology is a pathway to a new understanding of yourself and of the life on Earth around you.

CASE STUDY REVISITED Life on Earth—and Elsewhere?

Is there life on the moon? NASA was not taking any chances. When the *Apollo 11* astronauts who had spent 2.5 hours on the lunar surface splashed down in the ocean on July 24, 1969, a decontamination specialist met them and had them don biological isolation suits while still in the *Apollo 11* module. After the astronauts left the spacecraft, he sterilized the outsides of their isolation suits and the hatch of the spacecraft with disinfectant. The astronauts then stayed in a mobile decontamination unit aboard the recovery ship for four days, until they reached the Johnson Space Center in Houston, Texas. Here they remained in quarantine for an additional three weeks.

No foreign microorganisms were found on the astronauts or on the moon rocks they carried back with them. The only microbes found on the moon were discovered by *Apollo 12* astronauts in November 1969. Visiting the unmanned spacecraft *Surveyor 3* that had landed on the moon in 1967, they collected material from inside *Surveyor 3* in a sterile container. From this sample, scientists back on Earth cultured bacteria of the genus *Streptococcus*; ironically, this resident of the human mouth, nose, and throat may have been deposited by a NASA technician who sneezed as he assembled the spacecraft before it was launched. Normally residing in the warm, moist conditions inside the human body, these amazing microbes had, for two years, survived the vacuum of outer space and temperatures as low as −273 °C.

Astronomers estimate that there may be billions of Earth-like planets in the universe. Thus, the probability is very high that life has evolved elsewhere, although the likelihood of intelligent life is far less certain—and hotly debated. But as an intelligent species, we have hardly begun to understand the diversity, the complexity, and incredible versatility of life on our own home planet.

Consider This: In the late 1970s and 1980s, Dr. James Lovelock, a British chemist, published the controversial and provocative "Gaia hypothesis" (named after the Greek goddess who is said to have brought forth the living world from chaos). Lovelock suggested the living and nonliving components of Earth together constitute a super-organism—an enormous living thing. He noted that the interconnectedness among all forms of life and their environment and the way that living things modify their nonliving surroundings create a sort of homeostasis that helps maintain conditions conducive to life. Research Lovelock's Gaia hypothesis, either in the library or on the Internet, and discuss how the definition of *life* given in this chapter would need to be changed to accommodate his ideas. Do you believe that Gaia is a useful hypothesis? Is it testable? Should it be elevated to the status of a scientific theory? Explain.

Links to Life: The Life Around Us

The next time you walk across campus, look at the astonishing array of creatures thriving in a place as domesticated as a college campus. During the right seasons, you will undoubtedly pass beds of flowers and see honeybees or butterflies flitting among them, gathering the sweet nectar that powers their flight.

As you observe life, think about the "why" behind what you see. The plants' green color is due to a unique molecule, chlorophyll, that traps specific wavelengths of solar energy and uses them to power the life of the plant and synthesize the sugar in the nectar gathered by the bees and butterflies. Showy flowers evolved to entice insects to the energy-rich nectar. Why? If you look carefully at a bee, you may see yellow pollen clinging to its legs or to the hairs coating its body. The plants "use" the insects to fertilize each other, and both benefit. The sugar in nectar is assembled by chemical reactions that combine carbon dioxide and water, releasing oxygen as a waste product. So as you breathe out air rich in carbon dioxide, you are nourishing the plants with your "waste gas." Conversely, with each breath you take, you are inhaling the life-sustaining "waste gas" from the plants around you: oxygen. Wherever you look, if you look in the right way, you'll see evidence of the interdependence of living things, and you will never take life on Earth for granted.

Summary of Key Concepts

1.1 What Are the Characteristics of Living Things?

Organisms possess the following characteristics: their structure is complex and organized; they maintain homeostasis; they grow; they acquire energy and materials from the environment; they respond to stimuli; they reproduce; and they have the capacity to evolve.

1.2 How Do Scientists Categorize the Diversity of Life?

Organisms can be grouped into three major categories, called domains: Archaea, Bacteria, and Eukarya. Within the Eukarya are three kingdoms, Fungi, Plantae, and Animalia, and a group of unicellular eukaryotes known collectively as "protists." Features used to classify organisms include the type of cell(s) the organism possesses, the number of cells in each organism, and the mode of acquiring energy: *Cell Type*: The genetic material of eukaryotic cells is enclosed within a membrane-bound nucleus. Prokaryotic cells do not have a nucleus. *Cell Number*: Organisms may consist of a single cell (unicellular) or of many cells bound together and working cooperatively (multicellular). *Energy Acquisition*: Most autotrophic organisms obtain energy by capturing and storing the energy of sunlight in energy-rich molecules by means of photosynthesis. Heterotrophic organisms usually obtain energy by eating energy-rich molecules (food) synthesized in the bodies of other organisms. The food may be eaten in large chunks and broken down (ingestion) or may be absorbed molecule by molecule from the environment (absorption). The features of the domains and kingdoms are summarized in Table 1-1.

1.3 What Is the Science of Biology?
Biology is based on the scientific principles of natural causality, uniformity in space and time, and common perception. These principles are assumptions that cannot be directly proved but are validated by experience. Knowledge in biology is acquired through the application of the scientific method. First, an observation is made. Then a hypothesis is formulated that suggests a natural cause for the observation. The hypothesis is used to predict the outcome of further observations or experiments. A conclusion is then drawn about the validity of the hypothesis. A scientific theory is a general explanation of natural phenomena, developed through extensive and reproducible experiments and observations.

1.4 Evolution: The Unifying Theory of Biology
Evolution is the theory that modern organisms descended, with modification, from preexisting life-forms. Evolution occurs as a consequence of genetic variation among members of a population, caused by mutation, inheritance of those variations by offspring, and natural selection of the variations that best adapt an organism to its environment.

1.5 How Does Knowledge of Biology Illuminate Everyday Life?
The more you know about living things, the more fascinating they become!

Key Terms

adaptation *p. 13*
atom *p. 2*
autotroph *p. 8*
biodiversity *p. 13*
biosphere *p. 4*
cell *p. 4*
community *p. 4*
conclusion *p. 9*
control *p. 9*
cytoplasm *p. 4*
deductive reasoning *p. 12*
deoxyribonucleic acid (DNA) *p. 6*
domain *p. 6*
ecosystem *p. 4*
element *p. 2*
energy *p. 5*
eukaryotic *p. 8*
evolution *p. 6*
experiment *p. 9*
gene *p. 4*
heterotroph *p. 8*
homeostasis *p. 4*
hypothesis *p. 9*
inductive reasoning *p. 12*
kingdom *p. 6*
metabolism *p. 5*
molecule *p. 4*
multicellular *p. 8*
mutation *p. 6*
natural causality *p. 8*
natural selection *p. 6*
nucleus *p. 8*
nutrient *p. 5*
observation *p. 9*
organ *p. 4*
organelle *p. 4*
organic molecule *p. 4*
organism *p. 4*
organ system *p. 4*
photosynthesis *p. 5*
plasma membrane *p. 4*
population *p. 4*
prokaryotic *p. 8*
scientific method *p. 9*
scientific theory *p. 11*
species *p. 4*
subatomic particle *p. 4*
tissue *p. 4*
unicellular *p. 8*
variable *p. 9*

Thinking Through the Concepts

To take a multiple-choice quiz with feedback on the contents of this chapter, visit http://www.prenhall.com/audesirk7. *Log in to the Web site selected by your instructor and navigate to the Self Test section for this chapter.*

? Review Questions

1. What are the differences between a salt crystal and a tree? Which is living? How do you know? How would you test your knowledge? What controls would you use?
2. What is the difference between a scientific theory and a hypothesis? Explain how each is used by scientists. Why do scientists refer to basic principles as "theories," not "facts"?
3. Define and explain the terms *natural selection*, *evolution*, *mutation*, *creationism*, and *population*.
4. Starting with the cell, list the hierarchy of organization of life, briefly explaining each level.
5. Define *homeostasis*. Why must organisms continuously acquire energy and materials from the external environment to maintain homeostasis?
6. Describe the scientific method. In what ways do you use the scientific method in everyday life?
7. What is evolution? Briefly describe how evolution occurs.

Applying the Concepts

1. Review the properties of life, and then discuss whether humans are unique.
2. Design an experiment to test the effects of a new dog food, "Super Dog," on the thickness and water-shedding properties of the coats of golden retrievers. Include all the parts of a scientific experiment. Design objective methods to assess coat thickness and water-shedding ability.
3. Science is based on principles, including uniformity in space and time and common perception. Assume that humans encounter intelligent beings from a planet in another galaxy

who evolved under very different conditions. Discuss the two principles mentioned above and how they would affect the nature of scientific observations on the different planets and communications about these observations.

4. Identify two different types of organisms that you have seen interacting, for example, a caterpillar on a plant such as a milkweed, or a beetle in a flower. Now, form a single, simple hypothesis about this interaction. Use the scientific method and your imagination to design an experiment that tests this hypothesis. Be sure to identify variables and control for them.

For More Information

Attenborough, D. *Life on Earth*. Boston: Little, Brown, 1979. Gorgeously illustrated and beautifully written introduction to the diversity of life on Earth; the inspiration for our title.

Dawkins, R. *The Blind Watchmaker*. New York: W. W. Norton & Co., 1986. An engagingly written description of the process of evolution, which Dawkins compares to a blind watchmaker.

Ehrlich, P. R. *The Machinery of Nature*. New York: Simon & Schuster, 1986. Using layperson's terms, a foremost ecologist and author explains the science of ecology and the biological rationale for environmental concern.

Leopold, A. *A Sand County Almanac*. New York: Oxford University Press, 1949 (reprinted in 1989). A classic by a natural philosopher; provides an eloquent foundation for the conservation ethic.

Swain, R. B. *Earthly Pleasures*. New York: Charles Scribner's Sons, 1981. Insightful essays stress the interrelatedness and diversity of life.

Thomas, L. *The Medusa and the Snail*. New York: Bantam Books, 1980, and *The Lives of a Cell*, 1973. The late physician, researcher, and philosopher Lewis Thomas shares his awe of the living world in a series of delightful essays.

Wilson, E. O. *The Diversity of Life*. New York: W. W. Norton & Co., 1992. A celebration of the diversity of life, how it evolved, and how humans are impacting it. Wilson's writings have won two Pulitzer prizes.

Media Activities

To access a Media Activity visit http://www.prenhall.com/audesirk7. *Log in to the Web site selected by your instructor, navigate to this chapter, and select the appropriate Media Activity number.*

1.1 Defining Life

Estimated time: 15 minutes

Explore the characteristics of life and the way we use the scientific method to ask questions about life.

1.2 Experimental Design

Estimated time: 10 minutes

This activity will introduce you to the scientific method, which is at the center of all scientific studies.

1.3 Web Investigation: Life on Earth—and Elsewhere?

Estimated time: 10 minutes

The Internet can be a wonderful source of information for the novice scientist. Unfortunately, not all of the information on the World Wide Web is correct. In this exercise you will learn how to evaluate a Web site.

UNIT ONE

The Life of a Cell

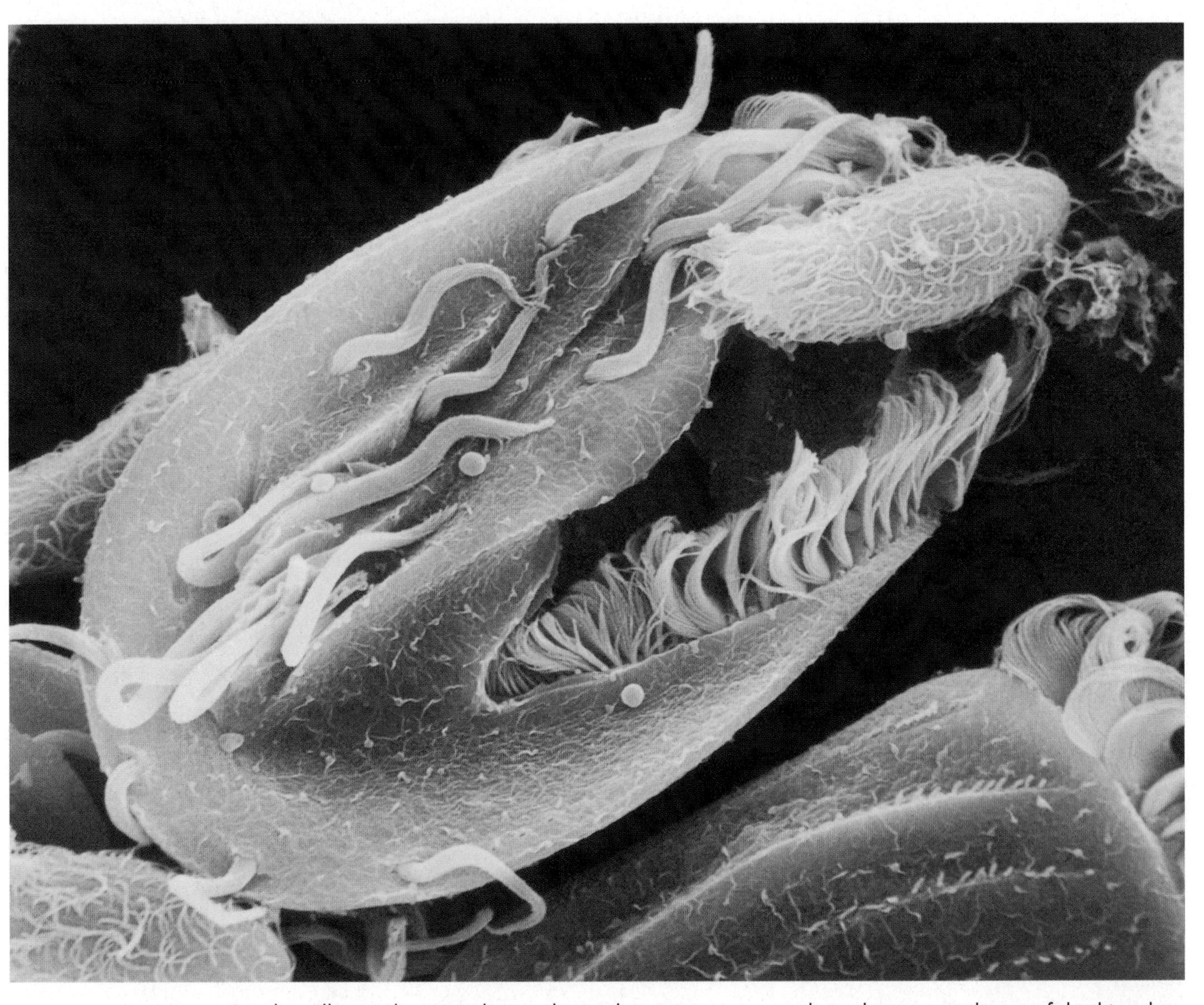

Single cells can be complex, independent organisms, such as these two ciliates of the kingdom *Protista*. A large *Euplotes* (about 300 μm in length) prepares to eat a much smaller *Paramecium*. Both are covered with cilia, short, beating, hairlike structures used to move and to ingest prey.

CHAPTER

2 Atoms, Molecules, and Life

The basilisk lizard and this child learning to ice skate have something in common: both are "walking on water."

AT A GLANCE

CASE STUDY Walking on Water

If you travel to the American tropics—from Mexico to Ecuador—you might discover a young basilisk lizard hiding among the leaves, searching for insects. Curious to learn more about this unusual animal, you might move too close to the basilisk and startle it. As it runs away, moving upright on its strong hind legs, the basilisk might encounter a small pond or a large puddle; instead of avoiding the water, it will stride across the water's still surface! A miracle? Hardly. Natural selection has endowed the basilisk with both speed and specialized feet that allow it to exploit one of several unique properties of water: surface tension. Water molecules tend to stick together. With care, you can float a paper clip in a bowl of water, but it will sink instantly in alcohol, which has far less surface tension than water.

Frozen water has unique properties that make ice skating both fun and feasible. First, ice is slippery; moving on it requires knowing how to use a whole different set of reflexes, a skill that this child is fortunate to be learning at a young age. Second, ice forms on the top of ponds and lakes, not on the bottom; in other words, it floats. Have you ever wondered why? Most other liquids freeze into denser solids. If the skating pond were filled with oil, the frozen oil would sink to the bottom. This child and the lizard are unknowingly exploiting different and unique properties of water in its liquid and solid phases.

Life almost certainly arose in water, and all the diverse molecules that make up living organisms function in a watery environment. But how are water molecules formed? How do water molecules interact with each other and with other forms of matter? What properties give water surface tension, and cause it to expand and become slippery when it freezes?

2.1 What Are Atoms?

Atoms, the Basic Structural Units of Matter, Are Composed of Still Smaller Particles

If you cut a diamond (a form of carbon) into pieces, each piece would still be carbon. If you could make finer and finer divisions of these pieces, you would eventually produce a pile of carbon atoms. **Atoms** are the fundamental structural units of matter. Atoms themselves, however, are composed of a central **atomic nucleus** (often called simply the *nucleus*; plural, *nuclei*—don't confuse it with the nucleus of a cell!). The nucleus contains two types of subatomic particles of equal weight: positively charged **protons** and uncharged **neutrons**. Other subatomic particles, called **electrons**, orbit the atomic nucleus (Fig. 2-1). Electrons are lighter, negatively charged particles. An atom has an equal number of electrons and protons and is therefore electrically neutral.

There are 92 types of atoms that occur naturally, each forming the structural unit of a different element. An **element** is a substance that can neither be broken down nor converted to other substances by ordinary chemical means. The number of protons in the nucleus—called the **atomic number**—is characteristic of each element. For example, every hydrogen atom has one proton in its nucleus; every carbon atom has six protons; and every oxygen atom has eight. Each element has unique chemical properties based on the number and configuration of its subatomic particles. Some elements, such as oxygen and hydrogen, are gases at room temperature; others, such as lead, are extremely dense solids. Most elements are quite rare, and relatively few are essential to life on Earth. Table 2-1 lists the most common elements in the

Media Activity
2.1 Interactive Atoms

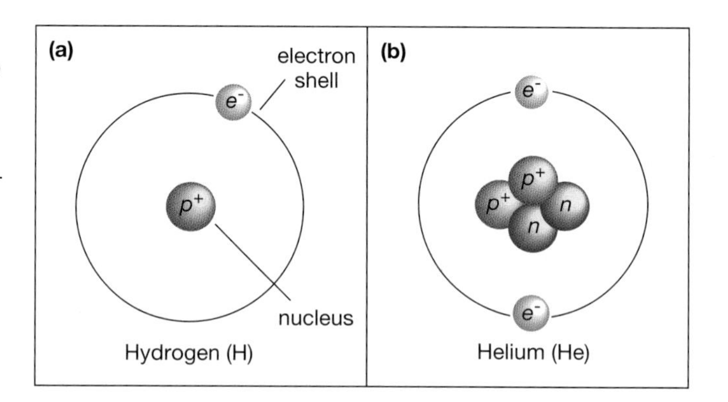

FIGURE 2-1 Atomic models Structural representations of the two smallest atoms, **(a)** hydrogen and **(b)** helium. In these simplified models, the electrons are represented as miniature planets, circling in specific orbits around a nucleus that contains protons and neutrons.

Table 2-1 Common Elements Important in Living Organisms

Element	Symbol	Atomic Number[a]	Percent in Universe[b]	Percent in Earth[b]	Percent in Human Body[b]
Hydrogen	H	1	91	0.14	9.5
Helium	He	2	9	Trace	Trace
Carbon	C	6	0.02	0.03	18.5
Nitrogen	N	7	0.04	Trace	3.3
Oxygen	O	8	0.06	47	65
Sodium	Na	11	Trace	2.8	0.2
Magnesium	Mg	12	Trace	2.1	0.1
Phosphorus	P	15	Trace	0.07	1
Sulfur	S	16	Trace	0.03	0.3
Chlorine	Cl	17	Trace	0.01	0.2
Potassium	K	19	Trace	2.6	0.4
Calcium	Ca	20	Trace	3.6	1.5
Iron	Fe	26	Trace	5	Trace

[a]Atomic number = number of protons in the atomic nucleus.

[b]Approximate percentage of atoms of this element, by weight, in the universe, in Earth's crust, and in the human body.

SCIENTIFIC INQUIRY Radioactivity in Research

How do biologists know that DNA is the genetic material of cells (Chapter 9)? How do paleontologists measure the ages of fossils (Chapter 17)? How do botanists know how plants transport sugars made in their leaves during photosynthesis to other parts of the plant (Chapter 24)? These discoveries, and many more, have been possible only through the use of radioactive isotopes. During *radioactive decay*, the process by which a radioactive isotope spontaneously breaks apart, an isotope emits particles that can be detected with devices such as Geiger counters.

A particularly fascinating and medically important use of radioactive isotopes is *positron emission tomography*, or *PET scans* (Fig. E2-1). In one common application of PET scans, a subject is given a sugar, glucose, that has been labeled with (attached to) a harmless radioactive isotope of fluorine. As the isotope decays, it emits two bursts of energy that travel in opposite directions. Detectors in a ring around the subject's head capture the emissions in that particular plane, recording the nearly simultaneous arrival times of the two energy bursts from each decaying particle. A powerful computer then calculates the location within the brain where the decay occurred and generates a color-coded map of the frequency of decays within a given "slice" of the brain. The more active a brain region is, the more glucose it uses as an energy source, and the more radioactivity is concentrated there. For example, tumor cells divide rapidly and use large amounts of glucose; thus they show up in PET scans as "hot spots" (Fig. E2-1c). Normal brain regions activated by a specific mental task (for example, a math problem) will also have higher glucose demands that can be detected by PET scans. Thus, physicians can use PET to locate brain problems, while researchers can use it to identify the locations of normal brain functions.

The development of PET scans required close cooperation among biologists and physicians (who recognized the need for brain scanning and can interpret the data), chemists (who developed and synthesized the radioactive probes), physicists (who interpreted the nature of isotopes and their energy emissions), and engineers (who designed and built the computers and other electronic components). Continued teamwork among scientists from different fields promises further advances in the fundamental understanding of biological processes as well as more practical applications such as the PET scanner.

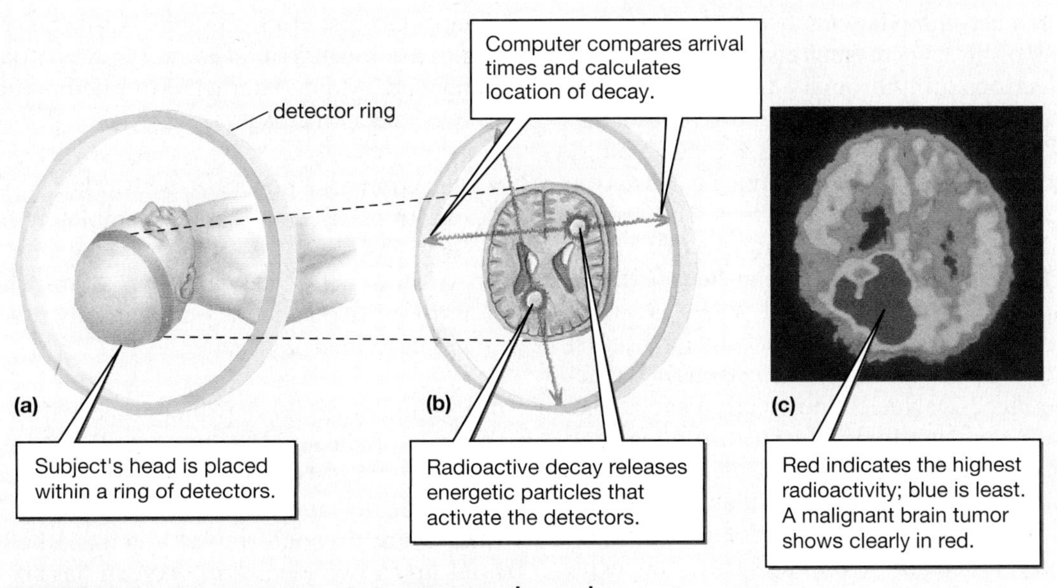
Computer compares arrival times and calculates location of decay.
detector ring
(a)
(b)
(c)
Subject's head is placed within a ring of detectors.
Radioactive decay releases energetic particles that activate the detectors.
Red indicates the highest radioactivity; blue is least. A malignant brain tumor shows clearly in red.

FIGURE E2-1 How positron emission tomography works

universe, on Earth, and in the human body. Notice how differently these elements are distributed.

Atoms of the same element may have different numbers of neutrons. When this occurs, the atoms are called **isotopes** of each other. Some, but not all, isotopes are **radioactive**; that is, they spontaneously break apart, forming different atoms and releasing energy in the process. Radioactive isotopes are extremely useful tools for studying biological processes (see "Scientific Inquiry: Radioactivity in Research").

Electrons Orbit the Nucleus at Fixed Distances, Forming Electron Shells That Correspond to Different Energy Levels

As you may know from experimenting with magnets, like poles repel each other and opposite poles attract each other. In a similar way, electrons, which are negatively charged, repel one another but are attracted to the positively charged protons of the nucleus. However, because of their mutual repulsion, only limited numbers of electrons can occupy the space closest to the nucleus.

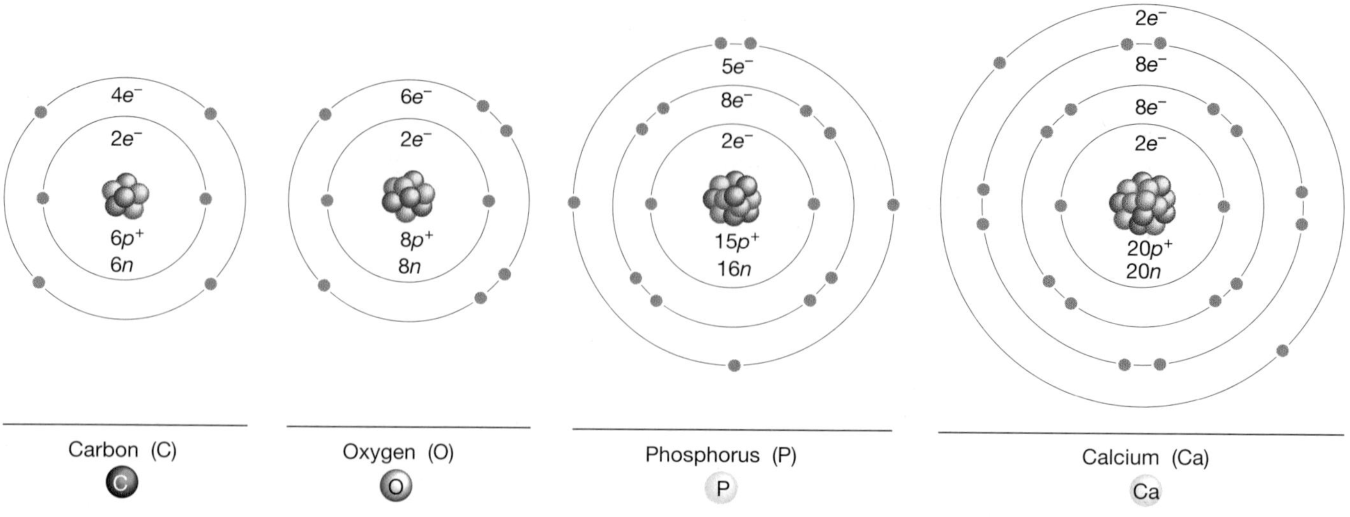

FIGURE 2-2 Electron shells in atoms
Most biologically important atoms have at least two shells of electrons. The first shell, closest to the nucleus, can hold two electrons; the next shell holds a maximum of eight electrons. More-distant shells can also hold eight electrons each. **QUESTION** Why do biologically active atoms have outer shells that are not full?

Large atoms can accommodate many electrons because their electrons orbit at increasing distances from their nuclei. The electrons orbit through a three-dimensional space; the orbits, which correspond to different energy levels, are called **electron shells** (see Fig. 2-1; Fig. 2-2).

The electron shell closest to the atomic nucleus is the smallest and can hold only two electrons. The second shell can hold up to eight electrons. The electrons in an atom usually fill the shell closest to the nucleus first, and then begin to occupy the next shell. Thus, a carbon atom, with six electrons, has two electrons in the first shell, closest to the nucleus, and four electrons in its second shell (see Fig. 2-2). Nuclei and electron shells play complementary roles in atoms. Nuclei (assuming they are not radioactive) provide stability, while the electron shells allow interactions, or *bonds*, with other atoms. Nuclei resist disturbance by outside forces. Ordinary sources of energy, such as heat, electricity, and light, hardly affect them at all. Because its nucleus is stable, a carbon atom remains carbon whether it is part of a diamond, carbon dioxide, or sugar. Electron shells, however, are dynamic; as you will soon see, atoms bond with one another by gaining, losing, or sharing electrons.

2.2 How Do Atoms Interact to Form Molecules?

Atoms Interact with Other Atoms When There Are Vacancies in Their Outermost Electron Shells

A **molecule** consists of two or more atoms of the same or different elements, held together by interactions among their outermost electron shells. A substance whose molecules are formed of different types of atoms is called a **compound**. Atoms interact with one another according to two basic principles:

- An atom will not react with other atoms when its outermost electron shell is completely full or empty. Such an atom is described as being *inert*.
- An atom will react with other atoms when its outermost electron shell is only partially full. Such atoms are described as *reactive*.

To demonstrate these principles, consider three types of atoms: hydrogen, helium (see Fig. 2-1), and oxygen (see Fig. 2-2). Hydrogen (the smallest atom) has one proton in its nucleus and one electron in its single (and therefore outermost) electron shell, which can hold up to two electrons. Oxygen has six electrons in its outer shell, which can hold eight. In contrast, helium has two protons in its nucleus, and two electrons fill its single electron shell. Therefore, we can predict that hydrogen and oxygen atoms, with partially empty outer shells, should be reactive; helium atoms, with full shells, should be stable. We might further predict that hydrogen and oxygen atoms could gain stability by reacting with each other. The single electrons from each of two hydrogen atoms would fill the outer shell of an oxygen atom, forming H_2O—otherwise known as water (see Fig. 2-4b). As we predicted, hydrogen *can* react readily with oxygen; in fact, the reaction is an explosive one. The space shuttle and other rockets use liquid hydrogen as fuel to power lift-off. The hydrogen fuel reacts with oxygen, releasing water as a by-product. In contrast, helium, with its full outer shell, is almost completely inert.

An atom with an outermost electron shell that is partially full can gain stability by losing electrons (emptying the shell completely), gaining electrons (filling the shell), or sharing electrons with another atom (allowing both atoms to behave as though they had full outer shells). The results of losing, gaining, and sharing electrons are **chemical bonds**, or *attractive forces* that hold atoms together in molecules. Each element has chemical bonding properties that arise from the configuration of electrons in its outer shell. **Chemical reactions**, the making and breaking of chemical bonds to form new substances, are essential for the maintenance of life and for the working of modern society. Whether they occur in a plant cell as it captures solar energy, your brain as it forms new memories, or your car's engine as it guzzles gas, chemical reactions consist of making new chemical bonds and/or breaking existing ones. There are three major types of chemical bonds: ionic bonds, covalent bonds, and hydrogen bonds.

Charged Atoms Called *Ions* Interact to Form Ionic Bonds

Both atoms that have an almost empty outermost electron shell and atoms that have an almost full outermost shell can become stable by losing electrons (emptying their outermost shell) or by gaining electrons (filling their outermost shell). The formation of table salt (sodium chloride) demonstrates this principle. Sodium (Na) has only one electron in its outermost electron shell, and chlorine (Cl) has seven electrons in its outer shell—one electron short of a full shell (Fig. 2-3a). Sodium, therefore, can become stable by losing the electron from its outer shell to chlorine, leaving that shell empty; chlorine then fills its outer shell by gaining the electron. Atoms that have lost or gained electrons, altering the balance between protons and electrons, are *charged*. These charged atoms are called **ions**. To form sodium chloride, sodium loses an electron and thereby becomes a positively charged sodium ion ($Na+$); chlorine picks up an electron and becomes a negatively charged chloride ion ($Cl-$). The two ions are held together by **ionic bonds**: the electrical attraction between positively and negatively charged ions (Fig. 2-3b). The ionic bonds between sodium and chloride ions form crystals containing a repeating, orderly arrangement of the two ions; we call this substance "table salt" (Fig. 2-3c). Ionic bonds are not unbreakable. As we will see later, water can easily break ionic bonds.

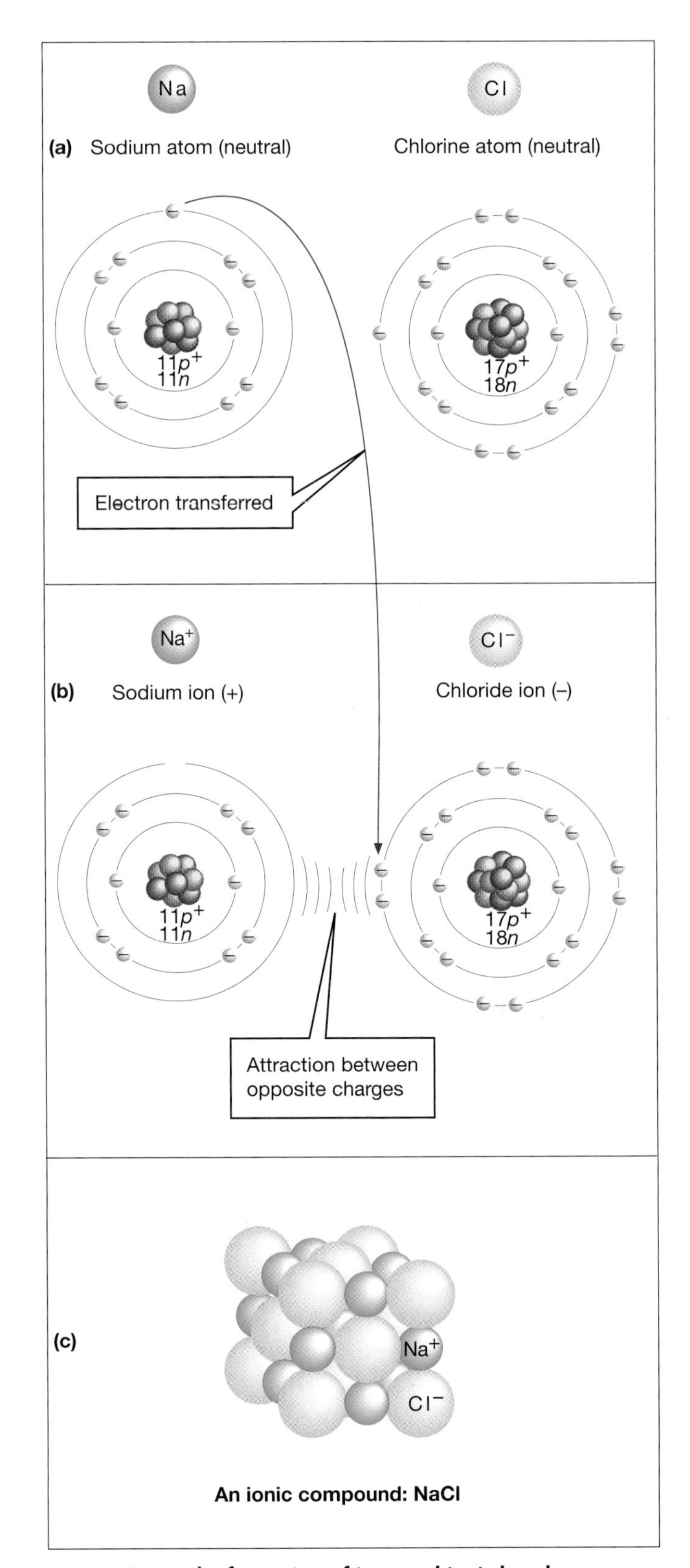

FIGURE 2-3 The formation of ions and ionic bonds
(a) Sodium has only one electron in its outer electron shell; chlorine has seven. **(b)** Sodium can become stable by losing an electron, and chlorine can become stable by gaining an electron. Sodium becomes a positively charged ion and chlorine a negatively charged ion. **(c)** Because oppositely charged particles attract one another, the resulting sodium ions (Na^+) and chloride ions (Cl^-) nestle closely together in a crystal of salt, NaCl.

Uncharged Atoms Can Become Stable by Sharing Electrons, Forming Covalent Bonds

An atom with a partially full outermost electron shell can also become stable by sharing electrons with another atom, forming a **covalent bond**. Consider the hydrogen atom, which has one electron in a shell that can hold two. A hydrogen atom can become reasonably stable if it shares its single electron with another hydrogen atom, forming a molecule of hydrogen gas, H_2 (Fig. 2-4a). Because the two hydrogen atoms are identical, neither nucleus can exert more attraction and capture the other's electron. So the two electrons orbit around both nuclei for equal amounts of time, forming a single covalent bond; each hydrogen atom behaves almost as if it had two electrons in its shell. Two oxygen atoms also share electrons equally, producing a molecule of oxygen gas, O_2, with a double covalent bond (see Fig. 2-4a). In such a bond, each atom contributes two electrons. If atoms share three pairs of electrons, a triple covalent bond is formed, as in nitrogen gas, N_2.

All covalent bonds are stronger than ionic bonds, but some are stronger than others, depending on the atoms involved (Table 2-2). Some covalent bonds, such as

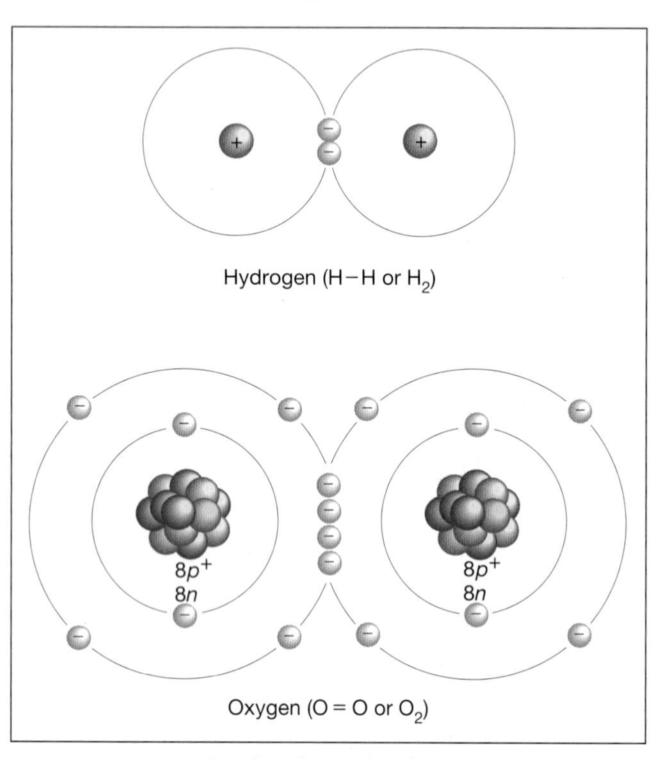

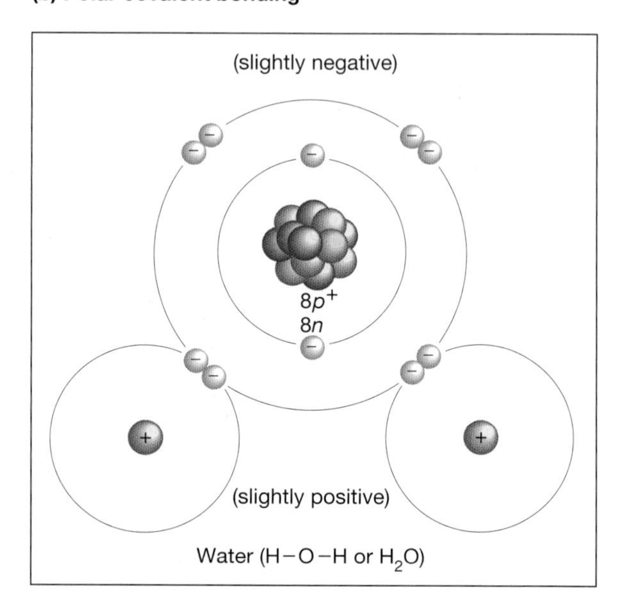

FIGURE 2-4 Covalent bonds involve shared electrons
(a) In hydrogen gas (top), an electron from each hydrogen atom is shared, forming a single nonpolar covalent bond. In oxygen gas (bottom), two oxygen atoms share four electrons, forming a double nonpolar covalent bond. **(b)** Oxygen lacks two electrons to fill its outer shell, so oxygen can form polar covalent bonds with two hydrogen atoms, creating water. Oxygen exerts a greater pull on the electrons than does hydrogen, so the "oxygen end" of the molecule has a slight negative charge and the "hydrogen end" has a slight positive charge. **QUESTION** In water's polar bonds, why is oxygen's pull on electrons greater than hydrogen's?

Table 2-2 Chemical Bonds

Type of Bond	Bond Forms:
Weak Bonds: allow interactions between individual atoms or molecules	
Ionic bonds	Between positive and negative ions
Hydrogen bonds	Between a hydrogen atom involved in a polar covalent bond and another atom involved in a polar covalent bond
Hydrophobic interactions	Because interactions between water molecules exclude hydrophobic molecules
Strong Bonds: hold atoms together within molecules	
Covalent bonds	By the sharing of electron pairs; equal sharing produces nonpolar covalent bonds; unequal sharing produces polar covalent bonds

those in water (H_2O; Fig. 2-4b) and carbon dioxide (CO_2), are extremely stable—that is, it takes a lot of energy to break the bonds. Other bonds, such as those in oxygen gas (see Fig. 2-4a) or gasoline, are less stable and come apart more easily. When a chemical reaction occurs in which less stable bonds are broken and more stable bonds are formed (such as burning gasoline with oxygen to form carbon dioxide and water), energy is released, as we will discuss in Chapter 6.

Most Biological Molecules Utilize Covalent Bonding

Covalent bonds are crucial to life on Earth. Since biological molecules must function in a watery environment in which ionic bonds rapidly dissociate, the atoms in most biological molecules—such as those found in proteins, sugars, and cellulose—are joined by covalent bonds. Hydrogen, carbon, oxygen, nitrogen, phosphorus, and sulfur are the most common atoms found in biological molecules. Hydrogen can form a covalent bond with one other atom; oxygen and sulfur with two other atoms; nitrogen with three; and phosphorus and carbon with up to four (Table 2-3). (Phosphorus is unusual; although it has only three spaces in its outer shell, it can form up to five covalent bonds with up to four other atoms.) This diversity of bonding arrangements allows the construction of biological molecules that have enormous variety and complexity.

Electron Sharing Determines Whether a Covalent Bond Is Nonpolar or Polar

In hydrogen gas, the two nuclei are identical and the shared electrons spend equal time near each nucleus. Therefore, not only is the molecule as a whole electrically neutral, but each end, or *pole*, of the molecule is also electrically neutral. Such an electrically symmetrical bond is called a **nonpolar covalent bond**, and the compound formed with such a nonpolar bond is a *nonpolar molecule*; examples include hydrogen (H_2) and oxygen (O_2) (see Fig. 2-4a). But electron sharing in covalent bonds is not always equal. In many molecules, one nucleus has a larger positive charge than the other, and therefore attracts the electrons more strongly. This situation produces a **polar covalent bond**. Although the molecule as a whole is electrically neutral, it has charged parts: the atom that attracts the electrons more strongly has a slightly negative charge (the negative pole of the molecule), and the other atom has a slightly positive charge (the positive pole). In water, for example, oxygen attracts electrons more strongly than does hydrogen, so the oxygen end of a water molecule is negative and each hydrogen is positive (see Fig. 2-4b). Water, with its charged ends, is an example of a *polar molecule*.

Free Radicals Are Highly Reactive and Can Damage Cells

Some reactions, particularly those that occur in cells as they process energy, give rise to molecules that have atoms (often oxygen atoms) with one or more unpaired electrons in their outer shells. This type of molecule, called a **free radical**, is very unstable and reacts readily with nearby molecules, capturing an electron to complete its outer shell. But by stealing an electron from the molecule it attacks, it creates a new free radical and begins a chain reaction that can lead to the destruction of biological molecules crucial to life. Cell death caused by free radicals contributes to a variety of human ailments, including heart disease and nervous system disorders such as Alzheimer's disease. By damaging genetic material, free radicals may also cause some forms of cancer. The gradual deterioration that accompanies aging is believed by many scientists to result (at least in part) from free radical damage that accumulates over a lifetime of exposure. Materials from the environment, including radiation (such as from the sun

Table 2-3 Bonding Patterns of Atoms Commonly Found in Biological Molecules

Atom	Capacity of Outer Electron Shell	Electrons in Outer Shell	Number of Covalent Bonds Usually Formed	Common Bonding Patterns
Hydrogen	2	1	1	—H
Carbon	8	4	4	—C— (four single bonds); —C= (two single, one double); =C= ; —C≡
Nitrogen	8	5	3	—N— (three single bonds); —N= ; N≡
Oxygen	8	6	2	—O— ; O=
Phosphorus	8	5	5	—P= (three single, one double)
Sulfur	8	6	2	—S—

and X-rays), chemicals in automobile exhaust, and industrial metals (such as mercury and lead) can also enter our bodies and generate free radicals. Fortunately, some molecules, called **antioxidants**, react with free radicals and render them harmless. Our bodies synthesize several antioxidants, and others can be obtained from a healthy diet. Vitamins E and C are antioxidants, as are a variety of substances found in fruits and vegetables. To learn about other sources of antioxidants, see "Links to Life: Health Food?"

Media Activity 2.2 Water and Life

Hydrogen Bonds Are Weaker Electrical Attractions Between or Within Molecules with Polar Covalent Bonds

Because of the polar nature of their covalent bonds, nearby water molecules attract one another. The partially negatively charged oxygens of water molecules attract the partially positively charged hydrogens of nearby water molecules. This electrical attraction is called a **hydrogen bond** (Fig. 2-5). As we will see shortly, hydrogen bonds between molecules give water several unusual properties that are essential to life on Earth.

Hydrogen bonds are also common and important in biological molecules. They occur whenever polar covalent bonds produce slightly negative and slightly positive charges that attract one another. Both nitrogen and oxygen atoms attract electrons more strongly than do hydrogen atoms. Therefore, the nitrogen or oxygen pole of a nitrogen–hydrogen or oxygen–hydrogen bond is slightly negative, and the hydrogen pole is slightly positive. The resulting polar parts of the molecules can form hydrogen bonds with water, with other biological molecules, or with polar parts of the same molecule. Although individual hydrogen bonds are quite weak, many of them working together are quite strong. As we will see in Chapter 3, hydrogen bonds play crucial roles in shaping the three-dimensional structures of proteins. In Chapter 9, you'll discover their importance in DNA.

2.3 Why Is Water So Important to Life?

As naturalist Loren Eiseley eloquently stated, "If there is magic on this planet, it is contained in water." Water is extraordinarily abundant on Earth, has unusual properties, and is so essential to life that it merits special consideration. Life is very likely to have arisen in the waters of the primeval Earth. Living organisms still contain about 60% to 90% water, and all life depends intimately on its properties. Why is water so crucial to life?

Water Interacts with Many Other Molecules

Water is involved in many of the chemical reactions that occur in living cells. The oxygen that green plants release into the air is derived from water during photosynthesis. When manufacturing a protein, fat, nucleic acid, or sugar, your body produces water; conversely, when you digest proteins, fats, and sugars in the foods you eat, water is used in the reactions. But why is water so important in biological chemical reactions?

Water is an extremely good **solvent**—that is, it is capable of dissolving a wide range of substances, including protein, salts, and sugars. Water or other solvents containing dissolved substances are called *solutions*. Recall that a crystal of table salt is held together by the electrical at-

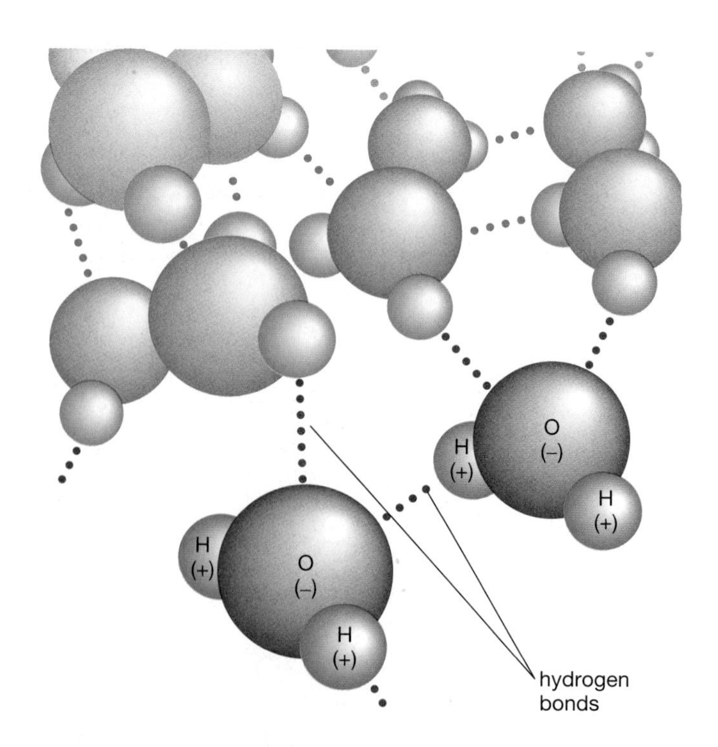

FIGURE 2-5 Hydrogen bonds The partial charges on different parts of water molecules produce weak attractive forces called *hydrogen bonds* (dotted lines) between the oxygen and hydrogen atoms in adjacent water molecules.

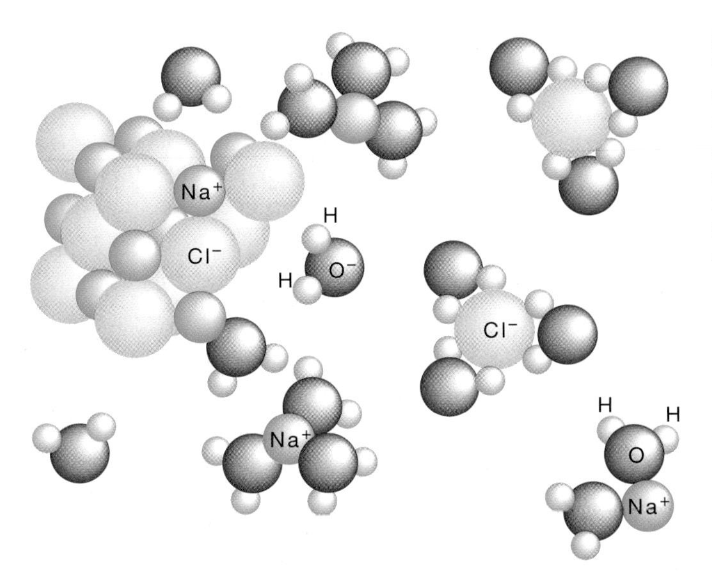

FIGURE 2-6 Water as a solvent
When a salt crystal is dropped into water, the water surrounds the sodium and chloride ions with oppositely charged poles of its molecules. Thus insulated from the attractiveness of other molecules of salt, the ions disperse, and the whole crystal gradually dissolves.

traction between positively charged sodium ions and negatively charged chloride ions (see Fig. 2-3c). Because water is a polar molecule, it has positive and negative poles. If a salt crystal is dropped into water, the positively charged hydrogen ends of the water molecules will be attracted to and will surround the negatively charged chloride ions, and the negatively charged oxygen poles of water molecules will surround the positively charged sodium ions. As water molecules enclose the sodium and chloride ions and shield them from interacting with each other, the ions separate from the crystal and drift away in the water—thus, the salt dissolves (Fig. 2-6).

Water also dissolves molecules held together by polar covalent bonds. Its positive and negative poles are attracted to oppositely charged regions of dissolving molecules. Ions and polar molecules are termed **hydrophilic** (Greek for "water-loving") because of their electrical attraction for water molecules. Many biological molecules, including sugars and amino acids, are hydrophilic and dissolve readily in water (Fig. 2-7). Water also dissolves gases such as oxygen and carbon dioxide. The fish swimming below the ice on a frozen lake rely on oxygen that dissolved before the ice formed, and they release CO_2 into solution in the water. By dissolving such a wide variety of molecules, the watery substance inside a cell provides a suitable environment for the countless chemical reactions essential to life.

Molecules that are uncharged and nonpolar, such as fats and oils, usually do not dissolve in water and hence are called **hydrophobic** ("water-fearing"). Nevertheless, water has an important effect on such molecules. Oils, for example, form globules when spilled into water. Oil molecules in water disrupt the hydrogen bonding among adjacent water molecules. When oil molecules encounter one another in water, their nonpolar surfaces nestle closely together, surrounded by water molecules that form hydrogen bonds with one another but not with the oil. To separate again, the oil molecules would have to break apart the hydrogen bonds that link surrounding water molecules. Thus, the oil molecules remain together, forming a glistening droplet that floats on the water's surface. The tendency of oil molecules to clump together is

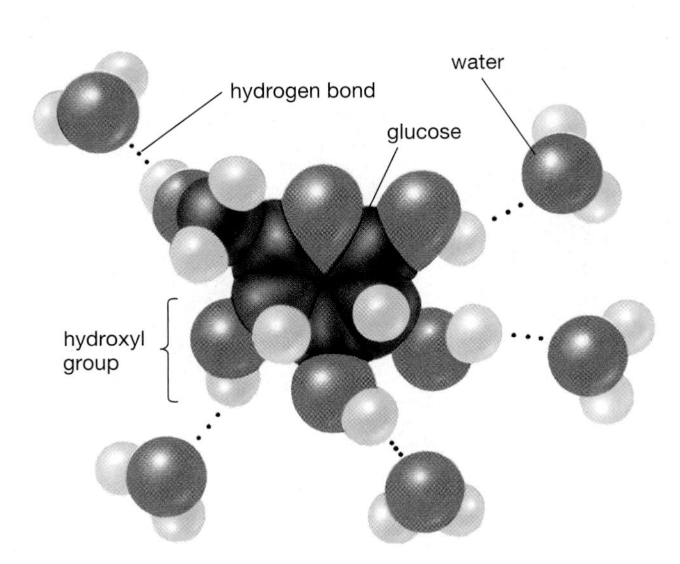

FIGURE 2-7 Water dissolves many biological molecules
Many biological molecules dissolve in water because they have polar parts—for example, OH^- (hydroxyl) groups—that can form hydrogen bonds with water molecules. Here, sugar dissolves because hydrogen bonds form between the hydroxyl groups on a glucose molecule (a simple sugar) and surrounding water molecules.
QUESTION Why is it important to human physiology that sugars dissolve easily in water?

described as a **hydrophobic interaction**. As we will discuss in Chapter 4, the membranes of living cells owe much of their structure to hydrophobic interactions.

Water Molecules Tend to Stick Together

In addition to interacting with other molecules, water molecules interact with each other. Because hydrogen bonds interconnect individual water molecules, liquid water has high **cohesion**—that is, water molecules have a tendency to stick together. Cohesion among water molecules at the water's surface produces **surface tension**, the tendency for the water surface to resist being broken. If you've ever experienced the slap and sting of a belly flop into a swimming pool, you've discovered firsthand the power of surface tension. Surface tension can support fallen leaves, some spiders and water insects (Fig. 2-8a), and even a running basilisk lizard.

A crucial role of cohesion in water occurs in the life of land plants. Since a plant absorbs water through its roots, how does the water reach the aboveground parts, especially if the plant is a 100-meter-tall redwood (Fig. 2-8b)? As we will see in Chapter 24, water molecules are pulled up by the leaves, filling tiny tubes that connect the leaves, stem, and roots. Water molecules that evaporate from the leaves pull water up the tubes, much like a chain being pulled up from the top. The system works because the hydrogen bonds among water molecules are stronger than the weight of the water in the tubes (even 100 meters' worth); thus, the water "chain" doesn't break. Without the cohesion of water, there would be no land plants as we know them, and terrestrial life would undoubtedly have evolved quite differently. The sting of a belly flop, the ability of a lizard to run on water, and the movement of water up a tree are all made possible by the hydrogen bonds between water molecules.

Water exhibits another property, *adhesion*, or a tendency to stick to polar surfaces having slight charges that attract polar water molecules. Adhesion helps water move within small spaces, such as the thin tubes in plants that carry water from roots to leaves. If you stick the end of a narrow glass tube into water, the water will move a short distance up the tube. Put some water in a narrow glass bud vase or test tube and you'll see that the upper surface is curved; water pulls itself up the sides of the glass by its adhesion to the surface of the glass and by the cohesion among water molecules.

Water-Based Solutions Can Be Acidic, Basic, or Neutral

Although water is generally regarded as a stable compound, a small fraction of water molecules are ionized—that is, broken apart into hydrogen ions (H^+) and hydroxide ions (OH^-):

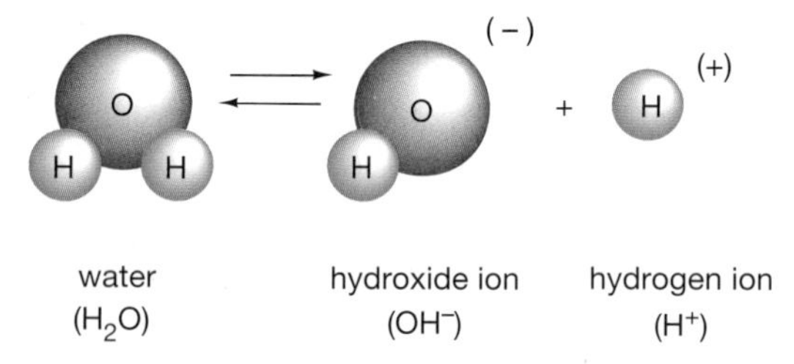

A hydroxide ion is negatively charged because it has gained an electron from the hydrogen atom. By losing

(a)

(b)

FIGURE 2-8 Cohesion among water molecules
(a) Buoyed by surface tension, the fishing spider rushes across the surface of a pond to capture an insect. **(b)** In giant redwoods, cohesion holds water molecules together in continuous strands from the roots to the topmost leaves as high as 300 feet above the ground.

an electron, the hydrogen atom is converted into a positively charged hydrogen ion. Pure water contains equal concentrations of hydrogen ions and hydroxide ions.

In many solutions, however, the concentrations of H^+ and OH^- are not the same. If the concentration of H^+ exceeds the concentration of OH^-, the solution is **acidic**. An **acid** is a substance that releases hydrogen ions when it is dissolved in water. When hydrochloric acid (HCl), for example, is added to pure water, almost all of the HCl molecules separate into H^+ and Cl^-. Therefore, the concentration of H^+ greatly exceeds the concentration of OH^-, and the resulting solution is acidic. (Many acidic substances, such as lemon juice and vinegar, have a sour taste. This is because the sour-taste receptors on your tongue are specialized to respond to the excess of H^+.)

If the concentration of OH^- is greater, the solution is **basic**. A **base** is a substance that combines with hydrogen ions, reducing their number. If, for instance, sodium hydroxide (NaOH) is added to water, the NaOH molecules separate into Na^+ and OH^-. The OH^- ions combine with H^+, reducing their number and creating a basic solution.

The degree of acidity is expressed on the **pH scale** (Fig. 2-9), in which neutrality (equal numbers of H^+ and OH^-) is indicated by the number 7. Pure water, with equal concentrations of H^+ and OH^-, has a pH of 7. Acids have a pH below 7; bases have a pH above 7. Each unit on the pH scale represents a tenfold change in the concentration of H^+. Thus, a cola drink (pH = 3) has a concentration of H^+ 10,000 times that of water (pH = 7)—no wonder it is so bad for your teeth!

A Buffer Helps Maintain a Solution at a Relatively Constant pH

In most mammals, including humans, both the cell interior (cytoplasm) and the fluids that bathe the cell are nearly neutral (pH about 7.3 to 7.4). Small increases or decreases in pH may cause drastic changes in both the structure and function of biological molecules, leading to the death of cells or entire organisms. Nevertheless, living cells seethe with chemical reactions that take up or give off H^+. How, then, does the pH remain constant overall? The answer lies in the many buffers found in living organisms. A **buffer** is a compound that tends to maintain a solution at a constant pH by accepting or releasing H^+ in response to small changes in H^+ concentration. If the H^+ concentration rises, buffers combine with them; if the H^+ concentration falls, buffers release H^+. Thus, the original concentration of H^+ is maintained. Common buffers in living organisms include bicarbonate (HCO_3^-) and phosphate ($H_2PO_4^-$ and HPO_4^{2-}), both of which can accept or release H^+ depending on the circumstances. If the blood becomes too acidic, for example, bicarbonate accepts H^+ to form carbonic acid:

$$\underset{\text{(bicarbonate)}}{HCO_3^-} + \underset{\text{(hydrogen ion)}}{H^+} \rightarrow \underset{\text{(carbonic acid)}}{H_2CO_3}$$

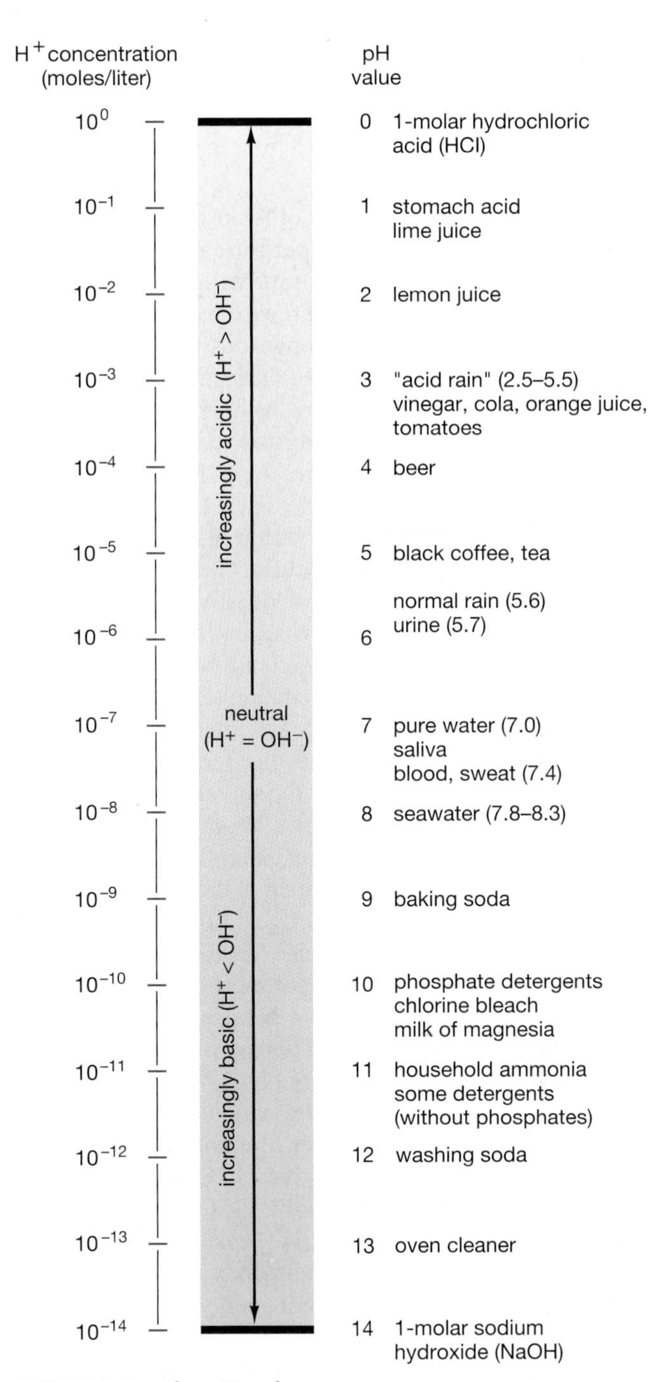

FIGURE 2-9 The pH scale
The pH scale expresses the concentration of hydrogen ions in a solution on a scale of 0 (very acidic) to 14 (very basic). Each unit of change on the scale represents a tenfold change in the concentration of hydrogen ions. Lemon juice, for example, is about 10 times more acidic than orange juice, and the most severe acid rains in the northeastern United States are almost 1000 times more acidic than normal rainfall. Except for the inside of your stomach, nearly all the fluids in your body are finely adjusted to a pH of about 7.4. The color-coding shown corresponds to a common pH indicator dye, bromthymol blue.

If the blood becomes too basic, carbonic acid liberates hydrogen ions, which combine with the excess hydroxide ions, forming water:

$$\underset{\text{(carbonic acid)}}{H_2CO_3} + \underset{\text{(hydroxide ion)}}{OH^-} \rightarrow \underset{\text{(bicarbonate)}}{HCO_3^-} + \underset{\text{(water)}}{H_2O}$$

In either case, the result is that the blood pH remains near its normal value.

Water Moderates the Effects of Temperature Changes

Your body and the bodies of other organisms can survive only within a limited temperature range. As we will see in Chapter 6, high temperatures may damage enzymes that guide the chemical reactions essential to life. Low temperatures are also dangerous because enzyme action slows as the temperature drops. Subfreezing temperatures within the body are usually lethal, because spearlike ice crystals can rupture cells. Fortunately, water has important properties that moderate the effects of temperature changes. These properties help keep the bodies of organisms within tolerable temperature limits. Also, large lakes and the oceans have a moderating effect on the climate of nearby land, making it warmer in winter and cooler in summer.

Temperature reflects the speed of molecules; the higher the temperature, the greater their average speed. Generally, if heat energy enters a system, the molecules of that system move more rapidly, and the temperature of the system rises. As you may recall, individual water molecules are weakly linked to one another by hydrogen bonds (see Fig. 2-5). When heat enters a watery system such as a lake or a living cell, much of the heat energy initially goes into breaking hydrogen bonds rather than speeding up individual molecules. Thus, it requires more energy to heat water than to heat most other substances. For example, 1 **calorie** of energy will raise the temperature of 1 gram of water by 1 °C, whereas only 0.6 calorie is needed to heat 1 gram of alcohol by 1 °C , 0.2 calorie to heat table salt, and only 0.02 calorie to heat a gram of rock such as granite or marble. Thus, the energy required to heat a pound (a pint) of water only 1 °C would raise the temperature of a pound of rock by 50 °C. For this reason, if a lizard wants to warm up, it will seek out a rock rather than a puddle. Because the human body is mostly water, a sunbather can absorb a lot of heat energy without sending his or her body temperature soaring—and many hot sunbathers can jump into a swimming pool to cool off without raising the temperature of the water very much. The energy required to heat a gram of a substance by 1 °C is called its *specific heat*. Because of its polar nature and hydrogen bonding, water has a very high specific heat, and therefore moderates temperature changes.

Water also moderates the effects of high temperatures because it takes a great deal of heat (539 calories per gram) to convert liquid water to water vapor. This, too, is due to the polar nature of water molecules and the resulting hydrogen bonds that interconnect them. For a water molecule to evaporate, it must absorb sufficient energy to make it move quickly enough to break all the hydrogen bonds that hold it to nearby water molecules. Only the fastest-moving water molecules, carrying the most energy, can break their hydrogen bonds and escape into the air as water vapor. The remaining liquid is cooled by the loss of these high-energy molecules. As children romp through a sprinkler on a hot summer day, water coats their bodies. Heat energy is transferred from their skin to the water and from the water to the vapor as the water evaporates. Evaporating just 1 gram of water cools 539 grams of a person's body by 1 °C, making water a very effective coolant. This also means that evaporating perspiration produces a considerable loss of heat without much loss of water. The heat required to vaporize water is called its *heat of vaporization*—water's heat of vaporization is one of the highest known.

Finally, water moderates the effects of low temperatures because an unusually large amount of energy must be removed from molecules of liquid water before they form the precise crystal arrangement of ice (see the in-text figure in the next section). As a result, water freezes more slowly than many other liquids at a given temperature and loses more heat to the environment in the process. This property of a substance is called its *heat of fusion*; water's heat of fusion is very high.

Water Forms an Unusual Solid: Ice

Water will become a solid after prolonged exposure to temperatures below its freezing point. But even solid water is unusual. Most liquids become denser when they solidify; therefore, the solid sinks. Ice is unique because it is less dense than liquid water. The regular arrangement of water molecules in ice crystals (below right) keeps them farther apart than they are in the liquid phase (below left), where they jumble more closely together; thus, ice is less dense than water.

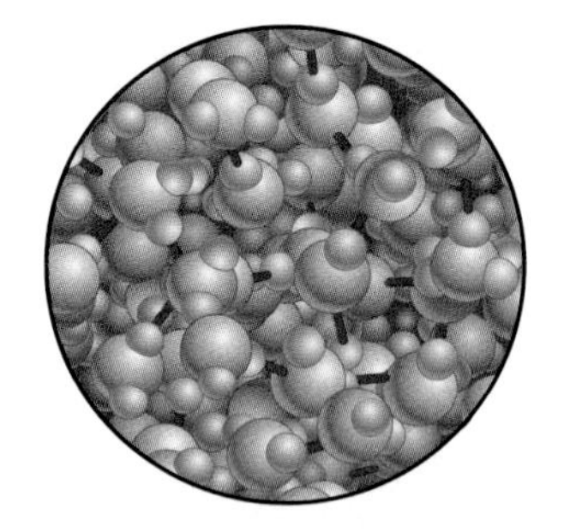

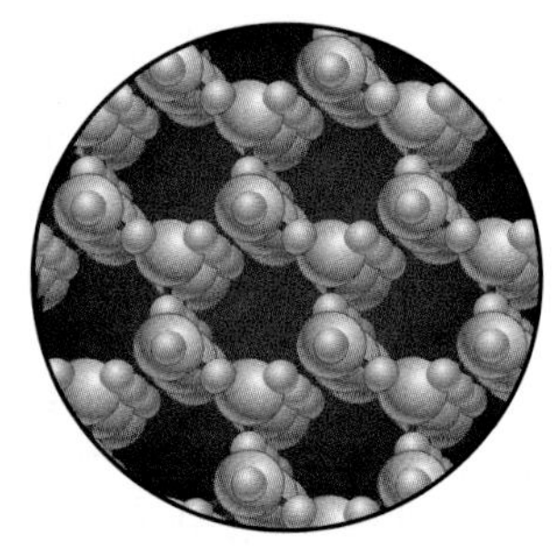

When a pond or lake starts to freeze in winter, the ice floats on top, forming an insulating layer that delays the freezing of the rest of the water and provides a slippery surface for skaters. This insulation allows fish and other lake residents to survive in the liquid water below. If ice were to sink, many ponds and lakes around the world would freeze solid from the bottom up during the winter, killing plants, fish and other underwater organisms.

CASE STUDY REVISITED Walking on Water

Most water-walkers are extremely lightweight insects—a 4-ounce basilisk is probably the heaviest animal able to support itself on its feet while moving across water. When the leaping lizard slaps its foot down hard on the water's surface, resistance caused by surface tension spreads special fringes on the lizard's toes, creating a larger surface area. As the lizard propels itself forward, stroking downward and backward, each of its fringed feet traps and pushes an air bubble behind it. Trapped between the water's surface tension and the lizard's foot, the air pocket acts as a momentary flotation device, providing support for a fraction of a second before the other foot splashes down ahead and repeats the process.

Ice skaters, meanwhile, exploit the floating property of frozen water. Below their skates, an entire community of lake-dwellers is insulated and protected. But why is ice so slippery? Surprisingly, scientist are not sure. They know that the water molecules in ice crystals are only loosely bonded to one another. Some speculate that the molecules on the icy surface readily move past one another when a solid slides over them, acting like molecular ball bearings. Others hypothesize that slipperiness is due to yet another unique property of ice: when ice is compressed, it melts. Perhaps under pressure from skates (or shoes, or tires) a microscopically thin layer of water forms and lubricates the surface of the ice.

Consider This: Many of water's unique properties are the result of its polar covalent bonds, which allow water molecules to form hydrogen bonds with each other. What if water molecules used *nonpolar* covalent bonds? What might the implications be? Using the information in this chapter, make a list of ways in which this might affect the properties of water and life in general.

Links to Life: Health Food?

Fruits and vegetables—particularly those with yellow, orange, and red coloring—contain not only vitamins C and E, but other antioxidants as well. But did you know that chocolate (Fig. E2-2)—sometimes described as "sinfully delicious" and often a source of guilt for those who indulge (or overindulge) in it—contains antioxidants, and thus might be a type of health food? Although it is extremely difficult to do controlled studies on the effects of antioxidants in the human diet, there is evidence that diets high in antioxidants may be beneficial. For example, the low incidence of heart disease among the French (many of whom who eat a relatively high-fat diet) has been attributed in part to antioxidants in wine, which the French consume regularly. They also eat considerably more fruits and vegetables than Americans (with the notable exception of potatoes—"french fries" are consumed far more regularly in the United States than in France). Antioxidant supplements abound in nutrition catalogs and in grocery and health food stores. Now, amazingly, researchers have given us an excuse to eat chocolate and feel good about it: cocoa powder (the dark, bitter powder made from the seeds inside cacao pods; see Fig. E2-2) contains high concentrations of *flavenoids*, which are powerful antioxidants (chemically related to those found in wine). No studies have yet been done to determine whether high consumption of chocolate actually reduces the risk of cancer or heart disease, but there will certainly be no shortage of volunteers for this research. Although becoming fat by eating too much chocolate candy could counteract any positive effects of the cocoa powder itself, slim "chocoholics" have reason to relax and enjoy.

FIGURE E2-2 Chocolate
Cocoa powder is derived from cacao beans (inset), which grow on trees in tropical regions of the Americas.

Summary of Key Concepts

2.1 What Are Atoms?

An element is a substance that can neither be broken down nor converted to different substances by ordinary chemical means. The smallest possible particle of an element is the atom, which is itself composed of a central nucleus, containing protons and neutrons, and electrons outside the nucleus. All atoms of a given element have the same number of protons, an amount different from the number of protons in the atoms of every other element. Electrons orbit the nucleus in electron shells at specific distances from the nucleus. Each shell contain a fixed maximum number of electrons. The chemical reactivity of an atom depends on the number of electrons in its outermost electron shell; an atom is most stable, and therefore least reactive, when its outermost shell is either completely full or empty.

2.2 How Do Atoms Interact to Form Molecules?

Atoms may combine to form molecules. The forces holding atoms together in molecules are called *chemical bonds*. Atoms that have lost or gained electrons are negatively or positively charged particles called *ions*. Ionic bonds are electrical attractions between charged ions, holding them together in crystals. When two atoms share electrons, covalent bonds form. In a nonpolar covalent bond, the two atoms share electrons equally. In a polar covalent bond, one atom may attract the electron more strongly than the other atom does; in this case, the strongly attracting atom bears a slightly negative charge, and the weakly attracting atom bears a slightly positive charge. Some polar covalent bonds give rise to hydrogen bonding, the attraction between charged regions of individual polar molecules or distant parts of a large polar molecule.

2.3 Why Is Water So Important to Life?

Water interacts with many other molecules, and dissolves many polar and charged substances. Water forces nonpolar substances, such as fat, to assume certain types of physical organization. Water participates in chemical reactions. Water molecules cohere to each other using hydrogen bonds. Because of its high specific heat, heat of vaporization, and heat of fusion, water helps maintain a fairly stable temperature in spite of wide temperature fluctuations in the environment.

Key Terms

acid *p. 31*
acidic *p. 31*
antioxidant *p. 28*
atom *p. 22*
atomic nucleus *p. 22*
atomic number *p. 22*
base *p. 31*
basic *p. 31*
buffer *p. 31*
calorie *p. 32*
chemical bond *p. 25*
chemical reaction *p. 25*
cohesion *p. 30*
compound *p. 24*
covalent bond *p. 25*
electron *p. 22*
electron shell *p. 24*
element *p. 22*
free radical *p. 27*
hydrogen bond *p. 28*
hydrophilic *p. 29*
hydrophobic *p. 29*
hydrophobic interaction *p. 30*
ion *p. 25*
ionic bond *p. 25*
isotope *p. 23*
molecule *p. 24*
neutron *p. 22*
nonpolar covalent bond *p. 27*
pH scale *p. 31*
polar covalent bond *p. 27*
proton *p. 22*
radioactive *p. 23*
solvent *p. 28*
surface tension *p. 30*

Thinking Through the Concepts

To take a multiple-choice quiz with feedback on the contents of this chapter, visit http://www.prenhall.com/audesirk7. *Log in to the Web site selected by your instructor and navigate to the Self Test section for this chapter.*

? Review Questions

1. What are the six most abundant elements that occur in living organisms?
2. Distinguish among atoms and molecules; elements and compounds; and protons, neutrons, and electrons.
3. Compare and contrast covalent bonds and ionic bonds.
4. Why can water absorb a great amount of heat with little increase in its temperature?
5. Describe how water dissolves a salt. How does this phenomenon compare with the effect of water on a hydrophobic substance such as corn oil?
6. Define *acid*, *base*, and *buffer*. How do buffers reduce changes in pH when hydrogen ions or hydroxide ions are added to a solution? Why is this phenomenon important in organisms?

Applying the Concepts

1. Many over-the-counter substances sold to bring relief from "acid stomach" or "heartburn" consist of calcium carbonate. Why do they work?
2. Fats and oils do not dissolve in water; polar and ionic molecules dissolve easily in water. Detergents and soaps help clean by dispersing fats and oils in water so that they can be rinsed away. From your knowledge of the structure of water and the hydrophobic nature of fats, what general chemical structures (for example, polar or nonpolar parts) must a soap or detergent have, and why?
3. What would the effects be for aquatic life if the density of ice were greater than that of liquid water? What would be the impact on terrestrial organisms?
4. How does sweating help you regulate your body temperature? Why do you feel hotter and more uncomfortable on a hot, humid day than on a hot, dry day?
5. Free radicals are commonly formed when animals use oxygen to metabolize sugar to make high-energy molecules. Ross Hardison, a researcher at Pennsylvania State University, stated eloquently: "Keeping oxygen under control while using it in energy production has been one of the great compromises struck in the evolution of life on Earth." What did he mean by this? (You may wish to revisit this question after reading Chapter 8.)

For More Information

Atkins, P. W. *Molecules*. New York: Scientific American Library, 1987. A layperson's introduction to atoms and molecules, with superb illustrations.

Eiseley, L. *The Immense Journey*. New York: Vintage Books, 1957. A delightful series of essays by a gifted naturalist and writer.

Glasheen, J. W., and McMahon, T. A. "Running on Water." *Scientific American*, September 1997. Answers the question, "How does the basilisk lizard run on water?"

Raloff, J. "Chocolate Hearts." *Science News,* March 18, 2000. Describes recent research indicating that chocolate is high in antioxidants.

Storey, K. B., and Storey, J. M. "Frozen and Alive." *Scientific American*, December 1990. By triggering ice formation here, suppressing it there, and loading up their cells with antifreeze molecules, some animals, including certain lizards and frogs, can survive with 60% of their body water frozen solid.

Woodley, R. "The Physics of Ice." *Discover*, June 1999. Ice is such a complicated solid that researchers are still unsure of exactly why it acts in the ways it does.

Media Activities

To access a Media Activity visit http://www.prenhall.com/audesirk7. *Log in to the Web site selected by your instructor, navigate to this chapter, and select the appropriate Media Activity number.*

2.1 Interactive Atoms

Estimated time: 5 minutes

An introduction to the basic chemistry needed to understand biological structures and processes.

2.2 Water and Life

Estimated time: 5 minutes

Explore the properties of water and why this molecule is essential for life.

2.3 Web Investigation: Walking on Water

Estimated time: 10 minutes

Water is central to life on our planet. This activity explores how the chemistry of water plays a central role in many different aspects of everyday life.

CHAPTER

3 Biological Molecules

Products from "nonfat" potato chips to diet sodas may contain artificially modified molecules that substitute for fats and sugar.

AT A GLANCE

CASE STUDY Improving on Nature?

Protein, carbohydrate, fat, sugar. Most people know that these are the building blocks of our bodies (and all other organisms on Earth). The special properties of biological molecules help make living bodies different from nonliving objects. As our understanding of these molecules increases, so do efforts to modify them. Why are we so anxious to improve on nature's handiwork?

In societies blessed with an overabundance of food, obesity is a serious health problem. One goal of food scientists is to modify biological molecules to make them noncaloric. Sugar is a prime candidate; several artificial sweeteners such as saccharin, aspartame (Nutrasweet™), and sucralose (Splenda™) add a sweet taste to foods, while providing few or no calories.

Government agencies such as the U.S. Food and Drug Administration (FDA) require exhaustive testing to demonstrate the safety of synthetic food products. Olestra (Olean™), an artificial oil substitute, was approved by the FDA in 1996 for use in snack foods. Olestra mimics the culinary properties of natural oils but is completely indigestible, ensuring that potato chips made with olestra have no fat calories and far fewer total calories than normal chips. It also means that some people's digestive systems might have difficulty coping with large quantities of this fake fat. To convince the FDA of its safety, olestra's developer, Procter & Gamble, submitted results from more than 150 studies on humans and animals.

How are these "nonbiological molecules" made? How can you, as an educated consumer, decide whether you want to add tasty "nonfat" potato chips or artificially sweetened diet cola to your shopping cart? For starters, you need to understand real biological molecules, the building blocks of life.

3.1 Why Is Carbon So Important in Biological Molecules?

You have probably seen "organic" fruits and vegetables in your grocery store. To a chemist, this phrase is redundant; all produce is organic because it is formed from biological molecules. In chemistry, the term **organic** describes molecules that have a carbon skeleton and also contain some hydrogen atoms. The word *organic* is derived from the ability of living *organisms* to synthesize and use this general type of molecule. **Inorganic** molecules include carbon dioxide and all molecules without carbon, such as water.

Although organisms are united by the common structures and functions of their organic molecules, the tremendous variety of organic molecules contributes to the diversity of structures within single organisms and even within individual cells. This vast array of organic molecules is possible because the carbon atom is so versatile. A carbon atom has four electrons in its outermost shell, with room for eight. Therefore, carbon atoms are able to form many bonds. They become stable by sharing four electrons with other atoms, forming up to four single covalent bonds or fewer double or triple covalent bonds. Molecules with many carbon atoms can assume complex shapes, including chains, branches, and rings—the basis for an amazing diversity of molecules.

Organic molecules are much more than just complicated skeletons of carbon atoms, however. Attached to the carbon backbone are **functional groups**, groups of atoms that determine the characteristics and chemical reactivity of the molecules. Functional groups are far less stable than the carbon backbone and are more likely to participate in chemical reactions. The common functional groups found in organic molecules are shown in Table 3-1.

The similarity among organic molecules from all forms of life is a consequence of two main features: the use of the same set of functional groups in virtually all organic molecules in all types of organisms, and the "modular approach" of synthesizing large organic molecules.

3.2 How Are Organic Molecules Synthesized?

In principle, there are two ways to manufacture a large, complex molecule: by combining atom after atom following an extremely detailed blueprint, or by preassembling smaller molecules and hooking them together. Just as trains are made by coupling engines to various train cars, life also takes a "modular approach." Small organic molecules (for example, sugars) are used as subunits that combine to synthesize longer molecules (for example, starches)—like cars in a train. The individual subunits are often called **monomers** (from Greek words meaning "one

Table 3-1 Important Functional Groups in Biological Molecules

Group	Structure	Properties	Types of Molecules
Hydrogen (—H)	—H	Polar or nonpolar, depending on which atom hydrogen is bonded to; involved in condensation and hydrolysis	Almost all organic molecules
Hydroxyl (—OH)	—O—H	Polar; involved in condensation and hydrolysis	Carbohydrates, nucleic acids, alcohols, some acids, and steroids
Carboxyl (—COOH)	—C(=O)—O—H	Acidic; negatively charged when H^+ dissociates; involved in peptide bonds	Amino acids, fatty acids
Amino (—NH_2)	—N(—H)—H	Basic; may bond an additional H^+, becoming positively charged; involved in peptide bonds	Amino acids, nucleic acids
Phosphate (—H_2PO_4)	—O—P(=O)(—O—H)—O—H	Acidic; up to two negative charges when H^+ dissociates; links nucleotides in nucleic acids; energy-carrier group in ATP	Nucleic acids, phospholipids
Methyl (—CH_3)	—C(—H)(—H)—H	Nonpolar; tends to make molecules hydrophobic	Many organic molecules; especially common in lipids

Media Activity
3.1 Structure of Biological Molecules

Table 3-2 The Principal Biological Molecules

Class of Molecule	Principal Subtypes (subunits in parentheses)	Example	Function
Carbohydrate: Normally contains carbon, oxygen, and hydrogen, in the approximate formula $(CH_2O)_n$	*(Monosaccharide*: Simple sugar)	Glucose	Important energy source for cells; subunit of polysaccharides
	Disaccharide: Two monosaccharides bonded together	Sucrose	Principal sugar transported throughout bodies of land plants
	Polysaccharide: Many monosaccharides (normally glucose) bonded together	Starch	Energy storage in plants
		Glycogen	Energy storage in animals
		Cellulose	Structural material in plants
Lipid: Contains high proportion of carbon and hydrogen; usually nonpolar and insoluble in water	*Triglyceride*: Three fatty acids bonded to glycerol	Oil, fat	Energy storage in animals, some plants
	Wax: Variable numbers of fatty acids bonded to long-chain alcohol	Waxes in plant cuticle	Waterproof covering on leaves and stems of land plants
	Phospholipid: Polar phosphate group and two fatty acids bonded to glycerol	Phosphatidylcholine	Common component of membranes in cells
	Steroid: Four fused rings of carbon atoms with functional groups attached	Cholesterol	Common component of membranes of eukaryotic cells; precursor for other steroids such as testosterone, bile salts
Protein: Chains of amino acids; contains carbon, hydrogen, oxygen, nitrogen, and sulfur	*(Amino acids)*	Keratin	Helical protein, principal component of hair
		Silk	Protein produced by silk moths and spiders
		Hemoglobin	Globular protein composed of four subunit peptides; transport of oxygen in vertebrate blood
Nucleic acid: Made of nucleotide subunits; may consist of a single nucleotide or long chain of nucleotides	*Long-chain nucleic acids*	Deoxyribonucleic acid (DNA)	Genetic material of all living cells
		Ribonucleic acid (RNA)	Genetic material of some viruses; in living cells, essential in transfer of genetic information from DNA to protein
	(Single nucleotides)	Adenosine triphosphate (ATP)	Principal short-term energy-carrier molecule in cells
		Cyclic adenosine monophosphate (cyclic AMP)	Intracellular messenger

Media Activity
3.2 Functions of Macromolecules

part"); long chains of monomers are called **polymers** ("many parts").

Biological Molecules Are Joined Together or Broken Apart by Removing or Adding Water

You have already learned some of the reasons that water is so important to life, but water also plays a central role in reactions that break down biological molecules to liberate subunits that the body can use. In addition, when complex biological molecules are synthesized in the body, water is often produced as a by-product.

The subunits that make up large biological molecules almost always join together by means of a chemical reaction called **dehydration synthesis** (literally, "to form by removing water"). In dehydration synthesis, a hydrogen ion (H^+) is removed from one subunit and a hydroxyl group (OH^-) is removed from a second subunit, creating openings in the outer electron shells of the two subunits. These openings are filled when the subunits share electrons, creating a covalent bond that links them. The free hydrogen and hydroxyl ions then combine to form a molecule of water (H_2O):

HO OH + HO OH → HO O OH + H O H

Dehydration synthesis

The reverse reaction, **hydrolysis** ("to break apart with water"), splits the molecule back into its original subunits:

HO O OH + H O H → HO OH + HO OH

Hydrolysis

Considering how complicated living things are, it might surprise you that nearly all biological molecules fall into one of only four general categories: carbohydrates, lipids, proteins, or nucleic acids (Table 3-2).

3.3 What Are Carbohydrates?

Carbohydrate molecules are composed of carbon, hydrogen, and oxygen in the approximate ratio of 1:2:1. This formula explains the origin of the name *carbohydrate*, which literally means "carbon plus water." All carbohydrates are either small, water-soluble **sugars** or polymers of sugar, such as starch. If a carbohydrate consists of just one sugar molecule, it is called a **monosaccharide** (Greek for "single sugar"). When two monosaccharides are linked, they form a **disaccharide** ("two sugars"), and a polymer of many monosaccharides is called a **polysaccharide** ("many sugars").

Carbohydrates are important energy sources for most organisms. Consider a breakfast that includes blueberry pancakes, syrup, and orange juice. The pancakes consist mainly of carbohydrates stored in the seeds of wheat or other grains. The sugar that sweetens the syrup, blueberries, and orange juice is also stored by plants as an energy source. Other carbohydrates, such as cellulose and similar molecules, provide structural support for individual cells or even for the entire bodies of organisms as diverse as plants, fungi, bacteria, and insects.

Most small carbohydrates are soluble in water. As in water molecules, the O–H bond in a hydroxyl group is polar, because oxygen attracts electrons more strongly than does hydrogen. Hydrogen bonds between water molecules and the polar hydroxyl groups of the carbohydrate keep the carbohydrate dissolved in a solution (see Fig. 2-7).

There Are Several Monosaccharides with Slightly Different Structures

Monosaccharides usually have a backbone of three to seven carbon atoms. Most of these carbon atoms have both a hydrogen (–H) and a hydroxyl group (–OH) attached to them, so carbohydrates generally have the approximate chemical formula $(CH_2O)_n$ (*n* is the number of carbons in the backbone.) When dissolved in water, such as in the cytoplasm of a cell, the carbon backbone of a sugar usually forms a ring. Sugars in ring form can link together to make disaccharides (see Fig. 3-2) and polysaccharides. Figure 3-1 shows various ways that the chemical structure of the sugar glucose might be depicted.

Glucose is the most common monosaccharide in living organisms and is a subunit of most polysaccharides. Glucose has six carbons, so its chemical formula is $C_6H_{12}O_6$. Many organisms synthesize other monosaccharides that have the same chemical formula as glucose but slightly different structures. These include *fructose* (the "corn sugar" found in corn syrup and many fruits) and *galactose* (part of lactose, or "milk sugar"), shown below:

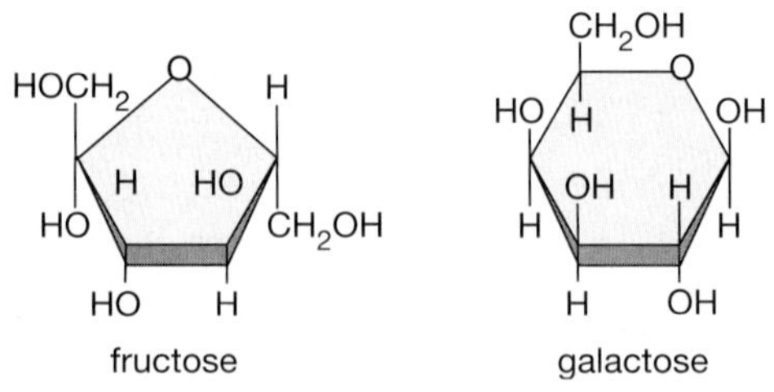

Other common monosaccharides, such as *ribose* and *deoxyribose*, have five carbons:

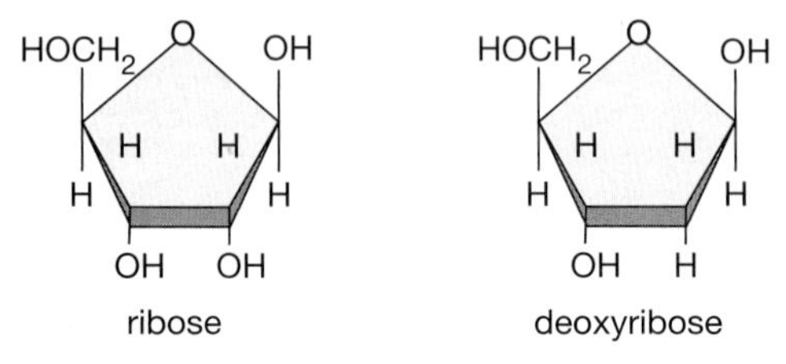

Ribose and deoxyribose are parts of the genetic molecules *deoxyribonucleic acid* (DNA, which stores the genetic code) and *ribonucleic acid* (RNA, which directs protein synthesis), respectively. You will learn more about these molecules in Chapters 9 and 10.

Disaccharides Consist of Two Single Sugars Linked by Dehydration Synthesis

Monosaccharides, especially glucose and its relatives, have a short life span in a cell. Most are either broken

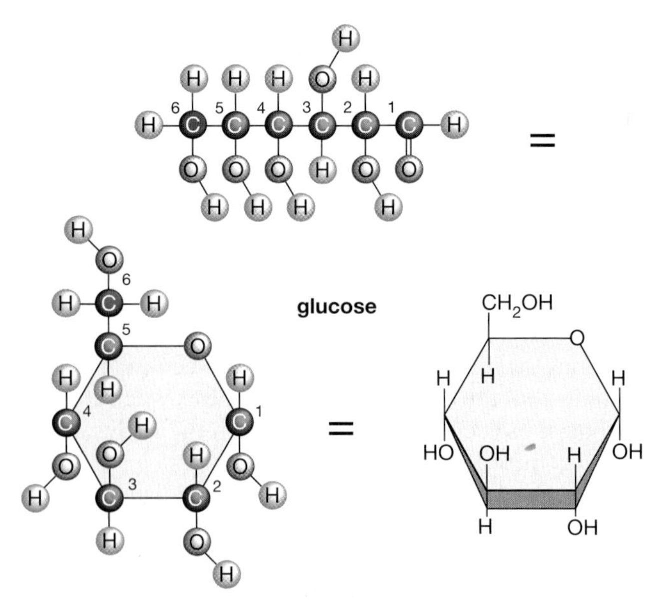

FIGURE 3-1 Glucose structure
Chemists can represent the same molecule in a variety of ways; glucose is shown here in linear (straight) form and in its ring configuration (shown two different ways), which it assumes when dissolved in water.

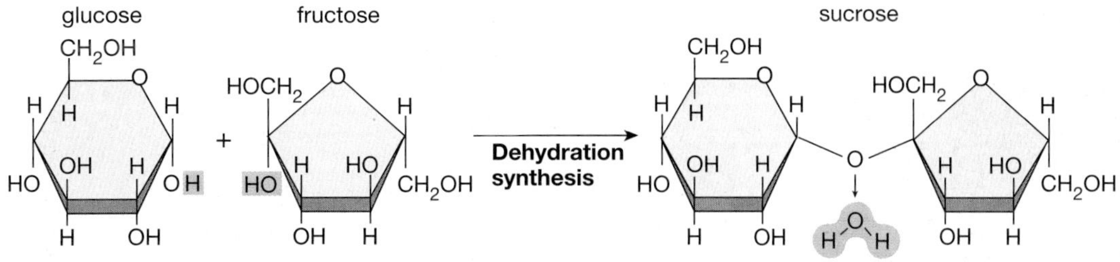

FIGURE 3-2 Synthesis of a disaccharide
The disaccharide sucrose is synthesized by a dehydration synthesis reaction in which a hydrogen (–H) is removed from glucose and a hydroxyl group (–OH) is removed from fructose. A water molecule (H–O–H) is formed in the process, leaving the two monosaccharide rings joined by single bonds to the remaining oxygen atom. Hydrolysis of sucrose is just the reverse of its synthesis, as water is split and added to the monosaccharides.

down to free their chemical energy for use in various cellular activities or are linked by dehydration synthesis to form disaccharides or polysaccharides (Fig. 3-2). Disaccharides are often used for short-term energy storage, especially in plants, and are present in many foods we eat. Perhaps you had pancakes and coffee with cream and sugar at breakfast. Common disaccharides include **sucrose** (glucose plus fructose), which forms the sugar you stirred into your coffee; **lactose** (milk sugar: glucose plus galactose), found in the milk you poured in your coffee; and **maltose** (glucose plus glucose, which will form in your digestive tract as you break down the starch in your pancakes). When energy is required, the disaccharides are broken apart into their monosaccharide subunits by hydrolysis (see in-text figure on pg. 39).

Polysaccharides Are Chains of Single Sugars

Try chewing a cracker for a long time to allow enzymes in your saliva to break it down into its component sugars. Does it taste sweeter the longer you chew? It should; monosaccharides (usually glucose) are joined together into polysaccharides to form **starch** (in plants; Fig. 3-3) or **glycogen** (in animals) for long-term energy storage. Starch is commonly formed in roots and seeds—in your cracker, from wheat seeds. Starch may occur as coiled, unbranched chains of up to 1000 glucose subunits or, more commonly, as huge branched chains of up to half a million glucose monomers. Glycogen, stored as an energy source in the liver and muscles of animals (including ourselves), is generally much smaller

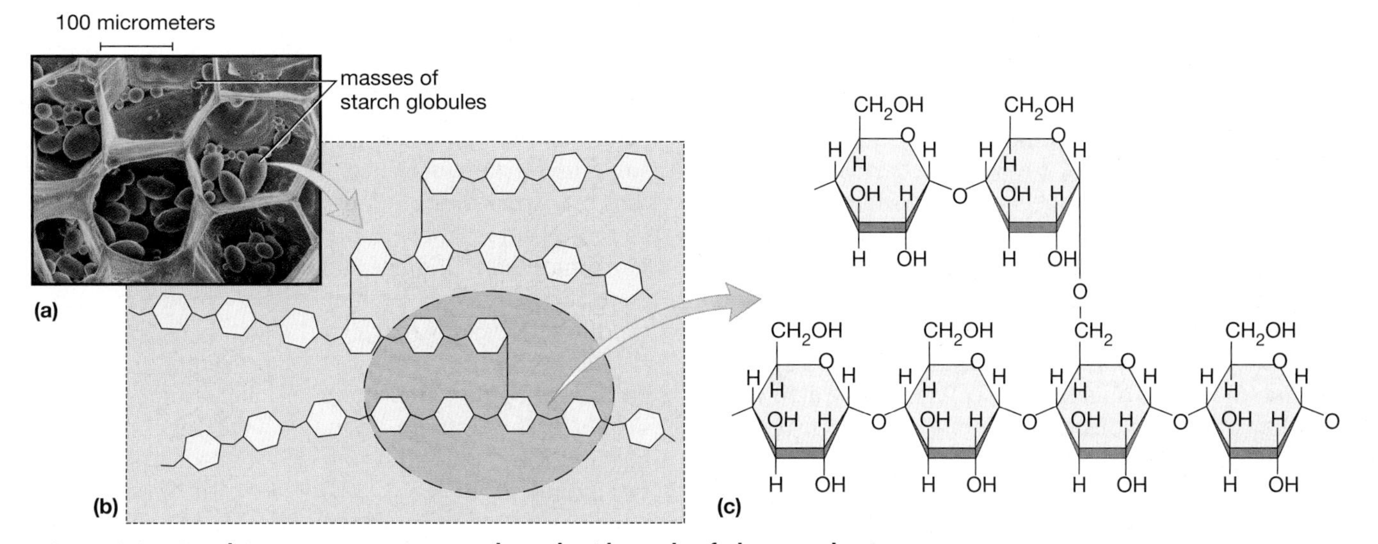

FIGURE 3-3 Starch is an energy-storage polysaccharide made of glucose subunits
(a) Starch globules inside individual potato cells. Most plants synthesize starch, which forms water-insoluble globules consisting of many starch molecules and makes up the bulk of this potato. **(b)** A small portion of a single starch molecule. Starch commonly occurs as branched chains of up to half a million glucose subunits. **(c)** The precise structure of the blue highlighted portion of the starch molecule in (b). Note the linkage between the individual glucose subunits for comparison with cellulose (Fig. 3-4).

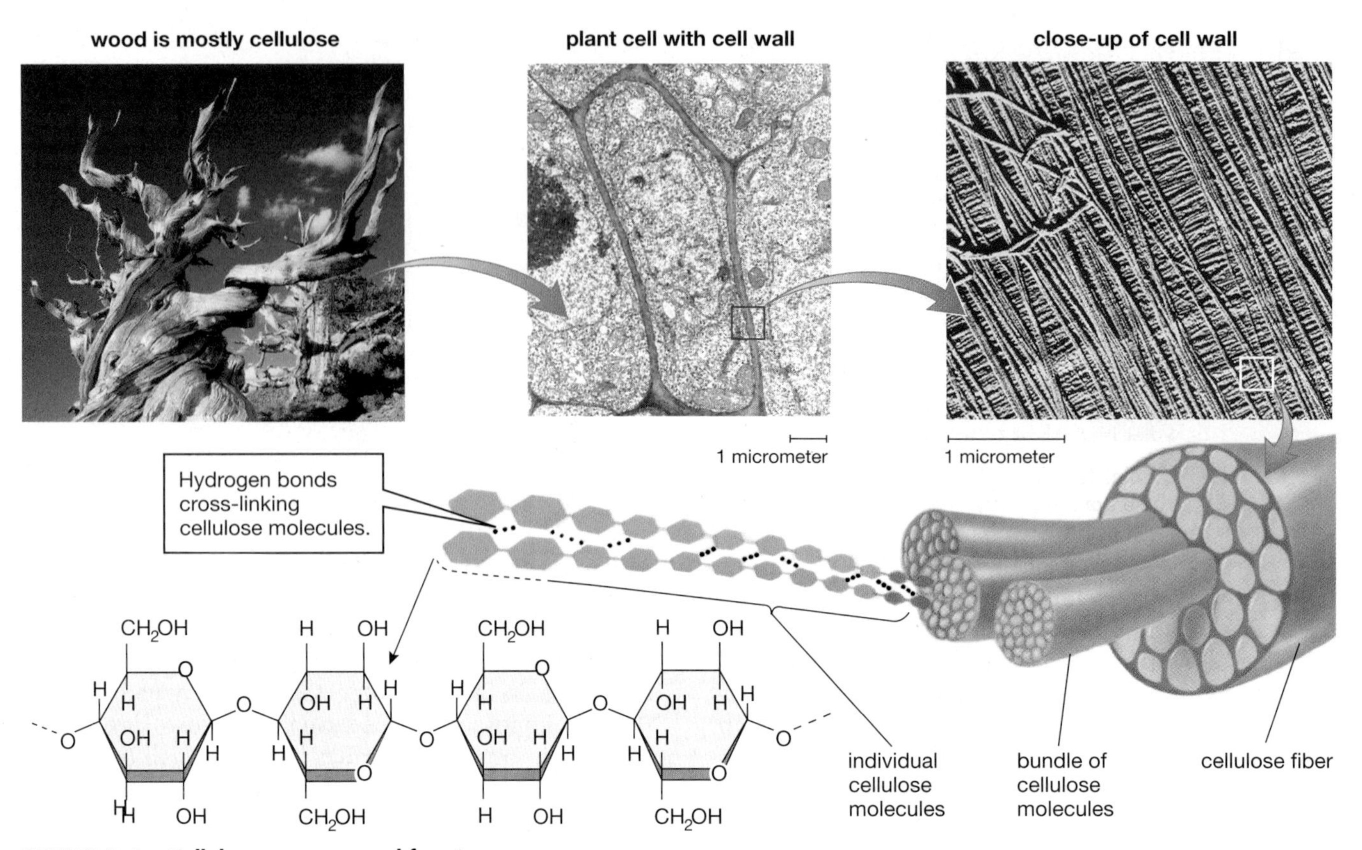

FIGURE 3-4 Cellulose structure and function
Cellulose is composed of glucose subunits. Compare with Fig. 3-3c and note that every other glucose molecule in cellulose is "upside down." Unlike starch, cellulose has great structural strength, due to the difference in bonding and to the arrangement of parallel molecules of cellulose into long, cross-linked fibers. Plant cells often lay down cellulose fibers in layers that run at angles to each other, resulting in resistance to tearing in both directions. The final product can be incredibly tough, as testified by this 3000-year-old bristlecone pine in California's White Mountains.
QUESTION Many types of plastic are composed of molecules derived from cellulose, but engineers are working hard to develop plastics based on starch molecules. Why might starch-based plastics be an improvement over existing types of plastic?

than starch and has branches every 10 to 12 glucose subunits. Its many small branches probably make it easier to split off the glucose subunits for quick energy release.

Many organisms also use polysaccharides as structural materials. One of the most important structural polysaccharides is **cellulose**, which makes up most of the cell walls of plants, 98% of the fluffy white bolls of a cotton plant, and about half the bulk of a tree trunk (Fig. 3-4). When you picture the vast fields and forests that blanket much of our planet, it may not surprise you that there is probably more cellulose on Earth than all other organic molecules combined. Ecologists estimate that about one trillion tons of cellulose are synthesized each year.

Cellulose, like starch, is a polymer of glucose. However, while most animals can easily digest starch, only a few microbes—such as those in the digestive tracts of cows or termites—can digest cellulose. Why is this the case, given that both starch and cellulose consist of glucose? The orientation of the bonds between subunits is different in the two polysaccharides. In cellulose, every other glucose is "upside down" (compare Fig. 3-3c with Fig. 3-4). This bond orientation prevents animals' digestive enzymes from attacking the bonds between glucose subunits. Certain microbes, however, synthesize enzymes that can break these bonds, and can consume cellulose as food. But for most animals, cellulose is *roughage* or *fiber*, material that passes undigested through the digestive tract. While it is valuable in preventing constipation, we don't derive any nutrients from it.

The hard outer coverings (exoskeletons) of insects, crabs, and spiders are made of **chitin**, a polysaccharide in which the glucose subunits have been chemically modified by the addition of a nitrogen-containing functional group (Fig. 3-5). Interestingly, chitin also stiffens the cell walls of many fungi. Bacterial cell walls contain other types of modified polysaccharides, as do the lubricating fluids in our joints and the transparent corneas of our eyes.

Many other molecules—including *mucus*, some chemical messengers called *hormones*, and many molecules in the plasma membrane that surrounds each

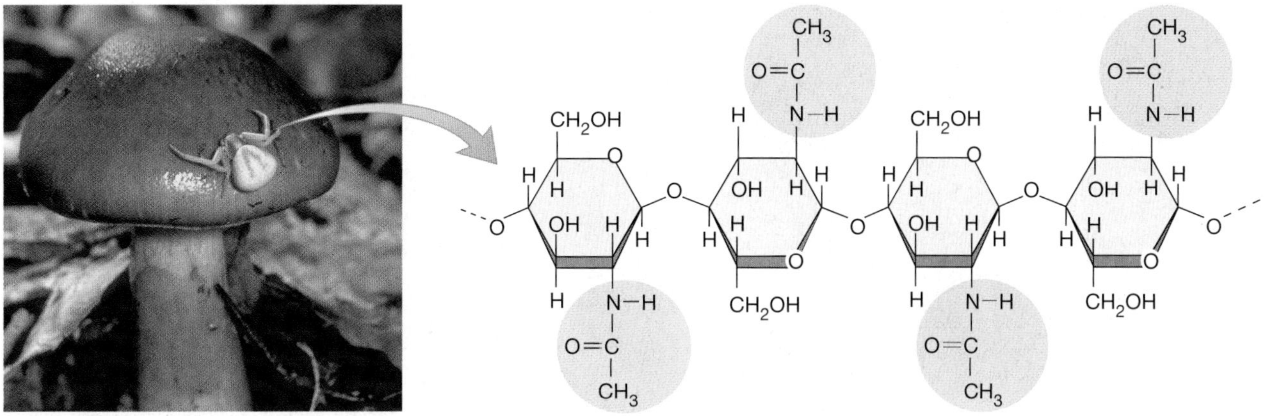

FIGURE 3-5 Chitin: A unique polysaccharide
Chitin has the same bonding configuration of glucose molecules as does cellulose, but chitin glucose subunits replace one of the hydroxyl groups with a nitrogen-containing functional group (yellow). Tough, slightly flexible chitin supports the otherwise soft bodies of arthropods (insects, spiders, and their relatives) and certain fungi.

cell—consist, in part, of carbohydrate. Perhaps the most interesting of these molecules are the nucleic acids (discussed later in this chapter) that carry hereditary information.

3.4 What Are Lipids?

Lipids form a diverse group of molecules with two important features. First, lipids contain large regions composed almost entirely of hydrogen and carbon, with nonpolar carbon–carbon or carbon–hydrogen bonds. Second, these nonpolar regions make lipids hydrophobic and insoluble in water. Lipids serve a wide variety of functions. Some are energy-storage molecules, some form waterproof coverings on plant or animal bodies, some make up the bulk of all the membranes of a cell, and still others are hormones.

Lipids are classified into three major groups: first, oils, fats, and waxes, which are similar in structure and contain only carbon, hydrogen, and oxygen; second, phospholipids, which are structurally similar to oils but also contain phosphorus and nitrogen; and third, the "fused-ring" family of steroids.

Oils, Fats, and Waxes Are Lipids Containing Only Carbon, Hydrogen, and Oxygen

Oils, fats, and waxes are related in three ways: they contain only carbon, hydrogen, and oxygen; they contain one or more **fatty acid** subunits, which are long chains of carbon and hydrogen with a *carboxyl group* (–COOH) at one end; and they usually do not have ring structures. **Fats** and **oils** are formed by dehydration synthesis from three fatty acid subunits and one molecule of **glycerol**, a short, three-carbon molecule with one hydroxyl group (–OH) per carbon (Fig. 3-6). This structure gives fats and oils their chemical name, **triglycerides**. Notice that a double bond between two carbons in the fatty acid subunit creates a kink in the chain.

Fats and oils possess a high concentration of chemical energy: about 9.3 Calories per gram, compared with 4.1 Calories per gram for sugars and proteins. (A *Calorie* with a capital *C* equals 1000 "small c" calories; the Calorie is used in measuring the energy content of foods.) Because fats are so calorie-dense, fat substitutes such as olestra may be especially appealing to dieters. Fats and oils are used for long-term energy storage in both plants and animals. For example, bears feast during

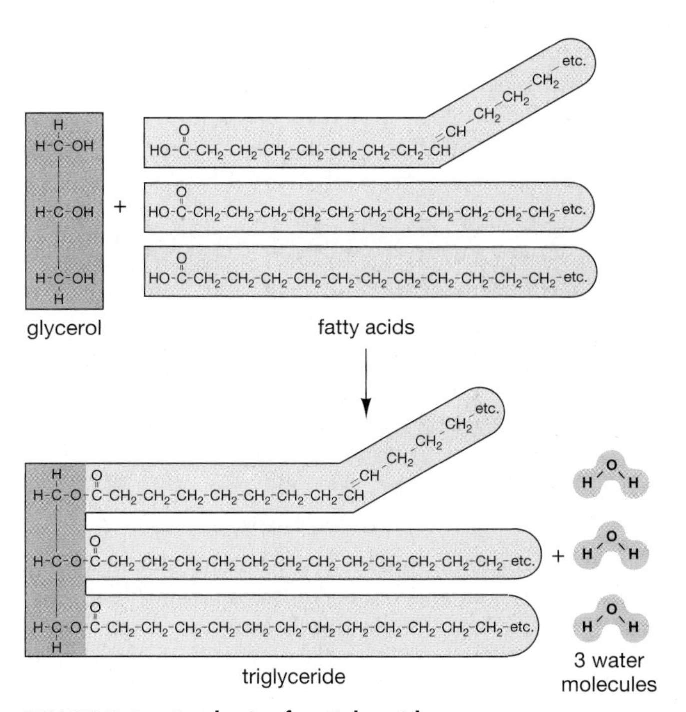

FIGURE 3-6 Synthesis of a triglyceride
Dehydration synthesis links a single glycerol molecule with three fatty acids to form a triglyceride.

(a) Fat

(b) Wax

FIGURE 3-7 Lipids
(a) A European brown bear ready to hibernate. If this bear stored the same amount of energy in carbohydrates instead of fat, she probably would be unable to walk! **(b)** Wax is a highly saturated lipid that remains very firm at normal outdoor temperatures. Its rigidity allows it to be used to form the strong but thin-walled hexagons of this honeycomb.

summer and fall to put on fat to tide them over during their winter hibernation (Fig. 3-7a). Because fats store the same energy with less weight than do carbohydrates, fat is an efficient way for animals to store energy.

The difference between a fat (such as beef fat), which is a solid at room temperature, and an oil (such as those used to make potato chips or french fries) lies in their fatty acids. Fats have fatty acids with all single bonds in their carbon chains. Hydrogens occupy all the other bond positions on the carbons. The resulting fatty acid is said to be **saturated**, because it is "saturated" with hydrogens—it has as many hydrogens as possible. Lacking double bonds between carbons, the carbon chain of the fatty acid is straight. The saturated fatty acids of fats (such as the beef fat molecule illustrated below) can nestle closely together, forming solid lumps at room temperature:

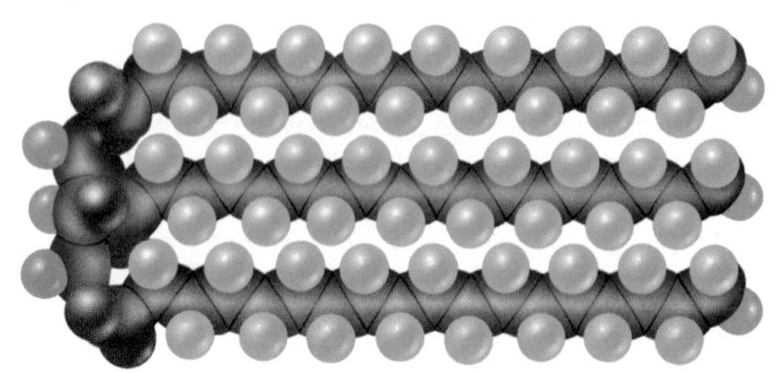

Beef fat (saturated)

Most of the saturated fat in the human diet, including butter, bacon fat, and the fat on steak, comes from animals. If there are double bonds between some of the carbons, and consequently fewer hydrogens, the fatty acid is said to be **unsaturated**. Oils have mostly unsaturated fatty acids. We get most of our unsaturated oils from the seeds of plants, where they are stored for the plants' developing embryos. Corn oil, peanut oil, and canola oil are all examples. The double bonds in the unsaturated fatty acids of oils produce kinks in the fatty acid chains, as illustrated by the linseed oil molecule:

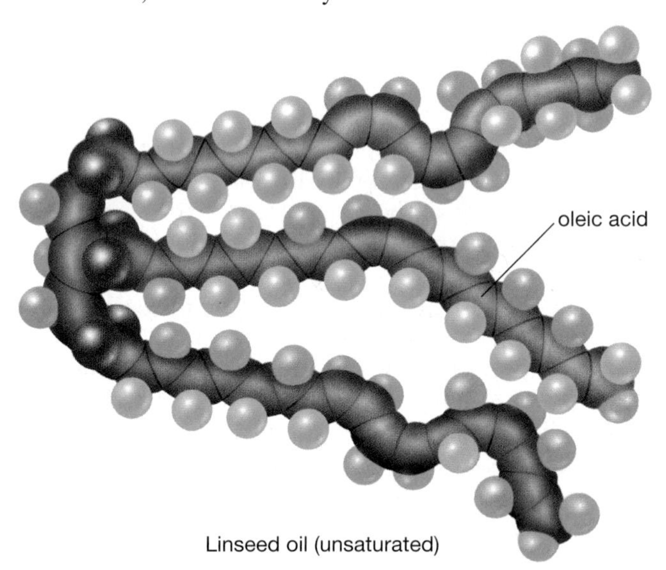

Linseed oil (unsaturated)

The kinks keep oil molecules apart; as a result, oil is liquid at room temperature. An oil can be converted to a fat by breaking some of the double bonds between carbons, replacing them with single bonds, and adding hydrogens to the remaining bond positions. The resulting substance is the "hydrogenated oil" listed in the ingredients on a box of margarine, which allows the margarine to be solid at room temperature. The partial hydrogenation process results in a configuration of double and single bonds, called the *trans* configuration, that is rare in nature. In this configuration, the carbon chain bends in a zig-zag shape that allows adjacent fatty acids to stack together: the "zigs" of one strand nestle within the "zags" of adjacent strands. This tight packing resembles the packing that occurs between the straight chain fatty

HEALTH WATCH Cholesterol—Friend and Foe

Cholesterol is a steroid with a bad reputation. Why are so many products advertised as "cholesterol free" or "low in cholesterol"? After all, cholesterol is a crucial component of cell membranes. It is also the raw material for the production of bile (which helps us digest fats), vitamin D, and both male and female sex hormones.

Although it is crucial to life, medical researchers have found that individuals with excessively high levels of cholesterol in their blood are at increased risk for heart attacks and strokes. Unfortunately, cholesterol builds up "silently" and without warning. A person may not know that anything is wrong until he or she actually suffers a heart attack. Cholesterol contributes to the formation of obstructions in arteries, called *plaques*, which in turn can promote the formation of blood clots. These clots can break loose and block an artery supplying the heart muscle, causing a heart attack, or can block a vessel to the brain, causing a stroke. (Plaque formation is described in more detail in "Health Watch: Matters of the Heart" in Chapter 28.)

Where does cholesterol come from? It comes from animal-derived foods; it is essentially nonexistent in plants. Egg yolks are a particularly rich source; sausages, bacon, whole milk, and butter contain it as well. Have you ever heard of "good" and "bad" cholesterol? Because cholesterol molecules are nonpolar, they do not dissolve in blood (which is mostly water). Thus, cholesterol molecules are transported through blood in packets surrounded by special carrier molecules called *lipoproteins* (phospholipids plus proteins). Cholesterol in high-density lipoproteins ("HDL cholesterol") is the good kind; these packets transport cholesterol to the liver, where it is removed from circulation and further metabolized (used in bile synthesis, for example). HDL cholesterol has more protein and less lipid, and is called "high-density" because protein is more dense than lipid. Cholesterol in low-density lipoprotein packets ("LDL cholesterol" with less protein and more lipid) is the bad kind; in this form, cholesterol circulates to cells throughout the body and can be deposited in artery walls. A high ratio of HDL ("good") to LDL ("bad") is correlated with reduced risk of heart disease. A complete cholesterol screening test will distinguish between these two forms in your blood.

Perhaps you've also heard about trans fatty acids as dietary villains. Margarine can be made from unsaturated vegetable oil that has been artificially hardened by partial saturation, a process that creates trans fatty acids. Research suggests that these trans fatty acids are not metabolized normally and can both increase LDL cholesterol and decrease HDL cholesterol. Some epidemiological studies have found a correlation between greater consumption of trans fats and a higher risk of heart disease. In response to these concerns, margarine manufacturers and some fast-food chains are reducing the amount of trans fats in their products.

Animals, including people, synthesize much of their own cholesterol from other lipids. Because of genetic differences, some people's bodies manufacture more than others. People with high cholesterol (about 25% of all adults in the United States) can sometimes reduce their levels by eating a diet low in both cholesterol and saturated fats. For people with dangerous levels of cholesterol who are unable to reduce it adequately by changing their diets, doctors often prescribe cholesterol-reducing drugs. These drugs work by interfering with cholesterol synthesis.

acids in saturated fats, and allows the *trans fats* to assemble into a solid, as do saturated fats. These artificially produced trans fats are now found in many commercial food products, from margarine to cookies and crackers to french fries. Recently, however, researchers have become concerned about our consumption of trans fats; see "Health Watch: Cholesterol—Friend and Foe."

Although **waxes** are chemically similar to fats, they are not a food source; we and most other animals do not have the appropriate enzymes to break them down. Waxes are highly saturated and therefore solid at normal outdoor temperatures. Waxes form a waterproof coating over the leaves and stems of land plants. Animals synthesize waxes as waterproofing for mammalian fur and insect exoskeletons and, in a few cases, to build elaborate structures such as beehives (Fig. 3-7b).

Phospholipids Have Water-Soluble "Heads" and Water-Insoluble "Tails"

The plasma membrane that surrounds each cell contains several types of **phospholipids**. A phospholipid is similar to an oil, except that one of the three fatty acids is replaced by a phosphate group with a short, polar functional group (typically containing nitrogen) attached to the end (Fig. 3-8). Unlike the two fatty acid

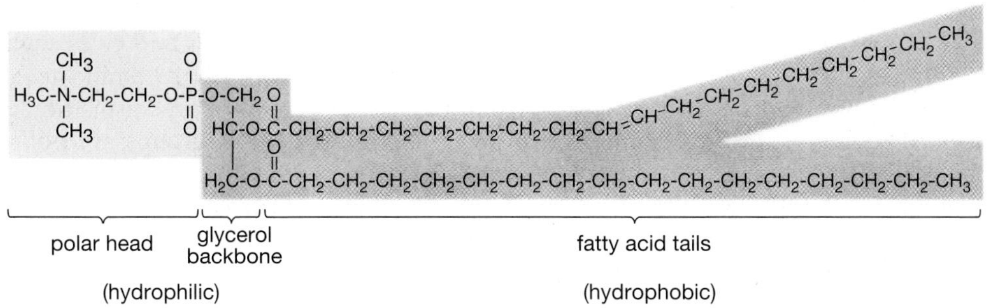

polar head (hydrophilic)
glycerol backbone
fatty acid tails (hydrophobic)

FIGURE 3-8 Phospholipids Phospholipids have only two fatty acid tails attached to the glycerol backbone. The third position is occupied by a polar head consisting of a phosphate group ($-PO_4^-$) to which a second (typically nitrogen-containing) functional group is attached. The phosphate group is negatively charged, and the nitrogen-containing group is positively charged.

FIGURE 3-9 Steroids
Steroids are synthesized from cholesterol. All steroids have almost the same molecular structure (note the colored, fused carbon rings in all). Differences in steroid function result from differences in functional groups attached to the rings. Notice the similarity between the male sex hormone testosterone and the female sex hormone estradiol (estrogen). **QUESTION** Why are steroid hormones able to act by binding with molecules inside the cell nucleus, while other types of hormones (i.e., not steroids) act only by interacting with molecules on the outside of the cell membrane?

"tails," which are insoluble in water, the phosphate–nitrogen "head" is polar, and so is water soluble. Thus, a phospholipid has two dissimilar ends: a hydrophilic head attached to hydrophobic tails. As you will see in Chapter 4, this dual nature of phospholipids is crucial to the structure and function of the plasma membrane.

Steroids Consist of Four Carbon Rings Fused Together

Steroids are structurally different from all other lipids. All steroids are composed of four rings of carbon fused together with various functional groups protruding from them (Fig. 3-9). One type of steroid is *cholesterol*; an egg yolk supplies more than half of your recommended daily allowance. Cholesterol is a vital component of the membranes of animal cells and is also used by cells to synthesize other steroids, including male and female sex hormones, hormones that regulate salt levels, and bile that assists in fat digestion.

3.5 What Are Proteins?

Proteins are molecules composed of one or more chains of *amino acids*. Proteins perform many functions; this diversity of function is made possible by the diversity of protein structures (Table 3-3). **Enzymes** are important proteins that guide almost all chemical reactions that occur inside cells, as you will learn in Chapter 6. Be-

Table 3-3 Functions of Proteins

Function	Example
Structure	Collagen in skin; keratin in hair, nails, horns
Movement	Actin and myosin in muscle
Defense	Antibodies in bloodstream
Storage	Albumin in egg white
Signaling	Growth hormone in bloodstream
Catalyzing reactions	Enzymes (Examples: amylase digests carbohydrates; ATP synthase makes ATP)

cause each enzyme assists only one or a few specific reactions, most cells contain hundreds of different enzymes. Other types of proteins are used for structural purposes, such as *elastin*, which gives skin its elasticity; *keratin*, the principal protein of hair, horns, nails, scales, and feathers; and the silk of spider webs and silk moth cocoons (Fig. 3-10). Still other proteins provide a source of amino acids for developing young animals, such as *albumin* protein in egg white and *casein* protein in milk. The protein hemoglobin transports oxygen in the blood, while contractile proteins, such as those that comprise muscle, allow for cell movement. Some hormones are proteins, including insulin and growth hormone, as are *antibodies* (which help fight disease and infection) and many poisons produced by animals, such as rattlesnake venom.

Proteins Are Formed from Chains of Amino Acids

Proteins are polymers of **amino acids**. All amino acids have the same fundamental structure, consisting of a central carbon bonded to four different functional groups: a nitrogen-containing *amino group* ($-NH_2$); a carboxyl, or carboxylic acid, group (–COOH); a hydrogen; and a variable group (represented by the letter R):

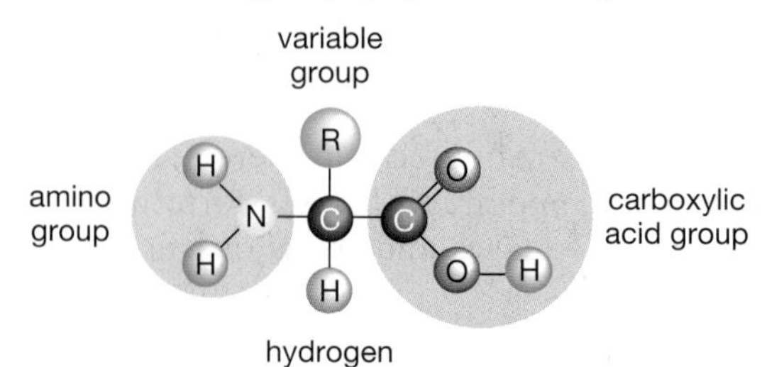

The R group differs among amino acids and gives each its distinctive properties (Fig. 3-11). Twenty amino acids are commonly found in the proteins of organisms. Some amino acids are hydrophilic; their R groups are polar and soluble in water. Others are hydrophobic, with nonpolar R groups that are insoluble in water. Another type of amino acid, cysteine, has sulfur in its R group and can form covalent bonds with other cysteines; these bonds

(a) Hair

(b) Horn

(c) Silk

FIGURE 3-10 Structural proteins
Common structural proteins include the keratin of **(a)** hair and **(b)** horn, and **(c)** the silk protein in a spider web.

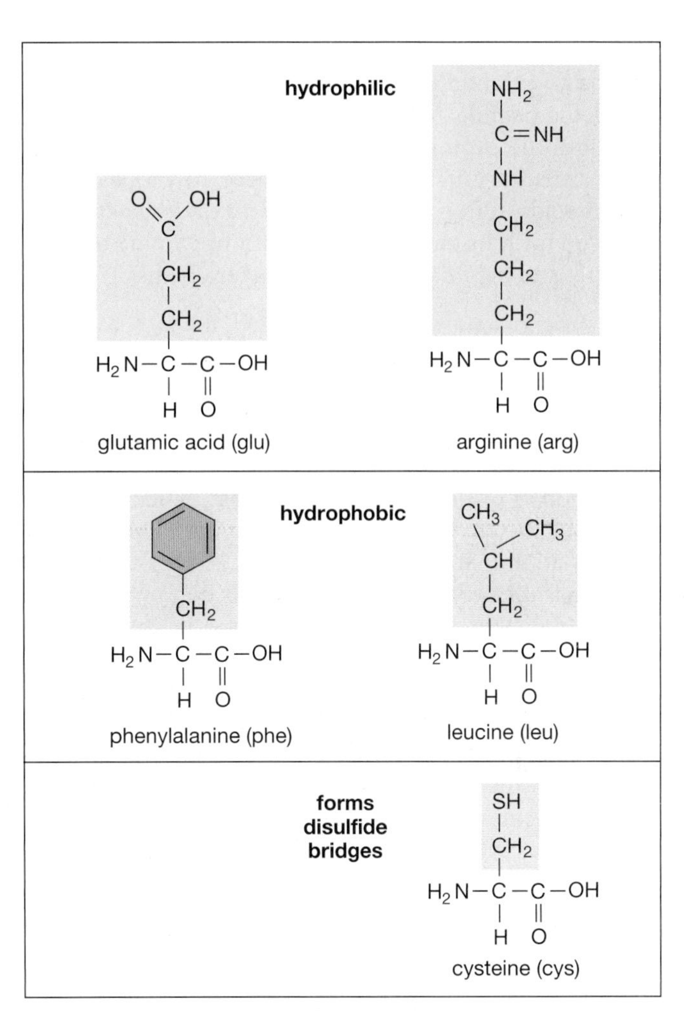

FIGURE 3-11 Amino acid diversity
The diversity of amino acid structures is a consequence of the variable R group (colored blue), which may be hydrophilic or hydrophobic. Cysteine is in a class by itself. Two cysteines in distant parts of a protein molecule can form a covalent bond between their sulfur atoms, creating a disulfide bridge that brings the cysteines very close together and bends the protein chain or links adjacent chains.

are called **disulfide bridges**. Disulfide bridges can link different polypeptide chains together or connect different parts of the same polypeptide chain, causing the protein to bend or fold.

Amino acids differ in their chemical and physical properties—size, water solubility, electrical charge—because of their different R groups. Therefore, the exact sequence of amino acids dictates the function of each protein: whether it is water-soluble and whether it is an enzyme, a hormone, or a structural protein. Scrambled sequences of amino acids are useless; in some cases, just one wrong amino acid can cause a protein to function incorrectly.

Amino Acids Are Joined to Form Chains by Dehydration Synthesis

Like lipids and polysaccharides, proteins are formed by dehydration synthesis. The nitrogen in the amino group ($-NH_2$) of one amino acid is joined to the carbon in the carboxyl group ($-COOH$) of another amino acid by a single covalent bond (Fig. 3-12). This bond is called a **peptide bond**, and the resulting chain of two amino acids is called a **peptide**. More amino acids are added, one by one, until the protein is complete. Amino acid chains in living cells vary in length from three to thousands of amino acids. Often, the word *protein* or *polypeptide* is reserved for long chains—say, 50 or more amino acids in length—and *peptide* is used for shorter chains.

A Protein Can Have Up to Four Levels of Structure

Proteins are highly organized molecules that come in a variety of shapes. Biologists recognize four levels of organization in protein structure. A single molecule of hemoglobin, the oxygen-carrying protein in red blood cells, exhibits all four structural levels (Fig. 3-13). The **primary structure** is the sequence of amino acids that make up the protein (see Fig. 3-13a). This sequence is specified by genes within molecules of DNA. Different types of proteins have different sequences of amino acids.

Polypeptide chains typically acquire one or both of two simple, repeating **secondary structures**. You may recall that hydrogen bonds can form between parts of polar molecules that have slight negative and slight positive charges, which attract one another (see Chapter 2). Hydrogen bonds between parts of amino acids give rise to the secondary structures of proteins. Many proteins, such as keratin and subunits of the hemoglobin molecule (see Fig. 3-13b) have a coiled, springlike secondary structure called a **helix**. Hydrogen bonds between the relatively negative oxygen of the –C=O and the relatively positive hydrogen of the –N–H amino acid groups hold the turns of the coils together. Other proteins, such as silk, consist of polypeptide chains that repeatedly fold back upon themselves, with hydrogen bonds holding adjacent segments of the polypeptide together in a **pleated sheet** arrangement (Fig. 3-14).

In addition to their secondary structures, proteins assume complex, three-dimensional **tertiary structures** that determine the final configuration of the polypeptide (see Fig. 3-13c). Probably the most important influence on the tertiary structure of a protein is its cellular environment—specifically, whether the protein is dissolved in the watery cytoplasm within a cell, in the lipids of cellular membranes, or spanning the membrane and thus straddling the two environments. Hydrophilic amino acids can form hydrogen bonds with nearby water molecules, whereas hydrophobic amino acids cannot. Therefore, a protein dissolved in water folds, exposing its hydrophilic amino acids to the watery environment outside, and causing its hydrophobic amino acids to cluster together in the center of the molecule. Disulfide bridges can also contribute to tertiary structure, by linking cysteine amino acids from different regions of the polypeptide. In keratin, disulfide bridges within individual helical polypeptides can distort them, creating a tertiary structure that makes the hair kinky or curly (see "Links to Life: A Hairy Subject"):

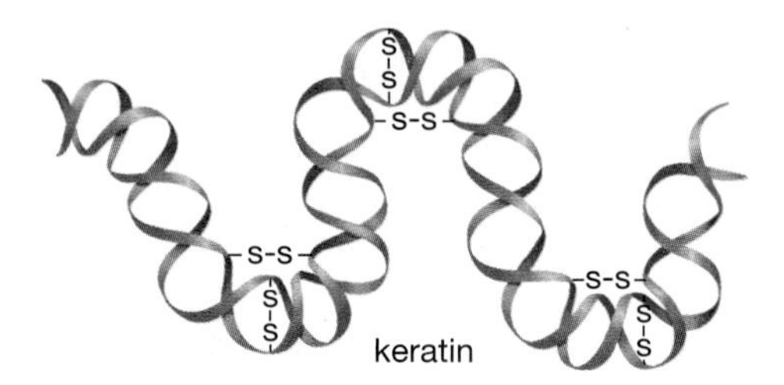

Individual polypeptides may sometimes be linked together, forming the fourth level of protein organization, called **quaternary structure**. Hemoglobin consists of four polypeptide chains (two pairs of very similar peptides) held together by hydrogen bonds (see Fig. 3-13d). Each of the four peptides holds an iron-containing organic molecule called a *heme* (the red disks in Fig. 3-13c, d), which can bind one molecule of oxygen.

The Functions of Proteins Are Linked to Their Three-Dimensional Structures

Within a protein, the exact type, position, and number of amino acids bearing specific R groups determine both the structure of the protein and its biological function. In any given protein, however, some amino acids are more important than others. In hemoglobin, for example, certain amino acids bearing specific R groups must be present in precisely the right places to hold the iron-containing heme group that binds oxygen. Some of the other amino acids are interchangeable if they are functionally equivalent.

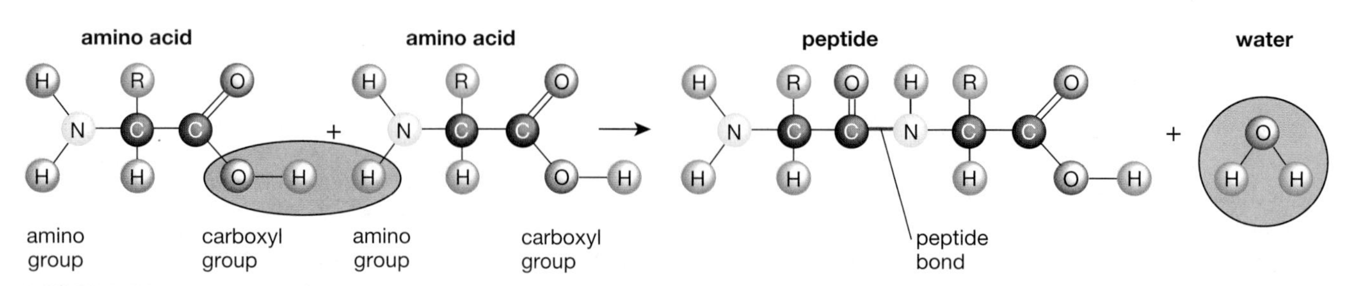

FIGURE 3-12 Protein synthesis
In protein synthesis, a dehydration synthesis joins the carbon of the carboxyl acid group of one amino acid to the nitrogen of the amino group of a second amino acid. The resulting covalent bond is called a *peptide bond*.

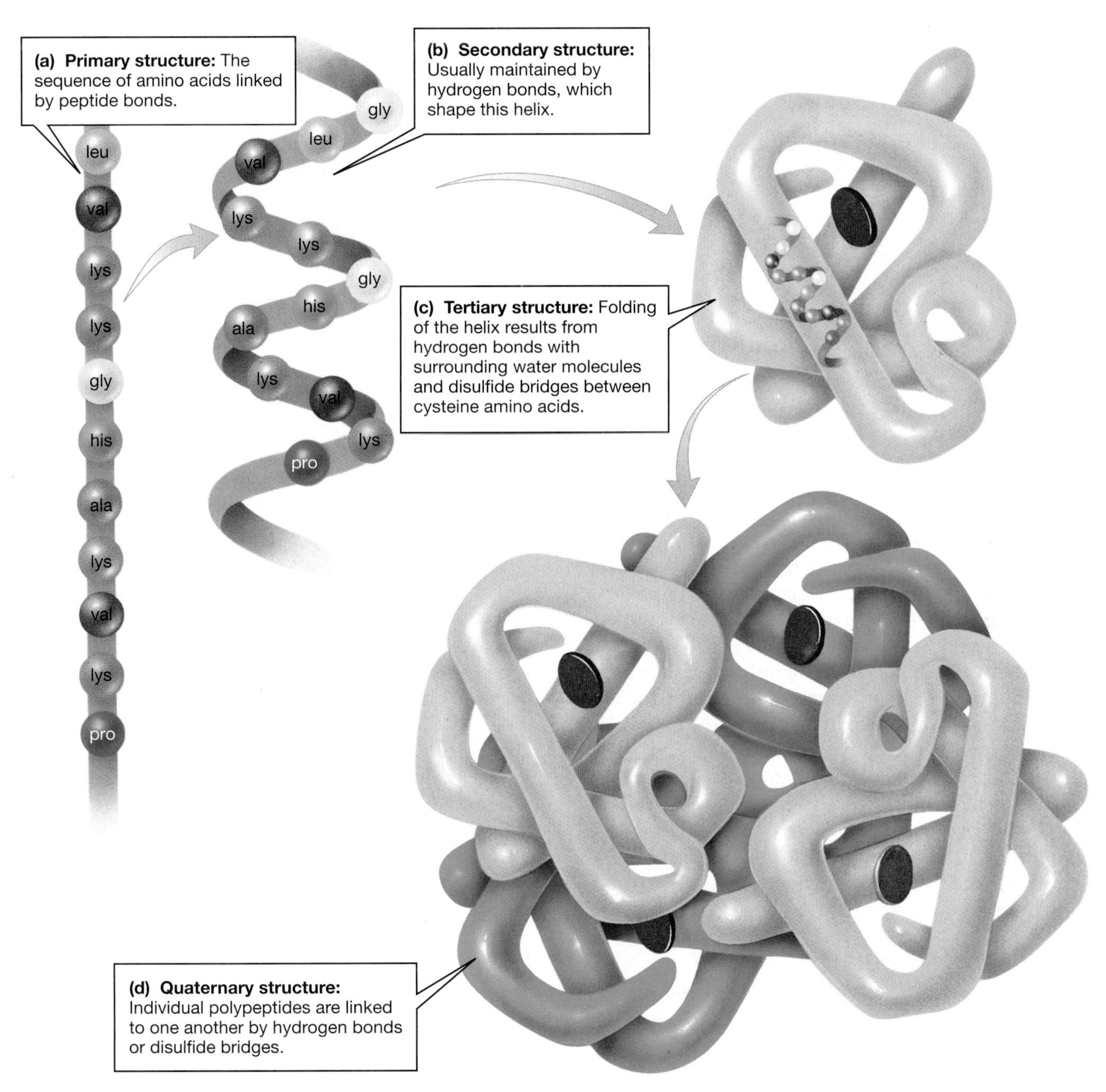

(a) Primary structure: The sequence of amino acids linked by peptide bonds.
leu
val
lys
lys
gly
his
ala
lys
val
lys
pro
(b) Secondary structure: Usually maintained by hydrogen bonds, which shape this helix.
gly
leu
val
lys
lys
gly
his
ala
lys
val
lys
pro
(c) Tertiary structure: Folding of the helix results from hydrogen bonds with surrounding water molecules and disulfide bridges between cysteine amino acids.
(d) Quaternary structure: Individual polypeptides are linked to one another by hydrogen bonds or disulfide bridges.

FIGURE 3-13 The four levels of protein structure
Levels of protein structure are represented here by hemoglobin, the oxygen-carrying protein in red blood cells (the red discs represent the iron-containing heme group that binds oxygen). All levels of protein structure are determined by the amino acid sequence of the protein, interactions among the R groups of the amino acids, and interactions between the R groups and their surroundings. **QUESTION** Why do most proteins, when heated, lose their ability to function?

For instance, the amino acids on the outside of a hemoglobin molecule serve mostly to keep it dissolved in the cytoplasm of a red blood cell. Therefore, as long as they are hydrophilic, changes in amino acids may not alter the function of the protein. As you will see in Chapter 12, however, replacing a hydrophilic amino acid with a hydrophobic amino acid can have catastrophic effects on the solubility of the hemoglobin molecule. In fact, such a substitution is the molecular cause of a painful and sometimes life-threatening disorder called sickle-cell anemia.

For an amino acid to be in the proper location within a protein, the amino acids must be in their proper sequence. Likewise, the protein must have the correct secondary and tertiary structures so that the amino acid is correctly positioned within the protein. When the secondary and tertiary structures of a protein are altered

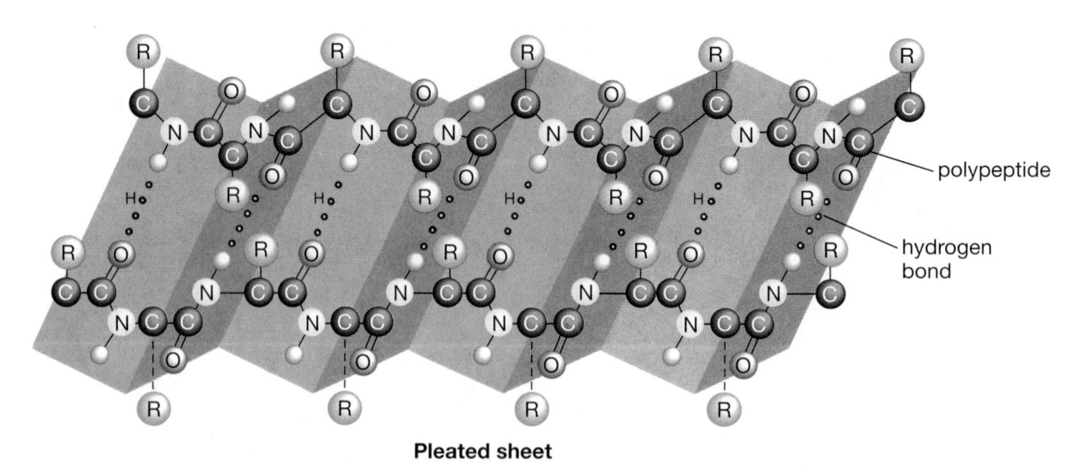

FIGURE 3-14 The pleated sheet is an example of protein secondary structure
In a pleated sheet, a single polypeptide chain is folded back upon itself repeatedly (connecting portions of the chain have been omitted). Adjacent segments of the folded polypeptide are linked by hydrogen bonds (dotted lines), creating a sheet-like configuration. The R groups (green) project alternately above and below the sheet. Despite its accordion-pleated appearance, produced by bonding patterns between adjacent amino acids, each peptide chain is in a fully extended state and cannot easily be stretched farther. For this reason, pleated sheet proteins such as silk are not elastic.

(leaving the peptide bonds between amino acids intact), the protein is said to be **denatured**, and it will no longer perform its function. There are many ways to denature protein. In a fried egg, for example, the heat of the frying pan denatures the albumin protein in the egg white, causing its appearance to change from clear to white and its texture to change from liquid to solid. Sterilization using heat or ultraviolet rays denatures the proteins of bacteria or viruses and causes them to lose their function. Salty, acidic solutions also denature proteins—dill pickles are preserved in this way.

3.6 What Are Nucleic Acids?

Nucleic acids are long chains of similar but not identical subunits called **nucleotides.** All nucleotides have a three-part structure: a five-carbon sugar (ribose or deoxyribose), a phosphate group, and a nitrogen-containing base that differs among nucleotides:

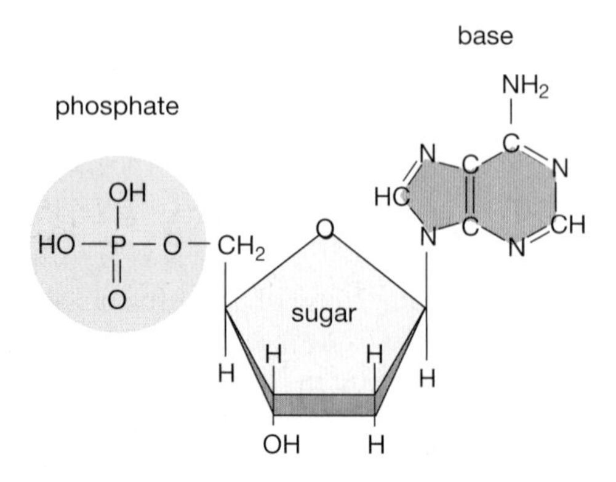

There are two types of nucleotides: ribose nucleotides (containing the sugar ribose) and deoxyribose nucleotides (containing the sugar deoxyribose). Deoxyribose nucleotides bond to four types of nitrogen-containing bases (adenine, guanine, cytosine, and thymine). Similarly, ribose nucleotides bond to four types of bases (adenine, guanine, cytosine, and uracil instead of thymine).

Nucleotides may be strung together in long chains, forming nucleic acids with the phosphate group of one nucleotide covalently bonded to the sugar of another:

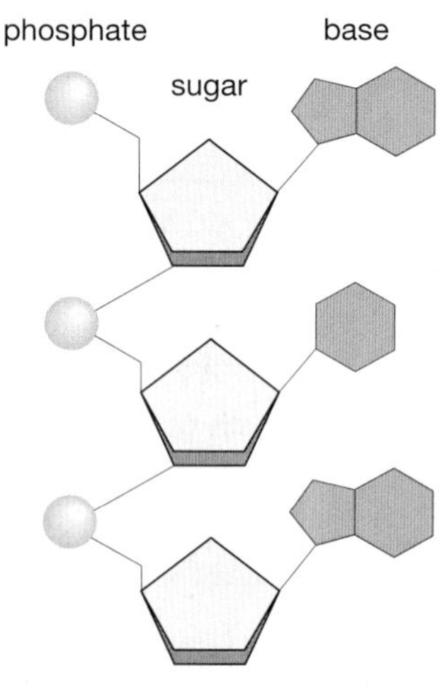

There are two types of nucleic acids: *deoxyribonucleic acid (DNA)* and *ribonucleic acid (RNA)*.

DNA and RNA, the Molecules of Heredity, Are Nucleic Acids

Deoxyribose nucleotides form chains millions of units long called **deoxyribonucleic acid**, or **DNA**. DNA is found in the chromosomes of all living things. Its se-

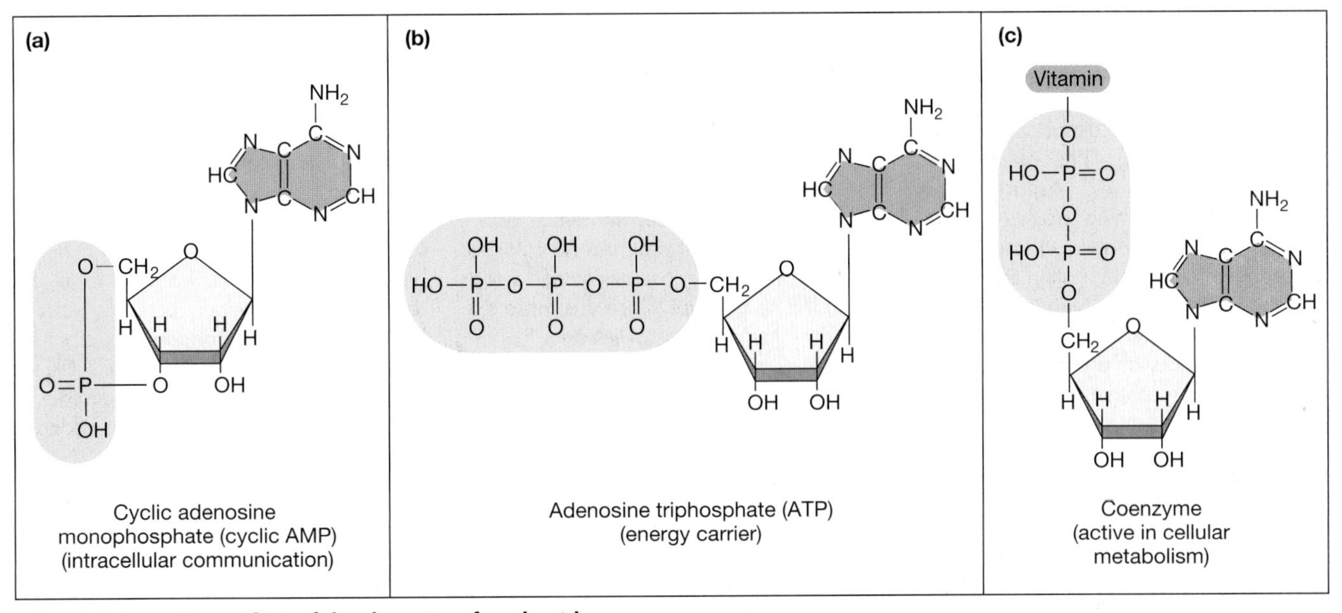

FIGURE 3-15 A sampling of the diversity of nucleotides
Individual nucleotides, constructed of a sugar, a phosphate group, and a nitrogen-containing base, are often modified by the addition of different functional groups and serve a variety of cellular functions.

quence of nucleotides, like the dots and dashes of a biological Morse code, spells out the genetic information needed to construct the proteins of each organism. Chains of ribose nucleotides, called **ribonucleic acid**, or **RNA**, are copied from the central repository of DNA in the nucleus of each cell. The RNA carries DNA's genetic code into the cell's cytoplasm and directs the synthesis of proteins. (DNA and RNA are described in detail in Chapters 9 and 10.)

Other Nucleotides Act as Intracellular Messengers, Energy Carriers, or Coenzymes

Not all nucleotides are part of nucleic acids. Some exist singly in the cell or occur as parts of other molecules. **Cyclic nucleotides**, such as *cyclic adenosine monophosphate* (*cyclic AMP*; Fig. 3-15a), are intracellular messengers that carry chemical signals from the plasma membrane to other molecules in the cell. Cyclic AMP is synthesized when certain hormones bind to other proteins in the plasma membrane; it then initiates biochemical reactions in the cytoplasm or nucleus that determine the effect of the hormone.

Some nucleotides have additional phosphate groups. These diphosphate and triphosphate nucleotides, such as **adenosine triphosphate** (**ATP**; Fig. 3-15b), are unstable molecules that carry energy, stored in bonds between the phosphate groups, from place to place within a cell. They capture energy where it is produced (during photosynthesis, for example) and release it to drive energy-demanding reactions elsewhere (say, to synthesize a protein). Other nucleotides (NAD+ and FAD) are known as "electron carriers" and transport energy in the form of high energy electrons. You will learn more about electron carriers and their role in cellular metabolism in Chapters 7 and 8.

Finally, certain nucleotides assist enzymes in their role of promoting and guiding chemical reactions. These nucleotides are called **coenzymes**. Most coenzymes consist of a nucleotide combined with a vitamin (Fig. 3-15c). You will learn more about energy-carrier nucleotides, enzymes, and coenzymes in Chapter 6, where we discuss energy release and use in the cell.

CASE STUDY REVISITED Improving on Nature?

The artificial sweetener aspartame carries an important warning: "Phenylketonurics: contains phenylalanine." Aspartame is composed of two amino acids: aspartic acid and phenylalanine. (Researchers aren't sure why this combination excites the same receptors on our tongues as does sugar.) You now know that amino acids are the building blocks of proteins. Phenylketonurics lack an enzyme that converts phenylalanine into tyrosine, another amino acid. Thus, phenyketonurics have abnormally high levels of phenylalanine, which is toxic to the developing nervous system and can cause brain damage and mental retardation. A pregnant woman with phenylketonuria must restrict her intake of this amino acid to avoid harming her developing child. Newborns are immediately tested for phenylalanine levels in urine, and they are put on a diet low in phenylalanine if they have this metabolic disorder.

The newly approved sucralose is made from sucrose that has been modified so that three of its hydroxyl groups are replaced with chlorine atoms. Sucralose activates our sweet taste receptors 600 times as effectively as sucrose, but our enzymes cannot digest it, so it provides no calories. Look for more sucralose-sweetened products in the near future, since it is more stable than other artificial sweeteners.

To understand olestra, recall that oils combine a glycerol backbone with three fatty acid chains. Olestra, however, contains a sucrose backbone with six to eight fatty acids attached. Apparently, the large number of fatty acid chains prevents digestive enzymes from reaching the digestible sucrose backbone of the olestra molecule. Since the molecule is not broken into absorbable fragments, it is excreted unchanged. Results of exhaustive tests on olestra left two concerns. First, since the oil passes undigested through the intestine, it acts as an intestinal lubricant and can have a laxative effect on people who consume large quantities. Second, olestra can reduce absorption of fat-soluble vitamins in food by dissolving the vitamins and carrying them out of the body. Before approving olestra, the FDA required that foods containing it be supplemented with fat-soluble vitamins and be identified with a warning label that reads: "This product contains olestra. Olestra may cause abdominal cramping and loose stools. Olestra inhibits the absorption of some vitamins and nutrients. Vitamins A, D, E, and K have been added."

Consider This: Some experts argue that the advantages of being able to eat appealing food while limiting fat and sugar consumption makes artificial sweeteners and fake oils and fats worthwhile. Others contend that people should select natural, nutritious foods that are already low in sugar and fats and shun these artificial dieting aids. Think of arguments on both sides of this issue. You might want to look at the case study in Chapter 30 for more about obesity.

Links to Life: A Hairy Subject

Pull out a long strand of hair—either your own or a friend's. You know now that hair is composed mostly of a helical protein called *keratin*. Notice the root or follicle that was embedded in the scalp. Living cells in the hair follicle produce new keratin at the rate of 10 turns of the protein helix every second. The keratin proteins in a hair entwine around one another, held together by disulfide bridges. If you pull gently on the hair, you will find it to be both strong and stretchy. When hair stretches, the hydrogen bonds within the keratin helix are broken, allowing the helix to be extended. Most of the covalent disulfide bonds, in contrast, are distorted by stretching but do not break. When tension is released, these disulfide bridges return the hair to its normal length, and the hydrogen bonds re-form. Now wet the hair and notice how limp it becomes. In wet hair, the hydrogen bonds of the helices are replaced by hydrogen bonds between the amino acids and the water molecules surrounding them, so the helices collapse. Notice that the hair is now easier to stretch, as well. If you roll the wet hair around a small rod and allow it to dry, the hydrogen bonds will re-form in slightly different places, holding the hair in a curl. However, the slightest moisture (even humid air) allows these hydrogen bonds to rearrange into their natural configuration.

If your hair is naturally curly (because of the particular sequence of amino acids specified by your genes), the disulfide bridges within and between the individual keratin helices form at locations that bend the keratin molecules, producing a curl (in straight hair, the disulfide bridges occur in places that do not distort the keratin). When straight hair is given a "permanent wave," two lotions are applied. The first lotion breaks disulfide bonds. After the hair is rolled tightly onto curlers, a second solution is applied which re-forms the disulfide bridges. The new disulfide bridges reconnect the keratin helices at new positions determined by the curler. These new bridges are more or less permanent, transforming genetically straight hair into "biochemically curly" hair.

Summary of Key Concepts

3.1 Why Is Carbon So Important in Biological Molecules?

Organic molecules are so diverse because the carbon atom is able to form many types of bonds. This ability, in turn, allows organic molecules (molecules with a backbone of carbon and hydrogen atoms) to form many complex shapes, including chains, branches, and rings. The presence of functional groups, shown in Table 3-1, produces further diversity among biological molecules.

3.2 How Are Organic Molecules Synthesized?

Most large biological molecules are polymers synthesized by linking together many smaller subunits, or monomers. The subunits are connected by covalent bonds through dehydration synthesis; the chains may be broken apart by hydrolysis reactions. The most important organic molecules fall into one of four classes: carbohydrates, lipids, proteins, and nucleic acids. Their major characteristics are summarized in Table 3-2.

3.3 What Are Carbohydrates?

Carbohydrates include sugars, starches, chitin, and cellulose. Sugars (monosaccharides and disaccharides) are used for temporary storage of energy and for the construction of other molecules. Starches and glycogen are polysaccharides that provide longer-term energy storage in plants and animals, respectively. Cellulose forms the cell walls of plants, and chitin strengthens the exoskeletons of many invertebrates and many types of fungi. Other types of polysaccharides form the cell walls of bacteria.

3.4 What Are Lipids?

Lipids are nonpolar, water-insoluble molecules of diverse chemical structure. Lipids include oils, fats, waxes, phospholipids, and steroids. Lipids are used for energy storage (oils and fats), as waterproofing for the outside of many plants and animals (waxes), as the principal component of cellular membranes (phospholipids), and as hormones (steroids).

3.5 What Are Proteins?

Proteins are chains of amino acids that possess primary, secondary, tertiary and sometimes quaternary structure. Both the structure and the function of a protein are determined by the sequence of amino acids in the chain, and how these amino acids interact with their surroundings and with each other. Proteins can be enzymes (which guide chemical reactions), structural molecules (hair, horn), hormones (insulin), or transport molecules (hemoglobin).

3.6 What Are Nucleic Acids?

The nucleic acid molecules deoxyribonucleic acid (DNA) and ribonucleic acid (RNA) are chains of nucleotides. Each nucleotide is composed of a phosphate group, a sugar group, and a nitrogen-containing base. Molecules formed from single nucleotides include intracellular messengers (cyclic AMP), energy-carrier molecules (ATP), and coenzymes.

Key Terms

adenosine triphosphate (ATP) *p. 51*
amino acid *p. 46*
carbohydrate *p. 40*
cellulose *p. 42*
chitin *p. 42*
coenzyme *p. 51*
cyclic nucleotide *p. 51*
dehydration synthesis *p. 39*
denatured *p. 50*
deoxyribonucleic acid (DNA) *p. 50*
disaccharide *p. 40*
disulfide bridge *p. 47*
enzyme *p. 46*
fat *p. 43*
fatty acid *p. 43*
functional group *p. 38*
glucose *p. 40*
glycerol *p. 43*
glycogen *p. 41*
helix *p. 48*
hydrolysis *p. 39*
inorganic *p. 38*
lactose *p. 41*
lipid *p. 43*
maltose *p. 41*
monomer *p. 38*
monosaccharide *p. 40*
nucleic acid *p. 50*
nucleotide *p. 50*
oil *p. 43*
organic *p. 38*
peptide *p. 48*
peptide bond *p. 48*
phospholipid *p. 45*
pleated sheet *p. 48*
polymer *p. 39*
polysaccharide *p. 40*
primary structure *p. 48*
protein *p. 46*
quaternary structure *p. 48*
ribonucleic acid (RNA) *p. 51*
saturated *p. 44*
secondary structure *p. 48*
starch *p. 41*
steroid *p. 46*
sucrose *p. 41*
sugar *p. 40*
tertiary structure *p. 48*
triglyceride *p. 43*
unsaturated *p. 44*
wax *p. 45*

Thinking Through the Concepts

To take a multiple-choice quiz with feedback on the contents of this chapter, visit http://www.prenhall.com/audesirk7. *Log in to the Web site selected by your instructor and navigate to the Self Test section for this chapter.*

? Review Questions

1. Which elements are common components of biological molecules?
2. List the four principal types of biological molecules and give an example of each.
3. What roles do nucleotides play in living organisms?
4. One way to convert corn oil to margarine (a solid at room temperature) is to add hydrogen atoms, decreasing the number of double bonds in the molecules of oil. What is this process called? Why does it work?
5. Describe and compare dehydration synthesis and hydrolysis. Give an example of a substance formed by each chemical reaction, and describe the specific reaction in each instance.
6. Distinguish among the following: monosaccharide, disaccharide, and polysaccharide. Give two examples of each and their functions.
7. Describe the synthesis of a protein from amino acids. Then describe primary, secondary, tertiary, and quaternary structures of a protein.
8. Most structurally supportive materials in plants and animals are polymers of special sorts. Where would we find cellulose? Chitin? In what way(s) are these two polymers similar? Different?
9. Which kinds of bonds or bridges between keratin molecules are altered when hair is (a) wet and allowed to dry on curlers and (b) given a permanent wave?

Applying the Concepts

1. A preview question for Chapter 4: In Chapter 2, you learned that hydrophobic molecules tend to cluster when immersed in water. In this chapter, you discovered that a phospholipid has a hydrophilic head and hydrophobic tails. What do you think would be the configuration of phospholipids that are immersed in water?
2. Many birds must store large amounts of energy to power flight during migration. Which type of organic molecule would be the most advantageous for energy storage? Why?
3. Do you remember the nuclear accident at Chernobyl in 1986? A scientist suspects that the food in a nearby ecosystem may have been contaminated with radioactive nitrogen over a period of months. Which substances in plants and animals could be examined for radioactivity to test his hypothesis?
4. Fat contains twice as many calories per unit weight as carbohydrate does, so fat is an efficient way for animals, who must move about, to store energy. Compare the way fat and carbohydrates interact with water, and explain why this interaction also gives fat an advantage for weight-efficient energy storage.

For More Information

Burdick, A., "Cement on the Half Shell." *Discover*, February 2003. Mussels produce a protein polymer that is waterproof and incredibly strong.

Gorman J. "Trans Fat." *Science News*, November 10, 2001. Reviews the structure and origin of trans fats and studies that implicate it in heart disease.

Hill, J. W., and Kolb, D. K. *Chemistry for Changing Times*. 10th ed. Upper Saddle River, NJ: Prentice Hall, 2004. A chemistry textbook for nonscience majors that is both clearly readable and thoroughly enjoyable.

King, J., Haase-Pettingell, C., and Gossard, D. "Protein Folding and Misfolding." *American Scientist*, September–October 2002. Protein folding holds the key to diverse functions.

Kunzig, R., "Arachnomania." *Discover*, September 2001. Researchers work to unravel the mystery of spider silk, and develop a process to synthesize it.

Radetsky, P. "Kim's Coils." *Discover*, June 1995. Biochemist Peter Kim is uncovering the relationship between structure and function in the elaborate coiling of proteins.

Sharon, N., and Lis, H. "Carbohydrates in Cell Recognition." *Scientific American*, January 1993. Sugar-protein complexes on the surfaces of cells regulate cell identification and interaction between cells.

Media Activities

To access a Media Activity visit http://www.prenhall.com/audesirk7. *Log in to the Web site selected by your instructor, navigate to this chapter, and select the appropriate Media Activity number.*

3.1 Structure of Biological Molecules

Estimated time: 10 minutes

Review the chemical structures of the most important biological macromolecules.

3.2 Functions of Macromolecules

Estimated time: 5 minutes

Explore the function of the major macromolecules.

3.3 Web Investigation: Improving on Nature?

Estimated time: 10 minutes

Whereas most artificial sweeteners, food colors, and fat replacers can be used for any food product, olestra is currently approved by the U.S. Food and Drug Administration (FDA) for use in "savory snacks" (e.g., potato chips) only. Some experts say that it is safe. Others think the side effects outweigh the benefits. In this exercise we will examine the pros and cons of the issue.

CHAPTER

4 Cell Membrane Structure and Function

A rattlesnake prepares to strike.

AT A GLANCE

CASE STUDY Vicious Venoms

Eager to explore their new environs, Karl and Mark, freshmen roommates at a university in southern California, drove to a trailhead in the Mojave Desert. Karl kidded Mark about his cellular phone—what kind of a wilderness experience could they have with a phone along? Mark joked about the large field guide, *Desert Flora and Fauna*, weighing down Karl's pack. Competitive and athletic, after several miles of hiking they spotted a rocky bluff and raced each other to the top. The sudden exertion in the hot sun made Karl feel momentarily lightheaded, and he reached for a rocky outcropping to steady himself. Expecting solid stone, he gasped at the feel of thick, scaly coils writhing under his hand. A sudden unmistakable warning rattle was followed almost immediately by an intense burning pain at the base of his thumb. Karl's yell brought Mark running. Seeing the huge snake slithering back into the protection of a crevice, Mark helped Karl, who was now feeling dizzy and nauseated, to lie down and then quickly rummaged in his pack for his cellular phone. The 911 dispatcher told them to wait quietly for a medical evacuation helicopter. By the time they heard the chopper, they had used Karl's field guide to identify the rattler as a Western Diamondback. Before he reached the hospital, a large bruised-looking area was spreading over Karl's hand, his blood pressure had dropped, and the paramedics were administering oxygen because he was gasping for air. Why was his hand bruising rapidly? Why did Karl feel short of breath? What treatment will Karl receive?

4.1 How Is the Structure of a Membrane Related to Its Function?

The Plasma Membrane Isolates the Cell While Allowing Communication with Its Surroundings

In Chapter 1, we defined the *cell* as the smallest unit of life. Each cell is surrounded by a thin **plasma membrane** that acts as a gatekeeper, allowing only specific substances in or out and passing chemical messages from the external environment to the cell's interior. As gatekeeper, the plasma membrane performs three general functions:

- It selectively isolates the cell's contents from the external environment.
- It regulates the exchange of essential substances between the cell's contents and the external environment.
- It communicates with other cells.

These are formidable tasks for a structure so thin that 10,000 plasma membranes stacked atop one another would scarcely equal the thickness of this page. The key to membrane function lies in its structure. Membranes are not simply homogeneous sheets; they are complex, heterogeneous structures whose different parts perform very distinct functions, and they change dynamically in response to their surroundings.

Most cells have internal membranes as well as plasma membranes that surround them. These internal membranes form compartments in which specialized biochemical activities can occur. All the membranes of a cell have a similar basic structure: proteins floating in a double layer of lipids. Lipids are responsible for the isolating function of membranes, whereas proteins regulate the exchange of substances and communication with the environment. Although this chapter focuses on the plasma membrane, much of the information presented here also applies to other cellular membranes.

Membranes Are "Fluid Mosaics" in Which Proteins Move Within Layers of Lipids

The **fluid mosaic model** of cellular membranes was developed in 1972 by cell biologists S. J. Singer and G. L. Nicolson. According to this model, a membrane, when viewed from above, looks something like a lumpy, constantly shifting mosaic of tiles (Fig. 4-1). A double layer of phospholipids forms a viscous, fluid "grout" for the mosaic; assorted proteins are the "tiles," which can move about within the phospholipid layers. Thus, although the components within the plasma membrane remain relatively constant, the overall distribution of proteins and various types of phospholipids can change over time. As strange as this model may seem, it captures something of the dynamic quality of real membranes. Now let's look more closely at the structure of membranes.

The Phospholipid Bilayer Is the Fluid Portion of the Membrane

As you learned in Chapter 3, a phospholipid consists of two very different parts: a polar, hydrophilic head (attracted to water) and a pair of nonpolar, hydrophobic tails (repelled by water). Notice that the double bond (which makes the lipid unsaturated) creates a kink in the tail that helps keep the membrane fluid at lower temperatures:

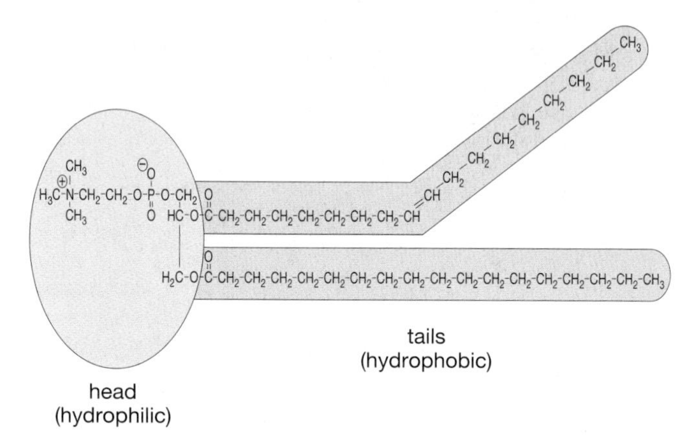

All cells are surrounded by a watery medium. Single-celled organisms may live in fresh water or in the ocean, while animal cells are bathed in a weakly salty *extracellular fluid* that filters out of the blood. The **cytoplasm** consists of all of a cell's internal contents (including all the organelles except the nucleus, in eukaryotes); cytoplasm is mostly water. Plasma membranes separate the watery cytoplasm from its watery external environment, and similar membranes surround watery compartments within the cell. Under these conditions, phospholipids spontaneously arrange themselves into a double layer called a **phospholipid bilayer**, in which the hydrophilic heads form the outer borders and the hydrophobic tails "hide" inside:

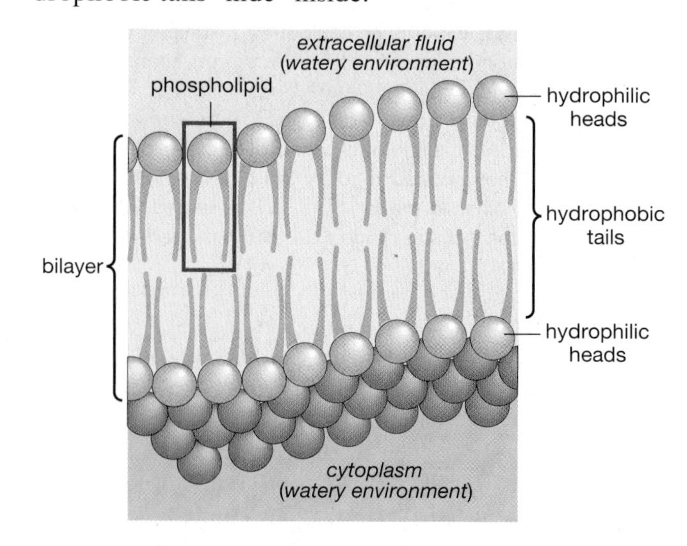

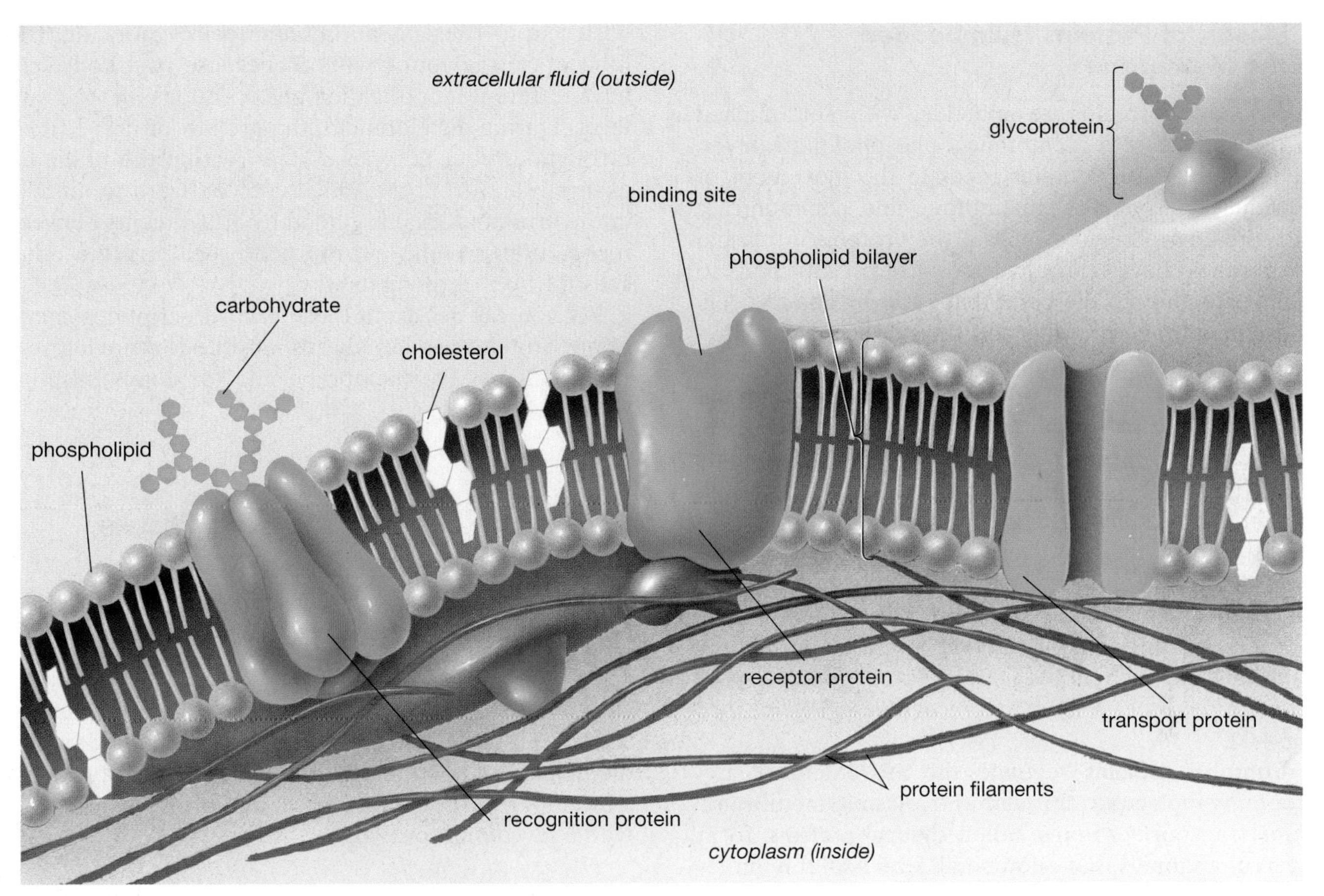

FIGURE 4-1 The plasma membrane is a fluid mosaic
The plasma membrane is a bilayer of phospholipids in which various proteins are embedded. Many proteins have carbohydrates attached to them, forming glycoproteins.

Media Activity
4.1 Membrane Structure and Transport

Hydrogen bonds can form between water and the phospholipid heads, so the hydrophilic heads face both the cytoplasm and the extracellular fluid, forming the outer layer of the bilayer. Hydrophobic interactions (see Chapter 2) cause the phospholipid tails to hide inside the bilayer. Because individual phospholipid molecules are not bonded to one another and the lipid tails contain saturated bonds, this double layer is quite fluid; individual phospholipids move about easily within each layer.

Most biological molecules, including salts, amino acids, and sugars, are polar and water soluble, and so are hydrophilic. In fact, most substances that contact a cell are water soluble—hydrophilic—and so cannot easily pass through the nonpolar, hydrophobic fatty acid tails of the phospholipid bilayer. The phospholipid bilayer is largely responsible for the first of the three membrane functions listed earlier: selectively isolating the cell's contents from the external environment. The isolation is not complete, however. As we will describe later, very small molecules, such as water and uncharged, lipid-soluble molecules can pass relatively freely through the phospholipid bilayer. Some of the most devastating effects of snake venom occur because it contains an enzyme that breaks down phospholipids and thereby destroys cell membranes.

In most animal cells, the phospholipid bilayer of membranes also contains cholesterol. Some cellular membranes have just a few cholesterol molecules; others have as many cholesterol molecules as they do phospholipids. Cholesterol affects membrane structure and function in several ways: it makes the bilayer stronger, more flexible but less fluid, and less permeable to water-soluble substances such as ions or monosaccharides.

The flexible, somewhat fluid nature of the bilayer is very important for membrane function. As you breathe, move your eyes, and turn the pages of this book, cells in your body change shape. If plasma membranes were stiff instead of flexible, cells would break open and die. Further, as you will learn in Chapter 5, membranes within eukaryotic cells are in constant motion. Membrane-enclosed compartments ferry substances into the cell, carry materials within the cell, and expel them to the outside, merging membranes in the process. This flow and merger of membranes is made possible by the fluid nature of the lipid bilayer.

A Mosaic of Proteins Is Embedded in the Membrane

Thousands of proteins are embedded within or attached to the surface of a membrane's phospholipid bilayer. Collectively, these proteins regulate the movement of substances through the membrane and communicate with the environment. Many of the proteins in plasma membranes have carbohydrate groups attached to them, especially to the parts that stick outside the cell. These membrane proteins and their attached carbohydrates are called **glycoproteins**.

Many membrane proteins can move about within the relatively fluid phospholipid bilayer. Others, however, are anchored in place to a network of protein filaments within the cytoplasm. The attachments between plasma membrane proteins and the underlying protein filaments produce the characteristic shapes of animal cells, from the dimpled discs of red blood cells to the elaborate branching of nerve cells.

There are three major categories of membrane proteins, each of which serves a different function: *transport proteins*, *receptor proteins*, and *recognition proteins* (see Fig. 4-1).

Transport proteins regulate the movement of hydrophilic molecules through the plasma membrane. Some transport proteins, called **channel proteins**, form pores or channels that allow small water-soluble molecules to pass through the membrane (see Fig. 4-1). Every plasma membrane bears a large assortment of channel proteins that are selective for specific ions such as potassium (K^+), sodium (Na^+), and calcium (Ca^{2+}). Other transport proteins, called **carrier proteins**, have binding sites that can temporarily attach to specific molecules on one side of the membrane. The protein then changes shape, in some cases through the use of cellular energy, and moves the molecule across the membrane. In the next section, we will discuss the mechanisms whereby transport proteins move molecules across the membrane.

Receptor proteins trigger cellular responses when specific molecules in the extracellular fluid, such as hormones or nutrients, bind to them. Most cells bear dozens of types of receptors on their plasma membranes. When activated by the appropriate molecule, some receptors set off elaborate sequences of cellular changes, such as increased metabolic rate, cell division, movement toward a nutrient source, or secretion of hormones. Other receptors act like gates on channel proteins; activating the receptor opens the gates, allowing ions to flow through the channels. For example, receptors allow nerve cells in your brain to communicate with one another (see Chapter 34).

Recognition proteins, many of which are glycoproteins, serve as identification tags and cell-surface attachment sites. The cells of your immune system, for example, recognize a bacterium as a foreign invader and target it for destruction. These same immune cells ignore the trillions of cells in your own body because your body cells have different identification glycoproteins on their surfaces. During development, the growth of nerve fibers throughout your nervous system (within the brain, for example, and from the spinal cord outward to muscles in the arms and legs) is guided by attachments between recognition proteins on the nerve cell and the other cells it traverses along its way.

As you can see from these brief descriptions, membrane proteins are largely responsible for moving substances across the membrane and for communicating with other cells.

4.2 How Do Substances Move Across Membranes?

Molecules in Fluids Move in Response to Gradients

Because the plasma membrane separates the fluid in the cell's cytoplasm from its fluid extracellular environment, let's begin our study of membrane transport with a brief look at the characteristics of fluids. We must start with a few definitions:

- A **fluid** is any substance that can move or change shape in response to external forces without breaking apart; for example, liquids and gases are fluids.
- The **concentration** of molecules in a fluid is the number of molecules in a given unit of volume.
- A **gradient** is a physical difference in properties such as temperature, pressure, electrical charge, or concentration of a particular substance between two adjoining regions. Basic principles of physics tell us that, over time, gradients tend to equalize unless energy is supplied to maintain them. Gradients of concentration, pressure, and electrical charge cause ions or molecules to move from one region to the other in a manner that tends to equalize the difference. Gradients in temperature cause a flow of energy from the higher-temperature region to the lower-temperature region. Cells frequently generate or encounter gradients across their membranes, particularly gradients of ions and various molecules in solution within their cytoplasm and their watery surroundings.

To understand how concentration gradients influence the movement of molecules or ions within a fluid, consider a sugar cube dissolving in coffee, or perfume molecules moving from an open bottle into the air. These substances are moving in response to a **concentration gradient**, a difference in concentration of those substances between one region and another. What causes this movement? The individual molecules in a fluid move continuously, bouncing off one another in random directions. Over time, these random movements will produce a net movement of molecules from regions of high concentration to regions of low concentration, a process called **diffusion**. Making an analogy with gravity, we will refer to such movements as going "down" the concentration gradient. If there are no factors opposing this movement, such as electrical charge, pressure differences, or physical barriers, the movement of molecules from regions of high to low concentration will continue until the substance is evenly dispersed throughout the fluid or the air. In this evenly dispersed state, called a *dynamic equilibrium*, the concentration gradient no longer exists. Molecules continue their random movements and collisions (the *dynamic* aspect), but there is no longer any change in concentration occurring; the substance has reached an *equilibrium* with its surroundings.

To watch diffusion in action, place a drop of food coloring in a glass of water, and check its progress every few minutes. With time, the drop will seem to spread out and become paler, until eventually, even without stirring, the entire glass of water will be uniformly faintly colored. Molecules of dye, simply owing to random motion, move out into the water from the region of high dye concentration into the surrounding water where the dye concentration is low (Fig. 4-2). Simultaneously, random motion causes some water molecules to enter the dye droplet, and the net movement of water is from the high water concentration outside the drop into the lower water concentration inside the drop. At first, there is a very steep concentration gradient, and the dye diffuses rapidly. As the concentration differences lessen, the dye diffuses more and more slowly. In other words, the greater the concentration gradient, the faster the rate of diffusion. However, as long as the concentration of dye within the expanding drop is greater than the concentration of dye in the rest of the glass, the net movement of dye will continue until it becomes uniformly dispersed in the water. Then, with no concentration gradient of either dye or water, diffusion stops. Individual molecules still move about randomly within the glass, but no changes occur in concentration of either water or dye. A dynamic equilibrium has been established.

As you can appreciate from this simple experiment, diffusion cannot move molecules rapidly over long distances. Although the drop of dye immediately begins to diffuse into the water, it will take many minutes or hours for the dye to disperse uniformly. As you will learn in Chapter 5, the slow rate of diffusion over long distances is one of the reasons that cells are small.

Summing Up

The Principles of Diffusion

- Diffusion is the net movement of molecules down a gradient from high to low concentration.
- The greater the concentration gradient, the faster the rate of diffusion.
- If no other processes intervene, diffusion will continue until the concentration gradient is eliminated.
- Diffusion cannot move molecules rapidly over long distances.

Movement Across Membranes Occurs by Both Passive and Active Transport

There are significant concentration gradients of ions and molecules across the plasma membranes of every cell because the cytoplasm of a cell is very different from the extracellular fluid. In its role as gatekeeper of the cell, the plasma membrane provides for two types of

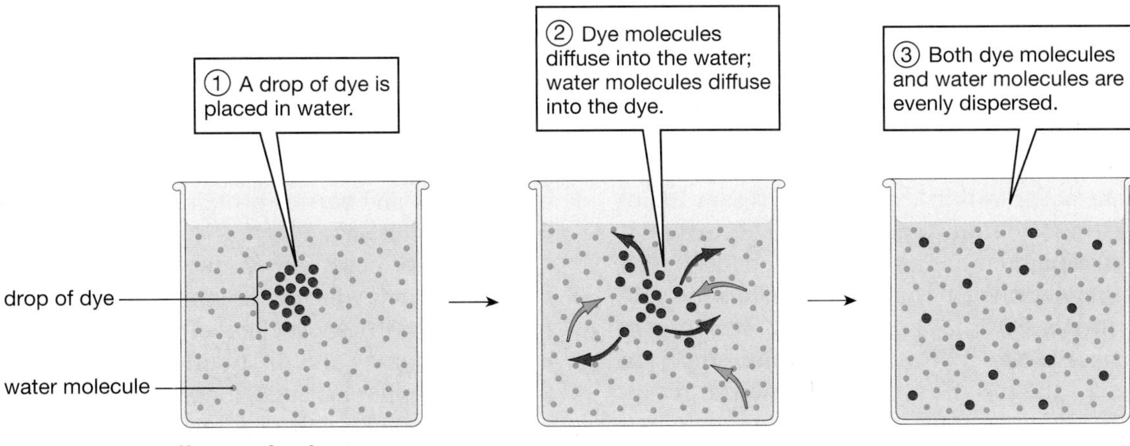

FIGURE 4-2 Diffusion of a dye in water

Table 4-1 Transport Across Membranes

Passive transport	Movement of substances across a membrane, going down a gradient of concentration, pressure, or electrical charge. Does not require the cell to expend energy.
Simple diffusion	Diffusion of water, dissolved gases, or lipid-soluble molecules through the phospholipid bilayer of a membrane.
Facilitated diffusion	Diffusion of (usually water-soluble) molecules through a channel or carrier protein.
Osmosis	Diffusion of water across a selectively permeable membrane—that is, a membrane that is more permeable to water than to dissolved molecules.
Energy-requiring transport	Movement of substances into or out of a cell using cellular energy.
Active transport	Movement of individual small molecules or ions through membrane-spanning proteins, using cellular energy, usually ATP.
Endocytosis	Movement of large particles, including large molecules or entire microorganisms, into a cell by engulfing extracellular material, as the plasma membrane forms membrane-bound sacs that enter the cytoplasm.
Exocytosis	Movement of materials out of a cell by enclosing the material in a membranous sac that moves to the cell surface, fuses with the plasma membrane, and opens to the outside, allowing its contents to diffuse away.

movement: *passive transport* and *energy-requiring transport* (Table 4-1). The movement of molecules directly through a cell membrane using energy is described as *active transport*.

During **passive transport**, substances move into or out of cells down concentration gradients. This movement by itself requires no expenditure of energy, since the concentration gradients provide the potential energy that drives the movement and controls its direction, either into or out of the cell. The lipids and protein pores of the plasma membrane regulate which molecules can cross, but they do not influence the direction of movement.

During **active transport**, the cell uses energy to move substances against a concentration gradient. In this case, transport proteins do control the direction of movement. The difference between passive and active transports can be compared to riding a bike. If you don't pedal, you can go only downhill, as in passive transport. However, if you put enough energy into pedaling, you can go uphill as well, as in active transport.

Passive Transport Includes Simple Diffusion, Facilitated Diffusion, and Osmosis

Plasma Membranes Are Selectively Permeable to Diffusion of Molecules

Diffusion can occur within a fluid or across a membrane separating two fluid compartments. Many molecules cross plasma membranes by diffusion, driven by differences between their concentration in the cytoplasm and in the external environment. Because of the properties of the plasma membrane, different molecules cross the plasma membrane at different locations and at different rates. Therefore, plasma membranes are said to be **selectively permeable**—that is, they allow some molecules to pass through, or *permeate*, but prevent other molecules from passing. A barrier that prevents the passage of all molecules is said to be *impermeable*.

Some Molecules Move Across Membranes by Simple Diffusion

Lipid-soluble molecules such as ethyl alcohol, vitamins A and E, and steroid hormones easily diffuse across the phospholipid bilayer, as do very small molecules, including water and dissolved gases such as oxygen and carbon dioxide. This process is called **simple diffusion** (Fig. 4-3a). Generally, the rate of simple diffusion is a function of the concentration gradient across the membrane, the size of the molecule, and how easily it dissolves in lipids (its lipid *solubility*); large concentration gradients, small molecule size, and high lipid solubility all increase the rate of simple diffusion.

Other Molecules Cross the Membrane by Facilitated Diffusion, with the Help of Membrane Transport Proteins

Most water-soluble molecules, such as ions (for example, K^+, Na^+, and Ca^{2+}), amino acids, and monosaccharides, cannot move through the phospholipid bilayer on their own. These molecules can diffuse across only with the aid of one of two types of transport proteins: channel proteins and carrier proteins. This process is called **facilitated diffusion**.

Channel proteins form pores, or channels, in the lipid bilayer through which certain ions can cross the membrane (Fig. 4-3b). Most channel proteins have a specific interior diameter and distribution of electrical charges that allow only particular ions to pass through. Nerve cells, for example, have separate channels for sodium ions, potassium ions, and calcium ions.

(a) Simple diffusion

(extracellular fluid)

lipid-soluble molecules (O_2, CO_2, H_2O)

(cytoplasm)

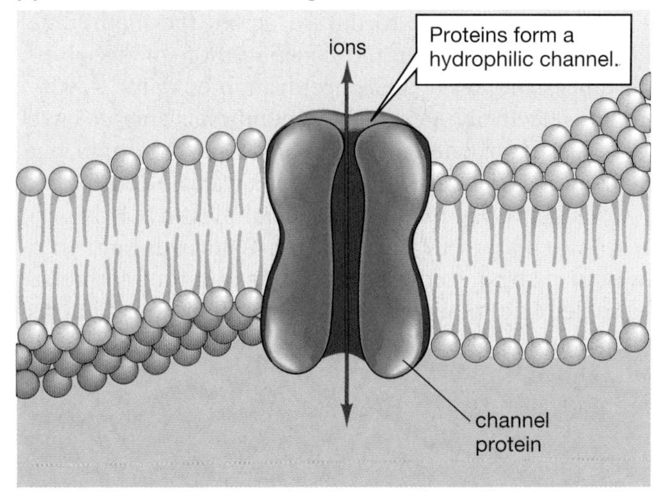

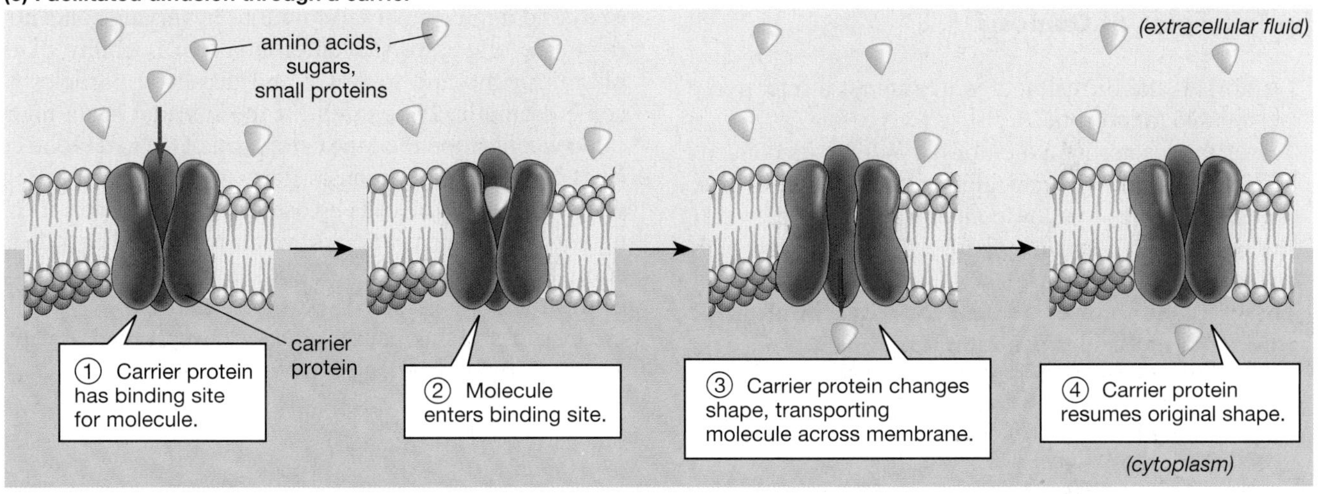

FIGURE 4-3 Diffusion through the plasma membrane
(a) Simple diffusion: gases such as oxygen and carbon dioxide and lipid-soluble molecules can diffuse directly through the phospholipids. **(b)** Facilitated diffusion through a channel: protein channels (pores) allow passage of some water-soluble molecules, principally ions, that cannot diffuse directly through the bilayer. **(c)** Facilitated diffusion through a carrier. **EXERCISE** Imagine an experiment that measures the initial rate of diffusion into cells placed in sucrose solutions of various different concentrations. Sketch a graph (initial diffusion rate versus solution concentration) that shows the result expected if diffusion is simple, and a graph that shows the result expected for facilitated diffusion.

Carrier proteins bind specific molecules from the cytoplasm or extracellular fluid, such as particular amino acids, sugars, or small proteins. Binding triggers a change in the shape of the carrier that allows the molecules to pass through the protein and across the plasma membrane. Facilitated diffusion occurs through carrier proteins that do not use cellular energy. These carrier proteins move molecules only down their concentration gradients (Fig. 4-3c).

Because they must rely on carrier proteins, molecules that cross the membrane by facilitated diffusion usually do so more slowly than do those that cross by simple diffusion through the lipid bilayer.

Osmosis Is the Diffusion of Water Across Membranes

Water also diffuses from regions of high water concentration to regions of low water concentration. However, the diffusion of water across selectively permeable membranes has such dramatic and important effects on cells that we refer to it by a special name: **osmosis**.

What do we mean when we describe a solution as having a "high water concentration" or a "low water concentration"? The answer is simple: pure water has the highest water concentration. Any substance added to pure water displaces some of the water molecules; the resulting solution will have a lower water content than pure water. In addition, dissolved substances often form weak bonds with some of the water molecules, making these water molecules unavailable to diffuse across the membrane (Fig. 4-4a). The higher the concentration of dissolved substances, the lower the concentration of water. A very simple selectively permeable membrane might have pores just large enough for water to pass through but small enough to be impermeable to sugar molecules.

Consider a bag made of a special plastic that is permeable to water but not to sugar. What will happen if we pour a sugar solution in the bag and then immerse the sealed bag in pure water? The principles of osmosis tell us that the bag will swell. If it is weak enough, it will burst (Fig. 4-4b).

Summing Up
The Principles of Osmosis

- Osmosis is the diffusion of water across a selectively permeable membrane.
- Water moves across a membrane—down its concentration gradient—from a high concentration of free water molecules to a low concentration of free water molecules (or down a pressure gradient from high pressure to low pressure).
- Dissolved substances reduce the concentration of free water molecules in a solution.

Osmosis Across the Plasma Membrane Plays an Important Role in the Lives of Cells

Most plasma membranes are highly permeable to water. Because all cells contain dissolved molecules such as salts, proteins, and sugars, the flow of water across the plasma membrane depends on the concentration of water in the liquid that bathes the cells. The extracellular fluid of animals is usually **isotonic** ("having the same strength") to the cytoplasmic fluid within each cell; that is, the concentration of water inside is the same as that outside the cells, so there is no net tendency for water to enter or leave the cells (Fig. 4-5a). Note that the types of dissolved particles are seldom the same inside and outside the cells, but the total concentration of all dissolved particles is equal; therefore, the water concentration inside is equal to that outside the cells.

For example, if red blood cells are taken out of the body and immersed in salt solutions of varying concentrations, the effects of the differential permeability of the plasma membrane to water and dissolved particles become dramatically apparent. If the solution has a higher salt concentration than the cytoplasm of the red blood cell (that is, if the solution has a lower water concentration), water will leave the cells by osmosis. The cells will shrivel up until the concentrations of water inside and outside become equal (Fig. 4-5b). Solutions that have a higher concentration of dissolved particles than does a cell's cytoplasm, and thus cause water to leave the cell by osmosis, are termed **hypertonic** ("having greater strength").

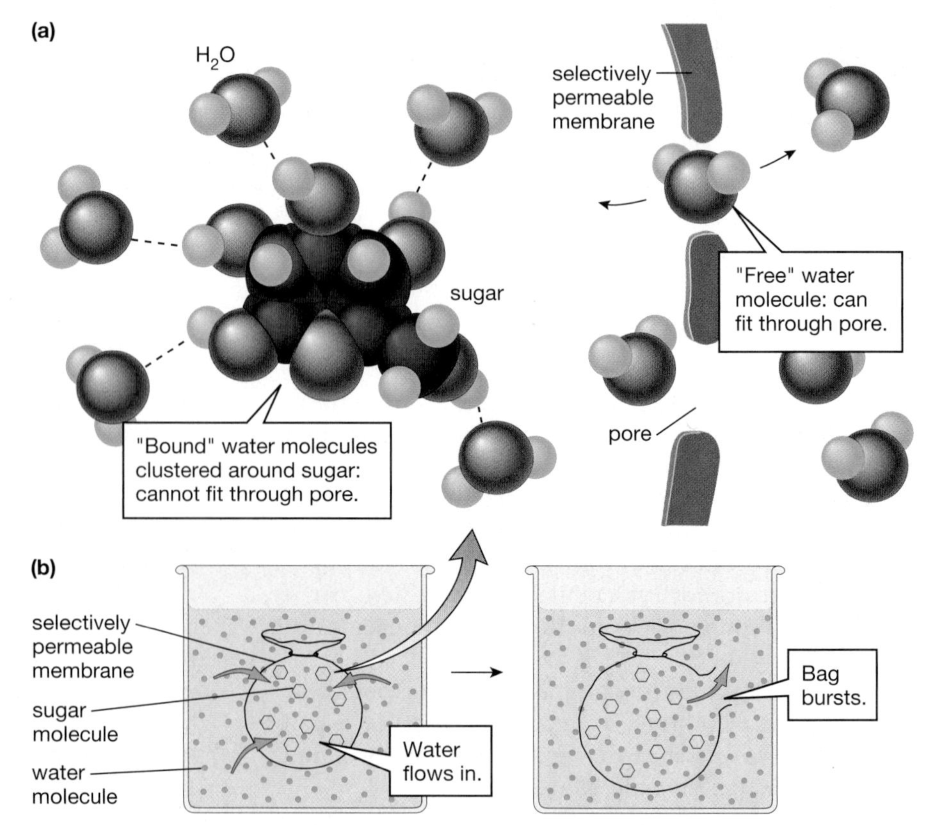

FIGURE 4-4 Osmosis
(a) Membrane pores allow "free" water molecules to pass through, but sugar molecules are too large. "Bound" water molecules, attracted to the sugars by hydrogen bonds, are also prevented from passing through the pore. **(b)** A bag is made of a membrane selectively permeable to free water molecules (white dots) but not to larger molecules, such as sugar (yellow hexagons) or water molecules held to the sugars by hydrogen bonds. If the bag is filled with a sugar solution and suspended in pure water, free water molecules will diffuse down their concentration gradient from the high concentration of water outside the bag to the lower concentration of water inside the bag. The bag will swell and may burst as water enters. **QUESTION** Imagine a container of glucose solution, divided into two compartments (A and B) by a membrane that is permeable to water and glucose but not to sucrose. If some sucrose is added to compartment A, how will the contents of compartment B change?

Media Activity
4.2 Osmosis

Conversely, if the solution has little or no salt, water will enter the cells, causing them to swell (Fig. 4-5c). If red blood cells are placed in pure water, they will swell and eventually burst. Solutions that have a lower concentration of dissolved particles than a cell's cytoplasm, and thus cause water to enter the cell by osmosis, are called **hypotonic** ("having lesser strength"). This process is why your fingers wrinkle after a long bath. It may appear that your fingers are shrinking, but they're not. Instead, water is diffusing into the outer skin cells of your fingers, swelling them more rapidly than the cells underneath and causing wrinkles.

The swelling caused by osmosis can have considerable effects on cells. As we will see in Chapter 5, protists that live in fresh water, such as *Paramecium*, have special structures called *contractile vacuoles* to eliminate the water that continuously leaks in. In contrast, water entry into *central vacuoles* of plant cells helps support the plant. Osmosis across plasma membranes is crucial to the functioning of many biological systems, including water uptake by plant roots (Chapter 24), absorption of dietary water from the intestine (Chapter 30), and reabsorption of water and minerals in kidneys (Chapter 31).

Active Transport Uses Energy to Move Molecules Against Their Concentration Gradients

All cells need to move some materials "uphill" across their plasma membranes, against concentration gradients. For example, every cell requires some nutrients that are less concentrated in the environment than in the cell's cytoplasm; diffusion would cause the cell to lose, not gain, these nutrients. Other substances, such as sodium and calcium ions in your brain cells, must be maintained at much lower concentrations inside the cells than in the extracellular fluid. When these ions diffuse into the cells, they must be pumped out again against their concentration gradients.

In active transport, membrane proteins use cellular energy to move individual molecules or ions across the

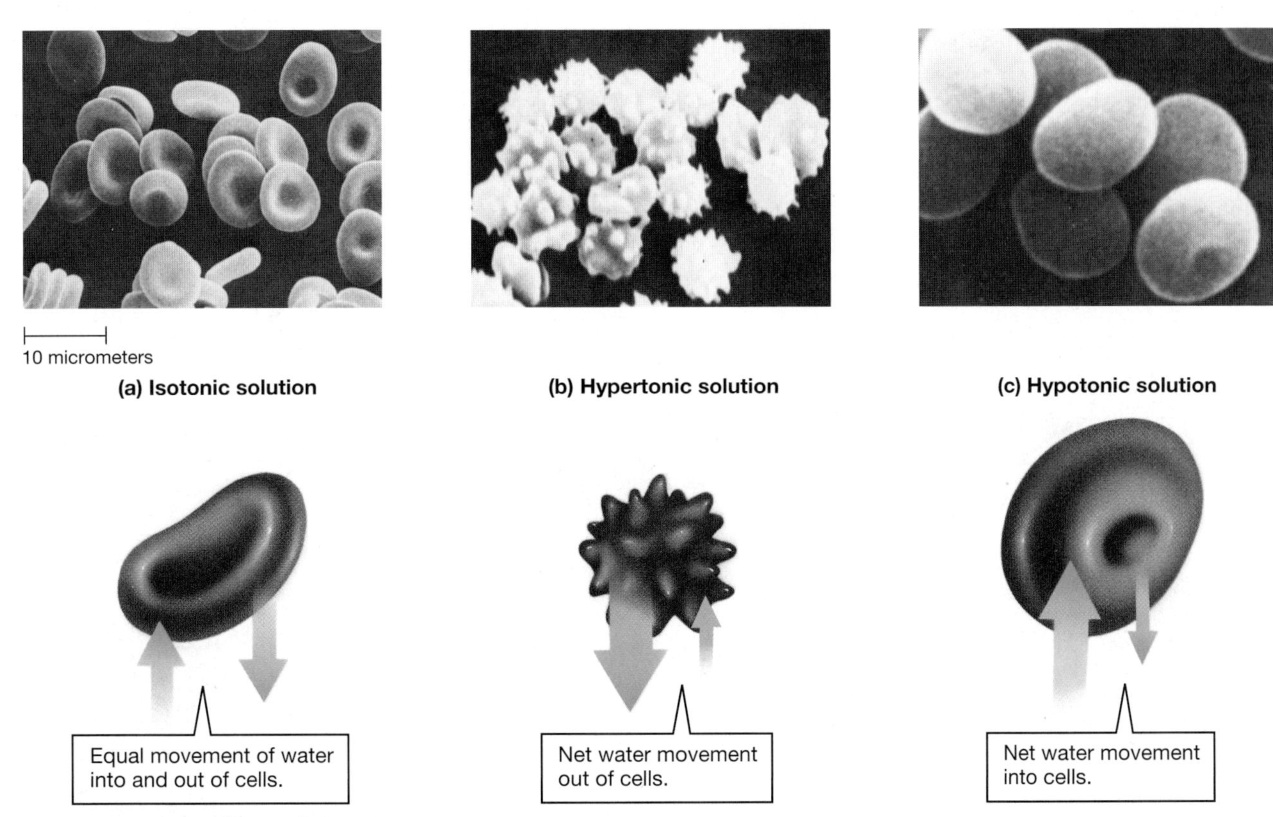
10 micrometers
(a) Isotonic solution
(b) Hypertonic solution
(c) Hypotonic solution
Equal movement of water into and out of cells.
Net water movement out of cells.
Net water movement into cells.

FIGURE 4-5 The effects of osmosis
(a) If red blood cells are immersed in an isotonic salt solution, which has the same concentration of dissolved substances as the blood cells do, there is no net movement of water across the plasma membrane. The red blood cells keep their characteristic dimpled disk shape. **(b)** A hypertonic solution, with too much salt, causes water to leave the cells, shriveling them up. **(c)** A hypotonic solution, with less salt than is in the cells, causes water to enter, and the cells swell. **QUESTION** All freshwater fish swim in a solution that is hypotonic to the fluid inside their bodies. Why don't freshwater fish swell up and burst?

plasma membrane, usually against their concentration gradient (Fig. 4-6). These active-transport proteins span the width of the membrane and have two active sites. One active site (which may be either on the face of the plasma membrane in contact with the cytoplasm or on the face in contact with the extracellular fluid, depending on the transport protein) binds a particular molecule or ion, such as a calcium ion. The second site (always on the inside of the membrane) binds an energy-carrier molecule, usually adenosine triphosphate (ATP). The ATP donates energy to the protein, causing it to change shape and move the calcium ion across the membrane. Active-transport proteins are often called *pumps*—in an analogy to water pumps—because they use energy to move ions or molecules "uphill" against a concentration gradient. We will see that plasma membrane pumps are vital in mineral uptake by plants (Chapter 24), mineral absorption in your intestines (Chapter 30), and maintaining concentration gradients essential to nerve cell functioning (Chapter 34).

Cells Engulf Particles or Fluids by Endocytosis

Many cells acquire or expel particles or substances that are too large to diffuse across a membrane regardless of concentration gradients. Cells have evolved several processes that use cellular energy to move materials into or out of the cell. Cells can acquire fluids or particles from their extracellular environment, especially large proteins or entire microorganisms such as bacteria, by a process called **endocytosis** (Greek for "into the cell"). During endocytosis, the plasma membrane engulfs the fluid droplet or particle and pinches off a membranous sac called a **vesicle**—with the fluid or particle inside—into the cytoplasm (Fig. 4-7). We can distinguish three types of endocytosis on the basis of the size of the particle acquired and the method of acquisition: *pinocytosis*, *receptor-mediated endocytosis*, and *phagocytosis*.

Pinocytosis Moves Liquids into the Cell

In **pinocytosis** ("cell drinking"), a very small patch of plasma membrane dimples inward as it surrounds extracellular fluid and buds off into the cytoplasm as a tiny vesicle (see Fig. 4-7a). Pinocytosis moves a droplet of extracellular fluid, contained within the dimpling patch of membrane, into the cell. Therefore, the cell acquires materials in the same concentration as in the extracellular fluid.

Receptor-Mediated Endocytosis Moves Specific Molecules into the Cell

Cells take up certain molecules or complexes of molecules (packets containing protein and cholesterol, for example) by the process of **receptor-mediated endocytosis** (see Fig. 4-7b). This process can selectively concentrate specific molecules inside a cell. Most plasma membranes bear many receptor proteins on their outside surfaces, each with a binding site for a particular molecule. In some cases, the receptors accumulate in depressions on the plasma membrane called *coated pits* (Fig. 4-8). If the right molecule contacts a receptor protein in one of these coated pits, the molecule attaches to the binding site. The coated pit deepens into a U-shaped pocket that eventually pinches off into the cytoplasm as a vesicle. Both the receptor with its bound molecules and a bit of extracellular fluid move into the cytoplasm within the vesicle.

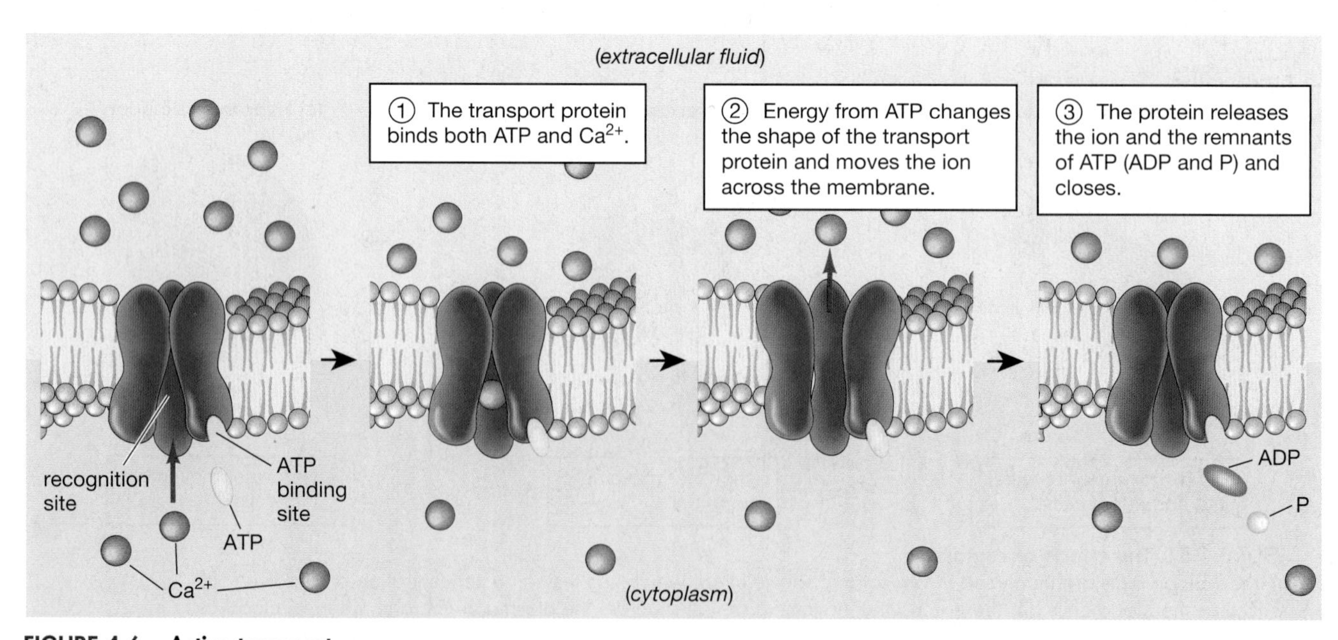

FIGURE 4-6 Active transport
Active transport uses cellular energy to move molecules across the plasma membrane, often against a concentration gradient. A transport protein (blue) has an ATP binding site and a recognition site for the molecules to be transported, in this case calcium ions (Ca^{2+}).

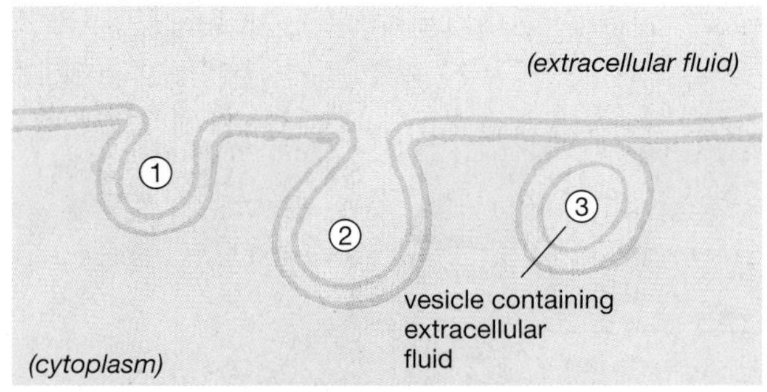

(1) A dimple forms in the plasma membrane, which (2) deepens and surrounds the extracellular fluid. (3) The membrane encloses the extracellular fluid, forming a vesicle.

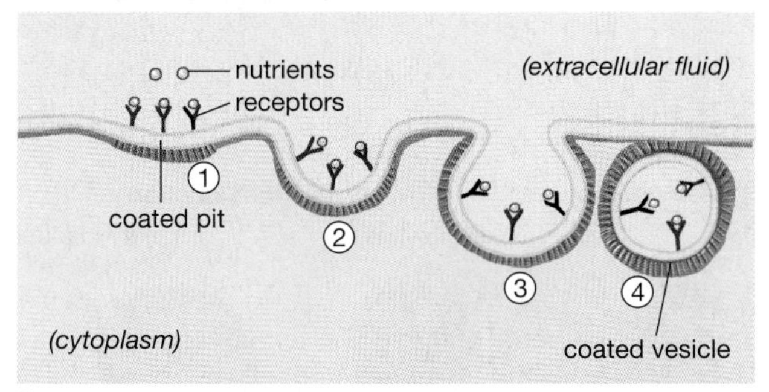

(1) Receptor proteins for specific molecules or complexes of molecules are localized at coated pit sites. (2) The receptors bind the molecules and the membrane dimples inward. (3) The coated pit region of the membrane encloses the receptor-bound molecules. (4) A vesicle ("coated vesicle") containing the bound molecules is released into the cytoplasm.

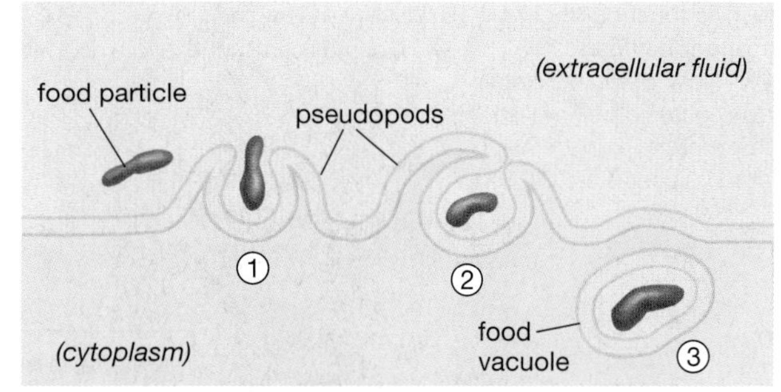

(1) The plasma membrane extends pseudopods toward an extracellular particle (for example, food). (2) The ends of the pseudopods fuse, encircling the particle. (3) A vesicle called a food vacuole is formed containing the engulfed particle.

FIGURE 4-7 Three types of endocytosis
QUESTION Compare and contrast receptor-mediated endocytosis with active transport.

Phagocytosis Moves Large Particles into the Cell

Cells use **phagocytosis** (which means "cell eating") to pick up large particles, including whole microorganisms (see Fig. 4-7c). When the freshwater protist *Amoeba*, for example, senses a tasty *Paramecium*, *Amoeba* extends parts of its surface membrane. These membrane extensions are called *pseudopods* (Latin for "false foot"). The pseudopod ends fuse around the luckless *Paramecium*, and the prey is carried into the interior of the *Amoeba* inside a vesicle—called a *food vacuole*—for digestion. Like *Amoeba*, white blood cells also use phagocytosis and intracellular digestion to engulf and destroy invading bacteria.

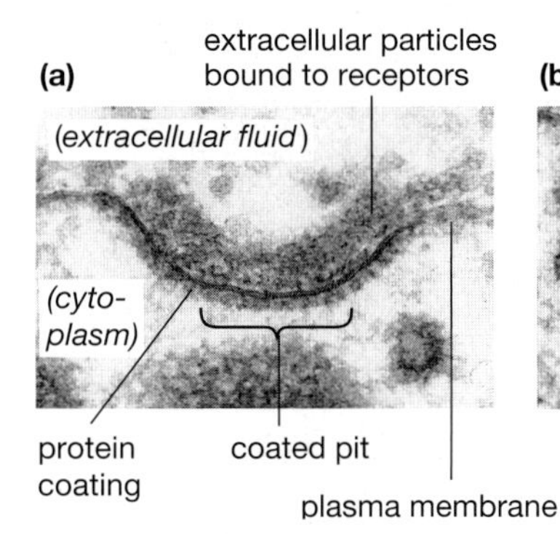

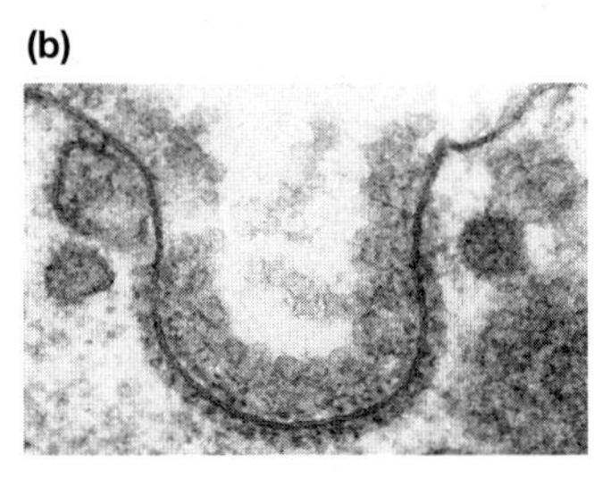

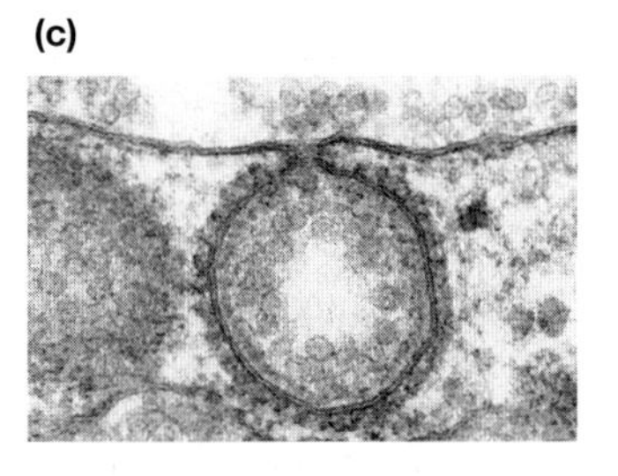

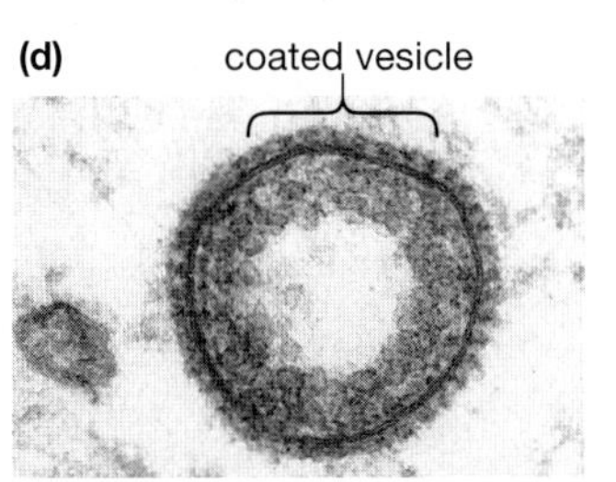

FIGURE 4-8 Receptor-mediated endocytosis
These electron micrographs illustrate the sequence of events in receptor-mediated endocytosis. **(a)** The shallow depression in the plasma membrane is coated on the inside with a protein (dark, fuzzy substance in the micrographs) and bears receptor proteins on the outside (not visible). **(b, c)** The pit deepens and **(d)** eventually pinches off as a coated vesicle. The protein coating is recycled back to the plasma membrane.

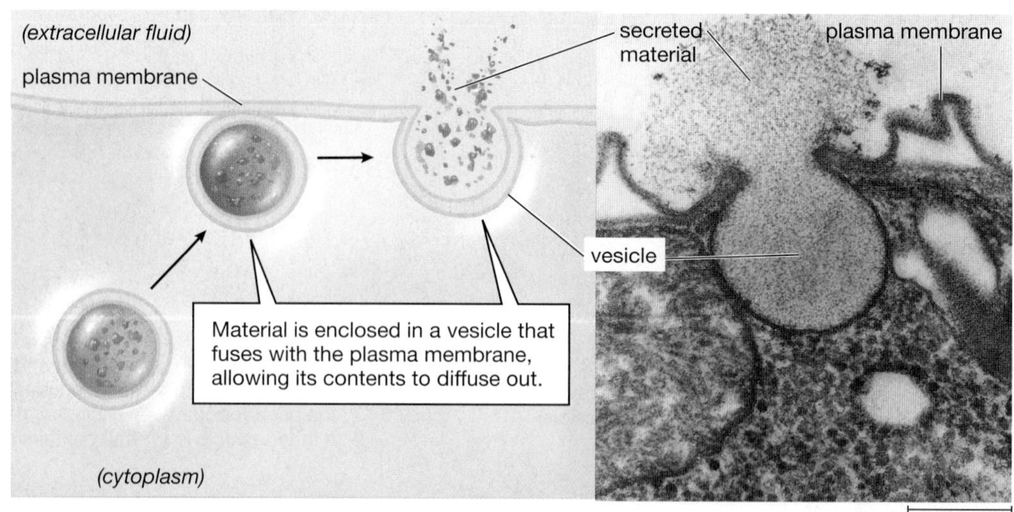

FIGURE 4-9 Exocytosis
Exocytosis is functionally the reverse of endocytosis.
QUESTION How does exocytosis differ from diffusion of materials out of a cell?

Exocytosis Moves Material Out of the Cell

Cells often use energy to accomplish the reverse of endocytosis, a process called **exocytosis** (Greek for "out of the cell"), to dispose of unwanted materials such as the waste products of digestion, or to secrete materials such as hormones into the extracellular fluid (Fig. 4-9). During exocytosis, a membrane-enclosed vesicle carrying material to be expelled moves to the cell surface, where the vesicle's membrane fuses with the cell's plasma membrane. The vesicle then opens to the extracellular fluid, and its contents diffuse out.

4.3 How Are Cell Surfaces Specialized?

Various Specialized Junctions Allow Cells to Connect and Communicate

In multicellular organisms, plasma membranes hold together clusters of cells and provide avenues through which cells communicate with their neighbors. Depending on the organism and the cell type, four types of connection may occur between cells: *desmosomes*, *tight junctions*, *gap junctions*, and *plasmodesmata.* While plasmodesmata are restricted to plant cells, some animal cells possess all three of the other types of junctions.

Desmosomes Attach Cells Together

As you know, animals tend to be flexible, mobile organisms. Many of an animal's tissues are stretched, compressed, and bent as the animal moves. Cells in the skin, intestine, urinary bladder, and other organs must adhere firmly to one another to avoid tearing under the stresses of movement. Such animal tissues have junctions called **desmosomes**, which hold adjacent cells together (Fig. 4-10a). In a desmosome, the membranes of adjacent cells are held together by proteins and carbohydrates. Protein filaments attached to the insides of the

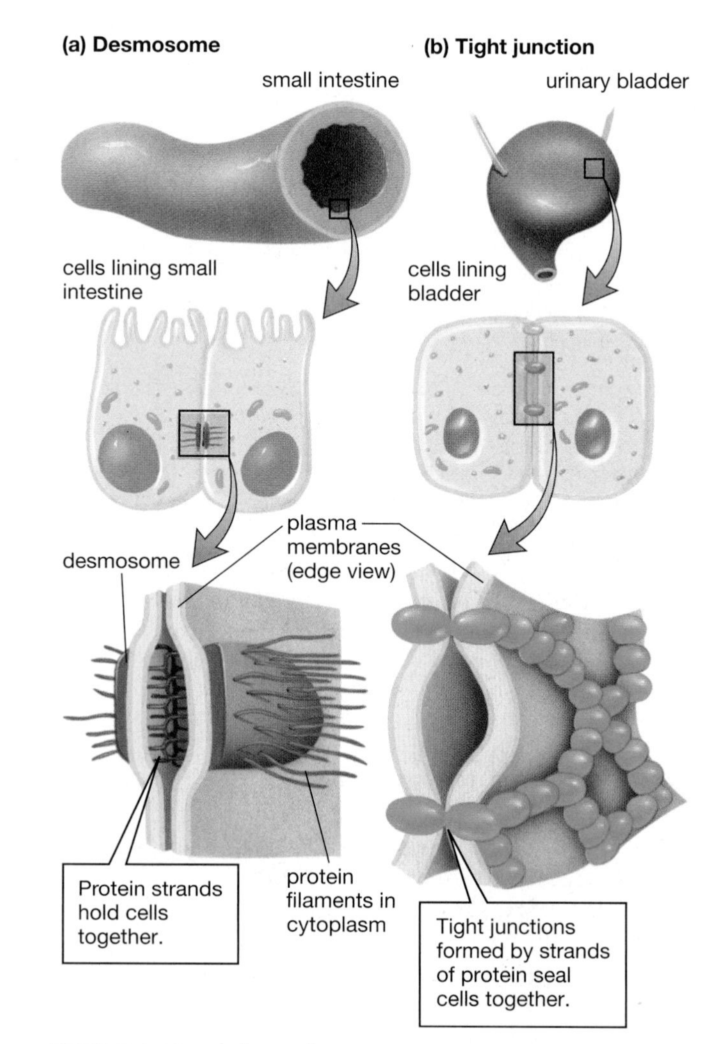

FIGURE 4-10 Cell attachment structures
(a) Cells lining the small intestine are attached by desmosomes. Protein filaments bound to the inside surface of each desmosome extend into the cytoplasm and attach to other filaments inside the cell, strengthening the connection between cells. **(b)** Tight junctions prevent leakage between cells, as in those of the urinary bladder.

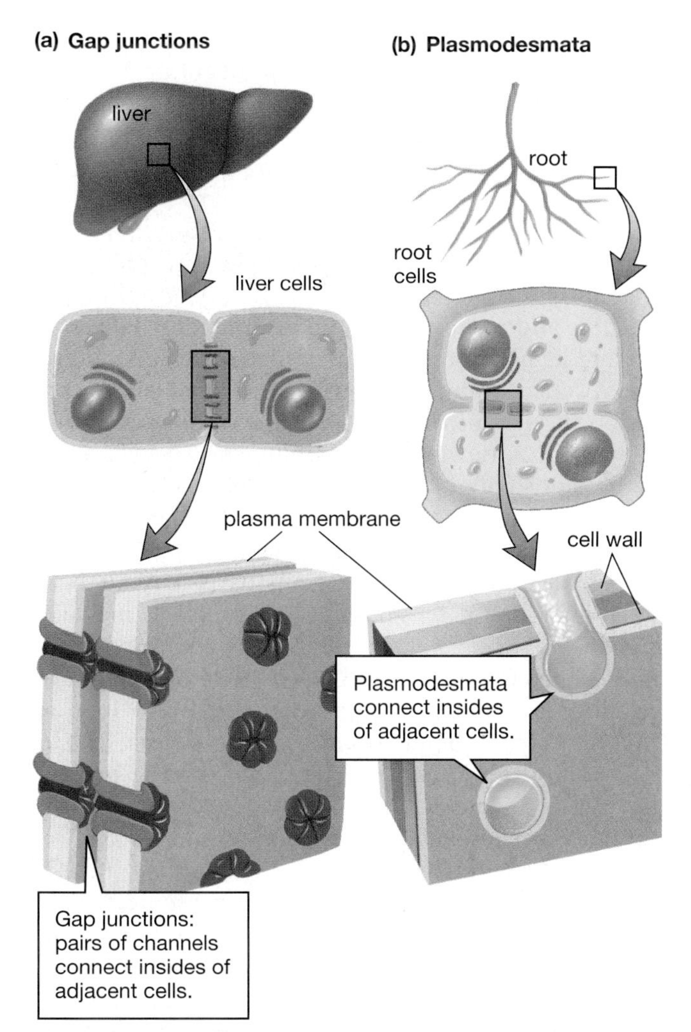

FIGURE 4-11 Cell communication structures
(a) Gap junctions, such as those between cells of the liver, contain cell-to-cell channels that interconnect the cytoplasm of adjacent cells. **(b)** Plant cells are interconnected by plasmodesmata, which form cytoplasmic connections through the walls of adjacent cells.

desmosomes extend into the interior of each cell, further strengthening the attachment.

Tight Junctions Make the Cell Leakproof

The animal body contains many tubes and sacs that must hold their contents without leaking; a leaky urinary bladder or intestine would spell disaster for the rest of the body. Spaces between the cells that line such tubes or sacs are sealed with strands of protein to form **tight junctions** (Fig. 4-10b). The membranes of adjacent cells nearly fuse along a series of ridges, effectively forming leakproof gaskets between cells. Continuous tight junctions sealing each cell to its neighbors prevent molecules from escaping between cells.

Gap Junctions and Plasmodesmata Allow Communication Between Cells

Multicellular organisms must coordinate the actions of their component cells. In animals, most cells that contact other cells—in other words, nearly all cells in the body—communicate through protein channels that

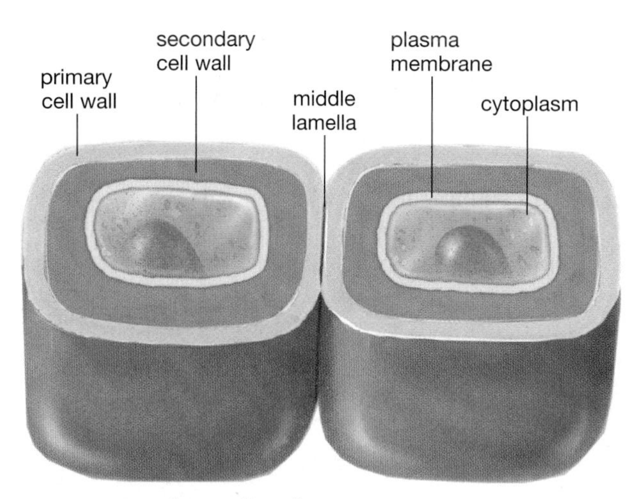

FIGURE 4-12 Plant cell walls
Each plant cell secretes cellulose and other carbohydrates to form a cell wall outside the plasma membrane. Many plant cells also produce secondary cell walls. The middle lamella separates adjacent plant cells.

connect the insides of adjacent cells. These cell-to-cell channels are called **gap junctions** (Fig. 4-11a). Hormones, nutrients, ions, and even electrical signals can pass through the channels at gap junctions.

Virtually all of the living cells of plants are connected to one another by **plasmodesmata**. Plasmodesmata are openings in the walls of adjacent plant cells, lined with plasma membrane and filled with cytoplasm. Plasmodesmata create continuous cytoplasmic bridges between the insides of adjacent cells (Fig. 4-11b). Many plant cells have thousands of plasmodesmata, allowing water, nutrients, and hormones to pass quite freely from one cell to another.

Some Cells Are Supported by Cell Walls

The outer surfaces of the cells of bacteria, plants, fungi, and some protists are covered with nonliving, typically stiff coatings called **cell walls**. Single-celled organisms that live in the ocean, for example, may have outer coverings of cellulose, protein, or glassy silica that protect the delicate plasma membrane. Plant cell walls are composed of cellulose and other *polysaccharides*, whereas fungal cell walls are made of polysaccharides and the modified polysaccharide chitin. (Polysaccharides and chitin are described in Chapter 3.) Bacterial cell walls have a chitin-like framework to which short chains of amino acids and other molecules are attached.

Cell walls are produced by the cells they surround. Plant cells secrete cellulose through their plasma membranes, forming the *primary cell wall*. Many plant cells later secrete more cellulose and other polysaccharides beneath the primary wall to form a thick *secondary cell wall*, pushing the primary cell wall away from the plasma membrane. In some plant cells, the secondary wall becomes thicker than the cell inside it. The primary cell walls of adjacent cells are joined by the *middle lamella*, a layer made primarily of *pectin* (Fig. 4-12). Pectin is the polysaccharide that makes jelly solidify.

Cell walls support and protect otherwise fragile cells. For example, cell walls allow plants and mushrooms to resist the forces of gravity and wind and to stand erect on land. Tree trunks, composed almost entirely of cellulose and other materials laid down over the years and capable of supporting impressive loads, are the ultimate proof of cell wall strength.

Although strong, cell walls are usually porous, permitting easy passage of small molecules such as minerals, water, oxygen, carbon dioxide, amino acids, and sugars. (Otherwise, the cell within would soon die.) However, the structure that really governs the interactions that occur between a cell and its external environment is the plasma membrane.

EVOLUTIONARY CONNECTIONS

Caribou Legs and Membrane Diversity

The membranes of all cells are similar in structure, reflecting the common evolutionary heritage of all life on Earth. Membrane function varies tremendously from organism to organism, however, and even from cell to cell within a single organism. This diversity arises largely from the different proteins and phospholipids in the membrane, which have evolved under different environmental pressures.

Our discussion of membranes emphasized the unique functions of membrane proteins. Consequently, you may think that the phospholipids are just a waterproof place for the proteins to reside. This isn't quite true, as we can see by examining the plasma membrane phospholipids in cells of the legs of caribou, animals that live in very cold regions of North America (Fig. 4-13). During the long arctic winters, temperatures plummet far below freezing. For caribou to keep their legs and feet really warm would waste precious energy. Fortunately, these conditions have favored the evolution of specialized arrangements of arteries and veins in caribou legs that allow the temperature of their lower legs to drop almost to freezing (0 °C), thus conserving body heat. The upper legs and main trunk of the body, in contrast, remain at about 37 °C. To remain fluid at these radically different temperatures requires the phospholipids in the membranes of cells in the upper legs of caribou to be very different from those near the hooves.

Remember, the membrane of a cell needs to be somewhat fluid to allow the membrane proteins to move to sites where they are needed. The fluidity of a membrane is a function of the fatty acid tails of its phospholipids: unsaturated fatty acids (with some double bonds between carbons) remain more fluid at lower temperatures than do saturated fatty acids (see Chapter 3). Caribou have a range of fatty acids in the phospholipids of the cells in their legs. The membranes of cells near the chilly hoof have lots of unsaturated fatty acids, whereas the membranes of cells near the warmer trunk have more saturated ones. This arrangement gives the plasma membranes throughout the leg the proper fluidity despite great differences in temperature.

As important as the phospholipids are, the membrane proteins play the major roles in determining cell function and in governing the interactions between a cell and its neighbors. Protein molecules can be altered as a result of mutations that change their amino acid composition, which may alter their shapes and, consequently, their functions. Over billions of years, an incredible diversity of proteins has evolved. Every nerve cell in your body, for instance, has membrane proteins essential for producing electrical signals and conducting them along the nerves to various parts of your body. Other membrane proteins receive chemical messages from neighboring nerve cells or from hormones and other chemicals in the blood. Each cell in the brain has a specific set of membrane proteins that allow it to respond to some stimuli while ignoring others. In fact, your ability to read this page depends on the proteins that reside in the membranes of your light-detecting cells and in the membranes of your brain cells.

Throughout this text, we will return many times to the concepts of membrane structure presented in this chapter. Understanding the diversity of membrane lipids and proteins is the key to understanding not just the isolated cell, but also entire organs, which function as they do largely because of the properties of the membranes of their component cells.

FIGURE 4-13 Caribou browse on the frozen Alaskan tundra The lipid composition of the membranes in the cells of a caribou's legs varies with distance from the animal's trunk. Unsaturated phospholipids predominate in the lower leg; more saturated phospholipids are found in the upper leg.

CASE STUDY REVISITED Vicious Venoms

Snake venoms are complex brews of poisonous proteins. The venom of the Western Diamondback rattlesnake is rich in enzymes called *phospholipases*, which break down the phospholipids of the plasma membranes of cells, causing the cells to rupture and die. Cell death blackens the tissue around the bite. In the bloodstream, the phospholipases attack red blood cells, reducing the oxygen-carrying capacity of the blood and causing the victim to become short of breath. Once carried to muscles, phospholipases also attack muscle cell membranes, exposing the complex contractile proteins within the muscle cells to attack by protein-digesting enzymes in the venom. Proteins that strengthen blood vessels also come under attack, and hemorrhaging may result if the vessels rupture. The venom also contains enzymes that attack blood-clotting proteins, making the hemorrhaging more dangerous. When internal bleeding is combined with loss of the oxygen-carrying red blood cells, it is no wonder the victim is short of breath and may go into shock.

Karl was extremely lucky that Mark had brought his cell phone. Trying to hike back to the car would have quickly spread the venom throughout his body, and the added delay would have decreased his chances of survival. Fortunately, Mark was able to get expert advice immediately: keep Karl lying down—since he was suffering from shock—and as immobile as possible until help arrives. Because they had identified the snake, the hospital had the correct antivenin waiting. Antivenin contains proteins that bind to and neutralize the various toxins in the snake venom. The Western Diamondback, with its half-inch-long fangs and large stores of venom, is responsible for more U.S. snakebite deaths than any other species. To keep matters in perspective, however, of the 7000 to 8000 people bitten by poisonous snakes in the United States each year, only 10 to 15 people are killed by the venom.

Consider This: Phospholipases and other digestive enzymes are found in animal (including human) digestive tracts as well as in snake venom. Why do we have phospholipases in our digestive tracts? Since snakes swallow their prey whole, what are two very different roles for the phospholipases and other enzymes in snake venom?

Links to Life: Gradients Keep You Going

One of the critical functions of membranes is to regulate the gradients that occur within cells and between cells and their surroundings. Each cell must maintain gradients for hundreds of different ions and molecules—a cell without gradients is dead. But since it's very difficult to appreciate gradients across the membranes of cells you can't even see, let's relate them to how the body works.

You stay alive, in part, because you eat and breathe. Eating results in gradients of nutrients such as sugars and amino acids between the digested food in your small intestine and the cells lining the intestine. These gradients, along with specialized membrane proteins, help the food molecules move into your intestinal cells, then into your bloodstream, and finally into cells throughout your body. Gradients of gases between the air in your lungs and the cells lining your lungs allow oxygen to diffuse in and carbon dioxide to diffuse out.

Nerve cells can produce nerve impulses (electrical signals) only because they maintain gradients of several different ions—some higher outside, some higher inside. To produce a nerve impulse, the nerve cell membrane changes its selective permeability and allows the gradients of specific ions to run down just a bit (like discharging a battery to create a flow of electricity in your flashlight). Then the nerve cell expends energy to restore the gradient (recharging its battery) so it can continue to make more impulses. These signals in your brain allow you to read this page, understand it, and remember it—and think about dinner at the same time!

Nerve cells also communicate with muscles to control everything you do. But muscles could not respond to neural signals without ion gradients of their own, as well as selective permeability that allows specific ions to flow down their concentration gradients, triggering muscles to contract. Heart muscle contractions, in turn, create pressure gradients that drive blood—and its life-sustaining oxygen and nutrients—throughout your body. The bottom line? Be grateful for gradients!

Summary of Key Concepts

4.1 How Is the Structure of a Membrane Related to Its Function?

The plasma membrane has three major functions: it selectively isolates the cytoplasm from the external environment; it regulates the flow of materials into and out of the cell; and it communicates with other cells. The membrane consists of a bilayer of phospholipids in which a variety of proteins are embedded. There are three major categories of membrane proteins: transport proteins, which regulate the movement of most water-soluble substances through the membrane; receptor proteins, which bind molecules in the external environment, triggering changes in the metabolism of the cell; and recognition proteins, which serve as identification tags and attachment sites.

4.2 How Do Substances Move Across Membranes?
Diffusion is the movement of particles from regions of higher concentration to regions of lower concentration. In simple diffusion, water, dissolved gases, and lipid-soluble molecules diffuse through the phospholipid bilayer. In facilitated diffusion, water-soluble molecules cross the membrane through protein channels or with the assistance of protein carriers. In both cases, molecules move down their concentration gradients, and cellular energy is not required.

Osmosis is the diffusion of water across a selectively permeable membrane down its concentration gradient. Dissolved substances decrease the concentration of free water molecules. Osmosis does not require cellular energy.

Several types of transport require energy. In active transport, protein carriers in the membrane use cellular energy (ATP) to drive the movement of molecules across the plasma membrane, usually against concentration gradients. Large molecules (for example, proteins), particles of food, microorganisms, and extracellular fluid may be acquired by endocytosis, either by pinocytosis, receptor-mediated endocytosis, or phagocytosis. The secretion of substances such as hormones and the excretion of wastes from a cell are accomplished by exocytosis.

4.3 How Are Cell Surfaces Specialized?
Cells may be connected to one another by a variety of junctions. Desmosomes attach cells firmly to one another, preventing a tissue from tearing during movement or stretching. Tight junctions seal off the spaces between adjacent cells, leakproofing organs such as the urinary bladder. Gap junctions in animals and plasmodesmata in plants interconnect the cytoplasm of adjacent cells.

The outer surfaces of some protist cells and of each bacterial, plant, and fungal cell are surrounded by a rigid cell wall outside the plasma membrane. The cell wall, produced by the cell that it surrounds, protects and supports that cell.

Key Terms

active transport *p. 62*
carrier protein *p. 60*
cell wall *p. 69*
channel protein *p. 60*
concentration *p. 60*
concentration gradient *p. 60*
cytoplasm *p. 58*
desmosome *p. 68*
diffusion *p. 61*
endocytosis *p. 66*
exocytosis *p. 68*
facilitated diffusion *p. 62*
fluid *p. 60*
fluid mosaic model *p. 58*
gap junction *p. 69*
glycoprotein *p. 60*
gradient *p. 60*
hypertonic *p. 64*
hypotonic *p. 65*
isotonic *p. 64*
osmosis *p. 63*
passive transport *p. 62*
phagocytosis *p. 67*
phospholipid bilayer *p. 58*
pinocytosis *p. 66*
plasma membrane *p. 58*
plasmodesmata *p. 69*
receptor-mediated endocytosis *p. 66*
receptor protein *p. 60*
recognition protein *p. 60*
selectively permeable *p. 62*
simple diffusion *p. 62*
tight junction *p. 69*
transport protein *p. 60*
vesicle *p. 66*

Thinking Through the Concepts

To take a multiple-choice quiz with feedback on the contents of this chapter, visit http://www.prenhall.com/audesirk7. *Log in to the Web site selected by your instructor and navigate to the Self Test section for this chapter.*

? Review Questions

1. Describe and diagram the structure of a plasma membrane. What are the two principal types of molecules in plasma membranes? What are the four principal functions of plasma membranes?
2. What are the three categories of proteins commonly found in plasma membranes, and what is the function of each?
3. Define *diffusion*, and compare that process to osmosis. How do these two processes help plant leaves remain firm?
4. Define *hypotonic*, *hypertonic*, and *isotonic*. What would be the fate of an animal cell immersed in each of the three types of solution?
5. Describe the following types of transport processes: simple diffusion, facilitated diffusion, active transport, pinocytosis, receptor-mediated endocytosis, phagocytosis, and exocytosis.
6. Name four types of cell-to-cell junctions and the function of each. Which junctions allow communication between the interiors of adjacent cells?

Applying the Concepts

1. Different cells have somewhat different plasma membranes. The plasma membrane of a *Paramecium*, for example, is only about 1% as permeable to water as the plasma membrane of a human red blood cell. Referring back to our discussion of the effects of osmosis on red blood cells and the role of contractile vacuoles in *Paramecium,* what do you think is the function of the low water permeability of *Paramecium*? What molecular differences do you think might account for this low water permeability?
2. A preview question for Chapter 32: The integrity of the plasma membrane is essential for cellular survival. Could the immune system utilize this fact to destroy foreign cells that have invaded the body? How might cells of the immune system disrupt membranes of foreign cells? (Two hints: virtually all cells can secrete proteins, and some proteins form pores in membranes.)
3. A preview question for Chapter 24: Plant roots take up minerals (inorganic ions such as potassium) that are dissolved in the water of the soil. The concentration of such ions is usually much lower in the soil water than in the cytoplasm of root cells. Design the plasma membrane of a hypothetical mineral-absorbing cell, with special reference to mineral-permeable channel proteins and mineral-transporting active transport proteins. Justify your choice of channels and active-transport proteins.
4. Red blood cells will swell up and burst when placed in a hypotonic solution such as pure water. Why don't we swell up and burst when we swim in water that is hypotonic to our cells and body fluids?

For More Information

Kunzig, R., "They Love the Pressure." *Discover*, August 2001. Living at depths that exert a pressure of 15,000 pounds per square inch necessitates more membranes in deep-sea dwellers.

McNeil, P. L. "Cell Wounding and Healing." *American Scientist*, May–June 1991. The fluidity of plasma membrane phospholipids makes cells able to withstand minor damage.

Rothman, J. E., and Orci, L. "Budding Vesicles in Living Cells." *Scientific American*, March 1996. Membranes within cells form small containers called *vesicles*, which transport materials inside the cell. Researchers are discovering the mechanisms by which these containers are formed.

Sharon, N., and Lis, H. "Carbohydrates in Cell Recognition." *Scientific American*, January 1993. Carbohydrates, usually attached to proteins as part of glycoproteins, identify cells, serve as parts of receptors in hormone binding, and regulate the attachment and movement of cells.

Media Activities

To access a Media Activity visit http://www.prenhall.com/audesirk7. *Log in to the Web site selected by your instructor, navigate to this chapter, and select the appropriate Media Activity number.*

4.1 Membrane Structure and Transport

Estimated time: 10 minutes

Explore how the cell membrane controls what enters and leaves a cell.

4.2 Osmosis

Estimated time: 5 minutes

Explore how osmosis affects cells.

4.3 Web Investigation: Vicious Venoms

Estimated time: 10 minutes

Many people are afraid of snakes. A person bitten by a poisonous snake might die without anti-venom treatment, yet venom components can also be life-saving drugs. This exercise takes a closer look at snakebites and venom.

CHAPTER

5 Cell Structure and Function

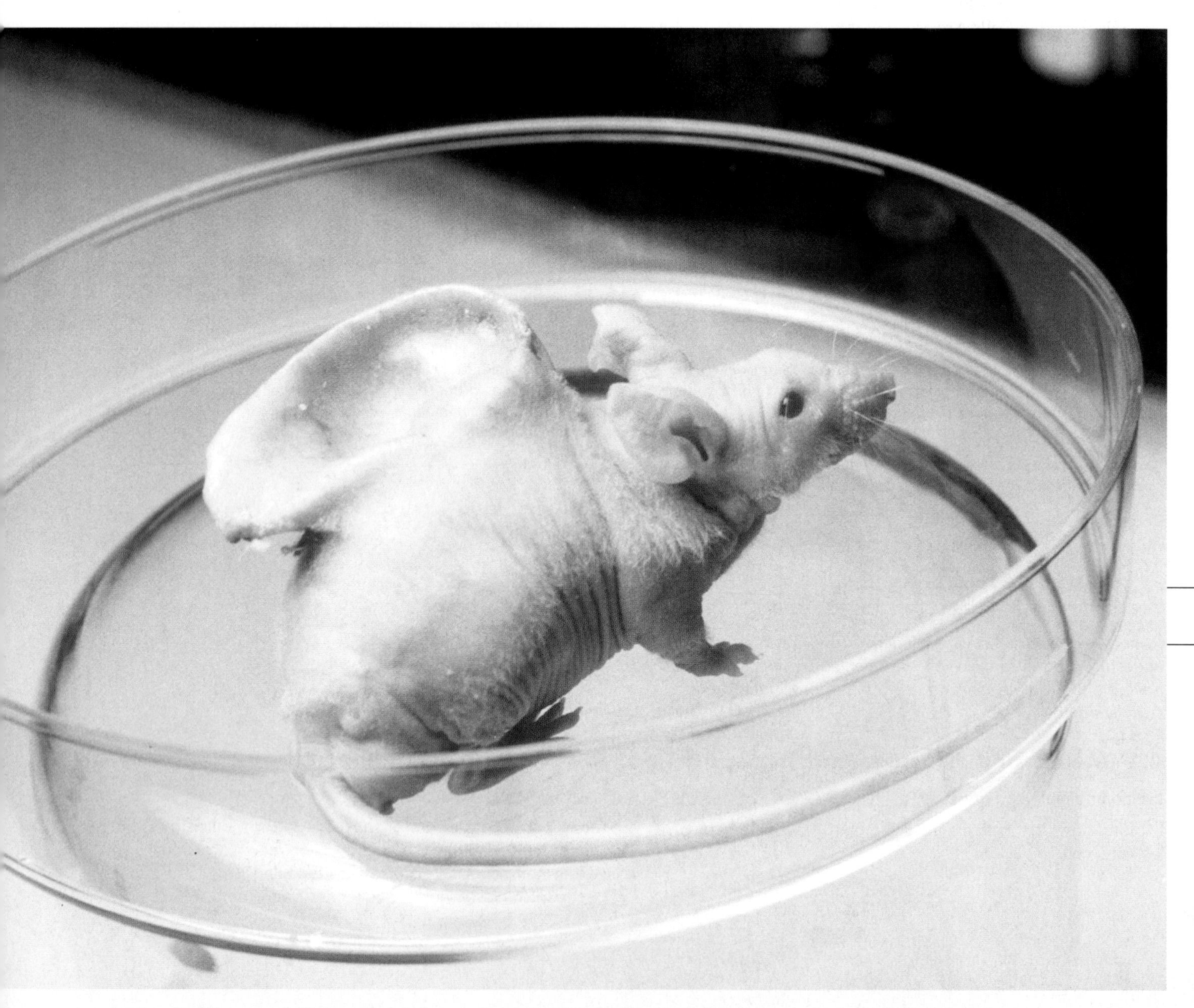

An "ear" of human cartilage cells has been transplanted under the skin of this mouse. But why is the mouse hairless? Researchers used a special strain of mutant mouse with an extremely weak immune system. The immune system of a normal mouse would immediately attack and reject the human cells as if they were foreign disease organisms. The defective stretch of DNA (mutant gene) that weakens the immune system of this type of mouse eliminates its hair as well.

AT A GLANCE

CASE STUDY Spare Parts for Human Bodies

Raul Mercia lost part of his thumb in an industrial accident, but he was lucky—doctors grew a new one for him. After the accident, doctors harvested some of his bone cells and kept them alive in the laboratory. Later, surgeons grafted a piece of coral shaped like a thumb bone onto Mercia's damaged digit. His cultured bone cells were then seeded onto the surface of the coral, where they grew and gradually replaced the biodegradable coral scaffolding.

Mercia was one of the first people to benefit from a new bioengineering technology in which doctors and researchers attempt to grow new body parts in the laboratory. The field made a leap forward in 1997, when Dr. Jay Vacanti of Massachusetts General Hospital in Boston produced a mouse with a human ear growing on its back. It wasn't a real ear, but an artificial ear made of human cartilage; researchers used special biodegradable plastic to construct an ear-shaped scaffold, onto which they seeded cartilage cells. They then grafted the resulting structure under the skin on a mouse's back, where the cartilage cells were nurtured by the mouse's circulatory system. The human cells reproduced and gradually replaced the plastic as it degraded, producing a living, ear-shaped piece of human cartilage (the same tissue that supports the real outer ear). In the future, artificial ears might be grown directly on people with missing or deformed ears.

The success of the rather bizarre experiment demonstrates our increasing power to manipulate cells, the fundamental units of life. All living things are constructed of cells, including tissues and organs that can be damaged by injury or disease. If scientists can shape cells into a living ear, might they someday be able to sculpt cells into working livers, kidneys, and lungs?

5.1 What Are the Basic Features of Cells?

All Living Things Are Composed of One or More Cells

In the late 1850s, Austrian pathologist Rudolf Virchow wrote, "Every animal appears as a sum of vital units, each of which bears in itself the complete characteristics of life." Furthermore, Virchow predicted, "All cells come from cells." The three principles of modern cell theory evolved directly from Virchow's statements:

- Every living organism is made up of one or more cells.
- The smallest living organisms are single cells, and cells are the functional units of multicellular organisms.
- All cells arise from preexisting cells.

All living things, from microscopic bacteria to a mighty oak tree to the human body, are composed of cells. Whereas each bacterium consists of a single, relatively simple cell, the human body consists of trillions of complex cells, each specialized to perform a specific function. To survive, all cells must obtain energy and nutrients from their environment. They must synthesize a variety of proteins and other molecules necessary for their growth and repair, and they must eliminate wastes. Many cells need to interact with other cells. To ensure the continuity of life, cells must also reproduce. These activities are accomplished by specialized parts of each cell, described in the following sections.

All Cells Share Certain Common Features

In spite of their diversity, cells share certain common features, which are summarized in Table 5-1 and described in the following sections.

Table 5-1 Common Features of All Cells

Molecular components	Protein, amino acids, lipids, carbohydrates, sugars, nucleotides, DNA, RNA
Structural components	Plasma membrane, cytoplasm, ribosomes
Metabolism	Extracts energy and nutrients from the environment; uses energy and nutrients to build, repair, and replace cellular parts

The Plasma Membrane Encloses the Cell and Mediates Interactions Between the Cell and Its Environment

As described in Chapter 4, the *plasma membrane* consists of a phospholipid bilayer in which a variety of proteins are embedded. The plasma membrane performs three major functions:

- It isolates the cell's contents from the external environment.
- It regulates the flow of materials into and out of the cell; for example, it acquires nutrients and expels wastes.
- It allows interaction with other cells.

All Cells Use DNA as a Hereditary Blueprint and Contain Cytoplasm

Each cell contains genetic material, an inherited blueprint that stores the instructions for making the other parts of the cell and for producing new cells. The genetic material in cells is **deoxyribonucleic acid (DNA)**. During cell division, "parent cells" pass exact copies of their DNA to their newly formed offspring, often called "daughter cells."

The **cytoplasm** consists of all the material inside the plasma membrane, but outside the region of the cell that contains DNA. The fluid portion of the cytoplasm in prokaryotic and eukaryotic cells contains water, salts, and an assortment of organic molecules, including proteins, lipids, carbohydrates, salts, sugars, amino acids, and nucleotides (see Chapter 3). Most of the cell's metabolic activities—the biochemical reactions that support life—occur in the cell cytoplasm. Protein synthesis is one example. This complex process takes place on special structures called *ribosomes*, located in the cytoplasm of all cells. The many types of proteins synthesized by cells include enzymes that allow metabolic reactions to occur, as we will see in Chapter 6.

All Cells Obtain Energy and Nutrients from Their Environment

To maintain their incredible complexity, all cells must continuously acquire and expend energy. As we explain in Chapters 6, 7, and 8, essentially all of the energy powering life on Earth originates in sunlight. Cells that can harness this energy directly and incorporate it into high-energy molecules provide energy for nearly all other forms of life. The building blocks of biological molecules, such as carbon, nitrogen, oxygen, and a variety of minerals, ultimately come from the environment—the air, water, rocks, and other forms of life. All cells obtain the materials to generate the molecules of life, and the energy to power this synthesis, from their living and nonliving environment.

Cell Function Limits Cell Size

Most cells are small, ranging from about 1 to 100 micrometers (millionths of a meter) in diameter (Fig. 5-1). Because cells are so small, they were not discovered until the invention of the microscope. Ever since seeing the first cells in the late 1600s, scientists have devised increasingly sophisticated ways of observing them, as described in "Scientific Inquiry: The Search for the Cell."

Why are most cells small, and why do large organisms consist of many cells rather than one large cell? The answer lies in the need for cells to exchange nutrients and wastes with their external environment

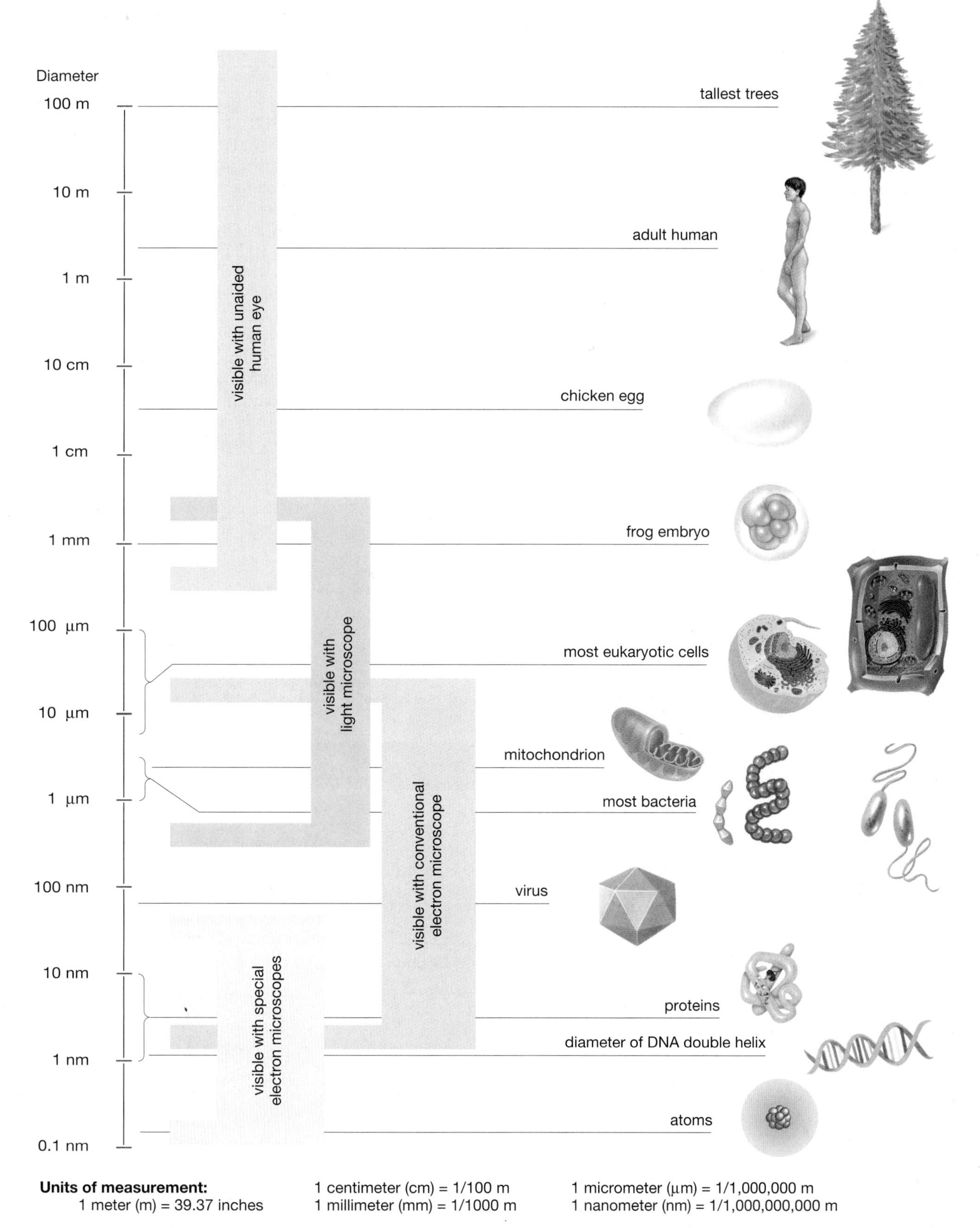

Units of measurement:
1 meter (m) = 39.37 inches
1 centimeter (cm) = 1/100 m
1 millimeter (mm) = 1/1000 m
1 micrometer (µm) = 1/1,000,000 m
1 nanometer (nm) = 1/1,000,000,000 m

FIGURE 5-1 Relative sizes
Dimensions commonly encountered in biology range from about 100 meters (the height of the tallest redwoods) through a few micrometers (the diameter of most cells) to a few nanometers (the diameter of many large molecules). Note that in the metric system (used almost exclusively in science and in many regions of the world), separate names are given to dimensions that differ by factors of 10, 100, and 1000.

SCIENTIFIC INQUIRY The Search for the Cell

Human understanding of the cellular nature of life came slowly. In 1665, English scientist and inventor Robert Hooke reported observations with a primitive microscope. He aimed his instrument at an "exceeding thin . . . piece of Cork" and saw "a great many little Boxes" (Fig. E5-1a). Hooke called the boxes "cells," because he thought they resembled the tiny rooms, or cells, occupied by monks. Cork comes from the dry outer bark of the Mediterranean oak. Hooke wrote that in the living oak and other plants, "These cells [are] fill'd with juices."

In the 1670s, Dutch microscopist Antoni van Leeuwenhoek was constructing his own simple microscopes and observing a previously unknown world. A self-taught amateur scientist, his descriptions of myriad "animalcules" (his term for protists) in rain, pond, and well water caused quite an uproar, because in those days, water was consumed without treatment. He made careful observations of an enormous range of microscopic specimens, including blood cells, sperm, and the eggs of small insects such as weevils, aphids, and fleas. His discoveries struck a blow to the then-common belief in spontaneous generation; at that time, fleas were believed to emerge spontaneously from sand or dust, as were weevils from grain! Although they appeared much more primitive than Hooke's microscopes, Leeuwenhoek's microscopes provided a much clearer images and higher magnification (Fig. E5-1b).

More than a century passed before biologists began to understand the role of cells in life on Earth. Microscopists first noted that many plants consist entirely of cells. The thick wall surrounding all plant cells, first observed by Hooke, made their observations easier. Animal cells, however, escaped notice until the 1830s, when German zoologist Theodor Schwann saw that cartilage contains cells that "exactly resemble [the cells of] plants." In 1839, after studying cells for years, Schwann was confident enough to publish his *cell theory*, calling cells the elementary particles of both plants and animals. By the mid-1800s, German botanist Matthias Schleiden further refined science's view of cells when he wrote: "It is . . . easy to perceive that the vital process of the individual cells must form the first, absolutely indispensable fundamental basis [of life]."

(a) 17th century microscope

(b) Leeuwenhoek's microscope

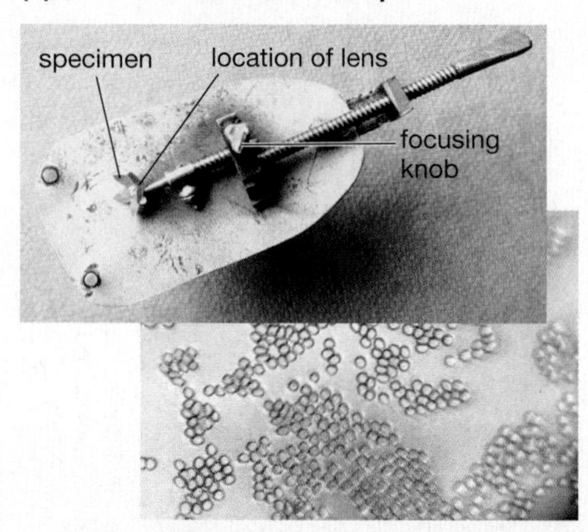

blood cells photographed through Leeuwenhoek's microscope

(c) Electron microscope

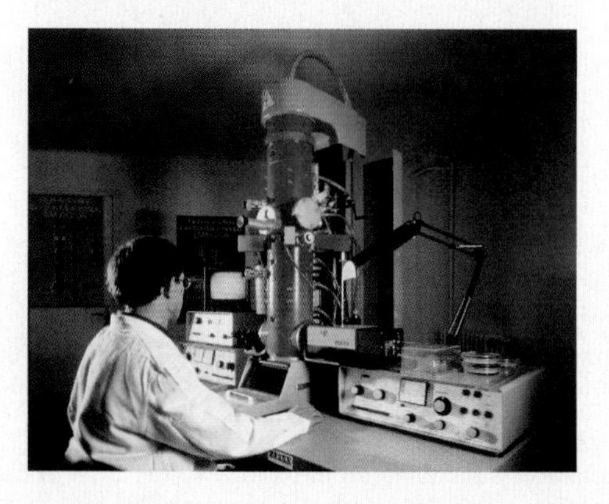

FIGURE E5-1 Microscopes yesterday and today
(a) Robert Hooke's drawings of cork cells, as he viewed them with an early light microscope similar to the one shown here. Only the cell walls remain. **(b)** One of Leeuwenhoek's microscopes, and a photograph of blood cells taken through a Leeuwenhoek microscope. The specimen is viewed through a tiny hole just underneath the lens. **(c)** A scanning electron microscope, which creates a three-dimensional image of a specimen's surface.

Ever since the pioneering efforts of Robert Hooke and Antoni van Leeuwenhoek, biologists, physicists, and engineers have collaborated in the development of a variety of advanced microscopes to view the cell and its components:

Light microscopes use lenses, usually made of glass, to focus and magnify light rays that either pass through or bounce off a specimen. Light microscopes provide a wide range of images, depending on how the specimen is illuminated and whether it has been stained (Fig. E5-2a). The *resolving power* of light microscopes—that is, the smallest structure that can be seen—is about 1 micrometer (a millionth of a meter).

Electron microscopes use beams of electrons instead of light, which are focused by magnetic fields rather than by lenses. Some types of electron microscopes can resolve structures as small as a few nanometers (billionths of a meter). *Transmission electron microscopes* (TEMs) pass electrons through a thin specimen and can reveal the details of interior cell structure, including organelles and plasma membranes (Fig. E5-2b). *Scanning electron microscopes* (SEMs; Fig. E5-1c) bounce electrons off specimens that have been coated with metals and provide three-dimensional images. SEMs can be used to view the surface details of structures that range in size from entire insects down to cells and even organelles (Fig. E5-2c,d).

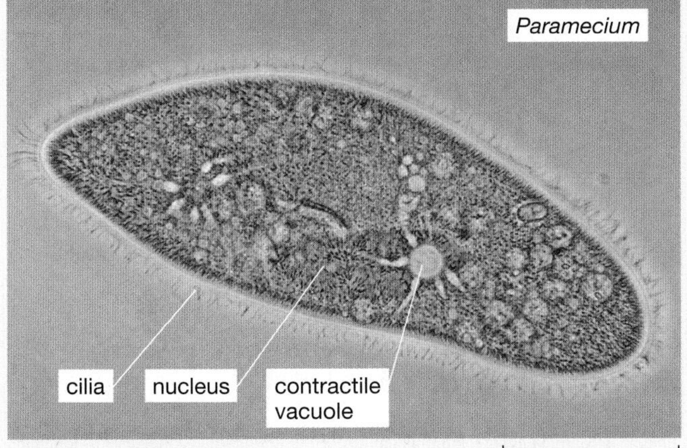

(a) Light microscope

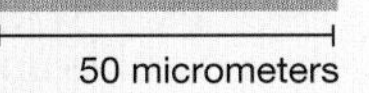

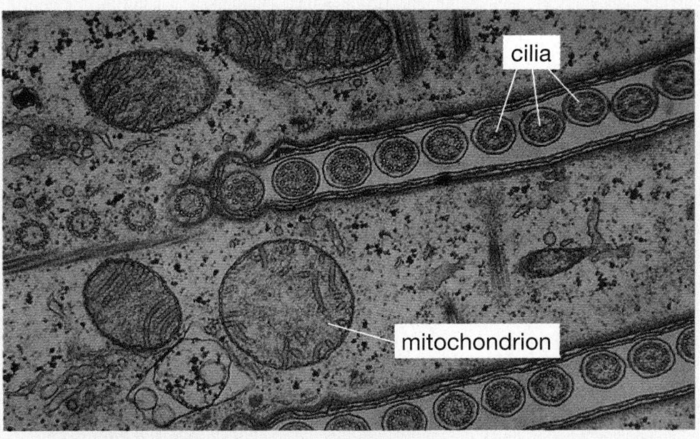

(b) Transmission electron microscope 0.5 micrometers

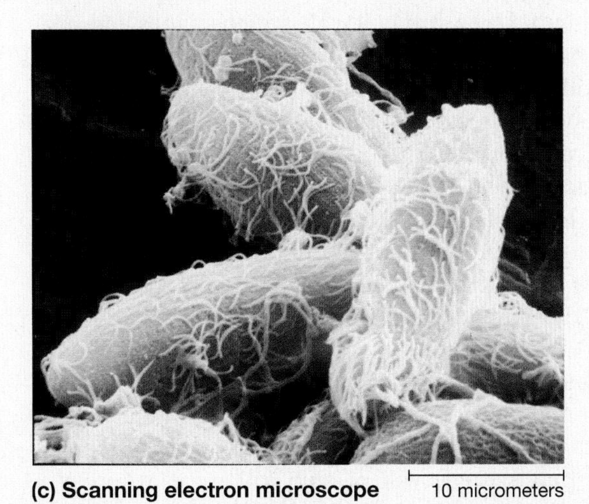

(c) Scanning electron microscope 10 micrometers

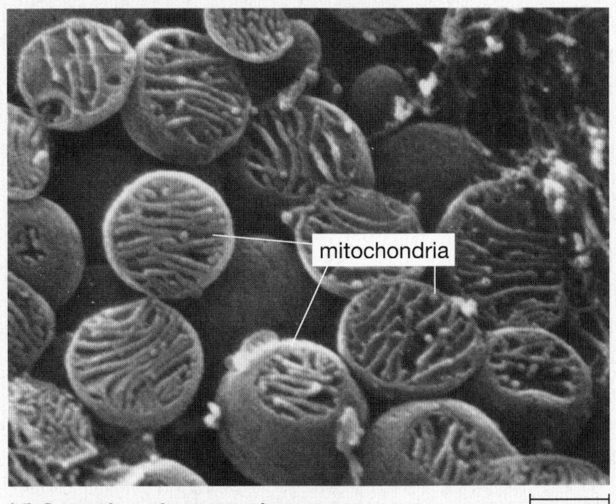

(d) Scanning electron microscope 5 micrometers

FIGURE E5-2 A comparison of microscope images
(a) A living *Paramecium* (a single-celled freshwater protist) viewed through a light microscope. **(b)** A false-color TEM photo of a *Paramecium*. **(c)** A SEM photo of a section of *Paramecium* showing sections of mitochondria and of the cilia that cover this protist. **(d)** An SEM photo at much higher magnification, showing mitochondria (many of which are sliced open) within the cytoplasm.

through the plasma membrane. As you learned in Chapter 4, many nutrients and wastes move into, through, and out of cells by *diffusion*, the movement of molecules from places of high concentration of those molecules to places of low concentration. As a roughly spherical cell becomes larger, its innermost regions become farther away from the membrane. Diffusion is a slow process. For example, in a hypothetical giant cell 8.5 inches (20 centimeters) in diameter, oxygen molecules would take more than 200 days to diffuse to the center of the cell, by which time the cell would be long dead for lack of oxygen. Also, as a cell enlarges, its volume increases more rapidly than does its surface area. For example, a cell that triples its radius becomes about 27 times greater in volume but only nine times greater in surface area. In contrast, a cluster of small cells of the same total volume maintains a large surface area:

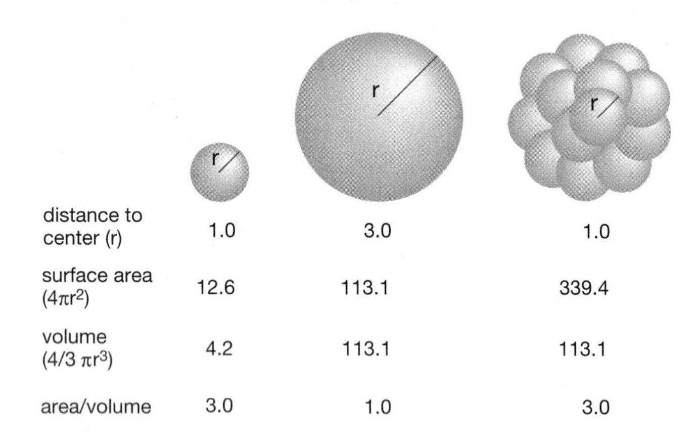

distance to center (r)	1.0	3.0	1.0
surface area ($4\pi r^2$)	12.6	113.1	339.4
volume ($4/3\ \pi r^3$)	4.2	113.1	113.1
area/volume	3.0	1.0	3.0

Thus, a larger cell has a greater need for the exchange of nutrients and wastes with the environment but a relatively smaller expanse of plasma membrane through which to make these exchanges. In a very large, roughly spherical cell, the surface area of the membrane would be too small to keep up with the cell's metabolic needs. These constraints limit the size of cells. However, some cells, such as neurons and muscle cells, can grow larger because they have an elongated shape that increases their membrane surface area.

There Are Two Basic Types of Cells: Prokaryotic and Eukaryotic

All forms of life are composed of only two fundamentally different types of cells. The first type, which includes the bacteria and archaeans, is called **prokaryotic** (Greek for "before the nucleus"; Fig. 5-2). The second type of cell, is called **eukaryotic**, (Greek for "true nucleus"; see Figs. 5-3 and 5-4). Eukaryotic cells, which almost certainly evolved from prokaryotic cells, make up the bodies of protists, plants, fungi, and animals. As their names imply, one striking difference between prokaryotic cells and eukaryotic cells is that the genetic material of eukaryotic cells is contained within a membrane-enclosed nucleus. In contrast, the genetic material of prokaryotic

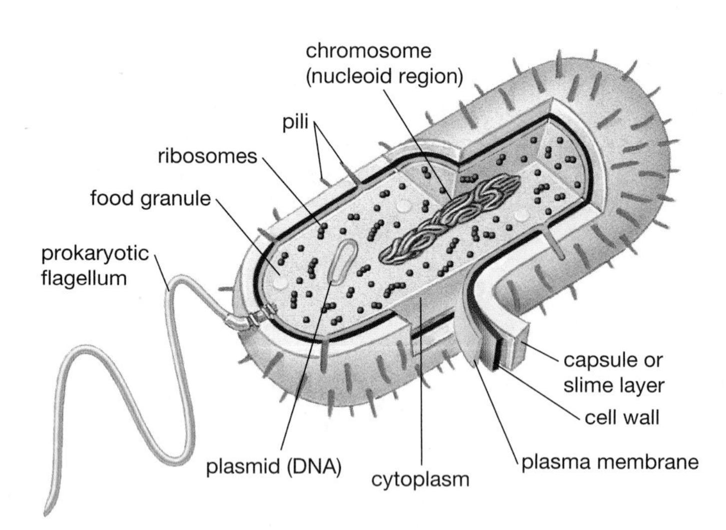

FIGURE 5-2 A generalized prokaryotic cell

cells is not enclosed within a membrane. In the following sections, we discuss the features of prokaryotic and eukaryotic cells.

5.2 What Are the Major Features of Prokaryotic Cells?

Prokaryotic Cells Are Small and Possess Specialized Surface Features

Most prokaryotic cells, or bacteria, are very small (less than 5 micrometers long), with a simple internal structure, as compared to eukaryotic cells (see Fig. 5-2). Most are surrounded by a stiff cell wall, which shapes and protects the bacterial cell. Several types of antibiotics, including penicillin, fight bacterial infections by interfering with cell wall synthesis, causing the bacteria to rupture. Some bacteria can move, propelled by flagella (different from eukaryotic flagella). Bacteria that infect other organisms, such as those that cause tooth decay and others that cause diarrhea, possess surface features that help them adhere to specific host tissues, such as the surface of a tooth or the lining of the small intestine. These surface features include *capsules* and *slime layers*, which are polysaccharide coatings that some bacteria secrete outside their cell walls. For example, in its normal encapsulated form, *Streptococcus pneumoniae* (which causes bacterial pneumonia) will readily kill mice, whereas a mutant form that lacks the capsule is harmless. *Pili* and *fimbriae* are proteins that project outward from the bacterial wall and also help certain infectious bacteria adhere to host tissues. Pili are also used to exchange genetic material (DNA) between bacterial cells.

Prokaryotic Cells Have Fewer Specialized Structures Within Their Cytoplasm

The cytoplasm of most prokaryotic cells is rather homogeneous in appearance compared to eukaryotic cells (although some photosynthetic bacteria have elaborate internal membranes). Prokaryotic cells generally have a single, circular *chromosome* consisting of a long strand of DNA that carries all the essential genetic information for the cell. The chromosome is usually coiled and localized in the region of the cell called the **nucleoid**. The nucleoid is not separated from the rest of the cytoplasm by a membrane. Most prokaryotic cells also contain small rings of DNA called *plasmids*, which are located outside the nucleoid. Plasmids usually carry genes that give the cell special properties; for example, some disease-causing bacteria possess plasmids that allow them to inactivate antibiotics, making them much more difficult to kill.

Prokaryotic cells lack nuclei and the other membrane-enclosed *organelles* that eukaryotic cells possess. Nonetheless, some prokaryotic cells use membranes to organize the molecules responsible for a series of biochemical reactions, situating the molecules in a particular sequence to promote the reactions in the necessary order. Photosynthetic bacteria possess internal membranes, in which light-capturing proteins and enzymes that catalyze the synthesis of high-energy molecules are embedded in a specific order. In prokaryotic cells, reactions that harvest energy from the breakdown of sugars are catalyzed by enzymes that may be either localized along the inner plasma membrane or free in the cytoplasm. Bacterial cytoplasm contains ribosomes, structures composed of a combination of proteins and **ribonucleic acid (RNA)**; as mentioned earlier, proteins are synthesized on ribosomes. The cytoplasm also may contain *food granules* that store energy-rich materials, such as glycogen. Table 5-2 compares prokaryotic cells with the eukaryotic cells of plants and animals. The diversity and specialized structures of bacteria are covered in more detail in Chapter 19.

Table 5-2 Cell Structures, Their Functions, and Their Distribution in Living Cells

Structure	Function	Prokaryotes	Plants	Animals
Cell surface				
Cell wall	Protects, supports cell	Present	Present	Absent
Plasma membrane	Isolates cell contents from environment; regulates movement of materials into and out of cell; communicates with other cells	Present	Present	Present
Organization of genetic material				
Genetic material	Encodes information needed to construct cell and control cellular activity	DNA	DNA	DNA
Chromosomes	Contain and control use of DNA	Single, circular, no proteins	Many, linear, with proteins	Many, linear, with proteins
Nucleus	Membrane-bound container for chromosomes	Absent	Present	Present
Nuclear envelope	Encloses nucleus; regulates movement of materials into and out of nucleus	Absent	Present	Present
Nucleolus	Synthesizes ribosomes	Absent	Present	Present
Cytoplasmic structures				
Mitochondria	Produce energy by aerobic metabolism	Absent	Present	Present
Chloroplasts	Perform photosynthesis	Absent	Present	Absent
Ribosomes	Provide site of protein synthesis	Present	Present	Present
Endoplasmic reticulum	Synthesizes membrane components and lipids	Absent	Present	Present
Golgi complex	Modifies and packages proteins and lipids; synthesizes carbohydrates	Absent	Present	Present
Lysosomes	Contain intracellular digestive enzymes	Absent	Present	Present
Plastids	Store food, pigments	Absent	Present	Absent
Central vacuole	Contains water and wastes; provides turgor pressure to support cell	Absent	Present	Absent
Other vesicles and vacuoles	Contain food obtained through phagocytosis; contain secretory products	Absent	Present (some)	Present
Cytoskeleton	Gives shape and support to cell; positions and moves cell parts	Absent	Present	Present
Centrioles	Synthesize microtubules of cilia and flagella; may produce spindle in animal cells	Absent	Absent (in most)	Present
Cilia and flagella	Move cell through fluid or move fluid past cell surface	Present[a]	Absent (in most)	Present

[a] Many prokaryotes have structures called *flagella*, but these are not made of microtubules and move in a fundamentally different way than do eukaryotic cilia or flagella.

5.3 What Are the Major Features of Eukaryotic Cells?

Eukaryotic Cells Contain Organelles

Eukaryotic cells differ from prokaryotic cells in many ways. For example, they are usually larger than prokaryotic cells—typically more than 10 micrometers in diameter. However, the cytoplasm of eukaryotic cells also houses a variety of **organelles**, membrane-enclosed structures that perform specific functions within the cell. The **cytoskeleton**, a network of protein fibers, gives shape and organization to the cytoplasm of eukaryotic cells. Many of the organelles are attached to the cytoskeleton.

Eukaryotic cells, however, are not all alike. Figures 5-3 and 5-4 illustrate the structures that are found in animal and plant cells, respectively, although few individual cells possess all the features shown in either drawing. Each type of cell has a few unique organelles not found in the other. Plant cells, for example, contain chloroplasts, plastids, and a central vacuole not present in animal cells, while only animal cells possess centrioles. You may want to refer to these illustrations as we describe the structures of the cell in more detail. The major components of eukaryotic cells (see Table 5-2) are described in more detail in the following sections.

The Nucleus Is the Control Center of the Eukaryotic Cell

Deoxyribonucleic acid (DNA) is the genetic material of all living cells. A cell's DNA stores all the information needed to construct the cell and direct the countless chemical reactions necessary for life and reproduction.

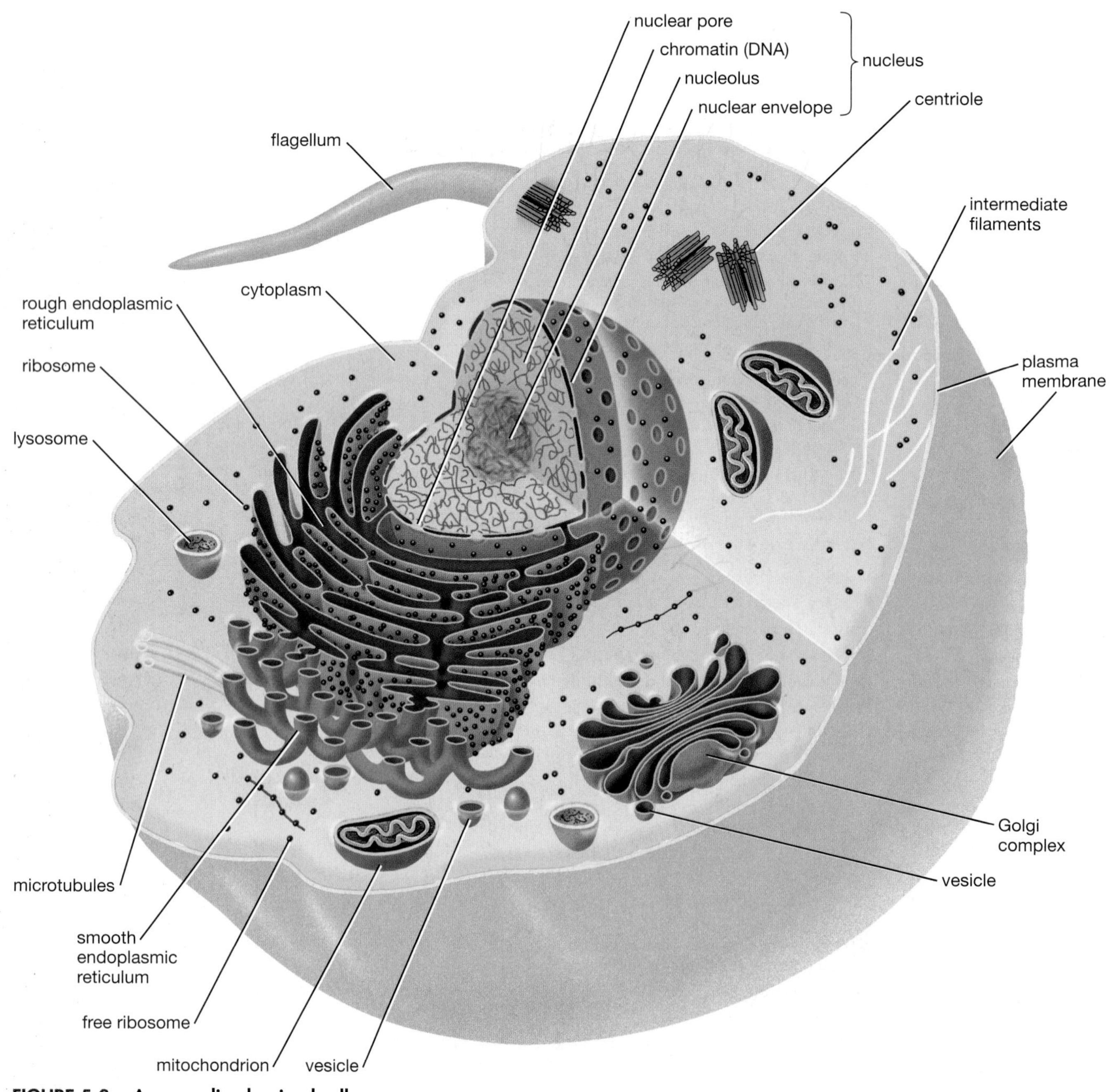

FIGURE 5-3 A generalized animal cell

Media Activity
5.1 Cell Structure

The genetic information in DNA is used selectively by the cell, depending on its stage of development and its environmental conditions. In eukaryotic cells, DNA is housed within the nucleus.

The **nucleus** is an organelle, usually the largest in the cell, that consists of three readily distinguishable parts (Fig. 5-5). The **nuclear envelope** separates the nuclear material from the cytoplasm. Inside the nuclear envelope, the nucleus contains a granular-looking material called **chromatin** and a darker region called the **nucleolus**, which are described in the following sections.

The Nuclear Envelope Allows Selective Exchange of Materials

The nucleus is isolated from the rest of the cell by a nuclear envelope that consists of a double membrane. The membrane is perforated with tiny membrane-lined channels called *nuclear pores*. Water, ions, and small molecules such as ATP can pass freely through the pores, but the passage of large molecules, particularly proteins, pieces of ribosomes, and RNA, is regulated by specialized "gatekeeper proteins" that line each nuclear pore. Ribosomes stud the outer nuclear membrane, which is continuous with membranes of the rough endoplasmic reticulum described later (see Figs. 5-3 and 5-4).

Chromatin Contains DNA, Which Codes for the Synthesis of Proteins

Because the nucleus is highly colored by common stains used in light microscopy, early microscopists, having no knowledge of its function, named the nuclear material *chromatin*, meaning "colored substance." Biologists have

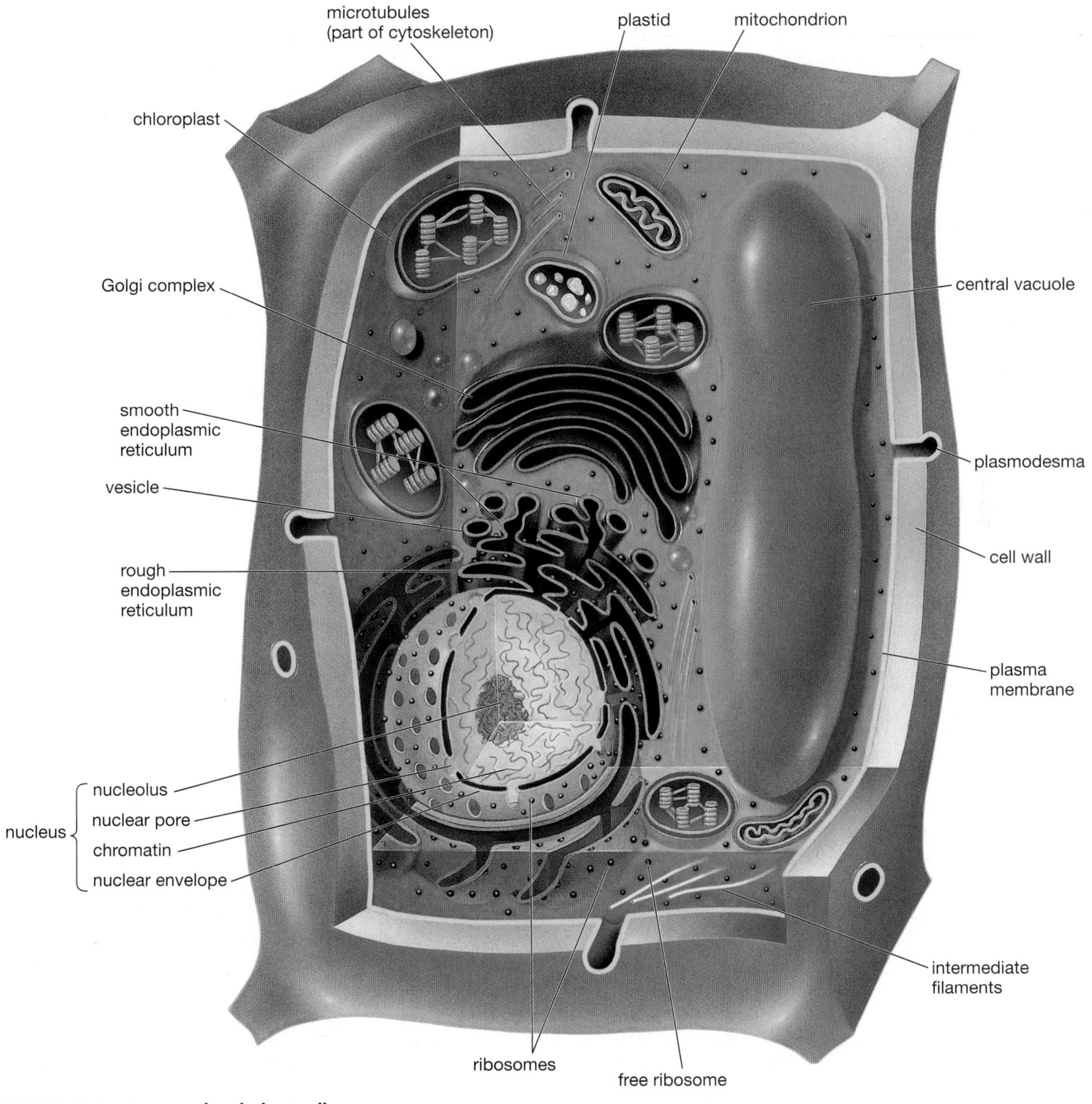

FIGURE 5-4 A generalized plant cell
QUESTION Of the nucleus, ribosome, chloroplast, and mitochondrion, which appeared earliest in the history of life?

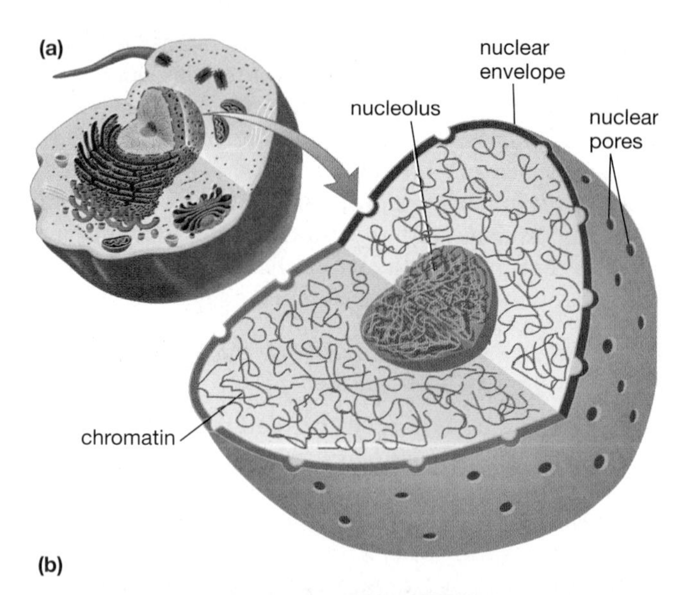

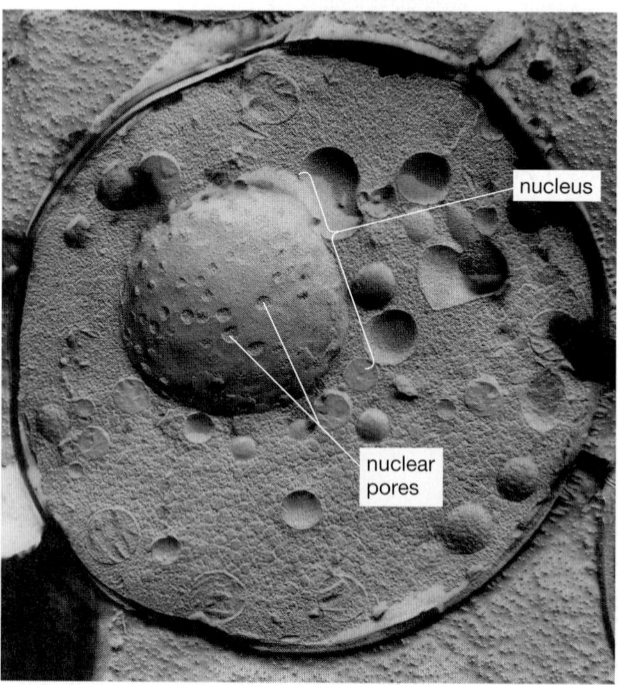

FIGURE 5-5 The nucleus
(a) The nucleus is bounded by a double outer membrane. Inside are chromatin and a nucleolus. **(b)** An electron micrograph of a yeast cell that has been frozen and broken open to reveal its internal structures. The large nucleus, with nuclear pores penetrating its nuclear membrane, is clearly visible.

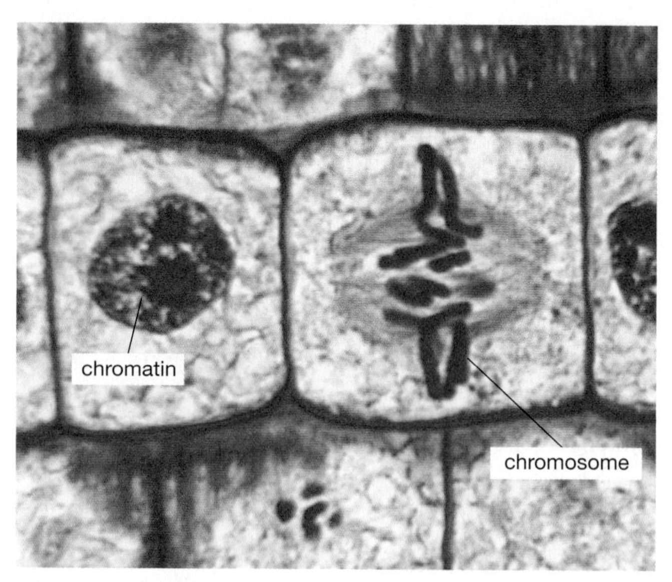

FIGURE 5-6 Chromosomes
Chromosomes, seen here in a light micrograph of a dividing cell (on the right) in an onion root tip, contain the same material (DNA and proteins) as the chromatin seen in adjacent nondividing cells, but in a more compact state. **QUESTION** Why doesn't chromatin remain in its condensed form throughout the cell cycle?

since learned that chromatin consists of DNA associated with proteins. Eukaryotic DNA and its associated proteins form long strands called **chromosomes** ("colored bodies"). When cells divide, each chromosome coils upon itself, becoming thicker and shorter. The resulting "condensed" chromosomes are easily visible even with light microscopes (Fig. 5-6).

Chemical reactions within the cell that are responsible for growth and repair, nutrient and energy acquisition and use, and reproduction are governed by the information encoded in DNA. Because most chemical reactions occur in the cytoplasm, molecules carrying genetic information must be exchanged between the DNA in the nucleus and the cytoplasm. To accomplish this, genetic information is copied from DNA into molecules of RNA, which move through the pores of the nuclear envelope into the cytoplasm. This information is then used to direct the synthesis of cellular proteins. These proteins include enzymes, which catalyze and regulate chemical reactions; membrane proteins, which govern interactions between the cell and its environment; and a variety of structural proteins. Some of these proteins pass from the cytoplasm into the nucleus and regulate the transfer of information from DNA to RNA, depending on what is happening in the cytoplasm and in the extracellular environment. We will take a closer look at these processes in Chapter 10.

The Nucleolus Is the Site of Ribosome Assembly

Most eukaryotic nuclei have one or more darkly staining regions called *nucleoli* ("little nuclei"; one nucleolus is shown in Fig. 5-5a). The nucleolus consists of ribosomal RNA, proteins, ribosomes in various stages of synthesis, and DNA (bearing genes that specify the blueprint for ribosomal RNA).

Nucleoli are the sites of ribosome synthesis. A **ribosome** is a small particle composed of RNA and proteins that serves as a kind of "workbench" for the synthesis of proteins. Just as a workbench can be used to construct many different objects, any ribosome can be used to synthesize any of the thousands of proteins made by a cell. In electron micrographs, ribosomes appear as dark granules, either distributed in the cytoplasm (Fig. 5-7) or clustered along the membranes of the nuclear envelope and the endoplasmic reticulum (see Fig. 5-8).

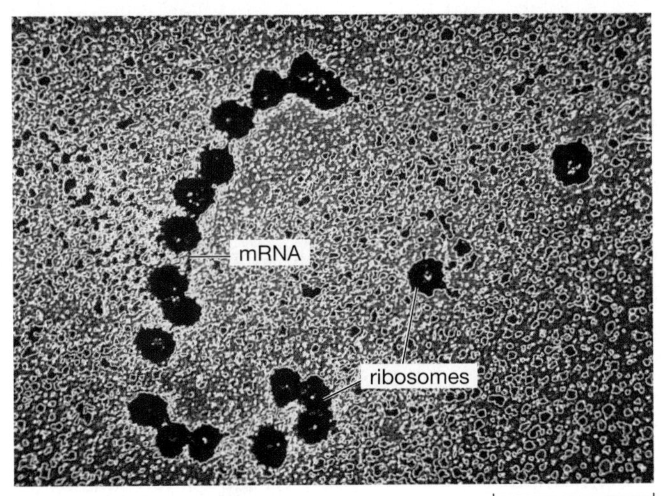

FIGURE 5-7 Ribosomes
Ribosomes may be found free in the cytoplasm either singly or strung along messenger RNA molecules as they participate in protein synthesis. Ribosomes also stud the rough endoplasmic reticulum, giving it a rough appearance and allowing the synthesis of proteins within the ER.

Eukaryotic Cells Contain a Complex System of Membranes

All eukaryotic cells have an elaborate system of membranes that enclose the cell and create internal compartments, allowing a huge variety of processes to occur within the cytoplasm. This membrane system is composed of the plasma membrane and membranous parts of several organelles, including the endoplasmic reticulum, nuclear envelope, Golgi complex, and a variety of membrane-enclosed sacs such as lysosomes. Membranes throughout the cell are fundamentally similar in composition and can merge with one another. Thus, the various parts of the cell's membrane system can exchange membrane material with one another and can also transfer membranous materials to different compartments for different types of processing (see Fig. 5-10).

The Plasma Membrane Both Isolates the Cell and Allows Selective Interactions Between the Cell and Its Environment

The plasma membrane forms the outer boundary of the living part of a cell, enclosing the cytoplasm. It is a marvelously complex structure that must perform the seemingly contradictory functions of separating the cytoplasm of the cell from the outside environment while enabling the transport of selected substances into or out of the cell. In plants, fungi, and some protists, a *cell wall* is secreted through the plasma membrane and forms an outer, protective coating. Plasma membrane and eukaryotic cell wall structure and function are discussed in Chapter 4.

The Endoplasmic Reticulum Forms Membrane-Enclosed Channels Within the Cytoplasm

The **endoplasmic reticulum (ER)** is a series of interconnected membrane-enclosed tubes and channels in the cytoplasm (Fig. 5-8); the ER membrane is continuous with the nuclear membrane. Eukaryotic cells have two forms of ER: rough and smooth. Numerous ribosomes stud the outside of the *rough endoplasmic reticulum*; in contrast, *smooth endoplasmic reticulum* lacks ribosomes.

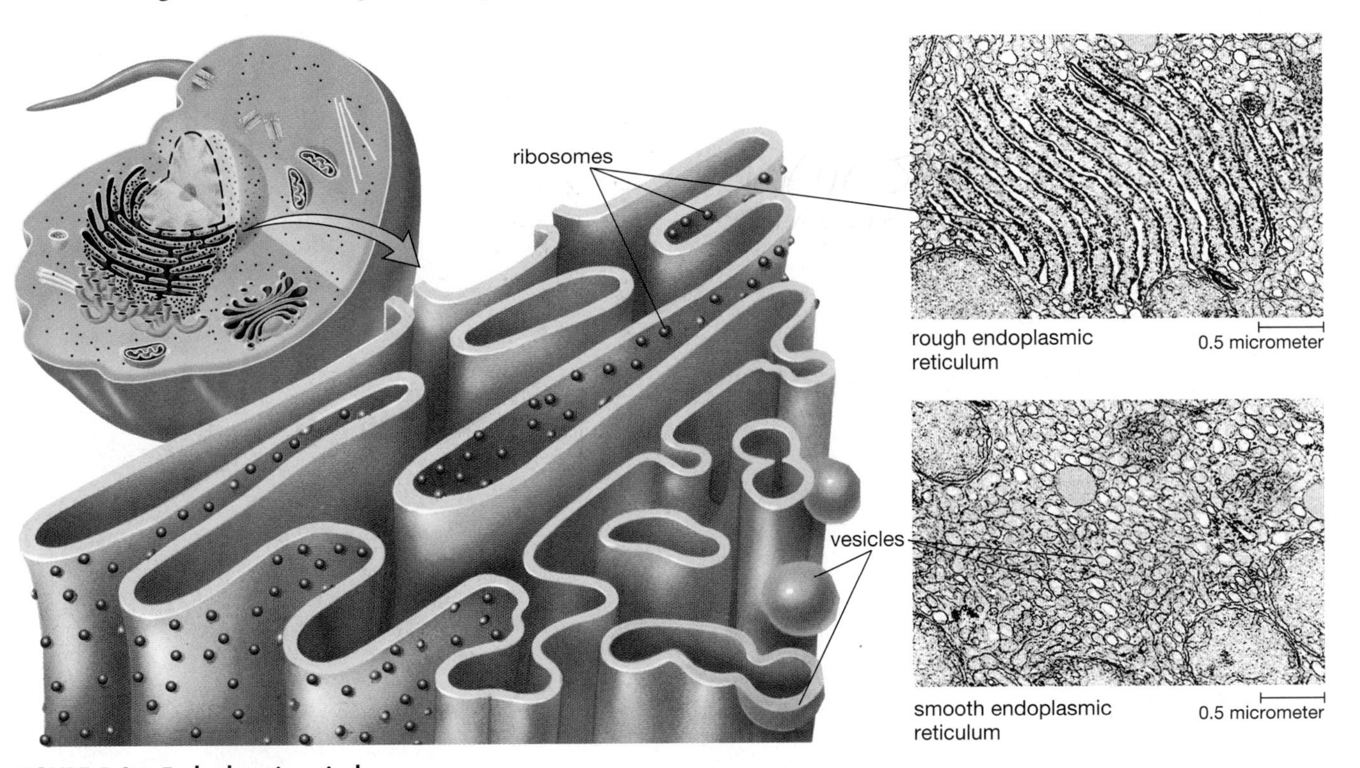

FIGURE 5-8 Endoplasmic reticulum
There are two types of endoplasmic reticulum: rough ER and smooth ER. Although in electron micrographs the ER looks like a series of tubes and sacs, it is actually a maze of folded sheets and interlocking channels. In many cells, the rough and smooth ER are thought to be continuous, as depicted in the drawing. Ribosomes (black) stud the cytoplasmic face of the rough ER membrane.

The different structures of smooth and rough ER reflect different functions. Enzymes embedded in the membranes of the smooth ER synthesize lipids, including the phospholipids and cholesterol used in membrane formation. Smooth ER in liver cells contains enzymes that detoxify harmful drugs and metabolic by-products. In some cells, the smooth ER synthesizes other types of lipids as well, such as the steroid hormones testosterone and estrogen, which are produced in the reproductive organs of mammals.

The ribosomes on the outside of rough ER are used to synthesize both proteins and phospholipids, and can produce new membrane that becomes incorporated into the ER. Continuous production of new ER membrane is important because membrane from the ER is constantly being budded off and transported to the Golgi complex, lysosomes, and the plasma membrane.

Ribosomes on rough ER also manufacture the proteins, such as digestive enzymes and protein hormones (for example, insulin), that some secretory cells export into their surroundings. As these proteins are synthesized, they are inserted through the ER membrane into the ER interior. The proteins synthesized for secretion or use within the cell move through the ER channels and accumulate in pockets. These pockets then bud off, forming membrane-bound sacs called **vesicles** that carry their protein cargo to the Golgi complex.

The Golgi Complex Sorts, Chemically Alters, and Packages Important Molecules

The **Golgi complex** (named for the Italian physician and cell biologist Camillo Golgi, who discovered it in the late 1800s) is a specialized set of membranes derived from the endoplasmic reticulum that looks very much like a stack of flattened sacs (Fig. 5-9). Vesicles from the ER fuse with one face of the Golgi complex, adding their membranes to the Golgi complex and emptying their contents into the Golgi sacs. Other vesicles bud off the Golgi complex on the opposite face of the stack, carrying away specific proteins, lipids, and other complex molecules. The Golgi complex performs the following three major functions:

- The Golgi separates proteins and lipids received from the ER according to their destinations; for example, it separates digestive enzymes that are bound for lysosomes from hormones that the cell will secrete.
- The Golgi modifies some molecules; for instance, it adds sugars to proteins to make glycoproteins.
- The Golgi packages these materials into vesicles that are then transported to other parts of the cell or to the plasma membrane for export.

The Travels of a Secreted Protein

To understand how the membranous organelles work together (Fig. 5-10), let's look at the secretion of an antibody. An antibody is a protein, secreted by a type of

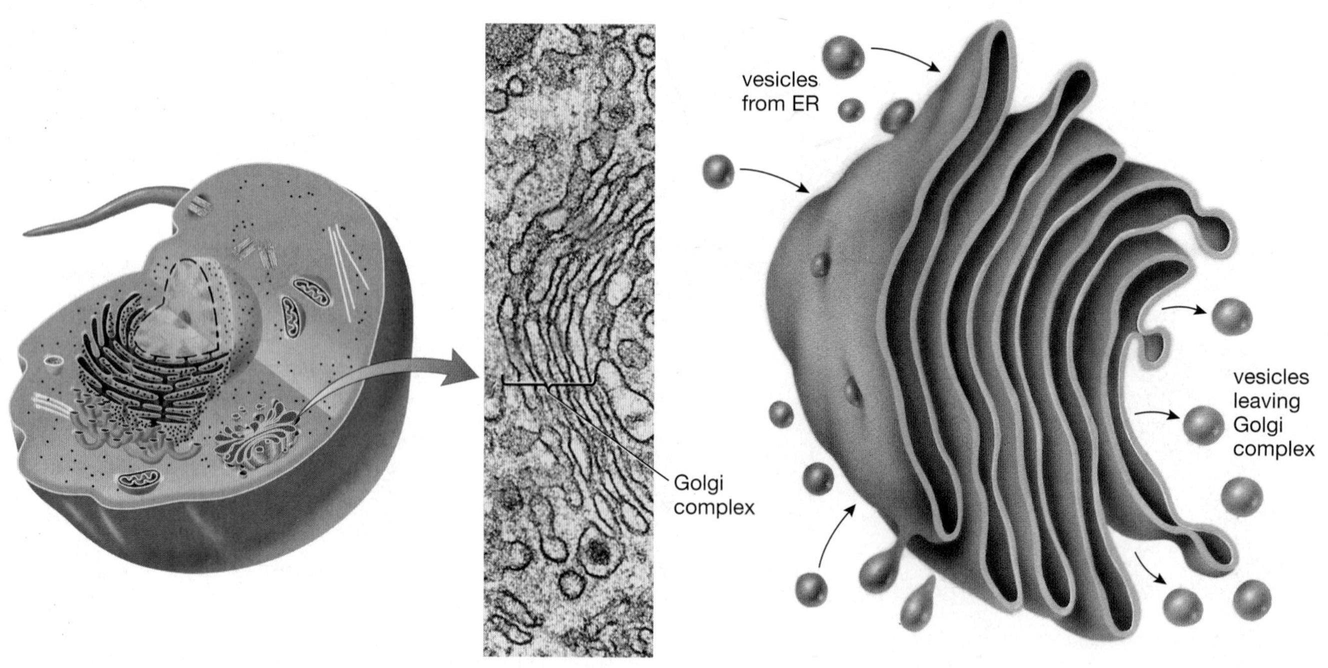

FIGURE 5-9 The Golgi complex
The Golgi complex is a stack of flat membranous sacs derived from the endoplasmic reticulum. Vesicles constantly bud off from and fuse with the Golgi and ER, transporting material from the ER to the Golgi and back again, and from the Golgi to plasma membrane, lysosomes, and vesicles.

white blood cell, that binds to foreign invaders (such as bacteria) and helps destroy them. Antibody proteins are synthesized on ribosomes of the rough ER and then packaged into vesicles formed from the ER membrane. These vesicles travel to the Golgi, where the membranes fuse, releasing the protein into the Golgi complex. Here, carbohydrates are attached to the protein, which is then repackaged into vesicles formed from Golgi membrane. The vesicle containing the completed antibody travels to the plasma membrane and fuses with it, releasing the antibody outside the cell, where it will make its way into the bloodstream to help defend the body against infection.

Lysosomes Serve as the Cell's Digestive System

Some of the proteins manufactured in the ER and sent to the Golgi complex are intracellular digestive enzymes that can break proteins, fats, and carbohydrates into their component subunits. In the Golgi, these enzymes are packaged in membranous vesicles called **lysosomes** (see Fig. 5-10). One major function of lysosomes is to digest food particles, which range from individual proteins to complete microorganisms.

As you have seen in Chapter 4, many cells "eat" by *phagocytosis*—that is, by engulfing extracellular particles with extensions of the plasma membrane. The food particles are then moved into the cytoplasm, enclosed within membranous sacs called **food vacuoles**. Lysosomes recognize these food vacuoles and fuse with them. The contents of the two vesicles mix, and the lysosomal enzymes digest the food into amino acids, monosaccharides, fatty acids, and other small molecules. These simple molecules then diffuse out of the lysosome and into the cytoplasm to nourish the cell. Cell biologists continue to search for the key to how lysosomes recognize these food vacuoles.

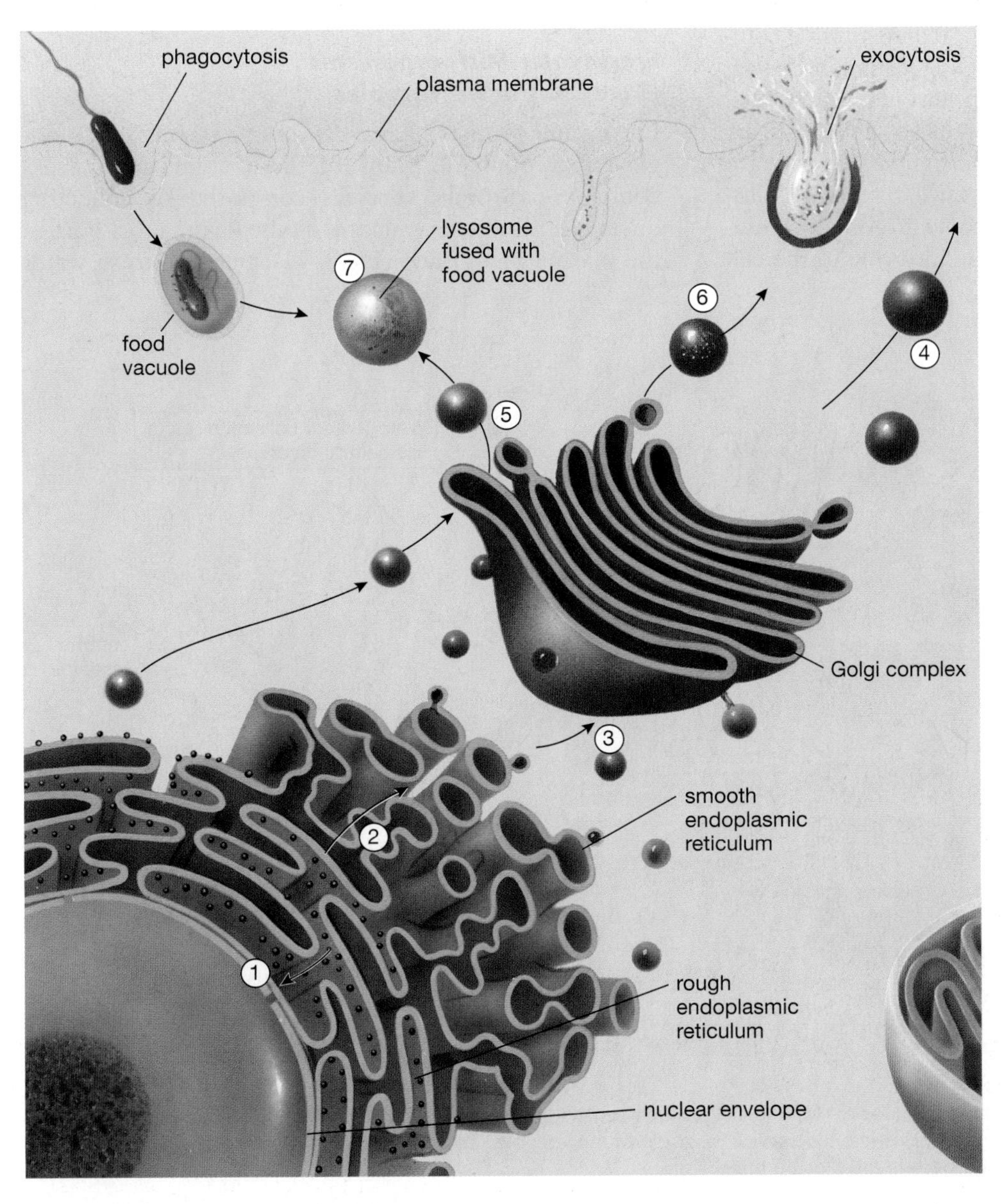

FIGURE 5-10 The flow of membrane within the cell
Membrane is synthesized by the ER. ① Some membrane moves inward to form new nuclear envelope, some moves outward to form ② smooth ER and ③ new Golgi membrane. From the Golgi, membrane buds off to form ④ new plasma membrane and ⑤ membrane surrounding organelles such as lysosomes. Some proteins synthesized in the rough ER are modified in the smooth ER and travel in membrane vesicles to the Golgi, where they are further modified and sorted. ⑥ Some of these proteins are packaged in vesicles bound for the plasma membrane, where they will be secreted from the cell. Some are enzymes packaged in lysosomes (⑤). Lysosomes may fuse with food vacuoles ⑦, allowing intracellular digestion of food particles.

Media Activity
5.2 Membrane Traffic

Lysosomes also digest excess cellular membranes and defective or malfunctioning organelles. The cell encloses them in vesicles made of membrane from the ER, which then fuse with lysosomes. Digestive enzymes within the lysosome enable the cell to recycle valuable materials from the defunct organelles. Scientists are still researching the question of how the cell identifies organelles that have outlived their usefulness.

Membrane Synthesized in the Endoplasmic Reticulum Flows Through the Membrane System of the Cell

The nuclear envelope, rough and smooth ER, Golgi complex, lysosomes, food vacuoles, and the plasma membrane all form an integrated membrane system (see Fig. 5-10). Membrane is synthesized in the ER and flows back and forth between these structures in a precisely determined manner. As an example, let's look at the movement of materials destined for inclusion in the plasma membrane. The ER synthesizes the phospholipids and proteins that make up the plasma membrane and buds off a vesicle whose membrane includes these plasma membrane components. The vesicle fuses with the Golgi complex. Plasma membrane material continues on through the Golgi, where it may be modified—for example, by adding sugars to make glycoproteins or *glycolipids* (lipids to which a sugar is attached). Eventually, the vesicle containing plasma membrane material becomes "outward-bound," budding off the far side of the Golgi and moving to the cell surface. The vesicle fuses with the plasma membrane, replenishing or enlarging it.

The Golgi complex processes and packages all membrane-enclosed materials produced by the cell. Many of the vesicles pinched off from the Golgi contain secretory products (for example, hormones) that are released outside the cell. Lysosomes contain digestive enzymes and often fuse with food vacuoles to carry out intracellular digestion. How these diverse materials are recognized, purified, modified, separated, and individually packaged remains a challenge to cell biologists.

Vacuoles Serve Many Functions, Including Water Regulation, Support, and Storage

Most cells contain one or more **vacuoles**—fluid-filled sacs surrounded by a single membrane. Some, such as the food vacuoles that form during phagocytosis, are only temporary. However, many cells contain permanent vacuoles that have important roles in maintaining the integrity of the cell, most notably by regulating the cell's water content.

Freshwater Microorganisms Have Contractile Vacuoles

Freshwater protists such as *Paramecium* consist of a single eukaryotic cell. Many of these organisms possess complex **contractile vacuoles** composed of collecting ducts, a central reservoir, and a tube leading to a pore in the plasma membrane (Fig. 5-11). Because fresh water

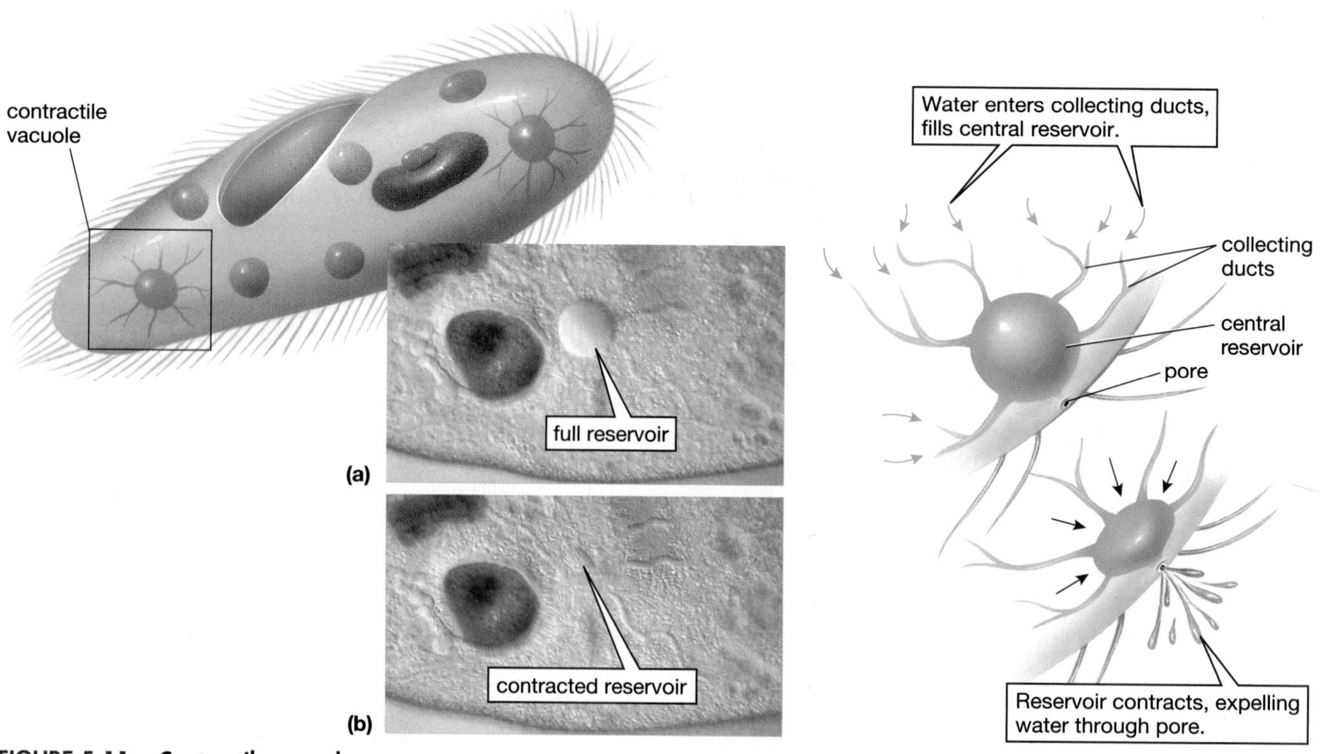

FIGURE 5-11 Contractile vacuoles
Many freshwater protists contain contractile vacuoles. **(a)** Water constantly enters the cell by osmosis. In the cell, water is taken up by collecting ducts and drains into the central reservoir of the vacuole. **(b)** When full, the reservoir contracts, expelling the water through a pore in the plasma membrane.

is hypotonic to the cytoplasm of these organisms, water constantly enters the cell by osmosis. The increasing volume of incoming water might soon burst the fragile creature if it did not have a mechanism to excrete the water. Cellular energy is used to pump salts from the cytoplasm of the protist into collecting ducts. Water follows by osmosis and drains into the central reservoir. When the reservoir is full, it contracts, squirting the water out through a pore in the plasma membrane.

Plant Cells Have Central Vacuoles

Three-quarters or more of the volume of many plant cells is occupied by a large **central vacuole** (Fig. 5-12, top; see also Fig. 5-4). The central vacuole has several functions. Filled mostly with water, the central vacuole is involved in the cell's water balance. It also provides a dump site for hazardous wastes, which plant cells often cannot excrete. Some plant cells store extremely poisonous substances, such as sulfuric acid, in their vacuoles, which deter animals from munching on the otherwise tasty leaves. Vacuoles may also store sugars and amino acids not immediately needed by the cell. Blue or purple pigments stored in central vacuoles are responsible for the colors of many flowers.

These dissolved substances make the vacuole contents hypertonic to the cell cytoplasm, which in turn is

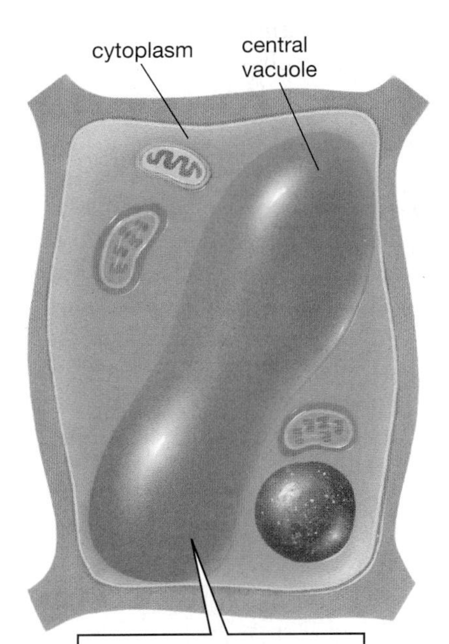

cytoplasm
central vacuole
When water is plentiful, it fills the central vacuole, pushes the cytoplasm against the cell wall, and helps maintain the cell's shape.

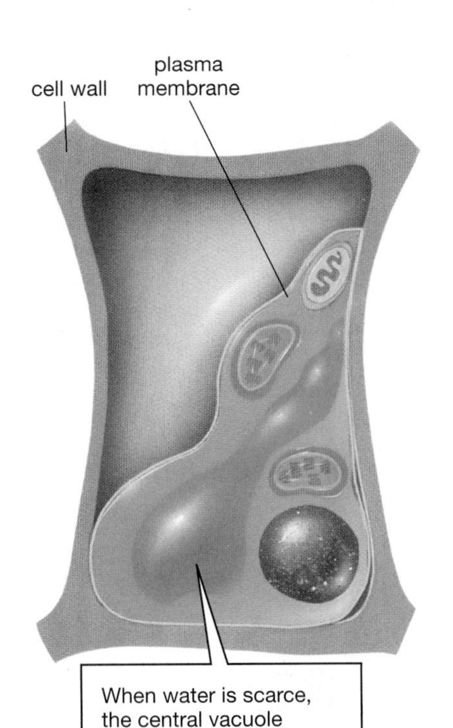

cell wall
plasma membrane
When water is scarce, the central vacuole shrinks and the cell wall is unsupported.

Water pressure supports the leaves of this impatiens plant.

Deprived of the support from water, the plant wilts.

FIGURE 5-12 The central vacuole and turgor pressure in plant cells
QUESTION If a plant cell is placed in distilled water (with no solutes), will the cell eventually burst? What about an animal cell?

usually hypertonic to the extracellular fluid that bathes the cells. Water therefore enters the vacuole by osmosis, which tends to make it swell. The pressure of the water within the vacuole, called *turgor pressure*, pushes the fluid portion of the cytoplasm up against the cell wall with considerable force (see Fig. 5-12). Cell walls are usually somewhat flexible, so both the overall shape and the rigidity of the cell depend on turgor pressure within the cell. Turgor pressure thus provides support for the nonwoody parts of plants. If you forget to water your houseplants, the central vacuoles and cytoplasm lose water and the cells shrink away from their cell walls. Just as a balloon goes limp when its air leaks out, so too the plant droops as its cells lose turgor pressure.

Mitochondria Extract Energy from Food Molecules, and Chloroplasts Capture Solar Energy

Every cell requires a continuous supply of energy to manufacture complex molecules and structures, to acquire nutrients from the environment and excrete waste materials, to move, and to reproduce. All eukaryotic cells have **mitochondria** that convert energy stored in sugar to ATP. Plant cells also have **chloroplasts**, which can capture energy directly from sunlight and store it in sugar molecules.

Biologists believe that both mitochondria and chloroplasts evolved from prokaryotic bacteria that took up residence long ago within the cytoplasm of cells ancestral to eukaryotic cells, a process called *endosymbiosis* (literally, "living together inside"). Mitochondria and chloroplasts are similar to each other in many ways. Both are about 1 to 5 micrometers in diameter and are surrounded by a double membrane. Both have assemblies of enzymes that synthesize ATP, although the assemblies are used in a very different manner in chloroplasts than in mitochondria. Finally, both have characteristics—such as their own DNA and ribosomes as well as their general size and shape—that seem to be remnants of their origin as free-living prokaryotic cells. The **endosymbiont hypothesis** of mitochondrial and chloroplast evolution is discussed in more detail in Chapter 17.

Mitochondria Use Energy Stored in Food Molecules to Produce ATP

All eukaryotic cells have mitochondria, which are sometimes called the "powerhouses of the cell" because they extract energy from food molecules and store it in the high-energy bonds of ATP. As you will see in Chapter 8, different amounts of energy can be released from a food molecule, depending on how it is metabolized. The breakdown of food molecules begins with enzymes in the fluid portion of the cytoplasm and does not use oxygen. This **anaerobic** (without oxygen) metabolism does not convert very much food energy into ATP energy. Mitochondria enable a eukaryotic cell to utilize oxygen to further break down high-energy molecules. These

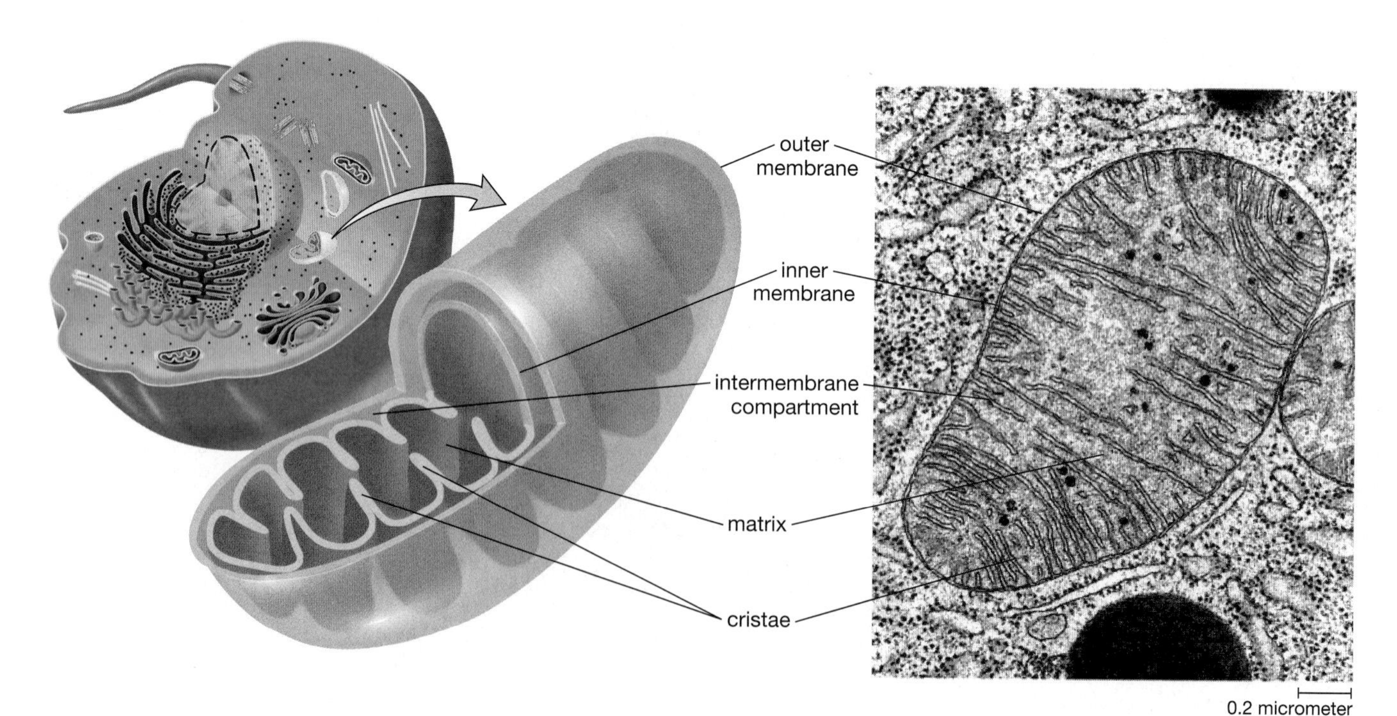

FIGURE 5-13 A mitochondrion
Mitochondria consist of a pair of membranes enclosing two fluid compartments: the intermembrane compartment between the outer and inner membranes, and the matrix within the inner membrane. The outer membrane is smooth, but the inner membrane forms deep folds called *cristae*.

aerobic (with oxygen) reactions generate energy much more effectively than the anaerobic reactions; 18 or 19 times more ATP is generated by aerobic metabolism in the mitochondria than by anaerobic metabolism in the cytoplasm. Not surprisingly, mitochondria are found in large numbers in metabolically active cells, such as muscle, and are less abundant in cells that are less metabolically active, such as those of bone and cartilage.

Mitochondria are round, oval, or tubular sacs made of a pair of membranes (Fig. 5-13). Although the outer mitochondrial membrane is smooth, the inner membrane loops back and forth to form deep folds called *cristae* (singular, *crista*, meaning "crest"). As a result, the mitochondrial membranes enclose two fluid-filled spaces: the *intermembrane compartment* between the inner and outer membranes and the *matrix*, or inner compartment, within the inner membrane. Some of the reactions of food metabolism occur in the fluid matrix contained within the inner membrane; the rest are conducted by a series of enzymes attached to the membranes of the cristae within the intermembrane compartment. The role of mitochondria in energy production is described in detail in Chapter 8.

Chloroplasts Are the Sites of Photosynthesis

If chloroplasts didn't exist, you wouldn't be reading this; none of the eukaryotic life-forms that dominate Earth today would exist, as we discuss in Chapter 7. Photosynthesis in the eukaryotic cells of plants and in photosynthetic protists occurs in chloroplasts. Chloroplasts (Fig. 5-14) are specialized organelles surrounded by a double membrane. The inner membrane of the chloroplast encloses a fluid called the *stroma*. Within the stroma are interconnected stacks of hollow membranous sacs. The individual sacs are called *thylakoids*, and a stack of sacs is a *granum* (plural, *grana*).

The thylakoid membranes contain the green pigment molecule **chlorophyll** (which gives plants their green color) as well as other pigment molecules. During photosynthesis, chlorophyll captures the energy of sunlight and transfers it to other molecules in the thylakoid membranes. These molecules in turn transfer the energy to ATP and other energy-carrier molecules. The energy carriers diffuse into the stroma, where their energy is used to drive the synthesis of sugar from carbon dioxide and water.

Plants Use Plastids for Storage

Chloroplasts are highly specialized **plastids**, which are organelles found only in plants and photosynthetic protists. Plastids are surrounded by a double membrane and serve a variety of functions. Plants and photosynthetic protists use nonchloroplast types of plastids as storage containers for various molecules, including pigments that give ripe fruits their yellow, orange, or red

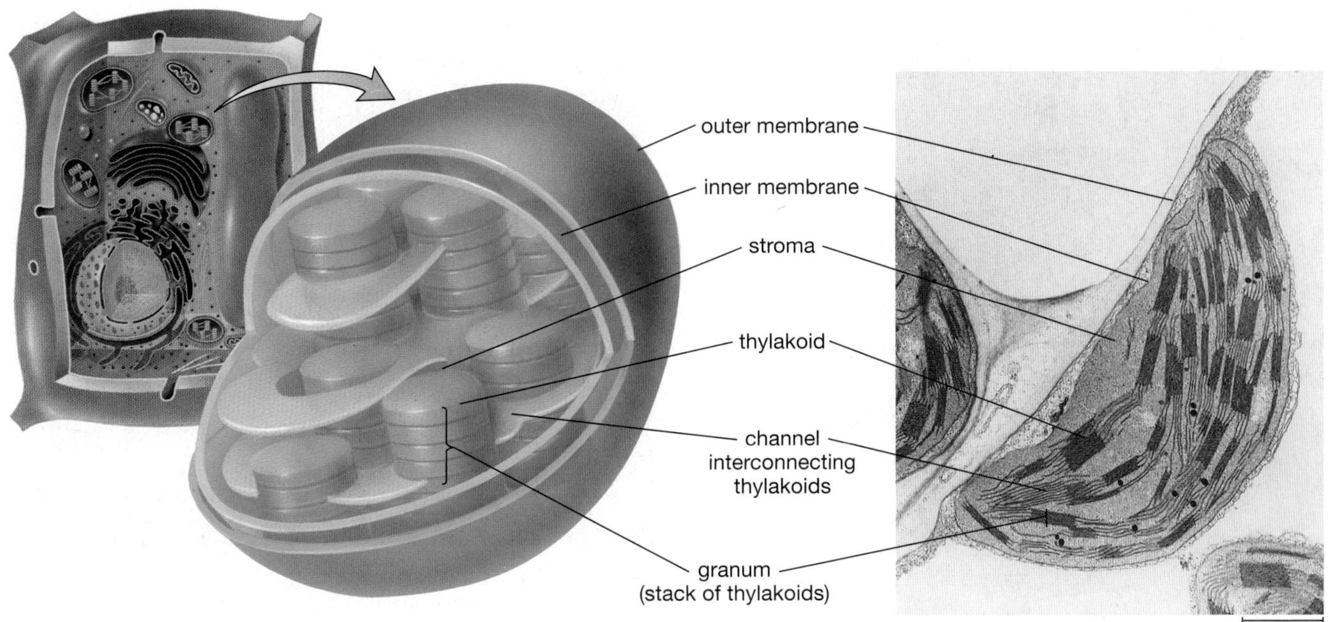

FIGURE 5-14 A chloroplast
Chloroplasts are surrounded by a double membrane, although the inner membrane is not usually visible in electron micrographs. The fluid stroma is enclosed by the inner membrane; within the stroma are stacks of thylakoid sacs called grana. Chlorophyll is embedded in the membranes of the thylakoids.

colors. In plants that continue growing from one year to the next, plastids store photosynthetic products from the summer for use during the following winter and spring. Most plants convert the sugars made during photosynthesis into starch, which is also stored in plastids (Fig. 5-15). Potatoes, for example, are masses of cells, each stuffed with starch-filled plastids.

The Cytoskeleton Provides Shape, Support, and Movement

Organelles do not drift about the cytoplasm haphazardly; most are attached to a network of protein fibers called the *cytoskeleton* (Fig. 5-16). Even individual enzymes, which are often a part of complex metabolic pathways, may be fastened in sequence to the cytoskeleton, so that molecules can be passed from one enzyme to the next in the correct order for a particular chemical transformation. Several types of protein fibers, including thin **microfilaments**, medium-sized **intermediate filaments**, and thick **microtubules**, make up the cytoskeleton.

plastid
starch globules
0.5 micrometer

FIGURE 5-15 A plastid
Plastids, found in the cells of plants and photosynthetic protists, are organelles surrounded by a double outer membrane. Chloroplasts are the most familiar plastids; other types store various materials, such as the starch filling these plastids in potato cells.

The cytoskeleton performs the following important functions:

- *Cell shape.* In cells without cell walls, the cytoskeleton, especially networks of intermediate filaments, determines the shape of the cell.
- *Cell movement.* The assembly, disassembly, and sliding of microfilaments and microtubules cause cell

(a)

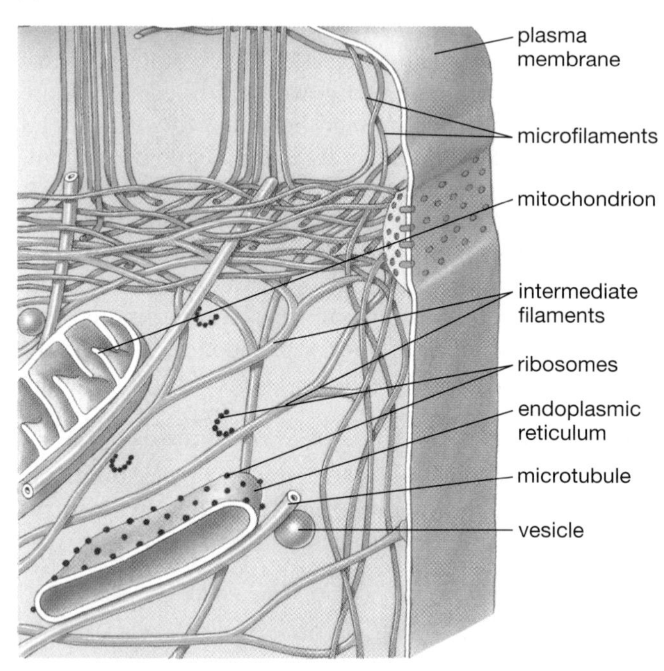

(b)

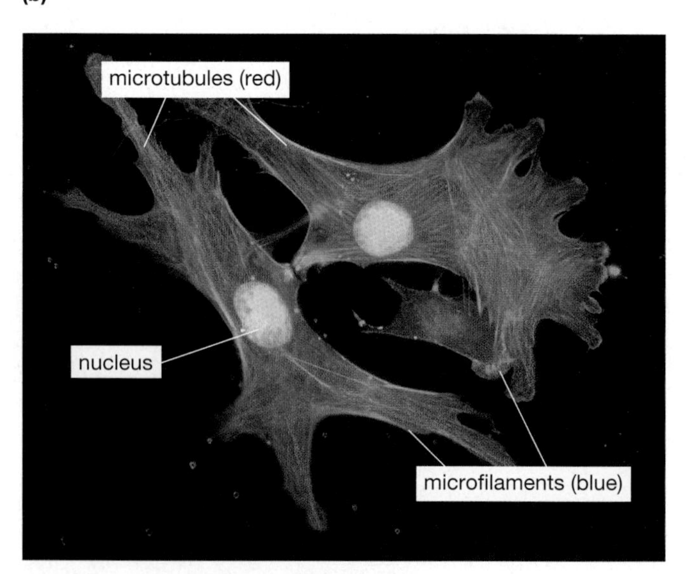

FIGURE 5-16 The cytoskeleton
(a) Eukaryotic cells are given shape and organization by the cytoskeleton, which consists of three types of proteins: microtubules, intermediate filaments, and microfilaments. **(b)** This cell from the lining of a cow artery has been treated with fluorescent stains to reveal microtubules and microfilaments, as well as the nucleus.

movement. Cell movement includes both the "crawling" of white blood cells, the contraction of muscle cells, and the migration and shape changes that occur during the development of multicellular organisms.

- *Organelle movement.* Microtubules and microfilaments move organelles from place to place within a cell. For example, microfilaments attach to vesicles formed during *endocytosis* (see Chapter 4), when large particles are engulfed by the plasma membrane and pull the vesicles into the cell. Vesicles budded off the ER and Golgi complex are probably guided by the cytoskeleton as well.
- *Cell division.* Microtubules and microfilaments are essential to cell division in eukaryotic cells. First, when eukaryotic nuclei divide, microtubules move the chromosomes into the daughter nuclei. Second, in animal cells, division of the cytoplasm of a single parent cell into two new daughter cells results from the contraction of a ring of microfilaments that pinch the "waist" of the parent cell around the middle. In addition, *centrioles* (see Fig. 5-3), which play a role in animal cell division, are composed of microtubules. Cell division will be covered in detail in Chapter 11.

Cilia and Flagella Move the Cell Through Fluid or Move Fluid Past the Cell

Both **cilia** (Latin for "eyelash") and **flagella** ("whip") are slender extensions of the plasma membrane. Each cilium and flagellum contains a ring of nine fused pairs of microtubules, with an unfused pair of microtubules in the center of the ring (forming what is called a "*9 + 2*" *arrangement*; Fig. 5-17). The microtubules span the length of the cilium or flagellum. This pattern of microtubules is produced by a centriole located in the cytoplasm just beneath the plasma membrane. A **centriole** is a short, barrel-shaped ring consisting of nine microtubule triplets, with no microtubules in the center (forming a "*9 + 0*" *arrangement*). Centrioles move to the plasma membrane and provide a center for the formation of cilia or flagella; in this situation, two of the members of each triplet give rise to pairs of microtubules in the cilium or flagellum (see Fig. 5-17). After it begins forming the cilium or flagellum, the centriole is referred to as a **basal body**, because it is located at the base of these structures, anchoring them to the plasma membrane.

Tiny "arms" of protein attach neighboring pairs of microtubules in cilia and flagella. When these arms flex, they slide one pair of microtubules past the adjacent pairs, causing the cilia or flagellum to move. The energy of ATP powers the movement of the protein arms during microtubule sliding. Cilia and flagella often move almost continuously and, consequently, require enormous supplies of ATP that are generated by the mitochondria usually found in abundance near the basal bodies.

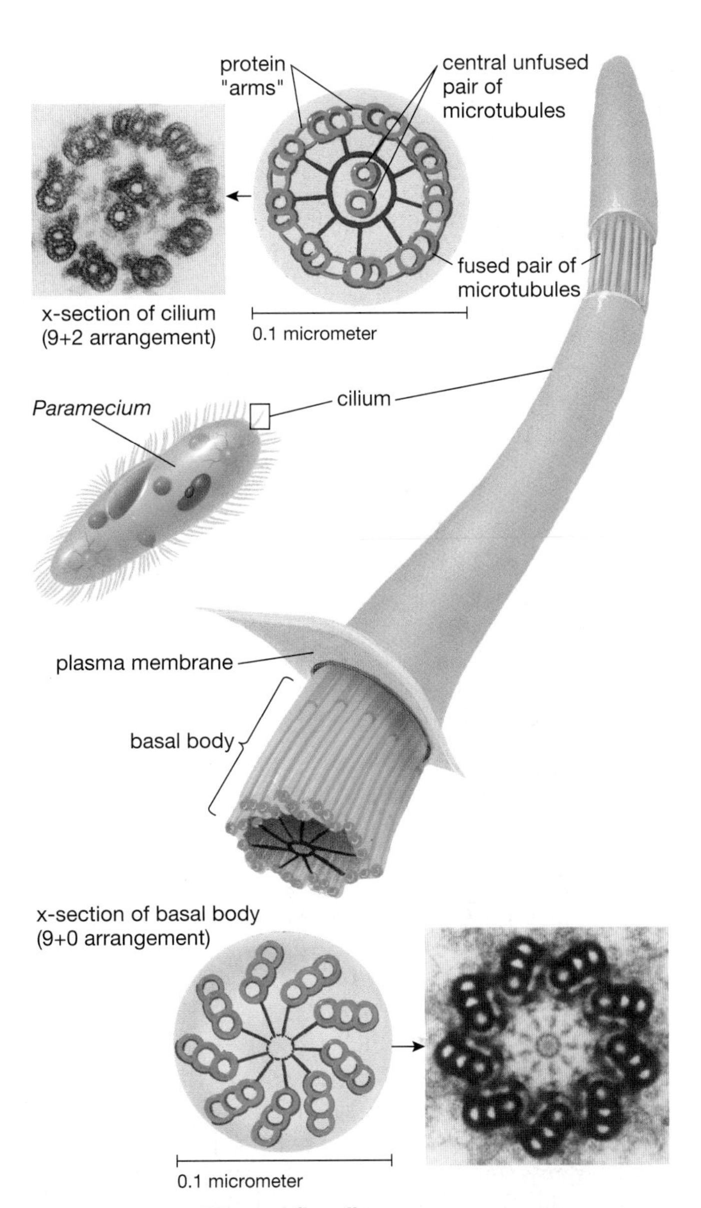

FIGURE 5-17 Cilia and flagella
Both cilia and flagella contain microtubules arranged in an outer ring of nine fused pairs of microtubules surrounding a central unfused pair (a 9 + 2 arrangement). The nine outer pairs have "arms" made of protein that interact with adjacent pairs to provide the force for bending. Cilia and flagella arise from basal bodies formed from centrioles located just beneath the plasma membrane. The fused pairs of microtubules arise from these basal bodies, which have nine fused triplets of microtubules and no central microtubules ("9 + 0").

The main differences between cilia and flagella lie in their length, number, and the direction of the force they generate. In general, cilia are shorter (about 10 to 25 micrometers long) and more numerous than flagella. They provide force in a direction parallel to the plasma membrane, like the oars in a canoe. This is accomplished using

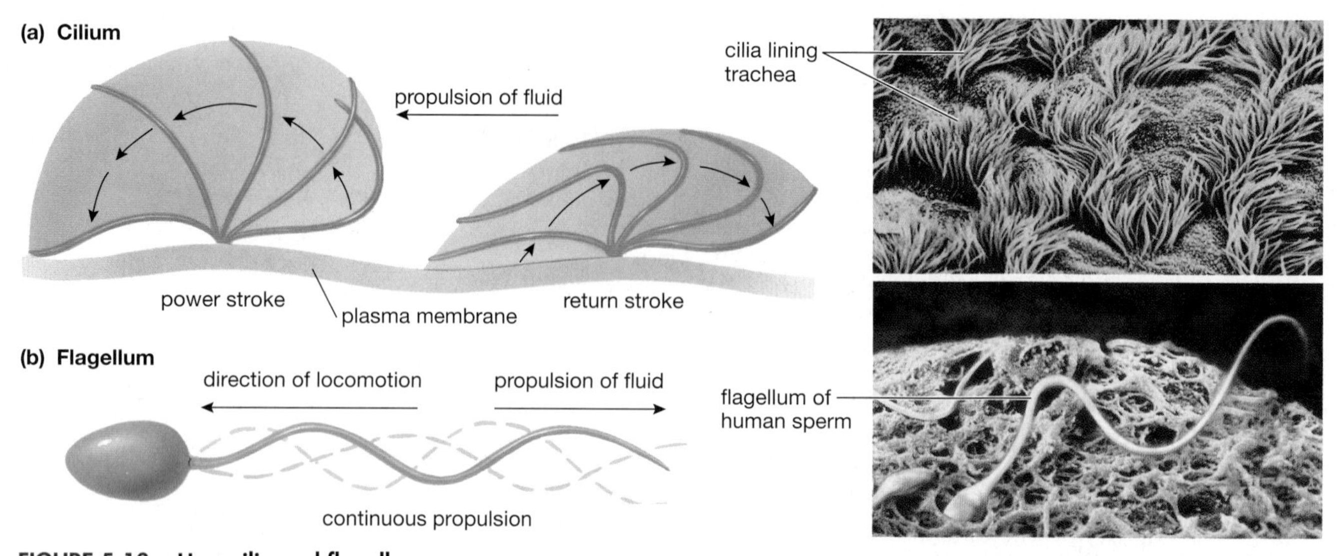

FIGURE 5-18 How cilia and flagella move
(a) (left) Cilia usually "row," providing a force of movement parallel to the plasma membrane. Their movement resembles the arms of a swimmer doing the breast stroke. (right) SEM photo of cilia lining the trachea (which conducts air to the lungs); these cilia sweep out mucus and trapped particles. **(b)** (left) Flagella move in a wavelike motion, providing continuous propulsion perpendicular to the plasma membrane. In this way, a flagellum attached to a sperm can move the sperm straight ahead. (right) A human sperm cell on the surface of a human egg cell.

a "rowing" motion (Fig. 5-18a). Flagella are longer (50 to 75 micrometers), usually fewer in number, and provide force perpendicular to the plasma membrane, like the engine on a motorboat (Fig. 5-18b).

Some unicellular organisms, such as *Paramecium* and *Euglena*, use cilia or flagella to move about. Most animal sperm also rely on flagella for movement (see Fig. 5-18b). In small multicellular animals, cilia are occasionally used for locomotion, moving the animal through water. Many small aquatic invertebrates, for example, swim by the coordinated beating of rows of cilia, like the oars of an ancient Roman galley ship. More commonly, however, cilia move fluids and suspended particles past a surface. Ciliated cells line such diverse structures as the gills of oysters (where they move water rich in food and oxygen over the gills), the oviducts of female mammals (where they move an egg through fluid from the ovary to the uterus), and the respiratory tracts of most land vertebrates (where they clear mucus that carries debris and microorganisms from the windpipe and lungs; see Fig 5-18a).

Prokaryotic cells may also bear slender protrusions that undulate or spin, thereby enabling the cell to move about. However, the "flagella" of prokaryotic cells do not contain microtubules and have no evolutionary relationship to eukaryotic flagella or cilia.

CASE STUDY REVISITED Spare Parts for Human Bodies

Raul Mercia's thumb remains one of the few bioengineered human tissue transplants. In the lab, however, scientists are growing skin, cartilage, bone, heart valves, and breast tissue on plastic scaffolds, and implanting some of these artificial tissues into experimental animals. Researchers continue to refine tissue culturing techniques and develop better scaffolding materials and hope to eventually duplicate entire organs. In one experiment, researchers removed bladders from beagle dogs and cultured cells from the muscle and lining of the bladder on plastic structures the size and shape of beagle bladders. They then transplanted the resulting artificial bladders back into the beagles, where they performed normally and became the first functioning bioengineered organs.

A major challenge in growing new organs is that, unlike the bladder, most are relatively thick, and it is difficult to deliver nutrients to the innermost cells. To solve this problem, Dr. Vacanti has teamed up with a microengineering expert to devise a bioengineered liver with its own blood supply. The team created a plastic cast of a liver's blood vessels by injecting liquid plastic into the vessels and, after the plastic hardened, dissolving away the surrounding tissue. They then produced a 3-D computer image of the plastic blood vessel network and used it to create a mold for scaffolding material. This material will be seeded with at least seven different types of cells that form the bulk of the liver. The network of blood vessels is represented by channels of varying size penetrating the scaffold framework. The researchers will inject these channels with blood vessel cells, which will hopefully line the channels and eventually form new vessels. Since the complexity of the project is staggering, it is unlikely that any of the 20,000 people currently awaiting liver transplants in the United States will benefit from this research. In the future, however, bioengineered organs could save hundreds of thousands of lives worldwide each year.

bio ETHICS **Consider This:** The research illustrated in the chapter-opening photo enraged some animal rights activists, many of whom would also be very disturbed by the bioengineering research in beagles. Virtually all modern medicinal drugs and medical procedures were developed using animal research. Where do you stand on this subject? Should some types of animals, or some types of animal experimentation, be off-limits? If you take the position that animals should not be used in research, would you be willing to forego all medical treatments that were developed with the help of animal research?

Links to Life: Unwanted Guests

In the late 1600s, Antoni van Leeuwenhoek scraped white matter from between his teeth and viewed it through a microscope that he had constructed himself. To his consternation, he saw millions of cells that he called "animalcules": microscopic single-celled organisms that we now recognize as bacteria. Annoyed at the presence of these life forms in his mouth, he attempted to kill them with vinegar and hot coffee—with little success. The warm, moist environment of the human mouth, particularly the crevices of the teeth and gums, is an ideal habitat for a variety of bacteria. Soon after a tooth is cleaned, glycoprotein from saliva begins to form a thin film on its surface. This film helps several species of bacteria attach to the surface of the tooth. Each bacterium divides repeatedly to form a colony of offspring. Some forms of bacteria produce slime layers that help them and others adhere to the tooth. Thick layers of bacteria, slime, and glycoproteins make up the white substance—called plaque—that Leeuwenhoek scraped from his teeth. Sugar in foods and beverages nourishes the bacteria, which break down the sugar into lactic acid. The acid, in turn, eats away at the tooth enamel, producing tiny crevices in which the bacteria multiply further, eventually producing a cavity. Fluoride in toothpaste and drinking water can help prevent cavities by incorporating fluoride into the enamel, making it more resistant to breakdown by acids released by decay bacteria. So, although he didn't know why, Leeuwenhoek was right to be concerned by the "animalcules" in his mouth!

Summary of Key Concepts

5.1 What Are the Basic Features of Cells?

The principles of the cell theory are:

- Every living organism is made up of one or more cells.
- The smallest living organisms are single cells, and cells are the functional units of multicellular organisms.
- All cells arise from preexisting cells.

Cells are limited in size because they must exchange materials with their surroundings by diffusion. Because diffusion is relatively slow, the interior of the cell must never be too far from the plasma membrane, and the plasma membrane must have a large surface area through which materials can diffuse relative to the volume of its cytoplasm. Both of these constraints limit the size of cells.

All cells are either prokaryotic or eukaryotic. Prokaryotic cells, or bacteria, are small and relatively simple in structure. More complex eukaryotic cells make up all other forms of life: protists, plants, fungi, and animals.

5.2 What Are the Major Features of Prokaryotic Cells?

Prokaryotic cells are generally very small with a relatively simple internal structure. Most are surrounded by a relatively stiff cell wall. The cytoplasm of prokaryotic cells lacks membrane-enclosed organelles (although some photosynthetic bacteria have elaborate internal membranes). A single, circular strand of DNA is found in the nucleoid. Table 5-2 compares prokaryotic cells to the eukaryotic cells of plants and animals.

5.3 What Are the Major Features of Eukaryotic Cells?

Genetic material (DNA) is contained within the nucleus, which is bounded by the double membrane of the nuclear envelope. Pores in the nuclear envelope regulate the movement of molecules between nucleus and cytoplasm. The genetic material of eukaryotic cells is organized into linear strands called *chromosomes*, which consist of DNA and proteins. The nucleolus consists of ribosomal RNA and ribosomal proteins, as well as the genes that code for ribosome synthesis. Ribosomes are particles of ribosomal RNA and protein that are the sites of protein synthesis.

The membrane system of a cell consists of the plasma membrane, endoplasmic reticulum (ER), Golgi complex, and vesicles derived from these membranes. Endoplasmic reticulum with ribosomes, called rough ER, manufactures many cellular proteins. Endoplasmic reticulum without ribosomes, called smooth ER, manufactures lipids. The ER is the site of all membrane synthesis within the cell. The Golgi complex is a series of membranous sacs derived from the ER. The Golgi complex processes and modifies materials synthesized in the rough and smooth ER. Some substances in the Golgi are packaged into vesicles for transport elsewhere in the cell. Lysosomes are vesicles that contain digestive enzymes, which digest food particles and defective organelles.

All eukaryotic cells contain mitochondria, which are organelles that use oxygen to complete the metabolism of food molecules, capturing much of their energy as ATP. Cells of plants and some protists contain plastids, including chloroplasts that capture the energy of sunlight during photosynthesis, enabling the cells to manufacture organic molecules, particularly sugars, from simple inorganic molecules. Both mitochondria and chloroplasts probably originated from bacteria. Storage plastids store pigments or starch.

Many eukaryotic cells contain sacs, called *vacuoles*, that are bounded by a single membrane and that store food or wastes, excrete water, or support the cell. Some protists have contractile vacuoles, which collect and expel water. Plants use central vacuoles to support the cell as well as to store wastes and toxic materials.

The cytoskeleton organizes and gives shape to eukaryotic cells and moves and anchors organelles. The cytoskeleton is composed of microfilaments, intermediate filaments, and microtubules. Cilia and flagella are whiplike extensions of the plasma membrane that contain microtubules in a characteristic pattern. These structures move fluids past the cell or move the cell through its fluid environment.

Study Note

Figures 5-3 and 5-4 illustrate the overall structure of animal and plant cells, respectively. Table 5-2 lists the principal organelles, their functions, and their occurrence in prokaryotic cells, animal cells, and plant cells.

Key Terms

aerobic *p. 91*
anaerobic *p. 90*
basal body *p. 93*
central vacuole *p. 89*
centriole *p. 93*
chlorophyll *p. 91*
chloroplast *p. 90*
chromatin *p. 83*
chromosome *p. 84*
cilium *p. 93*
contractile vacuole *p. 88*
cytoplasm *p. 76*
cytoskeleton *p. 82*
deoxyribonucleic acid (DNA) *p. 76*
endoplasmic reticulum (ER) *p. 85*
endosymbiont hypothesis *p. 90*
eukaryotic *p. 80*
flagellum *p. 93*
food vacuole *p. 87*
Golgi complex *p. 86*
intermediate filament *p. 92*
lysosome *p. 87*
microfilament *p. 92*
microtubule *p. 92*
mitochondrion *p. 90*
nuclear envelope *p. 83*
nucleoid *p. 81*
nucleolus *p. 83*
nucleus *p. 83*
organelle *p. 82*
plastid *p. 91*
prokaryotic *p. 80*
ribonucleic acid (RNA) *p. 81*
ribosome *p. 84*
vacuole *p. 88*
vesicle *p. 86*

Thinking Through the Concepts

To take a multiple-choice quiz with feedback on the contents of this chapter, visit http://www.prenhall.com/audesirk7. *Log in to the Web site selected by your instructor and navigate to the Self Test section for this chapter.*

? Review Questions

1. Diagram "typical" prokaryotic and eukaryotic cells, and describe their important similarities and differences.
2. Which organelles are common to both plant and animal cells, and which are unique to each?
3. Define *stroma* and *matrix*.
4. Describe the nucleus, including the nuclear envelope, chromatin, chromosomes, DNA, and nucleolus.
5. What are the functions of mitochondria and chloroplasts? Why do scientists believe that these organelles arose from prokaryotic cells?
6. What is the function of ribosomes? Where in the cell are they typically found? Are they limited to eukaryotic cells?
7. Describe the structure and function of the endoplasmic reticulum and Golgi complex.
8. How are lysosomes formed? What is their function?
9. Diagram the structure of cilia and flagella.

Applying the Concepts

1. If samples of muscle tissue were taken from the legs of a world-class marathon runner and a typical couch potato, which would you expect to have a higher density of mitochondria? Why? What about a muscle biopsy from the biceps of a weight lifter?
2. One of the functions of the cytoskeleton in animal cells is to give shape to the cell. Plant cells have a fairly rigid cell wall surrounding the plasma membrane. Does this mean that a cytoskeleton is unnecessary for a plant cell? Defend your answer in terms of other functions of the cytoskeleton.
3. Most cells are very small. What physical and metabolic constraints limit cell size? What problems would an enormous cell encounter? What adaptations might help a very large cell survive?

For More Information

de Duve, C. "The Birth of Complex Cells." *Scientific American*, April 1996. Describes the mechanisms by which the first eukaryotic cells were produced from prokaryotic ancestors.

Ford, B. J. "The Earliest Views." *Scientific American*, April 1998. The author used the original microscopes of Antoni van Leeuwenhoek to see the microscopic world as Leeuwenhoek saw it. Photographic images taken through these early and very primitive instruments reveal remarkable detail.

Goodsell, D. S. *The Machinery of Life*. New York: Springer, 1993. Wonderful, drawn-to-scale images of the organelles and molecules of the cell.

Hoppert, M., and Mayer, F. "Prokaryotes." *American Scientist*, November/December 1999. These relatively simple cells actually possess a great deal of internal organization.

Ingber, D. E. "The Architecture of Life." *Scientific American*, January 1998. Counteracting forces stabilize the design of organic structures, from carbon compounds to the cytoskeleton-reinforced architecture of the cell.

Kiester, E., Jr. "A Bug in the System."*Discover*, February 1991. Mitochondria may be the descendants of bacteria that live within our cells. Because mitochondria are essential for human life, defects in mitochondrial genes can cause some devastating diseases.

Murray, M. "Life on the Move."*Discover*, March 1991. Many cells, including *Amoeba* and white blood cells, crawl about by means of tiny protein "motors" to manipulate their cytoskeleton.

Media Activities

To access a Media Activity visit http://www.prenhall.com/audesirk7. *Log in to the Web site selected by your instructor, navigate to this chapter, and select the appropriate Media Activity number.*

5.1 Cell Structure

Estimated time: 5 minutes

Explore the structure of prokaryotic and eukaryotic cells.

5.2 Membrane Traffic

Estimated time: 5 minutes

This animation presents the details of membrane traffic in the cell, including transport of proteins, phagocytosis, and exocytosis.

5.3 Web Investigation: Spare Parts for Human Bodies

Estimated time: 10 minutes

Review some of the research scientists are conducting to create a cell from non-living components.

CHAPTER

6 Energy Flow in the Life of a Cell

The bodies of these runners efficiently convert energy stored in fats and carbohydrates to the energy of movement and heat. Their pounding footsteps noticeably shake the Verrazano Bridge during the New York Marathon.

AT A GLANCE

CASE STUDY Energy Unleashed

Medical staff members at the Memorial Sloan Kettering Cancer Center were puzzled by a dark spot on an X-ray of Betty Lawson, a 59-year-old woman who had just checked in for breast cancer treatment. Their confusion turned to admiration when they discovered it was a small medal that she had earned for completing the New York Marathon earlier that day—she had forgotten and left it on under her gown. "Cancer changes the way you look at life," she said. "It's like you've been asleep to all the possibilities of life, and cancer wakes you up." Betty described other remarkable people she encountered on the 26-mile journey: a participant from Britain supporting wildlife conservation ran dressed as a rhino; a 91-year-old man ran at a slow shuffle; a fireman ran in full fire-fighting gear to honor his fallen comrades; a man without a leg completed the race on his crutches; blind participants relied on sighted runners to guide them.

The 20,000 runners in the New York Marathon collectively convert more than 50 million Calories of energy into motion, heating their bodies, heating the air around them, and shaking the Verrazano Bridge. Once finished, they douse their overheated bodies with water and refuel on high-energy snacks. Finally, cars, buses, and airplanes—burning vast quantities of fuel and releasing enormous amounts of heat—carry the runners back to their homes throughout the world.

What exactly is energy? Do our bodies use it according to the same principles that govern energy use in the engines of cars and airplanes? Why do our bodies generate heat, and why do we give off more heat when exercising than when watching TV? Why are runners encouraged to drink water during a race—and is it possible to drink too much?

We often talk about "burning" Calories; setting sugar on fire involves similar overall reactions as allowing your body to "burn" sugar you've eaten. In both cases, oxygen combines with sugar to produce carbon dioxide, water, and heat. However, our bodies aren't roasted when we metabolize food, and some of the energy is captured in other molecules that can be used to power muscle movement or a huge variety of processes within our cells. How do we control the breakdown of high-energy molecules to produce useful energy? In this chapter, we discuss the physical laws that govern energy flow in the universe, how this energy flow governs chemical reactions, and how the chemical reactions within living cells are controlled by the cell's own molecules.

6.1 What Is Energy?

Energy can be defined simply as the capacity to do work, including synthesizing molecules, moving objects, and generating heat and light. There are two types of energy: *kinetic energy* and *potential energy*. Both of these, in turn, exist in many different forms. **Kinetic energy**, or energy of movement, includes light (movement of photons), heat (movement of molecules), electricity (movement of electrically charged particles), and movement of large objects. **Potential energy**, or stored energy, includes chemical energy stored in the bonds that hold atoms together in molecules, electrical energy stored in a battery, and positional energy stored in a diver poised to spring (Fig. 6-1). Under the right conditions, kinetic energy can be transformed into potential energy, and vice versa. For example, a diver converts kinetic energy of movement into potential energy of position when he climbs up to the platform; when he jumps off, the potential energy is converted back into kinetic energy.

To understand energy flow, we need to know two things: the quantity of available energy and the usefulness of the energy. These are described in the laws of thermodynamics.

FIGURE 6-1 From potential to kinetic energy
Perched atop the platform, the body of the diver has potential energy, because the heights of the platform and the pool are different. As he dives, the potential energy is converted to the kinetic energy of motion of the diver's body. Finally, some of this kinetic energy is transferred to the water, which itself is set in motion.

Media Activity
6.1 Energy and Chemical Reactions

The Laws of Thermodynamics Describe the Basic Properties of Energy

The **laws of thermodynamics** define the basic properties and behavior of energy. The **first law of thermodynamics** states that, assuming there is no influx of energy, the total amount of energy within a given system remains constant. Although nuclear reactions convert matter into energy, energy can neither be created nor destroyed by ordinary processes. Energy can, however, change form—for example, from chemical energy to heat energy. Therefore, the first law is often called the *law of conservation of energy*. When you drive a car, you convert the potential chemical energy of gasoline into the kinetic energy of movement and heat. The total amount of energy remains the same, although it has changed in form. Likewise, a runner converts the potential chemical energy from food into the same total amount of kinetic energy of movement plus heat.

The **second law of thermodynamics** states that when energy is converted from one form to another, the amount of useful energy decreases. Put another way, the second law states that all spontaneous changes result in a more uniform distribution of energy, reducing the energy differences that are essential for doing work; energy is spontaneously converted from more useful into less useful forms.

To illustrate the second law, let's examine a car engine burning gasoline. The kinetic energy of the moving vehicle represents only 25% or less of the chemical energy originally contained in the gasoline. According to the first law, the total amount of energy remains constant; where is the "missing" energy? The burning gas not only moved the car but also heated up the engine, the exhaust system, and the air around the car. The friction of tires on the pavement slightly heated the road as well, so, as the first law dictates, no energy is missing. However, energy released as heat is in a less usable form; it merely increased the random movement of molecules in the engine, the road, and the air.

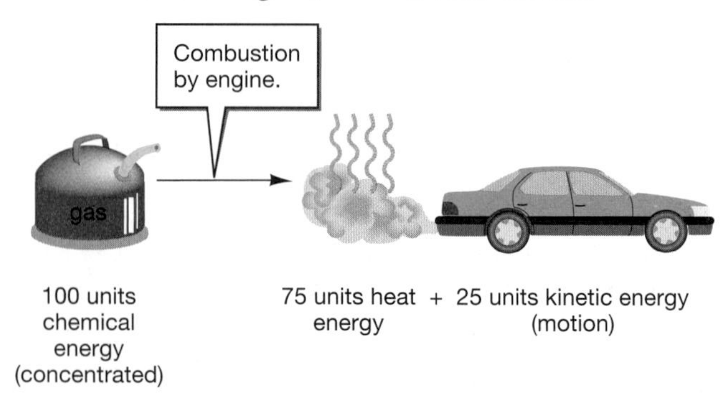

Similarly, the energy of heat that runners liberate to the air when food "burns" in their cells cannot be harnessed to allow them to run farther or faster. Thus, the second law tells us that no energy conversion process—including those that occur in the body—is 100% efficient.

The second law of thermodynamics also reveals something about the organization of matter. Regions of concentrated energy tend to be highly ordered. The eight carbon atoms in a single molecule of gasoline have a much more orderly arrangement than do the carbon atoms of the eight separate, randomly moving molecules of carbon dioxide that are formed when the gasoline burns. The same is true for the glycogen molecules stored in a runner's muscles, which are converted into simpler water and carbon dioxide as they are used by the muscles to provide energy. Therefore, we can also phrase the second law in terms of the organization of matter: unless energy is added to the system, processes that proceed spontaneously result in an increase in randomness and disorder. This tendency toward loss of orderliness and high-level energy and an increase in randomness, disorder, and low-level energy is called **entropy**. We all experience entropy in our homes. Frequent inputs of energy are required to keep debris confined to the trash can, newspapers to folded stacks, books to their shelves, and clothes to drawers and closets. Without our energetic cleaning and organizing efforts, these items tend to end up in their lowest-energy state—a state of disorder. When G. Evelyn Hutchinson said, "Disorder spreads through the universe, and life alone battles against it," he made an eloquent reference to entropy and the second law of thermodynamics.

Living Things Use the Energy of Sunlight to Create the Low-Entropy Conditions Characteristic of Life

If you think about the second law of thermodynamics, you may wonder how life can exist at all. If chemical reactions, including those inside living cells, cause the amount of unusable energy to increase, and if matter tends toward increasing randomness and disorder, how can organisms accumulate the concentrated energy and precisely ordered molecules that characterize living things? The answer is that nuclear reactions in the sun produce energy in the form of sunlight, a process that also produces vast increases in entropy. Living things on Earth use a continuous input of solar energy to synthesize complex molecules and maintain orderly structures—to "battle against disorder." The highly organized, low-entropy systems that characterize life do not violate the second law; they are achieved at the expense of an enormous loss of usable energy from the sun. The entropy of the solar system as a whole constantly increases.

6.2 How Does Energy Flow in Chemical Reactions?

A **chemical reaction** is a process that forms or breaks the chemical bonds that hold atoms together. Chemical reactions convert one set of chemical substances, the **reactants**, into another set, the **products**. All chemical reactions are either exergonic or endergonic. A reaction is **exergonic** (Greek for "energy out") if the reactants contain more energy than the products. Consequently, the reaction releases energy:

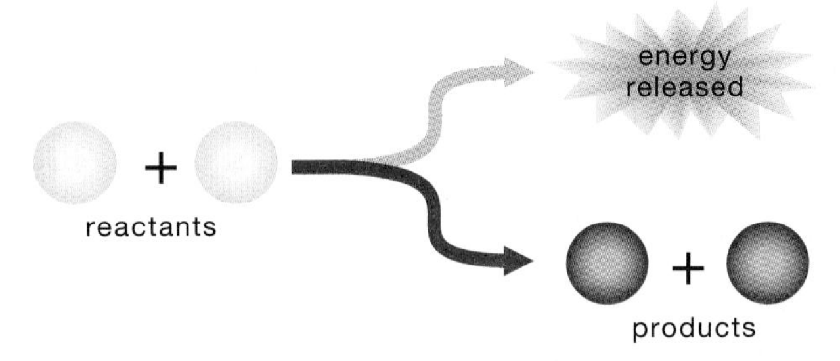

Conversely, a reaction is **endergonic** (Greek for "energy in") if the products contain more energy than the reactants. According to the second law of thermodynamics, endergonic reactions require an influx of energy from an outside source:

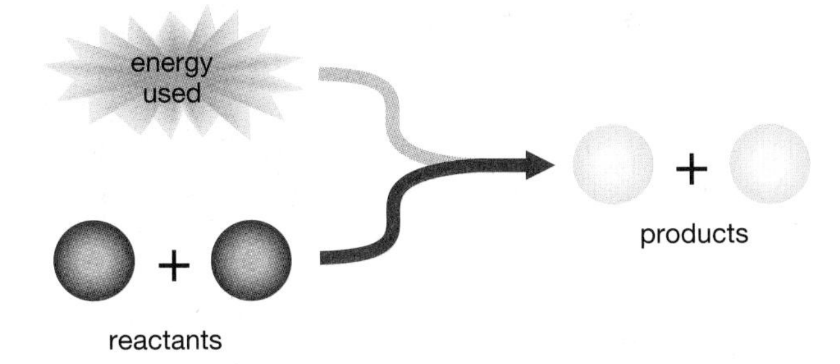

Let's look at two processes to illustrate each type of reaction: burning sugar and photosynthesis.

Exergonic Reactions Release Energy

In an exergonic reaction, the reactants contain more energy than the products. The sugar that the runners' bodies use as fuel contains more energy than the carbon dioxide and water that are produced when that sugar breaks down. The extra energy is liberated as muscular movement and heat. Sugar can also be burned, as any cook can tell you. When sugar is burned by a flame, similar to the way it is burned in the body, it reacts with oxygen (O_2) to produce carbon dioxide (CO_2) and water (H_2O) and release energy, as described by this equation:

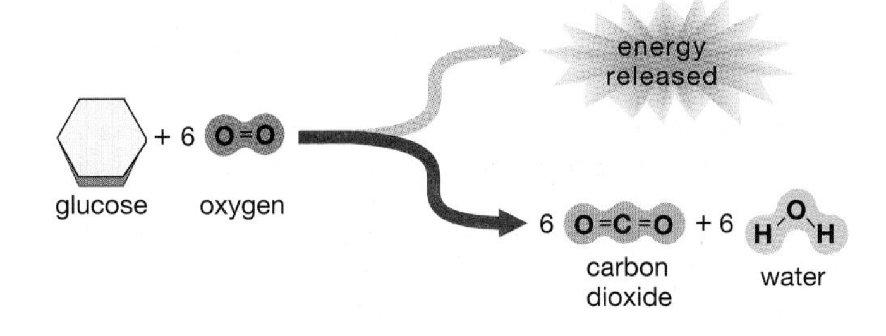

This reaction illustrates two important concepts, diagrammed in Figure 6-2a. First, molecules of sugar contain much more energy than molecules of carbon dioxide and water, so the reaction releases energy. Energy release allows exergonic reactions to occur without a net input of energy. Once ignited, sugar will continue to burn spontaneously. It may be helpful to think of exergonic reactions as running "downhill," from high energy to low energy.

Although burning sugar releases energy, a spoonful of sugar doesn't burst into flames by itself. This observation leads to the second important concept: all chemical reactions require an initial input of energy to get started, something like giving a rock poised at the top of a hill a push to start it rolling down. In chemical reactions, this initial energy input or "push" is called the **activation energy**. Chemical reactions require activation energy to get started because a shell of negatively charged electrons surrounds atoms and molecules. For two molecules to react with each other, their electron shells must be forced together, despite their mutual electrical repulsion; this process requires energy. The usual source of activation energy is the kinetic energy of movement. Molecules moving with sufficient speed collide hard enough to force their electron shells to mingle and react. Because molecules move faster as the temperature increases, most chemical reactions occur more readily at high temperatures. The initial heat provided by a match setting sugar on fire allows these reactions to begin. The combination of sugar with oxygen then releases enough of its own heat to sustain the reaction. Now, think about lighting a match; where does the heat to start that reaction come from?

Endergonic Reactions Require an Input of Energy

In contrast to what happens when sugar or a match is burned, many reactions in living systems result in products that contain more energy than the reactants. Sugar, produced by photosynthetic organisms, contains far more energy than the carbon dioxide and water from which it was formed. The protein in a muscle cell contains more energy than the individual amino acids that were joined together to synthesize it. In other words, synthesizing complex biological molecules requires an input of energy (Fig. 6-2b). As we will see in Chapter 7, photosynthesis in green plants takes low-energy water and carbon dioxide and produces oxygen and high-energy sugar from them:

Photosynthesis

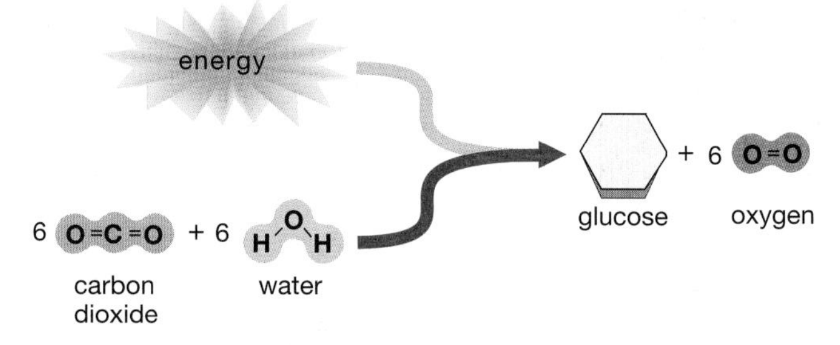

We might call endergonic reactions "uphill" reactions, or going from low energy to high energy, like pushing our rock back to the top of the hill. Photosynthesis requires energy, which photosynthetic organisms (plants, some protists, and some bacteria) obtain from sunlight. But where do we humans and other animals get the energy to synthesize muscle protein and other complex biological molecules?

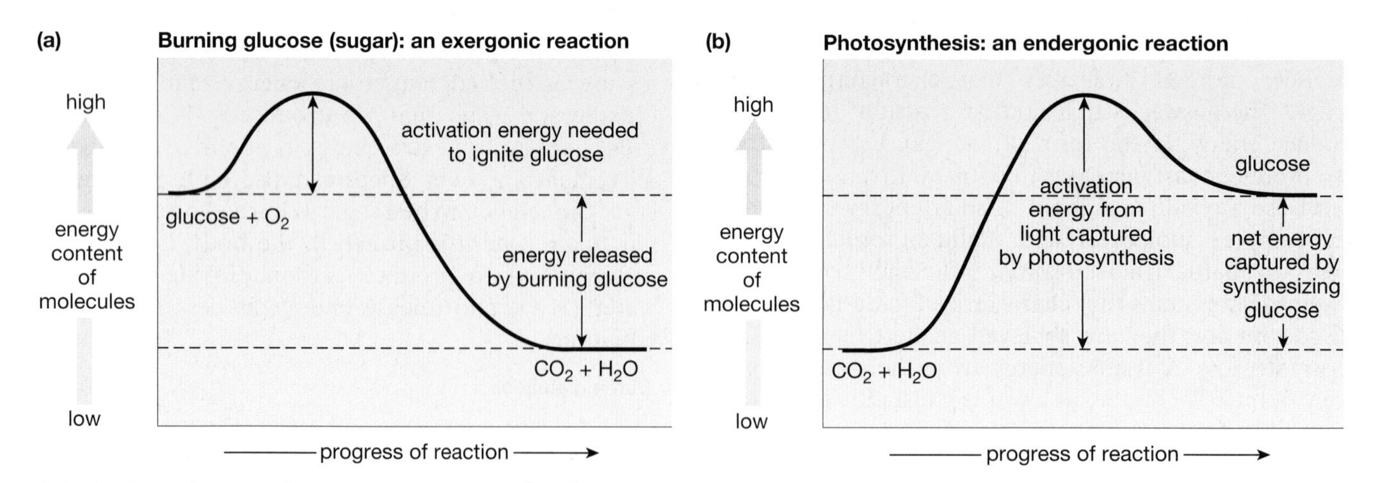

FIGURE 6-2 Energy relations in exergonic and endergonic reactions
(a) An exergonic ("downhill") reaction, such as the burning of sugar, proceeds from high-energy reactants (here, glucose) to low-energy products (CO_2 and H_2O). The energy difference between the chemical bonds of the reactants and products is released as heat. To start the reaction, however, an initial input of energy—the activation energy—is required. **(b)** An endergonic ("uphill") reaction, such as photosynthesis, proceeds from low-energy reactants (CO_2 and H_2O) to high-energy products (glucose) and therefore requires a net input of energy, in this case from sunlight. **QUESTION** In addition to heat and sunlight, what are some other potential sources of activation energy?

Coupled Reactions Link Exergonic and Endergonic Reactions

Endergonic reactions require energy from other sources; they obtain this energy from exergonic, or energy-releasing, reactions. In a **coupled reaction** (Fig. 6-3), an exergonic reaction provides the energy needed to drive an endergonic reaction. When you drive a car, the exergonic reaction of burning gasoline provides the energy for the endergonic reaction of starting a stationary car into motion and keeping it going; in the process, much energy is lost as heat. Photosynthesis is another coupled reaction. In photosynthesis, the exergonic reaction occurs in the sun, and the endergonic reaction occurs in the plant. Most of the energy liberated by the sun is lost as heat, so the second law of thermodynamics still applies: usable energy decreases.

Living organisms constantly use the energy given off by exergonic reactions (such as the chemical breakdown of sugar) to drive essential endergonic reactions (such as brain activity, muscular contraction and other types of movement, or the synthesis of complex molecules), as shown in Figure 6-3. Endergonic reactions cannot occur unless, somewhere in the body, an exergonic reaction has already happened to provide the energy to drive them. Further, since some energy is lost as heat every time it is transformed, the energy provided by exergonic reactions must exceed that needed to drive endergonic reactions. The exergonic and endergonic parts of coupled reactions often occur in different places, so there also must be some way to transfer the energy from the exergonic reaction that releases energy to the endergonic reaction that requires it. In coupled reactions occurring within cells, energy is usually transferred from place to place by *energy-carrier* molecules, the most common of which is ATP.

6.3 How Is Cellular Energy Carried Between Coupled Reactions?

As we saw earlier, cells couple reactions so that the energy released by exergonic reactions is used to drive endergonic reactions. In the case of a runner, the breakdown of a sugar (glucose) releases energy; this energy release is coupled to energy-consuming reactions that cause muscle contraction. But glucose cannot be used directly for muscle contraction. Instead, the energy from glucose must be transferred to an **energy-carrier molecule**, which provides the muscle protein with energy to contract. Energy carriers work something like rechargeable batteries; they pick up an energy charge at an exergonic reaction, move to another location within the cell, and release the energy to drive an endergonic reaction. Because energy-carrier molecules are unstable, they are used only for temporary energy transfer within cells. They are not used to transfer energy from cell to cell, nor are they used for long-term energy storage. Muscles store energy in glycogen, a stable molecule that consists of chains of glucose molecules. When energy is needed—for example, when the marathon begins—the glycogen in the body is broken down by enzymes first to glucose, then to carbon dioxide and water. The energy is then captured and transferred to muscle protein molecules by energy-carrier molecules such as ATP.

ATP Is the Principal Energy Carrier in Cells

Several exergonic reactions in cells produce **adenosine triphosphate**, or **ATP**, the most common energy-carrier molecule in cells. By providing energy for a wide variety of endergonic reactions, ATP serves as a common

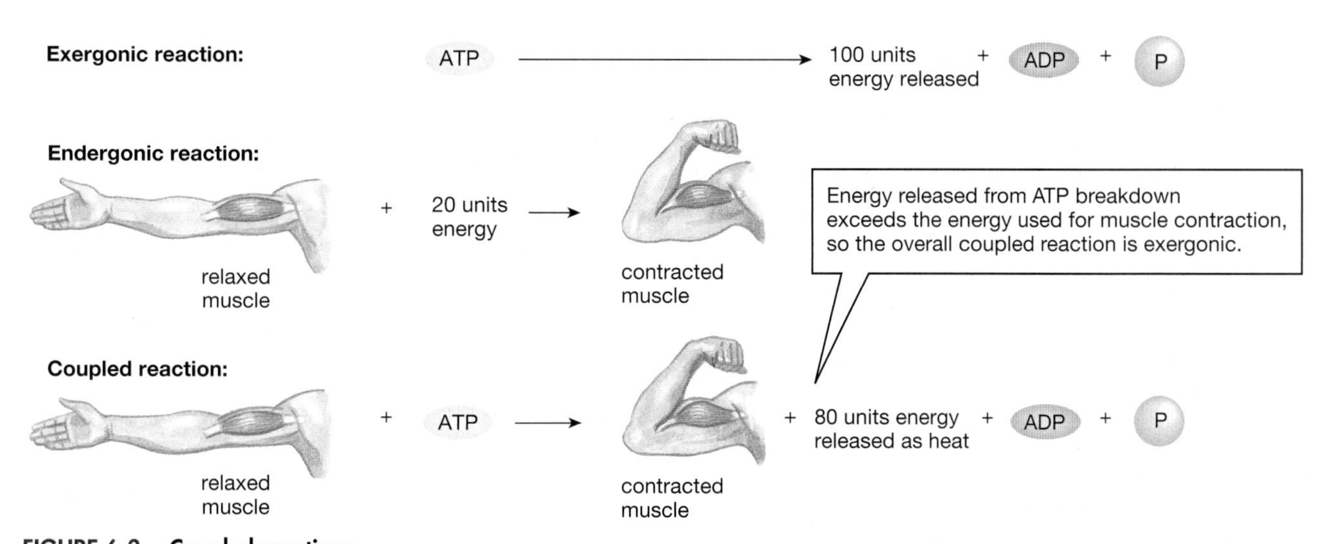

FIGURE 6-3 Coupled reactions
Muscle movement requires an endergonic reaction coupled to the exergonic reaction of ATP breakdown; the overall reaction is exergonic. (Energy units are arbitrary.)

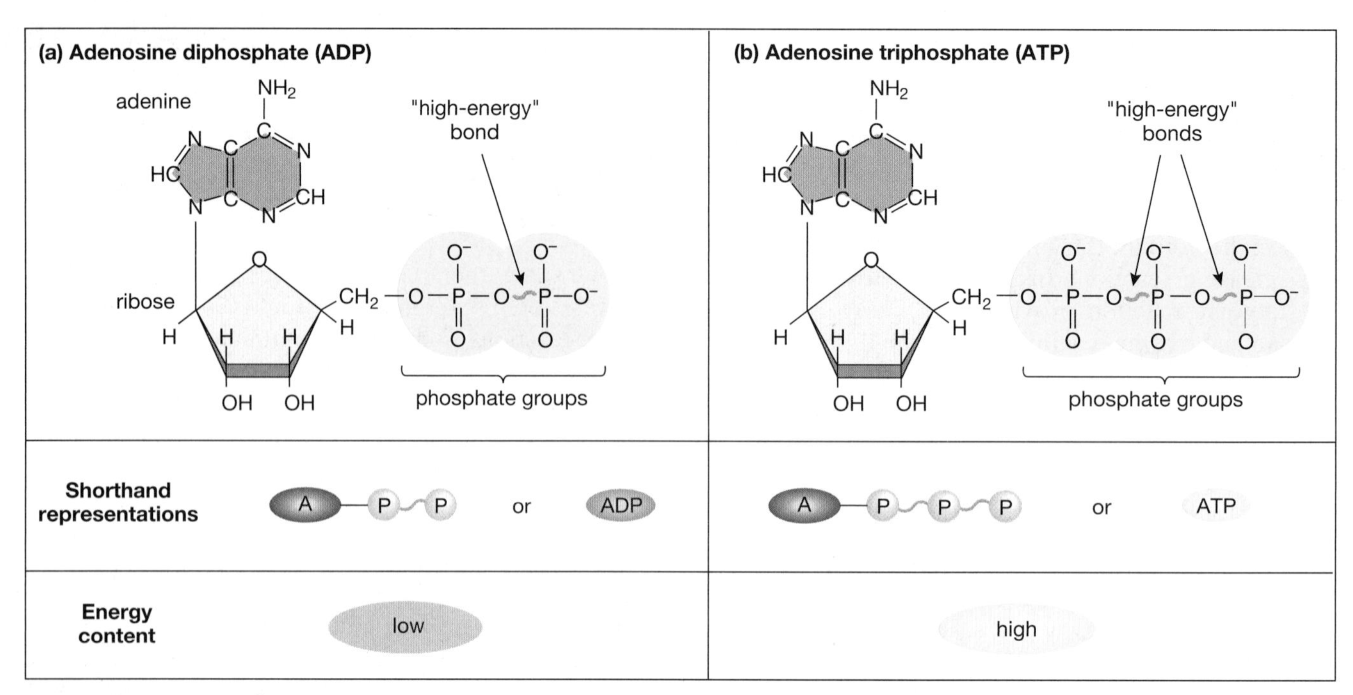

FIGURE 6-4 ADP and ATP
A phosphate group is added to **(a)** ADP (adenosine diphosphate) to make **(b)** ATP (adenosine triphosphate). In most cases, only the last phosphate group and its high-energy bond are used to carry energy and transfer it to endergonic reactions within a cell. **QUESTION** Why does conversion of ATP to ADP release energy for cellular work?

Media Activity
6.2 Energy and Life

currency of energy transfer. For this reason, it is sometimes called the "energy currency" of living cells. As you learned in Chapter 3, ATP is a nucleotide composed of the nitrogen-containing base adenine, the sugar ribose, and three phosphate groups (Fig. 6-4). Energy released in cells during glucose breakdown is used to synthesize ATP from **adenosine diphosphate (ADP)** and phosphate:

ATP synthesis: Energy is stored in ATP

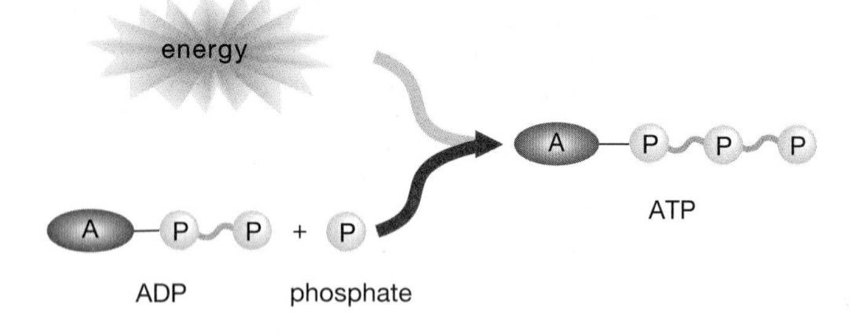

ATP stores this energy within its chemical bonds and can carry the energy to sites in the cell that perform energy-requiring reactions, such as the synthesis of proteins or muscle contraction (see Fig. 6-3). The ATP is then broken down to form ADP and phosphate:

ATP breakdown: Energy of ATP is released

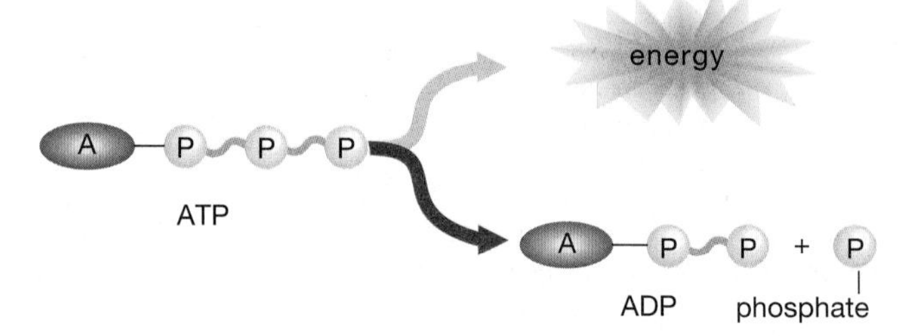

During these energy transfers, some heat is given off at each stage, and there is an overall loss of usable energy (Fig. 6-5). Heat, generated as a by-product of every biochemical transformation, is used by warm-blooded animals to maintain a high body temperature. By speeding up biochemical reactions, this elevated body temperature allows them to move faster and respond more quickly to stimuli than if their body temperatures were lower.

ATP is admirably suited to carrying energy within cells. The bonds joining the last two phosphate groups of ATP to the rest of the molecule (sometimes called *high-energy bonds*) require a large amount of energy to form, so considerable energy can be trapped from exergonic reactions by synthesizing ATP molecules. ATP is also unstable; it readily releases its energy in the presence of appropriate enzymes. Under most circumstances, only the bond joining the last phosphate group (the one joining phosphate to ADP to form ATP) carries energy from exergonic to endergonic reactions.

The life span of an ATP molecule in a living cell is very short because this energy carrier is continuously formed, broken down to ADP and phosphate, and resynthesized. If the molecules of ATP that you use just sitting at your desk all day could be captured (instead of recycled), they would weigh 40 kg—nearly 90 pounds of ATP! A marathon runner may use a pound of ATP every minute. (The ADP must be quickly converted back to ATP, or it would be a very brief run.) As you can see, ATP is *not* a long-term energy-storage molecule. More stable molecules, such as glycogen or fat, can store energy for hours, days, or (in the case of fat) even years.

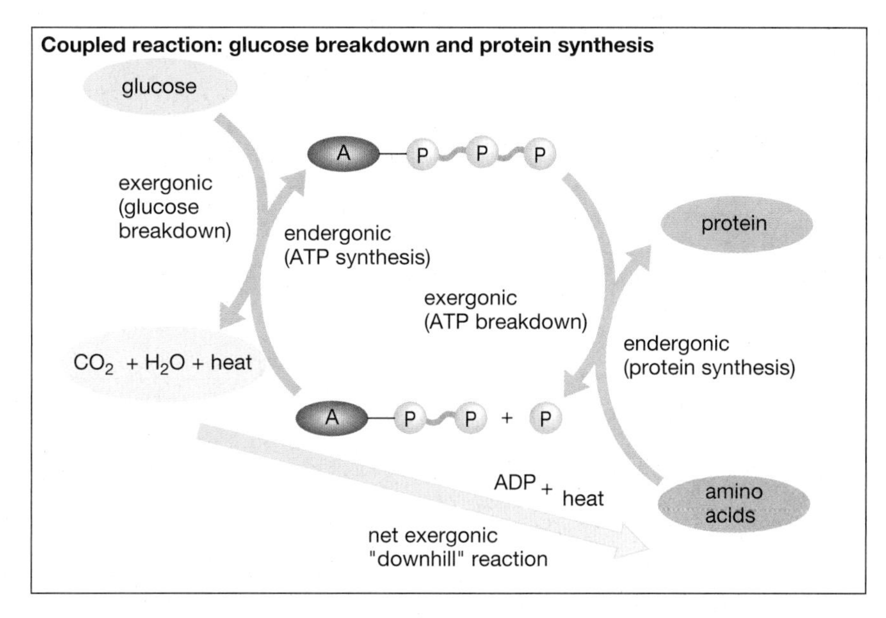

FIGURE 6-5 Coupled reactions within living cells
Exergonic reactions (such as glucose breakdown) drive the endergonic reaction of synthesizing ATP from ADP. The ATP molecule then moves to a part of the cell where the energy from the breakdown of ATP is needed to drive an essential endergonic reaction (such as protein synthesis). The ADP and phosphate are recycled back to ATP via endergonic reactions. The overall reaction is exergonic, or "downhill": more energy is released by the exergonic reaction than is needed to drive the endergonic reaction.

Electron Carriers Also Transport Energy Within Cells

In addition to ATP, other carrier molecules can also transport energy within a cell. In some exergonic reactions, including both glucose metabolism and the light-capturing stage of photosynthesis, some energy is transferred to electrons. These energetic electrons (in some cases, along with hydrogen atoms) are captured by **electron carriers** (Fig. 6-6). Common electron carriers include *nicotinamide adenine dinucleotide* (NAD^+) and its relative *flavin adenine dinucleotide* (FAD). Loaded electron carriers then donate the electrons, along with their energy, to other molecules. You will learn more about electron carriers and their role in cellular metabolism in Chapters 7 and 8.

6.4 How Do Cells Control Their Metabolic Reactions?

Cells are miniature, incredibly complex chemical factories. The **metabolism** of a cell is the sum of all its chemical reactions. Many of these reactions are linked in

Media Activity
6.3 Enzymes

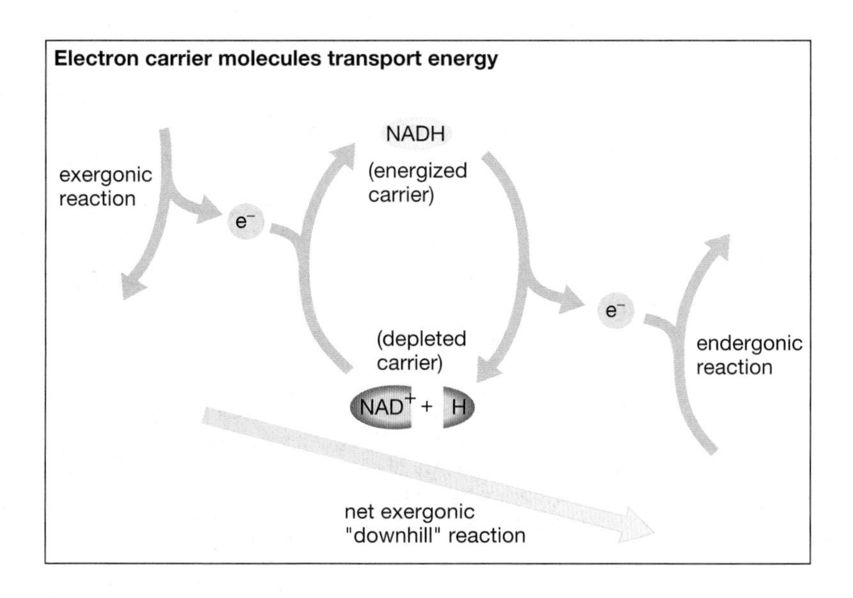

FIGURE 6-6 Electron carriers
Electron-carrier molecules such as NAD^+ pick up electrons generated by exergonic reactions and hold them in high-energy outer electron shells. Hydrogen atoms are often picked up simultaneously. The electron is then deposited, energy and all, with another molecule to drive an endergonic reaction, typically the synthesis of ATP.

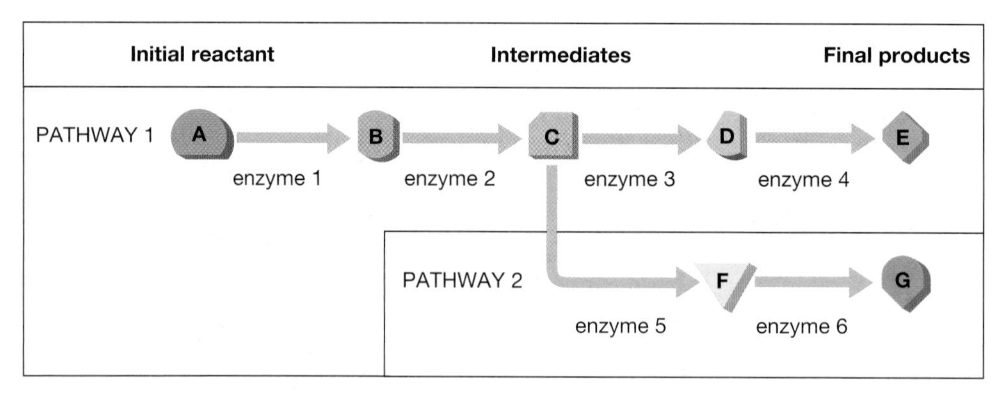

FIGURE 6-7 Simplified view of metabolic pathways
The original reactant molecule, A, undergoes a series of reactions, each catalyzed by a specific enzyme. The product of each reaction serves as the reactant for the next reaction in the pathway. Metabolic pathways are commonly interconnected such that the product of a step in one pathway may serve as a reactant for the next reaction in that pathway or for a reaction in another pathway.

sequences called **metabolic pathways** (Fig. 6-7). Photosynthesis (Chapter 7) is one such pathway. *Glycolysis*, the series of reactions that begin the digestion of glucose (Chapter 8), is another example. Different metabolic pathways may utilize the same molecules; as a result, all the reactions and all the molecules of a cell are interconnected in a single, enormously complicated metabolic pathway.

The chemical reactions in a cell are governed by the same laws of thermodynamics that control other reactions. How, then, do orderly metabolic pathways arise? The biochemistry of cells is finely tuned in three ways:

- Cells regulate chemical reactions through the use of proteins called *enzymes*.
- Cells couple reactions together, powering energy-requiring endergonic reactions with the energy released by exergonic reactions.
- Cells synthesize energy-carrier molecules that capture energy from exergonic reactions and transport it to endergonic reactions.

At Body Temperatures, Spontaneous Reactions Proceed Too Slowly to Sustain Life

In general, the speed at which a reaction occurs is determined by its activation energy, that is, how much energy is required to start the reaction (see Fig. 6-2). Reactions with low activation energies can proceed swiftly at body temperature, whereas reactions with high activation energies, such as gasoline combining with oxygen, are practically nonexistent at similar temperatures. Most reactions can be accelerated by raising the temperature, thereby increasing the speed of molecules.

The reaction of sugar with oxygen to yield carbon dioxide and water is exergonic, but it has an enormous activation energy. The heat of a match flame can cause sugar and oxygen molecules to collide violently enough to produce a vigorous and ungoverned burning. At the temperatures found in living organisms, however, sugar and many other energetic molecules would almost never break down spontaneously and give up their energy. It is the enzymes produced by living cells that allow sugar to be an important energy source for life on Earth. Let's see how enzymes and other *catalysts* influence chemical reactions.

Catalysts Reduce Activation Energy

Catalysts are molecules that speed up the rate of a reaction without themselves being used up or permanently altered. Catalysts speed up reactions by reducing the activation energy (Fig. 6-8). As an example of catalytic action, let's consider the catalytic converters in many

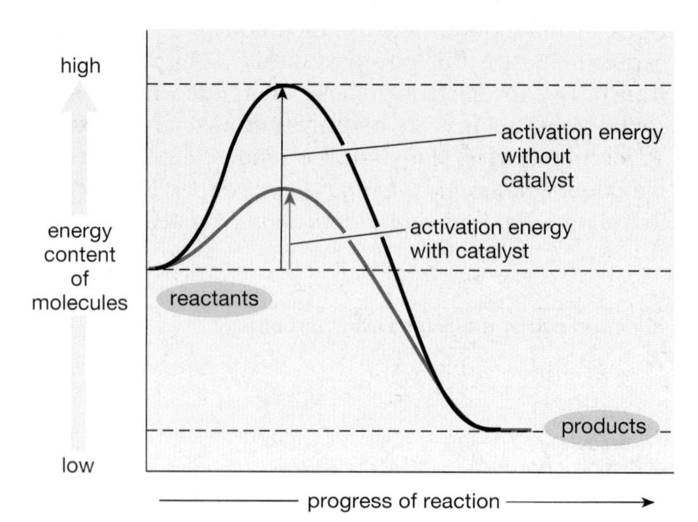

FIGURE 6-8 Catalysts lower activation energy, increasing the rate of reactions
A high activation energy (black curve) means that reactant molecules must collide very forcefully in order to react. Only very fast-moving molecules will collide hard enough to react, so reactions with high activation energies proceed slowly at low temperatures, where most molecules move relatively slowly. Catalysts lower the activation energy of a reaction (red curve), so a much higher proportion of molecules move fast enough to react when they collide. Therefore, the reaction proceeds much more rapidly. **QUESTION** Can a catalyst make a non-spontaneous reaction occur spontaneously?

automobile engines. When gasoline is burned completely, the final products are carbon dioxide and water:

$$2\ C_8H_{18} + 25\ O_2 \rightarrow 16\ CO_2 + 18\ H_2O + \text{energy}$$
(octane)

However, flaws in the combustion process generate other substances, including poisonous carbon monoxide (CO). Carbon monoxide reacts spontaneously but slowly with oxygen in the air to form carbon dioxide:

$$2\ CO + O_2 \rightarrow 2\ CO_2 + \text{energy}$$

In large cities, however, the spontaneous reaction of CO with O_2 can't keep pace with the vast amount of CO emitted into the atmosphere, and unhealthy levels of carbon monoxide accumulate. Enter the catalytic converter. Platinum catalysts in the converter hasten the conversion of CO to CO_2, thereby reducing air pollution.

Note three important principles about all catalysts:

- Catalysts speed up reactions.
- Catalysts can speed up only those reactions that would occur spontaneously, but at a much slower rate.
- Catalysts are not consumed in the reactions they promote. No matter how many reactions they accelerate, the catalysts themselves are not permanently changed.

Enzymes Are Biological Catalysts

Enzymes are biological catalysts, usually proteins, synthesized by living organisms. Enzymes use chemical means to orient, distort, and reconfigure molecules into new combinations in small, discrete steps. Enzymes possess the characteristics of catalysts that we just described. But enzymes have two additional attributes that set them apart from nonbiological catalysts:

- Enzymes are usually very specific, catalyzing at most only a few types of chemical reactions. In most cases, an enzyme catalyzes a single reaction that involves one or two specific molecules but leaves similar molecules untouched. (You may recall from Chapter 3, for example, that animals have enzymes that break apart starch molecules but leave cellulose intact, despite the fact that both starch and cellulose are composed of glucose subunits.)
- Enzyme activity is often regulated—that is, enhanced or suppressed—by the molecules whose reactions they catalyze.

The Structure of Enzymes Allows Them to Catalyze Specific Reactions

Why are enzymes specific, and how are they regulated? Enzyme function is intimately related to enzyme structure. Enzymes are proteins with complex three-dimensional shapes. Each enzyme has a "pocket," called the **active site**, into which reactant molecules, called **substrates**, can enter.

The active site of each enzyme has a distinctive shape and distribution of electrical charge that is complementary to those of its substrate. Because the enzyme and its substrate must fit together, only certain molecules can enter the active site. For example, several enzymes are required to completely digest all the proteins we eat, because each enzyme breaks apart only a specific sequence of amino acids.

How does an enzyme catalyze a reaction? First, both the shape and the charge of the active site force substrates to enter the enzyme in specific orientations (Fig. 6-9, step ①). Second, when substrates enter the active site, both the substrate and active site change shape (step ②). Certain amino acids within the part of the protein that forms the active site may temporarily bond with atoms of the substrates, or electrical interactions between the amino acids in the active site and substrates may distort the chemical bonds within the substrates. The combination of substrate selectivity, substrate orientation, temporary chemical bonds, and bond distortion promotes the specific chemical reaction catalyzed by a particular enzyme. When the final reaction between the substrates is finished, the product(s) no longer fit properly into the active site and are expelled (step ③). The temporary changes in shape, charge, and bonding patterns within the enzyme revert to their original configuration, and the enzyme is ready to accept another set of substrates (back to step ①).

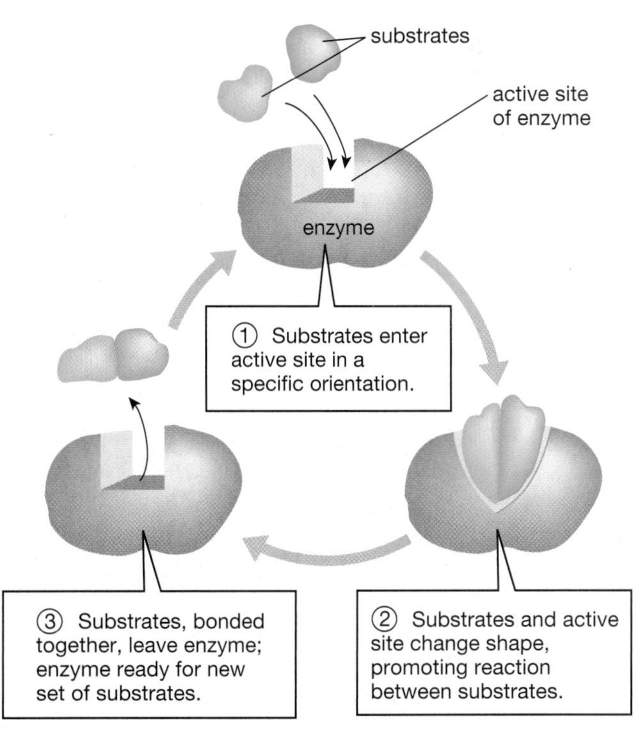

FIGURE 6-9 The cycle of enzyme–substrate interactions
QUESTION How would you modify reaction conditions if you wanted to increase the rate at which an enzyme-catalyzed reaction produced its product?

How do enzymes speed up the rate of chemical reactions? Enzymes are precisely regulated and promote only very specific reactions. The breakdown or synthesis of a molecule within a cell usually occurs in many discrete steps, each catalyzed by a different enzyme (see Fig. 6-7), which lowers the activation energy for its particular reaction (see Fig. 6-8). Just as carving a staircase into a cliff allows the cliff to be climbed—one small step at a time—this series of low activation energy steps allow the overall reaction to surmount its otherwise high activation energy "cliff," so the reaction can proceed at body temperature.

Cells Regulate the Amount and the Activity of Their Enzymes

To be useful, the metabolic reactions within cells must be carefully controlled; they must occur at the proper rate and with the proper timing. Cells have evolved many ways of regulating enzyme activity, including the following:

- Cells regulate the synthesis of enzymes to meet their changing needs. Reactions can occur only if the necessary enzymes are available.
- Cells synthesize some enzymes in inactive form and activate them only when needed. An example is the protein-digesting enzyme pepsin described in the following section. Pepsin doesn't digest the proteins of the cells that produce it because it is secreted in an inactive form with its active site blocked. In the acidic environment of the stomach, the blocking portion of the protein is removed, and the enzyme becomes active.
- Cells inhibit enzymes when adequate amounts of the enzyme's product are available. In **feedback inhibition**, the activity of an enzyme is inhibited by its own product or by a subsequent product produced farther along in a metabolic pathway (Fig. 6-10). For example, assume that an enzyme catalyzes the conversion of one amino acid to another. The concentration is regulated automatically by feedback inhibition if the end-product amino acid inhibits the enzyme once it reaches sufficiently high concentrations.
- Certain enzymes are subject to **allosteric regulation**. In allosteric regulation, enzyme action is enhanced or inhibited by small organic molecules that act as regulators. The regulator is neither the substrate nor the product of the enzyme, but may be the end product of a series of reactions in which the enzyme is involved. A regulator molecule will bind to a special *allosteric regulatory site* on the enzyme (Fig. 6-11a). As a result of this binding, the active site of the enzyme is changed (*allosteric* literally means "other shape"), and the enzyme may become either more or less able to bind its substrates (Fig. 6-11b). The specific enzyme and specific regulator molecule determine whether allosteric regulation increases or decreases enzyme activity. Allosteric regulation is one mechanism of feedback inhibition (see Fig. 6-10).
- In some cases, two or more molecules that are somewhat similar in structure compete for the active site of an enzyme, a situation called **competitive inhibition** (Fig. 6-11c). Some poisons are competitive inhibitors that keep an enzyme from breaking down its normal substrate. In another scenario, two types of alcohol, ethanol (the normal substrate, found in alcoholic beverages) and methanol (a poison), compete for the active site of the enzyme *alcohol dehydrogenase*. Methanol breakdown by this enzyme produces formaldehyde, which can cause blindness. Taking advantage of competitive inhibition, doctors administer ethanol to methanol-poisoning victims. By competing with methanol for the active site, ethanol blocks formaldehyde production.

The Activity of Enzymes Is Influenced by Their Environment

Enzymes, which are proteins, have very complex three-dimensional structures that are required for their proper function, but that are also sensitive to environmental

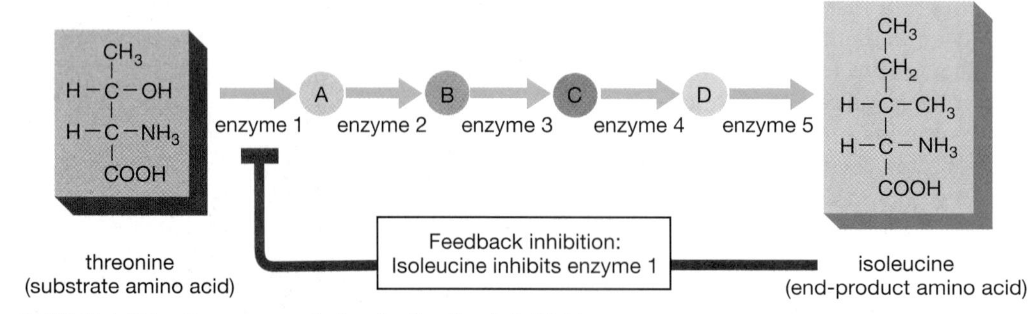

FIGURE 6-10 Enzyme regulation by feedback inhibition
In this example, the first enzyme in the metabolic pathway that converts threonine (an amino acid substrate) to isoleucine (an amino acid product) is inhibited by high concentrations of isoleucine. If a cell lacks isoleucine, the first enzyme is not inhibited, and the pathway proceeds rapidly. As isoleucine concentrations build up, the isoleucine binds to the first enzyme and gradually shuts down the pathway. When concentrations of isoleucine drop and fewer molecules are available to inhibit the enzyme, the pathway resumes production.

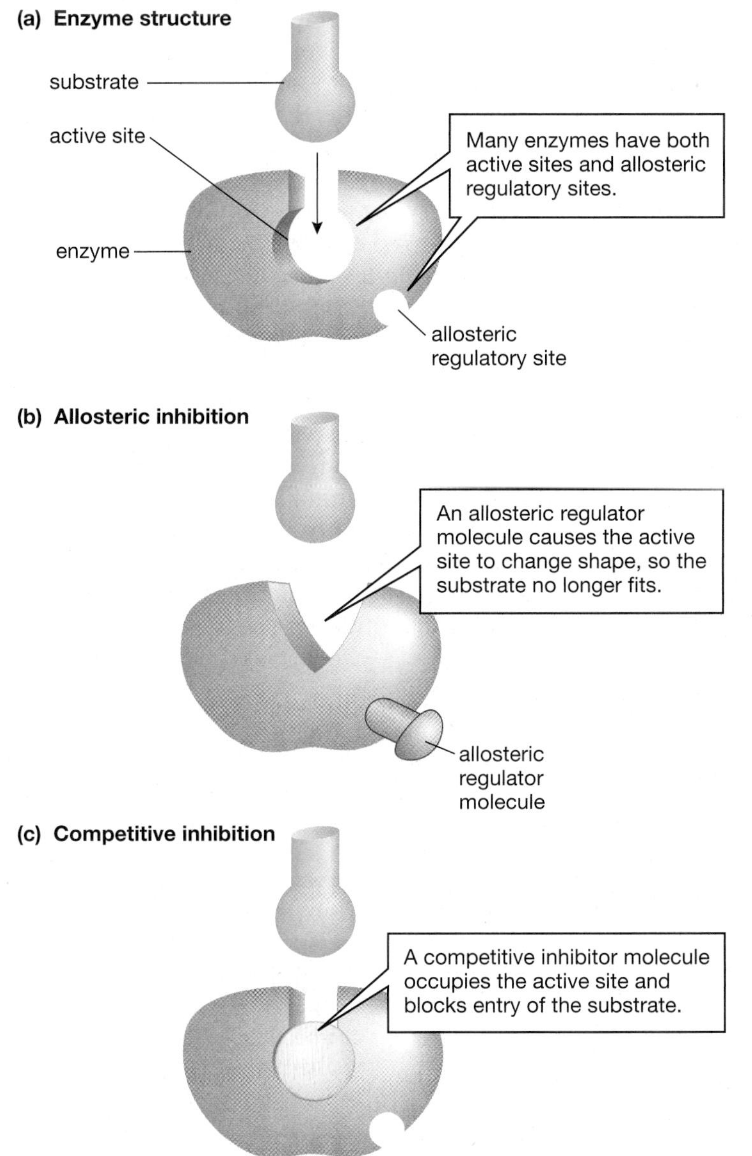
(a) Enzyme structure
substrate
active site
enzyme
Many enzymes have both active sites and allosteric regulatory sites.
allosteric regulatory site
(b) Allosteric inhibition
An allosteric regulator molecule causes the active site to change shape, so the substrate no longer fits.
allosteric regulator molecule
(c) Competitive inhibition
A competitive inhibitor molecule occupies the active site and blocks entry of the substrate.

FIGURE 6-11 Enzyme regulation by allosteric regulation and competitive inhibition
(a) Many enzymes have an active site and an allosteric regulatory site on different parts of the molecule. **(b)** When enzymes are inhibited by allosteric regulation, binding by a regulator molecule alters the active site so the enzyme is less compatible with its substrate. **(c)** During competitive inhibition, a molecule somewhat similar to the substrate fits into the active site and blocks entry of the substrate.

conditions. Recall from Chapter 3 that much of the three-dimensional structure of proteins is the result of hydrogen bonds between partially charged amino acids. These bonds can be altered by their chemical and physical surroundings. Each enzyme has evolved to function optimally at a particular pH, salt concentration, and temperature. Some also require the presence of other molecules called *coenzymes*, typically derived from water-soluble vitamins, in order to function.

Most enzymes function optimally at a pH between 6 and 8, the level found in most body fluids and maintained within cells. An exception is the protein-digesting enzyme *pepsin*. Pepsin is converted from an inactive to an active form by the acidic conditions within the stomach (pH of 2; see Fig. 2-9). At this pH, the excess of hydrogen ions causes hydrogen to attach to certain locations on the enzyme, altering its configuration and exposing the active site. In proteins that function best at neutral pH (pH 7), an acidic environment would distort the enzyme's structure and destroy its normal function.

Before the advent of refrigeration, foods such as meat were commonly preserved by using concentrated salt solutions. These solutions kill most bacteria, partly by interfering with enzyme function. Salts dissociate into ions, which form bonds with enzymes that interfere with the enzymes' normal three-dimensional structure, thus destroying enzyme activity. Dill pickles are very well preserved in a vinegar-salt solution, which combines both salty and acidic conditions. Organisms that live in highly salty environments, as you might predict, have enzymes whose configuration *depends* on the presence of salt ions.

Temperature also affects the rate of enzyme-catalyzed reactions. Because molecules move more rapidly at higher temperatures, their random movements are more likely to bring them into contact with the active site of an appropriate enzyme. Thus, these reactions are accelerated by moderately higher temperatures and slowed by lower temperatures. Stories abound of people who have fallen into frozen lakes for extended periods and survived. In one true example, a boy who fell through the ice was rescued and survived unharmed after 20 minutes under water. Although at normal body temperature the brain dies after about 4 minutes without oxygen, the boy's body temperature was lowered by the icy water, slowing his metabolic reactions and drastically reducing his need for oxygen. In contrast, when temperatures rise too high, the hydrogen bonds that regulate enzyme shape may be broken apart by the excessive molecular motion. Think of the protein in egg white and how it is completely altered in color and texture by cooking. Far lower temperatures than those required to fry an egg can still be too hot to allow enzymes to function properly.

Some enzymes require helper molecules called **coenzymes** to function. These organic molecules bind to the enzyme and interact with the substrate molecule. Coenzymes help weaken the bonds of the substrate, allowing it to react with an enzyme. Many water-soluble vitamins (such as the B vitamins) are essential to humans because they are used by the body to synthesize coenzymes.

In summary, the ability of an enzyme to catalyze reactions is controlled by many factors, including the amount of active enzyme, levels of allosteric regulator molecules, the concentration of inhibitor molecules, the concentration of substrates, pH, temperature, ionic environment, and, in some cases, the presence of coenzymes. In a healthy cell, the interactions among these molecules and precisely regulated environmental conditions maintain suitable concentrations of both substrates and products. Although runners and all other organisms use sugar as fuel, we life-forms burn it in a controlled fashion, using enzymes rather than flames. The enzymes allow our cells to break down sugar in many steps, each liberating small, manageable quantities of energy.

CASE STUDY REVISITED Energy Unleashed

Life generates heat, as dictated by the laws of thermodynamics and the energy demands of every living cell from bacteria to those in the energetic muscles of mammals (including marathon runners). In marathoners, for example, as ATP is broken down to power muscle contraction, some of the energy is converted to kinetic form as movement, and some is lost as heat. As muscles rapidly and repeatedly contract, they also warm up from friction as they slide past one another. You learned in Chapter 2 that, because water has one of the highest heats of vaporization of any molecule, we have evolved to use sweat (which is almost entirely water) to cool our bodies. Marathoners can lose considerable amounts of water through sweating during the race, and they risk overheating if they don't replenish it. Water stations are located at frequent intervals along the course so runners can "drink and run." But is it possible to drink too much water? Tragically, a healthy 28-year-old woman died from drinking too much water in the 2002 Boston Marathon. As discussed in Chapter 4, water moves readily through biological membranes, so it will enter the bloodstream and body cells whether you need it or not. Although the kidneys filter excess water from the blood, in extreme cases they are unable to keep up. The nervous system is particularly vulnerable. The brain can swell as excess water moves by osmosis into brain cells, and excess water also disrupts the precisely regulated ion concentrations that allow nerve cells to function.

Consider This: Runners must prevent overheating while avoiding excess water consumption. First, explain why proper enzyme function is disrupted by high temperatures. Then explain why Boston Marathon officials are considering placing scales near the end of the racecourse, on which runners can weigh themselves. For this to work, where else should scales be available?

Links to Life: Lack of an Enzyme Leads to Lactose Intolerance

Is it difficult for you to imagine life without milk, ice cream, or even pizza? Although some people consider these foods to be staples of the U.S. diet, they are not enjoyed by most of the world's population. Why? About 75% of people worldwide, including 25% of people in the United States, lose the ability to digest lactose, or "milk sugar," in early childhood. Roughly 75% of African Americans, Hispanics, and Native Americans, as well as 90% of Asian Americans, are *lactose intolerant.* From an evolutionary perspective, this makes perfect sense. The lactose enzyme (called lactase), is found in the small intestines of all normal young children. After weaning in early childhood, milk—the main source of lactose—was no longer available to our early ancestors. Since it takes energy to synthesize enzymes, losing the ability to synthesize an enzyme that is no longer needed is an adaptive advantage. However, a relatively small proportion of people, primarily those of northern European descent, retained the ability to digest lactose, raised cattle for their milk, and made dairy products a regular part of their diet.

When people who lack the enzyme lactase consume milk products, the undigested lactose draws water into the intestines by osmosis and also feeds intestinal bacteria that produce gas. The combination of excess water and gas leads to abdominal pain, bloating, diarrhea, and flatulence—a high price to pay for indulging in pizza or ice cream! Most people who are lactose intolerant do not need to avoid milk products altogether. Some produce enough lactase to tolerate a few servings daily. Aged cheeses (such as cheddar) and yogurt with live bacteria have relatively little lactose because the bacteria break it down. Lactase supplements can also be consumed along with dairy products. But compared to other consequences of enzyme deficiency, the inability to tolerate milk is a relatively minor inconvenience. These biological catalysts are crucial to all aspects of life, so mutations that render enzymes nonfunctional may prevent an embryo from developing at all, or cause life-threatening disorders.

Summary of Key Concepts

6.1 What Is Energy?

Energy is the capacity to do work. Kinetic energy is the energy of movement (light, heat, electricity, movement of large particles). Potential energy is stored energy (chemical energy, positional energy). The flow of energy among atoms and molecules obeys the laws of thermodynamics. The first law of thermodynamics states that, assuming there is no influx of energy, the total amount of energy remains constant, although it may change in form. The second law of thermodynamics states that any use of energy causes a decrease in the quantity of concentrated, useful energy and an increase in the randomness and disorder of matter. Entropy is a measure of disorder within a system.

6.2 How Does Energy Flow in Chemical Reactions?

Chemical reactions fall into two categories. In exergonic reactions, the reactant molecules have more energy than do the product molecules, so the reaction releases energy. In endergonic reactions, the reactants have less energy than do the products, so the reaction requires an input of energy. Exergonic reactions can occur spontaneously, but all reactions, including exergonic ones, require an initial input of energy (the activation energy) to overcome electrical repulsions between reactant molecules. Exergonic and endergonic reactions may be coupled such that the energy liberated by an exergonic reaction drives the endergonic reaction. Organisms couple exergonic reactions, such as light-energy capture or sugar metabolism, with endergonic reactions, such as the synthesis of organic molecules.

6.3 How Is Cellular Energy Carried Between Coupled Reactions?

Energy released by chemical reactions within a cell is captured and transported within the cell by energy-carrier molecules, such as ATP and electron carriers. These molecules are the major means by which cells couple exergonic and endergonic reactions that occur at different places in the cell.

6.4 How Do Cells Control Their Metabolic Reactions?

Cellular reactions are linked in interconnected sequences called metabolic pathways. The biochemistry of cells is regulated in three ways: first, through the use of protein catalysts called enzymes; second, by coupling endergonic with exergonic reactions; and third, through the use of energy-carrier molecules that transfer energy within cells.

High activation energies slow many reactions, even exergonic ones, to an imperceptible rate under normal environmental conditions. Catalysts lower the activation energy and thereby speed up chemical reactions without being permanently changed themselves. Organisms synthesize protein catalysts called enzymes that promote one or a few specific reactions. The reactants temporarily bind to the active site of the enzyme, making it easier to form the new chemical bonds of the products. Enzyme action is regulated in many ways, including by altering the rate of enzyme synthesis, by activating previously inactive enzymes, by feedback inhibition, by allosteric regulation, and by competitive inhibition. Environmental conditions including pH, salt concentration, and temperature can promote or inhibit enzyme function, by altering their three-dimensional structure.

Key Terms

activation energy *p. 102*
active site *p. 107*
adenosine diphosphate (ADP) *p. 104*
adenosine triphosphate (ATP) *p. 103*
allosteric regulation *p. 108*
catalyst *p. 106*
chemical reaction *p. 101*
coenzyme *p. 110*
competitive inhibition *p. 108*
coupled reaction *p. 103*
electron carrier *p. 105*
endergonic *p. 101*
energy *p. 100*
energy-carrier molecule *p. 103*
entropy *p. 101*
enzyme *p. 107*
exergonic *p. 101*
feedback inhibition *p. 108*
first law of thermodynamics *p. 100*
kinetic energy *p. 100*
laws of thermodynamics *p. 100*
metabolic pathway *p. 106*
metabolism *p. 105*
potential energy *p. 100*
product *p. 101*
reactant *p. 101*
second law of thermodynamics *p. 100*
substrate *p. 107*

Thinking Through the Concepts

To take a multiple-choice quiz with feedback on the contents of this chapter, visit http://www.prenhall.com/audesirk7. *Log in to the Web site selected by your instructor and navigate to the Self Test section for this chapter.*

? Review Questions

1. Explain why organisms do not violate the second law of thermodynamics. What is the ultimate energy source for most forms of life on Earth?
2. Define *metabolism*, and explain how reactions can be coupled to one another.
3. What is activation energy? How do catalysts affect activation energy? How does this change the rate of reactions?
4. Describe some exergonic and endergonic reactions that occur in plants and animals very regularly.
5. Describe the structure and function of enzymes. How is enzyme activity regulated?

Applying the Concepts

1. A preview question for ecology (Unit Six): When a brown bear eats a salmon, does the bear acquire all the energy contained in the body of the fish? Why or why not? What implications do you think this answer would have for the relative abundance (by weight) of predators and their prey? Does the second law of thermodynamics help explain the book title *Why Big Fierce Animals Are Rare*?
2. Many people in sub-Saharan Africa have experienced the effects of malnutrition and starvation, but the very young are most severely affected. Some individuals suffer permanent disability even if food is provided later. How could a lack of food intake interfere with functions of individual cells and tissues? Which tissues are likely to suffer the most irreversible damage?
3. As you learned in Chapter 3, the subunits of virtually all organic molecules are joined by condensation reactions and can be broken apart by hydrolysis reactions. Why, then, does your digestive system produce separate enzymes to digest proteins, fats, and carbohydrates—in fact, several of each type?
4. A preview question for evolution (Unit Three): Suppose someone tried to refute the existence of evolution with the following argument: "According to evolutionary theory, organisms have increased in complexity through time. However, evolution of increased biological complexity contradicts the second law of thermodynamics. Therefore, evolution is impossible." How would you respond to this argument in support of evolution?

For More Information

Farid, R. S. "Enzymes Heat Up." *Science News*, May 9, 1998. Scientists explore new ways to synthesize enzymes that will function at high temperatures.

Fenn, J. *Engines, Energy, and Entropy*. New York: W. H. Freeman, 1982. Elegantly simple introduction to the laws of thermodynamics and their relationship to everyday life.

Madigan, M. T., and Narrs, B. L. "Extremophiles." *Scientific American*, April 1997. Industrial processes are benefiting from an understanding of the molecules that allow certain microbes to thrive under highly acidic, salty, or extremely hot conditions that would denature most proteins.

Wu, C. "Hot-Blooded Proteins." *Science News*, May 9, 1998. Bacteria that thrive in near-boiling conditions have special enzymes that allow them to function at these extreme temperatures.

Media Activities

To access a Media Activity visit http://www.prenhall.com/audesirk7. *Log in to the Web site selected by your instructor, navigate to this chapter, and select the appropriate Media Activity number.*

6.1 Energy and Chemical Reactions

Estimated time: 5 minutes

Explore the basic concepts of thermodynamics and how they govern the life of a cell.

6.2 Energy and Life

Estimated time: 10 minutes

Explore the importance of ATP as the currency for work in the cell.

6.3 Enzymes

Estimated time: 5 minutes

Visualize the structure of an enzyme and explore how an enzyme's structure relates to its catalytic function.

6.4 Web Investigation: Energy Unleashed

Estimated time: 10 minutes

In this exercise, we will use online calorie calculators to explore how different factors affect energy needs in the human body.

CHAPTER

7 Capturing Solar Energy: Photosynthesis

As *Tyrannosaurus* threatens *Triceratops*, a giant meteor and its fragments streak toward Earth. Both hunter and hunted will die in this cataclysmic event. Such an event may have driven 70% of all life-forms to extinction 65 million years ago.

AT A GLANCE

CASE STUDY Did the Dinosaurs Die from Lack of Sunlight?

It is summer in the year 65,000,000 B.C. The Cretaceous period is about to reach an abrupt and catastrophic end. On an Earth where the continents we know as the Americas are largely submerged by shallow seas, a 20-foot *Triceratops* grazes in lush, tropical vegetation in what is now southern California. From the cover of dense forest nearby, a 45-foot, 8-ton *Tyrannosaurus rex* watches and waits. Suddenly, a deafening roar startles the animals, and they gaze upward to see a fireball ripping through the blue sky. A meteorite 6 miles in diameter has entered the atmosphere and is about to irrevocably alter life on Earth. Although any creatures that witnessed the meteorite strike Earth were immediately incinerated by the blast wave from the impact, the aftereffects were experienced by plants and animals all over the world. As it plowed into the ocean at the tip of the Yucatan Peninsula, the meteorite dug a crater a mile deep and 120 miles wide. The force of its impact sent trillions of tons of debris from Earth's crust and from the meteorite itself rocketing into the atmosphere. The heat of the blast caused fires that may have charred 25% of all the vegetation on land. Ashes, smoke, and dust obliterated the sun. Earth was plunged into a night that lasted for months. What would happen if the sun were obscured for months on end? Why is sunlight so important? Could a meteorite have been responsible for ending the rule of the dinosaurs?

7.1 What Is Photosynthesis?

At least 2 billion years ago, some cells, through chance mistakes (mutations) in their genetic makeup, acquired the ability to harness the energy in sunlight. These cells combined simple inorganic molecules—carbon dioxide and water—into more complex organic molecules such as glucose. In the process of *photosynthesis*, the cells captured a small fraction of the sunlight's energy, storing it as chemical energy in complex organic molecules. Exploiting this new source of energy without competition, early photosynthetic cells filled the seas, releasing oxygen as a by-product. A new element in the atmosphere, free oxygen was harmful to many organisms. But the endless variation produced by random genetic mistakes eventually gave rise to some cells that could survive in oxygen and, later, to cells that made use of oxygen to break down glucose in a new, more efficient process: *cellular respiration*. Today, most forms of life on Earth—including you—depend on the sugar produced by photosynthetic organisms as an energy source and release the energy from that sugar through cellular respiration, using the photosynthetic by-product oxygen (Fig. 7-1). In Chapter 8, we will examine the processes by which nearly all living things break down the sugary energy-storage molecules produced by photosynthesis and reclaim the energy to power other metabolic reactions. Sunlight powers practically all life on Earth and is captured only by photosynthesis.

Starting with the simple molecules of carbon dioxide (CO_2) and water (H_2O), **photosynthesis** converts the energy of sunlight into chemical energy stored in the bonds of glucose ($C_6H_{12}O_6$) and releases oxygen (O_2). The simplest overall chemical reaction for photosynthesis is:

$$6\,CO_2 + 6\,H_2O + \text{light energy} \longrightarrow C_6H_{12}O_6 + 6\,O_2$$

Photosynthesis occurs in eukaryotic plants, algae, and certain types of prokaryotes, all of which are described as *autotrophs* (literally, "self-feeders"). In this chapter, we will limit our discussion of photosynthesis to plants, particularly land plants. Photosynthesis in plants takes place within chloroplasts, most of which are contained within leaf cells. Let us begin, then, with a brief look at the structures of both leaves and chloroplasts. Chapter 24 examines leaf structure and function more thoroughly.

Leaves and Chloroplasts Are Adaptations for Photosynthesis

The leaves of most land plants are only a few cells thick; their structure is elegantly adapted to the demands of photosynthesis (Fig. 7-2). The flattened shape of leaves exposes a large surface area to the sun, and their thinness ensures that sunlight can penetrate them to reach the light-trapping chloroplasts inside. Both the upper and lower surfaces of a leaf consist of a layer of transparent cells, the *epidermis*. The outer surface of both epidermal layers is covered by the *cuticle*, a waxy, waterproof covering that reduces the evaporation of water from the leaf. A leaf obtains CO_2 for photosynthesis from the air; adjustable pores in the epidermis, called **stomata** (singular, *stoma*; Greek for "mouth"), open and close at appropriate times to admit air carrying CO_2. Inside the leaf are a few layers

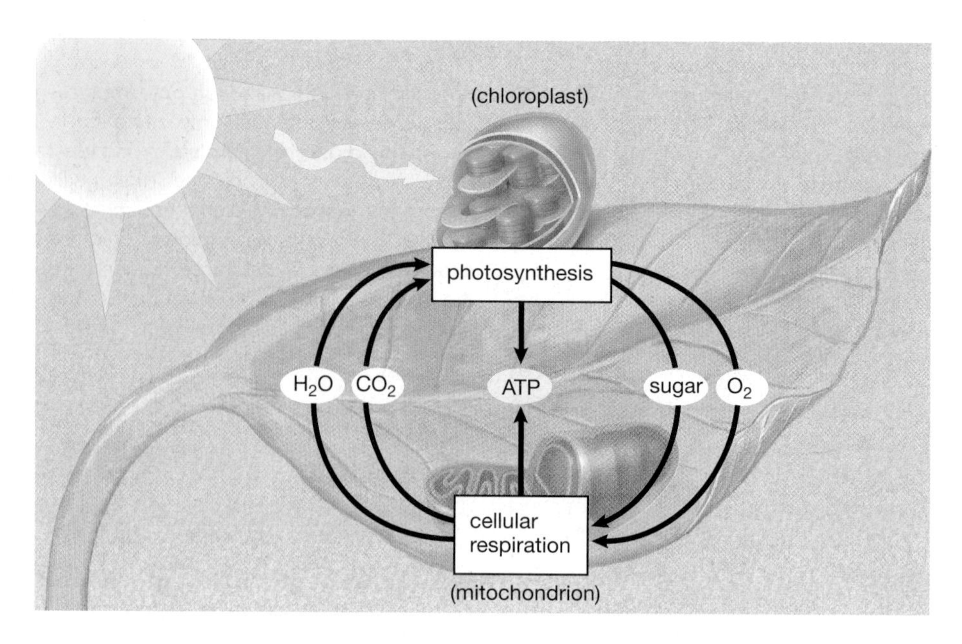

FIGURE 7-1 Interconnections between photosynthesis and cellular respiration Chloroplasts in green plants use the energy of sunlight to synthesize high-energy carbon compounds such as glucose from low-energy molecules of water and carbon dioxide. Plants themselves, and other organisms that eat plants or one another, extract energy from these organic molecules by cellular respiration, yielding water and carbon dioxide once again. This energy in turn drives all the reactions of life.

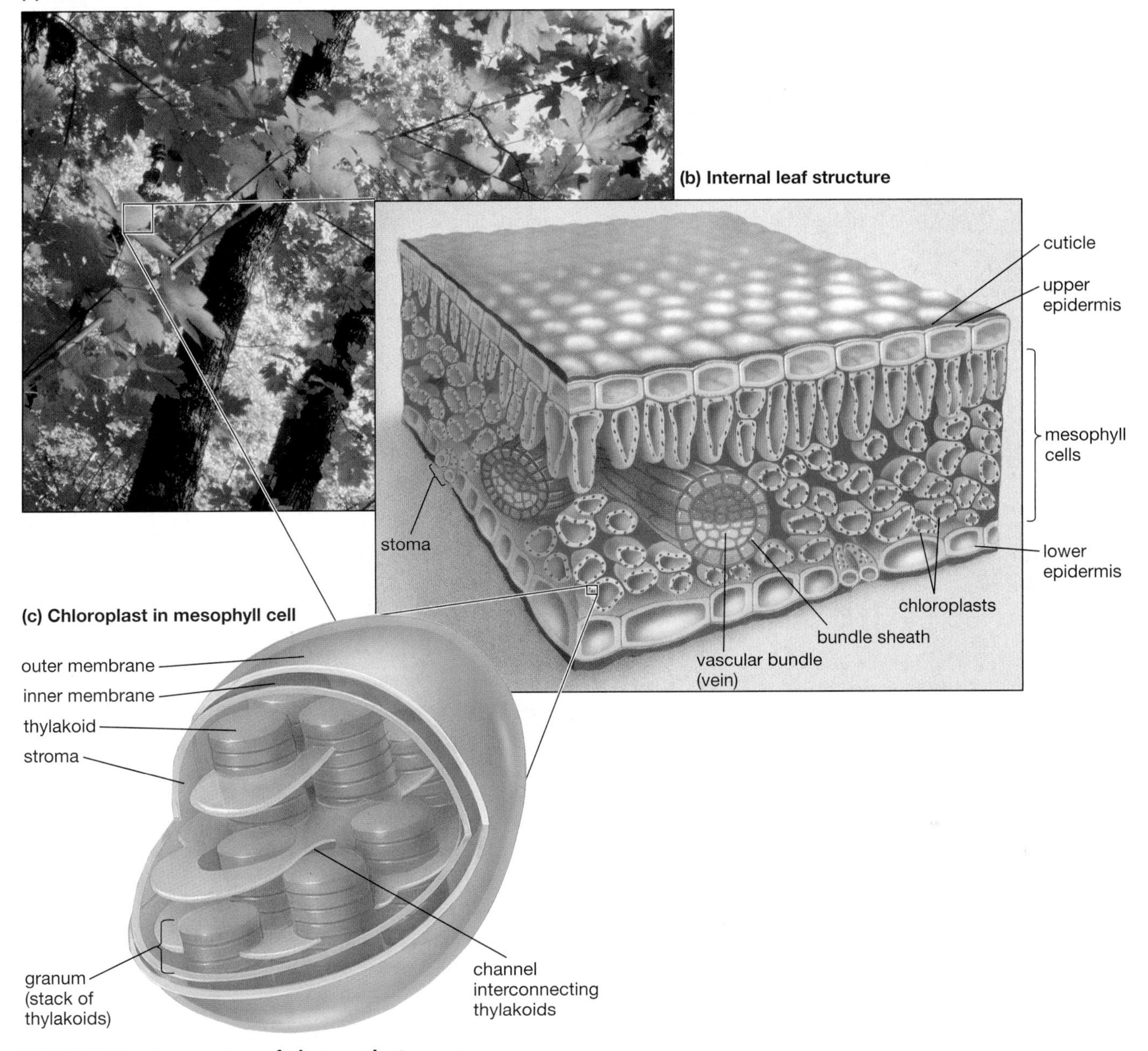

FIGURE 7-2 An overview of photosynthetic structures
(a) Photosynthesis occurs primarily in the leaves of land plants. **(b)** A section of a leaf, showing mesophyll cells where chloroplasts are concentrated and the waterproof cuticle that coats the leaf's upper epidermis. **(c)** A single chloroplast, showing the stroma and thylakoids where photosynthesis occurs.

of cells collectively called *mesophyll* (which means "middle of the leaf"). The mesophyll cells contain the vast majority of a leaf's chloroplasts (see Fig. 7-2), and consequently photosynthesis occurs principally in these cells. *Vascular bundles*, or veins, supply water and minerals to the mesophyll cells and carry the sugars they produce to other parts of the plant.

A single mesophyll cell can have from 40 to 200 chloroplasts, which are sufficiently small that 2000 of them lined up would just span your thumbnail. As we discussed in Chapter 5, chloroplasts are organelles that consist of a double outer membrane enclosing a semifluid medium, the **stroma** (see Fig. 7-2). Embedded in the stroma are disk-shaped, interconnected membranous sacs called **thylakoids**. In most chloroplasts, the thylakoids are piled atop one another in stacks called **grana** (singular, *granum*). The chemical reactions of photosynthesis that depend on light (*light-dependent reactions*) occur within the membranes of the thylakoids, while the photosynthetic reactions that can continue for a time in darkness (*light-independent reactions*) occur in the surrounding stroma.

Photosynthesis Consists of Light-Dependent and Light-Independent Reactions

The simple overall chemical summary of photosynthesis obscures the fact that photosynthesis actually involves dozens of enzymes catalyzing dozens of individual reactions. These reactions can be classified as light-dependent reactions and light-independent reactions. Each group of reactions occurs within a different region of the chloroplast; the two are linked by energy-carrier molecules.

- In **light-dependent reactions**, chlorophyll and other molecules embedded in the membranes of the thylakoids capture sunlight energy and convert some of it into the chemical energy stored in energy-carrier molecules (ATP and NADPH). Oxygen gas is released as a by-product.
- In **light-independent reactions**, enzymes in the stroma use the chemical energy of the carrier molecules to drive the synthesis of glucose or other organic molecules.

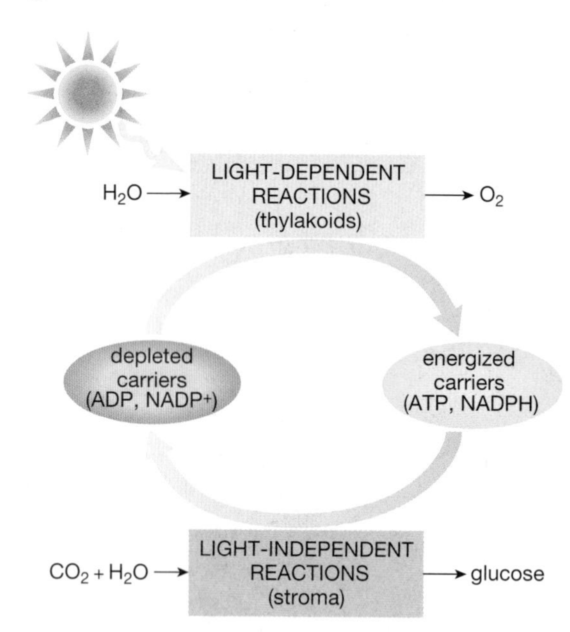

Media Activity 7.1 Properties of Light

7.2 Light-Dependent Reactions: How Is Light Energy Converted to Chemical Energy?

The light-dependent reactions capture the energy of sunlight, storing it as chemical energy in two different energy-carrier molecules: the familiar energy carrier ATP (*adenosine triphosphate*) and the high-energy electron carrier *NADPH* (*nicotinamide adenine dinucleotide phosphate*). The chemical energy stored in these carrier molecules will then be used to power the synthesis of high-energy storage molecules, such as glucose, during the light-independent reactions.

During Photosynthesis, Light Is First Captured by Pigments in Chloroplasts

The sun emits energy in a broad spectrum of electromagnetic radiation. The *electromagnetic spectrum* ranges from short-wavelength gamma rays, through ultraviolet, visible, and infrared light, to very long-wavelength radio waves (Fig. 7-3a). Light and the other types of radiation are composed of individual packets of energy called **photons**. The energy of a photon corresponds to its wavelength: short-wavelength photons are very energetic, whereas longer-wavelength photons have lower energies. Visible light consists of wavelengths with energies that are strong enough to alter the shape of certain pigment molecules (such as those in chloroplasts) but weak enough not to damage crucial molecules such as DNA. It is no coincidence that these wavelengths (with "just the right amount" of energy) not only power photosynthesis, but also stimulate the pigments in our eyes and allow us to see the world around us.

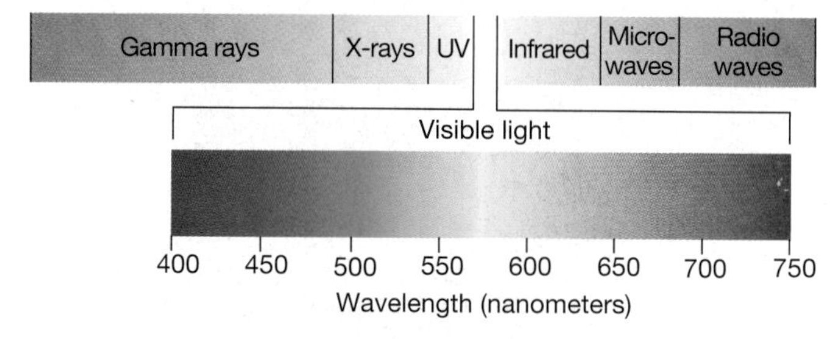

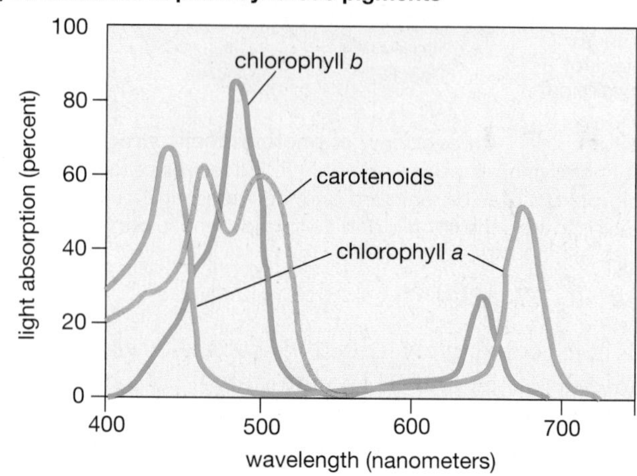

FIGURE 7-3 Light, chloroplast pigments, and photosynthesis **(a)** Visible light, a small part of the electromagnetic spectrum (top line), consists of wavelengths that correspond to the colors of the rainbow. **(b)** Chlorophyll (blue and green curves) strongly absorbs violet, blue, and red light. Carotenoids (orange curve) absorb blue and green wavelengths. **QUESTION** Based on the information in this graph, what color are carotenoids? What color is phycocyanin?

One of three events occurs when light strikes an object such as a leaf: the light is either *absorbed* (captured), *reflected* (bounced back), or *transmitted* (passed through). Light that is absorbed can heat up the object or drive biological processes, such as photosynthesis. Light that is reflected or transmitted reaches our eyes and gives an object its color.

Chloroplasts contain various pigment molecules (so called because they reflect specific colors to our eyes) that absorb different wavelengths of light. **Chlorophyll**, the key light-capturing pigment molecule in chloroplasts, strongly absorbs violet, blue, and red light but reflects green, thus giving green leaves their color (Fig. 7-3b). Chloroplasts also contain other molecules, called *accessory pigments*, that absorb additional wavelengths of light energy and transfer it to *chlorophyll a*. Some accessory pigments are actually slightly different forms of green chlorophyll; in land plants, chlorophyll a is the main light-capturing pigment, while *chlorophyll b* serves as an accessory pigment. **Carotenoids** are accessory pigments found in all chloroplasts. They absorb blue and green light and thus appear mostly yellow, orange, or red, since they reflect these wavelengths to our eyes (see Fig. 7-3b).

Although carotenoids (particularly yellow carotenoids) are present in all leaves, their color is usually masked by the green chlorophyll. In the autumn, as leaves begin to die, chlorophyll breaks down before carotenoids do, revealing the bright yellow carotenoids as fall colors. (Red fall colors are primarily anthocyanins, which are not involved in photosynthesis and are synthesized by some leaves in the fall; you'll learn more about these pigments in Chapter 24.) You may have heard of the carotenoid called beta-carotene, which is found in chloroplasts, and also produces the orange color of vegetables such as carrots. This plant pigment is the principal source of vitamin A for animals, including ourselves. In a beautiful symmetry, vitamin A, in turn, forms the visual pigment that captures light in our eyes. Thus, carotenoids not only capture light energy in plants, but, indirectly, in animals as well.

The Light-Dependent Reactions Occur Within the Thylakoid Membranes

The thylakoid membranes contain highly organized assemblies of proteins, chlorophyll, and accessory pigment molecules, called **photosystems**. Each thylakoid contains thousands of copies of each of two photosystems, *photosystem I* (PSI) and *photosystem II* (PS II), both of which are activated by light and work simultaneously. Each photosystem contains roughly 250 to 400 chlorophyll and accessory pigment molecules. These pigments absorb light and pass its energy to a pair of specialized chlorophyll molecules within a small region of the photosystem called the **reaction center**. The reaction-center chlorophyll molecules are located adjacent to an **electron transport chain** (ETC), which is a series or "chain" of electron carrier molecules embedded in the thylakoid membrane, as illustrated below. Note that each photosystem is associated with a different electron transport chain.

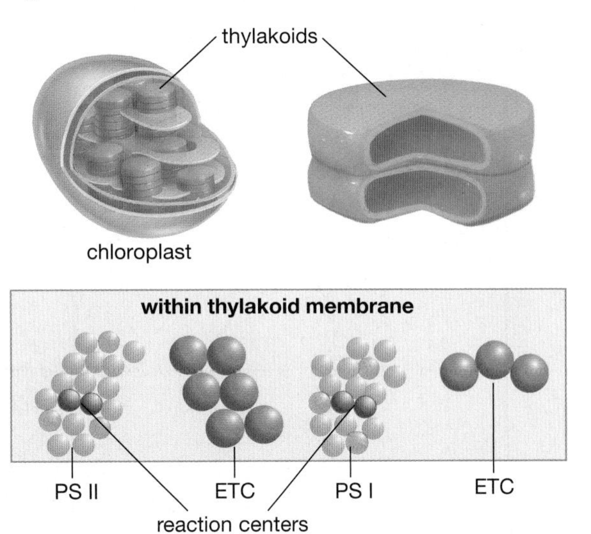

When the reaction-center chlorophyll molecules receive energy from the surrounding pigment molecules, an electron from each of the two reaction-center chlorophylls absorbs the energy. These "energized electrons" leave the chlorophyll molecules and jump over to the electron transport chain, where they are passed along from one carrier molecule to the next, losing energy as they go. At certain transfer points along the electron transport chain, the energy released from the electrons is captured and used to synthesize ATP from ADP plus phosphate or NADPH from $NADP^+$ plus H^+. (*NADP* is the electron carrier NAD, described in Chapter 6, plus a phosphate group.) With this overall scheme in mind, let's look more closely at the actual sequence of events in the light-dependent reactions (Fig. 7-4).

Photosystem II Generates ATP

For historical reasons, the photosystems are numbered "backward." The process of capturing light energy is most easily understood by starting with photosystem II and following the events initiated by the capture of two photons of light. The light-dependent reactions begin when photons of light are absorbed by photosystem II (step ① in Fig. 7-4). Light energy passes from molecule to molecule until it reaches the reaction center, where it boosts an electron out of each of the two chlorophyll molecules (step ②). The first electron carrier of the adjacent electron transport chain instantly accepts these two energized electrons (step ③). The electrons move along the chain from one carrier molecule to the next, releasing energy as they go; since no energy transfer is 100% efficient, some of the energy is lost as heat at each step. But some of the energy is captured and used to pump hydrogen ions (H^+) across the thylakoid membrane into the compartment within the

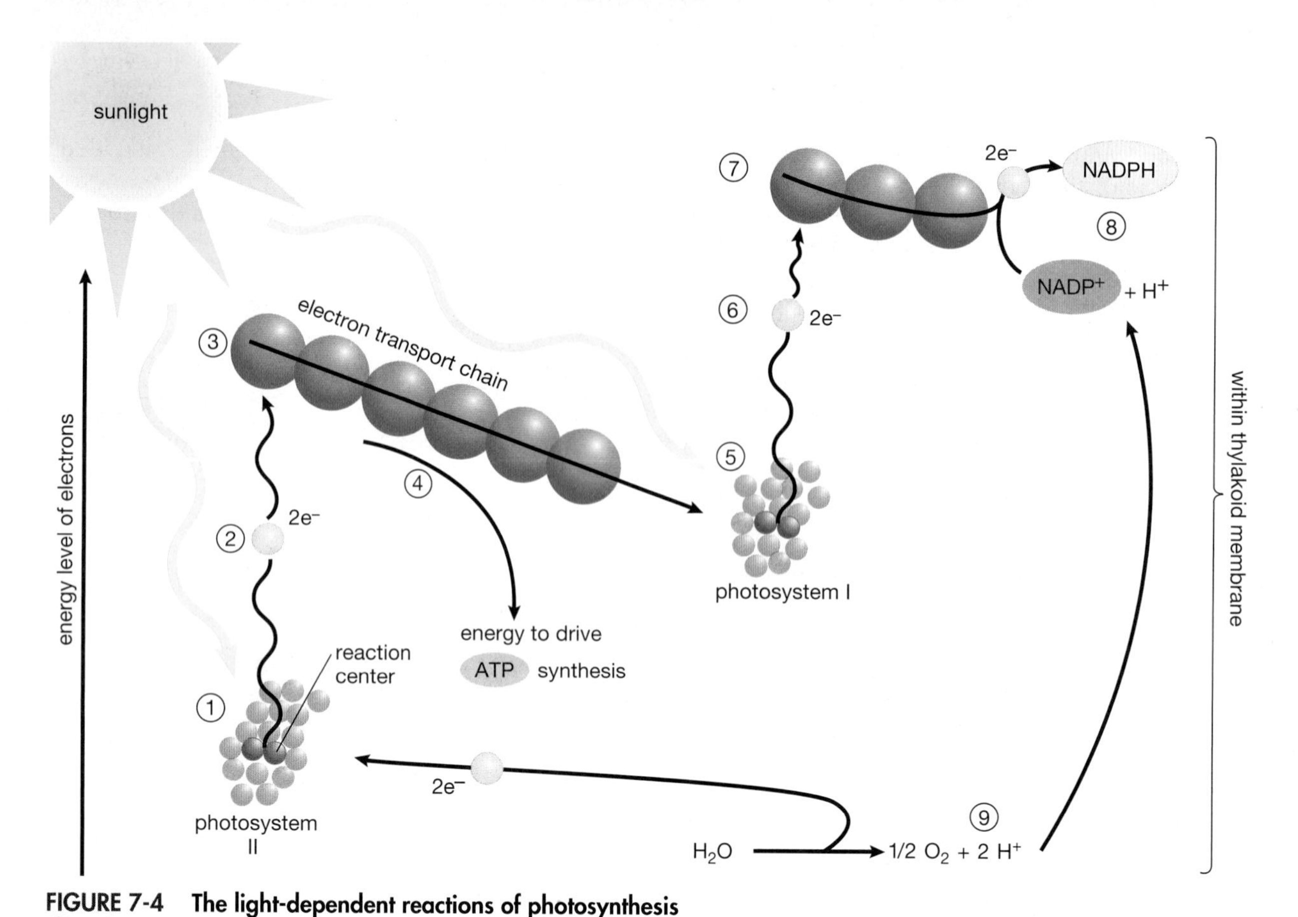

FIGURE 7-4 The light-dependent reactions of photosynthesis
① Light is absorbed by photosystem II, and the energy is passed to electrons in the reaction-center chlorophyll molecules. ② Energized electrons leave the reaction center. ③ The electrons move into the adjacent electron transport chain. ④ The chain passes the electrons along, and some of their energy is used to drive ATP synthesis by chemiosmosis. Energy-depleted electrons replace those lost by photosystem I. ⑤ Light strikes photosystem I, and the energy is passed to electrons in the reaction-center chlorophyll molecules. ⑥ Energized electrons leave the reaction center. ⑦ The electrons move into the electron transport chain. ⑧ The energetic electrons from photosystem I are captured in molecules of NADPH. ⑨ The electrons lost from the reaction center of photosystem II are replaced by electrons obtained from splitting water, a reaction that also releases oxygen, and H^+ used to form NADPH. **QUESTION** If these reactions produce ATP and NADPH, then why do plant cells need mitochondria?

thylakoid, creating a H^+ ion concentration gradient across the thylakoid membrane. The energy used to create this gradient is then harnessed to drive the synthesis of ATP, using a process called **chemiosmosis** (step ④). The essay "A Closer Look: Chemiosmosis—ATP Synthesis in Chloroplasts" describes chemiosmosis in more detail.

Photosystem I Generates NADPH

Meanwhile, light rays have also been striking the pigment molecules of photosystem I (step ⑤ in Fig. 7-4). Energy from the light photons is captured by these pigment molecules and funneled to the two reaction center chlorophyll molecules, which then eject high energy electrons (step ⑥). These electrons jump to the electron transport chain associated with photosystem I (step ⑦). The energized electrons ejected from photosystem I move through the adjacent, shorter electron transport chain and are finally transferred to the electron carrier $NADP^+$. The energy carrier molecule NADPH is formed when each $NADP^+$ molecule picks up two energetic electrons and one hydrogen ion (step ⑧); the hydrogen ion is obtained from splitting water (step ⑨). Both $NADP^+$ and NADPH are water-soluble molecules dissolved in the chloroplast stroma. The reaction-center chlorophylls of photosystem I immediately replace their lost electrons by obtaining the energy-depleted electrons from the final electron carrier of the electron transport chain fed by photosystem II.

Splitting Water Maintains the Flow of Electrons Through the Photosystems

Overall, electrons flow from the reaction center of photosystem II, through its nearby electron transport chain, to the reaction center of photosystem I, through its nearby electron transport chain, and finally form NADPH. To sustain this one-way flow of electrons, the reaction center of photosystem II must be continuously supplied with new electrons to replace the ones it gives up. These replacement electrons come from water (step

A CLOSER LOOK Chemiosmosis—ATP Synthesis in Chloroplasts

In the electron transport chain associated with photosystem II, energetic electrons move from carrier to carrier. The electron transfers do not directly drive ATP synthesis; rather, the energy released during the transfers is used to generate a concentration gradient of hydrogen ions across the thylakoid membrane. In a separate reaction, the energy stored in this gradient then powers ATP synthesis.

In the first step of the light-dependent reactions of photosynthesis, a photon strikes the light-harvesting complex of photosystem II; the energy is absorbed and passed to the reaction-center chlorophyll. There, an electron absorbs the energy and is ejected out of the chlorophyll molecule. Within a billionth of a second, the electron is captured by the first electron carrier of the adjacent electron transport chain.

As the electron moves from one carrier molecule to the next, it loses energy at each transfer. The energy released from the exergonic reaction of these electron transfers is used to power active transport of hydrogen ions across the thylakoid membrane from the stroma into the thylakoid interior:

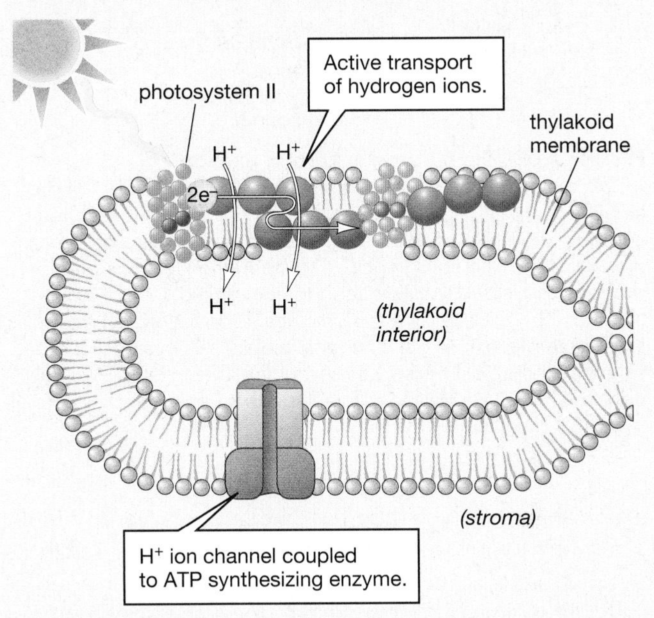

This transport increases the concentration of hydrogen ions (and therefore also the positive charge) inside the thylakoid, creating a gradient of both hydrogen ions and positive charge across the thylakoid membrane. Using energy to create a gradient of charge and ion concentration is very much like charging a rechargeable battery. The thylakoid membrane does not allow hydrogen ions to leak out, except at specific protein channels that are coupled to ATP-synthesizing enzymes. When hydrogen ions flow through these channels, down their gradients of charge and concentration, the energy released drives the synthesis of ATP:

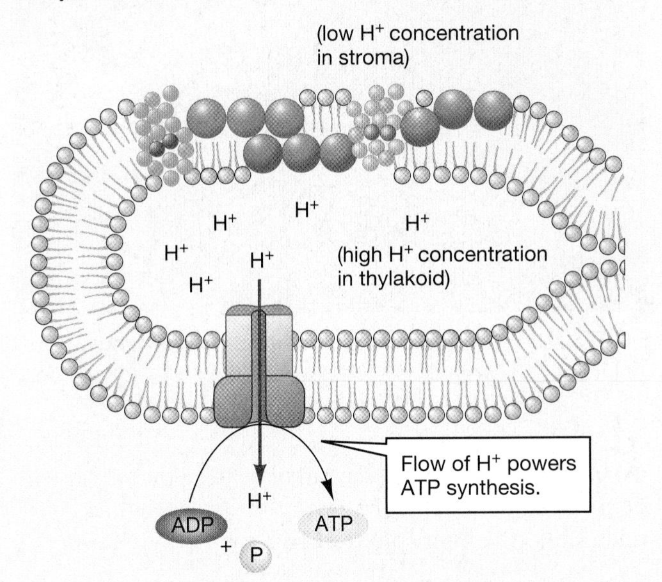

How does a gradient of hydrogen ions generate energy to drive ATP synthesis? Compare the hydrogen ion gradient across the thylakoid membrane to water stored behind a dam at a hydroelectric plant. When specific gates within the dam are opened, the rapid flow of water is directed through turbines that are turned by the rushing water and convert this energy of motion into electrical energy: they generate electricity. The thylakoid operates in an analogous way. Hydrogen ions in the thylakoid interior can move down their gradients into the stroma only through the ATP-synthesizing hydrogen ion channels. Something like flowing water, this flow of hydrogen ions through specific channels can do work: it drives ATP synthesis from ADP plus phosphate. About one ATP molecule is synthesized for every three hydrogen ions that pass through the channel.

Scientists are still investigating exactly how the ATP-synthesizing proton channel works. However, this general mechanism of ATP synthesis was first proposed in 1961 by British biochemist Peter Mitchell, who called it *chemiosmosis*. Chemiosmosis has been shown to be the mechanism of ATP generation in chloroplasts, mitochondria (as we will see in Chapter 8), and bacteria. For his brilliant hypothesis, Mitchell was awarded the Nobel Prize for Chemistry in 1978.

Media Activity 7.3 Chemiosmosis

⑨ in Fig. 7-4). In a series of reactions that scientists are still deciphering, the reaction-center chlorophylls of photosystem II attract electrons from water molecules within the thylakoid compartment, causing the water molecules to split apart:

$$H_2O \longrightarrow {}^1\!/_2 O_2 + 2H^+ + 2e^-$$

For every two photons captured by photosystem II, two electrons are boosted out of the reaction-center chlorophyll and are replaced by the two electrons obtained by splitting one water molecule. The loss of two electrons from water generates two hydrogen ions (H^+), which are used to form NADPH. As water molecules are split, their oxygen atoms combine to form molecules of

FIGURE 7-5 **Oxygen is a by-product of photosynthesis**
The bubbles released by the leaves of this aquatic plant (*Elodea*) are composed of oxygen, a by-product of photosynthesis.

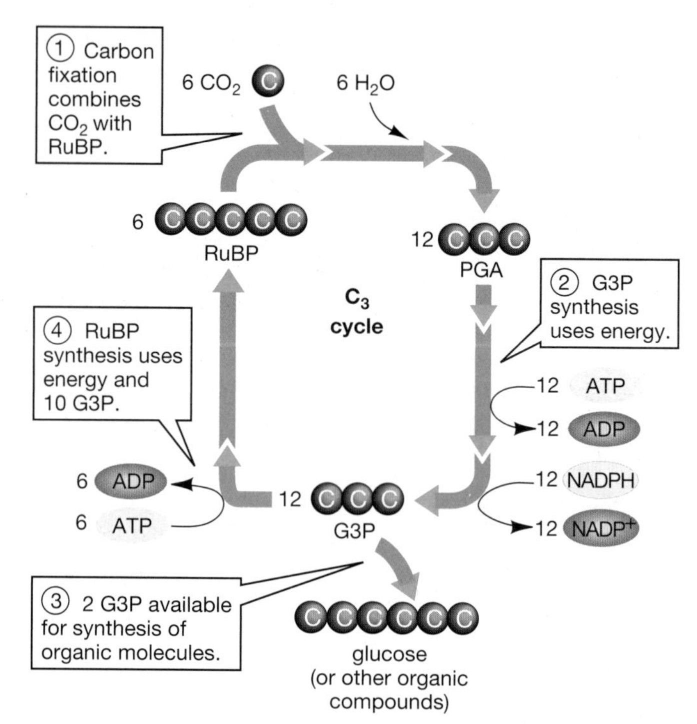

FIGURE 7-6 **The C_3 cycle of carbon fixation**
① Six molecules of RuBP react with 6 molecules of CO_2 and 6 molecules of H_2O to form 12 molecules of PGA. This reaction is carbon fixation, the capture of carbon from CO_2 into organic molecules. ② The energy of 12 ATPs and the electrons and hydrogens of 12 NADPHs are used to convert the 12 PGA molecules to 12 G3Ps. ③ Two G3P molecules are available to synthesize glucose or other organic molecules. This occurs outside the chloroplast and is not part of the C3 cycle. ④ Energy from 6 ATPs is used to rearrange 10 G3Ps into 6 RuBPs, completing one turn of the C_3 cycle.

oxygen gas (O_2). The oxygen may be used directly by the plant in its own cellular respiration (see Chapter 8) or released to the atmosphere (Fig. 7-5).

Summing Up

Light-Dependent Reactions

Chlorophyll and carotenoid pigments of photosystem II absorb light that is used to energize and eject electrons from the reaction-center chlorophyll molecules. The electrons are passed along the adjacent electron transport chain, where they release energy. Some of the energy is used to create a hydrogen ion gradient across the thylakoid membrane that drives ATP synthesis. The "electron-deprived" reaction-center chlorophylls of photosystem II replace their electrons by splitting water molecules. The resulting H^+ is used in NADPH, and oxygen gas is generated as a by-product. Light is also absorbed by photosystem I, which ejects energized electrons from its reaction-center chlorophylls. The electrons are picked up by the adjacent electron transport chain, and their energy is captured in NADPH. Electrons lost from the reaction center of photosystem I are replaced by those from the electron transport chain associated with photosystem II. The products of the light-dependent reactions are NADPH, ATP, and O_2.

7.3 Light-Independent Reactions: How Is Chemical Energy Stored in Glucose Molecules?

The ATP and NADPH synthesized during the light-dependent reactions are dissolved in the fluid stroma that surrounds the thylakoids. There, they provide the energy to power the synthesis of glucose from carbon dioxide and water, a process that requires enzymes that are also dissolved in the stroma. The reactions that eventually produce glucose are called the light-independent reactions because they can occur independently of light as long as ATP and NADPH are available. However, these high-energy molecules required for glucose synthesis are available only if they have been recharged by light. Thus, any event that reduces light availability (such as the dust, smoke, and ash that would accompany a meteorite collision with Earth) also reduces the availability of these high-energy compounds and consequently decreases the plant's ability to synthesize food for itself.

The C_3 Cycle Captures Carbon Dioxide

The process of capturing six carbon dioxide molecules from the air and using them to synthesize the six-carbon sugar glucose occurs in a set of reactions known as either the **Calvin-Benson cycle** (after its discoverers) or the **C_3 cycle** (three-carbon cycle; some of the important molecules in the cycle have three carbon atoms in them) (Fig. 7-6). The C_3 cycle requires CO_2 (usually from the air); a CO_2-capturing sugar, *ribulose bisphosphate* (RuBP); enzymes to catalyze all the reactions; and energy in the form of ATP and NADPH (provided by the light-dependent reactions).

The C_3 cycle is best understood if we mentally divide it into the following three parts: carbon fixation, the synthesis of *glyceraldehyde-3-phosphate* (G3P), and finally, the regeneration of RuBP. Keep track of the number of carbon atoms as you follow the process in Figure 7-6.

1. *Carbon fixation*. Acquiring carbon dioxide and incorporating it into a larger organic molecule is called **carbon fixation**. The C_3 cycle begins (and ends) with RuBP, a five-carbon sugar. Enzymes combine six RuBP molecules with six CO_2 molecules to form six molecules of an extremely unstable six-carbon compound (not illustrated). This compound spontaneously reacts with water to form 12 three-carbon molecules of *phosphoglyceric acid* (PGA), whose three carbons give the C_3 cycle its name (step ① in Fig. 7-6).
2. *Synthesis of G3P*. In a series of enzyme-catalyzed reactions, energy donated by ATP and NADPH (generated during the light-dependent reactions) is used to convert PGA to G3P (step ②).
3. *Regeneration of RuBP*. Through a series of reactions requiring ATP energy, 10 of the 12 molecules of G3P (10×3 carbons) regenerate the six molecules of RuBP (6×5 carbons; step ③) used at the start of carbon fixation (step ④). The remaining two molecules of G3P will be used to synthesize glucose (step ③).

Carbon Fixed During the C_3 Cycle Is Used to Synthesize Glucose

Because the C_3 cycle starts with RuBP, adds carbon from CO_2, and ends one "cycle" with RuBP, there is carbon left over from the captured CO_2. Using the simplest "carbon accounting" numbers shown in Figure 7-6, if you start and end one passage through the cycle with six molecules of RuBP, two molecules of G3P are left over. In light-independent reactions that occur outside the C_3 cycle, these two G3P molecules (three carbons each) are combined to form one molecule of glucose (six carbons). Later, glucose molecules may be broken down during cellular respiration. Most are modified and linked to form sucrose (table sugar; a storage molecule), or linked together in chains to form starch (a storage molecule) or cellulose (a major component of cell walls). Most of the synthesis of glucose from G3P and the subsequent synthesis of more-complex molecules from glucose occurs outside the chloroplast.

Summing Up
Light-Independent Reactions

For the synthesis of one molecule of glucose through the C_3 cycle, six molecules of CO_2 are captured by six molecules of RuBP. A series of reactions driven by energy from ATP and NADPH (obtained from the light-dependent reactions) produces 12 molecules of G3P. Two G3P molecules join to become one molecule of glucose. ATP energy is used to regenerate six RuBP molecules from the remaining 10 G3P molecules. Light-independent reactions generate glucose and depleted energy carriers (ADP and $NADP^+$) that will be recharged during light-dependent reactions.

7.4 What Is the Relationship Between Light-Dependent and Light-Independent Reactions?

Figure 7-7 illustrates the relationship between light-dependent and light-independent reactions, placing each in its appropriate location within the chloroplast. Both

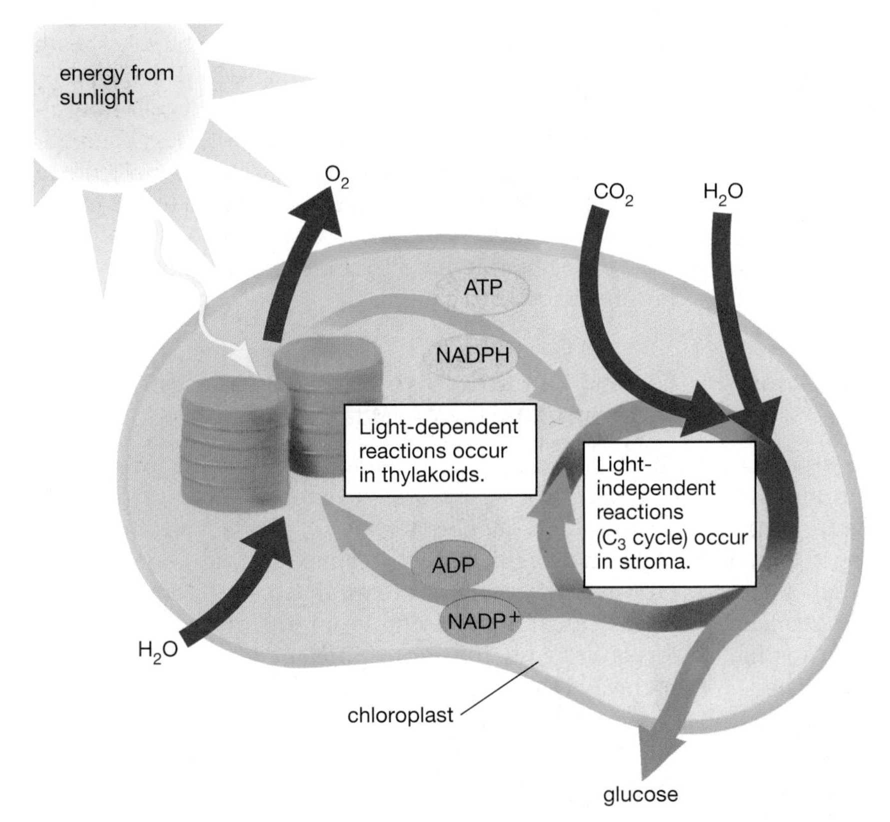

FIGURE 7-7 A summary diagram of photosynthesis
The light-dependent reactions in the thylakoids convert the energy of sunlight into the chemical energy of ATP and NADPH. Part of the sunlight energy is also used to split H_2O, forming O_2. In the stroma, the light-independent reactions use the energy of ATP and NADPH to convert CO_2 and H_2O to glucose. The depleted carriers, ADP and $NADP^+$, return to the thylakoids to be recharged by the light-dependent reactions.
QUESTION Could a plant survive in an oxygen-free atmosphere?

Figure 7-7 and the in-text figure on page 118 stress the interdependence of these two sets of reactions in the overall process of photosynthesis. Simply put, the "photo" part of photosynthesis refers to the capture of light energy by the light-dependent reactions. The "synthesis" part of photosynthesis refers to the synthesis of glucose that occurs during the light-independent reactions, using the energy captured by the light-dependent reactions. Stated in more detail, the light-dependent reactions in the membranes of the thylakoids use light energy to "charge up" the energy-carrier molecules ADP and $NADP^+$ to form ATP and NADPH. These energized carriers move to the stroma, where their energy is used to drive glucose synthesis by means of the light-independent reactions, which are catalyzed by enzymes dissolved in the stroma. The depleted carriers, ADP and $NADP^+$, then return to the light-dependent reactions to be recharged into ATP and NADPH.

7.5 Water, CO_2, and the C_4 Pathway

Photosynthesis requires light and carbon dioxide. Therefore, an ideal leaf should have a large surface area to intercept lots of sunlight and be very porous to allow lots of CO_2 to enter the leaf from the air. For land plants, however, being porous to air also allows water to evaporate easily from the leaf. Loss of water from leaves is a prime cause of stress to land plants, and may even be fatal.

Many plants have evolved leaves with features that represent a compromise between obtaining adequate light energy and CO_2 and reducing water loss. These leaves have a large surface area for intercepting light, a waterproof coating to reduce evaporation, and leaves with adjustable pores (the stomata) through which air carrying CO_2 can diffuse (see Fig. 7-2). When water supplies are adequate, the stomata open, letting in CO_2. If the plant is in danger of drying out, the stomata close. Closing the stomata reduces evaporation but has two disadvantages: it reduces CO_2 intake, and it also limits the ability of the leaf to release O_2, a by-product of photosynthesis.

When Stomata Are Closed to Conserve Water, Wasteful Photorespiration Occurs

What happens to carbon fixation when the stomata close, CO_2 levels drop, and O_2 levels rise? Unfortunately, the enzyme that catalyzes the reaction of RuBP with CO_2 is not very selective; it can cause either CO_2 or O_2 to combine with RuBP (Fig. 7-8a). When O_2 (rather than CO_2) is combined with RuBP, a wasteful process called **photorespiration** occurs. During photorespiration (as in cellular respiration), O_2 is used up and CO_2 is generated. But unlike cellular respiration, photorespiration does not produce any useful cellular energy, and it prevents the light-independent reactions from synthesizing glucose. Thus, photorespiration undermines the plant's ability to fix carbon.

Some photorespiration occurs all the time, even under the best of conditions. During hot, dry weather, the stomata seldom open; CO_2 from the air can't get in, and the O_2 generated by photosynthesis can't get out. In this situation, with CO_2 levels low and O_2 levels high, photorespiration dominates (see Fig. 7-8a). Plants, especially seedlings, may die during hot, dry weather because they are unable to capture enough energy in glucose to meet their metabolic needs.

C_4 Plants Reduce Photorespiration by Means of a Two-Stage Carbon-Fixation Process

Some plants, described as *C_4 plants*, have evolved a way to reduce photorespiration and boost photosynthesis during dry weather. In the leaves of C_3 plants, almost all of the chloroplasts reside in the mesophyll cells; their *bundle sheath cells* (which surround the vascular bundles in both C_3 and C_4 plants) lack chloroplasts. In C_4 plants (such as corn and crabgrass) that thrive in relatively hot and dry conditions, both mesophyll cells and bundle sheath cells contain chloroplasts (Fig. 7-8b). These C_4 plants use a two-stage carbon-fixation pathway known as the **C_4 pathway**.

In C_4 plants, the mesophyll cells contain a three-carbon molecule called *phosphoenolpyruvate* (PEP) instead of RuBP. The CO_2 reacts with PEP to form four-carbon intermediate molecules (for which C_4 plants are named). The reaction between CO_2 and PEP is catalyzed by an enzyme that is highly specific for CO_2 and is not hindered by high O_2 concentrations. A 4-carbon molecule is used to shuttle carbon from mesophyll cells into bundle sheath cells, where it breaks down, releasing CO_2. The high CO_2 concentration created in the bundle sheath cells now allows the regular C_3 cycle to proceed with less competition from oxygen. The remnant of the shuttle molecule (a 3-carbon molecule called pyruvate) returns to the mesophyll cells. Back in the mesophyll cells, ATP energy is used to regenerate the PEP molecule from pyruvate, allowing the cycle to continue.

C_3 and C_4 Plants Are Each Adapted to Different Environmental Conditions

Plants using the C_4 process to fix carbon are locked into this pathway, which uses up more energy to produce glucose than the C_3 pathway. C_4 plants have an advantage when light energy is abundant but water is not. However, if water is plentiful (so the stomata of C_3 plants can stay open and let in lots of CO_2) or if light levels are low, the more efficient C_3 carbon fixation pathway is advantageous to the plant.

Consequently, C_4 plants thrive in deserts and in hotter, drier regions of temperate climates, where light energy is plentiful but water is scarce. Plants using C_4

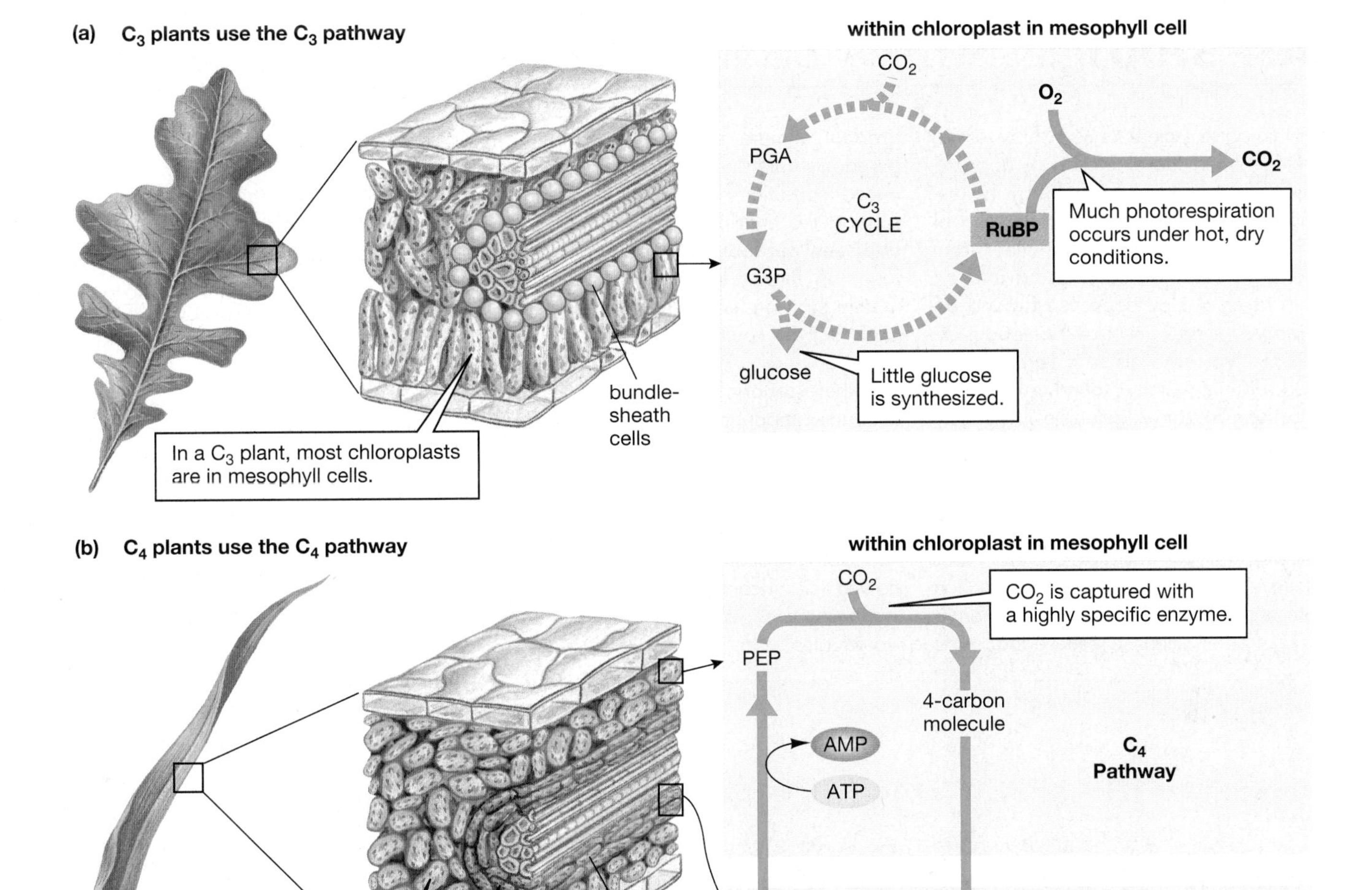

FIGURE 7-8 Comparison of C_3 and C_4 plants
(a) In C_3 plants, only the mesophyll cells carry out photosynthesis. All carbon fixation occurs by the C_3 pathway. With low CO_2 and high O_2 levels, photorespiration dominates in C_3 plants, because the enzyme that should catalyze the RuBP plus CO_2 reaction catalyzes the RuBP plus O_2 reaction instead. **(b)** In C_4 plants, both the mesophyll cells and bundle sheath cells contain chloroplasts and participate in photosynthesis. CO_2 is combined with PEP by a more selective enzyme, and the carbon is shuttled into bundle sheath cells by a four-carbon molecule, which releases CO_2 into the bundle sheath cells. Higher CO_2 levels allow efficient carbon fixation (with little photorespiration) in the C_3 pathway of the bundle sheath cells. Notice that the regeneration of PEP requires energy from ATP.
QUESTION Why do C_3 plants have an advantage over C_4 plants under conditions that are not hot and dry?

photosynthesis include corn, sugarcane, sorghum, some grasses (including crabgrass), and some types of thistles. The C_3 plants, which include most trees, grains such as wheat, oats, and rice, and grasses such as Kentucky bluegrass, have the advantage in cool, wet, cloudy climates, because the C_3 pathway is more energy-efficient. These differing adaptations explain why your spring lawn of lush Kentucky bluegrass (a C_3 plant) may be taken over by spiky crabgrass (a C_4 plant) during a long, hot, dry summer.

CASE STUDY REVISITED Did the Dinosaurs Die from Lack of Sunlight?

Paleontologists (scientists who study fossils) have identified the extinction of approximately 70% of all living species by the disappearance of their fossils at the end of the Cretaceous period. In sites from around the globe, researchers have found a thin layer of clay deposited around 65 million years ago; the clay has about 30 times the typical levels of a rare element called *iridium*, which is found in high concentrations in some meteorites. The clay also contains soot, such as would have been deposited in the aftermath of massive fires. Did a meteorite wipe out the dinosaurs? Many scientists believe it did. Certainly the evidence of an enormous meteorite impact, dated to 65 million years ago, is clear on the Yucatan Peninsula. But other scientists believe that more gradual climate changes, possibly triggered by intense volcanic activity, produced conditions that would no longer support the enormous reptiles. Volcanoes also spew out soot and ash, and iridium is found in higher levels in Earth's molten mantle than on its surface, so furious volcanic activity could also explain the iridium layer.

Either scenario would significantly reduce the amount of sunlight and immediately impact the rate of photosynthesis. Large herbivores (plant-eaters), such as *Triceratops*, which might have needed to consume hundreds of pounds of vegetation daily, would suffer if plant growth slowed significantly. Predators such as *Tyrannosaurus*, which fed on herbivores, would also suffer. In the Cretaceous as now, sunlight captured by photosynthesis powers all the dominant forms of life on Earth—interrupting this vital flow of energy would be catastrophic.

Consider This: Design an experiment to test the effects of light-blocking atmospheric pollutants (such as dust, smoke, and ash) on photosynthesis. Assume that you have lots of time, greenhouse space, equipment, money, and student helpers. What might you measure to determine the amount of photosynthesis that occurred under different light conditions? How would you determine which wavelengths of light are filtered out by these different types of pollutants? How would you test which adaptations in plants would help them survive under low-light conditions?

Links to Life: You Owe Your Life to Plants

As you study the details of photosynthesis, it is easy to become bogged down in the complexity of it all, losing sight of why photosynthesis is worth studying. The bottom line is that, without photosynthesis, you wouldn't be here to be perplexed by it—and neither would any other living thing that you encounter in a typical day. Over two billion years ago, when, in the words of poet Robinson Jeffers, the first bacteria "invented chlorophyll and ate sunlight," they sparked a revolution in the evolution of life on Earth. Capturing solar energy and using water as a source of electrons liberated oxygen into the primordial atmosphere for the first time. For many nonphotosynthetic organisms, this was a disaster. Oxygen is a highly reactive molecule that readily combines with and destroys biological molecules. The single-celled organisms that first encountered an oxygen-containing atmosphere had two choices: hide or evolve protective mechanisms. Descendants of those bacteria that hid from oxygen in the primordial ooze still survive today, and oxygen is still deadly to them. The others evolved cellular machinery to harness the reactive power of oxygen, using it to derive more energy from food molecules such as the glucose produced during photosynthesis. These efficient, oxygen-loving organisms quickly dominated the Earth and gradually evolved into the myriad creatures that inhabit it today, most of which would quickly die if deprived of oxygen.

Not only do we rely on the oxygen produced by photosynthesis, but all the energy in the food we eat originates in plants, which captured it from sunlight. Even if you revel in a diet of double bacon cheeseburgers and fried chicken, the energy stored in these animal fats and proteins ultimately came from the animals' food—plants. Even if you eat a carnivorous tuna fish, you can still trace the food chain (and thus the energy) that supported the tuna back to photosynthetic organisms in the ocean. Thus, photosynthesis gives us both our food and the oxygen that we need to "burn" it. Have you thanked a plant today?

Summary of Key Concepts

7.1 What Is Photosynthesis?

Photosynthesis captures the energy of sunlight and uses it to convert the inorganic molecules of carbon dioxide and water into high-energy organic molecules such as glucose. In plants, photosynthesis takes place in the chloroplasts, using two major reaction sequences: the light-dependent reactions and the light-independent reactions.

7.2 Light-Dependent Reactions: How Is Light Energy Converted to Chemical Energy?

The light-dependent reactions occur in the thylakoids. Light excites electrons in chlorophyll molecules and transfers the energetic electrons to electron transport chains. The energy of these electrons drives three processes:

- *Photosystem II generates ATP.* Some of the energy from the electrons is used to pump hydrogen ions into the thylakoids. The hydrogen ion concentration is therefore higher inside the thylakoids than in the stroma outside. Hydrogen ions move down this concentration gradient through ATP-synthesizing enzymes in the thylakoid membranes, providing the energy to drive ATP synthesis.
- *Photosystem I generates NADPH.* Some of the energy, in the form of energetic electrons, is added to electron-carrier molecules of $NADP^+$ to make the highly energetic carrier NADPH.
- *Splitting water maintains the flow of electrons through the photosystems.* Some of the energy is used to split water, generating electrons, hydrogen ions, and oxygen.

7.3 Light-Independent Reactions: How Is Chemical Energy Stored in Glucose Molecules?

In the stroma of the chloroplasts, both ATP and NADPH provide the energy that drives the synthesis of glucose from CO_2 and H_2O. The light-independent reactions begin with a cycle of chemical reactions called the Calvin-Benson, or C_3, cycle. The C_3 cycle has three major parts:

1. *Carbon fixation.* Carbon dioxide and water combine with ribulose bisphosphate (RuBP) to form phosphoglyceric acid (PGA).
2. *Synthesis of G3P.* PGA is converted to glyceraldehyde-3-phosphate (G3P), using energy from ATP and NADPH. G3P may be used to synthesize organic molecules, such as glucose.
3. *Regeneration of RuBP.* Ten molecules of G3P are used to regenerate six molecules of RuBP, again using ATP energy.

Light-independent reactions continue with the synthesis of glucose and other carbohydrates including sucrose, starch, and cellulose. These reactions occur primarily outside of the chloroplast.

7.4 What Is the Relationship Between Light-Dependent and Light-Independent Reactions?

The light-dependent reactions produce the energy carrier ATP and the electron carrier NADPH. Energy from these carriers is used in the synthesis of organic molecules during the light-independent reactions. The depleted carriers, ADP and $NADP^+$, return to the light-dependent reactions for recharging.

7.5 Water, CO_2, and the C_4 Pathway

The enzyme that catalyzes the reaction between RuBP and CO_2 may also catalyze a reaction, called photorespiration, between RuBP and O_2. If CO_2 concentrations drop too low or if O_2 concentrations rise too high, wasteful photorespiration, which prevents carbon fixation and does not generate ATP, may exceed carbon fixation. C4 plants have evolved an additional step for carbon fixation that minimizes photorespiration. In the mesophyll cells of these C_4 plants, CO_2 combines with phosphoenolpyruvic acid (PEP) to form a four-carbon molecule, which is modified and transported into adjacent bundle sheath cells, where it releases CO_2, thereby maintaining a high CO_2 concentration in those cells. This CO_2 is then fixed using the C_3 cycle.

Key Terms

C_3 cycle *p. 122*
C_4 pathway *p. 124*
Calvin-Benson cycle *p. 122*
carbon fixation *p. 123*
carotenoids *p. 119*
chemiosmosis *p. 120*
chlorophyll *p. 119*
electron transport chain *p. 119*
grana *p. 117*
light-dependent reactions *p. 118*
light-independent reactions *p. 118*
photon *p. 118*
photorespiration *p. 124*
photosynthesis *p. 116*
photosystems *p. 119*
reaction center *p. 119*
stomata *p. 116*
stroma *p. 117*
thylakoid *p. 117*

Thinking Through the Concepts

To take a multiple-choice quiz with feedback on the contents of this chapter, visit http://www.prenhall.com/audesirk7. *Log in to the Web site selected by your instructor and navigate to the Self Test section for this chapter.*

? Review Questions

1. Write the overall equation for photosynthesis. Does the overall equation differ between C_3 and C_4 plants?
2. Draw a diagram of a chloroplast, and label it. Explain specifically how chloroplast structure is related to its function.
3. Briefly describe the light-dependent and light-independent reactions. In what part of the chloroplast does each occur?
4. What is the difference between carbon fixation in C_3 and in C_4 plants? Under what conditions does each mechanism of carbon fixation work most effectively?
5. Describe the process of chemiosmosis in chloroplasts, tracing the flow of energy from sunlight to ATP.

Applying the Concepts

1. Many lawns and golf courses are planted with bluegrass, a C_3 plant. In the spring, the bluegrass grows luxuriously. In the summer, crabgrass, a weed and a C_4 plant, often appears and spreads rapidly. Explain this sequence of events, given the normal weather conditions of spring and summer and the characteristics of C_3 versus C_4 plants.
2. Suppose an experiment is performed in which plant I is supplied with normal carbon dioxide but with water that contains radioactive oxygen atoms. Plant II is supplied with normal water but with carbon dioxide that contains radioactive oxygen atoms. Each plant is allowed to perform photosynthesis, and the oxygen gas and sugars produced are tested for radioactivity. Which plant would you expect to produce radioactive sugars, and which plant would you expect to produce radioactive oxygen gas? Why?
3. You continuously monitor the photosynthetic oxygen production from the leaf of a plant illuminated by white light. Explain what will happen (and why) if you place (a) red, (b) blue, and (c) green filters between the light source and the leaf.
4. A plant is placed in a CO_2-free atmosphere in bright light. Will the light-dependent reactions continue to generate ATP and NADPH indefinitely? Explain how you reached your conclusion.
5. You are called before the Ways and Means Committee of the House of Representatives to explain why the U.S. Department of Agriculture should continue to fund photosynthesis research. How would you justify the expense of producing, by genetic engineering, the enzyme that catalyzes the reaction of RuBP with CO_2 and prevents RuBP from reacting with oxygen as well as CO_2? What are the potential applied benefits of this research?

For More Information

Bazzazz, F. A., and Fajer, E. D. "Plant Life in a CO_2-Rich World." *Scientific American*, January 1992. Burning fossil fuels is increasing CO_2 levels in the atmosphere (see Chapter 41). This increase could tip the balance between C_3 and C_4 plants.

Govindjee, and Coleman, W. J. "How Plants Make Oxygen." *Scientific American*, February 1990. The generation of oxygen during photosynthesis is just beginning to be understood.

Grodzinski, B. "Plant Nutrition and Growth Regulation by CO_2 Enrichment." *BioScience*, 1992. How higher CO_2 levels influence plant metabolism.

Hall, D. O., and Rao, K. K. *Photosynthesis*. 5th ed. New York: Cambridge University Press, 1994. An excellent short book recommended to any student interested in finding out more about photosynthesis.

Hinkle, P. C., and McCarthy, R. E. "How Cells Make ATP." *Scientific American*, March 1978. A good explanation of chemiosmosis, which is a difficult concept for many students.

Monastersky, R. "Children of the C_4 World." *Science News*, January 3, 1998. What role did a shift in global vegetation toward C_4 photosynthesis play in the evolution of humans?

Mooney, H. A., Drake, B. G., Luxmoore, R. J., Oechel, W. C., and Pitelka, L. F. "Predicting Ecosystems' Responses to Elevated CO_2 Concentrations." *BioScience*, 1994. What effects will CO_2 enrichment of the atmosphere due to human activities have on ecosystems?

Zimmer, C. "The Processing Plant." *Discover*, September 1995. Describes organisms inhabiting the watery digestive chamber of the pitcher plant, which is both photosynthetic and carnivorous.

Media Activities

To access a Media Activity visit http://www.prenhall.com/audesirk7. *Log in to the Web site selected by your instructor, navigate to this chapter, and select the appropriate Media Activity number.*

7.1 Properties of Light

Estimated time: 5 minutes

Explore the properties of light and how light is captured by chloroplasts to power photosynthesis.

7.2 Photosynthesis

Estimated time: 10 minutes

Explore the process of photosynthesis in detail.

7.3 Chemiosmosis

Estimated time: 5 minutes

Review an elegant test of the chemiosmotic hypothesis, which successfully demonstrated that a proton gradient can lead to ATP production.

7.4 Web Investigation: Did the Dinosaurs Die from Lack of Sunlight?

Estimated time: 10 minutes

This exercise will examine the causes of some modern extinctions in the hope that they might shed some light on the other mass extinctions, including the one that ended the age of dinosaurs.

CHAPTER

8 Harvesting Energy: Glycolysis and Cellular Respiration

With wings beating 60 times per second, the ruby-throated hummingbird has a metabolic rate 50 times that of a human. The muscles of its wings are packed with mitochondria, which supply the ATP needed to meet the bird's energy demands.

AT A GLANCE

CASE STUDY The Flight of the Hummingbird

When a broadtailed hummingbird crashed into the glass door of Susan Heriford's home in Colorado, the unfortunate event had a happy ending. She picked up the bird and (unaware that hummingbirds can't walk) placed it in the grass. When it recovered enough to fly, it flew to her shoulder. The bird adopted the Heriford family, including their pet dog and rabbit, as companions. It perched on Susan's hand to eat, followed her family on hikes, and slept on a branch outside their bedroom window. In September, the bird migrated south into Mexico, and Susan wondered if she would ever see it again. The following summer, "Buddy" returned, as friendly as ever, and to the Herifords' delight, fathered a new generation of hummingbirds.

Hummingbirds have extraordinary energy demands. Their wings are a blur, beating 60 times per second, "burning" calories at a rate 50 times that of an average human. Hummingbirds must eat frequently, sipping the sugar-rich nectar of flowers for energy and munching on small flying insects for protein. The rufous hummingbird makes the longest migration, from the Alaskan coast to central Mexico. As it passes through Colorado, it brazenly chases broadtails like Buddy from their favorite flowers and feeders. For some hummers, migration is a very dangerous undertaking. For example, ruby-throated hummingbirds fly continuously over 620 miles of open sea, traveling across the Gulf of Mexico from the southeastern United States to Mexico and Central America.

In Chapter 7, we described how plants trap the energy of sunlight and store it in sugar, some of which becomes nectar that feeds hummingbirds. How does a hummingbird extract the energy and convert it to muscular movement that powers flight? How does the ruby-throated hummingbird store enough energy from the sugar it eats to make it across the Gulf of Mexico?

8.1 How Is Glucose Metabolized?

Most cells can metabolize a variety of organic molecules to produce ATP. We will focus on the metabolism of glucose for three reasons. First, virtually all cells metabolize glucose for energy at least part of the time. Some, such as the nerve cells in your brain, rely almost entirely on glucose as a source of energy. Second, glucose metabolism is less complex than the metabolism of most other organic molecules. Finally, when using other organic molecules as energy sources, cells usually first convert the molecules to glucose or other compounds that enter the pathways of glucose metabolism (see "Health Watch: Why Can You Get Fat by Eating Sugar?").

As you learned in Chapter 7, photosynthetic organisms capture and store the energy of sunlight in glucose. During glucose breakdown, that solar energy is released and used to make ATP. The chemical equations for glucose formation by photosynthesis and for the complete metabolism of glucose back to CO_2 and H_2O (the original reactants in photosynthesis) are almost perfectly symmetrical:

Photosynthesis:

$$6\ CO_2 + 6\ H_2O + \text{sunlight} + \text{energy} \rightarrow C_6H_{12}O_6 + 6\ O_2$$

Complete Glucose Metabolism:

$$C_6H_{12}O_6 + 6\ O_2 \rightarrow 6\ CO_2 + 6\ H_2O + \text{chemical and heat energies}$$

This symmetry might lead you to assume that a cell can convert all of the chemical energy contained in a glucose molecule to high-energy bonds of ATP. Unfortunately, according to the second law of thermodynamics, "you can't break even." In other words, the conversion of energy into different forms always results in the decrease of the amount of concentrated, useful energy. In fact, over half of the energy listed on the right-hand side of the glucose metabolism equation is heat energy, not the chemical energy of ATP. Nevertheless, a cell can extract and store a great deal of chemical energy, in the form of ATP, from glucose if the glucose molecule is completely broken down to CO_2 and H_2O. Cells are actually extremely efficient at capturing chemical energy, trapping about 40% of the energy in glucose. If cells were as inefficient as our electric motors or gasoline engines (25% or less), animals would need to eat voraciously to remain active. Under these circumstances, a hummingbird would almost certainly be unable to get enough food to complete its long migration.

Figure 8-1 summarizes the major steps of glucose metabolism in eukaryotic cells. The first stage, *glycolysis*, does not require oxygen and proceeds in exactly the same way under both aerobic (with oxygen) and anaerobic (without oxygen) conditions. Glycolysis splits apart a single glucose molecule (a six-carbon sugar) into two three-carbon molecules of *pyruvate*. This splitting releases a small fraction of the chemical energy stored in the glucose, some of which is used to generate a net output of two ATP molecules. The presence of oxygen becomes an issue only in the processes that follow glycolysis. Under anaerobic conditions, the pyruvate is usually converted by fermentation into lactate or ethanol. Fermentation does not produce more ATP energy. Both glycolysis and fermentation occur in the fluid portion of the cytoplasm.

The pyruvate produced by glycolysis may also enter the mitochondria. There, if oxygen is available, cellular respiration uses oxygen to break pyruvate down completely to carbon dioxide and water, generating an additional 34 or 36 ATP molecules (the amount differs from cell to cell). The extra ATP produced by cellular respiration is so important to most organisms that if anything interferes with its production, such as cyanide poison or lack of oxygen, death occurs quickly.

8.2 How Is the Energy in Glucose Captured During Glycolysis?

The initial reactions that break down glucose without the use of oxygen are collectively called **glycolysis** (in Greek, "to break apart a sweet"). Glycolysis is believed to be one of the most ancient of all biochemical pathways because it is used by every living creature on the planet. This sequence of reactions occurs in the fluid portion of the cytoplasm and splits one molecule of glucose into two molecules of pyruvate. Glycolysis produces relatively little energy: only two molecules of ATP and two molecules of the electron carrier NADH. But without it, most forms of life would rapidly be extinguished. Reduced to its essentials, glycolysis consists of two major parts (each with several steps): ① glucose activation and ② energy harvest (Fig. 8-2).

Glycolysis Breaks Down Glucose to Pyruvate, Releasing Chemical Energy

Before glucose is broken down to release its energy, it must be activated—a process that actually uses up energy. During glucose activation, a molecule of glucose undergoes two enzyme-catalyzed reactions, each of which uses ATP energy (Fig. 8-2). These reactions convert a relatively stable glucose molecule into a highly unstable "activated" molecule of *fructose bisphosphate*. Fructose is a sugar molecule similar to glucose; *bisphosphate* refers to the two phosphate groups acquired from the ATP molecules. The considerable energy from ATP that is stored in the bonds linking the phosphate groups to the sugar make fructose bisphosphate an unstable molecule. Although forming fructose bisphosphate costs the cell two ATP molecules, this initial consumption of energy is necessary to produce greater energy returns later.

In the energy-harvesting steps, fructose bisphosphate splits apart into two three-carbon molecules of glyceraldehyde 3-phosphate (G3P, see Fig. 8-2; in Chapter 7, we encountered G3P in the C_3 cycle of photosynthesis).

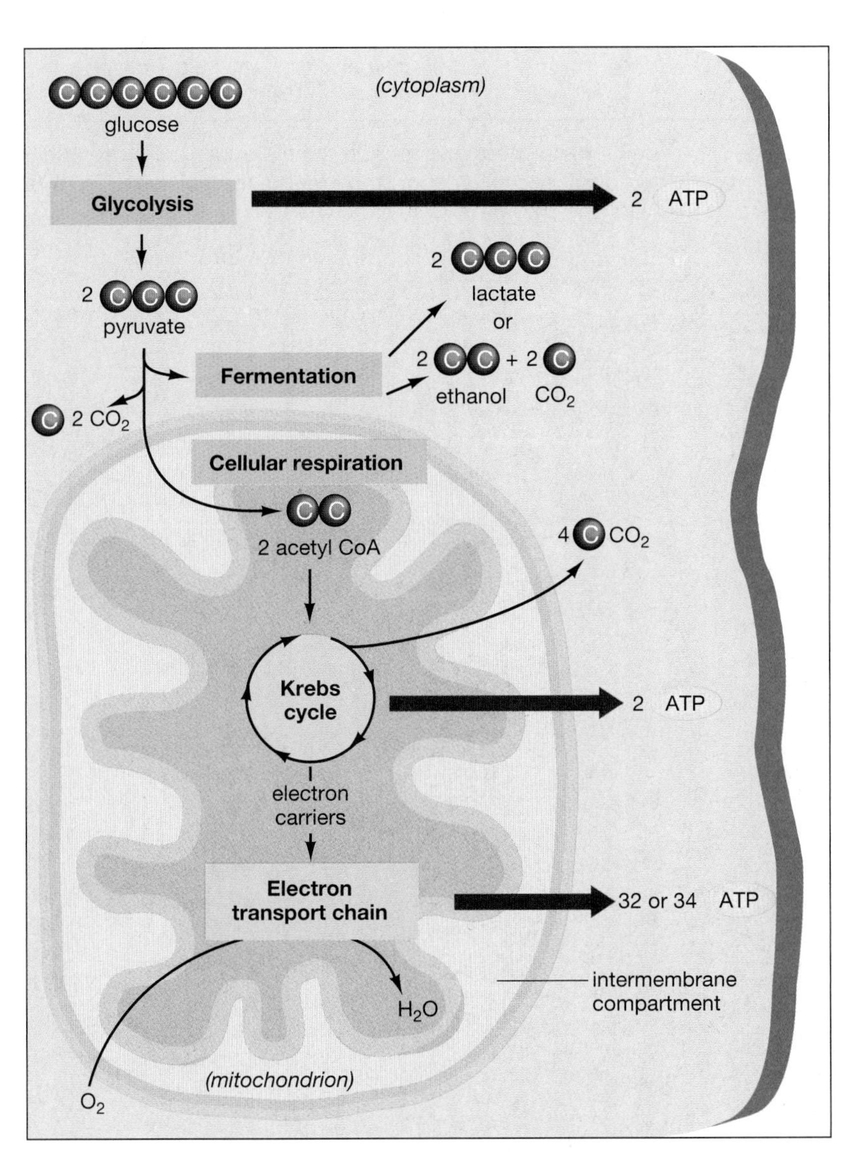

FIGURE 8-1 A summary of glucose metabolism Refer to this diagram as we progress through the reactions of glycolysis (in the fluid portion of the cytoplasm) and cellular respiration (in the mitochondria). The breakdown of glucose occurs in stages, with energy captured in ATP along the way. Most ATP is produced in the mitochondria.

Media Activity 8.1 Glucose Metabolism

Each G3P molecule, which retains a phosphate with its high energy bond, then goes through a series of reactions that convert it to pyruvate. During these reactions, two ATP are generated for each G3P, for a total of four ATPs. But because two ATPs were used up to activate the glucose molecule in the first place, there is a net gain of only two ATPs per glucose molecule. At another step along the way from G3P to pyruvate, two high-energy electrons and a hydrogen ion are added to the "empty" electron carrier NAD^+ to make the "energized" carrier molecule NADH (this is similar to the electron carrier $NADP^+$ used in photosynthesis). Two G3P molecules

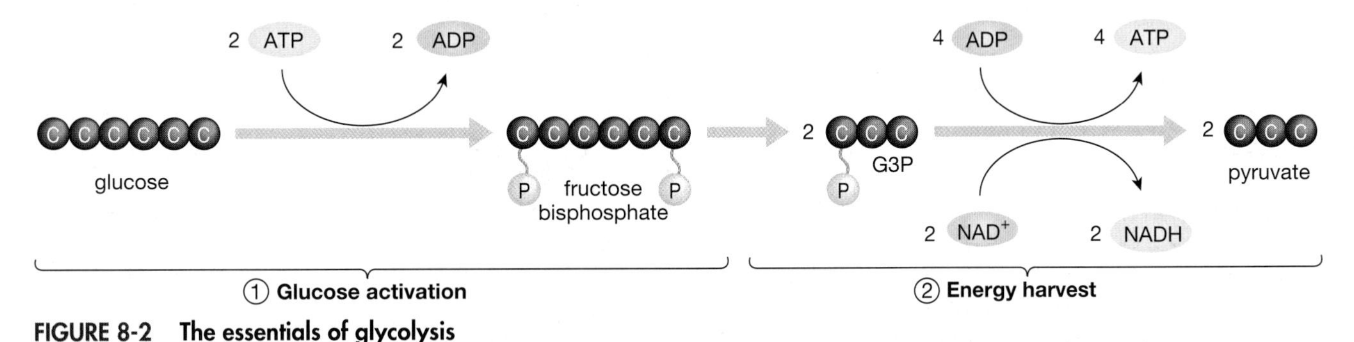

FIGURE 8-2 The essentials of glycolysis
① Glucose activation: The energy of two ATP molecules is used to convert glucose to the highly reactive fructose bisphosphate, which splits into two reactive molecules of G3P. ② Energy harvest: The two G3P molecules undergo a series of reactions that generate four ATP and two NADH molecules. Thus, glycolysis results in a net production of two ATP and two NADH molecules per glucose molecule.

A CLOSER LOOK Glycolysis

Glycolysis is a series of enzyme-catalyzed reactions that break down a single molecule of glucose into two molecules of pyruvate. To help you follow the reactions, we show only the "carbon skeletons" of glucose and the molecules produced during glycolysis. Each blue arrow represents a reaction catalyzed by at least one enzyme.

① A glucose molecule is energized by the addition of a high-energy phosphate from ATP.

② The molecule is slightly rearranged, forming fructose.

③ A second phosphate is added from another ATP.

④ The resulting molecule, fructose-1,6-bisphosphate, is split into two three-carbon molecules, one DHAP (dihydroxacetone phosphate) and one G3P. Each has one phosphate attached.

⑤ DHAP rearranges into G3P. From now on, there are two molecules of G3P going through the identical reactions.

⑥ Each G3P undergoes two almost-simultaneous reactions. Two electrons and a hydrogen ion are donated to NAD^+ to make the energized carrier NADH, and an inorganic phosphate (P) is attached to the carbon skeleton with a high-energy bond. The resulting molecules of 1,3-bisphosphoglycerate have two high-energy phosphates.

⑦ One phosphate from each bisphosphoglycerate is transferred to ADP to form ATP, for a net of two ATPs. This transfer compensates for the initial two ATPs used in glucose activation.

⑧ After another rearrangement, the second phosphate from each phosphoenolpyruvate is transferred to ADP to form ATP, leaving pyruvate as the final product of glycolysis. There is a net profit of two ATPs from each glucose molecule.

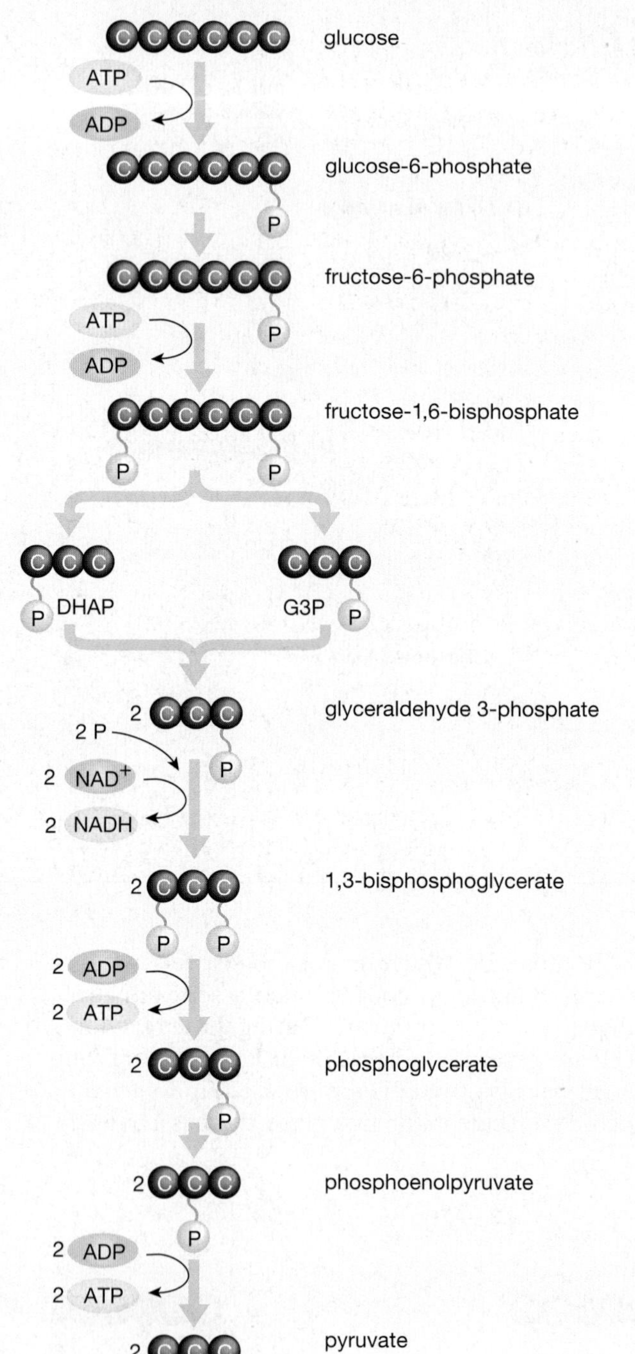

FIGURE E8-1 Glycolysis

(a)

(b)

FIGURE 8-3 Fermentation
(a) During a sprint, a runner's respiratory and circulatory systems cannot supply oxygen to her leg muscles fast enough to keep up with the demand for energy, so glycolysis must provide some of the ATP. In muscles, lactic acid fermentation follows glycolysis when oxygen is unavailable. **(b)** Bread rises as CO_2 is liberated by fermenting yeast, which converts glucose to ethanol. The dough on the left rose to the level on the right in a few hours. **QUESTION** Some species of bacteria use aerobic respiration and other species use anaerobic (fermenting) respiration. In an oxygen-rich environment, would either type be at a competitive advantage? What about in an oxygen-poor environment?

are produced per glucose molecule, so two NADH carrier molecules are formed when those G3P molecules are converted to pyruvate. For a complete description of glycolysis, see "A Closer Look: Glycolysis."

Summing Up
Glycolysis

Each molecule of glucose is broken down to two molecules of pyruvate. During these reactions, a net of two ATP molecules and two NADH electron carriers are formed.

Carrier molecules such as NAD^+ capture energy by accepting high-energy electrons. Carriers can transport these electrons to sites where their energy is used to form ATP. One major difference between anaerobic and aerobic glucose breakdown is the way in which these high-energy electrons are used. In the absence of oxygen, pyruvate accepts electrons from NADH, producing ethanol or lactate; this process is called **fermentation**. During cellular respiration, which occurs in the presence of oxygen, oxygen becomes the electron acceptor, allowing the pyruvate to be fully broken down and its energy harvested as ATP.

Some Cells Ferment Pyruvate to Form Lactate

The earliest forms of life appeared under anaerobic conditions (before the evolution of oxygen-liberating photosynthesis) and probably relied on glycolysis for energy production. Many microorganisms still thrive in places where oxygen is rare or absent, such as in the stomach and intestines of animals, deep in soil, or in bogs and marshes. Even some of our own body cells (and those of all animals) must cope without oxygen for brief periods of time. Under anaerobic conditions, NADH production is not used as a method of energy capture; it is actually a way of getting rid of the hydrogen ions and electrons produced during the breakdown of glucose to pyruvate. But this disposal method poses a problem for the cell, because NAD^+ is used up as it accepts electrons and hydrogen ions to become NADH. Without a way to regenerate NAD^+ and to dispose of the electrons and hydrogen ions, glycolysis would stop once the supply of NAD^+ was exhausted.

Fermentation solves this problem by enabling pyruvate to act as the final acceptor of electrons and hydrogen ions from NADH. Thus, NAD^+ is regenerated for use in further glycolysis. There are two main types of fermentation; one type converts pyruvate to lactate, and the other converts pyruvate to carbon dioxide and ethanol.

Fermentation to lactate occurs in your muscles when you exercise vigorously, such as when you race to class after you've overslept, or in the muscles of a runner sprinting through the finish line (Fig. 8-3a). You may hear lactate called "lactic acid"; lactate is

HEALTH WATCH Why Can You Get Fat by Eating Sugar?

As you know, humans do not live by glucose alone. Nor does the typical diet contain exactly the required amounts of each nutrient. Accordingly, the cells of the human body seethe with biochemical reactions, synthesizing one amino acid from another, making fats from carbohydrates, and channeling surplus organic molecules of all types into energy storage or release. Let's look at two examples of these metabolic transformations: the production of ATP from fats and proteins, and the synthesis of fats from sugars.

HOW ARE FATS AND PROTEINS METABOLIZED?

Even the leanest people have some fat in their bodies. During fasting or starvation, the body mobilizes these fat reserves for ATP synthesis; even the bare maintenance of life requires a continuous supply of ATP, and seeking out new food sources demands even more energy. Fat metabolism flows directly into the pathways of glucose metabolism.

Chapter 3 described the structure of a fat: three fatty acids connected to a glycerol backbone. In fat metabolism, the bonds between the fatty acids and glycerol are hydrolyzed (broken into subunits by the addition of water). The glycerol part of a fat, after activation by ATP, feeds directly into the middle of the glycolysis pathway (Fig. E8-2). The fatty acids are transported into the mitochondria, where enzymes in the inner membrane and matrix chop them up into acetyl groups. These groups attach to CoA to form acetyl CoA, which enters the Krebs cycle.

In individuals who are starving (a situation in which muscle protein is broken down to provide energy) or who are on a high-protein diet, amino acids can be used to produce energy. First, the amino acids are converted to pyruvate, acetyl CoA, or the compounds of the Krebs cycle. These molecules then proceed through the remaining stages of cellular respiration, yielding amounts of ATP that vary with their point of entry into the pathway.

HOW IS FAT SYNTHESIZED FROM SUGAR?

The body, in addition to having developed ways of coping with fasting or starvation, has also evolved strategies for coping with situations in which food intake exceeds current energy needs. The sugars and starches in corn flakes, candy bars, or the nectar of flowers can be converted into fats for energy storage. Complex sugars, such as starches and sucrose, are first hydrolyzed into their monosaccharide subunits (see Chapter 3). The monosaccharides are broken down to pyruvate and converted to acetyl CoA. If the cell needs ATP, the acetyl CoA will enter the Krebs cycle. If the cell has plenty of ATP, acetyl CoA will be used to make fatty acids by a series of reactions that are essentially the reverse of fatty-acid breakdown. Thus, the hummingbird can double its body weight by eating sugar and getting fat before it flies south. In humans, the liver synthesizes fatty acids, but fat storage is relegated to fat cells, with their all-too-familiar distribution in the body, particularly around the waist and hips. Acetyl CoA and other intermediate molecules of glucose breakdown can also be used in the synthesis of amino acids.

Energy use, fat storage, and nutrient intake are usually precisely balanced. Where the balance point lies, however, varies from person to person. Some people seem able to eat excessively without ever storing much fat; other people crave high-calorie foods even when they have an excess of fat stores. From an evolutionary perspective, overeating during times of easy food availability is highly adaptive behavior. During times of famine (which were common during our evolutionary history) heavier people are more likely to survive, while leaner individuals succumb to starvation. Only recently (from an evolutionary standpoint) have people in societies such as ours had continuous access to high-calorie food. Under these conditions, the drive to eat and the adaptation of storing excess food as fat leads to obesity, a growing health problem in the United States.

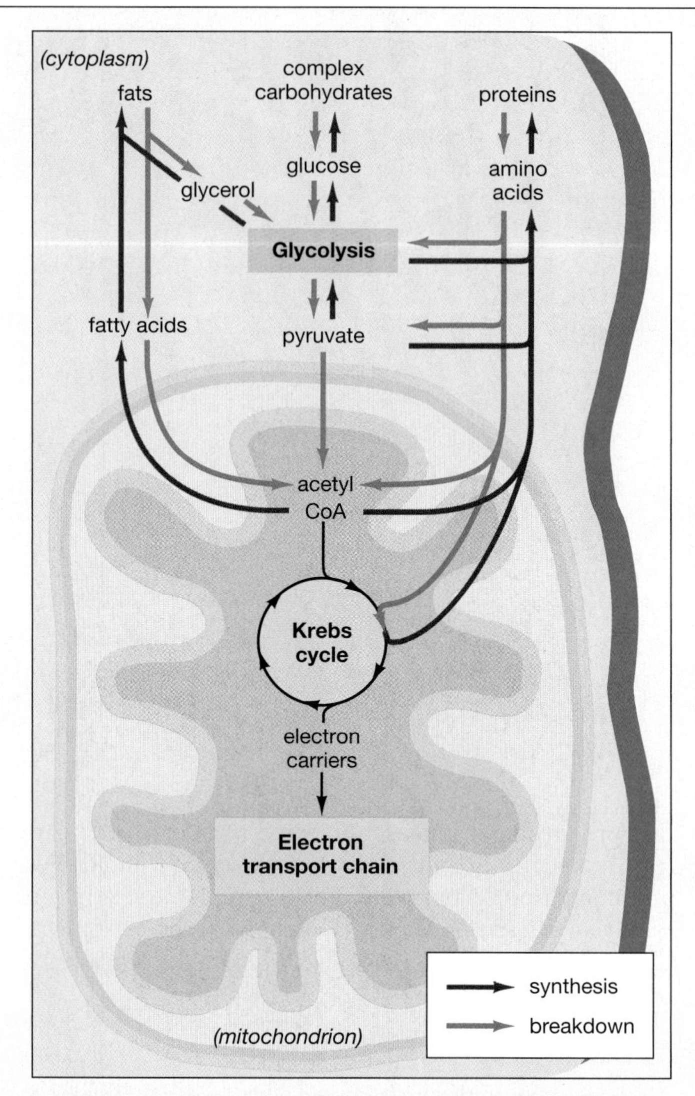

FIGURE E8-2 How various nutrients yield energy and can be interconverted
Metabolic pathways allow interconversion of fats, proteins, and carbohydrates via intermediate molecules formed along the same pathways that break down glucose. Blue arrows show breakdown of these substances to provide energy. Red arrows show that these molecules may also be synthesized when there is an excess of the intermediates.

the ionized form of lactic acid that is in solution in the cytoplasm. Even though working muscles need lots of ATP, and cellular respiration generates much more ATP than does glycolysis, cellular respiration is limited by the organism's ability to provide oxygen (by breathing, for example). While you exercise vigorously, you may not be able to get enough air into your lungs and enough oxygen into your blood to supply your muscles with sufficient oxygen to allow cellular respiration to meet all their energy needs. When deprived of adequate oxygen, your muscles do not immediately stop working. Instead, glycolysis continues for a short time, providing its meager two ATP molecules per glucose and generating both pyruvate and NADH. Then, to regenerate NAD^+, muscle cells ferment pyruvate molecules to lactate, using electrons and hydrogen ions from NADH:

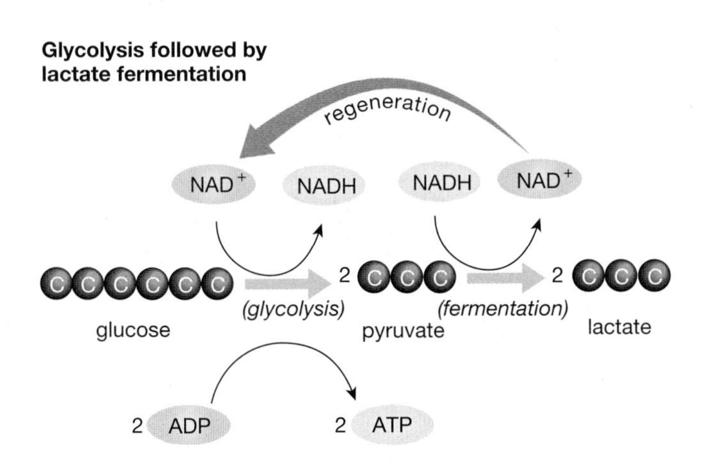

As you rest, breathing rapidly after your sprint, oxygen once more becomes available, and the lactate is converted back to pyruvate. Interestingly, this conversion occurs in the liver (to which the lactate is carried by the bloodstream) rather than in the muscle cells, because muscle cells lack the necessary enzymes. Some of this pyruvate is then broken down by cellular respiration into carbon dioxide and water, capturing additional energy.

Various microorganisms, including the bacteria that produce yogurt, sour cream, and cheese, also use lactate fermentation. As you may know, acids taste sour; thus, lactate (lactic acid) contributes to the distinctive tastes of these foods. Some microorganisms lack the enzymes for cellular respiration; these will ferment glucose even when oxygen is present, and some of them are actually poisoned by oxygen.

Other Cells Ferment Pyruvate to Alcohol

Many microorganisms use another type of fermentation to regenerate NAD^+ under anaerobic conditions: *alcoholic fermentation*. These organisms produce ethanol and CO_2 (rather than lactate) from pyruvate, using hydrogen ions and electrons from NADH:

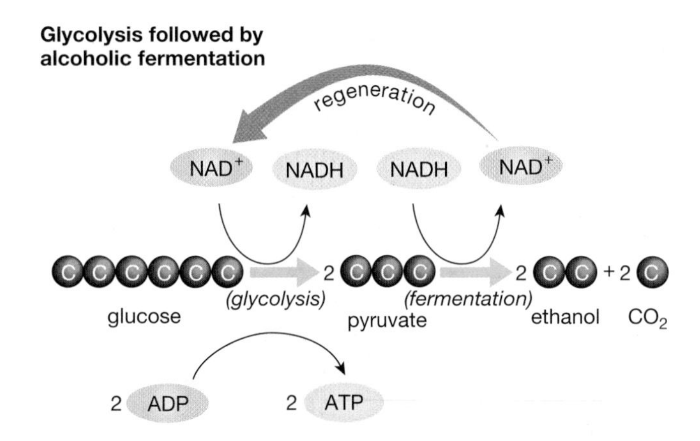

For thousands of years, people have taken advantage of this process to produce certain foods and beverages. Sparkling wines, such as champagne, are bottled while the yeasts are still alive and fermenting, trapping both the alcohol and the CO_2. When the cork is removed, the pressurized CO_2 is released, sometimes explosively. Baker's yeast in bread dough produces CO_2, making the bread rise; the alcohol generated by the yeast evaporates while the bread is baking (Fig. 8-3b). For more on alcoholic fermentation, see "Links to Life: A Jug of Wine, a Loaf of Bread . . ."

8.3 How Does Cellular Respiration Capture Additional Energy from Glucose?

Cellular respiration is a series of reactions, occurring under aerobic conditions, in which large amounts of ATP are produced. During cellular respiration, the pyruvate produced by glycolysis is broken down to carbon dioxide and water. The final reactions of cellular respiration require oxygen because oxygen acts as the final acceptor of electrons.

In eukaryotic cells, cellular respiration occurs in the mitochondria, which are sometimes called the "powerhouses of the cell." Recall from Chapter 5 that a mitochondrion has two membranes that produce two compartments: an inner compartment that is enclosed by the inner membrane and contains the fluid **matrix**, and an **intermembrane compartment** between the two membranes (see Fig. 8-4). The ATP produced during cellular respiration is generated by enzyme-catalyzed reactions in the matrix, by electron transport proteins in the inner membrane, and by the movement of hydrogen ions through ATP-synthesizing proteins in the inner membrane.

Figure 8-4 summarizes the main events of cellular respiration (the numbers in the figure refer to the steps below):

① The two molecules of pyruvate produced by glycolysis are transported across both mitochondrial membranes and into the matrix.

② Each pyruvate is split into CO_2 and a two-carbon acetyl group, which enters the *Krebs cycle* (discussed in the next section). The Krebs cycle releases the remaining carbons as CO_2, produces one ATP from each pyruvate, and donates energetic electrons to several electron-carrier molecules.

③ The electron carriers donate their energetic electrons to the electron transport chain of the inner membrane. There the energy of the electrons is used to transport H^+ from the matrix to the intermembrane compartment. At the end of the chain,

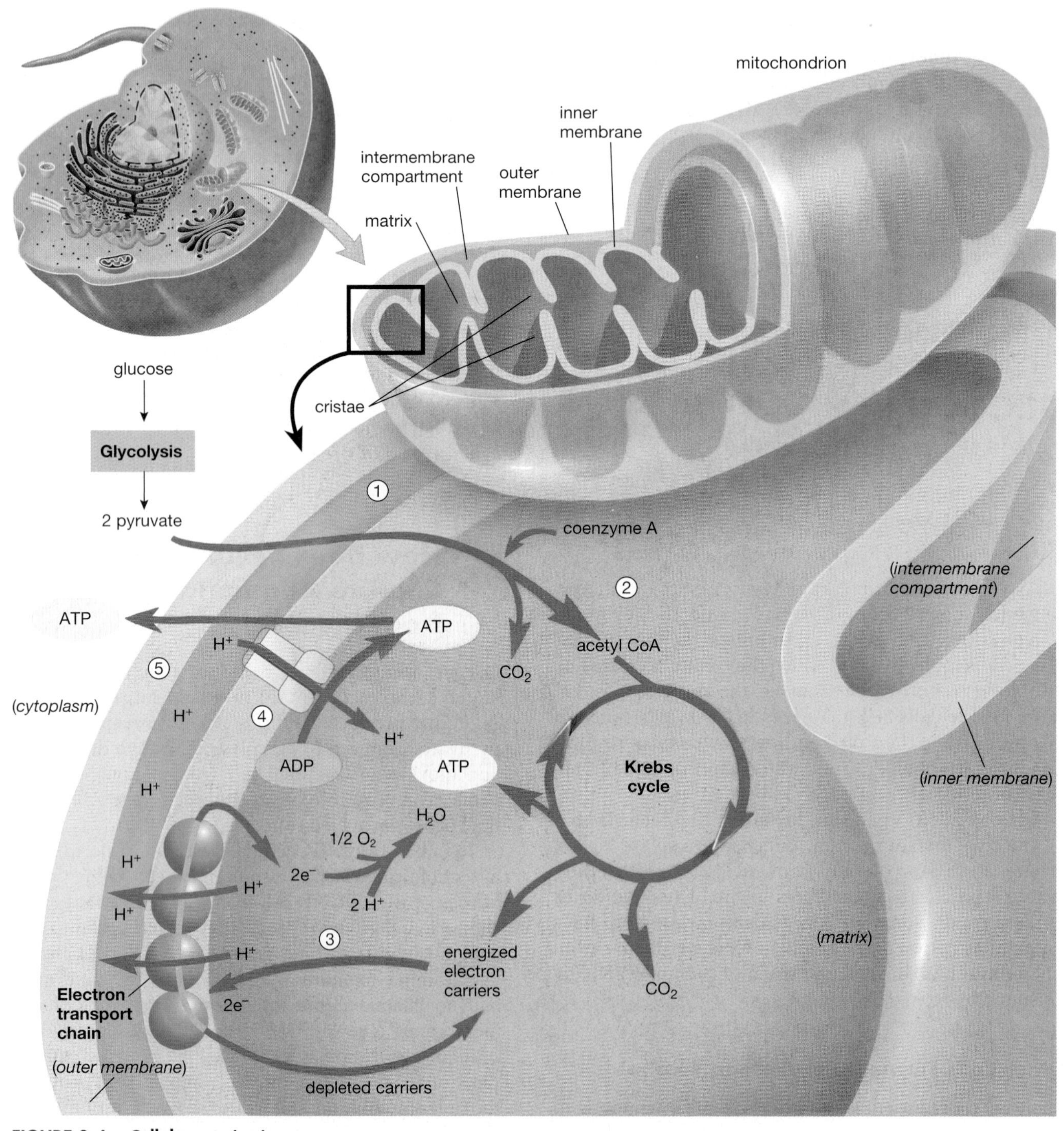

FIGURE 8-4 Cellular respiration
Cellular respiration occurs in mitochondria, whose structure accommodates the compartmentalized reactions that occur there. Note that the glycolysis that precedes cellular respiration occurs outside the mitochondrion in the cytoplasmic fluid.

Table 8-1 Summary of Glycolysis and Cellular Respiration of a Molecule of Glucose

Process	Location	Reactions	Electron Carriers Formed	ATP Yield (per glucose molecule)
Glycolysis	Fluid cytoplasm	Glucose broken down into two pyruvates	2 NADH	2 ATP
Cellular Respiration	Mitochondria			
Acetyl CoA formation	Matrix	Pyruvate combined with CoA to form acetyl CoA and CO_2	2 NADH	
Krebs cycle	Matrix	Acetyl group of acetyl CoA metabolized to two CO_2	6 NADH, 2 $FADH_2$	2 ATP
Electron transport	Inner membrane and intermembrane compartment	Energetic electrons from NADH and $FADH_2$ used to create a H^+ gradient, which is used to synthesize ATP		32 or 34 ATP*

* Glycolysis produces two NADH molecules in the fluid portion of the cytoplasm. The electrons from these two NADH molecules must be transported into the matrix before they can enter the electron transport chain. In most eukaryotic cells, the energy of 1 ATP molecule is used to transport the electrons from each NADH molecule into the matrix. Thus, the 2 "glycolytic NADH" molecules net only 2 ATPs, not the usual 3, during electron transport. The heart and liver cells of mammals, however, use a different transport chain, one that does not consume ATP. In these cells, the 2 NADH molecules produced during glycolysis net 3 ATPs apiece, just as the "mitochondrial NADH" molecules do.

the electrons combine with O_2 and H^+ to form H_2O. Depleted carriers are reused in the Krebs cycle.

④ In chemiosmosis, the hydrogen ion gradient created by the electron transport chain discharges through ATP-synthesizing enzymes in the inner membrane, and the energy is used to produce ATP.

⑤ ATP is transported out of the mitochondrion into the fluid of the cytoplasm, where it provides energy for cellular activities.

We have already discussed glycolysis; now let's look a little more closely at the processes of cellular respiration in the mitochondria. Glycolysis and cellular respiration are summarized in Table 8-1.

Pyruvate Is Transported to the Mitochondrial Matrix, Where It Is Broken Down via the Krebs Cycle

Recall that pyruvate is the end product of glycolysis and that it is synthesized in the fluid portion of the cytoplasm. The pyruvate diffuses down its concentration gradient into the mitochondria through pores in the mitochondrial membranes until it reaches the mitochondrial matrix, where it is used in cellular respiration.

In the matrix, pyruvate reacts with a molecule called *coenzyme A* (Fig. 8-5, step ①). Each pyruvate is split into CO_2 and a two-carbon molecule called an *acetyl group*, which immediately attaches to coenzyme A (CoA) and

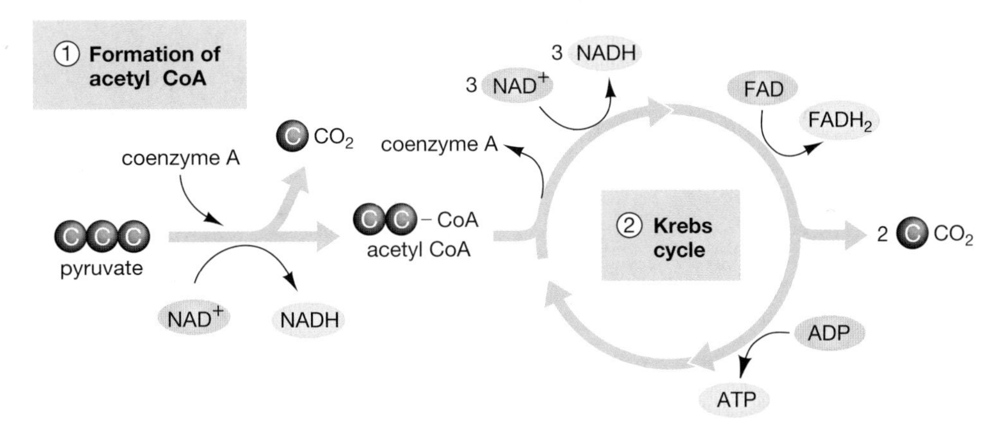

FIGURE 8-5 The reactions in the mitochondrial matrix
① Pyruvate reacts with CoA, forming CO_2 and acetyl CoA. During this reaction, an energetic electron is added to NAD^+ to form NADH. When acetyl CoA enters the Krebs cycle ②, CoA is released. One set of the reactions in the cycle produces three NADH, one $FADH_2$, two CO_2, and one ATP for each acetyl CoA. Because each glucose molecule yields two pyruvates, the total energy harvest per glucose molecule in the matrix is two ATP, eight NADH, and two $FADH_2$.

A CLOSER LOOK The Mitochondrial Matrix Reactions

Mitochondrial matrix reactions occur in two stages: the formation of acetyl coenzyme A, and the Krebs cycle. Recall that glycolysis produces two pyruvates from each glucose molecule, so each set of matrix reactions occurs twice during the metabolism of a single glucose molecule.

First Stage: Formation of Acetyl Coenzyme A

Pyruvate is split to form CO_2 and an acetyl group. The acetyl group attaches to CoA to form acetyl CoA. Simultaneously, NAD^+ receives two electrons and a hydrogen ion to make NADH. The acetyl CoA enters the second stage of the matrix reactions.

Second Stage: The Krebs Cycle

① Acetyl CoA donates its acetyl group to oxaloacetate to make citrate. CoA is released.

② Citrate is rearranged to form isocitrate.

③ Isocitrate loses a carbon to CO_2 forming α-ketoglutarate; NADH is formed from NAD^+.

④ Alpha-ketoglutarate loses a carbon to CO_2 forming succinate; NADH is formed from NAD^+ and additional energy is stored in ATP. (By this stage in the mitochondrial matrix reactions, all three carbons of the original pyruvate have been released as CO_2.)

⑤ Succinate is converted to fumarate, and the electron carrier FAD is charged up to $FADH_2$.

⑥ Fumarate is converted to malate.

⑦ Malate is converted to oxaloacetate, and NADH is formed from NAD^+.

The Krebs cycle produces two CO_2 and three NADH, one $FADH_2$, and one ATP per acetyl CoA. The formation of each acetyl CoA generates an additional CO_2 and an NADH. Overall, the mitochondrial matrix reactions produce four NADH and one $FADH_2$ and three CO_2 for each pyruvate supplied by glycolysis. Since each glucose molecule produces two pyruvates, the mitochondrial matrix reactions will generate a total of eight NADH and two $FADH_2$ per glucose molecule. These high-energy electron carriers will release their high-energy electrons to the electron transport chain of the inner membrane, where the energy of the electrons will be used to synthesize more ATP by chemiosmosis.

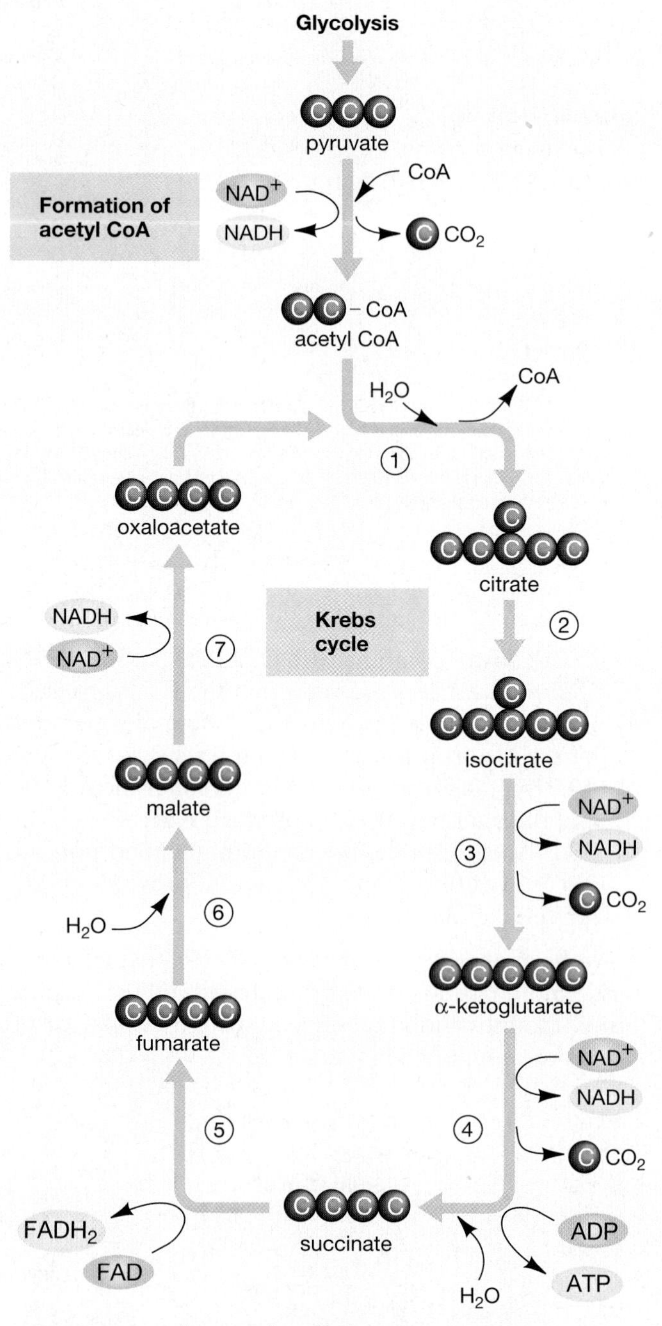

FIGURE E8-3 The mitochondrial matrix reactions

forms an acetyl–coenzyme A complex (*acetyl CoA* for short). During this reaction, two energetic electrons and a hydrogen ion are transferred to NAD^+, forming NADH.

The next stages of the reaction form a cyclic pathway known as the **Krebs cycle**, named after its discoverer, Hans Krebs, a biochemist who won the Nobel Prize for this work in 1953. The Krebs cycle is also called the *citric acid cycle* because citrate (the ionized form of citric acid) is the first molecule produced in the cycle. During the Krebs cycle (Fig. 8-5, step ②), each acetyl CoA briefly combines with a molecule of *oxaloacetate*. The two-carbon acetyl group is donated to the four-carbon oxaloacetate to form the six-carbon *citrate*. Coenzyme A is released once again; like an enzyme, coenzyme A is not permanently altered during these reactions and is reused many times. Mitochondrial enzymes then lead each citrate through a number of rearrangements that regenerate the oxaloacetate, give off two CO_2 molecules, and capture most of the energy of the acetyl group as one ATP and four electron carriers—one $FADH_2$ (flavin adenine dinucleotide) and three NADH.

The essay "A Closer Look: The Mitochondrial Matrix Reactions" shows the complete set of reactions that occur in the mitochondrial matrix, from acetyl CoA formation through the Krebs cycle.

Summing Up

The Mitochondrial Matrix Reactions

The synthesis of acetyl CoA produces one CO_2 and one NADH per pyruvate. The Krebs cycle produces two CO_2, one ATP, three NADH, and one $FADH_2$ per acetyl CoA. Therefore, at the conclusion of the matrix reactions, the two pyruvates that are produced from a single glucose molecule have been completely broken down by the addition of oxygen to form six CO_2 molecules. In the process, two ATPs, eight NADH, and two $FADH_2$ electron carriers have been produced.

Energetic Electrons Produced by the Krebs Cycle Are Carried to Electron Transport Chains in the Inner Mitochondrial Membrane

At this point, the cell has gained only four ATP molecules from the original glucose molecule: two during glycolysis and two during the Krebs cycle. The cell has, however, captured many energetic electrons in carrier molecules: 2 NADH during glycolysis, plus 8 more NADH and 2 $FADH_2$ from the matrix reactions, for a total of 10 NADH and 2 $FADH_2$. The carriers deposit their electrons in **electron transport chains** located in the inner mitochondrial membrane (Fig. 8-6). These electron transport chains are similar in function to those embedded in the thylakoid membrane of chloroplasts: the energetic electrons move from molecule to molecule along the transport chains. Energy released by the electrons during these transfers is used to pump hydrogen ions from the matrix across the inner membrane and into the intermembrane compartment during *chemiosmosis* (discussed in the next section).

Finally, at the end of the electron transport chain, oxygen and hydrogen ions accept the energetically depleted electrons: two electrons, one oxygen atom, and two hydrogen ions combine to form water. This step clears out the transport chain, leaving it ready to carry more electrons. Without oxygen, the electrons would "pile up" in the electron transport chain, and hydrogen ions would not be pumped across the inner membrane. The hydrogen ion gradient would soon dissipate, and ATP synthesis would stop.

Chemiosmosis Captures Energy Stored in a Hydrogen Ion Gradient and Produces ATP

Why pump hydrogen ions across a membrane? Hydrogen ion pumping across the inner membrane generates a large H^+ concentration gradient—that is, a high concentration of hydrogen ions in the intermembrane compartment and a low concentration in the matrix. Recall from Chapter 6 that, according to the second law of thermodynamics, energy must be expended to produce this nonuniform distribution of hydrogen ions, sort of like pumping water up into an elevated storage tank. Energy is released when the hydrogen ions are allowed to move down their concentration gradient—like opening the valves of the storage tank and allowing the water to rush out. This energy can be captured because the inner membrane is impermeable to hydrogen ions except at protein channels that are part of ATP-synthesizing enzymes. In the process of **chemiosmosis**, hydrogen ions move down their concentration gradient from the intermembrane compartment to the matrix through these ATP-synthesizing enzymes. The flow of hydrogen ions provides the energy to synthesize 32 to 34 molecules of ATP for each molecule of glucose by combining ADP (adenosine diphosphate) and phosphate. "A Closer Look: Chemiosmosis in Mitochondria" examines chemiosmosis in more detail.

The ATP that was synthesized in the matrix during chemiosmosis is transported across the inner membrane from the matrix to the intermembrane compartment. It then diffuses out of the mitochondrion to the

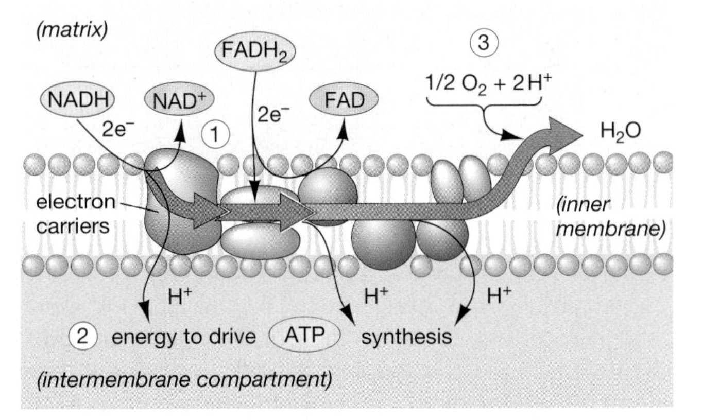

FIGURE 8-6 The electron transport chain of mitochondria ① NADH and $FADH_2$ donate their energetic electrons to the carriers of the transport chain. ② As the electrons pass through the transport chain, some of their energy is used to pump hydrogen ions from the matrix into the intermembrane compartment. This creates a hydrogen ion gradient that is used to drive ATP synthesis. ③ At the end of the electron transport chain, the energy-depleted electrons combine with oxygen and hydrogen ions in the matrix to form water. **QUESTION** How would the rate of ATP production be affected by the absence of oxygen?

A CLOSER LOOK Chemiosmosis in Mitochondria

ATP synthesis in mitochondria is similar to the process of chemiosmosis described for chloroplasts in Chapter 7. The inner membrane of a mitochondrion has an electron transport chain that functions similarly to the one in the thylakoids. Further, the intermembrane compartment between the outer and inner membranes of a mitochondrion is analogous to the interior of a thylakoid.

Anatomically, the arrangement in mitochondria looks like this:

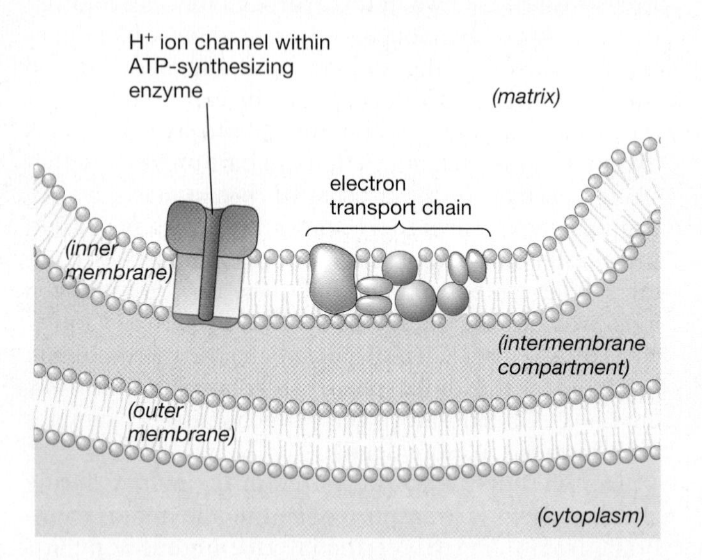

The electron carriers formed during glycolysis and the Krebs cycle—NADH and $FADH_2$—deposit their electrons with the electron transport chain of the inner membrane. (For clarity, $FADH_2$ is not shown in the illustration.) As they pass through the electron transport chain, the electrons provide the energy to pump hydrogen ions (H^+) across the inner membrane, from the matrix to the intermembrane compartment:

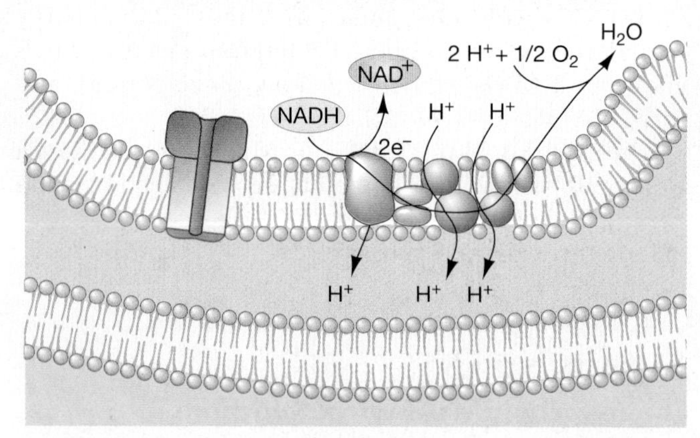

This pumping process increases the H^+ concentration in the intermembrane compartment and decreases the H^+ concentration in the matrix; therefore, a H^+ gradient is produced across the inner membrane. Like the thylakoid membrane of a chloroplast, the inner membrane of a mitochondrion is permeable to H^+ only at channels that are coupled with ATP-synthesizing enzymes. The movement of hydrogen ions down their concentration gradient through these channels drives ATP synthesis:

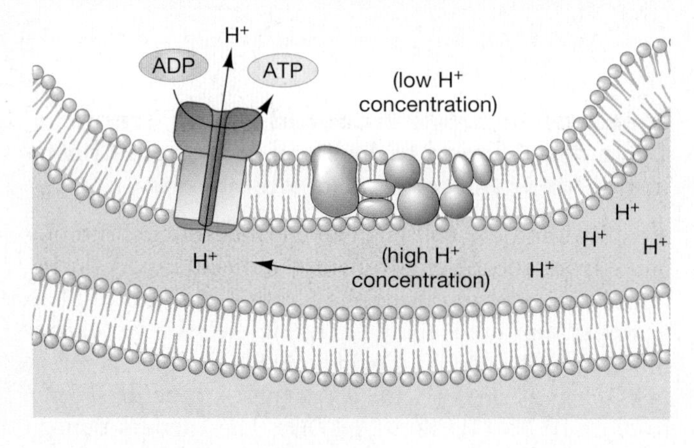

surrounding cytoplasm through the outer membrane, which is very permeable to ATP. These ATP molecules provide most of the energy needed by the cell. ADP simultaneously diffuses from the fluid of the cytoplasm across the outer membrane and is transported across the inner membrane to the matrix, replenishing the supply of ADP.

Summing Up

Electron Transport and Chemiosmosis

Electrons from the electron carriers NADH and $FADH_2$ enter the electron transport chain of the inner mitochondrial membrane. Here their energy is used to generate a hydrogen ion gradient across the inner membrane. The movement of hydrogen ions down their gradient through the pores of ATP-synthesizing enzymes drives the synthesis of 32 to 34 molecules of ATP. At the end of the electron transport chain, two electrons combine with one oxygen atom and two hydrogen ions to form water.

Glycolysis and Cellular Respiration Influence the Way Organisms Function

Many students believe that the details of glycolysis and cellular respiration are hard to learn and don't really help them understand the living world around them. But have you ever read a murder mystery and wondered how cyanide could kill a person almost instantly?

Cyanide reacts with one of the proteins in the electron transport chain, immediately blocking the movement of electrons through the sequence of electron transfer molecules and bringing cellular respiration to a screeching halt. Even under normal conditions, metabolic processes within individual cells have enormous impacts on the functioning of the entire organism. For a hummingbird to beat its wings, for your brain to process the information you are reading, for your hand to turn the pages of this book, cells require a continuous supply of energy. As an extreme example, let's consider Olympic track events.

Humans, like hummingbirds, must regulate energy reserves and energy use. Why is the average speed of the 5000-meter run in the Olympics slower than that of the 100-meter dash? During the dash, or during the sprint across the finish line of a marathon, runners' leg muscles use more ATP than cellular respiration can supply, because their bodies cannot deliver enough oxygen to keep up with the demand. Glycolysis and lactate fermentation can keep the muscles supplied with ATP for a short time, but soon the effects of lactate buildup (along with several other factors) cause fatigue and cramps. Although runners can do a 100-meter dash anaerobically, distance runners must pace themselves, using cellular respiration to power their muscles for most of the race and saving the anaerobic sprint for the finish.

Marathon runners face somewhat the same dilemma that migrating hummingbirds do. A marathon may require 3000 kilocalories of stored energy, with cellular respiration supplying nearly all the ATP. Marathoners train by running 50 or 100 miles a week, not so much to build up their leg muscles as to build up the capacity of their respiratory and circulatory systems to deliver enough oxygen to their muscles. An efficient transport of oxygen to the cells is necessary to support the cellular respiration that such vigorous exercise demands.

As you can see, sustaining life depends on efficiently obtaining, storing, and using energy. By gaining an understanding of the principles of cellular respiration, you can more fully appreciate the energy-related adaptations of living organisms.

CASE STUDY REVISITED The Flight of the Hummingbird

To fly more than 600 miles over open water, the ruby-throated hummingbird must store a great deal of energy. Hummers store the highest-energy molecules possible—fat—and extract the maximum usable energy during flight. A ruby-throated hummingbird weighs 0.11 to 0.16 ounces (2 to 3 grams, about as much as a penny) before it puts on weight for migration; it adds as much as 2 grams of fat in late summer, nearly doubling its weight. Recall from Chapter 3 that fats contain more than twice as much energy per unit weight as do proteins or carbohydrates. If a hummingbird stored glycogen or protein for energy, it would be too heavy to lift off. Even so, the hummer must still generate every ATP molecule possible out of each fat molecule. The hummer that just makes it to Guatemala on 2 grams of fat by using cellular respiration would collapse before reaching the Gulf Coast if it used lactate fermentation instead. Fortunately, the cells of a hummingbird's flight muscles are packed with mitochondria, so each cell is capable of producing large quantities of ATP. Furthermore, the hummingbird's lungs are exquisitely designed to extract oxygen from the air even while the bird exhales. Thus, even during strenuous flight, cellular respiration never falters for lack of oxygen.

Consider This: Some flowers are adapted to attract hummingbirds. They store relatively large quantities of nectar at the end of long, tubular compartments that bees cannot reach. The flowers are often red or orange (colors that attract hummingbirds). Look at the chapter-opening photo and form a hypothesis about why some plants have evolved to "waste" so much energy by putting their sugar into nectar. Now form a hypothesis as to why flowers that attract hummingbirds with large amounts of nectar prevent bees from reaching the nectar by their shape (if you are stumped by this, you may find some ideas in Chapter 25.)

Links to Life: A Jug of Wine, a Loaf of Bread . . .

Persian poet Omar Khayyam (1048–1122) described his vision of paradise on Earth as "A Jug of Wine, a Loaf of Bread—and Thou Beside me . . . " In fact, yeast's ability to ferment the sugars in fruit to form alcohol has been recognized and exploited by people for millennia; historical evidence suggests that wine and beer were commercially produced at least 5000 years ago. While the original fermenting yeasts were likely the wild types that colonize the surface of grapes, most wine makers now use specific species of yeast that have a higher tolerance for alcohol and produce fewer undesirable by-products. Yeasts (single-celled fungi) will engage in cellular respiration if oxygen is available, but switch to alcoholic fermentation if they run out of oxygen. But as you now know, carbon dioxide is also a by-product of alcoholic fermentation; thus, wine must be fermented in containers that allow the carbon dioxide to leave (so the containers don't explode) but prevent the entry of air (so cellular respiration doesn't occur). As the alcohol level rises, it becomes toxic to yeasts and other microorganisms; hence, alcohol can be used as a sterilizing agent. Even the most alcohol-tolerant yeast is killed by 15% alcohol, and most table wines have an alcohol around 10% to 12%. Sparking wines and champagne are made by adding additional yeast and sugar just before the wine is bottled, so that final fermentation occurs in the sealed bottle, trapping carbon dioxide. Fermentation is also a key element of bread production. Because yeasts are unable to break down starches, and bread is make from flour (a starch usually derived from wheat kernels), sugar must be added to allow the yeast to grow. Kneading the bread dough changes its texture—making it more elastic—and redistributes the yeast cells evenly throughout. Each cell will then multiply, ferment, and release CO_2, which is trapped within the bread dough. The CO_2 forms tiny pockets of gas that cause the dough to rise and give bread its evenly porous texture.

Summary of Key Concepts

8.1 How Is Glucose Metabolized?

Cells produce usable energy by breaking down glucose into lower-energy compounds and capturing some of the released energy as ATP. In glycolysis, glucose is metabolized in the fluid portion of the cytoplasm to form two molecules of pyruvate, generating two ATP molecules. In the absence of oxygen, pyruvate is converted by fermentation to lactate or ethanol and CO_2. If oxygen is available, the pyruvates are metabolized to release CO_2 and H_2O through cellular respiration in the mitochondria, generating much more ATP than does fermentation.

8.2 How Is the Energy in Glucose Captured During Glycolysis?

During glycolysis, a molecule of glucose is activated by the addition of phosphates from two ATP molecules to form fructose bisphosphate. In a series of reactions, the fructose bisphosphate is broken down into two molecules of pyruvate. These reactions produce four ATP molecules and two NADH electron carriers. Because two ATP were used in the activation steps, the net yield from glycolysis is two ATP and two NADH. Glycolysis, in addition to providing a small yield of ATP, uses up NAD^+ to produce NADH. Once the cell's supply of NAD^+ is consumed, glycolysis must stop. NADH may be regenerated by fermentation, with no additional ATP gain, or by cellular respiration, which also produces additional ATP.

8.3 How Does Cellular Respiration Capture Additional Energy from Glucose?

If oxygen is available, cellular respiration can occur. The pyruvates are transported into the matrix of the mitochondria. In the matrix, each pyruvate reacts with coenzyme A to form acetyl CoA plus CO_2. One NADH is also formed at this step. The two-carbon acetyl group of acetyl CoA enters the Krebs cycle, which releases the remaining 2 carbons as CO_2. One ATP, 3 NADH, and 1 $FADH_2$ are also formed for each acetyl group that goes through the cycle. At this point, each glucose molecule has produced 4 ATP (2 from glycolysis and 1 from each acetyl CoA during the Krebs cycle), 10 NADH (2 from glycolysis, 1 from each pyruvate during the formation of acetyl CoA, and 3 from each acetyl CoA during the Krebs cycle), and 2 $FADH_2$ (1 from each acetyl CoA during the Krebs cycle).

The NADH and $FADH_2$ deliver their energetic electrons to the proteins of the electron transport chain embedded in the inner mitochondrial membrane. The energy of the electrons is used to pump hydrogen ions across the inner membrane from the matrix to the intermembrane compartment. At the end of the electron transport chain, the depleted electrons combine with hydrogen ions and oxygen to form water. This is the oxygen-requiring step of cellular respiration. During chemiosmosis, the hydrogen ion gradient created by the electron transport chain is used to produce ATP, as the hydrogen ions diffuse back across the inner membrane through channels in ATP-synthesizing enzymes. Electron transport and chemiosmosis yield 32 or 34 additional ATP, for a net yield of 36 or 38 ATP per glucose molecule.

Study Note:

Figure 8-1 and Table 8-1 summarize the locations, major mechanisms, and overall energy harvest for the complete metabolism of glucose from glycolysis through cellular respiration.

Key Terms

cellular respiration *p. 137*
chemiosmosis *p. 141*
electron transport chain *p. 141*
fermentation *p. 135*
glycolysis *p. 132*
intermembrane compartment *p. 137*
Krebs cycle *p. 140*
matrix *p. 137*

Thinking Through the Concepts

To take a multiple-choice quiz with feedback on the contents of this chapter, visit http://www.prenhall.com/audesirk7. *Log in to the Web site selected by your instructor and navigate to the Self Test section for this chapter.*

? Review Questions

1. Starting with glucose ($C_6H_{12}O_6$), write the overall reactions for (a) aerobic respiration and (b) fermentation in yeast.
2. Draw a labeled diagram of a mitochondrion, and explain how its structure relates to its function.
3. What role do the following play in respiratory metabolism: (a) glycolysis, (b) mitochondrial matrix, (c) inner membrane of mitochondria, (d) fermentation, and (e) NAD^+?
4. Outline the major steps in (a) aerobic and (b) anaerobic respiration, indicating the sites of ATP production. What is the overall energy harvest (in terms of ATP molecules generated per glucose molecule) for each?
5. Describe the Krebs cycle. In what form is most of the energy captured?
6. Describe the mitochondrial electron transport chain and the process of chemiosmosis.
7. Why is oxygen necessary for cellular respiration to occur?
8. Compare the structure of chloroplasts (described in Chapter 7) to that of mitochondria, and describe how the similarities in structure relate to similarities in function. Also describe any differences in structure and function between chloroplasts and mitochondria.

Applying the Concepts

1. Some years ago a freight train overturned, spilling a load of grain. Because the grain was unusable, it was buried in the embankment. Although there is no shortage of other food, the local bear population has created a nuisance by continually uncovering the grain. Yeasts are common in the soil. What do you think has happened to the grain to make the bears do this, and how is it related to human cultural evolution?
2. In detective novels, "the odor of bitter almonds" is the telltale clue to murder by cyanide poisoning. Cyanide works by attacking the enzyme that transfers electrons from the respiratory electron transport chain to O_2. Why is it not possible for the victim to survive by using anaerobic respiration? Why is cyanide poisoning almost immediately fatal?
3. More than a century ago, French biochemist Louis Pasteur described a phenomenon, now called "the Pasteur effect," in the wine-making process. He observed that in a sealed container of grape juice and yeast, the yeast will consume the sugar very slowly as long as oxygen remains in the container. As soon as the oxygen is gone, however, the rate of sugar consumption by the yeast increases greatly, and the alcohol content in the container rises. Discuss the Pasteur effect on the basis of what you know about aerobic and anaerobic cellular respiration.
4. Some species of bacteria that live at the surface of sediment on the bottom of lakes are facultative anaerobes; that is, they are capable of either aerobic or anaerobic respiration. How will their metabolism change during the summer when the deep water becomes anoxic (deoxygenated)? If the bacteria continue to grow at the same rate, will glycolysis increase, decrease, or remain the same after the lake becomes anoxic? Explain why.
5. The dumping of large amounts of raw sewage into rivers or lakes typically leads to massive fish kills, although sewage itself is not toxic to fish. Similar fish kills also occur in shallow lakes that become covered in ice during the winter. What kills the fish? How might you reduce fish mortality after raw sewage is accidentally released into a small pond containing large bass?
6. Different cells respire at different rates. Explain why. How could you predict the relative respiratory rates of different tissues in a fish by microscopic examination of cells?
7. Imagine that a starving cell reached the stage where every bit of its ATP was depleted and converted to ADP plus phosphate. If that cell were placed in fresh nutrient broth at this point, would it recover and survive? Explain your answer based on what you know of glucose breakdown.

For More Information

Calder, W. A. "Red-Hot Hummers." *Nature Conservancy,* March/April 1998. Beautiful photos and lively writing describe the trials of the tiny rufous hummingbird as it fuels up for its long migration south.

McCarty, R. E. "H+-ATPases in Oxidative and Photosynthetic Phosphorylation." *BioScience*, January 1985. A description of the structure and function of the ATP-synthesizing enzymes in mitochondria and chloroplasts.

Nelson, M., Burgess, T. L., Alling, A., Alverez-Romo, N., Dempster, W. F., Walford, R. L., and Allen, J. P. "Using a Closed Ecological System to Study Earth's Biosphere." *BioScience*, 1993. An artificial ecosystem allows scientists to learn more about how natural ecosystems function.

Media Activities

To access a Media Activity visit http://www.prenhall.com/audesirk7. *Log in to the Web site selected by your instructor, navigate to this chapter, and select the appropriate Media Activity number.*

8.1 Glucose Metabolism

Estimated time: 10 minutes

Explore the mechanisms cells used to derive energy from the sugar glucose.

8.2 Interactions between Photosynthesis and Respiration

Estimated time: 10 minutes

In this exercise you will review how photosynthesis and respiration interact in a plant.

8.3 Web Investigation: The Flight of the Hummingbird

Estimated time: 10 minutes

Explore how the feeding behaviors and diet of hummingbirds are perfectly adapted to their energy needs.

UNIT TWO

Inheritance

Inheritance provides for both similarity and difference. All dogs share many similarities because their genes are nearly identical. The enormous variety of body size, fur length and color, and proportion among breeds results from tiny differences in their genes.

CHAPTER

9 DNA: The Molecule of Heredity

Within the double helix structure of DNA lies the explanation of the link between sunbathing and an increased risk of skin cancer.

AT A GLANCE

CASE STUDY Sunshine Perils

Given Rachel's hectic schedule as a college junior, skin cancer was the last thing on her mind. Compared to varsity swimming, her studies, and her part-time job, the bumpy, black mole on her back didn't seem important. Rachel would have ignored it completely, but her swim coach asked her to have it checked by a physician. So, she scheduled an appointment with her family doctor to have the mole removed. The doctor told her that he could remove the mole in his office and that the wound should heal in time for the next swim meet, leaving almost no scar.

After the appointment, Rachel didn't think about the mole at all. However, her doctor called back three days later. Following his general policy, he had sent the tissue to a laboratory for examination; the diagnosis was a type of cancer called *melanoma*.

Melanoma is a relatively common skin cancer that usually begins in pigmented cells in the inner parts of the skin. The cancer can then spread to other parts of the body, including internal organs. The resulting disease is challenging to treat and frequently deadly. The American Dermatology Association estimates that more than 54,000 people in the United States will be diagnosed with melanoma this year, with over 8000 deaths. Worse yet, the incidence of melanoma is rising by about 4% a year, faster than any other cancer. It is now the most common cancer in people between 25 and 29 years of age. Many physicians suspect that the cause may be increased exposure to sunlight.

Why do cancers form? How can sunlight cause cancer? To answer these questions, we need to understand the basic structure of DNA, the hereditary molecule.

9.1 How Did Scientists Discover That Genes Are Made of DNA?

Just 70 years ago, no one knew that *deoxyribonucleic acid,* or **DNA**, is the molecule that carries the blueprints for all forms of life on Earth. We now know that the "molecular instructions" in DNA direct the life of each cell in an organism. DNA also enables organisms, or cells within an organism, to transmit information accurately from one generation to the next. The discovery of how DNA carries life's blueprints was one of the greatest achievements of 20th-century biology.

The discovery of DNA's structure and how it worked was the culmination of decades of research by hundreds of scientists, although it is often credited to a handful of Nobel Prize winners. By the late 1800s, scientists had learned that heritable information exists in discrete units called **genes**. However, they could not provide a precise definition of a gene. Scientists merely knew that genes determine many of the heritable differences among individuals within a species. For example, genes for flower color determine whether roses are red, pink, yellow, or white. Studies of dividing cells provided strong evidence that genes are located in threadlike structures within cells, called **chromosomes**. Scientists also discovered that chromosomes consist of DNA and protein, indicating that genes are made of one or both of these molecules. For the first half of the 20th century, most scientists thought that genes were made of protein. However, experiments using bacteria showed that genes are actually composed of DNA.

Transformed Bacteria Revealed the Link Between Genes and DNA

In the late 1920s, a British researcher named Frederick Griffith was trying to make a vaccine to prevent bacterial pneumonia, a major cause of death at that time.

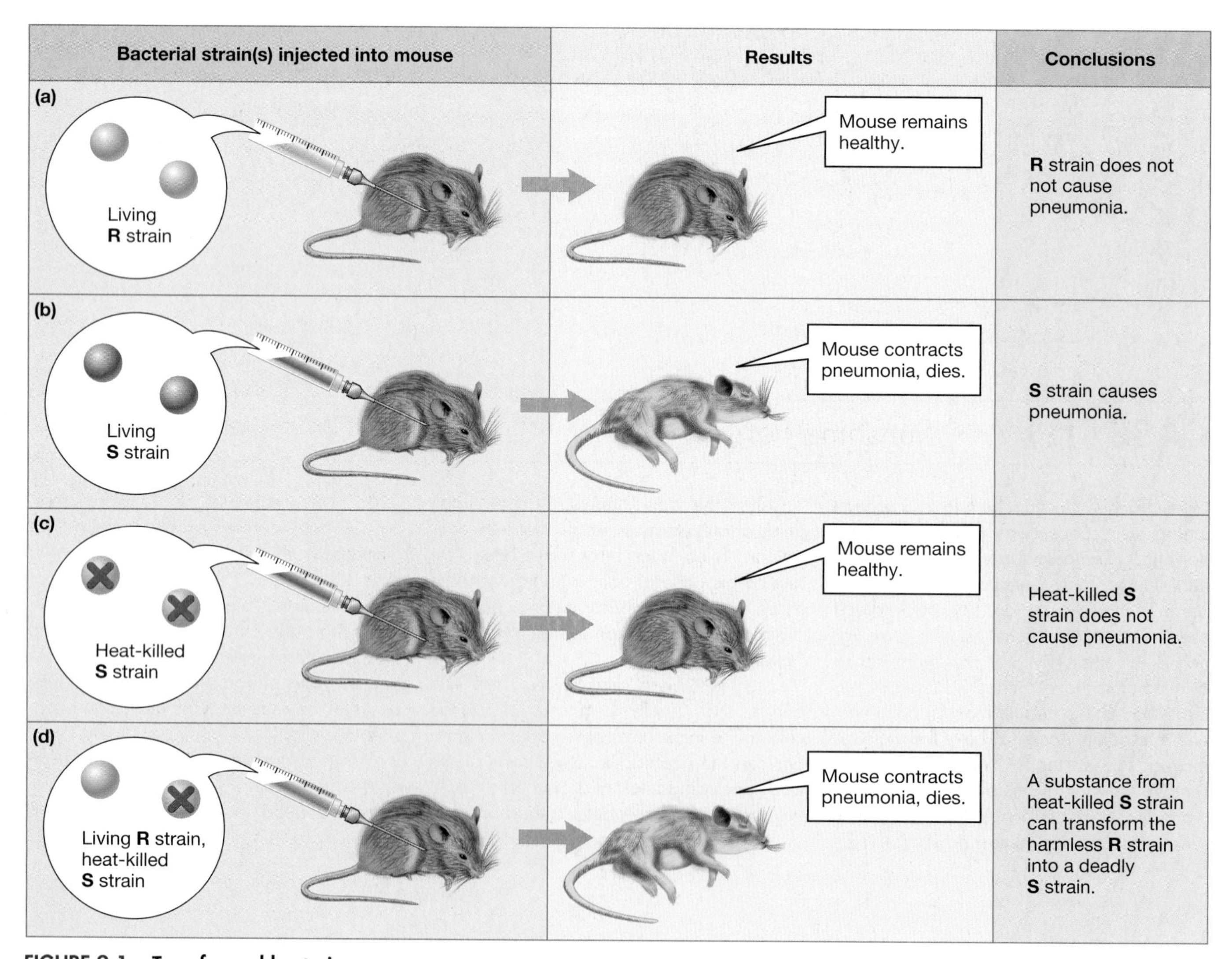

FIGURE 9-1 Transformed bacteria
Griffith's discovery that bacteria can be transformed from harmless to deadly laid the groundwork for the discovery that DNA contains genes. **QUESTION** If Griffith had used malaria, which is caused by a single-celled eukaryote, instead of pneumonia, would his experiment have had the same result?

Making vaccines against many infectious bacteria is very difficult (for example, the vaccines available against anthrax are neither completely safe nor completely effective), but this was not known back in the 1920s. Some antibacterial vaccines consist of a weakened strain of the bacteria that doesn't cause illness. Injecting this weakened but living strain into an animal may promote immunity against the disease-causing strains. Other vaccines use disease-causing (virulent) bacteria that have been killed by exposure to heat or chemicals. Griffith was trying to make a vaccine using two strains of the *Streptococcus pneumoniae* bacterium. One strain, R, did not cause pneumonia when injected into mice (Fig. 9-1a). The other strain, S, was deadly when injected, causing pneumonia and killing the mice in a day or two (Fig. 9-1b). As expected, when the S strain was killed and injected into mice, it did not cause disease (Fig. 9-1c). Unfortunately, neither the live R strain nor the killed S strain provided immunity against live S-strain bacteria.

Griffith also tried mixing living R-strain bacteria together with heat-killed S-strain bacteria and injecting the mixture into mice (Fig. 9-1d). Because neither of these bacterial strains causes pneumonia on its own, he expected the mice to remain healthy. To his surprise, the mice sickened and died. When he autopsied the mice, he recovered *living* S-strain bacteria from them. The simplest interpretation of these results is that some substance in the heat-killed S-strain changed the living but harmless R-strain bacteria into the deadly S-strain, a process he called *transformation*. The transformed S-strain cells then multiplied and caused pneumonia.

Griffith never discovered an effective pneumonia vaccine, so in that sense his experiments were a failure (in fact, an effective and safe vaccine against most forms of *Streptococcus pneumoniae* was not developed until a few years ago). However, Griffith's experiments marked a turning point in our understanding of genetics because other researchers suspected that the substance that causes transformation might be the long-sought molecule of heredity. In the 1940s, Oswald Avery, Colin MacLeod, and Maclyn McCarty of Rockefeller University purified the molecule from S-strain bacteria that could transform the R-strain into the S-strain. It was DNA.

This discovery helps us interpret the results of Griffith's experiments. Heating the S-strain cells killed them but did not completely destroy their DNA. When the killed S-strain bacteria were mixed in a test tube with living R-strain bacteria, fragments of DNA from the dead S-strain cells entered into some of the R-strain cells. If these fragments of DNA contained the genes needed to cause disease, an R-strain cell was transformed into an S-strain cell. Thus, Avery, MacLeod, and McCarty concluded that genes are made of DNA. However, many years of additional research were needed to convince everyone of DNA's central role in heredity.

9.2 What Is the Structure of DNA?

Knowing that genes are made of DNA does not answer critical questions about inheritance: How does DNA encode genetic information? How is DNA duplicated so that information can be accurately passed from one cell to its daughter cells (see Chapter 11 for more information about cell reproduction)? The secrets of DNA function, and therefore of heredity itself, can be found in the three-dimensional structure of the DNA molecule.

DNA Is Composed of Four Nucleotides

As you learned in Chapter 3, the DNA of every organism on Earth is composed of four small subunits called **nucleotides**. Each nucleotide in DNA consists of three parts: a phosphate group, a sugar called *deoxyribose*, and one of four possible nitrogen-containing **bases**—**adenine (A)**, **guanine (G)**, **thymine (T)**, or **cytosine (C)**.

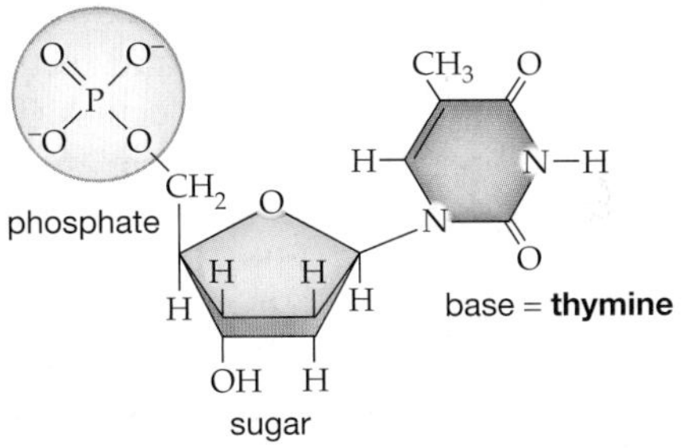

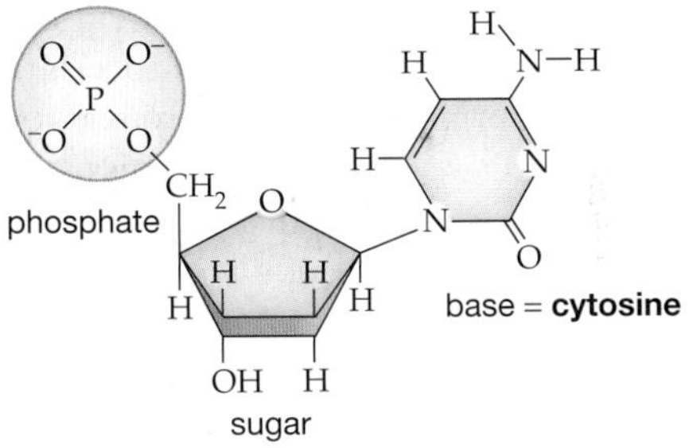

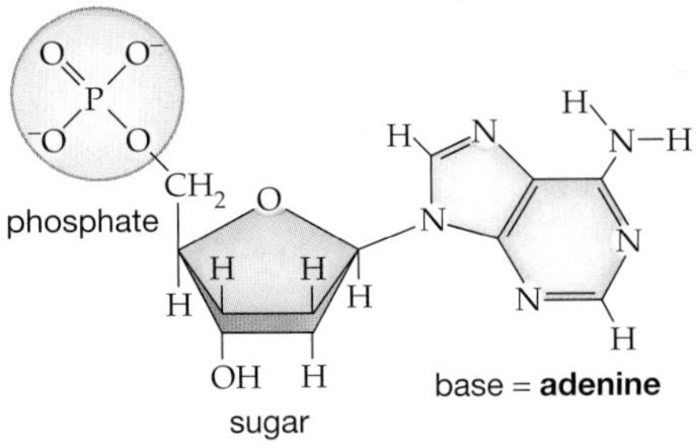

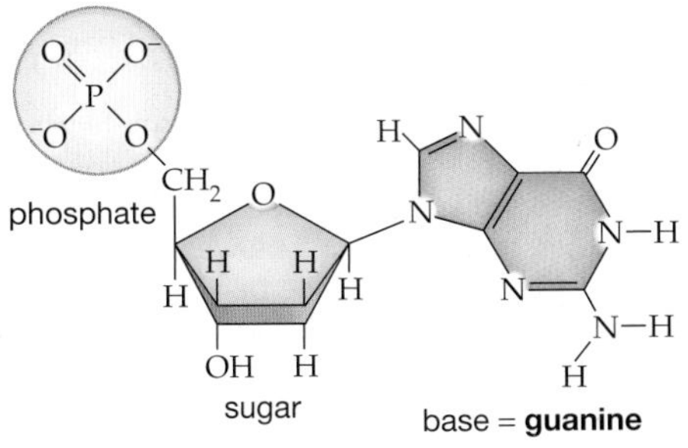

In the 1940s, when biochemist Erwin Chargaff of Columbia University analyzed the amounts of the four nucleotides in DNA from organisms as diverse as bacteria, sea urchins, fish, and humans, he found a curious consistency. The DNA of any given species contains *equal amounts of adenine and thymine*, as well as *equal amounts of guanine and cytosine*. Chargaff's observation certainly seemed significant, but it would be almost another decade before anyone figured out what it meant about DNA structure.

DNA Is a Double Helix of Two Nucleotide Strands

Determining the structure of any biological molecule is no simple task, even for scientists today. Nevertheless, in the late 1940s, several scientists began to study DNA, hoping to learn more about its structure. British scientists Maurice Wilkins and Rosalind Franklin used X-ray diffraction to study DNA structure. They bombarded crystals of purified DNA with X-rays and recorded how the X-rays bounced off the DNA molecules (Fig. 9-2a). As you can see, the resulting "diffraction" pattern does not provide a direct picture of DNA structure. However, experts like Wilkins and Franklin (Fig. 9-2b) could extract a lot of information about DNA from the pattern. First, a molecule of DNA is long and thin, with a uniform diameter of 2 nanometers (2 billionths of a meter). Second, DNA is helical; that is, it is twisted like a corkscrew. Third, the DNA molecule consists of repeating subunits.

These chemical and X-ray diffraction data did not provide enough information for researchers to work out the structure of DNA; some good guesses were also needed. Combining Wilkins and Franklin's data with a knowledge of how complex organic molecules bond together and an intuition that "important biological objects come in pairs," James Watson and Francis Crick proposed a model for the structure of DNA. (See "Scientific Inquiry: The Discovery of the Double Helix," for more information.) They suggested that the DNA molecule consists of two separate DNA polymers of linked nucleotides, called strands (Fig. 9-3). Within each DNA strand, the phosphate group of one nucleotide bonds to the sugar of the next nucleotide in the strand. This bonding pattern produces a "backbone" of alternating, covalently bonded sugars and phosphates. The nucleotide bases protrude from this **sugar-phosphate backbone**. All of the nucleotides within a single DNA strand are oriented in the same direction. Therefore, the two ends of a DNA strand differ; one end has a "free" or unbonded sugar (see Fig. 9-3a), and the other end has a "free" or unbonded phosphate (see Fig. 9-3a; think of a long line of cars stopped on a one-way street at night; the cars' headlights always point forward, and their taillights always point backward).

Hydrogen Bonds Between Complementary Bases Hold the Two DNA Strands Together

Watson and Crick proposed that two DNA strands are held together by hydrogen bonds that form between the protruding bases of the individual DNA strands (see Fig. 9-3a). These bonds give DNA a ladder-like structure, with the sugar-phosphate backbones on the outside (forming the uprights of the ladder) and the nucleotide bases on the inside (forming the rungs of the ladder). However, the DNA strands are not straight. Instead, they are twisted about each other to form a **double helix**, resembling a ladder twisted lengthwise into the shape of a circular staircase (see Fig. 9-3b). Further, the two strands in a DNA double helix are oriented in opposite directions (again, imagine an evening traffic jam, this time on a two-lane highway; a traffic helicopter pilot overhead would see only the headlights on cars in, say, the northbound lanes, and only taillights in the southbound lanes).

Take a closer look at the pairs of hydrogen-bonded bases that form each rung of the double helix ladder. Notice that adenine forms hydrogen bonds only with

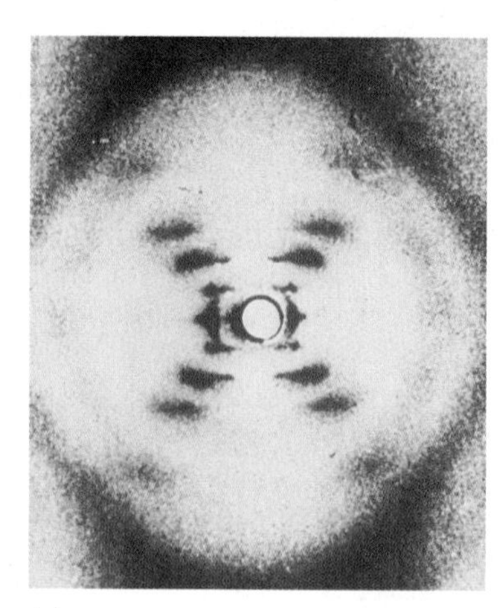

(a)

(b)

FIGURE 9-2 X-ray diffraction studies of DNA taken by Rosalind Franklin
(a) The X formed of dark spots is characteristic of helical molecules such as DNA. Measurements of various aspects of the pattern indicate the dimensions of the DNA helix; for example, the distance between the dark spots corresponds to the distance between turns of the helix. **(b)** Rosalind Franklin published about 40 scientific papers before her untimely death in 1958 at the age of 37.

Media Activity
9.1 DNA Structure

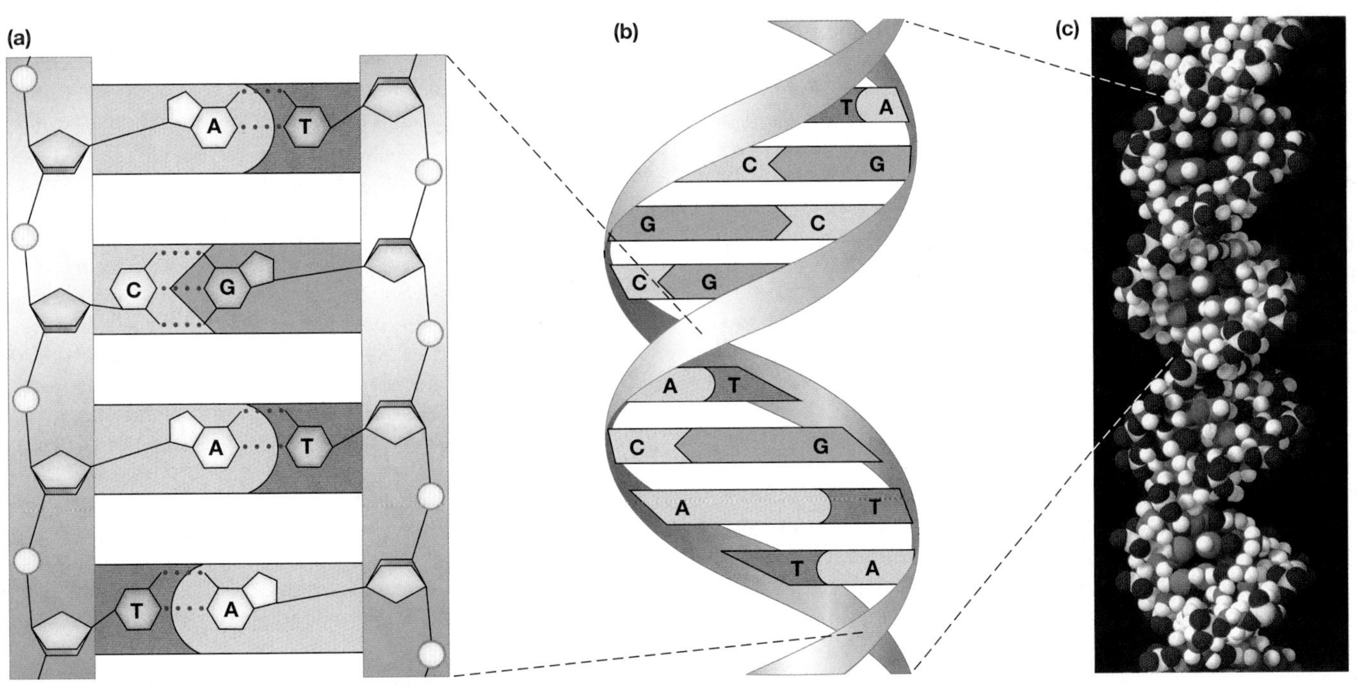

FIGURE 9-3 The Watson-Crick model of DNA structure
(a) Hydrogen bonding between complementary base pairs holds the two strands of DNA together. Three hydrogen bonds hold guanine to cytosine; two hydrogen bonds hold adenine to thymine. (b) Strands of DNA wind about each other in a double helix, like a twisted ladder, with the sugar-phosphate backbone forming the uprights and the complementary base pairs forming the rungs. (c) A space-filling model of DNA structure. **QUESTION** Which are harder to break, A–T bonds or C–G bonds?

thymine and that guanine forms hydrogen bonds only with cytosine (see Fig. 9-3a,b). These A–T and G–C pairs are called **complementary base pairs**, and their presence explains Chargaff's results—that the DNA of a given species contains equal amounts of adenine and thymine, as well as equal amounts of cytosine and guanine. Because an A in one DNA strand always pairs with a T in the other strand, the amount of A always equals the amount of T. Similarly, because a G in one strand always pairs with a C in the other DNA strand, the amount of G always equals the amount of C. Finally, look at the sizes of the bases: adenine and guanine are large, whereas thymine and cytosine are small. Because the double helix has only A–T and G–C pairs, all the rungs of the DNA ladder are the same width. Therefore, the double helix has a constant diameter, just as the X-ray diffraction pattern predicted.

The structure of DNA was solved. On March 7, 1953, at the Eagle Pub in Cambridge, England, Francis Crick proclaimed to the lunchtime crowd, "We have discovered the secret of life." This claim was not far from the truth. Although further data would be needed to confirm the details, within just a few years, the DNA model revolutionized biology, from genetics to medicine. As we will see in later chapters, the revolution continues today.

The Order of Nucleotides in DNA Can Encode Vast Amounts of Information

Look again at the elegant structure of DNA shown in Figure 9-3. Can you see why many scientists had trouble believing that DNA could be the carrier of genetic information? Consider the many characteristics of just one organism. How can the color of a bird's feathers, the size and shape of its beak, the ability to make a nest, its song, and its ability to migrate all be determined by a molecule with just four simple parts?

The answer is that it's not the *number* of different subunits but their *sequence* that's important. Within a DNA strand, the four types of bases can be arranged in any linear order, and this sequence is what encodes genetic information. An analogy might help: you don't need a lot of unique letters to make up a language. English has 26 letters, but Hawaiian has only 12, and the binary language of computers uses only two "letters" (0 and 1, or "on" and "off"). Nevertheless, all three can spell out thousands of different words. A stretch of DNA that is just 10 nucleotides long can have more than a million possible sequences of the four bases. Because an organism has millions (in bacteria) to billions (in plants or animals) of nucleotides, DNA molecules can encode a staggering amount of information.

SCIENTIFIC INQUIRY The Discovery of the Double Helix

In the early 1950s, many biologists realized that the key to understanding inheritance lay in the structure of DNA. They also knew that whoever deduced the correct structure of DNA would receive recognition, possibly the Nobel Prize. Linus Pauling of Caltech was considered the person most likely to solve the mystery of DNA structure. Pauling probably knew more about the chemistry of large organic molecules than any person alive. Like Rosalind Franklin and Maurice Wilkins, Pauling was an expert in X-ray diffraction techniques. In 1950, he used these techniques to show that many proteins were coiled into single-stranded helices (see Chapter 3). Pauling, however, had two important handicaps. First, he had concentrated on protein research for years and therefore had little data about DNA. Second, he was active in the peace movement. At that time, some government officials, including Senator Joseph McCarthy, considered such activity to be potentially subversive and threatening to national security. This latter handicap may have proved decisive.

The second most likely competitors were Wilkins and Franklin, the British scientists who had set out to determine the structure of DNA by using X-ray diffraction patterns. In fact, they were the only scientists who had good data about the general shape of the DNA molecule. Unfortunately for them, their methodical approach was also slow.

The door was open for the eventual discoverers of the double helix, James Watson and Francis Crick, two scientists with neither Pauling's tremendous understanding of chemical bonds nor Franklin and Wilkins' expertise in X-ray analysis. Watson and Crick did no experiments in the ordinary sense of the word; instead, they spent their time thinking about DNA, trying to construct a molecular model that made sense and fit the data. Because they were working in England and because Wilkins was very open about his and Franklin's data, Watson and Crick were familiar with all the X-ray information relating to DNA. This information was just what Pauling lacked. Because of Pauling's presumed subversive tendencies, the U.S. State Department refused to issue him a passport to leave the United States, so he could neither attend meetings at which Wilkins presented the X-ray data nor visit England to talk with Franklin and Wilkins directly. Watson and Crick knew that Pauling was working on DNA structure and were terrified that he would beat them to it. In his book, *The Double Helix*, Watson recounts his belief that, had Pauling seen the X-ray pictures, "in a week at most, Linus would have [had] the structure."

You might be thinking, "But wait just a minute! That's not fair. If the goal of science is to advance knowledge, then everyone should have access to all the data. If Pauling was the best, he should have discovered the double helix first." Maybe so. But after all, scientists are people, too. Although virtually all scientists want to see the advancement and benefit of humanity, each individual also wants to be the one responsible for that advancement and receive the credit and glory. And so, Linus Pauling remained in the dark about the X-ray data and was beaten to the correct structure (Fig. E9-1). When Watson and Crick discovered the double helix structure of DNA, Watson described it in a letter to Max Delbruck, a friend and adviser at Caltech. He asked Delbruck not to reveal the contents of the letter to Pauling until their structure was formally published. Delbruck, perhaps more of a model scientist, firmly believed that scientific discoveries belong in the public domain and promptly told Pauling about it. With the class of a great scientist and a great person, Pauling graciously congratulated Watson and Crick on their brilliant solution to the DNA structure. The race was over.

FIGURE E9-1 The discovery of DNA
James Watson and Francis Crick with a model of DNA structure.

Of course, to make sense, both the letters of a language and the bases of DNA must be in the correct sequence. Just as "friend" and "fiend" mean different things and "fliend" doesn't mean anything, different sequences of bases in DNA may encode very different pieces of information, or no information at all. For example, think back to the Case Study at the beginning of this chapter; Rachel's melanoma is a case of sunlight changing a "friendly" gene to a "fiendish" one.

In Chapter 10, we will discover how the information in DNA is used to produce the structures of living cells. In the remainder of this chapter, we will examine how DNA is replicated during cell division to ensure accurate copying of this genetic information.

9.3 How Does DNA Replication Ensure Genetic Constancy During Cell Division?

The Replication of DNA Is a Critical Event in a Cell's Life

In the 1850s, Austrian pathologist Rudolf Virchow realized that "all cells come from [preexisting] cells." All of the trillions of cells of your body are the offspring (usually called *daughter cells*) of other cells, going all the way back to when you were a fertilized egg. Moreover, nearly every cell of your body contains identical genetic information—the same genetic information present in that fertilized egg. To accomplish this, cells reproduce by a complex process of cell division that produces two daughter cells from a single parental cell (we will learn more about this in Chapter 11). Each daughter cell receives a nearly perfect copy of the parent cell's genetic information. Consequently, at an early stage of cell division, the parent cell must synthesize two exact copies of its DNA through a process known as **DNA replication**. Many cells in an adult human never divide at all and therefore do not replicate their DNA. In most of the millions of cells that *do* divide, initiation of DNA replication irreversibly commits the cell to division. If a cell tries to replicate its DNA without stockpiling enough raw materials or energy to complete the process, it can die. Therefore, the timing of replication is carefully regulated, ensuring that DNA replication does not begin unless the cell is ready to divide. These controls also ensure that the cell's DNA is replicated *exactly* one time prior to each cell division.

Once the "decision" is made to divide, the cell replicates its DNA. Recall that DNA is a component of chromosomes. Each chromosome contains a single DNA double helix. DNA replication produces two identical double helices, one of which will be passed to each of the new daughter cells, as we will see in Chapter 11.

DNA Replication Produces Two DNA Double Helices, Each with One Old Strand and One New Strand

How does a cell accurately copy its DNA? In their paper describing DNA structure, Watson and Crick included one of the greatest understatements in all of science: "It has not escaped our notice that the specific [base] pairing we have postulated immediately suggests a possible copying mechanism for the genetic material." In fact, base pairing is the foundation of DNA replication. Remember, the rules for base pairing are that an adenine on one strand must pair with a thymine on the other strand, and a cytosine must pair with a guanine. If one strand reads ATG, for example, then the other strand must read TAC. Therefore, the base sequence of each strand contains all of the information needed to replicate the other strand.

Conceptually, DNA replication is quite simple (Fig. 9-4). Enzymes called **DNA helicases** pull apart the parental DNA double helix, so that the bases of the two DNA strands no longer form base pairs with one another. Now DNA strands complementary to the two parental strands must be synthesized. Other enzymes, called **DNA polymerases**, move along each separated parental DNA strand, matching bases on the strand with complementary **free nucleotides**. DNA polymerase also connects these free nucleotides with one another to form two new DNA strands, each complementary to one of the parental DNA strands. Thus, if a parental DNA strand reads TAG, DNA polymerase will synthesize a new DNA strand with the complementary sequence ATC. For more information on how DNA is replicated, refer to "A Closer Look: DNA Replication."

When replication is complete, one parental DNA strand and its newly synthesized, complementary daughter DNA strand wind together into one double helix. At the same time, the other parental strand and its

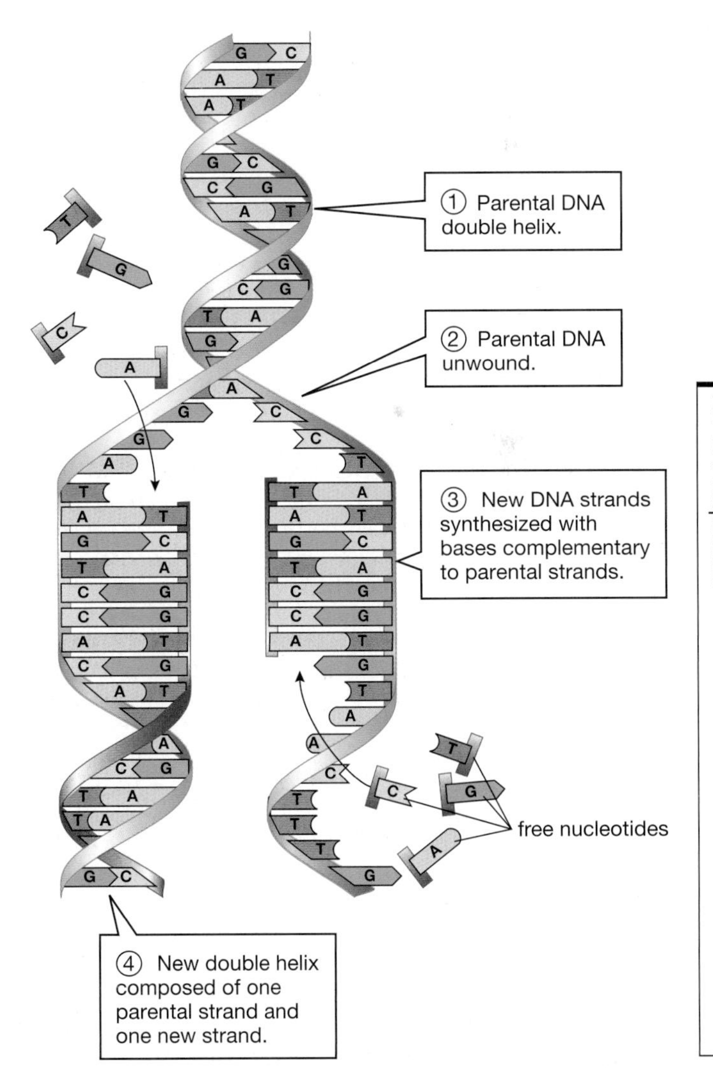
① Parental DNA double helix.
② Parental DNA unwound.
③ New DNA strands synthesized with bases complementary to parental strands.
free nucleotides
④ New double helix composed of one parental strand and one new strand.

FIGURE 9-4 Basic features of DNA replication
During replication, the two strands of the parental DNA double helix separate. Free nucleotides that are complementary to those in each strand are joined to make new daughter strands. Each parental strand and its new daughter strand then form a new double helix.

Media Activity
9.2 DNA Replication

A CLOSER LOOK DNA Replication

Media Activity
9.2 DNA Replication

DNA replication involves three major actions (Fig. E9-2). First, the DNA double helix must be opened up so that the base sequence can be "read." Then new DNA strands with base sequences complementary to the two original strands must be synthesized. In eukaryotic cells, these new DNA strands are synthesized in fairly short pieces. Therefore, the third step in DNA replication is to stitch the pieces together to form a continuous strand of DNA. Each step is carried out by a distinct set of enzymes.

DNA HELICASE SEPARATES THE PARENTAL DNA STRANDS

Acting in concert with several other enzymes, *DNA helicase* ("an enzyme that breaks apart the double helix") breaks the hydrogen bonds between complementary base pairs that hold the two parental DNA strands together. This separates and unwinds the parental double helix, forming a replication "bubble" (Fig. E9-2a). Within the replication bubble, the nucleotide bases of the parental DNA strands are no longer paired with one another. Each replication bubble contains two replication "forks" where the two parental DNA strands have not yet been unwound.

To help visualize this process, imagine that you are driving down an undivided two-lane road; each lane represents a single strand of a DNA double helix. The two DNA strands of the double helix point in opposite directions (see Fig. 9-3), just as cars in each lane of the road travel in opposite directions. A replication bubble is analogous to the road opening up, with a wide median separating the two lanes of the road. A little farther down the road, the median disappears, and the road once again becomes undivided. The places where the median begins and disappears are the "forks."

Eukaryotic chromosomes are so long that many DNA helicase enzymes open up many replication bubbles simultaneously, so that all of the DNA can be replicated in a reasonable length of time. The bubbles grow as DNA replication progresses and merge when they contact one another.

DNA POLYMERASE SYNTHESIZES NEW DNA STRANDS

Replication bubbles are essential because they allow a second enzyme, *DNA polymerase* ("an enzyme that makes a DNA polymer"), to gain access to the bases of each DNA strand (Fig. E9-2b). At each replication fork, DNA polymerase and other enzymes synthesize two new DNA strands that are complementary to the two parental strands. During this process, DNA polymerase recognizes an unpaired nucleotide base in the parental strand and matches it up with a free nucleotide that has the correct complementary base. For example, DNA polymerase pairs up an exposed adenine base in the parental strand with a thymine base in a free nucleotide. Then, DNA polymerase catalyzes the formation of new covalent bonds, linking the phosphate of the incoming free nucleotide to the sugar of the previously added nucleotide in the growing daughter strand. In this way, DNA polymerase synthesizes the sugar-phosphate backbone of the daughter strand.

SEGMENTS OF DNA ARE JOINED TOGETHER BY DNA LIGASE

DNA polymerase always moves toward the "free sugar" end of a single DNA strand. Because the two strands of the parental DNA double helix are oriented in opposite directions, the new complementary DNA strands will also be synthesized in opposite directions (Fig. E9-2b). Returning to our road analogy, a DNA polymerase enzyme stays on its own side of the DNA road, driving in the direction of the free sugar end of the strand.

Now picture how DNA helicase and DNA polymerase work together (Fig. E9-2c). DNA helicase "lands" on the double helix and moves along, unwinding the double helix and separating the strands. Because the two DNA strands run in opposite directions, as a DNA helicase enzyme moves toward the free sugar end of one strand, it is simultaneously moving toward the free phosphate end of the other strand. Now visualize two DNA polymerases "landing" on the two separated strands of DNA. One DNA polymerase (call it polymerase #1) can follow behind the helicase toward the free sugar end and can synthesize a continuous, complete new DNA strand. On the other strand, however, DNA polymerase #2 moves away from the helicase, and therefore can synthesize only part of a new DNA strand. As the helicase continues to unwind more of the double helix, additional DNA polymerases (#3, #4, and so on) must land on this strand and will in turn synthesize more pieces of DNA.

In this way, multiple DNA polymerases synthesize pieces of DNA of varying lengths, as many as 10 million pieces for a single human chromosome. How are all of these pieces sewn together? This is the job of the third major enzyme, **DNA ligase** ("an enzyme that ties DNA together"; Fig. E9-2d). Many DNA ligase enzymes stitch the fragments of DNA together until each daughter strand consists of one long, continuous DNA polymer.

FIGURE E9-2 Details of DNA replication
(a) DNA helicase separates the parental strands to form a replication bubble. **(b)** DNA polymerase synthesizes new pieces of DNA. **(c)** DNA helicase and DNA polymerase move along a replication bubble. **(d)** DNA ligase joins the small DNA segments into a single daughter strand. **QUESTION** During synthesis, why doesn't DNA polymerase move away from the replication fork on both strands?

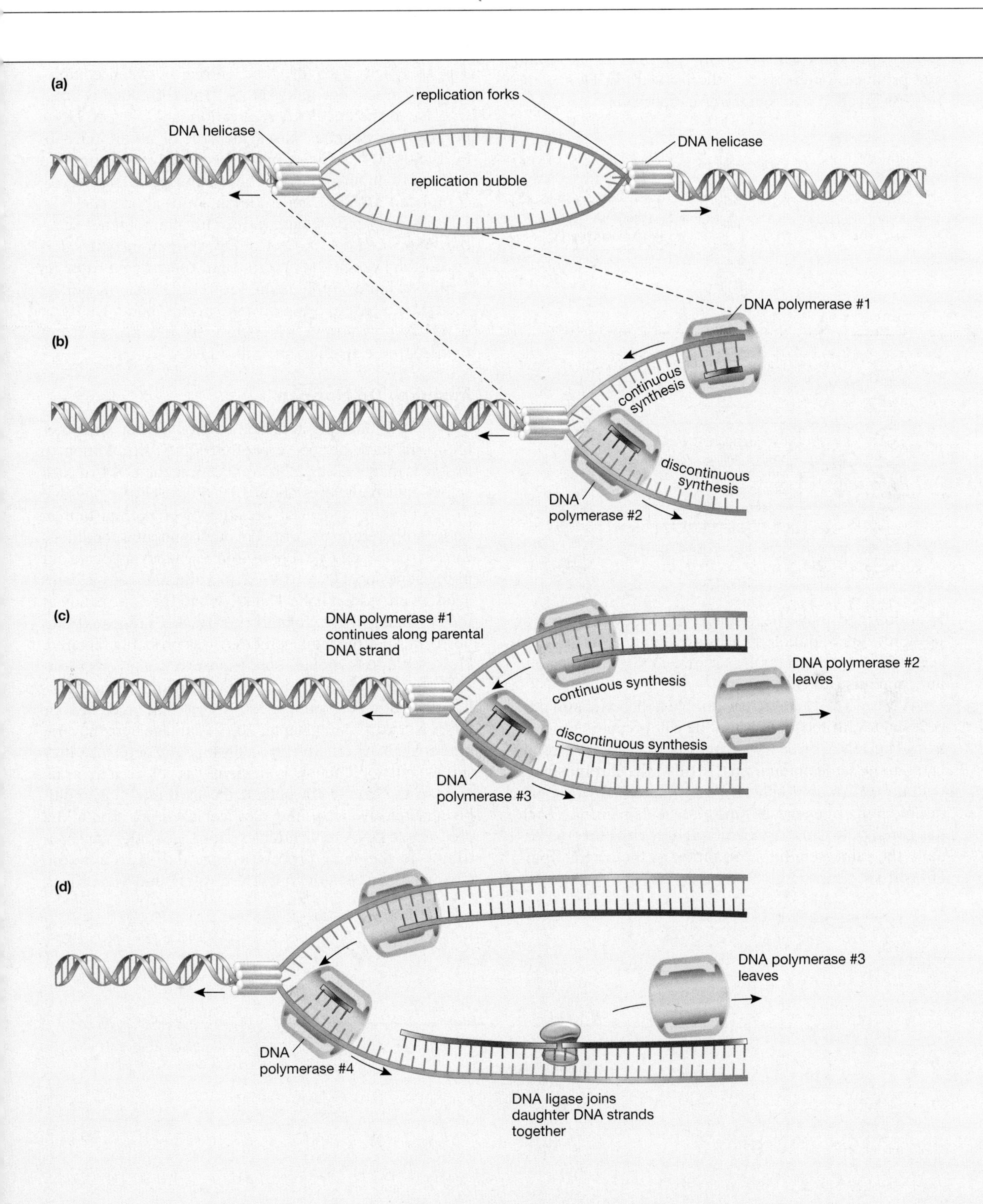
(a)
replication forks
DNA helicase
DNA helicase
replication bubble
(b)
DNA polymerase #1
continuous synthesis
discontinuous synthesis
DNA polymerase #2
(c)
DNA polymerase #1 continues along parental DNA strand
DNA polymerase #2 leaves
continuous synthesis
discontinuous synthesis
DNA polymerase #3
(d)
DNA polymerase #3 leaves
DNA polymerase #4
DNA ligase joins daughter DNA strands together

daughter strand wind together into a second double helix. In forming a new double helix, the process of DNA replication conserves one parental DNA strand and produces one newly synthesized strand. Hence, the process is called **semiconservative replication**.

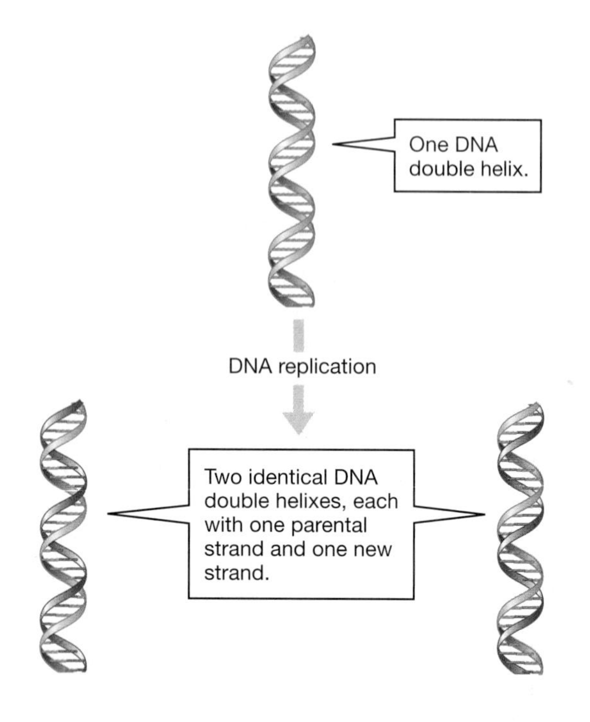

One DNA double helix.
DNA replication
Two identical DNA double helixes, each with one parental strand and one new strand.

If no mistakes have been made, the base sequences of both new DNA double helices are identical to the base sequence of the original, parental DNA double helix and, of course, to each other.

At this point, the two new double helices are still part of a single chromosome while the cell prepares for division. The DNA of every chromosome in the cell replicates in the same manner, so that all of the chromosomes contain two double helices. When the cell divides, one double helix of every chromosome is delivered to each daughter cell. Thus, the two daughter cells receive exactly the same genetic information as the original parent cell.

Proofreading Produces Almost Error-Free Replication of DNA

The specificity of hydrogen bonding between complementary base pairs makes DNA replication highly accurate. Nevertheless, DNA replication isn't perfect. DNA polymerase matches bases incorrectly about once in every 10,000 base pairs, partly because replication is so fast (up to about 700 nucleotides per second). However, completed DNA strands contain only about one mistake in every *billion* base pairs. This phenomenal accuracy is ensured by a variety of DNA repair enzymes that "proofread" each daughter strand during and after its synthesis. For example, some forms of DNA polymerase recognize a base-pairing mistake as it is made. This type of DNA polymerase will pause, fix the mistake, and then continue synthesizing more DNA.

Mistakes Do Happen

Despite this amazing accuracy, neither we nor any other life-forms have perfect, error-free DNA. In addition to mistakes made during DNA replication, the DNA in each cell in your body loses about 10,000 bases every day, due to spontaneous chemical breakdown at normal human body temperatures. A variety of environmental conditions can also damage DNA. For example, whenever you go out in the sunshine, DNA in some of your skin cells is damaged by ultraviolet light. Think back to the Case Study at the beginning of the chapter; after Rachel's repeated sunburns, most of the damaged DNA was probably repaired. However, after enough sun exposure, some errors inevitably remained. A cell with DNA damage may function normally. For example, it turns out that much of your DNA is "extra" (see Chapter 10), so damage here may not matter at all. Alternatively, a cell with damaged DNA may survive but not function as efficiently as before; it may die; or, as in the case of Rachel's skin cells, it may start multiplying explosively, forming a cancer. Deterioration in the accuracy of DNA replication as people get older may contribute to the aging process, a topic that is explored in Chapter 10 in "Health Watch: Sex, Aging, and Mutations."

CASE STUDY REVISITED Sunshine Perils

Ultraviolet rays in sunlight penetrate into the skin with just the right energy to damage the bases in DNA. For example, if UV rays hit a DNA strand with two adjacent thymines, the thymines may become linked together, preventing them from forming hydrogen bonds with adenines on the opposite DNA strand. In most cases, DNA repair enzymes snip out the offending thymine pair, DNA polymerase fills in the gap with normal thymines, and DNA ligase attaches the new little piece of DNA to the rest of the strand.

After a sunburn, many skin cells sustain more DNA damage than they can repair. Severely damaged cells commit a kind of cellular suicide. That's why you peel after a bad sunburn; the dead cells are being sloughed off. If a cell is not lethally injured, but the DNA of a gene involved in controlling cell division is damaged, the progeny of the damaged cell may begin to divide uncontrollably, starting a cancer. Without DNA repair, even brief exposure to sunlight would rapidly cause skin cancer. Rachel would probably have developed melanoma much earlier, except for constant repair jobs on her DNA.

As a child, Rachel had many sunburns, which may have caused her cancer. Being badly sunburned as a kid about triples the risk of developing melanoma. People with darker skin—which burns less easily—are about 15 times less likely to develop melanomas, though they are not totally protected.

Because her melanoma was caught at an early stage, Rachel's prognosis for total recovery is good. You, too, can learn to spot potential problems before they become deadly. Recognizing a possible melanoma is as easy as "ABCD": Examine your moles for *A*symmetry, irregular *B*order, irregular *C*olor, or a *D*iameter larger than the eraser on the end of a pencil. Check out this text's Web site for some tips on recognizing possible melanomas, and have any suspicious spots promptly examined by your doctor.

Consider This: DNA damage doesn't only cause cancer. Many scientists hypothesize that DNA damage may also be responsible for some aspects of aging. Why do you think that DNA damage might lead to certain characteristics of aging, such as wrinkles, thinning hair, and slower healing? Assuming this hypothesis is true, what actions might people take to slow the aging process?

Links to Life: *Tyrannosaurus rex* Reborn?

Remember *Jurassic Park*? Scientists found a mosquito entombed in amber, delicately inserted a needle into its stomach, and sucked out some dinosaur blood cells that the mosquito had dined on just before its demise. They extracted the cells' DNA and—*voila!*—made a *T. rex*. In principle, it seems plausible. After all, DNA contains all the genetic information needed to manufacture an entire tyrannosaurus, doesn't it? So if you could find *T. rex* DNA, why couldn't you make one?

Two reasons, actually. The first is the decay rate of DNA. DNA spontaneously breaks down and therefore needs constant repair. Seventy million years is a *lo-o-ong* time for DNA to decay without repair. DNA discovered in a perfectly preserved, quick-frozen mammoth from 10,000 years ago might be fairly intact. But *T.* rex DNA inside a 70 million-year-old mosquito? No way.

The second problem is determining which genes to use when. Nearly every cell in your body contains all the genes needed to replicate you, but your liver doesn't grow hair and your brain doesn't make teeth. During development, each cell of your body uses the genetic information in specific genes, at specific times, to produce a human being. As we'll see in Chapter 37, molecules in the fertilized egg begin the process of selecting which genes to turn on and which to turn off. Maybe elephant eggs could regulate mammoth DNA, but it's not likely that lizard or crocodile eggs could properly control the DNA of a *T. rex*.

It's really too bad that we'll never see a live *T. rex*, but you probably wouldn't want to meet one up close and personal, anyway . . .

Summary of Key Concepts

9.1 How Did Scientists Discover That Genes Are Made of DNA?

By the turn of the century, scientists knew that genes must be made of either protein or DNA. Studies by Griffith showed that genes can be transferred from one bacterial strain into another. This transfer could transform the bacterial strain from harmless to deadly. Avery, MacLeod, and McCarty showed that DNA was the molecule that could transform bacteria. Thus, genes must be made of DNA.

9.2 What Is the Structure of DNA?

DNA is composed of subunits called nucleotides, linked together into long strands. Each nucleotide consists of a phosphate group, the five-carbon sugar deoxyribose, and a nitrogen-containing base. Four types of bases occur in DNA: adenine, guanine, thymine, and cytosine. Within DNA, two nucleotide strands wind about one another to form a double helix. Within each strand, the sugar of one nucleotide is linked to the phosphate of the next nucleotide, forming a sugar-phosphate "backbone" on each side of the double helix. The nucleotide bases of each strand pair up in the middle of the helix, held together by hydrogen bonds. Only specific pairs of bases, called complementary base pairs, can bond together in the helix: adenine bonds with thymine, and guanine bonds with cytosine.

9.3 How Does DNA Replication Ensure Genetic Constancy During Cell Division?

When cells reproduce, they must replicate their DNA so that each daughter cell receives all of the original genetic information. During DNA replication, enzymes unwind the two parental DNA strands. The enzyme DNA polymerase binds to each parental DNA strand, selects free nucleotides with complementary bases, and links the nucleotides together to form new DNA strands. The sequence of nucleotides in each newly formed strand is complementary to the sequence of a parental strand. Replication is semiconservative because, when DNA replication is complete, both new DNA double helices consist of one parental DNA strand and one newly synthesized, complementary strand. The two new DNA double helices are therefore duplicates of the parental DNA double helix. DNA polymerase and other repair enzymes "proofread" the DNA, minimizing the number of mistakes during replication.

Key Terms

adenine (A) *p. 151*
bases *p. 151*
chromosome *p. 150*
complementary base pairs *p. 153*
cytosine (C) *p. 151*
DNA *p. 150*
DNA helicase *p. 155*
DNA ligase *p. 156*
DNA polymerase *p. 155*
DNA replication *p. 155*
double helix *p. 152*
free nucleotides *p. 155*
gene *p. 150*
guanine (G) *p. 151*
nucleotides *p. 151*
semiconservative replication *p. 158*
sugar-phosphate backbone *p. 152*
thymine (T) *p. 151*

Thinking Through the Concepts

To take a multiple-choice quiz with feedback on the contents of this chapter, visit http://www.prenhall.com/audesirk7. *Log in to the Web site selected by your instructor and navigate to the Self Test section for this chapter.*

? Review Questions

1. Draw the general structure of a nucleotide. Which parts are identical in all nucleotides, and which can vary?
2. Name the four types of nitrogen-containing bases found in DNA.
3. Which bases are complementary to one another? How are they held together in the double helix of DNA?
4. Describe the structure of DNA. Where are the bases, sugars, and phosphates in the structure?
5. Describe the process of DNA replication.

Applying the Concepts

1. As you learned in "Scientific Inquiry: The Discovery of the Double Helix," scientists in different laboratories often compete with one another to make new discoveries. Do you think this competition helps promote scientific discoveries? Sometimes researchers in different laboratories collaborate with one another. What advantages does collaboration offer over competition? What factors might provide barriers to collaboration and lead to competition?
2. Genetic information is encoded in the sequence of nucleotides in DNA. Let's suppose that the nucleotide sequence on one strand of a double helix encodes the information needed to synthesize a hemoglobin molecule. Do you think that the sequence of nucleotides on the other strand of the double helix also encodes useful information? Why? (An analogy might help. Suppose that English were a "complementary language," with letters at opposite ends of the alphabet complementary to one another [that is, A is complementary to Z, B to Y, C to X, etc.]. Would a sentence composed of letters complementary to "To be or not to be?" make sense?) Finally, why do you think DNA is double-stranded?
3. Today, scientific advances are being made at an astounding rate, and nowhere is this more evident than in our understanding of the biology of heredity. Using DNA as a starting point, do you believe that there are limits to the knowledge people should acquire? Defend your answer.

For More Information

Crick, F. *What Mad Pursuit: A Personal View of Scientific Discovery.* New York: Basic Books, 1998. Another view of the race to determine the structure of DNA, by Francis Crick himself.

Gibbs, W. W. "Peeking and Poking at DNA." *Scientific American (Explorations),* March 31, 1997. An update of new techniques for studying the DNA molecule, such as atomic force microscopy.

Judson, H. F. *The Eighth Day of Creation.* Cold Spring Harbor, NY: Cold Spring Harbor Laboratory Press, 1993. A very readable historical perspective on the development of genetics.

Mirsky, A. E. "The Discovery of DNA." *Scientific American,* June 1968. The early history of DNA research.

Olby, R. *The Path to the Double Helix.* Seattle: University of Washington Press, 1975. A historical perspective on the development of genetics in the twentieth century.

Radman, M., and Wagner, R. "The High Fidelity of DNA Duplication." *Scientific American,* August 1988. Faithful duplication of chromosomes requires both reasonably accurate initial replication of DNA sequences and final proofreading.

Rennie, J. "DNA's New Twists." *Scientific American,* March 1993. A reprise of new information on DNA structure and function.

Watson, J. D. *The Double Helix.* New York: Atheneum, 1968. If you still believe the Hollywood images that scientists are either maniacs or cold-blooded, logical machines, be sure to read this book. Although hardly models for the behavior of future scientists, Watson and Crick are certainly human enough!

Weinberg, R. "How Cancer Arises." *Scientific American,* September 1996. An overview of the molecular basis of cancer: mutations in DNA.

Wheelwright, J. "Bad Genes, Good Drugs." *Discover,* April 2002. The human genome project provides insights into genetic disorders and possible treatments.

Media Activities

To access a Media Activity visit http://www.prenhall.com/audesirk7. *Log in to the Web site selected by your instructor, navigate to this chapter, and select the appropriate Media Activity number.*

9.1 DNA Structure

Estimated time: 5 minutes

The structure of the DNA double helix is the key to the ability of this molecule to store and transmit genetic information. In this activity, you will explore the structure of the DNA molecule.

9.2 DNA Replication

Estimated time: 5 minutes

This activity explores the process of DNA replication, which is crucial for the organism growth and reproduction.

9.3 Web Investigation: Sunshine Perils

Estimated time: 20 minutes

Skin cancer is now the most common cancer in the United States. What is skin cancer? Who is most at risk? What are the treatment alternatives? How can it be prevented? This exercise explores the biochemical, medical, and psychological features of skin cancers.

CHAPTER

10 Gene Expression and Regulation

Many of the basic differences in the body structures of males and females can ultimately be traced to differences in a single gene.

AT A GLANCE

CASE STUDY Boy or Girl?

"Is it a boy or a girl?" When we hear about the birth of a baby, this is often the first thing we want to know. For many of us, the response will send us shopping for gifts: frilly, pink, and cuddly for girls; solid, blue, and sporty for boys. Such differences in what is considered "appropriate" for boys and girls reflect societal ideals rather than biological facts. However, boys and girls clearly have many physical differences that are biologically determined. An obvious difference is the presence of a vagina in girls and a penis in boys. Additional physical differences become obvious as children grow into adults. Men are usually heavier, stronger, and grow beards. Women are usually smaller, less muscular, and grow breasts. However, the genes of men and women do not differ so dramatically; human males and females have almost exactly the same genes. In fact, boys have all of the genes needed to make female genitalia, and girls have all of the genes needed to make male genitalia. In boys, the action of a single gene activates the male developmental pathway and deactivates the female developmental pathway. Without this gene we would all be physically female. How can a single gene determine something as complex as the sex of a human being? To solve this mystery, you will need to learn more about how genes work.

10.1 How Are Genes and Proteins Related?

Information, by itself, doesn't *do* anything. For example, a blueprint may describe the structure of a house in great detail, but unless that information is translated into action, no house will ever be built. Likewise, although the base sequence of DNA, the "molecular blueprint" of every cell, contains an incredible amount of information, DNA cannot carry out any action on its own. So how does DNA determine whether you are male or female, or have brown or blue eyes?

Proteins are a cell's "molecular workers." Every cell contains a particular set of proteins. The activities of these proteins control the cell's shape, function, and reproduction, as well as the synthesis of lipids, carbohydrates, and nucleic acids. Therefore, there must be a flow of information from the DNA of a cell's **genes** to the proteins that actually carry out the cell's functions.

Most Genes Contain the Information for the Synthesis of a Single Protein

Cells synthesize molecules in a series of linked steps called *biochemical pathways.* Each step in a biochemical pathway is catalyzed by an enzyme. (Recall from Chapters 3 and 6 that enzymes are proteins that catalyze a chemical reaction.) Within a biochemical pathway, the product produced by one enzyme becomes the substrate of the next enzyme in the pathway, like a molecular assembly line (refer back to Fig. 6-7). Genes must encode the information needed to produce these biochemical pathways, but how? The study of the biochemical pathways of a common bread mold, *Neurospora crassa*, revealed the relationship between an organism's genes and its enzymes.

Although we normally encounter *Neurospora* growing on stale bread, it can survive on a much simpler diet. All it needs is an energy source such as sugar, a few minerals, and vitamin B_6. Therefore, *Neurospora* has all the enzymes needed to make virtually all of its own organic molecules, including all of the amino acids. (In contrast, we humans cannot synthesize many vitamins or 9 of the 20 common amino acids; we must obtain these from our food.) Individual *Neurospora*, like any organism, may have mutations (usually defects) in some of its genes. In the 1940s, geneticists George Beadle and Edward Tatum used mutant *Neurospora* to test the hypothesis that many of an organism's genes encode the information needed to synthesize enzymes. If this hypothesis is true, a mutation in a specific gene might disrupt the synthesis of a specific enzyme. Without this enzyme, one of the mold's biochemical pathways wouldn't function properly. The mold would be unable to synthesize some of the organic molecules that it needs to survive, such as certain amino acids. These mutant *Neurospora* would be able to grow on a simple medium of sugar, minerals, and vitamin B_6 only if the missing organic molecules were added to the medium.

Beadle and Tatum isolated several strains of *Neurospora*, each with a single defective gene. Some of these mutant strains could grow if the simple medium were supplemented with the amino acid arginine. Arginine is synthesized from citrulline, which in turn is synthesized from ornithine (Fig. 10-1b). Mutant A could grow only if supplemented with arginine, but not if supplemented with citrulline or ornithine (Fig. 10-1a). Therefore, this strain had a defect in the enzyme that converts citrulline to arginine. Mutant B could grow if the medium were supplemented with either arginine or citrulline, but not with ornithine (see Fig. 10-1a). This mutant had a defect in the enzyme that converts ornithine to citrulline. Because a mutation in a single gene affected only a single enzyme within a single biochemical pathway, Beadle and Tatum concluded that one gene encodes the information for one enzyme. The importance of this observation was recognized in 1958 with a Nobel Prize, which was shared by Joshua Lederberg, one of Tatum's students.

Almost all enzymes are proteins, but many of the proteins in a cell are not enzymes. For example, keratin is a structural protein in hair and nails, but it does not catalyze any chemical reactions. In addition, many enzymes are composed of more than one protein subunit. For example, DNA polymerase is composed of more than a dozen proteins. Thus, the "one gene, one enzyme" relationship proposed by Beadle and Tatum was later clarified to "one gene, one protein." (As you know from Chapter 3, a protein is a chain of amino acids joined by peptide bonds. Depending on the length of the chain, proteins may be called peptides [short chains] or polypeptides [long chains]. In this text, we usually call any chain of amino acids, regardless of length, a protein.) There are exceptions to the "one gene, one protein" rule, including several in which the final product of a gene isn't protein, but a nucleic acid—called *ribonucleic acid*—described in the next section. Nevertheless, as a generalization, most genes encode the information for a single protein.

DNA Provides Instructions for Protein Synthesis via RNA Intermediaries

The DNA of a eukaryotic cell is housed in the nucleus, but protein synthesis occurs on ribosomes in the cytoplasm (see Chapter 5). Therefore, DNA cannot directly guide protein synthesis. There must be an intermediary, a molecule that carries the information from DNA in the nucleus to the ribosomes in the cytoplasm. This molecule is **ribonucleic acid**, or **RNA**.

RNA is similar to DNA but differs structurally in three respects: RNA is normally single-stranded; RNA has the sugar ribose (instead of deoxyribose) in its backbone; and RNA has the base uracil instead of the base thymine found in DNA (Table 10-1).

DNA codes for synthesis of three major types of RNA: **messenger RNA (mRNA)**, **transfer RNA (tRNA)**, and

(a) Growth characteristics of normal and mutant *Neurospora* on simple medium with different supplements show that defects in a single gene lead to defects in a single enzyme.

		Supplements Added to Medium				
		none	ornithine	citrulline	arginine	**Conclusions**
Normal *Neurospora*						Normal *Neurospora* can synthesize arginine, citrulline, and ornithine.
Mutants with single gene defect	A					Mutant A grows only if arginine is added. It cannot synthesize arginine because it has a defect in enzyme 2; gene A is needed for synthesis of arginine.
	B					Mutant B grows if either arginine or citrulline are added. It cannot synthesize arginine because it has a defect in enzyme 1. Gene B is needed for synthesis of citrulline.

(b) The biochemical pathway for synthesis of the amino acid arginine involves two steps, each catalyzed by a different enzyme.

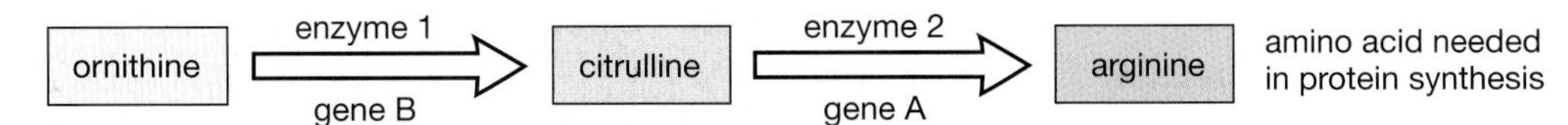

FIGURE 10-1 Beadle and Tatum's experiments with *Neurospora* mutants
QUESTION What result would you expect for a mutant that lacks an enzyme needed to produce ornithine?

Table 10-1 A Comparison of DNA and RNA

	DNA	RNA	
Strands	2	1	
Sugar	deoxyribose	ribose	
Types of Bases	adenine (A), thymine (T) cytosine (C), guanine (G)	adenine (A), uracil (U) cytosine (C), guanine (G)	
Base Pairs	DNA:DNA	RNA:DNA	RNA:RNA
	A–T	A–T	A–U
	T–A	U–A	U–A
	C–G	C–G	C–G
	G–C	G–C	G–C
Function	Contains genes; sequence of bases in most genes determines the amino acid sequence of a protein	**Messenger RNA (mRNA):** carries the code for a protein-coding gene from DNA to ribosomes **Ribosomal RNA (rRNA):** combines with proteins to form ribosomes, the structures that link amino acids to form a protein **Transfer RNA (tRNA):** carries amino acids to the ribosomes	

(a) Messenger RNA (mRNA)

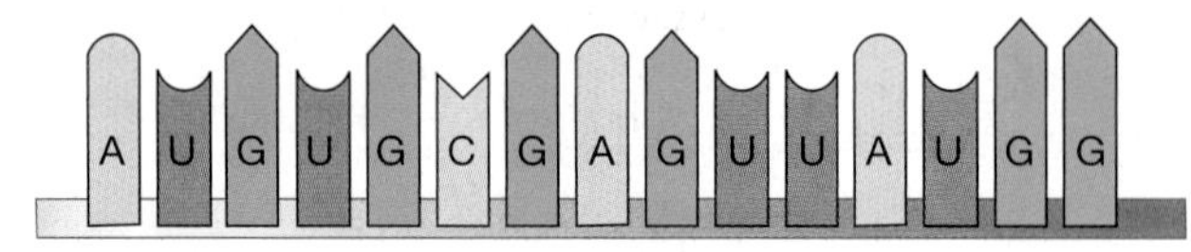

(b) Ribosome: contains ribosomal RNA (rRNA)

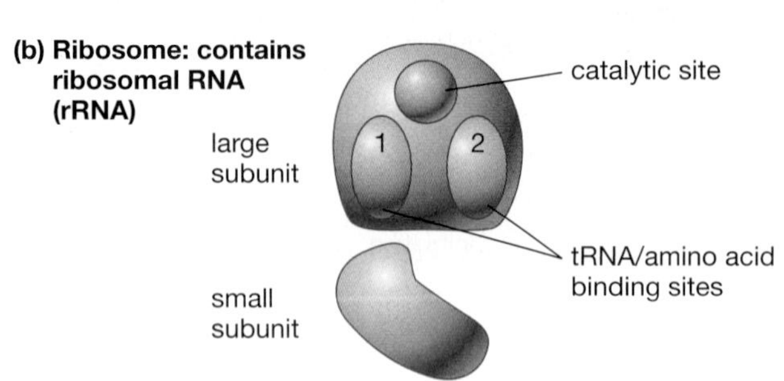

catalytic site
large subunit
1
2
tRNA/amino acid binding sites
small subunit
rRNA combines with proteins to form ribosomes. The small subunit binds mRNA. The large subunit binds tRNA and catalyzes peptide bond formation between amino acids during protein synthesis.

(c) Transfer RNA (tRNA)

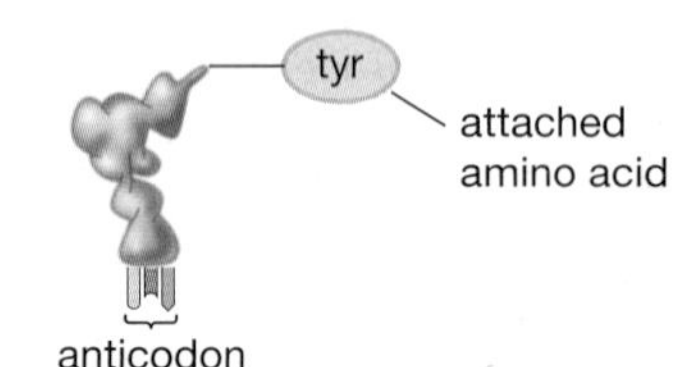

tyr
attached amino acid
anticodon
Each tRNA carries a specific amino acid to a ribosome during protein synthesis. The anticodon of tRNA pairs with a codon of mRNA, ensuring that the correct amino acid is incorporated into the protein.

FIGURE 10-2 Cells synthesize three major types of RNA

ribosomal RNA (rRNA) (Fig. 10-2). All these RNA molecules are involved in converting the nucleotide sequence of genes into the amino acid sequence of proteins. We will examine their functions in more detail shortly.

Overview: Genetic Information Is Transcribed into RNA, Then Translated into Protein

Information from DNA is used to direct the synthesis of proteins in a two-step process (Fig. 10-3 and Table 10-2):

1. During RNA synthesis, or **transcription** (see Fig. 10-3a), the information contained in the DNA of a specific gene is copied into RNA, either messenger RNA (mRNA), transfer RNA (tRNA), or ribosomal RNA (rRNA). Thus, a gene is a segment of DNA that can be copied, or transcribed, into RNA. Transcription is catalyzed by an enzyme, RNA polymerase, and occurs in the nucleus.
2. During protein synthesis, or **translation** (see Fig. 10-3b), tRNA and rRNA, together with proteins, use the nucleotide sequence in an mRNA molecule to synthesize the specific amino acid sequence of a protein. Ribosomal RNA combines with dozens of proteins to form a complex structure called a **ribosome**. Translation is catalyzed by ribosomes and occurs in the cytoplasm.

It is easy to confuse the terms *transcription* and *translation*. Comparing their common English meanings with their biological meanings may help you understand the difference. In everyday English, to *transcribe* means to make a written copy of something, almost always in the same language. In an American courtroom, for example, verbal testimony is transcribed into a written copy, and both the testimony of the witnesses and the transcriptions are in English. In biology, *transcription* is the process of copying information from DNA to RNA using the common "language" of nucleotides. In contrast, the common English meaning of *translation* is to convert words from one language to a different language; similarly, in biology, *translation* means to convert information from the "nucleotide language" of RNA to the "amino acid language" of proteins.

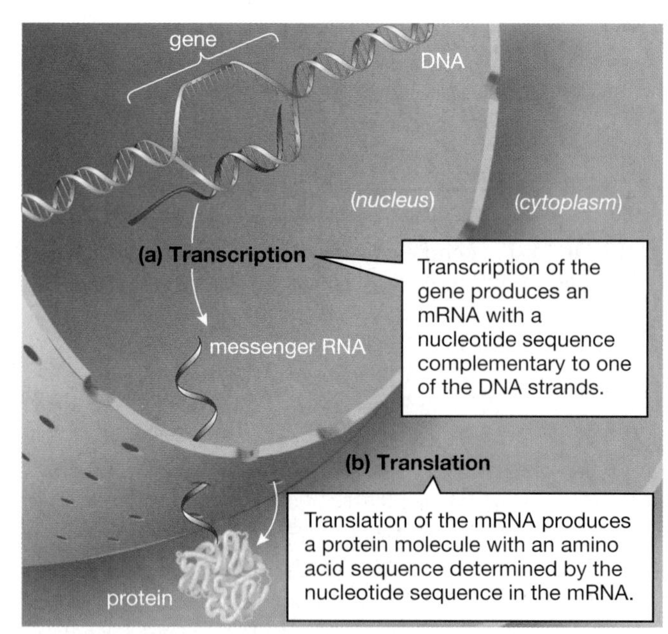

FIGURE 10-3 Genetic information flows from DNA to RNA to protein
(a) During transcription, the nucleotide sequence in a gene specifies the nucleotide sequence in a complementary RNA molecule. For protein-encoding genes, the product is an mRNA molecule that exits from the nucleus and enters the cytoplasm. **(b)** During translation, the sequence in an mRNA molecule specifies the amino acid sequence in a protein.

Table 10-2 Processes Involved in the Use and Inheritance of Genetic Information

Process	Information for Process	Product	Major Enzyme or Structure Involved in Process	Type of Base Pairing Required
Transcription (synthesis of RNA)	Short segment of one DNA strand	One RNA molecule (mRNA, tRNA, rRNA)	RNA polymerase	DNA with RNA: DNA bases pair with RNA bases in new RNA molecule.
Translation (synthesis of protein)	mRNA	One protein molecule	Ribosome (also requires tRNA)	mRNA with tRNA: Codon in mRNA forms base pairs with anticodon in tRNA.
Replication (synthesis of DNA; occurs only before cells divide)	Entire length of both DNA strands	Two DNA double helices (each with one old and one new strand)	DNA polymerase	DNA with DNA: DNA bases of each parental strand pair with DNA bases in the newly synthesized strands.

To understand the molecular mechanisms involved in the flow of information from DNA to RNA to protein, geneticists first had to break the language barrier: how does the language of nucleotide sequences in DNA and messenger RNA translate into the language of amino acid sequences in proteins? This translation relies on a "dictionary" called the genetic code.

The Genetic Code Uses Three Bases to Specify an Amino Acid

The **genetic code** translates the sequence of bases in nucleic acids into the sequence of amino acids in proteins. But which combinations of bases code for which amino acids? Both DNA and RNA contain four different bases: A, T (or U in RNA), G, and C (see Table 10-1). However, proteins are made of 20 different amino acids. Therefore, one base cannot code for one amino acid because there are simply not enough different types of bases. The genetic code must rely on a short sequence of bases to encode each amino acid. If a sequence of two bases codes for an amino acid, there would be 16 possible combinations, which still isn't enough to code for all 20 amino acids. A three-base sequence, however, gives 64 possible combinations of bases, which is more than enough. Under the assumption that nature operates as economically as possible, biologists hypothesized that the genetic code must be triplet: Three bases specify a single amino acid. In 1961, Francis Crick and three co-workers demonstrated that this hypothesis is correct (see "Scientific Inquiry: Cracking the Genetic Code" on the Web for more information).

For any language to be understood, its users must know what the words mean, where words start and stop, and where sentences begin and end. To decipher the "words" of the genetic code, researchers ground up bacteria and isolated the components needed to synthesize proteins. To this mixture, they added artificial mRNA, which allowed them to control what "words" were to be translated. They could then see which amino acids were incorporated into the resulting proteins. For example, an mRNA strand composed entirely of uracil (UUUUUUUU . . .) directed the mixture to synthesize a protein composed solely of the amino acid phenylalanine. Therefore, the triplet UUU must specify phenylalanine. Because the genetic code was deciphered by using these artificial mRNAs, it is usually written in terms of the base triplets in mRNA (rather than in DNA) that code for each amino acid (Table 10-3). These mRNA triplets are called **codons**.

Table 10-3 The Genetic Code (Codons of mRNA)

First Base	Second Base: U		Second Base: C		Second Base: A		Second Base: G		Third Base
U	UUU	Phenylalanine (Phe)	UCU	Serine (Ser)	UAU	Tyrosine (Tyr)	UGU	Cysteine (Cys)	U
U	UUC	Phenylalanine	UCC	Serine	UAC	Tyrosine	UGC	Cysteine	C
U	UUA	Leucine (Leu)	UCA	Serine	UAA	Stop	UGA	Stop	A
U	UUG	Leucine	UCG	Serine	UAG	Stop	UGG	Tryptophan (Trp)	G
C	CUU	Leucine	CCU	Proline (Pro)	CAU	Histidine (His)	CGU	Arginine (Arg)	U
C	CUC	Leucine	CCC	Proline	CAC	Histidine	CGC	Arginine	C
C	CUA	Leucine	CCA	Proline	CAA	Glutamine (Gln)	CGA	Arginine	A
C	CUG	Leucine	CCG	Proline	CAG	Glutamine	CGG	Arginine	G
A	AUU	Isoleucine (Ile)	ACU	Threonine (Thr)	AAU	Asparagine (Asp)	AGU	Serine (Ser)	U
A	AUC	Isoleucine	ACC	Threonine	AAC	Asparagine	AGC	Serine	C
A	AUA	Isoleucine	ACA	Threonine	AAA	Lysine (Lys)	AGA	Arginine (Arg)	A
A	AUG	Methionine (Met)	ACG	Threonine	AAG	Lysine	AGG	Arginine	G
G	GUU	Valine (Val)	GCU	Alanine (Ala)	GAU	Aspartic acid (Asp)	GGU	Glycine (Gly)	U
G	GUC	Valine	GCC	Alanine	GAC	Aspartic acid	GGC	Glycine	C
G	GUA	Valine	GCA	Alanine	GAA	Glutamic acid (Glu)	GGA	Glycine	A
G	GUG	Valine	GCG	Alanine	GAG	Glutamic acid	GGG	Glycine	G

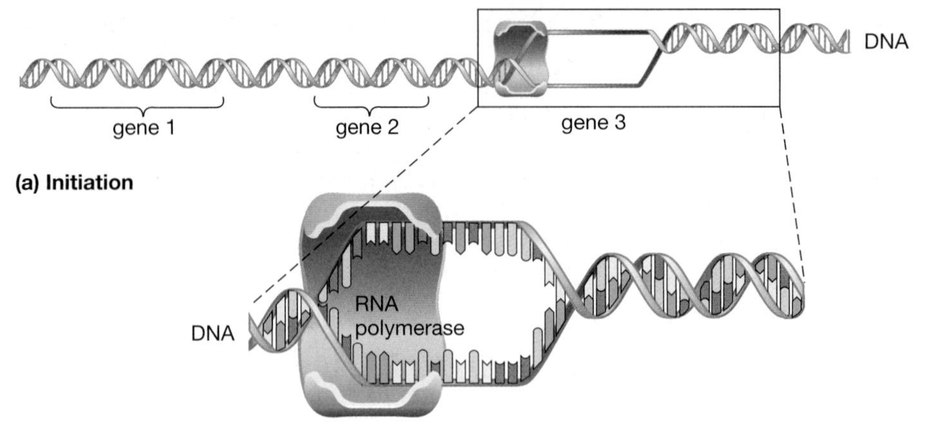

(a) Initiation

RNA polymerase binds to the promoter region of DNA near the beginning of a gene, separating the double helix near the promoter.

(b) Elongation

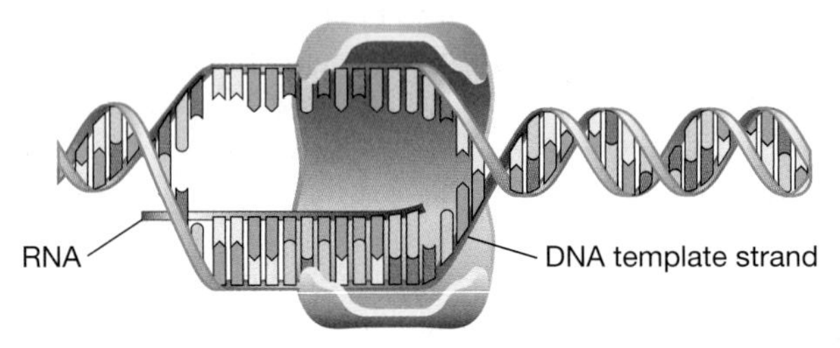

RNA polymerase travels along the DNA template strand, catalyzing the addition of ribose nucleotides into an RNA molecule. The nucleotides in the RNA are complementary to the template strand of the DNA.

(c) Termination

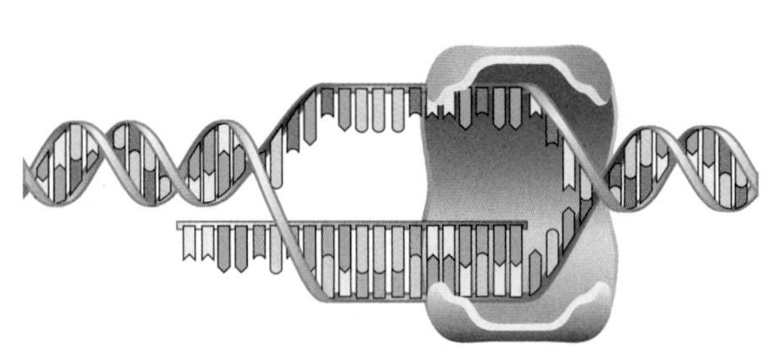

At the end of a gene, RNA polymerase encounters a sequence of DNA called a termination signal. RNA polymerase detaches from the DNA and releases the RNA molecule.

(d) Conclusion of transcription

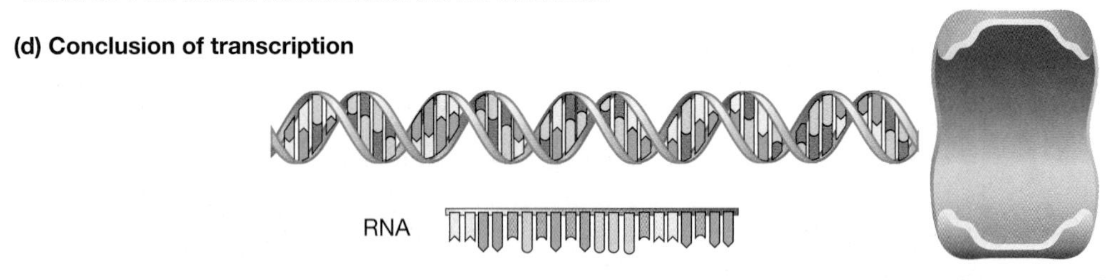

After termination, the DNA completely rewinds into a double helix. The RNA molecule is free to move from the nucleus to the cytoplasm for translation, and RNA polymerase may move to another gene and begin transcription once again.

Media Activity
10.1 Transcription

FIGURE 10-4 Transcription is the synthesis of RNA from instructions in DNA
A gene is a segment of a chromosome's DNA. One of the DNA strands will serve as the template for the synthesis of an RNA molecule with bases complementary to the bases in the DNA strand. **QUESTION** If the other DNA strand of this molecule were the template strand, in which direction would the RNA polymerase travel?

What about punctuation? Given that one mRNA molecule may contain hundreds or even thousands of bases, how does the cell recognize where codons start and stop, and where the code for an entire protein starts and stops? All proteins originally begin with the same amino acid, methionine (though it may be removed after the protein is synthesized). Methionine is specified by the codon AUG, which is known as the **start codon**. Three codons—UAG, UAA, and UGA—are **stop codons**. When the ribosome encounters a stop codon, it releases both the newly synthesized protein and the mRNA. Because all codons consist of three bases, and the beginning and end of a protein are specified, then punctuation ("spaces") between codons is unnecessary. Why? Consider what would happen if English used only three-letter words: a sentence such as THEDOGSAWTHECAT would be perfectly understandable, even without spaces between the words.

Because the genetic code has three stop codons, 61 nucleotide triplets remain to specify only 20 amino acids. Therefore, most amino acids are specified by several different codons. For example, six different codons specify leucine (see Table 10-3), so whether UUA or CUG is present in the mRNA sequence, ribosomes add leucine to the growing amino acid chain. However, each codon specifies one, and only one, amino acid.

10.2 How Is Information in a Gene Transcribed into RNA?

We can view transcription as a process consisting of (1) *initiation*, (2) *elongation*, and (3) *termination*. These three steps correspond to the three major parts of most genes in both eukaryotes and prokaryotes: (1) a *promoter* region at the beginning of the gene, where transcription is started, or initiated; (2) the "body" of the gene where elongation of the RNA strand occurs; and (3) a termination signal at the end of the gene, where RNA synthesis ceases, or terminates.

Initiation of Transcription Occurs When RNA Polymerase Binds to the Promoter of a Gene

The enzyme **RNA polymerase** synthesizes RNA. To initiate transcription, RNA polymerase must first locate the beginning of a gene. Near the beginning of every gene is an untranscribed sequence of DNA bases called the **promoter**, often one or more repetitions of the sequence TATA. When RNA polymerase binds to the promoter region of a gene, the DNA double helix at the beginning of the gene unwinds and transcription begins (Fig. 10-4a). As we will see shortly, the binding of RNA polymerase to the promoter of a specific gene, and therefore the transcription of that gene, may be blocked or enhanced, depending on conditions inside the cell.

Elongation Proceeds Until RNA Polymerase Reaches a Termination Signal

RNA polymerase then travels down one of the DNA strands, called the **template strand**, synthesizing a single strand of RNA with bases complementary to those in the DNA (Fig. 10-4b). Base-pairing between RNA and DNA is the same as between two strands of DNA, except that uracil in RNA pairs with adenine in DNA (see Table 10-1).

After about 10 nucleotides have been added to the growing RNA chain, the first nucleotides in the RNA molecule separate from the DNA template strand. This separation allows the two DNA strands to rewind into a double helix (Fig. 10-4b,c). Thus, as transcription continues to elongate the RNA molecule, one end of the RNA drifts away from the DNA; RNA polymerase keeps the other end temporarily attached to the DNA template strand (Figs. 10-4b and 10-5).

RNA polymerase continues along the template strand of the gene until it reaches a sequence of DNA bases known as the *termination signal*. At this point, RNA polymerase releases the completed RNA molecule and detaches from the DNA (Fig. 10-4c). The RNA polymerase is then free to bind to another promoter and synthesize another RNA molecule (Fig. 10-4d).

Transcription Is Selective

The complete human genome contains about 25,000 to 30,000 genes, most of which are transcribed only in certain cells at certain times. The transcription of genes into RNA is selective in two major ways. First, in any cell, transcription normally copies the DNA of only selected genes into RNA. Some genes are transcribed in all cells, because they encode proteins or RNA molecules that are essential for the life of any cell. For example, all cells need to synthesize proteins, so they all transcribe tRNA genes, rRNA genes, and genes for ribosomal proteins. Other genes are transcribed exclusively in certain types of cells. For example, even though every cell in your body contains the gene for insulin, a protein hormone that regulates glucose uptake, that gene is transcribed only in certain cells in your pancreas. How do cells regulate which genes are transcribed?

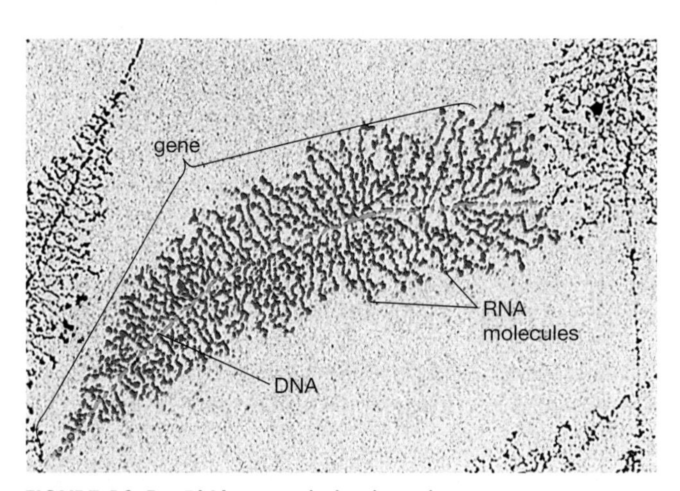

FIGURE 10-5 RNA transcription in action
This electron micrograph shows the progress of RNA transcription in the egg of an African clawed toad. In each treelike structure, the central "trunk" is DNA and the "branches" are RNA molecules. A series of RNA polymerase molecules are traveling down the DNA, synthesizing RNA as they go. The beginning of the gene is on the left. The short RNA molecules on the left have just begun to be synthesized; the long RNA molecules on the right are almost finished.

Proteins binding to "control regions" of DNA, near the promoter of a specific gene, block or enhance the binding of RNA polymerase. Thus, transcription is restricted to only those genes needed by that particular type of cell at that particular time of its life.

We have already encountered a second way in which transcription is selective. With rare exceptions, transcription copies only one of the two strands of DNA into RNA. Why? To answer this question, let's consider a protein-coding gene. Remember, the two strands of DNA are complementary, not identical. The sequence of bases on one strand codes for synthesis of a complementary mRNA that can be translated into a functional protein. This DNA strand is called the *template strand* because it is the template, or pattern, from which the complementary mRNA strand is made. The sequence of bases on the other DNA strand has a different base sequence, one that is not likely to code for a useful protein. As you know, a single chromosome contains many genes. The templates for some of its genes may reside on one of the two strands of its DNA double helix, and the templates for other genes may be located on the other strand.

10.3 How Is the Base Sequence of a Messenger RNA Molecule Translated into Protein?

As their names suggest, each type of RNA has a specific role in protein synthesis.

Messenger RNA Carries the Code for Protein Synthesis from the Nucleus to the Cytoplasm

All RNA is produced by transcription of DNA, but only mRNA carries the code for the amino acid sequence of a protein (see Fig. 10-2a). In eukaryotic cells, mRNA molecules are synthesized in the nucleus and enter the cytoplasm through the pores in the nuclear envelope. In the cytoplasm, mRNA binds to ribosomes, which synthesize a protein specified by the mRNA base sequence. The gene itself remains safely stored in the nucleus, like a valuable document in a library, while mRNA, like a "molecular photocopy," carries the information to the cytoplasm to be used in protein synthesis.

Ribosomes Consist of Two Subunits, Each Composed of Ribosomal RNA and Protein

Ribosomes, the structures that carry out translation, are composed of rRNA and many different proteins. Each ribosome is composed of two subunits—one small and one large. The small subunit has binding sites for mRNA, a "start" (methionine) tRNA, and several other proteins that collectively make up the "initiation complex." The large subunit has binding sites for two tRNA molecules and a catalytic site for joining together the amino acids attached to the tRNA molecules. Unless they are actively synthesizing proteins, the two subunits remain separate (see Fig. 10-2b). During protein synthesis, the small and large subunits come together and sandwich an mRNA molecule between them.

Transfer RNA Molecules Decode the Sequence of Bases in mRNA into the Amino Acid Sequence of a Protein

Delivery of the appropriate amino acids to the ribosome for incorporation into the growing protein chain depends on the activity of tRNA. Each cell synthesizes many different types of tRNA, one (sometimes several) for each amino acid. Twenty enzymes in the cytoplasm, one for each amino acid, recognize the tRNA molecules and attach the correct amino acid to one end (see Fig. 10-2c).

The ability of tRNA to deliver the proper amino acid depends on specific base-pairing between tRNA and mRNA. Each tRNA has three exposed bases, called the **anticodon**, which form base pairs with the mRNA codon. For example, the mRNA codon AUG forms base pairs with the anticodon UAC of a tRNA that has the amino acid methionine attached to its end. The ribosome can then incorporate methionine into a growing protein chain.

During Translation, mRNA, tRNA, and Ribosomes Cooperate to Synthesize Proteins

Now that we have introduced the major molecules involved in translation, let's look at the actual events (Fig. 10-6). Like transcription, translation has three steps: (1) *initiation* of protein synthesis, (2) *elongation* of the protein chain, and (3) *termination*.

Initiation: Protein Synthesis Begins When tRNA and mRNA Bind to a Ribosome

The first AUG codon in a eukaryotic mRNA sequence specifies the start of translation. Because AUG also codes for methionine, all newly synthesized proteins begin with this amino acid. An "initiation complex," which contains the small ribosomal subunit, a methionine tRNA, and several other proteins, binds to the end of an mRNA molecule with the AUG start codon (Fig. 10-6a). The AUG codon in the mRNA forms base pairs with the UAC anticodon of the methionine tRNA (Fig. 10-6b). The large ribosomal subunit then attaches to the small subunit, sandwiching the mRNA between the two subunits and holding the methionine tRNA in its first tRNA binding site (Fig. 10-6c). The ribosome is now fully assembled and ready to begin translation.

Elongation and Termination: Protein Synthesis Proceeds One Amino Acid at a Time Until a Stop Codon Is Reached

The assembled ribosome covers about 30 nucleotides of the mRNA. It holds two mRNA codons in alignment with the two tRNA binding sites of the large subunit. A second tRNA, with an anticodon complementary to the

Initiation:

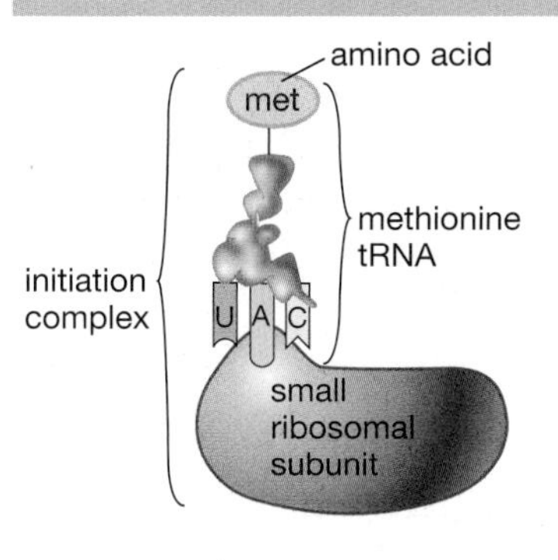

(a) A tRNA with an attached methionine amino acid binds to a small ribosomal subunit, forming an initiation complex.

(b) The initiation complex binds to an mRNA molecule. The methionine (met) tRNA anticodon (UAC) base-pairs with the start codon (AUG) of the mRNA.

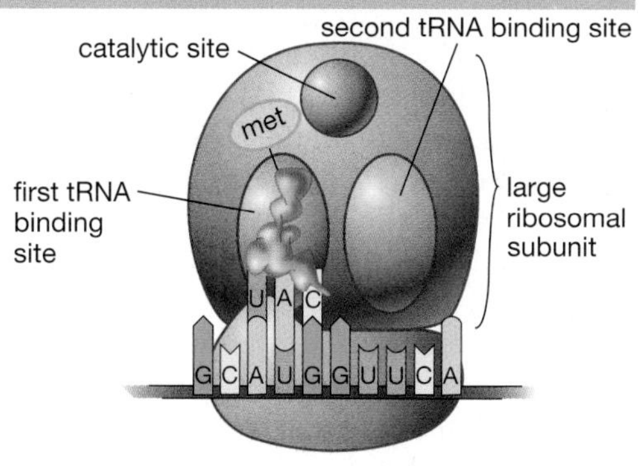

(c) The large ribosomal subunit binds to the small subunit. The methionine tRNA binds to the first tRNA site on the large subunit.

Elongation:

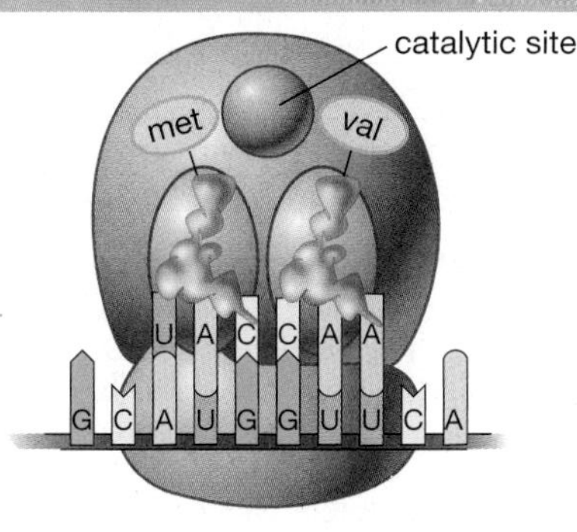

(d) The second codon of mRNA (GUU) base-pairs with the anticodon (CAA) of a second tRNA carrying the amino acid valine (val). This tRNA binds to the second tRNA site on the large subunit.

(e) The catalytic site on the large subunit catalyzes the formation of a peptide bond linking the amino acids methionine and valine. The two amino acids are now attached to the tRNA in the second binding position.

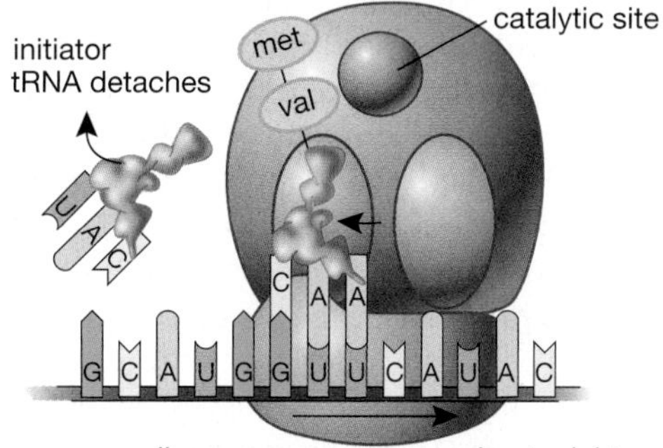

(f) The "empty" tRNA is released and the ribosome moves down the mRNA, one codon to the right. The tRNA that is attached to the two amino acids is now in the first tRNA binding site and the second tRNA binding site is empty.

Termination:

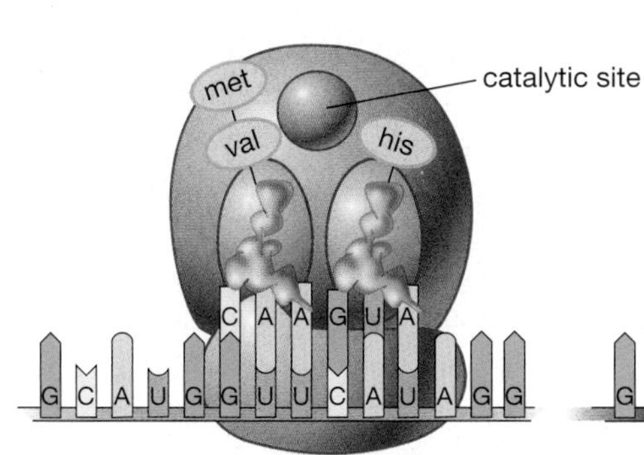

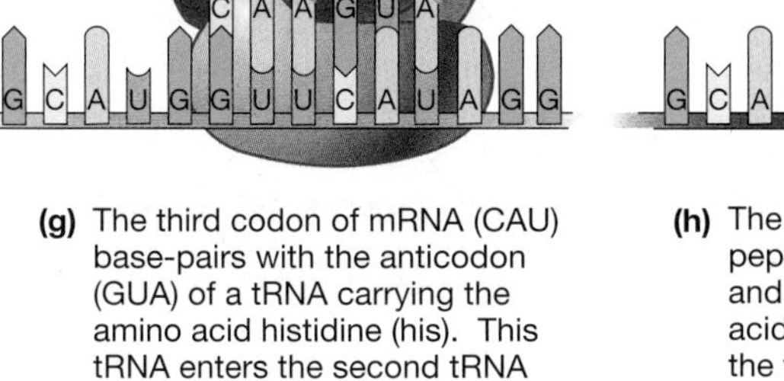

(g) The third codon of mRNA (CAU) base-pairs with the anticodon (GUA) of a tRNA carrying the amino acid histidine (his). This tRNA enters the second tRNA binding site on the large subunit.

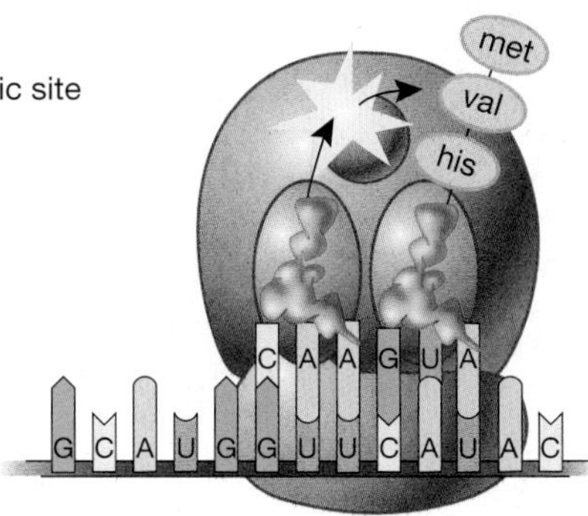

(h) The catalytic site forms a new peptide bond between valine and histidine. A three-amino-acid chain is now attached to the tRNA in the second binding site. The tRNA in the first site leaves, and the ribosome moves one codon over on the mRNA.

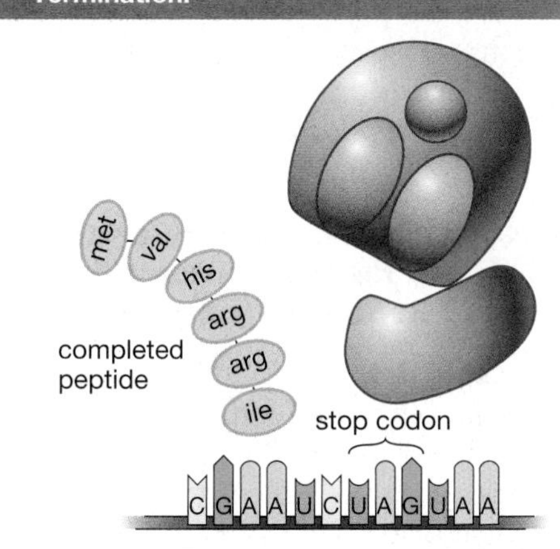

(i) This process repeats until a stop codon is reached; the mRNA and the completed peptide are released from the ribosome, and the subunits separate.

FIGURE 10-6 Translation is the process of protein synthesis
Protein synthesis, or translation, decodes the base sequence of an mRNA into the amino acid sequence of a protein.
QUESTION Examine panel (i) above. If mutations changed all of the guanine molecules visible in the mRNA sequence shown here to uracil, how would translated peptide differ from the one shown?

Media Activity
10.2 Translation

second codon of the mRNA, moves into the second tRNA binding site on the large subunit (Fig. 10-6d). The amino acids attached to the two tRNAs are now side by side. The catalytic site of the large subunit breaks the bond holding the first amino acid (methionine) to its tRNA and forms a peptide bond between this amino acid and the amino acid attached to the second tRNA (Fig. 10-6e). Interestingly, ribosomal RNA, and not one of the proteins of the large subunit, catalyzes the formation of the peptide bond. Therefore, this "enzymatic RNA" is often called a "ribozyme."

After the peptide bond is formed, the first tRNA is "empty," and the second tRNA carries a two-amino-acid chain. The ribosome then releases the empty tRNA and shifts to the next codon on the mRNA molecule (Fig. 10-6f). The tRNA holding the elongating chain of amino acids also shifts, moving from the second to the first binding site of the ribosome. A new tRNA, with an anticodon complementary to the third codon of the mRNA, binds to the empty second site (Fig. 10-6g). The catalytic site on the large subunit now links the third amino acid onto the growing protein chain (Fig. 10-6h). The "empty" tRNA leaves the ribosome, the ribosome shifts to the next codon on the mRNA, and the process repeats, one codon at a time.

A stop codon in the mRNA molecule signals the ribosome to terminate protein synthesis. Stop codons do not bind to tRNA. Instead, special proteins bind to the ribosome when it encounters a stop codon, forcing the ribosome to release the finished protein chain and the mRNA (Fig. 10-6i). The ribosome disassembles into its large and small subunits, which can then be used to translate another mRNA.

How fast does translation occur? Under optimum conditions, a ribosome can synthesize 5 to 15 peptide bonds per second. Most proteins are 100 to 200 amino acids long and thus can be synthesized in less than a minute.

Recap: Decoding the Sequence of Bases in DNA into the Sequence of Amino Acids in Protein Requires Transcription and Translation

We can now understand how a cell decodes the genetic information stored in its DNA to synthesize a protein. Each step involves the pairing of complementary bases and requires the action of a variety of proteins and enzymes. Follow these steps in Figure 10-7:

(a) With a few exceptions, such as the genes for tRNA and rRNA, each gene codes for a single protein.

(b) Transcription of a protein-coding gene produces an mRNA molecule that is complementary to one DNA strand of the gene. Starting from the first AUG, each codon within the mRNA is a sequence of three bases that specifies an amino acid or a "stop."

(c) Enzymes in the cytoplasm attach the appropriate amino acid to each tRNA based on the tRNA's anticodon.

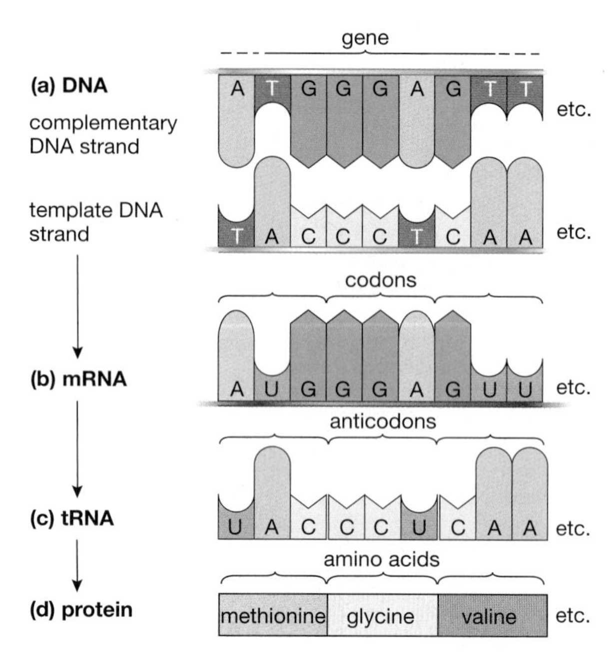

FIGURE 10-7 Complementary base pairing is critical to decode genetic information
(a) DNA contains two strands: the template strand is used by RNA polymerase to synthesize an RNA molecule. **(b)** Bases in the template strand of DNA are transcribed into a complementary mRNA. Codons are sequences of three bases that specify an amino acid or a stop during protein synthesis. **(c)** Unless it is a stop codon, each mRNA codon forms base pairs with the anticodon of a tRNA molecule that carries a specific amino acid. **(d)** The amino acids borne by the tRNAs are joined together to form a protein.

(d) During translation, tRNAs carry their attached amino acids to the ribosome. The appropriate amino acid is selected based on the complementary base pairs formed between the bases in the mRNA codon and the bases in the tRNA anticodon. The ribosome then links the amino acids together in sequence to form a protein.

This "decoding chain," moving from DNA bases to mRNA codons to tRNA anticodons to amino acids, results in the synthesis of a protein with a specific amino acid sequence. The amino acid sequence is ultimately determined by the base sequence within a single gene.

10.4 How Do Mutations in DNA Affect the Function of Genes?

Up to now, we have emphasized the accuracy of DNA replication, the transcription of RNA, and the translation of mRNA into proteins. However, nothing alive is perfect. Mistakes can occur in any of these processes. A single faulty mRNA molecule or a single defective protein doesn't normally affect a cell very much, because there are many correct copies of these molecules in the cell that can perform the proper cellular functions.

However, a faulty copy of a *gene* may be much more serious because all of the proteins synthesized according to the base sequence of the defective gene may themselves be defective. Changes in the sequence of bases in DNA, often resulting in a defective gene, are called **mutations**. We explore two cases in which single mutations cause far-reaching results in "Health Watch: Sex, Aging, and Mutations."

Mutations may occur in a variety of ways, including the following:

- There may be a mistake in base pairing during replication as a cell prepares for cell division. Despite the best efforts of proofreading enzymes, a few mistakes occur in each DNA replication.
- Occasionally, bases change spontaneously due to random movement of atoms in the DNA molecule.
- Certain chemicals (such as aflatoxins, synthesized by molds that live on grain and peanuts) and some types of radiation (such as X-rays and ultraviolet rays in sunlight) increase the frequency of base-pairing errors during replication, or even induce changes in DNA composition between replications (see the Case Study in Chapter 9).

Random changes in DNA sequence are usually detrimental to the functioning of the gene's products, much as randomly changing words in the middle of Shakespeare's *Hamlet* would probably interrupt the flow of the play. Some mutations, however, have no effect or—in very rare instances—are even beneficial, as you will learn later in this chapter.

Mutations Result from Nucleotide Substitutions, Insertions, or Deletions

During replication, a pair of bases is occasionally mismatched. Usually, repair enzymes recognize the mismatch, cut out the incorrect nucleotide, and replace it with a nucleotide that bears a complementary base. Sometimes, however, the enzymes replace the *correct* nucleotide instead of the incorrect one. The resulting base pair is complementary, but it is the wrong pair of nucleotides. These **nucleotide substitutions** are also called **point mutations**, because individual nucleotides in the DNA sequence are changed. There are two other types of mutations. An **insertion mutation** occurs when one or more new nucleotide pairs are inserted into a gene. A **deletion mutation** occurs when one or more nucleotide pairs are removed from a gene.

Mutations May Have a Variety of Effects on Protein Structure and Function

If a mutation occurs in cells whose offspring become gametes (sperm or eggs), the mutation may be passed on to future generations. But how does a change in base sequence affect the organism that inherits the mutated DNA? Deletions and insertions of one or two nucleotides can have particularly catastrophic effects on a gene because all of the codons that follow the deletion or insertion will be altered. (Think of our sample English sentence, THEDOGSAWTHECAT. Deleting or inserting a letter (deleting the first E, for example) means that all of the following three-letter words will be nonsense, such as THD OGS AWT HEC AT.) The protein synthesized from an mRNA containing such a mutation will almost always be nonfunctional.

Nucleotide substitutions within a protein-coding gene can produce at least four different outcomes (Table 10-4). As a concrete example, let's consider mutations that occur in the gene encoding beta-globin, one of the subunits of hemoglobin, the oxygen-carrying protein in red blood cells. The other type of subunit in hemoglobin is called alpha-globin; a normal hemoglobin molecule consists of two alpha and two beta subunits. In all but the last example, we will consider the results of mutations that occur in the sixth codon (CTC in DNA, GAG in mRNA), which specifies glutamic acid, a charged, hydrophilic, water-soluble amino acid.

Table 10-4 Effects of Mutations in the Hemoglobin Gene

	DNA (template strand)	mRNA	Amino Acid	Properties of Amino Acid	Functional Effect on Protein	Disease
Original codon 6	CTC	GAG	Glutamic acid	Hydrophilic	Normal protein function	None
Mutation 1	CT**T**	GA**A**	Glutamic acid	Hydrophilic	Neutral, normal protein function	None
Mutation 2	**G**TC	**C**AG	Glutamine	Hydrophilic	Neutral, normal protein function	None
Mutation 3	C**A**C	G**U**G	Valine	Hydrophobic	Loses water solubility; compromises protein function	Sickle-cell anemia
Original codon 17	TTC	AAG	Lysine	Hydrophilic	Normal protein function	None
Mutation 4	**A**TC	**U**AG	Stop codon	Ends translation after amino acid 16	Synthesizes only part of protein; eliminates protein function	Beta-thalassemia

- *The protein may be unchanged.* Remember that most amino acids can be encoded by several different codons. If a mutation changes the beta-globin DNA base sequence from CTC to CTT, this sequence still codes for glutamic acid. Therefore, the protein synthesized from the mutated gene remains the same even though the DNA sequence is different.
- *The new protein may be equivalent to the original one.* Many proteins have regions whose exact amino acid sequence is relatively unimportant. For example, in beta-globin, the amino acids on the outside of the protein must be hydrophilic to keep the protein dissolved in the cytoplasm of red blood cells. Exactly *which* hydrophilic amino acids are on the outside doesn't matter much. For example, a family in the Japanese town of Machida was found to contain a mutation from CTC to GTC, replacing glutamic acid (hydrophilic) with glutamine (also hydrophilic). Hemoglobin containing this mutant beta-globin protein—known as *Hemoglobin Machida*—appears to function well. Mutations such as the ones in Hemoglobin Machida and in the previous example are called **neutral mutations** because they do not detectably change the function of the encoded protein.
- *Protein function may be changed by an altered amino acid sequence.* A mutation from CTC to CAC replaces glutamic acid (hydrophilic) with valine (hydrophobic). This substitution is the genetic defect that causes sickle-cell anemia (see Chapter 12, page 227). The valines on the outside of the hemoglobin molecules cause them to clump together, distorting the shape of the red blood cells. These changes can cause serious illness.
- *Protein function may be destroyed by a premature stop codon.* A particularly catastrophic mutation occasionally occurs in the seventeenth codon of the beta-globin gene (TTC in DNA, AAG in mRNA). This codon specifies the amino acid lysine. A mutation from TTC to ATC (UAG in mRNA) results in a stop codon, halting translation of beta-globin mRNA before the protein is completed. People who inherit this mutant gene from both their mother and their father do not synthesize any functional beta-globin protein; they manufacture hemoglobin consisting entirely of alpha-globin subunits. This "pure alpha" hemoglobin does not bind oxygen very well. This condition, called beta-thalassemia, can be fatal unless treated with blood transfusions.

Mutations Provide the Raw Material for Evolution

If mutations in gametes are not lethal, they may be passed on to future generations. In humans, mutation rates range from about 1 in every 100,000 gametes to 1 in 1,000,000 gametes. For reference, a man releases about 300 to 400 million sperm per ejaculation, so each ejaculate contains about 600 sperm with new mutations. Although most mutations are neutral or potentially harmful, they are essential for evolution, because these random changes in DNA sequence are the ultimate source of all genetic variation. New base sequences undergo natural selection as organisms compete to survive and reproduce. Occasionally, a mutation proves beneficial in an organism's interactions with its environment. Through reproduction over time, the mutant base sequence may spread throughout the population and become common, as organisms that possess it may outcompete rivals bearing the original, unmutated base sequence. This process is described in detail in Unit Three.

10.5 How Are Genes Regulated?

As we mentioned in the Case Study at the beginning of this chapter, boys have all of the genes needed to make female genitalia. Obviously, these genes are not normally used as a boy matures. Why not? We'll return to this particular example at the end of the chapter. First, we need to look at how cells control transcription (which genes are used to make mRNA in a given cell), translation (how much protein is made from a particular type of mRNA), and the activity of proteins (how well a protein actually works as an enzyme and how long the protein lasts in a cell).

Proper Regulation of Gene Expression Is Critical for an Organism's Development and Health

Recent estimates suggest that the human genome contains about 25,000 to 30,000 genes. Each of these genes is present in most of your body cells, but individual cells *express* (transcribe and translate) only a small fraction of them—namely, those genes that are appropriate for the function of the particular cell type. Muscle cells, for example, synthesize large amounts of the contractile proteins actin and myosin, but do not synthesize insulin or hair proteins.

Expression of a gene also changes over time, depending on the body's needs from moment to moment. For example, during pregnancy, milk-producing cells in a woman's breasts multiply tremendously. Immediately after the baby's birth, these cells begin expressing the gene encoding casein, the major protein in milk. This change in gene expression allows the mother to produce large amounts of protein-rich milk to feed her new baby. Gene expression also varies from individual to individual. For example, a human male does not express the casein gene. However, he passes a copy of this gene to his daughters, who will express the gene if they become pregnant.

An organism's environment also helps determine which genes are transcribed. For example, in birds living in temperate climates, the longer days of spring stimulate the sex organs (testes or ovaries) to enlarge. The sex organs, in turn, produce sex hormones that cause the birds to produce eggs and sperm, and to sing, mate, and build nests. The proliferation of cells in the sex organs, the production of hormones by these cells, and the effects of those hormones on other cells throughout the body all result, directly or indirectly, from changes in gene expression.

Expression of genetic information by a cell is a multistep process, beginning with transcription of DNA and commonly ending with a protein that performs a particular function. Regulation of gene expression can occur at any of these steps, which are illustrated in Figure 10-8:

① *Cells can control the frequency at which an individual gene is transcribed.* Transcription of genes is turned on or off according to the demand for the protein (or RNA) product that they encode. Thus, whether a specific gene is transcribed and the frequency of its transcription depend on the type of cell and on the metabolic activity of both the cell and the organism as a whole. We consider this aspect of regulation in greater detail in the next section. Additionally, transcription in eukaryotes is complicated by the fact that most eukaryotic genes contain long stretches of nucleotides that do not code for protein; following transcription, these nucleotides must be removed from the newly synthesized RNA molecule to create a functional mRNA molecule. This process, described in "A Closer Look: Introns, Exons, and Splicing," can also be regulated to control the amount of mRNA produced from a particular gene.

② *Different messenger RNAs may be translated at different rates.* Messenger RNAs vary in their stability and in the rate at which they are translated into protein. Some mRNAs are long-lasting and are translated into protein many times. Others are translated only a few times before they are degraded. Furthermore, depending on metabolic requirements, a cell may completely block the translation of certain mRNAs until a signal activates their translation. This block is especially important in egg cells, which contain thousands of stored mRNAs that are not translated unless the egg is fertilized.

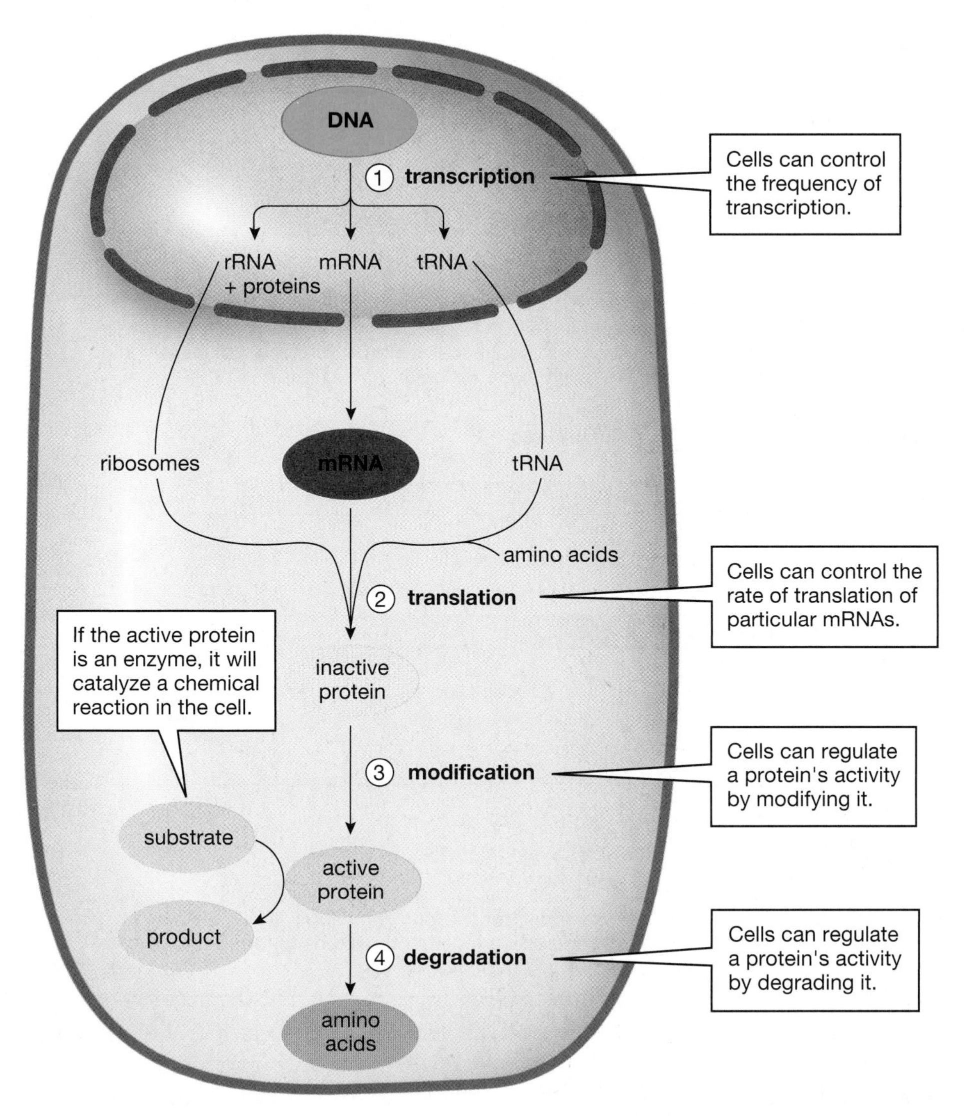

FIGURE 10-8 An overview of information flow in a cell, from gene transcription to chemical reactions catalyzed by enzymes

A CLOSER LOOK Introns, Exons, and Splicing

In the 1970s, molecular geneticists discovered that most eukaryotic protein-coding genes have much more DNA than is needed to encode the amino acids of the proteins. These genes consist of two or more segments with nucleotide sequences that encode for a protein, interrupted by other nucleotide sequences that are not translated into amino acids. The researchers called the coding segments **exons**, because they are ***ex***pressed in protein, and the noncoding segments **introns**, because they ***int***ervene between the exons (Fig. E10-1a). Most eukaryotic genes have introns; in fact, the gene that codes for a type of connective tissue in chickens has about 50 introns!

Transcription of a eukaryotic gene produces a very long RNA strand, which starts before the first exon and ends after the last exon (Fig. E10-1b). This RNA contains both introns and exons. To convert this RNA molecule into true mRNA, containing only the exons needed to code for the protein, enzymes in the nucleus precisely cut the RNA molecule apart at the junctions between intron and exon sequences, splice together the protein-coding segments (exons), and discard the rest.

Why are eukaryotic genes split up like this? Gene fragmentation appears to serve at least two functions. The first function is to allow a cell to produce multiple proteins from a single gene by splicing exons together in different ways. Rats, for example, have a gene that is transcribed in both the thyroid and the brain. In the thyroid, one splicing arrangement results in the synthesis of a hormone called calcitonin, which helps regulate calcium concentrations in the blood. In the brain, a different splicing arrangement results in the synthesis of a short protein that is probably used as a chemical messenger for communication between brain cells.

The second function of interrupted genes is more speculative, but is supported by some good experimental evidence: fragmented genes may provide a quick and efficient way for eukaryotes to evolve new proteins with new functions. Chromosomes sometimes break apart, and their parts may reattach to different chromosomes. If the breaks occur within the noncoding introns of genes, exons may be moved intact from one chromosome to another. Most such errors would be harmful. But some of these shuffled exons might code for a protein subunit that has a specific function (binding ATP, for example). In rare instances, adding this subunit to an existing gene may cause the gene to code for a new protein with useful functions. The accidental exchange of exons among genes produces new eukaryotic genes that will, on occasion, enhance the survival and reproduction of the organism that carries them.

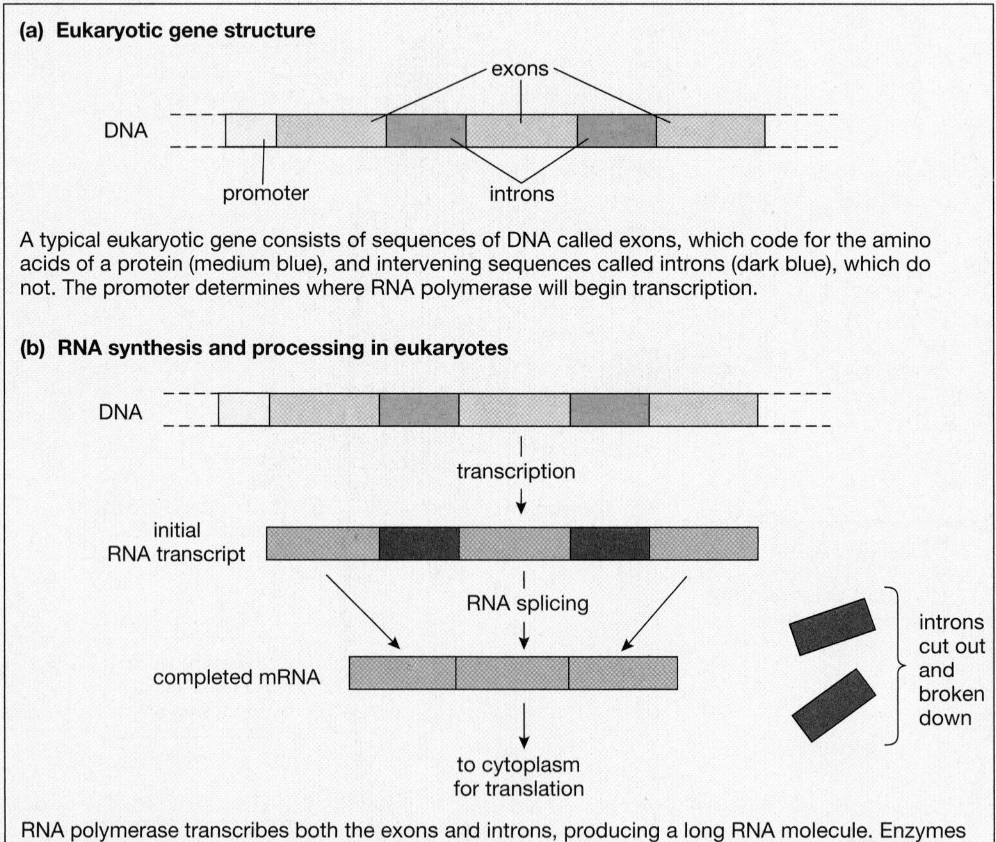

(a) Eukaryotic gene structure
exons
DNA
promoter
introns
A typical eukaryotic gene consists of sequences of DNA called exons, which code for the amino acids of a protein (medium blue), and intervening sequences called introns (dark blue), which do not. The promoter determines where RNA polymerase will begin transcription.
(b) RNA synthesis and processing in eukaryotes
DNA
transcription
initial RNA transcript
RNA splicing
completed mRNA
introns cut out and broken down
to cytoplasm for translation
RNA polymerase transcribes both the exons and introns, producing a long RNA molecule. Enzymes in the nucleus then cut out the RNA introns and splice together the exons to form the true mRNA, which moves out of the nucleus and is translated on the ribosomes.

FIGURE E10-1 Eukaryotic genes contain introns and exons

③ *Proteins may require modification before they can carry out their functions.* Many proteins must be modified before they become active. For instance, the protein-digesting enzymes produced by cells in your stomach wall and pancreas are initially synthesized in an inactive form, which prevents the enzymes from digesting the very cells that produce them. After these inactive forms are secreted into the digestive tract, portions of the enzymes are snipped out to unveil the enzyme's active site. Other modifications, such as adding or removing phosphate groups, can temporarily activate or inactivate a protein's function, allowing second-to-second control of the protein's activity.

④ *The life span of a protein can be regulated.* Most proteins have a limited life span within the cell. By preventing or promoting a protein's degradation, a cell can rapidly adjust the amount of a particular protein within it.

Eukaryotic Cells May Regulate the Transcription of Individual Genes, Regions of Chromosomes, or Entire Chromosomes

In eukaryotic cells, transcriptional regulation can operate on at least three levels: the individual gene, regions of chromosomes, or entire chromosomes.

Regulatory Proteins That Bind to the Gene's Promoter Alter the Transcription of Individual Genes

One of the best-known examples of transcriptional regulation is the role that the sex hormone, estrogen, plays in controlling egg production in birds. The gene for albumin, the major protein in egg whites, is not transcribed in the winter when birds are not breeding and estrogen levels are low. During the breeding season, the ovaries in female birds release estrogen, which enters the cells in the oviduct and binds to a receptor protein. The estrogen–receptor complex then attaches to a DNA control region near the promoter of the albumin gene. This attachment makes it easier for RNA polymerase to bind to the promoter and initiate transcription of mRNA. The mRNA is then translated into large amounts of albumin. Similar activation of gene transcription by steroid hormones occurs in other animals, including humans. (See Fig. E33-1b for more details of hormonal control of gene transcription.) The importance of hormonal regulation of transcription during development is illustrated by genetic defects in which receptors for sex hormones are nonfunctional (see "Health Watch: Sex, Aging, and Mutations"). In such cases, cells are unable to respond to the hormone, short-circuiting critical events in sexual development.

Some Regions of Chromosomes Are Condensed and Not Normally Transcribed

Certain parts of eukaryotic chromosomes are in a highly condensed, compact state in which most of the DNA seems to be inaccessible to RNA polymerase. Some of these regions are structural parts of chromosomes that don't contain genes. Other tightly condensed regions contain functional genes that are not currently being transcribed. When the product of a gene is needed, the portion of the chromosome containing that gene becomes "decondensed"—loosened so that the nucleotide sequence is accessible to RNA polymerase and transcription can occur.

Entire Chromosomes May Be Inactivated, Thereby Preventing Transcription

In some cases an entire chromosome may be condensed, making it largely inaccessible to RNA polymerase. An example occurs in the sex chromosomes of female mammals. Male mammals usually have an X and a Y chromosome (XY) and females usually have two X chromosomes (XX). As a consequence, females have the capacity to synthesize mRNA from genes on their two X chromosomes, while males, with only one X chromosome, may produce only half as much. To balance the expression of genes on the female's X chromosomes, one of her X chromosomes is condensed into a tight mass. In a light microscope, this inactivated, condensed X chromosome shows up in the nucleus as a dark spot called a **Barr body**, named after its discoverer, Murray Barr (Fig. 10-9). To prevent men from unfairly competing against women in the Olympics, officials attempt to

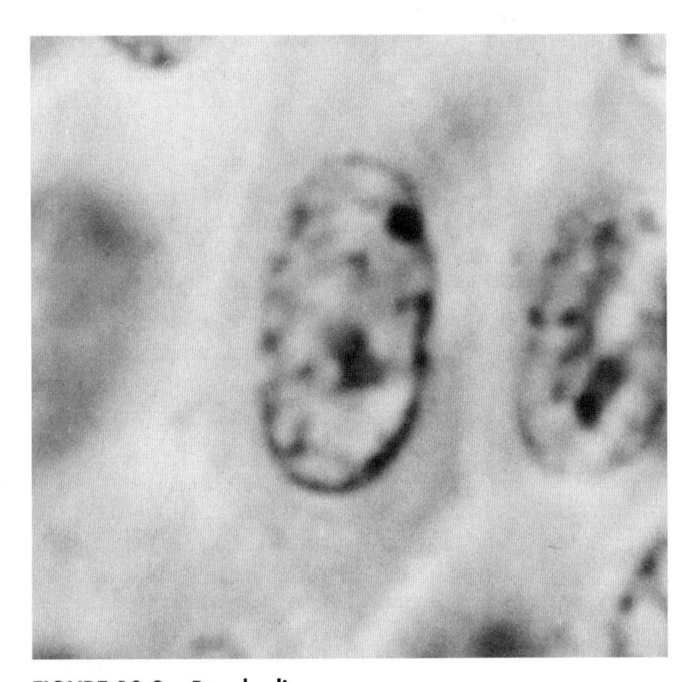

FIGURE 10-9 Barr bodies
The black spot on the upper right side of the nucleus is an inactivated X chromosome called a Barr body, usually found only in the cells of female mammals.

HEALTH WATCH Sex, Aging, and Mutations

Sometime in her early- to mid-teens, a girl usually goes through puberty: her breasts swell, her hips widen, and she begins to menstruate. In rare instances, however, a girl may develop all of the outward signs of womanhood, but does not menstruate. Eventually, when it becomes clear that she isn't merely developing a bit late, she reports this symptom to her physician, who may take a blood sample to do a chromosome test. In some cases, the chromosome test gives what might seem to be an impossible result: the girl's sex chromosomes are XY, a combination that would normally give rise to a boy. The reason she has not begun to menstruate is that she lacks ovaries and a uterus, but instead has testes that have remained inside her abdominal cavity. She has about the same concentrations of *androgens* (male sex hormones, such as testosterone) circulating in her blood as would be found in a boy her age. In fact, androgens, produced by the testes, have been present since early in her development. The problem is that her cells cannot respond to them—a rare condition called *androgen insensitivity*. This condition was a problem for Maria Jose Martinez Patino, an outstanding Spanish athlete who reached the Olympics, only to be barred from the hurdles competition because her cells lacked Barr bodies, which are normally present in females. After three years of struggle, the fact that she had developed as a female was finally recognized, and she was allowed to compete with others of her gender.

Many features of male development, including the formation of a penis, the descent of the testes into sacs outside the body cavity, and sexual characteristics that develop at puberty, such as a beard and increased muscle mass, occur because various body cells are responding to male sex hormones produced by the testes. In normal males, many body cells have androgen receptor proteins. When these proteins bind male hormones such as testosterone, the hormone–receptor complex attaches to the control regions of specific genes and influences their transcription into mRNA. The mRNA molecules serve as templates for the synthesis of proteins that contribute to maleness.

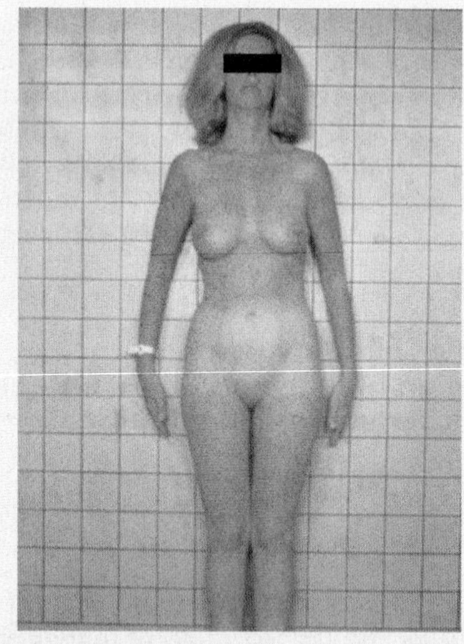

FIGURE E10-2 Androgen insensitivity leads to female features This individual has an X and a Y chromosome. She has testes that produce testosterone, but a mutation in her androgen receptor genes makes her cells unable to respond to testosterone, resulting in her female appearance.

verify that athletes who compete in women's events are truly female by performing a gene-based sex test. Women who "pass" the test are given a gender certification card, a requirement for a female athlete's participation in many competitions. One type of sex test, used as recently as the 1996 Olympics in Atlanta, checks the athlete's cells for the presence of Barr bodies. This test caused major problems for a female hurdler from Spain, Maria Jose Martinez Patino, when no Barr bodies were found in her cells. Learn more about her story in "Health Watch: Sex, Aging, and Mutations."

Usually, fairly large clusters of cells (all descended from a common "ancestral" cell during development) have the same X chromosome inactivated. As a result, the bodies of female mammals (including women) are composed of patches of cells in which one of the X chromosomes is active, and patches of cells in which the other X chromosome is active. The results of this phenomenon are easily observed in calico cats (Fig. 10-10). The X chromosome of a cat contains a gene encoding an enzyme that produces fur pigment. This gene comes in two versions, one producing orange fur and the other producing black fur. If one X chromosome in a female cat has the orange version of the fur color gene and the other X chromosome has the black version, the cat will have patches of orange and black fur. These patches represent areas of skin that developed from cells in the early embryo in which different X chromosomes were inactivated. Calico coloring is almost exclusively found in female cats. Because male cats have only one X chromosome, which is active in all of their cells, normal male cats can have black fur or orange fur, but not both.

In different groups of cells, the androgen receptor–testosterone complex influences gene transcription in different ways, giving rise to a wide range of male characteristics. Like other proteins, androgen receptors are coded by specific genes (interestingly, the gene encoding the androgen receptor protein is on the X chromosome). Even though genetically a male with both X and Y chromosomes, a person with a mutant androgen receptor gene will be unable to make functional androgen receptor proteins, and therefore will be unable to respond to the testosterone that the testes produce. Thus, a change in the nucleotide sequence of a single gene, causing a single type of defective protein to be produced, can cause a person who is genetically male to look and feel like a woman (Fig. E10-2).

A second type of mutation provides clues to why people age. Why will your hair whiten, skin wrinkle, joints ache, and eyes cloud as you become elderly? A small number of individuals carry a defective gene that causes *Werner syndrome*, which causes a type of premature aging (Fig. E10-3). People with this disorder typically die of aging-related conditions by age 50. Recent research has localized the mutations in most people with Werner syndrome to a gene that codes for an enzyme involved in DNA replication. As you have seen, the accurate replication of DNA is crucial to the production of normally functioning cells. If a mutation interferes with the ability of enzymes to replicate DNA accurately and to proofread and repair errors in DNA, then mutations will accumulate rapidly in cells throughout the body.

The fact that an overall increase in mutations caused by defective replication enzymes produces symptoms of old age provides support for one hypothesis to explain many of the symptoms of normal aging. During a typical long (say, 80-year) life span, mutations gradually accumulate because of mistakes in DNA replication and environmentally induced DNA damage. Eventually, these mutations interfere with nearly every aspect of body functioning and contribute to death from "old age."

Disorders such as androgen insensitivity and Werner syndrome provide profound insights into the impact of mutations, the function of specific genes and their protein products, the ways hormones regulate gene transcription, and even the mystery of aging.

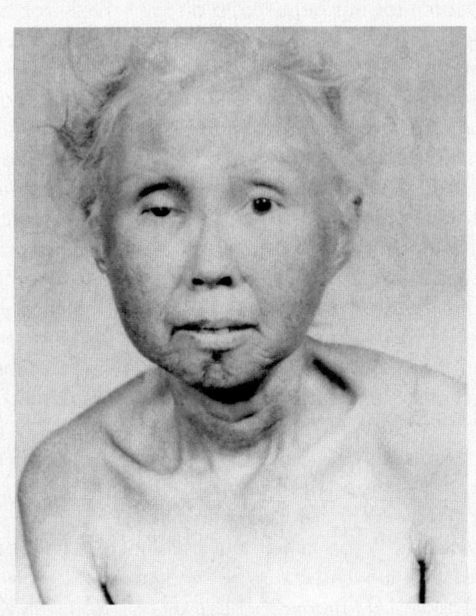

FIGURE E10-3 A 48-year-old woman with Werner syndrome This condition, most common among people of Japanese ancestry, is the result of a mutation that interferes with proper DNA replication, increasing the incidence of mutations throughout the body.

FIGURE 10-10 Inactivation of the X chromosome regulates gene expression This female calico cat carries a gene for orange fur on one X chromosome and a gene for black fur on her other X chromosome. Inactivation of different X chromosomes produces the black and orange patches. The white color is due to an entirely different gene that prevents pigment formation altogether. **QUESTION** Most orange cats are male. Why?

CASE STUDY REVISITED Boy or Girl?

How does knowing about transcription and translation help us understand the physical differences between males and females? By the 1930s, biologists knew that one or more genes on the Y chromosome were essential for determining whether a mammal would develop into a male or a female. In 1990, a search for this gene led to discovery of the SRY gene, for "sex-determining region on the Y chromosome." The SRY gene is found in all male mammals, including humans. Experiments with mice demonstrated its importance in sex determination. If a mouse embryo with two X chromosomes is given a copy of SRY, the embryo develops male characteristics: it will have testes and a penis, and behave like a male mouse. (These XX male mice are sterile, however, because other genes located on the Y chromosome are needed for production of functional sperm.) Mouse embryos that lack an SRY gene develop as females, regardless of whether they have two X chromosomes or an X and a Y. The conclusion: male (XY) mammals have all the genes needed to be female but usually aren't, because they have an SRY gene. Likewise, female (XX) mammals have all of the genes needed to be male, but because they *don't* have an SRY gene, they are usually female.

How does the SRY gene exert such an enormous effect on a mammal's characteristics? Even though it is critical for sex determination, the SRY gene is transcribed only for a short time in embryonic development, and only in cells that will become the testes. The SRY gene is then permanently inactivated for the rest of the animal's life. In the short time that it is transcribed, the protein encoded by the SRY gene apparently stimulates the expression of many other genes, whose protein products are essential to testicular development. Once formed, the testes in the embryo secrete testosterone, which activates other genes, leading to development of the penis and scrotum. Boy or girl? It depends on the carefully regulated expression of many genes, with a single gene, SRY, serving as the initial genetic switch to activate male development.

Consider This: We have briefly described two different ways in which a person with XY sex chromosomes can develop as a female: the Y chromosome may have a defective SRY gene, or the X chromosome may have a defective androgen receptor gene. Suppose a 16-year-old girl is tearful and frightened because she has never menstruated, and she asks a physician what is wrong with her. The doctor orders a chromosome test, and perhaps hormone tests as well, and discovers that, in fact, she has both X and Y chromosomes, but either has androgen insensitivity or lacks a functional SRY gene. What should the physician tell her? Clearly, she has to be told that she has no uterus, will never menstruate, and cannot bear children. But beyond that, what? To most people, a person with two X chromosomes is a female, and one with one X and one Y chromosome is a male, and that's that. Should the doctor tell her that she is genetically male although physiologically female? What will this do to her self-image and her psychological health? What would *you* do? To see how one physician handled this dilemma, see "The Curse of the Garcias," by Robert Marion, in *Discover* magazine, December 2000.

Links to Life: Genetics, Evolution, and Medicine

All life on Earth is related through evolution, sometimes closely (dogs and foxes), sometimes distantly (bacteria and people). As you know, mutations occur constantly, usually at a very low rate. Distantly related organisms may have shared a common ancestor millions of years ago. A lot of mutations may have occurred since then, so that the genes of these organisms may now differ by many nucleotides. Medicine takes advantage of these differences to develop antibiotics to treat bacterial infections.

Streptomycin and neomycin, commonly prescribed antibiotics, kill certain bacteria by binding to a specific sequence of RNA in the small subunits of the bacterial ribosomes, inhibiting protein synthesis. Without adequate protein synthesis, the bacteria die. Patients infected by these bacteria don't die, however, because the small subunits of their eukaryotic ribosomes have a different nucleotide sequence than the bacteria's prokaryotic ribosomes do.

You have probably heard of *antibiotic resistance*, in which bacteria that are frequently exposed to antibiotics evolve defenses against those antibiotics. Bacteria evolve resistance against streptomycin and related antibiotics rather rapidly. Why? It's actually pretty straightforward. If eukaryotic ribosomes are insensitive to streptomycin, then eukaryotic ribosomes must function perfectly well with a different RNA sequence than prokaryotic ribosomes have. Bacteria that are resistant to some of streptomycin's chemical relatives have a mutation that changes just a single nucleotide in their ribosomal RNA from adenine to guanine, which is precisely the nucleotide found at the comparable position in eukaryotic ribosomal RNA.

Genetics, mutations, the mechanisms of protein synthesis, and evolution are important not only to biologists, but to physicians, too. In fact, a whole discipline of medicine called evolutionary medicine has arisen that uses the evolutionary relationships between people and microbes to help fight disease.

Summary of Key Concepts

10.1 How Are Genes and Proteins Related?

Genes are segments of DNA that can be transcribed into RNA and, for most genes, translated into protein. Transcription produces the three types of RNA needed for translation: messenger RNA (mRNA), transfer RNA (tRNA), and ribosomal RNA (rRNA). During translation, tRNA and rRNA collaborate with enzymes and other proteins to decode the sequence of bases in mRNA and produce a protein with the amino acid sequence specified by the gene. The genetic code consists of codons, sequences of three bases in mRNA that specify either an amino acid in the protein chain or the end of protein synthesis (stop codons).

10.2 How Is Information in a Gene Transcribed into RNA?

Within an individual cell, only certain genes are transcribed. When the cell requires the product of a gene, RNA polymerase binds to the promoter region of the gene and synthesizes a single strand of RNA. This RNA is complementary to the template strand in the gene's DNA double helix.

10.3 How Is the Base Sequence in a Messenger RNA Molecule Translated into Protein?

In eukaryotes, mRNA carries the genetic information from the nucleus to the cytoplasm, where ribosomes can use this information to synthesize a protein. Ribosomes contain rRNA and proteins organized into large and small subunits. These subunits come together at the first AUG codon of the mRNA molecule to form the complete protein-synthesizing machine. tRNAs deliver the appropriate amino acids to the ribosome for incorporation into the growing protein. Which tRNA binds, and consequently which amino acid is delivered, depends on base pairing between the anticodon of the tRNA and the codon of the mRNA. Two tRNAs, each carrying an amino acid, bind simultaneously to ribosome; the large subunit catalyzes the formation of peptide bonds between the amino acids. As each new amino acid is attached, one tRNA detaches, and the ribosome moves over one codon, binding to another tRNA that carries the next amino acid specified by mRNA. Addition of amino acids to the growing protein continues until a stop codon is reached, signaling the ribosome to disassemble and to release both the mRNA and the newly formed protein.

10.4 How Do Mutations in DNA Affect the Function of Genes?

A mutation is a change in the nucleotide sequence of a gene. Mutations can be caused by mistakes in base pairing during replication, by chemical agents, and by environmental factors such as radiation. Common types of mutations include changes in a base pair (point mutations) and insertions or deletions of base pairs. Mutations may be neutral or harmful, but in rare cases a mutation will promote better adaptation to the environment and thus will be favored by natural selection.

10.5 How Are Genes Regulated?

The expression of a gene requires that it be transcribed and translated, and the resulting protein perform some action within the cell. Which genes are expressed in a cell at any given time is regulated by the function of the cell, the developmental stage of the organism, and the environment. Control of gene regulation can occur at many steps. The amount of mRNA synthesized from a particular gene can be regulated by increasing or decreasing the rate of its transcription, as well as by the stability of the mRNA itself. Rates of translation of mRNAs can also be regulated. Regulation of transcription and translation affects how many protein molecules are produced from a particular gene. Even after they are synthesized, many proteins must be modified before they can function. In addition to regulation of individual genes, cells can regulate transcription of groups of genes. For example, entire chromosomes or parts of chromosomes may be condensed and inaccessible to RNA polymerase, whereas other portions may be expanded, allowing transcription to occur.

Key Terms

anticodon *p. 170*
Barr body *p. 177*
codon *p. 167*
deletion mutation *p. 173*
exon *p. 176*
gene *p. 164*
genetic code *p. 167*
insertion mutation *p. 173*
intron *p. 176*
messenger RNA (mRNA) *p. 164*
mutation *p. 173*
neutral mutation *p. 174*
nucleotide substitution *p. 173*
point mutation *p. 173*
promoter *p. 169*
ribonucleic acid (RNA) *p. 164*
ribosomal RNA (rRNA) *p. 166*
ribosome *p. 166*
RNA polymerase *p. 169*
start codon *p. 168*
stop codon *p. 168*
template strand *p. 169*
transcription *p. 166*
transfer RNA (tRNA) *p. 164*
translation *p. 166*

Thinking Through the Concepts

To take a multiple-choice quiz with feedback on the contents of this chapter, visit http://www.prenhall.com/audesirk7. *Log in to the Web site selected by your instructor and navigate to the Self Test section for this chapter.*

? Review Questions

1. How does RNA differ from DNA?
2. What are the three types of RNA? What is the function of each?
3. Define the following terms: *genetic code*; *codon*; *anticodon*. What is the relationship among the bases in DNA, the codons of mRNA, and the anticodons of tRNA?
4. How is mRNA formed from a eukaryotic gene?
5. Diagram and describe protein synthesis.
6. Explain how complementary base pairing is involved in both transcription and translation.
7. Describe some mechanisms of gene regulation.
8. Define *mutation*, and give one example of how a mutation might occur. Would you expect most mutations to be beneficial or harmful? Explain your answer.

Applying the Concepts

1. As you know, women who can bear children have two X chromosomes, and men who can father children have an X and a Y. Therefore, a boy normally inherits an X chromosome from his mother and a Y from his father, while a girl inherits an X from both mother and father. As we described in this chapter, a person with XY sex chromosomes can develop as a female if her Y chromosome has a defective SRY gene, or if her X chromosome has a defective androgen receptor gene. Are these genetic abnormalities inherited from a person's mother or father? Why? Can the defective SRY condition be passed down through many generations? What about androgen insensitivity? Why? (If you're not sure how these conditions might be inherited, refer to Chapters 11 and 12 for more information.)
2. Although they are rare, male calico cats do occur. In fact, about 1 in 3000 calico cats is a male. Can you come up with an explanation of the origin of male calico cats? Most male calico cats are infertile. Why? Using your knowledge of chromosomes, genes, and mutations, what might be the differences between infertile and fertile male calico cats?
3. As you have learned in this chapter, many factors influence gene expression, including hormones. The use of anabolic steroids and growth hormones among athletes has created controversy in recent years. Hormones certainly affect gene expression, but, in the broadest sense, so do vitamins and foods. What do you think are appropriate guidelines for the use of hormones? Should athletes take steroids or growth hormones? Should children at risk of being unusually short be given growth hormones? Should parents be allowed to request growth hormones for a child of normal height in the hope of producing a future basketball player?

For More Information

Grunstein, M. "Histones as Regulators of Genes." *Scientific American*, October 1992. Histones are proteins associated with DNA in eukaryotic chromosomes. Once thought to be merely a scaffold for DNA, they are actually important in gene regulation.

Marion, R. "The Curse of the Garcias." *Discover*, December 2000. How one physician diagnosed and counseled a patient with androgen insensitivity.

Nirenberg, M. W. "The Genetic Code: II." *Scientific American*, March 1963. Nirenberg describes some of the experiments in which he deciphered much of the genetic code.

Tjian, R. "Molecular Machines That Control Genes." *Scientific American*, February 1995. Complexes of proteins regulate which genes are transcribed in a cell and therefore help determine the cell's structure and function.

Travis, J. "Biology's Periodic Table." *Science News*, March 22, 1997. A brief synopsis of the Human Genome Project.

Media Activities

To access a Media Activity visit http://www.prenhall.com/audesirk7. *Log in to the Web site selected by your instructor, navigate to this chapter, and select the appropriate Media Activity number.*

10.1 Transcription

Estimated time: 10 minutes

The genetic information that is stored in the DNA found in the nucleus of a cell must be conveyed to the cytoplasm, where it can be translated into a specific sequence of amino acids. The intermediate molecule is RNA, synthesized by the process of transcription. This tutorial investigates the steps involved in transcription.

10.2 Translation

Estimated time: 10 minutes

Transcription provides a messenger RNA transcript of the gene. Now it is the task of the cell to take the information encoded in this messenger RNA and translate it into a specific sequence of amino acids. This process occurs on the ribosomes and involves a number of steps illustrated in this tutorial.

10.3 Web Investigation: Boy or Girl?

Estimated time: 10 minutes

Everyone knows that chromosomes determine gender. Baby girls have two X chromosomes (XX) while baby boys are XY. But what about individuals with XO, XXY, or XYY genomes? Why are there female babies with XY karyotypes (chromosome sets)? Gender biology is much more complicated than one might think.

CHAPTER

11 The Continuity of Life: Cellular Reproduction

CC the cloned cat (right) and her genetic donor, Rainbow (left), have very different fur colors and patterns even though they share exactly the same genes. Inset: CC actually resulted from a preliminary experiment leading to the ultimate goal of cloning a dog, Missy.

CASE STUDY Cloning Conundrum

You've probably seen newspaper photos showing CC, a gray and white kitten, peering out of a beaker. CC is the first cloned cat, and the first tangible result of the Missyplicity Project. In 1997, John Sperling (founder of the University of Phoenix) and Joan Hawthorne realized that their beloved spayed mutt, Missy, was getting old, and they didn't want to lose her exceptional gene endowment. Hoping to obtain another dog with those genes, Sperling donated $3.7 million to Texas A&M University to develop the technology needed to clone dogs. The first fruit of the Missyplicity Project was a cat, not a dog, because cats are easier to clone. Even so, CC was the only successful birth out of 87 cloned embryos (current methods seem to have a higher success rate). There have been many attempts to clone Missy, but so far none have succeeded. Meanwhile, Missy died on July 6, 2002, at the age of 15. Many of her cells remain frozen, awaiting future cloning efforts.

Why can't Sperling and Hawthorne replace Missy the old-fashioned way? Missy was a Border Collie/Siberian Husky mix. What if they bred another Siberian Husky with a Border Collie? You probably already know that the puppies in a Siberian Husky/Border Collie litter wouldn't all be exactly the same, and that none of the puppies would be a precise duplicate of Missy. But why not? For that matter, if the Missyplicity Project succeeds, would Missy's clone really be *exactly* the same as the original? To answer these questions, we'll need to explore the mechanisms and consequences of cellular reproduction.

11.1 What Is the Role of Cellular Reproduction in the Lives of Individual Cells and Entire Organisms?

The **cell cycle** is the sequence of activities that occurs from one *cell division* to the next. When a cell divides, it must provide its offspring (usually called "daughter cells") with genetic information (DNA) and the other cellular components it needs, such as mitochondria, ribosomes, and endoplasmic reticulum. Much of this text is devoted to the activities of cells when they are not dividing. Here, however, we will focus on the mechanisms of cell division and the role of cell division in the lives of single cells and multicellular organisms.

Reproduction in which offspring are formed from a single parent, rather than through the union of gametes (sperm and egg) from two parents, is called **asexual reproduction**. Single-celled organisms, including *Paramecium* in ponds (Fig. 11-1a) and yeast in rising bread (Fig. 11-1b), reproduce asexually by cell division—each cell cycle results in two new organisms from each preexisting cell. Asexual reproduction isn't confined to single-celled organisms, however. You, too, have been involved in asexual reproduction throughout your life—or at least your cells have. Since your conception as a single fertilized egg, asexual reproduction via cell division has produced all of the trillions of cells in your body, and continues every day in many organs, such as your skin and intestines.

Entire multicellular organisms may also reproduce by asexual reproduction. Like its relative the sea anemone, a *Hydra* reproduces by growing a small replica of itself, called a bud, on its body (Fig. 11-1c). Eventually, the bud is able to live independently and separates from its par-

FIGURE 11-1 Cell division in eukaryotes enables asexual reproduction
(a) In unicellular microorganisms, such as the protist *Paramecium,* cell division produces two new, independent organisms. **(b)** Yeast, a unicellular fungus, reproduces by cell division. **(c)** *Hydra,* a freshwater relative of the sea anemone, grows a miniature replica of itself (a bud) on its side. When fully developed, the bud breaks off and assumes independent life. **(d)** Trees in an aspen grove are often genetically identical. Each tree grows up from the roots of a single ancestral tree. This photo shows three separate groves near Aspen, Colorado. In fall, the appearance of their leaves shows the genetic identity within a grove and the genetic difference between groves.

ent. Many plants and fungi reproduce both asexually and sexually. The beautiful aspen groves of Colorado, Utah, and New Mexico (Fig. 11-1d) develop asexually from shoots growing up from the root system of a single parent tree. Although the grove seems to be a population of separate trees, it can be considered to be a single individual whose multiple trunks are interconnected by a common root system. Aspen can also reproduce by seeds, which are made through sexual reproduction.

Both prokaryotic and eukaryotic cells have cell cycles that include growth, DNA replication, and cell division. Because of the structural and functional differences between these two cell types, prokaryotic and eukaryotic cell cycles differ considerably.

The Prokaryotic Cell Cycle Consists of Growth and Binary Fission

With sufficient nutrients and favorable temperatures, many prokaryotic cells are usually either dividing or getting ready to divide. The cell cycle consists of a relatively long period of growth—during which the cell also replicates its DNA—followed by rapid cell division (Fig. 11-2a).

Cell division in prokaryotic cells is known as **binary fission**, which means "splitting in two." The prokaryotic chromosome is usually a circle of DNA, attached at one point to the plasma membrane (Fig. 11-2b, ①). During the long "growth phase" of the prokaryotic cell cycle, the DNA is replicated, producing two identical chromosomes which become attached to the plasma membrane at nearby, but separate, points (Fig. 11-2b, ②). The cell increases in size both during and after DNA replication. As the cell grows, the plasma membrane between the attachment points of the chromosomes enlarges, pushing them apart (Fig. 11-2b, ③). When the cell has approximately doubled in size, the plasma membrane around the middle of the cell rapidly grows inward between the two DNA attachment sites (Fig. 11-2b, ④). Fusion of the plasma membrane along the equator of the cell completes binary fission, producing two daughter cells, each containing one of the chromosomes (Fig. 11-2b, ⑤). Because DNA replication produces two identical DNA molecules (except for the occasional mutation), the two daughter cells are genetically identical to one another and to the parent cell.

Under ideal conditions, binary fission in prokaryotes occurs rapidly. For example, the common intestinal bacterium *Escherichia coli* can grow, replicate its DNA, and divide in about 20 minutes. Luckily, the environment in our intestines is usually not ideal for bacteria growth; otherwise, the bacteria would soon outweigh the rest of our bodies!

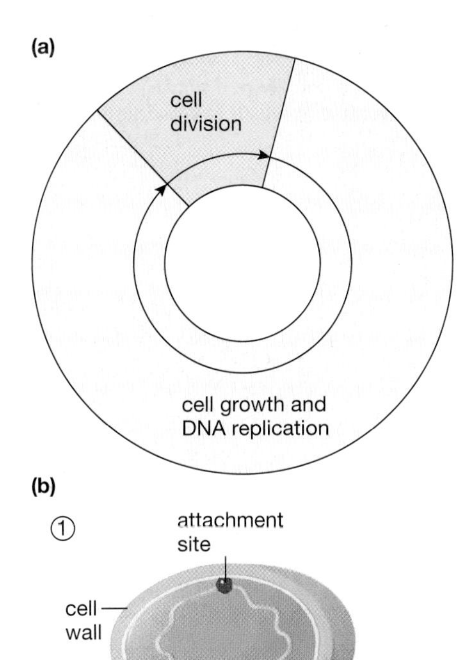

The circular DNA double helix is attached to the plasma membrane at one point.

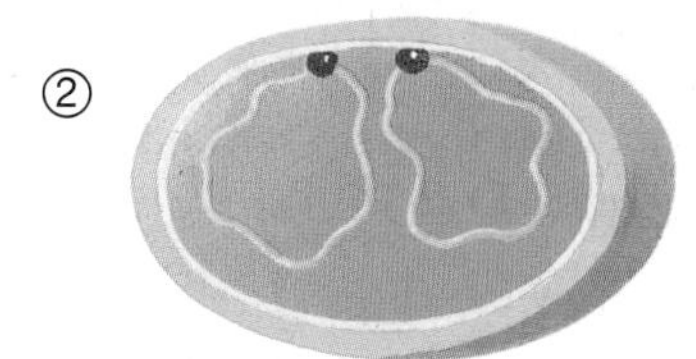

The DNA replicates and the two DNA double helices attach to the plasma membrane at nearby points.

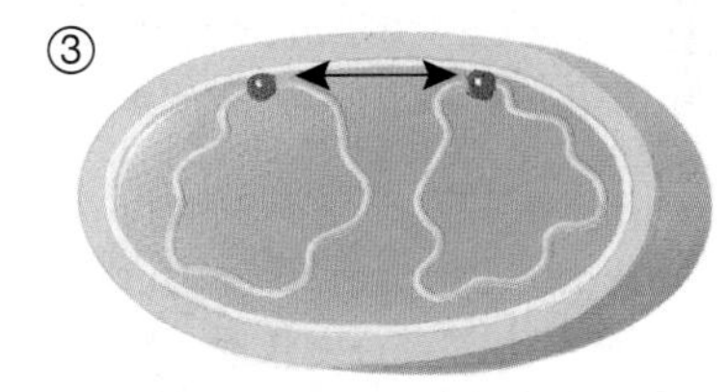

New plasma membrane is added between the attachment points, pushing them further apart.

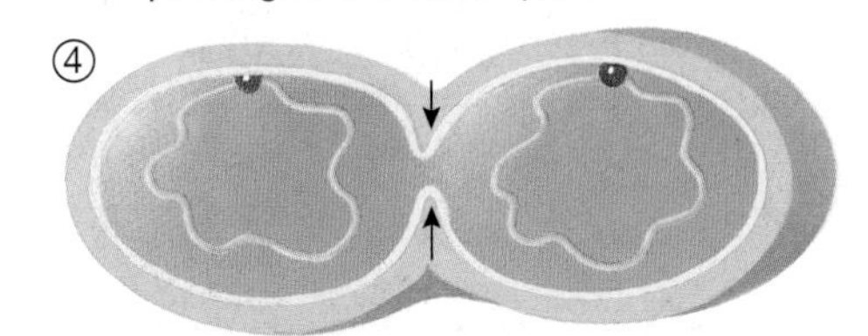

The plasma membrane grows inward at the middle of the cell.

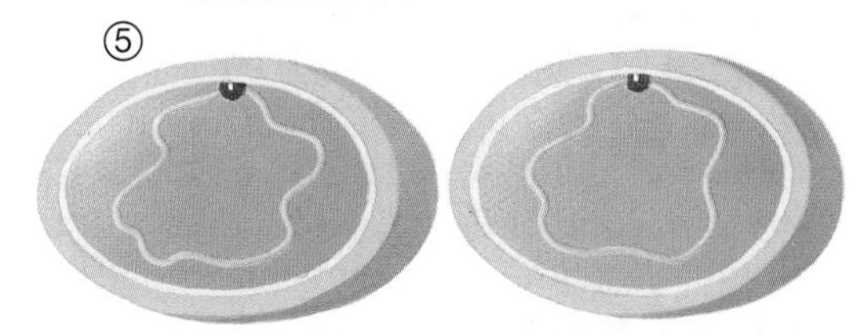

The parent cell divides into two daughter cells.

FIGURE 11-2 The prokaryotic cell cycle
(a) The prokaryotic cell cycle consists of growth and DNA replication, followed by binary fission. **(b)** Binary fission in prokaryotic cells.

FIGURE 11-3 The eukaryotic cell cycle
The eukaryotic cell cycle consists of interphase and mitotic cell division. Some cells enter the G_0 phase and may not divide again. **EXERCISE** Give an example of an adult animal body tissue in which many cells are likely to be in the cell division phase of the cell cycle, and an example in which few dividing cells would be found.

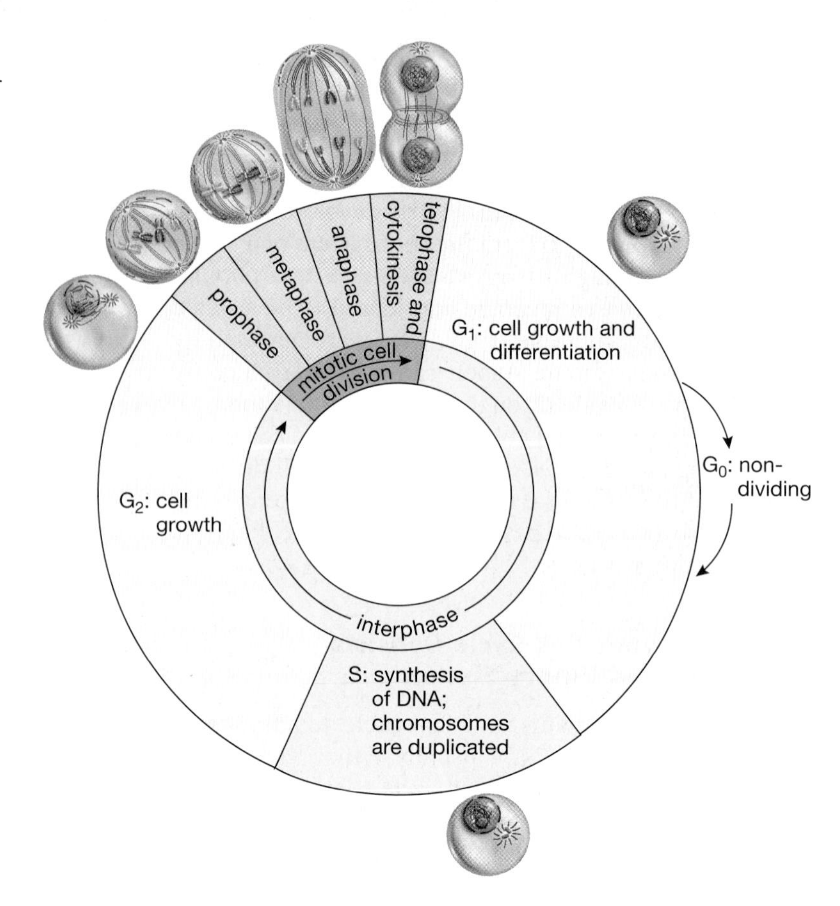

Media Activity
11.1 Cell Division in Humans

The Eukaryotic Cell Cycle Consists of Interphase and Cell Division

The eukaryotic cell cycle (Fig. 11-3) is somewhat more complex than the prokaryotic cell cycle. Newly formed cells usually acquire nutrients from their environment, synthesize additional cellular components, and grow larger. After a variable amount of time—depending on the organism, the type of cell, and the nutrients available—the cell may divide. Each daughter cell may then enter another cell cycle, producing additional cells. Some newly formed cells, however, divide only if they receive signals, such as growth hormones, that cause them to enter another cell cycle. Still other cells may exit the cell cycle completely and never divide again. In humans, cells in the bone marrow and skin may divide as frequently as once a day. At the other extreme, most nerve and muscle cells never divide after they mature; if one of these cells dies, it is not replaced.

During Interphase, the Eukaryotic Cell Grows in Size and Replicates Its DNA

The eukaryotic cell cycle is divided into two major phases: interphase and cell division (see Fig. 11-3). During **interphase**, the cell acquires nutrients from its environment, grows, and duplicates its chromosomes. During **cell division**, one copy of each chromosome and usually about half the cytoplasm (including mitochondria, ribosomes, and other organelles) are parceled out into each of the two daughter cells.

Most eukaryotic cells spend the majority of their time in interphase, preparing for cell division. For example, some cells in human skin, which divide about once a day, spend roughly 22 hours in interphase. Interphase itself contains three subphases: G_1 (gap or growth phase 1), S (DNA synthesis), and G_2 (gap or growth phase 2). To explore these stages, let us consider a newly formed daughter cell. This cell enters the G_1 portion of interphase, during which it acquires or synthesizes the materials needed for cell division. If the cell grows to a proper size and receives the necessary signals, it enters the *S* phase, which is when DNA *synthesis* occurs. After replicating its DNA, the cell completes its growth in the G_2 phase before dividing.

During the G_1 phase, the cell is sensitive to internal and external signals that help the cell "decide" whether to divide. If that decision is positive, the cell progresses through the rest of the cell cycle. Depending on the cell type and external signals, a cell can also exit from the cell cycle during G_1 and enter into a phase known as G_0. Cells in G_0 are alive and metabolically active. They may even grow in size, but they do not replicate their DNA or divide. This phase is also the time when many cells specialize, or **differentiate**. Muscle cells fill with the contractile proteins myosin and actin; some cells of the immune system become packed with endoplasmic reticulum to produce massive amounts of antibodies; and nerve cells grow long processes that allow them to connect with other cells. Many differentiated cells, including most of those in your heart muscle, eyes, and brain, remain in G_0 for your entire life.

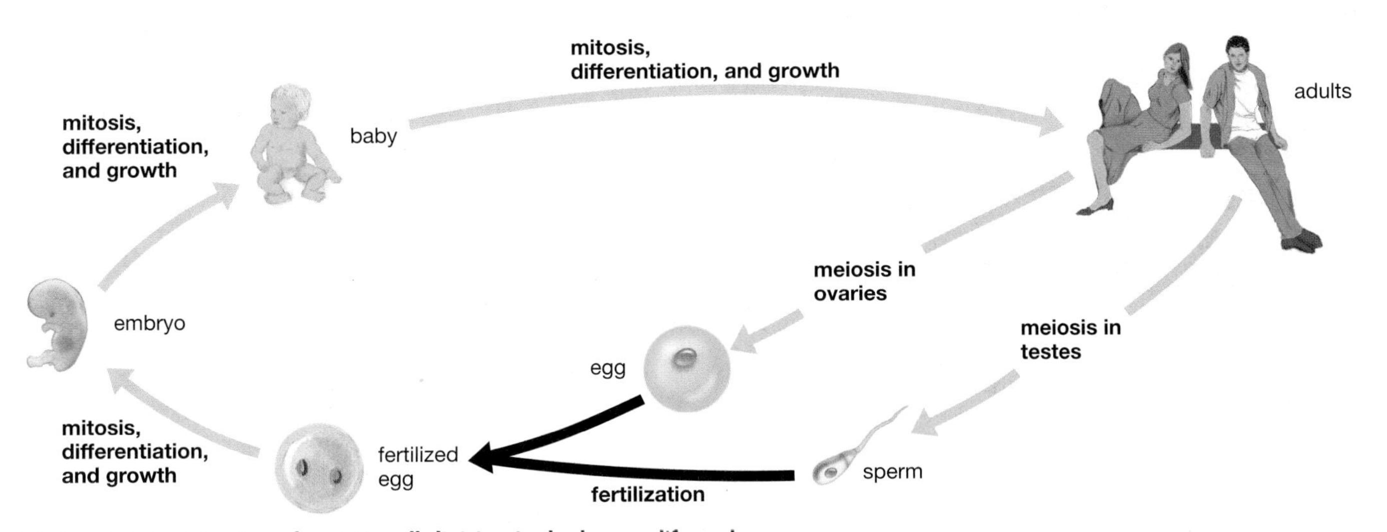

FIGURE 11-4 Mitotic and meiotic cell division in the human life cycle
Within ovaries, meiotic cell division produces eggs; within testes, meiotic cell division produces sperm. Fusion of egg and sperm produce a fertilized egg that develops into an adult by numerous mitotic cell divisions and differentiation of the resulting cells.

There Are Two Types of Cell Division in Eukaryotic Cells: Mitotic Cell Division and Meiotic Cell Division

Eukaryotic cells may undergo one of two evolutionarily related, but very different, types of cell division: *mitotic cell division* and *meiotic cell division*. **Mitotic cell division** consists of nuclear division (called **mitosis**) followed by cytoplasmic division (called **cytokinesis**). The word *mitosis* comes from the Greek for "thread"; during mitosis, chromosomes condense and appear as thin, threadlike structures when viewed through a light microscope. Cytokinesis (from the Greek words for "cell movement") is the process by which the cytoplasm is divided between the two daughter cells. As we will see later in this chapter, mitosis gives each daughter nucleus one copy of the parent cell's replicated chromosomes, and cytokinesis usually places one of these nuclei into each daughter cell. Hence, mitotic cell division typically produces two daughter cells that are genetically identical to each other and to the parent cell, and usually contain about equal amounts of cytoplasm.

Mitotic cell division takes place in all types of eukaryotic organisms. It is the mechanism of asexual reproduction in eukaryotic cells—including unicellular organisms such as yeast, *Amoeba*, and *Paramecium*—and in multicellular organisms such as *Hydra* and aspens. Finally, mitotic cell division is crucially important in multicellular organisms, even when the entire organism does not reproduce asexually.

In the life of any multicellular organism, mitotic cell division followed by differentiation of the daughter cells allows a fertilized egg to grow into an adult with perhaps trillions of specialized cells. Mitotic cell division also allows an organism to maintain its tissues, many of which require frequent replacement. For example, the cells of your stomach lining, which are constantly exposed to acid and digestive enzymes, survive only about 3 days. Without mitotic cell division to replace these short-lived cells, your body would soon be unable to function properly. Mitotic cell divisions also allow the body to repair itself, or sometimes even regenerate parts following injury.

Mitotic cell division also plays a role in biotechnology. Mitosis produced the nuclei used to create CC the cat (see this chapter's case study) and Dolly the cloned sheep, which you will read about in "Scientific Inquiry: Carbon Copies: Cloning in Nature and the Laboratory" later in the chapter. Because mitosis usually produces daughter cells that are genetically identical to the parent cell, CC and Dolly are genetically identical to their respective "nuclear donors" (the animals that provided the nuclei for each cloning procedure). In mammals, asexual reproduction—by cloning—occurs only in the laboratory.

Mitotic cell division also gives rise to "stem cells." These cells, which are found in both embryos and adults, may produce a wide variety of differentiated cell types, such as nerve cells, immune system cells, or muscle cells. We will discuss the medical applications and ethical implications of cloning in Chapter 13 and of stem cells in Chapter 37.

Meiotic cell division is a prerequisite for **sexual reproduction** in all eukaryotic organisms. In animals, meiotic cell division occurs only in ovaries and testes. The process of meiotic cell division involves a specialized nuclear division called **meiosis** and two rounds of cytokinesis to produce four daughter cells that can become **gametes** (eggs or sperm). Gametes carry half of the genetic material of the parent. Thus, the cells produced by meiotic cell division are not genetically identical to each other *or* to the original cell. During sexual reproduction, fusion of two gametes, one from each parent, reconstitutes a full complement of genetic material, forming a genetically unique offspring that is similar to both parents, but identical to neither (Fig. 11-4).

We will examine the events of mitosis and meiosis shortly. However, to understand the mechanisms of mitosis and meiosis, and their genetic and evolutionary significance, we first need to explore how DNA is packaged into eukaryotic chromosomes.

11.2 How Is DNA in Eukaryotic Cells Organized into Chromosomes?

The Eukaryotic Chromosome Consists of a Linear DNA Double Helix Bound to Proteins

Fitting all the DNA of a eukaryotic cell into the nucleus is no trivial task. If it were laid end to end, the total DNA in a single cell in your body would be about 6 feet long, and this DNA must fit into a nucleus that is at least a million times smaller! The degree of DNA compaction, or condensation, varies at each stage of the cell cycle. During most of a cell's life, much of the DNA is extended, making it readily accessible for transcription. In this extended state, individual **chromosomes**, which consist of a single DNA double helix and many associated proteins (Fig. 11-5), are too thin to be visible in light microscopes. Cell division, however, requires that the chromosomes be sorted out and moved into two daughter nuclei. Just as thread is easier to organize when it is wound onto spools, sorting and transporting chromosomes is easier when they are condensed and shortened. During cell division, proteins fold up the DNA of each chromosome into compact structures that can be seen in a light microscope.

How are chromosomes and genes related? Recall that genes are sequences of DNA from hundreds to thousands of nucleotides long. A single DNA double helix may contain hundreds or even thousands of genes, arranged in a particular linear order along the DNA strands. Each gene occupies a specific place, or **locus**, on a specific chromosome.

Chromosomes vary in length, and therefore in the number of genes they contain. The largest human chromosome, chromosome 1, contains approximately 3000 genes, whereas one of the smallest human chromosomes, chromosome 22, contains only about 600 genes.

In addition to genes, every chromosome has specialized regions that are crucial to its structure and function: two telomeres and one centromere. The two ends of a chromosome consist of repeated nucleotide sequences called **telomeres** ("end body" in Greek), which are essential for chromosome stability. Without telomeres, the ends of chromosomes might be removed by DNA repair enzymes, or the ends of two or more chromosomes might become connected, forming long, unwieldy structures that probably could not be distributed properly to the daughter nuclei during cell division.

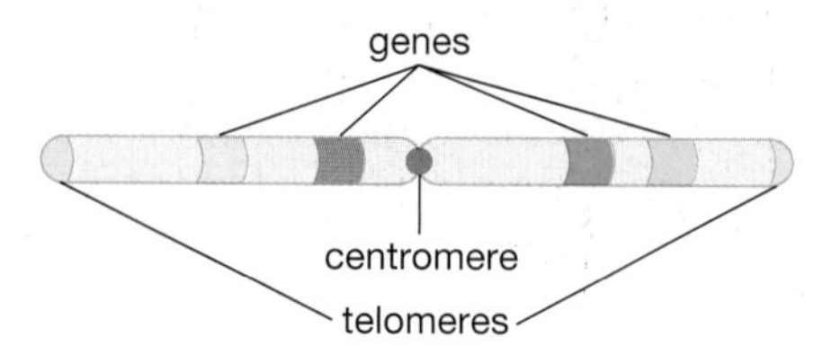

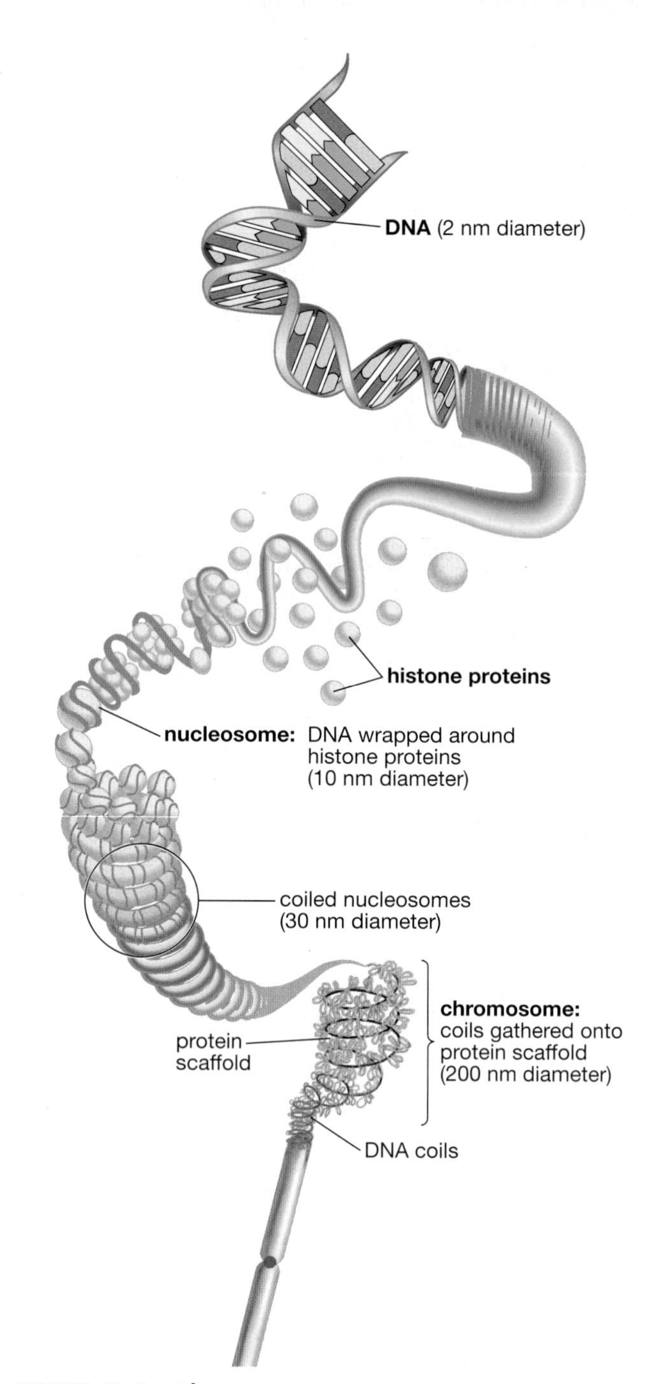

FIGURE 11-5 Chromosome structure
A eukaryotic chromosome contains a single, linear DNA double helix (top), which, in humans, is about 14 to 73 millimeters (mm) long and 2 nanometers (nm) in diameter. The DNA is wound around proteins called histones, forming nucleosomes (middle); this reduces the length by about a factor of 6. Other proteins coil up adjacent nucleosomes, much like a Slinky toy, reducing the length by another factor of 6 or 7. The coils of DNA and their associated proteins are attached in loops to still larger coils of protein "scaffolding" to complete the chromosome (bottom). All of this wrapping, coiling, and looping makes the extended interphase chromosome roughly 1000 times shorter than the DNA molecule it contains. Still other proteins produce about another 10-fold condensation during cell division (see Fig. 11-6).

At the time it condenses, the DNA within each chromosome has already replicated, forming two DNA double helices that remain attached to each other at the **centromere**. Although *centromere* means "middle body," a chromosome's centromere can be located almost any-

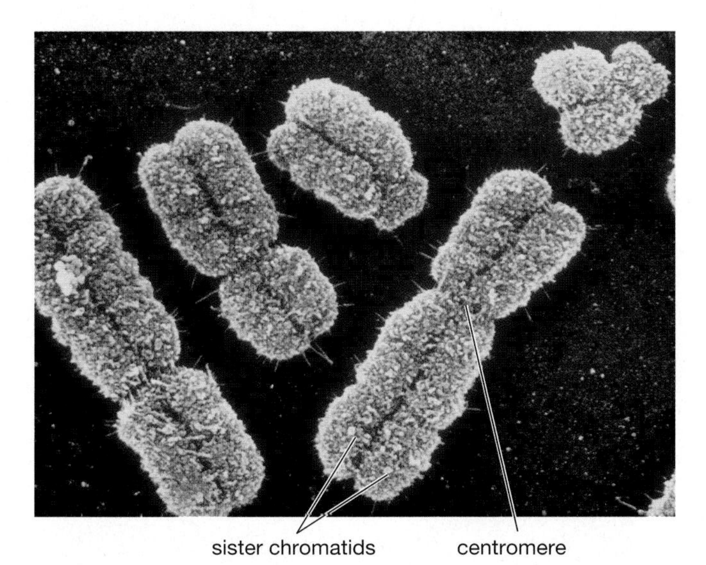

FIGURE 11-6 Human chromosomes during mitosis
The DNA and associated proteins in these duplicated human chromosomes have coiled up into the thick, short sister chromatids attached at the centromere. Each visible strand of "texture" is a loop of DNA. During cell division, the condensed chromosomes are about 5 to 20 micrometers long. At other times, the chromosomes uncoil until they are about 10,000 to 40,000 micrometers long.

where along the DNA double helix. While the two chromosomes are attached at their centromeres, we refer to each attached chromosome as a sister **chromatid**. Thus, DNA replication produces a **duplicated chromosome** with two identical sister chromatids (Fig. 11-6).

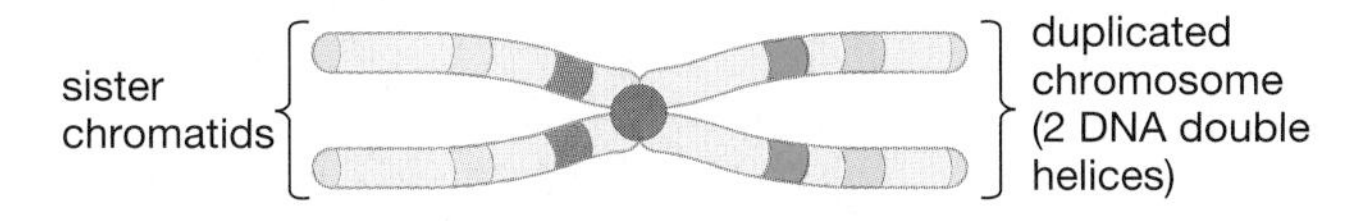

During mitotic cell division, the two sister chromatids separate, and each chromatid becomes an independent chromosome that is delivered to one of the two daughter cells.

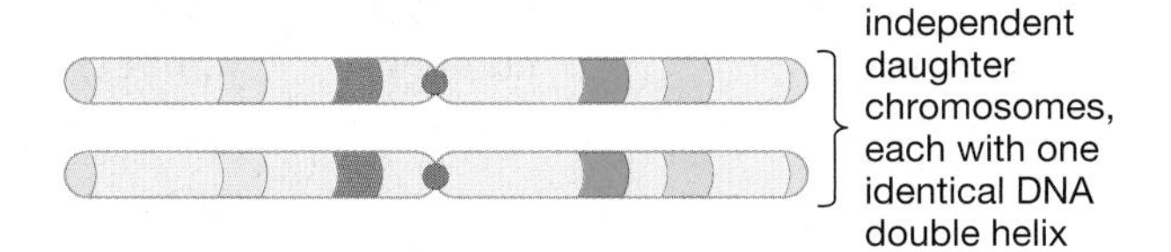

Eukaryotic Chromosomes Usually Occur in Homologous Pairs with Similar Genetic Information

The chromosomes of each eukaryotic species have characteristic shapes, sizes, and staining patterns (Fig. 11-7). When we view an entire set of stained chromosomes from a single cell (the **karyotype**), it becomes clear that the nonreproductive cells of many organisms, including humans, contain pairs of chromosomes. With one exception that we will discuss shortly, both members of each pair are the same length and have the same staining pattern. This similarity in size, shape, and staining occurs because each chromosome in a pair carries the same genes arranged in the same order. Chromosomes that contain the same genes are called *homologous chromosomes* or **homologues**, from Greek words that mean "to say the same thing." Cells with pairs of homologous chromosomes are called **diploid**, meaning "double."

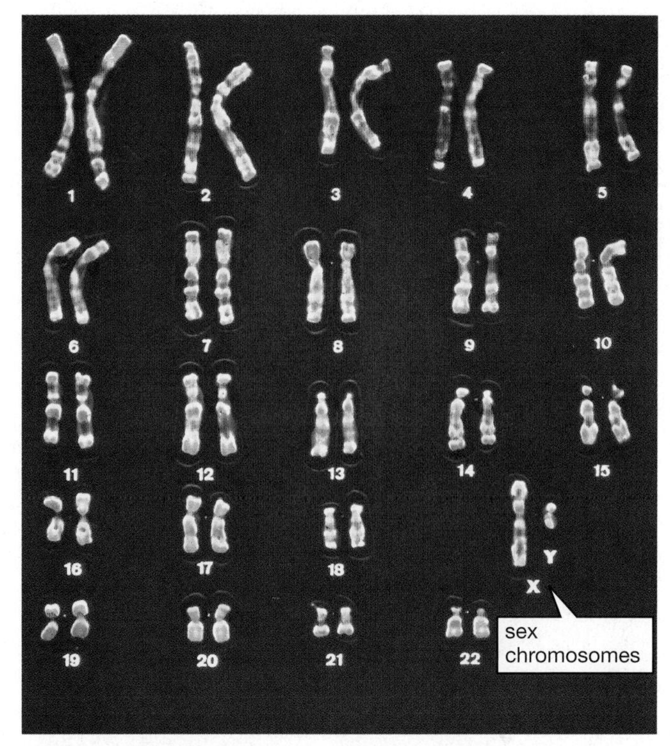

FIGURE 11-7 The karyotype of a human male
Staining and photographing the entire set of duplicated chromosomes within a single cell produces a karyotype. Pictures of the individual chromosomes are cut out and arranged in descending order of size. The chromosome pairs (homologues) are similar in both size and staining pattern and have similar genetic material. Chromosomes 1 through 22 are the autosomes; the X and Y chromosomes are the sex chromosomes. Notice that the Y chromosome is much smaller than the X chromosome. If this were a female karyotype, it would have two X chromosomes.

Let's consider a human skin cell. Although it has 46 chromosomes, it does not have 46 completely *different* chromosomes. The cell has two copies of chromosome 1, two copies of chromosome 2, and so on, up through chromosome 22. These chromosomes, which have similar appearance, similar genetic composition, and are paired in diploid cells of both sexes, are called **autosomes**. The cell also has two **sex chromosomes**: two X chromosomes or an X and a Y chromosome. The X and Y chromosomes are quite different in size (see Fig. 11-7) and genetic composition. Thus, sex chromosomes are an exception to the rule that homologous chromosomes contain the same genes. However, as we shall see later on, the X and Y chromosomes behave like homologues during meiotic cell division, and so the X and Y are considered as a pair in our "chromosomal bookkeeping."

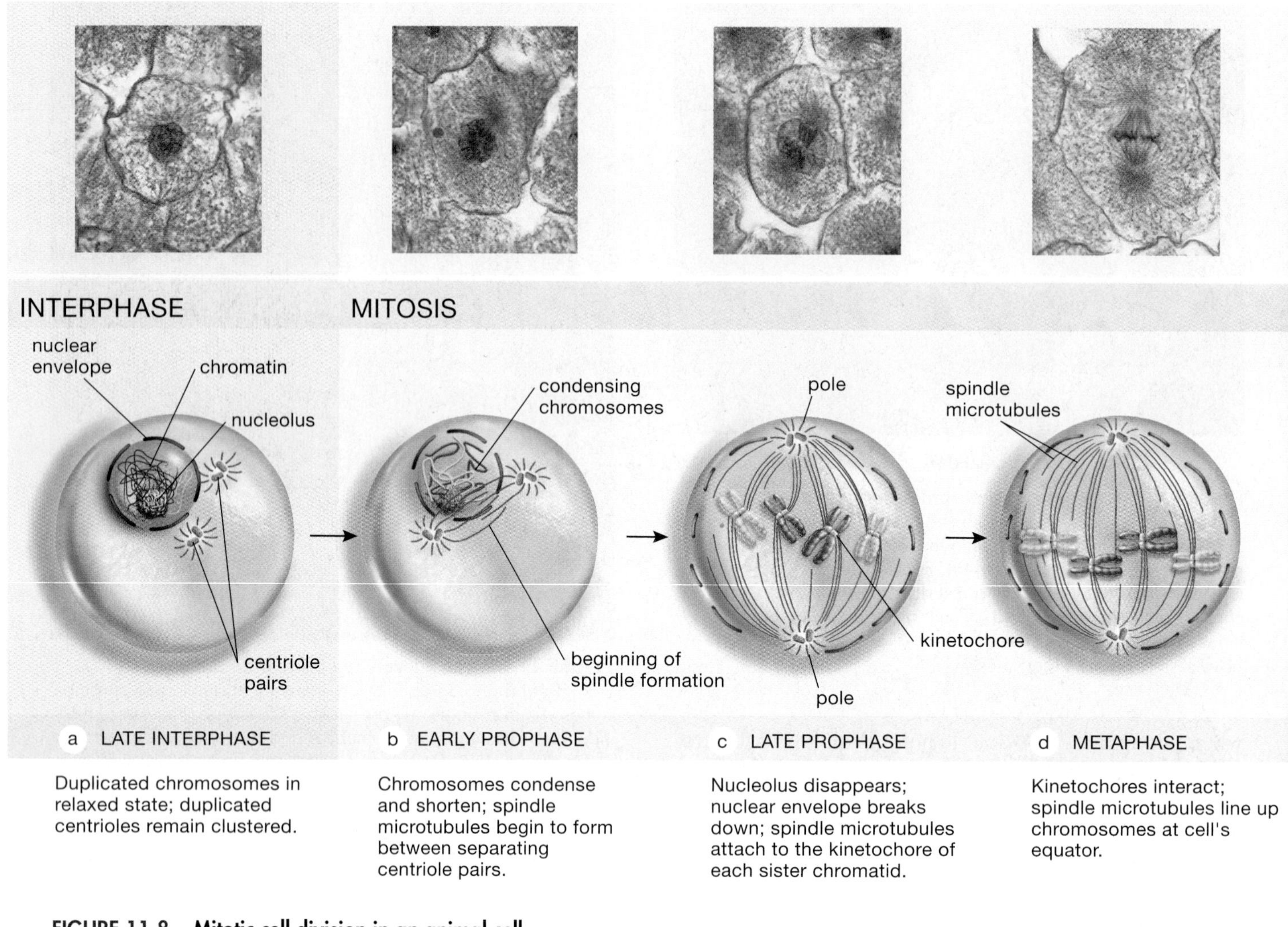
INTERPHASE
MITOSIS
nuclear envelope
chromatin
nucleolus
centriole pairs
condensing chromosomes
beginning of spindle formation
pole
kinetochore
pole
spindle microtubules
a LATE INTERPHASE
b EARLY PROPHASE
c LATE PROPHASE
d METAPHASE
Duplicated chromosomes in relaxed state; duplicated centrioles remain clustered.
Chromosomes condense and shorten; spindle microtubules begin to form between separating centriole pairs.
Nucleolus disappears; nuclear envelope breaks down; spindle microtubules attach to the kinetochore of each sister chromatid.
Kinetochores interact; spindle microtubules line up chromosomes at cell's equator.

FIGURE 11-8 Mitotic cell division in an animal cell
QUESTION What would the consequences be if one set of sister chromatids failed to separate at anaphase?

Most cells within our bodies are diploid. However, during sexual reproduction, cells in the ovaries or testes undergo meiotic cell division to produce gametes (sperm or eggs) that contain only one member of each pair of autosomes and one of the two sex chromosomes. Cells that contain only one of each type of chromosome are called **haploid** (meaning "half"). In humans, a haploid cell contains one each of the 22 autosomes, plus either an X or Y sex chromosome, for a total of 23 chromosomes. (Think of a haploid cell as one that contains *half* the diploid number of chromosomes, or one of each type of chromosome. A diploid cell contains two of each type of chromosome.) When a sperm fertilizes an egg, fusion of the two haploid cells produces a diploid cell with two copies of each type of chromosome.

In biological shorthand, the number of different types of chromosomes in a species is called the *haploid number* and is designated n. For humans, $n = 23$ because we have 23 different types of chromosomes (autosomes 1 to 22 plus one sex chromosome). Diploid cells contain $2n$ chromosomes. Thus, each human nonreproductive cell has 46 (2×23) chromosomes.

Every species has a specific number of chromosomes in its cells. As this table illustrates, the number of chromosomes varies tremendously between species.

Organism	n (haploid number)	$2n$ (diploid number)
Human	23	46
Gorilla, chimpanzee	24	48
Dog	39	78
Cat	19	38
Shrimp	127	254
Fruitfly	4	8
Pea	7	14
Potato	24	48
Sweet potato	45	90

Not all organisms are diploid. The bread mold *Neurospora*, for example, has haploid cells for most of its life cycle. Some plants, on the other hand, have more than two copies of each type of chromosome, with $4n$, $6n$, or even more chromosomes per cell.

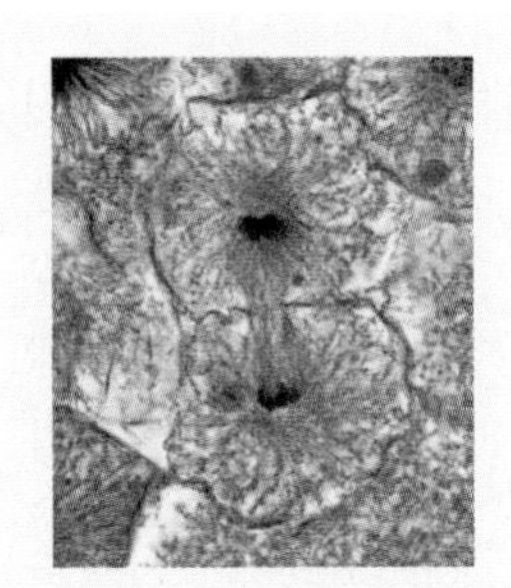

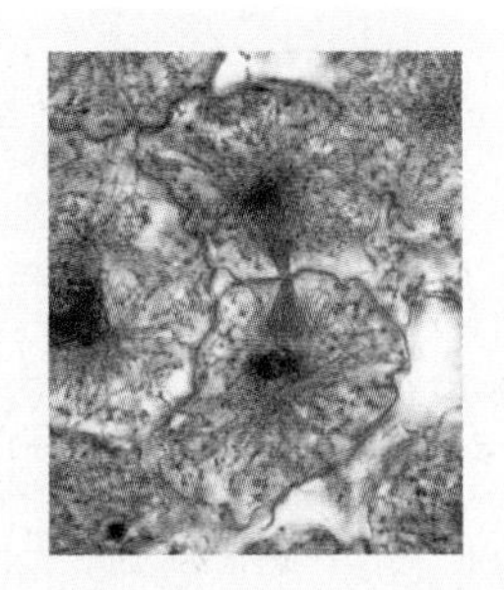

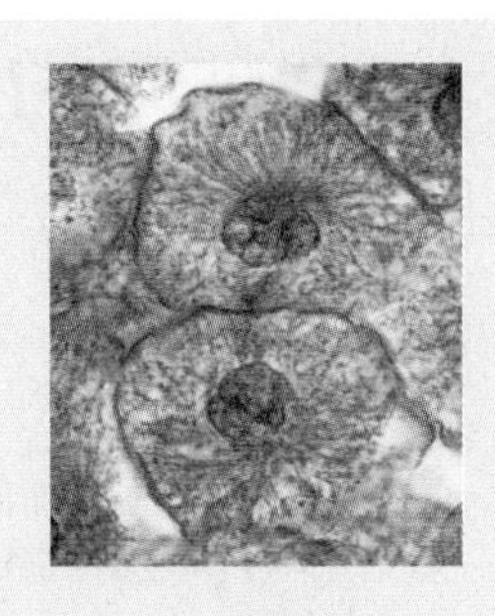

INTERPHASE

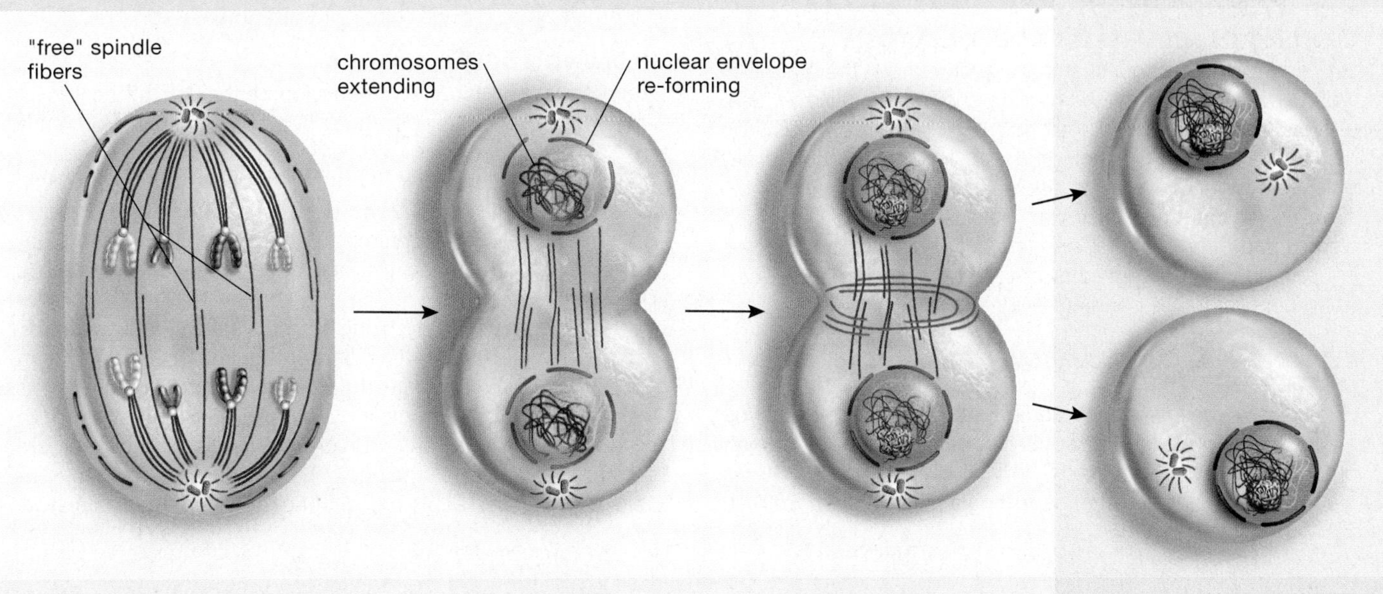

e ANAPHASE

Sister chromatids separate and move to opposite poles of the cell; spindle microtubules push poles apart.

f TELOPHASE

One set of chromosomes reaches each pole and relaxes into extended state; nuclear envelopes start to form around each set; spindle microtubules begin to disappear.

g CYTOKINESIS

Cell divides in two; each daughter cell receives one nucleus and about half of the cytoplasm.

h INTERPHASE OF DAUGHTER CELLS

Spindles disappear, intact nuclear envelopes form, chromosomes extend completely, and the nucleolus reappears.

11.3 How Do Cells Reproduce by Mitotic Cell Division?

As we described earlier, mitotic cell division (Fig. 11-8) consists of mitosis (nuclear division) and cytokinesis (cytoplasmic division). After interphase, when the cell's chromosomes have been replicated and all other necessary preparations for division have been made, mitotic cell division can occur. We will discuss mitosis and cytokinesis separately, even though they may overlap in time to some extent.

For convenience, biologists divide mitosis into four phases, based on the appearance and behavior of the chromosomes: (1) *prophase*, (2) *metaphase*, (3) *anaphase*, and (4) *telophase*. As with most biological processes, however, these phases are not really discrete events. Rather, they form a continuum, each phase merging into the next.

During Prophase, the Chromosomes Condense and the Spindle Microtubules Form and Attach to the Chromosomes

The first phase of mitosis is called **prophase** (meaning "the stage before" in Greek). During prophase, three major events occur: (1) the duplicated chromosomes condense, (2) the spindle microtubules form, and (3) the chromosomes are captured by the spindle (Fig. 11-8b,c).

Recall that chromosome duplication occurs during the S phase of interphase. Therefore, when mitosis begins, each chromosome already consists of two sister chromatids attached to one another at the centromere. During prophase, the duplicated chromosomes coil up and condense. In addition, the nucleolus, a structure within the nucleus where ribosomes assemble, disappears.

After the duplicated chromosomes condense, the **spindle microtubules** begin to assemble. In all eukaryotic cells, the proper movement of chromosomes during

Media Activity
11.2 Cell Cycle and Mitosis

SCIENTIFIC INQUIRY Carbon Copies: Cloning in Nature and the Lab

The word "cloning" usually brings to mind images of Dolly the sheep, CC the cat, or even *Star Wars: Attack of the Clones*, but nature has been quietly cloning for hundreds of millions of years. Everyone knows what **cloning** is: the creation of one or more individual organisms (**clones**) that are genetically identical to a preexisting individual. How are clones produced, either in nature or in the lab? Why is cloning such a hot—and controversial—topic in the news? And why is cloning included in a chapter on cell division?

CLONING IN NATURE: THE ROLE OF MITOSIS

Let's address the last question first. As you know, there are two types of cell division: mitotic division and meiotic division. Sexual reproduction relies on meiotic cell division, the production of gametes, and fertilization, and usually produces genetically unique offspring. In contrast, asexual reproduction (see Fig. 11-1) relies on mitotic cell division. Because mitotic cell division creates daughter cells that are genetically identical to the parent cell, offspring produced by asexual reproduction are genetically identical to their parents—clones.

CLONING PLANTS: A FAMILIAR APPLICATION IN AGRICULTURE

Humans have been in the cloning business a lot longer than you might think. For example, consider navel oranges, which don't produce seeds. Without seeds, how do they reproduce? Navel orange trees are propagated by cutting a piece of stem from an adult navel tree and grafting it onto the top of the root of a seedling orange tree, usually of a different type. (Why would the seedling usually not be a navel?) Therefore, the cells of the above-ground, fruit-bearing parts of the resulting tree are clones of the original navel orange stem. All navel oranges apparently originated from a single mutant bud of an orange tree discovered in Brazil in the early 1800s, and propagated asexually ever since. Three navel orange trees were brought from Brazil to Riverside, California, in the 1870s. (One of them is still there!) All American navels are clones of these three trees.

CLONING ADULT MAMMALS

Animal cloning isn't a recent development either. In the 1950s, John Gurdon and his colleagues inserted a nucleus from early frog embryos into eggs, and some of the resulting cells developed into complete frogs. By the 1990s, several labs had been able to clone mammals using embryonic nuclei, but it wasn't until 1996 that Dr. Ian Wilmut of the Roslin Institute in Edinburgh, Scotland, cloned the first adult mammal, the famous Dolly (Fig. E11-1).

Why is it important to clone an adult animal? In agriculture, it is usually worthwhile to clone only adults, because only in adults can we see the traits that we wish to propagate (such as milk and meat production in cows, or speed and strength in horses). Cloning an adult would produce "offspring" that are genetically identical to the adult. Therefore, insofar as the valuable traits of the adult are genetically determined, all of its

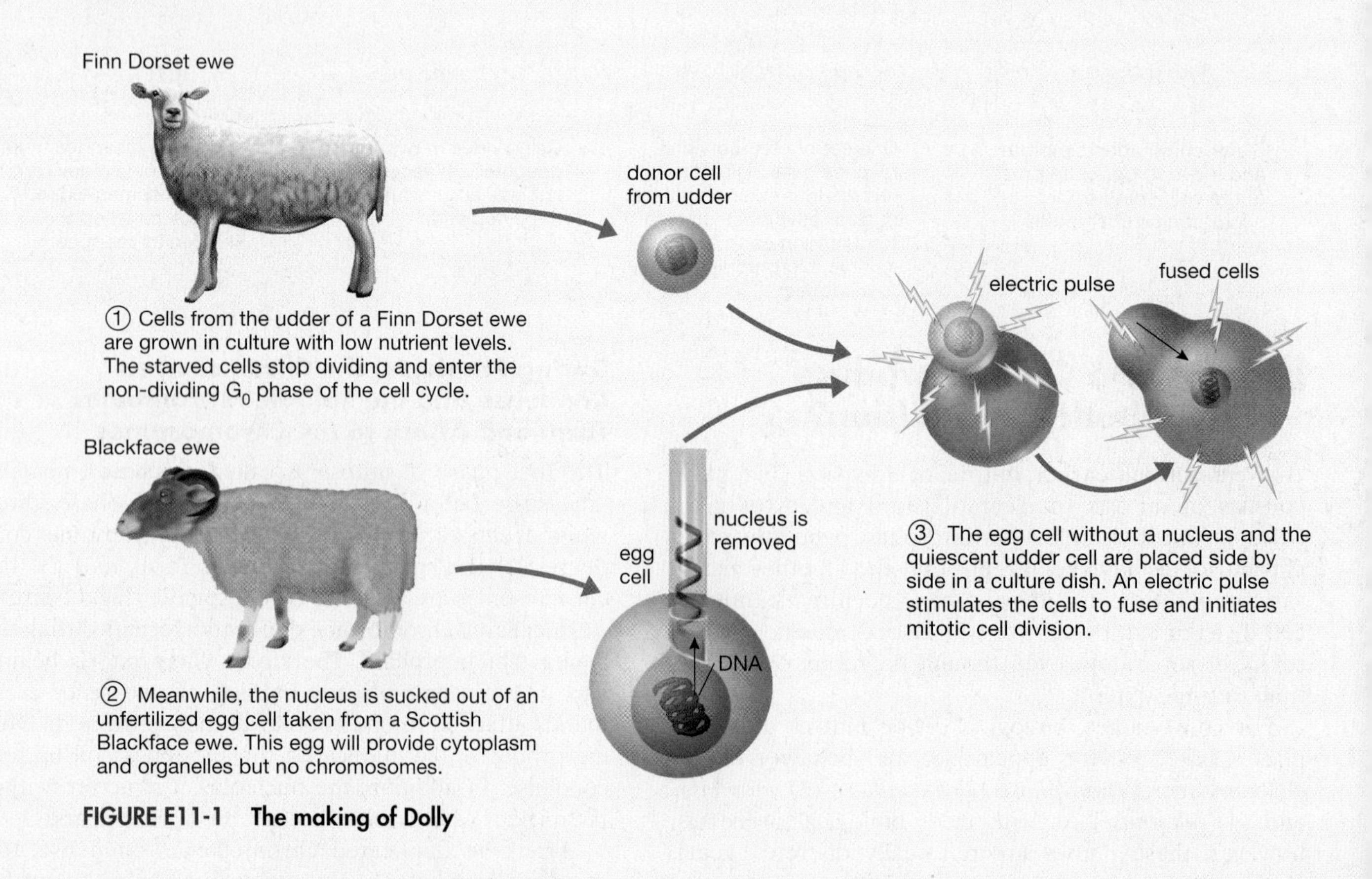

Finn Dorset ewe
donor cell from udder
electric pulse
fused cells
① Cells from the udder of a Finn Dorset ewe are grown in culture with low nutrient levels. The starved cells stop dividing and enter the non-dividing G_0 phase of the cell cycle.
Blackface ewe
nucleus is removed
egg cell
DNA
③ The egg cell without a nucleus and the quiescent udder cell are placed side by side in a culture dish. An electric pulse stimulates the cells to fuse and initiates mitotic cell division.
② Meanwhile, the nucleus is sucked out of an unfertilized egg cell taken from a Scottish Blackface ewe. This egg will provide cytoplasm and organelles but no chromosomes.

FIGURE E11-1 The making of Dolly

clones would also express those same valuable traits. Cloning of embryos would usually not be useful, because the embryonic cells would have been produced by sexual reproduction in the first place, and normally no one could tell if the embryo had any especially desirable traits.

For some medical applications, too, cloning adults is essential. Suppose that a pharmaceutical company genetically engineered (see Chapter 13) a cow that secreted a valuable molecule, such as an antibiotic, in its milk. These techniques are extremely expensive and somewhat hit-or-miss, so the company may successfully produce only one profitable cow. This cow could then be cloned, creating a whole herd of antibiotic-producing cows. Cloned cows that produce more milk or meat and pigs tailored to be organ donors for humans already exist.

Cloning might also help rescue endangered species, many of which don't reproduce well in zoos. As Richard Adams of Texas A&M, home of the Missyplicity Project, put it, "You could repopulate the world [with an endangered species] in a matter of a couple of years. Cloning is not a trivial pursuit."

CLONING: AN IMPERFECT TECHNOLOGY

Unfortunately, cloning mammals is inefficient and beset with difficulties. An egg is subjected to severe trauma when its nucleus is sucked out or destroyed, and a new nucleus is inserted (see Fig. E11-1). Often, the egg may simply die. Molecules in the cytoplasm that are needed to control development may be lost or moved to the wrong places, so that even if the egg survives and divides, it may not develop properly. If the eggs develop into viable embryos, the embryos must then be implanted into the uterus of a surrogate mother. Many clones die or are aborted during gestation, often with serious or fatal consequences for the surrogate mother. Even if the clone survives gestation and birth, it may have defects, commonly a deformed head, lungs, or heart. Given the high failure rate—it took 277 tries to produce Dolly, and 87 to make CC—cloning mammals is an expensive proposition.

To make things even more problematic, "successful" clones often have hidden defects. Dolly, for example, seemed to have "middle-aged" chromosomes. Remember the telomeres on the ends of chromosomes? At each mitotic cell division, the telomeres get a little shorter, and it appears that cells may die—or at least no longer divide—when their telomeres get too short. Dolly was born with short telomeres, as if she were already over 3 years old (not all cloned mammals have short telomeres, so there is hope that this problem can be solved). Dolly also developed arthritis when she was $5\frac{1}{2}$ and was euthanized with a serious lung disease when she was $6\frac{1}{2}$, both relatively young for a sheep, although no one can say for sure if these health problems occurred because she was a clone.

THE FUTURE OF CLONING

As a technology, cloning shows great promise. As the process becomes more routine, however, it will bring ethical questions. While hardly anyone objects to navel oranges and few would refuse antibiotics or other medicinal products from cloned livestock, many think that cloning pets is, at best, a frivolous luxury, and might even be a form of animal abuse. We will return to the ethics of cloning, especially cloning people, in the Case Study Revisited at the end of this chapter.

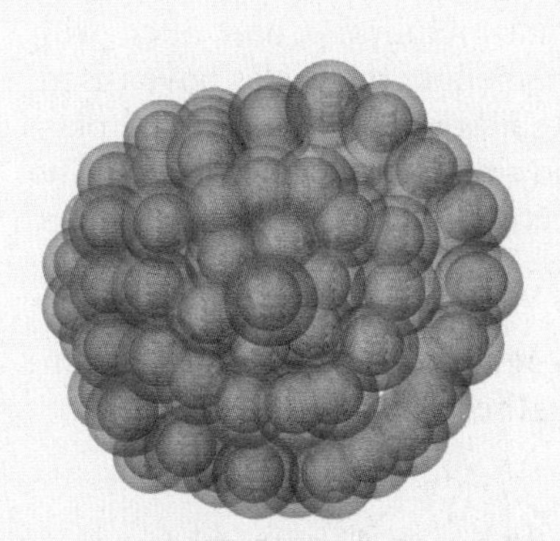

④ The cell divides, forming an embryo that consists of a hollow ball of cells.

⑤ The ball of cells is implanted into the uterus of another Blackface ewe.

⑥ The Blackface ewe gives birth to Dolly, a female Finn Dorset lamb, a genetic twin of the Finn Dorset ewe.

mitosis depends on these spindle microtubules. In animal cells, the spindle microtubules originate from a region in which a pair of microtubule-containing **centrioles** is located. During interphase, a new pair of centrioles forms near the previously existing pair. During prophase, the centriole pairs migrate to opposite sides of the nucleus. Each centriole pair serves as a central point from which the spindle microtubules radiate, both inward toward the nucleus and outward toward the plasma membrane. These points are called *spindle poles*. Though the cells of plants, fungi, and many algae do not contain centrioles, they nevertheless form functional spindles during mitotic cell division.

As the spindle microtubules form into a complete basket around the nucleus, the nuclear envelope disintegrates, releasing the duplicated chromosomes. At the centromere, each sister chromatid has a protein-containing structure called a **kinetochore** that serves as an attachment site for the ends of spindle microtubules. In each duplicated chromosome, the kinetochore of one sister chromatid binds to the ends of spindle microtubules leading to one pole of the cell, while the kinetochore of the other sister chromatid binds to spindle microtubules leading to the opposite pole of the cell (Fig. 11-8c) When the sister chromatids separate later in mitosis, the newly independent chromosomes will move along the spindle microtubules to opposite poles. Some spindle microtubules do not attach to chromosomes; rather, they have free ends that overlap along the cell's equator. As we will see, these unattached spindle microtubules will push the two spindle poles apart later in mitosis.

During Metaphase, the Chromosomes Align Along the Equator of the Cell

At the end of prophase, the two kinetochores of each duplicated chromosome are connected to spindle microtubules leading to opposite poles of the cell. As a result, each duplicated chromosome is connected to both spindle poles. During **metaphase** (the "middle stage"), the two kinetochores on a duplicated chromosome engage in a "tug of war." During this process, the microtubules lengthen and shorten, until each chromosome lines up along the equator of the cell, with one kinetochore facing each pole (Fig. 11-8d).

During Anaphase, Sister Chromatids Separate and Are Pulled to Opposite Poles of the Cell

At the beginning of **anaphase** (Fig. 11-8e), the sister chromatids separate, becoming independent daughter chromosomes. This separation allows "motor proteins" in the kinetochores to pull the chromosomes poleward along the spindle microtubules. One of the two daughter chromosomes derived from each original parental chromosome moves to each pole of the cell. As the kinetochores tow their chromosomes toward the poles, the unattached spindle microtubules interact and lengthen to push the poles of the cell apart, forcing the cell into an oval shape (see Fig. 11-8e). Because the daughter chromosomes are identical copies of the parental chromosomes, each cluster of chromosomes that forms on opposite poles of the cell contains one copy of every chromosome that was in the parent cell.

During Telophase, Nuclear Envelopes Form Around Both Groups of Chromosomes

When the chromosomes reach the poles, **telophase** (the "end stage") begins (Fig. 11-8f). The spindle microtubules disintegrate, and a nuclear envelope forms around each group of chromosomes. The chromosomes revert to their extended state, and the nucleoli reappear. In most cells, cytokinesis occurs during telophase, separating each daughter nucleus into a separate cell (Fig. 11-8g).

During Cytokinesis, the Cytoplasm Is Divided Between Two Daughter Cells

In animal cells, microfilaments attached to the plasma membrane form a ring around the equator of the cell. During cytokinesis, the ring contracts and constricts the cell's equator, much like pulling the drawstring on a pair of sweatpants tightens the waist when pulled. Eventually the "waist" constricts completely, dividing the cytoplasm into two new daughter cells (Fig. 11-9).

Cytokinesis in plant cells is quite different, perhaps because their stiff cell walls make it impossible to divide one cell into two by pinching at the waist. Instead, carbohydrate-filled vesicles, which bud off the Golgi complex, line up along the cell's equator between the two nuclei (Fig. 11-10). The vesicles fuse, producing a structure called the **cell plate**, which is shaped like a flattened sac, surrounded by plasma membrane, and filled with sticky carbohydrates. When enough vesicles have fused, the edges of the cell plate merge with the original plasma membrane around the circumference of the cell. The carbohydrate formerly contained in the vesicles remains between the plasma membranes as part of the cell wall.

Following cytokinesis, eukaryotic cells enter G_1 of interphase, thus completing the cell cycle (Fig. 11-8h).

11.4 Why Do So Many Organisms Reproduce Sexually?

The largest organism known on Earth is a mushroom whose underground, branching filaments extend through 2200 acres of soil in eastern Oregon (see Chapter 20). This organism was produced almost entirely by mitotic cell division. Clearly, asexual reproduction via mitotic cell division must work pretty well! Why, then, have nearly all known forms of life evolved ways of sexual reproduction? Mitosis can only produce

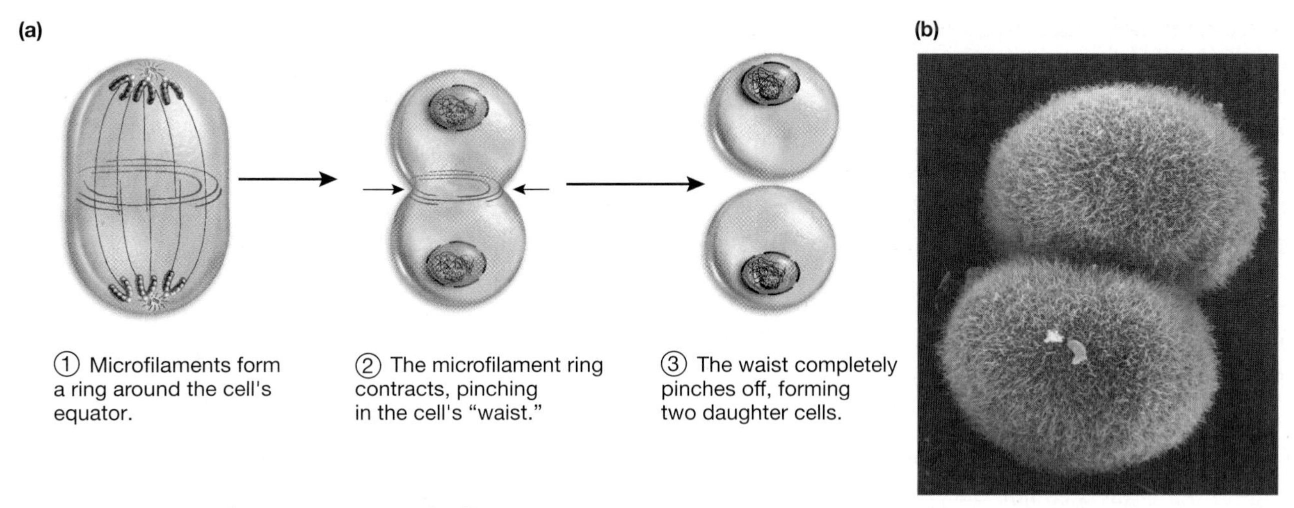

(a)
(b)
① Microfilaments form a ring around the cell's equator.
② The microfilament ring contracts, pinching in the cell's "waist."
③ The waist completely pinches off, forming two daughter cells.

FIGURE 11-9 Cytokinesis in an animal cell
(a) A ring of microfilaments just beneath the plasma membrane contracts around the equator of the cell, pinching it in two. **(b)** This scanning electron micrograph of cytokinesis shows the two daughter cells nearly separated.

clones of genetically identical offspring. In contrast, sexual reproduction shuffles genes to produce genetically unique offspring. The nearly universal presence of sexual reproduction provides evidence for the tremendous evolutionary advantage that DNA exchange among individuals confers on a species.

Mutations in DNA Are the Ultimate Source of Genetic Variability

As we saw in Chapter 10, the fidelity of DNA replication and proofreading minimizes errors during DNA replication, but changes in DNA base sequences do occur, producing mutations. Although most mutations are either neutral or harmful, they are also the raw material for evolution. Bacteria are different from bison—and you are different from your ancestors—because of differences in DNA sequence that originally arose as mutations. Mutations in gametes may be passed to offspring and become a part of the genetic makeup of the species. Such mutations form **alleles**, alternate forms of a given gene that may produce differences in structure or function, such as black, brown, or blond hair in humans, or different mating calls in frogs. As we saw earlier, most eukaryotic organisms are diploid, containing pairs of homologous chromosomes. Homologous chromosomes have the same genes, but each homologue may have the same alleles of some genes and different alleles of other genes.

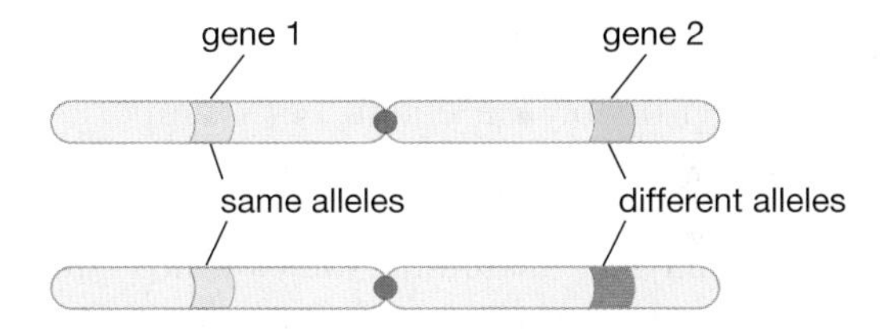

We'll explore the consequences of having paired genes—and more than one allele of each gene—in the next chapter.

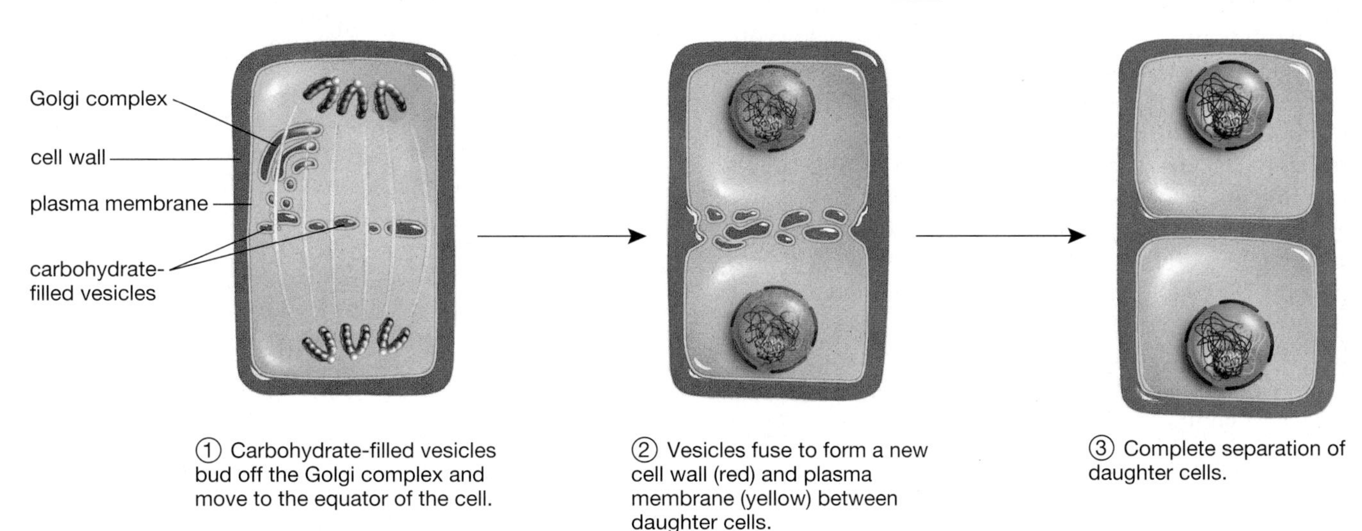

FIGURE 11-10 Cytokinesis in a plant cell

Sexual Reproduction May Combine Different Parental Alleles in a Single Offspring

To illustrate how sexual reproduction promotes genetic variability, let's consider a familiar organism: the domestic cat. As a result of generations of selective breeding by cat fanciers, a lot is known about cat genetics. For example, cat fur comes in two basic lengths, long and short, controlled by two alleles of a single gene. Cats may also have the typical number of toes (4 on each paw) or have extra toes (a condition called polydactyly). Toe number is also controlled by two alleles of a single gene (a different gene than the one that determines fur length). Let's suppose you have short-haired, 4-toed cats and long-haired, polydactylous cats, but you'd like to breed short-haired, polydactylous cats. You could breed each type of cat only to similar cats, wait for a mutation in either the fur-length gene or the toe-number gene, and hope that you live for 10,000 years and get lucky. Or, you could mate short-haired, 4-toed cats with long-haired, polydactylous cats. After a few generations, sexual recombination will produce some short-haired, polydactylous cats.

From an evolutionary perspective, making short-haired, polydactylous cats isn't very useful, but you can probably imagine many traits in wild animals or plants that would be useful only when combined. For example, camouflage coloration can help an animal avoid predation only if it stays still when it sees a predator. Both camouflaged animals that constantly jump around and brightly colored animals that freeze when a predator appears will probably be eaten. Let's suppose that a ground-nesting bird of a given species evolves better camouflage color, while another evolves more effective "freezing" behavior. Combining the two through sexual reproduction might produce offspring that are able to avoid predation better than either parent. Combining useful, genetically determined traits is one reason why sexual reproduction is so nearly ubiquitous in nature.

How does sexual reproduction combine traits from two parents in a single offspring? The first eukaryotic cells to evolve, about 1 billion to 1.5 billion years ago, were probably haploid, with only one copy of each chromosome. Relatively early on, two evolutionary events probably occurred in single-celled eukaryotic organisms that allowed them to shuffle and recombine genetic information. First, two haploid (parental) cells fused, resulting in a diploid cell with two copies of each chromosome. This cell could reproduce by mitotic cell division, producing diploid daughter cells. Second, this diploid cell evolved a variation in the process of cell division called meiotic cell division. Meiotic cell division produces haploid cells, each containing one copy of each chromosome. In animals, these haploid cells usually become gametes. A haploid sperm from animal A might contain alleles contributing to camouflage coloration, and a haploid egg from animal B might contain alleles that favor freezing at the first sign of a predator. Fusion of these gametes would produce an animal with camouflage coloration that also becomes motionless when a predator approaches.

11.5 How Does Meiotic Cell Division Produce Haploid Cells?

Meiosis Separates Homologous Chromosomes, Producing Haploid Daughter Nuclei

The key to sexual reproduction in eukaryotes is meiosis, the production of haploid nuclei with unpaired chromosomes from diploid parent nuclei with paired chromosomes. In meiotic cell division (meiosis followed by cytokinesis), each daughter cell receives one member of each pair of homologous chromosomes. Therefore, meiosis (from a Greek word meaning "to diminish") reduces the number of chromosomes in a diploid cell by half. For example, each diploid cell in your body contains 23 *pairs* of chromosomes; meiotic cell division produces sperm or eggs with 23 chromosomes, one from each pair.

Meiosis evolved from mitosis, so many of the structures and events of meiosis are similar or identical to those of mitosis. However, meiotic cell division differs from mitotic cell division in a major way: during meiotic cell division, the cell undergoes one round of DNA replication followed by two nuclear divisions. One round of DNA replication produces two chromatids in each duplicated chromosome. Because diploid cells have pairs of homologous chromosomes—with two chromatids per homologue—a single round of DNA replication creates four chromatids for each type of chromosome:

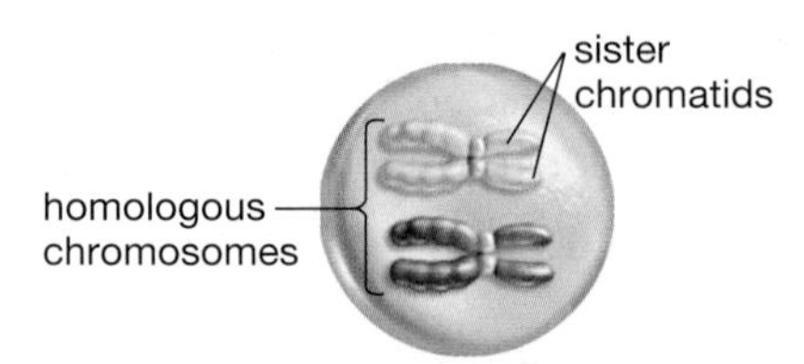

The first division of meiosis (called *meiosis I*) separates the pairs of homologues and sends one of each pair into each of two daughter nuclei, producing two haploid nuclei. Each homologue, however, still consists of two chromatids:

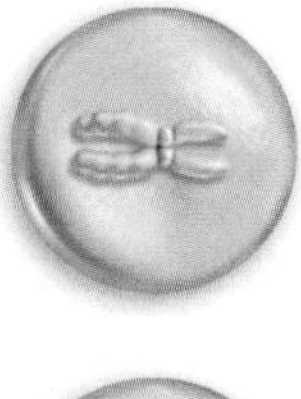

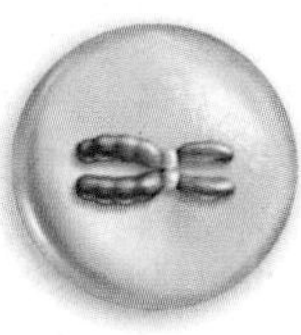

A second division (called *meiosis II*) separates the chromatids and parcels one chromatid into each of two more daughter nuclei. Therefore, at the end of meiosis, there are four haploid daughter nuclei, each with one copy of each homologous chromosome. Because each nucleus is usually contained in a different cell, meiosis normally produces four haploid cells from a single diploid parent cell:

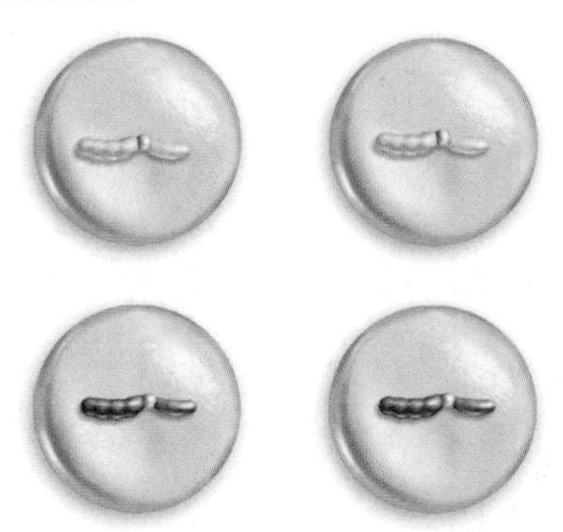

We'll explore the stages of meiosis in more detail in the following sections.

Meiotic Cell Division Followed by Fusion of Gametes Keeps the Chromosome Number Constant from Generation to Generation

Why is meiotic cell division so important to sexual reproduction? Consider what would happen if gametes were diploid, like the rest of the cells of the parent organism, with two copies of each homologous chromosome. Fertilization would result in a cell with four copies of each homologue, giving the offspring twice as many chromosomes as its parents. After a few generations, the cells of the offspring would have enormous amounts of DNA. On the other hand, when a haploid sperm fuses with a haploid egg, the resulting offspring are diploid, just like their parents:

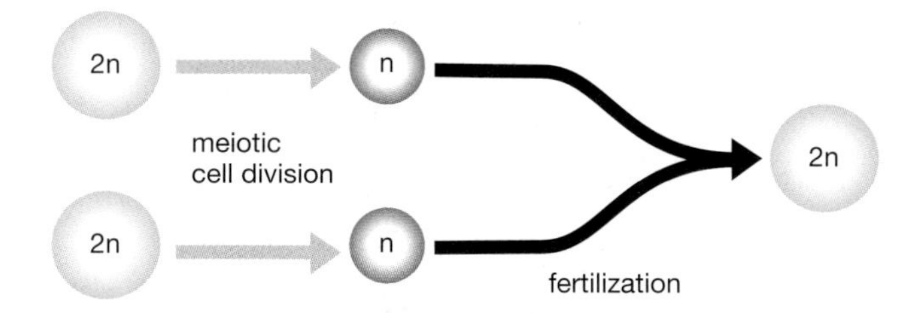

Meiosis I Separates Homologous Chromosomes into Two Haploid Daughter Nuclei

The phases of meiosis have the same names as the roughly equivalent phases in mitosis, followed by a I or II to distinguish the two nuclear divisions that occur in meiosis. In the descriptions that follow, we assume that cytokinesis accompanies the nuclear divisions. Meiosis begins with chromosome duplication. As in mitosis, the sister chromatids of each chromosome remain attached to one another at the centromere.

During Prophase I, Homologous Chromosomes Pair Up and Exchange DNA

During mitosis, homologous chromosomes move completely independently of each other. In contrast, during *prophase I* of meiosis, homologous chromosomes line up side by side and exchange segments of DNA (Fig. 11-11a and 11-12a, p. 200). We'll call one homologue the "maternal chromosome" and the other the "paternal chromosome," because one was originally inherited from the organism's mother and the other from the organism's father. During prophase I, proteins bind the maternal and paternal homologues together so that they match up exactly along their entire length, much like closing a zipper (Fig. 11-12b). In addition, enzyme complexes assemble at several places along the paired chromosomes (Fig. 11-12c). These enzymes cut through the DNA backbones within the chromosomes and graft the broken DNA ends together again, usually joining the maternal DNA to the paternal DNA and vice versa. This joining creates crosses, or **chiasmata** (singular, chiasma), where the maternal and paternal chromosomes intertwine (Fig. 11-12d). In human cells, each pair of homologues usually forms two or three chiasmata in prophase I. Eventually, the enzyme complexes detach from the chromosomes, and the protein zippers that held homologues together disassemble. Nevertheless, the homologues remain together, held by the chiasmata (Fig. 11-12e).

This exchange of DNA between maternal and paternal chromosomes at chiasmata is called **crossing over**. If the chromosomes had different alleles, then the formation of the chiasmata creates slight genetic differences in both chromosomes. The result of crossing over, then, is genetic **recombination**: the formation of new combinations of alleles on a chromosome.

As in mitosis, the spindle microtubules begin to assemble outside the nucleus during prophase I. Near the end of prophase I, the nuclear envelope breaks down and the spindle microtubules capture the chromosomes by attaching to their kinetochores.

During Metaphase I, Paired Homologous Chromosomes Line Up at the Equator of the Cell

During *metaphase I*, interactions between the kinetochores and the spindle microtubules move the paired homologues to the equator of the cell (Fig. 11-11b). Unlike mitosis, in which *individual* duplicated chromosomes line up along the equator, *homologous pairs* of duplicated chromosomes line up along the equator during metaphase I of meiosis.

The key to understanding meiosis lies in the way the duplicated chromosomes line up during metaphase I. Before going farther, then, let's look more closely at the differences in chromosome attachment to spindle microtubules in mitosis versus meiosis I. First, in mitosis, the homologues attach independently to the spindle. In meiosis I, the homologues remain associated with each other via chiasmata, attaching to the spindle as a unit

Media Activity
11.3 Meiosis

MEIOSIS I

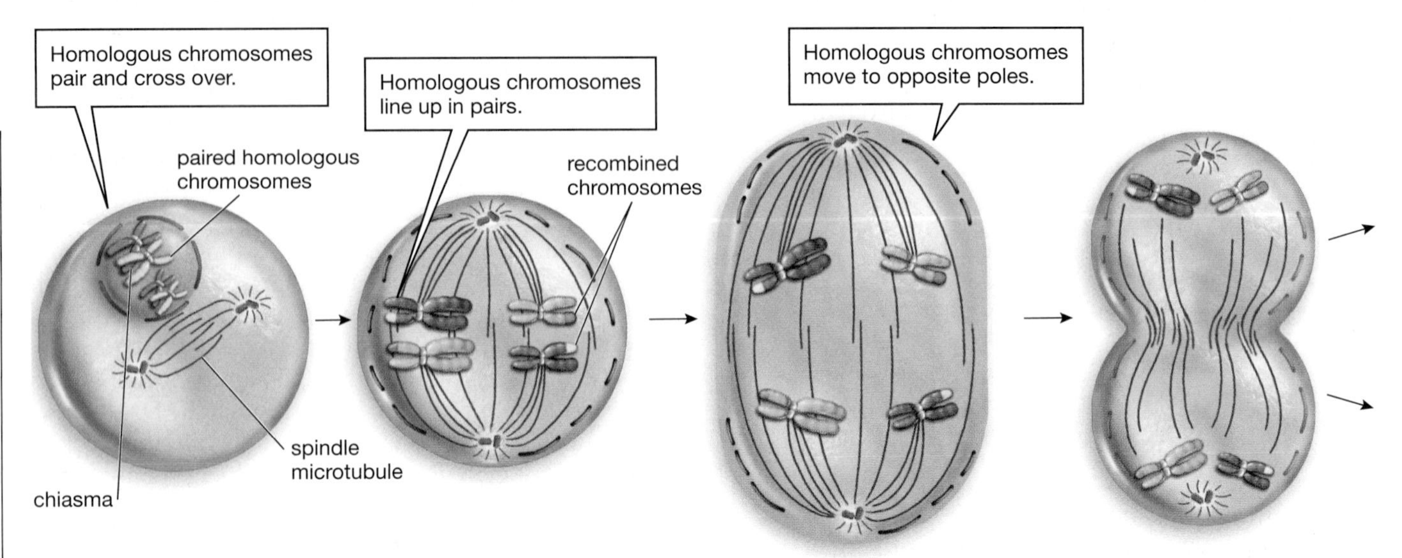

(a) Prophase I. Duplicated chromosomes condense. Homologous chromosomes pair up and chiasmata occur as chromatids of homologues exchange parts. The nuclear envelope disintegrates, and spindle microtubules form.

(b) Metaphase I. Paired homologous chromosomes line up along the equator of the cell. One homologue of each pair faces each pole of the cell and attaches to spindle microtubules via its kinetochore (blue).

(c) Anaphase I. Homologues separate, one member of each pair going to each pole of the cell. Sister chromatids do not separate.

(d) Telophase I. Spindle microtubules disappear. Two clusters of chromosomes have formed, each containing one member of each pair of homologues. The daughter nuclei are therefore haploid. Cytokinesis commonly occurs at this stage. There is little or no interphase between meiosis I and meiosis II.

FIGURE 11-11 Meiotic cell division in an animal cell
In meiotic cell division (meiosis and cytokinesis), the homologous chromosomes of a diploid cell are separated, producing four haploid daughter cells. Each daughter cell contains one member of each pair of parental homologous chromosomes. In these diagrams, two pairs of homologous chromosomes are shown, large and small. The yellow chromosomes are from one parent (for example, the father), and the violet chromosomes are from the other parent (for example, the mother). **QUESTION** What would the consequences be (for the resulting gametes) if one pair of homologues failed to separate at anaphase I? What if meiosis I was normal, but a pair of sister chromatids failed to separate at anaphase II?

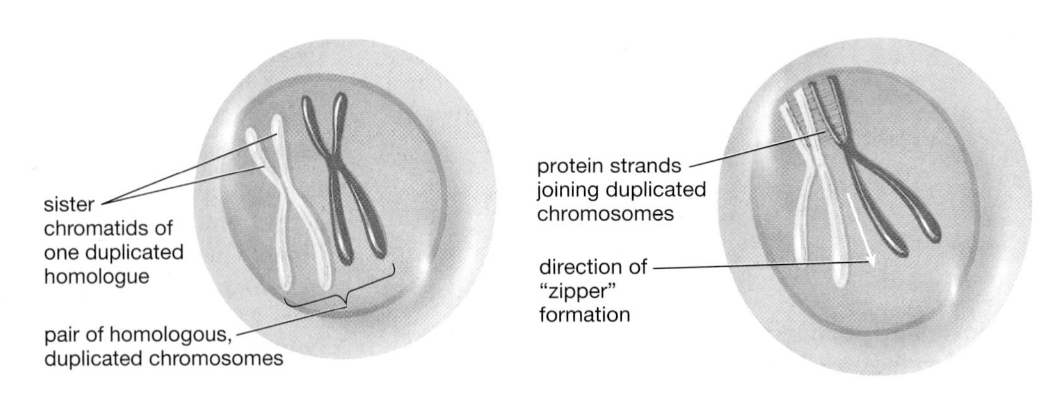

(a) Duplicated homologous chromosomes pair up side by side.

(b) Protein strands "zip" the homologous chromosomes together.

FIGURE 11-12 The mechanism of crossing over

MEIOSIS II

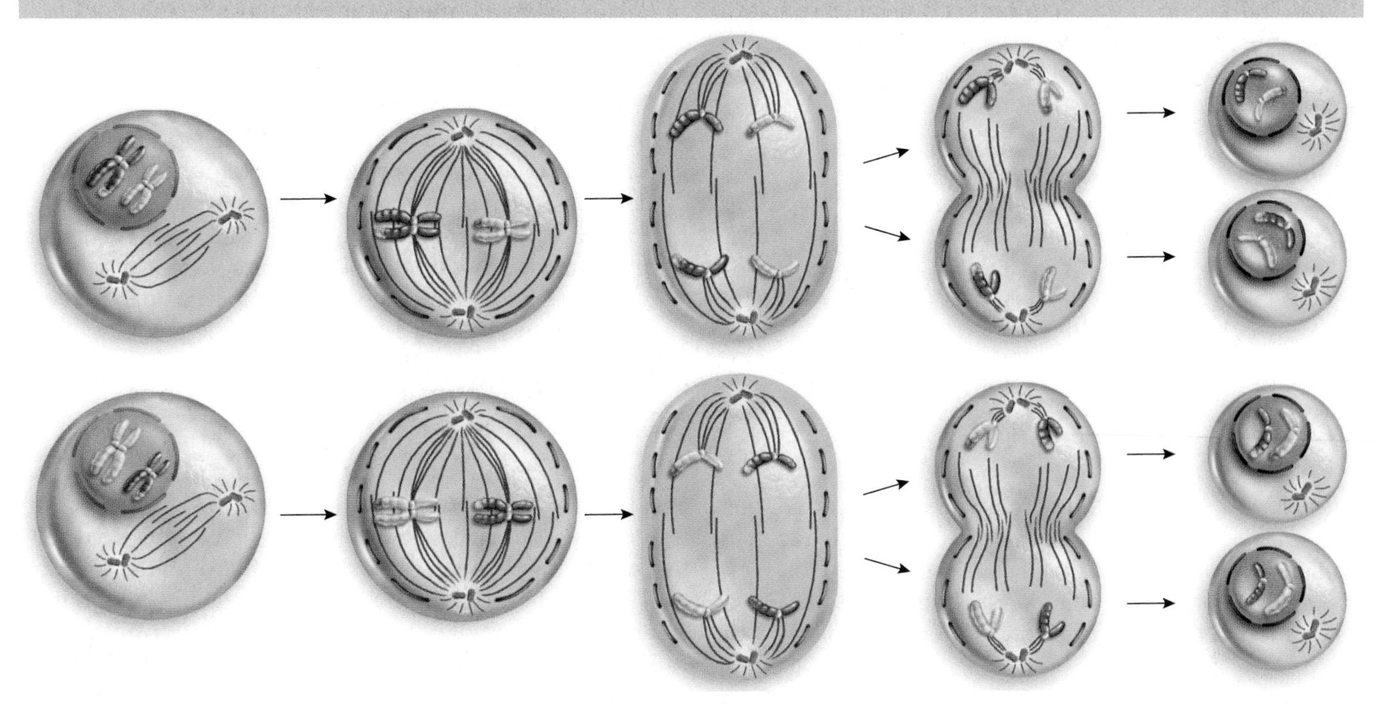

(e) Prophase II. If chromosomes have relaxed after telophase I, they recondense. Spindle microtubules re-form and attach to the sister chromatids.

(f) Metaphase II. Chromosomes line up along the equator, with sister chromatids of each chromosome attached to spindle microtubules that lead to opposite poles.

(g) Anaphase II. Chromatids separate into independent daughter chromosomes, one former chromatid moving toward each pole.

(h) Telophase II. Chromosomes finish moving to opposite poles. Nuclear envelopes re-form, and the chromosomes become extended again (not shown here).

(i) Four haploid cells. Cytokinesis results in four haploid cells, each containing one member of each pair of homologous chromosomes (shown here in condensed state).

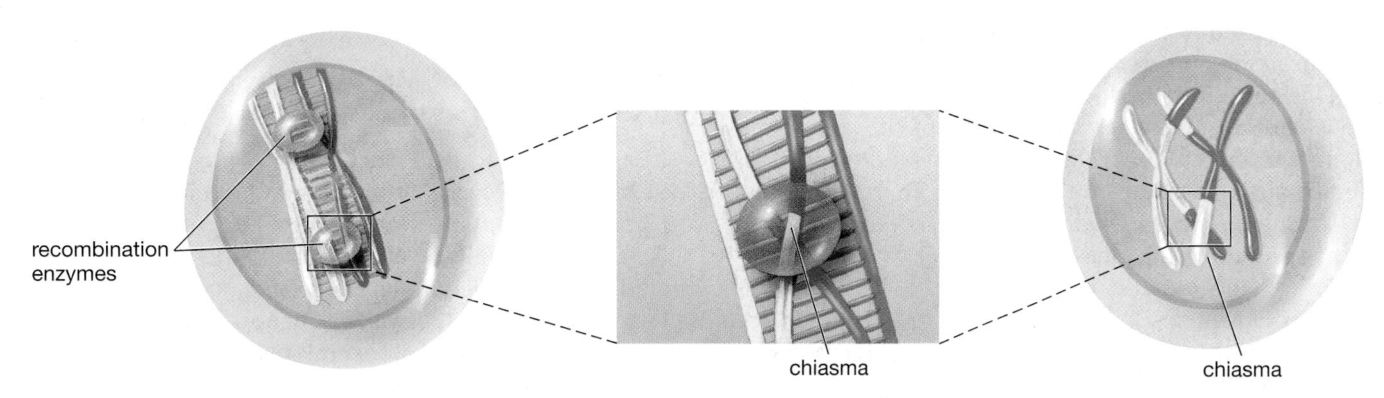

(c) Recombination enzymes bind to the joined chromosomes.

(d) Recombination enzymes snip chromatids apart and reattach the free ends. Chiasmata (the sites of crossing over) form when one end of the paternal chromatid (yellow) attaches to the other end of a maternal chromatid (purple).

(e) Recombination enzymes and protein zippers leave. Chiasmata remain, helping to hold homologous chromosomes together.

containing both the maternal and paternal homologues. Second, in mitosis, the duplicated chromosome has two functional kinetochores, one on each sister chromatid. Both kinetochores attach to spindle microtubules, so that each sister chromatid is attached to microtubules that pull toward opposite poles:

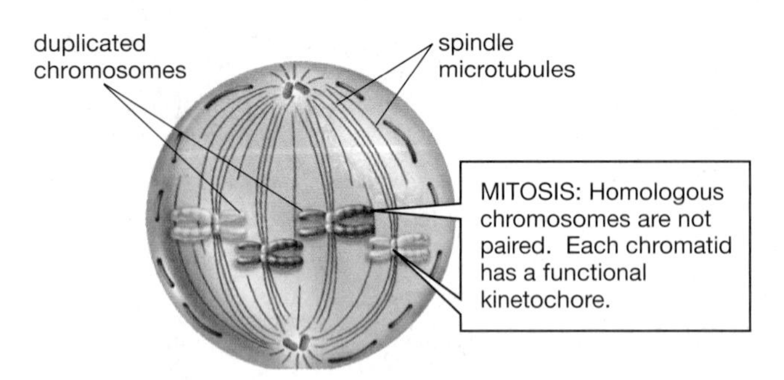

In meiosis I, each duplicated chromosome has only one functional kinetochore, so both sister chromatids attach to spindle microtubules leading to the same pole. However, the chromosomes of a homologous pair attach to spindle microtubules that pull them toward opposite poles:

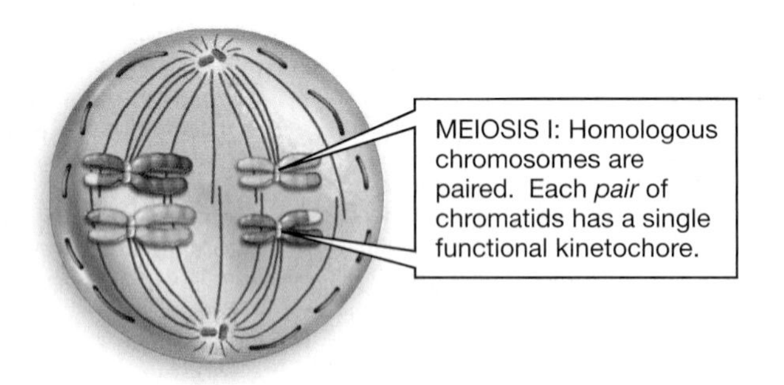

MEIOSIS I: Homologous chromosomes are paired. Each pair of chromatids has a single functional kinetochore.

These differences in attachment explain what will happen at anaphase. In mitosis, the *sister chromatids separate* and move to opposite poles. In contrast, in meiosis I the sister chromatids of each duplicated chromosome remain attached to each other and move to the same pole, but the *homologues separate* and move to opposite poles.

During meiosis I, which member of a pair of homologous chromosomes faces which pole of the cell is random. The maternal chromosome may face "north" for some pairs and "south" for other pairs. This randomness (also called *independent assortment*), together with genetic recombination caused by crossing over, is responsible for the genetic diversity of the haploid cells produced by meiosis.

During Anaphase I, Homologous Chromosomes Separate

In *anaphase I*, the homologues separate from one another and are towed by their kinetochores to opposite poles of the cell (Fig. 11-11c). One duplicated chromosome of each homologous pair (still consisting of two sister chromatids) moves to each pole of the dividing cell. At the end of anaphase I, the cluster of chromosomes at each pole contains one member of each pair of homologous chromosomes. Therefore, each cluster contains the haploid number of chromosomes.

During Telophase I, Two Haploid Clusters of Duplicated Chromosomes Form

In *telophase I*, the spindle microtubules disappear. Cytokinesis commonly occurs during telophase I (Fig. 11-11d), and nuclear envelopes may reappear. Telophase I is usually followed immediately by meiosis II, with little or no intervening interphase. It is important to remember that the chromosomes do not replicate between meiosis I and meiosis II.

Meiosis II Separates Sister Chromatids into Four Daughter Nuclei

During meiosis II, the sister chromatids of each duplicated chromosome separate in a process that is virtually identical to mitosis, though it takes place in haploid cells. During *prophase II*, the spindle microtubules re-form (Fig. 11-11e). The duplicated chromosomes attach individually to spindle microtubules as they do in mitosis. Each chromatid contains a functional kinetochore, allowing each sister chromatid in a duplicated chromosome to attach to spindle microtubules extending to opposite poles of the cell. During *metaphase II*, the duplicated chromosomes line up at the cell's equator (Fig. 11-11f). During *anaphase II*, the sister chromatids separate and are towed to opposite poles (Fig. 11-11g). *Telophase II* and cytokinesis conclude meiosis II as nuclear envelopes re-form, the chromosomes relax into their extended state, and the cytoplasm divides (Fig. 11-11h). Commonly, both daughter cells produced in meiosis I undergo meiosis II, producing a total of four haploid cells from the original parental diploid cell (Fig. 11-11i).

Now that we have covered all of the processes in detail, examine Table 11-1 to review and compare mitotic and meiotic cell division.

The Life Cycles of Most Organisms Include Both Meiosis and Mitosis

It may be easy to understand where mitosis and meiosis fit into the life cycles of humans and other animals (see Fig. 11-4). However, as you explore the diversity of life on Earth in later units, you will find a variety of life cycles among single-celled organisms, plants, and fungi. Despite this variability, the life cycles of almost all eukaryotic organisms have a common overall pattern. First, two haploid cells fuse, bringing together genes from two parent organisms and endowing the resulting diploid cell with new gene combinations. Second, at some point in the life cycle, meiosis occurs, recreating haploid cells. Third, mitosis of either haploid or diploid cells (or both) results in the growth of multicellular bodies and/or asexual reproduction. The seemingly vast differences in the life cycles of, say, fungi, ferns, and humans are due to variations in

Table 11-1 A Comparison of Mitotic and Meiotic Cell Divisions in Animal Cells

Feature	Mitotic Cell Division	Meiotic Cell Division
Cells in which it occurs	Body cells	Gamete-producing cells
Final chromosome number	Diploid—2*n*; two copies of each type of chromosome (homologous pairs)	Haploid—1*n*; one member of each homologous pair
Number of daughter cells	Two, identical to the parent cell and to each other	Four, containing recombined chromosomes due to crossing over
Number of cell divisions per DNA replication	One	Two
Function in animals	Development, growth, repair and maintenance of tissues, asexual reproduction	Gamete production for sexual reproduction

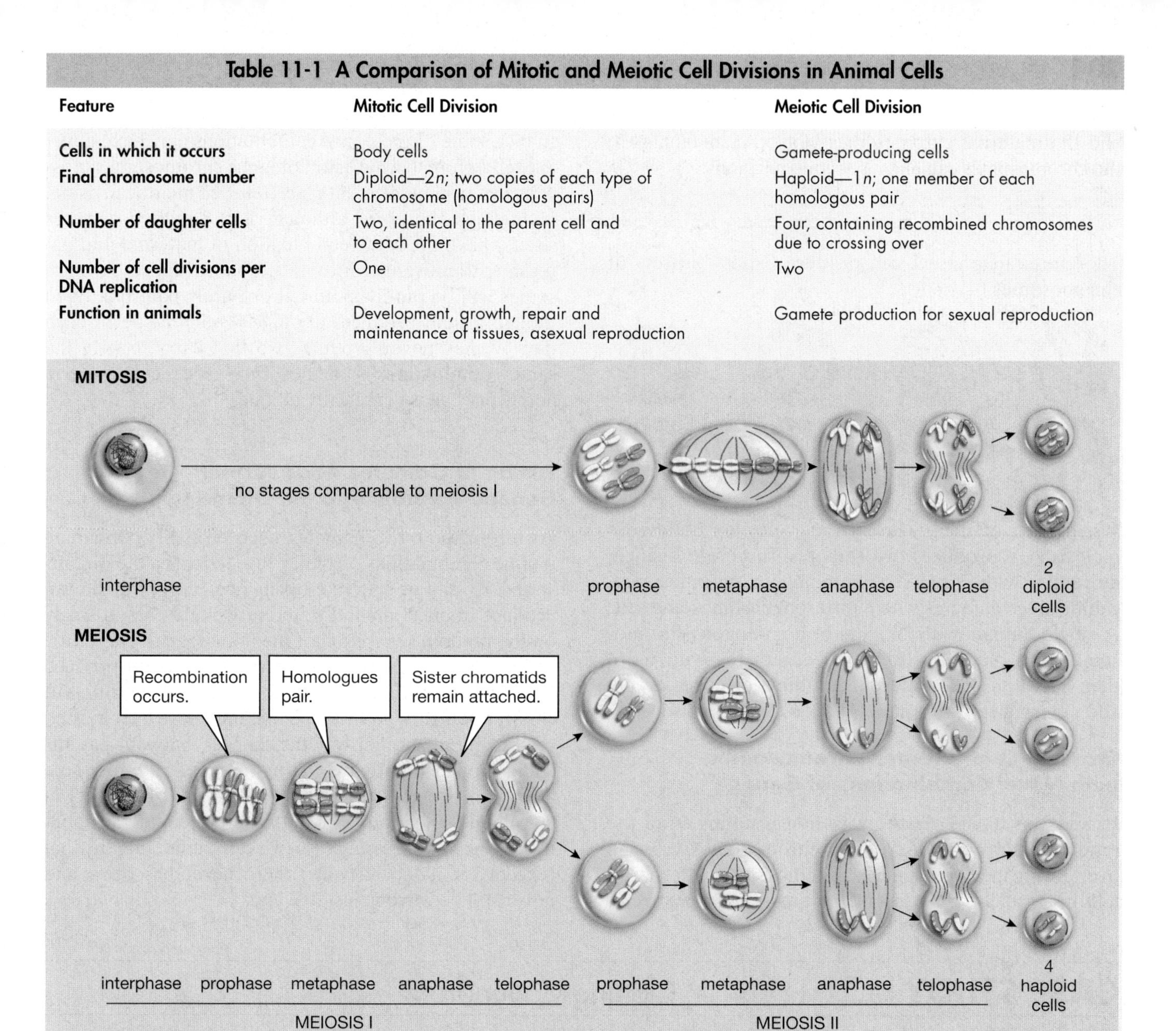

In these diagrams, comparable phases are aligned. In both mitosis and meiosis, chromosomes are replicated during interphase. Meiosis I, with the pairing of homologous chromosomes, formation of chiasmata, exchange of chromosome parts, and separation of homologues to form haploid daughter nuclei, has no counterpart in mitosis. Meiosis II, however, is similar to mitosis.

the parts of the life cycle in which mitosis and meiosis occur, and in the relative proportions of the life cycle spent in the diploid and haploid states.

11.6 How Do Meiosis and Sexual Reproduction Produce Genetic Variability?

Shuffling of Homologues Creates Novel Combinations of Chromosomes

Genetic variability among organisms is essential for survival and reproduction in a changing environment, and therefore for evolution. Mutations occurring randomly over millions of years are the original sources of genetic variability within the populations of organisms that exist today. However, mutations are rare events. Therefore, the genetic variability that occurs from one generation to the next results almost entirely from meiosis and sexual reproduction.

How does meiosis produce genetic diversity? One mechanism is the random distribution of maternal and paternal homologues to the daughter cells at meiosis I. Remember, at metaphase I the paired homologues line up at the cell's equator. In each pair of homologues, the maternal chromosome faces one pole and the paternal chromosome faces the opposite pole, but which homologue faces which pole is random.

Let's consider meiosis in mosquitoes, which have three pairs of homologous chromosomes ($n = 3, 2n = 6$). For the sake of simplicity, we'll represent these chromosomes as large, medium, and small. To keep track of the homologues, let's color-code the paternal chromosomes yellow

Media Activity
11.4 How Meiosis Produces Genetic Variability

and the maternal chromosomes violet. At metaphase I, the chromosomes can align in 4 configurations:

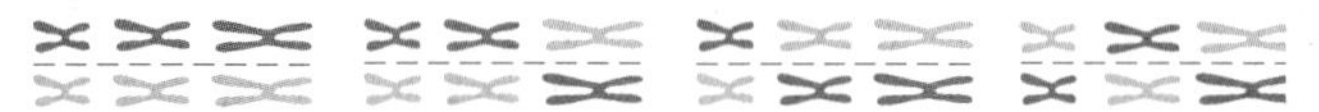

Therefore, anaphase I can produce 8 possible sets of chromosomes ($2^3 = 8$):

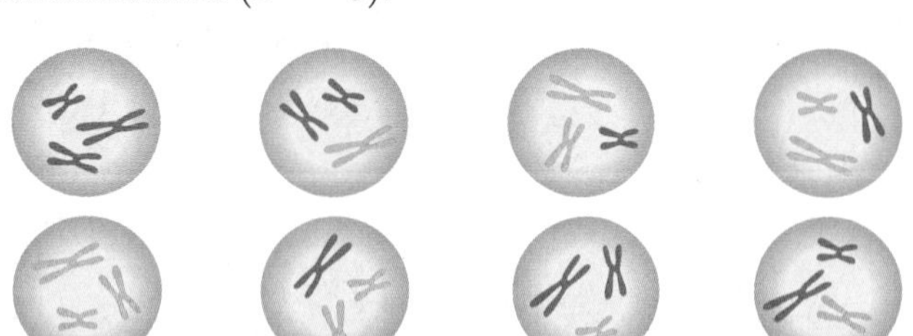

When each of these chromosome clusters undergoes meiosis II, it produces two gametes. Therefore, a single mosquito, with 3 pairs of homologous chromosomes, can produce gametes with 8 different chromosome sets. A single human, with 23 pairs of homologous chromosomes, can theoretically produce gametes with more than 8 million (2^{23}) different combinations of maternal and paternal chromosomes.

Crossing Over Creates Chromosomes with Novel Combinations of Genes

In addition to the genetic variation resulting from the random assortment of parental chromosomes, crossing over during meiosis produces chromosomes with combinations of alleles that differ from those of either parent. In fact, some of these new combinations may have never existed before, because homologous chromosomes cross over in new and different places at each meiotic division. In humans, therefore, although 1 in 8 million gametes should have the same combination of maternal and paternal chromosomes, in reality, none of those chromosomes will be purely maternal or purely paternal. Even though a man produces about 100 million sperm each day, he may never produce two that carry exactly the same combinations of alleles. In essence, every sperm and every egg is genetically unique.

Fusion of Gametes Adds Further Genetic Variability to the Offspring

At fertilization, two gametes, each probably containing unique combinations of alleles, fuse to form a diploid offspring. Even if we ignore crossing over, every human can produce about 8 million different gametes based solely on the random separation of the homologues. Therefore, fusion of gametes from just two people could produce 8 million × 8 million, or 64 trillion, genetically different children, which is far more people than have ever existed on Earth! Put another way, the chances that your parents could produce another child that is genetically the same as you are about 1/8,000,000 × 1/8,000,000, or about 1 in 64 trillion! When we factor in the almost endless variability produced by crossing over, we can confidently say that (except for identical twins) there never has been, and never will be, anyone just like you.

CASE STUDY REVISITED Cloning Conundrum

Why are no two puppies in a litter ever truly identical? Dogs have 39 pairs of homologous chromosomes, so the shuffling of homologues during meiosis can result in 2^{39}—or 550 billion—genetically different gametes in a single dog. Even if gametes from Missy's parents had been preserved and used for *in vitro* fertilization, there would be only one chance in 550 billion × 550 billion that they could produce another puppy genetically identical to Missy. When crossing over is added to the mix, well . . .

Mitosis, however, *does* produce genetically identical daughter cells. To clone Missy, the nucleus would be removed from a fertilized dog egg and replaced with a nucleus from one of Missy's cells. The resulting egg—containing Missy's chromosomes and cytoplasm from a different dog—would be implanted in a surrogate mother's uterus. Because all of Missy's cells are genetically identical, wouldn't a Missy clone really be another Missy?

Not exactly. First, mutations occur at a slow, but inexorable, rate. Thus, the DNA in the nucleus taken from Missy would be at least a few nucleotides different from the DNA in the fertilized egg that developed into the original Missy. Second, all the mitochondria in the egg that would become Missy's clone would be from the egg donor, not from Missy. Because mitochondria have their own DNA that encodes some of their own proteins, metabolic differences might make the energy level, life span, or disease susceptibility of Missy's clone different from Missy's. Third, the cytoplasm in the cloned egg would differ from the cytoplasm in the egg that became Missy. Factors in the cytoplasm regulate the transcription and translation of genes, which might affect the clone's development. Finally, the uterus in which Missy's clone develops, her surrogate mother's behavior, and the environment in which she is raised would all differ from the environment in which Missy developed. Nevertheless, the clone would still be more like Missy than any other dog would be. Would Missy's clone be a better pet for Sperling and Hawthorne than any other dog would be? Only the future (perhaps) can tell.

Consider This: What about human cloning? Early in 2003, there were claims that two cloned children had been born (although, as of mid-2003, this had not been confirmed). Assuming that the technology exists to clone people, would it be a good idea? CC the cat was the only survivor of 87 embryos implanted in 8 different surrogate mothers. A cloned Brahma bull, Second Chance, was in intensive care for two weeks with respiratory and cardiovascular disorders. With today's limited technology, it's hard to argue with Rudolf Jaenisch of the Whitehead Institute for Biomedical Research, who stated that human cloning is "just criminal." But what about in 2010, or 2050, if the technology is perfected? What do you think?

Links to Life: Cancer—Cell Division Run Amok

Mitotic cell division is essential for the development of multicellular organisms from single fertilized eggs, as well as for routine maintenance of certain body parts, such as the skin and the lining of the digestive tract. Unfortunately, uncontrolled cell division is a menace to life: cancer. Normally, cell division is regulated by a balance between growth-stimulating and growth-inhibiting factors. Mutations and certain types of viruses (which insert some of their DNA into the host cell's chromosomes) may either stimulate cell division inappropriately or reduce a cell's sensitivity to growth-inhibiting factors. These cells are usually eliminated by the immune system (see Chapter 32), but occasionally a renegade cell survives and reproduces. Because mitotic cell division faithfully transmits genetic information from cell to cell, all daughter cells of the original cancerous cell will themselves be cancerous.

Why does medical science, which has conquered smallpox, measles, and a host of other diseases, have such a difficult time curing cancer? Both normal and cancerous cells uses the same machinery for cell division, so treatments that slow down the multiplication of cancer cells also inhibit the maintenance of essential body parts, such as the stomach, intestine, and blood cells. Truly effective and *selective* treatments for cancer must target cell division only in malignant (cancerous) cells. Although great strides have been made in the fight to cure cancer, this daunting task is still far from being accomplished.

Summary of Key Concepts

11.1 What Is the Role of Cellular Reproduction in the Lives of Individual Cells and Entire Organisms?

The prokaryotic cell cycle consists of growth, DNA replication, and division by binary fission. The eukaryotic cell cycle consists of interphase and cell division. During interphase, the cell grows and duplicates its chromosomes. Interphase is divided into G_1 (growth phase 1), S (DNA synthesis), and G_2 (growth phase 2). During G_1, some cells may exit the cell cycle to enter a nondividing state called G_0. Cells may remain permanently in G_0, or may be induced to reenter the cell cycle. Eukaryotic cells can divide by mitotic or meiotic cell division.

Mitotic cell division consists of two processes: (1) mitosis (nuclear division) and (2) cytokinesis (cytoplasmic division). Mitosis parcels out one copy of every chromosome into two separate nuclei, and cytokinesis subsequently encloses each nucleus in a separate cell, producing two genetically identical daughter cells. Mitotic cell division of a fertilized egg produces genetically identical cells that grow and differentiate into an embryo and, eventually, an adult. Mitotic cell division also maintains body tissues and repairs damage to some organs. Asexual reproduction is based on mitotic cell division, resulting in formation of clones that are genetically identical to the parent.

Meiotic cell division produces haploid cells, which have only half of the parent's DNA. Fusion of haploid gametes creates a fertilized egg that has a different genetic makeup from either parent, which then grows and develops via mitotic cell division.

11.2 How Is DNA in Eukaryotic Cells Organized into Chromosomes?

Each chromosome in a eukaryotic cell consists of a single DNA double helix and proteins that organize the DNA. During cell growth, the chromosomes are extended and accessible for use by enzymes that read their genetic instructions. During cell division, the chromosomes condense into short, thick structures. Eukaryotic cells typically contain pairs of chromosomes called homologues, which appear virtually identical because they carry the same genes with similar nucleotide sequences. Cells with pairs of homologous chromosomes are diploid. Cells with only one member of each chromosome pair are haploid.

11.3 How Do Cells Reproduce by Mitotic Cell Division?

The chromosomes are duplicated during interphase, prior to mitosis. The two identical copies, called chromatids, remain attached to one another at the centromere during the early stages of mitosis. Mitosis consists of four phases (see Fig. 11-8), usually followed by cytokinesis.

1. **Prophase:** The chromosomes condense and their kinetochores attach to the spindle microtubules that form at this time.
2. **Metaphase:** The chromosomes move to the equator of the cell.
3. **Anaphase:** The two chromatids of each duplicated chromosome separate and are pulled along the spindle microtubules to opposite poles of the cell.
4. **Telophase:** The chromosomes relax into their extended state and nuclear envelopes re-form around each new daughter nucleus.
5. **Cytokinesis:** Cytokinesis normally occurs at the end of telophase and divides the cytoplasm into approximately equal halves, each containing a nucleus. In animal cells, a ring of microfilaments pinches the plasma membrane in along the equator. In plant cells, new plasma membrane forms along the equator by the fusion of vesicles produced by the Golgi complex.

11.4 Why Do So Many Organisms Reproduce Sexually?

Genetic differences among organisms originate as mutations, which, when preserved within a species, produce alternate forms of genes, called alleles. Alleles in different individuals of a species may be combined in offspring through sexual reproduction, creating variation among the offspring and potentially improving their likelihood of surviving and reproducing in their turn.

11.5 How Does Meiotic Cell Division Produce Haploid Cells?

Meiosis separates homologous chromosomes and produces haploid cells with only one homologue from each pair. During interphase before meiosis, chromosomes are duplicated. The cell then undergoes two specialized cell divisions—meiosis I and meiosis II—to produce four haploid daughter cells.

Meiosis I: During prophase I, homologous duplicated chromosomes, each consisting of two chromatids, pair up and exchange parts by crossing over. During metaphase I, homologues move together as pairs to the cell's equator, one member of each pair facing opposite poles of the cell. Homologous chromosomes separate during anaphase I, and two nuclei form during telophase I. Each daughter nucleus receives only one member of each pair of homologues, and is therefore haploid. The sister chromatids remain attached to each other throughout meiosis I.

Meiosis II: Meiosis II usually occurs in both daughter nuclei and resembles mitosis in a haploid cell. The duplicated chromosomes move to the cell's equator during metaphase II. The two chromatids of each chromosome separate and move to opposite poles of the cell during anaphase II. This second division produces four haploid nuclei. Cytokinesis normally occurs during or shortly after telophase II, producing four haploid cells.

11.6 How Do Meiosis and Sexual Reproduction Produce Genetic Variability?

The random shuffling of homologous maternal and paternal chromosomes creates new chromosome combinations. Crossing over creates chromosomes with allele combinations that may have never occurred before on single chromosomes. Because of the separation of homologues and crossing over, a parent probably never produces any two gametes that are completely identical. The fusion of two such genetically unique gametes adds further genetic variability to the offspring.

Key Terms

allele *p. 197*
anaphase *p. 196*
asexual reproduction *p. 186*
autosome *p. 191*
binary fission *p. 187*
cell cycle *p. 186*
cell division *p. 188*
cell plate *p. 196*
centriole *p. 196*
centromere *p. 190*
chiasma (chiasmata) *p. 199*
chromatid *p. 191*
chromosome *p. 190*
clone *p. 194*
cloning *p. 194*
crossing over *p. 199*
cytokinesis *p. 189*
differentiation *p. 188*
diploid *p. 191*
duplicated chromosome *p. 191*
gamete *p. 189*
haploid *p. 192*
homologue *p. 191*
interphase *p. 188*
karyotype *p. 191*
kinetochore *p. 196*
locus *p. 190*
meiosis *p. 189*
meiotic cell division *p. 189*
metaphase *p. 196*
mitosis *p. 189*
mitotic cell division *p. 189*
prophase *p. 193*
recombination *p. 199*
sex chromosome *p. 191*
sexual reproduction *p. 189*
spindle microtubule *p. 193*
telomere *p. 190*
telophase *p. 196*

Thinking Through the Concepts

To take a multiple-choice quiz with feedback on the contents of this chapter, visit http://www.prenhall.com/audesirk7. *Log in to the Web site selected by your instructor and navigate to the Self Test section for this chapter.*

? Review Questions

1. Diagram and describe the eukaryotic cell cycle. Name the various phases, and briefly describe the events that occur during each. What is the role of the cell cycle in a human?
2. Define *mitosis* and *cytokinesis*. What changes in cell structure result when cytokinesis does not occur after mitosis?
3. Diagram the stages of mitosis. How does mitosis ensure that each daughter nucleus receives a full set of chromosomes?
4. Define the following terms: *homologous chromosome, centromere, kinetochore, chromatid, diploid, haploid.*
5. Describe and compare the process of cytokinesis in animal cells and in plant cells.
6. Diagram the events of meiosis. At which stage do homologous chromosomes separate?
7. Describe homologue pairing and crossing over. At which stage of meiosis do they occur? Name two functions of chiasmata.
8. In what ways are mitosis and meiosis similar? In what ways are they different?
9. Describe how meiosis provides for genetic variability. If an animal had a haploid number of 2 (no sex chromosomes), how many genetically different types of gametes could it produce? (Assume no crossing over.) If it had a haploid number of 5?

Applying the Concepts

1. Most nerve cells in the adult human central nervous system, as well as heart muscle cells, remain in the G_0 portion of interphase. In contrast, cells lining the inside of the small intestine divide frequently. Discuss this difference in terms of why damage to the nervous system and heart muscle cells (such as caused by a stroke or heart attack) is so dangerous. What do you think might happen to tissues such as the intestinal lining if some disorder or drug blocked mitosis in all cells of the body?
2. Cancer cells divide out of control. Side effects of chemotherapy and radiation therapy that fight cancers include loss of hair and of the gastrointestinal lining, producing severe nausea. Note that cells in hair follicles and intestinal lining divide frequently. What can you infer about the mechanisms of these treatments? What would you look for in an improved cancer therapy?
3. Some animal species can reproduce either asexually or sexually, depending on the state of the environment. Asexual reproduction tends to occur in stable, favorable environments; sexual reproduction is more common in unstable and/or unfavorable circumstances. Discuss the advantages or disadvantages this behavior might have on survival of the species in an evolutionary sense or on survival of individuals.

For More Information

Axtman, K. "Quietly, Animal Cloning Speeds Onward." *Christian Science Monitor*, October 23, 2001. A discussion of the successes and failures of mammalian cloning.

Grant, M. C. "The Trembling Giant." *Discover*, October 1993. Aspen groves are really single individuals: huge, slowly spreading from the roots of the original parent tree, and potentially almost immortal.

Lanza, R. P., Dresser, B. L., and Damiani, P. "Cloning Noah's Ark." *Scientific American,* November 2000. Cloning rare and endangered species may offer hope of preventing extinction.

Leutwyler, K. "Turning Back the Strands of Time." *Scientific American (Explorations),* February 2, 1998. A brief discussion of telomeres, the repeating DNA regions at the ends of chromosomes.

Nash, M. "The Age of Cloning." *Time*, March 10, 1997. Well-written and well-illustrated description of the technique and implications of using an adult cell to clone a sheep.

Travis, J. "A Fantastical Experiment." *Science News*, April 5, 1997. A clear description of the cloning of Dolly the sheep and some of its implications.

Weiss, R. "Human Cloning Bid Stirs Experts' Anger." *Washington Post*, March 7, 2001. A group of pioneers in animal cloning discuss the ethical implications of Zavos and Antinori's attempts to clone humans.

Wilmut, I. "Cloning for Medicine." *Scientific American,* December 1998. Explanation of why cloning experiments might have medical applications.

Media Activities

To access a Media Activity visit http://www.prenhall.com/audesirk7. *Log in to the Web site selected by your instructor, navigate to this chapter, and select the appropriate Media Activity number.*

11.1 Cell Division in Humans

Estimated time: 5 minutes

This animation provides an overview of the cell cycle in humans.

11.2 Cell Cycle and Mitosis

Estimated time: 10 minutes

This activity describes the process of mitosis, and will help you to understand the steps into which mitosis is divided.

11.3 Meiosis

Estimated time: 10 minutes

In order for the fusion of sperm and egg to produce a cell with the correct number of chromosomes, the sperm and egg must each have one half the number of chromosomes as an adult cell. To accomplish this, certain cells in the body undergo a type of division, called meiosis, which reduces the chromosome number. This activity will lead you through the steps of meiosis.

11.4 How Meiosis Produces Genetic Variability

Estimated time: 5 minutes

Everyone is unique. How does the process of meiosis produce the amazing diversity we see around us? In this activity review the processes the produce such great genetic diversity.

11.5 Web Investigation: Cloning Conundrum

Estimated time: 20 minutes

The Missyplicity Project has raised questions about ethics and science. Explore some of these questions and find possible answers in this web investigation.

CHAPTER

12 Patterns of Inheritance

Olympic silver medallist Flo Hyman was struck down by Marfan syndrome at the height of her career.

AT A GLANCE

CASE STUDY Marfan Syndrome

What did volleyball star Flo Hyman, President Abraham Lincoln, and the Egyptian pharaoh Akhenaten have in common? They all had a genetic disorder called Marfan syndrome, which affects about one in 5000 people. People with Marfan syndrome are typically tall and slender, with unusually long limbs and large hands and feet. For some, these characteristics led to fame and fortune.

Flo Hyman, graceful, athletic, and over six feet tall, was one of the best woman volleyball players of all time. A star of the 1984 silver medal American Olympic volleyball team, Hyman later joined a professional Japanese team. In 1986, taken out of a game for a short breather, she died while sitting quietly on the bench. An autopsy showed that Hyman died from a ruptured aorta, the massive artery that carries blood from the heart to most of the body. Why did Hyman's aorta break? What does a weak aorta have in common with tallness and large hands?

Marfan syndrome is caused by a mutation in the gene that encodes a protein called fibrillin, which forms long fibers that give elasticity and strength to connective tissue. Many parts of your body contain connective tissue, including tendons, ligaments, and artery walls. Fibrillin also apparently acts as a "growth factor," stimulating the growth of various structures, probably including bone. Although no one knows for sure, it is likely that some mutations enhance the growth-promoting function of fibrillin, causing people with Marfan syndrome to grow tall and lanky. Defective fibrillin molecules also weaken artery walls, sometimes with tragic consequences.

As we described in Chapter 11, diploid organisms, including people, generally have two copies of each gene, one on each homologous chromosome. One defective copy of the fibrillin gene is enough to cause Marfan syndrome. Further, the children of a person with Marfan syndrome have a 50% chance of inheriting the disease. To understand these patterns of inheritance, we must go back in time to a monastery in Moravia and visit the garden of Gregor Mendel.

12.1 How Did Gregor Mendel Lay the Foundation for Modern Genetics?

Before settling down as a monk in the monastery of St. Thomas in Brünn (now Brno, in the Moravian part of the Czech Republic), Gregor Mendel (Fig. 12-1) attended the University of Vienna for two years. He studied many subjects, including botany and mathematics. This training proved crucial to his later experiments, which would become the foundation for the modern science of genetics. At St. Thomas, Mendel carried out both his monastic duties and a groundbreaking series of experiments on inheritance in the common edible pea. Although Mendel worked without knowledge of genes or chromosomes, we can more easily follow his experiments after a brief look at some modern genetic concepts.

A gene's specific physical location on a chromosome is called its **locus** (plural, *loci*) (Fig. 12-2). Homologous chromosomes carry the same **genes** at the same loci. Although the nucleotide sequence at a given gene locus is *similar* on homologous chromosomes, the sequence may not be *identical*. These differences in nucleotide sequences at the same gene locus on two homologous chromosomes produce alternate versions of the gene, called **alleles**. Human A, B, and O blood types, for example, are produced by the three different alleles of the blood type gene.

If both homologous chromosomes in an organism have the *same* allele at a given gene locus, the organism is said to be **homozygous** at that gene locus. (*Homozygous* comes from Greek words meaning "same pair.") For example, the chromosomes in Figure 12-2 are homozygous at the loci for both the M and D genes. If two homologous chromosomes have *different* alleles at a given gene locus, the organism is **heterozygous** ("different pair") at that locus and is sometimes called a **hybrid**. The chromosomes in Figure 12-2 are heterozygous at the locus for the Bk gene. Recall from Chapter 11 that, during meiosis, homologous chromosomes are separated, so each gamete receives one member of each pair of homologous chromosomes. As a result, every gamete has only one allele for each gene. Therefore, all the gametes produced by an organism that is homozygous at a particular gene locus contain the same allele. Gametes produced by an organism that is heterozygous at the same gene locus are of two kinds: half of the gametes contain one allele, and half contain the other allele.

FIGURE 12-1 Gregor Mendel
A portrait of Mendel, painted in about 1888, after he had completed his pioneering genetics experiments.

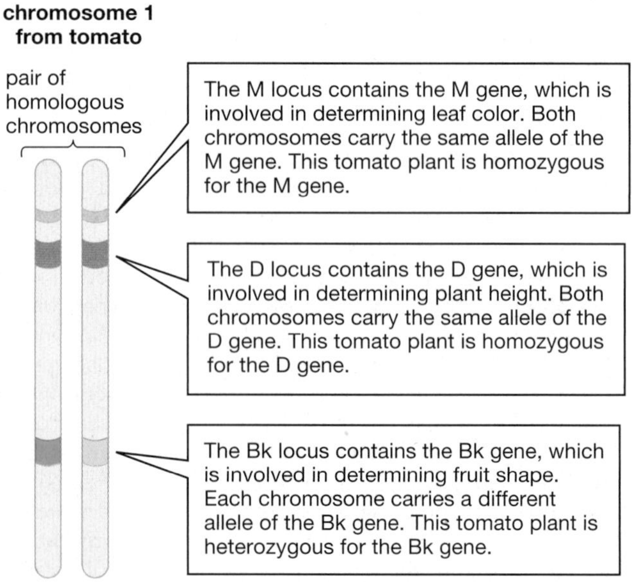

chromosome 1 from tomato

pair of homologous chromosomes

The M locus contains the M gene, which is involved in determining leaf color. Both chromosomes carry the same allele of the M gene. This tomato plant is homozygous for the M gene.

The D locus contains the D gene, which is involved in determining plant height. Both chromosomes carry the same allele of the D gene. This tomato plant is homozygous for the D gene.

The Bk locus contains the Bk gene, which is involved in determining fruit shape. Each chromosome carries a different allele of the Bk gene. This tomato plant is heterozygous for the Bk gene.

FIGURE 12-2 The relationships among genes, alleles, and chromosomes
Each homologous chromosome carries the same set of genes. Each gene is located at the same relative position, or locus, on its chromosome. Differences in nucleotide sequences at the same gene locus produce different alleles of the gene. Diploid organisms have two alleles of each gene.

Interestingly, patterns of inheritance and many essential facts about genes, alleles, and the distribution of alleles in gametes and zygotes during sexual reproduction were deduced by Gregor Mendel long before DNA, chromosomes, or meiosis had been discovered. Because his experiments are succinct, elegant examples of science in action, let's follow Mendel's paths of discovery.

Doing It Right: The Secrets of Mendel's Success

There are three key steps to any successful experiment in biology: choosing the right organism with which to work, designing and performing the experiment correctly, and analyzing the data properly. Mendel was the first geneticist to complete all three steps.

Mendel's choice of the edible pea as an experimental subject was critical to the success of his experiments. Stamens, the male reproductive structures of a flower, produce pollen. Each pollen grain contains male gametes, or *sperm*. Pollination allows sperm to fertilize the female gamete, or *egg*, located within the ovary at the base of the carpel, which is the female reproductive structure of the flower. The petals of a pea flower enclose all of the flower structures, normally preventing another flower's pollen from entering (Fig. 12-3). Instead, each pea flower normally supplies its own pollen, so the egg cells in each flower are fertilized by sperm from the pollen of the same flower. This process is called **self-fertilization**. In the edible pea, flower color is controlled by a single gene. If a pea plant is homozygous for this gene, all of its offspring will have the same flower color, which will be the same as the parent plant. Such plants are called **true-breeding**. Even in Mendel's time, commercial seed dealers sold many types of true-breeding pea varieties.

intact pea flower

flower dissected to show reproductive structures

carpel (female, produces eggs)

stamen (male, produces pollen)

FIGURE 12-3 Flowers of the edible pea
In the intact pea flower (left), the lower petals form a container enclosing the reproductive structures—the stamens (male) and carpel (female). Pollen normally cannot enter the flower from outside, so peas usually self-fertilize. If the flower is opened (right), it can be cross-pollinated by hand.

Although peas normally self-fertilize, plant breeders can mate two plants by hand, causing **cross-fertilization**. Breeders pull apart the petals and remove the stamens, preventing self-fertilization. By dusting the sticky end of the carpel with pollen from the plants they have selected, breeders can control fertilization. In this way, two plants can be mated to see what types of offspring they produce.

Mendel's experimental design was simple, but brilliant. Rather than looking at the entire plant in all of its complexity, Mendel chose to study individual characteristics (usually called *traits*) that had unmistakably different forms, such as white versus purple flowers. He also worked with one trait at a time.

Mendel followed the inheritance of these traits for several generations, counting the numbers of offspring with each type of trait. By analyzing these numbers, the basic patterns of inheritance became clear. Numerical analysis was something of an innovation in Mendel's time, but today, statistics is an essential tool in virtually every field of biology.

12.2 How Are Single Traits Inherited?

Media Activity
12.1 Monohybrid Crosses

Mendel raised pea plants that were true-breeding for different forms of a single trait and cross-fertilized them. He saved the resulting hybrid seeds and grew them the following year to determine their characteristics.

In one of these experiments, Mendel cross-fertilized a white-flowered pea plant with a purple-flowered one. This was the *parental generation*, denoted by the letter *P*. When he grew the resulting seeds, he found that all the first-generation offspring (the "first filial," or F_1, generation) produced purple flowers:

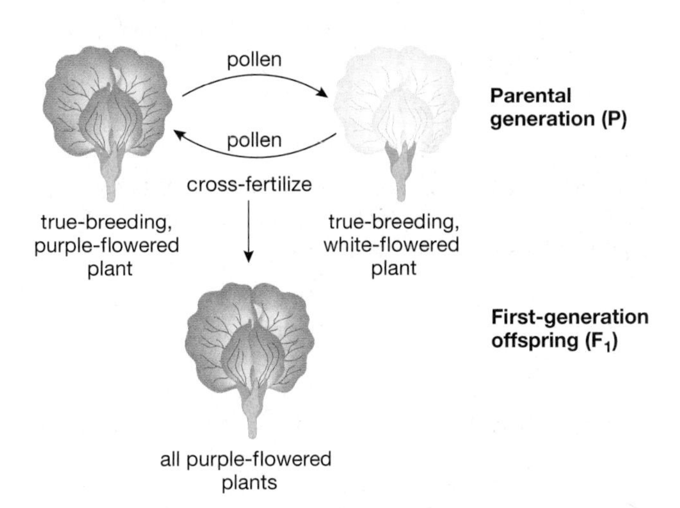

What had happened to the white color? The flowers of the hybrids were just as purple as their parent. The white color seemed to have disappeared in the F_1 offspring.

Mendel then allowed the F_1 flowers to self-fertilize, collected the seeds, and planted them the next spring. In the second generation (F_2), about three-fourths of the plants had purple flowers and one-fourth had white flowers:

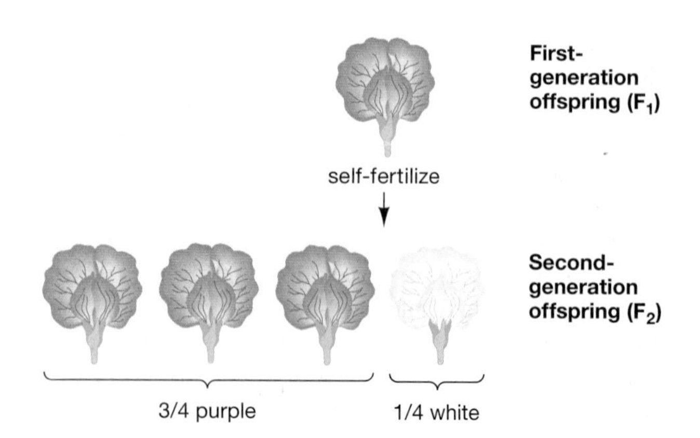

The exact numbers were 705 purple and 224 white, or a ratio of about 3 purple to 1 white. This result showed that the gene that produced white flowers had not disappeared but had only been "hidden."

Mendel allowed the F_2 plants to self-fertilize and produce yet a third (F_3) generation. He found that all the white-flowered F_2 plants produced white-flowered offspring; that is, they were true-breeding. For as many generations as he had time and patience to raise, white-flowered parents always gave rise to white-flowered offspring. The purple-flowered F_2 plants were of two types: About $\frac{1}{3}$ of these were true-breeding for purple; the remaining $\frac{2}{3}$ were hybrids that produced both purple- and white-flowered offspring, again in the ratio of 3 to 1. Therefore, the F_2 generation included $\frac{1}{4}$ true-breeding purple plants, $\frac{1}{2}$ hybrid purple, and $\frac{1}{4}$ true-breeding white.

The Inheritance of Dominant and Recessive Alleles on Homologous Chromosomes Can Explain the Results of Mendel's Crosses

Mendel's results, supplemented by our knowledge of genes and homologous chromosomes, allow us to develop a five-part hypothesis:

- Each trait is determined by pairs of discrete physical units, which we now call genes. Each individual has two alleles for a given gene, such as the gene that determines flower color. One allele of the gene is present on each homologous chromosome. True-breeding peas with white flowers have different alleles of the "flower-color" gene than true-breeding purple-flowered peas.
- The pairs of genes on homologous chromosomes separate from each other during gamete formation, so each gamete receives only one allele of an organism's pair of genes. This conclusion is known as Mendel's **law of segregation**: the two alleles of a gene segregate from one another at meiosis. When a sperm fertilizes an egg, the resulting offspring receives one allele from the father and one from the mother.
- Which allele becomes included in a gamete is determined by chance. This randomness occurs because the separation of homologous chromosomes during meiosis is random.
- When two different alleles are present in an organism, one (the **dominant** allele) may mask the expression of the other (the **recessive** allele). The dominant allele does not, however, alter the physical presence of the recessive allele. Each allele, whether dominant or recessive, is passed into the individual's gametes. In Mendel's experiments with flower color, the allele for purple flowers is dominant, and the allele for white flowers is recessive.
- True-breeding organisms have two of the same alleles for a given gene; these organisms are homozygous. All of the gametes from a homozygous individual have the same allele for that gene:

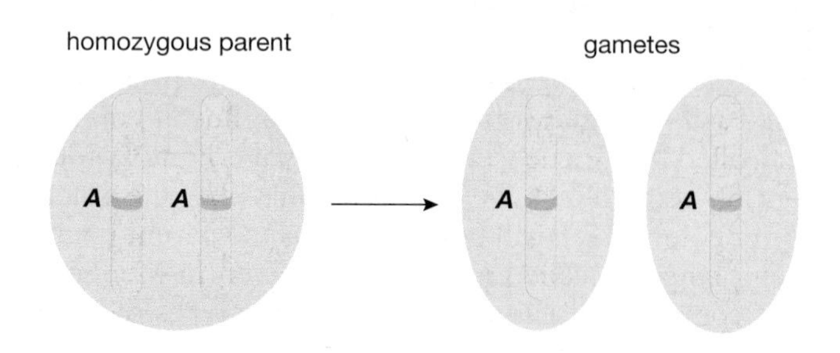

Hybrids have two different alleles for that gene; they are heterozygous. Half of the organism's gametes contain one allele for that gene; half contain the other allele. A heterozygous organism, with two different alleles of gene "A," produces equal numbers of gametes with each of the two alleles:

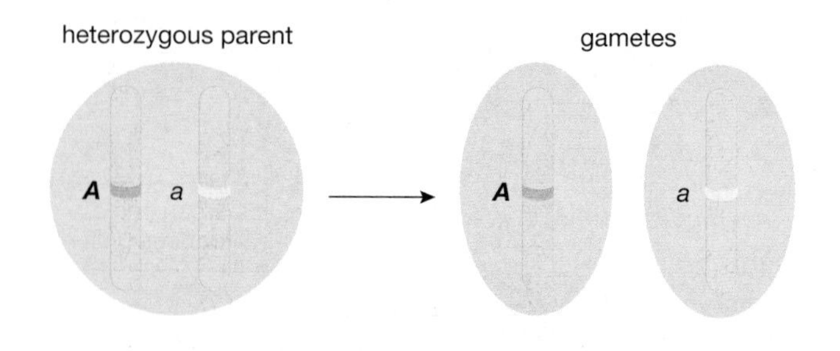

Let's see how this hypothesis explains the results of Mendel's experiments with flower color. Using letters to represent the different alleles, we will assign the uppercase letter P to the allele for purple (dominant) and the lowercase letter p to the allele for white (recessive). (By Mendel's convention, the dominant allele is represented by a capital letter.) A true-breeding (homozygous) purple-flowered plant has two alleles for purple flowers (PP), whereas a white-flowered plant has two alleles for white flowers (pp). All the sperm and eggs produced by a PP plant carry the P allele; all the sperm and eggs of a pp plant carry the p allele:

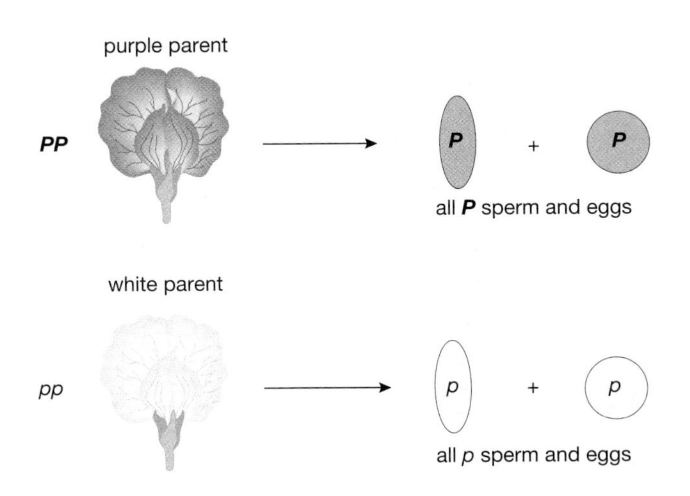

The F_1 hybrid offspring are produced when P sperm fertilize p eggs or when p sperm fertilize P eggs. In either case, the F_1 offspring are Pp. Because P is dominant to p, all the offspring are purple:

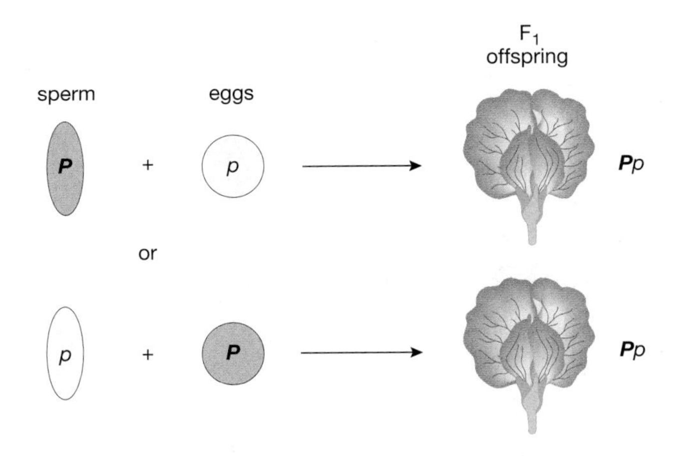

Each gamete produced by a heterozygous Pp plant has an equal chance of receiving either the P allele or the p allele. That is, the hybrid plant produces equal numbers of P and p sperm and equal numbers of P and p eggs. When a Pp plant self-fertilizes, each type of sperm has an equal chance of fertilizing each type of egg:

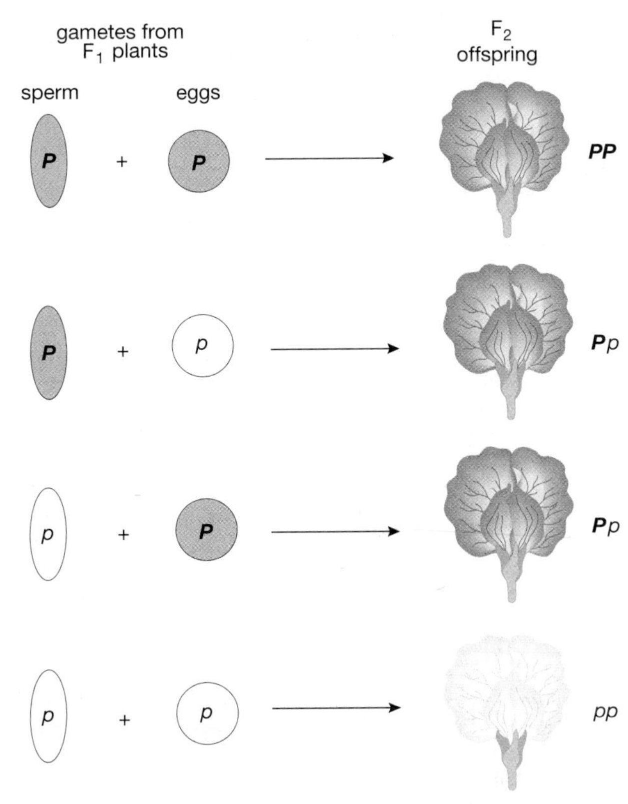

Therefore, three types of offspring can be produced: PP, Pp, and pp. The three types occur in the approximate proportions of $\frac{1}{4}PP$, $\frac{1}{2}Pp$, and $\frac{1}{4}pp$.

The actual combination of alleles carried by an organism (for example, PP or Pp) is its **genotype**. The organism's traits, including its outward appearance, behavior, digestive enzymes, blood type, or any other observable or measurable feature, make up its **phenotype**. As we have seen, plants with either the PP or the Pp genotype make purple flowers. Thus, even though they have different genotypes, they have the same phenotype. Therefore, the F_2 generation consists of three genotypes ($\frac{1}{4}PP$, $\frac{1}{2}Pp$, and $\frac{1}{4}pp$) but only two phenotypes ($\frac{3}{4}$ purple and $\frac{1}{4}$ white).

The **Punnett square method**, named after R. C. Punnett, a famous geneticist of the early 1900s, is a convenient way to predict the genotypes and phenotypes of offspring. Figure 12-4 (p. 214) shows how to use a Punnett square to determine the proportions of offspring that arise from the self-fertilization of a flower that is heterozygous for color (or the proportions of offspring produced by breeding any two organisms that are heterozygous for a single trait). This figure also provides the fractions that allow you to calculate the same outcomes based on probability theory. As you use these "genetic bookkeeping" techniques, keep in mind that, in a real experiment, the offspring will occur only in *approximately* the predicted proportions. Some sperm do not fertilize an egg, and some eggs are not fertilized by any sperm, so the alleles in these gametes never show up in the next generation. Let's consider an example. Each time a baby is conceived, it has a 50:50 chance of

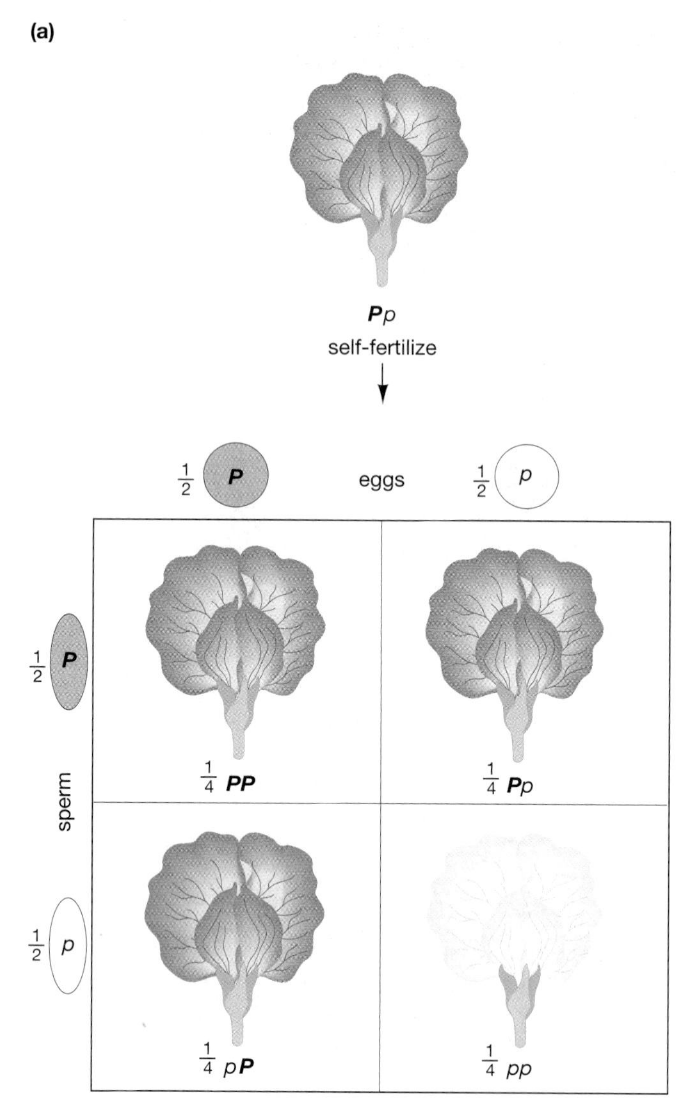

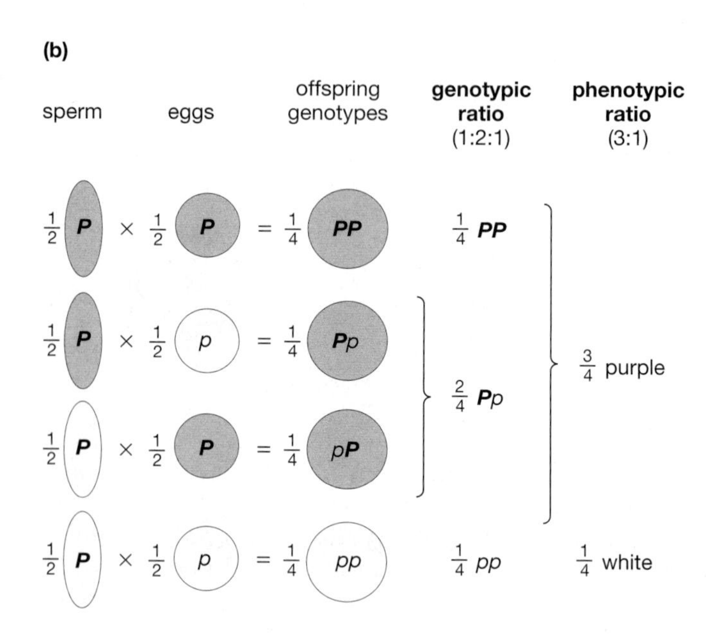

FIGURE 12-4 Determining the outcome of a single-trait cross **(a)** The Punnett square allows you to predict both genotypes and phenotypes of specific crosses; here we use it for a cross between plants that are heterozygous for a single trait, flower color.
(1) Assign letters to the different alleles; use uppercase for dominant and lowercase for recessive.
(2) Determine all the types of genetically different gametes that can be produced by the male and female parents.
(3) Draw the Punnett square, with each row and column labeled with one of the possible genotypes of sperm and eggs, respectively. (We have included the fractions of these genotypes with each label.)
(4) Fill in the genotype of the offspring in each box by combining the genotype of sperm in its row with the genotype of the egg in its column. (Multiply the fraction of sperm of each type in the row headers by the fraction of eggs of each type in the column headers.)
(5) Count the number of offspring with each genotype. (Note that *Pp* is the same as *pP*.)
(6) Convert the number of offspring of each genotype to a fraction of the total number of offspring. In this example, out of four fertilizations, only one is predicted to produce the *pp* genotype, so $\frac{1}{4}$ of the total number of offspring produced by this cross is predicted to be white. To determine phenotypic fractions, add the fractions of genotypes that would produce a given phenotype. For example, purple flowers are produced by $\frac{1}{4}$ *PP* + $\frac{1}{4}$ *Pp* + $\frac{1}{4}$ *pP*, for a total of $\frac{3}{4}$ of the offspring.
(b) Probability theory can also be used to predict the outcome of a single-trait cross. Determine the fractions of eggs and sperm of each genotype, and multiply these fractions together to calculate the fraction of offspring of each genotype. When two genotypes produce the same phenotype (e.g., *Pp* and *pP*), add the fractions of each genotype to determine the phenotypic fraction.

becoming a boy or a girl. However, many families with two children do not have one girl and one boy. The 50:50 ratio of girls to boys only occurs if we average the genders of the children in many families.

Mendel's Hypothesis Can Be Used to Predict the Outcome of New Types of Single-Trait Crosses

You have probably recognized that Mendel used the scientific method: observing results and formulating a hypothesis based on them. The scientific method has a third step: using the hypothesis to predict the results of other experiments and seeing if those experiments support or refute the hypothesis. For example, if the hybrid F_1 flowers have one allele for purple and one for white (*Pp*), then Mendel could predict the outcome of cross-fertilizing these *Pp* plants with homozygous recessive white plants (*pp*). Can you? Mendel predicted that there would be equal numbers of *Pp* (purple) and *pp* (white) offspring, and this is indeed what he found.

This type of experiment also has practical uses. Cross-fertilization of an organism with a dominant phenotype (in this case, a purple flower) but an unknown genotype with a homozygous recessive organism (a white flower) tests whether the organism with the dominant phenotype is homozygous or heterozygous; logically enough, this is called a **test cross**. When crossed with a homozygous recessive (*pp*), a homozygous dominant (*PP*) produces all phenotypically dominant offspring, whereas a heterozygous dominant (*Pp*) yields offspring with both dominant and recessive phenotypes in a 1:1 ratio:

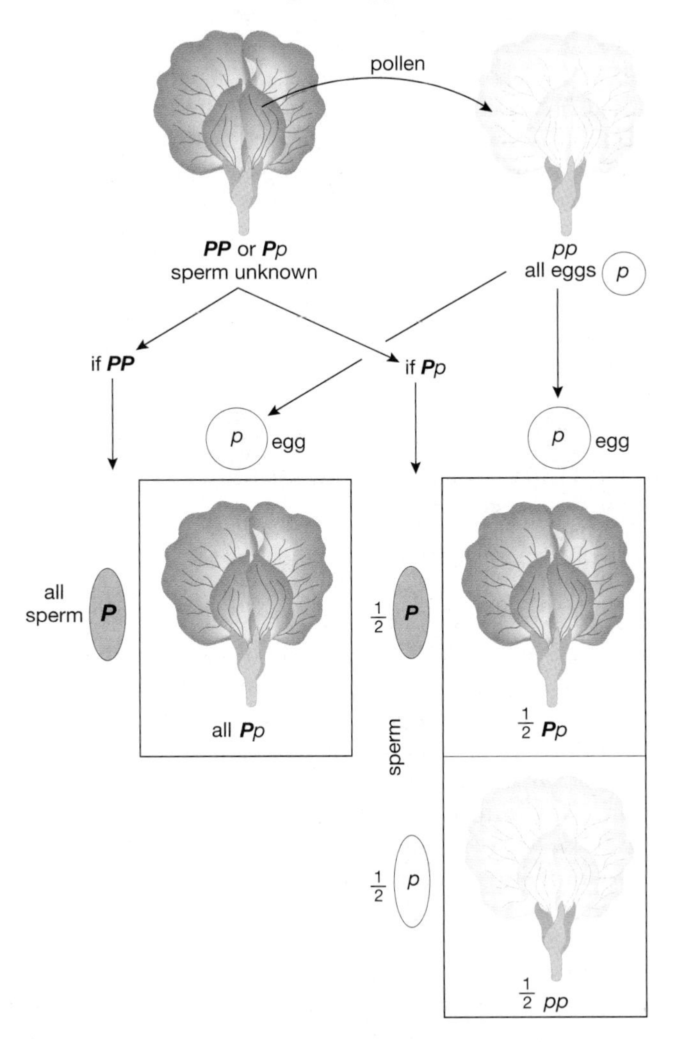

Referring back to the Case Study, when a person with Marfan syndrome marries a person without the syndrome, their children have a 50% chance of inheriting the condition. Do you think that Marfan syndrome is inherited as a dominant or as a recessive allele? Why? Check your reasoning in the Case Study Revisited at the end of the chapter.

12.3 How Are Multiple Traits on Different Chromosomes Inherited?

Mendel Hypothesized That Genes on Different Chromosomes Are Inherited Independently

Having determined the **inheritance** modes for single traits, Mendel then turned to the more complex question of multiple traits in peas (Fig. 12-5). He began by crossbreeding plants that differed in two traits—for example, seed color (yellow or green) and seed shape (smooth or wrinkled). From other crosses of plants with these traits, Mendel already knew that the smooth allele of the seed shape gene (*S*) is dominant to the wrinkled allele (*s*). In addition, the yellow allele of the seed color gene (*Y*) is dominant to the green allele (*y*). He crossed a true-breeding plant with smooth, yellow seeds (*SSYY*) to a true-breeding plant with wrinkled, green seeds (*ssyy*). All the F_1 offspring, therefore, were genotypically *SsYy*. They also all had the same phenotype: smooth, yellow seeds. Allowing these F_1 plants to self-fertilize, Mendel found that the F_2 generation consisted of 315 plants with smooth, yellow seeds, 101 with wrinkled, yellow seeds, 108 with smooth, green seeds, and 32 with wrinkled, green seeds, a ratio of about 9:3:3:1. The F_2 generations produced from other crosses of gametes that were heterozygous for two traits had similar phenotypic ratios.

Trait	Dominant form	Recessive form
Seed shape	smooth	wrinkled
Seed color	yellow	green
Pod shape	inflated	constricted
Pod color	green	yellow
Flower color	purple	white
Flower location	at leaf junctions	at tips of branches
Plant size	tall (1.8 to 2 meters)	dwarf (0.2 to 0.4 meters)

FIGURE 12-5 Traits of pea plants that Mendel studied

Mendel realized that these results could be explained if the genes for seed color and seed shape are inherited independently of each other and do not influence each other during gamete formation. For each trait, $\frac{3}{4}$ of the offspring should show the dominant phenotype (*SS* and *Ss* genotypes) and $\frac{1}{4}$ should show the recessive trait (*ss*), producing a 3:1 ratio. This is just what happened in Mendel's experiment. There were 423 plants with smooth seeds (of either color) and 133 plants with wrinkled seeds

Media Activity
12.2 Dihybrid Crosses

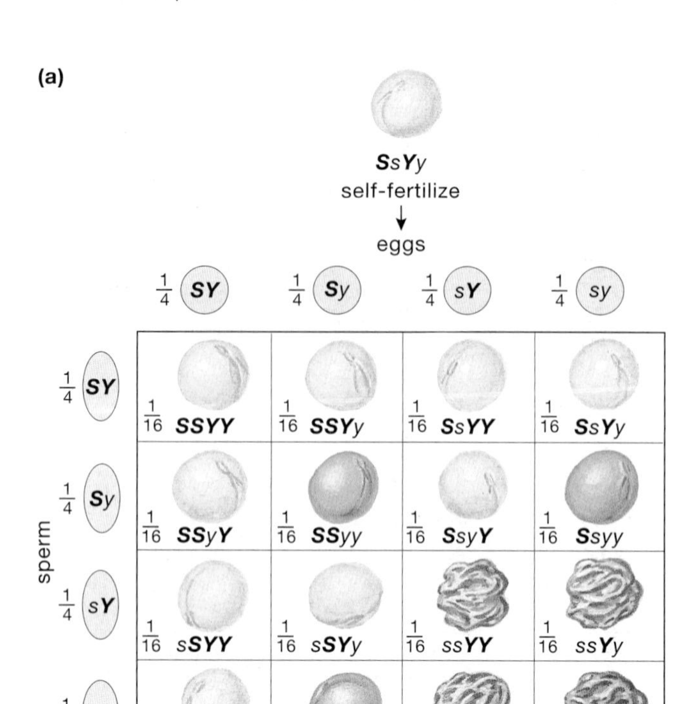

(b)

seed shape		seed color		phenotypic ratio (9:3:3:1)
$\frac{3}{4}$ smooth	×	$\frac{3}{4}$ yellow	=	$\frac{9}{16}$ smooth yellow
$\frac{3}{4}$ smooth	×	$\frac{1}{4}$ green	=	$\frac{3}{16}$ smooth green
$\frac{1}{4}$ wrinkled	×	$\frac{3}{4}$ yellow	=	$\frac{3}{16}$ wrinkled yellow
$\frac{1}{4}$ wrinkled	×	$\frac{1}{4}$ green	=	$\frac{1}{16}$ wrinkled green

FIGURE 12-6 Predicting genotypes and phenotypes for a cross between gametes that are heterozygous for two traits In pea seeds, yellow (*Y*) is dominant to green (*y*), and smooth (*S*) is dominant to wrinkled (*s*). **(a)** Punnett square analysis. In this cross, an individual heterozygous for both traits self-fertilizes. Note that the Punnett square predicts both the frequencies of combinations of traits ($\frac{9}{16}$ smooth yellow, $\frac{3}{16}$ smooth green, $\frac{3}{16}$ wrinkled yellow, $\frac{1}{16}$ wrinkled green) and the frequencies of individual traits ($\frac{3}{4}$ yellow, $\frac{1}{4}$ green, $\frac{3}{4}$ smooth, and $\frac{1}{4}$ wrinkled). **(b)** Probability theory states that the probability of two independent events is the product (multiplication) of their individual probabilities. Seed *shape* is independent of seed *color*. Therefore, multiplying the independent probabilities of the genotypes or phenotypes for each trait produces the predicted frequencies for the combined genotypes or phenotypes of the offspring. These ratios are identical to those generated by the Punnett square.
EXERCISE Use Punnett squares to determine if the genotype of a plant bearing yellow smooth seeds can be revealed by a test cross with a plant bearing green wrinkled seeds.

(about 3:1); in this same group of plants, 416 produced yellow seeds (of either shape) and 140 produced green seeds (also about 3:1). Figure 12-6 shows how a Punnett square or probability can be used to determine the outcome of a cross between organisms that are heterozygous for two traits, and how two independent 3:1 ratios combine to produce an overall 9:3:3:1 ratio.

The independent inheritance of two or more distinct traits is called the **law of independent assortment**, which states that the alleles of one gene may be distributed to gametes independently of the alleles for other genes. Independent assortment will occur when the traits being studied are controlled by genes on different pairs of homologous chromosomes. Why? From Chapter 11, recall the movement of chromosomes during meiosis. When paired homologous chromosomes line up during metaphase I, which homologue faces which pole of the cell is random, and the orientation of one homologous pair does not influence other pairs. Therefore, when the homologues separate during anaphase I, which allele of a gene on homologous pair 1 moves "north" does not affect which allele of a gene on homologous pair 2 moves "north"; that is, the alleles of genes on different chromosomes are distributed, or "assorted," independently (Fig. 12-7).

In an Unprepared World, Genius May Go Unrecognized

In 1865, Gregor Mendel presented his theories of inheritance to the Brünn Society for the Study of Natural Science, and they were published the following year. His paper did not mark the beginning of genetics. In fact, it didn't make any impression at all on the study of biology during his lifetime. Mendel's experiments, which eventually spawned one of the most important scientific theories in all of biology, simply vanished from the scene. Apparently, very few biologists read his paper, and those who did failed to recognize its significance or discounted it because it contradicted prevailing ideas of inheritance.

It was not until 1900 that three biologists—Carl Correns, Hugo de Vries, and Erich Tschermak—working independently and knowing nothing of Mendel's work, rediscovered the principles of inheritance. No doubt to their intense disappointment, when they searched the scientific literature before publishing their results, they found that Mendel had scooped them more than 30 years earlier. To their credit, they graciously acknowledged the important work of the Augustinian monk, who had died in 1884.

12.4 How Are Genes Located on the Same Chromosome Inherited?

Gregor Mendel knew nothing about the physical nature of genes or chromosomes. Much later, when scientists discovered that chromosomes are the vehicles of inheritance, it became obvious that there are many more traits (and therefore many more genes) than there are chromosomes. As you may recall from Chapter 11,

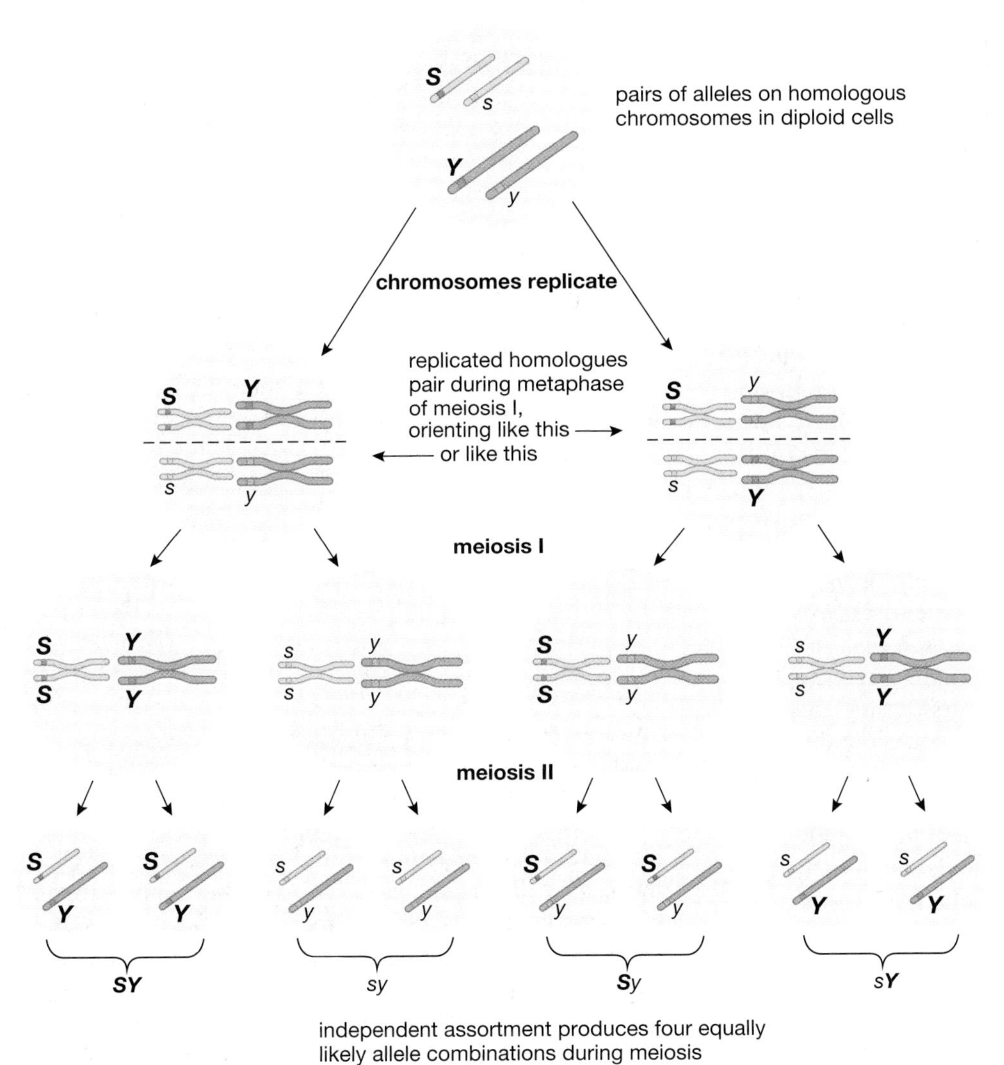

FIGURE 12-7 Independent assortment of alleles Chromosome movements during meiosis produce independent assortment of alleles of two different genes. Each combination of alleles is equally likely to occur. Therefore, an F_1 plant would produce gametes in the predicted proportions $\frac{1}{4}$ *SY*, $\frac{1}{4}$ *sy*, $\frac{1}{4}$ *sY*, and $\frac{1}{4}$ *Sy*.

genes are parts of chromosomes, and each chromosome contains many genes. These facts have important implications for inheritance.

Genes on the Same Chromosome Tend to Be Inherited Together

If chromosomes assort independently during meiosis I, then only genes located on *different chromosomes* will assort independently into gametes. In contrast, genes on the *same chromosome* tend to be inherited together. Genetic **linkage** is the inheritance of certain genes as a group because they are on the same chromosome. Alleles of linked genes tend to be inherited together and do not assort independently. One of the first pairs of linked genes to be discovered was found in the sweet pea, a different species from Mendel's garden pea. In sweet peas, the gene for flower color and the gene for pollen grain shape are carried on the same chromosome. Thus, the alleles for these genes normally assort *together* into gametes during meiosis and are inherited together as a result.

Consider a heterozygous sweet pea plant with purple flowers and long pollen that has the following chromosomes:

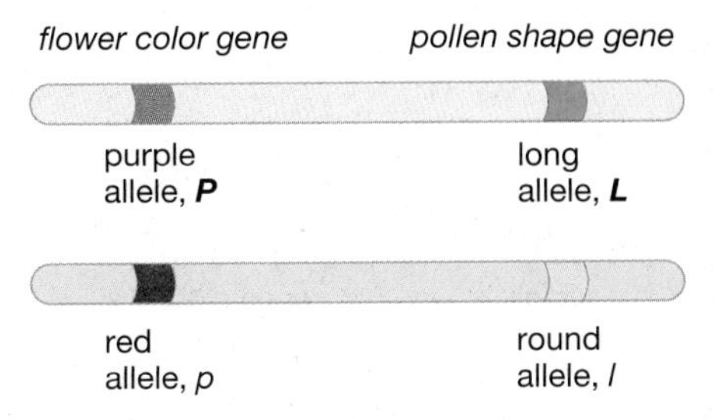

Note that the purple allele of the flower color gene and the long allele of the pollen shape gene are located on one homologous chromosome. The red allele of the flower color gene and the round allele of the pollen shape gene are located on the other homologue. Therefore, the gametes produced by this sweet pea plant are likely to have *either* purple and long alleles *or* red and round alleles. This pattern of inheritance breaks the law of independent assortment, because the alleles for flower color and pollen shape do *not* segregate independently of one another into the gametes, but tend to stay together through meiosis.

Recombination Can Create New Combinations of Linked Alleles

Although they tend to be inherited together, genes on the same chromosome do not always stay together. In the sweet pea cross just described, for example, the F_2 generation will commonly include a few plants in which the genes for flower color and pollen shape have been inherited as if they were not linked. That is, some of the offspring will have purple flowers and round pollen, and some will have red flowers and long pollen. How can this be?

As you learned in Chapter 11, during prophase I of meiosis, homologous chromosomes sometimes exchange parts, a process called **crossing over** (see Fig. 11-12). In most chromosomes, at least one exchange between each homologous chromosome pair occurs during each meiotic cell division. The exchange of corresponding segments of DNA during crossing over produces new allele combinations on both homologous chromosomes. Then, when homologous chromosomes separate at anaphase I, the chromosomes that each haploid daughter cell receives will have different sets of alleles from those of the parent cell.

Crossing over during meiosis explains the appearance of new combinations of alleles that were previously linked. Let's revisit the sweet pea during the early stages of meiosis I, when the chromosomes have duplicated and homologous chromosomes are paired up:

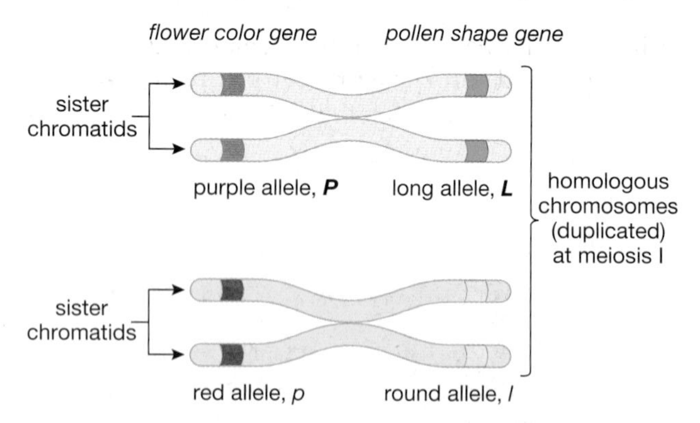

Each homologous chromosome will have one or more regions where crossing over will occur. Imagine that one cross-over occurs between the genes for flower color and pollen shape:

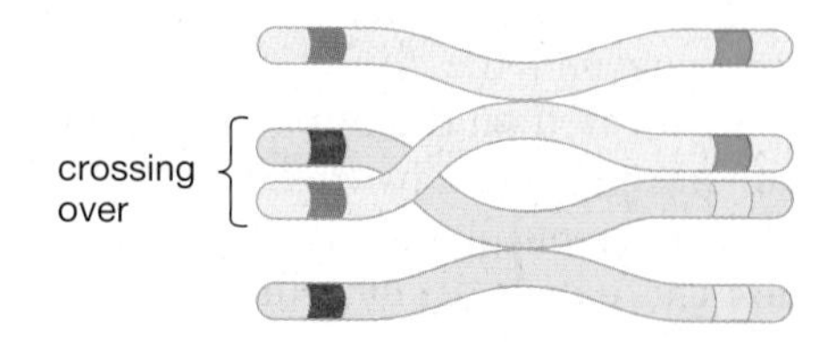

At anaphase I, the separated homologous chromosomes will have this gene composition:

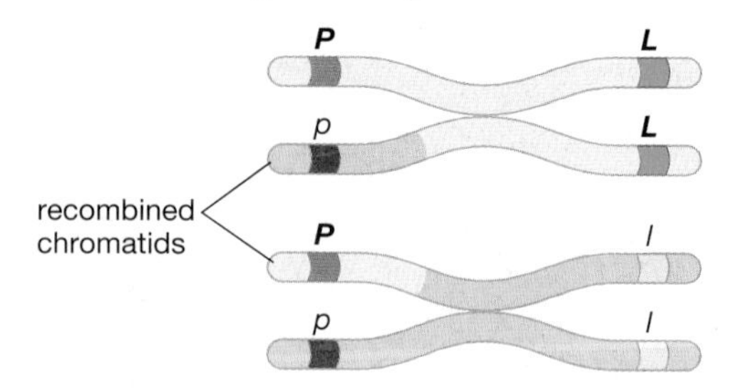

Four types of chromosomes are then distributed to the haploid daughter cells during meiosis II:

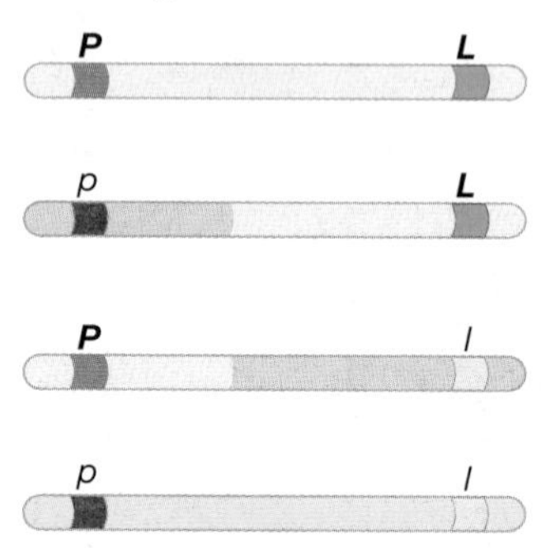

Therefore, some gametes will be produced with each of four chromosome configurations: *PL* and *pl* (the original parental types), and *Pl* and *pL* (recombined chromosomes). By exchanging DNA between homologous chromosomes, this **genetic recombination** creates new combinations of alleles. If a sperm with a *Pl* chromosome fertilizes an egg with a *pl* chromosome, the offspring plant will have purple flowers (*Pp*) and round pollen (*ll*). If a sperm with a *pL* chromosome fertilizes an egg with a *pl* chromosome, then the offspring will have red flowers (*pp*) and long pollen (*Ll*).

Not surprisingly, the farther apart the genes are on a chromosome, the more likely it is that crossing over will occur between them. In fact, if two genes are really far apart, crossing over occurs so often that they seem to be independently assorted, the same as if they were on different chromosomes. When Gregor Mendel discovered independent assortment, he was not only clever and careful, he was also lucky. The seven traits that he studied were controlled by genes on only four different chromosomes; he observed independent assortment because the genes that were on the same chromosome were far apart.

12.5 How Is Sex Determined, and How Are Sex-Linked Genes Inherited?

In mammals and many insects, males have the same number of chromosomes as females do, but one "pair," the **sex chromosomes**, is very different in appearance and genetic composition. Females have two identical sex chromosomes, called *X chromosomes*, whereas

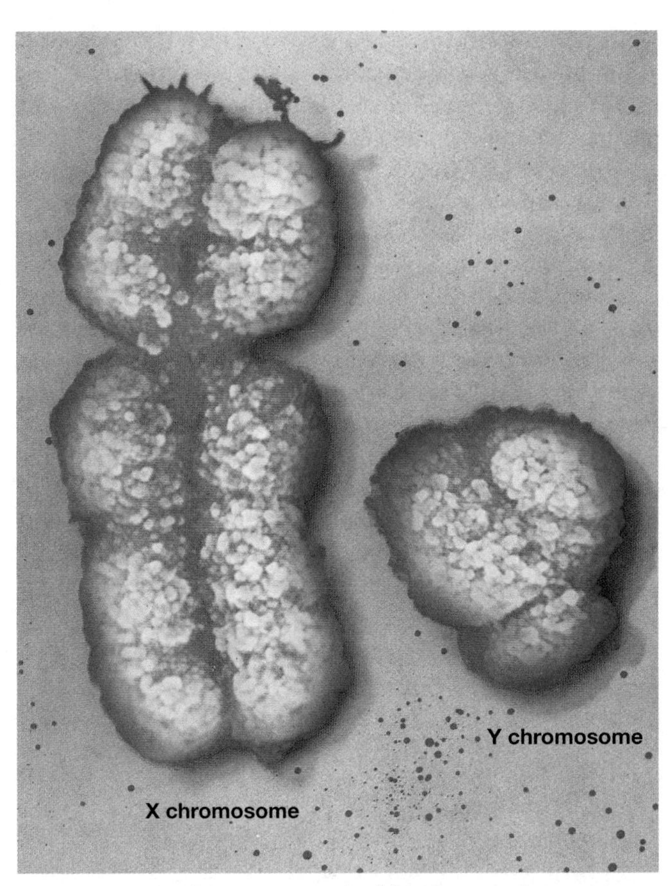

FIGURE 12-8 Photomicrograph of human sex chromosomes Notice the small size of the Y chromosome, which carries relatively few genes.

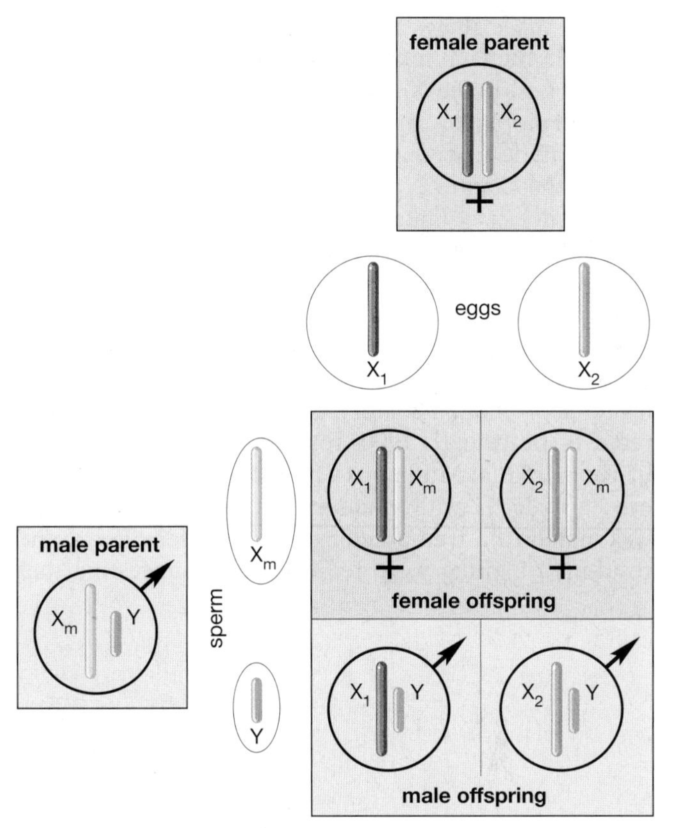

FIGURE 12-9 Sex determination in mammals Male offspring receive their Y chromosome from the father; female offspring receive the father's X chromosome (labeled X_m). Both male and female offspring receive an X chromosome (either X_1 or X_2) from the mother.

males have one X chromosome and one *Y chromosome* (Fig. 12-8). Although the Y chromosome normally carries far fewer genes than does the X chromosome, a small part of both sex chromosomes is homologous. As a result, the X and Y chromosomes pair up during prophase of meiosis I and separate during anaphase I. The other chromosomes, which occur in pairs that have identical appearance in both males and females, are called **autosomes**. Numbers of chromosomes vary tremendously among species, but there is only one pair of sex chromosomes. For example, the fruit fly *Drosophila* has 4 pairs of chromosomes (3 pairs of autosomes and 1 pair of sex chromosomes), humans have 23 pairs (22 pairs of autosomes and 1 pair of sex chromosomes), and dogs have 39 pairs (38 pairs of autosomes and 1 pair of sex chromosomes).

For organisms in which males are XY and females are XX, the sex chromosome carried by the sperm determines the sex of the offspring (Fig. 12-9). During sperm formation, the sex chromosomes segregate, and each sperm receives either the X or the Y chromosome (plus one member of each pair of autosomes). The sex chromosomes also segregate during egg formation, but because females have two X chromosomes, every egg receives one X chromosome (along with one member of each pair of autosomes). An offspring is male if an egg is fertilized by a Y-bearing sperm or female if an egg is fertilized by an X-bearing sperm.

Sex-Linked Genes Are Found Only on the X or Only on the Y Chromosome

Genes that are on one sex chromosome but not on the other are said to be **sex-linked**. In many animals, the Y chromosome carries only a few genes. As of late 2003, only 78 genes have been found on the Y chromosome; many of these play a role in male reproduction. In contrast, the X chromosome probably contains over 1000 genes, few of which have a specific role in female reproduction. Most have no counterpart on the Y chromosome, and encode traits that are important in both sexes, such as color vision, blood clotting, and certain structural proteins in muscles. Because they have two X chromosomes, females can be either homozygous or heterozygous for genes on the X chromosome, and dominant versus recessive relationships among alleles will be expressed. Males, in contrast, fully express all the alleles they have on their single X chromosome, whether those alleles are dominant or recessive. For this reason, in humans, most cases of recessive traits encoded by genes on the X chromosome, such as color blindness, hemophilia, and certain types of muscular dystrophy, occur in males. We will return to this concept later in the chapter.

How does sex linkage affect inheritance? Let's look at the first example of sex linkage to be discovered, the inheritance of eye color in the fruit fly *Drosophila*. Because these flies are small, reproduce rapidly, are easy to raise in the laboratory, and have few chromosomes, *Drosophila* have been favored subjects for genetics studies for more than a century. Normally, *Drosophila* have red eyes. In the early 1900s, researchers in the laboratory of Thomas Hunt Morgan at Columbia University discovered a male fly with white eyes. This white-eyed male was mated to a true-breeding, red-eyed female. All the resulting offspring were red-eyed flies, suggesting that white eye color (r) is recessive to red (R). The F_2 generation, however, was a surprise: there were nearly equal numbers of red-eyed males and white-eyed males, but no females had white eyes! A test cross of the F_1 red-eyed females to the original white-eyed male yielded roughly equal numbers of red-eyed and white-eyed males and females.

From these data, could you figure out how eye color is inherited? Morgan made the brilliant hypothesis that *the gene for eye color must be located on the X chromosome and that the Y chromosome has no corresponding gene* (Fig. 12-10). In the F_1 generation, both male and female offspring received an X chromosome, with its R allele for red eyes, from their mother. The F_1 males received a Y chromosome from their father with no allele for eye color, and so the males had an R- genotype, and red-eyed phenotype. (The "–" means that the Y chromosome doesn't have an eye color gene.) The F_1 fe-

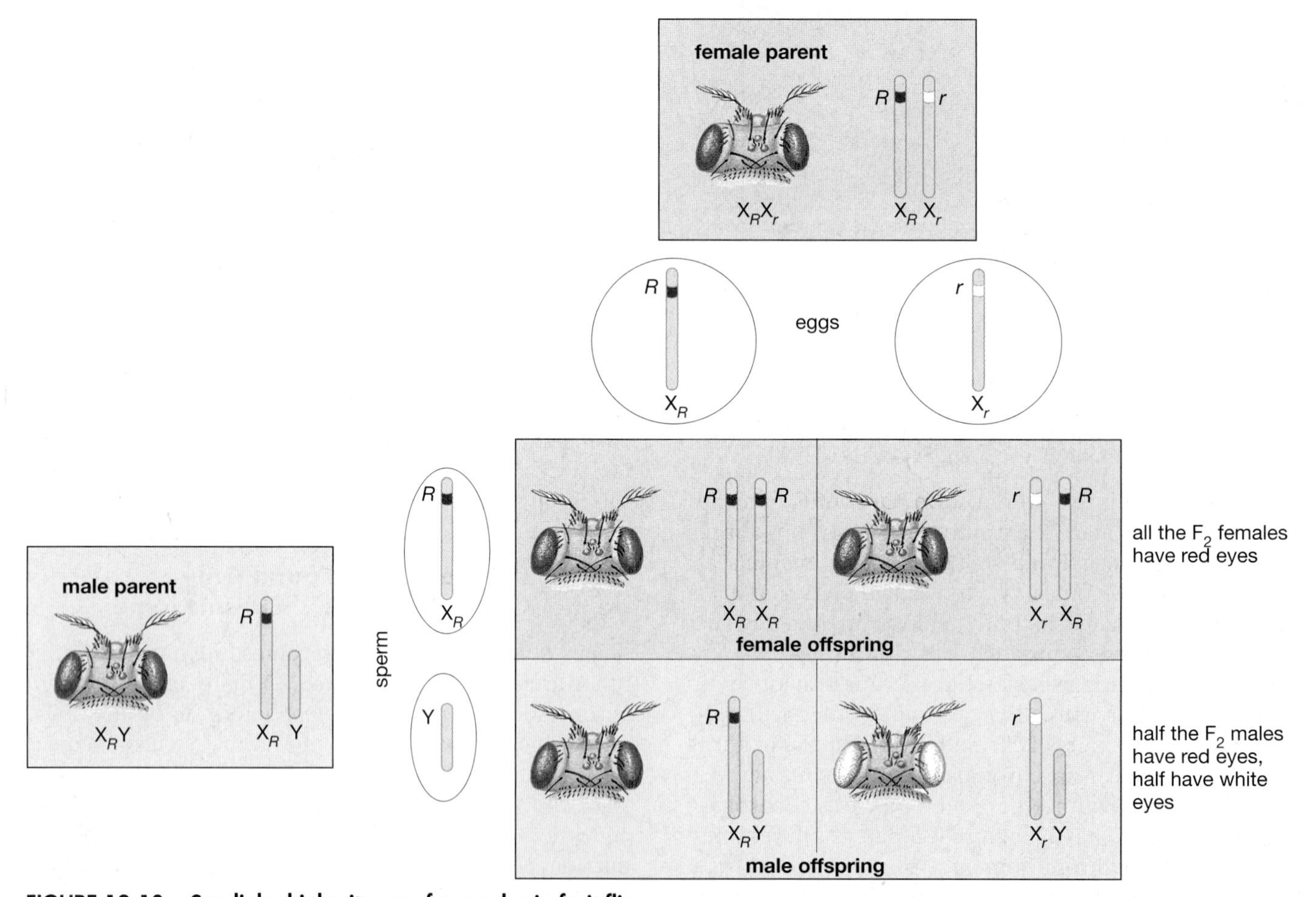

FIGURE 12-10 Sex-linked inheritance of eye color in fruit flies
The gene for eye color is located on the X chromosome; the Y chromosome does *not* contain an eye color gene. Red (R) is dominant to white (r). When a white-eyed male is mated to a homozygous red-eyed female, all the offspring have red eyes: F_1 females are heterozygous, receiving the r allele from their father and the R allele from their mother, while male offspring receive only the R allele from their mother. In the F_2 generation, the single R allele of the F_1 male parent is passed on to his daughters, so all the F_2 daughters have red eyes. The F_2 sons receive a Y chromosome from their father, and either the R or r allele on the X chromosome from their mother, so half the sons have white eyes and half have red eyes. **QUESTION** What would the phenotypic ratios in the offspring of this cross be if the alleles for eye color showed incomplete dominance, as shown in Fig. 12-11?

males received the father's X chromosome with its r allele, so the females had an Rr genotype and red-eyed phenotype. Thus, both male and female F_1 offspring had red eyes.

Crossing two F_1 flies, R- × Rr, resulted in an F_2 generation with the chromosome distribution shown in Figure 12-10. All the F_2 females received one X chromosome from their F_1 male parent, with its R allele, and therefore had red eyes. All the F_2 males inherited their single X chromosome from their F_1 female parent, which was heterozygous for eye color (Rr). So the F_2 males had a 50:50 chance of receiving an X chromosome with the R allele or one with the r allele. *With no corresponding gene on the Y chromosome, the F_2 males displayed the phenotype determined by the allele on the X chromosome.* Therefore, half the F_2 males had red eyes, and half had white eyes.

12.6 Do the Mendelian Rules of Inheritance Apply to All Traits?

In our discussion of patterns of inheritance thus far, we have made some major simplifying assumptions: that each trait is completely controlled by a single gene, that there are only two possible alleles of each gene, and that one allele is completely dominant to the other, recessive, allele. Most traits, however, are influenced in more varied and subtle ways than this.

Incomplete Dominance: The Phenotype of Heterozygotes Is Intermediate Between the Phenotypes of the Homozygotes

When one allele is completely dominant over a second allele, heterozygotes with one dominant allele have the same phenotype as homozygotes with two dominant alleles. However, relationships between alleles are not always this simple. When the heterozygous phenotype is intermediate between the two homozygous phenotypes, the pattern of inheritance is called **incomplete dominance**. In snapdragons, for example, crossing homozygous red-flowered plants (RR) with homozygous white-flowered ones ($R'R'$) produces F_1 hybrids that do not have red flowers. Instead, the F_1 (RR') flowers are pink. (In genetic nomenclature, incompletely dominant alleles are often given uppercase symbols with a superscript to denote the different alleles. In the inheritance of flower color in snapdragons, the two alleles are denoted as R and R'.) Although the phenotypes appear blended, the alleles remain unchanged from generation to generation. In the F_2 generation, the red and white colors of the homozygotes are as pure as ever (Fig. 12-11). The F_2 offspring include about $\frac{1}{4}$ red (RR), $\frac{1}{2}$ pink (RR') and $\frac{1}{4}$ white ($R'R'$) flowers.

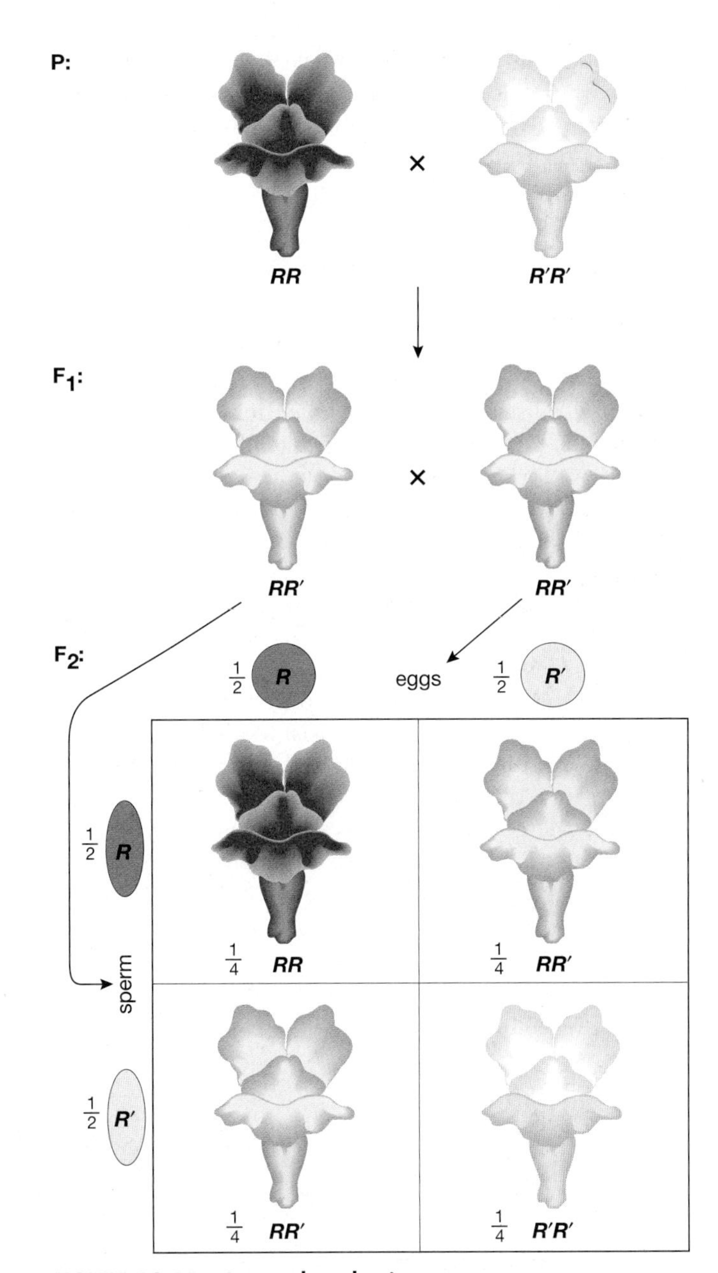

FIGURE 12-11 Incomplete dominance
The inheritance of flower color in snapdragons is an example of incomplete dominance. In such cases, we will use capital letters for both alleles, here R and R'. Hybrids RR' have pink flowers, whereas the homozygotes are red (RR) or white $R'R'$. Because heterozygotes can be distinguished from homozygous dominants, the distribution of phenotypes in the F_2 generation ($\frac{1}{4}$ red: $\frac{1}{2}$ pink: $\frac{1}{4}$ white) is the same as the distribution of genotypes ($\frac{1}{4}$ RR: $\frac{1}{2}$ RR': $\frac{1}{4}$ $R'R'$). **QUESTION** Is it possible for plant breeders to develop a true-breeding pink-flowered snapdragon?

What causes incomplete dominance? Recall that most genes encode the information for production of a protein. The different alleles of a gene may encode proteins with different functional capacity. In snapdragons, the R allele codes for an enzyme that catalyzes the formation of red pigment; the R' allele codes for a defective, nonfunctional enzyme. Plants with the RR genotype produce lots of red pigment and have red flowers, but those with the $R'R'$ genotype produce no pigment and have white flowers. Heterozygotes, with only one copy of the *R* allele, produce an intermediate amount of red pigment, so the resulting flowers are pink.

A Single Gene May Have Multiple Alleles

Alleles arise through mutation, and the same gene in different individuals may have different mutations, each producing a new allele. Therefore, although an *individual* can have at most two different alleles, a *species* may have **multiple alleles** of many of its genes. One eye-color gene in *Drosophila*, for example, has more than a thousand alleles. Depending on how they are combined, eye color in fruit flies can be white, yellow, orange, pink, brown, or red. There are hundreds of alleles for both Marfan syndrome and cystic fibrosis (see "Scientific Inquiry: Cystic Fibrosis"), each of which arose as a new mutation.

Human blood types are examples of multiple alleles of a single gene, with an added twist to the pattern of inheritance. The blood types A, B, AB, and O arise as a result of three different alleles (for simplicity, we will designate them *A, B,* and *o*) of a single gene located on chromosome 9. This gene codes for an enzyme responsible for adding sugar molecules to the ends of glycoproteins that protrude from the surfaces of red blood cells. Alleles *A* and *B* code for enzymes that add different end sugars to the glycoproteins (we will call the resulting molecules glycoproteins *A* and *B*, respectively). Allele *o* codes for a nonfunctional enzyme that doesn't add any sugar molecule. A person may have one of six genotypes: *AA, BB, AB, Ao, Bo,* or *oo* (Table 12-1). Alleles *A* and *B* are dominant to *o*. Therefore, people with genotypes *AA* or *Ao* have only type A glycoproteins and have type A blood. Those with genotypes *BB* or *Bo* synthesize only type B glycoproteins and have type B blood. Homozygous recessive *oo* individuals lack both types of glycoproteins and have type O blood. In people with type AB blood, both enzymes are present, so the plasma membranes of their red blood cells have both A and B glycoproteins. When heterozygotes express phenotypes of *both* of the homozygotes (in this case, both A and B glycoproteins), the pattern of inheritance is called **codominance**, and alleles are said to be *codominant* to one another.

People make antibodies to the type of glycoprotein(s) that they lack. These antibodies are proteins in blood plasma that bind to foreign glycoproteins by recognizing different end-sugar molecules. The antibodies cause red blood cells that bear foreign glycoproteins to clump together and to rupture. The resulting clumps and fragments can clog small blood vessels and damage vital organs such as the brain, heart, lungs, or kidneys. This means that blood type must be determined and matched carefully before a blood transfusion is made.

Type O blood, lacking any end sugars, is not attacked by antibodies in A, B, or AB blood, so it can be transfused safely to all other blood types. (The antibodies present in transfused blood become too diluted to cause

Table 12-1 Human Blood Group Characteristics

Blood Type	Genotype	Red Blood Cells	Has Plasma Antibodies to:	Can Receive Blood from:	Can Donate Blood to:	Frequency in U.S.
A	*AA* or *Ao*	A glycoprotein	B glycoprotein	A or O (no blood with B glycoprotein)	A or AB	40%
B	*BB* or *Bo*	B glycoprotein	A glycoprotein	B or O (no blood with A glycoprotein)	B or AB	10%
AB	*AB*	Both A and B glycoproteins	Neither A nor B glycoprotein	AB, A, B, O (universal recipient)	AB	4%
O	*oo*	Neither A nor B glycoprotein	Both A and B glycoproteins	O (no blood with A or B glycoprotein)	O, AB, A, B (universal donor)	46%

SCIENTIFIC INQUIRY Cystic Fibrosis

"Woe to that child which when kissed on the forehead tastes salty. He is bewitched and soon must die."

—17th-century English saying

This adage is based on a remarkably accurate diagnostic tool for the most common recessive genetic disorder in the United States and Europe—cystic fibrosis. About 30,000 Americans, 3,000 Canadians, and 20,000 Europeans have cystic fibrosis. The story of this disease is a blend of physiology, medicine, and both Mendelian and molecular genetics.

Let's begin with a child's salty forehead. Sweat cools the body by evaporating from the skin, and is mostly water. However, sweat also contains a lot of salt (sodium chloride) when it is first secreted, about as much as in blood and extracellular fluid. As sweat moves through tubes connecting the secreting cells with the surface of the skin, most of the salt is reclaimed if you are perspiring slowly enough. How? Pumps in the plasma membranes of the cells lining the tubes transport negatively charged chloride ions out of the sweat and return them to the extracellular fluid. Positively charged sodium ions follow along by electrical attraction. Cystic fibrosis is caused by defective chloride pumps: salt stays in the sweat, so the skin tastes salty.

Salty sweat isn't very harmful, but unfortunately the cells lining the lungs have the same chloride pumps. In the lungs, the pumps move chloride onto the surface of the airways. As you may recall from Chapter 4, water "follows" ions by osmosis, so the secreted sodium and chloride ions cause water to move to the airway surfaces. Some cells of the airways also secrete mucus. Ideally, the water dilutes the mucus, so that the fluid on the airway surfaces is thin and watery. Why does this matter? The mucus, along with bacteria and debris that it traps, is swept out of the lungs by cilia on the cells. In cystic fibrosis, reduced chloride secretion means that not much water reaches the airway surfaces, so the mucus is thick and the cilia can't move it very well. The mucus clogs the airways, and bacteria remain in the lungs, causing frequent infections. Even if a person survives the infections, the lungs usually become permanently damaged. Mucus also builds up in the stomach and intestines, reducing the absorption of nutrients and causing malnutrition. Before modern medical care, most people with cystic fibrosis died by age 4 or 5; even now, the average life span is only about 30 to 35 years.

Mutations in the *CTFR* gene, which encodes the protein of the chloride pump, cause cystic fibrosis. Researchers have identified nearly 1,000 mutations in this gene. Some mutations introduce a stop codon into the middle of the mRNA, which cuts off translation before the pump protein is completed; others change the amino acid sequence in ways that reduce the pumping rate. The most common mutation prevents the pump protein from moving through the endoplasmic reticulum and Golgi apparatus to the plasma membrane. Overall, about 1 American in 30 carries one of these mutations.

Why is cystic fibrosis a recessive trait? People who are heterozygous, with one normal *CFTR* allele and one copy of any of these mutations, produce enough chloride pump proteins to provide adequate chloride transport. Therefore, they are phenotypically normal—that is, they produce watery secretions in their lungs and do not develop cystic fibrosis. Someone with two defective alleles will have no functioning chloride pumps and will develop the disease.

Can anything be done to prevent, cure, or control the symptoms of cystic fibrosis? Because it is a genetic disorder, the only way to prevent the disease is to prevent the birth of affected infants. However, people usually don't know whether they are carriers, and therefore don't know whether their children may inherit the disease. Treatments that reduce lung damage include physical manipulation to drain the lungs, medicines that open the airways (similar to those used by people with asthma), and frequent, even continuous, administration of antibiotics (Fig. E12-1). Unfortunately, these treatments merely postpone inevitable damage to the lungs, intestine, pancreas, and other organs.

This may change in the next few years. Today, medical labs can identify carriers by a blood test and homozygous recessive embryos by prenatal diagnosis. Soon, children with cystic fibrosis may be cured or helped by one of several gene therapies now under development. We will explore these applications of biotechnology in the next chapter.

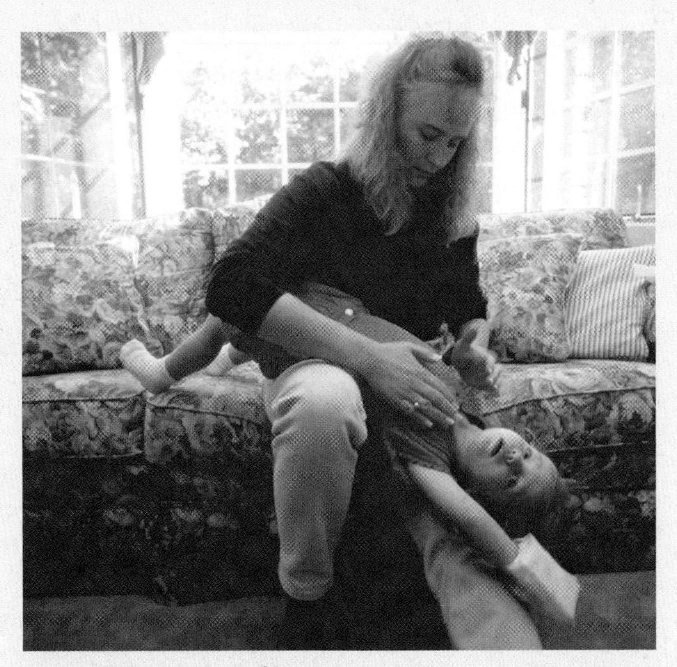

FIGURE E12-1 Cystic fibrosis
A child is treated for cystic fibrosis. Gentle pounding on the chest and back while the child is held upside-down helps dislodge mucus from the lungs. A device on the child's wrist injects antibiotics into a vein. These treatments combat the numerous lung infections to which cystic fibrosis patients are vulnerable.

problems.) People with type O blood are called "universal donors." But O blood carries antibodies to both A and B glycoproteins, so type O individuals can receive transfusions of only type O blood. Can you predict the blood type of people called "universal recipients"? Table 12-1 (p. 222) summarizes blood types and transfusion characteristics.

Many Traits Are Influenced by Several Genes

If you look around your class, you are likely to see people of varied heights, skin colors, and body builds. Traits such as these are not governed by single genes but are influenced by interactions among two or more genes, as well as by interactions with the environment. Many traits, such as human height, weight, eye color, and skin color, may have several phenotypes or even seemingly continuous variation that cannot be split up into convenient, easily defined categories. This is an example of **polygenic inheritance**, a form of inheritance in which the interaction of two or more genes contributes to a single phenotype.

The color of the grains of some types of wheat provides a simple example of polygenic inheritance. Red grain color is determined by two genes, each with two incompletely dominant alleles. Let's call one gene R_1 and the other R_2. Each gene has two alleles, R_1 and R_1', and R_2 and R_2'. The R alleles cause one "unit" of red pigment to be synthesized in the grains, while the R′ alleles don't produce any pigment at all. If there were only one gene, say R_1, then this would be simple incomplete dominance, like the snapdragon flowers of Fig. 12-11, with red, pink, and white grains. Because there are two genes, however, there are five possible combinations of alleles, producing five shades of red: 4 R alleles (darkest red), 3 R and 1 R′ alleles, 2 R and 2 R′ alleles, 1 R and 3 R′ alleles, and 4 R′ alleles (white). Crossing two wheat plants, both heterozygous for both genes, produces offspring with the full range of colors (Fig. 12-12).

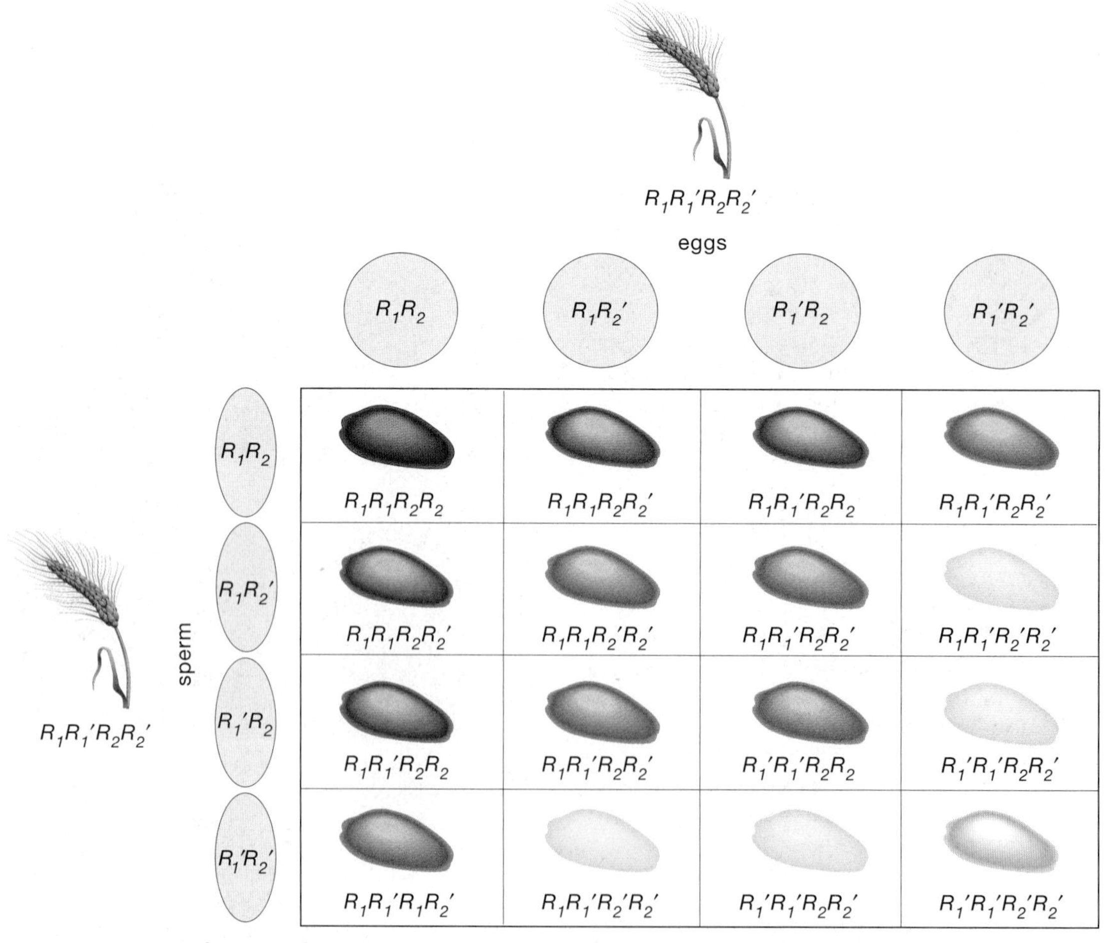

FIGURE 12-12 Polygenic inheritance of grain color in wheat
At least two separate genes, each with two incompletely dominant alleles, determine the color of wheat grains. A cross between two wheat plants, both heterozygous for both genes, produces five colors of offspring.

As you might imagine, the more genes that contribute to a single trait, the greater the number of phenotypes and the finer the distinctions among them. When four or more pairs of genes contribute to a trait, differences between phenotypes are small, and it is extremely difficult to classify the phenotypes reliably. For example, at least three or four genes control skin pigmentation in people. Exposure to the sun further alters skin color, with the result that humans show virtually continuous variation from very dark to very light skin.

Single Genes Typically Have Multiple Effects on Phenotype

We have just seen that a single phenotype may result from the interaction of several genes. The reverse is also true: single genes commonly have multiple phenotypic effects, a phenomenon called **pleiotropy**. A good example is the SRY gene, discovered on the Y chromosome in 1990. The SRY gene (short for "sex-determining region of the Y chromosome") codes for a protein that activates other genes; those genes in turn code for proteins that switch on male development in an embryo. Under the influence of the genes activated by the SRY protein, sex organs develop into testes. The testes, in turn, secrete sex hormones that stimulate the development of both internal and external male reproductive structures, such as the epididymis, seminal vesicles, prostate gland, penis, and scrotum. The SRY gene is described more fully in the Chapter 10 Case Study, "Boy or Girl?"

The Environment Influences the Expression of Genes

An organism is not just the sum of its genes. In addition to its genotype, the environment in which an organism lives profoundly affects its phenotype. A striking example of environmental effects on gene action occurs in the Himalayan rabbit, which, like the Siamese cat, has pale body fur but black ears, nose, tail, and feet (Fig. 12-13). The Himalayan rabbit actually has the genotype for black fur all over its body. However, the enzyme that produces the black pigment is inactive at temperatures above about 34 °C (93 °F). At typical ambient temperatures, extremities such as the ears and feet are cooler than the rest of the body, so black pigment can be produced there. Because the rabbit's main body surface is warmer than 34 °C, the fur there is pale.

Most environmental influences are more complicated and subtle. The complexity of environmental influences is particularly true of human characteristics. The polygenic trait of skin color is modified by the environmental effects of sun exposure. Height, another polygenic trait, can be reduced by poor nutrition.

Intelligence, too, has both genetic and environmental components. Dozens of studies have compared IQ levels in people of varying family relationships. Even when separated at birth and raised in different environments, identical twins achieve similar scores on IQ tests, although not as similar as identical twins who have been raised together. Brothers and sisters who are not twins differ more than do twins, but are still fairly similar. Thus, the more genetically related two people are, the more similar their IQ scores. These findings indicate that IQ testing capability has a genetic component. However, unrelated people who have been raised together as children (for example, adoptees) show more similarity on IQ tests than do unrelated people reared apart. Therefore, *both heredity and environment play major roles in the development of various mental abilities* and almost certainly other personality traits as well.

The interactions between complex genetic systems and varied environmental conditions can create a continuum of phenotypes that defies analysis into genetic and environmental components. The human generation time is long, and the number of offspring per couple is small. Add to these factors the countless subtle ways in which people respond to their environments, and you can see why it may be exceedingly difficult to determine the precise genetic basis of complex human traits such as intelligence or musical or athletic ability.

12.7 How Are Human Genetic Disorders Investigated?

Because experimental crosses with humans are out of the question, human geneticists search medical, historical, and family records to study past crosses. Records extending across several generations can be arranged in

FIGURE 12-13 Environmental influence on phenotype The expression of the gene for black fur in the Himalayan rabbit is a simple case of interaction between genotype and environment producing a particular phenotype. The gene for black fur is expressed in cool areas (nose, ears, and feet).

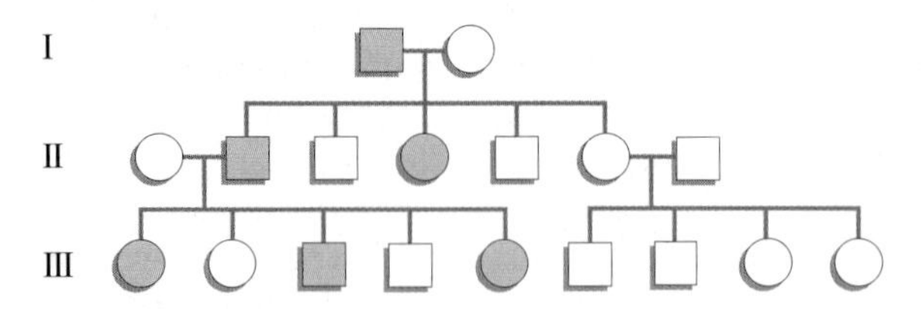

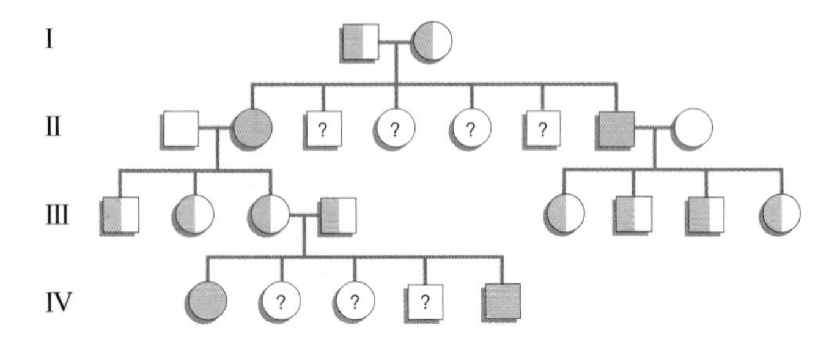

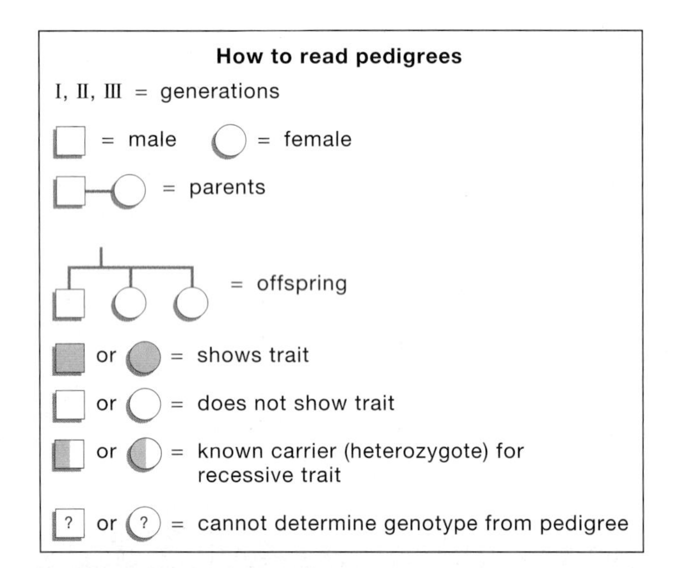

FIGURE 12-14 Family pedigrees
(a) A pedigree for a dominant trait. Note that any offspring showing a dominant trait *must* have at least one parent with the trait (see Figs. 12-4 and 12-6). **(b)** A pedigree for a recessive trait. Any individual showing a recessive trait must be homozygous recessive. If that person's parents did not show the trait, then *both* of the parents must be heterozygotes (carriers). Note that the genotype cannot be determined for some offspring, who may be either carriers or homozygous dominants.

the form of family **pedigrees**, diagrams that show the genetic relationships among a set of related individuals (Fig. 12-14). Careful analysis of pedigrees can reveal whether a particular trait is inherited in a dominant, recessive, or sex-linked pattern. Since the mid-1960s, analysis of human pedigrees, combined with molecular genetic technology, has produced great strides in understanding human genetic diseases. For instance, geneticists now know the genes responsible for dozens of inherited diseases, such as sickle-cell anemia, Marfan syndrome, and cystic fibrosis. Research in molecular genetics has increased our ability to predict genetic diseases and perhaps even to cure them, a topic we explore further in Chapter 13.

12.8 How Are Human Disorders Caused by Single Genes Inherited?

Many common human traits, such as freckles, long eyelashes, cleft chin, and widow's peak hairline, are inherited in a simple Mendelian fashion; that is, each trait appears to be controlled by a single gene with a dominant and a recessive allele. Here, we will concentrate on a few examples of medically important genetic disorders and the ways in which they are transmitted from one generation to the next.

Some Human Genetic Disorders Are Caused by Recessive Alleles

The human body depends on the integrated actions of thousands of enzymes and other proteins. A mutation in an allele of the gene coding for one of these enzymes can impair or destroy its function. However, the presence of one normal allele may generate enough functional enzyme or other protein to enable heterozygotes to be phenotypically indistinguishable from homozygotes with two normal alleles. Therefore, for many genes, a normal allele that encodes a functional protein is dominant to a mutant allele that encodes a nonfunctional protein. Put another way, a mutant allele of these genes is recessive to a normal allele. Thus, an abnormal phenotype occurs only in individuals who inherit two copies of the mutant allele. Cystic fibrosis, which affects about 30,000 Americans, is a recessive disease of this type (see "Scientific Inquiry: Cystic Fibrosis").

Heterozygous individuals are **carriers** of a recessive genetic trait: they are phenotypically dominant but can pass on their recessive allele to their offspring. Geneticists estimate that each of us carries recessive alleles of 5 to 15 genes that would cause serious genetic defects in homozygotes. Every time we have a child, there is a 50:50 chance that we will pass on the defective allele. This is usually harmless, because an unrelated man and woman are unlikely to possess a defective allele in the *same* gene, so they are unlikely to produce a child who is homozygous recessive for a genetic disease. Related couples, however (especially first cousins or closer), have inherited some of their genes from recent common ancestors, and so are more likely to carry a defective allele of the same gene. If they are heterozygous for the *same* defective recessive allele, such couples have a 1 in 4 chance of having a child affected by the genetic disorder (see Fig. 12-14).

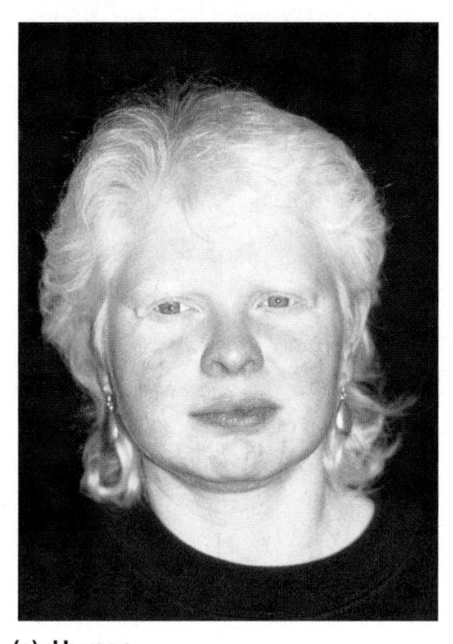

(a) Human

(b) Rattlesnake

(c) Wallaby

FIGURE 12-15 Albinism
Albinism is controlled by a single, recessive allele. Melanin is found throughout the animal kingdom, so albinos of many species have been observed. The female wallaby, having mated with a normally pigmented male, carries a normally colored offspring in her pouch.

Albinism Results from a Defect in Melanin Production

An enzyme called tyrosinase is needed to produce melanin, the dark pigment in skin cells. The gene that encodes tyrosinase is called *TYR*. If an individual is homozygous for a mutant *TYR* allele that encodes a defective tyrosinase enzyme, albinism results (Fig. 12-15). Albinism in humans and other mammals is manifested as white skin and hair and pink eyes (without melanin in the iris, you can see the color of the blood vessels in the retina).

Sickle-Cell Anemia Is Caused by a Defective Allele for Hemoglobin Synthesis

Sickle-cell anemia, a recessive disease in which defective hemoglobin is produced, results from a specific mutation in the hemoglobin gene. The hemoglobin protein, which gives red blood cells their color, transports oxygen in the blood. In sickle-cell anemia, the substitution of one nucleotide results in a single incorrect amino acid at a crucial position in hemoglobin, altering the properties of the hemoglobin molecule. Under conditions of low oxygen (such as in muscles during exercise), masses of hemoglobin molecules in each red blood cell clump together. The clumps force the red blood cell out of its normal disk shape (Fig. 12-16a) into a longer, sickle shape (Fig. 12-16b). The sickled cells are more fragile than normal red blood cells, making them likely to break; they also tend to aggregate, clogging capillaries. Tissues "downstream" of the block do not receive enough oxygen or have their wastes removed. This lack of blood flow can cause

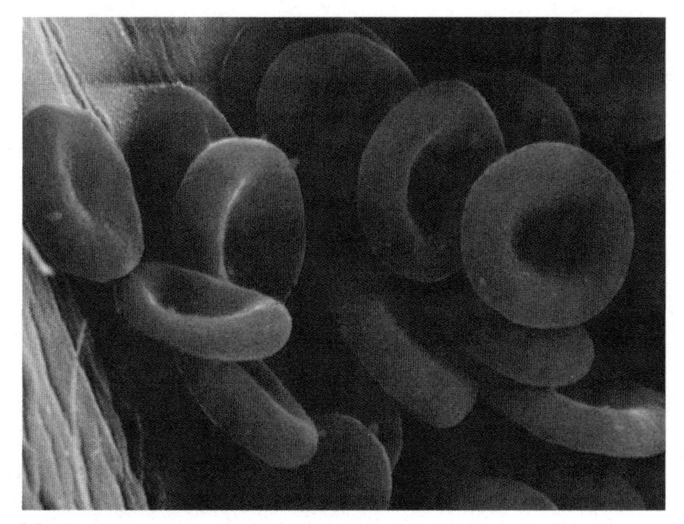

(a)

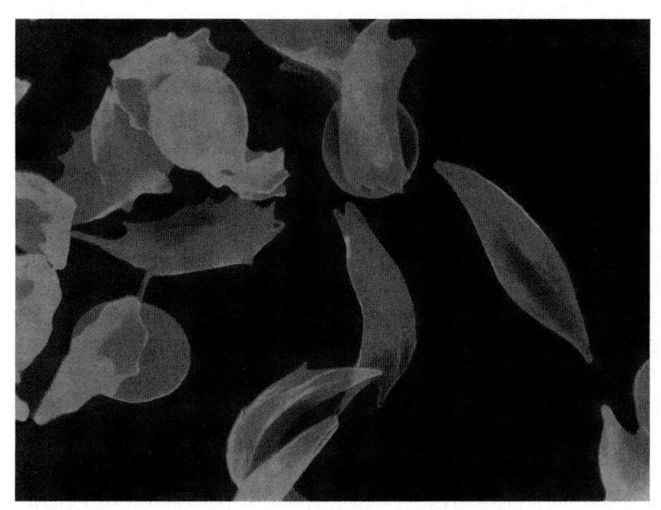

(b)

FIGURE 12-16 Sickle-cell anemia
(a) Normal red blood cells are disc-shaped with indented centers. **(b)** Sickled red blood cells in a person with sickle-cell anemia occur when blood oxygen is low. In this shape they are fragile and tend to clump together, clogging capillaries.

(a)

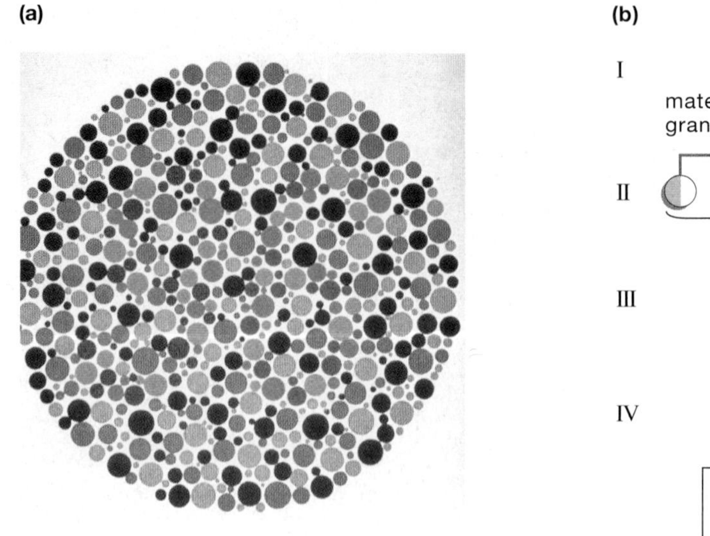

(b)

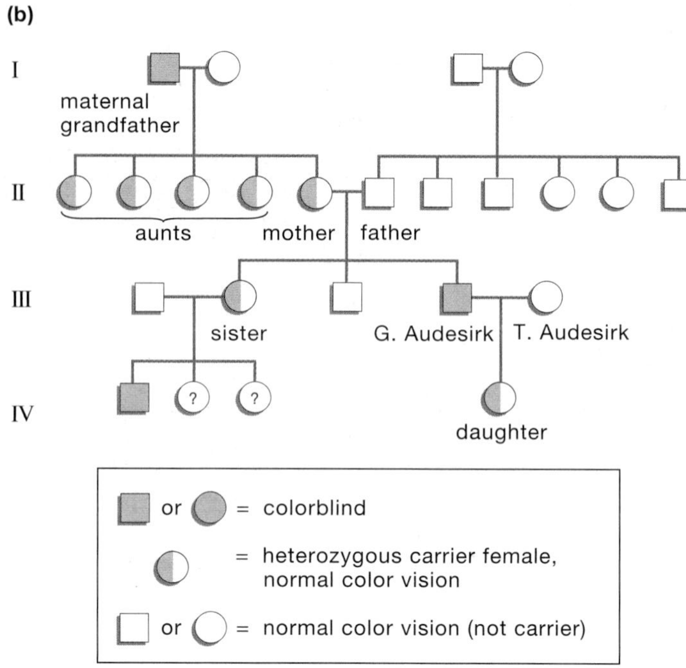

FIGURE 12-17 Color blindness, a sex-linked recessive trait
(a) This figure, called an *Ishihara chart* after its inventor, distinguishes color-vision defects. People with red-deficient vision see a 6, and those with green-deficient vision see a 9. People with normal color vision see 96. **(b)** Pedigree of one of the authors (G. Audesirk), showing sex-linked inheritance of red-green color blindness. Both the author and his maternal grandfather are color deficient; his mother and her four sisters carry the trait but have normal color vision. This patterns of more-common phenotypic expression in males and transmission from affected male to carrier female to affected male are typical of sex-linked recessive traits.

pain, especially in joints. Paralyzing strokes can occur if blocks occur in blood vessels in the brain; the condition can also cause anemia because so many red blood cells are destroyed. Although heterozygotes have about half normal and half abnormal hemoglobin, they usually have few sickled cells and are not disabled by the disease; in fact, many world-class athletes are heterozygotes for the sickle-cell allele.

About 8% of the African-American population is heterozygous for sickle-cell anemia, reflecting a genetic legacy of African origins. In some regions of Africa, 15% to 20% of the population carries the allele. The prevalence of the sickle-cell allele in Africa is explained by the fact that heterozygotes have some resistance to the parasite that causes malaria. We will explore this benefit further in Chapter 15. Individuals of Mediterranean, Middle Eastern, Central or South American, and East Indian ancestry also have an increased risk of carrying the allele responsible for sickle-cell anemia.

Heterozygous carriers of the sickle-cell allele can be detected by a blood test, but until fairly recently there was no way to learn whether a fetus was homozygous or heterozygous. If two heterozygous carriers have children, every conception will have a 1 in 4 chance of producing a child who is homozygous for the sickle-cell allele and consequently will have sickle-cell anemia. Modern DNA techniques can distinguish the normal hemoglobin allele from the sickle-cell allele, and analysis of fetal cells allows medical geneticists to diagnose sickle-cell anemia in fetuses. These methods are described in Chapter 13.

Some Human Genetic Disorders Are Caused by Dominant Alleles

Many normal physical traits, including cleft chin and freckles, are inherited as dominant traits. Many serious genetic diseases are also caused by dominant alleles. For dominant diseases to be passed on to offspring, at least one parent must suffer from the disease; thus, at least some people with dominant diseases must remain healthy enough to grow up and have children. Alternately, the dominant allele may result from a new mutation in a normal individual's eggs or sperm (this frequently occurs in Marfan syndrome—the fibrillin gene is huge, so it suffers mutations fairly often). In this situation, neither parent has the disease.

How can a mutant allele be dominant to the normal allele? Some dominant alleles produce an abnormal protein that interferes with the function of the normal one. For example, some proteins must link together into long chains in order to perform their function in the cell. The abnormal protein may enter into a chain but prevent addition of new protein "links." These short-

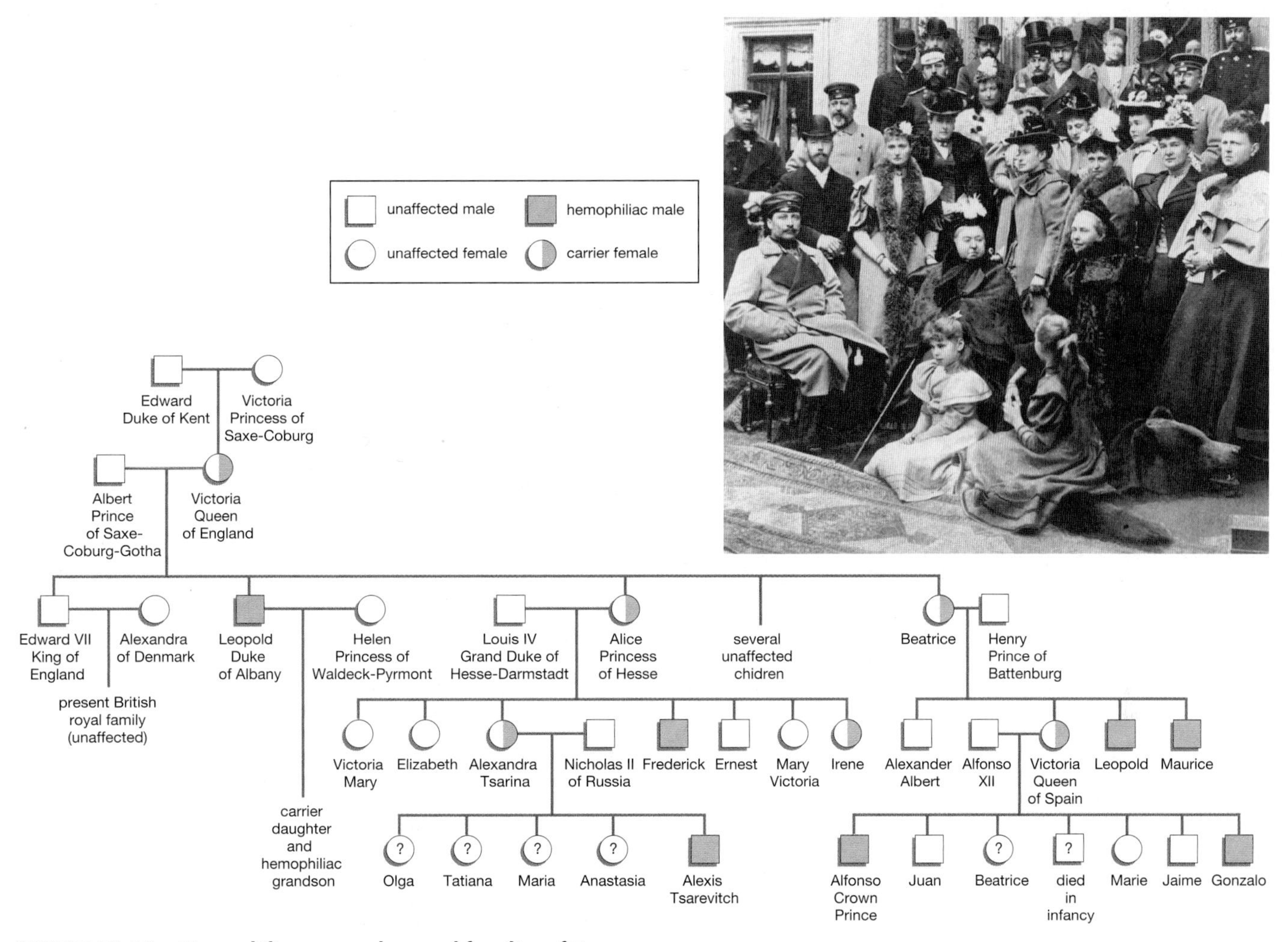

FIGURE 12-18 Hemophilia among the royal families of Europe
A famous genetic pedigree involves the transmission of sex-linked hemophilia from Queen Victoria of England (seated center front, with cane, 1885) to her offspring and eventually to virtually every royal house in Europe. Because Victoria's ancestors were free of hemophilia, the hemophilia allele must have arisen as a mutation either in Victoria herself or in one of her parents (or as a result of marital infidelity). Extensive intermarriage among royalty spread Victoria's hemophilia allele throughout Europe. Her most famous hemophiliac descendant was great-grandson Alexis, Tsarevitch (crown prince) of Russia. The Tsarina Alexandra (Victoria's granddaughter) believed that only the monk Rasputin could control Alexis's bleeding. Rasputin may actually have used hypnosis to cause Alexis to cut off circulation to bleeding areas by muscular contraction. The influence that Rasputin had over the imperial family may have contributed to the downfall of the tsar during the Russian Revolution. In any event, hemophilia was not the cause of Alexis's death; he was killed with the rest of this family by the Bolsheviks (Communists) in 1918.

ened chain fragments may be unable to properly perform a needed function. Other dominant alleles may encode proteins that carry out new, toxic reactions. Finally, dominant alleles may encode a protein that is overactive, performing its function at inappropriate times and places.

Some Human Genetic Disorders Are Sex-Linked

As we described earlier, the X chromosome contains many genes that have no counterpart on the Y chromosome. Because males have only one X chromosome, they have only one allele for each of these genes. This single allele will be expressed with no possibility of its activity being "hidden" by expression of another allele.

A son receives his X chromosome from his mother and passes it only to his daughters. Thus, sex-linked disorders caused by a recessive allele have a unique pattern of inheritance. Such disorders appear far more frequently in males and typically skip generations: an affected male passes the trait on to a phenotypically normal, carrier daughter, who in turn bears affected sons. The most familiar genetic defects due to recessive alleles of X-chromosome genes are red-green color blindness (Fig. 12-17) and **hemophilia** (Fig. 12-18). Hemophilia is caused by a recessive allele on the X chromosome that results in a deficiency in one of the proteins needed for blood clotting. People with hemophilia bruise easily and may bleed extensively from wounds or mild damage to internal structures. Hemophiliacs often have anemia due to blood loss. Nevertheless, even before modern treatment with clotting factors, some hemophiliac males survived to pass on their defective allele to their daughters, who carried the allele and could pass it to their sons.

12.9 How Do Errors in Chromosome Number Affect Humans?

In Chapter 11, we examined the intricate mechanisms of meiosis, which ensure that each sperm and egg receive only one chromosome from each homologous pair. Not surprisingly, this elaborate dance of the chromosomes occasionally misses a step, resulting in gametes that have too many or too few chromosomes. Such errors in meiosis, called **nondisjunction**, can affect the number of either sex chromosomes or autosomes. Most embryos that arise from the fusion of gametes with abnormal chromosome numbers spontaneously abort, accounting for 20% to 50% of all miscarriages, but some embryos with abnormal chromosome number survive to birth or beyond.

Some Genetic Disorders Are Caused by Abnormal Numbers of Sex Chromosomes

Because the X and Y chromosomes pair up during meiosis, sperm usually carry either an X or a Y chromosome. Nondisjunction of sex chromosomes in males produces sperm with 22 autosomes, but either no sex chromosome (often called "O" sperm), or two sex chromosomes (the sperm may be XX, YY, or XY, depending on whether the nondisjunction occurred in meiosis I or II). Nondisjunction of the sex chromosomes in females produces O or XX eggs instead of eggs with one X chromosome. When normal gametes fuse with these defective sperm or eggs, the zygotes have normal numbers of autosomes but abnormal numbers of sex chromosomes (Table 12-2). The most common abnormalities are XO, XXX, XXY, and XYY. (Genes on the X chromosome are essential to survival, so any embryo without at least one X chromosome spontaneously aborts very early in development.)

Turner Syndrome (XO)

About one in every 3000 phenotypically female babies has only one X chromosome, a condition known as **Turner syndrome**. Thus, approximately 48,000 people in the U.S. have Turner syndrome. At puberty, hormone deficiencies prevent XO females from menstruating or developing secondary sexual characteristics, such as enlarged breasts. Treatment with estrogen promotes physical development. However, because most women with Turner syndrome lack mature eggs, hormone treatment does not make it possible for them to bear children. Additional characteristics of women with Turner syndrome include short stature, folds of skin around the neck, and increased risk of cardiovascular disease, kidney defects, and hearing loss. Because women with Turner syndrome have only one X chromosome, they display X-linked recessive disorders, such as hemophilia and color blindness, much more frequently than do XX women.

The differences between XO and XX women suggest that the inactivation of one X chromosome in XX females (see Chapter 10) is not complete. Otherwise, XX women, with only one "active" X chromosome, and XO women, with only one X chromosome, should have

Table 12-2 Effects of Nondisjunction of the Sex Chromosomes During Meiosis

Nondisjunction in Father

Sex Chromosomes of Defective Sperm	Sex Chromosomes of Normal Egg	Sex Chromosomes of Offspring	Phenotype
O (none)	X	XO	Female—Turner syndrome
XX	X	XXX	Female—Trisomy X
YY	X	XYY	Male—XYY male
XY	X	XXY	Male—Klinefelter syndrome

Nondisjunction in Mother

Sex Chromosomes of Normal Sperm	Sex Chromosomes of Defective Egg	Sex Chromosomes of Offspring	Phenotype
X	O (none)	XO	Female—Turner syndrome
Y	O (none)	YO	Dies as embryo
X	XX	XXX	Female—Trisomy X
Y	XX	XXY	Male—Klinefelter syndrome

EARTH WATCH Modern Medicines from Wild Genes

What do new medicines and biodiversity have in common? Each of the 5 to 10 million species on Earth has a different genetic makeup. Their genes code for different proteins; different protein enzymes in turn synthesize different molecules. Each species has evolved adaptations that foster survival and reproduction in a specific environment—forests, soils, hot springs, or the internal organs of other organisms that they infect, to name just a few. An organism's unique genes and adaptations can also benefit humans; the examples below are the contributions of only a few of the millions of species on Earth.

MEDICINES FROM THE WILD

Many medicines are derived from, or were originally discovered in, plants. For example, chemicals from the rosy periwinkle (Fig. E12-2a) are used against leukemia; Taxol® from the Pacific yew is used to treat ovarian cancer; and morphine and related painkillers were originally derived from the opium poppy. You may think that all of the useful plant medicines have already been discovered, or that modern chemists can discover and synthesize in the lab any new medicines that we might need. Far from it. In 1987, John Burley of Harvard's Arnold Arboretum brought back samples of a rainforest tree named *Calophyllum lanigerum* from the Malaysian island of Borneo (Fig. E12-2b). Scientists at the National Cancer Institute discovered that an extract of the fruits and twigs of the tree inhibits replication of the AIDS virus in cell cultures. When Burley returned to Borneo to collect more, he found that the tree he had sampled had been cut down, and that other local *Calophyllum* trees were of a different species with only weak anti-viral activity. Fortunately, the same species of tree was protected in the Singapore Botanic Garden in Malaysia. Researchers were able to isolate and synthesize the active component, a molecule called calanolide. Calanolide inhibits replication of several strains of AIDS virus that are resistant to current drugs, so it might become an important addition to the arsenal of anti-AIDS drugs. Clinical trials in human AIDS patients began in 2002.

Combretastatin, a chemical derived from the bark of *Combretum caffrum*, the South African Cape bushwillow, destroys newly formed capillaries, such as those that nourish tumors. Therefore, it may be able to slow the growth of cancer. Several other plant-derived chemicals are also in clinical trials as anticancer drugs.

Plants aren't the only source of potential "wild medicines." For example, chemicals from a shell-less marine snail, a tunicate (sea squirt), and a sponge are also in clinical trials as potential therapies against cancer. The vast majority of antibacterial drugs come from fungi or other bacteria. Several new medicines derived from fungi are under development in the race to stay ahead of drug-resistant strains of bacteria (see Chapter 15 for more information on the evolution of drug resistance).

WILD PLACES AND WILD GENES

Some potential medicines are from common organisms, while others are from rare plants and animals. Our reserves of wild genes are diminishing rapidly, largely due to the pressures of human population and development. Most organisms on Earth—particularly small organisms such as fungi and bacteria—have never even been scientifically named or described, much less studied. Sadly, we are in danger of losing many species without ever having gotten to know them. With the destruction of wild habitats, countless species and genetic varieties are disappearing rapidly. Preserving the genes of plants, animals, fungi, and microbes is an important, but little recognized, reason to preserve wilderness and save endangered species, for once a species becomes extinct, its genes are lost forever. Genes, with all their various alleles, represent one of our most valuable and irreplaceable natural resources.

(a)

(b)

FIGURE E12-2 Medicinal plants
(a) The rosy periwinkle provides several anticancer drugs.
(b) *Calophyllum lanigerum* is the source of drugs that show great promise for treating AIDS.

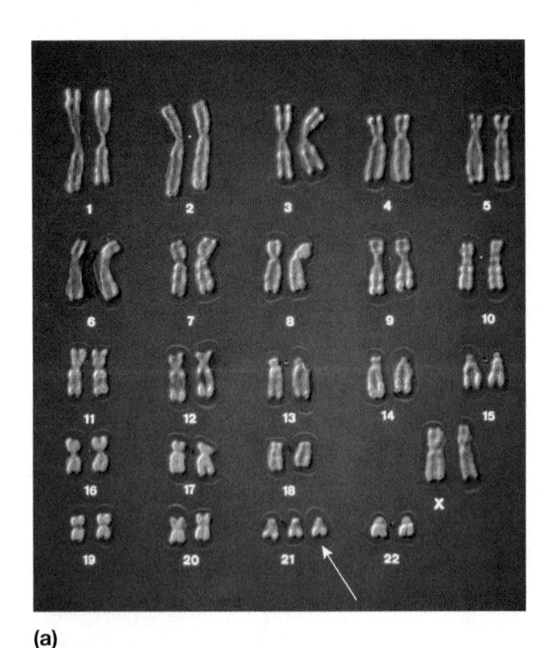

(a)

(b)

FIGURE 12-19 Trisomy 21, or Down syndrome
(a) This karyotype of a Down syndrome child reveals three copies of chromosome 21. **(b)** These girls have the relaxed mouth and distinctively shaped eyes typical of Down syndrome.

identical features. In fact, some two hundred genes on the inactivated X chromosome are functional in XX females, preventing the development of Turner syndrome.

Trisomy X (XXX)

About one in every 1000 women has three X chromosomes. Most such women have no detectable defects, except for a tendency to be tall and a higher incidence of below-normal intelligence. Unlike women with Turner syndrome, most **trisomy X** women are fertile and, interestingly enough, almost always bear normal XX and XY children. Some unknown mechanism must operate during meiosis to prevent an extra X chromosome from being included in their eggs.

Klinefelter Syndrome (XXY)

About one male in every 1000 is born with two X chromosomes and one Y chromosome. Most of these men go through life never realizing that they have an extra X chromosome. However, at puberty, some show mixed secondary sexual characteristics, including partial breast development, broadening of the hips, and small testes. These symptoms are known as **Klinefelter syndrome**. XXY men are usually infertile because of low sperm count, but are not impotent. They are usually diagnosed when the man and his partner seek medical help when they are unable to conceive a baby.

XYY Males

Another common type of sex chromosome abnormality is XYY, occurring in about one male in every 1000. You might expect that an extra Y chromosome, which has few active genes, would not make very much difference, and this seems to be true in most cases. However, XYY males usually have high levels of testosterone, often have severe acne, and are tall (about two-thirds of XYY males are over 6 feet tall, compared with the average male height of 5 feet 9 inches). Some appear to have slightly lower scores on IQ tests than their XY brothers. Debates continue about whether XYY males are genetically predisposed to violence. One estimate is that an XYY individual has a 24 times higher risk of behavioral problems or criminal activity. However, most of these studies involved only a few XYY individuals, and many used questionable statistics. In several countries, men accused of murder have attempted to use their XYY makeup as a defense, much like the insanity plea. They were not acquitted; only a tiny percentage of XYY males ever commits any sort of crime, so an extra Y chromosome certainly doesn't enforce a life of violence.

Some Genetic Disorders Are Caused by Abnormal Numbers of Autosomes

Nondisjunction of the autosomes may also occur, producing eggs or sperm missing an autosome or with two copies of an autosome. Fusion with a normal gamete (bearing one copy of each autosome) leads to an embryo with either one or three copies of the affected autosome. Embryos that have only one copy of any of the autosomes abort so early in development that the woman never knows she was pregnant. Embryos with three copies of an autosome (trisomy) also usually spontaneously abort. However, a small fraction of embryos with three copies of chromosomes 13, 18, or 21 survive to birth. In the case of trisomy 21, the child may live into adulthood.

Trisomy 21 (Down Syndrome)

In about 1 of every 900 births, the child inherits an extra copy of chromosome 21, a condition called **trisomy 21** or **Down syndrome**. Children with Down syndrome have several distinctive physical characteristics, including weak muscle tone, a small mouth held partially open because it cannot accommodate the tongue, and distinctively shaped eyelids (Fig. 12-19). Much more serious defects include low resistance to infectious diseases, heart malformations, and varying degrees of mental retardation, often severe.

The frequency of nondisjunction increases with the age of the parents, especially the mother (Fig. 12-20). Nondisjunction in sperm accounts for about a quarter of thc cases of Down syndrome, and there is a small increase in these defective sperm with increasing age of the father. Since the 1970s, it has become more common for couples to delay having children, increasing the probability of trisomy 21. Trisomy can be diagnosed before birth by examining the chromosomes of fetal cells (see "Health Watch: Prenatal Genetic Screening" in Chapter 13).

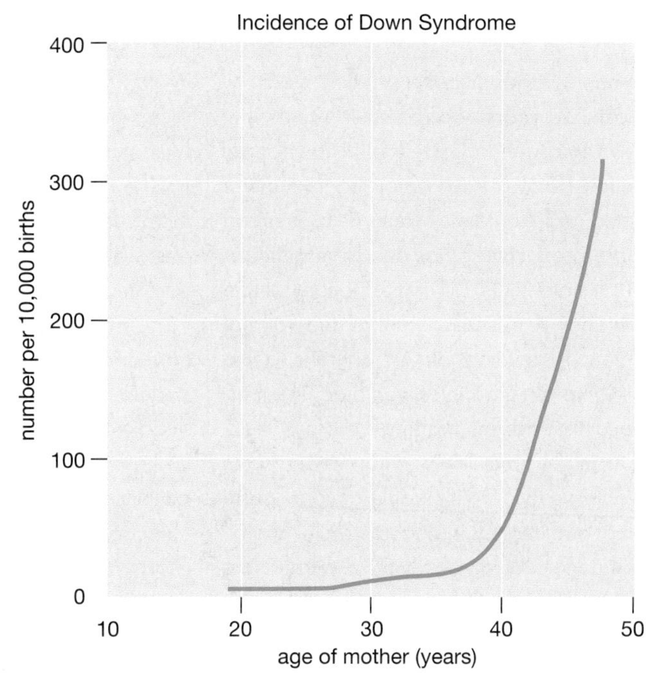

FIGURE 12-20 Down syndrome frequency increases with maternal age
The increase in frequency of Down syndrome after maternal age 35 is quite dramatic.

CASE STUDY REVISITED Marfan Syndrome

Recall that one defective copy of the fibrillin gene causes Marfan syndrome, and that the children of a person with Marfan syndrome have a 50% chance of inheriting the disease. You may recognize this pattern of inheritance—the allele that causes Marfan syndrome is dominant to the normal allele. People with two normal fibrillin alleles will produce only normal fibrillin protein, while people who inherit one mutated fibrillin allele and one normal allele will produce both normal and defective fibrillin. The mutant fibrillin will stimulate extra bone growth and weaken connective tissue, causing the typical symptoms of Marfan syndrome.

If you think carefully about this, you may conclude that Marfan syndrome should actually be inherited as an incompletely dominant allele. After all, if having half normal and half mutant fibrillin weakens connective tissue, shouldn't having nothing but mutant fibrillin be even worse? You're probably right. However, for all practical purposes, everyone with Marfan syndrome is heterozygous. Why? To produce a child homozygous for defective fibrillin, two heterozygotes must mate and have a child. The incidence of Marfan syndrome is about 1 in 5000, so the chances that two people with Marfan syndrome will mate is $1/5000 \times 1/5000$, or 1 in 25 million. Even then, only one child in four would be homozygous for defective fibrillin. It is also likely that an embryo homozygous for defective fibrillin will die very early in development. Therefore, we may never know if Marfan syndrome is inherited as a simple dominant or as an incompletely dominant allele.

Finally, you may wonder how anyone can be certain that Lincoln and Akhenaten had Marfan syndrome. You're right again. Nobody really knows; the "diagnosis" is based on pictures and descriptions. For example, Akhenaten was the only pharaoh to be depicted with long head, arms, and legs, all typical of Marfan syndrome. One visitor to the White House is said to have remarked to Lincoln, "Mr. President, what long legs you have!" Lincoln reportedly replied, "Just long enough to reach the ground, Madam." Although other genetic conditions can contribute to tall stature, the fact that these men were otherwise healthy, successful, and fertile points to Marfan syndrome.

Consider This: Marfan syndrome can't be detected in an embryo by a simple biochemical test (yet). Most cystic fibrosis mutations, however, can easily be detected, in both carriers and homozygotes, in adults, children, and embryos. A few years ago, some states considered mandatory testing of couples for cystic fibrosis as part of the application for a marriage license. If two carriers marry, each of their children has a 25% chance of having the disorder. While there is no cure, there will probably be better treatments within a few years. Do you think that carrier screening should be mandatory? If you and your spouse were both carriers, would you seek prenatal diagnosis of an embryo? What would you do if your embryo were destined to be born with cystic fibrosis?

Links to Life: Mendel, Mosquitoes, and Malaria

When Gregor Mendel studied peas almost 150 years ago, he couldn't possibly have foreseen today's applications of genetics. He may have predicted that understanding inheritance would help us breed more productive crops and farm animals. But what would he have thought of the October 4, 2002, issue of the journal *Science*, devoted almost entirely to the genome of *Anopheles gambiae*, the mosquito that carries malaria? His first reaction might have been, and maybe yours would be too, "Why would anyone waste time and money doing *that*?" In fact, if someone could genetically manipulate mosquitoes so that the malaria parasite couldn't reproduce in them, and could replace wild mosquitoes with resistant ones, billions of people might be ecstatic about it. The World Health Organization estimates that there are about 300 million cases of malaria each year.

A more immediate possibility is the development of better mosquito repellents. For some mosquitoes, there's nothing like a dame—or a guy. Put a human in the middle of a herd of cattle and release one of these mosquitoes, and they'll ignore the cattle and home right in on the person. How does the mosquito do that? By scent. Mosquitoes detect odors with proteins in the plasma membranes of cells on their antennae. And where do these proteins come from? Right—from instructions in genes. Researchers will use the mosquito genome to find the odor-sensing genes, use the nucleotide sequence to determine the amino acid sequence of the proteins, and use the amino acid sequence to determine the proteins' structures. Then they will try to design effective, nontoxic repellents (so the mosquitoes can't find people) and attractants (to lure the mosquitoes into traps).

If this ever comes to pass, it will all have started with a monk in Moravia.

Summary of Key Concepts

12.1 How Did Gregor Mendel Lay the Foundation for Modern Genetics?

Homologous chromosomes carry the same genes located at the same loci, but the genes at a particular locus can exist in alternate forms called *alleles*. An organism whose homologous chromosomes carry the same allele at a locus is homozygous for that particular gene. If the alleles at a locus differ, the organism is heterozygous for that gene. Gregor Mendel deduced many principles of inheritance in the mid-1800s, before the discovery of DNA, genes, chromosomes, or meiosis. He did this by choosing an appropriate experimental subject, designing his experiments carefully, following progeny for several generations, and analyzing his data statistically.

12.2 How Are Single Traits Inherited?

A trait is an observable or measurable feature of the organism's phenotype, such as eye color or blood type. Traits are inherited in particular patterns that depend on the types of alleles that parents pass on to their offspring. Each parent provides its offspring with one copy of every gene, so that the offspring inherits a pair of alleles for every gene. The combination of alleles in the offspring determines whether it displays a particular trait. Dominant alleles mask the expression of recessive alleles. The masking of recessive alleles can result in organisms with the same phenotype but different genotypes. That is, organisms with two dominant alleles (homozygous dominant) have the same phenotype as do organisms with one dominant and one recessive allele (heterozygous). Because each allele segregates randomly during meiosis, we can use the laws of probability to predict the relative proportions of offspring with a particular trait.

12.3 How Are Multiple Traits on Different Chromosomes Inherited?

If the genes for two traits are located on separate chromosomes, they will be assorted independently of one another into the egg or sperm. Thus, crossing two organisms that are heterozygous at two loci on separate chromosomes produces offspring with 10 different genotypes. If the alleles are typical dominant and recessive alleles, these progeny will display only 4 different phenotypes.

12.4 How Are Genes Located on the Same Chromosome Inherited?

Genes located on the same chromosome are linked to one another (encoded on the same DNA double helix) and they tend to be inherited together. Unless the alleles are separated by chromosomal recombination, the two alleles will be passed together to the offspring.

12.5 How Is Sex Determined, and How Are Sex-Linked Genes Inherited?

In many animals, sex is determined by sex chromosomes, often designated X and Y. The rest of the chromosomes, identical in the two sexes, are called *autosomes*. In many animals, females have two X chromosomes, whereas males have one X and one Y chromosome. The Y chromosome has many fewer genes than the X chromosome. Because males have only one copy of most X chromosome genes, recessive traits on the X chromosome are more likely to be phenotypically expressed in males.

12.6 Do the Mendelian Rules of Inheritance Apply to All Traits?

Not all inheritance follows the simple dominant–recessive pattern:

- In incomplete dominance, heterozygotes have a phenotype that is intermediate between the two homozygous phenotypes.
- Codominance occurs when two types of proteins, each encoded by a different allele at a single locus, both contribute to the phenotype.

- Many traits are determined by several different genes at different loci that contribute to the phenotype, a phenomenon called *polygenic inheritance*.
- Many genes have multiple effects on the organism's phenotype (pleiotropy).
- The environment influences the phenotypic expression of most, if not all, traits.

12.7 How Are Human Genetic Disorders Investigated?
The genetics of humans is similar to the genetics of other animals, except that experimental crosses are not feasible. Analysis of family pedigrees and, more recently, molecular genetic techniques must be used to determine the mode of inheritance of human traits.

12.8 How Are Human Disorders Caused by Single Genes Inherited?
Some genetic disorders are inherited as recessive traits; therefore, only homozygous recessive persons show symptoms of the disease. Heterozygotes are called *carriers*; they carry the recessive allele but do not express the trait. Some other diseases are inherited as simple dominant traits. In such cases, only one copy of the dominant allele is needed to cause the disease symptoms. The human Y chromosome bears few genes other than those that determine male reproductive attributes; therefore, men phenotypically display whichever allele they carry on their single X chromosome, a phenomenon called *sex-linked inheritance*.

12.9 How Do Errors in Chromosome Number Affect Humans?
Errors in meiosis can result in gametes with abnormal numbers of sex chromosomes or autosomes. Many people with abnormal numbers of sex chromosomes have distinguishing physical characteristics. Abnormal numbers of autosomes typically lead to spontaneous abortion early in pregnancy. In rare instances, the fetus may survive to birth, but severe mental and physical deficiencies always occur, as is the case with Down syndrome (trisomy 21). The likelihood of abnormal numbers of chromosomes increases with increasing age of the mother, and, to a lesser extent, the father.

Key Terms

allele *p. 210*
autosome *p. 219*
carrier *p. 226*
codominance *p. 222*
cross-fertilization *p. 211*
crossing over *p. 218*
dominant *p. 212*
Down syndrome *p. 233*
gene *p. 210*
genetic recombination *p. 218*
genotype *p. 213*
hemophilia *p. 229*
heterozygous *p. 210*
homozygous *p. 210*
hybrid *p. 210*
incomplete dominance *p. 221*
inheritance *p. 215*
Klinefelter syndrome *p. 232*
law of independent assortment *p. 216*
law of segregation *p. 212*
linkage *p. 217*
locus *p. 210*
multiple alleles *p. 222*
nondisjunction *p. 230*
pedigree *p. 226*
phenotype *p. 213*
pleiotropy *p. 225*
polygenic inheritance *p. 224*
Punnett square method *p. 213*
recessive *p. 212*
self-fertilization *p. 211*
sex chromosome *p. 218*
sex-linked *p. 219*
sickle-cell anemia *p. 227*
test cross *p. 215*
trisomy 21 *p. 233*
trisomy X *p. 232*
true-breeding *p. 211*
Turner syndrome *p. 230*

Thinking Through the Concepts

To take a multiple-choice quiz with feedback on the contents of this chapter, visit http://www.prenhall.com/audesirk7. *Log in to the Web site selected by your instructor and navigate to the Self Test section for this chapter.*

? Review Questions

1. Define the following terms: *gene, allele, dominant, recessive, true-breeding, homozygous, heterozygous, cross-fertilization*, and *self-fertilization*.
2. Explain why genes located on the same chromosome are said to be linked. Why do alleles of linked genes sometimes separate during meiosis?
3. Define *polygenic inheritance*. Why does polygenic inheritance sometimes allow parents to produce offspring that are notably different in eye or skin color than either parent?
4. What is sex linkage? In mammals, which sex would be most likely to show recessive sex-linked traits?
5. What is the difference between a phenotype and a genotype? Does knowledge of an organism's phenotype always allow you to determine the genotype? What type of experiment would you perform to determine the genotype of a phenotypically dominant individual?
6. In the pedigree of part (a) of Fig. 12-14, do you think that the individuals showing the trait are homozygous or heterozygous? How can you tell from the pedigree? In part (b), what information would you need to determine if the "?" individuals are carriers or homozygous dominants?
7. Define *nondisjunction*, and describe the common syndromes caused by nondisjunction of sex chromosomes and autosomes.

Applying the Concepts

1. Sometimes the term *gene* is used rather casually. Compare and contrast use of the terms *allele* and *locus* as alternatives to *gene.*
2. Using the information in the chapter, explain why AB individuals are referred to as "universal recipients" in terms of blood transfusions and why people with type O blood are called "universal donors."
3. Mendel's numbers seemed almost too perfect to be real; some believe he may have cheated a bit on his data. Perhaps he continued to collect data until the numbers matched his predicted ratios, then stopped. Recently, there has been much publicity over violations of scientific ethics, including researchers' plagiarizing others' work, using other scientists' methods to develop lucrative patents, or just plain fabricating data. How important an issue is this for society? What are the boundaries of ethical scientific behavior? How should the scientific community or society "police" scientists? What punishments would be appropriate for violations of scientific ethics?
4. Although American society has been described as a "melting pot," people often engage in "assortative mating," in which they marry others of similar height, socioeconomic status, race, and IQ. Discuss the consequences to society of assortative mating among humans. Would society be better off if people mated more randomly? Explain.
5. *Eugenics* is the term applied to the notion that the human condition might be improved by improving the human genome. Do you think there are both good and bad sides to eugenics? What examples can you think of to back up your stand? What would a eugenicist think of the medical advances that have ameliorated the problems of hemophilia?

Genetics Problems

(Note: An extensive group of genetics problems, with answers, can be found in the Study Guide.)

1. In certain cattle, hair color can be red (homozygous RR), white (homozygous $R'R'$), or roan (a mixture of red and white hairs, heterozygous RR').
   a. When a red bull is mated to a white cow, what genotypes and phenotypes of offspring could be obtained?
   b. If one of the offspring in (a) were mated to a white cow, what genotypes and phenotypes of offspring could be produced? In what proportion?
2. The palomino horse is golden in color. Unfortunately for horse fanciers, palominos do not breed true. In a series of matings between palominos, the following offspring were obtained:

   65 palominos, 32 cream-colored,
   34 chestnut (reddish brown)

   What is the probable mode of inheritance of palomino coloration?
3. In the edible pea, tall (T) is dominant to short (t), and green pods (G) are dominant to yellow pods (g). List the types of gametes and offspring that would be produced in the following crosses:
   a. $TtGg \times TtGg$
   b. $TtGg \times TTGG$
   c. $TtGg \times Ttgg$
4. In tomatoes, round fruit (R) is dominant to long fruit (r), and smooth skin (S) is dominant to fuzzy skin (s). A true-breeding round, smooth tomato ($RRSS$) was cross-bred with a true-breeding long, fuzzy tomato ($rrss$). All the F_1 offspring were round and smooth ($RrSs$). When these F_1 plants were bred, the following F_2 generation was obtained:

   Round, smooth: 43 Long, fuzzy: 13

   Are the genes for skin texture and fruit shape likely to be on the same chromosome or on different chromosomes? Explain your answer.
5. In the tomatoes of problem 4, an F_1 offspring ($RrSs$) was mated with a homozygous recessive ($rrss$). The following offspring were obtained:

   Round, smooth: 583 Round, fuzzy: 21
   Long, fuzzy: 602 Long, smooth: 16

   What is the most likely explanation for this distribution of phenotypes?
6. In humans, hair color is controlled by two interacting genes. The same pigment, melanin, is present in both brown-haired and blond-haired people, but brown hair has much more of it. Brown hair (B) is dominant to blond (b). Whether any melanin can be synthesized depends on another gene. The dominant form (M) allows melanin synthesis; the recessive form (m) prevents melanin synthesis. Homozygous recessives (mm) are albino. What will be the expected proportions of phenotypes in the children of the following parents?
   a. $BBMM \times BbMm$
   b. $BbMm \times BbMm$
   c. $BbMm \times bbmm$
7. In humans, one of the genes determining color vision is located on the X chromosome. The dominant form (C) produces normal color vision; red-green color blindness (c) is recessive. If a man with normal color vision marries a color-blind woman, what is the probability of their having a color-blind son? A color-blind daughter?
8. In the couple described in problem 7, the woman gives birth to a color-blind but otherwise normal daughter. The husband sues for a divorce on the grounds of adultery. Will his case stand up in court? Explain your answer.

Answers to Genetics Problems

1. a. A red bull (RR) is mated to a white cow ($R' R'$). The bull will produce all R sperm; the cow will produce all R' eggs. All the offspring will be RR' and will have roan hair (codominance).

   b. A roan bull (RR') is mated to a white cow ($R' R'$). The bull produces half R and half R' sperm; the cow produces R' eggs. Using the Punnett square method:

sperm \ eggs	R'
R	$R R'$
R'	$R' R'$

Using probabilities:

sperm	egg	offspring
$\frac{1}{2}R$	R'	$\frac{1}{2}RR'$
$\frac{1}{2}R'$	R'	$\frac{1}{2}R'R'$

The predicted offspring will be $\frac{1}{2}RR'$ (roan) and $\frac{1}{2}R'R'$ (white).

2. The offspring occur in three types, classifiable as dark (chestnut), light (cream), and intermediate (palomino). This distribution suggests incomplete dominance, with the alleles for chestnut (C) combining with the allele for cream (C') to produce palomino heterozygotes (CC'). We can test this hypothesis by examining the offspring numbers. There are approximately $\frac{1}{4}$ chestnut (CC), $\frac{1}{2}$ palomino (CC'), and $\frac{1}{4}$ cream ($C' C'$).
   If palominos are heterozygotes, we would expect the cross $CC' \times CC'$ to yield $\frac{1}{4}CC$, $\frac{1}{2}CC'$, and $\frac{1}{4}C'C'$. Our hypothesis is supported.

3. a. $TtGg \times TtGg$. This is a "standard" cross for differences in two traits. Both parents produce TG, Tg, tG, and tg gametes. The expected proportions of offspring are $\frac{9}{16}$ tall green, $\frac{3}{16}$ tall yellow, $\frac{3}{16}$ short green, $\frac{1}{16}$ short yellow.

   b. $TtGg \times TTGG$. In this cross, the heterozygous parent produces TG, Tg, tG, and tg gametes. However, the homozygous dominant parent can produce only TG gametes. Therefore, all offspring will receive at least one T allele for tallness and one G allele for green pods, and thus all the offspring will be tall with green pods.

   c. $TtGg \times Ttgg$. The second parent will produce two types of gametes, Tg and tg. Using a Punnett square:

sperm \ eggs	Tg	tg
TG	$TTGg$	$TtGg$
Tg	$TTgg$	$Ttgg$
tG	$TtGg$	$ttGg$
tg	$Ttgg$	$ttgg$

The expected proportions of offspring are $\frac{3}{8}$ tall green, $\frac{3}{8}$ tall yellow, $\frac{1}{8}$ short green, $\frac{1}{8}$ short yellow.

4. If the genes are on separate chromosomes—that is, assort independently—then this would be a typical two-trait cross with expected offspring of all four types (about $\frac{9}{16}$ round smooth, $\frac{3}{16}$ round fuzzy, $\frac{3}{16}$ long smooth, and $\frac{1}{16}$ long fuzzy). However, only the parental combinations show up in the F_2 offspring, indicating that the genes are on the same chromosome.

5. The genes are on the same chromosome and are quite close together. On rare occasions, crossing over occurs between the two genes, producing recombination of the alleles.

6. a. $BBMM$ (brown) × $BbMm$ (brown). The first parent can produce only BM gametes, so all offspring will receive at least one dominant allele for each gene. Therefore, all offspring will have brown hair.

   b. $BbMm$ (brown) × $BbMm$ (brown). Both parents can produce four types of gametes: BM, Bm, bM, and bm. Filling in the Punnett square:

sperm \ eggs	BM	Bm	bM	bm
BM	$BBMM$	$BBMm$	$BbMM$	$BbMm$
Bm	$BBMm$	$BBmm$	$BbMm$	$Bbmm$
bM	$BbMM$	$BbMm$	$bbMM$	$bbMm$
bm	$BbMm$	$Bbmm$	$bbMm$	$bbmm$

All mm offspring are albino, so we get the expected proportions $\frac{9}{16}$ brown-haired, $\frac{3}{16}$ blond-haired, $\frac{4}{16}$ albino.

   c. $BbMm$ (brown) × $bbmm$ (albino):

sperm \ eggs	bm
BM	$BbMm$
Bm	$Bbmm$
bM	$bbMm$
bm	$bbmm$

The expected proportions of offspring are $\frac{1}{4}$ brown-haired, $\frac{1}{4}$ blond-haired, $\frac{1}{2}$ albino.

7. A man with normal color vision is CY (remember, the Y chromosome does not have the gene for color vision). His color-blind wife is cc. Their expected offspring will be:

sperm \ eggs	c
C	Cc
Y	cY

We therefore expect that all the daughters will have normal color vision and all the sons will be color-blind.

8. The husband should win his case. All his daughters must receive one X chromosome, with the C allele, from him and therefore should have normal color vision. If his wife gives birth to a color-blind daughter, her husband cannot be the father (unless there was a new mutation for color blindness in his sperm line, which is very unlikely).

For More Information

Kahn, P. "Gene Hunters Close in on Elusive Prey." *Science*, March 8, 1996. Using pedigrees and biotechnology, researchers are uncovering the causes of diseases that involve complex interactions between genes and the environment.

McGue, M. "The Democracy of the Genes." *Nature*, July 1997. The environment has a more important role in the development of intelligence than previously thought.

National Institutes of Health. "Facts About Cystic Fibrosis." November 1995. On the World Wide Web at www.nhlbi.nih.gov/health/public/lung/other/cf.htm.

National Institutes of Health. "Questions and Answers About Marfan Syndrome." October 2001. On the World Wide Web at www.niams.nih.gov/hi/topics/marfan/marfan.htm.

O'Brien, S. J. "A Role for Molecular Genetics in Biological Conservation." *Proceedings of the National Academy of Sciences*, June 1994, Vol. 91. Habitat destruction has reduced many wild populations to tiny remnants of their former size. Genetic analysis is used to determine how much loss of genetic diversity has occurred due to inbreeding.

Sapienza, C. "Parental Imprinting of Genes." *Scientific American*, October 1990. It is not quite true that all genes are equal, regardless of whether they have been inherited from mother or father. In some cases, which parent a gene comes from greatly alters its expression in the offspring.

Stern, C., and Sherwood, E. R. *The Origin of Genetics: A Mendel Source Book*. San Francisco: W. H. Freeman, 1966. There is no substitute for the real thing, in this case a translation of Mendel's original paper to the Brünn Society.

Wivel, N. A., and Walters, L. "Germ-line Gene Modification and Disease Prevention: Some Medical and Ethical Perspectives." *Science* 1993. If we can fix a defective allele in an egg or zygote, should we?

Media Activities

To access a Media Activity visit http://www.prenhall.com/audesirk7. *Log in to the Web site selected by your instructor, navigate to this chapter, and select the appropriate Media Activity number.*

12.1 Monohybrid Crosses

Estimated time: 10 minutes

This tutorial will recreate one of Mendel's early experiments and show that a Punnett square can be used to determine how probability figures into genetic analysis.

12.2 Dihybrid Crosses

Estimated time: 10 minutes

In this tutorial, examine a set of crosses, focusing on two separate genetic traits. See how probability can help explain this situation also.

12.3 Web Investigation: Marfan Syndrome

Estimated time: 10 minutes

Marfan syndrome, a genetic disorder, afflicts one in every 3000 to 5000 people of all races and ethnicity. Marfan syndrome is one of many genetic disorders associated with a specific gene product, and the cellular and biochemical mechanisms underlying its many effects are relatively well understood. Explore Marfan syndrome in more detail in this investigation.

CHAPTER

13 Biotechnology

After spending many years in prison, Eddie Lloyd (left) and Jimmy Ray Bromgard (right) were proven innocent through the use of DNA fingerprinting.

AT A GLANCE

CASE STUDY Guilty or Innocent?

In 1984, a 16-year-old girl was brutally murdered in Detroit. Later, Eddie Lloyd, a patient in a mental hospital, wrote to the police, offering suggestions about how to solve the murder. The police investigated Lloyd's possible involvement, and, under police interrogation, he confessed to the crime. Because Lloyd would not permit his attorney to plead insanity, the only thing that saved him from execution was that Michigan had repealed the death penalty.

In 1987, an 8-year-old girl was repeatedly raped by an unknown intruder in her own bedroom in Montana. A composite sketch made by a forensic artist led police to Jimmy Ray Bromgard. The little girl said that Bromgard looked like her assailant, but she wasn't sure. Hairs left at the scene of the crime were similar to Bromgard's, and an expert witness testified that there was less than one chance in 10,000 that the hairs were not Bromgard's. Jimmy Ray Bromgard was convicted and sentenced to 40 years in prison.

These cases may appear to be flawless examples of justice in action. However, both Bromgard and Lloyd were innocent, the victims of shoddy investigation, inadequate defense attorneys, expert witnesses who were anything but, and, in Lloyd's case, possible police misconduct. After 15 and 17 years in prison, respectively, Bromgard and Lloyd are now free men, thanks to the professors and students of the Innocence Project at the Benjamin Cardozo School of Law at Yeshiva University and the science of biotechnology.

You may have already guessed how the Innocence Project was able to prove Bromgard and Lloyd's innocence: DNA evidence. In this chapter, we'll investigate techniques of biotechnology that now pervade so much of modern life, including forensics in the courtroom, prenatal diagnosis and treatment of inherited disorders, and genetically modified crops and livestock.

13.1 What Is Biotechnology?

If you look up the term **biotechnology** in the dictionary, you will find that the definition includes almost any kind of applied biology, with a modern emphasis on genetic engineering, recombinant DNA technology, and associated techniques for analyzing biological molecules, especially DNA and protein. Some traditional applications of biotechnology are thousands of years old. Prehistoric civilizations in Egypt and the Near East used yeast to produce beer and wine over 10,000 years ago. People have selectively bred plants and animals for just as long. Squash fragments 8000 to 10,000 years old, preserved in a dry cave in Mexico, have larger seeds and thicker rinds than those of wild squash, providing evidence of selective breeding by humans. Prehistoric art and animal remains suggest that the domestication and selective breeding of dogs, sheep, goats, pigs, and camels also began about 10,000 years ago.

Selective breeding continues to be an important tool in biotechnology. However, modern biotechnology also commonly uses **genetic engineering**, the modification of genetic material to achieve specific goals. Genetically engineered cells or organisms may have genes deleted, added, or changed. Major goals of genetic engineering include:

- Learning more about cellular processes, including inheritance and gene expression
- Providing better understanding and treatment of diseases, particularly genetic disorders
- Generating economic and social benefits, including efficient production of valuable biological molecules and improved plants and animals for agriculture

A key tool in genetic engineering is **recombinant DNA**. Recombinant DNA contains genes or portions of genes from different organisms, often from different species. Large amounts of recombinant DNA can be grown in bacteria, viruses, or yeast and then transferred into other species, including animals and plants. These plants and animals, which express DNA that has been modified or derived from another species, are called **transgenic**, or sometimes **genetically modified organisms (GMOs)**. Since its development in the 1970s, recombinant DNA technology has grown explosively, providing new methods, applications, and possibilities for genetic engineering. Today, most research labs involved in analysis of cell structure, genetics, the molecular basis of disease, and evolution routinely use recombinant DNA technology in their experiments. Many genetically engineered products are preferred over those that were previously available, including human insulin and enzymes used for making cheese.

Modern biotechnology also includes many methods of manipulating DNA, whether or not the DNA is subsequently put into a cell or an organism. For example, determining the nucleotide sequence of specific pieces of DNA is crucial to forensic science and the diagnosis of inherited disorders.

In this chapter, we will examine modern biotechnology, using four major themes to explain both the methods themselves and their applications: recombinant DNA mechanisms found in nature, mostly in bacteria and viruses; biotechnology in criminal forensics, principally DNA matching; biotechnology in agriculture, specifically the production of transgenic plants and animals; and biotechnology in medicine, focusing on the diagnosis and treatment of inherited disorders.

13.2 How Does DNA Recombination Occur in Nature?

Most of us tend to think that a species' genetic makeup is relatively stable: a given species has a certain number of genes, perhaps with several alleles of each gene. Occasionally, new alleles may arise through new mutations. Crossing over, separation of homologous chromosomes, and fusion of sperm and egg during sexual reproduction shuffle alleles around, but that's about it. However, many natural processes can transfer DNA from one organism to another, sometimes even organisms of different species. Many recombinant DNA technologies used in the laboratory are based on these naturally occurring mechanisms of DNA recombination.

Transformation May Combine DNA from Different Bacterial Species

Bacteria can undergo several types of recombination (Fig. 13-1). **Transformation** enables bacteria to pick up DNA from the environment (Fig. 13-1b). The DNA may be part of the chromosome from another bacterium, even from another species. You may recall from Chapter 9 that living, nonvirulent pneumonia bacteria can pick up genes from heat-killed virulent bacteria, allowing the formerly harmless bacteria to cause pneumonia (see Figure 9-1). Unraveling the mechanism of bacterial transformation was an important step leading to the discovery that DNA is the genetic material.

Transformation may also occur when bacteria pick up tiny circular DNA molecules called **plasmids** (Fig. 13-1c). Many types of bacteria contain plasmids ranging in size from about 1000 to 100,000 nucleotides long. For comparison, the *E. coli* chromosome is around 4,600,000 nu-

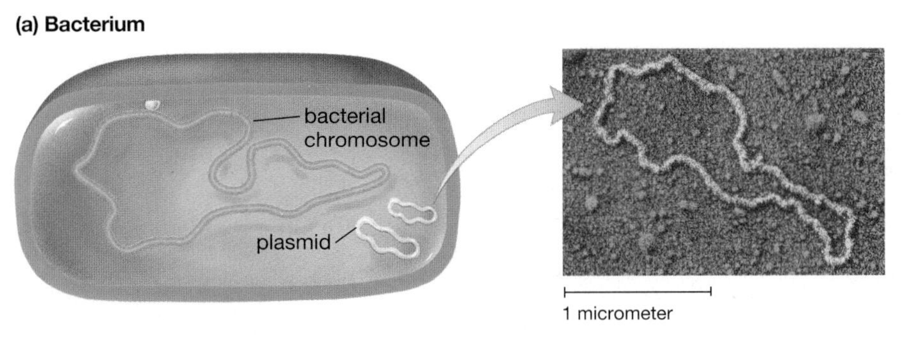

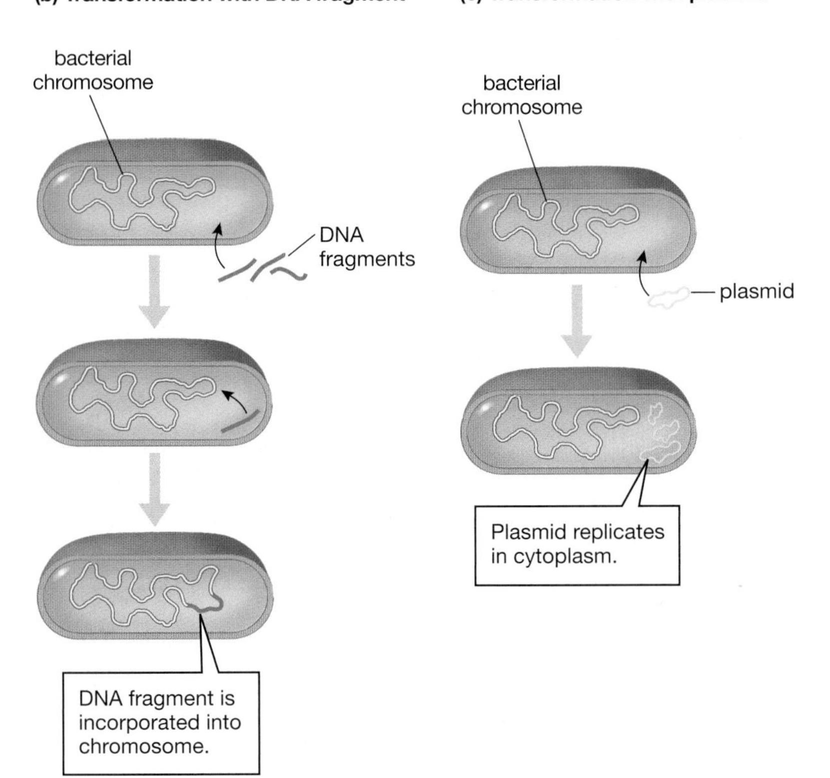

FIGURE 13-1 Recombination in bacteria **(a)** In addition to their large circular chromosome, bacteria commonly possess small rings of DNA called plasmids, which often carry additional useful genes. Bacterial transformation occurs when living bacteria take up **(b)** fragments of chromosomes or **(c)** plasmids.

Media Activity
13.1 Genetic Recombination in Bacteria

cleotides long. A single bacterium may contain dozens or even hundreds of copies of a plasmid. When the bacterium dies, it releases these plasmids into the environment, where they can be picked up by other bacteria of the same or different species. In addition, living bacteria can often pass plasmids directly to other living bacteria. Passing plasmids from bacteria to yeast may also occur, moving genes from a prokaryotic cell to a eukaryotic cell!

What use are plasmids? A bacterium's chromosome contains all the genes the cell normally needs for basic survival. However, genes carried by plasmids often allow the bacteria that carry them to grow in novel environments. Some plasmids contain genes that allow bacteria to metabolize unusual energy sources, such as petroleum. Other plasmids carry genes that cause disease symptoms, such as diarrhea, in the animal or other organism that the bacterium infects. (Diarrhea may benefit the bacterium, enabling it to spread and infect new hosts.) Still other plasmids carry genes that enable bacteria to grow in the presence of antibiotics, such as penicillin. In environments where antibiotic use is high, particularly in hospitals, bacteria carrying these antibiotic-resistance plasmids can quickly spread among patients and health care workers, making antibiotic-resistant infections a serious problem.

Viruses May Transfer DNA Between Bacteria and Between Eukaryotic Species

Viruses, which are often little more than genetic material encased in a protein coat, transfer their genetic material into cells during infection. Within the infected cell, viral genes replicate and direct the synthesis of viral proteins. The replicated genes and viral proteins assemble inside the cell, forming new viruses that are

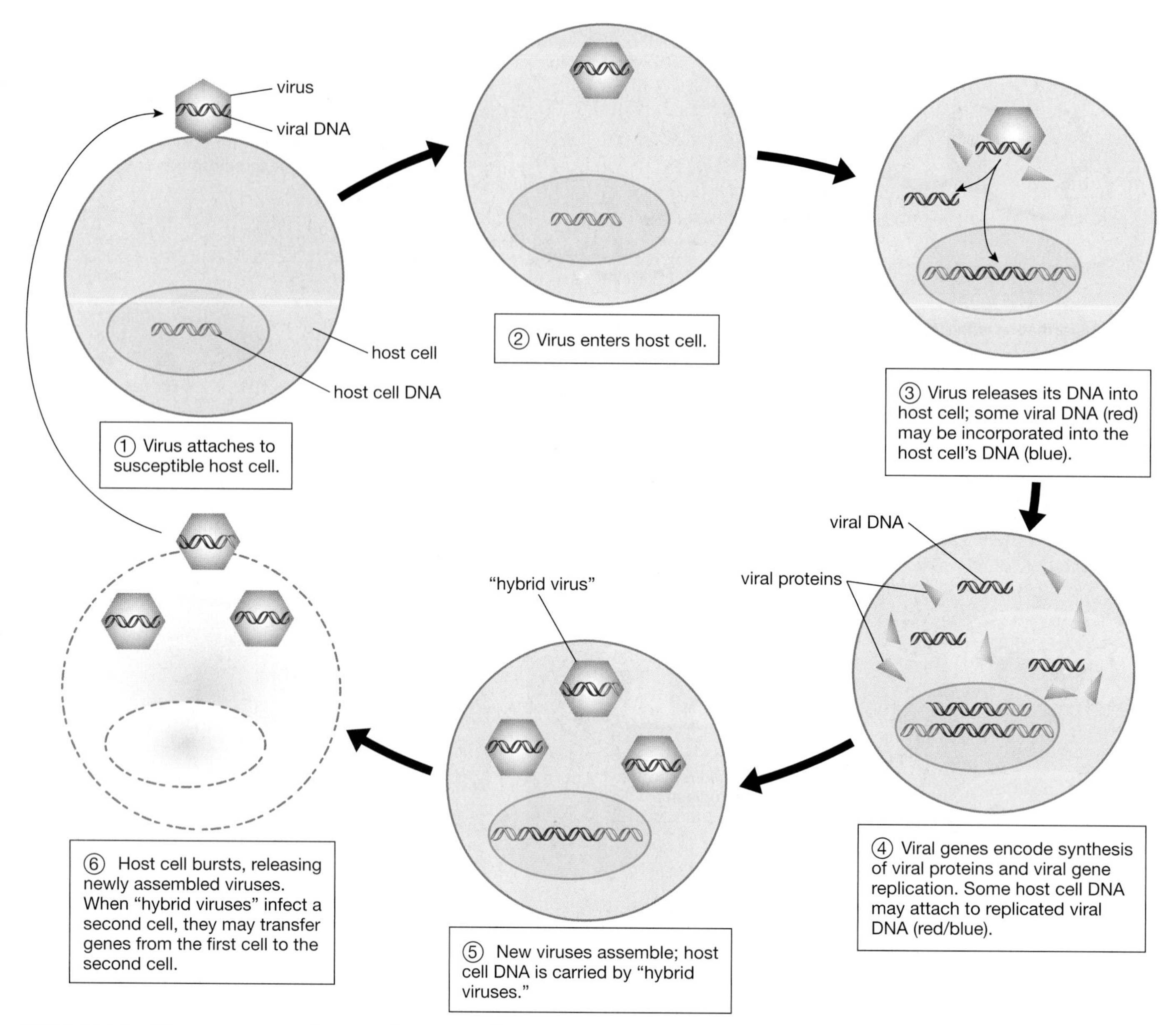
virus
viral DNA
host cell
host cell DNA
① Virus attaches to susceptible host cell.
② Virus enters host cell.
③ Virus releases its DNA into host cell; some viral DNA (red) may be incorporated into the host cell's DNA (blue).
viral DNA
viral proteins
④ Viral genes encode synthesis of viral proteins and viral gene replication. Some host cell DNA may attach to replicated viral DNA (red/blue).
"hybrid virus"
⑤ New viruses assemble; host cell DNA is carried by "hybrid viruses."
⑥ Host cell bursts, releasing newly assembled viruses. When "hybrid viruses" infect a second cell, they may transfer genes from the first cell to the second cell.

FIGURE 13-2 Viruses may transfer genes between cells

released to infect new cells (Fig. 13-2). Most viruses infect and replicate only in the cells of specific bacterial, animal, or plant species. For example, the canine distemper virus, which causes a frequently fatal disease in dogs, usually infects only dogs, raccoons, otters, and related species. Other viruses may infect species quite unrelated to each other; for example, rabies infects dogs, skunks, bats, and humans, whereas influenza infects birds, pigs, and humans.

During many viral infections, viral DNA sequences become incorporated into one of the host cell's chromosomes (see Fig. 13-2). The viral DNA may remain there for days, months, or even years. The cell replicates the incorporated viral DNA with its own DNA every time it divides. When new viruses are produced, the host's genes may be incorporated into the viral genome, creating a recombinant virus. When such viruses infect new cells, they may also transfer a portion of the previous host cell's DNA to the new cell. Occasionally, viruses may even cross species barriers. For example, canine distemper, which usually doesn't infect cats, has killed thousands of lions in Africa. During such cross-species infections, the new host may acquire genes that originally belonged to an unrelated species.

13.3 How Is Biotechnology Used in Forensics?

As with any technology, the applications of DNA biotechnology vary, depending on the goals of those who use it. Forensic scientists need to identify victims and criminals; biotechnology firms need to identify specific genes and insert them into organisms such as bacteria, cattle, or crop plants; and biomedical firms and physicians need to detect defective alleles and, ideally, devise ways to fix them or to insert normally functioning alleles into patients. First, we will discuss a few fundamental technologies that are common to nearly all DNA manipulations, using their application to forensic DNA analysis as a specific example. Later, we will investigate how biotechnology is used in agriculture and medicine.

When the Innocence Project investigated the Jimmy Ray Bromgard case, project workers needed to determine whether semen samples collected from the rape victim in 1987 came from Bromgard. They hired a DNA forensics laboratory to compare the semen samples with Bromgard's DNA. Now, the bits of DNA left in a 15-year-old sample were probably not in perfect shape. Even if intact DNA could be obtained, how could the lab determine whether the DNA samples matched? The technicians used two techniques that have become commonplace in virtually all DNA labs. First, they amplified the DNA so that they had enough material to analyze. Then they determined whether the DNA from the semen samples matched Bromgard's own DNA. Let's explore these two techniques.

The Polymerase Chain Reaction Amplifies DNA

Developed by Kary B. Mullis of the Cetus Corporation in 1986, the **polymerase chain reaction (PCR)** produces virtually unlimited amounts of DNA. Further, PCR can be used to amplify selected pieces of DNA, if desired. PCR is so crucial to molecular biology that it earned Mullis a share in the Nobel Prize for Chemistry in 1993. Let's see how PCR amplifies a specific piece of DNA (Fig. 13-3).

When we discussed DNA replication in Chapter 10, we omitted some of its real-life complexity. One of the things we did not discuss is crucial to PCR: by itself, DNA polymerase doesn't know where to start copying a strand of DNA. When a DNA double helix is unwound, enzymes put a little piece of complementary RNA, called a *primer*, on each strand. DNA polymerase recognizes this "primed" region of DNA as the place to start replicating the rest of the DNA strand.

In PCR, the nucleotide sequence of the beginning and the end of the DNA segment to be amplified must be known. A DNA synthesizer is used to make two sets of DNA fragments, one complementary to the beginning of one strand of the DNA segment and one complementary to the beginning of the other strand. These primers will "tell" DNA polymerase where to start working.

In a small test tube, DNA is mixed with primers, free nucleotides, and a special DNA polymerase, isolated from microbes that live in hot springs (see "Scientific Inquiry: Hot Springs and Hot Science"). PCR involves the following steps, which are repeated for as many cycles as are needed to generate enough copies of the DNA segment. PCR synthesizes DNA in a geometric progression ($1 \rightarrow 2 \rightarrow 4 \rightarrow 8$, etc.), so 20 PCR cycles

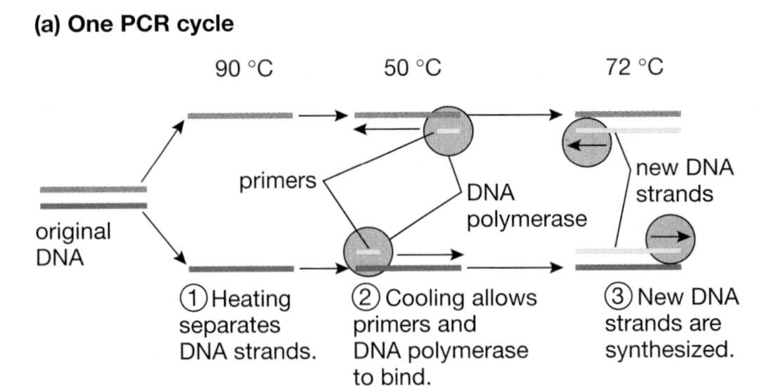

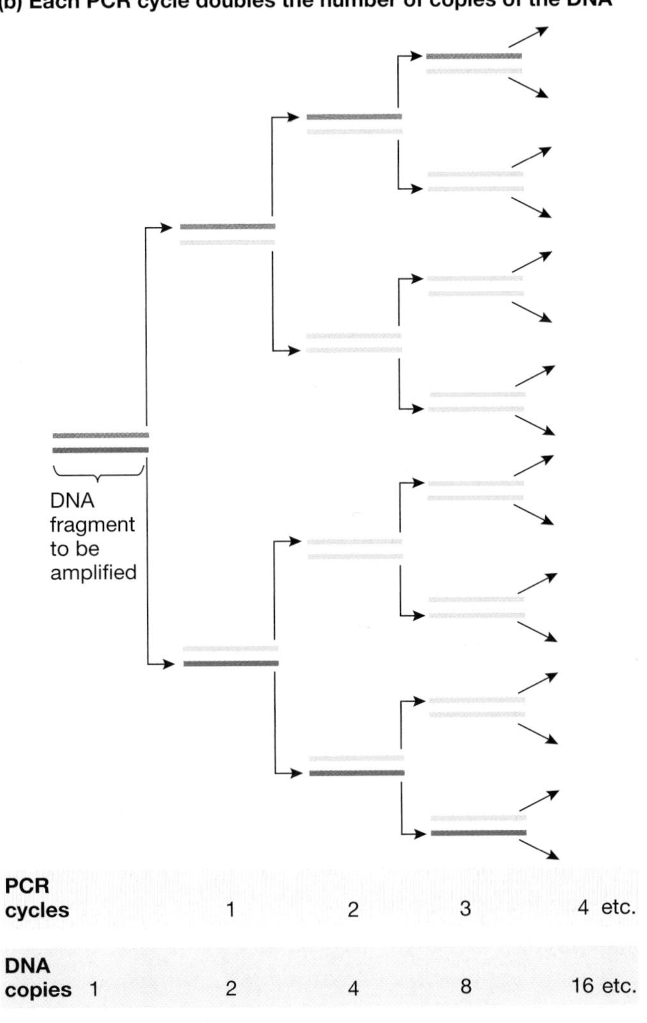

FIGURE 13-3 PCR copies a specific DNA sequence
The polymerase chain reaction consists of a series of 20 to 30 cycles of heating and cooling. After each cycle, the amount of target DNA doubles. After just 20 cycles, a million copies of the target DNA have been synthesized. **QUESTION** Why are primers necessary for PCR?

Media Activity 13.2 Polymerase Chain Reaction (PCR)

SCIENTIFIC INQUIRY Hot Springs and Hot Science

At a hot spring, such as those found in Yellowstone National Park, water literally boils out of the ground, gradually cooling as it flows to the nearest stream (Fig. E13-1). You might think that such springs, scalding hot and often containing poisonous metals and sulfur compounds, must be lifeless. However, closer examination often reveals a diversity of microorganisms, each adapted to a different temperature zone in the hot spring. In 1966, Thomas Brock of the University of Wisconsin discovered *Thermus aquaticus*, a bacterium that lives in water as hot as 80 °C (176 °F), in a Yellowstone hot spring.

When Kary Mullis first developed the polymerase chain reaction, he encountered a major technical difficulty. The DNA solution must be heated almost to boiling to separate the double helix into single strands, then cooled so DNA polymerase can synthesize new DNA; this process must be repeated over and over again. Ordinary DNA polymerase, like most proteins, is ruined, or *denatured*, by high temperatures. Therefore, new DNA polymerase had to be added after every heat cycle, which was expensive and labor-intensive. Enter *Thermus aquaticus*. Like other organisms, it replicates its DNA when it reproduces. But because it lives in hot springs, it has a particularly heat-resistant DNA polymerase. When DNA polymerase from *T. aquaticus* is used in PCR, it needs to be added to the DNA solution only once, at the start of the reaction.

FIGURE E13-1 Thomas Brock surveys Mushroom Spring
The colors in hot springs arise from minerals dissolved in the water and from various types of microbes that live at different temperatures.

make about 1 million copies, and a little over 30 cycles make 1 billion copies.

1. The test tube is heated to 90–95 °C. High temperatures break the hydrogen bonds between complementary bases, separating the DNA into single strands.
2. The temperature is lowered to about 50 °C, which allows the two primers to form complementary base pairs with the original DNA strands.
3. The temperature is raised to 70–72 °C. DNA polymerase, directed by the primers, uses the free nucleotides to make copies of the DNA segment bounded by the primers.
4. This cycle is repeated as many times as desired.

Using appropriate mixtures of primers, free nucleotides, and DNA polymerase, a PCR machine automatically runs heating and cooling cycles over and over again. Each cycle takes only a few minutes, so PCR can produce billions of copies of a gene or DNA segment in a single afternoon, starting, if necessary, from a single molecule of DNA. The DNA is then available for forensics, cloning, making transgenic organisms, or many other purposes.

The Choice of Primers Determines Which Segments of DNA Are Amplified

How would a forensics laboratory know which primers to use? After years of painstaking work, forensics experts have found that small, repeating segments of DNA called short tandem repeats (STRs) can be used to identify people with astonishing accuracy (Fig. 13-4). STRs are repeating sequences of DNA, 2 to 5 nucleotides long, that are scattered throughout the human genome, usually in *introns*, parts of genes that do *not* code for proteins (see Fig. E10-3). There may be as few as 5 or 6 copies of an individual STR in some people, and as many as 14 or 15 copies in others. Why so much variability? STRs probably have no biological function. As a result, DNA replication errors that lead to different numbers of STR repeats are not selected against during evolution, and variability has accumulated over evolutionary time. In 1999, British and American law enforcement agencies agreed to use a set of 10 to 13

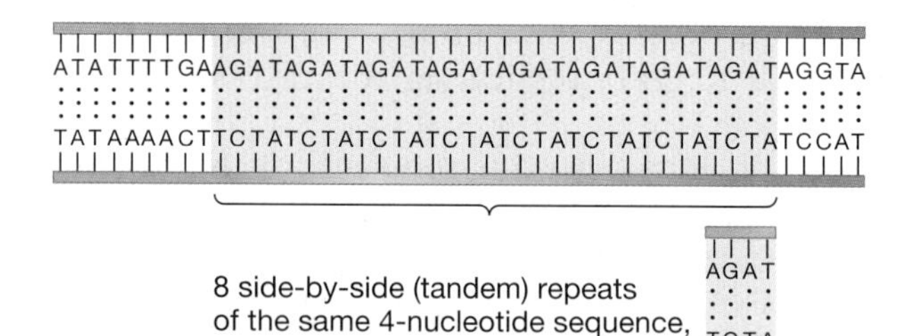

FIGURE 13-4 Short tandem repeats are common in noncoding regions of DNA
This STR, called D5, is not part of any known gene. The sequence AGAT may be repeated from 7 to 13 times in different individuals.

STRs, each 4 nucleotides long, that vary greatly among individuals. A perfect match of 10 STRs in a suspect's DNA and the DNA found at a crime scene means that there is less than one chance in a trillion that the two DNA samples did not come from the same person; with all 13 STRs, the chance is less than one in 200 quadrillion. What's more, the DNA around STRs doesn't seem to degrade very fast, so even old DNA samples, such as those in the Bromgard case, usually have STRs that are mostly intact.

Forensics labs use PCR primers that amplify only the DNA immediately surrounding the STRs. For example, one common STR repeats 7 to 13 times in different people. When a sample of DNA from a crime scene is amplified with the correct PCR primers, it produces DNA segments from 119 to 143 nucleotides long (some "extra" DNA on either end of the STRs is also amplified by the procedure). How does the lab determine how many repeats occurred in their DNA samples?

Gel Electrophoresis Separates and Identifies DNA Segments

Modern forensics labs use sophisticated and expensive machines to determine the number of times STRs repeat in their samples. Most of these machines, however, are based on two methods that are used in molecular biology labs around the world: first, separating the DNA by size, and second, labeling specific DNA segments of interest.

Mixtures of DNA pieces can be separated by a technique called **gel electrophoresis** (Fig. 13-5). First, the mixture of DNA fragments is loaded into shallow grooves, or wells, in a slab of agarose, a carbohydrate purified from certain types of seaweed (see Fig. 13-5a). Agarose is one of several materials that can form a *gel*, which is simply a meshwork of fibers with holes of various sizes between the fibers. The gel is put into a chamber with electrodes connected to each end. One electrode is made positive and the other negative; therefore, current will flow between the electrodes *through the gel*. The phosphate groups in the backbones of DNA are negatively charged. When electrical current flows through the gel, the negatively charged DNA fragments move toward the positively charged electrode. Because smaller fragments slip through the holes in the gel more easily than larger fragments, they move more rapidly toward the positively charged electrode. Eventually the DNA fragments are separated by size, forming distinct bands on the gel (see Fig. 13-5b).

Unfortunately, the DNA bands are usually invisible. There are simple stains that will make them visible, but this is often not very useful in either forensics or medicine. Why not? There may be many DNA fragments of approximately the same size; for example, several STRs with the same numbers of repeats might be mixed together in the same band. How can a technician identify a *specific* STR? How does *nature* identify sequences of DNA? Right, by base-pairing! Usually, the two strands of the DNA double helix are separated during gel electrophoresis; this allows pieces of synthetic DNA, called **DNA probes**, to base-pair with specific DNA fragments in the sample. DNA probes are short pieces of single-stranded DNA that are complementary to the nucleotide

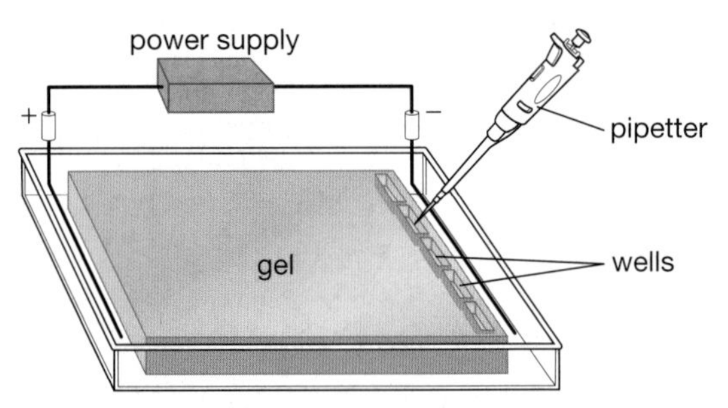

(a) DNA samples are pipetted into wells (shallow slots) in the gel. Electrical current is sent through the gel (negative at end with wells, positive at opposite end.)

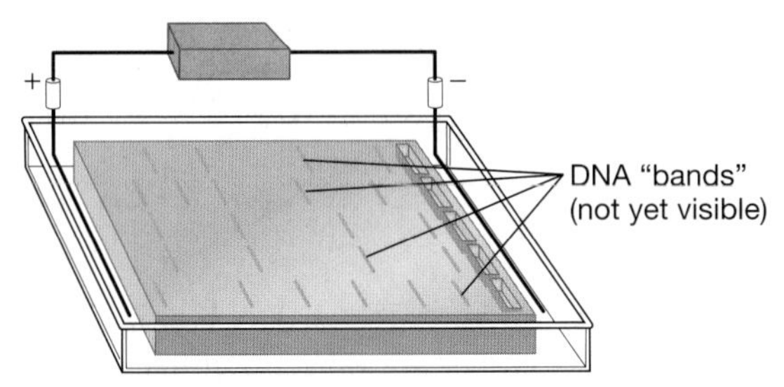

(b) Electrical current moves DNA segments through the gel. Smaller pieces of DNA move farther toward the positive electrode.

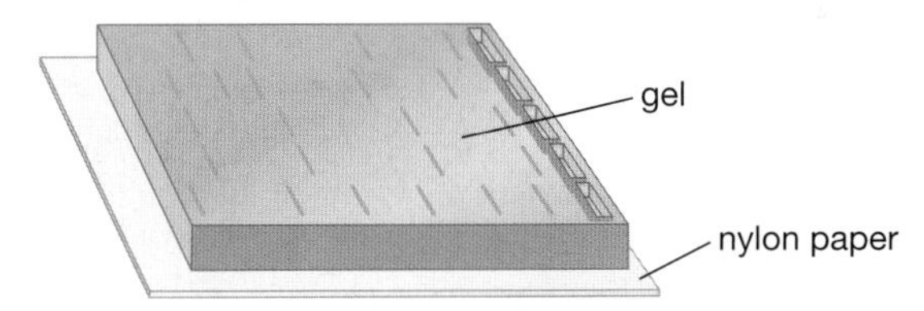

(c) Gel is placed on special nylon "paper." Electrical current drives DNA out of gel onto nylon.

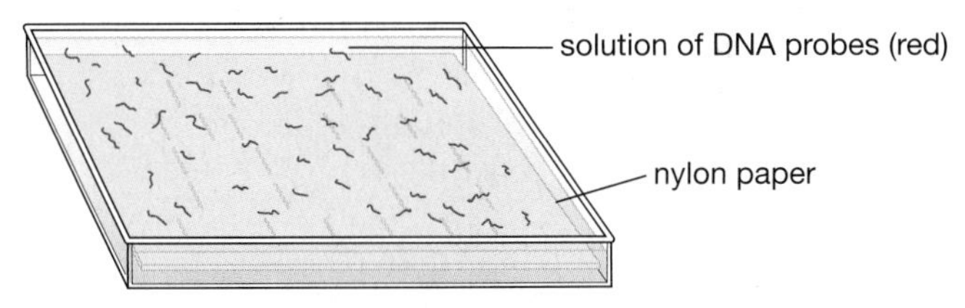

(d) Nylon paper with DNA is bathed in a solution of labeled DNA probes (red) that are complementary to specific DNA segments in the original DNA sample.

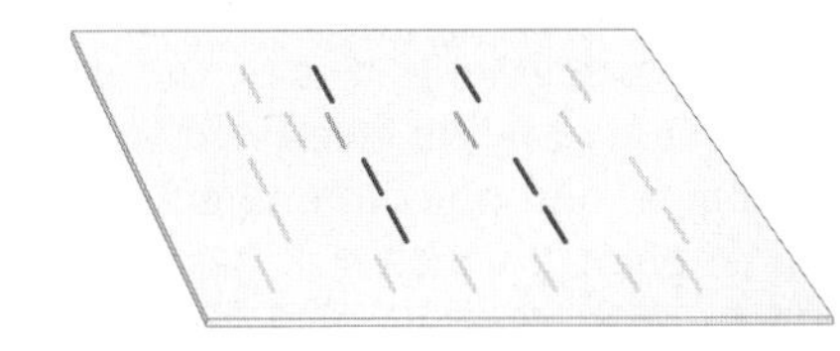

(e) Complementary DNA segments are labeled by probes (red bands).

FIGURE 13-5 Gel electrophoresis is used to separate and identify segments of DNA

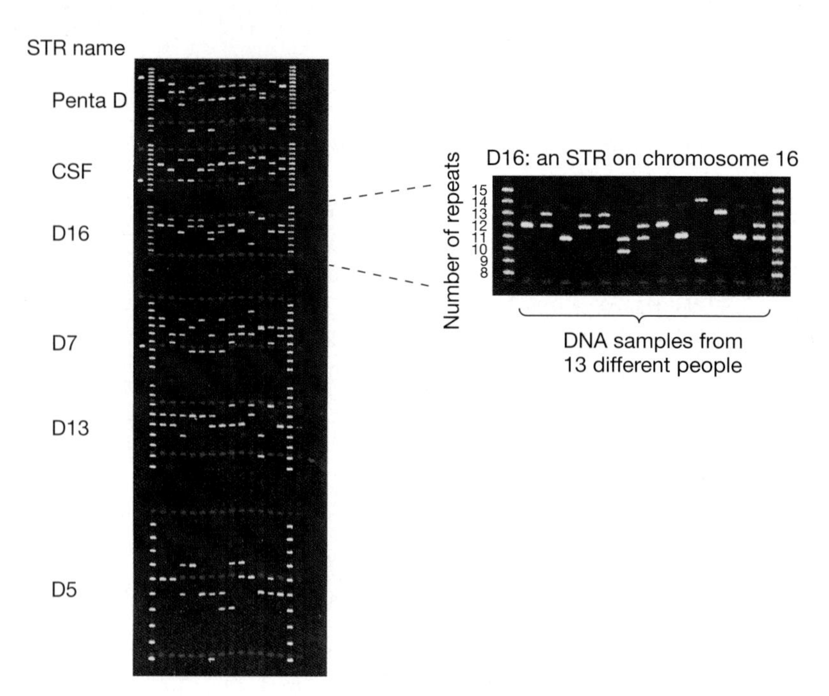

FIGURE 13-6 DNA fingerprinting The lengths of short tandem repeats of DNA form characteristic patterns on a gel; this gel displays six different STRs (Penta D, CSF, etc.). The evenly-spaced yellow-green bands on the far left and far right sides of the gel show the number of repeats of the individual STRs. DNA samples from 13 different people were run between these standards, resulting in one or two bands per vertical lane. For example, in the enlargement of the D16 STR on the right, the first person's DNA has 12 repeats, the second person's has 13 and 12, the third has 11, and so on. Although some people have the same number of repeats of some STRs, none has the same number of repeats of all the STRs. *(Photo courtesy of Dr. Margaret Kline, National Institute of Standards and Technology)* **QUESTION** On any individual's DNA fingerprint, a given STR always displays either one or two bands. Further, single bands are always about twice as bright as each band of a pair. For example, in the D16 STR on the right, the single bands of the first and third DNA samples are twice as bright as the pairs of bands of the second, fourth and fifth samples. Why?

sequence of a given STR (or any other DNA of interest in the gel). The DNA probes are labeled, either by radioactivity or by attaching one of several colored molecules to them. Therefore, a given DNA probe will label certain DNA sequences, and not others:

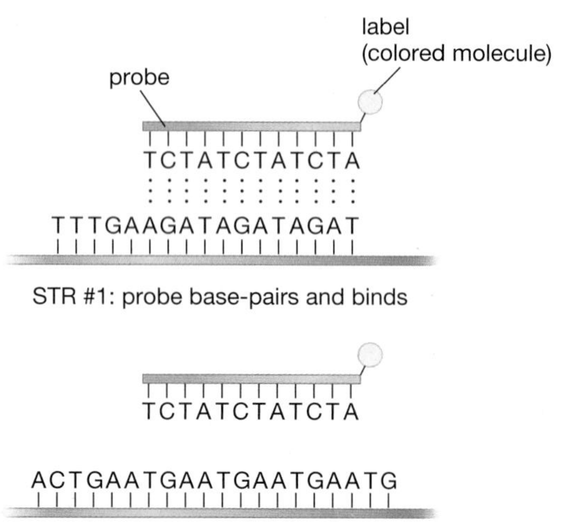

When the gel is finished running, the technician transfers the single-stranded DNA segments out of the gel and onto a piece of paper made of nylon (see Fig. 13-5c). Then, the paper is bathed in a solution containing a specific DNA probe (see Fig. 13-5d), which will base-pair with, and therefore bind to, only a specific STR, making this STR visible (see Fig. 13-5e). (Labeling the DNA fragments with radioactive or colored DNA probes is standard procedure in most research applications. In modern forensic analysis, however, the STRs are usually directly labeled with colored molecules during the PCR reaction; therefore, the STRs are already visible in the gel and do not have to be visualized with DNA probes.)

Until the early 1990s, forensic technicians ran DNA samples from a crime scene and from various suspects side-by-side on a gel, to see which suspect, if any, had DNA matching that found at the scene. In modern STR analysis, however, the suspect and crime scene DNA samples can be run on different gels, in different states or countries, and even years apart. Why? DNA samples run on STR gels produce a pattern, called a **DNA fingerprint** (Fig. 13-6), which is coded by recording the number of repeats for all the STR genes. The numbers and positions of the bands on the gel are determined by the numbers of repeats of each STR. Because an STR is part of a gene, each person has two copies of each STR—one on each homologous chromosome in every pair. Each of the two copies of the "STR genes" might have the same number of repeats (the person would be homozygous for that STR gene) or different numbers of repeats (the person would be heterozygous for that gene). For example, the first person in Fig. 13-6 is heterozygous for Penta D: the gel has two bands, with 9 repeats in one allele and 14 repeats in the other allele. That person is homozygous for CSF and D16, so the gel has single bands of 11 and 12 repeats, respectively.

In many states, anyone convicted of certain crimes (assault, burglary, attempted murder, etc.) must give a blood sample. Using the standard set of STRs, technicians determine the criminal's DNA fingerprint, encode the fingerprint, and send it off to state and national databases, where it will be permanently kept on file. When a crime has been committed, DNA collected at the scene and DNA from various suspects will be "DNA fingerprinted" and compared. In addition, because all forensics labs use the same STRs, the crime scene DNA code can be compared with the millions of DNA fingerprints on file in state and national databases. If the number of repeats of all the STRs in a suspect's DNA is the same as the crime scene DNA, then the odds are overwhelming that the crime scene DNA was left by the suspect. If they differ, then the DNA cannot belong to the suspect. If there aren't any matches, the crime scene DNA will remain on file. Sometimes, years later, a newly convicted

Table 13-1 Genetically Engineered Crops with USDA Approval

Genetically Engineered Trait	Potential Advantage	Examples of Bioengineered Crops Receiving USDA Approval Between 1992 and 2002
Resistance to herbicide	Application of herbicide kills weeds, but not crop plants, producing higher crop yields.	beet, canola, corn, cotton, flax, potato, rice, soybean, tomato
Resistance to pests	Crop plants suffer less damage from insects, producing higher crop yields.	corn, cotton, potato, soybean
Resistance to disease	Plants are less prone to infection by viruses, bacteria, or fungi, producing higher crop yields.	papaya, potato, squash
Sterile	Transgenic plants cannot cross with wild varieties, making them safer for the environment and more economically productive for the seed companies that produce them.	chicory, corn
Altered oil content	Oils can be made healthier for human consumption or can be made similar to more expensive oils (such as palm or coconut).	canola, soybean
Altered ripening	Fruits can be more easily shipped with less damage, producing higher returns for the farmer.	tomato

criminal's DNA fingerprint will match an archived crime scene fingerprint, and a "cold case" will be solved (see the Case Study Revisited at the end of this chapter).

13.4 How Is Biotechnology Used in Agriculture?

Many Crops Are Genetically Modified

Currently, almost all genetically modified organisms used in agriculture are plants. In 2002, about 34% of the corn, 71% of the cotton, and 75% of the soybeans grown in the United States were transgenic; that is, they contained genes from other species (Table 13-1). Crops are most commonly modified to improve their resistance to insects and herbicides.

There are several types of herbicide-resistant transgenic crops. Many herbicides kill plants by inhibiting an enzyme that is used by plants, fungi, and some bacteria—but not animals—to synthesize amino acids such as tyrosine, tryptophan, and phenylalanine. Without these amino acids, the plants die because they cannot synthesize proteins. Many herbicide-resistant transgenic crops have been given a bacterial gene that encodes an enzyme that functions even in the presence of herbicides, so the plants continue to synthesize normal amounts of amino acids and proteins. Herbicide-resistant crops allow farmers to kill weeds without harming their crops. Less competition from weeds means more water, nutrients, and light for the crops, hence larger harvests.

To promote insect resistance, many crops have been given a gene, called Bt, from the bacterium *Bacillus thuringiensis*. The protein encoded by the Bt gene damages the digestive tract of insects (but not mammals). Transgenic Bt crops therefore suffer far less damage from insects, and farmers can apply much less pesticide to their fields.

How would a seed company go about making a transgenic plant? Let's examine the process, using insect-resistant Bt plants as an example.

The Desired Gene Is Cloned

Cloning a gene usually involves two tasks: obtaining the gene and inserting it into a plasmid so that huge numbers of copies of the gene can be made.

There are two common ways of obtaining a gene. One is to isolate the gene from the organism that makes it. For a long time, this was the only practical method. More recently, biotechnologists can often synthesize the gene—or a modified version of it—in the lab, using PCR or DNA synthesizers.

Once the gene has been obtained, why insert it into a plasmid? Plasmids, small circles of DNA in bacteria, are replicated when the bacteria multiply. Therefore, once the desired gene has been inserted into a plasmid, producing huge numbers of copies of the gene is as simple as raising lots of bacteria. Inserting the gene into a plasmid also allows it to be separated from the bacteria fairly easily; this results in partial purification of the gene, now free of the DNA of the bacterial chromosome. Finally, plasmids may then be taken up by other bacteria (this is important when making transgenic Bt plants) or injected directly into animal eggs.

Restriction Enzymes Cut DNA at Specific Nucleotide Sequences

Without a special class of enzymes called **restriction enzymes**, inserting a gene into a plasmid would be a nearly hopeless task. Scores of restriction enzymes have been isolated from a wide variety of bacteria. Each enzyme cuts DNA at a specific nucleotide sequence. Many restriction enzymes cut straight across a double helix of DNA. Others make a staggered cut, snipping the DNA in a different location on each of the two strands, so that single-stranded sections hang off the ends of the DNA. Because these single-stranded regions can base-pair with, and thus stick to, other single-stranded pieces of DNA with complementary bases, they are commonly called "sticky ends":

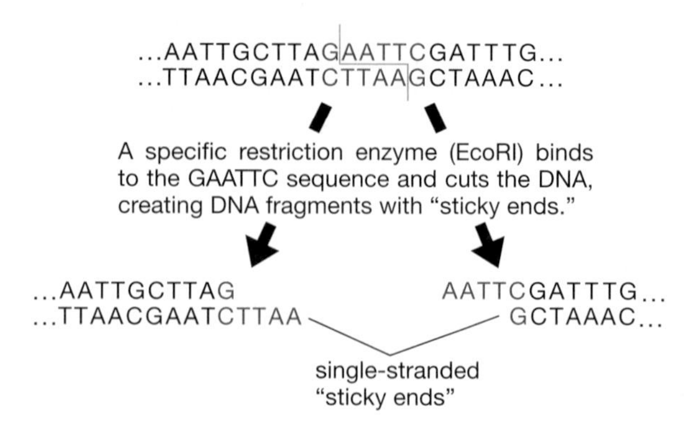
...AATTGCTTAGAATTCGATTTG...
...TTAACGAATCTTAAGCTAAAC...
A specific restriction enzyme (EcoRI) binds to the GAATTC sequence and cuts the DNA, creating DNA fragments with "sticky ends."
...AATTGCTTAG
...TTAACGAATCTTAA
AATTCGATTTG...
GCTAAAC...
single-stranded "sticky ends"

Cutting Two Pieces of DNA with the Same Restriction Enzyme Allows the Pieces to Be Joined Together

To insert the Bt gene into a plasmid, the same restriction enzyme is used to cut the DNA on either side of the Bt gene and to split open the circle of the plasmid (Fig. 13-7a). Therefore, the ends of the Bt gene and the opened-up plasmid both have complementary nucleotides in their "sticky ends" and can base-pair with each other. When the cut Bt genes and plasmids are mixed together, some of the Bt genes will be temporarily inserted between the ends of the plasmids. DNA ligase (see Chapter 9) permanently bonds the Bt genes into the plasmid (Fig. 13-7b). Bacteria are then transformed with the plasmids (Fig. 13-7c). By manipulating the plasmids and bacteria appropriately, biotechnologists can isolate and grow only the bacteria with the desired plasmid.

Plasmids Are Used to Insert the Bt Gene Into a Plant

The bacterium *Agrobacterium tumefaciens*, which contains a specialized plasmid called the Ti plasmid, can infect many plant species. When the bacterium infects a plant cell, the Ti plasmid inserts its DNA into one of the plant cell's chromosomes. Thereafter, every time that plant cell divides, it replicates the Ti plasmid DNA as well; all of its daughter cells inherit the Ti DNA. (Genes on the Ti plasmid cause plant tumors; however, biotechnologists have learned how to make "disabled" Ti plasmids that are harmless.) To make insect-resistant plants, Bt genes are inserted into disabled Ti plasmids. *A. tumefaciens* bacteria take up the plasmids and infect plant cells in culture (Fig. 13-7d). The modified Ti plasmids insert the Bt gene into the plant cells' chromosomes, so that the plant cells now permanently carry the Bt gene (Fig. 13-7e). Appropriate hormonal treatments stimulate the transgenic plant cells to divide and differentiate into entire plants. These plants are bred to one another or to other plants to create commercially valuable crop plants that resist insect attack (Fig. 13-8).

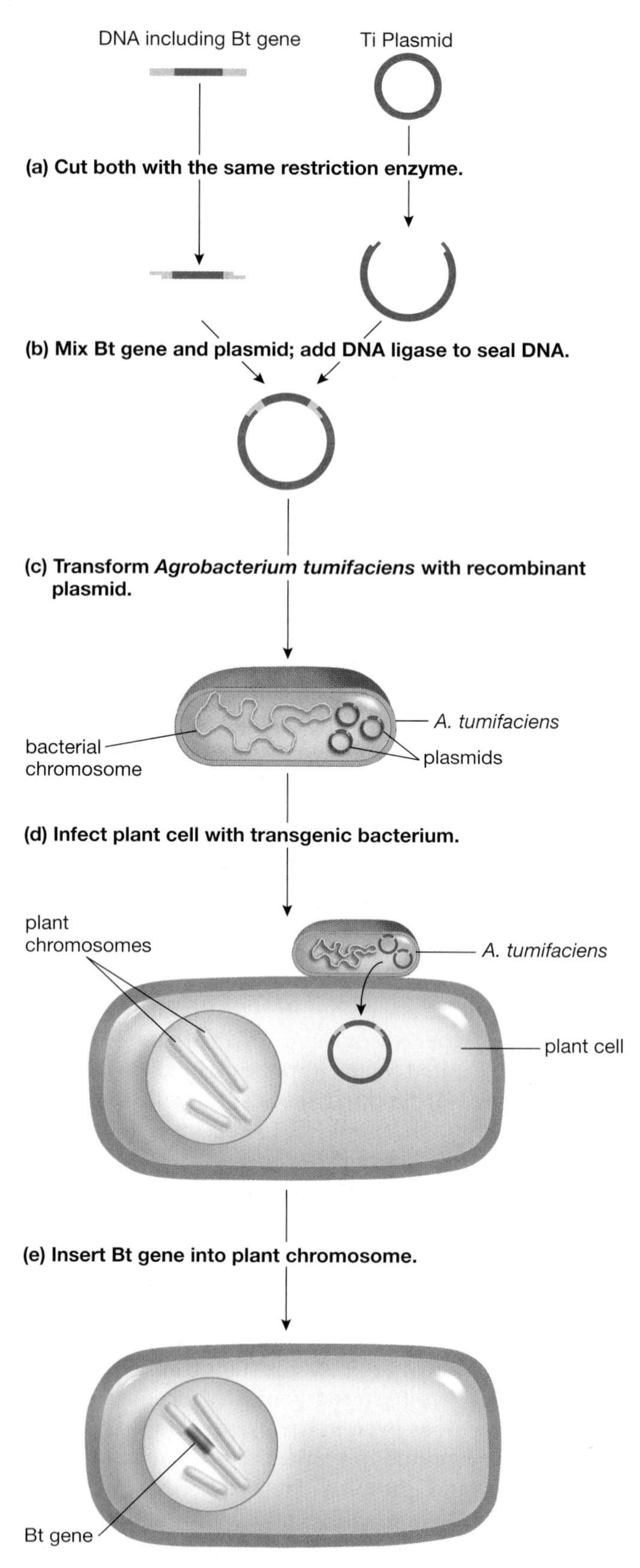
DNA including Bt gene
Ti Plasmid
(a) Cut both with the same restriction enzyme.
(b) Mix Bt gene and plasmid; add DNA ligase to seal DNA.
(c) Transform Agrobacterium tumifaciens with recombinant plasmid.
bacterial chromosome
A. tumifaciens
plasmids
(d) Infect plant cell with transgenic bacterium.
plant chromosomes
A. tumifaciens
plant cell
(e) Insert Bt gene into plant chromosome.
Bt gene

FIGURE 13-7 Using *Agrobacterium tumefaciens* to insert the Bt gene into plants

Genetically Modified Plants May Be Used to Produce Medicines

Similar techniques can be used to insert medically useful genes into plants. For example, a plant could be engineered to produce harmless proteins that are normally

FIGURE 13-8 Bt plants resist insect attack
The Bt gene was added to NewLeaf® potatoes (middle row) using *Agrobacterium tumefaciens* as a vector. These potatoes resist attack by Colorado potato beetles and consequently are much healthier than nontransgenic plants (right and left rows).
QUESTION How might herbicide-resistant crops help reduce topsoil erosion?

found in disease-causing bacteria or viruses. If these proteins resisted digestion in the stomach and small intestine, simply eating such plants could act as a vaccination against the disease organisms. Potatoes have been engineered to produce proteins that might act as vaccines against hepatitis B, the common intestinal bacteria *E. coli*, and a virus that causes diarrhea. But to avoid denaturing the proteins, the potatoes must be eaten raw, which isn't a very appealing prospect. Therefore, researchers are working to produce "vaccine bananas" that might protect against these and other diseases.

It might also be possible to engineer plants to produce human antibodies that would combat various diseases. As you will learn in Chapter 32, when a microbe invades your body, it takes several days for your immune system to respond and produce enough antibodies to overcome the infection. Meanwhile, you feel terrible, and might even die if the disease is serious enough. A direct injection of large quantities of the right antibodies might be able to cure the disease almost instantly. Plants have already been developed that produce antibodies against the genital herpes virus and some of the bacteria that cause tooth decay. Researchers hope that "plantibodies" might be produced very cheaply, making such therapies available to rich and poor alike.

Genetically Modified Animals May Be Useful in Agriculture and Medicine

Unlike plants, it is very difficult to make whole animals, especially vertebrates, from single cells in culture dishes. Therefore, making transgenic animals usually involves injecting the desired DNA, often incorporated into a disabled virus, into a fertilized egg. The egg is usually allowed to divide a few times in culture before being implanted into a surrogate mother. If the offspring are healthy and express the foreign gene, they are then bred together to produce homozygous transgenic organisms. So far, it has proven difficult to produce commercially valuable transgenic livestock. For example, pigs with extra growth-hormone genes indeed grow faster, but they also suffer from arthritis, ulcers, and sterility. Apparently healthy cows with extra genes for making milk proteins have been developed, but commercial quantities of transgenic milk are still 5 to 10 years away. Several types of fish with added growth-hormone genes grow much faster than wild fish and do not display any obvious ill effects. However, whether "fish farms" should be allowed to grow these fish remains controversial, principally because of concerns about what would happen if they escaped into the wild (see "Earth Watch: Bonanza from the Sea or Frankenfish?").

Because medicines are generally much more valuable than meat, many researchers are developing animals that will produce medicines, such as human antibodies or other essential proteins. For example, there are sheep whose milk contains a protein, alpha-1-antitrypsin, which may prove valuable in treating cystic fibrosis.

13.5 How Is Biotechnology Used for Medical Diagnosis and Treatment?

Media Activity
13.3 Human Genome Sequencing

Many people suffer from inherited disorders, including sickle-cell anemia, Marfan syndrome, and cystic fibrosis, to name a few. For over a decade, biotechnology has been routinely used to diagnose some inherited disorders. Potential parents can learn if they are carriers of a genetic disorder, and an embryo can be diagnosed early in a pregnancy (see "Health Watch: Prenatal Genetic Screening"). More recently, medical researchers have begun to use biotechnology to attempt to cure, or at least treat, genetic diseases.

DNA Technology Can Be Used to Diagnose Inherited Disorders

By definition, a person inherits a genetic disease because he or she inherits one or more dysfunctional alleles. Defective alleles differ from normal, functional alleles because of differences in nucleotide sequence. Two methods are currently used to find out if a person carries a normal allele or a malfunctioning allele.

Restriction Enzymes May Cut Different Alleles of a Gene at Different Locations

Remember that restriction enzymes cut DNA only at specific nucleotide sequences. Because chromosomes are so large, a restriction enzyme usually cuts the DNA of a chromosome in many places, producing many *restriction fragments*. What if two homologous chromosomes have

EARTH WATCH Bonanza from the Sea or Frankenfish?

Nutritional experts extol the benefits of eating fish. Not only are fish a good source of protein, but the fatty acids found in some species may protect against heart disease. With burgeoning human populations, the United Nations Food and Agriculture Organization predicts that the demand for fish might double by 2040. However, many fish populations are already declining due to overfishing; in 40 years, there almost certainly won't be enough wild fish to feed everyone.

Aquaculture—raising fish in freshwater ponds or enclosures in the oceans—has the potential to produce huge amounts of "domestic" fish. Just as ranching and farming have replaced hunting and gathering on land, aquaculture might do the same in the sea, thereby feeding more people and relieving pressures on wild fish populations. Farmers raise cattle, pigs, and chickens (instead of deer, wart hogs, or pheasants) because domestic animals have been bred for centuries for docility, fast weight gain, and high reproductive rate. Aquaculturists are just starting to develop more efficient strains of fish, using genetic engineering rather than centuries of artificial selection.

Perhaps the most famous engineered fish is a strain of Atlantic salmon produced by Aqua Bounty Farms. Aqua Bounty constructed a new "growth gene" for its Atlantic salmon, combining the growth-hormone gene from Chinook salmon (a Pacific Ocean species) with promoter sequences that cause the gene to be transcribed all year round. Aqua Bounty salmon, therefore, produce more growth hormone and grow as much as six times faster than their wild relatives, reaching marketable size a full year earlier than wild Atlantic salmon (Fig. E13-2).

Before Aqua Bounty salmon can be farmed commercially, several questions must be answered. First, are Aqua Bounty salmon safe to eat? Almost certainly, the answer is yes. People eat Chinook salmon—and, therefore, its growth hormone—with no ill effects. Second, are Aqua Bounty salmon safe to raise? The answer to this question is probably also yes. Salmon, engineered or not, require food to grow. The faster the Aqua Bounty salmon grow, the more they eat. The more they eat, the more uneaten food scraps and feces float away in the water. Large numbers of fast-growing salmon have the potential to produce massive water pollution, but this can almost certainly be solved with enough ingenuity and money.

The third question is more complicated: are Aqua Bounty salmon safe for the environment? In other words, what would happen if they escaped into the wild? There's a good chance that bioengineered salmon would outcompete their wild relatives and replace them in the ocean. Are they likely to escape, and if they did, would it matter?

Escape is really only a matter of time. For example, in December 2000, a huge storm destroyed the steel pens of a fish farming operation in Machias Bay in Maine. At least 100,000 fish swam away into the ocean. Sooner or later, engineered fish would enter the wild. Is that bad? They're still Atlantic salmon, aren't they? Maybe, but maybe not. Wild populations of most species vary greatly. If engineered salmon replaced wild fish, the gene pool of the salmon might shrink to a puddle, leaving them susceptible to changing environmental conditions. Even if the salmon are not adversely affected, what about other species? Voracious engineered salmon might eat a lot more than wild salmon, leaving less for other sea life. Whole food webs might be altered, or even disintegrate.

To prevent this, Aqua Bounty plans to breed the fish inland, sterilize the offspring, and ship the sterilized fish to pens along the coast to be grown for market. Aqua Bounty's sterilization procedure is 100% effective in small batches in the lab. Will it be 100% effective in commercial operation? If Aqua Bounty can't prove that sterilization is 100% effective in huge batches, should fisheries take the risk?

In 2002, the U.S. National Research Council investigated the safety of genetically modified animals. The NRC decided that the risk to people from eating transgenic animals is minimal, but that the risk to the environment is unknown but potentially large. A lost transgenic cow could be easily caught, but what about fish? In response to such concerns, the state of Washington banned transgenic fish in 2002. How do you think that governments should balance the probable benefits of increasing a valuable food supply with possible damage to the environment?

FIGURE E13-2
Transgenic salmon (bottom) grow much faster than their wild relatives (top).

different alleles of several genes, and some alleles have nucleotide sequences that *can* be cut by a restriction enzyme, while others have nucleotide sequences that *cannot* be cut by the enzyme? The result will be a mixture of DNA segments of various lengths. These are called **restriction fragment length polymorphisms (RFLPs)**. This rather daunting phrase simply means that *restriction* enzymes have cut DNA into *fragments* that vary in *length*, and that homologous chromosomes (from the same person or from different people) may differ (or be *polymorphic*) in the lengths of the fragments. Why is this useful? First, if different people have different RFLPs, this can be used to identify DNA samples. In fact, in the early 1990s, before STRs became the gold standard in DNA forensics, RFLPs were used to determine if DNA from a crime scene matched the DNA of a suspect. Second, with diligent research and a little luck, medically important alleles can sometimes be identified by differences in the lengths of the restriction fragments produced by cutting with a specific restriction enzyme.

RFLP analysis has become the standard technique to diagnose sickle-cell anemia, even in an embryo. You may recall that sickle-cell anemia is caused by a point mutation in which thymine replaces adenine near the beginning of the globin gene. This causes a hydrophobic amino acid (valine) to be placed in the globin protein instead of a hydrophilic amino acid (glutamic acid; see p. 173 in Chapter 10). The hydrophobic valines cause hemoglobin molecules to clump together, distorting and weakening the red blood cells.

One restriction enzyme, called MstII, cuts DNA in about the middle of both the normal and sickle-cell alleles. It also cuts DNA just outside of both alleles. However, the normal globin allele, but *not* the sickle-cell allele, is also cut in a third location (Fig. 13-9a). How can the one unique cut be identified? A DNA probe is synthesized that is complementary to the part of the globin allele spanning the site of the unique cut. When sickle-cell DNA is cut with MstII and run on a gel, this probe labels a single large band (Fig. 13-9b). When normal DNA is cut with MstII, the probe labels two bands, one small and one not quite as large as the sickle-cell band. Someone who is homozygous for the normal globin allele will have two bands; someone who is homozygous for the sickle-cell allele will have one band, and a heterozygote will have three bands. The genotypes of parents, children, and fetuses can be determined by this simple test.

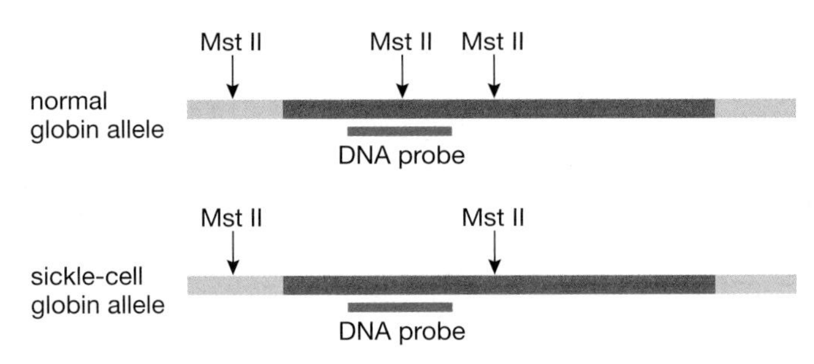

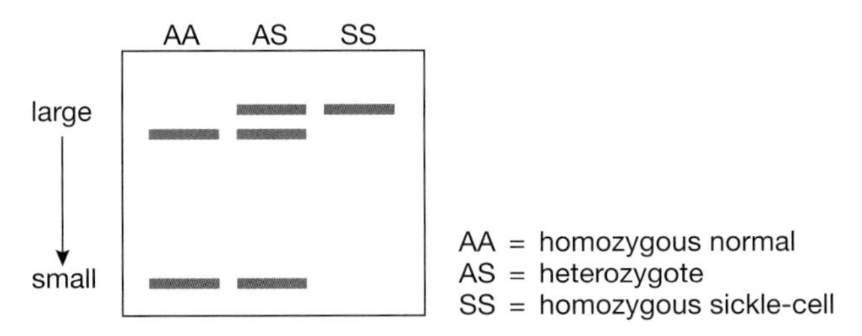

FIGURE 13-9 Diagnosing sickle-cell anemia with restriction enzymes
(a) The normal globin allele and the sickle-cell allele (both shown in red) are cut in half by the restriction enzyme MstII (far right arrow). The normal allele is also cut in another, unique location (middle arrow). Finally, regardless of which allele is present, the chromosome is cut somewhat ahead of the globin gene locus (far left arrow). A DNA probe (blue) is synthesized that is complementary to DNA on both sides of the unique cut site. Therefore, the probe will label two pieces of DNA from the normal allele but only a single piece of the sickle-cell allele.
(b) The cut DNA is run on a gel and made visible with the DNA probe. The large piece of DNA of the sickle-cell allele is close to the beginning of the gel, while the smaller pieces of the normal allele run farther into the gel.

Different Alleles May Bind to Different DNA Probes

In Chapter 12, we briefly discussed the genetics of cystic fibrosis, a disease caused by a defect in a protein that normally pumps chloride ions across cell membranes. Each of over 1000 different alleles—all at the same gene locus—encodes a slightly different defective chloride pump. People with either one or two normal alleles synthesize enough functioning chloride pumps so that they do not develop cystic fibrosis. People with two defective alleles (they may be the same or different alleles) do not synthesize any functional chloride pumps and develop cystic fibrosis. Therefore, the disease is inherited as a simple recessive trait.

How can anyone hope to diagnose a disorder caused by 1000 different alleles? Most of these alleles are extremely rare; only 32 alleles account for about 90% of the cases of cystic fibrosis. Still, 32 alleles are a lot. Although one could probably find restriction enzymes that would cut most of these alleles differently from the normal allele, testing would involve dozens of different enzymes, producing dozens of different patterns of DNA pieces that would need to be run on dozens of different gels. The cost would be astronomical.

Fortunately, each allele has a different nucleotide sequence. Therefore, one strand of each allele will form perfect base-pairs only with its own complementary strand. Several companies now produce cystic fibrosis "arrays," which are pieces of specialized filter paper to which segments of single-stranded DNA are bound. Each piece of DNA is complementary to a different one of the many cystic fibrosis alleles (Fig. 13-10). A person's DNA is cut into small pieces, separated into single strands, and labeled. The array is then bathed in the resulting solution of labeled DNA fragments. Under the right conditions, only

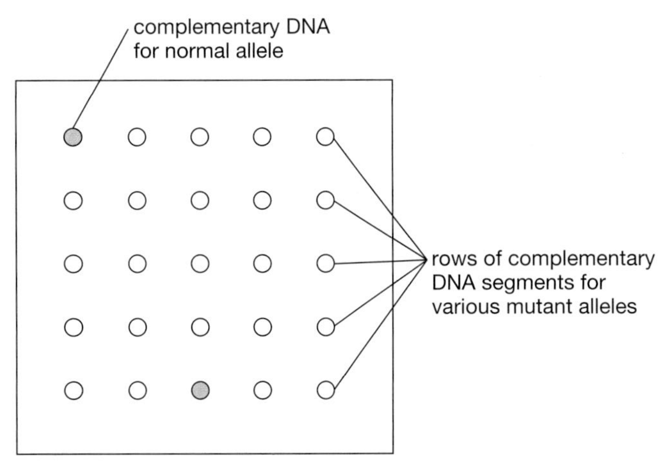

FIGURE 13-10 Diagnosing cystic fibrosis with a DNA array
DNA from a patient is cut into small pieces, separated into single strands, and labeled (blue, in this diagram). A cystic fibrosis screening array is bathed in this solution of labeled DNA. Each cystic fibrosis allele can bind to only one specific piece of complementary DNA on the array. In this simplified diagram, the patient has one normal allele (upper left) and one defective allele (middle bottom).

a perfect complementary strand of the person's DNA will bind to any given spot of DNA on the array; even a single "wrong" base will keep the person's DNA from binding. Depending on the number of different alleles represented on the array, up to 95% of all cases of cystic fibrosis can be diagnosed by this method.

DNA Technology Can Be Used to Treat Disease

Several therapeutically important proteins are now routinely made in bacteria, using technology similar to that used when making a transgenic plant: restriction enzymes are used to splice appropriate genes into plasmids, and bacteria are then transformed with these plasmids. The first human protein made by this recombinant DNA technology was insulin. Prior to 1982, when recombinant human insulin was first licensed for use, the insulin needed by diabetics was extracted from the pancreases of cattle or pigs slaughtered for meat. Although the insulin from these animals is very similar to human insulin, the slight differences cause an allergic reaction in about 5% of diabetics. Recombinant human insulin does not cause allergic reactions.

Media Activity
13.4 Manufacturing Human Growth Hormone

Other human proteins, such as growth hormone and clotting factors, can also be also produced in transgenic bacteria. Before recombinant DNA technology, some of these proteins were obtained from either human blood or human cadavers; these sources are expensive and sometimes dangerous. Human blood can be contaminated by HIV, the virus that causes AIDS. Cadavers may also contain several hard-to-diagnose infectious diseases, such as Creutzfeld-Jacob syndrome, in which an abnormal protein can be passed from the tissues of an infected cadaver to a patient and cause irreversible, fatal brain degeneration. Engineered proteins grown in bacteria circumvent these dangers. Some of the categories of human proteins produced by recombinant DNA technology are listed in Table 13-2.

These proteins, although tremendously helpful and often lifesaving, do not *cure* inherited disorders; they merely treat the symptoms. Often, as in the case of insulin-dependent diabetes, a patient may need to take the protein for his or her entire life. These proteins are soluble molecules, usually found dissolved in the blood, which often act as signaling molecules that tell cells how to regulate their metabolism. Insulin, for example, is released into the bloodstream and travels all over the body, instructing cells—such as liver and muscle cells—to take up glucose from the blood. Imagine how much better it would be if a diabetic could regain the ability to synthesize and release his or her *own* insulin, rather than taking daily injections. Nevertheless, diabetes sufferers are comparatively fortunate; in some diseases, such as cystic fibrosis, the defective molecule is an integral part of the patient's cells and cannot be replaced by a simple pill or injection.

Biotechnology offers the potential to treat diseases such as cystic fibrosis and possibly cure diseases such as diabetes, although progress has been painfully slow thus far. Let's look at two specific examples of how these advances may treat, or even cure, devastating illnesses.

Using Biotechnology to Treat Cystic Fibrosis

Cystic fibrosis causes devastating effects in the lungs, where the lack of chloride transport causes the usually thin, watery fluid lining the airways to become thick and

Table 13-2 Examples of Currently Used Products Produced by Recombinant DNA Methods

Type of Product	Purpose	Example: Product	Example: Year Approved	Example: Genetic Engineering
Human hormones	Used in treatment of diabetes, growth deficiency	Humulin™ (human insulin)	1982	Human gene inserted into bacteria
Human cytokines (regulate immune system function)	Used in bone marrow transplants and to treat cancers and viral infections, including hepatitis and genital warts	Leukine™ (granulocyte-macrophage colony stimulating factor)	1991	Human gene inserted into yeast
Antibodies (immune system proteins)	Used to fight infections, cancers, diabetes, organ rejection, and multiple sclerosis	Herceptin™ (antibodies to HER2 protein, expressed at high levels in some breast cancer cells)	1998	Recombinant antibody genes inserted into cultured hamster cell line
Viral proteins	Used to generate vaccines against viral diseases and for diagnosing viral infections	Energiz-B™ (Hepatitis B vaccine)	1989	Viral gene inserted into yeast
Enzymes	Used in treatment of heart attacks, cystic fibrosis, and other diseases, and production of cheeses and detergents	Activase™ (tissue plasminogen activator)	1987	Human gene inserted into cultured hamster cell line

clogged with mucus (see "Scientific Inquiry: Cystic Fibrosis" in Chapter 12). Several research groups are developing methods to deliver the allele for normal chloride pumps to the cells of the lungs, get them to synthesize functioning chloride pumps, and insert these pump proteins into their plasma membranes. Although different laboratories use slightly different methods, all involve inserting the DNA of the normal allele into a virus. When a virus infects a cell, it releases its genetic material into the cytoplasm of the cell, and uses the cell's own metabolism to transcribe the viral genes and make new viral proteins (see Fig. 13-2).

To treat cystic fibrosis, researchers first disable a suitable virus to ensure that the treatment doesn't cause yet another disease. Cold viruses are often used because they normally infect cells of the respiratory tract. The DNA of the normal chloride pump allele is then inserted into the DNA of the virus. The recombinant viruses are suspended in a solution and sprayed into the patient's nose or dripped directly into the lungs through a nasal tube. If all goes well, the viruses enter cells of the lungs and release the normal chloride pump allele into the cells. The cells then manufacture normal proteins, insert them into their plasma membranes, and successfully transport chloride into the fluid lining the lungs. The clinical trials underway for such treatments have been reasonably successful, but for only a few weeks. In all likelihood, the patients' immune systems see the viruses as undesirable invaders and mount attacks that eliminate them—and the helpful genes they carry—from the patients' bodies. Because lung cells are continually replaced over time, a single dose "wears off" as the modified cells die. Long-term success will probably require either inhibiting the patients' immune systems, which introduces its own set of problems, or designing delivery systems that can "hide" from the immune system.

Using Biotechnology to Cure Severe Combined Immune Deficiency

Like the cells of the lung, the vast majority of cells in the human body eventually die and are replaced by new cells. In many cases, the new cells come from special populations of cells called **stem cells**. When they divide, stem cells give rise to daughter cells that can differentiate into several different types of mature cells. In the brain, for example, stem cells give rise to several types of nerve cells, as well as multiple types of non-nervous support cells. It is possible that some stem cells, under the right conditions in the laboratory, might be able to give rise to *any* cell type of the entire body! We will discuss this potential of stem cells in Chapter 37. For now, we will look at a more limited function of stem cells in the body: producing or replacing cells of just one or two types.

All the cells of the immune system (mostly white blood cells) originate in the bone marrow. Some go on to produce antibodies; others kill cells that have been infected by viruses; and still others regulate the actions of these other cells. As mature cells die, they are replaced by new cells that arise from division of stem cells in the bone marrow. Severe combined immune deficiency (SCID) is a rare disorder in which a child fails to develop an immune system. About 1 in 80,000 children is born with some form of SCID. Infections that would be trivial in a normal child become life threatening. In some cases, if the child has an unaffected relative with a similar genetic makeup, a bone marrow transplant from the healthy relative can give the child functioning stem cells, so that he or she can develop a functioning immune system. Most SCID victims, however, die before their first birthday.

Although there are several forms of SCID, most are recessive, single-gene defects. In some cases, children are homozygous recessive for a defective allele that normally codes for an enzyme called adenosine deaminase. In 1990, the first test of human gene therapy was performed on such a SCID patient, 4-year-old Ashanti DeSilva. Some of her white blood cells were removed, genetically altered with a virus containing a functional version of her defective allele, and then returned to her bloodstream. Now a teenager, Ashanti is healthy, with a reasonably functional immune system. However, as the altered white blood cells die, they must be replaced with new ones, so Ashanti needs repeated treatments. She is also given regular intramuscular injections of a form of adenosine deaminase, which makes it difficult to assess the exact benefits of the gene therapy.

About half of the cases of SCID (over 1000 per year) have a different cause than Ashanti's. These children have X-linked SCID. Because the gene is carried on the X chromosome, the affected children are boys (see Chapter 12). As you will see in Chapter 32, white blood cells communicate with one another via hormonelike molecules called cytokines or interleukins. X-linked SCID is caused by a recessive allele that encodes a defective receptor on the surfaces of certain white blood cells. Without this receptor, white blood cells can secrete cytokines but cannot respond to them. The immune system cannot function without this communication. Because X-linked SCID is almost always fatal, bone marrow stem cells are relatively accessible, and the defect is simple and well-understood, X-linked SCID has become a prime target for gene therapy teams from several countries.

Bone marrow stem cells are removed from an affected child and infected with a virus carrying an allele coding for a normal cytokine receptor. The virus inserts the normal receptor allele into the stem cells' DNA. The "cured" stem cells are then injected back into the child's marrow. As these stem cells divide, some of their daughter cells mature into functioning cells of the immune system, while others remain in the bone marrow and continue dividing throughout the patient's lifetime. As many as nine children may have been permanently cured by this procedure, and several others have shown significant but incomplete immune system development. Unfortunately, the treatment is not without danger. In early 2002, one of the treated children developed leukemia. After intensive study, researchers concluded that DNA from the virus had inserted itself into a gene that, when overactive, causes leukemia. Possibly, the leukemia was triggered by a combination of genetic predisposition in the child and the gene therapy. The Recombinant DNA Advisory Committee of the National Institutes of Health concluded that the risks were very small, and that the participating U.S. teams could continue the therapy. However, a second case of leukemia occurred later in 2002, at least temporarily shutting down not only the SCID therapy, but dozens of other gene therapy trials employing similar viral techniques.

SCID gene therapy illustrates both the promise and peril of current techniques. Most children with SCID cannot live a normal life, and usually do not live at all, without gene therapy. Is gene therapy worth the risk of leukemia? Gene therapies, like most other biotechnologies, are still in their infancy. Difficult decisions must be made by governments, scientists, physicians, patients, and, in the case of affected children, parents. Science may, perhaps, be able to quantify the risks and benefits of such therapies, but who should decide how much risk is too much?

13.6 What Are the Major Ethical Issues Surrounding Modern Biotechnology?

Modern biotechnology offers the promise—some would say the threat—of greatly changing our lives and the lives of many other organisms on Earth. As Spiderman noted, "With great power comes great responsibility." Is humanity capable of handling the responsibility of biotechnology? Enormous controversy swirls around many applications of biotechnology. Here we will explore two such controversies: the use of genetically modified organisms in agriculture, and prospects for genetically modifying human beings.

Should Genetically Modified Organisms Be Permitted in Agriculture?

The aims of "traditional" and "modern" agricultural biotechnology are the same: to modify the genetic makeup of living organisms to make them more useful. However, there are three principal differences. First, traditional biotechnology is usually slow; many generations of selective breeding are necessary before significantly useful new traits appear in plants or animals. Genetic engineering, in contrast, can potentially introduce massive genetic changes in a single generation. Second, traditional biotechnology almost always recombines genetic material from the same, or at least very closely related, species, while genetic engineering can recombine DNA from very different species in one organism. Finally, traditional biotechnologists had no way to manipulate the DNA sequence of genes themselves. Genetic engineering, however, can produce new genes never before seen on Earth.

Transgenic crops have clear advantages for farmers. Herbicide-resistant crops allow farmers to rid their fields of weeds (which may reduce harvests by 10% or more) at virtually any stage of crop growth through the use of powerful, nonselective herbicides. Insect-resistant crops decrease the need to apply synthetic pesticides, saving the cost of the pesticides themselves, tractor fuel, and labor. Because transgenic crops produce larger harvests at less cost, these savings may be passed along to the consumer.

In addition to potentially lower prices, specific types of transgenic crops might have other benefits for the consumer. In many developing countries, rice is the primary food source. Unfortunately, with diets consisting mostly of rice, many people in these nations do not obtain enough vitamin A or the yellow pigment beta-carotene (the precursor to vitamin A). As a result, about a million children in Asia, Africa, and Latin America die each year, and another 250,000 children become blind. As a possible solution to this devastating problem, scientists recently developed "golden rice." Normal rice is tan (it can be made white by removing the husk and bran). To create golden rice, scientists inserted a gene from daffodils into rice, which caused the rice to produce beta-carotene. It was hoped that golden rice could protect children in developing countries from vitamin A deficiency. As it turns out, as of early 2003, the promise of golden rice remained unfulfilled for several reasons. First, the current strains don't make enough beta-carotene. Second, the original golden rice strains grow well only in certain areas; for worldwide use, the genes must be inserted into local varieties of rice. Third, beta-carotene is efficiently absorbed by the digestive tract only if the diet contains appreciable amounts of fat, which is usually not the case in developing countries. Finally, most people who might benefit from this transgenic technology are too poor to afford it.

Nevertheless, scientists are working to increase the beta-carotene content of golden rice. Further, the biotech company Zeneca and multiple patent holders have given the technology to research centers in the Philippines, India, China, and Vietnam for free, with the hope that they will modify native rice varieties for local use. Will this rice be "golden"? Only the future can provide the answer.

Regardless of potential monetary or health benefits, many people strenuously object to transgenic crops or livestock. There are two principal scientific objections to the use of genetically modified organisms (GMOs) in agriculture: they may be hazardous to human health, and they may be dangerous to the environment.

Are Foods from GMOs Dangerous to Eat?

The first argument against transgenic foods is that they may be dangerous to the people who eat them. In most cases, this is not a major concern. For example, tests have shown that the Bt protein used to make insect-resistant plants is not toxic to mammals, and should not prove a danger to human health. The Flavr Savr™ tomato, which lacks an enzyme that makes tomatoes get soft as they ripen (and therefore bruise easily during shipment), doesn't taste very good and has disappeared from grocery shelves, but it didn't make people sick. Transgenic fish that produce extra growth hormone are also very unlikely to be hazardous to eat, because growth hormone is also made in the human body. If growth-enhanced livestock are ever developed, they will simply have more meat, composed of exactly the same proteins that exist in nontransgenic animals, so they shouldn't be dangerous either.

Another possible danger is that people might be allergic to genetically modified plants. A few years ago, researchers inserted a gene from Brazil nuts into soybeans in an attempt to improve the balance of amino acids in soybean protein. It was discovered, however, that people allergic to Brazil nuts would probably also be allergic to the transgenic soybeans and eat them without suspecting that they might cause an allergic reaction. Needless to say, this transgenic soybean never made it to the farm. StarLink™ corn, which contained a genetically modified Bt protein, was not approved for human consumption because the engineered Bt protein is digested less readily than normal Bt protein and might cause allergic reactions. The U.S. Food and Drug Administration now monitors all new transgenic crop plants for allergenic potential. Interestingly, several groups of researchers are using genetic engineering to try to make nonallergenic peanuts (for consumption by people who are allergic to normal peanuts) by modifying or removing the genes that encode the most allergenic peanut proteins.

In 2003, the U.S. Society of Toxicology studied the risks of genetically modified plants and concluded that current transgenic plants pose no significant dangers to human health. The Society also recognized that past safety does not guarantee future safety, and recommended continued testing and evaluation of all new genetically modified plants.

Are GMOs Hazardous to the Environment?

The environmental effects of GMOs are much more problematic. Because the genes for herbicide resistance or pest resistance are incorporated into the genome of the transgenic crop, these genes will be in its pollen, too. A farmer cannot control where pollen from a transgenic crop will go; wind might carry pollen miles from the farmer's field. In some instances, this probably doesn't matter very much. In the United States, for example, there are no wild relatives of wheat, so pollen from transgenic wheat would be highly unlikely to spread resistance genes to wild plants. Many food crops, such as oats, wheat, and barley, originated in Eastern Europe and the Middle East; in these regions, many weedlike relatives of food crops exist in the wild. Suppose these plants interbred with transgenic crops and became resistant to herbicides or pests. Would herbicide resistance make them significant weed problems for agriculture? Would they displace other native plants in the wild because insects would be less likely to eat them? Even if transgenic crops have no close relatives in the wild, bacteria and viruses can carry genes from one plant to another, even between unrelated species. Might such lateral transfer spread unwanted genes into wild plant populations? No one knows the answers to these questions.

In 2002, a committee of the U.S. National Academy of Sciences studied the potential impact of transgenic crops on the environment. The committee pointed out that crops modified by both traditional breeding methods and recombinant DNA technologies have the potential to cause major changes in the environment. In addition, the committee found that the United States does not have an adequate system for monitoring changes in ecosystems that might be caused by transgenic crops. The committee recommended more thorough screening of transgenic plants before they are used commercially, as well as sustained ecological monitoring of both the agricultural and natural environments after commercialization.

What about transgenic animals? Unlike pollen, most domesticated animals, such as cattle or sheep, are relatively immobile. Further, most have few wild relatives with which they might exchange genes, so the dangers to natural ecosystems appear minimal. However, some transgenic animals, especially fish, may pose significant potential threats (see "Earth Watch: Bonanza from the Sea or Frankenfish?").

Should a Human Genome Be Changed by Biotechnology?

Many ethical implications of human applications of biotechnology are fundamentally the same as those connected with other medical procedures. For example, long before biotechnology enabled prenatal testing for cystic fibrosis or sickle-cell anemia, trisomy 21 (Down syndrome) could be diagnosed in embryos by simply counting the chromosomes in cells taken from the amniotic fluid (see "Health Watch: Prenatal Genetic Screening"). Whether parents should use such information as a basis for therapeutic abortion or to prepare to care for the affected child is an ethical issue that generates considerable debate. Other ethical concerns, however, have arisen purely as a result of advances in biotechnology; for example, should people be allowed to select the genomes of their offspring, or, even more controversially, should they be allowed to *change* the genomes?

Several years ago, a girl in Colorado was born with Franconi anemia, a genetic disorder that causes not only anemia but also skeletal abnormalities, such as missing thumbs. It is fatal without a bone marrow transplant. Her parents wanted another child—a very special child, without Franconi anemia, who could serve as a donor for their daughter. They went to Yury Verlinsky of the Reproductive Genetics Institute for help. Verlinsky used the parents' sperm and eggs to create dozens of embryos in culture. The embryos were then tested for the genetic defect and for tissue compatibility with the couple's daughter. Verlinsky chose an embryo with the desired genotype and implanted it into the mother's uterus. Nine months later, a son was born. Blood from his umbilical cord provided cells to transplant into his sister's bone marrow. Today, both children are healthy. Was this an appropriate use of genetic screening? Should dozens of embryos be created, knowing that the vast majority will be discarded? Is this ethical if it is the only way to save the life of another child? Assuming that it's possible someday, would this be an ethical method of selecting embryos that would grow up to be bigger or stronger football players?

Today's technology allows physicians only to select among existing embryos, not to change their genomes. But technologies do exist to alter the genomes of, for example, bone marrow stem cells in attempts to cure SCID. Soon, it may become possible to change the genes of fertilized eggs (Fig. 13-11). If such techniques were used to fix SCID or cystic fibrosis, would they be ethical? What about improving prospective athletes? If and when the technology is developed to cure diseases, it will be difficult to prevent it from being used for nonmedical purposes. Who will determine what is an appropriate use, and what is trivial vanity?

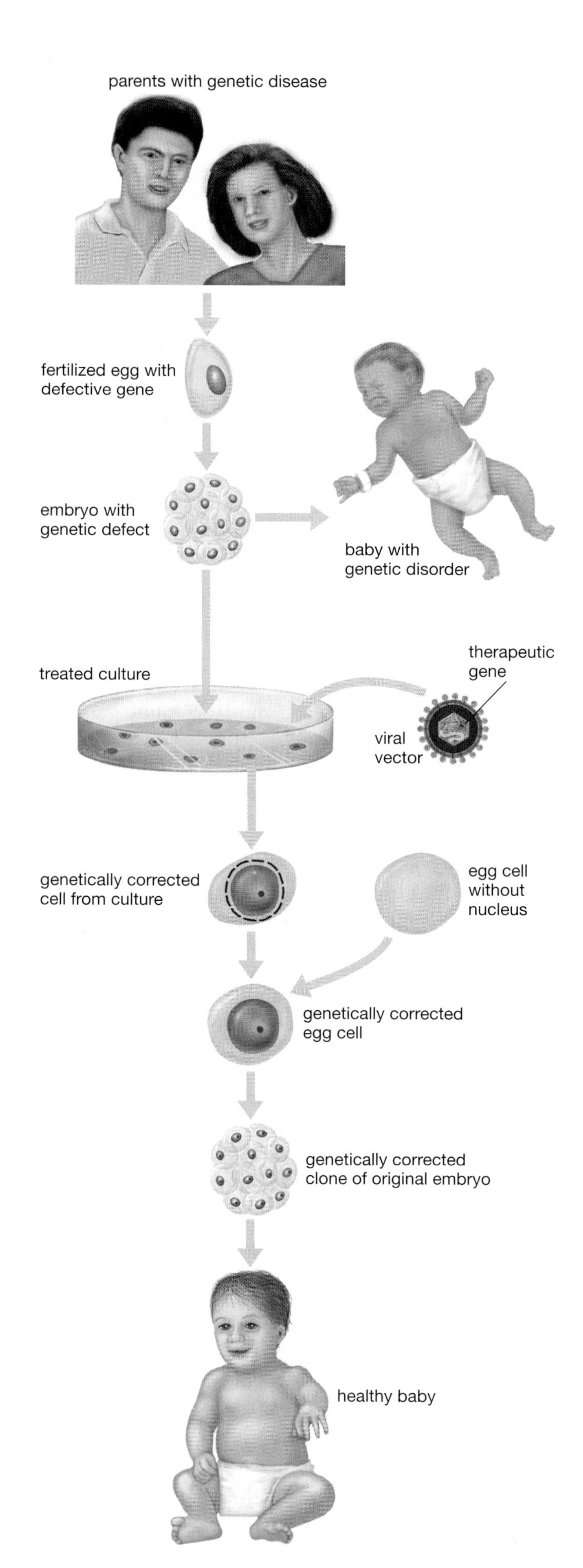

FIGURE 13-11 Human cloning technology might allow permanent correction of genetic defects
In this process, human embryos are derived from eggs fertilized in culture dishes using sperm and eggs from a man and woman, one or both of whom have a genetic disorder. When an embryo containing a defective gene grows into a small cluster of cells, a single cell would be removed from the embryo and the defective gene replaced using an appropriate vector. The repaired nucleus could then be implanted into another egg (taken from the same woman) whose nucleus had been removed. The repaired egg cell would then be implanted in the woman's uterus for normal development.

HEALTH WATCH Prenatal Genetic Screening

Prenatal diagnosis of a variety of genetic disorders, including cystic fibrosis, sickle-cell anemia, and Down syndrome, requires samples of fetal cells or chemicals produced by the fetus. Presently, two main techniques are used to obtain these samples: *amniocentesis* and *chorionic villus sampling*. A technique for analyzing fetal cells from maternal blood is also under development. Several types of diagnostic tests can be performed on the samples.

AMNIOCENTESIS

The human fetus, like all animal embryos, develops in a watery environment. As you'll see in Chapter 37, a waterproof membrane called the *amnion* surrounds the fetus and holds the fluid. As the fetus develops, it sheds some of its own cells into the fluid, which is called *amniotic fluid*. When a fetus is 16 weeks or older, amniotic fluid can be collected safely by a procedure called **amniocentesis**. A physician determines the position of the fetus by ultrasound scanning, inserts a sterilized needle through the abdominal wall, the uterus, and amnion, and withdraws 10 to 20 milliliters of fluid (Fig. E13-3). Biochemical analysis may be performed on the fluid immediately, but there are very few cells in the sample. For most analyses, such as karyotyping for Down syndrome, the cells must first be allowed to multiply in culture. After a week or two, there are normally enough cells to work with.

CHORIONIC VILLUS SAMPLING

The *chorion* is a membrane that is produced by the fetus and becomes part of the placenta. The chorion produces many small projections, called *villi*. In **chorionic villus sampling (CVS)**, a physician inserts a small tube into the uterus through the mother's vagina and suctions off a few fetal villi for analysis (see Fig. E13-3); this does not harm the fetus. CVS has two major advantages over amniocentesis. First, it can be done much earlier in pregnancy—as early as the eighth week. This is especially important if the woman is contemplating a therapeutic abortion if the fetus has a major defect. Second, the sample contains a much higher concentration of fetal cells than can be obtained by amniocentesis, so analyses can be performed immediately. However, chorionic cells tend to be more likely to have abnormal numbers of chromosomes (even when the fetus is normal), which complicates karyotyping. CVS also appears to have slightly greater risks than amniocentesis. Finally, CVS cannot detect certain disorders, such as spina bifida. For these reasons, CVS is much less commonly performed than amniocentesis.

FETAL CELLS FROM MATERNAL BLOOD

A tiny number of fetal cells cross the placenta and enter the mother's bloodstream as early as the sixth week of pregnancy. Separating fetal cells (perhaps as few as one per milliliter of blood) from the huge numbers of maternal cells is challenging, but can be done. Several companies now offer paternity testing from maternal blood, but practical genetic screening for inherited disorders seems to be several years in the future.

ANALYZING THE SAMPLES

Several types of analyses can be performed on amniotic fluid or fetal cells (see Fig. E13-3). Biochemical analysis is used to determine the concentration of chemicals in the amniotic fluid. For example, many metabolic disorders can be detected by a low concentration of enzymes that normally catalyze specific metabolic pathways or by the abnormal accumulation of precursors or by-products. Analysis of the chromosomes of the fetal cells can show whether all the chromosomes are present, whether there are too many or too few of some, and whether any chromosomes show structural abnormalities.

Recombinant DNA techniques can be used to analyze the DNA of fetal cells to detect many defective alleles, such as those for cystic fibrosis or sickle-cell anemia. Prior to the development of PCR, fetal cells typically had to be grown in culture for as long as 2 weeks before cells had multiplied sufficiently. Now, the second step in prenatal diagnosis is to extract the DNA from a few cells and to use PCR to amplify the region containing the gene of interest. After a few hours, enough DNA is available for techniques such as RFLP analysis, which is used to detect the allele that causes sickle-cell anemia (see Fig. 13-9). If the infant is homozygous for the sickle-cell allele, some therapeutic measures can be taken. In particular, regular doses of penicillin greatly reduce bacterial infections that otherwise kill about 15% of homozygous children. Further, knowing that a child has the disorder ensures correct diagnosis and rapid treatment, should "sickling crises" occur.

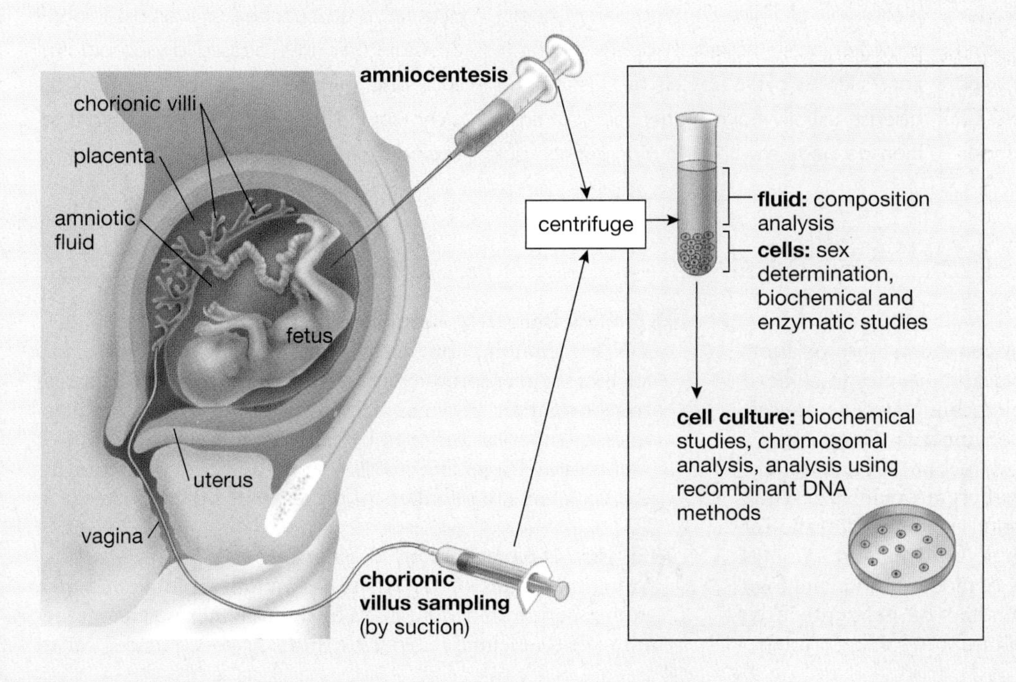

FIGURE E13-3 Prenatal cell sampling techniques
Two methods of obtaining fetal cell samples—amniocentesis and chorionic villus sampling—and some of the tests performed on the fetal cells.

CASE STUDY REVISITED Guilty or Innocent?

The Innocence Project is strictly a science-based effort—no courtroom theatrics, no picking away at fine points of law and procedure. The project takes cases only when molecular evidence exists that may exonerate the accused. Without the rapid advances in DNA technology in the past decade, Jimmy Bromgard, Eddie Lloyd, and dozens of other people convicted of serious crimes would still be in prison. Some innocent people might even have been executed.

Police and district attorneys also use DNA technology as an investigative tool. In 1990, three elderly women were raped in Goldsboro, North Carolina; two were murdered. DNA evidence indicated that all three crimes were committed by the same assailant, known only as the "Night Stalker." Over the years, the FBI and many state agencies have slowly built up DNA databases of criminals, each identified by "DNA fingerprints" of short tandem repeats (STRs). In 2001, the Goldsboro police created a DNA fingerprint profile of the Night Stalker from the evidence they had carefully stored for over a decade. They sent the profile to the North Carolina DNA database and discovered a match. Faced with indisputable DNA evidence, the Stalker confessed. He is now in prison.

Consider This: Who are the "heroes" in these stories? There are the obvious ones, of course—the professors and law students of the Innocence Project and the members of the Goldsboro Police Department. But what about Thomas Brock, who discovered *Thermus aquaticus* and its unusual lifestyle in Yellowstone hot springs (see "Scientific Inquiry: Hot Springs and Hot Science")? Or molecular biologist Kary Mullis, who discovered PCR? Or the hundreds of biologists, chemists, and mathematicians who, over the last century, developed procedures for gel electrophoresis, labeling DNA, and statistical analysis of sample matching?

Scientists often say that science is worthwhile for its own sake, and that it is difficult or impossible to predict which discoveries will lead to the greatest benefits for humanity. Nonscientists, when asked to pay the costs of scientific projects, are sometimes skeptical of such claims. How do you think public monetary support of science should be allocated? Forty years ago, would *you* have voted to give Thomas Brock public funds to see what types of organisms lived in hot springs?

Links to Life: Biotechnology, Privacy, and You

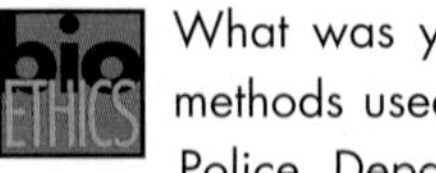

What was your reaction to the methods used by the Goldsboro Police Department to find the Night Stalker? Most people feel that matching the DNA from the crime scenes to the Night Stalker's DNA fingerprint files is a superb example of science and law enforcement working together to protect the public. Some, however, worry that the ever-increasing amount of personal information in government—and possibly industrial—files might be abused if placed in the wrong hands. Of course, the Night Stalker's DNA fingerprints were only on file because he had committed other serious crimes. How severe a crime do you think warrants DNA fingerprinting? Murder? Assault? Breaking and entering? Should *everyone's* DNA be on file? Every criminal has a first offense; perhaps more would be caught if there were a universal DNA file.

Others worry about the possible exploitation of biotechnology by health providers or employers. People with conditions such as cystic fibrosis or Huntington's disease usually have higher medical costs than do people without major genetic diseases. Medical care for a patient with cystic fibrosis, for example, is about $40,000 per year. Should health insurance rates reflect known genetic diseases? If our ability to understand and predict genetic disorders improves, should employers be able to discriminate against people who are genetically predisposed to potentially costly conditions or behaviors, such as schizophrenia, breast cancer, or even risk-taking? Several states have passed laws forbidding such discrimination. How do you think society should handle the information revealed by human biotechnology?

Summary of Key Concepts

13.1 What Is Biotechnology?

Biotechnology is any industrial or commercial use or alteration of organisms, cells, or biological molecules to achieve specific practical goals. Modern biotechnology generates altered genetic material via genetic engineering. Genetic engineering frequently involves the production of recombinant DNA by combining DNA from different organisms. When DNA is transferred from one organism to another, the recipients are called transgenic or genetically modified organisms (GMOs). Major applications of modern biotechnology include increasing our understanding of gene function, treating disease, improving agriculture, and solving crimes.

13.2 How Does DNA Recombination Occur in Nature?

DNA recombination occurs naturally through processes such as sexual reproduction (during crossing over); bacterial transformation, in which bacteria acquire DNA from plasmids or other bacteria; and viral infection, in which viruses incorporate fragments of DNA from their hosts and transfer the fragments to members of the same or other species.

13.3 How Is Biotechnology Used in Forensics?

Small quantities of DNA, such as might be obtained at a crime scene, can be amplified by the polymerase chain reaction (PCR) technique. Specific short DNA sequences, called short tandem repeats (STR), are amplified, labeled with colored molecules, and separated by gel electrophoresis. The pattern of STRs is unique to each individual, and can be used to match DNA found at a crime scene with DNA from suspects.

13.4 How Is Biotechnology Used in Agriculture?

Many crop plants have been modified by the addition of genes that promote herbicide resistance or pest resistance. The most common procedure uses restriction enzymes to insert the gene into a plasmid from the bacterium *Agrobacterium tumefaciens*. The genetically modified plasmid is then used to transform the bacteria, which are allowed to infect plant cells. The plasmid inserts the new gene into one of the plant chromosomes. Using cell culture, entire plants are grown from the transgenic cells, and eventually planted commercially. Plants may also be modified to produce human proteins, vaccines, or antibodies. Transgenic animals may be produced as well, with properties such as faster growth, increased production of valuable products such as milk, or the ability to produce human proteins, vaccines, or antibodies.

13.5 How Is Biotechnology Used for Medical Diagnosis and Treatment?

Biotechnology may be used to diagnose genetic disorders such as sickle-cell anemia or cystic fibrosis. For example, in the diagnosis of sickle-cell anemia, restriction enzymes cut normal and defective globin alleles in different locations. The resulting DNA fragments of different lengths may then be separated and identified by gel electrophoresis. In the diagnosis of cystic fibrosis, DNA probes complementary to various cystic fibrosis alleles are placed on a DNA array. Base-pairing of a patient's DNA to specific probes on the array identifies which alleles are present in the patient.

Inherited diseases are caused by defective alleles of crucial genes. Genetic engineering may be used to insert functional alleles of these genes into normal cells, into stem cells, or even into eggs to correct the genetic disorder.

13.6 What Are the Major Ethical Issues Surrounding Modern Biotechnology?

The use of genetically modified organisms in agriculture is controversial for two major reasons: consumer safety and environmental protection. In general, GMOs contain proteins that are harmless to mammals, are readily digested, or are already found in other foods. The transfer of potentially allergenic proteins to normally nonallergenic foods can be avoided by thorough testing. Environmental effects of GMOs are more difficult to predict. It is possible that foreign genes, such as those for pest resistance or herbicide resistance, might be transferred to wild plants, with resulting damage to agriculture and/or disruption of ecosystems. If they escape, highly mobile transgenic animals might displace their wild relatives.

Genetically selecting or modifying human embryos is highly controversial. As technologies improve, society may be faced with decisions about the extent to which parents should be allowed to correct or enhance the genomes of their children.

Key Terms

amniocentesis *p. 259*
biotechnology *p. 242*
chorionic villus sampling (CVS) *p. 259*
DNA fingerprinting *p. 248*
DNA probe *p. 247*
gel electrophoresis *p. 247*
genetic engineering *p. 242*
genetically modified organism (GMO) *p. 242*
plasmid *p. 242*
polymerase chain reaction (PCR) *p. 245*
recombinant DNA *p. 242*
restriction enzyme *p. 249*
restriction fragment length polymorphism (RFLP) *p. 252*
stem cell *p. 255*
transformation *p. 242*
transgenic *p. 242*

Thinking Through the Concepts

To take a multiple-choice quiz with feedback on the contents of this chapter, visit http://www.prenhall.com/audesirk7. *Log in to the Web site selected by your instructor and navigate to the Self Test section for this chapter.*

? Review Questions

1. Describe three natural forms of genetic recombination, and discuss the similarities and differences between recombinant DNA technology and these natural forms of genetic recombination.
2. What is a plasmid? How are plasmids involved in bacterial transformation?
3. What is a restriction enzyme? How can restriction enzymes be used to splice a piece of human DNA within a plasmid?
4. What is a short tandem repeat? How are short tandem repeats used in forensics?
5. Describe the polymerase chain reaction technique.
6. Describe several uses of genetic engineering in agriculture.
7. Describe several uses of genetic engineering in human medicine.
8. Describe amniocentesis and chorionic villus sampling, including the advantages and disadvantages of each. What are their medical uses?

Applying the Concepts

1. Discuss the ethical issues that surround the release of bioengineered organisms (plants, animals, or bacteria) into the environment. What could go wrong? What precautions might prevent the problems you listed from occurring? What benefits do you think would justify the risks?
2. Do you think that using recombinant DNA technologies to change the genetic composition of a human egg cell is ever justified? If so, what restrictions should be placed on such a use?
3. If you were contemplating having a child, would you want both yourself and your spouse tested for the cystic fibrosis gene? If both of you were carriers, how would you deal with this decision?
4. As you may know, many insects have evolved resistance to common pesticides. Do you think that insects might evolve resistance to Bt crops? If this is a risk, do you think that Bt crops should be planted anyway? Why?
5. Why are boys the victims of X-linked SCID? Would you *ever* expect to find a girl born with X-linked SCID? Why?

For More Information

Brown, K. "Seeds of Concern." *Scientific American*, April 2001. What are the environmental risks of genetically modified crops?

Gibbs, W. W. "Plantibodies." *Scientific American*, November 1997. Clinical trials using human antibodies engineered into corn began in 1998; the target was tooth decay.

Gura, T. "New Genes Boost Rice Nutrients," *Science*, August 1999. Explanation of how rice was genetically engineered to produce vitamin A precursors.

Hoplin, K. "The Risks on the Table." *Scientific American*, April 2001. Describes the controversies over whether GM crops are safe to eat.

Langridge, W. H. R. "Edible Vaccines." *Scientific American*, September 2000. Plants may be developed to produce vaccines or treatments for diseases.

Marvier, M. "Ecology of Transgenic Crops." *American Scientist*, March/April 2001. A thoughtful article that weighs the benefits, risks, and uncertainties of bioengineered crops.

Miller, R. V. "Bacterial Gene Swapping in Nature." *Scientific American*, January 1998. How likely is it that genes introduced into bioengineered organisms might be inadvertently transferred to wild organisms?

Palevitz, B. A. "Society Honors Golden Rice Inventor." *The Scientist*, August 2001. Ingo Potrykus, one of the key researchers involved in making golden rice, describes the motives, triumphs, and turmoils.

Scientific American, June 1997. A special issue devoted to the prospects of human gene therapy.

Stokstad, E. "Engineered Fish: Friend or Foe of the Environment." *Science*, September 2002. Describes the Aqua Bounty transgenic Atlantic salmon and efforts to ensure its environmental safety.

Weidensaul, S. "Raising the Dead." *Audubon*, May-June 2002. Can cloning be used to resurrect extinct species from museum specimens? Don Colgan is trying to recreate the Tasmanian marsupial wolf.

Wheelwright, J. "Body, Cure Thyself." *Discover*, March 2002. Gene therapy holds great promise, but it is turning out to be more difficult, and perhaps more dangerous, than expected.

Media Activities

To access a Media Activity visit http://www.prenhall.com/audesirk7. *Log in to the Web site selected by your instructor, navigate to this chapter, and select the appropriate Media Activity number.*

13.1 Genetic Recombination in Bacteria

Estimated time: 5 minutes

We often think of DNA and the genes it contains as being static and unchanging. In fact, it is anything but static. This activity will demonstrate some of the different ways that genetic recombination can occur in cells, both prokaryotic and eukaryotic.

13.2 Polymerase Chain Reaction (PCR)

Estimated time: 5 minutes

In this activity, you'll see the process of polymerase chain reaction (PCR), which allows minute amounts of DNA to be amplified into large amounts.

13.3 Human Genome Sequencing

Estimated time: 5 minutes

One strategy for sequencing the genome called map-based sequencing is the primary subject of this activity.

13.4 Manufacturing Human Growth Hormone

Estimated time: 5 minutes

In this activity, you'll see how human growth hormone can be produced cheaply and safely using recombinant DNA techniques.

13.5 Web Investigation: Guilty or Innocent?

Estimated time: 10 minutes

We usually think of molecular forensic evidence such as genetic "fingerprints" being used to convict a suspect. But the same molecular techniques are being used to free those wrongly convicted. The Innocence Project is a nonprofit legal clinic whose sole purpose is identifying innocent people who were wrongly convicted of serious crimes. Learn about The Innocence Project and molecular forensic evidence in this activity.

UNIT THREE

Evolution and Diversity of Life

The ghostly grandeur of ancient bones evokes images of a lost world. Fossil remnants of extinct creatures, such as this Triceratops *dinosaur skeleton, provide clues to the biologists who attempt to reconstruct the history of life.*

CHAPTER

14 Principles of Evolution

Newly discovered fossils of feathered dinosaurs such as *Caudipteryx* (shown here in an artist's reconstruction) provide powerful evidence that today's birds descended from dinosaur ancestors.

AT A GLANCE

CASE STUDY A Missing Link Unearthed

As the twentieth century drew to an end, the little village of Sihetun in northeastern China became the unlikely setting for some of history's most breathtaking fossil discoveries. Working in dusty quarries around the village, Chinese fossil hunters extracted the exquisitely preserved remains of some previously undiscovered types of dinosaurs. New dinosaur discoveries are always cause for celebration, but the new Chinese specimens displayed a distinctive feature that propelled them to superstar status: they had feathers. Along the margin of these clearly dinosaurian fossil skeletons, impressions of what appeared to be feathers were plainly visible. For the first time, scientists had solid evidence that some dinosaurs had feathers.

In the years since the discovery of these fossils, many additional feathered dinosaurs have been unearthed, including some with especially numerous and beautifully preserved fossil feathers. The steady accumulation of such fossils has been profoundly exciting for paleontologists (scientists who study fossils), and has convinced most of them to accept the longstanding but controversial idea that birds are descended from dinosaurs. Can it really be true that the little birds that flit about in our backyards today are actually the surviving descendents of extinct dinosaurs? It seems so; the existence of feathered dinosaurs provides very persuasive evidence that modern birds arose from a branch of the dinosaur family tree—a branch whose members had feathers. That is, the fossils demonstrate that today's birds arose through **evolution**, the process by which the characteristics of a population of organisms change over time.

14.1 How Did Evolutionary Thought Evolve?

Modern biology is based on our understanding that life has evolved, but this fundamental principle was not recognized by early scientists. The main ideas of evolutionary biology were widely accepted only after the publication of Charles Darwin's work in the late 19th century, but the intellectual foundation on which these ideas rest developed gradually over the centuries before Darwin's time.

Early Biological Thought Did Not Include the Concept of Evolution

Pre-Darwinian science, heavily influenced by theology, held that all organisms were created simultaneously by God, and that each distinct life-form remained fixed and unchanging from the moment of its creation. This explanation of how life's diversity arose was elegantly expressed by ancient Greek philosophers, especially Plato and Aristotle. Plato (427–347 B.C.) proposed that each object on Earth was merely a temporary reflection of its divinely inspired "ideal form." Plato's student Aristotle (384–322 B.C.) categorized all organisms into a linear hierarchy that he called the "ladder of Nature."

These ideas formed the intellectual basis for the view that each type of organism has a form that is permanently fixed. This view reigned unchallenged for nearly 2000 years. By the eighteenth century, however, several lines of newly emerging evidence began to erode the dominance of this static view of creation.

FIGURE 14-1 The Grand Canyon of the Colorado River
Layer upon layer of sedimentary rock forms the walls of the Grand Canyon. The layers formed from the accumulation of sediments over more than a billion years.

Exploration of New Lands Revealed a Staggering Diversity of Life

As early European naturalists explored the newly discovered lands of Africa, Asia, and America, they found that the number of species, or different types of organisms, was much greater than anyone had suspected. They also observed that some species closely resembled one another, yet differed in some characteristics. These observations led some naturalists to consider that perhaps species could change after all. Some of the similar species might have developed from a common ancestor.

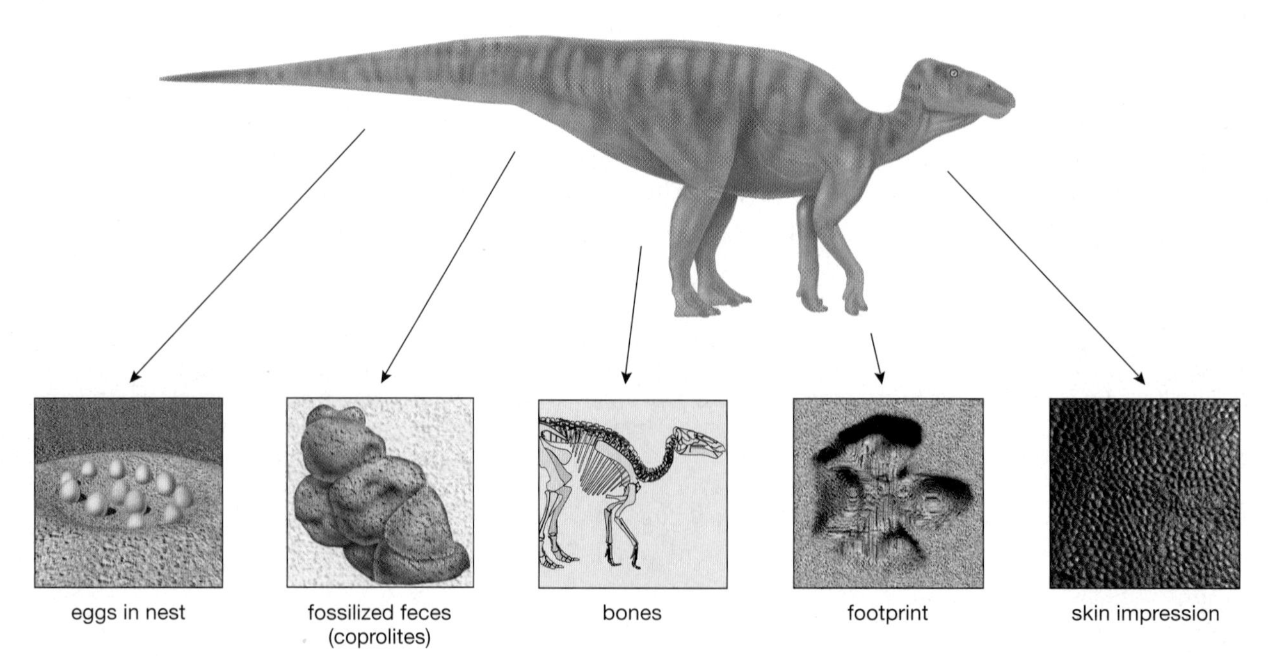

FIGURE 14-2 Types of fossils
Any part or trace of an organism that is preserved in rock or sediments is a fossil.

Fossil Discoveries Showed That Life Had Changed Over Time

As these new lands were explored, excavations for roads, mines, and canals revealed that many rocks occur in layers, formed when layers of sediment were deposited, with each new layer settling over the older layers (Fig. 14-1). Further exploration uncovered fragments that resembled parts of living organisms embedded within these layers. At first, these **fossils** were thought to be ordinary rocks that wind, water, or people had worked into lifelike forms. As more and more fossils were discovered, however, it became obvious that they were the remains or impressions of plants or animals that had died long ago, and which had been changed into or in some way preserved in rock (Fig. 14-2).

The way in which fossils are distributed in rock is also revealing. After studying fossils carefully, British surveyor William Smith (1769–1839) realized that certain fossils were always found in the same layers of rock. Further, the organization of fossils and rock layers was consistent: fossil type A could always be found in a rock layer directly above an older layer containing fossil type B, which in turn rested atop a still older layer containing fossil type C, and so on.

Fossil remains also showed a remarkable progression. Most fossils found in the oldest and deepest layers were very different from modern organisms; the resemblance to modern organisms gradually increased in progressively younger rocks. Many of the fossils were from plant or animal species that had gone *extinct*—that is, no members of the species still lived on Earth (Fig. 14-3).

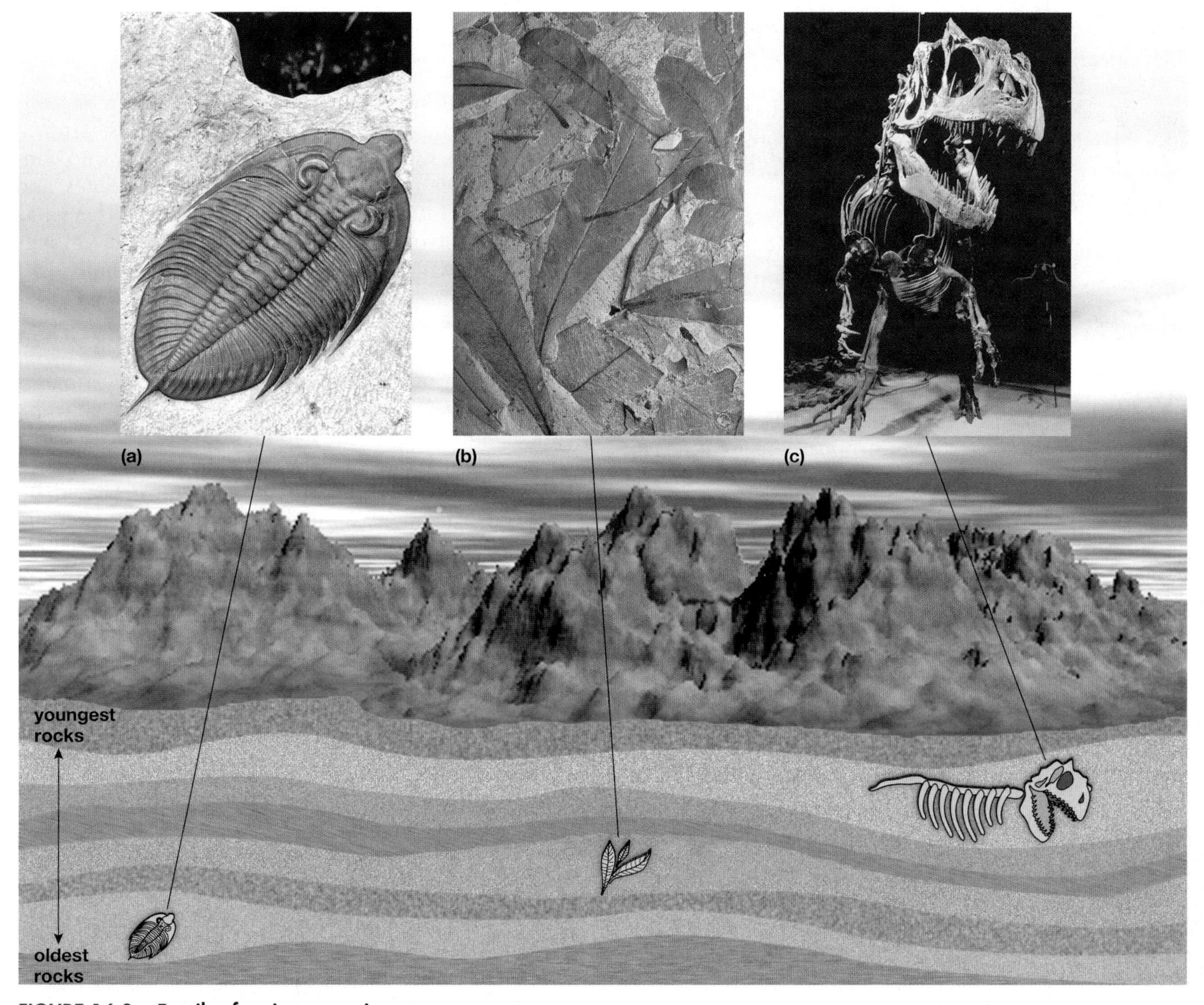

FIGURE 14-3 Fossils of extinct organisms
Fossils provide strong support for the idea that today's organisms were not created all at once, but arose over time by the process of evolution. If all species were created simultaneously, we would not expect **(a)** trilobites to be found in older rock layers than **(b)** seed ferns, which in turn would not be found deeper than **(c)** dinosaurs, such as *Allosaurus*. Trilobites became extinct about 230 million years ago, seed ferns about 150 million years ago, and dinosaurs 65 million years ago.

Putting these facts together, scientists came to the conclusion that different types of organisms had lived at various times in the past.

Some Scientists Devised Nonevolutionary Explanations for Fossils

What did this newfound richness of organisms mean? To account for the multitude of species, both living and extinct, while preserving the notion of creation by God, Georges Cuvier (1769–1832) proposed the theory of **catastrophism**. Cuvier, a French paleontologist, hypothesized that a vast supply of species was created initially. Successive catastrophes (such as the Great Flood described in the Bible) produced layers of rock and destroyed many species, fossilizing some of their remains in the process. The organisms of the modern world, he theorized, represent species that survived the catastrophes.

If Cuvier's belief that modern species survived from an original Creation were correct, then many individuals of modern, surviving species should also have died in the ancient catastrophes. Surely some of these would have been fossilized as well, so even the oldest rock layers should contain fossils of present-day species. Contrary to Cuvier's hypothesis, however, the vast majority of fossils are of extinct species. To account for this observation, French geologist Louis Agassiz (1807–1873) proposed that new creations after each catastrophe produced new and different species, and that modern species therefore result from the most recent creation. The fossils known at that time forced Agassiz to include at least 50 separate catastrophes and creations in his hypothesis!

A Few Scientists Speculated That Life Had Evolved

Some scientists could not readily accept that God would create so many types of organisms, only to let thousands of species go extinct. French naturalist Georges Louis LeClerc (1707–1788), who was known by the title Comte de Buffon, proposed one alternative explanation. Buffon suggested that perhaps the original Creation resulted in a relatively small number of founding species, and that some modern species had been "conceived by Nature and produced by Time"—that is, they had evolved through natural processes. Most people were not convinced by his claims. First, Buffon could not provide a mechanism whereby nature could "conceive" new species. Second, at that time, no one thought that Earth was old enough to allow sufficient time for the "production" of new species.

Geology Provided Evidence That Earth Is Exceedingly Old

In the early 1700s, few scientists suspected that Earth could be more than a few thousand years old. At that time, the age of Earth had been determined only by using religious texts (such as the Bible), which were considered to be historically accurate. For example, scholars who counted the number of generations described in the Old Testament estimated that Earth could not have been more than 4000 to 6000 years old. An Earth this young posed problems for the idea that life had evolved. For example, ancient writers such as Aristotle described organisms identical to those still present in Europe more than 2000 years later, such as wolves, deer, and lions. If organisms changed so little over that time, and Earth was only a few thousand years older than Aristotle, how could the entire range of modern species (plus extinct fossil species) have had time to arise?

One explanation was that, perhaps, Earth *was* old enough to allow for the development of new species. Such was the view of geologists James Hutton (1726–1797) and Charles Lyell (1797–1875), who contemplated the forces of wind, water, earthquakes, and volcanism. Hutton and Lyell concluded that there was no need to invoke catastrophes to explain the findings of geology. After all, layers of sediment normally drift to the bottoms of rivers and lakes, and lava flows create new rocks in areas of volcanic activity. Why, then, should we assume that layers of rock are evidence of anything but ordinary natural processes, occurring repeatedly over long periods of time? This concept, called **uniformitarianism**, had profound implications. If slow, natural processes alone produced layers of rock thousands of feet thick, then Earth must be old indeed, possibly many millions of years old. Hutton and Lyell, in fact, concluded that Earth was eternal: "No Vestige of a Beginning, no Prospect of an End," in Hutton's words. (Modern geologists estimate that Earth is about 4.5 billion years old; see "Scientific Inquiry: How Do We Know How Old a Fossil Is?" in Chapter 17.) Thus, Hutton and Lyell's reasoning provided the time for evolution to occur. But there was still no convincing mechanism.

Some Pre-Darwin Biologists Proposed Mechanisms for Evolution

One of the first scientists to propose a mechanism for evolution was French biologist Jean Baptiste Lamarck (1744–1829). Lamarck was impressed by the progression of fossil types through rock layers. He observed that older fossils tended to be simpler, whereas younger fossils tended to be more complex and more like existing organisms. In 1801, Lamarck hypothesized that organisms evolved through the **inheritance of acquired characteristics**, a process in which the bodies of living organisms are modified through the use or disuse of different parts; these modifications are then inherited by offspring. Why would bodies be modified? Lamarck proposed that all organisms possess an innate drive for perfection. For example, ancestral giraffes may have tried to gain access to additional food by stretching upward to feed on leaves that grew high up in trees, and their necks became slightly longer as a result. Their offspring would then inherit these longer necks, and then stretch their own necks farther still to reach even higher leaves. Eventually, this process would produce modern giraffes with very long necks indeed.

Today, we understand the mechanics of inheritance and can see that Lamarck's proposed evolutionary mechanism could not work as he described it. Acquired characteristics are not inherited; if a prospective father pumps iron, his children won't be born with large muscles. In Lamarck's time, however, the principles of inheritance had not yet been discovered; Gregor Mendel (see Chapter 12) was born only a few years before Lamarck's death. Nevertheless, Lamarck's insight—that inheritance plays an important role in evolution—greatly influenced the later biologists who discovered the key mechanism of evolution.

Darwin and Wallace Proposed a Mechanism of Evolution

By the mid-1800s, a growing number of biologists had concluded that modern species had evolved from earlier ones. But by what means? In 1858, Charles Darwin and Alfred Russel Wallace, working separately, provided convincing evidence that evolution was driven by a simple yet powerful process.

Although their social and educational backgrounds were very different, Darwin and Wallace had some similar experiences that would influence their ideas. Both had traveled extensively in the tropics and had studied the plants and animals living there. Both found that some species differed only in a few fairly subtle, but ecologically important, features (Fig. 14-4). Both were familiar with the fossils that had been discovered, which

(a) Large ground finch, beak suited to large seeds

(b) Small ground finch, beak suited to small seeds

(c) Warbler finch, beak suited to insects

(d) Vegetarian tree finch, beak suited to leaves

FIGURE 14-4 Darwin's finches, residents of the Galapagos Islands Darwin studied a group of closely related species of finches on the Galapagos Islands. Each species specializes in eating a different type of food and has a beak of characteristic size and shape, because natural selection has favored the individuals best suited to exploit each food source efficiently. Aside from the differences in their beaks, the finches are quite similar.

SCIENTIFIC INQUIRY Charles Darwin—Nature Was His Laboratory

Like many students, Charles Darwin excelled only in subjects that intrigued him. Although his father was a physician, Darwin was uninterested in medicine and unable to stand the sight of surgery. He eventually obtained a degree in theology from Cambridge University, although theology too was of minor interest to him. What he really liked to do was to tramp over the hills, observing plants and animals, collecting new specimens, scrutinizing their structures, and categorizing them.

In 1831, when Darwin was 22 years old (Fig. E14-1), he secured a position as "gentleman companion" to Captain Robert Fitzroy of the ship *H.M.S Beagle*. The *Beagle* soon embarked on a 5-year surveying expedition along the coastline of South America and then around the world. In addition to his duties as companion to the captain, Darwin also served as the expedition's official naturalist, whose task was to observe and collect geological and biological specimens. The *Beagle* sailed to South America, and made many stops along its coast. There Darwin observed the plants and animals of the Tropics and was stunned by the greater diversity of species compared with Europe. Although he had boarded the *Beagle* convinced of the permanence of species, his experiences soon led him to doubt this. He discovered a snake with rudimentary hind limbs, calling it "the passage by which Nature joins the lizards to the snakes." Another snake he encountered vibrated its tail like a rattlesnake but had no rattles and therefore made no noise. Similarly, Darwin noticed that penguins used their wings to paddle through the water rather than fly through the air. If a creator had individually created each animal in its present form, to suit its present environment, what could be the purpose behind these makeshift arrangements?

Perhaps the most significant stopover of the voyage was the month spent on the Galapagos Islands off the northwestern coast of South America. There, Darwin found huge tortoises; in fact, *galapagos* means "tortoise" in Spanish. Different islands were home to distinctively different types of tortoises (Fig. E14-2). Darwin also found several types of finches; as with the tortoises, different islands had slightly different finches. Could the differences in these organisms have arisen after they became isolated from one another on separate islands? The diversity of tortoises and finches haunted him for years afterward.

FIGURE E14-1 A painting of Charles Darwin as a young man

showed a trend of increasing complexity through time. Finally, both were aware of Hutton and Lyell's proposition that Earth is extremely ancient. These facts suggested to both Darwin and Wallace that species change over time. Both men sought a mechanism that might cause such evolutionary change.

In 1858, Darwin and Wallace each described the same mechanism for evolution in remarkably similar papers that were presented to the Linnaean Society in London. Like Gregor Mendel's manuscript on the principles of genetics, their papers had little impact. In fact, the secretary of the society wrote in his annual report that nothing very interesting had happened that year. Fortunately, the next year Darwin published his monumental book *On the Origin of Species by Means of Natural Selection*, which attracted a great deal of attention to the new theory.

14.2 How Does Natural Selection Work?

Darwin and Wallace concluded that life's huge variety of excellent designs arose by a process of descent with modification, in which individuals in each generation differ slightly from the members of the preceding generation. Over long stretches of time, these small differences accumulate to produce major transformations. The chain of logic leading to this powerful conclusion turns out to be surprisingly simple and straightforward. It is based on four observations and the conclusions that can be drawn from them.

Observation 1: A natural **population**, or all the individuals of one species in a particular area, has the potential to grow rapidly, because organisms can produce far

In 1836, Darwin returned to England and became established as one of the foremost naturalists of his time. But the problem of how isolated populations come to differ from each other gnawed constantly on his mind. Part of the solution came to him from an unlikely source: the writings of an English economist and clergyman, Thomas Malthus. In his *Essay on Population*, Malthus wrote, "It may safely be pronounced, therefore, that [human] population, when unchecked, goes on doubling itself every 25 years, or increases in a geometrical ratio." Darwin realized that a similar principle holds true for plant and animal populations. In fact, most organisms can reproduce much more rapidly than can humans (consider rabbits, dandelions, and houseflies) and consequently could produce overwhelming populations in short order. Nonetheless, the world is not chest-deep in rabbits, dandelions, or flies: natural populations do not grow "unchecked" but tend to remain approximately constant in size. Clearly, vast numbers of individuals must die in each generation, and most must not reproduce.

From his experience as a naturalist, Darwin realized that members of a species typically differ from one another. Further, which individuals die without reproducing in each generation is not arbitrary, but depends to some extent on the structures and abilities of the organisms. This observation was the source of the theory of evolution by natural selection. As Darwin's colleague Alfred Wallace put it, "Those which, year by year, survived this terrible destruction must be, on the whole, those which have some little superiority enabling them to escape each special form of death to which the great majority succumbed." Here you see the origin of the expression "survival of the fittest." That "little superiority" that confers greater fitness might be better resistance to cold, more efficient digestion, or any of hundreds of other advantages, some very subtle. Everything now fell into place. Darwin wrote, "It at once struck me that under these circumstances favorable variations would tend to be preserved, and unfavorable ones to be destroyed." If a favorable variation were inheritable, then the entire species would eventually consist of individuals possessing the favorable trait. With the continual appearance of new variations (due, as we now know, to mutations), which in turn are subject to further natural selection, "the result . . . would be the formation of new species. Here, then, I had at last got a theory by which to work."

FIGURE E14-2 One species of Galapogos island tortoise

When Darwin finally published *On the Origin of Species* in 1859, his evidence had become truly overwhelming. Although its full implications would not be realized for decades, Darwin's theory of evolution by natural selection has become a unifying concept for virtually all of biology.

more offspring than are required merely to replace the parents.

Observation 2: Nevertheless, the number of individuals in a natural population tends to remain relatively constant over time.

Conclusion 1: Therefore, more organisms must be born than survive long enough to reproduce. If some individuals fail to survive, it must also be true that organisms compete to survive and reproduce. In each generation, many individuals must die young. Even among those that survive, many must fail to reproduce, produce few offspring, or produce less-fit offspring that, in turn, fail to survive and reproduce.

Observation 3: Individual members of a population differ from one another in many respects, including their ability to obtain resources, withstand environmental extremes, and escape predators.

Conclusion 2: These differences among individuals help determine which individuals survive and reproduce most successfully, thereby leaving the most offspring. This process, by which those individuals whose traits are most advantageous leave a larger number of offspring, is known as **natural selection**.

Observation 4: At least some of the variation in traits that affect survival and reproduction is due to differences that may be passed from parent to offspring.

Conclusion 3: Because the individuals that are best suited to their environment leave more offspring, the traits (and underlying genes) of these individuals are passed to a larger proportion of the individuals in subsequent generations. Over many generations, this differential, or unequal, reproduction among individuals with different genetic makeup changes the overall genetic composition

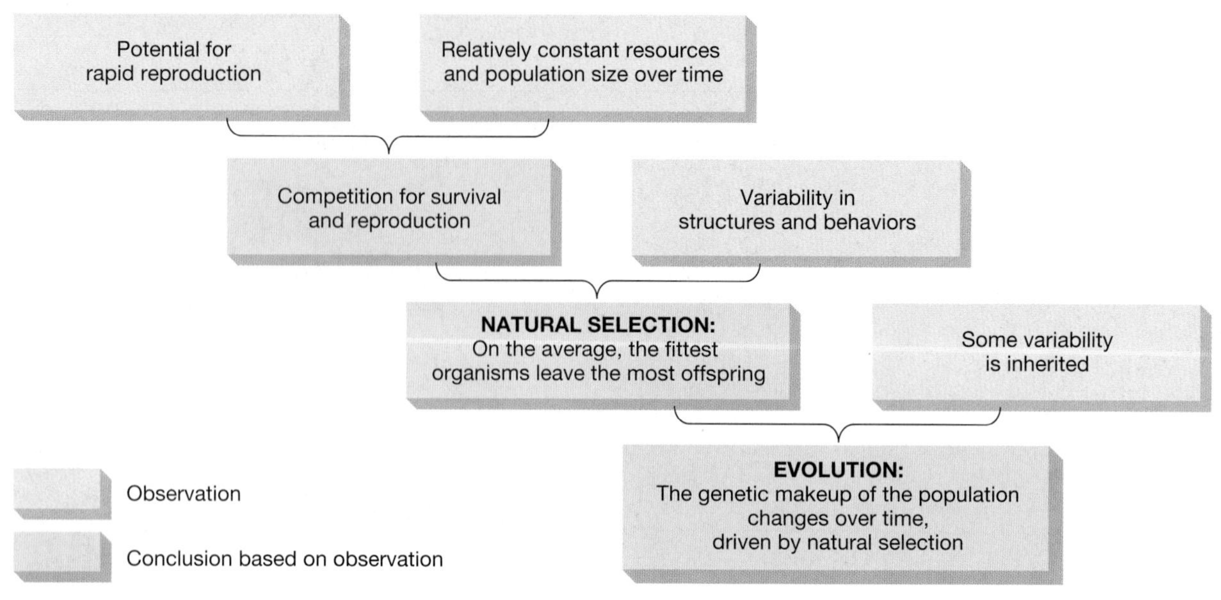

FIGURE 14-5 A flowchart of evolutionary reasoning
This chart is based on the hypotheses of Darwin and Wallace but incorporates ideas from modern genetics.
QUESTION Is sexual reproduction necessary to generate the variability in structures and behaviors needed for natural selection?

of the population. This process is evolution by natural selection.

Figure 14-5 charts these observations and conclusions, and "Scientific Inquiry: Charles Darwin—Nature Was His Laboratory" describes the observations that led to Darwin's understanding of how organisms evolve.

Modern Genetics Confirmed Darwin's Assumption of Inheritance

As you know, the principles of genetics had not yet been discovered when Darwin published *On the Origin of Species*. Therefore, Darwin and Wallace's "Observation 4" was an untested assumption—and thus a weakness in their theory at the time of its publication. Mendel's later work, of course, showed that Darwin's assumption that particular traits can be passed to offspring was correct. In addition, we now also know that the variations in natural populations arise purely by chance, as a result of random mutations in DNA (see Chapters 9 and 10). New variations can be good, bad, or neutral; there is no mechanism to ensure that favorable variations arise.

Natural Selection Modifies Populations Over Time

How might the natural selection of chance variations change the makeup of a species? In *On the Origin of Species*, Darwin proposed the following example: "Let us take the case of a wolf, which preys on various animals, securing [them] by . . . fleetness. . . . The swiftest and slimmest wolves would have the best chance of surviving, and so be preserved or selected. . . . Now if any slight innate change of habit or structure benefited an individual wolf, it would have the best chance of surviving and of leaving offspring. Some of its young would probably inherit the same habits or structure, and by the repetition of this process, a new variety might be formed." The same logic applies to the wolf's prey; the fastest or most alert would be most likely to avoid predation and would pass on these traits to its offspring. Notice that natural selection acts on individuals within a population. Over generations, the population changes as the percentage of individuals inheriting favorable traits increases. An individual cannot evolve, but a population can.

Although it is easier to understand how natural selection would cause changes *within* a species, under the right circumstances, the same principles might produce entirely *new* species. We will discuss the circumstances that might give rise to new species in Chapter 16.

14.3 How Do We Know That Evolution Has Occurred?

Virtually all biologists consider evolution to be a fact. Why? Because an overwhelming body of evidence—from fossils, comparative anatomy (the study of how body structures differ among species), embryology, and biochemistry and genetics—permits no other conclusion.

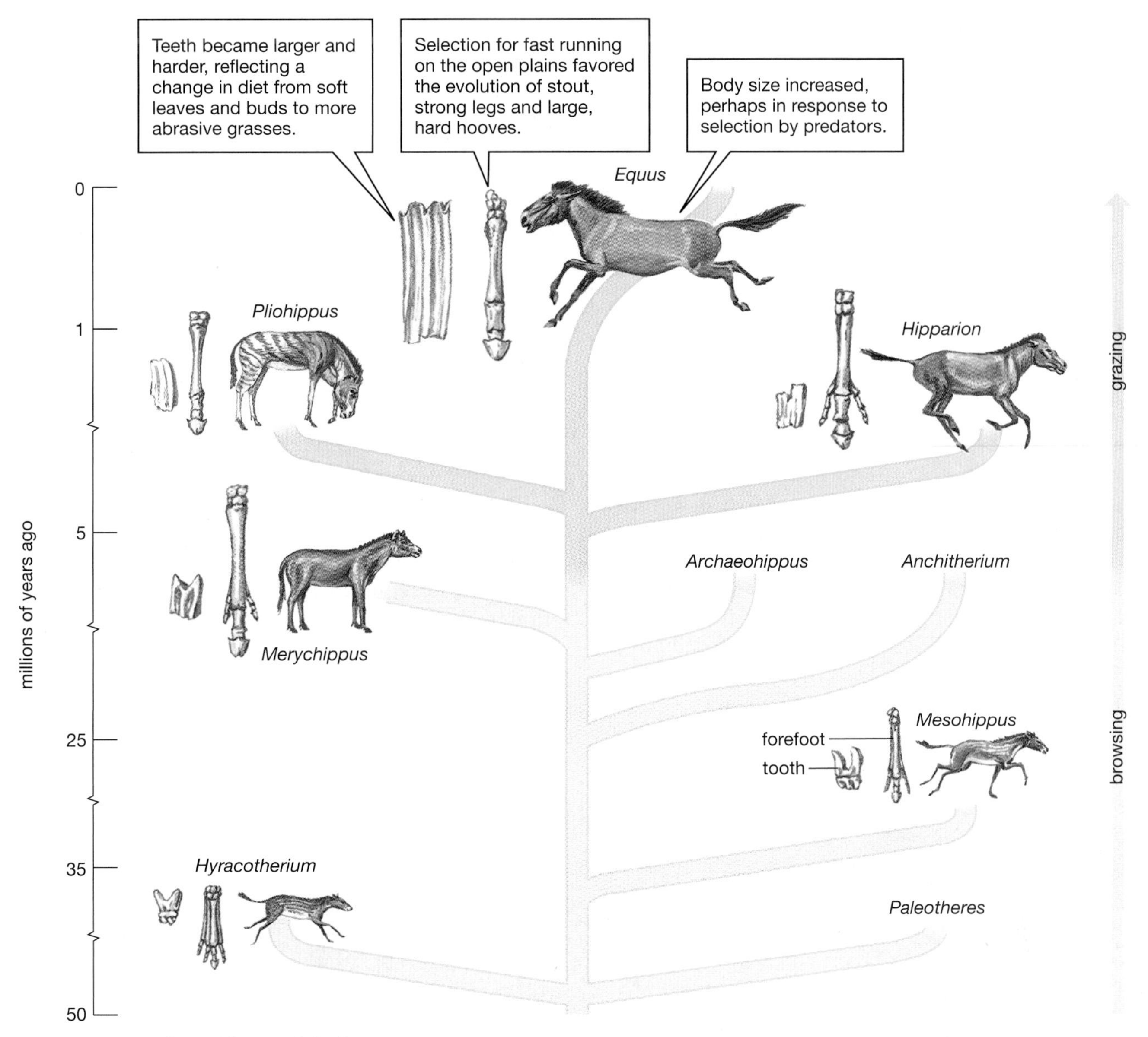

FIGURE 14-6 The evolution of the horse
Over the past 50 million years, horses have evolved from small woodland browsers to large plains-dwelling grazers. Three major changes include size, leg anatomy, and tooth anatomy. **QUESTION** The fossil history of some kinds of modern organisms, such as sharks and crocodiles, shows that their structure and appearance have changed very little over hundreds of millions of years. Is this evidence that such organisms have not evolved over that time?

Fossils Provide Evidence of Evolutionary Change Over Time

If it is true that many fossils are the remains of species ancestral to modern species, we might expect to find many progressive series of fossils that start with an ancient, primitive organism, progress through several intermediate stages, and culminate in the modern species. Such series have indeed been found. The best-known is probably the fossil horse series (Fig. 14-6), though series of fossil giraffes, elephants, whales, and mollusks also show the evolution of body structures over time. These fossil series suggest that new species gradually evolved from, and replaced, previous species. Certain sequences of fossil snails have such slight gradations in body structure between successive rock layers that paleontologists cannot easily determine where one species leaves off and the next one begins. Fossils also document large-scale evolutionary transitions. For example, the link between dinosaurs and birds was confirmed by fossils of dinosaurs with feathers, as described in this chapter's Case Study.

Media Activity
14.1 Analogous and Homologous Structures

Comparative Anatomy Gives Evidence of Descent with Modification

Fossils provide snapshots of the past that allow biologists to trace evolutionary changes, but careful examination of today's organisms can also uncover evolution's story. Comparing the bodies of organisms of different species can reveal similarities that can be explained only by shared ancestry, and differences that could result only from evolutionary change during descent from a common ancestor. In this way, the study of comparative anatomy has supplied strong evidence that different species are linked by common evolutionary heritage.

Homologous Structures Provide Evidence of Common Ancestry

The same body structure may be modified by evolution to serve different functions in different species. Birds and mammals, for example, use their forelimbs for many purposes, including flying, swimming, running over varied terrain, and grasping objects, such as branches and tools. Despite this enormous diversity of function, the internal anatomy of all bird and mammal forelimbs is remarkably similar (Fig. 14-7). It seems unlikely that, if each animal had been created separately, the same bone arrangement would be used to serve such diverse functions. This similarity is exactly what we would expect, however, if bird and mammal forelimbs were derived from a common ancestor. Through natural selection, each has been modified and now performs a particular function. Such internally similar structures are called **homologous structures**, meaning that they have the same evolutionary origin despite any differences in current function or appearance. Homologous structures can be found not only among living organisms, but also in comparisons between living organisms and extinct ones. For example, the feathers of birds and those of feathered dinosaurs are homologous structures that reveal common ancestry, even though dinosaurs did not use their feathers for flight, as birds do.

Functionless Structures Are Inherited from Ancestors

Evolution by natural selection also helps explain the curious existence of **vestigial structures** that serve no apparent purpose. Examples of vestigial structures include

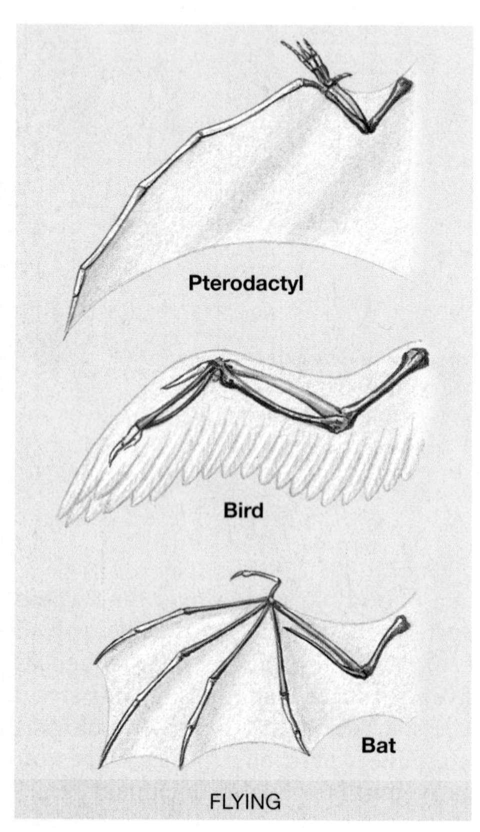

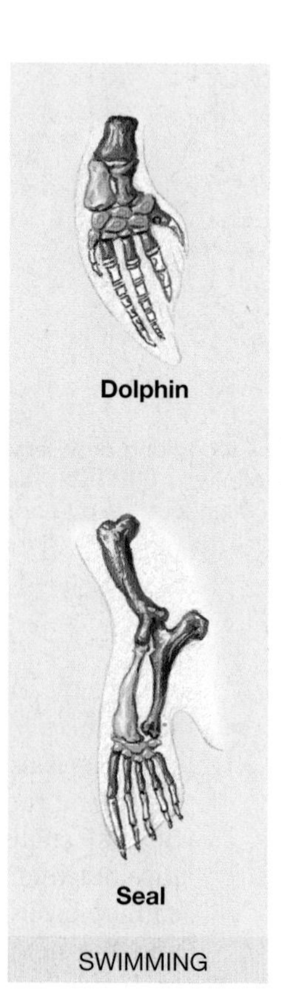

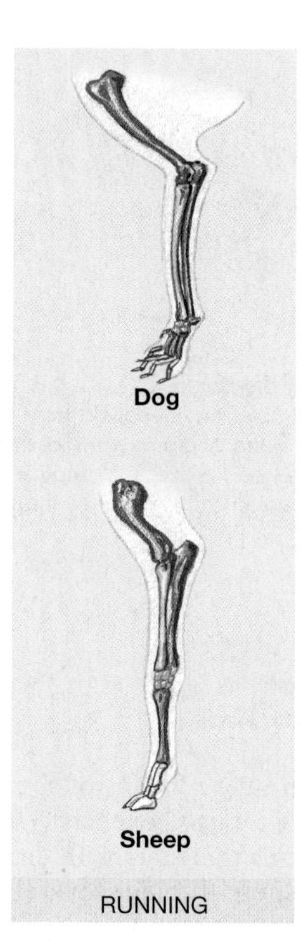

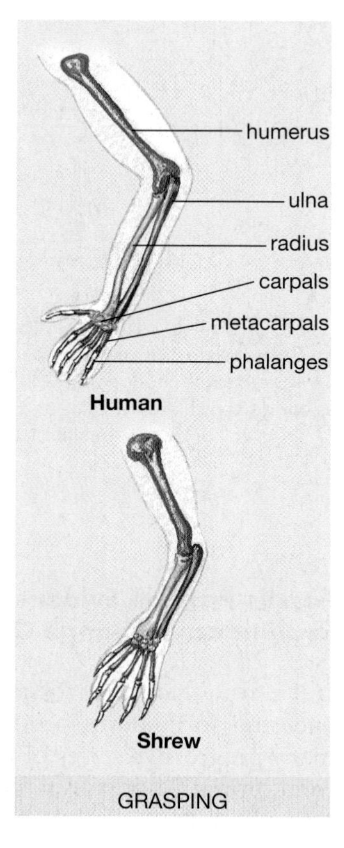

FIGURE 14-7 Homologous structures
Despite wide differences in function, the forelimbs of all these animals contain the same set of bones, inherited through evolution from a common ancestor. The different colors of the bones highlight the correspondences among the various species.

molar teeth in vampire bats (which live on a diet of blood and don't chew their food, and therefore do not need teeth designed for chewing) and pelvic bones in whales and certain snakes (which lack legs and therefore do not need pelvic bones to support them; Fig. 14-8). Both of these vestigial structures are clearly homologous to structures found in—and used by—other vertebrates. Their continued existence in animals that have no use for them is best explained as a sort of "evolutionary baggage." For example, the ancestral mammals from which whales evolved had four legs and a well-developed set of pelvic bones. Whales do not have hind legs, yet they have small pelvic and leg bones embedded in their sides. During whale evolution, the loss of hind legs was advantageous, as it better streamlined the body for movement through water. The result is the modern whale with small, unused pelvic bones.

Some Anatomical Similarities Result from Evolution in Similar Environments

The study of comparative anatomy has demonstrated the shared ancestry of life by identifying a host of homologous structures that different species have inherited from common ancestors. However, comparative anatomists have also identified many anatomical similarities that do not stem from common ancestry. Instead, these similarities stem from **convergent evolution**, in which natural selection causes nonhomologous structures that serve similar functions to resemble one another. For example, both birds and insects have wings, but the two types of wings are not derived from a structure that both birds and insects inherited from a common ancestor. Instead, bird and insect wings evolved from two different, nonhomologous structures that eventually gave rise to superficially similar structures. Because natural selection favored flight in both birds and insects, the two groups evolved superficially similar structures that are useful for flight. Such outwardly similar but nonhomologous structures are called **analogous structures** (Fig. 14-9). Analogous structures typically have very different internal anatomies, because the parts did not develop from a common ancestral structure.

Embryological Similarity Suggests Common Ancestry

In the early 1800s, German embryologist Karl von Baer noted that all vertebrate embryos look quite similar to

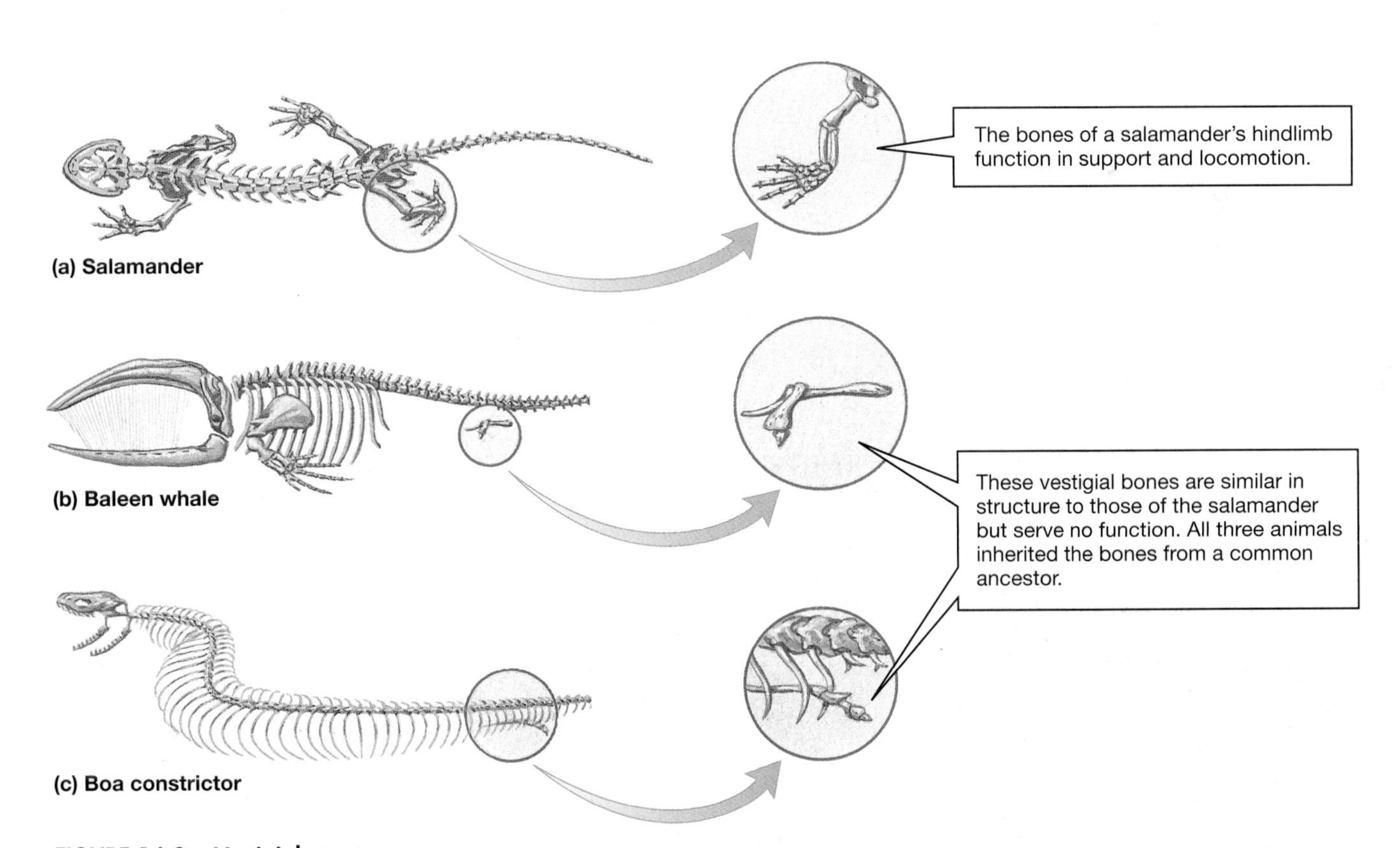

(a) Salamander
The bones of a salamander's hindlimb function in support and locomotion.
(b) Baleen whale
These vestigial bones are similar in structure to those of the salamander but serve no function. All three animals inherited the bones from a common ancestor.
(c) Boa constrictor

FIGURE 14-8 Vestigial structures
Many organisms have vestigial structures that serve no apparent function. The **(a)** salamander, **(b)** whale, and **(c)** snake all inherited hindlimb bones from a common ancestor; the bones remain functional in the salamander but are vestigial in the whale and snake. **EXERCISE** Compile a list of human vestigial structures. For each structure, name the corresponding homologous structure in a nonhuman species.

(a) (b) (c) (d)

FIGURE 14-9 Analogous structures
Convergent evolution can produce outwardly similar structures that differ anatomically. The wings of **(a)** insects and **(b)** birds and the sleek, streamlined shapes of **(c)** seals and **(d)** penguins are examples of such analogous structures. **QUESTION** Are a peacock's tail (see Fig. 15-7) and a dog's tail homologous structures or analogous structures?

one another early in their development (Fig. 14-10). In their early embryonic stages, fish, turtles, chickens, mice, and humans all develop tails and gill slits. Only fish go on to develop gills, and only fish, turtles, and mice retain substantial tails. Why do vertebrates that are so different have similar developmental stages? The only plausible explanation is that ancestral vertebrates possessed genes that directed the development of gills and tails; these genes are still present in all of their descendants. In fish, these genes are active throughout development, resulting in adults with fully developed tails and gills. In humans and chickens, they

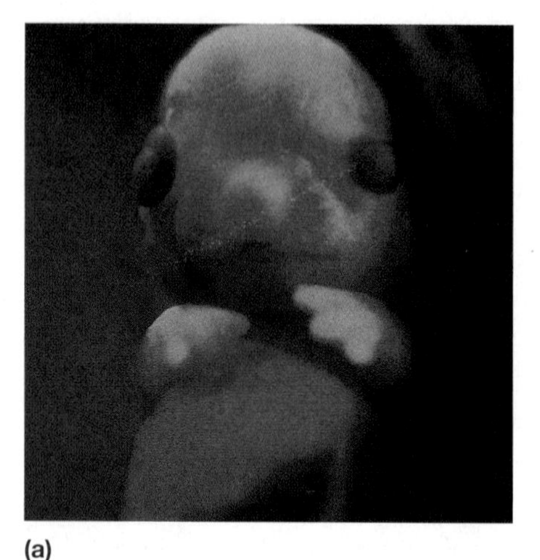

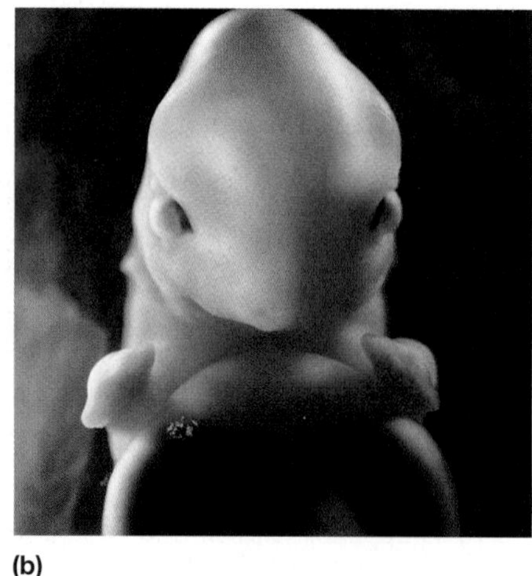

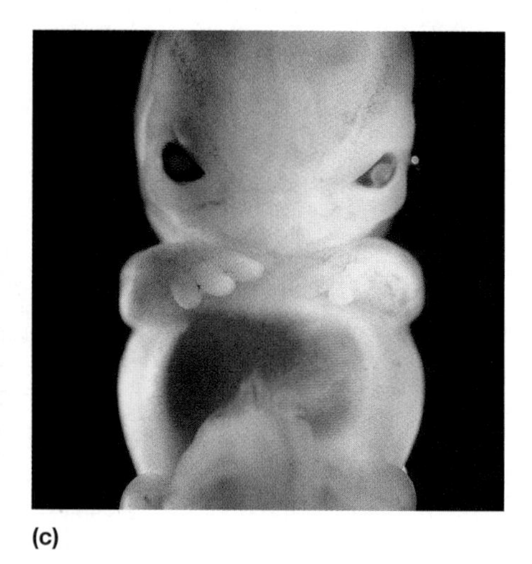

(a) (b) (c)

FIGURE 14-10 Embryological stages reveal evolutionary relationships
The early embryonic stages of a **(a)** lemur, **(b)** pig, and **(c)** human show strikingly similar anatomical features.

are active only during early developmental stages, and the structures whose development they direct are lost or inconspicuous in adults.

Modern Biochemical and Genetic Analyses Reveal Relatedness Among Diverse Organisms

Biochemistry and molecular biology provide striking evidence of the evolutionary relatedness of all organisms. At the most fundamental biochemical levels, all living cells are very similar. For example, all cells have DNA as the carrier of genetic information; all use RNA, ribosomes, and approximately the same genetic code to translate that genetic information into proteins; all use roughly the same set of 20 amino acids to build proteins; and all use ATP as an intracellular energy carrier. The most plausible explanation for such widespread sharing of such complex and specific biochemical traits is that they arose in the common ancestor of all living things, from which all of today's organisms inherited them.

14.4 What Is the Evidence That Populations Evolve by Natural Selection?

We have seen that evidence of evolution is vast and comes from a variety of sources. But what is the evidence that evolution occurs by the mechanism of natural selection?

Controlled Breeding Modifies Organisms

One line of evidence supporting evolution by natural selection is **artificial selection**, the breeding of domestic plants and animals to produce specific desirable features. The various breeds of dogs provide a striking example of artificial selection (Fig. 14-11). Dogs descended from wolves, and even today, the two will readily cross-breed. With few exceptions, however, modern dogs do not resemble wolves. Some breeds are so different from one another that they would be considered separate species if they were found in the wild. Humans produced these radically different dogs in a few thousand years by doing nothing more than repeatedly selecting individuals with desirable traits for breeding. Therefore, it is quite plausible that natural selection could, by an analogous process acting over hundreds of millions of years, produce the spectrum of living organisms. Darwin was very impressed by the connection between artificial selection and natural selection, and devoted a chapter of *Origin of Species* to the topic.

Evolution by Natural Selection Occurs Today

The logic of natural selection gives us no reason to believe that evolutionary change is limited to the past. After all, inherited variation and competition for access to resources continue today. If Darwin and Wallace were correct that those conditions inevitably lead to evolution by natural selection, then scientists ought to be able to detect evolutionary change as it occurs. And they have. Next, we consider some examples that give us a glimpse of natural selection at work.

When Fewer Predators Are Present, Brighter Coloration Can Evolve

On the island of Trinidad, guppies live in streams that are also inhabited by several species of larger, predatory fish that frequently dine on guppies. In upstream portions of these streams, however, the water is too shallow

Media Activity
14.2 Natural Selection for Antibiotic Resistance

(a)

(b)

FIGURE 14-11 Dog diversity illustrates artificial selection
A comparison of **(a)** the ancestral dog (the gray wolf, *Canis lupus*) and **(b)** various breeds of modern dogs. Artificial selection by humans has caused a great divergence in size and shape of dogs in only a few thousand years.

for the predators, and guppies are free of danger from predators. When scientists compared male guppies in an upstream area to those in a downstream area, they found that the upstream guppies were much more brightly colored than the downstream guppies. The scientists knew that the upstream population was composed of guppies that had found their way up into the shallower waters many generations earlier. The difference in coloration between the two populations is explained by the sexual preferences of female guppies. The females prefer to mate with the most brightly colored males, giving the brightest males a large reproductive advantage. Therefore, in predator-free areas, male guppies with bright colors have more offspring. Bright color, however, makes guppies more conspicuous to predators, and therefore more likely to be eaten. Thus, predators act as agents of natural selection by eliminating the bright-colored males before they can reproduce. In deeper areas where predators are common, the duller males have the advantage and produce more offspring. The color difference between the upstream and downstream guppy populations is a direct result of natural selection. (Researchers confirmed this conclusion by introducing some predatory fish to previously predator-free upstream areas. As expected, the male guppies in those areas evolved to become less colorful within a few generations.)

Natural Selection Can Lead to Pesticide Resistance

Natural selection is also evident in numerous instances of insect pests evolving resistance to the pesticides with which we try to control them. For example, a few decades ago, Florida homeowners were dismayed to realize that roaches were ignoring a formerly effective poison bait called Combat®. Researchers discovered that the bait had acted as an agent of natural selection. Roaches that liked it were consistently killed; those that survived inherited a rare mutation that caused them to dislike glucose, a type of sugar found in the corn syrup used as bait in Combat. By the time researchers identified the problem in the early 1990s, the formerly rare mutation had become common in Florida's urban roach population.

Unfortunately, the evolution of pesticide resistance in insects is a common example of natural selection in action. Such resistance has been documented in more than 500 species of crop-damaging insects, and virtually every pesticide has fostered the evolution of resistance in at least one insect species. We pay a heavy price for this evolutionary phenomenon; the additional pesticides that U.S. farmers apply in their attempts to control resistant insects cost almost $2 billion each year, and add millions of tons of poisons to Earth's soil and water.

Experiments Can Demonstrate Natural Selection

In addition to observing natural selection in the wild, scientists have also devised numerous experiments that confirm the action of natural selection. For example, one group of evolutionary biologists released small groups of *Anolis sagrei* lizards onto 14 small Bahamian islands that were previously uninhabited by lizards. The original lizards came from a population on Staniel Cay, an island with tall vegetation, including plenty of trees. In contrast, the islands to which the small lizard colonies were introduced had few or no trees and were covered mainly with small shrubs and other low-growing plants.

When the biologists returned to the islands 14 years after releasing the lizards, they found that the original colonies had given rise to thriving populations of hundreds of individuals. On all 14 of the experimental islands, lizards had legs that were shorter and thinner than those of the lizards from the original source population on Staniel Cay. In just over a decade, it appeared, the lizard populations had changed in response to new environments.

Why had the new lizard populations evolved shorter, thinner legs? Long legs allow greater speed for escaping predators, but shorter legs allow for more agility and maneuverability on narrow surfaces. So, natural selection favors legs that are as long and thick as possible while still allowing sufficient maneuverability. When the lizards were moved from an environment with thick-branched trees to an environment with only thin-branched bushes, the individuals with formerly favorable long legs were at a disadvantage. In the new environment, more agile, shorter-legged individuals were better able to escape predators and survive to produce a greater number of offspring. Thus, members of subsequent generations would have shorter legs on average.

Selection Acts on Random Variation to Favor the Traits That Work Best in Particular Environments

Two important points underlie the evolutionary changes described in the examples above:

- *The variations on which natural selection works are produced by chance mutations.* The bright coloration of Trinidadian guppies, distaste for glucose of Florida cockroaches, and shorter legs of Bahamian lizards were not *produced* by female mating preferences, poisoned corn syrup, or thinner branches The mutations that produced each of these beneficial traits arose spontaneously.
- *Natural selection selects for organisms that are best adapted to a particular environment.* Natural selection is not a mechanism for producing ever-greater degrees of perfection. Natural selection does not select for the "best" in any absolute sense, but only in the context of a particular environment, which varies from place to place and may change over time. A trait that is advantageous under one set of conditions may become disadvantageous if conditions change. For example, in the presence of poisoned corn syrup, a distaste for glucose yields an

advantage to a cockroach, but under natural conditions, avoiding glucose would cause the insect to bypass good sources of food.

14.5 A Postscript by Charles Darwin

"It is interesting to contemplate an entangled bank, clothed with many plants of many kinds, with birds singing on the bushes, with various insects flitting about, and with worms crawling through the damp earth, and to reflect that these elaborately constructed forms . . . have all been produced by laws acting around us. These laws, taken in the highest sense, being Growth with Reproduction; Inheritance [and] Variability; a Ratio of Increase so high as to lead to a Struggle for Life, and as a consequence to Natural Selection, entailing Divergence of Character and Extinction of less-improved forms. . . . There is grandeur in this view of life, with its several powers, having been originally breathed into a few forms or into one; and that, whilst this planet has gone cycling on according to the fixed law of gravity, from so simple a beginning endless forms most beautiful and most wonderful have been, and are being, evolved."

These are the concluding sentences of Darwin's *On the Origin of Species.*

CASE STUDY REVISITED A Missing Link Unearthed

Even though multiple examples of fossilized feathered dinosaurs have been discovered, the controversy over the evolutionary origin of birds has not been completely resolved. Some skeptical paleontologists question whether the new fossils provide truly conclusive evidence that dinosaurs gave rise to birds. For one thing, say the skeptics, the fossils of the feathered dinosaurs are much younger (by about 30 million years) than the fossils of *Archaeopteryx*, the oldest known bird, so these feathered dinosaur species could not have been the ancestors of birds. To the doubters, this anomaly suggests that the fossils represent not feathered dinosaurs, but rather birds that had lost the power of flight (as have some modern birds, such as ostriches and penguins). According to this view, birds more likely descended from a different, nondinosaur group of ancient reptiles, one that is represented in pre-*Archaeopteryx* fossils.

Most paleontologists, however, shrug off the question of the new fossils' age. To them, the comparatively young age of the fossils means nothing more than that some species of feathered dinosaurs persisted after birds first evolved from earlier dinosaurs (much as apes and humans both persist today, millions of years after a common ancestor gave rise to both groups). To proponents of the bird–dinosaur connection, feathered dinosaurs of any age demonstrate conclusively that pre-bird dinosaurs evolved feathers and that these feathered dinosaurs later gave rise to birds.

Consider This: An alternative hypothesis to explain the existence of feathered dinosaurs is convergent evolution. Is it possible that dinosaurs and birds are unrelated, and that each group separately evolved feathers? What is the likelihood of convergent evolution of feathers? What evidence might, if found, cause you to reject the conclusion that birds are descended from dinosaurs and accept the hypothesis of convergent evolution?

Links to Life: People Promote High-Speed Evolution

You probably don't think of yourself as a major engine of evolution. Nonetheless, as you go about the routines of your daily life, you are contributing to what is perhaps today's most significant cause of rapid evolutionary change. Human activity has changed Earth's environments tremendously. The biological logic of evolution, spelled out so clearly by Darwin, tells us that environmental change leads inevitably to natural selection. Humans have thus become a major agent of natural selection. Unfortunately, many of the evolutionary changes we have caused have turned out to be bad news. Our liberal use of pesticides has selected for resistant pests that frustrate efforts to protect our food supply. By overmedicating ourselves with antibiotics and other drugs, we have selected for resistant "supergerms" and diseases that are ever more difficult to treat. Heavy fishing in the world's oceans has selected for ever smaller fish (that can slip through nets more easily) and has reduced our ability to extract food from the sea. The rapid evolutionary changes caused by our technology now threaten us. Only strategies based on a sound understanding of the principles of evolution will be capable of solving them.

Summary of Key Concepts

14.1 How Did Evolutionary Thought Evolve?
Historically, the most common explanation for the origin of species was the divine creation of each species in its present form; species were believed to remain unchanged after their creation. This view was challenged by evidence from fossils, geology, and biological exploration of the Tropics. Since the middle of the nineteenth century, scientists have realized that species originate and evolve by the operation of natural processes that change the genetic makeup of populations.

14.2 How Does Natural Selection Work?
Charles Darwin and Alfred Russel Wallace independently proposed the theory of evolution by natural selection. Their theory can be concisely expressed as three conclusions based on four observations. These are summarized in modern biological terms in Figure 14-5.

14.3 How Do We Know That Evolution Has Occurred?
Many lines of evidence indicate that evolution has occurred, including the following:
- Fossils of ancient species tend to be simpler in form than modern species. Sequences of fossils have been discovered that show a graded series of changes in form. Both of these observations would be expected if modern species evolved from older species.
- Species thought to be related to a common ancestor through evolution show many similar anatomical structures. Examples include the forelimbs of amphibians, reptiles, birds, and mammals.
- Stages in early embryological development are quite similar among very different types of vertebrates.
- Similarities in such biochemical traits as the use of DNA as the carrier of genetic information support the notion of descent of related species through evolution from common ancestors.

14.4 What Is the Evidence That Populations Evolve by Natural Selection?
Similarly, many lines of evidence indicate that natural selection is the chief mechanism driving changes in the characteristics of species over time, including the following:
- Inheritable traits have been changed rapidly in populations of domestic animals and plants by selectively breeding organisms with desired features (artificial selection). The immense variations in species produced in a few thousand years of artificial selection by humans makes it seem likely that much larger changes could be wrought by hundreds of millions of years of natural selection.
- Evolution can be observed today. Both natural and human activities drastically change the environment over short periods of time. Characteristics of species have been observed to change significantly in response to such environmental changes.

Key Terms

analogous structures *p. 275*
artificial selection *p. 277*
catastrophism *p. 268*
convergent evolution *p. 275*
evolution *p. 265*
fossil *p. 267*
homologous structures *p. 274*
inheritance of acquired characteristics *p. 269*
natural selection *p. 271*
population *p. 270*
uniformitarianism *p. 268*
vestigial structure *p. 274*

Thinking Through the Concepts

To take a multiple-choice quiz with feedback on the contents of this chapter, visit http://www.prenhall.com/audesirk7. *Log in to the Web site selected by your instructor and navigate to the Self Test section for this chapter.*

? Review Questions

1. Selection acts on individuals, but only populations evolve. Explain why this is true.
2. Distinguish between catastrophism and uniformitarianism. How did these hypotheses contribute to the development of modern evolutionary theory?
3. Describe Lamarck's theory of inheritance of acquired characteristics. Why is it invalid?
4. What is natural selection? Describe how natural selection might have caused differential reproduction among the ancestors of a fast-swimming predatory fish (such as the barracuda).
5. Describe how evolution occurs through the interactions among the following: the reproductive potential of a species, the normally constant size of natural populations, variation among individuals of a species, natural selection, and inheritance.
6. What is convergent evolution? Give an example.
7. How do biochemistry and molecular genetics contribute to the evidence that evolution occurred?

Applying the Concepts

1. Does evolution through natural selection produce "better" organisms in an absolute sense? Are we climbing the "ladder of Nature"? Defend your answer.
2. Both the study of fossils and the idea of divine creation have had an impact on evolutionary thought. Discuss why one is considered scientific endeavor and the other is not.
3. In evolutionary terms, "success" can be defined in many different ways. What are the most successful organisms you can think of in terms of (a) persistence over time, (b) sheer numbers of individuals alive now, (c) numbers of species, and (d) geographical range?
4. In what sense are humans currently acting as agents of selection on other species? Name some organisms that are *favored* by the environmental changes humans cause.
5. Darwin and Wallace's discovery of natural selection is one of the great revolutions in scientific thought. Some scientific revolutions spill over and affect the development of philosophy and religion. Is this true of evolution? Does (or should) the idea of evolution by natural selection affect the way humans view their place in the world?

For More Information

Appleman, P. (ed.). *Darwin, a Norton Critical Edition.* New York: W. W. Norton, 2001. An excellent collection of early and modern writing about evolution, including excerpts from the *Origin of Species.*

Darwin, C. *On the Origin of Species by Means of Natural Selection.* Garden City, NY: Doubleday, 1960 (originally published in 1859). An impressive array of evidence amassed to convince a skeptical world.

Dennet, D. *Darwin's Dangerous Idea.* New York: Simon & Schuster, 1995. A philosopher's view of Darwinian ideas and their application to the world outside biology. A thought-provoking book that seems to have inspired admiration and condemnation in roughly equal proportions.

Eiseley, L. C. "Charles Darwin." *Scientific American,* February 1956. An essay on the life of Darwin by one of his foremost American biographers. Even if you need no introduction to Darwin, read this as an introduction to Eiseley, author of many marvelous essays.

Gould, S. J. *Ever Since Darwin,* 1977; *The Panda's Thumb,* 1980; and *The Flamingo's Smile,* 1985. New York: W. W. Norton. A series of witty, imaginative, and informative essays about evolution, mostly from *Natural History* magazine.

Paul, G. *Dinosaurs of the Air: The Evolution and Loss of Flight in Dinosaurs and Birds.* Baltimore: Johns Hopkins University Press, 2002. A clear and entertaining presentation of the evidence that birds descended from dinosaurs.

Weiner, J. "Evolution Made Visible." *Science,* January 6, 1995. A clear summary of modern evidence for evolution in action.

Media Activities

To access a Media Activity visit http://www.prenhall.com/audesirk7. *Log in to the Web site selected by your instructor, navigate to this chapter, and select the appropriate Media Activity number.*

14.1 Analogous and Homologous Structures

Estimated time: 5 minutes

This exercise will help you to learn the difference between structures that are similar because of common origin or because of similar evolutionary pressures.

14.2 Natural Selection for Antibiotic Resistance

Estimated time: 5 minutes

This activity demonstrates the mechanism of natural selection using the example of evolution of an antibiotic resistance trait in a population of the bacteria that cause the human disease tuberculosis.

14.3 Web Investigation: A Missing Link Unearthed

Estimated time: 10 minutes

Finding fossils and analyzing the fossil record is hard work. Sometimes the results are quite unexpected. This exercise will explore the fossil record of birds.

CHAPTER

15 How Organisms Evolve

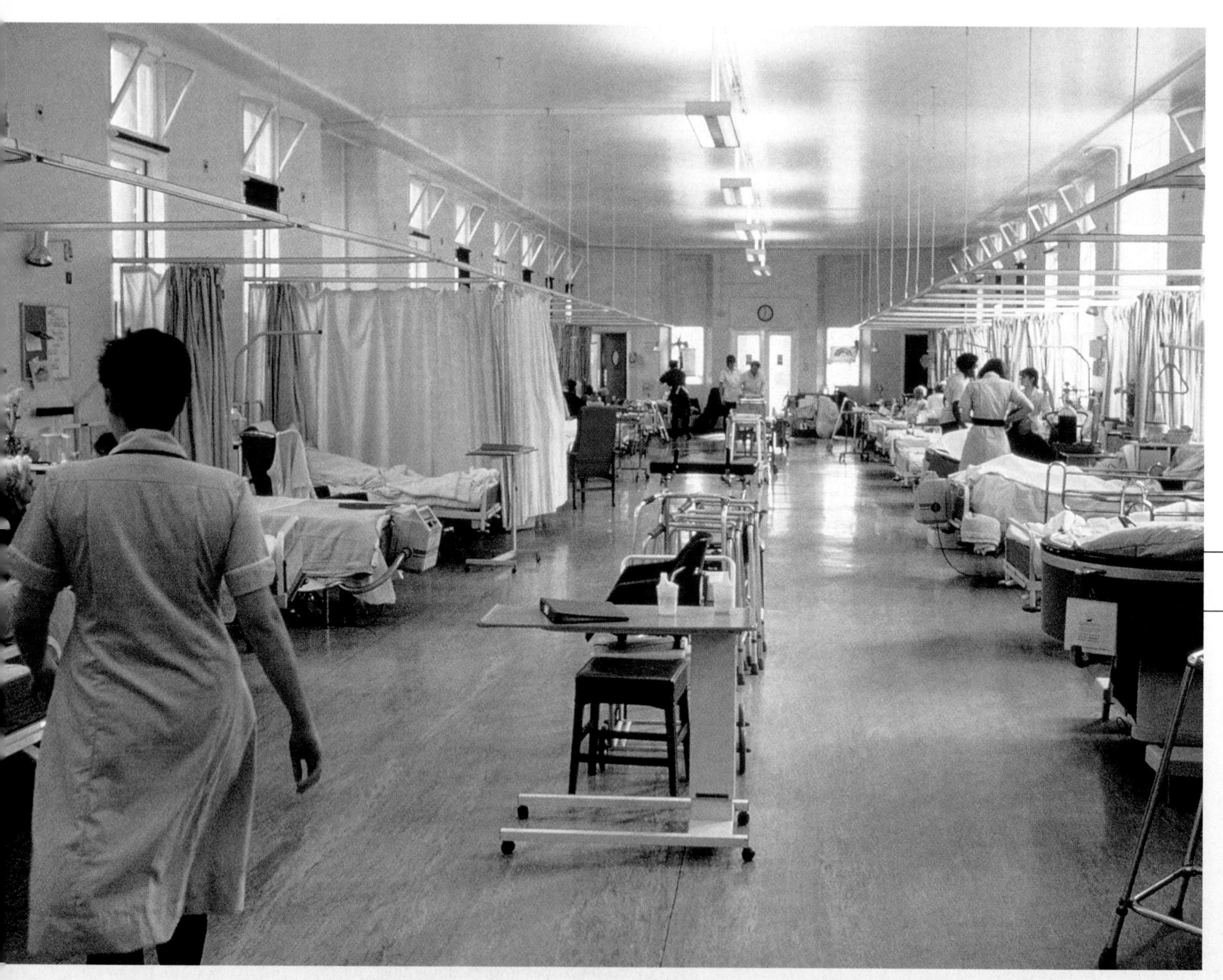

We think of hospitals as places in which to seek protection from disease, but they also foster the evolution of drug-resistant supergerms.

AT A GLANCE

CASE STUDY Evolution of a Menace

As she watched Jim leave the clinic, Dr. Lawson sighed and shook her head sadly. Jim, a middle-aged homeless man who was being treated for tuberculosis, was unlikely to recover. Lab tests had revealed that the bacteria infecting Jim were resistant to four different antibiotic drugs commonly used to treat the disease. Multidrug-resistant tuberculosis is extremely difficult to treat; it can be cured only by long-term treatment with a combination of several drugs. Even after such treatment, some cases prove to be incurable, and Dr. Lawson knew that Jim's treatment was especially unlikely to succeed. Jim lived on the streets, often struggling to survive, and was not likely to complete the treatment: five pills each day for two years. To make matters worse, Jim would no doubt sleep in overpopulated shelters, keep warm in crowded subway stations, and perhaps even spend time in packed city jails. In such places, he would likely pass tuberculosis bacteria to others, thereby hastening the spread of his multidrug-resistant strain.

Multidrug-resistant tuberculosis is a frightening and increasingly widespread threat to public health in many parts of the world, including the United States. Drug resistance is also becoming common in other dangerous bacteria as well, including those that cause food poisoning, blood poisoning, dysentery, pneumonia, gonorrhea, meningitis, and urinary tract infections. We are experiencing a global onslaught of resistant "supergerms," and are facing the specter of diseases that cannot be cured, even by our best medicines. In order to understand how this crisis arose and to devise a strategy to resolve it, we must have a clear understanding of the mechanisms by which populations evolve.

15.1 How Are Populations, Genes, and Evolution Related?

The changes that we see in an individual organism as it grows and develops are not evolutionary changes. Instead, evolutionary changes are those that occur from generation to generation and cause descendants to differ from their ancestors. Furthermore, we can't detect evolutionary change across generations by looking at a single set of parents and offspring. For example, if you observed that a six-foot-tall man had an adult son who stood five feet tall, could you conclude that humans were evolving to become shorter? Obviously not. Rather, if you wanted to learn about evolutionary change in human height, you would begin by measuring many humans for many generations to see if the average height changes over time. Clearly, evolution is a property not of individuals but of populations; a **population** is a group that includes all the members of a species living in a given area.

The recognition that evolution is a population-level phenomenon was one of Darwin's key insights. But populations are composed of individuals, and the fates of individuals determine which characteristics will be passed to descendant populations. In this fashion, inheritance provides the link between the lives of individual organisms and the evolution of populations. We will therefore begin our discussion of the processes of evolution by reviewing some principles of genetics as they apply to individuals, then extend those principles to the genetics of populations.

Genes and the Environment Interact to Determine Traits

Each cell of every organism contains genetic information encoded in the DNA of its chromosomes. Recall that a *gene* is a segment of DNA located at a particular place on a chromosome. The sequence of nucleotides in a gene encodes the sequence of amino acids in a protein, usually an enzyme that catalyzes a particular reaction in the cell. At a given gene's location, different members of a species may have slightly different nucleotide sequences, called *alleles*, as we learned in Chapter 12. Different alleles generate different forms of the same enzyme. In this way, various alleles of the gene that influences eye color in humans, for example, help produce eyes that are brown, or blue, or green, and so on. In any population of organisms, there are usually two or more alleles of each gene. An individual of a diploid species whose alleles of a particular gene are both the same is *homozygous* for that gene, and an individual with different alleles for that gene is *heterozygous*. The specific alleles borne on an organism's chromosomes (its *genotype*) interact with the environment to influence the development of its physical and behavioral traits (its *phenotype*).

Let's illustrate these principles with an example from Unit Two. A pea flower is colored purple because a chemical reaction in its petals converts a colorless molecule to a purple pigment. When we say that a pea plant has the allele for purple flowers, we mean that a particular stretch of DNA on one of its chromosomes contains a sequence of nucleotides that codes for the enzyme that catalyzes this reaction. A pea with the allele for white flowers has a different sequence of nucleotides at the same chromosomal position. The enzyme coded by this different sequence cannot produce purple pigment. If a pea is homozygous for the white allele, its flowers produce no pigment and thus are white.

The Gene Pool Is the Sum of the Genes in a Population

We can often improve our understanding of a subject by looking at it from more than one perspective. Looking at the process of evolution from the point of view of a gene has proven to be an enormously effective study tool. In particular, evolutionary biologists have made excellent use of the tools of a branch of genetics called population genetics, which deals with the frequency, distribution, and inheritance of alleles in populations. To take advantage of this powerful aid to understanding evolution, you will need to learn a few of the basic concepts of population genetics.

Population geneticists define the **gene pool** as the sum of all the genes in a population. In other words, the gene pool consists of all the alleles of all the genes in all the individuals of a population. Each particular gene can also be considered to have its own gene pool, which consists of all the alleles of that specific gene in a population. If we added up all the copies of each allele of that gene in all the individuals in a population, we could determine the relative proportion of each allele, a number called the **allele frequency**. For example, a population of 100 pea plants would contain 200 alleles of the gene that controls flower color (because pea plants are diploid). If 50 of those 200 alleles were of the type that codes for white flowers, then we would say that the frequency of that allele in the population is 0.25 (or 25%), because 50/200 = 0.25.

Evolution Is the Change Over Time of Allele Frequencies Within a Population

A casual observer might define evolution on the basis of changes in the outward appearance or behaviors of the members of a population. A population geneticist, however, looks at a population and sees a gene pool divided into the packages that we call individual organisms. Any outward changes that we observe in the individuals that make up the population can also be viewed as the visible expression of underlying changes to the gene pool. A population geneticist, therefore, defines evolution as changes in allele frequencies in a gene pool over time. Evolution is a change in the genetic makeup of populations over generations.

The Equilibrium Population Is a Hypothetical Population That Does Not Evolve

It is easier to understand what causes populations to evolve if the characteristics of a population that would *not* evolve are considered first. In 1908, English mathematician Godfrey H. Hardy and German physician Wilhelm Weinberg independently developed a simple mathematical model now known as the **Hardy-Weinberg principle**. This model showed that under certain conditions, allele frequencies and genotype frequencies in a population will remain constant no matter how many generations pass. In other words, this population does not evolve. Population geneticists use the term **equilibrium population** for this idealized, evolution-free population in which allele frequencies will not change as long as the following conditions are met:

- There must be no mutation.
- There must be no **gene flow** between populations; that is, there must be no movement of alleles into or out of the population (as would be caused, for example, by the movement of organisms into or out of the population).
- The population must be very large.
- All mating must be random, with no tendency for certain genotypes to mate with specific other genotypes.
- There must be no natural selection; that is, all genotypes must reproduce with equal success.

Under these conditions, allele frequencies within a population will remain the same indefinitely. If one or more of these conditions is violated, allele frequencies may change: the population will evolve.

As you might expect, few if any natural populations are truly in equilibrium. What, then, is the importance of the Hardy-Weinberg principle? The Hardy-Weinberg conditions are useful starting points for studying the mechanisms of evolution. In the following sections, we will examine each of the conditions, show that natural populations often fail to meet them, and illustrate the consequences of such failures. In this way, we can better understand both the inevitability of evolution and the forces that drive evolutionary change.

15.2 What Causes Evolution?

Population genetics theory predicts that the Hardy-Weinberg equilibrium can be disturbed by deviations from any of its five conditions. Therefore, we might predict five major causes of evolutionary change: mutation, gene flow, small population size, nonrandom mating, and natural selection.

Mutations Are the Source of Genetic Variability

A population remains in genetic equilibrium only if there are no **mutations** (changes in DNA sequence), but mutations are inevitable. Although cells have efficient mechanisms that protect the integrity of their genes, some changes in nucleotide sequence slip past the checking and repair systems. If a cell that produces gametes has an unrepaired mutation, it may be passed to an offspring and enter the gene pool of a population.

Mutations Are Rare but Important

How significant is mutation in changing the gene pool of a population? Mutations are rare; for example, only one out of every 100,000 to 1,000,000 human gametes carries a mutation for a given gene. Therefore, mutation by itself is not a major force in evolution. However, mutations are the source of new alleles, new variations on which other evolutionary processes can work. As such, they are the foundation of evolutionary change. Without mutations there would be no evolution and no diversity among organisms.

Mutations Are Not Goal-Directed

A mutation does not arise as a result of, or in anticipation of, environmental necessities (Fig. 15-1). A mutation simply happens, and may in turn produce a change in a structure or function of the organism. Whether that change is helpful, harmful, or neutral, now or in the future, depends on environmental conditions over which the organism has little or no control. The mutation provides a potential for evolutionary change. Other forces, especially natural selection, may act on that potential to spread the mutation through the population or to eliminate it from the population.

Gene Flow Between Populations Changes Allele Frequencies

When individuals move from one population to another and interbreed at the new location, alleles are transferred from one gene pool to another. This movement of alleles, or gene flow, between populations alters the distribution of alleles among populations. Baboons, for example, live in social groupings called *troops*. Within each troop, all the females mate with a handful of dominant males. Juvenile males usually leave the troop. If they are lucky, they join and perhaps even become dominant in another troop. Thus, the male offspring of one troop carry genes to the gene pools of other troops.

Movement of breeding organisms between populations has two significant effects. First, gene flow spreads

Media Activity
15.1 Agents of Change

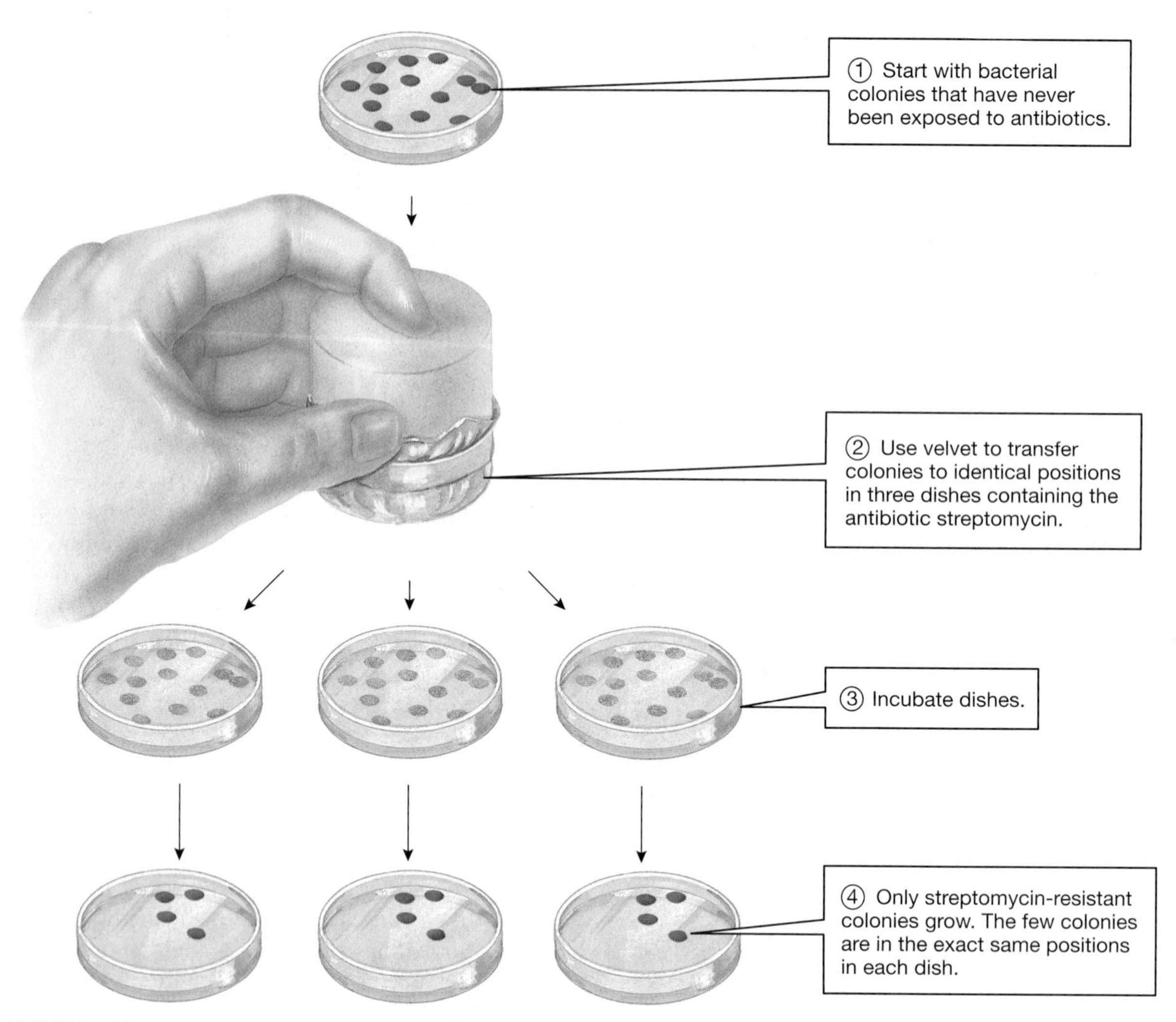
① Start with bacterial colonies that have never been exposed to antibiotics.
② Use velvet to transfer colonies to identical positions in three dishes containing the antibiotic streptomycin.
③ Incubate dishes.
④ Only streptomycin-resistant colonies grow. The few colonies are in the exact same positions in each dish.

FIGURE 15-1 Mutations occur spontaneously
This experiment demonstrates that mutations occur spontaneously and not in response to environmental pressures. When bacterial colonies that have never been exposed to antibiotics are exposed to the antibiotic streptomycin, only a few colonies grow. The observation that these surviving colonies grow in the exact same positions in all dishes shows that the mutations for resistance to streptomycin were present in the original dish before exposure to the environmental pressure, streptomycin. **QUESTION** If it were true that mutations *do* occur in response to the presence of antibiotic, how would the result of this experiment have differed from the actual result?

advantageous alleles throughout a species. If a new, beneficial allele arises by mutation and spreads through one population of a species, gene flow can carry the new allele to other populations. The second important effect of gene flow is that it prevents the development of large differences in allele frequencies among different populations of a species. If migrants constantly carry alleles back and forth among populations, the genetic compositions of the different populations are less likely to diverge. If gene flow between populations of a species is blocked, the resulting genetic differences may grow so large that one of the populations becomes a new species, a process we will discuss in Chapter 16.

Allele Frequencies May Drift in Small Populations

To remain in equilibrium, a population must be so large that chance events have no impact on its overall genetic makeup. Disaster may befall even the fittest organism. A maple seed that falls into a pond never sprouts; the yearling deer and elk killed by the eruption of Mount St. Helens left no descendants. If a population is sufficiently large, chance events are unlikely to alter its genetic composition; such events would interfere equally with the reproduction of organisms of all genotypes. In a small population, however, certain alleles may be carried by only a few organisms. Chance events could eliminate some or all of such alleles from the population, altering its genetic makeup. This process by which chance events change allele frequencies in a small population is called **genetic drift**.

Population Size Matters

To see how genetic drift works, imagine two populations of amoebas in which each amoeba is either red or blue, and color is controlled by two alleles (*A* and *a*) of a gene. Half the amoebas in each of our two populations are red, and half are blue. One population, however, has only four individuals in it, while the other has 10,000.

Now let's picture reproduction in our imaginary populations. Let's select at random half of the individuals in each population and allow them to reproduce by binary fission. Each reproducing amoeba splits in half to yield two amoebas, each of which is the same color as the parent. In the large population, 5000 amoebas reproduce, yielding a new generation of 10,000. What are the chances that all 10,000 members of the new generation will be red? Just about nil. In fact, it would be extremely unlikely for even 3000 amoebas to be red or for 7000 to be red. The most likely outcome is that about half will be red and half blue, just as in the original population. In this large population, then, we would not expect a major change in allele frequencies from generation to generation. One way to test this expectation is to write a computer program that simulates how the allele frequencies of the alleles could change over generations. Figure 15-2a shows that, after four runs of such a simulation, the frequency of allele *A*, encoding red color, remains close to 0.5, consistent with the expectation that half of the amoebas would be red.

In the small population, the situation is different. Only two amoebas reproduce, and there is a 25% chance that both reproducers will be red. (This outcome is as likely as flipping two coins and having both come up heads.) If only red amoebas reproduce, then the next generation will consist entirely of red amoebas—a relatively likely outcome. It is thus possible for the allele for blue color to disappear from the population within a single generation. Figure 15-2b shows the fate of allele *A* in four runs of a simulation of our small population. In one of the four runs, allele *A* reaches a frequency of 1.0 (100 percent) in the second generation, meaning that all the amoebas in the third and following generations are red. In another run, the frequency of *A* drifts to zero in the third generation; subsequently, the population is all blue. Thus, one of the two amoeba phenotypes disappeared in half of the simulations.

A Population Bottleneck Is an Example of Genetic Drift

Two causes of genetic drift, a *population bottleneck* and the *founder effect*, further illustrate the consequences that small population size may have on the allele frequencies of a species. In a **population bottleneck**, the size of a population is drastically reduced—as a result of a natural catastrophe or overhunting, for example. Only a few individuals, then, are available to contribute genes

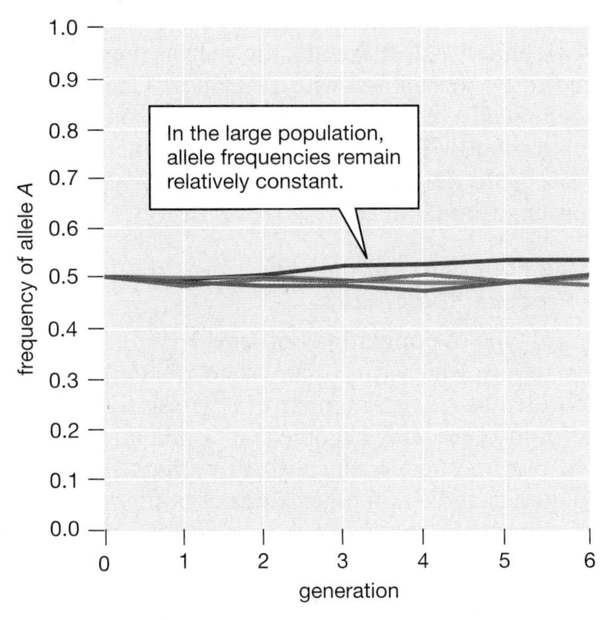

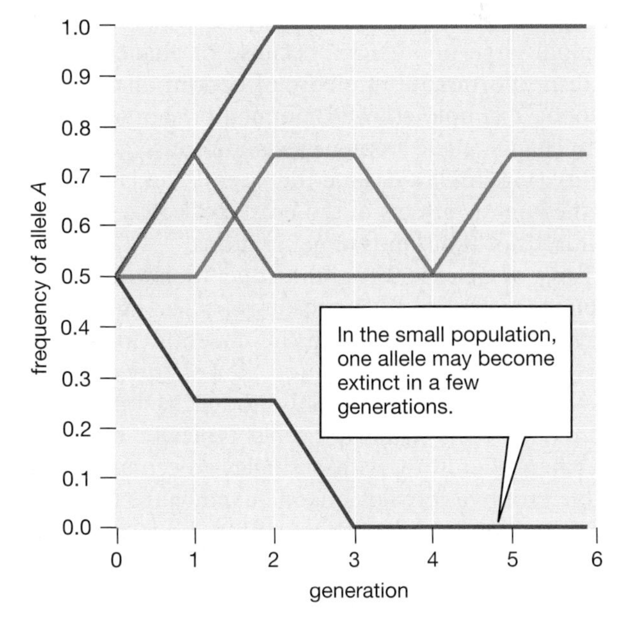

FIGURE 15-2 The effect of population size on genetic drift
Each colored line represents one computer simulation of the change over time in the frequency of allele *A* in a **(a)** large or **(b)** small population in which two alleles, *A* and *a*, were initially present in equal proportions, and in which randomly chosen individuals reproduced. **EXERCISE** Sketch a graph that shows the result you would predict if the simulation were run four times with a population size of 20.

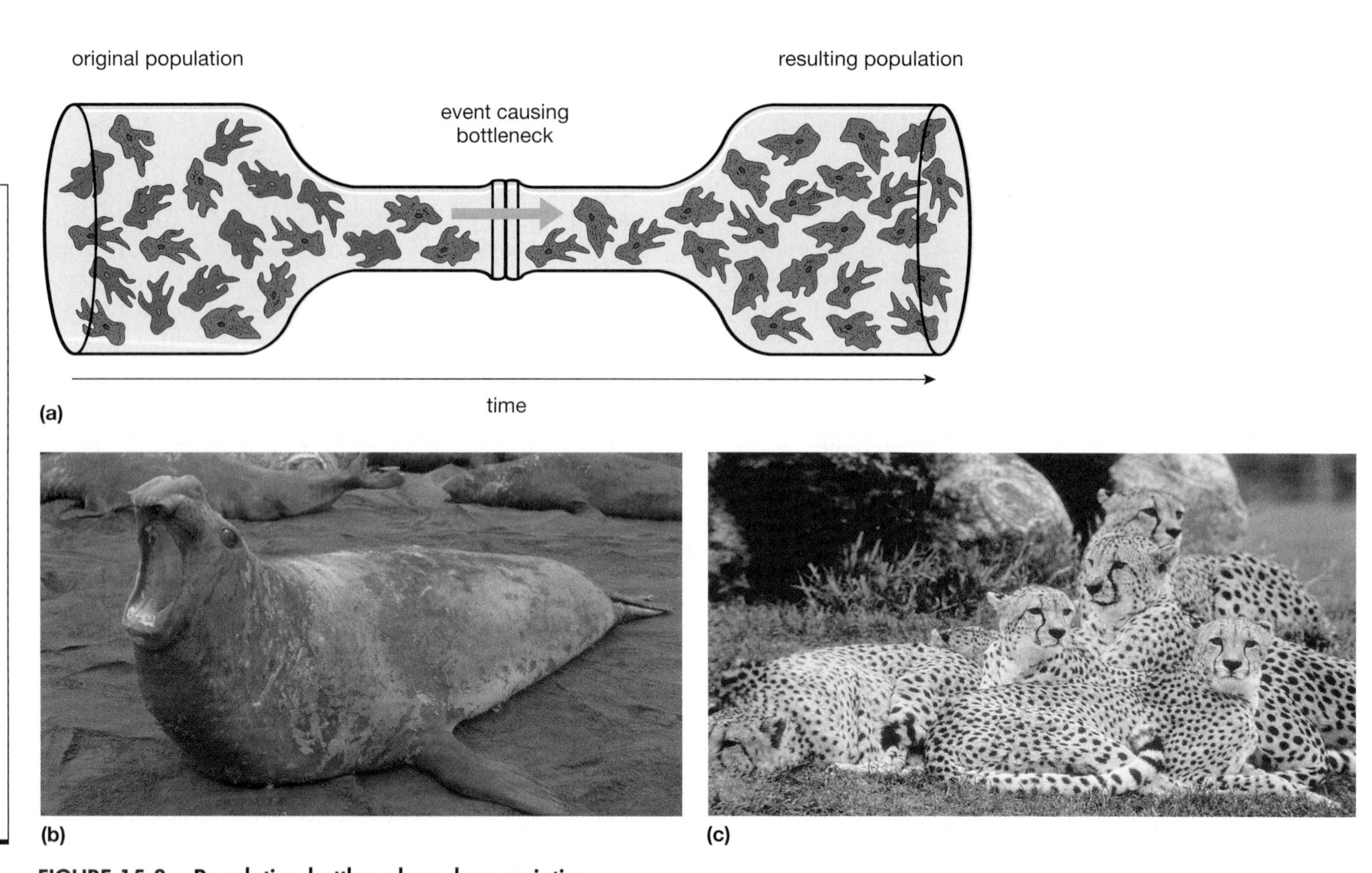

Media Activity
15.2 The Bottleneck Effect

FIGURE 15-3 Population bottlenecks reduce variation
(a) A population bottleneck may drastically reduce genetic and phenotypic variation because the few organisms that survive may carry similar sets of alleles. Both **(b)** the northern elephant seal and **(c)** the cheetah passed through population bottlenecks in the recent past, resulting in an almost total loss of genetic diversity. **QUESTION** If a population grows large again after a bottleneck, genetic diversity will eventually increase. Why?

to the next generation. Allele frequencies in this small group may be very different from the frequencies of the original large population, because of chance inclusion of disproportionate numbers of certain alleles. As our amoeba example showed, population bottlenecks can both change allele frequencies and reduce genetic variability (Fig. 15-3a). Even if the population later increases, the genetic effects of the bottleneck may remain for hundreds or thousands of generations.

Loss of genetic variability due to bottlenecks has been documented in numerous species, including the northern elephant seal and the cheetah (Fig. 15-3b,c). The elephant seal was hunted almost to extinction in the 1800s; by the 1890s, only about 20 still survived. Dominant male elephant seals typically monopolize breeding; therefore, with a single male mating with a stable group of females, one male may have fathered all the offspring at this extreme bottleneck point. Since then, elephant seals have increased in number to about 30,000 individuals, but biochemical analysis shows that all northern elephant seals are genetically almost identical. Other species of seals, whose populations have always remained large, exhibit much more genetic variability. The rescue of the northern elephant seal from extinction is rightly regarded as a triumph of conservation. With very little genetic variation, however, the elephant seal has much less potential to evolve in response to environmental changes. No matter how many elephant seals there are, the species must be considered to be threatened with extinction. Cheetahs are also genetically uniform, although the reason for their population bottleneck is unknown. Consequently, cheetahs too could be gravely threatened by small changes in their environment.

Isolated Founding Populations May Produce Bottlenecks

A special type of population bottleneck is caused by the **founder effect**, which can occur when isolated colonies are founded by a small number of organisms. A flock of birds, for instance, may become lost during migration or may be blown off course by a storm, perhaps settling on an isolated island. As in other kinds of bottlenecks, only a few individuals (the founders) contribute genes to the future population. If the founders remain isolated for a long period of time, a sizable new population may arise that differs greatly from the original population. For example, a set of genetic defects known as Ellis–van Creveld syndrome (Fig. 15-4) is far more common among the Amish inhabitants of Lancaster County, Pennsylvania, than among the general population. Today's Lancaster County Amish are descended from only 200 or so eigh-

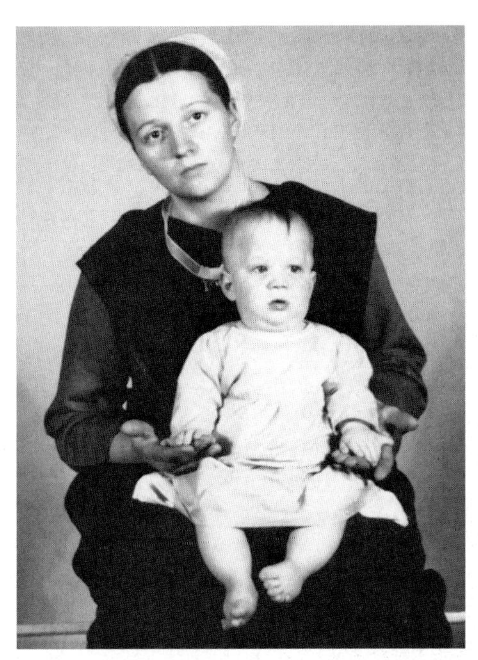

FIGURE 15-4 A human example of the founder effect
An Amish woman with her child, who suffers from a set of genetic defects known as Ellis–van Creveld syndrome (short arms and legs, extra fingers, and, in some cases, heart defects). The founder effect accounts for the prevalence of Ellis–van Creveld syndrome among the Amish residents of Lancaster County, Pennsylvania.

teenth-century immigrants, and one couple among these immigrants is known to have carried the Ellis–van Creveld allele. In such a small founder population, this single occurrence meant that the allele was carried by a comparatively high proportion of the Amish founder population (1 or 2 carriers out of 200, versus perhaps 1 in 1000 in the general population). This high initial allele frequency, combined with subsequent genetic drift, has led to extraordinarily high levels of Ellis–van Creveld syndrome among this Amish group.

Mating Within a Population Is Almost Never Random

Nonrandom mating by itself will not alter allele frequencies in a population. Nonetheless, it can have large effects on the distribution of different genotypes, and thus on the distribution of phenotypes, in the population. Certain genotypes may become more common, which can affect the direction of natural selection.

The effects of nonrandom mating can be important, because organisms seldom mate strictly randomly. For example, most animals have limited mobility and are most likely to mate with nearby members of their species. Further, they may choose to mate with certain individuals of their species rather than with others. The snow goose is a case in point. Individuals of this species come in two "color phases"; some snow geese are white, while others are blue-gray (Fig. 15-5). Although both white and blue-gray geese belong to the same species, mate choice is not random with respect to color. The

FIGURE 15-5 Nonrandom mating among snow geese
Snow geese, which have either white plumage or blue-gray plumage, are most likely to mate with other birds of the same color.

birds exhibit a strong tendency to mate with a partner of the same color. This preference for mates that are similar is known as *assortative mating*.

Another common form of nonrandom mating takes place in animal species in which only a few dominant males gain reproductive access to females. In these species, which include elephant seals, deer, elk, baboons, and bighorn sheep, all of the females in a population are fertilized by a small number of males. Fertilization typically follows some sort of contest among males, which may involve showing off with loud sounds or flashy colors, threatening gestures, or actual combat (Fig. 15-6).

In many animal species, mating is not random because one sex, normally the female, controls mate selection and is quite picky about choosing a breeding

FIGURE 15-6 Competition between males favors evolution of structures for ritual combat
Two male bighorn sheep spar during the fall mating season. In many species, the losers of such contests are unlikely to mate, while winners enjoy tremendous reproductive success.
QUESTION Imagine that you studied a population of bighorn sheep and were able to identify the father and mother of each lamb born. Would the difference in number of offspring between the most reproductively successful adult and the least successful one be greater for males or for females?

FIGURE 15-7 Peahens are attracted to the peacock's showy tail
The ancestors of today's peahens were apparently picky when deciding on males with which to mate, favoring males with longer and more colorful tails.

partner. Males display their virtues, such as the bright plumage of a peacock (Fig. 15-7) or the resource-rich territory of a songbird. A female evaluates the males and chooses her mate. We will explore this phenomenon in more detail later in this chapter.

All Genotypes Are Not Equally Beneficial

In a hypothetical equilibrium population, individuals of all genotypes survive and reproduce equally well—that is, no genotype has an advantage over the others. This condition, however, is probably met only rarely—if ever—in real populations. Even though some alleles are neutral, in the sense that organisms possessing any of several alleles are equally likely to survive and reproduce, this is clearly not true of all alleles in all environments. Any time an allele confers "some little superiority" (in the words of the naturalist Alfred Russel Wallace), natural selection favors the individuals who possess it; that is, those individuals have higher reproductive success. This phenomenon is illustrated by an example concerning an antibiotic drug.

Antibiotic Resistance Evolves by Natural Selection

The antibiotic penicillin first came into widespread use during World War II, when it was used to combat infections in wounded soldiers. Suppose that an infantryman suffers a gunshot wound in his arm; while the soldier recovers in the field hospital, his wound becomes infected by *Staphylococcus* bacteria. A medic sizes up the situation and resolves to treat the wounded soldier with an intravenous drip of penicillin. As the antibiotic courses through the soldier's blood vessels, millions of staph bacteria die before they can reproduce. A few bacteria, however, carry a rare allele that codes for an enzyme that destroys any penicillin coming in contact with the bacterial cell. (This allele is a variant of a gene that normally codes for an enzyme that breaks down the bacterium's waste products.) The bacteria carrying this rare allele are able to survive and reproduce, and their offspring inherit the penicillin-destroying allele. After a few generations, the frequency of the penicillin-destroying allele soars to nearly 100%, while the frequency of the normal, waste-processing allele declines to near zero. As a result of natural selection imposed by the antibiotic's killing power, the population of staph bacteria within the soldier's body has evolved. The gene pool of the staph population has changed, and natural selection, in the form of bacterial destruction by penicillin, has caused the change.

Penicillin Resistance Illustrates Key Points about Evolution

Natural selection does not cause genetic changes in individuals. In the above example, the allele for penicillin resistance arose spontaneously, long before penicillin was dripped into the soldier's vein. Penicillin did not cause resistance to appear; its presence merely caused bacteria with penicillin-destroying alleles to be favored over bacteria with waste-processing alleles.

Natural selection acts on individuals, but it is populations that are changed by evolution. The agent of natural selection, penicillin, acted on individual staph bacteria. As a result, some individuals reproduced and some did not. However, it was the population as a whole that evolved as its allele frequencies changed.

Evolution is change in allele frequencies of a population, owing to unequal reproductive success among organisms bearing different alleles. In evolutionary terminology, the **fitness** of an organism is measured by its reproductive success. In our example, the penicillin-resistant bacteria had greater fitness than the normal bacteria because the resistant bacteria produced greater numbers of viable (able to survive) offspring.

Evolutionary changes are not "good" or "progressive" in any absolute sense. The resistant bacteria were favored only because of the presence of penicillin in the soldier's body. At another time, when the environment of the soldier's body did not contain penicillin, the resistant bacteria may have been at a disadvantage relative to other bacteria that could process waste more effectively. Similarly, the long necks of male giraffes are helpful when the animals battle to establish dominance, but

EARTH WATCH Endangered Species: From Gene Pools to Gene Puddles

Since the Endangered Species Act was passed in 1973, the United States has had an official policy of protecting rare species (Fig. E15-1). The ultimate goal of the act, however, is not just to protect endangered species, but also to restore them. As one U.S. Fish and Wildlife Service official said, "The goal is to get species *off* the list." A species that has been officially listed as endangered can be removed from the list if government biologists determine that its population has grown large enough that it is no longer in danger of extinction from unpredictable events, such as a drought or a disease epidemic. If a species' population reaches this critical size, it is no longer legally "endangered," and it loses the legal protections provided by the act.

Unfortunately, a population that has become small enough to warrant endangered status is likely to undergo evolutionary changes that increase its chances of going extinct. One problem is that, in small populations, mating choices are limited and a high proportion of matings may be between close relatives. This inbreeding increases the odds that offspring will be homozygous for harmful recessive alleles, and these less-fit individuals may die before reproducing, further reducing the size of the population. The greatest threat to small populations, however, stems from their inevitable loss of genetic diversity. From our discussion of population bottlenecks, it is apparent that, when populations shrink to very small sizes, many of the alleles that were present in the original population will not be represented in the gene pool of the remnant population. Furthermore, we have seen that genetic drift in small populations will cause many of the surviving alleles to subsequently disappear permanently from the population (see Fig. 15-2b). Because genetic drift is a random process, many of the lost alleles will be advantageous ones that were previously favored by selection. Inevitably, the number of different alleles in the population grows ever smaller. As ecologist Thomas Foose aptly put it, "gene pools are being converted into gene puddles." Even if the size of an endangered population eventually begins to grow, the damage has already been done; lost genetic diversity is regained only very slowly.

Why does it matter if a population's genetic diversity is low? There are two main risks. First, the fitness of the population as a whole is reduced by the loss of advantageous alleles that underlie adaptive traits. A less-fit population is unlikely to thrive. Second, a genetically impoverished population lacks the variation that will allow it to adapt when environmental conditions change. When the environment changes, as it inevitably will, a genetically uniform species is less likely to contain individuals well suited to survive and reproduce under the new conditions. A species unable to adapt to changing conditions is at very high risk of extinction.

What can be done to preserve the genetic diversity of species? The best solution, of course, is to preserve plenty of diverse types of habitat so that species never become endangered. The human population, however, has grown so large and has appropriated so large a share of Earth's resources that this solution is impossible in many places. For many species, the only solution is to ensure that areas of preserved habitat are large enough to hold populations of sufficient size to contain most of a threatened species' total genetic diversity. If, however, circumstances dictate that preserved areas are small, it is important that the small areas be linked by corridors of the appropriate habitat, so that gene flow among populations in the small preserves can increase the spread of new and beneficial alleles.

bio ETHICS

ETHICAL CONSIDERATIONS

Does it matter that human activities are causing species to go extinct? Some bioethecists argue that, because humans have the power to extinguish species, we have an ethical obligation to protect the interests of all of the planet's inhabitants. In this view, it is unethical to allow any species to go extinct. For those who believe in the sanctity of other species, the biodiversity crisis poses profound ethical dilemmas. In many cases, the habitat destruction that endangers other species also helps make space for the farmland, housing, and workplaces needed by our growing human population. How can we reconcile the conflict between valid human needs and the needs of endangered species? Furthermore, it is becoming clear that, even with the best of intentions, we cannot save all of the species currently threatened with extinction. The resources available to preserve and manage protected habitats are limited, and we must make choices that will allow some species to survive while others perish. If all species are precious, how can we make such terrible choices? Who should decide which species will live and which will die, and what criteria should be used?

FIGURE E15-1 Endangered by habitat destruction
The bighorn sheep is among the species designated "endangered" by the U.S. government.

(a)

(b)

FIGURE 15-8 A compromise between opposing pressures
(a) A male giraffe with a long neck is at a definite advantage in combat to establish dominance. **(b)** But a giraffe's long neck forces it to assume an extremely awkward and vulnerable position when drinking. Thus, drinking and male–male contests place opposing evolutionary pressures on neck length.

are a hindrance to drinking (Fig. 15-8). The length of male giraffe necks represents an evolutionary compromise between the advantage of being able to win contests with other males and the disadvantage of vulnerability while drinking water. (The necks of female giraffes are long—though not as long as male necks—because successful males pass the alleles for long necks to daughters as well as sons.)

15.3 How Does Natural Selection Work?

Natural selection is not the *only* evolutionary force. As we have seen, mutation provides initial variability in heritable traits. The chance effects of genetic drift may change allele frequencies, even spawn new species. Further, evolutionary biologists are now beginning to appreciate that the history of life on Earth has been shaped, in part, by the power of random catastrophes—massively destructive events that exterminate thriving and failing species alike. Nevertheless, natural selection shapes the evolution of populations as they adapt to their changing environment; therefore, we will examine natural selection in more detail.

Media Activity
15.4 Natural Selection at Work: Alpine Skypilots

Natural Selection Stems from Unequal Reproduction

To most people, the phrase **natural selection** is synonymous with "survival of the fittest." Natural selection may evoke images of wolves chasing caribou, or of lions snarling angrily in competition over a zebra carcass. Natural selection, however, is not about survival alone; it is also about reproduction. It is certainly true that if an organism is to reproduce, it must survive long enough to do so. In some cases, it is also true that a longer-lived organism has more chances to reproduce. But no organism lives forever, and the only way that its alleles continue into the future is through successful reproduction. When an organism dies without reproducing, its alleles die with it. The organism that reproduces lives on, in a sense, through the alleles that it has passed to its offspring. Therefore, although evolutionary biologists often discuss survival—partly because survival is usually easier to observe than reproduction—the main issue of natural selection is *differences in reproduction*: individuals bearing certain alleles leave more offspring (who inherit those alleles) than do other individuals with different alleles.

Natural Selection Acts on Phenotypes

Although we have defined evolution as changes in the genetic composition of a population, it is important to recognize that natural selection cannot act directly on the genotypes of individual organisms. Rather, natural selection acts on phenotypes, the structures and behaviors displayed by the members of a population. This selection on phenotypes, however, inevitably affects the genotypes present in a population, because phenotypes and genotypes are closely tied. For example, we know that a pea plant's height is strongly influenced by the plant's alleles of certain genes. If a population of pea plants encountered environmental conditions that favored taller plants, then taller plants would leave more offspring. These offspring would carry the alleles that contributed to their parents' height. Thus, if natural selection favors a particular phenotype, it will necessarily also favor the underlying genotype.

Some Phenotypes Reproduce More Successfully Than Others

As we have seen, natural selection simply means that some phenotypes reproduce more successfully than others. This simple process is such a powerful agent of change because only the "best" phenotypes pass traits to subsequent generations. But what makes a phenotype the "best"? Successful phenotypes are those that have the best adaptations to their particular environment. **Adaptations** are characteristics that help an individual survive and reproduce.

An Environment Has Nonliving and Living Components

Individual organisms must cope with an environment that includes not only physical factors but also the other organisms with which the individual interacts. The nonliving (*abiotic*) component of the environment includes factors such as climate, the availability of water, and the concentration of minerals in the soil. The nonliving environment establishes the "bottom line" requirements that an organism must fulfill to survive and reproduce. However, many of the adaptations that we see in modern organisms have arisen because of interactions with other organisms—the living (*biotic*) component of the environment. As Darwin wrote, "The structure of every organic being is related . . . to that of all other organic beings, with which it comes into competition for food or residence, or from which it has to escape, or on which it preys." A simple example illustrates this concept.

Imagine a buffalo grass plant sprouting in a small patch of soil in the plains of eastern Wyoming. Its roots must be able to take up enough water and minerals for growth and reproduction, and to that extent it must be adapted to its abiotic environment. But even in the dry prairies of Wyoming, this requirement is relatively trivial, provided that the plant is alone and protected in its square yard of soil. In reality, however, many other plants—other buffalo grass plants as well as other grasses, sagebrush bushes, and annual wildflowers—also sprout in that same patch of soil. If our buffalo grass is to survive, it must compete with the other plants for resources. Its long, deep roots and efficient methods of mineral uptake have evolved not so much because the plains are dry, but mostly because the buffalo grass must share the dry prairies with other plants. Further, buffalo grass must also coexist with animals that wish to eat it, such as the cattle that graze the prairie (and the bison that grazed it in the past). As a result, buffalo grass is extremely tough; silica compounds reinforce its leaves, an adaptation that discourages grazing. Over time, tougher, hard-to-eat plants survived better and reproduced more than did less-tough plants—another adaptation to the biotic environment.

Competition Acts as an Agent of Selection

As the buffalo grass example shows, one of the major agents of natural selection in the biotic environment is **competition** with other organisms for scarce resources. Competition for resources is most intense among members of the same species. As Darwin wrote in *On the Origin of Species*, "The struggle almost invariably will be most severe between the individuals of the same species, for they frequent the same districts, require the same food, and are exposed to the same dangers." In other words, no competing organism has such similar requirements for survival as does another member of the same species. Different species may also compete for the same resources, although generally to a lesser extent than do individuals within a species.

Both Predator and Prey Act as Agents of Selection

When two species interact extensively, each exerts strong selection pressures on the other. When one evolves a new feature or modifies an old one, the other typically evolves new adaptations in response. This constant, mutual feedback between two species is called **coevolution**. Perhaps the most familiar form of coevolution is found in predator–prey relationships.

Predation includes any situation in which one organism eats another. In some instances, coevolution between predators (those who do the eating) and prey (those who are eaten) is a sort of "biological arms race," with each side evolving new adaptations in response to "escalations" by the other. Darwin used the example of wolves and deer: Wolf predation selects against slow or careless deer, thus leaving faster, more-alert deer to reproduce and continue the species. In their turn, alert, swift deer select against slow, clumsy wolves, because such predators cannot acquire enough food.

Sexual Selection Favors Traits That Help an Organism Mate

In many animal species, males have conspicuous features such as bright colors, long feathers or fins, or elaborate antlers. Males may also exhibit bizarre courtship behaviors or sing loud, complex songs. These extravagant features typically play a role in mating, but they also seem to be at odds with efficient survival and reproduction. Exaggerated ornaments and displays may help males gain access to females, but also make the males more vulnerable to predators. Darwin was intrigued by this apparent contradiction. He coined the term **sexual selection** to describe the special kind of selection that acts on traits that help an animal acquire a mate.

Darwin recognized that sexual selection could be driven either by sexual contests among males or by female preference for particular male phenotypes. Male–male competition for access to females can favor the evolution of features that provide an advantage in fights or ritual displays of aggression (see Fig. 15-6). Female mate choice provides a second source of sexual selection. In animal species in which females actively choose their mates from among males, females often seem to prefer males with the most elaborate ornaments or most extravagant displays (see Fig. 15-7). Why? A popular hypothesis is that male structures, colors, and displays that do not enhance survival might instead provide a female with an outward sign of a male's condition. Only a vigorous, energetic male can survive when burdened with conspicuous coloration or a large tail that might make him more vulnerable to predators. Conversely, males that are sick or under parasitic attack appear dull and frumpy compared with healthy males. A female that chooses the brightest, most ornamented male is also choosing the healthiest, most vigorous male. By doing so, she gains fitness if, for example, the most vigorous male provides superior parental care or if he carries alleles for disease resistance that will be inherited by offspring and help ensure their survival. Females thus gain a reproductive advantage by choosing the most highly ornamented males, and the traits (including the exaggerated male ornament) of these flashy males are more likely to be passed to subsequent generations.

Selection Can Influence Populations in Three Ways

Natural selection and sexual selection can lead to different patterns of evolutionary change. Evolutionary biologists group these patterns into three categories (Fig. 15-9).

- **Directional selection** favors individuals with an extreme value of a trait and selects against both average individuals and individuals at the opposite extreme. For example, directional selection might favor small size and select against both average and large individuals in a population.
- **Stabilizing selection** favors individuals with the average value of a trait (for example, intermediate body size) and selects against individuals with extreme values.
- **Disruptive selection** favors individuals at both extremes of a trait (for example, both large and small body sizes) and selects against individuals with intermediate values.

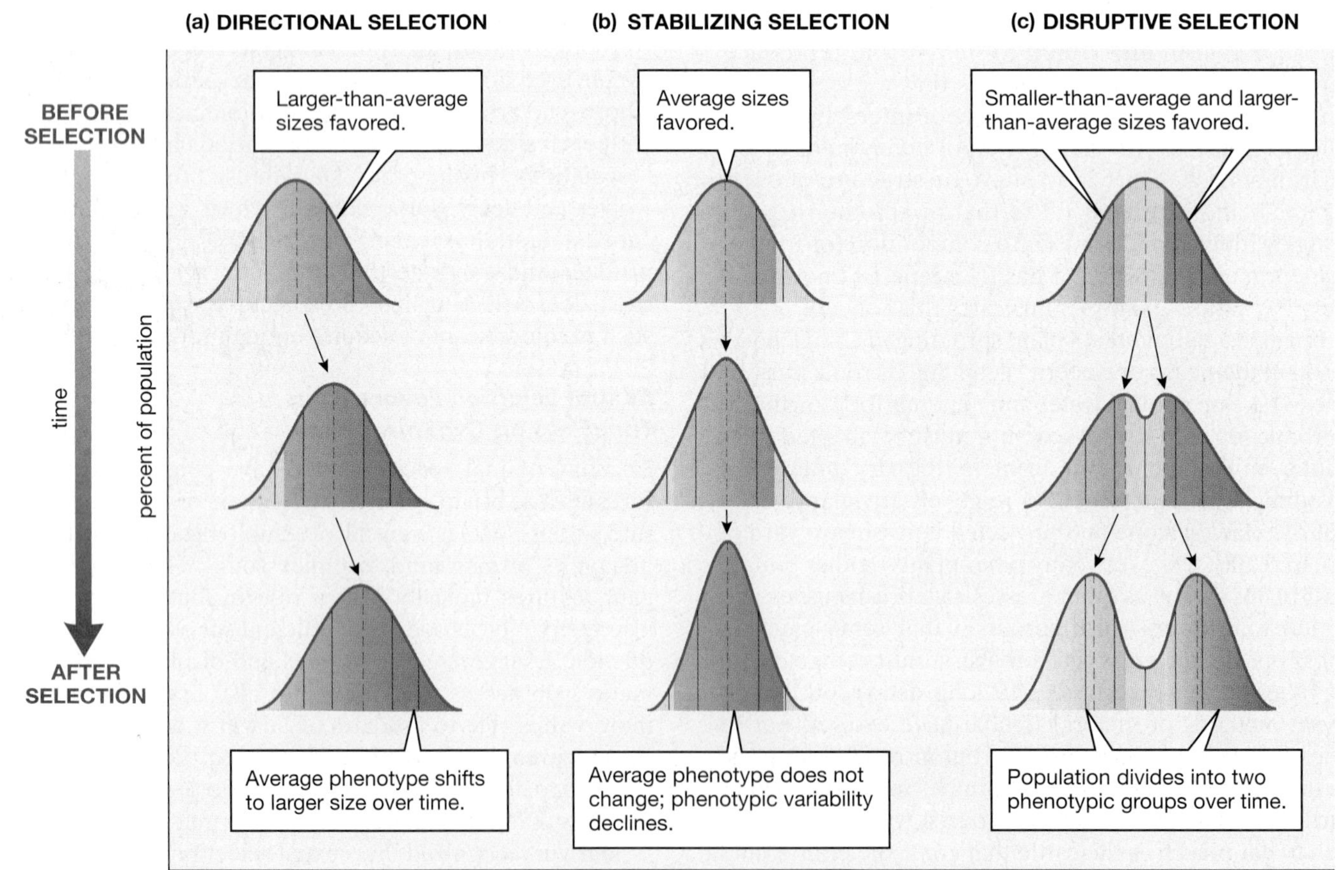
(a) DIRECTIONAL SELECTION
(b) STABILIZING SELECTION
(c) DISRUPTIVE SELECTION
BEFORE SELECTION
AFTER SELECTION
time
percent of population
Larger-than-average sizes favored.
Average sizes favored.
Smaller-than-average and larger-than-average sizes favored.
Average phenotype shifts to larger size over time.
Average phenotype does not change; phenotypic variability declines.
Population divides into two phenotypic groups over time.
range of a particular characteristic (size, color, etc.)

FIGURE 15-9 Three ways that selection affects a population over time
A graphical illustration of three ways natural and/or sexual selection, acting on a normal distribution of phenotypes, can affect a population over time. In all graphs, the beige areas represent individuals that are selected against—that is, do not reproduce as successfully as do the individuals in the purple range. **QUESTION** When selection is directional, is there any limit to how extreme the trait under selection will become? Why or why not?

Media Activity
15.3 Three Modes of Natural Selection

Directional Selection Shifts Character Traits in a Specific Direction

If environmental conditions change in a consistent way, a species may respond by evolving in a consistent direction. For example, if the climate becomes colder, mammal species may evolve thicker fur. The evolution of antibiotic resistance in bacteria is an example of directional selection (see Fig. 15-9a): when antibiotics are present in a bacterial species' environment, individuals with greater resistance reproduce more prolifically than do individuals with less resistance.

Stabilizing Selection Acts Against Individuals Who Deviate Too Far from the Average

Directional selection can't go on forever. What happens once a species is well adapted to a particular environment? If the environment in unchanging, most new variations that appear will be harmful. Under these conditions, we expect species to be subject to stabilizing selection, which favors the survival and reproduction of average individuals (see Fig. 15-9b). Stabilizing selection commonly occurs when a trait is under opposing environmental pressures from two different sources. For example, among lizards of the genus *Aristelliger*, the smallest lizards have a hard time defending territories, but the largest lizards are more likely to be eaten by owls. As a result, *Aristelliger* lizards are under stabilizing selection that favors intermediate body size.

It is widely assumed that many traits are under stabilizing selection. Although the long necks of giraffes probably originated under directional sexual selection for advantage in combat among males, they are probably now under stabilizing selection, as a compromise between the advantages of being able to win contests and the disadvantage of being vulnerable while drinking water (see Fig. 15-8).

Disruptive Selection Adapts Individuals Within a Population to Different Habitats

Disruptive selection (see Fig. 15-9c) may occur when a population inhabits an area with more than one type of useful resource. In this situation, the most adaptive characteristics may be different for each type of resource. For example, the food source of the black-bellied seedcracker (a small, seed-eating bird found in the forests of Africa) includes both hard seeds and soft seeds. Cracking hard seeds requires a large, stout beak, but a smaller, pointier beak is a more efficient tool for processing soft seeds. Consequently, black-bellied seedeaters have beaks in one of two sizes. A bird may have a large beak or small beak, but very few birds have a medium-sized beak; individuals with intermediate-sized beaks have a lower survival rate than individuals with either large or small beaks. Disruptive selection in black-bellied seedcrackers thus favors birds with large beaks and birds with small beaks, but not those with medium-sized beaks.

Black-bellied seedeaters represent an example of *balanced polymorphism*, in which two or more phenotypes are maintained in a population. In many cases of balanced polymorphism, multiple phenotypes persist because each is favored by a separate environmental force. For example, consider two different forms of hemoglobin that are present in some African human populations. In these populations, the hemoglobin molecules of people who are homozygous for a particular allele produce defective hemoglobin that clumps up into long chains, which distort and weaken red blood cells. This distortion causes a serious illness known as sickle-cell anemia, which can kill its victims. Before the advent of modern medicine, people homozygous for the sickle-cell allele were unlikely to survive long enough to reproduce. So why hasn't natural selection eliminated the allele?

Far from being eliminated, the sickle-cell allele is present in nearly half the population in some areas of Africa. The persistence of the allele seems to be the result of counterbalancing selection that favors heterozygous carriers of the allele. Heterozygotes, who have one allele for defective hemoglobin and one allele for normal hemoglobin, suffer from mild anemia but also exhibit increased resistance to malaria, a deadly disease affecting red blood cells that is widespread in equatorial Africa. In areas of Africa with high risk of malaria infection, heterozygotes must have survived and reproduced more successfully than either type of homozygote. As a result, both the normal hemoglobin allele and the sickle-cell allele have been preserved (Fig. 15-10).

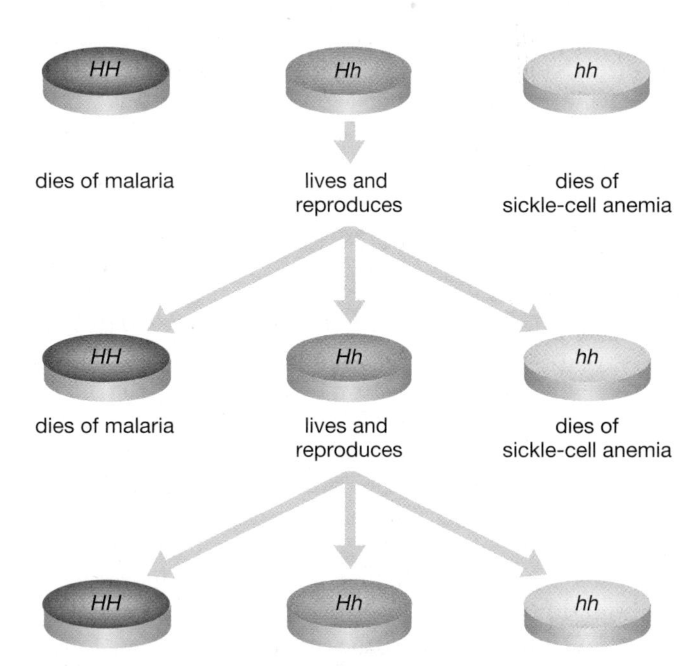

FIGURE 15-10 A balanced polymorphism
Natural selection may preserve two or more phenotypes in a population. In some African human populations, selection for malaria-resistant heterozygotes ensures that the alleles for both normal (*H*) and sickle-cell (*h*) hemoglobin persist in the population. **QUESTION** What will happen to the *h* allele in malaria-free environments?

EVOLUTIONARY CONNECTIONS

Knowing Your Relatives: Kin Selection and Altruism

Evolution is often portrayed as a bloody and vicious fight for survival. This portrayal, however, is not the complete picture. Although competitive and predatory interactions do influence the evolution of most species, cooperation and even self-sacrifice may be favored by selection. **Altruism** refers to any behavior that endangers an individual organism or reduces its reproductive success but benefits other members of its species. Altruistic behaviors are common in the animal kingdom. A male scrub jay helps feed and defend the nestlings of another mated pair of jays instead of seeking out a territory and mate of his own (Fig. 15-11); female worker bees forego reproduction and devote their lives to raising the offspring of the hive queen; and young male baboons scout around the edges of the troop, even though doing so increases their danger of predation by leopards.

FIGURE 15-11 Altruism by a helper at the nest
A yearling Florida scrub jay helps feed and defend his younger siblings rather than build a nest of his own.

From an evolutionary viewpoint, how can we explain altruistic behavior? Surely, a mutation that causes altruistic behavior would quickly disappear from a population, because individuals with the mutation would die or fail to reproduce because of their self-sacrificing behaviors. Maybe, or maybe not.

To understand the evolution of altruism, we will need to introduce a new concept: *inclusive fitness*. The inclusive fitness of an individual is determined by the individual's success at contributing its own alleles to the next generation. This contribution has two components: a direct contribution from producing offspring and an indirect contribution gained by helping relatives produce more offspring than would otherwise have been possible.

To see how altruism might increase the inclusive fitness of an individual, let's consider the Florida scrub jay. Year-old jays rarely mate and reproduce. Instead, these yearlings remain at their parents' nest and help feed and protect the parents' next brood.

Altruistic yearlings forego reproduction for at least one year; some die from predation or accidents and never reproduce at all. How, then, can this behavior be adaptive? It all has to do with the reproductive options available to young Florida scrub jays. In their particular ecological setting, suitable breeding habitat is in very short supply. Long-term studies have shown that a young jay that leaves its home territory is fairly unlikely to be successful at finding a territory and raising a brood. But if a young bird stays on its home territory, it may eventually inherit the territory or carve out a smaller area for itself. For this reason, selection has favored the behavior of "staying home." Still, this doesn't explain why the stay-at-home jays expend valuable energy to help raise their younger siblings. Answering *that* question requires us to recognize that, on average, a diploid, sexually reproducing animal shares 50% of its alleles with each sibling. A scrub jay—or any other animal—is related to its siblings just as closely as it would be to its own offspring. Therefore, if a yearling jay's help at the nest enables the survival of a sibling who would not otherwise have survived, the helper has accomplished as much in fitness terms as if it had raised an offspring of its own. As long as the average extra (indirect) fitness that results from helping is greater than the average (direct) fitness that would have resulted from an attempt at independent breeding, the altruistic helping behavior is the most adaptive behavior. This phenomenon, whereby the actions of an individual increase the survival or reproductive success of its relatives, is called **kin selection.**

As this example suggests, kin selection can favor the evolution of altruism if the altruistic behavior benefits relatives that also bear the alleles that help cause altruistic behavior. In the case of the scrub jays, it's easy to see how young jays can be sure that they help only relatives. Many biologists, however, have objected to kin selection as a plausible explanation of other instances of altruistic behaviors, arguing that many animals cannot evaluate degrees of relatedness. Two findings seem to address this objection. First, many social groups, including wolf packs and baboon troops, are actually family groups. Therefore, an animal in such a group would not have to be able to specifically identify its relatives for its altruistic behaviors to benefit them the most. Second, many animals, including birds, monkeys, tadpoles (Fig. 15-12), bees, and even tunicate larvae, can indeed identify relatives. Given the choice between relatives and strangers, these animals preferentially associate with their relatives, even if they were separated at birth and have never seen those relatives before. If animals selectively form related groups, altruistic behaviors will most likely benefit relatives. Although it is not the only mechanism of natural selection, kin selection has been a powerful environmental force in the evolution of altruism in many species, probably including humans.

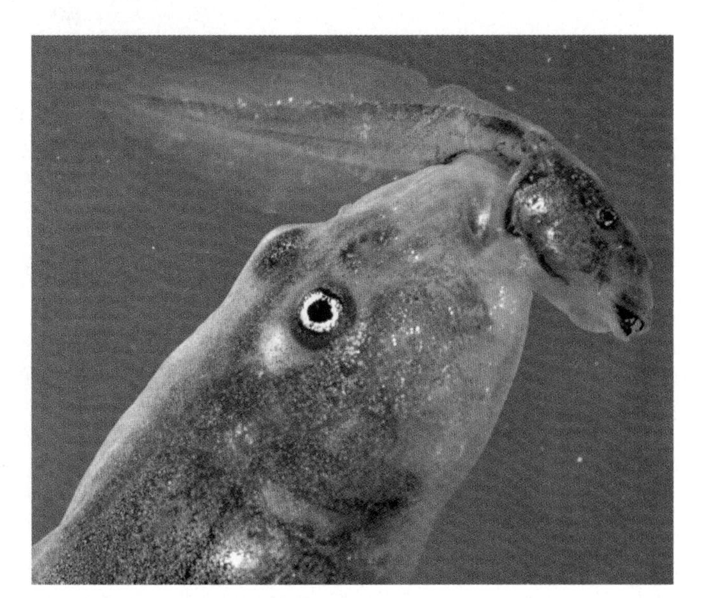

FIGURE 15-12 Cannibalistic animals don't eat close relatives
Spadefoot toad tadpoles, found in transient water holes of the Arizona desert, are cannibalistic. Many of their prey, however, are released unharmed after being tasted briefly. Researchers have discovered that the tadpoles distinguish, and spit out, their own brothers and sisters, preferring to eat unrelated members of their own species.

CASE STUDY REVISITED Evolution of a Menace

The evolution of antibiotic resistance in populations of bacteria—such as the bacteria that cause multidrug-resistant tuberculosis—is a direct consequence of natural selection applied by antibiotic drugs. When a population of disease-causing bacteria begins to grow in a human body, physicians try to halt population growth by introducing an antibiotic drug to the bacteria's environment. Although many bacteria are killed, some surviving bacteria have genomes with a mutant allele that confers resistance. Bacteria carrying the "resistance allele" produce a disproportionately large share of offspring, which inherit the allele. Soon, resistant bacteria predominate within the population. The resistant bacteria get even more of a boost when the presence of antibiotics is inconsistent, as occurs when a tuberculosis patient neglects to take his or her medicine. During antibiotic-free periods, the populations of resistant bacteria can grow very rapidly and spread to new hosts.

By introducing massive quantities of antibiotics into the bacteria's environment, humans have accelerated the pace of the evolution of antibiotic resistance. Each year, U.S. physicians write more than 100 million prescriptions for antibiotics; the Centers for Disease Control estimates that about half of these prescriptions are unnecessary.

Although medical use and misuse of antibiotics is the most important source of natural selection for antibiotic resistance, antibiotics also pervade the environment outside our bodies. Our food supply, especially meat, contains a portion of the 20 million pounds of antibiotics that are fed to farm animals each year. In addition, Earth's soils and water are laced with antibiotics that enter the environment through human and animal wastes, and from the antibacterial soaps and cleansers that are now routinely used in many households and workplaces. As a result of this massive alteration of the environment, resistant bacteria are now found not only in hospitals and the bodies of sick people but are also widespread in our food, water, and soil. Susceptible bacteria are under constant attack, and resistant strains have little competition. In our fight against disease, we have rashly overlooked some basic principles of evolutionary biology and are now paying a heavy price.

Consider This: Because natural selection acts only on existing variation among phenotypes, antibiotic resistance could not evolve if bacteria in natural populations did not already carry alleles that help them resist attack by antibiotic chemicals. Why are such alleles present (albeit at low levels) in bacterial populations? (Hint: Almost all medically useful antibiotics were originally derived from fungi or bacteria.) Conversely, if resistance alleles are beneficial, why are they rare in natural populations of bacteria?

Links to Life: Confining the Contagious

A great way to prevent natural selection for multidrug resistance is to eliminate variation in bacterial phenotypes, preferably by killing every disease-causing bacterium in a victim's body. But if a tuberculosis patient fails to take antibiotics until all of the bacteria in his or her body are completely wiped out, the goal of eliminating variation is not achieved—and the surviving bacteria are especially likely to be resistant. A noncompliant patient thus becomes a factory that produces multidrug-resistant strains, and does additional damage by helping spread those strains to other people. How can we contain the threat to public health posed by uncooperative tuberculosis victims? One solution is to round them up and detain them in a location where their daily treatment can be closely supervised and monitored. In many states, tuberculosis patients may be detained and held against their will if they fail to follow a treatment regimen for their disease. In essence, the freedom of such patients is sacrificed to protect society from the negative effects of their behavior. Nonetheless, the laws that permit such detention reflect sound evolutionary thinking. Do you believe that detention of uncooperative tuberculosis victims is justified?

Summary of Key Concepts

15.1 How Are Populations, Genes, and Evolution Related?

Evolution is change in frequencies of alleles in a population's gene pool. Allele frequencies in a population will remain constant over generations only if the following conditions are met: there is no mutation; there is no gene flow; the population is very large; all mating is random; and all genotypes reproduce equally well (that is, there is no natural selection). These conditions are rarely, if ever, met in nature. Understanding what happens when they are not met helps reveal the mechanisms of evolution.

15.2 What Causes Evolution?

- Mutations are random, undirected changes in DNA composition. Although most mutations are neutral or harmful to the organism, some prove advantageous in certain environments. Mutations are rare and do not change allele frequencies very much, but they provide the raw material for evolution.
- Gene flow is the movement of alleles between different populations of a species. Gene flow tends to reduce differences in the genetic composition of different populations.
- In any population, chance events kill or prevent reproduction by some of the individuals. If the population is small, chance events may eliminate a disproportionate number of individuals who bear a particular allele, thereby greatly changing the allele frequency in the population: This is genetic drift.
- Nonrandom mating, such as assortative mating, can change the distribution of genotypes in a population.
- The survival and reproduction of organisms are influenced by their phenotypes. Because phenotype depends at least partly on genotype, natural selection tends to favor the reproduction of certain alleles at the expense of others.

15.3 How Does Natural Selection Work?

Natural selection is driven by differences in reproductive success among different genotypes. Natural selection proceeds from the interactions of organisms with both the biotic and abiotic parts of their environments. When two or more species exert mutual environmental pressures on each other for long periods of time, both of them evolve in response. Such coevolution can result from any type of relationship between organisms, including competition and predation. Phenotypes that help organisms mate can evolve by sexual selection.

Key Terms

adaptation *p. 293*
allele frequency *p. 284*
altruism *p. 296*
coevolution *p. 293*
competition *p. 293*
directional selection *p. 294*
disruptive selection *p. 294*
equilibrium population *p. 285*
fitness *p. 290*
founder effect *p. 288*
gene flow *p. 285*
gene pool *p. 284*
genetic drift *p. 287*
Hardy-Weinberg principle *p. 285*
kin selection *p. 295*
mutation *p. 285*
natural selection *p. 292*
population *p. 284*
population bottleneck *p. 287*
predation *p. 293*
sexual selection *p. 293*
stabilizing selection *p. 296*

Thinking Through the Concepts

To take a multiple-choice quiz with feedback on the contents of this chapter, visit http://www.prenhall.com/audesirk7. *Log in to the Web site selected by your instructor and navigate to the Self Test section for this chapter.*

? Review Questions

1. What is a gene pool? How would you determine the allele frequencies in a gene pool?
2. Define *equilibrium population*. Outline the conditions that must be met for a population to stay in genetic equilibrium.
3. How does population size affect the likelihood of changes in allele frequencies by chance alone? Can significant changes in allele frequencies (that is, evolution) occur as a result of genetic drift?
4. If you measured the allele frequencies of a gene and found large differences from the proportions predicted by the Hardy-Weinberg principle, would that prove that natural selection is occurring in the population you are studying? Review the conditions that lead to an equilibrium population, and explain your answer.
5. People like to say that "you can't prove a negative." Study the experiment in Figure 15-1 again, and comment on what it demonstrates.
6. Describe the three ways in which natural selection can affect a population over time. Which way(s) is (are) most likely to occur in stable environments, and which way(s) might occur in rapidly changing environments?
7. What is sexual selection? How is sexual selection similar to and different from other forms of natural selection?
8. Define *kin selection* and *inclusive fitness*. Can these concepts help explain the evolution of altruism?

Applying the Concepts

1. In North America, the average height of adult humans has been increasing steadily for decades. Is directional selection occurring? What data would justify your answer?
2. Malaria is rare in North America. In populations of African Americans, what would you predict is happening to the frequency of the hemoglobin allele that leads to sickling in red blood cells? How would you go about determining whether your prediction is true?
3. By the 1940s, the whooping crane population had been reduced to fewer than 50 individuals. Thanks to conservation measures, its numbers are now increasing. What special evolutionary problems do whooping cranes have now that they have passed through a population bottleneck?
4. In many countries, conservationists are trying to design national park systems so that "islands" of natural area (the big parks) are connected by thin "corridors" of undisturbed habitat. The idea is that this arrangement will allow animals and plants to migrate between refuges. Why would such migration be important?
5. A preview question for Chapter 16: A species is all the populations of organisms that potentially interbreed with one another but that are reproductively isolated from (cannot interbreed with) other populations. Using the five conditions of the Hardy-Weinberg principle as a starting point, what factors do you think would be important in the splitting of a single ancestral species into two modern species?

For More Information

Allison, A. C. "Sickle Cells and Evolution." *Scientific American*, August 1956. The story of the interaction between sickle-cell anemia and malaria in Africa.

Dawkins, R. *Climbing Mount Improbable*. New York: Norton, 1996. An eloquent book-length tribute to the power of natural selection to design intricate adaptations. The chapter on the evolution of the eye is an instant classic.

Dugatkin, L. A., and Godin, J. J. "How Females Choose Their Mates." *Scientific American*, April 1998. A discussion of the role of female mate choice in sexual selection.

Fellman, B. "To Eat or Not to Eat." *National Wildlife*, February–March 1995. How animals with altruistic behaviors identify their relatives.

Levy, S. B. "The Challenge of Antibiotic Resistance." *Scientific American*, March 1998. An excellent summary of the public health implications of antibiotic resistance. Also discusses some strategies for ameliorating the problem.

O'Brien, S. J., Wildt, D. E., and Bush, M. "The Cheetah in Peril." *Scientific American*, May 1986. According to molecular and immunological studies, a population bottleneck has reduced the genetic variation of the world's cheetahs almost to zero.

Palumbi, S. R. *The Evolution Explosion*. New York: Norton, 2001. An evolutionary biologist explores cases of rapid evolution caused by humans, including antibiotic resistance, pesticide resistance, and the evolution of the virus that causes AIDS.

Rennie, J. "Fifteen Answers to Creationist Nonsense." *Scientific American*, July 2002. A summary of some common misconceptions espoused by creationists, and the scientific response to them.

Media Activities

To access a Media Activity visit http://www.prenhall.com/audesirk7. *Log in to the Web site selected by your instructor, navigate to this chapter, and select the appropriate Media Activity number.*

15.1 Agents of Change

Estimated time: 10 minutes

Natural selection is one agent of microevolution. Explore four other agents that can change allele frequencies in a population.

15.2 The Bottleneck Effect

Estimated time: 5 minutes

Genetic drift is one of the major mechanisms of evolutionary change and has significant implications for small natural populations. This activity illustrates the bottleneck effect, which involves genetic drift, in a population of amoebas.

15.3 Three Modes of Natural Selection

Estimated time: 5 minutes

Natural selection can produce three different effects on the genetic variation of a population. These three modes known as directional, stabilizing, and disruptive selection are demonstrated in this activity.

15.4 Natural Selection at Work: Alpine Skypilots

Estimated time: 10 minutes

The size of alpine skypilot flowers differs between the bare alpine and forested sub-alpine environments. What factors contribute to this difference? It is natural selection or some other factor at work? This activity will allow you to walk through Candace Galen's experiments as she uncovered this mystery.

15.5 Web Investigation: Evolution of a Menace

Estimated time: 15 minutes

Antibiotic resistance is reaching crisis proportions. Diseases that were all but eradicated only a few years ago are back and resistant to all known treatments. This exercise will take a brief look at the basic biology, medical repercussions, and social implications of bacterial antibiotic resistance.

CHAPTER

16 The Origin of Species

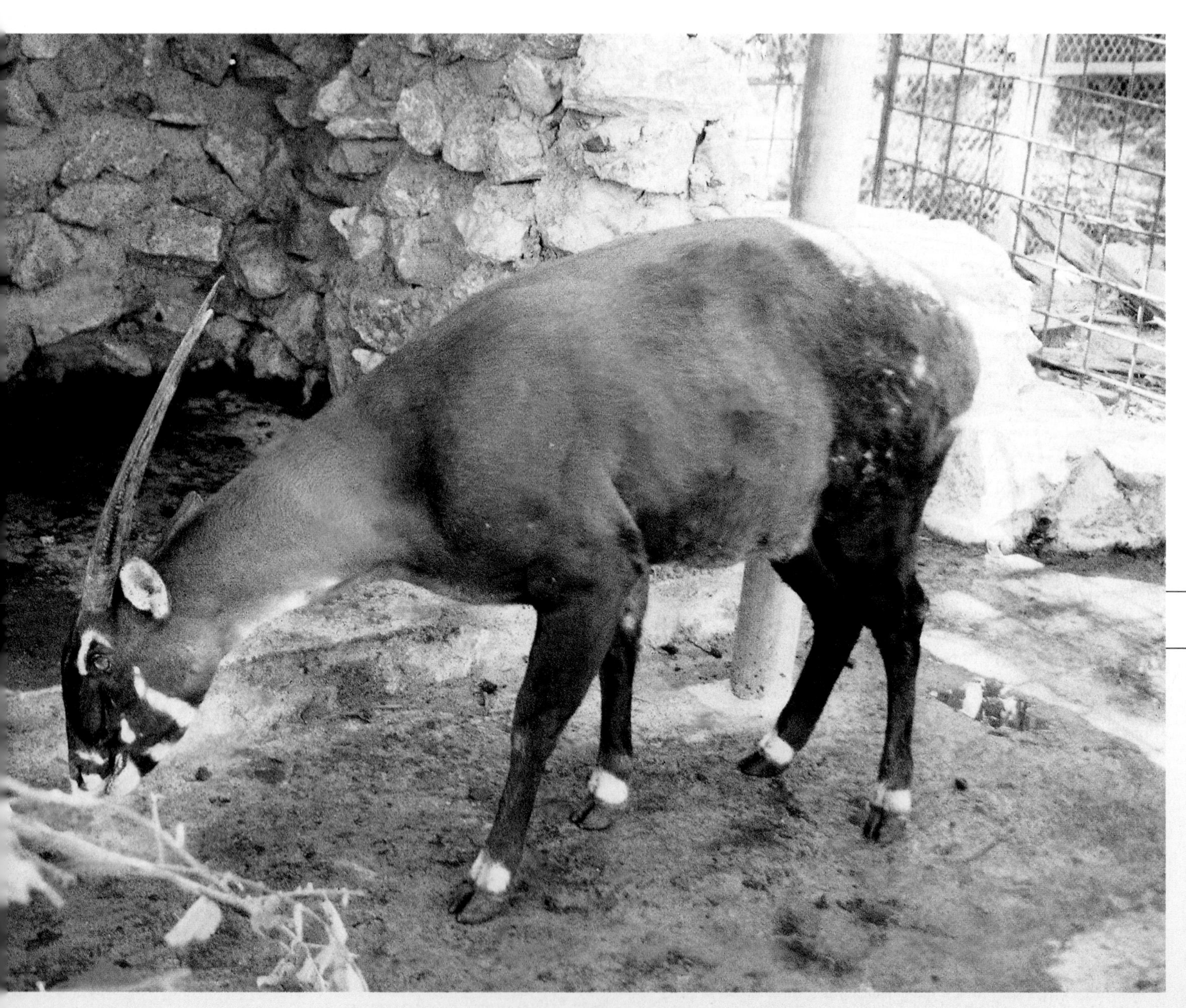

The saola, unknown to science until 1992, is one of a number of previously undiscovered species recently found in the mountains of Vietnam. The area's distinctive assemblage of species probably arose during a past period of geographic isolation.

AT A GLANCE

CASE STUDY Lost World

The steep, rain-drenched slopes of Vietnam's Annamite Mountains are remote and forbidding, cloaked in tropical mists that lend an air of mystery and concealment to the forested mountains. As it turns out, this remote refuge conceals a most astonishing biological surprise: the saola, a hoofed, horned mammal that was unknown to science until the early 1990s. The discovery of a new species of large mammal at this late date was a complete shock. After centuries of human exploration and exploitation in every corner of the world's forests, deserts, and savannas, scientists were certain that no large-sized mammal species could have escaped detection. As long ago as 1812, French naturalist Georges Cuvier wrote that "there is little hope of discovering new species of large quadrupeds." And yet, the saola, 3 feet high at the shoulder, weighing up to 200 pounds and sporting 20-inch black horns, remained hidden in Annamite Mountain forests, outside the realm of scientific knowledge until 1992 (though local tribespeople had apparently been hunting the creature for some time).

Since the discovery of the saola, scientists have described several additional new (if smaller-sized) mammal species, including the giant muntjac (also known as the barking deer) and a strange rabbit that has short ears and a brown-striped coat. This wave of discoveries has revealed the Vietnamese mountains to be a kind of lost world of animals. Isolated by inhospitable terrain and the wars fought in Vietnam during the last century, the animals of the Annamite Mountains remained unknown to scientists. In the face of increased scientific attention, however, this lost world has become increasingly well-known, and the curious biologist may wonder why these wonderfully unfamiliar species are concentrated in this particular part of the world. But before we can fully consider that question, we will need to explore the evolutionary process by which new species arise.

Media Activity
16.1 What Is Speciation?

16.1 What Is a Species?

Although Darwin brilliantly explained how evolution shapes complex, well-designed organisms, his ideas did not fully explain life's diversity. In particular, the mechanism of natural selection cannot by itself explain how living things came to be divided into groups, with each group distinctly different from all other groups. When we look at big cats, we don't see a continuous array of different tiger phenotypes that gradually changes into a lion phenotype. Lions and tigers are separate, distinct entities with no overlap. Each distinct group is known as a species.

Biologists Need a Clear Definition of Species

Before we can study the origin of species, we must first clarify our definition of the term. Throughout most of human history, "species" was a poorly defined concept. In pre-Darwinian Europe, the word "species" simply referred to one of the "kinds" produced by the biblical Creation. In this view, humans could not possibly know the criteria of the creator, but could only attempt to distinguish among species on the basis of visible differences in structure. In fact, "species" is Latin for "appearance."

On a coarse scale, it is easy to use quick visual comparisons to distinguish species. For example, warblers are clearly different from eagles, which are obviously different from ducks. But it is far more difficult to distinguish among different species of warblers, eagles, or ducks. How do scientists make these finer distinctions?

Species Are Groups of Interbreeding Populations

Today, biologists define a **species** as a group of populations that evolves independently; each species follows a separate evolutionary path because alleles do not move between the gene pools of different species. This definition, however, does not clearly state the standard by which such evolutionary independence is judged. The most widely used standard defines species as "groups of actually or potentially interbreeding natural populations, which are reproductively isolated from other such groups." This definition, known as the *biological species concept*, is based on the observation that reproductive isolation (no successful breeding outside the group) ensures evolutionary independence.

The biological species concept has at least two major limitations. First, because the definition is based on patterns of sexual reproduction, it does not help us discern species boundaries among asexually reproducing organisms. Second, it is not always practical or even possible to directly observe whether members of two different groups interbreed. Thus, a biologist who wishes to determine if a group of organisms is a separate species must often make the determination without knowing for sure if group members breed with organisms outside the group.

Despite the limitations of the biological species concept, most biologists accept it as the best criterion for identifying species of sexually reproducing organisms. Nonetheless, scientists who study bacteria and other organisms that mainly reproduce asexually must use alternate species definitions. Even some biologists who study sexually reproducing organisms prefer alternate species definitions that do not depend on a property (reproductive isolation) that can be difficult to measure. Several such alternatives to the biological species concept have been proposed; one that has gained many adherents is described in Chapter 18 (page 352).

Appearance Can Be Misleading

Biologists have found that organisms that appear to be very similar sometimes belong to different species. Conversely, differences in appearance do not always mean that two populations belong to different species. For example, bird field guides published in the 1970s list the myrtle warbler and Audubon's warbler (Fig. 16-1) as distinct species; these birds differ in geographic range and in the color of their throat feathers. More recently, scientists decided that these birds are local varieties of the same species. The reason: where their ranges overlap, these warblers interbreed, and the offspring are just as vigorous and fertile as their parents.

16.2 How Do New Species Form?

Despite his exhaustive exploration of the process of natural selection, Charles Darwin never proposed a complete mechanism of **speciation**, the process by which new species form. One scientist who did play a large role in describing the process of speciation was Ernst Mayr of Harvard University, an ornithologist (a scientist who studies birds) and a pivotal figure in the history of evolutionary biology. Mayr developed the species definition given above. He was also among the first to recognize that speciation depends on two factors acting on a pair of populations: isolation and genetic divergence.

- *Isolation of populations.* If individuals move freely between two populations, interbreeding and the resulting gene flow will cause changes in one population to soon become widespread in the other as well. Thus, two populations cannot grow increasingly different unless something happens to block gene flow between them. Speciation depends on isolation.
- *Genetic divergence of populations.* It is not sufficient for two populations simply to be isolated. They will become separate species only if, during the period of isolation, they evolve sufficiently large genetic differences. The differences must be large enough that, if the isolated populations are reunited, they can no longer interbreed and produce vigorous, fertile off-

(a)

(b)

FIGURE 16-1 Members of a species may differ in appearance
(a) The myrtle warbler and **(b)** Audubon's warbler are members of the same species.

spring. Such differences can arise by chance (genetic drift), especially if at least one of the isolated populations is small (see Chapter 15). Large genetic differences can also arise through natural selection, if the isolated populations experience different environmental conditions.

Speciation has seldom been observed in the wild. Nevertheless, evolutionary biologists have synthesized theories, observations, and experiments to devise hypothetical mechanisms for the origin of new species. These mechanisms fall into two broad categories: **allopatric speciation**, in which two populations are geographically separated from one another, and **sympatric speciation**, in which two populations share the same geographical area (Fig. 16-2).

Geographic Separation of a Population Can Lead to Allopatric Speciation

New species can arise by allopatric speciation when an impassible barrier physically separates different parts of a population. Physical separation could occur if, for example, some members of a population of land-dwelling organisms drifted, swam, or flew to a remote oceanic island. Populations of water-dwelling organisms might be split when geological processes such as volcanism or continental drift create new land barriers that subdivide previously continuous seas or lakes. Geological change can also divide terrestrial populations (Fig. 16-3). Portions of populations can become stranded in patches of suitable habitat that become isolated by climate shifts. You can probably imagine many other scenarios that could lead to the geographic subdivision of a population.

If two populations become geographically isolated for any reason, there will be no gene flow between them. If the pressures of natural selection differ in the separate locations, then the populations may accumulate genetic differences. Alternatively, genetic differences may arise if one or more of the separated populations is small enough for genetic drift to occur, which may be especially likely in the aftermath of a founder event (in which a few individuals become isolated from the main body of the species). In either case, genetic differences between the separated populations may eventually become large enough to make interbreeding impossible. At that point, the two populations will have become separate species. Most evolutionary biologists believe that geographic isolation followed by allopatric speciation has been the most common source of new species, especially among animals.

Ecological Isolation of a Population Can Lead to Sympatric Speciation

Only genetic isolation—limited gene flow—is required for speciation, so new species can arise by sympatric speciation when populations become isolated without geographic separation. If, for example a geographical area contains two distinct types of habitats (each with distinct food sources, places to raise young, and so on), different members of a single species may begin to specialize in one habitat or the other. If conditions are right, natural selection in the two different habitats may lead to the evolution of different traits in the two groups. Eventually, these differences may become large enough to prevent members of the two

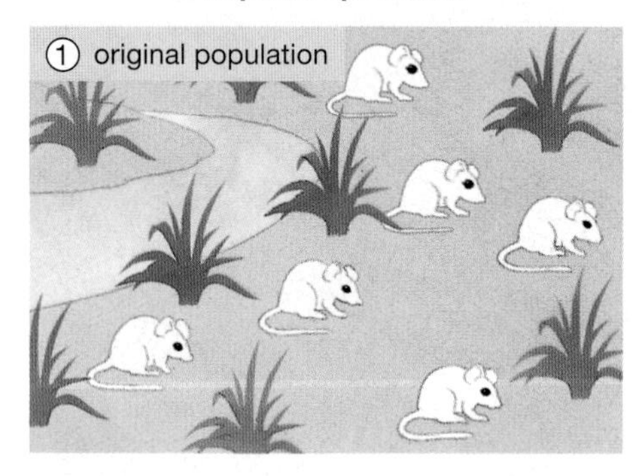

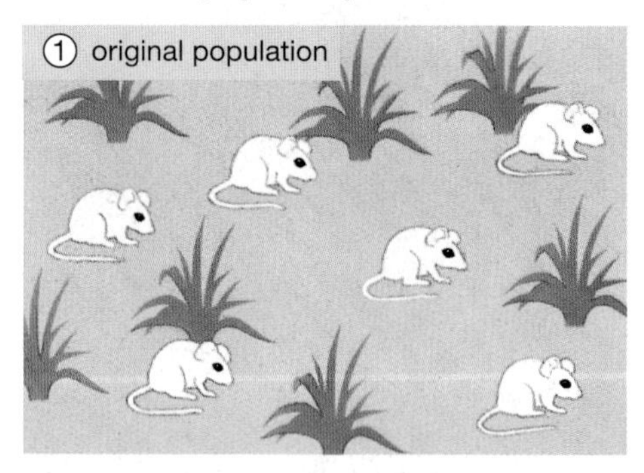

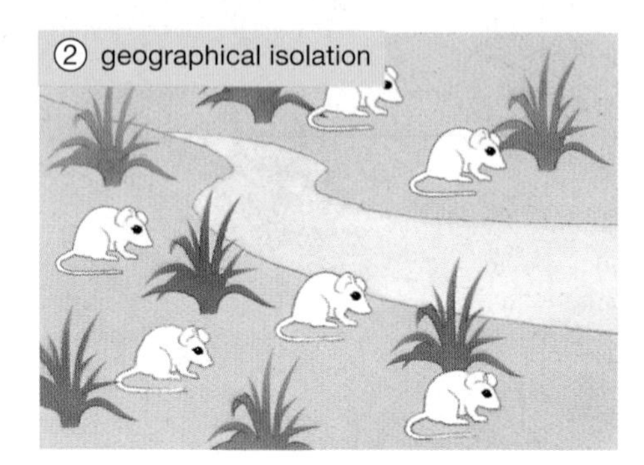

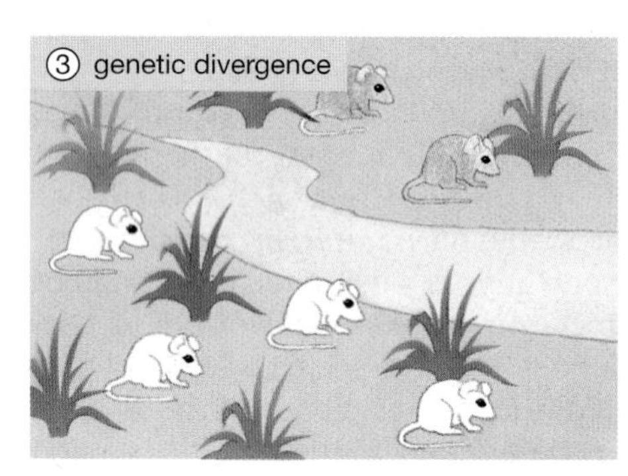

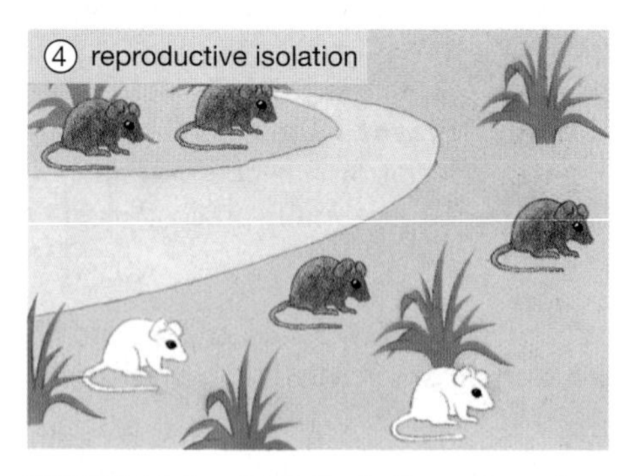

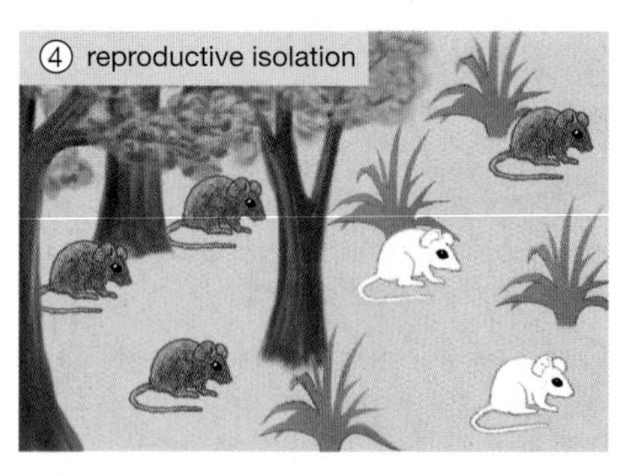

Media Activity 16.3 Sympatric Speciation

Media Activity 16.2 Allopatric Speciation

FIGURE 16-2 Models of allopatric and sympatric speciation
The main difference between allopatric (left column) and sympatric (right column) speciation is the mechanism by which genetic isolation arises. ① Both cases begin with a single species (white mice) occupying a relatively homogeneous habitat. ② In allopatric speciation, the population is separated into two isolated populations by an impassable geographical barrier (here, a river changing course). In sympatric speciation, there are no physical barriers to movement, but different parts of the population become restricted to different habitats. ③ In both cases, genetic drift and/or different environmental pressures cause the two populations to diverge genetically (tan vs. white mice). ④ The isolated populations may later come into contact. If the genetic differences between the two populations have become large enough to prevent inbreeding, then the two populations constitute separate species (brown vs. white mice). **EXERCISE** Make a list of events or processes that could cause geographical or ecological subdivision of a population. Are the items on your list sufficient to account for formation of the millions of species that have inhabited Earth?

(a) (b)

FIGURE 16-3 Geographical isolation
To determine if these two squirrels are members of different species, we must know if they are "actually or potentially interbreeding." Unfortunately, it is hard to tell, because **(a)** the Kaibab squirrel lives only on the north rim of the Grand Canyon and **(b)** the Abert squirrel lives only on the south rim. The two populations are geographically isolated but still quite similar. Have they diverged enough since their separation to be considered separate species? On the basis of our current knowledge, it is impossible to say.

groups from interbreeding, and the formerly single species will have split into two species. Such a split seems to be taking place right before biologists' eyes, so to speak, in the case of the fruit fly *Rhagoletis pomonella* (Fig. 16-4).

Rhagoletis is a parasite of the American hawthorn tree. This fly lays its eggs in the hawthorn's fruit; when the maggots hatch, they eat the fruit. About 150 years ago, entomologists (scientists who study insects) noticed that *Rhagoletis* had begun to infest apple trees, which were introduced into North America from Europe. Today, it appears that *Rhagoletis* is splitting into two species, one that breeds on apples and one that sticks to hawthorns. The two groups have evolved substantial genetic differences, some of which—such as those that affect the timing of emergence of the adult flies—are important for survival on a particular host plant.

The two kinds of flies will become two species only if they maintain reproductive separation. Apple trees and hawthorns typically grow in the same areas, and flies, after all, can fly. So why don't apple-flies and hawthorn-flies interbreed and cancel out any incipient genetic differences? First, female flies usually lay their eggs in the same type of fruit in which they developed. Males also tend to prefer the same type of fruit in which they developed. Therefore, apple-liking males will encounter and mate with apple-liking females. Second, apples mature two or three weeks later than does hawthorn fruit, and the two types of flies emerge with a timing appropriate for their chosen host fruit. Thus, the two varieties of flies have very little chance of meeting. Although some interbreeding between the two types of flies occurs, it seems they are well on their way to speciation. Will they make it? Entomologist Guy Bush suggests, "Check back with me in a few thousand years."

FIGURE 16-4 Sympatric isolation
Two sympatric populations of the fruit fly species *Rhagoletis pomonella* may be evolving into two separate, reproductively isolated species. **QUESTION** How might future scientists test whether the current *R. pomonella* has become two species?

Changes in Chromosome Number Can Lead to Sympatric Speciation

In some instances, new species can arise nearly instantaneously through changes in chromosome number. A common speciation mechanism in plants is **polyploidy**

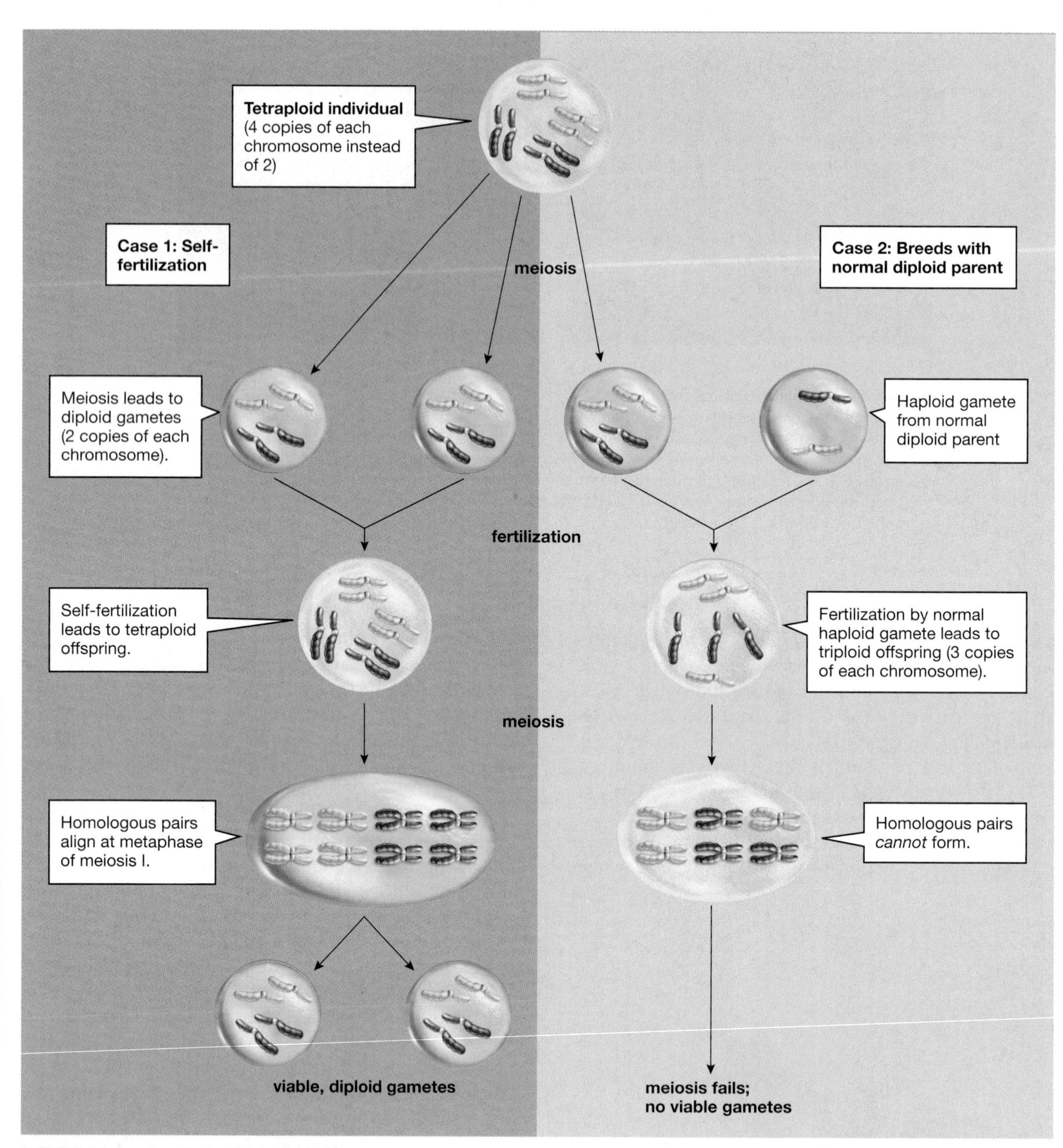

FIGURE 16-5 Speciation by polyploidy
A tetraploid mutant can successfully self-fertilize (or can interbreed with other tetraploid individuals) to yield a new generation of tetraploids, but mating between tetraploids and normal diploid individuals will yield only sterile offspring. Tetraploid mutants are thus reproductively isolated from their diploid ancestors and may constitute a new species.

Media Activity
16.4 Speciation by Polyploidy

(Fig. 16-5), the acquisition of multiple copies of each chromosome. As you may remember from Chapter 11, most plants and animals have paired chromosomes and are described as diploid. Occasionally, especially in plants, a fertilized egg duplicates its chromosomes but doesn't divide into two daughter cells. The resulting cell thus becomes *tetraploid*, with four copies of each chromosome. If all of the subsequent cell divisions are normal, this tetraploid zygote will develop into a plant with tetraploid cells. Most tetraploid plants are vigorous and healthy, and many can successfully complete meiosis to form viable gametes. The gametes, however, are diploid (meiosis normally produces haploid gametes from diploid cells). These diploid gametes can fuse with other diploid gametes to produce new tetraploid offspring, so it is no problem for tetraploids to interbreed with other tetraploids of that species or to self-fertilize (as many plants do).

If, however, a tetraploid interbreeds with a diploid individual from the "parental" species, the outcome is not so successful. For example, if a diploid sperm from a tetraploid plant fertilizes a haploid egg cell of the parental species, the offspring will be *triploid*, with three copies of each chromosome. Many triploid individuals experience problems during growth and development. Even if the triploid offspring develops normally, it will be sterile: when a triploid cell attempts to undergo meiosis, the odd number of chromosomes makes chromosome-pairing impossible. Meiosis fails, and viable gametes are not formed. Because the offspring of diploid–tetraploid matings are inevitably sterile, tetraploid plants and their diploid parents form distinct reproductive communities that cannot interbreed successfully. A new species can form in a single generation.

Why is speciation by polyploidy common in plants but not in animals? Many plants can either self-fertilize or reproduce asexually, or both. If a tetraploid plant self-fertilizes, then its offspring will also be tetraploid. Asexual offspring, of course, are genetically identical to the parent and are also tetraploid. In either case, the new tetraploid plant may perpetuate itself and form a new species. Most animals, however, cannot self-fertilize or reproduce asexually. Therefore, if an animal produced a tetraploid offspring, the offspring would have to mate with a member of the diploid parental species, producing all triploid offspring. The triploid offspring would almost certainly be sterile. Speciation by polyploidy is extremely common in plants; in fact, nearly half of all species of flowering plants are polyploid, and many of them are tetraploid.

Change Over Time Within a Species Can Cause Apparent "Speciation" in the Fossil Record

The mechanisms of speciation and reproductive isolation that we have described lead to forking branches in the *evolutionary tree* of life (Fig. 16-6), as one species splits into two. This kind of branching is a key source of evolutionary change, but changes *within* a species over time are also important. As generations pass and evolutionary innovations accumulate, the members of a species may come to be very different from their distant ancestors, even if no speciation takes place—that is, even if a new species doesn't form.

When a biologist encounters two populations of living organisms, he or she can, if the populations live in the same place, devise a test to see if the two populations are reproductively isolated and, therefore, separate species. For a paleontologist (a scientist who studies fossils), however, things are not so simple. Fossils cannot breed, so it is difficult to determine whether they were reproductively isolated from other fossils.

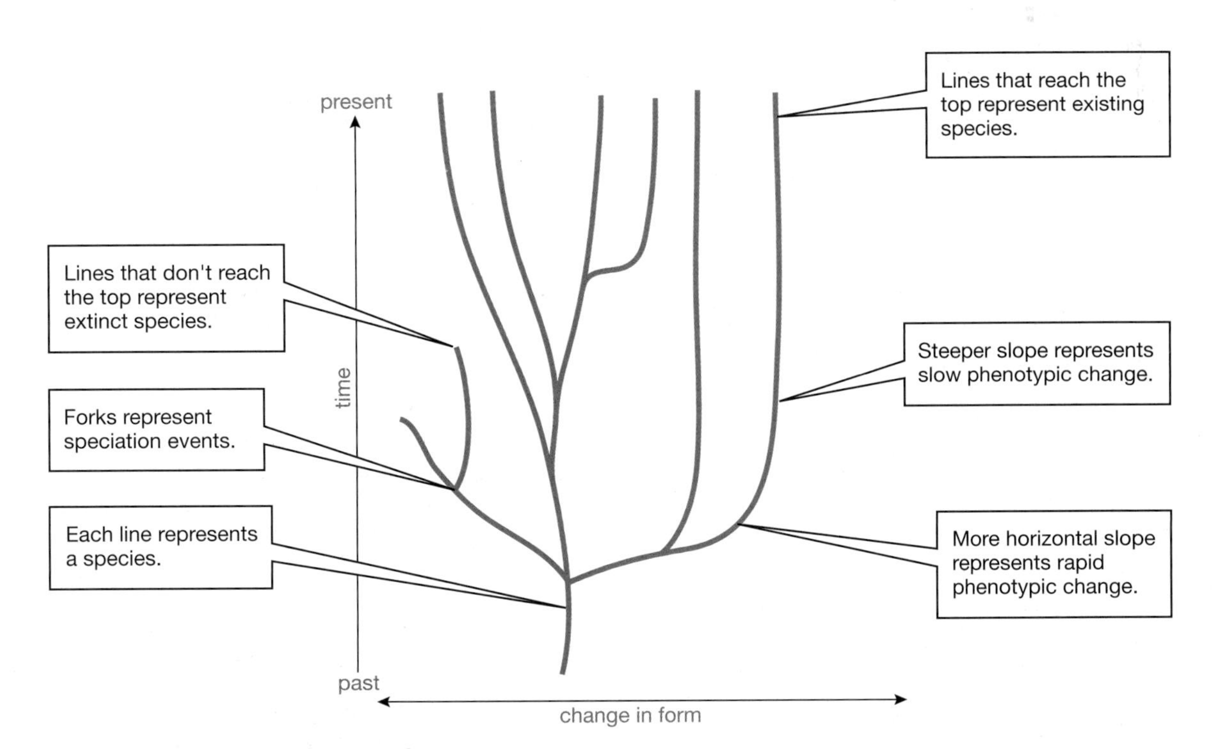

FIGURE 16-6 Interpreting an evolutionary tree
Evolutionary history is often represented by an evolutionary tree, a graph in which the vertical axis plots time and the horizontal axis stands for change in form (the greater the horizontal distance between points, the more phenotypic differences there are).

Furthermore, different fossil organisms may be found in different rock layers that were laid down thousands or even millions of years apart, and it is typically impossible to know if any branching, speciation events occurred between an older organism and a younger one. For these reasons, paleontologists must typically assign extinct organisms to species without reference to the biological-species concept. When comparing differences between extinct organisms, it is often impossible to tell if the differences arose due to branching evolution or to change within a single line of descent. Given the futility of applying the biological-species concept to fossils, many paleontologists choose to use a different system in which it is considered acceptable to assign different species names to anatomically different fossils, even if the two fossils may simply represent different time points on a single evolutionary branch.

Under Some Conditions, Many New Species May Arise

In some cases, many new species arise in a relatively short time. This process, called **adaptive radiation**, can occur when populations of one species invade a variety of new habitats and evolve in response to the differing environmental pressures in those habitats. Adaptive radiation has occurred many times and in many groups of organisms, typically when species encounter a wide variety of unoccupied habitats. For example, episodes of adaptive radiation took place when the ancestors of Darwin's finches colonized the Galapagos Islands, when marsupial mammals first invaded Australia, and when an ancestral cichlid fish species arrived at Lake Malawi (Fig. 16-7). In these examples, the invading species faced no competitors except other members of their own species; all the available habitats and food sources were rapidly exploited by new species that evolved from the original invaders.

FIGURE 16-7 Adaptive radiation
More than 300 species of cichlid fishes inhabit Lake Malawi in East Africa. These species are found nowhere else, and all of them descended from a single ancestral population within a million years. This dramatic adaptive radiation has led to a collection of closely related species with an array of adaptations for exploiting the many different food sources in the lake.
QUESTION Did the Lake Malawi cichlids arise by allopatric or sympatric speciation?

16.3 How Is Reproductive Isolation Between Species Maintained?

The process of speciation depends on the evolution of mechanisms that prevent interbreeding. Genetic divergence during the period of isolation is a necessary condition for the origin of a new species. It is not, however, a sufficient condition unless it includes changes that would block reproduction between the two separated groups if they were to come into contact again. If members of a population are unable to interbreed with members of other populations, the population is said to be in **reproductive isolation**.

The structural and behavioral modifications that prevent interbreeding and maintain reproductive isolation are called **isolating mechanisms**. Isolating mechanisms give a clear benefit to individuals. Any individual that mates with a member of another species will probably produce unfit or sterile offspring, thereby wasting its reproductive effort and contributing nothing to future generations. Thus, there is evolutionary pressure to avoid mating across species boundaries. Mechanisms that prevent mating between species are called **premating isolating mechanisms**.

When premating isolating mechanisms fail or have not yet evolved, members of different species may mate. If, however, all resulting hybrid offspring die during development, the two species are still reproductively isolated from one another. In some cases, however, viable hybrid offspring are produced. Even so, if these hybrids are less fit than their parents or are themselves infertile, the two species may still remain separate, with little or no gene flow between them. Mechanisms that prevent the formation of vigorous, fertile hybrids between species are called **postmating isolating mechanisms**.

Premating Isolating Mechanisms Prevent Mating Between Species

Reproductive isolation can be maintained by a variety of different mechanisms; those that prevent mating attempts are especially effective. Next, we discuss the most important types of such premating isolating mechanisms.

Members of Different Species May Be Prevented from Meeting

Members of different species cannot mate if they never get near one another. As we have already seen, geographical isolation typically provides the conditions for speciation in the first place. However, we cannot determine whether geographically separated populations are actually distinct species. Should the physical barrier separating the two populations disappear (for example, if an intervening river changes course), the reunited populations might interbreed freely and not be separate species after all. If they cannot interbreed, then other mechanisms, such as those considered below, must have developed during their isolation. Geographical isolation, therefore, is usually considered to be a mechanism that allows new species to form rather than a mechanism that maintains reproductive isolation between species.

Different Species May Occupy Different Habitats

Two populations that use different resources may spend time in different habitats within the same general area and thus exhibit ecological isolation. White-crowned and white-throated sparrows, for example, have extensively overlapping ranges. The white-throated sparrow, however, frequents dense thickets, whereas the white-crowned sparrow inhabits fields and meadows, seldom penetrating far into dense growth. The two species may coexist within a few hundred yards of one another and yet seldom meet during the breeding season. A more dramatic example is provided by the more than 750 species of fig wasp (Fig. 16-8). Each species of fig wasp breeds in (and pollinates) the fruits of a particular species of fig, and each fig species hosts one and only one species of pollinating wasp. Although such ecological isolation may slow down interbreeding, it seems unlikely that it could prevent gene flow entirely. Normally, other mechanisms also contribute to reproductive isolation.

FIGURE 16-8 Ecological isolation
This female fig wasp is carrying fertilized eggs from a mating that took place within a fig. She will find another fig of the same species, enter it through a pore, lay eggs, and die. Her offspring will hatch, develop, and mate within the fig. Because each species of fig wasp reproduces only in its own particular fig species, each wasp species is reproductively isolated.

Different Species May Breed at Different Times

Even if two species occupy similar habitats, they cannot mate if they have different breeding seasons, a phenomenon called temporal (time-related) isolation. Bishop pines and Monterey pines grow together near Monterey on the California coast (Fig. 16-9). Viable hybrids have been produced between these two species in the laboratory. In the wild, however, the two species release their sperm-containing pollen and have eggs ready to receive the pollen at different times; the Monterey pine releases pollen in early spring, the bishop pine in summer. Therefore, the two species never interbreed under natural conditions.

Different Species May Have Different Courtship Rituals

Among animals, the elaborate courtship displays that so enthrall human observers not only serve as recognition and evaluation signals between male and female; they also prevent mating with members of other species. Signals and behaviors that differ from species to species create behavioral isolation. The striking colors and calls of male songbirds, for example, may attract females of their own species, but females of other species treat

FIGURE 16-9 Temporal isolation
Bishop pines, such as these, and Monterey pines coexist in nature. In the laboratory they produce fertile hybrids. In the wild, however, they do not interbreed, because they release pollen at different times of the year.

them with the utmost indifference. Among frogs, males are often impressively indiscriminate, jumping on every female in sight, regardless of species, when the spirit moves them. Females, however, approach only male frogs that croak the "ribbet" appropriate to their species. If they do find themselves in an unwanted embrace, they utter the "release call," which causes the male to let go. As a result, few hybrids are produced.

Species' Differing Sexual Organs May Foil Mating Attempts

In many cases, mating between different species is physically impossible. For example, in animal species with internal fertilization (in which the sperm is deposited inside the female's reproductive tract), the male and female sexual organs may simply not fit together. Among plants, differences in flower size or structure may prevent pollen transfer between species because the differing flowers may attract different pollinators. Isolating mechanisms of this type are called mechanical incompatibilities.

Postmating Isolating Mechanisms Limit Hybrid Offspring

Premating isolation sometimes fails; members of different species mate, and the sperm of one species may reach the egg of another species. Such matings, however, often fail to produce vigorous, fertile hybrid offspring, due to postmating isolating mechanisms.

One Species' Sperm May Fail To Fertilize Another Species' Eggs

Even if a male inseminates a female of a different species, his sperm may not be able to fertilize her eggs, an isolating mechanism called gametic incompatibility. For example, the fluids of the female reproductive tract may weaken or kill sperm of other species. Among plants, chemical incompatibility may prevent the germination of pollen from one species that lands on the stigma (pollen-catching structure) of the flower of another species.

Hybrid Offspring May Survive Poorly

If cross-species fertilization does occur, the resulting hybrid may be unable to survive, a situation called hybrid inviability. The genetic instructions directing development of the two species may be so different that hybrids abort early in development. If the hybrid does survive, it may display behaviors that are mixtures of the two parental types. In attempting to do some things the way species *A* does them and other things the way species *B* does them, the hybrid may be hopelessly uncoordinated and therefore unable to reproduce. Hybrids between certain species of lovebirds, for example, have great difficulty learning to carry nest materials during flight and probably could not reproduce in the wild.

Table 16-1 Mechanisms of Reproductive Isolation

Premating Isolating Mechanisms: factors that prevent organisms of two populations from mating

- **Geographical isolation:** The populations cannot interbreed because a physical barrier separates them.
- **Ecological isolation:** The populations do not interbreed even if they are within the same area because they occupy different habitats.
- **Temporal isolation:** The populations cannot interbreed because they breed at different times.
- **Behavioral isolation:** The populations do not interbreed because they have different courtship and mating rituals.
- **Mechanical incompatibility:** The populations cannot interbreed because their reproductive structures are incompatible.

Postmating Isolating Mechanisms: factors that prevent organisms of two populations from producing vigorous, fertile offspring after mating

- **Gametic incompatibility:** Sperm from one population cannot fertilize eggs of another population.
- **Hybrid inviability:** Hybrid offspring fail to survive to maturity.
- **Hybrid infertility:** Hybrid offspring are sterile or have reduced fertility.

Hybrid Offspring May Be Infertile

Most animal hybrids, such as the mule (a cross between a horse and a donkey) or the liger (a zoo-based cross between a male lion and a female tiger), are sterile. Hybrid infertility prevents hybrids from passing on their genetic material to offspring. A common reason for such infertility is the failure of chromosomes to pair properly during meiosis, so eggs and sperm never develop.

Table 16-1 reviews the different types of isolating mechanisms.

16.4 What Causes Extinction?

Every living organism must eventually die, and the same is true of species. Just like individuals, species are "born" (through the process of speciation), persist for some period of time, and then perish. The ultimate fate of any species is **extinction**, the death of the last of its members. In fact, at least 99.9% of all the species that have ever existed are now extinct. The natural course of evolution, as revealed by fossils, is continual turnover of species as new ones arise and old ones go extinct.

The immediate cause of extinction is probably always environmental change, either in the living or the nonliving parts of the environment. Two major environmental factors that may drive a species to extinction are competition among species and habitat destruction.

FIGURE 16-10 Very localized distribution can endanger a species
The Devil's Hole pupfish is found in only one spring-fed water hole in the Nevada desert. This and other isolated small populations are at high risk of extinction.

FIGURE 16-11 Extreme specialization places species at risk
The Everglades kite feeds exclusively on the apple snail, found in swamps of the southeastern United States. Such behavioral specialization renders the kite extremely vulnerable to any environmental change that may exterminate its single species of prey.
QUESTION If specialization puts a species at risk for extinction, how could this hazardous trait have evolved?

Localized Distribution and Overspecialization Make Species Vulnerable in Changing Environments

Species vary widely in their range of distribution and, hence, in their vulnerability to extinction. Some species, such as herring gulls, white-tailed deer, and humans, inhabit entire continents or even the whole Earth; others, such as the Devil's Hole pupfish (Fig. 16-10), have extremely limited ranges. Obviously, if a species inhabits only a very small area, any disturbance of that area could easily result in extinction. If Devil's Hole dries up due to a drought or well-drilling nearby, its pupfish will immediately vanish. Conversely, wide-ranging species normally do not succumb to local environmental catastrophes.

Another factor that may make a species vulnerable to extinction is overspecialization. Each species evolves adaptations that help it survive and reproduce in its environment. In some cases, these adaptations include specializations that favor survival in a particular and limited set of environmental conditions. The Everglades kite, for example, is a bird of prey that feeds only on the apple snail, a freshwater snail (Fig. 16-11). As the swamps of the American Southeast are drained for farms and developments, the snail population shrinks. If the snail disappears, the kite will surely go extinct along with it.

Interactions with Other Organisms May Drive a Species to Extinction

As described earlier, interactions such as competition and predation serve as agents of natural selection. In some cases, these same interactions can lead to extinction rather than to adaptation.

Organisms compete for limited resources in all environments. If a species' competitors evolve superior adaptations and the species doesn't evolve fast enough to keep up, it may become extinct. A particularly striking example of extinction through competition occurred in South America, beginning about 2.5 million years ago. At that time, the isthmus of Panama rose above sea level and formed a land bridge between North America and South America. After the previously separated continents were connected, the mammal species that had evolved in isolation on each continent were able to mix. Many species did indeed expand their ranges, as North American mammals moved southward and South American mammals moved northward. As they moved, each species encountered resident species that occupied the same kinds of habitats and exploited the same kinds of resources. The ultimate result of the ensuing competition was that the North American species diversified and underwent an adaptive radiation that displaced the vast majority of the South American species, many of which went extinct. Clearly, evolution had bestowed on the North American species some (as yet unknown) set of adaptations that enabled their descendants to exploit resources more efficiently and effectively than their South American counterparts.

Habitat Change and Destruction Are the Leading Causes of Extinction

Habitat change, both contemporary and prehistoric, is the single greatest cause of extinctions. Present-day habitat destruction due to human activities is proceeding at a frightening pace. Many biologists believe that we are presently in the midst of the fastest-paced and most widespread episode of species extinction in the history of life. Loss of tropical forests is especially devastating to species diversity. As many as half the species presently on Earth may be lost over the next 50 years as

EARTH WATCH Hybridization and Extinction

The main cause of extinction is environmental change, especially habitat destruction. Some species with small populations, however, are also threatened by a less obvious danger: hybridization. Although premating isolating mechanisms ensure that, for the most part, members of one species cannot interbreed with members of a different species, matings between members of different species are nonetheless possible. Between-species matings and the resulting hybrid offspring are especially common in birds and plants.

How can hybrid mating be dangerous to endangered species? Recall that postmating isolating mechanisms ensure that, in most cases, hybrid offspring will survive poorly and may even be sterile. Now picture what happens when contact between two species produces hybrids, and one of the species has a much smaller population than the other. If the hybrid offspring fail to survive and reproduce, the numbers of each species will decline, but the decline will have a proportionally larger impact on the small population. When the more abundant species moves into the range of the rare species, the impact on the rare species can be severe. Even if the hybrid offspring survive well, high numbers of hybrids can overwhelm the rare species, essentially absorbing the rare species into the abundant species.

Damage from hybridization is most likely to occur when formerly isolated small populations come into contact with larger populations of a closely related species. For example, the plant *Clarkia lingulata* is extremely rare, known to exist only in two sites in the Sierra Nevada mountains of California. Unfortunately, it readily hybridizes with its abundant relative *Clarkia biloba* to produce sterile hybrid offspring. Because several populations of *biloba* grow near the *lingulata* populations, extinction by hybridization is a real possibility for this rare species.

Human activities often cause contact between an endangered species and a more abundant species with which it can hybridize. For example, the rare Hawaiian duck, found only on the Hawaiian Islands, hybridizes freely with mallard ducks, a nonnative species introduced to Hawaii by hunters in search of new game species. Similarly, the endangered Ethiopian wolf (Fig. E16-1) is threatened by interbreeding with feral domestic dogs, and the endangered European wildcat is at risk from hybridization with domestic cats. In these cases and others, a species first declined in number due to habitat destruction, then became vulnerable to further damage by hybridization with a more numerous species that was present as a result of human activities.

FIGURE E16-1 Ethiopian wolves
Fewer than 500 Ethiopian wolves remain. Among the threats to their continued existence is hybridization with wild dogs.

the tropical forests that contain them are cut for timber and to clear land for cattle and crops. We will discuss extinctions due to prehistoric habitat change in Chapter 17.

EVOLUTIONARY CONNECTIONS

Scientists Don't Doubt Evolution

In the popular press, conflicts among evolutionary biologists are sometimes seen as conflicts about evolution itself. We occasionally read statements implying that new theories are overthrowing Darwin's and casting doubt on the reality of evolution. Nothing could be farther from the truth. Despite some disagreements about the details of the evolutionary process, biologists unanimously agree that evolution occurred in the past and is still occurring today. The only argument is about the relative importance of the various mechanisms of evolutionary change in the history of life on Earth, their pace, and which forces were most important in shaping the evolution of particular species. Meanwhile, wolves still tend to catch the slowest caribou, small populations still undergo genetic drift, and habitats still change or disappear. Evolution continues, still generating, in Darwin's words, "endless forms most beautiful."

CASE STUDY REVISITED Lost World

One possible explanation for the distinctive collection of species found in the Annamite Mountains of Vietnam lies in the geological history of the region. During the ice ages that have occurred repeatedly over the past million years or so, the area covered by tropical forests must have shrunk dramatically. Organisms that depended on the forests for survival would have been restricted to any remaining "islands" of forest, isolated from their fellows in other, distant patches of forest. What is now the Annamite Mountain region may well have been an isolated forest during periods of glacial advance. As we learned in this chapter, this kind of isolation can set the stage for allopatric speciation and may have created the conditions that gave rise to the saola, giant muntjac, striped rabbit, and other unique denizens of Vietnamese forests.

Ironically, we have discovered the lost world of Vietnamese animals at a moment when that world is in grave danger of disappearing. Economic development in Vietnam has brought logging and mining to ever more remote regions of the country, and Annamite Mountain forests are being cleared at an unprecedented rate. The increasing local human population means that local animals are hunted heavily; most of our knowledge of the saola comes from carcasses found in local markets. All of the newly discovered mammals of Vietnam are quite rare, seen only infrequently even by local hunters. Fortunately, the Vietnamese government has established a number of national parks and nature preserves in key areas. Only time will tell if these measures are sufficient to ensure the survival of the mysterious mammals of the Annamites.

Consider This: The All Species Foundation is a nonprofit organization that promotes the goal of finding and naming all of Earth's undiscovered species within the next 25 years. According to the foundation, this task "deserves to be one of the great scientific goals of the new century." The foundation estimates the cost of the job at between $700 and $2000 per species, with perhaps millions of undiscovered species remaining to found. Do you think the search for undiscovered species should continue? What value or benefit to humans does the search for new species provide?

Links to Life: Biological Vanity Plates

Looking for a special gift for a friend or loved one? Why not name a species after him or her? Or, for that matter, name one after yourself! Thanks to the BIOPAT project (www.biopat.de), anyone with $3000 can be immortalized in the Latin name of a newly discovered plant or animal.

Typically, the scientist who discovers and describes a new species is entitled to choose its Latin name. Scientists usually choose a name that describes a trait of the species or perhaps the location where it was found. Sometimes, however, more whimsical choices are made. For example, a recently discovered snail was named *Bufonaria borisbeckeri*, in honor of the German tennis player Boris Becker, and a frog was named *Hyla stingi* after the British rock star.

If you donate money to the BIOPAT project, the name of a new species will be entirely up to you. In return for a contribution that supports efforts to discover and conserve endangered species, the people at BIOPAT will offer you a selection of newly discovered but unnamed species. You can choose your species and pick a name, which is then given an appropriate Latin ending and published in a scientific journal. Your chosen name becomes the official, recognized scientific name of the new species.

Summary of Key Concepts

16.1 What Is a Species?

According to the biological-species concept, a species is defined as all the populations of organisms that are potentially capable of interbreeding under natural conditions and that are reproductively isolated from other populations.

16.2 How Do New Species Form?

Speciation, the formation of new species, takes place when gene flow between two populations is reduced or eliminated and the populations diverge genetically. Most commonly, speciation follows geographical isolation and subsequent genetic divergence of the separated populations through genetic drift or natural selection.

16.3 How Is Reproductive Isolation Between Species Maintained?

Reproductive isolation between species may be maintained by one or more of several mechanisms, known as premating isolating mechanisms and postmating isolating mechanisms. Premating isolating mechanisms include geographical isolation, ecological isolation, temporal isolation, behavioral isolation, and mechanical incompatibility. Postmating isolating mechanisms include gametic incompatibility, hybrid inviability, and hybrid infertility.

16.4 What Causes Extinction?

Factors that can lead to extinction, or the death of all the members of a species, include overspecialization, competition among species, and habitat destruction.

Key Terms

adaptive radiation *p. 308*
allopatric speciation *p. 303*
extinction *p. 310*
isolating mechanism *p. 308*
polyploidy *p. 305*
postmating isolating mechanism *p. 308*
premating isolating mechanism *p. 314*
reproductive isolation *p. 308*
speciation *p. 302*
species *p. 302*
sympatric speciation *p. 303*

Thinking Through the Concepts

To take a multiple-choice quiz with feedback on the contents of this chapter, visit http://www.prenhall.com/audesirk7. *Log in to the Web site selected by your instructor and navigate to the Self Test section for this chapter.*

? Review Questions

1. Define the following terms: *species*, *speciation*, *allopatric speciation*, and *sympatric speciation*. Explain how allopatric and sympatric speciation might work, and give a hypothetical example of each.
2. Many of the oak tree species in central and eastern North America hybridize (interbreed). Are they "true species"?
3. Review the material on the possibility of sympatric speciation in *Rhagoletis* varieties that breed on apples or hawthorns. What types of genotypic, phenotypic, or behavioral data would convince you that the two forms have become separate species?
4. A drug called colchicine affects the mitotic spindle fibers and prevents cell division after the chromosomes have doubled at the start of meiosis. Describe how you would use colchicine to produce a new polyploid species of your favorite garden flower.
5. What are the two major types of reproductive isolating mechanisms? Give examples of each type and describe how they work.

Applying the Concepts

1. The biological-species concept has no meaning with regard to asexual organisms, and it is difficult to apply to extinct organisms that we know only as fossils. Can you devise a meaningful, useful species definition that would apply in all situations?
2. Seedless varieties of fruits and vegetables, created by breeders, are triploid. Explain why they are seedless.
3. Why do you suppose there are so many *endemic* species—that is, species found nowhere else—on islands? Why has the overwhelming majority of recent extinctions occurred on islands?
4. A biologist you've met claims that the fact that humans are pushing other species into small, isolated populations is good for biodiversity because these are the conditions that lead to new speciation events. Comment.
5. Southern Wisconsin is home to several populations of gray squirrels (*Sciurus carolinensis*) with black fur. Design a study to determine if they are actually separate species.
6. It is difficult to gather data on speciation events in the past or to perform interesting experiments about the process of speciation. Does this difficulty make the study of speciation "unscientific"? Should we abandon the study of speciation?

For More Information

Eldredge, N. *Fossils: The Evolution and Extinction of Species*. New York: Harry N. Abrams, 1991. A nicely illustrated exploration of a paleontologist's approach to examining and interpreting the past, including speciation events.

Levin, D. A. "Hybridization and Extinction." *American Scientist*, May-June 2002. A discussion of the impact of interbreeding on the conservation of rare species.

Quammen, D. *The Song of the Dodo*. New York: Scribner, 1996. Beautifully written exposition of the biology of islands. Read this book to understand why islands are known as "natural laboratories of speciation."

Schilthuizen, M. *Frogs, Flies, and Dandelions: Speciation—The Evolution of New Species.* Oxford: Oxford University Press, 2001. A readable and entertaining summary of the latest biological thought on species and speciation.

Wilson, E. O. *The Diversity of Life*. New York: W. W. Norton, 1992. Elegant description of how species arise, how they disappear, and why we should preserve them.

Media Activities

To access a Media Activity visit http://www.prenhall.com/audesirk7. *Log in to the Web site selected by your instructor, navigate to this chapter, and select the appropriate Media Activity number.*

16.1 What Is Speciation?

Estimated time: 2 minutes

The animation provides a brief overview of the concept of speciation through reproductive isolation.

16.2 Allopatric Speciation

Estimated time: 5 minutes

This activity demonstrates how a single population may be separated geographically into two populations, and one or both may evolve into new species.

16.3 Sympatric Speciation

Estimated time: 5 minutes

This activity demonstrates how a single population may split into two different populations, without being separated geographically, and one or both may evolve into new species.

16.4 Speciation by Polyploidy

Estimated time: 5 minutes

In this activity, explore how new plant species can form through changes in ploidy, which is the number of chromosome sets in an organism.

16.5 Web Investigation: Lost World

Estimated time: 20 minutes

Is the great age of exploration over? Think again. Thousands of new organisms, ecosystems, and other natural wonders remain to be discovered. This exercise looks at a few recently discovered animal species and how they were found.

CHAPTER

17 The History of Life

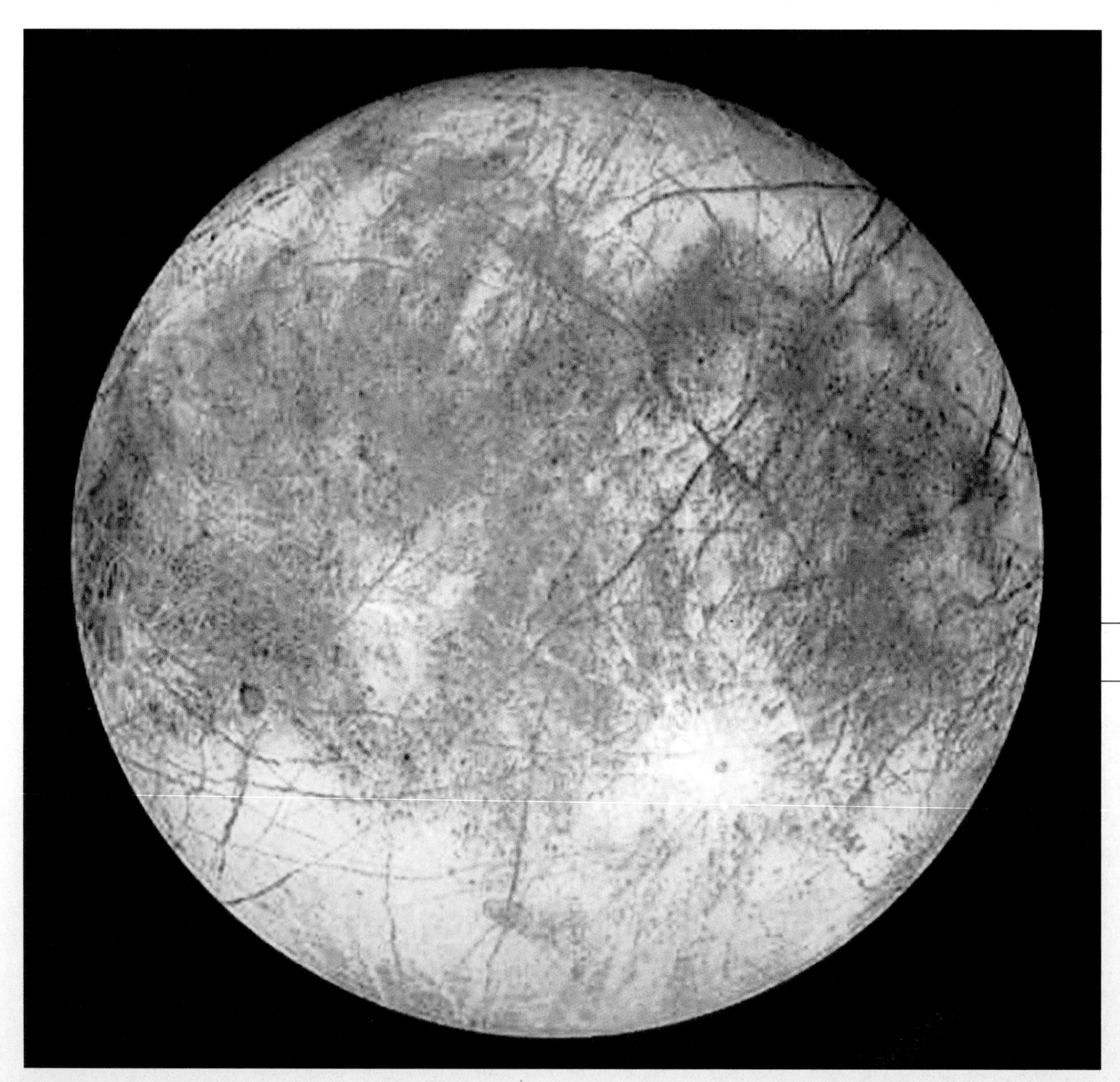

The icy surface of Europa, one of Jupiter's moons, may conceal a liquid ocean.
Could this extraterrestrial ocean harbor life?

AT A GLANCE

CASE STUDY Life on a Frozen Moon?

On the cold, dark surface of Europa, all is silent. On the horizon, the huge orange and purple mass of Jupiter looms. High overhead, a spidery metal spacecraft speeds across the sky, its instruments whirring. Within moments it is gone, hurtling back into space.

Back on Earth, NASA scientists gather around a computer screen. They gaze intently at the screen, examining images of Europa that have just been beamed home by *Galileo,* an unmanned spacecraft. Why would scientists be so interested in pictures from one of Jupiter's many moons? Because Europa is one of the celestial objects deemed most likely to hold extraterrestrial life.

Europa wasn't always considered a possible home to extraterrestrial life. Data gathered in the 1970s showed its surface to be entirely covered with frozen water; this icy world did not appear to be particularly hospitable to life. More recently, however, the *Galileo* spacecraft has flown quite close to Europa's surface and has produced a wealth of detailed pictures. These images, together with other data gathered by *Galileo*, have fostered a new outlook on the frozen moon.

The most recent images of Europa show that its frozen surface is littered with huge cracks, humps, and crevices. These deformities indicate that the ice sheet is moving and shifting; careful analyses by scientists suggest that the observed pattern of cracks would be expected if the ice is resting atop liquid water. This seemingly cold and lifeless world may actually conceal a huge, liquid ocean, warmed by heat from the moon's rocky core and hidden deep beneath a vast sheet of ice.

What is the significance of this finding? Water is essential to life; all forms of life on Earth require water to survive. According to our current understanding of the origin of life on Earth, water is a key component of the conditions under which life can arise. The discovery of Europa's shifting ice sheets is our first indication that liquid water may exist elsewhere in the solar system. If life could arise in the watery environment of early Earth, why not in the watery environment of Europa?

17.1 How Did Life Begin?

How and when did life first appear on Earth? Just a few centuries ago, this question would have been considered trivial. Although no one knew how life *first* arose, people thought that new living things appeared all the time, through **spontaneous generation** from nonliving matter and from other, unrelated forms of life. In 1609 a French botanist wrote, "There is a tree ... frequently observed in Scotland. From this tree leaves are falling; upon one side they strike the water and slowly turn into fishes, upon the other they strike the land and turn into birds." Medieval writings abound with similar observations and delightful recipes for creating life—even human beings. Microorganisms were thought to arise spontaneously from broth, maggots from meat, and mice from mixtures of sweaty shirts and wheat.

Experiments Refuted Spontaneous Generation

In 1668, Italian physician Francesco Redi disproved the maggots-from-meat hypothesis simply by keeping flies—whose eggs hatch into maggots—away from uncontaminated meat (see Chapter 1). In the mid-1800s, Louis Pasteur in France and John Tyndall in England disproved the broth-to-microorganism idea (Fig. 17-1). Although their work effectively demolished the notion of spontaneous generation, it did not address the question of how life on Earth originated in the first place. Or, as biochemist Stanley Miller put it, "Pasteur never proved it didn't happen once, he only showed that it doesn't happen all the time."

Chemical Evolution Preceded and Gave Rise to Life

For almost half a century, the subject lay dormant. Eventually, however, biologists returned to the question of the origin of life. In the 1920s and 1930s, Alexander Oparin in Russia and John B. S. Haldane in England noted that the oxygen-rich atmosphere now present on Earth would not have permitted the spontaneous formation of the complex organic molecules necessary for life. Oxygen reacts readily with other molecules, disrupting chemical bonds; thus, an oxygen-rich environment favors the presence of simple molecules. Oparin and Haldane speculated that the atmosphere of the young Earth must have contained very little oxygen gas and that, under such atmospheric conditions, complex organic molecules could have arisen through ordinary chemical reactions. Some kinds of molecules could persist in the lifeless environment of early Earth better than others, and would therefore have become more common over time. This chemical version of the "survival of the fittest" is called *prebiotic* (meaning "before life") evolution. In the scenario envisioned by Oparin and Haldane, prebiotic chemical evolution gave rise to progressively more complex molecules and eventually to living organisms.

Organic Molecules Can Form Spontaneously Under Prebiotic Conditions

Inspired by the ideas of Oparin and Haldane, Stanley Miller and Harold Urey of the University of Chicago set out in 1953 to simulate prebiotic evolution in the laboratory. They knew that, based on the chemical composition of the rocks that formed early in Earth's history, geochemists had concluded that the early atmosphere probably contained virtually no oxygen gas, but did contain other substances, including methane, ammonia, hydrogen, and water vapor. Miller and Urey simulated the oxygen-free atmosphere of early Earth by mixing these components in a flask. An electrical discharge mimicked the intense energy of early Earth's lightning storms. In this experimental microcosm, the researchers found that simple organic molecules appeared after just a few days (Fig. 17-2).

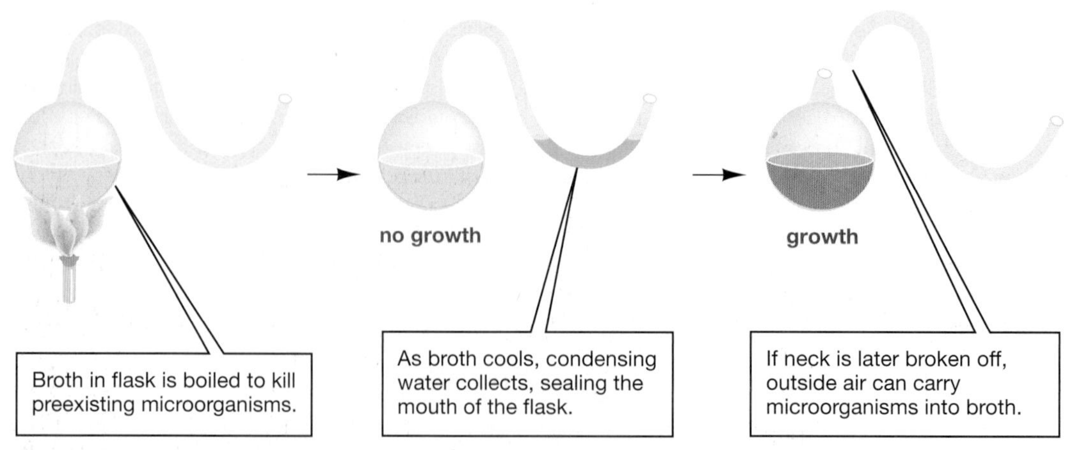

no growth
growth
Broth in flask is boiled to kill preexisting microorganisms.
As broth cools, condensing water collects, sealing the mouth of the flask.
If neck is later broken off, outside air can carry microorganisms into broth.

FIGURE 17-1 Spontaneous generation refuted
Louis Pasteur's experiment disproving the spontaneous generation of microorganisms in broth.

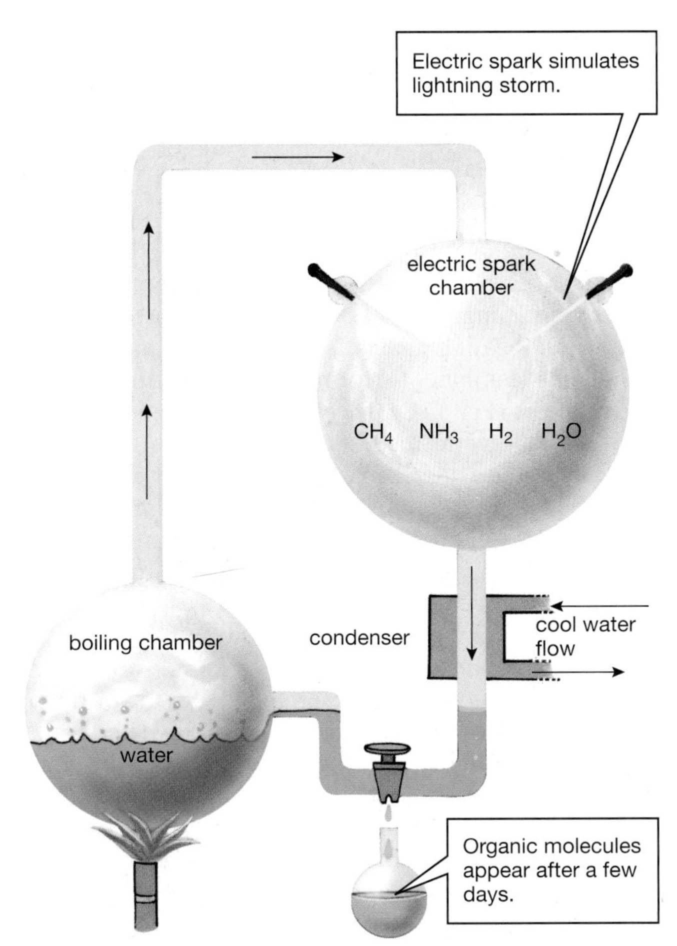

FIGURE 17-2 The experimental apparatus of Stanley Miller and Harold Urey
Life's earliest stages left no fossils, so evolutionary historians have pursued a strategy of re-creating in the laboratory the conditions that may have prevailed on early Earth. The mixture of gases in the spark chamber simulates Earth's early atmosphere. **QUESTION** How would the experiment's result change if O_2 were included in the spark chamber?

Similar experiments by Miller and others have produced amino acids, short proteins, nucleotides, adenosine triphosphate (ATP), and other molecules characteristic of living things. Interestingly, the exact composition of the "atmosphere" used in these experiments is unimportant, provided that hydrogen, carbon, and nitrogen are available and that oxygen gas is excluded. Similarly, a variety of energy sources, including ultraviolet light, electrical discharge, and heat, are all about equally effective. Even though we may never know exactly what the earliest atmosphere was like, we can be confident that organic molecules formed spontaneously on early Earth. Additional organic molecules probably arrived from space when meteorites and comets crashed into Earth's surface. (Analysis of present-day meteorites recovered from impact craters on Earth has revealed that some meteorites contain relatively high concentrations of amino acids and other simple organic molecules similar to those generated in the Miller-Urey experiment.)

Organic Molecules Probably Accumulated Under Prebiotic Conditions

Prebiotic synthesis was not very efficient or very fast. Nonetheless, in a few hundred million years, large quantities of organic molecules accumulated in early Earth's oceans. Today, most organic molecules have short lives; they are either digested by living organisms or they react with atmospheric oxygen. Early Earth, however, lacked both life and free oxygen, so molecules would not have been exposed to these threats. Still, the prebiotic molecules must have been threatened by the sun's high-energy ultraviolet (UV) radiation, because early Earth lacked an ozone layer. The ozone layer is a region high in today's atmosphere that is enriched with ozone (O_3) molecules, which absorb some of the sun's UV light before it reaches Earth's surface. Before the ozone layer formed, UV bombardment, which can break apart organic molecules, must have been fierce. Some places, however, such as those beneath rock ledges or at the bottoms of even fairly shallow seas, would have been protected from UV radiation. In these locations, organic molecules may have accumulated.

Organic Molecules May Have Become Concentrated in Tidal Pools

In the next stage of prebiotic evolution, simple molecules combined to form larger molecules. The chemical reactions that formed these larger molecules required that the reacting molecules be packed closely together. Scientists have proposed several mechanisms by which the requisite high concentrations of molecules might have been achieved on early Earth. One possibility is that waves crashing onto the shore during high tides filled shallow pools at the ocean's edge with water. Afterward, during low tides, some of the water in the pools might have evaporated, concentrating the dissolved substances. After enough cycles of refilling and evaporation, these pools could have contained a highly concentrated "primordial soup" in which spontaneous chemical reactions between simple molecules could generate complex organic molecules. These molecules could then have become the building blocks of the first living organisms.

RNA May Have Been the First Self-Reproducing Molecule

Although all living organisms use DNA to encode and store genetic information, it is unlikely that DNA was the earliest informational molecule. DNA can reproduce itself only with the help of large, complex protein enzymes, but the instructions for building these enzymes are encoded in DNA itself. Thus, the origin of

DNA's role as life's information storage molecule poses a "chicken and egg" puzzle: DNA requires proteins, but those proteins require DNA. It is thus difficult to construct a plausible scenario for the origin of self-replicating DNA from prebiotic molecules. It is therefore likely that the current DNA-based system of information storage evolved from an earlier system.

A prime candidate for the first self-replicating informational molecule is RNA. In the 1980s, Thomas Cech and Sidney Altman, working with the single-celled organism *Tetrahymena*, discovered a cellular reaction that was catalyzed not by a protein, but by a small RNA molecule. Because this special RNA molecule performed a function previously thought to be performed only by protein enzymes, Cech and Altman decided to give their catalytic RNA molecule the name **ribozyme**. In the years since their discovery, researchers have found dozens of naturally occurring ribozymes that catalyze a variety of different reactions, including cutting other RNA molecules and splicing together different RNA fragments. Ribozymes have also been found in the protein manufacturing machinery of cells, where they help catalyze the attachment of amino acid molecules to growing proteins. In addition, researchers have been able to synthesize different ribozymes in the laboratory, including some that can catalyze the replication of small RNA molecules.

The discovery that RNA molecules can act as catalysts for diverse reactions, including RNA replication, provides support for the hypothesis that life arose in an "RNA world." According to this view, the current era of DNA-based life was preceded by one in which RNA served as both the information-carrying genetic molecule and as the enzyme catalyst for its own replication. This RNA world may have emerged after hundreds of millions of years of prebiotic chemical synthesis, during which RNA nucleotides would have been among the molecules synthesized. After reaching a sufficiently high concentration, the nucleotides probably bonded together to form short RNA chains. Let's suppose that, purely by chance, one of these RNA chains was a ribozyme that could catalyze the production of copies of itself. This first self-reproducing ribozyme probably wasn't very good at its job and produced copies with lots of errors. These mistakes were the first mutations. Like modern mutations, most undoubtedly ruined the catalytic abilities of the "daughter molecules," but a few may have been improvements. Such improvements set the stage for the evolution of RNA molecules, as variant ribozymes with increased speed and accuracy of replication reproduced faster, making more copies of themselves and displacing less efficient molecules. Molecular evolution in the RNA world proceeded until, by some still unknown chain of events, RNA gradually receded into its present role as an intermediary between DNA and protein enzymes.

Membrane-Like Microspheres May Have Enclosed Ribozymes

Self-replicating molecules alone do not constitute life; these molecules must be contained within some kind of enclosing membrane. The precursors of the earliest biological membranes may have been simple structures that formed spontaneously from purely physical, mechanical processes. For example, chemists have shown that if water containing proteins and lipids is agitated to simulate waves beating against ancient shores, the proteins and lipids combine to form hollow structures called **microspheres** (Fig. 17-3). These hollow balls resemble living cells in several respects. They have a well-defined outer boundary that separates their internal contents from the external solution. If the composition of the microsphere is right, a "membrane" forms that is remarkably similar in appearance to a real cell membrane. Under certain conditions, microspheres can absorb material from the external solution, grow, and even divide.

If a microsphere happened to surround the right ribozymes, it would form something resembling a living cell. We could call it a **protocell**, structurally similar to a cell but not a living thing. In the protocell, ribozymes and any other enclosed molecules would have been protected from free-roaming ribozymes in the primordial soup. Nucleotides and other small molecules might have diffused across the membrane and been used to synthesize new ribozymes and other complex molecules. After sufficient growth, the microsphere may have divided, with a few copies of the ribozymes becoming incorporated into each daughter microsphere. If this occurred, the path to the evolution of the first cells would be nearly at its end. Was there a particular moment when a nonliving protocell gave rise to something alive? Probably

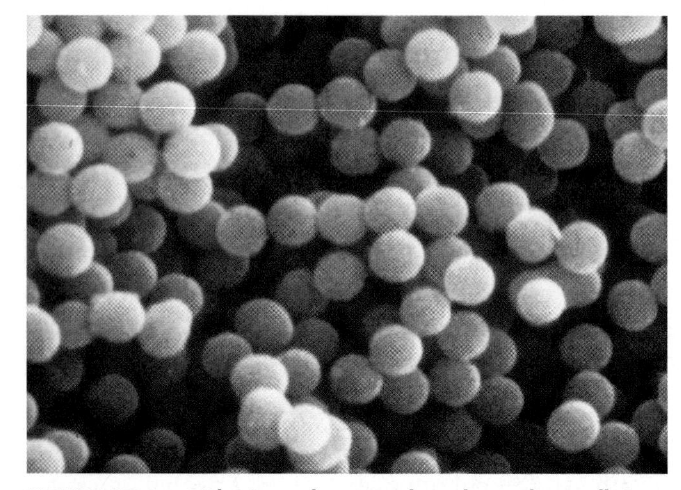

FIGURE 17-3 Did microspheres enclose the earliest cells? Cell-like microspheres can be formed by agitating proteins and lipids in a liquid medium. Each microsphere is this photo is about 5 micrometers (μm) in diameter.

not. Like most evolutionary transitions, the change from protocell to living cell was a continuous process, with no sharp boundary between one state and the next.

But Did All This Happen?

The above scenario, although plausible and consistent with many research findings, is by no means certain. One of the most striking aspects of origin-of-life research is the great diversity of assumptions, experiments, and contradictory hypotheses (Iris Fry's *The Emergence of Life on Earth*, cited in the "For More Information" section at the end of this chapter, offers a taste of these controversies). Researchers disagree about whether life arose in quiet pools, in the sea, in moist films on the surfaces of clay or iron pyrite (fool's gold), or in furiously hot deep-sea vents. (Proponents of the deep-sea vent hypothesis suggest that similar vents might keep Europa's seas thawed, providing conditions for an independent origin of life there.) A few researchers even argue that life arrived on Earth from space. Can we draw any conclusions from the research conducted so far? No one knows for sure, but we can make a few observations.

First, the experiments of Miller and others show that amino acids, nucleotides, and other organic molecules, along with simple membrane-like structures, would have formed in abundance on early Earth. Second, chemical evolution had long periods of time and huge areas of the Earth available to it. Given sufficient time and a sufficiently large pool of reactant molecules, even extremely rare events can occur many times. So, even if prebiotic evolution yielded only simple molecules, the earliest catalysts were not very efficient, and the earliest membranes were simple, the vast expanses of available time and space would have increased the likelihood of each small step on the path from primordial soup to living cell.

Most biologists accept that the origin of life was probably an inevitable consequence of the working of natural laws. We should emphasize, however, that this proposition cannot be definitively tested. The origin of life left no record, and researchers exploring this mystery can proceed only by developing a hypothetical scenario and then conducting laboratory investigations to determine if the scenario's steps are chemically and biologically possible and plausible.

17.2 What Were the Earliest Organisms Like?

When Earth first formed about 4.5 billion years ago, it was quite hot. A multitude of meteorites smashed into the forming planet, and the kinetic energy of these extraterrestrial rocks was converted into heat on impact. Still more heat was released by the decay of radioactive atoms. The rock composing Earth melted, and heavier elements such as iron and nickel sank to the center of the planet, where they remain molten today. It must have taken hundreds of millions of years for Earth to cool enough to allow water to exist as a liquid. Nonetheless, it appears that life arose in fairly short order once liquid water was available.

The oldest fossil organisms found so far are in rocks that are about 3.5 billion years old; their age was determined using radiometric dating techniques (see "Scientific Inquiry: How Do We Know How Old a Fossil Is?"). Chemical traces in older rocks have led some paleontologists to believe that life is even older, perhaps as old as 3.9 billion years. The period in which life began is known as the Precambrian era. This interval was designated by geologists and paleontologists, who have devised a hierarchical naming system of eras, periods, and epochs to delineate the immense span of geological time (Table 17-1).

The First Organisms Were Anaerobic Prokaryotes

The first cells to arise in Earth's oceans were **prokaryotes**; that is, their genetic material was not separated from the rest of the cell within a nucleus. These cells probably obtained nutrients and energy by absorbing organic molecules from their environment. There was no oxygen gas in the atmosphere, so the cells must have metabolized the organic molecules anaerobically. You may recall from Chapter 8 that anaerobic metabolism yields only small amounts of energy.

Thus, the earliest cells were primitive anaerobic bacteria. As these bacteria multiplied, they must have eventually used up the organic molecules produced by prebiotic chemical reactions. Simpler molecules, such as carbon dioxide and water, would still have been very abundant, as was energy in the form of sunlight. What was lacking, then, was not materials or energy itself, but energetic molecules—molecules in which energy is stored in chemical bonds.

Some Organisms Evolved the Ability to Capture the Sun's Energy

Eventually, some cells evolved the ability to use the energy of sunlight to drive the synthesis of complex, high-energy molecules from simpler molecules; in other words, photosynthesis appeared. Photosynthesis requires a source of hydrogen, and the very earliest photosynthetic bacteria probably used hydrogen sulfide gas dissolved in water for this purpose (much as today's purple photosynthetic bacteria do). Eventually, however, Earth's supply of hydrogen sulfide (which is produced mainly by volcanoes) must have run low. The shortage of

Table 17-1 The History of Life on Earth

Era	Period	Epoch	Millions of Years Ago*	Major Events
Precambrian			4600	Origin of solar system and Earth.
			4000–3900	Appearance of first rocks on Earth.
			3900–3500	First living cells (prokaryotes).
			3500	Origin of photosynthesis (in cyanobacteria).
			2200	Accumulation of free oxygen in atmosphere.
			2000–1700	First eukaryotes.
			By 1000	First multicellular organisms.
			About 1000	First animals (soft-bodied marine invertebrates).
Paleozoic	Cambrian		544–505	Primitive marine algae flourish; origin of most marine invertebrate types; first fishes.
	Ordovician		505–440	Invertebrates, especially arthropods and mollusks, dominant in sea; first fungi.
	Silurian		440–410	Many fishes, trilobites, mollusks in sea; first vascular plants; invasion of land by plants; invasion of land by arthropods.
	Devonian		410–360	Fishes and trilobites flourish in sea; first amphibians and insects; first seeds and pollen.
	Carboniferous		360–286	Swamp forests of tree ferns and club mosses; first conifers; dominance of amphibians; numerous insects; first reptiles.
	Permian		286–245	Massive marine extinctions, including last of trilobites; flourishing of reptiles and decline of amphibians; aggregation of continents into one land mass, Pangaea.
Mesozoic	Triassic		245–208	First mammals and dinosaurs; forests of gymnosperms and tree ferns; beginning of breakup of Pangaea.
	Jurassic		208–146	Dominance of dinosaurs and conifers; first birds; continents partially separated.
	Cretaceous		146–65	Flowering plants appear and become dominant; mass extinctions of marine life and some terrestrial life, including last dinosaurs; modern continents well separated.
Cenozoic	Tertiary	Paleocene	65–54	Widespread flourishing of birds, mammals, insects, and flowering plants; shifting of continents into modern positions; mild climate at beginning of period, with extensive mountain building and cooling toward end.
		Eocene	54–38	
		Oligocene	38–23	
		Miocene	23–5	
		Pliocene	5–1.8	
	Quaternary	Pleistocene	1.8–0.01	Evolution of genus *Homo*; repeated glaciations in Northern Hemisphere; extinction of many giant mammals.
		Recent	0.01–present	

*From University of California Museum of Paleontology, April 2000.

hydrogen sulfide set the stage for the evolution of photosynthetic bacteria that were able to use the planet's most abundant source of hydrogen: water (H_2O).

Photosynthesis Increased the Amount of Oxygen in the Atmosphere

Water-based photosynthesis converts water and carbon dioxide to energetic molecules of sugar, releasing oxygen as a by-product. The emergence of this new method for capturing energy introduced significant amounts of oxygen gas to the atmosphere for the first time. At first, the newly liberated oxygen was quickly consumed by reactions with other molecules in the atmosphere and in Earth's crust, or surface layer. One especially common reactive atom in the crust was iron, and much of the new oxygen combined with iron atoms to form huge deposits of iron oxide (also known as rust).

After all the accessible iron had turned to rust, the concentration of oxygen gas in the atmosphere began to increase. Chemical analysis of rocks suggests that significant amounts of oxygen first appeared in the atmosphere about 2.2 billion years ago, produced by bacteria that were probably very similar to modern cyanobacteria (you will undoubtedly breathe in some oxygen molecules today that were expelled 2 billion years ago by one of these early cyanobacteria). Atmospheric oxygen levels increased steadily until they reached a stable level about 1.5 billion years ago. Since that time, the proportion of oxygen in the atmosphere has been nearly constant, as the amount of oxygen released by photosynthesis worldwide is neatly balanced by the amount that is consumed by aerobic respiration.

SCIENTIFIC INQUIRY How Do We Know How Old a Fossil Is?

Early geologists could date rock layers and their accompanying fossils only in a *relative* way: fossils found in deeper layers of rock were generally older than those found in shallower layers. With the discovery of radioactivity, it became possible to determine *absolute* dates, within certain limits of uncertainty. The nuclei of radioactive elements spontaneously break down, or decay, into other elements. For example, carbon-14 (usually written ^{14}C) decays by emitting an electron to become nitrogen-14 (^{14}N) Each radioactive element decays at a rate that is independent of temperature, pressure, or the chemical compound of which the element is a part. The time it takes for half of a radioactive element's nuclei to decay at this characteristic rate is called its *half-life*. The half-life of ^{14}C, for example, is 5730 years.

How are radioactive elements used in determining the age of rocks? If we know the rate of decay and measure the proportion of decayed nuclei to undecayed nuclei, we can estimate how much time has passed since these radioactive elements became trapped in the rock. This process is called *radiometric dating*. A particularly straightforward dating technique measures the decay of potassium-40 (^{40}K), which has a half-life of about 1.25 billion years, into argon-40 (^{40}Ar). Potassium is a very reactive element commonly found in volcanic rocks such as granite and basalt. Argon, however, is an unreactive gas. Let's suppose that a volcano erupts with a massive lava flow, covering the countryside. All the ^{40}Ar, being a gas, will bubble out of the molten lava, so when the lava first cools and solidifies into rock, it will not contain any ^{40}Ar. Meanwhile, any ^{40}K present in the hardened lava will decay to ^{40}Ar, with half of the ^{40}K decaying every 1.25 billion years. This ^{40}Ar gas will be trapped in the rock. A geologist could take a sample of the rock and determine the proportion of ^{40}K to ^{40}Ar (Fig. E17-1). If the analysis finds equal amounts of the two elements, the geologist will conclude that the lava hardened 1.25 billion years ago. With appropriate care, such age estimates are quite reliable. If a fossil is found beneath a lava flow dated at, say, 500 million years, then we know that the fossil is at least that old.

As some radioactive elements decay, they can even give an estimate of the age of the solar system. Analysis of uranium, which decays to lead, has shown that the oldest meteorites and moon rocks collected by astronauts are about 4.6 billion years old.

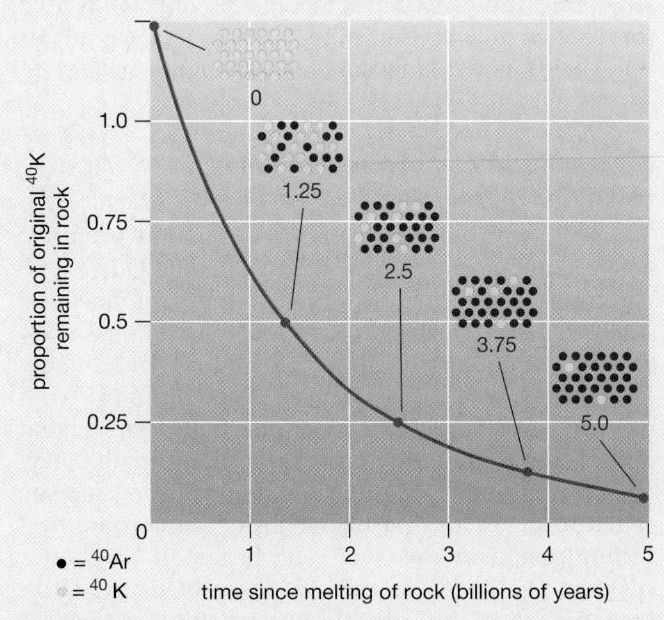

FIGURE E17-1 The relationship between time and the decay of radioactive ^{40}K to ^{40}Ar

EXERCISE Uranium-235 decays to lead-207 with a half-life of 713 million years. If you analyze a rock and find that in contains uranium-235 and lead-207 in a ratio of 3:1, how old is the rock?

Aerobic Metabolism Arose in Response to the Oxygen Crisis

Oxygen is potentially very dangerous to living things, because it reacts with and destroys organic molecules. Many of today's anaerobic bacteria perish when exposed to what is, for them, a deadly poison: oxygen. The accumulation of oxygen in the atmosphere of early Earth probably exterminated many organisms and fostered the evolution of cellular mechanisms for detoxifying oxygen. This crisis for evolving life also provided the environmental pressure for the next great advance in the "Age of Microbes": the ability to use oxygen in metabolism. This ability not only provides a defense against the chemical action of oxygen, but actually channels oxygen's destructive power through aerobic respiration to generate useful energy for the cell. Because the amount of energy available to a cell is vastly increased when oxygen is used to metabolize food molecules, aerobic cells had a significant selective advantage.

Some Organisms Acquired Membrane-Enclosed Organelles

Hordes of bacteria would offer a rich food supply to any organism that could eat them. There is no fossil evidence of the first predatory cells to roam the seas, but paleobiologists speculate that once a suitable prey population (such as these bacteria) appeared, predation would have evolved quickly. These early predators were specialized prokaryotic cells, with larger size and flexible membranes that allowed them to engulf whole bacteria as prey. According to the most widely accepted

Media Activity 17.1 Endosymbiosis

hypothesis, these predators were otherwise quite primitive, being capable of neither photosynthesis nor aerobic metabolism. Although they could capture large food particles, namely bacteria, they metabolized the food inefficiently. Nonetheless, by about 1.7 billion years ago, one such predator probably gave rise to the first eukaryotic cell.

As you know, the cells of **eukaryotes** differ from prokaryotic cells in that they have a nucleus, which contains the cell's genetic material, and other distinctive structures. These structures include the organelles used for energy metabolism: mitochondria and (in plants only) chloroplasts. How did these organelles evolve?

Mitochondria and Chloroplasts May Have Arisen from Engulfed Bacteria

The **endosymbiont hypothesis** proposes that primitive cells acquired the precursors of mitochondria and chloroplasts by engulfing certain types of bacteria. These cells and the bacteria trapped inside them (*endo* means "within") gradually entered into a *symbiotic* relationship, a close association between different types of organisms over an extended time. Let's suppose that an anaerobic predatory cell captured an aerobic bacterium for food, as it often did, but for some reason failed to digest this particular prey. The aerobic bacterium remained alive and well. In fact, it was better off than ever, because the cytoplasm of its predator/host was chock-full of half-digested food molecules, the remnants of anaerobic metabolism. The aerobe absorbed these molecules and used oxygen to metabolize them, thereby gaining enormous amounts of energy. So abundant were the aerobe's food resources, and so bountiful its energy production, that the aerobe must have leaked energy, probably in the form of ATP or similar molecules, back into its host's cytoplasm. The anaerobic predatory cell, with its symbiotic bacteria, could now metabolize food aerobically, gaining a great advantage over other anaerobic cells and leaving a greater number of offspring. Eventually, the endosymbiotic bacterium lost its ability to live independently of its host, and the mitochondrion was born (Fig. 17-4, ① and ②).

One of these successful new cellular partnerships must have managed a second feat: it captured a photosynthetic cyanobacterium and similarly failed to digest its prey. The cyanobacterium flourished in its new host and gradually evolved into the first chloroplast (Fig. 17-4, ③ and ④). Other eukaryotic organelles may have also originated through endosymbiosis. Many biologists believe that cilia, flagella, centrioles, and microtubules may all have evolved from a symbiosis between a spirilla-like bacterium (a form of bacterium with an elongated corkscrew shape) and an early eukaryotic cell.

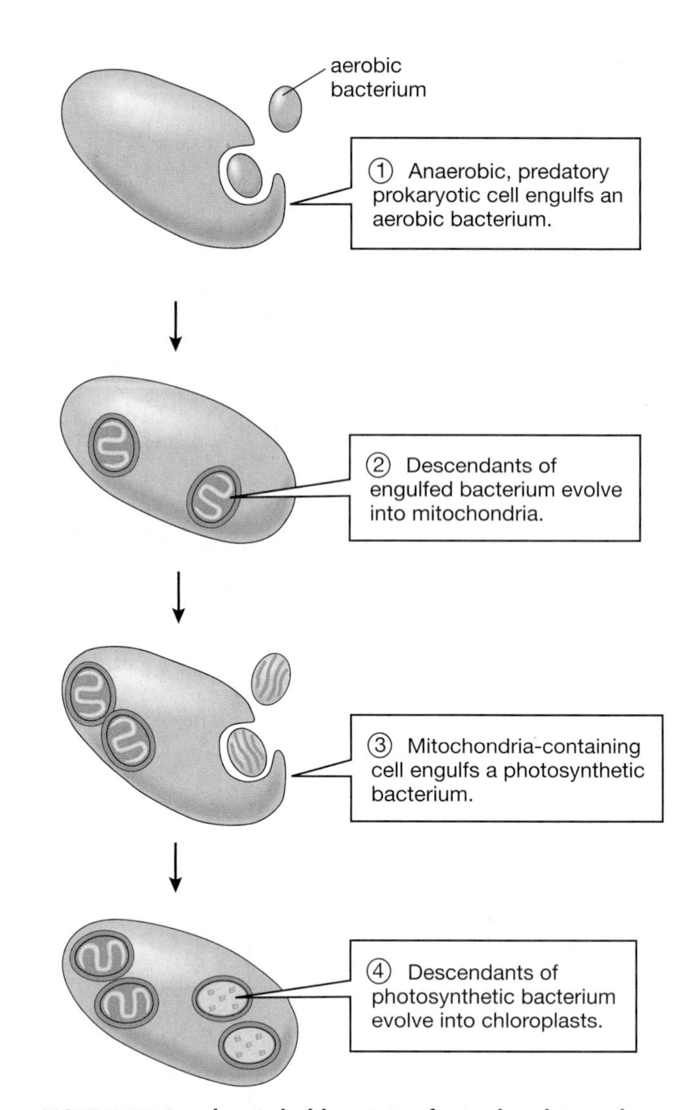

FIGURE 17-4 The probable origin of mitochondria and chloroplasts in eukaryotic cells

QUESTION Scientists have identified a living bacterium believed to be descended from the endosymbiont that gave rise to mitochondria. Would you expect the DNA sequence of this modern bacterium to be most similar to the sequence of DNA from a plant chloroplast, an animal cell nucleus, or a plant mitochondrion?

Evidence for the Endosymbiont Hypothesis Is Strong

Several types of evidence support the endosymbiont hypothesis. A particularly compelling line of evidence is the many distinctive biochemical features shared by eukaryotic organelles and living bacteria. In addition, mitochondria, chloroplasts, and centrioles each contain their own minute supply of DNA, which many researchers interpret as remnants of the DNA originally contained within the engulfed bacteria. Another kind of support comes from *living intermediates*, organisms alive today that are similar to their hypothetical ancestors and thus help demonstrate the plausibility of a pro-

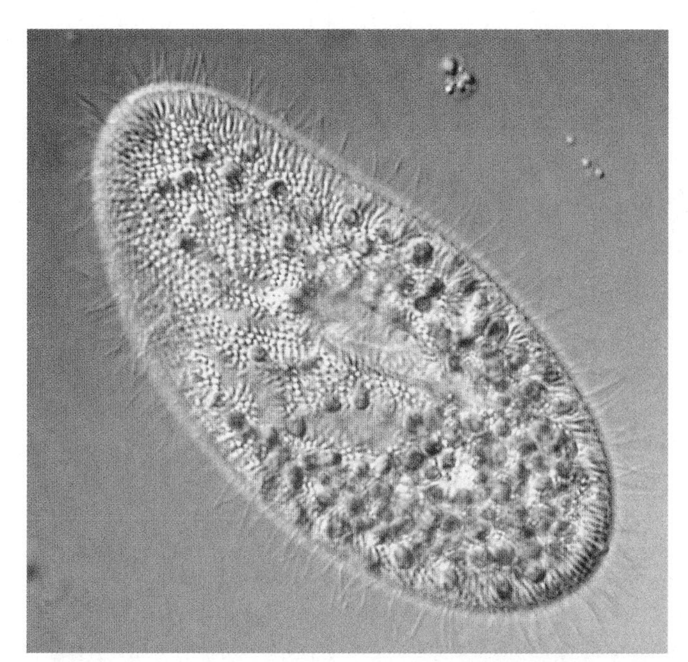

FIGURE 17-5 Symbiosis within a modern cell
The ancestors of the chloroplasts in today's plant cells may have resembled *Chlorella*, the green, photosynthetic, single-celled algae living symbiotically within the cytoplasm of the *Paramecium* pictured here.

posed evolutionary pathway. For example, the amoeba *Pelomyxa palustris* lacks mitochondria, but hosts a permanent population of aerobic bacteria that carry out much the same role. Similarly, a variety of corals, some clams, a few snails, and at least one species of *Paramecium* harbor a permanent collection of photosynthetic algae in their cells (Fig. 17-5). These examples of modern cells that host bacterial endosymbionts suggest that we have no reason to doubt that similar symbiotic associations could have occurred almost 2 billion years ago and led to the first eukaryotic cells.

17.3 What Were the Earliest Multicellular Organisms Like?

Once predation had evolved, increased size became an advantage. In the marine environments to which life was restricted, a larger cell could more easily engulf a smaller cell, while also being more difficult for other predatory cells to ingest. Most larger organisms can also move faster than smaller ones, making successful predation and escape more likely. But enormous single cells have problems. Oxygen and nutrients going into the cell and waste products going out must diffuse through the plasma membrane. As mentioned in Chapter 5, the larger a cell becomes, the less membrane surface area is available per unit volume of cytoplasm, making diffusion more difficult. An organism larger than a millimeter or so in diameter can survive only in one of two ways. First, it can have a low metabolic rate so that it doesn't need much oxygen or produce much carbon dioxide. This seems to work for certain very large single-celled algae. Alternatively, an organism can be multicellular—that is, consist of many small cells packaged into a larger, unified body.

Some Algae Became Multicellular

The oldest fossils of multicellular organisms are about 1 billion years old and include impressions of the first multicellular algae, which arose from single-celled eukaryotic cells containing chloroplasts. Multicellularity would have provided at least two advantages for these seaweeds. First, large, many-celled algae would have been difficult for single-celled predators to swallow. Second, specialization of cells would have conferred the potential for staying in one place in the brightly lit waters of the shoreline, as rootlike structures burrowed in sand or clutched onto rocks, while leaflike structures floated above in the sunlight. The green, brown, and red algae lining our shores today—some, such as brown kelp, more than 200 feet long—are the descendants of these early multicellular algae.

Animal Diversity Arose in the Precambrian Era

In addition to fossil algae, billion-year-old rocks have yielded fossil traces of animal tracks and burrows. This evidence of early animal life notwithstanding, fossils of animal bodies first appear in Precambrian rocks laid down between 610 million and 544 million years ago. Some of these ancient invertebrate animals (animals lacking backbones) are quite different in appearance from any animals that appear in later fossil layers, and may represent types of animals that left no descendants. Others fossils in these rock layers, however, appear to be ancestors of today's animals. Ancestral sponges and jellyfish appear in the oldest layers, followed later by ancestors of worms, mollusks, and arthropods. The full range of modern invertebrate animals, however, does not appear in the fossil record until the Cambrian period, marking the beginning of the Paleozoic era, about 544 million years ago. (The phrase "fossil record" is a shorthand reference to the entire collection of all fossil evidence that has been found to date.) These Cambrian fossils reveal an adaptive radiation (see Chapter 16) that had already yielded a diverse array of complex body plans; almost all of the animal phyla that are present on Earth today were already present in the early Cambrian. The apparently sudden appearance of so many different kinds of animals suggests that the earlier evolutionary history that produced such an impressive

FIGURE 17-6 Diversity of ocean life during the Silurian period
(a) Life characteristic of the oceans during the Silurian period, 440 million to 410 million years ago. Among the most common fossils from that time are **(b)** trilobites and their predators, such as nautiloids and **(c)** ammonites. This **(d)** living *Nautilus* is very similar in structure to the Silurian nautiloids, showing that a successful body plan may exist virtually unchanged for hundreds of millions of years.

range of different animal forms is not preserved in the fossil record.

The early diversification of animals was probably driven in part by the emergence of predatory lifestyles. Coevolution of predator and prey led to the evolution of new features in many kinds of animals. By the Silurian period (440 million to 410 million years ago), mud-skimming, armored trilobites were preyed on by ammonites and the chambered nautilus, which still survives in almost unchanged form in deep Pacific waters (Fig. 17-6). Many animals of this era were more mobile than their evolutionary predecessors. Predators gain an advantage by being able to travel over wide areas in search of suitable prey, and the ability to make a speedy escape is an advantage for prey. The evolution of efficient movement was often associated with the evolution of greater sensory capabilities and more complex nervous systems. Senses for detecting touch, chemicals, and light became highly developed, along with nervous systems capable of handling the sensory information and directing appropriate behaviors.

About 530 million years ago, one group of animals—the fishes—developed a new form of support for the body: an internal skeleton. These early fishes were inconspicuous members of the ocean community, but by 400 million years ago fishes were a diverse and prominent group. By and large, the fishes proved to be faster than the invertebrates, with more-acute senses and larger brains. Eventually, they became the dominant predators of the open seas.

17.4 How Did Life Invade the Land?

A compelling subplot in the long tale of life's history is the story of life's invasion of land after more than 3 billion years of a strictly watery existence. In moving to solid ground, organisms had many obstacles to overcome. Life in the sea provides buoyant support, but on land, an organism must bear its weight against the crushing force of gravity. The sea provides ready access to life-sustaining water, but a terrestrial organism must find adequate water. Sea-dwelling plants and animals can reproduce by means of mobile sperm and/or eggs that swim to each other through the water, but the gametes of land-dwellers must be protected from drying out.

Despite the obstacles to life on land, the vast empty spaces of the Paleozoic landmass represented a tremendous evolutionary opportunity. The potential rewards of terrestrial life were especially great for plants. Water strongly absorbs light, so even in the clearest water, photosynthesis is limited to the upper few hundred meters of depth—and usually much less. Out of the water, the dazzling brightness of the sun permits rapid photosynthesis. Furthermore, terrestrial soils are rich storehouses of nutrients, whereas seawater tends to be low in certain nutrients, particularly nitrogen and phosphorus. Finally, the Paleozoic sea swarmed with plant-eating animals, but the land was devoid of animal life. The plants that first colonized the land would have had ample sunlight, untouched nutrient sources, and no predators.

Some Plants Became Adapted to Life on Dry Land

In moist soils at the water's edge, a few small green algae began to grow, taking advantage of the sunlight and nutrients. They didn't have large bodies to support against the force of gravity, and, living in the thin layer of water on the soil, they could easily obtain water. About 400 million years ago, some of these algae gave rise to the first multicellular land plants. Initially simple, low-growing forms, land plants rapidly developed solutions to two of the main difficulties of plant life on land: obtaining and conserving water and staying upright despite gravity and winds. Waterproof coatings on aboveground parts reduced water loss by evaporation, and rootlike structures delved into the soil, mining water and minerals. Specialized cells formed tubes called vascular tissues to conduct water from roots to leaves. Extra-thick walls surrounding certain cells enabled stems to stand erect.

Primitive Land Plants Retained Swimming Sperm and Required Water to Reproduce

Reproduction out of water presented challenges. Like animals, plants produce sperm and eggs, which must be able to meet to produce the next generation. The first land plants had swimming sperm, presumably much like those of some of today's marine algae (some of which have swimming eggs as well). Consequently, the earliest plants were restricted to swamps and marshes, where the sperm and eggs could be released into the water, or to areas with abundant rainfall, where the ground would occasionally be covered with water. Later, plants with swimming sperm prospered during periods in which the climate was warm and moist. For example, the Carboniferous period (360 million to 286 million years ago) was characterized by vast forests of giant tree ferns and club mosses (Fig. 17-7). The coal we mine today is derived from the fossilized remains of those forests.

Media Activity
17.2 Evolutionary Timescales

FIGURE 17-7 The swamp forest of the Carboniferous period
The treelike plants in this artist's reconstruction are tree ferns and giant club mosses, most species of which are now extinct. **QUESTION** Why are today's ferns and club mosses so small in comparison to their giant ancestors?

Seed Plants Encased Sperm in Pollen Grains

Meanwhile, some plants inhabiting drier regions had evolved a means of reproduction that no longer depended on water. The eggs of these plants were retained on the parent plant, and the sperm were encased in drought-resistant pollen grains that blew on the wind from plant to plant. When the pollen grains landed near an egg, they released sperm cells directly into living tissue, eliminating the need for water. The fertilized egg remained on the parent plant, where it developed inside a seed, which provided protection and nutrients for the developing embryo within.

The earliest seed-bearing plants appeared in the late Devonian period (375 million years ago) and produced their seeds along branches, without any specialized structures to hold them. By the middle of the Carboniferous period, however, a new kind of seed-bearing plant had arisen. These plants, called **conifers**, protected their developing seeds inside cones. Conifers, which did not depend on water for reproduction, flourished and spread during the Permian period (286 to 245 million years ago), when mountains rose, swamps drained, and the climate became much drier. The good fortune of the conifers, however, was not shared by the tree ferns and giant club mosses, which, with their swimming sperm, largely went extinct.

Flowering Plants Enticed Animals to Carry Pollen

About 140 million years ago, during the Cretaceous period, the flowering plants appeared, having evolved from a group of conifer-like plants. Many flowering plants are pollinated by insects and other animals, a mode of pollination that seems to have conferred an evolutionary advantage. Flower pollination by animals can be far more efficient than pollination by wind; wind-pollinated plants must produce an enormous amount of pollen because the vast majority of pollen grains fail to reach their target. Flowering plants also evolved other advantages, including more rapid reproduction and, in some cases, much more rapid growth. Today, flowering plants dominate the land, except in cold northern regions, where conifers still prevail.

Some Animals Became Adapted to Life on Dry Land

Soon after land plants evolved, providing potential food sources for other organisms, animals emerged from the sea. The first animals to move onto land were **arthropods** (the group that today includes insects, spiders, scorpions, centipedes, and crabs). Why arthropods? The answer seems to be that they already possessed certain structures that, purely by chance, were suited to life on land. Foremost among these structures was an external skeleton, or **exoskeleton**, a hard covering surrounding the body, such as the shell of a lobster or crab. Exoskeletons are both waterproof and strong enough to support a small animal against the force of gravity.

For millions of years, arthropods had the land and its plants to themselves, and for tens of millions of years more, they were the dominant land animals. Dragonflies with a wingspan of 28 inches (70 centimeters) flew among the Carboniferous tree ferns, while millipedes 6.5 feet (2 meters) long munched their way across the swampy forest floor. Eventually, however, the arthropods' splendid isolation came to an end.

Amphibians Evolved from Lobefin Fishes

About 400 million years ago, a group of Silurian fishes called the lobefins appeared, probably in fresh water. **Lobefins** had two important features that would later enable their descendants to colonize land: stout, fleshy fins with which they crawled about on the bottoms of shallow, quiet waters, and an outpouching of the digestive tract that could be filled with air, like a primitive lung. One group of lobefins colonized very shallow ponds and streams, which shrank during droughts and whose water often became oxygen-poor. By taking air into their lungs, these lobefins could obtain oxygen anyway. Some began to use their fins to crawl from pond to pond in search of prey or water, as some modern fish do today (Fig. 17-8). Some of the lobefin's descendants even evolved limbs with toes on them, though the possessors of these earliest legs were water-dwellers and probably used their limbs mainly to move about in shallow water.

The benefits of feeding on land and moving from pool to pool favored the evolution of a group of animals that could stay out of water for longer periods and that could move about more effectively on land. With improvements in their lungs and legs, **amphibians** evolved from lobefins, first appearing in the fossil record about 350 million years ago. To an amphibian, the Carboniferous swamp forests were a kind of paradise: no predators to speak of, abundant prey, and a warm, moist climate.

FIGURE 17-8 A fish that walks on land
Some modern fishes, such as this mudskipper, walk on land. Like the ancient lobefin fishes that gave rise to amphibians, mudskippers use their strong pectoral fins to move across dry areas in their swampy habitats. **QUESTION** Does the mudskipper's ability to walk on land constitute evidence that lobefin fishes were the ancestors of amphibians?

As with the insects and millipedes, some amphibians evolved gigantic size, including salamanders more than 10 feet (3 meters) long.

Despite their success, the early amphibians were not fully adapted to life on land. Their lungs were simple sacs without very much surface area, so they had to obtain some of their oxygen through their skin. Therefore, their skin had to be kept moist, which restricted them to swampy habitats where they wouldn't dry out. Further, amphibian sperm and eggs could not survive in dry surroundings and had to be deposited in watery environments. So, although amphibians could move about on land, they could not stray too far from the water's edge. As with the tree ferns and club mosses, when the climate turned dry at the beginning of the Permian period about 286 million years ago, amphibians declined.

Reptiles Evolved from Amphibians

As the conifers were evolving on the fringes of the swamp forests, a group of amphibians was also evolving adaptations to drier conditions. These amphibians ultimately gave rise to the **reptiles**, which had three major adaptations to life on land. First, reptiles evolved shelled, waterproof eggs that enclosed a supply of water for the developing embryo. Thus, eggs could be laid on land without the reptiles' having to venture back to the dangerous swamps full of fish and amphibian predators. Second, ancestral reptiles evolved scaly, waterproof skin that helped prevent the loss of body water to the dry air. Finally, reptiles evolved improved lungs that were able to provide the entire oxygen supply for an active animal. As the climate dried during the Permian period, reptiles became the dominant land vertebrates, relegating amphibians to the swampy backwaters where most remain today.

A few tens of millions of years later, the climate returned to more moist and equable conditions. This period saw the evolution of some very large reptiles, in particular, the dinosaurs. The variety of dinosaur forms was enormous—from predators (Fig. 17-9) to plant eaters, from those that dominated the land, to others that took to the air, to still others that returned to the sea. Dinosaurs were among the most successful animals ever, if we consider persistence to be a measure of success. They flourished for more than 100 million years, until the last dinosaurs went extinct about 65 million years ago. No one is certain why they died out, but the aftereffects of a gigantic meteorite's impact with Earth seems to have been the final blow (as discussed in the following section).

Even during the age of dinosaurs, many reptiles remained quite small. One major difficulty faced by small reptiles is maintaining a high body temperature. A warm body is helpful to an active animal, because warmth increases the efficiency of the nervous system and muscles. But a warm body loses heat to the environment unless the air is also warm. Heat loss is a bigger problem for smaller animals, which have a larger surface area per unit of weight than do larger animals. Many species of small reptiles cope with the heat-loss problem by retaining slow metabolisms and developing lifestyles in which they remain active only when the air is sufficiently warm. Two groups of small reptiles, however, independently followed a different evolutionary pathway: they developed insulation. One group evolved feathers, and another evolved hair.

FIGURE 17-9 A reconstruction of a Cretaceous forest
By the Cretaceous period, flowering plants dominated terrestrial vegetation. Dinosaurs, such as the predatory pack of 6-foot-long *Velociraptors* shown here, were the preeminent land animals. Although small by dinosaur standards, velociraptors were formidable predators with great running speed, sharp teeth, and deadly, sickle-like claws on their hind feet.

Reptiles Gave Rise to Both Birds and Mammals

In ancestral birds, insulating feathers helped retain body heat. Consequently, these animals could be active in cool habitats and during the night, when their scaly relatives became sluggish. Later, some ancestral birds evolved longer, stronger feathers on their forelimbs, perhaps under selection for better ability to glide from trees or to jump after insect prey. Ultimately, feathers evolved into structures capable of supporting powered flight. Fully developed, flight-capable feathers are present in 150-million-year-old fossils, so the earlier insulating structures that eventually developed into flight feathers must have been present well before that time.

The earliest **mammals** coexisted with the dinosaurs but were small creatures, probably living in trees and being active mostly at night. When the dinosaurs went extinct, mammals colonized the habitats left empty by the extinctions. Mammal species prospered, diversifying into their modern array of forms.

Unlike birds, which retained the reptilian habit of laying eggs, mammals evolved live birth and the ability to feed their young with secretions of the mammary (milk-producing) glands. Ancestral mammals also developed hair, which provided insulation. Because the uterus, mammary glands, and hair do not fossilize, we may never know when these structures first appeared, or what their intermediate forms looked like. Recently, however, a team of paleontologists found bits of fossilized hair preserved in coprolites—fossilized animal feces. These coprolites, found in the Gobi Desert of China, were deposited by an anonymous predator 55 million years ago, so mammals have presumably had hair at least that long.

17.5 What Role Has Extinction Played in the History of Life?

If there is a moral to the great tale of life's history, it is that nothing lasts forever. The story of life can be read as a long series of evolutionary dynasties, with each new dominant group rising, ruling the land or the seas for a time, and, inevitably, falling into decline and extinction. Dinosaurs are the most famous of these fallen dynasties, but the list of extinct groups known only from fossils is impressively long. Despite the inevitability of extinction, however, the overall trend has been for species to arise at a faster rate than they disappear, so the number of different species on Earth has tended to increase over time.

Evolutionary History Has Been Marked by Periodic Mass Extinctions

Over much of life's history, dynastic succession has proceeded in a steady, relentless manner. This slow and steady turnover of species, however, has been interrupted by episodes of **mass extinction** (Fig. 17-10). These mass extinctions are characterized by the relatively sudden disappearance of a wide variety of species over a large part of Earth. In the most catastrophic episodes of extinction, more than half the planet's species disappeared. The worst episode of all, which occurred 245 million years ago at the end of the Permian period, wiped out more than 90% of the world's species, and life came perilously close to disappearing altogether.

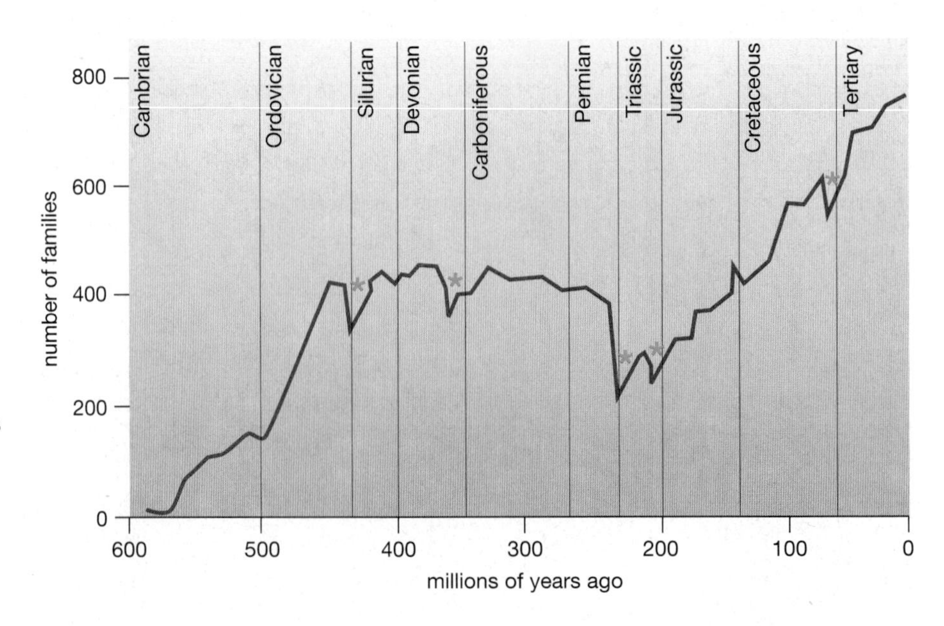

FIGURE 17-10 Mass extinctions This graph plots the number of marine-animal groups against time, as reconstructed from the fossil record. Note the general trend toward an increasing number of groups, punctuated by periods of sometimes rapid extinction. Five of these declines, marked by asterisks, are so steep that they qualify as catastrophic mass extinctions. **QUESTION** If extinction is the ultimate fate of all species, how can the total number of species have increased over time?

Climate Change Contributed to Mass Extinctions

Mass extinctions have had a profound impact on the course of life's history, repeatedly redrawing the picture of life's diversity. What could have caused such dramatic changes in the fortunes of so many species? Many evolutionary biologists believe that changes in climate must have played an important role. When the climate changes drastically, as it has done many times over the course of Earth's history, organisms that were adapted for survival in one climate may be unable to survive in a new, different climate. In particular, at times when warm climates gave way to drier, colder climates with more variable temperatures, species may have gone extinct after failing to adapt to the harsh new conditions.

One cause of climate change is the changing positions of continents. Earth's surface is divided into portions called plates, which include the continents and the seafloor. The solid plates slowly move above a viscous but fluid layer; this movement is called **plate tectonics.** As a plate wanders, its position may change in latitude (Fig. 17-11), thus changing the climate on its surface. For example, 350 million years ago much of North America was located at or near the equator, an area characterized by consistently warm and wet tropical weather. But plate tectonics carried the continent up into temperate and arctic regions. As a result, the tropical climate was replaced by a regime of seasonal changes, cooler temperatures, and less rainfall. Plate tectonics continues today; the Atlantic Ocean, for example, widens by a few centimeters each year.

Catastrophic Events May Have Caused the Worst Mass Extinctions

Geological data indicate that most mass extinction events coincided with periods of climate change. To many scientists, however, the rapidity of mass extinctions suggests that the slow process of climate change could not, by itself, be responsible for such large-scale disappearances of species. Perhaps more-sudden events also play a role. For example, catastrophic geological events, such as massive volcanic eruptions, could have had devastating effects. Geologists have found evidence

Media Activity
17.3 Plate Tectonics

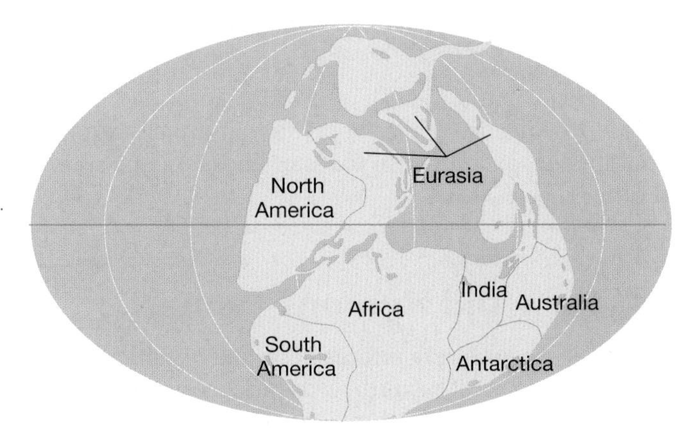

(a) 340 million years ago

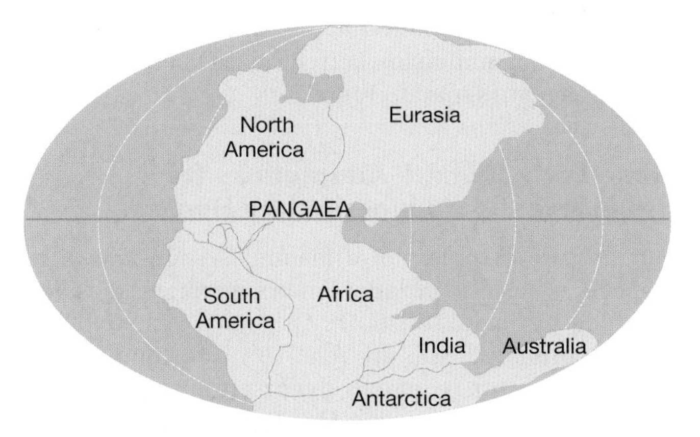

(b) 225 million years ago

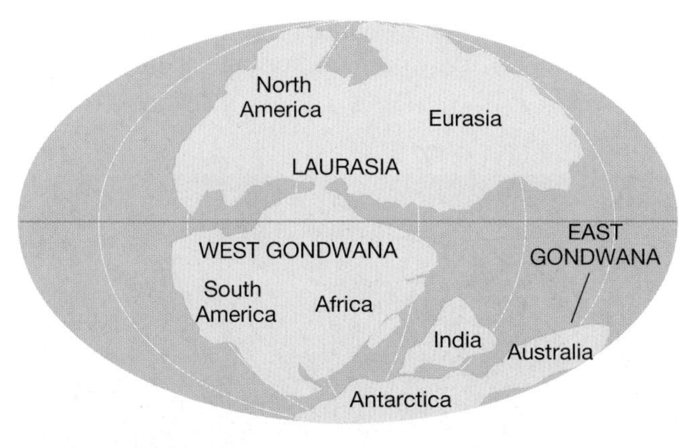

(c) 135 million years ago

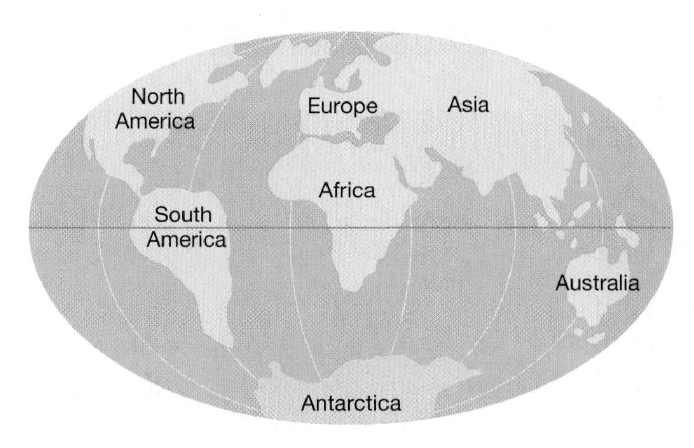

(d) Present

FIGURE 17-11 Continental drift from plate tectonics
The continents are passengers on plates moving on Earth's surface as a result of plate tectonics. **(a)** About 340 million years ago, much of what is now North America was positioned at the equator. **(b)** All the plates eventually fused together into one gigantic landmass, which geologists call Pangaea. **(c)** Gradually Pangaea broke up into Laurasia and Gondwanaland, which itself eventually broke up into West and East Gondwana. **(d)** Further plate motion eventually resulted in the modern positions of the continents.

of past volcanic eruptions so huge that they make the 1980 Mount St. Helens explosion in Washington look like a firecracker by comparison. Even such gigantic eruptions, however, would directly affect only a relatively small portion of Earth's surface.

The search for the causes of mass extinctions took a fascinating turn in the early 1980s, when Luis and Walter Alvarez proposed that the extinction event of 65 million years ago, which wiped out the dinosaurs and many other species, was caused by the impact of a huge meteorite. The Alvarezes' idea was met with great skepticism when it was first introduced, but geological research since that time has generated a great deal of evidence that a massive impact did indeed occur 65 million years ago. In fact, researchers have identified the Chicxulub crater, a 100-mile-wide crater buried beneath the Yucatan Peninsula of Mexico, as the impact site of a giant meteorite, 10 miles in diameter, that collided with Earth at the time that dinosaurs disappeared.

Could this immense meteorite strike have caused the mass extinction that coincided with it? No one knows for sure, but scientists suggest that such a massive impact would have thrown so much debris into the atmosphere that the entire planet would have been plunged into darkness for a period of years. With little light reaching the planet, temperatures would have dropped precipitously, and the photosynthetic capture of energy (upon which all life ultimately depends) would have declined drastically. This worldwide "impact winter" would have spelled doom for the dinosaurs and a host of other species.

17.6 How Did Humans Evolve?

Scientists are intensely interested in the origin and evolution of humans. The outline of human evolution that we present in this section is a synthesis of current thought on the subject, but it is speculative; the fossil evidence of human evolution is comparatively scarce. Paleontologists disagree about the interpretation of the fossil evidence, and many current ideas may have to be revised as new fossils are found.

Some Early Primate Adaptations for Life in Trees Were Inherited by Humans

Humans are members of a mammal group known as **primates**, which also includes lemurs, monkeys, and apes. The oldest primate fossils are 55 million years old, but because primate fossils are relatively rare compared to those of many other animals, the first primates probably arose considerably earlier but left no fossil record. Early primates probably fed on fruits and leaves and were adapted for life in the trees. Many modern primates retain the tree-dwelling lifestyle of their ancestors (Fig. 17-12). The common heritage of humans and

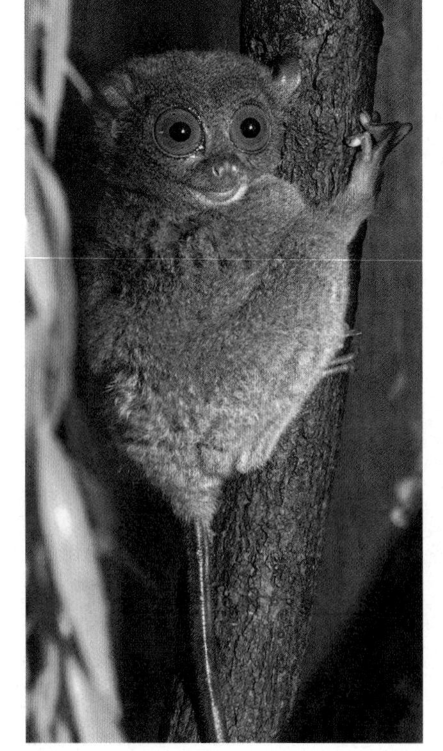
(a)

(b)

(c)

FIGURE 17-12 Representative primates
The **(a)** tarsier, **(b)** lemur, and **(c)** lion-tail macaque monkey all have relatively flat faces, with forward-looking eyes providing binocular vision. All also have color vision and grasping hands. These features, retained from the earliest primates, are shared by humans.

other primates is reflected in a set of physical characteristics that was present in the earliest primates and that persists in many modern primates, including humans.

Binocular Vision Provided Early Primates with Accurate Depth Perception

One of the earliest primate adaptations seems to have been large, forward-facing eyes (see Fig. 17-12). Jumping from branch to branch is risky business unless an animal can accurately judge where the next branch is located. Accurate depth perception was made possible by binocular vision, provided by forward-facing eyes with overlapping fields of view. Another key adaptation was color vision. We cannot, of course, tell if a fossil animal had color vision, but since modern primates have excellent color vision, it seems reasonable to assume that earlier primates did too. Many primates feed on fruit, and color vision helps the detection of ripe fruit among a bounty of green leaves.

Early Primates Had Grasping Hands

Early primates had long, grasping fingers that could wrap around and hold onto tree limbs. This adaptation to tree-dwelling was the basis for later evolution of human hands that could perform both a *precision grip* (used by modern humans for delicate maneuvers such as manipulating small objects, writing, and sewing) and a *power grip* (used for powerful actions, such as swinging a club and thrusting with a spear).

A Large Brain Facilitated Hand–Eye Coordination and Complex Social Interactions

Primates have brains that are larger, relative to their body size, than almost all other animals. No one really knows for certain which environmental forces favored the evolution of large brains. It seems reasonable, however, that controlling and coordinating rapid locomotion through trees, dexterous movements of the hands, and binocular, color vision would be facilitated by increased brain power. Most primates also have fairly complex social systems, which probably require relatively high intelligence. If sociality promoted increased survival and reproduction, then there would have been environmental pressures for the evolution of larger brains.

The Oldest Hominid Fossils Are from Africa

Based on comparisons of DNA from modern chimps, gorillas, and humans, researchers estimate that the **hominid** line (humans and their fossil relatives) diverged from the ape lineage sometime between 5 million and 8 million years ago. The fossil record, however, suggests that the split must have occurred at the early end of that range; paleontologists working in the African country of Chad in 2002 discovered fossils of a hominid, *Sahelanthropus tchadensis*, that lived more that 6 million years ago (Fig. 17-13). *Sahelanthropus* is clearly a hominid, as it shares several anatomical features with later members of the group. But because this oldest known member of our family also exhibits other features that are more characteristic of apes, it may represent a point on our family tree that is close to the split between apes and hominids.

In addition to *Sahelanthropus,* two other hominid species, *Ardipithecus ramidus* and *Orrorin tugenensis*, are known from fossils appearing in rocks that are between 4 million and 6 million years old. Our knowledge of these hominids is limited, as only a few specimens have been found so far, mostly in recent discoveries that typically include only small portions of skeletons. A more extensive record of early hominid evolution does not begin until about 4 million years ago. That date marks the beginning of the fossil record of the genus *Australopithecus* (Fig. 17-14), a group of African hominid species with brains larger than those of their prehominid forebears but still much smaller than those of modern humans.

The Earliest Australopithecines Could Stand and Walk Upright

Although the earliest australopithecines (as the various species of *Australopithecus* are collectively known) had legs that were shorter, relative to their height, than those of modern humans, their knee joints allowed

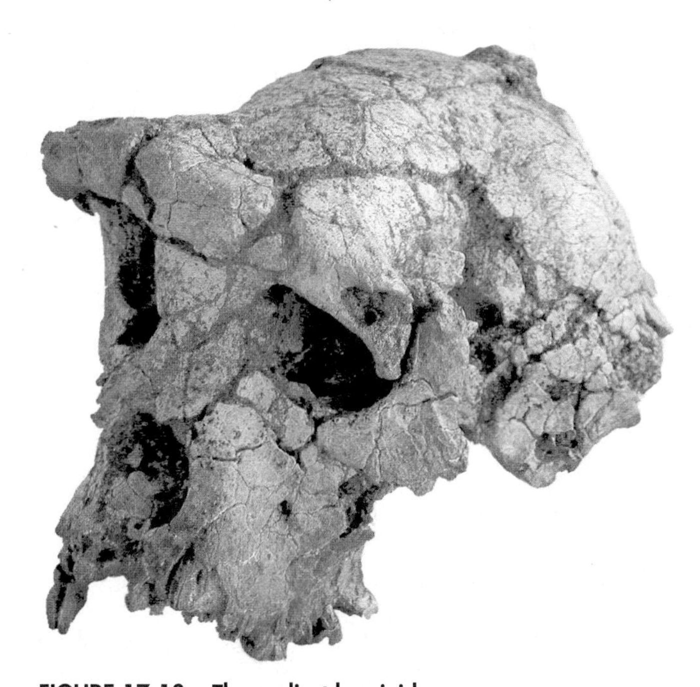

FIGURE 17-13 The earliest hominid
This nearly complete skull of *Sahelanthropus tchadensis*, which is more than 6 million years old, is the oldest hominid fossil yet found.

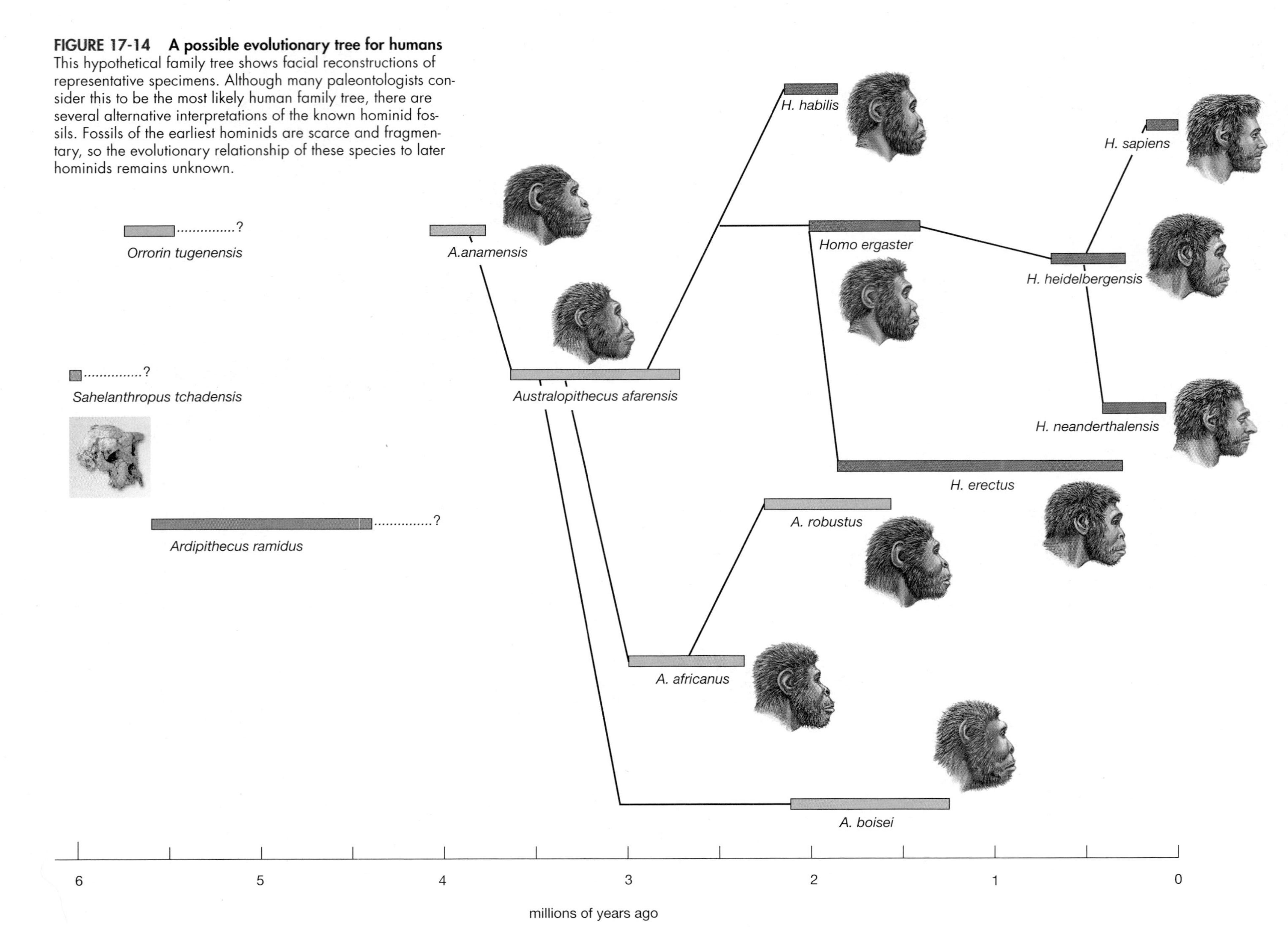

FIGURE 17-14 A possible evolutionary tree for humans This hypothetical family tree shows facial reconstructions of representative specimens. Although many paleontologists consider this to be the most likely human family tree, there are several alternative interpretations of the known hominid fossils. Fossils of the earliest hominids are scarce and fragmentary, so the evolutionary relationship of these species to later hominids remains unknown.

them to straighten their legs fully, permitting efficient bipedal (upright, two-legged) locomotion. Footprints almost 4 million years old, discovered in Tanzania by anthropologist Mary Leakey, show that even the earliest australopithecines could, and at least sometimes did, walk upright. Upright posture may have evolved even earlier. The discoverers of *Sahelanthropus* and *Orrorin* argue that the leg and foot bones of these earliest hominids have characteristics that indicate bipedal locomotion, but this conclusion will remain speculative until more complete skeletons of these species are found.

The reasons for the evolution of bipedal locomotion among the early hominids remain poorly understood. Perhaps hominids that could stand upright gained an advantage in gathering or carrying food in their forest habitat. Whatever its cause, the early evolution of upright posture was extremely important in the evolutionary history of hominids, because it freed their hands from use in walking. Later hominids were thus able to carry weapons, manipulate tools, and eventually achieve the cultural revolutions produced by modern *Homo sapiens*.

Several Species of *Australopithecus* Emerged in Africa

The oldest australopithecine species, represented by fossilized teeth, skull fragments, and arm bones, was first unearthed near an ancient lake bed in Kenya from sediments between 3.9 million and 4.1 million years old. It was named *Australopithecus anamensis* by its discoverers. The second most ancient australopithecine, called *Australopithecus afarensis*, was discovered in the Afar region of Ethiopia in 1974. Fossil remains of this species as old as 3.9 million years have been unearthed. Later, the *A. afarensis* line apparently gave rise to at least two distinct forms: small, omnivorous species such as *A. africanus* (which was similar to *A. afarensis* in size and eating habits), and larger, herbivorous species such as *A. robustus* and *A. boisei*. All of the australopithecine species had apparently gone extinct by 1.2 million years ago, but one of them first gave rise to a new branch of the hominid family tree, the genus *Homo*.

The Genus *Homo* Diverged from the Australopithecines 2.5 Million Years Ago

Hominids that are sufficiently similar to modern humans to be placed in the genus *Homo* first appear in African fossils that are about 2.5 million years old. Among the earliest African *Homo* fossils are *H. habilis* (see Fig. 17-14), a species whose bodies and brains were larger than those of the australopithecines, but which retained the apelike long arms and short legs of their australopithecine ancestors. In contrast, the skeletal anatomy of *H. ergaster*, a species whose fossils first appear 2 million years ago, has limb proportions more like those of modern humans. This species is believed by many paleoanthropologists (scientists who study human origins) to be on the evolutionary branch that led ultimately to our own species, *H. sapiens*. In this view, *H. ergaster* was the common ancestor of two distinct branches of hominids. The first branch led to *H. erectus*, the first hominid species to leave Africa. The second branch from *H. ergaster* ultimately led to *H. heidelbergensis*, some of which migrated to Europe and gave rise to the Neanderthals, *H. neanderthalensis*. Meanwhile, back in Africa, another branch split off from the *H. heidelbergensis* lineage. This branch ultimately became *H. sapiens*, modern humans.

The Evolution of *Homo* Was Accompanied by Advances in Tool Technology

Hominid evolution is closely tied to the development of tools, a hallmark of hominid behavior. The oldest tools discovered so far were found in 2.5-million-year-old East African rocks, dating them to the same time as the early emergence of the genus *Homo*. Early *Homo*, whose molar teeth were much smaller than those of its australopithecine ancestors, might have first used stone tools to break and crush tough foods that were hard to chew. Hominids constructed their earliest tools by striking rocks against one another to chip off fragments and leave a sharp edge behind. Over the next several hundred thousand years, tool-making techniques in Africa gradually became more advanced. By 1.7 million years ago, tools were more sophisticated, with flakes chipped symmetrically from both sides of a rock to form "biface" tools ranging from hand axes—used for cutting and chopping—to points probably used on spears (Fig. 17-15a,b). *Homo ergaster* and other bearers of these weapons presumably ate meat, probably acquired from both hunting and scavenging for the remains of prey killed by other predators. Biface tools were carried to Europe at least 600,000 years ago by migrating populations of *H. heidelbergensis*, and the Neanderthal descendants of these emigrants took stone-tool construction to new heights of skill and delicacy (Fig. 17-15c).

Neanderthals Had Large Brains and Excellent Tools

Neanderthals first appeared in the European fossil record about 150,000 years ago; by about 70,000 years ago, they had spread throughout Europe and western Asia. By 30,000 years ago, however, Neanderthals were extinct. Contrary to the popular image of a hulking, stoop-shouldered "caveman," Neanderthals were quite similar to modern humans in many ways. Although more heavily muscled, Neanderthals walked fully erect, were dexterous enough to manufacture finely crafted stone tools, and had brains that, on average, were slightly larger than those of modern humans. Many European Neanderthal fossils show heavy brow ridges and a broad, flat skull; others, particularly from areas around the eastern shores of the Mediterranean Sea, are somewhat more

FIGURE 17-15 Representative hominid tools
(a) *Homo habilis* produced only fairly crude chopping tools called *hand axes*, usually unchipped on one end to hold in the hand. **(b)** *Homo ergaster* manufactured much finer tools. The tools were typically sharp all the way around the stone; at least some of these blades were probably tied to spears rather than held in the hand. **(c)** Neanderthal tools were works of art, with extremely sharp edges made by flaking off tiny bits of stone. In comparing these weapons, note the progressive increase in the number of flakes taken off the blades and the corresponding decrease in flake size. Smaller, more numerous flakes produce a sharper blade and suggest either more insight into tool making, more patience, finer control of hand movements, or perhaps all three.

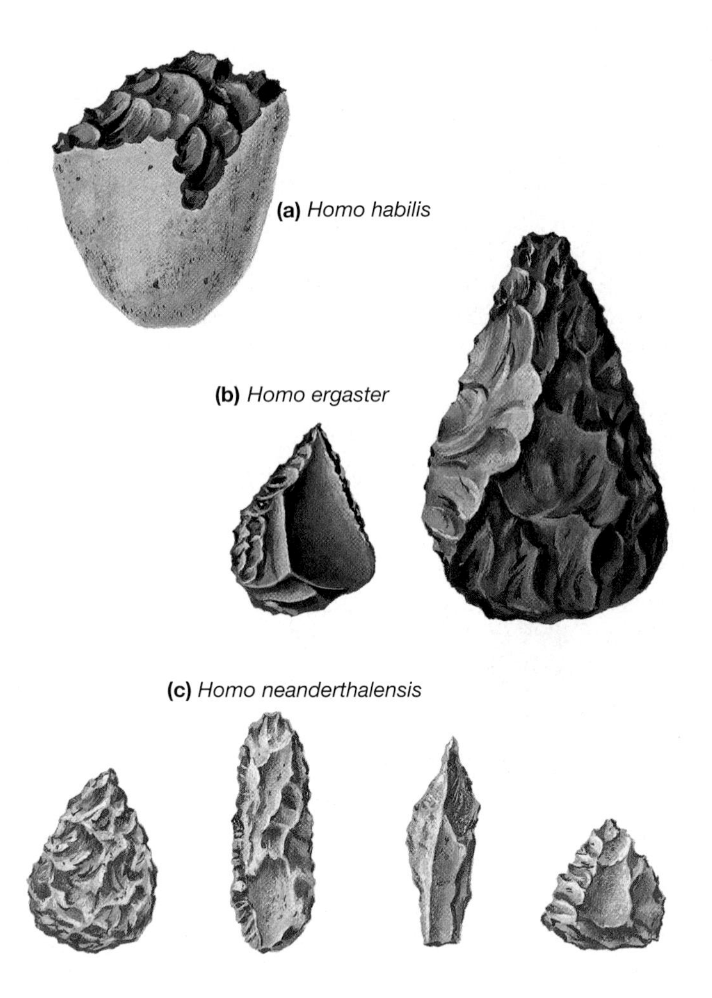

physically similar to *H. sapiens*. Despite the physical and technological similarities between *H. neanderthalensis* and *H. sapiens*, there is no solid archeological evidence that Neanderthals ever developed an advanced culture that included such characteristically human endeavors as art, music, and rituals. Some anthropologists argue that, because their skeletal anatomy shows that they were physically capable of making the sounds required for speech, Neanderthals might have acquired language. This interpretation of Neanderthal anatomy, however, is not unanimously accepted. In general, the available evidence of the Neanderthal way of life is limited and open to different interpretations, and anthropologists are engaged in a sometimes heated debate about how advanced Neanderthal culture became.

Though some anthropologists argue that Neanderthals were simply a variety of *H. sapiens*, most agree that Neanderthals were a separate species. Dramatic evidence in support of this hypothesis has come from two groups of researchers who were able to isolate and analyze ancient DNA from two different 30,000-year-old Neanderthal skeletons. The researchers determined the nucleotide sequence of a Neanderthal gene and compared it to the same gene from a large number of different modern humans from various parts of the world. The Neanderthal gene sequence was very different from that of modern humans, indicating that the evolutionary branch leading to Neanderthals diverged from the ancestral human line hundreds of thousands of years prior to the emergence of modern *H. sapiens*.

Modern Humans Emerged Only 150,000 Years Ago

The fossil record shows that anatomically modern humans appeared in Africa about 150,000 years ago. The location of these fossils suggests that *Homo sapiens* originated in Africa, but most of our knowledge about our own early history comes from European and Middle Eastern fossil *H. sapiens* collectively known as Cro-Magnons (after the district in France in which their remains were first discovered). Cro-Magnons appeared about 90,000 years ago, and had domed heads, smooth brows, and prominent chins (just like us). Their tools were precision instruments similar to the stone tools used until recently in many parts of the world. Behaviorally, Cro-Magnons seem to have been similar to, but more sophisticated than, Neanderthals. Artifacts from 30,000-year-old Cro-Magnon archeological sites include elegant bone flutes, graceful carved ivory sculptures, and evidence of elaborate burial ceremonies (Fig. 17-16). Perhaps the most remarkable accomplishment of

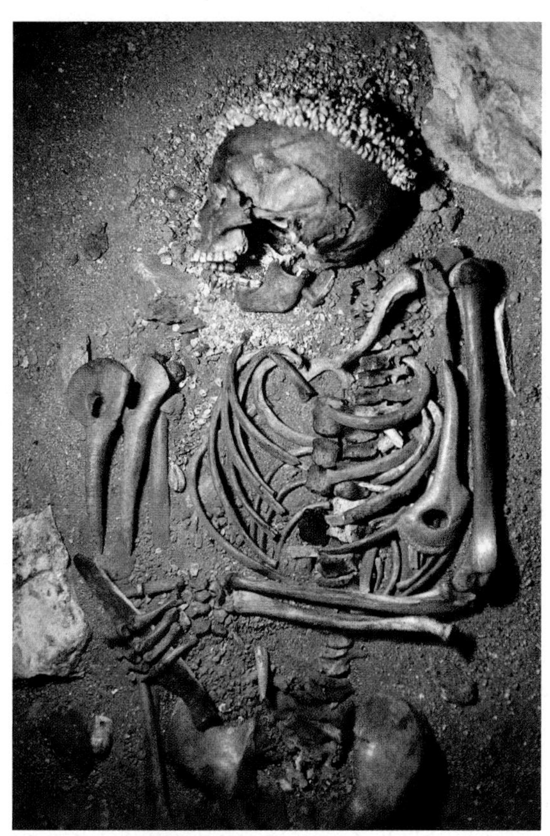

FIGURE 17-16 Paleolithic burial
This 24,000-year-old grave shows evidence that Cro-Magnon people ritualistically buried their dead. The body was covered with a dye known as red ocher, then buried wearing a head-dress made of snail shells and with a flint tool in its hand.

FIGURE 17-17 The art of Cro-Magnon people
Cave paintings by Cro-Magnons have been remarkably preserved by the relatively constant underground conditions of a cave in Lascaux, France.

Cro-Magnons is the magnificent art left in caves in places such as Altamira in Spain and Lascaux and Chauvet in France (Fig. 17-17). The oldest cave paintings so far found are more than 30,000 years old, and even the oldest ones make use of sophisticated artistic techniques. No one knows exactly why these paintings were made, but they attest to minds as fully human as our own.

Cro-Magnons and Neanderthals Lived Side by Side

Cro-Magnons coexisted with Neanderthals in Europe and the Middle East for perhaps as many as 50,000 years before the Neanderthals disappeared. Some researchers believe that Cro-Magnons interbred extensively with Neanderthals, so Neanderthals were essentially absorbed into the human genetic mainstream. Other scientists disagree, citing mounting evidence such as the fossil DNA described earlier, and suggest that later-arriving Cro-Magnons simply overran and displaced the less-well-adapted Neanderthals. Neither hypothesis, however, does a good job of explaining how the two kinds of hominids managed to occupy the same geographical areas for such a long time. The persistence in one area of two similar but distinct groups for tens of thousands of years seems inconsistent with both interbreeding *and* direct competition. Perhaps the competition between *H. neanderthalensis* and *H. sapiens* was indirect, so that the two species were able to coexist for a time in the same habitat, until the superior ability of *H. sapiens* to exploit the available resources slowly drove Neanderthals to extinction.

Several Waves of Hominids Emigrated from Africa

The human family tree is rooted in Africa, but hominids found their way out of Africa on numerous occasions. For example, *H. erectus* reached tropical Asia almost 2 million years ago and apparently thrived there, eventually spreading across Asia. Similarly, *H. heidelbergensis* made it to Europe at least 780,000 years ago. It is increasingly clear that the genus *Homo* made repeated long-distance emigrations, beginning as soon as sufficiently capable limb anatomy evolved. What is less clear is how all this wandering is related to the origin of modern *H. sapiens*. According to the "African replacement" hypothesis (the basis of the scenario outlined earlier), *H. sapiens* emerged in Africa and dispersed less than

150,000 years ago, spreading into the Near East, Europe, and Asia and replacing all other hominids (Fig. 17-18a). But some paleoanthropologists believe that populations of *H. sapiens* evolved simultaneously in many regions from the already widespread populations of *H. erectus*. According to this "multiregional origin" hypothesis, continued migrations and interbreeding among *H. erectus* populations in different regions of the world maintained them as a single species as they gradually evolved into *H. sapiens* (Fig. 17-18b). Although an increasing number of studies of modern human DNA support the African replacement model of the origin of our species, both hypotheses are consistent with the fossil record. Therefore, the question remains unsettled.

The Evolutionary Origin of Large Brains May Be Related to Meat Consumption

The main physical features that distinguish us from our closest relatives, the apes, are our upright posture and large, highly developed brains. As described earlier, upright posture arose very early in hominid evolution, and hominids walked upright for several million years before large-brained *Homo* species arose. What circumstances might have caused the evolution of increased brain size? Many explanations have been proposed, but little direct evidence is available; hypotheses about the evolutionary origins of large brains are necessarily speculative.

One proposed explanation for the origin of large brains suggests that they evolved in response to increasingly complex social interactions. In particular, fossil evidence suggests that, beginning about two million years ago, hominid social life began to include a new type of activity: the cooperative hunting of large game. The resulting access to significant amounts of meat must have fostered a need to develop methods for distributing this valuable, limited resource among group members. Some anthropologists hypothesize that the individuals best able to manage this social interaction would have been more successful at gaining a large share of meat and using their share to their own advantage. Perhaps this social management was best accomplished by individuals with larger, more powerful brains, and natural selection therefore favored such individuals. Observations of chimpanzee societies have shown that the distribution of group-hunted meat often involves intricate social interactions in which meat is used to form alliances, repay favors, gain access to sexual partners, placate rivals, and so on. Perhaps the mental skill required to plan, assess, and remember such interactions was the driving force behind the evolution of our large, clever brains.

The Evolutionary Origin of Human Behavior Is Highly Speculative

Even after the evolution of comparatively large brains in species such as *H. erectus*, more than a million years passed before the origin of modern humans and their extremely large brains. And even after the first appearance of modern *H. sapiens*, another 100,000 years passed before the appearance of any archeological evidence of the distinctively human characteristics that were made possible by a large brain: language, abstract thought, and advanced culture. The evolutionary origin of these human traits is another unresolved question, in part because direct evidence of the transition to advanced culture may never be found; early humans capable of language and symbolic thought would not necessarily have created artifacts that indicated these capabilities. We can uncover some clues by studying our ape relatives, which possess less-complex versions of many human behaviors and mental processes; their behavior might resemble that of ancestral hominids. Nonetheless, the late, seemingly rapid origin of advanced human culture remains a puzzle.

The Cultural Evolution of Humans Now Far Outpaces Biological Evolution

In recent millennia, human evolution has come to be dominated by *cultural evolution,* the transmission of learned behaviors from generation to generation. Our recent evolutionary success, for example, was engendered not so much by new physical adaptations as by a series of cultural and technological revolutions. The first such revolution was the development of tools, which began with the early hominids. Tools increased the efficiency with which food and shelter could be acquired and thus increased the number of individuals that could survive within a given ecosystem. About 10,000 years ago, human culture underwent a second revolution as people discovered how to grow crops and domesticate animals. This agricultural revolution dramatically increased the amount of food that could be extracted from the environment, and the human population surged, increasing from about 5 million at the dawn of agriculture to around 750 million by 1750. The subsequent industrial revolution gave rise to the modern economy and its attendant improvements in public health. Longer lives and lower infant mortality led to truly explosive population growth, and today we are more than 6 billion strong.

Human cultural evolution and the accompanying increases in human population have had profound effects on the continuing biological evolution of other lifeforms. Our agile hands and minds have transformed much of Earth's terrestrial and aquatic habitats. Humans have become the single overwhelming agent of natural selection. In the words of the late evolutionary biologist Stephen Jay Gould, "We have become, by the power of a glorious evolutionary accident called intelligence, the stewards of life's continuity on Earth. We did not ask for this role, but we cannot abjure it. We may not be suited for it, but here we are."

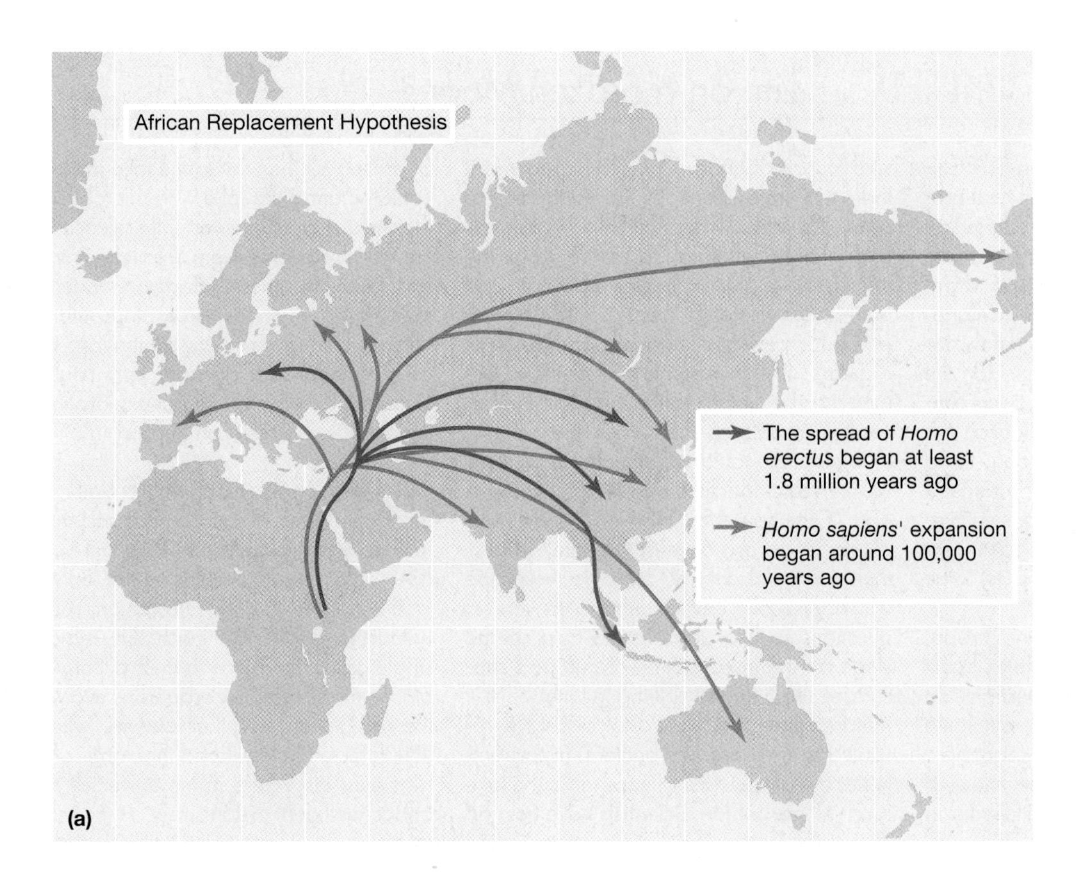

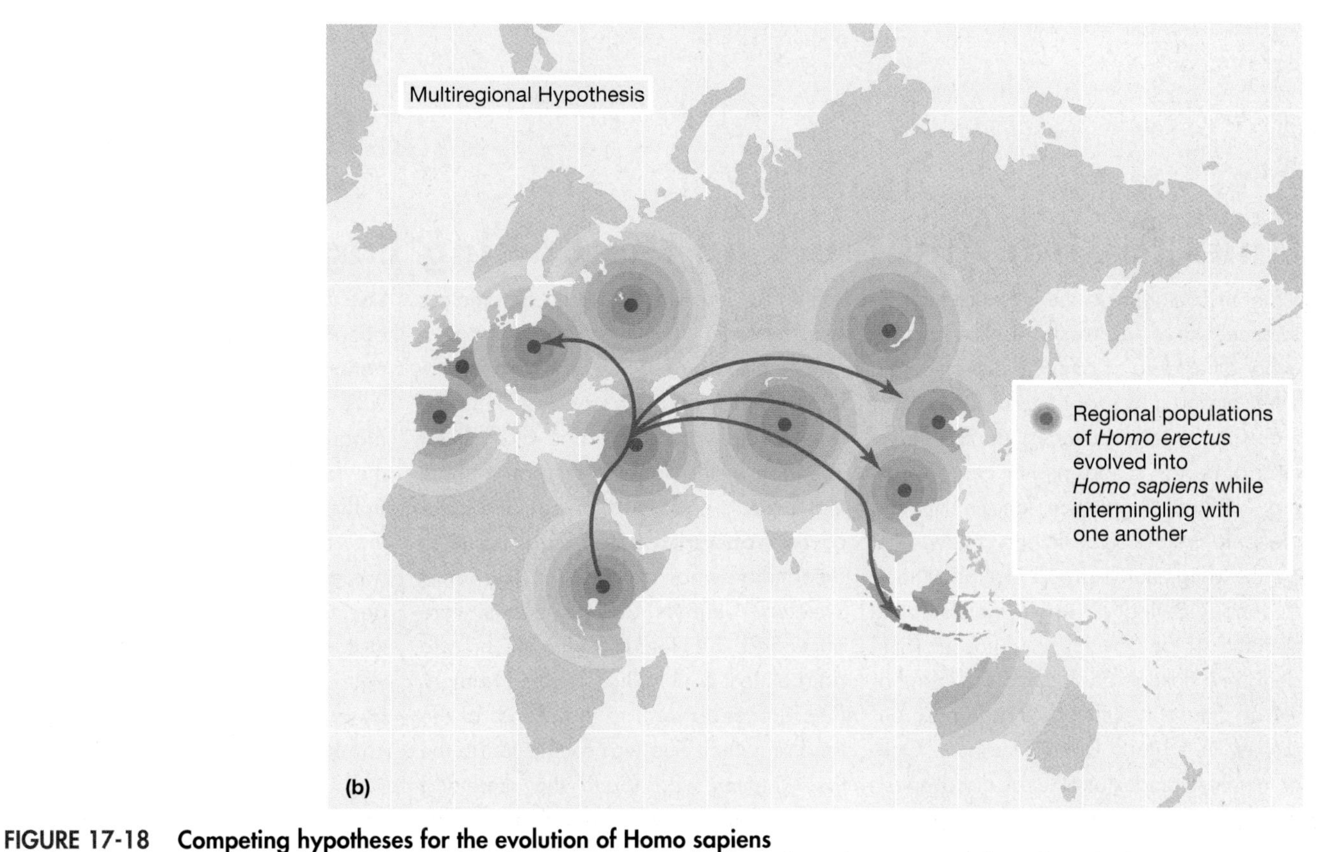

FIGURE 17-18 Competing hypotheses for the evolution of Homo sapiens
(a) The "African replacement" hypothesis suggests that *H. sapiens* evolved in Africa, then migrated throughout the Near East, Europe, and Asia, displacing the other hominid species that were present in those regions. **(b)** The "multiregional" hypothesis suggests that populations of *H. sapiens* evolved in many regions simultaneously from the already widespread populations of *H. erectus*. **QUESTION** Paleontologists recently discovered fossil hominids with features characteristic of modern humans in 160,000-year-old sediments in Africa. Which hypothesis does this new evidence support?

CASE STUDY REVISITED Life on a Frozen Moon?

Is there really liquid water on Europa? Calculations estimating the amount of heat produced by the gravitational forces exerted on Europa by Jupiter's other moons indicate that the moon's interior should be warm enough to sustain a liquid ocean. Unfortunately, however, we may never know for sure. After years of delay, plans for the *Europa Orbiter* spacecraft have been cancelled. NASA had planned to launch the spacecraft into orbit around Europa, using it to make subsurface radar measurements that would determine whether liquid is present. Unfortunately, due to budget constraints and changing priorities, the ambitious plan was scrapped.

If liquid water *is* present on Europa, has life arisen in it? If so, what form has it assumed? Answers to these questions do not appear to be close at hand, but some tantalizing clues might be found right here on Earth. The potential source of such clues is Lake Vostok, a huge lake (similar in size to Lake Ontario) that is buried more than two miles beneath the Antarctic ice pack. The cold waters of Lake Vostok are under extremely high pressure from the immense weight of the ice above it and have been isolated from Earth's atmosphere for perhaps hundreds of thousands of years; the environment of these waters may thus be similar to the environment beneath Europa's ice sheet.

Biologists would very much like to know if Lake Vostok harbors any living organisms and, if so, how they survive. If, for example, bacteria are alive in the lake, what is their source of energy? How do they cope with their cold environment? Answers to questions such as these could help set the stage for the investigation of Europa's subsurface seas. Recent drilling to within a few hundred feet of Lake Vostok's surface revealed some preserved bacteria (apparently not still alive), raising hopes that the lake itself will harbor life. Scientists have held off of drilling all the way to the lake for fear of contaminating the lake with microbes from the surface or of inadvertently releasing unknown organisms from the lake into the surface environment. Efforts are under way to develop methods to explore the lake without risking cross-contamination. Once such methods are devised, they could be adapted for use in future exploration of similar environments on Europa.

Consider This: In addition to water, what other conditions would be necessary for life to have arisen on Europa? If we were somehow able to explore the entire universe, would we find that life is common, rare, or unique to Earth? On Earth, life remained single-celled for more than 2 billion years before more complex organisms evolved. If life *has* arisen on other planets, what do you think is the likelihood that large organisms exist elsewhere in the universe? What about intelligent organisms?

Links to Life: Amateurs on the Cutting Edge

Want to contribute to the advancement of science? Hunt for fossils. Unlike almost every other area of science, paleontology offers opportunities for nonprofessionals to make important discoveries. You will never hear of a high school student characterizing a genetic sequence for the Human Genome Project or designing a new anticancer drug, but a group of high school students did find a rare complete fossil skeleton of an *Allosaurus* dinosaur. Students in a college geology course found a set of dinosaur skeletons that is now on display in a North Dakota museum. One of the most famous of all fossil discoveries, the complete *Tyrannosaurus* skeleton known as Sue, was found by an amateur fossil hunter.

The discovery of Sue was a spectacular example of an amateur's contribution to science, but the find also called attention to the sometimes seamy financial aspects of fossil hunting. Sue's discovery sparked a fierce legal battle to determine who owned the dinosaur skeleton. Ultimately, Sue's bones were auctioned; the Field Museum in Chicago paid almost $13 million for the skeleton. (The proceeds went to the owner of the land on which Sue was found, not to the fossil hunter who found the skeleton.) Clearly, fossils have monetary as well as scientific value. One unfortunate consequence of fossils' monetary value is that national parks and other public and private lands around the world are plagued by fossil poachers. In addition, the fossil trade keeps many specimens in the hands of private collectors, where the valuable information they contain is not accessible to scientists. So, if you decide to try fossil hunting, make sure that you have permission to explore your collecting area. A course offered by a college or natural history museum can help you get started.

Summary of Key Concepts

17.1 How Did Life Begin?

Before life arose, lightning, ultraviolet light, and heat formed organic molecules from water and the components of primordial Earth's atmosphere. These molecules probably included nucleic acids, amino acids, short proteins, and lipids. By chance, some molecules of RNA may have had enzymatic properties, catalyzing the assembly of copies of themselves from nucleotides in Earth's waters. These may have been the forerunners of life. Protein-lipid microspheres enclosing these ribozymes may have formed the first protocells.

17.2 What Were the Earliest Organisms Like?

The oldest fossils, about 3.5 billion years old, are of prokaryotic cells that fed by absorbing organic molecules that had been synthesized in the environment. Because there was no free oxygen in the atmosphere, their energy metabolism must have been anaerobic. As the cells multiplied, they depleted the organic molecules that had been formed by prebiotic synthesis. Some cells developed the ability to synthesize their own food molecules by using simple inorganic molecules and the energy of sunlight. These earliest photosynthetic cells were probably ancestors of today's cyanobacteria.

Photosynthesis releases oxygen as a by-product; by about 2.2 billion years ago, significant amounts of free oxygen were accumulating in the atmosphere. Aerobic metabolism, which generates more cellular energy than does anaerobic metabolism, probably arose about this time.

Eukaryotic cells had evolved by about 1.7 billion years ago. The first eukaryotic cells probably arose as symbiotic associations between predatory prokaryotic cells and other bacteria. Mitochondria may have evolved from aerobic bacteria engulfed by predatory cells. Similarly, chloroplasts may have evolved from photosynthetic cyanobacteria.

17.3 What Were the Earliest Multicellular Organisms Like?

Multicellular organisms evolved from eukaryotic cells and first appeared, in the seas, about 1 billion years ago. Multicellularity offers several advantages, including greater size. In plants, increased size offered some protection from predation. Specialization of cells allowed plants to anchor themselves in the nutrient-rich, well-lit waters of the shore. For animals, multicellularity allowed more-efficient predation and more-effective escape from predators. These in turn provided environmental pressures for faster locomotion, improved senses, and greater intelligence.

17.4 How Did Life Invade the Land?

The first land organisms were probably algae. The first multicellular land plants appeared about 400 million years ago. Although life on land required special adaptations for support of the body, reproduction, and the acquisition, distribution, and retention of water, the land also offered abundant sunlight and protection from aquatic herbivores. Soon after land plants evolved, arthropods invaded the land. Absence of predators and abundant land plants for food probably facilitated the invasion of the land by animals.

The earliest land vertebrates evolved from lobefin fishes, which had leglike fins and a primitive lung. A group of lobefins evolved into the amphibians about 350 million years ago. Reptiles evolved from amphibians, with several further adaptations for life on land: waterproof eggs that could be laid on land, waterproof skin, and better lungs. Birds and mammals evolved independently from separate groups of reptiles. A major advance in the evolution of both birds and mammals was body insulation: feathers and hair.

17.5 What Role Has Extinction Played in the History of Life?

The history of life has been characterized by the constant turnover of species as some species go extinct and are replaced by new ones. Mass extinctions, in which large numbers of species disappear within a relatively short time, have occurred periodically. Mass extinctions were probably caused by some combination of climate changes and catastrophic events, such as volcanic eruptions and meteorite impacts.

17.6 How Did Humans Evolve?

One group of mammals evolved into the tree-dwelling primates, which were the ancestors of apes and humans. The human and ape lines diverged 7 million to 8 million years ago; the oldest known homind fossils are between 6 million and 7 million years old and were found in Africa. The first well-known hominid line, the australopithecines, arose in Africa about 4 million years ago. These hominids walked erect, had larger brains than did their forebears, and fashioned primitive tools. One group of australopithecines gave rise to a line of hominids in the genus *Homo*, which in turn gave rise to modern humans.

Key Terms

amphibian *p. 328*
arthropod *p. 328*
conifer *p. 328*
endosymbiont hypothesis *p. 324*
eukaryote *p. 324*
exoskeleton *p. 328*
hominid *p. 333*
lobefin *p. 328*
mammal *p. 330*
mass extinction *p. 330*
microsphere *p. 320*
plate tectonics *p. 331*
primate *p. 332*
prokaryote *p. 321*
protocell *p. 320*
reptile *p. 329*
ribozyme *p. 320*
spontaneous generation *p. 318*

Thinking Through the Concepts

To take a multiple-choice quiz with feedback on the contents of this chapter, visit http://www.prenhall.com/audesirk7. *Log in to the Web site selected by your instructor and navigate to the Self Test section for this chapter.*

? Review Questions

1. What is the evidence that life might have originated from nonliving matter on early Earth? What kind of evidence would you like to see before you would accept this hypothesis?
2. If the first cells with aerobic metabolism were so much more efficient at producing energy, why didn't they drive cells with only anaerobic metabolism to extinction?
3. Explain the endosymbiont hypothesis for the origin of chloroplasts and mitochondria.
4. Name two advantages of multicellularity for plants and two advantages for animals.
5. What advantages and disadvantages would terrestrial existence have had for the first plants to invade the land? For the first land animals?
6. Outline the major adaptations that emerged during the evolution of vertebrates, from fishes to amphibians to reptiles to birds and mammals. Explain how these adaptations increased the fitness of the various groups for life on land.
7. Outline the evolution of humans from early primates. Include in your discussion such features as binocular vision, grasping hands, bipedal locomotion, social living, tool making, and brain expansion.

Applying the Concepts

1. What is cultural evolution? Is cultural evolution more or less rapid than biological evolution? Why?
2. Do you think that studying our ancestors can shed light on the behavior of modern humans? Why?
3. A biologist would probably answer the age-old question, "What is life?" by saying, "The ability to self-replicate." Do you agree with this definition? If so, why? If not, how would you define life in biological terms?
4. Traditional definitions of humans have emphasized "the uniqueness of humans" because we possess language and use tools. But most animals can communicate with other individuals in sophisticated ways, and many vertebrates use tools to accomplish tasks. Pretend that you are a biologist from Mars, and write a taxonomic description of the species *Homo sapiens*.
5. Extinctions have occurred throughout the history of life on Earth. Why should we care if humans are causing a mass extinction event now?
6. The "African replacement" and "multiregional origin" hypotheses of the evolution of *Homo sapiens* make contrasting predictions about the extent and nature of genetic divergence among human races. One predicts that races are old and highly diverged genetically; the other predicts that races are young and little diverged genetically. What data would help you determine which hypothesis is closer to the truth?
7. In biological terms, what do you think was the most significant event in the history of life? Explain your answer.

For More Information

de Duve, C. "The Birth of Complex Cells." *Scientific American*, April 1996. Narrative describing the origin of complex eukaryotic cells by repeated instances of endosymbiosis.

Fenchel, T., and Finlay, B. J. "The Evolution of Life Without Oxygen." *American Scientist*, January/February 1994. Clues to the origin of the first eukaryotic cells are provided by symbiotic relationships of organisms in oxygen-free environments.

Fry, I. *The Emergence of Life on Earth: A Historical and Scientific Overview*. Brunswick, NJ: Rutgers University Press, 2000. A thorough review of research and hypotheses concerning the origin of life.

Hay, R. L., and Leakey, M. D. "The Fossil Footprints of Laetoli." *Scientific American*, February 1982. The footprints of a hominid family were discovered by Hay and Leakey in volcanic ash 3.5 million years old.

Maynard Smith, J., and Szathmary. E. *The Origins of Life: From the Birth of Life to the Origin of Language*. New York: Oxford University Press, 1999. A thought-provoking review of the major shifts that have occurred over the 3.5-billion-year history of life.

Monastersky, R. "The Rise of Life on Earth." *National Geographic*, March 1998. A beautifully illustrated and engaging description of current ideas and evidence of how life arose.

Pappalardo, R., Head, J., and Greeley, R. "The Hidden Ocean of Europa." *Scientific American,* October 1999. A summary of the evidence that Europa's icy surface conceals a liquid ocean; illustrated with photos from the *Galileo* mission.

Tattersall, I. "Once We Were Not Alone" *Scientific American*, January 2000. A overview of the evolutionary history that led to modern *Homo sapiens*, with illustrations of some of the hominids that preceded us.

Tattersall, I. "How We Came to Be Human" *Scientific American*, December 2001. A discussion of the evolutionary origin of language, art, symbolic representation, and other distinctictive traits of modern humans.

Ward, P. D. *The End of Evolution: On Mass Extinctions and the Preservation of Biodiversity.* New York: Bantam Books, 1994. An engaging first-person account of a paleontologist's investigation of the causes of mass extinction.

Media Activities

To access a Media Activity visit http://www.prenhall.com/audesirk7. *Log in to the Web site selected by your instructor, navigate to this chapter, and select the appropriate Media Activity number.*

17.1 Endosymbiosis

Estimated time: 5 minutes

This activity demonstrates how cellular symbiosis may have resulted in the complex organelles of mitochondria and chloroplasts.

17.2 Evolutionary Timescales

Estimated time: 5 minutes

In this activity, you will explore the evolutionary history of three important types of organisms: land plants, terrestrial vertebrate animals, and primates.

17.3 Plate Tectonics

Estimated time: 5 minutes

This activity demonstrates how the continents have moved over the course of time and points out some important effects of continental drift on the distribution and evolutionary histories of organisms.

17.4 Web Investigation: Life on a Frozen Moon?

Estimated time: 10 minutes

Since water is one of the basic elements of life as we know it, many astrobiologists believe that the best way to detect life in space is to look for liquid water. Recent observations suggest that there may be liquid water on Europa. This exercise will take a closer look at the fourth largest satellite of Jupiter.

CHAPTER

18 Systematics: Seeking Order Amidst Diversity

Biologists studying the evolutionary history of type 1 human immunodeficiency virus (HIV-1) discovered that the virus, which causes AIDS, probably originated in chimpanzees.

AT A GLANCE

CASE STUDY Origin of a Killer

One of the world's most frightening diseases is also one of its most mysterious. Acquired immune deficiency syndrome (AIDS) appeared seemingly out of nowhere, and when it was first recognized in the early 1980s, no one knew what caused it or where it came from. Scientists raced to solve the mystery and, within a few years, had identified the infectious agent that caused AIDS: human immunodeficiency virus (HIV). Once HIV had been identified, researchers turned their attention to the question of its origin.

Finding the source of HIV required an evolutionary approach. To ask, "Where did HIV come from?" is really to ask, "What kind of virus was the ancestor of HIV?" Biologists who examine questions of ancestry are known as *systematists*. Systematists strive to categorize organisms according to their evolutionary history, building classifications that accurately reflect the structure of the tree of life. When a systematist concludes that two species are closely related, it means that the two species share a recent common ancestor from which both species evolved.

The systematists who explored the ancestry of HIV discovered that its closest relatives are found not among other viruses that infect humans, but among those that infect monkeys and apes. In fact, the latest research on HIV's evolutionary history has concluded that the closest relative of HIV-1 (the type of HIV that is most responsible for the worldwide AIDS epidemic) is a virus strain that infects a particular chimp subspecies that inhabits a limited range in West Africa. Therefore, the ancestor of the virus that we now know as HIV-1 did not evolve from a preexisting human virus but must have somehow jumped from West African chimpanzees to humans.

Table 18-1 Classification of Selected Organisms, Reflecting Their Degree of Relatedness*

	Human	Chimpanzee	Wolf	Fruit Fly	Sequoia Tree	Sunflower
Domain	**Eukarya**	**Eukarya**	**Eukarya**	**Eukarya**	**Eukarya**	**Eukarya**
Kingdom	**Animalia**	**Animalia**	**Animalia**	**Animalia**	**Plantae**	**Plantae**
Phylum	**Chordata**	**Chordata**	**Chordata**	Arthropoda	Coniferophyta	Anthophyta
Class	**Mammalia**	**Mammalia**	**Mammalia**	Insecta	Coniferosida	Dicotyledoneae
Order	**Primates**	**Primates**	Carnivora	Diptera	Coniferales	Asterales
Family	Hominidae	Pongidae	Canidae	Drosophilidae	Taxodiaceae	Asteraceae
Genus	*Homo*	*Pan*	*Canis*	*Drosophila*	*Sequoiadendron*	*Helianthus*
Species	*sapiens*	*troglodytes*	*lupus*	*melanogaster*	*giganteum*	*annuus*

*Boldface categories are those that are shared by more than one of the organisms classified.
Genus and species names are always italicized or underlined.

Media Activity
18.1 Taxonomic Classification

18.1 How Are Organisms Named and Classified?

Systematics is the science of reconstructing **phylogeny**, or evolutionary history. As part of their effort to reveal the tree of life, systematists name organisms and place them into categories on the basis of their evolutionary relationships. There are eight major categories: **domain**, **kingdom**, **phylum**, **class**, **order**, **family**, **genus**, and **species**. These categories form a nested hierarchy in which each level includes all of the other levels below it. Each domain contains a number of kingdoms, each kingdom contains a number of phyla, each phylum includes a number of classes, each class includes a number of orders, and so on. As we move down the hierarchy, smaller and smaller groups are included. Each category is increasingly narrow and specifies groups whose common ancestor is increasingly recent. Table 18-1 describes some examples of classifications of specific organisms.

The **scientific name** of an organism is formed from the two smallest categories, the genus and the species. Each genus includes a group of very closely related species, and each species within a genus includes populations of organisms that can potentially interbreed under natural conditions. Thus, the genus *Sialia* (bluebirds) includes the eastern bluebird (*Sialia sialis*), the western bluebird (*Sialia mexicana*), and the mountain bluebird (*Sialia currucoides*)—very similar birds that normally do not interbreed (Fig. 18-1).

Each two-part scientific name is unique, so referring to an organism by its scientific name rules out any chance of ambiguity or confusion. For example, the bird *Gavia immer* is commonly known in North America as the common loon, in Great Britain as the northern diver, and by still other names in non-English-speaking countries. But the Latin scientific name *Gavia immer* is recognized by biologists worldwide, overcoming language barriers and allowing precise communication.

FIGURE 18-1 Three species of bluebird
Despite their obvious similarity, these three species of bluebird—from left to right, the eastern bluebird *(Sialia sialis)*, the western bluebird *(Sialia mexicana)*, and the mountain bluebird *(Sialia currucoides)*—remain distinct because they do not interbreed.

Note that, by convention, scientific names are always underlined or *italicized*. The first letter of the genus name is always capitalized, and the first letter of the species name is always lowercase. The species name is never used alone but is always paired with its genus name.

Classification Originated as a Hierarchy of Categories

Aristotle (384–322 B.C.) was among the first to attempt to formulate a logical, standardized language for naming living things. Based on characteristics such as structural complexity, behavior, and degree of development at birth, he classified about 500 organisms into 11 categories. Aristotle's categories formed a hierarchical structure, with each category more inclusive than the one beneath it, a concept that is still used today.

Building on this foundation more than 2000 years later, Swedish naturalist Carl von Linné (1707–1778)—who called himself Carolus Linnaeus, a latinized version of his given name—laid the groundwork for the modern classification system. He placed each organism into a series of hierarchically arranged categories on the basis of its resemblance to other life-forms, and he also introduced the scientific name composed of genus and species. Nearly 100 years later, Charles Darwin (1809–1882) published *On the Origin of Species*, which demonstrated that all organisms are connected by common ancestry. Biologists then began to recognize that the categories ought to reflect the pattern of evolutionary relatedness among organisms. The more categories two organisms share, the closer their evolutionary relationship.

Systematists Identify Features That Reveal Evolutionary Relationships

Systematists seek to reconstruct the tree of life, but they must do so without much direct knowledge of evolutionary history. Because they can't see into the past, they must infer the past as best they can, based on similarities among living organisms. Not just any similarity will do, however. Some observed similarities stem from convergent evolution in organisms that are not closely related; such similarities are not useful for inferring evolutionary history. Instead, systematists value the similarities that exist because two kinds of organisms both inherited the same characteristic from a common ancestor. Therefore, an important task is to distinguish informative similarities caused by common ancestry from similarities that result from convergent evolution. In the search for informative similarities, biologists look at many different kinds of characteristics.

Anatomy Plays a Key Role in Systematics

Historically, the most important and useful distinguishing characteristics have been anatomical. Systematists look carefully at similarities in external body structure (see Fig. 18-1) and in internal structures, such as skeletons and muscles. For example, homologous structures, such as the finger bones of dolphins, bats, seals, and humans (see Fig. 14-7), provide evidence of a common ancestor. To detect relationships between more closely related species, biologists may use microscopes to discern finer details—the number and shape of the "teeth" on the tonguelike radula of a snail, the shape and position of the bristles on a marine worm, or the external structure of pollen grains of a flowering plant (Fig. 18-2).

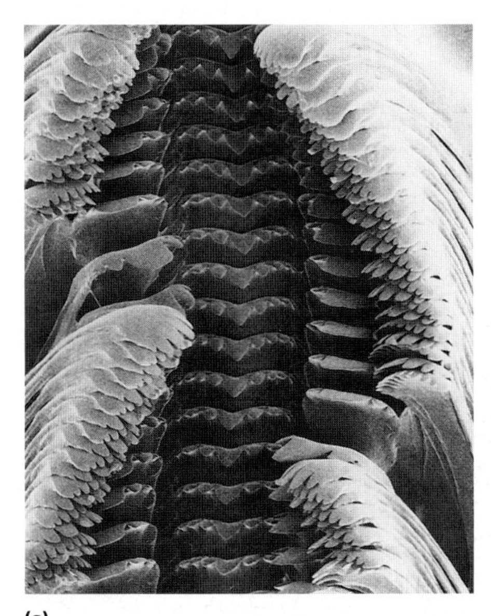
(a)

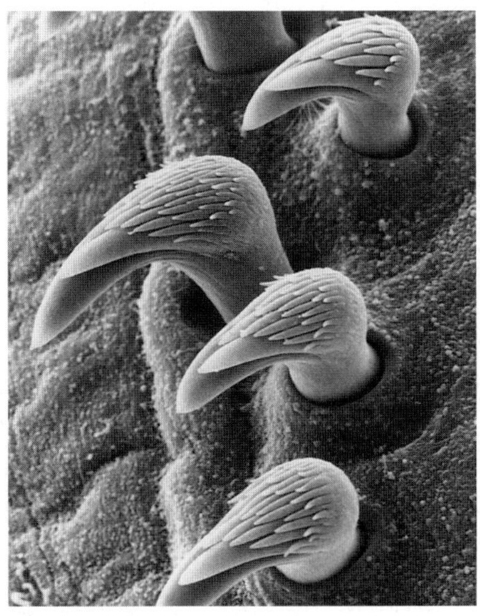
(b)

(c)

FIGURE 18-2 Microscopic structures may be used to classify organisms
(a) The "teeth" on a snail's tonguelike radula (a structure used in feeding), **(b)** the bristles on a marine worm, and **(c)** the shape and surface features of pollen grains are characteristics that are potentially useful in classification. Such finely detailed structures can reveal similarities between species that are not apparent in larger and more obvious structures.

Molecular Similarities Are Also Useful for Reconstructing Phylogeny

The anatomical characteristics shared by related organisms are expressions of underlying genetic similarities, so it stands to reason that evolutionary relationships among species must also be reflected in genetic similarities. Of course, direct genetic comparisons were not possible for most of the history of biology. In the past two decades, however, advances in the techniques of molecular genetics have revolutionized studies of evolutionary relationships. For the first time, the nucleotide sequence of DNA (that is, genotype), rather than phenotypic features such as appearance or behavior, can be used to investigate relatedness among different types of organisms. Genetic relatedness among organisms can also be evaluated by examining the structure of their chromosomes. Among the findings derived from this technique is that the chromosomes of chimpanzees and humans are extremely similar, showing that these two species are very closely related (Fig. 18-3). Some of the key methods and findings of genetic analysis are explored in "Scientific Inquiry: Molecular Genetics Reveals Evolutionary Relationships."

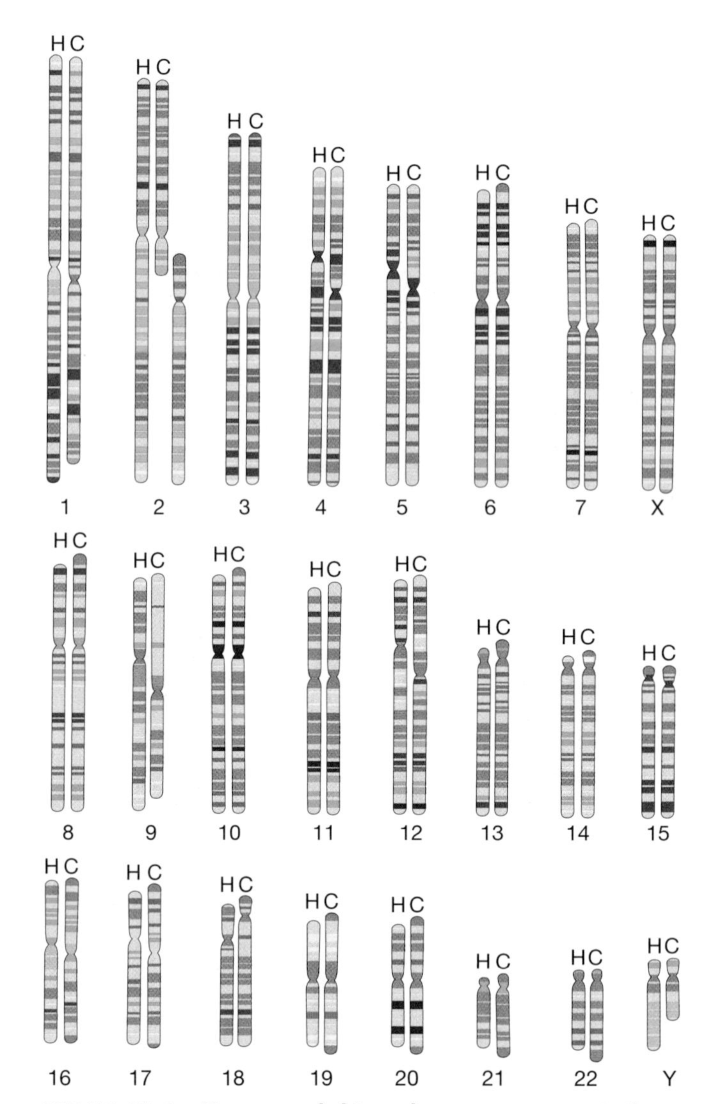

FIGURE 18-3 Human and chimp chromosomes are similar Chromosomes from different species can be compared by means of banding patterns that are revealed by staining. The comparison illustrated here, between human chromosomes (left member of each pair; H) and chimpanzee chromosomes (C), reveals that the two species are genetically very similar. In fact, it has been estimated that 99% of the two genomes is identical. The numbering system shown is that used for human chromosomes; note that human chromosome 2 corresponds to a combination of two chimp chromosomes.

18.2 What Are the Domains of Life?

Before 1970, all forms of life were classified into two kingdoms: Animalia and Plantae. All bacteria, fungi, and photosynthetic eukaryotes were considered to be plants, and all other organisms were classified as animals. As scientists learned more about fungi and microorganisms, however, it became apparent that the two-kingdom system oversimplified evolutionary history. To help rectify this problem, Robert H. Whittaker proposed a five-kingdom classification in 1969 that was eventually adopted by most biologists.

Media Activity 18.2 Tree of Life

The Five-Kingdom System Improved Classification

Whittaker's five-kingdom system placed all prokaryotic organisms into a single kingdom and divided the eukaryotes into four kingdoms. The designation of a separate kingdom (called Monera) for the prokaryotes reflected growing recognition that the evolutionary pathway of these tiny, single-celled organisms had diverged from that of the eukaryotes early in the history of life. Among the eukaryotes, the five-kingdom system recognized three kingdoms of multicellular organisms (Plantae, Fungi, and Animalia), and placed all of the remaining, mostly single-celled eukaryotes in a single kingdom (Protista).

Because it more accurately reflected current understanding of evolutionary history, the five-kingdom system was an improvement over the old two-kingdom system. As understanding continued to grow, however, our view of life's most fundamental categories needed yet another revision. The pioneering work of microbiologist Carl Woese has shown that biologists had overlooked a fundamental event in the early history of life, one that demands a new and more-accurate classification of life.

A Three-Domain System More Accurately Reflects Life's History

Woese and other biologists interested in the evolutionary history of microorganisms have studied the biochemistry of prokaryotic organisms. These re-

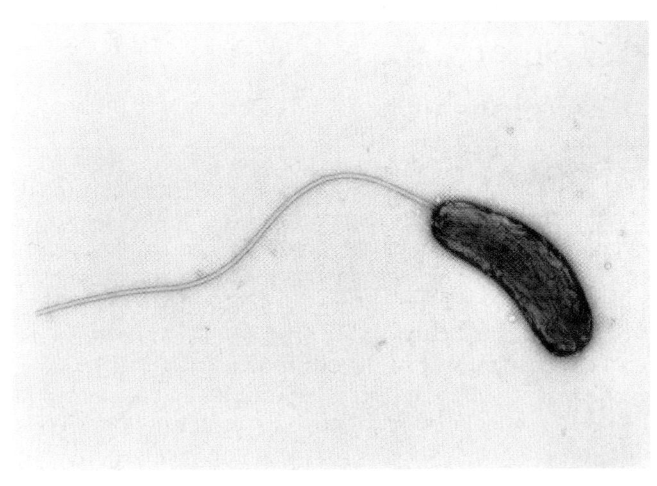

(a)

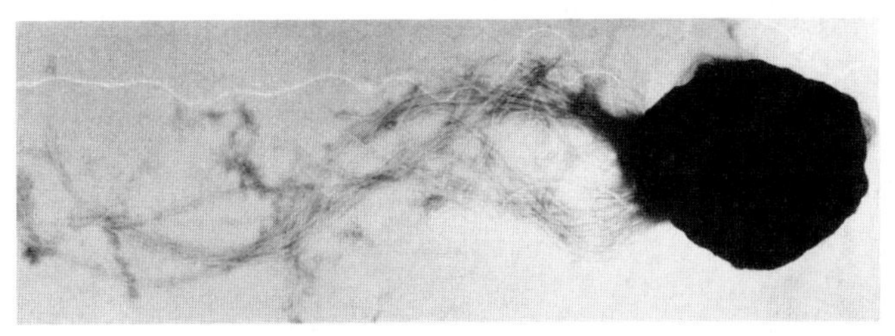

(b)

FIGURE 18-4 Two domains of prokaryotic organisms
Although similar in appearance, **(a)** *Vibrio cholerae* and **(b)** *Methanococcus jannaschi* are less closely related than a mushroom and an elephant. *Vibrio* is in the domain Bacteria, and *Methanococcus* is in Archaea.

searchers, focusing on nucleotide sequences of the RNA that is found in the organisms' ribosomes, discovered that the supposed kingdom Monera actually included two very different kinds of organisms. Woese has dubbed these two groups the Bacteria and the Archaea (Fig. 18-4).

Despite superficial similarities in their appearance under the microscope, the Bacteria and the Archaea are radically different. These two groups are no more closely related to one another than either is to any eukaryote. The tree of life split into three parts very early in the history of life, long before the appearance of plants, animals, and fungi. As a result of this new understanding, the five-kingdom system has been replaced by a classification that divides life into three domains: **Bacteria**, **Archaea**, and **Eukarya** (Fig. 18-5).

Kingdom-Level Classification Remains Unsettled

The move to a three-domain classification system has required systematists to reexamine the kingdoms within each domain, and the process of establishing kingdoms is still underway. If we accept that the striking differences between plants, animals, and fungi demand that each of these evolutionary lineages retains its kingdom status, then the logic of systematics requires that we also assign kingdom status to groups that branch off of the tree of life earlier than these three groups of multicellular eukaryotes. Following this logic, prokaryote systematists recognize about 15 kingdoms among the Bacteria and 3 or so kingdoms among the Archaea. Systematists also recognize additional kingdoms within the Eukarya, reflecting a number of very early evolutionary splits within the diverse array of single-celled eukaryotes formerly lumped together in the kingdom Protista. Systematists, however, have yet to reach a consensus about the precise definitions of new prokaryotic and eukaryotic kingdoms, though new information about the evolutionary history of single-celled organisms is emerging rapidly. Thus, kingdom-level classification is in a state of transition as systematists strive to incorporate the latest information.

This text's descriptions of the diversity of life—which appear in Chapters 19 through 23—sidestep the unsettled state of life's kingdoms. The prokaryotic domains Archaea and Bacteria are discussed without reference to kingdom-level relationships. Among the eukaryotes, fungi, plants, and animals are treated as distinct evolutionary units, while the generic term "protist" designates the diverse collection of eukaryotes that are not

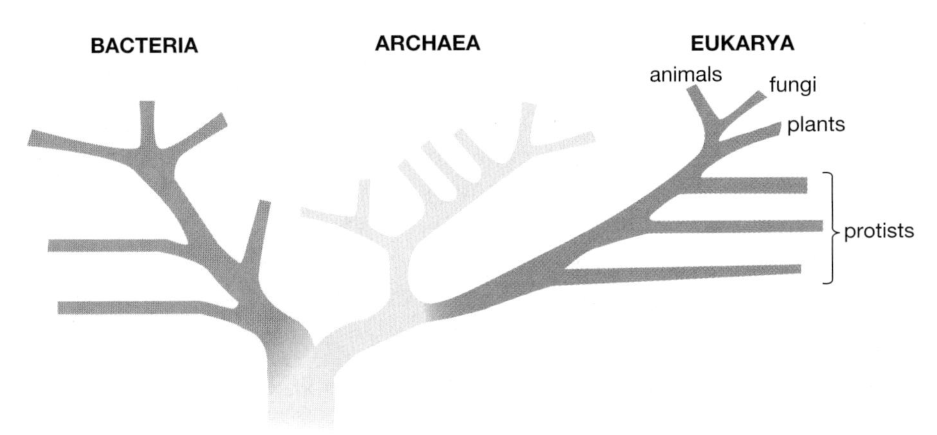

FIGURE 18-5 The tree of life
The three domains of life represent the three main "trunks" on the tree of life.

SCIENTIFIC INQUIRY Molecular Genetics Reveals Evolutionary Relationships

Evolution results from the accumulation of inherited changes in populations. Because DNA is the molecule of heredity, evolutionary changes must be reflected in changes in DNA. Systematists have long known that comparing DNA within a group of species would be a powerful method for inferring evolutionary relationships, but for most of the history of systematics, direct access to genetic information was nothing more than a dream. Today, however, **DNA sequencing**—determining the sequence of nucleotides in segments of DNA—is comparatively cheap, easy, and widely available. The *polymerase chain reaction* (PCR, see Chapter 13) allows systematists to easily accumulate large samples of DNA from organisms, and automated machinery makes sequence determination a comparatively simple task. Sequencing has rapidly become one of the primary tools for uncovering phylogeny.

The logic underlying molecular systematics is straightforward. It is based on the observation that when a single species divides into two species, the gene pool of each resulting species begins to accumulate mutations. The particular mutations in each species, however, will differ because the species are now evolving independently, with no gene flow between them. As time passes, more and more genetic differences accumulate. So, a systematist who has obtained DNA sequences from representatives of both species can compare the two species' nucleotide sequences at any given location in the genome. Fewer differences indicate more closely related organisms.

Putting the simple principles outlined above into practice usually involves more sophisticated thinking. For example, sequence comparisons become far more complex when a researcher wishes to assess relationships among, say, 20 or 30 species. Fortunately, mathematicians and computer programmers have devised some very clever computer-assisted methods for comparing large numbers of sequences and deriving the phylogeny that is most likely to account for the observed sequence differences.

Molecular systematists must also use care in choosing which segment of DNA to sequence. Different parts of the genome evolve at different rates, and it is crucial to sequence a DNA segment whose rate of change is well matched to the phylogenetic question at hand. In general, slowly evolving genes work best for comparing distantly related organisms, and rapidly changing portions of the genome are best for analyzing closer relationships. It is sometimes difficult to find any single gene that will yield sufficient information to provide an accurate picture of evolutionary change across the genome, so sequences from several different genes are often needed to construct reliable phylogenies, such as the one shown in Fig. E18-1.

Today, sequence data are piling up with unprecedented rapidity, and systematists have access to sequences from an ever-increasing number of species. The entire genomes of more than 100 species have been sequenced, and this number is expected to reach 1000 within a decade. The Human Genome Project has been completed, and our own DNA sequences are now a matter of public record. The revolution in molecular biology has fostered a great leap forward in our understanding of evolutionary history.

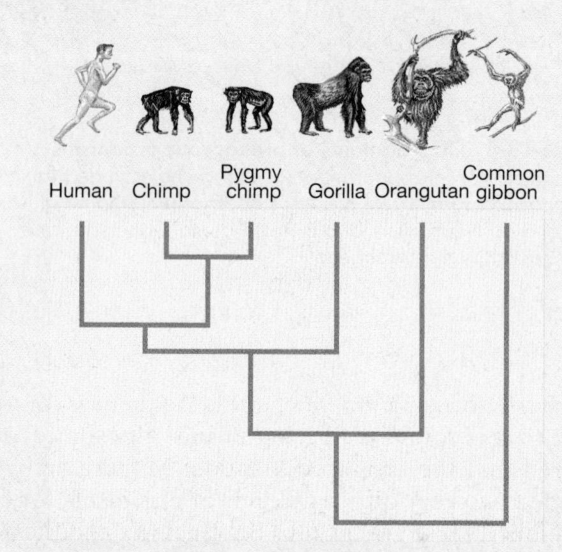

FIGURE E18-1 Relatedness can be determined by comparing DNA sequences
This evolutionary tree was derived from the nucleotide sequences of several different genes that are common to humans and apes.

members of these three kingdoms. Figure 18-6 shows the evolutionary relationships among some members of the domain Eukarya.

18.3 Why Do Classifications Change?

As the emergence of the three-domain system shows, the hypotheses of evolutionary relationships on which classification is based are subject to revision as new data emerge. Even domains and kingdoms, which represent deep, old branchings of the tree of life, must sometimes be rearranged. Such changes at the top levels of classification occur only rarely, but at the other end of the classification hierarchy, among species designations, revisions are more frequent.

Species Designations Change When New Information Is Discovered

As researchers uncover new information, systematists regularly propose changes in species-level classification. For example, until recently, systematists recognized two

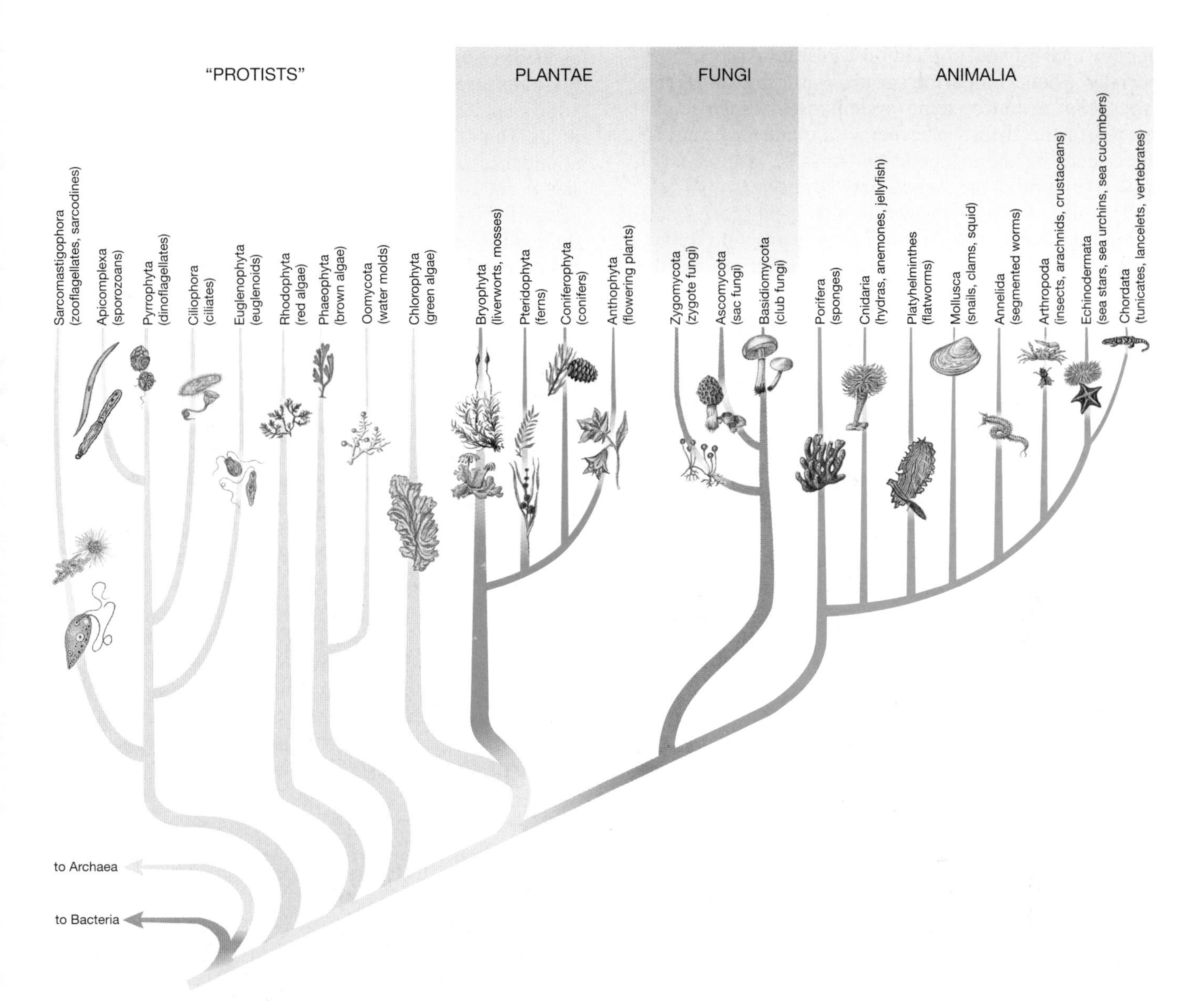

FIGURE 18-6 A closer look at the eukaryotic tree of life
Some of the major evolutionary lineages within the domain Eukarya are shown. The term "protist" refers to the many eukaryotes that are not plants, animals, or fungi.

species of elephant, the African elephant and the Indian elephant. Now, however, they recognize three elephant species; the former African elephant is now divided into two species, the savannah elephant and the forest elephant. Why the change? Genetic analysis of elephants in Africa revealed that there is little gene flow between forest-dwelling and savannah-dwelling elephants. The two groups of elephants are no more genetically similar than lions and tigers.

Another example is the red wolf, an inhabitant of the southern United States, which is currently listed as an endangered species. Researchers comparing red wolf DNA to that of gray wolves and coyotes have been unable to find any sequences that are unique to red wolves; all of the sequences examined so far match those of either gray wolves or coyotes. This DNA evidence strongly suggests that red wolves are actually hybrids between gray wolves and coyotes and may not be a distinct species after all.

The Biological Species Definition Can Be Difficult or Impossible to Apply

In some cases, systematists find themselves unable to say with certainty where one species ends and another begins. As discussed in Chapter 16, asexually reproducing organisms pose a particular challenge to systematists, because the criterion of interbreeding (the basis of the biological species definition that we have used in this text) cannot be used to distinguish among species. The irrelevance of this criterion in studies of asexual organisms leaves plenty of room for investigators to disagree about which asexual populations constitute a species, especially when comparing groups with similar phenotypes. For instance, some systematists recognize 200 species of the British blackberry (a plant that can produce seeds parthenogenetically—that is, without fertilization), but others recognize only 20 species.

The difficulty of applying the biological species definition to asexual organisms is a serious problem for systematists. After all, a significant portion of Earth's organisms reproduces without sex. Most bacteria, archaea, and protists, for example, reproduce asexually most of the time. Some systematists argue that we need a more universally applicable definition of species, one that won't exclude asexual organisms and that doesn't depend on the criterion of reproductive isolation.

The Phylogenetic Species Concept Offers an Alternative Definition

A number of alternative species definitions have been proposed over the history of evolutionary biology, but none has been sufficiently compelling to displace the biological species definition. One alternative definition, however, has been gaining adherents in recent years. The *phylogenetic species concept* defines a species as "the smallest diagnosable group that contains all the descendants of a single common ancestor." In other words, if we draw an evolutionary tree that describes the pattern of ancestry among a collection of organisms, each distinctive branch on the tree constitutes a separate species, regardless of whether the individuals represented by that branch can interbreed with individuals from other branches. As you might suspect, rigorous application of the phylogenetic species concept would vastly increase the number of different species recognized by systematists.

Proponents and opponents of the phylogenetic species concept are currently engaged in a vigorous debate about its merits. Perhaps one day it will replace the biological species concept as the "textbook definition" of species. In the meantime, classifications will continue to be debated and revised as systematists learn more and more about evolutionary relationships, particularly with the application of techniques derived from molecular biology. Although the precise evolutionary relationships of many organisms continue to elude us, classification is enormously helpful in ordering our thoughts and investigations into the diversity of life on Earth.

18.4 How Many Species Exist?

Scientists do not know even within an order of magnitude how many species share our world. Each year, between 7000 and 10,000 new species are named, most of them insects, many from tropical rain forests. The total number of named species is currently about 1.5 million. However, many scientists believe that 7 million to 10 million species may exist, and estimates range as high as 100 million. This total range of species diversity is known as **biodiversity**. Of all the species that have been identified thus far, about 5% are prokaryotes and protists. An additional 22% are plants and fungi, and the rest are animals. This distribution has little to do with the actual abundance of these organisms and a lot to do with the size of the organisms, how easy they are to classify, how accessible they are, and the number of scientists studying them. Historically, systematists have chiefly focused on large or conspicuous organisms in temperate regions, but biodiversity is greatest among small, inconspicuous organisms in the Tropics. In addition to the overlooked species on land and in shallow waters, an entire "continent" of species lies largely unexplored on the deep-sea floor. From the limited samples available, scientists estimate that hundreds of thousands of unknown species may reside there.

Although about 5000 species of prokaryotes have been described and named, prokaryotic diversity remains largely unexplored. Consider a study by Norwegian scientists, who analyzed DNA to count the number of different bacteria species present in a small sample of forest soil. To distinguish among species, the researchers arbitrarily defined bacterial DNA as coming from separate species if it differed by at least 30% from that of any other bacterial DNA in the sample. Using this criterion, they reported more than 4000 types of bacteria in their soil sample and an equal number of forms in a sample of shallow marine sediment.

FIGURE 18-7 The black-faced lion tamarin
Researchers estimate that no more than 260 individuals remain in the wild; captive breeding may be the black-faced lion tamarin's only hope for survival.

Our ignorance of the full extent of life's diversity adds a new dimension to the tragedy of the destruction of the tropical rain forests, discussed in Chapter 42. Although these forests cover only about 6% of Earth's land area, they are believed to be home to two-thirds of the world's existing species, most of which have never been studied or named. Because these forests are being destroyed so rapidly, Earth is losing many species that we will never even know existed! For example, in 1990, a new species of primate, the black-faced lion tamarin, was discovered in a small patch of dense rain forest on an island just off the east coast of Brazil (Fig. 18-7). Had the patch of forest been cut before this squirrel-sized monkey was discovered, its existence would have remained undocumented. At current rates of deforestation, most of the tropical rain forests, with their undescribed wealth of life, will be gone within the next century.

EVOLUTIONARY CONNECTIONS

Are Reptiles for Real?

The ever-changing nature of biological classification sometimes means that even the most entrenched and familiar categories must give way to new understandings of evolutionary history. Take reptiles, for example. Even schoolchildren are familiar with this well-known group of scaly animals and know that it includes snakes, lizards, turtles, alligators, and crocodiles. But many systematists would like to do away with the reptiles. Why? Because reptiles do not form what systematists call a *natural group.*

The goal of modern systematics is to devise classifications that accurately reflect evolutionary history. One characteristic of such classifications is that each designated group should contain only organisms that are more closely related to one another than to any organisms outside the group. So, for example, the members of family Canidae (which includes dogs, wolves, foxes, and coyotes) are more closely related to each other than to any member of any other family. Another way to state this principle is to say that each designated group should contain *all* of the living descendants of a common ancestor (Fig. 18-8a). In the terminology of systematics, such groups are said to be **monophyletic**.

The reptiles, however, are not a monophyletic group (Fig. 18-8b). As historically defined, the reptiles exclude the birds, which are now known to fall squarely within the

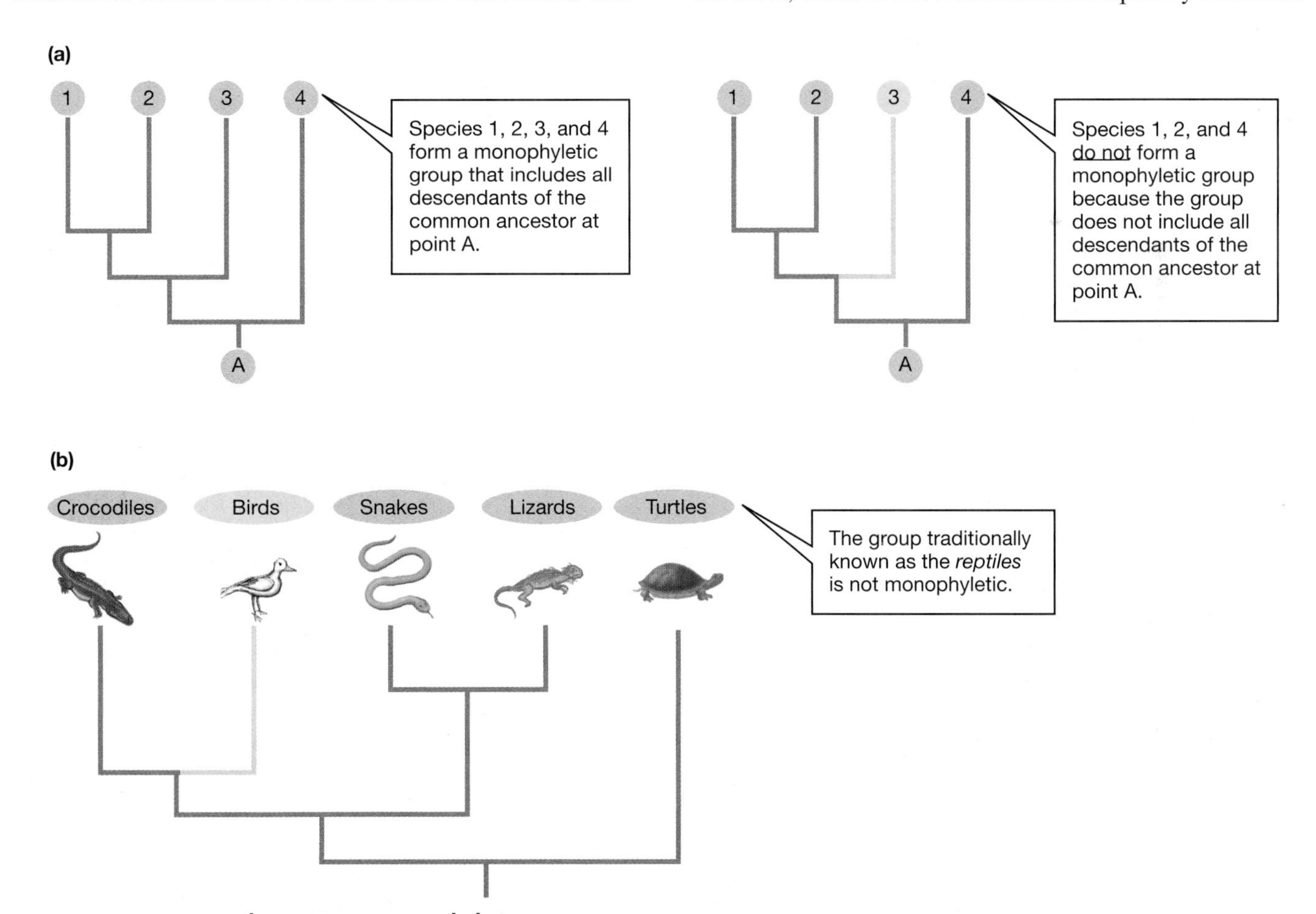

FIGURE 18-8 Reptiles are not a monophyletic group
Only groups that contain all of the descendants of a common ancestor are considered to be monophyletic groups.
EXERCISE Consider the following list of groups: (1) protists, (2) fungi, (3) great apes (chimpanzees, pygmy chimpanzees, gorillas, orangutans, and gibbons), (4) seedless plants (ferns, mosses, and liverworts), (5) prokaryotes (bacteria and archaea), (6) animals. Using Figures 18-5, 18-6, E18-1, and 21-1 for reference, identify the monophyletic groups on the list.

reptile family tree. The reptiles, therefore, do not include all living descendants of the common ancestor that gave rise to snakes, lizards, turtles, crocodilians, and birds. So systematists would prefer to dispense with the former Class Reptilia in favor of a scheme that names only monophyletic groups. The word *reptiles*, however, will most likely be with us for some time to come, if only because so many people (including systematists) are accustomed to using it. The word does, after all, provide a convenient way to describe a group of animals that share some interesting adaptations, even if that group isn't monophyletic.

CASE STUDY REVISITED Origin of a Killer

What evidence has persuaded evolutionary biologists that HIV originated in apes and monkeys? To understand the evolutionary thinking behind this conclusion, examine the evolutionary tree shown in Figure 18-9. This tree illustrates the phylogeny of HIV and its close relatives, as revealed by a comparison of RNA sequences among different viruses. Note the positions on the tree of the four human viruses (two strains of HIV-1 and two of HIV-2). One strain of HIV-1 is more closely related to a chimpanzee virus than to the other strain of HIV-1. Similarly, one strain of HIV-2 is more closely related to pig-tailed macaque SIV than to the other strain of HIV-2. Both HIV-1 and HIV-2 are more closely related to ape or monkey viruses than to one another.

The only way for the evolutionary history shown in the tree to have emerged is if viruses jumped between host species. If HIV had evolved strictly within human hosts, the human viruses would be each other's closest relatives. Because the human viruses do not cluster together on the phylogenetic tree, we can infer that cross-species infection occurred, probably on multiple occasions. The mainstream view is that transmission occurred in connection with human consumption of monkeys (HIV-2) and chimpanzees (HIV-1), but not all investigators agree with that assessment. Some feel that it is more likely that HIV was transferred via contaminated vaccines. Investigative journalist Edward Hooper, for example, has gathered evidence in support of the hypothesis that HIV-1 jumped to humans in an oral polio vaccine that was administered to more than a million people in Africa between 1957 and 1960. The vaccine may have been prepared using chimpanzee kidneys.

Consider This: Can understanding the evolutionary origin of HIV help researchers devise better ways to treat and control the spread of AIDS? How might such understanding influence strategies for treatment and prevention? More generally, how can evolutionary thinking help advance medical research?

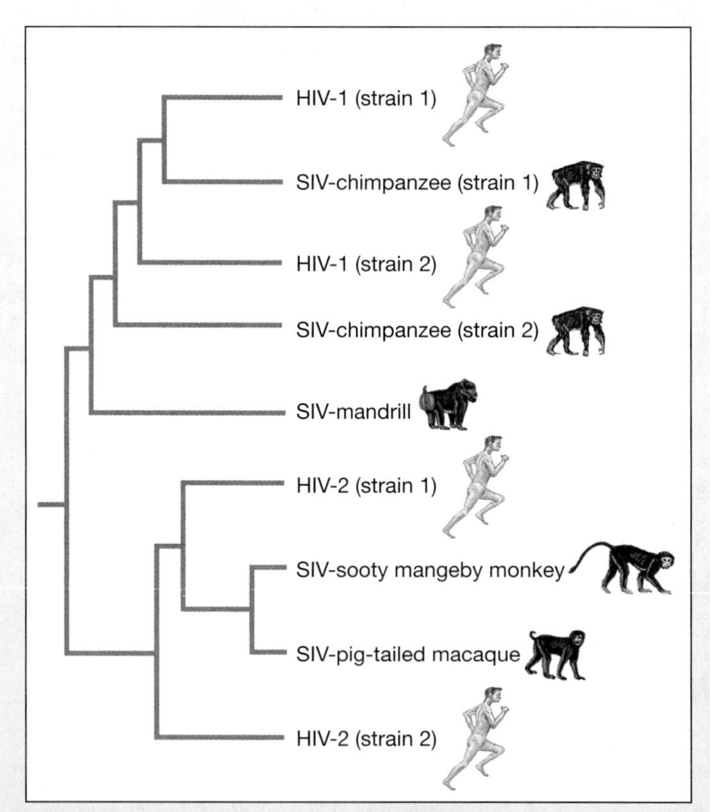

FIGURE 18-9 Evolutionary analysis helps reveal the origin of HIV
In this phylogeny of some immunodeficiency viruses, the viruses with human hosts do not cluster together. This lack of congruence between the evolutionary histories of the viruses and their host species suggests that the viruses must have jumped between host species. (Note that SIV stands for *simian immunodeficiency virus.*)

Links to Life: Small World

In light of humankind's intense curiosity about the origin of our species, it is not surprising that systematists have devoted a great deal of attention to uncovering the evolutionary history of *Homo sapiens.* Though much of this inquiry has focused on revealing the evolutionary connections between modern humans and the species to which we are most closely related, the techniques and methods of systematics have also been used to assess the evolutionary relationships among different populations within our species. Biologists have compared DNA sequences from human populations in many different parts of the world; different investigators have compared different portions of the human genome. As a result, a large amount of data has been gathered, and some interesting findings have emerged.

First, genetic divergence among human populations is very low compared to that of other animal species. For example, the range of genetic differences among all of Earth's humans is only one-tenth the size of the differences among the deer mice of North America (and many other species have even more genetic variability than deer mice). Clearly, all humans are genetically very similar, and the differences among different human populations are tiny.

It is also increasingly apparent that most of the human genetic variability that does exist can be found in African populations. The range of genetic differences found *within* sub-Saharan African populations is greater than the range between African populations and any non-African population. For many genes, all known variants are found in Africa, and no non-African population contains any distinctive variants; rather, non-African populations contain subsets of the African set. This finding strongly suggests that *Homo sapiens* originated in Africa, and that we have not lived anywhere else long enough to evolve very many differences from our African ancestors.

Summary of Key Concepts

18.1 How Are Organisms Named and Classified?

Organisms are classified and placed into hierarchical categories that reflect their evolutionary relationships. The eight major categories, in order of decreasing inclusiveness, are domain, kingdom, phylum, class, order, family, genus, and species. The scientific name of an organism is composed of its genus name and species name. Anatomical and molecular similarities among organisms are a measure of evolutionary relatedness.

18.2 What Are the Domains of Life?

The three domains of life, each representing one of three main branches of the tree of life, are Bacteria, Archaea, and Eukarya. Each domain contains a number of kingdoms, but the details of kingdom-level classification are in a period of transition and remain unsettled. Within the domain Eukarya, however, the kingdoms Fungi, Plantae, and Animalia are universally accepted as valid, monophyletic groups.

18.3 Why Do Classifications Change?

Classifications are subject to revision as new information is discovered. Species boundaries may be hard to define, particularly in the case of asexually reproducing species. However, systematics is essential for precise communication and contributes to our understanding of the evolutionary history of life.

18.4 How Many Species Exist?

Although only about 1.5 million species have been named, estimates of the total number of species range up to 100 million. New species are being identified at the rate of 7000 to 10,000 annually, mostly in tropical rain forests.

Key Terms

Archaea *p. 349*
Bacteria *p. 349*
biodiversity *p. 352*
class *p. 346*
DNA sequencing *p. 350*
domain *p. 346*
Eukarya *p. 349*
family *p. 346*
genus *p. 346*
kingdom *p. 346*
monophyletic *p. 353*
order *p. 346*
phylogeny *p. 346*
phylum *p. 346*
scientific name *p. 346*
species *p. 346*
systematics *p. 346*

Thinking Through the Concepts

To take a multiple-choice quiz with feedback on the contents of this chapter, visit http://www.prenhall.com/audesirk7. *Log in to the Web site selected by your instructor and navigate to the Self Test section for this chapter.*

? Review Questions

1. What contributions did Aristotle, Linnaeus, and Darwin each make to modern taxonomy?
2. What features would you study to determine whether a dolphin is more closely related to a fish or to a bear?
3. What techniques might you use to determine whether the extinct cave bear is more closely related to a grizzly bear or to a black bear?
4. Only a small fraction of the total number of species on Earth has been scientifically described. Why?
5. In England, "daddy long-legs" refers to a long-legged fly, but the same name refers to a spider-like animal in the United States. How do scientists attempt to avoid such confusion?

Applying the Concepts

1. There are many areas of disagreement about the classification of organisms. For example, there is no consensus about whether the red wolf is a distinct species or about how many kingdoms are within the domain Bacteria. What difference does it make whether biologists consider the red wolf a species, or into which kingdom a bacterial species falls? As Shakespeare put it, "What's in a name?"
2. The pressures created by human population growth and economic expansion place storehouses of biological diversity such as the Tropics in peril. The seriousness of the situation is clear when we consider that probably only 1 out of every 20 tropical species is known to science at present. What arguments can you make for preserving biological diversity in poor and developing countries, such as those in many areas of the Tropics? Does such preservation require that these countries sacrifice economic development? Suggest some solutions to the conflict between the growing demand for resources and the importance of conserving biodiversity.
3. During major floods, only the topmost branches of submerged trees may be visible above the water. If you were asked to sketch the branches below the surface of the water solely on the basis of the positions of the exposed tips, you would be attempting a reconstruction somewhat similar to the "family tree" by which taxonomists link various organisms according to their common ancestors (analogous to branching points). What sources of error do both exercises share? What advantages do modern taxonomists have?
4. The Florida panther, found only in the Florida Everglades, is currently classified as an endangered species, protecting it from human activities that could lead to its extinction. It has long been considered a subspecies of cougar (mountain lion), but recent mitochondrial DNA studies have shown that the Florida panther may actually be a hybrid between American and South American cougars. Should the Florida panther be protected by the Endangered Species Act?

For More Information

Avise, J. C. "Nature's Family Archives." *Natural History*, March 1989. Shows how evolutionary relationships can be determined by analyzing differences in DNA contained in mitochondria.

Gould, S. J. "What Is a Species?" *Discover*, December 1992. Discusses the difficulties of distinguishing separate species.

Mann, C., and Plummer, M. *Noah's Choice: The Future of Endangered Species*. New York: Knopf, 1995. A thought-provoking look at the hard choices we must make with regard to protecting biodiversity. Which species will we choose to preserve? What price are we willing to pay?

Margulis, L., and Sagan, D. *What Is Life?* London: Weidenfeld & Nicolson, 1995. A lavishly illustrated survey of life's diversity. Also includes an account of life's history and a meditation on the question posed by the title.

May, R. M. "How Many Species Inhabit the Earth?" *Scientific American*, October 1992. Although no one knows the precise answer to this question, an effective estimate is crucial to our effort to manage our biological resources.

Moffett, M. W. *The High Frontier: Exploring the Tropical Rainforest Canopy*. Cambridge, MA: Harvard University Press, 1994. The tremendous diversity of life in the rainforest treetops is only now becoming known to us. This book documents the unexpected and spectacular diversity of animals in the upper reaches of this endangered habitat.

Wilson, E. O. *The Diversity of Life*. Cambridge, MA: Harvard University Press, 1992. An outline of the processes that created the diversity of life and a discussion of the threats to that diversity and the steps required to preserve it.

Media Activities

To access a Media Activity visit http://www.prenhall.com/audesirk7. *Log in to the Web site selected by your instructor, navigate to this chapter, and select the appropriate Media Activity number.*

18.1 Taxonomic Classification

Estimated time: 5 minutes

Classification is a means of categorizing all organisms on Earth. This activity will help you understand the hierarchy of taxonomic groups used by biologists.

18.2 Tree of Life

Estimated time: 10 minutes

In this animation you will explore the Tree of Life, how molecular genetics are used to establish evolutionary relationships, and why there is so much disagreement about how the Tree of Life should be structured.

18.3 Web Investigation: Origin of a Killer

Estimated time: 20 minutes

The search for the origins of HIV involves the often conflicting activities of evolutionary biologists, epidemiologists, clinicians, and social activists. Alternate theories, intricate data analyses, and contradictory conclusions are common. This exercise will explore popular theories and recent developments in HIV studies.

CHAPTER

19 The Diversity of Viruses, Prokaryotes, and Protists

Workers prepare to decontaminate the Hart Office Building in Washington, D.C. after it was attacked with a biological weapon.

AT A GLANCE

CASE STUDY Agents of Death

In the fall of 2001, a longstanding fear became a terrible reality when residents of the United States were attacked with a biological weapon. The weapon, which killed five people and caused six others to become gravely ill, was simply a culture of bacteria. The bacteria had been placed in envelopes and mailed to the Hart Office Building in Washington, D.C. and to media company offices, where they were unknowingly inhaled by victims who opened the harmless-looking envelopes. The attack, though comparatively small, nonetheless dramatically illustrated the possibility and potential destructive power of a larger attack.

The bacterium used in the attack was *Bacillus anthracis*, which causes the disease anthrax. Anthrax normally infects domestic animals such as goats and sheep, but can also infect humans. The bacterium is a dangerous, often deadly infectious agent with properties that make it especially attractive to developers of biological weapons. Anthrax bacteria are easily isolated from infected animals, are cheap and easy to culture in large quantities, and, once produced, can be dried into a powder that remains viable for years. The powder is easily "weaponized" by packing it into a missile warhead or other delivery device, and a small volume of bacteria can infect a very large number of people. Areas contaminated with anthrax bacteria are very difficult to decontaminate.

Since the attack, it has become apparent that much of our ability to defend against such biological attacks depends on our understanding of the disease-causing microbes (as single-celled organisms are collectively known) that can be used as biological weapons. Scientific investigation of microbes can provide the knowledge required to detect an attack, destroy dangerous microbes in the environment, and prevent and treat infections. Fortunately, biologists already know quite a bit about microorganisms. In this chapter, we'll explore some of that knowledge.

19.1 What Are Viruses, Viroids, and Prions?

Viruses possess no membranes of their own, no ribosomes on which to make proteins, no cytoplasm, and no source of energy. They can reproduce only inside a **host** cell—the cell that a virus or other infectious agent infects. The simplicity of viruses makes it impossible to call them cells and, indeed, seems to place them outside the realm of living things.

A Virus Consists of a Molecule of DNA or RNA Surrounded by a Protein Coat

Virus particles are so small (0.05–0.2 micrometer in diameter) that they can be seen only under the enormous magnification of an electron microscope (Fig. 19-1). Viruses consist of two major parts: a molecule of hereditary material and a coat of protein surrounding the molecule. The hereditary molecule may be either DNA or RNA and may be single-stranded or double-stranded, linear or circular. The protein coat may be surrounded by an envelope formed from the plasma membrane of the host cell (Fig. 19-2a).

Viruses cannot grow or reproduce on their own, even if placed in a rich broth of nutrients at optimal temperature. They lack the complex cellular organization that these activities require. A virus's protein coat, however, is specialized to enable the virus to penetrate the cells of a specific host. After a virus enters a host cell, the viral genetic material takes command. The hijacked host cell is forced to use the instructions encoded in the viral genes to produce the components of new viruses. The pieces are rapidly assembled, and an army of new viruses bursts forth to invade and conquer neighboring cells (Fig. 19-2b; see "A Closer Look: Viruses—How the Nonliving Replicate").

Viruses Are Host-Specific

Each type of virus is specialized to attack a specific host cell (Fig. 19-3). As far as we know, no organism is immune to all viruses. Even bacteria fall victim to viral invaders; viruses that infect bacteria are called **bacteriophages** (Fig. 19-4). Bacteriophages may soon become important in treating diseases caused by bacteria, as many disease-causing bacteria have become increasingly resistant to antibiotics. Treatments based on bacteriophages could also take advantage of the viruses' specificity, attacking only the targeted bacteria and not the many other harmless or beneficial bacteria in the body.

In multicellular organisms such as plants and animals, different viruses specialize in attacking particular cell types. Viruses responsible for the common cold, for example, attack the membranes of the respiratory tract; those causing measles infect the skin, and the rabies virus attacks nerve cells. One type of herpes virus specializes in the mucous membranes of the mouth and lips, causing cold sores; a second type produces similar sores on or near the genitals. Herpes viruses take up

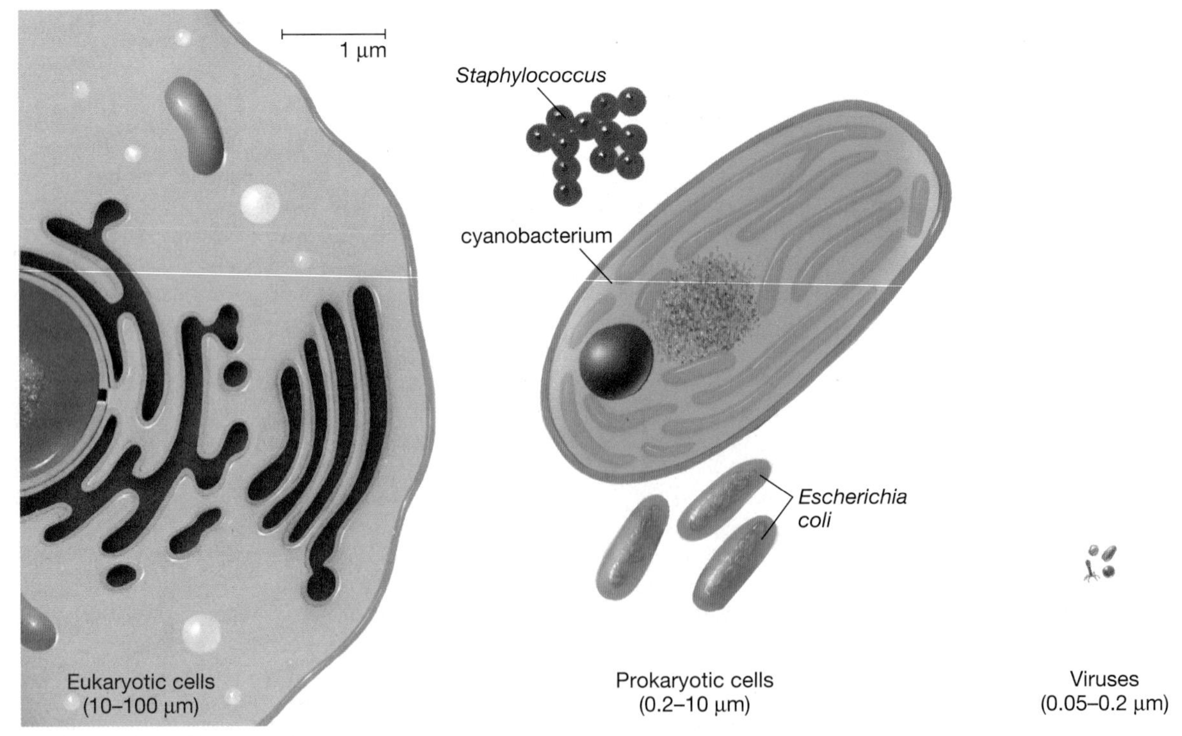

FIGURE 19-1 The sizes of microorganisms
The relative sizes of eukaryotic cells, prokaryotic cells, and viruses (1 μm = 1/1000 millimeter).

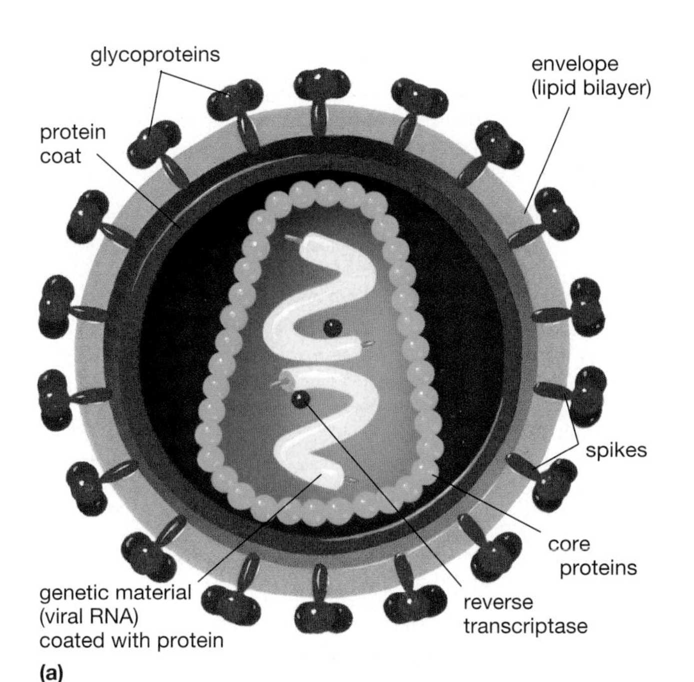

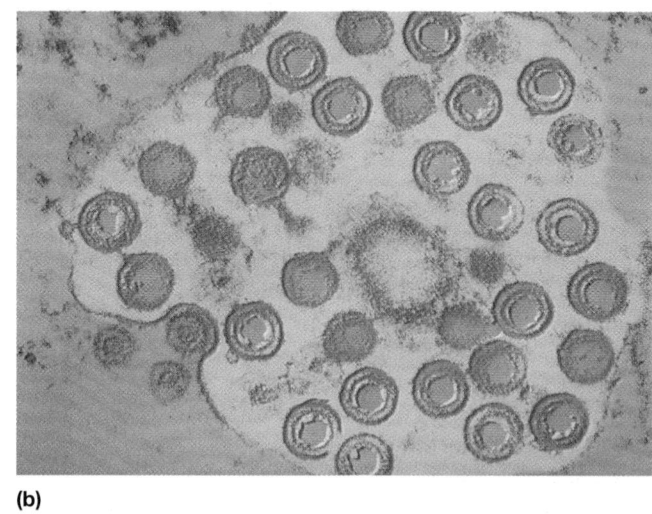

FIGURE 19-2 Viral structure and replication
(a) A cross section of the virus that causes AIDS. Inside, genetic material is surrounded by a protein coat and molecules of reverse transcriptase, an enzyme that catalyzes the transcription of DNA from the viral RNA template after the virus enters the host cell. This virus is among those that also have an outer envelope that is formed from the host cell's plasma membrane. Spikes made of glycoprotein (protein and carbohydrate) project from the envelope and help the virus attach to its host cell. **(b)** In this electron micrograph, herpes viruses are seen packed into an infected cell. **QUESTION** Why are viruses unable to replicate outside of a host cell?

permanent residence in the body, erupting periodically (typically during times of stress) as infectious sores. The devastating disease AIDS (acquired immune deficiency syndrome), which cripples the body's immune system, is caused by a virus that attacks a specific type of white blood cell that controls the body's immune response (as we will see in Chapter 32). Viruses have also been linked to some types of cancer, such as T-cell leukemia (a cancer of the white blood cells), liver cancer, and cervical cancer.

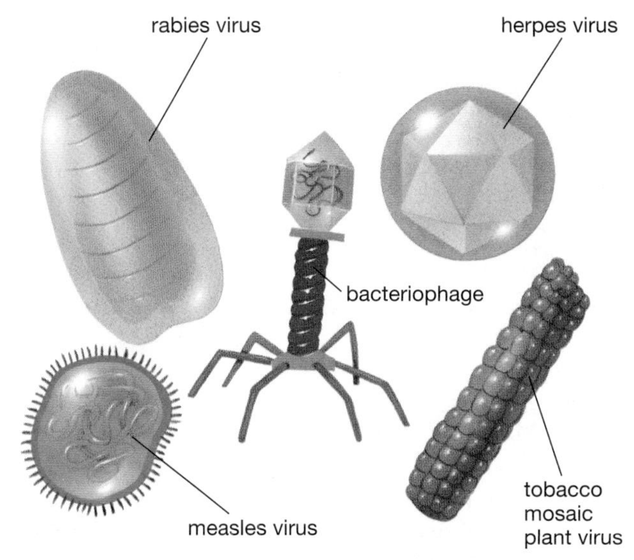

FIGURE 19-3 Viruses come in a variety of shapes
Viral shape is determined by the nature of the virus's protein coat. Viruses such as the rabies and herpes viruses are surrounded by an extra envelope derived from membranes of the host cell.

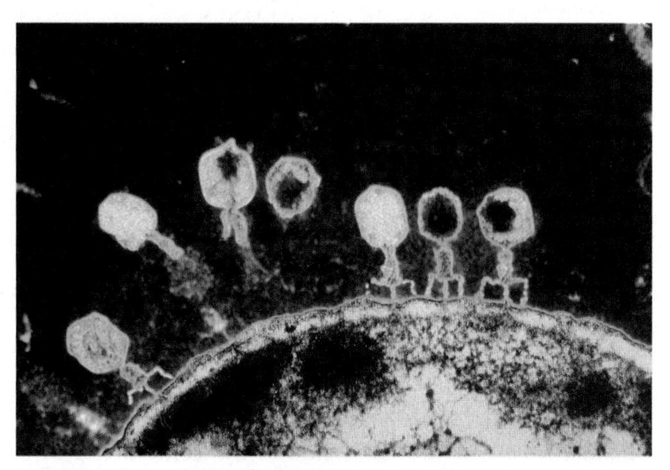

FIGURE 19-4 Some viruses infect bacteria
In this electron micrograph, bacteriophages are seen attacking a bacterium. They have injected their genetic material inside, leaving their protein coats clinging to the bacterial cell wall. The black objects inside the bacterium are newly forming viruses. **QUESTION** In biotechnology, viruses are often used to transfer genes from the cells of one species to the cells of another. Which properties of viruses make them useful for this purpose?

A CLOSER LOOK Viruses—How the Nonliving Replicate

Viruses multiply, or replicate, using their own genetic material, which—depending on the virus—consists of single-stranded or double-stranded RNA or DNA. This material serves as a template (or blueprint) for the viral proteins and genetic material required to make new viruses. Viral enzymes may participate in replication as well, but the overall process depends on the biochemical machinery that the host cell uses to make its own proteins.

Viral replication follows a general sequence:

1. **Penetration.** Viruses may be engulfed by their host cell (endocytosis). Some viruses have surface proteins that bind to receptors on the host cell's plasma membrane and stimulate endocytosis. Other viruses are coated with an envelope that can fuse with the host's membrane. The viral genetic material is then released into the cytoplasm.

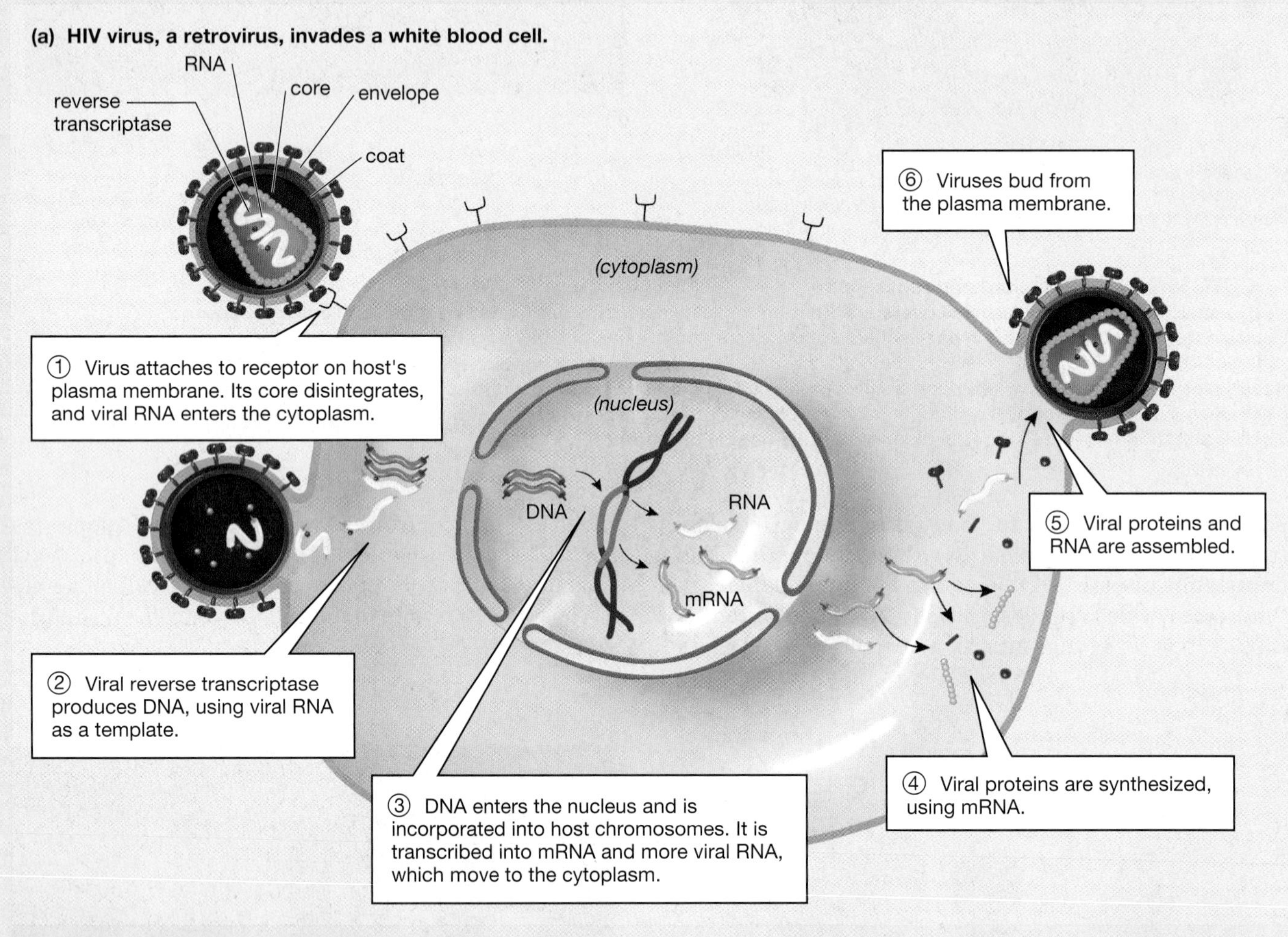
(a) HIV virus, a retrovirus, invades a white blood cell.
RNA
core
envelope
reverse transcriptase
coat
(cytoplasm)
(nucleus)
DNA
RNA
mRNA
① Virus attaches to receptor on host's plasma membrane. Its core disintegrates, and viral RNA enters the cytoplasm.
② Viral reverse transcriptase produces DNA, using viral RNA as a template.
③ DNA enters the nucleus and is incorporated into host chromosomes. It is transcribed into mRNA and more viral RNA, which move to the cytoplasm.
④ Viral proteins are synthesized, using mRNA.
⑤ Viral proteins and RNA are assembled.
⑥ Viruses bud from the plasma membrane.

Media Activity 19.1 Retrovirus Replication

FIGURE E19-1 How viruses replicate

Viral Infections Are Difficult to Treat

Because viruses are closely tied to the cellular machinery of their hosts, the illnesses they cause are difficult to treat. The antibiotics that are often effective against bacterial infections are useless against viruses, and antiviral agents may destroy host cells as well as viruses. Despite the difficulty of attacking viruses as they "hide" within cells, a number of antiviral drugs have been developed. Many of these drugs destroy or block the function of enzymes that the targeted virus requires for replication. In almost all cases, however, the benefits of antiviral drugs are limited, because viruses quickly evolve resistance to the drugs. Mutation rates are very high in viruses, in part because viruses lack mechanisms for correcting errors that occur during replication of DNA or RNA. It is thus almost inevitable that when a

2. **Replication.** The viral genetic material is copied many times.
3. **Transcription.** Viral genetic material is used as a blueprint to make messenger RNA (mRNA).
4. **Protein synthesis.** In the host cytoplasm, viral mRNA is used to synthesize viral proteins.
5. **Viral assembly.** The viral genetic material and enzymes are surrounded by their protein coat.
6. **Release.** Viruses emerge from the host cell by "budding" from the cell membrane or by bursting the cell.

Here, two types of viral life cycles are depicted. Figure E19-1a shows the *human immunodeficiency virus (HIV)*, the *retrovirus* that causes AIDS. Retroviruses use single-stranded RNA as a template to make double-stranded DNA by using a viral enzyme called *reverse transcriptase*. Many other retroviruses exist, and several cause cancers or tumors. Figure E19-1b shows the *herpes virus*, which contains double-stranded DNA that is transcribed into mRNA.

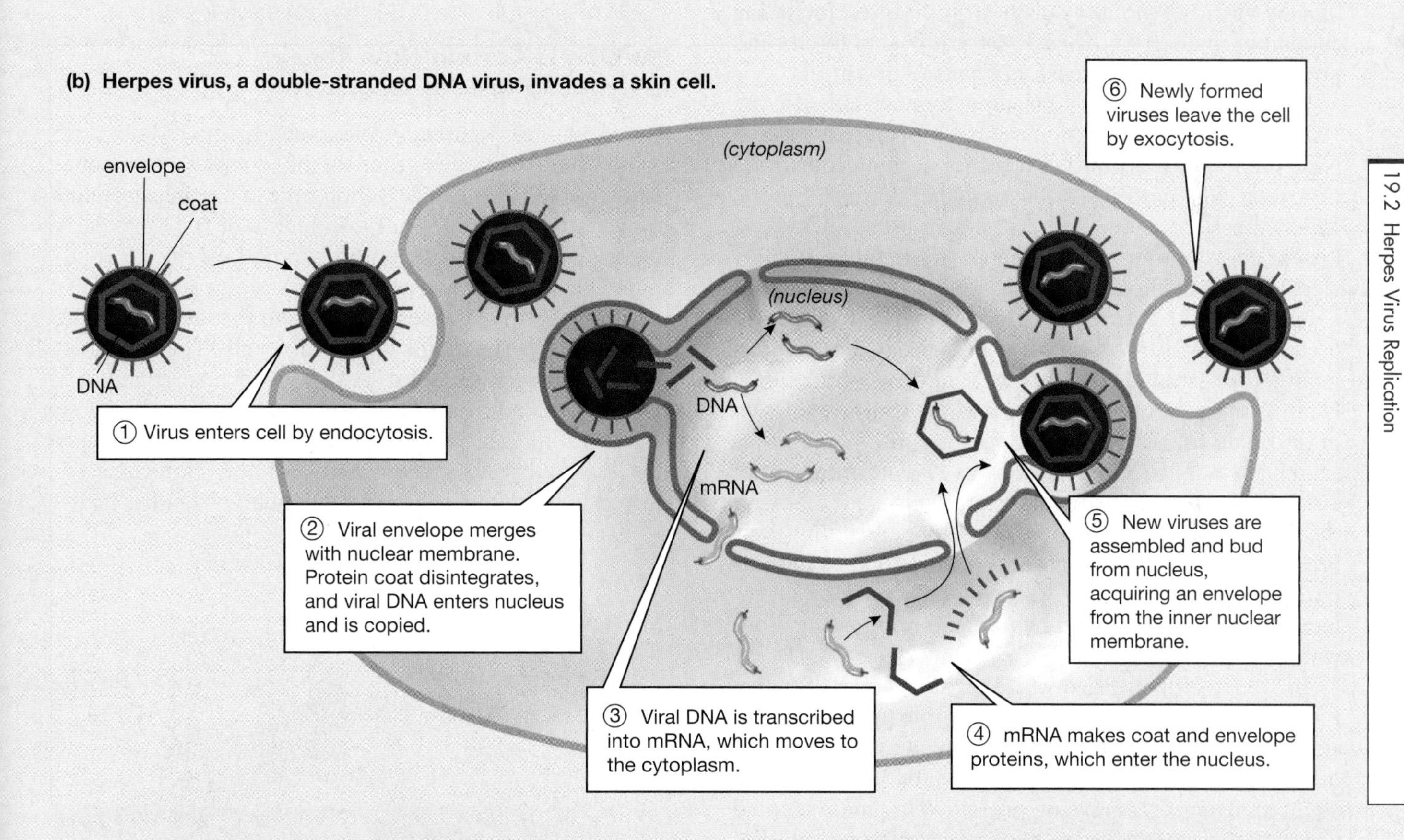
(b) Herpes virus, a double-stranded DNA virus, invades a skin cell.
envelope
coat
DNA
(cytoplasm)
(nucleus)
DNA
mRNA
① Virus enters cell by endocytosis.
② Viral envelope merges with nuclear membrane. Protein coat disintegrates, and viral DNA enters nucleus and is copied.
③ Viral DNA is transcribed into mRNA, which moves to the cytoplasm.
④ mRNA makes coat and envelope proteins, which enter the nucleus.
⑤ New viruses are assembled and bud from nucleus, acquiring an envelope from the inner nuclear membrane.
⑥ Newly formed viruses leave the cell by exocytosis.
Media Activity
19.2 Herpes Virus Replication

population of viruses is under attack by an antiviral drug, a mutation will arise that confers resistance to the drug. The resistant viruses prosper and replicate in great numbers, eventually spreading to new human hosts. Ultimately, resistant viruses predominate, and a formerly helpful antiviral drug is rendered ineffective.

The difficulty of treating viral infections makes the possibility of virus-based biological weapons troubling. Of particular concern is the smallpox virus. Smallpox has been eradicated as a naturally occurring disease, and the only known cultures of the virus are held at two well-guarded government labs, one in Russia and one in the United States; nevertheless, other samples may exist in unknown locations. Because of this possibility, plans to destroy the remaining stocks of the virus have been indefinitely deferred so that the stored viruses can be used in research to develop a more effective smallpox vaccine. Another potential threat is the virus that causes Ebola hemorrhagic fever, a serious disease of African origin that kills more than 90% of its victims. Ebola is especially worrisome, both as an emerging infectious disease and as a possible biological weapon, because there is currently no effective treatment and no vaccine to prevent infections.

Some Infectious Agents Are Even Simpler Than Viruses

Viroids are infectious particles that lack a protein coat, consisting of nothing more than short, circular strands of RNA. Despite their simplicity, viroids are able to enter the nucleus of a host cell and direct the synthesis of new viroids. About a dozen crop diseases, including cucumber pale fruit disease, avocado sunblotch, and potato spindle tuber disease, are caused by viroids.

Prions are even more puzzling than viroids. In the 1950s, physicians studying the Fore, a primitive tribe in New Guinea, were puzzled to observe numerous cases of a fatal degenerative disease of the nervous system, which the Fore called *kuru*. The symptoms of kuru—loss of coordination, dementia, and ultimately death—were similar to those of the rare but more widespread *Creutzfeldt-Jakob disease* in humans and of *scrapie* and *bovine spongiform encephalopathy*, diseases of domestic livestock (see "Health Watch: Mad Cows and Human Health"). Each of these diseases typically results in brain tissue that is spongy—riddled with holes. The researchers in New Guinea eventually determined that kuru was transmitted by ritual cannibalism; members of the Fore tribe honored their dead by consuming their brains. This practice has since stopped, and kuru has virtually disappeared. Clearly, kuru was caused by an infectious agent transmitted by infected brain tissue—but what was that agent?

In 1982, Nobel Prize-winning neurologist Stanley Prusiner published evidence that scrapie (and, by extension, kuru, Creutzfeldt-Jakob disease, and a number of other, similar afflictions) is caused by an infectious agent that consists only of protein. This idea seemed preposterous at the time, because most scientists believed that infectious agents must contain genetic material such as DNA or RNA in order to replicate. But Prusiner and his colleagues were able to isolate the infectious agent from scrapie-infected hamsters and demonstrate that it contained no nucleic acids. The researchers called these infectious protein particles *prions* (Fig. 19-5).

How can a protein replicate itself and be infectious? Not all researchers are convinced that this is possible. However, recent findings have sketched the outline of a possible mechanism for prion replication. It turns out that prions consist of a single protein produced by normal nerve cells. Some copies of this normal protein molecule, for reasons still poorly understood, become folded into the wrong shape and are thus transformed into infectious prions. Once present, prions can apparently induce other, normal copies of the protein molecule to become transformed into prions. Eventually, the concentration of prions in nerve tissue may get high enough to cause cell damage and degeneration. Why would a slight alteration to a normally benign protein turn it into a dangerous cell-killer? No one knows.

Another peculiarity of prion-caused diseases is that they can be inherited as well as transmitted by infection. Recent research has shown that certain small mutations in the gene that codes for "normal" prion protein increase the likelihood that the protein will fold into its abnormal form. If one of these mutations is genetically passed on to offspring, the tendency to develop a prion disease may be inherited.

No One Is Certain How These Infectious Particles Originated

The origin of viruses, viroids, and prions is obscure. Some scientists believe that the huge variety of mechanisms for self-replication among these particles reflects their status as evolutionary remnants of the very early history of life, before evolution settled on the more familiar large, double-stranded DNA molecules. Another possibility is that viruses, viroids, and prions may be the degenerate descendants of parasitic cells. These ancient parasites (organisms that live in or on host organisms, harming their hosts in the process) may have been so successful at exploiting their hosts that they eventually lost the ability to synthesize all of the molecules required for survival and became dependent on the host's

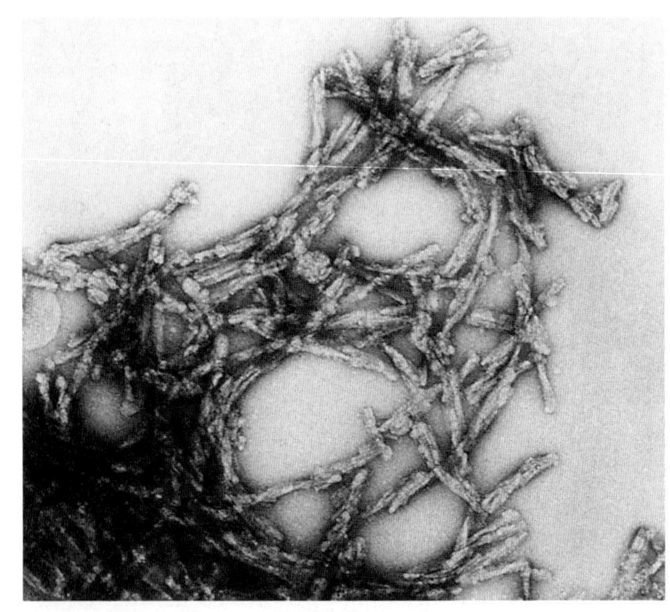

FIGURE 19-5 Prions: Puzzling proteins
A section from the brain of a hamster infected with scrapie contains fibrous clusters of prion proteins.

HEALTH WATCH Mad Cows and Human Health

In the past decade, more than 120 people have died of variant Creutzfeldt-Jakob disease (vCJD), a degenerative brain disease that leaves the brain tissue of its victims with a spongy texture. There is no treatment or cure for vCJD. Victims are typically young adults, and die within a year of their first symptoms. The disease is caused by the same infectious prion that causes bovine spongiform encephalopathy (BSE). (See page 364 for a description of prions.) Also known as "mad cow disease," BSE spread rapidly among cattle in Great Britain in the 1980s and early 1990s, mainly through the then-widespread practice of using the remains of slaughtered animals as feed for other cattle. The epidemic infected hundreds of thousands of cattle, and millions more were slaughtered in an attempt to limit the spread of the disease. Beef exports were halted, but not before mad cow disease had spread to most European countries and Japan.

The prion that causes both mad cow disease and vCJD passes from cow to human through the consumption of beef, and cannot be destroyed by cooking meat thoroughly. Unfortunately, widespread consumption of infected animals continued for more than a decade before the danger to humans was recognized. Because the symptoms of vCJD are not expressed until years after infection, and because the precise mechanism of infection remains poorly understood, it is impossible to guess how many people will eventually fall victim to the disease.

So far, cases of vCJD have been restricted to Europe (the only victim treated in America spent her childhood in England). Mad cow disease, however, recently appeared for the first time in the United States and in Canada. These first appearances of BSE have inspired fear that the disease will become widespread in North America, and ultimately infect humans as it has in Europe. Can this outcome be prevented? The U.S. currently bans beef imports from all countries in which BSE-infected cows have been found, and rules prohibit feeding cows with protein derived from other ruminants (the mammal group that includes cows, deer, and sheep). However, only about 12,500 cattle, out of more than 35 million slaughtered, are tested for BSE each year. Not everyone agrees that current precautions offer adequate protection, and controversy continues about whether more stringent regulation and testing is necessary.

One prion disease that *is* prevalent in North American is chronic wasting disease, a spongy-brain disease with symptoms similar to those of mad cow disease. Chronic wasting disease affects wild and captive elk and deer, species that are frequently eaten by humans. However, no case of transmission to humans has been documented, and experiments in which brain tissue from infected elk is injected into the brains of other species suggest that cross-species transmission is difficult. Nonetheless, some researchers urge caution, pointing to the long delay between infection and expression of prion diseases and noting the possibility that human infections may simply have gone undiagnosed, as the symptoms of spongy-brain prion diseases can resemble those of dementia or Alzheimer's disease.

biochemical machinery. Whatever the origin of these infectious particles, their success poses a continuing challenge to living things.

19.2 Which Organisms Make Up the Prokaryotic Domains—Bacteria and Archaea?

BACTERIA ARCHAEA EUKARYA
animals fungi plants protists

Two of life's three domains, Bacteria and Archaea, consist entirely of prokaryotes, single-celled microbes that lack organelles such as the nucleus, chloroplasts, and mitochondria. (See Chapter 5 for a comparison of prokaryotic and eukaryotic cells.) Both bacteria and archaea are normally very small, ranging from about 0.2 to 10 micrometers in diameter; in comparison, the diameters of eukaryotic cells range from about 10 to 100 micrometers. About 250,000 average-sized bacteria or archaea could congregate on the period at the end of this sentence, though a few species of bacteria are larger. The largest known bacterium is as much as 700 micrometers in diameter, making it visible to the naked eye.

Bacteria and Archaea Are Fundamentally Different

Bacteria and archaea are superficially similar in appearance under the microscope, but the ancient evolutionary separation between them is revealed by striking differences in their structural and biochemical features. For example, the cell walls of bacterial cells contain *peptidoglycan*, but the cell walls of archaea do not. Bacteria and archaea also differ in the structure and composition of plasma membranes, ribosomes, and RNA polymerases, and in basic processes such as transcription and translation.

Classification of Prokaryotes Within Each Domain Is Difficult

The sharp biochemical differences between archaea and bacteria make distinguishing the two domains a straightforward matter, but classification within each domain poses challenges. Prokaryotes are tiny, structurally simple, and do not exhibit the huge array of anatomical and developmental differences that are used to infer the evolutionary history of plants, animals, and other eukaryotes. Consequently, prokaryotes have been classified on the basis of such features as shape, means of locomotion, pigments, nutrient requirements, the appearance of colonies (groups of individual organisms descended from a single cell), and staining properties. For example, the **Gram stain**, a staining technique, distinguishes two types of cell-wall construction in bacteria; depending on the results of the stain, these bacteria are classified as either *gram-positive* or *gram-negative.*

In recent years, our understanding of the evolutionary history of the prokaryotic domains has been greatly expanded by comparisons of DNA and RNA nucleotide sequences. On the basis of this new information, some biologists now classify bacteria into 13 to 15 kingdoms and archaea into about 3 kingdoms. Prokaryote classification, however, is a rapidly changing field, and consensus on kingdom-level classification has proved elusive thus far. With new DNA sequence data being generated at a furious pace, and with new and distinctive types of bacteria and archaea being discovered and described on a regular basis, the revision of prokaryote classification schemes will likely continue to be dynamic for some time to come.

Prokaryotes Differ in Shape and Structure

The cell walls that surround prokaryotic cells give characteristic shapes to different types of bacteria and archaea. The most common shapes are rodlike, spherical, and corkscrew-shaped (Fig. 19-6). Some bacteria and archaea have **flagella** (singular, *flagellum*). These are simpler in structure than the eukaryotic flagella discussed in Chapter 5. Prokaryote flagella may appear singly at one end of a cell, in pairs (one at each end of the cell), as a tuft at one end of the cell (Fig. 19-7a), or scattered over the entire cell surface. Flagella can rotate rapidly, propelling the organism through its liquid environment. A unique wheel-like structure embedded in the bacterial membrane and cell wall allows the flagellum to rotate (Fig. 19-7b). Flagella allow bacteria to disperse into new habitats, migrate toward nutrients, and leave unfavorable environments.

Many Bacteria Form Films on Surfaces

The cell walls of some bacterial species are surrounded by sticky layers of protective slime, composed of polysaccharide or protein, which protects the bacteria and helps them adhere to surfaces. In many cases, slime-secreting bacteria of one or more species aggregate in colonies to form communities known as a *biofilms.* One familiar biofilm is dental plaque, which is formed by the bacteria that inhabit the mouth (Fig. 19-8). The protection afforded by biofilms helps defend the embedded bacteria against a variety of attacks, including those launched by antibiotics and disinfectants. As a result, biofilms formed by bacteria harmful to humans can be very difficult to eradicate. This is unfortunate, because the surfaces on which biofilms form include contact lenses, surgical sutures, and medical equipment such as catheters. In addition,

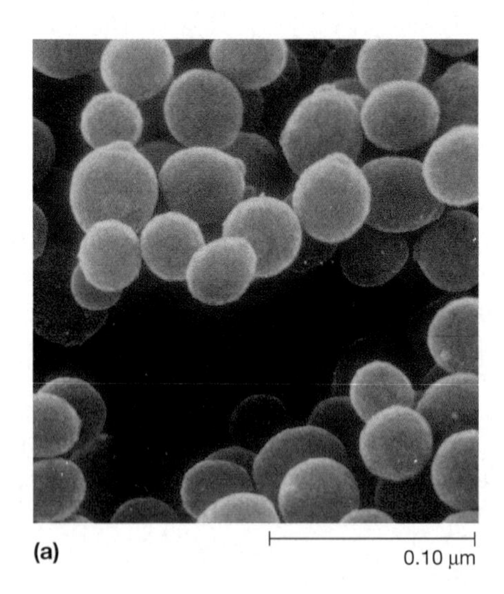

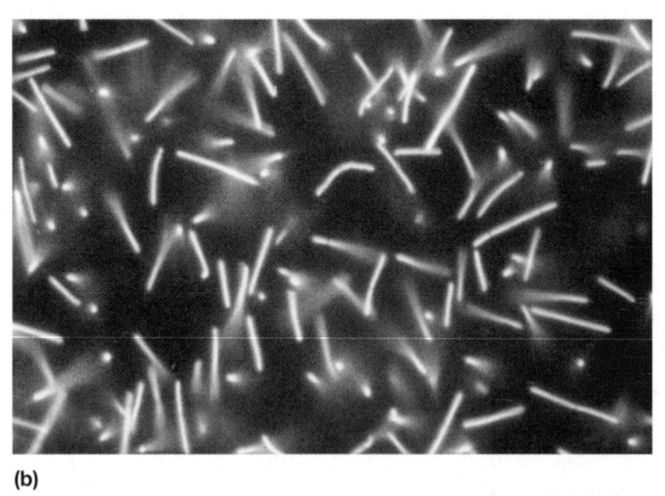

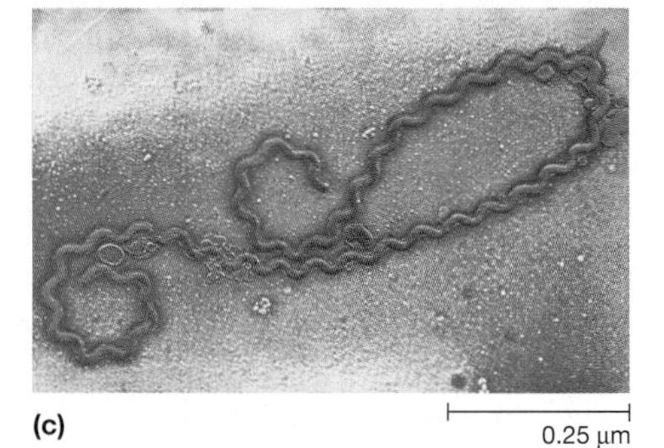

FIGURE 19-6 Three common prokaryote shapes **(a)** Spherical bacteria of the genus *Micrococcus*, **(b)** rod-shaped archaea of the genus *Methanopyrus*, and **(c)** corkscrew-shaped bacteria of the species *Leptospirosis interrogans.*

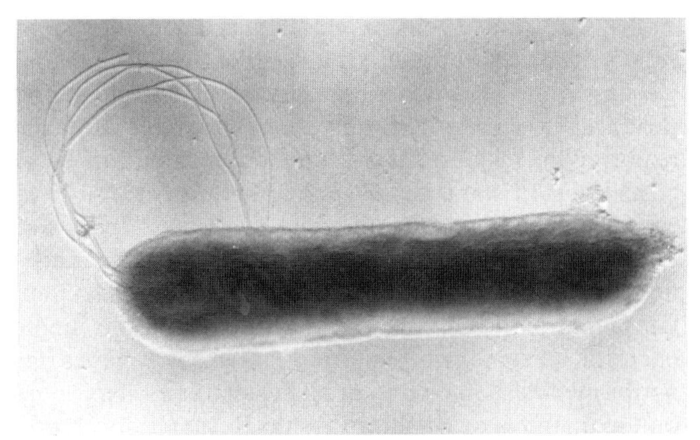

(a)

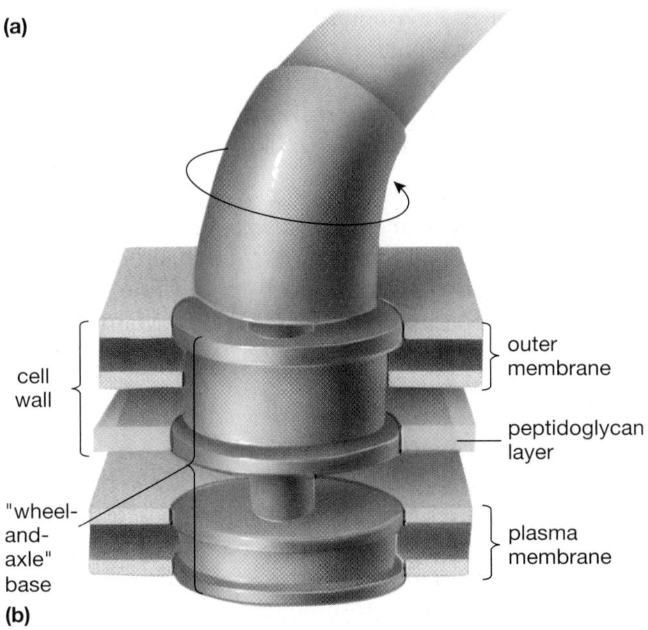

(b)

FIGURE 19-7 The prokaryote flagellum
(a) A flagellated archaean of the genus *Aquifex* uses its flagella to move toward favorable environments. **(b)** In bacteria, a unique "wheel-and-axle" arrangement anchors the flagellum within the cell wall and plasma membrane, enabling the flagellum to rotate rapidly.

many infections of the human body take the form of biofilms, including those responsible for tooth decay, gum disease, and ear infections.

Protective Endospores Allow Some Bacteria to Withstand Adverse Conditions

When environmental conditions become inhospitable, many rod-shaped bacteria form protective structures called **endospores**. An endospore, which forms inside a bacterium, contains genetic material and a few enzymes encased within a thick protective coat (Fig. 19-9). Metabolic activity ceases until the spore encounters favorable conditions, at which time metabolism resumes and the spore develops into an active bacterium.

Endospores are resistant to even extreme environmental conditions. Some can withstand boiling for an hour or more. Endospores are also able to survive for

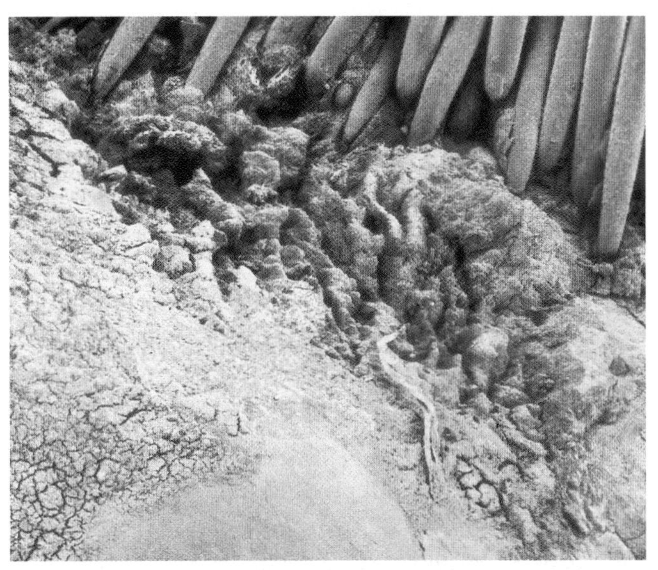

FIGURE 19-8 The cause of tooth decay
Bacteria in the human mouth form a slimy biofilm that helps them cling to tooth enamel and protects them from threats in the environment. The bacteria-laden biofilm can cause tooth decay unless it is removed by its chief antagonist, a toothbrush (seen here as green bristles).

extraordinarily long periods. In the most extreme example of longevity, scientists recently discovered bacterial spores that had been sealed inside rock for 250 million years. After being carefully extracted from their rocky "tomb," the spores were incubated in test tubes. Amazingly, live bacteria developed from the ancient spores, which were older than the oldest dinosaur fossils.

Endospores are one of the main reasons that the bacterial disease anthrax has become an agent of biological terrorism. The bacterium that causes anthrax forms endospores, which provide the means by which terrorists (or governments) can disperse the bacteria. The spores

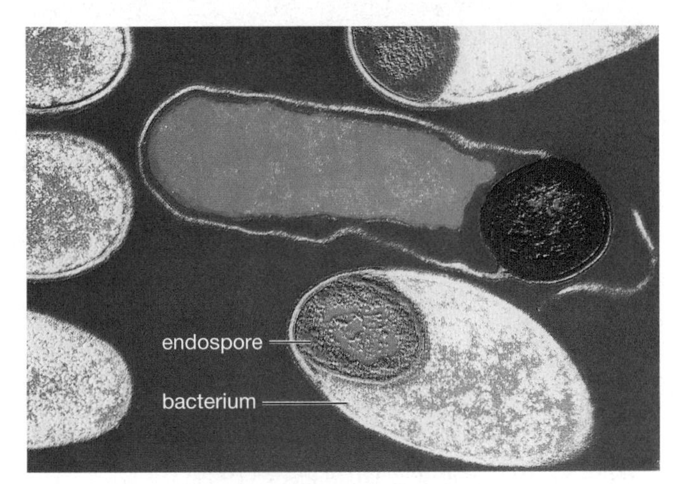

FIGURE 19-9 Spores protect some bacteria
Resistant endospores have formed inside bacteria of the genus *Clostridium*, which causes the potentially fatal food poisoning called botulism. **QUESTION** What might explain the observation that most species of endospore-forming bacteria live in soil?

can be stored indefinitely and can survive the harsh conditions they might encounter while traveling to their destination, including the stress of a missile launch and high-altitude travel. When they reach their target, the spores can survive dispersal into the atmosphere, remaining viable until inhaled by a potential victim.

Prokaryotes Reproduce by Binary Fission

Most prokaryotes reproduce asexually by a simple form of cell division called binary fission (see Chapter 11), which produces genetically identical copies of the original cell (Fig. 19-10). Under ideal conditions, a prokaryotic cell can divide about once every 20 minutes, potentially giving rise to sextillions (10^{21}) of offspring in a single day. This rapid reproduction allows prokaryotes to exploit temporary habitats, such as a mud puddle or warm pudding. Rapid reproduction also allows bacterial populations to evolve quickly. Recall that many mutations, the source of genetic variability, are the result of mistakes in DNA replication during cell division (see Chapter 10). Thus, the rapid, repeated cell division of prokaryotes provides ample opportunity for new mutations to arise and also allows mutations that enhance survival to spread quickly.

Prokaryotes May Exchange Genetic Material Without Reproducing

Media Activity 19.3 Bacterial Conjugation

Although prokaryote reproduction is generally asexual and does not involve genetic recombination, some bacteria and archaea nonetheless exchange genetic material. In these species, DNA is transferred from a donor to a recipient in a process called **conjugation**. The cell membranes of two conjugating prokaryotes fuse temporarily to form a cytoplasmic bridge across which DNA travels. In bacteria, donor cells may use specialized extensions called *sex pili* that attach to a recipient cell, drawing it closer to allow conjugation (Fig. 19-11). Conjugation produces new genetic combinations that may allow the resulting bacteria to survive under a greater variety of conditions. In some cases, genetic material may be exchanged even between individuals of different species.

DNA transferred during bacterial conjugation is contained within a structure called a **plasmid**, a small, circular DNA molecule that is separate from the single bacterial chromosome. Plasmids may carry genes for antibiotic resistance or even alleles of genes also found on the main bacterial chromosome. Researchers in molecular genetics have made extensive use of bacterial plasmids, as described in Chapter 13.

Prokaryotes Are Specialized for Specific Habitats

Prokaryotes occupy virtually every habitat, including those where extreme conditions prevent occupation by other forms of life. For example, some bacteria thrive in near-boiling environments, such as the hot springs of Yellowstone National Park (Fig. 19-12). Many archaea live in even hotter environments, including springs where the water actually boils or deep-ocean vents, where superheated water is spewed through cracks in Earth's crust at temperatures of up to 230 °F (110 °C). Prokaryotes can also survive at extremely high pressures, such as are found 1.7 miles (2.8 kilometers) below Earth's surface, where scientists recently discovered a new bacterial species. Bacteria and archaea are also found in very cold environments, such as Antarctic sea ice. Even extreme chemical conditions fail to impede invasion by prokaryotes. Thriving colonies of bacteria and

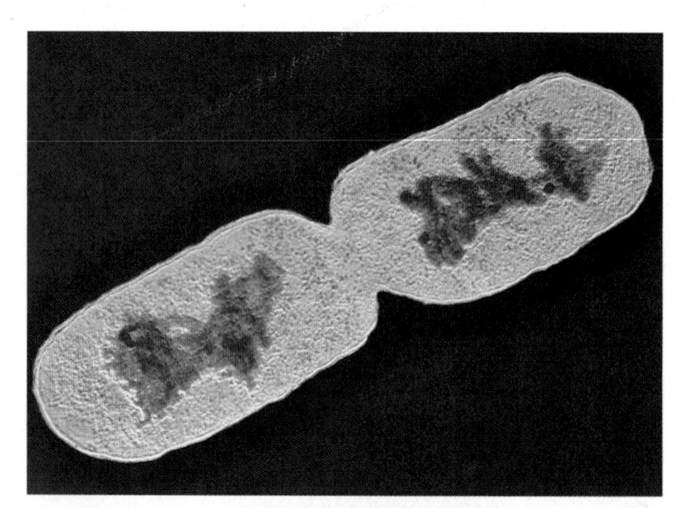

FIGURE 19-10 Reproduction in prokaryotes
Prokaryotic cells reproduce by binary fission. In this color-enhanced electron micrograph, an *Escherichia coli*, a normal component of the human intestine, is dividing. Red areas are genetic material. **QUESTION** What is the main advantage of binary fission, compared to sexual reproduction?

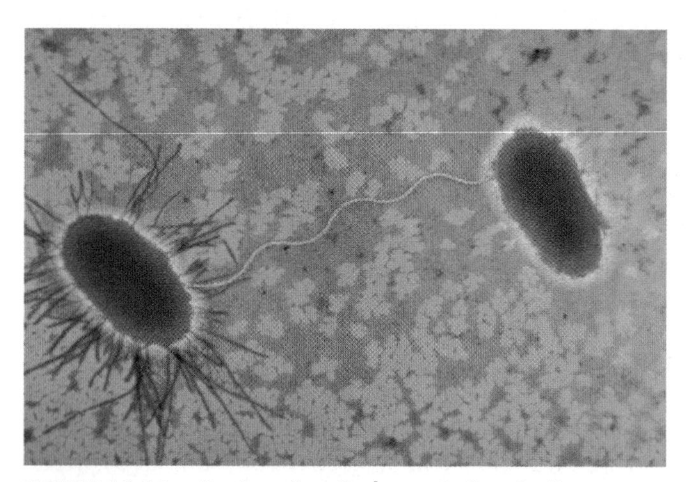

FIGURE 19-11 Conjugation: Prokaryotic "mating"
During conjugation, one prokaryote acts as a donor, transferring DNA to the recipient. In this photo, two *Escherichia coli* are connected by a long sex pilus. The sex pilus will retract, drawing the recipient bacterium (at right) to the donor bacterium. The donor bacterium is bristling with nonsex pili that help it attach to surfaces.

FIGURE 19-12 Some prokaryotes thrive in extreme conditions
Hot springs harbor bacteria and archaea that are both heat- and mineral-tolerant. Several species of cyanobacteria paint these hot springs in Yellowstone National Park with vivid colors; each is confined to a specific area determined by temperature range.
QUESTION Some of the enzymes that have important uses in molecular biology procedures are extracted from prokaryotes that live in hot springs. Can you guess why?

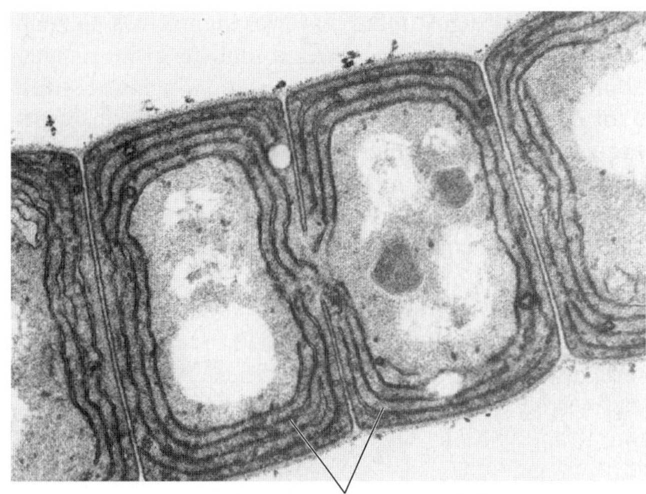

FIGURE 19-13 Cyanobacteria
Electron micrograph of a section through a cyanobacterial filament. Chlorophyll is located on the membranes visible within the cells.

archaea live in the Dead Sea, where a salt concentration seven times that of the oceans precludes all other life, and in waters that are as acidic as vinegar or as alkaline as household ammonia. Of course, rich bacterial communities also reside in a full range of more moderate habitats, including in and on the healthy human body. But an animal need not be healthy to harbor bacteria. Recently, a colony of bacteria was found dormant within the intestinal contents of a mammoth that had lain in a peat bog for 11,000 years.

No single species of prokaryote, however, is as versatile as these examples may suggest. In fact, most prokaryotes are specialists. One species of archaea that inhabits deep-sea vents, for example, grows optimally at 223 °F (106 °C) and stops growing altogether at temperatures below 194 °F (90 °C). Clearly, this species could not survive in a less-extreme habitat. Bacteria that live on the human body are also specialized; different species colonize the skin, the mouth, the respiratory tract, the large intestine, and the urogenital tract.

Prokaryotes Exhibit Diverse Metabolisms

Prokaryotes are able to colonize diverse habitats partly because they have evolved diverse methods of acquiring energy and nutrients from the environment. For example, unlike eukaryotes, many prokaryotes are **anaerobes**—their metabolisms do not require oxygen. The resulting ability to inhabit oxygen-free environments allows prokaryotes to exploit habitats that are off-limits to eukaryotes. Some anaerobes, such as many of the archaea found in hot springs and the bacterium that causes tetanus, are actually poisoned by oxygen. Others are opportunists, engaging in anaerobic respiration when oxygen is lacking and switching to aerobic respiration (a more efficient process) when oxygen becomes available. Many prokaryotes, of course, are strictly aerobic, and require oxygen at all times.

Whether aerobic or anaerobic, different prokaryote species can extract energy from an amazing array of different substances. Among the organic compounds on which prokaryotes subsist are not only the sugars, carbohydrates, fats, and proteins that we normally think of as foods, but also compounds that are inedible or even poisonous for humans, including petroleum, methane (the main component of natural gas), and solvents such as benzene and toluene. Prokaryotes can even metabolize inorganic molecules, including hydrogen, sulfur, ammonia, iron, and nitrite. The process of metabolizing inorganic molecules sometimes yields by-products that are useful to other organisms. For example, certain bacteria release sulfates or nitrates, crucial plant nutrients, into the soil.

Some species of bacteria, such as the *cyanobacteria* (Fig. 19-13) use photosynthesis to capture energy directly from sunlight. Like green plants, photosynthetic bacteria possess chlorophyll. Most species produce oxygen as a by-product of photosynthesis, but some, known as the sulfur bacteria, use hydrogen sulfide (H_2S) instead of water (H_2O) in photosynthesis, releasing sulfur instead of oxygen. No photosynthetic archaea are known.

Prokaryotes Perform Functions Important to Other Organisms

Many eukaryotic organisms depend on close associations with prokaryotes. For example, most animals that eat leaves, including cattle, rabbits, koalas, and deer, can't actually digest plant material themselves. Instead, they depend on certain bacteria that have the unusual ability to break down cellulose, the principal component of plant cell walls. Some of these bacteria live in the animals' digestive tracts, where they help liberate nutrients from plant fodder that the animals are unable to break down themselves. Without the bacteria, leaf-eating animals could not survive.

Prokaryotes also have important impacts on human nutrition. Many foods, including cheese, yogurt, and sauerkraut, are produced by the action of bacteria. Bacteria also inhabit your intestines. These bacteria feed on undigested food and synthesize such nutrients as vitamin K and vitamin B_{12}, which the human body absorbs.

Prokaryotes Capture the Nitrogen Needed by Plants

Humans could not live without plants, and plants are entirely dependent on bacteria. In particular, plants are unable to capture nitrogen from that element's most abundant reservoir, the atmosphere. Plants need nitrogen to grow. To acquire it, they depend on **nitrogen-fixing bacteria**, which live both in soil and in specialized nodules—small, rounded lumps on the roots of certain plants (legumes, which include alfalfa, soybeans, lupines, and clover; Fig. 19-14). The nitrogen-fixing bacteria capture nitrogen gas (N_2) from air trapped in the soil and combine it with hydrogen to produce ammonium (NH_4^+), a nitrogen-containing nutrient that plants can use directly.

Prokaryotes Are Nature's Recyclers

Prokaryotes play a crucial role in recycling waste. Many substances that we consider to be waste can serve as food for archaea and bacteria, most species of which obtain energy by breaking down complex organic (carbon-containing) molecules. The range of compounds attacked by prokaryotes is staggering. Nearly anything that human beings can synthesize, including detergents and the poisonous solvent benzene, can be destroyed by some prokaryote. The term "biodegradable" (meaning "broken down by living things") refers largely to the work of prokaryotes. Even oil is biodegradable. Soon after the tanker *Exxon Valdez* dumped 11 million gallons of crude oil into Prince William Sound in Alaska, researchers from Exxon sprayed oil-soaked beaches with a fertilizer that encouraged the growth of natural populations of oil-eating bacteria. Within 15 days, the oil deposits on these beaches were noticeably reduced in comparison to unsprayed areas. Prokaryotes are much less successful, however, at degrading most kinds of plastic. Therefore, the search for useful, economical plastics that are also biodegradable is a major focus of industrial research.

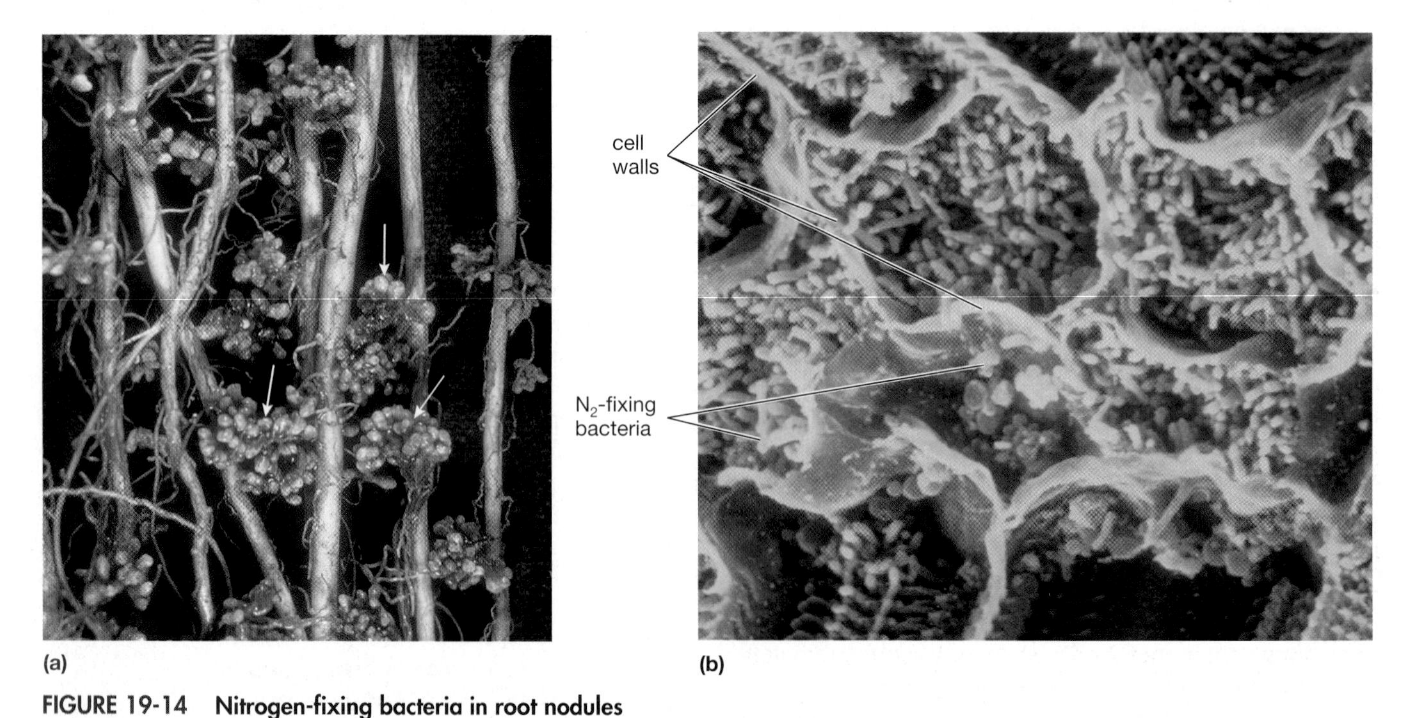

FIGURE 19-14 Nitrogen-fixing bacteria in root nodules
(a) Special chambers called nodules on the roots of a legume (alfalfa) provide a protected and constant environment for nitrogen-fixing bacteria. **(b)** This scanning electron micrograph shows the nitrogen-fixing bacteria inside cells within the nodules. **QUESTION** If all of Earth's nitrogen-fixing prokaryotes were to die suddenly, what would happen to the concentration of nitrogen gas in the atmosphere?

The appetite of some prokaryotes for nearly any organic compound is the key to their important role as decomposers in ecosystems. While feeding themselves, bacteria break down the waste products and dead bodies of plants and animals, thereby freeing nutrients for reuse. The recycling of nutrients provides the basis for continued life on Earth, as we will see in Chapter 41.

Some Bacteria Pose a Threat to Human Health

Despite the benefits some bacteria provide, the feeding habits of certain bacteria threaten our health and well-being. These **pathogenic** (disease-producing) bacteria synthesize toxic substances that cause disease symptoms. (So far, no pathogenic archaea have been identified.)

Some Anaerobic Bacteria Produce Dangerous Poisons

Some bacteria produce toxins that attack the nervous system. Examples of such pathogens include *Clostridium tetani*, which causes tetanus, and *Clostridium botulinum*, which causes botulism (a sometimes lethal food poisoning). Both of these bacterial species are anaerobic, surviving as spores until introduced into a favorable, oxygen-free environment. For example, tetanus bacteria can enter the human body through a puncture wound; if the wound is deep enough, they will be protected from contact with oxygen. As they multiply, the bacteria release their paralyzing poison into the bloodstream. For botulism bacteria, a sealed container of canned food that has been improperly sterilized may provide a haven. Thriving on the nutrients in the can, these anaerobes produce a toxin so potent that a single gram could kill 15 million people. Perhaps inevitably, this potent poison has caught the attention of biological weapon designers, who are presumed to have added it to their arsenals.

Humans Have Battled Bacterial Diseases Throughout History

Bacterial diseases have had a significant impact on human history. Perhaps the most infamous example is bubonic plague, or "Black Death," which killed 100 million people during the middle of the fourteenth century. In many parts of the world, one-third or more of the population died. Plague is caused by the highly infectious bacteria *Yersinia pestis*, which is spread by fleas that feed on infected rats and then move to human hosts. Although plague has not reemerged as a large-scale epidemic, about 2000 to 3000 people worldwide are still diagnosed with the disease each year.

Some bacterial pathogens seem to emerge suddenly. Lyme disease, for example, was unknown until 1975. This disease, named after the town of Old Lyme, Connecticut, where it was first described, is caused by the spiral-shaped bacterium *Borrelia burgdorferi*. The bacterium is carried by deer ticks, which transmit it to the humans they bite. At first, the symptoms resemble flu, with chills, fever, and body aches. If untreated, weeks or months later the victim may experience rashes, bouts of arthritis, and in some cases, abnormalities of the heart and nervous system. Both physicians and the general public are becoming more familiar with the disease, so more victims are receiving treatment before serious symptoms develop.

Perhaps the most frustrating pathogens are those that come back to haunt us long after we believed they were under control. Tuberculosis, a bacterial disease once almost vanquished in developed countries, is again on the rise in the United States and elsewhere. Two sexually transmitted bacterial diseases, gonorrhea and syphilis, have reached epidemic proportions around the globe. Cholera, a water-transmitted bacterial disease that flourishes when raw sewage contaminates drinking water or fishing areas, is under control in developed countries but remains a major killer in poorer parts of the world.

Some Common Bacterial Species Can Be Harmful

Some pathogenic bacteria are so widespread and ubiquitous that we cannot expect to ever be totally free of their damaging effects. For example, different species of the abundant streptococcus bacterium produce several diseases. One streptococcus causes strep throat. Another, *Streptococcus pneumoniae*, causes pneumonia by stimulating an allergic reaction that clogs the lungs with fluid. Yet another streptococcus has gained fame as the "flesh-eating bacterium." A small percentage of people who become infected with this bacterium experience severe symptoms, described luridly in tabloid newspapers with such headlines as "Killer Bug Ate My Face." About 800 Americans each year are victims of necrotizing fasciitis (as the "flesh-eating" infection is more properly known), and about 15% of these victims die. The streptococci enter through broken skin and produce toxins that either destroy flesh directly or stimulate an overwhelming and misdirected attack by the immune system against the body's own cells. A limb can be destroyed in hours, and in some cases, only amputation can halt the rapid tissue destruction. In other cases, these rare strep infections sweep through the body, causing death within a matter of days.

One of the most common bacterial inhabitants of the human digestive system, *Escherichia coli*, is also capable of doing harm. Different populations of *E. coli* may differ genetically, and some genetic differences can transform this normally benign species into a pathogen. One particularly notorious strain, known as O157:H7, infects about 70,000 Americans each year, about 60 of whom die from its effects. Most O157:H7 infections result from consumption of contaminated beef. About a third of the cattle in the United States carry O157:H7 in their intestinal tracts, and the bacteria can be transmitted to humans when a slaughterhouse inadvertently grinds some gut contents into hamburger. Once in a human digestive system, O157:H7 bacteria attach firmly to the wall of the intestine and begin to release a toxin that causes intestinal bleeding and that spreads to and damages other organs as well. The best defense against O157:H7 is to cook all meat thoroughly.

Most Bacteria Are Harmless

Although some bacteria assault the human body, most of the bacteria with which we share our bodies are harmless and many are beneficial. For example, the normal bacterial community in the vagina creates an environment that is hostile to infections by parasites such as yeasts. The bacteria that harmlessly inhabit our intestines are an important source of vitamin K. As the late physician, researcher, and author Lewis Thomas so aptly put it, "Pathogenicity is, in a sense, a highly skilled trade, and only a tiny minority of all the numberless tons of microbes on the Earth has ever been involved in it; most bacteria are busy with their own business, browsing and recycling the rest of life."

19.3 What Are Protists?

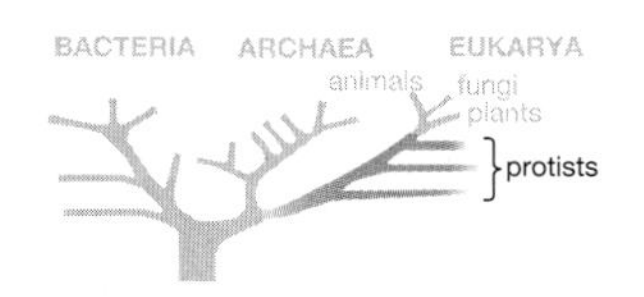

The third domain, Eukarya, includes all eukaryotic organisms. The most conspicuous Eukarya are members of the kingdoms Fungi, Plantae, and Animalia, which we will discuss in Chapters 20 through 23. The remaining eukaryotes constitute a diverse collection of evolutionary lineages collectively known as **protists** (Table 19-1). The term "protist" does not describe a true evolutionary unit united by shared features, but is a term of convenience that means "any eukaryote that is not a plant, animal, or fungus." About 60,000 protist species have been described.

Most Protists Are Single-Celled

Most protists are single-celled, and invisible to us as we go about our daily lives. If we could somehow shrink to their microscopic scale, we might be more impressed with their spectacular and beautiful forms, their varied and active lifestyles, their astonishingly diverse modes of repro-

Table 19-1 The Major Groups of Protists

Group	Subgroup	Locomotion	Nutrition	Representative Features	Representative Genus
Chromists	Water molds	Swim with flagella (gametes)	Heterotrophic bodies	Filamentous	*Plasmopara* (causes downy mildew)
	Diatoms	Glide along surfaces	Autotrophic; photosynthetic	Have silica shells; most marine	*Navicula* (glides toward light)
	Brown algae	Nonmotile	Autotrophic; photosynthetic	"Seaweeds" of temperate oceans	*Macrocystis* (forms kelp forests)
Alveolates	Dinoflagellates	Swim with two flagella	Autotrophic; photosynthetic	Many bioluminescent; often have cellulose	*Gonyaulax* (causes red tide)
	Apicomplexans	Nonmotile	Heterotrophic; all parasitic	Form infectious spores	*Plasmodium* (causes malaria)
	Ciliates	Swim with cilia	Heterotrophic	Most complex single cells	*Paramecium* (fast-moving pond-dweller)
Slime molds	Acellular slime molds	Sluglike mass oozes over surfaces	Heterotrophic	Form multinucleate plasmodium	*Physarum* (forms a large bright orange mass)
	Cellular slime molds	Amoeboid cells extend pseudopods; sluglike mass crawls over surfaces	Heterotrophic	Form pseudoplasmodium with individual amoeboid cells	*Dictyostelium* (often used in laboratory studies)
Euglenoids		Swim with one flagellum	Autotrophic; photosynthetic	Have an eyespot; all freshwater	*Euglena* (common pond-dweller)
Red algae		Nonmotile	Autotrophic; photosynthetic	Some deposit calcium carbonate; most marine	*Porphyra* (used as food in Japan)
Zooflagellates		Swim with flagella	Heterotrophic	Inhabit soil or water or may be parasitic	*Trypanosoma* (causes African sleeping sickness)
Pseudopod-users	Amoebas	Extend pseudopods	Heterotrophic	Both naked and shelled forms exist	*Amoeba* (common pond-dweller)
	Foraminifera Radiolarians Heliozoans				
Green algae		Swim with flagella (some species)	Autotrophic; photosynthetic	Closest relatives of land plants	*Ulva* (sea lettuce)

duction, and the structural and physiological innovations that are possible within the limits of a single cell. In reality, however, their small size makes them challenging to observe. A microscope and a good supply of patience are required to appreciate the majesty of protists.

Although most protists are single-celled, some are visible to the naked eye, and a few are genuinely large. Some larger protists are aggregations or colonies of single-celled individuals, while others are multicellular organisms.

Protists Use Diverse Modes of Reproduction and Nutrition

Most protists reproduce asexually by mitotic cell division (Fig. 19-15a), though many are also capable of sexual reproduction. Nonreproductive processes that combine the genetic material of different individuals are also common among protists (Fig. 19-15b). Protist reproduction, unlike that of plants and animals, never includes the formation and development of an embryo.

All three major modes of nutrition are represented among protists. Photosynthetic protists capture solar energy directly; predatory protists ingest their food; and parasitic protists can absorb nutrients from their surroundings. Past classifications of protists grouped species according to their mode of nutrition, but improved understanding of the evolutionary history of protists has revealed that the old categories did not accurately reflect phylogeny. Nonetheless, biologists still use terminology that refers to groups of protists that share particular characteristics but are not necessarily related. For example, photosynthetic protists are collectively known as **algae** (singular, alga), and single-celled, nonphotosynthetic protists are collectively known as **protozoa** (singular, protozoan).

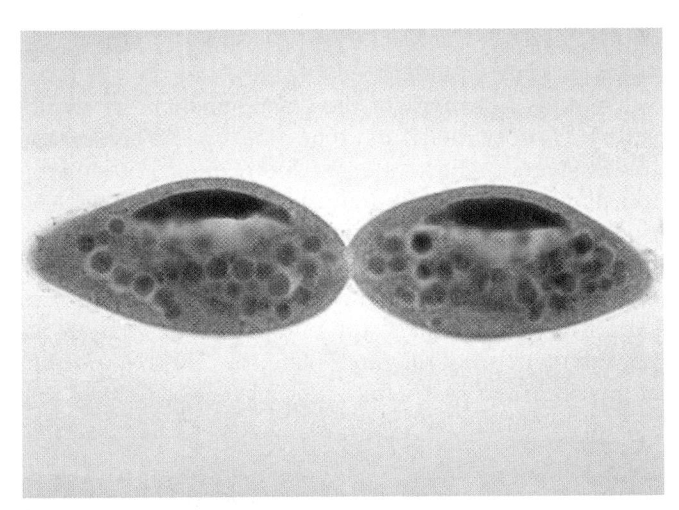

(a)

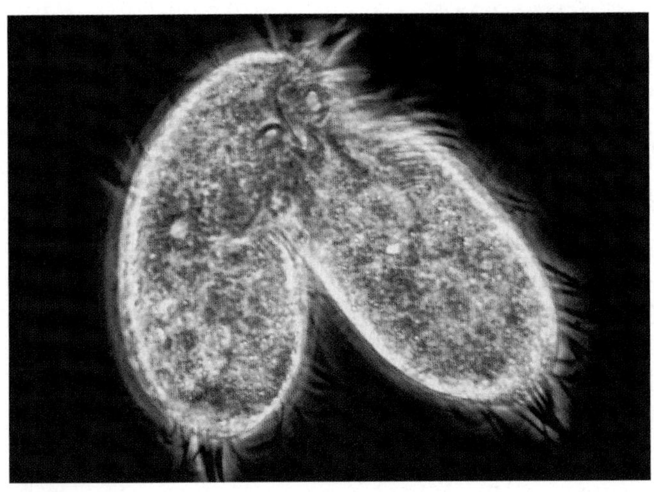

(b)

FIGURE 19-15 Protistan reproduction and gene exchange **(a)** *Paramecium*, a ciliate, reproduces asexually by cell division. **(b)** *Euplotes*, a ciliate, exchanges genetic material across a cytoplasmic bridge. **QUESTION** What do biologists mean when they say that sex and reproduction are uncoupled in most protists?

Protist Systematics Are in Transition

Genetic comparisons are helping systematists gain a better understanding of the evolutionary history of protist groups. Because systematists strive to devise classification systems that reflect evolutionary history, the new information has fostered a revision of protist classification. Some protist species, which had been previously grouped together on the basis of physical similarity, actually belong to independent evolutionary lineages that diverged from one another very early in the history of eukaryotes. Conversely, some protist groups bearing little physical resemblance to one another have been revealed to share a common ancestor, and thus have been classified together in new kingdoms. The process of revising protist classification, however, is far from complete. Thus, our understanding of the eukaryotic tree of life is still "under construction"; many of the branches are in place, but others await new information that will allow systematists to place them alongside their closest evolutionary relatives.

In the following sections, we'll explore a brief sampling of protist diversity.

The Chromists Include Photosynthetic and Nonphotosynthetic Organisms

The **chromists** form a group (designated as a kingdom by some systematists) whose shared ancestry was discovered through genetic comparison. All members of the group have fine, hairlike projections on their flagella (though in many chromists, flagella are present only at certain stages of the life cycle). Despite their shared evolutionary history, however, chromists display a wide range of different forms. Some are photosynthetic and some are not; most are single-celled, but some are multicellular. Three major chromist groups are the water molds, the diatoms, and the brown algae.

FIGURE 19-16 A parasitic water mold
Downy mildew, a plant disease caused by the water mold *Plasmopara*, nearly destroyed the French wine industry in the 1870s. **QUESTION** Although water molds are chromists, they look and function very much like fungi. What is the cause of this similarity?

FIGURE 19-17 Some representative diatoms
This photomicrograph illustrates the intricate, microscopic beauty and variety of the glassy walls of diatoms.

Water Molds Have Had Important Impacts on Humans

The **water molds**, or *oomycetes*, form a small group of protists, many of which are shaped as long filaments that aggregate to form cottony tufts. These tufts are superficially similar to structures produced by some fungi, but this resemblance is due to convergent evolution (see Chapter 14), not shared ancestry. Many water molds are decomposers that live in water and damp soil. Some species have profound economic impacts on humans; for example, a water mold causes a disease of grapes, known as *downy mildew* (Fig. 19-16). Its inadvertent introduction into France from the United States in the late 1870s nearly destroyed the French wine industry. Another oomycete has destroyed millions of avocado trees in California; still another is responsible for *late blight*, a devastating disease of potatoes. When accidentally introduced into Ireland about 1845, this protist destroyed nearly the entire potato crop, causing the devastating potato famine during which as many as 1 million people in Ireland starved and many more emigrated to the United States.

Diatoms Encase Themselves Within Glassy Walls

The **diatoms**, photosynthetic chromists found in both fresh and salt water, produce protective shells of *silica* (glass), some of exceptional beauty (Fig. 19-17). These shells consist of top and bottom halves that fit together like a pillbox or petri dish. Accumulations of diatoms' glassy walls over millions of years have produced fossil deposits of "diatomaceous earth" that may be hundreds of meters thick. This slightly abrasive substance is widely used in products such as toothpaste and metal polish.

Diatoms form part of the **phytoplankton**, the single-celled photosynthesizers that float passively in the upper layers of Earth's lakes and oceans. Phytoplankton play an immensely important ecological role. Marine phytoplankton account for nearly 70% of all the photosynthetic activity on Earth, absorbing carbon dioxide, recharging the atmosphere with oxygen, and supporting the complex web of aquatic life. Diatoms, as key components of the phytoplankton, are so important to marine food webs that they have been called the "pastures of the sea."

Brown Algae Dominate in Cool Coastal Waters

Though most photosynthetic protists, such as diatoms, are single-celled, some form multicellular aggregations that are commonly known as *seaweeds*. Although some seaweeds seem to resemble plants, they are not closely related to plants and lack many of the distinctive features of the plant kingdom. For example, none of the seaweeds have roots or shoots, and none form embryos during reproduction.

The chromists include one group of seaweeds, the brown algae. The brown algae are named for the brownish-yellow pigments that (in combination with green chlorophyll) increase the seaweed's light-gathering ability and produce a brown to olive-green color.

Almost all brown algae are marine. The group includes the dominant seaweed species that dwell along rocky shores in the temperate (cooler) oceans of the world, including the eastern and western coasts of the United States. Brown algae live in habitats ranging from nearshore, where they cling to rocks that are exposed at low tide, to far offshore. Several species use gas-filled floats to support their bodies (Fig. 19-18a). Some of the

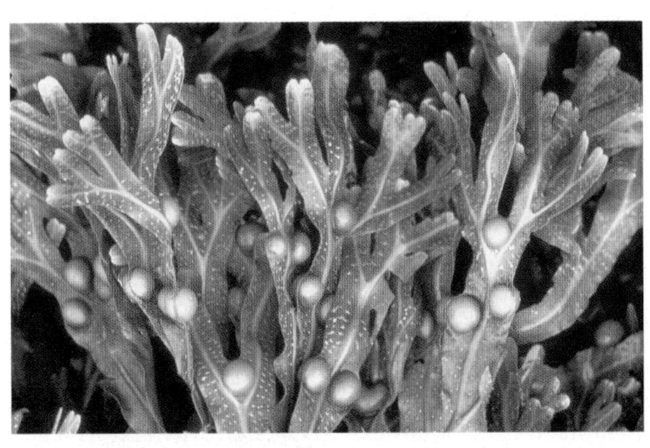

(a)

(b)

FIGURE 19-18 Brown algae, a multicellular protist
(a) *Fucus*, a genus found near shores, is shown here exposed at low tide. Notice the gas-filled floats, which provide buoyancy in water. **(b)** The giant kelp *Macrocystis* forms underwater forests off southern California.

giant kelp found along the Pacific coast reach heights of 325 feet (100 meters) and may grow more than 6 inches (15 centimeters) in a single day. With their dense growth and towering height (Fig. 19-18b), kelp form undersea forests that provide food, shelter, and breeding areas for marine animals.

The Alveolates Include Parasites, Predators, and Phytoplankton

The **alveolates** are single-celled organisms that have distinctive, small cavities beneath the surface of their cells. Like the chromists, the alveolates form a distinct lineage that may eventually be given kingdom status. Also like the chromists, the evolutionary link among the alveolates was long obscured by the variety of structures and ways of life among the group members, but was revealed by molecular comparisons. Some alveolates are photosynthetic, some are parasitic, and some are predatory. The major alveolate groups are the dinoflagellates, apicomplexans, and ciliates.

Dinoflagellates Swim by Means of Two Whiplike Flagella

Though most **dinoflagellates** are photosynthetic, there are also some nonphotosynthetic species. Dinoflagellates are named for the motion created by their two whiplike flagella (*dino* is Greek for "whirlpool"). One flagellum encircles the cell, and the second projects behind it. Some dinoflagellates are enclosed only by a cell membrane; others have cellulose walls that resemble armor plates (Fig. 19-19). Although some species live in fresh water, dinoflagellates are especially abundant in the ocean, where they are an important component of the phytoplankton and a food source for larger organisms. Many dinoflagellates are bioluminescent, producing a brilliant blue-green light when disturbed. Specialized dinoflagellates live within the tissues of corals, some clams, and even other protists, where they provide their hosts with nutrients from photosynthesis and remove carbon dioxide. Reef-building corals live only in the shallow, well-lit waters in which their embedded dinoflagellates can survive.

Warm water that is rich in nutrients may bring on a dinoflagellate population explosion. Dinoflagellates

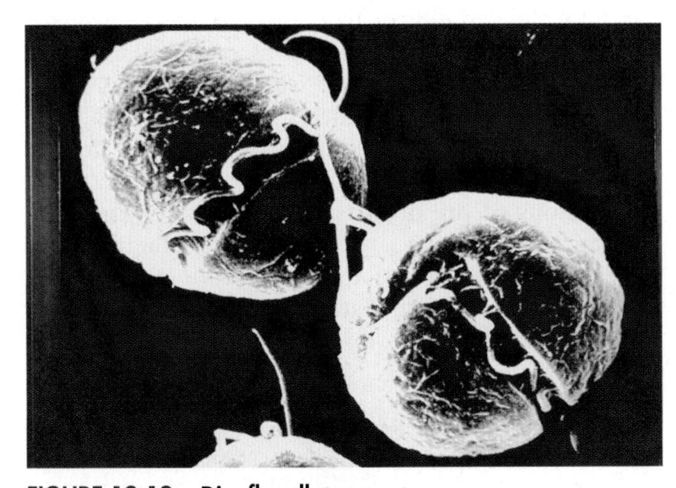

FIGURE 19-19 Dinoflagellates
Two dinoflagellates covered with protective cellulose armor. Visible on each is a flagellum in a groove that encircles the body.

FIGURE 19-20 A red tide
The explosive reproductive rate of certain dinoflagellates under the right environmental conditions can produce concentrations so great that their microscopic bodies dye the seawater red or brown, as in this bay in Mexico.

can become so numerous that the water is dyed red by the color of their bodies, causing a "red tide" (Fig. 19-20). During red tides, fish die by the thousands, suffocated by clogged gills or by the oxygen depletion that results from the decay of billions of dinoflagellates. One type of dinoflagellate, *Pfisteria*, even eats fish directly, first secreting chemicals that dissolve the victim's flesh. But dinoflagellate explosions can benefit oysters, mussels, and clams, which have a feast, filtering millions of the protists from the water and consuming them. In the process, however, their bodies accumulate concentrations of a nerve poison produced by the dinoflagellates. Humans who eat these mollusks may be stricken with potentially lethal paralytic shellfish poisoning.

Apicomplexans Are Parasitic and Have No Means of Locomotion

All **apicomplexans** (sometimes known as *sporozoans*) are parasitic, living inside the bodies and sometimes inside the individual cells of their hosts. They form infectious spores, resistant structures transmitted from one host to another through food, water, or the bite of an infected insect. As adults, apicomplexans have no means of locomotion. Many have complex life cycles, a common feature of parasites. A well-known example is the malarial parasite *Plasmodium* (Fig. 19-21). Parts of its life cycle are spent in the stomach, and later the salivary glands, of the female *Anopheles* mosquito. When the mosquito bites a human, it passes the *Plasmodium* to the unfortunate victim. The apicomplexan develops in the victim's liver, then enters the blood, where it reproduces rapidly in red blood cells. The release of large quantities of spores through the rupture of the blood cells causes the recurrent fever of malaria. Uninfected mosquitoes may acquire the parasite by feeding on the blood of a malaria victim, spreading the parasite when they bite another person.

Although the drug chloroquine kills the malarial parasite, drug-resistant populations of *Plasmodium* are, unfortunately, spreading rapidly throughout Africa, where the disease is prevalent. Programs to eradicate mosquitoes have failed because the mosquitoes rapidly evolve resistance to pesticides.

Ciliates Are the Most Complex of the Alveolates

Ciliates, which inhabit fresh and salt water, represent the peak of unicellular complexity. They possess many specialized organelles, including **cilia** (singular, cilium), the short hairlike outgrowths after which they are named. The cilia may cover the cell or may be localized. In the well-known freshwater genus *Paramecium*, rows of cilia cover the organism's entire body surface (Fig. 19-22). Their coordinated beating propels the cell through the water at a rate of one millimeter per second—a protistan speed record. Although only a single cell, *Paramecium* responds to its environment as if it had a well-developed nervous system. Confronted with a noxious chemical or a physical barrier, the cell immediately backs up by reversing the beating of its cilia and then proceeds in a new direction. Some ciliates, such as *Didinium*, are accomplished predators (Fig. 19-23).

Slime Molds Are Decomposers That Inhabit the Forest Floor

The *slime molds* are another distinctive protist lineage that could become its own kingdom when a consensus on eukaryote classification emerges. The physical form of slime molds seems to blur the boundary between a colony of different individuals and a single, multicellular individual. The life cycle of the slime mold consists of two phases: a mobile feeding stage and a stationary reproductive stage called a *fruiting body*. There are two main types of slime mold: acellular and cellular.

Acellular Slime Molds Form a Multinucleate Mass of Cytoplasm Called a Plasmodium

The **acellular**, or **plasmodial**, **slime molds** consist of a mass of cytoplasm that may spread thinly over an area of several square meters. Although the mass contains thousands

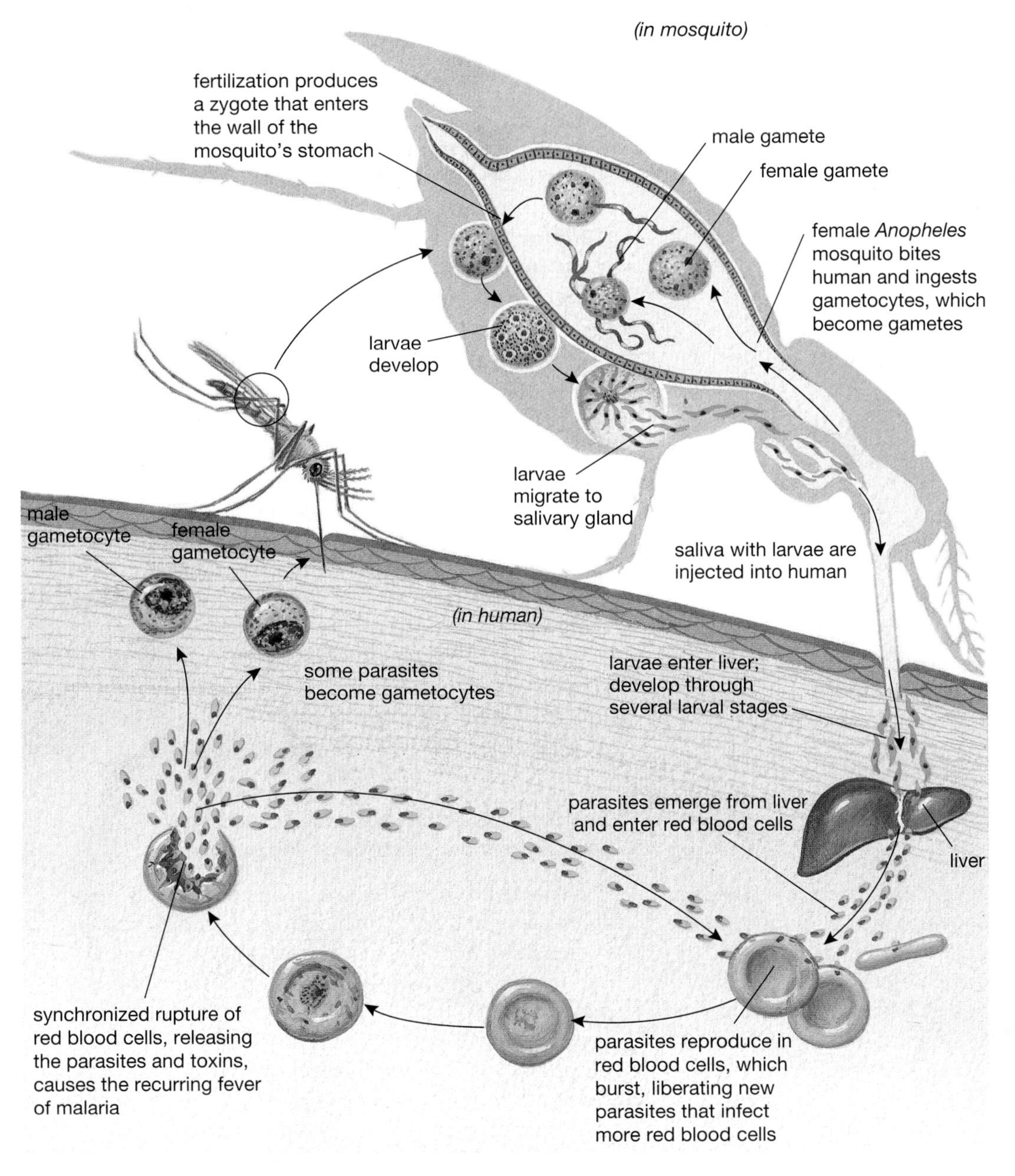
(in mosquito)
fertilization produces a zygote that enters the wall of the mosquito's stomach
male gamete
female gamete
female *Anopheles* mosquito bites human and ingests gametocytes, which become gametes
larvae develop
larvae migrate to salivary gland
saliva with larvae are injected into human
male gametocyte
female gametocyte
(in human)
some parasites become gametocytes
larvae enter liver; develop through several larval stages
parasites emerge from liver and enter red blood cells
liver
synchronized rupture of red blood cells, releasing the parasites and toxins, causes the recurring fever of malaria
parasites reproduce in red blood cells, which burst, liberating new parasites that infect more red blood cells

FIGURE 19-21 The life cycle of the malaria parasite

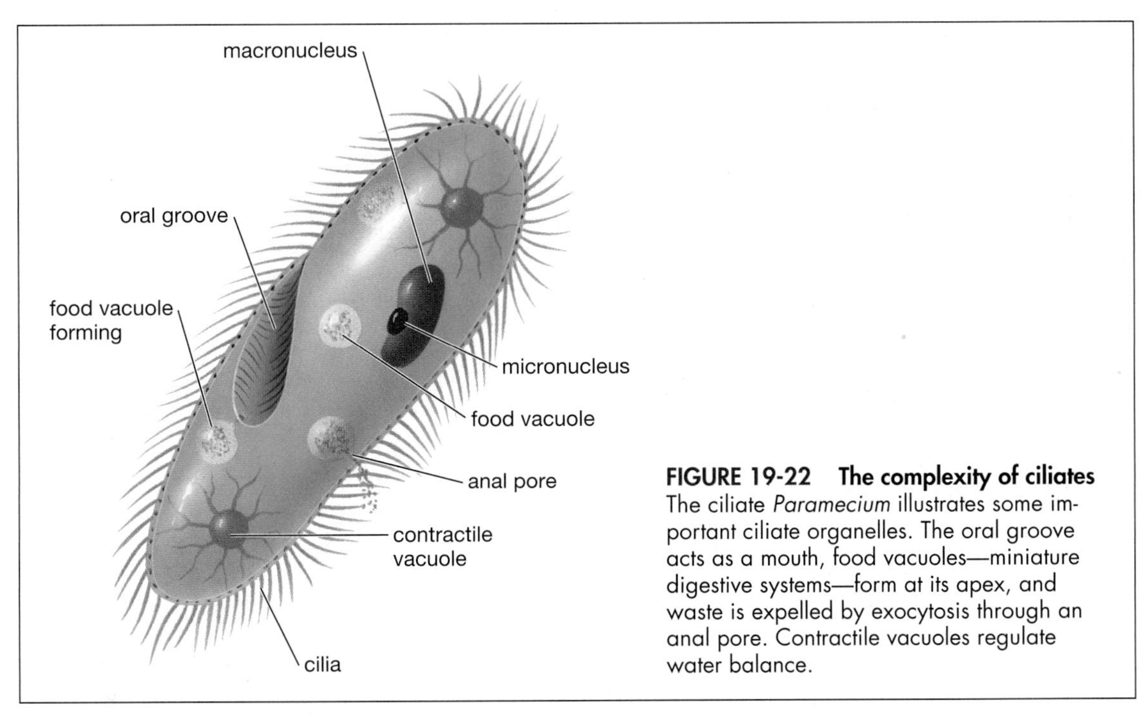
macronucleus
oral groove
food vacuole forming
micronucleus
food vacuole
anal pore
contractile vacuole
cilia

FIGURE 19-22 The complexity of ciliates The ciliate *Paramecium* illustrates some important ciliate organelles. The oral groove acts as a mouth, food vacuoles—miniature digestive systems—form at its apex, and waste is expelled by exocytosis through an anal pore. Contractile vacuoles regulate water balance.

FIGURE 19-23 A microscopic predator
In this scanning electron micrograph, the predatory ciliate *Didinium* attacks a *Paramecium*. Note that the cilia of *Didinium* are confined to two bands, whereas *Paramecium* has cilia over its entire body. Ultimately, the predator will engulf and consume its prey. This microscopic drama could take place on a pinpoint with room to spare.

of diploid nuclei, the nuclei are not confined in separate cells surrounded by plasma membranes, as in most multicellular organisms. This structure, called a **plasmodium**, explains why these protists are described as "acellular" (without cells). The plasmodium oozes through decaying leaves and rotting logs, engulfing food such as bacteria and particles of organic material. The mass may be bright yellow or orange—a large plasmodium can be rather startling (Fig. 19-24a). Dry conditions or starvation stimulate the plasmodium to form a fruiting body, on which haploid spores are produced (Fig. 19-24b). The spores are dispersed and germinate under favorable conditions, eventually giving rise to a new plasmodium.

Cellular Slime Molds Live as Independent Cells but Aggregate into a Pseudoplasmodium When Food Is Scarce

The **cellular slime molds** live in soil as independent haploid cells that move and feed by producing extensions called **pseudopods**. These extensions surround and engulf food such as bacteria. In the best-studied genus, *Dictyostelium*, individual cells release a chemical signal when food becomes scarce. This signal attracts nearby cells into a dense aggregation that forms a sluglike mass called a **pseudoplasmodium** ("false plasmodium") because, unlike a true plasmodium, it actually consists of individual cells (Fig. 19-25). The pseudoplasmodium then behaves like a multicellular organism. After crawling toward a source of light, the cells in the aggregation take on specific roles, forming a fruiting body. Haploid spores formed within the fruiting body are dispersed by wind and germinate directly into new single-celled individuals.

Euglenoids Lack a Rigid Covering and Swim by Means of Flagella

Euglenoids are single-celled protists that live mostly in fresh water and are named after the group's best-known representative, *Euglena* (Fig. 19-26), a complex single cell that moves about by whipping its flagellum through water. Many euglenoids are photosynthetic, but some species instead absorb or engulf food. Unlike diatoms or dinoflagellates, euglenoids lack a rigid outer covering, so some can move by wriggling as well as by whipping their flagella. Some euglenoids also possess simple light-sensing organelles consisting of a photoreceptor, called an

(a)

(b)

FIGURE 19-24 The acellular slime mold *Physarum*
(a) *Physarum* oozes over a stone on the damp forest floor. **(b)** When food becomes scarce, the mass differentiates into black fruiting bodies in which spores are formed.

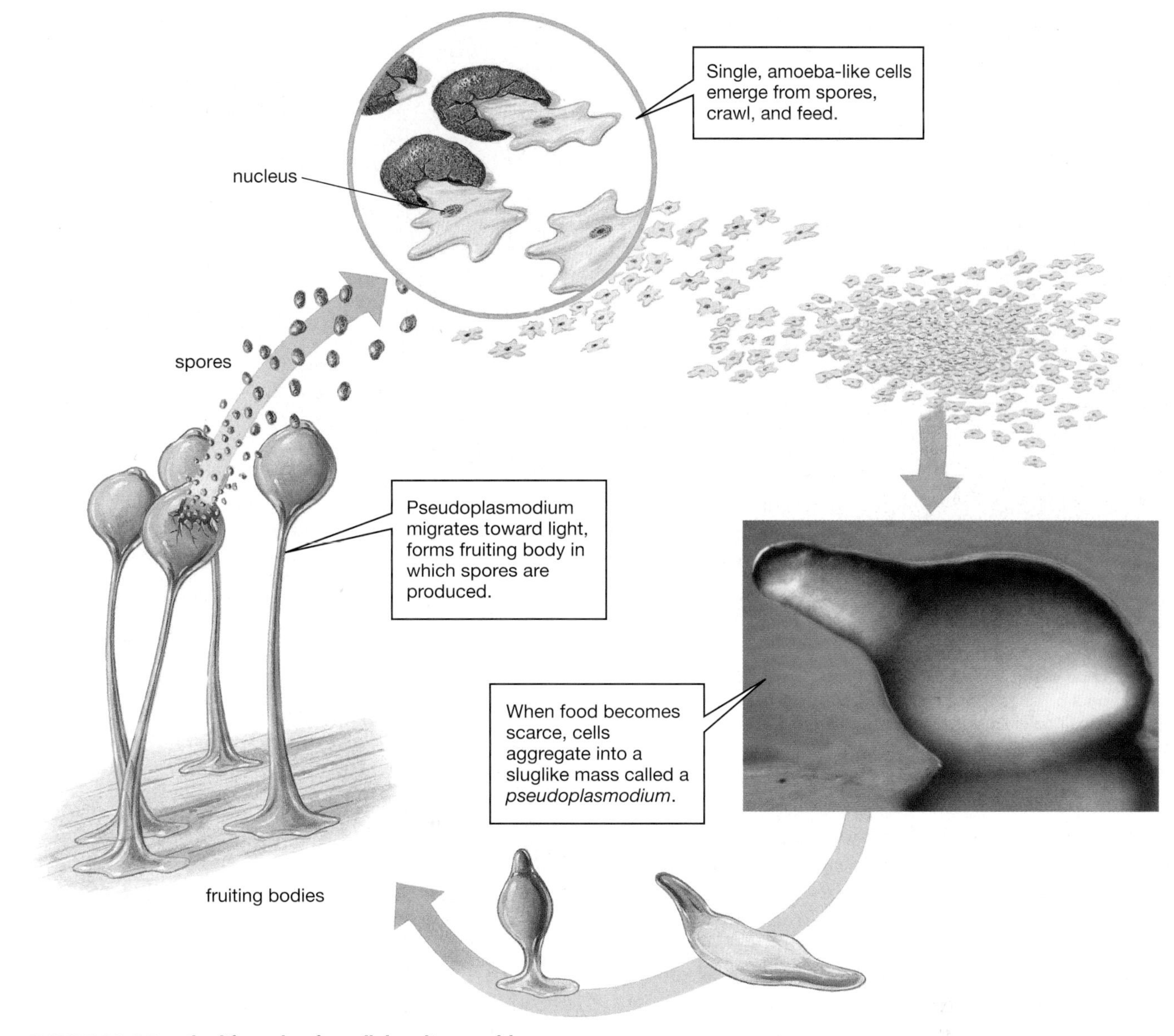
Single, amoeba-like cells emerge from spores, crawl, and feed.
nucleus
spores
Pseudoplasmodium migrates toward light, forms fruiting body in which spores are produced.
When food becomes scarce, cells aggregate into a sluglike mass called a *pseudoplasmodium*.
fruiting bodies

FIGURE 19-25 The life cycle of a cellular slime mold

eyespot, and an adjacent patch of pigment. The pigment shades the photoreceptor only when light strikes from certain directions, enabling the organism to determine the direction of the light source. Using information from the photoreceptor, the flagellum propels the protist toward light levels appropriate for photosynthesis.

Red Algae Live Primarily in Clear Tropical Oceans

The red algae are multicellular, photosynthetic seaweeds (Fig. 19-27). These protists range in color from bright red to nearly black, and derive their color from red pigments that mask their green chlorophyll. Red algae are found almost exclusively in marine environments. They dominate in deep, clear tropical waters, where their red pigments absorb the deeply penetrating blue-green light and transfer this light energy to chlorophyll, where it is used in photosynthesis.

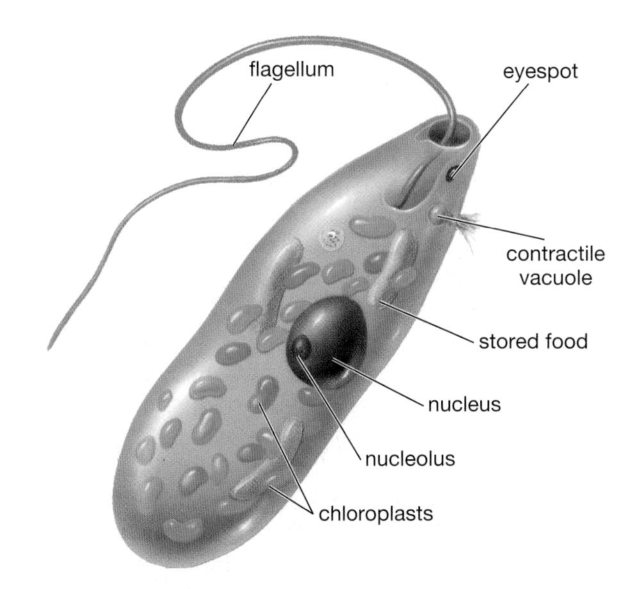
flagellum
eyespot
contractile vacuole
stored food
nucleus
nucleolus
chloroplasts

FIGURE 19-26 *Euglena*, a representative euglenoid
Euglena's elaborate single cell is packed with green chloroplasts, which will disappear if the protist is kept in darkness.

FIGURE 19-27 Red algae
Red coralline algae from the Pacific Ocean off California provide an anchoring site for other organisms. Coralline algae, which deposit calcium carbonate within their bodies, contribute to coral reefs in tropical waters.

Some species of red algae deposit calcium carbonate, which forms limestone, in their tissues and contribute to the formation of reefs. Other species are harvested for food in Asia. Red algae also contain certain gelatinous substances with commercial uses, including carrageenan (used as a stabilizing agent in products such as paints, cosmetics, and ice cream) and agar (a substrate for growing bacterial colonies in laboratories). However, the major importance of these and other algae lies in their photosynthetic ability; the energy they capture helps support nonphotosynthetic organisms in marine ecosystems.

Zooflagellates Possess Flagella

All **zooflagellates** possess at least one flagellum, which may propel the organism, sense the environment, or ensnare food. Many zooflagellates are free-living, inhabiting soil and water; others are *symbiotic*, living inside other organisms in a relationship that may be either mutually beneficial or parasitic. One symbiotic form can digest cellulose and lives in the gut of termites, where it helps them extract energy from wood. A more dangerous symbiotic zooflagellate in the genus *Trypanosoma* is responsible for African sleeping sickness, a potentially fatal disease (Fig. 19-28). Like many parasites, this organism has a complex life cycle, part of which is spent in the tsetse fly. While feeding on the blood of a mammal, the fly transmits the trypanosome to the mammal. The parasite then develops in the new host (which may be a human) and enters the bloodstream. It may then be ingested by another tsetse fly that bites the host, thus beginning a new cycle of infection.

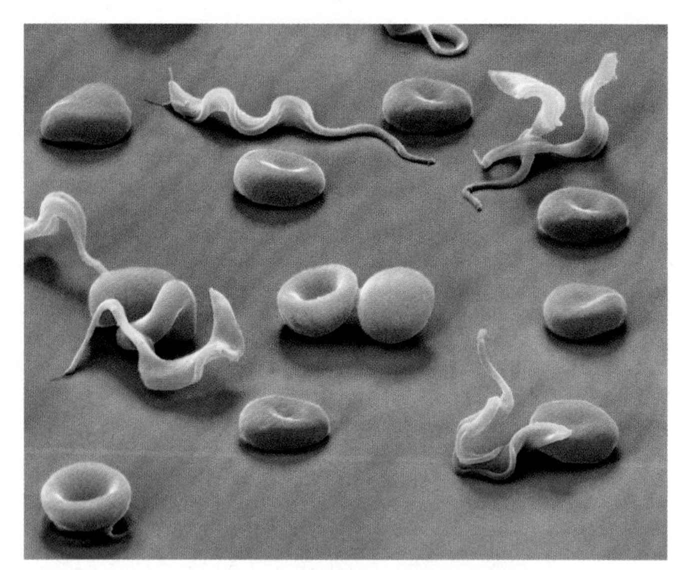

FIGURE 19-28 A disease-causing zooflagellate
This photomicrograph shows human blood that is heavily infested with the corkscrew-shaped, parasitic zooflagellate *Trypanosoma*, which causes African sleeping sickness. Note that the *Trypanosoma* are larger than red blood cells.

Another parasitic zooflagellate, *Giardia*, is an increasing problem in the United States, particularly to hikers who drink from apparently pure mountain streams. *Cysts* (tough structures that enclose the organism during one phase of its life cycle) of this flagellate are released in the feces of infected humans, dogs, or other animals; a single gram of feces may contain 300 million cysts. Once outside the animal's body, the cysts may enter freshwater streams and even community reservoirs. If a mammal drinks infected water, the cysts develop into the adult form (Fig. 19-29) in the small intestine of their mammalian host. In humans, infections can cause severe diarrhea, dehydration, nausea, vomit-

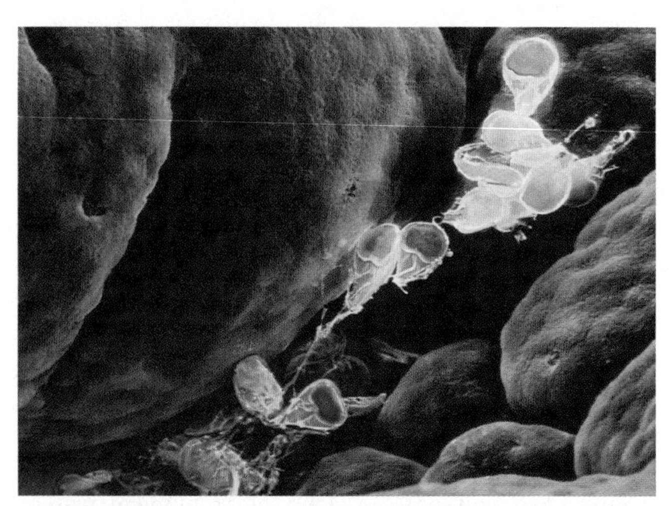

FIGURE 19-29 *Giardia:* The curse of campers
A zooflagellate (genus *Giardia*) that may infect drinking water, causing gastrointestinal disorders, is shown here in the human small intestine.

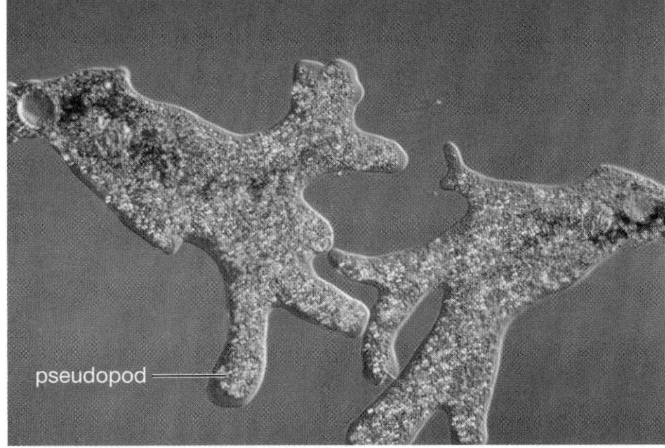

FIGURE 19-30 The amoeba
An amoeba uses cytoplasmic projections called *pseudopods* to move about and to capture prey.

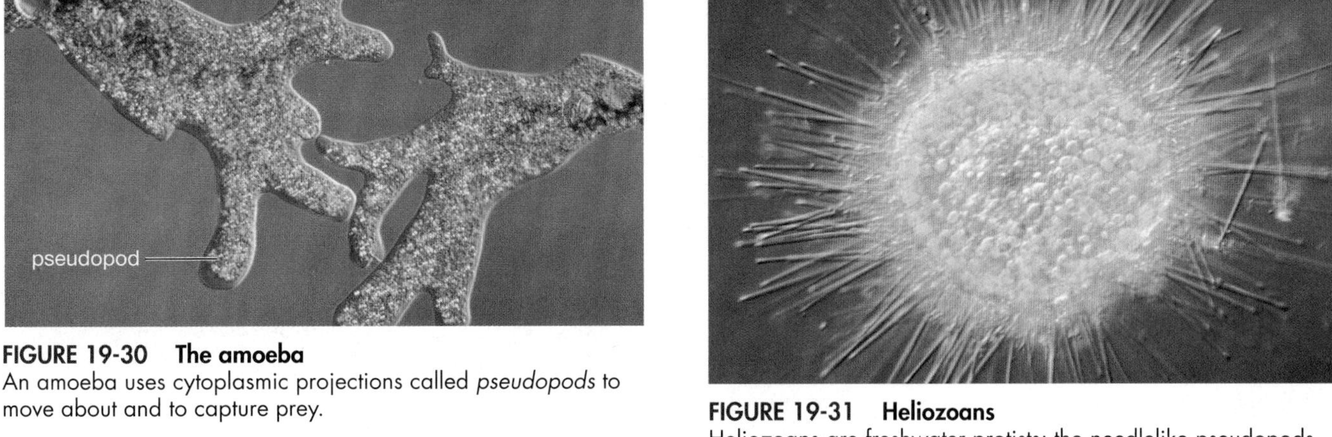

FIGURE 19-31 Heliozoans
Heliozoans are freshwater protists; the needlelike pseudopods are clearly visible in this specimen (genus *Acanthocystis*).

ing, and cramps. Fortunately, these infections can be cured with drugs, and deaths from *Giardia* infections are uncommon.

Various Protists Move by Means of Pseudopods

Like the slime molds, several other types of protist possess flexible plasma membranes that they can extend in any direction to form pseudopods, which are used for locomotion and for engulfing food (Fig. 19-30). Though these various protists all use pseudopods, they are probably not closely related to one another. But because a classification based on evolutionary history has not yet been developed for these groups, we'll discuss them together here.

Amoebas are common in freshwater lakes and ponds. Many amoebae are predators that stalk and engulf prey, but some species are parasites. One parasitic form causes amoebic dysentery, a disease that is prevalent in warm climates. The dysentery-causing amoeba multiplies in the intestinal wall, triggering severe diarrhea.

Heliozoans ("sun animals"), a striking form of freshwater protist, may be found floating in ponds or attached by stalks to an underwater plant or rock (Fig. 19-31). They have stiff, needlelike pseudopods, each of which is supported internally by a bundle of microtubules. Some heliozoans cover themselves with intricate and delicate shells of silica.

The **foraminiferans** and **radiolarians** are primarily marine protists that also produce beautiful shells. The shells of foraminiferans are constructed mostly of calcium carbonate (chalk; Fig. 19-32a); those of radiolarians are of silica (Fig. 19-32b). These elaborate shells are pierced by myriad openings through which pseudopods

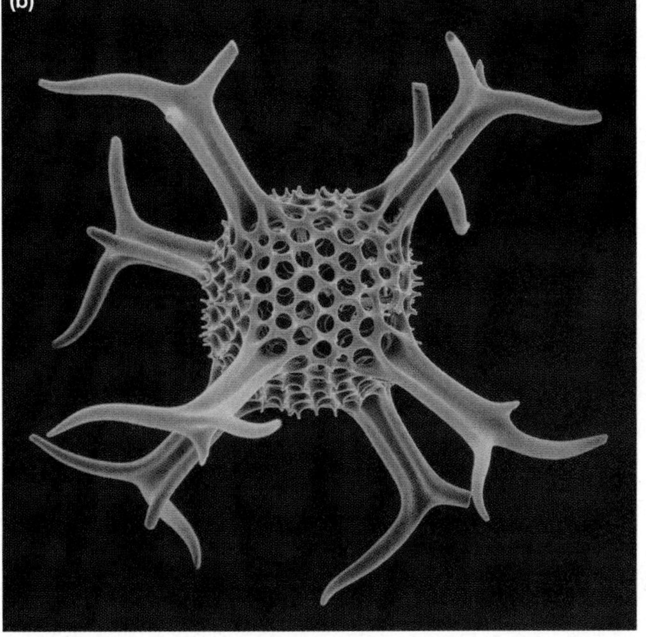

FIGURE 19-32 Foraminiferans and radiolarians
(a) The chalky shells of foraminiferans show numerous interior chambers. **(b)** The delicate, glassy shell of a radiolarian. Pseudopods, which sense the environment and capture food, extend out through the openings in the shell.

extend. The chalky shells of foraminiferans, accumulating over millions of years, have resulted in immense deposits of limestone such as those that form the famous White Cliffs of Dover, England.

Green Algae Live Mostly in Ponds and Lakes

The green algae, a large and diverse group of photosynthetic protists, include both multicellular and unicellular species. Most species live in freshwater ponds and lakes, but some live in the seas. Some green algae, such as *Spirogyra*, form thin filaments from long chains of cells (Fig. 19-33). Other species of green algae form colonies containing clusters of cells that are somewhat interdependent and constitute a structure intermediate between unicellular and multicellular forms. These colonies range from a few cells to a few thousand cells, as in species of *Volvox*. Most green algae are small, but some marine species are large. For example, the green alga *Ulva*, or sea lettuce, is similar in size to the leaves of its namesake.

The green algae are of special interest because, unlike other groups that contain multicellular, photosynthetic protists, green algae are closely related to plants. Plants share a common ancestor with some types of green algae, and many researchers believe that the very earliest plants were similar to today's multicellular green algae.

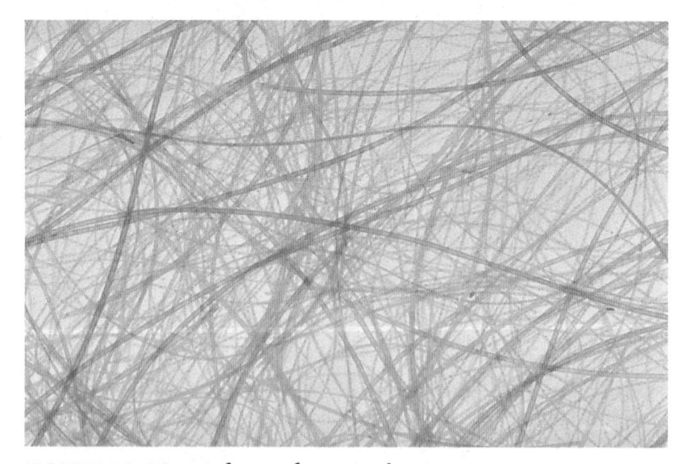

FIGURE 19-33 A form of green algae
Spirogyra is a filamentous green alga composed of strands only one cell thick.

EVOLUTIONARY CONNECTIONS

Our Unicellular Ancestors

Some of today's microbes are probably quite similar to the ancient species that ultimately gave rise to the complex multicellular organisms that are now the most conspicuous inhabitants of Earth. For example, the external appearance of many modern prokaryotes is basically indistinguishable from that of 3.5-billion-year-old fossilized cells. Similarly, the metabolism of today's anaerobic, heat-loving archaea is probably similar to the methods of energy acquisition used by Earth's earliest inhabitants, long before any oxygen entered the atmosphere. Likewise, modern purple sulfur bacteria and cyanobacteria are probably not too different from the first photosynthetic organisms that appeared more than 2 billion years ago.

Life might still consist solely of prokaryotic, single-celled organisms if protists, with their radical eukaryotic design, had not appeared nearly 2 billion years ago. As you learned in the discussion of the endosymbiont theory in Chapter 17, eukaryotic cells may have originated when one prokaryote, perhaps a bacterium capable of aerobic respiration, took up residence inside a partner, forming the first "mitochondrion." A separate but equally crucial merger may have occurred when a photosynthetic bacterium (probably resembling a cyanobacterium) took up residence within a nonphotosynthetic partner and became the first "chloroplast." The foundations of multicellularity were laid with the eukaryotic cell, whose intricacy allowed specialization of entire cells for specific functions within a multicellular aggregation. Thus, primitive protists, some absorbing nutrients from the environment, some photosynthesizing, and others consuming their food in chunks, almost certainly followed divergent evolutionary paths that led to the three multicellular kingdoms—the fungi, plants, and animals—that are the subjects of the following four chapters.

CASE STUDY REVISITED Agents of Death

Although anthrax is thought to be the most likely biological weapon, many other infectious agents also have the potential to become weapons. These include the viruses that cause smallpox and Ebola hemorrhagic fever, and the bacteria that cause plague. Evidence also exists that some countries are trying to use genetic engineering to "improve" pathogens: for example, by adding antibiotic-resistance genes to plague bacteria so that victims of an attack would be more difficult to treat and more likely to die.

Before 2001, humankind relied on politics, diplomacy, and widespread revulsion at the concept of biological warfare to protect us from its terrifying destructive potential. Now, however, it is painfully clear that humanity also includes people willing to unleash biological weapons. Unfortunately, little expertise is required to culture pathogenic bacteria or viruses, and the necessary supplies and equipment are easily acquired. Given the difficulty of preventing biological weapons from falling into the wrong hands, much current research is focused on developing tools to detect attacks and render them harmless.

It is not easy to detect a biological attack, as pathogens are invisible and symptoms may take hours or days to appear after an attack. Nonetheless, rapid detection is crucial if an effective response is to be mounted, and a variety of new detection technologies are being rapidly developed. Detectors must be able to distinguish released pathogens from the multitude of harmless microbes that ordinarily inhabit the air, water, and soil. One promising approach relies on sensors that incorporate living human immune cells that have been genetically altered to glow when receptor molecules in their cell membranes bind to a particular pathogen.

Once an attack is detected, the main task is to care for those who have been exposed. Thus, developing fast-acting, easily distributed postexposure treatments is a top priority for researchers. For example, biologists have intensively investigated the mechanism by which the toxin released by anthrax bacteria attacks and damages cells. Better understanding of this process has improved investigators' ability to block it, and has yielded several promising ideas for antidotes that could be used in conjunction with antibiotics as a treatment for anthrax exposure.

Consider This: The threat of biological attack has prompted a debate: should large numbers of people be immunized against potential agents of attack for which vaccinations exist? Mass vaccinations are costly and would inevitably cause some deaths due to occasional adverse reactions. Is the increased protection and peace of mind that would come with vaccinations worth the price?

Links to Life: Unwelcome Dinner Guests

Although the prospect of biological weapons is chilling, you are far more likely to encounter harmful microorganisms from a more mundane source: your food. The nutrients that you consume during meals and snacks can also provide sustenance for a wide variety of disease-causing bacteria and protists. Some of these invisible diners may accompany your lunch to your digestive tract and take up residence there, causing unpleasant symptoms. The Centers for Disease Control estimates that U.S. residents experience an astonishing 76 million cases of food-borne illness each year, resulting in 325,000 hospitalizations and 5200 deaths.

The most frequent culprits in food-borne diseases are bacteria. Species in the genera *Escherichia, Salmonella, Listeria, Streptococcus,* and *Campylocbacter* are responsible for an especially large number of illnesses, with *Campylocbacter* currently claiming the largest number of victims. Harmful protist parasites, especially apicomplexans such as *Crytosporidium* and *Cyclospora,* also make their way into human foods.

How can you protect yourself from the bacteria and protists that share our food supply? It's easy: clean, cook, and chill. Cleaning helps prevent the spread of pathogens. Wash your hands before preparing food, and wash all utensils and cutting boards after preparing each item. Thorough cooking is the best way to ensure that any bacteria and protists present in food are killed. Meats, in particular, must be thoroughly cooked; do *not* eat meat that is still pink inside. Fish should be cooked until it is opaque and flakes easily with a fork; cook eggs until both white and yolk are firm. Finally, keep food cold. Pathogens multiply most rapidly at temperatures between 40 °F and 140 °F. So get your groceries home from the store and into the refrigerator or freezer as quickly as possible. Don't leave cooked leftovers unrefrigerated for more than two hours. Thaw frozen foods in the refrigerator, not at room temperature. A little bit of attention to food safety can save you from unwelcome guests in your food.

Summary of Key Concepts

19.1 What Are Viruses, Viroids, and Prions?

Viruses are parasites consisting of a protein coat that surrounds genetic material. They are noncellular and unable to move, grow, or reproduce outside a living cell. They invade cells of a specific host and use the host cell's energy, enzymes, and ribosomes to produce more virus particles, which are liberated when the cell ruptures. Many viruses are pathogenic to humans, including those causing colds and flu, herpes, AIDS, and certain forms of cancer.

Viroids are short strands of RNA that can invade a host cell's nucleus and direct the synthesis of new viroids. To date, viroids are known to cause only certain diseases of plants.

Prions have been implicated in diseases of the nervous system, such as kuru, Creutzfeldt-Jakob disease, and scrapie. Prions are unique in that they lack genetic material. They are composed solely of mutated prion protein, which may act as an enzyme, catalyzing the formation of more prions from normal prion protein.

19.2 Which Organisms Make Up the Prokaryotic Domains—Bacteria and Archaea?

Members of the domains Bacteria and Archaea—the bacteria and archaea—are unicellular and prokaryotic. Archaea and bacteria are not closely related and differ in several fundamental features, including cell wall composition, ribosomal RNA sequence, and membrane lipid structure. A cell wall determines the characteristic shapes of prokaryotes: round, rodlike, or spiral. Certain types of bacteria can form spores that disperse widely and withstand inhospitable environmental conditions. Prokaryotes obtain energy in a variety of ways. Some, including the cyanobacteria, rely on photosynthesis. Others are chemosynthetic, breaking down inorganic molecules to obtain energy. Heterotrophic forms are capable of consuming a wide variety of organic compounds. Many are anaerobic, able to obtain energy from fermentation when oxygen is not available.

Some bacteria are pathogenic, causing disorders such as pneumonia, tetanus, botulism, and the sexually transmitted diseases gonorrhea and syphilis. Most bacteria, however, are harmless to humans and play important roles in natural ecosystems. Bacteria and archaea have colonized nearly every habitat on Earth, including hot, acidic, very salty, and anaerobic environments. Some live in the digestive tracts of ruminants and break down cellulose. Nitrogen-fixing bacteria enrich the soil and aid in plant growth; many others live off the dead bodies and wastes of other organisms, liberating nutrients for reuse.

19.3 What Are Protists?

Most protists are single, highly complex eukaryotic cells, but some form colonies and some, such as seaweeds, are multicellular. Photosynthetic protists form much of the phytoplankton, which plays a key ecological role. Protists exhibit diverse modes of nutrition, reproduction, and locomotion. Protist groups include the chromists (water molds, diatoms, and brown algae), the alveolates (dinoflagellates, apicomplexans, and ciliates), slime molds, euglenoids, red algae, zooflagellates, several groups of pseudopod-using organisms (amoebas, heliozoans, radiolarians, and foraminiferans), and green algae (the closest relatives of plants). Some protists, especially types of apicomplexans, cause human diseases. Others, especially certain water molds, are crop pests.

Key Terms

acellular slime mold *p. 376*
algae *p. 373*
alveolate *p. 375*
amoeba *p. 381*
anaerobe *p. 369*
apicomplexan *p. 376*
bacteriophage *p. 360*
cellular slime mold *p. 378*
chromist *p. 373*
cilia *p. 376*
ciliate *p. 376*
conjugation *p. 368*
diatom *p. 374*
dinoflagellate *p. 375*
endospore *p. 367*
euglenoid *p. 378*
flagellum *p. 366*
foraminiferan *p. 381*
Gram stain *p. 366*
heliozoan *p. 381*
host *p. 360*
nitrogen-fixing bacterium *p. 370*
pathogenic *p. 371*
phytoplankton *p. 374*
plasmid *p. 368*
plasmodial slime mold *p. 376*
plasmodium *p. 378*
prion *p. 364*
protist *p. 372*
protozoa *p. 373*
pseudoplasmodium *p. 378*
pseudopod *p. 378*
radiolarian *p. 381*
viroid *p. 364*
virus *p. 360*
water mold *p. 374*
zooflagellate *p. 380*

Thinking Through the Concepts

To take a multiple-choice quiz with feedback on the contents of this chapter, visit http://www.prenhall.com/audesirk7. *Log in to the Web site selected by your instructor and navigate to the Self Test section for this chapter.*

? Review Questions

1. Describe the structure of a typical virus. How do viruses replicate?
2. List the major differences between prokaryotes and protists.
3. Describe some of the ways in which bacteria obtain energy and nutrients.
4. What are nitrogen-fixing bacteria, and what role do they play in ecosystems?
5. What is an endospore? What is its function?
6. Describe some examples of bacterial symbiosis.
7. What is the importance of dinoflagellates in marine ecosystems? What happens when they reproduce rapidly?
8. What is the major ecological role played by single-celled algae?
9. What protist group consists entirely of parasitic forms?

Applying the Concepts

1. In some developing countries, antibiotics can be purchased without a prescription. Why do you think this is done? What biological consequences would you predict?
2. Before the discovery of prions, many (perhaps most) biologists would have agreed with the statement, "It is a fact that no infectious organism or particle can exist that lacks nucleic acid (such as DNA or RNA)." What lessons do prions teach us about nature, science, and scientific inquiry? You may wish to review Chapter 1 to help answer this question.
3. Recent research shows that ocean water off southern California has become 2 to 3 °F (1 to 1.5 °C) warmer over the past four decades, possibly due to the greenhouse effect. This warming has indirectly led to a depletion of nutrients in the water and thus a decline in photosynthetic protists such as diatoms. What effects is this warming likely to have on life in the oceans?
4. Argue for and against the statement, "Viruses are alive."
5. The internal structure of many protists is much more complex than that of cells of multicellular organisms. Does this mean that the protist is engaged in more complex activities than the multicellular organism? If not, why are protistan cells more complicated?
6. Why would the lives of multicellular animals be impossible if prokaryotic and protistan organisms did not exist?

For More Information

Costerton, J., and Stewart, P. "Battling Biofilms. *Scientific American*, July 2001. How biofilms form and how to fight them.

Madigan, M., and Marrs, B. "Extremophiles." *Scientific American*, April 1997. Prokaryotes that prosper under extreme conditions, and potential industrial uses of the enzymes that they use to do so.

Prusiner, S. "The Prion Diseases." *Scientific American*, January 1995. A description of prions and the research that led to their discovery, from the point of view of the most influential scientist in the field.

Raloff, J. "Taming Toxins." *Science News*, November 30, 2002. Describes a possible new strategy for combating red tides and other blooms of toxic dinoflagellates.

Young, J., and Collier, R. J. "Attacking Anthrax." *Scientific American*, March 2002. A summary of recent research that could help develop new techniques for detecting and treating anthrax.

Media Activities

To access a Media Activity visit http://www.prenhall.com/audesirk7. *Log in to the Web site selected by your instructor, navigate to this chapter, and select the appropriate Media Activity number.*

19.1 Retrovirus Replication

Estimated time: 5 minutes

The Closer Look essay in this chapter describes the structure of retroviruses and illustrates the process by which they reproduce. This activity demonstrates the uptake of a retrovirus (in this case HIV) by its host cell (a helper T cell) and the process by which new HIV components are produced and assembled into new viruses.

19.2 Herpes Virus Replication

Estimated time: 5 minutes

The herpes virus uses a different mechanism to invade human cells. Explore the details in this animation.

19.3 Bacterial Conjugation

Estimated time: 5 minutes

This animation shows how some prokaryotes are able to exchange genetic information through the process of conjugation.

19.4 Web Investigation: Agents of Death

Estimated time: 10 minutes

While biological weapons are not new, bacterial strains, delivery mechanisms, and toxins have become much more sophisticated. In this exercise we will study biological weapons from a biological, functional, and social perspective.

CHAPTER

20 The Diversity of Fungi

These honey mushrooms are part of the visible portion of the largest organism on Earth.

AT A GLANCE

CASE STUDY Humongous Fungus

What is the largest organism on Earth? A reasonable guess might be the world's largest animal, the blue whale, which can be 100 feet long and weigh 300,000 pounds. But the blue whale is dwarfed by the General Sherman tree, a giant sequoia specimen that is 275 feet high and whose weight is estimated at 6200 *tons*. Even these two behemoths, however, are pipsqueaks compared to the real record-holder, the fungus *Armillaria ostoyae*, also known as the honey mushroom. The largest known *Armillaria* is a specimen in Oregon that spreads over 2200 acres (about 3.4 square miles), and probably weighs even more than the General Sherman tree. Despite its huge size, no one has actually seen the monster fungus, as it is largely underground; its only aboveground parts are brown mushrooms that sprout occasionally from the creature's gigantic body. Just beneath the surface, however, the fungus spreads through the soil by means of long, string-like structures called rhizomorphs. These rhizomorphs extend until they encounter the tree roots on which *Armillaria* subsists, causing "root rot" that weakens or kills trees. This "root rot" provides aboveground evidence of *Armillaria*'s existence; the giant Oregon specimen was first identified by examining aerial photos to find forested areas with many dead trees.

How can researchers be sure that the Oregon fungus is truly one single individual and not many intertwined individuals? The strongest evidence is genetic. Researchers gathered *Armillaria* tissue samples from throughout the area thought to be inhabited by a single individual and compared DNA extracted from the samples. All were genetically identical, demonstrating that they came from the same individual.

It may seem strange that the world's largest organisms went unnoticed until very recently, but the lives of fungi typically take place outside of our view. Nonetheless, fungi play a fascinating role in human affairs. Read on to find out more about the inconspicuous but often influential members of kingdom Fungi.

20.1 What Are the Key Features of Fungi?

When you think of a fungus, you probably picture a mushroom. Mushrooms, however, are just temporary reproductive structures that extend from the main bodies of certain kinds of fungi. The body of almost all fungi is a **mycelium** (Fig. 20-1a), which is an interwoven mass of one-cell-thick, threadlike filaments called **hyphae** (singular, hypha; Fig. 20-1b,c). Depending on the species, hyphae either consist of single elongated cells with numerous nuclei or are subdivided—by partitions called **septa** (singular, septum)—into many cells, each containing from one to many nuclei. Pores in the septa allow cytoplasm to stream between cells, distributing nutrients. Like plant cells, fungal cells are surrounded by cell walls. Unlike plant cells, however, fungal cell walls are strengthened by *chitin*, the same substance found in the exoskeletons of arthropods.

Fungi cannot move. They compensate for this lack of mobility with filaments that can grow rapidly in any direction within a suitable environment. In this way, a fungal mycelium can quickly infuse itself into aging bread or cheese, beneath the bark of decaying logs, or into the soil. Periodically, the hyphae grow together and differentiate into reproductive structures that project above the surface beneath which the mycelium grows. These structures, including mushrooms, puffballs, and the powdery molds on unrefrigerated food, represent only a fraction of the complete fungal body, but are typically the only part of the fungus that we can easily see.

Fungi Obtain Their Nutrients from Other Organisms

Like animals, fungi survive by breaking down nutrients stored in the bodies or wastes of other organisms. Some fungi digest the bodies of dead organisms. Others are parasitic, feeding on living organisms and causing disease. Others live in close, mutually beneficial relationships with other organisms that provide food. There are even a few predatory fungi, which attack tiny worms in soil.

Unlike animals, fungi do not ingest food. Instead, they secrete enzymes that digest complex molecules outside their bodies, breaking down the molecules into smaller subunits that can be absorbed. Fungal filaments can penetrate deeply into a source of nutrients and are only one cell thick, presenting an enormous surface area through which to secrete enzymes and absorb nutrients. This mode of securing nutrition serves fungi well. Almost every biological material can be consumed by at least one fungal species, so nutritional support for fungi is likely to be present in nearly every terrestrial habitat.

Fungi Propagate by Spores

Unlike plants and animals, fungi do not form embryos. Instead, fungi reproduce by means of **spores**—tiny, lightweight reproductive packages that are extraordinarily mobile, even though most lack a means for self-propulsion. Spores are distributed far and wide as hitchhikers on the outside of animal bodies, as passengers inside the digestive systems of animals that have eaten them, or as airborne drifters, cast aloft by chance or shot into the atmosphere by elaborate reproductive

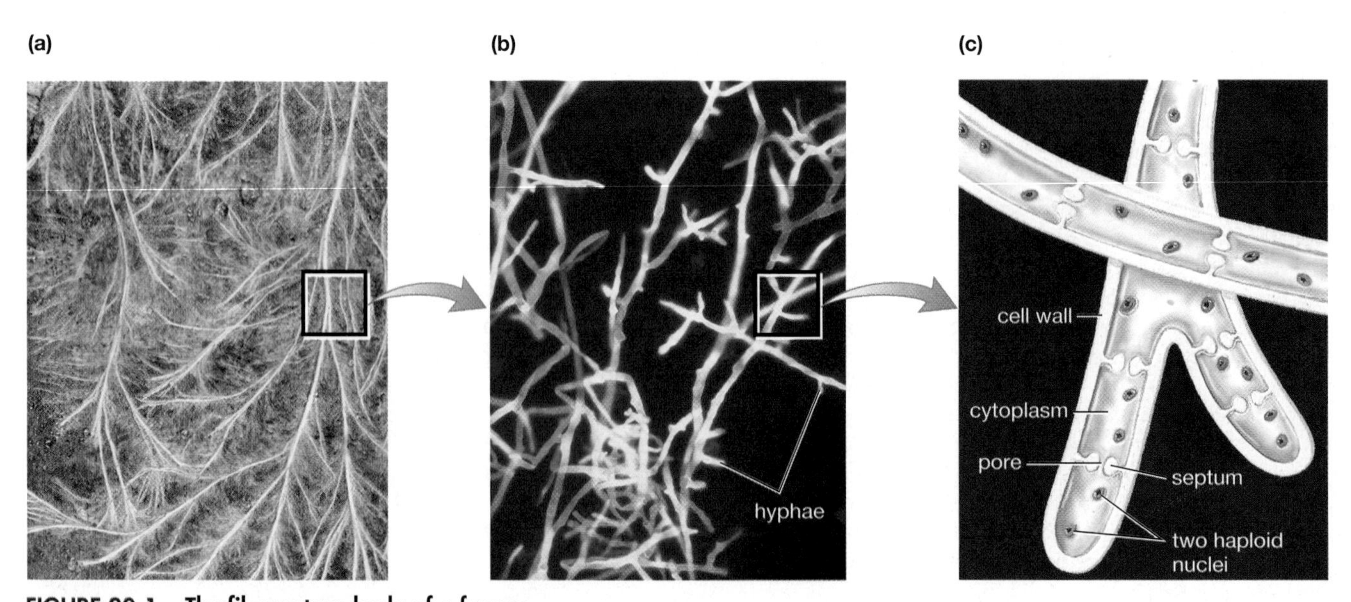

FIGURE 20-1 The filamentous body of a fungus
(a) A fungal mycelium spreads over decaying vegetation. The mycelium is composed of **(b)** a tangle of microscopic hyphae, only one cell thick, portrayed in cross section **(c)** to show their internal organization. **QUESTION** Which features of a fungus's body structure are adaptations related to its method of acquiring nutrients?

FIGURE 20-2 Some fungi can eject spores
A ripe earthstar mushroom, struck by a drop of water, releases a cloud of spores that will be dispersed by air currents.

structures (Fig. 20-2). Spores are often produced in great numbers (a single giant puffball may contain 5 trillion sexual spores; see Fig. 20-7a). The fungal combination of prodigious reproductive capacity and highly mobile spores ensures that fungi are ubiquitous in terrestrial environments and accounts for the inevitable growth of fungi on every uneaten sandwich and container of leftovers.

Most Fungi Can Reproduce Both Sexually and Asexually

In general, fungi are capable of both asexual and sexual reproduction. For the most part, fungi reproduce asexually by default under stable conditions, with sexual reproduction occurring mainly under conditions of environmental change or stress. Both asexual and sexual reproduction ordinarily involve the production of spores within special fruiting bodies that project above the mycelium.

The bodies and spores of fungi are haploid (contain only a single copy of each chromosome). A haploid mycelium produces haploid asexual spores by mitosis. If an asexual spore is deposited in a favorable location, it will begin mitotic divisions and develop into a new mycelium. This simple reproductive cycle results in the rapid production of genetically identical clones of the original mycelium.

Diploid structures form only during a brief period of the sexual portion of the fungal life cycle. Sexual reproduction begins when a filament of one mycelium comes into contact with a filament from a second mycelium that is of a different, but compatible, mating type (the different mating types of fungi are analogous to the different sexes of animals, except that there are often more than two mating types). If conditions are suitable, the two hyphae may fuse, so that nuclei from the two different hyphae share a common cell. This merger of hyphae is followed (immediately in some species, after some delay in others) by fusion of the two different haploid nuclei to form a diploid zygote. The zygote then undergoes meiosis to form haploid sexual spores. These spores are dispersed, germinate, and divide by mitosis to form new haploid mycelia. Unlike the cloned offspring of asexual spores, these sexually produced fungal bodies are genetically distinct from either parent.

20.2 How Are Fungi Classified?

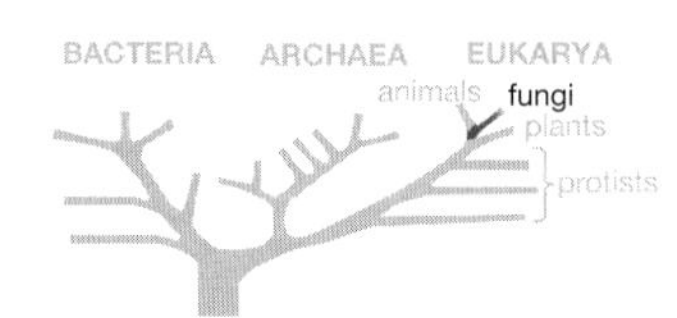

Although nearly 100,000 species of modern fungi have been described, biologists have only begun to comprehend the diversity of these organisms—at least 1000 additional species are described each year. Fungi belong to the following phyla: Chytridiomycota (chytrids), Zygomycota (zygote fungi), Ascomycota (sac fungi), and Basidiomycota (club fungi) (Table 20-1). Species that cannot be readily classified are placed, for convenience, in a group known as the *deuteromycetes* (imperfect fungi).

Media Activity
20.2 Classification of Fungi

The Chytrids Produce Swimming Spores

Unlike other types of fungi, most chytrids live in water. The chytrids (Fig. 20-3) are further distinguished from other fungi by their swimming spores, which require

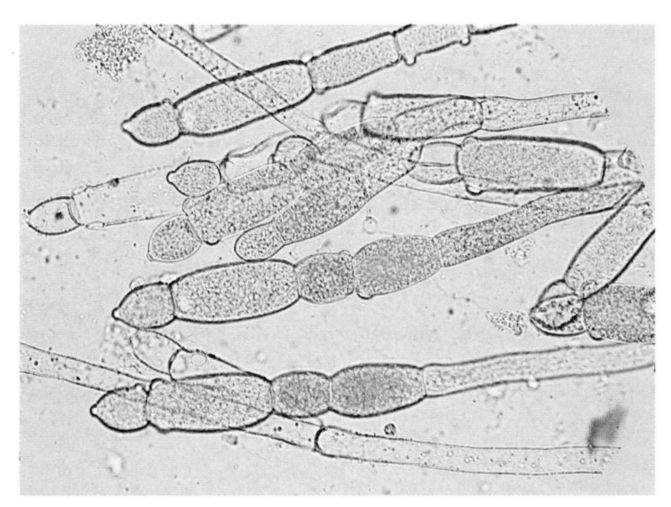

FIGURE 20-3 Chytrid filaments
These filaments of the chytrid fungus *Allomyces* are in the midst of sexual reproduction. The orange structures visible on many of the filaments will release male gametes; the clear structures will release female gametes. Chytrid gametes are flagellated, and these swimming reproductive structures aid dispersal of members of this mostly aquatic phylum.

Table 20-1 The Phyla of Fungi

Common Name (Phylum)	Reproductive Structures	Cellular Characteristics	Economic and Health Impacts	Representative Genera
Chytrids (Chytridiomycota)	Flagellated spores	Cell walls contain chitin; septa are absent	Contribute to decline of frog populations	*Batrachochytrium* (frog pathogen)
Zygote fungi (Zygomycota)	Produce sexual diploid zygospores	Cell walls contain chitin; septa are absent	Cause soft fruit rot and black bread mold	*Rhizopus* (causes black bread mold); *Pilobolus* (dung fungus)
Sac fungi (Ascomycota)	Sexual spores formed in saclike ascus	Cell walls contain chitin; septa are present	Cause molds on fruit; can damage textiles; cause Dutch elm disease and chestnut blight; include yeasts and morels	*Saccharomyces* (yeast); *Ophiostoma* (causes Dutch elm disease)
Club fungi (Basidiomycota)	Sexual reproduction involves production of haploid basidiospores on club-shaped basidia	Cell walls contain chitin; septa are present	Cause smuts and rusts on crops; include some edible mushrooms	*Amanita* (poisonous mushroom); *Polyporus* (shelf fungus)

water for dispersal (even soil-dwelling chytrids require a film of water for reproduction). A chytrid spore propels itself through the water by means of a single flagellum located on one end of the spore. No other fungus group has flagella.

Research by fungal systematists suggests that the chytrids form an ancient group that predates and gave rise to the other groups of modern fungi. This conclusion is bolstered by the fossil record, as the oldest known fossil fungi are chytrids that were found in rocks more than 600 million years old. Ancestral fungi may well have been similar in habit to today's aquatic and marine chytrids, so fungi (like plants and animals) probably originated in a watery environment before colonizing land.

Most chytrid species feed on dead aquatic plants or other debris in watery environments, but some species are parasites of plants or animals. One such parasitic chytrid is believed to be a major cause of the current worldwide die-off of frogs, which threatens many species and has apparently already caused the extinction of several. No one yet understands exactly why this fungal disease emerged as a major cause of death in frogs. One hypothesis is that frog populations under stress from pollution and other environmental challenges might be more susceptible to infection by chytrids.

The Zygote Fungi Can Reproduce by Forming Diploid Spores

The *zygomycetes*, also called the **zygote fungi**, generally live in soil or on decaying plant or animal material. The zygomycetes include species belonging to the genus *Rhizopus,* which cause the familiar annoyances of soft fruit rot and black bread mold. The life cycle of the black bread mold, which reproduces both asexually and sexually, is depicted in Figure 20-4. Asexual reproduction in zygote fungi is initiated by the formation of haploid spores in black spore cases called **sporangia**. These spores disperse through the air and, if they land on a suitable substrate (such as a piece of bread), germinate to form new haploid hyphae.

If two hyphae of different mating types of zygote fungi come into contact, sexual reproduction may ensue. The two hyphae "mate sexually," and their nuclei fuse to produce diploid **zygospores**: tough, resistant structures that give this group its name. Zygospores can remain dormant for long periods until environmental conditions are favorable for growth. Like asexually produced spores, zygospores disperse and germinate, but instead of producing new hyphae directly, they undergo meiosis. As a result, they form structures that bear haploid spores, which develop into new hyphae.

The Sac Fungi Form Spores in a Saclike Case

The *ascomycetes*, or **sac fungi**, also reproduce both asexually and sexually. Asexual spores of sac fungi are produced at the tips of specialized hyphae. During sexual reproduction, spores are produced by a complex sequence of events that begins when hyphae of two different mating types fuse. This sequence culminates in the formation of **asci** (singular, ascus), saclike cases that contain several spores and that give this phylum its name.

Some ascomycetes live in decaying forest vegetation and form either beautiful cup-shaped reproductive

FIGURE 20-4 The life cycle of a zygomycete
Top: During asexual reproduction in the black bread mold (genus *Rhizopus*), haploid spores, produced within sporangia, disperse and germinate on food such as bread. The resulting haploid hyphae may complete the asexual cycle by producing sporangia and spores. Bottom: During sexual reproduction, hyphae of different mating types (designated + and − on the bread) contact one another and fuse, producing a diploid zygospore. The zygospore undergoes meiosis and germinates, producing sporangia. The sporangia liberate haploid spores, which germinate into hyphae that can enter either the asexual or sexual cycle.

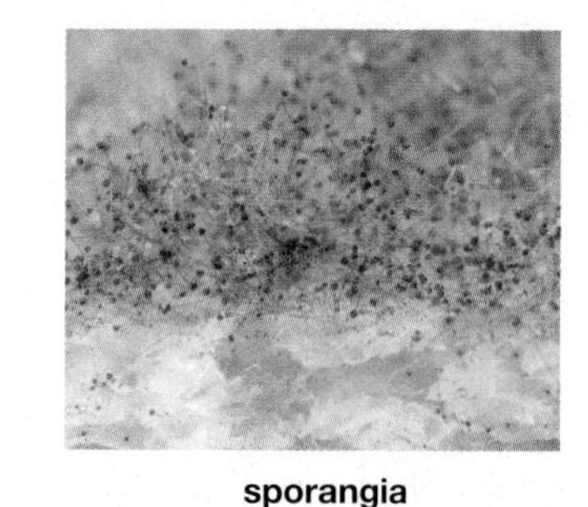

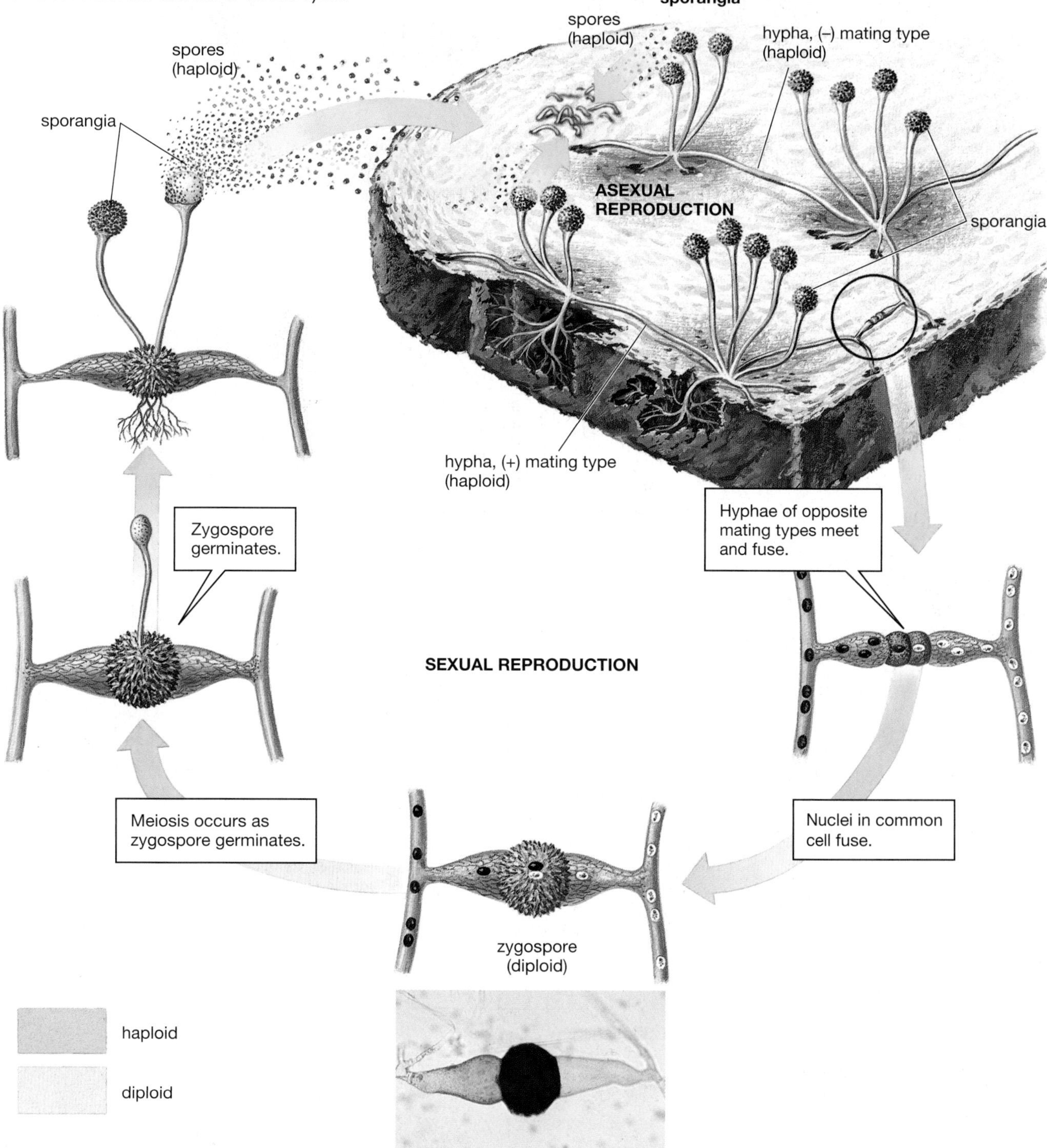

(a) (b)

FIGURE 20-5 Diverse ascomycetes
(a) The cup-shaped fruiting body of the scarlet cup fungus. **(b)** The morel, an edible delicacy. (Consult an expert before sampling any wild fungus—some are deadly!)

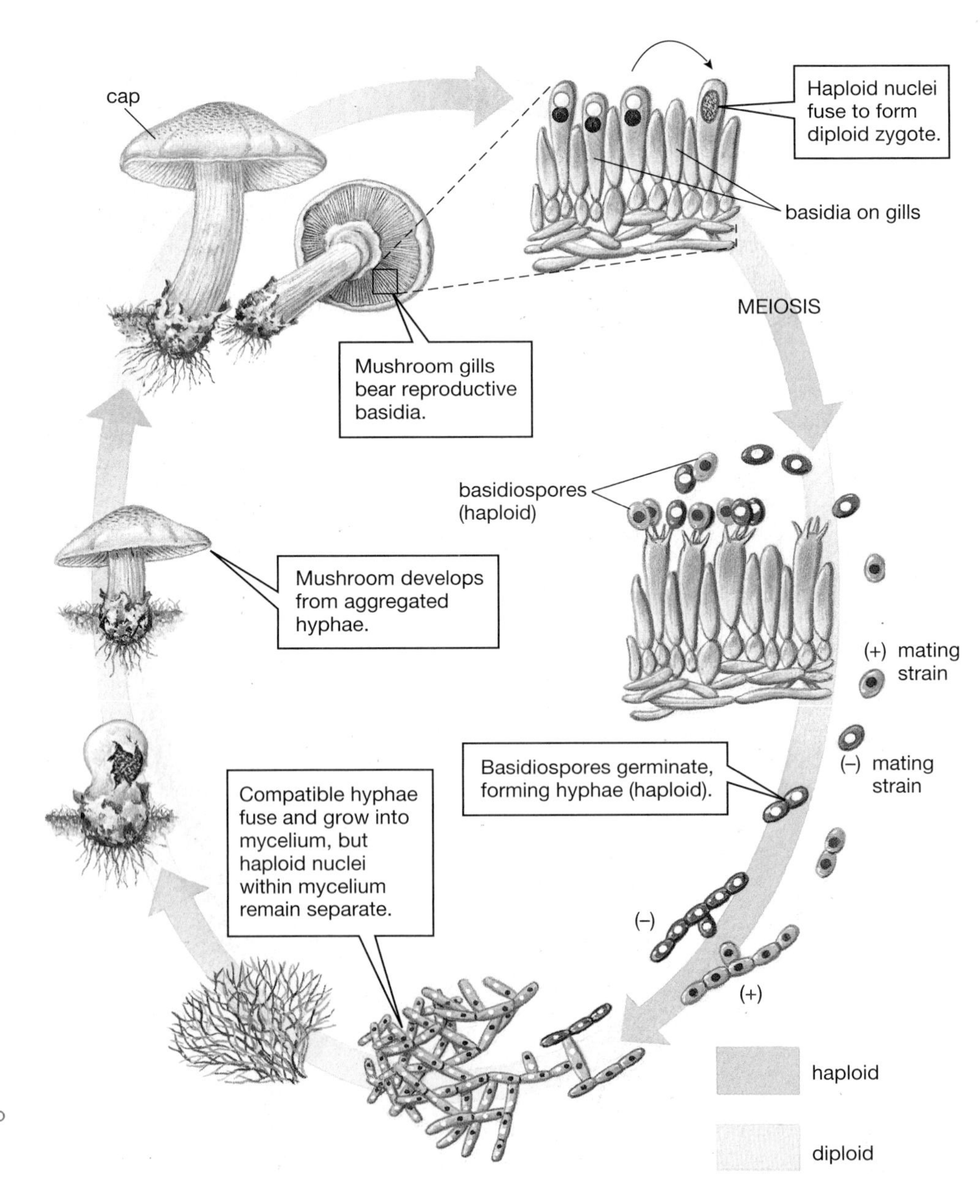
cap
Haploid nuclei fuse to form diploid zygote.
basidia on gills
MEIOSIS
Mushroom gills bear reproductive basidia.
basidiospores (haploid)
Mushroom develops from aggregated hyphae.
(+) mating strain
(–) mating strain
Basidiospores germinate, forming hyphae (haploid).
Compatible hyphae fuse and grow into mycelium, but haploid nuclei within mycelium remain separate.
(–)
(+)
haploid
diploid

FIGURE 20-6 The life cycle of a typical basidiomycete
QUESTION If two spores from the same sporangium each germinate and the resulting hyphae come into contact, can sexual reproduction follow?

(a)

(b)

(c)

FIGURE 20-7 Diverse basidiomycetes
(a) The giant puffball *Lycopedon giganteum* may produce up to 5 trillion spores. **(b)** Shelf fungi, the size of dessert plates, are conspicuous on trees. **(c)** The spores of stinkhorns are carried on the outside of a slimy cap that smells terrible to humans, but appeals to flies. The flies lay their eggs on the stinkhorn, and inadvertently disperse the spores that stick to their bodies. **QUESTION** Are the structures shown in these photos haploid or diploid?

structures (Fig. 20-5a) or corrugated, mushroomlike fruiting bodies called *morels* (Fig. 20-5b). This phylum also includes many of the colorful molds that attack stored food and destroy fruit and grain crops and other plants, as well as yeasts (some of the few unicellular fungi) and the species that produces penicillin, the first antibiotic.

The Club Fungi Produce Club-Shaped Reproductive Structures

Basidiomycetes are called the **club fungi** because they produce club-shaped reproductive structures. Members of this phylum typically reproduce sexually (Fig. 20-6); hyphae of different mating types fuse to form filaments in which each cell contains two nuclei, one from each parent. The nuclei do not themselves fuse until the formation of specialized, club-shaped diploid cells called **basidia** (singular, basidium). Basidia, in turn, give rise to haploid reproductive **basidiospores** by meiosis.

The formation of basidia and basidiospores takes place in special fruiting bodies, which are familiar to most of us as mushrooms, puffballs, shelf fungi, and stinkhorns (Fig. 20-7). These reproductive structures are actually dense aggregations of hyphae that emerge under proper conditions from a massive underground mycelium. On the undersides of mushrooms are leaflike gills on which basidia are produced. Basidiospores are released by the billions from the gills of mushrooms or through openings in the tops of puffballs and are dispersed by wind and water.

Falling on fertile ground, a mushroom basidiospore may germinate and form haploid hyphae. These hyphae grow outward from the original spore in a roughly circular pattern as the older hyphae in the center die. The subterranean body periodically sends up numerous

FIGURE 20-8 A mushroom fairy ring
Mushrooms emerge in a fairy ring from an underground fungal mycelium, growing outward from a central point where a single spore germinated, perhaps centuries ago.

mushrooms, which emerge in a ringlike pattern called a fairy ring (Fig. 20-8). The diameter of the fairy ring reveals the approximate age of the fungus—the wider the diameter, the older the underlying fungus. Some fairy rings are estimated to be 700 years old, and basidomycete mycelia can be even older than that. For example, the researchers who discovered the gigantic *Armillaria* in Oregon estimate that it took at least 2400 years to grow to its current size.

The Imperfect Fungi Are Species in Which Sexual Structures Have Not Been Observed

In many fungus species, researchers have never observed sexual reproduction or the special structures that result from sexual reproduction. Because fungi are categorized on the basis of their sexual structures (flagellated spores, zygospores, asci, and basidia), these species cannot be readily classified. Systematists place these unclassifiable forms into a group known as the *deuteromycetes,* or **imperfect fungi**. Unlike the chytrids, zygomycetes, ascomycetes, and basidiomycetes, the deuteromycetes do not form a natural evolutionary group of closely related organisms linked by descent from a common ancestor. Instead, the deuteromycetes are a category of convenience, a kind of "holding pen" for species that cannot be fit into one of the four fungal phyla. It is possible that some deuteromycetes truly lack a sexual stage, but most are believed to actually belong to one of the four main phyla. In fact, many former deuteromycetes have been placed into their appropriate phyla after sexual reproduction was observed.

Some Fungi Form Symbiotic Relationships

Many fungi live in direct contact with another species for a prolonged period. Such intimate, long-term relationships are known as *symbiosis.* In many cases, the fungal member of a symbiotic relationship is parasitic, and harms its host. But some symbiotic relationships are mutually beneficial.

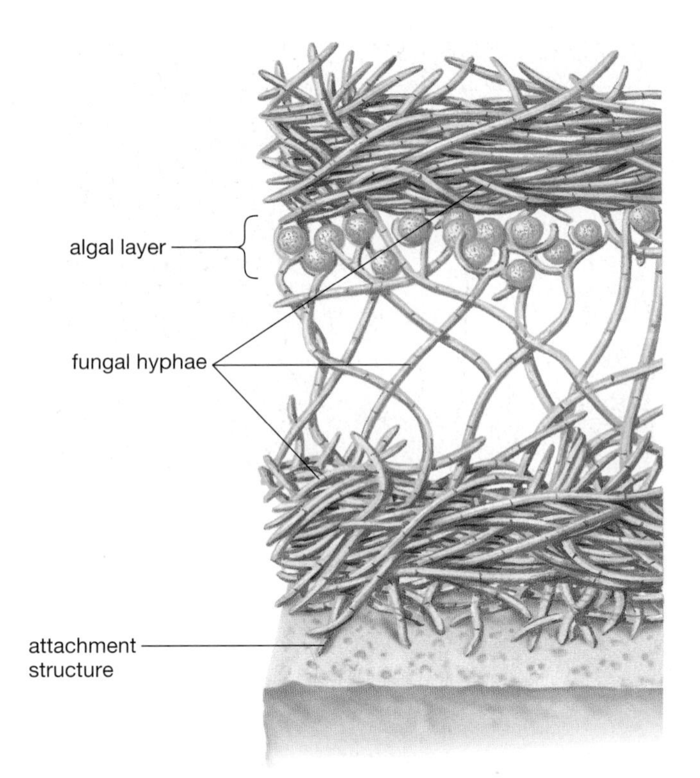

FIGURE 20-9 The lichen: A symbiotic partnership
Most lichens have a layered structure bounded on the top and bottom by an outer layer formed from fungal hyphae. The fungal hyphae emerge from the lower layer, forming attachments that anchor the lichen to a surface, such as a rock or a tree. An algal layer in which the alga and fungus grow in close association lies beneath the upper layer of hyphae.

Lichens Are Formed by Fungi That Live with Photosynthetic Algae or Bacteria

Lichens are symbiotic associations between fungi and single-celled green algae or cyanobacteria (Fig. 20-9). Lichens are sometimes described as "fungi that have learned to garden," because the fungal member of the partnership "tends" the photosynthetic algal or bacterial partner by providing shelter and protection from harsh conditions. In this protected environment, the photosynthetic member of the partnership uses solar energy to manufacture simple sugars, producing food for itself but also some excess food that is consumed by the fungus. In fact, the fungus often consumes the lion's share of the photosynthetic product (up to 90% in some species), leading some researchers to conclude that the symbiotic relationship in lichens is really much more one-sided than it is usually portrayed. This view was bolstered by the discovery that, in lichens that include algal symbionts, fungal hyphae actually penetrate the

(a)

(b)

FIGURE 20-10 Diverse lichens
(a) A colorful encrusting lichen, growing on dry rock, illustrates the tough independence of this symbiotic combination of fungus and algae. **(b)** A leafy lichen grows from a dead tree branch.

cell walls of the algae, as do the hyphae of fungi that parasitize plants.

Thousands of different fungal species (mostly ascomycetes) form lichens (Fig. 20-10), combining with one of a much smaller number of algal or bacterial species. Together, these organisms form a unit so tough and self-sufficient that lichens are among the first living things to colonize newly formed volcanic islands. Brightly colored lichens also invade other inhospitable habitats ranging from deserts to the Arctic, and can even grow on bare rock. Understandably, lichens in extreme environments grow very slowly; arctic colonies, for example, expand as slowly as 1 to 2 inches per 1000 years. Despite their slow growth, lichens can persist for long periods of time; some arctic lichens are more than 4000 years old.

Mycorrhizae Are Fungi Associated with Plant Roots

Mycorrhizae (singular, mycorrhiza) are important symbiotic associations between fungi and plant roots. More than 5000 species of mycorrhizal fungi (including representatives of all the major groups of fungi) can grow in intimate association with about 80% of plants that have roots, including most trees. These associations benefit both the plant and its fungal partner. The hyphae of mycorrhizal fungi surround the plant root and invade the root cells (Fig. 20-11). The fungus digests and absorbs minerals and organic nutrients from the soil, passing some of them directly into the root cells. The fungus also absorbs water and passes it to the plant—an advantage for plants in dry, sandy soils. In return, sugar produced photosynthetically by the plant is passed from the root to the fungus. Plants that participate in this unique relationship, especially those in poor soils, tend to grow larger and more vigorously than do those deprived of the fungus.

Some scientists believe that mycorrhizal associations may have been important in the invasion of land by plants more than 400 million years ago. Such a relationship between an aquatic fungus and a green alga (ancestral to terrestrial plants) could have helped the alga acquire the water and mineral nutrients it needed to survive out of water.

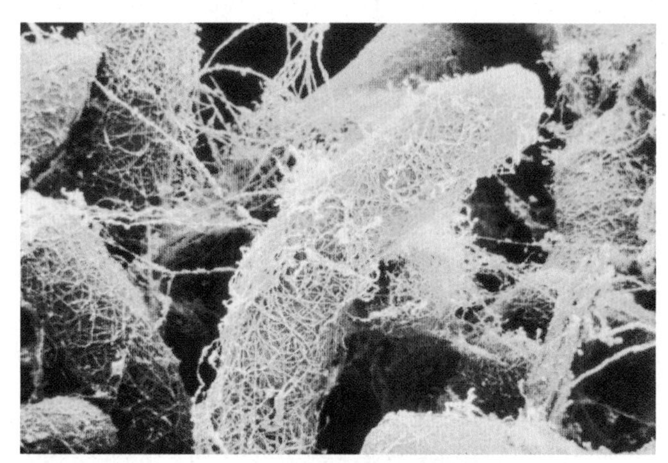

FIGURE 20-11 Mycorrhizae enhance plant growth
Hyphae of mycorrhizae entwining about the root of an aspen tree. Plants grow significantly better in a symbiotic association with these fungi, which help make nutrients and water available to the roots.

FIGURE 20-12 Corn smut
This basidiomycete pathogen destroys millions of dollars' worth of corn each year. Even a pest like corn smut has its admirers, though. In Mexico this fungus is known as *huitlacoche* and is considered to be a great delicacy.

FIGURE 20-13 A helpful fungal parasite
Although some parasitic fungi attack crops, others are used by farmers to control insect pests. Here, a *Cordyceps* species has killed a grasshopper.

20.3 How Do Fungi Affect Humans?

The average person gives little thought to fungi, except perhaps for an occasional, momentary appreciation for the mushrooms on a pizza. Nonetheless, fungi affect our lives in more ways than you might imagine.

Fungi Attack Plants That Are Important to People

Fungi cause the majority of plant diseases, and some of the plants that they infect are important to humans. In fact, fungal pathogens have a devastating effect on the world's food supply. Especially damaging are the basidiomycete plant pests descriptively called *rusts* and *smuts*, which cause billions of dollars' worth of damage to grain crops annually (Fig. 20-12). Fungal diseases also affect the appearance of our landscape. The American elm and the American chestnut, two tree species that were once prominent in many of America's parks, yards, and forests, were destroyed on a massive scale by the ascomycetes that cause Dutch elm disease and chestnut blight. Today, few people can recall the graceful forms of large elms and chestnuts, which are now almost entirely absent from the landscape.

Fungi continue to attack plant tissues long after they have been harvested for human use. To the dismay of homeowners, a host of different fungal species attack wood, causing it to rot. Some ascomycete molds secrete the enzymes cellulase and protease, which can cause significant damage to cotton and wool textiles, especially in warm, humid climates where molds flourish.

The fungal impact on agriculture and forestry is not entirely negative, however. Fungal parasites that attack insects and other arthropod pests can be an important ally in pest control (Fig. 20-13). Farmers who wish to reduce their dependence on toxic and expensive chemical pesticides are increasingly turning to biological methods of pest control, including the application of "fungal pesticides." Fungal pathogens are currently used to control termites, rice weevils, tent caterpillars, aphids, citrus mites, and other pests.

Fungi Cause Human Diseases

The kingdom Fungi includes parasitic species that attack humans directly. Some of the most familiar fungal diseases are caused by ascomycetes that attack the skin, resulting in athlete's foot, jock itch, and ringworm. These diseases, though unpleasant, are not life-threatening and can usually be treated with antifungal ointments. Prompt treatment can also control another common fungal disease: vaginal infections caused by the yeast *Candida albicans* (Fig. 20-14). Fungi can also infect the lungs if victims inhale spores of disease-causing fungal species, such as those that cause valley fever and histoplasmosis. Like other fungal infections, these diseases can, if promptly diagnosed, be controlled with antifungal drugs. If untreated, however, they can develop into serious systemic infections. Singer Bob Dylan, for instance, became gravely ill with histoplasmosis when a fungus infected the pericardial membrane surrounding his heart.

Fungi Can Produce Toxins

In addition to their role as agents of infectious disease, some fungi produce toxins that are dangerous to humans. Of particular concern are toxins produced by fungi that grow on grains and other foodstuffs that have

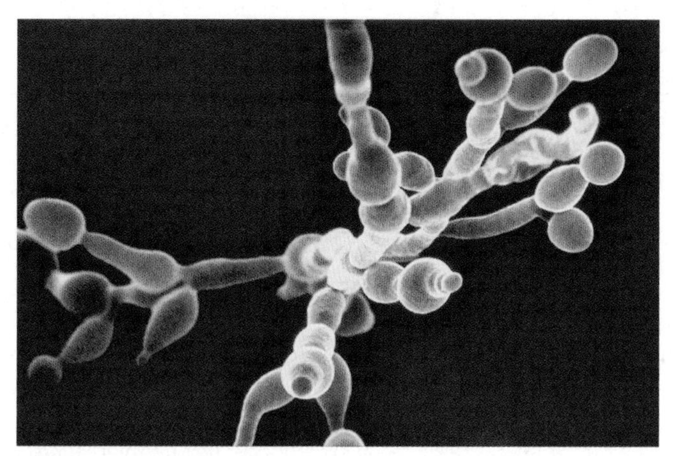

FIGURE 20-14 The unusual yeast
Yeasts are unusual, normally nonfilamentous ascomycetes that reproduce most commonly by budding. The yeast shown here is *Candida*, a common cause of vaginal infections.

been stored in too-moist conditions. For example, molds of the genus *Aspergillus* produce highly toxic, carcinogenic compounds known as aflatoxins. Some foods, such as peanuts, seem especially susceptible to attack by *Aspergillus*. Since aflatoxins were discovered in the 1960s, food growers and processors have developed methods for reducing the growth of *Aspergillus* in stored crops, so aflatoxins have been largely eliminated from the nation's peanut butter supply.

One infamous toxin-producing fungus is the ascomycete *Claviceps purpurea,* which infects rye plants and causes a disease known as ergot. This fungus produces several toxins, which can affect humans if infected rye is ground into flour and consumed. This happened frequently in northern Europe in the Middle Ages, with devastating effects. At that time, ergot poisoning was typically fatal, and victims experienced terrible symptoms before dying. One ergot toxin is a vasoconstrictor, which constricts blood vessels and reduces blood flow. The effect can be so extreme that gangrene develops and limbs actually shrivel and fall off. Other ergot toxins cause symptoms that include a burning sensation, vomiting, convulsive twitching, and vivid hallucinations. Today, new agricultural techniques have effectively eliminated ergot poisoning, but the hallucinogenic drug LSD, which is derived from a component of the ergot toxins, still remains as a legacy of this disease.

Many Antibiotics Are Derived from Fungi

Fungi have also had positive impacts on human health. The modern era of life-saving antibiotic medicines was ushered in by the discovery of penicillin, which is produced by an ascomycete mold (Fig. 20-15). Penicillin is still used, along with other fungi-derived antibiotics, to combat bacterial diseases. Other important drugs are also derived from fungi, including cyclosporin, which is used to suppress the immune response during organ transplants so that the body is less likely to reject the transplanted organs.

FIGURE 20-15 Penicillium
Penicillium growing on an orange. Reproductive structures, which coat the fruit's surface, are visible, while hyphae beneath draw nourishment from inside. The antibiotic penicillin was first isolated from this fungus. **QUESTION** Why do some fungi produce antibiotic chemicals?

Fungi Make Important Contributions to Gastronomy

Fungi make important contributions to human nutrition, the most obvious of which is represented by the fungi that we consume directly, including wild and cultivated basidiomycete mushrooms and ascomycetes, such as morels and the rare and prized truffle (see "Evolutionary Connections: Fungal Ingenuity—Pigs, Shotguns, and Nooses"). The role of fungi in cuisine, however, also has less-visible manifestations. For example, some of the world's most famous cheeses, including Roquefort, Camembert, Stilton, and Gorgonzola, gain their distinctive flavors from ascomycete molds that grow on them as they ripen. Perhaps the most important and pervasive fungal contributors to our food supply, however, are the single-celled ascomycetes (a few species are basidiomycetes) known as *yeasts*.

EARTH WATCH The Case of the Disappearing Mushrooms

Mycologists (scientists who study fungi) and gourmet cooks may seem to have little in common, but recently, they have been united by a common concern: mushrooms are rapidly declining in numbers, average size, and species diversity. Although the problem is most easily recognized in Europe, where people have been gathering wild mushrooms for centuries, American mycologists are also alarmed; the same decline may be occurring here as well. Why are mushrooms disappearing? Overhunting of edible mushrooms is not the culprit, because poisonous forms are equally affected. The loss is evident in all types of mature forests, so changing forest management practices could not be the cause. The most likely cause is air pollution, because the loss of mushrooms is greatest where the air contains the highest levels of ozone, sulfur, and nitrogen.

Although mycologists have not yet determined exactly how air pollution harms mushrooms, the evidence is clear. In Holland, for example, the average number of fungal species per 1000 square meters has dropped from 37 down to 12 over the past several decades. Twenty out of 60 fungal species surveyed in England are declining. Concern is intensified by the fact that the mushrooms most affected are those whose hyphae form mycorrhizal associations with tree roots. Trees with diminished mycorrhizae may have less resistance to periodic droughts or extreme cold spells. Because air pollution is also harming forests directly, the additional loss of mycorrhizae could be devastating.

Wine and Beer Are Made Using Yeasts

The discovery that yeasts could be harnessed to enliven our culinary experience is surely a key event in human history. Among the many foods and beverages that depend on yeasts for their production are bread, wine, and beer, which are consumed so widely that it is difficult to imagine a world without them. All derive their special qualities from fermentation by yeasts. Fermentation occurs when yeasts extract energy from sugar and, as by-products of the metabolic process, emit carbon dioxide and ethyl alcohol. As yeasts consume the fruit sugars in grape juice, the sugars are replaced by alcohol, and wine is the result. Eventually, the increasing concentration of alcohol kills the yeasts, ending fermentation. If the yeasts die before all available grape sugar is consumed, the wine will be sweet; if the sugar is exhausted, the wine will be dry.

Beer is brewed from grain (usually barley), but yeasts cannot effectively consume the carbohydrates that compose grain kernels. For the yeasts to do their work, the barley grains must have sprouted (recall that grains are actually seeds). Germination converts the kernels' carbohydrates to sugar, so the sprouted barley provides an excellent food source for the yeasts. As with wine, fermentation converts sugars to alcohol, but beer brewers capture the carbon dioxide by-product as well, giving the beer its characteristic bubbly carbonation.

Yeasts Make Bread Rise

In bread making, carbon dioxide is the crucial fermentation product. The yeasts added to bread dough do produce alcohol as well as carbon dioxide, but the alcohol evaporates during baking. In contrast, the carbon dioxide is trapped in the dough, where it forms the bubbles that give bread its light, airy texture. So the next time you're enjoying a slice of French bread with Camembert cheese and a nice glass of Chardonnay, or a slice of pizza and a cold bottle of your favorite brew, you might want to quietly give thanks to the yeasts. Our diets would certainly be a lot duller without the help we get from fungal partners.

Fungi Play a Crucial Ecological Role

No account of the fungi would be complete without mention of their fundamental ecological importance. The fungi are Earth's undertakers, consuming the dead of all kingdoms and returning their component substances to the ecosystems from which they came. The extracellular digestive activities of many fungi liberate nutrients such as carbon, nitrogen, and phosphorus compounds and minerals that can be used by plants. If fungi and bacteria were suddenly to disappear, the consequences would be disastrous. Nutrients would remain locked in the bodies of dead plants and animals, the recycling of nutrients would grind to a halt, soil fertility would rapidly decline, and waste and organic debris would accumulate. In short, ecosystems would collapse.

EVOLUTIONARY CONNECTIONS

Fungal Ingenuity—Pigs, Shotguns, and Nooses

Natural selection, operating over millions of years on the diverse forms of fungi, has produced some remarkable adaptations by which fungi disperse their spores and obtain nutrients.

The Rare, Delicious Truffle

Although many fungi are prized as food, none are as avidly sought as the truffle (Fig. 20-16). The finest Italian truffles may sell for as much as $1500 per pound. Truffles are the spore-containing structures of an ascomycete that forms a mycorrhizal association with the

FIGURE 20-16 The truffle
Truffles, rare ascomycetes (each about the size of a small apple), are a gastronomic delicacy.

roots of oak trees. Although it develops underground, the truffle has evolved an effective mechanism to entice animals to dig it up. As its spores ripen, the truffle releases an odor that animals, especially pigs, find appealing. When a pig follows the smell to the truffle, digs it up, and devours it, millions of spores are scattered to the winds. Traditionally, truffle collectors used muzzled pigs to hunt their quarry; a good truffle-pig can smell an underground truffle 150 feet (about 50 meters) away. Today, dogs are the most common assistants to truffle hunters.

The Shotgun Approach to Spore Dispersal

If you get close enough to a pile of horse manure to scrutinize it, you may observe the beautiful, delicate reproductive structures of the zygomycete *Pilobolus* (Fig. 20-17). Despite their daintiness, these are actually fungal shotguns. The clear bulbs, capped with sticky black spore cases, extend from hyphae that penetrate the dung. As the bulbs mature, the sugar concentration inside them increases, and water is drawn in by osmosis. Meanwhile, the bulb begins to weaken just below its cap. Suddenly, like an overinflated balloon, the bulb bursts and blows its spore-carrying top up to 3 feet (1 meter) away.

The airborne spores may well land in some leaves of grass; *Pilobolus* bends toward the light as it grows, thereby increasing the chances that its spores will be directed toward open pasture. Spores that adhere to grass remain there until consumed by a grazing herbivore, perhaps a horse. Later (likely some distance away), the horse will deposit a fresh pile of manure containing *Pilobolus* spores that have passed unharmed through its digestive tract. The spores germinate and growing hyphae penetrate the manure (a rich source of nutrients), ultimately sending up new projectiles to continue this ingenious cycle.

The Nematode Nemesis

Microscopic *nematodes* (roundworms) abound in rich soil, and fungi have evolved several fascinating forms of nematode-nabbing hyphae that allow them to exploit this rich source of protein. Some fungi produce sticky pods that adhere to passing nematodes and penetrate the worm body with hyphae, which then begin digesting the worm from within. One species shoots a microscopic harpoonlike spore into passing nematodes; the spore develops into a new fungus inside the worm. The fungal strangler *Arthrobotrys* produces nooses formed from three hyphal cells. When a nematode wanders into the noose, its contact with the inner parts of the noose stimulates the noose cells to swell with water (Fig. 20-18). In a fraction of a second, the hole constricts, trapping the worm. Fungal hyphae then penetrate and feast on their prey.

FIGURE 20-17 An explosive zygomycete
The delicate, translucent reproductive structures of the zygomycete *Pilobolus* literally blow their tops when ripe, dispersing the black caps with their payload of spores.

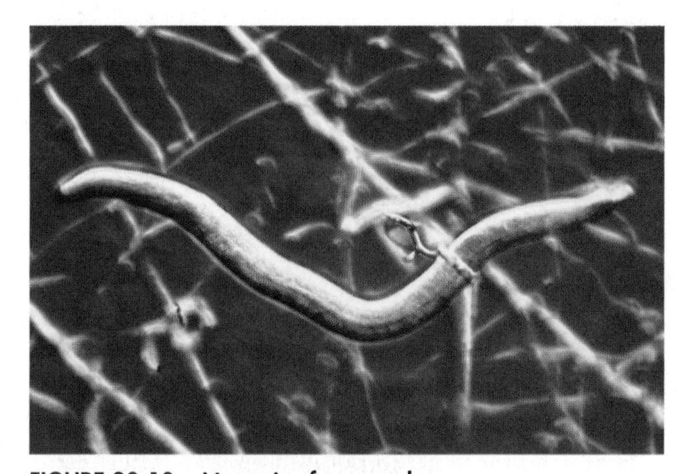

FIGURE 20-18 Nemesis of nematodes
Arthrobotrys, the nematode (roundworm) strangler, traps its prey in a nooselike modified hypha that swells when the inside of the loop is contacted.

CASE STUDY REVISITED Humongous Fungus

Why do *Armillaria* fungi grow so large? Their size is due in part to their ability to form rhizomorphs, which consist of hyphae bundled together inside a protective rind. The enclosed hyphae carry nutrients to the rhizomorphs, allowing them to extend long distances through nutrient-poor areas to reach new sources of food. The *Armillaria* fungus can thus grow beyond the boundaries of a particular food-rich area.

Another factor that may contribute to the gigantic size of the Oregon *Armillaria* is the climate in which it was found. In this dry region, fungal fruiting bodies form only rarely, so the colossal *Armillaria* rarely produces spores. In the absence of spores that might grow into new individuals, the existing individual faces little competition for resources, and is free to grow and fill an increasingly large area.

The discovery of the Oregon specimen is merely the latest chapter in a long-running, good-natured "fungus war" that began in 1992 with the discovery of the first humongous fungus, a 37-acre *Armillaria gallica* growing in Michigan. Since that initial landmark discovery, research groups in Michigan, Washington, and Oregon have engaged in a friendly competition to find the largest fungus. Will the current record holder ever be topped? Stay tuned.

Consider This: Because the entire Oregon *Armillaria* grew from a single spore, all of its cells are genetically identical. Its parts, however, are not all physiologically dependent on one another, so it is unlikely that any substances are transported through the entire 3.4-square-mile mycelium. And there is no continuous skin or bark or membrane that covers the entire mycelium and separates it from the environment as a unit. Is the fungus's genetic unity sufficient evidence for it to be considered a single individual, or is greater physiological integration required? Do you think that the claim of "world's largest organism" is valid?

Links to Life: Collect Carefully

In the early 1980s, doctors at a California hospital noticed a curious trend. Over a period of a few months, the number of patients admitted for treatment of poisoning had risen dramatically, and many of those admitted had died. What caused this sudden outbreak of poisoning? Further investigation revealed that, in almost all cases, the victims were recent immigrants from Laos or Cambodia. Struggling to adjust to their new country, they had been thrilled to find that California's forests contained mushrooms that looked just like the ones they had collected for food back in Asia. Unfortunately, the similarity was superficial; the mushrooms were, in fact, poisonous species. The immigrants' nostalgic pursuit of "comfort food" had tragic consequences.

In general, immigrants from countries where mushroom collecting is common have proved to be especially susceptible to poisoning by toxic mushrooms. But they are not the only victims. Each year, a number of small children, inexperienced collectors, and unlucky guests at gourmet dinners make unexpected trips to the hospital after eating poisonous wild mushrooms.

It can be fun and rewarding to collect wild mushrooms, which offer some of the richest and most complex flavors a human can experience. But if you decide to go collecting, be careful, because some of the deadliest poisons known to humankind are found in mushrooms. Especially noted for their poisons are certain species in the genus *Amanita*, which have evocative common names such as death cap and destroying angel. These names are apt, as even a single bite of one of these mushrooms can be lethal. Damage from *Amanita* toxins is most severe in the liver, where the toxins tend to accumulate. Often, a victim of *Amanita* poisoning can be saved only by undergoing a liver transplant. So be sure to protect your health by inviting an expert to join your mushroom-hunting expeditions.

Summary of Key Concepts

20.1 What Are the Key Features of Fungi?

Fungal bodies generally consist of filamentous hyphae, which are either multicellular or multinucleated and form large, intertwined networks called *mycelia.* Fungal nuclei are generally haploid. A cell wall of chitin surrounds fungal cells.

All fungi are heterotrophic, secreting digestive enzymes outside their bodies and absorbing the liberated nutrients.

Fungal reproduction is varied and complex. Asexual reproduction can occur either through fragmentation of the mycelium or through asexual spore formation. Sexual spores form after compatible haploid nuclei fuse to form a diploid zygote, which undergoes meiosis to form haploid sexual spores. Both asexual and sexual spores produce haploid mycelia through mitosis.

20.2 How Are Fungi Classified?

The major phyla of fungi and their characteristics are summarized in Table 20-1. A lichen is a symbiotic association between a fungus and algae or cyanobacteria. This self-sufficient combination can colonize bare rock. Mycorrhizae are associations between fungi and the roots of most vascular plants. The fungus derives photosynthetic nutrients from the plant roots and, in return, carries water and nutrients into the root from the surrounding soil.

20.3 How Do Fungi Affect Humans?

The majority of plant diseases are caused by parasitic fungi. Some parasitic fungi can help control insect crop pests. Others can cause human diseases, including ringworm, athlete's foot, and common vaginal infections. Some fungi produce toxins that can harm humans. Nonetheless, fungi add variety to the human food supply, and fermentation by fungi helps make wine, beer, and bread.

Fungi are extremely important decomposers in ecosystems. Their filamentous bodies penetrate rich soil and decaying organic material, liberating nutrients through extracellular digestion.

Key Terms

asci *p. 390*
basidia *p. 393*
basidiospore *p. 393*
club fungus *p. 393*
hyphae *p. 388*
imperfect fungus *p. 394*
lichen *p. 394*
mycelium *p. 388*
mycorrhizae *p. 395*
sac fungus *p. 390*
septa *p. 388*
sporangium *p. 390*
spore *p. 388*
zygospore *p. 390*
zygote fungus *p. 390*

Thinking Through the Concepts

To take a multiple-choice quiz with feedback on the contents of this chapter, visit http://www.prenhall.com/audesirk7. *Log in to the Web site selected by your instructor and navigate to the Self Test section for this chapter.*

? Review Questions

1. Describe the structure of the fungal body. How do fungal cells differ from most plant and animal cells?
2. What portion of the fungal body is represented by mushrooms, puffballs, and similar structures? Why are these structures elevated above the ground?
3. What two plant diseases, caused by parasitic fungi, have had an enormous impact on forests in the United States? In which division are these fungi found?
4. List some fungi that attack crops. To which phyla do they belong?
5. Describe asexual reproduction in fungi.
6. What is the major structural ingredient in fungal cell walls?
7. List the phyla of fungi, describe the feature that gives each its name, and give one example of each.
8. Describe how a fairy ring of mushrooms is produced. Why is the diameter related to its age?
9. Describe two symbiotic associations between fungi and organisms from other kingdoms. In each case, explain how each partner in these associations is affected.

Applying the Concepts

1. Dutch elm disease in the United States is caused by an *exotic*—that is, an organism (in this case, a fungus) introduced from another part of the world. What damage has this introduction done? What other fungal pests fall into this category? Why are parasitic fungi particularly likely to be transported out of their natural habitat? What can governments do to limit this importation?
2. The discovery of penicillin revolutionized the treatment of bacterial diseases. However, penicillin is now rarely prescribed. Why is this? *Hint:* Refer back to Chapter 15.
3. The discovery of penicillin was the result of a chance observation by an observant microbiologist, Alexander Fleming. How would you search systematically for new

antibiotics produced by fungi? Where would you look for these fungi?

4. Fossil evidence indicates that mycorrhizal associations between fungi and plant roots existed in the late Paleozoic era, when the invasion of land by plants began. This evidence suggests an important link between mycorrhizae and the successful invasion of land by plants. Why might mycorrhizae have been important fungi in the colonization of terrestrial habitats by plants?

5. General biology texts in the 1960s included fungi in the plant kingdom. Why do biologists no longer consider fungi as legitimate members of the plant kingdom?

6. What ecological consequences would occur if humans, using a new and deadly fungicide, destroyed all fungi on Earth?

For More Information

Angier, N. "A Stupid Cell with All the Answers." *Discover,* November 1986. Fascinating description of the uses of yeasts in molecular biology, including beautiful illustrations.

Barron, G. "Jekyll-Hyde Mushrooms." *Natural History*, March 1992. Fungi include a variety of forms; some absorb decaying plant material, and others prey on microscopic worms.

Dix, N. J., and Webster, J. *Fungal Ecology*. London: Chapman & Hall, 1995. A rather technical but comprehensive and readable account of fungal diversity that focuses on the roles of fungi in different ecological communities.

Hudler, G. W. *Magical Mushrooms, Mischievous Molds*. Princeton, NJ: Princeton University Press, 1998. Engaging treatment of the fungi, focusing on their importance in human affairs.

Kiester, E. "Prophets of Gloom." *Discover*, November 1991. Lichens can be used as indicators of air quality and provide evidence of deteriorating environmental conditions.

Radetsky, P. "The Yeast Within." *Discover*, March 1994. Describes the newly discovered ability of baker's yeast to form filaments and the possible implications of this for the study of yeasts that cause vaginal infections.

Schaechter, E. *In the Company of Mushrooms*. Cambridge: Harvard University Press, 1997. An accessible account of the world of mushrooms, written in a warm, personal style.

Strobel, G. A., and Lanier, G. N. "Dutch Elm Disease." *Scientific American*, August 1981. A description of the life cycle of this parasitic fungus and of techniques that limit its spread.

Vogel, S. "Taming the Wild Morel." *Discover*, May 1988. Describes the research that has allowed these rare delicacies to be cultivated in the laboratory.

Media Activities

To access a Media Activity visit http://www.prenhall.com/audesirk7. *Log in to the Web site selected by your instructor, navigate to this chapter, and select the appropriate Media Activity number.*

20.1 Fungi Structure and Reproduction

Estimated time: 5 minutes

Explore the major structures and reproductive life cycles of the fungi.

20.2 Classification of Fungi

Estimated time: 5 minutes

Explore the major divisions of the fungi and their distinguishing characteristics.

20.3 Web Investigation: Humongous Fungus

Estimated time: 10 minutes

The finding of giant fungi has provided new insights into how humans impact forests and the management of forests. Explore more about these giant fungi and what they are teaching us about the ecosystem.

CHAPTER

21 The Diversity of Plants

The huge, foul-smelling flower of the stinking corpse lily is a treat for visitors to Asian rain forests.

AT A GLANCE

CASE STUDY Queen of the Parasites

The flower of the stinking corpse lily makes a strong impression. For one thing, it's huge; a single flower may be 3 feet across. It also has a rather strange appearance, consisting largely of fleshy lobes that are almost fungus-like. But the thing that makes a stinking corpse lily almost impossible to ignore is its aroma, which has been described as "a penetrating smell more repulsive than any buffalo carcass in an advanced stage of decomposition." Though utterly revolting to humans, the smell is attractive to blowflies and other insects that normally feed on and lay their eggs in decaying flesh. When such insects visit a male stinking corpse lily, they may carry away pollen that can fertilize a nearby female flower.

Close examination of a stinking corpse lily reveals that it has no visible leaves, roots, or stems. In fact, it is a parasite, and its body is completely embedded in the tissue of its host, a vine of the genus *Tetrastigma*. Without leaves, the stinking corpse lily cannot produce any food of its own, but instead draws all of its nutrition from its host. The parasite becomes visible outside the body of its host only when one of its cabbage-shaped flower buds pushes through the surface of the host's stem and its gigantic, stinking flower opens for a week or so before shriveling and falling off. If a male and a female flower happen to be open simultaneously and close together, the female flower may be fertilized and produce seeds. A seed that is dispersed in animal droppings and happens to land on a *Tetrastigma* stem may germinate and penetrate a new host.

When you think of plants, you might first think of their most obvious feature: green leaves that capture solar energy by photosynthesis. It may seem odd, then, that this chapter about plants begins with a peculiar plant that does not photosynthesize. Oddities such as the stinking corpse lily, however, serve as reminders that evolution does not always follow a predictable pathway, and that even an adaptation as seemingly valuable as the ability to live on sunlight can be discarded.

Media Activity
21.1 Evolution of Plant Structure

21.1 What Is the Evolutionary Origin of Plants?

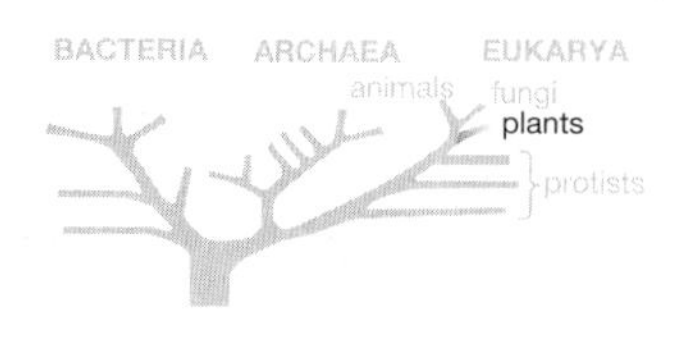

The ancestors of plants were photosynthetic protists, probably similar to today's algae. Like modern algae, the organisms that gave rise to plants presumably lacked true roots, stems, leaves, and complex reproductive structures such as flowers or cones. All of these features appeared later in the evolutionary history of plants (Fig. 21-1).

Green Algae Gave Rise to Plants

Of today's different groups of algae, green algae are probably most similar to ancestral plants. This supposition stems from the close phylogenetic relationship between the two groups. DNA comparisons have shown that green algae are plants' closest living relatives, and the hypothesis that plants evolved from green algal ancestors is supported by other evidence as well. For example, green algae and plants use the same type of chlorophyll and accessory pigments in photosynthesis. In addition, both plants and green algae store food as starch and have cell walls made of cellulose. In contrast, the photosynthetic pigments, food-storage molecules, and cell walls of other photosynthetic protists, such as the red algae and the brown algae, differ from those of plants.

The Ancestors of Plants Lived in Fresh Water

Most green algae live in fresh water, which suggests that the early evolutionary history of plants took place in freshwater habitats. In contrast to the nearly constant environmental conditions of the ocean, freshwater habitats are highly variable. Water temperature can fluctuate seasonally or even daily. Changing rates of rainfall and evaporation can cause frequent changes in the concentration of chemicals in the water, and can even cause an aquatic habitat to dry up periodically. Ancient freshwater green algae must have evolved characteristics that enabled them to withstand temperature extremes and periods of dryness. These adaptations to the difficulties of freshwater life provided a foundation that allowed the descendants of early algae to evolve the traits that made life on land possible.

21.2 What Are the Key Features of Plants?

Plants are multicellular, and most use photosynthesis to convert water and carbon dioxide to sugar. Neither multicellularity nor the ability to photosynthesize is unique

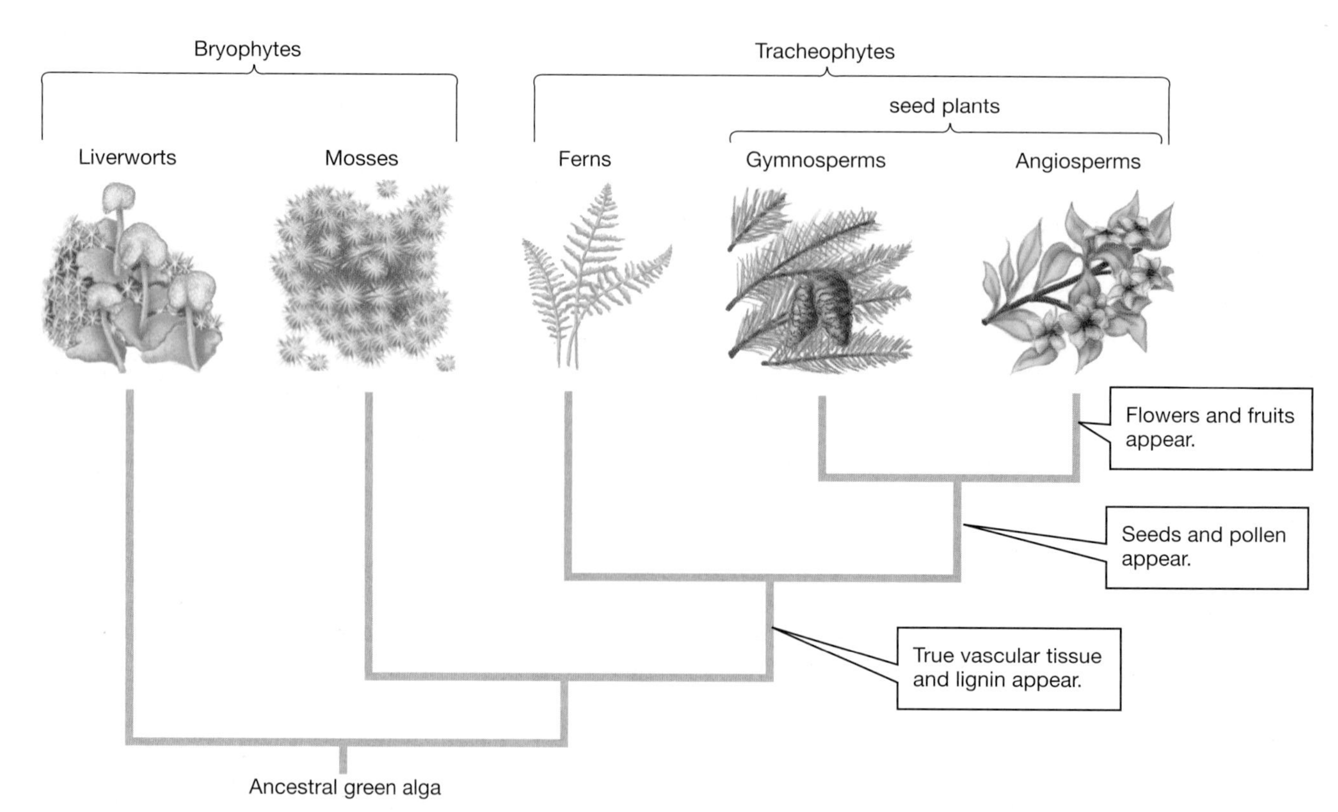

FIGURE 21-1 Evolutionary tree of some major plant groups

to plants, but the simultaneous appearance of these traits in a single organism is quite rare outside the plant kingdom. The most distinctive feature of plants, however, is their reproductive cycle.

Plants Have Alternating Multicellular Haploid and Diploid Generations

The plant life cycle is characterized by **alternation of generations** (Fig. 21-2), in which separate diploid and haploid generations alternate with one another. (Recall that a diploid organism has two sets of chromosomes; a haploid organism, one set.) In the diploid generation, the plant body consists of diploid cells and is known as the **sporophyte**. Certain cells of sporophytes undergo meiosis to produce haploid spores, which then divide by mitosis and develop into multicellular, haploid plants called **gametophytes**. The gametophytes ultimately produce male and female haploid gametes by mitosis. The gametes fuse to form diploid **zygotes**, which develop into a diploid sporophyte, and the cycle begins again.

Plants Have Multicellular, Dependent Embryos

In plants, zygotes develop into multicellular embryos that are retained within and receive nutrients from the tissues of the parent plant. A plant embryo is thus attached to and dependent on its parent as it grows and develops. Such multicellular, dependent embryos are not found among photosynthetic protists; they distinguish plants from their nearest relatives among the algae.

Plants Are Adapted to Life on Land

Most plants live on land. On land, the supportive buoyancy of water is missing, the body is not bathed in a nutrient solution, and the air tends to dry things out. In addition, gametes (sex cells) and zygotes (fertilized sex cells) cannot be carried by water currents or propelled by flagella, as they are in many water-dwelling organisms. Thus, life on land has favored the evolution in plants of structures that support the body and conserve water, vessels that transport water and nutrients to all parts of the plant, and processes that disperse gametes and zygotes by methods that are independent of water.

Plant Bodies Resist Gravity and Drying

Some of the key adaptations to life on land arose early in plant evolution and are now common to virtually all land plants. They include:

- Roots or rootlike structures that anchor the plant and/or absorb water and nutrients from the soil
- A waxy **cuticle** that covers the surfaces of leaves and stems and that limits the evaporation of water
- Pores called **stomata** (singular, stoma) in the leaves and stems that open to allow gas exchange but close when water is scarce, thus reducing the amount of water lost to evaporation

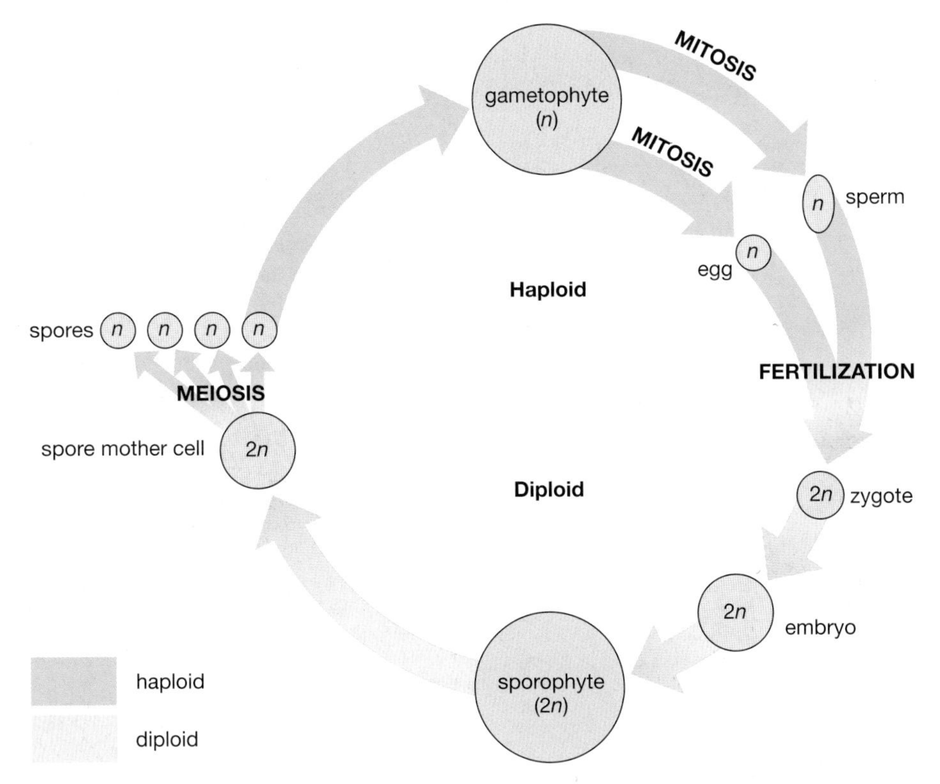

FIGURE 21-2 Alternation of generations in plants
As shown in this generalized depiction of a plant life cycle, a diploid sporophyte generation produces haploid spores through meiosis. The spores develop into a haploid gametophyte generation that produces haploid gametes by mitosis. The fusion of these gametes results in a diploid zygote that develops into the sporophyte plant.

Other key adaptations occurred somewhat later in the transition to terrestrial life, and are now widespread but not universal among plants (most nonvascular plants, a group described below, lack them):

- Conducting vessels that transport water and minerals upward from the roots and that move photosynthetic products from the leaves to the rest of the plant body
- The stiffening substance **lignin**, which is a rigid polymer that impregnates the conducting vessels and supports the plant body, helping the plant expose maximum surface area to sunlight

Plant Embryos Are Protected and Plant Sex Cells May Disperse Without Water

All plants protect developing embryos within parental tissues, but the most widespread groups of plants are characterized by especially well-protected and well-provisioned embryos and by waterless dispersal of sex cells. The key adaptations of these plant groups are *pollen*, *seeds*, and, in the flowering plants, *flowers* and *fruits*. Early seed plants produced dry, microscopic pollen grains that allowed wind, instead of water, to carry the male gametes. Seeds provided protection and nourishment for developing embryos and the potential for more effective dispersal. Later came the evolution of flowers, which enticed animal pollinators that delivered pollen more precisely than did wind. Fruits also attracted animal foragers, which consumed them and dispersed the indigestible seeds in their feces.

21.3 What Are the Major Groups of Plants?

Two major groups of land plants arose from ancient algal ancestors (Table 21-1). One group, the **bryophytes** (also called *nonvascular plants*), requires a moist environment to reproduce and thus straddles the boundary between aquatic and terrestrial life, much like the amphibians of the animal kingdom. The other group, the **vascular** plants (also called *tracheophytes*), has been able to colonize drier habitats.

Bryophytes Lack Conducting Structures

Bryophytes retain some characteristics of their algal ancestors: they lack true roots, leaves, and stems. They do possess rootlike anchoring structures called *rhizoids*, which bring water and nutrients into the plant body, but bryophytes are nonvascular—they lack well-developed structures for conducting water and nutrients. They must instead rely on slow diffusion or poorly developed conducting tissues to distribute water and other nutrients. As a result, their body size is limited. Their size is also limited by the absence of any stiffening agent in their bodies; without such material, they cannot overcome gravity to achieve much upward growth. Most bryophytes are less than 1 inch (2.5 centimeters) tall.

The most common bryophytes are the liverworts and mosses (Fig. 21-3), which are generally most abundant

Table 21-1 Features of the Major Plant Groups

Group	Subgroup	Relationship of Sporophyte and Gametophyte	Transfer of Reproductive Cells	Early Embryonic Development	Dispersal	Water and Nutrient Transport Structures
Bryophytes		Gametophyte dominant—sporophyte develops from zygote	Motile sperm swims to stationary egg retained on gametophyte	Occurs within archegonium of gametophyte	Haploid spores carried by wind	Absent
Vascular Plants	Ferns	Sporophyte dominant—develops from zygote retained on gametophyte	Motile sperm swims to stationary egg retained on gametophyte	Occurs within archegonium of gametophyte	Haploid spores carried by wind	Present
	Conifers	Sporophyte dominant—microscopic gametophyte develops within sporophyte	Wind-dispersed pollen carries sperm to stationary egg in cone	Occurs within a protective seed containing a food supply	Seeds containing diploid sporophyte embryo dispersed by wind or animals	Present
	Flowering plants	Sporophyte dominant—microscopic gametophyte develops within sporophyte	Pollen, dispersed by wind or animals, carries sperm to stationary egg within flower	Occurs within a protective seed containing a food supply; seed encased in fruit	Fruit, carrying seeds, dispersed by animals, wind, or water	Present

FIGURE 21-3 Bryophytes
Both plants shown here are less than $\frac{1}{2}$ inch (about 1 centimeter) in height. **(a)** Liverworts grow in moist, shaded areas. This female plant bears umbrella-like archegonia, which hold the eggs. Sperm must swim up the stalks through a film of water to fertilize the eggs. **(b)** Moss plants, showing the stalks that carry spore-bearing capsules. **QUESTION** Why are all bryophytes short?

in areas where moisture is plentiful. Some mosses, however, have a waterproof covering that retains moisture, preventing water loss. These mosses can survive in deserts, on bare rock, and in far northern and southern latitudes where humidity is low and liquid water is scarce for much of the year.

The Reproductive Structures of Bryophytes Are Protected

Among bryophytes' adaptations to terrestrial existence are their enclosed reproductive structures (Fig. 21-4). **Archegonia** (singular, archegonium), in which eggs develop, and **antheridia** (singular, antheridium), where sperm are formed, prevent the gametes from drying out. In some bryophyte species, both archegonia and antheridia are located on the same plant; in other species, each individual plant is either male or female.

In all bryophytes, the sperm must swim to the egg, which emits a chemical attractant, through a film of water. (Bryophytes that live in drier areas must time their reproduction to coincide with rains.) The fertilized egg is retained in the archegonium, where the embryo grows and matures into a small diploid sporophyte that remains attached to the parent gametophyte plant. At maturity, the sporophyte produces haploid spores by meiosis within a capsule. When the capsule is opened, spores are released and dispersed by the wind. If a spore lands in a suitable environment, it may develop into another haploid gametophyte plant.

Vascular Plants Have Conducting Vessels That Also Provide Support

Vascular plants are distinguished from bryophytes by specialized groups of conducting cells, called **vessels**, that are impregnated with the stiffening substance lignin and serve both supportive and conducting functions. Vessels allow vascular plants to grow taller than nonvascular plants, because of the extra support provided by lignin and because the conducting cells allow water and nutrients absorbed by the roots to move to the upper portions of the plant. Another difference between vascular plants and bryophytes is that, in vascular plants, the diploid sporophyte is the larger, more conspicuous generation; in nonvascular plants, the haploid gametophyte is more evident.

The vascular plants can be divided into two groups: the seedless vascular plants and the seed plants.

The Seedless Vascular Plants Include the Club Mosses, Horsetails, and Ferns

Like the bryophytes, seedless vascular plants have swimming sperm and require water for reproduction. As their name implies, they do not produce seeds; instead, they propagate by spores. The present-day seedless vascular plants—the club mosses, horsetails, and ferns—are much diminished in size compared to their ancestors, which dominated the landscape hundreds of millions of years ago. Today, the bodies of ancient seedless vascular plants—transformed by heat, pressure, and time—are burned as coal. The formerly dominant role of the seedless vascular plants has been assumed by the more-versatile seed plants.

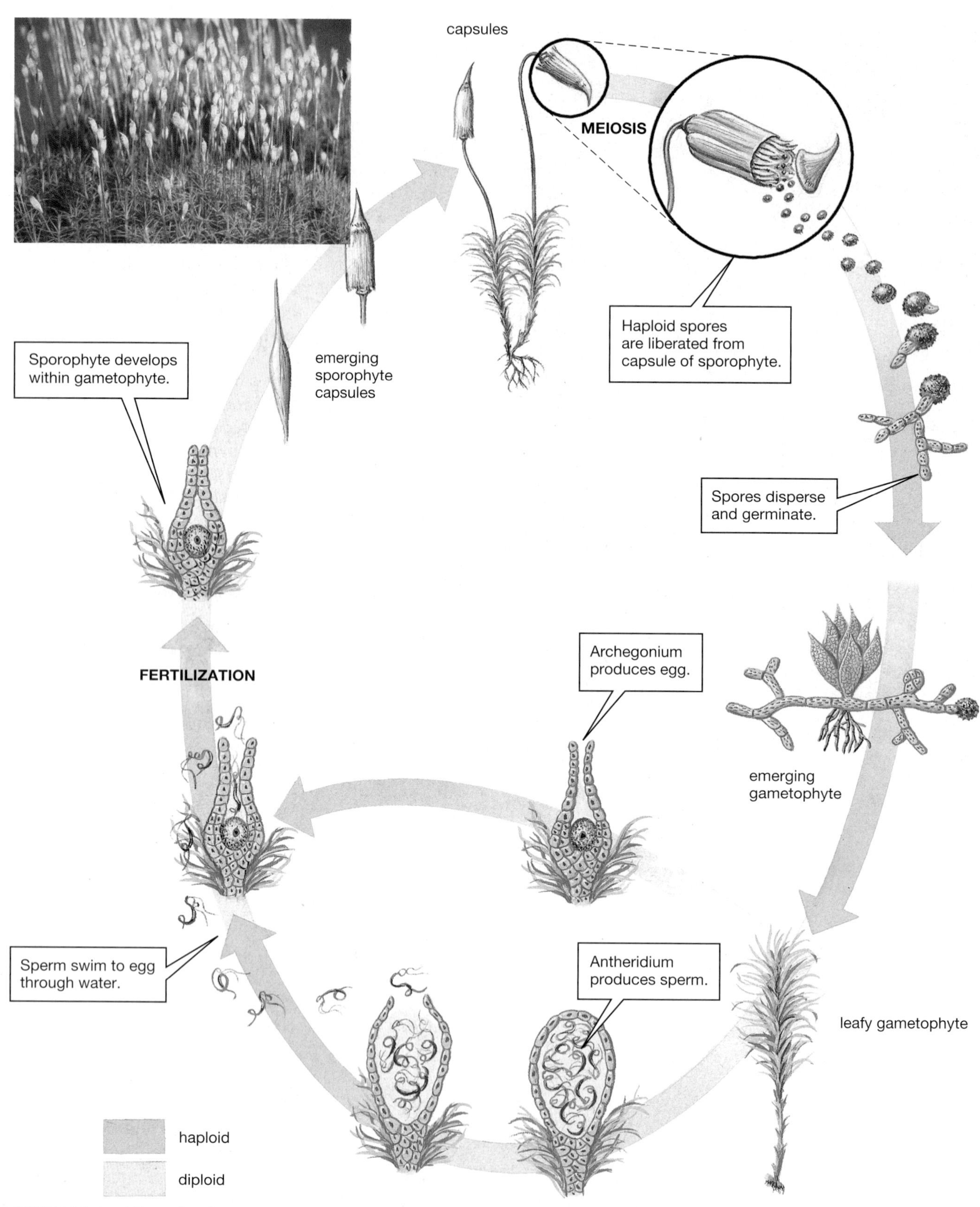
capsules
MEIOSIS
Haploid spores are liberated from capsule of sporophyte.
Spores disperse and germinate.
emerging gametophyte
leafy gametophyte
Antheridium produces sperm.
Archegonium produces egg.
Sperm swim to egg through water.
FERTILIZATION
Sporophyte develops within gametophyte.
emerging sporophyte capsules
haploid
diploid

FIGURE 21-4 Life cycle of a moss
The leafy green gametophyte (lower right) is the haploid generation that produces sperm and eggs. The sperm must swim through a film of water to the egg. The zygote develops into a stalked, diploid sporophyte that emerges from the gametophyte plant. The sporophyte is topped by a brown capsule in which haploid spores are produced by meiosis. These are dispersed and germinate, producing another green gametophyte generation. **(Inset)** Moss plants. The short, leafy green plants are haploid gametophytes; the reddish brown stalks are diploid sporophytes.

(a)

(c)

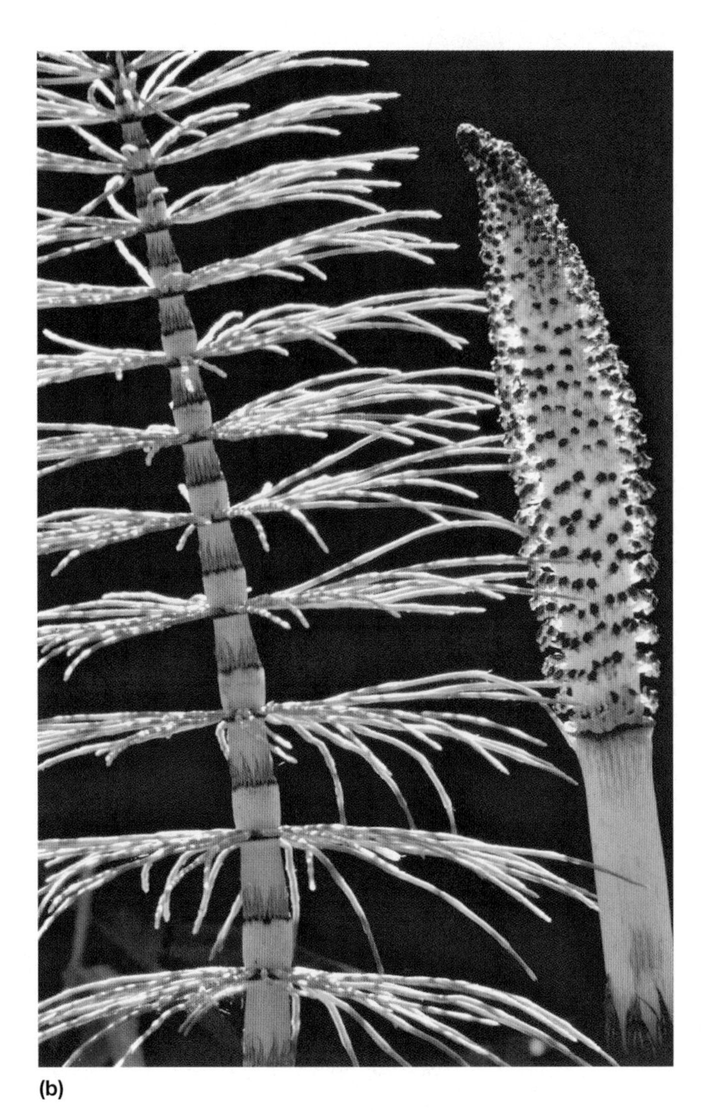

(b)

FIGURE 21-5 Some seedless vascular plants
Seedless vascular plants are found in moist woodland habitats. **(a)** The club mosses (sometimes called ground pines) grow in temperate forests. This specimen is releasing spores. **(b)** The giant horsetail extends long, narrow branches in a series of rosettes. Its leaves are insignificant scales. At right is a cone-shaped spore-forming structure.**(c)** The leaves of this deer fern are emerging from coiled fiddleheads. **QUESTION** In each of these photos, is the pictured structure a sporophyte or a gametophyte?

Club Mosses and Horsetails Are Small and Inconspicuous

The club mosses are now only a few inches in height (Fig. 21-5a). Their leaves are small and scalelike, resembling the leaflike structures of mosses. Club mosses of the genus *Lycopodium*, commonly known as ground pine, form a beautiful ground cover in some temperate coniferous and deciduous forests.

Modern horsetails belong to a single genus, *Equisetum*, that contains only 15 species, most less than 1 meter tall (Fig. 21-5b). The bushy branches of some species lend them the common name horsetails; the leaves are reduced to tiny scales on the branches. They are also called "scouring rushes" because they were used by early European settlers of North America to scour pots and floors. All species of *Equisetum* deposit large amounts of silica (glass) in their outer layer of cells, giving them an abrasive texture.

Ferns Are Broad-Leaved and More Diverse

The ferns, with 12,000 species, are the most diverse of the seedless vascular plants (Fig. 21-5c). In the Tropics, tree ferns still reach heights reminiscent of their ancestors from the Carboniferous period. Ferns are the only seedless vascular plants that have broad leaves.

In ferns, haploid spores are produced in structures called *sporangia* that form on special leaves of the

sporophyte
masses of sporangia
sporangium
leaf
stem
root
MEIOSIS
Sporophyte develops from gametophyte.
Haploid spores are liberated from sporangium.
Archegonium produces egg.
FERTILIZATION
Sperm swim to egg through water.
Spores disperse and germinate.
gametophyte
Antheridium produces sperm.
haploid
diploid

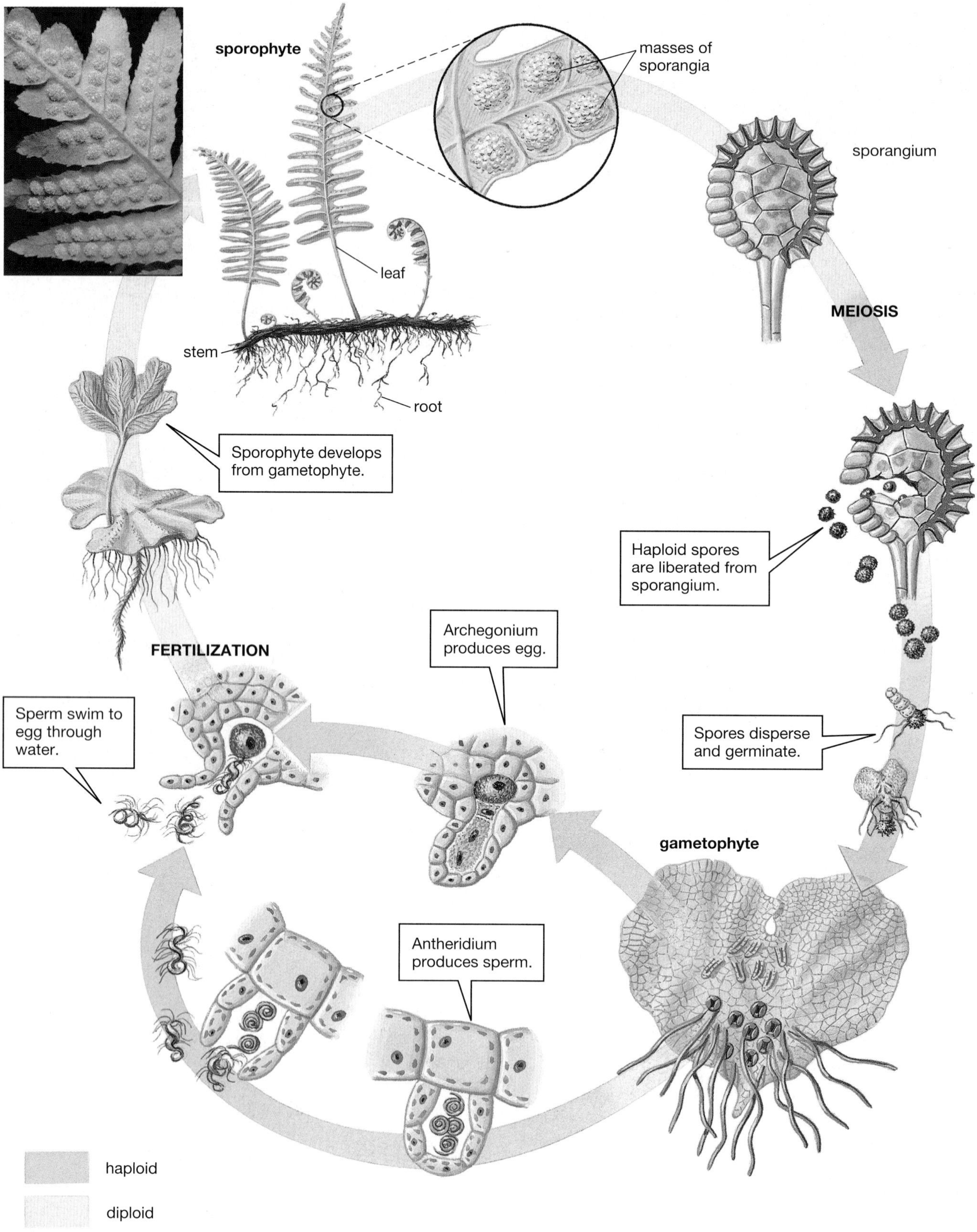

Media Activity
21.2 Fern Life Cycle

FIGURE 21-6 Life cycle of a fern
The dominant plant body (upper left) is the diploid sporophyte. Haploid spores, formed in sporangia located on the underside of certain leaves, are dispersed by the wind to germinate on the moist forest floor into inconspicuous haploid gametophyte plants. On the lower surface of these small, sheetlike gametophytes, male antheridia and female archegonia produce sperm and eggs. The sperm must swim to the egg, which remains in the archegonium. The zygote develops into the large sporophyte plant. **(Inset)** Underside of a fern leaf, showing clusters of sporangia.

sporophyte (Fig. 21-6). The spores are dispersed by the wind and give rise to tiny, haploid gametophyte plants, which produce sperm and eggs. The gametophyte generation retains two traits that are reminiscent of the bryophytes. First, the small gametophytes lack conducting vessels. Second, as in bryophytes, the sperm must swim through water to reach the egg.

The Seed Plants Dominate the Land, Aided by Two Important Adaptations: Pollen and Seeds

The seed plants are distinguished from bryophytes and seedless vascular plants by their production of pollen and seeds. **Pollen** grains are tiny structures that carry sperm-producing cells. Pollen grains are dispersed by wind or by animal pollinators such as bees. In this way, sperm move through the air to fertilize egg cells. Thus, the distribution of seed plants is not limited by the need for water through which sperm can swim to the egg; seed plants are fully adapted to life on dry land.

Analogous to the eggs of birds and reptiles, **seeds** consist of an embryonic plant, a supply of food for the embryo, and a protective outer coat (Fig. 21-7). The *seed coat* maintains the embryo in a state of suspended animation, or dormancy, until conditions are proper for growth. The stored food helps sustain the emerging plant until it develops roots and leaves and can make its own food by photosynthesis. Some seeds possess elaborate adaptations that allow them to be dispersed by wind, water, and animals.

In seed plants, gametophytes (which produce the sex cells) are greatly reduced in size. The female gametophyte is a small group of haploid cells that produces the egg. The male gametophyte is the pollen grain.

Seed plants are grouped into two general types: *gymnosperms*, which lack flowers, and *angiosperms*, the flowering plants.

Gymnosperms Are Nonflowering Seed Plants

Gymnosperms evolved earlier than the flowering plants. One group, the **conifers**, still dominates large areas of our planet. Conifers, whose 500 species include pines, firs, spruce, hemlocks, and cypresses, are

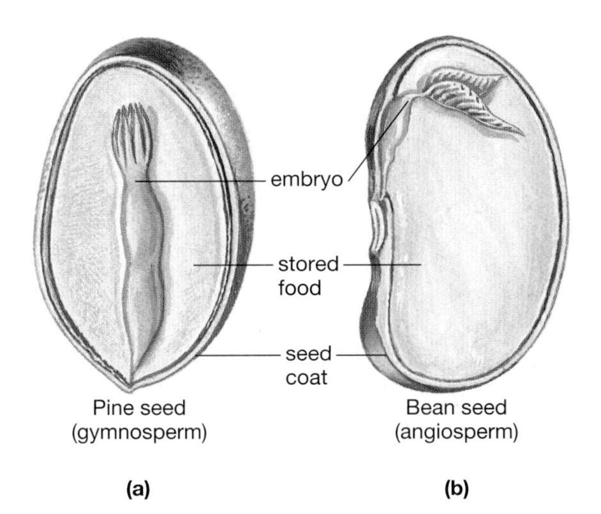

FIGURE 21-7 Seeds
Seeds from **(a)** a gymnosperm and **(b)** an angiosperm. Both consist of an embryonic plant and stored food confined within a seed coat. Seeds exhibit diverse adaptations for dispersal, including **(c)** the dandelion's tiny, tufted seeds that float in the air and **(d)** the massive, armored seeds (protected inside the fruit) of the coconut palm, which can survive prolonged immersion in seawater as they traverse oceans. **QUESTION** Can you think of some adaptations that help protect seeds from destruction by animal consumption?

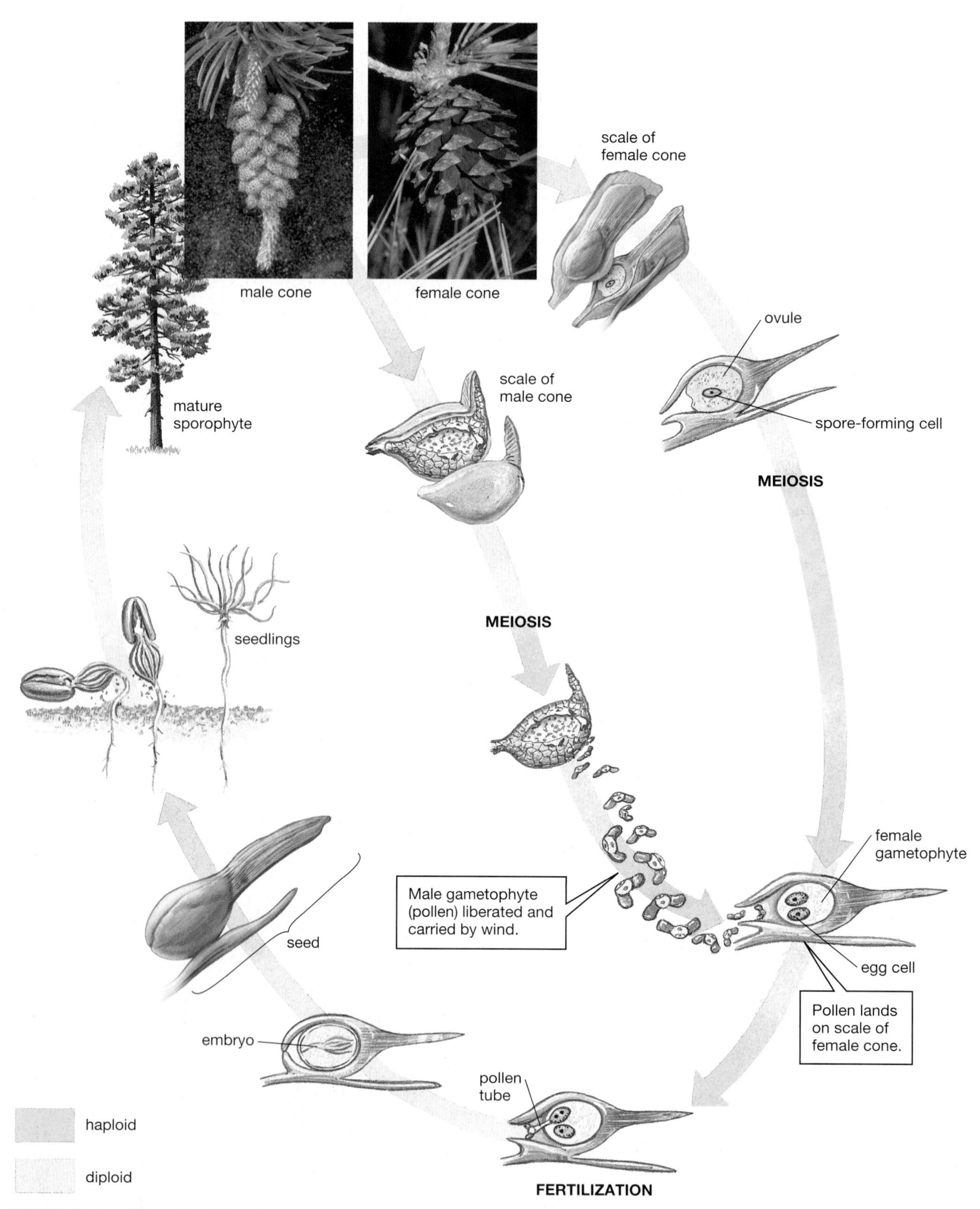

FIGURE 21-8 Life cycle of the pine
The pine tree is the sporophyte generation (upper left) and bears both male and female cones. Haploid female gametophytes develop within the scales of female cones and produce egg cells. Male cones produce pollen, the male gametophytes. A pollen grain, dispersed by the wind, may land on the scale of a female cone. It then grows a pollen tube that penetrates the female gametophyte and conducts sperm to the egg. The fertilized egg develops into an embryonic plant enclosed in a seed. The seed is eventually released from the cone, germinates, and grows into a sporophyte tree.

most abundant in the cold latitudes of the far north and at high elevations where conditions are dry. Not only is rainfall limited in these areas, but water in the soil remains frozen and unavailable during the long winters. Conifers are adapted to dry, cold conditions in three ways. First, conifers retain green leaves throughout the year, enabling these plants to continue photosynthesizing and growing slowly during times when most other plants become dormant. For this reason, conifers are often called "evergreens." Second, conifer leaves are actually thin needles covered with a thick, waterproof surface that minimizes evaporation. Finally, conifers produce an "antifreeze" in their sap that enables them to continue transporting nutrients in below-freezing temperatures. This substance gives them their fragrant "piney" scent.

Conifer Seeds Develop in Cones

Reproduction is similar in all conifers, so let's examine the reproductive cycle of a pine tree (Fig. 21-8). The tree itself is the diploid sporophyte, and develops both male and female cones. The male cones are relatively small (normally $\frac{3}{4}$ of an inch—about 2 centimeters—or less), delicate structures that release clouds of pollen during the reproductive season and then disintegrate. These clouds of pollen are immense; inevitably, some pollen grains land by chance on a female cone.

Each female cone consists of a series of woody scales arranged in a spiral around a central axis. At the base of each scale are two **ovules** (immature seeds), within which diploid spore cells form and undergo meiosis to produce haploid female gametophytes. These gametophytes then develop and produce egg cells. If a pollen grain from a male cone lands nearby, it sends out a pollen tube that slowly burrows into the female gametophyte. After nearly 14 months, the tube finally reaches the egg cell and releases the sperm that fertilize it. The fertilized egg becomes enclosed in a seed as it develops into a tiny embryonic plant. The seed is liberated when the cone matures and its scales separate.

Ginkgos and Cycads Are Less Abundant

Unlike the conifers, other gymnosperms, such as the ginkgos and cycads (Fig. 21-9), now occupy only a small fraction of their former range. Ginkgos were probably the first modern-day seed plants to evolve, becoming widespread during the Jurassic period, which began 208 million years ago. Today, however, they are represented by the single species *Ginkgo biloba*, the maidenhair tree. Ginkgo trees are either male or female; female trees bear foul-smelling, fleshy seeds the size of cherries (see Fig. 21-9a). Ginkgos have been maintained by cultivation, particularly in Asia; if not for this cultivation, they might be extinct today. Because they are more resistant to pollution than are most other trees, ginkgos (normally, the male trees) have been extensively planted in U.S. cities. Recently, the leaves of the ginkgo have gained attention as a herbal remedy that purportedly improves memory.

Cycads look like the large ferns from which they probably evolved (see Fig. 21-9b). Today there are approximately 160 species, most of which dwell in tropical or subtropical climates. Most cycads are about 3 feet (1 meter) in height, although some species can reach 65 feet (20 meters). Cycads grow slowly and live for a long time; one Australian specimen is estimated to be 5000 years old.

(a)

(b)

FIGURE 21-9 Two uncommon gymnosperms
(a) This ginkgo, or maidenhair tree, is female and bears fleshy seeds the size of large cherries. **(b)** A cycad. Common in the age of dinosaurs, these are now limited to about 160 species. Like ginkgos, cycads have separate sexes.

Angiosperms Are Flowering Seed Plants

Modern flowering plants, or **angiosperms**, have dominated Earth for more than 100 million years. The group is incredibly diverse, with more than 230,000 species. Angiosperms range in size from the diminutive duckweed (Fig. 21-10a) to the towering eucalyptus tree (Fig. 21-10b). From desert cactus to grasses to parasitic stinking corpse lilies, angiosperms rule over the plant kingdom.

Flowers Attract Pollinators

Three major adaptations have contributed to the enormous success of angiosperms: flowers, fruits, and broad leaves. **Flowers**, the structures in which both male and female gametophytes are formed, may have evolved when gymnosperm ancestors formed an association with animals (most likely insects) that carried their pollen from plant to plant. According to this scenario, the relationship between these ancient gymnosperms and their animal pollinators was so beneficial that natural selection favored the evolution of showy flowers that advertised the presence of pollen to insects and other animals (Fig. 21-10b,e). The animals benefited by eating some of the protein-rich pollen, and the plant benefited from the animals' unwitting transportation of pollen from plant to plant. With this animal assistance, many flowering plants no longer needed to produce prodigious quantities of pollen and send it flying on the fickle winds to ensure fertilization. Nonetheless, there are also many wind-pollinated angiosperms (Fig. 21-10c,d).

In the angiosperm life cycle (Fig. 21-11, p. 418), flowers develop on the dominant sporophyte plant. Male gametophytes (pollen) are formed inside a structure called the *anther*; the female gametophyte develops from an ovule within a part of the flower called the *ovary*. The egg, in turn, develops within the female gametophyte. Fertilization occurs when the pollen forms a tube through the *stigma*, a sticky pollen-catching structure of the flower, and bores into the ovule. There, the zygote develops into an embryo enclosed in a seed formed from the ovule.

Fruits Encourage Seed Dispersal

The ovary surrounding the seed of an angiosperm matures into a **fruit**, the second adaptation that has contributed to the success of angiosperms. Just as flowers encourage animals to transport pollen, so, too, many fruits entice animals to disperse seeds. If an animal eats a fruit, many of the enclosed seeds may pass through the animal's digestive tract unharmed, perhaps falling at a location suitable for germination. Not all fruits, however, depend on edibility for dispersal. Dog owners are well aware, for example, that some fruits (called *burrs*) disperse by clinging to animal fur. Others, such as the fruits of maple trees, form wings that carry the seed through the air. The variety of dispersal mechanisms made possible by various fruits have helped angiosperms invade nearly all possible terrestrial habitats.

Broad Leaves Capture More Sunlight

The third feature that gives angiosperms an advantage in warmer, wetter climates is broad leaves. When water is plentiful, as it is during the warm growing season of temperate and tropical climates, broad leaves provide an advantage by collecting more sunlight for photosynthesis. In regions with seasonal variation in growing conditions, many trees and shrubs drop their leaves during periods when water is in short supply, which reduces evaporative water loss (Fig. 21-12, p. 419). In temperate climates, such periods occur during the fall and winter, at which time most temperate angiosperm trees and shrubs drop their leaves. In the Tropics and subtropics, most angiosperms are evergreen, but species that inhabit certain tropical climates where periods of drought are common may drop their leaves to conserve water during the dry season.

The advantages of broad leaves are offset by some evolutionary costs. In particular, broad, tender leaves are much more appealing to herbivores than are the tough, waxy needles of conifers. As a result, angiosperms have developed a range of defenses against mammalian and insect herbivores. These adaptations include physical defenses such as thorns, spines, and resins that toughen the leaves. But the evolutionary struggle for survival has also led to a host of chemical defenses—compounds that make plant tissue poisonous or distasteful to potential predators. Many of the compounds responsible for chemical defense have properties that humans have exploited for medicinal and culinary uses. Medicines such as aspirin and codeine, stimulants such as nicotine and caffeine, and spicy flavors such as mustard and peppermint are all derived from angiosperm plants.

More Recently Evolved Plants Have Smaller Gametophytes

The evolutionary history of plants has been marked by a tendency for the sporophyte generation to become increasingly prominent, and for the longevity and size of the gametophyte generation to shrink (see Table 21-1). Thus, the earliest plants are believed to have been similar to today's nonvascular plants, which have a sporophyte that is smaller than the gametophyte and remains attached to it. In contrast, plants that originated somewhat later, such as ferns and the other seedless vascular plants, feature a life cycle in which the sporophyte is dominant, and the gametophyte is a much smaller, independent plant. Finally, in the most recently evolved group of plants, the seed plants, gametophytes are microscopic and barely recognizable as an alternate generation. These tiny gametophytes, however, still produce the eggs and sperm that unite to form the zygote that develops into the diploid sporophyte.

FIGURE 21-10 Angiosperms
(a) The smallest angiosperm is the duckweed, found floating on ponds. These specimens are about $\frac{1}{8}$ inch (3 millimeters) in diameter. **(b)** The largest angiosperms are eucalyptus trees, which can reach 325 feet (100 meters) in height. Both **(c)** grasses and many trees, such as **(d)** this birch, in which flowers are shown as buds (green) and blossoms (brown), have inconspicuous flowers and rely on wind for pollination. More conspicuous flowers, such as those on **(e)** this butterfly weed and on a eucalyptus tree (*b*, inset), entice insects and other animals that carry pollen between individual plants. **EXERCISE** List the advantages and disadvantages of wind pollination. Do the same for pollination by animals. Why do both types of pollination persist among the angiosperms?

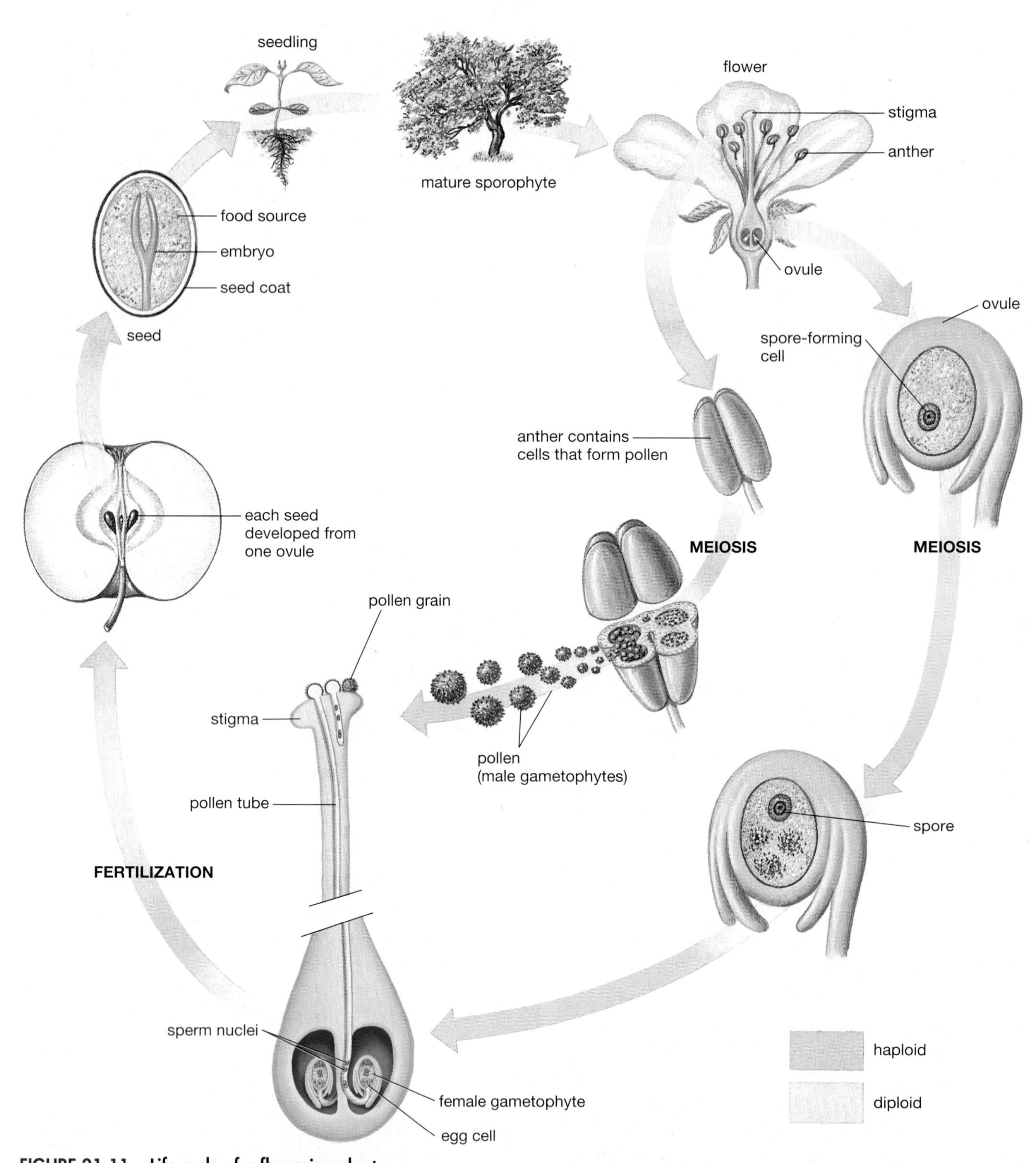

FIGURE 21-11 Life cycle of a flowering plant
The dominant plant body (upper right) is the diploid sporophyte, whose flowers normally produce both male and female gametophytes. Male gametophytes (pollen grains) are produced within anthers. The female gametophyte develops from a spore within the ovule, and contains one egg cell. A pollen grain that lands on a stigma grows a pollen tube that burrows down to the ovule and into the female gametophyte. There it releases its sperm, one of which fuses with the egg to form a zygote. The ovule gives rise to the seed, which contains the developing embryo and its food source. The seed is dispersed, germinates, and develops into a mature sporophyte.

FIGURE 21-12 Two ways of coping with the dryness of winter
The evergreen (a conifer) retains its needles throughout the year. The small surface area and heavy cuticle of the needles slow the loss of water through evaporation. In contrast, the aspen (an angiosperm) sheds its leaves each fall. The dying leaves turn brilliant shades of gold as pigments used to capture light energy for photosynthesis are exposed when the chlorophyll disintegrates.

CASE STUDY REVISITED Queen of the Parasites

The approximately 17 parasitic plant species of the genus *Rafflesia,* which includes the stinking corpse lily, are found in the moist forests of Southeast Asia, a habitat that is disappearing rapidly as forests are cleared for agriculture and development. The geographic range of the stinking corpse lily is limited to the dwindling forests of Malaysian peninsula and the Indonesian islands of Borneo and Sumatra; the species is rare and endangered. The government of Indonesia has established some parks and reserves that help protect the stinking corpse lily but—as is often the case in developing countries—a forest that is protected on paper may still be vulnerable in reality. Perhaps the best hope for the continued survival of the largest *Rafflesia* is the growing realization among the rural residents of Sumatra and Borneo that the spectacular, putrid-smelling flowers of the stinking corpse lily might lure interested tourists to their countries. Under an innovative conservation program that seeks to take advantage of this potential for ecotourism, people who live in the vicinity of the stinking corpse lily can become caretakers of the plants. These assigned caretakers watch over the plants and, in return, may charge a small fee to curious visitors. Thus, local inhabitants have been given an economic incentive to protect this rare parasitic plant.

Consider This: A parasitic lifestyle is unusual among plants, but it is not exactly rare. Fifteen different plant families contain parasitic species, and systematists estimate that parasitism has evolved at least nine different times over the evolutionary history of plants. Given the obvious benefits of photosynthesis, why has parasitism (which is often accompanied by loss of photosynthetic capability) evolved repeatedly in photosynthetic plants?

Links to Life: Our Friends the Angiosperms

Humans may dominate Earth, but we could not have achieved dominance without some help. Neither our explosive population growth nor our rapid technological advance would have been possible without our 10,000-year-old partnership with domesticated animal and plant species. Among domesticated plants, we have shown a special affinity for the angiosperms. Over generations of selective breeding, humans have modified the seeds, stems, roots, flowers, and fruits of favored angiosperm species to provide us with food, medicines, and shelter.

Our relationship with the angiosperms, however, seems to be based on something more profound than their ability to help us meet our material needs. Though we appreciate the practical value of wheat and wood, our most emotionally powerful connections with angiosperms are purely sensual. Many of life's pleasures come to us courtesy of our plant partners. We delight in the beauty and fragrance of flowers, and present them to others as symbols of our most sublime and inexpressible emotions. Quite a few of us spend hours of our leisure time tending gardens and lawns, for no reward other than the pleasure and satisfaction we derive from observing the fruits of our labor. In our homes, we reserve space not only for members of our families, but also for our houseplant companions. We feel compelled to line our streets with trees, and we seek refuge from the stress of daily life in parks with abundant plant life. Our mornings are enhanced by the aroma of coffee or tea, and our evenings by a nice glass of wine.

Clearly, angiosperms fill our desires as well as our needs. But what do plants get out of the deal? For the species fortunate enough to be accepted as our partners, we provide nutrients, water, protection from predators, and dispersal around the globe.

Summary of Key Concepts

21.1 What Is the Evolutionary Origin of Plants?
Photosynthetic protists, probably green algae, gave rise to the first plants. Ancestral plants were probably similar to modern multicellular green algae, which have photosynthetic pigments, starch molecules, and cell wall components—including cellulose—similar to those of plants. The freshwater heritage of green algae may have endowed them with qualities that enabled their descendants to invade land.

21.2 What Are the Key Features of Plants?
The kingdom Plantae consists of eukaryotic, photosynthetic, multicellular organisms. Unlike their green algae relatives, plants have multicellular, dependent embryos and exhibit alternation of generations in which a haploid gametophyte generation alternates with a diploid sporophyte generation.

Plants also have a number of key adaptations for a terrestrial existence: rootlike structures for anchorage and for absorption of water and nutrients; a waxy cuticle to slow the loss of water through evaporation; stomata that can open, allowing gas exchange, and that can also close, preventing water loss; conducting vessels to transport water and nutrients throughout the plant; and a stiffening substance, called *lignin*, to impregnate the vessels and support the plant body.

Plant reproductive structures suitable for life on land include a reduced male gametophyte (pollen) that allows wind to replace water in carrying sperm to eggs; seeds that nourish, protect, and help disperse developing embryos; flowers that attract animals, who carry pollen more precisely and efficiently than wind; and fruits that entice animals to disperse seeds.

21.3 What Are the Major Groups of Plants?
Two major groups of plants, bryophytes and vascular plants, arose from the ancient algal ancestors. Bryophytes, including the liverworts and mosses, are small, simple land plants that lack conducting vessels. Although some have adapted to dry areas, most live in moist habitats. Bryophyte reproduction requires water through which the sperm swim to the egg.

In vascular plants, a system of vessels—stiffened by lignin—conducts water and nutrients absorbed by the roots into the upper portions of the plant and supports the body as well. Owing to this support system, seedless vascular plants, including the club mosses, horsetails, and ferns, can grow larger than bryophytes. As in bryophytes, the sperm of seedless vascular plants must swim to the egg for sexual reproduction to occur, and the gametophyte lacks conducting vessels.

Vascular plants with seeds have two major additional adaptive features: pollen and seeds. Seed plants are often classified into two categories: gymnosperms and angiosperms. Gymnosperms include ginkgos, cycads, and the highly successful conifers. These plants were the first fully terrestrial plants to evolve. Their success on dry land is partially due to the evolution of the male gametophyte into the pollen grain. Pollen protects and transports the male gamete, eliminating the need for the sperm to swim to the egg. The seed, a protective resting structure containing an embryo and a supply of food, is a second important adaptation contributing to the success of seed plants.

Angiosperms, the flowering plants, dominate much of the land today. In addition to pollen and seeds, angiosperms also produce flowers and fruits. The flower allows angiosperms to utilize animals as pollinators. In contrast to wind, animals can in some cases carry pollen farther and with greater accuracy and less waste. Fruits may attract animal consumers, which incidentally disperse the seeds in their feces.

There has been a general evolutionary trend toward reduction of the haploid gametophyte, which is dominant in bryophytes but microscopic in seed plants.

Key Terms

alternation of generations *p. 407*
angiosperm *p. 416*
antheridium *p. 409*
archegonium *p. 409*
bryophyte *p. 408*
conifer *p. 413*
cuticle *p. 407*
flower *p. 416*
fruit *p. 416*
gametophyte *p. 407*
gymnosperm *p. 413*
lignin *p. 408*
ovule *p. 415*
pollen *p. 413*
seed *p. 413*
sporophyte *p. 407*
stomata *p. 407*
vascular *p. 408*
vessel *p. 409*
zygote *p. 407*

Thinking Through the Concepts

To take a multiple-choice quiz with feedback on the contents of this chapter, visit http://www.prenhall.com/audesirk7. *Log in to the Web site selected by your instructor and navigate to the Self Test section for this chapter.*

? Review Questions

1. What is meant by "alternation of generations"? What two generations are involved? How does each reproduce?
2. Explain the evolutionary changes in plant reproduction that adapted plants to increasingly dry environments.
3. Describe evolutionary trends in the life cycles of plants. Emphasize the relative sizes of the gametophyte and sporophyte.
4. From which algal group did green plants probably arise? Explain the evidence that supports this hypothesis.
5. List the structural adaptations necessary for the invasion of dry land by plants. Which of these adaptations are pos-

sessed by bryophytes? By ferns? By gymnosperms and angiosperms?

6. The number of species of flowering plants is greater than the number of species in the rest of the plant kingdom. What feature(s) are responsible for the enormous success of angiosperms? Explain why.

7. List the adaptations of gymnosperms that have helped them become the dominant trees in dry, cold climates.

8. What is a pollen grain? What role has it played in helping plants colonize dry land?

9. The majority of all plants are seed plants. What is the advantage of a seed? How do plants that lack seeds meet the needs served by seeds?

Applying the Concepts

1. You are a geneticist working for a firm that specializes in plant biotechnology. Explain what *specific* parts (fruit, seeds, stems, roots, etc.) of the following plants you would try to alter by genetic engineering, what changes you would try to make, and why: (a) corn, (b) tomatoes, (c) wheat, and (d) avocados.

2. Prior to the development of synthetic drugs, more than 80% of all medicines were of plant origin. Even today, indigenous tribes in remote Amazonian rain forests can provide a plant product to treat virtually any ailment. Herbal medicine is also widely and successfully practiced in China. Most of these drugs are unknown to the Western world. But the forests from which much of this plant material is obtained are being converted to agriculture. We are in danger of losing many of these potential drugs before they can be discovered. What steps can you suggest to preserve these natural resources while also allowing nations to direct their own economic development?

3. Only a few hundred of the hundreds of thousands of species in the plant kingdom have been domesticated for human use. One example is the almond. The domestic almond is nutritious and harmless, but its wild precursor can cause cyanide poisoning. The oak makes potentially nutritious seeds (acorns) that contain very bitter-tasting tannins. If we could breed the tannin out of acorns, they might become a delicacy. Why do you suppose we have failed to domesticate oaks?

For More Information

Diamond, J. "How to Tame a Wild Plant." *Discover*, September 1994. Cultivated plants have ecological and genetic properties that make them well suited for agriculture.

Doyle, J. "DNA, Phylogeny, and the Flowering of Plant Systematics." *BioScience*, June 1993. Chloroplast DNA is used in reconstructing evolutionary relationships, but there are difficulties with each of the methods in use.

Joyce, C. *Earthly Goods: Medicine-Hunting in the Rainforest.* Boston: Little, Brown, 1994. Science and adventure combine in this account of prospecting for new medicines and the people who do it.

Kaufman, P. B. *Plants—Their Biology and Importance.* New York: Harper & Row, 1989. Complete, readable coverage of all aspects of plant taxonomy, physiology, and evolution.

McClintock, J. "The Life, Death, and Life of a Tree. *Discover*, May 2002. The author describes the biology of California's majestic redwood trees and the threat they face from humanity's appetite for their wood.

Milot, V. "Blueprint for Conserving Plant Diversity." *BioScience,* June 1989. Points out the importance of preserving genetic diversity in endangered plant species.

Pollan, M. *The Botany of Desire.* New York: Random House, 2001. A literate look at the mutually beneficial relationship between humans and plants.

Media Activities

To access a Media Activity visit http://www.prenhall.com/audesirk7. *Log in to the Web site selected by your instructor, navigate to this chapter, and select the appropriate Media Activity number.*

21.1 Evolution of Plant Structure

Estimated time: 10 minutes

This animation illustrates the evolutionary changes that have occurred in plant structures and how the changes have allowed the plants to exploit different habitats.

21.2 Fern Life Cycle

Estimated time: 5 minutes

View the stages of and structures involved in the fern life cycle.

21.3 Web Investigation: Queen of Parasites

Estimated time: 10 minutes

Explore the complex relationships between plants and their pollinators.

CHAPTER

22 Animal Diversity I: Invertebrates

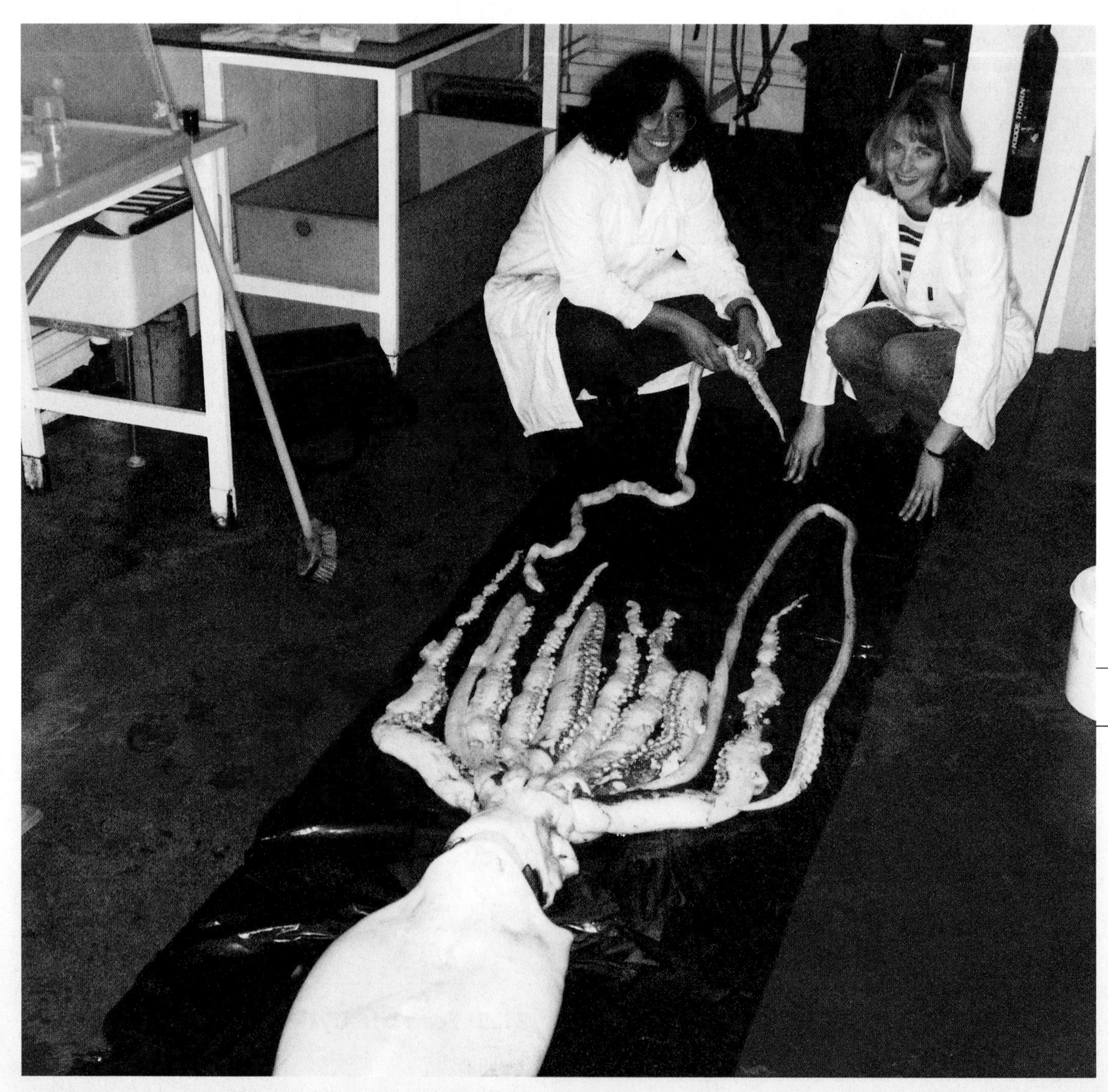

Although the giant squid is Earth's largest invertebrate animal, no human has ever observed a living giant squid in its natural habitat.

AT A GLANCE

CASE STUDY The Search for a Sea Monster

Everyone loves a good mystery, and a mystery involving a gigantic, fearsome predator is even better. Consider *Architeuthis*, the giant squid. It is the world's largest invertebrate animal, reaching lengths of 60 feet (18 meters) or more. Each of its huge eyes, the largest in the animal kingdom, can be as large as a human head. The squid's ten tentacles, two of which are longer than the others, are covered with powerful suckers. The suckers contain sharp, clawlike hooks to better grasp prey, which is then pulled toward the squid's mouth, where a heavily muscled beak tears the food apart. The giant squid is one of the most imposing organisms on Earth, yet we know nothing of its habits and lifestyle. What does it eat? Does it swim with its head elevated or pointed downward? How does it mate? Does it live alone or in groups? Even these basic questions about the behavior of the giant squid remain unanswered, because no one has ever seen an adult giant squid alive in its natural habitat, or even in an aquarium.

Our limited scientific knowledge of the giant squid comes entirely from specimens that have been found dead or dying, washed up on beaches, caught in fishermen's nets, or contained in the stomachs of sperm whales (which consume vast quantities of squid, including the occasional giant squid). More than 200 such specimens have been reported over the past century, and written accounts of squid corpses go back to the sixteenth century. Live squids, however, remain elusive because they inhabit deep ocean waters, beyond the reach of human divers.

Clyde Roper, a biologist at the Smithsonian Institution, has devoted much of his professional life to a quest to view and study live giant squids. We'll explore some of Dr. Roper's undersea search methods and his latest effort to find the giant squid at the close of this chapter, following a survey of the extraordinary diversity of invertebrate animal life.

Media Activity
22.1 Architecture of Animals

22.1 What Are the Key Features of Animals?

It is difficult to devise a concise definition of the term "animal." No single feature fully characterizes animals, so the group is defined by a list of characteristics. None of these characteristics is unique to animals, but when considered together, they distinguish animals from members of other kingdoms:

- Animals are multicellular.
- Animals are heterotrophic—they obtain their energy by consuming the bodies of other organisms.
- Animals typically reproduce sexually. Although animal species exhibit a tremendous diversity of reproductive styles, most are capable of sexual reproduction.
- Animal cells lack a cell wall.
- Animals are motile (able to move about) during some stage of their lives. Even the stationary sponges have a free-swimming *larval* stage (a juvenile form).
- Most animals are able to respond rapidly to external stimuli as a result of the activity of nerve cells, muscle tissue, or both.

22.2 Which Anatomical Features Mark Branch Points on the Animal Evolutionary Tree?

By the Cambrian period, which began 544 million years ago, most of the animal phyla that currently populate Earth were already present. Unfortunately, the pre-Cambrian fossil record is very sparse, and does not reveal the sequence in which the animal phyla arose. Therefore, animal systematists have looked to features of animal anatomy, embryological development, and DNA sequences for clues about the evolutionary history of animals. These investigations have shown that certain features mark major branching points on the animal evolutionary tree and represent milestones in

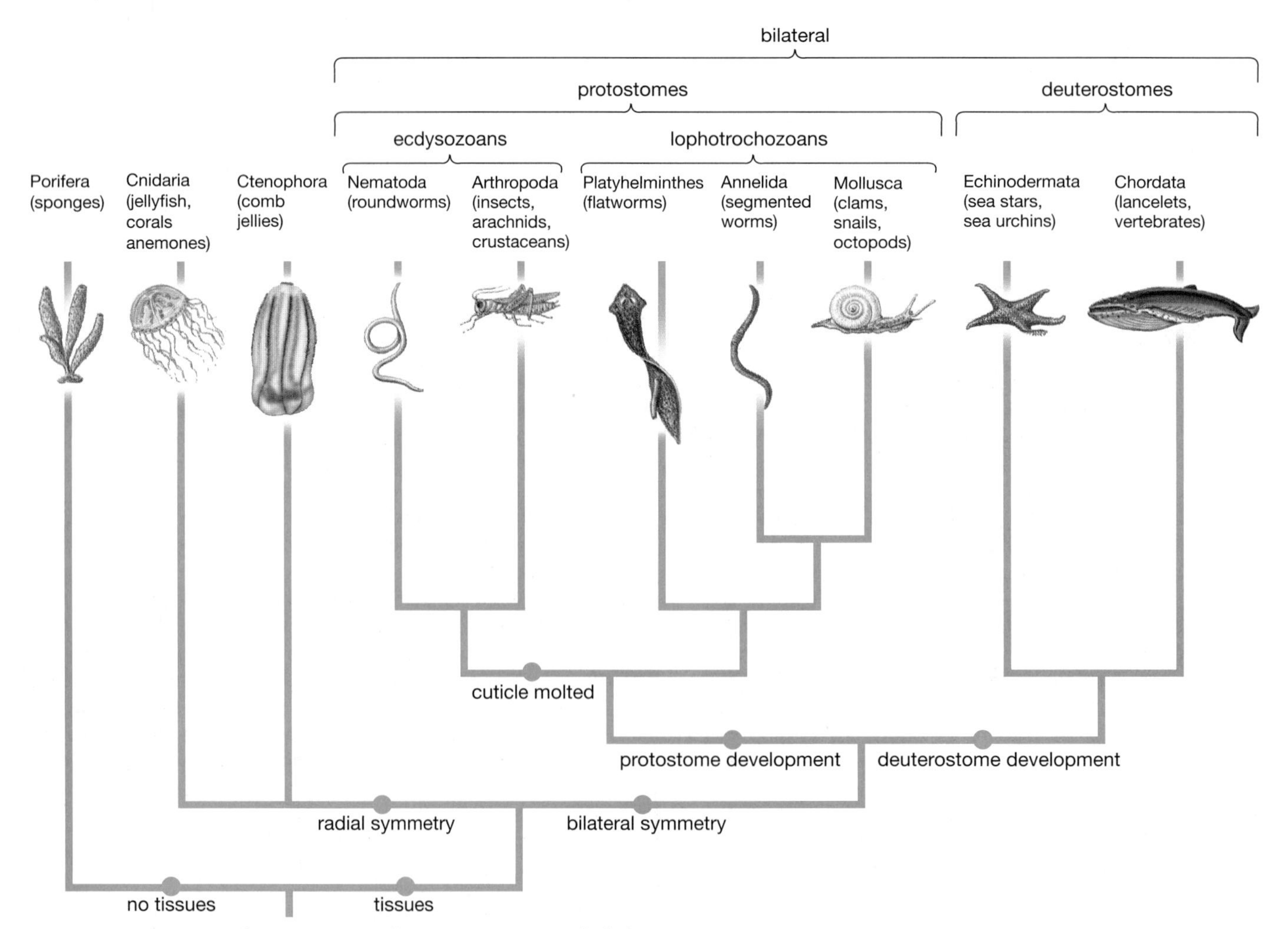

FIGURE 22-1 An evolutionary tree of some major animal phyla

the evolution of the different body plans of modern animals (Fig. 22-1). In the following sections, we will explore these evolutionary milestones and their legacies in the bodies of modern animals.

Lack of Tissues Separates Sponges from All Other Animals

One of the earliest major innovations in animal evolution was the appearance of **tissues**—groups of similar cells integrated into a functional unit, such as a muscle. Today, the bodies of almost all animals include tissues; the only animals that have retained the ancestral lack of tissues are the sponges. In sponges, individual cells may have specialized functions, but they act more or less independently and are not organized into true tissues. This unique feature of sponges suggests that the split between sponges and the evolutionary branch leading to all other animal phyla must have occurred very early in the history of animals. An ancient common ancestor without tissues gave rise to both the sponges and the remaining tissue-containing phyla.

Animals with Tissues Exhibit Either Radial or Bilateral Symmetry

The evolutionary advent of tissues coincided with the first appearance of body symmetry; all animals with true tissues also have symmetrical bodies. An animal is said to be symmetrical if it can be bisected along at least one plane such that the resulting halves are mirror images of one another. Note that, unlike the asymmetrical sponges, any symmetrical animal has an upper, or *dorsal*, surface and a lower, or *ventral*, surface.

The symmetrical, tissue-bearing animals can be divided into two groups, one containing animals that exhibit **radial symmetry** (Fig. 22-2a) and one with animals that exhibit **bilateral symmetry** (Fig. 22-2b). In radial symmetry, any plane through a central axis divides the object into roughly equal halves. In contrast, a bilaterally symmetrical animal can be divided into roughly mirror-image halves only along one particular plane through the central axis.

The difference between radially and bilaterally symmetrical animals reflects another major branching point in the animal evolutionary tree. This split separated the ancestors of the radially symmetrical cnidarians (jellyfish, anemones, and corals) and ctenophores (comb jellies) from the ancestors of the remaining animal phyla, all of which are bilaterally symmetrical.

Radially Symmetrical Animals Have Two Embryonic Tissue Layers; Bilaterally Symmetrical Animals Have Three

The distinction between radial and bilateral symmetry in animals is closely tied to a corresponding difference in the number of tissue layers, called germ layers, that arise during embryonic development. Animals with radial symmetry have two germ layers: an inner layer of **endoderm** (which lines most hollow organs) and an outer layer of **ectoderm** (which covers the body and lines its inner cavi-

(a) Radial symmetry

central axis

plane of symmetry

(b) Bilateral symmetry

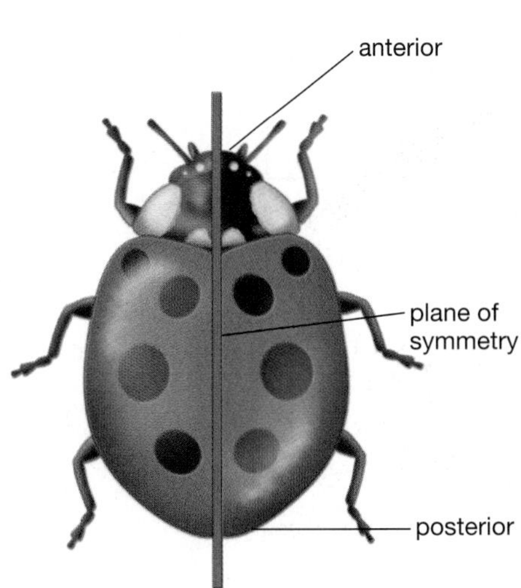

FIGURE 22-2 Body symmetry and cephalization
(a) Animals with radial symmetry lack a well-defined head. Any plane that passes through the central axis divides the body into mirror-image halves. **(b)** Animals with bilateral symmetry have an anterior head end and a posterior tail end. The body can be split into two mirror-image halves only along a particular plane that runs down the midline.

ties and nerve tissues). Bilaterally symmetrical animals add a third germ layer. Between the endoderm and ectoderm is a layer of **mesoderm** (which forms muscle and, when present, the circulatory and skeletal systems).

The parallel evolution of symmetry type and number of germ layers helps us make sense of the potentially puzzling case of the echinoderms (starfish, sea cucumbers, and sea urchins). Adult echinoderms are radially symmetrical, yet our evolutionary tree places them squarely within the bilaterally symmetrical group. It turns out that echinoderms have three germ layers, as well as several other characteristics (some described below) that unite them with the bilaterally symmetrical animals. So, the immediate ancestors of echinoderms must have been bilaterally symmetrical, and the group subsequently evolved radial symmetry (a case of convergent evolution; see page 275). Even now, however, larval echinoderms retain bilateral symmetry.

Bilateral Animals Have Heads

Radially symmetrical animals tend to either be *sessile* (fixed to one spot, as in sea anemones) or drift around on currents (as in jellyfish). Such animals may encounter food or threats from any direction, so a body that essentially "faces" all directions at once is advantageous. In contrast, most bilaterally symmetrical animals move under their own power in any given direction. Resources are most likely to be encountered by the part of the animal that is closest to the direction of movement. The evolution of bilateral symmetry was therefore accompanied by **cephalization**, the concentration of sensory organs and a brain in a defined head region. Cephalization produces an *anterior* (head) end, where sensory cells, sensory organs, clusters of nerve cells, and organs for ingesting food are concentrated. The other end of a cephalized animal is designated *posterior* and may feature a tail (see Fig. 22-2b).

Most Bilateral Animals Have Body Cavities

The members of many bilateral animal phyla have a fluid-filled cavity between the digestive tube (or gut, where food is digested and absorbed) and the outer body wall. In an animal with a body cavity, the gut and body wall are separated by a space, creating a "tube-within-a-tube" body plan. Body cavities are absent in radially symmetrical animals, so it is likely that this feature arose sometime after the split between radially and bilaterally symmetrical animals.

A body cavity can serve a variety of functions. In the earthworm it acts as a kind of skeleton, providing support for the body and a framework against which muscles can act. In other animals, internal organs are suspended within the fluid-filled cavity, which serves as a protective buffer between them and the outside world. A body cavity also allows organs such as the heart and digestive tract to move independently of the body wall. For example, your body cavity allows you to remain externally inactive after a meal even while your digestive tract churns energetically.

Body Cavity Structure Varies Among Phyla

The most widespread type of body cavity is a **coelom**, a fluid-filled cavity that is completely lined with a thin layer of tissue that develops from mesoderm (Fig. 22-3a). Phyla whose members have a coelom are called

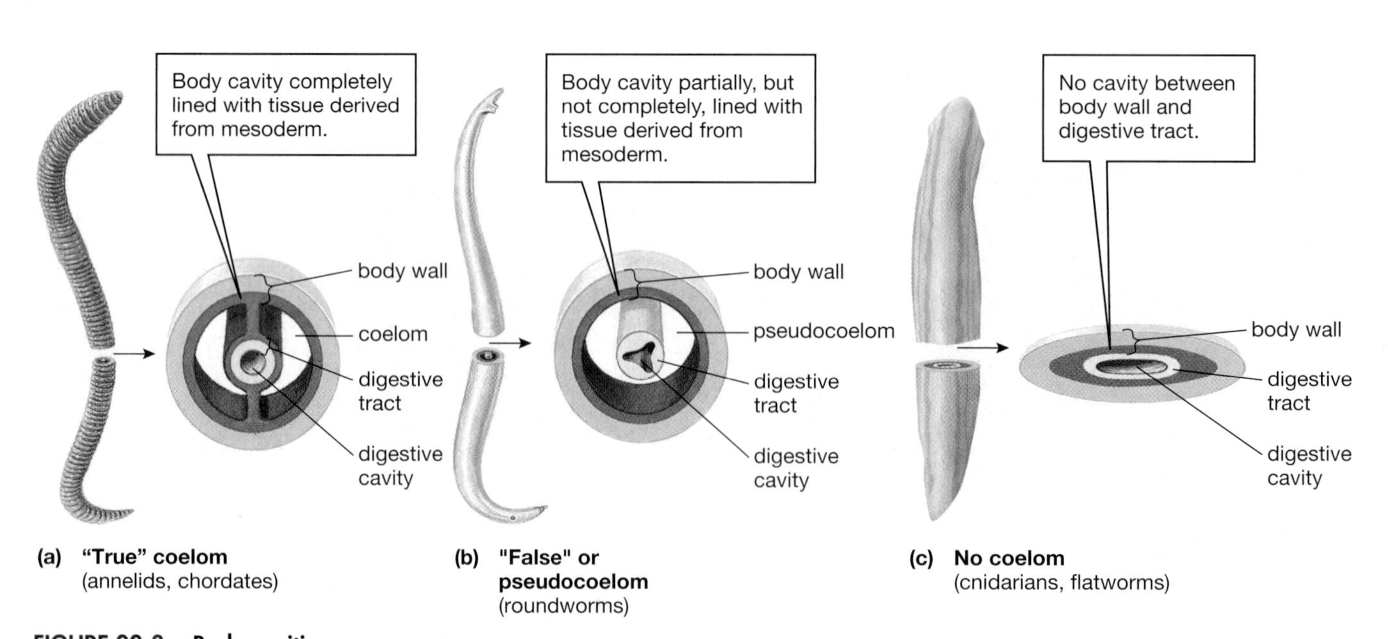

FIGURE 22-3 Body cavities
(a) Annelids have a true coelom. **(b)** Roundworms are pseudocoelomates. **(c)** Flatworms have no cavity between the body wall and digestive tract.

coelomates. The annelids (segmented worms), arthropods (insects, spiders, crustaceans), mollusks (clams and snails), echinoderms, and chordates (which include humans) are coelomate phyla.

Members of some phyla have a body cavity that is *not* completely surrounded by mesoderm-derived tissue. This type of cavity is known as a **pseudocoelom**, and phyla whose members have one are collectively known as *pseudocoelomates* (Fig. 22-3b). The roundworms (nematodes) are the largest pseudocoelomate group. Some phyla of bilateral animals have no body cavity at all and are known as *acoelomates*. For example, flatworms have no cavity between their gut and body wall; instead, the space is filled with solid tissue (Fig. 22-3c).

Simpler Body Cavities Evolved from Coelomate Body Plans

Because acoelomate and pseudocoelomate body plans appear to be more "primitive" than a coelomate body plan, the acoelomate and pseudocoelomate phyla were once presumed to represent a distinct lineage that diverged early in animal evolutionary history, before the origin of the coelom. Now, however, animal systematists recognize that the various aceolomate and pseudocoelomate phyla are not all closely related to one another, but instead form branches at various different points on the animal evolutionary tree (see Fig. 22-1). Thus, acoelomate and pseudocoelomate body plans are not evolutionary precursors of the coelom, but instead are modifications of it.

Bilateral Organisms Develop in One of Two Ways

Among the bilateral animal phyla, embryological development follows a variety of pathways. These diverse developmental pathways, however, can be grouped into two categories, known as **protostome** and **deuterostome** development. In protostome development, the body cavity forms within the space between the body wall and the digestive cavity. In deuterostome development, the body cavity forms as an outgrowth of the digestive cavity. The two types of development also differ in the pattern of cell division immediately after fertilization and in the method by which the mouth and anus are formed. Protostomes and deuterostomes represent distinct evolutionary branches within the bilateral animals. Annelids, arthropods, and mollusks exhibit protostome development; echinoderms and chordates are deuterostomes.

Protostomes Include Two Distinct Evolutionary Lines

The protostome animal phyla fall into two groups, which correspond to two different lineages that diverged early in the evolutionary history of protostomes. One group, the *ecdysozoans*, includes phyla such as the arthropods and roundworms, whose members have bodies covered by an outer layer that is periodically shed. The other group is known as the *lophotrochozoans* and includes phyla whose members have a special feeding structure called a lophophore, as well as phyla whose members pass through a particular type of developmental stage called a trochophore larva. The mollusks, annelids, and flatworms are examples of lophotrochozoan phyla.

22.3 What Are the Major Animal Phyla?

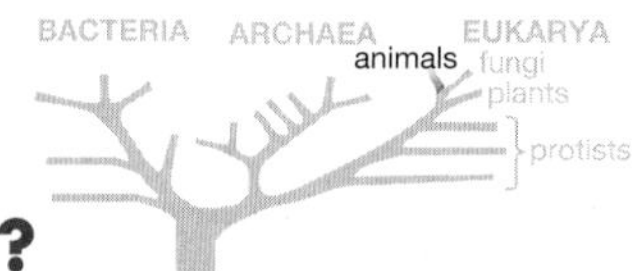

It's easy to overlook the differences among the multitude of small, boneless animals in the world. Even Carolus Linnaeus, the originator of modern biological classification, recognized only two phyla of animals without backbones (insects and worms). Today, however, biologists recognize about 27 phyla of animals, some of which are summarized in Table 22-1.

For convenience, biologists often place animals in one of two major categories: **vertebrates**, those with a backbone (or vertebral column), and **invertebrates**, those lacking a backbone. The vertebrates, which we will discuss in Chapter 23, are perhaps the most conspicuous animals from a human point of view, but less than 3% of all known animal species on Earth are vertebrates. The vast majority of animals are invertebrates.

The earliest animals probably originated from colonies of protists whose members had become specialized to perform distinct roles within the colonial body. In our survey of the invertebrate animals, we begin with the sponges, whose body plan most closely resembles the probable ancestral protozoan colonies.

Sponges Have a Simple Body Plan

Sponges (phylum Porifera) lack true tissues and organs. In some ways, a sponge resembles a colony of single-celled organisms, and a few biologists believe that sponges should be classified as the protist group most closely related to animals, rather than within the animal kingdom. The colony-like properties of sponges were revealed in an experiment performed by embryologist H. V. Wilson in 1907. Wilson mashed a sponge through a piece of silk, thereby breaking it apart into single cells and cell clusters. He then placed these tiny bits of sponge into seawater and waited for 3 weeks. By the end of the experiment, the cells had reaggregated into a functional sponge, demonstrating that individual sponge cells had been able to survive and function independently.

All sponges have a similar body plan. The body is perforated by numerous tiny pores, through which water enters, and by fewer, large openings (called *oscula*), through which it is expelled. Within the sponge, water travels through canals. During its passage, oxygen

Table 22-1 Comparison of the Major Animal Phyla

Common name (Phylum)		Sponges (Porifera)	Hydra, Anemones, Jellyfish (Cnidaria)	Flatworms (Platyhelminthes)	Segmented Worms (Annelida)
Body Plan	**Level of organization**	Cellular—lack tissues and organs	Tissue—lack organs	Organ system	Organ system
	Germ layers	Absent	Two	Three	Three
	Symmetry	Absent	Radial	Bilateral	Bilateral
	Cephalization	Absent	Absent	Present	Present
	Body cavity	Absent	Absent	Absent	Coelom
	Segmentation	Absent	Absent	Absent	Present
Internal Systems	**Digestive system**	Intracellular	Gastrovascular cavity; some intracellular	Gastrovascular cavity	Separate mouth and anus
	Circulatory system	Absent	Absent	Absent	Closed
	Respiratory system	Absent	Absent	Absent	Absent
	Excretory system (fluid regulation)	Absent	Absent	Canals with ciliated cells	Nephridia
	Nervous system	Absent	Nerve net	Head ganglia with longitudinal nerve cords	Head ganglia with paired ventral cords; ganglia in each segment
	Reproduction	Sexual; asexual (budding)	Sexual; asexual (budding)	Sexual (some hermaphroditic); asexual (body splits)	Sexual (some hermaphroditic)
	Support	Endoskeleton of spicules	Hydrostatic skeleton	Hydrostatic skeleton	Hydrostatic skeleton
	Number of known species	5000	9000	12,000	9000

is extracted, microorganisms are filtered out and taken into individual cells where they are digested, and wastes are released (Fig. 22-4).

Sponges have three major cell types, each with a specialized role. Flattened *epithelial cells* cover their outer body surfaces. Some epithelial cells are modified into *pore cells*, which surround pores, controlling their size and regulating the flow of water. The pores close when harmful substances are present. *Collar cells* maintain a flow of water through the sponge by beating flagella that extend into the inner canal. The collars that surround these flagella act as fine sieves, filtering out microorganisms that are then ingested by the cell. Some of the food is passed to the *amoeboid cells*. These cells roam freely between the epithelial and collar cells, digesting and distributing nutrients, producing reproductive cells, and secreting small skeletal projections called *spicules*.

Sponges come in a variety of shapes and sizes. Some species have a well-defined shape, but others grow free-form over underwater rocks (Fig. 22-5). The largest sponges can grow to more than 3 feet (1 meter)

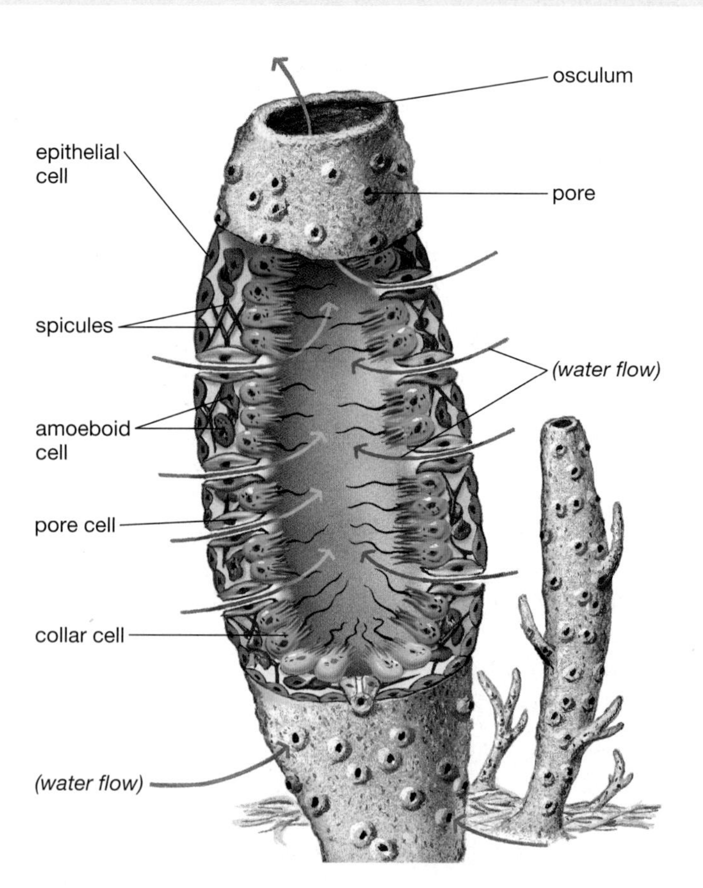

FIGURE 22-4 The body plan of sponges
Water enters through numerous tiny pores in the sponge body and exits through oscula. Microscopic food particles are filtered from the water.

Snails, Clams, Squid *(Mollusca)*	Insects, Arachnids, Crustaceans *(Arthropoda)*	Roundworms *(Nematoda)*	Sea Stars, Sea Urchins *(Echinodermata)*
Organ system	Organ system	Organ system	Organ system
Three	Three	Three	Three
Bilateral	Bilateral	Bilateral	Bilateral larvae, radial adults
Present	Present	Present	Absent
Coelom	Coelom	Pseudocoel	Coelom
Absent	Present	Absent	Absent
Separate mouth and anus	Separate mouth and anus	Separate mouth and anus	Separate mouth and anus (normally)
Open	Open	Absent	Absent
Gills, lungs	Tracheae, gills, or book lungs	Absent	Tube feet, skin gills, respiratory tree
Nephridia	Excretory glands resembling nephridia	Excretory gland	Absent
Well-developed brain in some cephalopods; several paired ganglia, most in the head; nerve network in body wall	Head ganglia with paired ventral nerve cords; ganglia in segments, some fused	Head ganglia with dorsal and ventral nerve cords	Head ganglia absent; nerve ring and radial nerves; nerve network in skin
Sexual (some hermaphroditic)	Normally sexual	Sexual (some hermaphroditic)	Sexual (some hermaphroditic); asexual by regeneration (rare)
Hydrostatic skeleton	Exoskeleton	Hydrostatic skeleton	Endoskeleton of plates beneath outer skin
50,000	1,000,000	12,000	6500

(a) (b) (c)

FIGURE 22-5 The diversity of sponges

Sponges come in a wide variety of sizes, shapes, and colors. Some, such as **(a)** this fire sponge, grow in a free-form pattern over undersea rocks. **(b)** Tiny appendages attach this tubular sponge to rocks, whereas **(c)** this reef sponge with flared tubular openings attaches to a coral reef. **QUESTION** Sponges are often described as the most "primitive" of animals. How can such a primitive organism have become so diverse and abundant?

in height. An internal skeleton composed of calcium carbonate (chalk), silica (glass), or protein spicules provides support for the body (see Fig. 22-4). Natural bath sponges, which are now largely replaced by factory-made cellulose imitations, are actually sponge skeletons.

All sponges live in water, mostly in saltwater environments. Adult sponges are generally sessile, attaching themselves to rocks or other underwater surfaces. Sponges may reproduce asexually by **budding**, in which the adult produces miniature versions of itself that drop off and assume an independent existence. Alternatively, they may reproduce sexually through the fusion of sperm and eggs. Fertilized eggs develop inside the adult into active larvae that escape through the openings in the sponge body. Water currents disperse the larvae to new areas, where they settle and develop into adult sponges.

Cnidarians Are Well-Armed Predators

Cnidarians (phylum Cnidaria)—jellyfish, sea anemones, corals, and hydrozoans—come in a bewildering and beautiful variety of forms (Fig. 22-6), all of which are actually variations on two basic body plans: the *polyp* (Fig. 22-7a) and the *medusa* (Fig. 22-7b). The generally tubular polyp is adapted to a life spent quietly attached to rocks. The polyp has *tentacles*, extensions that reach upward for grasping, stinging, and immobilizing prey. The bell-shaped body of the medusa ("jellyfish") floats in the water and is carried by currents, trailing its tentacles like multiple fishing lines. Both polyp and medusa develop from just two germ layers—the interior endoderm and the exterior mesoderm; between those layers is a jellylike *mesoglea*. Polyps and medusae are radially symmetrical, with body parts arranged in a circle around the mouth and digestive cavity (see Fig. 22-2a).

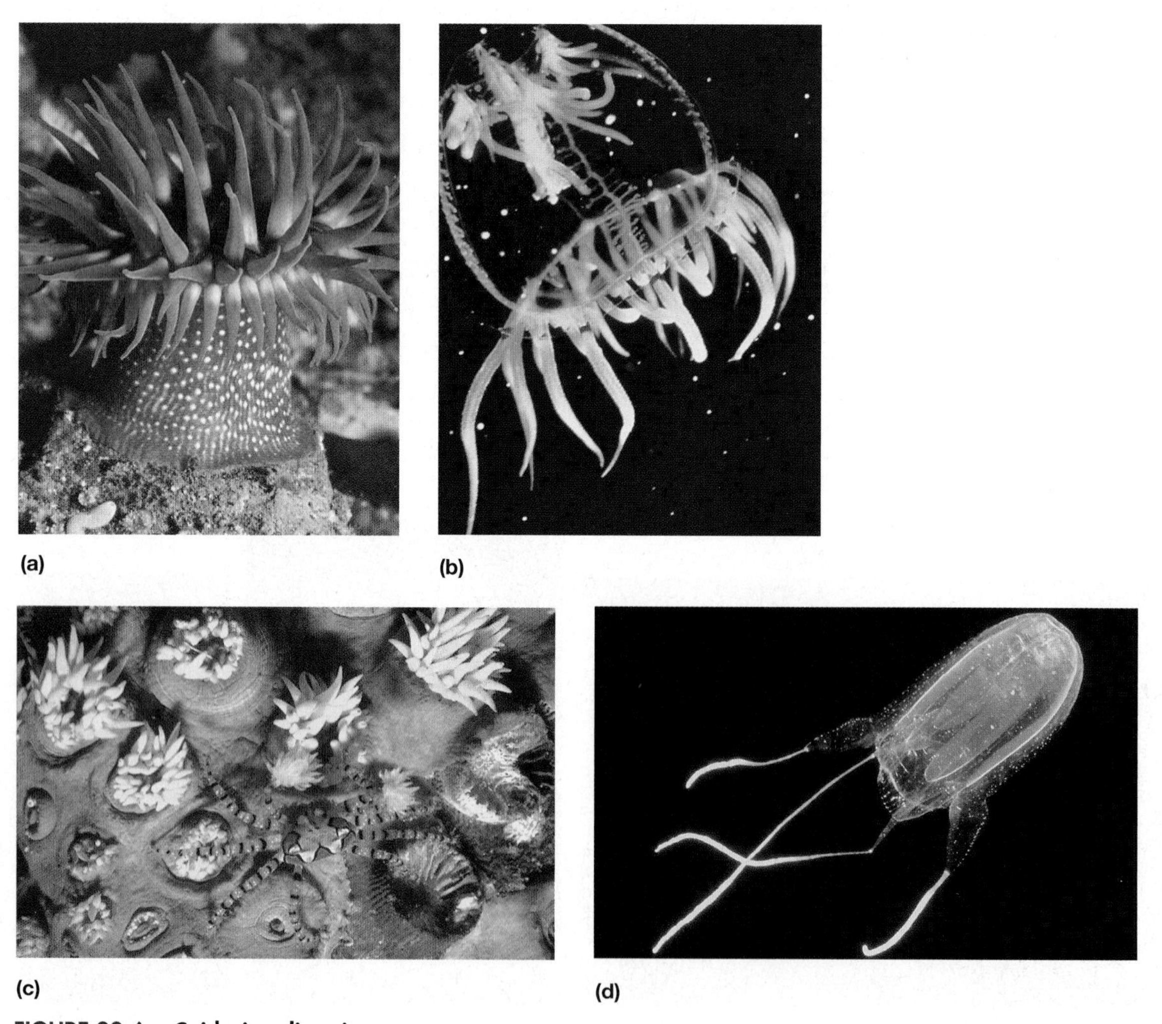

FIGURE 22-6 Cnidarian diversity
(a) A red-spotted anemone spreads its tentacles to capture prey. **(b)** A small medusa. **(c)** A close-up of coral reveals bright yellow polyps in various stages of tentacle extension. At the lower right, areas where the coral has died expose the calcium carbonate skeleton that supports the polyps and forms the reef. A strikingly patterned crab (an arthropod) sits atop the coral, holding tiny white anemones in its claws. Their stinging tentacles help protect the crab. **(d)** A sea wasp, a cnidarian whose stinging cells contain one of the most toxic of all known venoms. **QUESTION** In each of these photos, is the pictured organism a polyp or a medusa?

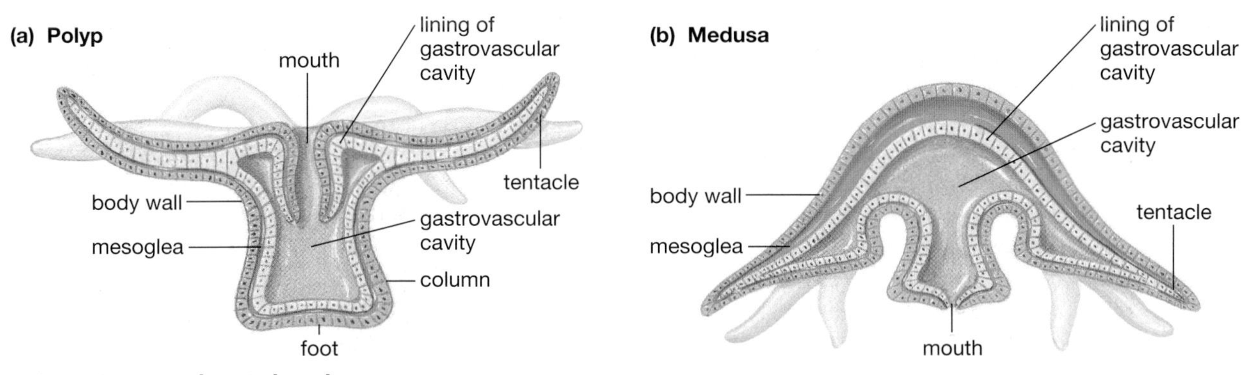

FIGURE 22-7 Polyp and medusa
(a) The polyp form is seen in hydra (see Fig. 22-8), sea anemones (Fig. 22-6a), and the individual polyps within a coral (Fig. 22-6c). **(b)** The medusa form, seen in the jellyfish (Fig. 22-6b), resembles an inverted polyp.

This arrangement of parts is well suited to these sessile or free-floating animals, as it enables them to respond to prey or threats from any direction.

The cells of cnidarians are organized into distinct tissues, including contractile tissue that acts like muscle. The nerve cells are organized into tissue called a *nerve net*, which branches through the body and controls the contractile tissue to bring about movement and feeding behavior. However, most cnidarians lack true organs and have no brain.

Cnidarian tentacles are armed with *cnidocytes*, cells containing structures that, when stimulated by contact, explosively inject poisonous or sticky filaments into prey (Fig. 22-8). The venom of some cnidarians can cause painful stings in humans unfortunate enough to come into contact with them, and the stings of a few jellyfish species can even be life-threatening. The most deadly of these species is the "sea wasp," *Chironex fleckeri*, which is found in the waters off northern Australia and southeast Asia. The amount of venom in a single sea wasp could kill up to 60 people, and the victim of a serious sting may die within minutes of being stung.

The function of stinging cells, of course, is not to sting human swimmers but to capture prey. Although all cnidarians are predatory, none hunt actively. Instead, they wait for their victims to blunder, by chance, into the grasp of their enveloping tentacles. Stung and firmly grasped, the prey is forced through an expansible mouth into a digestive sac, the *gastrovascular cavity*. Digestive enzymes secreted into this cavity break down some of the food, and further digestion occurs within the cells lining the cavity. Because the gastrovascular cavity has only a single opening, undigested material is expelled through the mouth when digestion is completed. Although this two-way traffic prevents continuous feeding, it is adequate to support the low energy demands of these animals.

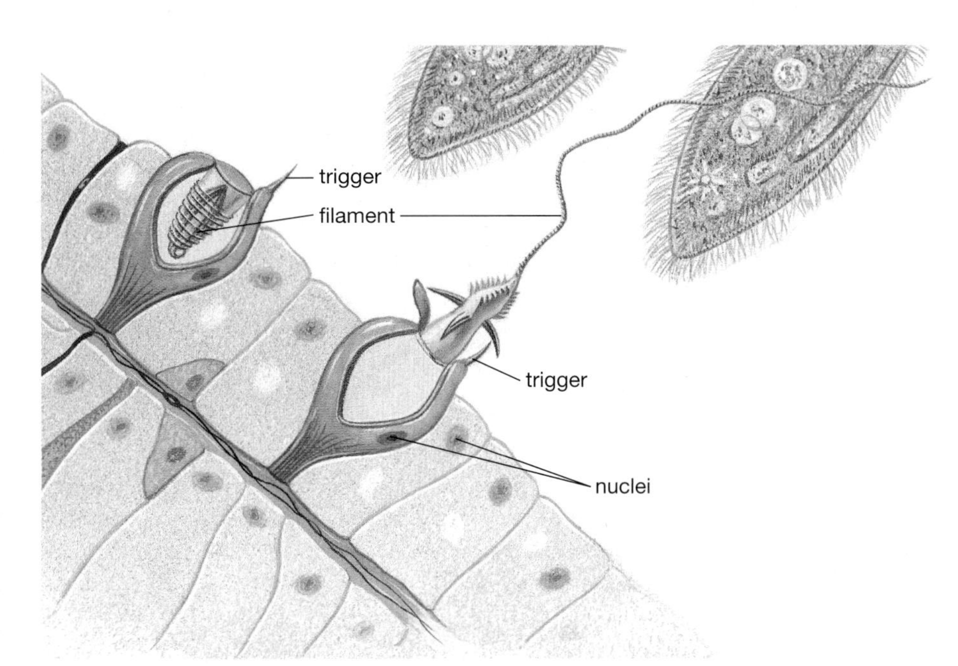

FIGURE 22-8 Cnidarian weaponry: the cnidocyte
At the slightest touch to the trigger of a special structure in their cnidocytes, cnidarians, such as this hydra, violently expel a poisoned filament.

Like sponges, cnidarians are confined to watery habitats, and most species are marine. One group of cnidarians, the corals, is of particular ecological importance (see Fig. 22-6c). Coral polyps form large colonies, and each member of the colony secretes a hard skeleton of calcium carbonate. The skeletons persist long after the organisms die, serving as a base to which other individuals may attach themselves. The cycle continues until, after thousands of years, massive coral reefs are formed. Corals are restricted to the warm, clear waters of the Tropics, where their reefs form undersea habitats that are the basis of an ecosystem of stunning diversity and unparalleled beauty (see Chapter 42).

Flatworms Have Organs but Lack Respiratory and Circulatory Systems

The *flatworms* (phylum Platyhelminthes) are bilaterally symmetrical, rather than radially symmetrical (see Fig. 22-2). This body plan and its accompanying cephalization foster active movement. Cephalized, bilaterally symmetrical animals possess an anterior end, which is the first part of a moving animal to encounter the environment ahead. Consequently, sense organs are concentrated in the anterior portion of the body, thus enhancing the animal's ability to respond appropriately to any stimuli it encounters (for example, eating food items and retreating from obstacles).

Unlike members of radially symmetrical phyla, flatworms have well-developed organs, in which tissues are grouped into functional units. For example, flatworms such as the freshwater planarians (Fig. 22-9) have sense organs, including eyespots that detect light and dark, and cells that respond to chemical and tactile stimuli. To process information, a flatworm has clusters of nerve cells called **ganglia** (singular, ganglion) in its head, forming a simple brain. Paired neural structures called **nerve cords** conduct nervous signals to and from the ganglia.

Despite the presence of some organs, flatworms lack respiratory and circulatory systems. In the absence of a respiratory system, gas exchange is accomplished by direct diffusion between body cells and the environment. This mode of respiration is possible because the small size and flat shape of flatworm bodies ensure that no body cell is very far from the surrounding environment. Without a circulatory system, nutrients move directly from the digestive tract to body cells. The digestive cavity has a branching structure (see Fig. 22-9a) that reaches all parts of the body and allows digested nutrients to diffuse into nearby cells. The digestive cavity has only one opening to the environment, so undigested waste must pass out through the same opening that also serves as a mouth.

Flatworms may be either parasitic or free-living. **Parasites** are organisms that live in or on the body of another organism, called a *host*, which is harmed as a result of the relationship; free-living organisms do not live in such intimate association with members of another species. Some parasitic flatworms can infect humans. For example, tapeworms can infect people who eat improperly cooked beef, pork, or fish that has been infected by the worms. Worm larvae form encapsulated resting structures, called *cysts*, in the muscles of these animals. The cysts hatch in the human digestive tract, where the young tapeworms attach themselves to the lining of the intestine. There they may grow to a length of more than 20 feet (7 meters), absorbing digested nutrients directly through their outer surface and eventually releasing packets of eggs that are shed in the host's feces. If pigs, cows, or fish eat food contaminated with infected human feces, the eggs hatch in the animal's digestive tract, releasing larvae that burrow into its muscles and form cysts, thereby continuing the infective cycle (Fig. 22-10).

Another group of parasitic flatworms is the *flukes.* Of these, the most devastating are liver flukes (common in Asia) and blood flukes, such as those of the genus *Schistosoma*, which cause the disease schistosomiasis. Like most parasites, flukes have a complex life cycle

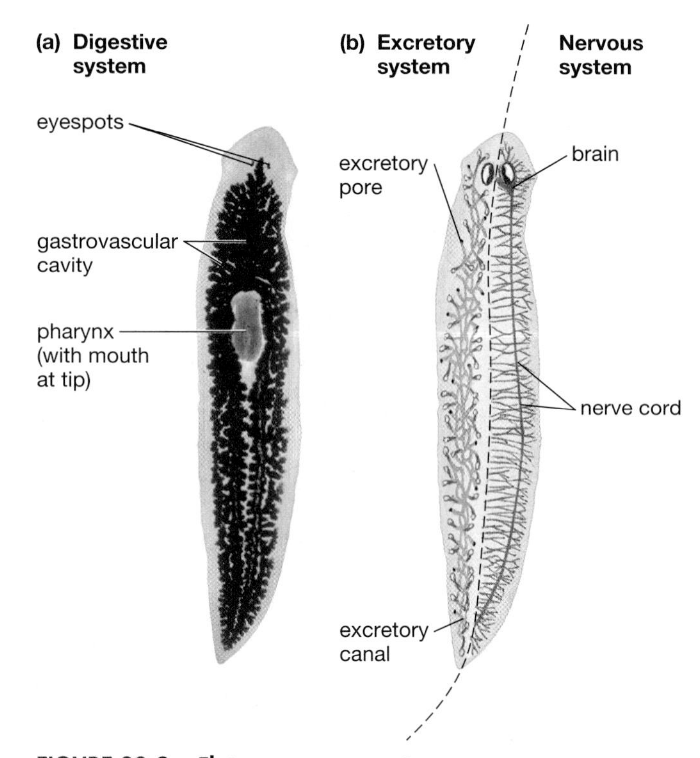

FIGURE 22-9 Flatworm organ systems
(a) The flatworm's elaborately branched digestive system, the centrally located ventral pharynx, and eyespots in the head are clearly visible. **(b)** (Left) The excretory system consists of branching tubes that conduct excess fluid to the outside through numerous pores. (Right) The nervous system of flatworms shows clear cephalization, with eyes and a brain composed of ganglia cells in a well-defined head. Ladderlike nerve cords carry signals through the rest of the body.

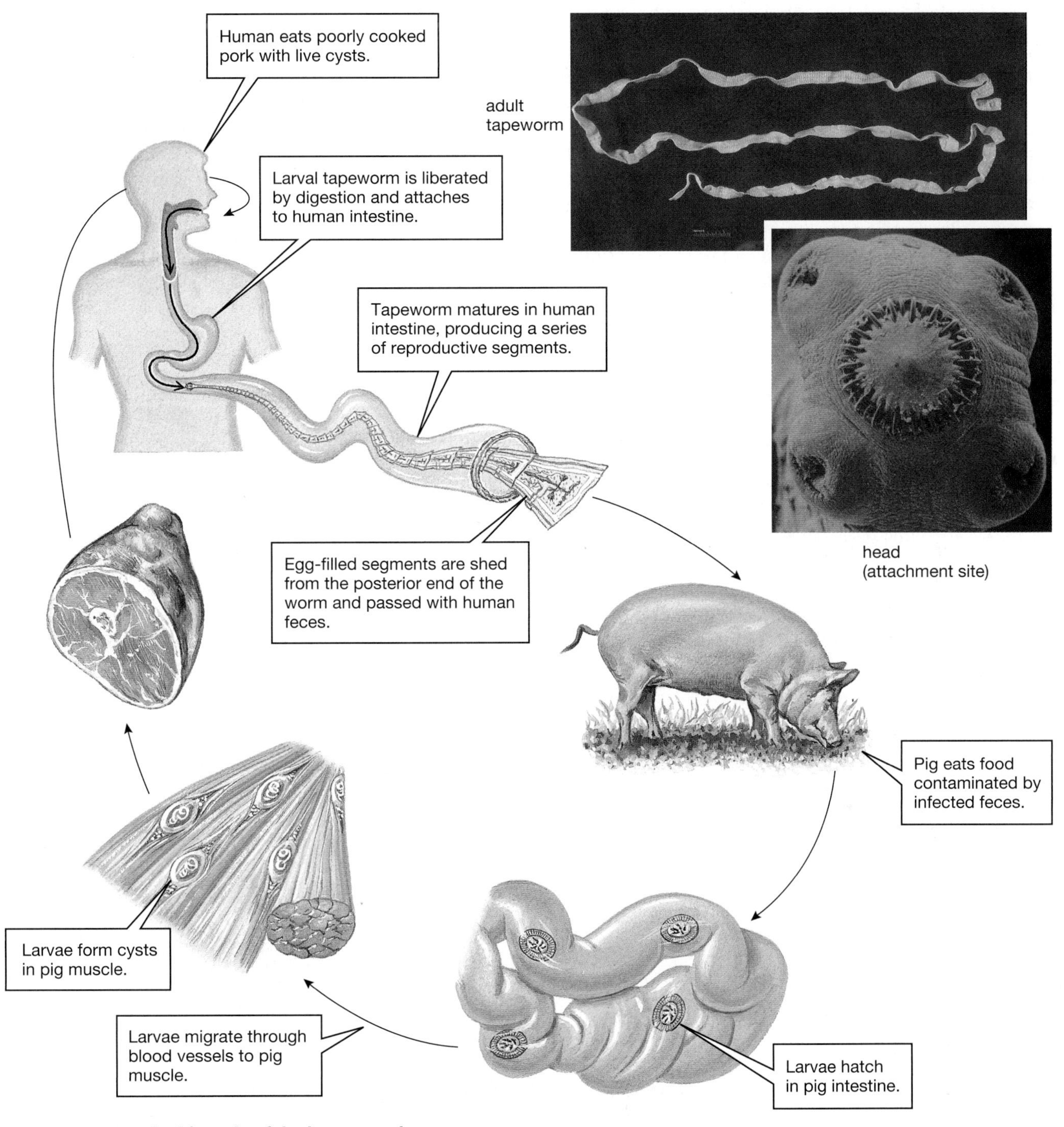

Human eats poorly cooked pork with live cysts.
adult tapeworm
Larval tapeworm is liberated by digestion and attaches to human intestine.
Tapeworm matures in human intestine, producing a series of reproductive segments.
Egg-filled segments are shed from the posterior end of the worm and passed with human feces.
head (attachment site)
Pig eats food contaminated by infected feces.
Larvae form cysts in pig muscle.
Larvae migrate through blood vessels to pig muscle.
Larvae hatch in pig intestine.

FIGURE 22-10 The life cycle of the human pork tapeworm
Each reproductive unit, or proglottid, is a self-contained reproductive factory that includes both male and female sex organs. **QUESTION** Why have tapeworms evolved a long, flat shape?

that includes an intermediate host (a snail, in the case of *Schistosoma*). Prevalent in Africa and parts of South America, schistosomiasis affects an estimated 200 million people worldwide. Its symptoms include diarrhea, anemia, and possible brain damage.

Flatworms can reproduce both sexually and asexually. Free-living forms may reproduce by cinching themselves around the middle until they separate into two halves, each of which regenerates its missing parts. All forms can reproduce sexually; most are **hermaphroditic**—that is, they possess both male and female sexual organs. This trait is a great advantage to parasitic forms, as each worm is able to reproduce through self-fertilization, even if it is the only individual present in its host.

Annelids Are Composed of Identical Segments

A prominent feature of the segmented worms, or *annelids* (phylum Annelida), is the division of the body into a series of repeating segments. Externally, these segments appear as ringlike depressions on the surface. Internally, most of the segments contain identical copies of nerves, excretory structures, and muscles. **Segmentation** is advantageous for locomotion, because the body compartments, each of which is controlled by separate muscles, collectively are capable of more complex movement. Another feature that distinguishes annelids from flatworms and roundworms is a fluid-filled true coelom between the body wall and the digestive tract (see Fig. 22-3a). The incompressible fluid in the coelom of many annelids is confined by the partitions between the segments and serves as a **hydrostatic skeleton**, a rigid framework against which muscles can act. The hydrostatic skeleton makes possible such feats as burrowing through soil.

Annelids have a well-developed **closed circulatory system** that distributes gases and nutrients throughout the body (see Chapter 28). In closed circulatory systems (including yours), blood remains confined to the heart and blood vessels. In the earthworm, for example, blood with oxygen-carrying hemoglobin is pumped through well-developed vessels by five pairs of "hearts" (Fig. 22-11). These hearts are actually short segments of specialized blood vessels that contract rhythmically. The blood is filtered and wastes are removed by excretory organs called *nephridia* (singular, nephridium), which are found in many of the segments. Nephridia resemble the individual tubules of the vertebrate kidney (see Chapter 31).

The annelid nervous system consists of a simple ganglionic brain in the head and a series of repeating paired segmental ganglia joined by a pair of ventral nerve cords that pass along the length of the body.

Annelids have a tubular gut that runs from the mouth to the anus. This kind of digestive tract, with two openings and a one-way digestive path, is much more efficient than the single-opening digestive systems of cnidarians and flatworms. Digestion in annelids occurs in a series of compartments, each specialized for a different phase of food processing (see Fig. 22-11). For example, in the earthworm, a muscular *pharynx* draws in food, consisting of bits of decaying plant and animal debris in soil. The food is conducted through the esophagus to a storage chamber, the *crop*, and then released slowly into the muscular *gizzard*, where it is ground into tiny particles by muscular contractions of the gizzard and the sharp-edged sand grains it contains. The food then passes into the intestine, where it is digested and nutrients are absorbed. Undigested food and soil exit through the anus.

The phylum Annelida includes three main subgroups, the *oligochaetes*, the *polychaetes*, and the *leeches*. The oligochaetes include the familiar earthworm and its relatives. Polychaetes live primarily in the ocean. Some polychaetes have paired fleshy paddles on most of their segments, used in locomotion. Others live in tubes secreted by the worms themselves, from which they project feathery gills that both exchange gases and sift the water for microscopic food (Fig. 22-12a,b). Leeches (Fig. 22-12c) live in fresh water or moist terrestrial habitats, and are either carnivorous or parasitic—some preying on smaller invertebrates, others sucking the blood of larger animals.

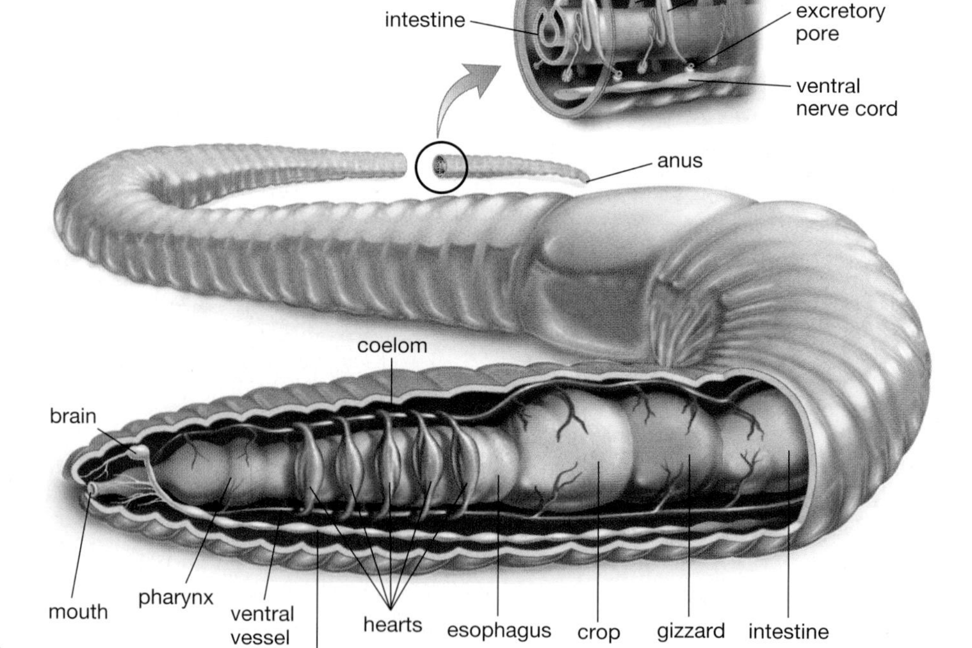

FIGURE 22-11 An annelid, the earthworm
This diagram shows an enlargement of segments, many of which are repeating similar units separated by partitions.
QUESTION What advantage does a digestive system with two openings have relative to digestive systems with only a single opening (like that of the flatworms)?

(a)

(b)

(c)

FIGURE 22-12 Diverse annelids
(a) A polychaete annelid projects brightly spiraling gills from a tube attached to rock. When the gills retract, the tube is covered by the trap door visible on the lower right. **(b)** The "fireworm" polychaete swims by using paddles on each segment. The bristles on each paddle can deliver a fiery sting. **(c)** This leech, a freshwater annelid, shows numerous segments. The sucker encircles its mouth, allowing it to attach to its prey. **QUESTION** Why does pouring salt on a leech harm it?

Most Mollusks Have Shells

In terms of number of species, the *mollusks* are second (albeit a distant second) only to the arthropods. Some mollusks protect their bodies with shells of calcium carbonate; others escape predation by moving swiftly or by having an unappealing taste. With the exception of some snails and slugs, mollusks are water dwellers. Mollusks have a *mantle*, an extension of the body wall that forms a chamber for the gills and, in shelled species, secretes the shell. Mollusks also have well-developed circulatory systems with a feature not seen in annelids: the **hemocoel**, or blood cavity. Blood empties into the hemocoel, where it bathes the internal organs directly. This arrangement, known as an **open circulatory system**, is also present in most arthropods. The nervous system, like that of annelids, consists of ganglia connected by nerves, but many more of the ganglia are concentrated in the brain. Reproduction is sexual; some species have separate sexes, and others are hermaphroditic. Although mollusks are enormously diverse, a simplified diagram of the body plan of a mollusk is shown in Figure 22-13.

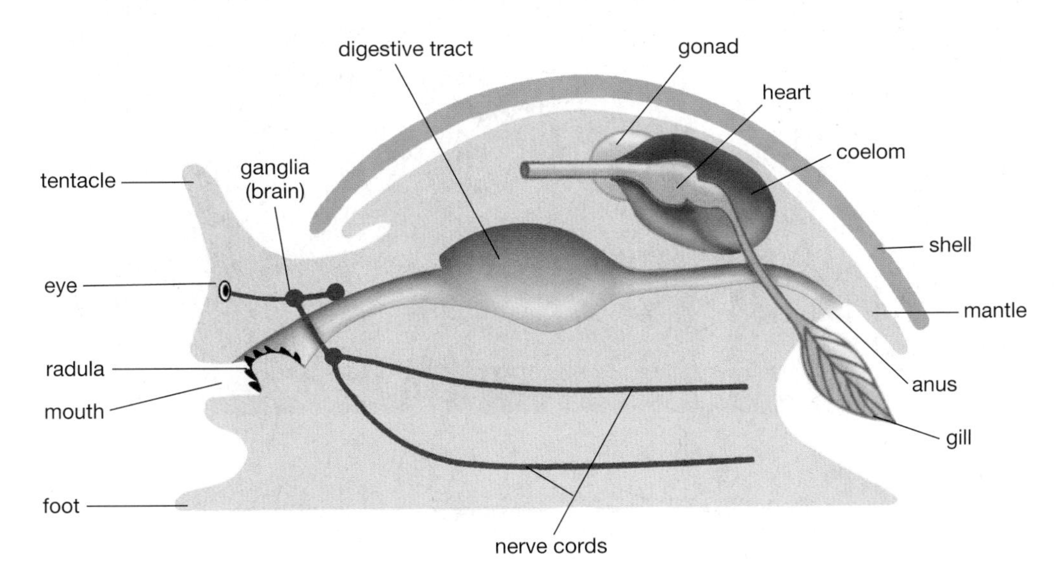

FIGURE 22-13 A generalized mollusk
The general body plan of a mollusk, showing the mantle, foot, gills, shell, radula, and other features that are seen in most (but not all) mollusk species.

(a)

(b)

FIGURE 22-14 The diversity of gastropod mollusks
(a) A Florida tree snail displays a brightly striped shell and eyes at the tip of stalks that retract instantly if touched. **(b)** Spanish shawl sea slugs prepare to mate. The brilliant colors of many sea slugs warn potential predators that they are distasteful.

Among the many classes of mollusks, we will discuss three in more detail: gastropods, cephalopods, and bivalves.

Gastropods Are One-Footed Crawlers

Snails and slugs—collectively known as *gastropods*—crawl on a muscular *foot*, and many are protected by shells that vary widely in form and color (Fig. 22-14a). Not all gastropods are shelled, however. Sea slugs, for example, lack shells, but their brilliant colors warn potential predators that they are poisonous or foul-tasting (Fig. 22-14b).

Gastropods feed with a *radula*, a flexible ribbon of tissue studded with spines that is used to scrape algae from rocks or to grasp larger plants or prey (see Fig. 22-13). Most snails use gills, typically enclosed in a cavity beneath the shell, for respiration. Gases can also diffuse readily through the skin of most gastropods; sea slugs rely on this mode of gas exchange. The few gastropod species that live in terrestrial habitats (including the destructive garden snails and slugs) use a simple lung for breathing.

Bivalves Are Filter Feeders

Included among the *bivalves* are scallops, oysters, mussels, and clams (Fig. 22-15). Members of the class lend exotic variety to the human diet and are important members of the nearshore marine community. Bivalves possess two shells connected by a flexible hinge. A strong muscle clamps the shells closed in response to danger; this muscle is what you are served when you order scallops in a restaurant.

Clams use a muscular foot for burrowing in sand or mud. In mussels, which live attached to rocks, the foot is

(a)

(b)

FIGURE 22-15 The diversity of bivalve mollusks
(a) This swimming scallop parts its hinged shells. The upper shell is covered with an encrusting sponge. **(b)** Mussels attach to rocks in dense aggregations exposed at low tide. White barnacles are attached to the mussel shells and surrounding rock.

smaller and helps secrete threads that anchor the animal to the rocks. Scallops lack a foot and move by a sort of jet propulsion achieved by flapping their shells together. Bivalves are filter feeders, using their gills as both respiratory and feeding structures. Water circulates over the gills, which are covered with a thin layer of mucus that traps microscopic food particles. Food is conveyed to the mouth by the beating of cilia on the gills. Probably because they filter-feed and do not move extensively, bivalves "lost their heads" as they evolved.

Cephalopods Are Marine Predators

The *cephalopods* include octopuses, nautiluses, cuttlefish, and squids (Fig. 22-16). The largest invertebrate, the giant squid, belongs to this group. All cephalopods are predatory carnivores, and all are marine. In these mollusks, the foot has evolved into tentacles with well-developed chemosensory abilities and suction disks for detecting and grasping prey. Prey grasped by the tentacles may be immobilized by a paralyzing venom in the saliva before being torn apart by beaklike jaws.

Cephalopods move rapidly by jet propulsion, which is generated by the forceful expulsion of water from the mantle cavity. Octopuses may also travel along the seafloor by using their tentacles like multiple undulating legs. The rapid movements and generally active lifestyles of cephalopods are made possible in part by their closed circulatory systems. Cephalopods are the only mollusks with closed circulation, which transports oxygen and nutrients more efficiently than open circulatory systems do.

Cephalopods have highly developed brains and sensory systems. The cephalopod eye rivals our own in complexity and exceeds it in efficiency of design. The cephalopod brain, especially that of the octopus, is (for an invertebrate brain) exceptionally large and complex. It is enclosed in a skull-like case of cartilage and endows the octopus with highly developed capabilities to learn and remember. In the laboratory, octopuses can rapidly learn to associate certain symbols with food and to open a screw-cap jar to obtain food.

Arthropods Are the Dominant Animals on Earth

In terms of both number of individuals and number of species, no other phylum comes close to the arthropods, which include insects, arachnids, myriopods, and crustaceans. Over 1 million arthropod species have been discovered, and scientists estimate that up to 9 million remain undescribed.

All arthropods have an **exoskeleton**, an external skeleton that encloses the arthropod body like a suit of armor. Secreted by the *epidermis* (the outer layer of skin), the exoskeleton is composed chiefly of protein and a polysaccharide called *chitin*. The external skeleton protects against predators and is responsible for arthropods'

(a)

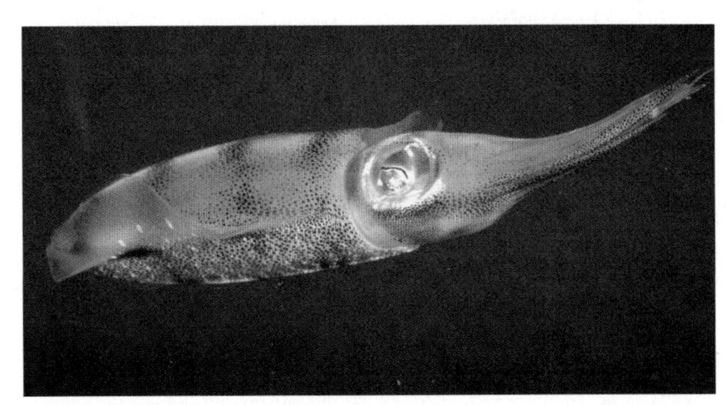

(b)

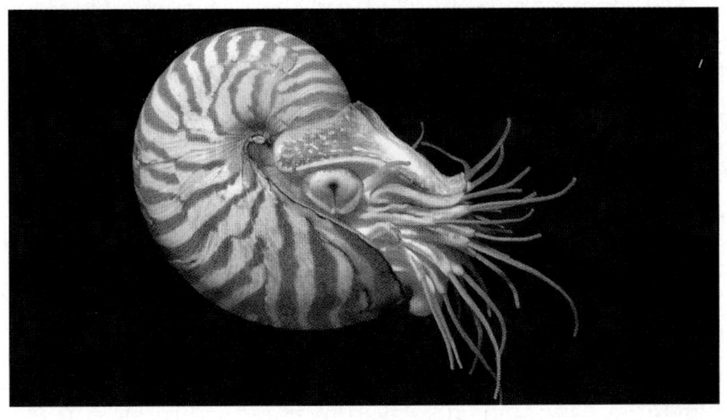

(c)

FIGURE 22-16 The diversity of cephalopod mollusks **(a)** An octopus can crawl rapidly by using its eight suckered tentacles, and can alter its color and skin texture to blend with its surroundings. In emergencies, this mollusk can jet backward by vigorously contracting its mantle. Octopuses and squid can emit clouds of dark purple ink to confuse pursuing predators. **(b)** The squid moves entirely by contracting its mantle to generate jet propulsion, which pushes the animal backward through the water. **(c)** The chambered nautilus secretes a shell with internal, gas-filled chambers that provide buoyancy. Note the well-developed eyes and the tentacles used to capture prey.

FIGURE 22-17 The exoskeleton allows precise movements
A garden spider, having immobilized its prey with a paralyzing venom, rapidly encases the prey in its web. Such dexterous manipulations are made possible by the exoskeleton and jointed appendages characteristic of arthropods.

FIGURE 22-18 The exoskeleton must be molted periodically
A newly emerged praying mantis (a predatory insect) hangs beside its outgrown exoskeleton (left).

greatly increased agility relative to their wormlike ancestors. The exoskeleton is thin and flexible in places, allowing movement of the paired, jointed appendages. By providing stiff but flexible appendages and rigid attachment sites for muscles, the exoskeleton makes possible the flight of the bumblebee and the intricate, delicate manipulations of the spider as it weaves its web (Fig. 22-17). The exoskeleton also contributed enormously to the arthropod invasion of dry terrestrial habitats by providing a watertight covering for delicate, moist tissues such as those used for gas exchange (arthropods were the earliest terrestrial animals; see Chapter 17).

The arthropod exoskeleton poses some unique problems. First, because it cannot expand as the animal grows, the exoskeleton must be periodically shed, or **molted**, and replaced with a larger one (Fig. 22-18). Molting uses energy and leaves the animal temporarily vulnerable until the new skeleton hardens. The exoskeleton is also heavy; its weight increases exponentially as the animal grows. It is no coincidence that the largest arthropods are crustaceans (crabs and lobsters), whose watery habitat supports much of their weight.

Arthropods are segmented, but their segments tend to be few and specialized for different functions such as sensing the environment, feeding, and movement (Fig. 22-19). For example, in insects, sensory and feeding structures are concentrated on the front segment, known as the *head*, and digestive structures are largely confined to the *abdomen*, which is the rear segment. Between the head and the abdomen is the *thorax*, the segment to which structures used in locomotion, such as wings and walking legs, are attached.

Efficient gas exchange is required to supply adequate oxygen to the muscles that allow the rapid flight, swimming, or running displayed by many arthropods. In aquatic forms such as crustaceans, gas exchange is accomplished by **gills** (thin, external respiratory membranes). In terrestrial arthropods, gas exchange is performed either by *tracheae* (singular, trachea; a network of narrow, branching respiratory tubes) or by *book lungs* (specialized respiratory structures of arach-

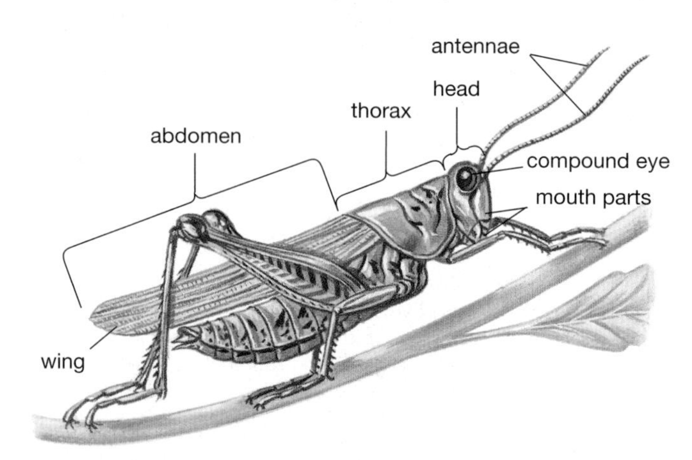

FIGURE 22-19 Segments are fused and specialized in insects
Insects, such as this grasshopper, show fusion and specialization of body segments into a distinct head, thorax, and abdomen. Segments are visible on the abdomen beneath the wings.

nids). Most arthropods have open circulatory systems, like those of mollusks, in which blood directly bathes the organs in a hemocoel.

Most arthropods possess a well-developed sensory system, including **compound eyes**, which have multiple light detectors (Fig. 22-20), and acute chemical and tactile senses. The arthropod nervous system is similar in plan to that of annelids, but is more complex. It consists of a brain, composed of fused ganglia, and a series of additional ganglia along the length of the body that are linked by a ventral nerve cord. The capacity for finely coordinated movement, combined with sophisticated sensory abilities and a well-developed nervous system, has allowed the evolution of complex behavior. In fact, the social behavior of some insect species, such as the honeybee, is as intricate and complex as any vertebrate behavior. Although many people associate communication and learning with vertebrates, both play important roles in insect societies.

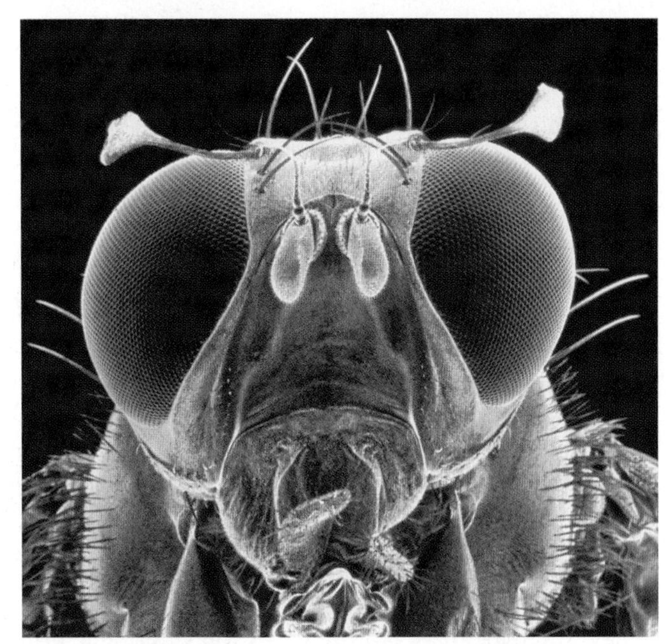

FIGURE 22-20 Arthropods possess compound eyes
This scanning electron micrograph shows the compound eye of a fruit fly. Compound eyes consist of an array of similar light-gathering and sensing elements whose orientation gives the arthropod a wide field of view. Insects have reasonably good image-forming ability and good color discrimination.

Insects Are the Only Flying Invertebrates

The number of described *insect* species is about 850,000, roughly three times the total number of known species in all other classes of animals combined (Fig. 22-21). Insects have a single pair of antennae, three pairs of legs,

FIGURE 22-21 The diversity of insects
(a) The rose aphid sucks sugar-rich juice from plants. **(b)** A mating pair of Hercules beetles. The large "horns" are found only on the male. **(c)** A June beetle displays its two pairs of wings as it comes in for a landing. The outer wings protect the abdomen and the inner wings, which are relatively thin and fragile. **(d)** Insects such as this locust can cause devastation of food crops and natural vegetation. **(e)** Caterpillars are larval forms of moths or butterflies. This caterpillar larva of the Australian fruit-sucking moth displays large eyespot patterns that may frighten potential predators, who mistake them for eyes of a large animal.

and, in most cases, two pairs of wings. Insects' capacity for flight distinguishes them from all other invertebrates and has contributed to their enormous success (see Fig. 22-21c). As anyone who has pursued a fly can testify, flight helps insects escape from predators. It also allows the insect to find widely dispersed food. For example, swarms of locusts (see Fig. 22-21d) can travel 200 miles a day in search of food; researchers tracked one swarm on a journey that totaled almost 3000 miles. Flight requires rapid and efficient gas exchange, which insects accomplish by means of tracheae. The network of tracheae conducts air to all parts of the body.

During development, insects undergo **metamorphosis**, a radical change from a juvenile body form to an adult body form. In insects with complete metamorphosis, the immature stage, called a **larva**, is worm-shaped (for example, the maggot of a housefly or the caterpillar of a moth or butterfly; see Fig. 22-21e). The larva hatches from an egg, grows by eating voraciously and shedding its exoskeleton several times, and then forms a nonfeeding stage called a **pupa**. Encased in an outer covering, the pupa metamorphoses, emerging in its adult winged form. The adults mate and lay eggs, continuing the cycle. Metamorphosis may include a change in diet as well as in shape, thereby eliminating competition for food between adults and juveniles and, in some cases, allowing the insect to exploit different foods when they are most available. For instance, a caterpillar that feeds on new green shoots in springtime metamorphoses into a butterfly that drinks nectar from the summer's blooming flowers. Some insects undergo a more gradual metamorphosis (called incomplete metamorphosis), hatching as young that bear some resemblance to the adult, then gradually acquiring more-adult features as they grow and molt.

Most Arachnids Are Predatory Meat Eaters

The *arachnids* include spiders, mites, ticks, and scorpions (Fig. 22-22). Members of the class Arachnida lack antennae and have eight walking legs. Most arachnids are carnivorous; many subsist on a liquid diet of blood or predigested prey. For example, spiders, the most numerous arachnids, first immobilize their prey with a paralyzing venom. They then inject digestive enzymes into the helpless victim (typically an insect) and suck in the resulting soup. Arachnids breathe by using either tracheae, book lungs (which are unique to arachnids), or both. In contrast to the compound eyes of insects and crustaceans, arachnids have simple eyes, each with a single lens. The eyes are particularly sensitive to movement, and in some species they probably can form images. Most spiders have eight eyes, which are placed to detect the movements of predators and prey over a wide field of view.

(a)

(b)

(c)

FIGURE 22-22 The diversity of arachnids
(a) The tarantula is one of the largest spiders but is relatively harmless. **(b)** Scorpions, found in warm climates such as the deserts of the southwestern United States, paralyze their prey with venom from a stinger at the tip of the abdomen. A few species can harm humans. **(c)** Ticks before (left) and after feeding on blood. The uninflated exoskeleton is flexible and folded, allowing the animal to become grotesquely bloated while feeding.

FIGURE 22-23 The diversity of myriapods
(a) Centipedes and **(b)** millipedes are common nocturnal arthropods. Each segment of a centipede's body holds one pair of legs, while each millipede segment has two pairs.

Myriapods Have Many Legs

The *myriapods* include the centipedes and millipedes, whose most prominent feature is an abundance of legs (Fig 22-23). Millipedes are especially notable in this regard; one species has 750 legs (375 pairs). Most millipede species, however, have between 100 and 300 legs. Centipedes are not quite as leggy; a typical species has around 70 legs, but many species have fewer. Both centipedes and millipedes have one pair of antennae. The legs and antennae of centipedes are longer and more delicate than those of millipedes. Myriapods have very simple eyes that detect light and dark but do not form images. In some species, the number of eyes can be high—up to 200. They respire by means of tracheae.

Myriapods inhabit terrestrial environments exclusively, living mostly in the soil or leaf litter or under logs and rocks. Centipedes are generally carnivorous, capturing prey (mostly other arthropods) with their frontmost legs, which are modified into sharp claws that inject poison into prey. Bites from large centipedes can be painful to humans. In contrast, most millipedes are not predators, but instead feed on decaying vegetation and other debris. When attacked, many millipedes defend themselves by secreting a foul-smelling, distasteful liquid.

Most Crustaceans Are Aquatic

The *crustaceans*, including crabs, crayfish, lobster, shrimp, and barnacles, live primarily in the water (Fig. 22-24). Crustaceans range in size from microscopic maxillopods

FIGURE 22-24 The diversity of crustaceans
(a) The microscopic waterflea is common in freshwater ponds. Notice the eggs developing within the body. **(b)** The sowbug, found in dark, moist places such as under rocks, leaves, and decaying logs, is one of the few crustaceans to invade the land successfully. **(c)** The hermit crab protects its soft abdomen by inhabiting an abandoned snail shell. **(d)** The goose-neck barnacle uses a tough, flexible stalk to anchor itself to rocks, boats, or even animals such as whales. Other types of barnacles attach with shells that resemble miniature volcanoes (see Fig. 22-15b). Early naturalists thought barnacles were mollusks until they observed barnacles' jointed legs (seen here extending into the water).

that live in the spaces between grains of sand to the largest of all arthropods, the Japanese crab, whose legs span nearly 12 feet (4 meters). Crustaceans have two pairs of sensory antennae, but the rest of their appendages are highly variable in form and number, depending on the habitat and lifestyle of the species. Most crustaceans have compound eyes similar to those of insects, and nearly all respire by means of gills.

Roundworms Are Abundant and Mostly Tiny

Although you may be blissfully unaware of their presence, *roundworms* (phylum Nematoda) are nearly everywhere. Roundworms, also called *nematodes*, have colonized nearly every habitat on Earth, and they play an important role in breaking down organic matter. They are extremely numerous; a single rotting apple may contain 100,000 roundworms. Billions thrive in each acre of topsoil. In addition, almost every plant and animal species hosts several parasitic nematode species.

In addition to being abundant and ubiquitous, roundworms are diverse. Although only about 12,000 roundworm species have been named, there may be as many as 500,000. Most, such as the one shown in Figure 22-25, are microscopic, but some parasitic forms reach a meter in length.

Nematodes have a rather simple body plan, featuring a tubular gut and a fluid-filled pseudocoelom that surrounds the organs and forms a hydrostatic skeleton. A tough, flexible, nonliving cuticle encloses and protects the thin, elongated body and is periodically molted. The molting of roundworms reveals that they share a common evolutionary heritage with arthropods and other ecdysozoan phyla. Sensory organs in the roundworm head transmit information to a simple "brain," composed of a nerve ring.

Like flatworms, nematodes lack circulatory and respiratory systems. Because most nematodes are extremely thin and have low energy requirements, diffusion suffices for gas exchange and distribution of nutrients. Most nematodes reproduce sexually, and the sexes are separate; the male (who is normally smaller) fertilizes the female by placing sperm inside her body.

During your life, you may become host to one of the 50 species of roundworms that infect humans. Most such worms are relatively harmless, but there are important exceptions. For example, hookworm larvae (found in soil) can bore into human feet, enter the bloodstream, and travel to the intestine, where they cause continuous bleeding. Another dangerous nematode parasite, *Trichinella*, causes the disease trichinosis. *Trichinella* worms can infect people who eat improperly cooked infected pork, which can contain up to 15,000 larval cysts per gram (Fig. 22-26a). The cysts hatch in the human digestive tract and invade blood vessels and muscles, causing bleeding and muscle damage.

Parasitic nematodes can also endanger domestic animals. Dogs, for example, are susceptible to heartworm,

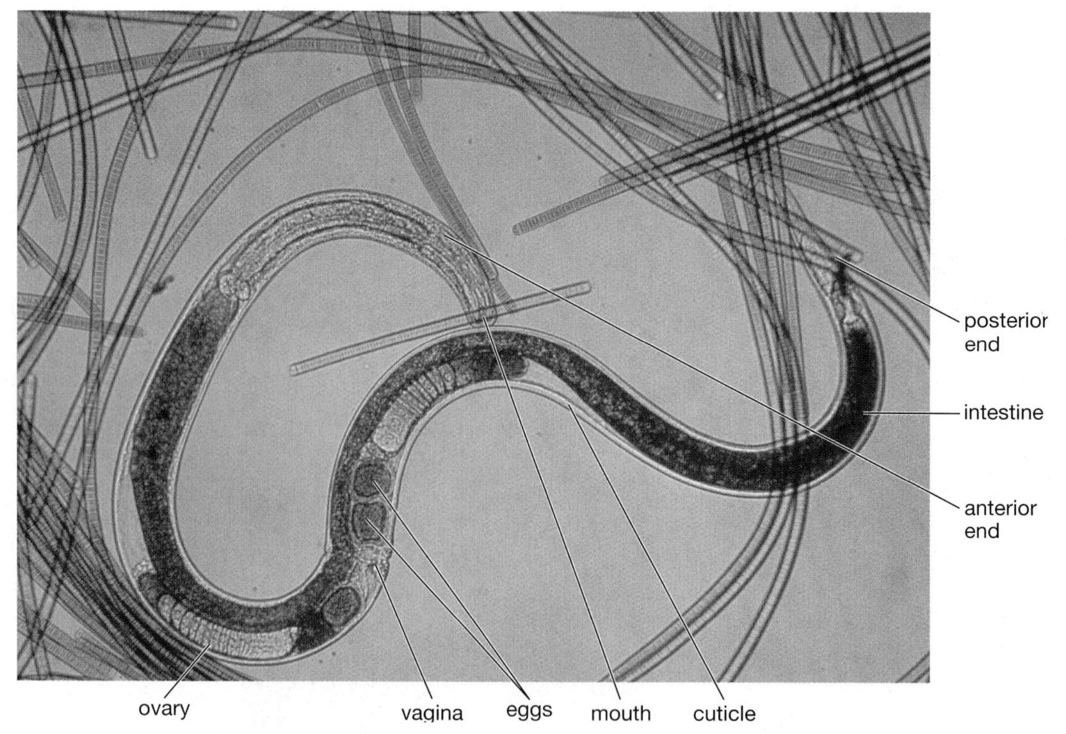

FIGURE 22-25 A freshwater nematode
Eggs can be seen inside this female freshwater nematode, which feeds on algae.

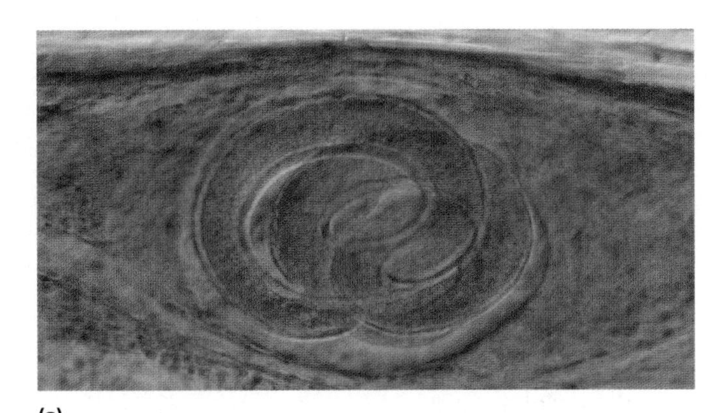

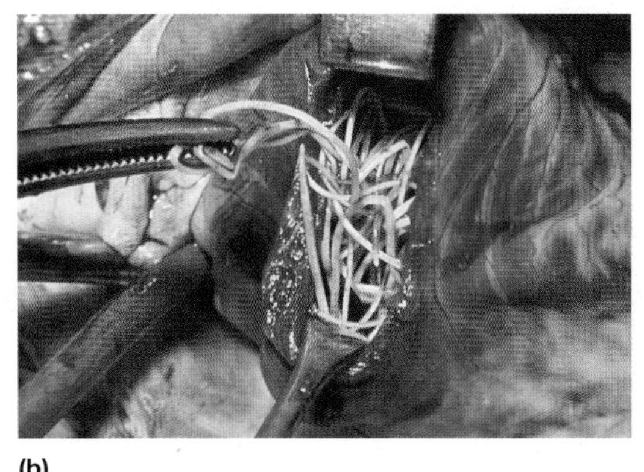

(a) (b)

FIGURE 22-26 Some parasitic nematodes
(a) Encysted larva of the *Trichinella* worm in the muscle tissue of a pig, where it may live for up to 20 years. **(b)** Adult heartworms in the heart of a dog. The juveniles are released into the bloodstream, where they may be ingested by mosquitoes and passed to another dog by the bite of an infected mosquito.

which is transmitted by mosquitoes (Fig. 22-26b). In the southern United States, and increasingly in other parts of the country, heartworm poses a severe threat to the health of unprotected pets.

Echinoderms Have a Calcium Carbonate Skeleton

Echinoderms (phylum Echinodermata) are found only in marine environments, and their common names tend to evoke their saltwater habitats: sand dollars, sea urchins, sea stars (or starfish), sea cucumbers, and sea lilies (Fig. 22-27). The name "echinoderm" (Greek, "hedgehog skin") stems from the bumps or spines that extend from the skin of most echinoderms. These spines are especially well developed in sea urchins and much reduced in sea stars and sea cucumbers. Echinoderm bumps and spines are actually extensions of an **endoskeleton** (internal skeleton) composed of plates of calcium carbonate that lie beneath the outer skin.

Echinoderms exhibit deuterostome development and are linked by common ancestry with the other deuterostome phyla, including the chordates (described below). Deuterostomes form a group of branches on the larger evolutionary tree of bilaterally symmetrical animals, but in echinoderms, bilateral symmetry is expressed only in embryos and free-swimming larvae. An adult echinoderm, in contrast, is radially symmetrical and lacks a head. This absence of cephalization is consistent with the sluggish or sessile existence of echinoderms. Most echinoderms move only very slowly as they feed on algae or small particles sifted from sand or water. Some echinoderms are slow-motion predators. Sea stars, for example, slowly pursue even slower-moving prey, such as bivalve mollusks.

Echinoderms move on numerous tiny *tube feet*, delicate cylindrical projections that extend from the lower surface of the body and terminate in a suction cup. Tube feet are part of a unique echinoderm feature, the *water-vascular system*, which functions in locomotion, respiration, and food capture (Fig. 22-28). Seawater enters

(a) (b) (c)

FIGURE 22-27 The diversity of echinoderms
(a) A sea cucumber feeds on debris in the sand. **(b)** The sea urchin's spines are actually projections of the internal skeleton. **(c)** The sea star has reduced spines and typically has five arms.

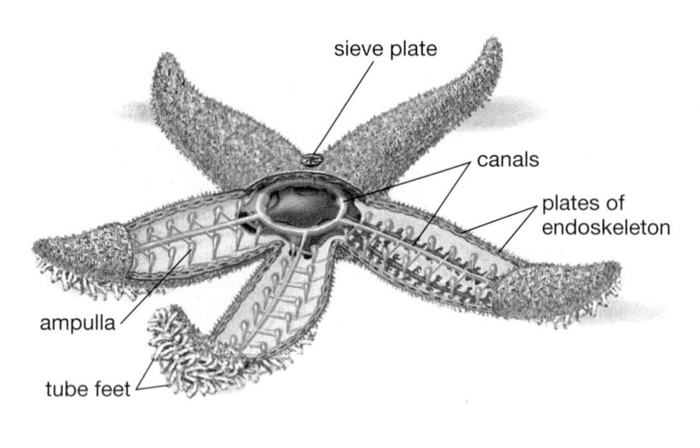

(a)

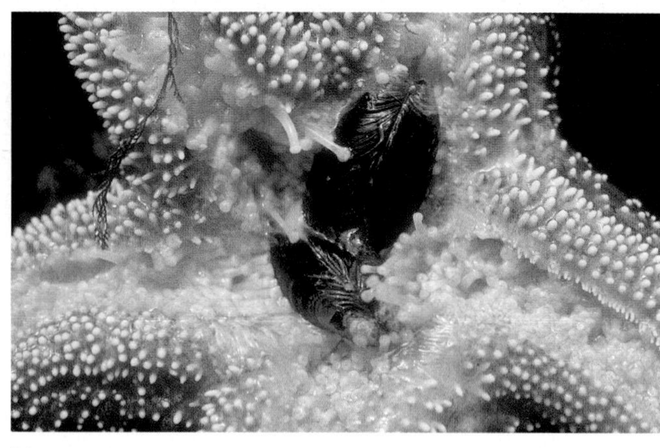

(b)

FIGURE 22-28 The water-vascular system of echinoderms
(a) Changing pressure inside the seawater-filled water-vascular system extends or retracts the tube feet. **(b)** The sea star often feeds on mollusks such as this mussel. A feeding sea star attaches numerous tube feet to the mussel's shells, exerting a relentless pull. Then, the sea star turns the delicate tissue of its stomach inside out, extending it through its centrally located ventral mouth. The stomach can fit through an opening in the bivalve shells which measures less than 1 millimeter. Once insinuated between the shells, the stomach tissue secretes digestive enzymes that weaken the mollusk, causing it to open further. Partially digested food is transported to the upper portion of the stomach, where digestion is completed.

through an opening (the *sieve plate*) on the animal's upper surface and is conducted through a circular central canal, from which branch a number of radial canals. These canals conduct water to the tube feet, each of which is controlled by a muscular squeeze bulb (*ampulla*). Contraction of the bulb forces water into the tube foot, causing it to extend. The suction cup may be pressed against the seafloor or a food object, to which it adheres tightly until pressure is released.

Echinoderms have a relatively simple nervous system with no distinct brain. Movements are loosely coordinated by a system consisting of a nerve ring that encircles the esophagus, radial nerves to the rest of the body, and a nerve network through the epidermis. In sea stars, simple receptors for light and chemicals are concentrated on the arm tips, and sensory cells are scattered over the skin. In some brittle star species, light receptors are associated with tiny lenses, smaller than the width of a human hair, that gather light and focus it on receptors. The optical quality of these "microlenses" is excellent, far superior to that of any human-created lens of comparable size.

Echinoderms lack a circulatory system, although movement of the fluid in their well-developed coelom serves this function. Gas exchange occurs through the tube feet and, in some forms, through numerous tiny "skin gills" that project through the epidermis. Most species have separate sexes and reproduce by shedding sperm and eggs into the water, where fertilization occurs.

Many echinoderms are able to regenerate lost body parts, and these regenerative powers are especially potent in sea stars. In fact, a single arm of a sea star is capable of developing into a whole animal, provided that part of the central body is attached to it. Before this ability was widely appreciated, mussel fishermen often tried to rid mussel beds of predatory sea stars by hacking them into pieces and throwing the pieces back. Needless to say, the strategy backfired.

The Chordates Include the Vertebrates

The phylum Chordata, which contains the vertebrate animals, also includes a few groups of invertebrates, such as the sea squirts and the lancelets. We will discuss these invertebrate chordates and their vertebrate relatives in Chapter 23.

CASE STUDY REVISITED The Search for a Sea Monster

Clyde Roper's search for the giant squid has led him to organize three major expeditions over the past few years. The first of these began in 1996, off the Azores Islands in the Atlantic Ocean. Because sperm whales are known to prey upon giant squid, Roper believed that the whales might lead him to the squids. To test this idea, he and his team affixed video cameras to sperm whales, allowing the scientists to see what the whales were seeing. These "crittercams" revealed a great deal of new information about sperm whale behavior but, alas, no footage of giant squids.

The next Roper-led expedition took place the following year in the Kaikoura Canyon, an area of very deep water (3300 feet, or 1000 meters) off the coast of New Zealand. This spot was chosen because deep-sea fishing boats had recently captured several giant squid in the vicinity. Cameras were again deployed on sperm whales, but this time the mobile cameras were supplemented by a stationary, baited camera and an unmanned, remote-controlled submarine. Again, however, a large investment of time, money, and equipment yielded no squid sightings.

In 1999, Roper assembled a team of scientists for a return to Kaikoura Canyon. This time, the group was able to use Deep Rover, a one-person submarine that could carry an observer to depths of 2200 feet. The scientists used Deep Rover to explore the canyon, following sperm whales in hopes that the huge mammals would lead them to giant squid.

Despite 30 days of trying, the expedition team once again failed to glimpse a giant squid. They did amass a wealth of data on a virtually unknown deep-sea ecosystem, but gathered no new clues about the mysterious "sea monster." The life of the giant squid remains a mystery. According to Roper, who is currently raising funds for another squid-searching expedition, "We probably know more about the dinosaurs than about the giant squid."

Consider This: Steve Shea, another scientist interested in the giant squid, captured a few juvenile giant squid in 2002. The tiny animals, only a few millimeters long, survived in captivity for just a few hours, but their identity as giant squid was confirmed by comparing their DNA to that of preserved adult specimens. Shea believes that with more research and experience, he could learn to raise the young animals to adulthood. Given that research funds are limited, which approach is better? Would we learn more from viewing wild adult squid in the depths of the ocean, or by capturing tiny juveniles from surface waters and figuring out how to raise them in the lab?

Links to Life: A Bloodsucker's Renaissance

Although invertebrate animals cause or transmit many human illnesses, some have been recruited to make a positive contribution to human health. Consider leeches, for example. For more than 2000 years, healers enlisted these parasitic annelids for treatment of almost every human illness or injury. For much of human medical history, treatment with leeches was based on the hope that the creatures would suck out the "tainted" blood that was believed to be the primary cause of disease. As the actual causes of disease were discovered, however, medical use of leeches declined. By the beginning of the twentieth century, leeches no longer had a place in the toolkit of modern medicine, and had become a symbol of the ignorance of an earlier age. Today, however, medical use of leeches is making a surprising comeback.

Currently, leeches are used to treat a surgical complication known as venous insufficiency. This complication is especially common in reconstructive surgery, such as reattaching a severed finger or repairing a disfigured face. In such cases, surgeons are often unable to reconnect all of the veins that would normally carry blood away from tissues. Eventually, new veins will grow, but in the meantime blood may accumulate in the repaired tissue. Unless the excess blood is removed, it will coagulate, causing clots that can deprive the tissue of the oxygen and nutrients it needs to live. Fortunately, leeches can help. Applied to the affected area, the leeches get right to work, making a small, painless incision and sucking blood into their stomachs. To aid them in their blood-removal task, the leeches' saliva contains a mixture of chemicals that dilate blood vessels and prevent blood from clotting. Although the chemical brew in the saliva is an adaptation that helps leeches consume blood more efficiently, it also helps the patient by promoting blood flow in the damaged tissue. In this way, leeches provide a painless, effective treatment for venous insufficiency, and have thereby regained their status as medical assistants to humanity.

Summary of Key Concepts

22.1 What Are the Key Features of Animals?

Animals are multicellular, sexually reproducing, heterotrophic organisms. Most can perceive and react rapidly to environmental stimuli and are motile at some stage in their lives. Their cells lack a cell wall.

22.2 Which Anatomical Features Mark Branch Points on the Animal Evolutionary Tree?

The earliest animals had no tissues, a feature retained by modern sponges. All other modern animals have tissues. Animals with tissues can be divided into radially symmetrical and bilaterally symmetrical groups. During embryonic development, radially symmetrical animals have two germ layers, bilaterally symmetrical animals have three. Bilaterally symmetrical animals also tend to have sense organs and clusters of neurons concentrated in the head, a process called *cephalization*. Bilateral phyla can be divided into two main groups, one of which undergoes protostome development, the other of which undergoes deuterostome development. Protostome phyla can in turn can be divided into ecdysozoans and lophotrochozoans. Some phyla of bilaterally symmetrical animals lack body cavities, but most have either pseudocoeloms or true coeloms.

22.3 What Are the Major Animal Phyla?

The bodies of sponges (phylum Porifera) are typically free-form in shape and are sessile. Sponges have relatively few types of cells. Despite the division of labor among the cell types, there is little coordination of activity. Sponges lack the muscles and nerves required for coordinated movement, and digestion occurs exclusively within the individual cells.

The hydra, anemones, and jellyfish (phylum Cnidaria) have tissues. A simple network of nerve cells directs the activity of contractile cells, allowing loosely coordinated movements. Digestion is extracellular, occurring in a central gastrovascular cavity with a single opening. Cnidarians exhibit radial symmetry, an adaptation to both the free-floating lifestyle of the medusa and the sedentary existence of the polyp.

Flatworms (phylum Platyhelminthes) have a distinct head with sensory organs and a simple brain. A system of canals forming a network through the body aids in excretion. They lack a body cavity.

The segmented worms (phylum Annelida) are the most complex of the worms, with a well-developed closed circulatory system and excretory organs that resemble the basic unit of the vertebrate kidney. The segmented worms have a compartmentalized digestive system, like that of vertebrates, which processes food in a sequence. Annelids also have a true coelom, a fluid-filled space between the body wall and the internal organs.

The snails, clams, and squid (phylum Mollusca) lack a skeleton; some forms protect the soft, moist, muscular body with a single shell (many gastropods and a few cephalopods) or a pair of hinged shells (the bivalves). The lack of a waterproof external covering limits this phylum to aquatic and moist terrestrial habitats. Although the body plan of gastropods and bivalves limits the complexity of their behavior, the cephalopod's tentacles are capable of precisely controlled movements. The octopus has the most complex brain and the best-developed learning capacity of any invertebrate.

Arthropods, the insects, arachnids, millipedes and centipedes, and crustaceans (phylum Arthropoda), are the most diverse and abundant organisms on Earth. They have invaded nearly every available terrestrial and aquatic habitat. Jointed appendages and well-developed nervous systems make possible complex, finely coordinated behavior. The exoskeleton (which conserves water and provides support) and specialized respiratory structures (which remain moist and protected) enable the insects and arachnids to inhabit dry land. The diversification of insects has been enhanced by their ability to fly. Crustaceans, which include the largest arthropods, are restricted to moist, usually aquatic habitats and respire by using gills.

The pseudocoelomate roundworms (phylum Nematoda) possess a separate mouth and anus and a cuticle layer that is molted.

The sea stars, sea urchins, and sea cucumbers (phylum Echinodermata) are an exclusively marine group. Like other complex invertebrates and chordates, echinoderm larvae are bilaterally symmetrical; however, the adults show radial symmetry. This, in addition to a primitive nervous system that lacks a definite brain, adapts them to a relatively sedentary existence. Echinoderm bodies are supported by a nonliving internal skeleton that sends projections through the skin. The water-vascular system, which functions in locomotion, feeding, and respiration, is a unique echinoderm feature.

The phylum Chordata includes two invertebrate groups, the lancelets and tunicates, as well as the vertebrates.

Key Terms

bilateral symmetry *p. 425*
budding *p. 430*
cephalization *p. 426*
closed circulatory system *p. 434*
coelom *p. 426*
compound eye *p. 439*
deuterostome *p. 427*
ectoderm *p. 425*
endoderm *p. 425*
endoskeleton *p. 443*
exoskeleton *p. 437*
ganglion *p. 432*
gill *p. 438*
hemocoel *p. 435*
hermaphroditic *p. 433*
hydrostatic skeleton *p. 434*
invertebrate *p. 427*
larva *p. 440*
mesoderm *p. 426*
metamorphosis *p. 440*
molt *p. 438*
nerve cord *p. 432*
open circulatory system *p. 435*
parasite *p. 432*
protostome *p. 427*
pseudocoelom *p. 427*
pupa *p. 440*
radial symmetry *p. 425*
segmentation *p. 434*
tissue *p. 425*
vertebrate *p. 427*

Thinking Through the Concepts

To take a multiple-choice quiz with feedback on the contents of this chapter, visit http://www.prenhall.com/audesirk7. *Log in to the Web site selected by your instructor and navigate to the Self Test section for this chapter.*

? Review Questions

1. List the distinguishing characteristics of each of the phyla discussed in this chapter, and give an example of each.
2. Briefly describe each of the following adaptations, and explain the adaptive significance of each: bilateral symmetry, cephalization, closed circulatory system, coelom, radial symmetry, segmentation.
3. Describe and compare respiratory systems in the three major arthropod classes.
4. Describe the advantages and disadvantages of the arthropod exoskeleton.
5. State in which of the three major mollusk classes each of the following characteristics is found:
   a. two hinged shells
   b. a radula
   c. tentacles
   d. some sessile members
   e. the best-developed brains
   f. numerous eyes
6. Give three functions of the water-vascular system of echinoderms.
7. To what lifestyle is radial symmetry an adaptation? Bilateral symmetry?

Applying the Concepts

1. The class Insecta is the largest taxon of animals on Earth. Its greatest diversity is in the Tropics, where habitat destruction and species extinction are occurring at an alarming rate. What biological, economic, and ethical arguments can you advance to persuade people and governments to preserve this biological diversity?
2. Discuss at least three ways in which the ability to fly has contributed to the success and diversity of insects.
3. Discuss and defend the attributes you would use to define biological success among animals. Are humans a biological success by these standards? Why?

For More Information

Adis, J., Zompro, O., Moombolah-Goagoses, E., and Marais, E. "Gladiators, A New Order of Insects." *Scientific American*, November 2002. An unusual insect, found fossilized in amber, is found to be a member of a previously unknown order. Later, living representatives of the new group were discovered in Africa.

Brusca, R. C., and Brusca, G. J. *Invertebrates*. Sunderland, MA: Sinauer, 1990. A thorough survey of the invertebrate animals in textbook format but readable and filled with beautiful and informative drawings.

Chadwick, D. H. "Planet of the Beetles." *National Geographic*, March 1998. The beauty and diversity of beetles, which comprise one-third of the world's insects, are described in text and photographs.

Conover, A. "Foreign Worm Alert." *Smithsonian*, August 2000. Escaped night crawlers, annelids imported for use as fishing bait, threaten North American ecosystems.

Hamner, W. "A Killer Down Under." *National Geographic*, August 1994. Among the most poisonous animals in the world is the box jellyfish, which lives off the coast of northern Australia.

Kunzig, R. "At Home with the Jellies." *Discover*, September 1997. An account of the biologists who study jellyfish, and some of their findings. Includes excellent photos.

Morell, V. "Life on a Grain of Sand." *Discover*, April 1995. The sand beneath shallow waters is home to an incredible range of microscopic creatures.

Rennie, J. "Living Together." *Scientific American*, January 1992. Most parasites do not kill their hosts, and the interaction of parasite and host provides some fascinating insights into the evolution of life on Earth.

Media Activities

To access a Media Activity visit http://www.prenhall.com/audesirk7. *Log in to the Web site selected by your instructor, navigate to this chapter, and select the appropriate Media Activity number.*

22.1 Architecture of Animals

Estimated time: 10 minutes

This chapter describes the anatomical features associated with the various animal phyla that indicate branching of the animal evolutionary tree. This activity will allow you to explore those features and the associated animal phyla.

22.2 Web Investigation: The Search For a Sea Monster

Estimated time: 10 minutes

Many fascinating creatures can be found in the oceans' depths. This exercise looks at the relatively limited information available about a few, better-known species.

CHAPTER

23 Animal Diversity II: Vertebrates

Would you be shocked to learn that dinosaurs still walked the Earth? The discovery of modern coelacanth fishes was no less surprising.

AT A GLANCE

CASE STUDY Fish Story

On December 22, 1938, Marjorie Courtney-Latimer received a phone call that would lead to one of the most spectacular discoveries in biological history. The call was from a local fisherman whom Courtney-Latimer, the curator of a small museum in South Africa, had asked to collect some fish specimens for the museum. His boat had returned from its most recent voyage and was waiting at the town dock. Dutifully, Courtney-Latimer went to the boat and began sorting through the fish that were strewn across the deck. Later, she wrote, "I noticed a blue fin sticking up from beneath the pile. I uncovered the specimen, and, behold, there appeared the most beautiful fish I had ever seen." In addition to its beauty, the fish had some odd features, including fins that were stumpy and lobed, unlike the fins of any other living species.

Courtney-Latimer did not recognize the strange fish, but she knew it was unusual. She tried to find a place to refrigerate it but, in her small town, she was unable to find a cold storage facility willing to store a fish. In the end, she was able to save only the skin. Undaunted, she made some drawings of the fish and used them to attempt an identification. To her amazement, the creature did not resemble any species known to inhabit the waters off South Africa, but did seem similar to members of a family of fishes known as coelacanths. The only problem with this assessment was that coelacanths were known only from fossils. The earliest coelacanth fossils were found in 400-million-year-old rocks and, as far as anyone knew, the group had been extinct for 80 million years!

Perplexed, Courtney-Latimer sent her drawings to J. L. B. Smith, a fish expert at Rhodes University. Smith was astounded when he saw the sketch, later writing that "a bomb seemed to burst in my brain." Although bitterly disappointed that the specimen's bones and internal organs had been lost, Smith arranged to view the preserved skin. Ultimately, he confirmed the astonishing news that coelacanths still swam Earth's waters.

23.1 What Are the Key Features of Chordates?

In terms of both number of species and number of individuals, Earth's animals are overwhelmingly boneless invertebrates. Nonetheless, when we think of animals we tend to think of vertebrates—fish, reptiles, amphibians, birds, and mammals. Our bias toward vertebrates arises in part because, compared to invertebrates, they are generally larger and more conspicuous; a person is simply more likely to notice a crow or a squirrel than a flatworm or a clam. But our affinity for vertebrates also stems from their similarity to us. We are, after all, vertebrates ourselves.

Humans are members of the phylum Chordata, which we share not only with birds and apes but also with some invertebrates, including the tunicates (sea squirts) and little fishlike creatures called lancelets. Though it may be hard to believe that humans have anything in common with sea squirts, several characteristics unite the vertebrate and invertebrate chordates. All chordates share deuterostome development (which is also characteristic of echinoderms; see Chapter 22) and are further united by four features that all possess at some stage of their lives:

- A **notochord**. A stiff but flexible rod that extends the length of the body and provides an attachment site for muscles.
- A dorsal, hollow **nerve cord**. Lying above the digestive tract, this hollow neural structure develops a thickening at its anterior end that becomes a brain.
- **Pharyngeal gill slits**. Located in the pharynx (the cavity behind the mouth), these may form functional respiratory openings or may appear only as grooves during an early stage of development.
- A **post-anal tail**. An extension of the body past the anus.

At first glance, this list may seem puzzling because, although humans are chordates, we seem to lack every feature except the nerve cord. But evolutionary relationships are sometimes seen most clearly during early stages of development; it is during our embryonic life that we develop—and lose—a notochord, gill slits, and a tail (Fig. 23-1).

Invertebrate Chordates Lack a Backbone

The invertebrate chordates lack the backbone that is the defining feature of vertebrates. These chordates comprise two groups of organisms, the *lancelets* and the *tunicates.* The small (2 inches, or about 5 centimeters, long) fishlike lancelet spends most of its time half-buried in the sandy sea bottom, filtering tiny food particles from the water. As can be seen in Figure 23-2a, all the typical chordate features are present in the adult lancelet.

The tunicates form a larger group of marine invertebrate chordates, which includes the sea squirts. It is difficult to imagine a less likely relative of humans than the sessile, filter-feeding, vaselike sea squirt (Fig. 23-2b). Its ability to move is limited to forceful contractions of its saclike body, which can send a jet of seawater into the face of anyone who plucks it from its undersea home; hence its common name, sea squirt. Adult sea squirts may be immobile, but their larvae swim actively and possess all the diagnostic chordate features (see Fig. 23-2b).

Vertebrates Have a Backbone

In **vertebrates**, the embryonic notochord is normally replaced during development by a backbone, or **vertebral column**. The vertebral column is composed of bone or **cartilage**, a tissue that resembles bone but is less brittle

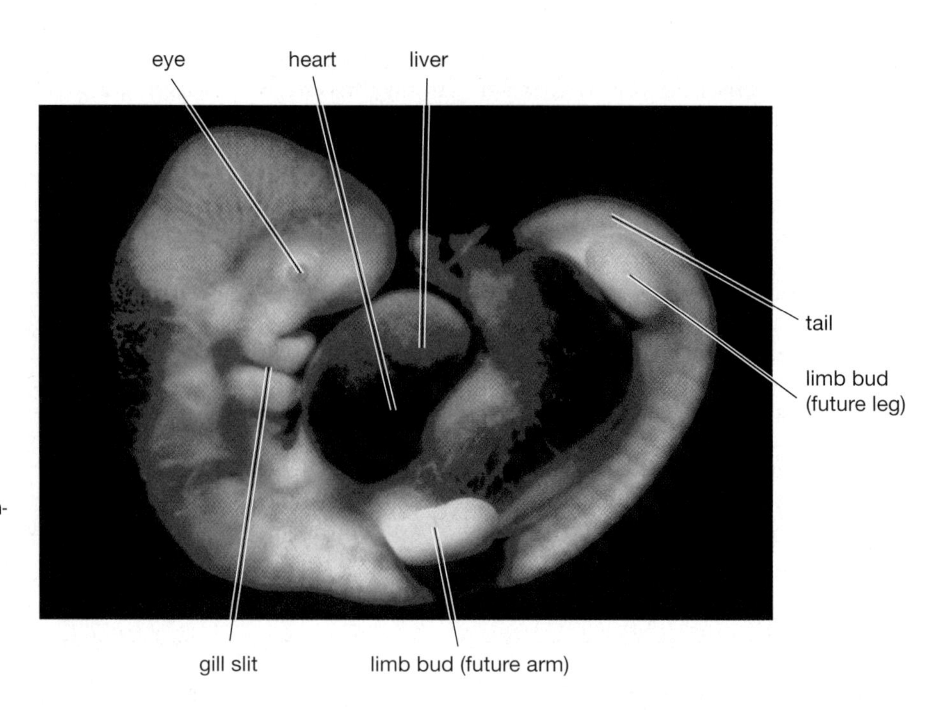

FIGURE 23-1 Chordate features in the human embryo
This 5-week-old human embryo is about 1 centimeter long and clearly shows a tail and external gill slits (more properly called grooves, since they do not penetrate the body wall). Although the tail will disappear completely, the gill grooves contribute to the formation of the lower jaw.

(a) Lancelet

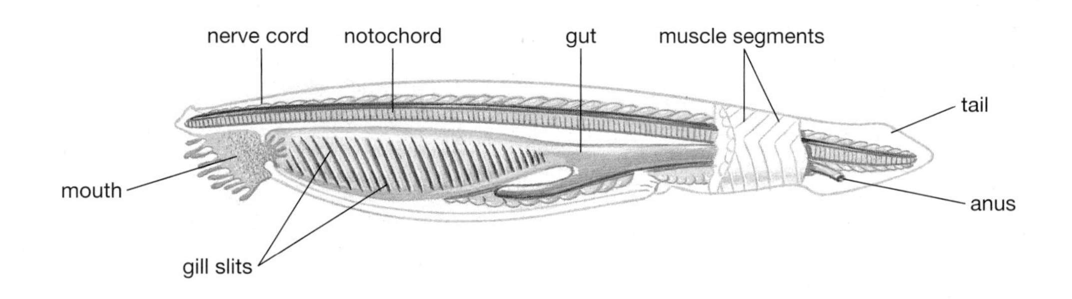

(b) Tunicate

mouth
atrial opening
tail
anus
attachment points
gill slits
gut
nerve cord
notochord
larva

water enters (mouth)
water exits
gill slits
heart
gut
gonad
adult

FIGURE 23-2 Invertebrate chordates
(a) A lancelet, a fishlike invertebrate chordate. The adult organism exhibits all the diagnostic features of chordates. **(b)** The sea squirt larva (left) also exhibits all the chordate features. The adult sea squirt (a type of tunicate, middle) has lost its tail and notochord and has assumed a sedentary life, as shown in the photo (right).

and more flexible. This column supports the body, offers attachment sites for muscles, and protects the delicate nerve cord and brain. It is also part of a living **endoskeleton** that can grow and repair itself. Because this internal skeleton provides support without the armorlike weight of the arthropod exoskeleton, it has allowed vertebrates to achieve great size and mobility and has contributed to their invasion of the land and the air.

Other adaptations have also contributed to vertebrates' successful invasion of most habitats. One is the presence of paired appendages. These first appeared as fins in fish, and serve as stabilizers for swimming. Over millions of years, some fins were modified by natural selection into legs that allowed animals to crawl onto dry land, and later into wings that allowed some to take to the air. Another adaptation that has contributed to the success of vertebrates is an increase in the size and complexity of their brains and sensory structures, which allows vertebrates to perceive their environment in detail and to respond to it in a great variety of ways.

23.2 What Are the Major Groups of Vertebrates?

The evolutionary ancestor of vertebrates was probably an organism similar to today's lancelets. The earliest known vertebrates, whose fossils were found in 530-million-year-old rocks, resembled lancelets but had brains, skulls, and eyes. Today, vertebrates include lampreys, cartilaginous fishes, bony fishes, amphibians, reptiles, birds, and mammals.

Some Vertebrates Lack Jaws

The mouths of the earliest vertebrates did not include jaws. The early history of vertebrates was characterized by an array of strange, now-extinct jawless fishes, many of which were protected by bony armor plates. Today, two groups of jawless fishes survive: the *hagfishes* (class Myxini) and the *lampreys* (class Petromyzontiformes). Although both hagfishes and lampreys have eel-shaped bodies and smooth, unscaled skin, the two groups represent distinct, early branches of the chordate evolutionary tree. The branch leading to modern hagfishes is the more ancient of the two.

Hagfishes Are Slimy Residents of the Ocean Floor

The hagfish body is stiffened by a notochord, but its "skeleton" is limited to a few small cartilage elements, one of which forms a rudimentary braincase. Because hagfishes lack skeletal elements that surround and protect the nerve chord, most systematists do not consider them to be vertebrates. Instead, they represent the chordate group that is most closely related to the vertebrates.

Hagfishes are exclusively marine (Fig. 23-3a). They live near the ocean floor, often burrowing in the mud, and feed primarily on worms. They will, however, eagerly attack dead and dying fish, using pincerlike teeth to burrow into their prey's body and consume the soft internal organs. Hagfishes are regarded with great disgust by some fishermen because they secrete massive quantities of slime as a defense against predators. Despite their well-deserved reputation as "slime balls of the sea," hagfishes are avidly pursued by many commercial fishermen, because the leather industry in some parts of the world provides a market for hagfish skin. Most leather items that purport to be "eel skin" are in fact made from tanned hagfish skin.

Some Lampreys Parasitize Fish

The spinal cord of lampreys is protected by segments of cartilage, so lampreys are considered to be true vertebrates. They live in both fresh and salt waters, but the marine forms must return to fresh water to spawn. Some lamprey species are parasitic. Parasitic lampreys have suckerlike mouths lined with teeth, which they use to attach themselves to larger fish (Fig. 23-3b). Using rasping teeth on its tongue, the lamprey excavates a hole in the host's body wall, through which it sucks blood and body fluids. Beginning in the 1920s, lampreys spread into the Great Lakes, where, in the absence of effective predators, they have multiplied prodigiously and greatly reduced commercial fish populations, including the lake trout. Vigorous measures to control the lamprey population have allowed some recovery of the other fish populations of the Great Lakes.

Jawed Fishes Rule Earth's Waters

About 425 million years ago, in the mid-Silurian period, jawless fishes that were ancestors of the lampreys and hagfishes gave rise to a group of fish that possessed an important new structure: jaws. Jaws allowed fish to grasp and chew their food, permitting them to exploit a much wider range of food sources than could jawless fish. Although the earliest forms of jawed fishes have been extinct for 230 million years, they gave rise to the two major classes of jawed fishes that survive today: the cartilaginous fishes (class Chondrichthyes) and the bony fishes (class Osteichthyes).

(a)

(b)

FIGURE 23-3 Jawless fishes
(a) Hagfishes live in communal burrows in mud, feeding on worms. **(b)** Some lampreys are parasitic, attaching to fish (such as this carp) with suckerlike mouths lined with rasping teeth (inset).

Cartilaginous Fishes Are Marine Predators

The class Chondrichthyes, whose name is derived from Greek words meaning "cartilage fishes," includes 625 marine species, among them the sharks, skates, and rays (Fig. 23-4). The *cartilaginous fishes* are graceful predators that lack any bone in their skeleton, which is formed entirely of cartilage. The body is protected by a leathery skin roughened by tiny scales. Members of this group respire using gills. Although some must swim to circulate water through their gills, most can pump water across their gills. Like all fishes, the cartilaginous fishes have two-chambered hearts.

Many shark species sport several rows of razor-sharp teeth; the back rows move forward as the front teeth are lost to age and use (see Fig. 23-4a). Although a few species consider us potential prey, most sharks are shy of humans. Skates and rays (see Fig. 23-4b) are also

(a)

(b)

FIGURE 23-4 Cartilaginous fishes
(a) A sand tiger shark displaying several rows of teeth. As outer teeth are lost, they are replaced by the new ones behind them. Both sharks and rays lack a swim bladder and tend to sink toward the bottom when they stop swimming. **(b)** The tropical blue-spotted sting ray swims by graceful undulations of lateral extensions of the body.

timid creatures, although some can inflict dangerous wounds with a spine near their tail, and others produce a powerful electrical shock that can stun their prey. Some cartilaginous fishes are very large. The whale shark, for example, can grow to more than 45 feet (15 meters) in length.

Bony Fishes Are the Most Diverse Vertebrates

Just as our size bias makes us overlook the most-diverse invertebrate groups, our habitat bias makes us overlook the most-diverse vertebrates. The most diverse and abundant vertebrates are not birds or the predominantly terrestrial mammals. Rather, the vertebrate diversity crown belongs to the lords of the oceans and fresh water, the *bony fishes* of the class Osteichthyes, whose skeletons are composed of bone rather than cartilage. From the snakelike moray eel to the bizarre, luminescent deep-sea fishes to the streamlined tuna, this enormously successful group has spread to nearly every possible watery habitat, both freshwater and marine (Fig. 23-5). Although about 17,000 species have been identified, scientists predict that perhaps nearly twice this number may exist if the undescribed species from deep waters and remote areas are considered. The

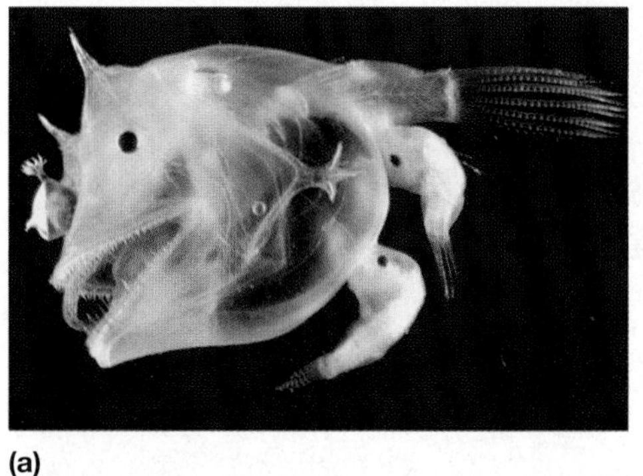

(a)

(b)

(c)

FIGURE 23-5 The diversity of the bony fishes
Bony fishes have colonized nearly every aquatic habitat. **(a)** This female deep-sea angler fish attracts prey with a living lure that projects just above her mouth. The fish is ghostly white; at the 6000 foot depth where anglers live, no light penetrates and thus colors are unnecessary. Male deep-sea angler fish are extremely small and remain attached to the female as permanent parasites, always available to fertilize her eggs. Two parasitic males can be seen attached to this female. **(b)** This tropical green moray eel lives in rocky crevices. A small fish (a banded cleaner goby) on its lower jaw eats parasites that cling to the moray's skin. **(c)** The tropical seahorse may anchor itself with its prehensile tail (adapted for grasping) while feeding on small crustaceans.
QUESTION In terms of water regulation (maintaining the proper amount of water in the body), how does the challenge faced by a freshwater fish differ from that faced by a saltwater fish?

EARTH WATCH Frogs in Peril

Frogs and toads have frequented Earth's ponds and swamps for nearly 150 million years, somehow surviving the Cretaceous catastrophe that extinguished the dinosaurs and so many other species about 65 million years ago. Their evolutionary longevity, however, appears to offer inadequate defense against the environmental changes wrought by human activities. Over the past two decades, herpetologists (biologists who study reptiles and amphibians) from around the world have documented an alarming decline in amphibian populations. Thousands of species of frogs, toads, and salamanders are dramatically decreasing in number, and many have apparently gone extinct.

This is not a localized phenomenon; population crashes have been reported from every part of the globe. Yosemite toads and yellow-legged frogs are disappearing from the mountains of California; tiger salamanders have been nearly wiped out in the Colorado Rockies. Leopard frogs, eagerly chased by children, are becoming rare in the United States. Logging destroys the habitats of amphibians from the Pacific Northwest to the Tropics (Fig. E23-1), but even amphibians in protected areas are dying. In the Monteverde Cloud Forest Preserve in Costa Rica, the golden toad (see Fig. 36-6) was common in the early 1980s but has not been seen since 1989. The gastric brooding frog of Australia fascinated biologists by swallowing its eggs, brooding them in its stomach, and later regurgitating fully formed offspring. This species was abundant and seemed safe in a national park. Suddenly, in 1980, the gastric brooding frog disappeared and hasn't been seen since.

The causes of the worldwide decline in amphibian diversity are not fully understood, but researchers have discovered that frogs and toads in many places are succumbing to infection by a pathogenic fungus. The fungus has been found in the skin of dead and dying frogs in widespread locations, including Australia, Central America, and the western United States. In those places, discovery of the fungus has coincided with massive frog and toad die-offs, and most herpetologists agree that the fungus is causing the deaths. It seems unlikely, however, that the fungus alone is responsible for the worldwide decline of amphibians. For one thing, die-offs have occurred in many places where the fungus has not been found. In addition, many herpetologists believe that the fungal epidemic would not have arisen if the frogs and toads had not first been weakened by other stresses. So, if the fungus is not doing all of the damage on its own, what are the other possible causes of amphibian decline? All of the most likely causes stem from human modification of the biosphere—the portion of Earth that sustains life.

Habitat destruction, especially the draining of wetland habitats that are especially hospitable to amphibian life, is one major cause of the decline. Amphibians are also vulnerable to toxic substances in the environment. For example, researchers found that frogs exposed to trace amounts of atrazine, a widely used herbicide that is found in virtually all fresh water in the United States, suffered severe damage to their reproductive tissues. The unique biology of amphibians makes them especially susceptible to poisons in the environment. Amphibian bodies at all stages of life are protected by only a thin, permeable skin that pollutants can easily penetrate. To make matters worse, the double life of many amphibians exposes their permeable skin to a wide range of aquatic and terrestrial habitats and to a correspondingly wide range of environmental toxins.

Amphibian eggs can also be damaged by ultraviolet (UV) light, according to research by Andrew Blaustein, an ecologist at Oregon State University. Blaustein demonstrated that the eggs of some species of frogs in the Pacific Northwest are sensitive to damage from UV light and that the most sensitive species are experiencing the most drastic declines. Unfortunately, many parts of Earth are subject to increasingly intense UV radiation levels, because atmospheric pollutants have caused a thinning of the protective ozone layer (as we will see in Chapter 41).

Another disturbing trend among frogs and toads is the increasing incidence of grotesquely deformed individuals. Researchers at the Environmental Protection Agency recently demonstrated that developing frogs exposed to current natural levels of UV light grow deformed limbs much more frequently than frogs protected from UV radiation. Other researchers have shown that deformities are more common in frogs exposed to low concentrations of commonly used pesticides. In addition, a growing body of evidence suggests that some deformities—especially the most common one, extra limbs—are caused by parasitic infections during embryonic development. Many frogs with extra legs are infested with a parasitic flatworm, and researchers have shown that tadpoles experimentally infected with the flatworm developed into deformed adults. Why have the parasites, which have long coexisted with frogs, suddenly begun causing so many deformities? A likely explanation is that exposure to UV radiation, pesticides, and herbicides weakens frogs' immune response. A compromised immune system leaves a developing tadpole more vulnerable to parasitic infection.

FIGURE E23-1 Amphibians in danger
The corroboree toad, shown here with its eggs, is rapidly declining in its native Australia. Tadpoles are developing within the eggs. The thin water-permeable and gas-permeable skin of the adult and the jellylike coating around the eggs make them vulnerable to both air and water pollutants.

Many scientists believe that the troubles of amphibians signal an overall deterioration of Earth's ability to support life. According to this line of reasoning, the highly sensitive amphibians are providing an early warning of environmental degradation that will eventually affect more-resistant organisms as well. Equally worrisome is the observation that amphibians are not just sensitive indicators of the health of the biosphere; they are also crucial components of many ecosystems. They may keep insect populations in check, in turn serving as food for larger carnivores. Their decline will further disrupt the balance of these delicate communities. Margaret Stewart, an ecologist at the State University of New York, Albany, aptly summarized the problem: "There's a famous saying among ecologists and environmentalists: 'Everything is related to everything else.' . . . You can't wipe out one large component of the system and not see dramatic changes in other parts of the system."

ocean's potential to conceal species was illustrated in dramatic fashion by the discovery of coelacanths, as described in this chapter's Case Study.

The ancestors of modern bony fish probably had lungs in addition to gills (the *swim bladder*, a sort of internal balloon that allows most bony fishes to float effortlessly at any level, probably evolved from these ancestral lungs). In addition, some groups of early bony fishes developed modified fleshy fins that, in an emergency, could be used as legs, allowing the fish to drag itself from a drying puddle to a deeper pool. We know from fossils that at least one species even evolved actual limbs, although the function of limbs in a water-dwelling organism is not well understood. Such ancestors ultimately gave rise to the first vertebrates to make the first tentative invasion of the land: the amphibians.

Amphibians Live A Double Life

The species of the class Amphibia straddle the boundary between aquatic and terrestrial existence (Fig. 23-6). The limbs of *amphibians* show varying degrees of adaptation to movement on land, from the belly-dragging crawl of salamanders to the long leaps of frogs and toads. A three-chambered heart (in contrast to the two-chambered heart of fishes) circulates blood more efficiently, and lungs replace gills in most adult forms. Amphibian lungs, however, are poorly developed and must be supplemented by the skin, which serves as an additional respiratory organ. This respiratory function requires that the skin remains moist, a constraint that greatly restricts the range of amphibian habitats on land.

(a)

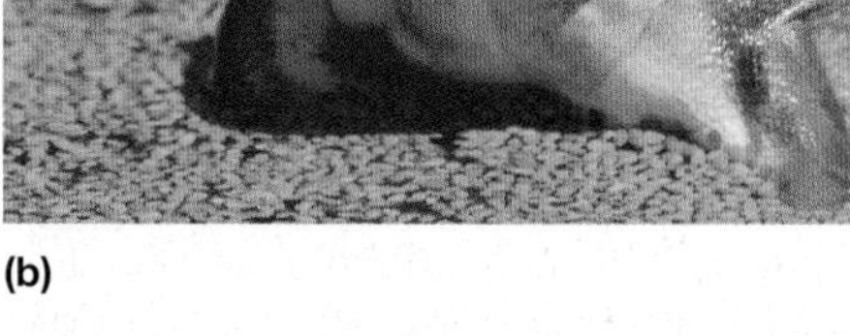

(b)

(c)

FIGURE 23-6 Amphibian means "double life"
The double life of amphibians is illustrated by the bullfrog's transition from **(a)** a completely aquatic larval tadpole to **(b)** an adult leading a semiterrestrial life. **(c)** The red salamander is restricted to moist habitats in the eastern United States. Salamanders hatch in a form that closely resembles the adult.
QUESTION What advantages might amphibians gain from their "double life"?

(a)

(b)

(c)

FIGURE 23-7 The diversity of reptiles
(a) The mountain king snake has a color pattern very similar to that of the poisonous coral snake, which potential predators avoid. This mimicry helps the harmless king snake elude predation. **(b)** The outward appearance of the American alligator, found in swampy areas of the South, is almost identical to that of 150-million-year-old fossil alligators. **(c)** The tortoises of the Galapagos Islands, Ecuador, may live to be more than 100 years old.

Amphibians are also tied to moist habitats by their breeding behavior, which requires water. Their fertilization is normally external and must therefore occur in water so that the sperm can swim to the eggs. The eggs must remain moist, as they are protected only by a jellylike coating that leaves them vulnerable to loss of water by evaporation. The mechanics of keeping the eggs moist varies considerably among different amphibian species, but in many species the requisite moisture is secured simply by laying the eggs in water. In some amphibian species, fertilized eggs develop into aquatic larvae such as the tadpoles of some frogs and toads. The dramatic transformation of these aquatic larvae into semiterrestrial adults gives the class Amphibia its name, which means "double life." Their double life and their thin, permeable skin have made amphibians particularly vulnerable to pollutants and to environmental degradation, as described in "Earth Watch: Frogs in Peril," on page 454.

Reptiles and Birds Are Adapted for Life on Land

The *reptiles* include the lizards and snakes (by far the most diverse of the modern groups) and the turtles, alligators, and crocodiles (Fig. 23-7). Reptiles evolved from an amphibian ancestor about 250 million years ago. Early reptiles—the dinosaurs—ruled the land for nearly 150 million years.

Reptiles Haves Scales and Shelled Eggs

Some reptiles, particularly desert dwellers such as tortoises and lizards, are completely independent of their aquatic origins. This independence was achieved through a series of adaptations, of which three are outstanding: a tough, scaly skin that resists water loss and protects the body; internal fertilization, in which the male deposits sperm within the female's body; and a shelled **amniote egg**, which can be buried in sand or dirt, far from water with its hungry predators. The shell prevents the egg from drying out on land. An internal membrane, the **amnion**, encloses the embryo in the watery environment that all developing animals require (Fig. 23-8).

In addition to these features, reptiles have more-efficient lungs than do earlier vertebrates, and do not use their skin as a respiratory organ. The three-chambered heart became modified, allowing better separation of oxygenated and deoxygenated blood, and the limbs and skeleton evolved features that provided better support and more efficient movement on land.

FIGURE 23-8 The amniote egg
An anole lizard struggles free of its egg. The amniote egg encapsulates the developing embryo in a liquid-filled membrane (the amnion), ensuring that development occurs in a watery environment, even if the egg is far from water.

(a)

(b)

(c)

FIGURE 23-9 The diversity of birds
(a) The delicate hummingbird beats its wings about 60 times per second and weighs about 0.15 ounce (4 grams). **(b)** This young frigate bird, a fish-eater from the Galapagos Islands, has nearly outgrown its nest. **(c)** The ostrich is the largest of all birds, weighing more than 300 pounds (136 kilograms); its eggs weigh more than 3 pounds (1500 grams). **QUESTION** Although the ancestor of all birds could fly, many bird species—such as the ostrich—cannot. Why do you suppose flightlessness has evolved repeatedly among birds?

Birds Are Feathered Reptiles

The birds are a very distinctive group of "reptiles" (Fig. 23-9). Although birds have traditionally been classified as a group separate from reptiles (class Aves *versus* class Reptilia), modern systematists have shown that birds are really a subset of an evolutionary group that includes both birds and the groups that have been traditionally designated as reptiles (see page 353 in Chapter 18 for a more complete explanation). The first birds appear in the fossil record roughly 150 million years ago (Fig. 23-10) and are distinguished from other reptiles by the presence of feathers, which are essentially a highly specialized version of reptilian body scales. Modern birds retain scales on their legs—a testimony to the ancestry they share with the rest of the reptiles.

FIGURE 23-10 *Archaeopteryx*, the "missing link" between reptiles and birds
The earliest known bird is *Archaeopteryx,* preserved here in 150-million-year-old limestone. Feathers, a feature unique to birds, are clearly visible, but the reptilian ancestry of birds is also apparent: like a modern reptile (but unlike a modern bird), *Archaeopteryx* had teeth, a tail, and claws.

Bird anatomy and physiology are dominated by adaptations that help meet the rigorous demands of flight. In particular, birds are exceptionally light for their size. Hollow bones reduce the weight of the skeleton to a fraction of that of other vertebrates, and many bones present in other reptiles are lost or fused with other bones in birds. Reproductive organs are considerably reduced in size during nonbreeding periods, and female birds possess only a single ovary, further minimizing weight. The shelled egg that contributed to the reptiles' success on land also frees the mother bird from carrying her developing offspring internally. Feathers form lightweight extensions to the wings and the tail, providing the lift and control required for flight as well as lightweight protection and insulation for the body. The nervous system of birds accommodates the special demands of flight with extraordinary coordination and balance combined with acute eyesight.

Birds are also able to maintain body temperatures high enough to allow their muscles and metabolic processes to operate at peak efficiency, supplying the power necessary to fly regardless of the outside temperature. This physiological ability to maintain an internal temperature that is usually higher than that of the surrounding environment is characteristic of both birds and mammals, which are therefore sometimes described as *warm-blooded.* In contrast, the body temperature of invertebrates, fishes, amphibians, and nonbird reptiles varies with the temperature of their environment, though these animals may exert some control of their body temperature by behavioral means (such as basking in the sun or seeking shade).

Warm-blooded animals such as birds have a high metabolic rate, which increases their demand for energy and requires efficient oxygenation of tissues. Therefore, birds must eat frequently and possess circulatory and respiratory adaptations that help meet the need for efficiency. A bird's heart has four chambers, thus preventing the mixing of oxygenated and deoxygenated blood (alligators and crocodiles also have four-chambered hearts). The respiratory system of birds is supplemented by air sacs that provide a continuous supply of oxygenated air to the lungs, even while the bird exhales.

Mammals Provide Milk to Their Offspring

One branch of the reptile evolutionary tree gave rise to a group that evolved hair and diverged to form the *mammals*. The mammals first appeared approximately 250 million years ago but did not diversify and assume terrestrial prominence until after the dinosaurs went extinct roughly 65 million years ago. Like birds, mammals are warm-blooded and have high metabolic rates. In most mammals, fur protects and insulates the warm body. Like birds, alligators, and crocodiles, mammals have four-chambered hearts that increase the amount of oxygen delivered to the tissues. Legs designed for running rather than crawling make many mammals fast and agile. In contrast to birds, whose bodies are almost uniformly molded to the requirements of flight, mammals have evolved a remarkable diversity of form. The bat, mole, impala, whale, seal, monkey, and cheetah exemplify the radiation of mammals into nearly all habitats, with bodies finely adapted to their varied lifestyles (Fig. 23-11).

(a)

(b)

(c)

(d)

FIGURE 23-11 The diversity of mammals
(a) A humpback whale gives its offspring a boost. **(b)** A bat, the only type of mammal capable of true flight, navigates at night by using a kind of sonar. Large ears help the animal detect echoes as its high-pitched cries bounce off nearby objects. **(c)** Mammals are named after the mammary glands with which females nurse their young, as illustrated by this mother cheetah. **(d)** The male orangutan can reach 75 kilograms (165 pounds). These gentle, intelligent apes occupy swamp forests in limited areas of the Tropics, but are endangered by hunting and habitat destruction.

Mammals are named for the milk-producing **mammary glands** used by all female members of this class to suckle their young (see Fig. 23-11c). In addition to these unique glands, the mammalian body is arrayed with sweat, scent, and sebaceous (oil-producing) glands, none of which are found in reptiles. With the exception of the egg-laying **monotremes**, such as the platypus and spiny anteater (Fig. 23-12a), mammalian embryos develop in the *uterus*, a muscular organ in the female reproductive tract. In one group of mammals, the **marsupials** (including opossums, koalas, and kangaroos), the period of uterine development is short and the young are born at a very immature stage of development. Immediately after birth, they crawl to a nipple, firmly grasp it, and, nourished by milk, complete their development. In most, but not all, marsupial species, this postbirth development takes place in a protective pouch (Fig. 23-12b). Most mammal species are **placental** mammals and retain their young in the uterus for a much longer period. "Placental" refers to the **placenta**, the uterine structure that functions in gas, nutrient, and waste exchange between the circulatory systems of the mother and embryo.

The mammalian nervous system has contributed significantly to the success of the mammals by making possible behavioral adaptation to changing and varied environments. The brain is more highly developed than that in any other class, endowing mammals with unparalleled curiosity and learning ability. The highly developed brain allows mammals to alter their behavior on the basis of experience and helps them survive in a changing environment. Relatively long periods of parental care after birth allow some mammals to learn extensively under parental guidance; humans and other primates are good examples of this behavior. In fact, the large brains of humans have been the major factor leading to human domination of Earth, as explored in the following section.

EVOLUTIONARY CONNECTIONS

Are Humans a Biological Success?

Physically, human beings are fairly unimpressive biological specimens. For such large animals, we are not very strong or very fast, and we lack natural weapons such as fangs or claws. It is the human brain, with its tremendously developed cerebral cortex, that truly sets us apart from other animals. Our brains give rise to our minds, which, in bursts of solitary brilliance and in the collective pursuit of common goals, have created wonders. No other animal could sculpt the graceful columns of the Parthenon, much less reflect on the beauty of this ancient Greek temple. We alone can eradicate smallpox

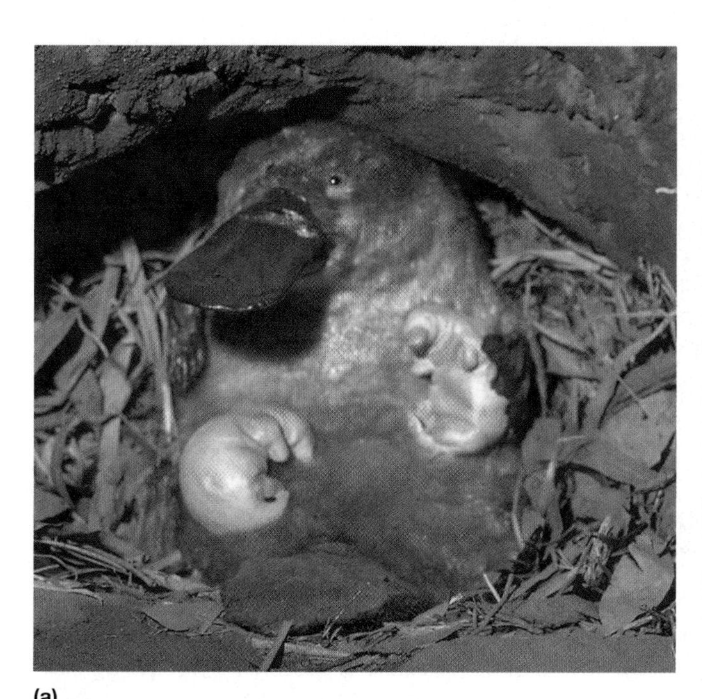

(a)

(b)

FIGURE 23-12 Nonplacental mammals
(a) Monotremes, such as this platypus from Australia, lay leathery eggs resembling those of reptiles. The newly hatched young obtain milk from slitlike openings in the mother's abdomen. **(b)** Marsupials, such as the wallaby, give birth to extremely immature young who immediately grasp a nipple and develop within the mother's protective pouch (inset).

and polio, domesticate other life-forms, penetrate space with rockets, and fly to the stars in our imaginations.

And yet, are we, as it appears at first glance, the most successful of all living things? The duration of human existence is a mere instant in the 3.5-billion-year span of life on Earth. But during the last 300 years, the human population has increased from 0.5 billion to 6 billion and now grows by a million people every 4 days. Is this a measure of our success? As we have expanded our range over the globe, we have driven at least 300 other species to extinction. Within your lifetime, the rapid destruction of tropical rain forests and other diverse habitats may wipe out millions of species of plants, invertebrates, and vertebrates, most of which we will never know. Many of our activities have altered the environment in ways that are hostile to life, including our own. Acid from power plants and automobiles rains down on the land, threatening our forests and lakes—and eroding the marble of the Parthenon. Deserts spread as land is stripped by overgrazing and the demand for firewood. Our aggressive tendencies, spurred by pressures of expanding wants and needs, their scope magnified by our technological prowess, have given us the capacity to destroy ourselves and most other life-forms as well.

The human mind is the source of our most pressing problems—and our greatest hope for solving them. Will we now devote our mental powers to reducing our impact, controlling our numbers, and preserving the ecosystems that sustain us and other forms of life? Are we a phenomenal biological success—or a brilliant catastrophe? Perhaps the next few centuries will tell.

CASE STUDY REVISITED Fish Story

After Marjorie Courtney-Latimer's discovery of the coelacanth, J. L. B. Smith dedicated himself to searching for more coelacanth specimens in the waters off South Africa. He didn't find one until 1952, when fishermen from the Comoro Islands, having seen leaflets that offered a reward for a coelacanth, contacted Smith with the news that they had one in their possession. Smith immediately booked a flight to the Comoros, and reportedly wept for joy upon holding the 88-pound coelacanth awaiting him.

In the years since Smith's trip, about 200 additional coelacanths have been caught by fishermen, mostly in waters around the Comoros but also around nearby Madagascar and off the coasts of Mozambique and South Africa. Scientists thought that the fish's range was restricted to this relatively small area in the western Indian Ocean, and it was therefore something of a shock when a few specimens were discovered in Indonesia, more than 6000 miles away. DNA tests showed that these Indonesian coelacanths were members of a second species.

Although the coelacanth specimens have revealed a great deal about the creatures' anatomy, their habits and behavior remain comparatively mysterious. Observations from research submarines suggest that coelacanths spend much of their time in caves and beneath rocky overhangs at depths of 300 to 1200 feet. Radio tracking suggests that they may venture into open water at night, presumably to forage for food. Almost all individuals observed (or caught) have been at least 3 feet long, which suggests that young fish must travel to sites away from the main adult populations to mature, though no such location has been discovered.

The known populations of coelacanths are small, consisting of a few hundred individuals, and appear to be declining. Part of this decline is due to fishing, though coelacanths are mostly caught accidentally by fishermen searching for more commercially desirable species. Conservation efforts in South Africa and the Comoros thus focus largely on introducing fishing methods that will reduce the chances of accidentally snaring a coelacanth.

Consider This: Many accounts of coelacanths refer to them as "living fossils," a term that is also applied to alligators, gingko trees, horseshoe crabs, and other species whose modern appearance matches that of ancient fossils. The implication of the living fossil designation is that these organisms have evolved very little over a very long period. Do you think this implication is accurate? Is it accurate to say that "living fossils" have evolved more slowly or undergone less evolutionary change than other species?

Links to Life: Do Animals Belong in Laboratories?

Vertebrate animals are the subjects of much laboratory research, in part because biologists, like most people, tend to be more interested in vertebrates than in other types of organisms. Much of the use of vertebrates in research, however, stems from their similarity to humans. Researchers often hope to answer questions about human biology by using information gained from experiments on rats, mice, dogs, monkeys, and other vertebrates. Many such experiments would be considered unethical if performed on humans. For example, it is impermissible to intentionally expose humans to disease-causing organisms, to inject humans with untested drugs, to perform experimental surgery on a healthy human, or to intentionally kill a human for research purposes. Nonetheless, such manipulations are routinely performed on laboratory animals.

Some observers and activists argue that nonhuman animals are entitled to protection against suffering inflicted by experimental research. In this view, humans have no right to subject members of other species to treatment that would be unethical if applied to humans, and the pursuit of experiments that cause animals to suffer cannot be justified. Many scientists, however, object strenuously to the assertion that research on animals is unethical, arguing that the advance of scientific knowledge, including that which leads to lifesaving treatments for humans, requires research on vertebrate animals.

Where do you stand? Is experimental research on animals always unethical? Or is it acceptable in some types of research, but unacceptable in others? Or are you satisfied with the current system, in which scientists are largely free to use animals in research, but the animals are protected by regulations intended to limit their suffering?

Summary of Key Concepts

23.1 What Are the Key Features of Chordates?

The phylum Chordata includes two invertebrate groups, the lancelets and tunicates, as well as the familiar vertebrates. All chordates possess a notochord, a dorsal hollow nerve cord, pharyngeal gill slits, and a post-anal tail at some stage in their development. The vertebrates are a subphylum of chordates and have a backbone, which is part of their living endoskeleton.

23.2 What Are the Major Groups of Vertebrates?

Hagfishes are jawless, eel-shaped chordates that lack a true backbone and, therefore, are not true vertebrates. Lampreys are jawless vertebrates; the best known lamprey species are parasites of fish.

All amphibians have legs, and most have simple lungs for breathing in air rather than in water. Most are confined to relatively damp terrestrial habitats by their need to keep their skin moist, their use of external fertilization, and their eggs and larvae, which develop in water.

Reptiles, with well-developed lungs, dry skin covered with relatively waterproof scales, internal fertilization, and an amniote egg with its own water supply, are well adapted to the driest terrestrial habitats.

Birds are also fully terrestrial and have additional adaptations that allow the muscles to respond rapidly regardless of the temperature of the environment, such as an elevated body temperature. The bird body is molded for flight, with feathers, hollow bones, efficient circulatory and respiratory systems, and well-developed eyes.

Mammals have insulating hair and give birth to live young that are nourished with milk. The mammalian nervous system is the most complex in the animal kingdom, providing mammals with enhanced learning ability that helps them adapt to changing environments.

Key Terms

amnion *p. 456*
amniote egg *p. 456*
cartilage *p. 450*
endoskeleton *p. 451*
mammary gland *p. 459*
marsupial *p. 459*
monotreme *p. 459*
nerve chord *p. 450*
notochord *p. 450*
pharyngeal gill slit *p. 450*
placenta *p. 459*
placental *p. 459*
post-anal tail *p. 450*
vertebral column *p. 450*
vertebrate *p. 450*

Thinking Through the Concepts

To take a multiple-choice quiz with feedback on the contents of this chapter, visit http://www.prenhall.com/audesirk7. *Log in to the Web site selected by your instructor and navigate to the Self Test section for this chapter.*

? Review Questions

1. Briefly describe each of the following adaptations, and explain the adaptive significance of each: vertebral column, jaws, limbs, amniote egg, feathers, placenta.
2. List the vertebrate groups that have each of the following:
   a. a skeleton of cartilage
   b. a two-chambered heart
   c. an amniote egg
   d. warm-bloodedness
   e. a four-chambered heart
   f. a placenta
   g. lungs supplemented by air sacs
3. List four distinguishing features of chordates.
4. Describe the ways in which amphibians are adapted to life on land. In what ways are amphibians still restricted to a watery or moist environment?
5. List the adaptations that distinguish reptiles from amphibians and help reptiles adapt to life in dry terrestrial environments.
6. List the adaptations of birds that contribute to their ability to fly.
7. How do mammals differ from birds, and what adaptations do they share?
8. How has the mammalian nervous system contributed to the success of mammals?

Applying the Concepts

1. Are hagfishes vertebrates or invertebrates? On which characteristics did you base your answer? Is it important to be able to place them in one category or the other? Why?
2. Is the decline of amphibian populations of concern to humans? What about the increase in frog deformities? Why is it important to understand the causes of these phenomena?
3. Discuss and defend the attributes you would use to define biological success among animals. Are humans a biological success by these standards? Why?

For More Information

Blaustein, A. R. "Amphibians in a Bad Light." *Natural History*, October 1994. Recent declines in amphibian population size and overall diversity are linked to possible harm from ultraviolet light that is penetrating a depleted ozone layer.

Blaustein, A., and Johnson, P. T. J. "Explaining Frog Deformities." *Scientific American*, February 2003. Dramatic increases in the occurrence of deformed frogs are caused by a parasite epidemic exacerbated by environmental degradation.

Diamond, J. "Stinking Birds and Burning Books." *Natural History*, October 1994. A description of a recently described species of bird (the pitohuis) and its peculiar chemical ecology.

Duellman, W. E. "Reproductive Strategies of Frogs." *Scientific American*, July 1992. Free-living tadpoles are only one way in which these amphibians progress from egg to adult.

Montgomery, S. "New Terror of the Deep." *International Wildlife*, July–August 1992. A description of the threat that overfishing by humans poses to shark populations.

Perkins, S. "The Latest Pisces of an Evolutionary Puzzle." *Science News*, May 5, 2001. A summary of recent research on coelacanths in their natural habitat.

Rahn, H., Ar, A., and Paganelli, C. V. "How Bird Eggs Breathe." *Scientific American*, February 1979. (Offprint No. 1420). A description of the remarkable adaptations of the amniote egg for gas exchange.

Media Activities

To access a Media Activity visit http://www.prenhall.com/audesirk7. *Log in to the Web site selected by your instructor, navigate to this chapter, and select the appropriate Media Activity number.*

23.1 Web Investigation: Fish Story

Estimated time: 10 minutes

Explore why scientists are interested in examining coelacanths.

UNIT FOUR

Plant Anatomy and Physiology

A prairie blanketed with Texas bluebonnets in the spring. Flowers such as these are actually elaborately modified leaves adapted to attract pollinators.

CHAPTER

24 Plant Anatomy and Nutrient Transport

Each fall color has a purpose, but scientists are still investigating the function of red anthocyanin pigments, which are synthesized just before the leaf drops.

CASE STUDY Why Do Leaves Turn Red in the Fall?

Every autumn, people flock to the deciduous forests of the northern United States—particularly New England—to enjoy the brilliant red, yellow, and orange colors of the leaves and the crisp, sunny weather that brings on this display. But these colors did not evolve to attract tourists! The yellow and orange pigments (carotenoids) are there all year, helping leaves trap sunlight for photosynthesis. As the leaves die and the predominant green chlorophyll pigment breaks down, these pigments are revealed. Of all the fall colors, the striking reds are the most mysterious. These are caused by anthocyanin pigments, which are synthesized just before the leaves drop in the fall and are not involved in photosynthesis.

After research spanning 200 years, scientists are finally beginning to obtain hard evidence to explain why leaves synthesize a new red pigment just as they are dying. For the past decade, plant physiologists David Lee and Kevin Gould have sought answers to this question in Harvard Forest, a Massachusetts nature sanctuary with unrivaled fall colors. Their research follows up on the observations and hypotheses of German botanists of the late 1800s, who noticed that a combination of low temperatures and intense light seems to stimulate anthocyanin production. Perhaps, as these early investigators suggested, the red pigments help warm the leaf and protect it from the damaging effects of too much sun. But why protect a dying leaf?

Media Activity
24.1 Plant Anatomy

24.1 How Are Plant Bodies Organized, and How Do They Grow?

Flowering Plants Consist of a Root System and a Shoot System

Flowering plants, or *angiosperms*, consist of two major regions, the *root system* and the *shoot system* (Fig. 24-1). The **root system** consists of all the roots of a plant. **Roots** are branched portions of the plant body that are usually embedded in soil. Plant roots serve six major functions: they anchor the plant in the ground; absorb water and minerals from the soil; store surplus sugars manufactured during photosynthesis; transport water, minerals, sugars, and hormones to and from the shoot; produce some hormones; and interact with soil fungi and microorganisms that provide nutrients to the plant.

The rest of the plant is the **shoot system**, which is usually located above ground. The shoot system consists of *leaves*, *buds*, and (in season) *flowers* and *fruits*; these are all borne on **stems**, which are typically branched. The functions of shoots include photosynthesis, mainly in leaves and young green stems; transport of materials among leaves, flowers, fruits, and roots; reproduction; and hormone synthesis.

Figure 24-2 illustrates the two broad groups of flowering plants, or angiosperms: the **monocots**, which include grasses, lilies, palms, and orchids; and the **dicots**, which include deciduous trees (those that drop their leaves in winter), bushes, and many garden flowers. These two broad groups are named after the first leaflike structures an angiosperm embryo produces: *cotyledons*, or *seed leaves*. Monocots and dicots differ in the structure of their flowers, leaves, vascular tissue, root pattern, and seeds. Don't worry about terms that are not yet familiar to you; for now, look over Figure 24-2 and refer to it as we examine the parts of flowering plants in more detail.

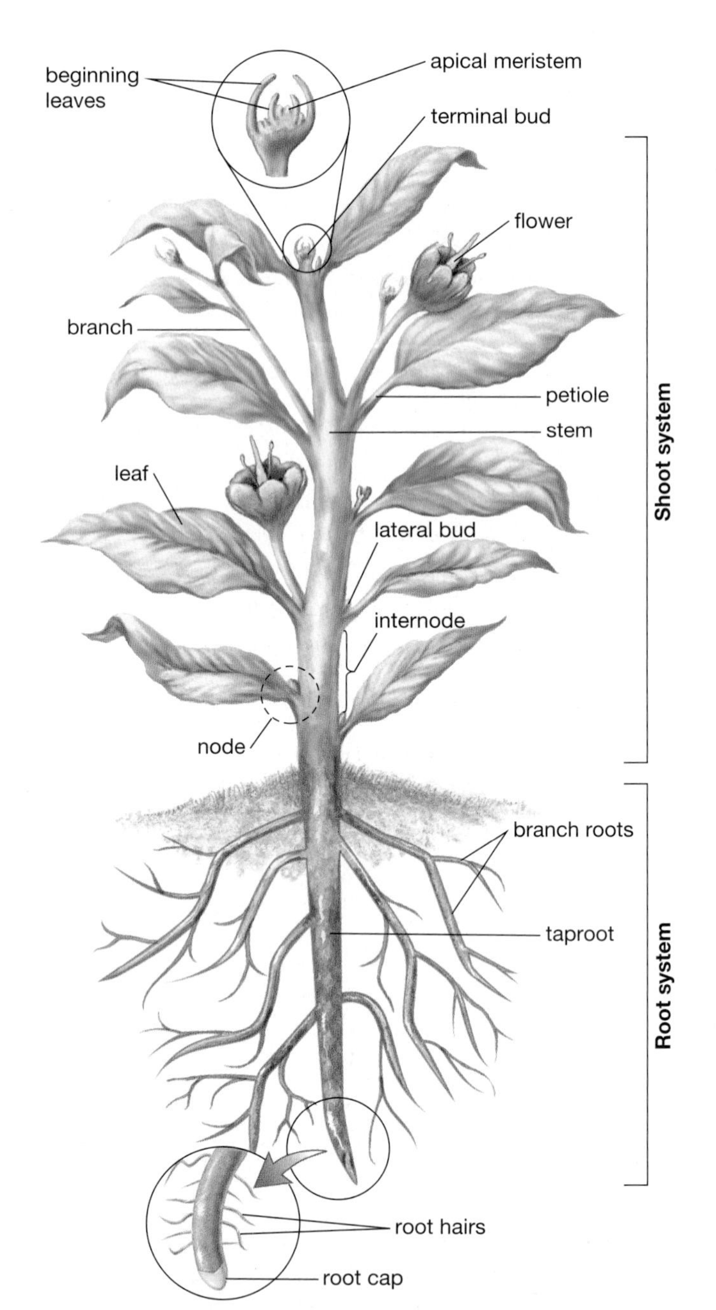

FIGURE 24-1 Flowering plant structure
A flowering plant consists of root and shoot systems. The highly branched root system includes microscopic root hairs that increase the surface area for nutrient absorption. The shoot system includes stems (from which branches grow), with buds and leaves. In the appropriate season, the shoot may bear flowers and fruit. Our model plant is a dicot.

During Plant Growth, Meristem Cells Give Rise to Differentiated Cells

Animals and plants develop in dramatically different ways. One difference is the timing and distribution of growth. As you grew from a baby to an adult, all parts of your body became larger. When you reached your adult height, you stopped growing (upward, at least!). In contrast, flowering plants grow throughout their lives, never reaching a stable "adult" body form. Moreover, most plants grow longer only at the tips of their branches and roots, and structures that developed earlier remain in exactly the same place; a swing tied to a tree branch does not move farther from the ground each year. Why do plants grow this way?

From the moment they sprout, plants are composites of two fundamentally different categories of cells: meristem cells and differentiated cells. Embryonic, undifferentiated **meristem cells** are capable of mitotic cell division. Some of the offspring, or daughter cells, of the dividing meristem cells lose the ability to divide and be-

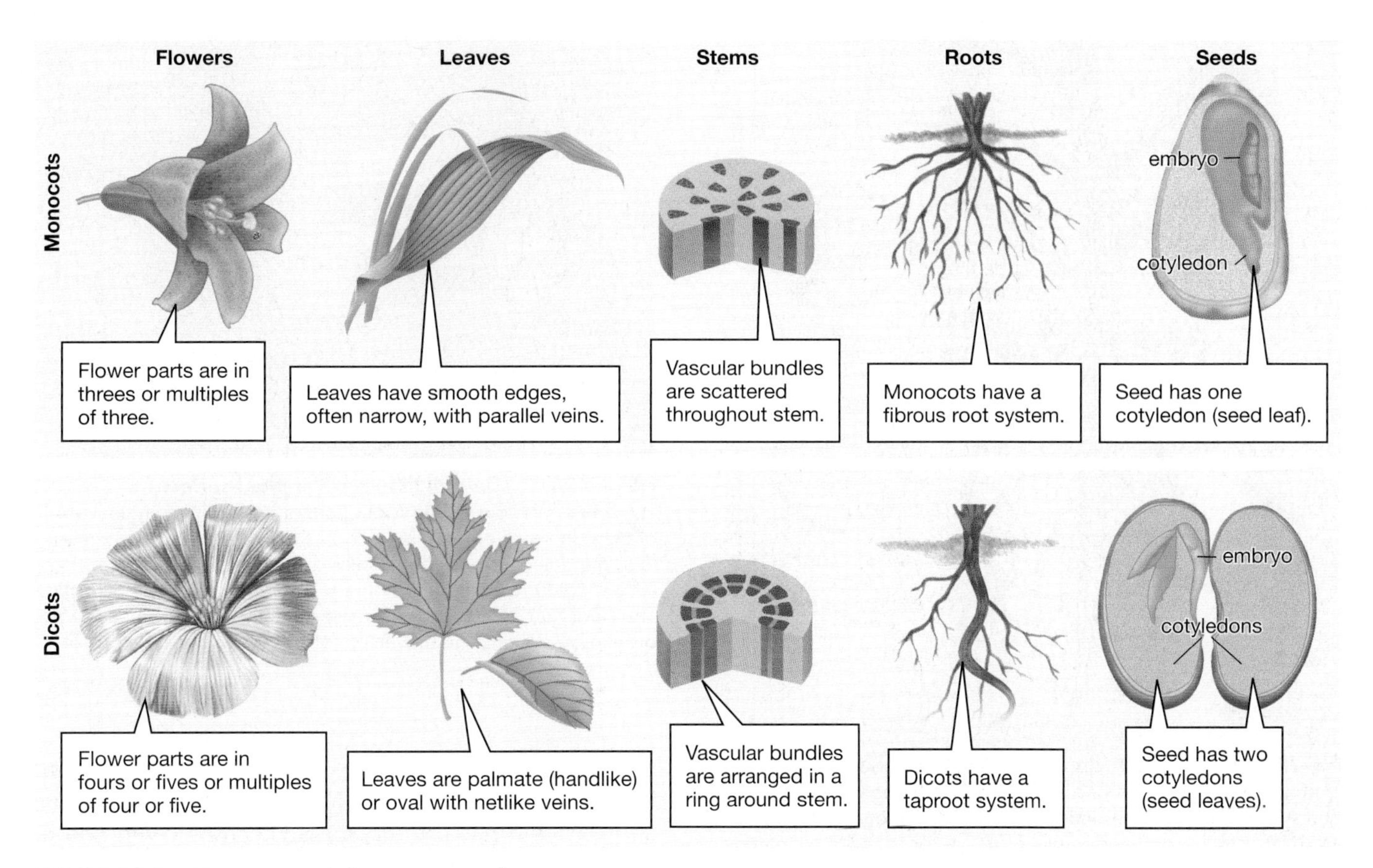

FIGURE 24-2 Monocots and dicots compared

come part of the nongrowing portions of the plant body. These **differentiated cells** have specialized structures and functions. Continued divisions of meristem cells, then, keep a plant growing throughout its life, whereas their differentiated daughter cells form more stable or permanent parts of the plant, such as mature leaves or the trunks of trees.

Plants grow through the division and differentiation of two major types of meristem cells: apical meristems and lateral meristems. **Apical meristems** ("tip meristems") are located at the ends of roots and shoots, including main stems and branches (see Figs. 24-1, 24-15). **Lateral meristems** ("side meristems"), also called *cambia* (singular, **cambium**), form cylinders that run parallel to the long axis of roots and stems (see Fig. 24-11).

Plant growth takes two forms: *primary growth* and *secondary growth*. **Primary growth** occurs by the mitotic cell division of apical meristem cells, followed by differentiation of the resulting daughter cells. This type of growth takes place in the growing tips of roots and shoots of plants. Primary growth is responsible both for increase in length and for the development of the specialized plant structures. The elongation of roots and shoots through primary growth allows them to enter new space from which to collect light, nutrients, and water. Primary growth also explains why your swing never gets any higher off the ground.

Secondary growth, which is responsible for increases in diameter, occurs by the division of lateral meristem cells and differentiation of their daughter cells. Secondary growth causes the stems and roots of most conifers (cone-bearing trees, or evergreens) and dicots to become thicker and woodier as they age. Although we will discuss secondary growth only in stems, keep in mind that secondary growth also occurs in roots.

24.2 What Are the Tissues and Cell Types of Plants?

The major structures of land plants, including roots, stems, and leaves, consist of three *tissue systems*—dermal, ground, and vascular—each containing more than one type of tissue. Each tissue, in turn, is composed of

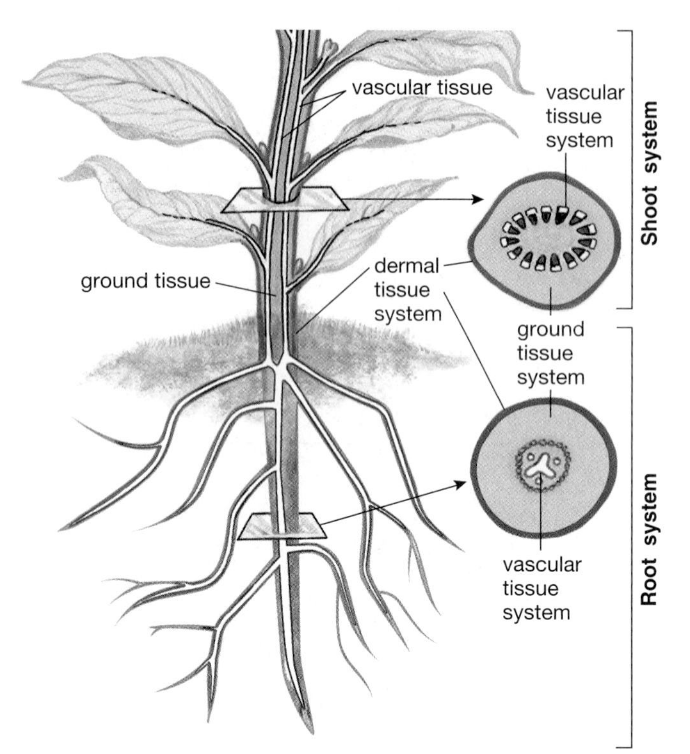

FIGURE 24-3 The structure of the root and shoot
Both the root and shoot of a flowering plant consist of three tissue systems: dermal, ground, and vascular tissue systems.
EXERCISE On this drawing, circle all of the locations at which primary growth occurs, and draw arrows pointing to some of the locations at which secondary growth occurs.

one or more specialized cell types (Fig. 24-3). The **dermal tissue system** covers the outer surfaces of the plant body. The **ground tissue system**, which consists of all nondermal and nonvascular tissues, makes up most of the body of young plants. Its functions include photosynthesis, support, and storage. The **vascular tissue system** transports water, minerals, sugars, and plant hormones throughout the plant.

Some flowering plants, described as *herbaceous*, are soft-bodied with flexible stems; herbaceous plants include lettuce, beans, and grasses. Such plants usually live only one year. Other plants, such as trees and bushes, are described as *woody*; most are perennial (living many years) and develop hard, thickened, woody stems as a result of secondary growth. As you will see, different types of tissue are present in herbaceous and woody plants.

The Dermal Tissue System Forms the Covering of the Plant Body

The dermal tissue system covers the outside of the plant body. There are two types of dermal tissue: *epidermal tissue* and *periderm*.

Epidermal tissue forms the **epidermis**, the outermost cell layer that covers the leaves, stems, and roots of all young plants (Fig. 24-4). Epidermal tissue also covers flowers, seeds, and fruit. In herbaceous plants, the epidermis forms the outer covering of the entire plant body throughout its life. The epidermal tissue of the aboveground parts of a plant is generally composed of tightly packed, thin-walled cells, covered with a waterproof, waxy **cuticle**. Secreted by the epidermal cells, the cuticle reduces the evaporation of water from the plant and helps protect it from the invasion of disease microorganisms. In contrast, the epidermal cells of roots are not covered with cuticle; a waterproof cuticle would prevent the absorption of water and minerals.

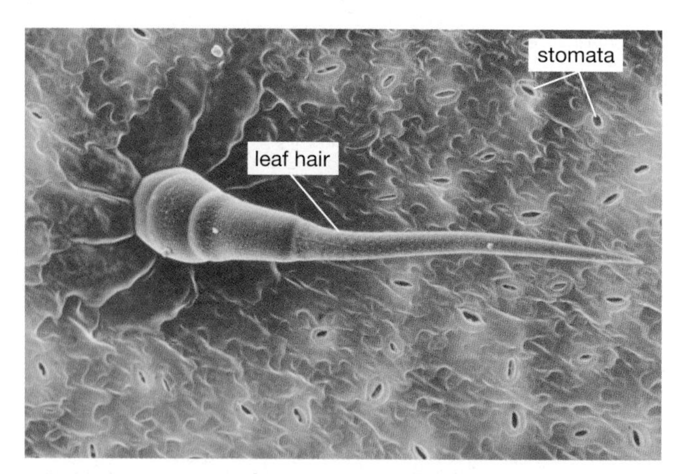

FIGURE 24-4 Dermal tissues cover plant surfaces
The epidermis of a young root or shoot is usually a single layer of cells. In shoot epidermis, such as the epidermis of this zinnia leaf, the outer surfaces of the cells are covered with cuticle, a waxy, waterproof coating that reduces the evaporation of water. The "leaf hair" protruding from the epidermis also reduces evaporation by slowing the movement of air across the leaf surface.

Some epidermal cells produce fine extensions called *hairs*. Many root epidermal cells bear **root hairs**, long projections that greatly increase the absorptive surface area of the root. Epidermal hairs on the stems and leaves of desert plants reduce evaporative water loss by reflecting sunlight and producing an unstirred layer of air near the plant's surface. In contrast, some tropical plants use their hairy leaves to capture and hold water.

Periderm replaces epidermal tissue on the roots and stems of woody plants as they age. This dermal tissue is composed primarily of **cork cells**, which have thick, waterproof walls and are dead at maturity. Periderm also includes the *cork cambium* that gives rise to cork cells (see Fig. 24-12). Cork cells form the protective outer layer of the bark of trees (see Fig. 24-13) and woody shrubs as well as the tough covering of their roots.

The Ground Tissue System Makes Up Most of the Young Plant Body

The ground tissue system, which makes up the bulk of a young plant, consists of all nondermal and nonvascular tissues. There are three types of ground tissue: *parenchyma*, *collenchyma*, and *sclerenchyma*.

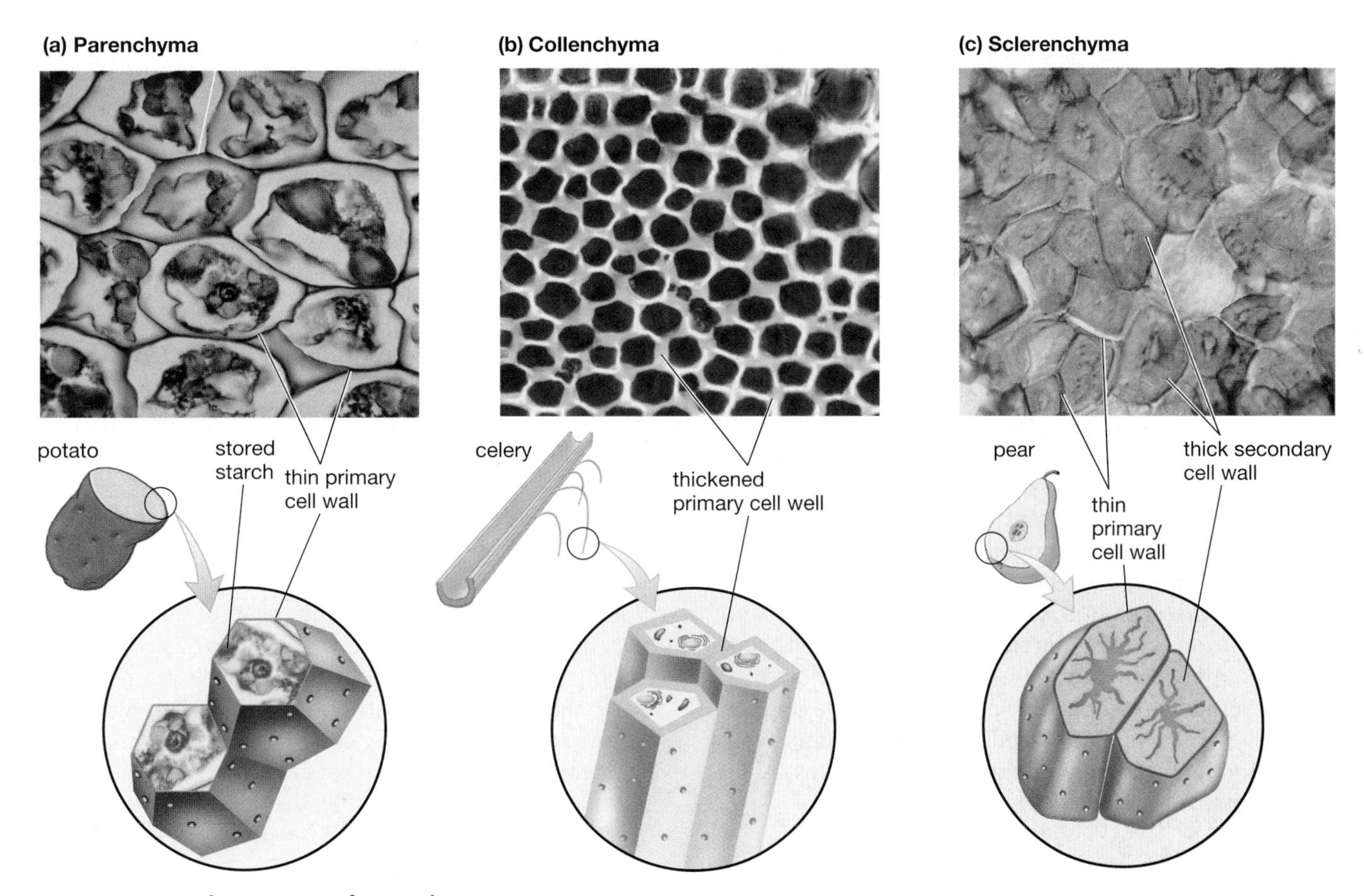

FIGURE 24-5 The structure of ground tissue
(a) Parenchyma cells are living and serve many functions. They have thin, flexible primary cell walls. These parenchyma cells are used for starch storage in a potato. **(b)** Collenchyma cells are living and have thickened, but somewhat flexible, primary walls. They help support the plant body (as seen in this celery stalk). **(c)** Sclerenchyma cells (such as these "stone cells" that give the meat of a pear its slightly gritty texture) have thick, rigid secondary cell walls, and die after they differentiate. Sclerenchyma cells also make up the shells of nuts and form fibers that support vascular tissue.

Parenchyma tissue is the most abundant of the ground tissues. Parenchyma cells are thin-walled and alive at maturity, and they typically carry out most of the metabolic activities of the plant (Fig. 24-5a). Depending on their location within the plant body, parenchyma cells have such diverse functions as photosynthesis, storage of sugars and starches, and secretion of hormones. Under the proper conditions, many parenchyma cells are capable of mitotic cell division. Roots that are adapted for storage, including carrots and sweet potatoes, are packed with parenchyma cells that store carbohydrates, such as starch and sugar. Starch-filled parenchyma cells pack white potatoes as well.

Collenchyma tissue consists of elongated, polygonal (many-sided) cells with unevenly thickened cell walls (Fig. 24-5b). Collenchyma cells are alive at maturity but generally cannot divide. Although strong, the cell walls of collenchyma are still somewhat flexible. In herbaceous plants and in the leaf stalks and young growing stems of all plants, collenchyma tissue is an important source of support. The strings in celery stalks, for example, are mostly collenchyma cells in association with vascular tissue.

Sclerenchyma tissue consists of cells with thick, hardened secondary cell walls (located between the outer primary wall and the plasma membrane) reinforced with the stiffening substance *lignin* (Fig. 24-5c). Like collenchyma, sclerenchyma cells support and strengthen the plant body; however, unlike collenchyma, they die after they differentiate. Their hardened cell walls then remain as a source of support. Sclerenchyma tissue can be found in many parts of the plant body, including xylem and phloem (described next). Sclerenchyma cells provide the fibers of hemp and jute, which are used for making rope. Other types of sclerenchyma cells form nut shells, the outer covering of peach pits, and the gritty texture of pears.

The Vascular Tissue System Consists of Xylem and Phloem

The vascular tissue system consists of two complex conducting tissues: *xylem* and *phloem*. The major role of each tissue is the transport of materials. Xylem transports water and minerals up from the roots to the rest of the plant; phloem conveys water, sugars, amino acids, and hormones throughout the plant.

Xylem Conducts Water and Dissolved Minerals from the Roots to the Rest of the Plant

Xylem conducts water and minerals from roots to shoots in tubes that are made from *tracheids* and *vessel elements* (Fig. 24-6). **Tracheids** are thin, tubelike cells with thick cell walls; their slanted ends resemble the tips of hypodermic needles. Tracheids are stacked atop one another with the slanted ends overlapping. The overlapping end walls contain **pits**, porous sections where secondary cell walls are absent. These pits allow water and minerals to pass from one tracheid to the next, or from a tracheid to an adjacent vessel element, by crossing only the thin, water-permeable primary cell wall.

Vessel elements also meet end to end, but they are larger in diameter than tracheids. Their ends may be either flat or overlapping and tapered. Perforations connect adjoining elements. In some cases, the ends of vessel elements completely disintegrate, leaving an open tube (see Fig. 24-6). Thus, vessel elements form wide-diameter, relatively unobstructed pipelines called **vessels** from root to leaf.

Most conifers have only tracheids; flowering plants usually have both tracheids and vessel elements. As these cells differentiate, they develop thick cell walls that help support the weight of the plant. In some trees (pines, for example), the bulk of the tree trunk consists of the thick cell walls of tracheids. The final step in the differentiation of both tracheids and vessel elements is death, when the cytoplasm and plasma membrane disintegrate, leaving behind a hollow tube of cell wall.

Phloem Conducts Water, Sugars, Amino Acids, and Hormones Throughout the Plant

Phloem carries water containing dissolved substances synthesized by the plant, including sugars, amino acids, and hormones. Among the supportive sclerenchyma cells of phloem are **sieve tubes**, constructed of a single strand of cells called **sieve-tube elements** (Fig. 24-7). As sieve-tube elements mature, most of their internal contents disintegrate, leaving behind only a thin layer of cytoplasm that lines the plasma membrane. At the ends of sieve-tube elements, where adjacent cells meet, holes form in the primary cell walls, creating structures called **sieve plates**. The interiors of adjacent sieve-tube elements are connected through the openings in the sieve plate. A continuous conducting system is forged by the linking of many sieve-tube elements end to end in this way.

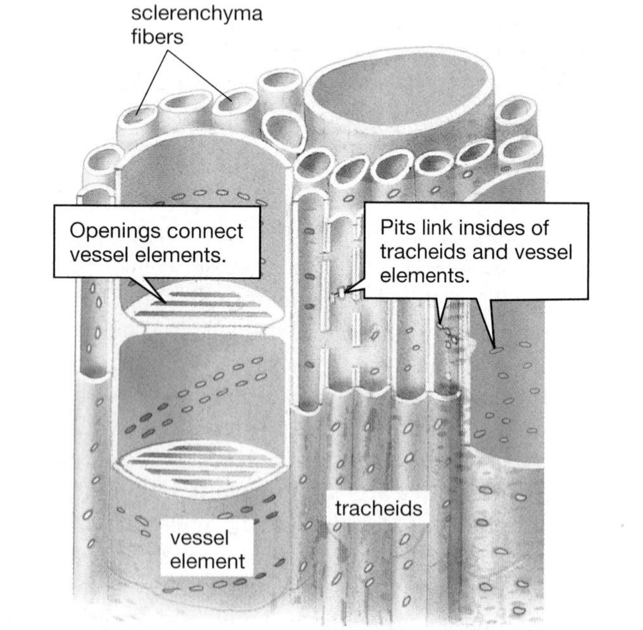

FIGURE 24-6 The structure of xylem
Xylem includes sclerenchyma fibers for support and two types of conducting cells: tracheids and vessel elements. Tracheids are thin with ends and sides connected by pits. The pits consist of a water-permeable primary cell wall that separates the interiors of adjoining cells. Vessel elements stacked atop one another form vessels. The end walls may be absent or have narrow openings. Pits in side walls interconnect tracheids and vessel elements, allowing water and dissolved minerals to flow between them.

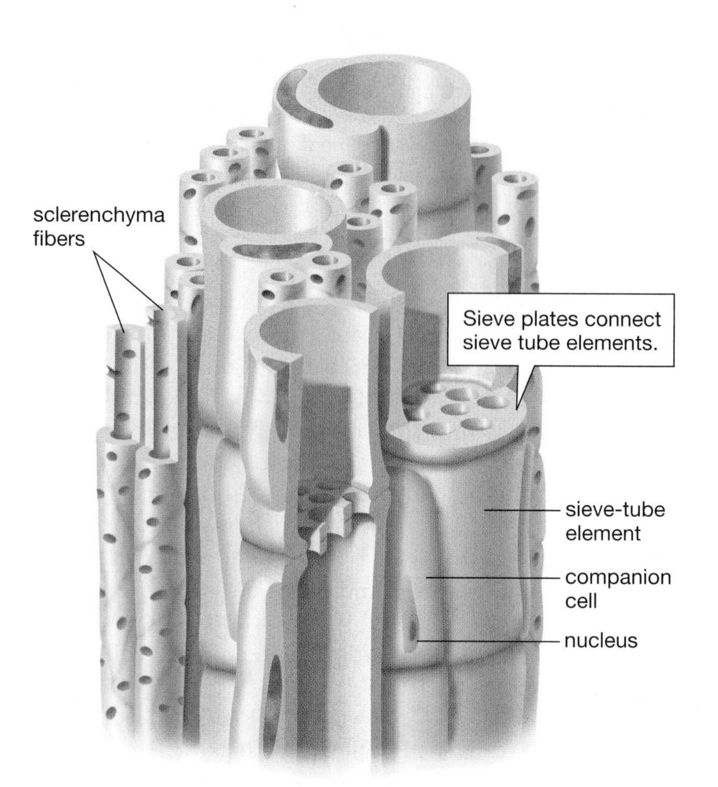

FIGURE 24-7 The structure of phloem
Phloem includes sclerenchyma fibers for support, sieve-tube elements, and companion cells. Sieve-tube elements, stacked end to end, form the conducting system of phloem. Where they join, membrane-lined pores allow fluid to pass. Each sieve-tube element has a companion cell that nourishes it and regulates its function.

A sieve-tube element has a plasma membrane, a few small mitochondria, and some endoplasmic reticulum. It is therefore considered to be alive, but it generally lacks ribosomes, Golgi complexes, and a nucleus. How, then, can a sieve-tube element remain alive? Each sieve-tube element is nourished by a smaller, adjacent **companion cell**. Companion cells are connected to sieve-tube elements by cytoplasm-filled channels called *plasmodesmata*, described in Chapter 4. The companion cells maintain the integrity of the sieve-tube elements by donating high-energy compounds and perhaps even by repairing the sieve-tube plasma membrane. As we will see later, companion cells also regulate the movement of sugars into and out of the sieve tubes.

24.3 Leaves: Nature's Solar Collectors

Only structures with chlorophyll are able to carry out photosynthesis and make valuable sugar out of widely available ingredients: sunlight, carbon dioxide, and water. **Leaves** are the major photosynthetic structures of most plants. Water is obtained from the soil and transported to the leaf through the xylem, and carbon dioxide (CO_2) must diffuse into the leaf from the air. An ideal leaf has a large surface area for gathering light and is porous to permit CO_2 to enter from the air to allow photosynthesis. However, a large, porous leaf would lose large amounts of water by evaporation, so the leaf must be reasonably waterproof as well. The leaves of flowering plants represent an elegant compromise among these conflicting demands (Fig. 24-8).

Leaves Have Two Major Parts: Blades and Petioles

A typical angiosperm leaf consists of a broad, flat portion, the **blade**, connected to the stem by a stalk called the **petiole** (see Fig. 24-1). The petiole positions the blade in space, usually orienting the leaf for maximum exposure to the sun. Inside the petiole are vascular tissues of xylem and phloem that are continuous with those in the stem, root, and blade. Within the blade, the vascular tissues branch into **vascular bundles**, or **veins**.

The leaf epidermis consists of a layer of nonphotosynthetic, transparent cells that secrete a waxy, waterproof cuticle on their outer surfaces that reduces evaporation. The epidermis and its cuticle are pierced by adjustable pores, the **stomata** (singular, *stoma*), which regulate the diffusion of CO_2 and water into and out of the leaf. Each stoma is surrounded by two sausage-shaped **guard cells**, which regulate the size of the opening into the interior of the leaf (see Fig. 24-8 and Fig. 24-23). Unlike the surrounding epidermal cells, guard cells contain chloroplasts and can carry out photosynthesis. As we will see, photosynthesis in the guard cells contributes to their ability to adjust the size of the pore.

Beneath the epidermis lies the **mesophyll** ("middle of the leaf"), which consists of loosely packed parenchyma cells. In many leaves, mesophyll cells are of two types: a layer of columnar *palisade cells* just beneath the upper epidermis, and a layer of irregularly shaped *spongy cells* above the lower epidermis. Both palisade and spongy cells contain chloroplasts; these cells perform most of the photosynthesis of the leaf. The openness of the leaf interior (see Fig. 24-8) allows CO_2 to diffuse easily to all the mesophyll cells. Vascular bundles, each containing both xylem and phloem, are embedded within the mesophyll,

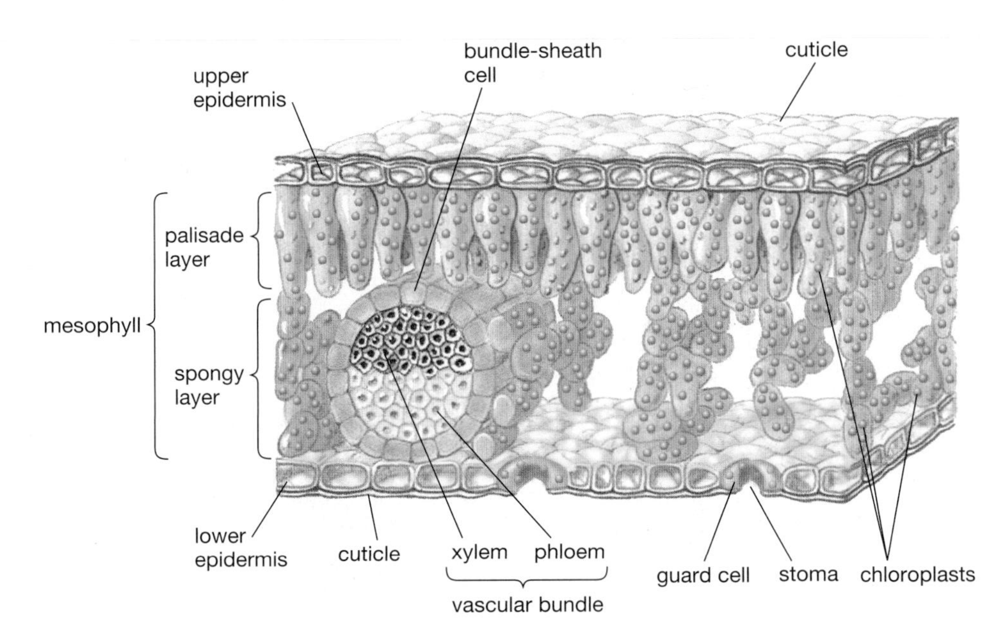

FIGURE 24-8 The structure of a typical dicot leaf
The cells of the epidermis lack chloroplasts and are transparent, allowing sunlight to penetrate to the chloroplast-containing mesophyll cells beneath. The stomata that pierce the epidermis and the loose, open arrangement of the mesophyll cells ensure that CO_2 can diffuse into the leaf from the air and reach all the photosynthetic cells.

with fine veins reaching very close to each photosynthetic cell. Thus, each mesophyll cell receives energy from sunlight transmitted through the clear epidermis; carbon dioxide from the air, diffusing through the stomata; and water from the xylem. The sugars it produces are carried to the rest of the plant by the phloem.

24.4 Stems: Reaching for the Light

The Stem Includes Four Types of Tissue

Like roots, stems develop from a small group of actively dividing cells, the apical meristem, which lies at the tip of the young shoot. The apical meristem is located within the **terminal bud**. The terminal bud consists of meristem tissue surrounded by developing leaves called **leaf primordia** (singular, *primordium*) at the tip of the shoot. The daughter cells of the apical meristem differentiate into the specialized cell types of stem, buds, leaves, and flowers.

As the shoot grows, small clusters of meristem cells are "left behind" at the surface of the stem. These meristem cells form the leaf primordia, which develop into the mature leaves unique to each species of plant. The meristem cells also produce **lateral buds**, which, under appropriate conditions, grow into branches. (We will discuss the growth of branches shortly.) Leaf primordia and lateral buds appear at characteristic locations, called **nodes**, on the stem; regions of stem between these nodes are called **internodes** (Fig. 24-9).

Most young stems are composed of four tissues: epidermis (dermal tissue), cortex (ground tissue), pith (ground tissue), and vascular tissues. As Figure 24-2 illustrates, monocots and dicots differ somewhat in the arrangement of vascular tissues. We will discuss only dicot stems here.

The Epidermis of the Stem Is Specialized to Retard Water Loss While Allowing Carbon Dioxide to Enter

In the stem and leaves, the epidermis is exposed to dry air, making it a potential pathway for water loss. Epidermal cells of the stem, unlike those of the root, secrete a waxy covering, the cuticle, that reduces evaporation of water. The cuticle also, however, reduces the diffusion of carbon dioxide and oxygen into and out of the plant. Hence, the epidermis is commonly perforated with stomata that regulate this exchange.

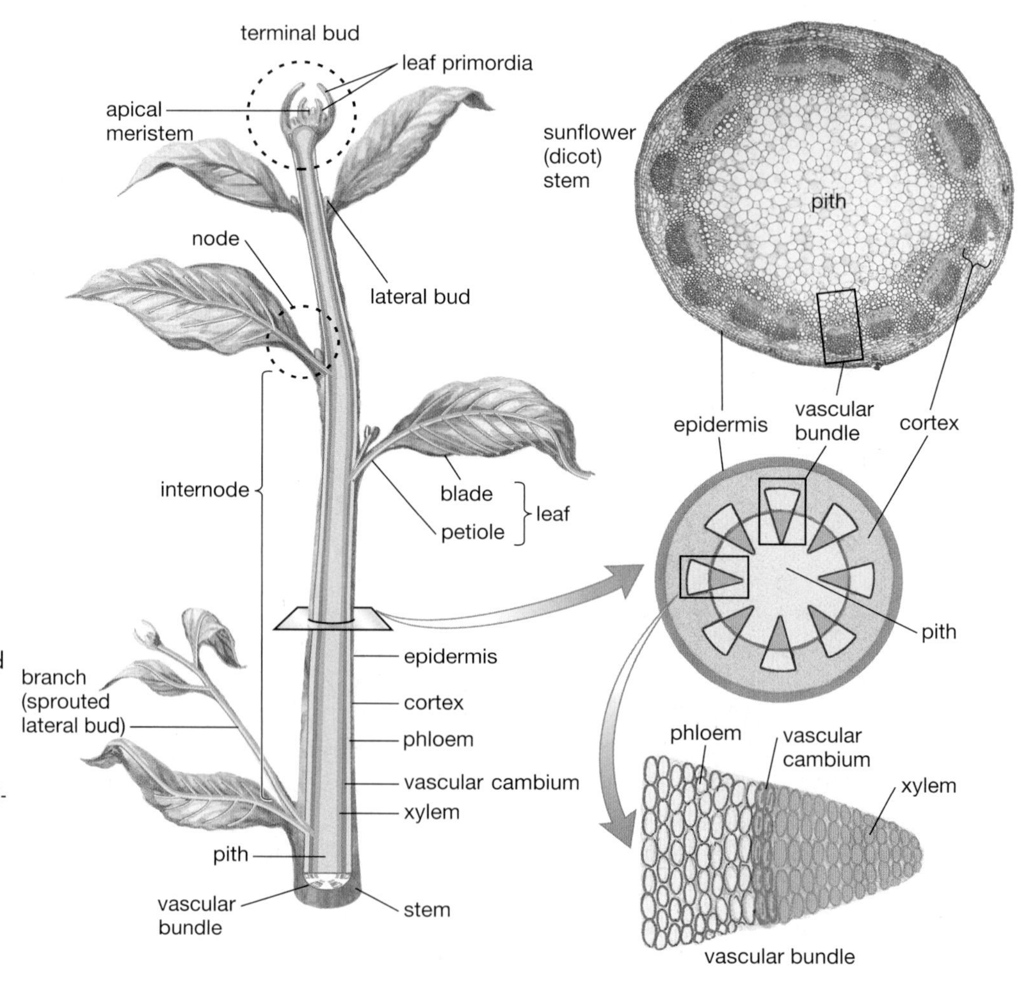

FIGURE 24-9 The structure of a young dicot stem
At the tip of the stem, the terminal bud includes the apical meristem and several leaf primordia, produced by the meristem. Other daughter cells of the apical meristem differentiate into epidermis, cortex, pith, and vascular tissues. Leaves and lateral buds are located at nodes; the naked stem between nodes is an internode. In cross section, vascular tissue forms a ring of vascular bundles in dicots such as the sunflower shown in the photomicrograph. As the stem grows, leaf primordia develop into leaves, and internodes elongate. A lateral bud (meristem tissue) remains between each leaf and the stem, and may sprout into a branch.

The Cortex and Pith Support the Stem, Store Food, and May Photosynthesize

Cortex (located between the epidermis and vascular tissues) and **pith** (inside the vascular tissues at the center of the stem) are similar in most respects; in fact, in some stems it is difficult to tell where cortex ends and pith begins. Cortex and pith perform three major functions: support, storage, and, in some cases, photosynthesis.

- *Support*. In very young stems, water filling the central vacuoles of cortex and pith cells causes turgor pressure (see Chapter 5). Turgor pressure stiffens the cells, much as air inflates a tire. If you forget to water your houseplants, their drooping tips show the importance of turgor pressure in keeping young stems erect. Somewhat older stems also have collenchyma or sclerenchyma cells with thickened cell walls, which don't depend on turgor pressure for strength.
- *Storage*. Parenchyma cells in both cortex and pith convert sugar into starch and store the starch as a food reserve.
- *Photosynthesis*. In many stems, the outer layers of cortex cells contain chloroplasts and carry out photosynthesis. In some desert plants such as cacti, in which the leaves are reduced or absent, the cortex of the stem is the only green photosynthetic part of the plant.

Vascular Tissues in Stems Transport Water, Dissolved Nutrients, and Hormones

Like those of roots, the vascular tissues of stems transport water, minerals, sugars, and hormones. Vascular tissues are continuous in root, stem, and leaf, interconnecting all the parts of the plant. The *primary xylem* and *primary phloem* found in young stems arise from the apical meristem. In young dicot stems, the primary xylem, **vascular cambium** (meristematic tissue that produces *secondary xylem* and *secondary phloem*), and primary phloem may form concentric cylinders or may appear as a ring of bundles running up the stem; each bundle contains both phloem and xylem (see Fig. 24-9). Secondary growth in dicot stems, discussed in the next section, always results in concentric cylinders of xylem and phloem.

Stem Branches Form from Lateral Buds Consisting of Meristem Cells

Branches grow from lateral buds. A lateral bud is a cluster of dormant meristem cells left behind by the apical meristem as the stem grows. Lateral buds are located at nodes, just above the attachment points of the leaves (see Fig. 24-9). When stimulated by the appropriate hormones (as we will see in Chapter 26), the meristem cells of a lateral bud activate and the bud sprouts, growing into a branch (Fig. 24-10). As the meristem cells divide, they release hormones that change the developmental fate of the cells between the bud and the vascular tissues of the stem. Parenchyma cells of the cortex differentiate into xylem and phloem, ultimately connecting with the main vascular systems in the stem. As the branch grows, it duplicates the development of the stem: it has an apical meristem at its tip and produces its own leaf primordia and lateral buds as it grows.

Secondary Growth Produces Thicker, Stronger Stems

In conifers and perennial dicots, stems may survive up `to hundreds of years, becoming thicker and stronger each year. This secondary growth in stem thickness

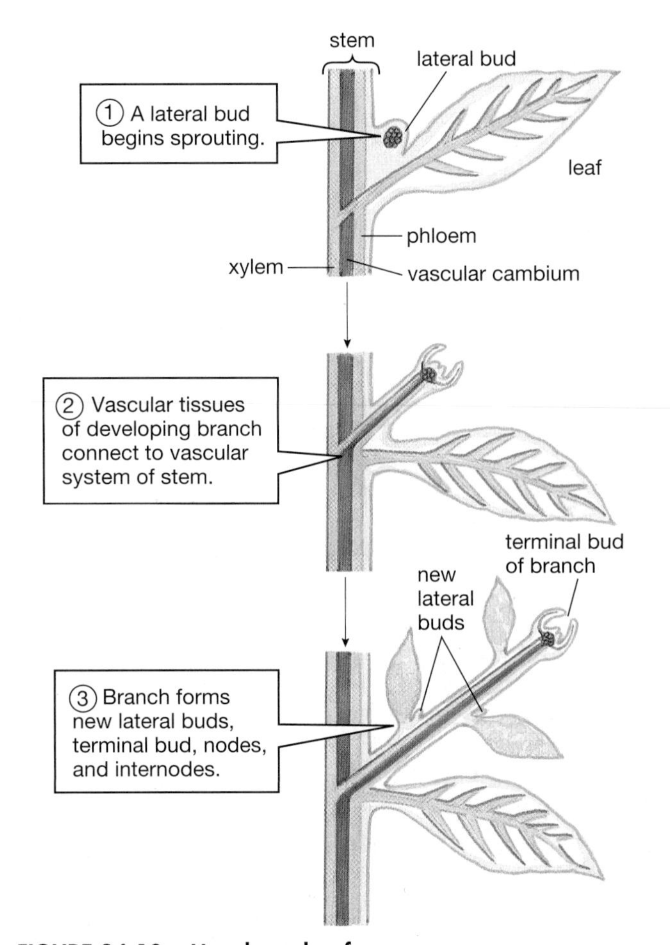

FIGURE 24-10 How branches form

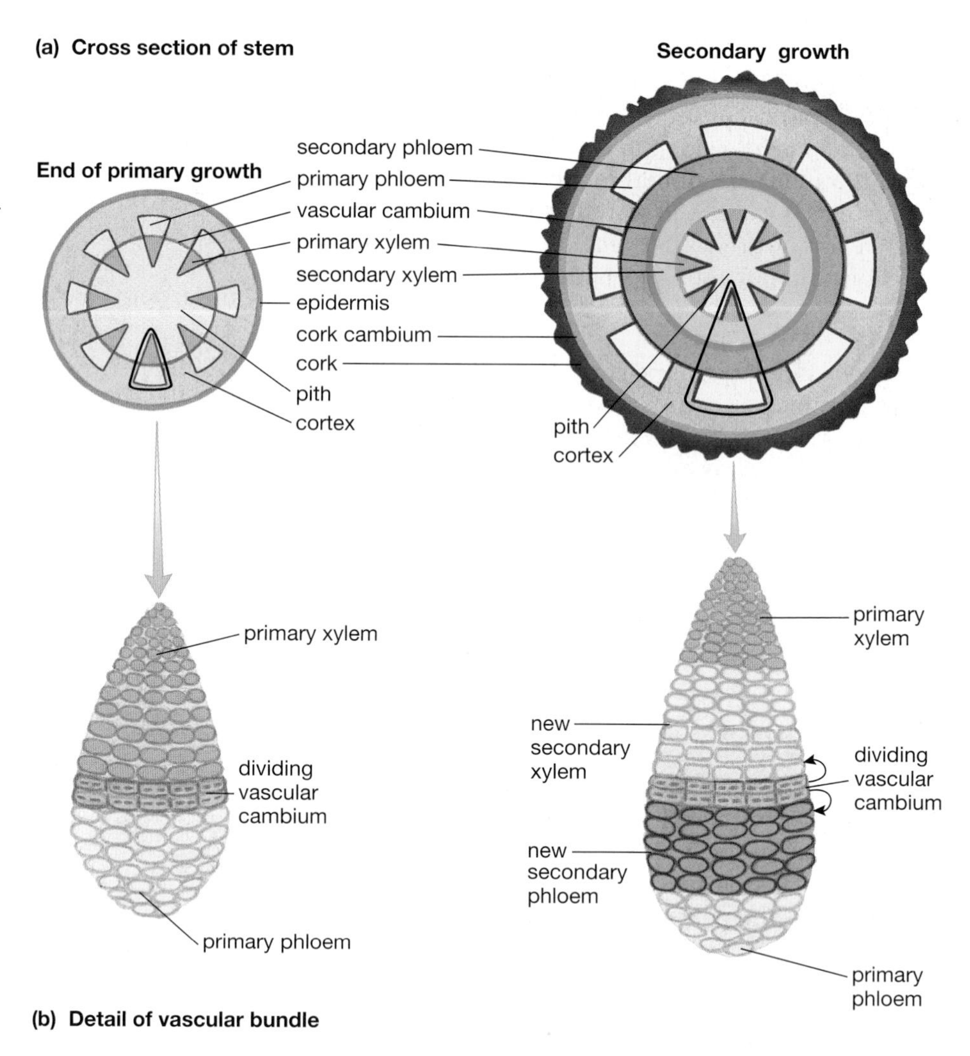

FIGURE 24-11 Secondary growth in a dicot stem
(a) Cross section of a dicot stem at the end of primary growth (left) and during early secondary growth (right). **(b)** A vascular bundle during secondary growth. Vascular cambium separates primary xylem and primary phloem. Division and differentiation of vascular cambium cells produce secondary xylem on the inside, and secondary phloem on the outside. Because xylem and pith already fill the inside of the stem, secondary xylem forces the outer tissues outward, increasing the diameter of the stem. Cork cambium produces cork cells that cover the outside of the stem.
QUESTION If a tree is "girdled" by removing a strip of bark completely around its trunk, it usually dies. Why?

results from cell division in two lateral meristems: the vascular cambium and *cork cambium* (Fig. 24-11).

Vascular Cambium Produces Secondary Xylem and Phloem

The vascular cambium is a cylinder of meristem cells located between the primary xylem and primary phloem. Daughter cells of the vascular cambium produced toward the inside of the stem differentiate into *secondary xylem*; those produced toward the outside of the stem differentiate into *secondary phloem* (see Fig. 24-11). Because the center of the stem is already filled with pith and primary xylem, newly formed secondary xylem pushes the vascular cambium and all outer tissues farther out, increasing the diameter of the stem. This secondary xylem, with its thick cell walls, forms the wood that makes up most of the trunk of a tree. Young xylem, called **sapwood** (located just inside the vascular cambium), transports water and minerals; older xylem, the **heartwood**, contributes only to the strength of the trunk, no longer carrying water and solutes. Heartwood serves as a collection site for metabolic wastes of the tree, such as gums, resins, and oils. These wastes increase the density of the heartwood, contribute to its darker color, and help the heartwood resist rotting.

Phloem cells are much weaker than xylem cells. As they die over time, the sieve-tube elements and companion cells are crushed between the hard xylem on the inside of the trunk and the tough cork on the outside. Only a thin strip of recently formed phloem remains alive and functioning.

In trees adapted to temperate latitudes, such as oaks and pines, cell division in the vascular cambium ceases during the cold of winter. In spring, the cambium cells divide, forming new xylem and phloem. The young cells grow by absorbing water and swelling, while the newly formed cell walls are still soft. As the cells mature, the cell walls thicken and harden, preventing further growth. Water is readily available in spring; therefore, young xylem cells swell considerably and are large when mature. As summer progresses and water becomes scarcer, new xylem cells absorb less water and consequently are smaller when they mature. As a result, tree trunks in cross section show a pattern of alternating pale regions (large cells formed in spring) and dark regions (small cells formed in summer), as shown in Fig. 24-12. This pattern forms the familiar **annual rings** of growth in temperate trees. You can determine the approximate age of a tree that has been cut by counting

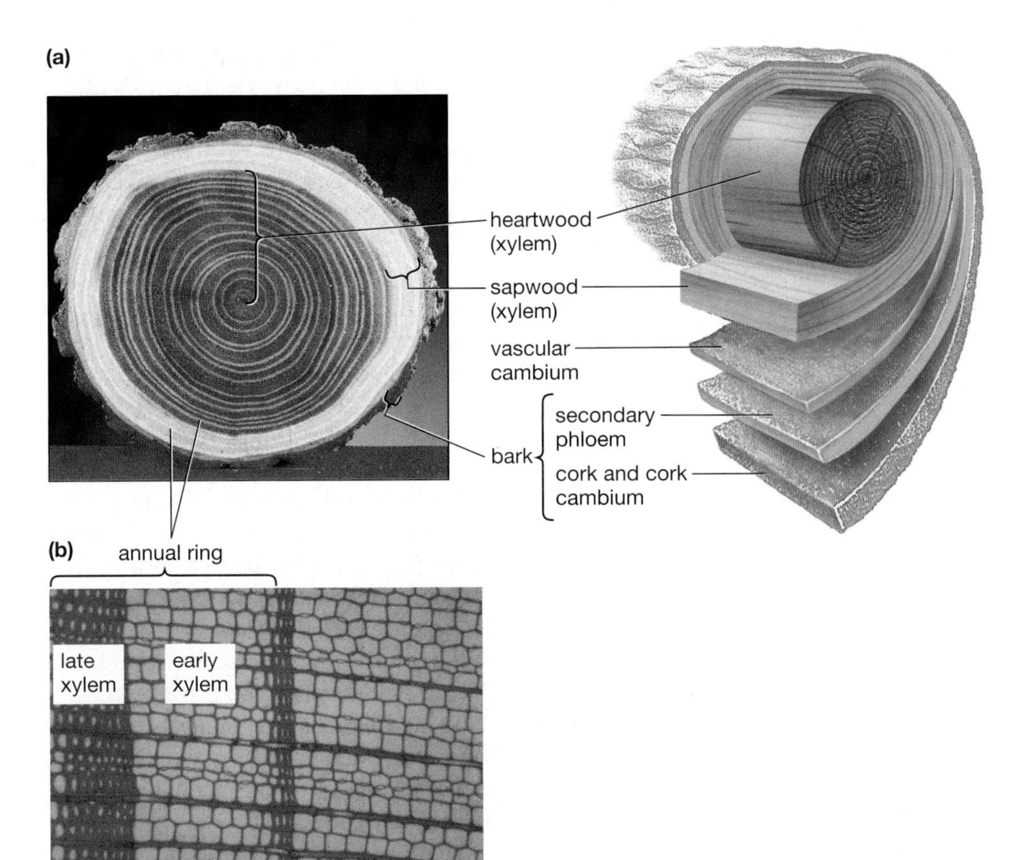

FIGURE 24-12 How annual tree rings are formed
(a) Tree rings are clearly visible in this section of tree trunk. The ratio of cell wall to "hole" (the now-empty interior of the cell) determines the color of the wood: Early wood, formed during the spring, with lots of hole, is pale; late wood, formed during the summer, with lots of wall, is dark. The water-transporting xylem of sapwood forms a lighter layer inside the bark. Xylem of the older heartwood, where the rings are most easily visible, no longer transports water and minerals. **(b)** As this micrograph shows, secondary xylem cells formed during the wet spring are large, whereas secondary xylem cells formed during the hotter, drier summer are small.

the dark growth rings. Scientists can also use the width of each ring to reconstruct past climate. Wet years produce more growth and wider rings. Using this technique and the rings in ancient trees, including a 1000-year-old cypress, researchers have constructed an 800-year record of climate in Virginia. They hypothesize that a 7-year drought from 1606 to 1612, recorded in tree rings, was responsible for the mysterious disappearance of the Virginia colony of Jamestown, which was founded 1607.

Secondary Growth Causes the Epidermis to Be Replaced by Woody Cork

Recall that epidermal cells are mature, differentiated cells that can no longer divide. Therefore, as new secondary xylem and phloem are added each year, enlarging the stem, the epidermis can't expand to keep up with the increasing circumference. The epidermis splits off and dies. Apparently stimulated by hormones, some parenchyma cells in the cortex become rejuvenated and form a new lateral meristem layer, the **cork cambium** (see Fig. 24-11). These cells divide, forming daughter cells toward the outside of the stem. These daughter cells, called *cork cells*, or simply cork, develop tough, waterproof cell walls that protect the trunk both from drying out and from physical damage. Cork cells die as they mature and may form a protective layer a half meter thick in some tree species, such as the fire-resistant sequoia (Fig. 24-13). As the trunk expands from year to year, the outermost layers of cork split apart or peel off, accommodating the growth. Corks used to plug bottles are made from the outermost layer of cork from a certain type of oak, the cork oak, which is carefully peeled off by harvesters. The cork of the cork oak breaks away from the cork cambium, so the tree is not

FIGURE 24-13 Cork forms the outer layer of bark
An ancient sequoia in the Sierra Nevada of California. The cork cambium of a sequoia produces new layers of cork each year, eventually producing a protective, fire-resistant outer covering half a meter or more thick. This massive cork layer contributes to the sequoia's great longevity; forest fires that kill lesser trees merely burn off a few inches of sequoia cork, leaving the tree unharmed. Small, blackened areas on this cork are from past fires.

harmed by stripping it off. The common term **bark** includes all the tissues outside the vascular cambium: phloem, cork cambium, and cork cells. The complete removal of a strip of bark all the way around a tree, called *girdling*, is invariably fatal to a tree because it severs the phloem. Without phloem, sugars synthesized in the leaves cannot reach the roots. Deprived of energy, the roots no longer take up water and minerals, causing the tree to die.

24.5 Roots: Anchorage, Absorption, and Storage

Primary Growth Causes Roots to Elongate and to Branch

As a seed sprouts, the **primary root**—the first root to develop—grows down into the soil. Many dicots, such as carrots and dandelions, develop a taproot system. A **taproot system** consists of the primary root and many smaller roots that grow out from its sides (Fig. 24-14a). In contrast, in monocots such as grasses and palms, the primary root soon dies off, replaced by many new roots that emerge from the base of the stem. These secondary roots are nearly equal in size, forming a **fibrous root system** (Fig. 24-14b).

In young roots of both taproot and fibrous root systems, divisions of the apical meristem give rise to four distinct regions (Fig. 24-15). At the very tip of the root, daughter cells produced on the lower portion of the apical meristem differentiate into the **root cap**. The root cap protects the apical meristem from being scraped off as the root pushes down between the rocky particles of the soil. Root-cap cells have thick cell walls and secrete a slimy lubricant that helps ease the root between soil particles. Nevertheless, root-cap cells wear away and must be continuously replaced by new cells from the meristem.

Daughter cells produced on the upper portion of the apical meristem differentiate into one of three parts: an outer envelope of epidermis, an adjacent internal layer of cortex, and a core called the *vascular cylinder* (see Fig. 24-15).

Root branches originate from the **pericycle** (the outermost layer of the vascular cylinder), a remnant of the apical meristem that retains the ability to divide. Under the influence of plant hormones, pericycle cells divide and form the apical meristem of a **branch root**, a root that branches off from an existing root (Fig. 24-16). Branch root development is similar to primary root development except that the branch must first break out through the cortex and epidermis of the primary root. It does so by both crushing the cells that lie in its path and by secreting enzymes that digest them away. Thus, the vascular tissues of the branch root connect with the vascular tissues of the primary root.

(a) (b)

FIGURE 24-14 **Typical root systems in dicots and monocots** (a) Dicots typically have a taproot system, consisting of a long central root with many smaller, secondary roots branching from it. (b) Monocots usually have a fibrous root system, with many roots of equal size.

The Epidermis of the Root Is Very Permeable to Water

The root's outermost covering of cells is the epidermis, which is in contact with the soil and any air or water trapped among the soil particles. The cell walls of the epidermal cells are highly water permeable. Therefore, water can penetrate into the interior of the root either by passing through the membranes of the epidermal cells or by passing between those cells through the porous cell walls. Many epidermal cells grow root hairs into the surrounding soil (Fig. 24-17). By increasing the root's surface area, root hairs increase its ability to absorb water and minerals. Root hairs may add dozens of square meters of surface area to the roots of even small plants.

Cortex Makes Up Much of the Interior of a Young Root

Cortex occupies most of the inside of a young root. The cortex consists of an outer mass of large, loosely packed parenchyma cells just beneath the epidermis and an inner layer of smaller, close-fitting cells that form a ring around the vascular cylinder called the **endodermis** (see Fig. 24-15). Sugars produced in the shoot by photosynthesis are transported down to the parenchyma cells of the cortex, where they are converted to starch and stored. These cells are particularly abundant in roots specialized for carbohydrate storage, such as the thick roots of carrots and dandelions.

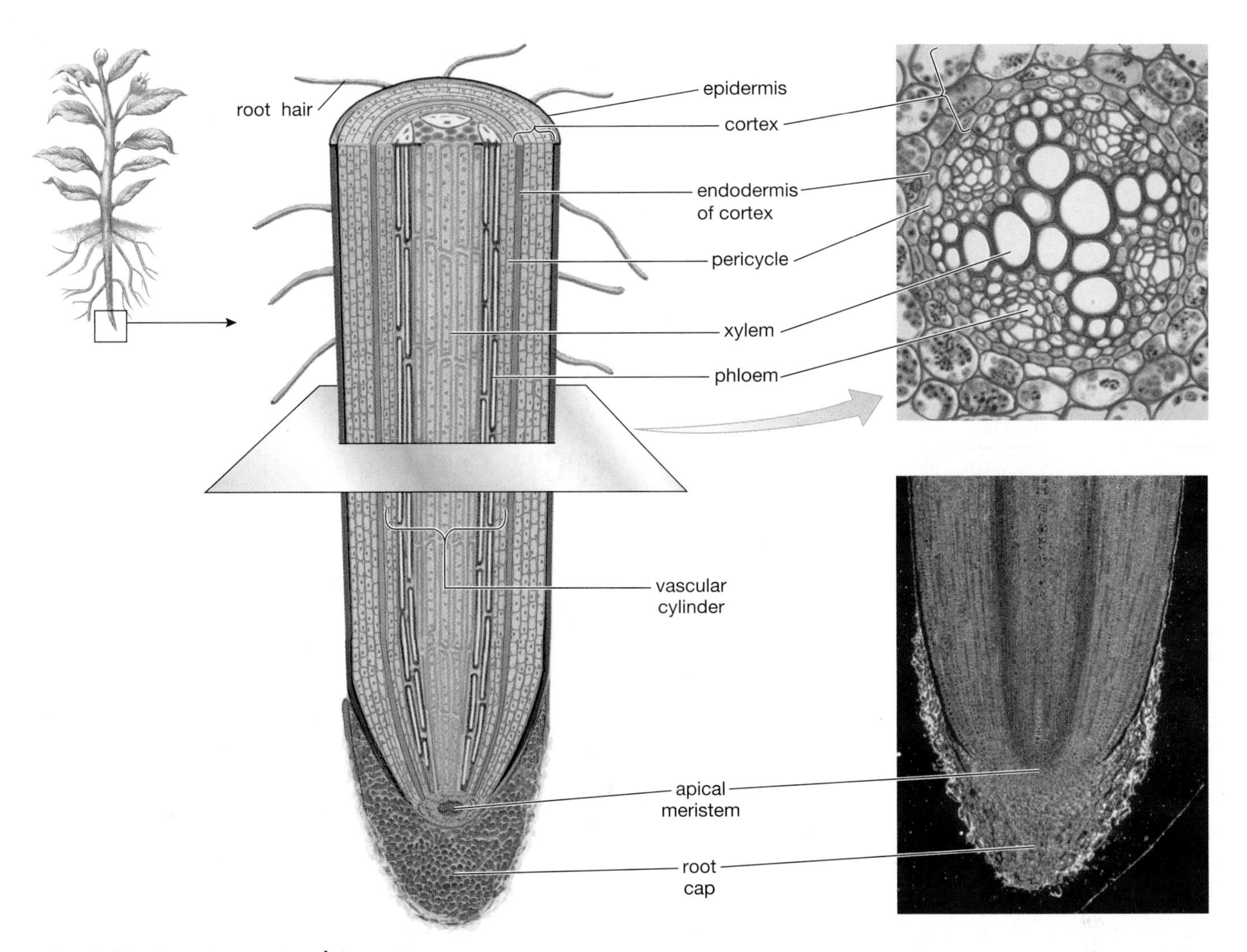

FIGURE 24-15 Primary growth in roots
Primary growth in roots results from cell divisions in the apical meristem near the root tip, which form the root cap, epidermis, vascular cylinder, and cortex. **QUESTION** Of the cells labeled in the cross-section photo at top right, which form ground tissue, which form vascular tissue, and which form epidermal tissue?

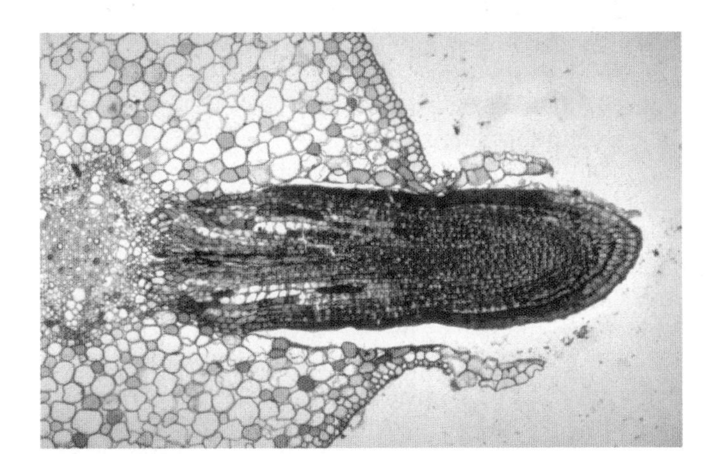

FIGURE 24-16 Branch roots
Branch roots emerge from the pericycle of a root. The central axis of this branch root is already differentiating into vascular tissue.

FIGURE 24-17 Root hairs
Root hairs, shown here in a sprouting radish, greatly increase a root's surface area for the absorption of water and minerals from the soil.

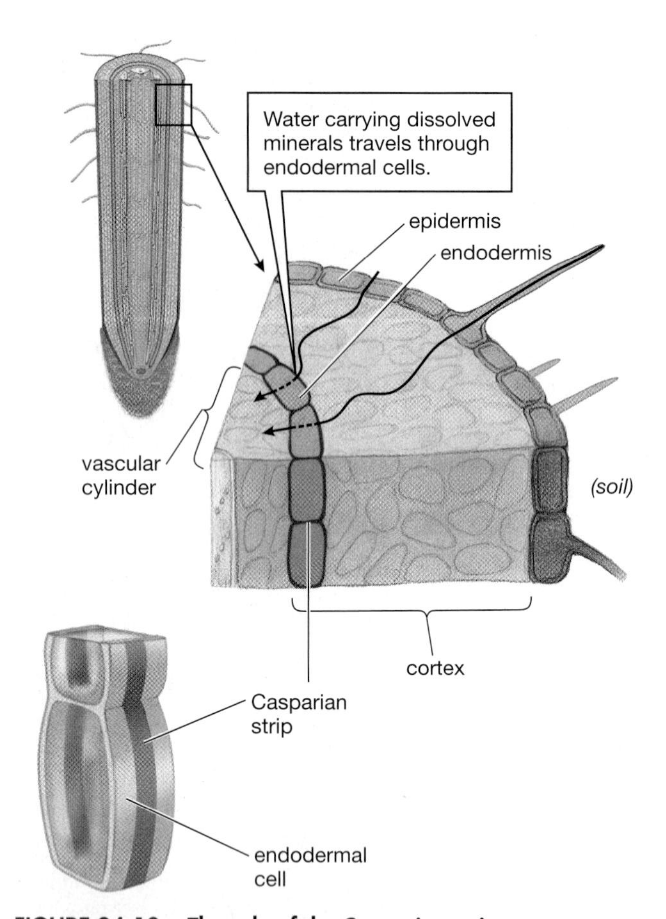

FIGURE 24-18 The role of the Casparian strip
The Casparian strip is a band of waterproof material joining the walls of endodermal cells. Encircling each cell, the Casparian strip forces water to move by osmosis through the cell membranes instead of flowing through the porous cell walls.
QUESTION What problem would arise in a root with no waterproof Casparian strip?

The cell wall of each endodermal cell contains a band of fatty, waterproof material called the **Casparian strip**. The Casparian strip prevents water from traveling through the cell walls between endodermal cells, but does not cover the cell surfaces that face the rest of the cortex or those that face the vascular cylinder. While water and dissolved minerals can travel *around* both epidermal and cortex parenchyma cells by moving through their porous cell walls, the Casparian strip forces water that enters the vascular cylinder to move by osmosis through the living membranes of the endodermal cells (Fig. 24-18).

The Vascular Cylinder Contains Conducting Tissues

The **vascular cylinder** contains the conducting tissues of xylem and phloem, which transport water and dissolved materials within the plant. The outermost layer of the vascular cylinder is the pericycle, which is located just inside the endoderm of the cortex. The pericycle cells receive water and minerals from the endodermal cells, and actively transport the minerals into the interior of the vascular cylinder (water follows by osmosis). This maintains the concentration gradient and keeps minerals moving by diffusion into the endoderm and pericycle. In the interior of the vascular cylinder, water carrying its high concentration of dissolved minerals flows through the cell walls of the nonliving xylem cells and up through the plant body. Phloem carries high-energy molecules derived from photosynthesis, such as sugars, down from the leaves to provide energy for root cells.

24.6 How Do Plants Acquire Nutrients?

Nutrients are elements essential to normal life that differ for different organisms. Plants, for example, require relatively large quantities of carbon (obtained from carbon dioxide), hydrogen (from water), oxygen (from air and water), phosphorus (from phosphate ions in soil), nitrogen (from nitrate and ammonium ions in soil), magnesium, calcium, and potassium (as ions in soil). Plants also require very small quantities of nutrients such as iron, chlorine, copper, manganese, zinc, boron, and molybdenum. Carbon dioxide and oxygen usually enter a plant by diffusion from the air into leaves, stem, and roots. Roots extract water and all other nutrients, collectively called **minerals**, from the soil.

Roots Acquire Minerals by a Four-Step Process

Soil consists of bits of pulverized rock, air, water, and organic matter. Although the rock particles and the organic matter contain essential nutrients, only minerals dissolved in the soil water are accessible to roots. The concentration of minerals in the soil water is very low, usually much lower than the concentration within plant cells and fluids—for example, the concentration of potassium (K^+) in root cells is at least 10 times greater than that in soil water, so diffusion cannot move potassium into the root. In general, most minerals are moved into a root against their concentration gradients by active transport. (Recall from Chapter 6 that the movement of molecules from areas of low concentration to areas of high concentration requires energy.) Sugar synthesized in the leaves by photosynthesis is transported through the phloem to the roots, where mitochondria in root cells use it to produce ATP (adenosine triphosphate) by cellular respiration. Some of this ATP drives the active transport of minerals. Because ATP production by mitochondria requires oxygen, soil must have some air spaces within it. Flooding (or overwatering) can kill plants by depriving their roots of oxygen.

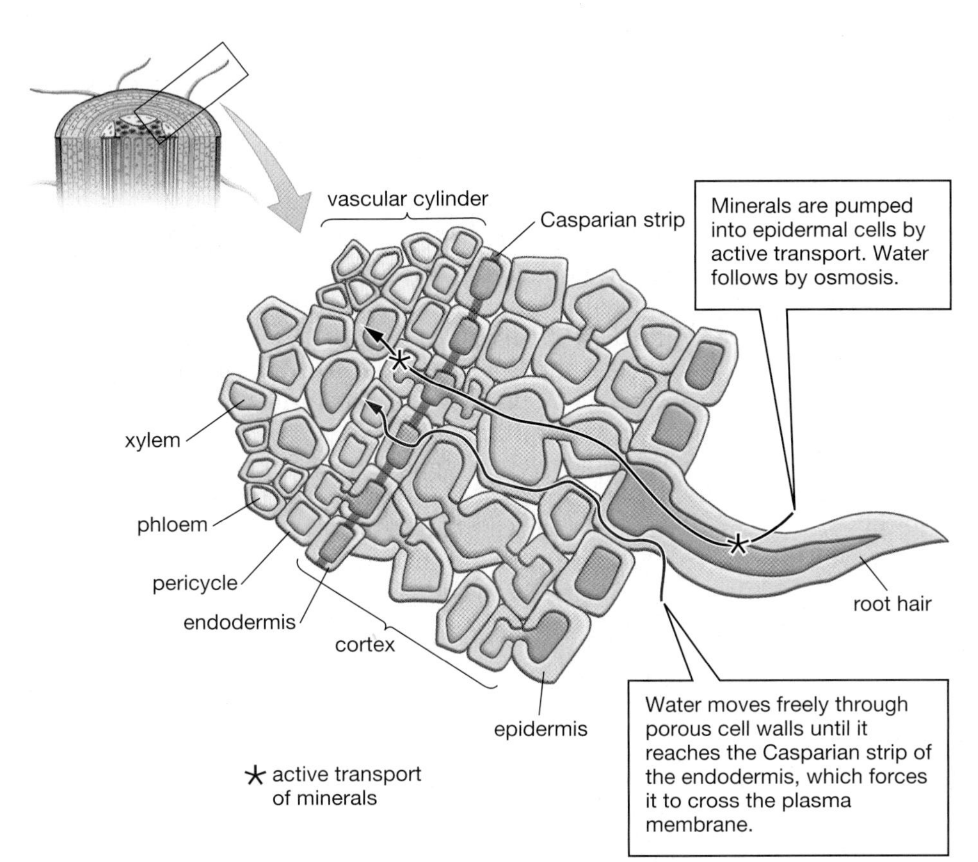
vascular cylinder
Casparian strip
Minerals are pumped into epidermal cells by active transport. Water follows by osmosis.
xylem
phloem
pericycle
endodermis
cortex
root hair
epidermis
Water moves freely through porous cell walls until it reaches the Casparian strip of the endodermis, which forces it to cross the plasma membrane.
* active transport of minerals

FIGURE 24-19 Mineral and water uptake by roots
Black arrow: Concentrated mineral solution follows a path through the interior of cells. Pericycle cells actively transport minerals into the interior of the vascular cylinder, maintaining a concentration gradient that keeps the minerals diffusing inward and water following by osmosis. **Blue arrow:** Water moves freely until it reaches the Casparian strip and must travel by osmosis through the endodermal cell membranes to enter the vascular cylinder. Sites of mineral active transport are indicated by asterisks (*).

Media Activity
24.3 Nutrient Uptake

Most mineral absorption by roots occurs in a four-step process (Fig. 24-19):

1. *Active transport into root hairs.* Root hairs projecting from the epidermal cells provide most of the surface area of the root and are in intimate contact with the soil water. The plasma membranes of the root hairs use the energy of ATP to transport minerals from the soil water, concentrating the minerals in the root hair cytoplasm.
2. *Diffusion through cytoplasm to pericycle cells.* The cytoplasm of adjacent living plant cells is interconnected by plasmodesmata. Minerals can diffuse through plasmodesmata from the epidermal cells into the cortex, endodermis, and pericycle cells.
3. *Active transport into the extracellular space of the vascular cylinder.* At the center of the vascular cylinder lies the xylem, into which the minerals must ultimately be transported. The tracheids and vessel elements of xylem are dead, without cytoplasm or plasma membrane—merely an outer skeleton of cell wall shot full of holes (see Fig. 24-6). Any minerals that enter the extracellular space surrounding the xylem can easily diffuse into the xylem cells through the holes in their walls. Therefore, pericycle cells actively transport minerals out of their own cytoplasm into the extracellular space around the xylem.
4. *Diffusion into the xylem.* The active transport of minerals into the extracellular space of the vascular cylinder increases the concentration of minerals in the extracellular space. This high concentration creates a gradient that promotes the diffusion of minerals from the extracellular space into the tracheids and vessel elements of the xylem.

You can now appreciate one of the functions of the waterproof Casparian strip that seals the spaces between the endodermal cells that surround the vascular cylinder. If the water and dissolved minerals that enter the vascular cylinder could flow back through the extracellular space *between* endodermal cells, minerals would leak back out of the extracellular space of the vascular cylinder as fast as they were pumped in. This leakage would waste the energy used to actively transport the minerals into the root and would reduce the concentration gradient that allows the minerals to diffuse into the conducting cells of the xylem. The Casparian strip, however, effectively leakproofs the vascular cylinder, retaining the concentrated mineral solution within the vascular cylinder.

Symbiotic Relationships Help Plants Acquire Nutrients

Many minerals are too scarce in soil water to support plant growth, although plenty of minerals may be bound up in the surrounding rock particles. One nutrient—nitrogen—is almost always in short supply in both rock particles and soil water. Scientists hypothesize that red fall colors are actually synthesized to help the plant recover nitrogen-containing compounds in its leaves before they drop in the fall (see "Case Study Revisited: Why Do Leaves Turn Red in the Fall?"). Most plants have evolved beneficial relationships with other organisms that help the plants acquire scarce minerals and nutrients such as nitrates and phosphates. Examples include root–fungus relationships, called *mycorrhizae*, and root–bacteria relationships formed in nodules of legumes.

Fungal Mycorrhizae Help Plants Acquire Minerals

Under normal conditions, water-soluble minerals are released very slowly from rock particles, and the chemical forms of the minerals may not be suitable for uptake by the plasma membranes of plant root cells. Most land plants form symbiotic relationships with fungi to form root–fungus complexes called **mycorrhizae**, which help the plant extract and absorb minerals. Fungal strands intertwine between the root cells and extend out into the soil (Fig. 24-20). In some way that is not yet understood, the fungus renders nutrients, such as phosphorus and certain minerals, accessible for uptake by the roots, perhaps by converting rock-bound minerals into simple soluble compounds that root plasma membranes can transport. The fungus, in return, receives sugars, amino acids, and vitamins from the plant. In this way, both the fungus and the plant can grow in places where neither could survive alone, including deserts and high-altitude, nutrient-scarce rocky soils.

Recent research has revealed that, in some forests, mycorrhizae form an immense underground web that interlinks trees—even trees of different species. This web of fungi transfers carbon compounds produced by one tree to another. Trees with access to abundant sunlight subsidize their shaded neighbors, with the mycorrhizae transferring photosynthetic products from the rich to the poor (like an underground Robin Hood). Researchers hypothesize that nutrient transfer among trees by mycorrhizae may be an important factor in the overall health of the forest, which in turn benefits the mycorrhizae.

Bacteria-Filled Nodules on the Roots of Legumes Help Those Plants Acquire Nitrogen

Since amino acids, nucleic acids, and chlorophyll all contain nitrogen, plants need large amounts of this element. Although nitrogen gas (N_2) makes up about 79% of the atmosphere, plants can take up nitrogen only through their roots, in the form of ammonium ions (NH_4^+) or nitrate ions (NO_3^-).

Although N_2 diffuses from the atmosphere into the air spaces in the soil, it cannot be used by plants. Plants don't have the enzymes needed to carry out **nitrogen fixation**, the conversion of N_2 into ammonium or nitrate. A variety of **nitrogen-fixing bacteria**, some of which live freely in the soil, do have these enzymes. However, nitrogen fixation is very costly, energetically speaking, using at least 12 ATPs per ammonium ion synthesized. Consequently, bacteria don't routinely manufacture a lot of extra ammonium and liberate it into the soil.

Some plants, particularly the **legumes** (such as peas, clover, and soybeans), enter into a mutually beneficial relationship with certain species of nitrogen-fixing bacteria. By secreting chemicals into the soil, legumes attract nitrogen-fixing bacteria to their roots. Once there, the bacteria enter the root hairs. The bacteria then digest channels through the cytoplasm of the epidermal cells and into underlying cortex cells. As both bacteria and their host cortex cells multiply, a **nodule**, or swelling that houses root–bacteria complexes (Fig. 24-21), forms, and a cooperative relationship develops. The plant transports sugars from its leaves down to the cortex for storage, just as it normally would. The bacteria within

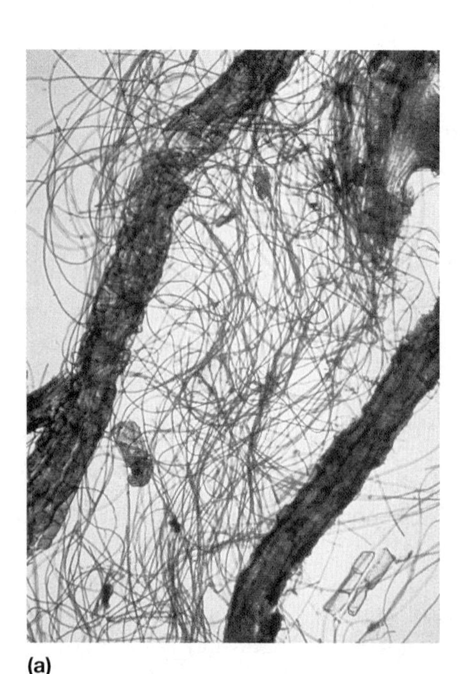

(a)

(b)

FIGURE 24-20 Mycorrhizae, a root–fungus symbiosis
(a) A tangled meshwork of fungal strands surrounds and penetrates into the root. **(b)** Seedlings growing under identical conditions with (right) and without (left) mycorrhizal fungi illustrate the importance of mycorrhizae in plant nutrition.
QUESTION Based on what you've learned about root function, which part of the root system might you expect to be infected by mycorrhizal fungi?

EARTH WATCH Plants Help Regulate the Distribution of Water

The distribution of plants on Earth is limited by both environmental factors and the adaptations of the plants. Probably the most important environmental factor influencing plant distribution is water: cacti inhabit deserts because they can withstand drought; orchids and mahogany trees need the frequent drenching rains of the rain forest. However, this relationship works both ways: plants, through transpiration, help regulate the amount and distribution of rainfall, soil water, and even river flow.

Consider the Amazon rain forest (Fig. E24-1). An acre of soil supports hundreds of towering trees, each bearing millions of leaves. The surface area of the leaves dwarfs the surface area of the soil, so up to 75% of all the water evaporating from the acre of forest is due to leaf transpiration. This transpiration raises the humidity of the air and causes rain to fall. In fact, about half of the water transpired from the leaves falls again as rain; thus, about one-third of the total rainfall is water recycled by transpiration. In a very real sense, the high humidity and frequent showers that the rain forest needs to survive are partly created by the forest itself! Large-scale cutting of the Amazon rain forest continues to destroy about 1% of the total forest each year. Trees are burned, releasing CO_2, and are replaced by crops to feed the growing population. Agricultural crops are not nearly as efficient at trapping and transpiring moisture as the trees they replaced. Consequently, runoff increases, humidity falls, and rainfall declines. Decreased rainfall not only slows regrowth of forest when the fields are abandoned, but also harms adjacent forests, gradually changing their composition to non-rain-forest vegetation. Using the rates of both deforestation and regrowth, researchers predict that unprotected portions of this spectacular ecosystem may be gone by the year 2150. But when the effects of reduced rainfall are factored in, the end comes much more swiftly, possibly by 2030. Although portions of the rain forest can be protected from logging, it will all be susceptible to the effects of climate change caused by the loss of transpiration in deforested areas.

FIGURE E24-1 The Amazon rain forest
The rainforest community helps mold its own environment.

The Monteverde cloud forest, which blankets the uppermost reaches of the Cordillera de Tilarán mountain range in Costa Rica, is entirely dependent on an almost constant shroud of fog. Transpiration from Costa Rica's lowland forests pumps moisture into the winds flowing up the mountain slopes; this moisture condenses and forms fog as the air cools. Scientists studying the Monteverde cloud forest have noticed an alarming trend: the clouds are lifting. Frog and toad populations have crashed in some areas, and birds characteristic of non-cloud forests are invading formerly clouded lowland areas. Why? A century of logging has eliminated over 80% of Costa Rica's lowland forests. Using satellite images, ecologist Robert Lawton and colleagues found that the deforested areas had relatively low cloud cover compared to nearby forests. Since the winds moving up the mountain slopes gain water from the lowlands, the researchers hypothesize that reduced transpiration has lowered the moisture content of the air, so it must rise farther before the clouds form. Mist-free conditions are becoming more common, and scientists warn that if clouds disappear for days at a time, the fragile ecosystem could collapse.

These examples show that plants wield an enormous influence on what we often consider to be nonliving aspects of the biosphere, such as humidity, rainfall, soil water, and river flow. These factors then feed back and influence all the life forms in the region. Human activities that alter plant cover can have far-reaching and unanticipated impacts on ecosystems.

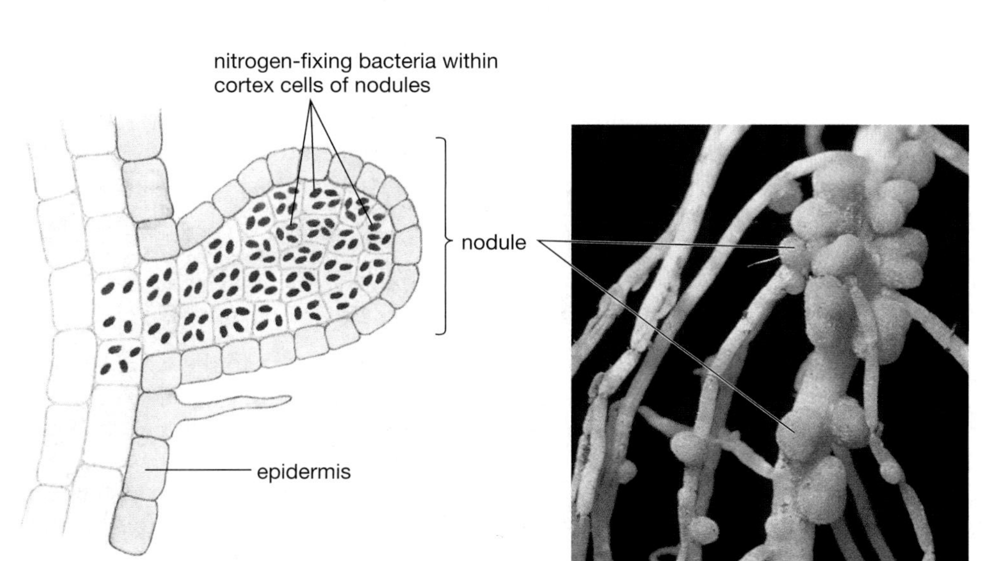

FIGURE 24-21 Nitrogen fixation in legumes
A diagram and photograph of root nodules in legumes that contain nitrogen-fixing bacteria.

the cortex cells take up the sugar and use its energy for all of their metabolic processes, including nitrogen fixation. The bacteria obtain so much energy that they produce more ammonium than they need. The surplus ammonium diffuses into the cytoplasm of their host cells, providing the plant with a steady supply of usable nitrogen. Surplus ammonium also diffuses into the surrounding soil, making it better able to support other types of plants. Farmers plant legumes not only for their commercial value but also to enrich the soil with ammonium for future crops.

Media Activity
24.4 Plant Transport Mechanisms

24.7 How Do Plants Acquire and Transport Water and Minerals?

Nearly 99% of the water absorbed by the roots of plants evaporates through the stomata of leaves and, to a lesser extent, through the stomata of stems in a process called **transpiration**. As you will see in the descriptions that follow, transpiration drives the movement of water through the plant body. To understand how water enters roots, it is first necessary to understand how transpiration, acting in conjunction with the properties of water, can pull water up through the xylem of the roots and stem and into the leaves. Once you understand this process, we will return to the question of how the water enters the roots.

Water Movement in Xylem Is Explained by the Cohesion–Tension Theory

After entering the root xylem, water and minerals still must be moved to the uppermost reaches of the plant. (In redwood trees, the distance may be more than 300 feet!) **Bulk flow** moves fluids up the xylem from root to stem and leaf in land plants. Because minerals are dissolved in water, they are passively carried along as the water flows upward. But how do plants overcome the force of gravity and make water flow upward? The *cohesion–tension theory* provides an explanation.

According to the **cohesion–tension theory**, water is pulled up the xylem, powered by transpiration—the evaporation of water from the leaves (Fig. 24-22). As its name suggests, this theory has two essential parts:

- *Cohesion*. Attraction among water molecules holds water together, forming a solid chainlike column within the xylem tubes.
- *Tension*. This "water chain" is pulled up the xylem; evaporation provides the necessary energy.

Let's briefly examine both.

Hydrogen Bonds Between Water Molecules Produce Cohesion

You may recall from Chapter 2 that water is a polar molecule, with the oxygen carrying a slight negative charge and the hydrogens carrying a slight positive charge. As a result, nearby water molecules attract one another, forming weak hydrogen bonds. Just as individually weak cotton threads together make the strong fabric of your jeans, the network of individually weak hydrogen bonds in water collectively produce a very high cohesion, or tendency to resist being separated. Experiments have demonstrated that the column of water within the xylem is at least as strong—and as unbreakable—as a steel wire of the same diameter. This is the "cohesion" part of the theory: hydrogen bonds among water molecules provide the cohesion that holds together a chain of water extending the entire height of the plant within the xylem. Supplementing the cohesion between water molecules is adhesion between water molecules and the walls of xylem. Attraction of water molecules to the cell walls of the thin xylem tubes helps the water creep upward, just as water is pulled upward into a very narrow glass tube.

Transpiration Produces the Tension That Pulls Water Upward

Transpiration provides the force for water movement—the "tension" part of the theory. As a leaf transpires, the concentration of water in the mesophyll drops. This drop causes water to move by osmosis from the xylem into the dehydrating mesophyll cells. Water molecules leaving the xylem are attached to other water molecules in the same xylem tube by hydrogen bonds. Therefore, when one water molecule leaves, it pulls adjacent water molecules up the xylem. As these water molecules move upward, other water molecules farther down the tube move up to replace them. This process continues all the way to the roots, where water in the extracellular space around the xylem is pulled in through the holes in the walls of vessel elements and tracheids. This upward and inward movement of water finally causes soil water to move through the endodermal cells into the vascular cylinder by osmosis. The force even extends out into the soil to pull water into the root. The force generated by the evaporation of water from the leaves, transmitted down the xylem to the roots, is so strong that water can be absorbed even from quite dry soils. But can the cohesion–tension theory explain the movement of water from soil to the topmost leaves of giant redwoods? Using a special apparatus, botanists have measured xylem water tensions strong enough to pull water up more than 600 feet (200 meters).

Water Enters Roots Mainly by Pressure Differences Created by Transpiration

Water takes two routes on its way to the vascular cylinder. First, water follows the minerals that are actively pumped into epidermal cells, moving into the epidermal cells by osmosis (black arrow in Fig. 24-19, p. 479). Second, water moves readily by bulk flow through the highly porous cell walls of the epidermis and outer layers of cortex, without penetrating cell membranes until it

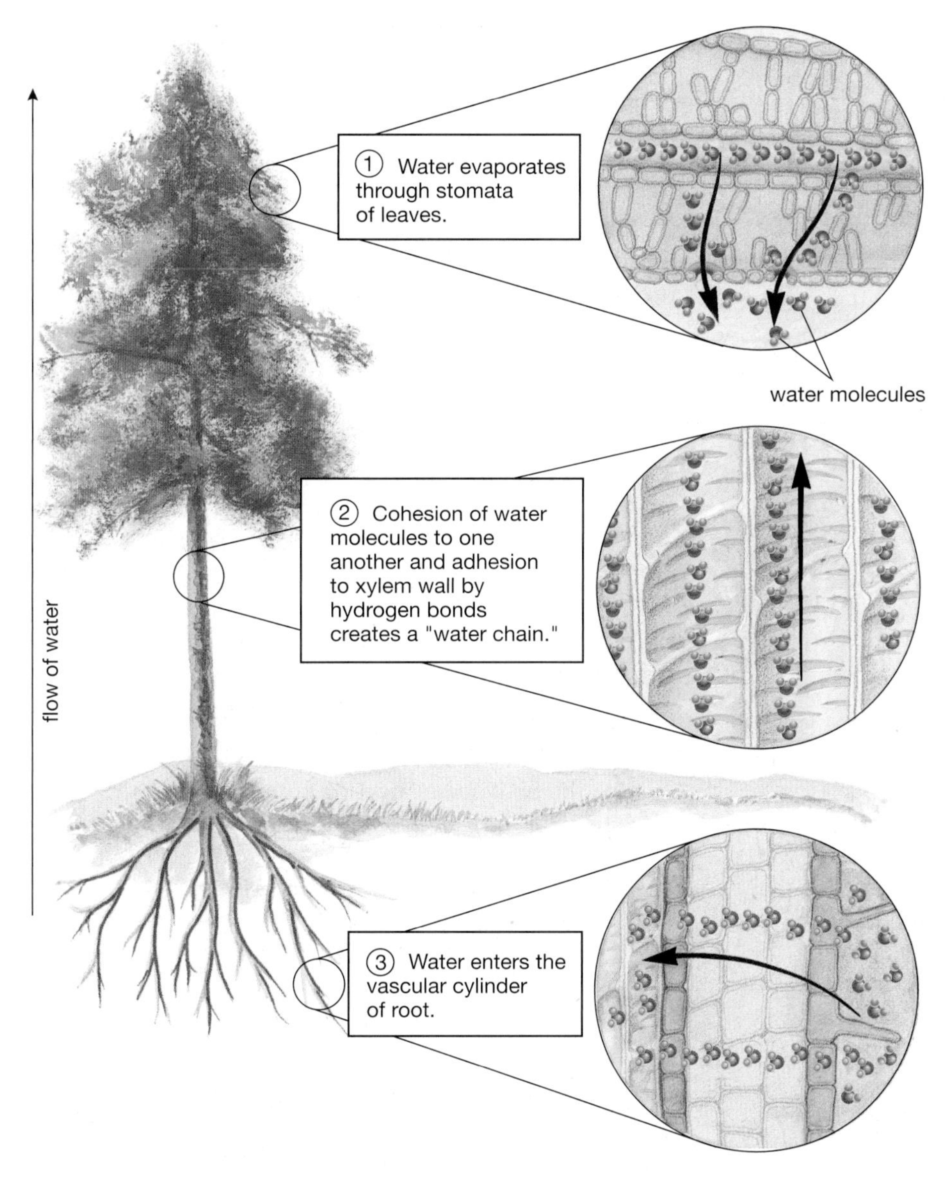
1 Water evaporates through stomata of leaves.
water molecules
2 Cohesion of water molecules to one another and adhesion to xylem wall by hydrogen bonds creates a "water chain."
3 Water enters the vascular cylinder of root.
flow of water

FIGURE 24-22 The cohesion–tension theory of water flow from root to leaf in xylem
(1) As water molecules evaporate out of the leaves through transpiration, other water molecules replace them from the xylem of the leaf veins. (2) As the top of the "water chain" is pulled up by evaporation, the rest of the chain, all the way down to the roots, comes along as well. (3) As the molecules of the water chain retreat up the xylem in the roots, the decreased water pressure within the root xylem and the surrounding extracellular space causes water to enter from the soil water, thus steadily replenishing the bottom of the chain.

reaches the endodermis (blue arrow in Fig. 24-19, p. 479). At the endodermis, the waterproof Casparian strip blocks further flow of water through the spaces between cells, so it is forced to cross the endodermal cell membrane by osmosis. After leaving the endodermal cells, water moves into the vascular cylinder, entering the tracheids and vessel elements of the xylem through their porous and pitted cell walls.

Movement of water into roots by osmosis due to the chemical gradient of minerals may be an important mechanism in certain plants when transpiration is low, such as at night when evaporation from the leaves is reduced. When the transpiration rate is high, such as during the day, the second mechanism dominates. Water moves by bulk flow into the epidermis and outer cortex layers of the root between the cells along a pathway created by their porous cell walls, then crosses the endodermal cells by osmosis. The driving force powering this bulk flow of water is lowered water pressure in the vascular cylinder created by water losses due to transpiration from the leaves. Since plants absorb most of their water during periods of high transpiration, this pressure-driven bulk flow mechanism of movement is the primary means by which water enters roots.

Summing Up

Water Transport in Xylem

Transpiration from leaves removes water from the top of a xylem tube. The transpired water is replaced by water from farther down the xylem tube, so water continues to move up the xylem tube by bulk flow. This upward flow removes water from the root xylem and the extracellular space surrounding it, promoting the movement of water by osmosis from the soil into the vascular cylinder of the root. The flow of water is unidirectional, from root to shoot, because only the shoot can transpire.

Adjustable Stomata Control the Rate of Transpiration

Although it provides the force that transports water and minerals to the leaves at the top of the plant, transpiration is also by far the largest source of water loss—a loss that may threaten the very survival of the plant, especially in hot, dry weather. Because most water transpires through the stomata of the leaves and stem, you might think that a plant could prevent water loss simply by closing its stomata. Don't forget, however, that photosynthesis requires carbon dioxide from the air, which diffuses into the leaf primarily through open stomata. Therefore, a plant must use its stomata to achieve a balance between carbon dioxide uptake and water loss.

A stoma consists of a central opening surrounded by two sausage-shaped, photosynthetic guard cells that regulate the size of the opening (Fig. 24-23). With some exceptions, stomata open during the day, when sunlight allows photosynthesis, and close at night, conserving water. They will also close in the sunlight if the plant is losing too much water. Plants whose leaves are oriented horizontally generally have more stomata on the shaded, lower surface than on the sunny, upper surface to reduce evaporation, as illustrated in Figure 24-8.

How do plants regulate stomatal opening and closing? Stomata open when the guard cells take up water and elongate, bowing outward and increasing the space between them. Stomata close when guard cells lose water and shrink, reducing the space between them. The entry of water, in turn, is regulated by changes in the potassium content of the guard cells. The plant opens its stomata by actively pumping potassium into the guard cells, causing water to follow by osmosis. When potassium leaves the guard cells, water follows by osmosis, and the stomata close.

Several factors regulate the potassium concentration inside guard cells. The three most important factors are availability of light, availability of carbon dioxide, and water levels within the leaf. These factors help the plant achieve a balance between the need to photosynthesize and the need to conserve water.

- *Light reception.* When light strikes special pigments within the guard cells, it triggers a series of reactions that cause potassium to be actively transported into the guard cells. Water follows by osmosis, and the stomata open. At night, when light is not present to activate the pigments, the potassium pumping stops. The "extra" potassium within the guard cells diffuses back out, and the stomata close, conserving water.
- *Carbon dioxide concentration.* Low CO_2 concentrations (such as those that occur during the day when photosynthesis exceeds cellular respiration) stimulate the active transport of potassium into the guard cells. This transport causes stomata to open and allows CO_2 to diffuse in. At night, cellular respiration in the absence of photosynthesis raises CO_2 levels, halting the inward transport of potassium and allowing the guard cells to close.

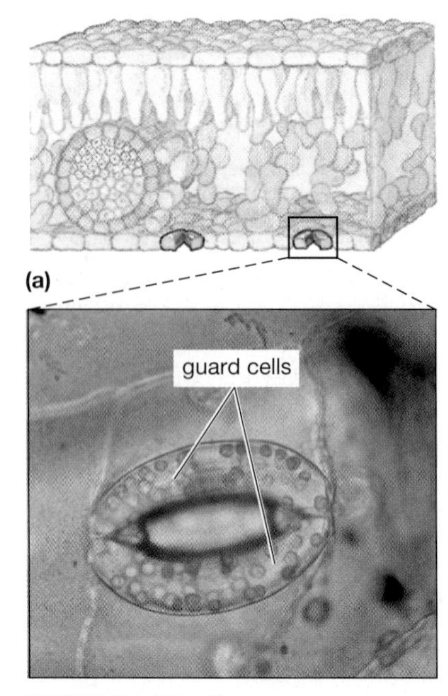

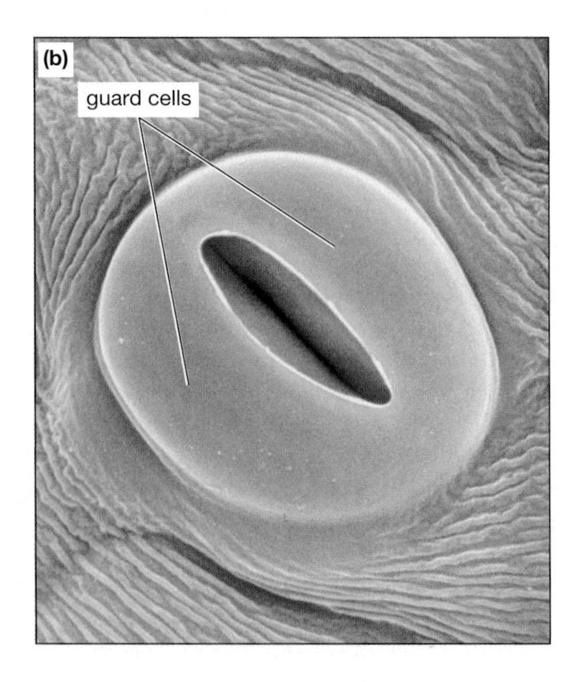

FIGURE 24-23 Stomata
Stomata seen through a **(a)** light microscope and **(b)** scanning electron microscope. In the light micrograph, note that the guard cells contain chloroplasts (the green ovals within the cells) but that the other epidermal cells do not. **QUESTION** When the stomata close, how is photosynthesis affected? How is the movement of water into the roots affected?

- *Water.* If a leaf loses water faster than it can be replaced with water from the xylem, it begins to wilt, and the mesophyll cells release a hormone called **abscisic acid**. This hormone strongly inhibits the active transport of potassium into the guard cells, overriding the stimulatory effects of light and low CO_2 levels, so potassium pumping stops. As potassium diffuses out of the guard cells, water follows by osmosis, the guard cells shrink, and the stomata close. As you might guess, when your house or garden plants are wilted, they are unable to carry out normal levels of photosynthesis.

24.8 How Do Plants Transport Sugars?

Sugars synthesized in the leaves must be moved to other parts of the plant, where they nourish nonphotosynthetic structures such as roots or flowers and can be stored in the cortex cells of the root and stem. Sugar transport is the function of phloem.

Botanists studying phloem contents have employed a most unlikely lab assistant: the aphid. *Aphids* are insects that feed on the fluid contained in phloem sieve tubes. An aphid inserts its *stylet*, a pointed, hollow tube, through the epidermis and cortex of a young stem into a sieve tube (Fig. 24-24). The aphid can then relax and let the plant do the work. The fluid in the sieve tubes is under pressure and flows through the stylet into the aphid's digestive tract. By cutting off the aphid but leaving its stylet in place, botanists have collected sieve-tube fluid and found that it consists mostly of water that is about 10% to almost 25% sucrose by weight. How is this concentrated sugar solution moved throughout the plant?

The Pressure-Flow Theory Explains Sugar Movement in Phloem

The movement of fluid in phloem is directed by sugar production and use. Any structure that actively synthesizes sugar is said to be a **source** away from which phloem fluid will be transported. Conversely, any structure that uses up sugar or converts sugar to starch is said to be a **sink** toward which phloem fluids will flow. A newly forming leaf will be a sink as it develops, with phloem flowing up into it from more mature leaves. When the leaf matures, it will photosynthesize and produce sugar, becoming a source for phloem flow to other newly developing leaves, to flowers or fruits, or to the roots. Therefore, fluid in phloem can move either up or down the plant, depending on the metabolic demands of the various parts of the plant at any given time.

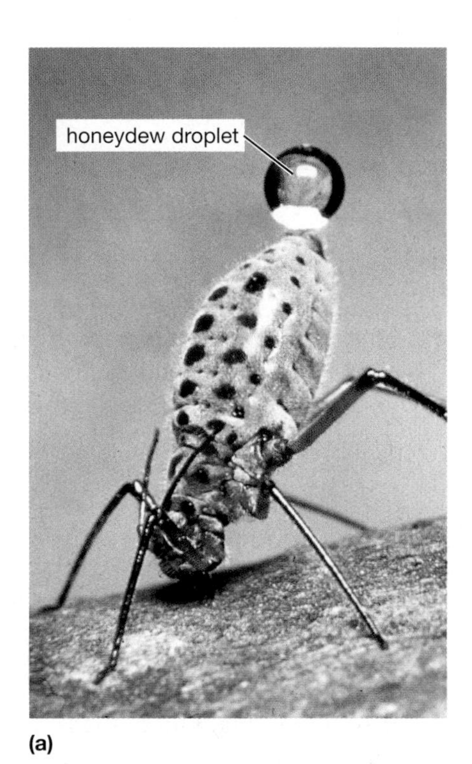

(a)

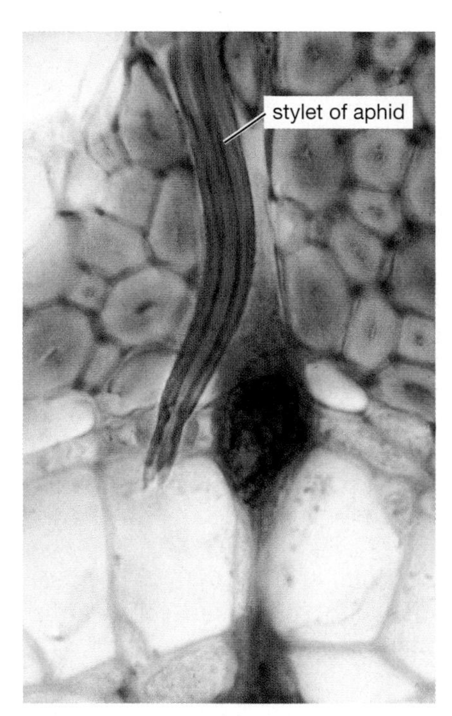

(b)

FIGURE 24-24 Aphids feed on the sugary fluid in phloem sieve tubes
(a) When an aphid pierces a sieve tube, pressure in the tube forces the fluid out of the phloem and into the aphid's digestive tract. The pressure can be so great that fluid is forced completely through the aphid and out its anus, as "honeydew." (This fluid is collected by certain species of ants that defend the aphids from predators in return for a diet of sweet honeydew.) **(b)** The flexible stylet of an aphid passes through many layers of cells to penetrate a sieve-tube element.

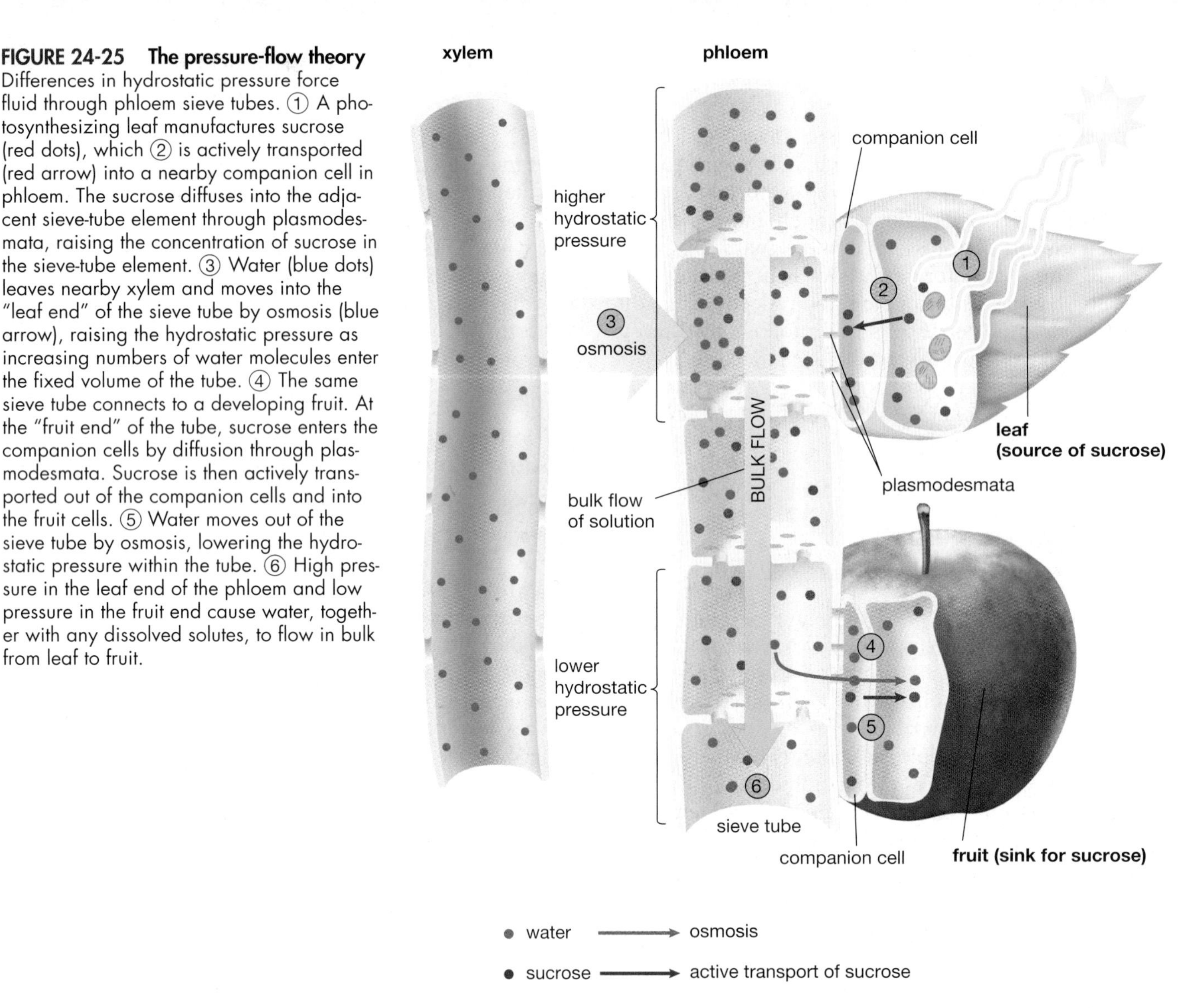

FIGURE 24-25 The pressure-flow theory Differences in hydrostatic pressure force fluid through phloem sieve tubes. ① A photosynthesizing leaf manufactures sucrose (red dots), which ② is actively transported (red arrow) into a nearby companion cell in phloem. The sucrose diffuses into the adjacent sieve-tube element through plasmodesmata, raising the concentration of sucrose in the sieve-tube element. ③ Water (blue dots) leaves nearby xylem and moves into the "leaf end" of the sieve tube by osmosis (blue arrow), raising the hydrostatic pressure as increasing numbers of water molecules enter the fixed volume of the tube. ④ The same sieve tube connects to a developing fruit. At the "fruit end" of the tube, sucrose enters the companion cells by diffusion through plasmodesmata. Sucrose is then actively transported out of the companion cells and into the fruit cells. ⑤ Water moves out of the sieve tube by osmosis, lowering the hydrostatic pressure within the tube. ⑥ High pressure in the leaf end of the phloem and low pressure in the fruit end cause water, together with any dissolved solutes, to flow in bulk from leaf to fruit.

The most widely accepted mechanism of sugar transport in phloem is the **pressure-flow theory** (Fig. 24-25), which relies on differences in hydrostatic pressure (water pressure) to move fluid through sieve tubes. Let's illustrate this theory by following sucrose movements from a mature leaf to a developing fruit:

① *Sucrose source: photosynthesis.* When a leaf is photosynthesizing rapidly, it manufactures a lot of glucose, much of which is converted to the larger sucrose molecule.

② *Phloem sieve-tube loading.* Much of this sucrose is actively transported into companion cells of the phloem through the leaf veins. This movement raises the concentration of sucrose within the companion cells, so sucrose then diffuses down its concentration gradient through plasmodesmata into adjacent sieve-tube elements. This diffusion, in turn, raises the sucrose concentration in the leaf sieve tube.

③ *Osmosis into the leaf sieve tube.* The high sucrose concentration in the leaf sieve tube lowers the water concentration in the sieve tube, causing water to enter the sieve tube by osmosis from nearby xylem. Hydrostatic pressure increases as more water molecules enter the tube.

④ *Sucrose sink: developing fruit.* Meanwhile, some distance away along the same sieve tube, sucrose is actively transported out of sieve-tube elements and companion cells into the cells of a fruit. The concentration of sugar in the fruit is raised, and the concentration of sugar in the sieve tubes is lowered.

⑤ *Osmosis out of the fruit sieve tube.* Water leaves the sieve tube by osmosis and follows the sugar into the fruit. Hydrostatic pressure drops within the tube.

⑥ *Bulk flow, driven by a hydrostatic pressure gradient.* Water moving by osmosis follows the sucrose into the sieve tube near the leaf, causing hydrostatic pressure to build up in the leaf portion of the sieve tube. Meanwhile, water entering the fruit by osmosis causes reduced hydrostatic pressure in the sieve tube near the fruit. In response to this pressure gradient, water moves by bulk flow from the leaf portion of the phloem into the fruit portion of the phloem, carrying the dissolved sugar.

FIGURE 24-26 Root adaptations
(a) Roots modified for nutrient storage include (left to right) beets, carrots, and radishes. Note that these are all dicot taproots. **(b)** This *Cattleya* orchid (a monocot) grows on a tree branch in the Amazon basin; its aerial roots dangle below the branch.

EVOLUTIONARY CONNECTIONS

Special Adaptations of Roots, Stems, and Leaves

Not all roots are sinuous fibers; not all stems are smooth and upright; and not all leaves are flat and fanlike. Just as evolution has modified the basic shape of the vertebrate forelimb to suit the demands of running, swimming, and flying, so too have plant parts become modified in response to environmental demands by the forces of natural selection. All plants, even the most ordinary, are adapted to their environments. The "typical" leaf of an oak or maple is just as much a special adaptation as a cactus spine or daffodil bulb. You may be surprised to learn that many familiar structures are derived from unlikely parts of a plant.

Some Specialized Roots Store Food; Others Photosynthesize

Roots have probably undergone fewer unusual modifications of their basic structure than have either stems or leaves. Some roots have extreme specializations for storage, such as the beet, carrot, or radish (Fig. 24-26a). Some of the most unusual root adaptations occur in certain orchids that grow perched on trees. A few of these aerial orchids have green, photosynthetic roots (Fig. 24-26b).

Some Specialized Stems Produce New Plants, Store Water or Food, or Produce Thorns or Climbing Tendrils

Many plants have modified stems that perform functions very different from merely raising leaves up to the light. Strawberries, for example, grow horizontal *runners* that snake out over the soil, sprouting new strawberry plants where nodes touch the soil (Fig. 24-27a). These new plants are connected to the "mother" plant, but once the plantlets form roots, they can live independently.

Some plants, such as the baobab tree (Fig. 24-27b), store water in aboveground stems. Many other plants store carbohydrates in underground stems. The common white potato is actually a storage stem; each "eye"

FIGURE 24-27 Stem adaptations
(a) The beach strawberry can reproduce using runners, horizontal stems that spread out over the surface of the sand. Where the node of a runner touches the soil, it sprouts roots and develops into a complete plant. **(b)** The baobab tree has an enormously expanded water-storing trunk, allowing it to survive in a dry climate. Some baobab trees have trunks so large that they have been hollowed out and used as small houses and, in one case, as a jail!

is a lateral bud, ready to send up a branch when conditions become favorable, using the energy stored in starch within the potato to power the growth of the branch. If you store potatoes for too long, you may find them sprouting branches in the refrigerator. Irises have horizontal underground stems called *rhizomes*, which store carbohydrates produced during the summer. Irises can be propagated by cutting up the rhizome; if it contains enough stored food, each piece with a node can generate a complete plant.

Many aboveground stems produce modified branches with special functions. One common branch adaptation is the *thorn*, generally growing from the usual branch location just above the attachment site of a leaf. Hard, sharp thorns discourage animals from dining on the branches. Some of the branches of grapes and Boston ivy are modified into grasping *tendrils*, which coil around trees, trellises, or buildings, providing an otherwise prostrate plant better access to sunlight.

Specialized Leaves May Conserve and Store Water, Store Food, or Even Capture Insects

The most important environmental factors affecting the growth of leaves are temperature and availability of light and water. For example, plants growing on the floor of a tropical rain forest have plenty of water year-round but very little light, owing to the deep shade cast by several layers of trees above them. Consequently, their leaves tend to be extremely large—an adaptation demanded by the low light level and permitted by the abundant water.

At the other extreme, deserts receive bright sunlight virtually every day but have limited water and scorching temperatures. Desert plants have evolved two strikingly different adaptations to this situation. Some have very thick leaves with large cells that store water from the infrequent rains against the inevitable long droughts (Fig. 24-28a). Such leaves are covered with a thick cuticle

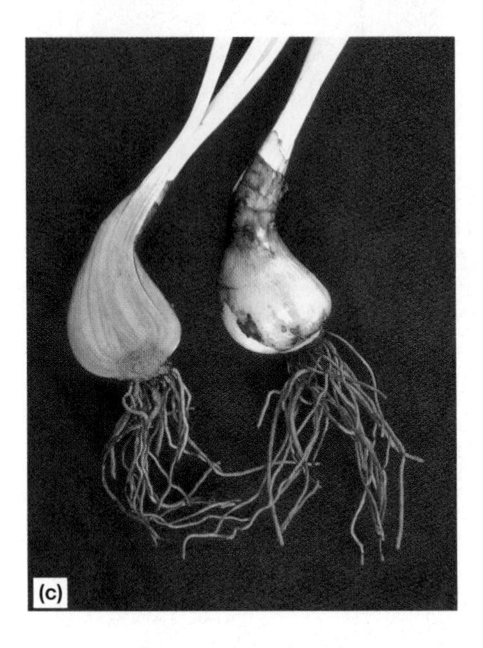

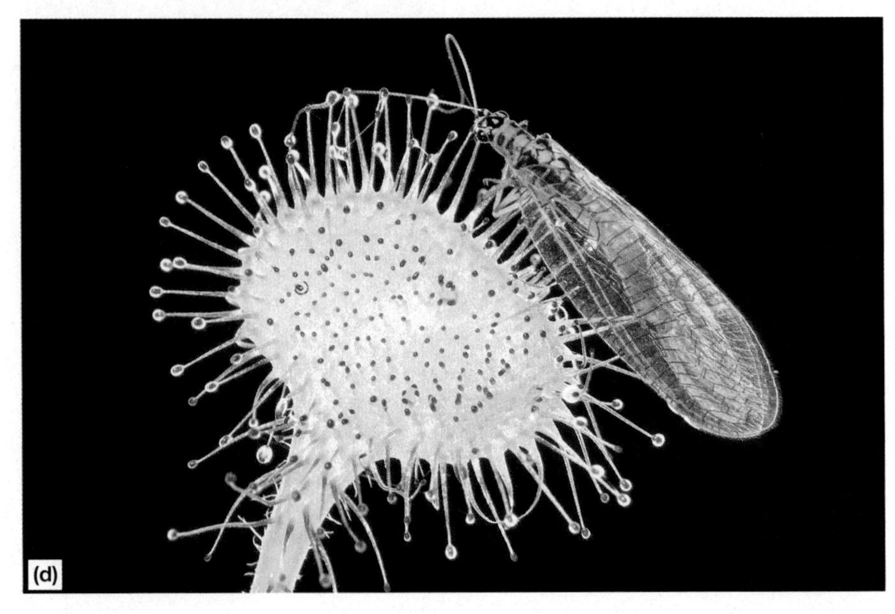

FIGURE 24-28 Leaf adaptations
(a) Some desert plants have evolved fleshy leaves that store water from the occasional rains. **(b)** Spines of desert cacti are non-photosynthetic leaves whose surface area has been minimized, reducing evaporation and protecting the plant from grazing animals. **(c)** Daffodil bulbs consist of short central stems surrounded by thick, water- and food-storing leaves. **(d)** A sundew grasps a lacewing fly with sticky, enzyme-laden hairs.

that greatly reduces the evaporation of water. Cacti use the opposite strategy, reducing the leaves to thin spines that protect the plant from herbivores and reduce water loss (Fig. 24-28b). Photosynthesis in cacti occurs in cortex cells of the green, water-storing stems.

Modified leaves in other plants function in ways unrelated to photosynthesis or water conservation. The common edible pea, for example, grasps fences, mailbox posts, or other plants with clinging tendrils. Unlike the tendrils of grapes, which are derived from branches, pea tendrils are slender, supple leaflets. Some plants, such as onions, daffodils, and tulips, use thick, fleshy leaves as storage organs. A daffodil bulb consists of a short stem bearing thick, overlapping leaves that store nutrients during the winter (Fig. 24-28c). Finally, a few plants have turned the table on the animals and have become predators. The protein-rich bodies of animals such as insects are excellent sources of nitrogen, if they can only be caught and digested—a formidable challenge for a plant. Carnivorous plants including Venus flytraps and sundews (Fig. 24-28d), have leaves that are modified into snares for trapping unwary insects. These leaves are adaptations evolved by plants that colonize nitrogen-poor soils. Bogs are prime habitat for carnivorous plants because their acidic soils create an environment that is hostile to nitrogen-fixing bacteria.

CASE STUDY REVISITED Why Do Leaves Turn Red in the Fall?

At the onset of autumn, as temperatures cool and light levels remain high, the metabolic rate of the leaf slows, rendering it unable to utilize all the light it absorbs. The excess light energy can damage chloroplasts and leaf cells and further reduce photosynthesis. In the laboratory, Lee and Gould exposed both red and green dogwood leaves to intense light. They found that the leaves containing more red anthocyanin pigment were much better protected from the effects of excess light energy than those lacking it. Intense sunlight hitting leaves also causes the production of free radicals—highly reactive molecules that can damage cellular components. Scientists now have evidence that anthocyanins reduce the formation of free radicals by absorbing wavelengths of sunlight energy that are not used in photosynthesis. Further, these versatile red molecules also act as antioxidants, reacting with any free radicals that are formed and rendering them harmless.

So why protect a dying leaf? Because both chlorophyll and carotenoids are rich in nitrogen, most of the plant's nitrogen is found in its leaves. To conserve this valuable nutrient, perennial plants salvage the nitrogen from dying leaves and pump it into woody tissues for storage over the winter. But this takes energy, which the plant derives from photosynthesis. Perhaps, Lee and Gould suggest, by protecting leaves during their final days, anthocyanin allows the plant to continue to photosynthesize for as long as possible and acquire the energy it needs to salvage its nitrogen for use in the coming spring.

Not all leaves turn red in the fall, and anthocyanins probably play a variety of roles in the leaves of different plant species. As researchers probe deeper into the question of why leaves turn red, we can simply delight in the autumn colors, and be glad that they exist.

Consider This: Researcher William Hoch investigated the redness of leaves from 74 species of plants native to regions with cold winter climates (in the northern United States and Canada) where temperatures plunge in the fall and milder climates (in coastal Europe). He found that the 41 species that produced the reddest leaves in autumn all came from the colder climates. What hypothesis does this support for the function of anthocyanins? Does it "prove" anything? Design another field survey that could provide further support for this role. Explain what you would expect if your survey supported the hypothesis.

Links to Life: From Pine to Pulp to Paper

If you're a typical student, you're probably surrounded by paper. About 90% of all the paper we use comes from trees. But how is it made? Paper is derived from wood pulp, which is made from the woody parts of conifers. At a paper mill, the bark is stripped off, and the rest of the trunk is ground up. The wood is mixed with water and treated with chemicals that dissolve the lignin. This results in a slurry of the tracheid fibers that comprise most of the conifer trunk. The slurry and dissolved lignin can then be dried to produce newsprint, which gradually becomes yellow and brittle due to the high content of lignin, a stiffening substance in wood. To make higher-grade paper, the lignin must be removed and the fiber bleached, which reduces the yield and greatly increases the costs. Clay, whitening agents, latex, and products derived from algae may also be added to produce white, glossy paper. Two to three tons of trees are required to make a ton of office paper. Recycling paper can conserve much of the energy and substantially reduce pollution that results from making new paper from trees. According to the Institute of Scrap Recycling Industries, recycling all the morning newspapers in the United States would save 41,000 trees every day. Currently, less than half of all paper used in the United States is recycled, and paper products make up 40% of municipal solid waste. Since the United States consumes roughly 100 million tons of paper annually, that's a lot of trees converted into trash!

Summary of Key Concepts

24.1 How Are Plant Bodies Organized, and How Do They Grow?

The body of a land plant consists of root and shoot. Roots are usually underground and have six functions: anchoring the plant in the soil; absorbing water and minerals from the soil; storing surplus photosynthetic products; transporting water, minerals, photosynthetic products, and hormones; producing some hormones; and interacting with soil fungi and microorganisms that provide nutrients. Shoots are generally located above ground and consist of stem, leaves, buds, and (in season) flowers and fruit. Shoot functions include photosynthesis, transport of materials, reproduction, and hormone synthesis.

Plant bodies are composed of two main classes of cells: meristem cells and differentiated cells. Meristem cells are undifferentiated cells that retain the capacity for mitotic cell division. Differentiated cells arise from divisions of meristem cells, become specialized for particular functions, and usually do not divide. Most meristem cells are located in apical meristems at the tips of roots and shoots and in lateral meristems in the shafts of roots and shoots. Primary growth (growth in length and differentiation of parts) results from the division and differentiation of cells from apical meristems; secondary growth (growth in diameter) results from the division and differentiation of cells from lateral meristems.

24.2 What Are the Tissues and Cell Types of Plants?

Plant bodies consist of three tissue systems: the dermal, ground, and vascular systems. The dermal tissue system forms the outer covering of the plant body. The dermal tissue system of leaves and of primary roots and stems is usually a single cell layer of epidermis. Dermal tissue after secondary growth is a multilayered covering of cork.

The ground tissue system consists of a variety of cell types, including parenchyma, collenchyma, and sclerenchyma. Most are involved in photosynthesis, support, or storage. Ground tissue makes up most of a young plant during primary growth. During secondary growth of stems and roots, ground tissue becomes an increasingly small part of the plant body.

The vascular tissue system consists of xylem, which transports water and minerals from the roots to the shoots, and phloem, which transports water, sugars, amino acids, and hormones throughout the plant body.

24.3 Leaves: Nature's Solar Collectors

Leaves are the main photosynthetic organs of plants. The blade of a leaf consists of a waterproof outer epidermis surrounding mesophyll cells, which have chloroplasts and carry out photosynthesis, and vascular bundles of xylem and phloem, which carry water, minerals, and photosynthetic products to and from the leaf. The epidermis is perforated by adjustable pores called stomata that regulate the exchange of gases and water.

24.4 Stems: Reaching for the Light

Primary growth in dicot stems results in a structure consisting of an outer, waterproof epidermis; supporting and photosynthetic cells of cortex beneath the epidermis; vascular tissues of xylem and phloem; and supporting and storage cells of pith at the center. Leaves and lateral buds are found at nodes along the surface of the stem. Under the proper hormonal conditions, lateral buds may sprout into a branch. Secondary growth in stems results from cell divisions in the vascular cambium and cork cambium. Vascular cambium produces secondary xylem and secondary phloem, increasing the stem's diameter. Cork cambium produces waterproof cork cells that cover the outside of the stem.

24.5 Roots: Anchorage, Absorption, and Storage

Primary growth in roots results in a structure consisting of an outer epidermis, an inner vascular cylinder of conducting tissues, and cortex between the two. The apical meristem near the tip of the root is protected by the root cap. Cells of the root epidermis absorb water and minerals from the soil. Root hairs are projections of epidermal cells that increase the surface area for absorption. Most cortex cells store surplus sugars (usually in the form of starch) produced by photosynthesis. The innermost layer of cortex cells is the endodermis, which controls the movement of water and minerals from the soil into the vascular cylinder. The vascular cylinder contains the conducting tissues—xylem and phloem.

24.6 How Do Plants Acquire Nutrients?

Most minerals are taken up from the soil water by active transport into the root hairs. These minerals diffuse into the root through plasmodesmata to the pericycle, just inside the vascular cylinder. There they are actively transported into the extracellular space of the vascular cylinder. The minerals diffuse from the extracellular space into the tracheids and vessel elements of xylem.

Many plants have fungi called *mycorrhizae* associated with their roots that help absorb soil nutrients. Nitrogen can be absorbed only as ammonium or nitrate, both of which are scarce in most soils. Legumes have evolved a cooperative relationship with nitrogen-fixing bacteria that invade legume roots. The plant provides the bacteria with sugars, and the bacteria use some of the energy in those sugars to convert atmospheric nitrogen to ammonium, which is then absorbed by the plant.

24.7 How Do Plants Acquire and Transport Water and Minerals?

The cohesion–tension theory explains xylem function: The cohesion of water molecules to one another by hydrogen bonds holds together the water within xylem tubes almost as if it were a solid chain. As water molecules evaporate from the leaves during transpiration, the hydrogen bonds pull other water molecules up the xylem to replace them. This movement is transmitted down the xylem to the root, where water loss from the vascular cylinder promotes water movement across the endodermis from the soil water by osmosis. Because the cells of the root epidermis and cortex are loosely packed and have porous walls, water in the soil has a continuous, uninterrupted pathway through the outer layers of the root to the waterproof layer of the Casparian strip between endodermal cells. Both mineral uptake and the upward movement of water in xylem, pulled by transpiration, contribute to a concentration gradient of water across the endodermal cells, with a higher concentration of free water molecules in the extracellular space outside the endodermis than in the extracellular space inside the endodermis.

Therefore, water moves by osmosis across the plasma membranes of the endodermal cells into the extracellular space of the vascular cylinder. The water pressure gradient caused by loss of water through transpiration is the primary force drawing water into the root.

24.8 How Do Plants Transport Sugars?

The pressure-flow theory explains sugar transport in phloem. Parts of the plant that synthesize sugar (for example, leaves) export sugar into the sieve tube. Increasing sugar concentrations attract water entry by osmosis, causing high hydrostatic pressure in that part of the phloem. Parts of the plant that consume sugar (for example, fruits) remove sugar from the sieve tube. The loss of sugar causes the loss of water by osmosis, resulting in low hydrostatic pressure. Water and dissolved sugar move by bulk flow in the sieve tube from areas of high to low pressure.

Key Terms

abscisic acid *p. 485*
annual ring *p. 474*
apical meristem *p. 467*
bark *p. 476*
blade *p. 471*
branch root *p. 476*
bulk flow *p. 482*
cambium *p. 467*
Casparian strip *p. 478*
cohesion–tension theory *p. 482*
collenchyma *p. 469*
companion cell *p. 471*
cork cambium *p. 475*
cork cell *p. 468*
cortex *p. 473*
cuticle *p. 468*
dermal tissue system *p. 468*
dicot *p. 466*
differentiated cell *p. 467*
endodermis *p. 476*
epidermal tissue *p. 468*
epidermis *p. 468*
fibrous root system *p. 476*
ground tissue system *p. 468*
guard cell *p. 471*
heartwood *p. 474*
internode *p. 472*
lateral bud *p. 472*
lateral meristem *p. 467*
leaf *p. 471*
leaf primordium *p. 472*
legume *p. 480*
meristem cell *p. 466*
mesophyll *p. 471*
mineral *p. 478*
monocot *p. 466*
mycorrhizae *p. 480*
nitrogen fixation *p. 480*
nitrogen-fixing bacterium *p. 480*
node *p. 472*
nodule *p. 480*
nutrient *p. 478*
parenchyma *p. 469*
pericycle *p. 476*
periderm *p. 468*
petiole *p. 471*
phloem *p. 470*
pit *p. 470*
pith *p. 473*
pressure-flow theory *p. 486*
primary growth *p. 467*
primary root *p. 476*
root *p. 466*
root cap *p. 476*
root hair *p. 468*
root system *p. 466*
sapwood *p. 474*
sclerenchyma *p. 469*
secondary growth *p. 467*
shoot system *p. 466*
sieve plate *p. 470*
sieve tube *p. 470*
sieve-tube element *p. 470*
sink *p. 485*
source *p. 485*
stem *p. 466*
stoma *p. 471*
taproot system *p. 476*
terminal bud *p. 472*
tracheid *p. 470*
transpiration *p. 482*
vascular bundle *p. 471*
vascular cambium *p. 473*
vascular cylinder *p. 478*
vascular tissue system *p. 468*
vein *p. 471*
vessel *p. 470*
vessel element *p. 470*
xylem *p. 470*

Thinking Through the Concepts

To take a multiple-choice quiz with feedback on the contents of this chapter, visit http://www.prenhall.com/audesirk7. *Log in to the Web site selected by your instructor and navigate to the Self Test section for this chapter.*

? Review Questions

1. Describe the locations and functions of the three tissue systems in land plants.
2. Distinguish between primary growth and secondary growth, and describe the cell types involved in each.
3. Distinguish between meristem cells and differentiated cells. Which meristems cause primary growth? Which ones form secondary growth? Where is each type located?
4. Diagram the internal structure of a root after primary growth, labeling and describing the function of epidermis, cortex, endodermis, pericycle, xylem, and phloem. What tissues are located in the vascular cylinder?
5. How do xylem and phloem differ?
6. What are the main functions of roots, stems, and leaves?
7. What types of cells form root hairs? What is the function of root hairs?
8. Diagram the internal structure of leaves. What structures regulate water loss and CO_2 absorption by a leaf?
9. What role does abscisic acid play in controlling the opening and closing of stomata? Describe the daily cycle of the opening and closing of guard cells. How are various environmental conditions involved in this process?
10. You and a friend carved your initials 5 feet above the ground on a tree on campus that was 40 feet tall. Now returning for your 25th reunion, you are ashamed of your actions, but you wonder if you will still find the damage on the tree, which is now 60 feet tall. How high above the ground should you look for your initials? Explain.

Applying the Concepts

1. A mutant form of aphid, the klutzphid, inserts its stylet into the vessel elements of xylem. What materials are found in the fluids of xylem? Could an aphid live on xylem fluid? Would xylem fluid flow into the aphid? Explain your answer.
2. One of the foremost goals of molecular botanists is to insert the genes for nitrogen fixation, or the ability to enter into symbiotic relationships with nitrogen-fixing bacteria, into crop plants such as corn or wheat (see Chapter 13). Why would the insertion of such genes be useful? What changes in farming practices would this technique allow?
3. We learned in Chapter 2 about the peculiar characteristics of water. Discuss several ways in which the evolution of vascular plants has been greatly influenced by water's special characteristics.
4. A major environmental problem is desertification, in which overgrazing by cattle or other animals results in too few plants in an area. Show how what you know about the movement of water through plants enables you to understand this process, in which there is less water in the atmosphere, less rain, and thus dry, desertlike conditions.
5. The tropical rain forest contains a large number of as-yet unidentified plants, many of which may have uses as medicines or food. If you were given the job of searching a particular portion of the rain forest for useful products, how would you use the information you gained from this chapter to help narrow your search? What kinds of plant tissues or organs would be most likely to contain such products?
6. The desert tends to have two types of plants with respect to their root systems—small grasses or herbs, and shrubs or small trees. The grasses and herbs typically form fibrous root systems. The shrubs and trees form taproot systems. What advantages can you think of for each system? How does each type of root allow for survival in a desert environment?
7. Grasses (monocots) form their primary meristem near the ground surface rather than at the tips of branches the way dicots do. How does this feature allow you to grow a lawn and mow it every week in the summer? What would happen if you had a dicot lawn and tried to mow it?
8. Discuss the structures and adaptations that might occur in the leaves of plants living in (a) dry, sunny habitats; (b) wet, sunny habitats; (c) dry, shady habitats; and (d) wet, shady habitats. Which of these habitats do you think would be most inhospitable (for example, in which habitat would it be most difficult to design a functioning leaf)?

For More Information

Baskin, Y. "Forests in the Gas." *Discover*, October 1994. As carbon dioxide increases in the atmosphere, plant relationships will be altered.

Lee, D. W., and Gould, K. S. "Why Leaves Turn Red." *American Scientist*, November/December 2002. Researchers describe studies that led to a hypothesis that red pigments protect dying leaves so they can help the plant conserve valuable nutrients.

Milius, S. "Why Turn Red?" *Science News*, October 16, 2002. Why do leaves about to fall expend the energy to synthesize a new red pigment?

Perkins, S. "Lowland Tree Loss Threatens Cloud Forests." *Science News*, October 20, 2001. Costa Rica's Monteverde cloud forest relies on almost continuous cloud cover. This is now threatened by deforestation of lowland trees whose transpiration humidifies the air.

Weiss, R. "When Plants Act Like Animals." *National Wildlife*, December 1994–January 1995. An engaging look at plant movements.

Zimmer, C. "The Processing Plant." *Discover*, September 1995. The purple pitcher plant supplements its diet by digesting flies, with the surprising help of insect larvae that thrive in the pitcher plant's "digestive chamber."

Zimmer, C. "The Web Below." *Discover*, November 1997. An underground web of mycorrhizae transfers nutrients between trees and helps maintain forest health.

Media Activities

To access a Media Activity visit http://www.prenhall.com/audesirk7. *Log in to the Web site selected by your instructor, navigate to this chapter, and select the appropriate Media Activity number.*

24.1 Plant Anatomy

Estimated time: 5 minutes

This tutorial allows you to explore the plant body. Click on the plant structure to zoom in on it and learn more details about its function.

24.2 Primary and Secondary Growth

Estimated time: 5 minutes

Consider the relationship between primary and secondary growth in a plant stem.

24.3 Nutrient Uptake

Estimated time: 5 minutes

Take a look at how roots take up nutrients from the soil.

24.4 Plant Transport Mechanisms

Estimated time: 5 minutes

Explore the structures and mechanisms plants use to transport water, minerals, and sugars throughout the plant body.

24.5 Web Investigations: Why Do Leaves Turn Red in the Fall?

Estimated time: 10 minutes

What does changing leaf color accomplish for trees? How are the colors developed? The first question is still under active investigation. In this activity, explore the answers to questions about the chemistry and environmental factors involved in fall leaf color.

CHAPTER

25 Plant Reproduction and Development

Although fire kills existing trees, it opens up sunny spaces and, through its smoke and heat, actually sets the stage for seed germination and forest renewal. Inset: The lotus seed holds the dormancy record at 1288 years.

AT A GLANCE

CASE STUDY Time Capsules, Smoke Signals, and Seed Sprouting

Years of drought in the United States have made forests more vulnerable to wildfires. Perhaps you have seen the desolate aftermath of a wildfire: thousands of acres of trees converted into blackened sticks. What might this charred landscape look like in several years? Think about the conditions created by a fire and its aftermath: intense heat and smoke followed by sunlight striking earth that had previously been in constant shade. The blackened soil contains mineral nutrients liberated by the fire from the bodies of trees where they had been stored. But wouldn't all vestiges of life be destroyed by the heat and smoke?

How long can seeds remain dormant? For over 1200 years, on a placid lake in northeastern China, Buddhist monks cultivated the lotus, a type of waterlily considered to be a sacred symbol of purity (inset). Over time, a series of earthquakes altered the landscape and caused the lake to dry up, revealing sediment that covered a treasure—thick-coated lotus seeds that had accumulated over the centuries. Jane Shen-Miller, a plant physiologist at UCLA, was delighted to receive a gift of these large marble-sized seeds from colleagues at the Beijing Institute of Botany. She later revisited the lake bed and observed that "looking for the seeds was like walking back into history."

While these lotus seeds are a natural time capsule, in 1879, Professor William Beal of Michigan State University buried 20 bottles, uncorked and upside down, several feet under the ground. Each bottle contained 50 seeds from each of 21 species of local plants. His goal was to determine how long seeds could remain dormant under the natural levels of darkness, moisture, and temperature found beneath their native soil. Knowing his experiment would outlast his lifetime, he left instructions for one bottle to be unearthed every five years, and the seeds to be exposed to conditions that were favorable for sprouting. Are any of them still viable? Can a lotus seed germinate after a millennium in a lake bed? How can a seed survive a forest fire?

25.1 What Are the Features of Plant Life Cycles?

Do plants engage in sexual activity? Certainly. Many plants can reproduce either sexually or asexually. Asexual reproduction in plants usually involves part of a single plant (such as a stem) giving rise to a new plant. Cells of the parent plant undergo mitotic cell division to produce their offspring asexually. Therefore, these offspring are genetically identical to the parent. In Chapter 24, you encountered several methods of asexual reproduction, including the spreading of runners by strawberries, bulb production by daffodils, and the sprouting of rhizomes by irises. Asexual reproduction can be highly effective, allowing offspring to colonize an entire area where their original parent plant found optimal conditions.

However, if an offspring is genetically identical to its parent, then it is only as well adapted to the environment as its parent. What if the environment changes? Most sexually produced offspring combine genes from both parents, and therefore may be endowed with traits that differ from those of either parent. This new combination of traits may help the offspring cope with a changing environment or invade slightly different habitats. As a result, most organisms, including plants, reproduce sexually at least some of the time.

During the animal life cycle, animals with diploid ($2n$) cells produce haploid (n) gametes (sperm or eggs) by the process of meiotic cell division (meiosis followed by cytoplasmic division). The gametes fuse to form a new diploid cell (a *zygote*) that develops into the adult organism through repeated mitotic cell divisions. The plant life cycle, however, is a bit more complex. Plants have two distinct, multicellular forms—one diploid and one haploid—that give rise to each other. For this reason, the plant life cycle is described as **alternation of generations**:

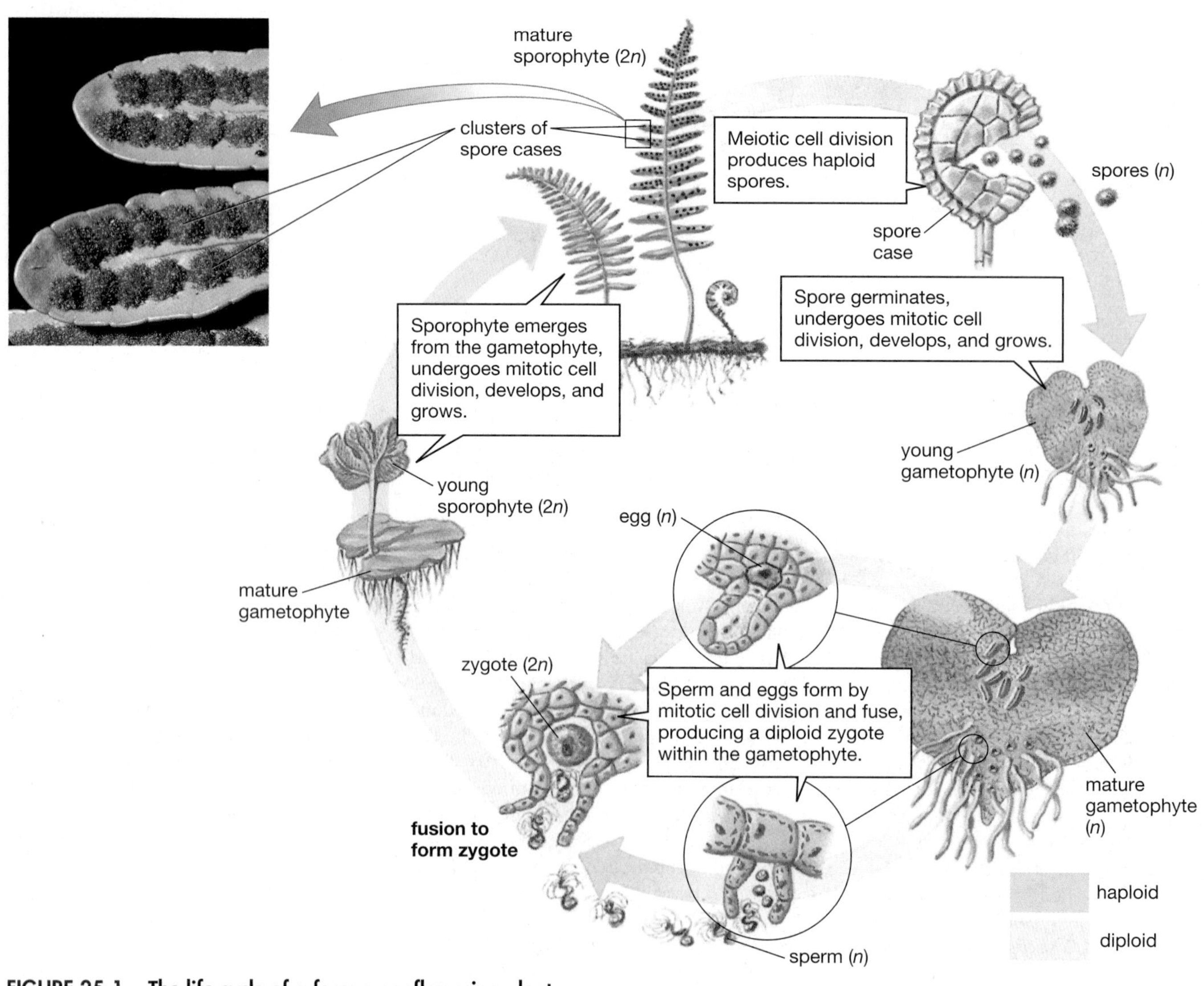

FIGURE 25-1 The life cycle of a fern: a nonflowering plant
Ferns illustrate the alternation-of-generations life cycle found in all plants, in which separate multicellular haploid and multicellular diploid organisms occur at different parts of the life cycle. n refers to the haploid state, $2n$ to the diploid state. When you see ferns, look on the undersides of their leaves; occasionally you will find clusters of brown sporangia (photo inset).

diploid plants (*sporophytes*) alternate with haploid plants (*gametophytes*).

Plants with the most easily visible alternating generations (mosses and ferns, for example) do not produce flowers. To illustrate the alternation of generations, let's examine the life cycle of a fern (Fig. 25-1), starting with the diploid adult form. This stage of the life cycle, the **sporophyte** ("spore plant" in Greek), bears reproductive cells. These cells undergo meiotic cell division to produce haploid cells that are **spores**, not gametes. Unlike gametes, spores do not fuse together to form a diploid cell. Instead, the spore is carried by wind or water from the parent leaf to the soil. There, the spore **germinates** (begins to grow and develop), undergoing repeated mitotic cell division to form a multicellular, haploid organism. This organism produces gametes and hence is called the **gametophyte** ("gamete plant" in Greek). Because its cells are haploid, the gametophyte can produce sperm and eggs without further meiosis. A single gametophyte usually produces both sperm and eggs, but typically at different times, thereby preventing self-fertilization. Sperm and egg fuse to form a fertilized egg, or **zygote**, that develops into a new diploid sporophyte plant.

Alternation of generations occurs in all plants. In primitive land plants, including mosses and ferns, the gametophyte is an independent plant. In ferns, it is smaller than the sporophyte, while in mosses it is much larger. The gametophyte liberates mobile sperm cells that reach an egg either by swimming through thin films of water that cover adjacent gametophytes or by being splashed by raindrops from one plant to the next. For this reason, ferns and mosses can reproduce only in moist habitats.

Many terrestrial habitats are relatively dry. In drier habitats, a plant can reproduce sexually by surrounding its sperm in a watertight package. This package can be transported to another plant, where the sperm are liberated directly into the egg-bearing structures of the second plant. The seed plants (both nonflowering and flowering plants) do just that. **Flowers** are the reproductive structures of flowering plants borne by the sporophyte generation. In flowering plants, two types of spores are formed by meiotic cell division within flowers (Fig. 25-2). These haploid spores develop into microscopic gametophytes within the flower; the gametophyte generation of flowering plants never assumes an independent existence in the soil. One type of spore, the *megaspore*,

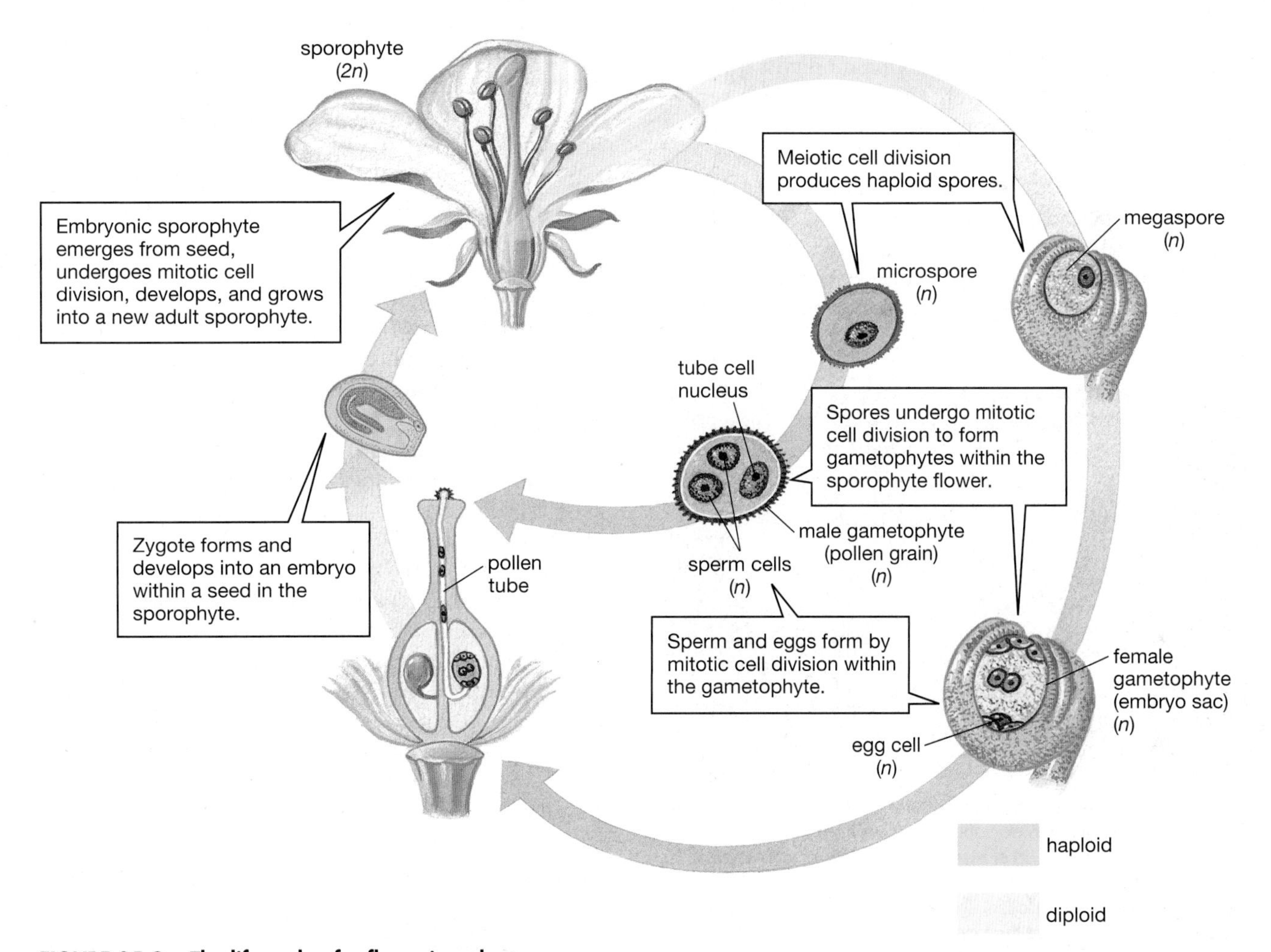

FIGURE 25-2 The life cycle of a flowering plant
Although this cycle shows the same basic stages as the life cycle of a fern (see Fig. 25-1), the haploid gametophyte stages are much smaller and cannot live independently of the diploid plant.

undergoes a few mitotic divisions and develops into the female gametophyte, a small cluster of cells permanently retained within the flower. The other type of spore, the *microspore*, develops into the male gametophyte: a tough, watertight **pollen grain** containing (usually) two sperm. The pollen grain drifts on the wind or is carried by an animal from one flower to another. On the recipient flower, the pollen grain elongates, forming a tube that burrows through the flower's tissues to the female gametophyte within. This miniature male gametophyte liberates its sperm inside the female gametophyte, where fertilization occurs. The zygote becomes enclosed in a drought-resistant **seed**. The seed, including an embryonic plant and a food reserve within a protective outer coating, may lie dormant (in a resting state) for months or years, waiting for conditions favorable for growth.

In the sections that follow, we examine sexual reproduction in flowering plants, from the evolution of the flower through the formation of the seed and the development of the new seedling.

25.2 How Did Flowers Evolve?

The flower is actually a sexual display that enhances a plant's reproductive success. By enticing animals to transfer pollen from one plant to another, flowers enable stationary plants to "court" distant members of their own species. This critical advantage has allowed the flowering plants to become the dominant plants on land.

The earliest seed plants were the gymnosperms, represented today mainly by conifers, a group that includes pines, firs, and spruces. As described in Chapter 21, conifers do not produce flowers; instead, they bear male and female gametophytes on separate cones. During early spring, the small male cones release millions of pollen grains that are carried widely by the wind (Fig. 25-3). So many grains are floating around that some, by chance, enter the pollen chambers located on the scales of the female cones, where they are captured by sticky coatings of sugars and resins. The pollen grains germinate and tunnel to the female gametophytes at the base of each scale (the individual woody extensions that make up the cone). Sperm are liberated and fertilize the eggs within each female gametophyte, and a new generation begins.

FIGURE 25-3 Conifers are wind-pollinated
Even slight breezes blow thick clouds of pollen from ripe male cones. Look for these "soft cones" in clusters near the ends of branches of pine, spruce, and fir trees, often in late spring. The cones disintegrate after releasing their pollen. The larger, woody cones are female and produce seeds at the base of each scale. **QUESTION** Compared to flowering plants pollinated by animals, what advantages do wind-pollinated plants have? What disadvantages?

Clearly, wind pollination is an inefficient operation, because most pollen grains do not reach their target. In a world of stationary plants and mobile animals, a gymnosperm that could entice an animal to carry its pollen from male to female cone would greatly enhance its reproductive rate and hence its evolutionary success. As it happens, gymnosperms and insects established just such a relationship about 150 million years ago.

Insects, especially beetles, are among the most abundant animals on Earth. They exploit nearly every possible food resource on land, including the reproductive parts of gymnosperms. About 150 million years ago, some beetles fed on both the protein-rich pollen of male cones and the sugar-rich secretions of female cones. Beetles can make quite a mess when they feed, and pollen feeders often wind up with pollen dusted all over their bodies. If the same beetle visited one plant and ate pollen, then wandered over to another plant of the same species to dine on the sugary secretions of a female cone, some of the loose pollen would quite likely rub off on the female cone.

Thus, the stage was set for the evolution of flowering plants. Efficient pollination by insects requires that a given insect visit several plants of the same species, pollinating them along the way. For the plants, two key adaptations were necessary. First, enough pollen or nectar (sugary secretions) must be produced within the reproductive structures so that insects will regularly visit them to feed. Second, the location and richness of these storehouses of pollen and nectar must be advertised to the insects, both to show them where to go and to entice them to specialize on that particular plant species. Any mutation that contributed to these adaptations would enhance the reproductive success of the plant that carried the mutation and would be favored by natural selection. By about 130 million years ago, flowers had evolved with exactly these adaptations. The advantages of flowers are so great that in today's temperate and tropical zones flowering plants are overwhelmingly dominant, and numerous animals—including bees, moths, butterflies, hummingbirds, and even some mammals—feed at and pollinate flowers (see "Evolutionary Connections: Adaptations for Pollination and Seed Dispersal"). Although flowers did not evolve to attract people, our nervous systems (for reasons that remain mysterious) are "wired" to appreciate their scents and beauty; we are lured to walk through flowery meadows for the pure pleasure of it.

Complete Flowers Have Four Major Parts

We have seen that flowers are the reproductive structures of flowering plants. Evolution commonly produces new structures by modifying old ones; flower parts are

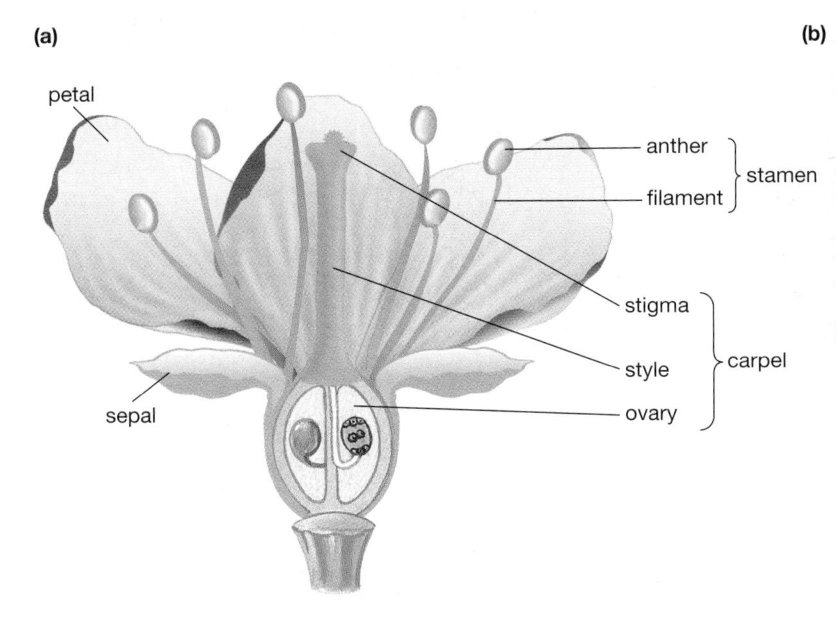

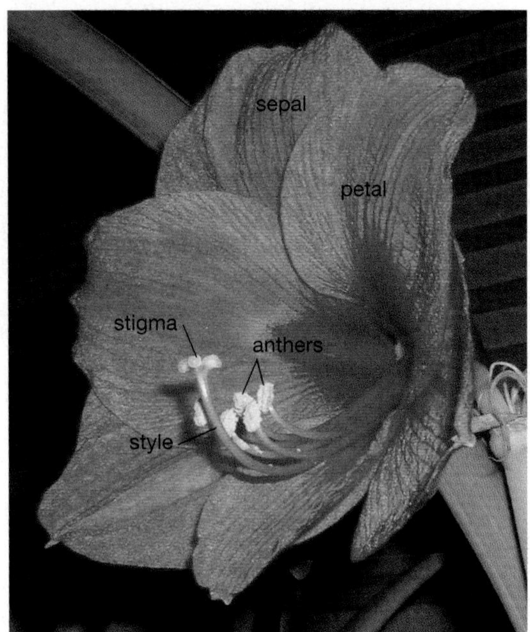

FIGURE 25-4 A complete flower
(a) A complete flower has four parts: sepals, petals, stamens (the male reproductive structures), and at least one carpel (the female reproductive structure). This drawing shows a complete dicot flower, with two ovules illustrated within the ovary. **(b)** The amaryllis is a complete monocot flower, with three sepals (virtually identical to the petals), three petals, six stamens, and three carpels (fused into a single structure). The anthers are well below the stigma, making self-pollination difficult.

actually highly modified leaves, shaped by mutation and natural selection into a form that enhances pollination. A **complete flower**, such as that of a petunia, rose, or lily, consists of a central axis that supports four successive sets of modified leaves (Fig. 25-4). These modified leaves form the *sepals*, *petals*, *stamens*, and *carpel*. The **sepals** are located at the base of the flower. In dicots, sepals are typically green and leaflike; in monocots, most sepals resemble the petals (see Fig. 25-4b). In either case, sepals surround and protect the flower bud as the remaining three structures develop. Just above the sepals are the **petals**, which are often brightly colored and fragrant, advertising the location of the flower.

The male reproductive structures, the **stamens**, are attached just above the petals. Each stamen typically consists of a slender **filament** that supports an **anther**, the structure that produces pollen. The female reproductive structure, the **carpel**, frequently occupies a central position in the flower. A generalized carpel is somewhat vase-shaped, with a sticky **stigma** for catching pollen mounted atop an elongated **style**. The style connects the stigma with the bulbous **ovary** (see Fig. 25-4a). Inside the ovary are one or more **ovules**, in which the female gametophytes develop. When mature, each ovule will become a seed, and the ovary will develop into a protective, adhesive, and/or edible enclosure, the **fruit**.

Incomplete flowers lack one or more of the four floral parts. For example, grass flowers (see Fig. 25-8) lack both petals and sepals. Other incomplete flowers lack either the male stamens or the female carpels. In such cases, the flowers are described as *imperfect*, as well as incomplete. This is not a value judgment; plant species with imperfect flowers are highly successful. They bear separate male and female flowers, sometimes on the same plant, as in the case of garden squash such as zucchini (Fig. 25-5), or on separate plants, as in the American holly, where flowers that produce the familiar red berries are only found on the "female" plants.

FIGURE 25-5 Some plants have separate male and female flowers
Plants of the squash family, such as these zucchinis, bear separate female (left) and male (right) flowers. Each plant initially produces only male flowers, ensuring some cross-pollination between plants that flower at slightly different times. Note the small zucchini (actually a fruit) forming at the base of the female flower. Zucchini fruits are produced only by female flowers.
QUESTION In species with separate male and female flowers on the same plant, why would natural selection favor individuals whose male and female flowers bloom at different times?

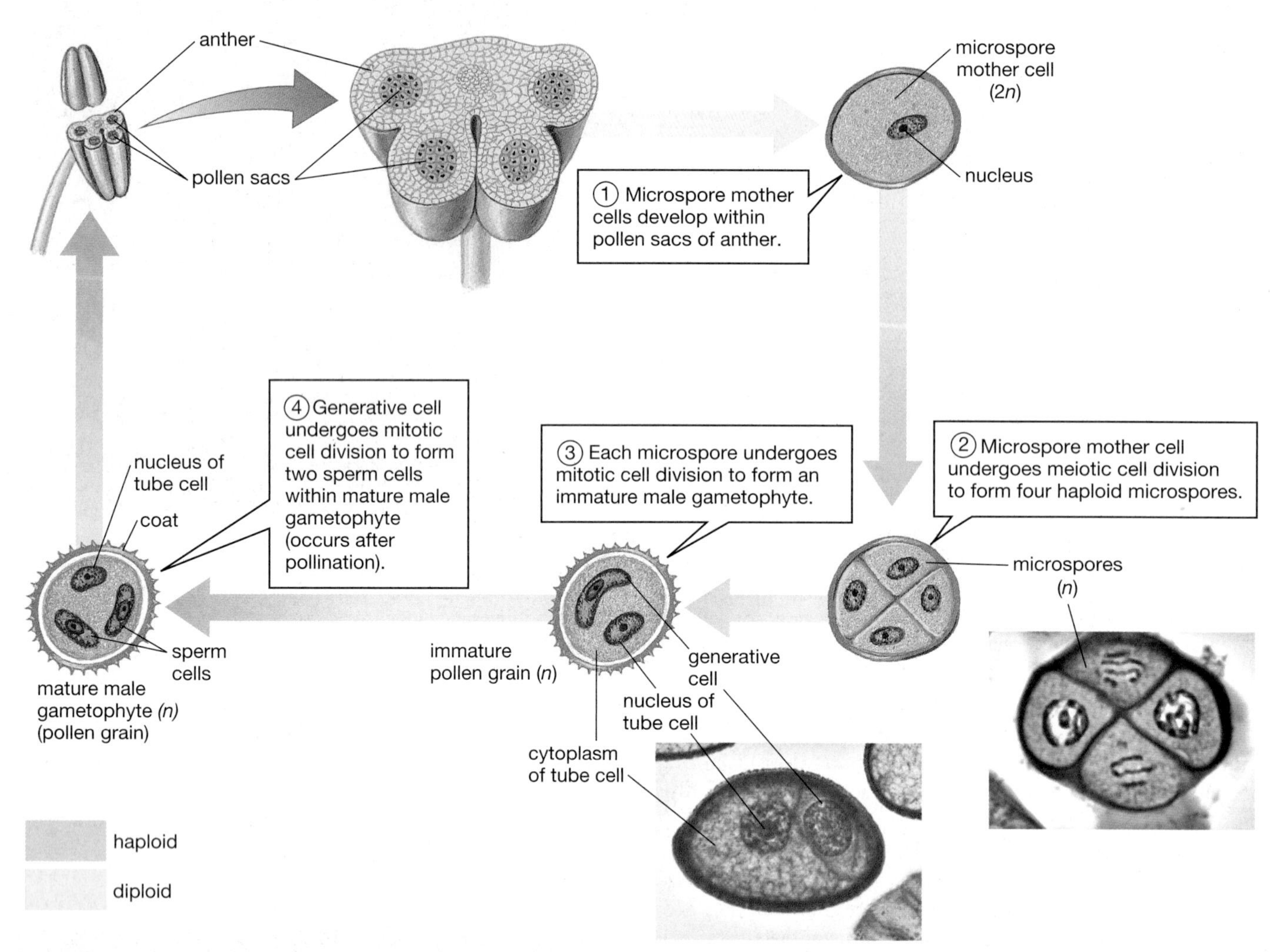

anther
pollen sacs
microspore mother cell (2n)
nucleus
① Microspore mother cells develop within pollen sacs of anther.
② Microspore mother cell undergoes meiotic cell division to form four haploid microspores.
microspores (n)
③ Each microspore undergoes mitotic cell division to form an immature male gametophyte.
immature pollen grain (n)
generative cell
nucleus of tube cell
cytoplasm of tube cell
④ Generative cell undergoes mitotic cell division to form two sperm cells within mature male gametophyte (occurs after pollination).
nucleus of tube cell
coat
sperm cells
mature male gametophyte (n) (pollen grain)
haploid
diploid

FIGURE 25-6 Male gametophyte (pollen) development

25.3 How Do Gametophytes Develop in Flowering Plants?

The familiar flowering plant seen in meadows, gardens, and farms is the diploid sporophyte, illustrated in Figure 25-2. The haploid gametophytes, the pollen grain (male) and the **embryo sac** (female), develop within the flowers of the sporophyte. Gametophytes of flowering plants are much smaller than those of ferns and mosses and cannot live independently of the sporophyte.

Pollen Is the Male Gametophyte

Pollen, the male gametophyte, develops within the anther of the sporophyte plant. Each anther consists of four chambers called pollen sacs (Fig. 25-6). Within each sac, hundreds to thousands of diploid **microspore mother cells** develop. Each microspore mother cell undergoes meiotic cell division (see Chapter 11) to produce four haploid **microspores**. Each microspore then undergoes one mitotic cell division to produce a haploid male gametophyte, or pollen grain. In many species, each immature pollen grain consists of only two cells: a large **tube cell** and a smaller **generative cell** that resides within the cytoplasm of the tube cell. As the pollen grain matures, the generative cell undergoes mitotic cell division and produces two haploid sperm cells (see Fig. 25-6). A tough surface coat develops around the pollen grain, often with an elaborate pattern of pits and protrusions specific to the plant species (Fig. 25-7). This coat protects the cells during their journey to the sometimes-distant female carpel.

When the pollen has matured, the pollen sacs of the anther split open. In wind-pollinated flowers, such as those of grasses (Fig. 25-8) and oaks, the pollen grains spill out and are widely distributed by wind currents; a few of those grains reach and pollinate other flowers of the same species. In animal-pollinated flowers, the pollen adheres weakly to the anther case until the pollinator comes along and brushes or picks it off.

The Embryo Sac Is the Female Gametophyte

Within an ovary, masses of cells differentiate into ovules. Each ovule consists of protective outer layers of cells called **integuments**, which surround a single, diploid **megaspore mother cell** (Fig. 25-9). The mother cell undergoes meiotic cell division to produce four large haploid

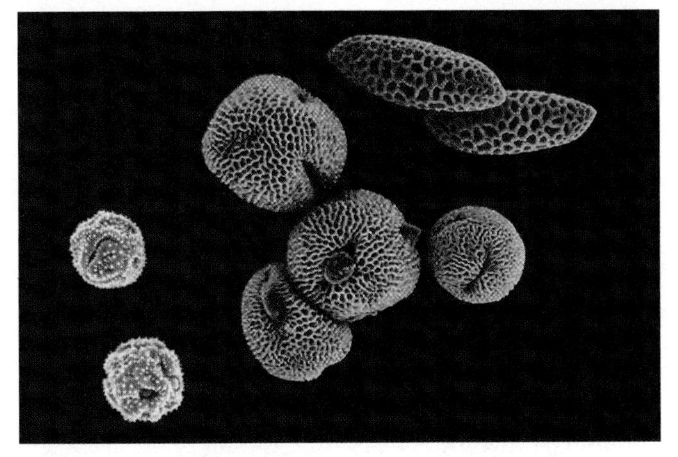

FIGURE 25-7 Pollen grains
The tough outer coats of many pollen grains are elaborately sculptured in species-specific shapes and patterns. The pollen grains in this false-color SEM photo are from a geranium (orange), a tiger lily (fuchsia), and a dandelion (yellow).

megaspores. Three megaspores degenerate, and one survives. This remaining megaspore undergoes an unusual set of mitotic divisions. Three nuclear divisions produce a total of eight haploid nuclei. Plasma membranes then divide the cytoplasm into seven (not eight) cells: three small cells at each end, each containing one nucleus, and one large cell containing two **polar nuclei**.

FIGURE 25-8 Wind-pollinated flowers
The flowers of grasses and many deciduous trees are wind-pollinated, with anthers (yellow structures hanging beneath the flowers) exposed to the wind. Petals are usually reduced or absent.

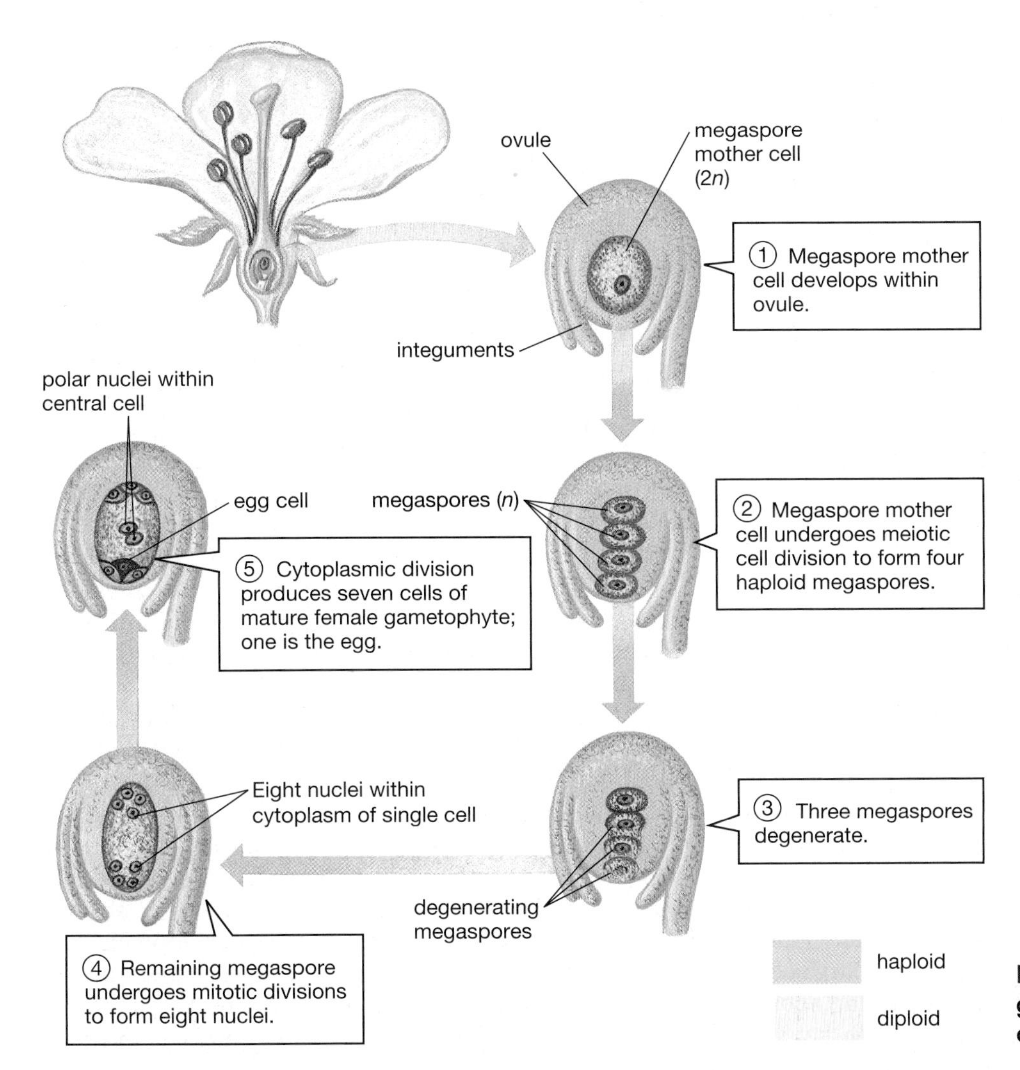

ovule
megaspore mother cell (2n)
① Megaspore mother cell develops within ovule.
integuments
polar nuclei within central cell
egg cell
megaspores (n)
② Megaspore mother cell undergoes meiotic cell division to form four haploid megaspores.
⑤ Cytoplasmic division produces seven cells of mature female gametophyte; one is the egg.
Eight nuclei within cytoplasm of single cell
③ Three megaspores degenerate.
degenerating megaspores
④ Remaining megaspore undergoes mitotic divisions to form eight nuclei.
haploid
diploid

FIGURE 25-9 Female gametophyte (embryo sac) development

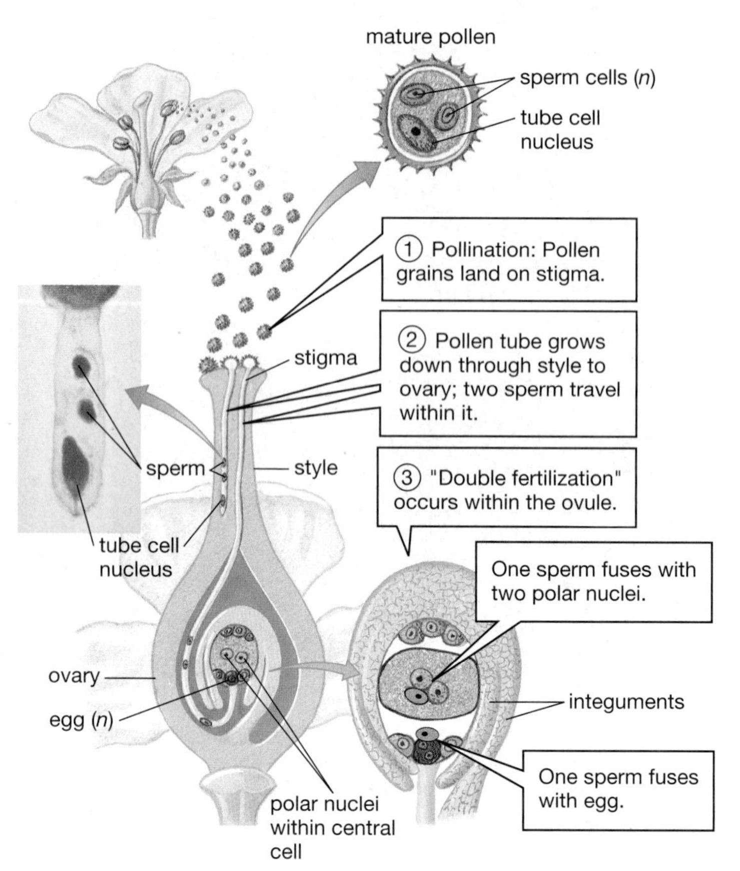

FIGURE 25-10 Pollination and fertilization of a flower
QUESTION How does double fertilization prevent plants from wasting energy?

This seven-celled organism, called the embryo sac (because it will contain the developing embryo), is the haploid female gametophyte. The **egg** is the central small cell at the bottom of the embryo sac, located near an opening in the integuments of the ovule.

25.4 How Does Pollination Lead to Fertilization?

When a pollen grain lands on the stigma of a flower of the same plant species, a remarkable series of events occurs (Fig. 25-10). The pollen grain absorbs water from the stigma. The tube cell elongates, growing down the style toward an ovule in the ovary. Meanwhile, the generative cell undergoes mitotic cell division to form two sperm cells.

If all goes well, the pollen tube reaches the pore in the integument of an ovule and breaks into the embryo sac. The tube's tip ruptures, releasing the two sperm. One sperm merges with the egg cell, a process called **fertilization**. Fertilization produces the diploid zygote that will develop into the embryo and eventually into a new sporophyte. The second sperm enters the large central cell, and its nucleus fuses with both polar nuclei, forming a triploid nucleus (containing three sets of chromosomes). Through repeated mitotic divisions, this cell will develop into the triploid ($3n$) **endosperm**, a food-storage tissue within the seed. **Double fertilization** describes the fusion of the egg with one sperm and the fusion of the polar nuclei with the second sperm, a process unique to flowering plants. The other five cells of the embryo sac degenerate soon after fertilization.

The distinction between pollination and fertilization is important. **Pollination** in flowering plants occurs when a pollen grain lands on a fully-developed stigma of a compatible plant; fertilization is the fusion of sperm and egg. Although pollination is necessary for fertilization, these are two separate events. For example, pollination will not lead to fertilization if the tube cell fails to grow properly, if the embryo sac has no egg, or if a sperm from another pollen grain has already fertilized the egg.

25.5 How Do Seeds and Fruits Develop?

Drawing on the resources of the parent plant, the embryo sac and the surrounding integuments of the ovule develop into a seed. The seed is surrounded by the ovary, which develops into a fruit (Fig. 25-11). Having already served their functions of attracting pollinators and producing pollen, petals and stamens shrivel and fall away as the fruit enlarges.

The Seed Develops from the Ovule and Embryo Sac

The integuments of the ovule develop into the **seed coat**, the outer covering of the seed. As we will see, in many plants the characteristics of the seed coat play a role in regulating when the seed will germinate. Meanwhile, within the integuments, two distinct developmental processes occur (Fig. 25-12a). First, the triploid endosperm cell divides rapidly. Its daughter cells absorb nutrients from the parent plant, forming a large, food-filled endosperm. Second, the zygote develops into the embryo. Both dicot and monocot embryos consist of an embryonic root and embryonic shoot, including the **cotyledons**, or seed leaves. The cotyledons absorb food molecules from the endosperm and transfer them to other parts of the embryo.

In dicots ("two cotyledons"), the cotyledons usually absorb most of the endosperm during seed development, so the mature seed is virtually filled with embryo (Fig. 25-12b, left). If you strip the seed coat from a pea or a bean (both dicots), you will find that the material inside splits easily into two parts; each is a cotyledon. In monocots ("one cotyledon"), the cotyledon absorbs some of the endosperm during seed development, but most of the endosperm remains in the mature seed to be used by the germinating seedling (Fig. 25-12b, right). Cultivated grains, including wheat, corn, and rice, are

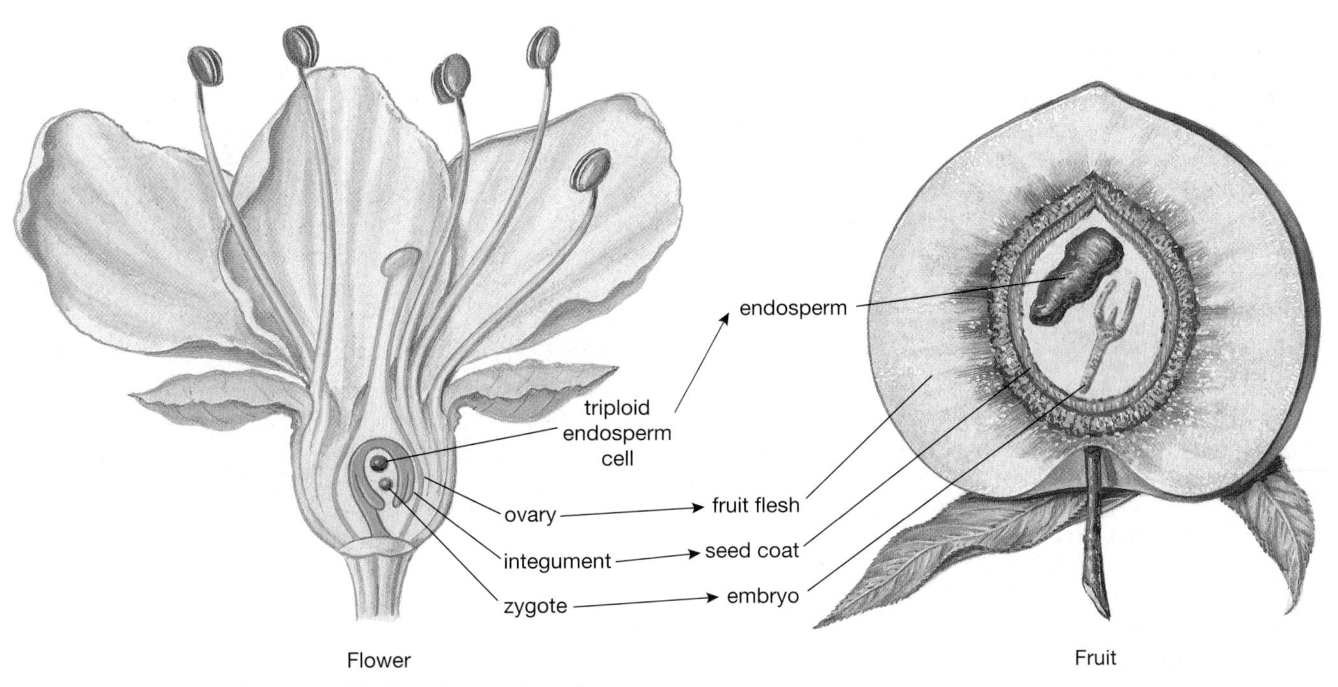

FIGURE 25-11 Development of fruit and seeds from flower parts
The fruit and seed coat are derived from the parent sporophyte plant. The ovary wall ripens into the fruit flesh, which may be soft and tasty, or hard, hooked, or tufted to facilitate dispersal. The integuments of each ovule, surrounding the embryo sac, form the seed coat. The triploid endosperm cell divides repeatedly, absorbs nutrients, and becomes the endosperm. The zygote, also within the seed, develops into the embryo.

Media Activity
25.2 Fruits and Seeds: Structure and Development

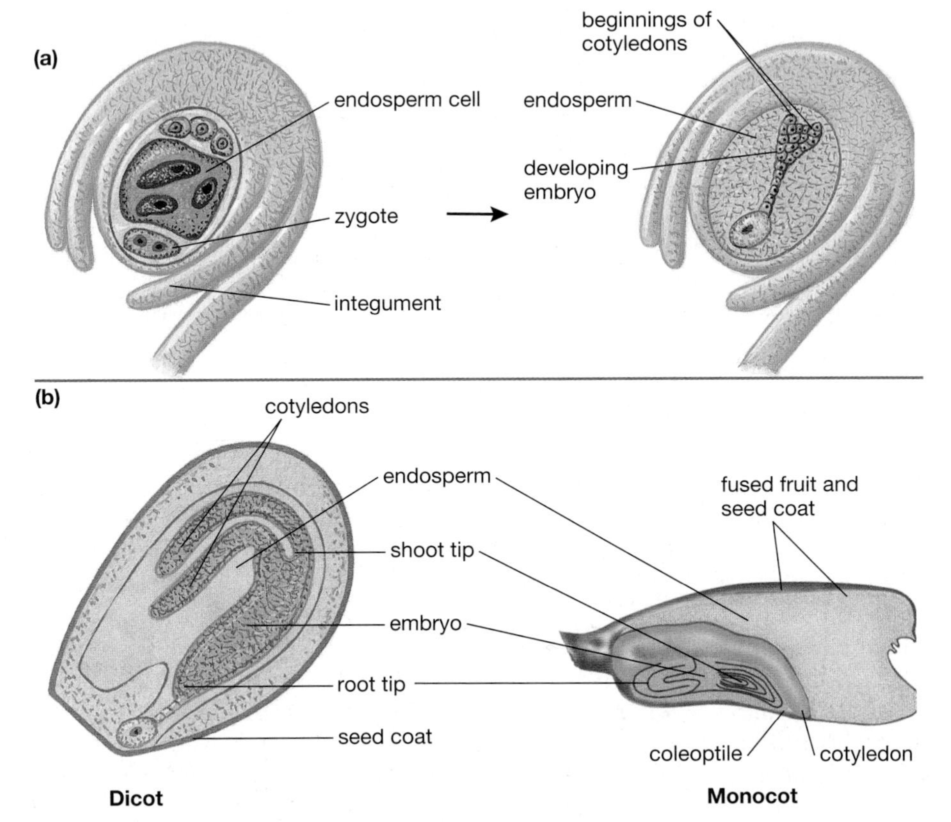

FIGURE 25-12 Seed development
(a) The endosperm develops first, absorbing nutrients from the parent plant. The embryo develops later, absorbing nutrients from the endosperm. **(b)** *(left)* The two cotyledons of dicots (such as this shepherd's purse) usually absorb most of the endosperm as the seed develops; hence, the mature seed is mostly cotyledon. *(right)* Monocot seeds (such as this corn kernel) retain large quantities of endosperm. As the seed germinates, the single cotyledon absorbs the food reserves of the endosperm and transfers them to the rest of the growing embryo.

Media Activity
25.2 Fruits and Seeds: Structure and Development

EARTH WATCH On Dodos, Bats, and Disrupted Ecosystems

Flowering plants dominate terrestrial ecosystems largely because of mutually beneficial relationships with animals that pollinate their flowers and disperse their seeds. Some plant and animal species, such as the yucca and yucca moths, have become totally dependent on one another. Within complex ecosystems, these mutually beneficial relationships sustain both plant and animal populations and ultimately the ecosystem itself.

The tambalacoque tree, like most of the native plants on the island of Mauritius in the Indian Ocean, is threatened. Tambalacoque trees produce large, edible fruit similar to peaches, with a pulpy outside surrounding a stone-hard pit. Today, the remaining trees produce healthy fruits that fall to the ground and rapidly rot in the tropical climate. The enclosed seeds are highly susceptible to destruction by fungal and bacterial infections, which are promoted by the rotting fruit. Before humans arrived, the island was home to the dodo (Fig. E25-1a). Early sailors found the large, slow dodos to be easy prey, and by 1681 they had hunted the dodo to extinction. Other native animals, including giant tortoises, large-billed parrots, and the giant skink (a large reptile), were also driven to extinction as humans introduced monkeys, pigs, and deer to Mauritius and destroyed natural habitats by clearing the land for farming. Scientists believe that some of these extinct animals ate the tambalacoque fruit before it had a chance to rot, thoroughly cleaning the seeds and thus protecting them from attack by fungi. The animals also dispersed the seeds throughout the island, ensuring that some of them reached habitats favorable for germination. As native animals are destroyed, plants that coevolved with them are also threatened, and natural ecosystems that evolved over millennia collapse.

In another example, on the island of Madagascar off the coast of Africa, researchers have identified more than 20 tree species that depend primarily on lemurs (tree-dwelling primates) for seed dispersal. But a burgeoning human population is displacing natural habitats, and lemurs are rapidly disappearing. Where lemurs are disappearing, so are these trees.

In many tropical forests, fruit-eating bats are the most important agents of seed dispersal (Fig. E25-1b). These bats may fly more than 20 miles each night, consuming up to twice their weight in fruit and defecating the seeds in flight. Biologist Donald Thomas discovered that after passing through a bat's digestive tract, nearly all the seeds germinated; in contrast, seeds planted directly from fruit had only a 10% germination rate. Today, in the tropical forests of southern Mexico, fruit-eating and dispersing animals such as monkeys, deer, and tapir have been overhunted. Tropical fruits are rotting on the forest floor or sending up doomed sprouts under the shade of their parents; dispersal has been dramatically reduced. As Alejandro Estrada of the University of Mexico put it, "The continued existence of tropical forests whose primates and . . . birds and bats have been shot is just as precarious as if their trees had been chainsawed and bulldozed." The web of interdependent life-forms linked by interactions forged over millennia of coevolution is fragile and easily disrupted. Only by understanding and preserving the complex and crucial interactions among plants and animals can we hope to conserve diverse, functioning ecosystems.

(a)

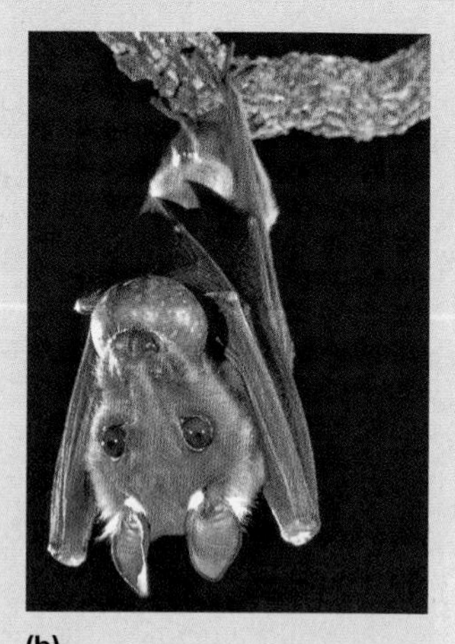

(b)

FIGURE E25-1 Animal seed dispersers are crucial to some ecosystems
(a) The dodo, driven to extinction in the late 1600s, probably helped disperse and promote germination of the tambalacoque trees of Mauritius. The trees are now seriously threatened and rarely germinate in the wild. **(b)** A Wahlberg's epauleted bat eats a ripe fig in Kenya. Without bats and other seed-dispersing animals, some types of tropical forest communities may not survive.

monocots. We take advantage of food stored by the plant for its embryo when we eat corn and rice or grind up the endosperm of the wheat seed to make flour. We also sometimes consume the embryo of the wheat seed in the form of "wheat germ."

The embryonic root develops at one end of the embryo. At the other end, the embryonic shoot is usually composed of two regions. Below the cotyledons, but above the root, is the **hypocotyl** ("hypo" is from the Greek, meaning "beneath" or "lower"); above the cotyledons, the shoot is called the **epicotyl** ("epi" means "above"). At the tip of the epicotyl lies the apical meristem of the shoot; its daughter cells differentiate into the specialized cell types of stem, leaves, and flowers. One or two developing leaves may already be present.

The Fruit Develops from the Ovary Wall

The ovary wall develops into a fruit (see Fig. 25-11); when you eat a fruit, you are consuming the ripened ovary of a plant. There are a bewildering variety of fruits, with outer layers that are variously fleshy, hard, winged, or even spiked like a medieval mace. The selective forces favoring the evolution of all fruits, however, are similar: fruits help disperse seeds to locations far from the parent plant. The burrs that attach to your socks during a walk in a meadow are likely specialized fruits that are using you as a mechanism to disperse their seeds (see "Earth Watch: On Dodos, Bats, and Disrupted Ecosystems" and "Evolutionary Connections: Adaptations for Pollination and Seed Dispersal").

Seed Dormancy Helps Ensure Germination at an Appropriate Time

All seeds need warmth and moisture to germinate. But many newly matured seeds will not germinate immediately, even under ideal conditions. Instead, they enter a period of **dormancy**, during which they will not germinate. Dormancy is usually marked by lowered metabolic activity and resistance to adverse environmental conditions.

Seed dormancy solves two problems, one intrinsic to the plant itself and one related to environmental factors. First, if a seed germinated while still enclosed in a fruit and hanging from a tree or vine, it might exhaust its food reserves before it ever touched the ground. Further, seedlings germinating within a fruit that contains many seeds would grow in a dense cluster, competing with one another for nutrients and light. Second, environmental conditions that are suitable for seedling growth (such as adequate moisture and temperatures) may not coincide with seed maturation. Seeds that mature in the late summer in temperate climates, for example, face the harsh winter to come. Spending the winter as a dormant seed is clearly preferable to freezing to death as a tender young sprout. In warm, moist, tropical regions, where environmental conditions are suitable for germination throughout the year, seed dormancy is much less common than in temperate regions. Refer to this chapter's Case Study for more information about seed dormancy.

Adequate moisture and proper temperature are almost universal prerequisites for germination. Beyond these, the seeds of many plant species require additional conditions for germination to occur. The requirements for maintaining and breaking out of dormancy are finely tuned to the plant's native environment and the mechanisms it uses for dispersal. The three most common requirements to break dormancy are drying, exposure to cold, and disruption of the seed coat.

- *Drying*. Many seeds that are housed in moist fruit must dry out before they are able to germinate. Drying prevents the seed from germinating while it is still within the fruit. Seeds that require drying often are dispersed by fruit-eating animals, which cannot digest the seeds. The seeds are excreted and exposed to air, where they dry out. Later, when temperature and moisture levels are favorable, they germinate.
- *Cold*. Seeds of many temperate and arctic plants will not germinate unless they are exposed to prolonged subfreezing temperatures, followed by sufficient warmth and moisture. This requirement ensures that seeds released in mild autumn weather do not immediately germinate, only to succumb to winter cold. Requiring a substantial cold spell guarantees that they will delay sprouting until the following spring.
- *Disruption of the seed coat*. The seed coat itself can be a barrier to seed germination; it may need to be weathered, or even partially digested, before germination can occur. Many seed coats are impermeable to water and oxygen, others confine the developing embryo and physically prevent growth, while still others contain chemicals that inhibit germination (as we will see in Chapter 26). In deserts, for example, years may go by without enough water for plants to germinate, grow, flower, and release more seeds. Therefore, the seeds must not sprout unless a rainfall is heavy enough to allow the plant to complete its entire life cycle, because it cannot rely on additional rain falling in time for later stages of growth. The seed coats of most desert plants have water-soluble chemicals that inhibit germination, and only a hard rainfall is sufficient to wash away enough of the inhibitors to allow sprouting.

25.6 How Do Seeds Germinate and Grow?

During germination, the embryo absorbs water, which makes it swell and burst its seed coat. The root is usually the first structure to emerge from the seed coat, growing rapidly and absorbing water and minerals from the soil. Much of the water is transported to cells in the shoot. As its cells elongate, the stem lengthens, pushing up through the soil.

The Shoot Tip Must Be Protected

The growing shoot faces a serious difficulty: it must push through the soil without scraping away the apical meristem and tender leaflets at its tip. Likewise, a root must also contend with tip abrasion, but its apical meristem is protected by a root cap (see Fig. 24-15 in Chapter 24). Shoots, however, spend most of their time in the air and do not develop permanent protective caps. Instead, germinating shoots have other mechanisms that cope with the abrasion of sprouting. In monocots, the **coleoptile**, a tough sheath, encloses the shoot tip like a glove around a finger (Fig. 25-13a). The coleoptile pushes aside the soil particles as it grows. Once out in the air, the coleoptile tip degenerates, allowing the tender shoot

FIGURE 25-13 Seed germination Stored food provides energy for seedling growth until photosynthesis can take over. The root grows rapidly, absorbing water and nutrients, and the delicate shoot tip is protected as it pushes upward through the soil. **(a)** In monocots, the tip is protected within a tough coleoptile. **(b)** In dicots, the hypocotyl (shown) or the epicotyl bends, forming a hook that emerges from the soil first, protecting the shoot tip. **QUESTION** How would shoot growth during germination be affected if a bean seed were allowed to germinate in a completely dark environment?

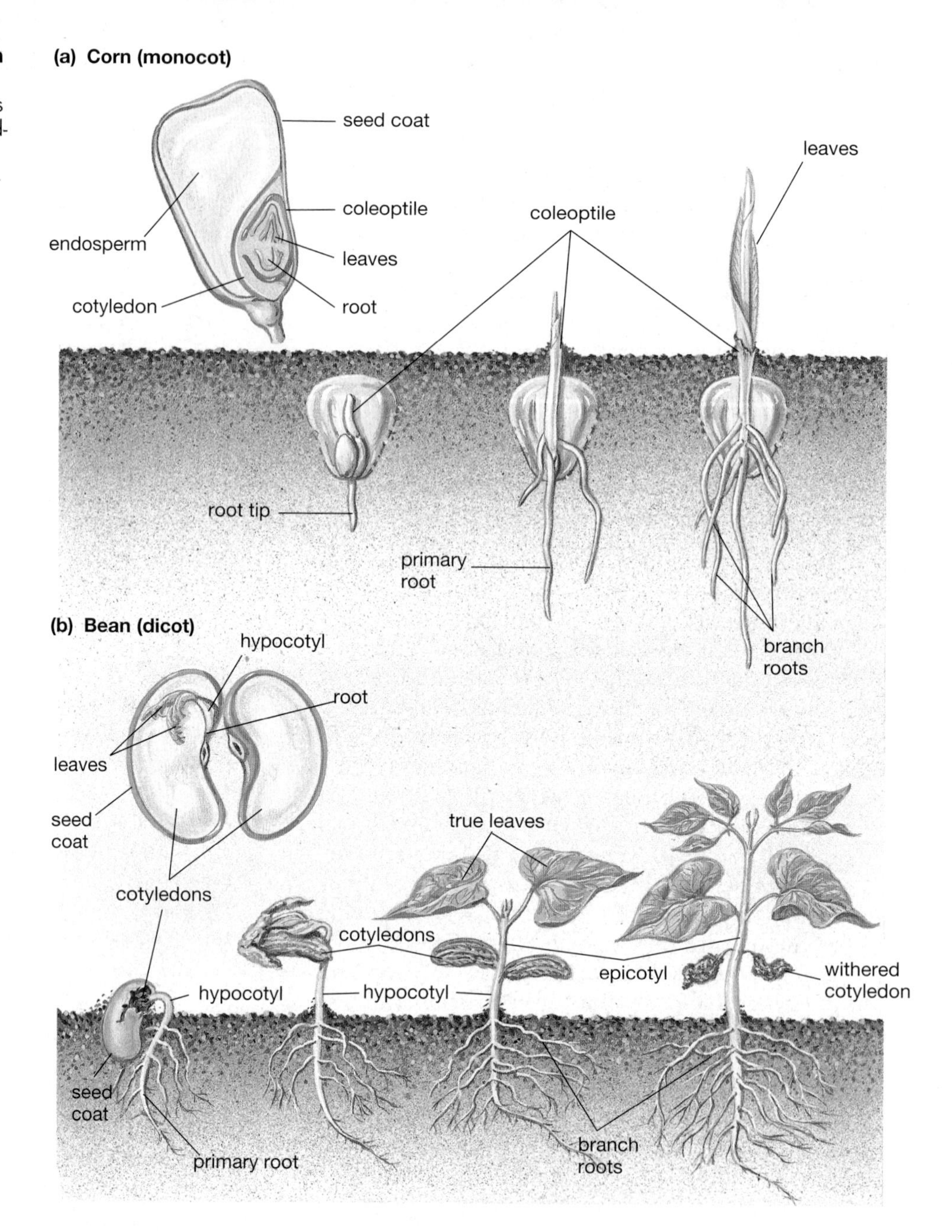

to emerge. Dicots do not have coleoptiles. Rather, the dicot shoot forms a hook in the hypocotyl or epicotyl (Fig. 25-13b). The bend of the hook, encased in epidermal cells with tough cell walls, forces its way up through the soil, clearing the path for the downward-pointing apical meristem with its delicate new leaves. The shoot immediately begins to straighten after it emerges, pointing its leaves toward the sunlight.

Cotyledons Nourish the Sprouting Seed

Food stored in the seed provides the energy for sprouting. Recall that the cotyledons of dicots absorb the endosperm while the seed develops, making them swollen with stored food. In dicots with hypocotyl hooks, such as members of the squash family, the elongating shoot carries the cotyledons out of the soil into the air. These aboveground cotyledons typically become green and photosynthetic and transfer both previously stored food and newly synthesized sugars to the shoot (Fig. 25-14). In dicots with epicotyl hooks, the cotyledons stay below the ground, shriveling up as the embryo absorbs their stored food. Monocots retain most of their food reserve in the endosperm until germination, when it is digested and absorbed by the cotyledon as the embryo grows. The cotyledon remains below ground in the remnants of the seed.

Controlling the Development of the Seedling

Once out in the air, the shoot rapidly spreads its leaves to the sun. Simultaneously, the root system delves into the soil. The apical meristem cells of shoot and root divide, giving rise to the mature structures discussed in Chapter 24. Eventually this plant will mature, flower, and set seed, renewing the cycle of life. The regulation of this cycle—why shoots grow upward while roots grow downward, and how plants produce flowers at the proper time of year—is described in Chapter 26.

FIGURE 25-14 Cotyledons nourish the developing plant
In the squash family, the cotyledons (the pair of smooth oval leaves) expand into photosynthetic leaves. The first true leaf (crinkled single leaf) expands a little later. Eventually, the cotyledons shrivel up and die.

EVOLUTIONARY CONNECTIONS

Adaptations for Pollination and Seed Dispersal

Coevolution Matches Plants and Pollinators

In many instances, plants and their pollinators have coevolved; that is, each has acted as an agent of natural selection on the other (see Chapter 15). Animal-pollinated flowers must attract useful pollinators and frustrate undesirable visitors who might eat nectar or pollen without fertilizing the flower. The animals must have behavioral and sensory capabilities that allow them to efficiently locate and identify flowers that can provide adequate nutrition, and extract the nectar or pollen. Animal-pollinated flowers can be loosely grouped into three categories, depending on the benefits (real or deceptive) that they offer to potential pollinators: food, sex, or a nursery. As you will see, some plants have evolved in ways that trick insects into pollinating them, and thus avoid expending energy to provide the insects with food.

Some Flowers Provide Food for Pollinators

Many flowers provide food for foraging animals such as beetles, bees, moths and butterflies, or hummingbirds. In return, the animals unwittingly distribute pollen from flower to flower.

We can thank the bees for most of the sweet-smelling flowers, because sweet "flowery" odors attract these pollinators (flowers produce and release their scents from a variety of structures that vary with the plant species). Bees also have good color vision, but they do not see exactly the same range of colors that humans do (Fig. 25-15a). To attract a bee from afar, bee-pollinated flowers must look brightly colored to a bee (Fig. 25-15b). Typically, these flowers are white, blue, yellow, or orange, and many have other markings—such as central spots or lines pointing toward the center—that reflect UV light.

(a)

human vision

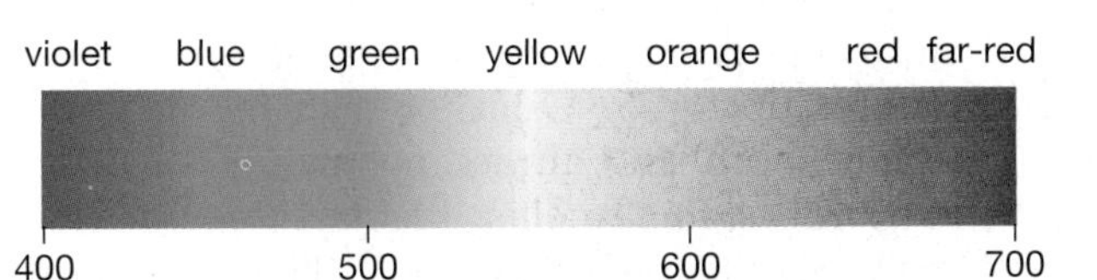

(b)

bee vision

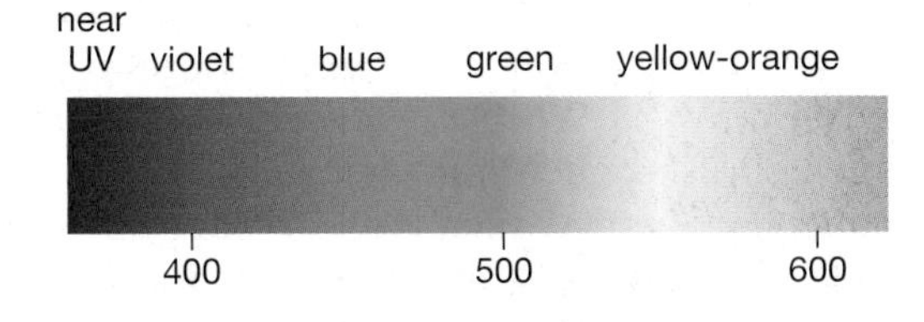

FIGURE 25-15 Ultraviolet patterns guide bees to nectar
The spectra of color vision for **(a)** humans and **(b)** bees overlap considerably in the blue, green, and yellow ranges but differ on the edges. Humans are sensitive to red, which bees do not perceive; bees can see UV light, which is invisible to the human eye. Many flowers photographed under **(a)** ordinary daylight and under **(b)** UV light show striking differences in color patterns. Bees can see the UV patterns that presumably lead them to the nectar- and pollen-containing centers of the flowers.

(a)

(b)

FIGURE 25-16 "Pollinating" a pollinator
(a) In Scotch broom flowers, the bee finds nectar near the junction of top and bottom petals. **(b)** The bee's weight deflects the bottom petals downward, causing curved, pollen-laden stamens to pop up and cover the bee's hairy back with pollen. The bee will carry the pollen to other Scotch broom flowers, leaving some on ready stigmas.

FIGURE 25-17 The corpse flower is overwhelming in size and scent
Scavenging insects such as flies and beetles are attracted to the odor of rotting meat produced by some of the largest flowers in the world. To date, only 11 corpse flowers (*Amorphophallus titanum,* a species native to Sumatra, also called "devil's tongue") have ever been coaxed into bloom in the United States; the first bloomed in 1937. Thousands of visitors flocked to see and smell the amazing specimen (nearly 6 feet tall) that bloomed in California's Huntington Botanical Gardens in August 1999. The "bloom" actually contains hundreds of separate male and female flowers, which (in Sumatra) are pollinated by large carrion beetles.

Bee-pollinated flowers have structural adaptations that help ensure pollen transfer. Many bee-pollinated flowers, such as nasturtiums and foxgloves, produce nectar at the bottom of a tube; in the Scotch broom flower, nectar forms in a crevice between enclosing petals (Fig. 25-16). Either pollen-laden stamens (in newly opened flowers) or the sticky stigma of the carpel (in older flowers) sticks out of the top of the tube or emerges from confinement when the bee's weight deflects the petals downward. When a bee visits a young flower, the stamens brush pollen onto her back (see Fig. 1-12 in Chapter 1). She may then visit an older flower and repeat her foraging behavior. This time, she leaves pollen behind on the stigma.

Many flowers adapted for moth and butterfly pollinators have tubes containing nectar that accommodate the long tongues of moths and butterflies. Flowers pollinated by night-flying moths open only in the evening; most are white and some give off strong, musky odors that help the moth locate the flower in the dark.

Beetles and flies often prefer to feed on animal material, and beetle-pollinated flowers typically smell like dung or rotting flesh, which attracts these scavenging insects. Their names, including "carrion flower," "skunk cabbage," and "corpse flower" (Fig. 25-17), describe their odor. The corpse flower and the skunk cabbage, common in U.S. swamps, are related. Both share an unusual adaptation: their flowers heat up! By metabolizing stored food, mostly fat, eastern skunk cabbage flowers can reach temperatures up to 25 °C higher than the air around them. The heat probably attracts pollinators and certainly helps broadcast foul-smelling scents. These flowers deceive their pollinators—they smell like nutrient-rich rotting meat, but offer no food at all.

Hummingbirds (see the opening photo of Chapter 8) are one of the few important vertebrate pollinators, although several mammals also visit flowers (Fig. 25-18). Since birds have notoriously poor senses of smell, hummingbird-pollinated flowers seldom synthesize fragrant chemicals. However, these flowers often produce more

(a)

(b)

FIGURE 25-18 Some vertebrate pollinators
(a) A tropical bat feeds at a cluster of tubular flowers with protruding stamens and stigma. As the bat hovers before the flower, the top of its head touches either the anthers or stigma or both, thus pollinating the flower. **(b)** As the honey possum stuffs its face into this flower, pollen adheres to its muzzle and whiskers. A visit to another flower will transfer the pollen. **QUESTION** Why have many plants that are pollinated by hummingbirds evolved flowers shaped like long, narrow tubes?

nectar than other flowers, because hummingbirds need more energy than insects and will favor flowers that provide it. These flowers must protect their large nectar supplies from insects that would become sated on the abundant sugar and fail to transfer pollen to another flower. Adaptations to hummingbird pollination include a deep, tubular shape that matches the long bills and tongues of hummers, and no place for insects to land and rest while dining. In addition, most hummingbird-pollinated flowers are red or orange. Red is particularly attractive to hummingbirds, but appears gray or black to a bee.

Sexy Deceptions Attract Pollinators

To pollinate their flowers, a few plants, most notably the orchids, take advantage of the mating drive and stereotyped behaviors of male wasps. Some orchid flowers mimic female wasps or bees both in scent (the orchids release a sexual attractant similar to that produced by the female insect) and shape (Fig. 25-19). The males land atop these "fake females" and attempt to copulate, but get only a packet of pollen for their efforts. As they repeat their attempts on other orchids of the same species, the pollen packet is transferred.

Some Plants Provide Nurseries for Pollinators

Perhaps the most elaborate relationships between plants and pollinators occur in a few cases in which insects fertilize a flower and then lay their eggs in the flower's ovary. This arrangement occurs between milkweeds and milkweed bugs, figs and certain wasps, and yuccas and yucca moths (Fig. 25-20). The yucca moth's remarkable behavior results in the pollination of yuccas and a well-stocked pantry for its own offspring. A female moth visits a yucca flower, collects pollen, and rolls it into a compact ball. The moth flies off with the pollen ball to another yucca flower, drills a hole in the ovary wall, and lays its eggs inside the ovary. Then it smears pollen from the pollen ball all over the stigma of the flower, performing this genetically programmed behavior flawlessly. By pollinating the yucca, the moth ensures that the plant will provide a supply of developing seeds for its caterpillar offspring. Because the caterpillars eat only a fraction of the seeds, the yucca also reproduces successfully. The mutual adaptation of yucca and yucca moth is so complete that neither can reproduce without the other.

Fruits Help Disperse Seeds

A plant species will be most successful if its seeds are dispersed far enough so that the young plants don't compete with the parents for light and nutrients. Plants will

FIGURE 25-19 Sexual deception promotes pollination
This male wasp is trying to copulate with an orchid flower. The result is successful reproduction—by the orchid, not the wasp!

FIGURE 25-20 A mutually dependent relationship
(a) Yuccas bloom on the dry plains of eastern Colorado in early summer. **(b)** A yucca moth places pollen on the stigma of a yucca flower.

also be more successful and widespread if its members send seeds to distant habitats. In flowering plants, seed dispersal is the function of fruits; a wide variety of fruits have evolved, each dispersing seeds in a different way.

Explosive Fruits Allow Shotgun Dispersal

A few plants develop explosive fruits that eject their seeds meters away from the parent plant. Mistletoes, common parasites of trees, produce fruits that shoot out sticky seeds. If one seed strikes a nearby tree, it sticks to the bark and germinates, sending rootlike fibers into the vascular tissues of its host, from which it draws its nourishment. Because the proper germination site for a mistletoe seed is not the ground but a tree limb, shooting the seeds, not dropping them, is clearly useful.

Lightweight Fruits Allow Wind Dispersal

Dandelions and maples (Fig. 25-21) produce lightweight fruits with surfaces that catch the wind. Each individual hairy tuft on a dandelion ball is a separate fruit. Each fruit typically contains a single small seed; having only one seed reduces weight and lets the fruit remain aloft longer. These featherweight fruits aid the seed in traveling away from the parent plant, from a few meters for maples to kilometers for a milkweed or dandelion on a windy day.

Floating Fruits Allow Water Dispersal

Many fruits can float on water for a time and may be dispersed by streams or rivers. The coconut fruit, however, is the ultimate floater. Round, buoyant, and watertight, the coconut drops off its parent palm, rolls to the sea, and floats for weeks or months until it washes ashore on some distant isle (Fig. 25-22). There it germinates, perhaps establishing a new coconut colony on a formerly barren island.

(a)

(b)

FIGURE 25-21 Wind-dispersed fruits
These fruits usually contain only one or two lightweight seeds. Some, such as **(a)** dandelions, have filamentous tufts that catch the breezes. Others, such as **(b)** maple fruits, are miniature glider–helicopters, whirling away from the tree as they fall. To see how the wings aid in seed dispersal, take two maple fruits and pluck the wings off one. Hold both fruits over your head and drop them; the wingless fruit will fall at your feet; the winged one will glide some distance away.

FIGURE 25-22 Water-dispersed fruit
After a long journey at sea, this coconut was washed high onto a beach by a storm. The large size and massive food reserves of coconuts are probably adaptations required for successful germination and seedling growth on barren, sandy beaches.

Clingy or Tasty Fruits Allow Animal Dispersal

The majority of fruits use animals as agents of seed dispersal. Two quite distinct strategies have evolved for dispersal by animals: grabbing an animal as it passes by, or enticing it to eat the fruit but not digest the seeds.

Anyone who takes a long-haired dog on a walk through a field in the fall knows about fruits that hitchhike on animal fur. Burdocks, burr clover, foxtails, and sticktights all develop fruits with prongs, hooks, spines, or adhesive hairs (Fig. 25-23). The parent plants hold these fruits very loosely, so even slight contact with fur pulls the fruit off of the plant and leaves it stuck on the animal. Some of these fruits may fall off the next time the animal brushes against a tree or rock or come out when the animal grooms its fur.

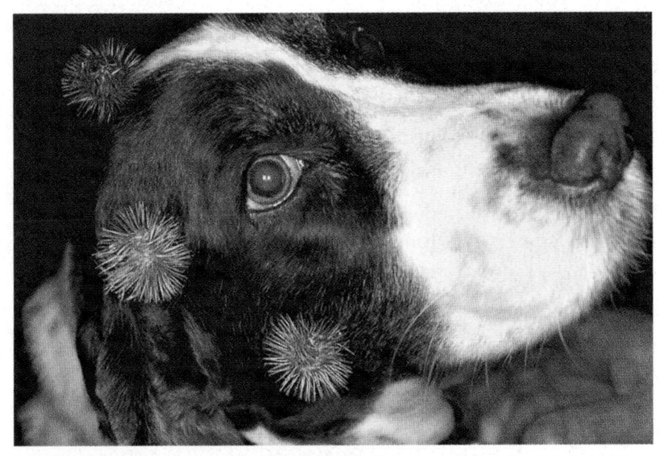

FIGURE 25-23 The cocklebur fruit uses hooked spines to hitch a ride on furry animals

Unlike hitchhiker fruits, edible fruits benefit both animal and plant. The plant stores sugars and tasty flavors in a fleshy fruit that surrounds the seeds, enticing hungry animals to eat the fruit (Fig. 25-24). Some fruits, such as peaches and plums, contain large, hard seeds that animals usually do not eat. After consuming the flesh of the fruit, the animals discard the seeds. Other fruits, including blackberries, raspberries, strawberries, and tomatoes, have small seeds that are swallowed along with the fruit flesh. The seeds then pass through the animal's digestive tract without harm. In some cases, passing through an animal's gut may even be essential to seed germination, by scraping or digesting away part of the seed coat. Recently, a graduate student in ecology discovered the secret of chili pepper seed dispersal. The burning flavor discourages local mammals, such as desert mice and pack rats, from eating the fruit but does not act as a deterrent to birds that can't taste it. The researcher discovered that the digestive tracts of mammals destroy the chili seeds, but passage through the digestive tract of birds increases the germination rate by a factor of three compared to seeds that just fall to the ground. Besides transport away from its parent plant, a seed that is swallowed and excreted benefits in another way: it ends up with its own supply of fertilizer!

FIGURE 25-24 The colors of ripe fruits attract animals
A bright red raspberry fruit has attracted a resplendent quetzal in Costa Rica. Only ripe fruits with mature seeds inside are sweet and brightly colored. Most unripe fruits are unpalatable: green, hard, and bitter. This too is an evolutionary adaptation. The immature seeds within unripe fruit may not survive passage through an animal's gut.

CASE STUDY REVISITED Time Capsules, Smoke Signals, and Seed Sprouting

Even a fire-blackened landscape will look quite green in several years. Some plant species are perfectly adapted to withstand regular fires. Their extremely hard seed coats crack in the fire's heat, allowing moisture to enter and stimulate germination. In many species, fire-adapted seedlings also survive better in the higher light levels that occur after a forest has burned. Jack pine is a good example of a fire-adapted tree. Its seedlings thrive in bright sunlight, and cones of the jack pine are sealed shut with resin and require high temperatures (such as those caused by fire) to open. The seeds remain dormant within the sealed cones for many years; half the seeds from a 20-year-old cone may remain alive and able to germinate. Recently, plant physiologists discovered that, in some species, chemical signals from smoke alone can stimulate germination. Studying whispering bells, a flowering plant characteristic of chaparral ecosystems that rely on frequent fires, researchers found that 30 seconds of exposure to smoke stimulated germination. Of the many chemicals in smoke, they isolated nitrogen dioxide as the trigger for germination. Now, they are trying to discover exactly how this volatile gas acts on the dormant seed.

Could a lotus seed from a plant lovingly tended by a Buddhist monk over a millennium ago still sprout? After Jane Shen-Miller filed through their tough seed coats, her lotus seeds sprouted in about 4 days. Using radiocarbon dating (described in Chapter 17), Shen-Miller discovered that one of her sprouts was nearly 1288 years old—a new record for seed dormancy. Meanwhile, researchers at Michigan State University, where Professor Beal buried his seed time capsules, continue to follow his instructions, although they now wait 20 years before opening the next bottle. Two of the original 21 species (a mallow and a mullein) still sprout, 120 years after they were buried.

Consider This: A century of fire suppression by people has left many woodlands far denser than their natural state. Thus, when fires do start, they tend to burn hotter and encompass larger areas than they would have when fires were more frequent. In cities such as Los Angeles, which is surrounded by a fire-dependent chaparral ecosystem (see Chapter 42), millions of cars burning fossil fuels have elevated nitrogen dioxide well above its natural levels. Explain why these situations could be harmful to fire-promoted regeneration. Design experiments in the lab or in the field—or list observations that could be made in nature—that would test the hypothesis that these situations may delay regeneration after a fire.

Links to Life: Walk Through a Meadow

"By the time I get to the wood, I am carrying all manner of seeds hooked in my coat or piercing my socks or sticking by ingenious devices to my shoestrings. I let them ride ... After all, who am I to contend against such ingenuity? It is obvious that nature, or some part of it in the shape of these seeds, has intentions beyond this field, and has made plans to travel with me."

Loren Eiseley in "*The Immense Journey*" (1957)

Imagine walking through a wildflower-strewn meadow. You've learned that plants have evolved showy flowers for the birds and bees—and for beetles, moths, and even bats. Perhaps the real wonder is that we, too, are attracted to flowers. But if you feel "sneezy" just imagining a walk through a grassy field in summer, you might have a special interest in pollen. Flowering grasses don't produce showy flowers, but in the right season you'll see that their "heads" are yellow with pollen. Are there any bees on the grasses? Based on what you now know about how grasses transfer pollen, you can understand why grass pollen is such a major contributor to "allergy season." Look closely for pollinators on flowers and notice what colors and shapes of flowers are being visited by bees or butterflies. Find a foraging bee and look at it closely—you may see pollen on its body. Notice how the anthers or the stigma of the flower contact the bee's body. You are witnessing a symbiotic relationship in action!

If you return to your field in the fall, you'll find seed pods in various shades of green and brown—some filled with seeds, others split open and empty. As you walk, your socks will soon be prickly with small burs and grass seeds that (to paraphrase Eiseley's colorful quote) have intentions beyond the meadow and have made plans to travel with you. Although they have no real plans or intentions, natural selection has favored these adaptations for travel. They are using you as an agent of seed dispersal. Where will you take them?

Summary of Key Concepts

25.1 What Are the Features of Plant Life Cycles?

The sexual life cycle of plants—alternation of generations—includes both a multicellular diploid form (the sporophyte generation) and a multicellular haploid form (the gametophyte generation).

In seed plants, the gametophyte stage is greatly reduced. The male gametophyte is the pollen grain, a drought-resistant structure that can be carried from plant to plant by wind or animals. The female gametophyte is also reduced and is retained within the body of the sporophyte stage. In this way, seed plants can reproduce independently of liquid water.

25.2 How Did Flowers Evolve?

Flowering plants evolved from gymnosperms. In gymnosperms, wind blows pollen from male cones to female cones, but wind pollination is inefficient. In many habitats, flowering plants enjoy a selective advantage over gymnosperms because many types of flowers attract insects that carry pollen from plant to plant.

Complete flowers consist of four parts: sepals, petals, stamens (male reproductive structures), and carpels (female reproductive structures). The sepals form the outer covering of the flower bud. Most petals (and in some cases, the sepals) are brightly colored and attract pollinators to the flower. The stamen consists of a filament that bears at its tip an anther, in which pollen (the male gametophyte) develops. The carpel consists of the ovary, in which one or more embryo sacs (female gametophytes) develop, and a style, which bears a sticky stigma, to which pollen adheres during pollination. Incomplete flowers lack one or more of the four floral parts. Incomplete flowers may lack either the male or the female reproductive structures, in which case they are described as "imperfect" flowers.

25.3 How Do Gametophytes Develop in Flowering Plants?

Pollen develops in the anthers. The diploid microspore mother cell undergoes meiotic cell division to produce four haploid microspores. Each of these divides using mitotic cell division to form haploid pollen grains. An immature pollen grain consists of two cells: a tube cell and a generative cell. The generative cell divides once to produce two sperm cells.

The embryo sac develops within the ovules of the ovary. A diploid megaspore mother cell undergoes meiotic cell division to form four haploid megaspores. Three of these degenerate; the fourth undergoes three sets of mitotic divisions to produce the eight nuclei of the embryo sac. These eight nuclei come to reside in only seven cells. One of these cells, with a single nucleus, is the egg cell; the other, a very large cell with two nuclei, is the primary endosperm cell. These two cells are involved in seed formation; the rest of the cells degenerate.

25.4 How Does Pollination Lead to Fertilization?

Pollination is the transfer of pollen from anther to stigma. When a pollen grain lands on a stigma, its tube cell grows through the style to the embryo sac. The generative cell divides to form two sperm cells that travel down the style within the tube cell, eventually entering the embryo sac. One sperm fuses with the egg to form a diploid zygote, which will give rise to the embryo. The other sperm fuses with the binucleate primary endosperm cell to produce a triploid cell. This cell will give rise to the endosperm, a food-storage tissue within the seed.

25.5 How Do Seeds and Fruits Develop?

The embryo consists of an embryonic root and embryonic shoot, including the cotyledon (one in monocots, two in dicots). Cotyledons absorb food from the endosperm and transfer it to the growing embryo. The seed is enclosed within a fruit, which develops from the ovary wall. The function of the fruit is to disperse the seeds away from the parent plant. Fruits may be juicy and edible, enclosing seeds that are adapted for passage through an animal digestive tract. Alternatively, fruits may have hooks that attach to animal fur, or wings that promote wind dispersal.

25.6 How Do Seeds Germinate and Grow?

Seed germination requires warmth and moisture. Energy for germination comes from food, stored in the endosperm, that is transferred to the embryo by the cotyledons. Seeds may remain dormant for some time after fruit ripening, particularly in temperate climates. Environmental conditions that favor seed germination include appropriate temperature and moisture, but some seeds also require other factors including initial drying, exposure to cold, or disruption of the seed coat.

Key Terms

alternation of generations *p. 496*
anther *p. 499*
carpel *p. 499*
coleoptile *p. 505*
complete flower *p. 499*
cotyledon *p. 502*
dormancy *p. 505*
double fertilization *p. 502*
egg *p. 502*
embryo sac *p. 500*
endosperm *p. 502*
epicotyl *p. 504*
fertilization *p. 502*
filament *p. 499*
flower *p. 497*
fruit *p. 499*
gametophyte *p. 497*
generative cell *p. 500*
germinate *p. 497*
hypocotyl *p. 504*
incomplete flower *p. 499*
integument *p. 500*
megaspore *p. 501*
megaspore mother cell *p. 500*
microspore *p. 500*
microspore mother cell *p. 500*
ovary *p. 499*
ovule *p. 499*
petal *p. 499*
polar nucleus *p. 501*
pollen grain *p. 498*
pollination *p. 502*
seed *p. 498*
seed coat *p. 502*
sepal *p. 499*
spore *p. 497*
sporophyte *p. 497*
stamen *p. 499*
stigma *p. 499*
style *p. 499*
tube cell *p. 500*
zygote *p. 497*

Thinking Through the Concepts

To take a multiple-choice quiz with feedback on the contents of this chapter, visit http://www.prenhall.com/audesirk7. *Log in to the Web site selected by your instructor and navigate to the Self Test section for this chapter.*

? Review Questions

1. Diagram the plant life cycle, comparing ferns with flowering plants. Which stages are haploid, and which are diploid? At which stage are gametes formed?
2. What are the advantages of the reduced gametophyte stages in flowering plants, compared with the more substantial gametophytes of ferns?
3. Diagram a complete flower. Where are the male and female gametophytes formed? What are these gametophytes called?
4. How does an egg develop within an embryo sac? How does this structure allow double fertilization to occur?
5. What does it mean when we say that pollen is the male gametophyte? How is pollen formed?
6. What are the parts of a seed, and how does each part contribute to the development of a seedling?
7. Describe the characteristics you would expect to find in flowers that are pollinated by the wind, beetles, bees, and hummingbirds, respectively. In each case, explain why.
8. What is the endosperm? From which cell of the embryo sac is it derived? Is endosperm more abundant in the mature seed of a dicot or of a monocot?
9. Describe three mechanisms whereby seed dormancy is broken in different types of seeds. How are these mechanisms related to the typical environment of the plant?
10. How do monocot and dicot seedlings protect the delicate shoot tip during seed germination?
11. Describe three types of fruits and the mechanisms whereby these fruit structures help disperse their seeds.

Applying the Concepts

1. A friend gives you some seeds to grow in your yard. When you plant some, nothing happens. What might you try to get the seed to germinate?
2. In areas where farms have been left uncultivated for several years, it is often possible to see certain kinds of trees growing in straight lines, which mark old fences where birds sat and deposited seeds they had eaten. Why are such seeds more likely to germinate than are those of the same species that have not passed through a bird's digestive tract? How might an anthropologist use such lines of trees to study past inhabitants of an area?
3. Charles Darwin once described a flower that produced nectar at the bottom of a tube 25 centimeters (10.5 inches) deep. He predicted that there must be a moth or other animal with a 25-centimeter-long tongue to match; he was right. Such specialization almost certainly means that this particular flower could be pollinated only by that specific moth. What are the advantages and disadvantages of such specialization?
4. Many plants that we call weeds were brought from another continent either accidentally or purposefully. In their new environment, they have few competitors or animal predators, so they tend to grow in such large numbers that they displace native plants. Think of several ways in which humans become involved in plant dispersal. To what degree do you think humans have changed the distributions of plants? In what ways is this change helpful to humans? In what ways is it a disadvantage?
5. In the Tropics, there are a number of plant–animal coevolutionary relationships in which both are dependent on the relationship. In light of the rapid rate of destruction of tropical ecosystems, how does this type of relationship leave both organisms particularly vulnerable to extinction? What political and economic problems might this rapid rate of extinction create?

For More Information

Brown, Kathryn. "Patience Yields Secrets of Seed Longevity." *Science*, March 9, 2001. William Beal's plant germination study continues over 120 years later.

Eiseley, L. "How Flowers Changed the World." *National Wildlife*, April/May 1996. The late philosopher/naturalist Loren Eiseley eloquently explains how the evolution of flowers has changed the history of life on Earth. Beautifully written and lavishly illustrated.

Fleming, T. H. "Cardon and the Night Visitors." *Natural History*, October 1994. A beautifully illustrated look at the way bats pollinate the world's largest cactus and may help determine the numbers of each of the plant's three sexual types.

Handel, S. N., and Beattie, A. J. "Seed Dispersal by Ants." *Scientific American*, August 1990. Many plants depend on ants for seed dispersal, even producing fat deposits on the outside of the seed as a lure for the ants. Dispersal is not the only benefit to the seed: Being "planted" in an anthill seems to provide an ideal environment for germination and growth as well.

Kearns, C. A., and Inouye, D. W. "Pollinators, Flowering Plants, and Conservation Biology." *Bioscience*, May 1997. To preserve plant species, we must preserve their pollinators and the web of life that sustains these pollinators.

Milius, S. "Attractive Tree ISO Lemur to Start a Family." *Science News*, July 31, 1999. Loss of lemurs on Madagascar threatens trees that rely on these primates to disperse their seeds.

Milius, S. "The Science of Big, Weird, Flowers." *Science News*, September 11, 1999. Many giant flowers deceive flies and beetles into pollinating them by smelling like carrion.

Mlot, C. "Where There's Smoke, There's Germination." *Science News*, May 31, 1997. Researchers have discovered that nitrogen dioxide, such as that released in smoke, is a potent stimulus for seed germination in some species.

Moore, P. D. "The Buzz About Pollination." *Nature*, November 7, 1996. The buzzing of bees may actually shake loose the pollen of certain specially adapted flowers, a kind of "sonic pollination."

Seymour, R. S., and Schultz-Motel, P. "Thermoregulating Lotus Flowers." *Nature*, September 26, 1996. The lotus flower generates a significant amount of heat and effectively regulates its own temperature. The heat may serve as an attractant to pollinators.

Sunquist, F. "Blessed Are the Fruit Eaters." *International Wildlife*, May/June 1992. Especially in the Tropics, fruit-eating mammals and birds are crucial to the dispersal and germination of seeds. Overhunting threatens not only the animals but also the plants that depend on them.

Media Activities

To access a Media Activity visit http://www.prenhall.com/audesirk7. *Log in to the Web site selected by your instructor, navigate to this chapter, and select the appropriate Media Activity number.*

25.1 Reproduction in Flowering Plants

Estimated time: 5 minutes

Explore the basic reproductive structures in flowering plants and learn their function.

25.2 Fruits and Seeds: Structure and Development

Estimated time: 5 minutes

Examine the basic structures of fruits and seeds and then explore the process of fruit and seed development.

25.3 Web Investigations: Time Capsules, Smoke Signals and Seed Sprouting

Estimated time: 15 minutes

In this activity you will explore the complex role of fires in the environment and some of the benefits of fire in world ecosystems.

CHAPTER

26 Plant Responses to the Environment

A Venus flytrap captures its prey.

AT A GLANCE

CASE STUDY Predacious Plants

In a marsh, plants are "hungry"—not for sunlight, but for nitrogen. Though lacking in this watery environment, nitrogen is abundant in the bodies of animals; thus, a variety of bog-dwelling plants have evolved carnivorous lifestyles that satisfy their needs.

An unsuspecting fly lands on a small, seemingly harmless plant, attracted by nectar lining the edges of the plant's clamshell-like leaves. As the fly explores a leaf, seeking food, the leaf—belonging to a Venus flytrap—suddenly snaps shut. Within a fraction of a second, spikes along the leaf's outer edges trap the hapless insect (inset). During the next four to five days, enzymes will digest the fly, and the leaf will absorb the nitrogen-containing molecules before the trap opens to attract its next victim.

Nearby, a lacewing insect lands on cluster of glistening, nectar-like droplets—only to find itself struggling helplessly in the sticky mass. To make matters worse, red tentacles bearing more balls of "glue" bend toward it, miring it hopelessly within a few seconds. The bending tentacles then transport the struggling insect toward the center of the leaf, where digestive enzymes attack its body and the leaf cells absorb the liberated nutrients. In this way, the sundew plant captures its meals (see Fig. 24-28d).

Beneath the surface of the marsh, still another drama unfolds. A bladderwort dangles hundreds of pear-shaped chambers in the water, which—depending on the species—range in size from pinheads to peas. Each of these "bladders" is sealed by a watertight trap door whose lower edge is fringed with bristles. A copepod, a crustacean related to shrimp but smaller than a pinhead, swims by, brushing the hairs. Within one-sixtieth of a second, the bladder traps the animal inside, where enzymes gradually kill and digest it. But how do these predacious plants move quickly enough to trap flying or swimming animals?

26.1 What Are Plant Hormones, and How Do They Act?

Plant cells, like those of animals, are miniature factories filled with diverse chemicals that allow plants to respond appropriately to their environment. Some convey messages within the plant, and others even communicate among individual plants. Animal physiologists have long recognized that chemicals called **hormones** are produced by cells in one location and transported to other parts of the body, where they exert specific effects. Comparably, plant-regulating chemicals are called **plant hormones**. So far, plant physiologists have identified five major classes of plant hormones: auxins, gibberellins, cytokinins, ethylene, and abscisic acid (Table 26-1). Several other types of hormones are also thought to exist. Each hormone can elicit a variety of responses from plant cells, depending on factors such as the type of target cell, the developmental stage of the plant, the concentration of the hormone, and the presence of other hormones. Further, the roles of some plant hormones vary among plant species. This should not be too surprising; after all, in animals, a single hormone can be involved in actions as diverse as a salmon's transition from fresh water to salt water, a frog's metamorphosis from tadpole to adult, and a snake's shedding of its skin (see "Evolutionary Connections: The Evolution of Hormones," in Chapter 33).

Auxins promote the elongation of cells in coleoptiles and other parts of the shoot; high concentrations cause cells to elongate (see "Scientific Inquiry: How Were Plant Hormones Discovered?"). In roots, which differ from stems in their response to auxin, low concentrations stimulate elongation, while slightly higher concentrations inhibit elongation. Both light and gravity affect the distribution of auxin in roots and shoots, so auxin plays a major role in both **phototropism** (growth toward light) and **gravitropism** (directional growth with respect to gravity). Auxin also affects many other aspects of plant development, such as the differentiation of conducting tissues (xylem and phloem) and the development of fruits. Auxin may also prevent sprouting by lateral buds. It stimulates root branching and can be used to cause stems of plants to grow roots, which is useful when growing a new plant from a cutting.

Gibberellins are a group of chemically similar molecules that, like auxin, promote the elongation of cells in stems. In some plants, gibberellins stimulate flowering, fruit development, seed germination, and bud sprouting.

Cytokinins promote cell division in many plant tissues; consequently, they stimulate the sprouting of buds and the development of fruit, seed endosperm, and the embryo. Cytokinins also stimulate plant metabolism, delaying the aging of plant parts, especially leaves.

Ethylene is an unusual plant hormone in that it is a gas at typical environmental temperatures. Ethylene is best known, and most commercially valuable, for its ability to cause fruit to ripen. It also stimulates the separation of cell walls into *abscission layers*, which, we will soon see, allow leaves, flowers, and fruit to drop off at the appropriate times without harming the plant. Ethylene's effects were first recognized in the 1800s, when gas lamps were installed in German cities. Residents observed that plants growing near leaky gas mains supplying the lamps grew abnormally and lost their leaves prematurely. In the early 1900s, a Russian plant physiologist tested all the components of the "illuminating gas" and discovered that ethylene was responsible for the effects on plants.

Abscisic acid helps plants withstand unfavorable environmental conditions. As you learned in Chapter 24, abscisic acid causes stomata to close when water availability is low. It inhibits the activity of gibberellins, thus helping maintain dormancy in buds and seeds at times when germination would be dangerous.

The rest of this chapter describes a year in the life of a plant, illustrating how hormones regulate its growth and development.

Table 26-1 Hormone Actions in Plants

Hormone	Functions
Abscisic acid	Closing of stomata; seed dormancy; bud dormancy
Auxins	Elongation of cells in coleoptiles and shoots; phototropism; gravitropism in shoots and roots; root growth and branching; apical dominance; development of vascular tissue; fruit development; retarding senescence in leaves and fruit; ethylene production in fruit
Cytokinins	Promotion of sprouting of lateral buds; prevention of leaf senescence; promotion of cell division; stimulation of fruit, endosperm, and embryo development
Ethylene	Ripening of fruit; abscission of fruits, flowers, and leaves; inhibition of stem elongation; formation of hook in dicot seedlings
Gibberellins	Germination of seeds and sprouting of buds; elongation of stems; stimulation of flowering; development of fruit

Media Activity 26.1 Hormones

26.2 How Do Hormones Regulate the Plant Life Cycle?

The life cycle of a plant results from a complex interplay between its genetic information and its environment. Hormones mediate many of the genetic determinants of growth and development, as well as nearly all responses to environmental factors. At each stage in its life cycle, a plant produces a distinctive set of hormones that interact with one another to direct the growth of the plant body.

Abscisic Acid Maintains Seed Dormancy; Gibberellin Stimulates Germination

As we pointed out in Chapter 25, a warm autumn day provides ideal germination conditions for a seed maturing within a juicy fruit, yet the seed remains dor-

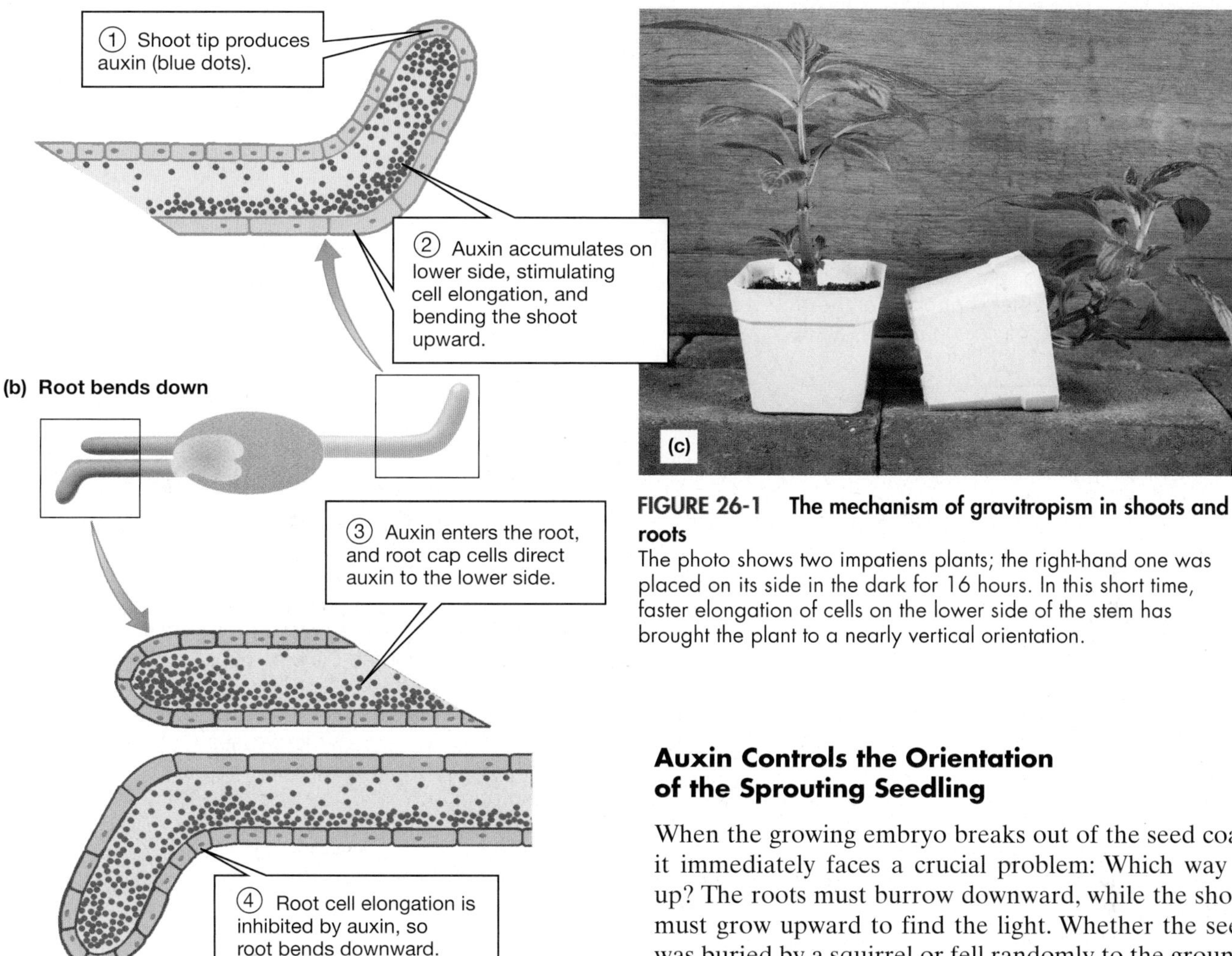
(a) Shoot bends upward
① Shoot tip produces auxin (blue dots).
② Auxin accumulates on lower side, stimulating cell elongation, and bending the shoot upward.
(b) Root bends down
③ Auxin enters the root, and root cap cells direct auxin to the lower side.
④ Root cell elongation is inhibited by auxin, so root bends downward.
root cap
(c)

FIGURE 26-1 The mechanism of gravitropism in shoots and roots
The photo shows two impatiens plants; the right-hand one was placed on its side in the dark for 16 hours. In this short time, faster elongation of cells on the lower side of the stem has brought the plant to a nearly vertical orientation.

mant until the following spring. In many seeds, abscisic acid enforces dormancy. Abscisic acid slows down the metabolism of the embryo within the seed, preventing its growth. The seeds of some desert plants contain high concentrations of abscisic acid; only a really hard rain can wash it away, freeing the embryo from its inhibitory effects and allowing the seed to germinate. Seeds of most high-latitude plants require a prolonged period of cold, such as a full winter season, to break dormancy; in these seeds, chilling induces the destruction of abscisic acid.

Germination is stimulated by other hormones, especially gibberellin. The same environmental conditions that cause the breakdown of abscisic acid also promote the synthesis of gibberellin. In germinating seeds, gibberellin initiates the synthesis of enzymes that digest the food reserves of the endosperm and cotyledons, making sugars, lipids, and amino acids available to the growing embryo.

Auxin Controls the Orientation of the Sprouting Seedling

When the growing embryo breaks out of the seed coat, it immediately faces a crucial problem: Which way is up? The roots must burrow downward, while the shoot must grow upward to find the light. Whether the seed was buried by a squirrel or fell randomly to the ground, it faces a high probability of being oriented "upside-down." Auxin controls the responses of both roots and shoots to light and gravity.

Auxin Stimulates Shoot Elongation Away from Gravity and Toward Light

Let's look at the growth of a shoot as it first emerges from the seed, buried underground. Auxin is synthesized in shoot tips, moves down the shaft of the stem, and stimulates cell elongation. If the stem is not exactly vertical, organelles in the cells of the stem detect the direction of gravity and cause auxin to accumulate on the stem's lower side (Fig. 26-1a). Therefore, the lower cells elongate rapidly, forcing the stem to bend upward. When the shoot tip is vertical, the auxin distribution becomes symmetrical. The stem then grows straight up, emerging from the soil into the light (Fig. 26-1c).

In addition to gravitropism, auxin also mediates phototropism. Ordinarily, the distribution of auxin caused by light is the same as the distribution caused by gravity, because the direction of brightest light (the sun) is roughly opposite that of gravity. For example, if a young shoot still buried underground is close enough to the surface that some light penetrates down to it, both light

Anyone who keeps houseplants on a windowsill knows that as they grow, the plants bend toward the window in response to the sunlight streaming in. More than a hundred years ago, Charles Darwin and his son Francis studied this phenomenon of growth toward the light, or phototropism.

FIRST, THE DARWINS DETERMINED THE DIRECTION OF INFORMATION TRANSFER

The Darwins illuminated grass coleoptiles (protective sheaths surrounding monocot seedlings) from various angles. They noted that a region of the coleoptile a few millimeters below the tip bent toward the light, causing the tip to point toward the light source. When they controlled for the presence of light by covering the tip of the coleoptile with an opaque cap, the coleoptile didn't bend.

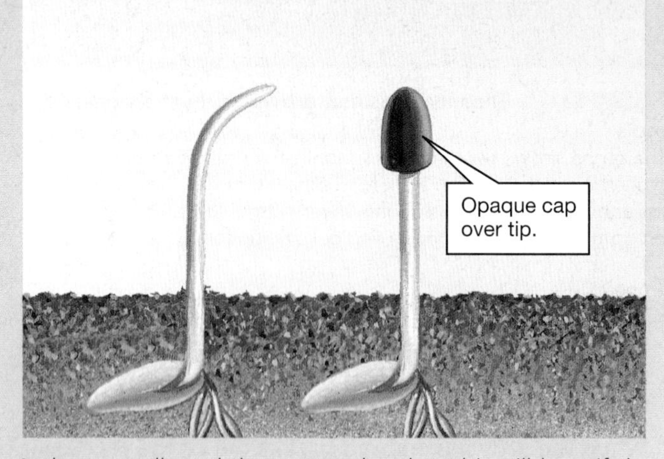

A clear cap allowed the stem to bend, and it still bent if the Darwins covered the bending region below the tip with an opaque sleeve.

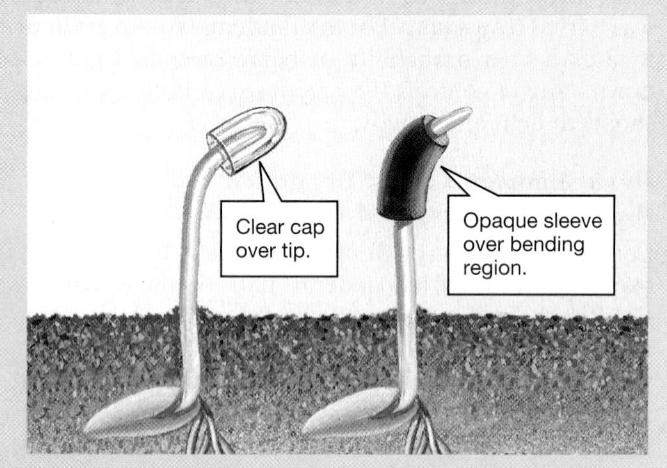

They concluded that the tip of the coleoptile perceives the direction of light and bending occurs farther down the coleoptile; therefore, the tip transmits information about the light direction down to the bending region. How does the coleoptile bend? Although the Darwins didn't know this, the coleoptile bends as a result of unequal elongation of cells on opposite sides of the shaft. The cells on the darker side elongate faster than those facing the light, bending the shaft toward the light. So information transmitted from the tip to the bending region causes unequal cell elongation.

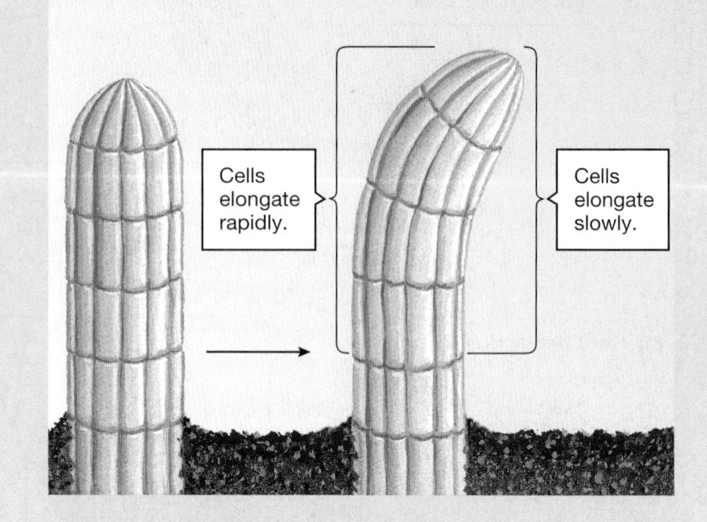

NEXT, PETER BOYSEN-JENSEN DEMONSTRATED THAT THE INFORMATION IS CHEMICAL IN NATURE

About 30 years after the Darwins' experiments, Peter Boysen-Jensen cut the tips off coleoptiles and found that the remaining stump neither elongated nor bent toward the light. If he replaced the tip and placed the patched-together coleoptile in the dark, it elongated straight up. In the light, it showed normal phototropism. When he inserted a thin layer of porous gelatin that prevented direct contact but permitted diffusion of substances between the severed tip and the stump, he still observed elongation and bending. In contrast, an impenetrable barrier eliminated these responses.

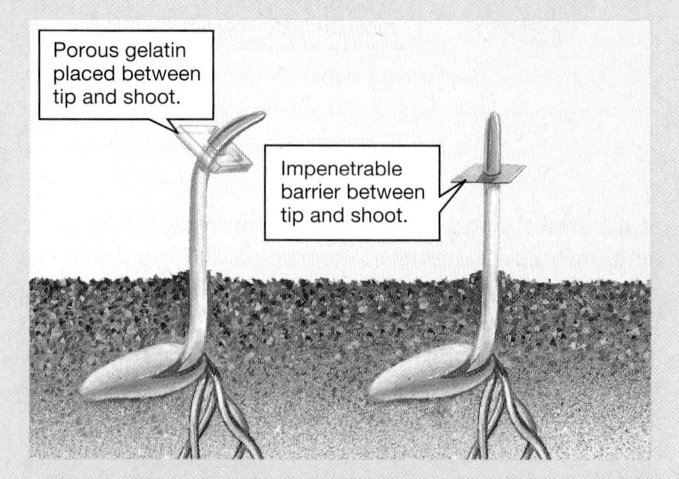

Porous gelatin placed between tip and shoot.
Impenetrable barrier between tip and shoot.

Boysen-Jensen concluded that a chemical is produced in the tip and moves down the shaft, causing cell elongation. In the dark, the chemical that causes the cells to elongate diffuses straight down from the tip and causes the coleoptile to elongate straight up. Presumably, light causes the chemical to become more concentrated on the "shady" far side of the shaft, so cells on the shady side elongate faster than do cells on the "sunny" near side, causing the shaft to bend toward the light.

FINALLY, THE CHEMICAL AUXIN WAS IDENTIFIED

The next step was to isolate and identify the chemical. In the 1920s, Frits Went devised a way to collect the elongation-promoting chemical. He cut off the tips of oat coleoptiles and placed them on a block of agar (a porous, gelatinous material) for a few hours. Went hypothesized that the chemical would migrate out of the coleoptiles into the agar.

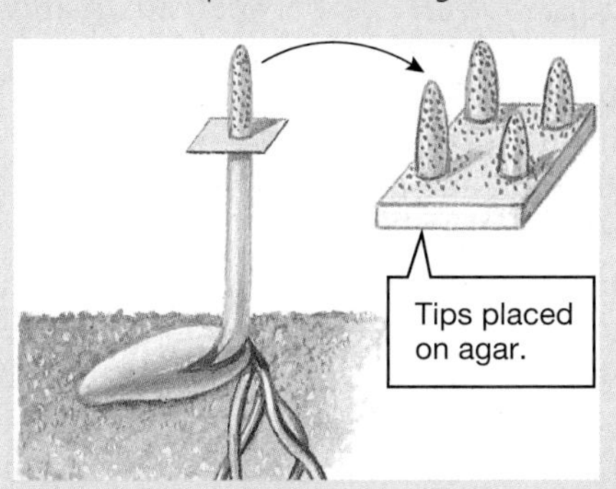
Tips placed
on agar.

He then cut up the agar, now presumably loaded with the chemical, and placed small pieces on the tops of coleoptile stumps growing in darkness. When he put a piece of agar squarely atop a stump, the stump elongated straight up. All the stump cells received equal amounts of the chemical and elongated at the same rate. If he placed a piece on one side of a cut stump, the stump would invariably bend away from the side with the agar.

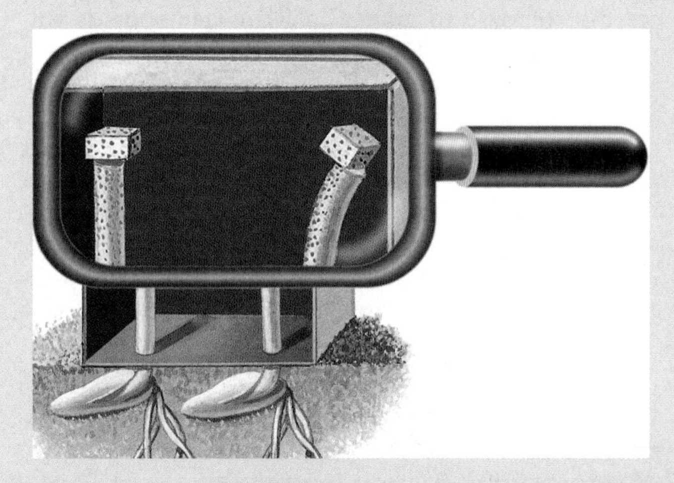

It was apparent that cells on the side under the agar received more of the chemical and were more stimulated to elongate. Went called the chemical *auxin*, from a Greek word meaning "to increase." Kenneth Thimann later purified auxin and determined its molecular structure.

and gravity cause auxin to be transported to the lower side of the shoot and promote upward bending. Thus, under normal conditions, gravitropism and phototropism add to each other's effects.

Auxin May Control the Direction of Root Growth

Gravitropism in roots is less well understood than it is in stems. According to one model, auxin controls the direction of root growth (Fig. 26-1b). Auxin is transported from the shoot down to the root. If the root is not vertical, the root cap senses the direction of gravity and causes the auxin to accumulate on the lower side of the root. Unlike shoots, in which moderate concentrations of auxin stimulate cell elongation, in roots these same concentrations of auxin inhibit cell elongation. Therefore, cell elongation in the lower side of the root, where auxin accumulates, is inhibited, whereas cell elongation remains unaffected in the upper side of the root. As a result, the root bends downward. When the root tip points directly downward, the auxin distribution becomes equal on all sides, and the root continues to grow straight downward. Note that auxin slows down but does not eliminate root cell elongation; thus, a vertical root continues to grow.

Plants May Sense Gravity by Means of Starch-Filled Plastids

Although auxin plays a major role in the unequal growth rates of cells that cause bending away from gravity (in shoots) and toward gravity (in roots), auxin itself does not detect gravity. Instead, specialized cells in stems and in root caps contain starch-filled plastids. By staining these plastids and observing them under the microscope, plant physiologists have discovered that they settle to the downward side of the cell within minutes when a plant is laid on its side (Fig. 26-2). The time it takes the plastids to settle is similar to the time it takes the root to

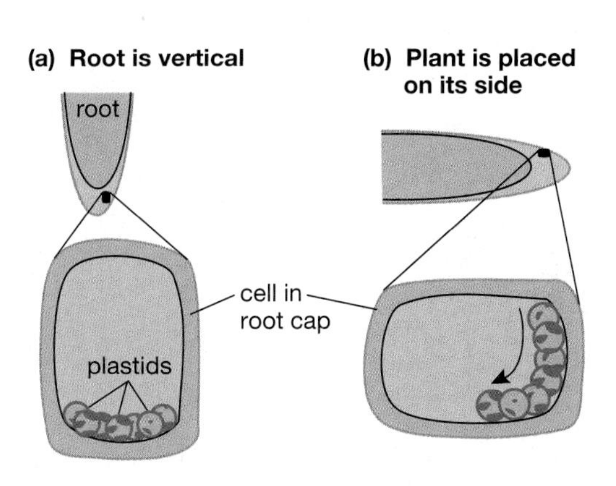
(a) Root is vertical
root
cell in
root cap
plastids
(b) Plant is placed
on its side

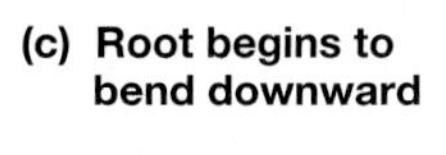
(c) Root begins to
bend downward

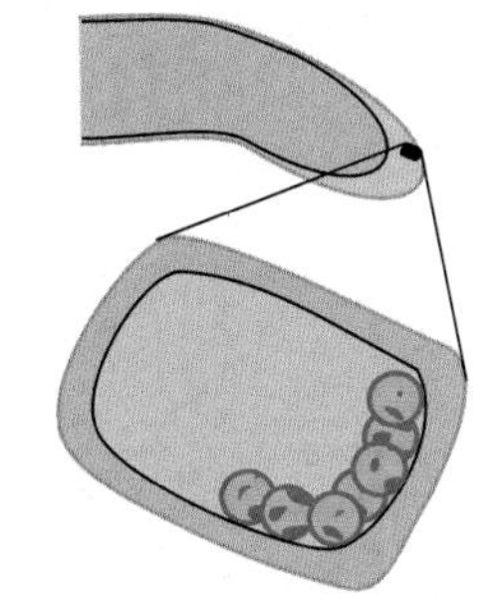

FIGURE 26-2 Starch-filled plastids allow plants to sense gravity
(a) Normal orientation of starch-filled plastids in a root cap cell. **(b)** When the plant is placed on its side (as in Figure 26-1c), the plastids begin tumbling downward, coming to rest on the new lower surface. **(c)** This change triggers the bending of the root downward.

begin its unequal cell elongation so that it once more grows downward. Exactly how the falling plastids initiate the response to gravity in stems and roots is still under investigation. One hypothesis is that the plastids are enmeshed in fibers of the cytoskeleton (see Chapter 5) that connect them to ion channels. As the plastids are pulled downward by gravity, the ion channels are pulled open, allowing movement of ions, such as calcium, into the cell. Thus, a mechanical stimulus (movement of plasmids in response to gravity) could be converted to a chemical stimulus. This chemical signal might initiate a series of reactions that cause auxin to accumulate on the downward side of the root cap.

The Genetically Determined Shape of the Mature Plant Is the Result of Interactions Among Hormones

As a plant grows, both its root and shoot develop branching patterns that are largely determined by its genetic heritage. For example, the stems of some plants, such as sunflowers, hardly branch at all; others, such as oaks and cottonwoods, branch profusely without any clear pattern; still others branch in very regular patterns, producing the conical shapes of firs and spruces.

The amount of growth in shoot and root systems must also be kept in balance. The shoot must be large enough to supply the roots with sugars, and the roots must be large enough to provide the shoot with water and minerals. Interactions between auxin and cytokinin regulate root and stem branching, thereby regulating the relative sizes of root and shoot systems.

Stem Branching Is Influenced by the Growing Tip of the Shoot

Gardeners know that pinching back the tip of a growing plant causes the plant to become bushier. The explanation for this is that the growing tip suppresses the sprouting of lateral buds to form branches, a phenomenon known as **apical dominance**. The control mechanism for the sprouting of lateral buds remains a subject of research. However, some evidence shows that the proper levels and ratios of auxin and cytokinin must be present (Fig. 26-3). Auxin is produced by the shoot tip (where it is found in highest concentration) and transported down the stem, gradually decreasing in concentration.

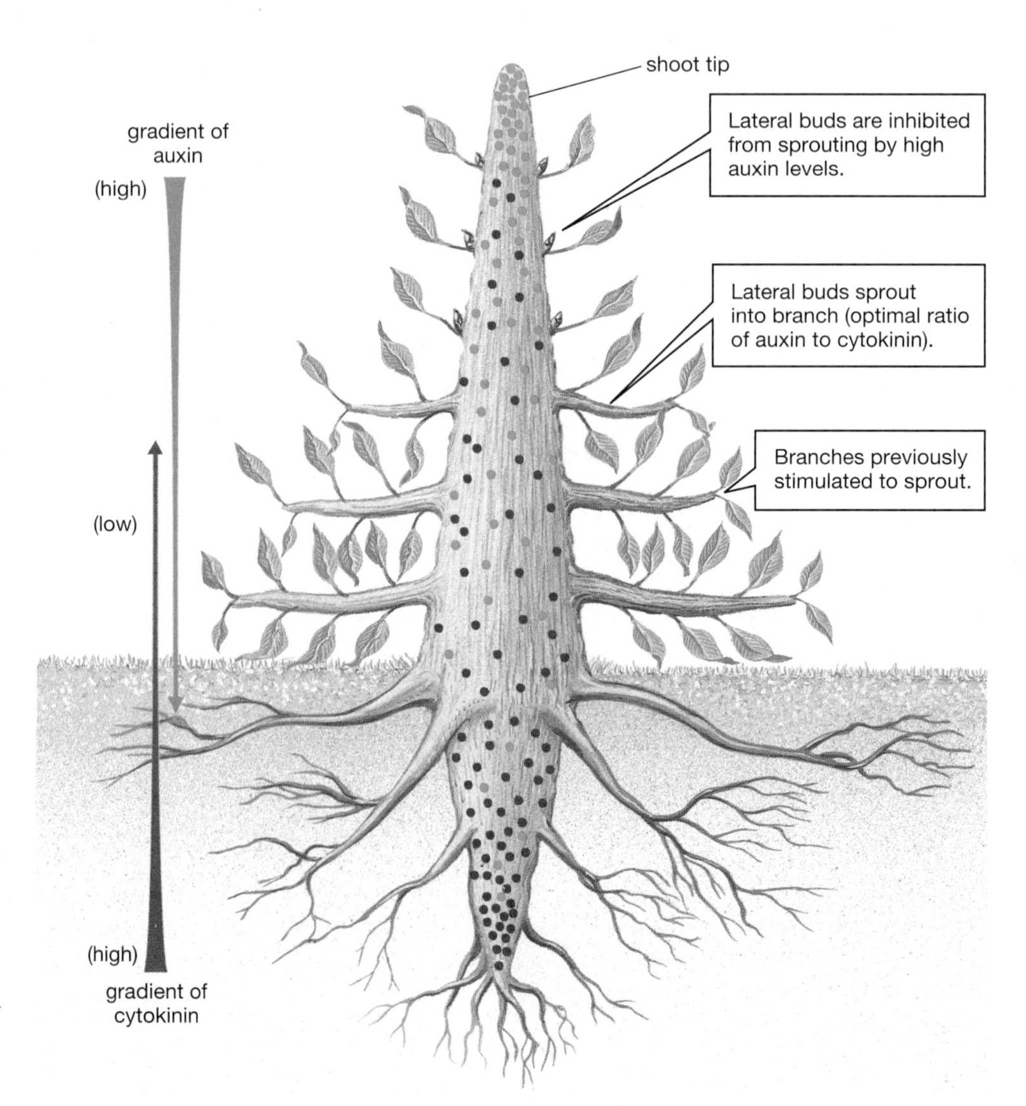
shoot tip
Lateral buds are inhibited from sprouting by high auxin levels.
Lateral buds sprout into branch (optimal ratio of auxin to cytokinin).
Branches previously stimulated to sprout.
gradient of auxin
(high)
(low)
(high)
gradient of cytokinin

FIGURE 26-3 The role of auxin and cytokinin in lateral bud sprouting
A simplified diagram of the interplay of auxin (blue dots) and cytokinin (red dots) in the control of sprouting of lateral buds. Auxin is produced by shoot tips and moves downward; cytokinin is produced by root tips and moves upward. **QUESTION** What result would you expect if a plant's shoot tip were removed and auxin were applied to the cut surface?

Cytokinin is produced by the root tips (where its concentration is highest) and is transported up through the roots and into the stem. Therefore, the relative concentrations of these two hormones will vary along the length of the stem. Buds at different positions will experience differing hormonal influences.

Auxin by itself appears to inhibit the sprouting of lateral buds, whereas auxin and cytokinin together stimulate bud sprouting. The lateral buds closest to the shoot tip receive enough auxin to inhibit their growth, and receive very little cytokinin because they are so far from the roots. Therefore, they remain dormant. Lower buds receive less auxin while receiving much more cytokinin. They are stimulated by optimal concentrations of both hormones, so they sprout (see Fig. 26-3). In many types of plants, this interaction between auxin and cytokinin produces an orderly progression of bud sprouting from the bottom to the top of the shoot. The exact ratio of cytokinin to auxin that promotes sprouting varies among species.

Auxin Stimulates Root Branching

Even in extremely low concentrations, auxin stimulates the branching of roots. As we described in Chapter 24, branch roots arise from the pericycle layer of the vascular cylinder. Auxin, transported down from the stem, stimulates pericycle cells to divide and form a branch root. Auxin also stimulates root development. Commercial auxin powder allows people to produce a new plant by dipping the cut end of a stem into the auxin and placing it in soil or water, where it will develop roots.

Gradients of Auxin and Cytokinin Create a Balance Between the Root and Shoot Systems

Through the interaction of auxin and cytokinin, the root and shoot systems regulate each other's growth. This interaction is important, because the roots and the shoots supply complementary nutrients. An enlarging root system synthesizes large amounts of cytokinin, which stimulates lateral buds to break dormancy and sprout. If the root system isn't keeping up with the growing shoot system, less cytokinin is produced. The sprouting of lateral buds is delayed, slowing the growth of the shoot system. Simultaneously, as the stem grows and branches, it produces more auxin, which stimulates root branching and growth. Thus, neither root nor shoot can outgrow the other, and the plant is adequately supplied with all its needs.

Daylength Controls Flowering

The timing of flowering and seed production is finely tuned to the physiology of the plant and the rigors of its environment. In temperate climates, plants must flower early enough so that their seeds can mature before the deadly frosts of autumn. Depending on how quickly the seed and fruit develop, flowering may occur in spring, as it does in oaks; in summer, as in lettuce; or even in autumn, as in asters.

What environmental cues do plants use to determine the seasons? Most cues, such as temperature or water availability, are quite variable: October can be warm, a late snow may fall in May, or the summer might be unusually cool and wet. The only reliable cue is daylength: longer days always mean that spring and summer are coming; shorter days foretell the onset of autumn and winter.

With respect to flowering, botanists classify plants as *day-neutral*, *long-day*, or *short-day* (Fig. 26-4). A **day-neutral plant** flowers as soon as it has sufficiently grown and developed, regardless of the length of the day. Day-neutral plants include tomatoes, corn, and snapdragons. Although the naming is traditional, long-day and short-day plants are better described as *short-night* and *long-night* plants, respectively, because their flowering actually depends on the duration of continuous darkness rather than daylength. A **short-night plant** flowers when the duration of darkness is shorter than a species-specific critical value. A **long-night plant** flowers when the duration of uninterrupted darkness is longer than a species-specific critical value. Thus, spinach is classified as a short-night plant because it flowers only if the night is shorter than 11 hours, and the cocklebur is a long-night plant because it flowers only if the night is longer than 8.5 hours (thus, both will flower with 10 hours of darkness).

Grafting a branch from a long-night cocklebur plant maintained on a long-night schedule onto a cocklebur plant grown in a short-night environment will induce flowering. In fact, plant physiologists have been able to induce flowering in the cocklebur by exposing a *single leaf* to long nights in a special chamber, while the rest of the plant continued to experience short nights. Clearly, a signal must travel from leaf to bud. Studies on a variety of plants suggest that still-unidentified substances can both trigger and inhibit flowering. These hypothetical substances, which may differ among plant species, are collectively called **florigens** (literally, "flower makers"). Duration of darkness is a crucial stimulus for these substances, but how do plants detect this?

Pigments Called Phytochromes Measure Darkness by Resetting the Biological Clock

To measure darkness, a plant needs two things: some sort of metabolic clock to measure time (the duration of darkness) and a light-detecting system to set the clock. Virtually all organisms have an internal **biological clock** that measures time even without environmental cues. However, environmental cues, particularly light, can reset the clock. In most organisms, the operating method of the biological clock is poorly understood.

The light-detecting system of plants is a pigment in the leaves called **phytochrome** (meaning "plant color"). Phytochrome occurs in two interchangeable forms.

FIGURE 26-4 The effects of night length on flowering

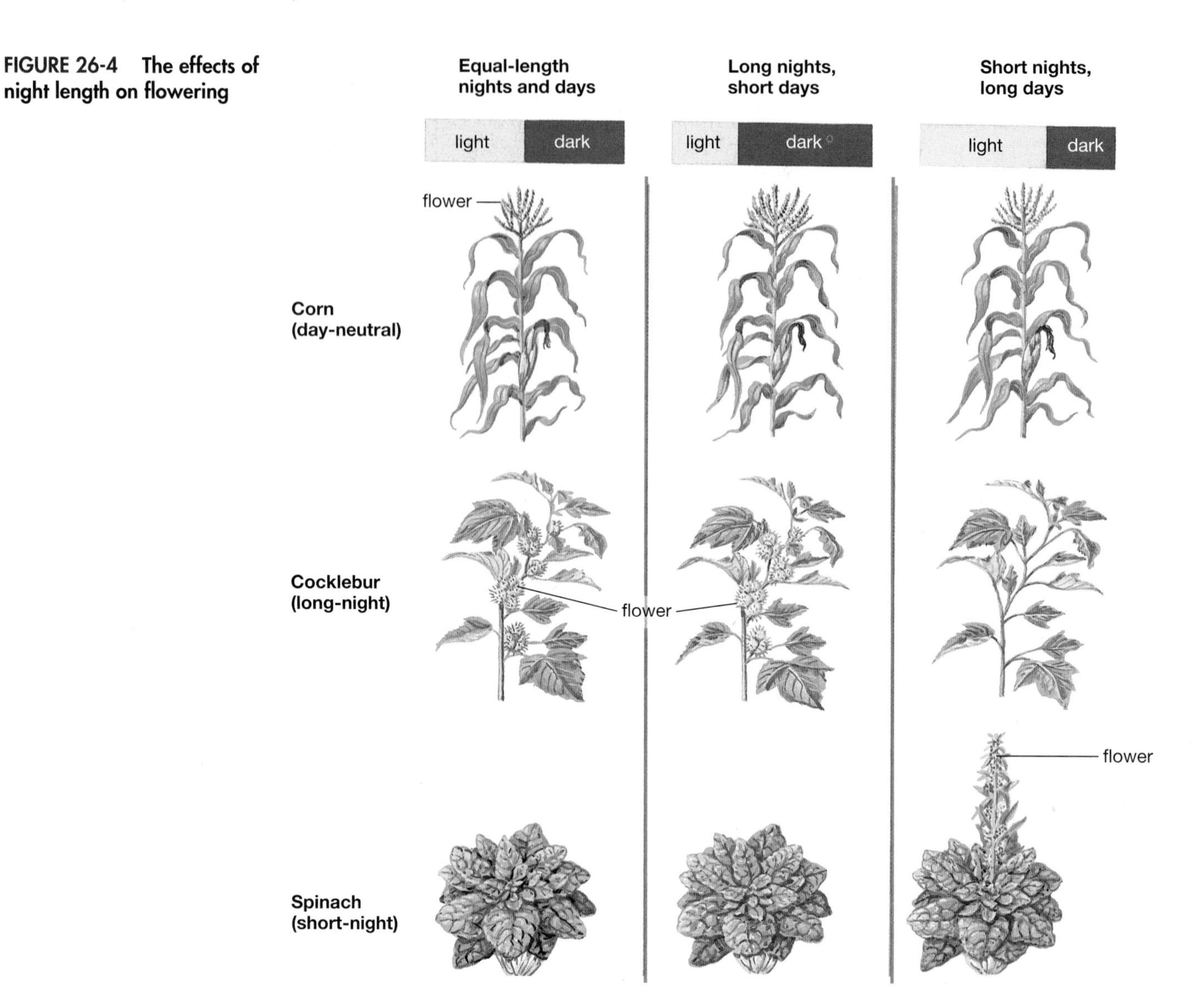

Media Activity
26.2 Plant Responses to Phytochrome

One form strongly absorbs red light (r) and is called P_r; the other form absorbs far-red light (fr; almost infrared) and is accordingly called P_{fr} (Fig. 26-5). In most plants, P_{fr} is the active form of phytochrome; that is, a suitable concentration of P_{fr} stimulates or inhibits physiological processes, such as flowering or setting the biological clock. The inactive form, P_r, has no effect on these same processes.

Phytochrome changes from one form to the other when the pigment absorbs light of the appropriate color: upon absorbing red light, P_r is converted into active P_{fr}; upon absorbing far-red light, P_{fr} is transformed back into inactive P_r. Daylight consists of all wavelengths of visible light, including both red and far-red. Therefore, during the day, a leaf contains both forms of phytochrome. In the dark, however, P_{fr} rapidly and spontaneously reverts to P_r.

Plants seem to use the phytochrome system and their internal biological clocks to detect the duration of darkness and light. Cockleburs, for example, flower under a lighting schedule of 16 hours of darkness and 8 hours of light. However, interrupting the middle of the dark period with just a minute or two of light prevents flowering. Thus, their flowering is controlled by the length of *continuous* darkness.

The color of the light used for the night flash is also important. A midnight flash of red light inhibits flowering, but a far-red flash allows flowering. This observation implicates phytochrome in the control of flowering. However, scientists are still researching how the response of phytochrome to light determines whether a plant will flower. How phytochromes influence the production of florigens is another area of active research.

Phytochrome Influences Other Responses of Plants to Their Environment

Phytochrome is involved in many plant responses; for example, P_{fr} inhibits the elongation of seedlings. Because P_{fr} reverts to P_r in the dark, seedlings germinating in the darkness of the soil contain no P_{fr} and consequently elongate very rapidly, emerging from the soil. Seedlings growing beneath other plants will be exposed

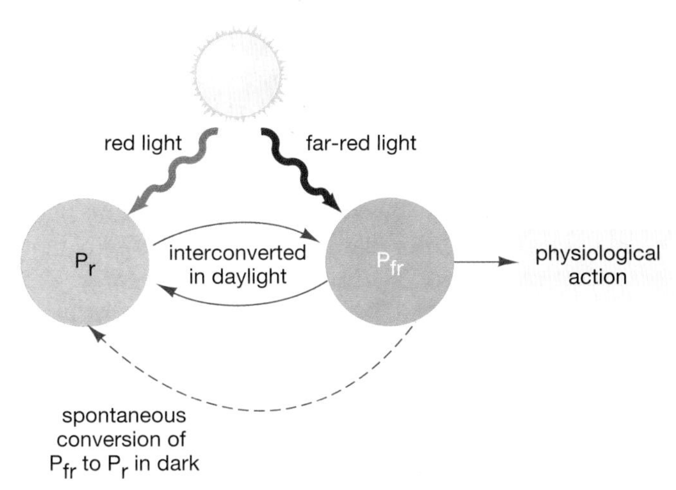

FIGURE 26-5 The light-sensitive pigment phytochrome
Phytochrome exists in two forms: inactive (P_r) and active P_{fr}. P_r is converted to P_{fr} by red light. In daylight, the two forms reach an equilibrium, with slightly more P_{fr}. P_{fr} may be converted to P_r by far-red light, and also reverts spontaneously to P_r in darkness.

FIGURE 26-6 Ripe fruit becomes attractive to animal seed dispersers
The prickly pear cactus fruit is green, hard, and bitter before it ripens, which discourages animals from eating it. After the seeds mature, the fruit becomes soft, red, and tasty, attracting animals such as this desert tortoise. The mature seeds are not harmed by the animal's digestive tract and are dispersed in the animal's feces. **QUESTION** Agricultural engineers have developed genetically modified tomato plants in which ethylene production is blocked. Why might such a plant be valuable to tomato growers?

largely to far-red light, because the green chlorophyll of the leaves above them absorb most of the red light but transmit the far-red. Far-red light converts P_{fr} to P_r, so shaded seedlings grow rapidly, which may bring them out of the shade. Once out in the sunlight, P_{fr} forms. The P_{fr} slows down elongation, which prevents the seedlings from becoming too spindly.

Other plant responses that are stimulated by P_{fr} include leaf growth, chlorophyll synthesis, and the straightening of the epicotyl or hypocotyl hook of dicot seedlings (see Chapter 25). As in the case of stem elongation, these responses are adaptations related to burial in the soil or shading by the leaves of other plants. For example, a newly germinating shoot needs to retain its protective bend while still in the soil (that is, in the dark) and straighten out only in the open air, where sunlight converts P_r to P_{fr}.

Hormones Coordinate the Development of Seeds and Fruit

When a flower is pollinated, auxin or gibberellin released by the pollen stimulates the ovary to begin developing into a fruit. If fertilization also occurs, the developing seeds release still more auxin or gibberellin (or both) into the surrounding ovary tissues. Cells of the ovary multiply and grow larger, commonly storing starches and other food materials, and producing a mature fruit. In this way, the plant coordinates the development of seeds and fruit.

Seeds and fruits acquire nutrients for growth and development from their parent plant. If the seed is separated from the parent too soon, it may not complete its development. Not surprisingly, seed maturation and fruit ripening are closely coordinated. Most unripe fruits are inconspicuously colored (usually green, like the rest of the plant), hard, bitter, and in some cases even poisonous. As a result, animals seldom eat unripe fruit.

When the seeds mature, the fruit ripens; it becomes sweeter as starches are converted to sugar, softer, and more brightly colored, making it more noticeable and attractive to animals (Fig. 26-6). Look around the produce section of your supermarket at all the brightly colored fruits—adapted to attract animal seed dispersers. Interestingly, gibberellin sprayed on fruit such as grapefruit causes the peel to remain tough and green, although the inside continues to ripen. Citrus growers in Florida can now use this technique to discourage fruit flies, which are attracted to the yellow color and must penetrate the ripening peel to lay their eggs.

Fruit ripening is frequently stimulated by ethylene, which also causes the breakdown of green chlorophyll, revealing the attractive pigments that signal a ripe fruit. Ethylene is synthesized by fruit cells in response to a surge of auxin that is released by the seeds (another mechanism by which seed and fruit development are coordinated). Because ethylene is a gas, many ripe fruits continually leak ethylene into the air. In nature, this probably doesn't make much difference. When you store fruit in a closed container, however, ethylene released from one fruit will hasten ripening in the rest.

Media Activity
26.3 Plant Response to Stimuli

The discovery of the role of ethylene in ripening revolutionized modern fruit and vegetable marketing. The gibberellin-sprayed green (but ripe) grapefruit described earlier will turn its normal yellow when exposed to ethylene. Bananas grown in Central America can be picked green and tough and shipped to North American markets. By exposing them to ethylene at their destination, grocers can market perfectly ripe fruit. Unfortunately, not all fruits ripen properly when separated from the plant. Strawberries, for example, do not ripen in response to ethylene and must be allowed to vine-ripen, so shipping the ripe, soft fruit to markets without damage can be a challenge. Although green tomatoes do ripen when gassed with ethylene, they never taste the same as those that ripen on the vine (see "Links to Life: The Quest for the Perfectly Ripe Tomato").

Senescence and Dormancy Prepare the Plant for Winter

In autumn, fruit ripens and drops to the ground, making it more available to animals who will eat it and disperse its seeds. For perennial broadleaf plants, leaves must be also be shed in autumn, because they would be a liability in winter: unable to photosynthesize but still allowing water to evaporate. Leaves, fruits, and flowers undergo a rapid aging called **senescence**.

Senescence is a complex process controlled by several different hormones. In most plants, healthy leaves and developing seeds produce auxin, which in turn helps maintain the health of the leaf or fruit. Simultaneously, the roots synthesize cytokinin, which is transported up the stem and out to the branches. Cytokinin also prevents senescence (a leaf plucked from a tree and floated in water in which cytokinin is dissolved may stay green for weeks). But as winter approaches, cytokinin production in roots slows, and fruits and leaves produce less auxin. Meanwhile, ethylene is released by both aging leaves and ripening fruit. Ethylene stimulates leaf senescence, during which proteins, starches, and chlorophyll are broken down to simple molecules that are transported to the roots and other permanent tissues of the plant for winter storage. The culmination of senescence is the formation of the **abscission layer** at the base of the petiole (Fig. 26-7). Ethylene stimulates this layer of thin-walled cells to produce an enzyme that digests their cell walls. When the petiole attachment site weakens sufficiently, the leaf or fruit falls.

Other changes also occur that prepare the plant for winter. New buds, rather than developing into leaves and branches as they would have during spring and summer, now become tightly wrapped up and dormant, waiting out the winter. Dormancy in buds, as in seeds, is enforced by abscisic acid. Metabolism slows to a crawl, and the plant enters its long winter "sleep," awaiting signals of warmth and longer spring days before "awakening" once again.

26.3 Can Plants Communicate and Move Rapidly?

Plants May Summon "Bodyguards" When Attacked

The hundred-million-year war between plants and their animal parasites and predators has led to the evolution of sophisticated plant defenses that surprise people who are accustomed to thinking of plants as passive, helpless organisms. Researchers studying how plants respond to attack by predators or disease-causing viruses have recently discovered that plants under attack help protect themselves—and sometimes neighboring plants as well—by releasing volatile chemicals into the air around them; a chemical "cry for help." Working with maize (a relative of corn), researchers discovered that in response to attack by hungry caterpillars, maize plants release a mixture of volatile chemicals. A wasp has evolved an attraction to these volatile chemicals, probably because they signal the presence of the caterpillars, which provide food for the wasp's offspring. The wasp lays its eggs in a caterpillar's body, where the larvae hatch and consume their host from the inside out. Scientists discovered that merely tearing the leaves of the maize plant will not elicit the chemical alarm signal; the attack on the plant must come from an actual caterpillar. A chemical called *volicitin* in caterpillar saliva causes maize to release the volatile chemicals (Fig. 26-8). Similarly, when spider

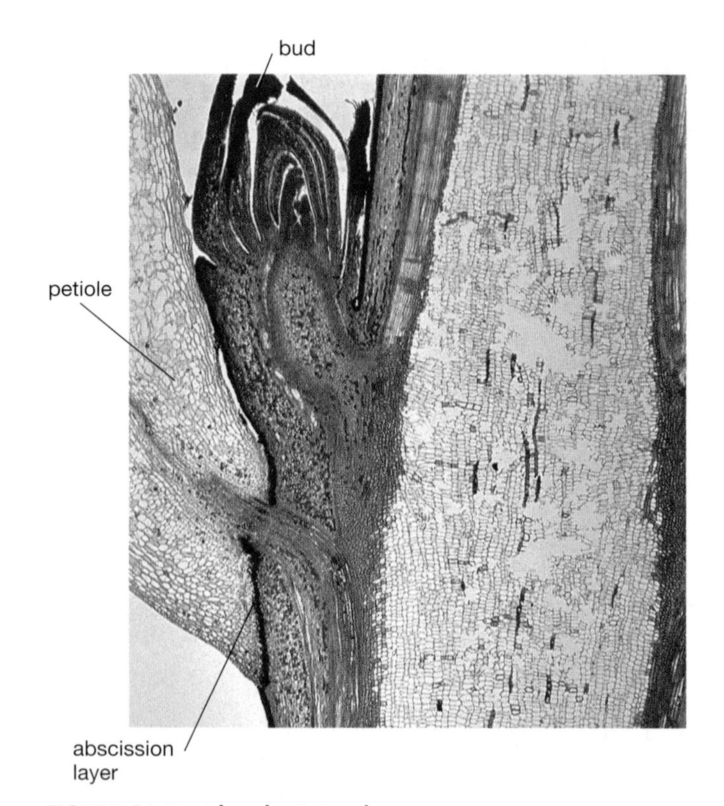

FIGURE 26-7 The abscission layer
This cross section shows the abscission layer forming at the base of a maple leaf. A new leaf bud is visible above the dying leaf.

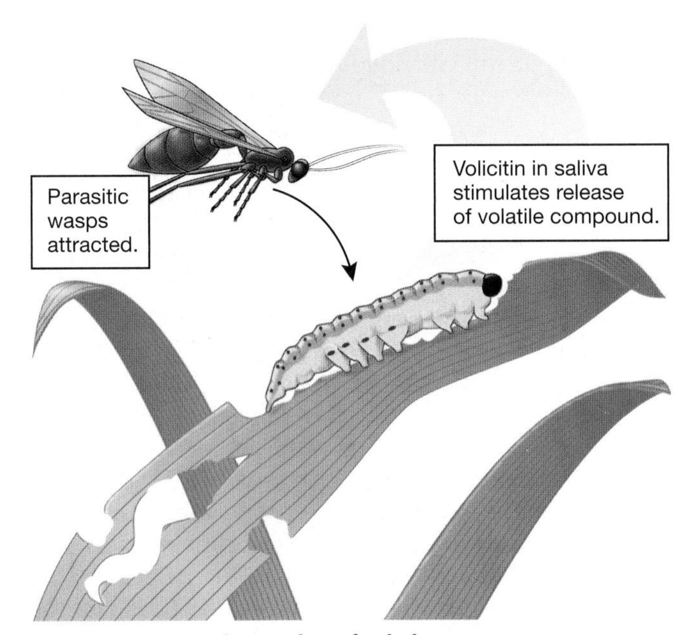
Parasitic
wasps
attracted.
Volicitin in saliva
stimulates release
of volatile compound.

FIGURE 26-8 A chemical cry for help

mites attack lima bean plants, the beans release a chemical that attracts another type of mite that preys on the spider mite. Wild tobacco plants attacked by hornworms (hawkmoth larvae) actually time their chemical defenses, releasing wasp-attracting chemicals during the day but releasing a blend of chemicals at night that seems to deter the night-flying hawk moths from laying eggs on the plants. As you contemplate these apparently ingenious strategies, keep in mind that they evolved gradually because natural selection favored them.

Plants May Warn Their Neighbors and Offspring of Attacks

If attacked plants can summon help, might their neighbors sense the message and prepare themselves for attack as well? Evidence is accumulating that many types of plants (including barley and willow, alder, and birch trees) warn members of their own species of attack; the warning boosts defenses in unattacked individuals, making them more able to defend themselves. This communication may also cross species lines. Wounded sagebrush release large amounts of volatile chemical signals; researchers have found that wild tobacco plants planted downwind from wounded sagebrush are less prone to insect damage than those planted downwind of unwounded sagebrush. Likewise, injured lima bean plants seem to trigger defenses in nearby cucumber plants.

Radish plants that are attacked by butterfly larvae increase their production of a bitter-tasting compound and defensive hairs on their leaves. When these better-defended plants reproduced, researchers found that their seedlings were less attractive to predators than seedlings from unattacked parent plants. This indicates that plants not only defend themselves, but may pass on a chemical signal within their seeds that triggers development of defenses in their offspring.

Plants also produce proteins that defend against diseases. Many plants produce *salicylic acid*, the compound from which aspirin is derived. Ilya Raskin and colleagues at Rutgers University found that tobacco plants infected with a plant virus produce large quantities of salicylic acid. The salicylic acid, in turn, activates an immune response in the plants, helping them fight off the viral attack. The plant also converts some of the salicylic acid to *methyl salicylate* (used to flavor "wintergreen" candy). This highly volatile compound diffuses into the air from virus-infected plant tissues and is absorbed by nearby plants. The plant's healthy neighbors reconvert the methyl salicylate to salicylic acid, enhancing their immune defenses and making them better able to resist the viral infection.

As Ilya Raskin explains, "Plants can't run away and they can't make . . . noises. But they are wonderful chemists." Perhaps human chemists, learning from the plants, will enable farmers of the future to protect their crops from predators and diseases using natural signaling substances instead of toxic pesticides.

Rapid-Fire Plant Responses

All plants are alive, but (as described in this chapter's case study) some are livelier than others. If you touch a sensitive plant (*Mimosa*), its leaflets immediately fold together while it petioles droop (Fig. 26-9). This rapid and dramatic movement probably discourages leaf-eaters. Watch a fly brush against the sensory hairs in a Venus flytrap, and you will see a response that is almost animal-like in its speed of movement. How do these plants perceive touch, and how do they move their leaves so rapidly? In this chapter, you have seen how hormones such as auxin can trigger the expansion of cells, but hormones don't carry signals fast enough to catch a fly or withdraw a leaf. In response to touch, both Venus flytraps and sensitive plants transmit electrical signals that resemble the nerve impulses of animals, thus permitting their animal-like speed of movement.

In the sensitive plant, electrical signals travel from the touched leaf to the petiole that attaches the leaf to the stem. As a signal travels, it causes specialized "motor cells" to increase their permeability to certain ions, including potassium (K^+). The motor cells are located at the base of each leaflet and in the petiole where it joins the stem. As ions flow out of the motor cells, water follows by osmosis. As the cells shrink from water loss, both the leaflets and the petioles rapidly droop.

The Venus flytrap (see the chapter opener photo) uses a different mechanism to control its movement. Each of the fringed trapping leaves of a Venus flytrap bears three sensory "hairs" on its inside surface. If a foraging insect, attracted to nectar secreted by the leaves, touches one hair twice in rapid succession or touches

(a)

(b)

FIGURE 26-9 A rapid response to touch
(a) A leaf of the sensitive plant (*Mimosa*) consists of an array of leaflets emerging from a central stalk, which is attached to the main stem by a short petiole. **(b)** Touching the causes the leaflets to close together and the petiole to droop.

two hairs simultaneously, the hairs initiate a change in electrical potential analogous to the action potential of animal nerve cells (see Chapter 34). The electrical potential sets off a rapid chain of events that causes the trap to close.

Researchers have discovered that the flytrap's leaf closes because of rapid expansion of cells. Flytrap leaves can be pictured most simply as two layers of cells—the outer and inner epidermis—as illustrated below. The electrical potential triggered by hair movement stimulates the cells of the outer layer to rapidly pump hydrogen ions (H^+) into their cell walls. Enzymes in the cell walls are activated by acidic conditions (created by a high concentration of H^+; see Chapter 2) and loosen the cellulose fibers of the walls. As the walls weaken, high osmotic pressure inside the cells causes them to absorb water from extracellular fluids and rapidly increase in size by about 25%. Because the outer layer expands while the inner layer does not, the leaf is pushed closed.

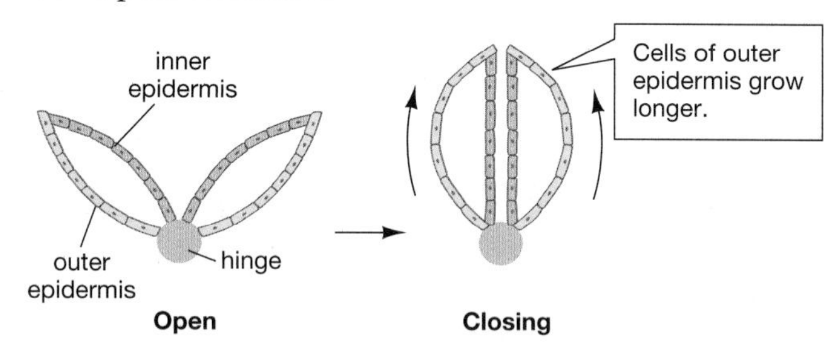

Although reopening the trap takes several hours, the fundamental mechanism is similar. The cells on the inside of the leaf expand, pushing apart the lobes of the trap. So much energy is used up by the hydrogen ion pumps that closing the trap consumes nearly one-third of all the ATP within the entire leaf. It is therefore very important that something digestible actually be in the leaf before it closes the trap! Although much is known about the mechanisms that produce movement in the Venus flytrap, mysteries still remain. How is touch transformed into an electrical stimulus by the sensory hairs? What is the nature of the electrical potential change? How does the electrical signal cause the cells to begin pumping hydrogen ions? As so often happens in biology, the answer to one question immediately poses several new ones.

CASE STUDY REVISITED Predacious Plants

You've learned how a Venus flytrap snaps shut and that sundew tentacles curl inward—but what about the world's speediest plant predator, the aquatic bladderwort? How does it trap the fast-swimming crustaceans that provide it with nitrogen? The answer is an evolutionary marvel. The door to its hollow bladder is hinged at the top, opens inward, and is sealed shut by sticky secretions. Any water that leaks in is extruded back out by glandular cells lining the bladder. This produces a slight vacuum inside that actually pulls the walls inward, giving them a concave shape. Stiff bristles guard the sealed trapdoor. If a small aquatic creature bumps the bristles, they act as levers, pushing the flexible door inward and breaking the seal. The instant the seal is broken, water is sucked inward by the vacuum inside, carrying the prey to death and eventual digestion.

Consider This: Many wetlands in the United States are threatened by runoff water from nearby farms, which may be heavily fertilized or rich in animal wastes. Carnivorous plants thrive in nitrogen-poor bogs partly because other species, which can't trap nitrogen-rich food, cannot compete with them. Explain why runoff from farms poses a threat to carnivorous plants in nearby wetlands.

FIGURE E26-1 A bladderwort snares its prey
(Inset) A copepod is trapped inside the bladder of a bladderwort (genus *Utricularia*).

Links to Life: The Quest for the Perfectly Ripe Tomato

If you've ever tasted a bright red, juicy, tangy-sweet garden-grown tomato, you may recall that it was nothing like the pale, tasteless slices you find on many fast-food burgers. Tomatoes present a challenge for commercial growers and shippers. If they are allowed to ripen on the vine, they become so soft that they are almost impossible to package and ship without damage. Thus, most tomatoes are picked while still green and relatively hard so they can withstand the rigors of shipping. They are then exposed to ethylene gas to stimulate ripening, but artificially ripened tomatoes aren't the same as those allowed to ripen on the plant.

In an attempt to solve this problem, scientists at Calgene created the Flavr Savr™ tomato, an early attempt at genetic modification. Researchers blocked a gene that codes for an enzyme that causes the ripening tomatoes to soften, allowing them to be shipped at a later stage of ripening. But the tomatoes were still more fragile than ideal for shipping; the costs of development were enormous; and (competing with cheaper, unmodified tomatoes) they failed to make a profit. Flavr Savr™ tomatoes disappeared from grocery shelves in 1997, three years after they were first produced.

In the past few years, researchers have learned more about the biochemistry of tomato ripening and the genes that control it, and have identified a "master gene" for tomato ripening. By blocking it, scientists hope to allow the tomato to stay on the vine longer, developing more taste before being shipped and later exposed to ethylene. Recently, researchers working to delay tomato ripening by inserting a gene from yeast accidentally produced a strain of tomato with three times the amount of lycopene, a type of carotenoid that gives tomatoes their red color and may have cancer-fighting and antioxidant properties. As of 2003, there were no genetically engineered tomatoes on the market, but who knows? Within a few years, you may find red ripe and juicy tomatoes—with flavor rivaling tomatoes grown in a home garden and with a higher level of antioxidants as well—on the shelves of your local supermarket.

Summary of Key Concepts

26.1 What Are Plant Hormones, and How Do They Act?

Plant hormones are chemicals that are produced by cells in one part of a plant body and transported to other parts of the plant, where they exert specific effects. The five major classes of plant hormones are auxins, gibberellins, cytokinins, ethylene, and abscisic acid. The functions of these hormones are summarized in Table 26-1.

26.2 How Do Hormones Regulate the Plant Life Cycle?

Dormancy in seeds is enforced by abscisic acid. Falling levels of abscisic acid and rising levels of gibberellin trigger germination. As the seedling grows, it shows differential growth with respect to the direction of light (phototropism) and gravity (gravitropism). Auxin mediates phototropism and gravitropism in shoots and gravitropism in roots. In shoots, auxin stimulates the elongation of cells. In roots, similar concentrations of auxin inhibit elongation. Plants apparently detect gravity by means of organelles called plastids.

Branching in stems results from the interplay of two hormones, auxin (produced in shoot tips and transported downward) and cytokinin (synthesized in roots and transported up the shoot). High concentrations of auxin inhibit the growth of lateral buds. An optimum concentration of both auxin and cytokinin stimulates the growth of lateral buds. Auxin also stimulates the growth of branch roots.

The timing of flowering is normally controlled by duration of darkness. Flowering is both stimulated and inhibited by hormones called florigens. Plants appear to detect light and darkness by changes in phytochrome, a pigment in the leaves. Plant processes influenced by phytochrome responses to light include flowering, straightening the epicotyl or hypocotyl hook, seedling elongation, leaf growth, and chlorophyll synthesis.

Developing seeds produce auxin, which diffuses into the surrounding ovary tissues and causes the production of a fruit. A surge of auxin as the seed matures stimulates fruit cells to release another hormone, ethylene, which causes the fruit to ripen. Ripening includes the conversion of starches to sugars, softening of the fruit, development of bright colors, and, commonly, the formation of an abscission layer at the base of the petiole.

Several changes prepare perennial plants of temperate zones for winter. Leaves and fruits undergo a rapid aging process called senescence, including the formation of an abscission layer. Senescence occurs as a result of a fall in levels of auxin and cytokinin and, perhaps, a rise in ethylene concentrations. Other parts of the plant, including buds, become dormant. Dormancy in buds is enforced by high concentrations of abscisic acid.

26.3 Can Plants Communicate and Move Rapidly?

Many plants under attack by insects release volatile chemicals into the air around them. Other insects that prey on the insect predators are attracted by these chemicals and attack the plant's predator, thus (indirectly) defending the plant. Volatile compounds released by injured or infected plants may also serve as signals to neighboring plants of both the same and different species. In response, the unattacked plants produce substances that help protect them from predation or infection.

A few plants, such as the sensitive plant and the Venus flytrap, can move their leaves rapidly. Touch sensors in the leaf cause electrical signals, which initiate a series of events involving ion movement. This ion movement causes specialized cells in critical hinge areas to quickly absorb or lose water. Changes in the size of these cells of leaves or petioles cause the leaf to move or the petiole to droop.

Key Terms

abscisic acid *p. 518*
abscission layer *p. 526*
apical dominance *p. 522*
auxin *p. 518*
biological clock *p. 523*
cytokinin *p. 518*
day-neutral plant *p. 523*
ethylene *p. 518*
florigen *p. 523*
gibberellin *p. 518*
gravitropism *p. 518*
hormone *p. 518*
long-night plant *p. 523*
phototropism *p. 518*
phytochrome *p. 523*
plant hormone *p. 518*
senescence *p. 526*
short-night plant *p. 523*

Thinking Through the Concepts

To take a multiple-choice quiz with feedback on the contents of this chapter, visit http://www.prenhall.com/audesirk7. *Log in to the Web site selected by your instructor and navigate to the Self Test section for this chapter.*

? Review Questions

1. What did the Darwins, Boysen-Jensen, and Went each contribute to our understanding of phototropism? Do their experiments truly prove that auxin is the hormone that controls phototropism? What other experiments would you like to see?
2. How do hormones interact to cause apical dominance? To control seed dormancy?
3. How can one hormone, an auxin, cause shoots to grow up and roots to grow down?
4. What is the phytochrome system? How do the two forms of phytochrome interact to help control the plant life cycle?
5. Which hormones cause fruit development? From where do these hormones come? Which hormone causes fruit ripening?
6. What is a biological clock?
7. Describe the role of phytochrome in stem elongation in seedlings that grow in the shade of other plants. What is the likely adaptive significance of this response?
8. What is apical dominance? How do auxin and cytokinin interact in determining the growth of lateral buds?
9. Which hormone(s) is (are) involved in leaf and fruit drop? In bud dormancy?
10. Describe one example of a chemical defense mechanism in plants.
11. Describe how a sensitive plant closes its leaves. Why might this behavior have evolved?

Applying the Concepts

1. Suppose you got a job in a greenhouse in which the owner was trying to start the flowering of chrysanthemums (a long-night plant) for Mother's Day. You accidentally turned on the light in the middle of the night. Would you be likely to lose your job? Why or why not? What would happen if you turned on the lights in the day?
2. A student reporting on a project said that one of her seeds did not grow properly because it was planted upside down so that it got confused and tried to grow down. Do you think the teacher accepted this explanation? Why or why not? Which plant hormone or hormones would be involved?
3. Agent Orange, a combination of two synthetic auxins, was used in Vietnam to defoliate the rain forest during the Vietnam War. When they are similar to natural growth hormones, how can synthetic auxins be used to harm or kill plants? What do you think would happen if natural auxins were used in excess quantities on plants?
4. Bean sprouts such as those you might eat in a salad have to be grown in the dark to form the long, yellowish stems that you see. We call such stems *etiolated*. If they are grown in the light, they will be short and green. Why do seedlings grow etiolated in the dark? Under what conditions does etiolation occur in nature? How do plant hormones enable these seedlings to form this shape?
5. Suppose that on July 4, you discover that both a long-night plant and a short-night plant have bloomed in your garden. Discuss how it is possible for both to bloom.
6. Suppose you work in a lab with a well-equipped greenhouse, healthy tomato plants, tomato hornworms by the dozen, and a supply of the parasitic wasps that attack tomato hornworms. Design a controlled study that will support or refute the hypothesis that tomato plants, like maize, can summon the wasps when attacked by a hornworm. Be sure to control for other types of attack.

For More Information

Farmer, E. E. "New Fatty Acid-Based Signals: A Lesson from the Plant World." *Science*, May 9, 1997. Describes the research leading to the discovery of volicitin, which attracts parasitic wasps to plants under attack by caterpillars.

Hansen, E. "Where Rocks Sing, Ants Swim, and Plants Eat Animals." *Discover*, October 2001. Researchers explore carnivorous plants in the wilds of Borneo.

Mlot, C. "Where There's Smoke, There's Germination." *Science News*, May 31, 1997. Researchers have recently discovered that nitrogen dioxide produced by fires can induce germination in plants that live in ecosystems where fires are common.

Moffatt, A. S. "How Plants Cope with Stress." *Science*, November 1, 1994. The hormone "systemin," similar to animal hormones, enables plants to respond to stress.

Russell, S. A. "Talking Plants." *Discover*, April 2002. A clear and engaging summary of research documenting chemical communication among plants.

Saunders, F. "Keep the Aspirin Flying." *Discover*, January 1998. Describes how plants use methyl salicylate to help nearby plants resist infection.

Media Activities

To access a Media Activity visit http://www.prenhall.com/audesirk7. *Log in to the Web site selected by your instructor, navigate to this chapter, and select the appropriate Media Activity number.*

26.1 Hormones

Estimated time: 5 minutes

In this activity you will view the basic principles of hormone structure, transport, and activity. See how hormones produce the effects of phototropism and gravitropism.

26.2 Plant Responses to Phytochrome

Estimated time: 5–15 minutes

In this simulation, experiment with the effects of daylength and different wavelengths of light on plant flowering.

26.3 Plant Response to Stimuli

Estimated time: 5 minutes

Explore some of the unique mechanisms used by plants to respond to outside stimuli.

26.4 Web Investigations: Predacious Plants

Estimated time: 15 minutes

Remember the plant Seymour in "The Little Shop of Horrors"? Now that's a carnivorous plant! Real carnivorous plants may not be quite as flashy, but they have evolved fascinating mechanisms for attracting, catching, and digesting their prey. This exercise takes a brief look at Seymour's "relatives."

UNIT FIVE

Animal Anatomy and Physiology

The animal body is an exquisite expression of the elegance with which evolution has linked form to function. All the animal body systems work in concert to maintain life.

CHAPTER

27 Homeostasis and the Organization of the Animal Body

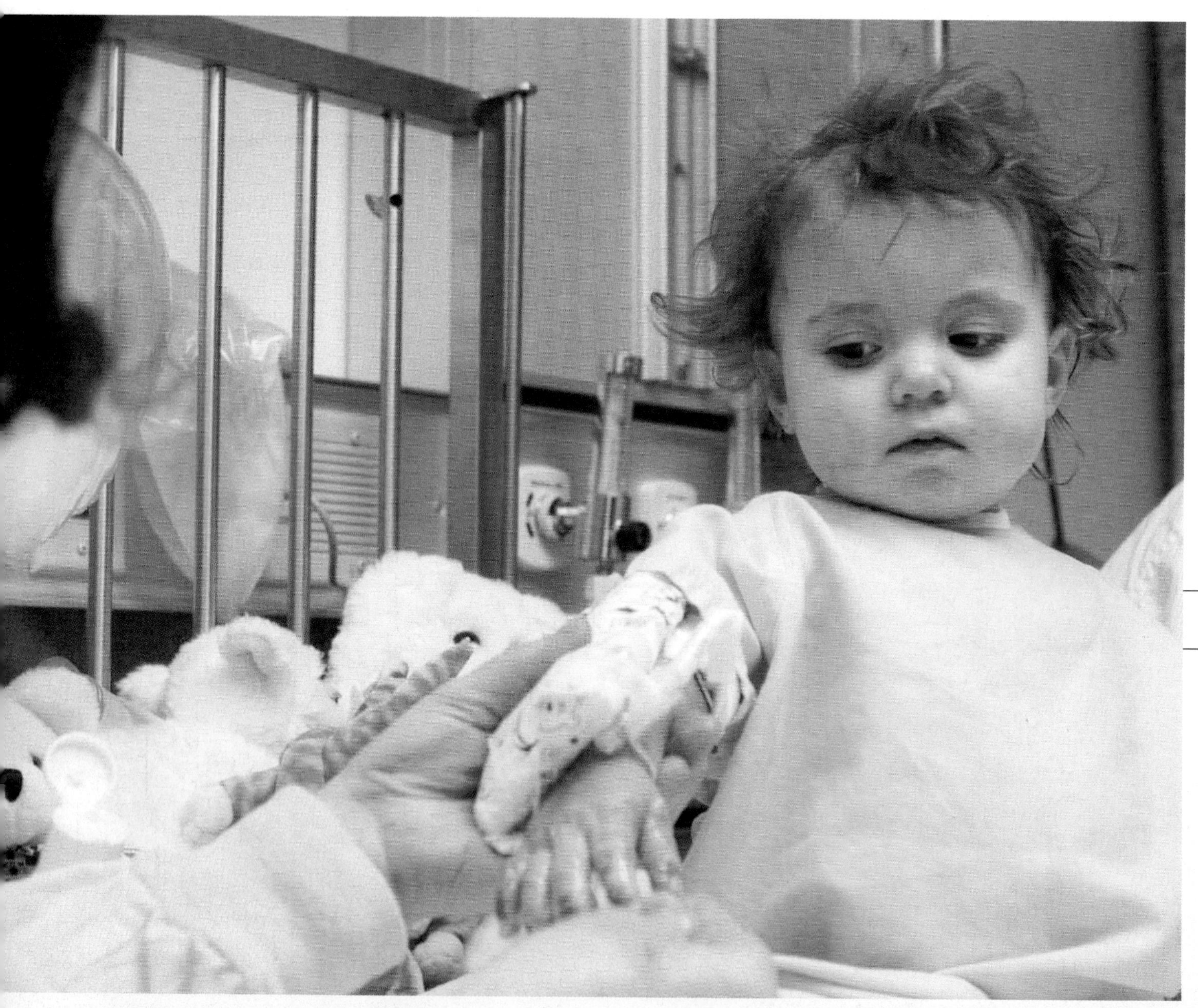

Caregivers tend to Erica Norby's frostbitten fingers in a hospital in Alberta, Canada.

AT A GLANCE

CASE STUDY From Nightmare to Medical Miracle

In Alberta, Canada, on a subzero morning in February, Leyla Nordby experienced a parent's worst nightmare. Just after 3 A.M., she awoke to discover that her 13-month-old toddler was missing from the bed they had been sharing. Screaming her child's name, Leyla ran through the quiet house to find the back door open. Tiny footprints led to a small figure lying face down in the snow. Erika Nordby was rigid as her frantic mother wrapped her in a blanket and dialed 911. Paramedics found her clinically dead; Erika was not breathing, lacked a heartbeat, and had a core body temperature of only 61 °F (normal is near 98.6 °F). Erika's stiffened jaw muscles prevented them from inserting a breathing tube. Performing CPR, the paramedics rushed her to the nearest hospital, where doctors wrapped her in a special blanket that circulated warm air over her body and gradually raised her temperature. Medical personnel continued CPR until, finally, two hours after she was discovered in the snow, Erika's heart began beating on its own. A day later, she was alert and hungrily sucking milk from a bottle. Her mother and doctors hailed her recovery as a medical miracle, but remained concerned that the lack of oxygen during her ordeal may have damaged her brain. The human body is well equipped to cope with drastic changes in the external environment, but there is always a risk of permanent injury in a life-threatening situation such as Erika's. The body fights a constant battle to maintain a state of internal constancy, or homeostasis, in the face of changing environmental conditions. How did Erika's body respond when her ability to maintain a constant internal temperature was overwhelmed by the relentless cold? How did she survive?

27.1 Homeostasis: How Do Animals Maintain Internal Constancy?

Whether you dive into a swimming pool, hike in the desert, or swim in the ocean, your cells remain isolated from outside conditions. They are bathed in extracellular fluid containing a very specific, complex mixture of dissolved substances that must be maintained regardless of conditions on the outside. Many animals have evolved elaborate physiological mechanisms that allow them to maintain precise internal conditions despite lifetimes spent in harsh environments. For example, desert dwellers such as the kangaroo rat have kidneys that allow them to conserve water, while freshwater animals such as the trout or frog must excrete copious amounts of water. Ocean-dwelling fish secrete excess salt from their gills. Because the cells of the animal body cannot survive if the internal environment deviates from a narrow range of acceptable states, cells devote a large portion of their energy to actions that keep the cellular environment stable.

This "constancy of the interior milieu" was first recognized by French physiologist Claude Bernard in the mid-nineteenth century. Later, in the 1920s, Walter B. Cannon coined the term **homeostasis** to describe the constancy of the body's internal environment. Although the word *homeostasis* (derived from Greek words meaning "to stay the same") implies a static, unchanging state, the internal environment in fact seethes with activity as the body continuously adjusts to internal and external changes.

The internal state of an animal body is better described as a *dynamic equilibrium*. Many physical and chemical changes do occur, but the net result of all this activity is that physical and chemical parameters are kept within the narrow range that cells require to function. These equilibrium conditions are maintained by mechanisms collectively known as *feedback systems*. *Negative feedback systems* counteract the effects of changes in the internal environment; less-common *positive feedback systems* reinforce changes when such reinforcement serves a physiological need.

Negative Feedback Reverses the Effects of Changes

The most important mechanism governing homeostasis is **negative feedback**, in which the response to a change is to counteract the change. In other words, an input stimulus causes an output response that "feeds back" to the initial input and decreases its effects. Because the initial change triggers a response that reverses its effects, the overall result is to return the system to its original condition. This kind of feedback is called "negative" because it reverses or negates the initial change.

A familiar example of negative feedback is your home thermostat (Fig. 27-1). In a thermostat, a stimulus (temperature dropping below a *set point*, the thermostat setting) is detected by a thermometer, which signals a control device that switches on a heater. The heater restores the temperature to the set point, and the heater is switched off. Continuously repeated on-off cycles keep your home's temperature near the set point. Note that the thermostat's negative feedback mechanism requires a *control center* with a set point, a *sensor* (the thermometer), and an *effector* (the furnace), which accomplishes the change.

Negative Feedback Maintains Body Temperature

How do people and other "warm-blooded" animals maintain their internal temperature despite extreme fluctuations in the temperature around them, such as those experienced in by Erika Nordby on a Canadian winter night? The set point in the temperature control system, which varies by only about 1 °F in healthy humans, is located in a control center in your *hypothalamus*, a region deep in the brain that controls many homeostatic responses (see Fig. 27-1). Nerve endings in your hypothalamus, abdomen, spinal cord, skin, and large veins act as temperature sensors and transmit this information to the hypothalamus. When body temperature drops, the hypothalamus activates various effector mechanisms that tend to raise body temperature. When normal body temperature is restored, sensors signal the hypothalamus to switch off these temperature control mechanisms. For example, Erika undoubtedly began shivering soon after she walked out the door into the snow. Shivering uses rapid, reflexive contractions of skeletal muscles to burn stored fuel and generate heat. The blood vessels supplying nonvital areas of her body (such as her face, hands, feet, and skin) became constricted, reducing heat loss and diverting warm blood to vital inner regions (brain, heart, and other internal organs). Finally, the hypothalamus initiated a series of chemical signals that raised her metabolic rate, generating more heat to maintain vital functions. But with very limited energy reserves, small body mass, and a large surface area, the small child's homeostatic control mechanisms were quickly overwhelmed by the subzero environment.

Negative feedback mechanisms abound in physiological systems. In the chapters that follow, you will find many examples of homeostatic control that operate by negative feedback, including the systems regulating blood oxygen content, water balance, blood sugar levels, and many other components of the "internal milieu."

Positive Feedback Drives Events to a Conclusion

When you first think about it, positive feedback is a rather frightening concept. In contrast to a negative feedback system, a change in a **positive feedback** system produces a response that intensifies the original change (see Fig. 27-1). The end result is that change tends to proceed in the same direction as the initial stimulus (rather than reversing to return to a set point). Positive feedback, as you can imagine, tends to create chain reactions

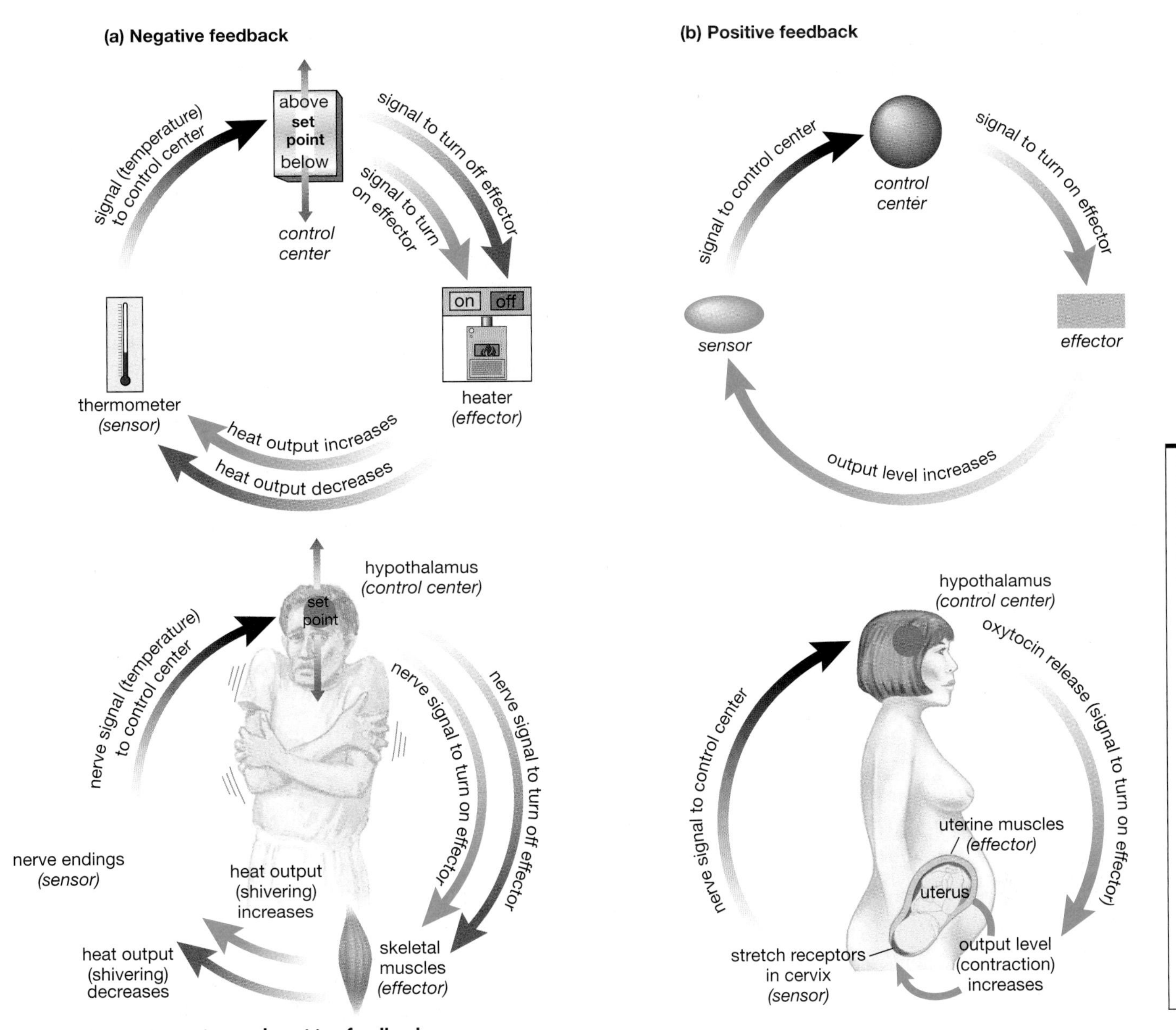

FIGURE 27-1 Negative and positive feedback
(top) Both our homes (left) and our bodies (right) use negative feedback to maintain appropriate temperatures. **(bottom)** Positive feedback is seen in this hypothetical model (left) and in childbirth (right). **QUESTION** What would happen if a cold, shivering mammal ingested a poison that destroyed all of the nerve endings that detect heat? More generally, what happens if the sensor in a negative feedback system loses its ability to receive signals?

Media Activity
27.1 Homeostasis

that must somehow be controlled. For example, in nuclear fission, each particle that is split from an atom triggers the splitting of another atom, the pieces of which trigger the fission of other atoms, and so on. When controlled, the chain reaction supplies nuclear power. When deliberately set out of control, it produces an atomic explosion. A familiar biological example of positive feedback is population growth (as we will see in Chapter 39); each offspring gives rise to still more offspring. Ecologist Paul Ehrlich coined the apt expression "population bomb" to describe unchecked population growth.

In physiological systems, events governed by positive feedback mechanisms are generally self-limiting and occur relatively infrequently. Positive feedback occurs, for example, during childbirth (see Fig. 27-1). The early contractions of labor begin to force the baby's head against the cervix, located at the base of the uterus; this pressure causes the cervix to dilate (open). Stretch-receptor neurons in the cervix respond to this expansion by signaling the hypothalamus, which responds by triggering the release of a hormone (oxytocin) that stimulates more and stronger uterine contractions. Stronger contractions create further pressure on the cervix, which in turn prompts the release of more hormones. The feedback cycle is finally terminated by the expulsion of the baby and its placenta.

The Body's Internal Systems Act in Concert

The systems of the animal body are all "team players" that work together in a coordinated manner to maintain a relatively constant internal environment. The job of regulating a multitude of factors throughout the body cannot be accomplished by a few independent feedback mechanisms. Instead, numerous mechanisms are constantly at work, responding to various stimuli that continuously change as the animal's activities and external environment change.

Fortunately, evolution has ensured that the various systems work together. For example, the systems that take substances into the body (for example, the digestive system) act in concert with those responsible for transporting substances within the body (such as the circulatory system) and those that remove substances from the body (such as the urinary system). This kind of coordinated action is possible because the body has mechanisms for sending signals from one part to another. Each cell is indirectly connected to all of the others by an elaborate network of blood vessels and nerves that can carry molecules and messages to the appropriate locations. For example, to return to our Case Study, when temperature receptors in Erika's skin sensed the subzero air, they sent the message, "Cold!" to her hypothalamus. Simultaneously, temperature sensors in the hypothalamus monitored her blood temperature as it began to drop rapidly. Cold signals sent from both of these sources to Erika's hypothalamus caused it to trigger shivering and constriction of blood vessels to the skin.

In order to maintain homeostasis, chemical signals are transported throughout the body, acting only on appropriate target cells that are specialized to receive and respond to specific signals. There are often many links in this chain of communication. Exposure to cold, for example, causes the hypothalamus to release chemical signals that travel in the bloodstream to a nearby gland (the pituitary). The pituitary gland then releases a hormone that triggers the thyroid gland to release a different hormone. This hormone acts to increase the body's metabolic rate, generating more heat. The hypothalamus stops releasing the chemical signal when its blood temperature receptors indicate that body temperature has been restored to normal. Thus, using a variety of routes and mechanisms, messages are carried from sensors to effectors and back again, allowing feedback mechanisms to maintain homeostasis.

27.2 How Is the Animal Body Organized?

The animal body is an engineering marvel. From simple, free-living cells, evolutionary change has produced astonishingly complex systems consisting of trillions of specialized cells that accomplish hundreds of functions simultaneously. The parts fit together with a degree of precision and integration of which human engineers can only dream. This complexity is based on a simple organizational hierarchy:

Cells → Tissues → Organs → Organ Systems

An example of this hierarchy is illustrated in Figure 27-2. You learned in Chapter 1 that cells are the building blocks of all life. The animal body incorporates cells into **tissues**; each tissue is composed of dozens to billions of structurally similar cells that act in concert to perform a particular function. Tissues are the building blocks of **organs**, discrete structures that perform complex functions. Examples of organs include the stomach, small intestine, kidneys, and urinary bladder. Organs, in turn, are organized into **organ systems**, groups of organs that function in a coordinated manner. For example, the digestive system is an organ system that includes the stomach, small intestine, large intestine, and other organs that work together to allow us to digest food and absorb nutrients from it. Major organ systems of vertebrates are illustrated in Table 27-1.

Animal Tissues Are Composed of Similar Cells That Perform a Specific Function

A tissue is composed of cells that are similar in structure and perform a specialized function. Tissue may also include extracellular components produced by these cells, as in the case of cartilage and bone. Here we present a brief overview of the four major categories of animal tissue and the major cell types that comprise these tissues: epithelial tissue, connective tissue, muscle tissue, and nerve tissue.

Epithelial Tissue Covers the Body, Lines Its Cavities, and Forms Glands

Epithelial cells are the body's gatekeepers, protecting and regulating the movement of substances into and out of the body. **Epithelial tissues** (also called the *epithelium*), bound to loose connective tissue (described below), form continuous sheets of cells called **membranes** (not to be confused with the plasma membrane that surrounds each cell). These membranes cover the body and line body cavities such as the mouth, the stomach, and the bladder. The epithelial tissue portion of membranes faces either the outside of the body (as in skin) or the inside of a cavity within the body. Membranes create barriers that either resist the movement of substances across them (such as in the skin) or allow the movement of specific substances across them (such as the lining of the small intestine). Epithelial tissues can serve as effective barriers because epithelial cells are packed closely together and connected to one another by several types of junctions (see Chapter 4). The structure of epithelial tissue is adapted to its function. For example, the epithelium lining the lungs, where the exchange of gas molecules takes

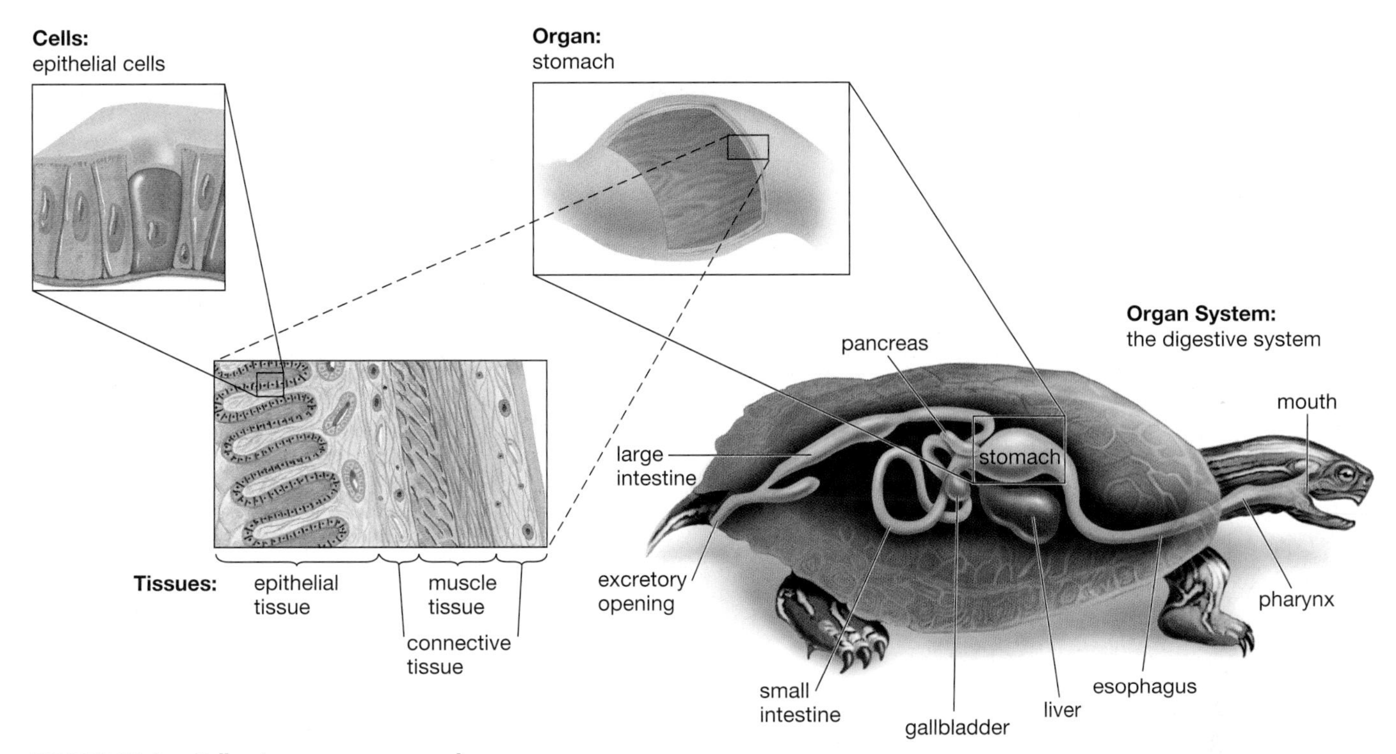

FIGURE 27-2 Cells, tissues, organs, and organ systems
The animal body is composed of cells, which make up tissues, which combine to form organs that work in harmony as organ systems.

place, consists of thin, flattened cells arranged in a single layer (Fig. 27-3a). Another form of epithelium consists of elongated cells, often with cilia, which are capable of secreting mucus (Fig. 27-3b). This type of epithelium is part of the membrane that lines the trachea that leads to the lungs. Here, the mucus traps dust particles, and the cilia transport them away from the lungs. This ciliated epithelium also lines tubes of the reproductive organs, where the cilia transport sex cells to their destinations (see Fig. 27-3b).

No blood vessels penetrate epithelial tissue; it is nourished by diffusion from capillaries (the smallest blood vessels, whose thin walls allow exchange of wastes and nutrients). These capillaries are embedded within connective tissue that lies beneath the epithelium.

Another important property of epithelial tissues is that they are continuously lost and replaced by mitotic cell division. For example, consider the abuse suffered by the epithelium that lines your mouth. Scalded by coffee and scraped by corn chips, it would be destroyed within a few days if it did not replace itself continuously. The stomach lining, abraded by food and attacked by acids and protein-digesting enzymes, is completely replaced every 2 to 3 days. Your skin's outer membrane, the epidermis (see Fig. 27-10), is renewed about twice a month.

During development, some epithelial tissues fold inward; their cells change shape and function to form **glands**, clusters of cells that are specialized to secrete (release) substances. Glands are classified into two broad categories: exocrine glands and endocrine glands. **Exocrine glands** remain connected to the epithelium by a passageway, or *duct*. Examples of exocrine glands are sweat

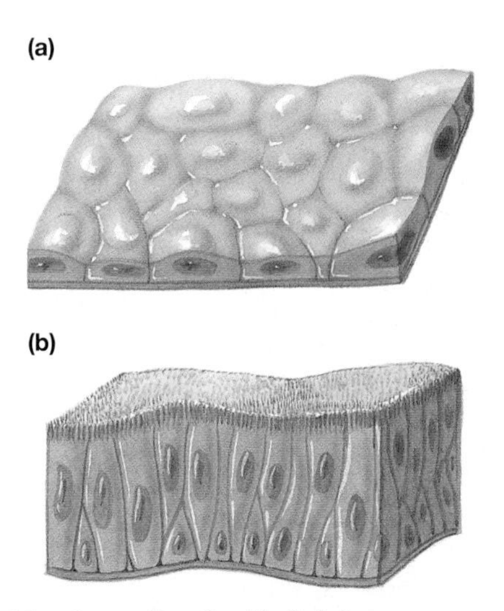

FIGURE 27-3 Examples of epithelial tissue
(a) Thin, flattened cells in a single layer form the epithelial tissue that lines lungs and blood vessels, where exchange of materials is important. **(b)** Elongated epithelial cells bearing cilia and capable of secreting mucus line the trachea and tubes of the reproductive organs.

Table 27-1 Major Vertebrate Organ Systems

Organ System	Major Structures	Physiological Role	Organ System	Major Structures	Physiological Role
Circulatory system 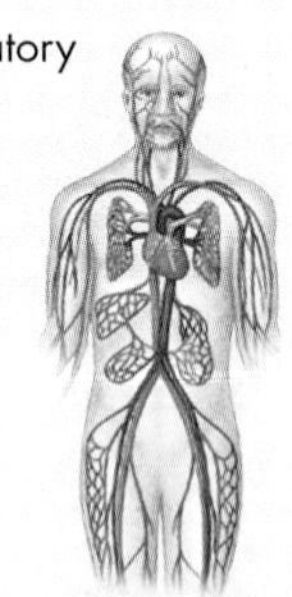	Heart, blood vessels, blood	Transports nutrients, gases, hormones, metabolic wastes; also assists in temperature control	Endocrine system 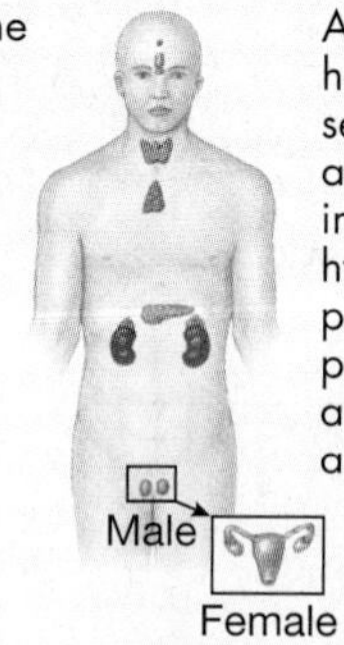	A variety of hormone-secreting glands and organs, including the hypothalamus, pituitary, thyroid, pancreas, adrenals, ovaries, and testes	Controls physiological processes, typically in conjunction with the nervous system
Lymphatic/immune system 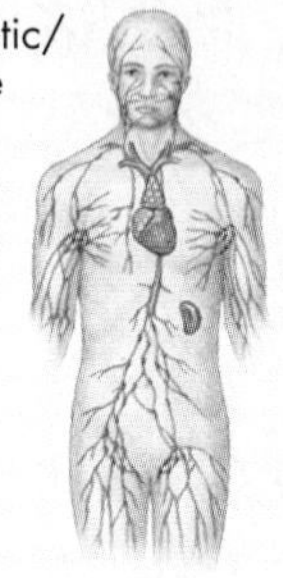	Lymph, lymph nodes and vessels, white blood cells	Carries fat and excess fluids to blood; destroys invading microbes	Nervous system 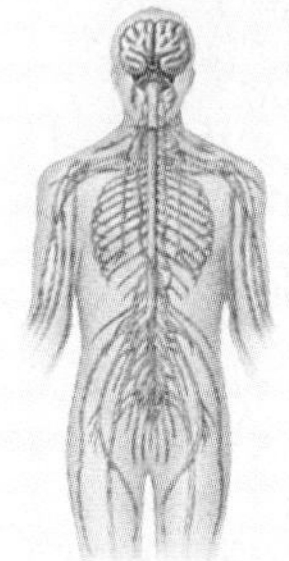	Brain, spinal cord, peripheral nerves	Controls physiological processes in conjunction with the endocrine system; senses the environment, directs behavior
Digestive system 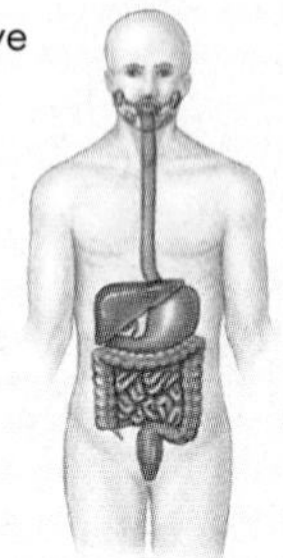	Mouth, esophagus, stomach, small and large intestines, glands producing digestive secretions	Supplies the body with nutrients that provide energy and materials for growth and maintenance	Muscular system 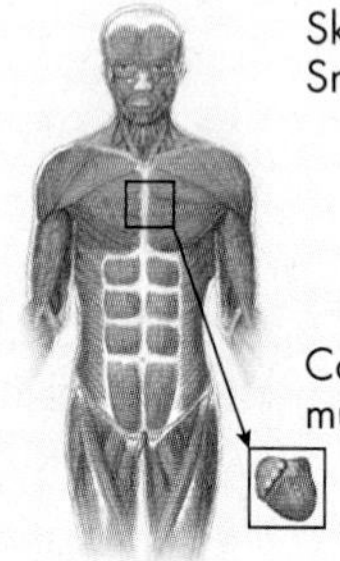	Skeletal muscle Smooth muscle Cardiac muscle	Moves the skeleton Controls movement of substances through hollow organs (digestive tract, large blood vessels) Initiates and implements heart contractions
Urinary system 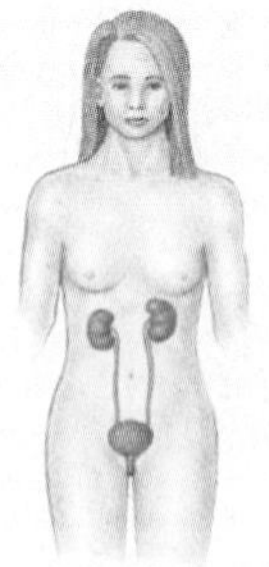	Kidneys, ureters, bladder, urethra	Maintains homeostatic conditions within bloodstream; filters out cellular wastes, certain toxins, and excess water and nutrients	Skeletal system 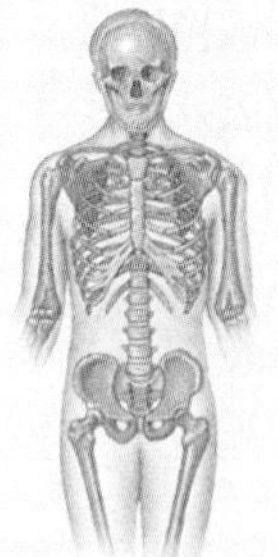	Bones, cartilage, tendons, ligaments	Provides support for the body, attachment sites for muscles, and protection for internal organs
Respiratory system 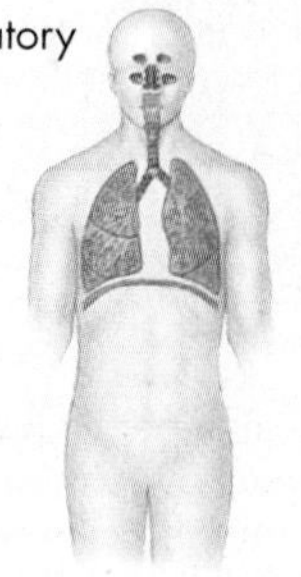	Nose, pharynx, trachea, lungs (mammals, birds, reptiles, amphibians), gills (fish and some amphibians)	Provides an area for gas exchange between the blood and the environment; allows oxygen acquisition and carbon dioxide elimination	Reproductive system 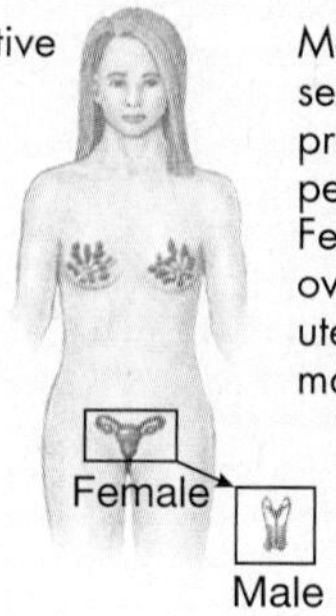	Males: testes, seminal vesicles, prostate gland, penis Female (mammal): ovaries, oviducts, uterus, vagina, mammary glands	Male: produces sperm, inseminates female Female (mammal): Produces egg cells, nurtures developing offspring

glands and *sebaceous* (oil-secreting) glands; both types are found in the skin and are derived from skin epithelium (see Fig. 27-10). Exocrine glands called *salivary glands* release saliva into the mouth, and still other exocrine glands line the stomach, where they secrete a protective layer of mucus. **Endocrine glands** become separated from the epithelium that produced them. Most products of endocrine glands are hormones, which are secreted into the extracellular fluid that surrounds the glands and then diffuse into nearby capillaries. Endocrine glands and their hormones are covered in detail in Chapter 33.

Connective Tissues Have Diverse Structures and Functions

Connective tissues serve mainly to support and bind other tissues. Most connective tissues include large quantities of extracellular substances, typically secreted by the connective tissue cells themselves. With the exception of blood and lymph, connective tissues are interwoven with flexible, fibrous strands of an extracellular protein called **collagen**, which is secreted by the cells. Connective tissues can be placed into three main categories, described below.

- *Loose connective tissue* combines with epithelial cells to form membranes. It underlies all epithelial tissue and contains capillaries and fluid-filled spaces that nourish the epithelium. Underlying the epidermis of the skin, for example, is connective tissue called the **dermis**, which is richly supplied with capillaries (see Fig. 27-10). Loose connective tissue contains a diffuse network of protein fibers, loosely woven though a clear, extracellular fluid with a syrup-like consistency. It surrounds, cushions and supports most organs of the body.
- *Fibrous connective tissue* includes **tendons** (which connect bones to muscles) and **ligaments** (which connect bones to bones). Fibrous connective tissue contains collagen fibers, which are densely packed in an orderly parallel arrangement—a design that gives tendons and ligaments their flexibility and tremendous strength.
- *Specialized connective tissues*, which include *cartilage, bone, fat, blood*, and *lymph*, are diverse in composition. **Cartilage** is flexible and resilient, consisting of widely spaced cells surrounded by a thick, nonliving matrix. This matrix is composed of collagen secreted by the cartilage cells (Fig. 27-4). Cartilage covers the ends of bones at joints, provides the supporting framework for the respiratory passages, supports the ear and nose, and forms shock-absorbing pads between the vertebrae. **Bone** (Fig. 27-5) resembles cartilage, but its matrix is hardened by deposits of calcium phosphate. Bone forms in concentric circles around a central canal, which contains a blood

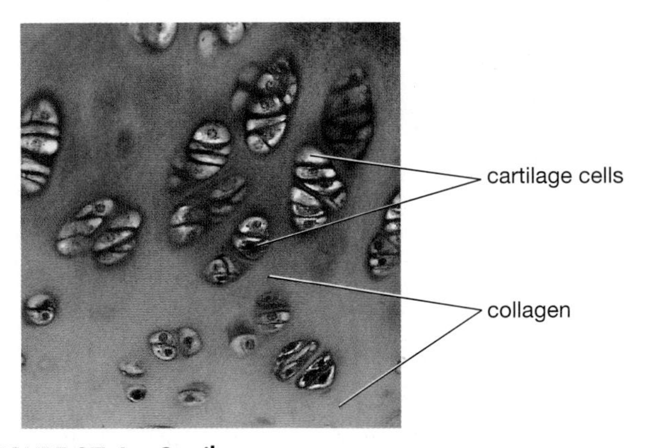

FIGURE 27-4 Cartilage
The cells of the cartilage are stained dark purple and surrounded by clear spaces. The homogeneous material stained pale purple is the matrix of collagen secreted by the cartilage cells.

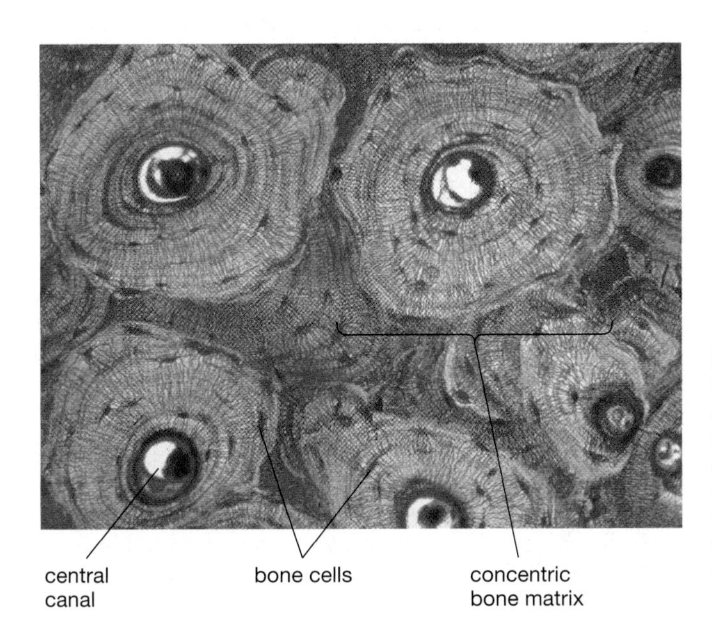

FIGURE 27-5 Bone
Concentric circles of bone (a specialized connective tissue), deposited around a central canal that contains a blood vessel, are clearly visible in this micrograph. Individual bone cells appear as dark spots trapped in small chambers within the hard matrix that the cells themselves deposit.

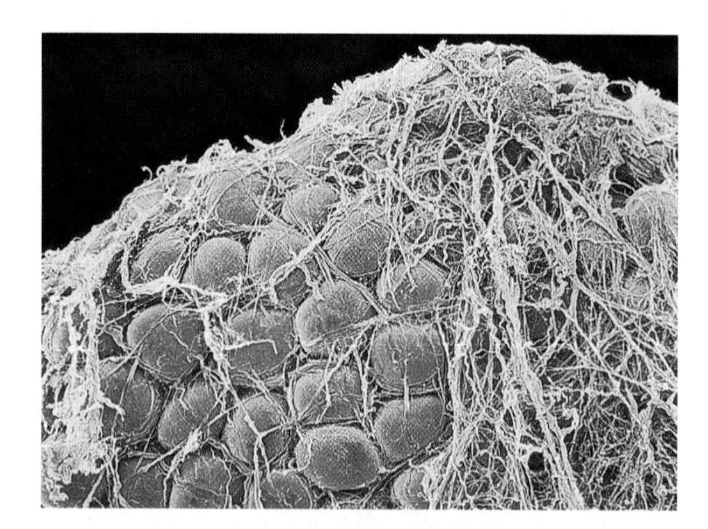

FIGURE 27-6 Adipose tissue
(a) Adipose tissue, shown here from a human abdomen, is specialized connective tissue made up almost exclusively of fat cells. A droplet of oil occupies most of the volume of the cell. Yellow fibrous connective tissue covers the cells. **(b)** A hooded seal at 4 days of age has doubled her birth weight by drinking mother's milk that consists of 61% fat, derived from the mother's own stores of blubber. At 100 pounds, she is almost too fat to move. The fat will feed and insulate the pup as the ice floes break up and she dives into icy water to learn to hunt and feed on her own. **QUESTION** Why are young mammals often fatter than adults?

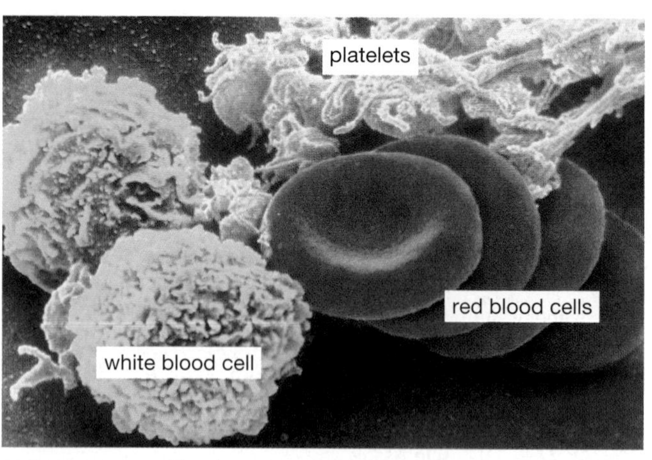

FIGURE 27-7 Blood
Blood contains three types of cellular components, shown in this color-enhanced scanning electron micrograph. The cells are suspended in plasma, which is also a component of this specialized connective tissue. **QUESTION** Why do the red blood cells of mammals lack cell nuclei?

vessel. (We will discuss cartilage and bone in depth in Chapter 35.) Fat cells, collectively called **adipose tissue** (Fig. 27-6a), are modified for long-term energy storage. Adipose tissue is especially important in the physiology of animals adapted to cold environments, because it not only stores energy but also serves as insulation (Fig. 27-6b). Although they are liquids, **blood** and **lymph** are considered connective tissues because they are composed largely of extracellular fluids. The cellular portion of blood consists of red blood cells (which transport oxygen), white blood cells (which fight infection), and cell fragments called *platelets* (which aid in blood clotting). These are all suspended in extracellular fluid called *plasma* (Fig. 27-7). Lymph consists largely of fluid that has leaked out of blood capillaries (the smallest of the blood vessels) and is carried back to the circulatory system within lymph vessels. You will learn more about blood and lymph in Chapter 28.

Muscle Tissue Has the Ability to Contract

The long, thin cells of muscle tissue (Fig. 27-8) contract (shorten) when stimulated, then relax passively. There are three types of muscle tissue: skeletal, cardiac, and smooth. **Skeletal muscle** is generally under voluntary, or conscious, control. As its name implies, its main function is to move the skeleton, as occurs when you walk or turn the pages of this text. **Cardiac muscle** is located only in the heart. Unlike skeletal muscle, it is spontaneously active and involuntary, that is, not under conscious control. Cardiac muscle cells are interconnected by gap junctions, through which electrical signals spread rapidly through the heart, stimulating the cardiac muscle cells to contract in a coordinated fashion. **Smooth muscle**, so named because it lacks the orderly arrangement of thick and thin filaments seen in cardiac and skeletal muscles, is embedded in the walls of the digestive tract, the uterus, the bladder, and large blood

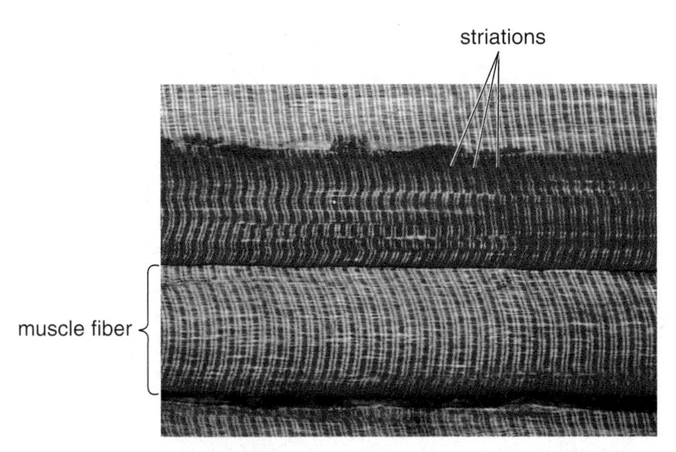

FIGURE 27-8 Muscle tissue consists of contractile cells called muscle fibers
A regular arrangement of fibrous proteins inside the muscle fibers of skeletal muscle gives this muscle tissue stripes or "striations" when viewed under a microscope.

vessels. It produces slow, sustained contractions that are mostly involuntary. Muscles and muscle contraction are covered in Chapter 35.

Nerve Tissue Is Specialized to Transmit Electrical Signals

You owe your ability to sense and respond to the world to **nerve tissue**, which makes up the brain, the spinal cord, and the nerves that travel from them to all parts of the body. Nerve tissue is composed of two types of cells: nerve cells, also called **neurons**, and glial cells. Neurons are specialized to generate electrical signals and to conduct these signals to other neurons, muscles, or glands (Fig. 27-9). **Glial cells** surround, support, electrically insulate, and protect neurons. Glial cells also regulate the composition of the extracellular fluid, allowing neurons to function optimally. We will discuss nerve tissue more fully in Chapter 34.

Organs Include Two or More Interacting Tissue Types

Organs are formed from at least two tissue types that function together. If an organ is hollow, such as the bladder or blood vessels, its interior is lined with epithelium underlain by connective tissue. Different organs have different types and proportions of glandular, muscular, and nervous tissues. Some organs, such as the skin, include all four of these tissue types. The structure of the skin is, in a general sense, representative of many organs. An outer epithelium is underlain by connective tissue that contains a blood supply, a nerve supply, muscle (in some cases), and glandular structures derived from the epithelium. In the following section, we will examine the components and functions of the skin.

The Skin Illustrates the Properties of Organs

Although we take our skin for granted, it is so important as a barrier against infection and water loss that large-scale destruction of skin, such as by extensive

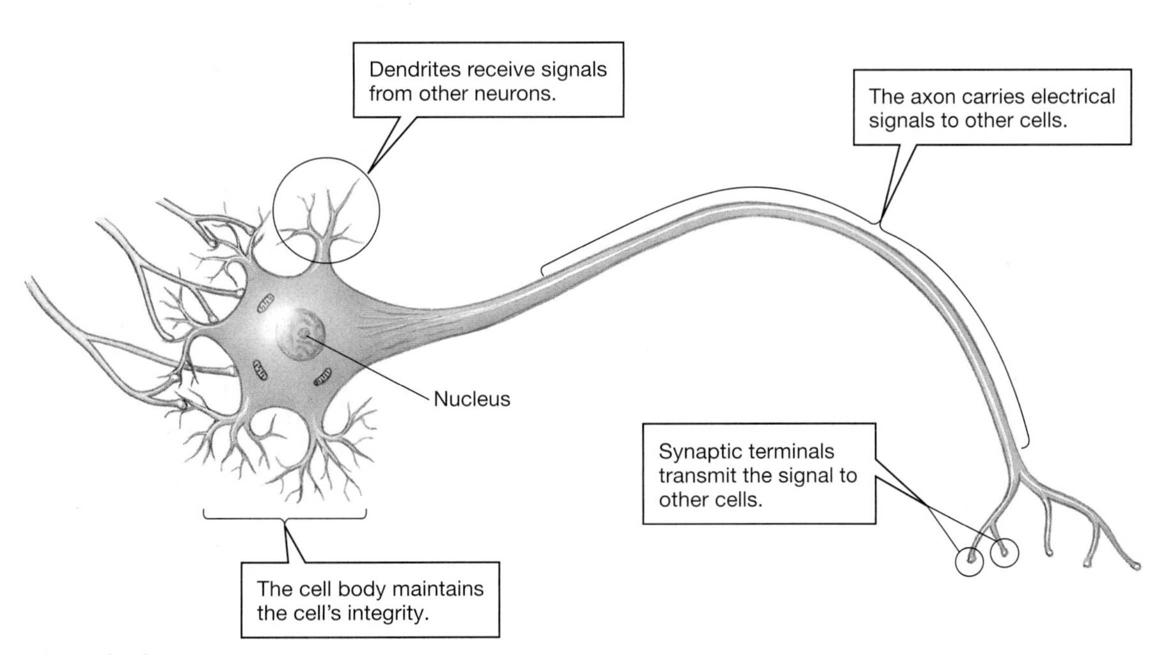

Dendrites receive signals from other neurons.
The axon carries electrical signals to other cells.
Nucleus
Synaptic terminals transmit the signal to other cells.
The cell body maintains the cell's integrity.

FIGURE 27-9 Nerve tissue
Nerve cells are specialized to transmit electrical signals, while glia regulate the extracellular environment that allows nerve cells to function.

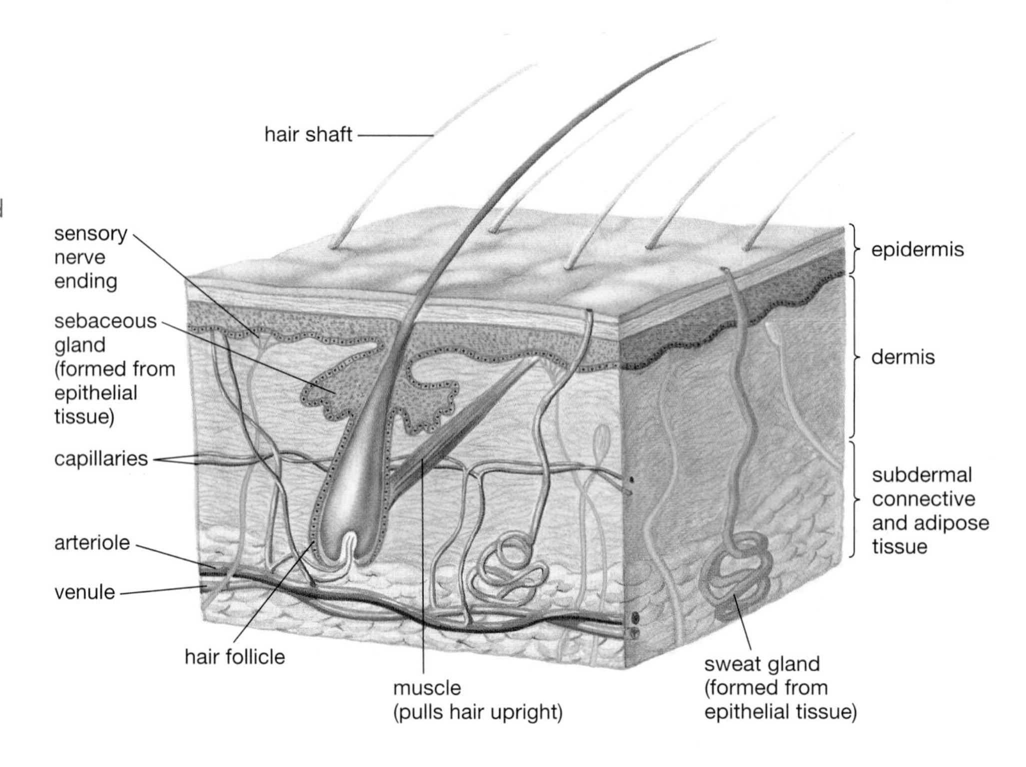

FIGURE 27-10 Skin
Mammalian skin, a representative organ, in cross section. The skin is a membrane containing embedded glands, muscles, and nerve cells. **QUESTION** Why is skin considered to be an organ but blood is considered to be a tissue?

burns, can prove fatal. The **epidermis**, or outer layer of the skin, is a specialized epithelial tissue (Fig. 27-10). It is covered by a protective layer of dead cells produced by the underlying living epidermal cells. These dead cells are packed with the protein *keratin*, which helps keep the skin both airtight and relatively waterproof.

Immediately beneath the epidermis lies a layer of connective tissue, the dermis. The loosely packed cells of the dermis are permeated by *arterioles* (small arteries). Arterioles feed blood pumped from the heart into a dense meshwork of capillaries that nourish both the dermal and epidermal tissue and empty into a network of *venules* (small veins) in the dermis. Loss of heat through the skin is precisely regulated by neurons controlling the degree of dilation (expansion) of the arterioles. When cooling is required, the arterioles dilate and flood the capillary beds with blod, thus releasing excess heat. When heat conservation is required, the arterioles supplying the skin capillaries are constricted; this is what caused frostbite in Erika's feet. Lymph vessels collect and carry off extracellular fluid within the dermis. Various sensory nerve endings responsive to temperature, touch, pressure, vibration, and pain are scattered throughout the dermis and epidermis and provide feedback to the nervous system.

The dermis is also packed with glands derived from epithelial tissue. Glands called **hair follicles** produce hair from protein-containing secretions. Sweat glands produce watery secretions that cool the skin and excrete substances such as salts and urea. *Sebaceous glands* secrete an oily substance (*sebum*) that lubricates the epithelium.

In addition to the epithelial, connective, and nerve tissues already mentioned, the skin also contains muscle tissue. Tiny muscles attached to the hair follicles can cause the hairs of the skin to "stand on end" in response to signals from motor neurons. Although this reaction is useless for retaining heat in humans, most mammals are able to increase the thickness of their insulating fur in cold weather by erecting individual hairs.

Organ Systems Consist of Two or More Interacting Organs

Organ systems consist of two or more individual organs (in some cases, located in different regions of the body) that work together, performing a common function. An example is the digestive system, in which the mouth, esophagus, stomach, intestines, and other organs that supply digestive enzymes, such as the liver and pancreas, all function together to convert food into nutrient molecules (see Fig. 27-2). The major organ systems of the vertebrate body and their representative organs and functions are listed in Table 27-1. The structure and physiology of these organ systems are the subjects of the remainder of the unit.

CASE STUDY REVISITED From Nightmare to Medical Miracle

Cold slows all biochemical reactions; Erika's metabolic rate (the rate at which biochemical reactions occurred in her body) at 61 °F may have been only about 25% of the normal rate. As Erika's body temperature dropped below 85 °F, the activity of her hypothalamus was depressed to the point where it lost its ability to control her body temperature. With these homeostatic mechanisms shut down, her temperature continued to drop rapidly.

Doctors believe that Erika's small size may have saved her, as it allowed her core temperature to drop rapidly. This reduced the demand for oxygen by her brain and other internal organs, which suffered no lasting damage. The only lasting sign of her ordeal was minor frostbite damage to the skin of one foot. Just over a month after she entered the hospital, Erika waved a cheerful good-bye to the caregivers who had become her friends as she recovered.

Consider This: If Erika were in a warm room, why would wrapping her in a thick blanket do more harm than good? Imagine snorkeling in the cold ocean waters and encountering a seal. As you watch it in fascination, you become thoroughly chilled, though the seal thrives in its environment. Describe the homeostatic mechanisms that will help restore your body temperature. Why can the seal remain in the ocean for hours while maintaining its body temperature?

Links to Life: Heat or Humidity?

It's not the heat, it's the humidity! Actually, the discomfort you feel on a hot muggy day results from both these factors. Meteorologists have developed a formula, the heat index, that generates an "apparent temperature" by taking humidity into account. For example, a temperature of 90 °F has a heat index of 100 °F at 60% humidity, but feels like only 85 °F at 10% humidity. Hot weather reduces your ability to dissipate your body's excess heat, while high humidity undermines your body's ability to cool itself by sweating. Heat and humidity together disrupt your body's attempts to maintain the narrow range of temperature that promotes homeostasis.

In extreme cases, excessive heat can lead to a deadly condition called hyperthermia or heat stroke. During heat stroke, the body's homeostatic mechanisms are overcome, and body temperature rises to 106 °F (41.1 °C) or higher, often accompanied by dehydration. Although anyone can succumb, the elderly and the very young are most susceptible because their ability to regulate body temperature is less efficient. Heat stroke leads to a failure of the mechanisms the body uses to regulate temperature: Sweating ceases, and the perception of thirst may be lost. Dozens of young children in the United States die every summer from being left in closed cars, where temperatures can reach lethal levels in as little as 15 minutes.

As you gripe about the heat and humidity this summer, keep in mind that your discomfort has evolved as a warning system that homeostasis is threatened. So run through a sprinkler, seek shade or air conditioning, or drink a tall cool glass of water!

Summary of Key Concepts

27.1 Homeostasis: How Do Animals Maintain Internal Constancy?

Homeostasis refers to the tendency of many physiological processes to maintain an organism's internal conditions within a narrow range that permits the continuation of life. These conditions are maintained through negative feedback, in which a change triggers a response that counteracts the change and restores conditions to a set point. Temperature regulation as well as many hormone systems use negative feedback to maintain homeostasis. Positive feedback, in which a change initiates events that intensify the change, occurs relatively rarely and is self-limiting. For example, the uterine contractions that lead to childbirth are driven by positive feedback. Within the animal body, multiple feedback mechanisms work in concert to maintain life.

27.2 How Is the Animal Body Organized?

The animal body is composed of organ systems consisting of two or more organs. Organs, in turn, are made up of tissues. A tissue is a group of cells and extracellular material that form a structural and functional unit and is specialized for a specific task. Animal tissues include epithelial, connective, muscle, and nerve tissue.

Epithelial tissue forms membranous coverings over internal and external body surfaces and also gives rise to glands. Connective tissue usually contains considerable extracellular material and includes dermal tissue, bone, cartilage, tendons, ligaments, fat, and blood. Muscle tissue is specialized for movement. There are three types of muscle tissue: skeletal, cardiac, and smooth. Nerve tissue, including neurons and glial cells, is specialized for the generation and conduction of electrical signals.

Organs include at least two tissue types that function together. Mammalian skin is a representative organ. The epidermis, an epithelial tissue, covers and protects the dermis beneath it. The dermis contains blood and lymph vessels, a variety of glands, and tiny muscles that erect the hairs. Animal organ systems include the digestive, urinary, immune, respiratory, circulatory/lymphatic, nervous, muscular, skeletal, endocrine, and reproductive systems, summarized in Table 27-1.

Key Terms

adipose tissue *p. 542*
blood *p. 542*
bone *p. 541*
cardiac muscle *p. 542*
cartilage *p. 541*
collagen *p. 541*
connective tissue *p. 541*
dermis *p. 541*
endocrine gland *p. 541*
epidermis *p. 544*
epithelial tissue *p. 538*
exocrine gland *p. 539*
gland *p. 539*
glial cell *p. 543*
hair follicle *p. 544*
homeostasis *p. 536*
ligament *p. 541*
lymph *p. 542*
membrane *p. 538*
negative feedback *p. 536*
nerve tissue *p. 543*
neuron *p. 543*
organ *p. 538*
organ system *p. 538*
positive feedback *p. 536*
skeletal muscle *p. 542*
smooth muscle *p. 542*
tendon *p. 541*
tissue *p. 538*

Thinking Through the Concepts

To take a multiple-choice quiz with feedback on the contents of this chapter, visit http://www.prenhall.com/audesirk7. *Log in to the Web site selected by your instructor and navigate to the Self Test section for this chapter.*

? Review Questions

1. Define *homeostasis,* and explain how negative feedback helps maintain it. Explain one example of homeostasis in the human body.
2. Explain positive feedback, and provide one physiological example. Explain why this type of feedback is relatively rare in physiological processes.
3. Explain what goes on in your body to restore temperature homeostasis when you become overheated by exercising on a hot, humid day.
4. Describe the structure and functions of epithelial tissue.
5. What property distinguishes connective tissue from all other tissue types? List five types of connective tissue, and briefly describe the function of each type.
6. Describe the skin, a representative organ. Include the various tissues that compose it and the role of each tissue.

Applying the Concepts

1. Why does life on land present more difficulties in maintaining homeostasis than does life in water? What made it evolutionarily advantageous for organisms to colonize dry land?
2. The majority of homeostatic regulatory mechanisms in animals do not require conscious control. Discuss several reasons why this type of regulation is more advantageous to the animal than is conscious regulation of homeostatic controls.
3. Third-degree burns are usually painless. Skin regenerates only from the edges of these wounds. Second-degree burns regenerate from cells located at the burn edges, in hair follicles, and in sweat glands. First-degree burns are painful but heal rapidly from undamaged epidermal cells. From this information, draw the depth of first-, second-, and third-degree burns on Figure 27-10.
4. A coroner dictates the following description during an autopsy: "The tissue I am looking at forms part of the fetal skeleton. The extracellular matrix appears transparent. Fibers of collagen are present but are small and evenly dispersed in the extracellular matrix. Chondrocytes appear in tiny spaces, *lacunae,* within the matrix. Blood vessels have not yet penetrated the matrix." What tissue is the coroner describing?
5. Imagine you are a health-care professional teaching a prenatal class for fathers. Design a real-world analogy with sensors, electrical currents, motors, and so on to illustrate feedback relationships involved in the initiation of labor that a layperson could understand.

For More Information

Bruemmer, F. "Five Days with Fat Hoods." *International Wildlife,* January–February 1999. The rapid growth and prodigious fat-storing ability of the hooded seal adapts it to maintaining homeostasis under the extreme conditions of the far north.

Nuland, S. *The Wisdom of the Body.* New York: Alfred A. Knopf, 1997. Human physiology as seen through a surgeon's eyes. A firsthand account of the beauty and power of the body's mechanisms for maintaining homeostasis.

Pool, R. "Saviors." *Discover*, May 1998. Many seriously ill individuals die while waiting for organ transplants. Thanks to genetic engineering, organ donors of the future may be raised on a farm.

Storey, K. B., and Storey, J. M. "Frozen and Alive." *Scientific American*, December 1990. Some animals have special adaptations that allow them to withstand freezing.

Media Activities

To access a Media Activity visit http://www.prenhall.com/audesirk7. *Log in to the Web site selected by your instructor, navigate to this chapter, and select the appropriate Media Activity number.*

27.1 Homeostasis

Estimated time: 5 minutes

In this activity, you will discover how feedback systems help maintain stable equilibrium in animals.

27.2 Web Investigation: From Nightmare to Medical Miracle

Estimated time: 25 minutes

In this investigation, we will explore the experience and serious life-threatening problems that arise when the human body is super-cooled.

CHAPTER

28 Circulation

At a memorial service, Darryl Kile's teammates watch him wind up for a pitch. (Inset) A plaque in this coronary artery (an artery supplying the heart) has stimulated the formation of a blood clot that is totally blocking the vessel and preventing blood from reaching a portion of the heart muscle. This blockage will cause a heart attack.

AT A GLANCE

CASE STUDY Sudden Death

On June 22, 2002, players for the St. Louis Cardinals prepared for their upcoming game against the Chicago Cubs. As game time approached, they were first puzzled, then concerned, by the unexplained absence of their pitcher, Darryl Kile. Their concern turned to shock and grief when 33-year-old Kile was found dead in his hotel room, apparently having died in his sleep. One of the country's top pitchers and noted for his exceptional curveball, Kile was an athlete in his prime. But an autopsy revealed that two of his three coronary arteries (arteries that supply blood to the heart itself) were 80% to 90% blocked by atherosclerosis, in which arteries are narrowed by fatty deposits called plaque (see the inset in the chapter opening photo). His heart was also enlarged, a result of its heroic efforts to force blood through the partially blocked arteries. Some inherited genetic traits favor plaque buildup, and can cause life-threatening levels of plaque to accumulate at a much younger age than in people without these risk factors. The fact that Kile's father died of a heart attack at the age of 44 suggests that Darryl Kile may have been in this high-risk group.

Atherosclerosis often begins in childhood. Millions of children in the United States have elevated blood cholesterol levels, exacerbated by high-fat diets and lack of exercise, but decades usually pass before the disease is recognized. How does the heart work? How it is threatened by atherosclerosis and hypertension? What treatments might have benefited Darryl Kile had he known about his condition?

Media Activity
28.1 Circulatory Systems

28.1 What Are the Major Features and Functions of Circulatory Systems?

Billions of years ago, the first cells were nurtured by the sea in which they evolved. The sea brought them nutrients, which diffused into the cells and washed away any wastes that diffused out. Today, microorganisms and some simple multicellular animals still rely almost exclusively on diffusion for the exchange of nutrients and wastes with the environment. Sponges, for example, circulate seawater through pores in their bodies, bringing the environment close to each cell. As larger, more complex animals evolved, individual cells became increasingly distant from the outside world. However, the constant demands of a cell require short diffusion distances so that adequate nutrients reach the cell and the cell isn't poisoned by its own wastes. With the evolution of the circulatory system, a sort of "internal sea" was created, serving the same purpose as the sea did for the first cells. This internal sea transports food and oxygen close to each cell and carries away wastes produced by the cells.

All circulatory systems have three major parts:

- A fluid, **blood**, that serves as a medium of transport.
- A system of channels, or **blood vessels**, that conduct the blood throughout the body.
- A pump, the **heart**, that keeps the blood circulating.

Animals Have Two Types of Circulatory Systems

Animals have one of two major types of circulatory systems: open and closed. **Open circulatory systems** are present in many invertebrates, including arthropods—such as crustaceans, spiders, and insects—and mollusks, such as snails and clams. An animal with an open circulatory system has one or more hearts, a network of blood vessels, and a large open space within the body called a **hemocoel** (Fig. 28-1a). The heart pumps blood through vessels that release the blood into the hemocoel. Within the hemocoel (which may occupy 20% to 40% of the animal's body volume), tissues and internal organs are directly bathed in blood. The vessels also deliver blood back to the heart. For example, when the hearts of a grasshopper contract, valves in the hearts are pressed shut, forcing the blood to travel out through the vessels to the hemocoel. When the hearts relax, blood is drawn back into them through openings guarded by valves.

Closed circulatory systems are present in some invertebrates, including the earthworm (Fig. 28-1b) and very active mollusks (such as squid and octopuses). Closed circulatory systems are also a characteristic of all vertebrates, including humans. In closed circulatory systems, the blood (whose volume is only 5% to 10% of body volume) is confined to the heart and a continuous series of blood vessels. Closed circulatory systems allow more rapid blood flow, more efficient transport of wastes and nutrients, and higher blood pressure than is possible in open systems. In the earthworm, five contractile vessels serve as hearts, pumping blood through major vessels from which smaller vessels branch.

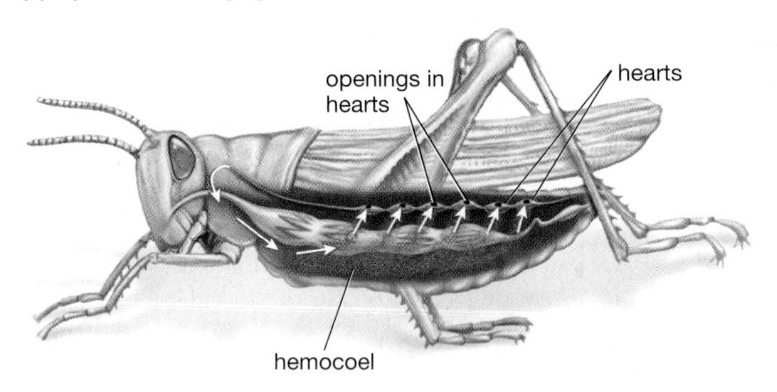

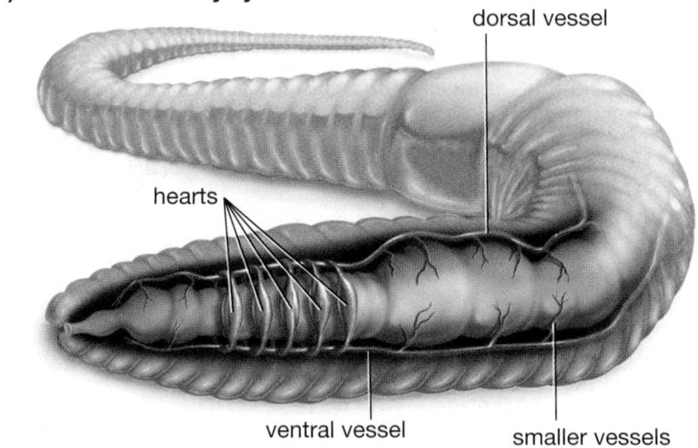

FIGURE 28-1 Open and closed circulatory systems
(a) In the open circulatory system of insects and other arthropods, a series of hearts pumps blood through vessels into the hemocoel, where blood directly bathes the other organs. Although shown in red for visibility, insect blood lacks hemoglobin and is almost clear, sometimes with a greenish tint. **(b)** In a closed circulatory system, blood remains confined to the heart(s) and the blood vessels. In the earthworm, five contractile vessels serve as hearts and pump blood through major ventral and dorsal vessels, from which smaller vessels branch. Earthworm blood, like ours, contains red hemglobin.

The Vertebrate Circulatory System Has Many Diverse Functions

The circulatory system supports all the other organ systems in the body. The circulatory systems of humans and other vertebrates perform the following functions:

- Transport oxygen from the lungs or gills to the tissues, and transport carbon dioxide from the tissues to the lungs or gills.
- Distribute nutrients from the digestive system to all body cells.

- Transport waste products and toxic substances to the liver (where many of them are detoxified) and kidneys for excretion.
- Distribute hormones from the glands and organs that produce them to the tissues upon which they act.
- Regulate body temperature, which is achieved partly by adjustments in blood flow.
- Prevent blood loss by means of the clotting mechanism.
- Protect the body from bacteria and viruses by circulating antibodies and white blood cells.

In the following sections we examine the three parts of the circulatory system: the heart, the blood, and the vessels, with an emphasis on the human system. Finally, we describe the lymphatic system, which works closely with the circulatory system.

28.2 How Does the Vertebrate Heart Work?

Increasingly Complex and Efficient Hearts Have Arisen During Vertebrate Evolution

No circulatory system can operate without a dependable pump. Blood must be moved through the body continuously throughout an animal's life; the vertebrate heart consists of muscular chambers capable of strong contractions. Chambers called **atria** (singular, **atrium**) collect blood. Atrial contractions send blood into the **ventricles**, chambers whose contractions circulate blood through the body. During the course of vertebrate evolution, the heart has become increasingly more complex, with more separation between oxygenated blood (which has picked up oxygen from the lungs or gills) and deoxygenated blood (which, in passing through body tissues, has lost oxygen).

The hearts of fishes, the first vertebrates to evolve, consist of two contractile chambers: a single atrium that empties into a single ventricle (Fig. 28-2a). Blood pumped from the ventricle passes first through the gill *capillaries*, thin-walled vessels where blood picks up oxygen and gives off carbon dioxide. The blood then travels to the rest of the body, delivering oxygen to the tissues and picking up carbon dioxide in the body capillaries.

Over evolutionary time, as fish gave rise to amphibians and amphibians to reptiles, a three-chambered heart evolved, consisting of two atria and one ventricle (Fig. 28-2b). In the three-chambered hearts of amphibians and most reptiles, deoxygenated blood from the body is delivered into the right atrium, while blood from the lungs travels into the left atrium. Both atria empty into the single ventricle. Although some mixing does occur, the deoxygenated blood tends to remain in the right portion of the ventricle and is pumped into vessels that enter the lungs, while most of the oxygenated blood remains in the left portion of the ventricle and is pumped to the rest of the body. This separation is enhanced in reptiles by a partial wall between the right and left portions of the ventricle.

Warm-blooded birds and mammals have high metabolic demands and require more efficient delivery of oxygen to their tissues than do cold-blooded animals. That demand is met by the four-chambered heart (Fig. 28-2c). Separate right and left ventricles isolate oxygenated from deoxygenated blood, ensuring that blood reaching the tissues has the highest possible oxygen content.

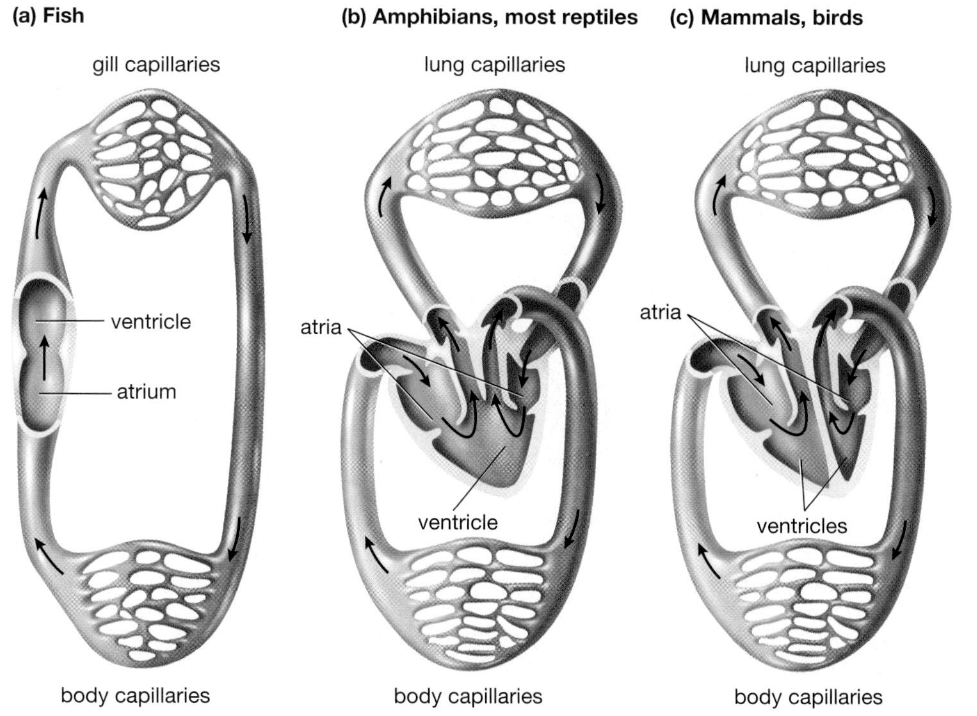

(a) Fish

gill capillaries

ventricle

atrium

body capillaries

(b) Amphibians, most reptiles

lung capillaries

atria

ventricle

body capillaries

(c) Mammals, birds

lung capillaries

atria

ventricles

body capillaries

FIGURE 28-2 The evolution of the vertebrate heart
(a) The earliest vertebrate heart is illustrated by the two-chambered heart of fishes. **(b)** Amphibians and most reptiles have hearts with two atria, from which blood empties into a single ventricle. Many reptiles have a partial wall down the middle of the ventricle. **(c)** The hearts of birds and mammals are actually two separate pumps that prevent mixing of oxygenated and deoxygenated blood. Note that in this and in subsequent illustrations, oxygenated blood is depicted as bright red, while deoxgenated blood is colored blue.

The Vertebrate Heart Consists of Muscular Chambers

Human hearts (and those of other mammals and birds) can be considered as two separate pumps, each with two chambers. In each pump, an atrium receives and briefly stores the blood before passing it to a ventricle that propels it through the body (Fig. 28-3). One pump, consisting of the right atrium and right ventricle, deals with deoxygenated blood. The right atrium receives oxygen-depleted blood from the body through two large **veins** (vessels that carry blood toward the heart): the *superior vena cava* and the *inferior vena cava*. After being filled with blood, the right atrium contracts, forcing the blood into the right ventricle. Contraction of the right ventricle sends the oxygen-depleted blood to the lungs via pulmonary **arteries** (vessels that carry blood away from the heart). The other pump, consisting of the left atrium and ventricle, deals with oxygenated blood. Oxygen-rich blood from the lungs enters the left atrium through pulmonary veins and is then squeezed into the left ventricle. Strong contractions of the left ventricle, the heart's most muscular chamber, send the oxygenated blood coursing out through a major artery, the *aorta*, to the rest of the body.

The Coordinated Contractions of Atria and Ventricles Produce the Cardiac Cycle

The human heart beats about 100,000 times each day. During each beat, the two atria first contract in synchrony, emptying their contents into the ventricles. A fraction of a second later, the two ventricles contract simultaneously, forcing blood into arteries that exit the heart. Both atria and ventricles then relax briefly before this **cardiac cycle** repeats (Fig. 28-4). At a normal resting heart rate, the cardiac cycle lasts just under 1 second. The cardiac cycle is involved in blood pressure measurement (Fig. 28-5); *systolic pressure* (the higher of the two readings) is measured during ventricular contraction and *diastolic pressure* is measured between contractions.

Valves Maintain the Direction of Blood Flow, and Electrical Impulses Coordinate the Sequence of Contractions

When the ventricles contract, blood must be directed out through the arteries and not back up into the atria. Then, once blood has entered the arteries, it must be prevented from flowing back as the heart relaxes. This directionality of blood flow is maintained by one-way

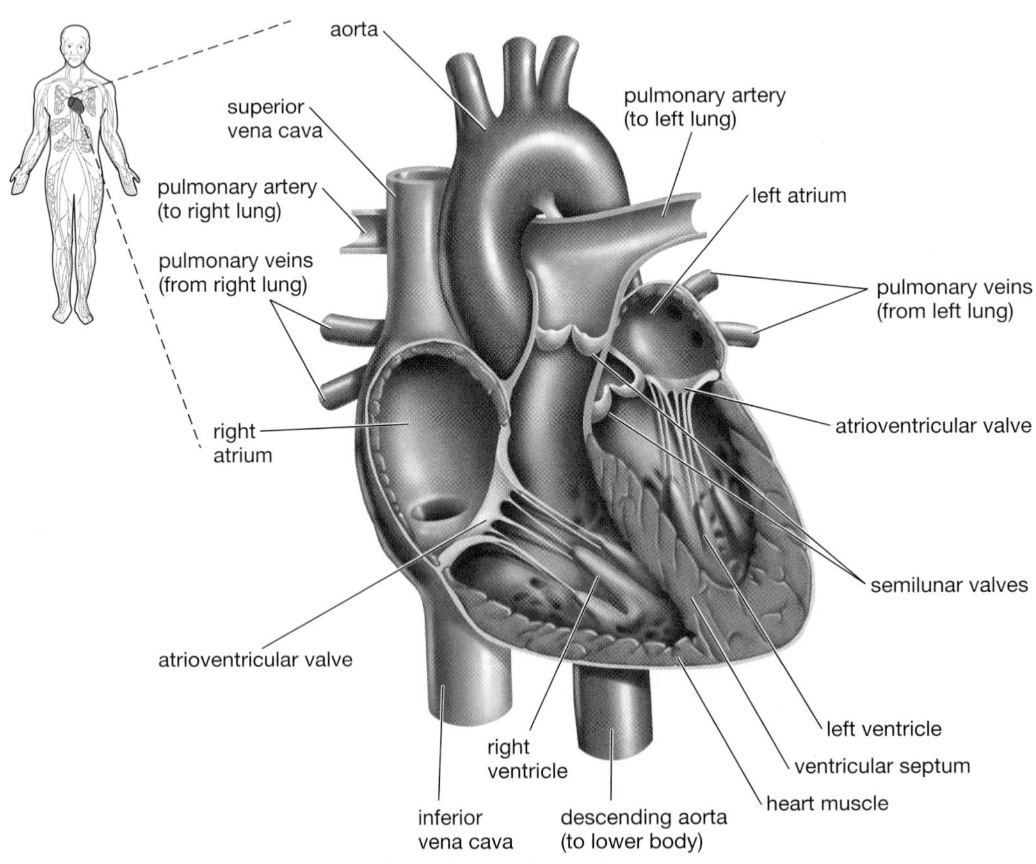

FIGURE 28-3 The human heart and its valves and vessels
The heart is drawn as if it were in a body facing you, so that right and left appear reversed. Note the thickened walls of the left ventricle, which pumps blood much farther through the body than does the right ventricle, which propels blood to the lungs. One-way semilunar valves separate the aorta from the left ventricle, and the pulmonary artery from the right ventricle. Atrioventricular valves separate the atria and ventricles.

Media Activity
28.2 Heart Structure and Function

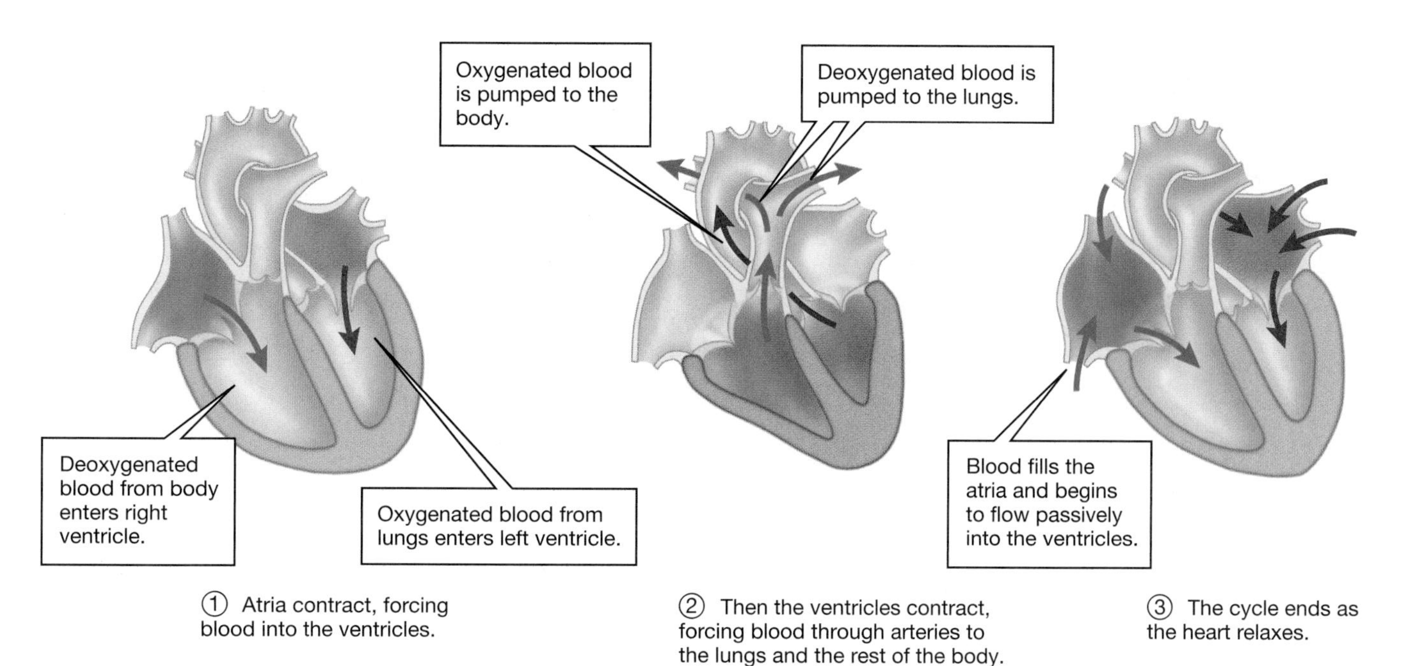
Oxygenated blood is pumped to the body.
Deoxygenated blood is pumped to the lungs.
Deoxygenated blood from body enters right ventricle.
Oxygenated blood from lungs enters left ventricle.
Blood fills the atria and begins to flow passively into the ventricles.
① Atria contract, forcing blood into the ventricles.
② Then the ventricles contract, forcing blood through arteries to the lungs and the rest of the body.
③ The cycle ends as the heart relaxes.

FIGURE 28-4 The cardiac cycle
QUESTION How would the cardiac cycle be altered if the right atrioventricular valve were damaged badly enough to prevent it from functioning?

Media Activity 28.3 Cardiac Cycle

valves (see Figs. 28-3 and 28-4). Pressure in one direction opens them easily, but reverse pressure forces them closed. **Atrioventricular valves** allow blood to flow from the atria into the ventricles (but not the reverse), and **semilunar valves** allow blood to enter the pulmonary artery and the aorta when the ventricles contract (and prevent it from returning as the ventricles relax). You may notice that no valves separate the left and right atria from the pulmonary veins and the vena cavas, respectively. In fact, when the atria contract, some blood does flow backward into these veins, but since the ventricles fill adequately in spite of this, no valves have evolved here.

The contraction of the heart is initiated and coordinated by a **pacemaker**, a cluster of specialized heart muscle cells that produce spontaneous electrical signals at a regular rate. These electrical signals are transmitted among the heart muscle cells and stimulate them to contract. The heart's primary pacemaker is the **sinoatrial (SA) node**, located in the upper wall of the right atrium

Media Activity 28.4 Blood Pressure

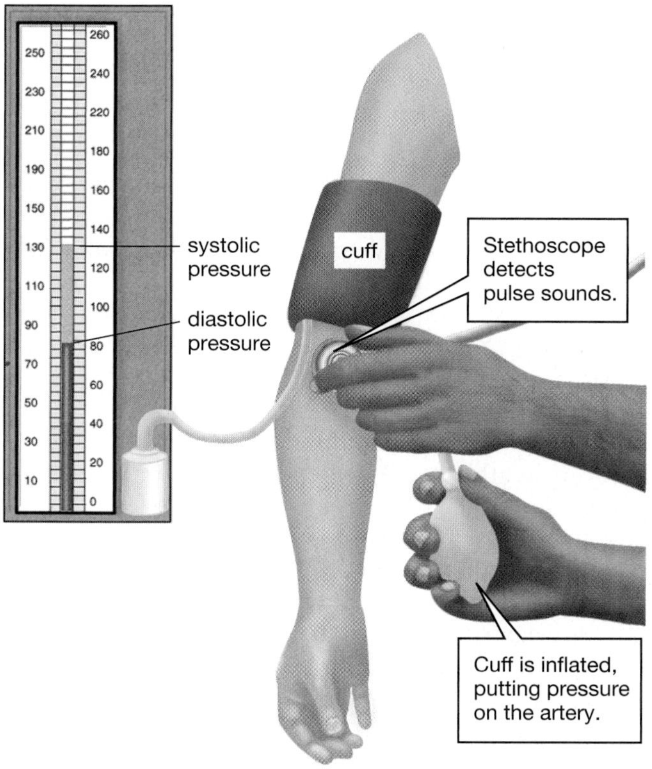

FIGURE 28-5 Measuring blood pressure
First, the cuff is inflated until its pressure closes off the arm's main artery; this pressure is then gradually reduced. When the pulse is first detected by the stethoscope, it indicates that the contractions of the left ventricle have overcome the pressure in the cuff and blood is flowing. This is the upper (and higher) reading: the systolic pressure. Next, cuff pressure is reduced until no pulse is audible, indicating that blood is flowing continuously through the artery. In other words, the pressure *between* ventricular contractions is adequate to overcome the cuff pressure. This is the lower reading: the diastolic pressure. The numbers are in millimeters of mercury, a standard measure of pressure also used in barometers. **EXERCISE** Sketch a graph that shows how blood pressure inside an artery changes during the cardiac cycle. On the graph, label the points that correspond to the systolic and diastolic blood pressure measured by the blood pressure cuff.

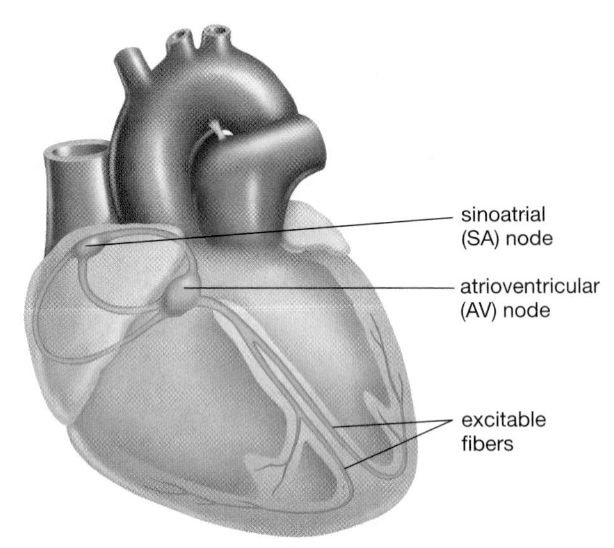

FIGURE 28-6 The heart's pacemaker and its connections
The sinoatrial (SA) node serves as the heart's pacemaker. The signal to contract spreads from the SA node through the muscle fibers of both atria, finally exciting the atrioventricular (AV) node in the lower right atrium. The AV node then transmits the signal to contract to the ventricular muscle.

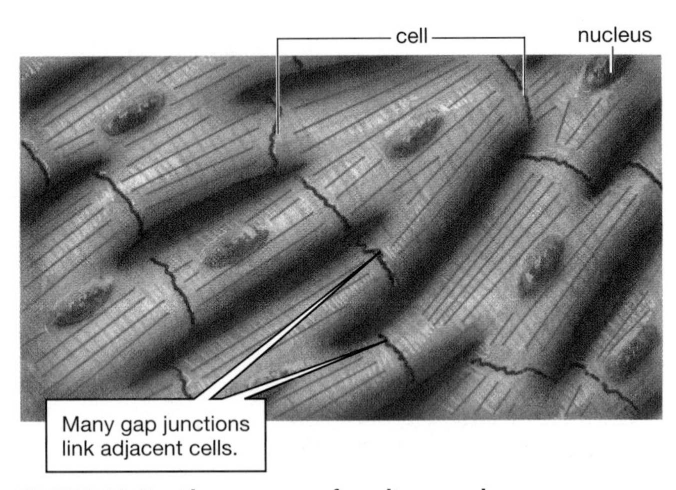

FIGURE 28-7 The structure of cardiac muscle
Cardiac muscle cells are branched. Adjacent plasma membranes meet in folded areas that are densely packed with gap junctions, which connect the interiors of adjacent cells. This arrangement allows direct transmission of electrical signals between the cells, coordinating their contractions. **QUESTION** If a muscle is repeatedly exercised, it increases in size. Why is the resting heart rate of a well-conditioned athlete slower than the heart rate of a less active person?

(Fig. 28-6). Individual cardiac muscle cells communicate directly with one another through gap junctions (Fig. 28-7). These connecting pores, introduced in Chapter 4, allow the electrical signals to pass freely and rapidly from cells near the pacemaker to adjoining muscle cells.

During the cardiac cycle, the atria contract first and empty their contents into the ventricles, then refill while the ventricles contract. Thus, there must be a delay between the atrial and the ventricular contractions. How is this accomplished? Starting at the SA node, an electrical impulse creates a wave of contraction that sweeps through the muscles of the right and left atria, which contract in smooth synchrony. The signal then reaches a barrier of inexcitable tissue between the atria and the ventricles. Here, the excitation is channeled through the **atrioventricular (AV) node**, a small mass of specialized muscle cells located in the floor of the right atrium (see Fig. 28-6). The impulse is slowed at the AV node, postponing the ventricular contraction until about 0.1 second after the atria contract. This delay gives the atria time to complete the transfer of blood into the ventricles before ventricular contraction begins. From the AV node, the signal to contract spreads to the base of the two ventricles along tracts of excitable fibers, shown in Fig. 28-6. The impulse then travels rapidly from these fibers through the communicating muscle fibers, causing the ventricles to contract in unison from the base upward, forcing blood up into the major arteries.

A variety of disorders can interfere with the complex series of events that produces the normal cardiac cycle. When the pacemaker fails, or if other areas of the heart become more excitable and usurp the pacemaker's role, uncoordinated, irregular, weak contractions called fibrillation may occur. Fibrillation of the ventricles is fatal if not reversed, because blood cannot be pumped by the quivering muscle. A defibrillating machine applies a jolt of electricity to the heart, synchronizing the contraction of the ventricular muscle cells and sometimes allowing the pacemaker to resume its normal coordinating function. An electrocardiogram (ECG), which measures the electrical activity underlying the contractions of the heart chambers, is described at the end of the chapter in "Links to Life: Charting Heartbeats."

The Nervous System and Hormones Influence Heart Rate

Your heart rate is finely tuned to your activity level, whether you are running to class or basking in the sun. On its own, the SA node pacemaker would maintain a steady rhythm of about 100 beats per minute. However, nerve impulses and hormones significantly alter heart rate. In a resting individual, activity of the parasympathetic nervous system, which regulates body systems during periods of rest (see Chapter 34) slows heart rate to about 70 beats per minute (this resting rate is typically lower in athletes). When exercise or stress creates a demand for greater blood flow to the muscles, the sympathetic nervous system, which prepares the body for emergency action, accelerates the heart rate. Simultaneously, the hormone epinephrine (also known as adrenaline) increases the heart rate while mobilizing the body to respond to threatening or exciting events. For example, when astronauts landed on the moon, their heart rates were more than 170 beats per minute, even though they were sitting in their spacecraft.

28.3 What Is Blood?

Blood, which has been called the "river of life," transports dissolved nutrients, gases, hormones, and wastes through the body. Blood has two major components: a fluid called **plasma**, and cellular components—*red blood cells, white blood cells*, and *platelets*—that are suspended in the plasma. The cellular components of blood typically account for 40% to 45% of its volume; the other 55% to 60% is plasma. The average human has 5 to 6 liters of blood, constituting about 8% of total body weight.

Plasma Is Primarily Water in Which Proteins, Salts, Nutrients, and Wastes Are Dissolved

Plasma, which is straw-colored, is about 90% water. Dissolved in the plasma are proteins, hormones, nutrients (glucose, vitamins, amino acids, lipids), gases (carbon dioxide, oxygen), salts (sodium, calcium, potassium, magnesium), and wastes, such as urea. Plasma proteins are the most abundant of these dissolved substances. The three major plasma proteins are *albumins*, which help maintain the blood's osmotic pressure (which controls the flow of water across plasma membranes); *globulins*, which transport nutrients and play a role in the immune system; and *fibrinogen*, which is important in blood clotting and is discussed later in this chapter.

Red Blood Cells Carry Oxygen from the Lungs to the Tissues

The most abundant cells in the blood are those that carry oxygen, called red blood cells or **erythrocytes**. Each cubic millimeter of blood (a small droplet) contains about 5 million erythrocytes, which make up about 99% of all blood cells and constitute about 40% of the total blood volume in females and 45% in males. A red blood cell resembles a ball of clay squeezed between a thumb and forefinger (Fig. 28-8). This shape, which results when the cell loses its nucleus during development, provides a larger surface area than would a spherical cell of the same volume, and increases the cell's ability to absorb and release oxygen through its plasma membrane.

The red color of erythrocytes is caused by the pigment **hemoglobin** (Fig. 28-9). This large, iron-containing protein accounts for about one-third of the weight of each red blood cell and carries about 97% of the blood's oxygen. One hemoglobin molecule can bind and carry up to four molecules of oxygen, permitting blood to hold far more oxygen than would be possible if it were dissolved in plasma. Hemoglobin takes on a bright cherry-red color when it binds oxygen, and becomes a deeper red when oxygen is released. Since deoxygenated blood is found in veins, which appear bluish when viewed through the skin, the color conventions in most diagrams depict arteries as red and veins as blue. Hemoglobin binds loosely to oxygen, picking it up in the capillaries of the lungs—where oxygen concentration is high—and releasing it in other tissues of the body where oxygen concentration is lower. After releasing its oxygen, some of the hemoglobin picks up carbon dioxide from the tissues for transport back to the lungs. Blood's ability to transport a large amount of oxygen to tissues is difficult to duplicate with artificial fluids. Nonetheless, because blood, like hearts, can be in short supply, researchers are experimenting with a variety of blood substitutes to use in

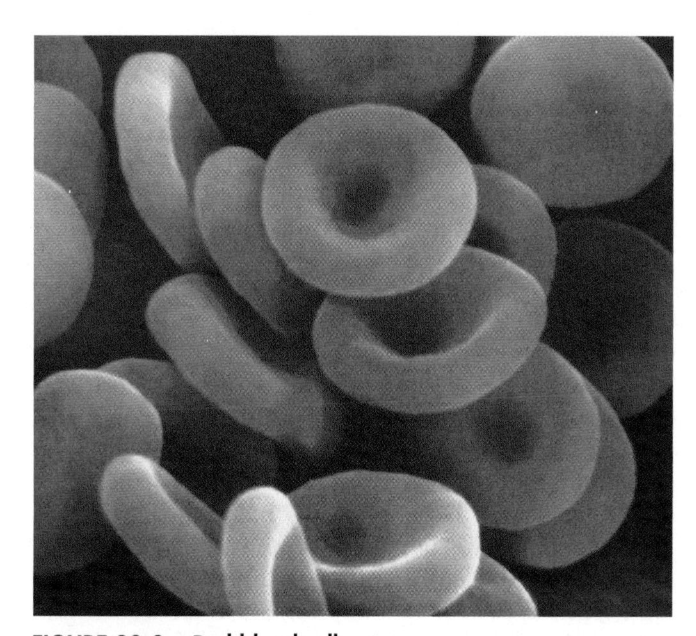

FIGURE 28-8 Red blood cells
This false-color scanning electron micrograph clearly shows the biconcave disk shape of red blood cells.

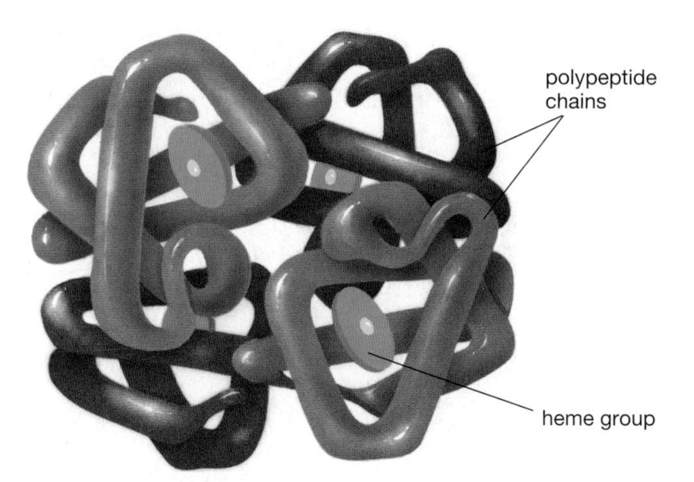

FIGURE 28-9 Hemoglobin
A molecule of hemoglobin is composed of four polypeptide chains (two pairs of similar chains), each surrounding a heme group. **QUESTION** An insufficient amount of iron in a person's diet can cause anemia, a condition in which the blood contains too few red blood cells to keep the body's tissues supplied with adequate oxygen. Why does dietary iron deficiency cause anemia?

SCIENTIFIC INQUIRY Artificial Blood? Artificial Vessels? Artificial Hearts?

Every minute, someone in the United States dies of coronary heart disease. Research continues on a number of fronts to combat cardiovascular disorders, which are the leading causes of death in the United States. For example, thousands of patients whose vessels are clogged by atherosclerosis might benefit from bypass surgery, but do not have vessels suitable for grafting. In an effort to help these patients, researchers have taken collagen from pigs and cows, molded it into tubes, and either placed it in a nutrient broth with cells from blood vessels or grafted it directly into the vessels of experimental animals. Under the right conditions, living blood vessel cells invade and cover these tubes to form somewhat weak but (at least temporarily) functional vessels. Researchers at Duke University Medical Center induced layers of smooth muscle and endothelial cells from blood vessels to grow on a polymer tube while it was subjected to pressure pulses from a pump, mimicking the conditions under which arteries develop. After the polymer broke down as expected, the remaining artificial vessels were muscular and strong and functioned for about a month when grafted into pigs.

Meanwhile, blood supplies often fail to meet the need for blood transfusions, a problem that is worsening as the population ages. Researchers are using several approaches to make artificial blood with the oxygen-delivering properties of real blood. Some are using perfluorocarbons (PFCs), chemicals that can carry oxygen and carbon dioxide much as blood does (Fig. E28-1), trying to create PFC-based blood substitutes that are nontoxic, do not build up in body tissues, and exchange appropriate amounts of oxygen and carbon dioxide. Meanwhile, others are working to modify real hemoglobin so it is available in large quantities for transfusion. Hemoglobin extracted from outdated human blood or cow's blood (which are usually discarded) is being treated to function outside a red blood cell in artificial blood that may be marketed soon. Recently, researchers have genetically engineered bacteria to produce human hemoglobin that is stable outside the red blood cell for use in a blood substitute.

FIGURE E28-1 Breathing fluid?
After a moment of wide-eyed shock, a mouse submerged in this oxygen-carrying PFC solution begins to breathe the liquid as it would air—with no ill effects. Researchers hope that PFCs can be made to carry oxygen safely within the human circulatory system.

For some people suffering from severe heart disease, heart transplants are the only hope of survival, but while 4000 people await heart donors in the United States, only 2500 hearts are available for transplant each year (for more information, refer to the Case Study "Xenotransplants" in Chapter 28 on this text's Web site). Researchers are currently working with "bridging devices"—pumps that can keep patients alive while they wait for

transfusions (see "Scientific Inquiry: Artificial Blood? Artificial Vessels? Artificial Hearts?"). The role of blood in gas exchange is discussed further in Chapter 29.

Red Blood Cells Have a Relatively Short Life Span

Red blood cells are formed in the *bone marrow*, the soft interior portion of certain bones found in the chest, upper arms, upper legs, and hips. During their development, mammalian red blood cells lose their nuclei. Without a nucleus, the cell cannot divide or synthesize new enzymes and other cellular components coded by genetic material. Thus, erythrocytes are short-lived, with an average life span of about 4 months. Every second, more than 2 million red blood cells die and are replaced by new ones from the bone marrow. Dead or damaged red blood cells are removed from circulation, primarily in the liver and spleen, and broken down to release their iron. Blood carries the salvaged iron back to the bone marrow, where it is used to make more hemoglobin and packaged into new red blood cells. Although this recycling process is efficient, small amounts of iron are excreted daily and must be replenished by the diet. Bleeding from injury or menstruation also tends to deplete iron stores.

Negative Feedback Regulates Red Blood Cell Numbers

The number of red blood cells in the blood determines how much oxygen it can carry; this number is maintained by a negative feedback system that involves the hormone **erythropoietin**. Erythropoietin is produced by the kidneys and released into the blood in response to oxygen deficiency. This lack of oxygen may be caused by a loss of blood, insufficient production of hemoglobin, high altitude (where less oxygen is available), or lung disease that interferes with gas exchange in the lungs. Erythropoietin stimulates the rapid production of new

donors after their hearts fail. In 2001 researchers began human trials with the artificial AbioCor heart, implanting it in patients who probably had less than a month to live and were deemed too ill for a transplant (Fig. E28-2). One of these early transplant pioneers survived for over a year. Another bridging device, the *left ventricular assist device* (LVAD), diverts blood from the failing left ventricle to a pump in the abdomen or chest. Clinical trials have shown that patients implanted with LVADs survive significantly longer than those receiving the best standard medical treatment without the device. Using another approach entirely, some researchers and biotechnology firms are working toward the goal of using stem cells to generate new heart muscle, which might one day help failing hearts regenerate.

Although the manufacturers of the AbioCor heart provide patients with a 13-page consent form, medical ethicists debate whether patients faced with almost certain death are able to give truly informed consent about any procedure that might prolong their lives. Transplants and other procedures with uncertain outcomes prompt many complex ethical questions from both researchers and patients. Should clinical trials begin with the most desperately ill patients, who have the least to lose, or are these patients the least likely to demonstrate the device's true potential? If the LVAD proves more effective than an artificial heart, should patients who might otherwise receive LVADs be recruited into trials of new artificial hearts that might (or might not) function better? Finally, bridging devices that prolong the lives of patients awaiting transplants will inevitably increase the demand for the already limited supply of donor hearts. Should patients on bridging devices (whose hearts have already failed) have priority to receive this limited resource, or would people whose disease is less advanced be more likely to benefit from a transplant? Alternatively, should bridging device research and human trials be continued with the goal of providing a survival rate comparable to that of transplants, so that the bridging devices of today might become the permanent replacement hearts of the future?

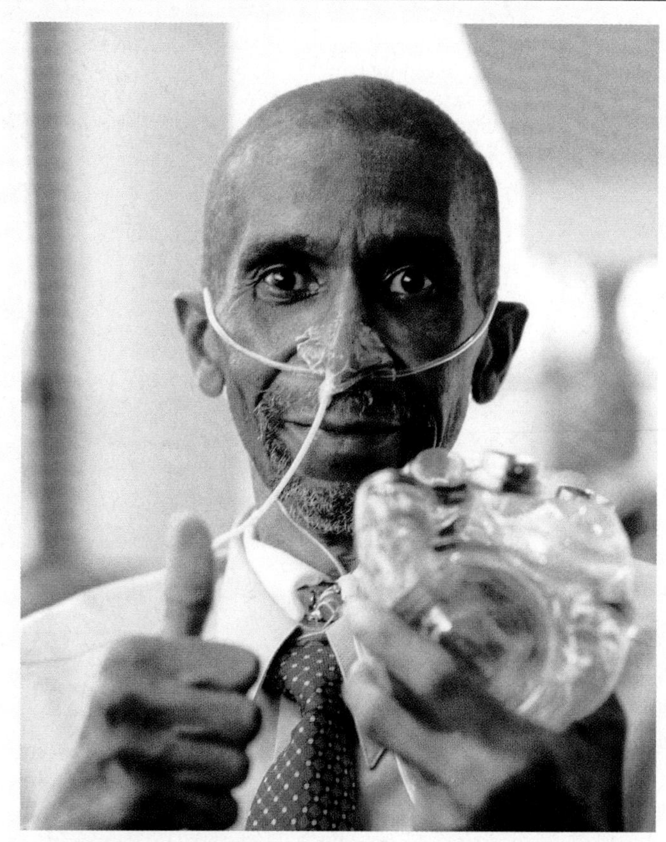

FIGURE E28-2 The AbioCor heart
Fifty-nine-year-old Robert Tools, the first recipient of the AbioCor heart, is shown holding the device. He survived about four months with his artificial heart before suffering a fatal stroke.

red blood cells by the bone marrow. When adequate oxygen levels are restored, erythropoietin production declines and the rate of red blood cell production returns to normal (Fig. 28-10).

Blood Type Is Determined by Specific Proteins on Red Blood Cell Membranes

You inherit your blood type, and if you ever need a transfusion, getting the proper blood type can be a matter of life and death. Blood is classified as type A, B, AB, or O, depending on the presence or absence of specific glycoproteins (designated A and B) on the plasma membranes of red blood cells. Because A blood carries antibodies that attack the proteins on B blood (and vice versa), transfusion of B blood into an individual with A blood (or A blood into a type B individual) could be fatal. We discussed the genetics and properties of these major blood types in Chapter 12 (see Table 12-1).

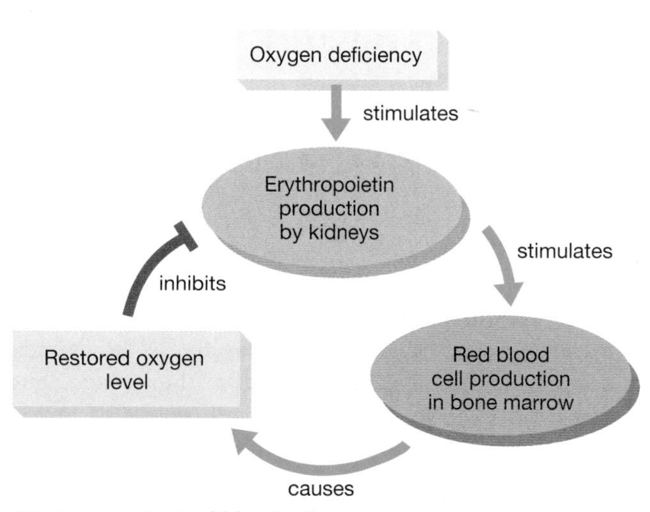

FIGURE 28-10 Red blood cell regulation by negative feedback
QUESTION Some endurance althetes cheat by "blood-doping": injecting large doses of erythropoietin. Why does this provide a competitive advantage?

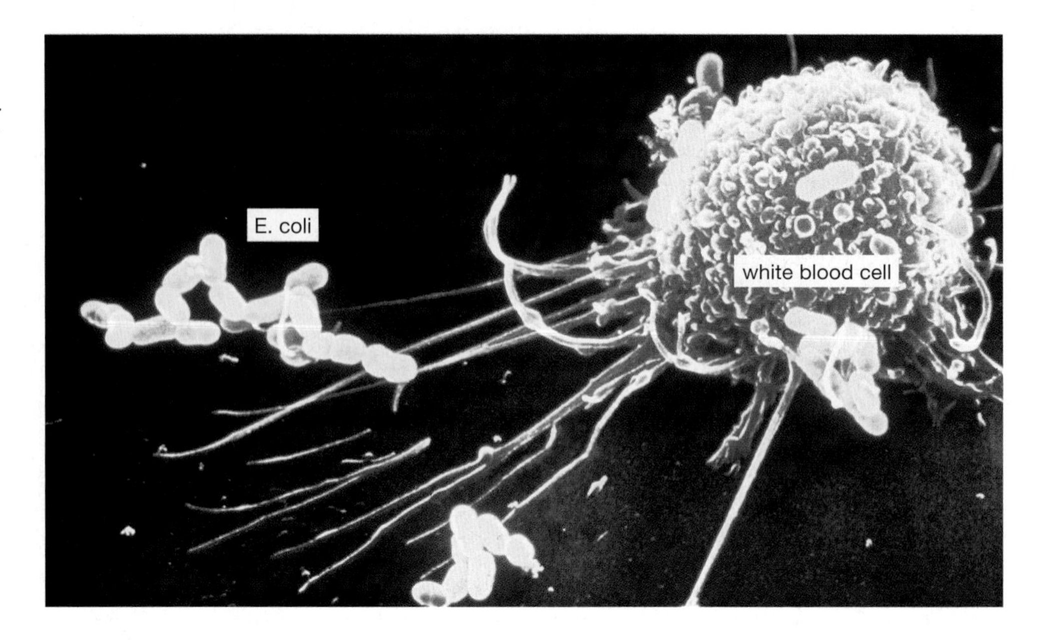

FIGURE 28-11 A white blood cell attacks bacteria
These bacteria are *Escherichia coli*. Some forms of these intestinal bacteria can cause disease if they enter the bloodstream.

Another type of protein on red blood cells is the **Rh factor**. If this protein is present, blood is described as *Rh-positive*; if it is absent, the blood is *Rh-negative*. The presence or absence of the Rh factor is important in transfusions and in pregnancy. If an Rh-negative woman has children with an Rh-positive man, half of her children are likely to be Rh-positive, because Rh-positive blood is a dominant genetic trait. The first Rh-positive child will trigger antibody production in the mother's blood, usually without noticeable ill effects. Subsequent Rh-positive children, however, could be born with erythroblastosis fetalis. In this condition, the mother's Rh antibodies invade the fetus and attack its red blood cells, causing severe anemia in the newborn. Fortunately, this condition can now be prevented with injections of a substance that prevents the pregnant woman from forming Rh antibodies.

White Blood Cells Help Defend the Body Against Disease

There are five common types of white blood cells, or **leukocytes**, which together make up less than 1% of all blood cells. All white blood cells are derived from cells that originate in bone marrow. Most white blood cells function in some way to protect the body against microbes (such as bacteria and viruses) that cause disease, and they use the circulatory system to travel to the site of invasion. Some travel through capillaries to wounds where bacteria have gained entry, then ooze out through narrow openings in the capillary walls. After leaving the capillaries, some change into amoeba-like cells called **macrophages** (literally "big eaters") that engulf foreign particles, including bacteria and cancer cells (Fig. 28-11). They typically die in the process, and their dead bodies accumulate and contribute to a white substance, called *pus*, often seen at infection sites. The **lymphocytes**, described in Chapter 32, are another type of white blood cell. They are responsible for the production of antibodies that help provide immunity against disease. Cells that give rise to lymphocytes migrate from the bone marrow through the bloodstream to tissues of the lymphatic system, such as the thymus, spleen, and lymph nodes, described later in this chapter.

Platelets Are Cell Fragments That Aid in Blood Clotting

Platelets, which are crucial to blood clotting, are pieces of large cells called *megakaryocytes*. Megakaryocytes remain in the bone marrow, where they pinch off membrane-enclosed chunks of their cytoplasm to form platelets (Fig. 28-12). The platelets then enter the blood and play a central role in blood clotting. Like red blood cells, platelets lack a nucleus, but their life span is even shorter—about 10 to 12 days.

Blood clotting is a complex process that keeps animals from bleeding to death, not only from trauma, but also from normal wear and tear on the body. Clotting (Fig. 28-13a) begins when blood comes in contact with injured tissue, for example, a break in the blood vessel wall. The ruptured surface causes platelets to adhere and partially block the opening. Both the adhering platelets and the ruptured cells release a variety of substances, initiating complex sequences of reactions among circulating plasma proteins. One important outcome of these chemical reactions is the production of the enzyme **thrombin**, which catalyzes the conversion of the plasma protein **fibrinogen** into insoluble strands of protein called **fibrin**. Fibrin proteins adhere to one another, forming a fibrous network on top of the aggregated platelets. This protein web traps red blood cells and more platelets (Fig. 28-13b), increasing the density of the clot. Platelets adhering to the fibrous mass send out sticky projections that grip one another. Within about half an hour, the platelets

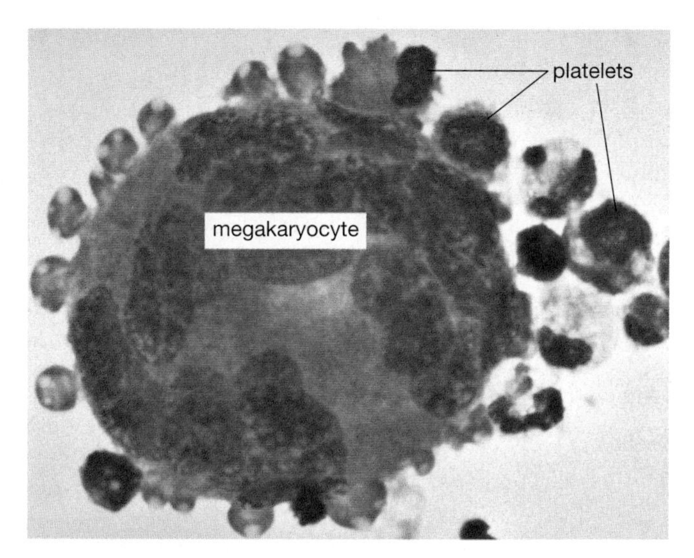
platelets
megakaryocyte

FIGURE 28-12 The production of platelets
Here, dozens of membrane-enclosed pieces of cytoplasm called *platelets* are budding off a single megakaryocyte.

contract, pulling the fibrin web tighter and forcing liquid out. This action creates a denser, stronger clot (on the skin it is called a *scab*) and also constricts the wound, pulling the damaged surfaces closer together in a way that promotes healing.

Despite blood's clotting ability, each year in the United States tens of thousands of people bleed to death from gunshots and other forms of trauma. U.S. Army and Red Cross researchers have developed bandages impregnated with large quantities of thrombin and fibrinogen that both physically and chemically staunch the flow of blood. To reduce costs and safety concerns from the use of human blood products, researchers have recently genetically engineered cows that secrete large quantities of human fibrinogen into their milk. This and other animal research holds promise for an abundant and safe supply of these vital proteins in the near future.

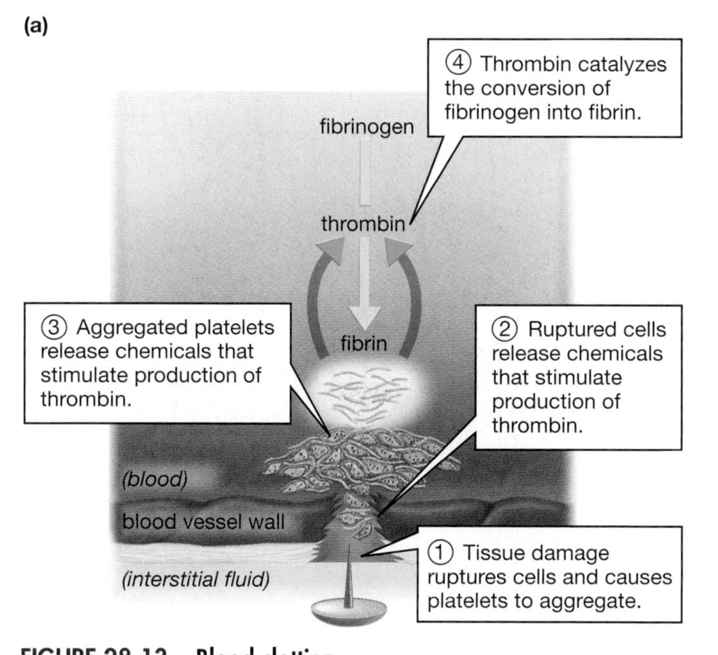
(a)
④ Thrombin catalyzes the conversion of fibrinogen into fibrin.
fibrinogen
thrombin
③ Aggregated platelets release chemicals that stimulate production of thrombin.
fibrin
② Ruptured cells release chemicals that stimulate production of thrombin.
(blood)
blood vessel wall
(interstitial fluid)
① Tissue damage ruptures cells and causes platelets to aggregate.

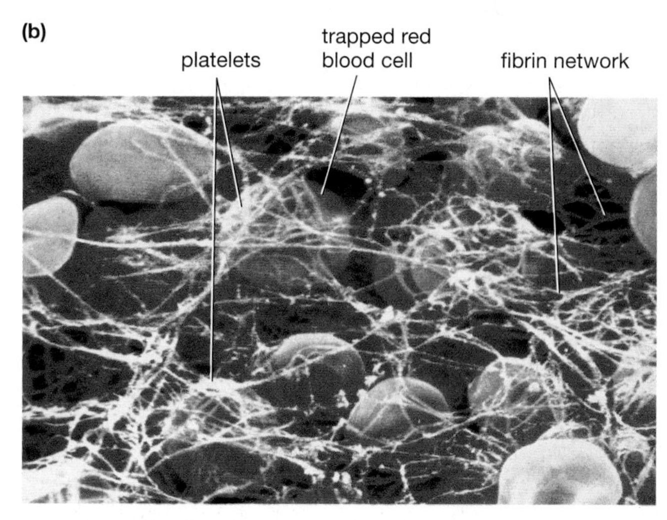
(b)
platelets
trapped red blood cell
fibrin network

FIGURE 28-13 Blood clotting
(a) Injured tissue and adhering platelets cause a complex series of biochemical reactions among blood proteins. These reactions produce thrombin, which catalyzes the conversion of fibrinogen to insoluble fibrin strands.
(b) Threadlike fibrin proteins produce a tangled sticky mass that traps red blood cells and eventually forms a clot.
QUESTION A common cause of hemophilia, a disease in which victims' blood does not clot properly, is a genetic defect that prevents production of a single protein. How might such a simple problem prevent clotting?

28.4 What Are the Types and Functions of Blood Vessels?

The blood flows in well-defined channels called *blood vessels*. Some of the major blood vessels of the human circulatory system are diagrammed in Figure 28-14. As it leaves the heart, blood travels from arteries to *arterioles* to capillaries to *venules* to veins, which return it to the heart. These vessels are shown in Figure 28-15.

Arteries and Arterioles Are Thick-Walled Vessels That Carry Blood Away from the Heart

Arteries carry blood away from the heart. These vessels have thick walls embedded with smooth muscle and elastic tissue (see Fig. 28-15). With each surge of blood from the ventricles, the arteries expand slightly, like thick-walled balloons. As their elastic walls recoil between heartbeats, the arteries actually help pump the blood and maintain its steady flow through the smaller vessels. Arteries branch into vessels of smaller diameter called **arterioles**, which play a major role in determining how blood is distributed within the body, as described later.

Capillaries Are Microscopic Vessels That Allow the Blood and Body Cells to Exchange Nutrients and Wastes

The entire circulatory system is an elaborate device that allows each cell of the body to exchange nutrients and wastes by diffusion. Arterioles conduct blood to **capillaries**, the tiniest of all vessels, where the actual process of diffusion occurs. Here, wastes, nutrients, gases, and hormones are exchanged between the blood and the body cells. Capillaries are well-adapted to their role of exchange, with walls only one cell thick. Most nutrients, oxygen, and carbon dioxide diffuse readily through capillary plasma membranes. Salts and small charged molecules (including some small proteins) move through fluid-filled spaces between adjacent capillary cells. The high pressure within the capillaries that branch from arterioles causes fluid to leak continuously from the blood plasma into the spaces that surround the capillaries and tissues. This fluid, known as the **interstitial fluid**, consists

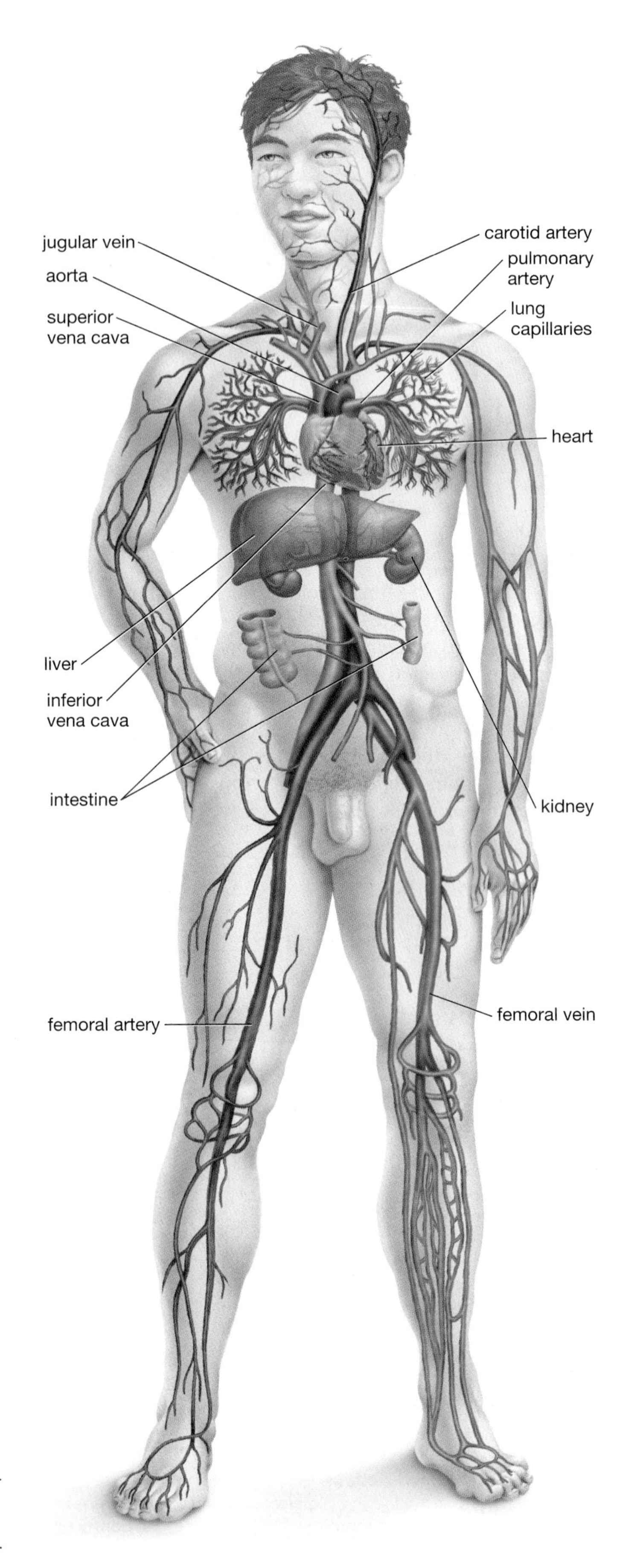

FIGURE 28-14 The human circulatory system
Most veins (right) carry deoxygenated blood to the heart, and most arteries (left) conduct oxygenated blood away from the heart. The pulmonary veins (which carry oxygenated blood) and arteries (which carry deoxygenated blood) are exceptions. Organs such as the liver, kidneys, and intestine are illustrated for reference. All organs (with the exception of the lungs, where the pattern is reversed) receive blood from arteries, send it back via veins, and are nourished by microscopic capillaries (the lung capillaries illustrated here are greatly enlarged).

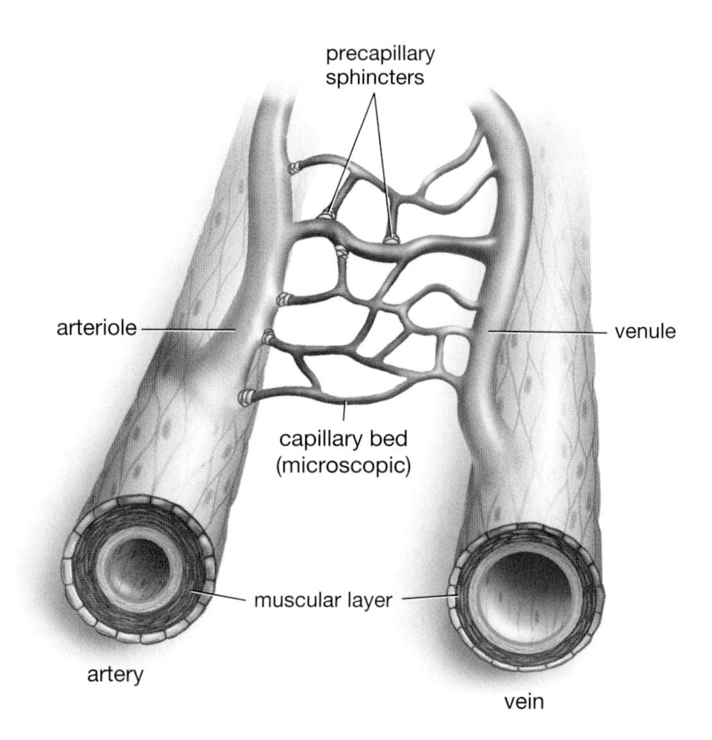
precapillary sphincters
arteriole
venule
capillary bed (microscopic)
muscular layer
artery
vein

FIGURE 28-15 Structures and interconnections of blood vessels
Arteries and arterioles are more muscular than veins and venules. Oxygenated blood moves from arteries to arterioles to capillaries. Capillaries, which have walls only one cell thick, empty deoxygenated blood into venules, which empty into veins. Precapillary sphincters regulate the movement of blood from arterioles into capillaries.

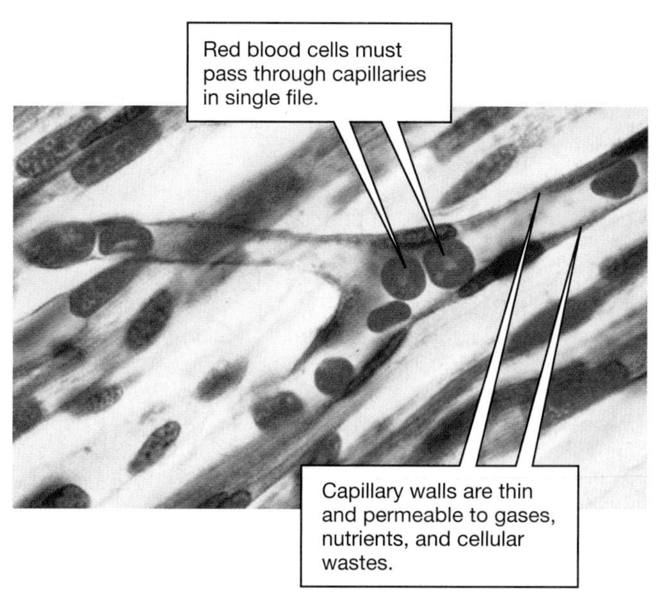
Red blood cells must pass through capillaries in single file.
Capillary walls are thin and permeable to gases, nutrients, and cellular wastes.

FIGURE 28-16 Red blood cells travel single file through a capillary
QUESTION Why does oxygen move out of capillaries in body tissues while carbon dioxide moves in (instead of the other way around)?

primarily of water containing dissolved nutrients, hormones, gases, wastes, and small proteins from the blood. Large plasma proteins (such as albumins), red blood cells, and platelets are unable to leave the capillaries because they are too large to fit through plasma membrane channels. The exchange of materials between capillary blood and nearby cells occurs through this interstitial fluid, which bathes nearly all the body's cells. Approximately 8 gallons (24 liters) of interstitial fluid is pressure-filtered from the blood each day through the walls of capillaries at the arteriole end of the capillary network. Pressure within the capillaries drops as blood travels toward the venules, and the high osmotic pressure of the blood that remains inside the capillaries (due to the presence of albumins and other large proteins) draws water back into the vessels by osmosis as blood approaches the venous end of the capillaries. As water moves into the capillaries and the blood within becomes more dilute, dissolved substances in the interstitial fluid tend to diffuse back into the capillaries as well. Thus, much of the interstitial fluid (about 85%) is restored to the bloodstream through the capillary walls on the venous side of the capillary network. As you will learn later in this chapter, the lymphatic system restores the remaining fluid (see Fig. 28-19).

Capillaries are so narrow that red blood cells must pass through them in single file (Fig. 28-16). Consequently, all blood is sure to pass close to the capillary walls, where the exchange of gases, nutrients, and wastes occurs with the surrounding interstitial fluid. Capillaries are so numerous that most body cells are no more than 100 micrometers (about as thick as four pages of this book) from a capillary; your body contains about 50,000 miles (80,600 kilometers) of capillaries, enough to encircle the globe twice! The speed of blood flow drops very quickly as blood moves through this narrow, lengthy capillary network, allowing more time for diffusion to occur. These factors combine to facilitate the exchange of materials by diffusion.

Veins and Venules Carry Blood Back to the Heart

After picking up carbon dioxide and other cellular wastes from cells, capillary blood drains into larger vessels called **venules**, which empty into still larger veins (see Fig. 28-15). Veins provide a low-resistance pathway that conducts blood back toward the heart. The walls of veins are thinner, less muscular, and more expandable than those of arteries, although both contain a layer of smooth muscle. Because blood pressure in the veins is low, contractions of skeletal muscle during exercise and breathing help return blood to the heart by squeezing the veins and forcing blood through them.

When veins are compressed, why isn't blood forced away from the heart as well as toward it? Veins are

HEALTH WATCH Matters of the Heart

Disorders of the heart and blood vessels, called *cardiovascular disorders*, are the leading cause of death in the United States, killing nearly 1 million Americans annually—and no wonder. Your heart is expected to contract vigorously more than 2.5 billion times during your lifetime without once stopping to rest, forcing blood through vessels whose total length would encircle the globe twice. Since these vessels may become constricted, weakened, or clogged, the cardiovascular system is a prime candidate for malfunction.

Hypertension Strains the Heart

High blood pressure, also called **hypertension**, is caused by the constriction of arterioles, causing resistance to blood flow and strain on the heart. For most of the 50 million Americans with this condition, the cause is unknown. An approximate borderline reading for high blood pressure is 140/90. The strain on the heart caused by hypertension may cause it to increase in size, but its own blood supply may not increase proportionately. The heart muscle is then inadequately supplied with blood, especially during exercise. Lack of sufficient oxygen to the heart can cause chest pain called **angina**. High blood pressure also contributes to "hardening of the arteries," or *atherosclerosis*, described below. In conjunction with hardened arteries, hypertension can lead to the rupture of an artery and internal bleeding. The rupture of vessels supplying the brain causes a **stroke**, in which brain function in the area deprived of blood and of the vital oxygen and nutrients the blood carries is lost.

Mild hypertension is sometimes alleviated by weight reduction, exercise, stress management, and (for some individuals) reduction of dietary salt. For severe cases, doctors may prescribe drugs to reduce fluid in the body, reduce the heart rate, and expand the arteries and arterioles.

Atherosclerosis Obstructs Blood Vessels

Atherosclerosis (derived from the Greek *athero*, meaning "gruel" or "paste," and *scleros*, "hard") causes the walls of the large arteries to thicken and lose their elasticity. This is caused by deposits called **plaques** within the wall of the artery between the smooth muscle and the inner lining. Plaques are composed largely of *LDL-cholesterol*, which is cholesterol bound to a carrier molecule called *low-density lipoprotein* (LDL). If LDL cholesterol levels are too high, this "bad" form of cholesterol accumulates within arterial walls. (In contrast, cholesterol bound to *high-density lipoprotein* (HDL) is metabolized or excreted and hence is often called "good" cholesterol. Cholesterol is discussed in "Health Watch: Cholesterol—Friend and Foe" in Chapter 3.)

The accumulation of LDL-cholesterol triggers inflammation—usually a response to infection—and brings macrophages to the site. These ingest large quantities of cholesterol and accumulate within the vessel, enlarging the plaque. Smooth muscle cells from the arterial wall migrate and form a tough, fibrous cap over the enlarging plaque. Most heart attacks occur when the fibrous cap ruptures, exposing the blood to clot-promoting factors within the plaque. The clot further obstructs the artery and may completely block it (see Chapter 28 opener photo, inset), or it may be swept away to a narrower part of the artery where it lodges and prevents blood flow. Arterial clots are responsible for the most serious consequences of atherosclerosis: heart attacks and strokes.

About 1.1 million people suffer **heart attacks** in the U.S. each year, and over half a million people die from them. A heart attack occurs when one of the coronary arteries is blocked (coronary arteries supply the heart muscle itself; one with blockage is shown in Fig. E28-3). Deprived of nutrients and oxygen, the heart muscle supplied by the blocked artery rapidly and painfully dies. Although heart attacks are the major cause of death from atherosclerosis, this disease causes plaques and clots to form in arteries throughout the body. If a clot or plaque obstructs an artery that supplies the brain, it can cause a stroke, with results similar to those caused by a ruptured artery.

As with hypertension, the exact cause of atherosclerosis is unclear, but it is promoted by hypertension, cigarette smoking, genetic predisposition, obesity, diabetes, lack of exercise, and high blood cholesterol levels, particularly LDL-cholesterol. Traditional treatment of atherosclerosis includes the use of drugs or changes in diet and lifestyle to lower blood pressure and blood cholesterol levels. Plaques are sometimes squashed

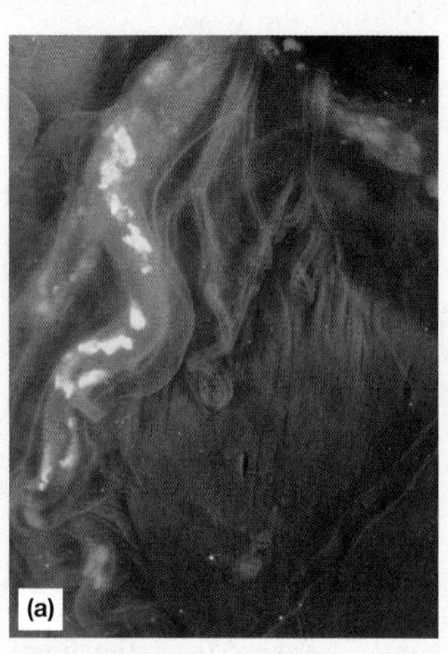

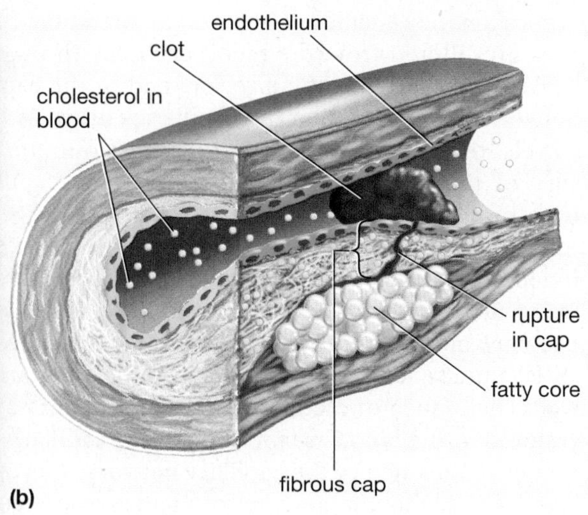

FIGURE E28-3 Plaques clog arteries
(a) In this remarkable photo of coronary arteries, plaques are seen in glowing yellow. **(b)** Plaque accumulates and stimulates a blood clot to form in an artery.

by inserting a tiny balloon, which is inflated to flatten the plaque, shaved off the artery wall with a miniature drill, or destroyed by a laser. After removing the plaque, a wire mesh tube or *stent* is often inserted into the artery to help keep it open. *Coronary bypass surgery* bypasses obstructed coronary arteries with segments of vein (usually obtained from the patient's leg) or artery (often from the patient's forearm). Researchers are working to perfect artificial vessels (see "Scientific Inquiry: Artificial Blood? Artificial Vessels? Artificial Hearts?") so the patient's own vessels can be spared. Patients who are not good candidates for coronary bypass operations can often benefit from *transmyocardial laser revascularization*, in which a laser beam is used to shoot 15 to 30 small (1 millimeter in diameter) channels through the ventricular walls. Blood clots form on the outside of the ventricle, keeping blood from leaking out. As the heart beats, blood from inside the ventricles flows in and out of the channels, supplying oxygen to the ventricular muscle and partially replacing the function of coronary arteries clogged by plaques.

If a heart attack occurs, rapid treatment can minimize the damage and significantly increase the victim's chances of survival. Blood clots in coronary arteries or in the brain are commonly dissolved by injecting substances that stimulate the production of an enzyme that breaks down fibrin, the protein that binds the clot together. Although heart disease is still the leading cause of death in the United States, steady progress in treatment has significantly reduced the rate of early deaths from atherosclerosis.

Note: Statistics in this essay are from the American Heart Association, "2002 Heart and Stroke Statistical Update." Some numbers are rounded from those provided.

equipped with one-way valves that allow blood to flow only toward the heart (Fig. 28-17). When you sit or stand for long periods, the lack of muscular activity allows blood to accumulate in the veins of the lower legs. This is why you may find your feet swollen after a long airplane flight. Long periods of inactivity can contribute to the formation of varicose veins, in which the valves become stretched and weakened and the veins become permanently swollen.

If blood pressure should fall—for instance, after extensive bleeding—veins can help restore it. In such cases, the sympathetic nervous system (which prepares the body for emergency action) automatically stimulates contraction of the smooth muscles in the vein walls. This action decreases the internal volume of the veins and raises blood pressure, speeding up the return of blood to the heart.

Arterioles Control the Distribution of Blood Flow

Muscular arteriole walls are under the influence of nerves, hormones, and chemicals produced by nearby tissues. Therefore, arterioles contract and relax in response to changing needs of the tissues and organs they supply. In a suspenseful novel, you might read, "The blood drained from her face as she beheld the gruesome sight"; the heroine is experiencing the constriction of the arterioles that supply her skin with blood. In threatening situations, the sympathetic nervous system stimulates the smooth muscle of the arterioles to contract. This contraction raises blood pressure overall, but selective constriction redirects blood to the heart and muscles, where it may be needed for vigorous action, and away from the skin, where it is less essential.

On a hot summer day, however, you become flushed as arterioles in the skin expand and bring more blood to the skin capillaries. Bringing this blood closer to the surface

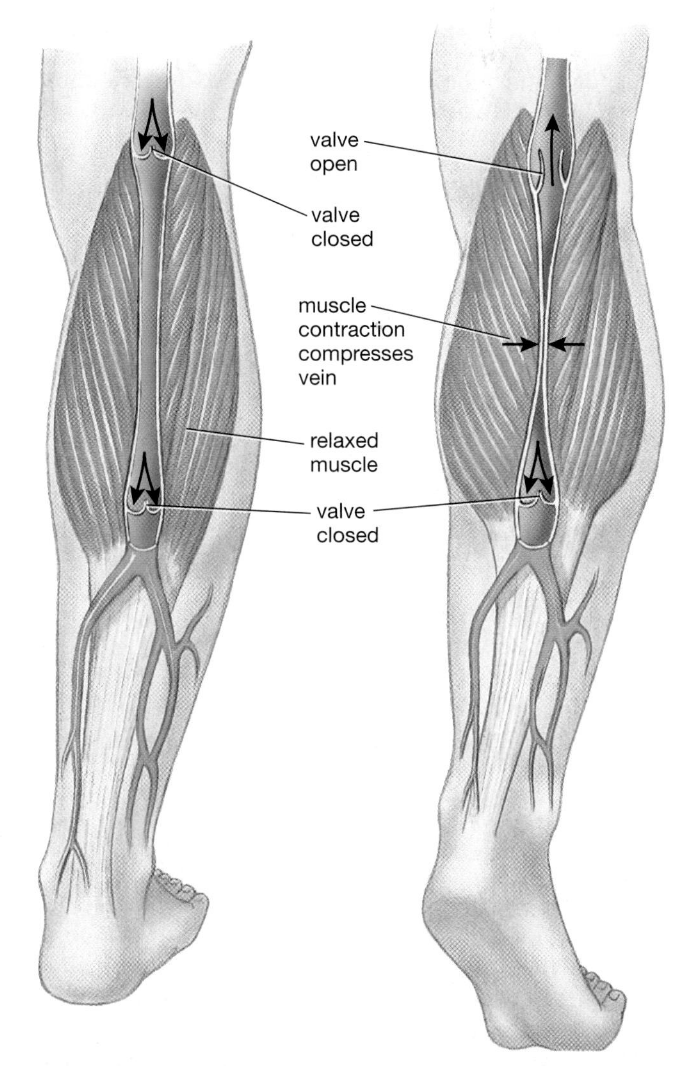

FIGURE 28-17 Valves direct blood flow in veins
Veins and venules have one-way valves that keep blood flowing in the proper direction. When the vein is compressed by nearby muscles, these valves allow blood to flow toward the heart, but clamp shut to prevent backflow.

enables the body to dissipate excess heat to the outside, thus maintaining a relatively constant internal temperature. In extremely cold weather, fingers and toes can become frostbitten because the arterioles that supply blood to the extremities constrict. Blood is shunted to vital organs, such as the heart and brain, which cannot function properly if their temperature drops. By minimizing blood flow to heat-radiating extremities, the body conserves heat.

The flow of blood in capillaries is regulated by tiny rings of smooth muscle (called **precapillary sphincters**), which surround the junctions between arterioles and capillaries (see Fig. 28-15). These open and close in response to local changes that signal the needs of nearby tissues. For example, the accumulation of carbon dioxide, lactic acid, or other cellular wastes signals the need for increased blood flow to the tissues. These signals cause the precapillary sphincters, as well as the muscles in the walls of nearby arterioles, to relax, allowing more blood to flow through the capillaries.

What happens when blood vessels rupture, are narrowed by deposits of cholesterol, or are blocked by clots, damming the "river of life"? We explore these questions in "Health Watch: Matters of the Heart."

28.5 How Does the Lymphatic System Work with the Circulatory System?

The **lymphatic system** consists of a network of lymph capillaries and larger vessels that empty into the circulatory system, numerous small *lymph nodes*, patches of lymphocyte-rich connective tissue (including the *tonsils*), and two additional organs: the *thymus* and the *spleen* (Fig. 28-18). Although not strictly part of the circulatory system, the lymphatic system is closely associated with it. The lymphatic system serves to:

- Return excess fluid and dissolved substances that leak from the capillaries to the bloodstream.
- Transport fats from the small intestine to the bloodstream.
- Defend the body by exposing bacteria and viruses to white blood cells.

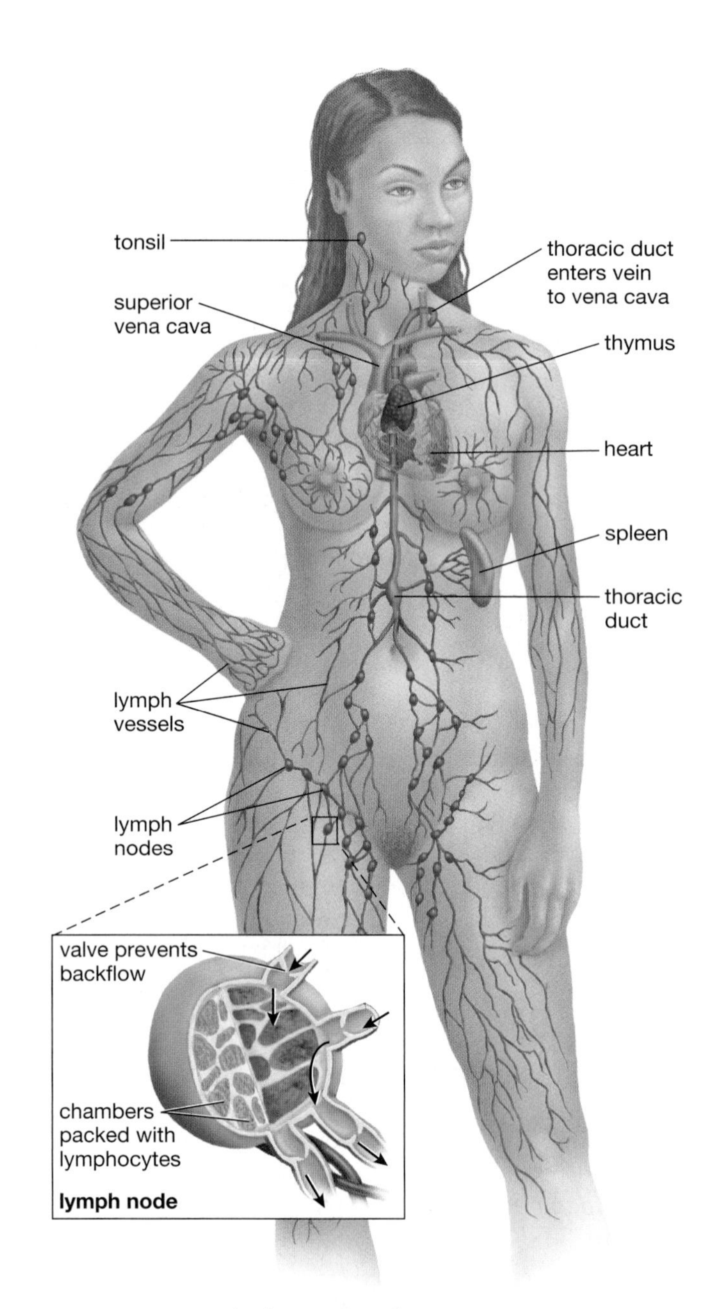

FIGURE 28-18 The human lymphatic system
Lymph vessels, lymph nodes, and two auxiliary lymph organs, the thymus and spleen. Lymph is returned to the circulatory system by way of the thoracic duct. (Inset) A cross section of a lymph node. The node is filled with channels lined with white blood cells that remove foreign matter from the lymph.

Lymphatic Vessels Resemble the Veins and Capillaries of the Circulatory System

Like blood capillaries, *lymph capillaries* form a complex network of microscopically narrow, thin-walled vessels into which substances move readily. Lymph capillary walls are composed of cells with openings between them that act as one-way valves. These openings allow relatively large particles, along with fluid, to be carried into the lymph capillary. Unlike blood capillaries, which form a continuous connected network, lymph capillaries "dead-end" in the body's tissues (Fig. 28-19). Materials collected by the lymph capillaries flow into larger lymph vessels. These lymph vessels have somewhat muscular walls, but, as in veins, most of the impetus for lymph flow comes from the contraction of nearby muscles, such as those used in breathing and walking. As in blood veins, the direction of flow is regulated by one-way valves (Fig. 28-20).

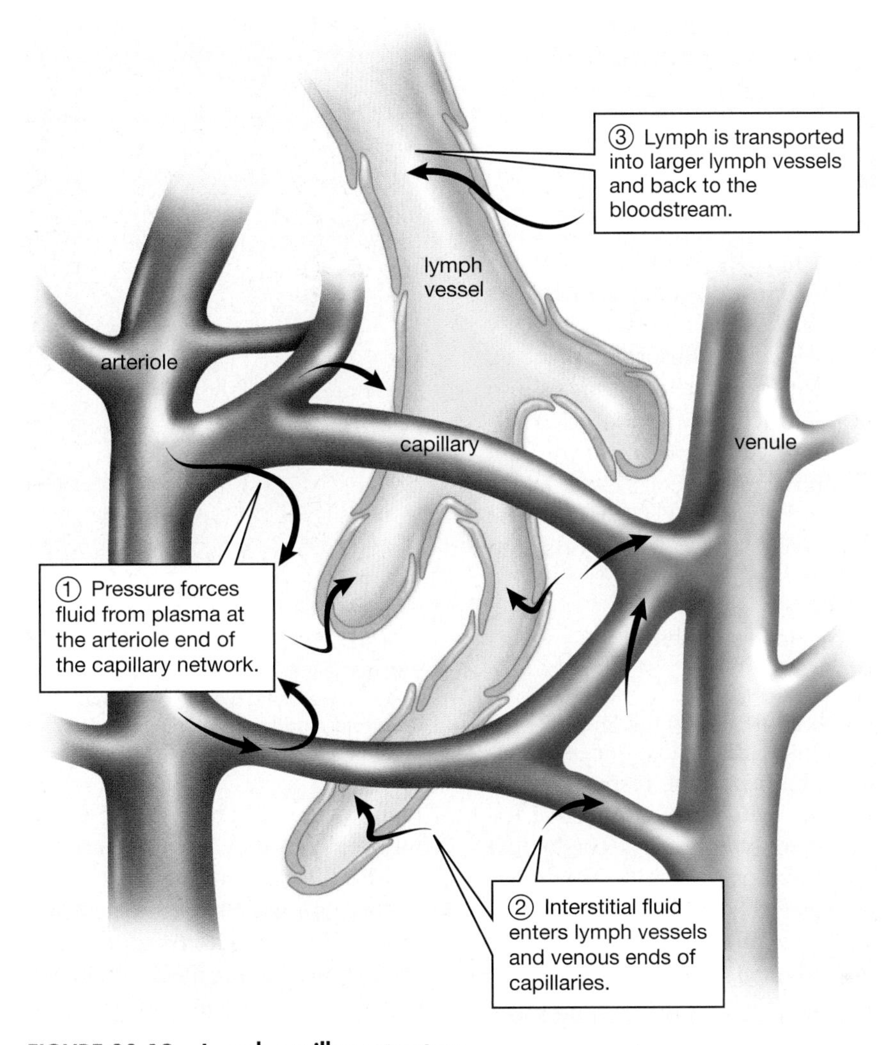

③ Lymph is transported into larger lymph vessels and back to the bloodstream.
lymph vessel
arteriole
capillary
venule
① Pressure forces fluid from plasma at the arteriole end of the capillary network.
② Interstitial fluid enters lymph vessels and venous ends of capillaries.

FIGURE 28-19 Lymph capillary structure
Lymph capillaries end blindly in the body tissues, where pressure from the accumulation of interstitial fluid forces the fluid into the lymph capillaries as well as into the venous side of the capillary network.

The Lymphatic System Returns Fluids to the Blood

As described earlier, dissolved substances are exchanged between the capillaries and body cells by means of interstitial fluid (pressure-filtered from blood plasma through capillary walls), which bathes nearly all the body's cells. In an average person, about 3 or 4 liters more fluid leaves the blood capillaries than is reabsorbed by them each day. One function of the lymphatic system is to return this excess fluid and its dissolved molecules to the blood. As interstitial fluid accumulates, its pressure forces the fluid through the openings between the cells in the lymph capillaries (see Fig. 28-19). The lymphatic system transports this fluid, now called **lymph**, back to the circulatory system. The importance of the lymphatic system in returning fluid to the bloodstream is illustrated by the condition

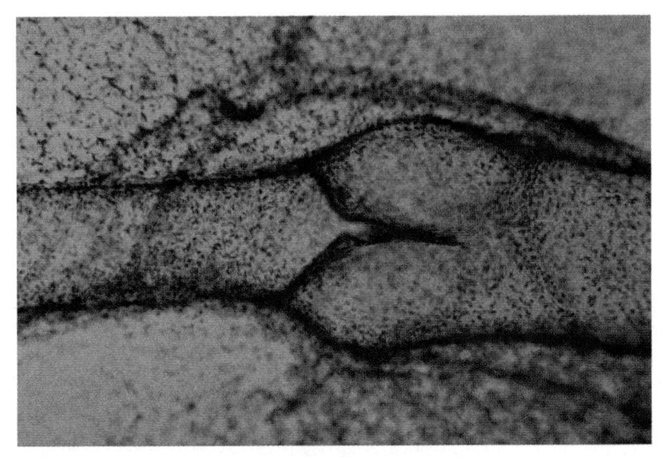

FIGURE 28-20 A valve in a lymph vessel
Like blood-carrying veins, lymph vessels have internal one-way valves that direct the flow of lymph toward the large veins into which they empty.

known as *elephantiasis* (Fig. 28-21). This disfiguring disorder is caused by a parasitic roundworm that colonizes lymphatic vessels, scarring them and preventing them from draining off excess fluid.

The Lymphatic System Transports Fats from the Small Intestine to the Blood

After a fatty meal, fat-transporting particles may make up 1% of the lymphatic fluid. How does this happen? As you will learn in Chapter 30, the small intestine is richly supplied with lymph capillaries. After absorbing digested fats, intestinal cells release fat-transporting particles into the interstitial fluid. These particles are too large to diffuse into blood capillaries, but can easily move through the openings between lymph capillary cells. Once in the lymph, they are dumped into veins that merge into the superior vena cava, a large vein that carries blood into the heart.

The Lymphatic System Helps Defend the Body Against Disease

In addition to its other roles, the lymphatic system also helps defend the body against foreign invaders, such as bacteria and viruses. In the linings of the respiratory, digestive, and urinary tracts are patches of connective tissue containing large numbers of lymphocytes. The largest of these patches are the **tonsils**, located in the cavity behind the mouth. The lymph vessels are interrupted periodically by kidney-bean-shaped structures about 1 inch (2.5 centimeters) long called **lymph nodes** (see Fig. 28-18). Lymph is forced through spaces within the lymph nodes; these spaces are lined with masses of macrophages. Lymphocytes are also produced in the lymph nodes. Both macrophages and lymphocytes recognize and destroy foreign particles, such as bacteria and viruses, and are killed in the process. The painful swelling of lymph nodes that accompanies certain diseases (mumps is an extreme example) is largely a result of the accumulation of dead

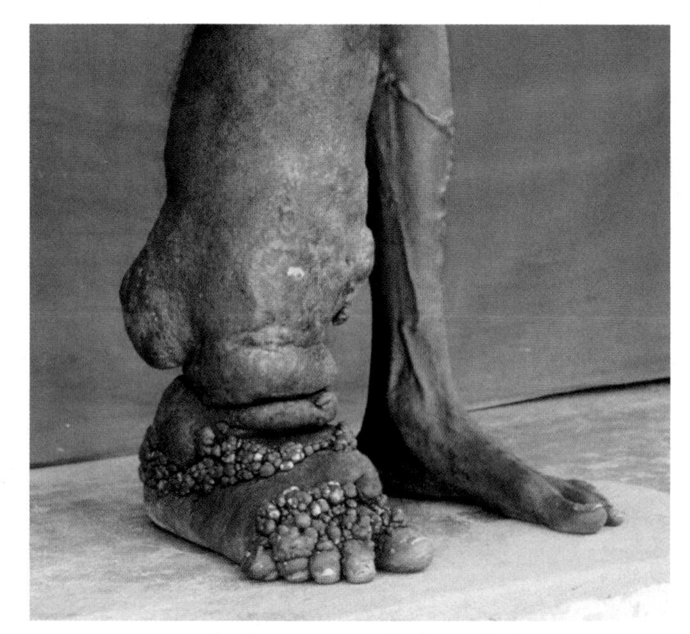

FIGURE 28-21 Elephantiasis results from blocked lymphatic vessels
When a parasitic worm scars lymph vessels, preventing fluid from returning to the bloodstream, the affected area can become massively swollen.

lymphocytes and macrophages, as well as the dead virus-infested cells they have engulfed.

The thymus and the spleen are often considered part of the lymphatic system (see Fig. 28-18). The **thymus**, an organ that produces lymphocytes, is located beneath the breastbone slightly above the heart. The thymus is particularly active in infants and young children, but decreases in size and importance in early adulthood. The **spleen**, another lymphocyte-producing organ, is located on the left side of the abdominal cavity, between the stomach and diaphragm. Just as the lymph nodes filter lymph, the spleen filters blood, exposing it to macrophages and lymphocytes that destroy foreign particles and aged red blood cells.

CASE STUDY REVISITED Sudden Death

Darryl Kile may have been entirely unaware of the insidious progress of atherosclerosis that threatened and suddenly claimed his life. Although extremely narrowed coronary arteries can cause chest pain (angina) and warn of impending trouble, only about 17% of heart attacks are caused by the plaque buildup itself. Most come without warning when the fibrous cap that contains the plaque within the artery wall breaks open and stimulates the formation of a blood clot within the artery (see inset in the chapter opener photo). Such clots can break off suddenly and be carried to a narrower part of the artery where they completely block blood flow. In the United States, about 3000 young people between the ages of 15 and 34 die of heart attacks each year. In some individuals, genetic factors contribute to exceptionally high levels of LDL cholesterol, hypertension, and atherosclerosis. But with lifestyle adjustments and the help of drugs that lower blood pressure, reduce LDL cholesterol, and decrease inflammation, even people with special risk factors may lead long, active lives and avoid the tragedy that befell Darryl Kile—and his family, friends, and teammates.

Consider This: Do risk factors such as high cholesterol (total and LDL), hypertension, or angina run in your family? Has anyone in your immediate family died prematurely of heart disease? It is never too early to find out! Even in the absence of a family history, a complete physical exam that includes a blood lipid evaluation and an ECG (see "Links to Life: Charting Heartbeats") can help you evaluate your own risks and take action to help safeguard your cardiovascular health.

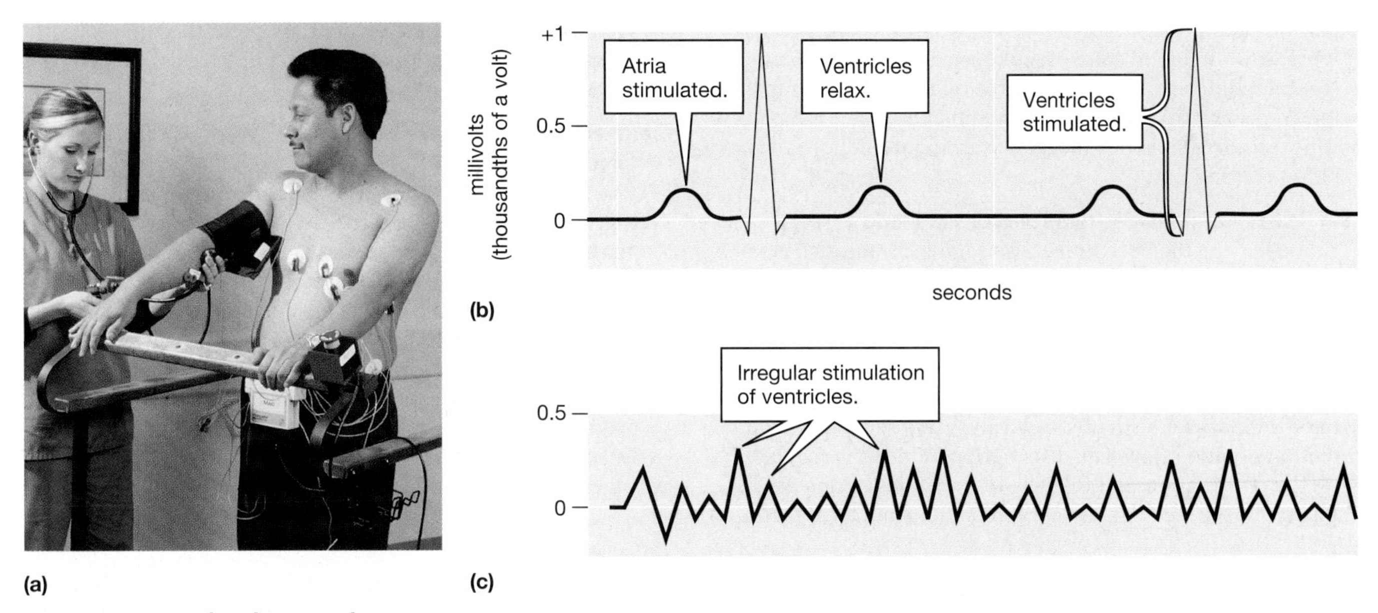

FIGURE E28-4 The electrocardiogram
(a) A patient is wired for an ECG. **(b)** A normal ECG readout. **(c)** Ventricular fibrillation, which is fatal if not reversed.
QUESTION In the normal ECG readout, what causes the sharp peaks? (Refer to the description of control of heart rhythm.)

Links to Life: Charting Heartbeats

If you've ever had a complete physical exam, it probably included an ECG (electrocardiogram), a simple yet powerful tool used to detect signs of heart malfunction. The electrical activity that drives the cardiac cycle (see Fig. E28-4b) is strong enough to be measured by electrodes placed on the skin. Normal ECGs record characteristic waves of electrical activity that give rise to contractions of specific heart chambers. Figure E28-4 shows the placement of electrodes on the body, an example of a healthy ECG, and ventricular fibrillation, in which ventricles quiver and are unable to pump blood. During your physical exam, your ECG will be read by a trained clinician who can detect subtle irregularities in wave shape, relative wave sizes, or the intervals between waves that may reflect damage to the cardiac muscle or the pacemaker.

Summary of Key Concepts

28.1 What Are the Major Features and Functions of Circulatory Systems?

Circulatory systems transport blood rich in dissolved nutrients and oxygen close to each cell, where nutrients can be released and wastes absorbed by diffusion. All circulatory systems have three major parts: blood, a fluid; vessels, a system of channels to conduct the blood; and a heart, a pump to circulate the blood. Invertebrates have open or closed circulatory systems; nearly all vertebrates have closed systems. In open systems, blood is pumped by a heart into a hemocoel, where it directly bathes internal organs. In closed systems, the blood is confined to the heart and blood vessels.

28.2 How Does the Vertebrate Heart Work?

Vertebrate circulatory systems transport gases, hormones, and wastes; distribute nutrients; help regulate body temperature; and defend the body against disease.

The vertebrate heart evolved from two chambers in fishes, to three in amphibians and most reptiles, to four in birds and mammals. In the four-chambered heart, blood is pumped separately to the lungs and through the body, maintaining complete separation of oxygenated and deoxygenated blood. Deoxygenated blood is collected from the body in the right atrium and passed to the right ventricle, which pumps it to the lungs. Oxygenated blood from the lungs enters the left atrium, passes to the left ventricle, and is pumped to the rest of the body.

The cardiac cycle consists of two stages: (1) atrial contraction, followed by (2) ventricular contraction. The direction of blood flow is maintained by valves within the heart. The contractions of the heart are initiated and coordinated by the sinoatrial node, the heart's pacemaker. Heart rate can be modified by the nervous system and by hormones such as epinephrine.

28.3 What Is Blood?

Blood is composed of both fluid and cellular materials. The fluid plasma consists of water that contains proteins, hormones, nutrients, gases, and wastes. Red blood cells, or erythrocytes, are packed with iron-containing protein called hemoglobin, which carries oxygen. Their numbers are regulated by the hormone erythropoietin. Proteins on their plasma membranes determine blood type. There are five types of white blood cells, or leukocytes, that fight infection (for more information, see the table on this book's companion Web site, http://www.prenhall.com/audesirk7). Platelets, which are fragments of megakaryocytes, are important for blood clotting.

28.4 What Are the Types and Functions of Blood Vessels?
Blood leaving the heart travels (in sequence) through arteries, arterioles, capillaries, venules, veins, and then back to the heart. Each vessel is specialized for its role. Elastic, muscular arteries help pump the blood. The thin-walled capillaries allow the exchange of materials between the body cells and the blood. Veins provide a path of low resistance back to the heart, with one-way valves that maintain the direction of blood flow. The distribution of blood is regulated by the constriction and dilation of arterioles under the influence of the sympathetic nervous system and local factors such as the amount of carbon dioxide in the tissues. Local factors also regulate precapillary sphincters, which control blood flow to the capillaries.

28.5 How Does the Lymphatic System Work with the Circulatory System?
The human lymphatic system consists of lymphatic vessels, tonsils, lymph nodes, and the thymus and spleen. The lymphatic system removes excess interstitial fluid that leaks through blood capillary walls. It transports fats to the bloodstream from the small intestine and fights infection by filtering the lymph through lymph nodes, where white blood cells ingest foreign invaders, such as viruses and bacteria. The thymus, which is most active in young children, produces lymphocytes that function in immunity. The spleen filters blood past macrophages and lymphocytes, which remove bacteria and damaged blood cells.

Key Terms

angina *p. 562*
arteriole *p. 560*
artery *p. 552*
atherosclerosis *p. 562*
atrioventricular (AV) node *p. 554*
atrioventricular valve *p. 553*
atrium *p. 551*
blood *p. 550*
blood clotting *p. 558*
blood vessel *p. 550*
capillary *p. 560*
cardiac cycle *p. 552*
closed circulatory system *p. 550*
erythrocyte *p. 555*
erythropoietin *p. 556*
fibrin *p. 558*
fibrinogen *p. 558*
heart *p. 550*
heart attack *p. 562*
hemocoel *p. 550*
hemoglobin *p. 555*
hypertension *p. 562*
interstitial fluid *p. 560*
leukocyte *p. 558*
lymph *p. 565*
lymphatic system *p. 564*
lymph node *p. 566*
lymphocyte *p. 558*
macrophage *p. 558*
open circulatory system *p. 550*
pacemaker *p. 553*
plaque *p. 562*
plasma *p. 555*
platelet *p. 558*
precapillary sphincter *p. 564*
Rh factor *p. 558*
semilunar valve *p. 553*
sinoatrial (SA) node *p. 553*
spleen *p. 566*
stroke *p. 562*
thrombin *p. 558*
thymus *p. 566*
tonsil *p. 566*
vein *p. 552*
ventricle *p. 551*
venule *p. 561*

Thinking Through the Concepts

To take a multiple-choice quiz with feedback on the contents of this chapter, visit http://www.prenhall.com/audesirk7. *Log in to the Web site selected by your instructor and navigate to the Self Test section for this chapter.*

? Review Questions

1. Trace the flow of blood through the circulatory system, starting and ending with the right atrium.
2. List three types of blood cells and describe their principal functions.
3. What are five functions of the vertebrate circulatory system?
4. In what way do veins and lymph vessels resemble one another? Describe how fluid is transported in each of these vessels.
5. Describe three important functions of the lymphatic system.
6. Distinguish among plasma, interstitial fluid, and lymph.
7. Describe veins, capillaries, and arteries, noting their similarities and differences.
8. Trace the evolution of the vertebrate heart from two chambers to four chambers.
9. Explain in detail what causes the vertebrate heart to beat.
10. Describe the cardiac cycle, and relate the contractions of the atria and ventricles to the two readings taken during the measurement of blood pressure.
11. Describe how the number of red blood cells is regulated by a negative feedback system.
12. Describe the formation of an atherosclerotic plaque. What are the risks associated with atherosclerosis?

Applying the Concepts

1. Discuss the steps you can take now and in the future to reduce your risks of developing heart disease.
2. Discuss why a four-chambered heart is much more efficient than a two-chambered heart in delivering oxygenated blood to the various body parts. What evolutionary changes in the lifestyles of organisms selected for the evolution of the four-chambered heart?
3. Discuss the ethical, medical, and technological implications of the use of "bridging devices" such as the artificial heart.
4. Considering the prevalence of cardiovascular disease and the high, increasing costs of treating it, certain treatments may not be available to all who might benefit from them. What factors would you take into account if you had to ration cardiovascular procedures, such as heart transplants?
5. Joe, a 45-year-old executive of a major corporation, has been diagnosed with mild hypertension. What treatments or lifestyle changes might Joe's physician recommend? If Joe's hypertension becomes more severe, what treatments might Joe's physician use? Should Joe be concerned about mild hypertension? Explain your answer.

For More Information

Ditlea, S. "The Trials of an Artificial Heart." *Scientific American*, July 2002. Describes the workings and clinical trials of artificial hearts and other bridging devices, with a discussion of ethical considerations.

Gibbons, R., and associates. "Waiting for Organ Transplantation." *Science*, January 14, 2000. Statistics about the need, availability, and waiting time for organ transplantation suggest that new sources of organs are needed.

Jain, R. K., and Carmeliet, P. F. "Vessels of Death or Life." *Scientific American*, December 2001. By learning to manipulate angiogenesis (the formation of blood vessels), researchers may find ways to thwart tumors or bring more blood to the heart.

Libby, A. "Atherosclerosis: The New View." *Scientific American*, May 2002. Beautifully illustrated description of plaque formation with an emphasis on the role of inflammation in heart disease.

Radetsky, P. "The Mother of All Blood Cells." *Discover*, March 1995. Discusses Irving Weissman's discovery of a "stem cell" that gives rise to all other types of blood cells.

Seppa, N. "Secondary Smoke Carries High Price." *Science News*, January 17, 1998. New research suggests that smoking causes atherosclerosis, that the damage persists after a smoker quits, and that exposure to secondhand smoke significantly increases the buildup of plaque in the carotid artery of nonsmokers.

Wang, L. "Blood Relatives." *Science News*, March 31, 2001. Describes a variety of approaches and problems in making artificial blood, and the most likely prospects for the near future.

Wu, C. "Engineered Blood Vessel Is Only Human." *Science News*, January 17, 1998. Researchers have constructed an experimental blood vessel composed entirely of human tissue.

Media Activities

To access a Media Activity visit http://www.prenhall.com/audesirk7. *Log in to the Web site selected by your instructor, navigate to this chapter, and select the appropriate Media Activity number.*

28.1 Circulatory Systems

Estimated time: 10 minutes

Explore the structure, function, and evolution of circulatory systems in animals.

28.2 Heart Structure and Function

Estimated time: 5 minutes

Explore the various structures and function of the vertebrate heart.

28.3 The Cardiac Cycle

Estimated time: 5 minutes

View a brief animation of the steps of the cardiac cycle.

28.4 Blood Pressure

Estimated time: 10 minutes

See what blood pressure is, learn how to measure it yourself, and explore the effects of various chemicals on blood pressure.

28.5 Web Investigation: Sudden Death

Estimated time: 20 minutes

In this investigation, you will explore the nature of Coronary Artery Disease (CAD) and uncover some of the risk factors that may have contributed to Darryl Kile's death.

CHAPTER

29 Respiration

These students say they will quit smoking later. Only one in five will succeed.

AT A GLANCE

CASE STUDY Lives Up in Smoke

College students cluster just outside the building doors, lighting up between classes. Coughs punctuate their conversation. Like most adult smokers, many of these students picked up the habit in high school. "I smoke because I like smoking, and when I don't want to anymore, that's when I'll quit," says a 17-year-old high school student. Another has watched her relatives become addicted to cigarettes but smokes because of peer pressure and to relieve stress. "It scares me because . . . a lot of people get cancer. I don't want to die early," she says, lighting up and inhaling deeply. Every day in the United States, roughly 3000 teenagers light up their first cigarette; for many, this is the beginning a life-long struggle with addiction. Surveys cited by the American Lung Association indicate that about 35% of high school seniors are smokers, and more than 20% smoke daily. Encouragingly, these figures have dropped in recent years, following a steady and alarming increase during the 1990s. Statistically, young smokers will suffer more frequent and severe respiratory infections, and smoking may retard the growth of their lungs. Most of these students are aware of the dangers but say they will stop "when the time comes." Are these new smokers likely to quit? What are their chances of dying from smoking? How does smoking interfere with the respiratory system from the bronchi to the bloodstream? You can find answers to these questions at the end of the chapter and in "Health Watch: Smoking—A Life and Breath Decision."

29.1 Why Exchange Gases?

Late again! Sprinting up two flights of stairs to your classroom, you feel your calves "burning." Remembering Chapter 8, you think "Ah ha! That's lactic acid building up—my muscle cells are fermenting glucose because they can't get enough oxygen for cellular respiration." As you slip into your seat, quietly panting and feeling your heart pounding, the discomfort eases. Less exertion coupled with rapid breathing ensures that adequate oxygen is now available; the lactic acid is being reconverted to pyruvate and then broken down into carbon dioxide (CO_2) and water, while providing additional energy.

You are experiencing firsthand the relationship between cellular respiration and the act of breathing, which is also called *respiration*. Each cell in your body (and in the bodies of all other organisms) must continuously expend energy to maintain itself. When you call on your muscles to carry you upstairs quickly, the demands are extreme. As cellular respiration converts the energy in nutrients (such as sugar) into ATP that can be used by body cells, the process requires a steady supply of oxygen and generates carbon dioxide as a waste product. The rapid beating of your heart as you relax after your sprint upstairs reminds you that the circulatory system works in close harmony with the respiratory system. It extracts oxygen from air in your lungs, carries it within diffusing distance of each cell, then picks up carbon dioxide for release in the lungs.

How does breathing help support cellular respiration? What does the inside of a lung look like, and how is it adapted for gas exchange? Why aren't our lungs outside our bodies, where they would be directly exposed to the air? How do aquatic animals respire? In this chapter, we will explore the specialized structures of respiratory systems and how they work.

29.2 What Are Some Evolutionary Adaptations for Gas Exchange?

Gas exchange in all organisms ultimately relies on diffusion. Cellular respiration depletes O_2 and increases CO_2 levels, creating concentration gradients that favor the diffusion of carbon dioxide out of cells and the diffusion of oxygen into them. Although animal respiratory systems are amazingly diverse, they all share three features that facilitate diffusion: First, the respiratory surface must remain moist, because gases must be dissolved in water when they diffuse into or out of cells. Second, cells lining respiratory surfaces are very thin, a feature that facilitates diffusion of gases through them. Third, the respiratory system must have a large surface area in contact with the environment to allow adequate gas exchange.

In the following sections we will examine some diverse respiratory systems in animals, each shaped by the environment in which it evolved. Notice how each type of system meets the demand for a large, moist surface that allows gas exchange.

(a) Flatworm

(b) Jellyfish

(c) Sponge

FIGURE 29-1 Some animals lack specialized respiratory structures
Most animals that lack a respiratory system have low metabolic demands and a large, moist body surface. **(a)** The flattened body of this marine flatworm exchanges gases with the water. **(b)** The cells in the bell-shaped body of a jellyfish have a low metabolic rate, and seawater flowing in and out of the bell during swimming allows adequate gas exchange. **(c)** Flagellated cells draw currents of water through numerous openings in the body of the sponge. These currents carry microscopic food particles and allow the cells to exchange gases with the water.

Some Animals in Moist Environments Lack Specialized Respiratory Structures

Some animals that live in moist environments are able to exchange gases without specialized respiratory structures. The outside of their bodies provides an adequate surface area for the diffusion of gases. If the body is extremely small and elongated, as in microscopic roundworms, gases need to diffuse only a short distance to reach all cells of the body. Alternatively, an animal's body may be thin and flattened, producing a large surface area for diffusion. In flatworms, most cells are close to the moist skin through which gases can diffuse (Fig. 29-1a).

If energy demands are sufficiently low, the relatively slow rate of gas exchange by diffusion may suffice even for a larger, thicker-bodied organism. For example, jellyfish can be quite large, but cells that are far from the surface are relatively inert and require little oxygen (Fig. 29-1b).

Another adaptation for gas exchange involves bringing the watery environment close to all body cells, allowing direct exchange of gases between the body cells and water. Sponges, for example, circulate seawater through channels within their bodies, enabling it to come close to all of their cells (Fig. 29-1c).

Some animals combine a large skin surface—through which diffusion occurs—with a well-developed circulatory system. For example, in the earthworm, gases diffuse through the moist skin and are distributed throughout the body by an efficient circulatory system (see Fig. 28-1b). Blood in skin capillaries rapidly carries off oxygen that has diffused through the skin, maintaining a concentration gradient that favors the inward diffusion of oxygen. The worm's elongated shape ensures a large skin surface relative to its internal volume. This system is quite effective, since a worm's sluggish metabolism demands relatively little oxygen. To remain effective as a gas-exchange organ, the skin must stay moist; a dry earthworm will suffocate.

Respiratory Systems Facilitate Gas Exchange by Diffusion

Most animals have evolved specialized respiratory systems that interface closely with their circulatory systems to exchange gases between the cells and the environment. The transfer of gases from the environment to the blood, and then to the cells and back again, usually occurs in stages that alternate bulk flow with diffusion. During **bulk flow**, fluids or gases move in bulk through relatively large spaces, from areas of higher pressure to areas of lower pressure. Bulk flow contrasts with diffusion, in which molecules move individually from areas of higher concentration to areas of lower concentration (see Chapter 4). In general, gas exchange in respiratory systems occurs in the following stages, illustrated for the mammalian system in Figure 29-2:

1. Air or water, relatively high in oxygen and low in carbon dioxide, is moved past a respiratory surface by bulk flow; this is usually facilitated by muscular movements.
2. Oxygen and carbon dioxide are exchanged through the respiratory surface by diffusion; oxygen diffuses into the capillaries of the circulatory system and carbon dioxide diffuses out, both along their concentration gradients.
3. Gases are transported between the respiratory system and the tissues by the bulk flow of blood as it is pumped throughout the body by the heart.
4. Gases are exchanged between the tissues and the circulatory system by diffusion. At the tissues, oxygen diffuses out of the capillaries and carbon dioxide diffuses into them along their concentration gradients.

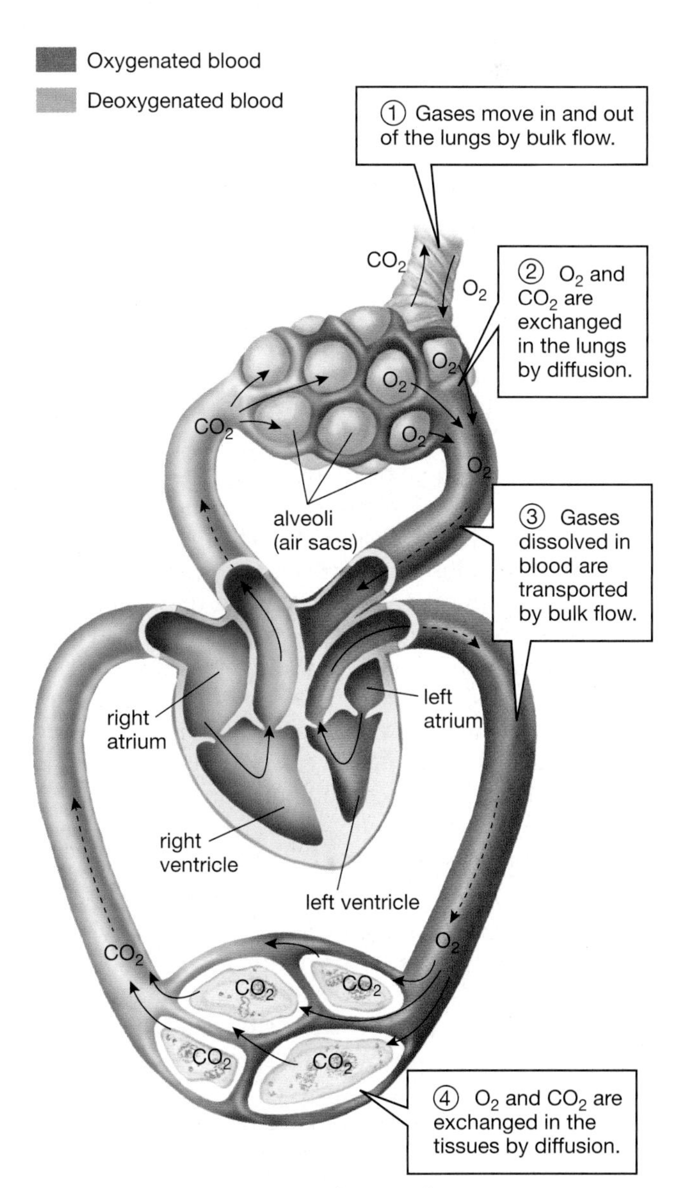

Oxygenated blood
Deoxygenated blood
① Gases move in and out of the lungs by bulk flow.
CO_2
O_2
② O_2 and CO_2 are exchanged in the lungs by diffusion.
alveoli (air sacs)
③ Gases dissolved in blood are transported by bulk flow.
right atrium
left atrium
right ventricle
left ventricle
④ O_2 and CO_2 are exchanged in the tissues by diffusion.

FIGURE 29-2 An overview of gas exchange

Media Activity 29.1 Gas Exchange

Gills Facilitate Gas Exchange in Aquatic Environments

Gills are the respiratory structures of many aquatic animals. The simplest type of gill, found in certain mollusks and amphibians, consists of numerous projections of the body surface into the surrounding water. In general, gills are elaborately branched or folded to maximize their surface area. In some animals, gill size is determined by the availability of oxygen in the surrounding water. For example, salamanders living in stagnant water (which has little opportunity to mix with air) have larger gills than do those living in well-aerated water. Gills have a dense profusion of capillaries just beneath their delicate outer membranes. These capillaries bring blood close to the surface, where gas exchange occurs.

The surface of a fish's body protects the delicate gill membranes from nibbling predators. Fish create a continuous current over their gills by pumping water into their mouths and ejecting it through openings just behind the gills (Fig. 29-3). They can increase the flow of water by swimming with their mouths open; some fast swimmers, such as mackerel, tuna, and a few (but not all) sharks, rely heavily on swimming to ventilate their gills. Gills are useless out of water because they collapse and dry out in air. Therefore, before animals could inhabit dry land, they needed to evolve respiratory organs with both support and protection from drying.

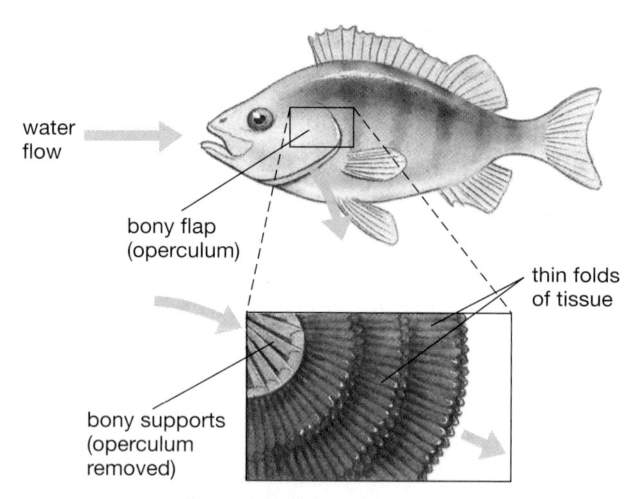

FIGURE 29-3 Gills exchange gases with water
The gill reaches its greatest complexity in the fish, where it is made of thin folds of tissue and is protected within the body, here under a bony flap. A one-way flow of water is maintained over the gill by pumping water through the mouth and out the opening beneath the flap.

Terrestrial Animals Have Internal Respiratory Structures

Terrestrial (land-dwelling) animals live in air, which has a far higher oxygen concentration than does water. However, extracting oxygen from dry air presents special challenges. All respiratory surfaces need to remain moist, because gases must be dissolved in water to diffuse across membranes. But if moist respiratory surfaces were on the outside of the body, they would lose water continuously by evaporation and would tend to dry out. Thus, land animals have evolved structures in which respiratory surfaces are moistened, supported, and protected from drying. Natural selection has produced a variety of terrestrial respiratory structures, including tracheae in insects and lungs in vertebrates.

Insects Respire by Means of Tracheae

Insects use a system of elaborately branching internal tubes called **tracheae**, which convey air directly to the body cells. Reinforced with chitin (which also supports the insect's external skeleton), tracheae subdivide and branch into smaller channels (*tracheoles*) that penetrate the body tissues and allow gas exchange (Fig. 29-4). Each body cell is close to a tracheole, minimizing diffusion distances. Air enters the tracheae through a series of openings called **spiracles**, located along each side of the abdomen. Spiracles have valves that allow them to be opened or closed. Some large insects use muscular pumping movements of the abdomen to enhance air movement through the tracheae.

Most Terrestrial Vertebrates Respire by Means of Lungs

Lungs are chambers containing moist respiratory surfaces that are protected within the body, where water loss is minimized and the body wall provides support. The first vertebrate lung probably appeared in a freshwater fish and consisted of an outpocketing of the digestive tract. Gas exchange in this simple lung helped the fish survive in stagnant water, in which oxygen is scarce. Amphibians, which straddle the boundary between aquatic and terrestrial life,

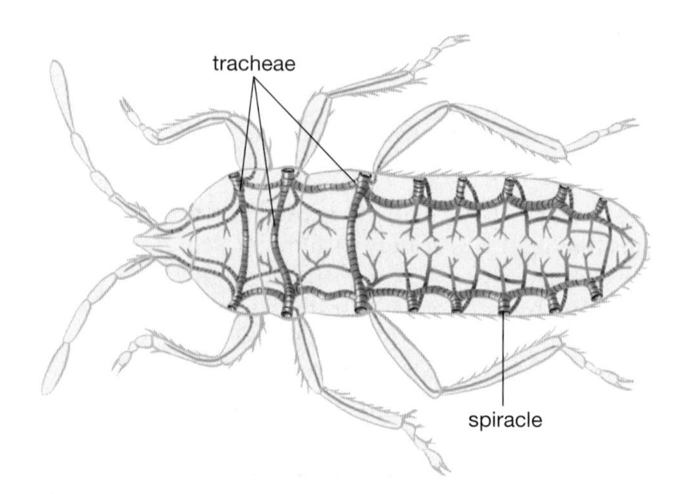

FIGURE 29-4 Insects breathe via tracheae
The tracheae of insects, such as this beetle, branch intricately throughout the body and open to the air through spiracles in the abdominal wall. **QUESTION** Why are the respiratory organs of an insect (its tracheae) so numerous compared to the respiratory organs of a mammal or bird (its lungs)?

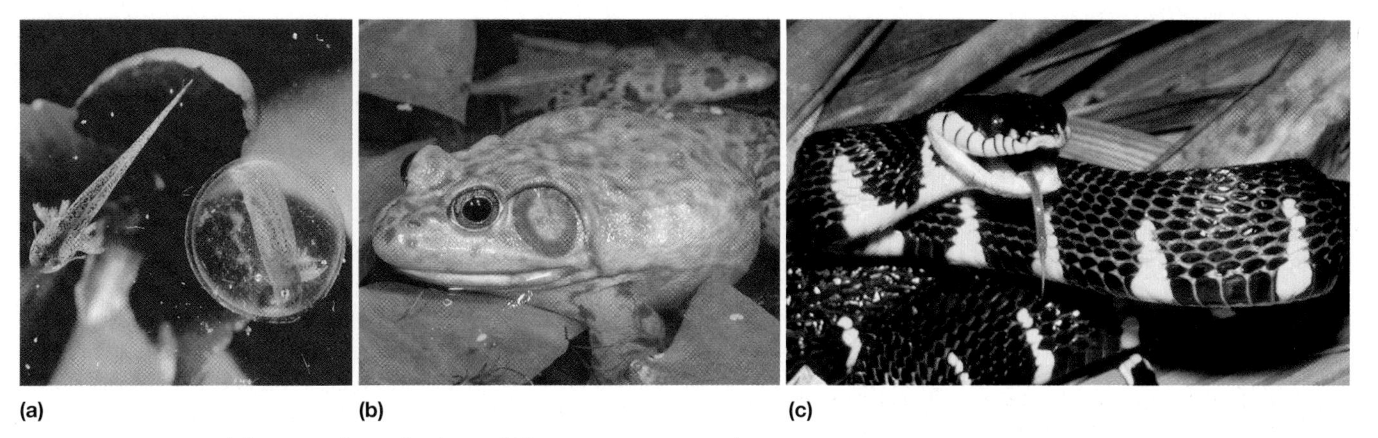

(a) (b) (c)

FIGURE 29-5 Amphibians and reptiles have different respiratory adaptations
(a) The bullfrog, an amphibian, begins life as a fully aquatic tadpole with feathery external gills that will later become enclosed in a protective chamber. **(b)** During metamorphosis into an air-breathing adult frog, the gills are lost and replaced by simple saclike lungs. In both tadpole and adult, gas exchange also occurs by diffusion through the skin, which must be kept moist to function as a respiratory surface. **(c)** The fully terrestrial reptile, such as this mangrove snake, is covered with dry scales that restrict gas exchange through the skin. Reptilian lungs are more efficient than are those of amphibians. **QUESTION** How do the respiratory adaptations of amphibians influence the range of habitats in which amphibians are found?

may use gills in the larval stage and lungs in their more terrestrial adult form. For example, the purely aquatic tadpole exchanges its gills for lungs as it develops into a more terrestrial frog (Fig. 29-5a,b). Frogs and salamanders also use their moist skin as a supplemental respiratory surface.

The scales of reptiles (Fig. 29-5c) reduce the loss of water through the skin and allow reptiles to survive in dry environments. But scales also reduce the diffusion of gases through the skin, so the lungs of reptiles are better developed than those of amphibians.

Birds and mammals are exclusively lung breathers. The bird lung has evolved special adaptations that allow extremely efficient gas exchange. These adaptations support the enormous energy demands of flight, sometimes at thousands of feet up, where oxygen is scarce. As a bird breathes in, it draws air through its lungs, where oxygen is extracted, and simultaneously pulls air into air sacs, some of which are located beyond the lungs (Fig. 29-6a). As the bird breathes out, oxygenated air from the air sacs is forced back through the lungs, allowing the bird to extract oxygen even as it exhales. In contrast to other vertebrate lungs, which are saclike, bird lungs are filled with hollow, thin-walled tubes called *parabronchi*, which allow air to pass through them in both directions (Fig. 29-6b).

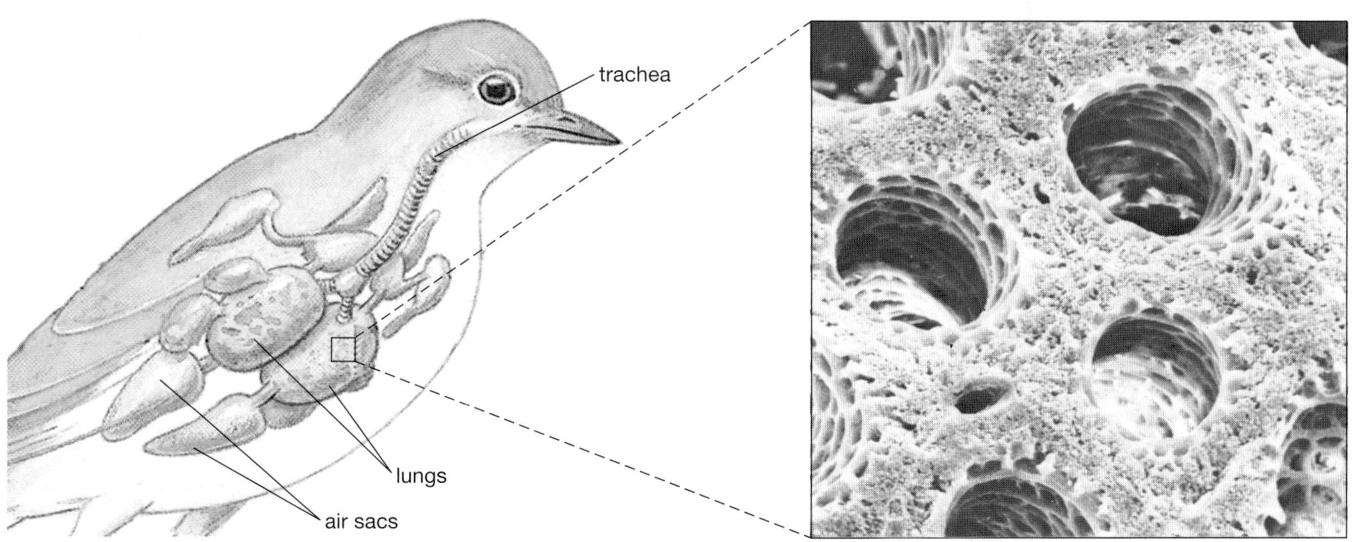

FIGURE 29-6 The bird respiratory system is extremely efficient
(a) Birds have air sacs in addition to their lungs that allow more efficient gas exchange. **(b)** In birds, tubular gas-exchange organs called *parabronchi* allow air to flow entirely through the lung to and from the air sacs.

29.3 How Does the Human Respiratory System Work?

The respiratory system in humans and other lung-breathing vertebrates can be divided into two parts: the **conducting portion** and the **gas-exchange portion**. The conducting portion consists of a series of passageways that carry air into and out of the gas-exchange portion, where gases are exchanged with the blood in tiny sacs in the lungs.

The Conducting Portion of the Respiratory System Carries Air to the Lungs

The conducting portion brings air to the lungs and also contains the apparatus that makes speaking possible. Air enters through the nose or the mouth, passes through the nasal cavity or oral cavity into a common chamber, the **pharynx**, and then travels through the **larynx,** or "voice box" (Fig. 29-7). The opening to the larynx is guarded by the *epiglottis*, a flap of tissue supported by cartilage. During normal breathing, the epiglottis is tilted upward, as shown in Figure 29-7, allowing air to flow freely into the larynx. During swallowing, the epiglottis tilts downward and covers the larynx, directing substances into the esophagus instead. If an individual attempts to inhale and swallow at the same time, this reflex may fail and food can become lodged in the larynx, blocking air from entering the lungs. What should you do if you see this happen? The *Heimlich maneuver* (Fig. 29-8) is easy to perform and has saved countless lives.

Within the larynx are the **vocal cords**, bands of elastic tissue controlled by muscles. Muscular contractions can cause the vocal cords to partially obstruct the opening within the larynx. Exhaled air causes them to vibrate, producing the tones of speech or song. Stretching the cords changes the pitch of the tones, which can be articulated into words by movements of the tongue and lips.

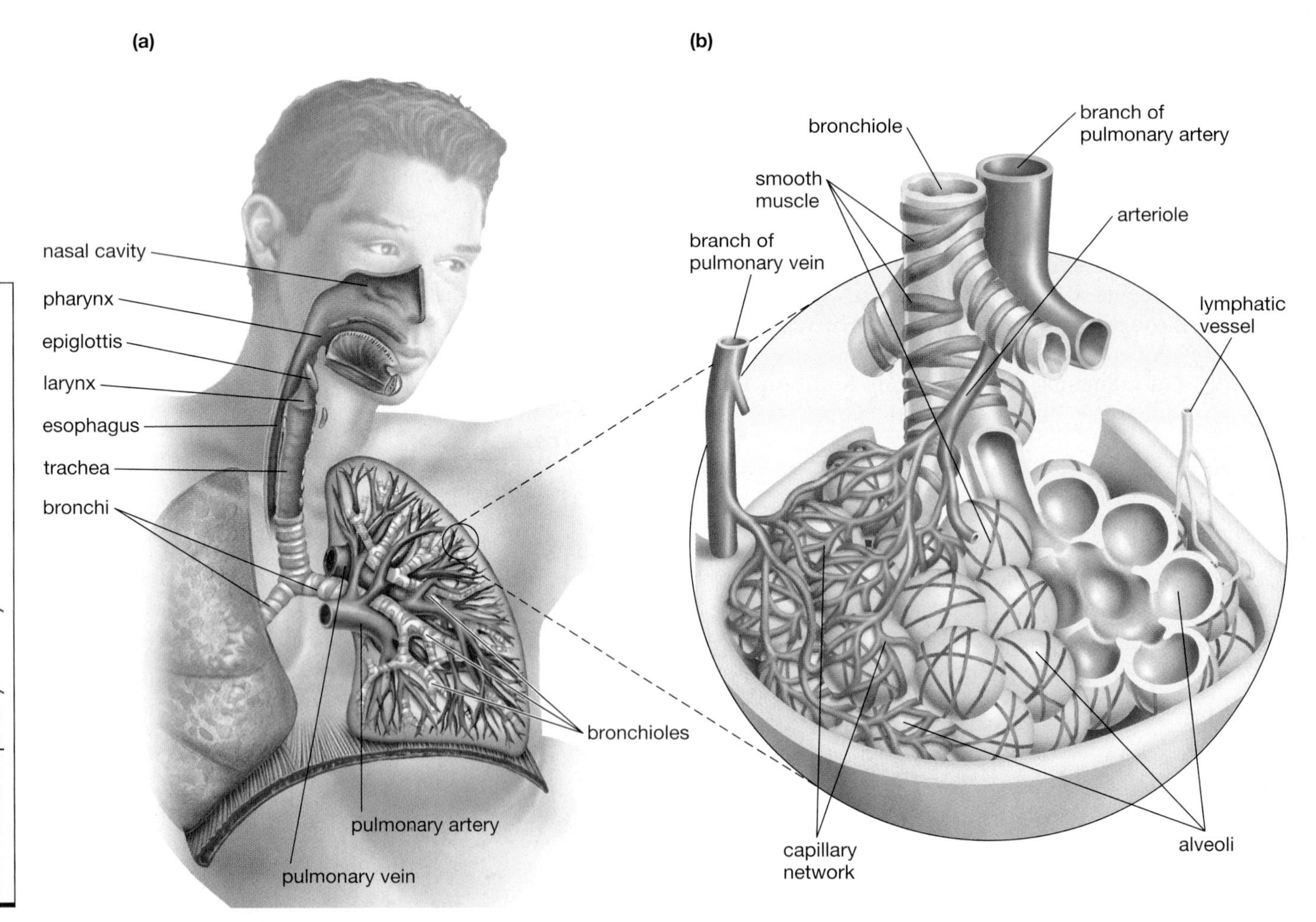

Media Activity
29.2 Human Respiratory Anatomy

FIGURE 29-7 The human respiratory system
(a) Air enters through the nasal cavity and mouth and passes through the pharynx and the larynx into the trachea. The epiglottis prevents food from going down the trachea. The trachea splits into two large branches, the bronchi, which lead into the two lungs. The smaller branches of the bronchi, the bronchioles, lead to the microscopic alveoli (which are enmeshed in capillaries), where gas exchange occurs. The pulmonary artery carries deoxygenated blood (in blue) to the lungs; the pulmonary vein carries oxygenated blood (in red) back to the heart. **(b)** Close-up of alveoli (their interiors are shown in the cut-away section) and their surrounding capillaries. **QUESTION** The nasal passages of mammals tend to follow long, complicated, winding pathways. Why are nasal passages so convoluted?

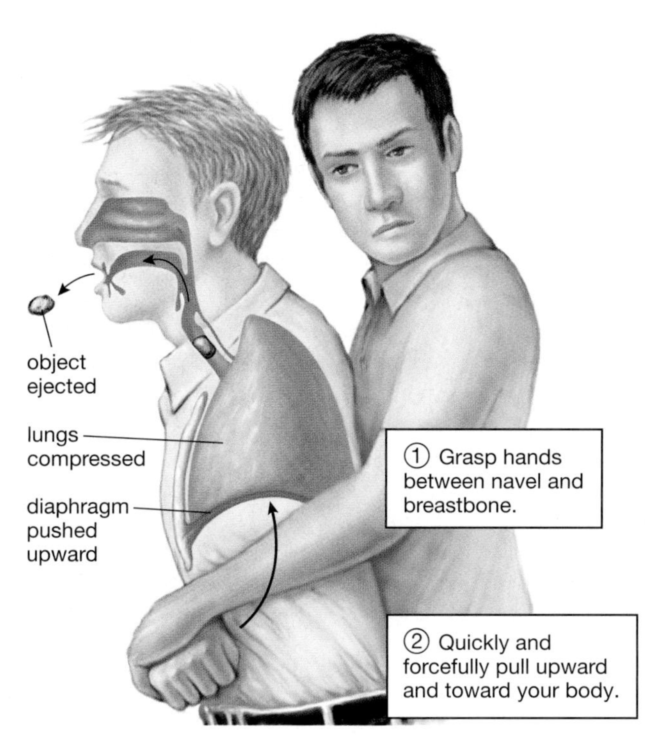

FIGURE 29-8 The Heimlich maneuver can save lives
If a person is choking on food or another object and is unable to breathe, the Heimlich maneuver pushes upward on the victim's diaphragm and forces air out of his or her lungs, possibly dislodging the object. Repeat if necessary.

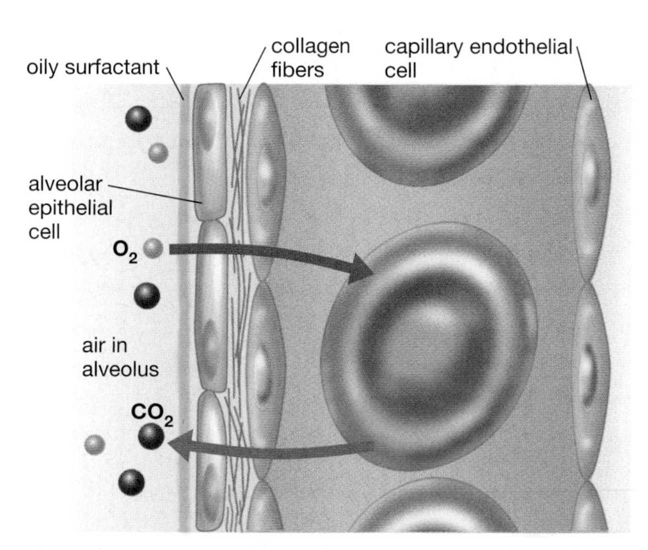

FIGURE 29-9 Gas exchange between alveoli and capillaries
The alveoli and capillary walls are only one cell thick and are very close to one another, with cells coated in a thin layer of fluid. This allows gases to dissolve and diffuse easily between the lungs and circulatory system.

Inhaled air travels past the larynx into the **trachea**, a flexible tube whose walls are reinforced with semicircular bands of stiff cartilage. Within the chest, the trachea splits into two large branches called **bronchi** (singular, *bronchus*), one leading to each lung. Inside the lung, each bronchus branches repeatedly into ever-smaller tubes called **bronchioles**. Bronchioles lead finally to the microscopic **alveoli** (singular, *alveolus*), tiny air sacs where gas exchange occurs (see Fig. 29-7). During its passage through the conducting system, air is warmed and moistened. Much of the dust and bacteria it carries is trapped in mucus secreted by cells that line the respiratory passages. The mucus, with its trapped debris, is continuously swept upward toward the pharynx by cilia that line the bronchioles, bronchi, and trachea. Upon reaching the pharynx, the mucus is coughed up or swallowed. Smoking interferes with this cleansing process by paralyzing the cilia (see "Health Watch: Smoking—A Life and Breath Decision"). *Asthma* occurs when muscle tissue in the bronchioles becomes hyperexcitable and mucus production increases, usually due to an allergy to an airborne substance, such as pollen. The bronchioles spasm and reduce the diameter of the airway. The asthma victim has particular difficulty breathing out through the constricted bronchioles, since the reduced air volume caused by exhalation allows the mucus-clogged passageways to collapse more readily.

Gas Exchange Occurs in the Alveoli

The lung provides an enormous moist surface for gas exchange. The dense network of bronchioles conducts air to tiny structures, the alveoli, which cluster about the end of each bronchiole like a bunch of grapes. In a typical adult, the two lungs combined have approximately 300 million alveoli. These microscopic (0.2 millimeter in diameter) chambers give magnified lung tissue the appearance of a pink sponge (see Fig. E29-2). Alveoli provide an extensive surface area for diffusion, totaling about 143 square meters (which is about 80 times the skin surface area of a human adult). The alveoli are enmeshed in a network of capillaries that cover about 85% of their surface (see Fig. 29-7b). The walls of the alveoli consist of a single layer of epithelial cells, and form the innermost portion of the *respiratory membrane*. The respiratory membrane, across which gas exchange occurs, consists of the alveolar epithelium and the layer of endothelial cells that forms the innermost wall of each capillary, which are fused together by collagen fibers. Because both the alveolar wall and the adjacent capillary walls are only one cell thick, diffusion distance for gases between the blood and the air is minimized (Fig. 29-9). The alveoli are lined with a thin layer of fluid containing an oily secretion (called a *surfactant*) that reduces surface tension and prevents the alveoli from collapsing during exhalation. The lung disease *emphysema,* most cases of which are caused by cigarette smoking, causes the alveoli to rupture, severely reducing the area available for gas exchange (emphysema is discussed in greater detail in "Health Watch: Smoking—A Life and Breath Decision").

HEALTH WATCH Smoking—A Life and Breath Decision

An estimated 440,000 people in the United States die of smoking-related diseases each year, and the cost of health care and loss of productivity due to smoking-related ailments costs more than $100 billion annually. More than 150,000 individuals die of lung cancer annually, and the American Cancer Society estimates that 87% of those deaths are due to smoking. The rest of the smoking-related deaths are from emphysema, chronic bronchitis, heart disease, stroke, and other forms of cancer. Altogether, one out of every five deaths in the United States can be attributed to smoking-related causes.

Tobacco smoke has a dramatic impact on the human respiratory tract. As smoke is inhaled, toxic substances such as nicotine and sulfur dioxide paralyze the cilia that line the respiratory tract; a single cigarette can inactivate them for a full hour. Because these ciliary sweepers remove inhaled particles, smoking inhibits them just when they are most needed. The visible portion of cigarette smoke consists of billions of microscopic carbon particles. Adhering to these particles are about 200 different toxic substances, of which more than a dozen are known or probable *carcinogens* (cancer-causing substances). With the cilia incapacitated, the particles stick to the walls of the respiratory tract and enter the lungs.

Cigarette smoke also impairs the white blood cells that defend the respiratory tract by engulfing foreign particles and bacteria. Consequently, still more bacteria, dust, and smoke particles enter the lungs. In response to the irritation of cigarette smoke, the respiratory tract produces more mucus, another method of trapping foreign particles. But without the cilia sweeping it along, the mucus builds up and can obstruct the airways; the familiar "smoker's cough" is an attempt to expel it. Microscopic smoke particles accumulate in the alveoli over the years until the lungs of a heavy smoker are literally blackened. The longer the delicate tissues of the lungs are exposed to the carcinogens on the trapped particles, the greater the chance that cancer will develop (Fig. E29-1).

Each year, approximately 124,000 people in the United States die from *chronic obstructive pulmonary disease (COPD),* which is the fourth leading cause of mortality (data from National Center for Health Statistics, 1999). COPD is actually a catch-all term for disorders including chronic bronchitis and emphysema, which generally occur together; smoking is the leading contributing factor to each of these diseases. COPD victims frequently suffer from asthma as well. **Chronic bronchitis** is a persistent lung infection characterized by coughing, swelling of the lining of the respiratory tract, an increase in mucus production, and a decrease in the number and activity of cilia. The result is a decrease in air flow to the alveoli. **Emphysema** occurs when toxic substances in cigarette smoke cause the body to produce substances that lead to brittle and ruptured alveoli. The lung (under magnification) gradually loses its normal sponge-like appearance (Fig. E29-2a) and more closely resembles blackened Swiss cheese (Fig. E29-2b). The loss of the alveoli, where gas exchange occurs, leads to oxygen deprivation of all body tissues. In an individual with emphysema, breathing becomes labored and grows increasingly difficult, sometimes leading to death.

Carbon monoxide, present in high levels in cigarette smoke, binds tenaciously to red blood cells in place of oxygen. This binding reduces the blood's oxygen-carrying capacity and thereby increases the workload on the heart. Chronic bronchitis and emphysema compound this problem. Smoking also promotes *atherosclerosis,* or thickening of the arterial walls by fatty deposits that can lead to heart attacks (see Chapter 28). As a result, smokers are 70% more likely than nonsmokers to die of heart disease. Smokers' wounds and broken bones take longer to heal, and the skin of smokers often develops premature wrinkles. The carbon monoxide in cigarette smoke may also contribute to the reproductive problems experienced by women who smoke, because it deprives the developing fetus of oxygen. These complications include an increased incidence of infertility, miscarriage, lower-birth-weight babies, and, later, more learning and behavioral problems in children.

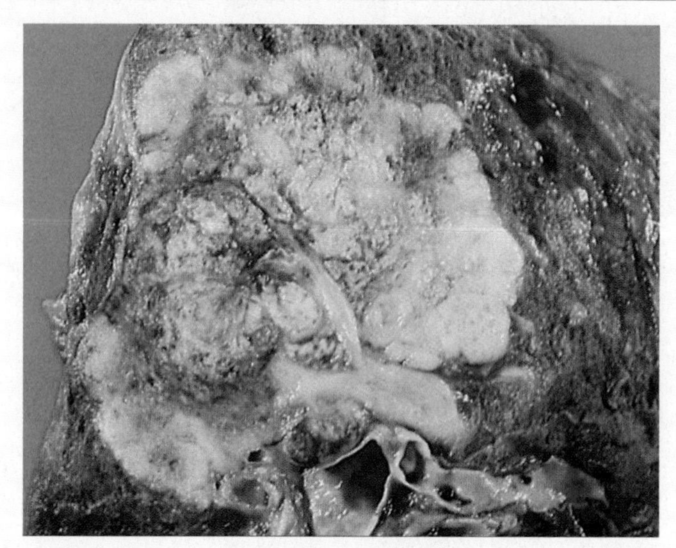

FIGURE E29-1 Smoking causes lung cancer
A lung cancer tumor is visible as a large, pale mass; the lung tissue surrounding it is blackened by trapped smoke particles.

"Passive smoking," or breathing secondhand smoke, poses real health hazards for both children and adults. Researchers have concluded that children whose parents smoke are more likely to contract bronchitis, pneumonia, ear infections, coughs, and colds and to have decreased lung capacity. Children who grow up with smokers are more likely to develop asthma and allergies; for children with asthma, the number and severity of asthma attacks are increased by secondhand smoke. Among adults, studies have concluded that nonsmoking spouses of smokers face a 30% higher risk of both heart attack and lung cancer than do spouses of nonsmokers. A recent study links even relatively infrequent exposure to secondhand smoke with an increased risk of atherosclerosis. Government agencies report that secondhand smoke is responsible for an estimated 3000 lung-cancer deaths and 37,000 deaths from heart disease in nonsmokers in the United States each year.

For smokers who quit, healing begins immediately and the chances of heart attack, lung cancer, and numerous other smoking-related illnesses gradually diminish (see "Links to Life: Quitters Are Winners").

Non Smoker

(a)

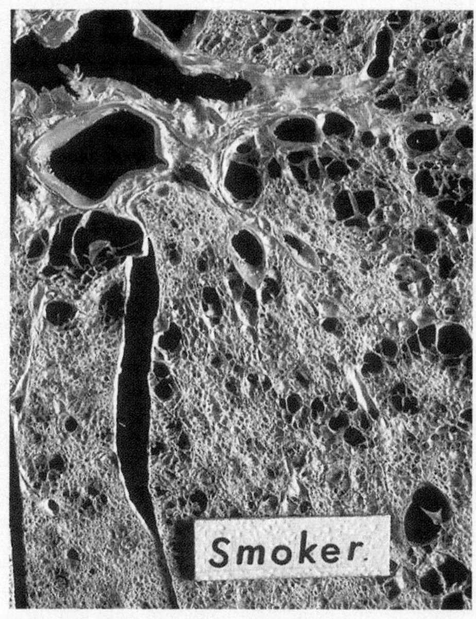
Smoker

(b)

FIGURE E29-2 Smoking causes emphysema
(a) Normal lung tissue from a nonsmoker, seen in cross section, has nearly invisible small openings, the alveoli, surrounded by healthy tissue. **(b)** The lung of a smoker suffering from emphysema is full of large holes, each caused by the rupture of hundreds of alveoli.

After the blood circulates through the body tissues, it is pumped to the lungs by the heart. The incoming blood surrounding the alveoli is low in oxygen (because the body cells have used it up) and high in carbon dioxide (which is released by the cells; see Fig. 29-2). Carbon dioxide diffuses out of the blood, where its concentration is high, into the air in the alveoli, where its concentration is lower (see Fig. 29-9). Carbon dioxide concentration in the blood is especially high after heavy exertion—for example, if you sprinted up the stairs on your way to class. As carbon dioxide diffuses into the air in the alveoli, oxygen diffuses from the air, where its concentration is high, into the blood, where its concentration is low. Blood from the lungs, now oxygenated and purged of carbon dioxide, returns to the heart, which pumps it to the body tissues. In the tissues, oxygen diffuses into the cells because the concentration of oxygen is lower in the cells than in the blood.

Oxygen and Carbon Dioxide Are Transported Using Different Mechanisms

In the blood, oxygen binds loosely and reversibly with **hemoglobin**, a large, iron-containing protein in the red blood cells, as described in Chapter 28 (see Fig. 28-9). Each hemoglobin molecule can bind up to four oxygen molecules (eight oxygen atoms). Nearly all the oxygen carried by the blood is bound to hemoglobin. By removing oxygen from solution in the plasma, hemoglobin maintains a concentration gradient that favors the diffusion of oxygen from the air into the blood (Fig. 29-10a). Thanks to hemoglobin, our blood can carry about 70 times as much oxygen as it could if the oxygen were simply dissolved in the plasma.

(a) Transport of oxygen ( )
① O_2 diffuses through lung capillary wall.
③ O_2 diffuses through tissue capillary walls.
② O_2 is carried to tissues bound to hemoglobin.
hemoglobin
lung side
body cell side
(b) Transport of carbon dioxide ( )
dissolved in plasma
HCO_3^- as bicarbonate
bound to hemoglobin
lung side
body cell side
② CO_2 is carried to lungs.
③ CO_2 diffuses through lung capillary wall.
① CO_2 diffuses through tissue capillary walls.

FIGURE 29-10 The chemistry and mechanism of gas exchange

Media Activity
29.3 Oxygen and Carbon Dioxide Transport

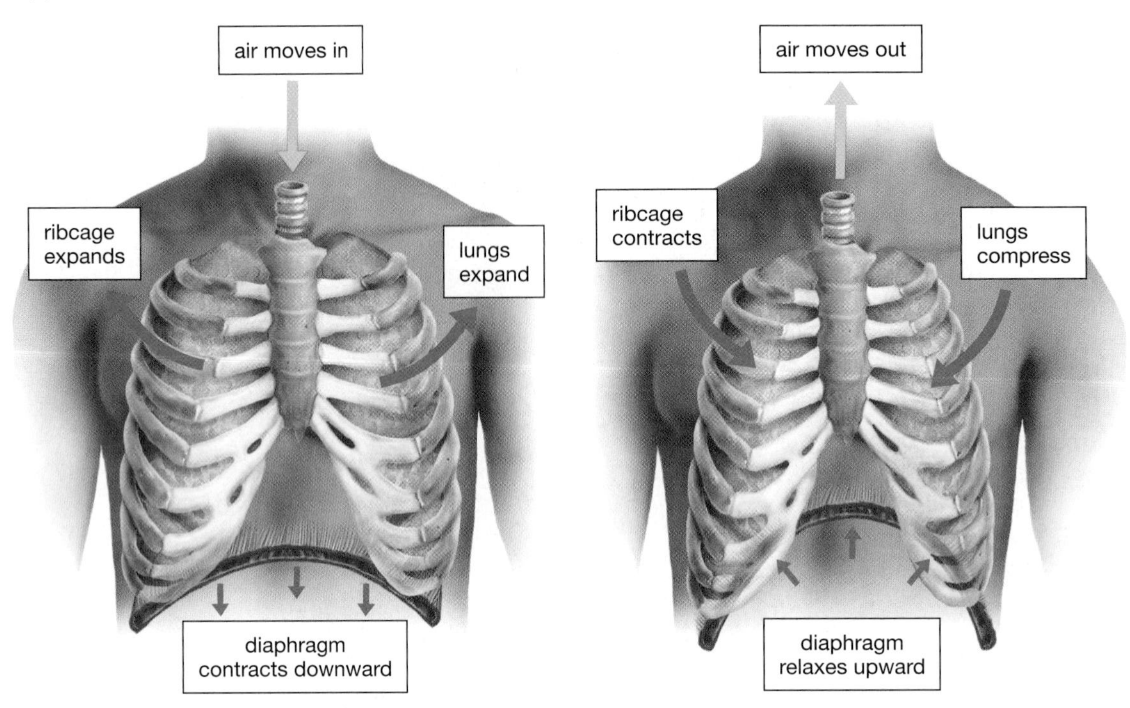

FIGURE 29-11 The mechanics of breathing
(a) During inhalation, rhythmic nerve impulses from the brain stimulate the diaphragm to contract (pulling it downward) and the muscles surrounding the ribs to contract (moving them up and outward). The result is an increase in the size of the chest cavity, causing air to rush in. **(b)** Relaxation of these muscles (exhalation) allows the diaphragm to dome upward and the rib cage to collapse, forcing air out of the lungs. **QUESTION** Imagine that a person living at sea level travels to a high mountaintop, and there inhales with a muscular contraction of exactly the same strength as had been her habit while at rest in her sea-level home. Would the resulting inhalation contain a volume of air that was larger, smaller, or the same as at sea level? Why?

As hemoglobin binds oxygen, the protein undergoes a slight change in shape, which alters its color. Oxygenated blood is a bright cherry-red; deoxygenated blood is dark maroon-red and appears bluish through the skin.

Carbon dioxide (CO_2) is transported in three different ways. Approximately 70% of the CO_2 reacts with water to form bicarbonate ion (HCO_3^-), which then diffuses into the plasma (Fig. 29-10b). About 20% of the CO_2 binds to hemoglobin (which has released its O_2 to the tissues) for its return trip to the lungs. The remaining 10% stays dissolved in the plasma as CO_2. The production of bicarbonate ion and the binding of CO_2 to hemoglobin reduce the concentration of dissolved CO_2 in the blood and increase the gradient for CO_2 to flow from the body cells into the blood.

Carbon monoxide (CO) is a toxic gas produced by combustion, such as that which occurs in engines, furnaces, and cigarettes when the fuel is not completely burned to form carbon dioxide. CO can be deadly because, at high levels, it binds to hemoglobin with more than 200 times more tenacity than oxygen. Humans will die from breathing air with as little as 0.1% CO. Hemoglobin bound to CO is bright red (like oxygen-bound hemoglobin), but it is incapable of transporting oxygen. While most victims of asphyxiation have bluish lips and nail beds because their hemoglobin is deoxygenated, the lips and nail beds of victims of carbon monoxide poisoning (such as could occur from breathing car exhaust in a closed space) are brighter red than normal.

Air Is Inhaled Actively and Exhaled Passively

Our ability to breathe depends on an airtight chest cavity—if the chest is punctured, the lungs may collapse. Outside the lungs, the chest cavity is bounded by neck muscles and connective tissue on top and by the dome-shaped, muscular **diaphragm** on the bottom.

Breathing occurs in two stages: (1) **inhalation**, when air is actively drawn into the lungs, and (2) **exhalation**, when it is passively expelled from the lungs. Inhalation is accomplished by enlarging the chest cavity. To do so, the diaphragm muscle contracts, drawing the diaphragm downward. The rib muscles also contract, lifting the ribs up and outward (Fig. 29-11). When the chest cavity is expanded, the lungs expand with it, because a vacuum holds them tightly against the inner wall of the chest. As the lungs expand, their increased volume creates a partial vacuum that draws air into the lungs.

Exhalation occurs automatically when the muscles that cause inhalation are relaxed. As it relaxes, the diaphragm domes upward; at the same time, the ribs fall down and inward, decreasing the size of the chest cavity and forcing air out of the lungs. Additional air can be forced out by contracting the abdominal muscles. After exhalation, the lungs still contain air, which helps prevent the thin alveoli from collapsing and fills the spaces within the conducting portion of the respiratory system. A typical breath moves about 500 milliliters (1 pint) of "new" air into the respiratory system. Of this, only about 350 milliliters reaches the alveoli for gas exchange; the rest remains in the conducting portion. During exercise, your deeper breathing can move several times more air through your respiratory system.

Breathing Rate Is Controlled by the Respiratory Center of the Brain

Imagine having to think about every breath. Fortunately, breathing occurs rhythmically and automatically without conscious thought. But, unlike the heart muscle, the muscles used in breathing are not self-activating; each contraction is stimulated by impulses from nerve cells. These impulses originate in the **respiratory center**, which is located in the medulla, a portion of the brain just above the spinal cord. Nerve cells in the respiratory center generate cyclic bursts of impulses that cause the contraction (followed by passive relaxation) of the respiratory muscles.

The respiratory center receives input from several sources and adjusts breathing rate and volume to meet the body's changing needs. The respiratory rate is regulated to maintain a constant level of carbon dioxide in the blood, as monitored by carbon dioxide receptors in the medulla. For example, an elevated level of carbon dioxide caused by an increase in cellular activity, such as that experienced by your muscle cells when you run up stairs, signals a need for more oxygen, causing the receptors to stimulate an increase in the rate and depth of breathing. These receptors are extremely sensitive; an increase in carbon dioxide of only 0.3% can double the breathing rate.

The respiratory rate is much less sensitive to changes in oxygen concentration because normal breathing supplies an overabundance of oxygen. But if blood oxygen levels fall drastically, receptors in the aorta and carotid arteries stimulate the respiratory center. Surprisingly, when you begin strenuous activity, such as running, an increase in breathing rate actually *precedes* any change in blood gas levels. Apparently, when higher brain centers activate muscles during heavy exercise, they simultaneously stimulate the respiratory center to increase breathing rate. Breathing is then "fine-tuned" by the receptors that monitor carbon dioxide concentrations.

CASE STUDY REVISITED Lives Up in Smoke

Nicotine is a powerfully addictive drug, as likely to lead to addiction as cocaine or heroin. For this reason, sometimes the only way to quit smoking is to not start smoking at all. Researchers at the University of Massachusetts Medical School have evidence that teenagers become addicted to smoking far more easily than was commonly believed. A recent 30-month study of nearly 700 students who started smoking at ages 12 and 13 found that 40% reported symptoms of addiction: craving cigarettes and feeling nervous, irritable, anxious, or less able to concentrate when they were unable to smoke. Half of those reporting symptoms of addiction smoked only once a week, and one-third smoked only once a month. After becoming regular smokers, only one out of five will successfully quit, and one out of every five who continue will eventually die from a smoking-related illness—the legacy of a decision made while they were barely out of childhood.

Consider This: A school for seventh and eighth graders asks you to develop a lesson plan designed to discourage students from smoking. The program you are designing can occupy two school days and can also include a field trip. What subjects would you cover, and in what order? Ideally, what "props" would you use? What discussions would you lead? What type of field trip might you organize?

Links to Life: Quitters Are Winners

Have you or has someone you know quit smoking? The American Lung Association has some encouraging news—a timeline of improvements experienced by former smokers as their bodies begin to recover, starting from the last cigarette puffed. If you stop smoking, after 20 minutes your blood pressure and pulse rate drop. After 8 hours the level of carbon monoxide in your blood drops, while oxygen increases to normal levels. After 24 hours your chance of a heart attack decreases. After 48 hours you regain more ability to smell and taste your food. From 2 weeks to 3 months later you are able to exercise more easily as circulation and respiratory function improve. From 1 to 9 months later you cough less, and have less sinus congestion and more energy. After 1 year you have about half the risk of coronary heart disease as does a smoker. After 5 years your risk of a stroke begins to drop. After 10 years your risk of lung cancer may be half that of a continuing smoker, and your risk of cancers of the pancreas, kidney, bladder, esophagus, throat, and mouth also decline. After 15 years your risk of death from all smoking-related causes is nearly as low as that of people who have never smoked. (Modified from "What Are the Benefits of Quitting Smoking" published on-line by the American Lung Association; http://www.lungusa.org)

Summary of Key Concepts

29.1 Why Exchange Gases?

The respiratory system supports cellular respiration. Oxygen-rich air is inhaled and supplies oxygen to the blood, which carries it to cells throughout the body. Blood also picks up CO_2 (a product of cellular respiration) from body cells and transports it to the lungs, where it is released into the atmosphere.

29.2 What Are Some Evolutionary Adaptations for Gas Exchange?

The exchange of oxygen and carbon dioxide between the body and the environment by diffusion across a moist surface is made possible by respiration. In moist environments, animals with very small or flattened bodies may rely exclusively on diffusion through the body surface. Animals with low metabolic demands and/or well-developed circulatory systems may also lack specialized respiratory structures. Larger, more active animals have evolved specialized respiratory systems. Animals in aquatic environments have evolved gills, such as those of fish and many amphibians. On land, respiratory surfaces must be protected and moistened internally. This need has selected for the evolution of tracheae in insects and lungs in terrestrial vertebrates.

The transfer of gases between respiratory systems and tissues occurs in a series of stages that alternate bulk flow with diffusion. Air or water moves by bulk flow past the respiratory surface, and gases in blood are also carried by bulk flow. Gases move by diffusion across membranes between the respiratory system and the capillaries and between the capillaries and the tissues.

29.3 How Does the Human Respiratory System Work?

The human respiratory system consists of a conducting portion and a gas-exchange portion. Air passes first through the conducting portion, consisting of the nose and mouth, pharynx, larynx, trachea, bronchi, and bronchioles, and then into the gas-exchange portion, composed of alveoli (microscopic sacs). Blood, within a dense capillary network surrounding the alveoli, releases carbon dioxide and absorbs oxygen from the air.

Most of the oxygen in the blood is bound to hemoglobin within red blood cells. By removing oxygen from solution, hemoglobin maintains a favorable concentration gradient that allows oxygen to readily diffuse from the air into the blood. Hemoglobin then transports the oxygen to the body tissues, where it diffuses out along its concentration gradient. Carbon dioxide diffuses into the blood from the tissues and is transported in three ways: as bicarbonate; bound to hemoglobin; or dissolved in blood plasma.

Breathing involves actively drawing air into the lungs by contracting the diaphragm and the rib muscles, which expands the chest cavity. Relaxing these muscles causes the chest cavity to collapse, expelling the air. Respiration is controlled by nerve impulses that originate in the medulla's respiratory center. The respiration rate is modified by receptors, such as those in the medulla that monitor carbon dioxide levels in the blood.

Key Terms

alveoli *p. 577*
bronchiole *p. 577*
bronchi *p. 577*
bulk flow *p. 573*
chronic bronchitis *p. 578*
conducting portion *p. 576*
diaphragm *p. 580*
emphysema *p. 578*
exhalation *p. 580*
gas-exchange portion *p. 576*
gill *p. 574*
hemoglobin *p. 579*
inhalation *p. 580*
larynx *p. 576*
lung *p. 574*
pharynx *p. 576*
respiratory center *p. 71*
spiracle *p. 574*
trachea (in birds/ mammals) *p. 577*
tracheae (in insects) *p. 574*
vocal cord *p. 576*

Thinking Through the Concepts

To take a multiple-choice quiz with feedback on the contents of this chapter, visit http://www.prenhall.com/audesirk7. *Log in to the Web site selected by your instructor and navigate to the Self Test section for this chapter.*

? Review Questions

1. Describe three arthropod respiratory systems and two vertebrate respiratory systems.
2. Trace the route taken by air in the vertebrate respiratory system, listing the structures through which it flows and the point at which gas exchange occurs.
3. Explain some characteristics of animals in moist environments that may supplement respiratory systems or make them unnecessary.
4. How are human respiratory movements initiated? How are they modified, and why are these controls adaptive?
5. What events occur during human inhalation? Exhalation? Which of these is always an active process?
6. Trace the pathway of an oxygen molecule in the human body, starting with the nose and ending with a body cell.
7. Describe the effects of smoking on the human respiratory system.
8. Explain how bulk flow and diffusion interact to promote gas exchange between air and blood and between blood and tissues.
9. Compare carbon dioxide and oxygen transport in the blood. Include the source and destination of each.
10. Explain how the structure and arrangement of alveoli make them well suited for their role in gas exchange.

Applying the Concepts

1. Heart–lung transplants are performed in some cases, but donors are scarce. On the basis of your knowledge of the respiratory and circulatory systems and of lifestyle factors that might damage them, what criteria would you use in selecting a recipient for such a transplant?
2. Nicotine is a drug in tobacco that is responsible for several of the effects that smokers crave. Discuss the advantages and disadvantages of low-nicotine cigarettes.
3. Discuss why a brief exposure to carbon monoxide is much more dangerous than a brief exposure to carbon dioxide.
4. Describe several adaptations that might evolve to help members of a species of mammal respire better if the population began living continuously for many generations at very high altitudes.
5. Mary, a strong-willed 3-year-old, threatens to hold her breath until she dies if she doesn't get her way. Can she carry out her threat? Explain.

For More Information

Gibbs, W. "Breath of Fresh Liquid." *Scientific American,* February 1999. Perfusing the lungs with oxygen-carrying fluid containing perfluorocarbons may help patients with lung disease.

Harding, C. "Going to Extremes." *National Wildlife*, August/September 1993. Describes adaptations that allow diving animals to plunge to enormous depths without running out of oxygen.

Houston, C. "Mountain Sickness." *Scientific American*, October 1992. The mechanisms of potentially fatal altitude sickness are explained.

Platt, C. "Here, Breathe This Liquid." *Discover*, October 2001. Perfusing ice-cold, oxygen-enriched perfluorocarbons into the lungs might increase the chance for survival of patients whose hearts have stopped. After the heart is restarted, the chilled liquid would rapidly cool the body and reduce the brain's oxygen requirements and the damaging chemical reactions that occur in oxygen-starved brain tissue.

Seppa, N. "Secondary Smoke Carries High Price." *Science News*, January 17, 1998. Research suggests that smoking causes atherosclerosis, that the damage continues after a smoker quits, and that exposure to secondhand smoke significantly increases the buildup of plaque in the carotid artery of nonsmokers.

Media Activities

To access a Media Activity visit http://www.prenhall.com/audesirk7. *Log in to the Web site selected by your instructor, navigate to this chapter, and select the appropriate Media Activity number.*

29.1 Gas Exchange

Estimated time: 5 minutes

Watch an overview of the stages of gas exchange in the animal respiratory system.

29.2 Human Respiratory Anatomy

Estimated time: 5 minutes

Explore the structures and function of the human respiratory system.

29.3 Oxygen and Carbon Dioxide Transport

Estimated time: 5 minutes

Explore mechanisms for transporting oxygen and carbon dioxide to and from the respiratory system.

29.4 Web Investigation: Lives Up in Smoke

Estimated time: 10 minutes

This exercise will survey the scientific evidence for the harmful effects of environmental tobacco smoke.

CHAPTER

30 Nutrition and Digestion

More than half of the U.S. population is overweight. As obesity rates climb, researchers seek genetic causes and chemical cures.

AT A GLANCE

CASE STUDY Fat in the Family?

"I turn sideways in the mirror and gasp with horror," admits one resident of Texas, whom we'll call Chuck. "I stepped on the bathroom scales, and they broke. I couldn't see them over my gut anyway." Obesity is a growing epidemic in the United States. More than half of us are overweight; more than 20% are classified as *obese* (30% or more above ideal weight). The percentage of overweight people has doubled in the past two decades, and many have given up. "I would rather die than eat fat-free," says Chuck. Exercise? "Forget that. Who has time for it?"

Genetic research may fuel this sense of resignation. When an auto-parts store employee was fired for "poor job performance" (in spite of an exemplary record) after his weight climbed to 400 pounds, he sued. After an expert medical witness testified that a person's weight is controlled approximately 80% by genes, a jury agreed that the employee was a victim of "weight discrimination" and awarded him more than $1 million. The relative contributions of genetics and environment to the weight of any individual is an open question. However, researchers have found at least 130 different genes that influence weight in some way. The number of different combinations of these genes that people can inherit is staggering, suggesting thousands of different genetic tendencies that promote weight gain.

People differ genetically in how their bodies respond to exercise, how full they feel after eating, how much fat they store, and how many calories their bodies consume while lying still. "I look at the family pictures of my ancestors, and they are a bunch of pudgy people," Chuck muses. "I'm not going to fight it, because I can't win." Dieters face the dilemma that their bodies react to dieting as if they were encountering starvation—their metabolic rates go down, energy consumption plummets, and losing weight becomes even more difficult. Why do people and other animals have such a strong tendency to store fat? Should overweight people resign themselves to their "genetically fat fate?"

30.1 What Nutrients Do Animals Need?

All foods, whether they fall in the "vegetable" category or the "chocolate" category, contain nutrients that you need to survive. **Nutrition** is the process of acquiring and processing nutrients into a usable form. In animals, these **nutrients** are typically supplied by the diet. Nutrients fall into five major categories: lipids, carbohydrates, proteins, minerals, and vitamins. These substances provide the body with its basic needs, including energy and the raw materials to synthesize the molecules of life. These molecules include enzymes, structural proteins, genetic material, energy carriers, the calcium-based components of bone, and the lipid-based components of all cell membranes, to name just a few.

Cells Continuously Expend Energy, Which Is Derived from Nutrients and Measured in Calories

Cells rely on a continuous supply of energy to maintain their incredible complexity and their wide range of activities. Deprived of this energy, cells begin to die within seconds. Two types of nutrients provide most of the energy in the animal diet: carbohydrates and fats (in the United States, people on the average also get about 16% of their energy from protein). These molecules are broken down by digestion, and their subunits are metabolized during cellular respiration, releasing energy that is captured in adenosine triphosphate (ATP; see Chapter 8).

The energy in nutrients is measured in calories. A **calorie** is the amount of energy required to raise the temperature of 1 gram of water by 1 degree Celsius. The calorie content of foods is measured in units of 1000 calories (*kilocalories*), also known as **Calories** (with a capital *C*). The average human body at rest burns roughly 70 Calories per hour, but this value is influenced by body size, muscle mass, age, sex, and genetic factors. Exercise significantly boosts caloric requirements; well-trained athletes can temporarily raise their calorie consumption from a resting rate of about 1 Calorie per minute to nearly 20 Calories per minute during vigorous exercise (Table 30-1).

Lipids Include Fats, Phospholipids, and Cholesterol

Although they are sometimes viewed as the enemy in our overweight society, fats and other lipids are essential nutrients. Lipids are a diverse group of molecules that generally contain long chains of carbon atoms and are insoluble in water. The principal types of lipids are *triglycerides* (fats), *phospholipids*, and *cholesterol* (see Chapter 3). Triglycerides are used primarily as a source of energy. Phospholipids are important components of all cellular membranes. Cholesterol is used in the synthesis of cellular membranes, sex hormones, and bile (which aids in fat breakdown). Some animal species can synthesize all the lipid "building blocks" necessary to make the specialized lipids they need. Others must acquire specific types of lipid building blocks, called **essential fatty acids**, from their food. For example, humans are unable to synthesize linoleic acid (which is required for the synthesis of certain phospholipids), so we need to obtain this essential fatty acid from our diet.

Animals Store Energy as Fat

When an animal's diet provides more energy than is expended through metabolic activities, most of the excess carbohydrate, fat, or protein is converted to fat for storage. About 3600 Calories are stored in each pound of fat. Fats have two major advantages as energy-storage molecules. First, they are the most concentrated energy source, containing more than twice the energy per unit weight of either carbohydrates or proteins (about 9 Calories per gram for fats compared with about 4 per gram for proteins and carbohydrates). Second, lipids are

Table 30-1 Approximate Energy Consumed by a 150-Pound Person for Different Activities

		Time to "Work Off"			
Activity	Calories/ hr	500 Calories Cheeseburger	300 Calories Ice Cream Cone	70 Calories Apple	40 Calories 1 Cup Broccoli
Running (6 mph)	700	43 min	26 min	6 min	3 min
Cross-country skiing (moderate)	560	54 min	32 min	7.5 min	4 min
Roller skating	490	1 hr 1 min	37 min	8.6 min	5 min
Bicycling (11 mph)	420	1 hr 11 min	43 min	10 min	6 min
Walking (3 mph)	250	2 hr	1 hr 12 min	17 min	10 min
Frisbee® playing	210	2 hr 23 min	1 hr 26 min	20 min	11 min
Studying	100	5 hr	3 hr	42 min	24 min

hydrophobic—that is, they do not mix with water. Fat deposits, therefore, do not cause any extra accumulation of water in the body. For both these reasons, fats store more calories with less weight than do other molecules. Minimizing weight allows an animal to move faster (important for escaping predators and hunting prey) and to use less energy for movement (important when food supplies are limited). Since people evolved under the same food constraints as other animals, we have a strong tendency to eat when food is available, often in excess of our needs because we may require the energy later. Some modern societies now have access to almost unlimited high-calorie food. In this environment, our natural tendency to overeat can become a liability, and we need to exert considerable willpower to avoid becoming obese.

In animals that maintain an elevated body temperature, fat deposits do double duty by providing insulation as well as storing energy. Fat, which conducts heat at only one-third the rate of other body tissues, is typically stored in a layer beneath the skin where it insulates the body. Birds (especially flightless birds such as penguins) and mammals who live in polar climates or in cold ocean waters are particularly dependent on this insulating layer, which reduces the amount of energy they must expend to keep warm (Fig. 30-1).

Carbohydrates, Including Sugars and Starches, Are a Source of Quick Energy

Athletes sometimes engage in "carbo-loading"—gorging on pasta or other carbohydrates—to build up reserves of quick energy for a sporting event. Carbohydrates consist of monosaccharide sugars (such as glucose), disaccharide sugars (such as sucrose), and longer chains of sugars called *polysaccharides* (see Chapter 3). Polysaccharides include starches, the principal energy-storage material of plants; *glycogen*, a short-term energy-storage molecule in animals; and *cellulose*, the major structural component of plant cell walls. Although cellulose is the most abundant carbohydrate on the planet, few species of animals are able to digest it, as described later. During digestion, carbohydrates are broken down into sugars and absorbed. Cells obtain most of their energy from a single sugar: glucose. Various metabolic transformations allow the body to derive glucose from carbohydrates, fats, and even amino acids.

FIGURE 30-1 Fat provides insulation
These walruses can withstand the icy waters of the polar seas because a thick layer of fat beneath the skin insulates them from the cold.

Animals, including humans, store the carbohydrate **glycogen** (a large, highly branched chain of glucose molecules) in the liver and muscles. Although humans can potentially store hundreds of pounds of fat, most of us store only a pound of glycogen. During exercise, such as running, the body draws on this store of glycogen as a source of quick energy. When the activity is prolonged, as in the case of a marathon runner, the stored glycogen can be almost totally depleted. The expression "hitting the wall" describes the extreme fatigue that long-distance runners may experience after exhausting their glycogen supply, often about 18 miles into a race. With glycogen stores exhausted, the body switches to fatty acids as a source of energy. Because fatty acids require more complex metabolic transformations to yield glucose, the body is only able to extract energy from them at about half the rate as for glycogen. A glycogen-depleted runner, even one with plenty of fat stores, may need to slow to almost a walking pace to compensate for this reduction in available energy. For this reason, marathon runners often drink solutions high in glucose during the course of a race.

Proteins, Composed of Amino Acids, Perform a Wide Range of Functions Within the Body

In the digestive tract, ingested protein is broken down into its amino acid subunits, most of which are used to synthesize new proteins. These proteins perform various roles in the body, acting as enzymes, receptors on cell membranes, oxygen transport molecules (hemoglobin), structural proteins (hair and nails), antibodies, and muscle proteins. Excess proteins are also broken down into amino acids, which can serve as an energy source, be converted into fat for storage, or be used to synthesize certain hormones and some neurotransmitters (chemicals used in communication between neurons). Each day, your body requires 20 to 30 grams of protein. Protein breakdown produces the waste product **urea**, which is filtered from the blood by the kidneys.

Humans can synthesize (from other amino acids) 11 of the 20 different amino acids used in proteins. The 9 amino acids that we cannot synthesize are called **essential amino acids**. Humans must get these essential amino acids from protein-rich foods such as meat, milk, eggs, corn, beans, and soybeans. Many plant proteins are deficient in some of the essential amino acids. Thus, to avoid protein deficiency, individuals on a vegetarian diet must include a variety of plants (for example, legumes, grains, and corn) whose proteins collectively provide all 9 of the essential amino acids. Protein deficiency can cause a variety of debilitating conditions, including kwashiorkor (Fig. 30-2a), which is most often encountered in poverty-stricken countries.

Minerals Are Elements Required by the Body

You may have seen a display of **minerals** in a museum—beautiful, brightly colored crystals. Minerals are elements that play crucial roles in animal nutrition (Table 30-2). Since no organism can manufacture minerals, they must be obtained through the diet, either from food or dissolved in drinking water. Minerals such as calcium, magnesium, and phosphorus are major constituents of bones and teeth. Sodium, calcium, and potassium are essential for muscle contraction and the conduction of nerve impulses. Iron is a central component of each hemoglobin molecule in blood, and iodine is found in hormones produced by the thyroid gland. We also require trace amounts of several other minerals including zinc and magnesium (both required for the function of some enzymes), copper (needed for hemoglobin synthesis), and chromium (used in the metabolism of sugar and fat).

Vitamins Are Required in Small Amounts and Play Many Roles in Metabolism

"Take your vitamins!" is a familiar refrain in many households with children. But why are vitamins so important? **Vitamins** are a diverse group of organic compounds that animals require in small amounts. In general, the body cannot synthesize vitamins (or cannot do so in adequate amounts), so they must be obtained from food. Our modern diet is now so different from the natural diet on which we evolved that many people find vitamin supplements a prudent way to compensate for possible suboptimal vitamin levels. For example, our skin can manufacture some vitamin D when it is exposed to sunlight, but most of us spend so much time indoors that we do not synthesize enough and must augment our natural supply with dietary sources or supplements. The vitamins considered essential in human nutrition are listed in Table 30-3.

Some vitamins, such as C and E, also function as *antioxidants*. As our cells generate and use energy, damaging molecules called *free radicals* are produced. These molecules react with and can damage DNA, in some cases causing cancer. Free radicals can also promote atherosclerosis; over a lifetime, they contribute to the deterioration of physiological functioning associated with aging. Antioxidants combine with free radicals to limit their damaging effects.

Water-Soluble Vitamins

Human vitamins are often grouped into two categories: water-soluble and fat-soluble. Water-soluble vitamins include vitamin C and the 8 compounds that make up the B-vitamin complex. Since these substances dissolve in the water of the blood plasma and are excreted by the kidneys, they are not stored in the body in any appreciable amounts. Therefore, the body's supply of these vitamins must be constantly replenished by diet. Most water-soluble vitamins work in conjunction with

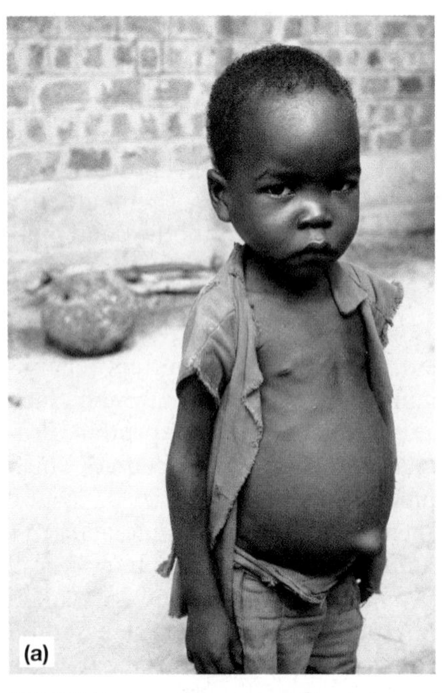

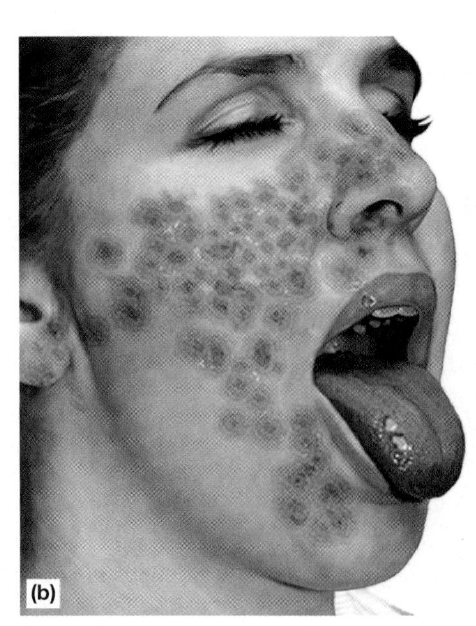

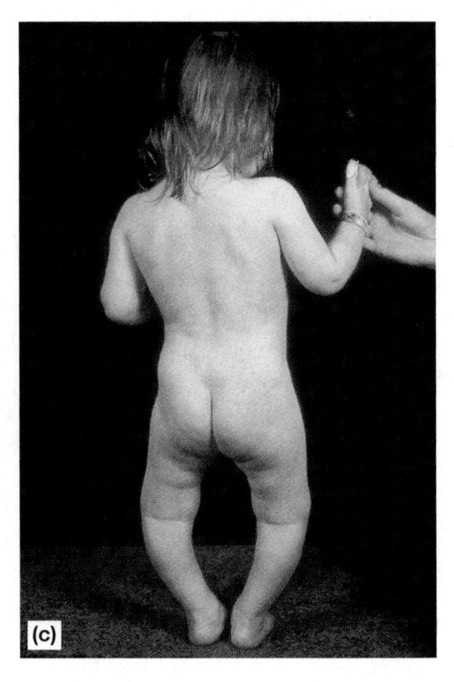

FIGURE 30-2 Symptoms of protein and vitamin deficiency
(a) Symptoms of kwashiorkor include a swollen abdomen and emaciated arms and legs; this condition is caused by protein deficiency. **(b)** Pellagra, characterized by scaly, reddish-brown skin lesions, is a symptom of niacin deficiency. **(c)** Rickets, which causes bone deformation, is a result of vitamin D deficiency.

Table 30-2 Minerals, Sources, and Functions for Humans

Mineral	Dietary Sources	Major Functions in Body	Deficiency Symptoms
Calcium	Milk, cheese, green vegetables, legumes	Bone and tooth formation Blood clotting Nerve impulse transmission	Stunted growth Rickets, osteoporosis Convulsions
Phosphorus	Milk, cheese, meat, poultry, grains	Bone and tooth formation Acid–base balance	Weakness Demineralization of bone Loss of calcium
Potassium	Meats, milk, fruits	Acid–base balance Body water balance Nerve function	Muscular weakness Paralysis
Chlorine	Table salt	Formation of gastric juice Acid–base balance	Muscle cramps Apathy Reduced appetite
Sodium	Table salt	Acid–base balance Body water balance Nerve function	Muscle cramps Apathy Reduced appetite
Magnesium	Whole grains, green leafy vegetables	Activation of enzymes in protein synthesis	Growth failure Behavioral disturbances Weakness, spasms
Iron	Eggs, meats, legumes, whole grains, green vegetables	Constituent of hemoglobin and enzymes involved in energy metabolism	Iron-deficiency anemia (weakness, reduced resistance to infection)
Fluorine	Fluoridated water, tea, seafood	Maintenance of teeth and probably bone structure	High frequency of tooth decay
Zinc	Widely distributed in foods	Constituent of enzymes involved in digestion	Growth failure Small sex glands
Iodine	Seafish and shellfish, dairy products, many vegetables, iodized salt	Constituent of thyroid hormones	Goiter
Chromium	Fruits, vegetables, whole grains	Metabolism of sugar and fats	Reduced glucose tolerance Elevated insulin in blood

enzymes to promote chemical reactions that supply energy or synthesize biological molecules. Because each vitamin participates in several metabolic processes, a deficiency of a single vitamin can have wide-ranging effects (see Table 30-3). For example, deficiency of the B-vitamin niacin causes the cracked, scaly skin of pellagra (Fig. 30-2b), as well as some nervous disorders. Folic acid is required to synthesize thymine, a component of DNA; folic acid deficiency impairs cell division throughout the body. As you might predict, it is particularly important for pregnant women to get enough folic acid to supply the rapidly growing fetus. Folic acid deficiency can also lead to a reduction in red blood cells and anemia. In order for folic acid to function properly, trace amounts of vitamin B_{12} are required. In the human diet, vitamin B_{12}—which is synthesized by bacteria—can be obtained only from eating animal protein, so strict vegetarians require supplements of this vitamin.

Fat-Soluble Vitamins

The fat-soluble vitamins A, D, E, and K have even more varied roles (See Table 30-3). Vitamin K, for example, helps regulate blood clotting. Vitamin A deficiency can lead to difficulty seeing at night because it is used to produce visual pigment (the light-capturing molecule in the retina of the eye). Vitamin D is required for normal bone formation; a deficiency can lead to rickets (Fig. 30-2c). U.S. researchers have recently discovered that many adult women (particularly African-Americans, whose darker skin cannot synthesize as much vitamin D in the presence of sunlight) have inadequate levels of this vitamin. Children born to vitamin D-deficient mothers are at particular risk for rickets, the prevalence of which is increasing in the United States. Fat-soluble vitamins can be stored in body fat and may accumulate in the body over time. For this reason, high doses of certain fat-soluble vitamins (vitamin A, for example) are toxic.

Nutritional Guidelines Help People Obtain a Balanced Diet

Most people in the United States are fortunate to live amid an abundance of food. However, the overwhelming diversity of foods available in a typical U.S. supermarket and the easy availability of "fast food" can lead to poor nutritional choices. To help people make informed choices, the U.S. government has developed recommendations and goals for the average U.S. citizen, which are summarized in Table 30-4.

Table 30-3 Vitamins, Sources, and Functions for Humans

Vitamin	Dietary Sources	Functions in Body	Deficiency Symptoms
Water soluble			
B-complex			
Vitamin B_1 (thiamin)	Milk, meat, bread	Coenzyme in metabolic reactions	Beriberi (muscle weakness, peripheral nerve changes, edema, heart failure)
Vitamin B_2 (riboflavin)	Widely distributed in foods	Constituent of coenzymes in energy metabolism	Reddened lips, cracks at corner of mouth, lesions of eye
Niacin	Liver, lean meats, grains, legumes	Constituent of two coenzymes in energy metabolism	Pellagra (skin and gastrointestinal lesions; nervous, mental disorders)
Vitamin B_6 (pyridoxine)	Meats, vegetables, whole-grain cereals	Coenzyme in amino acid metabolism	Irritability, convulsions, muscular twitching, dermatitis, kidney stones
Pantothenic acid	Milk, meat	Constituent of coenzyme A, with a role in energy metabolism	Fatigue, sleep disturbances, impaired coordination
Folic acid	Legumes, green vegetables, whole wheat	Coenzyme involved in nucleic and amino acid metabolism	Anemia, gastrointestinal disturbances, diarrhea, retarded growth, birth defects
Vitamin B_{12}	Meats, eggs, dairy products	Coenzyme in nucleic acid metabolism	Pernicious anemia, neurological disorders
Biotin	Legumes, vegetables, meats	Coenzymes required for fat synthesis, amino acid metabolism, and glycogen formation	Fatigue, depression, nausea, dermatitis, muscular pains
Others			
Choline	Egg yolk, liver, grains, legumes	Constituent of phospholipids, precursor of the neuro-transmitter acetylcholine	None reported in humans
Vitamin C (ascorbic acid)	Citrus fruits, tomatoes, green peppers	Maintenance of cartilage, bone, and dentin (hard tissue of teeth); collagen synthesis	Scurvy (degeneration of skin, teeth, gums, blood vessels; epithelial hemorrhages)
Fat soluble			
Vitamin A (retinol)	Beta-carotene in green, yellow, and red vegetables Retinol added to dairy products	Constituent of visual pigment Maintenance of epithelial tissues	Night blindness, permanent blindness
Vitamin D	Cod-liver oil, eggs, dairy products	Promotes bone growth and mineralization Increases calcium absorption	Rickets (bone deformities) in children; skeletal deterioration
Vitamin E (tocopherol)	Seeds, green leafy vegetables, margarines, shortenings	Antioxidant, prevents cellular damage	Possibly anemia
Vitamin K	Green leafy vegetables Product of intestinal bacteria	Important in blood clotting	Bleeding, internal hemorrhages

Further help is provided by the Food Guide Pyramid, designed by the U.S. Department of Agriculture, which illustrates the relative abundance of different food groups in an optimal diet (Fig. 30-3). Still more nutritional information can be found on the labeling required on commercially packaged foods. These labels provide complete information about calorie, fiber, fat, sugar, and vitamin content (Fig. 30-4). Some fast-food chains also make fliers available that list nutritional information about their products.

30.2 How Is Digestion Accomplished?

An Overview of Digestion

After a meal, you may hear your stomach gurgling and churning; these noises are generated during one of the several phases of digestion. **Digestion** is the process that physically grinds up and chemically breaks down food. The **digestive systems** of animals take in food and then digest its complex molecules

Table 30-4 Dietary Changes Advocated by the U.S. Government

	Percentage of Total Daily Energy Intake	
Dietary Component	**Average U.S. Diet**	**Dietary Goals**
Carbohydrates	46	45–65
Lipids	38	20–35
Proteins	16	10–35

Summary of Recommendations:

- Increase consumption of fruits, vegetables, and whole grains.
- Decrease consumption of refined sugars.
- Decrease consumption of fats, replace saturated with unsaturated fats, and minimize intake of trans fats (see Chapter 3).
- Decrease consumption of animal fats by selecting lean meats, poultry, and fish.
- Decrease consumption of high-cholesterol foods.
- Decrease consumption of salt and foods high in salt content.
- Decrease caloric intake to maintain desirable weight.
- Engage in a total of 1 hour of moderate physical activity (such as brisk walking) daily.

into simpler molecules that can be absorbed. Material that cannot be broken down or used is then expelled from the body.

Animals eat the bodies of other organisms, but these bodies may resist becoming food. The plant body, for example, supports each cell with a wall of indigestible cellulose. Animal bodies may be covered with equally indigestible fur, scales, or feathers. In addition, the complex lipids, proteins, and carbohydrates in food do not occur in a form that can be used directly. These nutrients must be broken down before they can be absorbed and distributed to the cells of the animal that has consumed them, where they recombine in unique ways. Different types of animals acquire nutrients with various types of digestive tracts, each finely tuned to meet the challenges of a unique diet and lifestyle. Amid this diversity, however, all digestive systems must accomplish certain tasks:

1. *Ingestion.* The food must be brought into the digestive tract through an opening, usually called a **mouth**.
2. *Mechanical breakdown.* The food must be physically broken down into smaller pieces. This is accomplished by gizzards or teeth as well as by the churning action of the digestive cavity itself. The particles produced by mechanical breakdown provide increased surface area, allowing digestive enzymes to attack them more effectively.

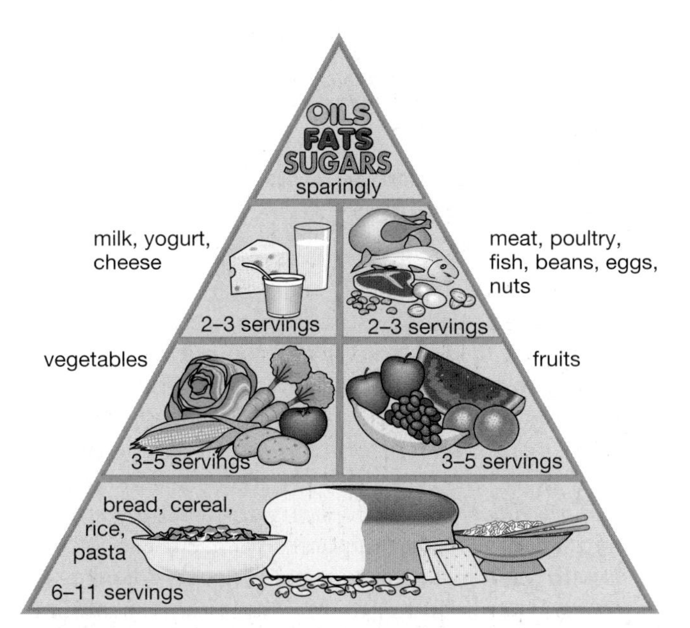

FIGURE 30-3 The Food Guide Pyramid
This chart shows the daily servings suggested by the U.S. Department of Agriculture.

NUTRITION FACTS

Serving Size	1 Cup (55 g)
Servings Per Package	8
Amount Per Serving	
Calories 210	Calories from Fat 0
	% Daily Value*
Total Fat 0 g	0%
Saturated Fat 0 g	0%
Cholesterol 0 mg	0%
Sodium 20 mg	1%
Total Carbohydrate 46 g	15%
Dietary Fiber 6 g	24%
Sugars 12 g	
Protein 6 g	
Vitamin A* •	Vitamin C 2%
Calcium 4% •	Iron 18%
Thiamin 38% •	

*Percent Daily Values are based on a 2,000 calorie diet. Your Daily Values may be higher or lower depending on your calorie needs:

	Calories	2,000	2,500
Total Fat	Less than	65 g	80 g
Saturated Fat	Less than	20 g	25 g
Cholesterol	Less than	300 mg	300 mg
Sodium	Less than	2,400 mg	2,400 mg
Total Carbohydrate		300 g	375 g
Dietary Fiber		25 g	30 g

Calories per gram:
Fat 9 • Carbohydrate 4 • Protein 4

FIGURE 30-4 Complete food labels
The U.S. government requires complete nutritional labeling of foods, as illustrated by this sample. The weight (in grams) of various nutrients—such as fat, cholesterol, and sodium—is shown as a percentage of the recommended daily allowance, assuming a 2000-Calorie diet. At the bottom of the label, the total recommended number of grams of these nutrients is listed for a 2000- and a 2500-Calorie diet. **QUESTION** Compare this food label to the recommendations in the Food Guide Pyramid (Fig. 30-3). According to the USDA, would eating this food be a wise nutritional choice?

HEALTH WATCH Eating Disorders—Betrayal of the Body

Natural selection has provided animals with strong drives to eat when nutrients are needed (and even when nutrients are not needed, if good food is available). In humans, these natural impulses, so crucial to health and well-being, can go terribly awry. In recent decades we have seen an increase in both overeating and *eating disorders*, ailments characterized by the disruption of normal eating behavior.

Eating disorders include two particularly debilitating illnesses, *anorexia nervosa* and *bulimia nervosa*. People with anorexia experience an intense fear of gaining weight, and they achieve extreme weight loss—often 30% or more below their normal weight—by eating very little food. Victims of this disorder also engage in self-induced vomiting, laxative intake, and excessive exercise. The consequences are disastrous. Anorexics become emaciated, losing both fat and muscle. This emaciation can in turn disrupt cardiac, digestive, endocrine, and reproductive functions. As many as 18% of anorexics die as a result of their disorder. Victims of bulimia, who maintain a normal weight, engage in binge eating, rapidly consuming huge amounts of food. After bingeing, bulimic individuals typically induce vomiting and may also take laxatives. The repeated vomiting damages the digestive tract and upsets the normal balance of salts in the blood, which can lead to heart disorders.

FIGURE E30-1 Eating disorders
The late Princess Diana made headlines when she announced that she had been suffering from bulimia.

More than 90% of diagnosed eating disorders occur in females (Fig. E30-1); the incidence of anorexia among U.S. women is about 1% or 2%. Among Caucasian females, the disorder usually starts between 12 and 20 years of age. Scientists don't understand the causes of eating disorders and have not identified a clear genetic link. Therefore, researchers suspect that environmental factors—especially societal pressures to be thin—contribute to these diseases, coupled with the effects of individual personality traits, such as a high need for achievement and acceptance. Anthropologist Anne Becker of Harvard University studied girls in the Fiji Islands, where 80% of women are overweight by U.S. standards. Shortly after U.S. shows glorifying slim actresses were introduced to the islands, the incidence of vomiting to control weight increased by a factor of five among teenaged Fijian girls. Dr. Becker's findings support the concept that social pressures play a major role in eating disorders.

Unfortunately, it is difficult to treat eating disorders. Victims are usually hospitalized to restore nutritional health and given counseling to deal with the disorder, but they often fail to recover completely. Antidepressant drugs are helpful in some cases, but are more likely to be effective for bulimia than for the more serious anorexia nervosa. Research into the hormones that control appetite may one day lead to more effective treatments for eating disorders. For example, researchers have discovered a class of hormones known as *orexins* (from the Greek word *orexis*, meaning "appetite"), which, when injected into mice, bind to receptors in the brain and cause food consumption to increase dramatically. Such new discoveries raise hopes that, as our understanding of the physiological control of appetite grows, we may yet be able to devise chemical treatments for eating disorders.

3. *Chemical breakdown.* The particles of food must be exposed to enzymes and other digestive fluids that break down large molecules into smaller subunits.
4. *Absorption.* The small subunits must be transported out of the digestive cavity and into cells.
5. *Elimination.* Indigestible materials must be expelled from the body.

In the following section, we will briefly explore some of the diverse mechanisms by which animal digestive systems accomplish these functions. From a simplified perspective, animals are machines for converting food into more animals, allowing them to renew their bodies and reproduce. Natural selection has favored a wide variety of behaviors and digestive adaptations that allow animals to acquire and digest food from almost every conceivable source.

Digestion Within Single Cells Occurs in the Sponges

Simple digestive systems can be as efficient as more complex ones if a comparatively small amount of energy is used to acquire and digest food. Sponges, for example, are sedentary feeders with no specialized digestive systems; they prosper with a system in which digestion occurs within individual cells (Fig. 30-5). Such **intracellular digestion** occurs after a cell has engulfed microscopic food particles. Once ingested by a cell, the food is enclosed in a **food vacuole**, a space surrounded by a membrane that serves as a temporary stomach. The vacuole fuses with **lysosomes**, which are small packets of digestive enzymes. Food is broken down within the vacuole into smaller molecules that can be absorbed into the cell cytoplasm. Undigested remnants remain in the vacuole,

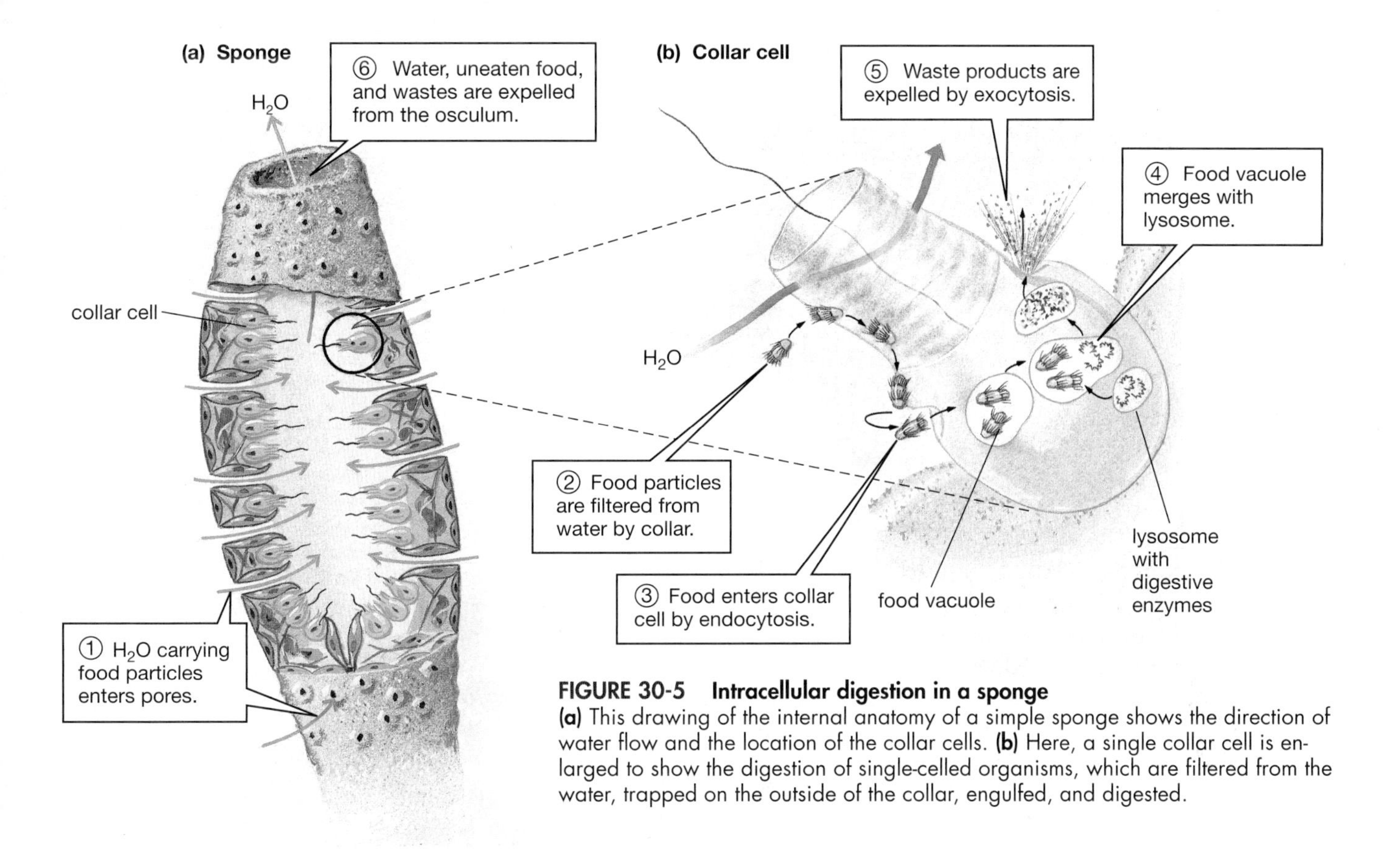
(a) Sponge
⑥ Water, uneaten food, and wastes are expelled from the osculum.
H_2O
collar cell
① H_2O carrying food particles enters pores.
(b) Collar cell
⑤ Waste products are expelled by exocytosis.
④ Food vacuole merges with lysosome.
H_2O
② Food particles are filtered from water by collar.
③ Food enters collar cell by endocytosis.
food vacuole
lysosome with digestive enzymes

FIGURE 30-5 Intracellular digestion in a sponge
(a) This drawing of the internal anatomy of a simple sponge shows the direction of water flow and the location of the collar cells. **(b)** Here, a single collar cell is enlarged to show the digestion of single-celled organisms, which are filtered from the water, trapped on the outside of the collar, engulfed, and digested.

which eventually expels its contents outside the cell. Intracellular digestion such as this is seen in single-celled protists and in the simplest animals. Sponges, for example, rely entirely on intracellular digestion. This process limits their menu to microscopic food particles, such as minute organisms (phytoplankton) that are filtered from the surrounding sea by means of the sievelike collar cells (see Chapter 22).

A Sac with One Opening Forms the Simplest Digestive System

Larger, more complex organisms evolved a chamber within the body where chunks of food are broken down by enzymes that act outside the cells. This process is called **extracellular digestion.** One of the simplest of these chambers is found in cnidarians, such as sea anemones, hydra, and jellyfish. As you learned in Chapter 22, these animals possess a digestive sac called a **gastrovascular cavity**, which has a single opening through which food is ingested and wastes are ejected (Fig. 30-6). Although it is

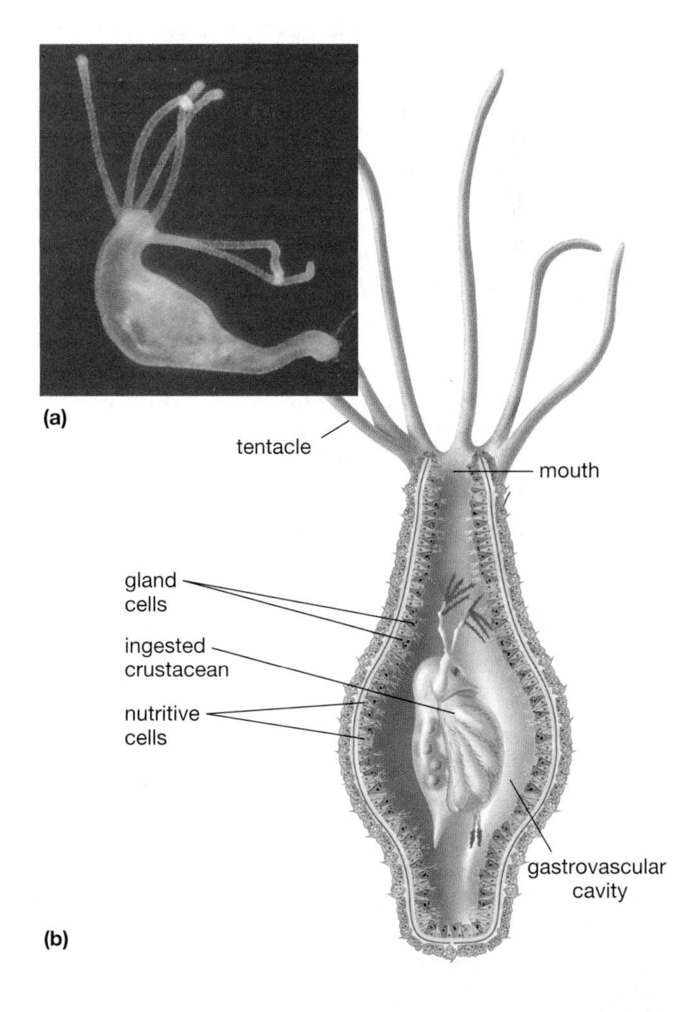
(a)
tentacle
mouth
gland cells
ingested crustacean
nutritive cells
gastrovascular cavity
(b)

FIGURE 30-6 Digestion in a sac
(a) A *Hydra* has captured and ingested a tiny crustacean.
(b) Within the gastrovascular cavity of a *Hydra*, gland cells secrete enzymes that digest the prey into smaller particles and nutrients. Elongated cells lining the cavity ingest and process these particles by intracellular digestion (see Fig. 30-5). Undigested waste is then expelled through the single opening.

FIGURE 30-7 The ruminant digestive system
Arrows trace the path of food through the digestive tract. **QUESTION** In addition to the ability to digest cellulose, what other nutritional benefits might ruminants gain by having microorganisms in their guts?

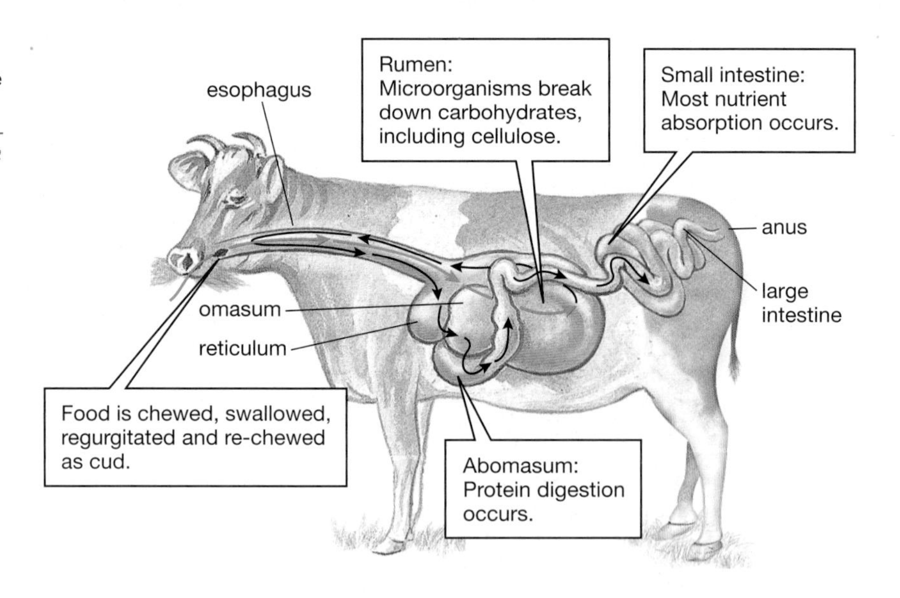
esophagus
Rumen: Microorganisms break down carbohydrates, including cellulose.
Small intestine: Most nutrient absorption occurs.
anus
large intestine
omasum
reticulum
Food is chewed, swallowed, regurgitated and re-chewed as cud.
Abomasum: Protein digestion occurs.

generally referred to as the mouth, this opening also serves as an anus. The animal's stinging tentacles capture food and escort it into the gastrovascular cavity, where enzymes break it down. Cells lining the cavity absorb the nutrients and engulf small food particles. Further digestion then occurs via intracellular digestion. The undigested remains are eventually expelled through the same opening by which the food entered.

Because the same chamber is used for digestion, absorption, and elimination of wastes, only one meal can be processed at a time. Thus, this type of digestive system is unsuited to active animals, which require frequent meals, and to animals whose food offers so little nutrition that they must feed continually. The needs of such animals are met by a digestive system that consists of a one-way tube with a series of compartments—such as the esophagus, stomach, and intestines—with an opening at each end.

Digestion in a Tube Allows Animals to Feed More Frequently

A tubular digestive tract allows the animal to eat frequently, since the incoming food does not interfere with any outgoing wastes. Most animals, including people and other vertebrates, earthworms, mollusks, arthropods, and echinoderms, have digestive systems that are basically tubes that begin with a mouth and end with an anus. Specialized regions within the tube process the food in an orderly sequence: first physically grinding it up, then enzymatically breaking it down, and finally absorbing the nutrients into the body. Animals with tubular digestive systems use extracellular digestion to break down their food.

Digestive Specializations

Specialized tubular digestive tracts allow different types of animals to eat a wide range of foods and to extract the maximum amount of nutrients from them. **Carnivores**, such as dogs and wolves, cats, and predatory birds, eat other animals. **Herbivores** eat only plants. These animals include seed-eating birds, grazing animals such as deer, horses, and cows, and many rodents such as mice. Animals such as humans, bears, and raccoons are called **omnivores** because they are adapted to digest all different types of food.

Special Adaptations Allow Ruminants to Digest Cellulose

The cellulose surrounding each plant cell is potentially one of the most abundant food energy sources on Earth; nevertheless, if people were restricted to a cow's diet of grass, we would soon starve. Although cellulose, like starch, consists of long chains of glucose molecules, it resists the attack of animal digestive enzymes (see Chapter 3). **Ruminant** animals—cows, sheep, goats, camels, and hippos, to name a few—have evolved elaborate digestive systems housing microorganisms that can break down cellulose. *Rumination*, or "cud-chewing," is the process of regurgitating food and rechewing it, and is one of several adaptations that enable these animals to digest tough plant material. Ruminant stomachs consist of several chambers (Fig. 30-7). The first chamber, the *rumen*, has evolved into a large fermentation vat which, in cows, can hold up to 200 liters (about 50 gallons). There, microorganisms—including many species of bacteria and ciliates—thrive. In addition to digesting plant sugars and starches, these microorganisms produce **cellulase**, an enzyme that breaks down cellulose into its component sugars. After being processed in the rumen, the plant material enters the reticulum and is formed into masses called *cud*. The cud is regurgitated, chewed, and swallowed into the rumen. The extra chewing exposes more of the cellulose and cell contents to the cellulase and other enzymes of the rumen's microorganisms for further digestion. Gradually, the partially digested plant material and microorganisms are released into the remaining chambers, traveling through the narrow *omasum* and then into the larger *abomasum*, where protein digestion occurs. Here the cow digests not only plant proteins, but also the microorganisms from its rumen. The cow then absorbs most of the products of digestion through the walls of

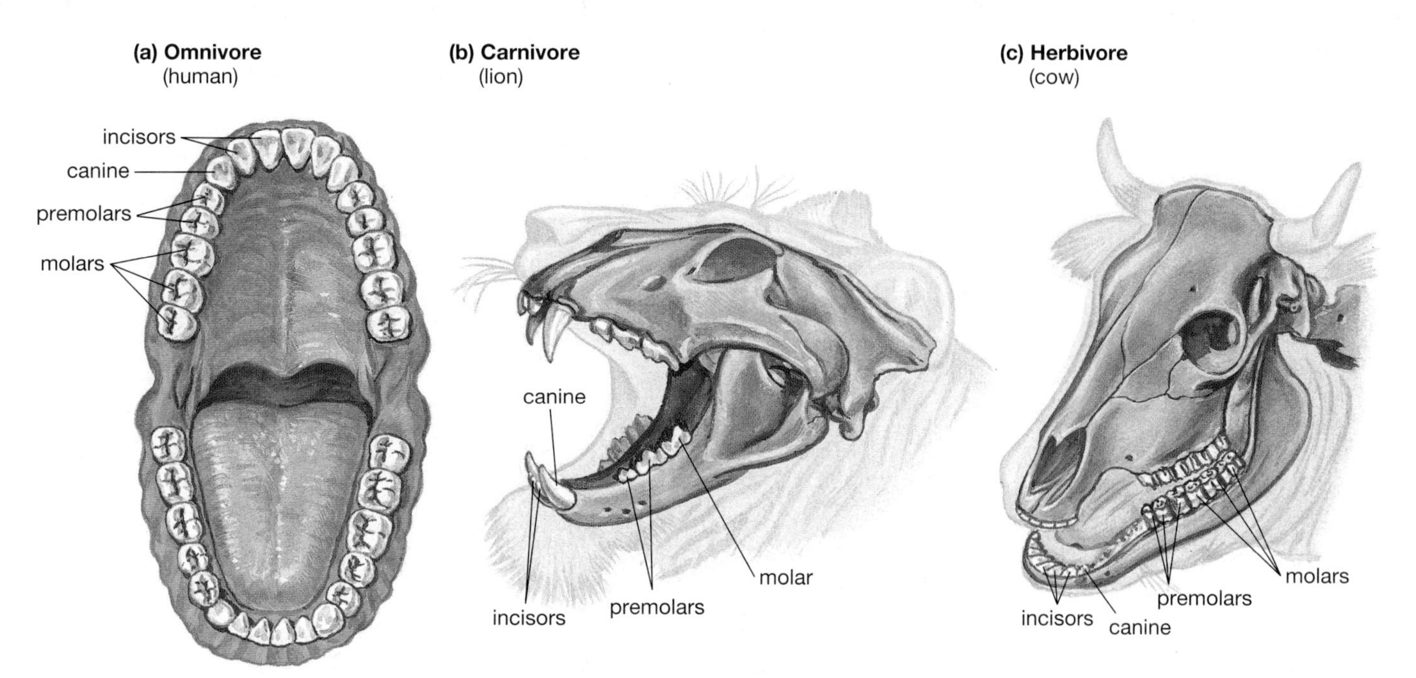

FIGURE 30-8 Teeth evolved to suit different diets
QUESTION The mouthparts of many animals that lack teeth have also evolved to suit their diets. For each of the following animals, briefly describe its mouthparts and explain how they help the animal eat a particular kind of food: mosquito (Fig. 19-21), butterfly, toad (Fig. 38-3), hummingbird (Fig. 23-9), eagle (Fig. 39-1).

its small intestine, but absorbs any remaining water through the large intestine.

Intestinal Length Is Correlated with Diet

Since most digestion and absorption of nutrients occurs in the intestine, a longer intestine provides herbivores with more opportunity to extract nutrients from plants (whose cell walls are difficult to digest). In general, carnivores, whose diets are mainly protein (which is relatively east to digest) have shorter intestines than herbivores. This tendency is strikingly illustrated during the development of frogs. In its juvenile tadpole stage, the frog is an algae-eating herbivore with a long intestine. When it metamorphoses into a carnivorous (usually insect-eating) adult frog, its intestine shortens to about one-third of its previous length (you can see a frog and its tadpole larva in Fig. 29-5a,b).

Teeth Evolved to Accommodate Different Diets

Each type of animal has teeth that are uniquely suited to its diet. The varied, omnivorous diet of humans has fostered the evolution of flat incisors for biting, pointed canines for tearing, premolars for grinding, and molars for crushing and chewing (Fig. 30-8a). If you have a dog, look carefully in its mouth. Carnivores have modest incisors but greatly enlarged canines for stabbing and tearing flesh. They have a reduced set of molars and premolars with specialized sharp edges for shearing through tendon and bone (Fig. 30-8b). Herbivores have incisors designed for snipping leaves, and their canines—which are reduced in size and placed forward in the mouth—also help with that job. A herbivore's cheek is filled with a full set of wide, flat premolars and molars that grind up tough, cellulose-containing plants (Fig. 30-8c). Many herbivores have teeth that grow continuously throughout their lives to compensate for the wear.

Birds Have Gizzards for Grinding Food

Birds lack teeth, but most eat food that requires the equivalent of chewing, such as seeds or small bony mammals (for example, mice). Birds swallow their food whole, after which it passes through a muscular, tubular esophagus. In seed-eating birds, the food may be stored and softened by water in a large, expandable crop. The food then passes gradually into a highly specialized stomach (Fig. 30-9). The first portion of the stomach secretes digestive enzymes, while the next

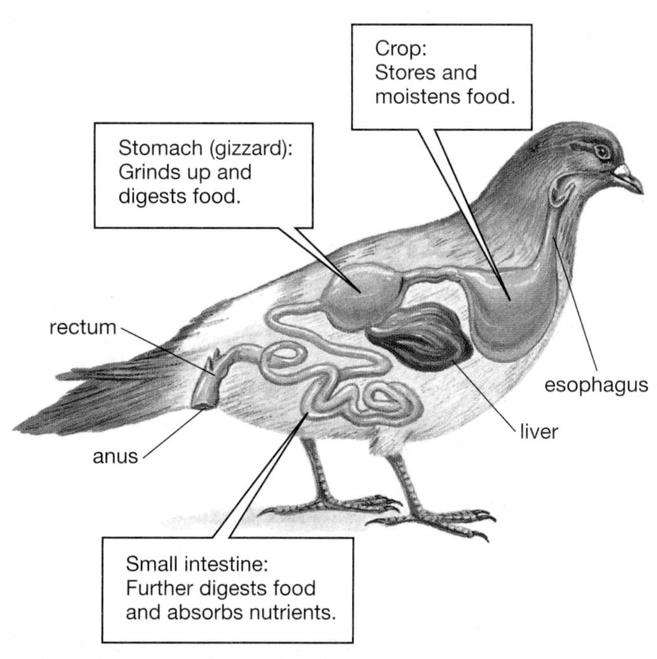

FIGURE 30-9 Bird digestive adaptations
QUESTION Why do birds lack teeth?

Media Activity
30.1 The Digestive System

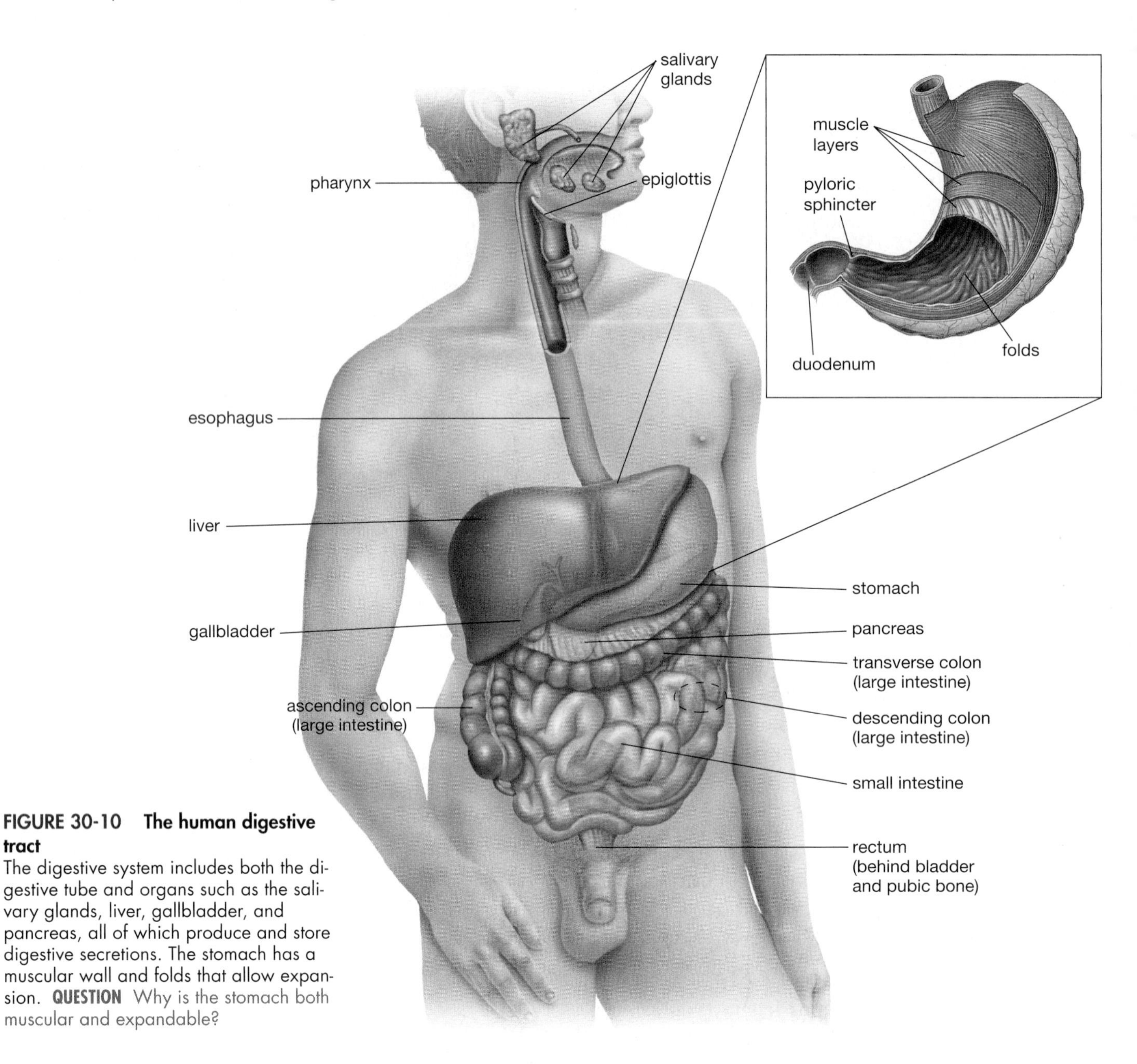

FIGURE 30-10 The human digestive tract
The digestive system includes both the digestive tube and organs such as the salivary glands, liver, gallbladder, and pancreas, all of which produce and store digestive secretions. The stomach has a muscular wall and folds that allow expansion. **QUESTION** Why is the stomach both muscular and expandable?

Media Activity
30.2 Physical and Chemical Digestion

portion is modified into a grinding gizzard—a thick-walled, muscular chamber with a hard lining. Many bird species swallow small, sharp-edged stones that lodge in the gizzard and act like teeth, crushing and grinding the food under pressure from the gizzard's muscular contractions. In carnivorous birds, the gizzard is smaller; bones, hair, and feathers of prey are trapped in the gizzard and regurgitated. Further digestion and most absorption of nutrients occurs in the small intestine.

30.3 How Do Humans Digest Food?

The human digestive system (Fig. 30-10) is adapted for processing the wide variety of different foods in our omnivorous diet, and provides a good example of the mammalian digestive system. Food travels in a continuous tube from mouth to anus; in the course of this circuitous route, it is subjected to a precisely orchestrated succession of digestive operations. By the time its passage is complete, the food has been chopped, mashed, mixed, churned, and bathed in a series of powerful chemicals. Everything of nutritional value has been extracted, and the residue is ejected. This sequential breakdown of food requires coordinated action from the variety of structures that make up the digestive system.

The Mechanical and Chemical Breakdown of Food Begins in the Mouth

You take a bite, your mouth waters, and you begin chewing. This begins both the mechanical and the chemical breakdown of food. In people and other mammals, the mechanical work is done mostly by teeth. *Incisors* at the front of the mouth snip off pieces of food, the pointed *canine* teeth beside them are useful for tearing the

Table 30-5 Digestive Structures and Secretions

Site of Digestion	Secretion	Source of Secretion	Role in Digestion
Mouth	Amylase	Salivary glands	Breaks down starch into disaccharides
	Mucus, water	Salivary glands	Lubricates, dissolves food
Stomach	Hydrochloric acid	Cells lining stomach	Allows pepsin to work, kills bacteria, solubilizes minerals
	Pepsin	Cells lining stomach	Breaks down proteins into large peptides
	Mucus	Cells lining stomach	Protects stomach
Small intestine	Sodium bicarbonate	Pancreas	Neutralizes acidic chyme from stomach
	Amylase	Pancreas	Breaks down starch into disaccharides
	Proteases	Pancreas	Breaks down proteins into large peptides
	Lipase	Pancreas	Breaks down lipids into fatty acids and glycerol
	Bile	Liver	Emulsifies lipids
	Peptidases	Small intestine	Split small peptides into amino acids
	Disaccharidases	Small intestine	Split disaccharides into monosaccharides

pieces apart, and the *premolars* and *molars* at the back of the mouth have flat surfaces for grinding food to a paste (see Fig. 30-8). In adult humans, 32 teeth of varying shapes and sizes cut and grind food into small pieces.

While the teeth pulverize the food, the first phase of chemical digestion occurs as three pairs of salivary glands pour out saliva in response to the smell, feel, taste, and (if you're hungry) even the thought of food. Saliva contains the digestive enzyme **amylase**, which begins the breakdown of starches into sugar (Table 30-5). Saliva has other functions as well. It contains a bacteria-killing enzyme and antibodies that help guard against infection. Saliva also lubricates the food to facilitate swallowing and dissolves some food molecules such as acids and sugars, carrying them to *taste buds* on the tongue. The taste buds are sensory receptors that help identify the type and quality of the food.

With the help of the muscular tongue, the food is manipulated into a mass and pressed backward into the **pharynx**, a muscular cavity connecting the mouth with the esophagus (Fig. 30-11a). Through the *larynx*, the pharynx also connects the nose and mouth with the *trachea*, which conducts air to the lungs. This arrangement occasionally causes problems, as anyone who has ever choked on a piece of food can attest. Normally, however, the swallowing reflex (triggered by food entering the pharynx) elevates the larynx so it meets the

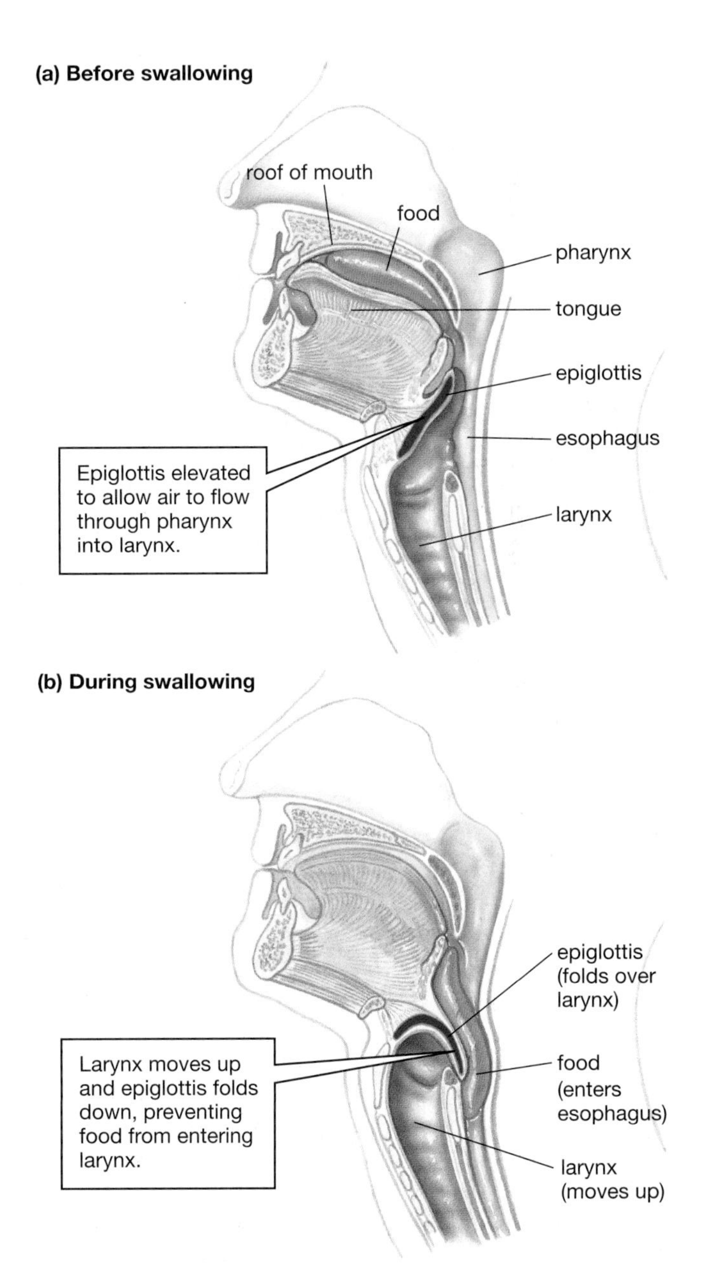

FIGURE 30-11 The challenge of swallowing
(a) Swallowing is complicated by the fact that both the esophagus (part of the digestive system) and the larynx (part of the respiratory system) open into the pharynx. **(b)** During swallowing, the larynx moves upward beneath a small flap of cartilage, the epiglottis. The epiglottis folds down over the larynx, sealing off the opening to the respiratory system and directing food down the esophagus instead.

HEALTH WATCH Ulcers: Digesting the Digestive Tract

About 1 out of every 10 Americans eventually develops an *ulcer*. Ulcers occur when localized areas of the tissue layers that line the stomach or upper portion of the small intestine become eroded. Ulcer victims can experience burning pain in the stomach area, vomiting and nausea, and in severe cases, blood may by passed with the feces, due to bleeding at the site of tissue destruction (Fig. E30-2). Until recently, the medical community believed that this destruction was caused mainly by overproduction of acid (thought to be stress-related), which led to the breakdown of the protective mucus barrier and destruction of tissue by acid and pepsin. Ulcers were treated with antacids and stress-reduction programs. Now, however, the Centers for Disease Control reports that the bacterium *H. pylori* causes about 90% of all ulcers, and that appropriate antibiotics (used in conjunction with acid-reducing medications) can cure most ulcers in 1 to 2 weeks.

What caused this recent change in our understanding of ulcers? In 1983 J. Robin Warren, an Australian pathologist, noticed that samples of inflamed stomach tissue were consistently infected with a spiral-shaped bacterium. Barry Marshall, a trainee in internal medicine, joined Warren in his attempts to isolate and culture the undescribed bacterium, which was later named *Helicobacter pylori*. The researchers then proposed that *H. pylori* caused the inflammation that could lead to ulcers. But the medical community was skeptical. How could bacteria survive, much less flourish, in the acidic, protein-digesting environment of the stomach? To prove their point, Marshall and another volunteer swallowed a batch of the bacteria and later provided samples of their own *H. pylori*-infected stomach tissue. More research and epidemiological studies supported Warren and Marshall's hypothesis and confirmed the results of Marshall's unusual experiment. Scientists now know that the bacterium seeks shelter beneath the protective layer of mucus that coats the stomach wall. The bacteria both weaken the mucus layer and increase stomach acid production, making the stomach lining more susceptible to attack by acid and protein-digesting enzymes. The inflammation caused by the infection also contributes to the destruction of tissue in the stomach and upper small intestine. Interestingly, although about half of the world's population harbors *H. pylori*, most infected people do not have ulcers or other obvious symptoms of infection.

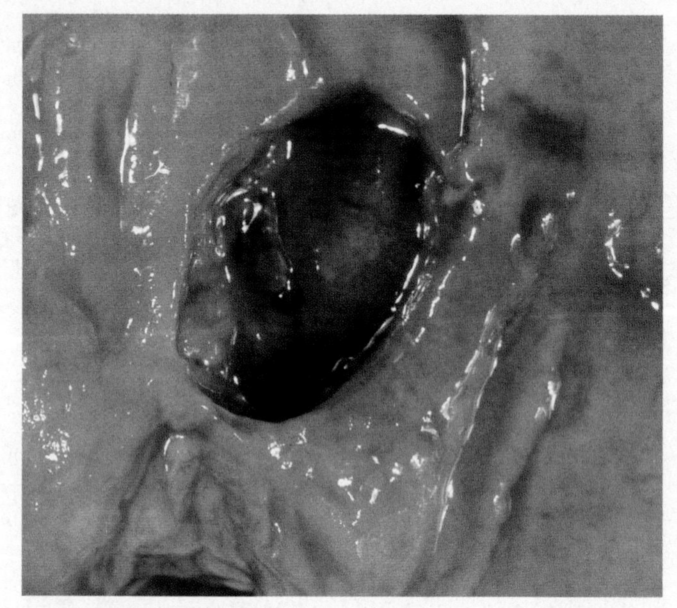

FIGURE E30-2 An ulcer
This ulcer was photographed through a fiber-optic viewing device called an endoscope.

Prolonged use of pain-killers such as aspirin or ibuprofen can also cause ulcers by interfering with the mechanisms that protect the stomach and intestinal cells from the action of acids and digestive enzymes. Other factors that may aggravate ulcers and slow their healing include smoking, caffeine, and possibly alcohol.

epiglottis, a flap of tissue that blocks off the respiratory passages. Food is thus directed into the esophagus rather than into the trachea (Fig. 30-11b).

The Esophagus Conducts Food to the Stomach

Swallowing forces food into the esophagus, a muscular tube that propels food from the mouth to the stomach. Muscles surrounding the esophagus produce a wave of contraction that begins just above the swallowed mass and progresses down the esophagus, forcing the food down toward the stomach. This muscular action, called **peristalsis**, also occurs in the stomach and intestines, where it propels food along the digestive tract. Peristalsis is so effective that a person can actually swallow when upside down. Mucus secreted by cells that line the esophagus helps protect it from abrasion and lubricates the food during its passage.

The **stomach** in humans is an expandable muscular sac capable of holding from 2 to 4 liters (as much as a gallon) of food and liquids. Food is retained in the stomach by a ring of circular muscle that separates the lower portion of the stomach from the upper *small intestine*. This muscle, called the **pyloric sphincter**, regulates the passage of food into the small intestine, as described later.

The stomach has three major functions. First, the stomach stores food and releases it gradually into the small intestine at a rate suitable for proper digestion and absorption. Folds in the stomach wall (see Fig. 30-10, inset) increase its capacity, allowing us to eat large, infrequent meals. Carnivores carry this ability to an extreme. A lion, for instance, may consume about 18 kilograms (40 pounds) of meat at one meal, then spend the next few days quietly digesting it. A second function of the stomach is to assist in the mechanical breakdown of food. In addition to peristalsis, its muscular walls undergo a variety of churning contractions that help break apart large pieces of food. The third function of the stomach is the chemical breakdown of food. Glands in the lining of the stomach secrete enzymes and other substances, including gastrin, hydrochloric acid (HCl), pepsinogen, and mucus. **Gastrin** (a hormone) stimulates the secretion of hydrochloric acid by specialized stomach cells. Other cells release *pepsinogen*, an inactive form of the protein-digesting enzyme *pepsin*. Pepsin is a **protease**, an enzyme that helps break proteins into shorter chains of amino acids called *peptides* (see Table 30-5). Pepsin is secreted in the form of pepsinogen to prevent it from digesting the very cells that produce it. Hydrochloric acid, which gives the fluid in the stomach a very acidic pH of 1 to 3, converts the pepsinogen into pepsin, which functions best in an acidic environment.

As you may have noticed, the stomach produces all the ingredients necessary to digest itself. Indeed, this is what happens when a person develops ulcers (see "Health Watch: Ulcers: Digesting the Digestive Tract"). However, cells lining the stomach normally produce a large quantity of thick mucus that coats the walls of the stomach and serves as a barrier to self-digestion. The protection is not perfect, however, and the cells lining the stomach are digested to some extent and must be replaced every few days.

Food in the stomach is gradually converted to a thick, acidic liquid called **chyme**, which consists of partially digested food and digestive secretions. Peristaltic waves then propel the chyme toward the small intestine. The pyloric sphincter allows only about a teaspoon of chyme to be expelled with each contraction, which occurs about every 20 seconds. Depending on the size of the meal, it takes 2 to 6 hours to empty the stomach completely. The continued churning movements of an empty stomach are felt as "hunger pangs."

Only a few substances, including water, some drugs, and alcohol, can enter the bloodstream through the stomach wall. Alcohol that is consumed when the stomach is empty is immediately absorbed into the bloodstream, with strong and rapid effects. Because food in the stomach slows alcohol absorption, the advice "never drink on an empty stomach" is based on sound physiological principles.

Most Digestion Occurs in the Small Intestine

The **small intestine** is narrow (about 1 to 2 inches in diameter in an adult human), but with a length of 10 feet, it is the longest part of the digestive tract. The small intestine functions to digest food into small molecules and to absorb these molecules into the bloodstream. The first role of the small intestine—digestion—is accomplished with the aid of digestive secretions from three sources: the liver, the pancreas, and the cells of the small intestine itself.

The Liver and Gallbladder Provide Bile, Important in Fat Breakdown

The **liver** is perhaps the most versatile organ in the body. The liver stores fats and carbohydrates for energy, regulates blood glucose levels, synthesizes blood proteins, stores iron and certain vitamins, converts toxic ammonia (released when amino acids are metabolized) into urea, and detoxifies harmful substances we ingest such as nicotine and alcohol. The role of the liver in digestion is to produce *bile*, a liquid stored and concentrated in the **gallbladder** and released into the small intestine through a tube called the *bile duct* (see Fig. 30-10).

Bile is a complex mixture composed of **bile salts**, water, other salts, and cholesterol. Bile salts are synthesized in the liver from cholesterol and amino acids. Although they assist in the breakdown of lipids, bile salts are not enzymes. Rather, they act as detergents or emulsifying agents, dispersing globs of fat in the chyme into microscopic particles. These particles expose a large surface area for attack by **lipases**, lipid-digesting enzymes produced by the pancreas.

The Pancreas Supplies Several Digestive Secretions to the Small Intestine

The **pancreas** lies in the loop between the stomach and small intestine (see Fig. 30-10). It consists of two major types of cells. One type produces hormones involved in blood sugar regulation (as we will see in Chapter 33), and the other produces a digestive secretion called **pancreatic juice**, which is released into the small intestine. Pancreatic juice neutralizes the acidic chyme and digests carbohydrates, lipids, and proteins. About 1 liter (1.06 quarts) of pancreatic juice is released into the small intestine each day. This secretion contains water, sodium bicarbonate, and several digestive enzymes (see Table 30-5). Sodium bicarbonate (the active ingredient in baking soda) neutralizes the acidic chyme in the small intestine, producing a slightly basic pH. Pancreatic digestive enzymes require a more basic (alkaline) pH to function properly, in contrast to the stomach's digestive enzymes, which require an acidic pH.

The pancreatic digestive enzymes break down three major types of nutrients: Amylase breaks down carbohydrates; lipases attack lipids; and several proteases break down proteins and peptides.

The Digestive Process Is Completed by Cells of the Intestinal Wall

The wall of the small intestine is studded with cells that are specialized to complete the digestive process and absorb the small molecules that result. These cells have various enzymes on their external membranes, which form the lining of the small intestine. The enzymes include *peptidases*, which complete the breakdown of peptides into amino acids, and *disaccharidases*, which break down disaccharides into monosaccharides (see Chapter 3). Small amounts of lipase digest lipids. Because these enzymes are actually embedded in the membranes of the cells that line the small intestine, this final phase of digestion occurs *as* the nutrients are being absorbed into the cell. Like the stomach, the

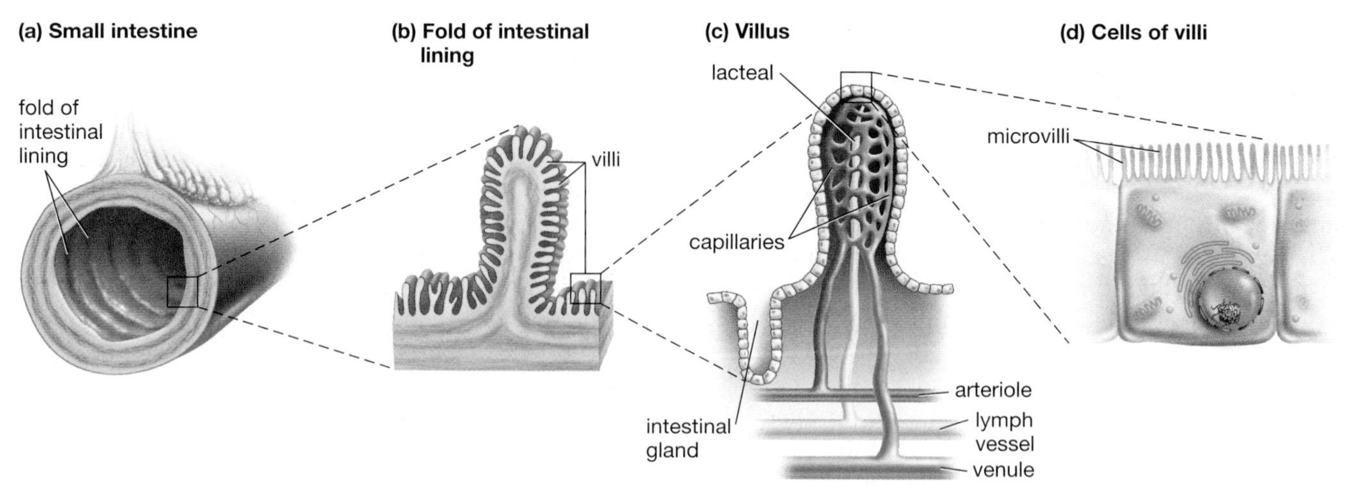

FIGURE 30-12 The small intestine
The folds of the small intestine maximize the surface area available to absorb nutrients. **(a)** Macroscopic folds in the intestinal lining are carpeted with **(b)** tiny projections called villi, which enclose **(c)** a network of capillaries and lymph vessels. **(d)** If we use a microscope to zoom in on one villus, we see that the epithelial cells on its surface are sheathed in plasma membranes with yet another level of microscopic projections, microvilli. **QUESTION** What might the anatomy of the digestive system be like if the folds, villi, and microvilli of the small intestine had not evolved?

small intestine is protected from digesting itself by large amounts of mucous secretions from specialized cells in its lining.

Most Absorption Occurs in the Small Intestine

The small intestine is not only the principal site of chemical digestion, but it is also the major site of nutrient **absorption** into the blood. The small intestine has numerous folds and projections that give it an internal surface area that is 600 times that of a smooth tube of the same length (Fig. 30-12). Minute, fingerlike projections called **villi** (literally, "shaggy hairs"; singular, *villus*) cover the entire folded surface of the intestinal wall. Villi, which range from 0.5 to 1.5 millimeters in length, give the intestinal lining a velvety appearance to the naked eye. They move gently back and forth in the chyme that passes through the intestine, increasing the exposure of the villi to the molecules to be digested and absorbed. Further, each individual cell of the villi bears a fringe of microscopic projections called **microvilli**. Taken together, these specializations of the lining of the small intestine give it a surface area of about 250 square meters (more than 2200 square feet; almost the size of a tennis court).

Unsynchronized contractions of the circular muscles of the intestine, called **segmentation movements**, slosh the chyme back and forth, bringing nutrients into contact with the absorptive surface of the small intestine. When absorption is complete, coordinated peristaltic waves conduct the leftovers into the *large intestine*.

Nutrients absorbed by the small intestine include water, monosaccharides, amino acids and short peptides, fatty acids produced by lipid digestion, vitamins, and minerals. The mechanisms by which this absorption occurs are varied and complex. In most cases, energy is expended to transport nutrients into the intestinal cells. (Water follows by osmosis, as described in Chapter 4.) The nutrients then diffuse out of the intestinal cells into the interstitial fluid, where they enter the bloodstream.

Each villus of the small intestine is provided with a rich supply of blood capillaries and a single lymph capillary, called a **lacteal**, to carry off the absorbed nutrients and distribute them throughout the body (see Fig. 30-12). Most of the nutrients enter the bloodstream through the capillaries, but fat subunits take a different route. After diffusing into the epithelial cells lining the small intestine, they are resynthesized into fats, coated with protein, and then released as particles into the interstitial fluid (see "A Closer Look: The Fate of Fats" at this text's Web site for Chapter 30). In this form, they enter the lymph vessels and are eventually delivered to the bloodstream when the lymph vessels empty into the veins.

Water Is Absorbed and Feces Are Formed in the Large Intestine

The **large intestine** in an adult human is about 5 feet long and about 3 inches in diameter, which is both wider and shorter than the small intestine. The large intestine has two parts: For most of its length it is called the **colon**, but its final 6-inch compartment is called the **rectum**. Into the large intestine flow the leftovers of digestion: a mixture of water, undigested fats and proteins, and indigestible fibers, such as the cell walls of vegetables and fruits. The large intestine contains a flourishing population of bacteria that thrive on these unabsorbed nutri-

Table 30-6 Some Important Digestive Hormones

Hormone	Site of Production	Stimulus for Production	Effect
Gastrin	Stomach	Food in mouth Peptides in stomach Distension of stomach	Stimulates acid secretion by cells in stomach
Secretin	Small intestine	Acid in small intestine	Stimulates bicarbonate production by pancreas and liver; increases bile output by liver
Cholecystokinin	Small intestine	Amino acids, fatty acids in small intestine	Stimulates secretion of pancreatic enzymes and release of bile by gallbladder
Gastric inhibitory peptide	Small intestine	Fatty acids and sugars in small intestine	Inhibits stomach movements and release of stomach acid

ents (although, among mammals, only ruminants harbor intestinal microorganisms that can digest cellulose). The intestinal bacteria earn their keep by synthesizing vitamin B_{12}, thiamin, riboflavin, and vitamin K. A typical human diet would be deficient in vitamin K without these helpful bacteria. Cells lining the large intestine absorb these vitamins as well as leftover water and salts.

After absorption is complete, any remaining material is compacted into semisolid **feces**. Feces consist of indigestible wastes and the dead bodies of bacteria (bacteria account for about one-third of the dry weight of feces). The feces are transported by peristaltic movements until they reach the rectum. Expansion of this chamber stimulates the urge to defecate. Although defecation is a reflex (as any new parent can attest), it is initiated voluntarily after about the age of two.

Digestion Is Controlled by the Nervous System and Hormones

The waiter places a chef salad in front of you, and you hungrily begin to eat. Without any conscious thought on your part, your body coordinates a complex series of events that converts the salad into nutrients circulating in your blood. As your mouth responds to the first bite, your stomach is alerted that food is on the way. The environments of the stomach and small intestine are maintained to suit the requirements of their respective enzymes (highly acidic in the stomach, slightly basic in the small intestine), and secretions into various parts of the digestive tract correspond with the arrival of food. Not surprisingly, the secretions and activity of the digestive tract (Table 30-6) are coordinated by both nerves and hormones.

The nervous system controls the initial phase of digestion, which involves responses to signals that originate in the head. These signals include the sight, smell, taste, and sometimes even the thought of food, as well as the muscular activity of chewing. In response to these stimuli, the salivary glands secrete saliva into the mouth while nervous signals to the stomach walls cause the secretion of acid and the hormone gastrin, which in turn stimulates further acid secretion. The concentration of acid is regulated by a negative feedback mechanism (see Chapter 27). When acid levels reach a certain point, they inhibit gastrin secretion, which inhibits further acid production.

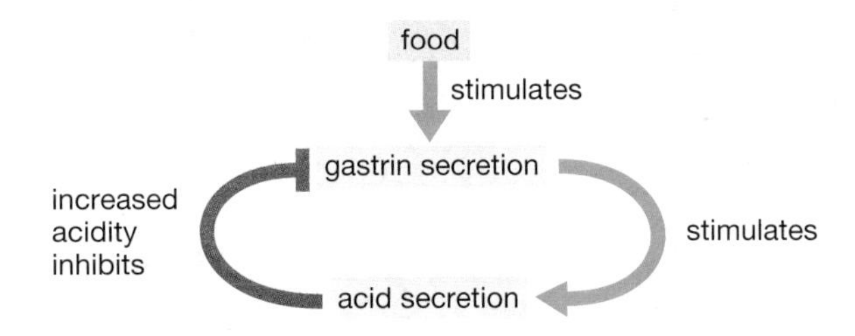

Food arriving in the stomach triggers the second phase of digestion. As the stomach wall is stretched, it produces a large quantity of mucus, which protects the stomach from digesting itself. The acidity of the stomach converts pepsinogen to its active form, pepsin, which begins protein digestion. However, the presence of protein in food tends to reduce the concentration of stomach acid. Thus, as protein is broken down, the acidity drops, and the release of gastrin is no longer inhibited; gastrin is released again and stimulates further acid production. The cells secreting stomach acid are also activated by the expansion of the stomach and by the presence of peptides produced by protein digestion.

As the liquid chyme is gradually released into the small intestine, its acidity stimulates the release of a second hormone, **secretin**, into the bloodstream by cells of the upper small intestine. Secretin causes the pancreas to pour sodium bicarbonate into the small intestine. Sodium bicarbonate neutralizes the acidity of the incoming chyme, creating an environment in which the pancreatic enzymes can function. A third hormone, **cholecystokinin**, is also produced by cells of the upper small intestine in response to the presence of chyme. This hormone stimulates the pancreas to release various digestive enzymes into the small intestine. It also stimulates the gallbladder to contract, squeezing bile through the bile duct to the small intestine. Bile assists in fat breakdown, as described earlier.

Gastric inhibitory peptide, a hormone secreted by the small intestine in response to the presence of fatty acids and sugars in chyme, inhibits acid production and peristalsis in the stomach. This inhibition slows the rate at which chyme is dumped into the small intestine, providing additional time for digestion and absorption to occur.

CASE STUDY REVISITED Fat in the Family?

Our weight is not determined by genes alone. While our genes haven't changed appreciably in the past 20 years, the number of overweight people in the United States has doubled. Although some people do have genetic tendencies that make it easier for them to gain weight and harder to lose it, everyone who is overweight has eaten more than his or her body needs. Early in this chapter, we discussed why animal bodies store fat. But most people in developed countries no longer need these adaptations, and delicious, inexpensive, calorie-laden foods beckon us from nearly every street corner. Must we succumb? Not necessarily. Researchers at the University of Pittsburgh School of Medicine and the University of Colorado Health Sciences Center are studying a group of 2800 people who were obese but reduced their weight and have maintained a weight loss of at least 30 pounds for more than 5 years. These individuals share certain characteristics: They typically exercise for at least an hour a day, weigh themselves regularly, and get about 24% of their daily calories from fat. Even Chuck is considering some changes: "There must be something between not caring at all and fighting it," he muses. "Maybe it's an alertness. . . . I am learning not to overeat."

Consider This: An obese man recently sued four prominent fast-food restaurants, claiming that their food, which he has been consuming regularly since the 1950s, is responsible for his obesity and two heart attacks. He claims that the high amounts of fat, cholesterol, salt, and sugar in the foods these chains serve lead to a variety of health problems, and that the fast-food chains should be made to pay the cost of treating these problems. If you were on a jury trying this case, how would you respond?

Links to Life: Are You Too Heavy?

A simple way to determine whether your weight might pose a health risk is by calculating your **body mass index** (BMI). The BMI uses your weight and height measurements to arrive at an estimate of body fat. The actual formula (using weight in kilograms and height in meters) is weight divided by height^2, but you can calculate the same number by multiplying your weight in pounds by 703, then dividing by your height in inches, squared. Alternatively, if you type in "body mass index" on your favorite Internet search engine, you will find many sites that calculate it for you. A BMI between 20 and 24 is considered healthy. A BMI between 25 and 30 indicates that you are overweight, and you are obese if your BMI exceeds 30. Obese people are at greater risk for heart disease, diabetes, stroke, and some forms of cancer. Good ways to lose weight include getting more exercise and eating only when you are hungry. Feeling stressed? Reach for your walking shoes instead of a candy bar!

Summary of Key Concepts

30.1 What Nutrients Do Animals Need?

Each type of animal has specific nutritional requirements. These requirements include molecules that can be broken down to liberate energy, such as lipids, carbohydrates, and proteins; chemical building blocks used to construct complex molecules, such as amino acids that can be linked together to form proteins; and minerals and vitamins that facilitate the diverse chemical reactions of metabolism.

30.2 How Is Digestion Accomplished?

Digestive systems must accomplish five tasks: ingestion, mechanical and chemical breakdown of food, absorption, and elimination of wastes. Digestive systems convert the complex molecules of the animal and plant bodies that have been eaten into simpler molecules that can be utilized. Animal digestion at its simplest is intracellular, as occurs within the individual cells of a sponge. Extracellular digestion, utilized by more complex animals, occurs within a body cavity. The simplest form is a saclike gastrovascular cavity in organisms such as flatworms and *Hydra*. Still more complex animals utilize a tubular compartment with specialized chambers where food is processed in a well-defined sequence.

30.3 How Do Humans Digest Food?

In humans, digestion begins in the mouth, where food is physically broken down by chewing, and chemical digestion is initiated by saliva. Food is then conducted to the stomach by peristaltic waves of the esophagus. In the acidic environment of the stomach, food is churned into smaller particles, and protein digestion begins. Gradually, the liquefied food, now called chyme, is released into the small intestine. There, it is neutralized by sodium bicarbonate from the pancreas. Secretions from the pancreas, liver, and the cells of the intestine itself complete the breakdown of proteins, fats, and carbohydrates. In the small intestine, the simple molecular products of digestion are absorbed into the bloodstream for distribution to the body cells. The large intestine absorbs the remaining water and converts indigestible material to feces, which are temporarily stored in the rectum and eliminated through the anus.

Digestion is regulated by the nervous system and hormones. The smell and taste of food and the action of chewing trigger the secretion of saliva in the mouth and the production of gastrin by the stomach. Gastrin stimulates stomach acid production. As chyme enters the small intestine, three additional hormones are produced by intestinal cells: secretin, which causes sodium bicarbonate production to neutralize the acidic chyme; cholecystokinin, which stimulates bile release and causes the pancreas to secrete digestive enzymes into the small intestine; and gastric inhibitory peptide, which inhibits acid production and peristalsis by the stomach. This inhibition slows the movement of food into the intestine.

Key Terms

absorption *p. 600*
amylase *p. 597*
bile *p. 599*
bile salt *p. 599*
body mass index *p. 602*
calorie *p. 586*
Calorie *p. 586*
carnivore *p. 594*
cellulase *p. 594*
cholecystokinin *p. 601*
chyme *p. 599*
colon *p. 600*
digestion *p. 590*
digestive system *p. 590*
epiglottis *p. 598*
essential amino acid *p. 587*
essential fatty acid *p. 586*
extracellular digestion *p. 593*
feces *p. 601*
food vacuole *p. 592*
gallbladder *p. 599*
gastric inhibitory peptide *p. 601*
gastrin *p. 599*
gastrovascular cavity *p. 593*
glycogen *p. 587*
herbivore *p. 594*
intracellular digestion *p. 592*
lacteal *p. 600*
large intestine *p. 600*
lipase *p. 599*
liver *p. 599*
lysosome *p. 592*
microvillus *p. 600*
mineral *p. 588*
mouth *p. 591*
nutrient *p. 586*
nutrition *p. 586*
omnivore *p. 594*
pancreas *p. 599*
pancreatic juice *p. 599*
peristalsis *p. 598*
pharynx *p. 597*
protease *p. 599*
pyloric sphincter *p. 598*
rectum *p. 600*
ruminant *p. 594*
secretin *p. 601*
segmentation movement *p. 600*
small intestine *p. 599*
stomach *p. 598*
urea *p. 587*
villi *p. 600*
vitamin *p. 588*

Thinking Through the Concepts

To take a multiple-choice quiz with feedback on the contents of this chapter, visit http://www.prenhall.com/audesirk7. *Log in to the Web site selected by your instructor and navigate to the Self Test section for this chapter.*

? Review Questions

1. List four general types of nutrients, and describe the role of each in nutrition.
2. Describe two different types of digestive tract specializations, including their function and relationship to the animal's diet.
3. List and describe the function of the three principal secretions of the stomach.
4. List the substances secreted into the small intestine, and describe the origin and function of each.
5. Name and describe the muscular movements that usher food through the human digestive tract.
6. Vitamin C is a vitamin for humans but not for dogs. Certain amino acids are essential for humans but not for plants. Explain.
7. Name four structural or functional adaptations of the human small intestine that ensure good digestion and absorption.
8. Describe protein digestion in the stomach and small intestine.

Applying the Concepts

1. The food label on a soup can shows that the product contains 10 grams of protein, 4 grams of carbohydrate, and 3 grams of fat. How many Calories are in this soup?
2. Stomach ulcers that resist antibiotic therapy are treated with several kinds of drugs. Anticholinergic drugs decrease the nerve signals to the stomach walls that are produced by the sight, smell, and taste of food. Antacids neutralize stomach acid. Why would it be inadvisable to take anticholinergics and antacids together?
3. Small birds have high metabolic rates, efficient digestive tracts, and high-calorie diets. Some birds consume an amount of food equivalent to 30% of their body weight every day. They rarely eat leaves or grass but often eat small stones. The bird's small intestine has an attached pancreas and liver. Using this information and Figure 30-9, explain how a bird's digestive tract is adapted to its lifestyle (foods consumed, flight, habitat, and so on).
4. Control of the human digestive tract involves several feedback loops and messages that coordinate activity in one chamber with those taking place in subsequent chambers. List the coordinating events you discovered in this chapter, in order, beginning with tasting, chewing,

and swallowing a piece of meat and ending with residue that enters the large intestine. What turns on and what shuts off each process?

5. Symbiotic protozoa in the digestive tracts of termites produce cellulase used by their hosts. In return, termites provide protozoa with food and shelter. Imagine that the human species is gradually invaded, over many generations, by symbiotic protozoa capable of producing cellulase. What evolutionary adaptive changes in body structure and function might occur simultaneously?

6. Trace a ham and cheese with lettuce sandwich through the human digestive system, discussing what happens to each part of the sandwich as it passes through each region of the digestive tract.

7. One of the common remedies for constipation (difficulty eliminating feces) is a laxative solution that contains magnesium salts. In the large intestine, magnesium salts are absorbed very slowly by the intestinal wall, remaining in the intestinal tract for long periods of time. Thus, the salts affect water movement in the large intestine. On the basis of this information, explain the laxative action of magnesium salts.

For More Information

Blaser, M. "The Bacteria Behind Ulcers." *Scientific American*, February 1996. Discoveries about the causes of ulcers are leading to new treatment strategies.

Diamond, J. "Dining with the Snakes." *Discover*, April 1994. Follow food as it is consumed and digested by a snake.

Mason, M. "Why Ulcers Run in Families." *Health*, September 1994. The story of the bacterium that causes most ulcers.

Moog, F. "The Lining of the Small Intestine." *Scientific American*, November 1981. A description of this intricate tissue, which is responsible for absorbing nutrients into the body.

Nuland, S. B. "The Beast in the Belly." *Discover*, February 1995. An interesting medical tale of bacterial disease, digestive enzymes, and food.

Pollen, M. "Power Steer." *New York Times Magazine*, March 31, 2002. Excellent description of how the cattle industry has caused major problems—including the development of antibiotic resistant bacteria—by ignoring the specialized ruminant digestive tract.

Willett, W. C. "Diet and Health: What Should We Eat?" *Science*, April 22, 1994. Summarizes studies that suggest that diet can play a major role in the prevention of disease.

Vogel, S. "Why We Get Fat." *Discover*, April 1999. Americans are getting fatter for many reasons—environmental, genetic, and behavioral.

Media Activities

To access a Media Activity visit http://www.prenhall.com/audesirk7. *Log in to the Web site selected by your instructor, navigate to this chapter, and select the appropriate Media Activity number.*

30.1 The Digestive System

Estimated time: 5 minutes

Explore the anatomy and function of the human digestive system.

30.2 Physical and Chemical Digestion

Estimated time: 5 minutes

See how major nutrients are broken down in each portion of the digestive system and transported to the body's cells.

30.3 Web Investigation: Fat in the Family

Estimated time: 10 minutes

This exercise will examine the roles of heredity, biochemistry, and behavior in obesity.

CHAPTER

31 The Urinary System

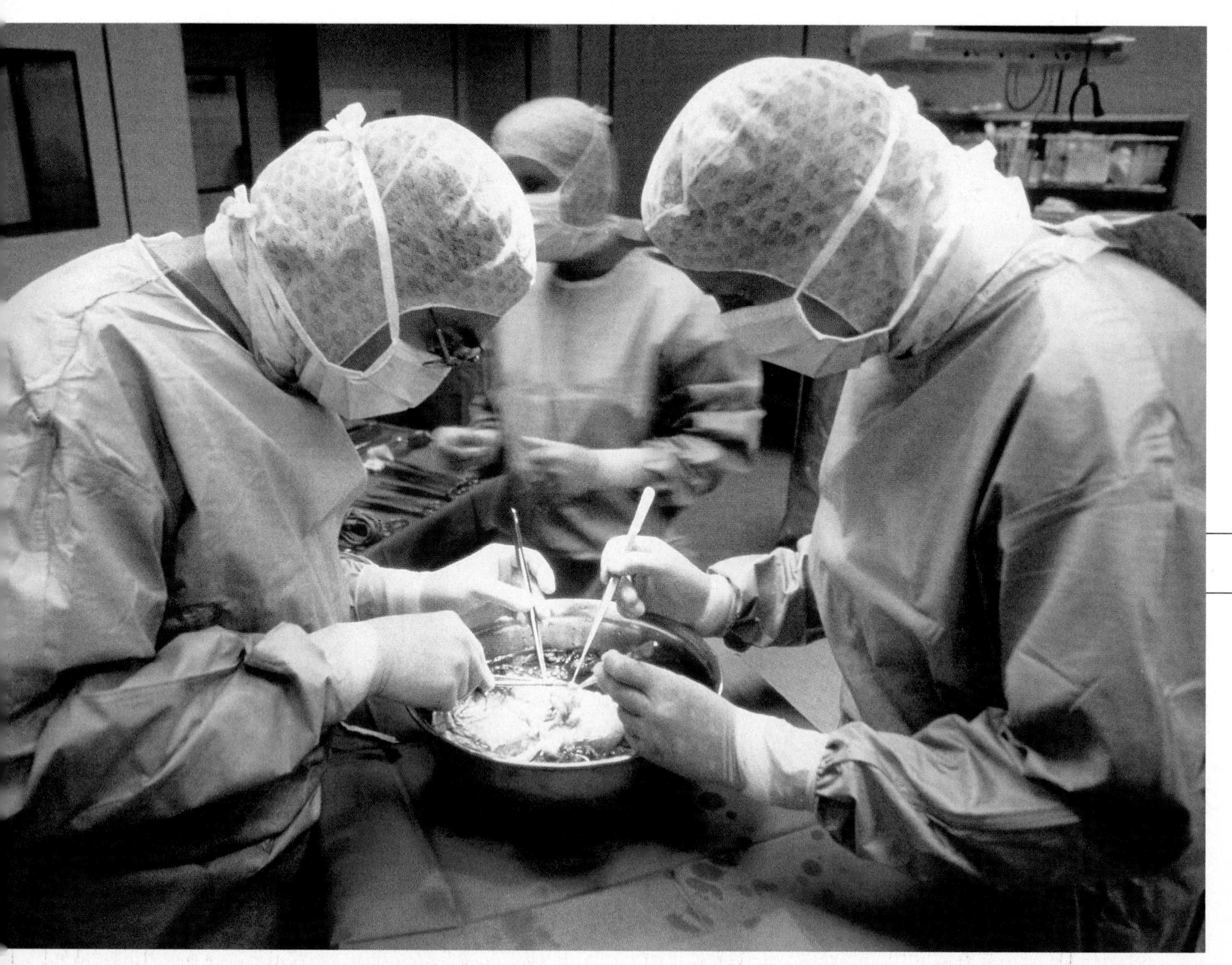

Surgeons prepare a donated kidney for transplant.

AT A GLANCE

CASE STUDY Family Ties

Kay Burt is a survivor. Her story began in 1966—at the dawn of kidney transplantation—and now involves three generations of her family. Born with congenitally small kidneys, Kay was dying by the time she was 14 years old. The primitive dialysis machines of the time could not keep her alive, and her weight dropped to only 57 pounds. When her doctor suggested a radical procedure—kidney transplantation—her father, who was a compatible donor for Kay, unhesitatingly offered one of his. At that time, kidney transplantation was in its early stages of development; Kay's was only the fifth kidney transplant in the entire state of Texas. Her surgery was a grueling 12-hour ordeal. When the transplanted kidney began to function normally in Kay's body, it seemed like a miracle. Kay was warned that she must never become pregnant; doctors feared that the developing baby could damage the transplanted organ or its delicate connections. Despite this advice, Kay gave birth to a healthy daughter, Cherry, five years later. After three decades of good health, Kay's life was threatened again in 1998. Just two weeks after her beloved father's death, his kidney, which had functioned in her body for 32 years, also failed, forcing Kay to seek further medical assistance.

Each year in the United States, tens of thousands of people lose kidney function. While about 20,000 receive kidney transplants from compatible donors, over 46,000 patients currently await kidney transplants and about 200,000 are kept alive by hemodialysis (more commonly referred to simply as "dialysis"). What is hemodialysis, and how does it work? What other "medical miracles" might help people with urinary system problems in the future? How have Kay and her family dealt with her new setback?

31.1 What Are the Basic Functions of Urinary Systems?

In this chapter we explore the workings of urinary systems and discover that they do far more than just produce urine. The **urinary system** serves many crucial functions relating to homeostasis (see Chapter 27), helping maintain the chemical composition of the blood and extracellular fluid within the narrow bounds required for cellular metabolism. One critical element in homeostasis is water balance. Why? If the volume of water inside body cells fluctuates too much, causing the chemicals dissolved in internal fluids to become too concentrated or too diluted, the chemical equilibrium of the cells will be disrupted, with disastrous consequences for the animal.

As with many physiological systems, the urinary system is a master of multitasking. An important function of the urinary system is **excretion**, a general term referring to the elimination of substances from the body. Excretion occurs through the respiratory system (carbon dioxide), the digestive system (undigested material), and the urinary system, which eliminates the nongaseous waste products of cellular metabolism (for example, the urea from amino acid metabolism), excess water, excesses of certain vitamins, and some drugs or drug-related by-products.

Whether we are looking at flatworms, fishes, or people, all urinary systems (often called *excretory systems*) perform similar functions using the same basic process: First, the blood or other fluid that bathes cells is filtered, removing water and small dissolved molecules; second, nutrients are selectively reabsorbed from the filtrate; and third, any remaining water and dissolved wastes are excreted from the body. Next, we will explore two of the many types of urinary systems that have evolved in invertebrates.

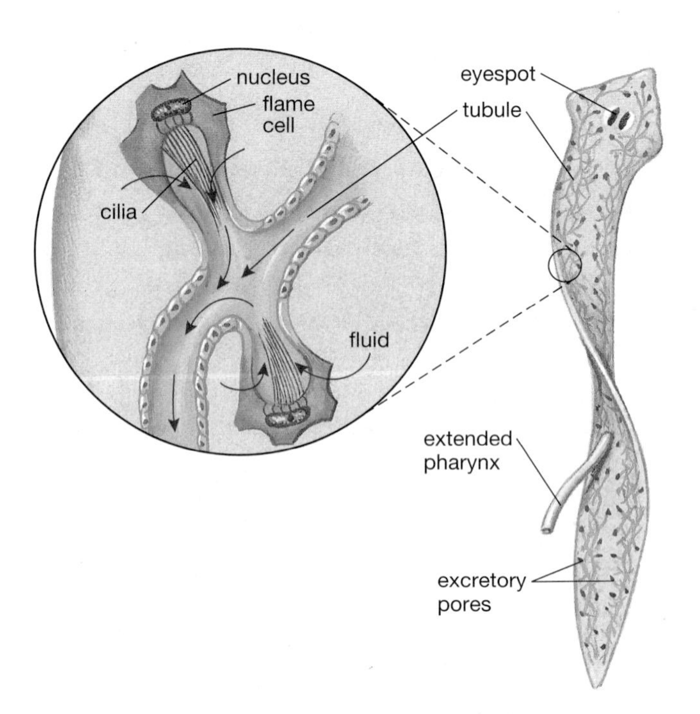

FIGURE 31-1 The simple excretory system of a flatworm Hollow flame cells direct excess water and dissolved wastes into a network of tubes. The beating cilia of the flame cells help circulate the fluid to excretory pores.

31.2 How Does Excretion Occur in Invertebrates?

Flame Cells Filter Fluids in Flatworms

The first specialized excretory structures to arise during the course of animal evolution were most likely **protonephridia** (literally "before the nephridium"; nephridia are described below). The flatworm, found under rocks in streams, still uses a protonephridial system, whose major function is to regulate water balance. The flatworm's protonephridia consist of a network of tubes that branch throughout the body (Fig. 31-1). At intervals, the branches end blindly in single-celled bulbs called **flame cells**, which derive their name from the tuft of beating cilia that extends into the hollow bulb. Under the microscope, each beating cluster of cilia resembles a flickering flame. Water and dissolved substances are filtered into the bulbs, where the beating cilia produce a current that conducts the fluid through the tubular network. Within the tubes, more waste products are added and some nutrients are reabsorbed. The water, carrying some dissolved waste products, eventually reaches one of numerous pores that release it to the outside. The shape of flatworms provides a large skin surface through which cellular waste products also leave by diffusion.

Nephridia Filter Fluid in Earthworms

Earthworms, mollusks, and several other invertebrates have simple kidney-like structures called **nephridia** (singular, *nephridium*). In the earthworm, fluid fills the body cavity, or coelom, that surrounds the internal organs. This coelomic fluid collects both wastes and nutrients from the blood and tissues. The fluid is conducted into a funnel-shaped opening called the **nephrostome** and is swept by cilia along a narrow, twisted tube (Fig. 31-2). There, salts and other dissolved nutrients are absorbed back into the blood, leaving behind water and wastes. The resulting urine is stored in an enlarged bladderlike portion of the nephridium and is then excreted through the **excretory pore**, an opening in the body wall. The earthworm body is composed of repeating segments, nearly every one of which contains its own pair of nephridia.

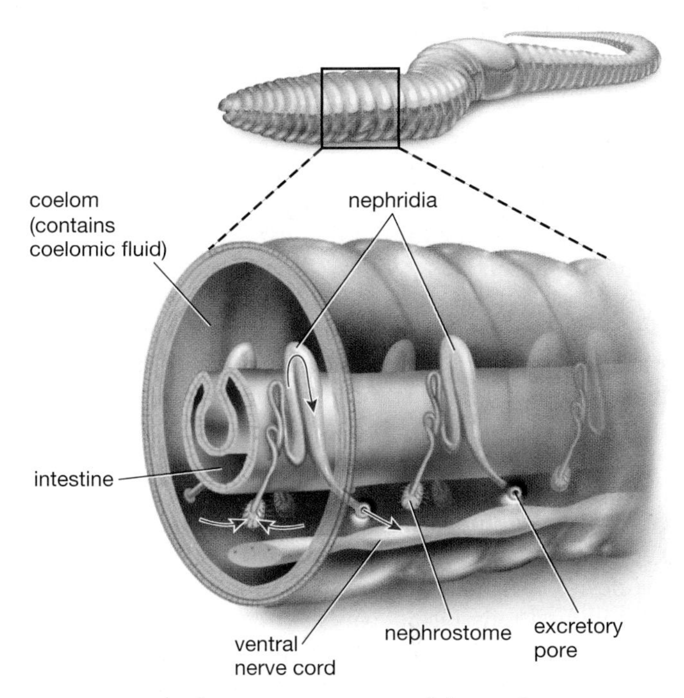

FIGURE 31-2 The excretory system of the earthworm
This system consists of structures called *nephridia*, one pair per segment. Coelomic fluid is drawn into the nephrostome, and urine is released through the excretory pore. Each nephridium resembles a vertebrate nephron.

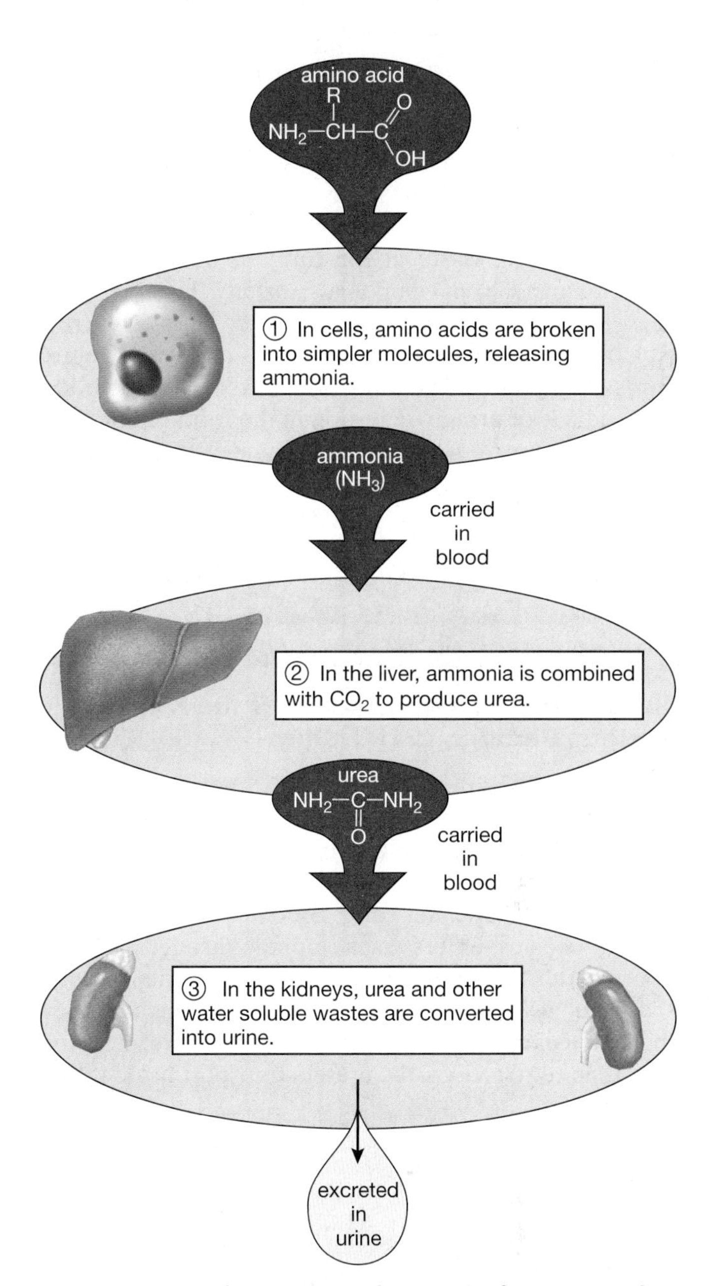

FIGURE 31-3 A flow diagram showing the formation and excretion of urea
QUESTION In some animals, ammonia is not converted to urea but instead circulates in the body until it is excreted. In what kinds of environments would you expect to find such animals?

31.3 What Are the Functions of Vertebrate Urinary Systems?

As mentioned at the beginning of the chapter, the urinary system plays a crucial role in **homeostasis**. Urinary systems face a challenge: Excreting dissolved wastes requires the elimination of water, but water may also need to be retained in order to maintain the proper water balance. Furthermore, water that contains wastes may also contain dissolved nutrients and salts that the body cannot afford to lose. How can wastes be excreted without undue loss of water and nutrients? The kidneys have evolved a complex internal structure and metabolic abilities that meet these challenges.

The **kidneys** are organs in which the fluid portion of the blood is collected. From this fluid, water and important nutrients are reabsorbed into the blood, while toxic substances, cellular waste products, and excess vitamins, salts, hormones, and water are left behind to be excreted as urine. The rest of the urinary system channels and stores urine until it is excreted from the body. The mammalian urinary system helps maintain homeostasis in several ways:

- Regulates blood levels of ions such as sodium, potassium, chloride, and calcium.
- Regulates the water content of the blood.
- Maintains proper pH of the blood.
- Retains important nutrients such as glucose and amino acids in the blood.
- Secretes hormones, such as *erythropoietin,* which stimulates red blood cell production.
- Eliminates cellular waste products such as *urea.*

Urea is a product of amino acid metabolism. As you may recall from Chapter 30, the digestive system breaks proteins into their amino acid building blocks, which are then absorbed. When amino acids are taken into cells, some are used directly to synthesize new proteins. Others have their amino ($-NH_2$) groups removed and are then used either as a source of energy or in the synthesis of new molecules. The amino groups are released as **ammonia** (NH_3) which is very toxic. In mammals, ammonia travels in the blood to the liver, where it is converted to urea, a far less toxic substance (Fig. 31-3).

Urea is filtered from the blood by the kidneys and excreted in **urine**, a fluid consisting of water, dissolved wastes, and some excess nutrients.

By excreting waste in the form of urea, mammals avoid damage from ammonia's toxicity. However, because urea is water-soluble, some water must be excreted along with urea, even if water loss is disadvantageous. Birds and reptiles avoid this problem; they excrete the by-products of protein digestion in the form of **uric acid**. Uric acid is not very soluble and can be excreted in crystalline form with very little loss of water.

31.4 What Are the Structures and Functions of the Human Urinary System?

edia Activity
1.1 Urinary System Anatomy

The Urinary System Consists of the Kidneys, Ureters, Bladder, and Urethra

Human kidneys are paired organs located on either side of the spinal column and extending slightly above the waist (Fig. 31-4). Each is approximately 5 inches long, 3 inches wide, and 1 inch thick, and resembles a kidney bean in both shape and color. Blood carrying dissolved cellular wastes enters each kidney through a **renal artery**. After the blood has been filtered, it exits through the **renal vein** (Fig. 31-5). Urine leaves each kidney through a narrow, muscular tube called the **ureter**. Using rhythmic contractions, the ureters transport urine to the *urinary bladder,* or simply **bladder**. This hollow, muscular chamber collects and stores the urine.

The walls of the bladder, which contain smooth muscle, are capable of considerable expansion. Urine is retained in the bladder by two sphincter muscles located at its base, just above the junction with the urethra. When the bladder becomes distended, receptors in the walls signal its condition and trigger reflexive contractions. The sphincter nearest the bladder, the *internal sphincter*, opens during this reflex. However, the lower *external sphincter* is under voluntary control, allowing the brain to suppress the reflex unless the bladder becomes overly full. The average adult bladder holds about 500 milliliters (approximately a pint) of urine, but the desire to urinate is triggered by considerably smaller accumulations. Urine exits the body via the **urethra**, a single narrow tube about 1.5 inches long in the female and about 8 inches long in the male (it is longer in the male because it extends the length of the penis).

Urine Is Formed in the Nephrons of the Kidneys

Each kidney contains a solid outer layer where urine is formed, which consists of the **renal cortex** overlying the **renal medulla**, and a subdivided inner chamber, called

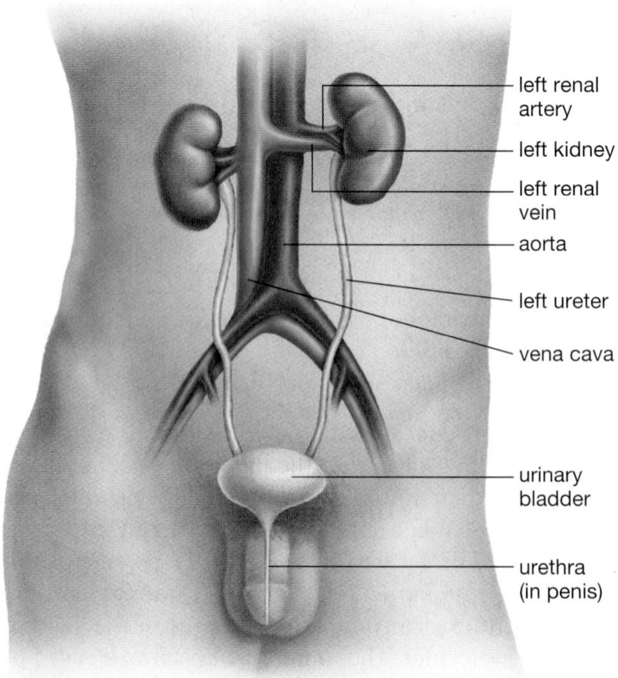

left renal artery
left kidney
left renal vein
aorta
left ureter
vena cava
urinary bladder
urethra (in penis)

FIGURE 31-4 The human urinary system
A diagrammatic view of the human urinary system and its blood supply.

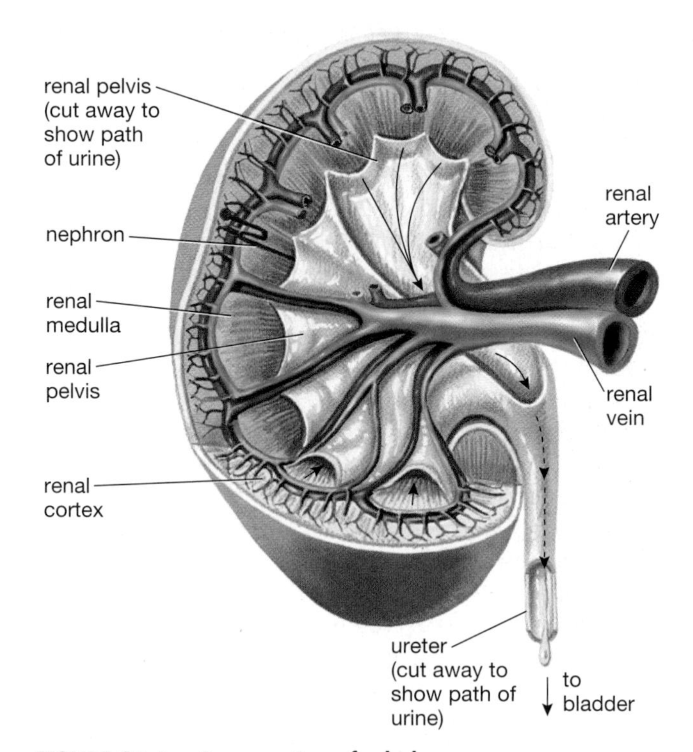

FIGURE 31-5 Cross section of a kidney
The renal artery and the renal vein branch extensively within the kidney. The two are connected by a highly permeable capillary network through which substances are exchanged between the blood and the nephrons. A nephron is drawn considerably larger than normal to show its orientation in the kidney. The renal pelvis is a collecting chamber that funnels urine out of the kidney.

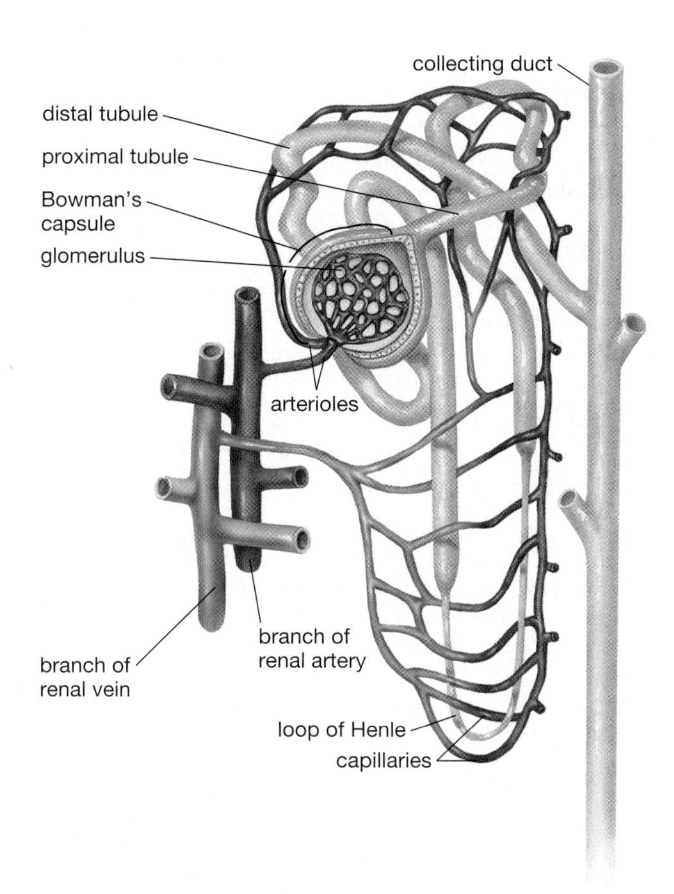

FIGURE 31-6 An individual nephron and its blood supply

the **renal pelvis**, that collects urine and funnels it into the ureter (see Fig. 31-5). Microscopic examination of the outer layer of the kidney reveals an array of tiny individual filters, or **nephrons**, which are richly supplied with blood vessels. More than 1 million nephrons, which evolved from and are somewhat similar to the nephridia described earlier, are packed into the outer layer of each human kidney. The nephron has three major parts: the **glomerulus**, a dense knot of capillaries where fluid from the blood is filtered into **Bowman's capsule**, a cuplike structure surrounding the glomerulus, which leads to a long, twisted **tubule** (from the Latin, "little tube"). The tubule is further subdivided into the *proximal tubule*, the *loop of Henle*, and the *distal tubule*. **Collecting ducts** within the medulla of the kidney collect fluid from many nephrons and conduct it into the renal pelvis (Figs. 31-6 and 31-7). The processes by which these structures produce urine and help maintain homeostasis are examined in the following sections and covered in more detail in "A Closer Look: The Nephron and Urine Formation."

Blood Is Filtered Through Capillaries in the Glomerulus

Nearly one-quarter of the volume of blood pumped by each heartbeat travels through the kidneys, which receive more than one liter of blood (over a quart) every minute. Each nephron receives blood from an arteriole that branches from the renal artery. Within a cup-shaped portion of the nephron—Bowman's capsule—the arteriole branches further into microscopic capillaries that intertwine in a mass called the glomerulus (see Fig. 31-6). The walls of the glomerular capillaries are extremely

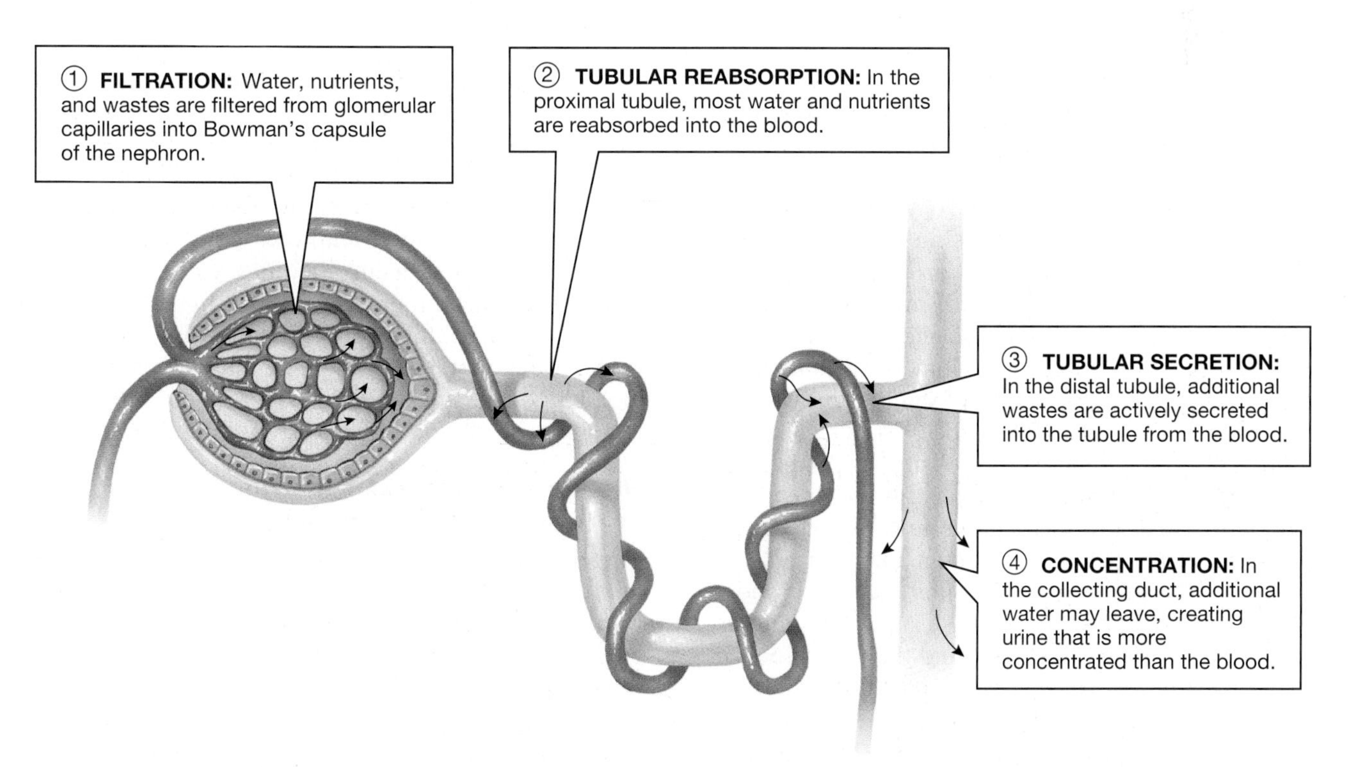

① FILTRATION: Water, nutrients, and wastes are filtered from glomerular capillaries into Bowman's capsule of the nephron.
② TUBULAR REABSORPTION: In the proximal tubule, most water and nutrients are reabsorbed into the blood.
③ TUBULAR SECRETION: In the distal tubule, additional wastes are actively secreted into the tubule from the blood.
④ CONCENTRATION: In the collecting duct, additional water may leave, creating urine that is more concentrated than the blood.

FIGURE 31-7 Urine formation in the nephron

A CLOSER LOOK The Nephron and Urine Formation

The complex structure of the nephron is finely adapted to its function. In Figure E31-1, the nephron is presented in diagram form to illustrate the process occurring in each part. The circled numbers in the illustration refer to the following descriptions:

① *Filtration*. Water and dissolved substances are forced out of the glomerular capillaries into Bowman's capsule and then funneled into the proximal tubule.

② *Tubular reabsorption*. In the proximal tubule, most of the important nutrients remaining in the filtrate are actively pumped out through the walls of the tubule to be reabsorbed into the blood. These nutrients include about 75% of salts, as well as amino acids, sugars, and vitamins. The proximal tubule is highly permeable to water, so water accompanies the nutrients, moving out of the tubule and back into the blood by osmosis.

③ The loop of Henle, which is unique to birds and mammals, is essential for the concentration of urine. It maintains a salt concentration gradient in the extracellular fluid that surrounds it, with the highest concentration at the bottom of the loop. The descending portion of the loop of Henle is very permeable to water but not to salt or other dissolved substances. As the filtrate passes through the descending portion, water leaves by osmosis as the concentration of the surrounding fluid increases.

④ The thin portion of the ascending loop of Henle is relatively impermeable to water and urea but is permeable to salt, which moves out of the filtrate by diffusion. Why? Although the osmotic concentrations inside and outside the tubule are about equal, at this portion of the loop, the urea level is higher outside and the salt level is higher inside. Thus, the concentration gradient favors the movement of salt outward. Because water cannot follow it, the filtrate now becomes less concentrated than its surroundings.

⑤ The thick portion of the ascending loop of Henle is also impermeable to water and urea. Here, salt is actively pumped out of the filtrate, leaving water and wastes behind.

⑥ The watery filtrate, low in salt but retaining wastes such as urea, now arrives at the distal portion of the tubule, where more salt is pumped out. Because this portion is permeable to water, water follows by osmosis. Tubular secretion occurs throughout the tubule but is especially active in the distal portion. There, substances such as K^+, H^+, NH_3, and some drugs and toxins are actively pumped into the tubule from the extracellular fluid.

⑦ By the time the filtrate reaches the collecting duct, very little salt is left, and about 99% of the water has been reabsorbed into the bloodstream. The collecting duct conducts the fluid, now called urine, down through the increasing concentration gradient created by the loop of Henle. The collecting duct is very permeable to water when antidiuretic hormone (ADH) is present, so water moves out by osmosis as the concentration of the external fluid increases. If ADH is absent, the collecting duct remains impermeable to water, and the urine stays dilute and watery.

⑧ The lower portion of the collecting duct is also permeable to urea. Therefore, as the filtrate moves farther down the collecting duct, some urea diffuses out, contributing to the osmotic concentration of the surrounding fluid. As water (when ADH is present) and urea move out, the concentration of dissolved wastes in the urine in the collecting duct approaches equilibrium with the high osmotic concentration of the external fluid.

permeable to water and small dissolved molecules, but they prevent the movement of blood cells and most large proteins, such as albumin and fat droplets. Beyond the glomerulus, the capillaries reunite to form an arteriole whose diameter is smaller than that of the incoming arteriole (you may notice that this is an exception to the usual rule in the circulatory system: arterioles → capillaries → venules). The difference in diameter between the larger incoming arteriole and the smaller outgoing arteriole creates pressure within the glomerulus that drives water carrying many dissolved substances through the very porous capillary walls. This process is called **filtration** (see Fig. 31-7, step ①, and Fig. E31-1, step ①), and the resulting fluid is called **filtrate**. The watery filtrate, resembling blood plasma without its plasma proteins, is collected in Bowman's capsule, then travels through the nephron.

After it passes through the glomerulus, the blood is more concentrated; the blood in the small outgoing arteriole contains about 20% less fluid than the blood in the larger incoming arteriole. Beyond the glomerulus, the outgoing arteriole again branches into smaller, highly porous capillaries that entwine around the length of the tubule (see Figs. 31-6 and 31-7). Carrying blood low in water and nutrients, the capillaries reacquire these substances from the filtrate as it passes through the tubule of the nephron, as described in the following sections. The capillaries eventually conduct the blood, with most of its water and nutrients restored, into venules that transport it away from the kidneys.

Diseases that prevent adequate filtration of blood can cause kidney failure, resulting in the need for dialysis, or, ideally, a kidney transplant. When filtration becomes inadequate, blood volume may increase (causing high blood pressure); regulation of salt and pH balance is disrupted; and toxic wastes are retained in the blood.

The Filtrate Is Converted to Urine in the Tubule of the Nephron

The filtrate collected in Bowman's capsule contains a mixture of both wastes and essential nutrients, including water. The tubule of the nephron will restore the nutrients and most of the water to the blood, while retaining wastes for elimination; in this way, the balance of water and nutrients required for homeostasis is maintained. This is accomplished by two processes: *tubular reabsorption* and *tubular secretion*.

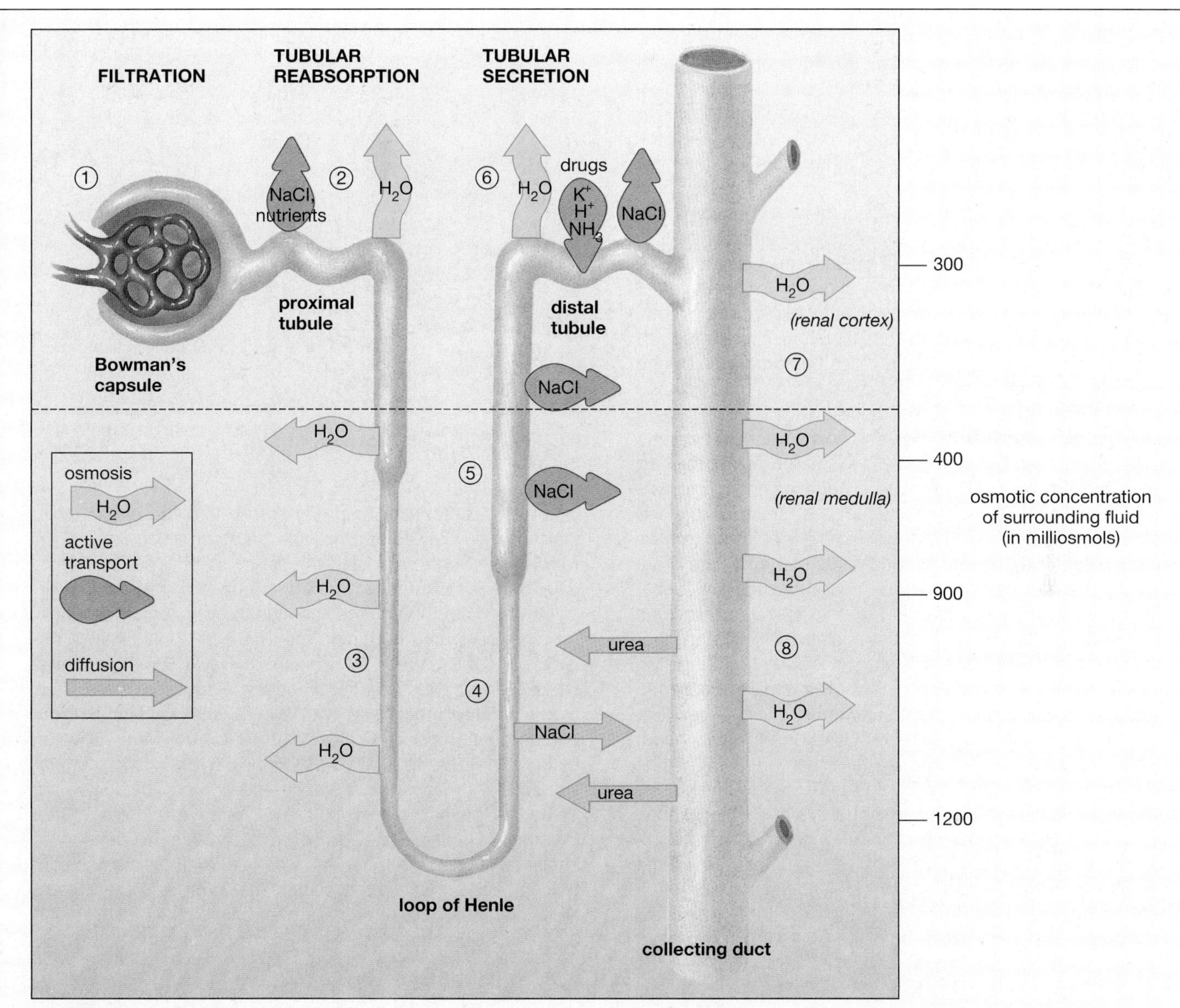

FIGURE E31-1 Details of urine formation
A single nephron, showing the movement of materials through different regions. The concentration of dissolved substances in the filtrate inside the nephron increases from top to bottom in this diagram. Outside the nephron, darker shades of blue represent higher concentrations of salts in the surrounding fluid. The dashed line marks the boundary between the renal cortex and the renal medulla.

Media Activity
31.2 Urine Formation

Tubular reabsorption occurs primarily in the proximal tubule, although water and other nutrients may be reabsorbed throughout the length of the tubule. During tubular reabsorption, tubule cells use active transport to pump nutrients such as salts, amino acids, and glucose from the filtrate, out of the tubule, and into the surrounding fluid. From the extracellular fluid, the nutrients diffuse into the entwining capillaries. Water follows the nutrients passively, moving out of the tubule and into the capillaries by osmosis (see Fig. 31-7, step ②, and Fig. E31-1, step ②). As it passes through the entire length of the tubule, 99% of the water that was initially filtered out is reabsorbed.

Tubular secretion occurs primarily in the distal tubule. During tubular secretion, wastes and excess substances still in the blood are actively secreted from extracellular fluid into the distal tubule by tubule cells (see Fig. 31-7, step ③, and Fig. E31-1, step ⑥). These wastes are added to the filtrate and become part of the urine. Substances secreted into the tubule for excretion include

HEALTH WATCH When the Kidneys Collapse

If the kidneys fail, death occurs rapidly, usually within two weeks. Each year in the United States, about 90,000 people die as a result of urinary system disease. The kidneys are vulnerable to attack from several sources. Overdoses of painkilling medicines and infections, particularly by intestinal bacteria that reach the kidney via the urethra, can harm the kidneys. However, the most common cause of kidney failure is diabetes, followed by high blood pressure; both of these disorders damage the glomerular capillaries. Kidney failure is typically treated by hemodialysis.

First used in 1945, hemodialysis operates on a simple principle: Substances will diffuse from areas of higher concentration to areas of lower concentration across an artificial permeable membrane. This process is called *dialysis*; therefore, the filtration of blood according to this principle is called **hemodialysis**. During hemodialysis, the patient's blood is diverted from the body and pumped through narrow tubing made of a cellophane membrane, which is suspended in dialyzing fluid. Like the glomerular capillaries, this membrane has pores too small to permit the passage of blood cells and large proteins, but large enough to pass small molecules such as water, sugar, salts, amino acids, and urea. The composition of the dialyzing fluid is adjusted to have normal blood levels of salts and nutrients and no waste products. This causes molecules whose concentrations are higher than normal in the patient's blood to diffuse into the dialyzing fluid. Urea, for example, is present in a relatively high concentration in the blood of a dialysis patient and is absent in the dialyzing fluid, so it diffuses out. The patient must remain attached to the dialysis machine for 4 to 6 hours, three times a week (Fig. E31-2). People in their 40s can be kept alive on dialysis for about 10 years or less, and their lives are far from normal. Their diet and fluid intake must be carefully regulated. Even so, their blood composition fluctuates and toxic substances reach higher than normal levels between sessions. *Peritoneal dialysis* is a less common technique in which dialyzing fluid is pumped or poured through a tube implanted directly into the abdominal cavity. The abdominal cavity is lined with a natural membrane called the *peritoneum*. Waste products from blood circulating in capillaries within the peritoneum gradually diffuse into the dialysis fluid, which is then drained out through the tube. A patient can perform this at home, replacing the dialysis fluid about four times daily, or can link his or her implanted tube to a machine that circulates the fluid through the abdominal cavity during the night.

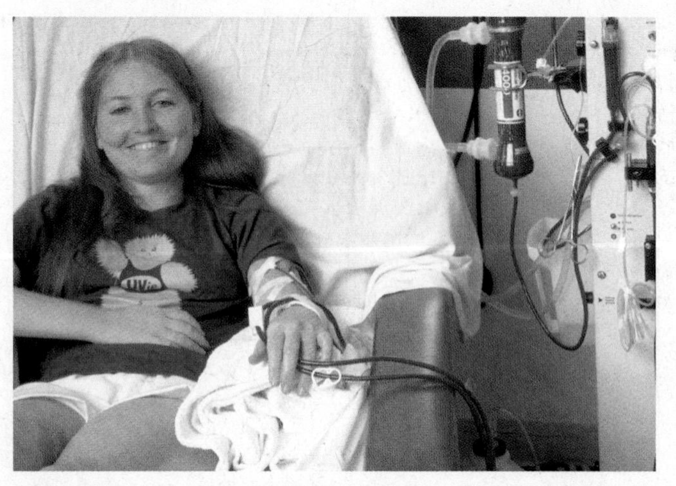

FIGURE E31-2 Patient on dialysis

What does the future hold for victims of urinary system diseases? At the University of Michigan, Dr. David Humes has developed a prototype "bioartificial kidney." Humes's device filters blood through tubes lined with living kidney cells taken from human donor kidneys. Clinical trials, which used the device on patients in intensive care whose kidneys were failing, demonstrated that the bioartificial kidney—used in conjunction with a hemodialysis machine—significantly improves survival rates. Eventually, Humes hopes to use this technology to develop an implantable bioartificial kidney. Large medical firms are also continuing work on *xenotransplantation*—a process that would allow people to receive kidneys from animals such as pigs that are genetically engineered to prevent tissue rejection.

---

hydrogen and potassium ions, ammonia, and many drugs. These substances often accumulate in the blood of kidney failure victims, a condition that can lead to death if not treated.

The Loop of Henle Allows Urine to Become Concentrated

The kidneys of mammals and birds are able to produce urine that has a higher concentration of dissolved materials than their blood. This ability to concentrate urine results from the structures of both the nephron and the collecting duct into which several nephrons empty (see Fig. 31-7, step ④ and Fig. E31-1, steps ⑦ and ⑧). Urine can become concentrated because there is an osmotic concentration gradient of salts and urea in the interstitial fluid that surrounds the loop of Henle and the collecting ducts. This gradient is produced by the loop of Henle within the renal medulla; the longer the loop, the higher the concentration gradient, since there is a greater length of tubule to transport salt out into the surrounding fluid (see Fig. E31-1, steps ④ and ⑤). The most concentrated fluid (with the greatest amount of dissolved substances and the least amount of water, about four times the concentration of the blood) surrounds the bottom of the loop. The urine in the collecting duct also passes through this osmotic gradient on its way to the renal pelvis. As the filtrate passes through the portion of the collecting duct surrounded by the osmotically concentrated fluid, additional water may leave the filtrate by osmosis through the walls of the collecting duct and be carried off by the surrounding capillaries, while wastes are left behind in the collecting duct. Therefore, as it moves through the collecting duct, the filtrate, now called *urine*, can reach an osmotic equilibrium, becoming as concentrated as the surrounding fluid. In humans, this can be more than four times the concentration of the blood. Because the rest of the excretory system does not allow water to enter or urea to diffuse out, the urine remains concentrated.

It is important to produce concentrated urine when water is scarce, and to produce dilute, watery urine when there is excess water in the blood. Whether the urine becomes concentrated depends on the permeability of the collecting duct to water, which in turn is controlled by the amount of *antidiuretic hormone* in the blood, as will be described shortly.

31.5 How Do Mammalian Kidneys Help Maintain Homeostasis?

Every drop of blood in your body passes through a kidney about 350 times a day; thus, the kidney is able to fine-tune the composition of the blood and thereby help maintain homeostasis. The importance of this task is illustrated by the fact that kidney failure results in death within a short time (see "Health Watch: When the Kidneys Collapse").

The Kidneys Regulate the Water Content of the Blood

One important function of the kidney is to regulate the water content of the blood (see Fig. 31-7). Human kidneys filter out about half a cup of fluid from the blood each minute. Without reabsorption of water, you would produce about 50 gallons of urine daily! Water reabsorption occurs passively by osmosis as the filtrate travels through the tubule and the collecting duct.

The amount of water reabsorbed into the blood is controlled by a negative feedback mechanism (see Chapter 27) that involves the amount of **antidiuretic hormone** (**ADH**; also called *vasopressin*) circulating in the blood. ADH is produced by cells in the hypothalamus and is released by the posterior pituitary gland (as we will describe in Chapter 33). This hormone increases the permeability of the distal tubule and the collecting duct to water, thereby allowing more water to be reabsorbed from the urine.

The release of ADH is regulated by receptor cells in the hypothalamus that monitor the osmotic concentration of the blood and by receptors in the heart that monitor blood volume. Let's look at an example: As a lost traveler staggers through the searing desert sun, perspiring copiously and losing water with every breath, dehydration occurs. His blood volume falls, and the osmotic concentration of his blood rises, triggering release of ADH. The ADH increases water reabsorption to conserve water, producing urine more concentrated than the blood (Fig. 31-8). Encountering an oasis, our traveler overindulges in the cool, clear water of a natural spring. His blood volume rises and its osmotic strength falls, triggering a decrease in his ADH output. Reduced ADH makes distal tubules and collecting

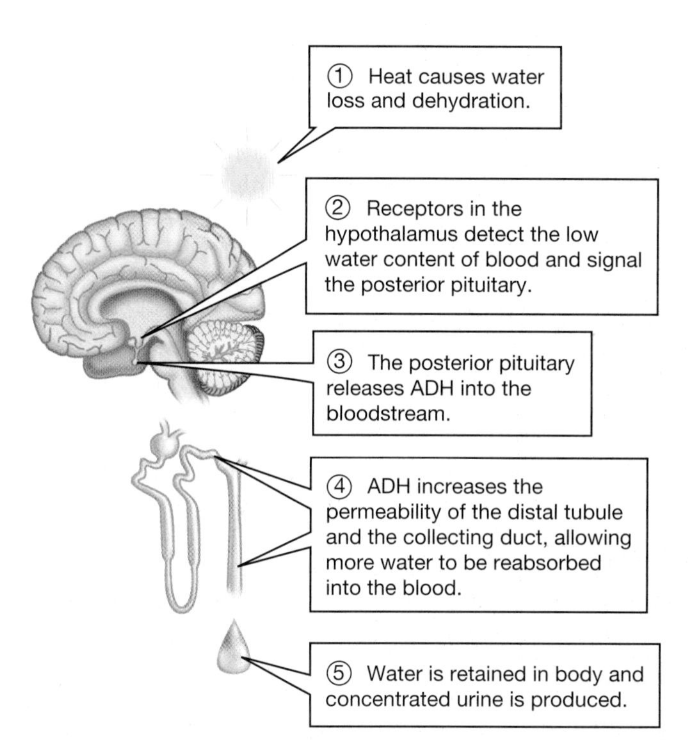

① Heat causes water loss and dehydration.
② Receptors in the hypothalamus detect the low water content of blood and signal the posterior pituitary.
③ The posterior pituitary releases ADH into the bloodstream.
④ ADH increases the permeability of the distal tubule and the collecting duct, allowing more water to be reabsorbed into the blood.
⑤ Water is retained in body and concentrated urine is produced.

FIGURE 31-8 Dehydration stimulates ADH release and water retention

QUESTION Alcohol inhibits the release of ADH. How does consumption of alcoholic beverages affect the body's water balance?

ducts less permeable to water, so less water is reabsorbed after the urine leaves the loop of Henle. He will now produce urine that is more dilute than the blood. In extreme cases, urine flow may exceed 1 liter (about 1 quart) per hour. As the proper water level in his blood is restored, the increased osmotic concentration of the blood and decreased blood volume will again stimulate some ADH release, thus maintaining homeostasis by keeping the blood water content within narrow limits.

Kidneys Release Hormones That Help Regulate Blood Pressure and Oxygen Levels

The kidneys produce two hormones that are extremely important in regulating blood pressure and the blood's oxygen-carrying capacity. When blood pressure falls, the kidneys release **renin** into the bloodstream. Renin acts as an enzyme, catalyzing the formation of a second hormone, **angiotensin**, from a protein that circulates in the blood. Angiotensin, in turn, causes arterioles to constrict, elevating blood pressure. The constriction of the arterioles that carry blood to the kidneys also reduces the rate of blood filtration, causing less water to be removed from the blood. Water retention causes an increase in blood volume and, consequently, an increase in blood pressure.

In response to low blood oxygen levels, the kidneys release a second hormone, **erythropoietin**, described in Chapter 28. Erythropoietin travels in the blood to the bone marrow, where it stimulates the marrow to produce more red blood cells, whose function is to transport oxygen. Kidney failure almost always leads to anemia because the kidneys do not produce enough erythropoeitin to stimulate adequate red blood cell production. Physicians may now give human erythropoietin made by recombinant DNA techniques (described in Chapter 13) to patients with anemia caused by kidney failure.

Kidneys Monitor and Regulate Dissolved Substances in the Blood

As the kidney filters the blood, it monitors and regulates blood composition and adjusts tubular secretion and reabsorption rates, maintaining a constant internal environment. Substances the kidney regulates, in addition to water, include nutrients such as glucose, amino acids, vitamins, urea, and a variety of ions, including sodium, potassium, chloride, and sulfate. The kidney maintains a constant blood pH by regulating the amount of hydrogen and sodium bicarbonate ions that are secreted into the urine. This remarkable organ also eliminates potentially harmful substances, including some drugs, food additives, pesticides, and toxic substances such as nicotine from cigarette smoke.

FIGURE 31-9 A well-adapted desert dweller
The desert kangaroo rat of the southwestern United States seldom drinks, partly because its long loops of Henle allow it to produce very concentrated urine.

Mammalian Kidneys Are Adapted to Diverse Environments

In addition to regulating water balance using hormones such as ADH, different types of mammals have kidneys with structures adapted to the availability of water in their natural habitats. Mammals that must conserve water do so by producing urine that is more concentrated than their blood. The degree of concentration that can be achieved is determined by the length of the loop of Henle. The longer the loop, the higher the salt concentration produced in the fluid that surrounds it. Higher salt concentration causes more water to move by osmosis out of the urine as it passes through the collecting duct, resulting in more concentrated urine. As you might predict, animals living in very dry climates have the longest loops of Henle, whereas those living in watery environments have relatively short loops.

The beaver, for example, has only short-looped nephrons and is unable to concentrate its urine to more than twice its plasma concentration. Human kidneys have a mixture of long-looped and short-looped nephrons and can concentrate urine to about four times the level of plasma concentration. The masters of urine concentration are desert rodents such as kangaroo rats, which can produce urine 14 times more concentrated than their plasma (Fig. 31-9). Kangaroo rats (as you might predict) have only very long-looped nephrons. Because they have an extraordinary ability to conserve water, they do not need to drink; instead, they rely entirely on water derived from their food.

Kidney structure is not the only factor that affects water balance in the mammalian body. The body's water content is also affected by food and water consumption, by evaporative water loss through the skin (for example, sweating), and by an animal's behavior. Kangaroo rats, for instance, have evolved behaviors that help them avoid unnecessary water loss while acquiring the greatest possible amount of water from their food. Kangaroo rats live in underground burrows and venture above ground only at night. This behavior minimizes their exposure to hot desert temperatures by day, reducing water loss by evaporation. During their nightly sojourns to the surface, kangaroo rats gather seeds to eat. But instead of eating them right away, the rats carry the seeds back to their burrows. Why? Because the air in the underground burrows is far more moist than the parched desert air on the surface, and a seed stored underground will absorb precious water before it is eaten. For the kangaroo rat, behaviors that help them conserve water make an important contribution to maintaining homeostasis, and thus maintaining life.

CASE STUDY REVISITED Family Ties

When Cherry realized her mother might be dependent on hemodialysis for the rest of her life, she immediately offered to donate a kidney. Kay's remarkable luck continued; her daughter, whom doctors had ordered her not to conceive, was a good match. Kay's second transplant operation, 32 years after her first, was another success. Kay Burt, who made history by receiving transplanted kidneys from both her father and her daughter, has been healthy since her second operation and continues to do well.

Kidneys are now routinely transplanted from both living and recently deceased donors. Fortunately, the number of living donors is on the rise. Recipients of kidneys from living donors have better long-term survival rates, and this increase may also help to reduce the number of people—currently about 3100—who die each year while waiting for a donor kidney. Some transplant centers are now using endoscopic surgery to remove kidneys from living donors. This technique, in which the surgeon views the operation through a tiny camera inserted into the body, allows the kidney to be extracted through an incision about 2.5 inches long—far shorter than the 9-inch incision of traditional methods. Endoscopic surgery dramatically reduces pain, hospitalization time, and overall recovery time for the donor.

Consider This: You read in your local paper that a family is looking for a kidney donor for their child. You have never met the family. Would you be willing to undergo compatibility testing, and, if you were found to be a compatible donor, consider donating a kidney?

Links to Life: Dehydrating Drinks

Those of you who consume alcoholic drinks (or strong coffee, for that matter) have undoubtedly noticed an interesting phenomenon: The volume of urine you produce after consuming these beverages seems to exceed the volume of liquid you drank. What's going on? Alcohol has wide-ranging effects on the body, one of which is that it inhibits the production of ADH. As you now know, ADH increases the permeability of the walls of the collecting ducts to water; without ADH, urine remains very dilute and watery. Certainly, you'd expect to produce urine after consuming a few beers; after all, beer contains a lot of water, which will reduce the release of ADH. But alcohol further suppresses ADH release, causing you to excrete more urine than necessary to restore water balance. So, ironically, having "too much to drink" can actually dehydrate you if the drinks are alcoholic. Dehydration contributes to the hangover often experienced by people who overindulge. You may have heard that strong coffee can help counteract a hangover. In fact, this is not a good strategy; the caffeine in coffee interferes with ADH's ability to promote water reabsorption through the collecting duct, dehydrating you even further. So, if you make the mistake of overindulging in alcohol, reach for some water or fruit juice—after a sober friend drives you home!

Summary of Key Concepts

31.1 What Are the Basic Functions of Urinary Systems?

All urinary systems (often called *excretory systems*) perform similar functions using the same basic process: First, the blood or other fluid that bathes cells is filtered, removing water and small dissolved molecules; second, nutrients are selectively reabsorbed from the filtrate; and third, any remaining water and dissolved wastes are excreted from the body.

31.2 How Does Excretion Occur in Invertebrates?

The simple excretory system of the flatworm consists of a network of tubules that branch through the body, collecting wastes and excess water for excretion through excretory pores. Many of the more complex invertebrates, including earthworms and mollusks, use nephridia, which resemble vertebrate nephrons, to filter the fluid that fills the body cavity. Wastes and excess water are released through excretory pores.

31.3 What Are the Functions of Vertebrate Urinary Systems?

Urinary systems eliminate cellular wastes and toxic substances, while retaining vital nutrients in the blood. Urinary systems play a crucial role in homeostasis, regulating the water content, the ion content, and the pH of the blood. They also secrete hormones such as erythropoietin, which stimulates red blood cell production.

31.4 What Are the Structures and Functions of the Human Urinary System?

The human urinary system consists of kidneys, ureters, bladder, and urethra. Kidneys produce urine, which is conducted by the ureters to the bladder, a storage organ. Distension of the muscular bladder wall triggers urination, during which urine passes out of the body through the urethra.

Each kidney consists of more than 1 million individual nephrons in an outer layer, which consists of the renal cortex overlying the renal medulla. Urine formed in the nephrons enters collecting ducts that empty into the renal pelvis, from which the urine is funneled into the ureter.

Each nephron is served by an arteriole that branches from the renal artery. The arteriole further branches into a mass of porous-walled capillaries called the *glomerulus*. There, water and dissolved substances are filtered from the blood by pressure. The filtrate is collected in the cup-shaped Bowman's capsule and conducted along the tubular portion of the nephron. During tubular reabsorption, nutrients are actively pumped out of the filtrate through the walls of the tubule. Nutrients then enter capillaries surrounding the tubule, and water follows by osmosis. Some wastes remain in the filtrate; others are actively pumped into the tubule by tubular secretion. The tubule forms the loop of Henle, which creates a salt concentration gradient surrounding it. After completing its passage through the tubule, the filtrate enters the collecting duct, which passes through the concentration gradient. Final passage of the filtrate through this gradient via the collecting duct allows the concentration of the urine.

31.5 How Do Mammalian Kidneys Help Maintain Homeostasis?

The kidneys are important organs of homeostasis. The water content of the blood is regulated by antidiuretic hormone (ADH), produced in the hypothalamus and released into the blood by the posterior pituitary gland. Dehydration stimulates the release of ADH, which increases water absorption into the blood through the distal tubule and the collecting duct. The kidneys also control blood pH, remove toxic substances, and regulate ions such as sodium, chloride, potassium, and sulfate. The kidneys excrete excess glucose, vitamins, and amino acids.

The kidneys secrete hormones, including renin and erythropoietin. Renin, released in response to low blood pressure, catalyzes the formation of angiotensin, which constricts arterioles and elevates blood pressure. Erythropoietin is released when the oxygen content of the blood is reduced, and stimulates the bone marrow to produce red blood cells.

Mammalian kidneys are adapted to the animal's environment. Animals such as beavers that live where water is abundant tend to have short loops of Henle and produce dilute urine, while desert animals have very long loops of Henle and can produce very concentrated urine.

Key Terms

ammonia *p. 609*
angiotensin *p. 615*
antidiuretic hormone (ADH) *p. 615*
bladder *p. 610*
Bowman's capsule *p. 611*
collecting duct *p. 611*
erythropoietin *p. 615*
excretion *p. 608*
excretory pore *p. 608*
filtrate *p. 612*
filtration *p. 612*
flame cell *p. 608*
glomerulus *p. 611*
hemodialysis *p. 614*
homeostasis *p. 609*
kidney *p. 609*
nephridium *p. 608*
nephron *p. 611*
nephrostome *p. 608*
protonephridium *p. 608*
renal artery *p. 610*
renal cortex *p. 610*
renal medulla *p. 610*
renal pelvis *p. 611*
renal vein *p. 610*
renin *p. 615*
tubular reabsorption *p. 613*
tubular secretion *p. 613*
tubule *p. 611*
urea *p. 609*
ureter *p. 610*
urethra *p. 610*
uric acid *p. 610*
urinary system *p. 608*
urine *p. 610*

Thinking Through the Concepts

To take a multiple-choice quiz with feedback on the contents of this chapter, visit http://www.prenhall.com/audesirk7. *Log in to the Web site selected by your instructor and navigate to the Self Test section for this chapter.*

? Review Questions

1. Explain the two major functions of excretory systems.
2. Trace a urea molecule from the bloodstream to the external environment.
3. What is the function of the loop of Henle? The collecting duct? Antidiuretic hormone?
4. Describe and compare the processes of filtration, tubular reabsorption, and tubular secretion.
5. Describe the role of the kidneys as organs of homeostasis.
6. Compare and contrast the excretory systems of humans, earthworms, and flatworms. In what general ways are they similar? How do they differ?

Applying the Concepts

1. Discuss the differences in function of the two major capillary beds in the kidneys: the glomerular capillaries and those surrounding the tubules.
2. Desert animals, such as the kangaroo rat, need to conserve water. These animals have larger kidneys than do animals that live in moist environments and thus need not conserve water. The larger kidneys allow for a greater distance between the glomerulus and the bottom of the loop of Henle. Discuss why this anatomical difference assists water conservation in the desert animals.
3. Some "quick weight loss" diets require the ingestion of much protein-rich food and the elimination of carbohydrates. Two side effects of such diets are increased thirst and increased urination. Explain the connections between the diets and their side effects.
4. Some employers require their employees to submit to urine tests before they can be employed and at random intervals during their employment. Refusal to take the test or failure to "pass" it could be grounds for termination. What is the purpose of a urine test? What types of employers might find such tests necessary? How would you feel if you had to undergo a urine test to obtain or keep a job? Explain your answers.

For More Information

Greenberg, A. (ed.). *Primer of Kidney Disease*. San Diego: Academic Press, 1998. An up-to-date, comprehensive review of kidney diseases, written by a variety of experts.

O'Brien, C. "Lucky Break for Kidney Disease Gene." *Science*, June 24, 1994. Discusses research on a dominant gene that causes kidney disease.

Plafrey, C., and Cossins, A. "Fishy Tales of Kidney Function." *Nature*, September 24, 1994. A description of research findings that have increased our understanding of how fish cope with the physiological challenges posed by watery and/or salty environments.

Media Activities

To access a Media Activity visit http://www.prenhall.com/audesirk7. *Log in to the Web site selected by your instructor, navigate to this chapter, and select the appropriate Media Activity number.*

31.1 Urinary System Anatomy

Estimated time: 5 minutes

Explore the structure and function of the urinary system.

31.2 Urine Formation

Estimated time: 10 minutes

Examine how the nephron cleans waste and toxins from the blood.

31.3 Web Investigation: Family Ties

Estimated time: 10 minutes

Each year in the U.S., tens of thousands of people lose kidney function. While about 20,000 receive kidney transplants from compatible donors, over 46,000 patients currently await kidney transplants and about 200,000 are kept alive by hemodialysis. In this investigation, explore what hemodialysis is and how it works. Also explore what other "medical miracles" might help people with urinary system problems in the future.

CHAPTER

32 Defenses Against Disease

A sneeze propels thousands of microscopic droplets out of the nose and mouth at a rate of 200 miles per hour! There may be as many as 100,000 bacteria released during a single sneeze; viruses are also spread in this manner.

AT A GLANCE

CASE STUDY Fighting the Flu

Have you ever been sitting in class when the person next to you lets loose with an explosive, full-bodied sneeze? As you draw in a breath to whisper "Gesundheit," you unwittingly inhale thousands of microscopic droplets laden with "germs," including common cold and flu viruses and bacteria from the flourishing populations that grow in everyone's nose and mouth. The symptoms of a cold and of some types of flu—sneezing, coughing, and a runny nose—are ideally suited to spread these viruses. In fact, natural selection favors viruses that cause symptoms that help them spread to other victims.

After class, your sneezing classmate admits he's been feeling achy and feverish. Remembering your own experiences with similar symptoms, you advise him to take some aspirin and go home to bed. You realize that your classmate's sneeze may have exposed you to the same virus that caused his illness. Remembering being debilitated by flu last year, you fervently hope you are now immune. Should you get a flu shot from your doctor? Was your suggestion regarding aspirin good advice? How will your body cope with the microbes you have unwittingly inhaled?

32.1 What Are the Basic Mechanisms of Defense Against Disease?

Our environment teems with microscopic *parasites*—organisms that live and reproduce on or within other life forms, doing harm in the process. These include viruses, bacteria, fungi, and protists that we will collectively call **microbes**. Residents of tropical regions face even greater challenges from parasitic protists (such as the species that causes malaria) and from a variety of parasitic worms, including hookworms and tapeworms. New microbial threats are appearing with disturbing regularity. Since the early 1980s, several viruses have emerged as threats to people, including HIV, Ebola virus, hantavirus, and (more recently) West Nile virus and SARS (severe acute respiratory syndrome). We have witnessed outbreaks of relatively rare but deadly strains of the common intestinal bacterium *Escherichia coli* and the respiratory bacterium *Streptococcus pyrogens* ("flesh-eating bacteria"). Meanwhile, hundreds of "cold virus" strains and constantly mutating strains of influenza (flu) virus continue to threaten us.

Media Activity 32.1 Defense Against Infectious Agents

Given the prevalence and diversity of disease-causing organisms, you might wonder, "Why don't we get sick more often?" Over evolutionary time, animals and their parasites have engaged in a constantly escalating battle. As animals evolve more sophisticated defense systems, parasites in turn evolve more effective tactics for penetrating those defenses. This "evolutionary arms race" has honed our defenses into a stunningly complex system designed to resist parasitic invasion and attack.

The body has evolved three major forms of protection against disease (Fig. 32-1). First, it provides *nonspecific external barriers* that prevent most disease-causing microbes from entering the body. These barriers are primarily anatomical structures and secretions such as skin, hair, cilia, and tears, saliva, and mucus. Such barriers cover the body and also line body cavities that are continuous with the outside; for example, the surfaces of the respiratory, digestive, and urogenital tracts. Second, if the barriers are breached, a variety of *nonspecific internal defenses* are called into action. Certain white blood cells engulf foreign particles or destroy infected cells. Chemicals released by damaged body cells and by white blood cells are responsible for inflammation and fever, which can also deter infection. Both the nonspecific barriers and the non-specific internal defenses operate regardless of the exact nature of the invader, repelling, killing, or neutralizing the threat. The third and final line of defense consists of the **immune response**, in which immune cells selectively destroy the particular toxin or microbe and "remember" the invader, allowing a faster response if it reappears in the future.

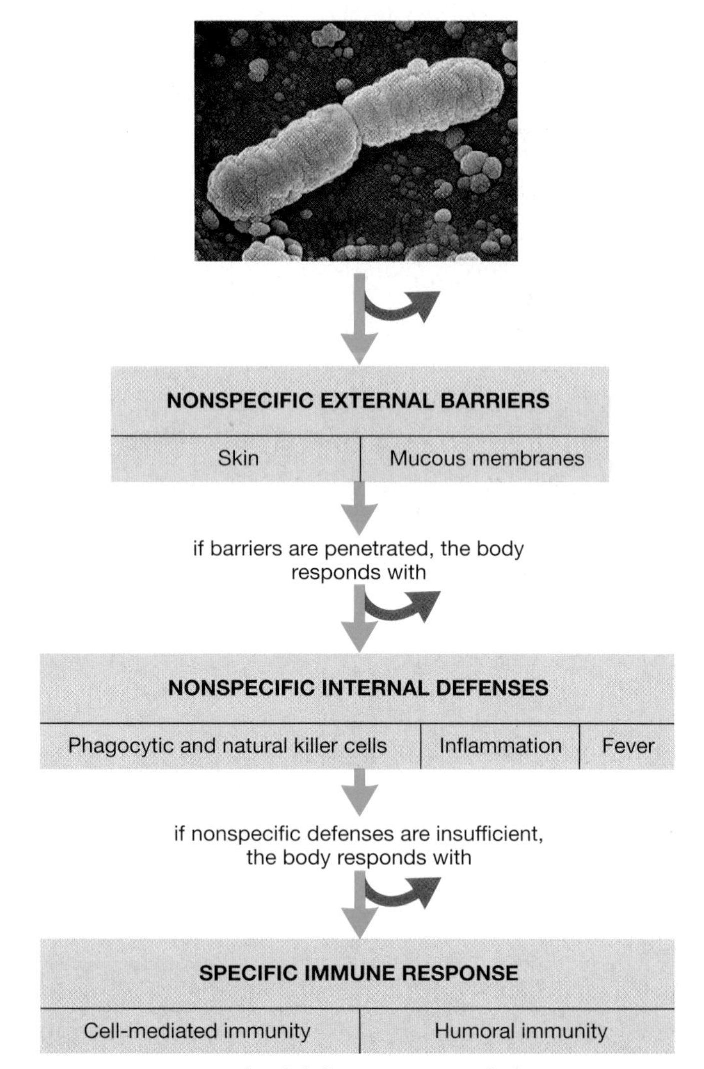

FIGURE 32-1 Levels of defense against infection

32.2 How Do Nonspecific Defenses Function?

Skin and Mucous Membranes Form External Barriers to Invasion

An ideal defense strategy prevents invaders from entering the body in the first place. In animal bodies, this first line of defense consists of the two surfaces with direct exposure to the environment: the *skin* and the *mucous membranes* of the digestive, respiratory, and urogenital tracts.

The Skin and Its Secretions Block Entry and Provide an Inhospitable Environment for Microbial Growth

Any virus and bacteria-laden droplets from your neighbor's sneeze that land on your skin will encounter an outer surface of dry, dead cells. Most microbes that come in contact with the skin do not obtain the water and nutrients they need to survive. The few bacteria and fungi that manage to gain a foothold on skin will usually be shed before they can do harm, because skin cells are constantly sloughed off and replaced. Secretions from sweat glands, sebaceous (oil) glands, and wax-secreting glands in the external ear canal all contain natural antibiotics that inhibit the growth of bacteria foreign to the body. These multiple defenses make the unbroken skin an extremely effective barrier against microbial invasion.

Antimicrobial Secretions, Mucus, and Ciliary Action Defend the Mucous Membranes Against Microbes

The warm, moist mucous membranes that surround the eye and line the digestive, respiratory, and urogenital tracts are more hospitable to microbes than the skin, but they also possess effective defense mechanisms. The mucus and tears secreted by these membranes contain antibacterial enzymes—called *lysozymes*—that destroy bacterial cell walls. The mucus also physically traps microbes that enter the body through the nose or mouth (Fig. 32-2). Cilia on the membranes lining the respiratory tract sweep up the mucus, microbes and all, until it is coughed or sneezed out of the body, or swallowed. If microbes are swallowed, they enter the stomach, where they encounter both extreme acidity and protein-digesting enzymes, both of which can kill many types of microbes. Farther along in the digestive tract, the intestine is inhabited by bacteria that are harmless to their intestinal habitat, but secrete substances that destroy invading foreign bacteria or fungi. In the urinary tract, the slight acidity of urine inhibits bacterial growth. In females, acidic secretions and mucus help protect the vagina. Despite these defenses, many disease-causing organisms enter the body through the mucous membranes; for example, many viruses enter your lungs when you inhale the fallout from your classmate's sneeze. Some of these will probably penetrate your vulnerable membranes. Let's look at what happens next.

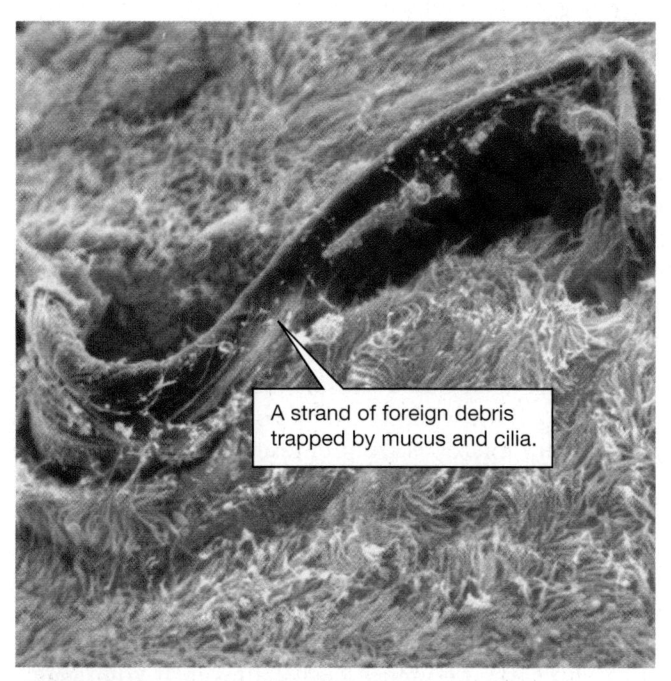

A strand of foreign debris trapped by mucus and cilia.

FIGURE 32-2 The protective function of mucus
The ciliated, mucus-coated epithelium lining the respiratory tract traps microbes and debris, then sweeps both mucus and foreign matter out of the body.

Nonspecific Internal Defenses Combat Disease

Invading parasites that penetrate the skin and mucous membranes encounter an array of internal defenses, some of which are nonspecific—that is, they do not target specific invaders. Nonspecific internal defenses fall into three main categories. First, the body has a standing army of **phagocytic cells**, which engulf and directly destroy microbes, and *natural killer cells*, which destroy cells of the body that have been infected by viruses. The steady trickle of microbes that passes through the body's external barriers is mostly mopped up by these cells. Second, an injury, with its combination of tissue damage and relatively massive invasion of microbes, provokes an *inflammatory response*. The inflammatory response simultaneously recruits phagocytic cells and natural killer cells and walls off the injured area, isolating the infected tissue from the rest of the body. Finally, if a population of microbes succeeds in establishing a major infection, the body may produce a **fever**, an elevated body temperature which both slows down microbial reproduction and enhances the body's own fighting abilities.

Phagocytic Cells and Natural Killer Cells Destroy Invading Microbes

The body contains several types of phagocytic white blood cells (also called *leukocytes*) that engulf and digest microbes rather indiscriminately. Two important types are **macrophages** and **neutrophils**. Both travel within the bloodstream, penetrate through capillary walls, and patrol the body's tissues. Macrophages (literally, "big eaters") and neutrophils ingest dead and dying cells, cellular debris, and microbes by phagocytosis; for example, they consume bacteria and foreign substances that penetrate the delicate mucous membranes of your

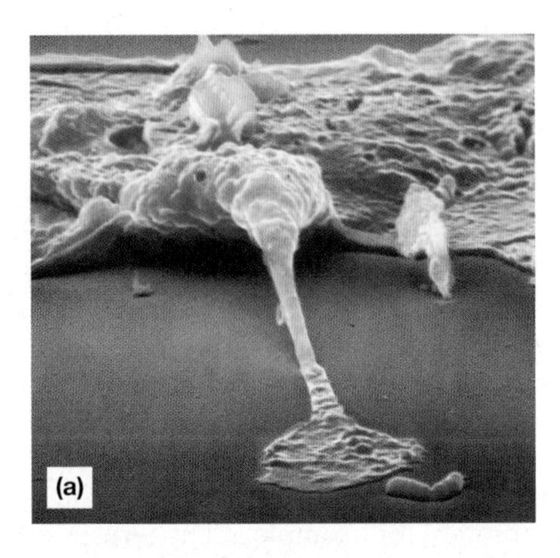

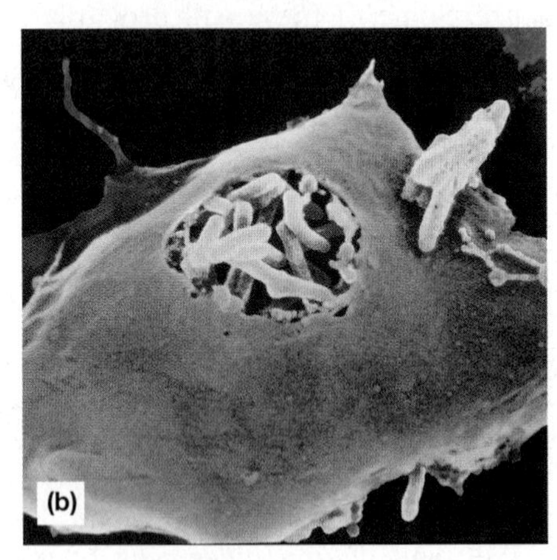

FIGURE 32-3 The attack of the macrophages
(a) A phagocyte reaches for a rod-shaped bacteria. **(b)** This macrophage has stuffed itself with bacteria, which are visible through a hole in its plasma membrane.

lungs (Fig. 32-3). Macrophages also act as housekeepers, scavenging dead and dying cells that are produced as the body continuously renews its tissues and suffers various traumas. If they were not engulfed, the leaking contents of these dead cells would damage healthy neighboring cells. As we will discuss later, macrophages also act as *antigen-presenting cells*—"presenting" parts of the microbe to cells involved in the immune response.

Natural killer cells are another class of white blood cells. In general, they strike at the body's own cells that have become cancerous or have been invaded by viruses, recognizing these cells by any abnormal molecules on their surfaces. For example, virus-infected cells acquire viral proteins on their surfaces, and cancerous cells also bear abnormal molecules within their outer cell membranes. Rather than engulf their victims, as neutrophils and macrophages do, natural killer cells secrete enzymes that attack the infected or cancerous cell, and also secrete proteins that open up holes in its membrane. Chewed on by enzymes and shot full of holes, the infected or cancerous target cell soon dies.

The Inflammatory Response Attracts Phagocytes to Infected or Injured Tissue

Whether you catch the flu from your classmate or accidentally get a splinter in your skin (Fig. 32-4), you will experience inflammation—an important component of the body's non-specific internal defense. The **inflammatory response**, which causes mucous membranes to become leaky and injured tissues to become warm, red, and swollen, has several functions: It attracts phagocytic cells to the area, promotes blood clotting, and causes pain that stimulates protective behaviors.

As shown in Figure 32-4, the inflammatory response can be initiated by cells damaged by infectious microbes or a wound. These cells release substances that, in turn, cause connective tissue cells called **mast cells** to release **histamine** (and other chemicals) into the wounded area. Histamine relaxes the smooth muscle that surrounds arterioles, causing increased blood flow, and also makes capillary walls leaky. Extra blood flowing through the leaky capillaries drives watery fluid filtered from the blood through capillary walls into the region around the wound, which becomes swollen, warm, and reddened (*inflammation* literally means "to set on fire"). Chemicals—released by wounded cells and mast cells, and produced by the microbes themselves—attract phagocytic macrophages and neutrophils. These squeeze out through the leaky capillary walls and ingest bacteria, dirt, and cellular debris caused by the injury (see Fig. 32-3). Macrophages release **cytokines**, chemical messengers secreted by cells that allow them to communicate with one another and with other body systems. These cytokines supplement histamine in making the capillaries leaky, attract more white blood cells to the area, and facilitate the cells' movement through the capillary walls.

The swollen, painful tissues in the throat of a flu victim and the fluid secretions that lead to sneezing, a "runny nose," and coughing are a direct result of the inflammatory response. In the case of a dirty wound (if the penetrating splinter, for example, is loaded with bacteria), *pus*—a thick whitish mixture of dead bacteria, tissue debris, and living and dead white blood cells—may accumulate. In addition, leaky capillaries and injured cells release chemicals that lead to blood clotting, which blocks damaged blood vessels (see Chapter 28). Clotting seals off the wound from the outside world and prevents more microbes from entering. Finally, sensations of pain, activated by swelling and chemicals released by the injured tissue, alert the injured person (or other animal) to protect the area from further damage.

Fever Combats Large-Scale Infections

If invaders breach other defenses and mount a large-scale infection, they may trigger a fever. If your sneezing classmate has the flu and you catch it, you will almost certainly develop a fever (a cold will rarely produce a fever). Although high fevers can be dangerous

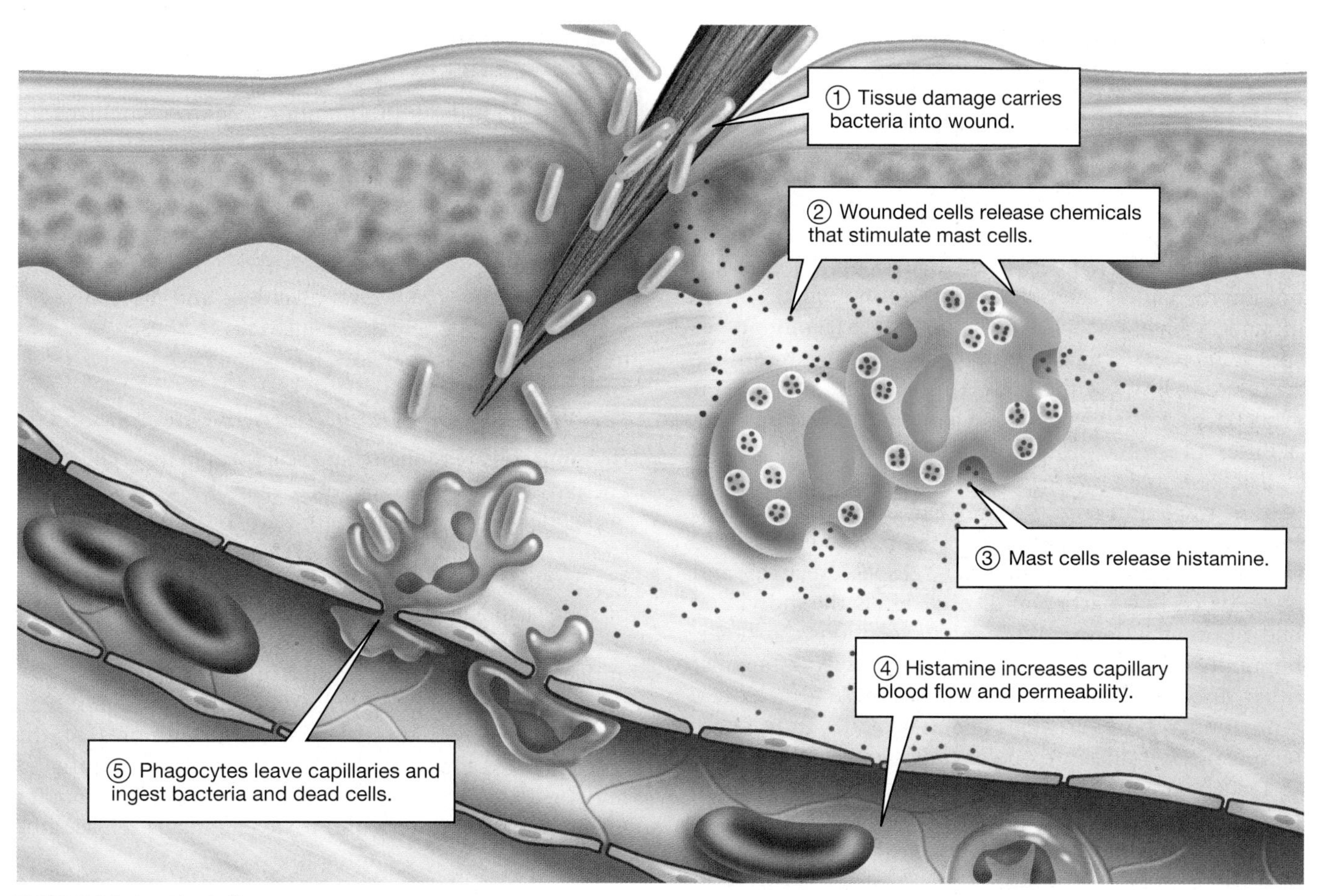
① Tissue damage carries bacteria into wound.
② Wounded cells release chemicals that stimulate mast cells.
③ Mast cells release histamine.
④ Histamine increases capillary blood flow and permeability.
⑤ Phagocytes leave capillaries and ingest bacteria and dead cells.

FIGURE 32-4 The inflammatory response

QUESTION Why can white blood cells, but not red blood cells, leave a capillary?

Media Activity 32.2 Inflammation

and even moderate ones can be very unpleasant, fever is actually part of the body's nonspecific response to infection. The onset of fever is controlled by the hypothalamus, the part of the brain that contains the temperature-sensing nerve cells that are the body's thermostat. In humans, this thermostat is set at about 37 °C (98.6 °F). Certain macrophages, as they respond to the infection, release a cytokine called *endogenous pyrogen* ("self-produced fire-maker"). This cytokine travels in the bloodstream to the hypothalamus and raises the thermostat's set point, triggering responses that increase body temperature. These responses include increased fat metabolism, constriction of surface blood vessels, and heat-producing behaviors such as shivering. Cytokines also cause other cells to reduce the concentration of iron in the blood.

Fever both enhances the body's normal defenses and harms invading microbes. In the case of bacterial infection, fever increases the activity of the phagocytic white blood cells that attack bacteria. Immune cells (discussed later in the chapter) multiply more rapidly and produce more antibodies. At the same time, feverish body temperatures of about 102 °F force many bacteria to reproduce more slowly and to require more iron for reproduction; the high temperature and reduced iron in the blood combine to help keep the numbers of bacteria in check.

Fever also helps fight viral infections by increasing the production of another cytokine called *interferon*. Some types of cells synthesize and release interferon after invasion by viruses; the interferon then travels to other cells and increases their resistance to viral attack. Because fever is a defense mechanism, it can be a mistake to attempt to reduce mild fevers. In one study, individuals with viral infections were treated either with aspirin (to reduce fever) or a *placebo* (an inactive substance that looks like the real drug, ensuring that the subjects don't know whether they have been given the drug). The subjects given aspirin had far more viruses in their noses and throats, and consequently sneezed and coughed out far more viruses, than did the subjects given the placebo. The defenses of subjects with fevers lowered by aspirin were not as effective at controlling infections, making these subjects much more infectious to other people. Because fevers are a sign of infection—and because high fevers are dangerous—be sure to seek advice from a health professional before treating them.

32.3 What Are the Key Characteristics of the Immune Response?

External barriers, phagocytic cells, natural killer cells, the inflammatory response, and fever are all *nonspecific* defenses, which prevent or overcome any microbial invasion of the body. Unfortunately, these nonspecific defenses are not impregnable. When they fail to do the job, the body mounts a highly specific and coordinated immune response directed against the particular organism that has successfully invaded the body. This response involves an army of specialized white blood cells that secrete an array of different chemicals and communicate in complex ways.

The essential features of the immune response to infection were recognized more than 2000 years ago by the Greek historian Thucydides. He observed that, occasionally, a person would contract a disease, recover, and never catch that particular disease again; the individual had become immune. With rare exceptions, however, immunity to one disease confers no protection against other diseases. Thus, the immune response attacks one type of microbe, overcomes it, and provides future protection against that microbe or toxin, but no others.

Let's assume the worst happens and you catch the flu from your classmate. The virus penetrates your nonspecific defenses and activates your immune response. Specialized white blood cells called **lymphocytes** are involved in this response. Lymphocytes are distributed throughout the body in the blood and lymph, and many are clustered in specific organs, particularly the thymus, lymph nodes, and spleen (described in Chapter 28). The lymphocytes that produce the immune response, the chemical antibodies they generate to target infectious microbes, and the organs in which the lymphocytes are produced and reside, collectively constitute the **immune system**.

The immune response arises from interactions among the various types of lymphocytes and the molecules that they produce. Table 32-1 provides a brief overview of the major cellular players and their roles in both the nonspecific defenses and specific immune responses.

The key cellular players in the immune response are two types of lymphocytes, called **B cells** and **T cells** (see Table 32-1). Like all white blood cells, B cells and T cells arise from lymphocyte precursor cells in the bone marrow (marrow is found in the interior of certain bones; see Chapter 35). Some of these lymphocyte precursors are released into the bloodstream and travel to the thymus, where they complete their differentiation into T (for *thymus*) cells. In contrast, B cells differentiate in the bone marrow. The two cell types play quite different roles in the immune response, but immune responses produced by both B cells and T cells consist of the same three fundamental steps. The cells of the immune system must first recognize the invader; second, launch an attack; and third, retain a memory of the invader to ward off future infections.

Cells of the Immune System Recognize the Invader

To understand how the immune system recognizes invading microbes and initiates a specific response to them, we must answer three related questions: How do the immune cells recognize foreign molecules? How can immune cells produce specific responses to so many different types of molecules? How do immune cells avoid mistaking the body's own cells and molecules for invaders? We'll explore the answers to these questions in the three sections below.

Foreign Invaders Exhibit Characteristic Antigens

Immune cells respond to specific foreign molecules that are particular to the invading microbe or toxin; these molecules serve as **antigens** (short for "*anti*body response *gen*erating"). In general, only large, complex molecules such as proteins, polysaccharides, and glycoproteins can act as antigens. Antigens are located on the surfaces of invading microbes or cancer cells, while viral antigens become incorporated into the plasma membranes of infected body cells. Viral or bacterial antigens are also exposed on the plasma membranes of the macrophages that engulf them. Other types of antigens may be dissolved in the blood or extracellular fluid; for example, snake venom and toxic chemicals released by bacteria are dissolved antigens.

Table 32-1 The Body's Cellular Armory

Mast Cells	Connective tissue cells that release histamine; important in the inflammatory response
Neutrophils	White blood cells that engulf invading microbes
Macrophages	White blood cells that engulf invading microbes and present antigens
Natural killer cells	White blood cells that destroy infected or cancerous cells
B cells	Lymphocytes that produce antibodies
Plasma cells	Offspring of B cells that secrete antibodies into the bloodstream
Memory B cells	Offspring of B cells that provide future immunity against invasion by the same antigen
T cells	Lymphocytes that regulate the immune response or kill certain types of cells
Cytotoxic T cells	Offspring of T cells that destroy specific targeted cells, including foreign eukaryotic cells, infected body cells, or cancerous body cells
Helper T cells	Offspring of T cells that stimulate immune responses by both B cells and cytotoxic T cells
Memory T cells	Offspring of T cells that provide future immunity against invasion by the same antigen

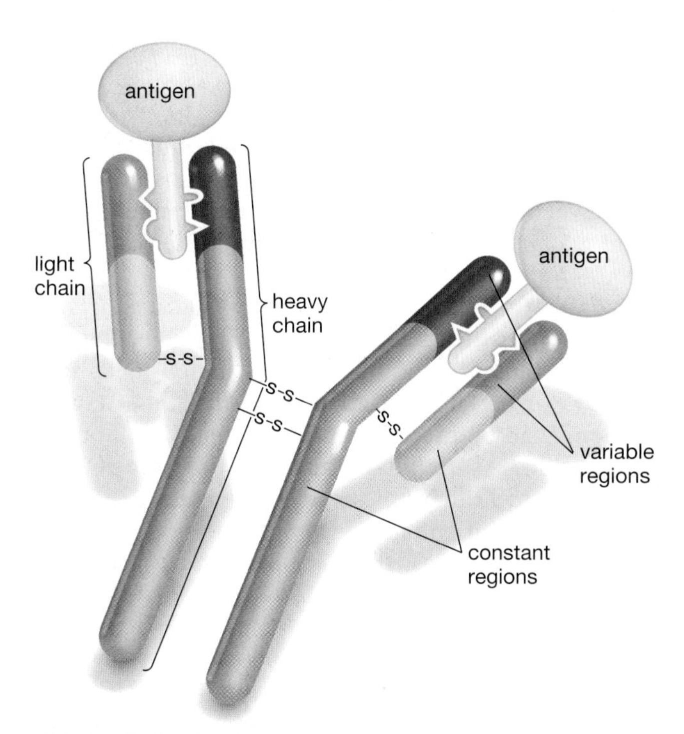

FIGURE 32-5 Antibody structure
Antibodies are proteins composed of two pairs of peptide chains (light chains and heavy chains) arranged like the letter Y. Constant regions on both chains form the stem of the Y; variable regions on the two chains form a specific binding site at the end of each arm of the Y. Different antibodies have different variable regions, forming unique binding sites. **QUESTION** Why do antibody molecules have both constant and variable regions?

(a) Antibody bound to the surface of a B cell recognizes microbes bearing foreign antigens.

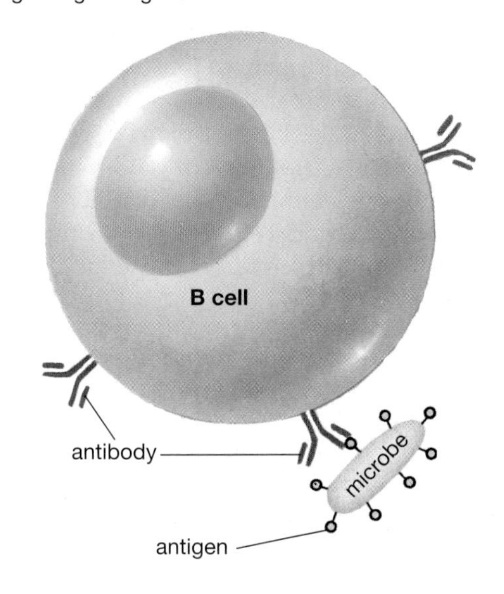

(b) Circulating antibodies bind to antigens on a microbe and promote phagocytosis by a macrophage.

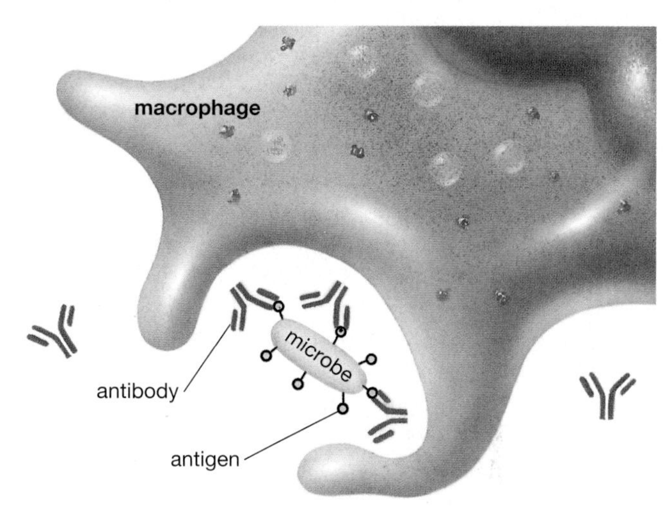

FIGURE 32-6 Some functions of antibodies

Antibodies and T-Cell Receptors Recognize and Bind to Foreign Antigens

The key to the immune cell's ability to recognize antigens lies in two types of large proteins: **antibodies**, which are produced by B cells, and **T-cell receptors**, produced by T cells. Both of these proteins recognize and bind to specific antigens.

Antibody proteins either remain attached to the surfaces of the B cells that produce them or are released and become dissolved in the blood plasma. Antibodies are Y-shaped molecules composed of two pairs of peptide chains: one pair of identical large (heavy) chains and one pair of identical small (light) chains (Fig. 32-5). Both heavy and light chains consist of a **constant region**, which is similar in all antibodies of the same type, and a **variable region**, which differs among individual antibodies. The combination of light and heavy chains results in an antibody with two functional parts: the "arms" and the "stem" of the Y. The variable regions (located at the tips of the arms of the antibody) form highly specific binding sites for antigens. These binding sites resemble the active sites of enzymes (see Chapter 6): Each binding site has a particular shape and electrical charge, so only certain molecules can fit in and bind. The binding sites are so specific that each antibody can bind only a few types of antigen molecules, sometimes only one.

There are five classes of true antibodies, each serving a different function in the defense of the body. (For more information, see "A Closer Look: Cellular Communication During the Immune Response" on this text's Web site.) The different types of antibodies are distinguished by small differences in the structure of the constant region—the stem (see Fig. 32-5). The stem determines the mechanism by which the antibody will act against invaders. For example, the stem of one type of antibody may attach the antibody to the plasma membrane of a B cell, while the stem of another type of antibody may bind to certain proteins in the blood, called *complement* proteins (described later), to promote the destruction of microbes. When the antibody stem remains attached to the B cell that produced it (Fig. 32-6a), the two arms of the antibody protrude outward from the B cell, sampling the blood and lymph for antigen molecules. When the arms of the antibody encounter an antigen with a compatible chemical structure, they bind to it. This binding triggers a response in the B cell that bears the antibody. Other antibodies circulate in the bloodstream (Fig. 32-6b), where they bind and neutralize toxins, or bind to microbes that bear complementary antigens. This binding then attracts phagocytic white blood cells that "eat" the microbes. These functions are described in more detail in the following sections.

Though T-cell receptors are different from antibodies, they are similar in some ways. T-cell receptors are found only on the surfaces of T cells and (unlike antibodies) are not released into the bloodstream. T-cell receptors consist of two peptide chains of about equal size. The ends of the two chains protrude from the T cell, forming highly specific binding sites for an antigen, just as the ends of the arms of an antibody molecule do. T-cell receptors trigger a response in the T cell only when they encounter antigen molecules borne on the membranes of cancerous or infected cells, or presented on the membranes of macrophages that have ingested the invading microbes. Unlike antibodies, the T-cell receptor proteins do not directly contribute to the destruction of invading microbes or toxic molecules.

Cells of the Immune System Can Recognize and Produce Specific Responses to Millions of Types of Molecules

During your lifetime, your body will be challenged by a multitude of different invaders. Your classmates may sneeze cold and flu viruses into the air you breathe. Weeds and trees release pollen that will find its way to your lungs. Your food may play host to a bacterial population or to toxins released by bacteria. You may drink water that contains the microbes that cause dysentery or the *Giardia* parasite. As you relax outdoors, a tick carrying the bacteria responsible for Lyme disease might bite you. There is no escape from these pervasive and persistent assaults. Fortunately, the immune system recognizes and responds to virtually all of the millions of antigens that an animal encounters (both natural and artificial), because immune cells produce millions of different antibodies and T-cell receptors, each capable of binding a different antigen. This amazing recognition ability is one of the main reasons that people and other animals don't get sick more often.

The mechanisms by which cells of the immune system recognize antigens are both complex and fascinating. Antibodies and T-cell receptors are, after all, proteins, and proteins are encoded by genes. It would seem that, to recognize millions of different antigens, each person would have to possess millions of different genes for antibodies and T-cell receptors. But there are only about 35,000 genes in the entire human genome. How, then, can a subset of these genes code for millions of antibodies and T-cell receptors? The answer lies in two distinct but complementary mechanisms that join forces to produce an enormous diversity of antibodies and T-cell receptors from a comparatively small number of genes. The following description explains how antibody diversity develops in B cells; similar phenomena produce a large diversity of T-cell receptors.

Antibodies Are Pieced Together from Fragments

There are no genes for entire antibody molecules. Instead, the human genome includes many genes that encode a variety of different antibody fragments; these fragments can be pieced together in many millions of combinations. As each individual B cell develops, it retains only a few randomly selected fragment genes (including both light and heavy chains with both constant and variable regions); the rest are cut out from the DNA as the cell matures (Fig. 32-7). Therefore, each B cell produces a single type of antibody—specified by the chance recombination of variable- and constant-region genes—that is different from the antibodies produced by other B cells (except its own daughter cells). The random selection of antibody-fragment genes from a large pool of choices yields a vast number of possible unique combinations. It may be helpful to think of antibody gene formation in terms of card playing. Each B cell is dealt a "hand" of two variable-region genes, one for the light chain and one for the heavy chain, randomly chosen from two large "decks" of genes. With each deck containing hundreds of "cards" (genes), virtually every cell will synthesize its own unique antibody.

The Immune System Does Not Design Antibodies or T-Cell Receptors Expressly to Bind Invading Antigens

The end result of mutation and gene recombination is that each B cell has its own particular antibody genes, different from those of most other B cells. At any time, the human body contains an army of perhaps 100 million different antibodies (and an even larger number of T-cell receptors), so antigens almost always encounter antibodies or receptors that will bind them. It is important to recognize that the immune cells do not "design" antibodies and T-cell receptors to fit invading antigens, as a tailor might design custom clothes for individual customers. Instead, the immune system randomly synthesizes millions of different antibodies and receptors. Like clothes in a department store, the array of antibodies and receptors is simply there, waiting. The fact that virtually every possible invading antigen binds to at least a few antibodies and receptors is due simply to the immense numbers of different antibodies and receptors present in the body. In our clothing analogy, given enough racks of clothing from which to choose, each of us will find something suitable that fits. The binding of antigen to antibody triggers changes in the B cells. These changes usually lead to the destruction of microbes bearing that antigen. We will examine these changes in more detail in a moment.

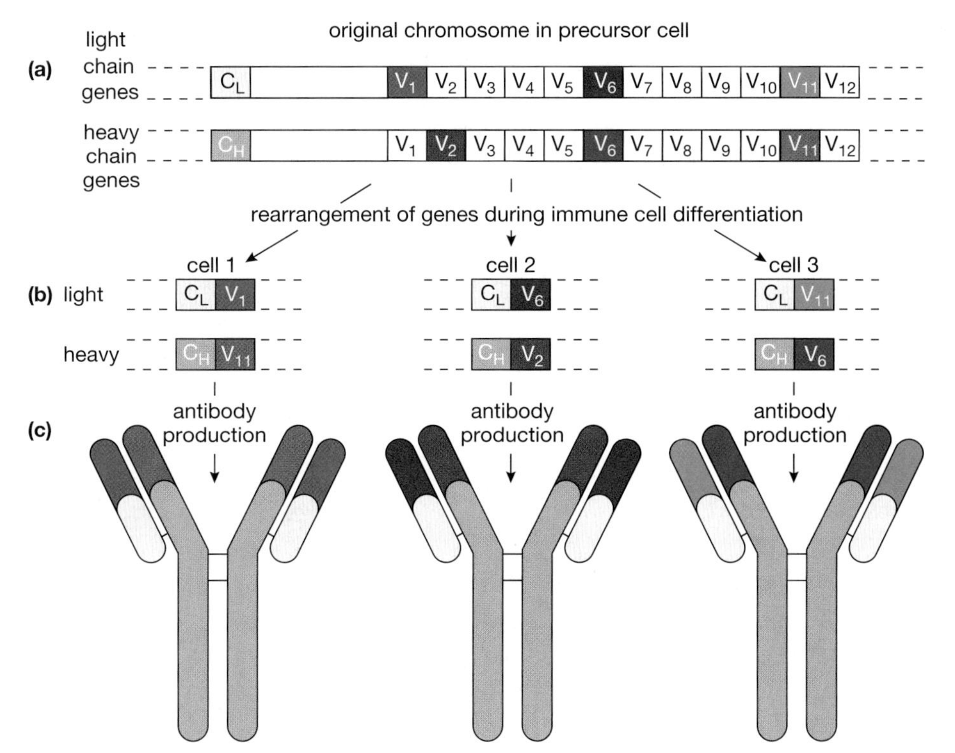

FIGURE 32-7 Recombination produces antibody genes
(a) Each precursor cell of the immune system contains at least one gene for the constant regions (C) of the light (C_L) and heavy (C_H) chains of antibodies and many genes for the variable regions (V_{1-12}). **(b)** During the development of each B cell, these genes are rearranged, moving one of the variable-region genes next to a constant-region gene. For each chain, each cell thus generates a "recombined antibody gene" that differs from the recombined antibody genes generated by other B cells. **(c)** The different antibodies synthesized by each B cell in (b).

THE IMMUNE SYSTEM DISTINGUISHES "SELF" FROM "NON-SELF" Why doesn't your immune system destroy your own cells? The surfaces of the body's own cells bear large proteins and polysaccharides, just as microbes do. Some of these proteins, collectively called the **major histocompatibility complex (MHC)**, are unique to each individual (except identical twins, who have the same genes and hence the same MHC proteins). Because your MHC proteins are different from those of everyone else, they act as foreign antigens in other people's bodies. This is why organ transplants are rejected. Physicians must find a donor whose MHC proteins are as similar as possible to those of the recipient, and also use drugs to suppress the recipient's immune system. If this is not done, the recipient's immune system recognizes MHC proteins on the donor's cells as foreign and destroys the transplanted tissue.

So why don't these "self" antigens arouse an individual's own immune system? The key seems to be the continuous presence of the body's own antigens while the immune cells mature. As an embryo develops, some newly differentiating immune cells do indeed produce antibodies or T-cell receptors that can bind the body's own proteins and polysaccharides, treating them as antigens. However, as soon as any *immature* immune cell contacts a molecule it treats like antigen, the immature immune cell is destroyed. Since immune cells in their immature state are unlikely to encounter any *foreign* antigens, only immune cells with antibodies to "self" are likely to encounter antigens. By destroying these cells, the immune system distinguishes "self" from "non-self" by retaining only those immune cells that do not respond to the body's own molecules.

Cells of the Immune System Launch an Attack

Once a body has been invaded, say by a flu virus, the immune cells launch two types of attack: humoral immunity and cell-mediated immunity. **Humoral immunity** is provided by B cells and the circulating antibodies they secrete into the bloodstream, which attack invaders before they can enter body cells. **Cell-mediated immunity** is produced by *cytotoxic T cells*, which attack invaders that have made their way into body cells. These two types of immune responses require considerable communication among the cell types (see "A Closer Look: Cellular Communication During the Immune Response," at this text's Web site for Chapter 32). Both humoral and cell-mediated immunity are stimulated by **helper T cells**. Helper T cells have T-cell receptors for microbial antigens. These antigens are presented either on the plasma membranes of macrophages that have engulfed the invaders or on the surfaces of infected cells. Upon binding an antigen, the helper T cells release cytokines that stimulate cell division and differentiation in both the B cells and cytotoxic T cells that respond to the same microbial invasion (see Fig. 32-11). In fact, very little immune response, either cell-mediated or humoral, can occur without the

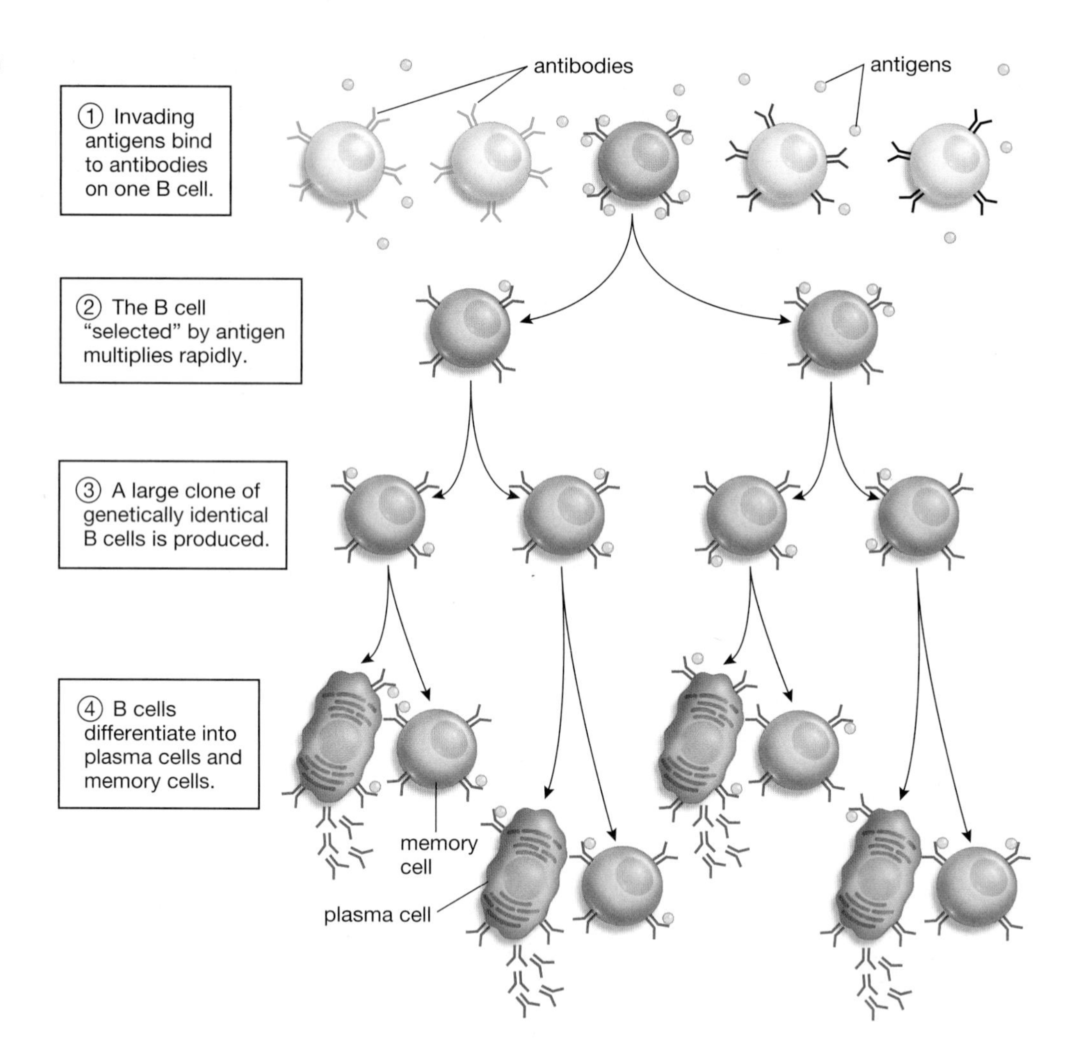

antibodies
antigens
① Invading antigens bind to antibodies on one B cell.
② The B cell "selected" by antigen multiplies rapidly.
③ A large clone of genetically identical B cells is produced.
④ B cells differentiate into plasma cells and memory cells.
memory cell
plasma cell

FIGURE 32-8 Clonal selection among B cells by invading antigens

Media Activity
32.3 Clonal Selection

chemical boost provided by helper T cells. That is why AIDS, which destroys helper T cells, is such a deadly disease. Although humoral and cell-mediated immunity are not completely independent, we consider them separately below for ease of understanding.

Humoral Immunity Is Produced by Antibodies Dissolved in Blood

Each B cell bears a specific type of antibody on its surface. When an infection occurs, the antibodies borne by a few B cells are able to bind to antigens on the invader. Antigen–antibody binding causes these B cells to divide rapidly. This process is called **clonal selection** because the resulting population of cells is composed of "clones" (genetically identical to the parent B cells) that have been "selected" to multiply by the presence of particular invading antigens (Fig. 32-8). The daughter cells differentiate into two cell types: **memory B cells** and **plasma cells.** Memory cells do not release antibodies but do play an important role in future immunity to the particular invader that stimulated their production (as described later). Plasma cells become enlarged and packed with rough endoplasmic reticulum, within which huge quantities of their own specific antibody proteins are synthesized (Fig. 32-9). These antibodies are then released into the bloodstream (hence the name "humoral" immunity; to the ancient Greeks, blood was one of the four "humors," or body fluids). Scientists have exploited the ability of plasma cells to produce large quantities of a specific type of antibody for a variety of medical uses; see "Scientific Inquiry: Designer Drugs Fight Disease."

Because antibodies circulate in the bloodstream, humoral immunity can defend only against invaders that are in the blood or extracellular fluid. Bacteria (most of which never enter the body's cells), bacterial toxins, and some fungi and protists are therefore vulnerable to the humoral immune response. Invaders that penetrate into the body's cells, as viruses do, are safe from antibody attack as long as they are within the cytoplasm of a body cell. Antibodies can attack viruses either during initial infection or after the viruses have replicated in one host cell, have ruptured it, and have been released into the body fluids.

Humoral immunity produced by antibodies in the blood destroys foreign chemicals or microbes in several

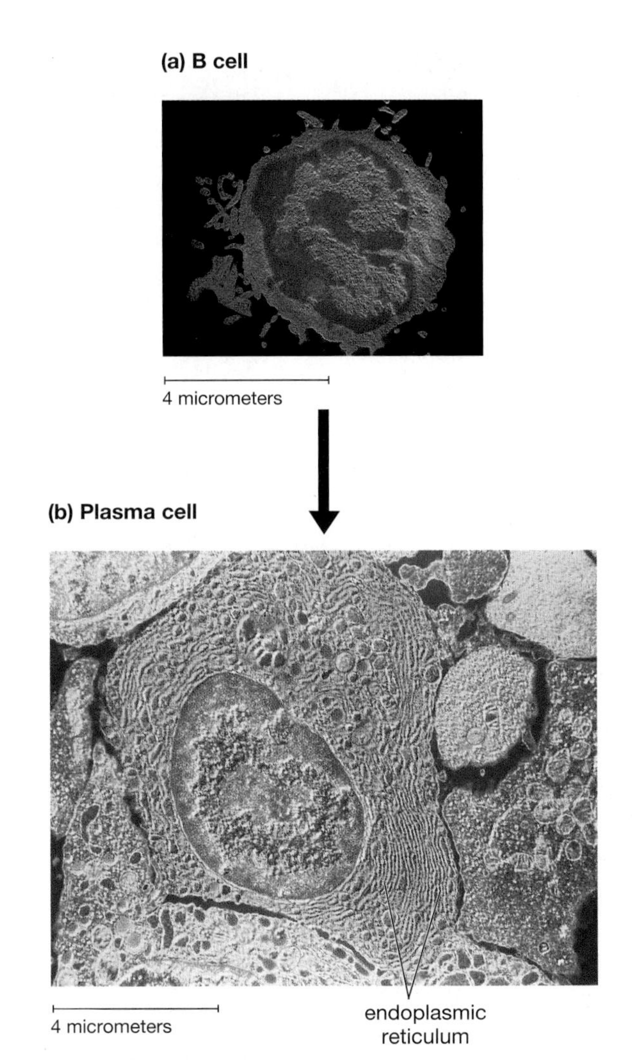

FIGURE 32-9 A B cell becomes a plasma cell
Colorized micrographs of B cells **(a)** before and **(b)** after conversion to plasma cells. The plasma cell is much larger than the B cell and is packed with rough endoplasmic reticulum that synthesizes antibodies.

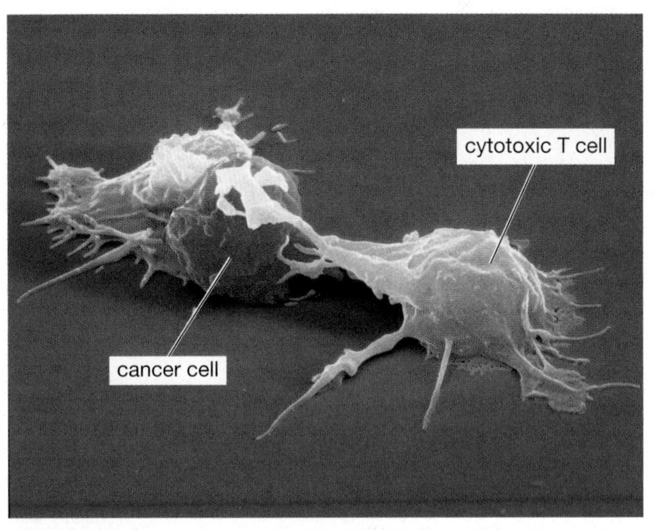

FIGURE 32-10 Cell-mediated immunity in action
Cytotoxic T cells attack a cancer cell in this false-color scanning electron micrograph. **QUESTION** A cancer tumor consists of the body's own cells, so why does the immune system attack cancer cells?

ways. First, the circulating antibodies can bind to a foreign molecule and render it inactive. Second, the antibodies may coat the surface of an invading cell, or third, each antibody may bind two microbes and cause them to clump together; both of these tactics encourage macrophages to engulf the microbes. Finally, the antibody–antigen complex on the surface of an invading cell may trigger a series of reactions with a group of blood proteins called the **complement system**. When these complement proteins bind to the antibody stems, a complex series of reactions is produced among the different members of the complement system. The end result is that the proteins attract phagocytic white blood cells to the site, promote phagocytosis of the foreign cells, or in some instances directly destroy the invaders by creating holes in their plasma membranes, much as natural killer cells do.

Humoral immunity works best against toxic chemicals, or bacterial, fungal, or parasitic microbes. Humoral immunity will also work against viruses as they enter the body or after they are released from infected cells. To fight the viral infection you contract from your sneezing classmate, however, you will also need the help of cell-mediated immune responses.

Cell-Mediated Immunity Is Produced by T Cells

How can cells of the immune system recognize body cells that are cancerous or harboring viruses? Why are transplanted organs sometimes rejected? These events occur because of cell-mediated immunity. Three types of T cells contribute to cell-mediated immunity: *helper T cells*, *cytotoxic T cells*, and *memory T cells*.

Recall that helper T cells bind antigens on the surfaces of infected or cancerous cells and produce cytokines that stimulate T-cell division and differentiation. Some then become **cytotoxic T cells**, which release proteins that disrupt the infected cell's plasma membrane. This attack is activated when T-cell receptors on the cytotoxic T cells' membranes bind to antigens on the surface of infected or cancerous cells (Fig. 32-10). The cytotoxic T cells then release proteins (similar to those produced by natural killer cells) that create large holes in the target cell's membranes. After the infection is over, some helper T cells persist and function as **memory T cells**. Like memory B cells, these help protect

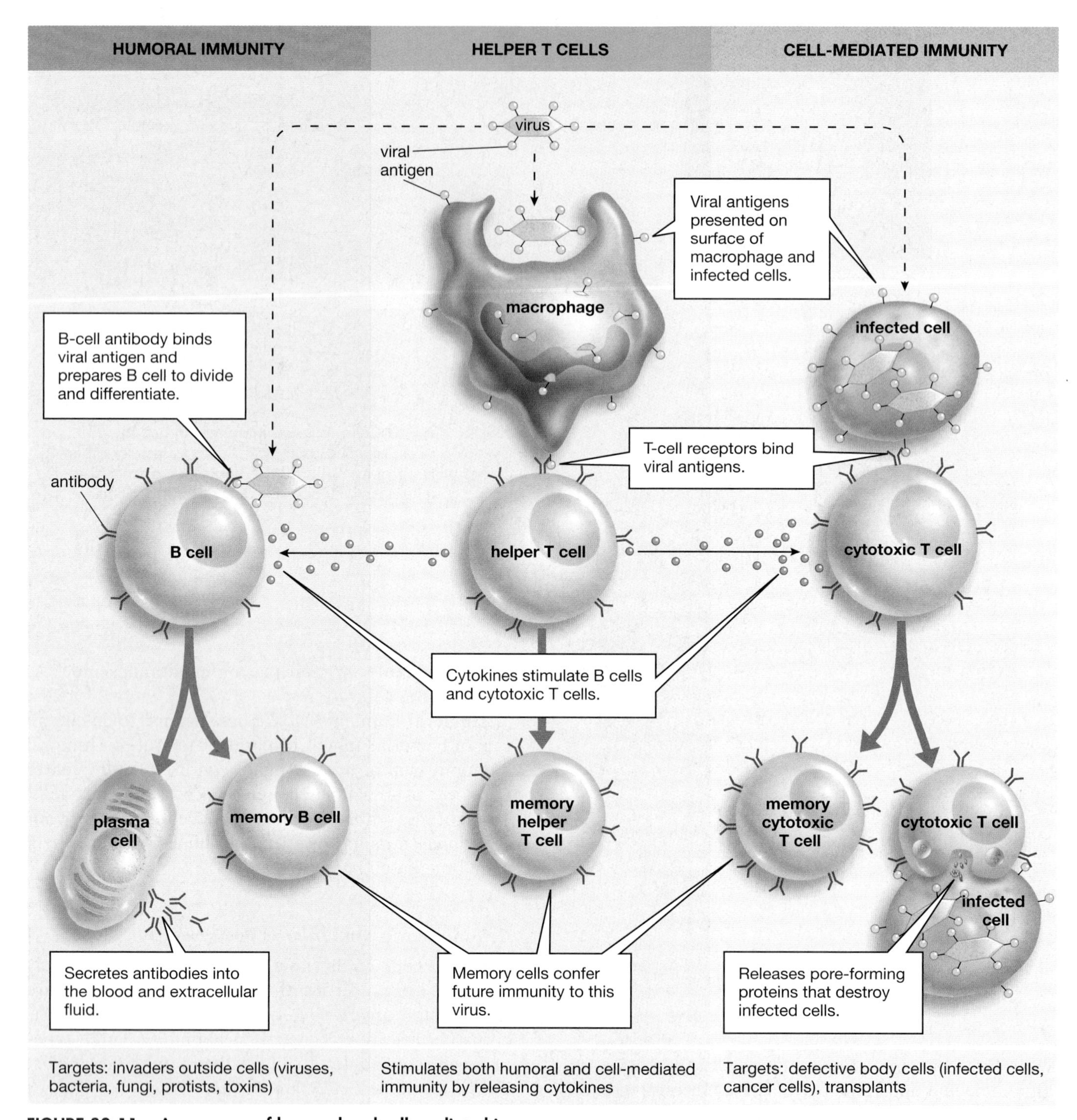
HUMORAL IMMUNITY
HELPER T CELLS
CELL-MEDIATED IMMUNITY
virus
viral antigen
Viral antigens presented on surface of macrophage and infected cells.
macrophage
infected cell
B-cell antibody binds viral antigen and prepares B cell to divide and differentiate.
T-cell receptors bind viral antigens.
antibody
B cell
helper T cell
cytotoxic T cell
Cytokines stimulate B cells and cytotoxic T cells.
plasma cell
memory B cell
memory helper T cell
memory cytotoxic T cell
cytotoxic T cell
infected cell
Secretes antibodies into the blood and extracellular fluid.
Memory cells confer future immunity to this virus.
Releases pore-forming proteins that destroy infected cells.
Targets: invaders outside cells (viruses, bacteria, fungi, protists, toxins)
Stimulates both humoral and cell-mediated immunity by releasing cytokines
Targets: defective body cells (infected cells, cancer cells), transplants

FIGURE 32-11 A summary of humoral and cell-mediated immune responses

the body against future infection. Figure 32-11 compares the humoral immune response with the cell-mediated immune response.

Cells of the Immune System Remember Past Victories

After you have recovered from a disease, you will remain immune to that particular strain of microbe for many years, perhaps a lifetime. Retaining immunity is the function of memory cells. Plasma cells and cytotoxic T cells directly fight disease-causing organisms, but they generally live only a few days. Each plasma or cytotoxic T cell, however, leaves behind hundreds or thousands of memory cells that may survive for many years. If the body is invaded by microbes to which the body has already mounted an immune response, the appropriate memory cells will recognize the invaders. These memory cells will then multiply rapidly and produce a second immune response by generating huge populations of

SCIENTIFIC INQUIRY Designer Drugs Fight Disease

In addition to their crucial role in defending the body, antibodies are valuable for many medical and scientific purposes. To acquire a supply of pure antibody, scientists inject an animal with the antigen that induces B cells to form plasma cells, which in turn produce the desired antibody. Since normal plasma cells cannot be grown outside the body, scientists induce them to merge with plasma tumor cells to form hybrid cells called *hybridomas*. Hybridomas proliferate under laboratory conditions and produce relatively large quantities of a specific antibody, known as *monoclonal antibodies*. Because a given antibody binds only to one specific type of antigen, antibodies are invaluable tools for finding and/or marking molecules in living animals or in dissected tissues. For example, pregnancy tests typically use an antibody that binds to a hormone released by a developing embryo.

For "therapeutic antibodies" to reach their full potential in the fight against disease, they must be produced in massive quantities. Modern genetic engineering techniques (see Chapter 13) are being used to introduce the genes for human antibodies into farm animals (engineered cows and goats now secrete antibodies in their milk) and even into corn and other plants. Scientists are experimenting with antibody therapies directed against a variety of invaders and toxins, including tooth decay bacteria, herpes viruses, cancer cells, and even addictive drugs. Cancer cell antibodies can be linked to radioactive particles or chemotherapy drugs, allowing them to be carried directly and specifically to the cancer. Sometimes called "magic bullets," there are several of these monoclonal antibody-based drugs now on the market, including those targeting breast cancer and leukemia. Antibodies to cocaine can bind to the cocaine molecule and eliminate its effects, potentially counteracting a drug overdose or making the drug unable to elicit a "high." Recently, drug company researchers have cloned genes for human T-cell receptors and introduced them into harmless bacteria that can produce the T cells in large quantities. Unlike natural T-cell receptors, these cloned versions can exist independently of the T cell. The drug companies hope to be able to produce "designer T-cell receptors" that are specific for particular cancers or infectious agents. As with monoclonal antibodies, the T-cell receptors could be used to ferry drugs or radioisotopes directly to infected or cancerous cells. The ability to mass-produce specifically designed antibodies and T-cell receptors opens up an exciting frontier in the prevention and cure of diseases—using an arsenal derived from the immune system's own weapons.

plasma cells (from memory B cells) and cytotoxic T cells (from memory T cells). Because so many memory cells are primed to respond, the body will be able to fend off a second attack before it gains a foothold (Fig. 32-12). Why then, do you repeatedly suffer from colds and flu? The problem, as described in "Health Watch: Can We Beat the Flu Bug?" is that each year's flu is caused by a different virus than the previous year's; additionally, so many cold viruses exist that you are likely to encounter different versions of these, as well.

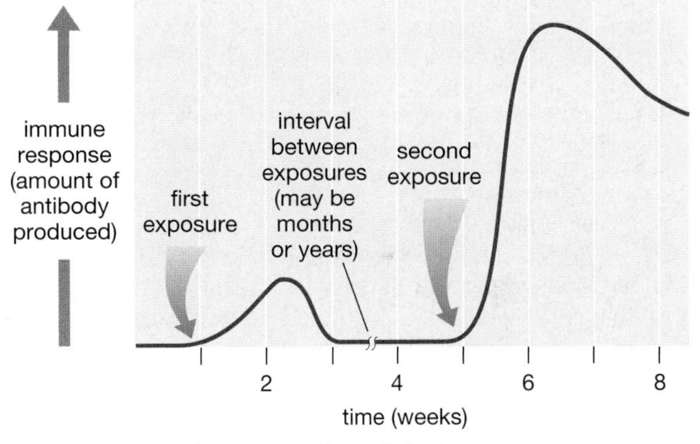

FIGURE 32-12 Memory cells and the immune response The immune response to the first exposure to a disease-causing organism is fairly slow and not very large, as B and T cells are selected and multiply. A second exposure activates memory cells formed during the first response, so the second response is both faster and larger.

32.4 How Does Medical Care Augment the Immune Response?

Antibiotics Slow Down Microbial Reproduction

The immune system may sound invincible, but as you know, this is not the case. If left untreated, many diseases kill their victims because—unfortunately—the body provides ideal conditions for the growth and reproduction of disease-causing microbes. Any infection, then, is a race between the invading microbes and the immune response. If the initial infection is massive or if the microbes produce particularly toxic by-products, the immune response may not fully activate before it is too late. Further, some microbes, such as the microscopic worm responsible for schistosomiasis and the protist responsible for malaria, have evolved ingenious defenses that help them evade the immune system. Schistosomiasis worms, which may live within an infected individual for up to two decades, fool the immune system by coating themselves with human proteins that are not perceived as antigens. The malaria protist undergoes three stages of its life cycle within humans, (see Chapter 19; Fig. 19-20) each stage exposes different antigens for too short a time for the immune system to launch an effective attack.

Antibiotics are drugs that retard the growth and multiplication of many invaders, including bacteria, fungi, and protists (but not viruses). Although antibiotics usually don't destroy every single microbe, they give the body's non-specific internal defenses and immune response

HEALTH WATCH Can We Beat the Flu Bug?

Every winter, a wave of *influenza*, or flu, sweeps across the world. Thousands of the elderly, the newborn, and those already suffering from illness perish, while hundreds of millions more suffer the respiratory distress, fever, and muscle aches of milder cases. Occasionally, devastating flu varieties appear, causing worldwide epidemics called *pandemics*. In the great flu pandemic of 1918, 20 million people worldwide died in one winter. In 1968 the Hong Kong flu infected 50 million people in the United States, causing 70,000 deaths in 6 weeks.

Flu Viruses

Flu is caused by several viruses that invade the cells of the respiratory tract, turning each cell into a factory for manufacturing new viruses. The outer surface of a flu virus is studded with proteins, some of which are recognized by the immune system as antigens. Most people survive the flu because their defenses eventually inactivate the viruses or kill virus-infected body cells. Why, then, don't our memory T and B cells make us immune to future outbreaks of flu?

The answer lies in a flu virus's amazing ability to change. Flu virus genes are made up of RNA, which lacks the proofreading mechanisms that reduce mutations in genes made of DNA (see Chapter 10). Therefore, flu RNA genes mutate rapidly. Four or five mutations in the same virus may alter the surface antigens enough that the immune system doesn't fully recognize the virus as the same old flu that was beaten off last year. Since your memory cells provide only partial protection, you may get the flu year after year.

Deadly New Strains

On rare occasions, dramatically new and deadly flu viruses appear and cause pandemics. These viruses carry entirely new antigens with distinctive structures that the human immune system has never before encountered. To respond, the immune system must start from scratch, selecting out entirely new lines of B cells and T cells to attack the intruder. In the meantime, the virus multiplies so rapidly that many individuals die or become so weakened that they contract some other fatal disease.

Where do the genes that encode these new flu antigens come from? Believe it or not, they come from viruses that infect birds and pigs. The intestinal tracts of birds, especially migratory waterfowl such as ducks, may carry viruses strikingly similar to human flu viruses, although the infected waterfowl show no symptoms. While human flu viruses aren't known to infect birds, both human and bird viruses can infect pigs, so occasionally, both viruses will meet within the same pig cell. Once in a great while, new viruses that spring from a doubly infected pig cell end up with a mixture of genes from human and bird viruses (Fig. E32-1). Some of these hybrid viruses combine the worst genes of each (at least from our perspective): From the human virus, the deadly new viruses pick up the genes needed to subvert human cellular metabolism to produce new viruses; from the bird virus, they pick up genes for new surface antigens. The hybrid viruses can move easily from pigs to humans, because pigs live near humans and, like people, pigs cough when they have the flu.

Most flu pandemics are thought to originate in China, where the feces from pigs and ducks are used to fertilize fish ponds. This is a very efficient farming practice, but unfortunately it also places pigs (ideal mixing vessels for flu viruses) in close proximity to humans and ducks. In Hong Kong in 1997, health officials made an alarming discovery—a new and fatal form of flu virus appeared in poultry and could be passed directly from the birds to humans. After 6 people died, all the chickens in Hong Kong's markets and farms were slaughtered. Fortunately, this virus had not yet evolved the capacity to spread from person to person; had it acquired this ability, a new and deadly pandemic might have swept the world.

Swatting the Flu Bug

Flu viruses mutate so rapidly that they may be impossible to eradicate, but we can fight back with flu shots. Flu shots are vaccinations; how is a vaccine against a mutating virus created? Each year, World Health Organization officials collect samples of mutated influenza virus from more than 100 sites throughout the world, and they identify three strains they believe are most likely to spread widely. These viruses are grown

enough time to finish the job. One problem with antibiotics, however, is that they are potent agents of natural selection (see Chapter 15). The occasional mutant microbe that is resistant to an antibiotic will pass on the gene(s) for resistance to its offspring. The result is that resistant mutants proliferate, whereas susceptible microbes die off. Eventually, many antibiotics become ineffective in treating diseases.

Until recently, little could be done about viral infections except to treat the symptoms and hope that the immune system would triumph. Now, a new class of drugs called *neuraminidase inhibitors* has entered the physician's arsenal. These antiviral drugs block the ability of newly formed viruses to escape the host cells in which they replicate. In the coming years, the immune system may receive powerful help from these drugs (see "Health Watch: Can We Beat the Flu Bug?").

Vaccinations Stimulate the Development of Memory Cells

As early as A.D. 1000, people in India, China, and Africa deliberately exposed themselves to mild cases of smallpox to acquire immunity to the disease. In 1796 the English surgeon Edward Jenner inoculated an 8-year-old boy with bacteria-laden material taken from cowpox lesions on the fingers of a dairymaid. The boy developed a small lesion and slight fever. A few months later, Jenner inoculated the

in fertilized chicken eggs, and viral proteins are then isolated and injected into people to stimulate an immune response that protects against subsequent infections by these strains. Flu shots are often quite effective, especially if the correct strains have been selected. For people who contract flu anyway, *neuraminidase inhibitors*, first marketed in the United States in 2000, can help. The viral enzyme neuraminidase is crucial to the viral life cycle because it allows newly formed viruses to escape from their host cells; this enzyme is conserved on all successful flu strains. Administered immediately after flu symptoms begin, the drugs block neuraminidase, interrupt the viral life cycle, and hasten recovery. Researchers hope that people will someday be tested for flu virus infections before symptoms occur, allowing very early treatment that might prevent the symptoms entirely. While we may never eradicate influenza, it may soon lose its ability to disrupt our lives.

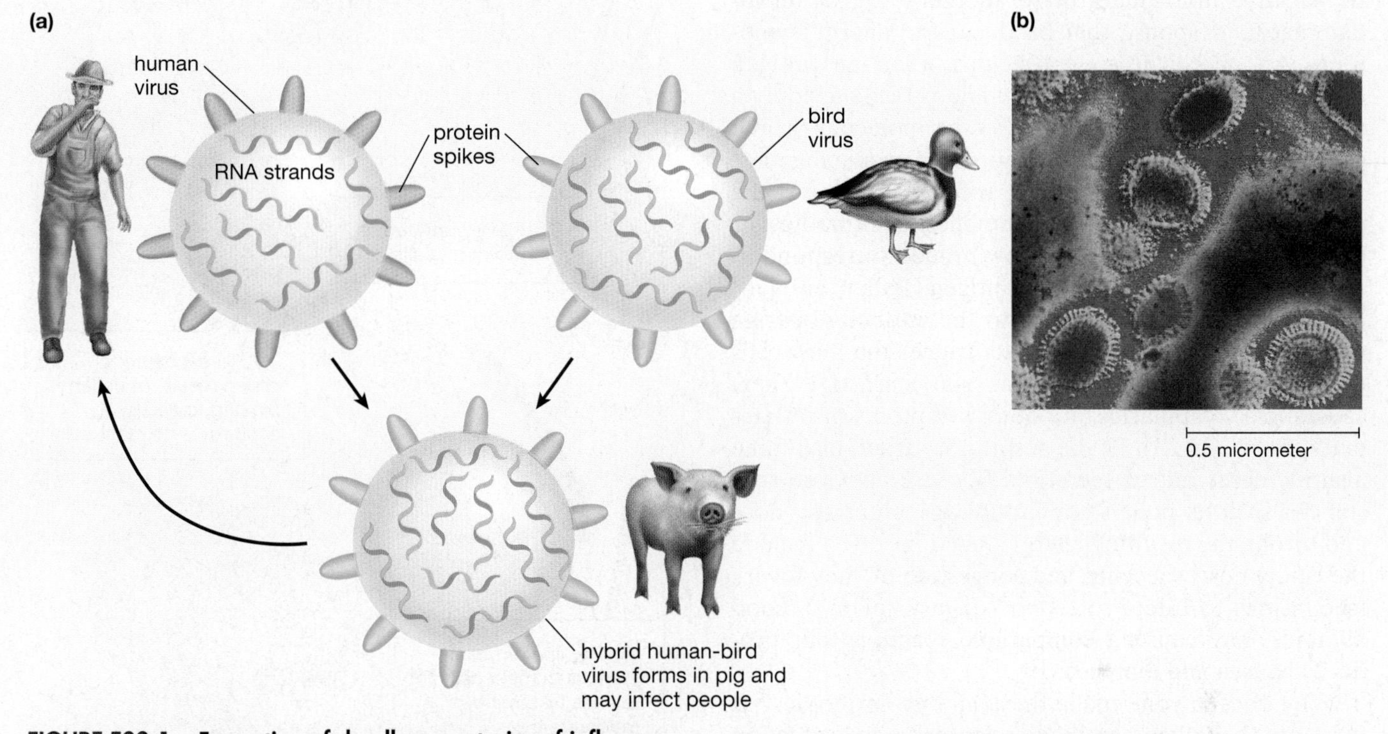

FIGURE E32-1 Formation of deadly new strains of influenza
(a) Rare recombination of genes from bird and human flu viruses can result in deadly new strains of flu. Note the protein "spikes" projecting from the virus coats in the drawings and in **(b)** the colorized transmission electron micrograph of flu viruses. The spikes attach to cells in the human respiratory system, helping the virus gain entry into the cells. If these spike proteins are derived from the bird virus, they may be totally new to the human immune system.

boy with material from a smallpox lesion, but the boy remained healthy. After repeating these results, Jenner published his findings in 1798. This discovery initiated the modern practice of **vaccination**, the injection of weakened or killed microbes to confer immunity. The word "vaccination" is derived from the Latin word for "cow," in reference to Jenner's pioneering efforts using cowpox. In the late 1800s, Louis Pasteur extended the use of vaccination to several other diseases by injecting weakened or dead microbes into healthy individuals. The weakened microbes do not cause disease (or at least not a severe case). However, they do bear antigens that elicit a vigorous immune response that produces an army of memory cells that confer immunity against subsequent exposure to the living, dangerous microbes. Today, many diseases, including polio, diphtheria, typhoid fever, and measles can be controlled through vaccination. Smallpox, one of the deadliest diseases of all, has been completely eradicated since 1980 as a result of a massive vaccination program sponsored by the World Health Organization. Two facilities, one in Siberia and one in Atlanta, legally store what health officials hope are the last smallpox viruses on Earth. One compelling reason to maintain these stocks is as a source of genetic material that will allow scientists to make additional smallpox vaccines. A generation of people has never been vaccinated against smallpox, and some authorities believe that "bioterrorists" may possess their own stocks of smallpox virus.

32.5 What Happens When the Immune System Malfunctions?

Allergies Are Misdirected Immune Responses

As your classmate sneezes nearby, you entertain a fleeting hope that he is merely allergic to something and not coming down with the flu. More than 35 million Americans suffer from **allergies**, immune responses to antigens that are not harmful. Common allergies include those to pollen, dust, mold spores, animal dander, and bee stings. In sensitive individuals, these allergens trigger an inflammatory response that produces the allergy symptoms. An allergic response begins when an antigen (such as pollen) enters the body and is recognized by a B cell carrying an antibody to some molecule on the pollen's surface. This B cell proliferates, producing plasma cells that pour out "allergy antibodies" against the pollen antigens. The stems of the allergy antibodies attach to mast cells, which are now primed to respond to subsequent exposures to the antigen. Pollen antigens that later encounter and bind to the antibody-bearing mast cells in the respiratory tract trigger the mast cells to release histamine, which (as you learned earlier) causes leaky capillaries and other symptoms of inflammation (Fig. 32-13). In the respiratory tract, histamine also increases mucus secretion. Because airborne substances such as pollen grains typically enter the nose and throat, the resulting allergic reactions often include the runny nose, sneezing, and congestion of "hay fever" (see "Links to Life: Protection Against Pollen"). Food allergies may cause a comparable reaction that produces nausea and diarrhea.

Why does anyone make these allergy antibodies? It turns out that allergy antibodies confer some protection against parasites, such as worms and the freshwater protist *Giardia*, which colonizes the gastrointestinal tract and causes nausea and diarrhea. Many parasites enter the body through the mouth or nose, moving into the respiratory or digestive tract and attaching to the lining of the nasal passages, throat, or intestine. Since the typical symptoms of allergies—increased mucus secretion, sneezing and coughing, or intestinal contractions causing diarrhea—could help dislodge and expel the parasites, the allergic response can be an important defense mechanism. The fact that the same type of antibodies may also target common, harmless substances in the environment is an unfortunate side effect of an otherwise helpful immune response.

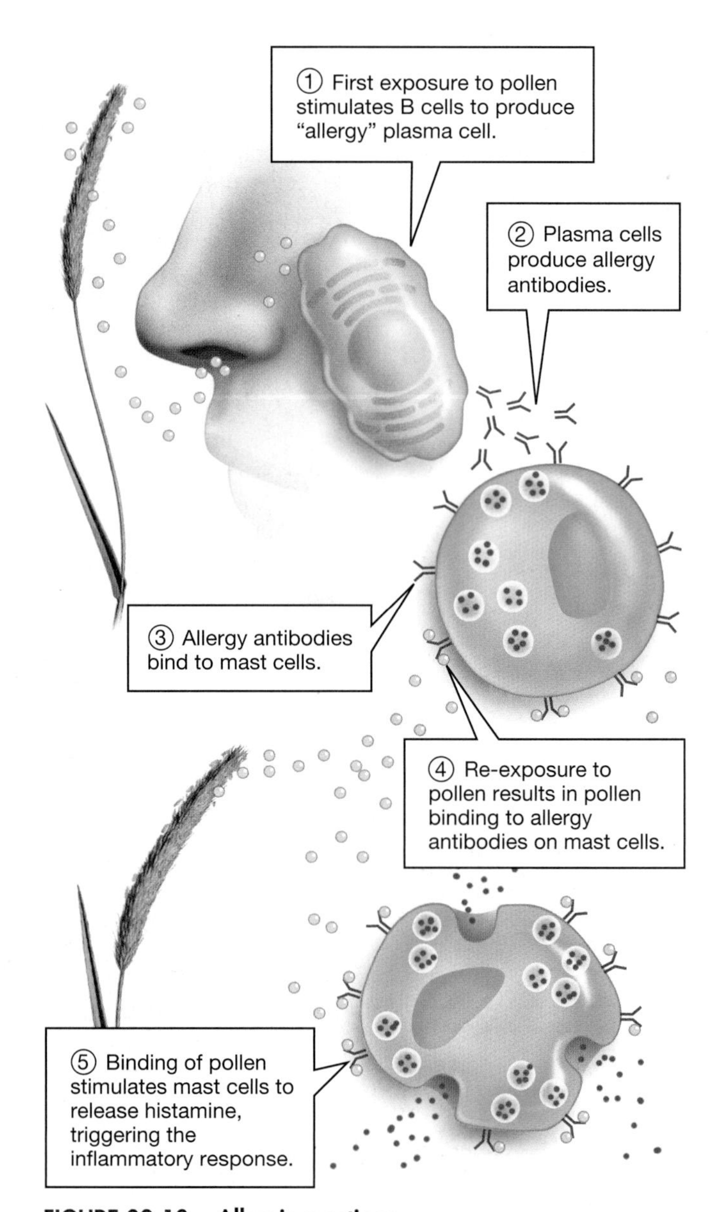

FIGURE 32-13 Allergic reactions

An Autoimmune Disease Is an Immune Response Against the Body's Own Molecules

Fortunately, our immune systems rarely mistake our own cells for invaders. Occasionally, however, something goes awry, and "anti-self" antibodies are produced. The reason for this is not understood; one hypothesis is that infection-causing cells occasionally bear antigens that resemble those of the individual's body. The antibodies produced by the immune system may then mistakenly attack some of the animal's own cells. The result is an **autoimmune disease**, in which the body mounts an immune response against a particular cell type that it should recognize as "self." Some types of anemia, for example, are caused by antibodies that destroy an individual's red blood cells. Many cases of juvenile-onset diabetes occur because the immune system attacks the insulin-secreting cells of the pancreas. Multiple sclerosis occurs when immune cells launch a misdirected attack against the insulating fatty sheath that coats parts of neurons in the brain and spinal cord. Rheumatoid arthritis is caused when the immune system attacks the cartilage of the joints of the body. Unfortunately, at present there is no known cure for autoimmune diseases. In some cases, replacement therapy can alleviate the symptoms—for instance, administering insulin to diabetics or blood transfusions to people with anemia. Alternatively, the autoimmune response can be suppressed with drugs. Immune suppression, however, also reduces immune responses to the everyday assaults of disease microbes, so this therapy has major drawbacks.

An Immune Deficiency Disease Results from the Inability to Mount an Effective Immune Response to Infection

David, the "bubble boy," lived all of his short, 12-year life in a germ-proof "bubble," isolated from direct contact with every unsterilized object, including other people. Occasionally, a child like David is born with **severe combined immune deficiency (SCID)**, a disorder in which few or no immune cells are formed due to a defective gene. A child with SCID may survive the first few months of postnatal life, protected by antibodies acquired from the mother during pregnancy or in her milk. Once these antibodies are lost, however, common bacterial infections can prove fatal. One form of therapy that has had remarkable success involves removing bone marrow (from which immune cells arise) from the child with SCID. In a culture dish, researchers infect the bone marrow cells with a virus that has been genetically engineered to carry a functional copy of the defective gene. The virus inserts its genetic material, including the functional gene, into the DNA of the marrow cells. The modified marrow cells are then injected into the bloodstream of the SCID victim. In successful cases, they return to the bone marrow, proliferate, and stimulate the formation of healthy immune cells. The most fortunate recipients of this experimental therapy have been able to resume normal lives.

AIDS Is a Devastating Immune Deficiency Disease

The most common and widespread immune deficiency disease is **acquired immune deficiency syndrome**, or **AIDS**. In 2003, the UN estimated that 3 million people died of AIDS and 5 million more became infected, bringing the total infected population to 40 million. Two viruses, named **human immunodeficiency viruses** 1 and 2 (HIV-1 and HIV-2), cause AIDS. These viruses undermine the immune system by infecting and destroying the helper T cells, which stimulate both the cell-mediated and humoral immune responses. AIDS does not kill people directly, but AIDS victims become increasingly susceptible to other diseases as their helper T-cell populations decline. Although AIDS was first recognized in 1981, genetic studies show that the virus almost certainly arose from viruses that have been infecting chimpanzees in Africa for as many as 100,000 years. Infected chimps show no signs of the disease. Preserved tissue samples taken from a man who lived in Africa, dating from 1957, contain fragments of the HIV genome that resemble the common ancestor of three modern strains of HIV and also resemble the ancestral chimp virus. Researchers believe that this ancestral virus mutated and jumped from chimps to humans between the mid-1940s and the early 1950s.

The Human Immunodeficiency Virus Is a Retrovirus That Infects and Destroys Helper T Cells

How does HIV wreak havoc on the human immune system? Both HIV-1 and HIV-2 are *retroviruses*—viruses that have RNA as their genetic material and reproduce by first transcribing their RNA into DNA, then inserting the DNA into the chromosomes of a host cell (retroviruses are described in Chapter 19). HIV consists of an outer envelope, which is taken from an infected cell's plasma membrane as the virus leaves the cell, and two protein layers, the innermost of which encloses the viral RNA and an enzyme called *reverse transcriptase* (Fig. 32-14). When HIV attacks a helper T cell, the virus's outer envelope binds to the plasma membrane of the cell and allows the virus to enter the cell. Once the virus is inside, its reverse transcriptase catalyzes the transcription of the virus's RNA genome into DNA (a process known as *reverse transcription*, because it reverses the "normal" DNA-to-RNA transcription in living cells). The resulting DNA then travels to the nucleus and is inserted into the T cell's genome, directing it to make more HIV particles. Early in the infection, the patient may have a fever, rash, muscle aches, headaches, and enlarged lymph nodes, as the immune system fights the infection. After about six months, the rate of viral replication slows and stabilizes at a relatively low level. Enough helper T cells remain that patients are able to resist disease, and they generally feel quite well. This state may persist for several years. Eventually, however, helper T cell levels begin to drop. When only 200 helper T cells per milliliter of blood remain (one-fourth the normal level), the patient is described as having AIDS. At this point, in untreated AIDS, HIV levels skyrocket; helper T cell numbers drop further; and the patient falls victim to a variety of other infections. The life expectancy for untreated AIDS victims is about 1 to 2 years.

The Human Immunodeficiency Virus Is Transmitted by Body Fluids

HIV cannot survive for very long outside the body. The virus can be transmitted only by the direct contact of broken skin or mucous membranes with virus-laden body fluids—including blood, semen, vaginal secretions, and breast milk. HIV infection can be spread by sexual activity, by sharing of needles among intravenous drug users, by blood transfusions (this is rare in developed countries since it became standard practice to screen all donated blood for anti-HIV antibodies), or from mother to child during pregnancy, childbirth, or breast feeding. Almost all people infected by HIV eventually develop AIDS.

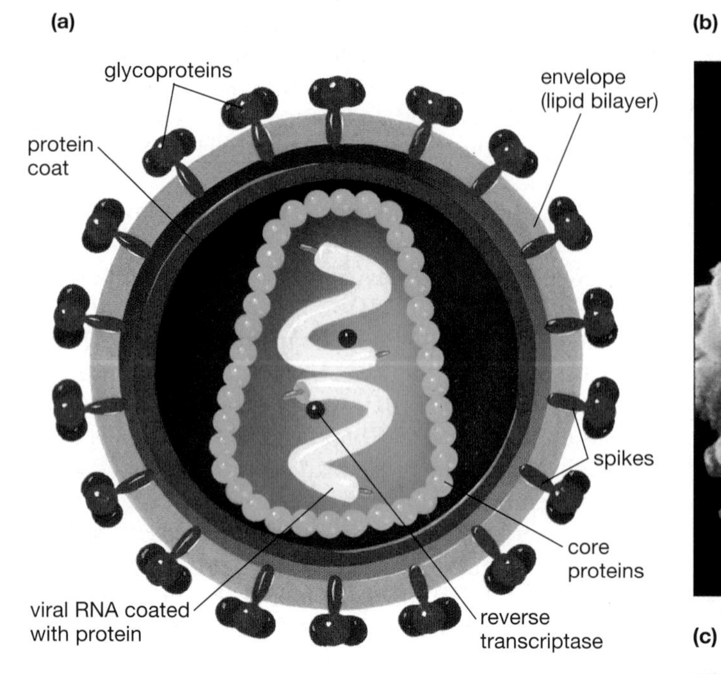

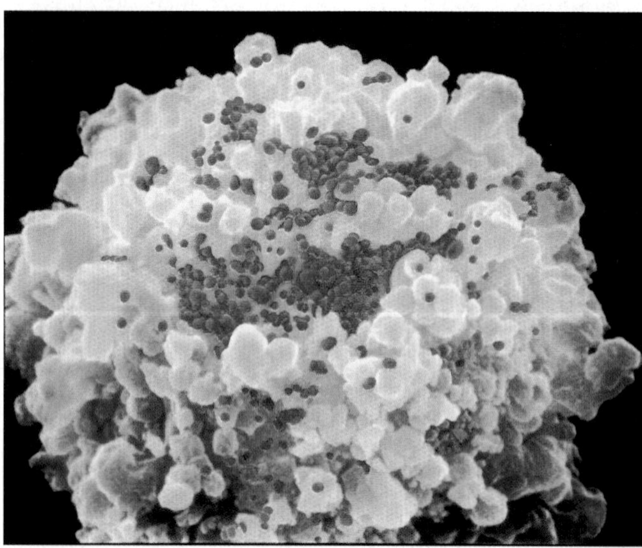

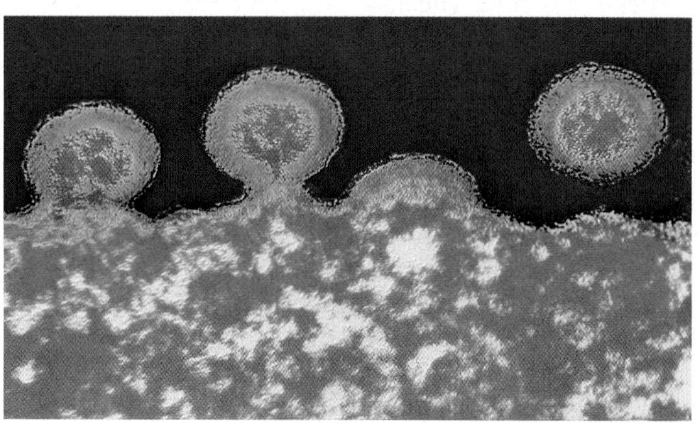

FIGURE 32-14 HIV
(a) HIV, the virus that causes AIDS, consists of an outer envelope, a protein coat, and an inner protein capsule, which contains RNA (the genetic material of HIV) and the enzyme reverse transcriptase. Glycoproteins protruding through the envelope attach to the plasma membranes of helper T cells. These proteins are potential targets for AIDS vaccines. **(b)** The red specks in this colorized scanning electron micrograph are HIV that have just emerged from the large, green helper T cell. **(c)** In this colorized transmission electron micrograph, HIV are emerging from the helper T cell and acquiring an outer envelope of plasma membrane (green) from the infected T cell.

Media Activity
32.4 Effects of HIV on the Immune System

THERE ARE PARTIALLY EFFECTIVE TREATMENTS, BUT NO CURES, FOR AIDS

For persons already infected with HIV, there are two categories of therapy. First, infections that occur because the impaired immune system is unable to fight them, called opportunistic infections—such as *Kaposi's sarcoma* (a deadly form of cancer affecting the skin) or *Pneumocystis carinii* pneumonia—can be treated as they would be in any patient. Second, new drugs that target retroviruses can disable HIV and slow the progress of AIDS, improving the quality and duration of life for AIDS patients. Some drugs block reverse transcriptase, while others block a protease that allows the viruses to assemble within the host cells. Used together, these drugs have tremendously increased the life expectancy of AIDS victims. Unfortunately, HIV can mutate into forms resistant to the drugs and in some patients, the drugs have severe side effects.

Clearly, the best solution would be to develop a vaccine against HIV. This is a major challenge, since the HIV vaccine would have to evoke a more effective immune response than does the HIV infection itself. Further, HIV has an incredible mutation rate, perhaps a thousand times faster than that of flu viruses; in fact, a single infected individual may harbor different strains of HIV in blood and in semen. Researchers are currently attempting to develop a vaccine that stimulates antibodies to a protein found in all HIV strains. This vaccine is in its early stages of animal research, but there are human clinical trials of many other types of potential AIDS vaccines underway in countries throughout the world.

Cancer Can Evade or Overwhelm the Immune Response

Cancer is one of the most dreaded words in the English language, and with good reason. Despite an encouraging 1% decline per year in the United States in recent years, cancer still accounts for 1 out of every 4 deaths, and is second only to heart disease as the leading cause of mortality. A sobering 40% of U.S. citizens will eventually contract cancer. Despite decades of intensive research, the quest for a cancer cure remains unfulfilled. Why can't we cure or prevent cancer?

A **tumor** is a population of cells that has escaped normal regulatory processes and grows at an abnormal rate. The cells in a *benign tumor*, or *polyp*, remain confined to one area, but the cells of a *malignant tumor* grow uncontrollably and spread to other areas of the body. As a malignant tumor grows, it uses increasing amounts of the body's energy and nutrient supplies, and literally squeezes out vital organs nearby. **Cancer** is a disease characterized by the unchecked growth of malignant tumor cells.

Unlike most other diseases, cancer is not a straightforward invasion of the body by a foreign organism. Although some cancers are triggered by viruses, cancer is essentially a failure of the mechanisms that control the growth of the body's own cells—a disease in which the body destroys itself. The mechanisms of regulation of cell division and the genetic mutations that cause cancer are discussed in Chapter 11.

The Immune System Defends Against Cancerous Cells

Cancer cells form in our bodies every day. We cannot avoid some carcinogens, such as gamma rays from the sun, radioactivity from the rocks beneath our feet, and naturally produced carcinogens in our food. Fortunately, natural killer cells and cytotoxic T cells screen the body for cancer cells and destroy nearly all of them before they have a chance to proliferate and spread. Cancer cells are, of course, "self" cells (the body's own cells), and the immune response does not respond to "self." How are cancer cells weeded out? It seems likely that the very processes that cause cancer also cause new and slightly different proteins to appear on the surfaces of cancer cells. Cytotoxic T cells encounter these new proteins, recognize them as "non-self" antigens, and destroy the cancer cells (see Fig. 32-10). However, some cancer cells may not bear antigens that allow the immune system to recognize them as foreign; thus, they evade detection. Even tumors that are attacked by immune cells may develop variant cell types that are resistant to immune attack. Some types of cancers actively suppress the immune system, and others simply grow so fast that the immune response can't keep up. If the immune system is thwarted by any of these means, the tumor grows and spreads. At this point, the individual's health depends on medical treatment. Unfortunately, the available treatments are only partially effective and have serious drawbacks. The rate of cure is increasing, but it is still scarcely a third of all cancers.

Medical Treatments for Cancer Depend on Distinguishing and Selectively Killing Cancerous Cells

The three main forms of cancer treatment are surgery, radiation, and chemotherapy. The surgical removal of the tumor is the first step in the treatment of many cancers. Unfortunately, the surgeon may not be able to see and remove very small patches of cancer cells that may extend from the main tumor, and surgery cannot be used to treat cancer that has begun to spread throughout the body. Alternatively, cancerous cells can be bombarded with radiation in an effort to kill them. Unlike surgery, radiation may destroy even microscopic clusters of cancer cells. Like surgery, however, radiation cannot be used to treat widespread cancers, because irradiating the whole body would damage a great deal of healthy tissue. Both surgery and radiation therapy can be traumatic and dangerous.

Chemotherapy, or drug treatment, is commonly used to supplement surgery and/or radiation or to treat cancers that cannot be treated with surgery or radiation. Chemotherapy drugs attack the machinery of cell division; because cancer cells divide much more frequently than do normal cells, the theory is that attacks on dividing cells will selectively kill cancer cells. Unfortunately, other cells of the body divide too, and chemotherapy inevitably also kills some healthy cells. Damage to dividing cells in patients' hair follicles and intestinal lining produce the well-known side effects of hair loss, nausea, and vomiting.

A tremendous amount of research has been devoted to the search for cancer treatments that are effective and have few unpleasant side effects. Developing a "cancer vaccine" is a high priority. Other approaches include developing therapies that stimulate the immune system to attack tumors, and developing antibodies or T-cell receptors that recognize tumor cells and could be used to deliver drugs or radioactive particles directly to tumor cells without affecting healthy cells (see "Scientific Inquiry: Designer Drugs Fight Disease"). Research and clinical trials continue, and cancer patients may soon reap the benefits of innovative new treatments.

Without constant surveillance by the body's defenses, it is unlikely that any of us would survive more than a few years. However, we can reduce our own chances of developing cancer by avoiding cigarette smoke and by eating a diet high in fruits and vegetables. Other behaviors that lower the risk of cancer include avoiding excessive alcohol consumption, using sunscreen, and getting regular exercise.

CASE STUDY REVISITED Fighting the Flu

You got your flu shot too late; it takes a few weeks for the flu shot to activate your specific immune system to recognize and remember the virus. A few days after intercepting your classmate's sneeze, your nose itches, your head aches, and you feel cold and exhausted. Not only that, your muscles ache and your skin is hypersensitive. What's going on? Most of the unpleasant symptoms of flu and other infections are caused by the body's nonspecific defenses. As macrophages ingest the virus, they release cytokines such as the pyrogen that causes fever. These same cytokines also sensitize pain endings throughout the body; suddenly your back aches, old injuries throb, and your skin protests as you change clothes. In ways that are not yet understood, cytokines also make you tired. Why do our bodies produce these symptoms? Biologists hypothesize that feeling tired, sensitive, and achy sent our ancestors to the safety of their caves and sends us to bed—to conserve energy for the battle against microscopic invaders that our bodies are waging.

When you finally recover from the flu, you'll be safe from reinfection this year—until next year's mutated form shows up. If flu shots are safe for you, consider getting one early in the flu season.

Consider This: It is likely that neuraminidase inhibitors will be heavily prescribed to fight flu in the coming years. Based on what you learned about flu viruses and about neuraminidase inhibitors and how they work, suggest a couple of different reasons why these drugs might continue to be effective against flu, avoiding the problem of resistance that is causing major setbacks in the use of antibiotics to fight bacterial infections. Only time will tell if this prediction is correct!

Links to Life: Protection Against Pollen

Are you one of the millions who fight "hay fever" every summer? You now know that hay fever is caused when your immune system reacts to certain pollens as if they were disease-causing organisms, mounting an inflammatory response against them. So your eyes water, your nose runs, and you sneeze a lot. You might be a candidate for a treatment called *allergy vaccination*, in which the immune system is trained to ignore pollen (or other allergens) as it should have been doing all along. This treatment consists of a series of injections of tiny amounts of the allergen (such as pollen, dog or cat dander, mold, or dust mites). Eventually, allergic symptoms subside, and the beneficial effects usually last at least 5 years. The immune system is so complex that researchers are not entirely certain how allergy vaccinations work. They agree, however, that three major changes occur during the course of the therapy. First, chemicals that promote inflammation are reduced. Second, the body produces antibodies to the allergen that are different from the "allergy antibodies" that promote inflammation. These nonallergy antibodies may bind the allergen and prevent it from interacting with the allergy antibodies on the mast cells, so they don't release histamine (see Fig. 32-13). Finally, the allergy antibodies to the allergen are eventually reduced. Although they involve a large number of injections, allergy vaccinations have helped many people with severe allergies resume a more active and comfortable life.

Summary of Key Concepts

32.1 What Are Basic Mechanisms of Defense Against Disease?

First, nonspecific external barriers, including anatomical structures and secretions such as skin, hair, cilia, and mucus, prevent disease-causing organisms from easily entering the body. Second, nonspecific internal defenses consisting of a variety of white blood cells destroy microbes, toxins, cancerous and infected body cells, and scavenge dead and dying cells. Finally, the specific immune response selectively destroys the particular toxin or microbe and "remembers" the invader, allowing a faster response if it reappears in the future.

32.2 How Do Nonspecific Defenses Function?

The skin and its secretions physically block the entry of microbes into the body and inhibit their growth. The mucous membranes of the respiratory and digestive tracts secrete antibiotic substances and mucus that traps microbes. If microbes do enter the body, they are engulfed by white blood cells. Natural killer cells secrete proteins that kill infected or cancerous cells. Injuries stimulate the inflammatory response, in which chemicals are released that attract phagocytic white blood cells, increase blood flow, and make capillaries leaky. Later, blood clots wall off the injury site. Fever is caused by endogenous pyrogens, chemicals released by white blood cells in response to infection. High temperatures inhibit bacterial growth and accelerate the immune response.

32.3 What Are the Key Characteristics of the Immune Response?

The immune response involves two types of immune cells or lymphocytes: B cells and T cells. B cells give rise to plasma cells, which secrete antibodies into the bloodstream, causing humoral immunity. Cytotoxic T cells destroy some microbes, cancer cells, and virus-infected cells on contact, causing cell-mediated immunity. Helper T cells stimulate both the humoral and cell-mediated immune responses. Immune responses have three steps: (1) recognition, (2) attack, and (3) memory.

First, antibodies (on B cells) and T-cell receptors (on T cells) recognize foreign antigens and trigger the immune response. Antibodies are Y-shaped proteins composed of a constant region and a variable region. Antigens are molecules that generate an antibody response. Antibodies both detect and actively destroy antigens. Each B cell synthesizes only one type of antibody, unique to that particular cell and its progeny. The diversity of antibodies arises from gene shuffling and the mutation of antibody genes during B-cell development. Each antibody has specific sites that bind only one or a few types of antigen. Normally, only foreign antigens are recognized by the B cells.

Second, antibodies attack the invaders. Antigens from an invader bind to and activate only those B and T cells with the complementary antibodies or T-cell receptors. In humoral immunity, B cells with the proper antibodies, stimulated by the presence of particular antigens, divide rapidly to produce plasma cells that synthesize massive quantities of the antibody. The circulating antibodies destroy antigens and antigen-bearing microbes. In cell-mediated immunity, T cells with the proper receptors bind antigens and divide rapidly. Cytotoxic T cells bind to antigens on microbes, infected cells, or cancer cells and then kill the cells. Helper T cells stimulate both the B-cell and cytotoxic-T-cell responses.

Finally, some progeny cells of both B and T cells are long-lived memory cells. If the same antigen reappears in the bloodstream, these memory cells are immediately activated, dividing rapidly and causing an immune response that is much faster and more effective than the original response.

32.4 How Does Medical Care Augment the Immune Response?

Antibiotics kill microbes or slow down their reproduction, thus allowing the body's defenses more time to respond and exterminate the invaders. Vaccinations are injections of antigens from disease organisms, in some cases the weakened or dead microbes themselves. An immune response is evoked by the antigens, providing memory and a rapid response should a real infection occur.

32.5 What Happens When the Immune System Malfunctions?

Allergies are immune responses to normally harmless foreign substances, such as pollen or dust. B cells treat these as antigens and produce "allergy antibodies" that bind to mast cells. When exposed to the antigen, the mast cells then release histamine, which causes a local inflammatory response. Autoimmune diseases arise when the immune system mistakes the body's own cells for foreign invaders and destroys them. Immune deficiency diseases occur when the immune system cannot respond strongly enough to ward off usually minor diseases.

Infection with human immunodeficiency viruses (HIV) nearly always leads to AIDS (acquired immune deficiency syndrome). These viruses invade and destroy helper T cells. Without helper T cells to stimulate the immune responses of B cells and cytotoxic T cells, an individual with AIDS is extremely susceptible to a wide assortment of infections, which are eventually fatal.

Cancer is a population of the body's cells that multiplies without control. Cancerous cells may be recognized as "different" by the immune system and destroyed by natural killer cells and cytotoxic T cells. A few evolve the capacity to evade the immune system; some attack immune cells; and others multiply too fast for the immune system to keep up. In these cases, cancer develops.

Key Terms

acquired immune deficiency syndrome (AIDS) *p. 637*
allergy *p. 636*
antibody *p. 627*
antigen *p. 626*
autoimmune disease *p. 636*
B cell *p. 626*
cancer *p. 639*
cell-mediated immunity *p. 629*
clonal selection *p. 630*
complement system *p. 631*
constant region *p. 627*
cytokine *p. 624*
cytotoxic T cell *p. 631*
fever *p. 623*
helper T cell *p. 629*
histamine *p. 624*
human immunodeficiency virus (HIV) *p. 637*
humoral immunity *p. 629*
immune response *p. 622*
immune system *p. 626*
inflammatory response *p. 624*
lymphocyte *p. 626*
macrophage *p. 623*
major histocompatibility complex (MHC) *p. 629*
mast cell *p. 624*
memory B cell *p. 630*
memory T cell *p. 631*
microbe *p. 622*
natural killer cell *p. 624*
neutrophil *p. 623*
phagocytic cell *p. 623*
plasma cell *p. 630*
severe combined immune deficiency (SCID) *p. 637*
T cell *p. 626*
T-cell receptor *p. 627*
tumor *p. 639*
vaccination *p. 635*
variable region *p. 627*

Thinking Through the Concepts

To take a multiple-choice quiz with feedback on the contents of this chapter, visit http://www.prenhall.com/audesirk7. *Log in to the Web site selected by your instructor and navigate to the Self Test section for this chapter.*

? Review Questions

1. List the human body's three lines of defense against invading microbes. Which are nonspecific (that is, act against all types of invaders), and which are specific (act only against a particular type of invader)? Explain your answers.
2. How do natural killer cells and cytotoxic T cells destroy their targets?
3. Describe humoral immunity and cell-mediated immunity. Include in your answer the types of immune cells involved in each, the location of antibodies and receptors that bind foreign antigens, and the mechanisms by which invading cells are destroyed.
4. How does the immune system construct so many different antibodies?
5. How does the body distinguish "self" from "non-self"?
6. Diagram the structure of an antibody. What parts bind to antigens? Why does each antibody bind only to a specific antigen?
7. What are memory cells? How do they contribute to long-lasting immunity to specific diseases?
8. What is a vaccination? How does it confer immunity to a disease?
9. Compare and contrast an inflammatory response with an allergic reaction from the standpoint of cells involved, substances produced, and symptoms experienced.
10. Distinguish between autoimmune diseases and immune deficiency diseases, and give one example of each.
11. Describe the causes and eventual outcome of AIDS. How do AIDS treatments work? How is HIV spread?
12. Why is cancer sometimes fatal?

Applying the Concepts

1. Why is it essential that antibodies and T-cell receptors bind only relatively large molecules (such as proteins) and not relatively small molecules (such as amino acids)?
2. Smallpox formerly killed about 30% of those infected. Due to a massive vaccination program, it was eradicated from society in 1977, and routine vaccination of schoolchildren was halted in 1972. There are only two known, legitimate stores of the smallpox virus on Earth. Now, due to concerns that there may be undeclared stocks of smallpox viruses in the hands of potential bioterrorists, there is a new impetus to vaccinate members of the military and health professions. Some people object because of concerns about the safety of the vaccine, which has rare, but potentially serious, side effects. Argue for and against mandatory vaccination for military and health workers.
3. The essay "Health Watch: Can We Beat the Flu Bug?" states that the flu virus is different each year. If that is true, what good does it do to get a "flu shot" each winter?
4. Organ transplant patients typically receive the drug cyclosporine. This drug inhibits the production of a cytokine that stimulates helper T cells to proliferate. How does cyclosporine prevent the rejection of transplanted organs? Some patients who received successful transplants many years ago are now developing various kinds of cancers. Propose a hypothesis to explain this phenomenon.

For More Information

Carpenter, S. "Modern Hygiene's Dirty Tricks." *Science News*, August 14, 1999. This article describes the controversial hypothesis that keeping our environment ultra-clean may create an imbalance in the immune system that predisposes people to allergies and asthma.

"Defeating AIDS: What Will It Take?" *Scientific American*, July 1998. A series of nine articles covering all aspects of AIDS.

Enserink, M., and Stone, R. "Dead Virus Walking." *Science*, March 15, 2002. Fear of bioterrorism has led to a revival of research on the two remaining stores of smallpox viruses.

Ezzell, C. "Hope in a Vial: Will There Be an AIDS Vaccine Anytime Soon?" *Scientific American*, June 2002. For nearly two decades, HIV has thwarted attempts to produce a vaccine.

Laver, W. G., Bischofberger, N., and Webster, R. G. "Disarming Flu Viruses." *Scientific American*, January 1999. Presents a clear description of how flu viruses work, how new strains arise, and the development of the new neuraminidase inhibitors that may stop them.

Lichtenstein, L. M. "Allergy and the Immune System." *Scientific American*, September 1993. Allergic responses, evolved as protection against parasites, turn against us when we respond violently to harmless pollen and foods.

Marrack, P., and Kappler, J. W. "How the Immune System Recognizes the Body." *Scientific American*, September 1993. How the immune system's potential weapons against the body are disarmed.

Paul, W. E. "Infectious Diseases and the Immune System." *Scientific American*, September 1993. Microbes and the immune system engage in constant evolutionary warfare.

Raloff, J. "A Rash of Kisses." *Science News*, July 20, 2002. Some people have such violent food allergies that the inflammatory response can be triggered by a kiss from someone who has consumed the allergen.

Sapolsky, R. M., "Why You Feel Crummy When You're Sick." *Discover*, July 1990. Why feeling bad is part of the body's defense mechanism.

Media Activities

To access a Media Activity visit http://www.prenhall.com/audesirk7. *Log in to the Web site selected by your instructor, navigate to this chapter, and select the appropriate Media Activity number.*

32.1 Defense Against Infectious Agents

Estimated time: 5 minutes

View the body's defenses against infectious agents.

32.2 Inflammation

Estimated time: 5 minutes

Review the details of the inflammatory response.

32.3 Clonal Selection

Estimated time: 10 minutes

Explore how B cells are activated to produce antibodies by foreign antigens.

32.4 Effects of HIV on the Immune System

Estimated time: 5 minutes

In this animation, you will explore how HIV enters host cells, how it replicates and destroys the host, and why the loss of HIV-infected cells compromises the immune response.

32.5 Web Investigation: Fighting the Flu

Estimated time: 10 minutes

In this investigation, examine why some individuals still get polio, rubella, and other diseases despite extensive vaccination programs.

CHAPTER

33 Chemical Control of the Animal Body: The Endocrine System

Canada's track superstar Ben Johnson was stripped of his 1988 Olympic gold medal for the 100-meter dash after he tested positive for anabolic steroids.

AT A GLANCE

CASE STUDY Losing on Artificial Hormones

With fractions of a second making the difference between winning and losing, competitive athletes are under enormous pressure to improve their performance—even though they already push the limits of human ability. Many, from world-class athletes to aspiring body builders on campus, turn to "anabolic steroids." Some athletes began to use steroids in the 1950s; it wasn't until the 1970s that some sports organizations began to ban their use. Despite increasing evidence that steroids pose significant health risks, steroid use has increased. In 2003, dozens of Olympic and professional athletes were accused of artificially boosting their performances by taking a new anabolic steroid. Called THG (tetrahydrogestrinone), the steroid is not detected by routine drug screening tests. The U.S. Food and Drug Administration described THG as a "purely synthetic designer steroid" and moved quickly to ban the drug.

Anabolic steroids, which resemble the male sex hormone testosterone, stimulate the body to increase the amount of lean muscle mass. Other hormones can also be used and abused to improve athletic performance. At the 2002 Olympics in Salt Lake City, Utah, cross-country skiers from Russia and Spain were disqualified when they tested positive for a new hormone mimic, darbepoetin. This drug, which is prescribed for victims of anemia, acts like erythropoietin, stimulating bone marrow to produce more red blood cells. By increasing the oxygen-carrying capacity of the blood, darbepoetin (or erythropoietin, which is also abused by athletes) can improve performance by 7% to 10%. But athletes take dangerous risks to gain this artificial advantage, risks that go well beyond the possible loss of their medals. As you read this chapter, notice how many different effects the same hormone can have throughout the body. Think about the possible effects of artificially increasing the amount of a hormone that stimulates red blood cell production or increasing the amount of testosterone-like steroids. What risks might accompany steroid use?

33.1 How Do Animal Cells Communicate?

In all multicellular organisms, individual cells must remain in continuous communication with one another (Table 33-1). In some specialized tissues, such as heart muscle, *gap junctions* directly link the insides of cells, allowing the flow of ions and electrical signals. More commonly, cells release chemical signaling molecules sometimes called "messenger molecules" that affect other cells, either nearby or distant. Like conversation at a crowded party, this communication is directed to specific "target" cells (the people you are talking to) and not to others (nearby people holding their own conversations). To ensure that the chemical message reaches only appropriate targets, cells have **receptors**, specialized protein molecules that bind only to specific chemical messengers. Receptors may be located either on the plasma membrane or inside target cells. Upon binding to its receptor, the chemical triggers some type of change within the target cell.

There are three general classes of messenger molecules, each utilizing a different distribution system: *Local hormones* diffuse through the extracellular fluid to cells in the immediate vicinity; *endocrine hormones* are released into the blood, where they are distributed to both nearby and distant cells; and *neurotransmitters* are released across a very narrow gap (the *synaptic cleft*) between a specialized region of the neuron and its target (see Table 33-1). This chapter deals with endocrine hormones and local hormones called *prostaglandins*. You will learn more about neurotransmitters in Chapter 34.

33.2 What Are the Characteristics of Animal Hormones?

Local Hormones Diffuse to Nearby Target Cells

Most—if not all—cells secrete **local hormones** into their immediate vicinity, such as the cytokines that allow immune cells to communicate (see Chapter 32). **Prostaglandins**, modified fatty acids synthesized by cells from membrane phospholipids, are another type of local hormone. Unlike most other hormones, which are synthesized by a limited number of specialized cells, prostaglandins are produced by many, perhaps all, cells of the body. Research on this diverse and potent family of compounds is still in its infancy; several prostaglandins are known, and probably a great many more await discovery. One prostaglandin causes arteries to constrict and stop the bleeding from newborn infants' umbilical cords. Another prostaglandin works in conjunction with oxytocin during labor, stimulating uterine contractions. Menstrual cramps are caused by the overproduction of uterine prostaglandins that stimulate uterine contractions. Some prostaglandins contribute to inflammation (such as occurs in arthritic joints) and stimulate pain receptors. Drugs such as aspirin and ibuprofen provide relief from these symptoms (as well as from menstrual cramps) by inhibiting prostaglandin synthesis. The use of local hormones such as prostaglandins to communicate with nearby cells is called "*paracrine communication*" (*para*, appropriately, means "beside"), whereas "endocrine communication" utilizes chemicals that travel within the bloodstream, often over considerable distances.

Table 33-1 How Cells Communicate

Communication	Chemical Messengers	Mechanism of Transmission	Examples
Direct	Ions, small molecules	Direct movement through gap junctions linking cytoplasm of adjacent cells	Ions flowing between cardiac muscle cells
Paracrine	Local hormones	Diffusion through extracellular fluid to nearby cells bearing receptors	Prostaglandins
Endocrine	Hormones	Carried in the bloodstream to near or distant cells bearing receptors	Insulin
Synaptic	Neurotransmitters	Diffusion from a neuron across a narrow synaptic cleft to a cell bearing receptors	Acetylcholine

Hormones of the Endocrine System Are Transported by the Circulatory System

Endocrine hormones are chemical messages produced by specialized cells; often they are released in response to some stimulus from inside or outside the body. Endocrine hormones are carried by the circulatory system and influence target cells bearing specific receptors for those hormones. The changes induced by hormonal messages may be prolonged and irreversible, as in puberty or the metamorphosis of a caterpillar into a butterfly. More typically, the induced changes are transient and reversible, helping control and regulate the interrelated physiological systems that compose the animal body. The regulation of the body requires communication; in animal bodies, hormones provide much of that communication. In fact, the **endocrine system**—consisting of hormones and the various cells that secrete and receive them—can be viewed as the "postal system of physiology," moving information and instructions between cells that may be some distance apart. Hormones are released by the cells of major endocrine glands and endocrine organs located throughout the body (Fig. 33-1).

There are three classes of vertebrate endocrine hormones: **peptide hormones**, made from chains of amino acids; **amino acid based hormones**, which are synthesized from single amino acids; and **steroid hormones**, which resemble cholesterol, from which most steroid hormones are synthesized. For more information about the chemical structures of hormones, see "A Closer Look: The Chemical Diversity of Vertebrate Hormones" in Chapter 33 on this text's Web site.

Hormones Bind to Specific Receptors on Target Cells

Because nearly all cells have a blood supply, a hormone released into the bloodstream will reach nearly every cell of the body. However, a given hormone acts only on **target cells** bearing receptors for that particular hormone molecule, thus exerting precise control; cells that

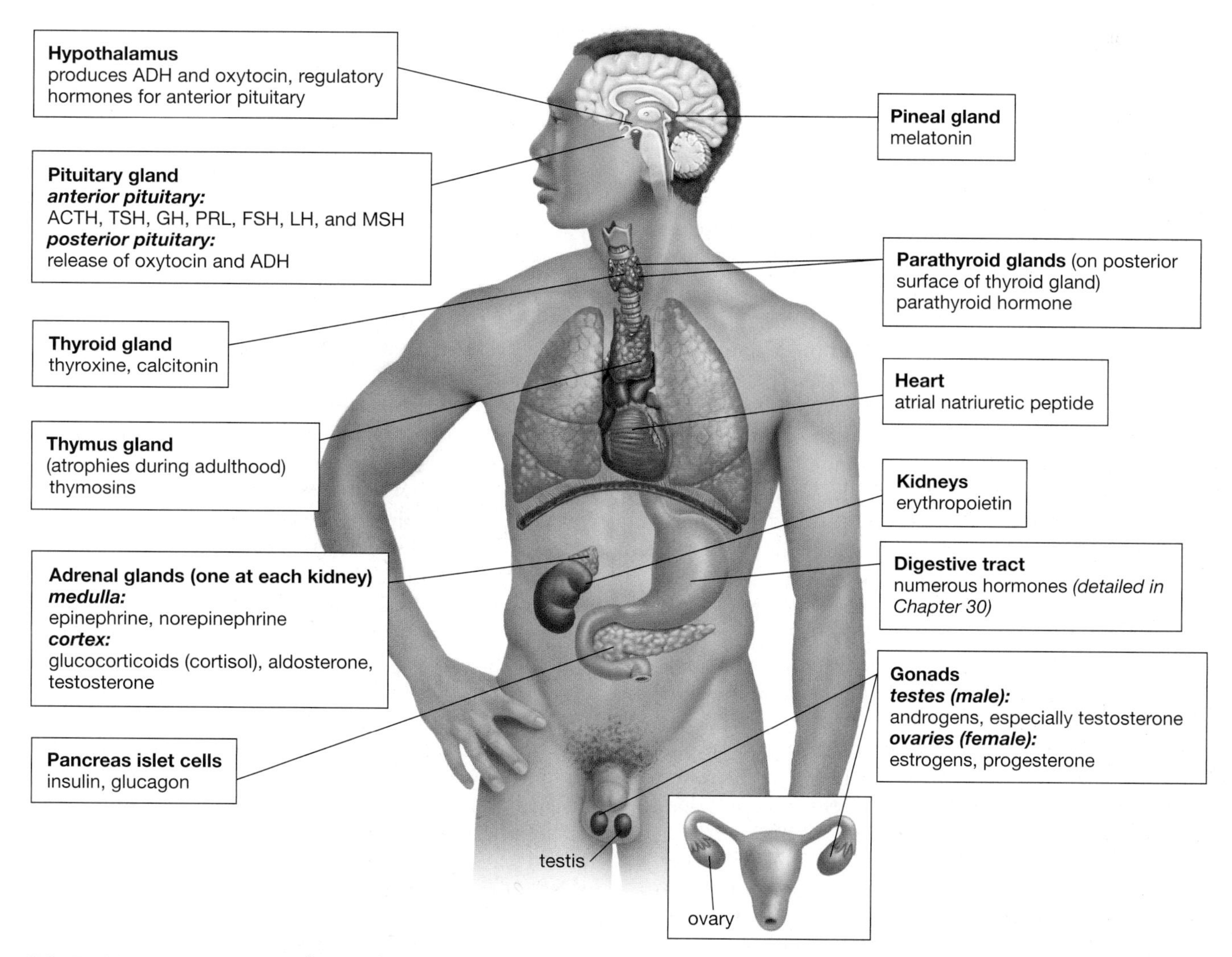

FIGURE 33-1 Major mammalian endocrine glands and their secretions

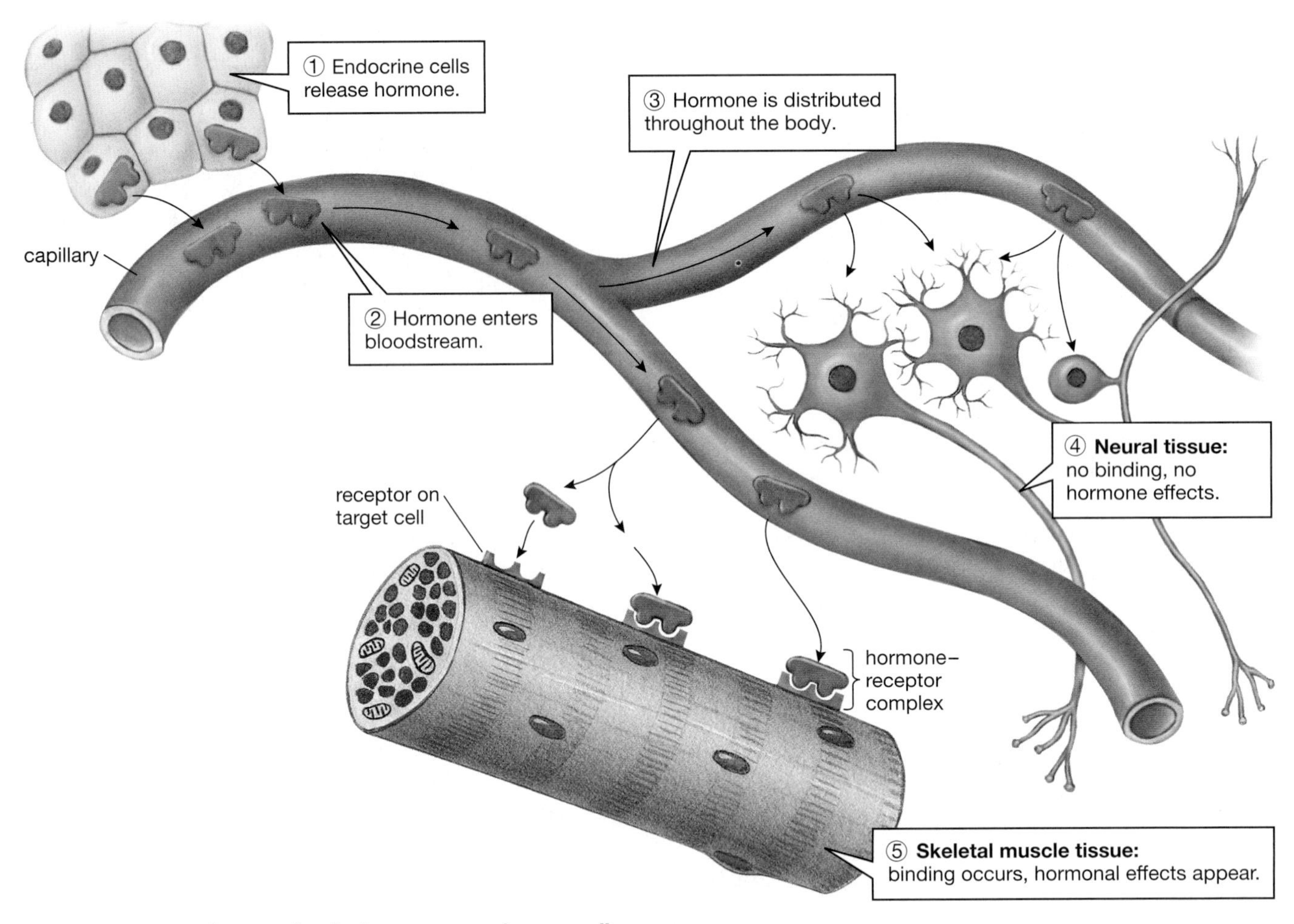
① Endocrine cells release hormone.
③ Hormone is distributed throughout the body.
capillary
② Hormone enters bloodstream.
④ Neural tissue: no binding, no hormone effects.
receptor on target cell
hormone–receptor complex
⑤ Skeletal muscle tissue: binding occurs, hormonal effects appear.

FIGURE 33-2 Endocrine glands, hormones, and target cells

lack the appropriate receptor will not respond to the hormonal message (Fig. 33-2). In addition, a given hormone may have several different effects, depending on the nature of the receptor on the target cell it contacts. Receptors for hormones are found in two general locations on target cells: on the plasma membrane and inside the cell, within the cytoplasm or the nucleus. The mechanisms by which hormone receptors cause changes in the target cell are described in more detail in "A Closer Look: How Hormones Influence Target Cells."

Many peptide and amino acid-based hormones are soluble in water but not in lipids; therefore, these hormones cannot penetrate plasma membranes, which are largely composed of phospholipids. Instead, the hormones bind to protein receptors on the target cell's plasma membrane (see Fig. E33-1a). These "membrane receptors" span the plasma membrane, so a hormone binding to the external portion of the receptor can cause a shape change on the portion of the receptor protein that protrudes into the cell. This change in configuration triggers reactions that produce a molecule, called a **second messenger**, inside the cell. The second messenger transfers the signal from the *first messenger*—the hormone—to other molecules within the cell, often initiating a series of biochemical reactions (see Fig. E33-1a). Although a variety of molecules can act as second messengers, in many cases hormones binding to receptors result in the conversion of ATP to **cyclic AMP** (cAMP), a nucleotide that regulates many cellular activities (see Chapter 3). Cyclic AMP, in its role of second messenger, initiates a chain of reactions inside the cell. Each reaction in the chain involves an increasing number of molecules, amplifying the original signal. The end result varies with the target cell; channels may be opened in the plasma membrane, or substances may be synthesized or secreted. For example, the hormone epinephrine (also called *adrenaline*) binds to membrane receptors on heart muscle, triggers cAMP formation, and starts a chain of molecular events that keeps calcium channels in the heart muscle open longer. The extra calcium that comes in causes a stronger contraction in the cardiac muscle. (You will learn more about calcium and muscle contraction in Chapter 35). This is one of several ways in which epinephrine helps your body prepare for emergency situations, as described later in this chapter.

In contrast, some hormones, such as the steroid hormones, are lipid soluble and are therefore able to pass through the plasma membrane. Once inside the cell, these hormones bind to receptors. Some of these receptors are in the nucleus, while others bind the hormone in the cytoplasm and transport it into the nucleus The receptor–hormone complex binds to DNA and stimulates particular genes to transcribe messenger RNA, which travels into the cytoplasm and directs the synthesis of proteins (see Fig. E33-1b). These hormones may take minutes or even days to exert their full effects. Researchers have also found receptors for steroid hormones on the plasma membrane, which make steroids extremely versatile in their signaling mechanisms.

Hormone Release Is Regulated by Feedback Mechanisms

If a hormone is to be useful for physiological control, there must be a way to turn its message both on and off. In animals, the switching mechanism usually involves negative feedback: The secretion of a hormone stimulates a response in target cells; the response then inhibits further secretion of the hormone.

Most hormones exert powerful effects on the body that could be harmful if prolonged; therefore, control of hormone release by negative feedback is especially important. Suppose you have jogged a few miles on a hot, sunny day and have lost a pint of water through perspiration. In response to the loss of water from your bloodstream, your pituitary gland releases *antidiuretic hormone* (ADH), which causes your kidneys to reabsorb water and to produce a very concentrated urine (see Chapter 31). However, if you arrive home and drink a quart of Gatorade®, you will more than replace the water you lost in sweat. Continued retention of excess water could raise blood pressure and strain the circulatory system. Negative feedback ensures that ADH secretion is turned off when the water content of the blood returns to normal, allowing the kidneys to begin eliminating the excess water. Look for more examples of negative feedback as you read through this chapter (see Fig. 31-8).

In a few cases, positive feedback stimulates hormone release, at least for a short time. For example, as described in Chapter 27, contractions of the uterus early in childbirth stretch the cervix and cause the posterior pituitary to release the hormone *oxytocin*. Oxytocin stimulates stronger contractions of the uterus, which cause more oxytocin to be released, creating a positive feedback cycle. Simultaneously, oxytocin causes uterine cells to release prostaglandins, which further enhance uterine contractions, another example of positive feedback. But even positive feedback systems also elicit negative feedback that limits their duration. In this case, the uterine contractions cause the eventual birth of the baby, eliminating the stretching of the cervix and disrupting the positive feedback cycle that sustained and enhanced the uterine contractions.

33.3 What Are the Structures and Hormones of the Mammalian Endocrine System?

Endocrinologists don't fully understand how animal hormones work. New hormones and new roles for known hormones are discovered nearly every year. The key functions of the major endocrine glands and endocrine organs, however, have been known for many years. Here we will focus on the endocrine functions of the hypothalamus–pituitary complex, the thyroid and parathyroid glands, the pancreas, the sex organs, and the adrenal glands (see Fig. 33-1). Table 33-2 lists these and other glands, their major hormones, and their principal functions.

Mammals Have Both Exocrine and Endocrine Glands

Although this chapter focuses on endocrine glands, you should be aware that glands come in two basic types: exocrine glands and endocrine glands. **Exocrine glands** produce secretions that are released outside the body (*exo* means "out of" in Greek) or into the digestive tract (a hollow tube continuous with the outside world). Exocrine gland secretions are released through tubes or openings called **ducts**. The exocrine glands include the sweat glands and oil-producing (sebaceous) glands of the skin, the tear-producing (lacrimal) glands of the eye, and the milk-producing (mammary) glands, as well as glands that produce digestive secretions, such as the salivary glands and some cells of the pancreas.

Endocrine glands, sometimes called *ductless glands,* release their hormones within the body (*endo* means "inside of"). An endocrine gland generally consists of clusters of hormone-producing cells embedded within a network of capillaries. The cells secrete their hormones into the extracellular fluid surrounding the capillaries (see Fig. 33-2). The hormones then enter the capillaries by diffusion and are carried throughout the body by the bloodstream.

The Hypothalamus Controls the Secretions of the Pituitary Gland

If the endocrine system is the body's postal service, then the hypothalamus is the main post office, and the pituitary gland is the administrative headquarters. Together, these structures coordinate the action of many key hormonal messaging systems. The **hypothalamus** is a part of the brain that contains clusters of specialized nerve cells called **neurosecretory cells**. Neurosecretory cells synthesize peptide hormones, store them, and release them when stimulated. The **pituitary gland** is a pea-sized gland that dangles from the hypothalamus by a stalk (see Fig. 33-1). Anatomically, the pituitary consists of two distinct

Table 33-2 Mammalian Endocrine Glands and Hormones

Endocrine Gland	Hormone	Type of Chemical	Principal Function
Hypothalamus (via posterior pituitary)	Antidiuretic hormone (ADH)	Peptide	Promotes reabsorption of water from kidneys; constricts arterioles
	Oxytocin	Peptide	In females, stimulates contraction of uterine muscles during childbirth, milk ejection, and maternal behaviors; in males, causes sperm ejection
Hypothalamus (to anterior pituitary)	Releasing and inhibiting hormones	Peptides	At least nine hormones; releasing hormones stimulate release of hormones from anterior pituitary; inhibiting hormones inhibit release of hormones from anterior pituitary
Anterior pituitary	Follicle-stimulating hormone (FSH)	Peptide	In females, stimulates growth of follicle, secretion of estrogen, and perhaps ovulation; in males, stimulates spermatogenesis
	Luteinizing hormone (LH)	Peptide	In females, stimulates ovulation, growth of corpus luteum, and secretion of estrogen and progesterone; in males, stimulates secretion of testosterone
	Thyroid-stimulating hormone (TSH)	Peptide	Stimulates thyroid to release thyroxine
	Adrenocorticotropic hormone (ACTH)	Peptide	Stimulates adrenal cortex to release hormones, especially glucocorticoids, such as cortisol
	Growth hormone (GH)	Peptide	Stimulates growth, protein synthesis, and fat metabolism; inhibits sugar metabolism
	Prolactin (PRL)	Peptide	Stimulates milk synthesis in and secretion from mammary glands
	Melanocyte-stimulating hormone (MSH)	Peptide	Promotes synthesis of brown skin pigment, melanin
Thyroid	Thyroxine	Amino acid derivative	Increases metabolic rate of most body cells; increases body temperature; regulates growth and development
	Calcitonin	Peptide	Inhibits release of calcium from bones
Parathyroid	Parathyroid hormone	Peptide	Stimulates release of calcium from bones; promotes absorption of calcium by intestines; promotes reabsorption of calcium by kidneys
Pancreas	Insulin	Peptide	Decreases blood glucose levels by increasing uptake of glucose into cells and converting glucose to glycogen, especially in liver; regulates fat metabolism
	Glucagon	Peptide	Converts glycogen to glucose, raising blood glucose levels
Ovaries[a]	Estrogen	Steroid	Causes development of female secondary sexual characteristics and maturation of eggs; promotes growth of uterine lining
	Progesterone	Steroid	Stimulates development of uterine lining and formation of placenta
Testes[a]	Testosterone	Steroid	Stimulates development of genitalia and male secondary sexual characteristics; stimulates spermatogenesis
Adrenal medulla	Epinephrine (adrenaline) and norepinephrine (noradrenaline)	Amino acid derivatives	Increase levels of sugar and fatty acids in blood; increase metabolic rate; increase rate and force of contractions of the heart; constrict some blood vessels
Adrenal cortex	Glucocorticoids (cortisol)	Steroid	Increase blood sugar; regulate sugar, lipid, and fat metabolism; anti-inflammatory effects
	Aldosterone	Steroid	Increases reabsorption of salt in kidney
	Testosterone	Steroid	Causes masculinization of body features, growth
		Other Sources of Hormones	
Pineal gland	Melatonin	Amino acid derivative	Regulates seasonal reproductive cycles and sleep–wake cycles; may regulate onset of puberty
Thymus	Thymosin	Peptide	Stimulates maturation of cells of immune system
Kidney	Renin	Peptide	Acts on blood proteins to produce hormone (angiotensin) that regulates blood pressure
	Erythropoietin	Peptide	Stimulates red blood cell synthesis in bone marrow
Heart	Atrial natriuretic peptide (ANP)	Peptide	Increases salt and water excretion by kidneys; lowers blood pressure
Digestive tract[b]	Secretin, gastrin, cholecystokinin, and others	Peptides	Control secretion of mucus, enzymes, and salts in digestive tract; regulate peristalsis
Fat cells	Leptin	Peptide	Regulates appetite; stimulates immune function; promotes blood vessel growth; required for onset of puberty

[a]See Chapters 36 and 37.
[b]See Chapter 30.

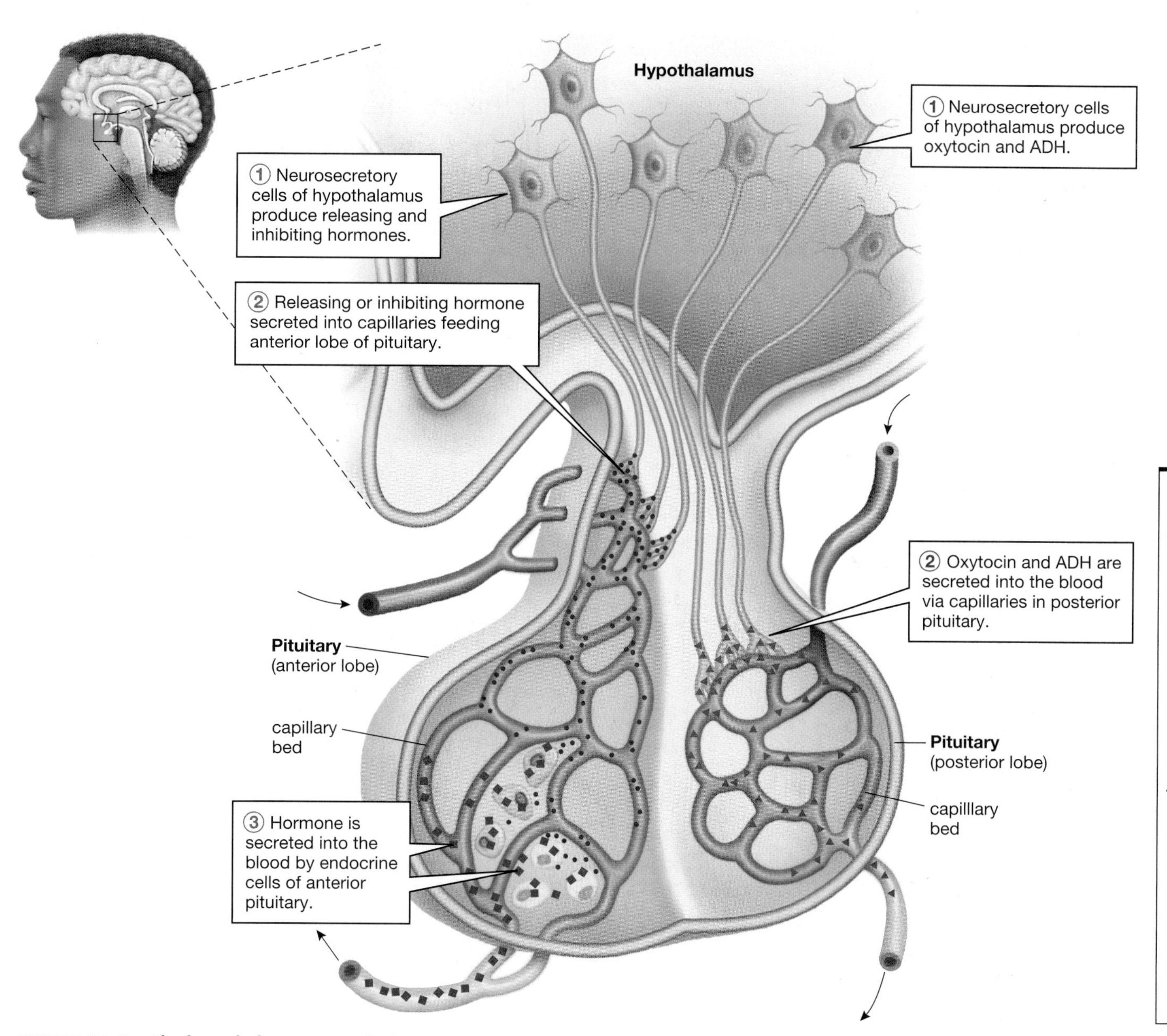

FIGURE 33-3 The hypothalamus controls the pituitary
Neurosecretory cells of the hypothalamus control hormone release in the anterior lobe of the pituitary by producing releasing or inhibiting hormones (left). These cells secrete their hormones into a capillary network that carries them to the anterior pituitary. There, each hormone stimulates endocrine cells with appropriate receptors, while leaving other types unaffected. The posterior lobe of the pituitary (right) is an extension of the hypothalamus. Neurosecretory cells in the hypothalamus have cell endings on a capillary bed in the posterior lobe of the pituitary, where the cells release oxytocin or antidiuretic hormone (ADH). **QUESTION** What advantage is gained by having nerve cells in the hypothalamus participate in controlling the release of pituitary hormones?

Media Activity
33.1 Hypothalamic Control of the Pituitary

parts: the **anterior pituitary** and the **posterior pituitary**. The hypothalamus controls the release of hormones from both parts. The anterior pituitary is a true endocrine gland, composed of several types of hormone-secreting cells enmeshed in a network of capillaries. The posterior pituitary, however, is derived from an outgrowth of the hypothalamus.

Hypothalamic Hormones Control the Anterior Pituitary

Neurosecretory cells of the hypothalamus produce at least nine peptide hormones that regulate the release of hormones from the anterior pituitary. These peptides are called **releasing hormones** or **inhibiting hormones**, depending on whether they stimulate or inhibit the release of a particular pituitary hormone (Fig. 33-3). Releasing and inhibiting hormones are synthesized in nerve cells in the hypothalamus, secreted into a capillary bed in the lower portion of the hypothalamus, and travel a short distance through blood vessels to a second capillary bed that surrounds the endocrine cells of the anterior pituitary. There, the releasers and inhibitors diffuse out of the capillaries and influence pituitary hormone secretion.

A CLOSER LOOK How Hormones Influence Target Cells

The effects of hormones on their target cells depend on the nature of the receptor and how the receptor influences biochemical pathways inside the target cell. Generally, peptide and amino acid hormones, which are not soluble in lipids, bind receptors on the outside of the target cell's plasma membrane (Fig. E33-1a). Such hormones usually stimulate formation of an intracellular "second messenger" molecule, such as the cAMP molecule illustrated here.

① A peptide or amino acid hormone binds to a receptor on the plasma membrane.

② Hormone–receptor binding activates an enzyme that catalyzes the synthesis of a second-messenger such as cAMP.

③ The second messenger activates other enzymes.

④ The activated enzymes promote specific cellular reactions, producing a variety of responses (depending on the hormone and the receptor), such as an increase in glucose production (induced by epinephrine) or an increase in estrogen synthesis (induced by luteinizing hormone).

FIGURE E33-1 Mechanisms of hormone action

(a) Amino acid and peptide hormones

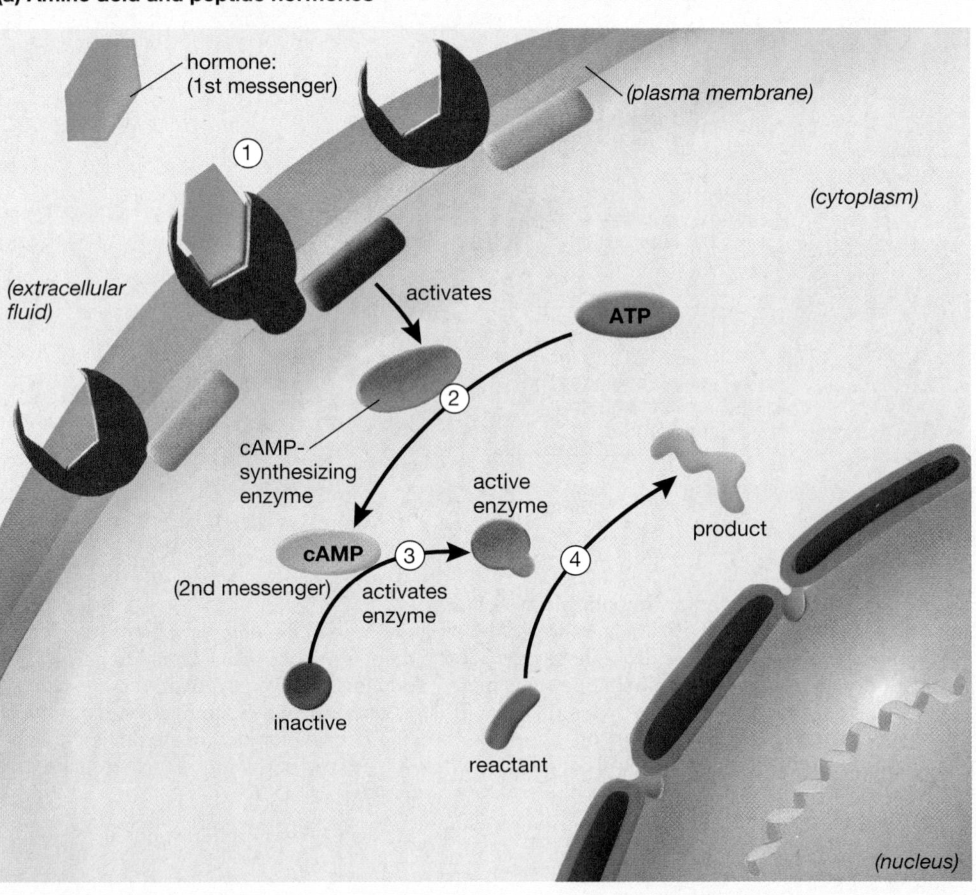

Media Activity
33.2 Modes of Action of Hormones

Because the releasing and inhibiting hormones are secreted very close to the anterior pituitary, they are produced only in minute amounts. Not surprisingly, they were extremely difficult to isolate and study. Andrew Schally and Roger Guillemin, U.S. endocrinologists who shared the Nobel Prize in Medicine in 1977 for characterizing several of these hormones, used the brains of millions of sheep and pigs (obtained from slaughterhouses) to extract enough releasing hormone for chemical analysis.

The Anterior Pituitary Produces and Releases a Variety of Hormones

The anterior pituitary produces several peptide hormones. Four of these regulate hormone production in other endocrine glands. **Follicle-stimulating hormone (FSH)** and **luteinizing hormone (LH)** stimulate the production of sperm and testosterone in males, and of eggs, estrogen, and progesterone in females. We will discuss the roles of FSH and LH further in Chapter 36. **Thyroid-stimulating hormone (TSH)** stimulates the thyroid gland to release its

Figure E33-1b illustrates a mechanism by which lipid-soluble steroid hormones may bind to receptors in the target cell nucleus and alter the transcription of proteins.

① Lipid-soluble steroid hormones diffuse through the plasma membrane and into the target cell.

② The hormone binds to a receptor in the nucleus or to a receptor in the cytoplasm that carries it into the nucleus.

③ The hormone–receptor complex binds to DNA and causes RNA polymerase to bind to a nearby promoter site for a specific gene.

④ RNA polymerase catalyzes the transcription of DNA into messenger RNA (mRNA).

⑤ The mRNA leaves the nucleus, then attaches to a ribosome and directs the synthesis of a specific protein product. In hens, for example, the steroid hormone estrogen promotes the transcription of the albumin gene, causing the synthesis of albumin (egg white), which is packaged in the egg as a food supply for the developing chick.

(b) Steroid hormones

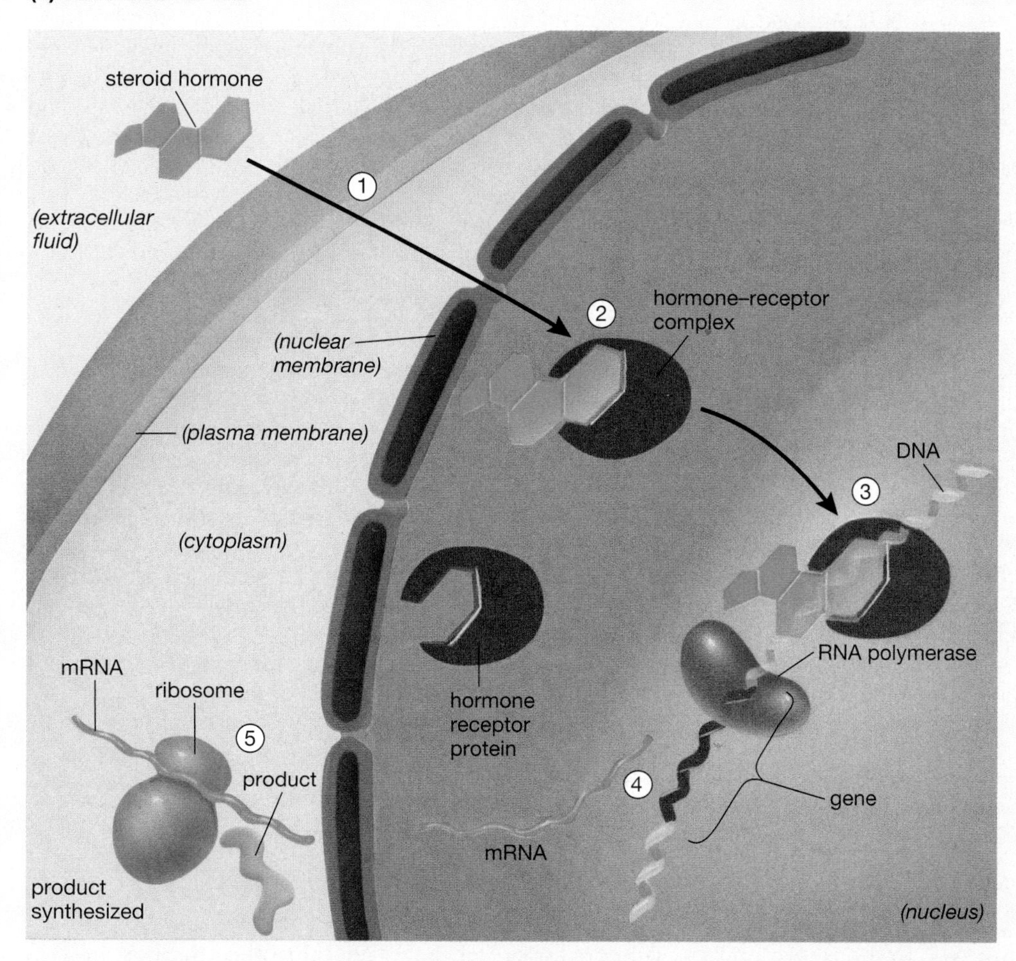

hormones, and **adrenocorticotropic hormone** (**ACTH**; "hormone that stimulates the adrenal cortex") causes the release of the hormone *cortisol* from the adrenal cortex. We will discuss the effects of thyroid and adrenal cortical hormones later in this chapter.

The remaining hormones of the anterior pituitary do not act on other endocrine glands. **Prolactin**, in conjunction with other hormones, stimulates mammary gland (milk-producing exocrine gland) development during pregnancy. **Melanocyte-stimulating hormone** **(MSH)** stimulates the synthesis of the skin pigment melanin. **Growth hormone** regulates the body's growth by acting on nearly all the body's cells—increasing protein synthesis, fat utilization, and the storage of carbohydrates. During maturation, it has a stimulatory effect on bone growth, which influences the ultimate size of the adult organism. Much of the normal variation in human height is due to differences in the secretion of growth hormone from the anterior pituitary. Too little growth hormone causes some cases of

EARTH WATCH Endocrine Deception

Human activities have introduced a vast number and amount of foreign substances into the environment, which are now found—in small quantities—in our water, air, and food. All of us are exposed to them daily. Some have disrupted the reproductive function of wildlife exposed at high levels. These compounds are extremely diverse in chemical structure and originate from a wide variety of sources, including pesticides (DDT, methoxychlor), plastics (bisphenol A, phthalates), detergents (nonylphenyl), and industrial processes (PCBs). Some, called **environmental estrogens**, mimic the effects of estrogen. Others block the effects of testosterone or either mimic or block reproductive hormones, depending on the site of action or the species. The effects of these **endocrine disrupters** have been determined through studies of laboratory animals, cell cultures, and wild animals that have been exposed in their natural habitats.

Scientists have identified a wide variety of harmful effects caused by endocrine disrupters, include feminization in males, masculinization in females, reproductive cancers, malformed sex organs, altered blood hormone levels, and reduced fertility. When a chemical plant near Lake Apopka in Florida released several known environmental estrogens into the water, wildlife biologists noted an alarming decline in the alligator population of the lake. Many eggs were not hatching. Males had high estrogen, low testosterone, smaller penises, and abnormal testes. Females typically had exceptionally high estrogen levels and abnormal ovaries. In a different study, researchers discovered that male freshwater fish downstream from sewage outlets in the United States and England produce an egg yolk protein normally found only in females. Researchers suspect that this is caused by natural human estrogen and synthetic estrogen used in birth control pills and excreted in urine by women; the synthetic estrogen estrinyl estadiol resists degradation and survives passage through wastewater treatment plants. There is increasing concern over two new forms of birth control: contraceptive patches and vaginal rings, which retain high levels of synthetic estrogen after use, and may be flushed down toilets or discarded into landfills, where the estrogen will be released into waterways or leach into groundwater.

Some of the most devastating endocrine disrupters, including DDT and PCBs, have been banned in developed countries (although they remain in our air, water, or soil), but many others are still widely used and are very persistent in the environment. New endocrine disrupters are discovered on a regular basis. Atrazine, the most widely used weed-killer in the United States, causes a tenfold decrease in testosterone as well as sexual abnormalities in frogs at concentrations that are common in streams and are well below the maximum EPA standard. Researchers compared sperm motility and numbers in men living in a semirural agricultural region to men from several urban areas. Both sperm numbers and motility were higher in New York and Minneapolis than in agricultural central Missouri. Researchers are now investigating a possible link between herbicides, which are widespread in streams and groundwater in the Midwest, and reduced human semen quality. Recently, PBDEs (polybromo diphenyl ethers), compounds that serve as flame-retardants in manufactured products including computers, plastics, carpet, and furniture, have leached into air, water, and the human food supply, and can be found in human breast milk. Though levels in humans are still very low, the effects on people are unknown; animal research has revealed toxicity very similar to PCBs, which cause nervous system damage and birth defects in both animals and humans. Both PCBs and PBDEs resemble thyroxine, and are believed to exert their toxic effects by disrupting thyroid function.

Although high levels of endocrine disrupters are known to be harmful, no one knows what effects long-term, low-level exposure to these substances will have on human and other animal populations, particularly during vulnerable early developmental stages. How many of the thousands of industrial chemicals we use act as endocrine disrupters? What are their mechanisms of action? What levels of exposure do various human and wildlife populations experience, and is there a threshold of exposure for toxic effects to occur? Does exposure to multiple endocrine disrupters produce synergistic ill-effects in people or in wildlife? Answers to these questions will allow us to formulate appropriate controls over the release of these chemicals into the environment. Unfortunately, these are difficult and complex questions—and as we seek answers, more and more of these chemicals are entering the environment.

dwarfism; too much can cause *gigantism* (Fig. 33-4). In adulthood, many bones lose their ability to lengthen, but growth hormone continues to be secreted throughout life, helping regulate protein, fat, and sugar metabolism.

A major advance in the treatment of pituitary dwarfism occurred when molecular biologists successfully inserted the gene for human growth hormone into bacteria, which churned out large quantities of the substance. Previously, the main commercial source of growth hormone was human cadavers, from which tiny amounts were extracted at great cost. Thanks to the new, cheaper source, many more children with underactive pituitary glands, who would previously have been extremely short, can achieve normal height.

The Posterior Pituitary Releases Hormones Produced by Cells in the Hypothalamus

The posterior pituitary contains the endings of two types of neurosecretory cells whose cell bodies are located in the hypothalamus. These neurosecretory cell endings are enmeshed in a capillary bed into which they release hormones to be carried into the bloodstream (see Fig. 33-3). Two peptide hormones are synthesized in the hypothalamus and released from the posterior pituitary: **antidiuretic hormone (ADH)** and **oxytocin**.

Antidiuretic hormone, which literally means "hormone that prevents urination," helps prevent dehydration. As you learned in Chapter 31, ADH increases the permeability to water of the collecting ducts of nephrons in the kidney, causing water to be reabsorbed from the

STAGE
10

FIGURE 33-4 When the anterior pituitary malfunctions An improperly functioning anterior pituitary produces either too much or too little growth hormone. Too little results in a particular type of dwarfism; too much causes gigantism. **QUESTION** Why is gigantism usually more difficult to treat than dwarfism?

urine and retained in the body. Interestingly, alcohol inhibits the release of ADH, greatly increasing urination, so a beer drinker may temporarily lose more fluid than he or she has taken in.

Oxytocin triggers the "milk letdown reflex" in nursing mothers by causing muscle tissue within the breasts (mammary glands) to contract as a result of the stimulation of breastfeeding. This reflex ejects milk from the saclike milk glands into the nipples (Fig. 33-5). Oxytocin also causes contractions of the muscles of the uterus during childbirth, helping expel the fetus from the womb.

Recent studies using laboratory animals indicate that oxytocin also has behavioral effects. In rats, for example, oxytocin injections cause virgin females to exhibit maternal behavior such as building a nest, licking pups, and retrieving pups that have strayed. Oxytocin may also have a

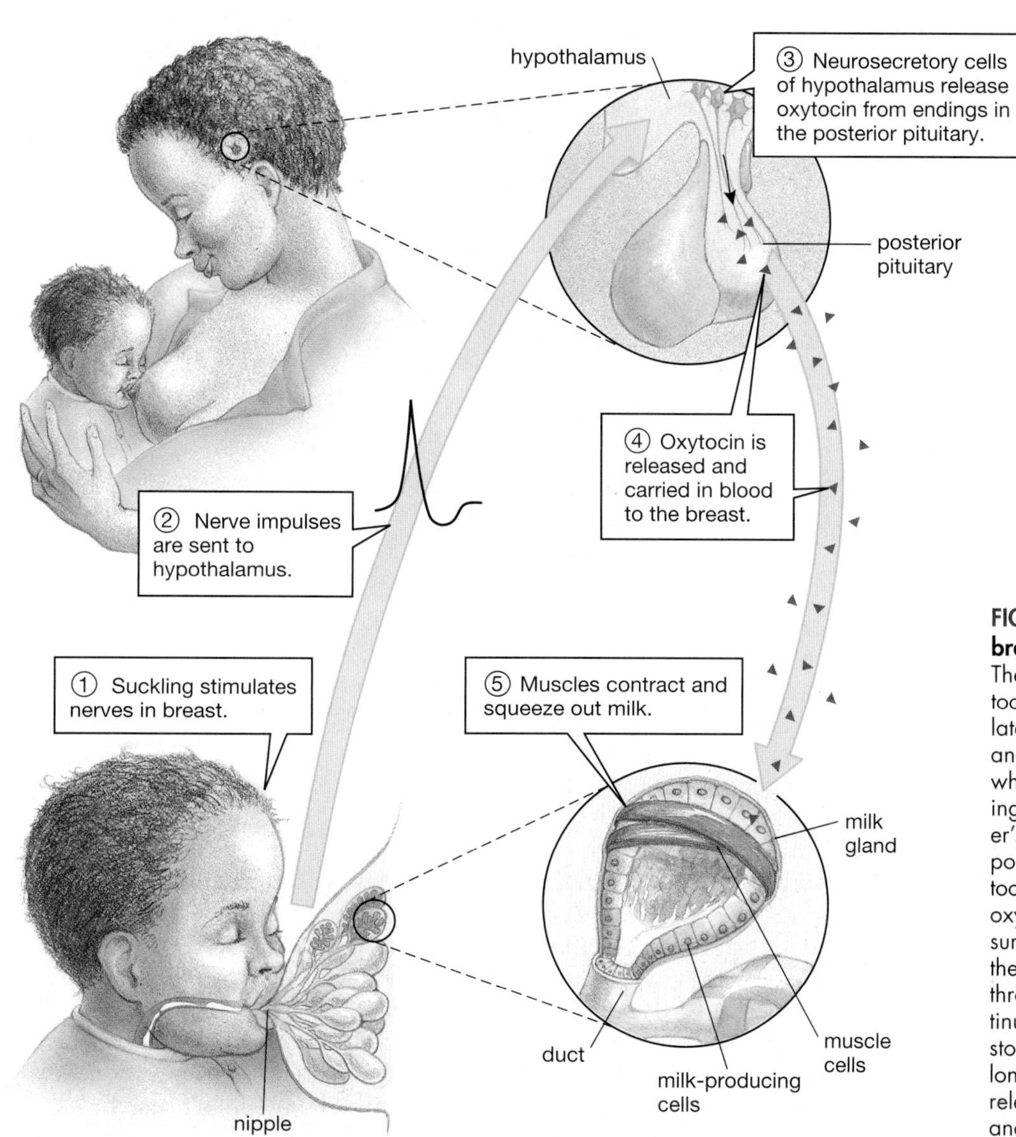
hypothalamus
③ Neurosecretory cells of hypothalamus release oxytocin from endings in the posterior pituitary.
posterior pituitary
④ Oxytocin is released and carried in blood to the breast.
② Nerve impulses are sent to hypothalamus.
① Suckling stimulates nerves in breast.
⑤ Muscles contract and squeeze out milk.
milk gland
duct
milk-producing cells
muscle cells
nipple

FIGURE 33-5 Hormones and breastfeeding The control of milk letdown by oxytocin during breastfeeding is regulated by feedback between a baby and its mother. Milk is expelled when suckling stimulates nerve endings that send a signal to the mother's hypothalamus, causing the posterior pituitary to secrete oxytocin into the bloodstream. When oxytocin reaches the muscles that surround the milk ducts, it causes them to contract and expel milk through the nipple. This cycle continues until the infant is full and stops suckling. With the nipple no longer being stimulated, oxytocin release stops, the muscles relax, and milk flow ceases.

role in male reproductive behavior. In several types of animals, oxytocin stimulates the contraction of muscles that surround the tubes that conduct sperm from the testes to the penis, causing ejaculation.

The Thyroid and Parathyroid Glands Influence Metabolism and Calcium Levels

Lying at the front of the neck, nestled just below the larynx (Fig. 33-6a), the **thyroid gland** produces two major hormones: thyroxine and calcitonin. **Thyroxine**, often referred to as "thyroid hormone," is an iodine-containing modified amino acid that raises the metabolic rate of most body cells. **Calcitonin** is a peptide important in calcium metabolism.

Thyroxine influences most of the cells in the body, elevating their metabolic rate and stimulating the synthesis of enzymes that break down glucose and provide energy. In adults, levels of thyroxine determine the overall metabolic rate—that is, the resting rate of cellular metabolism. Normal levels of thyroxine are required for mental alertness. Low thyroxine leads to mental and physical lethargy, decreased appetite, and intolerance to cold (the body generates less of its own heat when metabolic rate is low). Excess thyroxine leads to restlessness and irritability, increased appetite, and intolerance to heat.

In juvenile animals, including humans, thyroxine helps regulate growth by stimulating both metabolic rate and the development of the nervous system. Undersecretion of thyroid hormone early in life causes *cretinism,* a condition characterized by retardation in both mental and physical development. Fortunately, early diagnosis and thyroxine supplementation can reverse this condition. Conversely, oversecretion of thyroxine in developing vertebrates can trigger precocious development. In 1912, in one of the first demonstrations of hormone action, a physiologist discovered that thyroxine can induce early metamorphosis in tadpoles (see "Evolutionary Connections: The Evolution of Hormones").

Levels of thyroxine in the bloodstream are fine-tuned by negative feedback loops. Thyroxine release is stimulated by thyroid-stimulating hormone (TSH) from the anterior pituitary, which in turn is stimulated by a releasing hormone from the hypothalamus. The amount of TSH released from the pituitary is regulated by negative feedback. Adequate levels of thyroxine circulating in the bloodstream inhibit the secretion of both the releasing hormone (from the hypothalamus) and TSH (from the anterior pituitary), thus inhibiting further release of thyroxine from the thyroid gland (Fig. 33-7).

An iodine-deficient diet can reduce the production of thyroxine and trigger a feedback mechanism that acts to restore normal hormone levels by dramatically increasing the number of thyroxine-producing cells. This compensating mechanism leads to excessive growth of the thyroid gland; the enlarged gland may bulge from the neck, producing a condition called **goiter** (Fig. 33-6b).

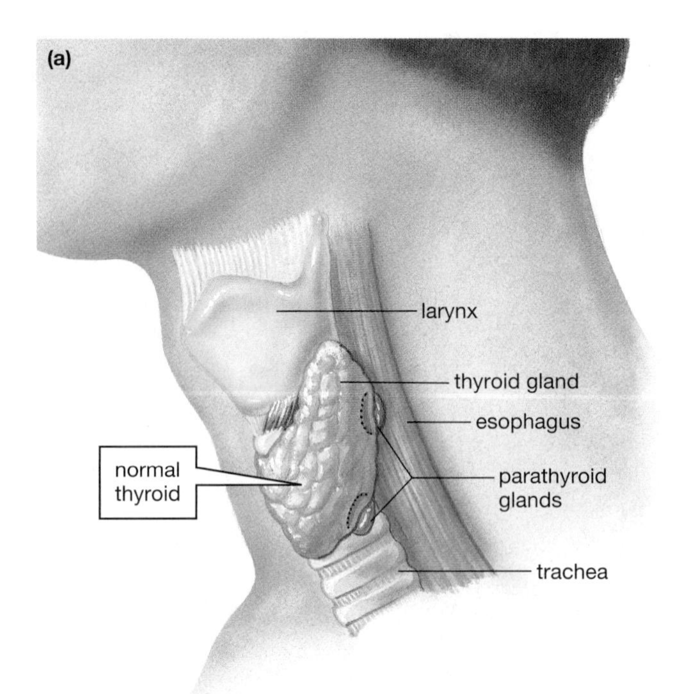

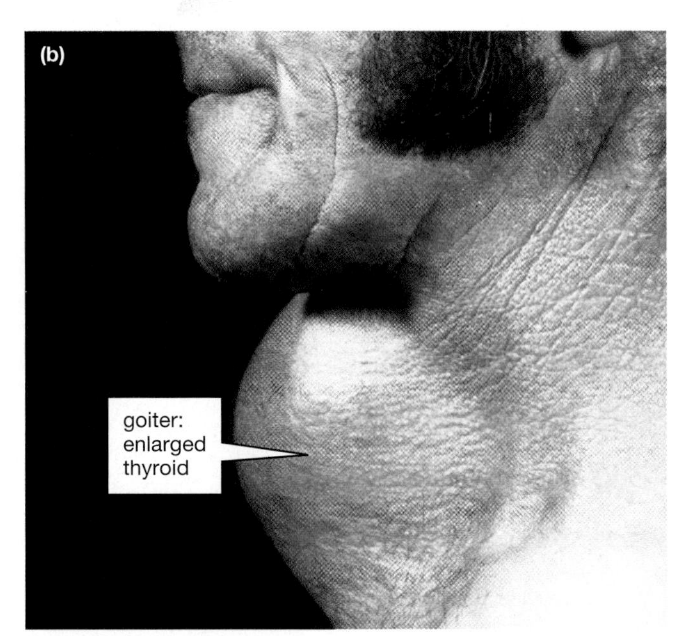

FIGURE 33-6 The thyroid and parathyroid glands
(a) The thyroid and parathyroid glands wrap around the front of the larynx in the neck. **(b)** Goiter, a condition in which the thyroid gland becomes greatly enlarged, is caused by an iodine-deficient diet.

Goiter was once common in some regions of the United States, but widespread use of iodized salt has now all but eliminated this condition in developed countries.

The four small disks of the **parathyroid glands** are embedded in the back of the thyroid gland (see Fig. 33-6a). The parathyroids secrete **parathyroid hormone**, which, along with calcitonin, controls the concentration of calcium in the blood and other body fluids. Calcium is essential for many processes, including nerve and muscle

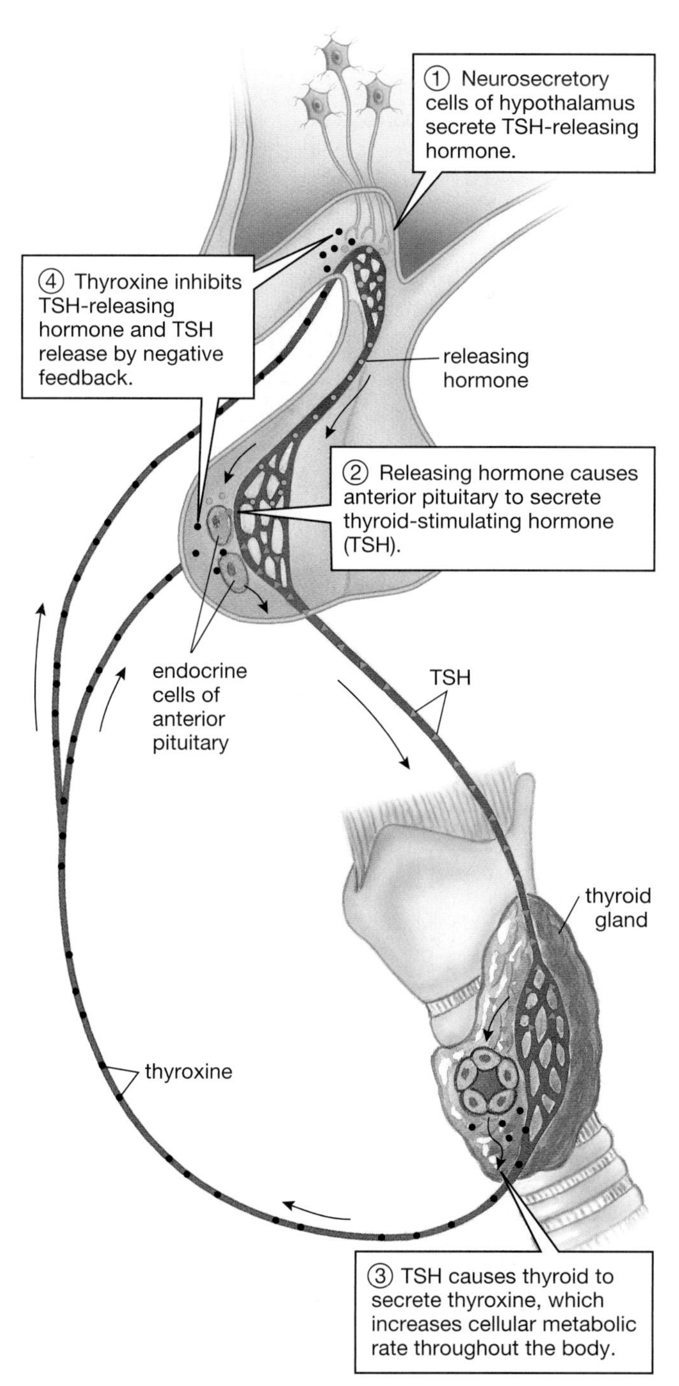
① Neurosecretory cells of hypothalamus secrete TSH-releasing hormone.
④ Thyroxine inhibits TSH-releasing hormone and TSH release by negative feedback.
releasing hormone
② Releasing hormone causes anterior pituitary to secrete thyroid-stimulating hormone (TSH).
endocrine cells of anterior pituitary
TSH
thyroid gland
thyroxine
③ TSH causes thyroid to secrete thyroxine, which increases cellular metabolic rate throughout the body.

FIGURE 33-7 Negative feedback in thyroid gland function
QUESTION A common test of thyroid gland function is to measure the amount of thyroid-stimulating hormone circulating in the blood. What would you conclude if such a test found an abnormally high level of TSH?

function, so the calcium concentration in body fluids must be kept within narrow limits. Parathyroid hormone and calcitonin regulate calcium absorption and release by the bones, which serve as a bank into which calcium can be deposited or withdrawn as necessary. In response to low blood calcium, the parathyroids release parathyroid hormone, which causes the release of calcium from bones. The parathyroids increase in size in pregnant and lactating women, thereby enhancing parathyroid hormone output and allowing the mother's body to meet the extra demands for calcium imposed by the developing fetus and, later, milk production. If blood calcium levels become too high, the thyroid releases calcitonin, which inhibits the release of calcium from bones.

The Pancreas Is Both an Exocrine and an Endocrine Gland

The **pancreas** is a gland that produces both exocrine and endocrine secretions. The exocrine portion synthesizes digestive secretions that are released into the *pancreatic duct* and flow into the small intestine (see Chapter 30). The endocrine portion consists of clusters of cells called **islet cells**, which produce peptide hormones. One type of islet cell produces the hormone **insulin**; another type produces the hormone **glucagon**.

Insulin and glucagon work in opposition to regulate carbohydrate and fat metabolism: Insulin reduces the blood glucose level; glucagon increases it (Fig. 33-8). Together, the two hormones help keep the blood glucose level nearly constant. When blood glucose rises (for example, after you've eaten), insulin is released. Insulin causes body cells to take up glucose and either metabolize it for energy or convert it to fat or *glycogen* (a polysaccharide made of long chains of glucose molecules) for storage. When blood glucose levels drop (for example, after you've skipped breakfast or have run a 10-kilometer race), glucagon is released. Glucagon activates an enzyme in the liver that breaks down glycogen (which is stored primarily in the liver), releasing glucose into the blood. Glucagon also promotes lipid breakdown, which releases fatty acids that can be metabolized for energy.

Lack of insulin production or the failure of target cells to respond to insulin results in **diabetes mellitus**. There are several causes of diabetes, but in all cases, blood glucose levels are high and fluctuate with food intake. For reasons that are not yet fully understood, diabetes causes a wide range of circulatory problems that can result in high blood pressure, atherosclerosis, and increased levels of LDL (bad) cholesterol. Diabetes damages the circulatory system, which in turn causes heart attacks, blindness, and kidney failure. Until recently, the insulin supplements required to treat diabetes were formulated from insulin extracted from the pancreases of cows and pigs obtained from slaughterhouses. Now, however, scientists have inserted the gene for human insulin into bacteria, which are cultivated to produce large quantities of human insulin.

The Sex Organs Secrete Steroid Hormones

The sex organs do far more than produce sperm or eggs. The **testes**, in males, and **ovaries**, in females, are also important endocrine organs. The testes secrete several steroid hormones, collectively called **androgens**. The most important of these is **testosterone**. The ovaries secrete two types of steroid hormones: **estrogen** and

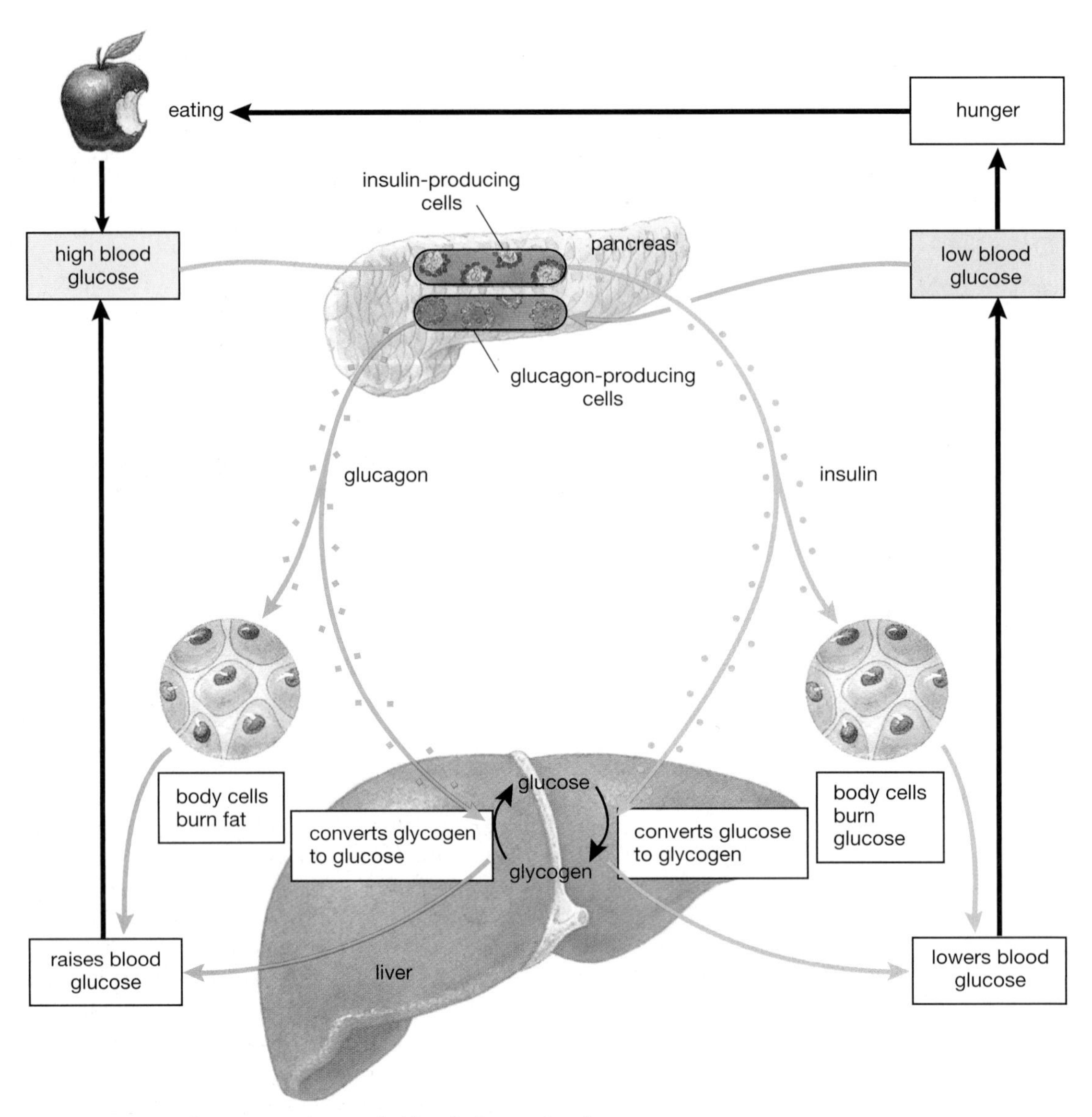

FIGURE 33-8 The pancreas controls blood glucose levels
The pancreatic islet cells contain two populations of hormone-producing cells, one producing insulin, the other producing glucagon. These two hormones cooperate in a two-part negative feedback loop to control blood glucose concentrations. High blood glucose stimulates the insulin-producing cells and inhibits the glucagon-producing cells; low blood glucose stimulates the glucagon-producing cells and inhibits the insulin-producing cells. This dual control quickly corrects high or low blood glucose levels. **QUESTION** How would control of blood glucose be affected in a person who was born with a mutation that prevented glucagon receptors from binding to glucagon?

progesterone. The roles of the sex hormones in sperm and egg production, the menstrual cycle, pregnancy, and development are discussed in Chapters 36 and 37.

The sex hormones also play a key role in *puberty,* the phase of life during which the reproductive systems of both sexes become mature and functional. This is accompanied by behavioral changes that make puberty such an interesting time for teenagers and their parents. Puberty begins when, for reasons not fully understood, the hypothalamus starts to secrete increasing amounts of releasing hormones that in turn stimulate the anterior pituitary to secrete more lutenizing hormone (LH) and follicle-stimulating hormone (FSH) into the bloodstream. Both LH and FSH stimulate target cells in the testes or ovaries to produce higher levels of sex hormones. The resulting increase in circulating sex hormones ultimately affects tissues throughout the body that bear the appropriate receptors. Both sexes develop pubic and underarm hair. Testosterone, secreted by the testes, stimulates the development of male secondary sexual attributes, including body and facial hair, muscle growth, and a larger larynx ("voice box"), which lowers the voice. Testosterone also promotes sperm cell production. Estrogen from the ovaries in females stimulates growth of mammary glands and maturation of the female reproductive system, including production of mature egg cells. Progesterone, secreted by the ovaries during pregnancy, prepares the reproductive tract to receive and nourish the fertilized egg.

Although there is a surge of sex hormone production at puberty, sex hormones are present from the fetal stage onward. They influence development in both sexes, and they continue to have an effect on both behavior and cognitive processes throughout life.

The Adrenal Glands Have Two Parts That Secrete Different Hormones

Imagine how your body feels when you are startled, afraid, or angry. These physical reactions are the result of hormones produced by the adrenal glands, which act in conjunction with your sympathetic nervous system (see Chapter 34). Like the pituitary gland and pancreas, the **adrenal glands** (Latin for "on the kidney") are two glands in one: the *adrenal medulla* and the *adrenal cortex* (Fig. 33-9). The **adrenal medulla** is located in the center of each gland (*medulla* means "marrow" in Latin). It consists of secretory cells derived during development from nervous tissue, and its hormone secretion is controlled directly by the nervous system. The adrenal medulla produces two hormones—**epinephrine** and, in much smaller quantities, **norepinephrine** (also called *adrenaline* and *noradrenaline,* respectively)—in response to stress. These hormones, which are amino-acid derivatives, prepare the body for emergency action. They increase the heart and respiratory rates, cause blood glucose levels to rise, and direct blood flow away from the digestive tract toward the brain and muscles. They also cause the air passages of the lungs to expand, allowing more efficient exchange of gases. For this reason, substances that mimic epinephrine are often administered to asthmatics, whose airways become constricted during an asthma attack. The adrenal medulla is activated by the sympathetic nervous system, which prepares the body to respond to emergencies, as we will describe in Chapter 34.

The outer layer of the adrenal gland forms the **adrenal cortex** (*cortex* is Latin for "bark"), which secretes three types of steroid hormones collectively called **glucocorticoids**. Of these, **cortisol** is secreted in the largest quantity. Glucocorticoid release is stimulated by ACTH from the anterior pituitary in response to a releasing hormone from the hypothalamus. Hormone levels are controlled by negative feedback; circulating glucocorticoids inhibit the release of both the hypothalamic releasing hormone and ACTH. Cortisol is released when the body is under stress for reasons such as trauma, infection, exposure to temperature extremes, or final exams. The effects of cortisol help the body cope with short-term stressors, raising blood glucose levels by stimulating glucose production and promoting the use of fats instead of glucose for energy production. Cortisol also inhibits the immune response—if you find yourself getting sick around exam time, this may be a contributing factor. Since the immune response itself can be quite debilitating (producing inflammation and fever), temporarily suppressing it can help an organism cope with more immediate and possibly life-threatening situations.

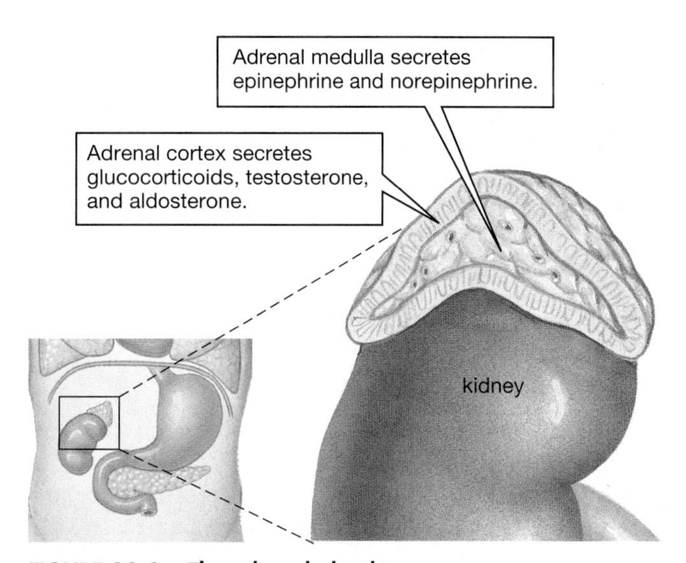

FIGURE 33-9 The adrenal glands
Atop each kidney sits an adrenal gland, a two-part gland composed of very dissimilar cells. The outer cortex consists of endocrine cells that secrete steroid hormones. The inner medulla, derived from nervous tissue during development, secretes epinephrine and norepinephrine.

You may have noticed that many different hormones are involved in glucose metabolism: thyroxine, insulin, glucagon, epinephrine, and the glucocorticoids. Why? The reason can probably be traced to a metabolic requirement of the brain. Although most body cells can produce energy from fats and proteins as well as from carbohydrates, brain cells can burn only glucose. Thus, blood glucose levels cannot be allowed to fall too low, or brain cells rapidly starve, leading to unconsciousness and death.

The adrenal cortex also secretes the hormone **aldosterone**, which regulates the sodium content of the blood. Sodium ions, derived from salt in the diet, are the most abundant positive ions in blood and extracellular fluid. The sodium ion gradient across plasma membranes (high outside, low inside) is a factor in many cellular events, including the production of electrical signals by nerve cells. If blood sodium falls, the adrenal cortex releases aldosterone, which causes the kidneys and sweat glands to retain sodium; this enables salt and other sources of dietary sodium to raise blood sodium levels, shutting off further aldosterone secretion (an example of negative feedback).

In both women and men, the adrenal cortex also produces the male sex hormone testosterone, although normally in much smaller amounts than produced by the testes. Tumors of the adrenal medulla can lead to excessive testosterone release, causing masculinization of women. Many of the "bearded ladies" who once appeared in circus sideshows probably had this condition.

Other Sources of Hormones Include the Pineal Gland, Thymus, Kidneys, Heart, Digestive Tract, and Fat Cells

The **pineal gland** is located between the two hemispheres of the brain, just above and behind the hypothalamus (see Fig. 33-1). Named for its resemblance to a pine cone, the pineal gland is smaller than a pea. In 1646 philosopher René Descartes described it as "the seat of the rational soul." Since then, scientists have learned more about this organ, but many of its functions are still poorly understood.

The pineal produces the hormone **melatonin**, an amino-acid derivative. Melatonin is secreted in a daily rhythm, which in mammals is regulated by the eyes. In some vertebrates, such as the frog, the pineal itself contains photoreceptive cells; the skull above it is thin, so the pineal can detect sunlight and thus daylength. By responding to daylengths characteristic of different seasons, the pineal appears to regulate the seasonal reproductive cycles of many mammals. Despite years of research, the function of melatonin and the pineal gland in humans is still unclear. One hypothesis is that the pineal gland and melatonin secretion influence sleep–wake cycles. Melatonin is sold as a sleeping aid, but its use is controversial. Improper pineal function may contribute to the depression that some people experience during the short days of winter.

The **thymus** is located in the chest cavity behind the breastbone, also called the *sternum* (see Fig. 33-1). In addition to producing white blood cells, the thymus produces the hormone **thymosin**, which stimulates the development of specialized white blood cells (T cells) that play an important role in the immune system (see Chapter 32). The thymus is extremely large in infants but, under the influence of sex hormones, begins decreasing in size after puberty.

The kidneys, which play a central role in maintaining body fluid homeostasis, are important endocrine organs as well. When the oxygen content of the blood drops, the kidneys produce the hormone **erythropoietin**, which increases red blood cell production (see Chapter 28). The kidneys also produce a second hormone, **renin**, in response to low blood pressure, such as that caused by bleeding. Renin is an enzyme that catalyzes the production of the hormone **angiotensin** from proteins in the blood. Angiotensin raises blood pressure by constricting arterioles. It also stimulates the release of aldosterone by the adrenal cortex, causing the kidneys to retain sodium, which in turn increases blood volume.

In 1981 a substance extracted from atrial tissue of the heart and injected into rats was found to cause an increase in the output of salt and water by the kidneys. Two years later, the active substance **atrial natriuretic peptide (ANP)** was described and its amino acid sequence determined. This peptide is released by cells in the atria when blood volume increases, causing extra distension of the heart. Atrial natriuretic peptide then reduces blood volume by decreasing the release of both ADH and aldosterone.

The stomach and small intestine produce a variety of peptide hormones that help regulate digestion. These hormones include **gastrin**, **secretin**, and **cholecystokinin**, discussed in Chapter 30.

Can fat be an endocrine organ? In 1995 researchers described the peptide hormone **leptin** (derived from a word meaning "slender"), which is released by adipose (fat) cells. Mice missing the gene for leptin became obese (Fig. 33-10), and leptin injections caused them to lose weight. The researchers hypothesized that adipose tissue, by releasing leptin, tells the body how much fat it has stored and therefore how much to eat. Unfortunately, trials of leptin as a human weight-loss aid have not been encouraging. Many obese people have high levels of leptin but seem to be relatively insensitive to it. However, researchers are discovering surprising new functions for leptin and are finding leptin receptors in unexpected places, such as blood vessels and white blood cells. Leptin appears to stimulate the growth of new capillaries and to speed wound healing. It also stimulates the immune system and is required for the onset of puberty and the development of secondary sexual characteristics.

Research continues to expand our understanding of the multiple effects of hormones and the wide variety of organs and cells that produce them. In time, this

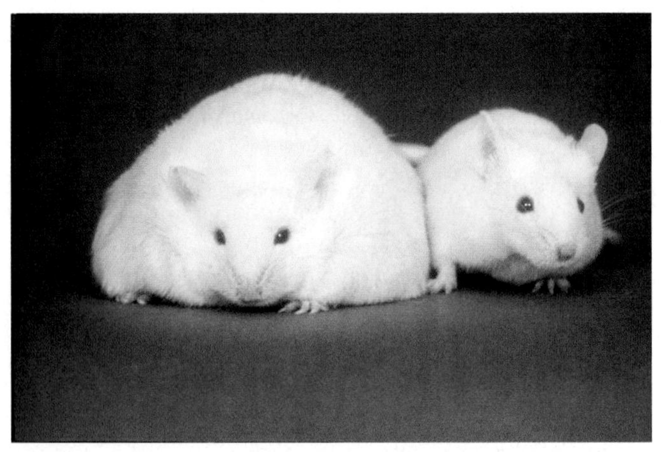

FIGURE 33-10 Leptin helps regulate body fat
The mouse on the left has been genetically engineered to lack the gene for the hormone leptin. **QUESTION** Obese mice that lack leptin lose weight when injected with the hormone, but humans with ample leptin in their blood can be obese. What might account for this discrepancy?

understanding will lead to a wealth of new medical treatments. It should also increase our respect for these substances and our recognition that any hormone we take might influence physiological systems throughout our body.

EVOLUTIONARY CONNECTIONS

The Evolution of Hormones

Not long ago, vertebrate endocrine systems were considered unique to our phylum, and endocrine chemicals were thought to have evolved expressly for their role in vertebrate physiology. In recent years, however, physiologists have discovered that hormones are evolutionarily ancient. Insulin, for example, is found not only in vertebrates but also in protists, fungi, and bacteria, although research has not yet determined the function of insulin in most of these organisms. Protists also manufacture ACTH, even though they have no adrenal glands to stimulate. Yeasts have receptors for estrogen but no ovaries. Thyroid hormones have been found in certain invertebrates, such as worms, insects, and mollusks, as well as in vertebrates. Even among vertebrates, the effects of chemically identical hormones, secreted by the same glands, can vary dramatically from organism to organism. Let's look briefly at the diverse effects that the thyroid hormone thyroxine has on several organisms.

Some fish undergo radical physiological changes during their lifetimes. A salmon, for example, begins life in fresh water, migrates to the ocean, and returns to fresh water to spawn. In the stream where the salmon hatched, fresh water tends to enter the fish's tissues by osmosis; in salt water, the fish tends to lose water, becoming dehydrated. The salmon's migrations, therefore, require complete revamping of salt and water control. In salmon, one of the functions of thyroxine is to produce the metabolic changes necessary to go back and forth from life in streams to life in the ocean.

In amphibians, thyroxine has the dramatic effect of triggering metamorphosis. In 1912, in one of the first demonstrations of the action of any hormone, tadpoles were fed minced horse thyroid. As a result, the tadpoles metamorphosed prematurely into miniature adult frogs (Fig. 33-11). In high mountain lakes in Mexico, where the water is deficient in the iodine needed to synthesize thyroxine, natural selection has produced one species of salamander that has the ability to reproduce while still in its juvenile form.

Thyroxine regulates the seasonal molting of most vertebrates. From snakes to birds to the family dog, surges of thyroxine stimulate the shedding of skin, feathers, or hair. In humans (who neither migrate regularly, metamorphose, nor molt), thyroxine regulates growth and metabolism.

The use of chemicals to regulate cellular activity is extremely ancient. The diversity of life on Earth rests on a conservative foundation: A relative handful of chemicals coordinates activities within single cells and among groups of cells. Life's diversity originated in part by changing the systems used to deliver the chemicals and by evolving new types of responses. Early in their evolution, animals developed a complement to hormonal communication that provides faster, more precise delivery of chemical messages: the nervous system. As we will explain in the next chapter, the nervous system permits rapid responses to environmental stimuli, flexibility in response options, and ultimately consciousness itself.

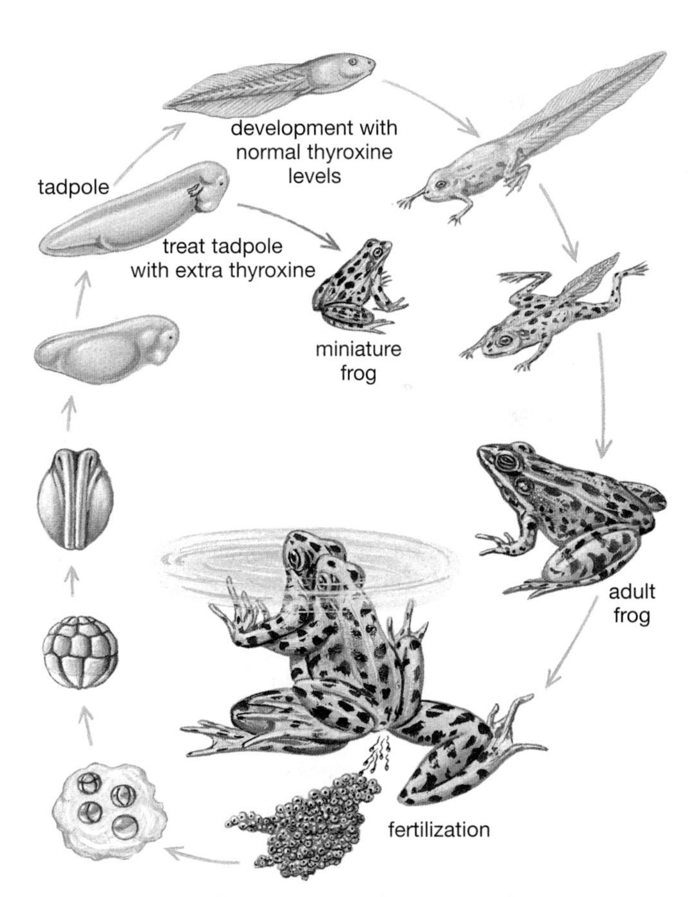

FIGURE 33-11 Thyroxine controls metamorphosis in amphibians
The life cycle of the frog begins with fertilization of the eggs (bottom). The fertilized eggs develop into an aquatic, fishlike tadpole, which grows and ultimately metamorphoses into an adult frog. Metamorphosis is triggered by a surge of thyroxine from the tadpole's thyroid gland. If injected with extra thyroxine, a young tadpole will metamorphose ahead of schedule into a miniature adult frog. **QUESTION** What would happen if tadpoles were given a drug that blocked the production of thyroxine?

CASE STUDY REVISITED Losing on Artificial Hormones

As with many hormones, anabolic steroids exert their effects throughout the body; a steroid user risks damage to the liver, circulatory system, reproductive system, and immune system. Liver damage can lead to jaundice and liver tumors. Steroid-induced high blood pressure and decreases in the "good" (HDL) form of cholesterol threaten the circulatory system because both of these effects are risk factors for heart attacks and strokes. In males, artificially increased levels of anabolic steroids create a negative feedback effect that can reduce the natural production of testosterone, reduce testicle size and sperm count, and cause the prostate to enlarge and interfere with urination. In females, anabolic steroids can interfere with menstrual periods and cause a more masculine distribution of body hair. For both sexes, anabolic steroids depress the immune system. Aggressiveness and mood swings have also been attributed to steroid use. Likewise, darbepoetin can be dangerous to a person whose blood already has a normal amount of red blood cells. Blood packed with too many cells has increased viscosity, forcing the heart (already working at near-maximum capacity in many athletes) to work even harder to pump the thickened blood. Although darbepoetin is so new that statistics are unavailable, erythropoietin supplementation is suspected in the deaths of about 20 competitive European cyclists since the late 1980s. What unearned glory could possibly be worth such risks?

Many individuals in the world of sports are not surprised when they learn that even elite athletes have used anabolic steroids and other artificial hormones. The practice is all too widespread, and evading detection has become part of the "game" for some members of this intensely competitive world.

Consider This: Given what you know about steroids and other hormones, do you think that high school and college athletes should be routinely tested, just as Olympic athletes are? Explain your answer.

Links to Life: Closer to a Cure for Diabetes

The worst kind of diabetes, Type I, occurs when a person's immune system attacks and kills the insulin-producing islet cells of the pancreas. This form of diabetes often strikes early in life; its victims rarely survive for a full life span, and their lives, filled with multiple daily blood tests and insulin injections, are never normal. For the one million people in the United States who suffer from this form of diabetes, a research team led by Dr. James Shapiro at the University of Alberta, Canada, has offered a ray of hope. These researchers removed pancreases from recently deceased donors and infused extracted islet cells into a vein feeding the livers of diabetes victims; there, the cells took up residence and began secreting insulin. Patients returned home the same day. In preliminary trials, 92% of transplant recipients had functioning islet cells, and 82% had become totally independent of insulin. Although the recipients have an immensely improved quality of life, they must continue to take immunosuppressant drugs to avoid rejection. Unfortunately, only about 3000 pancreases become available each year in the United States, and most successful islet transplants require two donor pancreases. The Canadian breakthrough will undoubtedly spur further research that may one day allow islet cells to be grown in culture, perhaps from the patient's own stem cells, offering freedom from the need for donors and immunosuppressant drugs, and hope to millions of diabetics worldwide.

Summary of Key Concepts

33.1 How Do Animal Cells Communicate?

Within multicelluar organisms, communication among cells occurs through gap junctions directly linking cells, by diffusion of chemicals to nearby cells (local hormones and neurotransmitters), and by transport of chemicals within the bloodstream (endocrine hormones). Extracellular chemical messengers act selectively on target cells that bear specific receptors for the chemical.

33.2 What Are the Characteristics of Animal Hormones?

Most cells release local hormones, such as prostaglandins, to communicate with nearby cells. The endocrine system is a collection of glands and organs that release endocrine hormones, which are transported in the bloodstream to other parts of the body, where they affect the activity of specific target cells bearing receptors for the hormones. Hormones are synthesized either from amino acids (amino-acid based hormones and peptides) or from lipids (steroid hormones).

Most hormones act on their target cells in one of two ways: Peptide hormones and amino acid based hormones bind to receptors on the surfaces of target cells and activate intracellular second messengers, such as cyclic AMP, which then alter the metabolism of the cell. Steroid hormones can bind to surface receptors or diffuse through the plasma membranes of their target cells and bind with receptors in the cytoplasm or nucleus. The hormone–receptor complex promotes the transcription of specific genes within the nucleus. Thyroid hormones also penetrate the plasma membrane but diffuse into the nucleus, where they bind to receptors associated with the chromosomes and influence gene transcription.

Hormone action is commonly regulated through negative feedback, a process in which a hormone causes changes that inhibit further secretion of that hormone.

33.3 What Are the Structures and Hormones of the Mammalian Endocrine System?

Many hormones are produced by endocrine glands, which are clusters of cells embedded within a network of capillaries. Hormones are secreted into the extracellular fluid and diffuse into capillaries. The major endocrine glands of the human body are the hypothalamus–pituitary complex, the thyroid and parathyroid glands, the pancreas, the sex organs, and the adrenal glands. The hormones released by these glands and their actions are summarized in Table 33-2. Other structures that produce hormones include the pineal gland, thymus, kidneys, heart, stomach and small intestine, and fat cells.

Key Terms

Thinking Through the Concepts

To take a multiple-choice quiz with feedback on the contents of this chapter, visit http://www.prenhall.com/audesirk7. *Log in to the Web site selected by your instructor and navigate to the Self Test section for this chapter.*

? Review Questions

1. What are the three types of molecules used as hormones in vertebrates? Give an example of each.
2. What is the difference between an endocrine gland and an exocrine gland? Which type releases hormones?
3. When peptide hormones attach to target cell receptors, what cellular events follow? How do steroid hormones behave?
4. Diagram the process of negative feedback, and give an example of it in the control of hormone action.
5. What are the major endocrine glands in the human body, and where are they located?
6. Describe the structure of the hypothalamus–pituitary complex. Which pituitary hormones are neurosecretory? What are their functions?
7. Describe how releasing hormones regulate the secretion of hormones by cells of the anterior pituitary. Name the hormones of the anterior pituitary, and give one function of each.
8. Describe how the hormones of the pancreas act together to regulate the concentration of glucose in the blood.
9. Compare the adrenal cortex and adrenal medulla by answering the following questions: Where are they located within the adrenal gland? What are their embryological origins? Which hormones do they produce? Which organs do their hormones target? What homeostatic processes regulate blood levels of the respective hormones?

Applying the Concepts

1. A student decides to do a science project on the effect of the thyroid gland on frog metamorphosis. She sets up three aquaria with tadpoles. She adds thyroxine to the water of one, the drug thiouracil to a second, and nothing to the third. Thiouracil reacts with thyroxine in tadpoles to produce an ineffective compound. Assuming that the student uses appropriate physiological concentrations, predict what will happen.
2. If you were obese, would you consider injections of leptin? What pros and cons can you think of? Defend your decision.
3. Suggest a hypothesis about the endocrine system to explain why many birds lay their eggs in the spring and why poultry farmers who produce eggs keep lights on at night.
4. Anabolic steroids, used by risk-taking athletes and bodybuilders, are chemically related to testosterone. They increase bone mass and muscle mass and seem to improve athletic performance. But anabolic steroids can cause liver problems, heart attacks, strokes, testicular atrophy, and personality changes in males. Females on anabolic steroids also have liver and circulatory problems. In addition, their voices deepen, their bodies develop more hair, and their menstrual cycles are disturbed. Explain how the same compound can produce different effects in males and females.
5. Some parents who are interested in college sports scholarships for their children are asking physicians to prescribe growth hormone treatments. Farmers also have an economic incentive to treat cows with growth hormone, which can now be produced in large quantities by genetic-engineering techniques. What biological and ethical problems do you foresee for parents, children, physicians, coaches, college scholarship boards, food consumers, farmers, the U.S. Food and Drug Administration, and biotechnology companies?
6. Argue for and against banning or restricting the use of common endocrine disrupters, such as plasticizers and certain pesticides. What compromises can you suggest?

For More Information

Atkinson, M., and MacLaren, N. "What Causes Diabetes?" *Scientific American*, July 1991. An in-depth look at what's behind a common but serious disease.

Beardsley, T. "Melatonin Mania." *Scientific American*, April 1996. The hormone melatonin has been touted as a treatment for ailments ranging from insomnia to cancer. What do scientists think of the evidence?

Christensen, D. "Transplanted Hopes." *Science News*, September 2, 2000. Describes successful pancreatic islet cell transplants into diabetic patients.

Gold, C. "Hormone Hell." *Discover*, September 1996. Environmental pollutants that mimic the chemical activity of hormones may be playing havoc with animal (including human) physiology.

McLachlan, J., and Arnold, S. "Environmental Estrogens." *American Scientist*, September–October 1996. A sober evaluation of the effects of estrogen-like substances in the environment, written by two of the leading researchers in the field.

Raloff, J. "Common Pollutants Undermine Masculinity." *Science News*, April 3, 1999. Several recent research articles report feminizing effects of estrogen disrupters in laboratory animals.

Sapolsky, R. "Testosterone Rules." *Discover*, March 1997. An engaging inquiry into the question of whether testosterone causes aggression.

Schubert, C. "Burned by Flame Retardants?" *Science News*, October 13, 2001. Flame retardants from TVs, computers, drapes, and sofas are building up in our bodies.

Zorpette, G. "All Doped Up—and Going for the Gold." *Scientific American*, May 2000. Describes the efforts to develop drug tests for erythropoietin and human growth hormone, both abused by competitive athletes.

Media Activities

To access a Media Activity visit http://www.prenhall.com/audesirk7. *Log in to the Web site selected by your instructor, navigate to this chapter, and select the appropriate Media Activity number.*

33.1 Hypothalamic Control of the Pituitary

Estimated time: 5 minutes

The hypothalamus and the pituitary gland work together to coordinate many actions of the endocrine system. In this activity, you will learn how the hypothalamus controls hormonal release by the pituitary gland.

33.2 Modes of Action of Hormones

Estimated time: 5 minutes

Explore two ways hormones act to effect cells throughout the body.

33.3 Web Investigation: Losing on Artificial Hormones

Estimated time: 30 minutes

Can a drug-free athlete win in the modern world? The modern pharmacopoeia is full of enticing performance-enhancing drugs, but the long-term effects can be devastating.

CHAPTER

34 The Nervous System and the Senses

Christopher Reeve discusses spinal cord injuries at the National Press Club in Washington, D.C. in December 1999.

AT A GLANCE

CASE STUDY From Tragedy to Triumph

Actor Christopher Reeve's real-life athletic abilities made him well suited to his movie role as Superman. But in 1995, that phase of his life ended abruptly at an equestrian jumping competition. Racing up to a jump, his horse suddenly balked, throwing Reeve headfirst onto the jumping rail. The impact shattered the first two vertebrae in his neck and crushed his spinal cord, interrupting the flow of signals from his brain to the rest of his body and back again. Unable to move or breathe, Reeve's life was saved by the fast action of paramedics, who began pumping oxygen into his lungs within three minutes of the fall. If one or two more minutes had elapsed, he would almost certainly have had irreversible brain damage from the lack of oxygen.

Completely paralyzed from the neck down and unable to breathe on his own, Reeve battled depression, finally making a conscious decision to find new ways to be productive and active. In 1999, Reeve began an intensive rehabilitation program involving *functional electrical stimulation* (FES). This procedure uses a computer program to send electrical signals to the nerves that activate specific muscles, mimicking the signals that the brain would normally send to allow the individual to perform an activity. Using FES, Reeve regularly pedals a stationary bicycle. In the years since his accident, he and his wife Dana have raised awareness and funds for research that may help people with spinal cord injuries regain the use of their bodies.

How are signals conducted from the brain to and from distant parts of our body? Why doesn't the spinal cord repair itself? What hopes can research offer to individuals with spinal cord injuries? What progress has Christopher Reeve made in the years since his accident?

34.1 What Are the Structures and Functions of Neurons?

Our study of the nervous system begins with the individual nerve cell, or **neuron**. As the fundamental unit of the nervous system, each neuron must perform four specialized functions:

1. Receive information from the internal or external environment or from other neurons.
2. Integrate the information it receives and produce an appropriate output signal.
3. Conduct the signal to its terminal ending, which may be some distance away.
4. Transmit the signal to other nerve cells or to glands or muscles.

Although neurons vary enormously in structure, a "typical" vertebrate neuron has four distinct structural regions that carry out the functions listed above: the *dendrites,* the *cell body,* the *axon,* and the *synaptic terminals* (Fig. 34-1). **Dendrites**, branched tendrils that extend outward from the nerve cell body, respond to signals from other neurons or from the external environment. Dendrites receive information; their branched form provides a large surface area for receiving signals. In neurons of the brain and spinal cord, dendrites respond to the chemical neurotransmitters released by other neurons. These dendrites have protein receptors in their membranes that bind specific neurotransmitters and, as a result, produce electrical signals. Dendrites of *sensory neurons* have special membrane adaptations that allow them to produce electrical signals in response to specific stimuli from the environment, such as pressure, odor, light, or heat.

Electrical signals travel down the dendrites and converge on the neuron's **cell body**, which serves as an integration center. In this role, the cell body adds up the various positive and negative signals from the dendrites. If the sum of all these signals is sufficiently positive, the neuron will produce an **action potential**, an electrical output signal. The cell body, containing the usual assortment of organelles, also carries on the routine activities common to most other body cells, such as synthesizing complex molecules and coordinating the metabolic activities of the cell.

In a typical neuron, a long, thin fiber called an **axon** extends outward from the cell body and conducts the electrical signal. Single axons may stretch from your spinal cord to your toes, a distance of about a meter (about 3 feet), making neurons the longest cells in the body. Axons are distribution lines, carrying action potentials from the cell body to the synaptic terminals, located at the far end of each axon. Axons are normally bundled together into **nerves**, much like bundles of wires in an electrical cable. However, unlike electrical power distribution cables (in which energy is lost along the way from power station to customer), the plasma membranes of axons conduct action potentials undiminished from the cell body to their synaptic terminals. Some axons are wrapped with insulation called **myelin**, which allows the electrical signal to be conducted more rapidly. Myelin is formed from non-neuronal cells that wrap around the axon (see Fig. 34-1). In vertebrates, axons bundled into nerves emerge from the brain and spinal cord and extend to all regions of the body.

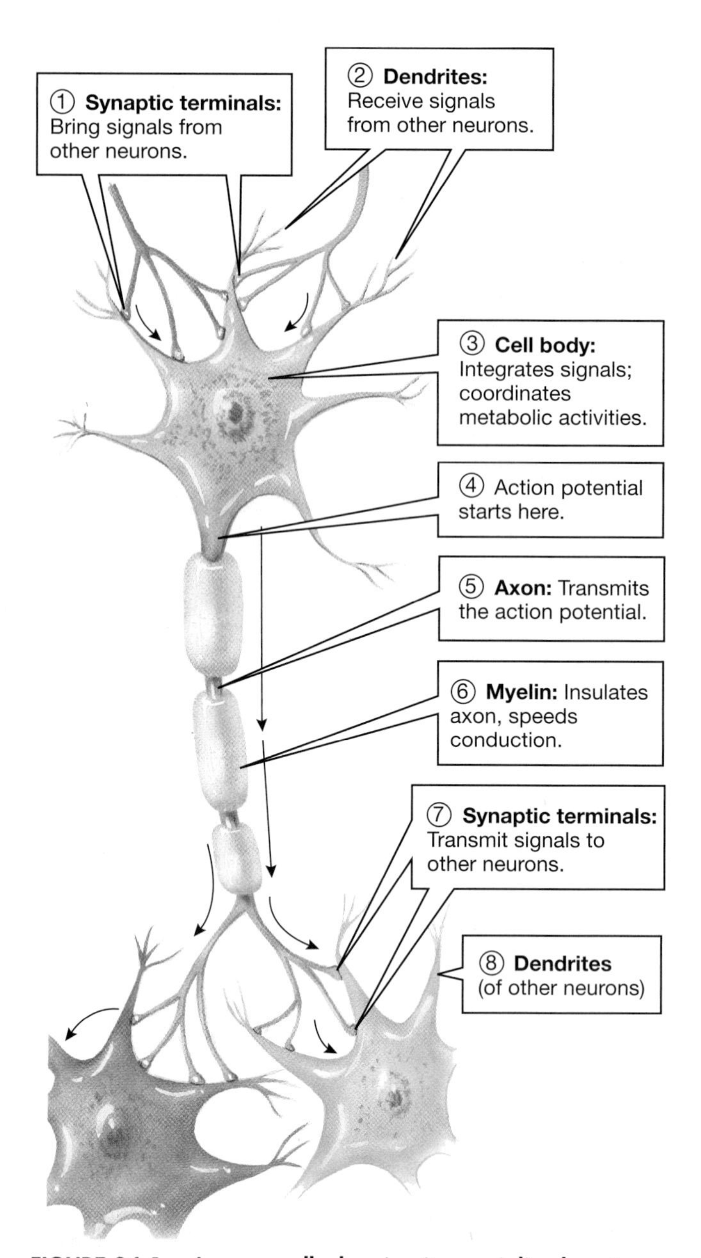

FIGURE 34-1 A nerve cell, showing its specialized parts and their functions

The transmission of the signal to other cells occurs at **synaptic terminals**, swellings at the branched endings of axons (see Fig. 34-1). Most synaptic terminals contain a **neurotransmitter**, a specific chemical released in response to an action potential reaching the terminal. The synaptic terminals of one neuron may communicate with a gland, a muscle cell, or the dendrites or cell body of a second neuron. The output of the first cell becomes the input to the second cell. The site at which synaptic terminals communicate with other cells is called the **synapse**.

34.2 How Is Neural Activity Produced and Transmitted?

Neurons Create Electrical Signals Across Their Membranes

In the early 1950s, biologists using the giant axon of a squid (a mollusk) developed ways to record electrical events inside individual neurons. The researchers found that unstimulated, inactive neurons maintain a constant electrical difference, or *potential,* across their plasma membranes, similar to that across the poles of a battery. As in a battery, the electrical potential across a neuron membrane stores energy. This potential, called the **resting potential**, is always negative inside the cell and ranges from −40 to −90 millivolts (mV; thousandths of a volt).

If the neuron is stimulated, either naturally or by an experimenter using an electrical current, the negative potential inside the neuron can be made either more or less negative. If the potential is made sufficiently less negative that it reaches a level called **threshold**, an action potential is triggered (Fig. 34-2). During an action potential, the neuron's potential rapidly rises to about +50 mV inside the cell. Action potentials last a few milliseconds (thousandths of a second) before the cell's negative resting potential is restored. The positive charge of the action potential flows rapidly down the axon to the synaptic terminal, where the signal is communicated to another cell at a synapse. In "A Closer Look: Ions and Electrical Signals," we examine these electrical potentials, the language of the nervous system.

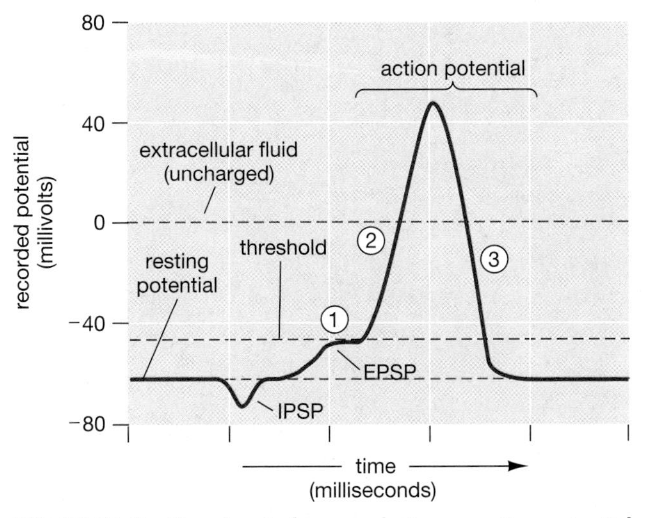

FIGURE 34-2 The electrical events during an action potential The resting potential of a nerve cell is about −60 millivolts with respect to the outside. (1) When a PSP brings the cell to threshold, membrane channels permeable to Na^+ open and Na^+ enters the cell, powered by diffusion and by electrical attraction; (2) the inside of the cell becomes positively charged. (3) Shortly thereafter, other membrane channels permeable to K^+ open, and K^+ leaves, driven by diffusion and electrical repulsion from the now-positive inside of the cell, until the resting potential is reestablished. **QUESTION** Imagine that the neuron cell body recorded in this figure simultaneously received three postsynaptic potentials: an IPSP that made the membrane potential 20 mV more negative, an EPSP that made the potential 5 mV less negative, and an EPSP that made the potential 15 mV less negative. Would an action potential be generated?

Neurons Communicate at Synapses

Once an action potential reaches the synaptic terminal of a neuron, the signal must be transmitted to another cell, such as another neuron, a muscle cell, or a gland. This transmission occurs at synapses, and for simplicity we will discuss only neuron-to-neuron synapses. The signals transmitted at synapses are called *postsynaptic potentials*.

At a synaptic terminal, an action potential encounters a synapse, where parts of two neurons are specialized to communicate with one another. A tiny gap separates the synaptic terminal of the first neuron, the **presynaptic neuron**, from the second neuron, or **postsynaptic neuron** (Fig. 34-3). Both the dendrites and the cell bodies of neurons are typically covered with synapses. The synapse includes the synaptic terminal of the presynaptic neuron, the gap, and the specialized membrane of the postsynaptic neuron, which contains receptors for neurotransmitters.

When an action potential reaches a synaptic terminal, the inside of the terminal becomes positively charged. This charge causes storage vesicles in the synaptic terminal to release neurotransmitter into the gap between the cells. The neurotransmitter molecules rapidly diffuse across the gap and briefly bind to receptors in the membrane of the postsynaptic neuron before diffusing away and being taken back into the presynaptic neuron (see Fig. 34-3).

Excitatory or Inhibitory Postsynaptic Potentials Are Produced at Synapses and Summate in the Cell Body

Receptor proteins in the postsynaptic membrane bind to a specific type of neurotransmitter. This binding causes specific types of ion channels in the postsynaptic membrane to open, allowing ions to flow across the plasma membrane along their concentration gradients. This flow of ions in the postsynaptic neuron causes a small, brief change in electrical charge called the **postsynaptic potential (PSP)**. The type of PSP depends on the type of channels opened and the movement of ions through them. *Excitatory postsynaptic potentials* (EPSPs) make the neuron less negative inside and more likely to produce an action potential. *Inhibitory postsynaptic potentials* (IPSPs) make the neuron more negative and less likely to produce an action potential (see Fig. 34-2). Some EPSPs open Na^+ channels, allowing positive Na^+ ions to flow into the neuron and bringing it closer to threshold. Some IPSPs open K^+ channels, allowing K^+ to flow out.

A CLOSER LOOK Ions and Electrical Signals

Media Activity
34.1 The Nervous System: Electrical Signals

HOW IS THE RESTING POTENTIAL GENERATED?

The resting potential is based on a balance between chemical and electrical gradients, maintained by active transport and a membrane that is selectively permeable to specific ions. The ions of the cytoplasm consist mainly of positively charged potassium ions (K^+) and large, negatively charged organic molecules such as proteins, which cannot leave the cell (Fig. E34-1). Outside the cell, the extracellular fluid contains positively charged sodium ions (Na^+) and negatively charged chloride ions (Cl^-). These concentration differences are maintained by a specialized membrane protein called a *sodium–potassium pump*, which simultaneously pumps K^+ into and Na^+ out of the cell.

In an unstimulated neuron, only K^+ can cross the plasma membrane by traveling through specific membrane proteins called *potassium channels* (shown in yellow). Although *sodium channels* (shown in purple) are also present, they remain closed in unstimulated neurons. Because the K^+ concentration is higher inside the cell than outside, that ion tends to diffuse out, leaving the negatively charged organic ions behind:

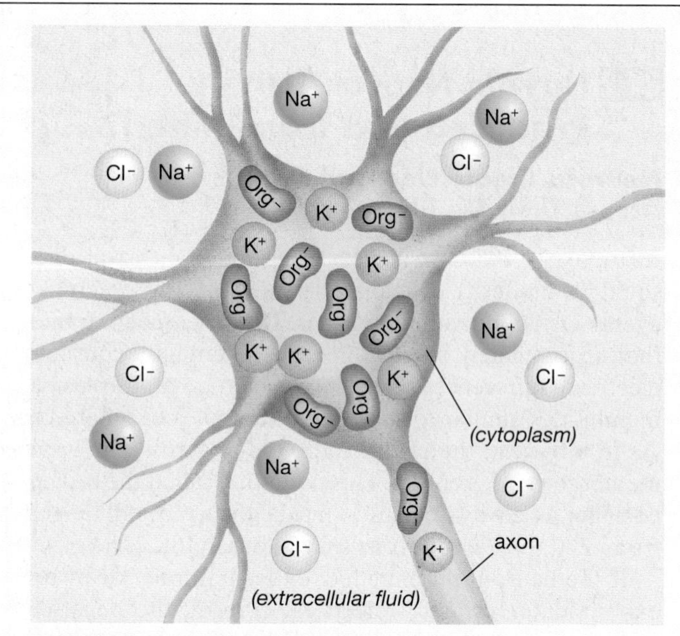

FIGURE E34-1 The neuron maintains ionic gradients
The ionic composition of a neuron's cytoplasm is significantly different from that of the extracellular fluid. The neuron maintains high concentrations of K^+ and large organic ions (Org^+); the extracellular fluid is high in Na^+ and Cl^-.

resting potential

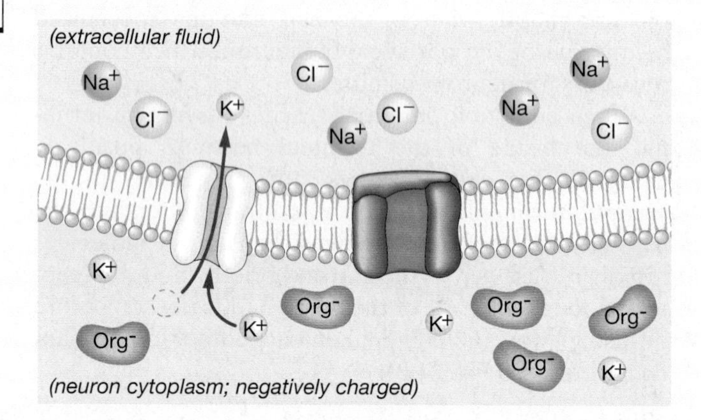

The outward diffusion of K^+ stops when the negative charge (which tends to pull the K^+ back inside the cell) becomes sufficiently large to counteract the diffusion gradient across the membrane. This negative charge at rest is the neuron's resting potential.

ACTION POTENTIALS CAN CARRY MESSAGES RAPIDLY OVER LONG DISTANCES

Neuron signals are carried long distances by action potentials. Action potentials occur if the PSPs from other neurons bring the potential inside the neuron (at a specific region where the axon joins the cell body) to threshold. At threshold, Na^+ channels (purple) are triggered to open, allowing a rapid influx of Na^+ (① in the figure to the right). Soon after the Na^+ channels open, they spontaneously close and a different type of K^+ channels (orange) are triggered to open by the positive charge inside the axon, allowing more K^+ to flow out of the cell and restore the negative resting potential (②):

action potential **resting potential restored**

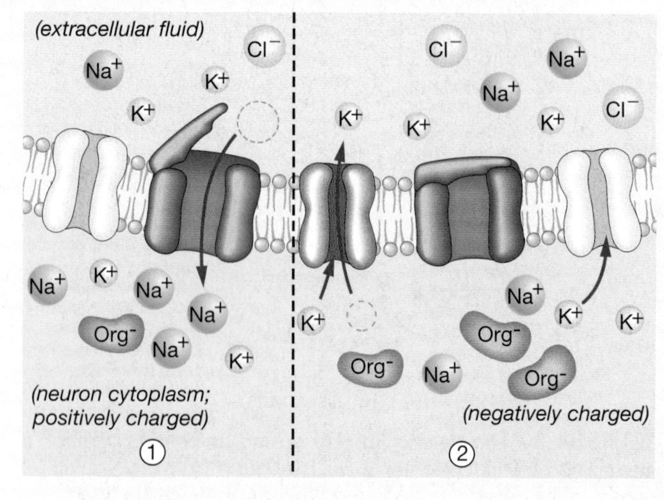

An action potential resembles a fast-moving wave of positive charge that travels, undiminished in size, along the axon to the synaptic terminal. The positive charge carried into the axon by Na^+ causes Na^+ channels farther along the axon to open. More

Na^+ can then flow in, more channels farther down the axon are opened, and the action potential continues along the axon:

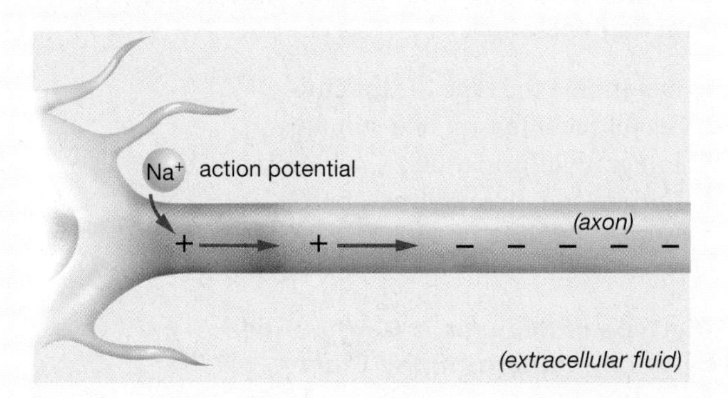

As the wave of positive charges passes a given point along the axon, the resting potential is restored as K^+ flows out:

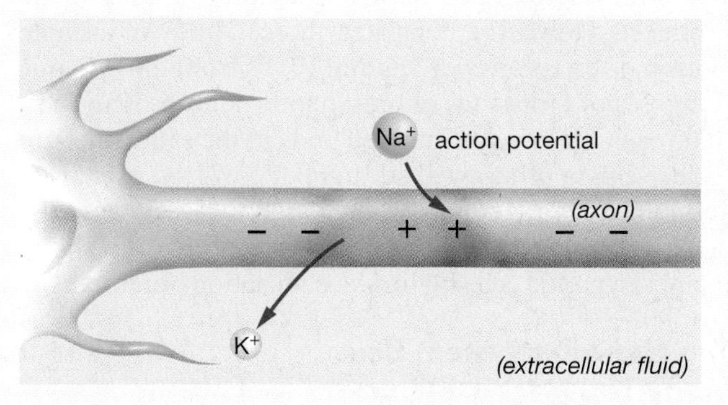

Action potentials are "all-or-none" phenomena. That is, if the neuron does not reach threshold, there will be no action potential, but if threshold is reached, a full-sized action potential will occur and travel the entire length of the axon. Only a tiny fraction of the total potassium and sodium in and around each neuron is exchanged during each action potential.

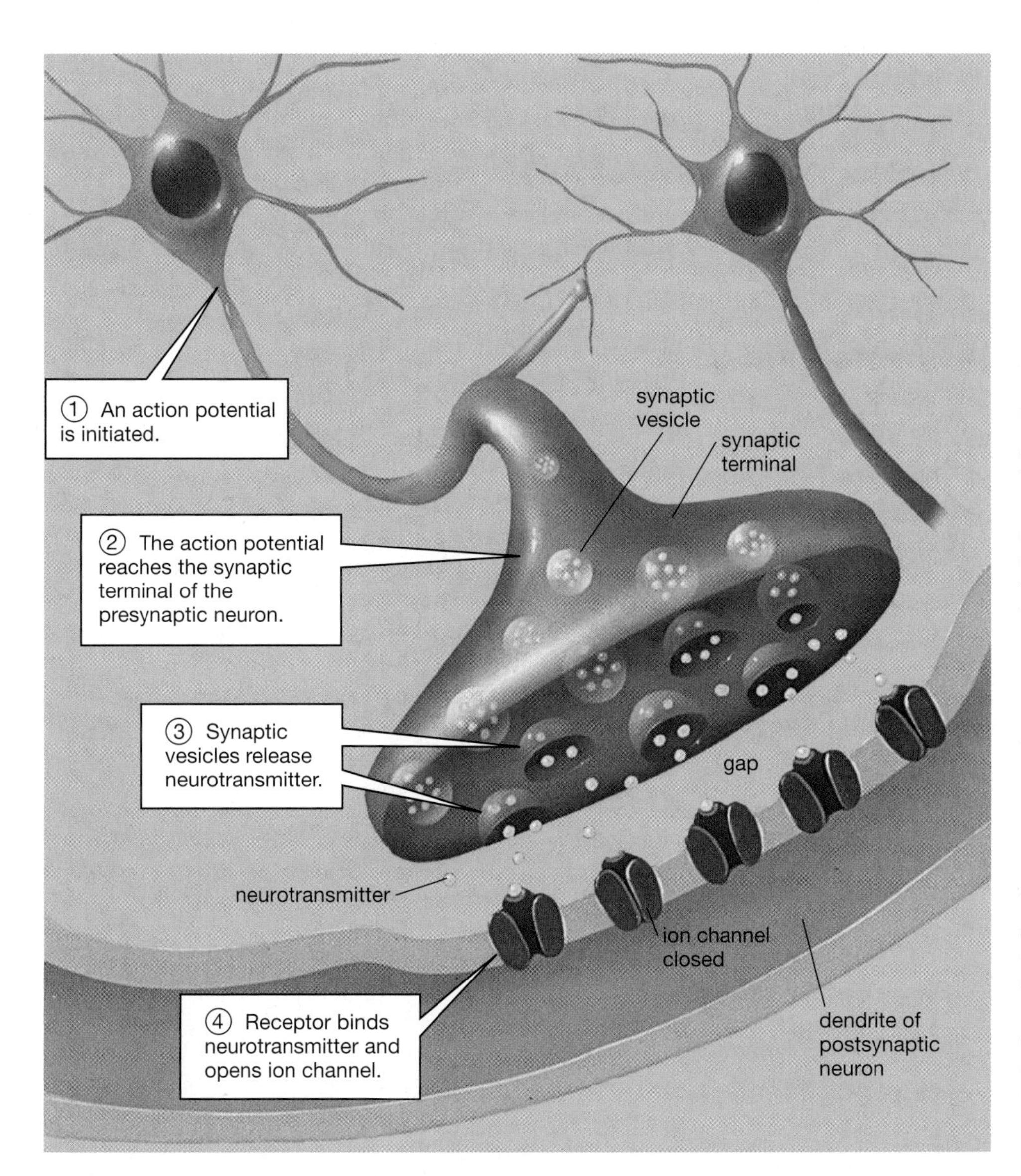

FIGURE 34-3 The structure and operation of the synapse
The synaptic terminal contains numerous neurotransmitter-filled vesicles. When an action potential enters the synaptic terminal, the vesicles release their neurotransmitter into the space between the neurons. The neurotransmitter diffuses rapidly across the gap, binds to postsynaptic receptors, and causes ion channels to open. Ions flow through these open channels, causing a postsynaptic potential in the postsynaptic cell. **QUESTION** Imagine an experiment in which the neurons pictured here are bathed in a solution containing a nerve poison. The presynaptic neuron is stimulated and produces an action potential, but this does not result in an action potential in the postsynaptic neuron. When the experimenter adds some neurotransmitter to the synapse, the postsynaptic neuron still produces no action potential. How does the poison act to disrupt nerve function?

Media Activity
34.2 The Nervous System: Synapse

Postsynaptic potentials are small, rapidly fading signals, but they travel far enough to reach the cell body. There, they determine whether an action potential will be produced. How? The dendrites and cell body of a single neuron often receive EPSPs and IPSPs from the synaptic terminals of thousands of presynaptic neurons. The PSPs that reach the postsynaptic cell body at the same time are "added up," a process called *summation*. If the excitatory and inhibitory potentials, when added together, raise the electrical potential inside the neuron above threshold, the postsynaptic cell will produce an action potential.

The Nervous System Uses Many Neurotransmitters

Over the past few decades, researchers have become increasingly aware that the brain is a teeming cauldron; its neurons synthesize and respond to a vast array of chemicals, including many of the hormones once thought unique to the endocrine system. For example, hormones that control digestive tract secretions are now known to be synthesized in the brain as well, where they influence appetite. Nitric oxide, a gas that lasts only a few seconds, was recently recognized as an important neurotransmitter. At least 50 neurotransmitters have been identified, and the list is growing. Table 34-1 lists a few well-known neurotransmitters and some of their functions; "Health Watch: Drugs, Diseases, and Neurotransmitters" further explores the role of neurotransmitters in addiction and in neurological diseases.

34.3 What Are Some General Features of Nervous Systems?

The individual neuron uses a language of action potentials, yet somehow this basic language allows even simple animals to perform a variety of complex behaviors. One key to the versatility of the nervous system is the presence of complex networks of neurons. These neural networks range from thousands to billions of cells. As in computers, small, simple elements can perform amazing feats when connected properly.

Information Processing Requires Four Basic Operations

Before we delve into the basic anatomy of nervous systems, we should first examine their operating principles. At a minimum, a nervous system must be able to perform four operations:

1. Determine the type of stimulus.
2. Signal the intensity of a stimulus.
3. Integrate information from many sources.
4. Initiate and direct the response.

Let's examine each of these operations.

The Type of Stimulus Is Distinguished by Wiring Patterns in the Brain

The nervous system must be able to identify the type of stimulus—for example, light, touch, or sound. All action potentials are similar; their properties tell us nothing about the kind of stimulus that elicited them. Instead, the nervous system monitors *which* neurons are producing action potentials. Thus, your brain interprets action potentials that occur in the axons of your optic nerves (originating in the eye and traveling to a specific area of the brain) as the sensation of light, action potentials in olfactory nerves (originating in receptors in the nose and traveling to a different region of the brain) as odors, and so on. For this reason, you have no trouble distinguishing the action potentials caused by music from those caused by the bitter taste of coffee.

Evidence of this wiring occurs when we activate specific sensory pathways in unnatural ways. The slight trauma of being poked in the eye may trigger action potentials in the optic nerve. Even though the stimulus is mechanical, your brain interprets all action potentials that occur in the optic nerve as light, so you "see stars." For the same reason, a blow to the head can make your ears "ring." A special subset of individuals have neural circuits that produce mixtures of sensations; for example, they may see vivid colors whenever they hear music. Learn more about this in "Links to Life: The Sensory Stew of Synesthesia."

Table 34-1 Some Important Neurotransmitters

Neurotransmitter	Location in Nervous System	Some Functions
Acetylcholine	Motor neuron-to-muscle synapse; autonomic nervous system, brain	Activates skeletal muscles; activates target organs of parasympathetic nervous system
Dopamine	Midbrain	Important in control of movement
Epinephrine (adrenaline)	Sympathetic nervous system	Activates target organs of sympathetic nervous system
Serotonin	Midbrain, pons, and medulla	Influences mood, sleep
Glutamate	Brain and spinal cord	Major excitatory neurotransmitter in CNS
Glycine	Spinal cord	Major inhibitory neurotransmitter in spinal cord
GABA (gamma amino butyric acid)	Throughout brain	Major inhibitory neurotransmitter in brain
Endorphins	Brain and spinal cord	Influence mood, reduce pain sensations
Nitric oxide	Brain	Important in forming memories

HEALTH WATCH Drugs, Diseases, and Neurotransmitters

Chances are, you know someone who is addicted. How can substances such as nicotine, alcohol, and cocaine exert such profound influence over people's lives? A big part of the answer lies in the effects of these drugs on neurotransmitters and in how the nervous system adapts to those insidious effects. Addictive drugs activate the "reward circuitry" of the brain, creating feelings of intense pleasure. Cocaine is a good example. Synapses in the brain that use the neurotransmitters *dopamine, serotonin,* or *norepinephrine* contribute to our energy level and our overall sense of well-being. Normally, the presynaptic neuron, after releasing one of these neurotransmitters, immediately starts pumping it back in, thus limiting its effects. Researchers have found that cocaine works by blocking this pump mechanism. The result? When a person takes cocaine, the neurotransmitters remain in their synapses much longer and reach higher levels than normal, so their effects are enhanced. The user feels euphoric and energetic until the brain attempts to restore the status quo; to reduce the impact of cocaine, the postsynaptic neuron decreases its number of receptors for these neurotransmitters. When fewer receptors are present, the high levels of neurotransmitter caused by cocaine are now *required* for the user to feel normal. When cocaine is withdrawn, the postsynaptic neurons are inadequately stimulated, and the user experiences an emotional "crash" that can be relieved only by more cocaine. Increasing amounts of the drug are required to produce the euphoric effects; the user has become an addict (Fig. E34-2).

Alcohol stimulates receptors for the neurotransmitter GABA (gamma amino butyric acid), enhancing inhibitory neuronal signals, and blocks receptors for glutamate, reducing excitatory signals. When a person drinks frequently, the brain compensates by decreasing GABA receptors and increasing glutamate receptors. Without alcohol, an alcoholic feels jittery and nervous—in short, overstimulated. In extreme cases, withdrawal from alcohol can cause convulsions. Nicotine and other components of cigarette smoke also interfere with normal synaptic transmission, producing a variety of addictive effects. To overcome addictions, drug users must undergo the misery caused by a nervous system that is deprived of a drug to which it has adapted. Although receptors eventually return to normal levels, for unknown reasons drug cravings often recur periodically.

You may also know someone with Parkinson's or Alzheimer's disease. Both are caused by the death of specific neurons in the brain and the loss of their neurotransmitters, which normally communicate with other neurons. In Parkinson's disease, dopamine-releasing neurons in the midbrain die, interfering with the complex control system that underlies smooth movements. Parkinson's patients experience tremors and have difficulty initiating movement. In Alzheimer's disease, neurons in the temporal lobes that produce the neurotransmitter *acetylcholine* die in large numbers. Memory loss is a prominent symptom of Alzheimer's.

The neurotransmitter serotonin acts in the brain and spinal cord. Too little serotonin can cause depression; the antidepressant Prozac® selectively blocks the re-uptake of serotonin into the presynaptic neuron, enhancing the neurotransmitter's effects. You may have heard of Ecstasy (MDMA), a relative of the stimulant drug amphetamine. This drug causes a temporary massive increase in serotonin in synapses. Users report feelings of pleasure, increased energy, heightened sensory awareness, and improved rapport with other people. Increasing evidence from both animal and human research suggests that Ecstasy users may incur long-term damage to serotonin-producing neurons, and may suffer from deficits in learning and memory. New studies with primates suggest that MDMA may also damage dopamine-producing neurons; if this is also true of humans, MDMA users may be particularly vulnerable to Parkinson's disease later in life.

The analgesic (pain-relieving) effects of plant-derived opiates, such as morphine, opium, codeine, and heroin, have been recognized for centuries. Since the brain has receptors that bind these molecules, researchers reasoned that perhaps these plant opiates resemble (then) unknown substances produced by the brain for which these receptors evolved. The search for such substances was rewarded in 1975 with the discovery of *opioids* (opiate-like substances); *endorphins* are a group of opioids. Certain opioids suppress pain in times of extreme stress, such as on a battlefield or a football field. Opioids released during strenuous exercise may account for the well-known "runner's high."

FIGURE E34-2 The neuron maintains ionic gradients
An addict experiences extreme physical and emotional distress without his drug because his nervous system has adapted to it.

The Intensity of a Stimulus Is Coded by the Frequency of Action Potentials

Because all action potentials are of roughly the same magnitude and duration, no information about the strength, or **intensity**, of a stimulus (for example, the loudness of a sound) can be encoded in a single action potential. Instead, intensity is coded in two other ways (Fig. 34-4). First, intensity can be signaled by the frequency of action potentials in a single neuron. The more intense the stimulus, the faster the neuron produces action potentials, or *fires*. Second, most nervous systems have many neurons that can respond to the same input. Stronger stimuli tend to excite more of these neurons, whereas weaker stimuli excite fewer. Thus, intensity can also be signaled by the number of similar neurons that fire at the same time. For example, the loud wail of a fire alarm activates many of your auditory neurons and causes them to fire action potentials very rapidly.

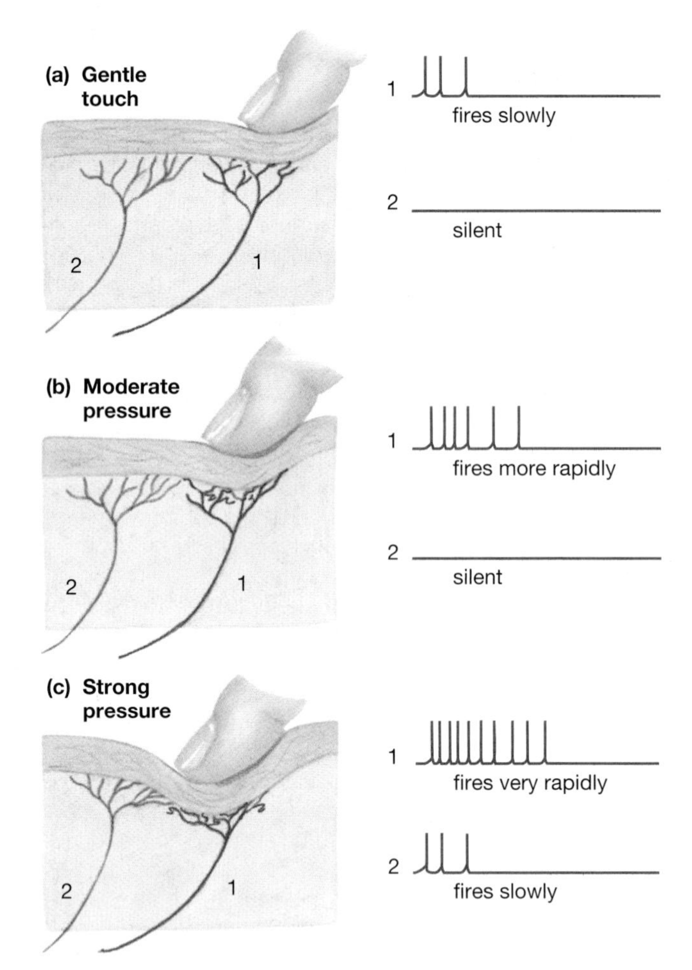

FIGURE 34-4 Signaling stimulus intensity
The intensity of a stimulus is signaled by the rate at which individual sensory neurons produce action potentials and by the number of sensory neurons that are activated. In this example, increasing pressure on the skin first causes faster firing, then causes an adjacent receptor to be activated. **QUESTION** In what way do skin areas that are especially sensitive to touch differ from less sensitive areas?

The Nervous System Processes Information from Many Sources Through Convergence

Your brain is continuously bombarded by sensory stimuli that originate both inside and outside the body. The brain must filter all these inputs, determine which ones are important, and decide how to respond. Nervous systems, like individual neurons, integrate information through **convergence**. In this process, many neurons funnel their signals to fewer neurons. For example, many sensory neurons may converge onto a smaller number of brain cells. Some of these brain cells act as "decision-making cells," adding up the postsynaptic potentials that result from the synaptic activity of the sensory neurons; depending on their relative strengths (and other internal factors such as hormones or metabolic activity), they produce appropriate outputs.

Divergence of Signals Allows Complex Responses

The output of the decision-making cells is responsible for initiating activity. The actions directed by the brain may involve many parts of the body. These actions require **divergence**, the flow of electrical signals from a relatively small number of decision-making cells onto many different neurons that control the activity of muscles or glands.

Neural Pathways Direct Behavior

Most behaviors are controlled by neuron-to-muscle pathways composed of four elements:

1. **Sensory neurons**, which respond to a stimulus, either internal or external to the body.
2. **Association neurons**, which receive signals from many sources, including sensory neurons, hormones, neurons that store memories, and many others. On the basis of this input, association neurons activate motor neurons.
3. **Motor neurons**, which receive instructions from association neurons and activate muscles or glands.
4. **Effectors**, usually muscles or glands that perform the response directed by the nervous system.

The Simplest Behavior Is the Reflex

The simplest type of behavior in animals is the **reflex**, a largely involuntary movement of a body part in response to a stimulus. Reflexes occur without involving conscious portions of the brain; many occur entirely within the spinal cord and peripheral neurons. Examples of human reflexes include the familiar knee-jerk and pain-withdrawal reflexes, both of which are produced by neurons in the spinal cord. The pain-withdrawal reflex, which moves a body part away from a painful stimulus such as a thumbtack, uses one of each of the three types of neurons and an effector (see Fig. 34-9). Although reflexes of this sort do not require the brain, other pathways inform the brain of pricked fingers and may trigger other, more complex behaviors (cursing, for example!).

Nearly all animals are capable of much more subtle and varied behavior than can be accounted for by simple reflexes. In principle, these more complex behaviors can be organized by *interconnected neural pathways,* in which several types of sensory input (along with memories, hormones, and other factors) converge on a set of association neurons. By integrating the postsynaptic potentials from several sources, the association neurons can "decide" what to do and can stimulate the motor neurons to direct the appropriate activity in muscles and glands.

Increasingly Complex Nervous Systems Are Increasingly Centralized

In the animal kingdom, there are really only two nervous system designs: a diffuse nervous system, such as that of cnidarians (*Hydra*, jellyfish, and their relatives; Fig. 34-5a), and a centralized nervous system, found to varying degrees in more complex organisms. Not surprisingly, nervous system design is highly correlated with the animal's lifestyle. Radially symmetrical cnidarians have no "front end," so there has been no evolutionary pressure to concentrate the senses in one place. A *Hydra* sits anchored to the seafloor, so prey or predators are equally likely to come from any direction. Cnidarian nervous systems are composed of a network of neurons, often called a **nerve net**, woven through the animal's tissues. Here and there we find a cluster of neurons, called a **ganglion** (plural, **ganglia**), but nothing resembling a real brain.

Almost all other animals are bilaterally symmetrical, with definite head and tail ends. Because the head is usually the first part of the body to encounter food, danger, and potential mates, it is advantageous to have sense organs concentrated there. Sizable ganglia evolved that integrate the information gathered by the senses and initiate appropriate action. Over evolutionary time, the major sense organs of animals with increasingly complex nervous systems became localized in the head and the ganglia became centralized into a brain. This trend, called **cephalization**, is clearly seen in different types of mollusks (Fig. 34-5b,c). Cephalization reaches its peak in the vertebrates, in which nearly all the cell bodies of the nervous system are localized in the brain and spinal cord.

34.4 How Is the Human Nervous System Organized?

The vertebrate nervous system, including that of humans, can be divided into two parts: central and peripheral. Each of these has further subdivisions (Fig. 34-6). The **central nervous system (CNS)** consists of a **brain** and a **spinal cord**, which extends down the dorsal part of the torso. The **peripheral nervous system (PNS)** consists of nerves that connect the central nervous system to the rest of the body.

The Peripheral Nervous System Links the Central Nervous System to the Body

The peripheral nervous system consists of **peripheral nerves**, which link the brain and spinal cord to the rest of the body, including the muscles, the sensory organs, and the organs of the digestive, respiratory, excretory, and circulatory systems. Within the peripheral nerves are axons of sensory neurons that bring sensory information *to* the central nervous system from all parts of the body. Peripheral nerves also contain the axons of motor neurons that carry signals *from* the central nervous system to the organs and muscles.

The motor portion of the peripheral nervous system can be subdivided into two parts: the **somatic nervous system** and the **autonomic nervous system**. Motor neurons

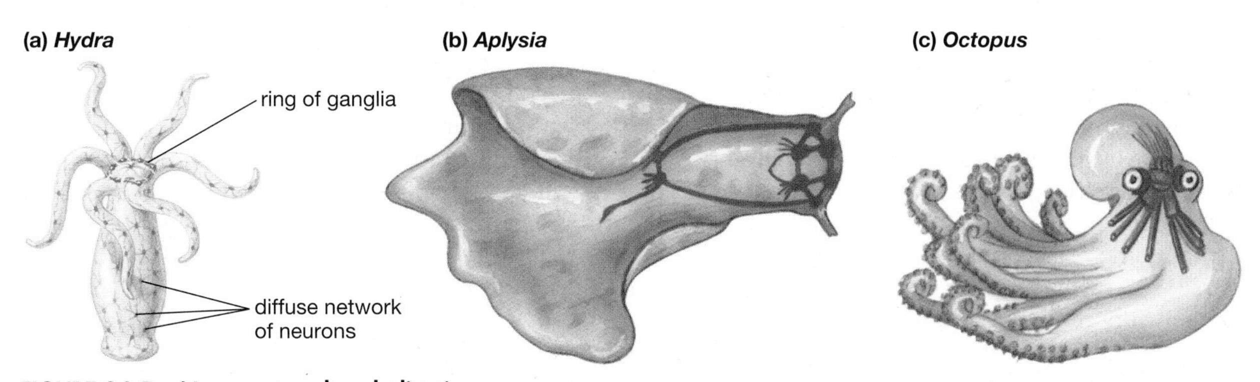

FIGURE 34-5 Nerve net and cephalization
(a) The diffuse nervous system of *Hydra* contains a few concentrations of neurons, particularly at the bases of tentacles, but no brain. Neural signals are conducted in virtually all directions throughout the body. **(b)** *Aplysia*, a shell-less marine snail, can crawl quite rapidly or even swim. Many of its neurons are aggregated into clusters of ganglia within the head. **(c)** Mollusk mobility and intelligence culminate in *Octopus*, with a large, complex brain and learning capabilities that rival those of some mammals.

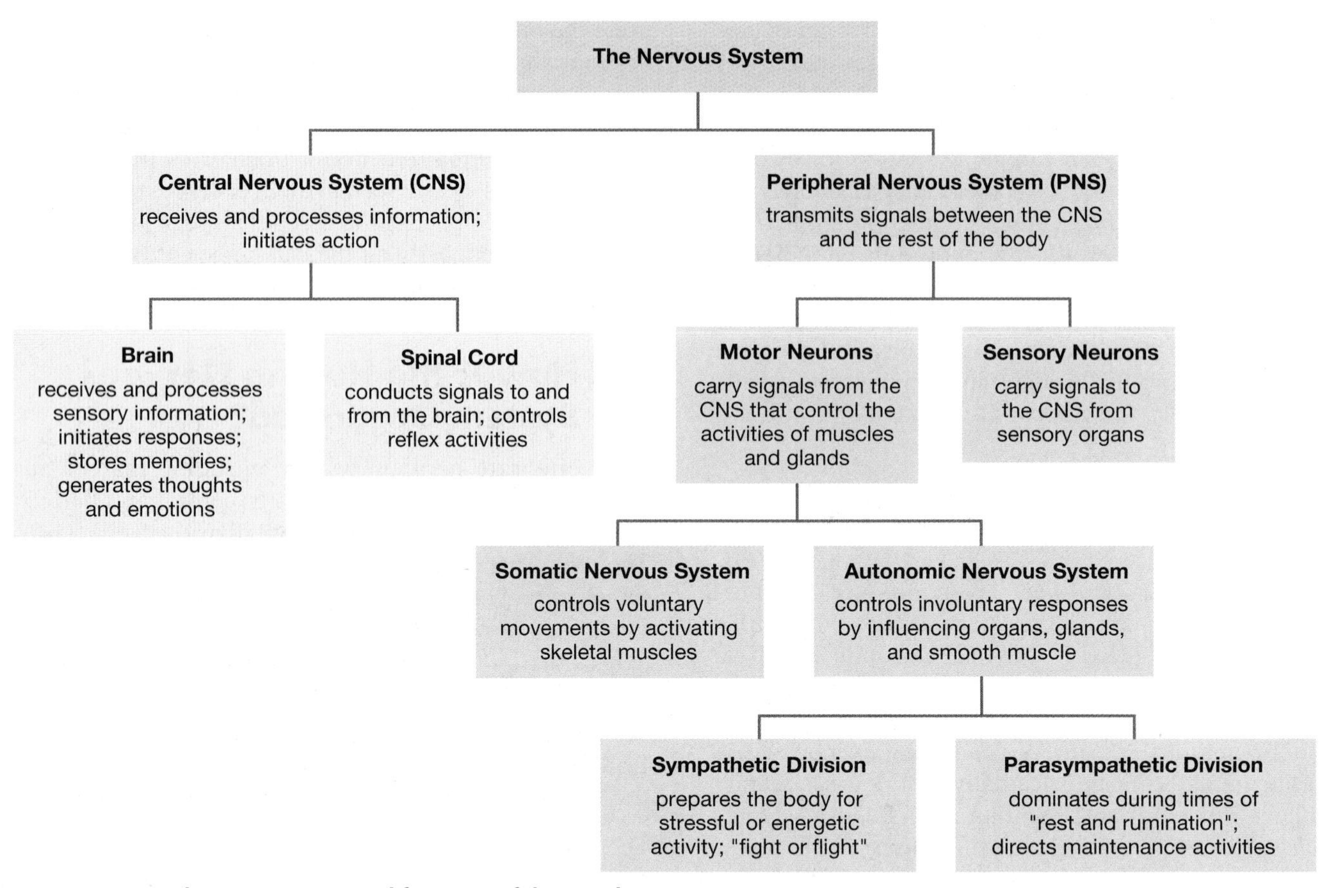

FIGURE 34-6 The organization and functions of the vertebrate nervous system

of the somatic nervous system form synapses on skeletal muscles and control voluntary movement. As you take notes, lift a coffee cup, or adjust your stereo, your somatic nervous system is in charge. The cell bodies of somatic motor neurons are located in the *gray matter* of the spinal cord (to be described shortly), and their axons go directly to the muscles they control. (Muscles and their control will be discussed in Chapter 35.)

Motor neurons of the autonomic nervous system control involuntary responses. They form synapses on the heart, smooth muscle, and glands. The autonomic nervous system is controlled primarily by the hypothalamus of the brain, described later in this chapter. It consists of two divisions: the **sympathetic division** and the **parasympathetic division** (Fig. 34-7). The two divisions of the autonomic nervous system generally make synaptic contacts with the same organs but usually produce opposite effects.

The sympathetic division releases the neurotransmitter norepinephrine (noradrenaline) onto its target organs, preparing the body for stressful or highly energetic activity, such as fighting, escaping, or taking an exam. During such "fight-or-flight" activities, the sympathetic nervous system curtails activity of the digestive tract, redirecting some of its blood supply to the muscles of the arms and legs. The heart rate accelerates. The pupils of the eyes open wider, admitting more light, and the air passages in the lungs expand, accommodating more air. These things may happen if you are suddenly called on in class to answer a question—especially if you don't know the answer!

The parasympathetic division, which releases acetylcholine onto its target organs, dominates during maintenance activities that can be carried on at leisure, often called "rest and rumination." Under its control, the digestive tract becomes active, the heart rate slows, and air passages in the lungs constrict. As you read this text, your parasympathetic nervous system is probably dominating your unconscious functions.

You may notice two differences in the organization of the sympathetic and parasympathetic divisions in Figure 34-7. First, although both divisions use a two-neuron pathway to carry messages to each target organ, in the sympathetic division, the synapse occurs in sympathetic ganglia near the spinal cord. In the parasympathetic division, the synapse occurs in smaller ganglia located on or very near each target organ. Second, their nerves originate at different levels of the central nervous system. Parasympathetic nerves emerge from the pons and medulla at the base of the brain, as well as from the lowest (sacral) region of the spinal cord. In contrast, all of the sympathetic nerves

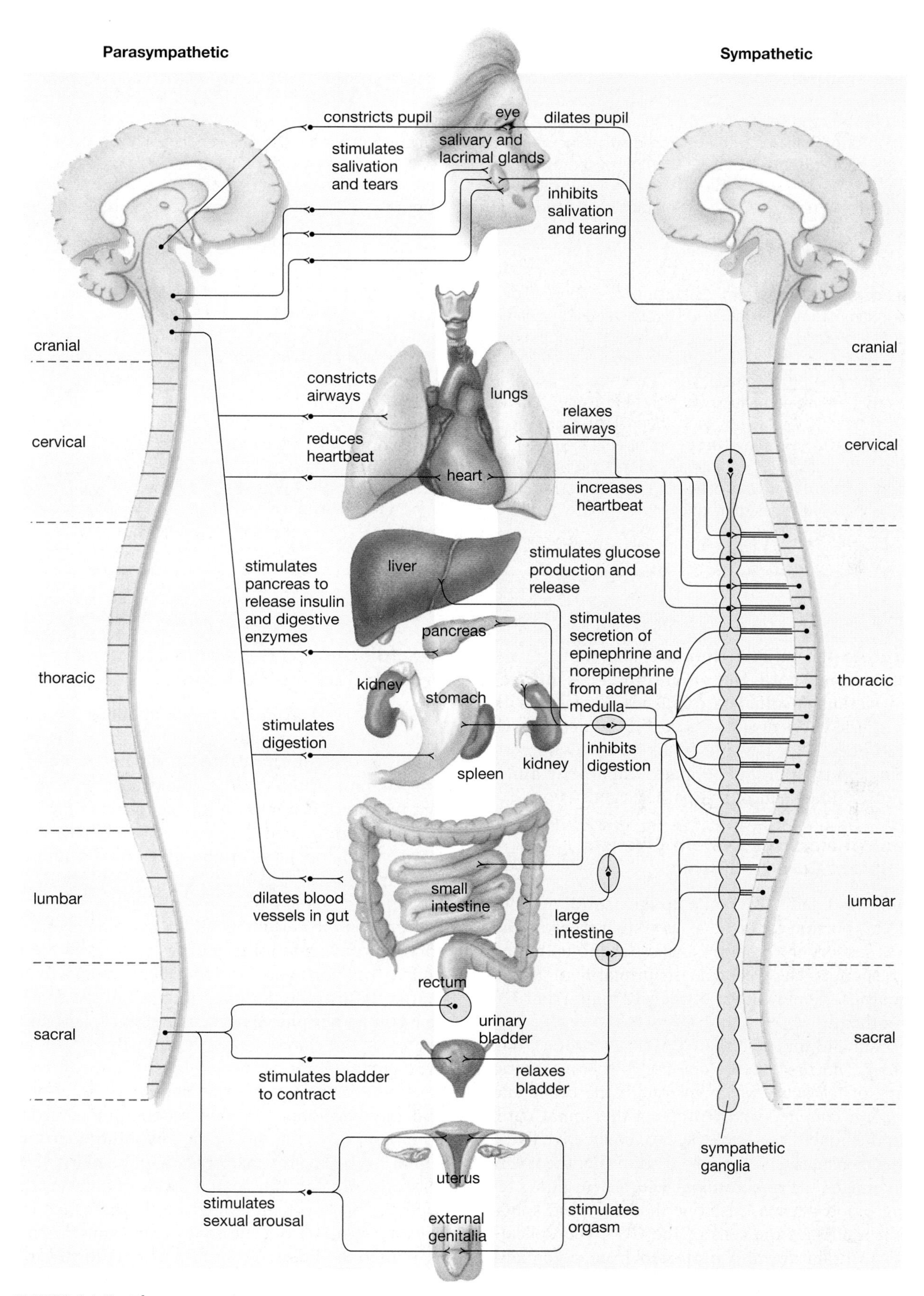

FIGURE 34-7 The autonomic nervous system
The autonomic nervous system has two divisions, sympathetic and parasympathetic, which supply nerves to many of the same organs but generally produce opposite effects. Activation of the autonomic nervous system is involuntarily commanded by signals from the hypothalamus.

HEALTH WATCH Healing the Spinal Cord

Why are spinal cord injuries so devastating? Unfortunately, injuries to the brain and spinal cord (the CNS) do not heal as easily as those in peripheral nerves. When the spinal cord is crushed, swelling and inflammation in the cord adds to the destruction of axons and cell bodies. Damaged cells release toxic substances, further enlarging the neuron death zone. As severed axons with living cell bodies attempt to regrow, their progress is thwarted by scar tissue that creates a physical barrier and secretes proteins that inhibit axon growth. To add insult to injury, the healthy insulating myelin cells in the CNS also produce proteins that deter axon regeneration, one of which is appropriately called "Nogo."

To counteract these setbacks, doctors now routinely administer anti-inflammatory medication to limit damage immediately after the trauma occurs. Researchers are developing antibodies to inactivate the Nogo protein and other molecules to prevent it from binding to receptors. To help regenerating axons find their way, scientists are experimenting with grafting tissue, such as myelin from the PNS, to promote and channel the regrowth of axons. Researchers have found that special olfactory receptor neurons, which regularly regenerate and form appropriate new connections in the brain, can direct this regrowth; rats with damaged spinal cords made dramatic recoveries after receiving transplants of these unique cells.

As you learned in Chapter 13, *stem cells* are cells that have not yet assumed their adult function, and so can be induced to differentiate into a variety of tissues when placed in the appropriate environment. In controlled studies, transplantation of stem cells into rodents with damaged spinal cords allowed the animals to regain some use of previously paralyzed limbs. Scientists had long believed that neurons in the mammalian brain were never replaced. In the late 1990s, however, researchers published striking new evidence that new neurons are produced in a specific area of the hippocampus at a rate of 500 to 1000 per day. In the future, doctors might be able to harvest a few of these adult neural stem cells from the brain of a patient with spinal cord damage, grow them in large numbers in culture, and transplant them into the same person at the site of spinal injury. In the appropriate environment, with inflammation controlled, Nogo inhibited, and perhaps guided by transplanted olfactory cells, these brain cells might differentiate into neurons that will bridge the gap in the damaged cord.

For more immediate relief, researchers are continuing to develop functional electrical stimulation (FES) computer programs that directly stimulate motor neurons to command muscles to perform useful functions; these programs are now available to assist paralyzed patients. There will be no quick or complete cures, but Christopher Reeve and other individuals with spinal cord injuries have reasons to be cautiously optimistic.

emerge from the middle (thoracic and lumbar) regions of the spinal cord. This organization has implications for victims of spinal cord injuries. A person with damage to the cervical spinal cord (in the neck), such as Christopher Reeve, would be likely to retain considerable ability to regulate parasympathetic functions, but would lose most sympathetic activation.

The Central Nervous System Consists of the Spinal Cord and Brain

The spinal cord and brain make up the central nervous system. This portion of the nervous system receives and processes sensory information, generates thoughts, and directs responses. The CNS consists primarily of association neurons—somewhere between 10 billion and 100 billion of them!

The brain and spinal cord are physically protected in three ways. The first line of defense is a bony armor, consisting of the *skull,* which surrounds the brain, and the *vertebral column,* which protects the spinal cord. Beneath the bones lies a triple layer of connective tissue called the **meninges** (see Fig. 34-11). Between the layers of the meninges, the **cerebrospinal fluid**, a clear liquid resembling blood plasma, cushions the brain and spinal cord as it nourishes the cells of the CNS. The delicate cells of the brain are also protected from potentially damaging chemicals that reach the bloodstream because walls of brain capillaries are far less permeable than capillaries in the rest of the body. This third line of defense is called the **blood–brain barrier**.

The Spinal Cord Is a Cable of Axons Protected by the Backbone

The spinal cord is a neural cable, about as thick as your little finger, that extends from the base of the brain to the lower back. It is protected by the bones of the vertebral column. Although the spinal cord is both strong and flexible, it is no match for the trauma of a headfirst fall off a horse or other types of violent trauma. (To learn what happens when the spinal cord is injured and what treatments are under study, see "Health Watch: Healing the Spinal Cord.")

Between the vertebrae, nerves carrying axons of sensory neurons and motor neurons arise from the dorsal and ventral portions of the spinal cord, respectively, and merge to form the peripheral nerves of the spinal cord (part of the peripheral nervous system; Fig. 34-8). In the center of the spinal cord is a butterfly-shaped area of **gray matter**. Gray matter consists of the cell bodies of several types of neurons that control voluntary muscles and the autonomic nervous system, plus neurons that communicate with the brain and other parts of the spinal cord. The gray matter is surrounded by **white matter**, containing myelin-coated axons of neurons that extend up or down the spinal cord. These axons carry sensory signals from internal organs, muscles, and the skin up to the brain. Axons also extend downward from the brain, carrying signals that direct the motor portions of the peripheral nervous system; the motor neurons of the spinal cord also control the muscles involved in conscious, voluntary activities—such as eating, writing, or

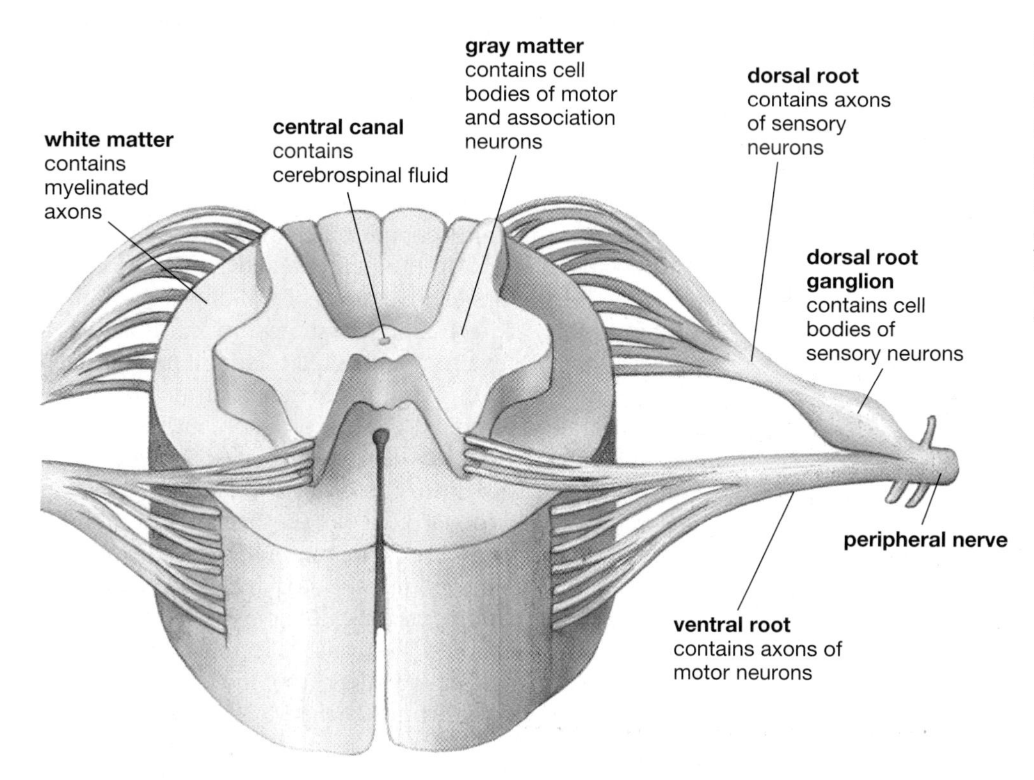

FIGURE 34-8 The spinal cord In cross section, the spinal cord has an outer region of myelinated axons (white matter) that travel to and from the brain and an inner, butterfly-shaped region of dendrites and the cell bodies of association and motor neurons (gray matter). The cell bodies of the sensory neurons are outside the cord in the dorsal root ganglion.

playing tennis—that are directly activated by motor portions of the brain. If the spinal cord is crushed or severed, preventing transmission of neural signals, the body below the injury feels numb, and motor signals from the brain cannot get through to command complex behaviors—even if the motor neurons and peripheral nerves and muscles remain intact.

In addition to relaying neural signals between the brain and the rest of the body, the spinal cord contains the neural pathways for certain simple behaviors, such as reflexes. Let's examine the simple pain-withdrawal reflex, which involves neurons of both the central and the peripheral nervous systems (Fig. 34-9). The cell bodies of the sensory neurons from the skin (in this case,

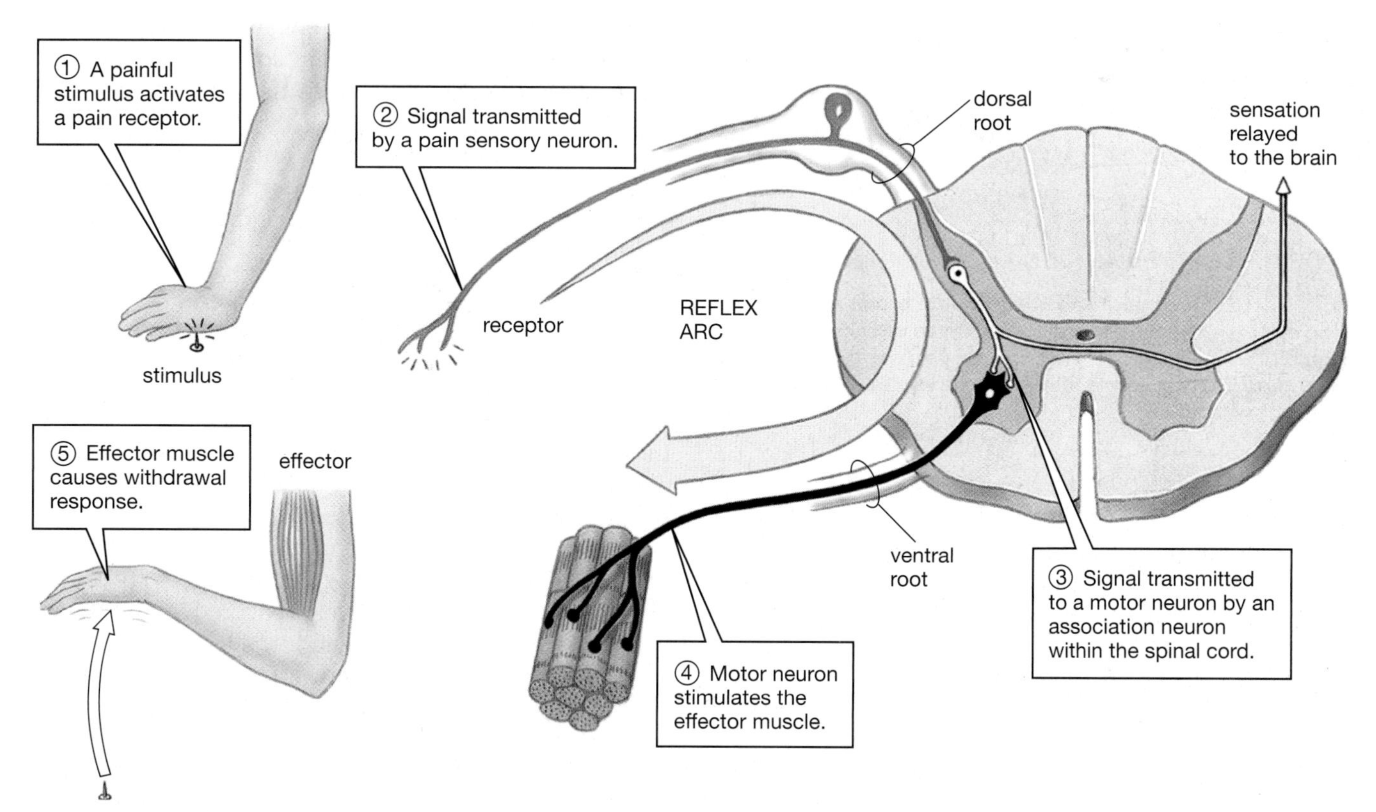

FIGURE 34-9 The pain-withdrawal reflex **QUESTION** Why does a paralyzed victim of a spinal cord injury, when pricked with a pin on a paralyzed part of his or her body, typically exhibit a normal pain-withdrawal reflex but feel no pain?

Media Activity 34.3 Reflex Arc

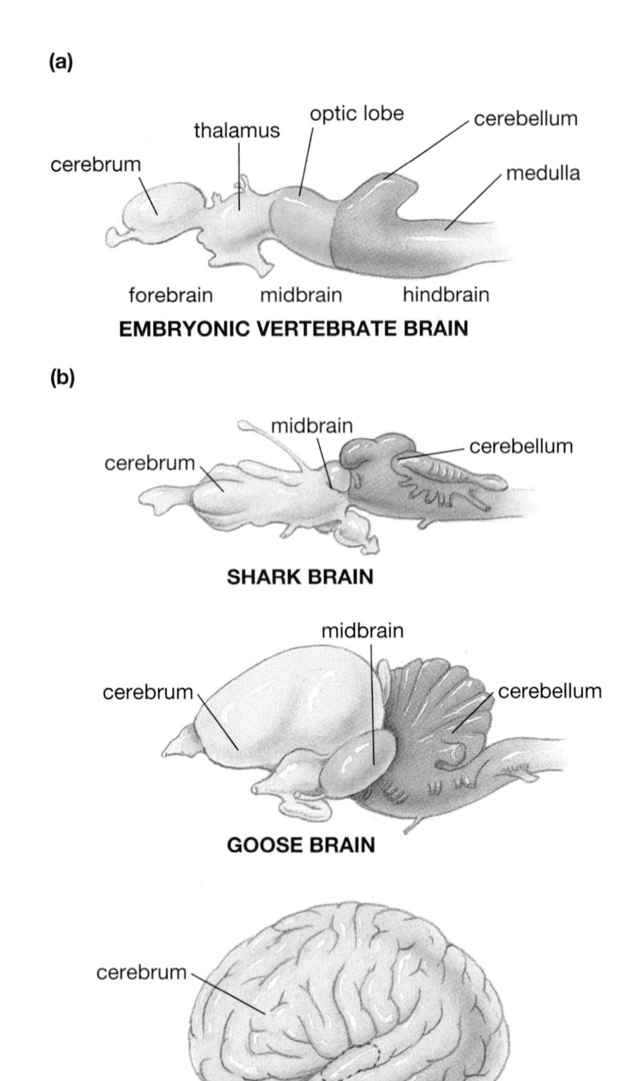

FIGURE 34-10 A comparison of vertebrate brains
(a) The embryonic vertebrate brain shows three distinct regions: the forebrain, midbrain, and hindbrain. **(b)** This basic structure persists in all adult vertebrate brains, but the relative size of the parts varies enormously.

signaling pain) are located in **dorsal root ganglia**, clusters of neurons on spinal nerves just outside the spinal cord. Both association neuron and motor neuron cell bodies are found in the gray matter in the center of the spinal cord. Association neurons not only form synapses on motor neurons, but also have axons that extend up to the brain. Signals carried along these axons alert the brain to the painful event. The brain, in turn, sends impulses down axons in the white matter to cells in the gray matter. These signals can modify spinal reflexes. With sufficient motivation, you can suppress the pain-withdrawal reflex; to rescue a child from a burning crib, for example, you could reach into the flames.

In addition to simple reflexes, the entire "software" for operating some fairly complex activities also resides within the spinal cord. All the neurons and interconnections needed for the basic movements of walking and running, for example, are contained in the spinal cord, allowing Christopher Reeve to "walk" on a treadmill in a harness. The advantage of this semi-independent arrangement between brain and spinal cord is probably an increase in speed and coordination, because messages do not have to travel all the way up to the brain and back down again merely to swing forward one leg (in the case of walking). The brain's role in these "semi-automatic" behaviors is to initiate, guide, and modify the activity of spinal motor neurons, based on conscious decisions (where are you going; how fast should you walk?). To maintain balance, the brain also uses sensory input from the muscles to command motor neurons to adjust the way the muscles move.

The Brain Consists of Many Parts Specialized for Specific Functions

All vertebrate brains have the same general structure, with major modifications corresponding to lifestyle and intelligence. Embryologically, the vertebrate brain begins as a simple tube that soon develops into three parts: the **hindbrain**, **midbrain**, and **forebrain** (Fig. 34-10a). Scientists believe that in the earliest vertebrates, these three anatomical divisions were also functional divisions: the hindbrain governed automatic behaviors such as breathing and heart rate, the midbrain controlled vision, and the forebrain dealt largely with the sense of smell. In nonmammalian vertebrates, the three divisions remain prominent. However, in adult mammals—particularly in humans—the brain regions are significantly modified. Some have been reduced in size, and others, especially the forebrain, greatly enlarged (Fig. 34-10b). A section through the midline of the human brain reveals many of its structural features, as shown in Figure 34-11.

The Hindbrain Includes the Medulla, Pons, and Cerebellum

In humans, the hindbrain is comprised of the *medulla*, the *pons*, and the *cerebellum* (see Fig. 34-11). In both structure and function, the **medulla** is very much like an enlarged extension of the spinal cord. Like the spinal cord, the medulla has neuron cell bodies at its center, surrounded by a layer of myelin-covered axons; it controls several automatic functions, such as breathing, heart rate, blood pressure, and swallowing. Certain neurons in the **pons**, located above the medulla, appear to influence transitions between sleep and wakefulness and between stages of sleep. Other neurons influence the rate and pattern of breathing. The **cerebellum** is crucial in coordinating movements of the body. It receives information from command centers in the conscious

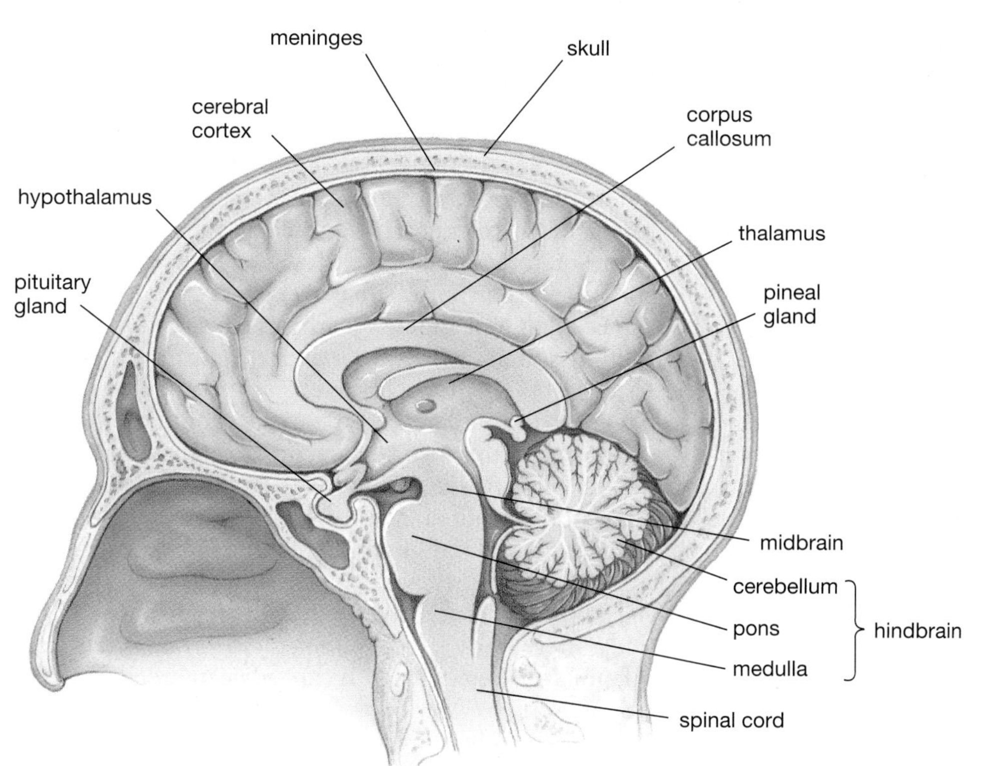

FIGURE 34-11 The human brain
A section through the midline of the human brain reveals some of its major structures.

Media Activity
34.4 Brain Structure

areas of the brain that control movement and also from position sensors in muscles and joints. By comparing information from these two sources, the cerebellum guides smooth, accurate motions and body position. The cerebellum is also involved in learning and memory storage for behaviors. As you take notes, your cerebellum instructs your brain about the order and timing of muscle movements in your hand. Not surprisingly, the cerebellum is largest in animals whose activities require fine coordination, such as in birds (see Fig. 34-10b), which engage in the complex activity of flight.

The Midbrain Contains the Reticular Formation

The midbrain is extremely reduced in humans (see Figs. 34-10b and 34-11). It contains an auditory relay center and a center that controls reflex movements of the eyes, as well as a portion of the **reticular formation**. The neurons of the reticular formation, an important relay center, extend all the way from the central core of the medulla into lower regions of the forebrain. It receives input from virtually every sense, from every part of the body, and from many areas of the brain as well. The reticular formation plays a role in sleep and wakefulness, emotion, muscle tone, and certain movements and reflexes. It filters sensory inputs before they reach the conscious regions of the brain, although the selectivity of the filtering seems to be set by higher brain centers, such as those that control conscious thought. Activities of the reticular formation allow you to read and concentrate in the presence of a variety of distracting stimuli, such as the music from your stereo and the smell of coffee. The fact that a mother wakens upon hearing the faint cry of her infant but sleeps through loud traffic noise outside her window testifies to the effectiveness of the reticular formation in screening inputs to the brain.

The Forebrain Includes the Thalamus, Limbic System, and Cerebral Cortex

The forebrain, also called the **cerebrum**, includes the *thalamus*, *limbic system*, and the *cerebral cortex*. In mammals, the cerebral cortex is much enlarged compared with that of fish, amphibians, and reptiles. This trend culminates in the human cerebral cortex (see Fig. 34-10b).

The **thalamus** is a complex relay station that channels sensory information from all parts of the body to the limbic system and cerebral cortex (Fig. 34-12). Signals

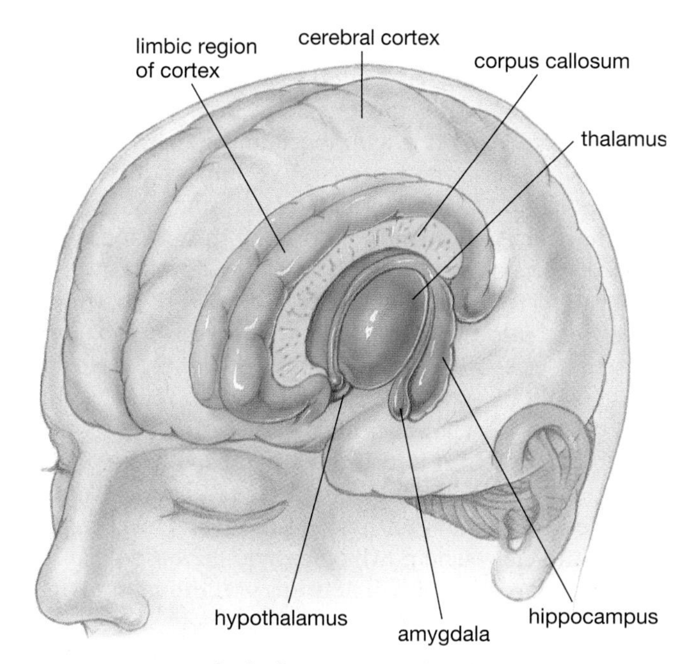

FIGURE 34-12 The limbic system and thalamus
The limbic system extends through several brain regions. The thalamus is a crucial relay center among the senses, the limbic system, and the cerebral cortex.

from the cerebellum and limbic system back to the cerebral cortex are also channeled through this busy thoroughfare. Anatomically, the **limbic system** is a diverse group of structures located in an arc between the thalamus and cerebral cortex (see Fig. 34-12). These structures work together to produce our most basic and primitive emotions, drives, and behaviors, including fear, rage, calm, hunger, thirst, pleasure, and sexual responses. Portions of the limbic system are also important in the formation of memories. The limbic system includes the *hypothalamus*, the *amygdala*, and the *hippocampus* as well as nearby regions of the cerebral cortex.

The **hypothalamus** (literally, "under the thalamus") contains many clusters of neurons. Some are neurosecretory cells that release hormones into the blood; others control the release of a variety of hormones from the pituitary gland (see Chapter 33). Other regions of the hypothalamus direct the activities of the autonomic nervous system. The hypothalamus, through its hormone production and neural connections, acts as a major coordinating center, maintaining homeostasis in a variety of ways: controlling body temperature, food intake, water balance, the menstrual cycle, and the sleep–wake cycle.

Clusters of neurons in the **amygdala** produce sensations of pleasure, fear, or sexual arousal when stimulated. Conscious humans whose amygdalas are electrically stimulated have reported feelings of rage or fear. Recent studies have revealed that damage to the amygdala early in life eliminates the ability both to feel fear and to recognize fearful facial expressions in other people.

The shape of the **hippocampus** as it curves around the thalamus inspired its name, which is derived from the Greek word meaning "seahorse." As in the amygdala and hypothalamus, behaviors that reflect a variety of emotions, including rage and sexual arousal, can be elicited by stimulating portions of the hippocampus. The hippocampus also plays an important role in the formation of long-term memory and is thus required for learning, discussed in more detail later in this chapter.

In humans, the largest part of the brain by far is the **cerebral cortex**, the outer layer of the forebrain. The cerebral cortex and underlying parts of the forebrain are divided into two halves, called **cerebral hemispheres**. These halves communicate with each other by means of a large band of axons, the **corpus callosum**. The cerebral cortex is the most sophisticated information-processing center known, but it is also the area of the brain about which scientists know the least. Tens of billions of neurons are packed into this thin surface layer. The cortex is folded into **convolutions**, which greatly increase its area. In the cortex, cell bodies of neurons predominate, giving this outer layer—in preserved specimens—a gray color; hence the term "gray matter." These neurons receive sensory information, process it, create memories for future use, direct voluntary movements, and allow us to plan and think in ways we cannot yet understand.

The cerebral cortex is divided into four anatomical regions: the *frontal*, *parietal*, *occipital*, and *temporal lobes* (Fig. 34-13). Functionally, the cortex contains *primary sensory areas,* regions where signals originating in sensory organs such as the eyes and ears are received and converted into subjective impressions—for instance, light and sound. Nearby *association areas* interpret the sounds as speech or music, for example, and the visual stimuli as recognizable objects or words on this page. Association areas also link the stimuli with memories stored in the cortex and generate commands to produce speech. Research has revealed that the association areas of the brain do not always have the same function in the right and left hemispheres (see Fig. 34-13).

Primary sensory areas in the parietal lobe interpret sensations of touch that originate in all parts of the body; these body parts are "mapped" in an orderly sequence (see Fig. 34-13). In an adjacent region of the frontal lobe, *primary motor areas* command movements in corresponding areas of the body by stimulating motor neurons that form synapses with muscles, allowing you to walk to class, throw a Frisbee®, or type a term paper. Like the primary sensory area, the primary motor area also has adjacent association areas, including the motor association area (the premotor area), which seems to be responsible for directing the motor area to produce more-complex movements. Behind the bones of the forehead lies another association area of the frontal lobe, which is important in complex reasoning functions such as decision making, predicting the consequences of actions, controlling aggression, and planning for the future.

Damage to the cortex from trauma, stroke, or a tumor results in specific deficits, such as problems with speech, difficulty reading, or the inability to sense or move specific parts of the body. Most brain cells of adults cannot be replaced, so if a brain region is destroyed, these deficits may be permanent. Fortunately, however, training can sometimes allow undamaged regions of the cortex to take control over and restore some of the lost functions.

34.5 How Does the Brain Produce the Mind?

Historically, people have always had difficulty reconciling the physical presence of a few pounds of soft material in the skull with the range of thoughts, emotions, and memories of the human mind. This "mind–brain problem" has occupied generations of philosophers and, more recently, neurobiologists. Beginning with observations of individuals with head injuries and progressing to sophisticated surgical, physiological, and biochemical experiments, the outlines of how the brain creates the mind are beginning to emerge. Here, we will be able to touch on only a few of the most fascinating features.

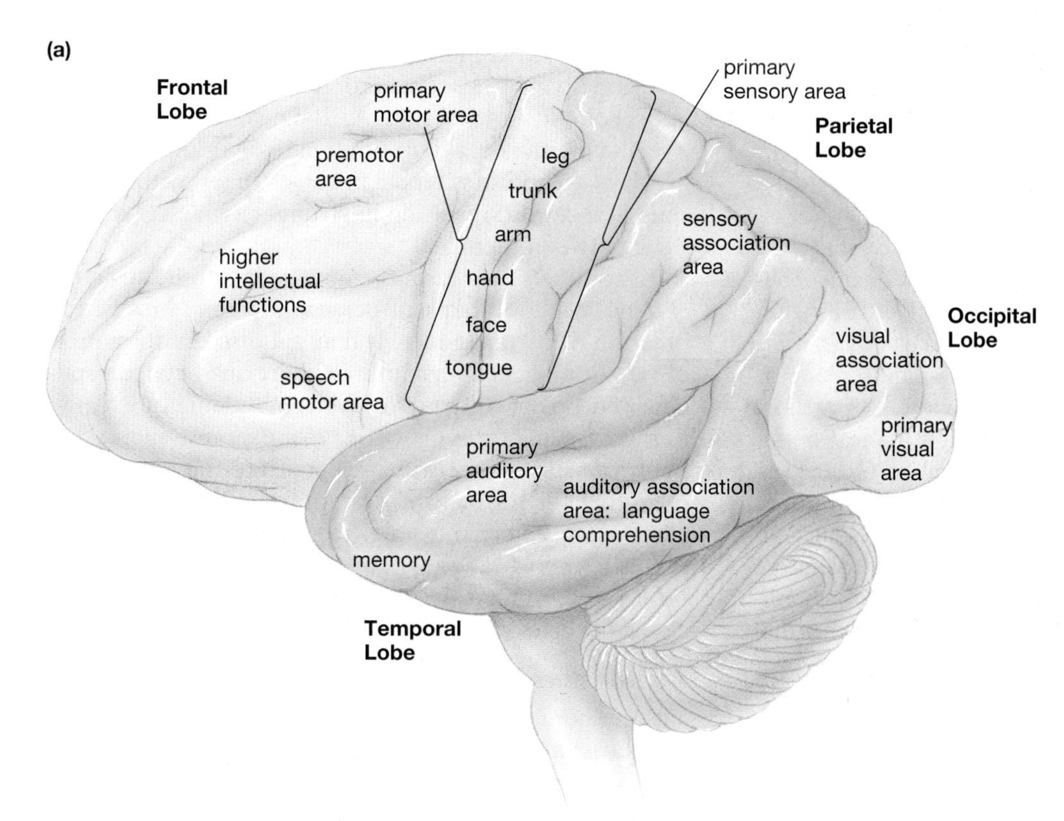

(b)

Left hemisphere

1. Controls right side of body
2. Input from right visual field, right ear, left nostril
3. Centers for language, mathematics

Right hemisphere

1. Controls left side of body
2. Input from left visual field, left ear, right nostril
3. Centers for spatial perception, music, creativity

FIGURE 34-13 The cerebral cortex
(a) A map of the human left cerebral cortex. A map of the right cerebral cortex would be similar, except that speech and language would not be as well developed. **(b)** The general distribution of abilities between the two hemispheres in most right-handed people.

The "Left Brain" and "Right Brain" Are Specialized for Different Functions

Although the cerebrum consists of two extremely similar-looking hemispheres, since the early 1900s it has been known that this symmetry does not extend to brain function. Much of what is known of the differences in hemisphere function comes from studies of accident or stroke victims with localized damage to one hemisphere, patients who have one hemisphere temporarily anesthetized, or those in whom the corpus callosum (which connects the two hemispheres) has been severed. This surgical procedure is still performed in certain cases of uncontrollable epilepsy, to prevent the spread of seizures from one hemisphere to the other. Because sensory input reaches both hemispheres, and each side guides appropriate responses based on its abilities, a person whose corpus callosum has been severed can function quite normally. As you read about hemispheric differences, remember that the brain performs as a coordinated unit. A person with an intact corpus callosum cannot use either side selectively; both hemispheres contribute in complex ways to our perceptions, thoughts, behavior, and abilities.

Beginning in the 1950s, Roger Sperry, a neurobiologist working at the California Institute of Technology, studied people whose hemispheres had been surgically separated by cutting the corpus callosum. In his studies, Sperry made use of the knowledge that axons within each optic tract (which are not severed by the surgery) follow a pathway that causes the left half of each visual field to be projected onto the right cerebral hemisphere and the right half to be projected onto the left hemisphere. Through an ingenious device that projected different images onto the left and right visual fields (thus sending different signals to each hemisphere), Sperry and other investigators gained more insight into the roles of the two hemispheres.

When Sperry projected an image of a nude figure onto just the left visual field, the subjects would blush and smile but would claim to have seen nothing, because the image had reached only the nonverbal right side of the brain. The same figure projected onto the right visual field was readily described verbally. These and later experiments have revealed that—in right-handed people—the left hemisphere is almost always dominant in speech, reading, writing, language comprehension, mathematical ability, and logical problem solving. The right side of the brain is superior to the left in musical skills, artistic ability, recognition of faces, spatial visualization, and the ability to recognize and express emotions. For his pioneering work, Sperry shared the Nobel Prize for Physiology in 1981.

Recent experiments, however, indicate that the left–right dichotomy is not as rigid as was once believed. An individual who has suffered a stroke that disrupted the blood supply to specific parts of the left hemisphere may lose the ability to speak, read, or write. Fortunately, these deficits can be at least partially overcome through training, even though the left hemisphere itself has not recovered. This suggests that the right hemisphere has some latent language capabilities. Interestingly, females have a slightly larger corpus callosum than males, suggesting a gender difference in the extent of interconnections between the two hemispheres. Imaging neural activity in the brains of normal subjects as they perform various mental tasks has provided further evidence of this difference. When subjects were asked to compare word lists for rhyming words, a specific region of the left cortex of male subjects became active, but in females, similar areas in *both* the left and right hemispheres were activated. (Further brain-imaging studies are described in "Scientific Inquiry: Peering into the 'Black Box.'").

Never let anyone tell you that you only use a small fraction of your brain! Although PET and fMRI images sometimes make it appear that only a small area of the brain is active, this is because activity of other neurons is subtracted out during the imaging process to show where neural activity has changed as a result of stimulation. At rest, your brain is responsible for an amazing 20% to 25% of your total energy consumption.

The Mechanisms of Learning and Memory Are Poorly Understood

Although hypotheses abound about the cellular mechanisms of learning and memory, we are a long way from understanding these phenomena. However, we do know a fair amount about short-term "working" memory, long-term memory, and some of the brain sites involved in learning, memory storage, and recall in mammals—particularly humans.

Memory May Be Brief or Long Lasting

Experiments show that learning occurs in two phases: initial **working memory** followed by **long-term memory**. For example, if you look up a number in the phone book, you will probably remember the number only long enough to dial it—this is working memory. But if you call the number frequently, eventually you will remember the number more or less permanently—an example of long-term memory.

Some working memory seems to be electrical in nature, involving the repeated activity of a particular neural circuit in the brain. As long as the circuit is active, the memory stays. In other cases, working memory involves temporary biochemical changes within neurons of a circuit, resulting in stronger synaptic connections between them.

In contrast, long-term memory seems to be structural—the result, perhaps, of persistent changes in the expression of certain genes. It may require the formation of new, long-lasting synaptic connections between specific neurons or the long-term strengthening of existing but weak synaptic connections (for example, increasing neurotransmitter release or increasing the number of receptors for the neurotransmitter). Working memory can be converted into long-term memory; this process seems to involve the hippocampus, which is believed to process new memories and then transfer them to the cerebral cortex for permanent storage.

The Temporal Lobes Are Important for Memory

The mechanisms of learning, memory storage, and memory retrieval are subjects of extensive research. Intense electrical activity occurs in the hippocampus—deep within the brain's temporal lobe—during learning, providing strong evidence that the two are closely linked. Even more striking are the results of hippocampal damage. An individual with two nonfunctioning hippocampi (one in each hemisphere) retains much of his or memory but is unable to learn new information after the loss. A patient whose hippocampi and associated brain structures were surgically removed in 1953 (in an attempt to control seizures) remains unable to recall his address or find his way home after many years at the same residence. He can entertain himself indefinitely by reading the same magazine over and over, and people whom he sees daily require reintroduction at each encounter. This example and others suggest that the hippocampus is responsible for transferring information from working memory to long-term memory.

The temporal lobes of the cerebral hemispheres seem to be important in the *retrieval,* or recall, of long-term memories. In a famous series of experiments in the 1940s, neurosurgeon Wilder Penfield electrically stimulated the temporal lobes of conscious patients undergoing brain surgery. The patients did not merely recall memories but felt that they were experiencing the past events right there in the operating room!

Insights on How the Brain Creates the Mind Come from Diverse Sources

Until about 100 years ago, the mind was a more appropriate subject for philosophers than for scientists, because tools for studying the brain did not yet exist. Through the first half of the twentieth century, the mind was treated by psychologists as a "black box" whose internal workings could be deduced only through the in-

SCIENTIFIC INQUIRY Peering into the "Black Box"

For most of human history, the brain has been a "black box"; its inputs and outputs were observable, but its internal workings were inherently unknowable. Now, however, new imaging techniques provide exciting insights into brain function. These include *PET* (positron emission tomography, described in Chapter 2) and *fMRI* (functional magnetic resonance imaging), which allow researchers to observe the brain in action.

Regions of the brain that are most active have higher energy demands; they utilize more glucose and attract a greater flow of oxygenated blood than do less active areas. In PET scans, scientists inject the subject with a radioactive substance, such as a radioactive form of glucose, and then monitor levels of radioactivity that reflect differences in metabolic rate. These are translated by computer into colors on cross-sectional images of the brain. By monitoring radioactivity while a specific task is performed, scientists can identify which parts of the brain are most active during that task. In contrast, fMRI detects differences in the way oxygenated and deoxygenated blood respond to a powerful magnetic field applied by an enormous electromagnet that surrounds the body. Active brain regions can be distinguished with fMRI without the use of radioactivity and over much shorter time spans than required by PET.

Using fMRI or PET, researchers can observe changes as the brain performs a specific reasoning task or responds to an odor or a visual or auditory stimulus. Through brain scans, scientists have confirmed that different aspects of the processing of language occur in distinct areas of the cerebral cortex (Fig. E34-3). Using fMRI, researchers analyzed the frontal lobe areas used in generating words in individuals who spoke two languages. In subjects who had grown up speaking two languages, the same region of the frontal lobe was used in speaking each language. Subjects who had learned a second language later in life used different but adjacent frontal lobe areas for the two languages. PET or fMRI scans can also precisely pinpoint damaged portions of the brain, such as the result of a stroke (Fig. E34-4a). We can also observe the contrast between disturbed brain functioning, as in Alzheimer's disease, and normal brain functioning (Fig. E34-4b).

FIGURE E34-3 Localization of language tasks
Changes in glucose utilization, measured by PET, reveal different cortical regions involved in different language-related tasks, on the basis of research by Dr. Marcus Raichle of the Washington University School of Medicine in St. Louis. The scale ranges from white (lowest) to red (highest).

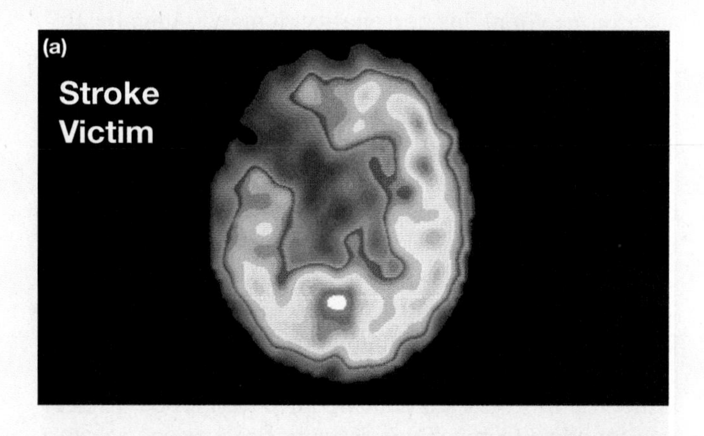

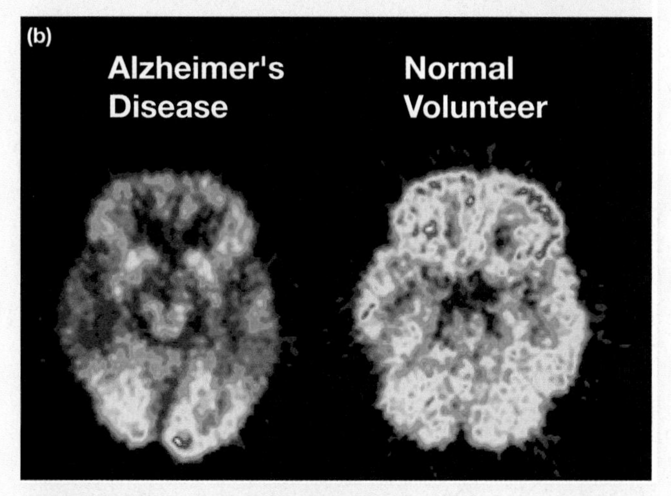

FIGURE E34-4 PET scans reveal malfunctioning brains
Metabolic activity decreases from red areas to yellow to green and to blue; black areas exhibit little or no metabolic activity. **(a)** As a result of stroke, part of the brain dies from lack of blood flow. The damaged region can be precisely located by the lack of neural activity in the upper-left region of this brain scan. **(b)** The brain of an individual with Alzheimer's disease is compared with that of a healthy elderly person. The top of each image corresponds to the front of the brain, just behind the forehead.

vestigation of how past and present experiences were interpreted and influenced behavior. New discoveries, however, are rapidly changing our views of the workings of the brain.

During recent decades, we have begun to understand the neural bases of at least some psychological phenomena. Many forms of mental illness, such as schizophrenia, manic depression, and autism, once thought to be due to childhood trauma or inept parenting, are now recognized as the result of neurotransmitter imbalances and/or structural abnormalities in the brain. Studies have also revealed a strong heritability factor (and hence, a biological basis) for traits that were once considered entirely learned or acquired, such as alcoholism or a tendency toward shyness.

A striking illustration of how the physical structure of the brain is related to personality was unwittingly provided by Phineas Gage in 1848. A railroad construction foreman, Gage was setting an explosive charge when it triggered prematurely. The blast blew a 13-pound steel rod that was more than a yard long through his skull, damaging both of his frontal lobes (Fig. 34-14). Although Gage survived for many years after his accident, his personality changed radically. Before the accident, Gage was conscientious, industrious, and well-liked. After his recovery, he became impetuous, profane, and incapable of working toward a goal. Subsequent research has implicated the frontal lobe in emotional expression, control of aggression, recognition of appropriate behavior, and the ability to work for delayed rewards.

Other sites of damage have revealed additional specializations. One patient with very localized damage to the left frontal lobe was unable to name fruits and vegetables (although he could name everything else). Other victims of brain damage have developed a selective inability to recognize faces or, in a recent case, to recognize any object that is *not* a face, suggesting that the brain has a region specialized to recognize faces that is separate from the region that allows it to recognize objects in general. Researchers have recorded individual neurons in the temporal lobes of monkeys that are selective for particular orientations of faces, such as a profile or a head-on view. The association area of the posterior parietal lobe seems to integrate information from several different senses, enabling an individual to recognize his or her own body and its relationship to the environment. People with damage to one of the parietal lobes sometimes lose track of one side of their body, and of objects on that side as well. For example, they may ignore food on one side of their plates or become lost because they don't recognize corridors on one side of the building. One patient did not recognize his own leg, and tried repeatedly to throw it out of his hospital bed. Fortunately, with time, other areas of the brain usually compensate for this damage, restoring more normal perception.

In the past, much of our understanding of the human mind–brain connection came from the study of victims of brain damage such as that caused by a stroke, trauma, tumor, or surgery. Typically, the exact extent of the damage remained unknown until revealed by autopsy. New techniques, such as the PET and fMRI scans, now permit insight into the functioning of normal, as well as diseased, brains (see "Scientific Inquiry: Peering into the 'Black Box'"). These and increasingly sophisticated techniques of the future will create ever-larger windows into the human brain and a clearer understanding of how the brain generates the human mind.

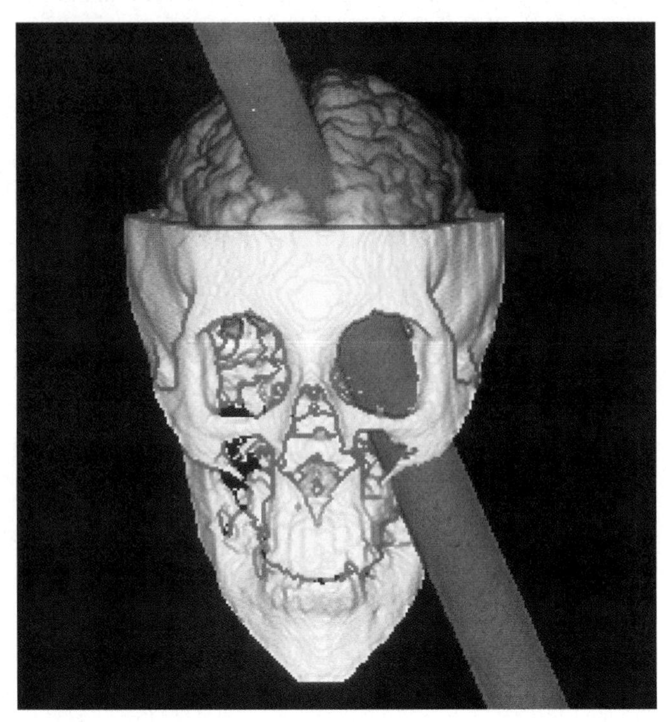

FIGURE 34-14 A revealing accident
Studies of the skull of Phineas Gage have enabled scientists to create this computer-generated reconstruction of the path taken by the steel rod that was blown through his head by an explosion.

34.6 How Do Sensory Receptors Work?

The word *receptor* is used in several contexts in biology. In the most general sense, a **receptor** is a structure that changes when it is acted on by a stimulus from its surroundings, causing a signal to be produced. A receptor may be a membrane protein that changes configuration when it binds a specific hormone or neurotransmitter, as discussed in previous chapters. Alternatively, a **sensory receptor** may be an entire specialized cell (typically, a neuron) that produces an electrical response to particular stimuli—that is, it translates sensory stimuli into the language of the nervous system. All sensory receptors produce electrical signals, but each receptor type is specialized to produce its signal only in response to a particular type of environmental stimulus. Some receptors, called *free nerve endings*, consist of branching dendrites of sensory neurons; other receptors have specialized structures that help them respond to a specific stimulus. Many sensory receptors are clustered into sensory organs, such as the eye, ear, skin, or tongue. Their electrical activity, after being processed by the brain, gives rise to the subjective perceptions of light, sound, touch, and taste that we describe as our "senses."

The stimulation of a sensory receptor causes an electrical signal called a **receptor potential**. Receptor potential amplitude varies with the intensity of the stimulus: the stronger the stimulus, the larger the receptor potential. Sensory receptors of different types influence postsynaptic neurons differently. In some sensory receptor neurons, a receptor potential will bring the cell above threshold and cause action potentials. In some very small sensory receptors, receptor potentials directly cause neurotransmitters to be released onto postsynaptic neurons, which in turn produce action potentials that travel to the central nervous system. A large positive receptor potential will cause a higher frequency of action potentials; as we learned earlier, that is how the brain interprets intensity. Sensory receptor cells are named after the stimulus to which they respond, as summarized in Table 34-2.

Sensory receptors are our links to the world around us. As Aristotle observed in the fourth century B.C., "Nothing is understood by the intellect which is not first perceived by the senses." In the following sections, we will focus on some senses that determine the way we, as humans, perceive the world.

34.7 How Is Sound Sensed?

Sound is produced by any vibrating object—a drum, a motor, vocal cords, or the speakers of a stereo. These vibrations are transmitted through air and intercepted by our ears, which are elaborate structures that detect the direction, pitch, and intensity of sound.

The Ear Captures, Transmits, and Converts Sound into Electrical Signals

The ear of humans and most other vertebrates consists of three parts: the outer, middle, and inner ear (Fig. 34-15a). The **outer ear** consists of the **external ear** and the **auditory canal**. The external ear, with its fleshy folds, modifies sound waves in ways that the brain uses to determine the location of the sound source. The air-filled auditory canal conducts the sound waves to the **middle ear**, consisting of the **tympanic membrane**, or *eardrum*;

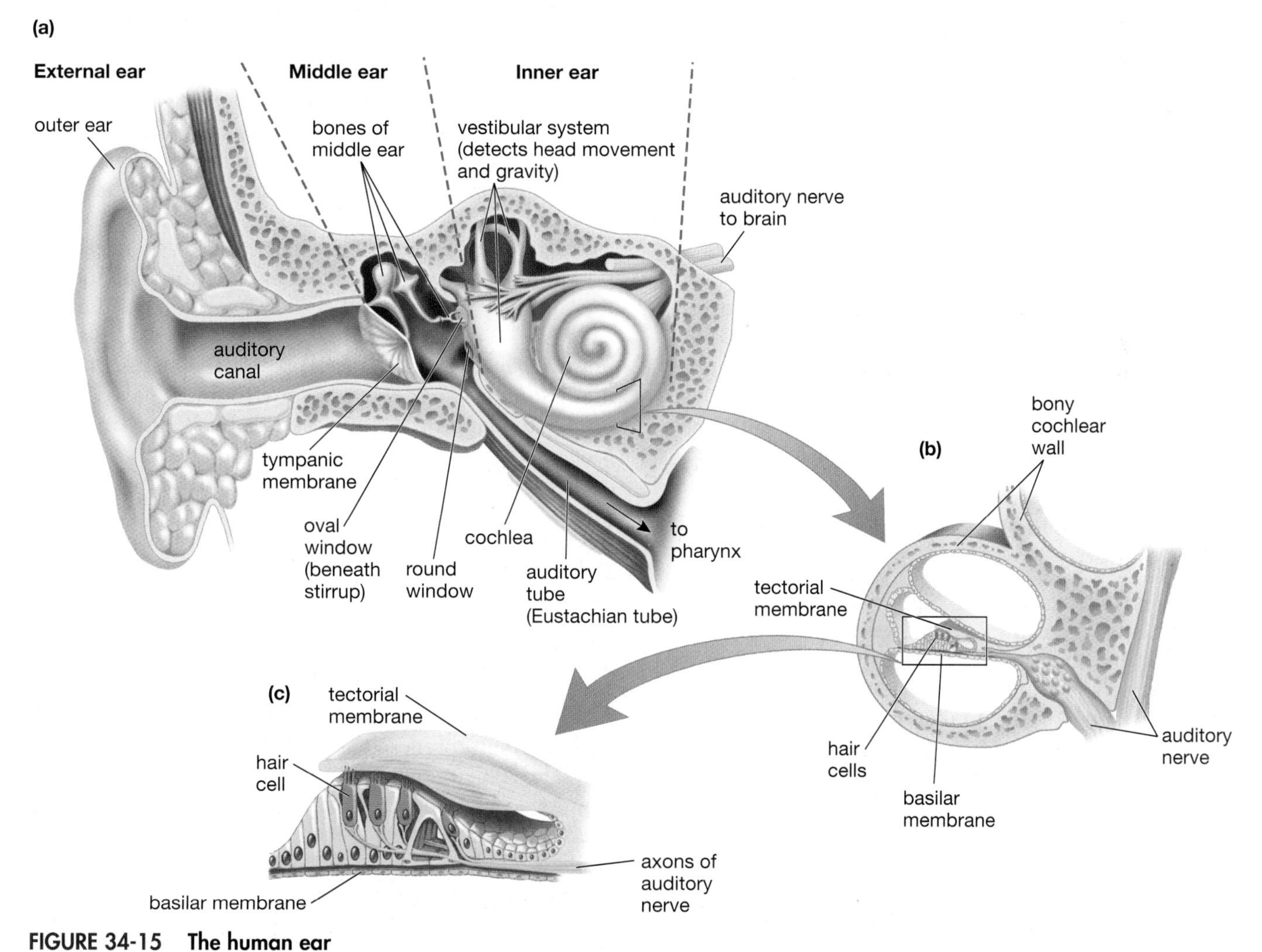

FIGURE 34-15 The human ear
(a) Overall anatomy of the ear. **(b)** In cross-section, the cochlea appears as three fluid-filled compartments; hair cells sit atop the basilar membrane in a central compartment. **(c)** The hairs of hair cells span the gap between the membranes in the central canal. Sound vibrations move the membranes relative to one another, bending the hairs and producing a receptor potential in the hair cells.

Table 34-2 Some Vertebrate Receptor Types

Type of Receptor	Sensory Cell Type	Stimulus	Location
Thermoreceptor	Free nerve ending	Heat, cold	Skin
Mechanoreceptor	Hair cell	Vibration, motion, gravity	Inner ear
	Specialized nerve endings and free nerve endings in skin (Pacinian corpuscle, Merkel's disc)	Vibration, pressure, touch	Skin
	Specialized nerve endings in muscles or joints (muscle spindle, Golgi tendon organ)	Stretch	Muscles, tendons
Photoreceptor	Rod, cone	Light	Retina of eye
Chemoreceptor	Olfactory receptor	Odor (airborne molecules)	Nasal cavity
	Taste receptor	Taste (waterborne molecules)	Tongue
Pain receptor	Free nerve ending	Chemicals released by tissue injury	Widespread in body

three tiny bones called the *hammer*, *anvil*, and *stirrup*; and the **auditory tube** (also called the *Eustachian tube*). The auditory tube connects the middle ear to the pharynx and equalizes the air pressure between the middle ear and the atmosphere. This tube may become swollen shut if you have a cold; if this happens, air pressure changes (such as those experienced during aircraft takeoff and landing) can be painful.

Within the middle ear, sound vibrates the tympanic membrane, which in turn vibrates the hammer, the anvil, and the stirrup. These bones transmit vibrations to the **inner ear**. The fluid-filled hollow bones of the inner ear form the spiral-shaped **cochlea** ("snail" in Latin) as well as structures of the *vestibular system* that detect head movement and the pull of gravity. The stirrup bone transmits vibrations to the fluid within the cochlea by vibrating a membrane in the cochlea called the *oval window*. The *round window* is a second membrane below the oval window; it allows fluid within the cochlea to shift back and forth as the stirrup bone vibrates the oval window.

Sound Is Converted into Electrical Signals in the Cochlea

The cochlea, in cross section, consists of three fluid-filled compartments. The central compartment houses the receptors and the supporting structures that activate them in response to sound vibrations. The floor of the central chamber consists of the **basilar membrane**, on top of which sit mechanoreceptors called **hair cells**. Hair cells have small cell bodies topped by hairlike projections that resemble stiff cilia. Some of these hairs are embedded in a gelatinous structure called the **tectorial membrane**, which protrudes into the central canal (Fig. 34-15b).

How do these structures allow the perception of sound? The oval window passes vibrations from the small bones of the middle ear to the fluid in the cochlea, which in turn vibrates the basilar membrane, causing it to move up and down. This movement bends the hairs of the hair cells, opening channels in the hair cell membrane that allow the flow of ions, producing receptor potentials. The receptor potentials cause the hair cells to release neurotransmitter onto neurons whose axons form the auditory nerve (see Fig. 34-15b). This causes action potentials within the auditory nerve, which are transmitted to auditory processing centers within the brain.

The cochlea also allows us to perceive *loudness* (the magnitude of sound vibrations) and *pitch* (the frequency of sound vibrations). Weak sounds cause small vibrations, which bend the hairs only slightly and result in a low frequency of action potentials in axons of the auditory nerve. Loud sounds cause large vibrations, which cause greater bending of the hairs and a larger receptor potential, producing high-frequency action potentials in the auditory nerve. Loud sounds can damage the hair cells (Fig. 34-16a), resulting in hearing loss, a fate suffered by many rock musicians and their fans. In fact, many sounds in our everyday environment have the potential to damage hearing, especially if they are prolonged (Fig. 34-16b).

The structure of the basilar membrane allows the perception of pitch. Humans can detect vibration frequencies from about 30 vibrations per second (very low pitched) to about 20,000 vibrations per second (very high pitched). The basilar membrane is stiff and narrow at the end near the oval window but more flexible and wider near the tip of the cochlea. This progressive change in structure causes each successive portion of the membrane to resonate or vibrate in synchrony with a particular frequency of sound. The brain interprets signals from receptors near the oval window as high-pitched sound, whereas signals from receptors located progressively closer to the tip of the cochlea are interpreted as increasingly lower in pitch.

34.8 How Is Light Sensed?

Animal vision varies in its ability to provide sharp, accurate representations of the real world; in fact, several types of eyes have evolved independently. All forms of vision, however, use *photoreceptors*. These sensory cells

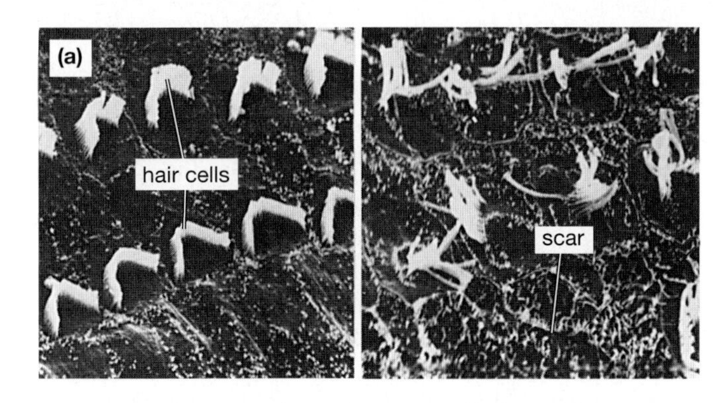

(b)

Ⓧ hours before hearing can be damaged
loudness range
jet takeoff (at 200 ft) (1/4)
rock concert
subway, stereo headphones (high volume) ②
motorcycle, lawn mower ⑧
urban street
normal talking
quiet background
0 20 40 60 80 100 120
decibels

FIGURE 34-16 Loud sounds can damage hair cells
(a) Scanning electron micrographs show the effect of intense sound on the hair cells of the inner ear. The hairs of hair cells in a normal guinea pig; hairs emerge from each receptor in a V-shaped pattern (left). After 24-hour exposure to a sound level approached by loud rock music (2000 vibrations per second at 120 decibels), many of the hairs are damaged or missing, leaving "scars" (right). Hair cells in humans do not regenerate, so such hearing loss is permanent. [SEMs by Robert S. Preston, courtesy of Professor J. E. Hawkins, Kresge Hearing Research Institute, University of Michigan Medical School] **(b)** Sound levels of everyday noises and their potential to damage hearing. Sound intensity is measured in *decibels* on a logarithmic scale; a 10-decibel sound is 10 times as loud as a 1-decibel sound, and a 20-decibel sound is 100 times as loud. You feel pain at sound intensities above 120 decibels. [Source: Deafness Research Foundation, National Institute on Deafness and Other Communication Disorders]

contain receptor molecules called **photopigments** (because they are colored), which absorb light and chemically change in the process. This chemical change alters ion channels in the receptor cell membrane, producing a receptor potential.

(a) Compound eyes

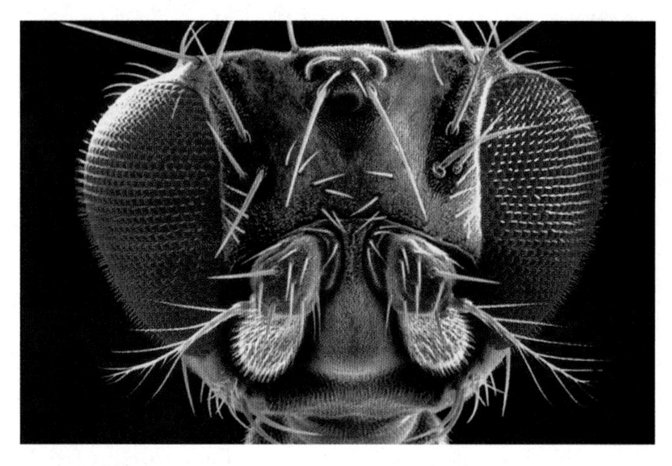

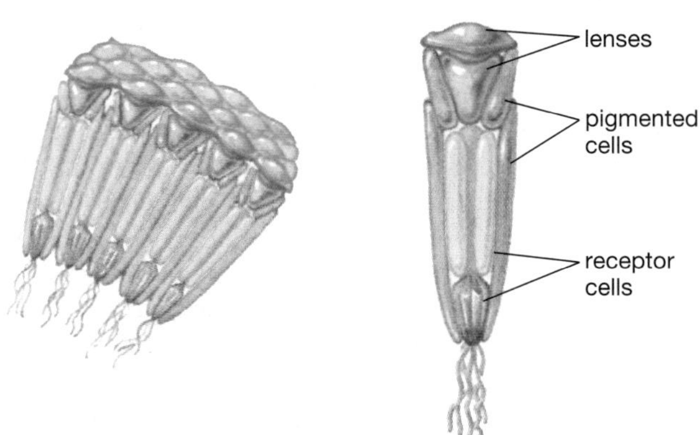

FIGURE 34-17 Compound eyes
(a) Scanning electron micrograph of the head of a fruit fly, showing a compound eye on each side of the head. **(b)** Each eye is made up of numerous light-receptive ommatidia. Within each ommatidium are several receptor cells, capped by a lens. Pigmented cells surrounding each ommatidium prevent the passage of light to adjacent receptors.

The Compound Eyes of Arthropods Produce a Mosaic Image

The arthropods (insects, spiders, and crustaceans) evolved **compound eyes**, which consist of a mosaic of many individual light-sensitive subunits called **ommatidia** (singular, **ommatidium**; Fig. 34-17). Each ommatidium functions as an on–off, bright–dim detector. Using a large number of ommatidia (up to 36,000 per eye in a dragonfly), most arthropods probably see a reasonably accurate—though grainy—image of the world. Compound eyes are excellent at detecting movement, an advantage in avoiding predators and in hunting. Many arthropods, such as bees and butterflies, also have good color perception.

The Mammalian Eye Collects, Focuses, and Transduces Light Waves

As you read this, light reflected from the page first encounters the **cornea**, a transparent covering over the front of the eyeball that collects light waves and begins to focus them. Behind the cornea, light passes through a chamber filled with a watery fluid called **aqueous humor**, which provides nourishment for both the lens and cornea. The amount of light entering the eye is adjusted by the **iris**, pigmented muscular tissue. The iris regulates the size of the **pupil**, a circular opening in the center of the iris. Light passing through the pupil encounters the **lens**, a structure that resembles a flattened sphere composed of transparent protein fibers. The lens is suspended behind the pupil by muscles that regulate its shape and allow fine focusing of the image. Behind the lens is a much larger chamber filled with the **vitreous humor**, a clear jellylike substance that allows light to pass freely while supporting and maintaining the shape of the eyeball (Fig. 34-18a).

After passing through the vitreous humor, light reaches the **retina**, a multilayered sheet of photoreceptors and neurons. There, the light energy is converted into electrical nerve impulses that are transmitted to the brain. Behind the retina is the **choroid**, a darkly pigmented tissue. The choroid's rich blood supply helps nourish the cells of the retina. Its dark pigment absorbs stray light whose reflection inside the eyeball would interfere with clear vision. Surrounding the outer portion of the eyeball is the **sclera**, a tough connective tissue layer that is visible as the white of the eye and is continuous with the cornea.

Driving along a country road at night, have you ever been startled by apparently disembodied glowing eyes? In vertebrates that are most active at dusk (such as deer), the choroid may be modified to reflect light rather than to absorb it. By reflecting light that escaped the photoreceptors during its initial passage, the choroid gives the receptors a second chance to detect it, maximizing the animal's ability to see in dim light. Reflective choroids give the eyes of these animals an eerie red or blue glow when bright light (such as that from car headlights) is reflected back through the wide-open pupil. Imagine how blindingly bright your headlights must be to these creatures!

The Adjustable Lens Allows Focusing of Both Distant and Nearby Objects

The visual image is focused most sharply on a small area of the retina called the **fovea**. Although focusing begins at the cornea, whose rounded contour bends light rays, the lens is responsible for final, sharp focusing. The shape of the lens is adjusted by its encircling muscle. When viewed from the side, the lens is either rounded, to focus on nearby objects, or flattened, to focus on distant objects (Fig. 34-19a).

If your eyeball is even a millimeter too long, you will be **nearsighted**—unable to focus on distant objects. **Farsighted** people, whose eyeballs are slightly too short, cannot focus on nearby objects. These conditions can be corrected by external lenses of the appropriate shape, either contact lenses or glasses (Fig. 34-19b,c). Both nearsightedness and farsightedness can also be correct-

(a) Anatomy of the human eye

(b) Layers of the retina

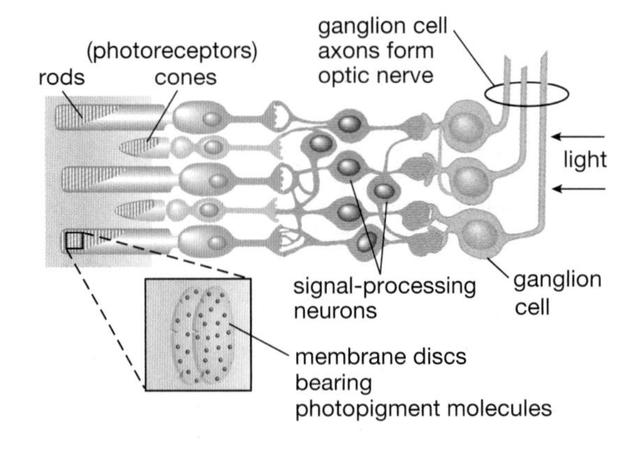

Media Activity
34.5 Vertebrate Eye

FIGURE 34-18 The human eye
(a) The anatomy of the human eye. **(b)** The human retina has rods and cones (photoreceptors), signal-processing neurons, and ganglion cells. Each rod and cone bears a long extension packed with membranes in which the light-sensitive molecules are embedded.

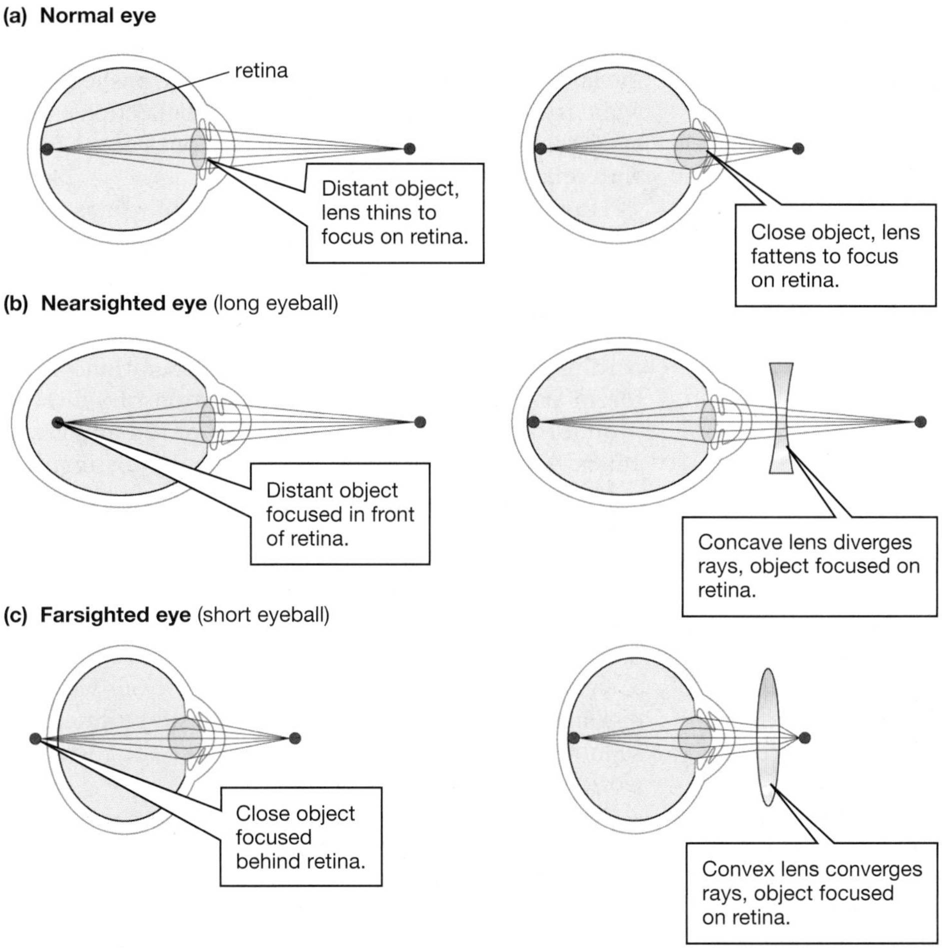

(a) Normal eye

retina

Distant object, lens thins to focus on retina.

Close object, lens fattens to focus on retina.

(b) Nearsighted eye (long eyeball)

Distant object focused in front of retina.

Concave lens diverges rays, object focused on retina.

(c) Farsighted eye (short eyeball)

Close object focused behind retina.

Convex lens converges rays, object focused on retina.

FIGURE 34-19 Focusing in the human eye
QUESTION Today, many nearsighted and farsighted people choose to correct their vision problems with laser surgery on their corneas rather than with corrective lenses. For nearsightedness and farsightedness, how should corneas be reshaped to correct the problem?

ed with laser surgery that reshapes the cornea, producing a new corneal curvature that acts much like a corrective lens. As people age, the lens within the eye stiffens, causing farsightedness because the lens can no longer curve enough to focus on nearby objects. By their mid-forties, most people require glasses for close work such as reading.

Light Striking the Retina Is Captured by Photoreceptors; The Signal Is Processed by Layers of Overlying Neurons

The vertebrate eye provides the sharpest vision in the animal kingdom, even though the complex, multilayered retina is, from an engineering perspective, "built backward." The photoreceptors, called **rods** and **cones** after their shapes, gather light at the rear of the retina (Fig. 34-18b). Between the receptors and incoming light are several layers of neurons that process the signals from the photoreceptors. The retinal layer nearest the vitreous humor consists of **ganglion cells**, whose axons make up the **optic nerve**. Ganglion cell axons must pass back through the retina to reach the brain at a location called the **blind spot** (Fig. 34-20; see also Fig. 34-18a).

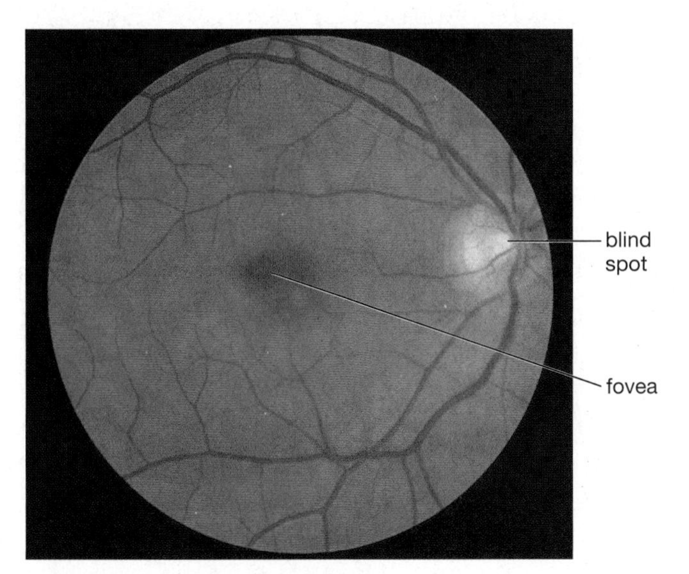

FIGURE 34-20 The human retina
A portion of the human retina, photographed through the cornea and lens of a living person. The blind spot and fovea are visible. Blood vessels supply oxygen and nutrients; notice that they are dense over the blind spot (where they won't interfere with vision) and scarce near the fovea.

This area lacks receptors; objects focused there seem to disappear. You can locate your blind spot by closing your left eye and focusing steadily on the star below with your right eye. Start with the book about a foot away and gradually move it closer. The spot will disappear when the image falls on your blind spot.

The receptor potential from the photoreceptors is processed by other retinal neurons in ways that enhance our ability to detect edges, movement, dim light, and changes in light intensity. The much-modified signal from the photoreceptors is finally converted to action potentials, which are carried to the brain by the optic nerve. In the brain, further processing ultimately results in the sensation of vision.

Rods and Cones Differ in Distribution and Light Sensitivity

Photoreception in both rods and cones begins with the absorption of light by photopigment molecules that are embedded in the plasma membranes of the photoreceptors (see Fig. 34-18b). Light hitting a photopigment molecule causes a change in the receptor membrane's permeability to ions, producing a receptor potential in the photoreceptor cell.

Although cones are located throughout the retina, they are concentrated in the fovea (there are no rods in the human fovea), where the lens focuses images most sharply (see Figs. 34-18a and 34-20). The fovea appears as a depression near the center of the retina, because the layers of signal-processing neurons are pushed aside while still retaining their synaptic connections. This arrangement allows light to reach the cones of the fovea with relatively little interference.

Human eyes have three varieties of cones, each containing a slightly different photopigment. Each type of photopigment is most strongly stimulated by a particular wavelength of light, corresponding roughly to red, green, or blue. The brain distinguishes color according to the relative intensity of stimulation of different cones. For example, the sensation of yellow is produced by fairly equal stimulation of red and green cones. About 3% of males have difficulty distinguishing red from green because they possess a defective gene for the red or green photopigment on the X chromosome.

Rods dominate in the peripheral portions of the retina. Rods, which have longer outer segments and far more photopigment than cones, are far more sensitive to light, so they are largely responsible for our vision in dim light. Unlike cones, rods do not distinguish colors. In moonlight, which is too dim to activate the cones, the world appears in shades of gray.

Not all animals have both rods and cones. Animals that are active almost entirely during the day (certain lizards, for example) may have all-cone retinas, whereas many nocturnal animals (such as rats and ferrets) and those dwelling in dimly lit habitats (such as deep-sea fishes) have mostly or entirely rods.

(a)

(b)

FIGURE 34-21 Eye position differs in predators and prey
(a) Most predators, such as this owl, and primates have eyes in front; both eyes can be focused on a target, providing binocular vision. **(b)** Most herbivorous prey animals, such as rabbits, have eyes at the sides, which allow them to scan for predators. **QUESTION** Why do some herbivorous or fruit-eating animals, such as monkeys and fruit bats, have both eyes in front?

Binocular Vision Allows Depth Perception

The placement of vertebrate eyes on the head is determined by the lifestyle of the animal. Predators and omnivores, such as humans, have both eyes facing forward (Fig. 34-21a), but most herbivores have one eye on each side of the head (Fig. 34-21b). The forward-facing eyes of predators and omnivores have slightly different but extensively overlapping visual fields. This **binocular vision** allows depth perception, the accurate judgment of the size and distance of an object from the eyes. These abilities are important to a cat about to pounce on a mouse or to a monkey leaping from branch to branch.

In contrast, the widely spaced eyes of herbivores have little overlap in their visual fields; some depth perception is sacrificed in favor of a nearly 360-degree field of view. This view allows these animals, who are frequently preyed on, to spot a predator approaching from any direction.

34.9 How Are Chemicals Sensed?

Through chemical senses that utilize chemoreceptors (see Table 34-2), animals find food, avoid poisonous materials, locate homes, find mates, and maintain homeostasis. Chemoreceptors in certain large blood vessels and in the hypothalamus of the brain monitor levels of crucial molecules such as sugar, water, and oxygen and carbon dioxide in the blood, and stimulate activities that maintain them within narrow limits. Terrestrial vertebrates have two separate senses for detecting chemicals outside the body: one for airborne molecules, called **smell** (*olfaction*), and one for chemicals dissolved in water or saliva, the sense of **taste** *(gustation)*.

The Ability to Smell Arises from Olfactory Receptors

In humans and most other vertebrates, receptors for olfaction are nerve cells located in a patch of mucus-covered epithelial tissue in the upper portion of each nasal cavity (Fig. 34-22). The human olfactory epithelium is small compared with that of many other mammals (dogs, for example), whose sense of smell is hundreds of times more acute than ours. Olfactory receptors have hairlike dendrites that protrude into the nasal cavity and are embedded in a layer of mucus. Odorous molecules in the air, such as those generated by coffee, diffuse into the mucus layer and bind with receptors on the dendrites.

Recent research suggests that there may be 1000 types of receptor proteins embedded in the olfactory dendrites. Each receptor protein is specialized to bind a particular type of molecule and stimulate the olfactory receptor to send a message to the brain.

Taste Receptors Are Located in Clusters on the Tongue

The human tongue bears about 10,000 **taste buds**, structures embedded in small bumps (called *papillae*) that

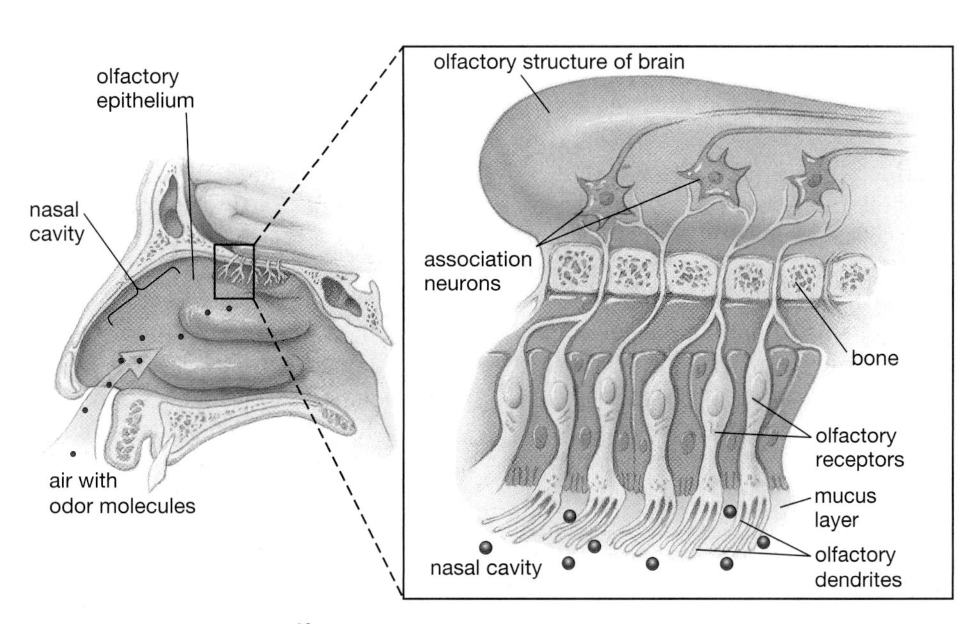

FIGURE 34-22 Human olfactory receptors
The receptors for olfaction in humans are neurons bearing microscopic hairlike projections that protrude into the nasal cavity. The projections are embedded in a mucus layer in which odorous molecules dissolve before contacting the receptors.

(a) The human tongue

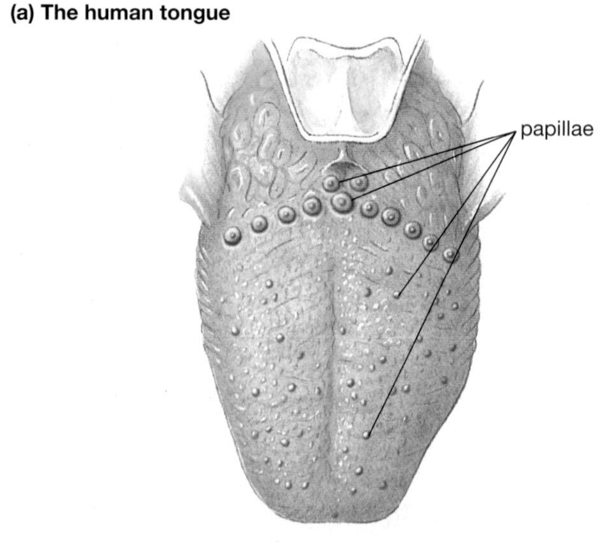

(b) Taste bud

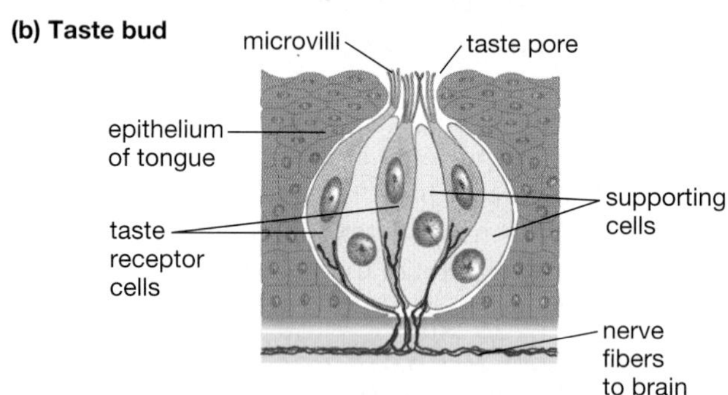

FIGURE 34-23 Human taste receptors
(a) The human tongue is covered with papillae, bumps in which taste buds are embedded. Small papillae are located on the front two-thirds of the tongue; larger ones with more taste buds are far in the back. **(b)** Each taste bud consists of supporting cells surrounding 60 to 80 taste receptor cells, whose microvilli protrude into the taste pore. The microvilli bear protein receptors that bind tasty molecules that enter through the taste pore, producing a receptor potential.

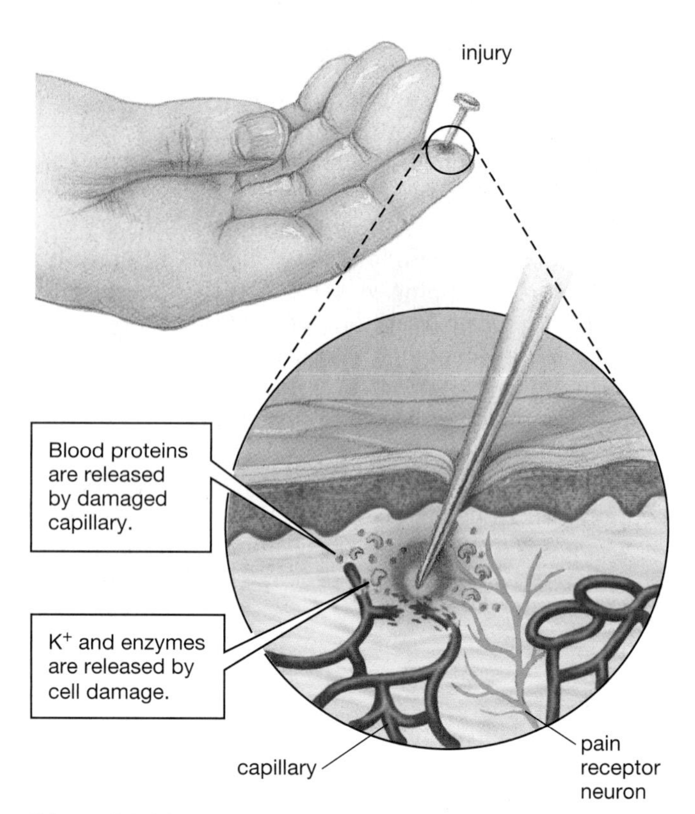

injury
Blood proteins are released by damaged capillary.
K+ and enzymes are released by cell damage.
capillary
pain receptor neuron

FIGURE 34-24 Pain perception
Pain perception is a specialized chemical sense. An injury damages both cells and blood vessels, releasing K^+, which activates pain receptor neurons. Damaged cells also release enzymes that convert certain blood proteins into bradykinin, which also stimulates pain-sensitive neurons.

cover the tongue's surface (Fig. 34-23a). Each taste bud consists of a cluster of 60 to 80 taste receptor cells surrounded by supporting cells in a small pit. The cells in the pit communicate with the mouth through a taste pore (Fig. 34-23b). Microvilli (thin membrane projections) of taste receptor cells protrude through the pore. Dissolved chemicals enter the pore and bind to receptor molecules on the microvilli, producing a receptor potential.

Four major types of taste receptors have long been known: sweet, sour, salty, and bitter; a fifth type, *umami* (a Japanese word loosely translated as "delicious"), has recently been identified. The umami receptor responds to glutamate, an amino acid that serves as a neurotransmitter; glutamate is found in MSG (monosodium glutamate), a seasoning that enhances the flavor of foods. Although it was formerly believed that taste buds for specific tastes were concentrated on specific areas of the tongue, recent research has shown they are more or less evenly distributed.

We perceive a great variety of tastes as a result of two mechanisms. First, a particular substance may stimulate two or more receptor types to different degrees, making it taste "salty–sweet," for example. Second and more important, a substance being tasted usually releases molecules into the air inside the mouth. These odorous molecules diffuse to the olfactory receptors, which contribute an odor component to the basic flavor. (Recall that the mouth and nasal passages are connected.)

To prove that what we call taste is really mostly smell, try holding your nose (and closing your eyes) while you eat different flavors of gourmet jelly beans. The flavors—from grape, cherry, and pear to buttered popcorn—will be indistinguishable sweet, sticky pastes. Likewise, when you have a bad cold, notice how usually tasty foods seem bland and unappealing, and how your coffee tastes merely bitter.

Pain Is a Specialized Chemical Sense

Whether you burn, cut, or crush a fingertip, you will feel the same sensation: pain. Most pain is produced by tissue damage. Researchers have found that pain perception is actually a special kind of chemical sense (Fig. 34-24).

(a)

(b)

FIGURE 34-25 Echolocation
(a) The enormous size and elaborate folds of a long-eared bat's external ears help it localize returning echoes. **(b)** The bottlenose porpoise focuses ultrasonic clicks by using the oil-filled sac in its forehead.
QUESTION Why do echolocating porpoises lack the large external ears that are so helpful to echolocating bats?

When cells and capillaries are damaged by a cut or a burn, for example, their contents flow into the extracellular fluid. The cell contents include potassium ions, which stimulate **pain receptors**. Damaged cells also release enzymes that convert certain blood proteins into a chemical called *bradykinin*, another stimulus that activates pain receptors. Each part of the body has a separate set of pain receptor neurons that provide input to particular brain cells. Hence, the brain can identify the location of the pain. Drugs that provide pain relief, such as morphine or Demerol®, block synapses in the pain pathways of the brain or spinal cord. As we mention in "Health Watch: Drugs, Diseases, and Neurotransmitters," the brain can modulate its perception of pain through its own narcotic-like endorphins.

EVOLUTIONARY CONNECTIONS

Uncommon Senses

We have reviewed the "common" senses of sound, sight, smell, taste, and pain. But if this text focused on bats, it undoubtedly would include a large section on *echolocation* and almost no coverage of vision! Here, we'll discuss a few of the "uncommon" senses that have evolved in response to different environments.

Echolocation

Some animals who hunt in darkness or murky water have evolved a type of sonar called **echolocation**, similar to the navigational system used by ships. Using echolocation, bats can navigate and hunt insect prey in total darkness. An echolocating bat emits pulses of noise at ultrasonic frequencies (higher than can be detected by the human ear) that bounce off nearby objects. The patterns of returning sound convey accurate information about the size, shape, surface texture, and location of objects in the environment. Little brown bats can detect wires only 1 millimeter thick from a distance of 2 meters (more than 6 feet). Several adaptations contribute to this remarkable sensitivity. The bat's enormous, elaborately folded outer ears collect the returning echoes and help the bat locate their source (Fig. 34-25a). As the bat emits its cry, muscles attached to the bones of the middle ear contract briefly, reducing the bones' vibrations and preventing the bat from being deafened by its own calls. The tympanic membrane and bones of the middle ear are exceptionally light and easily vibrated by the faint returning echoes.

Porpoises produce ultrasonic clicks within their nasal passages and emit them through the front of their head (Fig. 34-25b). There, a large, flexible, oil-filled sac directs the sound forward in a broad beam (for navigation) or a narrow beam (for prey location). An echolocating porpoise can locate a pea-sized object on the floor of its tank and can distinguish among species of fish. Porpoises may also use the narrowly focused beam to stun fish with a blast of sound, making them easier to capture.

Detecting Electrical Fields

Some fish, called weak electric fish, use electrical fields for **electrolocation** in much the same way that bats and porpoises use sound waves for echolocation. These fish

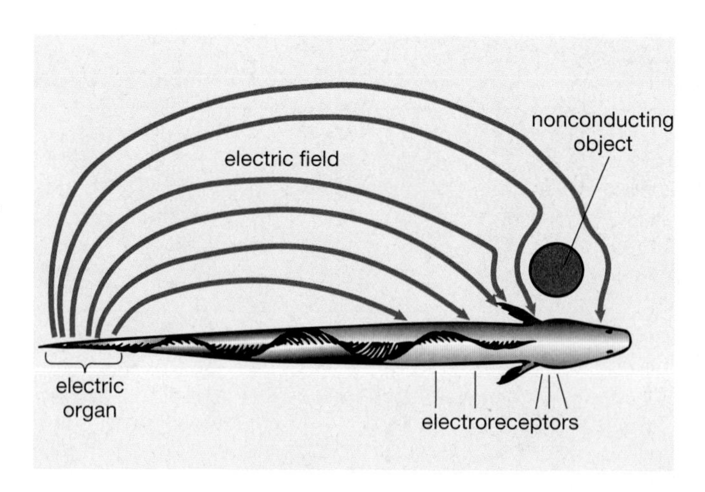

FIGURE 34-26 Electrolocation
A weak electric fish locates nearby objects by sensing distortions the objects produce in the its own electric field. The electric field, which surrounds both sides of the body, is generated by electric organs near the tail and detected by electroreceptors along the sides of the body.

produce high-frequency electrical signals from an electric organ just in front of their tails; they detect these signals via *electroreceptor* cells located along both sides of their bodies (Fig. 34-26). Objects near the fish distort the electrical field that surrounds the fish. This distortion is detected by the electroreceptors, which send an altered pattern of action potentials to the brain. The fish uses this information to detect and localize nearby objects. In this way, the platypus, whose bill is covered with electroreceptors (as well as sensitive touch receptors), can hunt for crayfish and tadpoles at night in murky freshwater ponds and streams.

Detecting Magnetic Fields

Homing pigeons are famous for their ability to fly home after being released some distance away. They can accurately locate their home roost even under cloudy skies in terrain with few landmarks. If a researcher straps a small magnet to a pigeon's back, the bird will find its way home successfully in sunny weather but will lose its way under overcast skies. Apparently, pigeons (and other birds that migrate long distances) can navigate either by the sun or, if the sun is hidden, by magnetic fields. In cloudy weather, the experimental magnets confuse the pigeon's internal magnetic compass. How do pigeons detect magnetic fields? Pigeons have deposits of magnetite (a magnetic iron compound) just beneath their skulls. These deposits may act as a built-in magnet that is used to tell direction.

Eels of eastern North America and western Europe swim out of streams and rivers into the Atlantic Ocean and migrate to the Sargasso Sea (near the West Indies) to spawn. Eels probably also use magnetic fields for navigation. The currents of the Gulf Stream, flowing from the Gulf of Mexico along the east coast of the United States and moving through Earth's magnetic field, generate an electrical field. The electrical field is roughly equivalent to that produced by a 1-volt battery with its poles more than 12 miles (7 km) apart. Researchers have discovered that eels can detect electrical fields as weak as that of a 1-volt battery with poles *more than 3000 miles* (5000 kilometers) apart. For an eel, finding the Gulf Stream must be a piece of cake!

CASE STUDY REVISITED From Tragedy to Triumph

Nearly 10,000 people, most young and otherwise healthy, become paralyzed from spinal cord injuries in the United States each year, joining the 250,000 people living with the disabling effects of these injuries. Research using laboratory animals and patients in experimental programs have shown that a walking reflex exists within the spinal cord. With the proper assistance, this automatic walking reflex, controlled entirely by neurons within the spinal cord, can be activated by electrodes implanted in the lower back. At the University of Arizona, a patient had been confined to a wheelchair for three years after partially severing his spinal cord: Now, after extensive treadmill therapy, he can walk nearly 300 yards on his own, aided only by electrical stimuli. Meanwhile, Christopher Reeve, with much more extensive damage, has astonished doctors with his progress. He has regained at least 50% normal sensation throughout most of his body; he can now detect heat and cold, distinguish a pinprick from a soft touch, and feel the embrace of his loved ones. Although still unable to stand or even sit up unaided, Reeve is beginning to move some of his joints, especially under weightless conditions provided by suspension in the swimming pool where he regularly works out. Stimulating electrodes recently implanted in his diaphragm may allow him to temporarily turn off his ventilator, inhale normally, and enjoy the scent of coffee and flowers for the first time in nearly a decade. The Christopher Reeve Paralysis center is funding proposals to bring more findings from animal research to clinical human trials.

Consider This: Whereas scientists are eagerly searching for ways to promote regeneration of damaged nerves in the CNS, evolution seems to have favored mechanisms that *inhibit* CNS (brain and spinal cord) regeneration after the nervous system matures (see "Health Watch: Healing the Spinal Cord"). Why hasn't natural selection favored such regrowth? Consider the following as you speculate: during development, the billions of neurons in the CNS form trillions of very precise synaptic connections that allow the enormously complex information-processing capabilities of the brain. As the nervous system matures, the capability for regrowth of injured axons diminishes in the CNS but not in the PNS. In the wild, are animals with major CNS injuries likely to survive? How does this situation apply to people?

Links to Life: The Sensory Stew of Synesthesia

A pager beeps and a woman exclaims, "Turn that thing down! It's red lightning bolts!" Taking a bite of roasted chicken, a young man muses, "The chicken's too round—it needs more points." Welcome to the unique world of the synesthete. For people with *synesthesia* (literally "mingling of the senses"), one stimulus activates more than one sensory area of the cortex, evoking multiple sensory impressions unique to that individual. For example, one synesthete sees specific colors associated with specific numbers, letters, syllables, and spoken words, thus converting an ordinary conversation or a black-and-white printed page into a fascinating palette of colors. Another perceives xylophone music as golden balls, while yet another describes electronic music as colored bars and green pyramids. In a recent fMRI study, synesthetes with "colored hearing" showed heightened activity in both the auditory cortex and the color-perception portion of the visual cortex when they listened to spoken words. In contrast, people without synesthesia had very little activation of the visual color center, even when they tried to visualize colors along with the words.

Synesthetes can experience almost any combination of sensory perceptions; the young man described earlier perceives tastes as tactile sensations, feeling different shapes and textures pressed against his face, arms, and hands while eating. A chocolate mint pie evokes cool columnar structures, while bitterness causes him to feel tendrils emerging from holes with his hands. Synesthetes are not merely people with vivid imaginations; their unique and lifelong sensory impressions are precisely duplicated whenever a specific stimulus is presented. Their unusual sensations are added to their "normal" sensory responses, rather than replacing them, creating a heightened sensory experience. Having experienced it all their lives, some do not even realize that their sensory world is unique. For this reason, although synesthesia is certainly rare, the number of people with synesthesia is impossible to estimate. Are you a synesthete?

Summary of Key Concepts

34.1 What Are the Structures and Functions of Neurons?

Nervous systems are composed of billions of individual cells called neurons. A neuron has four major specialized functions, which are reflected in its structure: Dendrites receive information from the environment or from other neurons. The cell body adds together electrical signals from the dendrites and from synapses on the cell body itself and "decides" whether to produce an action potential. (The cell body also coordinates the cell's metabolic activities.) The axon conducts the action potential to its output terminal: the synapse. Synaptic terminals transmit the signal to other nerve cells, glands, or muscles.

34.2 How Is Neural Activity Produced and Transmitted?

An unstimulated neuron maintains a negative resting potential inside the cell. Signals received from other neurons are small, rapidly fading changes in potential called postsynaptic potentials. Inhibitory and excitatory postsynaptic potentials (IPSPs and EPSPs) make the neuron less likely or more likely, respectively, to produce an action potential. If postsynaptic potentials, added together within the cell body, bring the neuron to threshold, an action potential will be triggered. The action potential is a wave of positive charge that travels, undiminished in magnitude, along the axon to the synaptic terminals.

A synapse, where two neurons communicate, consists of the synaptic terminal of the presynaptic neuron, a specialized region of the postsynaptic neuron, and the tiny gap between them. Neurotransmitters from the presynaptic neuron, released in response to an action potential, bind to receptors in the postsynaptic cell's plasma membrane, opening ion channels there. Ions flow, producing either an EPSP or an IPSP, depending on the type of ion channels opened.

There are many neurotransmitters. They are being intensively studied to explore their diverse roles in neurological disease, drug addiction, and all aspects of normal nervous system function.

34.3 What Are Some General Features of Nervous Systems?

Information processing in the nervous system requires four operations. The nervous system must (1) determine the type of stimulus, (2) signal the intensity of the stimulus, (3) integrate information from many sources, and (4) initiate and direct the response. The nervous system collects and processes sensory information from many sources. These sensory stimuli may come together (converge) on fewer neurons whose activity in response to the stimuli determines the action taken. The "decision" to act may then be transmitted to many more neurons (divergence), which direct the activity.

Neural pathways normally have four elements: (1) sensory neurons, (2) association neurons, (3) motor neurons, and (4) effectors. Overall, nervous systems consist of numerous interconnected neural pathways, which may be either diffuse or centralized.

34.4 How Is the Human Nervous System Organized?

The nervous system of humans and other vertebrates consists of the central nervous system and the peripheral nervous system. The peripheral nervous system is further subdivided into sensory and motor portions. The motor portion consists of the somatic nervous system (which controls voluntary movement) and the autonomic nervous system (which directs involuntary responses).

Within the central nervous system, the spinal cord contains neurons controlling voluntary muscles and the autonomic nervous system and neurons communicating with the brain and other parts of the spinal cord; axons leading to and from the brain; and neural pathways for reflexes and certain simple behaviors. The brain consists of three parts: the hindbrain, midbrain, and forebrain, each further subdivided into distinct regions.

The hindbrain in humans consists of the medulla and pons, which control involuntary functions (such as breathing), and the cerebellum, which coordinates complex motor activities (such as typing). In humans, the small midbrain contains the reticular formation, a filter and relay for sensory stimuli. The forebrain includes the thalamus, a sensory relay station that shuttles information to and from conscious centers in the forebrain. The diverse structures of the limbic system of the forebrain are involved in emotion, learning, and the control of instinctive behaviors such as sex, feeding, and aggression. The cerebral cortex of the forebrain is the center for information processing, memory, and initiation of voluntary actions. It includes primary sensory and motor areas and association areas that analyze sensory information and plan movements.

34.5 How Does the Brain Produce the Mind?

The cerebral hemispheres are specialized. In general, the left hemisphere dominates speech, reading, writing, language comprehension, mathematical ability, and logical problem solving. The right hemisphere specializes in recognizing faces and spatial relationships, artistic and musical abilities, and recognition and expression of emotions.

Memory takes two forms: short-term memory is electrical or chemical, while long-term memory probably involves structural changes that increase the effectiveness or number of synapses. The hippocampus is an important site for the transfer of information from short-term into long-term memory. The temporal lobes are important for memory recognition of objects and faces, and understanding language.

34.6 How Do Sensory Receptors Work?

Receptors convert signals from one form to another. Receptor cells are named after the stimulus to which they respond.

34.7 How Is Sound Sensed?

In the vertebrate ear, air vibrates the tympanic membrane, which transmits vibrations to the bones of the middle ear and then to the oval window of the fluid-filled cochlea. Within the cochlea, vibrations bend the hairs of hair cells, which are receptors located between the basilar and tectorial membranes. This bending produces receptor potentials in the hair cells that cause action potentials in the axons of the auditory nerve, which leads to the brain.

34.8 How Is Light Sensed?

In the vertebrate eye, light enters the cornea and passes through the pupil to the lens, which focuses an image on the fovea of the retina. Two types of photoreceptors, rods and cones, are located deep in the retina. They produce receptor potentials in response to light. These signals are processed through several layers of neurons in the retina and are translated into action potentials in the optic nerve, which leads to the brain. Rods are more abundant and more light-sensitive than cones, providing vision in dim light. Cones, which are concentrated in the fovea, provide color vision.

34.9 How Are Chemicals Sensed?

Terrestrial vertebrates detect chemicals in the external environment either by smell (olfaction) or by taste. Each olfactory or taste receptor cell type responds to only one or a few specific types of molecules, allowing discrimination among tastes and odors. Olfactory neurons of vertebrates are located in a tissue that lines the nasal cavity. Taste receptors are located in clusters called taste buds on the tongue. Pain is a special type of chemical sense in which sensory neurons respond to chemicals released by damaged cells.

Key Terms

action potential *p. 668*
amygdala *p. 682*
aqueous humor *p. 690*
association neuron *p. 674*
auditory canal *p. 687*
auditory tube *p. 688*
autonomic nervous system *p. 675*
axon *p. 668*
basilar membrane *p. 693*
binocular vision *p. 693*
blind spot *p. 691*
blood–brain barrier *p. 678*
brain *p. 675*
cell body *p. 668*
central nervous system *p. 675*
cephalization *p. 675*
cerebellum *p. 680*
cerebral cortex *p. 682*
cerebral hemisphere *p. 682*
cerebrospinal fluid *p. 678*
cerebrum *p. 681*
choroid *p. 690*
cochlea *p. 688*
compound eye *p. 689*
cone *p. 691*
convergence *p. 674*
convolution *p. 682*
cornea *p. 689*
corpus callosum *p. 682*
dendrite *p. 668*
divergence *p. 674*
dorsal root ganglion *p. 680*
echolocation *p. 695*
effector *p. 674*
electrolocation *p. 695*
external ear *p. 687*
farsighted *p. 690*
forebrain *p. 680*
fovea *p. 690*
ganglion *p. 675*
ganglion cell *p. 691*
gray matter *p. 678*
hair cell *p. 688*
hindbrain *p. 680*
hippocampus *p. 682*
hypothalamus *p. 682*
inner ear *p. 688*
intensity *p. 674*
iris *p. 690*
lens *p. 690*
limbic system *p. 682*
long-term memory *p. 684*
medulla *p. 680*
meninges *p. 678*
midbrain *p. 680*
middle ear *p. 687*
motor neuron *p. 674*
myelin *p. 668*
nearsighted *p. 690*
nerve *p. 668*
nerve net *p. 675*
neuron *p. 668*
neurotransmitter *p. 668*
ommatidium *p. 689*
optic nerve *p. 691*
outer ear *p. 687*
pain receptor *p. 695*
parasympathetic division *p. 676*
peripheral nerve *p. 675*
peripheral nervous system *p. 675*
photopigment *p. 689*
pons *p. 680*
postsynaptic neuron *p. 669*
postsynaptic potential (PSP) *p. 669*
presynaptic neuron *p. 669*
pupil *p. 690*
receptor *p. 686*
receptor potential *p. 686*
reflex *p. 674*
resting potential *p. 669*
reticular formation *p. 681*
retina *p. 690*
rod *p. 691*
sclera *p. 690*
sensory neuron *p. 674*
sensory receptor *p. 686*
somatic nervous system *p. 675*
smell *p. 693*
spinal cord *p. 675*
sympathetic division *p. 676*
synapse *p. 669*
synaptic terminal *p. 668*
taste *p. 693*
taste bud *p. 693*
tectorial membrane *p. 688*
thalamus *p. 681*
threshold *p. 669*
tympanic membrane *p. 687*
vitreous humor *p. 690*
white matter *p. 678*
working memory *p. 684*

Thinking Through the Concepts

To take a multiple-choice quiz with feedback on the contents of this chapter, visit http://www.prenhall.com/audesirk7. *Log in to the Web site selected by your instructor and navigate to the Self Test section for this chapter.*

? Review Questions

1. List four major parts of a neuron, and explain the specialized function of each part.
2. Diagram a synapse. How are signals transmitted from one neuron to another at a synapse?
3. How does the brain perceive the intensity of a stimulus? The type of stimulus?
4. What are the four elements of a simple nervous pathway? Describe how these elements function in the human pain-withdrawal reflex.
5. Draw a cross section of the spinal cord. What types of neurons are located in the spinal cord? Explain why severing the cord paralyzes the body below the level where it is severed.
6. Describe the functions of the following parts of the human brain: medulla, cerebellum, reticular formation, thalamus, limbic system, and cerebrum.
7. What structure connects the two cerebral hemispheres? Describe the evidence that the two hemispheres are specialized for distinct intellectual functions.
8. Distinguish between long-term memory and working memory.
9. What are the names of the specific receptors used for taste, vision, hearing, smell, and touch?
10. Why are we apparently able to distinguish hundreds of different flavors if we have only five types of taste receptors? How are we able to distinguish so many different odors?
11. Describe the structure and function of the various parts of the human ear by tracing a sound wave from the air outside the ear to the receptor cells.
12. How does the structure of the inner ear allow for the perception of pitch? Of sound intensity?
13. Diagram the overall structure of the human eye. Label the cornea, iris, lens, sclera, retina, and choroid. Describe the function of each structure.
14. How does the lens change shape to allow focusing of distant objects? What defect makes focusing on distant objects impossible, and what is this condition called? What type of lens can be used to correct it, and how does it do so?
15. List the similarities and differences between rods and cones.
16. Distinguish between taste and smell.
17. Describe how pain is signaled by tissue damage.

Applying the Concepts

1. Argue for or against the statement, "Consciousness by its nature is incomprehensible; the brain will never understand the mind."
2. In Parkinson's disease, which afflicts several million Americans, the cells that produce the neurotransmitter dopamine degenerate in a small part of the brain that is important in the control of movement. Some physicians have reported improvement after injecting cells taken from the same general brain region of an aborted fetus into appropriate parts of the brain of a Parkinson's patient. Discuss this type of surgery from as many viewpoints as possible: ethical, financial, practical, and so on. On the basis of your responses, is fetal transplant surgery the answer to curing Parkinson's disease?
3. If the axons of human spinal cord neurons were unmyelinated, would the spinal cord be larger or smaller? Would you move faster or slower? Explain your answer.
4. What is the adaptive value of reflexes? If Christopher Reeve accidentally pricked his toe on a thumbtack, would he withdraw his leg? Explain your answer.
5. Explain the statement, "Your sensory perceptions are purely a creation of your brain." Discuss the implications for communicating with other humans, with other animals, and with intelligent life elsewhere in the universe.
6. Corneal transplants can help restore vision and greatly improve the recipient's quality of life. What properties of the cornea make it an excellent candidate for transplantation? Suggest some ways in which society could improve the availability of corneal and other tissues for transplantation.

For More Information

Axel, R. "The Molecular Logic of Smell." *Scientific American*, October 1995. Describes research that uncovers some of the mechanisms by which the nose and brain decipher scents.

Beardsley, T. "The Machinery of Thought." *Scientific American*, August 1997. Using PET and functional MRI on monkeys and humans, researchers are learning more about where working memory resides.

Bower, B. "Creatures in the Brain." *Science News*, April 13, 1996. Using imaging techniques, scientists have discovered clues about how different regions of the brain are specialized for different concepts.

Damasio, A. R. "How the Brain Creates the Mind." *Scientific American*, December 1999. The author presents an intriguing hypothesis for how the sense of self emerges from the machinery of the brain.

Gazzaniga, M. S. "The Split Brain Revisited. " *Scientific American*, July 1998. More insights into hemispheric specializations by one of the pioneers of human split-brain studies.

Raichle, M. E. "Visualizing the Mind." *Scientific American*, April 1994. Brain imaging techniques partially open the "black box" of the mind.

Smith, D. V., and Margolskee, R. F. "Making Sense of Taste." *Scientific American*, March 2001. Scientists are beginning to unravel the mechanisms by which taste receptors respond to various tastes.

Wickelgren, I. "Animal Studies Raise Hopes for Spinal Cord Repair." *Science*, July 12, 2002. Reviews the most promising lines of research for spinal cord regeneration.

Wuethrich, B. "Getting Stupid. " *Discover*, March 2001. Heavy drinking may damage the brain, particularly in young people.

Media Activities

To access a Media Activity visit http://www.prenhall.com/audesirk7. *Log in to the Web site selected by your instructor, navigate to this chapter, and select the appropriate Media Activity number.*

34.1 The Nervous System: Electrical Signals

Estimated time: 10 minutes

Neurons send electrical signals called an action potential. Review the events that generate an action potential in a neuron.

34.2 The Nervous System: Synapses

Estimated time: 5 minutes

Review signals are transferred from neuron to neuron across the synapses.

34.3 Reflex Arc

Estimated time: 5 minutes

Watch how the reflex arc allows us to react quickly to pain.

34.4 Brain Structure

Estimated time: 5 minutes

Review the basic structures of the brain and their function.

34.5 The Vertebrate Eye

Estimated time: 5 minutes

Review the structure of the eye and how it captures light and sends visual signals to the brain.

34.6 Web Investigation: From Tragedy to Triumph

Estimated time: 30 minutes

At this time there is no cure for SCI, but there are several promising leads. Explore the types of injuries, ongoing research and what can be done now.

CHAPTER

35 Action and Support: The Muscles and Skeleton

Astronaut Millie Hughes-Fulford floats in the weightless environment of the life sciences module in the space shuttle *Columbia* during a 1991 mission. *Columbia* was lost in a tragic explosion on February 1, 2003.

AT A GLANCE

CASE STUDY Hidden Hazards of Space Travel

At this moment, astronauts from diverse cultures are circling Earth while living and working together in one of humankind's most remarkable cooperative ventures—the International Space Station. Unfortunately, this once-in-a-lifetime experience does not come without risk. In addition to the obvious dangers of space flight, these astronauts face a more insidious threat: the loss of both muscle and bone due to weightlessness. Although being weightless seems like fun, our bodies are not adapted for it. Humans (and other terrestrial species) evolved under the relentless pull of gravity, which strengthens our bones and muscles on a day-to-day basis. Astronauts routinely exercise to prevent muscle atrophy, but bone loss poses a more difficult dilemma. Research has shown that space travelers lose 0.5 to 2 percent of their total bone mass for every month they spend under weightless conditions. The loss is most pronounced in weight-bearing bones such as the lower portion of the vertebral column and the legs. Contrary to popular perception, bones are not merely dry scaffolding for the body; instead, they continuously change in response to the demands we place on them. In a process called "remodeling," bone thickens under stress and thins when stress is removed. How does remodeling happen? Does it change as we get older? What can be done to counteract bone loss in space?

35.1 An Introduction to the Muscular and Skeletal Systems

The system of muscles and skeleton that moves and supports the animal body is an engineering marvel. The flight of a bat, the pounce of a cat, the fluid grace of a ballet dancer, and the simple movements involved in walking to your classroom all depend on the same humble yet elegant mechanism. Muscle cells perform only one activity: they exert a force by contracting. Under the influence of natural selection, however, this simple unidirectional force has been applied to complex bony scaffolding that has been molded into structural elements such as wings, hands, and fins, and its action has come to be coordinated by the nervous system. The resulting capacity for movement gives animals the ability to search for food, seek out new habitats, flee from danger, and, on occasion, move in ways that we find awe-inspiring.

Muscles and skeletons also perform more mundane, yet crucial, functions. Pumping blood through the circulatory system, moving food through the digestive system, and breathing are some of the essential processes that depend on muscle contraction. The skeleton plays an equally essential role by opposing the force of gravity and providing a framework against which muscles exert force to move the body. Terrestrial organisms in particular depend on skeletal support to maintain their shapes. Without your skeleton, you'd be a formless, quivering mound of tissue.

Table 35-1 Location, Characteristics, and Functions of the Three Muscle Types

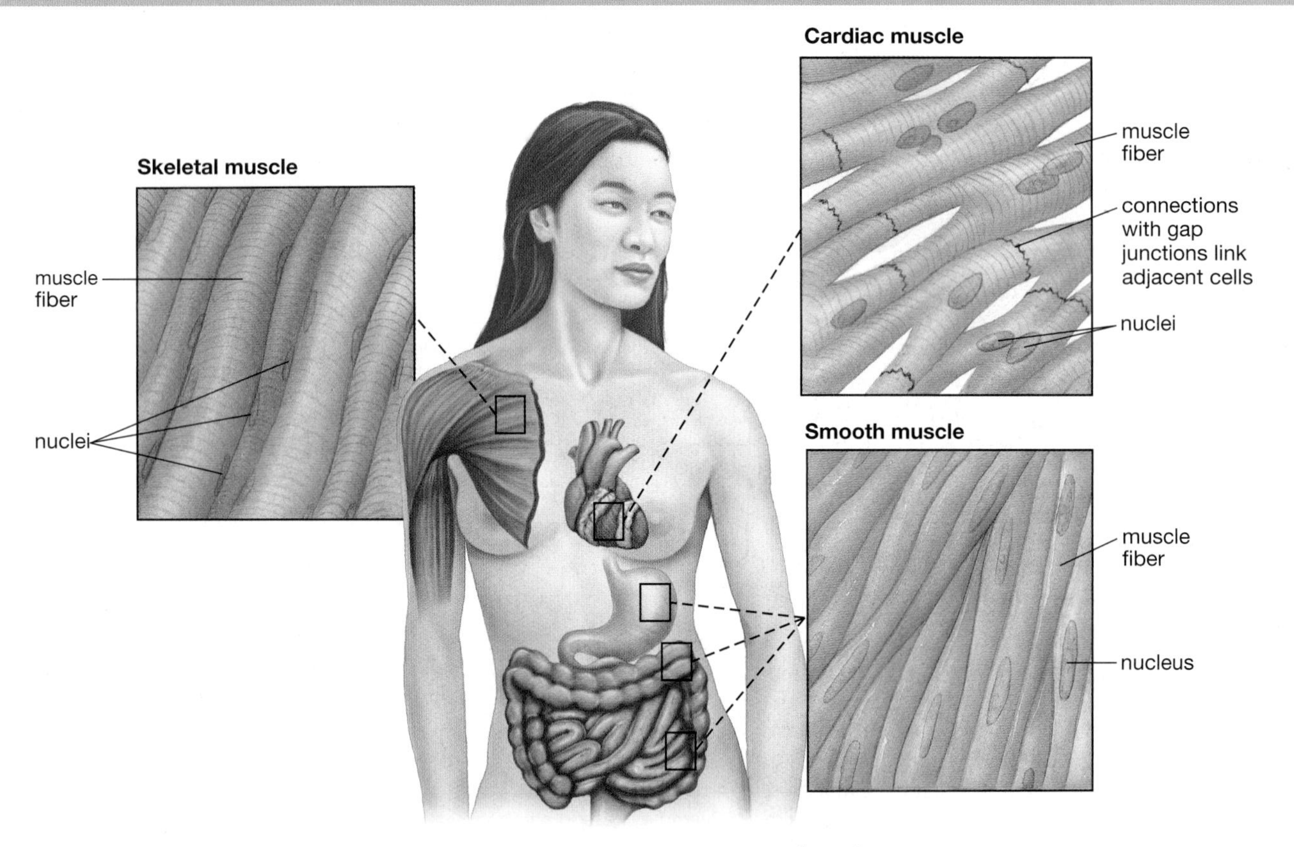

Property	Smooth	Cardiac	Skeletal
Muscle appearance	Nonstriated	Irregular striations	Regular striations
Cell shape	Spindle	Branched	Spindle
Number of nuclei	One per cell	One or more per cell	Many per cell
Speed of contraction	Slow	Intermediate	Slow to rapid
Contraction caused by	Spontaneous, stretch, nervous system, hormones	Spontaneous	Nervous system
Function	Controls movement of substances through hollow organs and tubes	Pumps blood	Moves the skeleton
Voluntary control	Usually no*	Usually no*	Yes

(Type of Muscle spans Smooth, Cardiac, Skeletal columns.)

*Smooth and cardiac muscles usually contract without conscious control. In some cases, however, their contractions may be initiated or modified voluntarily. For example, heart rate can be voluntarily slowed after biofeedback training, and bladder contractions are initiated consciously.

35.2 How Do Muscles Work?

All muscular work requires muscles to alternately contract and lengthen, but muscles are active only during contraction. The lengthening that follows contraction is passive, occurring when muscles relax and are stretched out by other forces. For example, a relaxed muscle may be lengthened by contractions of opposing muscles, the weight of a limb, or pressure from food on the muscular walls of the stomach.

Animals display a dazzling diversity of muscle function, adapted from a remarkable uniformity of muscle structure. Vertebrates have evolved three types of muscle: skeletal, cardiac, and smooth. All work on the same basic principles but differ in function, appearance, and control (Table 35-1). Invertebrate muscles closely resemble vertebrate muscles, but show a tremendous range of adaptations to their diverse lifestyles. For example, bivalves (mollusks with two shells, such as scallops and clams) possess a special type of smooth muscle that holds their shells tightly closed for hours using very little energy. In contrast to these sustained contractions, some flies have flight muscles that contract at rate of 1000 times per second. The sections below describe vertebrate muscles, using the human muscular system as an example.

Skeletal muscle, so named because it moves the skeleton, appears striped when viewed through a microscope and is often referred to as *striated* (meaning "striped"). Most skeletal muscle is under voluntary, or conscious, control. It can produce contractions ranging from quick twitches (as in blinking) to powerful, sustained tension (as in carrying an armload of textbooks). **Cardiac muscle** is so named because it is located only in the heart. It is spontaneously active, initiating its own contractions, but it is influenced by nerves and hormones. Like skeletal muscle, cardiac muscle is also striated. **Smooth muscle**, as its name suggests, lacks the orderly striations of skeletal and cardiac muscle. Smooth muscle lines the walls of the digestive tract and large blood vessels and produces slow, sustained contractions. These contractions are primarily involuntary; that is, they are not under conscious control.

The human body has some 700 different skeletal muscles, which make up 35% to 45% of its total weight. Our discussion of muscles begins with and emphasizes skeletal muscle, followed by a survey of cardiac and smooth muscle.

Skeletal Muscle Cell Structure and Function Are Closely Linked

In most eukaryotic cells, movement of organelles within the cell, changes in cell shape, and cellular locomotion depend on microfilaments constructed of the protein **actin** (see Chapter 6) interacting with strands of another protein, **myosin**. During movements, actin and myosin slide past one another, changing the shape of the cell. This evolutionarily ancient mechanism also enables animal muscle cells to contract.

Skeletal muscles are attached to the skeleton by tough, fibrous cords of connective tissue called **tendons**. Tendons are composed of fibers of collagen protein, bundled together much like small wires within a larger cable. Each muscle is encased in connective tissue and consists of bundles of muscle cells, as well as blood vessels and nerves (Fig. 35-1). Individual muscle cells, called **muscle fibers**, are among the largest cells in the

Media Activity
35.1 Muscle Structure

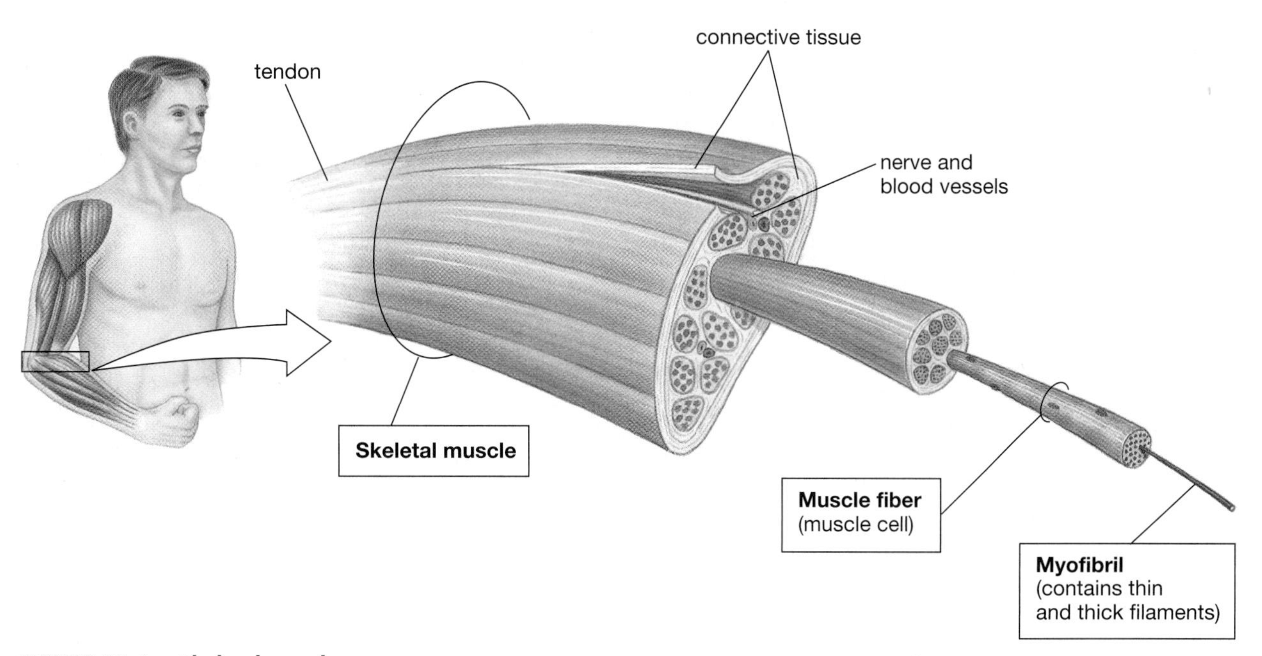

FIGURE 35-1 Skeletal muscle structure
A muscle is surrounded by connective tissue and attached to bones by tendons. It contains from a few to 1000 muscle cells called muscle fibers, often packaged into bundles within the muscle. Each fiber is packed with cylindrical subunits called myofibrils, which contain thick and thin filaments of protein.

human body. Ranging from 10 to 100 micrometers in diameter (a bit smaller than the period at the end of this sentence), each muscle fiber runs the entire length of the muscle, which may be as long as 35 centimeters (about 14 inches) in a human thigh. Muscle fibers are unusual in that they contain many nuclei located just beneath the cell's outer membrane; the largest skeletal muscle cells may contain several thousand nuclei. Each muscle fiber, in turn, contains many **myofibrils**, contractile cylinders composed primarily of actin and myosin, extending from one end of the fiber to the other. Each myofibril is surrounded by **sarcoplasmic reticulum**. Like the endoplasmic reticulum from which it is derived, sarcoplasmic reticulum consists of flattened, membrane-enclosed compartments (Fig. 35-2a). The fluid within the sarcoplasmic reticulum stores high concentrations of calcium ions, which play a key role in muscle contraction. Surrounding each muscle fiber is a plasma membrane which periodically tunnels deeply into the muscle fiber, forming membrane-lined channels called **T tubules** that are filled with extracellular fluid. T tubules form close connections with the sarcoplasmic reticulum, an association that is crucial to the control of muscle contraction, as described later.

Within each myofibril is a beautifully precise arrangement of actin and myosin filaments. The myofibrils are comprised of subunits called **sarcomeres**, which are aligned end to end along the length of the myofibril (Fig. 35-2b). Within each sarcomere, actin molecules (in association with two accessory proteins, *troponin* and *tropomyosin*) form the **thin filaments**. Suspended between the thin filaments are **thick filaments**. Thick filaments are composed of myosin protein, which is capable of linking temporarily to the thin filaments using small projections called **myosin heads** (Fig. 35-2c). The thin filaments are attached to fibrous protein bands called **Z lines**, which separate adjacent sarcomeres. The regular arrangement of thick and thin filaments within each myofibril gives the muscle fiber its striated appearance (see Table 35-1).

Media Activity
35.1 Muscle Structure

Muscle Contraction Results from Thick and Thin Filaments Sliding Past One Another

Muscle contraction is controlled by a process that is directly related to the molecular structure of the thin filament. The actin protein that makes up most of the thin filament is formed from a double chain of subunits, resembling a twisted double strand of pearls. Each subunit has a binding site for a myosin head. In a relaxed muscle cell, however, these binding sites are blocked by molecules of accessory proteins (see Fig. 35-2c). The accessory proteins prevent the myosin heads from attaching to the actin of the thin filament.

When a muscle contracts, the accessory proteins of the thin filament move aside, exposing binding sites on the actin. As soon as the sites are exposed, myosin heads bind the actin, temporarily linking the thick and thin filaments. Using energy from the splitting of adenosine triphosphate (ATP), the myosin heads repeatedly bend, release, and reattach farther along the thin filament, much like a sailor pulling in an anchor line hand over hand (Fig. 35-3a). The thin filaments are pulled past the

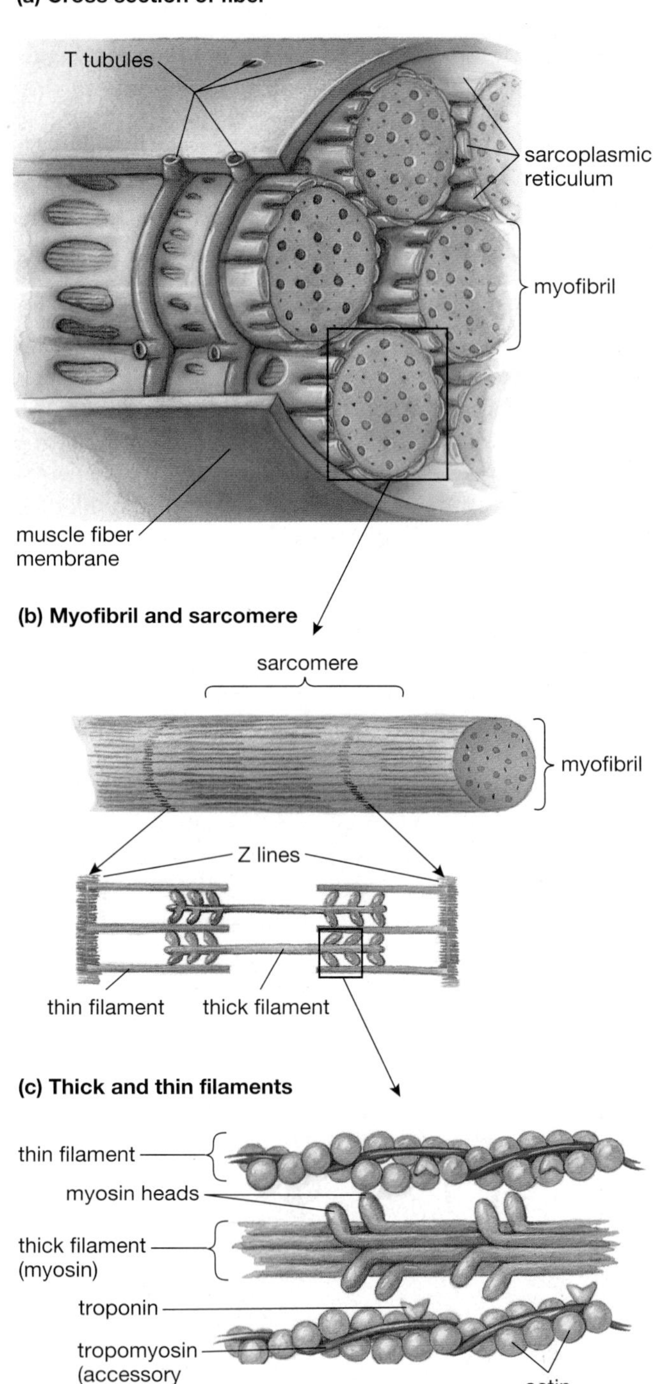

FIGURE 35-2 A skeletal muscle fiber
(a) Each muscle fiber is surrounded by plasma membrane that tunnels inside, forming T tubules. The sarcoplasmic reticulum surrounds each myofibril within the muscle cell. **(b)** Each myofibril consists of a series of subunits called sarcomeres, attached end to end by protein bands called Z lines. **(c)** Within each sarcomere are alternating thick and thin filaments, which can be connected by the myosin heads.

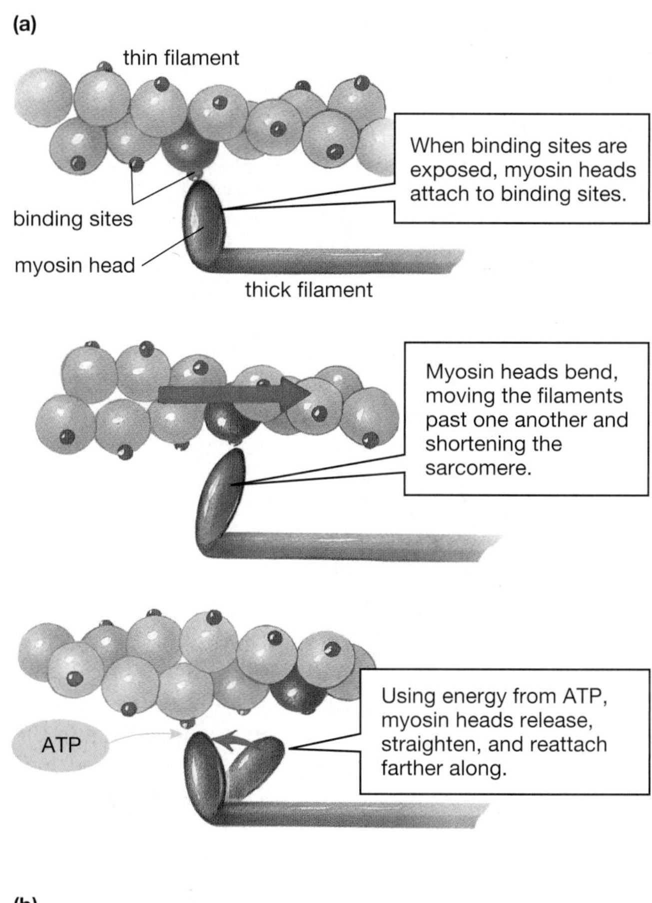

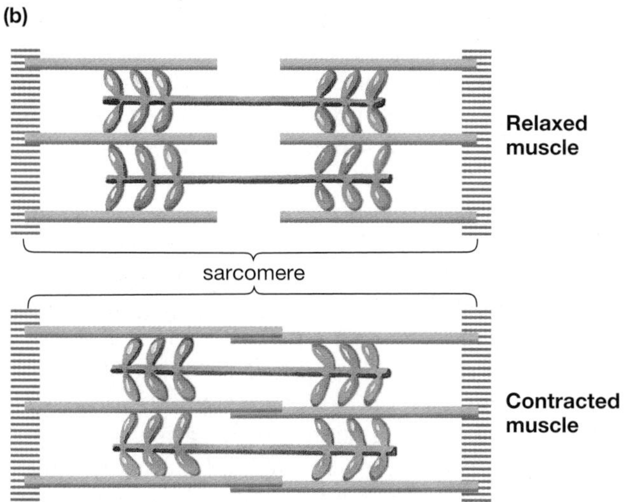

FIGURE 35-3 Muscle contraction
(a) Repeated cycles of myosin head attachment, bending, release, and reattachment result in muscle contraction. **(b)** Muscle contraction causes the thick and thin filaments to slide past one another toward the center of each sarcomere, shortening the muscle cell. **QUESTION** As the sarcomere shortens during muscle contraction, do the thick filaments shorten? Do the thin filaments?

thick filaments, shortening the sarcomere and contracting the muscle (Fig. 35-3b). Because the thick and thin filaments slide past one another during contraction, muscle contraction is described as using a *sliding-filament mechanism*; this mechanism is found throughout the animal kingdom.

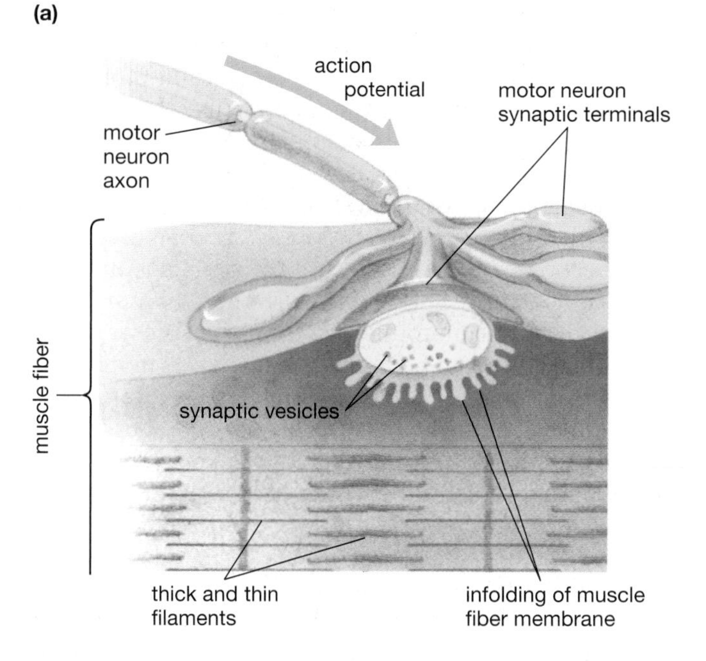

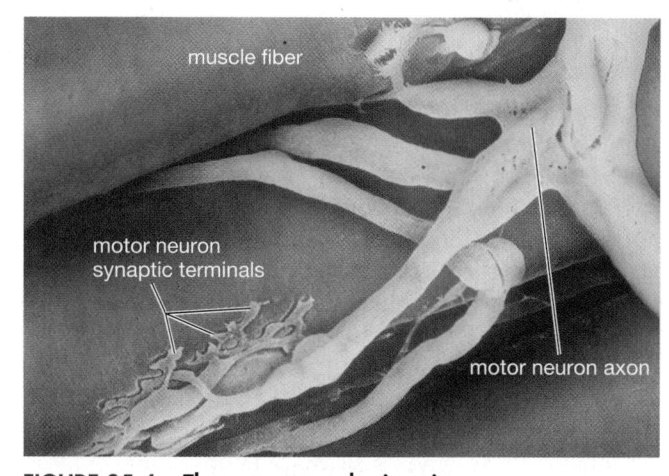

FIGURE 35-4 The neuromuscular junction
(a) Diagram of a neuromuscular junction in cross section. Action potentials in the motor neuron stimulate the muscle fiber membrane, which lies in folds beneath the terminal. **(b)** A scanning electron micrograph of motor neuron terminals that form synapses on muscle fibers.

Media Activity
35.2 Muscle Contraction

Skeletal Muscle Contraction Is Controlled by the Nervous System

Motor neurons activate skeletal muscles at specialized synapses called **neuromuscular junctions** (Fig. 35-4). Neuromuscular junctions differ from most other synapses in two important ways: they are always excitatory, never inhibitory; and every action potential in a motor neuron elicits an action potential in a muscle fiber, causing all its sarcomeres to contract—unlike a typical neuron, which needs the summation of multiple synaptic inputs to generate an action potential. The nervous system controls the strength and degree of muscle contraction by controlling the number of muscle fibers stimulated and the frequency of action potentials in each fiber. A single action potential doesn't cause a muscle cell to contract fully; this requires many action potentials in rapid succession. If rapid firing is prolonged, the muscle produces a sustained maximal contraction, such as you experience when you carry an armful of books like this one.

Most motor neurons have many synaptic terminals, each synapsing on a different muscle fiber; therefore, a single action potential will cause the simultaneous contraction of a cluster of muscle cells. The group of fibers on which a single motor neuron forms synapses is called a **motor unit**. The number of muscle fibers in a motor unit varies from muscle to muscle. Muscles used for large-scale movement, such as the thigh muscles you use when climbing stairs or the muscles in your back that help maintain your posture, can contain from hundreds to approximately 1000 muscle fibers in each motor unit. In muscles used for fine control of small body parts, such as those of the lips, eyes, and tongue, only a few muscle cells may be stimulated by each motor neuron. Therefore, the action potential fired by a single motor neuron can cause the contraction of a few muscle cells or many, depending on the size of the motor unit.

Muscle Contraction Depends on the Availability of Calcium Ions

An action potential in the muscle cell travels into the cell's interior by passing down the T tubules (see Fig. 35-2a). On reaching the sarcoplasmic reticulum, the action potential causes the release of calcium ions from the interior of the sarcoplasmic reticulum; these ions then flow into the cytoplasm surrounding the thick and thin filaments. Once in the cytoplasm, calcium ions bind to the small accessory protein (*troponin*) of the thin filament, changing its shape and pulling the larger protein (*tropomyosin*) away from the binding sites for myosin. As long as these binding sites are exposed and ATP is available, the myosin heads repeatedly bind, bend, release, and reattach, contracting the muscle fiber. As soon as the action potential is over, active transport proteins in the sarcoplasmic reticulum membrane pump the calcium ions back to their "holding area" inside the sarcoplasmic reticulum. When the calcium leaves the troponin, the accessory proteins return to a configuration that blocks the myosin binding sites, and the muscle fiber relaxes.

You've probably heard of rigor mortis, in which muscles become rigid (without contracting) hours after death. This occurs because the muscle cells run out of ATP and can no longer pump calcium back into the sarcoplasmic reticulum. As the calcium leaks out, it binds to troponin, which pulls tropomyosin off the binding sites and allows the myosin heads to bind the actin. Without ATP, no contraction occurs, but because calcium remains present the binding persists, making the muscles rigid. Rigor mortis passes many hours later as the muscle cells begin to decompose.

Muscle Contraction Requires a Steady Supply of Energy

Contracting muscles require a continuous supply of ATP, but a muscle's ATP stores are used up after only a few seconds of high-intensity exercise. Skeletal muscles also stock a large supply of creatine phosphate, an energy-storage molecule that quickly resynthesizes ATP from ADP (see Chapter 6), but this is also depleted within seconds. The exercising muscle then produces ATP via cellular respiration (see Chapter 8), burning glucose and fatty acids found in the bloodstream and in glycogen (long chains of glucose molecules) stored in the muscle. As you may recall, cellular respiration relies on an adequate supply of oxygen. During high-intensity exertion, muscles may not receive enough oxygen, even with the lungs and heart pumping at their maximum rates. In such cases, the inefficient process of glycolysis (see Chapter 8) can provide adequate levels of ATP. Glycolysis does not require oxygen, but it generates the by-product lactic acid, which builds up in overworked muscles and likely causes the burning sensation felt just before total exhaustion. As the body rests after exercise and heavy breathing restores oxygen levels, most of the lactic acid is transported in the bloodstream to the liver, where it is resynthesized into glucose, some of which is returned to the muscles and used to replenish glycogen stores.

Skeletal Muscle Responds to Exercise

Do bodybuilders have more muscle cells? Surprisingly, no. The total number of skeletal muscle fibers in an individual's body is established early in life, and although muscle fibers can be lost due to injury or normal aging processes, they are never replaced. They can, however, grow larger if more myofibrils are added. Repeated mechanical stresses on the muscle, such as occur during weightlifting, trigger the activation of genes that code for production of the actin and myosin that make up myofibrils. Intensive weight training can triple the size of a muscle; however, the old adage "use it or lose it" applies to both the muscular and skeletal systems. If you've ever broken a bone and had an arm or leg immobilized in a cast, you have probably seen firsthand that unused muscles lose mass over time. Inactivity, such as might occur during prolonged bed rest or the weightlessness of space flight, reduces or eliminates the need for the body to fight the forces of gravity, and takes a toll on muscles. People who have spinal cord injuries that prevent signals from reaching their muscles can help prevent the atrophy and eventual death of muscle fibers by electrically stimulating muscles or motor neurons.

Cardiac Muscle Powers the Heart

Cardiac muscle, located only in the heart, is both similar to and different from skeletal muscle (see Table 35-1). Although cardiac muscle fibers are smaller than most skeletal muscle cells and most possess only a single nucleus, cardiac muscle is striated, due to its regular arrangement of sarcomeres with their alternating thick and thin filaments. Like skeletal muscle, cardiac muscle contractions are induced when action potentials spread into the cell through the T tubules, causing calcium to be released from the sarcoplasmic reticulum; however, calcium also enters the cardiac muscle cytoplasm from the extracellular fluid. While skeletal muscle fibers contract only in response to action potentials that originate in motor neurons, cardiac muscle fibers can initiate their own contractions. This quality is particularly well-developed in the specialized cardiac muscle fibers of the *sinoatrial* (SA) *node,* which serves as the heart's pacemaker (see Chapter 28). Action potentials originating in the pacemaker spread rapidly through the heart, transmitted by specialized regions between cardiac muscle cells that contain numerous gap junctions, which allow action potentials to travel from one cell to the next, synchronizing their contractions. These connections also attach the muscle fibers together, allowing the force of contraction to pull adjacent cells. These connecting areas are absent in skeletal muscle, where each cell is individually stimulated by a branch of a motor neuron.

Smooth Muscle Produces Slow, Involuntary Contractions

Smooth muscle surrounds blood vessels and most hollow organs, including the uterus, bladder, and digestive tract. As its name suggests, smooth muscle is not striated; it lacks the regular arrangement of sarcomeres that characterizes skeletal and cardiac muscles (see Table 35-1). Like cardiac muscle fibers, smooth muscle cells contain only a single nucleus. Smooth muscle has less sarcoplasmic reticulum than skeletal muscle; as in cardiac muscle, calcium stored in the sarcoplasmic reticulum is supplemented by calcium influx from the extracellular fluid during an action potential. Smooth muscle cells are directly connected to one another by gap junctions, allowing synchronized contraction. Smooth muscle generally produces either slow, sustained contractions (such as the constriction of arteries that elevates blood pressure during times of stress) or slow, wavelike contractions (such as the peristaltic waves that move food through the digestive tract or expel a baby from the uterus during childbirth). Smooth muscle expands easily, as demonstrated by the bladder, the stomach, and the uterus. Smooth muscle contraction may be initiated by stretching, by hormones, by signals from the autonomic nervous system (see Chapter 34), or by some combination of these stimuli. Although contractions of the bladder can be initiated voluntarily, most smooth muscle contraction is involuntary.

35.3 What Does the Skeleton Do?

For most of us, the word *skeleton* conjures up the image of human bones. But skeletons are as diverse as any other structure in the animal kingdom and need not even be made of bone. A **skeleton** can be broadly defined as a supporting framework for the body. Within the animal kingdom, skeletons come in three radically different forms: *hydrostatic skeletons* (made of fluid), *exoskeletons* (on the outside of the animal), and *endoskeletons* (internal).

The **hydrostatic skeletons** of worms, mollusks, and cnidarians are the simplest, consisting of a fluid-filled sac (Fig. 35-5a). Fluid, which cannot be compressed, provides excellent support, but because fluid is formless, these animals rely on two layers of surrounding muscles in the body wall—one circular, the other longitudinal—to determine their shape. The wavelike movements of a burrowing earthworm, alternately extending to string-like thinness and then fattening as it contracts, provide an excellent illustration of the flexibility of hydrostatic skeletons.

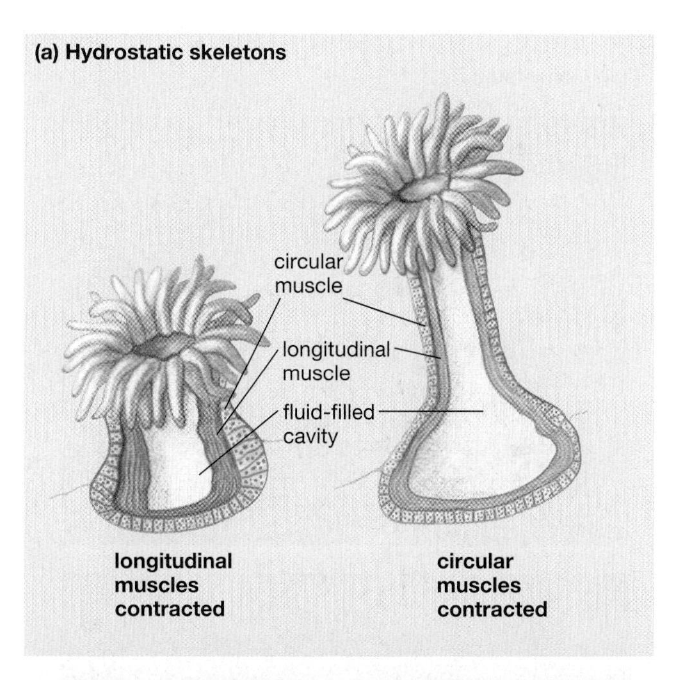

FIGURE 35-5a Not all skeletons are made of bone
A hydrostatic skeleton, such as in this sea anemone (art on top) and this earthworm (photo on bottom), is essentially a fluid-filled tube with muscular walls. Circular muscles form bands around the circumference of the tube, and longitudinal muscles run lengthwise. Because fluids are incompressible and can take any shape, if the circular muscles contract, the animal will become long and thin; if the longitudinal muscles contract, it will become short and fat.

Exoskeletons (literally, "outside skeletons") encase the bodies of arthropods (such as spiders, crustaceans, and insects). Exoskeletons vary tremendously in thickness and rigidity, from the thin flexible covering of many insects and spiders to the armorlike covering of many crustaceans (Fig. 35-5b). All exoskeletons are thin and flexible at the *joints* (moveable junctures of adjacent body parts), allowing complex and skillful movements such as those of a web-spinning spider. Exoskeletons and hydrostatic skeletons are discussed further in Chapter 22.

Endoskeletons, the internal skeletons of humans and other vertebrates, are found only in echinoderms and chordates (see Chapters 22 and 23). Although we think of internal skeletons as "normal," this is actually the least common type of skeleton in the animal kingdom.

(b) Exoskeletons

FIGURE 35-5b Not all skeletons are made of bone
Arthropods, such as this Sally Lightfoot crab (on top) and this cicada (on bottom, shown molting its exoskeleton), have armor-like skeletons on the outside of their bodies. Pairs of muscles move the exoskeleton at joints. **QUESTION** Why are thick, armor-like exoskeletons found mostly in water-dwelling animals, whereas land-dwelling insects and spiders tend to have thinner exoskeletons?

The Vertebrate Skeleton Serves Many Functions

The bony endoskeleton of humans and most vertebrates serves a wide variety of functions:

- The skeleton provides a rigid framework that supports the body and protects the internal organs. The central nervous system (CNS), for example, is almost completely enclosed within the skull and vertebral column; the rib cage protects the lungs and the heart with its major blood vessels.
- Vertebrates depend on bones for locomotion. Although muscle contractions provide the power, skeletal structures actually move the animal. Natural selection for efficient locomotion has produced the wonderfully designed wings, limbs, fins, and other complex skeletal structures that allow vertebrates to fly, run, swim, or slam dunk a basketball.
- Bones produce red blood cells, white blood cells, and platelets (see Chapter 28). In adults, these cells of the circulatory system are produced by *red bone marrow*, located in porous areas of bone in the sternum (breastbone), ribs, upper arms and legs, and hips.
- Bone serves as a storage site for calcium and phosphorus. Bone contains 99% of the calcium and 90% of the phosphorus in the human body. It absorbs and releases these minerals as needed, maintaining a constant concentration in the blood. *Yellow bone marrow*, dominated by fat cells, also stores energy reserves.
- The skeleton even participates in sensory transduction. As you may recall from Chapter 34, three tiny bones of the middle ear (hammer, anvil, and stirrup) transmit sound vibrations between the eardrum and the cochlea.

The 206 bones of the human skeleton can be placed in two categories: the axial skeleton and the appendicular skeleton. The **axial skeleton**, whose bones form the axis of the body, includes the bones of the head, vertebral column, and rib cage. The **appendicular skeleton**, whose bones form the appendages (extremities) and their attachments to the axial skeleton, includes the pectoral (shoulder) and pelvic (hip) girdles and the bones of the arms, legs, hands, and feet (Fig. 35-6).

35.4 Which Tissues Compose the Vertebrate Skeleton?

The vertebrate skeleton primarily consists of three types of tissue: *bone*, *cartilage*, and *ligaments*. Both cartilage and bone are rigid tissues that consist of living cells embedded in a matrix of collagen protein (see Chapter 27). Collagen also forms the tough bands of connective tissue called **ligaments**, which join bones at joints, allowing the

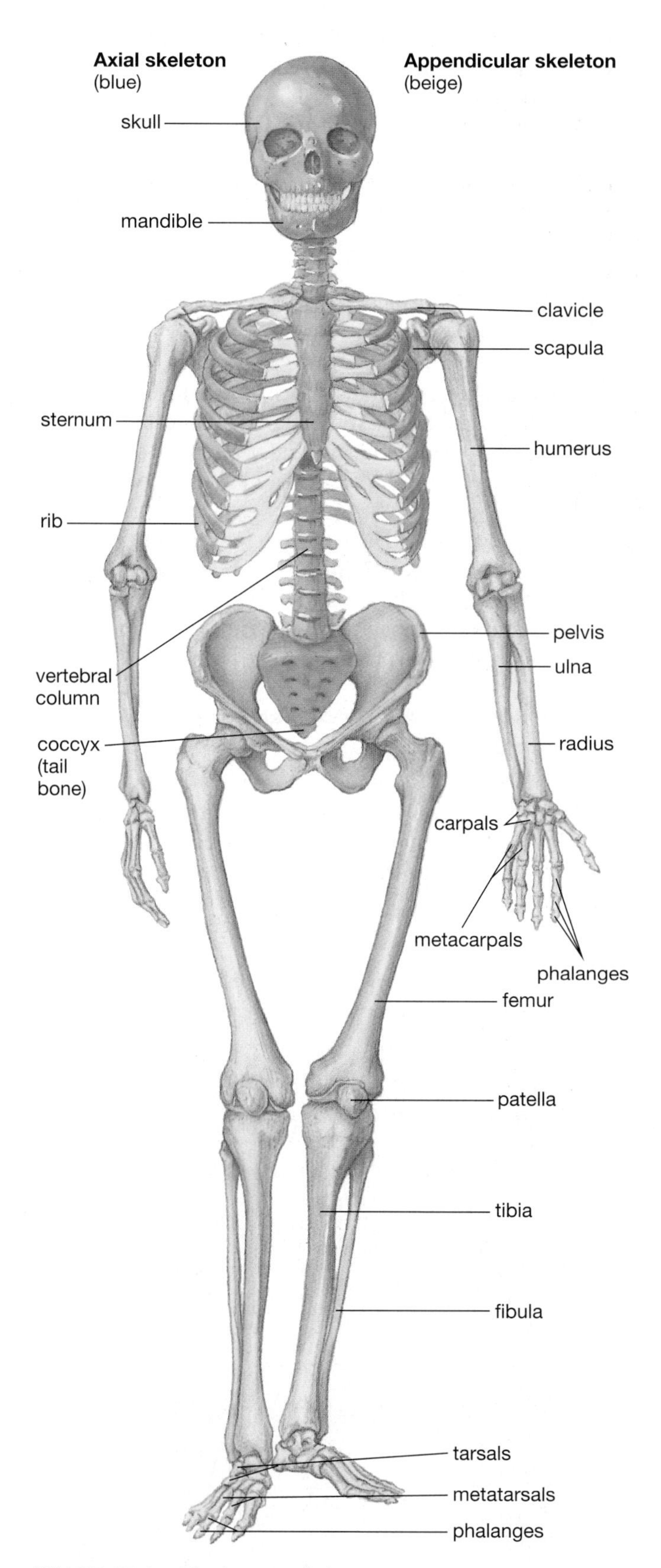

FIGURE 35-6 The human skeleton
The human skeleton, showing the axial skeleton (shaded in blue-gray) and the appendicular skeleton (bone color).

bones to move relative to one another. Ligaments are similar in structure to the tendons that connect muscles to bones.

Cartilage Provides Flexible Support and Connections

Cartilage plays many roles in the human skeleton. For example, during the development of the embryo, the skeleton forms first from cartilage that is later replaced by bone (Fig. 35-7). Cartilage also covers the ends of bones at joints, supports the flexible portion of the nose and external ears, connects the ribs to the sternum (breastbone), and provides the framework for the larynx, trachea, and bronchi of the respiratory system. In addition, it forms tough, shock-absorbing pads that cushion the knee joints and that form the **intervertebral discs** between the vertebrae of the backbone.

The living cells of cartilage are called **chondrocytes**. These cells secrete a flexible, elastic, noncellular matrix

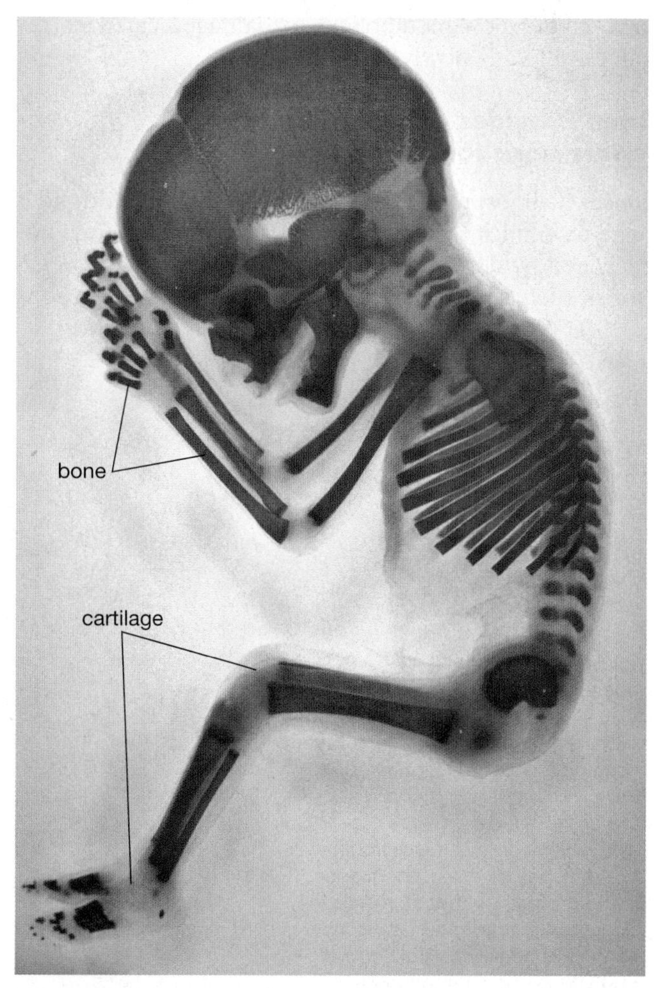

FIGURE 35-7 Bone replaces cartilage during development
In this 16-week-old human fetus, bone is stained magenta. The clear areas at the wrists, knees, ankles, elbows, and breastbone show cartilage that will later be replaced by bone.

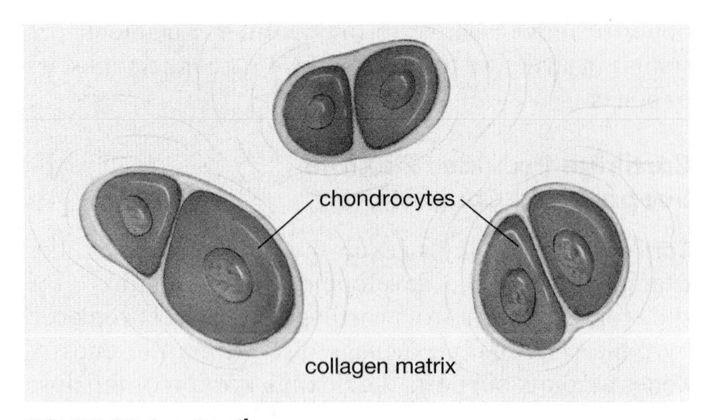

FIGURE 35-8 Cartilage
In cartilage, the chondrocytes, or cartilage cells, are embedded in an extracellular matrix of the protein collagen, which they secrete.

of collagen that surrounds them and forms the bulk of the cartilage (Fig. 35-8). No blood vessels penetrate cartilage; to exchange wastes and nutrients, chondrocytes must rely on the gradual diffusion of materials through the collagen matrix. As you might predict, cartilage cells have a very slow metabolic rate; damaged cartilage repairs itself very slowly, if at all.

Bone Provides a Strong, Rigid Framework for the Body

Bone is the most rigid form of connective tissue. Although bone resembles cartilage, the collagen fibers of bone are hardened by deposits of the mineral *calcium phosphate.* Bones, such as those supporting your arms and legs, consist of a hard outer shell of **compact bone**, with **spongy bone** in the interior (Fig. 35-9). Compact bone is dense and strong and provides an attachment site for muscle. Spongy bone is lightweight, rich in blood vessels, and highly porous. Bone marrow, where blood cells form, is found in cavities of spongy bone. In contrast to cartilage, bone is well supplied with blood capillaries.

There are three types of bone cells: **osteoblasts** (bone-forming cells), **osteocytes** (mature bone cells), and **osteoclasts** (bone-dissolving cells). Early in development, when bone replaces cartilage in the skeleton, osteoclasts invade and dissolve the cartilage; osteoblasts then replace it with bone.

As bones grow, osteoblasts form a thin layer covering the outside of the bone. The osteoblasts secrete a hardened matrix of bone and gradually become entrapped within it; matrix secretion ceases and the osteoblasts become mature osteocytes. Osteocytes are nourished by nearby capillaries and are connected to other osteocytes by thin extensions sent out by bone cells through narrow channels in the bone. Although unable to produce more bone, osteocytes may secrete substances that control the continuous remodeling of bone.

Bone Remodeling Allows Skeletal Repair and Adaptation to Stresses

Each year, 5% to 10% of all the bone in your body is dissolved away and replaced, a process called *bone remodeling.* This process allows your skeleton to subtly alter its shape in response to the demands placed on it. For example, bones that carry heavy loads or are subjected to

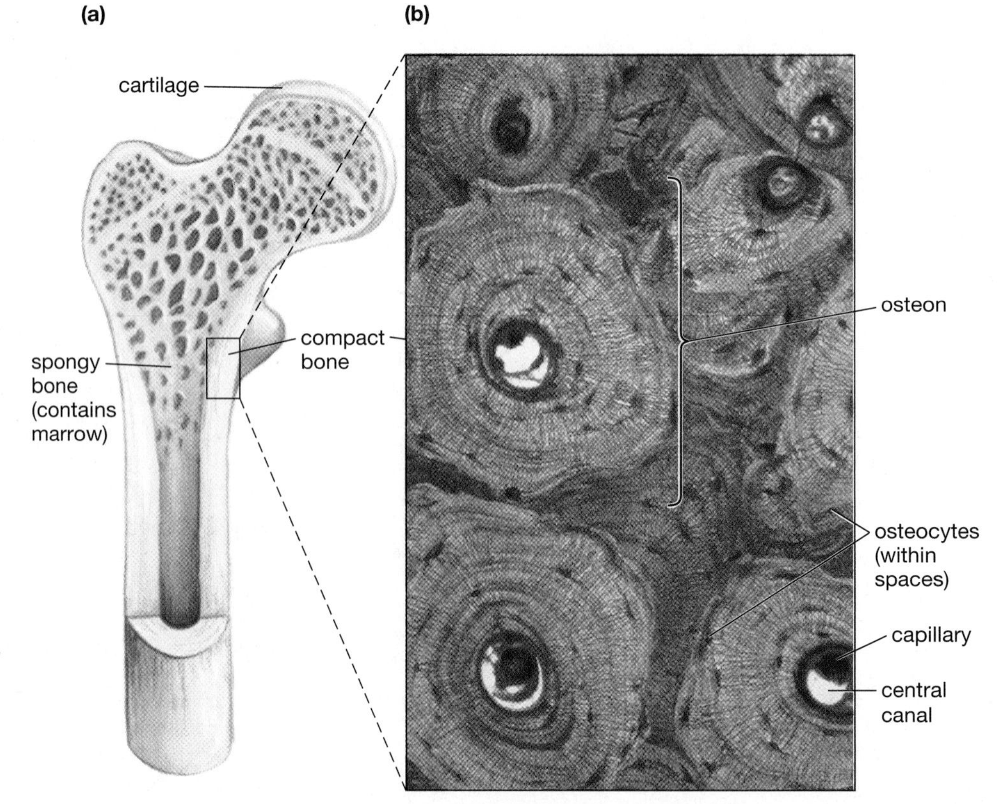

FIGURE 35-9 The structure of bone
(a) A typical bone, such as one in an arm or leg, consists of an outer layer of compact bone with spongy bone inside. For simplicity, blood vessels are not shown. **(b)** Osteons are clearly visible in this micrograph. Each includes a central canal containing a capillary that nourishes the osteocytes, embedded in the concentric rings of bone material.
QUESTION In some bone diseases, the collagen in the bone matrix is defective or present in abnormally small amounts, even though the calcium phosphate portion is normal. What would you predict to be the main symptom of such diseases?

HEALTH WATCH How Bones Heal

Bones are strong, but they're not indestructible. Too many of us have experienced the trauma and pain of a broken bone. We cringe at each recollection of the ill-fated mishap, the audible cracking sound, the dawning realization that the painful swelling and oddly bent-looking limb necessitate a trip to the emergency room. There, the attending physician coaxes the damaged bone back into its proper orientation and immobilizes it with a cast or splint. The rest of the healing process is up to the body's own repair mechanisms. Over the next 6 weeks or so, the body orchestrates a systematic restoration of the bone's strength and integrity.

The healing process begins when a large blood clot from vessels ruptured during the injury surrounds the break (Fig. E35-1, ①). Phagocytic cells and osteoclasts in the blood ingest and dissolve cellular debris and bone fragments. The fracture ruptures the *periosteum*, a thin layer of connective tissue that normally surrounds the bone and is rich in capillaries, osteoblasts, and osteoblast-forming cells. The osteoblasts, in conjunction with cartilage-forming cells, secrete a *callus*, a porous mass of bone and cartilage that surrounds the break (Fig. E35-1, ②). The callus replaces the original blood clot and holds the ends of the bones together while remodeling processes re-form the original shape of the bone. Once the callus is in place, osteoclasts, osteoblasts, and capillaries invade it. Nourished by the capillaries, osteoclasts break down cartilage while osteoblasts add new bone (Fig. E35-1, ③). Finally, osteoclasts remove excess bone, restoring the bone's original shape, but often leaving a slight thickening (Fig. E35-1, ④).

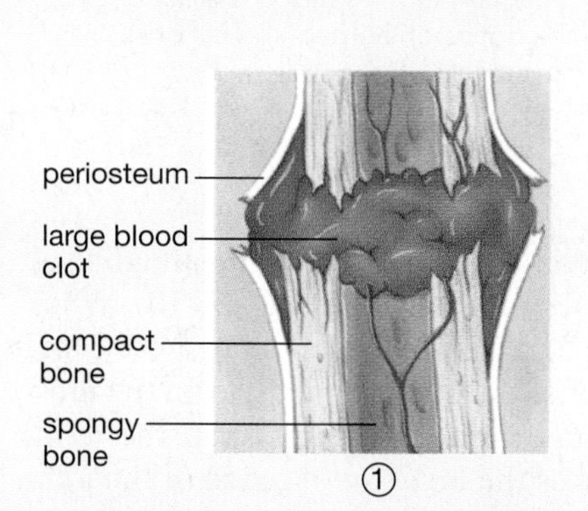

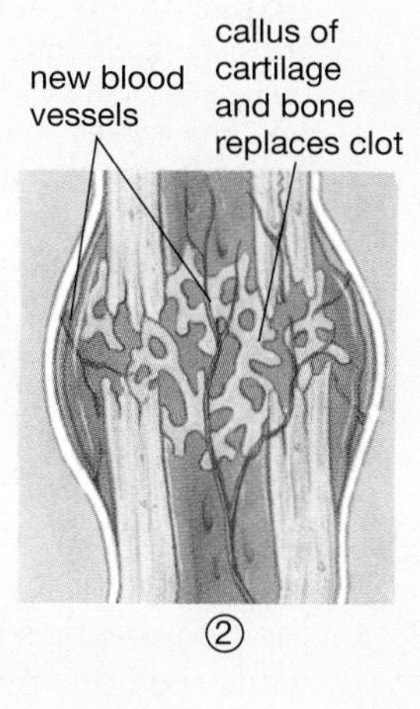

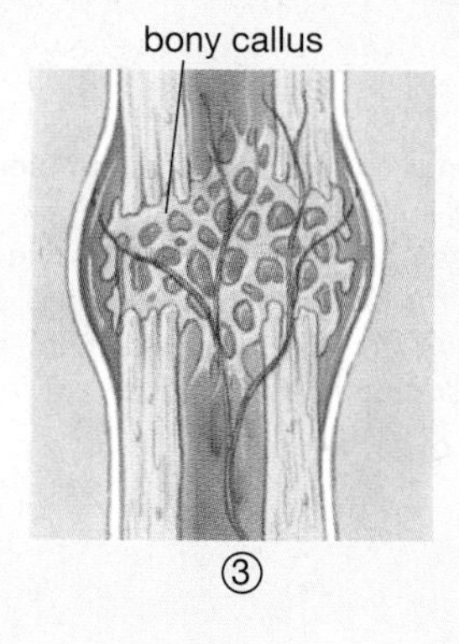

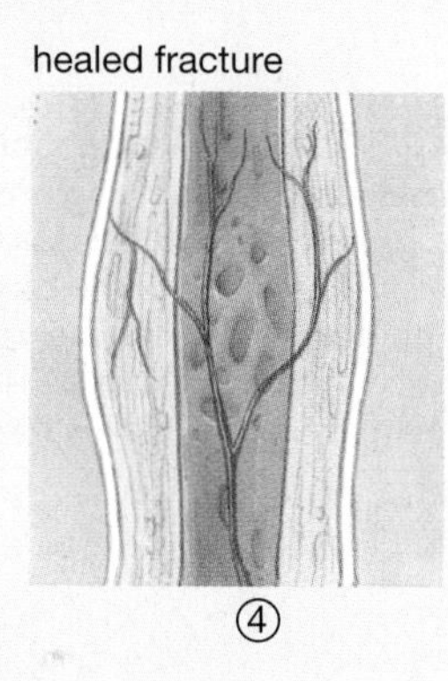

FIGURE E35-1 The steps in bone repair

extra stress become thicker, providing more strength and support. Archeologists excavating the skeletons of individuals buried in volcanic ash at Pompeii could identify archers because the bones of their right and left arms differed considerably in thickness. A professional tennis player may have 30% more bone mass in his or her playing arm. Actually, normal stress is a major factor in maintaining bone strength. The bones of a limb immobilized in a cast rapidly lose significant amounts of calcium. People confined to wheelchairs initially lose 1% to 2% of bone mass per month, although the loss eventually stops. Bed rest and space flight also reduce normal stresses on the bone, leading to bone loss.

Bone remodeling is the result of the coordinated activity of the osteoclasts that dissolve bone and the osteoblasts that rebuild it. Osteoclasts cling to the bone surface, secreting acids and enzymes that dissolve the hard matrix. Working in small groups, osteoclasts tunnel into the bone, creating channels. These channels are invaded by capillaries and osteoblasts. The osteoblasts fill the channel with concentric deposits of new bone matrix, leaving only a small opening for the capillary. As a result of this process, hard bone is made up of tightly packed units called **osteons** (also known as *Haversian systems*), each consisting of concentric layers of bone with embedded osteocytes. The concentric layers of bony material surround a *central canal*, through which a capillary passes (see Fig. 35-9). (Osteoclasts and osteocytes also play a crucial role in the repair of bone fractures, as described in "Health Watch: How Bones Heal.") The continuous turnover of bone also allows the body to maintain constant levels of calcium in the blood; calcium from bones is retained in the blood if blood calcium levels drop but is returned to bone if blood calcium is adequate or high. This process is regulated by hormones; calcitonin and parathormone cause bones to absorb calcium from the blood and release calcium into the blood, respectively (see Chapter 33). Early in life, the activity of osteoblasts outpaces that of osteoclasts, allowing bones to become larger and thicker as a child grows. In the aging body, however, the balance of power shifts to favor osteoclasts, and bones tend to become more fragile as a result. (See "Health Watch: Osteoporosis—When Bones Become Brittle.")

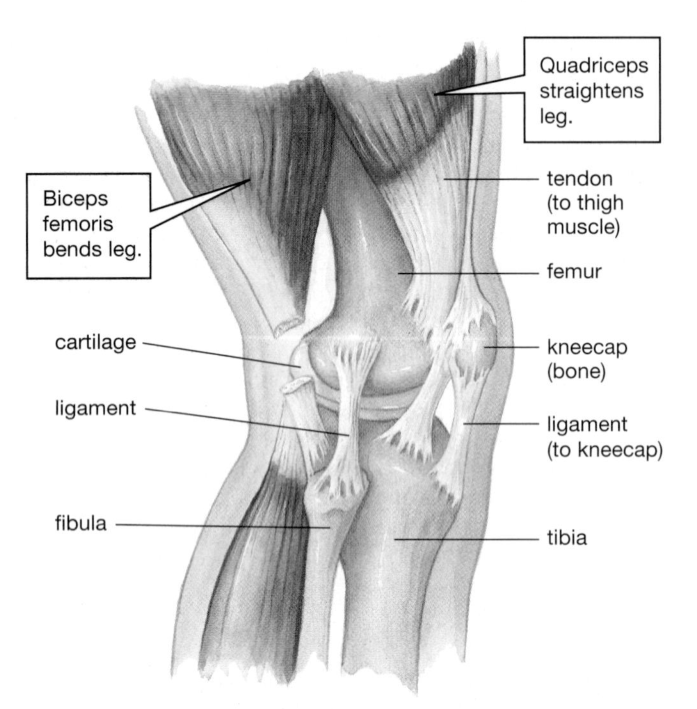

FIGURE 35-10 A hinge joint
The human knee showing antagonistic muscles (here, the biceps femoris and the quadriceps of the thigh), tendons, and ligaments. (The tendon of the biceps femoris has been cut for viewing purposes.) The complexity of this joint, coupled with the extreme stresses placed on it during activities such as jumping, running, or skiing, makes it very susceptible to injury.

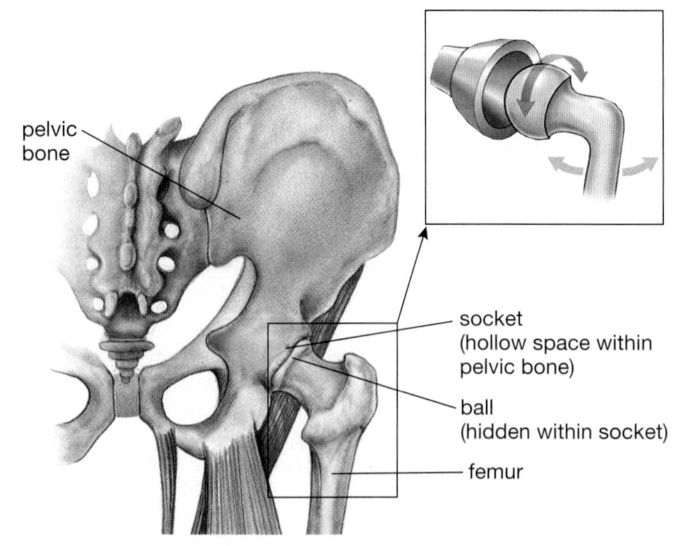

FIGURE 35-11 A ball-and-socket joint
The human hip joint consists of the rounded end of the femur (the ball; seen at the end of the femur) that fits into a cuplike depression (the socket; seen in the bones of the pelvis). This arrangement permits rotational movement.

35.5 How Does the Body Move?

In addition to providing support for the body, the skeleton facilitates movement by providing a framework that muscles can move. Movement of the skeleton is accomplished by the action of pairs of **antagonistic muscles**: one muscle actively contracts, causing the other to be passively extended (Fig. 35-10). Antagonistic muscles alter the configuration of the skeleton by causing movement around joints (in exoskeletons and vertebrate endoskeletons) or by altering the shape of the internal fluid (in hydrostatic skeletons; see Fig. 35-5a).

Muscles Move the Skeleton Around Flexible Joints

In vertebrates, bones act as levers that can be moved by the skeletal muscles to which they are attached. The bones typically move around **joints**, the points at which two bones meet. Not all joints are movable, but in those that move, the portion of each bone that forms the joint is coated with a layer of cartilage, whose smooth, resilient surface allows the bone surfaces to slide past one another during movement. On either side of a joint, skeletal muscles are attached to bones by tendons, and the bones themselves are joined at joints by ligaments (see Fig. 35-10).

Most skeletal muscles are arranged in antagonistic pairs on opposite sides of a joint (see Fig. 35-10). When one of the muscles in a pair contracts, it moves a bone and simultaneously stretches the opposing muscle. In the most common type of joints, skeletal muscles span the joint; their contraction moves the bone on one side of the joint while the bone on the other side of the joint remains in a fixed position. These joints, including the elbows, knees, and finger joints, are called **hinge joints**. Like a hinged door, these joints are movable in only two dimensions.

In hinge joints, pairs of muscles lie in roughly the same plane as the joint, and the members of the muscle pair are known as the **flexor** and the **extensor**. One end of each muscle, called the **origin**, is fixed to a relatively immovable bone on one side of the joint; the other end, the **insertion**, is attached to a mobile bone on the far side of the joint. When the flexor muscle contracts, it bends the joint; when the extensor muscle contracts, it straightens the joint. In Figure 35-10, for example, contraction of the biceps femoris bends the leg at the knee, while contraction of the quadriceps straightens it. Thus, alternate contractions of flexor and extensor muscles cause the movable bone to move back and forth at the joint.

Other joints, such as those of the hip and shoulder, are **ball-and-socket joints**, in which the rounded end of one bone fits into a hollow depression in another (Fig. 35-11). Ball-and-socket joints allow movement in several directions; compare the wide-ranging swinging of your upper arm or upper leg with the limited bending of your knee or elbow. The range of motion in ball-and-socket joints is made possible by at least two pairs of muscles, oriented at right angles to each other, that provide flexibility of movement.

HEALTH WATCH Osteoporosis—When Bones Become Brittle

As a human grows and matures, bone density increases steadily, reaching a peak at about age 35. After this point, however, the activity of osteoclasts exceeds that of osteoblasts, and bone density begins a slow, natural decline. Some bone loss is normal, but in people with **osteoporosis** (literally, "porous bones"; Fig. E35-2a), the loss is enough to weaken the bones, making them vulnerable to fractures and deformities. In many cases, the vertebrae of individuals with osteoporosis compress, causing a hunchbacked appearance (Fig. E35-2b). In extreme cases, simple activities such as lifting a shopping bag, opening a window, or sneezing can break a bone. Nearly one-third of women living to age 85 will fracture a hip weakened by osteoporosis.

Women are eight times as likely to suffer from osteoporosis as men. Why? The bones of women are about 30% less massive than those of men. Additionally, dietary calcium tends to be lower in women's diets than in men's, and women tend to consume even less calcium as they become older. As a result, when women begin to lose bone naturally, their bones may already be more fragile than they should be. Another factor unique to women is the hormone estrogen, which stimulates osteoblasts and helps maintain bone density. After menopause, estrogen production drops dramatically, and women may lose 3% to 5% of their bone mass each year for several years. Half of all women over age 65 are estimated to have some degree of osteoporosis. Alcoholism and smoking also contribute to bone loss and osteoporosis.

Bones thrive on moderate stress, but older people tend to be less active. Being inactive or bedridden (or weightless, as astronauts have discovered) results in rapid loss of bone calcium. Even in elderly people, weight-bearing exercise such as walking or dancing can reverse bone loss and even increase bone mass.

Some women who are at risk for osteoporosis choose medical therapy to help maintain bone density. The hormone calcitonin, administered as a nasal spray, has beneficial effects on bone deposition. A drug that mimics estrogen's effects on bone while blocking estrogen's effects on breast and uterine tissues is now available for use against osteoporosis. In the mid-1990s, molecular biologists engineered mice with extra copies of a newly discovered gene and observed that these mice had unusually thick bones. This gene codes for a protein (dubbed "OPG") that can shift the balance in favor of bone regeneration; OPG is being tested as an agent to prevent bone loss in postmenopausal women.

Someday astronauts in space and women at risk for osteoporosis may spend 10 to 20 minutes each day standing on a vibrating plate. Studies using rats, turkeys, and sheep have all reported impressive bone-strengthening as a result of vibrations barely perceptible to human volunteers. Preliminary work with postmenopausal women has been encouraging, and larger-scale human trials are currently underway. Researchers hypothesize that the minute stresses caused by the vibrations may activate bone formation much as do the everyday stresses imposed by muscle contractions.

Although medical intervention can partially reverse osteoporosis and slow its progression, there is currently no cure. Fortunately, much of the pain, incapacitation, and expense (estimated at $14 billion per year) caused by fractures due to osteoporosis can be prevented. The best way to prevent osteoporosis is a combination of regular exercise and adequate dietary or supplemental calcium and vitamin D (which is essential for deposition of calcium in bone). These measures will help ensure that bone mass is as high as possible before natural, age-related losses begin and will also minimize such losses in old age.

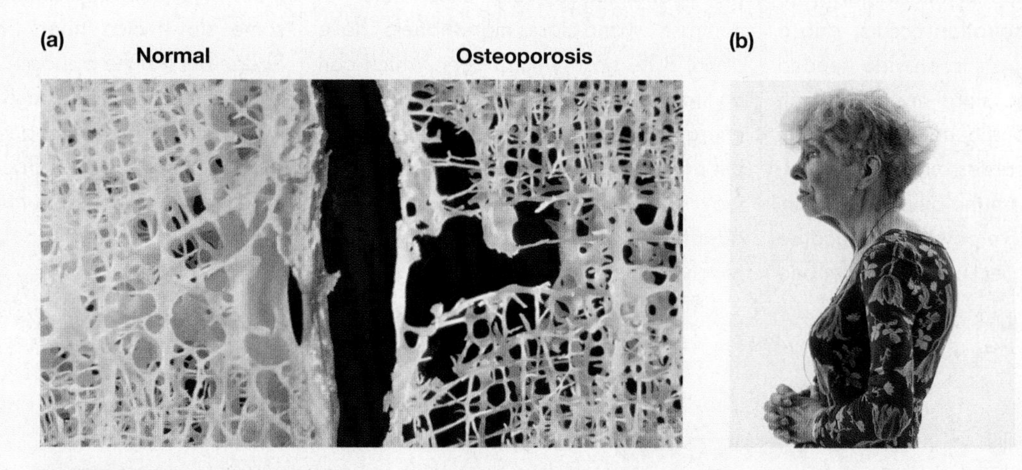

FIGURE E35-2 Osteoporosis
(a) Cross section of (left) a normal bone compared with (right) a bone from a woman with osteoporosis. **(b)** The devastating effects of osteoporosis extend beyond the obvious deformities, such as a hunchbacked appearance. Its victims are also at high risk for bone fractures.

CASE STUDY REVISITED Hidden Hazards of Space Travel

To prevent muscle loss during space flight, astronauts aboard the International Space Station exercise for about two hours daily, using a stationary bicycle and a treadmill. Bone loss, however, is more difficult to control. Researchers hypothesize that weightlessness favors the bone-dissolving activity of osteoclasts over the bone-strengthening actions of osteoblasts. Besides the effects this may have on bone strength, there are other hidden hazards. As unused bone dissolves, it releases extra calcium into the bloodstream, which may eventually be deposited as kidney stones. Since weightlessness will remain a fact of life for space travelers, scientists are trying to determine whether bone production can be stimulated chemically, perhaps by boosting enzymes or hormones that have already been shown to favor osteoblast production and activity. For example, scientists are testing the effects of the recently discovered bone-conserving protein OPG on mice taken on a 12-day shuttle mission to the International Space Station. To simulate prolonged weightlessness here on Earth, a team of international researchers recruited 28 hardy volunteers to spend 90 days confined to a bed (with a slight head-down tilt) to simulate the weightless conditions of space. One group was given a drug used to combat osteoporosis; a second group was required to exercise (head-down) three times weekly; and the third group served as a control. They will be given a battery of tests to determine the physiological effects of weightlessness on the muscular, skeletal, cardiovascular, and other physiological systems.

Consider This: Some people believe that research into space travel is a waste of money and talent, since there are many problems and unmet needs on Earth. How might a NASA scientist justify research on bone loss in space to a skeptical audience? Would you volunteer for a bed-rest study like the one described above? Why or why not?

Links to Life: Marathon Man

Next time you watch a track and field event, compare the legs of long-distance runners to those of sprinters. Although sprinters have larger calf muscles, marathoners have greater endurance. Why? Skeletal muscle fibers come in two basic types, fast-twitch and slow-twitch. Slow-twitch fibers contract more slowly. They have abundant mitochondria, in which cellular respiration occurs, and a plentiful blood supply to provide needed oxygen. They also store myoglobin, an oxygen-storing compound similar to the hemoglobin that carries oxygen in blood. Fast-twitch fibers, on the other hand, contract faster and more powerfully. Because they have a smaller blood supply and fewer mitochondria, they are better adapted to utilize glycolysis, which does not use oxygen and supplies ATP much more rapidly than cellular respiration. While a typical healthy adult has roughly even numbers of these two fiber types, champion sprinters have about 80% fast-twitch fibers, which allow bursts of amazing speed for a very short time. In contrast, world-class marathoners have about 80% slow-twitch fibers, which can contract for hours before exhausting the energy supplied by cellular respiration. But are marathoners or sprinters born this way, or do their muscles gradually adapt? The proportions of fast- and slow-twitch fibers vary greatly within the human population. Athletes with a high proportion of slow-twitch fibers are likely to excel in—and therefore gravitate to—endurance sports, while those with more fast-twitch fibers will find their niches in sports requiring bursts of energy. Can people alter their ratios of slow- to fast-twitch fibers? Researchers have found that very intensive training can convert some slow-twitch fibers into fast-twitch fibers, but not the opposite. Weight training, which selectively bulks up existing fast-twitch fibers, changes the relative masses of the different fiber types and is an effective way to train for sports requiring bursts of energy.

Summary of Key Concepts

35.1 An Introduction to the Muscular and Skeletal Systems

Muscles can only contract. Skeletal muscles work in close harmony with the skeleton, which provides a framework against which contraction can move the body. Contraction of other types of muscles pumps blood, moves air through the respiratory system, and moves food through the digestive tract.

35.2 How Do Muscles Work?

Skeletal muscle fibers (muscle cells) consist of subunits—myofibrils—surrounded by the sarcoplasmic reticulum. Sarcomeres within each myofibril are bounded by Z lines and include alternating thick filaments of myosin and thin filaments of actin and two accessory proteins (troponin and tropomyosin).

When stimulated by motor neurons at neuromuscular junctions, skeletal muscles produce action potentials. Calcium, stored in the sarcoplasmic reticulum, is released when an action potential invades the muscle fiber via the T tubules. Calcium causes the accessory proteins to move off myosin binding sites on the actin molecules of thin filaments. When the binding sites are exposed, myosin heads bind to them. Using ATP, the myosin heads bend, release, and reattach, sliding the filaments past each other and shortening the muscle fiber. Both the strength and the degree of muscle contractions are determined by the number of muscle fibers stimulated and the frequency of action potentials in each fiber. Rapid firing can result in maximum contraction.

Skeletal muscles rely on a steady supply of ATP derived primarily from glucose or fatty acids using cellular respiration, or from glycolysis which does not use oxygen and produces lactic acid as a by-product. In response to the stresses of maximal exertion, muscle fibers grow larger and stronger by adding myofibrils of actin and myosin, but do not increase in number.

Cardiac, or heart, muscle also consists of sarcomeres that contain alternating thick and thin filaments. Its cells tend to contract rhythmically and spontaneously, but these contractions are synchronized by electrical signals produced by specialized muscle fibers in the sinoatrial node. Cardiac muscle fibers are interconnected electrically by gap junctions that allow passage of electrical signals and coordinated contraction.

Smooth muscle lacks organized sarcomeres, but like cardiac muscle, its cells are electrically coupled by gap junctions. Smooth muscle surrounds hollow organs (uterus, digestive tract, bladder) and blood vessels, producing slow and sustained or rhythmic contractions, which are usually involuntary.

35.3 What Does the Skeleton Do?

Three types of skeletons are found in animals. Hydrostatic skeletons (in cnidarians, mollusks, and worms) consist of fluid confined in a chamber and surrounded by muscle. Exoskeletons, found in arthropods, are outer coverings with flexible joints. Endoskeletons, including the bony vertebrate skeleton, are found in both echinoderms and chordates.

The vertebrate skeleton provides support for the body, attachment sites for muscles, and protection for internal organs. Red blood cells, white blood cells, and platelets form in the marrow of bones. Bone acts as a storage site for calcium and phosphorus. The axial skeleton includes the skull, vertebral column, and rib cage. The appendicular skeleton consists of the pectoral and pelvic girdles and the bones of the arms, legs, hands, and feet.

35.4 Which Tissues Compose the Vertebrate Skeleton?

Cartilage is located at the ends of bones and forms pads in the knee joints and the intervertebral discs. It also supports the nose, ears, and respiratory passages. During embryological development, cartilage is the precursor of bone. Cartilage is formed by chondrocytes, which surround themselves with a matrix of fibrous collagen. Collagen also forms dense bands of tissue called ligaments that connect bones at movable joints, and tendons that connect muscles to bones.

Bone is formed by osteoblasts, which secrete a collagen matrix that becomes hardened by calcium phosphate. A typical bone consists of an outer shell of compact, hard bone, to which muscles are attached, and inner spongy bone, which provides a space for bone marrow. Remodeling of bone occurs continuously. Osteoclasts tunnel through the bone by means of acids and enzymes. Nourishing capillaries invade the tunnels, and osteoblasts fill the space with concentric layers of new bone, leaving a small central canal for the capillaries. This process produces osteons. Osteoblasts trapped within the bone are called osteocytes.

35.5 How Does the Body Move?

Skeletal muscles form antagonistic pairs that move the skeleton. In the vertebrate skeleton, movement occurs around joints, where bones are joined by ligaments. Muscles attach to bones on either side of the joint by tendons. The contraction of one muscle bends the joint and straightens its antagonistic muscle. At hinge joints, muscles are attached to the immovable bone at their origins; their insertions attach to the mobile bone. The contraction of the flexor muscle bends the joint; the contraction of its antagonistic extensor straightens it.

Key Terms

actin *p. 705*
antagonistic muscles *p. 714*
appendicular skeleton *p. 710*
axial skeleton *p. 710*
ball-and-socket joint *p. 714*
bone *p. 712*
cardiac muscle *p. 705*
cartilage *p. 711*
chondrocyte *p. 711*
compact bone *p. 712*
endoskeleton *p. 710*
exoskeleton *p. 710*
extensor *p. 714*
flexor *p. 714*
hinge joint *p. 714*
hydrostatic skeleton *p. 709*
insertion *p. 714*
intervertebral disc *p. 711*
joint *p. 714*
ligament *p. 710*
motor unit *p. 708*
muscle fiber *p. 705*
myofibril *p. 706*
myosin *p. 705*
myosin head *p. 706*
neuromuscular junction *p. 707*
origin *p. 714*
osteoblast *p. 712*
osteoclast *p. 712*
osteocyte *p. 712*
osteon *p. 713*
osteoporosis *p. 715*
sarcomere *p. 706*
sarcoplasmic reticulum *p. 706*
skeletal muscle *p. 705*
skeleton *p. 709*
smooth muscle *p. 705*
spongy bone *p. 712*
tendon *p. 705*
thick filament *p. 706*
thin filament *p. 706*
T tubule *p. 706*
Z line *p. 706*

Thinking Through the Concepts

To take a multiple-choice quiz with feedback on the contents of this chapter, visit http://www.prenhall.com/audesirk7. *Log in to the Web site selected by your instructor and navigate to the Self Test section for this chapter.*

? Review Questions

1. Sketch a relaxed muscle fiber containing a myofibril, sarcomeres, and thick and thin filaments. How would a contracted muscle fiber look by comparison?
2. Describe the process of skeletal muscle contraction, beginning with an action potential in a motor neuron and ending with the relaxation of the muscle. Your answer should include the following words: *neuromuscular junction, T tubule, sarcoplasmic reticulum, calcium, thin filaments, binding sites, thick filaments, sarcomere, Z line,* and *active transport.*
3. Explain the following two statements: muscles can only actively contract; muscle fibers lengthen passively.
4. What are the three types of skeletons found in animals? For one of them, describe how the muscles are arranged around the skeleton and how contractions of the muscles result in movement of the skeleton.
5. Compare the structure and function of the following pairs: spongy and compact bone, smooth and striated muscle, and cartilage and bone.
6. Explain the functions of osteoblasts, osteoclasts, and osteocytes.
7. How is cartilage converted to bone during embryonic development? Where is cartilage located in the body, and what functions does it serve?
8. Describe a hinge joint and how it is moved by antagonistic muscles.

Applying the Concepts

1. Discuss some of the problems that would result if the human heart were made of skeletal muscle instead of cardiac muscle.
2. Muscle fibers in individuals with Duchenne muscular dystrophy (DMD) lack a protein called *dystrophin*, which normally helps control calcium release from sarcoplasmic reticulum. Lack of dystrophin leads to a constant leaking of calcium ions, which activates an enzyme that dissolves muscle fibers. The gene that causes DMD is inherited as a sex-linked recessive gene. Women with this gene have a 50% chance of passing the disease to their male children and a 50% chance of passing the gene to their female children. Afflicted children gradually become unable to walk and die of respiratory problems as young adults. Tests have been developed that allow a woman to determine if she is a carrier and if her fetus has inherited the gene. What factors would make a woman a candidate for this test? If a woman discovers she carries this gene, what are her options with regard to having children? Discuss the ethical implications of these various options.
3. Myasthenia gravis is caused by the abnormal production of antibodies that bind to acetylcholine receptors on muscle cells and that eventually destroy the receptors. The disease causes muscles to become flaccid, weak, or paralyzed. Drugs, such as neostigmine, that inhibit the action of acetylcholinesterase (an enzyme that breaks down acetylcholine) are used to treat myasthenia gravis. How does neostigmine restore muscle activity?
4. Some insects would have a tough time flying if one nerve impulse was required for each muscle contraction. Gnats, for example, may beat their wings 1000 times a second. At such high frequencies, contraction is "myogenic," originating from the stretching caused by the contraction of antagonistic muscles. Also, insect flight muscle cells are filled with giant mitochondria. Suggest a mechanism to explain how myogenic contraction works inside cells. Explain the significance of giant mitochondria.
5. Human muscle cells contain a mixture of slow-twitch and fast-twitch fibers. Slow-twitch muscle cells break down ATP slowly; they contain many mitochondria and large amounts of myoglobin, a dark pigment that acts as a reservoir for oxygen. All fast-twitch muscle cells break down ATP rapidly and possess smaller amounts of myoglobin. The relative numbers of these fibers in different muscles is under genetic control. Use this information to explain the location of dark (myoglobin-containing) and white meat in birds.

For More Information

Andersen, J. L., Schjerling, P., and Saltin, B. "Muscle, Genes, and Athletic Performance." *Scientific American*, September 2000. Muscle physiology is influenced by both genes and training, and contributes to athletic ability.

Huyghe, P. "No Bone Unturned." *Discover*, December 1988. The story of a remarkable forensic anthropologist who uncovers the secrets hidden in unidentified skeletons.

Raloff, J. "Medicinal EMFs." *Science News*, November 13, 1999. Describes the use of electromagnetic fields and vibrations to heal bones and possibly help prevent osteoporosis.

Smith, K. K., and Kier, W. M. "Trunks, Tongues, and Tentacles: Moving with Skeletons of Muscle." *American Scientist*, January–February 1989. In certain organs of many animals, muscles provide support as well as movement.

Stossel, T. P. "The Machinery of Cell Crawling." *Scientific American*, September 1994. Cell movement relies on the orderly assembly and disassembly of scaffold proteins resembling those found in muscle.

Travis, J. "Boning Up." *Science News*, January 15, 2000. Researchers investigate a newly discovered gene whose protein shifts the balance between osteoblasts and osteoclasts.

Media Activities

To access a Media Activity visit http://www.prenhall.com/audesirk7. *Log in to the Web site selected by your instructor, navigate to this chapter, and select the appropriate Media Activity number.*

35.1 Muscle Structure

Estimated time: 5 minutes

In this animation you will explore the various levels of structure found within skeletal muscles.

35.2 Muscle Contraction

Estimated time: 10 minutes

The contraction of skeletal muscles is described as a sliding-filament mechanism. Here you will review the mechanism and the molecular processes that produce a muscle contraction.

35.3 Web Investigation: Hidden Hazards of Space Travel

Estimated time: 30 minutes

Take a closer look at the effects of weightlessness on the human body and get a new appreciation of the problems that astronauts encounter in space.

CHAPTER

36 Animal Reproduction

The frozen zoo consists of tissue samples and sex cells, often of endangered species, preserved in liquid nitrogen. (*Inset*) Born in 1990, this test-tube tiger is the result of the first successful use of *in vitro* fertilization in the fight to save endangered species.

AT A GLANCE

CASE STUDY The Frozen Zoo

In January 2000, the last living bucardo, a mountain goat native to the Spanish Pyrenees, was killed by a falling tree. Did another unique species vanish from the planet? Perhaps not. Some bucardo cells live on, having been cryopreserved (kept alive in a deeply frozen state) before the species' untimely extinction. The genetic blueprint of the bucardo now exists only in a "frozen zoo"—a collection of tissues, sperm, and eggs from a huge variety of species. Many of these species are endangered, but most are not yet extinct. The San Diego Zoo houses tissue samples from over 5400 animals, all stored in a single room in canisters of liquid nitrogen at −320 °F. There are a few dozen such frozen zoos on the planet, providing the raw material for a unique brand of wildlife conservation using "assisted reproductive technology" (ART).

This relatively new approach to wildlife conservation involves techniques such as artificial insemination, *in vitro* fertilization, interspecies embryo transfers (the use of surrogate mothers from different but related species), and even cloning. These techniques are surrounded by controversy and frustration, yet they have led to some inspiring success stories. Artificial insemination has become a cornerstone in efforts to save the black-footed ferret, giant panda, and cheetah, and *in vitro* fertilization (IVF, in which sperm and egg are joined in a dish) is also well-established. A major advantage of IVF is that it allows sperm from a male of an endangered species to be transported between continents, if necessary, to fertilize an appropriate female. This method eliminates the danger and trauma of transporting the animals themselves, and also circumvents the very real probability that, once together, the animals will refuse to mate. In a landmark success in April 1990, the first "test-tube" Siberian tiger was born (inset); only about 200 individuals of this rare Siberian subspecies remain in the wild. Ironically, the loss of species that people are trying to prevent by using assisted reproductive technology is due to the reproductive success of a single species—*homo sapiens*. The burgeoning human population threatens wildlife throughout the world, as our energy use modifies the global climate and we continue to usurp wildlife habitat for food, housing, and the extraction of natural resources.

How do animals, including people, reproduce? What options do we have for controlling reproduction? How does assisted reproductive technology help infertile couples have children? Are high-tech solutions for preserving endangered species justified?

36.1 How Do Animals Reproduce?

Animals reproduce either sexually or asexually. As you learned in Chapter 11, **sexual reproduction** requires the production of haploid gametes through meiosis. In a process called **fertilization**, two gametes—usually from separate parents—fuse and give rise to a diploid individual, which then divides mitotically to produce a diploid individual. Because the offspring receives genes from each of its two parents, its genome is not identical to that of either parent. In contrast, **asexual reproduction** involves only a single animal that produces offspring through repeated mitoses of cells in some part of its body. The offspring are, therefore, genetically identical to the parent.

Because humans and most other animals reproduce sexually, we tend to regard sexual reproduction as the normal and best method. In reality, however, asexual reproduction is far more efficient because there is no need to find and court a mate or fend off rivals; additionally, there is no waste of sperm and eggs that never meet. Not surprisingly, many animals reproduce asexually at least some of the time. Let's begin, then, with a brief survey of asexual reproduction among animals before we move on to sexual reproduction.

Asexual Reproduction Does Not Involve the Fusion of Sperm and Egg

Budding Produces a Miniature Version of the Adult

Many sponges and cnidarians, such as *Hydra* and some sea anemones, reproduce by **budding** (Fig. 36-1). A miniature version of the animal—a *bud*—grows directly on the body of the adult, drawing nourishment from its parent. When it has grown large enough, the bud breaks off and becomes an independent individual.

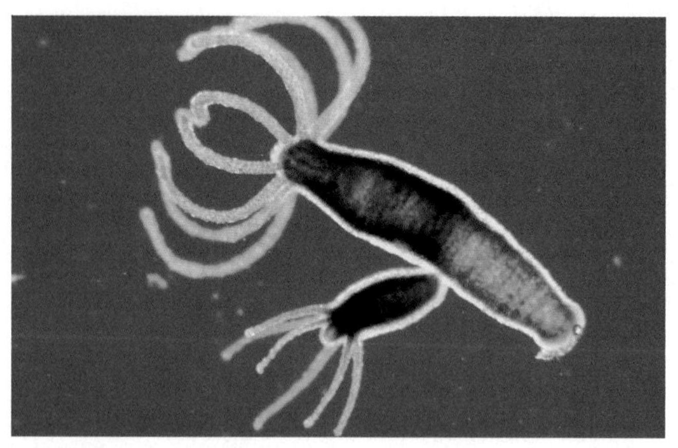

FIGURE 36-1 Budding
The offspring of some cnidarians, such as the *Hydra* shown here, grow as buds that appear as miniature adults sprouting from the body of the parent. The buds eventually break off as independent organisms. **QUESTION** Which type of cell division gives rise to the cells of the bud's body?

Fission Followed by Regeneration Can Produce a New Individual

Many animals are capable of **regeneration**, the ability to regrow lost body parts. For example, sea stars will regenerate an arm that is lost to an accident or predator. Even if multiple arms are lost, they can all be replaced as long as most of the central disc remains intact. However, the regenerative abilities of sea stars are not part of a reproductive strategy, because in most cases no new individuals are formed.

Nonetheless, some species do use regeneration for true reproduction; they reproduce by **fission**. Several annelid and flatworm species can reproduce by dividing into two or more pieces, each of which regenerates an entire body (Fig. 36-2). A few brittle star species routinely reproduce in a similar fashion; they split apart and each half regenerates a complete animal. Some coral species can divide lengthwise to produce two smaller but complete individuals. Among cnidarian medusae (that is, cnidarians with bell-shaped bodies; see Chapter 22), a few species can reproduce by a fission process in which the entire bell folds in half, splitting the stomach and other organs into two parts, and then divides into two individuals.

During Parthenogenesis, Eggs Develop Without Fertilization

The females of some animal species can reproduce by a process known as **parthenogenesis**, in which haploid egg cells divide mitotically and develop into adults without being fertilized. In some species, parthenogenetically produced offspring remain haploid. For example, male honeybees develop from unfertilized eggs and are haploid; their diploid sisters develop from fertilized eggs. Some fish, amphibians, and reptiles reproduce parthenogenetically, but restore the diploid number of chromosomes by duplicating all their chromosomes either before or after meiosis. All the resulting offspring are female.

Some species of fish, including relatives of the mollies and platys that are popular in tropical fish tanks, and some lizards, such as the whiptail of the southwestern United States and Mexico, have populations consisting entirely of parthenogenetically reproducing

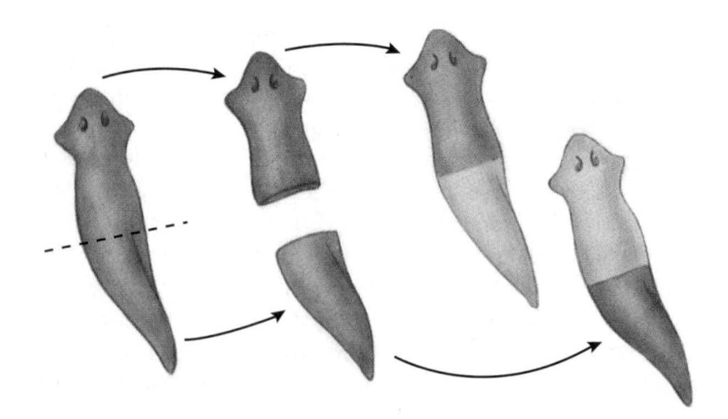

FIGURE 36-2 Fission followed by regeneration
Some flatworm species reproduce by dividing across the middle. Each offspring then regenerates the missing half of its body.

FIGURE 36-3 A female aphid gives live birth
In spring and early summer, when food is abundant, aphid females reproduce parthenogenetically; in fact, the females are born pregnant! In fall, they reproduce sexually. Aphids have the ability to exploit the advantages of both asexual reproduction (rapid population growth during times of abundant food, no energy spent in seeking a mate, no wasted gametes) and sexual reproduction (genetic recombination). **QUESTION** Why is fall the season in which aphids switch to sexual reproduction?

females. Still other animals, such as the aphid, can reproduce either sexually or parthenogenetically, depending on environmental factors such as the season of the year and the availability of food (Fig. 36-3).

Sexual Reproduction Requires the Union of Sperm and Egg

Given the obvious efficiency of asexual reproduction, no one is sure why sex evolved and became the dominant form of reproduction. Sex does have an important outcome: the genetic recombination that results from sexual reproduction creates novel genotypes—and therefore novel phenotypes—that are an important source of variation upon which natural selection may act.

In animals, sexual reproduction occurs when a haploid sperm fertilizes a haploid egg, generating a diploid offspring. In most animal species, an individual is either male or female. The sexes are defined by the type of gamete that each produces. Females produce **eggs**, which are large, nonmotile cells containing food reserves. Males produce small, motile **sperm**, which have almost no cytoplasm and no food reserves.

In other species, such as earthworms and many snails, single individuals produce both sperm and eggs. Such individuals are commonly called **hermaphrodites** (after Hermaphroditos, a male Greek god whose body merged with that of a female water nymph, producing a half-male and half-female being). In most hermaphroditic species, reproduction involves a mutual exchange of sperm between individuals. In some hermaphroditic species, however, individuals can fertilize their own eggs if a mate is unavailable. These animals, including tapeworms and many pond snails, are relatively immobile and may find themselves isolated from other members of their species. Obviously, the ability to fertilize oneself is advantageous under these circumstances.

For species with two distinct sexes and hermaphrodites that cannot self-fertilize, successful reproduction requires that sperm and eggs from different animals be brought together for fertilization. The union of sperm and egg is accomplished in a variety of ways, depending on the mobility of the animals and whether they breed in water or on land.

External Fertilization Occurs Outside the Parents' Bodies

In **external fertilization**, the union of the sperm and egg takes place outside the bodies of the parents. When animals breed in water, the parents release sperm and eggs into the water, through which the sperm swim to reach an egg. This process is called **spawning**. Because sperm and eggs are relatively short-lived, spawning animals must synchronize their reproductive behaviors, both *temporally* (male and female spawn at the same time) and *spatially* (male and female spawn in the same place). Synchronization may be achieved through chemical signals, courtship behaviors, environmental cues, or some combination of these factors.

Most spawning animals rely on environmental cues to some extent. Breeding usually occurs only during certain seasons of the year, and cues such as seasonal changes in day length typically stimulate the physiological changes that lead to readiness for breeding. More precise synchrony, however, is required to coordinate the actual release of sperm and egg. For example, many corals of Australia's Great Barrier Reef synchronize spawning by the phase of the moon. On the fourth or fifth night after the full moons of November and December, all the corals of a particular species on an entire reef release a blizzard of sperm and eggs into the water (Fig. 36-4).

FIGURE 36-4 Environmental cues may synchronize spawning
In the Great Barrier Reef of Australia, thousands of corals spawn simultaneously, creating this "blizzard" effect. Spawning in these corals is linked to the phase of the moon. (*Inset*) Close-up of a package of sperm and eggs erupting from a spawning hermaphroditic coral.

Other animals communicate their sexual readiness to one another by sending visual, acoustic, or chemical signals. Chemical signals are especially common among immobile or sluggish invertebrates, such as mussels and sea stars. These animals release chemical signals called **pheromones** into the water, where they are sensed by other members of the species. Usually, a female ready to spawn releases eggs and a sex pheromone into the water. Nearby males, detecting the mating pheromone, immediately release millions of sperm. The sperm themselves are lured by a chemical attractant released by the eggs in some, if not most, animals. Such "egg pheromones," which have been detected in animals as diverse as sea stars and humans, help ensure fertilization.

Synchronized timing alone does not guarantee efficient reproduction. Corals, sea stars, and mussels all waste enormous quantities of sperm and eggs because the gametes are released relatively far apart. In species of mobile animals, both temporal *and* spatial synchrony can be ensured by mating behaviors. Most fish, for example, have some form of courtship ritual in which the male and female come close together and release their gametes in the same place and at the same time. The courtship dances of Siamese fighting fish and the seahorse provide exquisite examples (Fig. 36-5). The female seahorse, laden with eggs, approaches the male and initiates an elaborate dance in which the partners approach, quiver, and nod heads before entwining their tails and lining up their bodies face to face. In an unusual reversal of sex roles, the female inserts a tube for depositing eggs into the pouch in the male's abdomen. As she injects her eggs into the pouch, the male releases a cloud of sperm from an opening just above the pouch, then seals the fertilized eggs into his pouch. The eggs eventually hatch within the pouch, and a few weeks later, the male later gives birth to perfect miniature seahorses (see Fig. 37-2a). Frogs and toads assume a characteristic mating pose called *amplexus* (Fig. 36-6). In shallow water near the edges of ponds and lakes, the male mounts the female and prods the sides of her abdomen. This stimulates her to release eggs, which he fertilizes by releasing a cloud of sperm above them. The golden toad shown here in amplexus was once abundant in the cloud forests of Costa Rica, but has not been seen since 1989. It is presumed to be extinct. Its disappearance, which scientists believe was caused by environmental changes due to global warming, was so sudden that no one had thought to preserve its genetic heritage in a frozen zoo; it is lost forever.

(a)

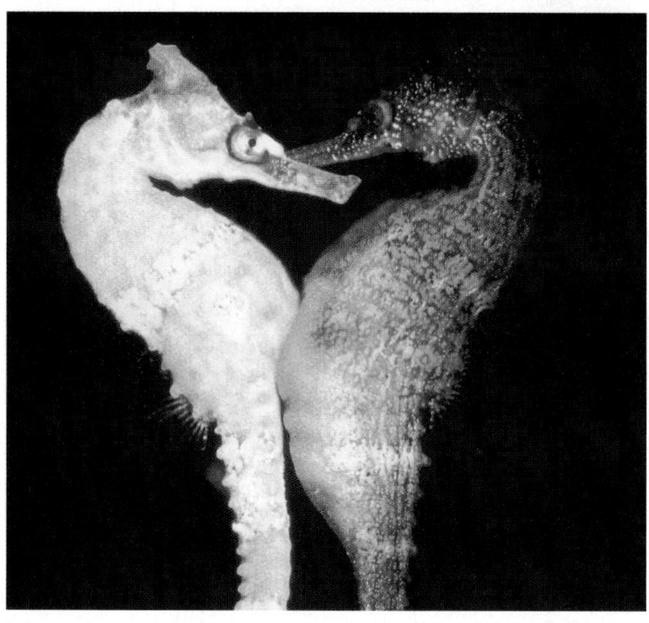

(b)

FIGURE 36-5 Courtship rituals synchronize release of sperm and eggs
(a) Courtship rituals among Siamese fighting fish *(Betta splendens)* ensure fertilization of the female's eggs, as male and female curl about one another, releasing sperm and eggs together. The male retrieves the eggs as they fall, spits them into his bubble nest (seen here as bubbles floating on the surface above him), and cares for the offspring during their first few weeks of life. **(b)** Spawning in the seahorse requires the male and female to orient their bodies so that the female can deposit her eggs in the male's pouch. **QUESTION** In addition to ensuring synchronized release of gametes, what other advantage is gained by performing courtship rituals?

Internal Fertilization Occurs Within the Female Body

In **internal fertilization**, sperm are taken into the body of the female, where the egg is fertilized. Internal fertilization is an important adaptation to terrestrial life: sperm must be bathed in fluid until they reach the eggs; on land, this liquid passage is best achieved inside the female's body. Even in aquatic environments, internal fertilization increases the likelihood of successful fertilization, because the sperm and eggs are confined together in a small space rather than relying on encounters within a large volume of water.

Internal fertilization usually occurs by **copulation**, the behavior by which the male deposits sperm directly into

FIGURE 36-6 Golden toads in amplexus
The smaller male clutches the female and stimulates her to release eggs. Golden toads are now believed to be extinct.

the reproductive tract of the female (Fig. 36-7). In a variation of internal fertilization, males of some species package their sperm in a container called a **spermatophore** (Greek for "sperm carrier"). In many spermatophore-producing species, including some scorpions, grasshoppers, and salamanders, no copulation occurs. The male simply drops a spermatophore on the ground, and if a female finds it, she fertilizes herself by inserting it into her reproductive cavity, where the enclosed sperm are liberated. Some male octopuses transfer a spermatophore to a cavity in the female's body using a special tentacle that breaks off and remains inside the cavity. Before this process was first observed, biologists believed that the detached tentacles inside the females were parasitic worms.

When animals must copulate to reproduce (or when a mating ritual is required for spawning), males may compete for access to copulations with females. This competition has driven the evolution of a wide variety of sexually selected structures and reproductive behaviors (see Chapters 15 and 38). One spectacular example of competition for access to mates occurs in the early spring of each year in the woods of western Canada. As the snows melt and the ground warms, male red-sided garter snakes emerge from the underground dens where they hibernate by the thousands. Later the females emerge, and a mating frenzy begins. In a sea of thousands of writhing snake bodies, each female attracts a crowd of dozens or even hundreds of males. Only one will copulate successfully.

Simply depositing sperm into the body of the female, however, does not guarantee fertilization; a mature egg must also be present. Many female snails and insects store sperm within their bodies for days or even months, thus ensuring a supply of sperm whenever eggs are ready. Among mammals, mating behavior must be synchronized. For example, copulation usually occurs only at certain seasons of the year or when the female signals readiness to mate. The season or signal typically

(a)

(b)

(c)

FIGURE 36-7 Internal fertilization is essential for reproduction on land
(a) Ladybugs mate on a dandelion flower. **(b)** South American tortoises must cope with confining shells. **(c)** King penguins mate comfortably in the snow.

coincides with **ovulation**, the release of the egg cell from the ovary. In a few mammals, such as rabbits, copulation triggers ovulation. Zoo scientists attempting to breed a rare female Sumatran rhino finally discovered that, in this species, ovulation is stimulated by courtship. A pair of rhino horns (which, ironically, are often sold to be ground up and eaten as an aphrodisiac) are valued at $50,000; therefore, the remaining 300 wild Sumatran rhinos are at risk from poachers as well as habitat loss. The first Sumatran rhino born in captivity in over a century was made possible by increased knowledge of their reproductive physiology.

36.2 How Does the Human Reproductive System Work?

Like other mammals, humans have separate sexes and reproduce sexually, with internal fertilization. The **gonads** of mammals are paired organs that produce sex cells—sperm and eggs. Although most mammal species reproduce only during certain seasons of the year and consequently produce sperm and eggs only at that time, human reproduction is not restricted by season. Men produce sperm more or less continuously, and women *ovulate* (release a mature egg cell) about once a month.

The Ability to Reproduce Begins at Puberty

Sexual maturation occurs at **puberty**, a stage of development characterized by rapid growth and the appearance of secondary sexual characteristics in both sexes. Although puberty generally begins in the early teens, it may occasionally start as early as age 8 or as late as age 15. During puberty in both sexes, brain maturation causes the hypothalamus to increase **gonadotropin-releasing hormone (GnRH)**, which stimulates the anterior pituitary to produce **luteinizing hormone (LH)** and **follicle-stimulating hormone (FSH)**. FSH and LH, in turn, stimulate the production of gametes (sperm or eggs). These hormones also stimulate the testes to produce more of the male sex hormone **testosterone** and the ovaries to produce more of the female sex hormone **estrogen**. In response to increases in testosterone, males develop secondary sexual char-

Table 36-1 The Human Male Reproductive Tract

Structure	Function
Testis (male gonad)	Produces sperm and testosterone
Epididymis and vas deferens (ducts)	Store sperm; conduct sperm from testes to penis
Urethra (duct)	Conducts semen from vas deferens and urine from urinary bladder to the tip of the penis
Penis	Deposits sperm in female reproductive tract
Seminal vesicles (glands)	Secrete fluid into semen
Prostate gland	Secretes fluids into semen
Bulbourethral glands	Secrete fluid into semen

Media Activity
36.1 Human Reproductive System

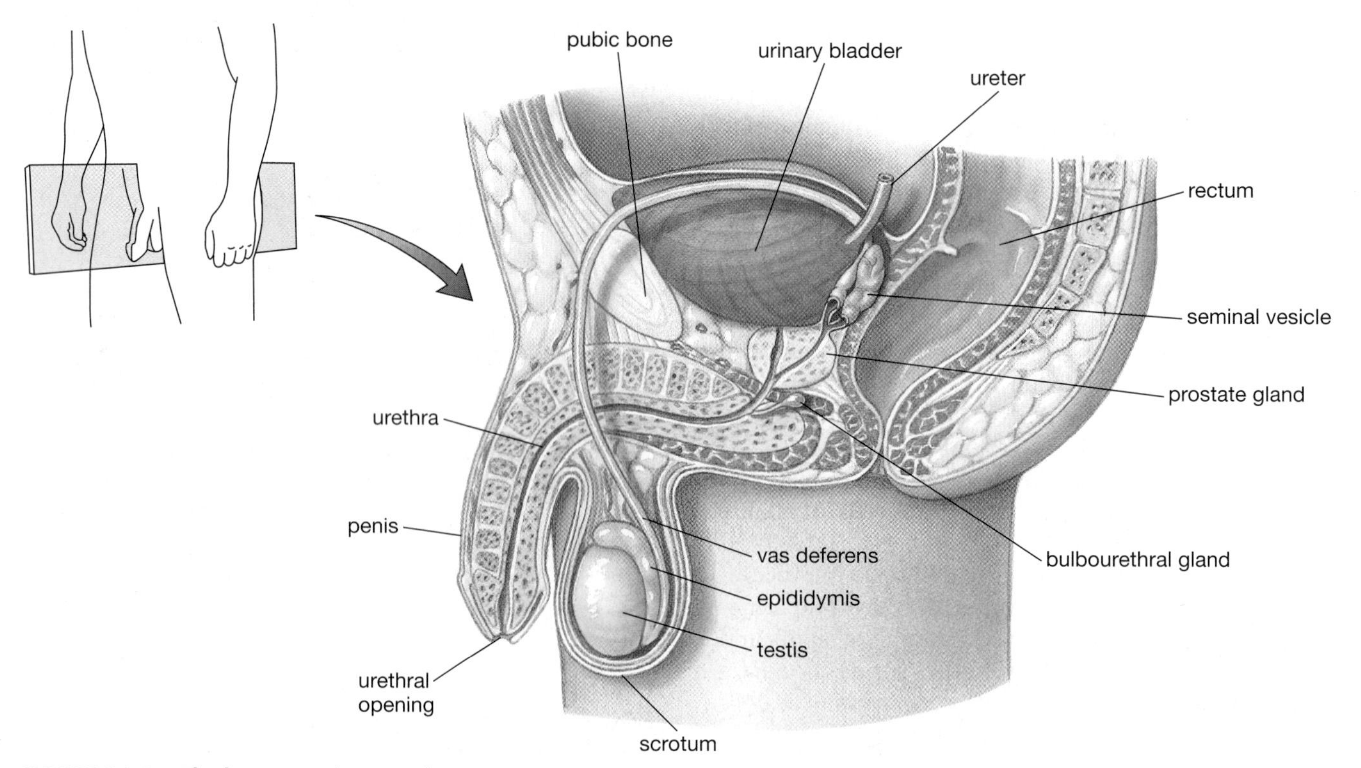

FIGURE 36-8 The human male reproductive tract
The male testes hang beneath the abdominal cavity in the scrotum. Sperm pass from the seminiferous tubules of a testis to the epididymis, then through the vas deferens and urethra to the tip of the penis. Along the way, fluids are added from the seminal vesicles, the bulbourethral glands, and the prostate gland.

acteristics: the **penis** and testes enlarge; pubic, underarm, and facial hair appears; the larynx enlarges (causing a deepening voice); and muscular development increases. In response to increased estrogen (and other hormones that surge at puberty), females develop enlarged breasts and pubic and underarm hair, and begin menstruating.

The Male Reproductive Tract Includes the Testes and Accessory Structures

The central structures of the male reproductive tract are the **testes** (singular, *testis*, which are the gonads that produce sperm. The male reproductive system (Table 36-1 and Fig. 36-8) also includes accessory structures that secrete substances that activate and nourish the sperm, store it, and conduct it to the female reproductive tract.

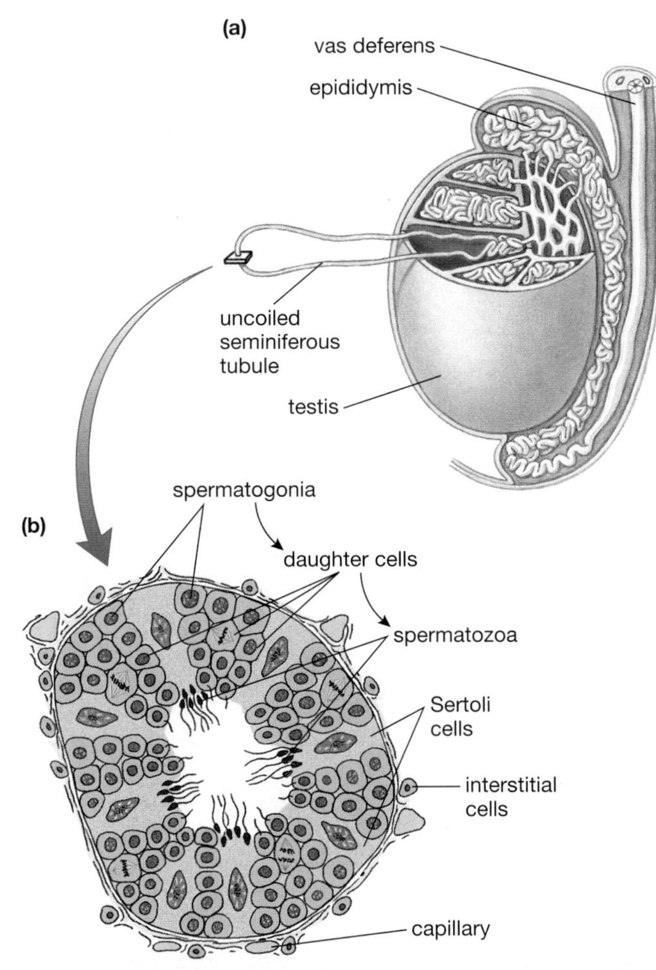

FIGURE 36-9 The structures involved in spermatogenesis
(a) A section of the testis, showing the seminiferous tubules, epididymis, and vas deferens. **(b)** Cross section of a seminiferous tubule. The walls of the seminiferous tubules are lined with Sertoli cells and spermatogonia. As spermatogonia undergo meiosis, the daughter cells move inward, enfolded in the Sertoli cells. There, they differentiate into sperm (spermatozoa), nourished by the Sertoli cells. Mature sperm are freed into the central cavity of the tubules for transport to the penis. Testosterone is produced by interstitial cells in the spaces between tubules.

Sperm Are Produced in the Testes

The testes, which produce both sperm and male sex hormones, are located in the **scrotum**, a pouch that hangs outside the main body cavity. This location keeps the testes about 4 °C cooler than the core of the body, providing the optimal temperature for sperm development. (Tight pants push the scrotum up against the body, raising the temperature of the testes. Some researchers believe that such clothing reduces sperm counts and decreases fertility. This is not, however, a reliable means of birth control!) Coiled, hollow **seminiferous tubules**, in which sperm are produced, nearly fill each testis (Fig. 36-9a). **Interstitial cells**, which synthesize the male hormone testosterone, are located in the spaces between the tubules.

Just inside the wall of each seminiferous tubule lie **spermatogonia** (singular, *spermatogonium*), the diploid cells from which the sperm eventually will arise, and the much larger **Sertoli cells** (Fig. 36-9b). Each time a spermatogonium divides, it takes one of two developmental paths. In the first developmental option, the cell may undergo mitosis. Mitosis ensures that the male has a steady supply of new spermatogonia throughout his life. Alternatively, the spermatogonium may undergo **spermatogenesis**—that is, it may experience a series of developmental events that leads to the production of haploid sperm (Fig. 36-10).

Spermatogenesis begins with the growth and differentiation of spermatogonia into **primary spermatocytes**, which are large diploid cells. The primary spermatocytes then undergo meiosis (see Chapter 11). At the end of meiosis I, each primary spermatocyte gives rise to two haploid **secondary spermatocytes**. Each secondary spermatocyte divides again, during meiosis II, to produce two **spermatids**; thus, each primary spermatocyte gives rise to a total of four spermatids. Spermatids undergo

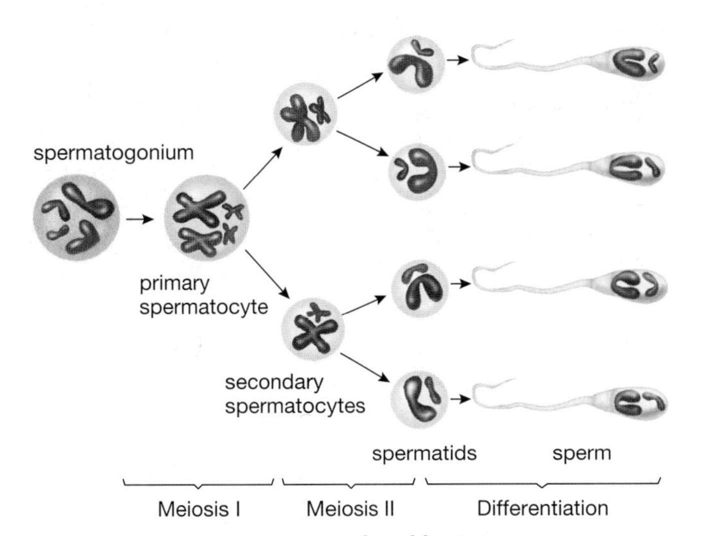

FIGURE 36-10 Sperm are produced by meiosis
Spermatogenesis is accomplished by meiotic divisions followed by differentiation, producing haploid sperm. Although 4 chromosomes are shown for clarity, in humans, the diploid number is 46 and the haploid number is 23.

radical rearrangements of their cellular components as they differentiate into sperm (see Fig. 36-10).

Sertoli cells regulate the process of spermatogenesis and nourish the developing sperm. The spermatogonia, spermatocytes, and spermatids are enfolded in the Sertoli cells. As spermatogenesis proceeds, they migrate up from the outermost edge of the seminiferous tubule to the central cavity of the tubule. The mature sperm (or *spermatozoa*, see Fig. 36-9b) are then liberated into the central cavity.

A human sperm (Fig. 36-11) is unlike any other cell of the body. Most of the cytoplasm disappears, leaving a haploid nucleus nearly filling the *head*. Atop the nucleus lies a specialized lysosome called the **acrosome**. The acrosome contains enzymes that will dissolve protective layers around the egg and enable the sperm to enter and fertilize it. Behind the head is the *midpiece*, which is packed with mitochondria. These organelles provide the energy needed to move the *tail*, which protrudes out the back. Whiplike movements of the tail, which is really a long flagellum, propel the sperm through the female reproductive tract.

In humans and other mammals, spermatogenesis does not begin until puberty, when GnRH from the hypothalamus stimulates the anterior pituitary to produce LH and FSH. Luteinizing hormone stimulates the interstitial cells of the testes to produce testosterone (Fig. 36-12). Testosterone, in combination with FSH, stimulates the Sertoli cells to promote spermatogenesis. Like many physiological processes, sperm production is regulated by negative feedback. Testosterone, while stimulating spermatogenesis, also inhibits both GnRH release by the

(a)

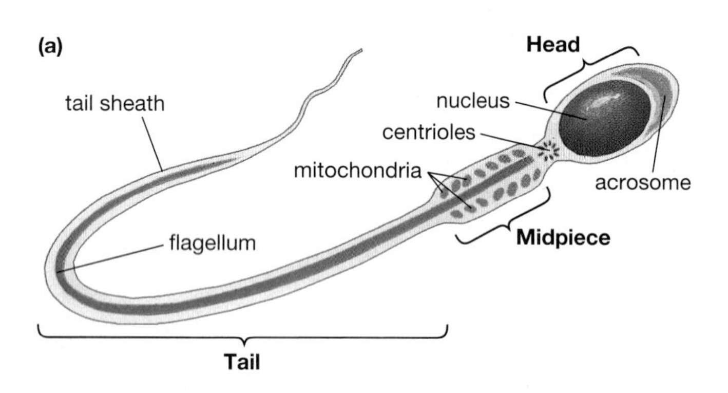

(b)

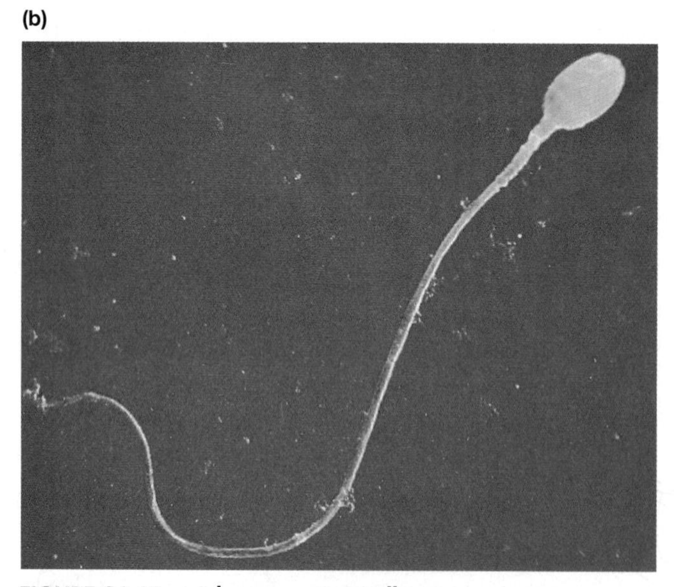

FIGURE 36-11 A human sperm cell
(a) A mature sperm is a cell equipped with only the essentials: a haploid nucleus, the acrosome (containing enzymes that will digest the barriers surrounding the egg), mitochondria for energy production, and a tail (a long flagellum) for locomotion. **(b)** A false-color electron micrograph of a human sperm. **QUESTION** The sperm competition hypothesis suggests that the evolution of sperm cells has been shaped by competition to fertilize the egg when several males have mated with the same female. Which features of sperm might have been shaped by sperm competition?

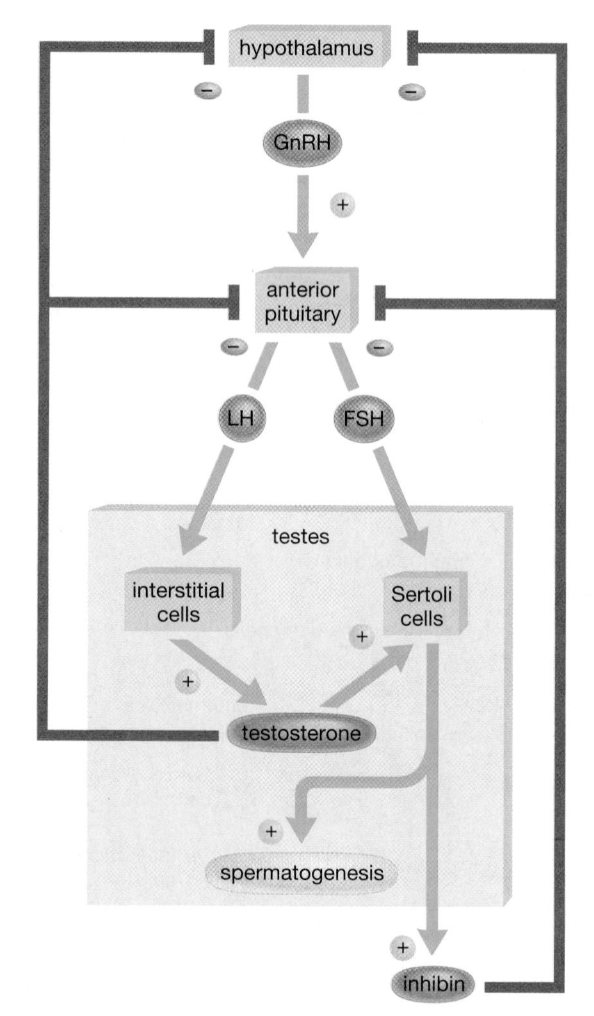

FIGURE 36-12 Hormonal control of spermatogenesis
GnRH from the hypothalamus stimulates the anterior pituitary to release LH and FSH. LH stimulates the interstitial cells to produce testosterone. Testosterone and FSH stimulate the Sertoli cells and the spermatogonia, causing spermatogenesis. Testosterone and chemicals produced during spermatogenesis inhibit further release of FSH and LH, forming a negative feedback loop that keeps the rate of spermatogenesis and the concentration of testosterone in the blood nearly constant. (+ stimulates; – inhibits) **QUESTION** Why do injections of testosterone suppress sperm production?

hypothalamus and LH and FSH release by the pituitary, limiting further testosterone production and sperm development. The Sertoli cells, when stimulated by FSH and testosterone, not only promote spermatogenesis, but secrete a hormone (*inhibin*) that also inhibits GnRH, LH, and FSH production (see Fig. 36-12). This feedback process maintains sperm production at relatively constant levels throughout the male reproductive life span.

Accessory Structures Produce Semen and Conduct the Sperm Outside the Body

The seminiferous tubules merge to form the **epididymis**, a long, continuous, folded tube (see Fig. 36-9a). The epididymis leads into the **vas deferens**, which leaves the scrotum and enters the abdominal cavity. Most of the hundreds of millions of sperm produced each day are stored in the vas deferens and epididymis. The vas deferens joins the **urethra**, which connects the bladder to the tip of the penis. This final common path is used, at different times, by both urine (during urination) and sperm (during ejaculation—a reflex caused by sexual stimulation that forces sperm out through the penis).

The fluid ejaculated from the penis, called **semen**, consists of sperm mixed with fluid secretions from three glands that empty into the vas deferens or urethra: the **seminal vesicles**, the **prostate gland**, and the **bulbourethral glands**. The secretions of these glands activate the sperm and provide energy that powers their swimming movements. The secretions also neutralize the acidic fluids of the vagina that inhibit bacterial growth but would also interfere with sperm (see Fig. 36-8 and Table 36-1).

Table 36-2 The Human Female Reproductive Tract

Structure	Function
Ovary (gonad)	Produces eggs, estrogen, and progesterone
Fimbria (opening of uterine tube)	Bears cilia that sweep egg into oviduct
Uterine tube	Conducts egg to uterus; site of fertilization
Uterus	Muscular chamber where fetus develops
Cervix	Closes off lower end of uterus
Vagina	Receptacle for semen; birth canal

The Female Reproductive Tract Includes the Ovaries and Accessory Structures

The female reproductive tract is almost entirely contained within the abdominal cavity (Table 36-2 and Fig. 36-13). It consists of paired gonads—the **ovaries**

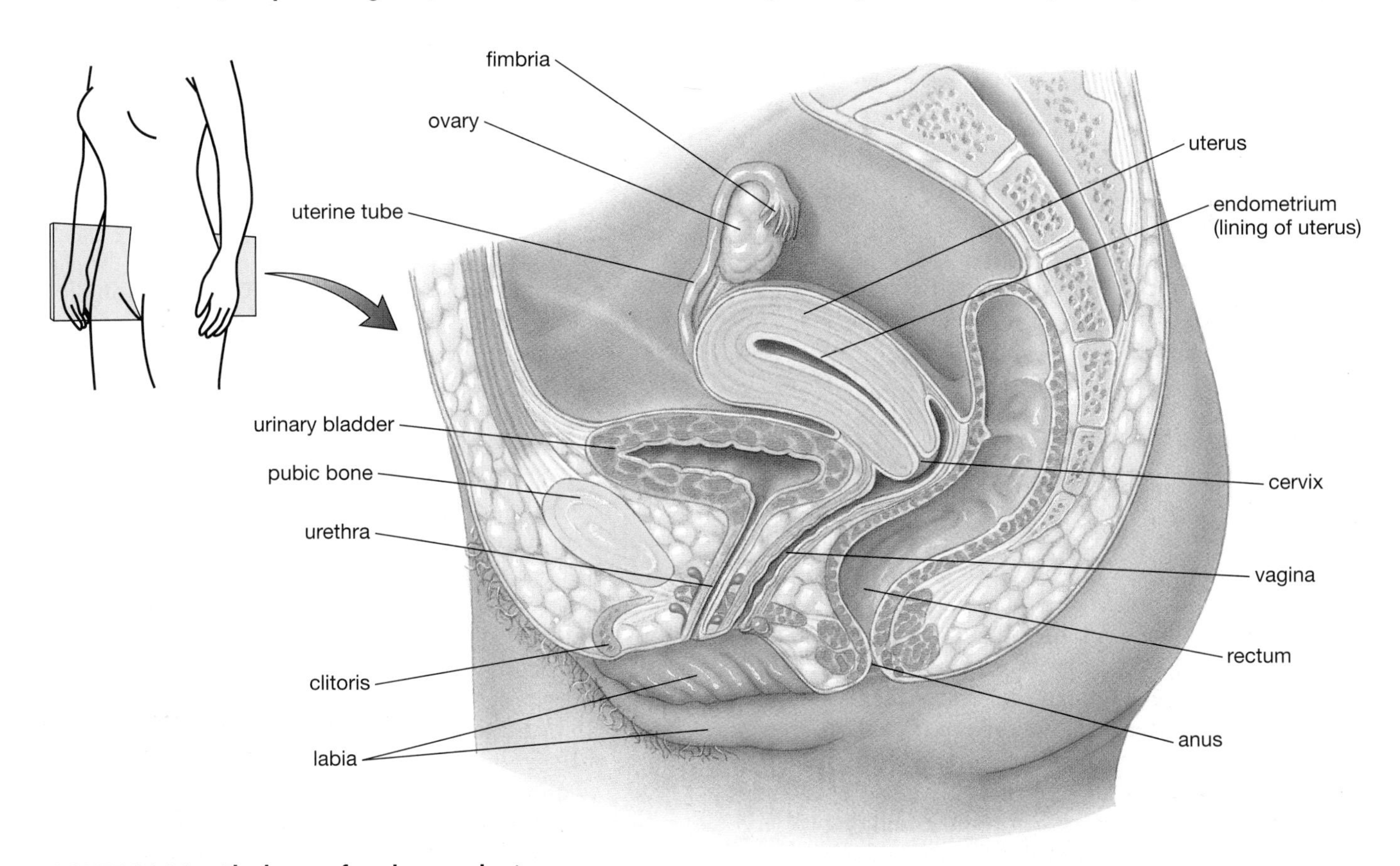

FIGURE 36-13 The human female reproductive tract
Eggs are produced in the ovaries and swept by cilia into the uterine tube. A male deposits sperm in the vagina, from which they move up through the cervix and uterus into the uterine tube. Sperm and egg usually meet in the uterine tube, where fertilization occurs and early stages of development occur. The early embryo embeds in the lining of the uterus, where development continues.

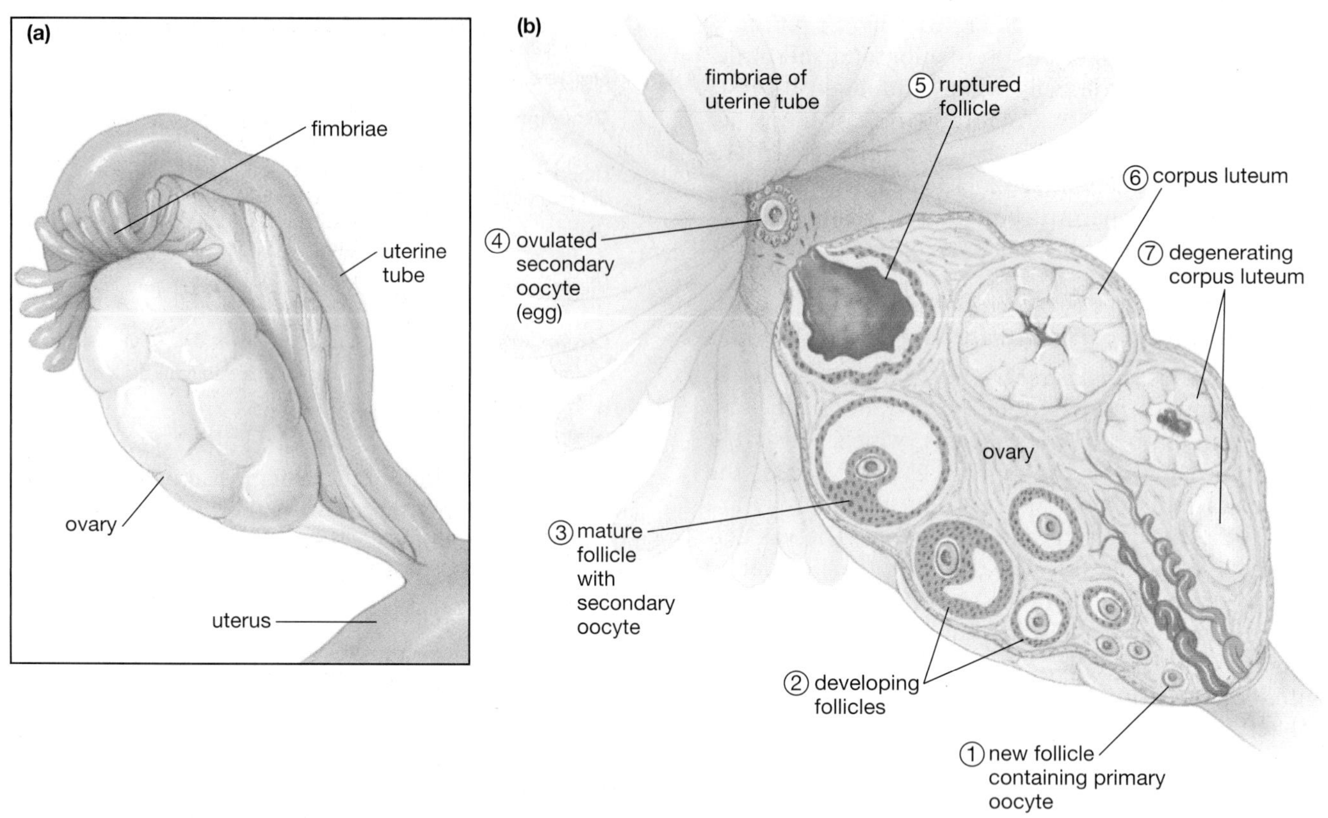

FIGURE 36-14 The structures involved in oogenesis
(a) External view of the ovary and uterine tube. **(b)** The development of follicles in an ovary, portrayed in a time sequence (clockwise from the lower right). ① A primary oocyte begins development within a follicle. ②, ③ The follicle grows, providing both hormones and nourishment for the enlarging oocyte. ④ At ovulation, the egg, surrounded by follicle cells, bursts through the ovary wall. ⑤, ⑥, ⑦ The remaining follicle cells develop into the corpus luteum, which secretes hormones. If fertilization does not occur, the corpus luteum breaks down after a few days.

(Fig. 36-14a)—and accessory structures that accept sperm, conduct the sperm to the egg, and nourish the developing **embryo**.

Eggs Are Produced in the Ovaries

Oogenesis, the formation of egg cells, begins in the developing ovaries of a female **fetus** (an embryo that is sufficiently developed to be recognizably human). This process starts with the formation of precursor egg cells called **oogonia** (singular, *oogonium*). By the end of the third month of fetal development, no oogonia remain, as they have all divided by mitosis and grown into **primary oocytes**. As fetal development continues, meiosis begins in all primary oocytes but is halted at prophase of meiosis I. By birth, a lifetime supply of primary oocytes is already in place, and no new ones will be generated later in life. The ovaries start out with about 2 million primary oocytes. Many primary oocytes die each day, until at puberty only about 400,000 remain. That number is more than enough, because only a few oocytes resume meiosis during each month of a woman's reproductive span (from puberty to *menopause* at about age 50).

Surrounding each oocyte is a layer of much smaller cells that both nourish the developing oocyte and secrete female sex hormones. Together, the oocyte and these accessory cells make up a **follicle** (Fig. 36-14b). Approximately once a month during a woman's reproductive years, she undergoes a *menstrual cycle*, which is described later in this chapter. During the menstrual cycle, pituitary hormones stimulate the development of a dozen or more follicles, although usually only one follicle matures completely. At this time, the primary oocyte completes its first meiotic division (which was halted during development), dividing into a single **secondary oocyte** and a **polar body**, which is little more than a discarded set of chromosomes (Fig. 36-15). Meanwhile, the small accessory cells of the follicle multiply and secrete estrogen. As the follicle matures, it grows, eventually erupting through the surface of the ovary and releasing the secondary oocyte, a process called *ovulation* (Fig. 36-16). The secondary oocyte then travels through the tube leading out of the ovary, called the **uterine tube** (also called the *oviduct* or *Fallopian tube*). For convenience, we will refer to the ovulated secondary oocyte as the *egg*. If the egg is fertilized, it

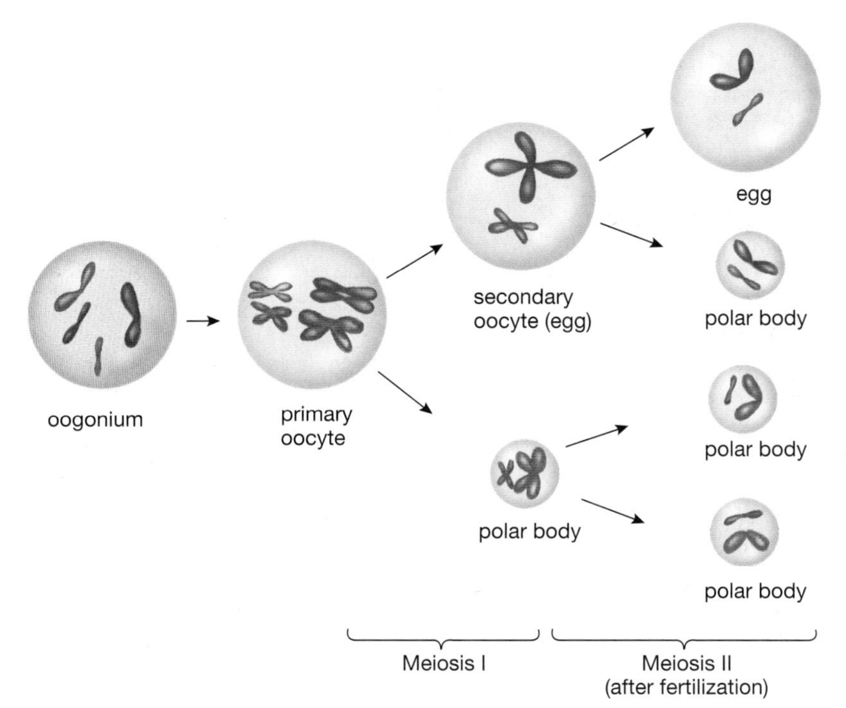

FIGURE 36-15 Egg cells are formed by meiosis The cellular stages of oogenesis. The oogonium enlarges to form the primary oocyte. At meiosis I, almost all the cytoplasm is included in one daughter cell, the secondary oocyte. The other daughter cell is a small polar body that contains chromosomes but little cytoplasm. At meiosis II, almost all the cytoplasm of the secondary oocyte is included in the egg, and a second small polar body discards the remaining "extra" chromosomes. The first polar body may also undergo the second meiotic division. In humans, meiosis II does not occur until a sperm penetrates into the egg.

may undergo the second meiotic division while in the uterine tube.

Some of the follicle cells accompany the egg, but most remain in the ovary. These cells enlarge and become glandular, forming the **corpus luteum** (see Fig. 36-14b). The corpus luteum secretes both estrogen and a second hormone, **progesterone**. If fertilization does not occur, the corpus luteum breaks down a few days later.

A human male is able to produce large numbers of sperm continuously. In contrast, a woman does not produce mature gametes (ovulate) unless her uterus is prepared to receive and nourish a fertilized egg. The coordination of ovulation with preparation of the uterus is accomplished by the **menstrual cycle**. The menstrual cycle, which is regulated by hormonal interactions among the hypothalamus, anterior pituitary gland, and ovary, is described in detail in "A Closer Look: Hormonal Control of the Menstrual Cycle."

Accessory Structures Include the Uterine Tubes, Uterus, and Vagina

Each ovary is adjacent to, but not attached to, its associated uterine tube (see Fig. 36-14a). The open end of the uterine tube is fringed with ciliated "fingers" called *fimbriae* which nearly surround the ovary. The cilia create a current that sweeps the egg into the mouth of the uterine tube. Fertilization usually occurs within the tube. The **zygote**, or fertilized egg, is swept down the uterine tube by beating cilia and released into the pear-shaped **uterus**, or *womb*. There it will develop for 9 months. The wall of the uterus has two layers that correspond to its dual functions of nourishing the developing embryo and childbirth. The inner lining, or **endometrium**, is richly supplied with blood vessels. This lining will form the mother's contribution to the **placenta**, the structure that transfers oxygen, carbon dioxide, nutrients, and wastes between fetus and mother, as we will see in Chapter 37. The outer muscular wall of the uterus gradually expands as the developing child grows, then contracts strongly during delivery, expelling the infant out into the world.

Developing follicles secrete estrogen, which stimulates the uterine lining to grow an extensive network of blood vessels and nutrient-producing glands. After ovulation, estrogen and progesterone released by the corpus luteum promote continued development of the endometrium. Thus, if an egg is fertilized, it encounters

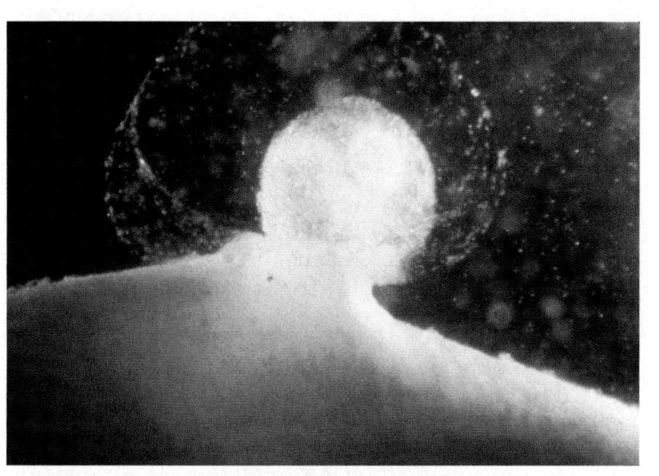

FIGURE 36-16 A follicle erupts from the ovary The mature follicle grows so large and is filled with so much fluid that it literally bursts through the ovary wall, releasing the secondary oocyte into the uterine tube.

A CLOSER LOOK Hormonal Control of the Menstrual Cycle

Our discussion of the menstrual cycle begins with the onset of menstruation, illustrated by the loss of the uterine lining as shown in the lowest panel of Fig. E36-1. Hormonally, the menstrual cycle is initiated by the spontaneous release of gonadotropin-releasing hormone (GnRH) by cells in the hypothalamus (top panel). This release occurs continuously unless it is suppressed by other hormones, notably progesterone. The cycle starts on day 1 (which immediately follows day 28 of the cycle) with an increase in GnRH. Follow the descriptions in the diagram by matching the numbers to Figure E36-1. Note that numbers are duplicated in the figure when the description applies to multiple panels.

① GnRH (top panel) stimulates the anterior pituitary (second panel) to release FSH (blue line) and LH (red line). The endometrium of the uterus is shed during menstruation (lowest panel).

② FSH initiates the development of several follicles, which secrete estrogen, within the ovaries. Under the combined influences of FSH, LH, and estrogen, the follicles grow and the primary oocyte within each follicle begins developing. Usually, only one follicle completes development each month.

③ As the maturing follicle grows, it secretes greater amounts of estrogen (purple line, fourth panel). This estrogen has three effects. First, it promotes the continued development of the follicle and its primary oocyte (third panel). Second, estrogen stimulates the growth of the endometrium of the uterus (lowest panel). Third, estrogen stimulates the hypothalamus to produce more GnRH.

④ The GnRH stimulates a surge of LH (and a smaller increase in FSH) at about the 14th day of the cycle. The surge of LH has three important consequences. First, it triggers the resumption of meiosis I in the oocyte, resulting in the formation of the secondary oocyte and the first polar body.

⑤ Second, the LH surge causes the final explosive growth of the follicle, culminating in ovulation, and third, it transforms the remnants of the follicle into the corpus luteum.

⑥ The corpus luteum secretes both estrogen (purple line) and progesterone (green line).

⑦ The combination of estrogen and progesterone inhibit GnRH production and reduce FSH and LH, preventing the development of more follicles. Simultaneously, estrogen and progesterone stimulate the endometrium to develop a network of blood vessels and nutrient-producing glands. The endometrium eventually becomes about 4 millimeters thick.

⑧ If pregnancy does not occur, the corpus luteum starts to disintegrate about 12 days after ovulation. This disintegration is caused by the corpus luteum itself, which secretes progesterone that shuts down LH secretion. Because the corpus luteum can persist only while it is stimulated by LH (or by a similar hormone released by the developing embryo, described below), it essentially induces its own destruction, a form of negative feedback.

⑨ With the corpus luteum gone, estrogen and progesterone levels plummet. Deprived of stimulation by estrogen and progesterone, the endometrium of the uterus dies within few days; its blood and tissue form the menstrual flow that defines the

Media Activity
36.2 Hormonal Control of the Menstrual Cycle

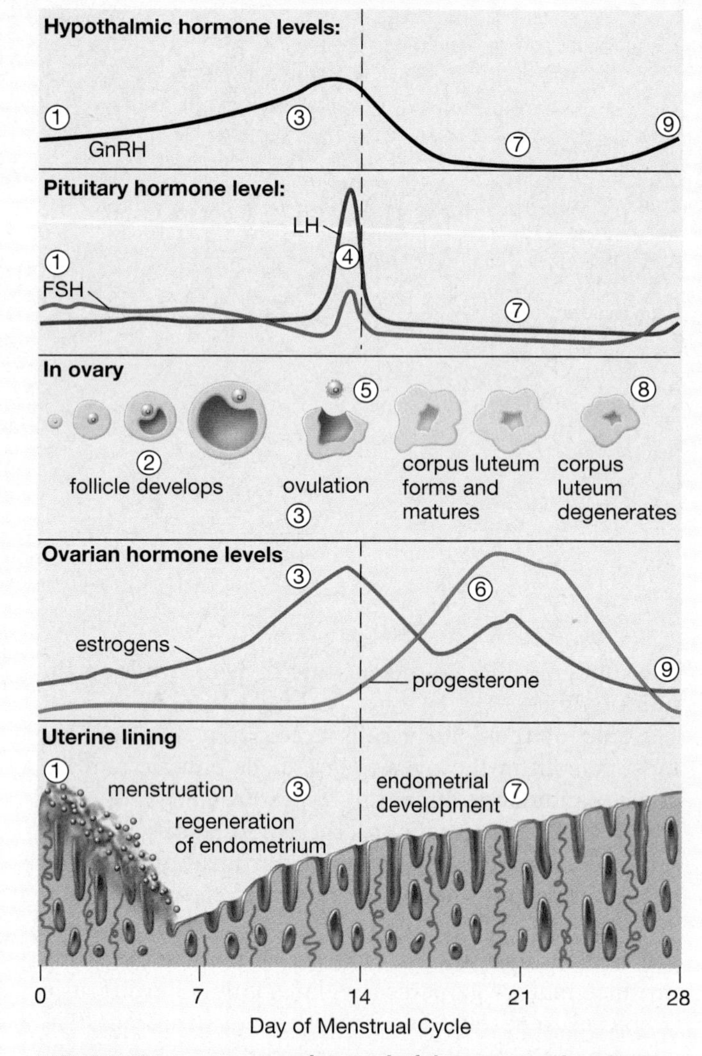

FIGURE E36-1 Hormonal control of the menstrual cycle The menstrual cycle is generated by interactions among the hormones of the hypothalamus, the anterior pituitary, and the ovaries. The circled numbers refer to the interactions discussed in the text.

a rich environment for growth. If the egg is not fertilized, however, the corpus luteum disintegrates, estrogen and progesterone levels fall, and the overgrown endometrium disintegrates as well. The uterus contracts, squeezing out the excess endometrial tissue (and sometimes causing menstrual cramps). The resulting flow of tissue and blood, resulting from the breakdown of endometrial blood vessels, is called **menstruation** (from the Latin *mensis,* meaning "month").

The outer end of the uterus is nearly closed off by the **cervix**, a ring of connective tissue. The cervix holds a developing baby in the uterus, expanding only at the onset

first day of the new cycle. The reduced level of circulating progesterone no longer inhibits the hypothalamus, so the spontaneous release of GnRH resumes. GnRH stimulates the release of FSH and LH (cycling back to step ①), initiating the development of a new set of follicles and restarting the cycle.

During pregnancy, the embryo itself prevents these changes from occurring. Shortly after the ball of cells formed by the dividing fertilized egg embeds itself in the endometrium, it starts secreting an LH-like hormone called *chorionic gonadotropin (CG)*. This hormone travels in the bloodstream to the ovary, where it prevents the breakdown of the corpus luteum. The corpus luteum continues to secrete estrogen and progesterone, and the uterine lining continues to grow, nourishing the embryo. The embryo releases so much CG that the hormone is excreted in the mother's urine; most pregnancy tests use the presence of CG in urine to determine pregnancy.

Although negative feedback regulates the levels of most hormones, the hormones of the menstrual cycle are regulated by both positive and negative feedback. During the first half of the cycle, FSH and LH stimulate estrogen production by the follicles. High levels of estrogen then *stimulate* the mid-cycle surge of FSH and LH release (positive feedback). During the second half of the cycle, estrogen and progesterone together *inhibit* the release of FSH and LH (negative feedback). The early positive feedback causes hormone concentrations to reach high levels; later, negative feedback shuts the system down again unless pregnancy intervenes.

of labor to permit passage of the child. Beyond the cervix is the **vagina**, which opens to the outside. The vagina serves both as the receptacle for the penis during intercourse and as the birth canal (see Fig. 36-13).

Copulation Allows Internal Fertilization

As terrestrial mammals, humans use internal fertilization to deposit sperm into the moist environment of the female's reproductive tract. During intercourse, the penis is inserted into the vagina, where it releases sperm. The sperm swim from the vagina through the cervix and into the uterus, finally entering the uterine tubes. If the female has ovulated within the past day or so, the sperm will meet an egg in one of the uterine tubes. Only one sperm can succeed in fertilizing the egg and begin the development of a new human being.

During Copulation, Sperm Are Deposited in the Female's Vagina

The male role in copulation begins with erection of the penis. Before erection, the penis is relaxed (flaccid), because the arterioles that supply it are constricted, allowing little blood flow (Fig. 36-17a). Under psychological and physical stimulation, the arterioles dilate and blood flows into spaces in the tissue within the penis. As these tissues swell, they squeeze off the veins that drain the penis (Fig. 36-17b). Blood pressure increases, causing an erection. After the penis is inserted into the vagina, movements further stimulate touch receptors on the penis, triggering ejaculation. *Ejaculation* occurs when muscles encircling the epididymis, vas deferens, and urethra contract, forcing semen out of the penis and into the vagina. On average, 3 or 4 milliliters of semen, containing roughly 300 million sperm, is ejaculated. *Male orgasm* causes ejaculation and feelings of intense pleasure and release.

In the female, sexual arousal causes increased blood flow to the vagina, to the external paired folds of tissue called the **labia** (singular, *labium*), and to the **clitoris**, a

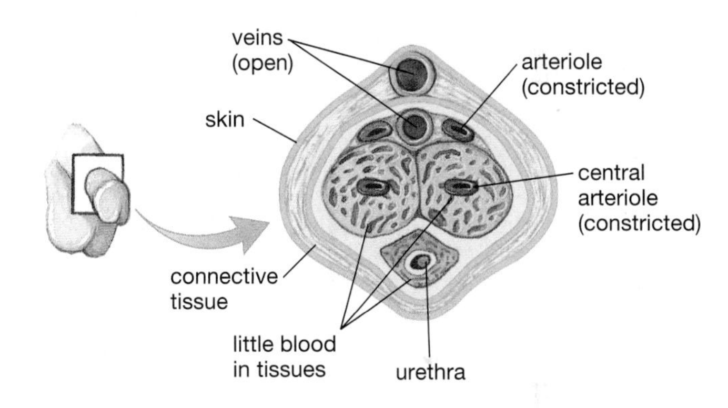

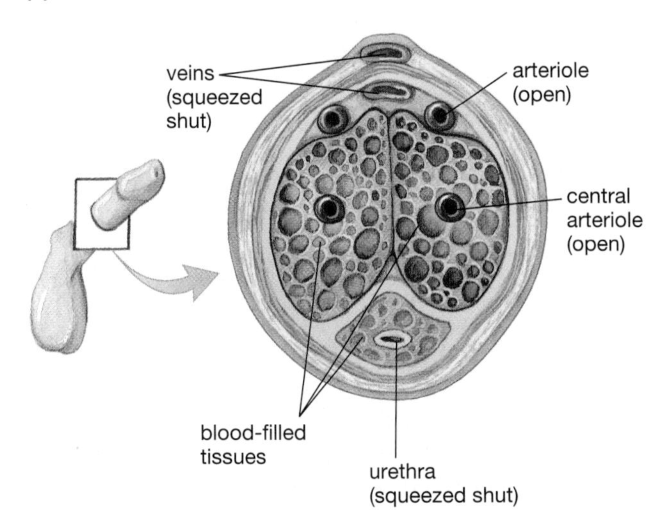

FIGURE 36-17 Changes in blood flow within the penis cause erection
(a) Smooth muscles encircling the arterioles leading into the penis are usually contracted, limiting blood flow. **(b)** During sexual excitement, these muscles relax, and blood flows into spaces within the penis. The swelling penis squeezes off the veins leaving the penis, increasing the blood pressure inside and causing the penis to become elongated and firm.

HEALTH WATCH Sexually Transmitted Diseases

Sexually transmitted diseases (STDs), caused by viruses, bacteria, protists, or arthropods that infect the sexual organs and reproductive tract, are a serious and growing health problem worldwide. As the name implies, these diseases are transmitted either exclusively or primarily through sexual contact. Here we discuss some of the more common of these diseases.

BACTERIAL INFECTIONS

Gonorrhea, an infection of the genital and urinary tract, is a very common STD in the United States. The causative bacterium is transmitted almost exclusively by intimate contact. It penetrates the membranes that line the urethra, anus, cervix, uterus, uterine tubes, and throat. In males, inflammation of the urethra results in painful urination and a discharge of pus from the penis. Some infected individuals do not seek treatment because they have very mild (or absent) symptoms; these people become carriers who can readily spread the disease. Gonorrhea can lead to infertility by blocking the uterine tubes with scar tissue. The disease can be treated with antibiotics. Infants born to infected mothers can acquire the bacterium during delivery. The bacterium attacks the eyes of newborns and was once a major cause of blindness. Today, most newborns are immediately given antibiotic eyedrops as a preventive measure to kill the bacterium.

Syphilis is caused by a spiral-shaped bacterium which enters the mucous membranes of the genitals, lips, anus, or breasts. It is readily killed by exposure to air and is spread only by intimate contact. Syphilis begins with a sore at the site of infection and can be cured with antibiotics. If untreated, syphilis bacteria spread through the body, damaging many organs including the skin, kidneys, heart, and brain, in some cases with fatal results. Syphilis can be transmitted to the fetus during pregnancy; and the skin, teeth, bones, liver, and central nervous system of the infant may be damaged.

Chlamydia causes inflammation of the urethra in males and the urethra and cervix in females. In many cases, there are no obvious symptoms, so the infection goes untreated and spreads. The chlamydia bacterium can infect and block the uterine tubes, resulting in sterility. Chlamydial infection can cause eye inflammation in infants born to infected mothers.

VIRAL INFECTIONS

Acquired immune deficiency syndrome, or **AIDS**, is caused by the human immunodeficiency virus (HIV), discussed in Chapter 32. Because the virus does not survive exposure to air, it is spread primarily by sexual activity and by contaminated blood and needles. HIV attacks the immune system, leaving the victim vulnerable to a variety of infections that almost invariably prove fatal. Children born to mothers with AIDS can become infected before or during birth, or during breastfeeding. There is no cure, but certain drug combinations, such as AZT and protease inhibitors, can prolong life.

Genital herpes causes painful blisters on the genitals and surrounding skin and is transmitted primarily when blisters are present. The herpes virus never leaves the body but resides in certain nerve cells, emerging unpredictably, possibly in response to stress. The first outbreak is the most serious; subsequent outbreaks produce fewer blisters and can be quite infrequent. Drugs that inhibit viral DNA replication can reduce the severity of outbreaks. A pregnant woman with an active case of genital herpes can transmit the virus to the developing fetus, possibly causing severe mental or physical disability or stillbirth. Herpes can also be transmitted from mother to infant if the infant contacts blisters during childbirth.

Genital warts are growths or bumps that appear on the external genitalia, in or around the vagina or anus, or on the cervix in females, and on the penis, scrotum, groin, or thighs in males. The warts, which are usually painless, are caused by the *human papillomavirus (HPV)*, which is transmitted by skin-to-skin contact during sex. The most common treatment is removal of the warts, typically by freezing them with liquid nitrogen. Antiviral drugs can also be effective. Like the herpes virus, the human papillomavirus resides inside cells and is therefore difficult to cure. In women, genital warts are considered to be a risk factor for certain types of cervical cancer.

PROTIST AND ARTHROPOD INFECTIONS

Trichomoniasis is caused by a flagellated protist that colonizes the mucous membranes lining the urinary tract and genitals of both males and females. Symptoms include a discharge caused by inflammation in response to the parasite. The protist is spread by intercourse but can also be acquired from contaminated clothing and toilet articles. Lengthy untreated infections can result in sterility.

Crab lice, also called *pubic lice*, are tiny arachnids (relatives of spiders) that live and lay their eggs in pubic hair. Their mouthparts are adapted for penetrating skin and sucking blood and body fluids, a process that causes severe itching. "Crabs" are not only irritating; they can also spread infectious diseases. They can be controlled through careful hygiene and chemical treatments.

---

small structure just in front of the vagina (see Fig. 36-13). The clitoris, which is derived from the same embryological tissue as the tip of the penis, becomes engorged with blood. Stimulation by the penis may result in *female orgasm*, a series of rhythmic contractions of the vagina and uterus accompanied by sensations of pleasure and release. Female orgasm is not necessary for fertilization.

The intimate contact involved in copulation creates a situation in which disease-causing organisms can readily be transmitted, as described in "Health Watch: Sexually Transmitted Diseases."

During Fertilization, the Sperm and Egg Nuclei Unite

Sperm and eggs live for only a few days if fertilization does not occur, so fertilization can succeed only if copulation occurs within a couple of days before or after ovulation. You may recall that when it leaves the ovary,

(a)

secondary oocyte (egg)

corona radiata

zona pellucida

(b)

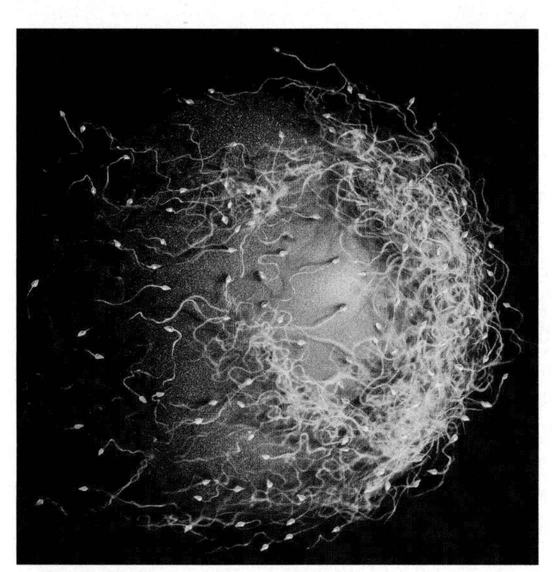

FIGURE 36-18 The secondary oocyte and fertilization
(a) A human secondary oocyte shortly after ovulation. Sperm must digest their way through the small follicular cells of the corona radiata and the clear zona pellucida to reach the oocyte itself. **(b)** Sperm surround the oocyte, attacking its defensive barriers. **QUESTION** Why is the oocyte so well protected by surrounding barriers?

the egg is surrounded by follicle cells. These cells, now called the **corona radiata**, and an inner jellylike layer **zona pellucida** ("clear area"), form a barrier between the sperm and the egg (Fig. 36-18a). Recent research suggests that the human egg releases a chemical attractant that lures the sperm toward it.

In the uterine tube, hundreds of sperm reach the egg and encircle the corona radiata, each sperm releasing enzymes from its acrosome (Fig. 36-18b). These enzymes weaken both the corona radiata and the zona pellucida, allowing the sperm to wriggle through to the egg. If there aren't enough sperm, an insufficient amount of enzyme is released, and none of the sperm will reach the egg. This may be the reason that natural selection has led to the ejaculation of so many sperm. Perhaps 1 in 100,000 reach the uterine tube, and 1 in 20 of those encounters the egg, so only a few hundred of the 300 million sperm that were ejaculated join in attacking the barriers around the egg.

When the first sperm finally contacts the surface of the egg, the plasma membranes of egg and sperm fuse, and the sperm's head is drawn into the egg's cytoplasm. As the sperm enters, it triggers two vital changes in the egg. Vesicles near the surface of the egg release chemicals into the zona pellucida, reinforcing it and preventing further sperm from entering the egg. The egg undergoes its second meiotic division, producing a haploid gamete at last. Fertilization occurs as the haploid nuclei of sperm and egg fuse, forming a diploid nucleus that contains all the genes of a new human being.

Anomalies in the male or female reproductive system can prevent fertilization. For example, a blocked uterine tube can prevent sperm from reaching the egg. Likewise, men who produce fewer than 20 million sperm per milliliter of semen (about one-fifth the normal amount) generally cannot fertilize a woman's egg during intercourse because too few sperm reach the egg. If the sperm are otherwise healthy, such men can father children by *artificial insemination*, in which a large quantity of their semen is injected directly into the uterine tube. Today, some couples seek high-technology help in the form of *in vitro* fertilization (see "Scientific Inquiry: High-Tech Reproduction").

36.3 How Can People Limit Fertility?

Natural selection favors individuals who reproduce successfully and whose offspring survive long enough to reproduce themselves. During most of human evolution, child mortality was high, and natural selection favored people who produced enough children to offset this high mortality rate. Today, most people no longer need to have many children to ensure that a few will survive to adulthood, but we nonetheless retain the reproductive drives with which evolution has endowed us. As a result, about 1.5 million new people are added to our increasingly overcrowded planet every week. Species are driven to extinction as we modify the world to accommodate our increasing numbers, so the control of birth rates has become an environmental necessity. On the individual level, birth control allows people to plan their families and provide the best opportunities for themselves and their children.

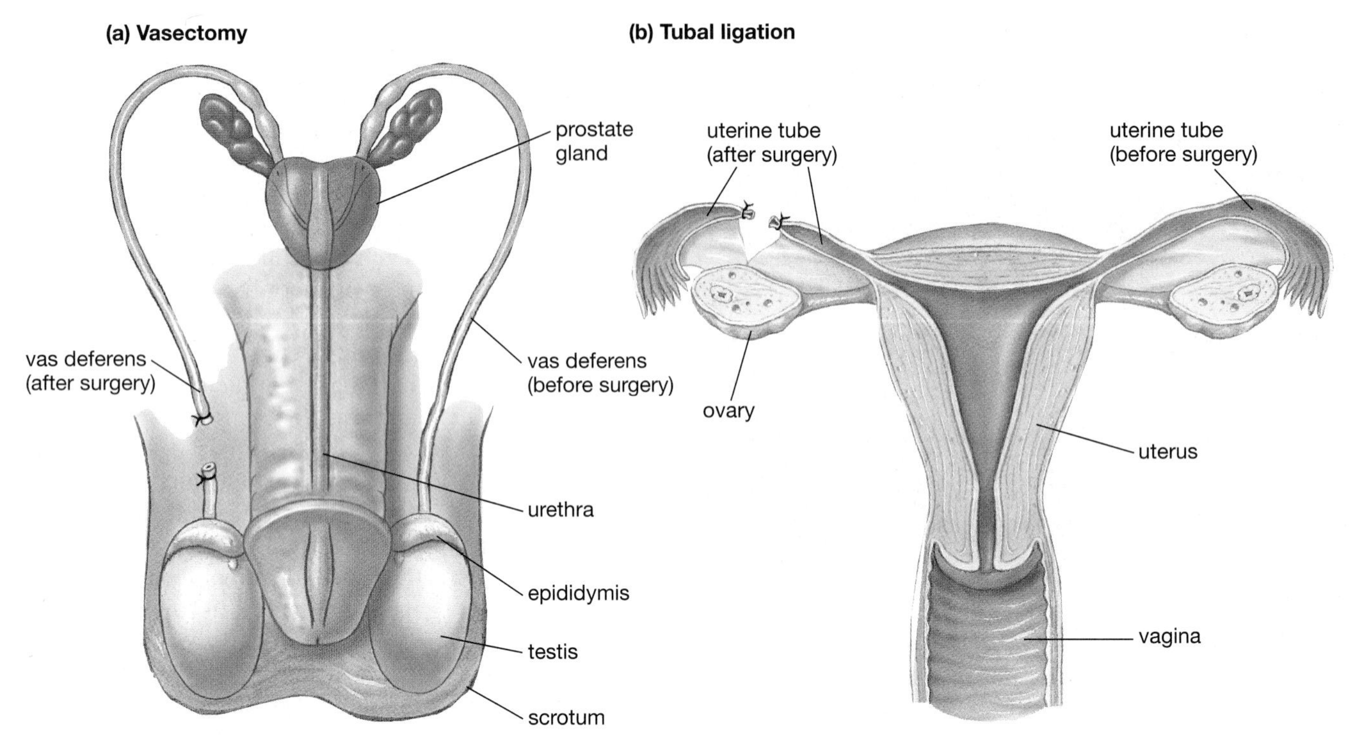

FIGURE 36-19 Sterilization
(a) Vasectomy involves removing a short segment of the male's vas deferens and tying off the cut ends. This prevents sperm from leaving the body. **(b)** Tubal ligation involves removing a short segment of the female's uterine tube and tying off the cut ends, preventing sperm from reaching the oocyte and preventing the oocyte from reaching the uterus.

Historically, limiting fertility has not been easy. In the past, women in some cultures have tried such inventive, if bizarre, techniques as swallowing froth from the mouth of a camel or placing crocodile dung in the vagina. Since the 1970s, however, several effective techniques have been developed for **contraception**, the prevention of pregnancy. The choice of a contraceptive should be made in consultation with a health professional who can provide more complete information.

Permanent Contraception Can Be Achieved Through Sterilization

In the long run, the most effortless method of contraception is **sterilization**, in which the pathways through which sperm or eggs must travel are interrupted (Fig. 36-19). In men, the vas deferens leading from each testis may be severed in an operation called a *vasectomy*. Sperm are still produced, but they cannot reach the penis during ejaculation. The surgery is performed under a local anesthetic, and vasectomy has no known effects on health or sexual performance. The slightly more complex operation of *tubal ligation* renders a woman infertile by cutting her uterine tubes. Ovulation still occurs, but sperm cannot travel to the egg, nor can the egg reach the uterus. Recently, the U.S. Food and Drug Administration approved a new sterilization device consisting of tiny springlike structures. They are inserted into each uterine tube using a tiny camera and flexible insertion device that are threaded up through the vagina and uterus. The procedure requires no incisions and only local anesthesia. The coil causes the uterine tube to form scar tissue that blocks passage of both sperm and eggs. Although sterilization is generally permanent, sometimes, in a delicate and expensive operation, a surgeon can reconnect the vas deferens or uterine tubes.

There Are Three General Approaches to Temporary Contraception

Temporary contraception techniques fall into three general categories: preventing ovulation, preventing sperm and egg from meeting, and preventing implantation in the uterus. Temporary forms of contraception are summarized in Table 36-3.

As you learned earlier in this chapter, ovulation is triggered by a mid-cycle surge of LH. An obvious way to prevent ovulation is to suppress LH release by providing a continuous supply of estrogen and progesterone. This is the basis for a variety of contraception techniques, the best known being the birth control pill. Several other delivery systems for estrogen and progesterone (generally in synthetic form) are now available, as described in Table 36-3.

Table 36-3 Temporary Contraceptive Techniques

Method	Technique and Mechanism	Failure Rate*	Protection from STD
Hormonal Methods: Prevent Ovulation			
Birth control pill	Pill containing either estrogen and synthetic progesterone (combination pill) or progesterone only (minipill). Taken daily.	0.1% to 3%	None
Ortho Evra® contraceptive patch	Skin patch containing synthetic estrogen and progesterone. Replaced weekly.	< 1%**	None
Depo-Provera®	Injection of synthetic progesterone that blocks ovulation for 3 months. Repeated at 3-month intervals.	0.3%	None
NuvaRing®	Flexible plastic ring impregnated with synthetic estrogen and progesterone. Inserted into vagina around the cervix, remains in place for 3 weeks, replaced every 4 weeks.	1% to 2%	None
Barrier Methods: Prevent Sperm and Egg from Meeting			
Abstinence	Deciding not to be sexually active.	0%	Good
Condom (male)	Thin, disposable latex sheath placed over penis just before intercourse. Prevents sperm from entering vagina. More effective with spermicide.	2% to 12%	Good; should be used with other methods of birth control for STD protection.
Condom (female)	Lubricated polyurethane pouch inserted into vagina, prevents sperm from entering cervix. More effective with spermicide.	5% to 25%	Partial
Today Sponge®	Domed disposable sponge impregnated with spermicide. Inserted in vagina, it works for 24 hours.	10% to 15%	Partial
Diaphragm/ Cervical cap	Reusable, flexible domed rubber-like barriers. Spermicide is placed within the dome, and the diaphragm (larger) or cap (smaller) is fitted over the cervix just before intercourse.	6% to 18%	Partial
Rhythm	Measuring body temperature and cervical mucus changes to estimate the time of ovulation and avoiding intercourse during the fertile period.	2% to 20% (rarely performed correctly)	None
Spermicide	Sperm-killing foam is placed in vagina prior to intercourse, forming a chemical barrier to sperm.	6% to 21%	Partial; may not protect against HIV
Methods That Prevent Implantation			
IUD (intrauterine device)	Small plastic device treated with hormones or copper and inserted through the cervix into the uterus.	0.6% to 2.6%	None
"Morning after" pill	Concentrated dose of the hormones in birth control pills, taken within 72 hours after intercourse.	25%	None

*Percentage of women becoming pregnant per year. The low and high numbers, respectively, indicate the differences between consistent and correct use and use in a more typical way that is not always consistent or correct.

**Preliminary findings report that the patch is as effective as the pill, and more likely to be used properly; however, women weighing more than 198 pounds may find it less effective.

There are several effective *barrier methods*, which prevent the encounter of sperm and egg. A variety of devices block the opening of the cervix, preventing entry of sperm. Alternatively, the male condom prevents sperm from being deposited in the vagina. A female condom is now also available that completely lines the vagina. Condoms are particularly effective against the spread of sexually transmitted diseases. Barrier devices are more effective when used with a *spermicide* (sperm-killing substance). Less-reliable techniques include the use of spermicides alone and the *rhythm method* (abstinence from intercourse during the ovulatory period of the menstrual cycle). In practice, the rhythm method has a high failure rate because of inaccuracies in determining the menstrual cycle, which usually varies somewhat from month to month. It can be

SCIENTIFIC INQUIRY High-Tech Reproduction

Though some assisted reproductive technologies may be used to save some animal species from extinction, most were developed with the intention of helping infertile couples have children. About 15,000 babies born in the United States each year are conceived in a shallow glass dish (*in vitro* fertilization literally means "fertilization in glass"). IVF is a complex and delicate procedure. First, the woman is given daily injections of drugs and/or hormones to stimulate multiple ovulations. Surgeons then insert a long, hollow needle into each ripe follicle within the ovary and suck out the oocyte. Usually, at least four oocytes are incubated in a dish with freshly collected sperm. In 48 hours, some of the oocytes will have been fertilized, begun dividing, and reached the eight-cell stage. A few of these early embryos are sucked into a tube and expelled very gently into the uterus. (Extra embryos can be frozen for later use in case the first attempt at implantation is unsuccessful.) Transplanting multiple embryos increases the success rate for implantation, but it also increases the probability of multiple births, which are much riskier than single births. The popularity of IVF, which can cost up to $15,000 per attempt, is a testimony to the strong biological drive to have children. With an average success rate of about 30%, a typical conception via IVF costs about $35,000.

With a new technique, *intracytoplasmic sperm injection* (ICSI), even men whose sperm are unable to swim to and fertilize an egg may be able to father their own children. In ICSI, immature sperm (or even sperm precursor cells) are extracted from the testes. Then, a tiny pipette is used to pierce the plasma membrane of an egg cell and inject a single sperm directly into the egg's cytoplasm (Fig. E36-2). In almost all cases, this procedure leads to a successful fertilization. Apparently, sperm are capable of fertilization very early in their developmental process, and further maturation mainly provides them with the ability to swim to the egg and penetrate its protective layers.

For women who do not ovulate regularly, fertility drugs cause multiple ovulations; as a result of this procedure, the rate of multiple births in the United States has quadrupled since 1971 (Fig. E36-3). In the world of assisted reproduction, a widow could be impregnated by her dead husband's cryopreserved sperm. Recently, twins were born to a woman whose ovaries could not produce eggs. Her children came from donor eggs that had been cryopreserved for two years. A woman can also put her uterus up for rent; as a surrogate mother, she may bear a child for a woman who has had a hysterectomy or who simply does not want to go through pregnancy. The egg and sperm that produced the fetus developing within the surrogate mother could come from the couple who has hired her, or alternatively, the egg, the sperm, or both gametes could come from unrelated individuals. Conceivably, a modern newborn could have up to five "parents."

Using assisted reproductive technology, parents can now strongly influence the sexes of their children. This can be important if the parents are carriers of sex-linked disorders, but some parents may simply seek to balance their families. Researchers have found that sperm carrying an X chromosome have 2.8% more DNA than sperm carrying a Y chromesome; this difference can be detected and used to sort the sperm. So far, they have been most successful at increasing the percentage of X sperm in a sample and increasing the probability that a couple will have a girl.

Although efforts to clone humans have received a great deal of publicity, human cloning (discussed in Chapter 11) is not a feasible form of assisted reproduction. Whether it will ever be safe or socially acceptable is a matter of intense debate.

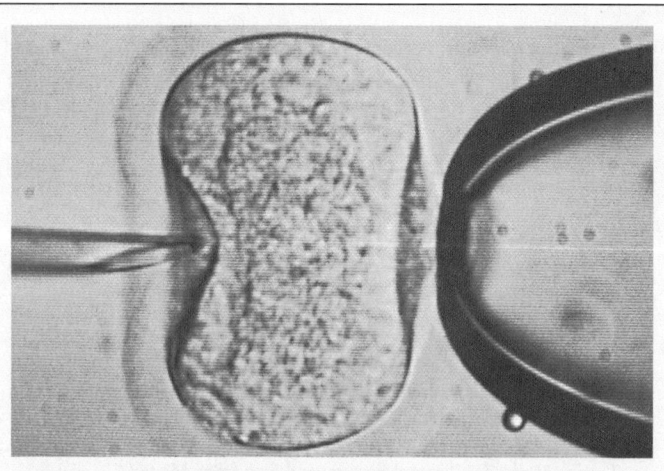

FIGURE E36-2 Piercing an egg to inject a sperm cell
An egg, stripped of its protective layers and held in place with a pipette, is injected with a single sperm cell that is placed directly into the egg's cytoplasm. **QUESTION** Why is the sperm injected into the egg, instead of merely being placed in contact with it?

FIGURE E36-3 Septuplets
The McCaugheys of Iowa had septuplets in 1997 after Mrs. McCaughey took fertility drugs.

made somewhat more reliable by carefully monitoring the slight rise in body temperature and changes in the discharge of mucus from the cervix that help predict ovulation. Withdrawal (the removal of the penis from the vagina before ejaculation) and douching (attempting to wash sperm out of the vagina before they have entered the uterus) are not reliable ways to avoid conception.

Even if an egg is fertilized, contraception can be achieved by preventing implantation of the early embryo (a hollow ball of cells) in the uterus. This type of contraception can be provided by an *intrauterine device (IUD)*, a small copper or plastic loop, squiggle, or shield that is inserted through the cervix and into the uterus by a physician. Another method of preventing implantation is the "morning after" pill, which contains hormones similar to birth control pills, only in larger dosages. Taken within 72 hours after intercourse, these pills prevent the early embryo from implanting.

Abortion Removes the Embryo from the Uterus

When contraception fails, pregnancy can be terminated by *abortion*. Abortion commonly involves dilating the cervix and removing the embryo and placenta by suction; most such abortions are performed during the first 3 months of pregnancy. Alternatively, abortion can be induced during the first 7 weeks of pregnancy by the drug RU-486 (mifepristone), which binds to progesterone receptors and blocks the actions of progesterone, which is essential to the maintenance of the uterine lining during pregnancy. Under the supervision of a physician, RU-486 is taken in pill form. Two days later, the physician administers a synthetic prostaglandin that induces uterine contractions, causing the embryo to be expelled. Ethical concerns, such as debate about when a fetus becomes a "person" and the relative merits of fetal rights versus maternal rights, make abortion controversial.

CASE STUDY REVISITED The Frozen Zoo

Although it is directed at saving endangered species, assisted reproductive technology (ART) does not appeal to all conservationists. Some contend that the only appropriate way to preserve a species is to maintain enough natural habitat to support a breeding population large enough to sustain itself and maintain reasonable genetic diversity. Proponents of ART agree, but they support high-tech efforts as a necessary adjunct to habitat preservation, particularly for critically endangered animals. Presiding over the frozen zoo in San Diego, geneticist Dr. Oliver Ryder explains: "[The frozen zoo] represents a genetic legacy. A DNA bank. In the future, scientists may have better tools, but they won't have access to more genes." Dr. Betsy Dresser, who heads the Audubon Center for Research of Endangered Species in New Orleans, describes ART as "a safety net": "If you freeze 200 or 300 embryos, that's enough to keep a population from going extinct." Dresser, who is working to develop interspecies embryo transfer techniques that will allow lions to serve as surrogate mothers for the endangered Siberian tiger, adds: "I don't want to see tigers just in textbooks someday. Nor do I want people a hundred years from now to look back and say, 'God, they had this technology and they just let these animals go extinct.'" Advocates of ART look forward to a future when habitat is restored and protected so that populations of critically endangered species that have been bred in zoos (while maintaining as much genetic diversity as possible) can be released to flourish and reproduce naturally in their native environment.

Consider This: Frozen tissue from the world's last remaining bucardo provides the only hope that Earth may again harbor this unique species. But only cloning will produce a new bucardo; any animals produced by cloning will be genetically identical, and may suffer from other problems that have plagued cloned mammals (see Chapter 11). Should scientists put the money and effort into trying to resurrect this species? Defend your answer.

Links to Life: A Male Contraceptive?

In humans and other mammalian species, child-rearing responsibility falls disproportionately on the mother (with some notable exceptions). Some women, therefore, wonder why females must take much of the responsibility for contraception as well. As you can see in Table 36-3, the only temporary methods of birth control available to men are the condom and abstinence. Actually, there is a sound biological reason for this. As John Amory, an endocrinologist at the University of Washington, puts it, "It's a lot easier to interfere with the ovulation of one egg per month than it is to interfere with the production of trillions of sperm per month." Nevertheless, efforts are well underway to bring a male hormonal contraceptive to the market. Researchers are testing an implant that gradually releases a synthetic version of progesterone, which—after about three months—brings sperm production to a halt by blocking FSH release. Unfortunately, the progesterone also stops testosterone production by blocking LH release and reduces the male sex drive. To counteract this, synthetic testosterone injections must be given every three months. Drug companies are hoping that the method will appeal to couples who are tired of using barrier methods and in which the woman is unable to (or no longer wishes to) use hormonal forms of contraception. Will men go for it? Only time will tell.

Summary of Key Concepts

36.1 How Do Animals Reproduce?

Animals reproduce either sexually or asexually. In sexual reproduction, haploid gametes, usually from two separate parents, unite and produce an offspring that is genetically different from either parent. In asexual reproduction, offspring tend to be genetically identical to the parent. Asexual reproduction can occur by budding, fission, or parthenogenesis.

During sexual reproduction, a female gamete—a large, nonmotile egg—fuses with the male gamete, a small, motile sperm. A few species of animals are hermaphroditic, producing both sperm and eggs, but most animals have separate male and female sexes. Fertilization, the union of sperm and egg, can occur outside the bodies of the animals (external fertilization) or inside the body of the female (internal fertilization). External fertilization must occur in water so that the sperm can swim to meet the egg. Internal fertilization generally occurs by copulation, in which the male deposits sperm directly into the female's reproductive tract.

36.2 How Does the Human Reproductive System Work?

The human male reproductive tract consists of paired testes, which produce sperm and testosterone, and accessory structures that conduct the sperm to the female's reproductive tract and secrete fluids that activate swimming by the sperm and provide energy. In human males, spermatogenesis and testosterone production are stimulated by FSH and LH, secreted by the anterior pituitary. Spermatogenesis and testosterone production are nearly continuous, beginning at puberty and lasting until death.

The human female reproductive tract consists of paired ovaries, which produce eggs as well as the hormones estrogen and progesterone, and accessory structures that conduct sperm to the egg and receive and nourish the embryo during prenatal development. In human females, oogenesis, hormone production, and development of the lining of the uterus vary in a monthly menstrual cycle. The cycle is controlled by hormones from the hypothalamus (GnRH), anterior pituitary (FSH and LH), and ovaries (estrogen and progesterone).

During copulation, the male ejaculates semen into the female's vagina. The sperm swim through the vagina and uterus into the uterine tube, where fertilization usually takes place. The unfertilized egg is surrounded by two barriers, the corona radiata and the zona pellucida. Enzymes released from the acrosomes in the heads of sperm digest these layers, permitting sperm to reach the egg. Only one sperm enters the egg and fertilizes it.

The ability to reproduce begins in puberty, when hypothalamic GnRH causes release of FSH and LH from the anterior pituitary. These, in turn, stimulate the sex glands to produce testosterone (male) and estrogen (female). These induce secondary sexual characteristics and the development of sperm and eggs.

36.3 How Can People Limit Fertility?

Contraception can be achieved by abstinence or by sterilization: severing the vas deferens in males (vasectomy) or the uterine tubes in females (tubal ligation). The uterine tubes may also be blocked by inserting a springlike device that causes scar tissue to form. Temporary contraception techniques include those that prevent ovulation by delivering estrogen and progesterone, for example, the pill, the contraceptive patch, the vaginal ring, and Depo Provera® injections. Barrier methods, which prevent sperm and egg from meeting, include the diaphragm, the cervical cap, the sponge, and the condom, accompanied by spermicide. Spermicide alone is less effective, while withdrawal and douching are unreliable. The rhythm method, which has a high failure rate, requires abstinence around the time of ovulation. Intrauterine devices may block sperm and also prevent implantation of the early embryo, as does the "morning after" pill. Abortion causes the expulsion of the developing embryo.

Key Terms

acquired immune deficiency syndrome (AIDS) *p. 734*
acrosome *p. 728*
asexual reproduction *p. 722*
budding *p. 722*
bulbourethral gland *p. 729*
cervix *p. 732*
chlamydia *p. 734*
clitoris *p. 733*
contraception *p. 736*
copulation *p. 724*
corona radiata *p. 735*
corpus luteum *p. 731*
crab lice *p. 734*
egg *p. 723*
embryo *p. 730*
endometrium *p. 731*
epididymis *p. 729*
estrogen *p. 726*
external fertilization *p. 723*
fertilization *p. 722*
fetus *p. 730*
fission *p. 722*
follicle *p. 730*
follicle-stimulating hormone (FSH) *p. 726*
genital herpes *p. 734*
genital wart *p. 734*
gonad *p. 726*
gonadotropin-releasing hormone (GnRH) *p. 726*
gonorrhea *p. 734*
hermaphrodite *p. 723*
internal fertilization *p. 724*
interstitial cell *p. 727*
labia *p. 733*
luteinizing hormone (LH) *p. 726*
menstrual cycle *p. 731*
menstruation *p. 732*
oogenesis *p. 730*
oogonia *p. 730*
ovary *p. 729*
ovulation *p. 726*
parthenogenesis *p. 722*
penis *p. 727*
pheromone *p. 724*
placenta *p. 731*

polar body *p. 730*
primary oocyte *p. 730*
primary spermatocyte *p. 727*
progesterone *p. 731*
prostate gland *p. 729*
puberty *p. 726*
regeneration *p. 722*
scrotum *p. 727*
secondary oocyte *p. 730*
secondary spermatocyte *p. 727*
semen *p. 729*
seminal vesicle *p. 729*
seminiferous tubule *p. 727*
Sertoli cells *p. 727*
sexually transmitted disease (STD) *p. 734*
sexual reproduction *p. 722*
spawning *p. 723*
sperm *p. 723*
spermatid *p. 727*
spermatogenesis *p. 727*
spermatogonia *p. 727*
spermatophore *p. 725*
sterilization *p. 736*
syphilis *p. 734*
testes *p. 727*
testosterone *p. 726*
trichomoniasis *p. 734*
urethra *p. 729*
uterine tube *p. 730*
uterus *p. 731*
vagina *p. 733*
vas deferens *p. 729*
zona pellucida *p. 735*
zygote *p. 731*

Thinking Through the Concepts

To take a multiple-choice quiz with feedback on the contents of this chapter, visit http://www.prenhall.com/audesirk7. *Log in to the Web site selected by your instructor and navigate to the Self Test section for this chapter.*

? Review Questions

1. List the advantages and disadvantages of asexual reproduction, sexual reproduction, external fertilization, and internal fertilization, including an example of an animal that uses each type.
2. Compare the structures of the egg and sperm. What structural modifications do sperm have that facilitate movement, energy use, and digestion?
3. What is the role of the corpus luteum in a menstrual cycle? In early pregnancy? What determines its survival after ovulation?
4. Construct a chart of common sexually transmitted diseases. List the disease's name, the cause (organism or virus), symptoms, and treatment.
5. List the structures, in order, through which a sperm passes on its way from the seminiferous tubules of the testis to the uterine tube of the female.
6. Name the three accessory glands of the male reproductive tract. What are the functions of the secretions they produce?
7. Diagram the menstrual cycle, and describe the interactions among hormones secreted by the pituitary gland and ovaries that produce the cycle.

Applying the Concepts

1. Discuss the most effective or appropriate method of birth control for each of the following couples: Couple A, who have intercourse three times a week but never want to have children; Couple B, who have intercourse once a month and may want to have children someday; and couple C, who have intercourse three times a week and want to have children someday.
2. *Pelvic endometriosis* is a relatively common disease in which bits of the endometrial lining find their way onto abdominal organs and respond in typical ways to hormones during a menstrual cycle. When the uterine lining bleeds during menstruation, so do these implants. Common treatments are oral contraceptives, Danazol™ (a compound that inhibits gonadotropins), and synthetic GnRH analogues that are, paradoxically, powerful inhibitors of FSH and LH. How does each of these compounds provide relief?
3. Would contraceptive drugs that block cell receptors for FSH be useful in males and/or females? Explain. What side effects would such drugs have?
4. Think of all the ways a couple can obtain a child, including *in vitro* fertilization using the couple's eggs and sperm, *in vitro* fertilization using a donor's sperm or egg, and insemination of a surrogate mother with sperm from the couple's husband. Think of some more. What ethical issues do these various options present?
5. Fertility drugs have greatly increased the incidence of multiple births. When more than two embryos share the uterus, the incidence of premature birth and developmental problems increase substantially. The costs of caring for multiple premature infants is staggering. When fertility drugs produce multiple embryos, the physician can selectively eliminate some of these embryos early in development, so the remaining few have a better chance to develop fully and normally. Given these facts, discuss the ethical implications of taking fertility drugs.

For More Information

Alexander, N. "Future Contraceptives." *Scientific American*, September 1995. A review of the techniques that may be used for contraception in the twenty-first century.

Crews, D. "Animal Sexuality." *Scientific American*, January 1994. Explores the wide range of mechanisms that control male and female sexual development among different types of animals.

Estabrook, B. "Staying Alive." *Wildlife Conservation*, June 2002. Assisted reproductive technology offers hope for saving endangered species.

Fackelmann, K. "It's a Girl." *Science News*, November 28, 1998. Sperm sorting can help parents select the sex of their children, particularly for girls.

Lanza, R. P., Dresser, B. L., and Damiani, P. "Cloning Noah's Ark." *Scientific American*, November 2000. For some endangered species, cloning may offer the best chance for survival.

Ness, E. "How to Breed a 2,000 Pound Rhino" *Discover*, November 2001. The endangered Sumatran rhino has given birth in captivity for the first time in over a century, and none too soon—only about 300 survive in the wild.

Riddle, J. M., and Estes, J. W. "Oral Contraceptives in Ancient and Medieval Times." *American Scientist*, May–June 1992. How did women control their fertility before modern medicine stepped in?

Wright, K. "Human in the Age of Mechanical Reproduction." *Discover*, May 1998. The child's question "Where do babies come from?" has become far more difficult to answer as technology has provided more than a dozen ways to have a child.

Wright, K. "Male Contraception." *Discover*, October 2002. Explores challenges and progress in developing a male contraceptive.

Media Activities

To access a Media Activity visit http://www.prenhall.com/audesirk7. *Log in to the Web site selected by your instructor, navigate to this chapter, and select the appropriate Media Activity number.*

36.1 Human Reproductive System

Estimated time: 10 minutes

In this activity, you will explore the structure and function of the human reproductive system and the formation of sperm in the male and an egg in the female.

36.2 Hormonal Control of the Menstrual Cycle

Estimated time: 10 minutes

Here you will explore the interplay of hormones and physical events during the course of the menstrual cycle.

36.3 Web Investigation: The Frozen Zoo

Estimated time: 30 minutes

Explore the science and controversy surrounding "Frozen Zoos" (also called Genome Resource Banks) as a tool for biologists to combat the extinction of endangered species.

CHAPTER

37 Animal Development

A pregnant woman's decisions can have a lifelong influence on the health, behavior, and physical attributes of her developing child.

AT A GLANCE

CASE STUDY Far-Reaching Choices

Maria is excited and nervous as she phones Nicole, a close friend from high school who took part in Maria's recent wedding. "My home pregnancy test showed up blue this morning—I think we did it!" she exclaims. Sounding overwhelmed, Nicole offers her congratulations, then adds, "Let's celebrate! Join me at Sudsy Smith's tonight. Just one drink, and—oh, I'm late for work! How about 5:30?" But recalling the smoky ambiance of Sudsy's and a warning she once noticed on a wine bottle, Maria suggests a local coffee shop instead. As she hangs up the phone, she considers her new responsibilities. What did the wine bottle say? Retrieving a beer from the refrigerator, she reads the small print on the side of the can: "Warning: According to the Surgeon General, women should not drink alcoholic beverages during pregnancy because of the risk of birth defects."

Contemplating the seeming miracle by which her single fertilized egg will become a dazzlingly complex individual, Maria vows to do everything she can to protect the health of her developing child. Her next phone call is to her doctor to schedule a prenatal visit. As Maria prepares a list of questions for her physician, she also plans a trip to the library to educate herself and her husband. The first question on her list is: "How might alcohol affect my developing child?" As a smoker, she also wonders if her pregnancy will finally give her the motivation she needs to quit for good. It suddenly occurs to her that even her favorite triple mocha espresso might not be the best choice to celebrate her pregnancy.

As you learn more about the amazing series of events that constitute development, keep Maria's investigation in mind. How could her home pregnancy test tell her she was pregnant after only a few weeks? What questions would you ask your doctor if you were an expectant mother or father? How would you change your lifestyle? If you were concerned about a friend who smoked or drank during pregnancy, how would you deal with the issue?

How does form arise from formlessness? Developmental biologists continue to learn more about how a single cell—the zygote formed from the fusion of sperm and egg—transforms itself into a complex organism. Because the cells of the embryo proliferate by mitosis, each cell has an identical genome. What chemical commands transform genetically uniform cells into the different components of bones, blood, and the brain? As scientists learn more, optimism grows that we might harness the ability to direct cellular differentiation, eventually developing techniques to replace damaged cells in sick or disabled individuals. Let's explore the types and stages of animal development, a little of what is known about cell differentiation, and ways in which foreign substances can interfere with this delicate process.

37.1 How Do Indirect and Direct Development Differ?

When we think of development, images of a newborn infant may come to mind. Certainly, their proportions are different, but babies are, in all important ways, miniature versions of adult humans. People and other mammals—as well as birds and reptiles—are all born as "miniature adults," developing via a process called **direct development.** For the vast majority of animal species, however, indirect development is the norm.

During Indirect Development, Animals Undergo a Radical Change in Body Form

In **indirect development**, a juvenile animal differs significantly from an adult of the same species and undergoes radical changes in body form during development, such as the transformation of a caterpillar into a butterfly. Indirect development occurs in most invertebrates, including insects and echinoderms, and in a few vertebrates—notably the amphibians. Animals with indirect development typically produce huge numbers of eggs, and each egg has only a small amount of food reserve called **yolk**. The yolk nourishes the developing embryo during its rapid transformation into a small, sexually immature form called a **larva** (Fig. 37-1). Because the yolk is small and the time spent as an embryo is relatively short, indirect development does not place great demands on the mother; therefore, many offspring can be produced.

Some larval animals not only look very different from adult animals but also occupy entirely different habitats. In addition, most larvae feed on different organisms than they will as adults. For instance, the aquatic larva of the dragonfly feeds on aquatic organisms such as tadpoles, but the adult dragonfly, which is terrestrial, feeds on insects (Fig. 37-1b). Eventually, the larvae undergo a revolution in body form, or **metamorphosis**, and become sexually mature adults.

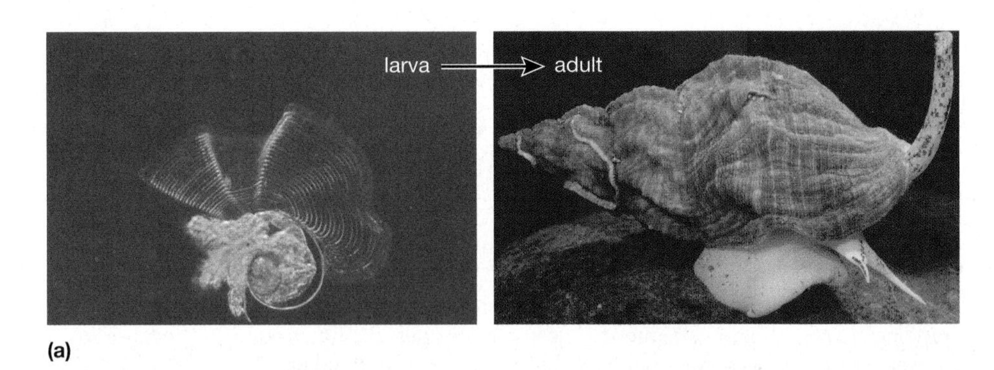

FIGURE 37-1 Indirect development **(a)** Many marine mollusks, such as this common whelk, undergo indirect development in which the nearly microscopic larva is very different from the adult in size, appearance, and lifestyle. **(b)** The larval dragonfly is aquatic and feeds on tadpoles (as shown here) and small fish, whereas the adult form is terrestrial and eats other insects. **EXERCISE** List and explain some advantages and disadvantages of indirect development.

FIGURE 37-2 Direct development
The offspring of animals with direct development closely resemble their parents from the moment of birth. All are nurtured either by egg yolk or by nutrients from the mother's blood. **(a)** A male seahorse gives birth to young that have developed from yolk-rich eggs placed in his pouch by the female. **(b)** Lizards hatch from large, yolk-filled eggs. **(c)** Snails hatch from small, yolk-rich eggs. **(d)** Mammalian mothers nourish their developing young within their bodies. **EXERCISE** List and explain some advantages and disadvantages of direct development.

Although we tend to regard the adult form as the "real animal" and the larval stage as "preparatory," most of the life span of some animals, especially insects, is spent as a larva. The adult may live for only a few days, reproducing frantically and in some cases not even eating. The mayfly, for example, metamorphoses from an aquatic larva that may spend a year or more feeding and growing. Emerging in huge swarms from freshwater streams, ponds, and lakes, adult mayflies live a few hours or, at most, a few days. The adults' sole occupation is to mate and lay eggs; their fragile dead bodies then accumulate in piles to be swept away by the wind.

Newborn Animals That Undergo Direct Development Resemble Miniature Adults

Other animals, including such diverse groups as land snails, reptiles, birds, and mammals, undergo direct development, in which the newborn animal is a miniature, but sexually immature, version of the adult (Fig. 37-2). As the young animal matures, it may grow much bigger, but it does not radically change its body form.

Juveniles of directly developing species are typically much larger than larvae and consequently need much more nourishment before emerging into the world. Two

Table 37-1 Vertebrate Embryonic Membranes

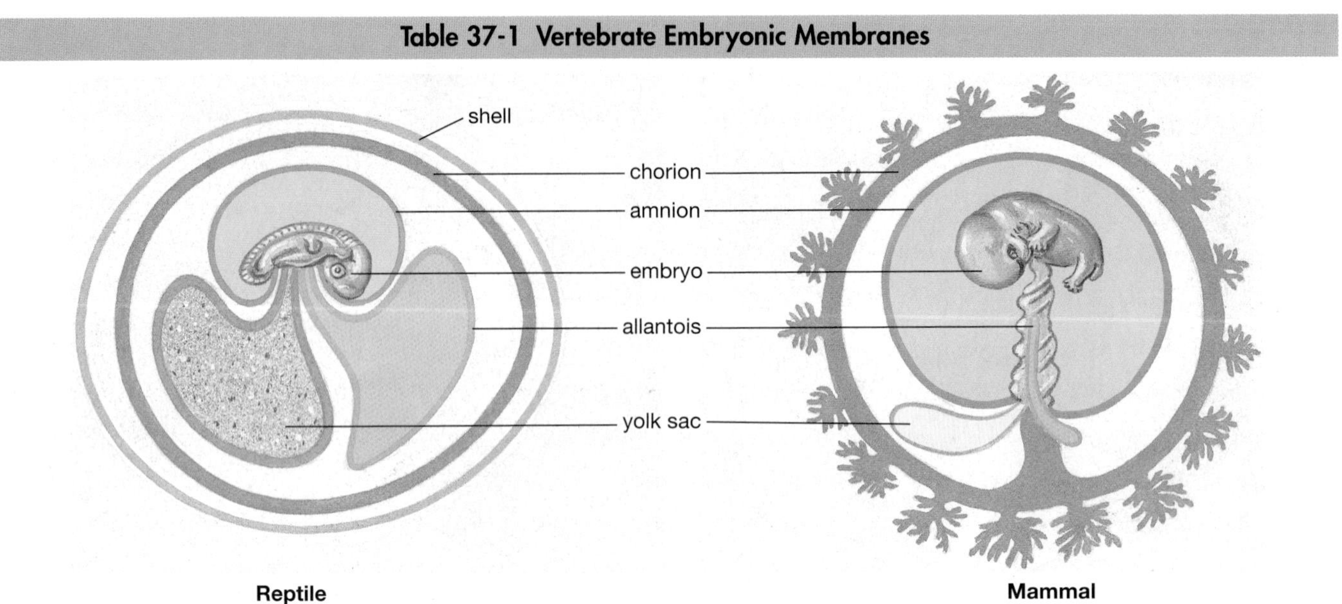

Membrane	Reptilian Embryo: Structure	Reptilian Embryo: Function	Mammalian Embryo: Structure	Mammalian Embryo: Function
Chorion	Membrane lining inside shell	Acts as respiratory surface; regulates exchange of gases and water between embryo and air	Fetal contribution to placenta	Provides surface for exchange of gases, nutrients, and wastes between embryo and mother
Amnion	Sac surrounding embryo	Encloses embryo in fluid	Sac surrounding embryo	Encloses embryo in fluid
Allantois	Sac connected to embryonic urinary tract; capillary-rich membrane lining inside of chorion	Stores wastes (especially urine); acts as respiratory surface	Provides blood vessels of umbilical cord	Carries blood between embryo and placenta
Yolk sac	Membrane surrounding yolk	Contains yolk as food; digests yolk and transfers nutrients to embryo	"Empty" membranous sac	Forms blood cells

strategies have evolved that meet the embryo's food requirement. Snails, reptiles, and birds produce eggs that contain relatively large amounts of yolk. An ostrich egg, for example, weighs several pounds. Mammals, some snakes, and a few fish have relatively little yolk in their eggs; instead, developing embryos are nourished within the body of the mother. Either way, providing food for directly developing embryos places great demands on the mother, in contrast to indirect development, and relatively few offspring are produced.

Reptiles, Birds, and Mammals Produce Similar Extraembryonic Membranes

Amphibians were the first vertebrates to live on land, but their reproduction remains tied to water, where they deposit their eggs and their larval offspring grow and metamorphose into adults. Fully terrestrial vertebrate life was not possible until the evolution of the shelled **amniotic egg**. This innovation, which encases the embryo in a protected, liquid-filled space, arose first in reptiles and persists today in that group and its descendants, birds and mammals. It allows these groups to complete their development into the adult form in their own "private pond." The amniotic egg is characterized by four membranes, called **extraembryonic membranes**: the *chorion*; the *amnion*; the *allantois*; and the *yolk sac*. The **chorion** lines the shell and exchanges oxygen and carbon dioxide between the embryo and the egg's external environment. The **amnion** encloses the embryo in a watery environment; the **allantois** surrounds and isolates wastes; and (in non-mammalian vertebrates) the **yolk sac** contains the stored food, or "egg yolk." Although mammalian eggs contain almost no yolk (the mammalian yolk sac is a source of blood cells), all four extraembryonic membranes still persist, remnants of the reptilian genetic program for development. Table 37-1 compares the structures and functions of these extraembryonic membranes in reptiles and mammals.

37.2 How Does Animal Development Proceed?

The transformation from fertilized egg—a single cell—to a multicellular, differentiated embryo is a beautiful, nearly magical process that has been carefully described for a number of animals. Actual development is a smoothly continuous process; the stages depicted in textbooks are just convenient "snapshots" (see "A Closer Look: Stages of Development" in Chapter 37 of this text's Web site). The initial stages of *cleavage*, *gastrulation*, *organogenesis*, and *growth* occur during embryonic life, in which nearly all the organs that will be present in the adult are formed. After birth, the animal typically undergoes further growth, achieves *sexual maturity* (at which time the animal may reproduce), *ages*, and finally dies. Let's examine these stages of development.

Cleavage Begins the Process

The formation of an embryo begins with **cleavage**, a series of mitotic divisions of the fertilized egg without an increase in size. Cleavage reduces cell size and distributes gene-regulating substances to the newly formed cells. An egg is a very large cell. Unlike most mitotic cell divisions—during which cells divide, grow, duplicate genetic material, then divide again—embryonic cell divisions during cleavage skip the growth phase. Consequently, as cleavage progresses, the available cytoplasm is split up into ever-smaller cells whose sizes approach those of cells in the adult organism. Finally, a solid ball of small cells, the **morula**, is formed. The morula is about the same size as the fertilized egg. As cleavage continues, a cavity opens within the morula, and its cells become the outer covering of a hollow, typically spherical structure called the **blastula**. The space inside the blastula is called the *blastocoel* (Fig. 37-3a).

The details of cleavage differ by species. The pattern is largely determined by the amount of yolk present, because yolk hinders cytokinesis (cytoplasmic division). The almost yolkless eggs of sea urchins divide symmetrically, but eggs with extremely large yolks, such as a hen's egg, don't divide all the way through. Nevertheless, a hollow blastula is always produced; in reptiles and birds, it is flattened on top of the yolk (and thus, not spherical).

Gastrulation Forms Three Tissue Layers

In the next step of development, an indentation called the **blastopore** forms on one side of the blastula. Blastula cells migrate in a continuous sheet in through the blastopore, much as if you punched in an underinflated basketball (Fig. 37-3b), to form three embryonic tissue layers (Table 37-2). The enlarging dimple is destined to become the digestive tract; the cells that line this cavity are now called **endoderm** (Greek for "inner skin"). The cells remaining on the outside will form the epidermis of the skin and the nervous system and are called

ectoderm mesoderm endoderm

(a) The blastula just before gastrulation.

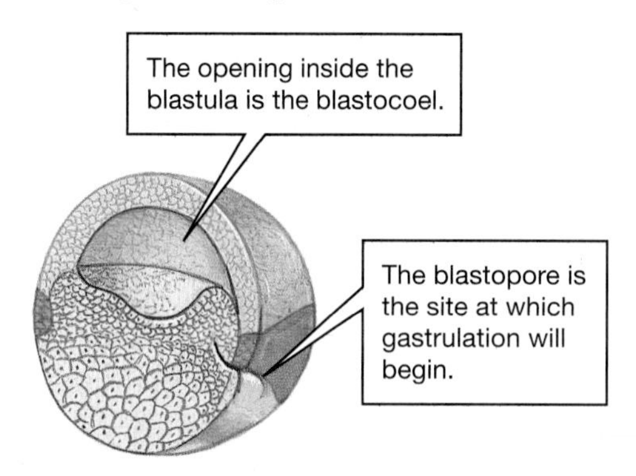

The opening inside the blastula is the blastocoel.
The blastopore is the site at which gastrulation will begin.

(b) Cells migrate at the start of gastrulation. Cells migrating in will form the endoderm and mesoderm layers of the gastrula; the cells remaining on the surface will form ectoderm.

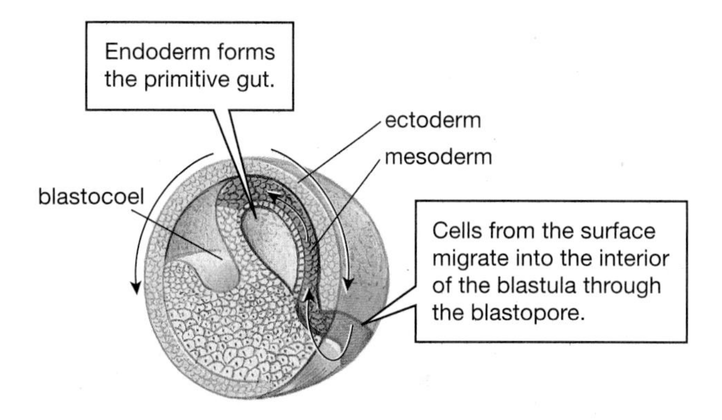

(c) Mesoderm differentiates.

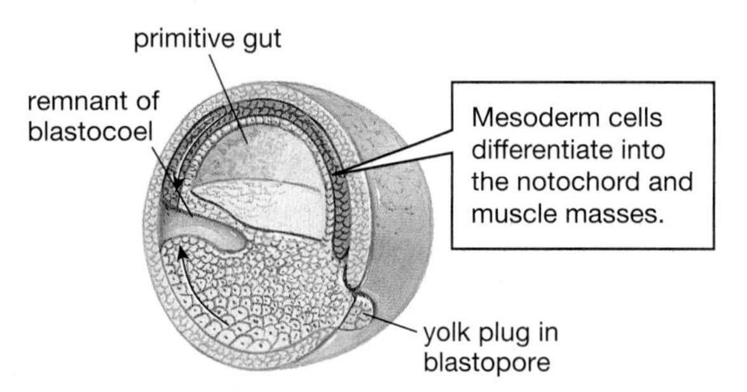

FIGURE 37-3 A blastula becomes a gastrula

Table 37-2 Derivation of Adult Tissues from Embryonic Cell Layers

Embryonic Layer	Adult Tissue
Ectoderm	Epidermis of skin; hair; lining of mouth and nose; glands of skin; nervous system
Mesoderm	Dermis of skin; muscle, skeleton; circulatory system; gonads; kidneys; outer layers of digestive and respiratory tracts
Endoderm	Lining of digestive and respiratory tracts; liver; pancreas

Media Activity
37.1 Stages of Animal Development

bio ETHICS SCIENTIFIC INQUIRY The Promise of Stem Cells

The ability of a single cell to give rise to the 200 or so types of cells in an adult organism is one of the wonders of the living world. Scientists have long known that every cell's nucleus contains the entire genetic blueprint for an organism, and that whether a cell becomes muscle, bone, or brain is determined by complex factors that shape the environment in which it develops and differentiates. If scientists could harness this potential, they might create new tissues from a person's own donor cells—tissues that wouldn't be rejected by the immune system—to replace those that are damaged or missing. For example, different types of neurons might be transplanted to bridge the gap in a crushed spinal cord or to replace cells that die from Parkinson's disease. Likewise, bone cells might fill in a shattered fracture, or heart muscle cells might replace those killed by a heart attack. In 1998, James Thompson and his co-workers at the University of Wisconsin reported the first success in isolating human **embryonic stem cells** and inducing them to grow in culture dishes. These embryonic stem cells differentiated into a wide variety of human tissues when they were transplanted under the skin of mice.

The process of creating embryonic stem cells as a source of replacement tissue, starting with a human blastocyst, is described in Figure E37-1. Once the stem cells start growing in culture, researchers must stimulate them to differentiate into the desired tissues. To accomplish this, they first need to determine what crucial chemical signals in the developing embryo carry the messages to produce specific tissues; then they must deliver these chemical messages to cells growing in culture dishes. Alternatively, it might be possible to introduce undifferentiated stem cells into a specific part of the body—say, the brain or the bone marrow—and let the normal chemical environment in that area guide differentiation. The ability to culture and differentiate human stem cells has generated a flurry of research and tremendous excitement in the scientific community. While researchers of human aging hope that worn-out organs might someday be replaced, biotech firms are scrambling to find new ways to produce these cells quickly and efficiently for use as tissue transplants—perhaps in the near future.

One vexing problem is that even cells derived from embryos elicit an immune response in recipients. Researchers are investigating the possibility of making stem cells by generating blastocyst clones of the individual to be treated, replacing the nucleus of a donor egg with a nucleus from one of the recipient's cells. If this *therapeutic cloning* technology were successful, the stem cells from the resulting blastocyst would carry the same genes as the recipient, and could potentially be cultivated into replacement tissues that would not be rejected.

Another approach uses *adult stem cells*. All of the cells in a blastocyst's inner cell mass are stem cells, but adult stem cells are more elusive and more difficult to culture. The most promising source of these cells appears to be bone marrow, where researchers have recently isolated what they call "multipotent adult progenitor cells." When the comparable mouse cells were injected into mice, the cells took up residence in several different organs, such as blood, liver, lungs, and intestines, and appeared to differentiate into their adopted host tissues. Using stem cells found in easily accessible tissues in an individual's own body would have tremendous advantages. This approach would circumvent not only ethical concerns but also problems of tissue rejection, since the immune system would not attack cells taken from the recipient's own body.

Legislators grapple with ethical concerns in embryonic stem cell research, since the blastocyst is an extremely early human embryo. Blastocysts, each consisting of about 100 cells, are frequently obtained from *in vitro* fertilization clinics where they would otherwise be destroyed after a successful pregnancy has been achieved. Therapeutic cloning to produce transplantable tissues that would not be rejected (using an adult cell nucleus) could, in theory, also be used to produce a human clone of the donor. Many scientists and laypeople favor therapeutic cloning in order to generate stem cells to correct various disabilities, but reject the concept of reproductive cloning.

There are other hurdles to overcome, including finding ways to ensure that transplanted stem cells do not grow out of control and form tumors, but the potential rewards of stem cell technology are far-reaching. In addition to their tremendous potential for transplantation, stem cells might one day be cultured in large quantities to screen for drugs that might produce birth defects and also to study the incredible complexity of the control of development.

---

ectoderm ("outer skin"). Meanwhile, some cells migrate between the endoderm and ectoderm, forming a third and final layer, the **mesoderm** ("middle skin"). Mesoderm gives rise to muscles, the skeleton (including the **notochord**, a supporting rod found at some stage in all chordates), and the circulatory system (Fig. 37-3c). This process of cell movement is called **gastrulation**, and the three-layered embryo that results is the **gastrula**.

Adult Structures Develop

Gradually, the ectoderm, mesoderm, and endoderm rearrange themselves into the organs characteristic of the animal's species (see Table 37-2) in a process called **organogenesis**. In some cases, adult structures are, in effect, "sculpted" by the death of excess cells produced during embryonic development. Some cells are programmed to die at precise times during development; this cell death is controlled by at least two mechanisms that function in different tissues. Some cells die during development unless they receive a "survival signal." Embryonic vertebrates, for example, have far more motor neurons in their spinal cords than do adult animals. Motor neurons survive only if they successfully form synapses with a skeletal muscle cell, which releases a chemical that prevents the death of its own motor neuron.

For other cells, the situation is just the reverse: some cells live unless they receive a "death signal" from other

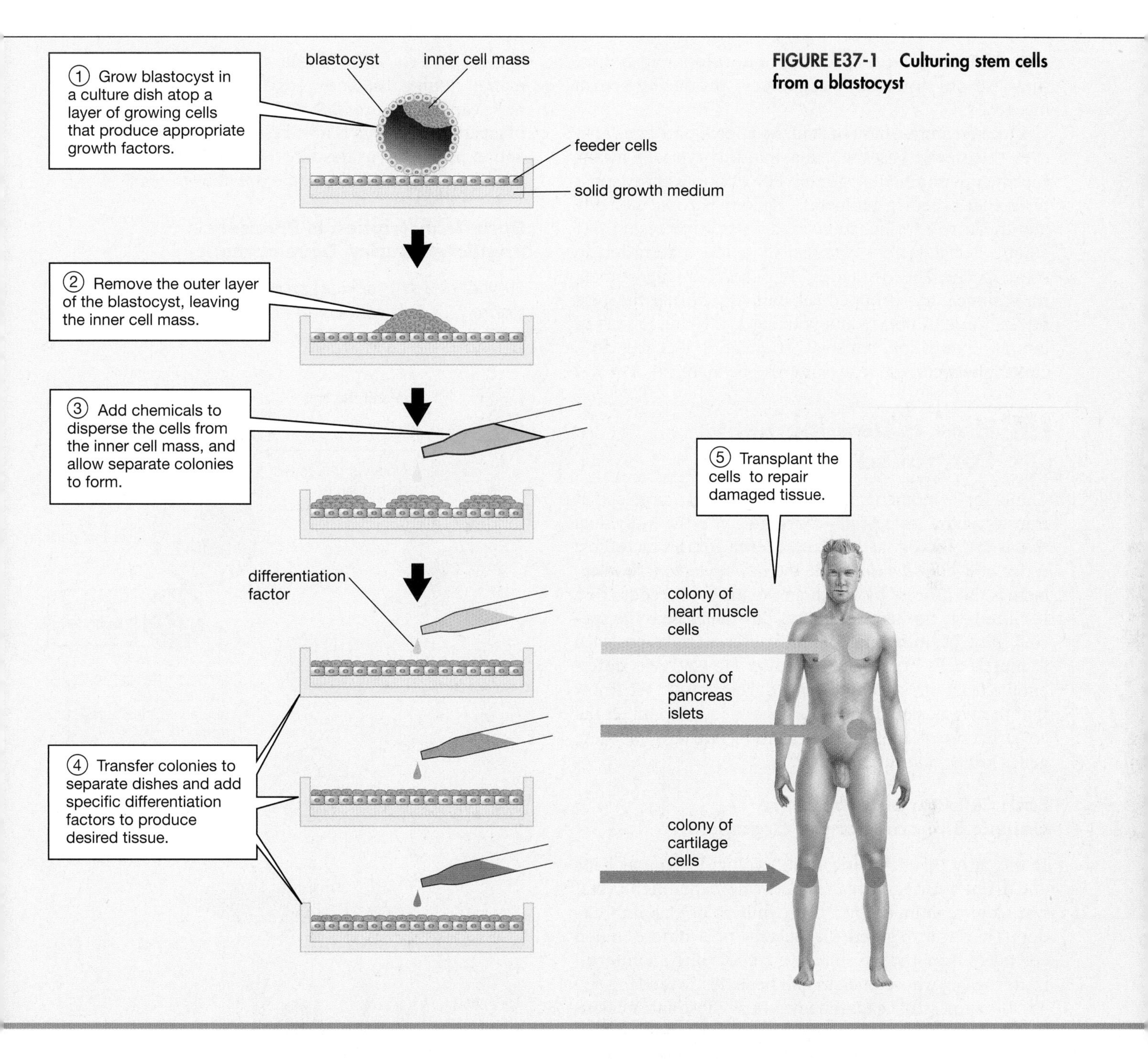

① Grow blastocyst in a culture dish atop a layer of growing cells that produce appropriate growth factors.
blastocyst
inner cell mass
feeder cells
solid growth medium
② Remove the outer layer of the blastocyst, leaving the inner cell mass.
③ Add chemicals to disperse the cells from the inner cell mass, and allow separate colonies to form.
⑤ Transplant the cells to repair damaged tissue.
differentiation factor
colony of heart muscle cells
colony of pancreas islets
④ Transfer colonies to separate dishes and add specific differentiation factors to produce desired tissue.
colony of cartilage cells

FIGURE E37-1 Culturing stem cells from a blastocyst

cells of the developing animal. Many embryonic structures disappear during development in this way. For example, all vertebrates pass through developmental stages in which they have tails and webbed hands and feet. In humans, these stages can be seen clearly in the 5-week-old human embryo (see Fig. 37-11a). Two weeks later, the cells of the webbing have died, revealing separate fingers, and the tail cells are dying, causing the tail to regress (see Fig. 37-11b). In frogs, the tail is lost during metamorphosis from its tadpole larva. Thyroid hormone, which triggers metamorphosis, stimulates cells in the tail to synthesize enzymes that digest the tail away. If the thyroid gland is surgically removed, the frog retains its tail.

Sexual Maturation Is Controlled by Genes and the Environment

Development does not stop at birth; animals continue to change throughout their lives. The period between birth and sexual maturity is one of very active growth; cells of all types increase in number, resulting in an increase in the size of the entire organism. Growth usually slows when the organism becomes sexually mature and stops relatively soon afterward.

Animals are not born with full reproductive capability; they become sexually mature at an age determined by interactions between genetic and environmental factors. Most vertebrates must undergo months to years of genetically regulated growth and development before they are physiologically capable of producing sperm or eggs.

Once an animal has reached the appropriate age, however, the precise onset of sexual maturity typically awaits specific environmental stimuli. For example, most songbirds that breed in temperate regions become sexually mature in the spring, stimulated by the increasing day length. Social factors can also influence maturation in some species. The average age of puberty among women, for instance, has dropped substantially during the past few centuries. Puberty at a younger age is due in part to improved nutrition, but social stimulation by early close contact between the sexes may also be involved.

37.3 How Is Development Controlled?

Media Activity 37.2 Control of Development

Think for a moment about the biological miracle that transformed a single cell—a zygote—into the individual that is you. Biologists use prosaic terms for this incredible series of events: *development* and *differentiation*. **Development** is the process by which an organism proceeds from fertilized egg through adulthood. **Differentiation** is the specialization of embryonic cells into different cell types, such as muscle cells, brain cells, and so on. How do cells differentiate from one another during development? We know that the zygote contains all the genes needed to direct the construction of the entire organism. Are any of these genes lost as cells differentiate?

Each Cell Contains the Entire Genetic Blueprint for the Organism

In the early 1950s, American embryologists Thomas King and Robert Briggs began pioneering experiments that were later continued by British embryologist John Gurdon. They transplanted the nucleus of a differentiated cell taken from the intestine of a tadpole into an unfertilized frog egg whose nucleus had been destroyed (Fig. 37-4). In successful experiments, the intestinal nucleus directed the frog egg's development into a normal tadpole, a feat that would have been impossible if genes were lost during differentiation. The experiments provided strong evidence that each differentiated cell in an animal contains all the genetic information needed for the development of the entire organism. Scientists now know that different types of cells differ not in which genes they *contain* but in which genes they *use*. In other words, cell types differ because different genes are activated, transcribed to messenger RNA, and translated into proteins. Does this experiment sound familiar? Think back to Chapter 11, in which you learned about the process by which a sheep (and since then, many other animals) was "cloned" to create Dolly; the nucleus of a single adult cell, injected into an egg from which the nucleus had been removed, grew and differentiated into a normal sheep. The frog experiments conducted half a century ago set the stage for modern breakthroughs in mammalian cloning. The knowledge that all cells retain the genetic capacity to produce all of the specialized structures of an adult organism is now being utilized in another new and exciting way, as described in "Scientific Inquiry: The Promise of Stem Cells (on the previous pages)."

Gene Transcription Is Precisely Regulated During Development

How does a cell "decide" to become bone, muscle, or intestine? In any cell at any time, only a portion of the cell's genes are used, or transcribed. (Recall from

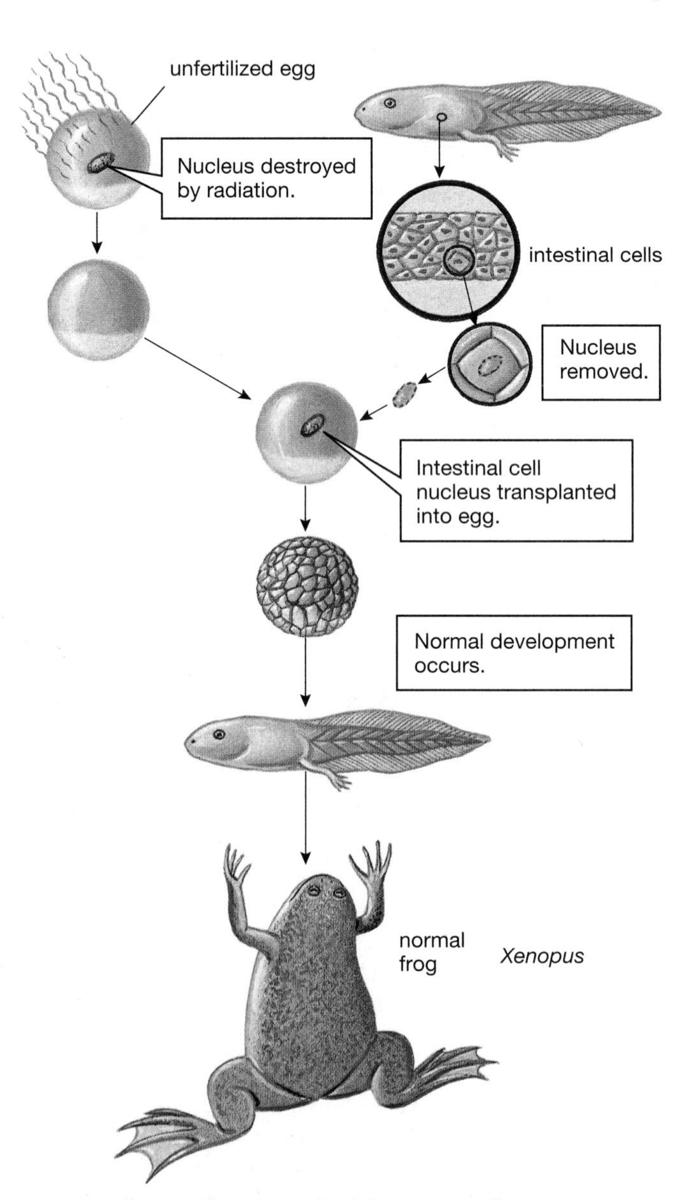

FIGURE 37-4 Cells retain all of their genes during differentiation
Researchers destroyed the nucleus of an unfertilized frog egg before transplanting the nucleus of a tadpole's intestinal cell into the egg. The resulting egg cells developed normally, demonstrating that intestinal cells retain all the genes necessary for the development of an entire organism. **QUESTION** In this experiment, would transplantation of a nucleus from any cell of an adult frog lead to normal development?

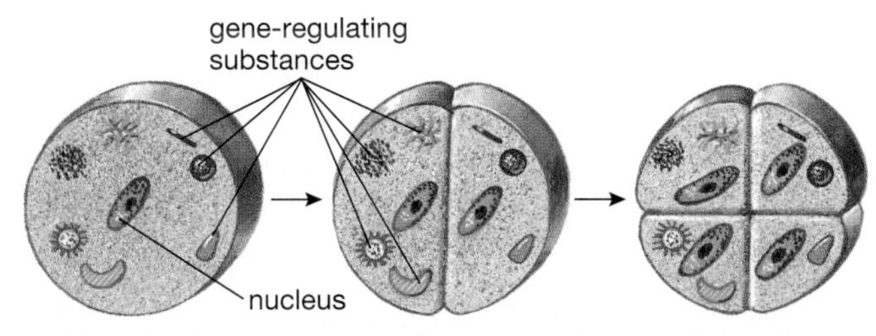
gene-regulating
substances
nucleus

FIGURE 37-5 Distribution of gene-regulating substances In some animals, gene-regulating substances are distributed in a precisely regulated manner during the division of a fertilized egg. As the egg divides, the daughter cells inherit different substances. **EXERCISE** Imagine that you are investigating an animal species whose developmental process is unknown. Design an experiment to test whether unevenly distributed gene-regulating substances in the egg cytoplasm are responsible for differentiation during embryonic development.

Chapter 10 that transcription is the production of messenger RNA using the gene as a blueprint.) The particular combination of genes that is transcribed in a cell largely determines the shape, structure, and activity of that cell. Thus, differentiation during development is accomplished by selectively activating transcription in different sets of genes. In general, the mechanism by which transcription is controlled involves regulatory molecules, typically proteins or proteins combined with activating substances such as steroid hormones, that travel to the nucleus and bind to the chromosomes (see Chapter 10). These substances bind to specific genes and can either block or promote their transcription.

In many invertebrates, various gene-regulating substances become concentrated in different places in the egg's cytoplasm as it develops. As the fertilized egg divides, each of its daughter cells receives different gene-regulating substances (Fig. 37-5), which then influence the fate of the daughter cells. In vertebrates (including mammals), however, gene-regulating substances are present in the fertilized egg but seem to be distributed evenly during early cleavage, producing cells that can each give rise to a complete individual if they are separated. For example, identical twins can be produced if the two cells formed by the first cleavage become separated. During later embryonic development (and throughout adult life), cells constantly receive chemical messages—including growth factors, hormones, and neurotransmitters—from other cells of the body. These chemical messages can alter the developmental fate of a cell by altering the transcription of genes and the activity of enzymes within the cell.

During gastrulation, the developmental fate of most of the embryo's cells is determined by chemical messages received from other cells, a process called **induction**. In amphibian embryos, special cells form at the site of dimpling as the blastula is transformed into the gastrula. This area, called the *dorsal lip of the blastopore*, controls the developmental fate of the cells around it, as Hans Spemann and Hilde Mangold demonstrated with well-designed transplantation experiments performed in the 1920s (Fig. 37-6).

Guided by chemical cues, cells migrate within the developing embryo (see Fig. 37-3). The process by which cells reach their appropriate positions in the developing

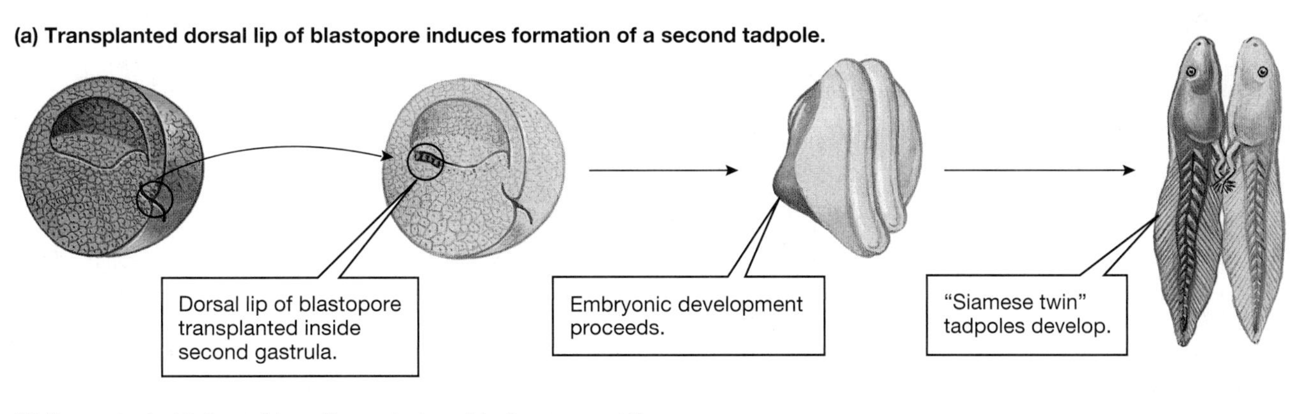
(a) Transplanted dorsal lip of blastopore induces formation of a second tadpole.
Dorsal lip of blastopore transplanted inside second gastrula.
Embryonic development proceeds.
"Siamese twin" tadpoles develop.

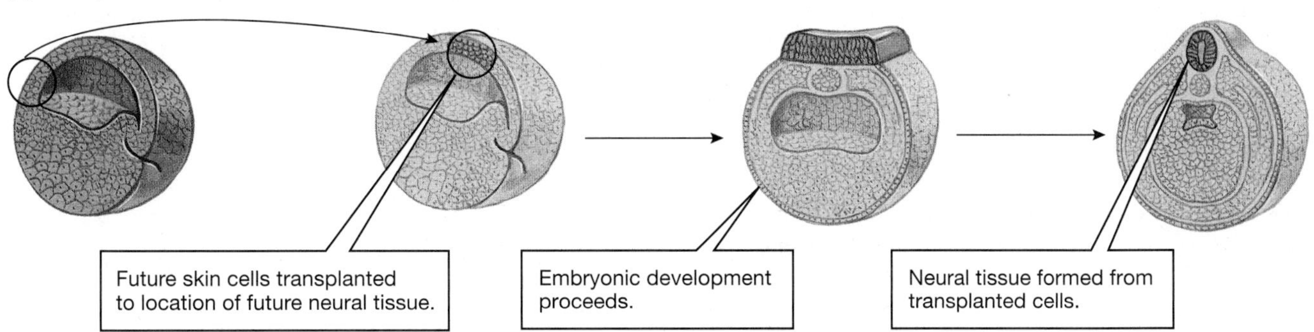
(b) Transplanted future skin cells are induced to form neural tissue.
Future skin cells transplanted to location of future neural tissue.
Embryonic development proceeds.
Neural tissue formed from transplanted cells.

FIGURE 37-6 Induction and its role in differentiation

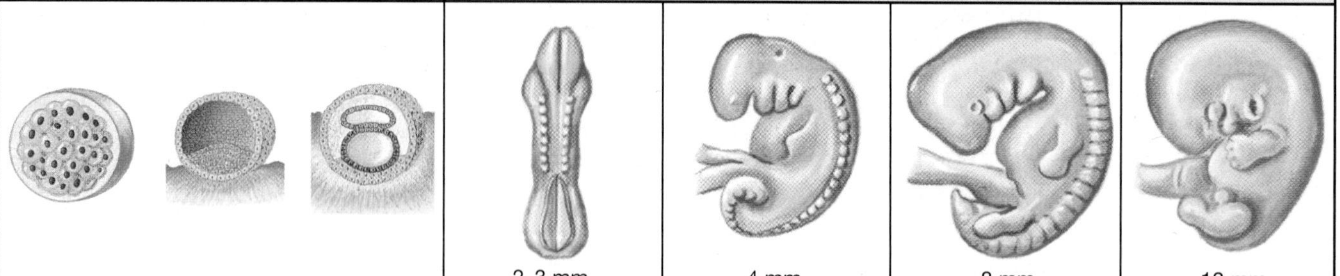

1 week	2 weeks	3 weeks	4 weeks	5 weeks	6 weeks
zygote to late blastula		embryo			
		2–3 mm	4 mm	8 mm	13 mm
Egg is fertilized to form zygote; zygote cleaves to form blastula which implants	Blastula burrows into endometrium; forms yolk sac, amnion, and embryonic disc	Gastrulation occurs; notochord and beginning of neural tube form; heart beats	Neural tube closes; arm buds, tail, and gill grooves form	Incipient eye parts—retina (as optic cup) and lens (as lens pits)—form; leg buds form; brain enlarges	Webbed fingers and external ear form; pigment appears in retina; tail and gill grooves disappearing

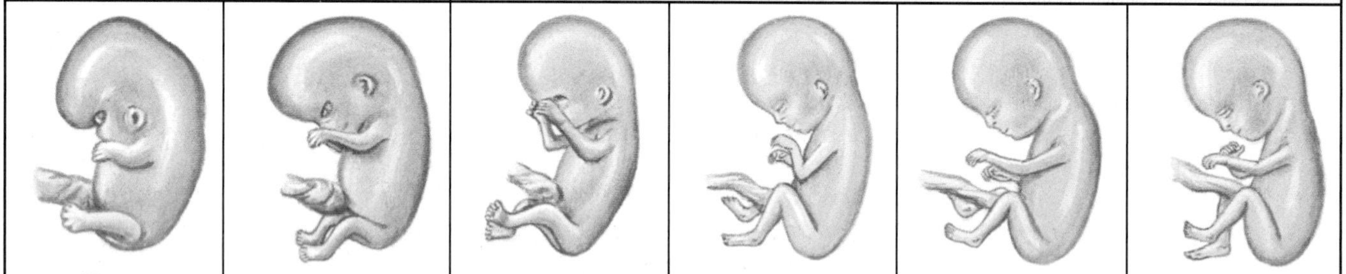

7 weeks	8 weeks	9 weeks	10 weeks	11 weeks	12 weeks
embryo		fetus			
18 mm	30 mm	50 mm	61 mm	73 mm	87 mm
Webbed toes form; bones begin to harden; back straightens; eyelids form	Upper limbs bend at elbows; genitalia begin to differentiate; fingers are distinct	Toes separate; eyelids develop; major parts of brain are present	Chin grows; nostrils separate; face appears human; genitals appear male or female	Well-defined neck appears; genitalia are complete; sucking reflex appears	

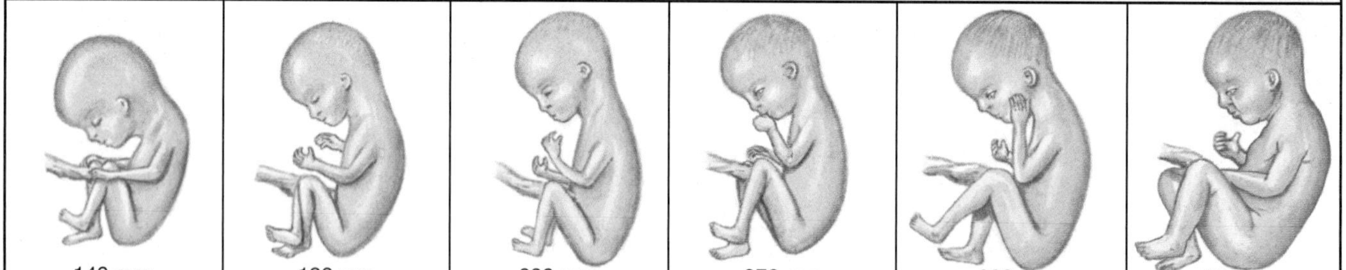

4 months	5 months	6 months	7 months	8 months	9 months
fetus					
140 mm	190 mm	230 mm	270 mm	300 mm	350 mm
Blood cells form; all major organs form; head and body hair appear; movements are felt by mother		Fetus may be viable if born; eyelids open; lungs and lung circulation develop; may suck thumb; fat deposited under skin		Fat deposits increase; body hair is lost; head hair is well developed; most senses are well developed; fetus turns head down in uterus	

for reference: 10 mm

FIGURE 37-7 Human embryonic development
A calendar of human embryonic development, from zygote to birth.

Media Activity
37.3 Human Development

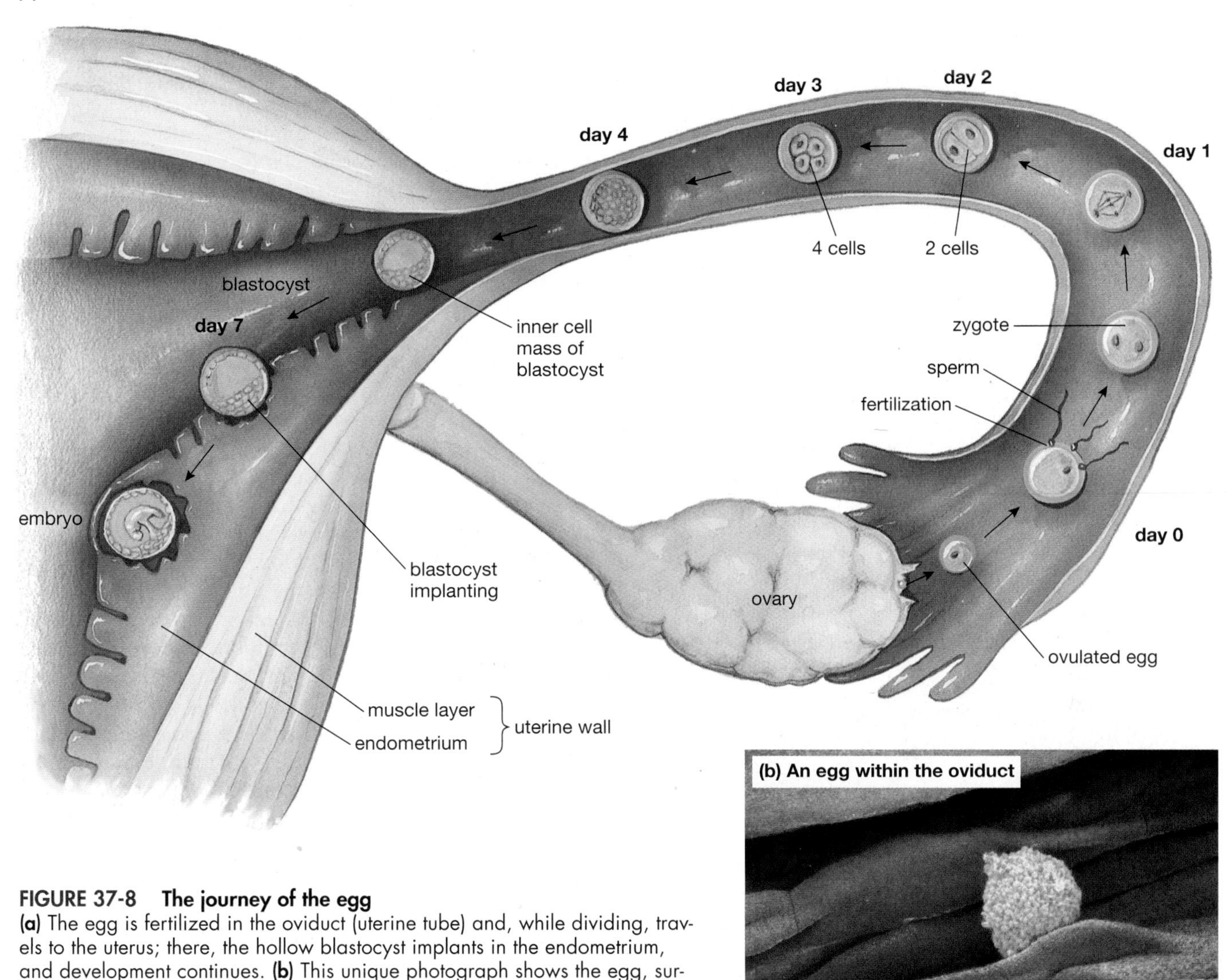

FIGURE 37-8 The journey of the egg
(a) The egg is fertilized in the oviduct (uterine tube) and, while dividing, travels to the uterus; there, the hollow blastocyst implants in the endometrium, and development continues. **(b)** This unique photograph shows the egg, surrounded by the follicle cells of the corona radiata, traveling along the oviduct toward the uterus. The egg emits chemicals that attract sperm, increasing its chances of being fertilized.

embryo—for example, in the spinal cord or an arm muscle—is a topic of active research. One such mechanism utilizes surface proteins associated with specific cell types. These proteins recognize chemical pathways laid out by other cells and cause the cells bearing the protein to migrate along these pathways. Although the exact mechanisms are not fully understood, the production of cell-type-specific proteins and of pathways along which cells bearing these specific proteins migrate depends on differential gene expression.

Researchers studying diverse developmental questions, from the formation of segments on a fruit fly to the development of the forelimb in chickens and frogs, have found that chemical gradients of regulatory proteins specify the fate of cells by causing appropriate genes to be expressed. For example, these chemical gradients (as might occur by diffusion from a concentrated source) can cause the structures from the shoulder to the fingers of a forelimb to develop in the proper order by causing different genes to become active at the appropriate times. This topic is explored in more detail in "A Closer Look: Homeobox Genes and the Control of Body Form" in Chapter 37 of this text's Web site.

37.4 How Do Humans Develop?

Human development is controlled by the same mechanisms that control the development of other animals. In fact, our development strongly reflects our evolutionary heritage. Figure 37-7 summarizes the stages of human embryonic development.

During the First 2 Months, Rapid Differentiation and Growth Occur

A human egg is usually fertilized in the mother's oviduct and undergoes a few cleavage divisions on its way to the uterus, a journey that takes about 4 days (Fig. 37-8a,b). By about 1 week after fertilization, the zygote has developed into a hollow ball of cells, known as the

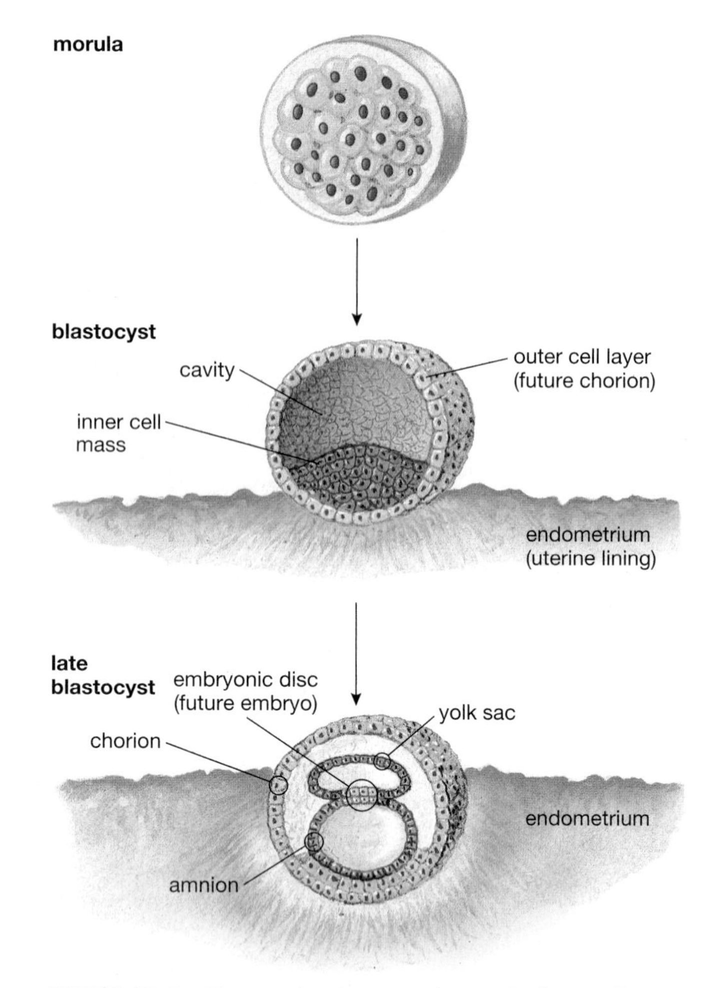

FIGURE 37-9 Human development during the first and second weeks
As it travels through the oviduct, the fertilized egg undergoes cleavage, forming a morula. In the uterus, the morula becomes a blastocyst (mammalian blastula) and implants in the uterine lining. As it burrows into the uterine lining, the outer cell layer forms the chorion, the embryonic contribution to the placenta. The inner cell mass forms the amnion, yolk sac, and the embryonic disc (future embryo).

blastocyst (the mammalian version of a blastula; see Fig. 37-8a). The blastocyst consists of a hollow ball of cells with a thicker **inner cell mass** on one side (Fig. 37-9; see also Fig. 37-8a). The sticky outer wall will adhere to the uterus and burrow into the endometrium, a process called **implantation**. This outer cell layer will first become the chorion and will later form the embryonic contribution to the placenta; the inner cell mass develops into the embryo and the three other extraembryonic membranes.

After implantation, the inner cell mass grows and splits, forming two fluid-filled sacs that are separated by a double layer of cells called the **embryonic disc** (Fig. 37-9). One sac, bounded by the amnion, forms the amniotic cavity. The amnion eventually grows around the embryo and contains the watery environment needed by all animal embryos. The yolk sac forms the second cavity (though in most mammals, it contains no yolk). At this stage, the embryonic disc consists of an upper layer of future ectoderm cells (on the side facing the amniotic cavity) and a lower layer of future endoderm cells (on the side facing the yolk sac). Gastrulation—which forms the three embryonic layers (ectoderm, mesoderm, and endoderm; see Table 37-2)—begins at the end of the second week after fertilization; many women, like Maria, first discover that they are pregnant at about this time. As you learned in Chapter 36, a woman's ovaries release an egg about halfway through the menstrual cycle. Failure to menstruate two weeks after ovulation is often a woman's first sign that she may have become pregnant.

During the third week of development, the embryo, enclosed in its amniotic sac, curls to form the future head ("head fold"; Fig. 37-10a). The chorion extends tiny fingers called **chorionic villi** into the endometrium of the uterus, and embryonic blood vessels invade the villi, carrying blood pumped by the embryonic heart, which has just begun to beat. As the embryo grows during the fourth week (Fig. 37-10b), the endoderm forms a tube—the embryonic gut—that will become the digestive tract. A *yolk stalk* connects the yolk sac with the embryonic gut (a reminder of yolk's role in nourishing the embryos of fish, birds, and reptiles). A rudimentary tail is visible. The embryo is connected to the chorion by the *body stalk* that includes the allantois and embryonic blood vessels. Within the embryo, ectoderm is forming structures that will become the brain and spinal cord. By the beginning of the fifth week, the embryo bulges into the cavity of the uterus and the placenta has become restricted to one side of the embryo (Fig. 37-10c). The embryo is completely surrounded by the amnion which is penetrated by the *umbilical stalk* (the future umbilical cord) that links the embryo to the placenta and exchanges wastes and nutrients. The umbilical cord forms from a merger of the yolk stalk and body stalk.

By the end of the sixth week, the embryo clearly displays its chordate ancestry (see Chapters 22 and 23), having developed a notochord, a prominent tail, and *gill grooves* (indentations behind the head that are homologous to the developing gills that fish and some amphibians retain as adults; Fig. 37-11a). These structures disappear as human development continues. The embryo already has the rudimentary beginnings of eyes, a beating heart, and separating fingers and toes on its tiny hands and feet (Fig. 37-11b). Especially notable at this stage is the rapid growth of the brain, which is nearly as large as the rest of the body. In fact, many of the structures of the adult brain are already recognizable.

As the second month draws to an end, nearly all the major organs have formed. The gonads appear and develop into testes or ovaries, depending on the presence or absence of the Y chromosome. Sex hormones—either testosterone from the testes or estrogen from the ovaries—are secreted. These hormones will affect the future development of the embryonic organs—not only

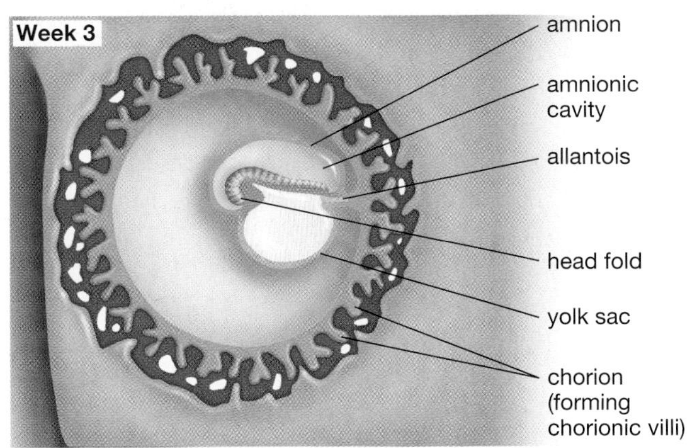

(a) The embryo is suspended in amniotic fluid. Chorionic villi carry embryonic blood vessels into the endometrium, beginning the formation of the placenta.

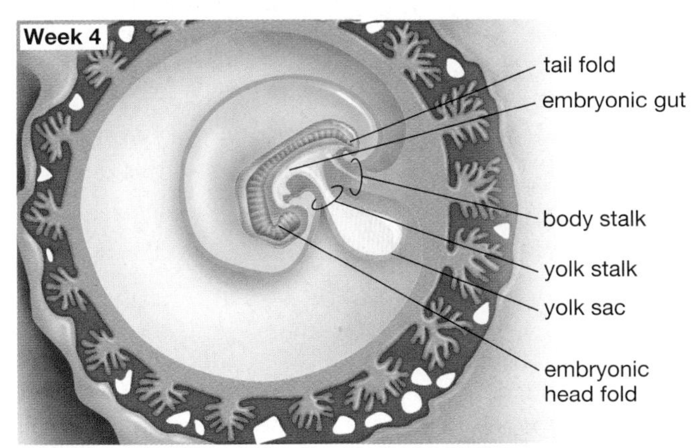

(b) Endoderm forms the embryonic gut (future digestive tract), which is connected to the yolk sac by the yolk stalk. The body stalk carries embryonic blood into the chorionic villi.

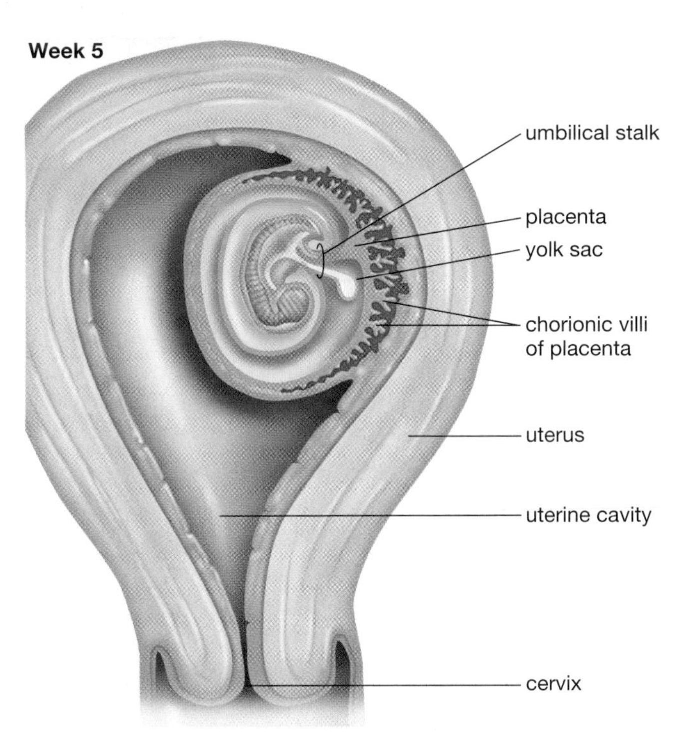

(c) The embryo bulges into the uterus, and the placenta is restricted to one side. The umbilical stalk (future umbilical cord) exchanges wastes and nutrients.

FIGURE 37-10 Human development during the third, fourth, and fifth weeks

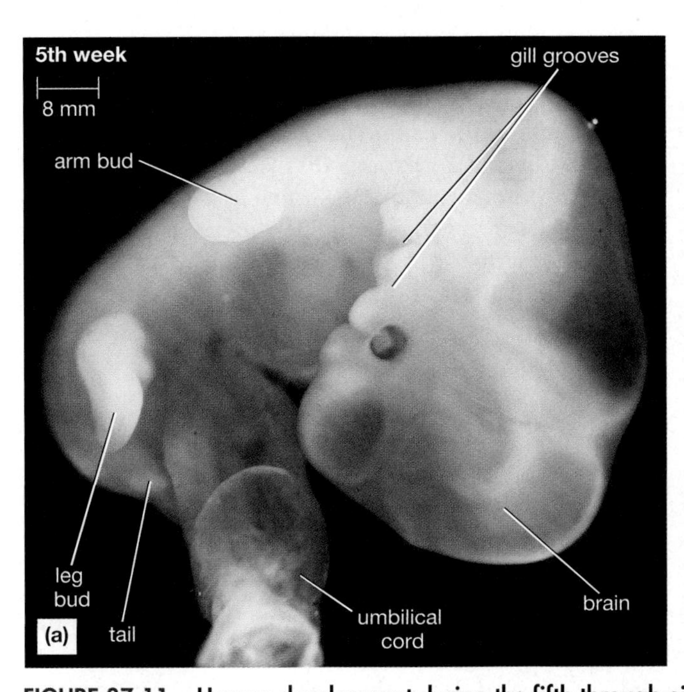

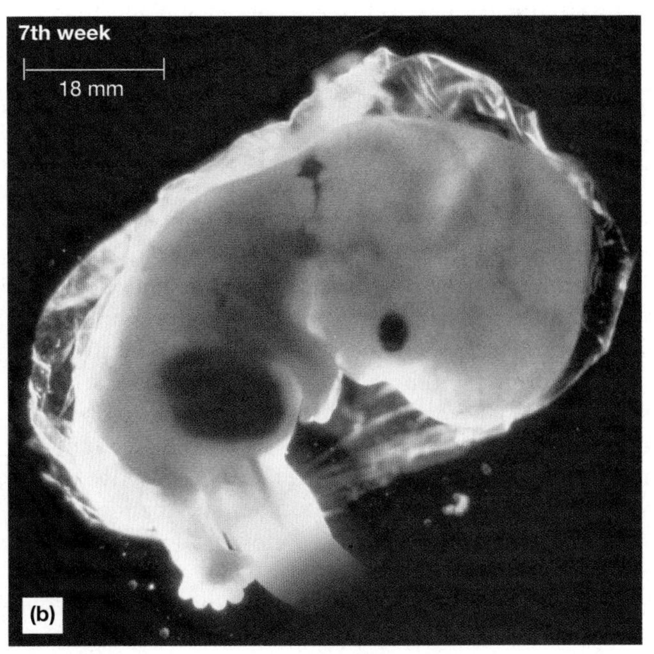

FIGURE 37-11 Human development during the fifth through eighth weeks
(a) At the end of the fifth week, the head comprises about half of the human embryo. The feet and hands have begun to develop digits. A tail and gill grooves are clearly visible. **(b)** By the seventh week, the human form has been more clearly defined by the selective death of cells that form the tail and connect the fingers and toes.

FIGURE 37-11 (continued) **Human development during the fifth through eighth weeks**
(c) At the end of the eighth week, the fetus is clearly human in appearance. Most of the major organs of the adult body have begun to develop.

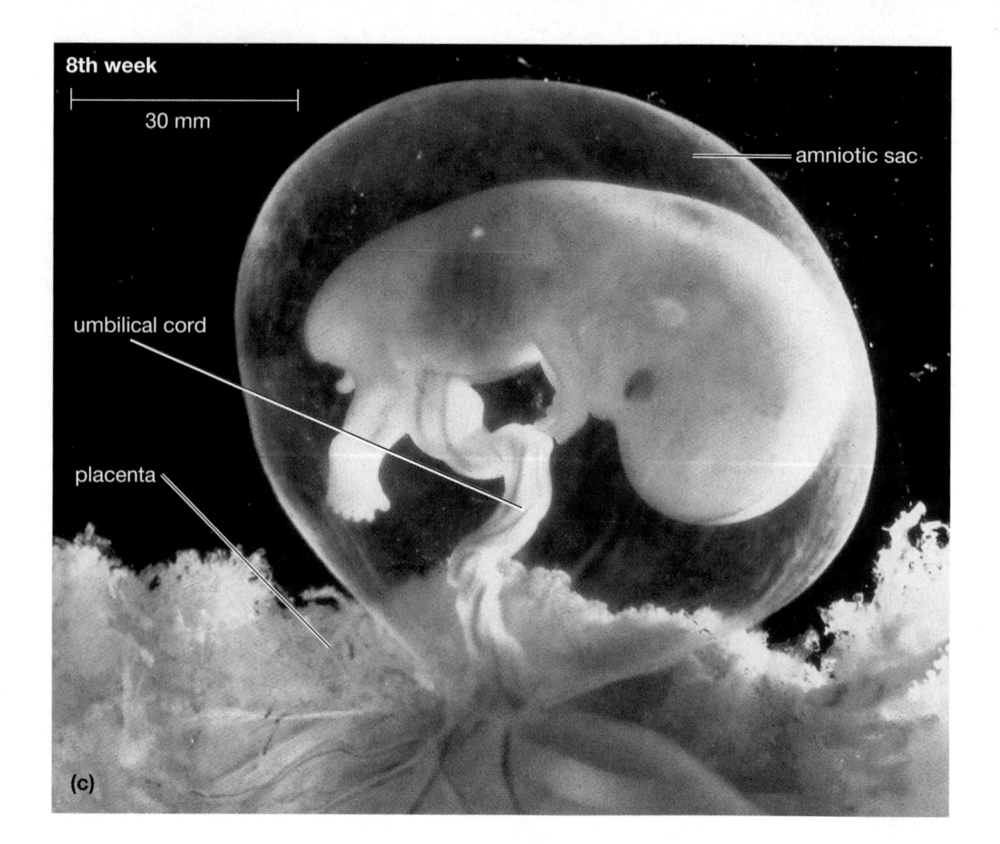

the reproductive organs, but also certain regions of the brain. After the second month of development, the embryo is called a **fetus**, and has taken on a generally human appearance (Fig. 37-11c).

The first 2 months of pregnancy are a time of extremely rapid differentiation and growth for the embryo, and also a time of considerable danger. Although the fetus is vulnerable throughout development, rapidly developing organs are the most sensitive to negative environmental effects, such as drugs (including alcohol and nicotine) and certain medications taken by the mother.

The Placenta Secretes Hormones and Exchanges Materials Between Mother and Embryo

During the first few weeks of pregnancy, embryonic cells burrow into the thickened lining of the uterus and obtain nutrients directly from the nearby cells of the endometrium. The outer cells of the embryo form the chorion, which penetrates the endometrium with fingerlike chorionic villi. During the third week, the **placenta** arises from this complex interweaving of tissues from the embryo and the endometrium. The placenta has two major functions: it secretes hormones, and it allows the selective exchange of materials between the mother and the fetus.

As it develops during the first 2 months of pregnancy, the placenta begins secreting estrogen and progesterone. Estrogen stimulates the growth of the mother's uterus and mammary glands; progesterone also stimulates the mammary glands and inhibits premature contractions of the uterus.

The placenta also regulates the exchange of materials between the blood of the mother and the blood of the fetus without allowing the two to mix. The chorionic villi contain a dense network of fetal capillaries and are bathed in pools of maternal blood (Fig. 37-12). This arrangement permits many small molecules to diffuse between fetal blood and maternal blood. Oxygen diffuses from maternal blood to fetal blood, and carbon dioxide from fetal blood to maternal blood. Nutrients, some aided by active transport, travel from mother to fetus. Fetal urea diffuses into the mother's blood, to be filtered out by the mother's kidneys.

While allowing exchange by diffusion, the membranes of the capillaries and chorionic villi act as barriers to the passage of some large proteins and most cells. Despite this barrier, some disease-causing organisms and many harmful chemicals—such as alcohol and nicotine, as Maria learned—can penetrate the placental barrier, as described in "Health Watch: The Placenta Provides Only Partial Protection."

Growth and Development Continue During the Last 7 Months

The fetus continues to grow and develop for another 7 months. Although the size of its body is "catching up" to the size of its head, the brain continues to develop rapidly and the head remains disproportionately large. Nearly every nerve cell ever formed during the entire human life span develops during embryonic life, one reason that the developing brain is such a sensitive target for drugs ingested during pregnancy. As the brain and spinal cord grow, they begin to generate specific types of behavior. As early as the third month of pregnancy, the fetus begins to move and respond to stimuli. Some instinctive behaviors appear, such as sucking,

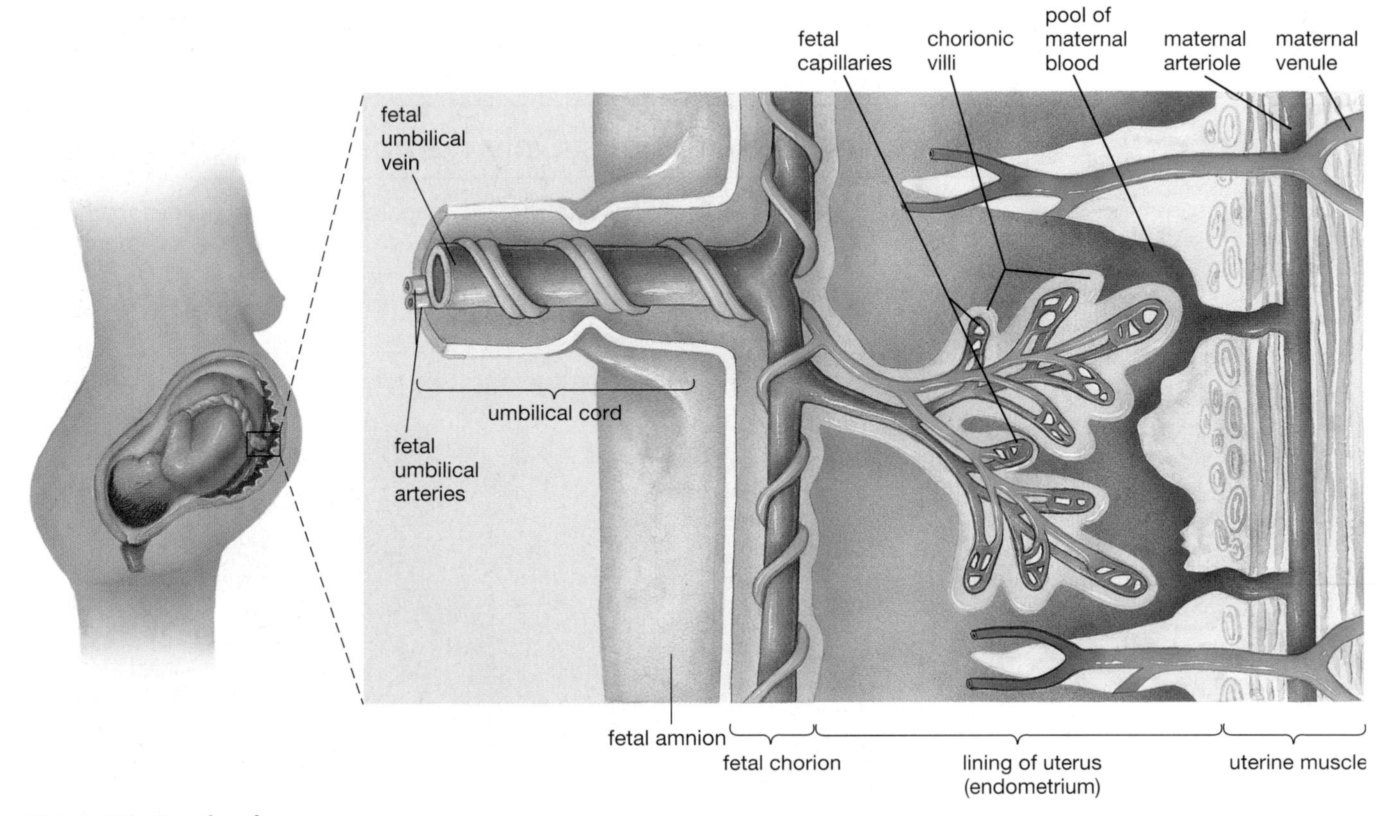

FIGURE 37-12 The placenta
The placenta is formed from both the chorion of the embryo and the endometrium of the mother. Capillaries of the endometrium break down, releasing blood that forms pools within the placenta. Chorionic villi containing embryonic capillaries extend into these pools of maternal blood. The placenta allows diffusion of oxygen, carbon dioxide, nutrients, and wastes between the fetal capillaries and the maternal blood pools, while keeping the fetal and maternal blood supply separate. Umbilical arteries carry deoxygenated blood from the fetus to the placenta, and umbilical veins carry oxygenated blood back to the fetus.
QUESTION Some mammals lack a placenta. What would you predict about the nature of development in nonplacental mammals?

which will have obvious importance soon after birth. Structures that the fetus will need when it emerges from the uterus, such as the lungs, stomach, intestine, and kidneys, enlarge and become functional, although they will not be used until after birth. Most fetuses 7 months or older can survive outside the womb with medical assistance, but larger and more mature fetuses have a much greater chance of survival.

Development Culminates in Labor and Delivery

Usually, during the last months of pregnancy, the fetus becomes positioned head downward in the uterus, with the crown of the skull resting against (and being held up by) the cervix. Normally, the process of birth begins around the end of the ninth month (Fig. 37-13). Birth

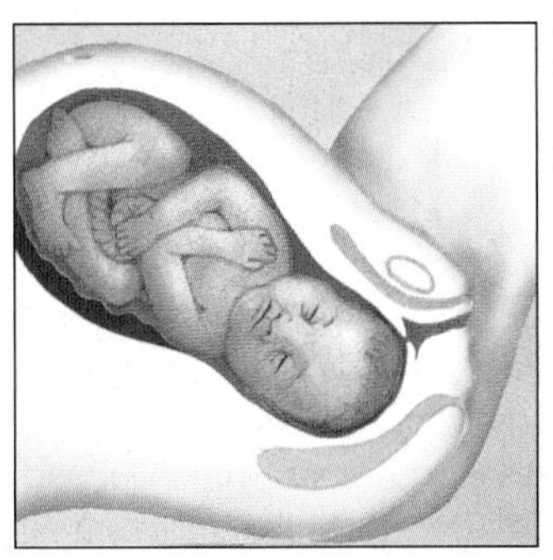

① The baby orients head downward, facing the mother's side. The cervix begins to thin and expand in diameter (dilate).

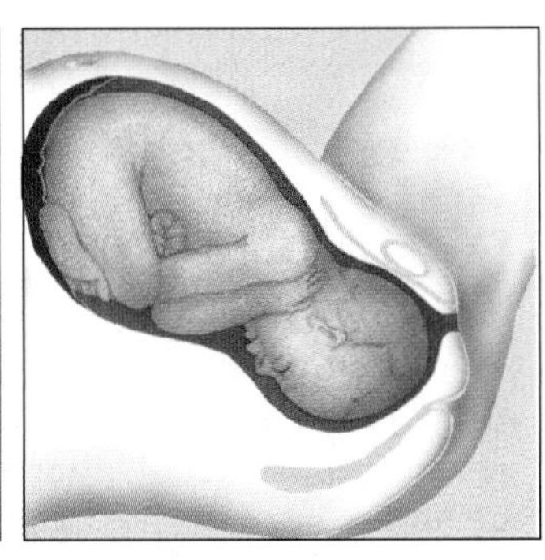

② The cervix dilates completely to 10 centimeters (almost 4 inches wide), and the baby's head enters the vagina, or birth canal. The baby rotates to face the mother's back.

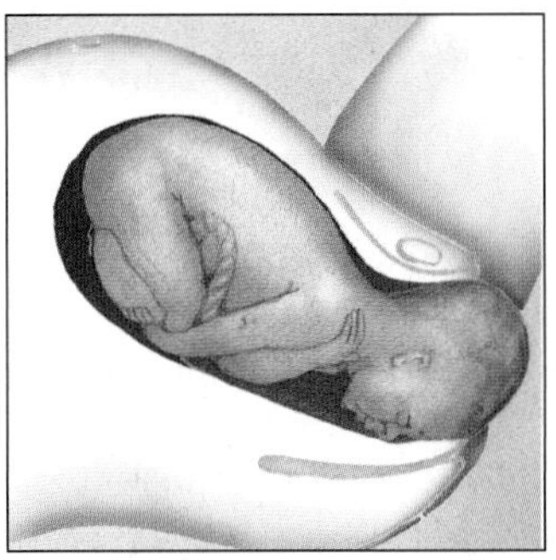

③ The baby's head emerges.

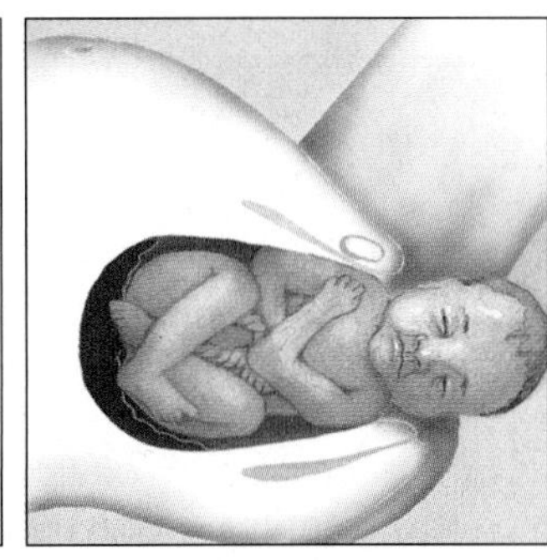

④ The baby rotates to the side once again as the shoulders emerge.

FIGURE 37-13 Delivery

HEALTH WATCH The Placenta Provides Only Partial Protection

Through the first half of the twentieth century, physicians assumed that the placenta protected the developing fetus from most harmful substances in maternal blood. We now know that this is far from true. In fact, most drugs (both medicinal and "recreational") and even some disease-producing organisms readily penetrate the placental barrier and affect the fetus.

Infections Can Cross the Placenta

The German measles virus can cross the placenta and attack the fetus, causing potentially severe retardation and other defects. As mentioned in Chapter 36, the virus causing genital herpes (during active outbreaks) and the bacterium causing syphilis can cause mental or physical defects in the developing fetus. The virus causing AIDS can also cross the placenta, so some infants are born with this deadly disease.

Drugs Readily Cross the Placenta

A tragic example of a drug that crosses the placenta is *thalidomide,* which was commonly prescribed in Europe in the late 1950s and early 1960s as a sedative and antinausea medication to combat morning sickness. Thalidomide's devastating effects on embryos were discovered only when many babies were born with missing or extremely abnormal limbs (Fig. E37-2). Thalidomide is gaining new respect as a treatment for a variety of disorders, including brain cancer, AIDS, rheumatoid arthritis, and leprosy, but it will never again be prescribed to pregnant women. In the late 1980s, the anti-acne drug Accutane® was found to cause gross deformities in babies born to women using it. Researchers now know that Accutane contains a substance that acts in a manner similar to natural regulatory molecules that control body formation in the developing embryo.

Although these are extreme examples, any drug, including aspirin, has the potential to harm the fetus, and any woman who thinks she may be pregnant should seek medical advice about any drugs that she takes.

The Effects of Smoking

Probably the most common toxic substances to which fetuses are exposed are those in cigarette smoke. Many women who smoke are so addicted that they don't stop smoking during pregnancy, thus exposing their developing children to nicotine, carbon monoxide, and a host of carcinogens. As a result, they have a higher incidence of miscarriages, and tend to give birth to smaller infants who have a higher risk of death shortly after

FIGURE E37-2 Drugs interfere with development
Children born to mothers who took the tranquilizer thalidomide during pregnancy cope heroically with missing and deformed limbs. A drug producing no obvious ill effects on the mother can have a dramatic impact on her developing fetus.

results from a complex interplay between uterine stretching caused by the growing fetus and fetal and maternal hormones that finally trigger **labor** (contractions of the uterus that result in *delivery*, the expulsion of the fetus from the uterus).

Unlike skeletal muscles, uterine muscles can contract spontaneously, and stretching enhances these contractions. As the baby grows, it stretches the uterine muscles, which occasionally contract weeks before delivery. The final trigger for labor is probably provided by the fetus. The near-term fetus produces steroid hormones that cause increased estrogen and prostaglandin production by the placenta and uterus. These hormones make the uterus even more likely to contract. When the combination of hormones and stretching activate the uterus beyond some critical point, strong contractions begin, signaling the onset of labor. As the contractions proceed, the baby's head pushes against the mother's cervix, making it expand in diameter (dilate). Stretch receptors in the walls of the cervix send signals to the hypothalamus, triggering oxytocin release. Under the dual stimulation of prostaglandin and oxytocin, the uterus contracts even more strongly. This positive feedback cycle is finally halted when the baby emerges from the vagina, or *birth canal.* The infant's head is so large that it can barely fit through the mother's pelvis. The skull is compressed into

birth. There is evidence that some children born to heavy smokers also suffer behavioral and intellectual impairment. Researchers analyzing the blood of infants born to mothers who smoke found that a potent carcinogen that causes mutations in DNA was passed to the infants through the placenta. Unfortunately, 61% of U.S. women who smoke continue to do so during pregnancy.

Fetal Alcohol Syndrome

The effects of alcohol on a developing fetus can be devastating. When a pregnant woman drinks, alcohol in the blood of her unborn child reaches a level as high as in her own blood. In spite of warnings on every wine and beer bottle, over 500,000 women in the United States continue to drink during pregnancy. Children born to women who drink even lightly during pregnancy tend to be smaller. A 2001 study that followed about 500 children for 6 years concluded that any consumption of alcohol during pregnancy was associated with a greater probability that the child would display anxiety, depression, and aggression compared to children whose mothers abstained from alcohol. Many children born to mothers who drink heavily on a regular basis (4 to 5 drinks per day or more) or go on alcoholic binges exhibit **fetal alcohol syndrome (FAS)**. Such children are mentally retarded and can be hyperactive and irritable. Children afflicted with FAS have small heads, abnormally small, improperly developed brains (Fig. E37-3), facial abnormalities, inhibited growth, and a higher-than-normal incidence of defects of the heart and other organs. The damage is irreversible. Research with rats has revealed learning deficits that persist to adulthood in rats born to mothers exposed to low levels of alcohol. Further, in rats, even a single 4-hour exposure to blood alcohol levels of 0.2% (such as might result from a single alcoholic binge for a person) during critical stages of brain development can cause massive death of fetal brain cells.

FAS is the single most common cause of mental retardation in the United States, with at least 4000 FAS infants born each year (1 in every 1000 births). It is likely that ten times as many children are born each year with *fetal alcohol effect*, a milder form of FAS that also impairs development. A woman who consumes one or two alcoholic drinks a day during the first 3 months of pregnancy also significantly increases her chances of having a miscarriage. Researchers have established no safe level of alcohol consumption during any phase of pregnancy. The U.S. Surgeon General advises pregnant women and those who are likely to become pregnant to avoid all alcohol consumption.

In summary, a pregnant woman should assume that any drugs she takes will find their way into the bloodstream of her developing infant. Women who are likely to become pregnant need to consider that crucial stages of development may occur before they even realize that they are pregnant, and take the same precautions as if they were pregnant. The mother's choices during this critical 9-month period can strongly influence her child's future well-being.

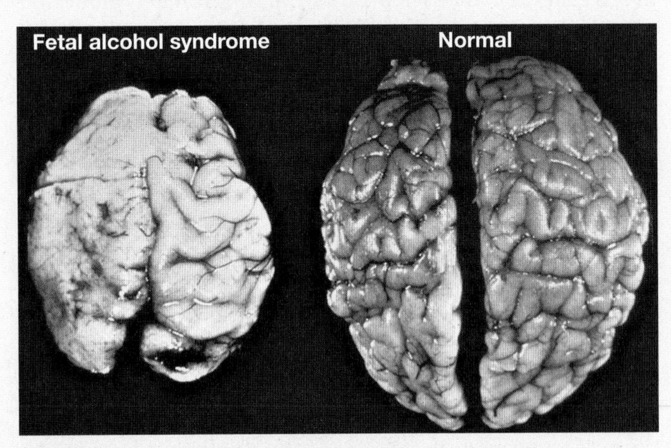

FIGURE E37-3 Alcohol impairs brain development
The brain of a child with fetal alcohol syndrome (left) and the brain of a normal child of the same age (right) show the devastating effects of alcohol on the developing brain.

a slightly conical shape as it passes through the vagina. The infant is in for a rude awakening. The uterus was soft, fluid-cushioned, and warm. All of a sudden, the baby must obtain oxygen and eliminate carbon dioxide on its own by breathing. It must regulate its own body temperature, and it must suckle to obtain food.

After a brief rest, uterine contractions resume, causing the uterus to shrink remarkably. During these contractions, the placenta is sheared from the uterine wall and expelled through the vagina as the *afterbirth*. The umbilical cord now releases prostaglandins that cause the muscles surrounding fetal blood vessels in the umbilical cord to contract and shut off blood flow. (Although tying off the umbilical cord is standard practice, it is not usually necessary; if it were, other mammals would not survive birth.) A new human being has been born.

Milk Secretion Is Stimulated by Hormones of Pregnancy

As the fetus grows, nourished by nutrients diffusing through the placenta, changes in the mother's breasts prepare her to continue nourishing her child after it is born. When pregnancy occurs, large quantities of estrogen and progesterone (acting together with several other hormones) stimulate **mammary glands**, milk-producing glands in the breasts, to grow, branch, and develop the capacity to secrete milk (Fig. 37-14). The mammary glands are arranged in a circle around the nipple; each gland has a milk duct that leads to the *nipple*, a projection of epithelial tissue. The actual secretion of milk, called **lactation**, is promoted by the pituitary hormone prolactin (see Chapter 33).

The level of prolactin rises steadily from about the fifth week of pregnancy until birth. Immediately after birth, estrogen and progesterone levels plummet, and prolactin takes over, stimulating the production of milk. Milk is released when the infant's suckling stimulates nerve endings in the nipples. The stimulated nerves send a signal to the hypothalamus, triggering an extra surge of prolactin and oxytocin from the pituitary. Oxytocin causes muscles surrounding the mammary glands to contract, ejecting the milk into the ducts that lead to the nipples (see Fig. 33-5).

During the first few days after birth, the mammary glands secrete a thin, yellowish fluid called **colostrum**. Colostrum is high in protein and contains antibodies that help protect the newborn against some diseases and are absorbed directly through the infant's intestine. Colostrum is gradually replaced by mature milk, which is higher in fat and milk sugar (lactose) and lower in protein.

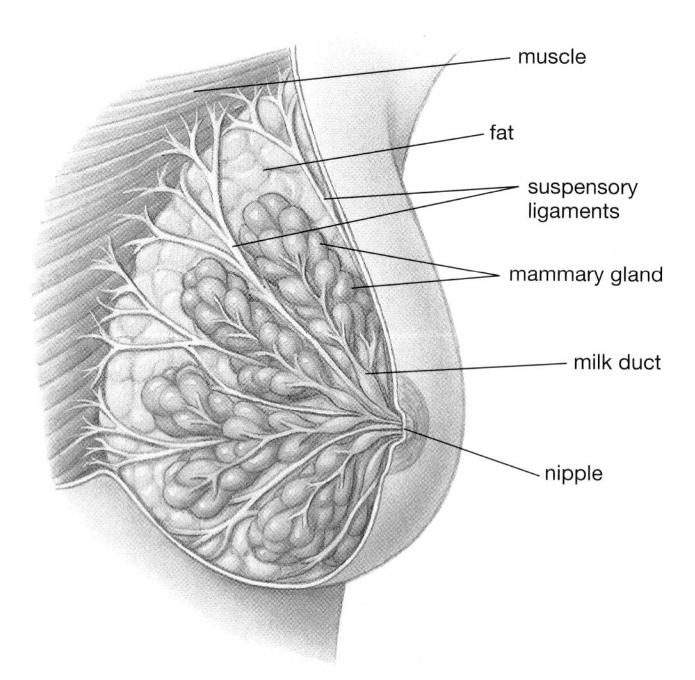

FIGURE 37-14 The structure of the mammary glands
During pregnancy, both fatty tissue and the milk-secreting glands and ducts increase in size.

Aging Is Inevitable

The aging process begins at the moment of birth. For thousands of years, people have attempted to delay aging and extend the human life span. The mythical notion of a "fountain of youth" has now been replaced by more modern elixirs. This plethora of pills, hormones, and diets all claim to delay aging, but they have only one thing in common: none has been demonstrated to work. The increase in life expectancy over the past 100 years is not the result of delayed aging, but of the reduction of premature death through improved sanitation, vaccines, and more sophisticated medical treatments.

What is **aging** (Fig. 37-15)? Researchers have defined aging as a gradual accumulation of random damage to essential biological molecules—particularly DNA—that begins at an early age. Eventually, the body's ability to repair or compensate for the damage is exceeded, impairing function at all levels, from cells to tissues to organs. Aging is manifested in many ways: muscle and bone mass are lost, skin elasticity decreases, reaction time slows, and senses such as vision and hearing become less acute. A less robust immune response renders the aging individual more vulnerable to disease.

Is death, both for cells and for entire organisms, a programmed part of life? From an evolutionary perspective, aging is best described as the result of neglect, rather than genetic programming. Natural selection preserves only those mechanisms that keep an organism alive and healthy while it is producing and nurturing its young. Repair mechanisms that extend longevity beyond this period are not selected for. Further, deleterious genes whose effects manifest themselves only after an organism has finished reproducing are passed along to offspring, and so can readily spread and accumulate within a population.

One factor that contributes to the cellular damage that accumulates with aging is the production of free radicals (see Chapter 2) that attack cellular components. Free radicals are generated as by-products of many crucial biochemical reactions, particularly those that harness energy, such as the Krebs cycle that produces ATP

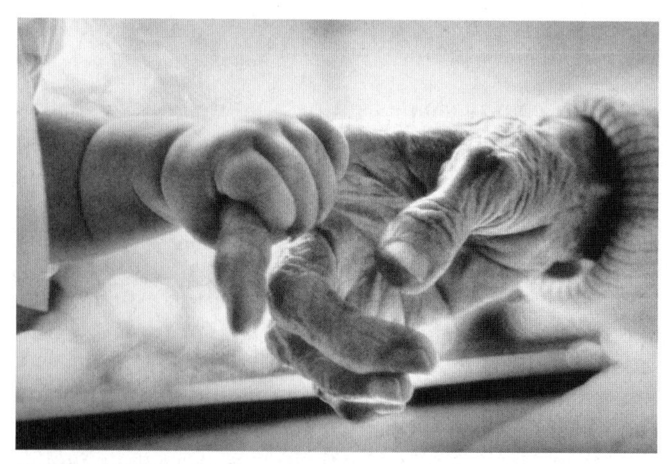

FIGURE 37-15 Youth vs. age

(see Chapter 8). It is ironic that, while an organism would die almost instantly if its energy production were halted, the same metabolic reactions that produce usable energy may eventually—indirectly—lead to an organism's death. The ability to repair damage, particularly to DNA, relies on enzymes that are also encoded by DNA. These enzymes themselves become less functional as mutations accumulate in the genes that code for them, dooming the organism to ever-increasing numbers of genetic mistakes and metabolic malfunctions.

Researchers have bred genetically simple animals such as fruit flies and roundworms to have increased life spans, and they are studying the genetic differences between individuals with normal life spans and those that live longer. Whether genetic changes made as a result of this knowledge can make a difference in the human life span remains to be seen. It is likely that hundreds of genetically determined biochemical pathways all play interrelated roles in longevity. Would extending the human life span even be desirable, given the health problems that accompany old age and the resource constraints of an overpopulated planet? No one knows, but we can be certain it will not happen soon.

CASE STUDY REVISITED Far-Reaching Choices

Maria's doctor emphasizes the incredible delicacy of the events of development and how easily they may be disrupted. At two weeks, Maria's child is a small ball of cells beginning to nestle in the endometrium of her uterus, growing in complexity as the embryonic disk (the future embryo) takes shape. Maria feels fortunate that she realized that she was pregnant early enough to change her behavior to protect these very early development stages. But how did Maria's home pregnancy test work? As soon as the blastocyst embeds itself in the endometrium, it begins secreting large quantities of chorionic gonadotropin (CG). This hormone maintains the corpus luteum and stimulates it to secrete other hormones that maintain and thicken the uterine lining, assuring the blastocyst of a safe haven and nourishment in the uterus. The CG is eventually excreted in the mother's urine. A home pregnancy test consists of an absorbent strip impregnated with antibodies to CG. When Maria dipped her home pregnancy test strip in a urine sample, the CG in her urine bound to the CG antibodies in the test strip, activating an enzyme that caused a colorless substance in the test strip to display a bright color, usually blue.

Must Maria give up her triple espresso? Her doctor recommended that she order a single espresso instead, and that she have no more than one a day. Caffeine readily reaches the developing child through the placenta, and there is evidence that high caffeine consumption during pregnancy may increase the risks of miscarriage, low birth weight, and premature delivery. Her physician also explained that drinking alcohol or smoking during pregnancy is likely to impair the development of her child in ways that range from subtle to devastating. Maria's doctor tells her that continuing to smoke during her pregnancy will increase her chances of miscarriage and of her child's having learning and behavioral problems later on. Further, she should avoid smoky places, which will not only tempt her to smoke but will also expose her and her developing child to secondhand smoke. Digging a half-empty pack of cigarettes from her purse, Maria tosses it in the wastebasket as she leaves the doctor's office.

bio ETHICS **Consider This:** Many health-care providers are not trained to recognize addictions and deal with the problems of addictions and pregnancy. The South Carolina Supreme Court has ruled that women who use illegal drugs during pregnancy may be charged with child abuse. How should society deal with the dilemma of pregnant women who are (perhaps unwittingly) damaging their unborn children? Should mothers who cause brain damage by drinking during pregnancy be charged with child abuse? Likewise, should fathers who smoke be charged with child abuse because of the dangerous effects of secondhand smoke? Based on what you now know about development, how would you deal with a friend who continued to smoke, drink, or take other drugs during pregnancy?

Links to Life: Why Is Human Childbirth So Difficult?

You've probably seen films of animals giving birth. Compared to the human ordeal, it seems painless, almost effortless. In fact, other primate species regularly give birth in trees, unassisted. Why are we so different? Researchers Karen Rosenberg and Wenda Trevathan have found that assisted birth is universal throughout the enormous range of human cultures. They and others hypothesize that the need for help giving birth is a result of two human traits: our large heads, and bipedalism (walking upright on two legs), which has altered the shape of the human pelvis. Apparently to accommodate the human baby's relatively huge head, the human birth canal differs from that of other primates in that it is widest from side to side (from the mother's perspective) at its entrance, and widest from front to back halfway through. Because the baby's head and shoulders just barely fit through the widest dimension, the infant must change positions as it travels through the canal, emerging somewhat backwards and sideways (see Fig. 37-13). Monkeys (whose heads are relatively smaller and whose birth canals are of uniform width throughout) are born with head and body facing forward, allowing the mother to reach down, help the infant emerge, and bring it up to her chest unassisted. Researchers postulate that as the brains (and heads) of our early ancestors increased dramatically in size, so did the difficulty and discomfort of human childbirth. Also, our large heads, coupled with the pelvic shape constraints imposed by bipedalism, necessitated the evolution of the twisting birth canal which causes the infant to emerge in an awkward position for the mother, favoring assisted birth. In March 2000, newspapers around the world featured photos of a young woman who had given birth in a tree in which she had been stranded for 4 days by raging floodwaters. But even she had help—her mother-in-law, stranded in the same tree, assisted her.

Summary of Key Concepts

37.1 How Do Indirect and Direct Development Differ?

Animals undergo either indirect or direct development. In indirect development, eggs (usually with relatively little yolk) hatch into larvae, which progress through a feeding stage and later undergo metamorphosis to become adults with notably different body forms. In direct development, the newborn animal is sexually immature but otherwise resembles a small adult. Animals with direct development tend to either produce large, yolk-filled eggs or nourish the developing embryo within the mother's body. In birds, reptiles, and mammals, extraembryonic membranes (the chorion, amnion, allantois, and yolk sac) encase the embryo in a fluid-filled space and regulate the exchange of nutrients and wastes between the embryo and its environment.

37.2 How Does Animal Development Proceed?

Animal development occurs in several stages. *Cleavage:* The fertilized egg undergoes cell divisions with little intervening growth, so the egg cytoplasm is partitioned into smaller cells. Cleavage divisions result in the formation of the morula, a solid ball of cells. A cavity then opens up within the morula, forming the blastula, a hollow ball of cells. *Gastrulation:* A dimple forms in the blastula, and cells migrate from the surface into the interior of the ball, eventually forming a three-layered gastrula. These three cell layers—ectoderm, mesoderm, and endoderm—give rise to all the adult tissues (see Table 37-2). *Organogenesis:* The cell layers of the gastrula form organs characteristic of the animal species. *Growth* and *sexual maturation:* The juvenile animal increases in size and achieves sexual maturity. *Aging:* Cells begin to function less efficiently, as damage to DNA and other cellular components accumulates; the cell's self-repair abilities deteriorate and eventually, the animal dies.

37.3 How Is Development Controlled?

All the cells of an animal body contain a full set of genetic information, yet cells are specialized for particular functions. During development, cells differentiate by stimulating and repressing the transcription of specific genes. Gene transcription is regulated in two ways. In some animals, gene-regulating substances in the egg cytoplasm are distributed in different proportions to different daughter cells during the first few cleavage divisions. Later in development, certain cells produce chemical messages that induce other cells to differentiate into particular cell types, a process called *induction*. Cells migrate within the developing embryo, a process requiring chemical communication among cells. Surface proteins associated with specific cell types recognize chemical pathways laid out by other cells and cause the cells bearing the protein to migrate along these pathways. Researchers have found that chemical gradients of regulatory proteins specify the fate of cells by causing appropriate genes to be expressed.

37.4 How Do Humans Develop?

A fertilized egg (zygote) develops into a hollow blastocyst and implants in the endometrium. The outer wall will become the chorion and will form the embryonic contribution to the placenta; the inner cell mass develops into the embryo and the three other extraembryonic membranes. During gastrulation, cells migrate and differentiate into ectoderm, mesoderm, and endoderm. During the third week, the endoderm forms a tube that will become the digestive tract; the heart begins to beat; and the rudiments of a nervous system appear. By the end of the second month, the major organs have formed, and the embryo—now called a fetus—appears human. In the next 7 months before birth, the fetus continues to grow, and the lungs, stomach, intestine, kidneys, and nervous system enlarge, develop, and become more functional. Human embryonic development, the stages of which are summarized in Fig. 37-7, follows the same principles as the development of other mammals. During pregnancy, mammary glands in the mother's breasts grow under the influence of estrogen, progesterone, and other hormones. After about 9 months, uterine contractions are triggered by a complex interplay of uterine stretch and prostaglandin and oxytocin release. As a result, the uterus expels the baby and then the placenta. After birth, prolactin and oxytocin, whose release is activated by the infant's suckling, trigger milk secretion.

Aging is the gradual accumulation of cellular damage (particularly to genetic material) over time that leads to loss of functionality of an organism and, eventually, to death.

Key Terms

aging *p. 762*
allantois *p. 748*
amnion *p. 748*
amniotic egg *p. 748*
blastocyst *p. 756*
blastopore *p. 749*
blastula *p. 749*
chorion *p. 748*
chorionic villus *p. 756*
cleavage *p. 749*
colostrum *p. 762*
development *p. 752*
differentiation *p. 752*
direct development *p. 746*
ectoderm *p. 750*
embryonic disc *p. 756*
embryonic stem cells *p. 750*
endoderm *p. 749*
extraembryonic membrane *p. 748*
fetal alcohol syndrome (FAS) *p. 761*
fetus *p. 758*
gastrula *p. 750*
gastrulation *p. 750*
implantation *p. 756*
indirect development *p. 746*
induction *p. 753*
inner cell mass *p. 756*
labor *p. 760*
lactation *p. 762*
larva *p. 746*
mammary glands *p. 761*
mesoderm *p. 750*
metamorphosis *p. 746*
morula *p. 749*
notochord *p. 750*
organogenesis *p. 750*
placenta *p. 758*
yolk *p. 746*
yolk sac *p. 748*

Thinking Through the Concepts

To take a multiple-choice quiz with feedback on the contents of this chapter, visit http://www.prenhall.com/audesirk7. *Log in to the Web site selected by your instructor and navigate to the Self Test section for this chapter.*

? Review Questions

1. Distinguish between indirect and direct development, and give examples of each.
2. Describe the structure and function of the four extraembryonic membranes found in reptiles and birds. Are these four present in placental mammals? In what ways are their roles similar in reptiles and birds as compared to mammals? How do they differ?
3. What is yolk? How does it influence cleavage?
4. What is gastrulation? Describe gastrulation in frogs and in humans.
5. Name two structures derived from each of the three embryonic tissue layers—endoderm, ectoderm, and mesoderm.
6. How does cell death contribute to development?
7. Describe the process of induction, and give two examples.
8. Define *differentiation.* How do cells differentiate; that is, how is it that adult cells express some but not all the genes of the fertilized egg?
9. What role do gradients of regulatory proteins play in animal development?
10. In humans, where does fertilization occur, and what stages of development occur before the fertilized egg reaches the uterus?
11. Describe how the human blastocyst gives rise to the embryo and its extraembryonic membranes.
12. Explain how the structure of the placenta prevents mixing of fetal and maternal blood while allowing the exchange of substances between the mother and the fetus.
13. Is the placenta an effective barrier against substances that can harm the fetus? Describe two types of harmful agents that can cross the placenta and their effects on the fetus.
14. How do changes in the breast prepare a mother to nurse her newborn? How do hormones influence these changes and stimulate milk production?
15. Describe the events that lead to the expulsion of the baby and the placenta from the uterus. Explain why this is an example of positive feedback.

Applying the Concepts

1. A researcher obtains two frog embryos at the gastrula stage of development. She carefully removes a cluster of cells from a location that she knows would normally become neural tube tissue and transplants it into the second gastrula in a location that would normally become skin. Does the second gastrula develop two neural tubes? Explain your answer.
2. Based on your knowledge of genetics (Unit II) and evolution (Unit III), explain why the human embryo passes through a developmental stage in which it has gill grooves and a tail.
3. Embryologists have used embryo fusion to produce *tetraparental* (four-parent) mice and have also produced "geeps" from goat and sheep embryos. The resulting bodies are patchworks of cells from both animals. Why does fusion succeed with very early embryos (four-cell to eight-cell stages) and fail when much older embryos are used?
4. If the nuclei of adult cells can be transplanted into eggs from which the nucleus has been removed to produce clones of the parent, is it theoretically possible to produce human clones? Would such clones yield offspring that are *exactly* identical to the parents who supplied the nuclei? Explain.
5. Based on your knowledge of cloning from this chapter and from Chapter 11, state as many arguments as possible for and against therapeutic cloning. Where do you stand?

For More Information

Caldwell, M. "How Does a Single Cell Become a Whole Body?" *Discover*, November 1992. Explores the miracles of development in layperson's terms.

Cibelli J. B., Lanza R. P., and West M. D. "The First Human Cloned Embryo." *Scientific American*, November 2001. Therapeutic cloning using the recipient's own cells can generate stem cells that will not undergo rejection by the immune system.

Holden, C., "Versatile Cells Against Intractable Disease." *Science*, July 26, 2002. Explores the potential of stem cells for combating neural disorders.

Nüsslein-Volhard, C. "Gradients That Organize Embryo Development." *Scientific American*, August 1996. A Nobel laureate describes developmental experiments that have revealed a great deal about how chemical gradients in embryos help guide cell differentiation.

Olshansky S. J., Hayfllick L., and Carnes B. A. "No Truth to the Fountain of Youth." *Scientific American*, June 2002. Anti-aging "remedies" abound—but none have been proven effective.

Pedersen, R. A. "Embryonic Stem Cells for Medicine." *Scientific American*, April 1999. Embryonic stem cells hold promise for regenerating damaged tissues.

Rose, M. R. "Can Human Aging Be Postponed?" *Scientific American*, December 1999. Aging can theoretically be postponed, but this will not happen soon.

Rosenberg K. W., and Trevathan, W. R. "The Evolution of Human Birth." *Scientific American*, November 2001. Explains why humans have such a difficult time delivering babies.

Smith, B. R. "Visualizing Human Embryos." *Scientific American*, March 1999. Magnetic resonance microscopy allows computerized reconstruction of three-dimensional structures within human embryos.

Taubes, G. "Ontongeny Recapitulated." *Discover*, May 1998. Will biotechnology allow us to turn on genes that normally become silent after development has been completed and that will provide the opportunity to regrow limbs or to replace dying brain cells?

Zigova, T., and Sanberg, P. R. "Neural Stem Cells for Brain Repair." *Science and Medicine*, September/October 1999. The discovery of stem cells in the adult brain that are capable of giving rise to all types of brain cells suggests that physicians may one day be able to repair damaged portions of the nervous system.

Media Activities

To access a Media Activity visit http://www.prenhall.com/audesirk7. *Log in to the Web site selected by your instructor, navigate to this chapter, and select the appropriate Media Activity number.*

37.1 Stages of Animal Development

Estimated time: 5 minutes

Explore the process of animal development and gastrulation in an animal embryo.

37.2 Control of Development

Estimated time: 5 minutes

Observe how genes, gene regulating substances, and proteins coordinate to control embryonic development.

37.3 Human Development

Estimated time: 10 minutes

Explore the key events that occur during each stage of human embryonic development.

37.4 Web Investigation: Far-Reaching Choices

Estimated time: 30 minutes

Explore how chemicals you are exposed to every day, such as caffeine, alcohol, cigarette smoke, over-the-counter drugs, formaldehyde from carpets, paint fumes, gas, and more, can have devastating effects on a developing fetus.

UNIT SIX

Behavior and Ecology

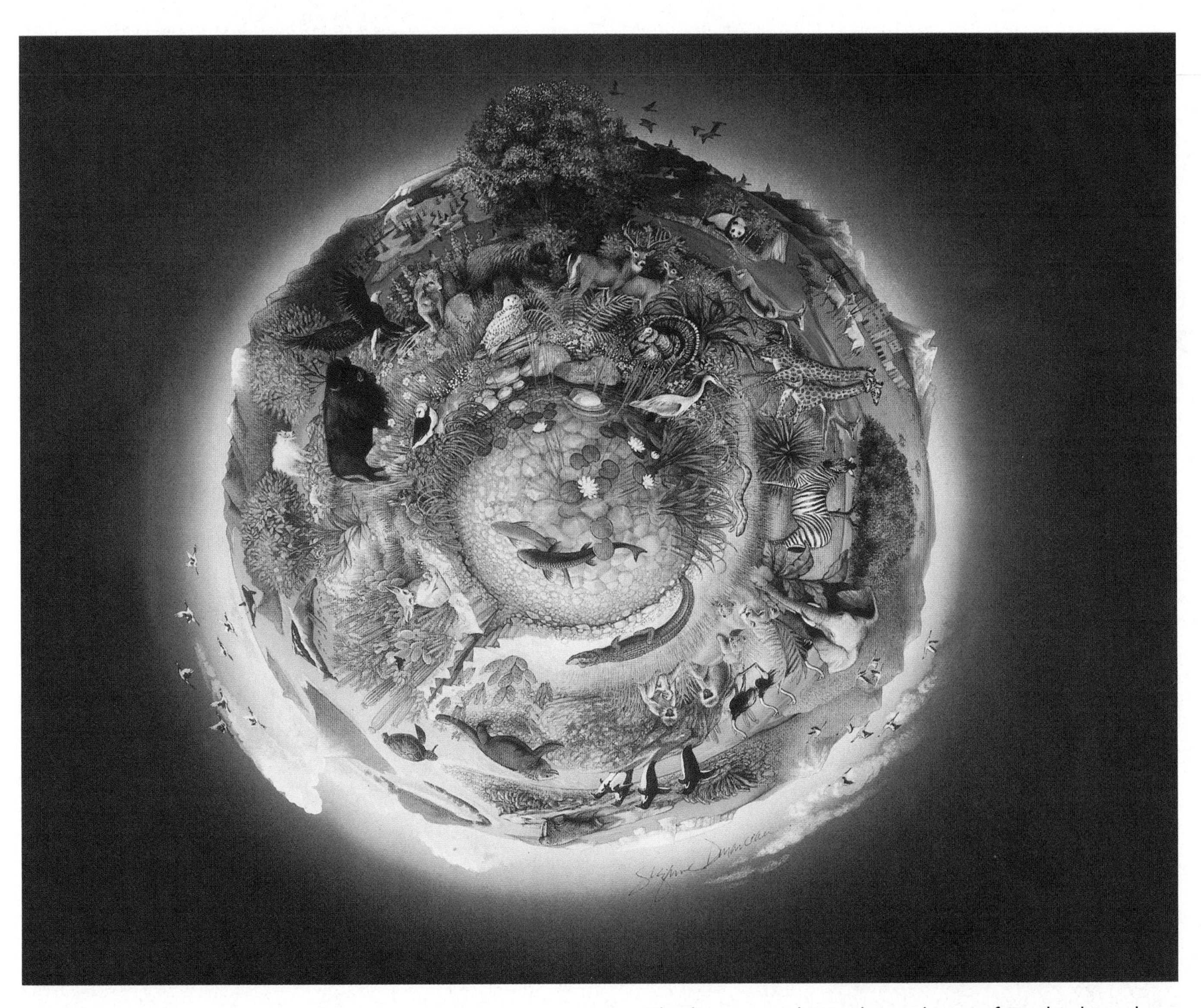

The beauty and interdependence of Earth's biosphere is illustrated in "Paradise," part one of "The Trilogy of the Earth" by Suzanne Duranceau/Illustratrice, Inc.

CHAPTER

38 Animal Behavior

Both this male scorpionfly and this male human are exceptionally attractive to females of their species. The secret to their sex appeal may be that both have highly symmetrical bodies.

AT A GLANCE

CASE STUDY Sex and Symmetry

For a male Japanese scorpionfly, finding a mate can be a real struggle. Female scorpionflies will mate only with males that can offer a tasty meal (usually a dead insect). Competition for dead insects is fierce. Typically, once a male finds an insect, he must defend it from other males. The competition for insects often erupts in bitter combat, characterized by repeated head-butting and grappling with sharp-pointed genital claspers.

In the competition to gain access to mates, not all male Japanese scorpionflies can be equally successful. The most successful males have qualities that help them defeat other males in combat and are especially attractive to females. Biologist Randy Thornhill found one quality in particular that accurately predicts the mating success of male scorpionflies: symmetry. In Thornhill's experiments and observations, the most successful males were those whose left and right wings were equal or nearly equal in length. Males with one wing longer than the other were less likely to win fights or to copulate; the greater the difference between the two wings, the lower the likelihood of success.

Thornhill's work with scorpionflies led him to wonder if the advantages of symmetry also extended to humans. Working with psychologist Steven Gangestad, he devised some fascinating studies that suggest that male symmetry does indeed play an important role in human sexual relationships. Women find symmetrical men more attractive. The preferences of human females resemble, at least in this one respect, those of female scorpionflies. One could almost say that insects and humans share a standard of beauty.

Media Activity
38.1 Observing Behavior: Homing in Digger Wasps

38.1 How Do Innate and Learned Behaviors Differ?

Behavior is any observable activity of a living animal. For example, a moth flies toward a bright light, a honey-bee flies toward a cup of sugar-water, and a housefly flies toward a piece of rotting meat. Bluebirds sing, wolves howl, and frogs croak. Mountain goats butt heads in ritual combat, chimpanzees groom one another, and ants attack a termite that approaches an anthill. Humans smoke cigarettes, play tennis, and plant gardens. Even the most casual observer sees many examples of animal behavior each day, and a careful observer can encounter a virtually limitless number of fascinating behaviors.

Innate Behaviors Can Be Performed Without Prior Experience

Innate behaviors are performed in reasonably complete form even the first time an animal of the appropriate age and motivational state encounters a particular stimulus. (The proper motivational state for feeding, for example, is hunger.) Scientists can demonstrate that a behavior is innate by depriving an animal of the opportunity to learn it. For example, red squirrels, which in the wild bury nuts in the fall for retrieval during the winter, can be raised from birth in a bare cage on a liquid diet, providing them with no experience of nuts, digging, or burying. Nonetheless, such a squirrel will, when presented with nuts for the first time, carry one to the corner of its cage and make covering and patting motions with its forefeet. Nut burying is therefore an innate behavior.

Innate behaviors can also be recognized by their occurrence immediately after birth, before any opportunity for learning presents itself. The cuckoo, for example, lays its eggs in the nest of another bird species, to be raised by the unwitting adoptive parent. Soon after a cuckoo egg hatches, the cuckoo chick performs the innate behavior of shoving the nest owner's eggs (or baby birds) out of the nest, eliminating its competitors for food (Fig. 38-1).

Learned Behaviors Are Modified by Experience

Natural selection may favor innate behaviors in many circumstances. For instance, it is clearly to the advantage of a herring gull chick to peck at its parent's bill as soon as possible after hatching, because pecking stimulates the parent to feed the chick. But in other circumstances, rigidly fixed behavior patterns may be less useful. For example, a male red-winged blackbird presented with a stuffed female blackbird will often attempt to copulate with the stuffed bird, a behavior that obviously will produce no offspring. In many situations, a certain degree of behavioral flexibility is advantageous.

The capacity to make changes in behavior on the basis of experience is called **learning**. This deceptively simple definition encompasses a vast array of different phenomena. A toad learns to avoid distasteful insects, a

(a)

(b)

FIGURE 38-1 Innate behavior
(a) The cuckoo chick evicts the eggs of its foster parents from the nest just hours after it hatches, before its eyes have opened. **(b)** The parents, responding to the stimulus of the cuckoo chick's wide-gaping mouth, feed the chick, unaware that it is not related to them. **QUESTION** The cuckoo chick benefits from its innate behavior, but the foster parent harms itself with its innate response to the cuckoo chick's begging. Why hasn't natural selection eliminated this disadvantageous innate behavior?

baby shrew learns which adult is its mother, a human learns to speak a language, and a sparrow learns to use the stars for navigation. Each of the many examples of animal learning represents the outcome of a unique evolutionary history, so learning is as diverse as animals themselves. Nonetheless, it can be useful to categorize types of learning, as long as we keep in mind that the categories are only rough guides; many examples of learning will not fit neatly into any category.

Habituation Is a Decline in Response to a Repeated Stimulus

A common form of simple learning is **habituation**, defined as a decline in response to a repeated stimulus. The ability to habituate prevents an animal from wasting its energy and attention on irrelevant stimuli. This form of learning is displayed by even the simplest animals. For example, a sea anemone will retract its tentacles when touched, but gradually stops retracting if touching is continued (Fig. 38-2).

The ability to habituate is clearly adaptive. If a sea anemone withdrew every time it was brushed by a strand of waving seaweed, the animal would waste a great deal of energy, and its retracted posture would prevent it from snaring food. Humans habituate to many stimuli; city dwellers habituate to nighttime traffic sounds as country dwellers do to choruses of crickets and tree frogs. Each may find the other's habitat unbearably noisy at first, but each habituates after a time.

Conditioning Is a Learned Association Between a Stimulus and a Response

A more complex form of learning is **trial-and-error learning**, in which animals acquire new and appropriate responses to stimuli through experience. Many animals are faced with naturally occurring rewards and punishments and can learn to modify their responses to them. For example, a hungry toad that captures a bee quickly learns to avoid future encounters with bees (Fig. 38-3). After only one experience with a stung tongue, a toad modifies its response to flying insects, ignoring bees (and even other insects that resemble them).

Trial-and-error learning is an important factor in the behavioral development of many animal species and often occurs during play and exploratory behavior (see "Evolutionary Connections: Why Do Animals Play?"). This type of learning also plays a key role in human behavior, for example, allowing a child to learn which foods taste good or bad, that a stove can be hot, and not to pull a cat's tail.

Some interesting properties of trial-and-error learning have been revealed by a laboratory technique known as **operant conditioning**. During operant conditioning, an animal learns to perform a behavior (such as pushing a lever or pecking a button) to receive a reward or to avoid punishment. This technique is most closely associated with the American comparative psychologist B. F. Skinner, who designed the "Skinner box" in which an animal is isolated and allowed to train itself. The box might contain a lever that, when pressed, ejects a food pellet. If the animal accidentally bumps the lever, a food reward appears. After a few such occurrences, the animal learns the connection between pressing the lever and receiving food and begins to press the lever repeatedly.

Operant conditioning has been used to train animals to perform tasks far more complex than pressing a lever, but perhaps the most interesting revelation fostered by the technique is that species differ in their propensity for learning particular associations. In particular, species seem to be predisposed to learn behaviors that are relevant to their own needs. For example, if

FIGURE 38-2 Habituation in a sea anemone

(a) A naive toad is presented with a bee.

(b) While trying to eat the bee, the toad is stung painfully on the tongue.

(c) Presented with a harmless robber fly, which resembles a bee, the toad cringes.

(d) The toad is presented with a dragonfly.

(e) The toad immediately eats the dragonfly, demonstrating that the learned aversion is specific to bees and insects resembling bees.

FIGURE 38-3 Trial-and-error learning in a toad

a rat is given a distinctively flavored food that has been mixed with a substance that makes the rat sick, the animal learns to avoid eating that food. In contrast, it is very difficult to train a rat to rear up on its hind legs in response to a particular sound or visual cue. The difference can be explained by asking which learning task is more likely to benefit a wild Norway rat (the species from which lab rats are descended). Clearly, an ability to avoid foods that induce illness is beneficial to an animal like the Norway rat, which is well known for eating a tremendous variety of foods. However, the animal gains no obvious benefit from learning to stand up in response to a noise. In general, the specific learning abilities of each species have evolved to support its particular mode of life.

Insight Is Problem Solving Without Trial and Error

In certain situations, animals seem able to solve problems suddenly, without the benefit of prior experience. This kind of sudden problem solving is sometimes called **insight learning**, because it seems at least superficially similar to the process by which humans mentally manipulate concepts to arrive at a solution. We cannot, of course, know for sure if nonhuman animals experience similar mental states when they solve problems.

In 1917, animal behaviorist Wolfgang Kohler showed that a hungry chimpanzee, without any training, could stack boxes to reach a banana suspended from the ceiling. This type of mental problem solving was once believed to be limited to very intelligent types of animals such as primates, but similar abilities may also be present in species that we tend to view as less intelligent. For example, R. Epstein and colleagues performed an experiment that showed that pigeons may be capable of insight learning. In the experiment, pigeons (whose wings had been clipped to prevent flight) were first trained to perform two unrelated tasks in return for food rewards. The tasks were to push a small box around the cage and to peck at a small plastic banana. Later, the trained birds were presented with a novel situation: a plastic banana that hung above their reach in a cage that also contained a small box. Many of the pigeons pushed the box to a position beneath the plastic banana and climbed atop the box to peck the faux fruit.

Apparently, a pigeon trained to execute the necessary physical movements can also solve the suspended banana problem.

There Is No Sharp Distinction Between Innate and Learned Behaviors

Although the terms "innate" and "learned" can be useful tools to help us describe and understand behaviors, these words also have the potential to lull us into an oversimplified view of animal behavior. In practice, no behavior is unambiguously innate or unequivocally learned; all behaviors are an intimate mixture of the two.

Seemingly Innate Behavior Can Be Modified by Experience

Behaviors that seem to be performed correctly on the first attempt without prior experience can later be modified by experience. For example, a newly hatched herring gull chick is able to peck at a red spot on its parent's beak (Fig. 38-4), an innate behavior that causes the parent to regurgitate food for the chick to eat. Biologist Niko Tinbergen studied this pecking behavior and found that the pecking response of very young chicks was triggered by the long, thin shape and red color of the parent's bill. In fact, when Tinbergen offered newly hatched chicks a thin, red rod with white stripes painted on it, they pecked at it more often than at a real beak. Within a few days, however, the chicks learned enough about the appearance of their parents that they began pecking more frequently at models more closely resembling the parents. After one week, the young gulls recognized their parents' appearance enough to prefer models of their own species to models of a closely related species. Eventually, the young birds learned to beg only from their own parents.

FIGURE 38-4 Innate behaviors can be modified by experience
A herring gull chick pecks at the red spot on its mother's bill, causing her to regurgitate food.

Habituation (a decline in response to a repeated stimulus) can also fine-tune an organism's innate responses to environmental stimuli. For example, young birds crouch down when a hawk flies over but ignore harmless birds such as geese. Early observers hypothesized that only the very specific shape of predatory birds provoked crouching. Using an ingenious model (Fig. 38-5), Niko Tinbergen and Konrad Lorenz (two of the founding fathers of **ethology**, the study of animal behavior) tested this hypothesis. When moved in one direction, the model resembled a goose and was ignored by the chicks. When its movement was reversed, however, the model resembled a hawk and elicited crouching behavior. Further research revealed that newborn chicks instinctively crouch when *any* object moves over their heads. Over time, their response habituates to things that soar by harmlessly and frequently, such as leaves, songbirds, and geese. Predators are much less common, and the novel shape of a hawk continues to elicit instinctive crouching. Thus, learning modifies the innate response, making it more advantageous.

Learning May Be Governed by Innate Constraints

Learning always occurs within boundaries that help increase the chances that only the appropriate behavior is acquired. For example, even though young robins hear the singing of sparrows, warblers, finches, and other bird species that share their nesting area, they do not imitate the songs of these other species. Instead, young robins learn only the songs of adult robins. The robin's ability

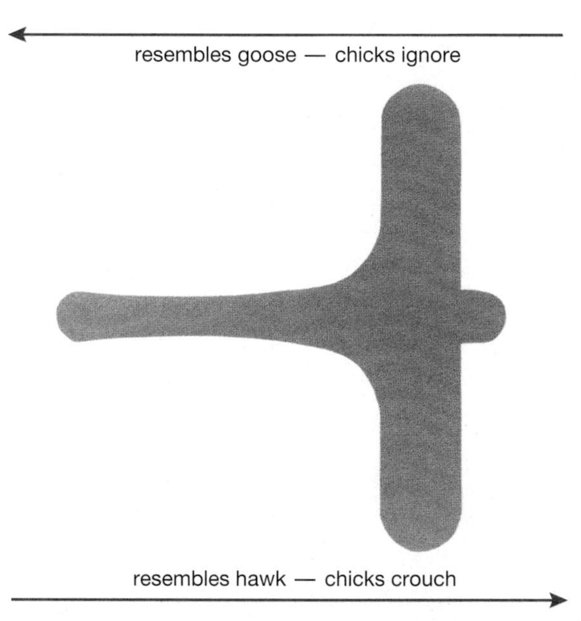

FIGURE 38-5 Habituation modifies innate responses
Konrad Lorenz and his student Niko Tinbergen used this model to investigate the response of chicks to the shape of objects flying overhead. Moving toward the right, the model resembles a predatory hawk, but when it moves left, it resembles a harmless goose.

to learn songs is limited to those of its own species, and the songs of other species are excluded from the learning process.

The innate constraints on learning are perhaps most strikingly illustrated by **imprinting**, a special form of learning in which an animal's nervous system is rigidly programmed to learn a certain thing only at a certain period of development. This causes strong associations to be formed during a particular stage, called a *sensitive period*, in the animal's life. During this stage, the animal is primed to learn specific information, which is then incorporated into behaviors that are not easily altered by further experience. Imprinting is best known in birds such as geese, ducks, and chickens. These birds learn to follow the animal or object that they most frequently encounter during an early sensitive period. In nature, a mother bird is the object most likely to be nearby during the sensitive period, so her offspring imprint on her. In the laboratory, however, these birds may imprint on a toy train or other moving object (Fig. 38-6). If given a choice, however, they select an adult of their own species.

All Behavior Arises Out of Interaction Between Genes and Environment

Many early ethologists saw innate behaviors as rigidly controlled by genetic factors and viewed learned behaviors as determined exclusively by an animal's environment. Today, however, this false dichotomy has given way to the realization that, just as no behavior is wholly innate or wholly learned, no behavior can be caused strictly by genes or strictly by the environment. Instead, all behavior develops out of an interaction between genes and the environment. The relative contributions of heredity and learning vary among animal species and among individuals.

FIGURE 38-6 Konrad Lorenz and imprinting
Konrad Lorenz, known as the "father of ethology," is followed by goslings that imprinted on him shortly after they hatched. They follow him as they would their mother.

The precise nature of the link among genes, environments, and behaviors is not well understood in most cases. The chain of events between the transcription of genes and the performance of a behavior may be so complex that we will never be able to decipher it fully. Nonetheless, a great deal of evidence demonstrates the existence of both genetic and environmental components in the development of behaviors. For example, consider bird migration. Even though it is well known that migratory birds must learn by experience how to navigate with celestial cues, this learning is not the only factor involved.

Bird Migration Behavior Has an Inherited Component

At the close of the summer, many birds disappear from their breeding habitats and head for their winter territory, which may be hundreds or even thousands of miles to the south. Many of these migrating birds are traveling for the first time, because they were hatched only a few months earlier. Amazingly, these naïve birds depart at the proper time, head in the proper direction, and locate the proper wintering location, even though they often do not simply follow more-experienced birds (which typically depart a few weeks in advance of the first-year birds). Somehow, these young birds execute a very difficult task the first time they try it. Thus, it seems that birds must be born with the ability to migrate; it must be "in their genes." Indeed, birds hatched and raised in indoor isolation still orient in the proper migratory direction when autumn comes, apparently without the need for any learning or experience.

The conclusion that birds must have a genetically controlled ability to migrate in the right direction has been further supported by hybridization experiments with blackcap warblers. This species breeds in Europe and migrates to Africa, but populations from different areas travel by different routes. Blackcaps from western Europe travel in a southwesterly direction to reach Africa, whereas birds from eastern Europe travel to the southeast (Fig. 38-7). If birds from the two populations are crossbred in captivity, however, the hybrid offspring exhibit migratory orientation due south, which is intermediate between the orientation of the two parents. This result suggests that parental genes—of which offspring inherit a mixture—influence migratory direction.

38.2 How Do Animals Communicate?

Animals frequently make information available for sharing. The sounds uttered, movements made, and chemicals emitted by animals can reveal their physical

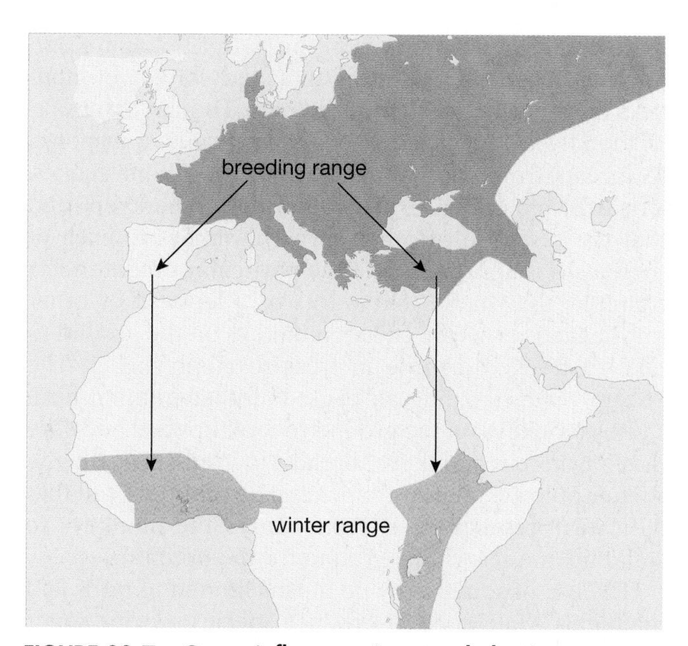

FIGURE 38-7 Genes influence migratory behavior
Blackcap warblers from western Europe begin their fall migration by flying in a southwesterly direction, but those from eastern Europe migrate to the southeast. If members of the two populations are crossbred in captivity, the hybrid offspring orient in a direction intermediate between the migratory directions of the parents: due south. **QUESTION** If young blackcap warblers from a wild population in western Europe were transported to eastern Europe and reared to adulthood in an outdoor environment, in which direction would you expect them to orient?

location, level of aggression, readiness to mate, and so on. If this information evokes a response from other individuals, and if that response tends to benefit the sender and the receiver, then a communication channel can form. **Communication** is defined as the production of a signal by one organism that causes another organism to change its behavior in a way beneficial to one or both.

Although animals of different species may communicate (picture a cat, its tail erect and bushy, hissing at a dog), most animals communicate only with members of their own species. Potential mates must communicate, as must parents and offspring. Members of the same species also compete directly with one another for food, space, and mates; communication is often used to resolve these conflicts.

The mechanisms by which animals communicate are astonishingly diverse and use all of the senses. In the following sections, we will look at communication by visual displays, sound, chemicals, and touch.

Visual Communication Is Most Effective over Short Distances

Animals with well-developed eyes, from insects to mammals, use visual signals to communicate. Visual signals can be *active,* in which a specific movement (such as baring fangs) or posture (such as lowering the head) conveys a message (Fig. 38-8). Alternatively, visual signals may be *passive,* in which case the size, shape, or color of the animal conveys important information, commonly about its sex and reproductive state. For example, when female mandrills become sexually receptive, they develop a large, brightly colored swelling on their buttocks (Fig. 38-9). Active and passive signals can

FIGURE 38-8 Active visual signals
The wolf signals aggression by lowering its head, ruffling the fur on its neck and along its back, facing its opponent with a direct stare, and exposing its fangs. These signals can vary in intensity, communicating different levels of aggression.

FIGURE 38-9 A passive visual signal
The female mandrill's colorfully swollen buttocks serve as a passive visual signal that she is fertile and ready to mate.

FIGURE 38-10 Active and passive visual signals combined The South American *Anolis* lizard raises his head high in the air (an active visual signal), revealing a brilliantly colored throat pouch (a passive visual signal) that warns others to keep their distance.

be combined, as illustrated by the lizard in Figure 38-10 and the courtship behavior of the three-spined stickleback fish (see Fig. 38-23).

Like all forms of communication, visual signals have both advantages and disadvantages. On the plus side, they are instantaneous, and active signals can be rapidly changed to convey a variety of messages in a short period. Visual communication is quiet and unlikely to alert distant predators, although the signaler does make itself conspicuous to those nearby. On the negative side, visual signals are generally ineffective in darkness and in dense vegetation, although female fireflies signal potential mates by using species-specific patterns of flashes. Finally, visual signals are limited to close-range communication.

Communication by Sound Is Effective over Longer Distances

The use of sound overcomes many of the shortcomings of visual displays. Like visual displays, sound signals reach receivers almost instantaneously. But unlike visual signals, sound can be transmitted through darkness, dense forests, and murky water. Sound signals can also be effective over longer distances than visual signals. For example, the low, rumbling calls of African elephants can be heard by elephants several miles away, and the songs of humpback whales are audible for hundreds of miles. Likewise, the howls of a wolf pack carry for miles on a still night. Even the small kangaroo rat produces a sound (by striking the desert floor with its hind feet) that is audible 150 feet (45 meters) away.

Sound signals are similar to visual displays in that they can be varied to convey rapidly changing messages. (Think of speech and the emotional nuances conveyed by the human voice during a conversation.) Changes in motivation can be signaled by a change in the loudness or pitch of the sound. An individual can convey different messages by variations in the pattern, volume, and pitch of the sound produced. In a study of vervet monkeys in Kenya in the 1960s, ethologist Thomas Struhsaker found that they produced different calls in response to threats from each of their major predators: snakes, leopards, and eagles. In 1980 other researchers reported that the response of other vervet monkeys to each of these calls is appropriate to the particular predator. For example, the "bark" that warns of a leopard or other four-legged carnivore causes monkeys on the ground to take to trees and those in trees to climb higher. The "rraup" call, signaling an eagle or other hunting bird, causes monkeys on the ground to look upward and take cover, whereas monkeys already in trees drop to the shelter of lower, denser branches. The "chutter" call that indicates presence of a snake causes the monkeys to stand up and search the ground for the predator.

The use of sound is by no means limited to birds and mammals. Male crickets produce species-specific songs that attract female crickets of the same species. The annoying whine of the female mosquito as she prepares to bite alerts nearby males that she may soon have the blood meal necessary for laying eggs. Male water striders vibrate their legs, sending species-specific patterns of vibrations through the water, attracting mates and repelling other males (Fig. 38-11). Ranging from these rather simple signals to the virtuoso performance of human language, sound is one of the most important forms of communication.

Chemical Messages Persist Longer but Are Hard to Vary

Chemical substances that are produced by individuals and influence the behavior of other members of the species are called **pheromones**. Pheromones can carry

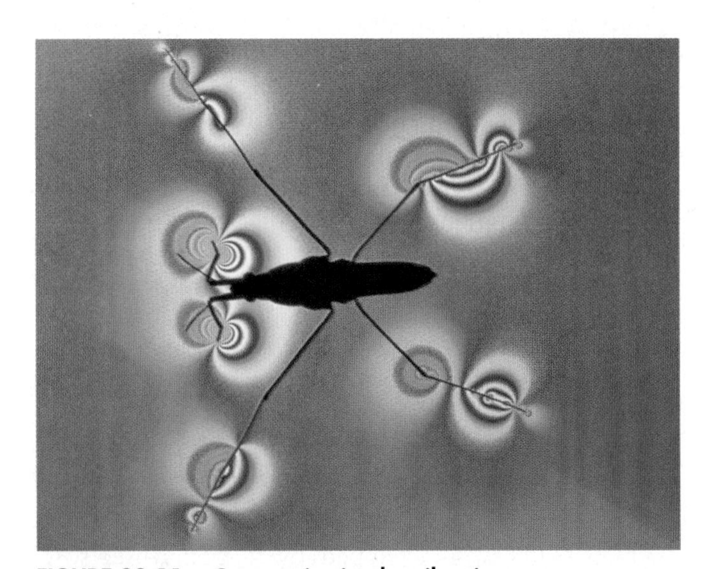

FIGURE 38-11 Communication by vibration The light-footed water strider relies on the surface tension of water to support its weight. By vibrating its legs, the water strider sends signals that radiate out over the surface of the water. These vibrations advertise the strider's species and sex to others nearby.

messages over long distances, and, unlike sound, take very little energy to produce. Pheromones may not even be detected by other species, whereas predators might be attracted to visual or sound signals. Like a signpost, a pheromone persists over time and can convey a message after the animal has departed. Wolf packs, hunting over areas as large as 385 square miles (1000 square kilometers), warn other packs of their presence by marking the boundaries of their travels with pheromone-containing urine. As anyone who has walked a dog can attest, the domesticated dog reveals its wolf ancestry by staking out its neighborhood with urine that carries the chemical message, "I live in this area."

Such communication requires animals to synthesize a different chemical for each message. As a result, chemical signaling systems communicate fewer different messages than do sight- or sound-based systems. In addition, pheromone signals cannot easily convey rapidly changing messages. Nonetheless, chemicals effectively convey a few simple but critical messages.

Many pheromones cause an immediate change in the behavior of the animal that detects them. For example, foraging termites that discover food lay a trail of pheromones from the food to the nest; other termites then follow the trail (Fig. 38-12). Pheromones can also stimulate physiological changes in the animal that detects them. For example, the queen honeybee produces a pheromone called *queen substance*, which prevents other females in the hive from becoming sexually mature. Similarly, mature males of some mouse species produce urine containing a pheromone that influences female reproductive physiology. The pheromone stimulates newly mature females to become fertile and sexually receptive. It will also cause a female mouse that is newly pregnant by another male to abort her litter and become sexually receptive to the new male.

Humans have harnessed the power of pheromones to combat insect pests; the sex-attractant pheromones of some agricultural pests, such as the Japanese beetle and the gypsy moth, have been successfully synthesized. These synthetic pheromones can be used to disrupt mating or to lure the insects into traps. Controlling pests with pheromones has major environmental advantages over conventional pesticides, which kill beneficial as well as harmful insects and foster the evolution of pesticide-resistant insects. In contrast, each pheromone is specific to a single species and does not promote the spread of resistance, because insects resistant to the attraction of their own pheromones do not reproduce successfully.

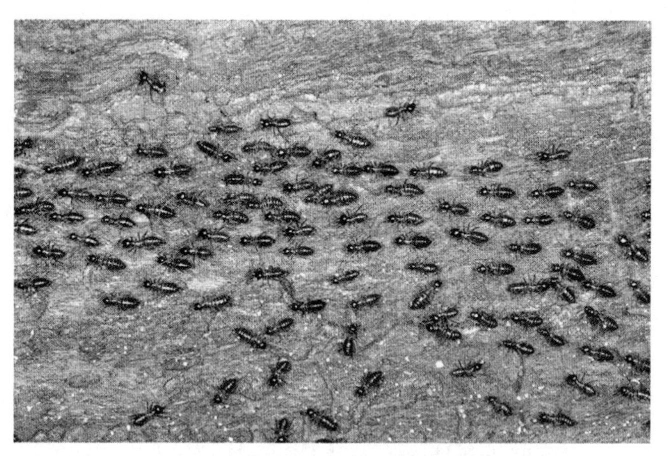

FIGURE 38-12 Communication by chemical messages
A trail of pheromones, secreted by termites from their own colony, orients foraging termites toward a source of food.

Communication by Touch Helps Establish Social Bonds

Communication by physical contact often serves to establish and maintain social bonds among group members. This function is especially apparent in humans and other primates, which have many gestures—including kissing, nuzzling, patting, petting, and grooming—with important social functions (Fig. 38-13a). Touch may even be essential to human well-being. For example, research has shown that when the limbs of premature human infants were

(a)

(b)

FIGURE 38-13 Communication by touch
(a) An adult olive baboon grooms a juvenile. Grooming both reinforces social relationships and removes debris and parasites from the fur. **(b)** Touch is also important in sexual communication. These land snails engage in courtship behavior that will culminate in mating.

stroked and moved for 45 minutes daily, the infants were more active, responsive, and emotionally stable, and gained weight more rapidly than did premature infants who received standard hospital treatment.

Communication by touch is not limited to primates, however. In many other mammal species, close physical contact helps cement the bond between parent and offspring. Species in which sexual activity is preceded or accompanied by physical contact can also be found across the animal kingdom (Fig. 38-13b).

38.3 How Do Animals Compete for Resources?

The Darwinian contest to survive and reproduce stems from the scarcity of resources relative to the reproductive potential of populations. The resulting competition underlies many of the most frequent types of interactions between animals.

Aggressive Behavior Helps Secure Resources

One of the most obvious manifestations of competition for resources such as food, space, or mates is **aggression**, or antagonistic behavior, between members of the same species. Although the expression "survival of the fittest" evokes images of the strongest animal emerging triumphantly from among the dead bodies of its competitors, in reality most aggressive encounters between members of the same species end without physical damage to the participants. Natural selection has favored the evolution of symbolic displays or rituals for resolving conflicts. During fighting, even the victorious animal can be injured and might not survive to pass on its genes. Aggressive *displays*, in contrast, allow the competitors to assess each other and acknowledge a winner on the basis of size, strength, and motivation, rather than on the wounds the competitor can inflict.

During aggressive displays, animals may exhibit weapons, such as claws and fangs (Fig. 38-14a), and often make themselves appear larger (Fig. 38-14b). Competitors often stand upright and erect their fur, feathers, ears, or fins (see Fig. 38-8). The displays are typically accompanied by intimidating sounds (growls, croaks, roars, chirps) whose loudness can help decide the winner. Fighting tends to be a last resort when displays fail to resolve a dispute.

In addition to aggressive visual and vocal displays, many animal species engage in ritualized combat. Deadly weapons may clash harmlessly (Fig. 38-15) or may not be used at all. In many cases these encounters involve shoving rather than slashing. The ritual thus allows contestants to assess the strength and the motivation of their rivals, and the loser slinks away in a submissive posture that minimizes the size of its body.

(a)

(b)

FIGURE 38-14 Aggressive displays
(a) Threat display of the male baboon. Despite the potentially lethal fangs so prominently displayed, aggressive encounters between baboons rarely cause injury. **(b)** The aggressive display of many male fish (here, striped grunts) includes elevating the fins and flaring the gill covers, thus making the body appear larger.

FIGURE 38-15 Displays of strength
Ritualized combat of fiddler crabs. Oversized claws, which could severely injure another animal, grasp harmlessly. Eventually one crab, sensing greater vigor in his opponent, retreats unharmed.

FIGURE 38-16 A dominance hierarchy
The dominance hierarchy of the male bighorn sheep is signaled by the size of the horns; these rams increase in status from right to left. The backward-curving horns, clearly not designed to inflict injury, are used in ritualized combat.

Dominance Hierarchies Help Manage Aggressive Interactions

Aggressive interactions use a lot of energy, can cause injury, and can disrupt other important tasks, such as finding food, watching for predators, or raising young. Thus, there are advantages to resolving conflicts with minimal aggression. In a **dominance hierarchy**, each animal establishes a rank that determines its access to resources. Although aggressive encounters occur frequently while the dominance hierarchy is being established, once each animal learns its place in the hierarchy, disputes are infrequent; the dominant individuals obtain most access to the resources needed for reproduction, including food, space, and mates. For example, domestic chickens, after squabbling, sort themselves into a reasonably stable "pecking order." Thereafter, all birds in the group defer to the dominant bird, all but the dominant bird give way to the second-most dominant, and so on. In wolf packs, one member of each sex is the dominant, or "alpha," individual, to whom all others of that sex are subordinate. Among male bighorn sheep, dominance is reflected in horn size (Fig. 38-16).

Perhaps the most thoroughly studied dominance hierarchy is that of chimpanzees. Ethologist Jane Goodall (Fig. 38-17) has devoted more than 30 years to meticulously observing chimpanzee behavior in the field at Gombe National Park in Tanzania, describing and documenting the animals' complex social organization. Chimps live in groups, and dominance hierarchies among males are a key aspect of their social life. Many males devote a significant amount of time to maintaining their position in the hierarchy, largely by way of an aggressive *charging display* in which a male rushes forward, throws rocks, leaps up to shake vegetation, and otherwise seeks to intimidate rival males. The advantages that accrue to dominant males, however, are not clear. According to Goodall, low-ranking males can still secure access to food and successful copulations (albeit not quite as easily as higher-ranking males). In Goodall's view, little evolutionary advantage accrues to a dominant male, and the function of chimpanzee dominance hierarchies remains unexplained.

Animals May Defend Territories That Contain Resources

In many animal species, competition for resources takes the form of **territoriality**, the defense of an area where important resources are located. The defended area may include places to mate, raise young, feed, or store food. Territorial animals generally restrict most or all of their activities to the defended area and advertise their presence there. Territories may be defended by males, females, a mated pair, or entire social groups (as in the defense of a nest by social insects). However, territorial behavior is most commonly seen in adult males, and territories are normally defended against members of the

FIGURE 38-17 Jane Goodall observing chimps at play

FIGURE 38-18 **A feeding territory**
Acorn woodpeckers live in communal groups that excavate acorn-sized holes in dead trees, stuffing the holes with green acorns for dining during the lean winter months. The group defends the trees vigorously against other groups of acorn woodpeckers and against acorn-eating birds of other species, such as jays.

same species, who compete most directly for the resources being protected. Territories are as diverse as the animals defending them. For example, a territory can be a tree where a woodpecker stores acorns (Fig. 38-18), small depressions in the sand used as nesting sites by cichlid fish, a hole in the sand that is home to a crab, or an area of forest providing food for a squirrel.

Territories Reduce Aggression

Acquiring and defending a territory requires considerable time and energy, yet territoriality is seen in animals as diverse as worms, arthropods, fish, birds, and mammals. The fact that organisms as distantly related as worms and humans independently evolved similar behavior suggests that territoriality provides some important advantages. Although the particular benefits depend on the species and the type of territory it defends, we can make some broad generalizations. First, as with dominance hierarchies, once a territory is established through aggressive interactions, relative peace prevails if boundaries are recognized and respected. The saying "good fences make good neighbors" also applies to nonhuman territories. One reason is that an animal is highly motivated to defend its territory and will often defeat even larger, stronger animals that attempt to invade it. Conversely, an animal outside its territory is much less secure and more easily defeated. This principle was demonstrated by Niko Tinbergen in an experiment using stickleback fish (Fig. 38-19).

Competition for Mates May Be Based on Territories

For males of many species, successful territorial defense has a direct impact on reproductive success. In these species, males defend territories, and females are attract-

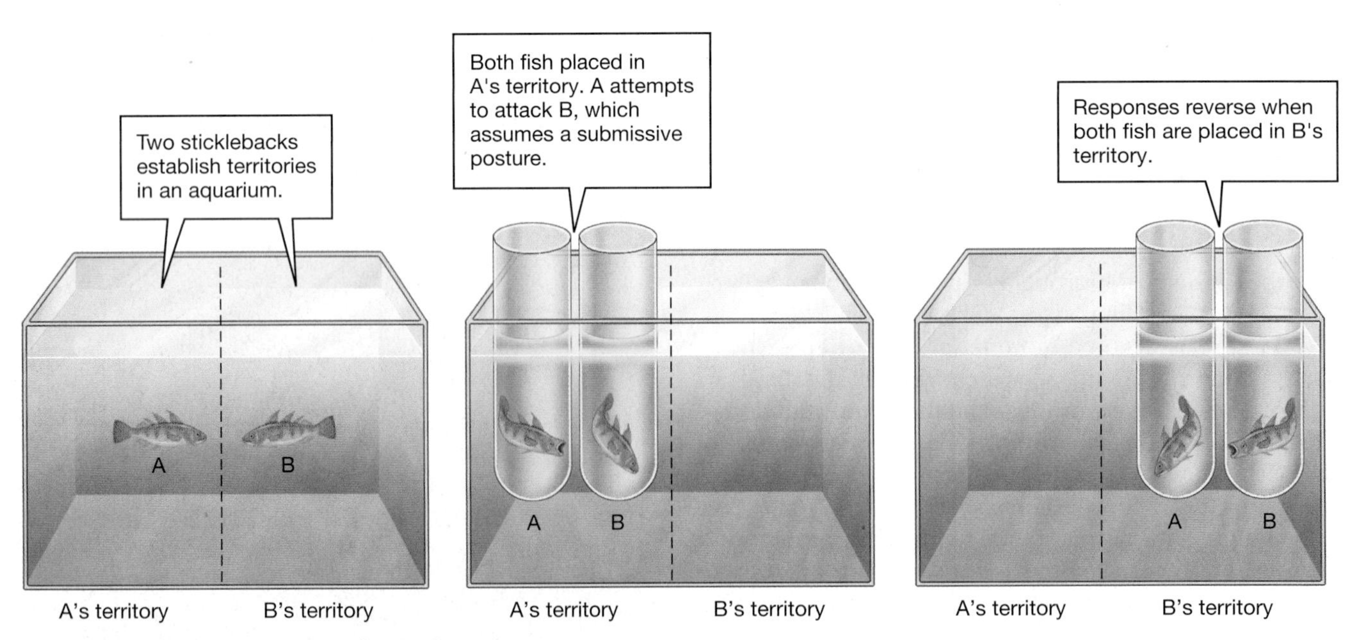
Two sticklebacks establish territories in an aquarium.
Both fish placed in A's territory. A attempts to attack B, which assumes a submissive posture.
Responses reverse when both fish are placed in B's territory.
A
B
A
B
A
B
A's territory
B's territory
A's territory
B's territory
A's territory
B's territory

FIGURE 38-19 **Territory ownership and aggression**
Niko Tinbergen's experiment demonstrating the effect of territory ownership on aggressive motivation.

ed to high-quality territories, which might have features such as large size, abundant food, and secure nesting areas. Males who successfully defend the best territories have the greatest chance of mating and passing on their genes. For example, experiments have shown that male stickleback fish that defend large territories are more successful in attracting mates than are males that defend small territories. Females that select males with the best territories increase their own reproductive success and pass their genetic traits (typically including their mate-selection preferences) to their offspring.

Animals Advertise Their Occupancy

Territories are advertised through sight, sound, and smell. If a territory is small enough, its owner's mere presence, reinforced by aggressive displays toward intruders, can provide sufficient defense. A mammal that has a territory but cannot always be present may use pheromones to scent-mark the boundaries of its territory. Male rabbits use pheromones secreted by glands in their chins and by anal glands to mark their territories. Hamsters rub the areas around their dens with secretions from special glands in their flanks.

Vocal displays are a common form of territorial advertisement. Male sea lions defend a strip of beach by swimming up and down in front of it, calling continuously. Male crickets produce a specific pattern of chirps to warn other males away from their burrows. Birdsong is a striking example of territorial defense. The husky trill of the male seaside sparrow is part of an aggressive display, warning other males to steer clear of his territory (Fig. 38-20). In fact, male sparrows that are unable to sing are unable to defend territories. The importance of singing to seaside sparrows' territorial defense was elegantly demonstrated by ornithologist M. Victoria McDonald, who captured territorial males and performed an operation that left them temporarily unable to sing but still able to utter the shorter and quieter signals in their vocal repertoires. The songless males were unable to defend territories or attract mates, but regained their lost territories when they recovered their singing ability.

FIGURE 38-20 Defense of a territory by song
A male seaside sparrow announces ownership of his territory.

38.4 How Do Animals Find Mates?

In many sexually reproducing animal species, mating involves copulation or other close contact between males and females. Before animals can successfully mate, however, they must identify one another as members of the same species, as members of the opposite sex, and as being sexually receptive. In many species, finding an appropriate potential partner is only the first step; often, the male must demonstrate his quality before the female will accept him as a mate. The need to fulfill all of these requirements has resulted in the evolution of a diverse and fascinating array of courtship behaviors.

Signals Encode Sex, Species, and Individual Quality

Individuals that waste energy and gametes by mating with members of the wrong sex or wrong species are at a disadvantage in the contest to reproduce. Thus, natural selection favors behaviors by which animals communicate their sex and species to potential mates.

Many Mating Signals Are Auditory

Animals often use sounds to advertise their sex and species. Consider the raucous nighttime chorus of male tree frogs, each singing a species-specific song. Male grasshoppers and crickets also advertise their sex and species by their calls, as does the female mosquito with her high-pitched whine.

Signals that advertise sex and species may also be used by potential mates in comparisons among rival suitors. For example, the male bellbird uses its deafening song to defend large territories and attract females from great distances. A female flies from one territory to another, alighting near each male in his tree. The male, beak gaping, leans directly over the flinching female and utters an earsplitting note. The female apparently endures this noise to compare the songs of the various males, perhaps choosing the loudest as a mate.

(a)

(b)

FIGURE 38-21 Sexual displays
(a) During courtship, the male gardener bowerbird builds a bower out of twigs and decorates it with colorful items that he gathers. **(b)** A male frigate bird inflates his scarlet throat pouch to attract passing females.
QUESTION The male bowerbird provides no protection, feeding, or other care to his mate or his offspring. Why, then, do females carefully compare the bowers of different males before choosing a mate?

Visual Mating Signals Are Also Common

Many species use visual displays for courting. The firefly, for example, flashes a message that identifies its sex and species. Male fence lizards bob their heads in a species-specific rhythm; females distinguish and prefer the rhythm of their own species. The elaborate construction projects of the male gardener bowerbird and the scarlet throat of the male frigate bird serve as flashy advertisements of sex, species, and male quality (Fig. 38-21). Sending these extravagant signals must be risky, as they make it much easier for predators to locate the sender. For males, the added risk is an evolutionary necessity, as females won't mate with males that lack the appropriate signal. Females, in contrast, typically do not need to attract males or assume the risk associated with a conspicuous signal, so in many species females are drab in comparison to males (Fig. 38-22).

The intertwined functions of sex recognition and species recognition, advertisement of individual quality, and synchronization of reproductive behavior commonly require a complex series of signals, both active and passive, by both sexes. Such signals are beautifully illustrated by the complex underwater "ballet" executed by the male and female three-spined stickleback fish (Fig. 38-23).

FIGURE 38-22 Sexual dimorphism in guppies
As in many animal species, the male guppy (left) is brighter and more colorful than the female. **QUESTION** Male guppies in streams with few predatory fish are brighter than males in streams with many predators. Why?

Chemical Signals Can Bring Mates Together

Pheromones can also play an important role in reproductive behavior. A sexually receptive female silk moth, for example, sits quietly and releases a chemical message so powerful that it can be detected by males up to 3 miles (5 kilometers) away. The exquisitely sensitive and selective receptors on the antennae of the male silk moth respond to just a few molecules of the substance, allowing him to travel upwind along a concentration gradient to find the female (Fig. 38-24a, p. 784).

Water is an excellent medium for dispersing chemical signals, and fish commonly use a combination of pheromones and elaborate courtship movements to ensure the synchronous release of gametes. Mammals, with their highly developed sense of smell, often rely on pheromones released by the female during her fertile periods to attract males (Fig. 38-24b).

(a) A male, inconspicuously colored, leaves the school of males and females to establish a breeding territory.

(b) As his belly takes on the red color of the breeding male, he displays aggressively at other red-bellied males, exposing his red underside.

(c) Having established a territory, the male begins nest construction by digging a shallow pit that he will fill with bits of algae cemented together by a sticky secretion from his kidneys.

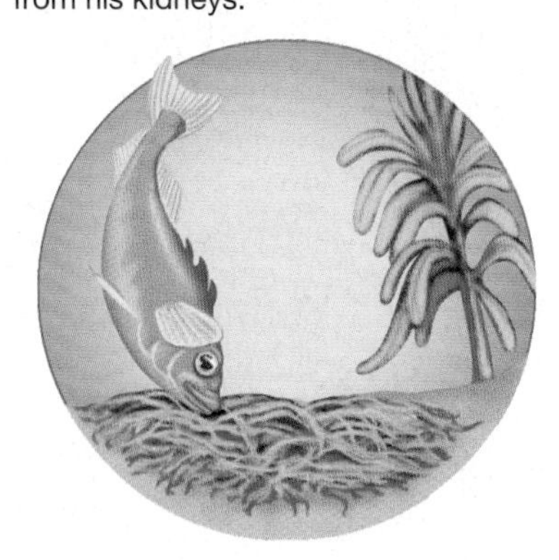

(d) After he tunnels through the nest to make a hole, his back begins to take on the blue courting color that makes him attractive to females.

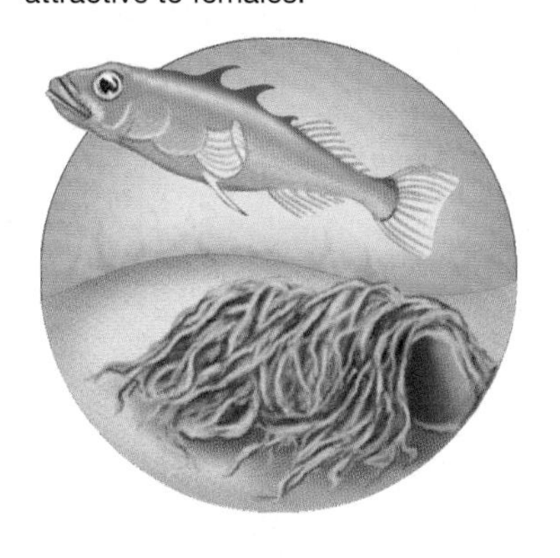

(e) An egg-carrying female displays her enlarged belly to him by assuming a head-up posture. Her swollen belly and his courting colors are passive visual displays.

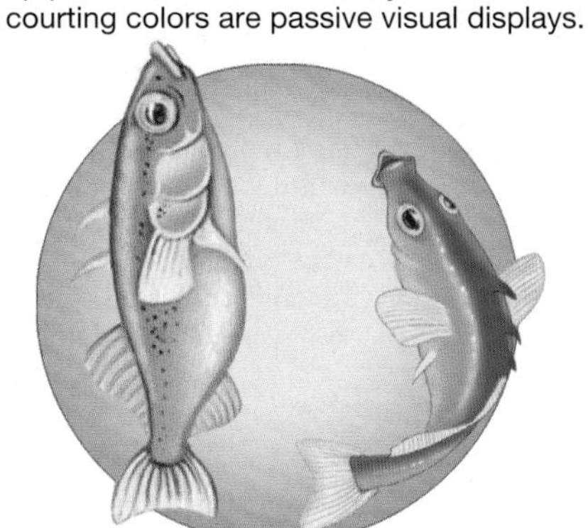

(f) Using a zigzag dance, he leads her to the nest.

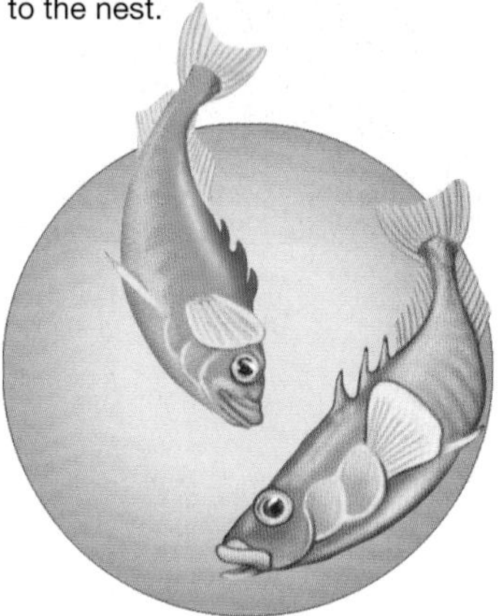

(g) After she enters, he stimulates her to release eggs by prodding at the base of her tail.

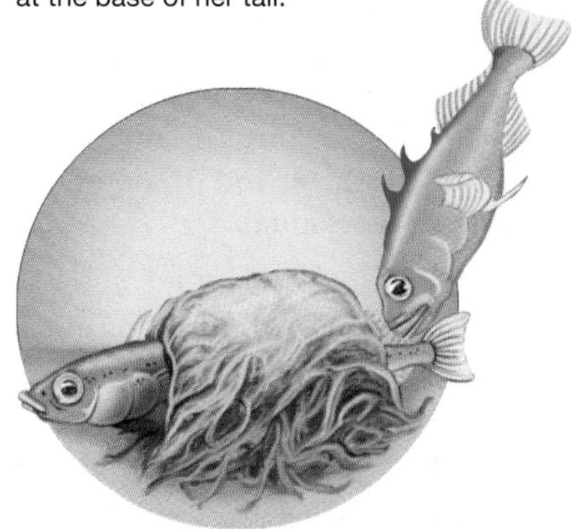

(h) He enters the nest as she leaves and deposits sperm, which fertilize the eggs.

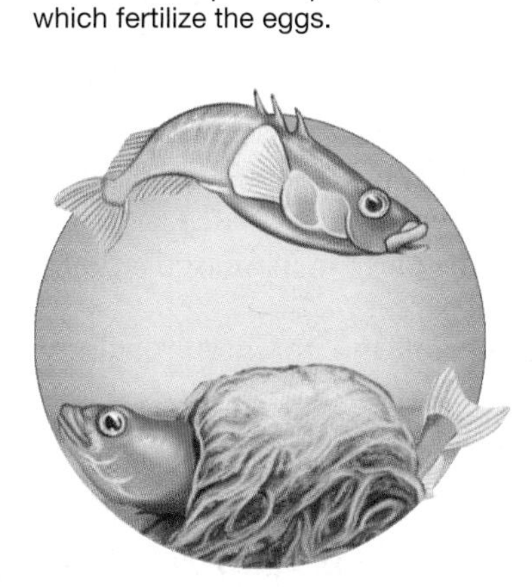

FIGURE 38-23 Courtship of the three-spined stickleback

38.5 What Kinds of Societies Do Animals Form?

Sociality is a widespread feature of animal life. Most animals interact at least a little with other members of their species; many spend the bulk of their lives in the company of others; and a few species have developed complex, highly structured societies. Social interaction can be cooperative or competitive, but is typically a mixture of the two.

Group Living Has Advantages and Disadvantages

Living in a group has both costs and benefits, and a species will not evolve social behavior unless the benefits of doing so outweigh the costs. On the negative side, social animals may encounter:

- Increased competition within the group for limited resources

(a)

(b)

FIGURE 38-24 Pheromone detectors
(a) Male moths find females not by sight but by following airborne pheromones released by females. These odors are sensed by receptors on the male's huge antennae, whose enormous surface area maximizes the chances of detecting the female scent. **(b)** When dogs meet, they typically sniff each other near the base of the tail. Scent glands located there broadcast information about the bearer's sex and interest in mating. **QUESTION** Females dogs use a pheromone to signal readiness to mate, but female mandrills (see Fig. 38-9) signal mating readiness with a visual signal. What differences would you predict between the two species' methods of searching for food?

- Increased risk of infection from contagious diseases
- Increased risk that offspring will be killed by other members of the group
- Increased risk of being spotted by predators

Benefits to social animals include:

- Increased abilities to detect, repel, and confuse predators
- Increased hunting efficiency or increased ability to spot localized food resources
- Advantages resulting from the potential for division of labor within the group
- Increased likelihood of finding mates

The Extent of Sociality Varies Among Species

The degree to which animals of the same species cooperate varies from one species to the next. Some types of animals, such as the mountain lion, are basically solitary; interactions between adults consist of brief aggressive encounters and mating. Other types of animals cooperate on the basis of changing needs. For example, the coyote is solitary when food is abundant but hunts in packs in the winter when food becomes scarce.

Loose social groups, such as pods of dolphins, schools of fish, flocks of birds, and herds of musk oxen (Fig. 38-25), can provide benefits. For example, the characteristic spacing of fish in schools or the V-pattern of geese in flight provides a hydrodynamic or aerodynamic advantage for each individual in the group, reducing the energy required for swimming or flying. Some biologists hypothesize that herds of antelope or schools of fish confuse predators; their myriad bodies make it difficult for the predator to focus on and pursue a single individual.

A Few Species Form Complex Societies

At the other end of the social spectrum are a few highly integrated cooperative societies, found primarily among insects and mammals. As you read the following section, you may notice that some cooperative societies are based on behavior that seems to sacrifice the individual for the good of the group. There are many examples: young, mature Florida scrub jays may remain at their parents' nest and help them raise subsequent broods instead of breeding; worker ants often die in defense of their nest; Belding ground squirrels may sacrifice their own lives to warn the rest of their group of an approaching predator. These behaviors are examples of **altruism**—behavior that decreases the reproductive success of one individual to benefit another.

FIGURE 38-25 Cooperation in loosely organized social groups A herd of musk oxen functions as a unit when threatened by predators such as wolves. Males form a circle, horns pointed outward, around the females and young.

Forming Groups with Relatives Fosters the Evolution of Altruism

How could altruistic behavior evolve? When individuals perform self-sacrificing deeds, why aren't the alleles that contribute to this behavior eliminated from the gene pool? One possibility is that other members of the group are close relatives of the altruistic individual. Because close relatives share alleles, the altruistic individual may promote the survival of its own alleles through behaviors that maximize the survival of its close relatives. If an altruistic behavior benefits close relatives, the penalty suffered by the altruist may be offset if the behavior helps its relatives pass more copies of the shared alleles to the next generation than the amount sacrificed by the altruist through its selfless behavior. This concept is called **kin selection**. Kin selection helps explain the self-sacrificing behaviors that contribute to the success of cooperative societies (see "Evolutionary Connections: Knowing Your Relatives: Kin Selection and the Evolution of Altruism" in Chapter 15). Cooperative behavior is illustrated in the following sections, which describe three examples of complex societies: one in an insect species, one in a fish species, and one in a mammal species.

Honeybees Live Together in Rigidly Structured Societies

Perhaps the most perplexing of all animal societies are those of bees, ants, and termites. Scientists have long struggled to explain the evolution of a social structure in which most individuals never breed, but instead labor intensively to feed and protect the offspring of a different individual. Whatever its evolutionary explanation, the intricate organization of a social insect colony makes a compelling story. In these communities, the individual is a mere cog in an intricate, smoothly running machine and could not survive by itself.

Individual social insects are born into one of several castes within the society. Each caste is a group of similar individuals that perform a specific function. Honeybees emerge from their larval stage into one of three major preordained roles. One role is *queen*. Only one queen is tolerated in a hive at any time. Her functions are to produce eggs (up to 1000 per day for a lifetime of 5 to 10 years) and regulate the lives of the workers. Male bees, called *drones,* serve merely as mates for the queen. Soon after the queen hatches, drones lured by her sex pheromones swarm around her, and she mates with as many as 15 of them. This relatively brief "orgy" supplies her with sperm that will last a lifetime, enough to fertilize more than 3 million eggs. Their sexual chore accomplished, the drones become superfluous and are eventually driven out of the hive or killed.

The hive is run by the third class of bees, sterile female *workers.* A worker's tasks are determined by her age and by conditions in the colony. A newly emerged worker starts life as a waitress, carrying food such as honey and pollen to the queen, to other workers, and to developing larvae. As she matures, special glands begin wax production, and she becomes a builder, constructing perfectly hexagonal cells of wax in which the queen deposits her eggs and the larvae develop. She also takes shifts as a maid, cleaning the hive and removing the dead, and as a guard, protecting the hive against intruders. Her final role in life is that of a forager, gathering pollen and nectar—food for the hive. She spends nearly half of her 2-month life in this role. Acting as a forager scout, she seeks new and rich sources of nectar and, if she finds one, returns to the hive and communicates its location to other foragers. She communicates by means

Media Activity
38.2 Communication in Honeybees

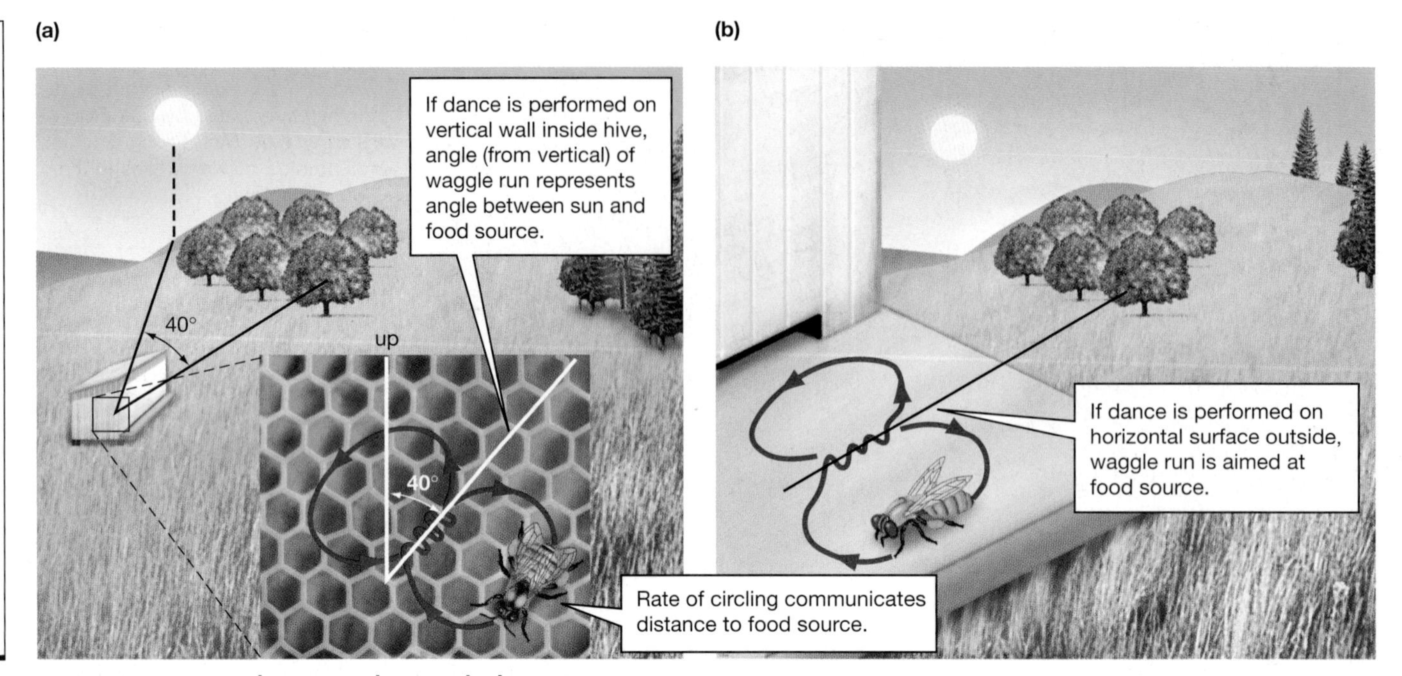
(a)
(b)
If dance is performed on vertical wall inside hive, angle (from vertical) of waggle run represents angle between sun and food source.
40°
up
40°
If dance is performed on horizontal surface outside, waggle run is aimed at food source.
Rate of circling communicates distance to food source.

FIGURE 38-26 Bee language: the waggle dance
A forager, returning from a rich source of nectar, performs a waggle dance that communicates the distance and direction of the food source as other foragers crowd around her, touching her with their antennae. The bee moves in a straight line while shaking her abdomen back and forth ("waggling") and buzzing her wings. She repeats this over and over in the same location, circling back in alternating directions.

of the **waggle dance**, an elegant form of symbolic expression (Fig. 38-26).

Pheromones play a major role in regulating the lives of social insects. Honeybee drones are drawn irresistibly to the queen's sex pheromone (*queen substance*), which she releases during her mating flights. Back at the hive, she uses the same substance to maintain her position as the only fertile female. The queen substance is licked off her body and passed among all the workers, rendering them sterile. The queen's presence and health are signaled by her continuing production of queen substance; a decrease in production (which occurs normally in the spring) alters the behavior of the workers. Almost immediately they begin building extra-large "royal cells." The workers feed the larvae that develop in these cells a special glandular secretion known as "royal jelly." This unique food alters the development of the growing larvae so that, instead of a worker, a new queen emerges from the royal cell. The old queen then leaves the hive, taking a swarm of workers with her to establish residence elsewhere. If more than one new queen emerges, a battle to the death ensues; the victorious queen takes over the hive.

Bullhead Catfish Form a Simple Vertebrate Society

The nervous systems of vertebrates are far more complex than those of insects, and we might therefore expect vertebrate societies to be proportionately more complex. With the exception of human society, however, they are not. Perhaps because the vertebrate brain *is* more complex, vertebrate societies tend to be simpler than those of the social insects such as honeybees, army ants, and termites. The robotic precision that makes complex insect societies possible is rarely found among vertebrates, which generally exhibit greater behavioral flexibility than do insects and therefore greater behavioral variation among individuals.

The social interactions of bullhead catfish, described by John Todd of the Woods Hole Oceanographic Institution in Massachusetts, provide a fascinating illustration of a vertebrate whose complex social interactions are based almost entirely on pheromones. Todd observed these nocturnal fish in large aquariums in the laboratory. He discovered that when a group was housed together, the fish staked out territories and established a dominance hierarchy, with the dominant fish defending the largest and best-protected area of the tank. Contests between tankmates consisted of open-mouthed aggressive displays. Once a fish became dominant, its aggressive displays caused the subordinate fish to flee. Actual violence occurred only when a stranger was introduced into a tank with a preestablished hierarchy. In this case, the established group exhibited cooperative behavior. The dominant fish allowed others to take refuge in his protected territory, then fought the intruder (Fig. 38-27). After defeating the newcomer, the dominant fish chased the others back out of his territory.

Todd discovered that blinding the bullheads did not cause any appreciable change in their social interactions. When their sense of smell was temporarily blocked, however, the fish did not recognize each other, interacting aggressively for weeks until their sense of smell returned. Both the status of an individual and a change in status were communicated by scent. If a dominant fish was removed from his tank and later returned, in most cases both his territory and his status were remembered and respected by his tankmates. But if he was removed and subjected to defeat in the tank of a more aggressive fish, his pheromones were somehow altered. Upon return to his home tank, he was attacked by his former subordinates.

When many newly caught fish are placed in the same tank, bullheads may form a dense and peaceful community that lacks territories or dominance hierarchies. Todd established such a community in one tank and placed a pair of aggressive rival fish in an adjacent tank. When water was pumped continuously from the "community tank" into the adjacent tank, the aggressive bullheads became peaceful, resuming their fighting only when the flow was stopped. Under the crowded conditions of the community tank, the fish apparently produced an "anti-aggression pheromone," minimizing conflict.

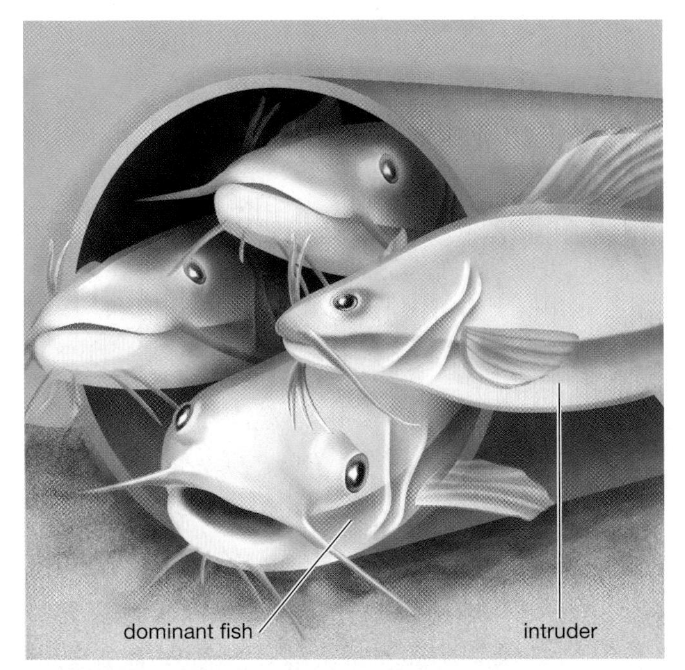

FIGURE 38-27 Cooperation among bullhead catfish
Three bullhead catfish occupy a section of pipe in a laboratory tank. The dominant fish is normally the exclusive occupant of the pipe, which is part of his territory. When an intruder is introduced into the tank, however, the dominant fish allows two subordinates to seek refuge in the pipe and reacts aggressively to the intruder.

Naked Mole Rats Form a Complex Vertebrate Society

Perhaps the most bizarre society among nonhuman mammals is that of the naked mole rat. These nearly blind, nearly hairless relatives of guinea pigs live in large underground colonies in southern Africa and exhibit a form of social organization not unlike that of an ant or termite colony. The colony is dominated by the queen, a single reproducing female to whom all other members are subordinate. The queen is the largest individual in the colony and maintains her status by aggressive behavior, particularly shoving. She prods and shoves lazy workers, stimulating them to become more active (Fig. 38-28). As in honeybee hives, there is a division of labor among the workers, in this case based on size. Small, young rats clean the tunnels, gather food, and tunnel. Tunnelers line up head to tail and pass excavated dirt along the completed tunnel to an opening. Just below the opening, a larger mole rat flings the dirt into the air, adding it to a cone-shaped mound. Biologists observing this behavior from the surface dubbed it "volcanoing." In addition to volcanoing, large mole rats defend the colony against predators and members of other colonies.

If another female begins to become fertile, the queen apparently senses changes in the estrogen levels of the subordinate female's urine. The queen then selectively shoves the would-be breeder, causing stress that prevents the rival from ovulating. Although all adult males are fertile, large males are more likely to mate with the queen than are small ones. When the queen dies, a few of the females gain weight and begin shoving one another. The aggression may escalate until a rival is killed. Ultimately, a single female becomes dominant. Her body lengthens; she assumes the queenship and begins

FIGURE 38-28 The naked mole rat queen
Encountering a lazy worker in one of the colony's underground tunnels, the queen shoves it and knocks it over to stimulate the worker to greater efforts.

to breed. Litters averaging 14 pups are produced about four times a year. During the first month, the queen nurses her pups, and the workers feed the queen. Then the workers begin feeding the pups solid food.

It is far from clear why naked mole rats, alone among mammals, have developed a social order in which almost all individuals sacrifice their reproductive potential. One likely contributing factor is that the queen produces the pups that grow up to form her colony, so all colony members are quite closely related. Another factor that may have favored the evolution of this communal social structure is the high cost of leaving the colony. It's unlikely that a single departing naked mole rat would be able to construct the elaborate underground burrow that the species inhabits.

38.6 Can Biology Explain Human Behavior?

The behaviors of humans, like those of all other animals, have an evolutionary history. Thus, the techniques and concepts of ethology can help us understand and explain human behavior. Human ethology, however, will remain less rigorous than animal ethology. We cannot treat people as laboratory animals, devising experiments that control and manipulate the factors that influence their attitudes and actions. In addition, some observers argue that human culture has been freed from the constraints of its evolutionary past for so long that we cannot explain our behavior in terms of biological evolution. Nevertheless, many scientists have taken an ethological, evolutionary approach to human behavior, and their work has had a major impact on our view of ourselves.

The Behavior of Newborn Infants Has a Large Innate Component

Because newborn infants have not had time to learn, we can assume that much of their behavior is innate. The rhythmic movement of an infant's head in search of its mother's breast is an innate behavior that is expressed in the first days after birth. Sucking, which can be observed even in a human fetus, is also innate (Fig. 38-29). Other behaviors seen in newborns—even premature infants—include grasping with the hands and feet and making walking movements when the body is supported. Another example is smiling, which can occur soon after birth. Initially, smiling can be induced by almost any object looming over the newborn. This initial indiscriminate response, however, is soon modified by experience. Infants up to 2 months old will smile in response to a stimulus consisting of two dark, eye-sized spots on a light background, which at that stage of development is a more potent stimulus for smiling than is an accurate representation of a human face. But as the child's development continues, learning and further development of the nervous system interact to limit the response to more-correct representations of a face.

Newborns in their first 3 days of life can be conditioned to produce certain sucking rhythms when their mother's voice is used as reinforcement. In an experiment performed by William Fifer of the New York State Psychiatric Institute, infants preferred their mothers' voices to other female voices, as indicated by their responses (Fig. 38-30). The infant's ability to learn his or her mother's voice and respond positively to it within days of birth has strong parallels to imprinting and may help initiate bonding with the mother.

Young Humans Acquire Language Easily

One of the most important insights to have emerged from studies of animal learning is that animals of a given species tend to have an inborn predilection for specific types of learning that are important to that species' mode of life. In humans, one such inborn predilection seems to be for the acquisition of language. Young children are able to acquire language rapidly and nearly effortlessly; they typically acquire a vocabulary of 28,000 words before the age of 8. Research suggests that we are born with a brain that is already primed for this early facility with language. For example, the human fetus begins responding to sounds dur-

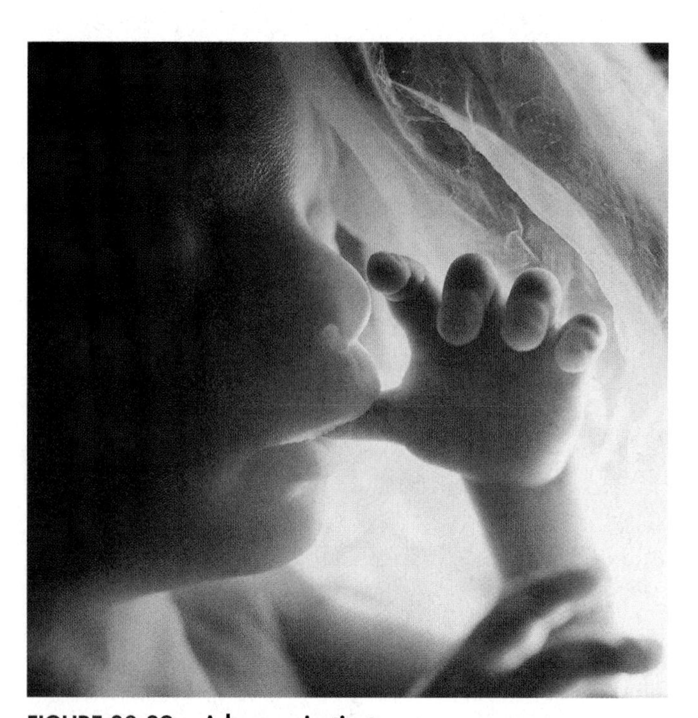

FIGURE 38-29 A human instinct
Thumb sucking is a difficult habit to discourage in young children, because sucking on appropriately sized objects is an instinctive, food-seeking behavior. This fetus sucks its thumb at about 4 ½ months of development.

ing the third trimester of pregnancy, and researchers have demonstrated that infants are able to distinguish among consonant sounds by 6 weeks after birth. In this demonstration, an infant responded to various consonant sounds by sucking on a pacifier that contained a force transducer to record the sucking rate. When one sound (such as "ba") was presented repeatedly, the infant became habituated and decreased her sucking rate. But when a new sound (such as "pa") was presented, sucking rate increased, revealing that the infant perceived the new sound as different.

Behaviors Shared by Diverse Cultures May Be Innate

Another way to study the innate bases of human behavior is to compare simple acts performed by people from diverse cultures. This comparative approach, pioneered by ethologist Irenaus Eibl-Eibesfeldt, has revealed several gestures that seem to form a universal, and therefore probably innate, human signaling system. Such gestures include facial expressions for pleasure, rage, and disdain, and greeting movements such as an upraised hand or the "eye flash" (in which the eyes are widely opened and the eyebrows rapidly elevated). The evolution of "hard-wiring" for these gestures presumably depended on the advantages that accrued to both senders and receivers from sharing information about the emotional state and intentions of the sender. A species-wide method of communication was perhaps especially important before the advent of language, and later remained useful during encounters between people who shared no common language.

Certain complex social behaviors are widespread among diverse cultures. For example, the incest taboo (avoidance of mating with close relatives) seems to be universal across human cultures (and even across many species of nonhuman primates). It seems unlikely, however, that a shared belief could be encoded in our genes. Some biologists have suggested that the taboo is instead a cultural expression of an evolved, adaptive behavior. According to this hypothesis, close contact among family members early in life suppresses sexual desire, an effect that arose due to the negative consequences of inbreeding (such as a higher incidence of genetic diseases). The hypothesis does not require us to assume an innate social belief, but rather proposes that we inherit a learning program that causes us to undergo a kind of imprinting early in life.

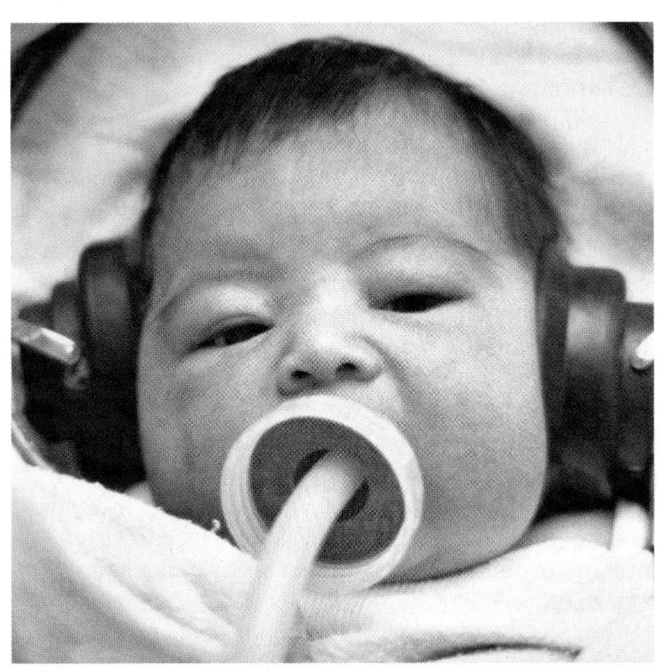

FIGURE 38-30 Newborns prefer their mother's voices Using a nipple connected to a computer that plays audio tapes, researcher William Fifer demonstrated that newborns can be conditioned to suck at specific rates for the privilege of listening to their own mothers' voices through headphones. For example, if the infant sucks faster than normal, her mother's voice is played; if she sucks more slowly, another woman's voice is played. Researchers found that infants easily learned and were willing to work hard at this task just to listen to their own mothers' voices, presumably because they had become used to her voice in the womb.

Humans May Respond to Pheromones

Although the main channels of human communication are through the eyes and ears, humans also seem to respond to chemical messages. The possible existence of human pheromones was hinted at in the early 1970s, when biologist Martha McClintock found that the menstrual cycles of roommates and close friends tended to become synchronized. McClintock suggested that the synchrony resulted from some chemical signal between the women, but almost 30 years passed before she and her colleagues uncovered more conclusive evidence that a pheromone was at work.

In 1998 McClintock's research group asked nine female volunteers to wear cotton pads in their armpits for 8 hours each day during their menstrual cycles. The pads were then disinfected with alcohol and swabbed above the upper lips of another set of 20 female subjects (who reported that they could detect no odors other than alcohol on the pads). The subjects were exposed to the pads in this way each day for two months, with half the group sniffing secretions from women in the early (pre-ovulation) part of the menstrual cycle, while the other half was exposed to secretions from later in the cycle (post-ovulation). Women exposed to early-cycle secretions had shorter-than-usual menstrual cycles, and women exposed to late-cycle secretions experienced delayed menstruation. Thus, it appears that women release different pheromones, with different effects on receivers, at different points in the menstrual cycle.

Although McClintock's experiment offers strong evidence for the existence of human pheromones, little else is known about chemical communication in humans. The actual molecules that caused the effects documented by McClintock remain unknown, as does their function (what benefit would a woman gain by influencing the menstrual cycles of other women?). Receptors for chemical messages have not yet been found in humans, and we don't know if the "menstrual pheromones" are the first known example of an important communication system or merely an isolated case of a vestigial ability. Despite the hopeful advertisements for sex-attraction pheromones on late-night television, chemical communication in humans is a scientific mystery awaiting a solution.

Studies of Twins Reveal Genetic Components of Behavior

Twins present an opportunity to examine the hypothesis that differences in human behavior are related to genetic differences. If a particular behavior is heavily influenced by genetic factors, we would expect to find similar expression of that behavior in *identical twins* (who arise from a single fertilized egg and have identical genes) but not in *fraternal twins* (who arise from two individual eggs and are no more similar genetically than other siblings). Data from twin studies, and from other within-family investigations, have tended to confirm the heritability of many human behavioral traits. These studies have documented a significant genetic component for traits such as activity level, alcoholism, sociability, anxiety, intelligence, dominance, and even political attitudes. On the basis of tests designed to measure many aspects of personality, identical twins are about two times more similar in personality than are fraternal twins.

The most fascinating twin findings come from observations of identical twins separated soon after birth, reared in different environments, and reunited for the first time as adults. Identical twins reared apart have been found to be as similar in personality as those reared together, indicating that the differences in their environments had little influence on their personality development. They have been found to share nearly identical taste in jewelry, clothing, humor, food, and names for children and pets. In some cases personal idiosyncrasies such as giggling, nail biting, drinking patterns, hypochondria, and mild phobias are shared by these separated twins.

Biological Investigation of Human Behavior Is Controversial

The field of human behavioral genetics is controversial, especially among nonscientists, because it challenges the long-held belief that environment is the most important determinant of human behavior. As discussed earlier in this chapter, we now recognize that all behavior has some genetic basis and that complex behavior in nonhuman animals typically combines elements of both innate and learned behaviors. Thus, it seems likely that our own behavior is influenced by both our evolutionary history and our cultural heritage. The debate over the relative importance of heredity and environment in determining human behavior continues and is unlikely ever to be fully resolved. Human ethology is not yet recognized as a rigorous science, and it will always be hampered because we can neither view ourselves with detached objectivity nor experiment with people as if they were laboratory rats. Despite these limitations, there is much to be learned about the interaction of learning and innate tendencies in humans.

EVOLUTIONARY CONNECTIONS

Why Do Animals Play?

Pigface, a giant 50-year-old African softshell turtle, spends hours each day batting a ball around his aquatic home in the National Zoo in Washington, D.C., to the delight of thousands of visitors and the puzzlement of behavioral biologists. Play has always been somewhat of a mystery. It has been observed in many birds and in most mammals, but, until zookeepers tossed Pigface a ball a few years ago, it had never been seen in animals as evolutionarily ancient as turtles.

Animals at play are fascinating. Pygmy hippopotamuses push one another, shake and toss their heads, splash in the water, and pirouette on their hind legs. Otters delight in elaborate acrobatics. Bottlenose dolphins balance fish on their snouts, throw objects, and carry them in their mouths while swimming. Baby vampire bats chase, wrestle, and slap each other with their wings. Even octopuses have been seen playing a game: pushing objects away from themselves and into a current, then waiting for the objects to drift back, only to push them back into the current to start the cycle over again.

Play can be solitary, typically consisting of a single animal manipulating an object, such as a cat with a ball of yarn, or the dolphin with its fish, or a macaque monkey making and playing with a snowball. Play can also be social. Often, young of the same species play together, but parents may join them (Fig. 38-31a). Social play typically includes chasing, fleeing, wrestling, kicking, and gentle biting (Fig. 38-31b,c).

Play seems to lack any clear immediate function, and is abandoned in favor of feeding, courtship, and escaping from danger. Young animals play more frequently than do adults. Play typically borrows movements from other behaviors (attacking, fleeing, stalking, and so on) and uses considerable energy. Also, play is potentially dangerous. Many young humans and other animals are

(a)

(b)

(c)

FIGURE 38-31 Young animals at play

injured, and some are killed, during play. In addition, play can distract an animal from the presence of danger while making it conspicuous to predators. So why do animals play?

The most logical conclusion is that play must have survival value and that natural selection has favored those individuals who engage in playful activities. One of the best explanations for the survival value of play is "practice theory," first proposed by K. Groos in 1898. He suggested that play allows young animals to gain experience in behaviors that they will use as adults. By performing these acts repeatedly in play, the animal practices skills that will later be important in hunting, fleeing, or social interactions.

Recent research supports and extends Groos's proposal. Play is most intense early in life, when the brain develops and crucial neural connections form. John Byers, a zoologist at the University of Idaho, has observed that animals with larger brains tend to be more playful than animals with smaller brains. Because larger brains are generally linked to increased learning ability, this relationship supports the idea that adult skills are learned during juvenile play. Watch children roughhousing or playing tag, and you will see how play fosters strength and coordination, and develops skills that might have helped our hunting ancestors survive. Quiet play with other children incorporating dolls, blocks, and other toys helps children prepare to interact socially, nurture their own children, and deal with the physical world.

Although Shakespeare tells us "play needs no excuse," there is good evidence that the tendency to play has evolved as an adaptive behavior in animals capable of learning. Play is quite literally "serious fun."

CASE STUDY REVISITED Sex and Symmetry

To assess whether male body symmetry is correlated with male "mating success" in humans, Randy Thornhill, Steven Gangestad, and their colleagues began by measuring symmetry in some young adult males. Each man's degree of symmetry was assessed by measurements of his ear length and the width of his foot, ankle, hand, wrist, elbow, and ear. From these measurements, the researchers derived an index that summarized the degree to which the size of these features differed between the right and left sides of the body.

The researchers found that, among the males in their sample, the most symmetrical men were judged (by a panel of female observers who were unaware of the nature of the study) to be more attractive than other men. A survey of the study subjects revealed that the more symmetrical men also tended to begin having sex earlier in life and to have had a larger number of different sexual partners. Apparently, a man's sexual activity and attractiveness to women are correlated with his symmetry.

Why would male body symmetry affect mating success? The most likely explanation is that symmetry is an indicator of good physical condition. As developmental biologists have long known, disruptions of normal embryological development can cause bodies to be asymmetrical, so a highly symmetrical body indicates healthy, normal development. It also may indicate a high-quality genotype that was able to overcome any disturbances (such as diseases or exposure to toxic substances) during development. Females that mate with individuals whose health and vitality are announced by their symmetrical bodies might have offspring that are similarly healthy and vital. Natural selection would thus favor females who chose to mate with symmetrical males, and that kind of mating behavior would come to predominate.

Consider This: What makes a person beautiful? Is our concept of human beauty essentially arbitrary, determined by cultural standards that vary among cultures and that can change over time, or are our perceptions of others' appearances hard-wired into our biological makeup, the product of our evolutionary heritage? As a scientist, how would you approach the problem of determining the source of our standards for beauty? What evidence would persuade you that beauty is a biological phenomenon, or that it is a cultural one?

Links to Life: Drunk Monkeys

Researchers hoping to understand the causes of complex human behaviors, such as alcoholism, often study animals that exhibit similar behaviors. A fascinating example of this approach is a long-term study of alcoholism in rhesus monkeys conducted at the National Institutes of Health. Researchers provide the monkeys with access to a sweet, alcohol-containing liquid for an hour each day. Some monkeys ignore the alcohol; some consume moderate amounts; and others imbibe so much alcohol that their ability to function normally is impaired, much like human alcoholics. What makes these monkeys become alcoholic?

Through experiments spanning almost two decades, the NIH researchers have made some fascinating discoveries. A tendency toward alcoholic behavior is tied to low levels of certain neurotransmitters, especially serotonin. Serotonin seems to act in the brain in part to "put the brakes" on impulsive behavior, and monkeys with low serotonin activity show little inclination to limit alcohol consumption. Monkeys with low serotonin tend to pass this trait to their offspring, so a tendency to alcoholic behavior is at least partly innate. But the researchers have found that social conditions also exert a powerful influence on monkey behavior. In particular, a monkey that is taken from its mother in infancy and raised in a group of same-age peers is far more likely to become an alcoholic adult than a monkey that spends the normal 7 months in its mother's care. Though low-serotonin monkeys may become alcoholic even if their upbringing is normal, and monkeys with especially high serotonin may resist the ill effects of a motherless infancy, social conditions in early life clearly have a crucial influence on adult behavior.

Summary of Key Concepts

38.1 How Do Innate and Learned Behaviors Differ?

Although all animal behavior is influenced by both genetic and environmental factors, biologists can distinguish between behaviors whose development is not highly dependent on external factors and behaviors that require more extensive environmental stimuli in order to develop. Behaviors in the first category are sometimes designated as innate and can be performed properly the first time an animal encounters the appropriate stimulus. Behavior that changes in response to input from an animal's social and physical environment is said to be learned. Learning can modify innate behavior to make it more appropriate.

Although the distinction between innate and learned behavior is conceptually useful, the distinction is not sharp in naturally occurring behaviors. Learning allows animals to modify innate responses so that they occur only with appropriate stimuli. Imprinting, a form of learning with innate constraints, is possible only at a certain time in an animal's development.

38.2 How Do Animals Communicate?
Communication allows animals of the same species to interact effectively in their quest for mates, food, shelter, and other resources. Animals communicate through visual signals, sound, chemicals (pheromones), and touch. Visual communication is quiet and can convey rapidly changing information. Visual signals are active (body movements) or passive (body shape and color). Sound communication can also convey rapidly changing information and is effective where vision is impossible. Pheromones can be detected after the sender has departed, conveying a message over time. Physical contact reinforces social bonds and is a part of mating.

38.3 How Do Animals Compete for Resources?
Although many competitive interactions are resolved through aggression, serious injuries are rare. Most aggressive encounters are settled by means of displays that communicate the motivation, size, and strength of the combatants.

Some species establish dominance hierarchies that minimize aggression. On the basis of initial aggressive encounters, each animal acquires a status in which it defers to more dominant individuals and dominates subordinates. When resources are limited, dominant animals obtain the largest share and are most likely to reproduce.

Territoriality, a behavior in which animals defend areas where important resources are located, also minimizes aggressive encounters. In general, territorial boundaries are respected, and the best-adapted individuals defend the richest territories and produce the most offspring.

38.4 How Do Animals Find Mates?
Successful reproduction requires that animals recognize the species, sex, and sexual receptivity of potential mates. In many species, animals also assess the quality of potential mates. These requirements have contributed to the evolution of sexual displays that use all forms of communication.

38.5 What Kinds of Societies Do Animals Form?
Social living has both advantages and disadvantages, and species vary in the degree to which their members cooperate. Some species form cooperative societies. The most rigid and highly organized are those of the social insects such as the honeybee, in which the members follow rigidly defined roles throughout life. These roles are maintained through both genetic programming and the influence of certain pheromones. Naked mole rats exhibit the most complex and rigid vertebrate social interactions, resembling those of social insects.

38.6 Can Biology Explain Human Behavior?
The degree to which human behavior is genetically influenced is highly controversial. Because we cannot freely experiment on humans, and because learning plays a major role in nearly all human behavior, investigators must rely on studies of newborn infants and comparative cultural studies, correlations between certain behaviors and physiology (which suggest a role for pheromones), and studies of identical and fraternal twins. Evidence is mounting that our genetic heritage plays a role in personality, intelligence, simple universal gestures, responses to certain stimuli, and the tendency to learn specific things such as language at particular stages of development.

Key Terms

aggression *p. 778*
altruism *p. 784*
behavior *p. 770*
communication *p. 775*
dominance hierarchy *p. 779*
ethology *p. 773*
habituation *p. 771*
imprinting *p. 774*
innate *p. 770*
insight learning *p. 772*
kin selection *p. 785*
learning *p. 770*
operant conditioning *p. 771*
pheromone *p. 776*
territoriality *p. 779*
trial-and-error learning *p. 771*
waggle dance *p. 786*

Thinking Through the Concepts

To take a multiple-choice quiz with feedback on the contents of this chapter, visit http://www.prenhall.com/audesirk7. *Log in to the Web site selected by your instructor and navigate to the Self Test section for this chapter.*

? Review Questions

1. Explain why neither *innate* nor *learned* adequately describes the behavior of a given organism.
2. Explain why animals play.
3. List four senses through which animals communicate, and give one example of each form of communication. After each, present both the advantages and disadvantages of that form of communication.
4. A bird will ignore a squirrel in its territory but will act aggressively toward a member of its own species. Explain why.
5. Why are most aggressive encounters among members of the same species relatively harmless?
6. Discuss advantages and disadvantages of group living.
7. In what ways do naked mole rat societies resemble those of the honeybee?

Applying the Concepts

1. Male mosquitoes orient toward the high-pitched whine of the female, and female mosquitoes—which suck blood—are attracted to the warmth, humidity, and carbon dioxide exuded by their prey. Using this information, design a mosquito trap or killer that exploits a mosquito's innate behaviors. Then design one for moths.
2. You raise honeybees but are new at the job. Trying to increase honey production, you introduce several queens into the hive. What is the likely outcome? What different things could you do to increase production?
3. Describe and give an example of a dominance hierarchy. What role does it play in social behavior? Give a human parallel, and describe its role in human society. Are the two roles similar? Why? Repeat this exercise for territorial behavior in humans and in another animal.
4. You are manager of an airport. Planes are being endangered by large numbers of flying birds, which can be sucked into the engines, disabling them. Without harming the birds, what might you do to discourage them from nesting and flying near the airport and its planes?

For More Information

Brown, S. L. "Animals at Play." *National Geographic*, December 1994. Clearly written description of why animals play, accompanied by beautiful photographs of wild animals at play.

de Waal, F. "The End of Nature Versus Nurture." *Scientific American*, December 1999. An eminent ethologist makes the case for taking a nuanced biological approach to understanding human behavior.

Dugatkin, L., and Godin, J. "How Females Choose Their Mates." *Scientific American*, April 1998. An entertaining look at how biologists study the processes by which females choose among available males.

Gould, J. L., and Gould, C. G. "The Instinct to Learn." *Science 81*, May 1981. Birds, bees, and perhaps even humans are genetically programmed to learn specific things at particular stages in life.

Kirchner, W. H., and Towne, W. F. "The Sensory Basis of the Honeybee Dance Language." *Scientific American*, June 1994. Combines a historical look at the work of Karl von Frisch with a modern update on recent research that has almost fully elucidated the mechanisms of honeybee communication during the waggle dance.

Lorenz, K. *King Solomon's Ring: New Light on Animal Ways.* New York: Thomas Y. Crowell, 1952. Beautifully written and filled with interesting anecdotes; provides important insights into early ethology.

Macdonald, D., and Brown, R. "The Smell of Success." *New Scientist*, May 1985. Describes the amazing diversity of mammalianpheromones.

Pinker, S. *The Language Instinct.* New York: William Morrow, 1994. An entertaining account of linguists' current understanding of how we develop the ability to use language and how that ability may have evolved.

Seeley, T. D. "The Honeybee Colony as a Superorganism." *American Scientist*, November–December 1989. The honeybee society exemplifies the concept of kin selection in which the society, rather than the individual, serves as a "vehicle for the survival of genes."

Sherman, P., and Alcock, J. *Exploring Animal Behavior.* Sunderland, MA: Sinauer, 1998. A collection of articles in which behavioral biologists describe their research and conclusions about the mechanisms, function, and evolution of behavior.

Sherman, P. W., Jarvis, J. U. M., and Braude, S. H. "Naked Mole Rats." *Scientific American*, August 1992. Describes the newly investigated society of the naked mole rat, a vertebrate whose social behavior resembles that of some social insects.

Weiner, J. *Time, Love, Memory: A Great Biologist and His Quest for the Origins of Behavior.* New York: Knopf, 2000. A wonderful science writer chronicles the study of the genetics of behavior, especially the work of a great pioneer of the field, Seymour Benzer.

Media Activities

To access a Media Activity visit http://www.prenhall.com/audesirk7. *Log in to the Web site selected by your instructor, navigate to this chapter, and select the appropriate Media Activity number.*

38.1 Observing Behavior: Homing in Digger Wasps

Estimated time: 15 minutes

Repeat some of Niko Tinbergen's famous experiments with digger wasps on innate versus learned behaviors.

38.2 Communication in Honeybees

Estimated time: 10 minutes

Social animals often have to communicate with each other to survive. This activity explores one of the most sophisticated and well-studied examples of communication in non-vertebrate animals, the dancing "language" of honeybees.

38.3 Web Investigation: Sex and Symmetry

Estimated time: 20 minutes

In his 1871 treatise *On the Origin of Species,* Charles Darwin not only proposed natural selection as a driving force for evolutionary change but also identified sexual selction as an important special case thereof. In this activity explore what characteristics are possible predictors of the fitness of a potential mate's genes.

CHAPTER

39 Population Growth and Regulation

Populations of white-footed mice, white-tailed deer, gypsy moth caterpillars, deer ticks, and acorns interact in complex ways in a forest ecosystem in the eastern United States. This forest is in the Pocono Mountains of Pennsylvania.

AT A GLANCE

CASE STUDY Acorns, Mice, Moths, Deer, and Disease

In the autumn of 1995, the hardwood forest in upstate New York was ablaze with fall colors. It was also alive with Girl Scouts industriously strewing acorns on the forest floor. But why bring acorns to an oak forest? The scouts were assisting researchers Clive Jones and Richard Ostfeld from the Institute of Ecosystem Studies in Millbrook, New York, who were monitoring several aspects of this forest ecosystem: the availability of acorns and the populations of white-footed mice, gypsy moths, deer, and black-legged deer ticks.

The oak trees didn't produce many acorns in 1995; in fact, oaks bear large numbers of acorns only every two to five years. Because white-footed mice feed on acorns, their populations fluctuate dramatically in response to acorn availability. The gypsy moth population also undergoes related fluctuations: Every decade or so, outbreaks of gypsy moth caterpillars—the larval form of this imported pest—strip the oak trees of almost all their leaves. Gypsy moth populations, in turn, are controlled by white-footed mice, which eat the pupae of the moths. The researchers were testing the hypothesis that an abundant acorn crop would lead to a large mouse population, and therefore a reduction in gypsy moths. By comparing mouse and moth populations in areas of forest artificially enriched with acorns to populations in nearby acorn-poor forests, the researchers tested this link. They also removed mice from selected areas to test the impact of reduced mouse predation on gypsy moths.

The biologists also monitored the population size of black-legged deer ticks because they can transmit Lyme disease to people. As adults, the ticks feed on deer, and blood-sated adult ticks drop from the deer and lay their eggs in the soil. Both deer and mice are attracted to acorns; when the tick eggs hatch, the larvae often attach themselves to nearby white-footed mice. Nearly all of these mice carry the bacterium that causes Lyme disease.

By manipulating the numbers of mice and acorns and monitoring the responses of these interacting populations, the researchers made some surprising and potentially important discoveries. What were they? Join us at the end of the chapter.

This chapter begins our study of **ecology** (derived from the Greek word *oikos*, "a place to live"). Ecology refers to the study of interrelationships between living things and their nonliving environment. The environment consists of the **abiotic** (nonliving) component—including soil, water, and weather—and the **biotic** (living) component, which includes all forms of life. The term **ecosystem** refers both to the nonliving environment and to all the organisms within a defined area, such as the eastern forest described in this chapter's Case Study. Within an ecosystem, all interacting populations of organisms—for example, the oak trees, mice, deer, ticks, moths, bacteria, and other forms of life in the forest—are described as a **community**.

What keeps natural populations from overpopulating their habitat and starving? What happens when different organisms compete for the same type of food, for space, or for other resources? Why has the human population continued to expand while other populations have fluctuated, remained stable, or declined? Look for answers to these questions in this chapter as we explore how populations grow and how population growth is controlled. In Chapters 40 through 42, we will proceed from *populations* to levels of increasing complexity: first to *communities* and the interactions within them, then to the organization of ecosystems, and finally to an exploration of the diversity of ecosystems that make up the *biosphere*, which encompasses all life on Earth.

39.1 How Does Population Size Change?

A **population** consists of all the members of a particular species that live within an ecosystem and can potentially interbreed. The mice, the oak trees, the gypsy moths, and the ticks in the Case Study each constitute a different population.

Studies of undisturbed ecosystems show that some populations tend to remain relatively stable in size over time, others fluctuate in a roughly cyclical pattern, and still others vary sporadically in response to complex environmental variables. In contrast to most nonhuman species, the global human population has exhibited steady growth for centuries. Let's examine how and why populations grow and then look at the forces that control this growth.

Three factors determine whether and how much the size of a population changes: births, deaths, and migration. Organisms join a population through birth or **immigration** (migration in) and leave it through death or **emigration** (migration out). A population remains stable if, on the average, as many individuals leave as join. Population growth occurs when the number of births plus immigrants exceeds the number of deaths plus emigrants. Populations decline when the reverse occurs. A simple equation for the change in population size within a given time period is as follows:

$$(\textbf{births} - \textbf{deaths}) + (\textbf{immigrants} - \textbf{emigrants}) = \textbf{change in population size}$$

In many natural populations, organisms moving in and out contribute relatively little to population change, making birth and death rates the primary factors that influence population growth.

The ultimate size of any population (if migration is not a factor) is the result of a balance between two major opposing factors that determine birth and death rates. The first factor is **biotic potential**, the maximum rate at which the population could increase, assuming ideal conditions that allow a maximum birth rate and minimum death rate.

Opposing this potential for growth are limits set by the living and nonliving environments; collectively, these limits are called **environmental resistance**. Environmental resistance is imposed by the availability of food and space, competition with other organisms, and certain interactions among species, such as predation and parasitism. Natural events such as storms, fires, freezing weather, floods, and droughts also come under this heading. Environmental resistance can both decrease birth rates and increase death rates. For example, a drought might kill plants directly. Drought would also harm animal populations that rely on these plants by reducing reproduction and increasing the number of deaths from starvation. The interaction between biotic potential and environmental resistance usually results in a balance between population size and available resources. To understand how populations grow and how their size is regulated, let's examine each of these forces in more detail.

Biotic Potential Can Produce Exponential Growth

Changes in population size (ignoring migration) are functions of the birth rate, the death rate, and the number of individuals in the original population. Rates of change in population size are often measured as the changes in these variables per individual during a given unit of time. For example, the birth rate may be expressed as the number of births per organism per year.

The **growth rate** (r) of a population is a measure of the change in population size per individual per unit of time. This value is determined by subtracting the death rate (d) from the birth rate (b):

$$\begin{array}{ccccc} b & - & d & = & r \\ \textbf{(birth rate)} & - & \textbf{(death rate)} & = & \textbf{(growth rate)} \end{array}$$

(If the death rate exceeds the birth rate, the growth rate will be negative and the population will decline; here we will focus on growing populations.) If we wish to calculate the annual growth rate of a human population of

Media Activity
39.1 Population Growth and Regulation

10,000 in which 1500 births and 500 deaths occur each year, we can use this simple equation:

$$\textbf{growth rate } r = \textbf{birth rate } b - \textbf{death rate } d$$

$$r = \frac{1500}{10,000} - \frac{500}{10,000} = 1000/10,000 = 1/10 = 0.10 = 10\%$$

or

$$r = 0.15 \text{ births per person per year} - 0.05 \text{ deaths per person per year}$$

$$= 0.10 \text{ (or } 10\%) \text{ increase per person per year}$$

To determine the number of individuals added to a population in a given time period, multiply the growth rate (r) by the original population size (N):

$$\textbf{population growth} = rN$$

In this example, population growth (rN) equals $0.10 \times 10,000 = 1000$ people in the first year. If this growth rate is constant, then the following year, r must be multiplied by an even larger population size: ($N + rN = 11,000$). Thus, in the second year, 1100 more individuals are added to the population, which further increases the number of individuals added in the third year, and so on.

This pattern of continuously accelerating increase in population size is **exponential growth**. During exponential growth, a population (over a given time period) grows by a fixed percentage of its size at the beginning of that time period. Thus, an increasing number of individuals is added to the population during each succeeding time period, causing population size to grow at an ever-accelerating pace. The graph of exponential population growth is often called a *J-shaped growth curve*, or a **J-curve**, after its shape. The population growth rate (r) also allows us to calculate the **doubling time** of the population, or the time it takes a population to double in size at its current rate of growth. Based on the calculation for exponential growth, the doubling time can be determined by dividing the constant 0.693 (about 0.7) by r. Using this calculation, our hypothetical population has a doubling time of about $0.7/0.1 = 7$ years.

Populations approximate exponential growth whenever births consistently exceed deaths. This occurs if, on the average, each individual produces more than one surviving offspring during its lifetime. Although the number of offspring an individual produces each year varies from millions (for an oyster) to one or fewer (for a human), each organism—whether working alone or as part of a sexually reproducing pair—has the potential to replace itself many times over during its lifetime. This high biotic potential evolved because it helps ensure that some offspring survive to bear their own young. Several factors influence biotic potential, including the following:

- The age at which the organism first reproduces
- The frequency at which reproduction occurs
- The average number of offspring produced each time
- The length of the organism's reproductive life span
- The death rate of individuals under ideal conditions

We will use examples in which these factors differ to illustrate the concept of exponential growth. The bacterium *Staphylococcus* (see Fig. 39-1a) normally resides harmlessly in and on the human body, where its population growth is restricted by environmental resistance. But in an ideal culture medium—such as warm custard—where *Staphylococcus* may accidentally be introduced, each bacterial cell can divide every 20 minutes; the population doubles three times each hour (in this case, threatening the consumer with food poisoning). The larger the population grows, the more cells there are to divide. The biotic potential of bacteria is so great that, hypothetically, the offspring of a single bacterium could coat Earth in a layer 7 feet deep within 48 hours!

In contrast, the golden eagle is a relatively long-lived, rather slowly reproducing species (see Fig. 39-1b). Let's assume that the golden eagle can live 30 years and reaches sexual maturity at age 4 years, and that each pair of eagles produces two offspring annually for the remaining 26 years (red line). Figure 39-1 compares the potential population growth of eagles to that of bacteria, assuming no deaths occur in either population during the time graphed. Although the time scale differs tremendously, notice that the shapes of the curves are virtually identical—both populations exhibit exponential growth. Figure 39-1b also shows what happens if eagle reproduction begins at age 6 years (green line) instead of 4. Exponential growth still occurs, but the time required to reach a particular size increases considerably. This result has important implications for the human population: Delayed childbearing significantly slows population growth. If each woman bears three children in her early teens, the population will grow much faster than if each woman bears five children after age 30.

So far we have looked only at birth rates. Even under ideal conditions, however, deaths are inevitable. Figure 39-2 compares three bacterial populations experiencing different death rates. Again, the shapes of the curves are the same; the population eventually approaches infinite size (assuming births exceed deaths), but the time required to reach a given population size is increased by increased mortality.

39.2 How Is Population Growth Regulated?

Exponential Growth Cannot Continue Indefinitely

In 1859 Charles Darwin wrote, "There is no exception to the rule that every organic being naturally increases at so high a rate that, if not destroyed, the Earth would

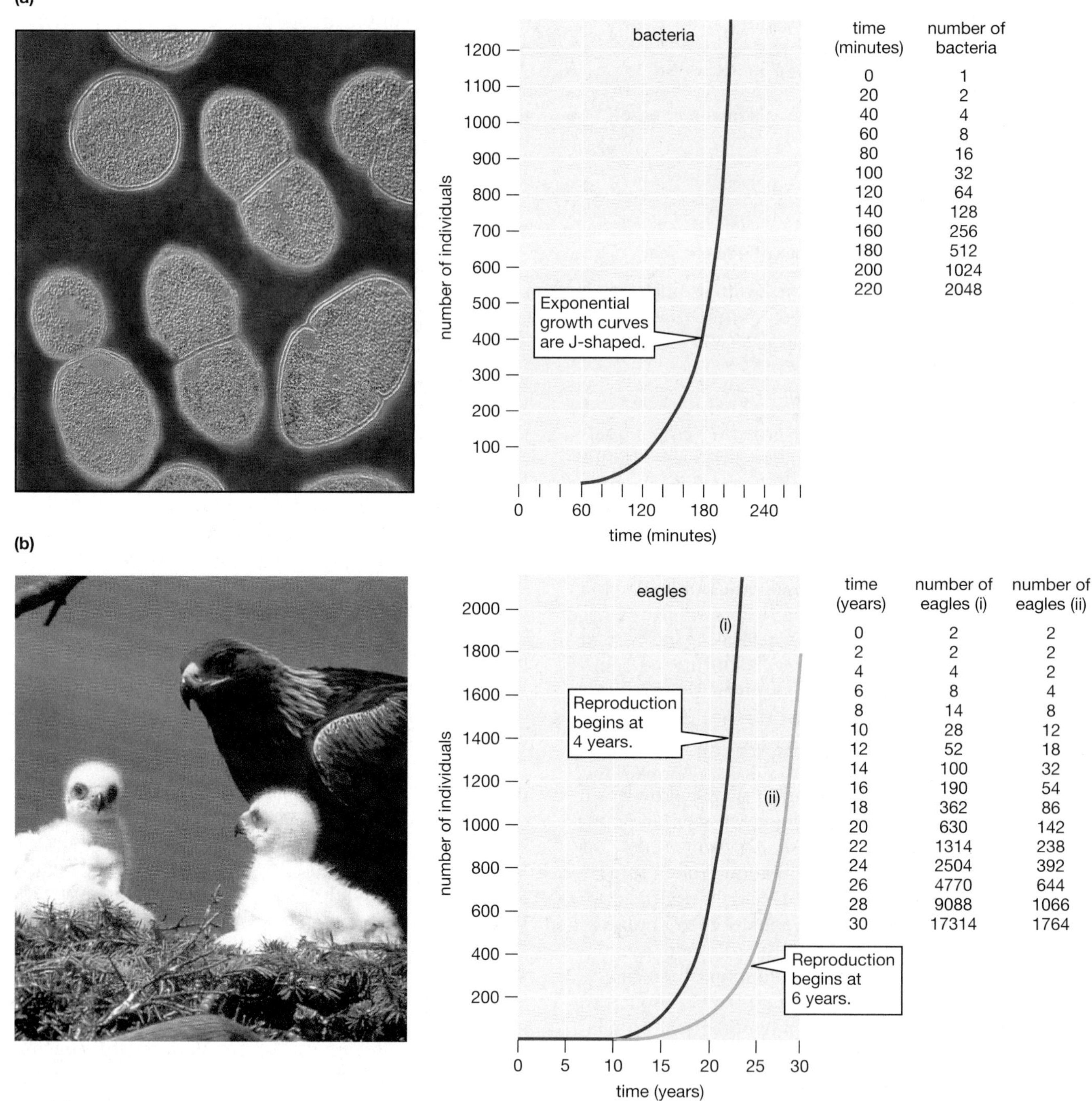

time (minutes)	number of bacteria
0	1
20	2
40	4
60	8
80	16
100	32
120	64
140	128
160	256
180	512
200	1024
220	2048

time (years)	number of eagles (i)	number of eagles (ii)
0	2	2
2	2	2
4	4	2
6	8	4
8	14	8
10	28	12
12	52	18
14	100	32
16	190	54
18	362	86
20	630	142
22	1314	238
24	2504	392
26	4770	644
28	9088	1066
30	17314	1764

FIGURE 39-1 Exponential growth curves are J-shaped
All such curves share a similar J shape; the major difference is the time scale. **(a)** Growth of a population of bacteria, starting with a single individual and with a doubling time of 20 minutes. **(b)** Growth of a population of eagles, starting with a single pair of hatchlings, for ages at first reproduction of 4 years (red line) and 6 years (green line). Notice in the table that after 26 years, the population of eagles that began reproducing at age 4 is nearly seven times that of the eagles that began reproducing at age 6.

soon be covered by the progeny of a single pair." But in nature, exponential growth occurs only under special circumstances and for a limited time. For example, exponential growth is observed in populations that undergo regular cycles in which rapid population growth is followed by a sudden, massive die-off. These **boom-and-bust cycles** occur in a variety of organisms for complex and varied reasons. Many short-lived, rapidly reproducing species—from algae to insects—have seasonal population cycles that are linked to predictable changes in rainfall, temperature, or nutrient availability (Fig. 39-3). In temperate climates, insect populations grow rapidly during the spring and summer, then crash with the killing hard frosts of winter. More complex factors pro-

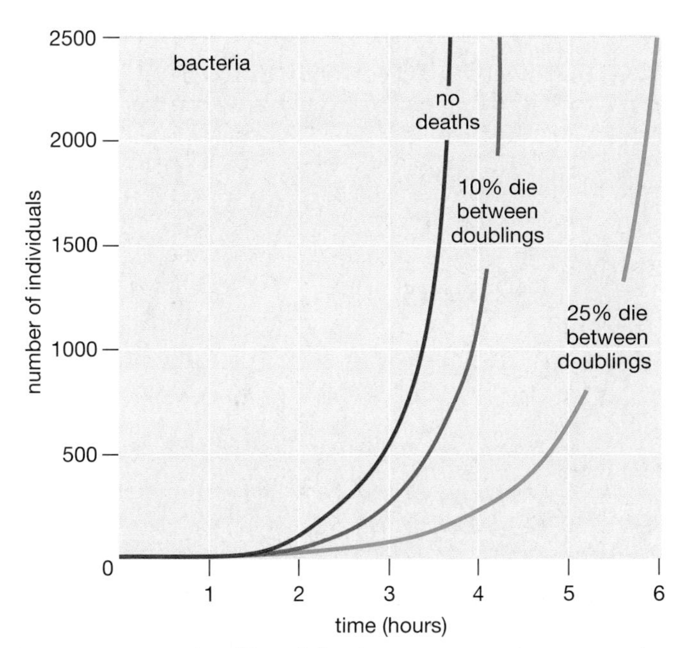

FIGURE 39-2 The effect of death rates on population growth
The graphs assume that a bacterial population doubles every 20 minutes. Notice that the population in which a quarter of the bacteria die every 20 minutes reaches 2500 only 2 hours and 20 minutes later than one in which no deaths occur.
QUESTION What would the death rate need to be for these populations to stabilize?

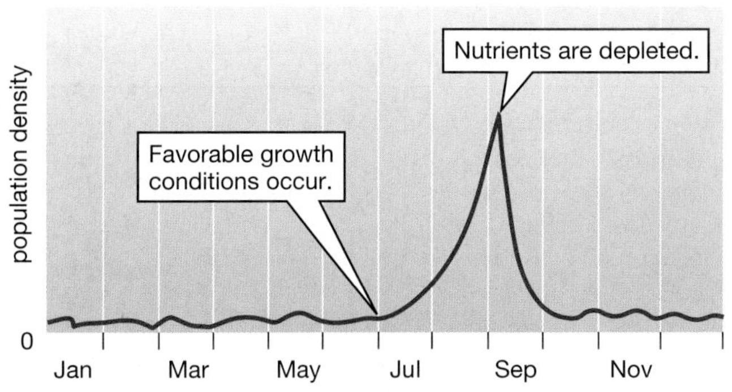

FIGURE 39-3 A boom-and-bust population cycle
Population density of cyanobacteria (blue-green algae) in an annual boom-and-bust cycle in a lake. Algae survive at a low level through the fall, winter, and spring. Early in July, conditions become favorable for growth, and exponential growth occurs through August. Nutrients soon become depleted, and the population "goes bust."

duce roughly 4-year cycles for small rodents such as voles and lemmings, and much longer population cycles in hares, muskrats, and grouse.

Lemming populations, for instance, may grow until the rodents overgraze their fragile arctic tundra ecosystem. Lack of food, increasing populations of predators, and social stress caused by overpopulation may all contribute to a sudden high mortality. Many deaths occur as waves of lemmings emigrate from regions of high population density. During these dramatic mass movements, lemmings are easy targets for predators. Many drown; they begin swimming when they encounter a body of water, including the ocean, but cannot make it all the way across. The reduced lemming population eventually contributes to a decline in predator numbers (see "Scientific Inquiry: Cycles in Predator and Prey Populations") and a recovery of the plant community on which the lemmings normally feed. These responses, in turn, set the stage for the next round of exponential growth in the lemming population (Fig. 39-4).

In populations that do not show boom-and-bust cycles, exponential growth may occur temporarily under special circumstances—for example, if population-controlling factors, such as predators or parasites, are eliminated or if the food supply is increased. This occurs in white-footed mouse populations when oak

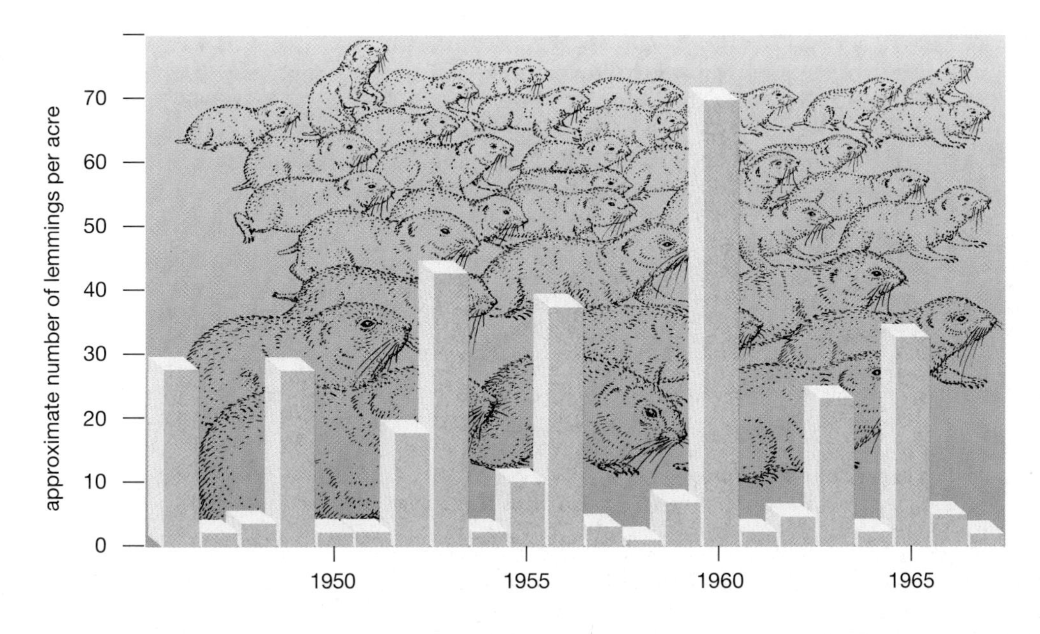

FIGURE 39-4 Lemming population cycles follow a boom-and-bust pattern
Lemming population density follows roughly a 4-year cycle (data from Point Barrow, Alaska).
QUESTION What factors might make the data in this graph somewhat erratic and irregular?

SCIENTIFIC INQUIRY Cycles in Predator and Prey Populations

If we assume that a certain prey species is eaten exclusively by a particular predator, it seems logical that both populations might show cyclic changes, with changes in the predator population size lagging behind changes in the prey population size. For example, a large hare population would provide abundant food for lynx and their offspring, which would then survive in large numbers. The increased lynx population would eat more hares, reducing the hare population. With fewer prey, fewer lynx would survive and reproduce, so the lynx population would decline a short time later.

Does this out-of-phase population cycle of predators and prey actually occur in nature? A classic example of such a cycle was demonstrated by using the ingenious method of counting all the pelts of northern Canada lynx and snowshoe hares purchased from trappers by the Hudson Bay Company between 1845 and 1935. The availability of pelts (which was assumed to reflect population size) showed dramatic, closely linked population cycles of these predators and their prey (Fig. E39-1). However, many uncontrolled variables could have influenced the relationship between hares and lynx. For example, hare populations sometimes fluctuate even without lynx present, possibly because, in the absence of predators, hares overshoot the carrying capacity of their environment and reduce their food supply. Further, lynx do not feed exclusively on hares but can eat a variety of small mammals. Density-independent environmental variables, such as exceptionally severe winters, also could have adversely affected both populations and produced similar cycles. Recently, researchers tested the hare–predator relationship more rigorously by fencing off 1-kilometer-square areas in northern Canada to exclude predators. The crash of the hare population was lessened either by providing extra food or by excluding predators, but by far the greatest success in preventing the crash of the hare population was achieved when researchers excluded predators *and* provided extra food, suggesting that both these factors may contribute to natural boom-and-bust cycles in hares.

To test the predator–prey cycle hypothesis in an even more controlled manner, investigators performed laboratory studies on populations of small predators and their prey. The study illustrated in Figure E39-2 involved tiny braconid wasp predators and their bean weevil prey. Abundant food was supplied to the weevils, no other food was available to the wasps, and other variables were carefully controlled. As predicted, the two populations showed regular cycles, with the predator population rising and falling slightly later than the prey population. The wasps lay their eggs on weevil larvae, which provide food for the newly hatched wasps. A large weevil population ensures a high survival rate for wasp offspring, increasing the predator population. Then, under intense predation pressure, the weevil population plummets, reducing the population size of the next generation of wasps. Reduced predation then allows the weevil population to increase rapidly, and so on.

It is highly unlikely that such a straightforward example will ever be found in nature, but this type of predator–prey interaction clearly can contribute to the fluctuations observed in many natural populations.

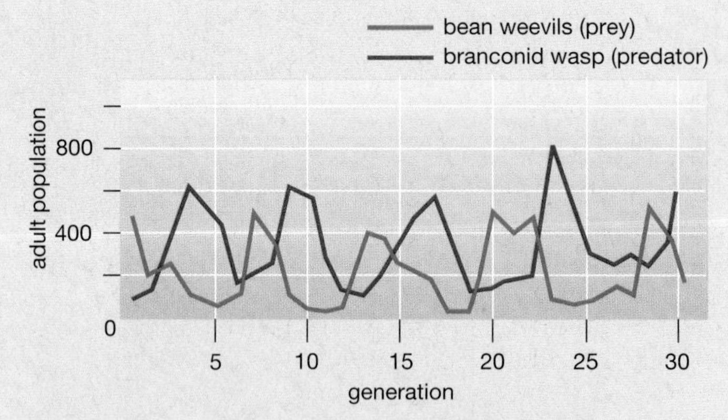

FIGURE E39-2 Experimental predator–prey cycles
Out-of-phase fluctuations in laboratory populations of the bean weevil and its braconid wasp predator.

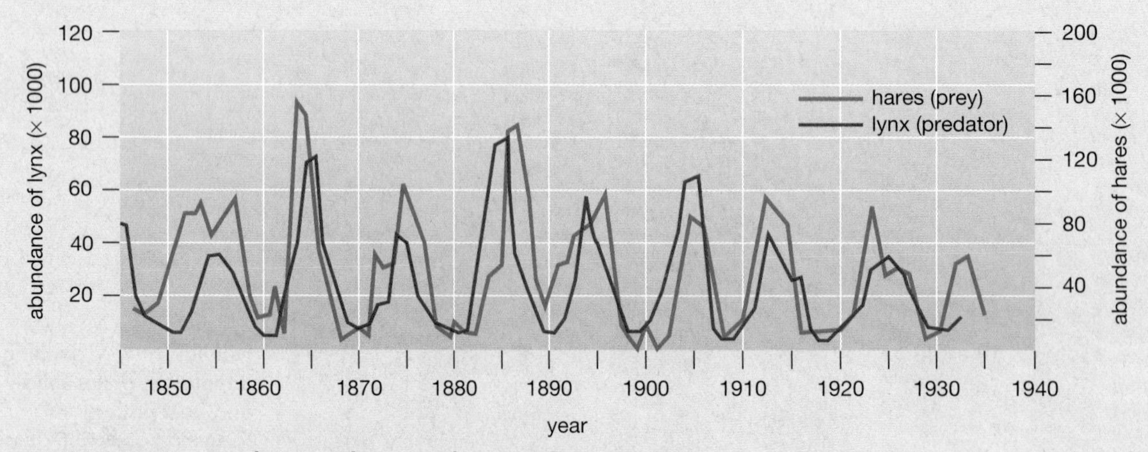

FIGURE E39-1 Population cycles in predators and prey
Populations of snowshoe hares and their lynx predators, graphed on the basis of the number of pelts received by the Hudson Bay Company.

trees produce large acorn crops. Exponential growth can also occur when individuals invade a new habitat with favorable conditions and little competition. For example, if a farm field is plowed, then abandoned, it provides an ideal habitat for weedy annual plants and perennial grasses, whose populations may increase exponentially at first. Invasion of new habitats and subsequent exponential growth also occurs frequently when people introduce foreign, or **exotic**, species into ecosystems, in many cases with damaging results; this is illustrated by the gypsy moth's devastation of oak forests. As you will learn in the next section, all populations that exhibit exponential growth must eventually either stabilize or "crash"—that is, precipitously decrease in size.

Environmental Resistance Limits Population Growth

As individuals join a population, competition for resources intensifies. For example, as plants invade an abandoned farm field and their populations grow, competition for space, water, sunlight, and soil nutrients increases until further expansion is impossible. In response to increases in populations of animals such as prairie dogs, predators such as hawks may increase in number or make this newly abundant prey a larger part of their diet. As crowding increases, parasites and diseases spread more readily, a process enhanced by weakness resulting from lack of food or from stress caused by adverse social interactions that can occur under crowded conditions. Crowded animals might emigrate to establish new populations, die at a younger age, or not reproduce as well. Consequently, after a period of exponential growth, the growth rate gradually declines and the population tends to stabilize at or below the maximum number the environment can sustain. A long-term state of equilibrium results, and the growth rate fluctuates around zero. In this equilibrium, the birth rate is balanced by the death rate, and population size is stabilized. This type of population growth, which is typical of long-lived organisms colonizing a new area, is represented graphically by an S-shaped growth curve, or **S-curve** (Fig. 39-5).

Populations may stabilize at a level at or below the **carrying capacity** of the ecosystem. The carrying capacity is the maximum population size that an ecosystem can sustain. It is determined primarily by the continuous availability of two types of resources. The first type of resources is *renewable*, that is, replenished by natural processes. Renewable resources include food and inorganic nutrients, water, and light (the energy source for plants). The second type of resource is *nonrenewable*; for nonhuman populations, the most important nonrenewable resource is space.

Organisms will starve if demands on renewable resources are too high. If space requirements are exceeded, animals may emigrate, but commonly to less-suitable areas where their death rate will be higher. Reproduction will decline, because animals may not find adequate breeding sites or the seeds of plants may not reach a suitable place to germinate. If a population exceeds the carrying capacity of its environment, excess demands on resources may damage ecosystems. This damage can reduce the carrying capacity and thus the ability of the ecosystem to sustain the population. As a result, the population either declines until the ecosystem recovers or the population is permanently reduced. For example, overgrazing by cattle on dry western grasslands has reduced the grass cover and allowed sagebrush (which cattle will not eat) to thrive. Once established, sagebrush replaces edible grasses and reduces the land's carrying capacity for cattle (see Fig. 42-18). Other dramatic cases of overgrazing have occurred

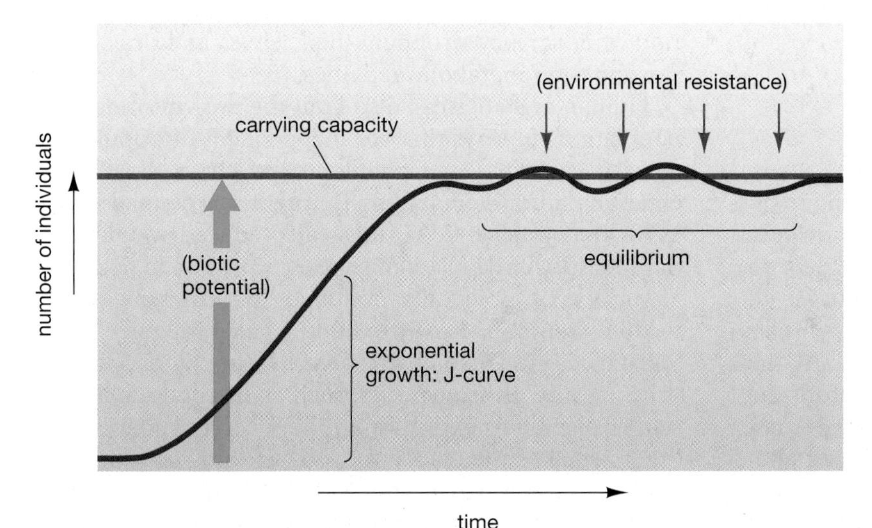

FIGURE 39-5 The S-curve of population growth
The population grows exponentially at first, then fluctuates around the carrying capacity. The growth is driven by biotic potential, but levels off owing to environmental resistance.

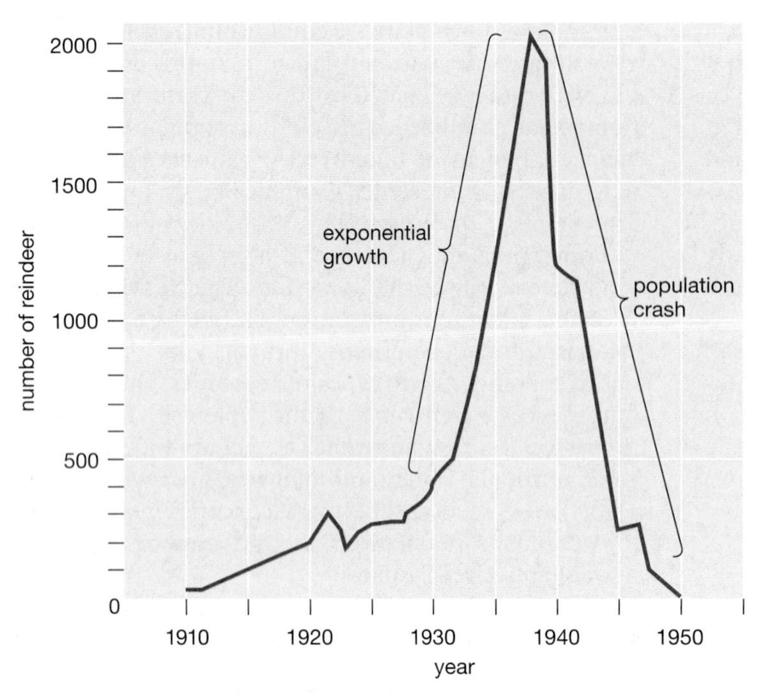

FIGURE 39-6 The effects of exceeding carrying capacity
Exceeding carrying capacity can damage an ecosystem, reducing its ability to support the population. In 1911, 25 reindeer were introduced onto one of the Pribilof Islands (St. Paul) in the Bering Sea off Alaska. Food was plentiful, and the reindeer encountered no predators on the island. The herd grew exponentially (note the initial J shape) until it reached 2000 reindeer in 1939. At this point, the small island was seriously overgrazed, food was scarce, and the population declined dramatically. By 1950, only eight reindeer remained.

when herbivores, such as reindeer, have been introduced onto islands without large predators (Fig. 39-6).

In nature, environmental resistance maintains populations at or below the carrying capacity of their environment. Factors of environmental resistance can be classified into two broad categories. **Density-independent** factors limit population size regardless of the population density (number of individuals per given area). **Density-dependent** factors increase in effectiveness as the population density increases. Note that food and space, the primary determinants of carrying capacity, are both density-dependent regulators of population size. In the following sections, we will look more closely at how population growth is controlled.

Density-Independent Factors Limit Populations Regardless of Their Density

Perhaps the most important natural density-independent factors are climate and weather. Hurricanes, droughts, floods, and fire can have profound effects on local populations, particularly those of small, short-lived species, regardless of population density. Many insects and annual plant populations are limited in size by the number of individuals that can be produced before the first hard freeze. Such populations typically do not reach the carrying capacity of their environment because density-independent factors intervene first. Weather is largely responsible for the boom-and-bust population cycles described earlier, and can cause significant variations within natural populations from year to year.

Organisms with a life span of one year or more have evolved various mechanisms to compensate for seasonal changes in weather, thereby circumventing these weather-related, density-independent population control factors. Many mammals, for example, develop thick coats and store fat for the winter; some also hibernate. Other animals, including many birds, migrate long distances to find food and a hospitable climate. Many trees and bushes survive the rigors of winter by entering a period of dormancy, dropping their leaves and drastically slowing their metabolic activities.

Human activities can also limit the growth of natural populations in ways that are independent of population density. Pesticides and pollutants can cause drastic declines in natural populations. Before it was banned in the 1970s, the pesticide DDT drastically reduced populations of predatory birds, including eagles, ospreys, and pelicans; a variety of pollutants continue to adversely affect wildlife (see "Earth Watch: Food Chains Magnify Toxic Substances" in Chapter 41). Overhunting by people has driven entire animal species, such as the dodo and passenger pigeon, to extinction, while severely endangering others, such as the black rhino. Overfishing continues to deplete once-abundant populations of nearly all commercially fished species, as discussed later in this chapter.

Density-Dependent Factors Become More Effective as Population Density Increases

For long-lived species, by far the most important elements of environmental resistance are density-dependent factors. Because they become increasingly effective as population density increases, density-dependent factors exert a negative feedback effect on population size. The larger the population grows, the more such factors are triggered that counteract this growth, and the more effective they become. Density-dependent factors include community interactions, such as predation and parasitism, as well as competition within the species or with members of other species. These factors are discussed below and in Chapter 40.

Predators and Parasites Exert Density-Dependent Controls on Populations

Both *predation* and *parasitism* involve one organism feeding on another and harming it in the process. Although the distinction is not always clear-cut, predation typically occurs when one organism, the **predator**, kills another, its **prey**, in order to eat it. Parasitism occurs when one organism, the **parasite**, lives on another, its **host** (usually a much larger organism) and feeds on the host's body without killing it—at least not immediately. While predators must kill their prey to feed, parasites benefit by having their hosts remain alive.

Predation includes wolves working together to kill an elk (Fig. 39-7), mice munching on acorns or gypsy moth pupae, and a Venus flytrap plant digesting an insect (described in Chapter 26). Predation becomes an increasingly important factor in population control as prey populations increase, simply because the more abundant the prey, the more often predators encounter them. Many predators will eat a variety of prey, depending on what is most abundant and easiest to find. Coyotes might eat more field mice when the mouse population is high but switch to eating more ground squirrels as the mouse population declines, thus allowing the mouse population to recover.

Predators can also exert density-dependent effects on their prey by increasing in number as their prey population grows. For instance, predators such as the arctic fox and snowy owl, which feed heavily on lemmings, regulate the number of offspring they produce according to the abundance of lemmings. The snowy owl might produce up to 13 chicks when lemmings are abundant but not reproduce at all in years when lemmings are scarce. In some cases, an increase in predators might cause a crash of the prey population, which in turn may result in a decline in the predator population. This pattern can result in out-of-phase **population cycles** of both predators and prey (see "Scientific Inquiry: Cycles in Predator and Prey Populations").

Some predators feed primarily on prey made vulnerable because their populations have exceeded the carrying capacity of their environment. Such prey may be weakened by lack of food or may be exposed because they cannot find appropriate shelter. In such cases, predation may maintain prey populations near a density that can be sustained by the resources of the ecosystem. In other cases, predators may maintain their prey at well below carrying capacity. A dramatic example of this phenomenon is the prickly pear cactus, which was introduced into Australia from Latin America. Lacking natural predators, its population grew exponentially and it spread uncontrollably, destroying millions of acres of valuable pasture and range land. Finally, in the 1920s, a cactus moth (a predator of the prickly pear) was imported from Argentina and released to feed on the cacti. Within a few years, the cacti were virtually eliminated. Today the moth continues to maintain its prey cacti at very low population densities, well below the carrying capacity of the ecosystem.

FIGURE 39-7 Predators help control prey populations
A pack of grey wolves has brought down an elk, who may have been weakened by old age or parasites.

In contrast to a predator, a parasite feeds on a larger organism without killing it immediately or directly. Examples include any organism that causes disease, including some bacteria (such as those causing Lyme disease), fungi, intestinal worms, ticks, and protists such as the malaria parasite. Insects that feed on plants without necessarily killing them—such as the gypsy moth—are also parasites. Parasitism is density-dependent. Most parasites have limited motility and spread more readily among hosts when their host population density is high. For example, plant diseases and pest insects spread readily through acres of densely planted crops, and childhood diseases spread rapidly through schools and daycare centers. Parasites influence population size by weakening their hosts and making them more susceptible to death from other causes. Although a parasite does not benefit from the death of its host, its activities may nonetheless kill the host; malaria, for example, kills over 1 million people annually. Infection by parasites can also contribute to deaths by predation, in that the parasites make the host organisms weaker and more vulnerable to predators.

Both predators and parasites can have beneficial effects on the prey population as a whole. Predators, parasites, and their prey *coevolve*; that is, they evolve together, each acting as an agent of natural selection on the other. Parasites and predators tend to destroy the least fit of the prey, leaving the better-adapted prey to reproduce. This results in a balance in which the prey population is regulated but not eliminated. The population balance in ecosystems can be destroyed when predators or parasites are introduced into regions in which they did not evolve, and in which local prey species have had no opportunity to evolve defenses against them through natural selection. An example is the prickly pear cactus in Australia, mentioned earlier (this topic will be discussed further in Chapter 40). The smallpox virus, inadvertently carried by traveling Europeans, caused heavy losses of life among native inhabitants of the continental U.S., Hawaii, and South America, and among the aborigines of Australia. In addition, introduced rats, snakes, and mongooses have exterminated many of Hawaii's native bird populations.

Competition for Resources Helps Control Populations

The resources that determine carrying capacity (space, nutrients, water, and light) are often limited relative to the demand for them. In other words, there may not be enough resources to support all the organisms that are produced. **Competition**, defined as the interaction among individuals who attempt to utilize a limited resource, limits population size in a density-dependent manner. There are two major forms of competition: **interspecific competition** (competition among individuals of different species; described further in Chapter 40), and **intraspecific competition** (competition among individuals of the same species). Because the needs of members of the same species for water and nutrients, shelter, breeding sites, light, and other resources are almost identical, intraspecific competition is more intense than interspecific competition.

Organisms have evolved several ways to deal with intraspecific competition. Some organisms, including most plants and many insects, engage in **scramble competition**, a kind of free-for-all with resources as the prize. For example, when a plant disperses its seeds in a small area, hundreds may germinate. However, as they grow, plants that germinated first begin to shade the smaller ones, while their more extensive root systems absorb most of the water; those that germinated later wither and die.

Many animals (and even a few plants) have evolved **contest competition**, which involves social or chemical interactions used to limit access to important resources. Territorial species—such as wolves, many fish, rabbits, and songbirds—defend an area that contains important resources such as food or sites adequate for raising young. When the population exceeds the available resources, only the best-adapted individuals are able to defend territories that adequately support their need for food and shelter. Those without territories may not reproduce (reducing the future population), or they may fail to obtain adequate food or shelter and become easy prey for predators.

As population densities increase and competition becomes more intense, some animals react by emigrating. Large numbers leave their homes to colonize new areas, and many, sometimes most, die in the quest. For example, the mass movements of lemmings, which in some instances end in marches into the sea, are apparently in response to overcrowding. Emigrating swarms of locusts plague the African continent, stripping all vegetation in their path (Fig. 39-8).

FIGURE 39-8 Emigration
In response to overcrowding and lack of food, locusts emigrate in swarms, devouring all the vegetation in their path.
QUESTION What benefits does mass emigration give to animals such as locusts or lemmings? Can you see any parallels in human emigration?

The size of a population at any given time is the result of complex interactions between density-independent and density-dependent forms of environmental resistance. For example, a stand of pines weakened by drought (a density-independent factor) may more readily fall victim to the pine bark beetle (a density-dependent parasite). Likewise, a herd of caribou weakened by hunger (density-dependent) and attacked by parasites (density-dependent) is more likely to be killed by an exceptionally cold winter (a density-independent factor). The demands of the growing human population are destroying the natural habitat of many animals, radically reducing their population size. Although bulldozing prairie dog towns to build shopping malls and cutting rain forest for agriculture reduces animal populations in a density-independent fashion, the end result is a reduced carrying capacity of the environment, which in turn exerts density-dependent limits on future population sizes.

Survivorship in Populations Follows Three Basic Patterns

Over time, populations show characteristic patterns of deaths or (more optimistically) survivorship. These patterns, called **survivorship curves**, are revealed when the number of individuals of each age is graphed against their age. Three types of survivorship curve—described as "late loss," "constant loss," and "early loss," according to the part of the life cycle during which most deaths occur—are shown in Figure 39-9. Survivorship curves reflect the number of offspring produced and the amount of parental care and protection that the offspring receive.

Late-loss populations produce *convex* survivorship curves. Such populations have relatively low juvenile death rates, and most individuals survive to old age. Late-loss survivorship curves are characteristic of humans and many other large and long-lived animals, such as Dall mountain sheep. These species produce relatively few offspring, which are protected by their parents during early life.

Populations with constant-loss survivorship curves have a fairly constant death rate; their survivorship graphs appear as a relatively straight line. In these populations, individuals have an equal chance of dying at any time during their life span. This phenomenon is seen in some birds such as gulls and the American robin, and in laboratory populations of organisms that reproduce asexually, such as hydra and bacteria.

Early-loss survivorship produces a *concave* curve and is characteristic of organisms that produce large numbers of offspring. These offspring receive little parental care; they are largely left to compete on their own. The death rate is initially very high among the offspring, but those that reach adulthood have a good chance of surviving to old age. Most invertebrates, most plants, and many fish exhibit such early-loss survivorship curves. Even some mammalian populations have early-loss survivorship curves: In some populations of black-tailed deer, 75% of the population dies within the first 10% of its life span.

FIGURE 39-9 Survivorship curves
Three types of survivorship curve are shown. Because the life spans differ, the percentages of survivors (rather than ages) are used.

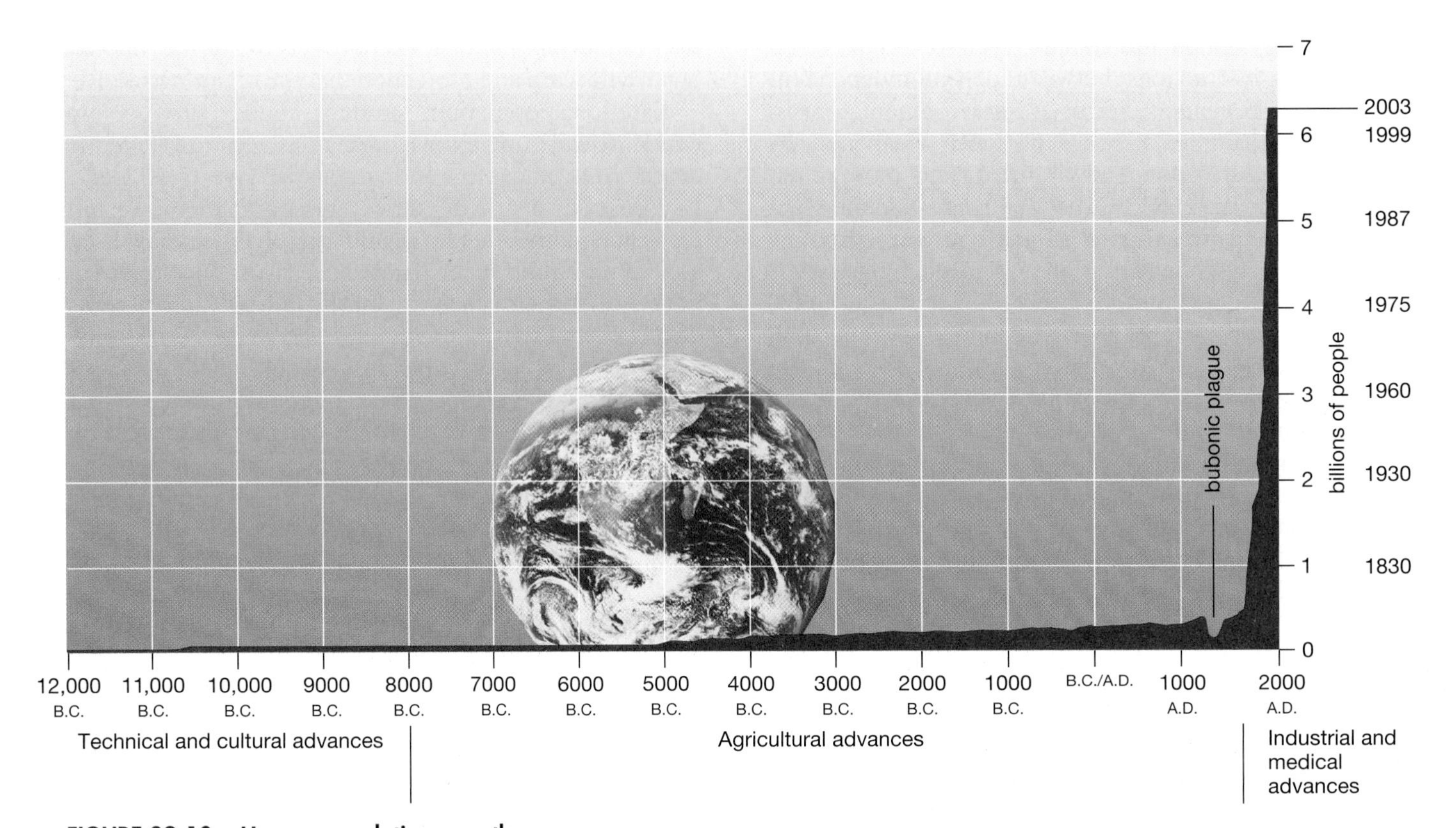

FIGURE 39-10 Human population growth
The human population from the Stone Age to the present has shown continued exponential growth as various advances overcame environmental resistance. Note the dip in the fourteenth century caused by the bubonic plague. Note also the steadily decreasing time intervals over which additional billions were added. (Inset) Earth is an island of life in a sea of emptiness; its space and resources are limited. **QUESTION** The human population continues to grow rapidly, but evidence suggests we have already exceeded Earth's carrying capacity at current levels of technology. What do you think this curve will look like when we reach the year 2500? 3000? Explain.

Media Activity
39.2 Human Population Growth and Regulation

39.3 How Is the Human Population Changing?

The Human Population Is Growing Rapidly

Compare the graph of human population growth in Figure 39-10 with the exponential growth curves in Figure 39-1. The time spans are different, but each has the J-shape characteristic of exponential growth. It took more than 1 million years for the human population to reach 1 billion, but the second billion was added in just 100 years, the third billion in 30 years, the fourth billion in 15 years, and the fifth and sixth billion were each added in 12 years. The fact that the sixth billion was not added in a shorter time than the fifth billion suggests that we may no longer be growing exponentially. Are we starting to enter the final bend of the S-shaped growth curve shown in Figure 39-5? No one knows. However, Earth's human population (currently over 6.3 billion) grows by about 1.3% or 82 million yearly—more than 224,000 people are added every day, nearly 1.6 million every week! Why hasn't environmental resistance put an end to our continued growth? What is the carrying capacity of the world for humans?

Like all populations, humans have encountered environmental resistance, but unlike nonhuman populations, we have responded to environmental resistance by devising ways to overcome it. As a result, the human population has grown for an unprecedented time span. To accommodate our growing numbers, we have altered the face of the globe. Human population growth has been spurred by a series of "revolutions," each of which circumvented environmental resistance and increased Earth's carrying capacity for people.

Technological Advances Have Increased Earth's Carrying Capacity for Humans

Primitive people produced a *technical and cultural revolution* when they discovered fire, invented tools and weapons, built shelters, and designed protective clothing. Tools and weapons allowed them to hunt more effectively and increased their food supply, while shelter and clothing increased the habitable areas of the globe.

Domesticated crops and animals supplanted hunting and gathering by about 8000 B.C. This *agricultural revolution* provided people with a larger, more dependable food supply, further increasing Earth's carrying capacity for humans. An increased food supply resulted in a longer life span and more childbearing years, but a high death rate from disease still restrained population growth.

Human population growth continued slowly for thousands of years until the *industrial–medical revolution* began in England in the mid-eighteenth century, spreading throughout Europe and North America in the nineteenth century. Medical advances dramatically decreased the death rate by reducing environmental resistance caused by disease. These advances included the discovery of bacteria and their role in infection, which led to the control of bacterial diseases through improved sanitation and the use of antibiotics. Viruses were also discovered, leading to the development of vaccines for diseases such as smallpox.

Today, the countries of the world are often described as either *developed* or *developing*. People in developed countries (including North America, Europe, Australia, New Zealand, and Japan) benefit from a relatively high standard of living, including access to modern technology and medicine (including contraception). Income is relatively high; education and employment opportunities are available to both sexes; and death rates from infectious diseases are relatively low. Yet fewer than 20% of the world's inhabitants live in developed countries. The majority of the people in developing countries (Central and South America, much of Asia, and Africa) lack many of these advantages.

In developed countries, the industrial–medical revolution resulted in an initial rise in population due to a decrease in death rates, which was followed by a decline in birth rates, resulting in a relatively stable population. This changing population dynamic, in which a stable population experiences rapid growth and then returns to a stable (although much larger) size, is called a **demographic transition**:

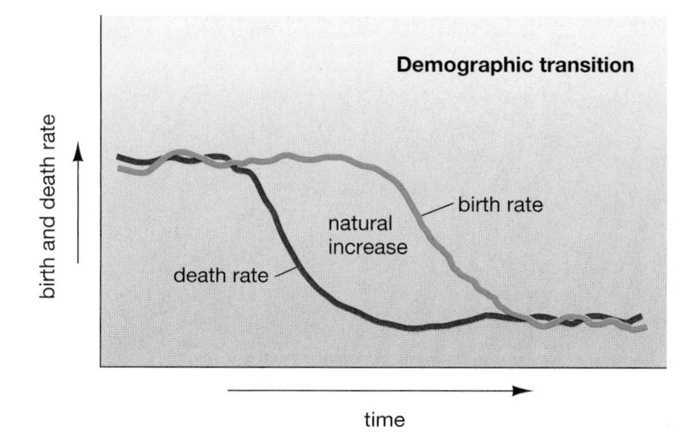

The decline in birth rates that ends the demographic transition is a result of many factors, including better education, increased availability of contraceptives, a shift to a primarily urban lifestyle, and employment of more women outside the home. In most developed countries, populations have more or less stabilized, and in some cases they are declining as people choose to have smaller families.

In developing countries, such as most of Central and South America, Asia (excluding China and Japan), and Africa (excepting those African countries devastated by the AIDS epidemic), medical advances have decreased death rates and increased the life span, but birth rates remain relatively high. These countries have begun but have not yet completed the demographic transition, in part because they have not benefited from the increase in wealth that has contributed to the decline in birth rates in developed countries. In developing nations, children may be the only support for elderly parents, may contribute significantly to the labor force, and can also be a source of social prestige. In Nigeria, the most populous country in Africa, only 9% of women use modern contraceptive methods; the average woman bears 5.8 children. Nigeria is suffering from loss of forests, soil erosion, and water pollution. With nearly half (44%) of its population of 134 million under age 15, continued population growth is a certainty.

Ironically, population growth in developing countries helps perpetuate the poverty and lack of education that tend to sustain high birth rates. The relationship of income and education to birth rate has been documented in the United States. Here, on the average, women who do not complete high school have twice as many children as do those with more than 4 years of college. Of the more than 6.3 billion people on Earth today, over 5 billion reside in developing countries. Encouragingly, birth rates in some developing countries have begun to decline and approach **replacement-level fertility (RLF)**, in which the adults of reproductive age have just enough children to replace themselves. Because not all children survive to maturity, RLF is slightly higher than 2. This decline has occurred because some governments are taking steps to encourage smaller families and increase access to contraceptive methods. But the prospects for population stabilization—*zero population growth*—in the near future are nil. For the year 2050, the United Nations predicts that the population will be approximately 8.9 billion and growing, with 7.7 billion people living in the developing nations:

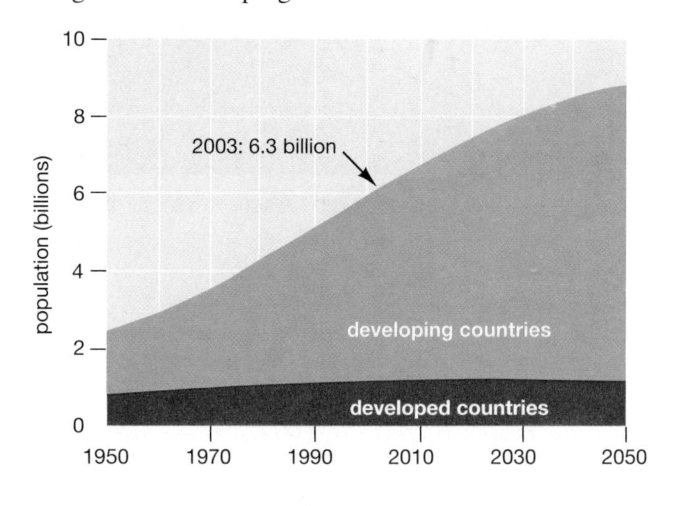

The Age Structure of a Population Predicts Its Future Growth

The **age structure** of a population is the distribution of males and females of each age group; it can be shown graphically in an age-structure diagram, in which the vertical axis represents age and the horizontal axis shows numbers of individuals in each age group, with males and females graphed on opposite sides. Age-structure diagrams all rise to a peak that reflects the maximum human life span, but the shape of the rest of the diagram reveals at-a-glance whether the population is expanding, stable, or shrinking. The shape is determined by the relative numbers of reproductive age adults (ages 15 to 45) and their children (ages 0 to 14). If the adults of reproductive age have just enough children to replace themselves, the population is at RLF. The age structure diagram of a population which has been at RLF for many years will have relatively straight sides. If the number of children exceeds the number of reproductive-age adults, the population is above RLF and is expanding; its age structure will have a pyramidal shape. In shrinking populations, there are fewer children than reproducing adults and the age-structure narrows at the base.

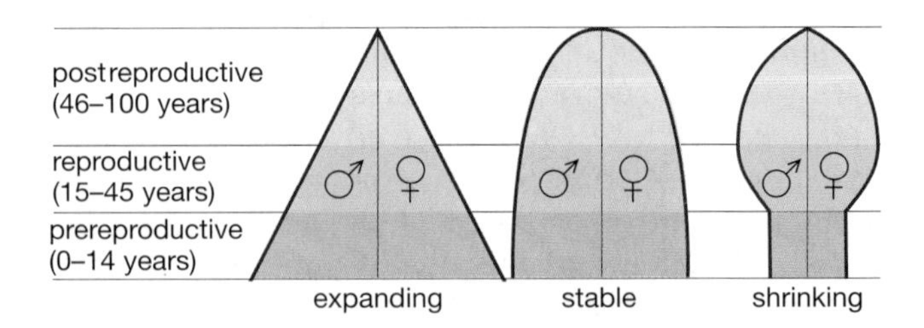

Figure 39-11 shows the average age structures for the populations of developed and developing countries. The

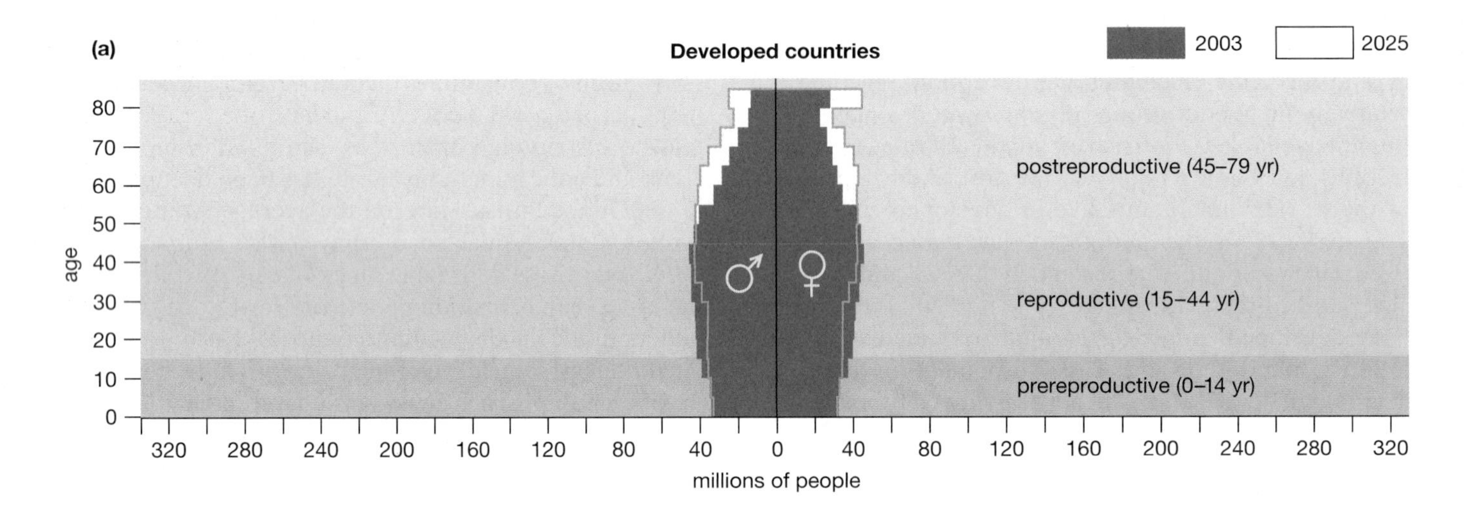

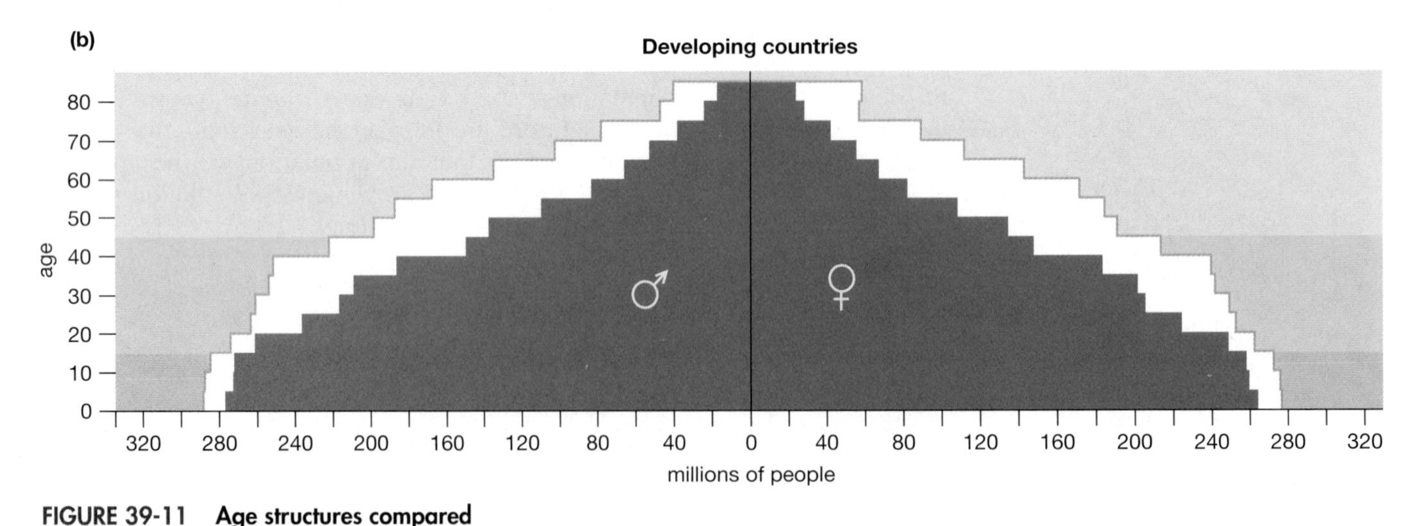

FIGURE 39-11 Age structures compared
(a) Developed countries and **(b)** developing countries. Compare these with the stylized diagrams for expanding and stable populations shown in the text. Notice that the predicted excess number of children over parents in developing countries is smaller in 2025 than in 2003, as these populations approach RLF. But as huge numbers of young people enter childbearing years, growth will continue. (*Note:* The uppermost bars include all individuals 80 years of age and above.) *Data provided by the U.S. Census Bureau.* **QUESTION** How does a fertility rate above RLF create a positive feedback effect in population growth?

outermost boundaries represent the projected population structure for the year 2025; the inner ones are actual values for 2003. In 2003, developing countries had an average annual *natural increase* (population growth based on the births minus deaths; this does not include migration) of 1.6%. In contrast, developed countries showed an average annual natural increase of 0.1%. Today, 97% of the growth of the human population occurs in the developing countries, where increasing numbers of people enter their reproductive years and give birth to an ever-increasing "base" of infants each year. Even if these countries were to reach RLF immediately, their population growth would continue for decades; the children of the large families of the recent past create a built-in momentum for population growth as they enter their reproductive years and begin having children—even if they have only 2 children each. This momentum fuels China's population growth of 0.6% annually, even though China's fertility rate of 1.7 is well below RLF. While less than 20% of a stable population is in the prereproductive age group, in many African nations well over 40% of the total population is younger than 15.

Figure 39-12 illustrates growth rates for various world regions. In Europe, the average annual change in population is −0.1%, with an average fertility rate of 1.4, substantially below RLF, as many women delay or forego having children for a variety of reasons relating both to family economics and lifestyle. This raises governmental concerns about the availability of future workers and taxpayers to support the increasing percentage of elderly people. Several European countries are considering incentives for couples to have children at an earlier age, which shortens the generation time and increases population. Meanwhile, Japan's 127 million people (about 43% of the entire U.S. population) inhabit an area about the size of Montana. Yet despite this tremendous population density, the Japanese government is concerned about their low fertility rate of 1.3 and provides a variety of subsidies that encourage larger families. Although a reduced and eventually stable population will ultimately offer tremendous benefits for both people and the biosphere that sustains them, current economic structures in countries throughout the world are based on growing populations. The difficult adjustments required by stabilizing or declining populations lead governments to adopt policies that encourage more childbearing and continued growth.

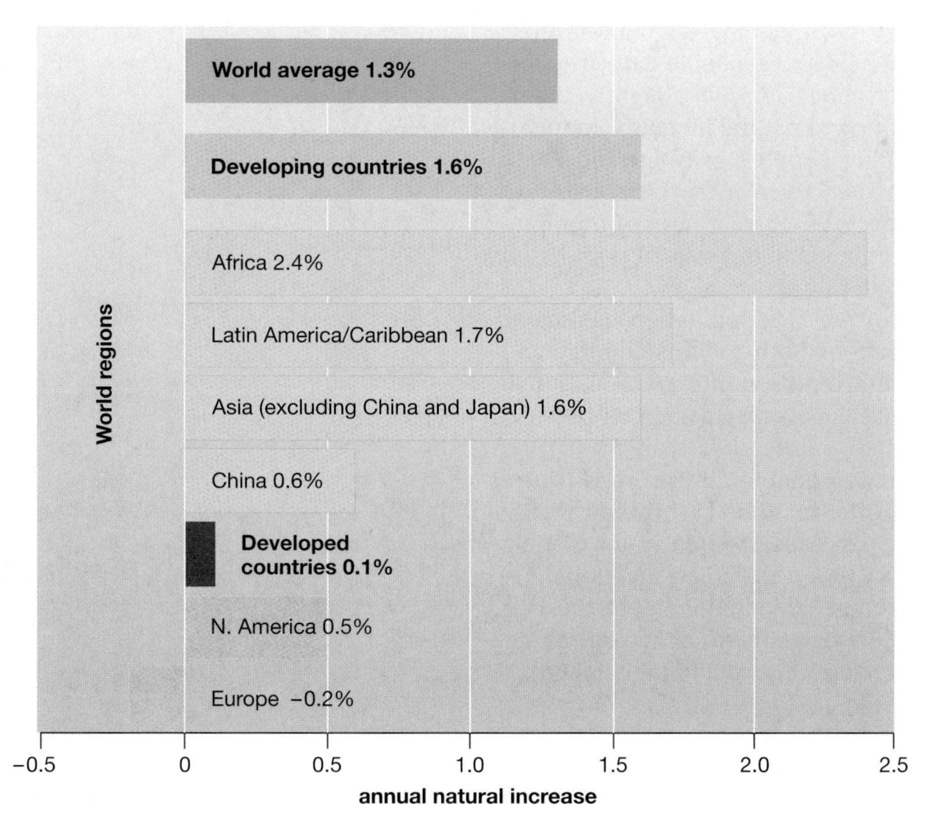

FIGURE 39-12 Population change by world regions
Growth rates shown are due to natural increase (births − deaths) expressed as the percentage increase per year for various regions of the world. These figures do not include immigration or emigration. *Data from the Population Reference Bureau's "World Population Data Sheet, 2003."* **QUESTION** Why are there such big differences between developed and developing countries?

EARTH WATCH Have We Exceeded Earth's Carrying Capacity?

In Côte d'Ivoire (Ivory Coast), a small country in West Africa, the government is waging a battle to protect some of its rapidly dwindling tropical rain forest from thousands of illegal hunters, farmers, and loggers. Officials burn the homes of the squatters, who immediately return and rebuild. One illegal resident is Sep Djekoule. "I have 10 children and we must eat," he explains. "The forest is where I can provide for my family, and everybody has that right." His words illustrate the conflict between population growth and environmental protection, between the desire to have many children and the ability to provide for them using Earth's finite resources. The middle level United Nations projection is that the human population will reach 8.9 billion by the year 2050, and will still be increasing. How many people can Earth support?

Humans have already increased carrying capacity by the use of technology. We have the potential to increase it further while reducing our destructive impact, for example by developing higher-yield crops, reducing erosion, conserving energy and water, reducing manufacturing wastes, and recycling far more paper and plastic, and the metals that we currently mine and discard. However, our ability to reproduce far exceeds our ability to increase Earth's carrying capacity.

Recently, a group of 11 scientists from around the world published a paper in the prestigious *Proceedings of the National Academy of Sciences*. In this rigorous assessment of humanity's impact on global ecosystems, the researchers compared the resource needs of the global human population with the ability of global ecosystems to provide them. They identify six types of

The United States Population Is Growing Rapidly

With a population of over 290 million, the United States (Fig. 39-13) is the fastest-growing developed country in the world. The natural increase of 0.6% is six times the average rate of developed countries. Between 2002 and 2003, the U.S. grew by 1.4%, adding 4.1 million more people: an increase of over 11,000 people daily. The fertility rate is currently about 2.03, just slightly below RLF (2.1). But in addition to natural increase, immigration adds about 1.1 million people legally and 350,000 to 500,000 illegally each year. Even if the U.S. birth rate falls substantially below RLF, current rates of immigration will guarantee continued high levels of population growth for the indefinite future.

The rapid growth of the U.S. population has major environmental implications both for this country and for the world. The average person in the U.S. uses five times as much energy as the average person worldwide; although U.S. residents comprise less than 5% of the world's population, they account for 25% of world energy use (see "Links to Life: Treading Lightly—How Big Is Your "Footprint"?"). Because people in developing countries have smaller "footprints" than the global average, the 4 million people added annually to the U.S. population have over 2.5 times the total ecological impact of the 18 million people added annually to India's population.

When and how will human numbers ever stabilize? How many people can Earth support? There are no certain answers to these questions, but in "Earth Watch: Have We Exceeded Earth's Carrying Capacity?" we explore them in more detail.

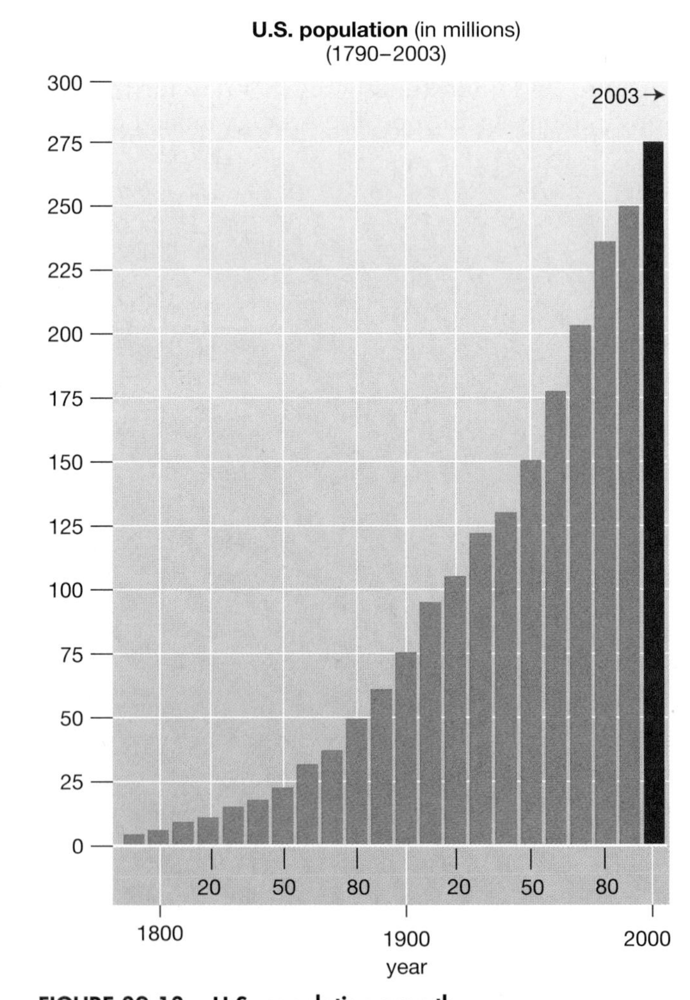

FIGURE 39-13 U.S. population growth
Since 1790, U.S. population growth has produced a J-shaped curve similar to that seen for exponential growth. Each bar reflects the date below it. **QUESTION** At what stage of the S-curve is the U.S. population? What factors do you think will cause it to stabilize, and when?

human activities that place demands on Earth's biological resources: growing crops, grazing livestock, fishing, harvesting timber, occupying space (for housing, roads, and industry), and burning fossil fuels (which places demands on the biosphere to absorb CO_2). The authors calculated the amount of biologically productive space needed to supply the demands of an average person at current levels of technology, an area they call an **ecological footprint**. They provide compelling evidence that the collective ecological footprint of Earth's 6.3 billion people already exceeds the available biologically productive space (4.7 acres per person) by about 20%. In other words, these calculations suggest that Earth's carrying capacity has already been considerably exceeded by the human population. Unfortunately, these estimates are conservative; they do not take into account the depletion of nonrenewable fossil fuels and underground freshwater stores, or the need to leave significant portions of the biosphere untouched to provide habitat for species other than people. Currently, only 3% of land has been set aside in such preserves, and ecologists estimate that we may need 10 times that much to adequately protect other species. When carrying capacity is exceeded, the biosphere is damaged and its ability to sustain the population (in this case, people) is reduced. There is compelling evidence that this is already occurring.

Stanford University biologist Peter Vitousek estimates that human activities have already reduced the productivity of Earth's forests and grasslands by 12%. Each year, overgrazing and deforestation further decreases the productivity of land, especially in developing countries (Fig. E39-3). In a world where nearly 780 million people are chronically undernourished (Fig. E39-3, inset), two-thirds of the world's agricultural land is suffering moderate to severe erosion. The quest for more agricultural land is leading to deforestation and attempts to farm land that is poorly suited for agriculture. These actions contribute to the destruction of tens of millions of acres of rain forest annually. As a result, some ecologists estimate that we are driving 50 to 150 undescribed species to extinction each day. Each year, the U.S. loses nearly half a million acres of productive farmland to development for homes, shopping malls, and roads. Already, most countries must import the grain they need from countries such as the U.S. As our growing population spreads onto our farmland, our future ability to export grain to help sustain other nations will diminish, while their needs increase. Worldwide, the amount of cropland per person has declined by half in the past 50 years.

In many developing countries, water supplies are badly polluted and underground water supplies, called *aquifers*, are depleted and not replaced. Both India and China are rapidly depleting aquifers to supply the needs of their growing cities and to irrigate their cropland. Currently, about 1.5 billion people lack access to safe drinking water, and this number is expected to double in the next 25 years.

The demand for wood in developing countries causes large areas to be deforested annually. This leads to erosion of precious topsoil, runoff of much-needed fresh water, and a decline in the ability of the land to regrow forests or support crops. The total world fish harvest peaked in the late 1980s and has been gradually declining since then, despite increased investments in fishing equipment, improved technology for finding fish, and increased harvests of smaller and less-desirable fish species. Almost 70% of commercial ocean fish populations have been fully exploited or overfished, and many formerly abundant fish populations, such as cod harvested off New England, Canada, and in the North Sea, have collapsed because of overfishing. Our present population, at its present level of technology, is clearly "overgrazing" the world ecosystem. As the 5 billion people of less-developed countries strive to increase their standard of living, the damage to Earth's ecosystems accelerates.

In estimating how many people Earth can—or should—support, it is important to recognize that people desire more from life than a minimum number of food calories daily. The standard of living in developed countries is already an unattainable luxury for most of Earth's inhabitants. Inevitably, the human population must stop growing. Either we must voluntarily reduce our birth rates, or various forces of environmental resistance, including disease and starvation, will eventually dramatically increase human death rates; the choice is ours. Hope for the future lies in recognizing the signs of "human overgrazing" and acting to reduce our population before we decimate our biodiversity and irrevocably damage the biosphere.

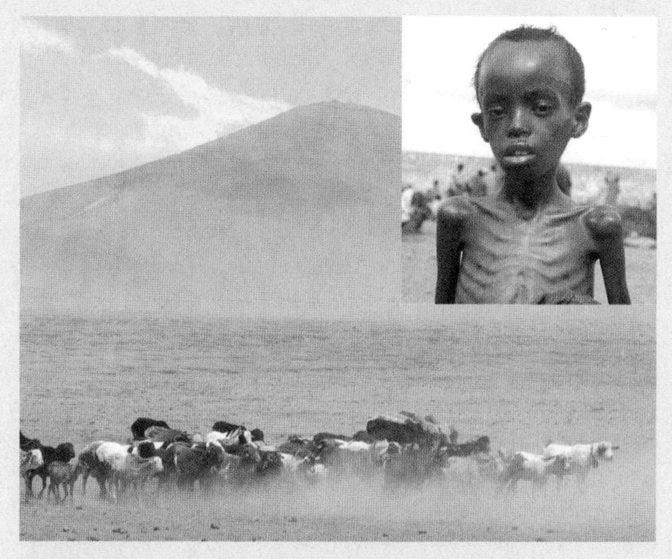

FIGURE E39-3 Deforestation can lead to the loss of productive land
Human activities, including overgrazing, deforestation, and poor agricultural practices, reduce the productivity of the land. (Inset) An expanding human population, coupled with a loss of productive land, can lead to tragedy.

CASE STUDY REVISITED Acorns, Mice, Moths, Deer, and Disease

From their carefully designed study, the ecologists concluded that an abundant acorn crop led to an enormous increase in the mouse population. Likewise, in areas where the researchers removed mice, the artificially reduced mouse population allowed far more gypsy moth larvae to survive. This established links among acorns, mice, and moths. Surprisingly, the number of larval ticks increased several times over in the acorn-enriched area. These were the offspring of parent ticks carried to the area by deer, which feed on acorns and were attracted to the bounty. The excess of larval ticks led to a 39% increase in the number of ticks feasting on mice and picking up the Lyme bacterium in the process. Looking back, the researchers noted that in 1994, the oak trees produced a bumper crop of acorns. They predicted that, as a result, 1995 would see extra-large populations of mice and tick larvae, and 1996 would see a large population of one-year-old larval ticks infected with Lyme bacteria. These pinhead-sized "yearling" ticks feed in the summer and pose the greatest threat to people. The prediction was accurate; in 1996, Lyme disease cases in Connecticut doubled compared to 1995. This research makes an important contribution to understanding how populations of oak trees, mice, deer, moths, and ticks interact within the forest ecosystem and alter the abundance of these two harmful parasites: gypsy moth caterpillars and the bacterium causing Lyme disease.

Consider This: A public health official suggests that treating oak trees with hormones that suppress the production of acorns would be a good way to reduce the incidence of Lyme disease. Explain her logic and design a study to test this hypothesis. Think about how a forester concerned with the health of the oak forest might respond to this suggestion. What might he or she say, and why?

Links to Life: Treading Lightly—How Big Is Your "Footprint"?

You now know that an "ecological footprint" measures a person's environmental impact. While animal populations tend to have the minimal footprints needed to sustain health and reproduce, human ecological footprints differ tremendously among different countries and among individuals within those countries. What determines the size of a person's footprint? If you search the Internet for "ecological footprint," you will find Web sites that further describe this concept, compare different countries, and allow you to calculate the size of your own footprint. You'll find that your energy use, the type of home you live in, and even the type of food you eat all influence your footprint.

Individuals in the U.S., on average, tread with larger footprints than any other country on Earth. U.S. footprints average 24 acres each, compared to the global average of 5.6 acres and the Earth's carrying capacity estimated at 4.7 acres per person. If all 6.3 billion people on Earth lived as extravagantly as the average U.S. citizen, we would need 5.4 Earths to supply their demands. People in the Netherlands and Canada also enjoy a comfortable standard of living, with footprints of 14 and 17, respectively. Nonetheless, we would need 3.8 Earths to support our current global population at the average Canadian living standard.

But, you might ask, what is wrong with eating meat or imported fruits, driving a car, or living in your own house with a big yard? In fact, there is nothing inherently wrong with any of these options; they have become ecologically harmful only because of humanity's ongoing failure to limit its population. Individuals must recognize that the decision to bear more than two children will result in more footprints treading planet Earth and fewer resources to go around. For example, if 1 billion people inhabited Earth, each could live in reasonable luxury without damaging the planet. Fewer footprints would also allow us to set aside enough untouched land for the continued survival and well-being of the myriad irreplaceable other species that provide Earth's wealth of biodiversity.

Summary of Key Concepts

39.1 How Does Population Size Change?

Individuals join populations through birth or immigration and leave through death or emigration. The ultimate size of a stable population results from interactions among biotic potential, the maximum possible growth rate, and environmental resistance (which limits population growth).

All organisms have the biotic potential to more than replace themselves over their lifetime, resulting in population growth. Populations tend to grow exponentially, with increasing numbers of individuals added during each successive time period. Populations cannot exhibit exponential growth for long; they either stabilize or undergo periodic boom-and-bust cycles as a result of environmental resistance.

39.2 How Is Population Growth Regulated?

Environmental resistance restrains population growth by increasing the death rate or decreasing the birth rate. The maximum size at which a population may be sustained indefinitely by an ecosystem is termed the carrying capacity, determined by limited resources such as space, nutrients, and light. Environmental resistance generally maintains populations at or below the carrying capacity.

Population growth is restrained by density-independent forms of environmental resistance (such as weather) and density-dependent forms of resistance (including predation, parasitism, and competition).

Populations show specific survivorship curves that describe the likelihood of survival at any given age. Late-loss (convex) curves are characteristic of long-lived species with few offspring, which receive parental care. Species with constant-loss curves have an equal chance of dying at any age. Early-loss (concave) curves are typical of organisms that produce numerous offspring, most of which die before reaching maturity.

39.3 How Is the Human Population Changing?

The human population has exhibited exponential growth for an unprecedented time, the result of a combination of high birth rates and technological, agricultural, industrial, and medical advances that have overcome several types of environmental resistance and increased Earth's carrying capacity for humans. Age-structure diagrams depict numbers of males and females in various age groups in different countries. Expanding populations have pyramidal age structures; stable populations show rather straight-sided age structures; and shrinking populations are illustrated by age structures that are constricted at the base.

Most of Earth's people live in developing countries with growing populations. Although birth rates have declined considerably in many places, momentum from previous high birth rates assures continued substantial population growth. The U.S. is the fastest-growing developed country, owing both to higher birth rates and higher immigration rates. Earth's carrying capacity for humans depends on many factors. Recently, scientists have calculated the amount of biologically productive space needed to supply the demands of an average person at current levels of technology. This "ecological footprint" provides evidence that the demands of Earth's 6.3 billion people already exceed the available biologically productive space. A steady decline in available resources also suggests that we have exceeded carrying capacity and are damaging our world ecosystem, decreasing its future ability to sustain us. As the U.S. population continues to surge, and as people in less-developed countries strive to increase their standard of living, the damage will be accelerated.

Key Terms

abiotic *p. 798*
age structure *p. 810*
biotic *p. 798*
biotic potential *p. 798*
boom-and-bust cycle *p. 800*
carrying capacity *p. 803*
community *p. 798*
competition *p. 806*
contest competition *p. 806*
demographic transition *p. 809*
density-dependent *p. 804*
density-independent *p. 804*
doubling time *p. 799*
ecological footprint *p. 813*
ecology *p. 798*
ecosystem *p. 798*
emigration *p. 798*
environmental resistance *p. 798*
exotic *p. 803*
exponential growth *p. 799*
growth rate *p. 798*
host *p. 805*
immigration *p. 798*
interspecific competition *p. 806*
intraspecific competition *p. 806*
J-curve *p. 799*
parasite *p. 805*
population *p. 798*
population cycle *p. 805*
predator *p. 805*
prey *p. 805*
replacement-level fertility (RLF) *p. 809*
scramble competition *p. 806*
S-curve *p. 803*
survivorship curve *p. 807*

Thinking Through the Concepts

To take a multiple-choice quiz with feedback on the contents of this chapter, visit http://www.prenhall.com/audesirk7. *Log in to the Web site selected by your instructor and navigate to the Self Test section for this chapter.*

? Review Questions

1. Define *biotic potential* and *environmental resistance*.
2. Draw the growth curve of a population before it encounters significant environmental resistance. What is the name of this type of growth, and what is its distinguishing characteristic?
3. Distinguish between density-independent and density-dependent forms of environmental resistance.
4. Describe (or draw a graph illustrating) what is likely to happen to a population that far exceeds the carrying capacity of its ecosystem. Explain your answer.
5. List three density-dependent forms of environmental resistance, and explain why each is density-dependent.
6. Distinguish between populations showing concave and convex survivorship curves. Which is characteristic of people living in the U.S., and why?
7. Given that the U.S. birth rate is currently slightly below replacement-level fertility, why is our population growing?
8. Discuss some reasons why making the transition from a growing to a stable population can be economically difficult.

Applying the Concepts

1. Explain natural selection in terms of biotic potential and environmental resistance.
2. The U.S. has a long history of accepting large numbers of immigrants. Discuss the implications of immigration for population stabilization.
3. What factors encourage rapid population growth in developing countries? What will it take to change this growth?
4. Contrast age structure in rapidly growing versus stable human populations. Why is there a momentum in population growth built into a population that is above RLF?
5. Why is the concept of carrying capacity difficult to apply to human populations? In reference to human population, should the concept be modified to include quality of life?
6. Search the Internet for "ecological footprint," and calculate your footprint using a questionnaire found on one of the resulting Web sites. For five of your daily activities, explain how and why each contributes to your ecological footprint.

For More Information

Kates, R. W. "Sustaining Life on the Earth." *Scientific American*, October 1994. Provides a prescription for an environmentally sustainable future.

Korpimaki, E., and Krebs, C. J. "Predation and Population Cycles of Small Mammals." *BioScience*, November 1996. A review of recent studies designed to evaluate cycles of predators and their prey.

Myers, N. "Biotic Holocaust." *International Wildlife*, March–April 1999. Human activities are causing extinction of species unprecedented since the disappearance of the dinosaurs. How can we reverse this trend?

Pauly, D., and Watson, R. "Counting the Last Fish." *Scientific American*, July 2003. Overfishing is causing the collapse of fisheries worldwide.

Potts, M. "The Unmet Need for Family Planning." *Scientific American*, January 2000. Reducing population growth and increasing the quality of life requires increased access to contraceptives in developing countries.

Sanz, C. "Summer of Danger: Lyme Disease." *Discover*, May 1999. Describes the interrelated populations that contribute to the spread of Lyme disease, and also discusses the new Lyme vaccine.

Wackernagel, M., et. al. "Tracking the Ecological Overshoot of the Human Economy." *Proceedings of the National Academy of Sciences*, Vol. 99, July 2002. A groundbreaking and conservative assessment of the human ecological footprint suggests that we have already surpassed Earth's ability to sustain our population at current living standards.

Wilson, E. O. "The Bottleneck." *Scientific American*, February 2002. Expanding human population combined with shrinking resources creates a bottleneck for humanity. This fascinating article by a National Medal of Science- and Pulitzer Prize-winning biologist compares and contrasts environmental and economic viewpoints.

Media Activities

To access a Media Activity visit http://www.prenhall.com/audesirk7. *Log in to the Web site selected by your instructor, navigate to this chapter, and select the appropriate Media Activity number.*

39.1 Population Growth and Regulation

Estimated time: 10 minutes

Ecologists describe the growth of populations using models, which can be very simple or very complicated equations or sets of equations. In this activity, you'll examine two types of models: density-independent growth and density-dependent growth. Both models are important in understanding the dynamics of populations in nature.

39.2 Human Population Growth and Regulation

Estimated time: 5 minutes

This activity allows you to explore some of the many parameters influencing human population growth and the potential issues associated with unrestrained growth.

39.3 Web Investigation: Acorns, Mice, Moths, Deer, and Disease

Estimated time: 10 minutes

It seems like a very straightforward experiment. Add acorns to forest tracts and observe the effects on gypsy moth and mouse populations. But behind this apparently simple protocol lies extensive biological knowledge. This exercise will explore the facts and theories used to design the Lyme disease experiment.

CHAPTER

40 Community Interactions

Workers blast jets of hot water at zebra mussels coating the interior of a Michigan water-treatment plant. (Inset) Zebra mussels smother a crayfish.

AT A GLANCE

CASE STUDY Invasion of the Zebra Mussels

In 1989 residents of Monroe, Michigan, a town on the Lake Erie shore, suddenly found themselves without water. Their schools, industries, and businesses were closed for two days while workers labored to fix the problem: Zebra mussels had clogged their water-treatment plant. The town's problem was not unique; at another treatment plant on Lake Erie, zebra mussel populations reached 600,000 per square yard (see the opening photo). Where did they come from?

Sometime in 1985 or 1986, a trading vessel bringing cargo from Europe discharged fresh water into Lake St. Clair, located between Lake Huron and Lake Erie at the border between Ontario and Michigan. The water, used for ballast during the ship's transatlantic voyage, contained stowaways—millions of zebra mussel larvae. Although these mollusks are native to the Caspian and Black Seas (large inland seas between Europe and Asia), they found ideal conditions in North America. Spreading throughout the Great Lakes and the Mississippi and Ohio River drainage systems, they have now reached as far south as New Orleans and as far west as Oklahoma. Each year, control efforts cost the U.S. about $5 billion.

Microscopic mussel larvae can be carried in currents for hundreds of miles. Using sticky threads, they attach themselves to nearly any underwater surface, including piers, pipes, machinery, underwater debris, boat hulls, and even sand and silt. Since they can survive out of water for days, mussels clinging to small boats may be portaged to other lakes and rivers, where they quickly move in. An adult female can produce 50,000 eggs each year, and the mussel menace has proven unstoppable. The mussels cover and suffocate other forms of shellfish, threatening many rare varieties with extinction. Think about the zebra mussel as you read about the community interactions that characterize healthy ecosystems. Why have these invaders been so enormously successful? Will anything control them?

40.1 Why Are Community Interactions Important?

An ecological **community** consists of all the interacting populations within an ecosystem; in other words, a community is the *biotic*, or living, component of an ecosystem. In the previous chapter, you learned that community interactions such as predation, parasitism, and competition help limit the size of populations. A community's interacting web of life tends to maintain a balance between resources and the numbers of individuals consuming them. When populations interact and influence each other's ability to survive and reproduce, they serve as agents of natural selection. For example, in killing prey that are easiest to catch, predators leave behind individuals with better defenses against predation. These individuals leave the most offspring, and over time their inherited characteristics increase within the prey population. Thus, as community interactions limit population size, they simultaneously shape the bodies and behaviors of the interacting populations. This process, by which two interacting species act as agents of natural selection on one another over evolutionary time, is called **coevolution**.

Media Activity
40.1 Competition

You have probably heard the expression "the balance of nature." This balance is the result of community interactions finely tuned to one another over evolutionary time. The sometimes fragile balance can be overturned when organisms are introduced into an ecosystem in which they did not evolve, as were the zebra mussels the Case Study. You will learn more about introduced species in "Earth Watch: Exotic Invaders."

The most important community interactions are competition, predation, parasitism, and mutualism. If we assume that each of these interactions involves two species, the types of interactions can be characterized according to whether each of the two species is harmed or benefits, as shown in Table 40-1. Over evolutionary time, these interactions have shaped the bodies and behaviors of organisms.

Table 40-1 Interactions Among Organisms

Type of Interaction	Effect on Organism A	Effect on Organism B
Competition between A and B	Harms	Harms
Predation by A on B	Benefits	Harms
Symbiosis		
Parasitism by A on B	Benefits	Harms
Commensalism of A with B	Benefits	No effect
Mutualism between A and B	Benefits	Benefits

40.2 What Are the Effects of Competition Among Species?

Community interactions may be harmful or beneficial to participating organisms (see Table 40-1). During **interspecific competition**, competition among members of different species, two or more species attempt to use the same limited resources, particularly food and/or space. Interspecific competition harms both species because access to resources is reduced. The intensity of interspecific competition depends on how similar the requirements of the species are. In other words, the degree of competition is proportional to the amount of overlap in the *ecological niches* of the competing species.

The Ecological Niche Defines the Place and Role of Each Species in Its Ecosystem

Although the word *niche* may call to mind a small cubbyhole, in ecology it means much more. Each species occupies a unique **ecological niche** that encompasses all aspects of its way of life, including its physical home or habitat. The primary habitat of a white-tailed deer, for example, is the eastern deciduous forest. In addition, the niche includes all the physical environmental factors necessary for survival, such as the range of temperatures under which the organism can survive, the amount of moisture it requires, the pH of the water or soil it may inhabit, the type of soil nutrients required, and the degree of shade it can tolerate. Although different species share many aspects of their niche with others, no two species ever occupy exactly the same ecological niche.

The ecological niche extends well beyond habitat. It also specifies how the species gets its supply of energy and materials—what might be called the species' role or "occupation" within its ecosystem. A species' predators, prey, and competitors, as well as its behaviors and interactions with other organisms, are considered elements of its niche.

Adaptations Reduce the Overlap of Ecological Niches Among Coexisting Species

Just as no two organisms can occupy exactly the same physical space at the same time, no two species can inhabit exactly the same ecological niche simultaneously and continuously. This important concept, often called the **competitive exclusion principle**, was formulated in 1934 by Russian biologist G. F. Gause. If two species with the same niche were placed together and forced to compete for limited resources, inevitably one would outcompete the other, and the less adapted of the two would die out. Gause used two species of the protist *Paramecium*—*P. aurelia* and *P. caudatum*—to demonstrate this principle. In laboratory flasks, both species thrived on bacteria and fed in the same parts of the

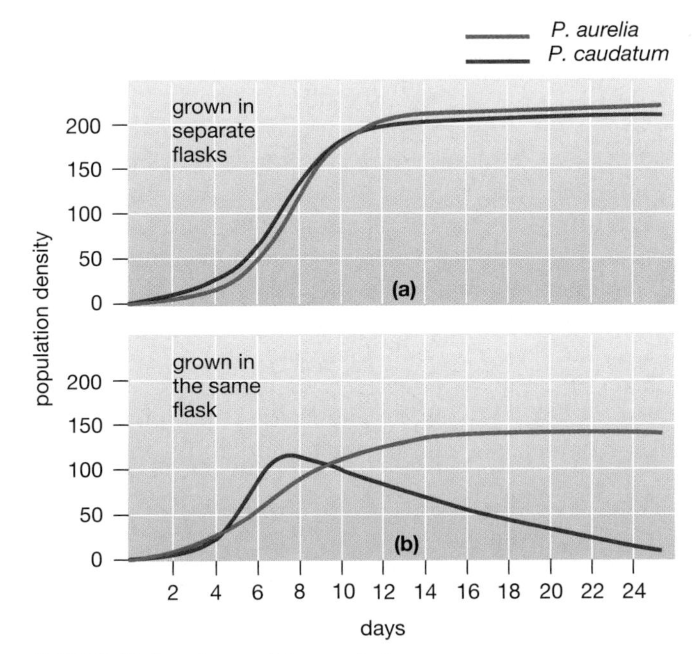

FIGURE 40-1 Competitive exclusion
(a) Raised separately with a constant food supply, both *Paramecium aurelia* and *P. caudatum* show the S-curve typical of a population that initially grows rapidly, then stabilizes. **(b)** Raised together and forced to occupy the same niche, *P. aurelia* consistently outcompetes *P. caudatum* and causes that population to die off. (Modified from G. F. Gause, *The Struggle for Existence*. Baltimore: Williams & Wilkins, 1934.)
QUESTION Exotic species, such as the zebra mussel, can pose a serious threat to native species. Explain how competitive exclusion could contribute to the threat posed by an exotic species.

flask. Grown separately, each population flourished (Fig. 40-1a). But when Gause placed the two species together in a flask, one always eliminated, or "competitively excluded," the other (Fig. 40-1b). Gause then repeated the experiment, replacing *P. caudatum* with a different species, *P. bursaria*, which tended to feed in a different part of the flask. In this case, the two species of *Paramecium* were able to coexist indefinitely because they occupied slightly different niches. Invaders such as zebra mussels have niches that overlap significantly with those of native species, such as freshwater clams, which they are able to outcompete.

Ecologist R. MacArthur tested Gause's laboratory findings under natural conditions by investigating five species of North American warbler. These birds all hunt for insects and nest in the same type of spruce tree. Although the niches of these birds appear to overlap considerably, MacArthur found that each species concentrates its search in specific areas of the tree, employs different hunting tactics, and nests at slightly different times. By partitioning a resource (their habitat), the warblers minimize the overlap of their niches and reduce competition among the different species (Fig. 40-2).

As MacArthur discovered, when two species with similar requirements coexist, they typically occupy a smaller niche than either would if it were by itself. This phenomenon, called **resource partitioning**, is an

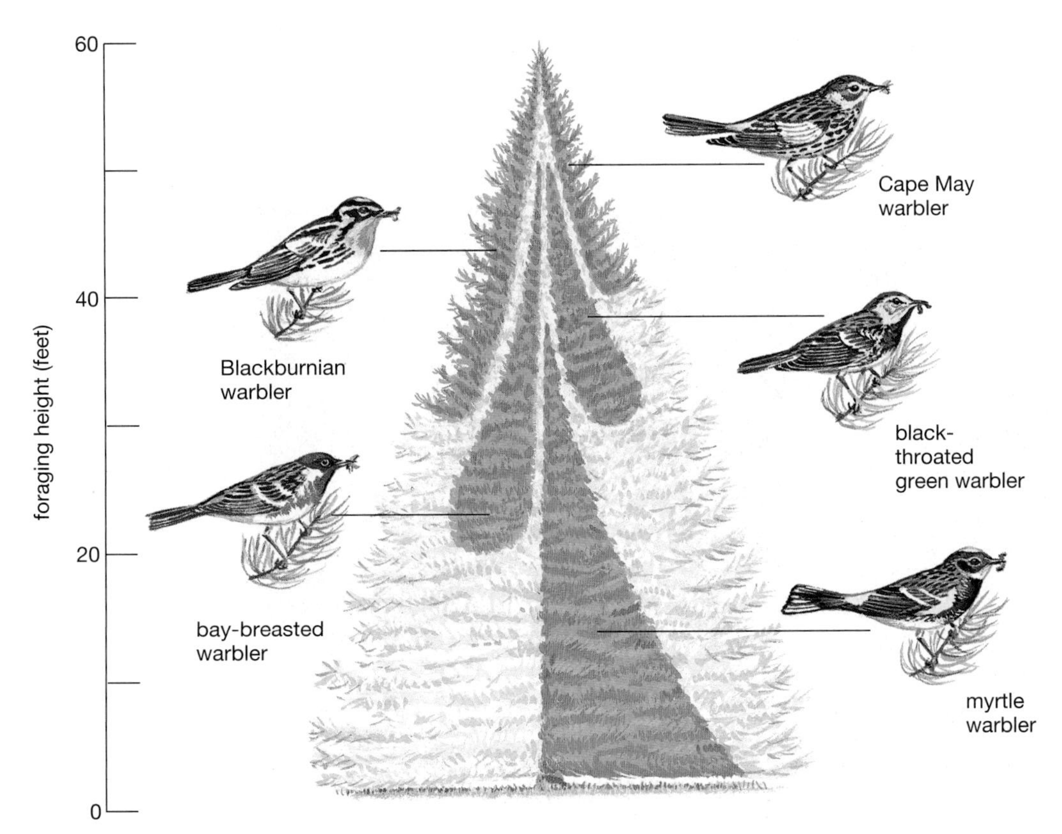

FIGURE 40-2 Resource partitioning
Each of these five insect-eating species of North American warblers searches for food in different regions of spruce trees. They reduce competition by occupying very similar, but not identical, niches.

evolutionary adaptation that reduces the harmful effects of interspecific competition. Resource partitioning is the outcome of the coevolution of species with extensive—but not total—niche overlap. Because natural selection favors individuals with fewer competitors, over evolutionary time the competing species develop physical and behavioral adaptations that minimize their competitive interactions. A dramatic example of resource partitioning was discovered by Charles Darwin among finches of the Galapagos Islands. The finches had evolved different bill sizes and shapes and different feeding behaviors that reduced the competition among them (described in Chapter 16).

Competition Helps Control Population Size and Distribution

Competition can occur both among individuals of the same species and among different species. Individuals of the same species have essentially identical requirements for resources and thus occupy the same ecological niche. For this reason, **intraspecific competition**—that is, competition among individuals of the same species—is a major factor in controlling population size. Although natural selection leads to a reduction of niche overlap among individuals of different species, those with similar niches still directly compete for limited resources. This interspecific competition may restrict the size and distribution of the competing populations.

A classic study of the effects of interspecific competition was performed by ecologist J. Connell, using barnacles (crustaceans that attach permanently to rocks and other surfaces; gray barnacles coat the rocks in Figure 40-13a). Barnacles of the genus *Chthamalus* share the rocky shores of Scotland with another genus, *Balanus*, and their niches overlap considerably. Both genera live in the **intertidal zone**, an area of the shore that is alternately covered and exposed by the tides. Connell found that *Chthamalus* dominates the upper shore and *Balanus* dominates the lower. When he scraped off *Balanus*, the *Chthamalus* population increased, spreading downward into the area its competitor had once inhabited. Where the habitat is appropriate for both genera, *Balanus* conquers because it is larger and grows faster. But *Chthamalus* tolerates drier conditions, so on the upper shore, where only high tides submerge the barnacles, it has a competitive advantage. As this example illustrates, interspecific competition can limit both the size and the distribution of the competing populations.

40.3 What Are the Results of Interactions Between Predators and Their Prey?

Predators kill and eat other organisms. Ecologists sometimes include herbivorous animals (animals that eat plants) in this general category because herbivores can have a major influence on the size and distribution of plant populations. Here we will define predation in its broadest sense—to include the grazing cow, the zebra mussel that is eating microscopic algae, the goby fish eating a zebra mussel, and the bat homing in on a moth (Fig. 40-3). Most predators are either larger than their prey or hunt collectively, as wolves do when bringing down an elk (see Fig. 39-7). Predators are generally less abundant than their prey; you will learn why in the next chapter.

(a)

(b)

FIGURE 40-3 Forms of predation
(a) A cow grazes on prairie grass. The tough stems of grass have evolved under predation pressure by herbivores. **(b)** A long-eared bat uses a sophisticated echolocation system to hunt moths, which have evolved special sound detectors and behaviors to avoid capture.
QUESTION Describe some other examples of coevolution of predators and prey.

Predator–Prey Interactions Shape Evolutionary Adaptations

To survive, predators must feed and prey must avoid becoming food. Therefore, predator and prey populations exert intense environmental pressure on one another, resulting in coevolution. As prey become more difficult to catch, predators must become more adept at hunting. Coevolution has endowed the mountain lion with tearing claws and given the fawn it hunts dappling spots and the behavior of lying still as it awaits its mother. It has produced the keen eyesight of the hawk, and the earthy camouflage coloration of its small mammal prey. Coevolution of predators and prey has also given rise to the poisons and bright colors of the poison arrow frog and the coral snake (see Figs. 40-7 and 40-8a). In the following sections, we examine a few of the evolutionary results of predator–prey interactions.

Some Predators and Prey Have Evolved Counteracting Behaviors

Bat and moth adaptations provide an excellent example of how both physical structures and behaviors are molded by coevolution. Most bats are nighttime hunters that navigate and locate prey by echolocation. They emit extremely high-frequency and high-intensity pulses of sound and, by analyzing the returning echoes, create an "image" of their surroundings. Under selection pressure from this specialized prey-location system, certain moths (a favored prey of bats) have evolved simple ears that are particularly sensitive to the frequencies used by echolocating bats. When they hear a bat, these moths take evasive action, flying erratically or dropping to the ground. The bats, in turn, have evolved the ability to counter this defense by switching the frequency of their sound pulses away from the moth's sensitivity range. Some moths have evolved a way to interfere with the bats' echolocation by producing their own high-frequency clicks. In response, when hunting a clicking moth, a bat may turn off its own sound pulses temporarily and zero in on the moth by following the moth's clicks.

Camouflage Conceals Both Predators and Their Prey

An old saying in detective novels is that the best hiding place may be right out in plain sight. Both predators and prey have evolved colors, patterns, and shapes that resemble their surroundings. Such disguises, called **camouflage**, render animals inconspicuous even when they are in plain sight (Fig. 40-4).

Some animals closely resemble specific objects such as leaves, twigs, seaweed, thorns, or even bird droppings (Fig. 40-5a–c). Camouflaged animals tend to remain motionless rather than to flee their predators; a fleeing "bird dropping" would be quite conspicuous! Whereas many camouflaged animals resemble plants, a few types of plants have evolved to resemble rocks, which are ignored by their herbivorous predators (Fig. 40-5d).

Predators who ambush are also aided by camouflage. For example, a spotted cheetah becomes inconspicuous in the grass as it watches for grazing mammals. The frogfish closely resembles the algae-covered rocks and algae

(a)

(b)

FIGURE 40-4 Camouflage by blending in
(a) The sand dab is a flat, bottom-dwelling ocean fish with a mottled color that closely resembles the sand on which it rests. **(b)** This nightjar bird on its nest in Central America is barely visible among the surrounding leaf litter.

(a) (b) (c) (d)

FIGURE 40-5 Camouflage by resembling specific objects
(a) A moth whose color and shape resemble a bird dropping sits motionless on a leaf. **(b)** The leafy sea dragon (an Australian "seahorse" fish) has evolved extensions of its body that mimic the algae in which it normally hides. **(c)** Florida treehopper insects avoid detection by resembling thorns on a branch. **(d)** This cactus of the American Southwest is appropriately called the "living rock cactus." **QUESTION** How could such camouflage evolve?

(a)

(b)

FIGURE 40-6 Camouflage assists predators
(a) As it waits for prey, a cheetah blends into the background of the grass. **(b)** Combining camouflage and aggressive mimicry, a frogfish waits in ambush, its camouflaged body matching the sponge-encrusted rock on which it rests. Above its mouth dangles a lure that closely resembles a small fish. The lure attracts small predators, who will find themselves to be prey.

FIGURE 40-7 Warning coloration
The South American poison arrow frog, with its poisonous skin, advertises its unpleasant taste with bright and contrasting color patterns.

on which it sits motionless, dangling a small lure from its upper lip (Fig. 40-6). Small fish notice only the lure and are engulfed as they approach it.

Bright Colors Often Warn of Danger

Some animals have evolved very differently, exhibiting bright **warning coloration** (Fig. 40-7; see Figs. 40-8 and 40-11). These animals are usually distasteful, and many are poisonous, such as the yellow jacket with its bright yellow and black stripes. Poisoning a predator is small consolation for an organism that has already been eaten; thus, the bright colors declare, "Eat me at your own risk!" One unpleasant experience is enough to teach predators to avoid these conspicuous prey.

Some Organisms Gain Protection Through Mimicry

Mimicry refers to a situation in which a species evolves to resemble something else, typically another type of organism. For example, once warning coloration evolved,

FIGURE 40-8 Warning mimicry
The warning coloration of **(a)** the poisonous coral snake is mimicked by **(b)** the harmless mountain king snake. Mutual mimicry of warning coloration by the distasteful monarch butterfly **(c)** and the equally distasteful viceroy **(d)** protects both species.

there arose a selective advantage for harmless animals to resemble poisonous ones. The deadly coral snake has brilliant warning coloration, and the harmless mountain king snake avoids predation by resembling the coral snake (Fig. 40-8a,b). By resembling each other, two distasteful species may each benefit from predators' painful experience with the other. For example, predators rapidly learn to avoid the conspicuous stripes on bees, hornets, and yellow jackets. Poisonous and distasteful monarch butterflies have wing patterns strikingly similar to those of equally unappealing viceroy butterflies (Fig. 40-8c,d). A common color pattern results in faster learning by predators—and less predation on all similarly colored species.

Some predators have evolved **aggressive mimicry**, a "wolf-in-sheep's-clothing" approach, in which they entice their prey to come close by resembling a harmless animal or part of the environment. For example, although the frogfish resembles the algae-covered rocks where it lurks, it dangles a wriggling lure that resembles a small fish just above its mouth (see Fig. 40-6b). A curious fish attracted to the lure is quickly swallowed.

A sophisticated variation on the theme of prey mimicking predators is seen in snowberry flies, which are hunted by territorial jumping spiders. When a fly spots an approaching spider, it spreads its wings, moving them back and forth in a jerky dance. Seeing this display, the spider will likely flee from the harmless fly. Why? Researchers have observed that the markings on the fly's wings closely resemble the legs of another jumping spider. The jerky movements of the fly mimic those of a jumping spider when it drives another spider from its territory (Fig. 40-9). Natural selection has finely tuned both the behavior and the appearance of the fly to avoid predation by jumping spiders.

Certain prey species use another form of mimicry: **startle coloration**. Several insects and even some vertebrates (such as the false-eyed frog) have evolved patterns of color that closely resemble the eyes of a much larger, and possibly dangerous, animal (Fig. 40-10). If a predator gets close, the prey suddenly flashes its eye-spots, startling the predator and allowing the prey to escape.

Some Predators and Prey Engage in Chemical Warfare

Both predators and prey have evolved a variety of toxic chemicals for attack and defense. The venom of spiders and poisonous snakes, such as the coral snake (see Fig. 40-8), serves both to paralyze prey and to deter its predators. Many plants also produce defensive toxins. For example, lupine plants, whose flowers grace both gardens and mountain meadows, produce chemicals called *alkaloids*, which deter attack by the blue butterfly, whose larvae feed on the lupine's buds. In fact, different individuals of the same species of lupine produce different forms of alkaloids, thus making it more difficult for the butterflies to evolve resistance to it.

Certain mollusks (including squid, octopus, and some sea slugs) emit clouds of ink when attacked. These colorful chemical "smoke screens" confuse predators and mask the prey's escape. A dramatic example of chemical defense is seen in the bombardier beetle. In response to the bite of an ant, the beetle releases secretions from special glands into an abdominal chamber. There, enzymes catalyze an explosive chemical reaction that shoots a toxic, boiling-hot spray onto the attacker (Fig. 40-11a).

Plants and Herbivores Have Coevolutionary Adaptations

Herbivores don't fit neatly into parasite or predator categories. A grazing horse or cow uproots and kills some grass, but most of the time acts more like a lawn mower, cropping but not killing the plants. Regardless of how we

(a)

(b)

FIGURE 40-9 Visual and behavioral mimicry
(a) In response to the approach of a jumping spider, **(b)** the snowberry fly spreads its wings, revealing a pattern that resembles spider legs. The fly enhances the effect by performing a jerky, side-to-side dance that resembles the leg-waving display of a jumping spider defending its territory.

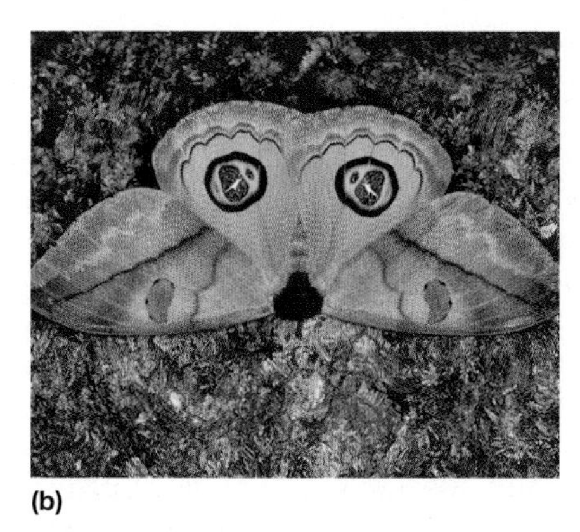
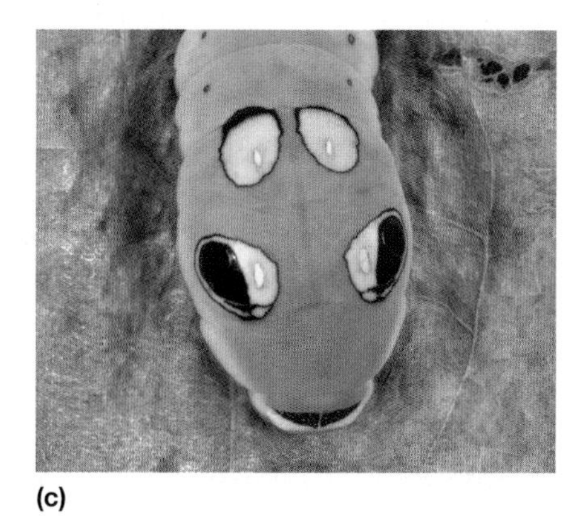

(a) (b) (c)

FIGURE 40-10 Startle coloration
(a) When threatened, the false-eyed frog raises its rump, which resembles the eyes of a larger predator. This startles the real predator, giving the frog a chance to flee. **(b)** The peacock moth from Trinidad is well camouflaged, but should a predator approach, it opens its wings to reveal spots resembling large eyes. **(c)** Predators of this caterpillar larva of the swallowtail butterfly are deterred by its resemblance to a snake. The caterpillar's head is the "snake's" nose.

categorize them, herbivores exert strong selective pressure on plants to avoid being eaten. Plants have evolved a variety of chemical adaptations that deter their herbivorous "predators." Many, such as the milkweed, synthesize toxic and distasteful chemicals. As plants evolved toxic chemicals for defense, certain insects evolved increasingly efficient ways to detoxify or even use the chemicals. The result is that nearly every toxic plant is eaten by at least one type of insect. For example, monarch butterflies lay their eggs on milkweed; when their larvae hatch, they consume the toxic plant (Fig. 40-11b). The caterpillars not only tolerate the milkweed poison but store it in their tissues as a defense against their own predators. The stored toxin is retained in the metamorphosed monarch butterfly (see Fig. 40-8c).

Grasses have evolved tough silicon (glassy) substances in their blades that make them difficult to chew, selecting for grazing animals with longer, harder teeth. Over an evolutionary time scale, as grasses evolved tougher blades that discouraged predation, horses evolved longer teeth with thicker enamel coatings that resist wear and abrasion from the tough grasses.

(a)

(b)

FIGURE 40-11 Chemical warfare
(a) The bombardier beetle sprays a hot toxic brew in response to a leg pinch. **(b)** A monarch caterpillar feeds on milkweed that contains a powerful toxin. Why do you think the caterpillar is colored with bright stripes instead of being green?
QUESTION Why is the caterpillar colored with bright stripes?

40.4 What Is Symbiosis?

Symbiosis, which literally means "living together," is the close interaction between organisms of different species for an extended time. Considered in its broadest sense, symbiosis includes parasitism, mutualism, and commensalism. Although one species always benefits in symbiotic relationships, the second species may be unaffected, harmed, or helped (see Table 40-1). **Commensalism** is a relationship in which one species benefits while the other is relatively unaffected. Barnacles that attach themselves to the skin of a whale, for example, get a free ride through nutrient-rich waters without harming the whale. In parasitism and mutualism, which affect both organisms, the participants act on each other as strong agents of natural selection. Here we will discuss these two forms of community interaction.

Parasitism Harms, but Does Not Immediately Kill, the Host

In parasitism, one organism benefits by feeding on another. **Parasites** live in or on their prey, which are called *hosts*, usually harming or weakening them but not immediately killing them. Although it is sometimes difficult to distinguish clearly between a predator and a parasite, parasites are generally much smaller and more numerous than their hosts. Familiar parasites include tapeworms, fleas, and the many types of disease-causing protozoa, bacteria, and viruses. Many parasites, particularly worms and protozoa, have complex life cycles involving two or more hosts (see Chapter 22). There are few parasitic vertebrates; the lamprey eel, which attaches itself to a host fish and sucks its blood, is a rare example.

The variety of infectious bacteria and viruses and the precision of the immune system that counters their attacks are evidence of the powerful forces of coevolution between parasitic microorganisms and their hosts. Consider the malaria parasite, which has provided strong environmental pressure for humans to carry a defective hemoglobin gene that causes sickle-cell anemia: The parasite can't survive in the affected red blood cells. In certain areas of Africa where malaria is common, up to 20% of the human population carries the sickle-cell gene.

Another example is *Trypanosoma*, a parasitic protozoan that causes both human sleeping sickness and a disease in cattle called *nagana*. African antelope, which coevolved with this parasite, are relatively unaffected by it. Most infected cattle, a more recently introduced species, suffer but survive infection if they have been bred in an infested area for many generations. Newly imported cattle, however, generally die if not treated.

In Mutualistic Interactions, Both Species Benefit

When two species interact in a way that benefits both, the relationship is called **mutualism**. If you see colored patches on rocks, they are probably lichens, a mutualistic association of an alga and a fungus (Fig. 40-12a). The fungus provides support and protection while deriving food from the photosynthetic alga, whose bright colors are actually light-trapping pigments. The mutualistic interactions between flowering plants and their pollinators are discussed in Chapter 25. Mutualistic associations also occur in the digestive tracts of cows and termites, where protists and bacteria find food and shelter while helping their hosts extract nutrients, as well as in our own intestines, where bacteria synthesize certain vitamins. The nitrogen-fixing bacteria inhabiting special chambers on the roots of legume plants are another important example. These bacteria obtain food and shelter from the plant and in return trap nitrogen in a form the plant can utilize. Some mutualistic partners have coevolved to the extent

(a)

(b)

FIGURE 40-12 Mutualism
(a) This brightly colored lichen growing on bare rock is a mutualistic relationship between an alga and a fungus. **(b)** The clownfish snuggles unharmed among the stinging tentacles of the anemone. Note the bright "warning" color of the clownfish. Although the fish itself is defenseless, the coloration may warn potential predators of the threat posed by the anemone.

SCIENTIFIC INQUIRY Ants and Acacias—An Advantageous Association

Daniel Janzen, a doctoral student at the University of Pennsylvania, was walking down a road in Veracruz, Mexico, when he saw a flying beetle alight on a thorny tree, only to be driven off by an ant. Looking more closely, he saw that the tree, a bull's-horn acacia, was covered with ants. A large ant colony of the genus *Pseudomyrmex* made its home inside the enlarged thorns of the plant; the ants easily excavated the thorns' soft pulpy interiors to provide shelter (Fig. E40-1).

To determine whether the ants were important to the tree, Janzen began stripping the thorns by hand until he found and removed the thorn that housed the ant queen, thus destroying the colony. Later, he turned to more efficient but dangerous methods, using an insecticide to eliminate all the ants on a large stand of acacias. Though the acacias were unharmed by the poison, Janzen became ill from it, and all the ants were killed. Within a year of spraying the insecticide, Janzen found nearly all the acacia trees dead, consumed by insects and other herbivores and shaded out by competing plants. The ground surrounding the trees, which the ants normally kept trimmed, was overgrown with vegetation. The trees were apparently dependent on their resident ants for survival.

Wondering if the ants could survive without the tree, Janzen painstakingly peeled ant-inhabited thorns off 100 acacia trees, suffering multiple stings in the process. Janzen housed each ant colony in a jar supplied with local non-acacia vegetation and insects for food, but all the ant colonies starved. Carefully examining the acacia, he found swollen structures filled with sweet syrup at the base of the leaves and protein-rich capsules on the leaf tips (Fig. E40-1, inset). Together, these materials provide a balanced diet for the ants.

Janzen's experiments strongly suggest that these species of ant and acacia have an obligatory mutualistic relationship: Neither can survive without the other. Of course, further observations are required to support this hypothesis. The fact that the ants starved in Janzen's jars did not rule out their possible survival elsewhere, but, in fact, this species of ant has never been found living independently. Similarly, the bull's-horn acacia has never been found without a resident ant colony. Thus, a chance observation followed by careful research led to the discovery of an important mutualistic association.

FIGURE E40-1 A mutualistic relationship
Yellow, protein-rich capsules produced at the tips of acacia leaves provide food for the resident ants. (Inset) A hole in the enlarged thorn of the bull's-horn acacia provides shelter for members of the ant colony. The ant entering the thorn is carrying a food capsule. As the ant colony grows, it invades more thorns.

that neither can survive alone. An example is the ant–acacia mutualism described in "Scientific Inquiry: Ants and Acacias—An Advantageous Association."

Mutualistic relationships involving vertebrates are rare and typically less intimate and extended. The clownfish, which is coated with a layer of protective mucus, takes shelter among the venomous tentacles of certain species of anemones. The anemone provides the fish with protection from predators. In return, the clownfish drives away other fish that eat anemones, cleans dirt and debris from its host, and may bring bits of food to its anemone (Fig. 40-12b).

40.5 How Do Keystone Species Influence Community Structure?

In some communities, a particular key species, called a **keystone species**, plays a major role in determining community structure—a role that is out of proportion to its abundance in the community. Removal of the keystone species dramatically alters the community. For example, in 1969 Robert Paine, an ecologist at the University of Washington, removed predatory starfish

(a)

(b)

FIGURE 40-13 Keystone species
(a) The starfish *Pisaster* is a keystone species along the rocky coast of the Pacific Northwest. **(b)** The elephant is a keystone species on the African savanna.

Media Activity
40.2 The Importance of Keystone Species

Pisaster (Fig. 40-13a) from sections of Washington's rocky intertidal coast. Mussels (two-shelled or "bivalve" mollusks that are a favored prey of *Pisaster*) became so abundant that they outcompeted algae and other invertebrates that normally coexist in intertidal communities. Another marine invertebrate, the lobster, may be a keystone species off the east coast of Canada. Overfishing of the lobster allowed its prey, sea urchins, to increase in numbers. The population explosion of sea urchins nearly eliminated certain types of algae on which the urchins prey, leaving large expanses of bare rock where a diverse community once existed. The sea otter appears to be a keystone species along the coast of western Alaska. Starting around 1990, observers noted an alarming decline in otter numbers, resulting in an increase in their sea urchin prey. This led to overgrazing by sea urchins on the kelp forests that provide critical undersea habitat for a variety of marine species. What killed the sea otters? Killer whales, which formerly fed primarily on seals and sea lions, were seen increasingly dining on sea otters, as their favored prey disappeared. Scientists hypothesize that the seal and sea lion decline, in turn, is a result of overfishing by humans in the North Pacific, depleting the food supply of these fish-eaters. In the African savanna, the African elephant is a keystone predator. By grazing on small trees and bushes, elephants prevent the encroachment of forests and help maintain the grassland community (Fig. 40-13b).

It is difficult to identify keystone species; doing so properly requires that the species be selectively removed and the community studied for several years before and after its removal. However, many ecological studies performed since the concept was introduced provide evidence that keystone species are important in a wide variety of communities. Why is it important to study keystone species? As human activities infringe on natural ecosystems, it becomes increasingly urgent to understand community interactions and to preserve species that are crucial to maintaining the natural community.

40.6 Succession: How Do Community Interactions Cause Change Over Time?

In a mature terrestrial ecosystem, the populations that make up the community interact with one another and with their nonliving environment in intricate ways. But this tangled web of life did not spring fully formed from bare rock or naked soil; rather, it emerged in stages over a long period, a process called succession. **Succession** is a structural change in a community and its nonliving environment over time. It is a kind of "community relay" in which assemblages of plants and animals replace one another in a sequence that is somewhat predictable.

Succession is preceded by a **disturbance**, an event that disrupts the ecosystem either by altering its community, its abiotic structure, or both. For example, beavers, landslides, or people may dam streams, causing marshes, ponds, or lakes to form. A landslide or avalanche may strip a swath of trees from a mountainside. Volcanoes may smother nearby ecosystems in ash, cover an ecosystem in solid rock from solidifying lava (Fig. 40-14a), or even form an entire new island. Fire is another common disturbance. Each disturbance sets the stage for succession. Volcanic eruptions, as in the case of Mount St. Helens, leave behind a nutrient-rich environment that encourages rapid invasion of new life (Fig. 40-14b). Forest fires, while destroying an existing community, also release nutrients and create conditions that favor rapid succession (Fig. 40-14c).

FIGURE 40-14 Succession in progress
Pairs of photographs illustrate primary and secondary succession. **(a)** Primary succession. Left: The Hawaiian volcano Mount Kilauea has erupted repeatedly since 1983, sending rivers of lava over the surrounding countryside. Right: A pioneer fern takes root in a crack in hardened lava. **(b)** Secondary succession. Left: On May 18, 1980, the explosion of Mount St. Helens in Washington State devastated the surrounding pine forest ecosystem. Right: Twenty years later, life abounds on the once-barren landscape. Because traces of the former ecosystem remained, this is an example of secondary succession. **(c)** Secondary succession. Left: In the summer of 1988, extensive fires swept through the forests of Yellowstone National Park in Wyoming. Right: Twenty-two years later, trees and flowering plants are thriving in the sunlight, and wildlife populations are rebounding as secondary succession occurs.
QUESTION People have suppressed fires for decades. What are the implications of fire suppression for forest ecosystems and succession?

The precise changes that occur during succession are as diverse as the environments in which succession occurs, but we can recognize certain general stages. In each case, succession is begun by a few hardy plant invaders called **pioneers**. The pioneers may alter the ecosystem in ways that favor competing plants, which gradually displace them. If allowed to continue, succession progresses to a diverse and relatively stable **climax community**. Alternatively, recurring disturbances maintain many communities in **subclimax** stages. Our discussion of succession will focus on plant communities, which dominate the landscape and provide both food and habitat for animals.

There Are Two Major Forms of Succession: Primary and Secondary

Succession takes two major forms: primary and secondary. During **primary succession**, a community gradually colonizes bare rock, sand, or a clear glacial pool where there is no trace of a previous community. This building of a community "from scratch" typically requires thousands or even tens of thousands of years. During **secondary succession**, a new community develops after an existing ecosystem is disturbed in a way that leaves traces of the previous community behind, such as soil and seeds. For this reason, secondary succession happens much more rapidly than does primary succession; it may take only a few centuries. In the following examples, we examine these processes in more detail.

Primary Succession Can Begin on Bare Rock

Figure 40-15 illustrates primary succession on Isle Royale, Michigan, an island in Lake Superior. Bare rock, such as that exposed by a retreating glacier or cooled from molten lava, liberates nutrients such as minerals by *weathering*. In weathering, cracks form as the rock alternately freezes and thaws, contracting and expanding. Chemical action such as acid rain further breaks down the surface.

Weathered rock provides a place for lichens, a pioneer species, to attach where there are no competitors and plenty of sunlight. Lichens can photosynthesize, and they obtain minerals by dissolving some of the rock with an acid they secrete. As the lichens spread over the rock, drought-resistant, sun-loving mosses begin growing in the cracks. Fortified by nutrients liberated by the lichens, the moss forms a dense mat that traps dust, tiny rock particles, and bits of organic debris. It may cover and kill the lichen that made its growth possible. As some mosses die each year, their bodies add to a growing nutrient base, and the living moss mat acts like a sponge, trapping moisture. Within the moss, seeds of larger plants, such as bluebell and yarrow, germinate. When these plants die, their bodies contribute to a growing layer of soil. As woody shrubs such as blueberry and juniper take advantage of the newly formed soil, the moss and remaining lichens may be shaded out and buried by decaying leaves and vegetation. Eventually, trees such as jack pine, blue spruce, and aspen take root in the deeper crevices, and the sun-loving shrubs are shaded out. Within the forest, shade-tolerant seedlings of taller or faster-growing trees, such as balsam fir, paper birch, and white spruce, thrive. In time they tower over and replace the original trees, which are intolerant of shade. After a thousand years or more, a tall climax forest thrives on what was once bare rock.

An Abandoned Farm Will Undergo Secondary Succession

Figure 40-16 illustrates secondary succession on an abandoned farm in the southeastern U.S. The pioneer species—fast-growing annual weeds such as crabgrass, ragweed, and sorrel—root in the rich soil already present

Media Activity
40.3 Primary Succession: Glacier Bay, Alaska

FIGURE 40-15 Primary succession Primary succession as it occurs on bare rock in upper Michigan.

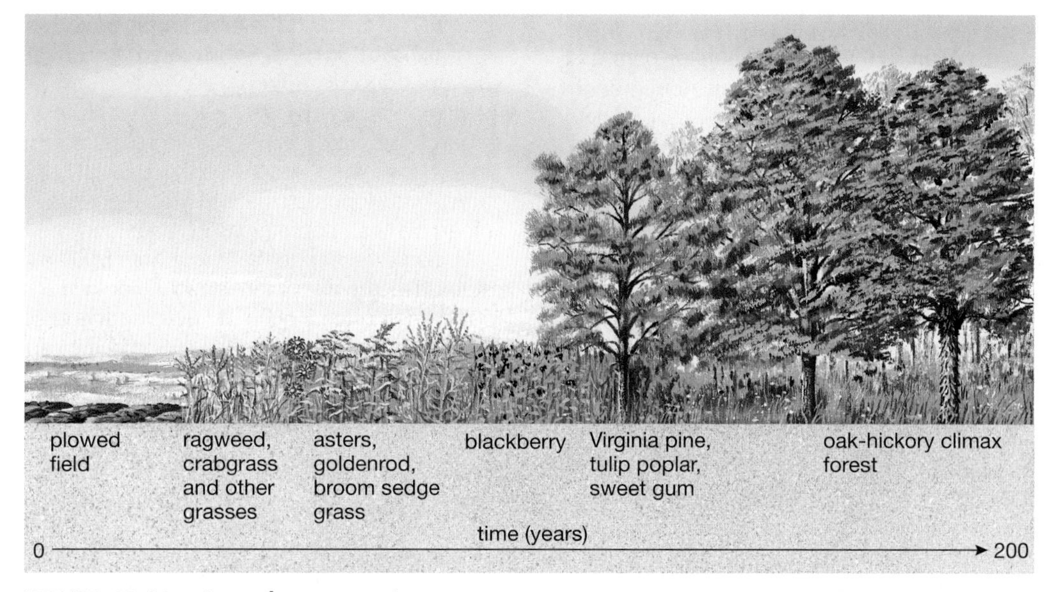

FIGURE 40-16 Secondary succession
Secondary succession as it occurs on a plowed, abandoned southeastern U.S. farm field.

and thrive in direct sunlight. They generally produce large numbers of easily dispersed seeds that help them colonize open spaces, but they don't compete well against longer-lived (perennial) species that gradually grow larger and shade out the pioneers. After a few years, perennial plants such as asters, goldenrod, and broom sedge grass invade, as do woody shrubs such as blackberry. These multiply rapidly and dominate for the next few decades. Eventually, they are replaced by pines and fast-growing deciduous trees, such as tulip poplar and sweet gum, which sprout from windblown seeds. These trees become prominent after about 25 years, and a pine forest dominates the field for the rest of the first century. Meanwhile, shade-resistant, slow-growing hardwoods such as oak and hickory take root beneath the pines. After the first century, they begin to tower over and shade the pines, which eventually die from lack of sun. A relatively stable climax forest dominated by oak and hickory is present by the end of the second century.

Succession Also Occurs in Ponds and Lakes

In freshwater ponds or lakes, succession occurs both from changes within the pond or lake and as a result of an influx of nutrients from outside the ecosystem. Sediments and nutrients carried in by runoff from the surrounding land have a particularly large impact on small freshwater lakes, ponds, and bogs, which gradually undergo succession to dry land (Fig. 40-17). In forests, meadows may be produced by lakes undergoing succession. As the lake fills in from the edges, grasses colonize the newly exposed soil. As the lake shrinks and the

(a)

(b)

FIGURE 40-17 Succession in a small freshwater pond
In small ponds, succession is speeded by an influx of materials from the surroundings. **(a)** In this small pond, dissolved minerals carried by runoff from the surroundings support aquatic plants, whose seeds or spores were carried in by the winds or by birds and other animals. **(b)** Over time, the decaying bodies of aquatic plants build up soil that provides anchorage for more terrestrial plants. Finally, the pond is entirely converted to dry land.

EARTH WATCH Exotic Invaders

Exotic species, or species introduced into an ecosystem where they did not evolve, sometimes find no predators or parasites in their new environment. Their unchecked population growth may seriously damage the ecosystem as they displace, outcompete, and prey on native species. Not all nonnative species become pests, but those that do often have high reproductive rates, effective means of moving into new habitats, and the ability to thrive under a relatively wide range of environmental conditions. Invasive plants may spread by runners as well as seeds, and some can form new plants from plant fragments. Invasive animals are usually not "picky eaters."

Both starlings and English sparrows have spread dramatically since their deliberate introduction into the eastern U.S. in the 1890s. Their success displaces native songbirds, with which they compete for nesting sites. Red fire ants from South America were accidentally introduced into Alabama on shiploads of lumber in the 1930s and have since spread throughout the South. Fire ants displace and may kill other insects, birds, and even small mammals. Their mounds can ruin farm fields, and their fiery stings and aggressive temperament can make backyards uninhabitable. Gypsy moths were introduced from Europe in 1866 and still pose a serious threat to fruit and forest trees in North America. Now a new invader, the Asian long-horned beetle, is devouring hardwood trees in the eastern and midwestern U.S. The beetle, which officials believe may pose the biggest threat to forests since the gypsy moth, arrived around 1996 in wooden pallets and boxes shipped from China.

Exotic plants also threaten natural communities. In the 1940s, the Japanese vine kudzu was planted extensively in the southern U.S. to control erosion. Today kudzu is a major pest and covers 7 million acres, where it overgrows and kills trees and underbrush and occasionally engulfs small houses (Fig. E40-2a). The water hyacinth, introduced from South America as an ornamental plant, now clogs about 2 million acres of lakes and waterways in the southern U.S., slowing boat traffic and displacing natural vegetation (Fig. E40-2b). Meanwhile, imported Old World climbing ferns, probably introduced by nurseries, are blanketing treetops throughout southern Florida and rapidly invading the Everglades, killing trees by depriving them of light.

The West Nile virus, endemic to Africa and the Middle East, was first recognized in the U.S. in 1999, when birds (particularly crows) began dying in large numbers in New York City. The virus replicates in birds, which are bitten by mosquitoes, who bite and infect more birds, people and a few other mammals including horses. By 2002 the West Nile virus had spread through most of the U.S., with cases reported in 44 states. Birds, horses, and people of the U.S. lack the immunity that comes with long association with the virus, and so are more vulnerable than populations in Africa and the Middle East.

Ecologists estimate that more than 6500 exotic species have established themselves in the U.S., driving 315 native species to the brink of extinction. By evading the checks and balances imposed by millennia of coevolution, exotic species are wreaking havoc on natural ecosystems throughout the world. Recently, wildlife officials have made cautious attempts to reestablish these checks and balances by importing predators or parasites to attack selected exotic species (a successful example, the importation of the cactus moth to Australia, was described in Chapter 39). In Australia, a dozen European rabbits introduced in the 1840s exploded in number to 300 million by 1996, wiping out native vegetation and competing for food with native species such as kangaroos. In 1996, officials released a virus deadly to rabbits at hundreds of sites across the continent. This controversial experiment has dramatically reduced rabbit populations in some areas, allowing native plants and animals to rebound. This type of control is fraught with danger, however, because introducing more exotic predators or parasites into an ecosystem can have unpredicted and possibly disastrous consequences for native species. In 1958, a large predatory Florida snail, the rosy wolf snail (Fig. E29-2c), was imported into Hawaii to feed on another exotic pest, the giant African snail, which was a menace to the native vegetation. The rosy wolf snail is now attacking many species of native snail, threatening them with extinction. In spite of the risks of importing these biocontrols, there often seems to be little alternative, since poisons intended for exotics will kill natives as well. United States Department of Agriculture scientists are planning to release a fly whose larvae feed selectively on fire ants and are researching the possibility of importing insects that will feed on the Old World climbing fern and a number of other invasive foreign weeds—hopefully without attacking native plants.

(a)

(b)

(c)

FIGURE E40-2 Exotic species
(a) The Japanese vine kudzu will rapidly cover entire trees and houses. **(b)** Water hyacinths, originally from South America, today clog waterways in the southern U.S. **(c)** Importing the rosy wolf snail proved disastrous for Hawaii's native snails, one of which is being attacked in this photo.

meadow grows, trees will encroach around the meadow's edges. If you return to a forest lake 20 years later, it probably will be a bit smaller.

Succession Culminates in the Climax Community

Succession ends with a relatively stable climax community, which perpetuates itself if it is not disturbed by external forces (such as fire, invasion of an introduced species, or human activities). The populations within a climax community have ecological niches that allow them to coexist without replacing one another. In general, climax communities have more species and more types of community interactions than do early stages of succession. The plant species that dominate climax communities are generally longer-living and tend to be larger than pioneer species; this trend is particularly evident in ecosystems where forest is the climax community.

In your travels, you have undoubtedly noticed that the type of climax community varies dramatically from one area to the next. For example, if you drive through Colorado, you will see a shortgrass prairie climax community on the eastern plains (in those rare areas where it has not been replaced by farms), pine-spruce forests in the mountains, tundra on the mountain summits, and sagebrush-dominated climax community in the western valleys. The exact nature of the climax community is determined by numerous geological and climatic variables, including temperature, rainfall, elevation, latitude, type of rock (which determines the type of nutrients available), and exposure to sun and wind. Natural events such as hurricanes, avalanches, and fires started by lightning may destroy sections of climax forest, reinitiating secondary succession and producing a patchwork of various successional stages within an ecosystem.

In many forests throughout the U.S., rangers allow fires set by lightning to run their course, recognizing that this natural process is important for the maintenance of the entire ecosystem. Fires liberate nutrients used by plants and also kill some (but usually not all) of the trees they engulf. Sunlight can then reach the forest floor, encouraging the growth of subclimax plants, which belong to a successional stage earlier than the climax stage. The combination of climax and subclimax regions within the ecosystem provides habitats for a larger number of species.

Human activities may dramatically alter the climax vegetation. Large stretches of grasslands in the western U.S., for example, are now dominated by sagebrush due to overgrazing. The grass that usually outcompetes sagebrush is selectively eaten by cattle, allowing the sagebrush to prosper.

Some Ecosystems Are Maintained in a Subclimax State

Some ecosystems are not allowed to reach the climax stage but are maintained in a subclimax stage by frequent disturbance. The tallgrass prairie that once covered northern Missouri and Illinois is a subclimax stage of an ecosystem whose climax community is deciduous forest. The prairie was maintained by periodic fires, some set by lightning and others deliberately set by Native Americans to increase grazing land for buffalo. Forest now encroaches, and limited prairie preserves are maintained by carefully managed burning.

Agriculture also depends on the artificial maintenance of carefully selected subclimax communities. Grains are specialized grasses characteristic of the early stages of succession, and much energy goes into preventing competitors (weeds and shrubs) from taking over. The suburban lawn is a painstakingly maintained subclimax ecosystem. Mowing (a disturbance) destroys woody invaders, and some suburbanites also use herbicides to selectively kill pioneers such as crabgrass and dandelions.

The study of succession is the study of variations in communities over time. The climax communities that form during succession are strongly influenced by climate and geography—the distribution of ecosystems in space. Deserts, grasslands, and deciduous forests are climax communities formed over broad geographical regions with similar environmental conditions. These extensive areas of characteristic plant communities are called **biomes**. Although the communities within the various biomes differ radically in the types of populations they support, communities worldwide are structured according to general rules. These principles of ecosystem structure, as well as some of the great biomes of the world, are described in the following chapters.

EVOLUTIONARY CONNECTIONS

Is Camouflage Splitting a Species?

Walking stick insects are aptly named; their elongated, camouflaged bodies blend in beautifully with the plants on which they feed and hide from predatory birds and lizards. In the Santa Ynez Mountains of California, a single species of walking stick (*Timena cristinae*) exhibits two distinct, genetically determined color patterns: green with a white stripe and solid green. Researcher Cristina Sandoval found that the striped form most often hides in and prefers to feed on chamise bushes, where it almost disappears among the needlelike leaves (Fig. 40-18, left). In contrast, she found the solid-colored form feeding mostly on wild blue lilac (Fig. 40-18, right), camouflaged among the lilac's solid green leaves.

Birds and lizards voraciously eat both colors of walking sticks. Therefore, striped forms that prefer striped leaves will be better camouflaged, allowing more to survive to reproduce and pass their plant preference on to their offspring. Solid green walking sticks that prefer solid leaves will have a similar survival advantage. Colleagues from Simon Frazer University in Canada and Sandoval brought both forms

FIGURE 40-18 Color variants of "walking sticks" prefer different plants
(left) The striped form of walking stick is well-hidden among the needlelike leaves of its preferred food, the chamise bush. **(right)** The solid-colored version of the same species blends well with the leaves of the wild lilac, which it prefers. This photo shows a mating pair. In the lab, the insects preferred to mate with others of the same color pattern.

into the laboratory and allowed them to mate. They observed that walking sticks from chamise plants preferred to mate with others from chamise plants, and those from wild lilacs preferred others from wild lilacs, indicating that natural selection had favored behavioral (as well as color) differences that accompanied the food preferences. This selective mating assures that the offspring will resemble the parents' host plant. Although the two color forms are still capable of interbreeding, the scientists hypothesize that they are observing the early stages of the splitting of a single species into two different species. Inherited traits that cause differently colored insects to resemble and prefer dining on different plant species create a type of *ecological isolation* (described in Chapter 16) in which the two color forms are unlikely to encounter one another, and unlikely to mate if they do. This sets the stage for further divergence of both physical and behavioral traits as the two forms encounter different selection pressures based on their preference for different plants.

CASE STUDY REVISITED Invasion of the Zebra Mussels

About 5 years after the zebra mussel arrived, scientists were pleased to see a native sponge growing on top of zebra mussels. Both sponges and mussels obtain food by filtering water and removing microscopic algae, so these species compete with one another for food. In some study areas, the number of mussels has declined, partly as a result of being smothered by sponges and partly from being eaten by another exotic species: the round goby. In 1990 a biologist at the University of Michigan found a round goby in the St. Clair River. The goby probably arrived the same way the mussels did and also originated in southeastern Europe. Recognizing one of its favorite prey, the 5-inch-long predator immediately began feasting on small zebra mussels and expanding its range into habitats already invaded by mussels; gobies are now found in all five Great Lakes.

Is this an accidental solution to the mussel problem? Unfortunately not. The gobies ignore the largest zebra mussels, so these continue to spawn. Further, the gobies are not picky eaters. In addition to mussels, they will eat the eggs and young of any other fish in their habitat, including smallmouth bass, walleye, perch, and sculpins. Researchers are now investigating ways to halt the goby's spread toward the Mississippi River. Meanwhile, zebra mussels are invading new waterways.

Consider This: Although the round goby was introduced accidentally, many exotic predators have been imported to control exotic pests, and some officials have even proposed importing exotic predators to control native pest species, such as grasshoppers. Discuss the implications of this approach for ecological communities and for native species. Describe the types of studies that should be conducted before any new predator is imported.

Links to Life: Frightening Flowerbeds

The hillsides of the California coast are now home to dense stands of a towering plumed grass. The word *exotic* perfectly describes this pampas grass, which reaches 9 feet in height and allows little else to grow. This South American native has become a favorite of North American gardeners. Each plume can produce tens of thousands of seeds that readily colonize sunny locations with exposed soil. Although recognized as an invasive weed, it is still surprisingly easy to purchase from nurseries, catalogs, Internet suppliers, and even from large discount department stores. The same its true for many exotics, because laws have not been enacted to ban their sale or import.

Recently, both Wal-Mart and Home Depot voluntarily stopped selling pampas grass in their California stores. Both retailers made the decision based on public input from concerned individuals. You, too, can help prevent the landscape from being transformed by invasive imports. Find out from your local conservation agency or the Internet what type of plants are considered invasive or noxious weeds in your area. If you see these for sale locally, explain the problem to the retailer and provide a printout of your findings. Ask nursery catalog companies and Internet retailers not to ship these species to any areas where they have been declared noxious weeds. Do you have a bit of time and want some healthy exercise? Many parks use volunteer labor to remove exotics. Finally, especially if you live near open space, be particularly careful about what you choose to plant in your garden or pond. When it comes to controlling exotic species, individuals can make a big difference—for better or for worse.

Summary of Key Concepts

40.1 Why Are Community Interactions Important?

Community interactions influence population size, and the interacting populations within communities act on one another as agents of natural selection. Thus, community interactions also shape the bodies and behaviors of members of the interacting populations.

40.2 What Are the Effects of Competition Among Species?

The ecological niche defines all aspects of a species' habitat and interactions with its living and nonliving environments. Each species occupies a unique ecological niche. Interspecific competition occurs when the niches of two populations within a community overlap. When two species with the same niche are forced—under laboratory conditions—to occupy the same ecological niche, one species always outcompetes the other. Species within natural communities have evolved in ways that avoid excessive niche overlap, with behavioral and physical adaptations that allow resource partitioning. Interspecific competition limits both the size and the distribution of competing populations.

40.3 What Are the Results of Interactions Between Predators and Their Prey?

Predators eat other organisms and are generally larger and less abundant than their prey. Predators and prey act as strong agents of selection on one another. Prey animals have evolved a variety of protective colorations that render them either inconspicuous (camouflage) or startling (startle coloration) to their predators. Some prey are poisonous and exhibit warning coloration by which they are readily recognized and avoided. The situation in which an animal has evolved to resemble another is called mimicry. Both predators and prey have evolved a variety of toxic chemicals for attack and defense. Plants that are preyed on have evolved elaborate defenses, ranging from poisons to thorns to overall toughness. These defenses, in turn, have selected for predators that can detoxify poisons, ignore thorns, and grind down tough tissues.

40.4 What Is Symbiosis?

Symbiotic relationships involve two species that interact closely over an extended time, and include parasitic, commensal, and mutualistic associations. In parasitism, the parasite feeds on a larger, less abundant host, usually harming it but not killing it immediately. In commensalism, one species benefits, typically by finding food more easily in the presence of the other species, which is not affected by the association. Mutualism benefits both symbiotic species.

40.5 How Do Keystone Species Influence Community Structure?

Keystone species have a greater influence on community structure than can be predicted by their numbers. Removal of a keystone species will radically alter community structure to an extent that would not be predicted on the basis of the species' abundance.

40.6 Succession: How Do Community Interactions Cause Change Over Time?

Succession is a progressive change over time in the types of populations that comprise a community. Primary succession, which may take thousands of years, occurs where no remnant of a previous community existed (such as on rock scraped bare by a glacier or cooled from molten lava, on a sand dune, or in a newly formed glacial lake). Secondary succession occurs much more rapidly because it builds on the remains of a disrupted community, such as an abandoned field or the aftermath of a fire. Secondary succession on land is initiated by fast-growing, readily dispersing pioneer plants, which are eventually replaced by longer-lived, generally larger and more shade-tolerant species. Uninterrupted succession ends with a climax community, which tends to be self-perpetuating unless acted on by outside forces, such as fire or human activities. Some ecosystems, including tallgrass prairie and farm fields, are maintained in relatively early stages of succession by periodic disruptions.

Key Terms

aggressive mimicry *p. 826*
biome *p. 835*
camouflage *p. 823*
climax community *p. 832*
coevolution *p. 820*
commensalism *p. 828*
community *p. 820*
competitive exclusion principle *p. 820*
disturbance *p. 830*
ecological niche *p. 820*
exotic species *p. 834*
interspecific competition *p. 820*
intertidal zone *p. 822*
intraspecific competition *p. 822*
keystone species *p. 829*
mimicry *p. 825*
mutualism *p. 828*
parasite *p. 828*
pioneer *p. 832*
primary succession *p. 832*
resource partitioning *p. 821*
secondary succession *p. 832*
startle coloration *p. 826*
subclimax *p. 832*
succession *p. 830*
symbiosis *p. 828*
warning coloration *p. 825*

Thinking Through the Concepts

To take a multiple-choice quiz with feedback on the contents of this chapter, visit http://www.prenhall.com/audesirk7. *Log in to the Web site selected by your instructor and navigate to the Self Test section for this chapter.*

? Review Questions

1. Define an ecological *community* and list three important types of community interactions.
2. Describe four very different ways in which specific plants and animals protect themselves from being eaten. In each, describe an adaptation that might evolve in predators of these species that would overcome their defenses.
3. List two important types of symbiosis; define and provide an example of each.
4. Which type of succession would occur on a clear-cut (a region in which all the trees have been removed by logging) in a national forest, and why?
5. List two subclimax and two climax communities. How do they differ?
6. Define *succession*, and explain why it occurs.

Applying the Concepts

1. Herbivorous animals that eat seeds are considered by some ecologists to be predators of plants, and herbivorous animals that eat leaves are considered to be parasites of plants. Discuss the validity of this classification scheme.
2. An interesting interspecific relationship exists between the tarantula spider and the tarantula hawk wasp. This wasp attacks tarantulas, paralyzing them with venom from their stingers. The wasp then lays eggs on the paralyzed spider. The eggs hatch, and the young eat the living, immobilized tissues of the spider. Discuss whether this relationship between spider and wasp exemplifies parasitism or predation.
3. An ecologist visiting an island finds two very closely related species of birds, one of which has a slightly larger bill than the other. Interpret this finding with respect to the competitive exclusion principle and the ecological niche, and explain both concepts.
4. Think about the case of the camouflaged frogfish and its prey. As the frogfish sits camouflaged on the ocean floor, wiggling its lure, a small fish approaches the lure and is eaten, while a very large predatory fish fails to notice the frogfish. Describe all the possible types of community interactions and adaptations these organisms have selected for. Remember that predators can also be prey and that community interactions are complex!
5. Design an experiment to determine whether the kangaroo is a keystone species in the Australian outback.
6. Why is it difficult to study succession? Suggest some ways you would approach this challenge for a few different ecosystems.

For More Information

Amos, W. H. "Hawaii's Volcanic Cradle of Life." *National Geographic*, July 1990. A naturalist explores succession on lava flows.

Enserink, M. "Biological Invaders Sweep In"; Kaiser, J. "Stemming the Tide of Invading Species"; and Malkoff, D. "Fighting Fire with Fire." *Science*, September 17, 1999. A series of articles covering the problems posed by exotic species.

Freindel, S. "If All the Trees Fall in the Forest ..." *Discover*, December 2002. The imported fungus responsible for chestnut blight killed 3.5 billion chestnuts in the 1940s. Now, a new exotic species of fungus threatens a variety of native trees, including live oaks and redwoods.

Gutin, J. C. "Purple Passion." *Discover*, August 1999. The exotic plant called purple loosestrife can grow 10 feet tall. Introduced to the U.S. East Coast 200 years ago, it is now rapidly spreading westward, threatening native species.

Harder, B. "Stemming the Tide." *Science News*, April 13, 2002. How can ships' ballast water be prevented from spreading exotic species such as the zebra mussel?

Power, M., et al. "Challenges in the Quest for Keystones." *Bioscience*, September 1996. A comprehensive review of the importance of keystone species and the challenges of studying them.

Withgott, J., "California Tries to Rub Out the Monster of the Lagoon." *Science*, March 22, 2002. The invasive tropical algae now blankets coastal areas in the Mediterranean and Australia, while California desperately tries to keep it from invading the U.S. West Coast.

Media Activities

To access a Media Activity visit http://www.prenhall.com/audesirk7. *Log in to the Web site selected by your instructor, navigate to this chapter, and select the appropriate Media Activity number.*

40.1 Competition

Estimated time: 10 minutes

Explore what competition is and the different types of competition between and within species.

40.2 The Importance of Keystone Species

Estimated time: 5 minutes

Explore how the loss of a keystone species affects the community in which it lives.

40.3 Primary Succession: Glacier Bay, Alaska

Estimated time: 10 minutes

Explore the process of succession in Glacier Bay, Alaska. View scientific data to determine how the pattern of succession by numerous tree species varies through the bay.

40.4 Web Investigation: Invasion of the Zebra Mussels

Estimated time: 30 minutes

As this chapter's Case Study in the textbook has pointed out, the zebra mussel has had profound effects on the economy and ecology of freshwater habitats in eastern and midwestern North America following its lightning-fast spread across the continent. In this activity, explore some more of the questions and research about this species.

CHAPTER

41 How Do Ecosystems Work?

A grizzly intercepts a salmon on its spawning journey as it struggles up a waterfall in an attempt to reach the same streambed where it hatched years earlier.

AT A GLANCE

CASE STUDY When the Salmon Return

Sockeye salmon of the Pacific Northwest have a remarkable life cycle. Hatching in shallow depressions in the gravel bed of a swiftly flowing stream, they follow the stream's path into ever-larger rivers that finally enter the ocean. Emerging into estuaries—wetlands where fresh water and salt water mix—the salmon's remarkable physiology allows the fish to adapt to the change to salt water before they reach the sea. The small percentage of young salmon that evade predators grow to adulthood, feeding on crustaceans and smaller fish. Years later, their bodies undergo another transformation. As they reach sexual maturity, a compelling instinctive drive—still poorly understood despite decades of research—lures them back to fresh water, but not any stream or river will do. The salmon swim along the coast (probably navigating by sensing Earth's magnetic field) until the unique scent of their home stream entices them to swim inland. Battling swift currents, leaping up small waterfalls, undulating over shallow sandbars and evading human traps, they carry their precious cargo of sperm and eggs back home to renew the cycle of life. This journey back to their birthplace is remarkable in another way. Nutrients almost always flow downstream, carried from the land into the ocean; the salmon, filled with muscle and fat acquired from feeding in the ocean, not only battle against the flow of the current in their upstream journey; they also reverse the usual movement of nutrients. What awaits the salmon at their journey's end? How does their journey impact the web of life upstream?

41.1 What Are the Pathways of Energy and Nutrients?

The activities of life, from the migration of the salmon to active transport of molecules through a cell membrane, are powered by the *energy* of sunlight. The molecules of life are constructed of chemical building blocks that are obtained as *nutrients* from the environment. Solar energy continuously bombards Earth, is used and transformed in the chemical reactions that power life, and is ultimately converted to heat energy that radiates back into space. Chemical nutrients, in contrast, remain on Earth. While they may change in form and distribution, and even be transported among different ecosys-

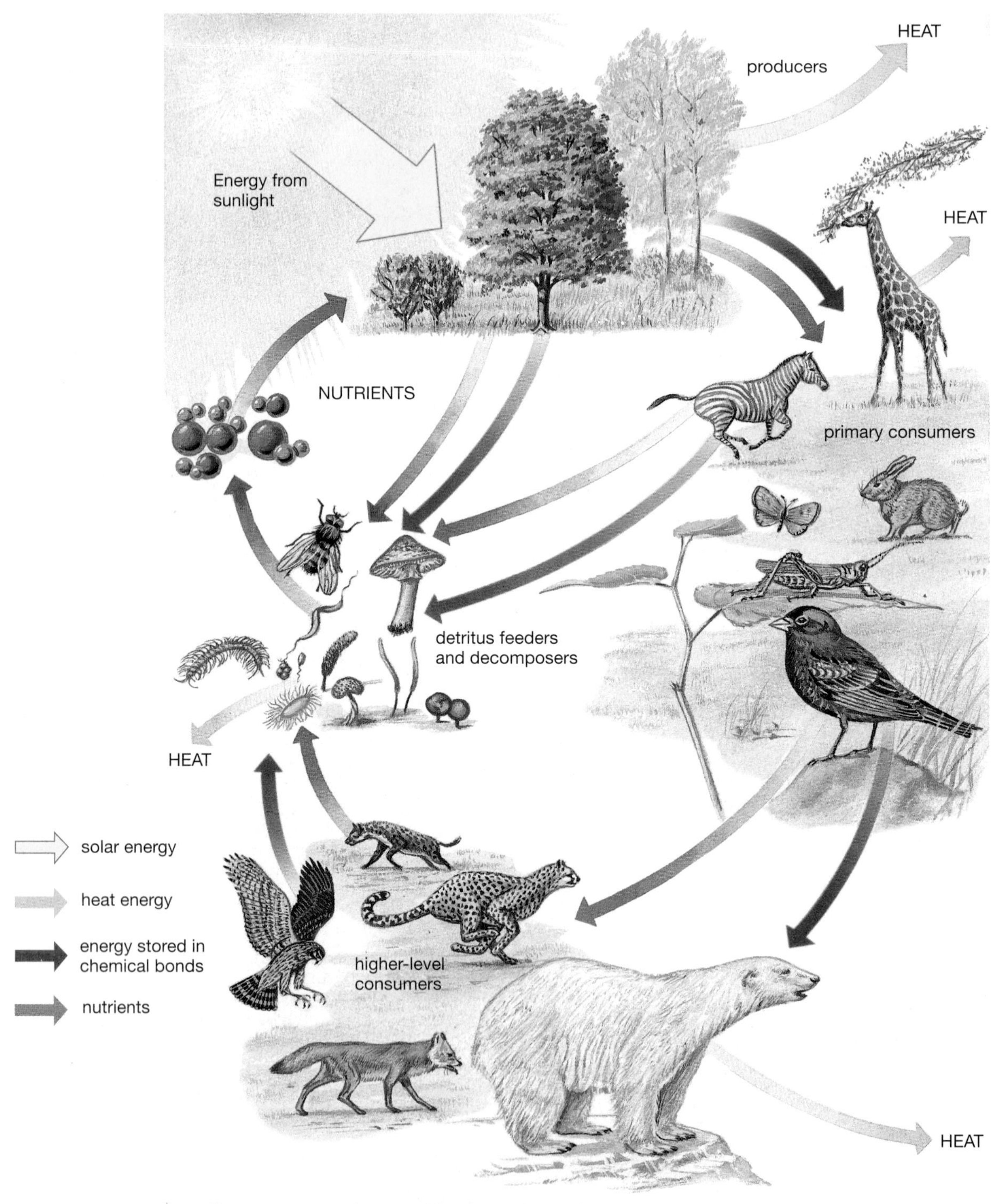

FIGURE 41-1 Energy flow, nutrient cycling, and feeding relationships in ecosystems
Note that nutrients (purple) neither enter nor leave the cycle. Energy (yellow), continuously supplied to producers as sunlight, is captured in chemical bonds and transferred through various levels of organisms (red). At each level, some energy is lost as heat (orange).

Media Activity
41.1 Ecology Models—Building a Food Web

tems, nutrients are constantly recycled. Thus, two basic laws underlie ecosystem function. First, energy moves through the communities within ecosystems in a continuous one-way flow, needing constant replenishment from an outside source, the sun. Second, nutrients constantly cycle and recycle within and among ecosystems (Fig. 41-1). These laws shape the complex interactions among populations within ecosystems, and between communities and their abiotic environment.

41.2 How Does Energy Flow Through Communities?

Energy Enters Communities Through Photosynthesis

Ninety-three million miles away, the sun fuses hydrogen molecules into helium molecules, releasing tremendous quantities of energy. A tiny fraction of this energy reaches Earth in the form of electromagnetic waves, including heat, light, and ultraviolet energy. Of the energy that reaches Earth, much is reflected by the atmosphere, clouds, and Earth's surface. Still more is absorbed as heat by Earth and its atmosphere, leaving only about 1% to power all life. Of this 1%, which reaches Earth's surface as light, green plants and other photosynthetic organisms capture 3% or less. The teeming life on this planet is thus supported by less than 0.03% of the energy reaching Earth from the sun.

During photosynthesis (described in detail in Chapter 7), pigments such as chlorophyll absorb specific wavelengths of sunlight. This solar energy is then used in reactions that store energy in chemical bonds, producing sugar and other high-energy molecules (Fig. 41-2). Photosynthetic organisms, from mighty oak trees to single-celled diatoms in the ocean, are called **autotrophs** (Greek, "self-feeders") or **producers**, because they produce food for themselves using nonliving nutrients and sunlight. In doing so, they directly or indirectly produce food for nearly all other forms of life as well. Organisms that cannot photosynthesize, called **heterotrophs** (Greek, "other-feeders") or **consumers**, must acquire energy and many of their nutrients prepackaged in the molecules that compose the bodies of other organisms.

The amount of life that a particular ecosystem can support is determined by the energy captured by the producers in that ecosystem. The energy that photosynthetic organisms store and make available to other members of the community over a given period is called **net primary productivity**. Net primary productivity can be measured in units of energy (calories) stored per unit area by autotrophs during a specified timespan. It is also measured as the **biomass**, or dry weight of organic material stored in producers that is added to the ecosystem per unit area over a specified time. The productivity of an ecosystem is influenced by many environmental variables, including the amount of nutrients available to the producers, the amount of sunlight reaching them, the availability of water, and the temperature. In the desert, for example, lack of water limits productivity; in the open ocean, light is limited in deep water and nutrients are limited in surface water. When resources are abundant, as in estuaries and tropical rain forests, productivity is high. Some average productivities for a variety of ecosystems are shown in Figure 41-3.

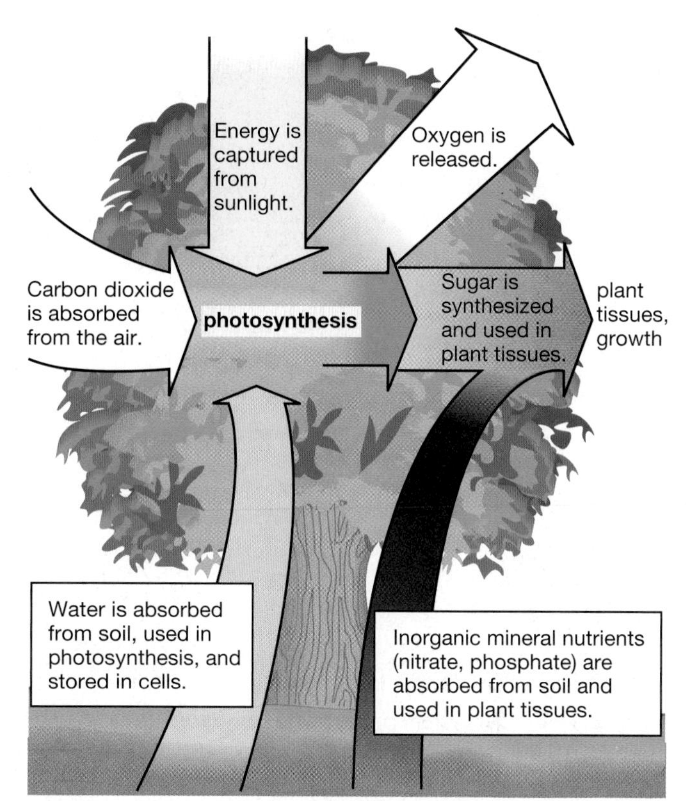

FIGURE 41-2 Primary productivity: photosynthesis Photosynthetic organisms capture solar energy. They also acquire inorganic nutrients from reservoirs and incorporate them into living tissue. Photosynthetic organisms ultimately provide all the energy and most of the nutrients for organisms in higher trophic levels.

Energy Is Passed from One Trophic Level to Another

Energy flows through communities from photosynthetic producers through several levels of consumers. Each category of organisms is called a **trophic level** (literally, "feeding level"). Producers—from redwood trees to cyanobacteria—form the first trophic level, obtaining their energy directly from sunlight (see Fig. 41-1). Consumers occupy several trophic levels. Some consumers feed directly and exclusively on producers, the most abundant living energy source in any ecosystem. These **herbivores** (meaning "plant eaters"), ranging from grasshoppers to giraffes, are also called **primary consumers**; they form the second trophic level. **Carnivores** (meaning "meat eaters")—such as the spider, eagle, and

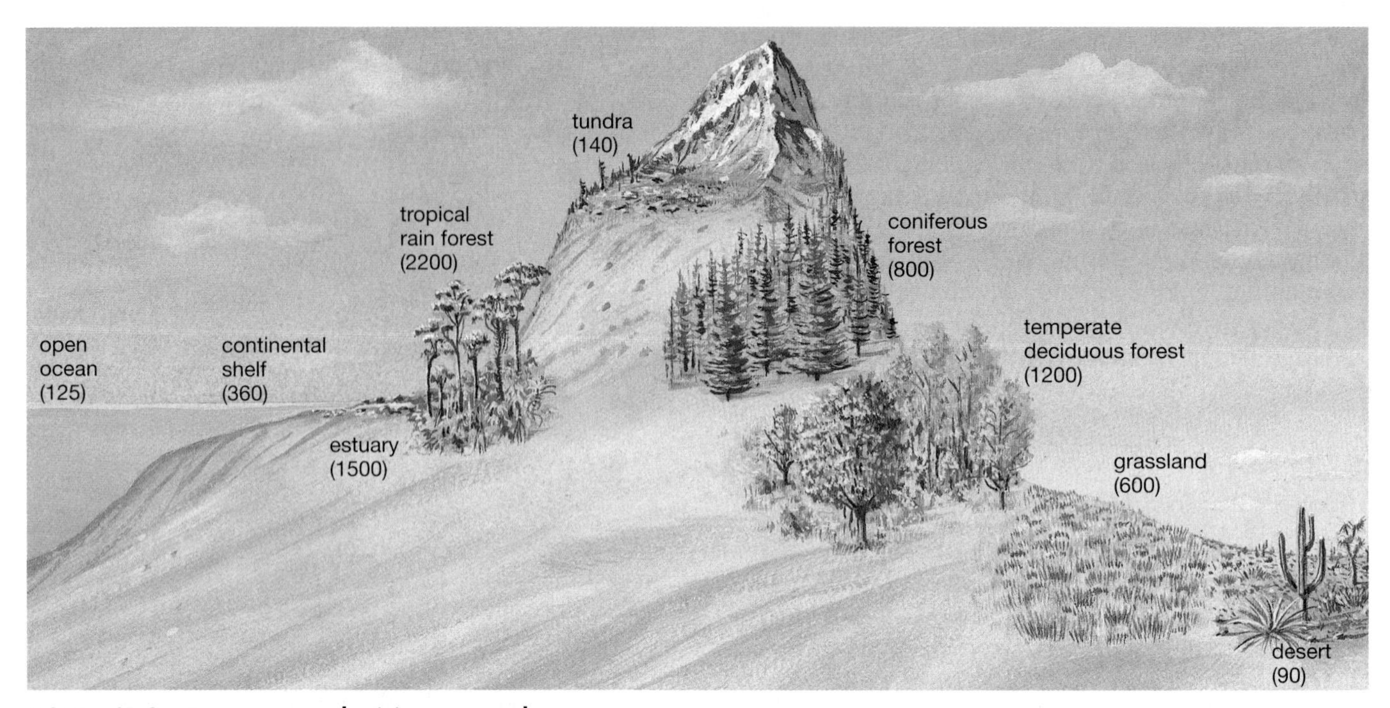

FIGURE 41-3 Ecosystem productivity compared
Average net primary productivity, in grams of organic material per square meter per year, of some terrestrial and aquatic ecosystems. Notice the enormous differences in productivity among the ecosystems. **QUESTION** What factors contribute to these differences in productivity?

wolf—are predators that feed primarily on primary consumers. Carnivores, also called **secondary consumers**, form the third trophic level. Some carnivores occasionally eat other carnivores; when doing so, they occupy the fourth trophic level, **tertiary consumers**.

Food Chains and Food Webs Describe the Feeding Relationships Within Communities

To illustrate who feeds on whom in a community, it is common to identify a representative of each trophic level that eats a representative of the level below it. This linear feeding relationship is called a **food chain**. As illustrated in Figure 41-4, different ecosystems have radically different food chains.

Natural communities, however, rarely contain well-defined groups of primary, secondary, and tertiary consumers. A **food web** shows many interconnecting food chains and more accurately describes the actual feeding relationships within a given community (Fig. 41-5, p. 846). Some animals, such as raccoons, bears, rats, and humans, are **omnivores** (Latin, "eating all")—that is, at different times they act as primary, secondary, and occasionally tertiary (third-level) consumers. Many carnivores will eat either herbivores or other carnivores, thus acting as secondary or tertiary consumers, respectively. An owl, for instance, is a secondary consumer when it eats a mouse, which feeds on plants, but a tertiary consumer when it eats a shrew, which feeds on insects. A shrew that eats a carnivorous insect is a tertiary consumer, and the owl that fed on the shrew is then a quaternary (fourth-level) consumer. When digesting a spider, a carnivorous plant, such as the sundew, can "tangle the web" hopelessly by serving simultaneously as a photosynthetic producer and a secondary consumer!

Detritus Feeders and Decomposers Release Nutrients for Reuse

Among the most important strands in the food web are the *detritus feeders* and *decomposers*. The **detritus feeders** are an army of small and often unnoticed animals and protists that live on the refuse of life: molted exoskeletons, fallen leaves, wastes, and dead bodies (*detritus* means "debris"). The network of detritus feeders is extremely complex, and includes earthworms, mites, protists, centipedes, some insects, a unique land-dwelling crustacean called a pillbug (or "rolypoly"), nematode worms, and even a few large vertebrates such as vultures. They consume dead organic matter, extract some of the energy stored within it, and excrete it in a further decomposed state. Their excretory products serve as food for other detritus feeders and for decomposers. The **decomposers** are primarily fungi and bacteria that digest food outside their bodies by secreting digestive enzymes into the environment. They absorb the nutrients they need and release the remaining nutrients. The black coating or gray fuzz you may notice on tomatoes and bread crusts left too long in your refrigerator are fungal decomposers hard at work. Thanks to detritus feeders and decomposers, most of the stored energy is eventually utilized.

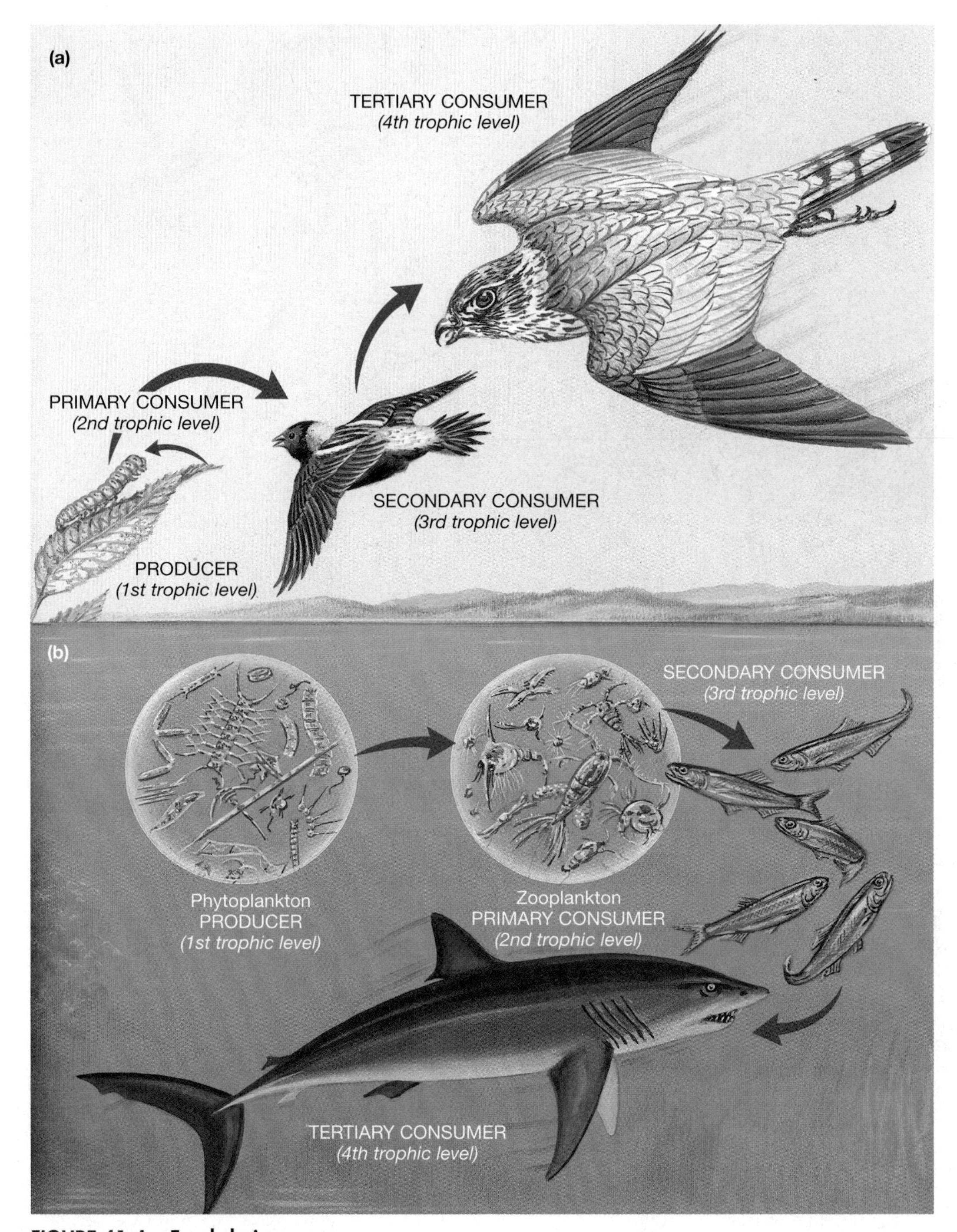

FIGURE 41-4 Food chains
(a) A simple terrestrial food chain. **(b)** A simple marine food chain.

Through the activities of detritus feeders and decomposers, the bodies and wastes of living organisms are reduced to simple molecules—such as carbon dioxide, water, minerals, and organic molecules—that return to the atmosphere, soil, and water. By liberating nutrients for reuse, detritus feeders and decomposers form a vital link in the nutrient cycles of ecosystems. In some ecosystems, such as deciduous forests, more energy passes through the detritus feeders and decomposers than through the primary, secondary, or tertiary consumers.

What would happen if detritus feeders and decomposers disappeared? This portion of the food web, although inconspicuous, is absolutely essential to life on Earth. Without it, communities would gradually be smothered by accumulated wastes and dead bodies. The nutrients stored in these bodies would be unavailable to enrich the soil. The quality of the soil would become poorer and poorer until plant life could no longer be sustained. With plants eliminated, energy would cease to enter the community; the higher trophic levels, including humans, would disappear as well.

FIGURE 41-5 A food web
A simple terrestrial food web on a short-grass prairie.

Energy Transfer Through Trophic Levels Is Inefficient

As we discussed in Chapter 6, a basic law of thermodynamics is that energy use is never completely efficient. For example, as your car burns gasoline, about 75% of the energy released is immediately lost as heat. This is also true in living systems. For example, splitting the chemical bonds of ATP to cause muscular contraction produces heat as a by-product; this is why walking briskly on a cold day will warm you. Small amounts of waste heat are produced by all the biochemical reactions that keep cells alive.

Energy transfer from one trophic level to the next is also quite inefficient. When a caterpillar (a primary consumer) eats the leaves of a tomato plant (a producer), only some of the solar energy originally trapped by the plant is available to the insect. Some energy was used by the plant for growth and maintenance, and more was lost as heat during these processes. Some energy was converted into the chemical bonds of molecules such as cellulose, which the caterpillar cannot digest. Therefore, only a fraction of the energy captured by the first trophic level is available to organisms in the second trophic level. The energy consumed by the caterpillar is in turn partially used to power crawling and the gnashing of mouthparts. Some is used to construct the indigestible exoskeleton, and much is given off as heat. All this energy is unavailable to the songbird in the third trophic level when it eats the caterpillar. The bird loses energy as body heat, uses more in flight, and converts a considerable amount into indigestible feathers, beak, and bone. All this energy will be unavailable to the hawk that catches it. A simplified model of energy flow through the trophic levels in a deciduous forest ecosystem is illustrated in Figure 41-6.

Energy Pyramids Illustrate Energy Transfer Between Trophic Levels

Studies of a variety of communities indicate that the net transfer of energy between trophic levels is roughly 10% efficient, although transfer among levels within different communities varies significantly. This means that, in general, the energy stored in primary consumers (herbivores) is only about 10% of the energy stored in the bodies of producers. In turn, the bodies of secondary consumers possess roughly 10% of the energy stored in

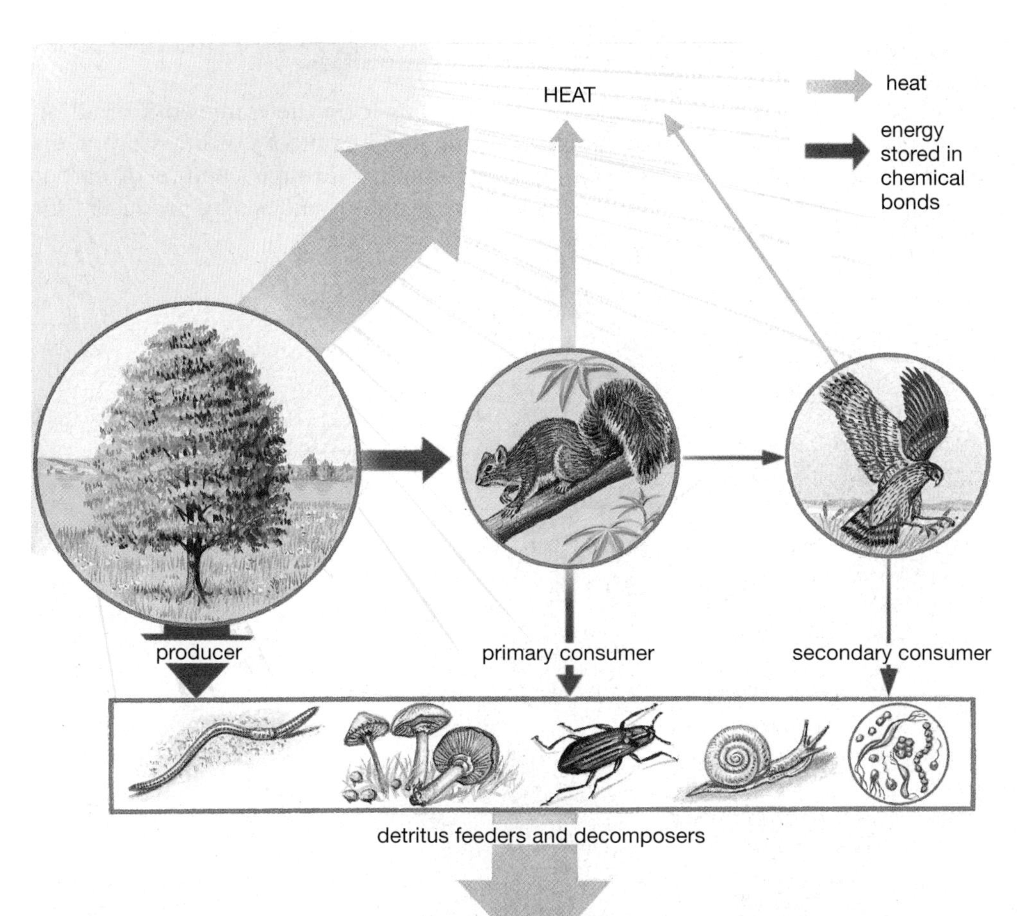

FIGURE 41-6 Energy transfer and loss
This diagram shows approximate amounts of energy transferred between trophic levels in the form of chemical energy (red) and lost in the form of heat (orange) in a forest community. The width of the arrows is roughly proportional to the quantity of energy transferred or lost. **QUESTION** Why is so much energy lost as heat? Explain this in terms of the second law of thermodynamics (introduced in Chapter 6) and relate it to the energy pyramid in Figure 41-7.

primary consumers. In other words, for every 100 calories of solar energy captured by grass, only about 10 calories are converted into herbivores, and only 1 calorie into carnivores. This inefficient energy transfer between trophic levels is called the "10% law." An **energy pyramid**, which shows maximum energy at the base and steadily diminishing amounts at higher levels, illustrates the energy relationships between trophic levels graphically (Fig. 41-7). Ecologists sometimes use biomass as a measure of the energy stored at each trophic level. Because the dry weight of organisms' bodies at each trophic level is roughly proportional to the amount of energy stored in the organisms at that level, a *biomass pyramid* for a given community often has the same general shape as its energy pyramid.

What does this mean for community structure? If you wander through an undisturbed ecosystem, you will notice that the predominant organisms are plants. Plants have the most energy available to them, because they trap it directly from sunlight. The most abundant animals will be those that feed on plants, and carnivores will be relatively rare. The inefficiency of energy transfer also has important implications for human food production. The lower the trophic level we utilize, the more food energy is available to us; in other words, far more people can be fed on grain than on meat.

An unfortunate side effect of the inefficiency of energy transfer, coupled with human production and release of toxic chemicals, is that certain persistent toxic chemicals become concentrated in the bodies of carnivores, including people, as described in "Earth Watch: Food Chains Magnify Toxic Substances."

41.3 How Do Nutrients Move Within and Among Ecosystems?

In contrast to the energy of sunlight, nutrients do not flow down onto Earth in a steady stream from above. Essentially the same pool of nutrients has been supporting life for more than 3 billion years. *Nutrients* are elements and small molecules that form the chemical building blocks of life. Some, called *macronutrients*, are required by organisms in large quantities. These include water, carbon, hydrogen, oxygen, nitrogen, phosphorus, sulfur, and calcium. *Micronutrients,* including zinc, molybdenum, iron, selenium, and iodine, are required only in trace quantities. **Nutrient cycles**, also called **biogeochemical cycles**, describe the pathways these substances follow as they move from communities to nonliving portions of ecosystems and back again to communities.

The sources and storage sites of nutrients are called **reservoirs**. The major reservoirs are generally in the nonliving, or abiotic, environment. For example, carbon has several major reservoirs: It is stored as carbon dioxide gas in the atmosphere, in dissolved form in oceans, in rock as limestone, and as fossil fuels underground. In the following section, we briefly describe the cycles of carbon, nitrogen, phosphorus, and water.

Carbon Cycles Through the Atmosphere, Oceans, and Communities

Chains of carbon atoms form the framework of all organic molecules, the building blocks of life. Carbon enters the living community through capture of carbon dioxide (CO_2) during photosynthesis by producers. On

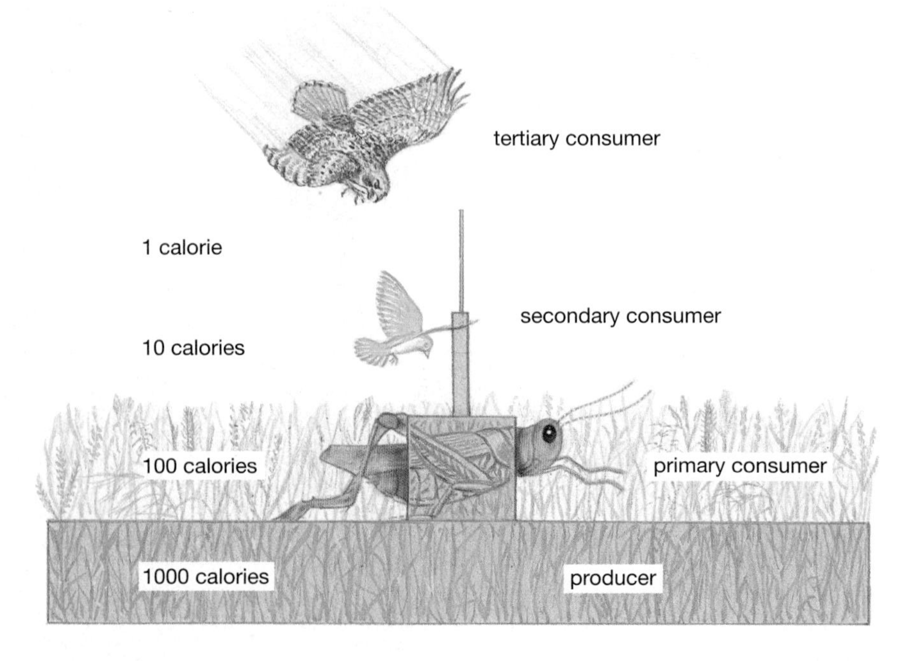

FIGURE 41-7 An energy pyramid for a prairie ecosystem
Each trophic level from producer to tertiary consumer stores less energy. The width of each rectangle is proportional to the energy stored at that trophic level. A biomass pyramid for this ecosystem would look quite similar.

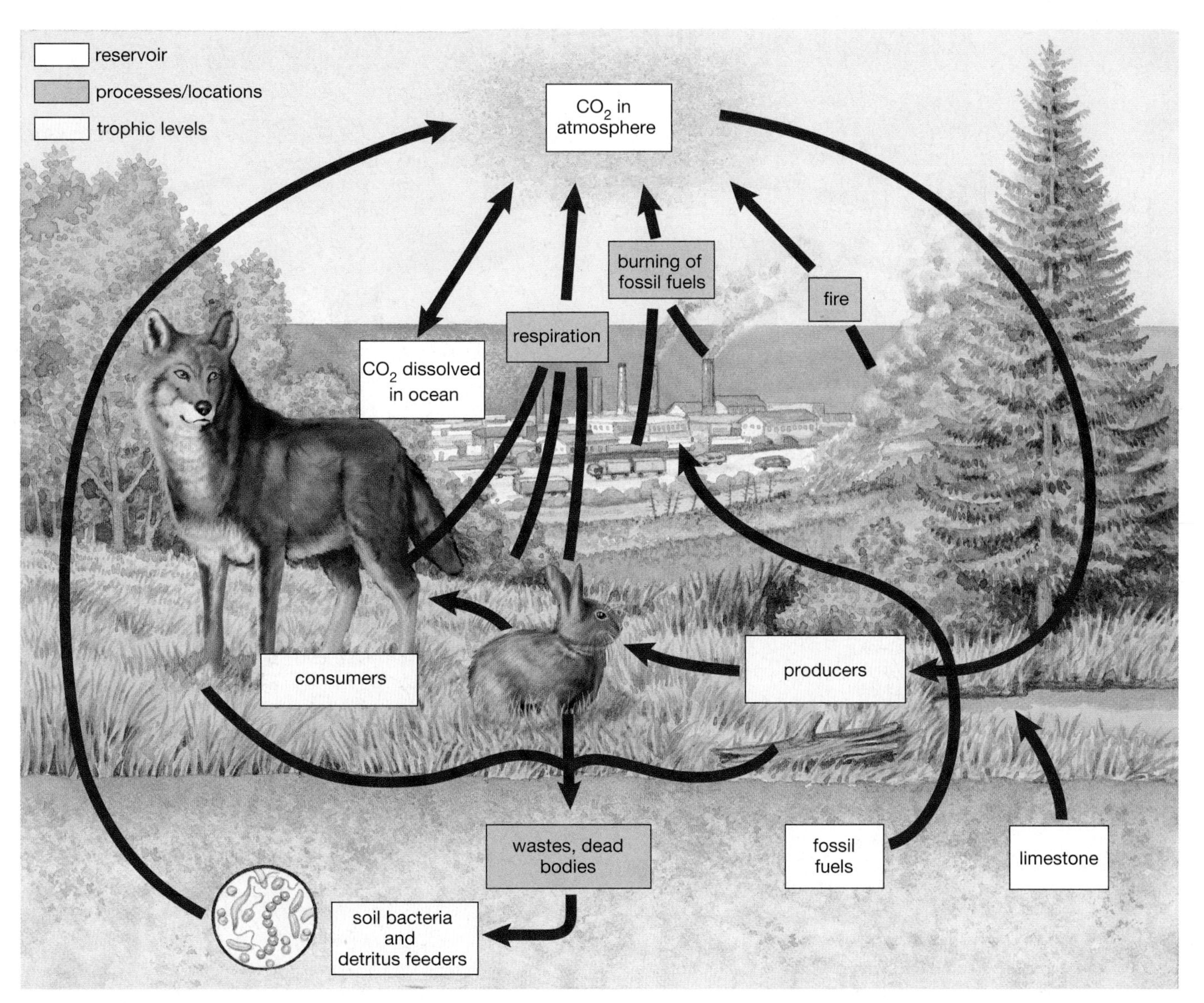

FIGURE 41-8 The carbon cycle
Carbon is stored in the atmosphere, in limestone, in the oceans, and in fossil fuels. Carbon dioxide is captured from the atmosphere during photosynthesis and passed up through the trophic levels. It is released during respiration from all trophic levels and by the burning of forests and fossil fuels.

Media Activity
41.2 The Global Carbon Cycle and Greenhouse Effect

land, producers acquire CO_2 from the atmosphere, where it represents a mere 0.036% of all atmospheric gases. Aquatic producers in the ocean, such as seaweeds and diatoms, find abundant CO_2 for photosynthesis dissolved in the water; in fact, far more CO_2 is stored in the oceans than in the atmosphere. Producers return some CO_2 to the atmosphere and ocean during cellular respiration and incorporate the rest into their bodies. Primary consumers, such as cows, shrimp, or tomato hornworms, eat the producers and acquire the carbon stored in their tissues. These herbivores also release some carbon through respiration and store the rest, which is sometimes consumed by organisms in higher trophic levels. All living things eventually die, and their bodies are broken down by detritus feeders and decomposers. Cellular respiration by these organisms returns CO_2 to the atmosphere and oceans. Carbon dioxide passes freely between these two great reservoirs (Fig. 41-8).

Some carbon cycles much more slowly. For example, mollusks and marine protists extract CO_2 dissolved in water and combine it with calcium to form calcium carbonate ($CaCO_3$), from which they construct their shells. After death, the shells of these organisms collect in undersea deposits, are buried, and may eventually be converted to limestone. Geological events may expose the limestone, which dissolves gradually as water runs over it, making the carbon available to living organisms once again.

Another long-term portion of the carbon cycle is the production of fossil fuels. **Fossil fuels** form from the remains of ancient forms of life. The carbon in the organic molecules of these prehistoric organisms was transformed by high temperatures and pressures over millions of years into coal, oil, and natural gas. The energy of prehistoric sunlight is also trapped in these fossil fuels, captured by autotrophs and then passed upward through various trophic levels before being sequestered

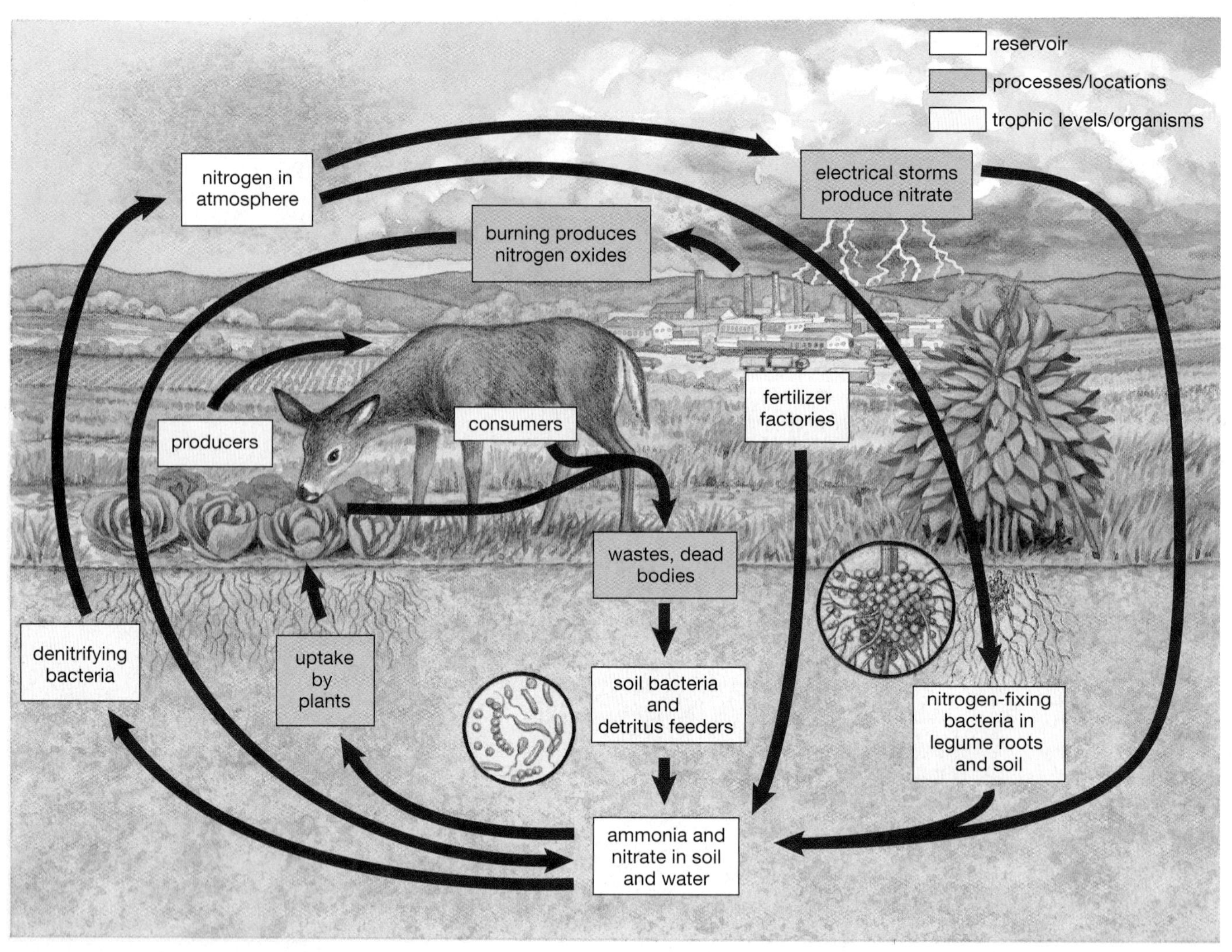

FIGURE 41-9 The nitrogen cycle
Atmospheric nitrogen is combined with oxygen to form nitrates by lightning or by burning and then is carried to Earth dissolved in rain. Nitrogen-fixing bacteria also produce ammonia from N_2. Nitrate and ammonia are synthesized for use in fertilizers. These are absorbed by plants and other producers and incorporated into biological molecules that are passed up through the trophic levels. Nitrate and ammonia are released by excretion or by decomposer bacteria into soil and water. Denitrifying bacteria convert these molecules back to atmospheric nitrogen, completing the cycle. **QUESTION** What incentives caused people to capture nitrogen from the air and pump it into the nitrogen cycle? What are some of the consequences of human augmentation of the nitrogen cycle?

in the high energy hydrocarbons that we burn today. When people burn fossil fuels to utilize this stored energy, CO_2 is released into the atmosphere. In addition to the burning of fossil fuels, human activities such as cutting and burning Earth's great forests (where much carbon is stored) also increase the amount of CO_2 in the atmosphere, as we will discuss later in this chapter.

The Major Reservoir for Nitrogen Is the Atmosphere

The atmosphere contains about 79% nitrogen gas (N_2) and is thus the major reservoir for this important nutrient. Nitrogen is a crucial component of proteins, many vitamins, and the nucleic acids DNA and RNA. Interestingly, neither plants nor animals can extract this gas from the atmosphere. Instead, plants must be supplied with nitrate (NO_3^-) or ammonia (NH_3). But how is atmospheric nitrogen converted to these molecules? Ammonia is synthesized by certain bacteria that engage in **nitrogen fixation**, a process that combines nitrogen with hydrogen. Some of these bacteria live in water and soil. Others have entered a symbiotic association with plants called *legumes* (including alfalfa, soybeans, clover, and peas), where they live in special swellings on the roots (see Chapter 19). Decomposer bacteria can also produce ammonia from the amino acids and urea found in dead bodies and wastes. Still other bacteria convert ammonia to nitrate.

Nitrogen is also combined with oxygen by nonbiological processes: electrical storms and the combustion of forests and fossil fuels. Plants incorporate the nitrogen from ammonia and nitrate into amino acids, proteins, nucleic acids, and vitamins. These nitrogen-containing molecules from the plant are eventually consumed by primary consumers, detritus feeders, or decomposers. As

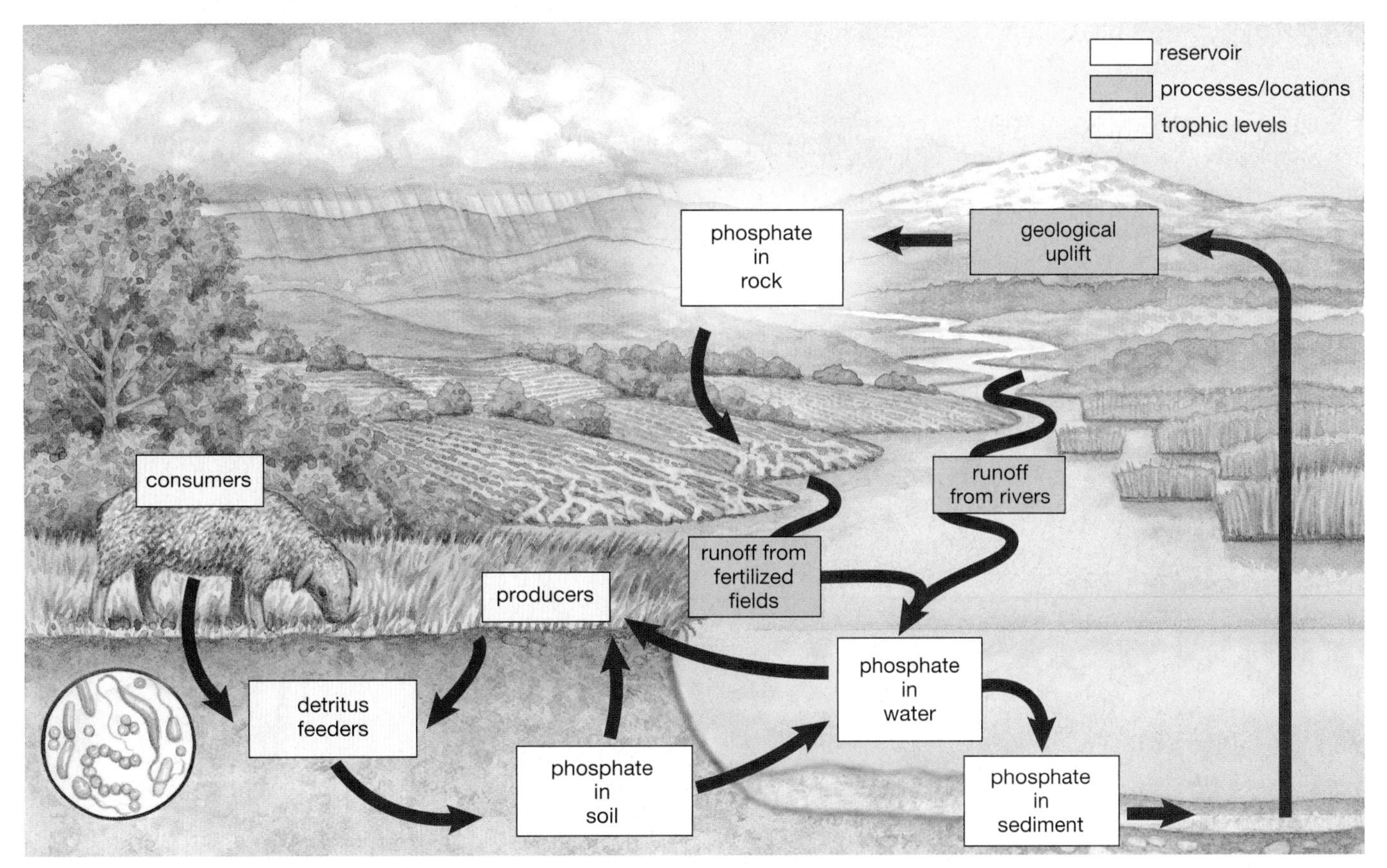

FIGURE 41-10 The phosphorus cycle
Phosphate dissolves from phosphate-rich rocks or from fertilizers and enters plants and other producers, where it is incorporated into biological molecules. These molecules are passed up through the trophic levels. Phosphate is excreted or returned to the soil and water by decomposer bacteria. It may then be reused by producers or eventually incorporated into rock.

it is passed through the food web, some of the nitrogen is released in wastes and dead bodies, which decomposer bacteria in soil or water convert back to nitrate and ammonia. This form of nitrogen is available to plants; nitrates and ammonia in soil and water constitute a second reservoir. The nitrogen cycle is completed by a continuous return of nitrogen to the atmosphere by *denitrifying bacteria*. These residents of wet soil, swamps, and estuaries break down nitrate, releasing nitrogen gas back to the atmosphere (Fig. 41-9).

In human-dominated ecosystems, such as farm fields, gardens, and lawns, ammonia and nitrate are supplied by chemical fertilizers. These fertilizers are produced by using the energy in fossil fuels to artificially "fix" atmospheric nitrogen. In fact, human activities now dominate the nitrogen cycle, creating serious environmental concerns. By burning fossil fuels, burning forests and draining wetlands that store nitrogen, cultivating legumes, and producing fertilizers by industrial processes and applying them extensively to crops, humans each year are estimated to convert over 140 million metric tons of nitrogen from its gaseous form to ammonia and nitrogen oxides (nitrogen combined with oxygen, including nitrate). By comparison, natural processes produce and utilize somewhere between 90 million and 140 million metric tons of nitrogen yearly. Runoff from fertilized farm fields into the Mississippi River is carried to the Gulf of Mexico, where scientists believe it is causing a 7000-square-mile "dead zone" that appears each summer. When they enter ecosystems, nitrogen oxides may change the composition of plant communities or destroy forests and freshwater communities by acidifying the environment, which we discuss later in this chapter.

The Phosphorus Cycle Has No Atmospheric Component

Phosphorus is a crucial component of biological molecules, including energy transfer molecules (ATP and NADP), nucleic acids, and the phospholipids of cell membranes. It is also a major component of vertebrate teeth and bones. In contrast to the cycles of carbon and nitrogen, the phosphorus cycle has no atmospheric component. The major storage site for phosphorus in ecosystems is rock, where it is bound to oxygen in the form of phosphate. As phosphate-rich rocks are exposed and eroded, rainwater dissolves the phosphate. Dissolved phosphate is readily absorbed through the roots of plants and by other autotrophs, such as photosynthetic protists and cyanobacteria, which incorporate it into biological molecules. From these producers, phosphorus is passed through food webs (Fig. 41-10). At each level, excess phosphate is excreted. Ultimately,

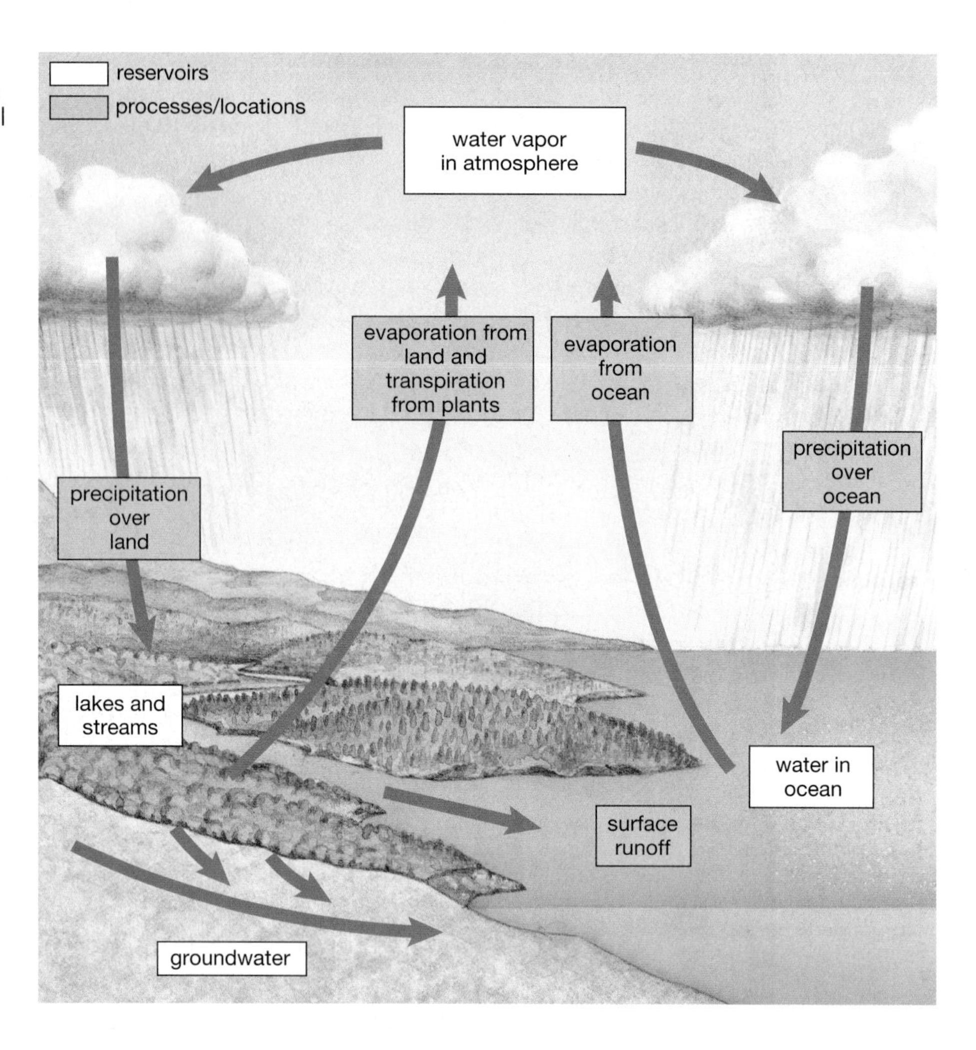

FIGURE 41-11 The hydrologic cycle The hydrologic cycle is the simplest of the nutrient cycles. Although water is essential to all life, most of its movement occurs independently of living things.

decomposers return the phosphorus that remains in dead bodies back to the soil and water in the form of phosphate. It may then be reabsorbed by autotrophs, or it may become bound to sediment and eventually reincorporated into rock.

Some of the phosphate dissolved in fresh water is carried to the oceans. Although much of this phosphate ends up in marine sediments, some is absorbed by marine producers and eventually incorporated into the bodies of invertebrates and fish. Some of these, in turn, are consumed by seabirds, which excrete large quantities of phosphorus back onto the land. At one time, the guano (droppings) deposited by seabirds along the western coast of South America was mined; it provided a major source of the world's phosphorus. Phosphate-rich rock is also mined, and the phosphate is incorporated into fertilizer. Soil that erodes from fertilized fields carries large quantities of phosphates into lakes, streams, and the ocean, where it stimulates the growth of producers. In lakes, phosphorus-rich runoff from land can stimulate such a rich growth of algae and bacteria that the natural community interactions of the lake are disrupted.

Most Water Remains Chemically Unchanged During the Hydrologic Cycle

The water cycle, or **hydrologic cycle** (Fig. 41-11), differs from most other nutrient cycles in that most water remains in the form of water throughout the cycle and is not used in the synthesis of new molecules. The major reservoir of water is the ocean, which covers about three-quarters of Earth's surface and contains more than 97% of Earth's water. Another 2% is trapped in ice, leaving only 1% available to the biosphere. The hydrologic cycle is driven by solar energy, which evaporates water, and by gravity, which draws it back to Earth in the form of precipitation (rain, snow, sleet, and dew). Most evaporation occurs from the oceans, and much water returns directly to them as precipitation. Water falling on land takes various paths. Some water is evaporated from the soil, lakes, and streams. A portion runs off the land back to the oceans, and a small amount enters underground reservoirs. Because the bodies of living things are roughly 70% water, some of the water in the hydrologic cycle enters the living communities of ecosystems. It is absorbed by the roots of plants; much of this water is evaporated back to the atmosphere from plants' leaves. A small amount is combined with carbon dioxide during photosynthesis to produce high-energy molecules. Eventually these molecules are broken down during cellular respiration, releasing water back to the environment. Consumers get water from their food or by drinking.

As the human population has grown, fresh water has become scarce in many regions of the world where the human demand for it is high. This scarcity limits the ability to grow crops; currently, 41% of the world's food is grown on irrigated cropland. Over the past few decades farmers have relied increasingly on aquifers—natural underground reservoirs—from which they pump water to irrigate their crops. Unfortunately, in many areas of the world—including China, India, Northern Africa, and the midwestern U.S.—this groundwater is being "mined" for agriculture; that is, it is removed faster than it is replenished. The High Plains aquifer, which extends from the Texas Panhandle north to South Dakota, has now lost over half of its water stores. Isotope analysis of some deep groundwater in the western U.S. suggests that this water fell as precipitation during the last ice age, about 15,000 years ago. In India, loss of groundwater places up to 25% of the grain harvest in jeopardy. Additionally, contaminated, untreated drinking water is a major problem in developing countries, where 1.2 billion people drink it. Impure water spreads diseases that kill millions of children each year.

FIGURE 41-12 A natural substance out of place
This bald eagle was killed by an oil spill off the coast of Alaska.

41.4 What Causes Acid Rain and Global Warming?

Many of the environmental problems that plague modern society have resulted from human interference in ecosystem function. Primitive peoples were sustained solely by the energy flowing from the sun, and they produced wastes that were readily taken back into the nutrient cycles. But as the population grew and technology increased, humans began to act more and more independently of these natural processes. We have mined substances—such as lead, arsenic, cadmium, mercury, uranium, and oil—that are foreign to natural ecosystems and toxic to many of the organisms in them (Fig. 41-12). In our factories, we synthesize substances never before found on Earth: pesticides, solvents, and a wide array of other industrial chemicals harmful to many forms of life. The Industrial Revolution, which began in earnest in the mid-nineteenth century, resulted in a tremendous increase in our reliance on energy from fossil fuels (rather than from sunlight) for heat, light, transportation, industry, and even agriculture. Modern industrial and agricultural processes often produce more nutrients than nutrient cycles can efficiently process. In the following sections, we describe two environmental problems of global proportion that are a direct result of human reliance on fossil fuels: acid rain and global warming.

Overloading the Nitrogen and Sulfur Cycles Causes Acid Rain

Each year, the U.S. discharges about 20 million tons of sulfur dioxide into the atmosphere, two-thirds of it from power plants burning coal or oil. The rest is largely a by-product of industrial boilers, smelters, and refineries. Although volcanoes and hot springs also release sulfur dioxide, human industrial activities account for 90% of the sulfur dioxide in the atmosphere. Twenty-four million tons of nitrogen oxides are also released from vehicles, power plants, and industry in the U.S. annually.

Excess production of these substances was identified in the late 1960s as the cause of a growing environmental threat: *acid rain*, more accurately called **acid deposition**. When combined with water vapor in the atmosphere, nitrogen oxides and sulfur dioxide are converted to nitric acid and sulfur dioxide, respectively. Days later, and often hundreds of miles from the source, these acids fall with rainwater, eating away at statues and buildings (Fig. 41-13), damaging trees and crops, and rendering lakes lifeless. Although both sulfuric and nitric acids form solutions in water vapor, sulfuric acid may also form particles that visibly cloud the air, even under dry conditions. In the U.S., the Northeast, Mid-Atlantic, Upper Midwest, and West regions, as well as the state of Florida, are the most vulnerable, because the rocks and soils that predominate there are not able to buffer the acidity.

In the U.S., amendments to the Clean Air Act in 1990 have resulted in substantial reductions in emissions of both sulfur dioxide and nitrogen oxides from power plants. Air quality and rain acidity has improved in some regions. However, nitrogen oxide emissions remain high, primarily because more gasoline is being burned by automobiles. Unfortunately, damaged ecosystems recover slowly. Many freshwater ecosystems in New York's Adirondack Mountains continue to grow more acidic, and high-elevation forests remain at risk throughout the U.S. Some Southeastern soils have become saturated with acid-releasing substances, and freshwater acid levels are increasing rapidly. Many scientists believe that considerable additional reductions in emissions will be needed to allow the recovery of damaged ecosystems.

FIGURE 41-13 Acid deposition is corrosive
This limestone statue at Rheims Cathedral in France is being dissolved by acid deposition.

Acid Deposition Damages Life in Lakes, Farms, and Forest

In the Adirondack Mountains, acid rain has made about 25% of all the lakes and ponds too acidic to support fish. But by the time the fish die, much of the food web that sustains them has been destroyed. Clams, snails, crayfish, and insect larvae die first, then amphibians, and finally fish. The result is a crystal-clear lake—beautiful but dead. The impact is not limited to aquatic organisms. Acid rain also interferes with the growth and yield of many farm crops by leeching out essential nutrients such as calcium and potassium and killing decomposer microorganisms, thus preventing the return of nutrients to the soil. Plants, poisoned and deprived of nutrients, become weak and vulnerable to infection and insect attack. High in the Green Mountains of Vermont, scientists have witnessed the death of about half of the red spruce and beech trees and one-third of the sugar maples since 1965. The snow, rain, and heavy fog that commonly cloak these eastern mountaintops are highly acidic. At a monitoring station atop Mount Mitchell in North Carolina, the pH of fog has been recorded at 2.9, more acidic than vinegar (Fig. 41-14).

Acid deposition increases the exposure of organisms to toxic metals, including aluminum, lead, mercury, and cadmium, which are far more soluble in acidified water than in water of neutral pH. Aluminum dissolved from rock may inhibit plant growth and kill fish. Drinking water in some households has been found to be dangerously contaminated with lead dissolved by acidic water from lead solder in old pipes. Fish in acidified water have been found to have dangerous levels of mercury in their bodies, which is subject to biological magnification as it is passed through trophic levels (see "Earth Watch: Food Chains Magnify Toxic Substances").

FIGURE 41-14 Acid deposition can destroy forests
Acid rain and fog have destroyed this forest atop Mount Mitchell in North Carolina.

EARTH WATCH Food Chains Magnify Toxic Substances

In the 1940s, the properties of the new insecticide DDT seemed close to miraculous. In the Tropics, DDT saved millions of lives by killing the mosquitoes that spread malaria. Increased crop yields resulting from DDT's destruction of insect pests saved millions more from starvation. Swiss chemist Paul Müller, the discoverer of its properties as a pesticide, was awarded the 1948 Nobel Prize for Medicine and Physiology, and people looked forward to a new age of freedom from insect pests. Little did they realize that the indiscriminate use of this pesticide was unraveling the complex web of life. For example, in the mid-1950s, the World Health Organization sprayed DDT on the island of Borneo to control malaria. A caterpillar that fed on the thatched roofs of houses was relatively unaffected, but a predatory wasp that fed on the caterpillar was destroyed. Eaten by the burgeoning caterpillar population, thatched roofs collapsed. Gecko lizards that ate poisoned insects accumulated high concentrations of DDT in their bodies. Both they, and the village cats that ate the geckos, died of DDT poisoning. With the cats eliminated, the rat population exploded. Villages were threatened with an outbreak of plague, carried by the uncontrolled rats. The outbreak was avoided by airlifting new cats to the villages.

In the U.S., wildlife biologists during the 1950s and 1960s witnessed an alarming decline in populations of several predatory birds, especially fish-eaters such as bald eagles, cormorants, ospreys, and brown pelicans. The decline pushed some, including the brown pelican and the bald eagle, close to extinction. The aquatic ecosystems supporting these birds had been sprayed with relatively low amounts of DDT to control insects. Scientists were amazed to find that predatory birds carried concentrations of DDT up to one million times greater than the concentration present in the water. This led to the discovery of **biological magnification**, the process by which toxic substances accumulate at increasingly high concentrations in progressively higher trophic levels.

DDT and other man-made substances that undergo biological magnification contain chlorine, which tends to stabilize molecules. All such substances have two properties that make them dangerous. First, decomposer organisms cannot readily break them down into harmless substances—that is, they are not **biodegradable**; second, they are not water soluble and tend to be stored in fat. Thus, they accumulate in the bodies of animals, rather than being broken down and excreted in the watery urine. Since the transfer of energy from lower to higher trophic levels is extremely inefficient, herbivores must eat large quantities of plant material (which may have been sprayed with pesticides), carnivores must eat many herbivores, and so on. Because these substances remain in the body, a predator accumulates the poison from its prey over many years. The Great Lakes have built up relatively high levels of bioaccumulating compounds, including DDT and PCBs (chlorinated compounds used in electrical transformers and cables), and mercury, which is extremely toxic to the nervous system and is released by coal-burning power plants, waste incinerators, mining activities, and some industrial processes. Bioaccumulating compounds often reach high levels in predatory fish and can pose a threat both to wildlife and to people who eat large quantities of fish. In the Great Lakes region, populations of fish-eating river otters have declined sharply, and a variety of fish-eating birds, including bald eagles, produce deformed offspring or eggs that never hatch (Fig. E41-1).

We must understand the properties of pollutants and the workings of food webs to reduce human health hazards as well as loss of wildlife. Many of the children born to women consuming PCB-contaminated fish from Lake Michigan exhibited slowed intellectual development. Exposure to high levels of pesticides and other persistent pollutants has been linked to certain forms of cancer, infertility, heart disease, and suppressed immune function in people. Humans often "feed high on the food chain." When we eat tuna or swordfish, for example, we act as tertiary or even quaternary consumers, so we are vulnerable to bioaccumulating substances. In addition, the long human life span provides more time for substances stored in our bodies to accumulate to toxic levels.

Today, although DDT, PCBs, and many other bioaccumulating pollutants are banned in the U.S., they persist from past usage; they are still used in developing countries and are transported worldwide by wind, water, and commerce. Inuit natives living north of the Arctic Circle, where pollutants might be expected to be minimal, have high levels of PCBs and chlordane (a banned pesticide similar to DDT) in their bodies from consuming the fat of whales and other marine predators. There is growing concern that a number of widely used synthetic chemicals share the tendencies of DDT to persist, accumulate, and interfere with hormone functions. These chemicals, called "endocrine disruptors," are released by plastics, detergents, and pesticides and as by-products of manufacturing processes (see "Earth Watch: Endocrine Deception" in Chapter 33).

FIGURE E41-1 The price of pollution
Deformities such as the twisted beak of this double-crested cormorant from Lake Michigan have been linked to bioaccumulating chemicals. Abnormalities of the reproductive and immune systems are also common in many types of organisms exposed to these pollutants. Predatory animals are especially vulnerable owing to biological magnification.

Interfering with the Carbon Cycle Contributes to Global Warming

Between 345 million and 280 million years ago, huge quantities of carbon were diverted from the carbon cycle when, under the warm, wet conditions of the Carboniferous period, the bodies of prehistoric organisms were buried in sediments, escaping decomposition. Over time, heat and pressure converted their bodies into fossil fuels such as coal, oil, and natural gas. Without human intervention, this carbon would have remained untouched. Beginning with the Industrial Revolution, however, we have increasingly relied on the energy stored in these fuels. As we burn them in our power plants, factories, and cars, we release CO_2 into the atmosphere. Since 1850, the CO_2 content of the atmosphere has increased from 280 parts per million (ppm) to 370 ppm, or more than 30%, creating the highest CO_2 concentrations in more than 410,000 years. That increase is continuing at the rate of 1.5 ppm yearly. Burning fossil fuels accounts for 80–85% of the CO_2 added to the atmosphere each year.

A second source of added atmospheric CO_2, accounting for 15–20% of CO_2 emissions, is **deforestation**, which eliminates tens of millions of forested acres annually. Deforestation is occurring principally in the Tropics, where rain forests are rapidly being converted to marginal agricultural land. The carbon stored in the massive trees in these forests returns to the atmosphere (primarily through burning) after they are cut.

Greenhouse Gases Trap Heat in the Atmosphere

Carbon dioxide still represents only a tiny fraction of Earth's atmosphere. However, atmospheric CO_2 acts something like the glass in a greenhouse; it allows solar energy to enter, then absorbs and holds that energy once it has been converted to heat (Fig. 41-15). Several other **greenhouse gases** share this property, including methane, chlorofluorocarbons (CFCs), water vapor, and nitrous oxide. The **greenhouse effect**, the ability of greenhouse gases to trap the sun's energy in a planet's atmosphere as heat, is a natural process. By keeping our atmosphere relatively warm, it allows life on Earth as

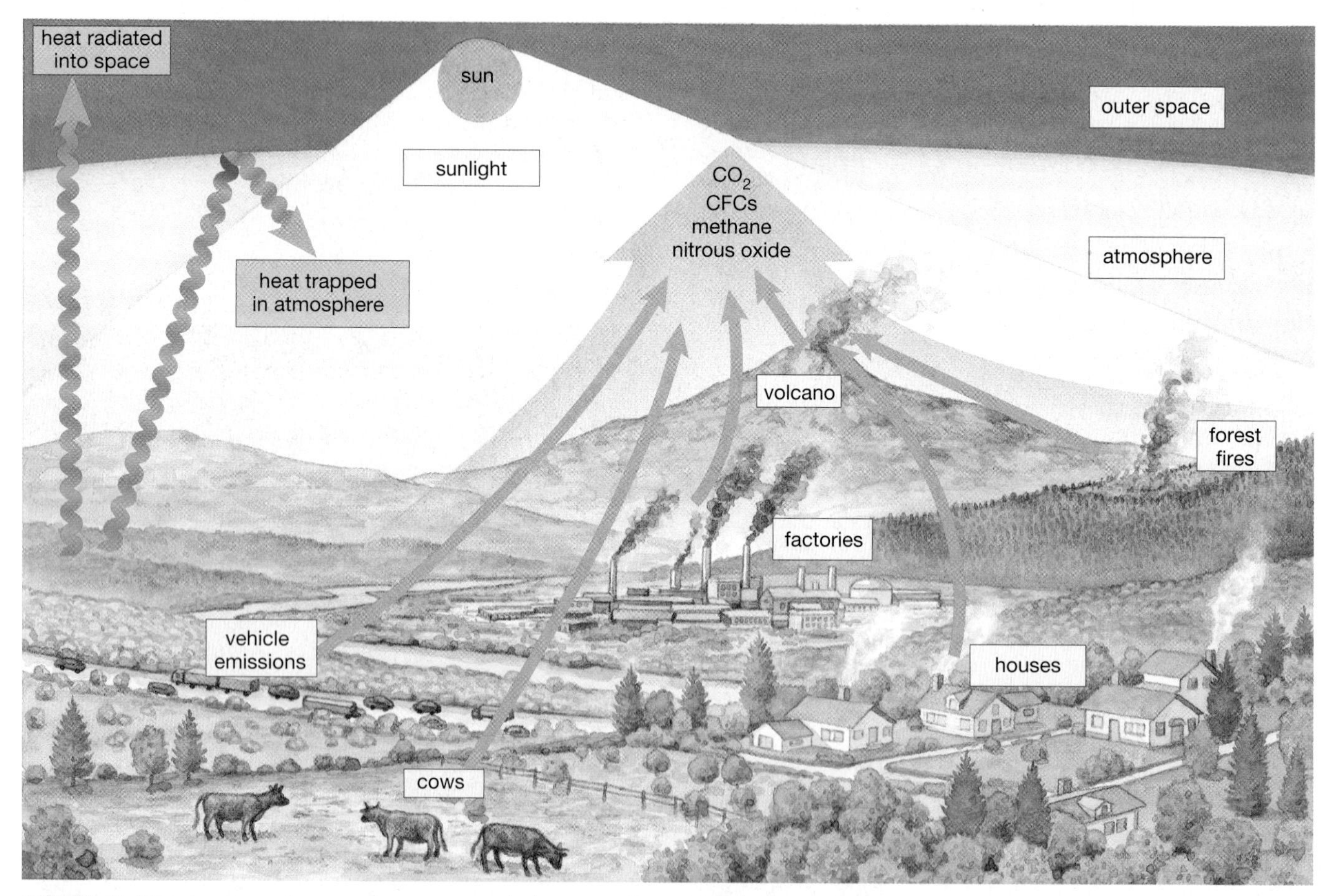

FIGURE 41-15 Increases in greenhouse gas emissions contribute to global warming
Incoming sunlight warms Earth's surface and is radiated back to the atmosphere. Greenhouse gases absorb some of this heat, trapping it in the atmosphere. Human activities have greatly increased levels of greenhouse gases, resulting in a gradual rise in average global temperatures. **QUESTION** Why do temperatures rise in an actual greenhouse? Why is this a good analogy for heat-trapping by greenhouse gases?

we know it. However, there is almost complete consensus among atmospheric scientists that human activities have amplified the natural greenhouse effect, producing a phenomenon called **global warming**.

Historical temperature records have revealed a global temperature increase paralleling the rise in atmospheric CO_2 (Fig. 41-16). In the last 100 years, average temperatures have risen about 1 °F. The 1980s were the warmest decade on record, until their record was topped by the 1990s; thus far, 2001, 2002, and 2003 have been the three hottest years on record.

Many interacting factors, such as increases in cloud cover caused by increased evaporation, possible increases in net primary productivity due to increased CO_2, and the uncertain capacity of the ocean to absorb CO_2, make future climate prediction difficult and imprecise. The United Nations-sponsored Intergovernmental Panel on Climate Change (IPCC), which consists of hundreds of scientists throughout the world, recently reported that "there is new and stronger evidence that most of the warming observed over the last 50 years is attributable to human activities" (Third Assessment Report, 2001). If greenhouse gas emissions are not curtailed, the IPCC predicts an increase of between 2.5 to 10.4 °F (1.4 to 5.8 °C) in average world surface air temperatures from 1990 to 2100, with temperatures rising more rapidly than in the past 10,000 years. To put this in perspective, average air temperatures during the peak of the last Ice Age (20,000 years ago) were only about 5 °C lower than at present. This rapid temperature rise is of particular concern because it is likely to exceed the rate at which natural selection can allow organisms to adapt. The temperature change will not be distributed evenly worldwide; U.S. temperatures are predicted to increase considerably faster than the global average.

Global Warming May Have Severe Consequences

As geochemist James White at the University of Colorado quipped, "If the Earth had an operating manual, the chapter on climate might begin with the caveat that the system has been adjusted at the factory for optimum comfort, so don't touch the dials." The consequences of global warming are only partially predictable and may be severe.

Throughout the world, ice is melting. Ice in the Arctic Sea and on Greenland's ice cap has shrunk to record-breaking lows. In Glacier National Park, where 150 glaciers once graced the mountainsides, only 35 remain, and scientists estimate that these may all disappear within the next 30 years. The glacier covering Mt. Everest has retreated three miles in the last 50 years. As polar ice caps and glaciers melt and ocean waters expand in response to atmospheric warming, sea levels will rise, threatening coastal cities and flooding coastal wetlands. Global warming is also predicted to increase the intensity of hurricanes, posing a further threat to coastal regions.

Even small temperature changes can dramatically alter the paths of major air and ocean currents, altering precipitation patterns in unpredictable ways. Some land might become too hot and dry for agriculture, while other areas might become warmer, wetter, and more productive. Overall, U.S. crop productivity might increase. As the world warms, however, experts predict longer, more severe droughts and more extremes in rainfall, leading to more frequent crop failure and flooding. Agricultural disruption could be disastrous for some nations that are already barely able to feed themselves.

The impact of global warming on forests could be profound. While overall forest growth in the U.S. may increase, species distributions will change. For example, sugar maples may disappear from northeastern U.S. forests, while southeastern forests could be partially replaced by grasslands. The IPCC notes that global warming will increase the range of tropical disease-carrying organisms, such as malaria-transmitting mosquitoes, with negative consequences for human health. Coral reefs, already stressed by human activities, are likely to suffer further damage from warmer waters.

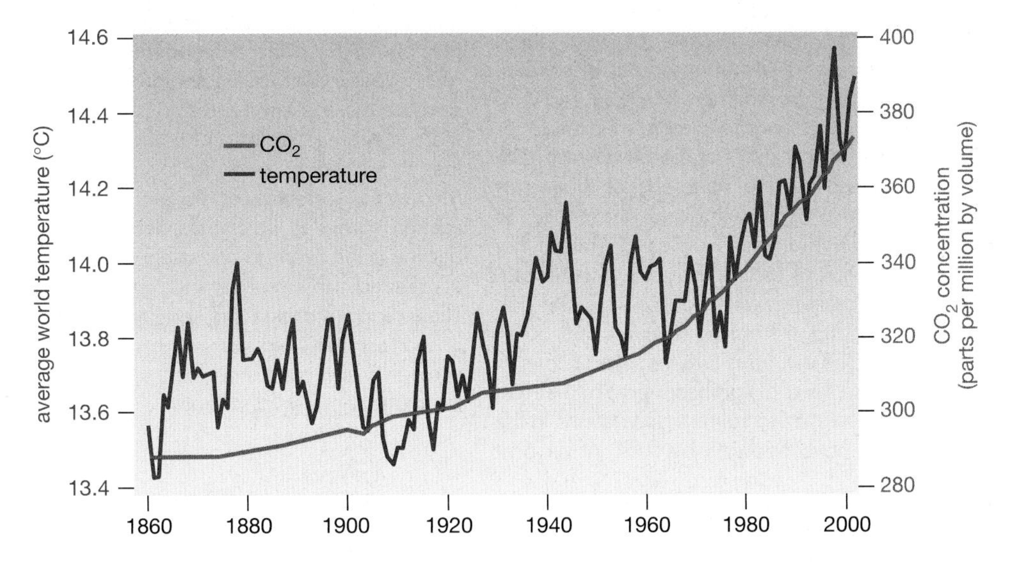

FIGURE 41-16 Global warming parallels CO_2 increases
The CO_2 concentration of the atmosphere (blue line) has increased steadily since 1860. The dashed portion of that curve represents measurements made from air trapped in ice cores; the solid portion reflects direct measurements made at Mauna Loa, Hawaii. Average global temperatures (red line) have also increased gradually, paralleling the increasing atmospheric CO_2. (With thanks to Drs. Kevin Trenberth and Jim Hurrell of the National Center for Atmospheric Research.)

Biologists worldwide are documenting changes related to warming. After analyzing plant data from sites throughout Europe, scientists at the University of Munich, Germany, concluded that the growing season has increased by more than 10 days over the past 41 years. Mexican Jays in southern Arizona are nesting 10 days earlier than they did in 1971. Many species of butterflies have shifted their ranges northward. While each individual report could be due to other factors, the cumulative weight of data from diverse sources worldwide provides strong evidence that warming-related biological changes have begun. In 2003 two different teams of scientists analyzed previously published data for over 1400 species, documenting plant and animal ranges and springtime events (such as flower blooming or bird nesting) over recent history. The data included diverse species from mollusks and mammals to grasses and trees. Both teams independently concluded, based on a shift of ranges toward the poles and an advancement of springtime events to earlier dates, that "climate change is already affecting living systems."

Our Decisions Make a Difference

What can we as individuals do to help reduce both acid rain and global warming? The total greenhouse gas emissions produced by the U.S. amount to 6.6 tons per person each year, more than any other country on Earth. With less than 5% of the world's population, the U.S. is responsible for more than 20% of world's greenhouse emissions. A car with a fuel efficiency of 20 miles per gallon releases 1 pound of CO_2 for every mile it travels, so we can substantially reduce emissions of CO_2 by using fuel-efficient vehicles, carpools, and public transportation. As electricity is generated in fossil fuel-fired power plants, so are tremendous quantities of CO_2, sulfur dioxide, and nitrogen oxides. We can support the efforts of utility companies to use renewable energy sources such as wind and solar power. To conserve electricity, we can purchase more efficient appliances, turn off unused computers and lights, and replace incandescent lighting with fluorescent light bulbs. Insulating and weatherproofing our homes, incorporating solar energy features into new homes, and planting deciduous trees near our houses to provide summer shade and winter sun will significantly reduce fuel consumption while cutting back on heating and air conditioning costs. Recycling is also a tremendous energy saver—for example, 95% of the energy used to produce an aluminum can from raw materials is conserved when the can is recycled instead of tossed in the trash. We can also support reforestation efforts to replace trees both in tropical rain forests and in our communities. Although continued global warming is inevitable, its rate can be reduced by human decisions.

CASE STUDY REVISITED When the Salmon Return

Researchers investigating the Sockeye salmon's return to an Alaskan stream witness an awesome sight. Hundreds of brilliant red bodies writhe in water so shallow it barely covers them. A female beats her tail, excavating a shallow depression in the gravel where she releases her coral-colored eggs; meanwhile, a male showers them with sperm. But after their long and strenuous migration, these adult salmon are dying. Their flesh is tattered, their muscles wasted, and the final act of reproduction saps the last of their energy. Soon the stream is clogged with dying, dead, and decomposing bodies—an abundance of nutrients unimaginable at any other time of the year. Eagles, grizzly bears, and gulls gather to gorge themselves on the fleeting bounty. Flies breed in the carcasses, feeding spiders, birds, and trout. Local mink populations have evolved breeding cycles around the event; females lactate just when the salmon provide them with abundant food. Isotope studies reveal that up to one-fourth of the nitrogen incorporated into the leaves of trees and shrubs near these streams comes from the bodies of salmon. Historically, researchers estimate that 500 million pounds of salmon migrated upstream in the U.S. Pacific Northwest each year, contributing hundreds of thousands of pounds of nitrogen and phosphorus to the Columbia River watershed alone. Now as a result of factors including overfishing, river damming, diversion of water for irrigation, runoff from agriculture, and pollution of the estuaries (where several salmon species spend a significant part of their life cycle), migratory salmon populations in the region have declined by 94% in the past century. The web of life that relied on the mighty annual upstream flow of nutrients has been disrupted.

Consider This: Some salmon populations have been so thoroughly depleted that they qualify for protection under the Endangered Species Act. Some people argue that because these salmon are also raised commercially in hatcheries and artificial ponds, they should not be afforded this protection. Meanwhile, researchers studying Chinook salmon raised in hatcheries noted a 25% decline in the average size of eggs of hatchery-reared fish over just four generations. These eggs produce smaller juvenile fish. Based on this information, explain why ecologists and conservationists are arguing for protection of wild salmon.

Links to Life: Making a Difference

This chapter discusses several ways in which individuals can reduce their energy consumption. But can one person's actions make a difference? Jonathan Foley of the University of Wisconsin thinks so. He is on the cutting edge of climatological research, having led a team that developed one of the first computer models of global climate change to consider the impact of biological systems and human land use (such as converting forests to cropland) on climate. In 1998, Jon and his wife Andrea, recognizing that greenhouse gas emissions and the resulting climate change can be significantly impacted by individual decisions and choices, made a choice of their own: to cut their family's energy use and carbon dioxide emissions in half. The Foleys and their young daughter lived in a 5 bedroom house 30 miles from their work; Jon and Andrea each used a separate car to commute about 60 miles daily. First, they moved to a smaller house much closer to work. A visitor to the Foleys' new home—warm and cozy in winter and cool in summer—would never realize how little energy it consumes. Cracks have been sealed and the attic insulated. Every appliance has been chosen for energy efficiency. Compact fluorescent bulbs, using 75% less energy than incandescents, shed light throughout. Decorative ceiling fans reduce the need for summer air conditioning. Solar collectors supply over two-thirds of the family's water heating needs, while low-emittance window glass lets sunlight in while reducing heat loss in winter. The Foleys can now ride bicycles or take the bus to work, but they also enjoy their Toyota Prius hybrid gas/electric car—which gets nearly 50 mpg in city driving. Have they reached their goals? Within two years of their resolution, the Foleys cut their energy use by roughly 65%. Foley says, "Cutting your greenhouse gas emissions doesn't have to be a 'sacrifice' at all. We have cut our emissions more than 50%, and we now have lower energy bills, a more comfortable house, more time to spend with our family, and a higher quality of life. Americans have a lot to gain by cutting fossil fuel use: reduced greenhouse gas emissions, improved air quality in our cities, less dependence on foreign oil supplies, and so on. This is a win-win scenario, so why not go for it?"

Summary of Key Concepts

41.1 What Are the Pathways of Energy and Nutrients?

Ecosystems are sustained by a continuous flow of energy from sunlight and a constant recycling of nutrients.

41.2 How Does Energy Flow Through Communities?

Energy enters the biotic portion of ecosystems when it is harnessed by autotrophs during photosynthesis. Net primary productivity is the amount of energy that autotrophs store in a given unit of area over a given period of time.

Trophic levels describe feeding relationships in ecosystems. Autotrophs are the producers, the lowest trophic level. Herbivores occupy the second level as primary consumers. Carnivores act as secondary consumers when they prey on herbivores and as tertiary or higher-level consumers when they eat other carnivores.

Feeding relationships in which each trophic level is represented by one organism are called *food chains*. In natural ecosystems, feeding relationships are far more complex and are described as food webs. Detritus feeders and decomposers, which digest dead bodies and wastes, use and release the energy stored in these substances and liberate nutrients for recycling. In general, only about 10% of the energy captured by organisms at one trophic level is converted to the bodies of organisms in the next higher level. The higher the trophic level, the less energy is available to sustain it. As a result, plants are more abundant than herbivores, and herbivores are more common than carnivores. The storage of energy at each trophic level is illustrated graphically as an energy pyramid. The energy pyramid leads to biological magnification, the process by which toxic substances accumulate in increasingly high concentrations in progressively higher trophic levels.

41.3 How Do Nutrients Move Within and Among Ecosystems?

A nutrient cycle depicts the movement of a particular nutrient from its reservoir (usually in the abiotic, or nonliving, portion of the ecosystem) through the biotic, or living, portion of the ecosystem and back to its reservoir, where it is again available to producers. Carbon reservoirs include the oceans, atmosphere, and fossil fuels. Carbon enters producers through photosynthesis. From autotrophs it is passed through the food web and released to the atmosphere as CO_2 during cellular respiration.

The major reservoir of nitrogen is the atmosphere. Nitrogen gas is converted by bacteria and human industrial processes into ammonia and nitrate, which can be used by plants. Nitrogen passes from producers to consumers and is returned to the environment through excretion and the activities of detritus feeders and decomposers.

The reservoir of phosphorus is in rocks as phosphate, which dissolves in rainwater. Phosphate is absorbed by photosynthetic organisms, then passed through food webs. Some is excreted, and the rest is returned to the soil and water by decomposers. Some is carried to the oceans, where it is deposited in marine sediments. Humans mine phosphate-rich rock to produce fertilizer.

The major reservoir of water is the oceans. Solar energy evaporates water, which returns to Earth as precipitation. Water flows into lakes and underground reservoirs and in rivers, which flow to the oceans. Water is absorbed directly by plants and animals and is also passed through food webs. A small amount is combined with CO_2 during photosynthesis to form high-energy molecules.

41.4 What Causes Acid Rain and Global Warming?
Environmental disruption occurs when human activities interfere with the natural functioning of ecosystems. Human industrial processes release toxic substances and produce more nutrients than nutrient cycles can efficiently process. Through massive consumption of fossil fuels, we have overloaded the natural cycles of carbon, sulfur, and nitrogen. The burning of fossil fuels has substantially increased atmospheric carbon dioxide, a greenhouse gas. This increase is correlated with increased global temperatures, leading nearly all atmospheric scientists to conclude that global warming is due to human industrial activities. Global warming is causing formerly permanent ice deposits to melt and is influencing the distribution and seasonal activities of wildlife. Scientists believe it will have a major impact on precipitation and weather patterns, with unpredictable results. Fossil fuel combustion also releases sulfur dioxide and nitrogen oxides. In the atmosphere, these are converted to sulfuric acid and nitric acid, which fall to Earth as acid deposition, including acid rain. Acidification of freshwater ecosystems has substantially reduced their ability to support life, particularly in the eastern U.S. At high elevations, acid deposition has significantly damaged many eastern forests and threatens other forests throughout the U.S.

Key Terms

acid deposition *p. 853*
autotroph *p. 843*
biodegradable *p. 855*
biogeochemical cycle *p. 848*
biological magnification *p. 855*
biomass *p. 843*
carnivore *p. 843*
consumer *p. 843*
decomposer *p. 844*
deforestation *p. 856*
detritus feeder *p. 844*
energy pyramid *p. 848*
food chain *p. 844*
food web *p. 844*
fossil fuel *p. 849*
global warming *p. 857*
greenhouse effect *p. 856*
greenhouse gas *p. 856*
herbivore *p. 843*
heterotroph *p. 843*
hydrologic cycle *p. 852*
net primary productivity *p. 843*
nitrogen fixation *p. 850*
nutrient cycle *p. 848*
omnivore *p. 844*
primary consumer *p. 843*
producer *p. 843*
reservoir *p. 848*
secondary consumer *p. 844*
tertiary consumer *p. 844*
trophic level *p. 843*

Thinking Through the Concepts

To take a multiple-choice quiz with feedback on the contents of this chapter, visit http://www.prenhall.com/audesirk7. *Log in to the Web site selected by your instructor and navigate to the Self Test section for this chapter.*

? Review Questions

1. What makes the flow of energy through ecosystems fundamentally different from the flow of nutrients?
2. What is an autotroph? What trophic level does it occupy, and what is its importance in ecosystems?
3. Define *primary productivity*. Would you predict higher productivity in a farm pond or an alpine lake? Defend your answer.
4. List the first three trophic levels. Among the consumers, which are most abundant? Why would you predict that there will be a greater biomass of plants than herbivores in any ecosystem? Relate your answer to the "10% law."
5. How do food chains and food webs differ? Which is the more accurate representation of actual feeding relationships in ecosystems?
6. Define *detritus feeders* and *decomposers*, and explain their importance in ecosystems.
7. Trace the movement of carbon from its reservoir through the biotic community and back to the reservoir. How have human activities altered the carbon cycle, and what are the implications for future climate?
8. Explain how nitrogen gets from the air to a plant.
9. Trace a phosphorus molecule from a phosphate-rich rock into the DNA of a carnivore. What makes the phosphorus cycle fundamentally different from the carbon and nitrogen cycles?
10. Trace the movement of a water molecule from the moment it leaves the ocean until it reaches a plant root and then a plant stoma before making its way back to the ocean.

Applying the Concepts

1. What could your college or university do to reduce its contribution to acid rain and global warming? Be specific and, if possible, offer practical alternatives to current practices.
2. Relate fossil fuel consumption to (a) the loss of aquatic life in lakes in the Northeast and Canada and (b) the lengthening of the growing season in Europe. Trace each step from the burning of gasoline in a car or power plant to the change in question.
3. Define and give an example of *biological magnification*. What qualities are present in materials that undergo biological magnification? In which trophic level are the problems worst, and why?
4. Discuss the contribution of population growth to (a) acid rain and (b) the greenhouse effect.
5. Describe what would happen to a population of deer if all predators were removed and hunting banned. Include effects on vegetation as well as on the deer population itself. Relate your answer to carrying capacity as discussed in Chapter 39.
6. Draw a food web incorporating salmon of the Pacific Northwest.

For More Information

Epstein, P. R. "Is Global Warming Harmful to Health?" *Scientific American*, August 2000. Global warming may increase the incidence of disease as the climate becomes more hospitable for disease-causing organisms and their vectors.

Krajick, K. "Long-Term Data Show Lingering Effects from Acid Rain." *Science*, April 13, 2001. Acid rain's damaging effects linger, and current control levels are inadequate to restore ecosystem health.

Lavendel, B. "GreenHouse." *Audubon*, March–April 2001. Climatologist Jonathan Foley's family cut energy use by 65%.

Mlot, C. "Tallying Nitrogen's Increasing Toll." *Science News*, February 15, 1997. Human activities now dominate the nitrogen cycle, with increasingly disruptive effects.

Moore, K. D., and Moore, J. W. "The Gift of Salmon." *Discover*, May 2003. Salmon migrating upstream to spawn and die reverse the usual travel of nutrients and help replenish nutrients carried downstream during the rest of the year.

Nickens, T. E. "North America's Fish Feel the Heat." *National Wildlife*, June/July 2002. Studies across North America suggest that global warming could dramatically alter the distribution and reproductive success of a variety of freshwater fish.

Perkins, S. "Crisis on Tap?" *Science News*, July 20, 2002. Earth's freshwater stores are being stretched to the limit by a burgeoning human population and pollution.

Wuethrich, B. "How Climate Change Alters Rhythms in the Wild." *Science*, February 4, 2000. The distribution, behavior, and ecology of many animal populations are already being influenced by global warming.

Media Activities

To access a Media Activity visit http://www.prenhall.com/audesirk7. *Log in to the Web site selected by your instructor, navigate to this chapter, and select the appropriate Media Activity number.*

41.1 Ecology Models—Building a Food Web

Estimated time: 10 minutes

Explore how energy and nutrients are transferred in a food web and the roles that organisms play within the ecosystem.

41.2 The Global Carbon Cycle and Greenhouse Effect

Estimated time: 10 minutes

Explore how carbon moves between the atmosphere, oceans, rocks, and living organisms. See how the accumulation of carbon in the atmosphere leads to a rise in the earth's temperature.

41.3 Web Investigation: When the Salmon Return

Estimated time: 30 minutes

This chapter's Case Study documents the plight of salmon populations in the Pacific Northwest and indicates how the Endangered Species Act has been used as a weapon to help preserve their dwindling wild stocks. It also outlines the importance of the salmon for the productivity of forests along the rivers and streams in which they spawn. In this activity explore more of the science and controversy surrounding salmon.

CHAPTER

42 Earth's Diverse Ecosystems

Kahindi Samson snares a butterfly. Insets: **(top)** A dark blue pansy butterfly. **(bottom)** Pupae are identified and sorted for shipment.

AT A GLANCE

CASE STUDY Wings of Hope

To help support and feed his five younger brothers and sisters, Kahindi Samson began sneaking into the Arabuko-Sokoke forest when he was 12 years old, snaring its endangered antelope and cutting the old growth trees that provide homes for the rare Sokoke Scops owl. This precious Kenyan forest is protected by the government as the largest remnant of coastal forest in East Africa and a final refuge for endangered birds and mammals that have been displaced by the growing human population. But to the farmers on surrounding land, the forest was the enemy, home to elephants and baboons that emerged at night to eat their crops. Most wanted to see the forest felled.

Ian Gordon, a butterfly ecologist, watched the poaching and tree-felling with alarm; Arabuko-Sokoke forest is home to 250 species of butterflies. Unwilling to stand by helplessly, Gordon founded Project Kipepeo, meaning "butterfly" in Swahili. His mission was to convince skeptical local farmers to grow butterflies instead of crops. Today, Samson enters the forest with a license and a butterfly net. In a cage outside his home, he places his catch of pregnant female butterflies. As their eggs hatch, Samson fattens the caterpillars on leaves he collects from the forest. Within a month, the caterpillars are ready to form pupae and be shipped to the U.S. and Europe. There they will hatch amidst lush tropical vegetation in butterfly gardens, where they will delight visitors who have never witnessed the splendor of rainforest butterflies. Samson is one of 550 local butterfly workers who now rely on the Arabuko-Sokoke forest for their livelihood and make a far better living than before. "We used to wish the forest would go away," says butterfly farmer Priscilla Kiti, "but these days we earn our living mostly in butterfly farming, so if they cut the forest, things are going to be very difficult."

Media Activity
42.1 Tropical Atmospheric Circulation and World Biomes

42.1 What Factors Influence Earth's Climate?

The distribution of life, particularly on land, is dramatically affected by both weather and climate. **Weather** refers to short-term fluctuations in temperature, humidity, cloud cover, wind, and precipitation in a region over periods of hours or days. **Climate**, in contrast, refers to patterns of weather that prevail from year to year and even century to century in a particular region. The amount of sunlight and water and the range of temperatures determine the climate of a given region. Whereas weather affects individual organisms, climate influences and limits the overall distribution of entire species.

Both Climate and Weather Are Driven by the Sun

Both climate and weather are driven by a great thermonuclear engine: the sun. Solar energy reaches Earth in a range of wavelengths—from short, high-energy ultraviolet (UV) rays, through visible light, to the longer infrared wavelengths that produce heat. The solar energy that reaches Earth drives the wind, the ocean currents, and the global water cycle. Before it reaches Earth's surface, however, sunlight is modified by the atmosphere. A layer relatively rich in ozone (O_3) is located in the middle atmosphere. This *ozone layer* absorbs much of the sun's high-energy UV radiation, which can damage biological molecules (see "Earth Watch: The Ozone Hole—A Puncture in Our Protective Shield"). Dust, water vapor, and clouds scatter light, reflecting some of the energy back into space. Carbon dioxide, water vapor, methane, and other *greenhouse gases* selectively absorb energy at infrared wavelengths, trapping heat in the atmosphere. Human activities have significantly increased levels of greenhouse gases, as described in Chapter 41.

Only about half the solar energy reaching the atmosphere actually strikes Earth's surface. Of this, a small fraction is immediately reflected back into space; another small fraction is captured by photosynthetic plants and microorganisms and used to power photosynthesis; and the rest is absorbed as heat. Eventually, nearly all the incoming solar energy is returned to space, either as light or as infrared radiation (heat). The solar energy temporarily absorbed and stored as heat by the atmosphere and by Earth's surface maintains Earth's relative warmth.

Many Physical Factors Also Influence Climate

Many physical factors influence climate. Among the most important are latitude, air currents, ocean currents, and the presence of mountains and irregularly shaped continents.

Latitude Influences the Angle at Which Sunlight Strikes Earth

Latitude, measured in degrees, is the distance north or south from the equator, which is located at 0° latitude. The amount of sunlight that strikes a given area of Earth's surface has a major effect on average yearly temperatures. At the equator, sunlight hits Earth's surface nearly at a right angle, making the weather there consistently warm. Farther north or south, the sun's rays strike Earth's surface at a greater slant. This angle spreads the same amount of sunlight over a larger area, producing lower overall temperatures (Fig. 42-1).

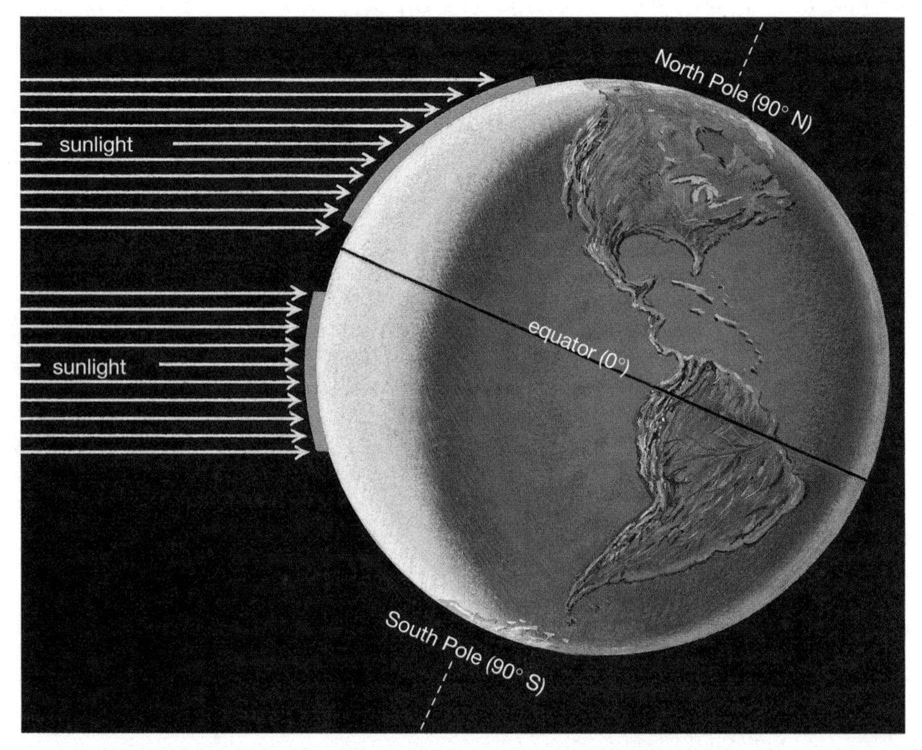

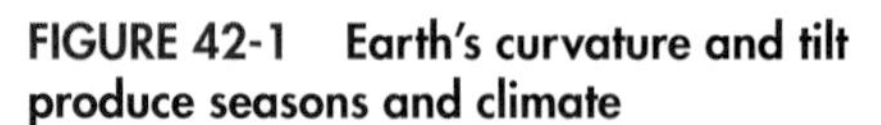

FIGURE 42-1 Earth's curvature and tilt produce seasons and climate
Near the equator, sunlight falls nearly perpendicularly on Earth's surface, concentrating its warmth over a relatively small area. Toward the poles, the same amount of sunlight is distributed over a much larger surface area. Temperatures are thus highest at the equator and lowest at the poles. The tilt of Earth on its axis causes seasonal variations in the directness of sunlight. Here, we see winter in the Northern Hemisphere and summer in the Southern Hemisphere. **QUESTION** Describe the seasons and daylength if Earth were not tilted on its axis. Would there still be a temperature gradient from the equator to the poles?

EARTH WATCH The Ozone Hole—A Puncture in Our Protective Shield

A small fraction of the radiant energy produced by the sun—*ultraviolet*, or UV, radiation—is so highly energetic that it can damage biological molecules. In small quantities, UV radiation helps human skin produce vitamin D and causes tanning in fair-skinned people. But in larger doses, UV causes sunburn and premature aging of skin, skin cancer, and cataracts, a condition in which the lens of the eye becomes cloudy.

Fortunately, most UV radiation is filtered out by ozone in the stratosphere, a layer of atmosphere extending from 10 to 50 kilometers (6 to 30 miles) above Earth. In pure form, *ozone* (O_3) is a bluish, explosive, and highly poisonous gas. In the stratosphere, the normal concentration of ozone is about 0.1 part per million (ppm), compared with 0.02 ppm in the lower atmosphere. This ozone-enriched layer is called the **ozone layer**. Ultraviolet light striking ozone and oxygen causes reactions that break down as well as regenerate ozone. In the process, the UV radiation is converted to heat, and the overall level of ozone remains reasonably constant—or so it did until humans intervened.

In 1985 British atmospheric scientists published the startling news that springtime levels of stratospheric ozone over Antarctica had declined by more than 40% since 1977. In the *ozone hole* over Antarctica, ozone now dips to about one-third of its predepletion levels (Fig. E42-1). Although ozone layer depletion is most severe over Antarctica, the ozone layer is somewhat reduced over most of the world, including nearly all of the continental U.S. Satellite data reveal that, since the early 1970s, UV radiation has risen by nearly 7% per decade in the Northern Hemisphere and by almost 10% per decade in the Southern Hemisphere. Epidemiological studies indicate that for every 1% increase in lifetime exposure to UV radiation, the lifetime risk of melanoma also increases by about 1%. But human health effects are only one cause for concern. Photosynthesis by phytoplankton, the producers for marine ecosystems, is reduced under the ozone hole above Antarctica. Some types of trees and farm crops are also harmed by increased UV radiation.

The ozone hole is caused by human production and release of chlorofluorocarbons (CFCs). Developed in 1928, these gases were widely used as coolants in refrigerators and air conditioners, as aerosol spray propellants, in the production of foam plastic, and as cleansers for electronic parts. CFCs are very stable and were considered safe. Their stability, however, has proved to be a major problem because they remain chemically unchanged as they gradually rise into the stratosphere. There, under intense bombardment by UV light, the CFCs break down, releasing chlorine atoms. Chlorine catalyzes the breakdown of ozone to oxygen gas (O_2) while remaining unchanged itself. Clouds over the Antarctic regions are composed of ice particles that provide a surface on which the reaction can occur.

Fortunately, we have taken the first steps toward "plugging" the ozone hole. In an almost unprecedented example of global cooperation and concern, talks beginning in 1985 led to treaties in 1990 and 1992 in which industrialized nations throughout the world agreed to phase out ozone-depleting chemicals rapidly, eliminating CFCs by 1996. These compounds are no longer used in foam plastic production, and CFC substitutes have been found for use in spray cans, refrigerators, and car air conditioners as well. Ground-level global atmospheric chlorine levels (an indicator of CFC use) peaked in 1994. By 1999, scientists had detected chlorine reductions in the stratosphere, and ozone depletion over mid-latitude regions had come to a standstill. In 2003, NASA satellite observations showed that, while ozone was still being destroyed, the rate of depletion had slowed. Because CFCs can persist for 50 to 100 years and take a decade or more to ascend into the stratosphere, current release of CFCs by developing countries—adding to the millions of tons already released by developed countries—will continue to erode the protective ozone shield. Recovery is decades away and will require further reductions in CFC release. In a spirit of continued cooperation, developed countries have recently pledged funding to help developing countries devise alternatives to CFCs that will allow them to cut their CFC production in half by 2005, providing more hope that our shield will eventually be restored.

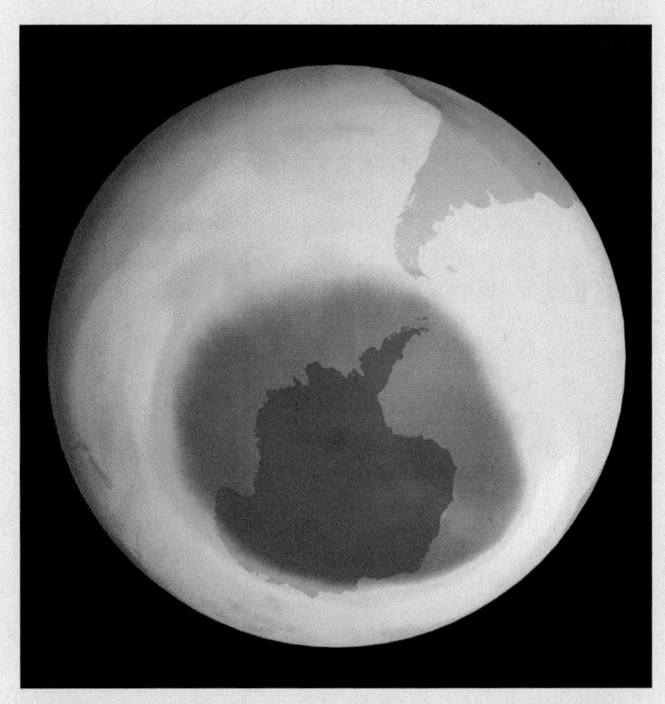

FIGURE E42-1 Satellite image of the Antarctic ozone hole The second largest ozone hole on record is seen in purple on this image based on satellite data from September, 2003. Notice that the hole nearly reaches the tip of South America. *(Image courtesy of NASA.)*

Earth is tilted on its axis as it makes its yearly trip around the sun. The higher latitudes experience considerable variation in the directness of sunlight throughout the year, resulting in pronounced seasons. When the Northern Hemisphere is tilted toward the sun, it receives sunlight most directly and experiences summer. During the Northern Hemisphere's winter, the Southern Hemisphere is closest to the sun (see Fig. 42-1). The tilting hardly affects the angle of the sun's rays as they strike the equator, so this region experiences little seasonal variation.

Air Currents Produce Broad Climatic Regions

Air currents are generated by Earth's rotation and by differences in temperature between different air masses. Warm air is less dense than cold air, so as the sun's direct rays fall on the equator, heated air rises there. The warm air near the equator is also laden with water evaporated by solar heat (Fig. 42-2a). As the water-saturated air rises, it cools somewhat. Cool air cannot retain as much moisture as can warm air, so water condenses from the rising air and falls as rain. The direct rays of the sun and the rainfall produced when warm, moist air rises and is cooled create a band around the equator called the *Tropics*. This region is both the warmest and the wettest on Earth. The cooler dry air then flows north and south from the equator. Near 30° N and 30° S, the cooled air is dense enough to sink. As it sinks, the air is warmed by heat radiated from Earth. By the time it reaches the surface, it is both warm and very dry. Not surprisingly, the major deserts of the world are found at these latitudes (Fig. 42-2a,b). This air then flows back toward the equator. Farther north and south, this general circulation pattern is repeated, dropping moisture at around 60° N and 60° S and creating extremely dry conditions at the North and South Poles.

Ocean Currents Moderate Near-Shore Climates

Ocean currents are driven by Earth's rotation, winds, and the direct heating of the water by the sun. Continents interrupt the currents, breaking them into roughly circular patterns called **gyres**. Gyres rotate clockwise in the Northern Hemisphere and counterclockwise in the Southern Hemisphere (Fig. 42-3). Because water both heats and cools more slowly than does land or air, ocean currents tend to moderate temperature extremes. Coastal areas, then, generally have less-variable climates than do areas near the center of continents. For example, a gyre in the Atlantic Ocean (the Gulf Stream) brings warm water from equatorial regions north along the eastern coast of North America, creating a warmer, moister climate than is found farther inland. It then carries the still-warm water farther north and east, warming the western coast of Europe before returning south.

Continents and Mountains Complicate Weather and Climate

If Earth's surface were uniform, climate zones would occur in bands corresponding to latitude. These zones would result from the interaction of temperature and rainfall, which are determined by the rise and fall of air masses (see Fig. 42-2a). The presence of irregularly shaped continents (which heat and cool relatively quickly) amidst oceans (which heat and cool more slowly) alters the flow of wind and water and contributes to the irregular distribution of ecosystems.

Variations in elevation within continents further complicate the situation. As elevation increases, the atmosphere becomes thinner and retains less heat.

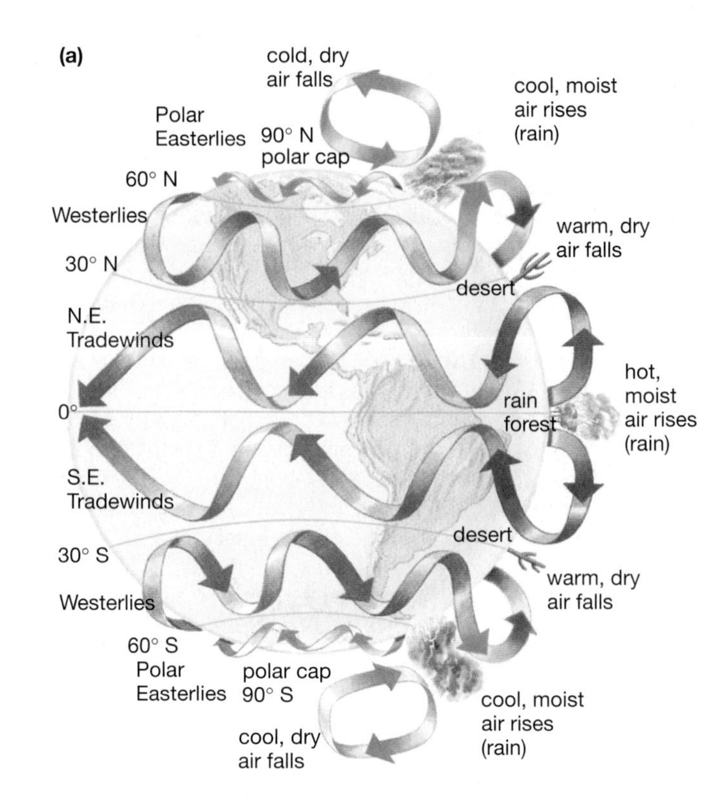

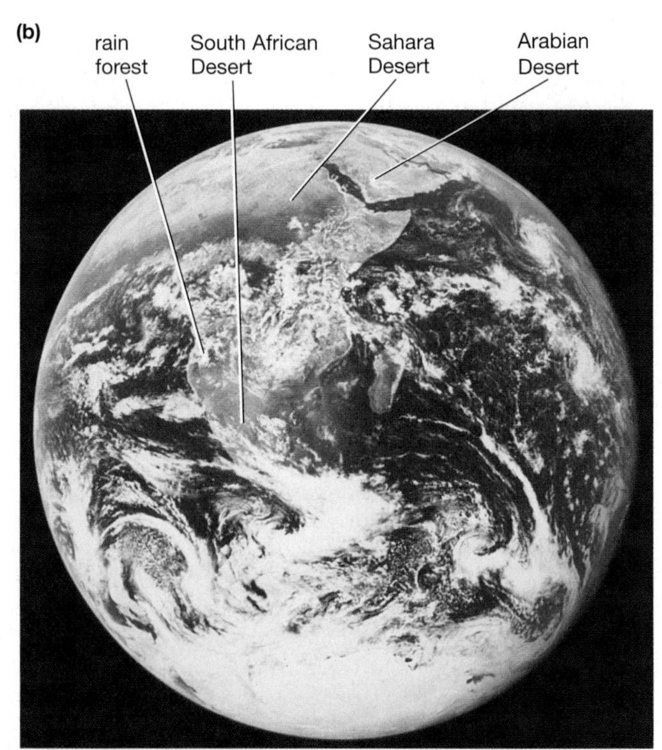

FIGURE 42-2 Distribution of air currents and climatic regions
(a) Rainfall is determined primarily by the distribution of temperatures and by Earth's rotation. The interaction of these two factors creates air currents that rise and fall predictably with latitude, producing broad climatic regions. **(b)** Some of these regions are visible in this photograph of the African continent taken from *Apollo 11*. Along the equator are heavy clouds that drop moisture on the central African rain forests. Note the lack of clouds over the Sahara and Arabian Deserts near 30° N and the South African Desert near 30° S.

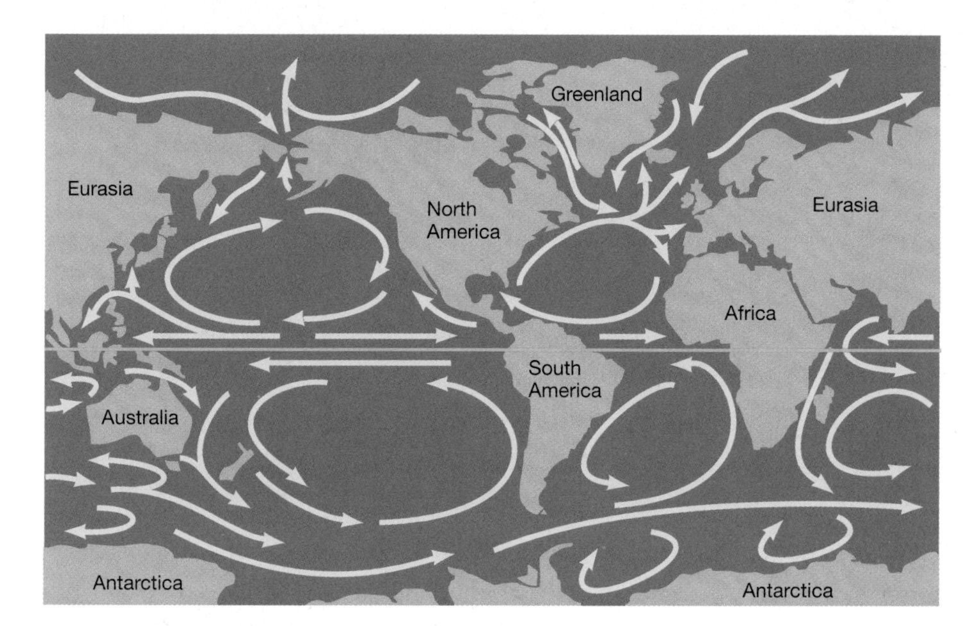

FIGURE 42-3 Ocean circulation patterns are called gyres
Gyres travel clockwise in the Northern Hemisphere and counterclockwise in the Southern Hemisphere. These currents tend to distribute warmth from the equator to northern and southern coastal areas.

The temperature drops approximately 3.5 °F (2 °C) for every 1000 foot (305 meter) elevation gain. This characteristic explains why snow-capped mountains are found even in the Tropics (Fig. 42-4).

Mountains also modify rainfall patterns. When water-laden air is forced to rise as it meets a mountain, it cools. Cooling reduces the air's ability to retain water, and the water condenses as rain or snow on the windward (near) side of the mountain. The cool, dry air is warmed again as it travels down the far side of the mountain and absorbs water from the land, creating a local dry area called a **rain shadow**. For example, mountain ranges such as the Sierra Nevada of the western U.S. wring the moisture from the westerly winds that come off the Pacific Ocean, leaving deserts in the rain shadow on their eastern sides (Fig. 42-5).

El Niño Periodically Disrupts the Usual Ocean-Atmosphere Interactions

The tropical western Pacific ocean typically harbors a huge pool of warm water that is pushed westward by the Northeast Tradewinds (see Fig. 41-2a). Water evaporating from this warm mass falls as rain over countries bordering the western Pacific, such as Indonesia

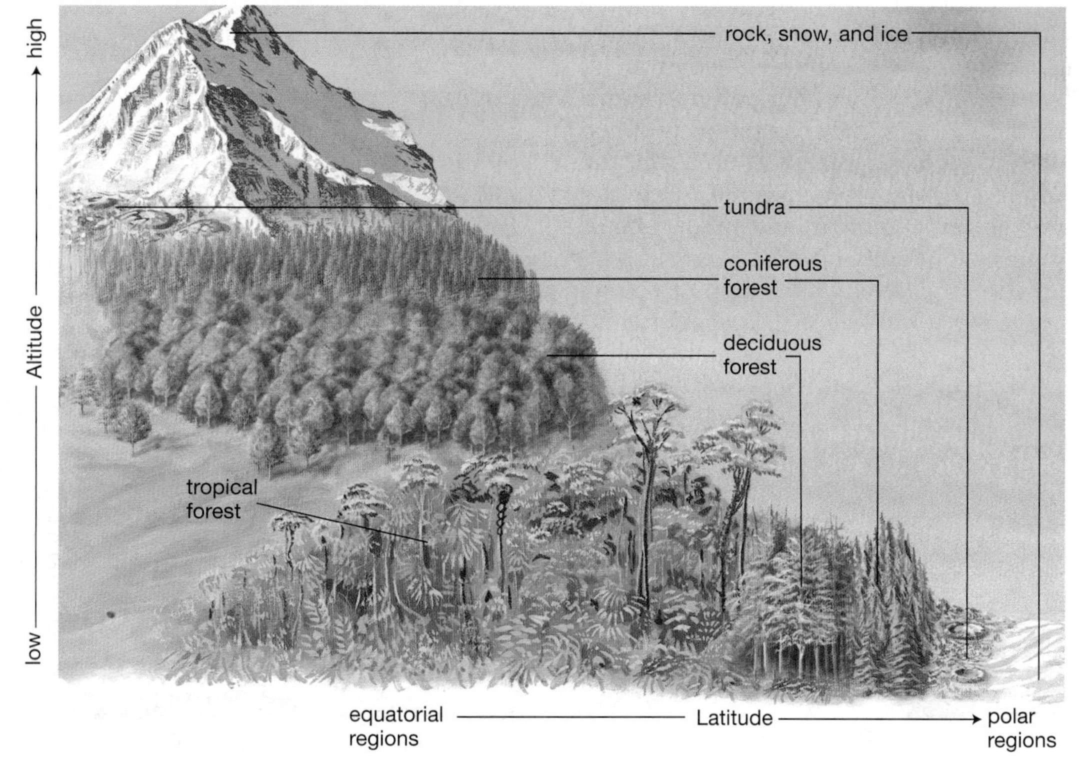

FIGURE 42-4 Effects of elevation on temperature
In terms of temperature, climbing a mountain is like going north; in both cases, increasingly cool temperatures produce a similar series of biomes.

FIGURE 42-5 The rain shadow of the Sierra Nevada
Here, rainfall is graphed against altitude. Westerly (eastward-moving) winds deposit moisture on a mountain's western slope, leaving a desert to the east. **QUESTION** What would this illustration look like between the equator and 30° south latitude?

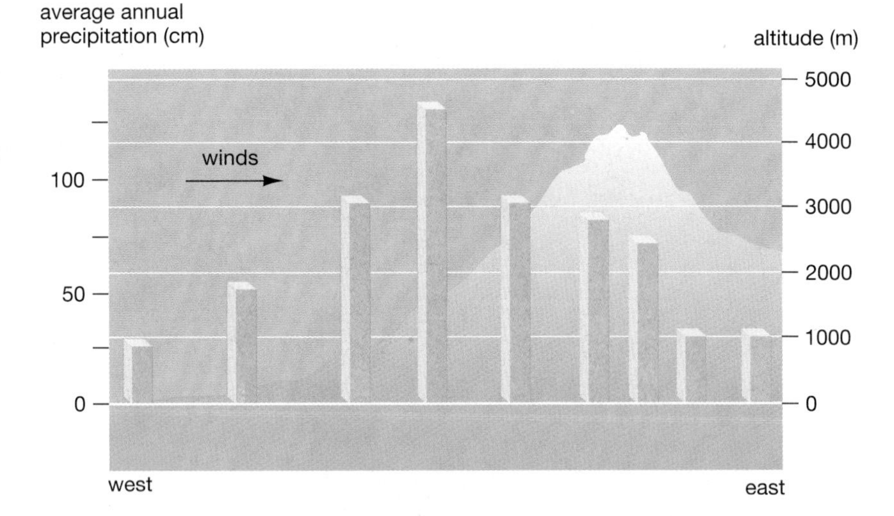

and Australia. To replace the westward-moving water, cooler water from the ocean depths, rich in nutrients, wells up along the western coast of South America, bringing Peru a rich harvest of fish. But for unknown reasons, at intervals of around 3 to 7 years, the tradewinds die down, and an **El Niño** (literally "boy child") occurs. Named by Peruvian fishermen in reference to the baby Jesus, El Niño brings rain to Peru during the normally arid December; the diminishing tradewinds allow warm water to spread back eastward across the Pacific until it reaches the western coast of South America. Meanwhile, drought plagues Indonesia and southern Africa. An El Niño of immense strength disrupted weather worldwide during 1997–1998, causing disastrous floods in both Peru and east Africa, droughts and associated fires in Indonesia, and massive deaths of seabirds, dolphins and other marine mammals as their supply of food fish was disrupted. In the U.S., Texas sizzled in a withering heat and drought, while Florida was deluged. As weather patterns return to normal after an El Niño, the interaction between ocean and atmosphere may overshoot, reversing the pattern and causing a **La Niña** ("girl child"). Researchers are working energetically to create computer models that will allow them to predict both El Niños and La Niñas well before they occur, as well as their effects on various regions of the world. Since these patterns interact with many other variables in the weather, this is a task of staggering complexity. Meanwhile, scientists debate global warming's effects on the phenomena; some predict that El Niños will become more frequent and severe.

42.2 What Conditions Does Life Require?

From the lichens on bare rock to the thermophilic (Greek for "heat-loving") algae in the hot springs of Yellowstone National Park to the bacteria thriving under the pressure-cooker conditions of a deep-sea vent, Earth teems with life. Underlying the diversity of habitats is the common ability to provide, to varying degrees, the four fundamental resources required for life:

- Nutrients from which to construct living tissue
- Energy to power that construction
- Liquid water to serve as a medium in which metabolic reactions occur
- Appropriate temperatures in which to carry out these processes

As we will see in the following sections, these resources are unevenly distributed. Their availability limits the types of organisms that can exist within the various terrestrial and aquatic ecosystems on Earth.

Ecosystems are extraordinarily diverse, yet clear patterns exist. The community that is characteristic of each ecosystem is dominated by organisms specifically adapted to particular environmental conditions. Variations in temperature and in the availability of light, water, and

FIGURE 42-6 Environmental demands mold physical characteristics
Evolution in response to similar environments has molded the bodies of **(a)** American cacti and **(b)** Canary Islands euphorbia into nearly identical shapes, although they are in different families. **QUESTION** Describe the similar selection pressures that operate on these two different families of plants.

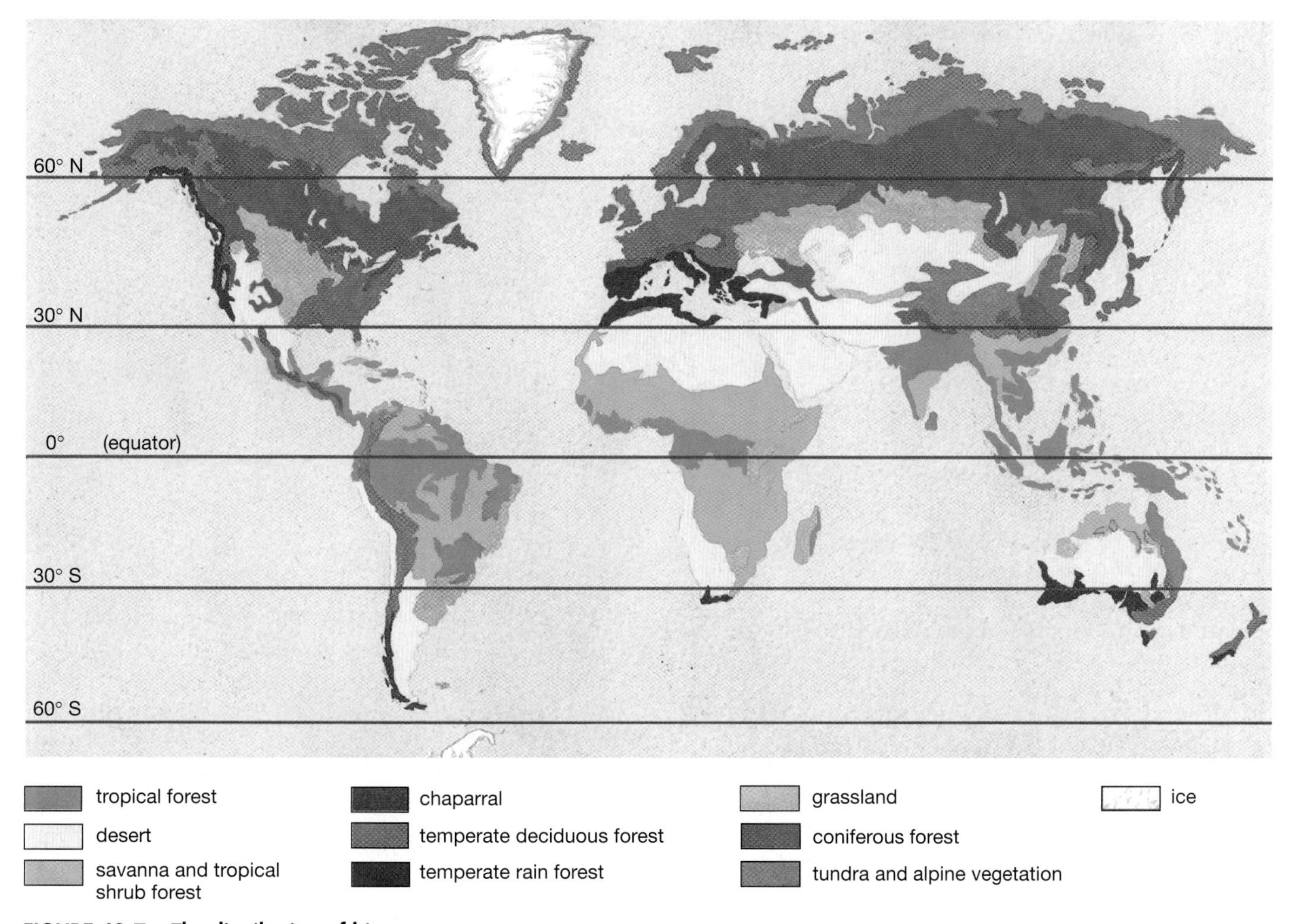

FIGURE 42-7 The distribution of biomes
Although mountain ranges and the sheer size of continents complicate the pattern of biomes, note the overall consistencies. Tundras and coniferous forests are in the northernmost parts of the Northern Hemisphere, whereas the deserts of Mexico, the Sahara, Saudi Arabia, South Africa, and Australia are located about 20° to 30° N and S.

nutrients shape the adaptations of organisms that inhabit an ecosystem. The desert community, for example, is dominated by plants adapted to heat and drought. The cacti of the Mojave Desert in the American Southwest are strikingly similar to the euphorbia of the deserts of Africa and of the nearby Canary Islands, although these plants are in separate families and are only distantly related genetically. Their spinelike leaves and thick, green, water-storing stems are adaptations for water conservation (Fig. 42-6). Likewise, the plants of the arctic tundra and those of the alpine tundra of the Rocky Mountains show growth patterns clearly recognizable as adaptations to a cold, dry, windy climate. Thus, in regions with similar environmental conditions, similar types of organisms are organized into similar types of communities.

42.3 How Is Life on Land Distributed?

Terrestrial organisms are restricted in their distribution largely by the availability of water and by temperature. Terrestrial ecosystems receive plenty of light, even on an overcast day, and the soil provides abundant nutrients. Water, however, is limited and very unevenly distributed, both in place and in time. Terrestrial organisms must be adapted to obtain water when it is available and to conserve it when it is scarce.

Like water, temperatures favorable to life are very unevenly distributed in place and time. At the South Pole, even in summer, the average temperature is usually well below freezing; not surprisingly, life is scarce there. Places such as central Alaska have favorable temperatures during only the summer, whereas the Tropics have a uniformly warm, moist climate in which life abounds.

Terrestrial Biomes Support Characteristic Plant Communities

Terrestrial communities are dominated and defined by their plant life. Because plants can't escape from drought, sun, or winter weather, they tend to be extremely well-adapted to the climate of a particular region. Large land areas with similar environmental conditions and characteristic plant communities are called **biomes** (Fig. 42-7). Biomes are generally named after the major type of vegetation found there. The predominant vegetation of each

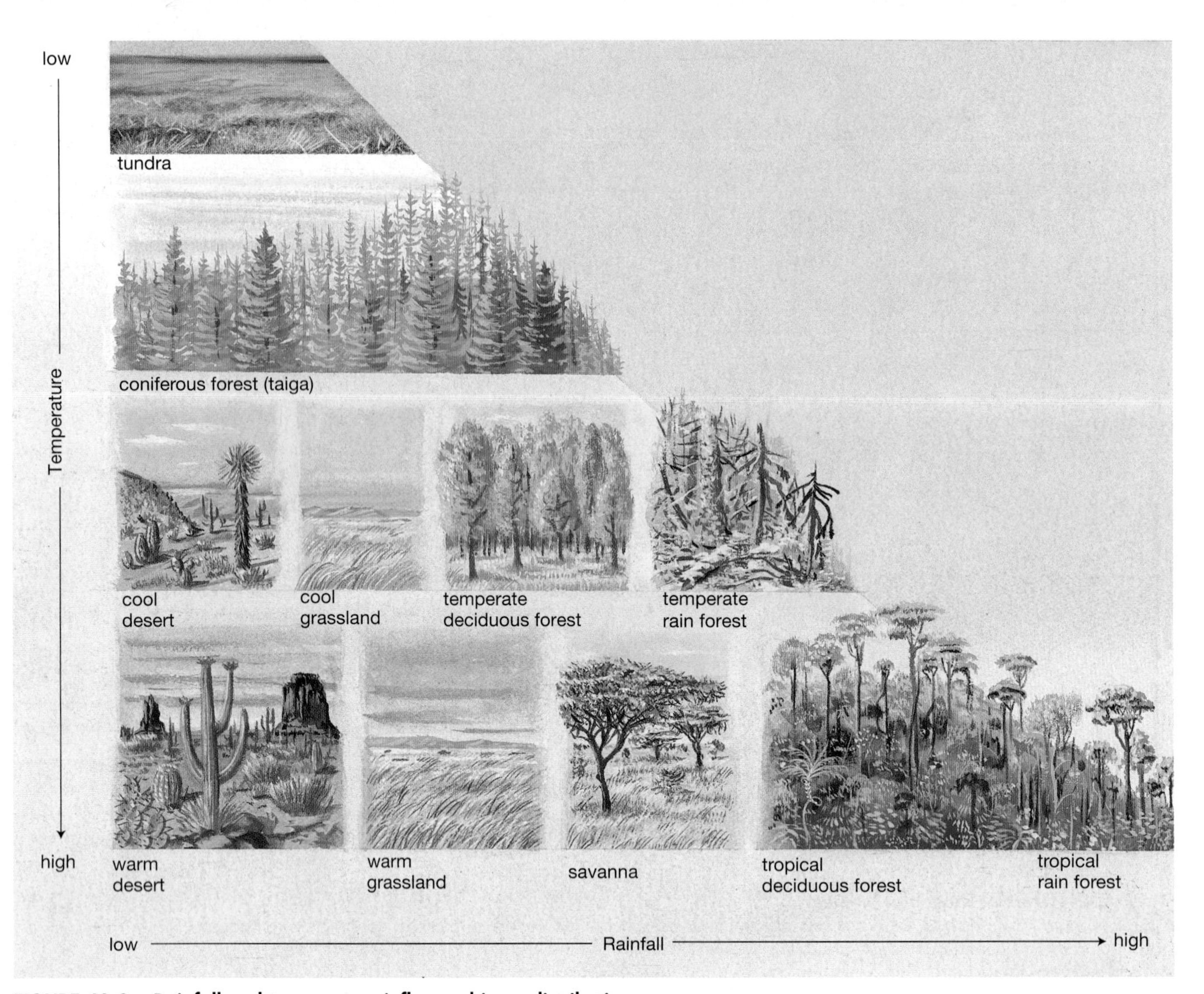

FIGURE 42-8 Rainfall and temperature influence biome distribution
Together, rainfall and temperature determine the soil moisture available to support plants.

biome is determined by the complex interplay of rainfall and temperature (Fig. 42-8). These factors determine the soil moisture available for plants to grow and to compensate for their evaporative water losses. In addition to the total amount of rainfall and the overall yearly average temperature, the variability of these factors over the year determines which plants can grow in an area. Arctic tundra plants, for example, must be adapted to marshy conditions in the early summer but cold and desertlike conditions for much of the rest of the year, when water is solidly frozen and unavailable. In the following sections we discuss the major biomes, beginning at the equator and working our way poleward. We also discuss some of the impacts of human activities on these biomes.

Tropical Rain Forests

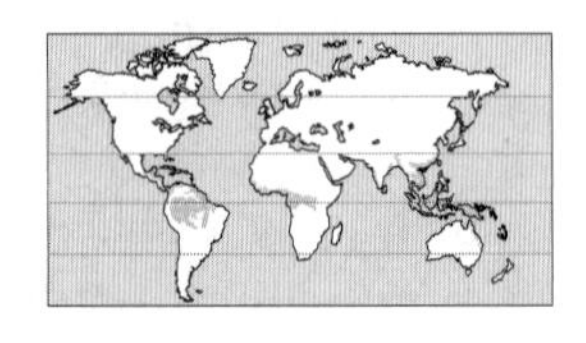

Near the equator, the temperature averages between 77 °F and 86 °F (25 °C and 30 °C) with little variation, and rainfall ranges from 100 to 160 inches (250 to 400 centimeters) annually. These evenly warm, evenly moist conditions combine to create the most diverse biome on Earth, the **tropical rain forest**, dominated by huge broadleaf evergreen trees (Fig. 42-9). Extensive rain forests are found in Central and South America, Africa, and Southeast Asia.

Biodiversity refers to the total number of species within an ecosystem and the resulting complexity of interactions among them; in short, it defines the biological "richness" of an ecosystem. Rain forests have the highest biodiversity of any ecosystem on Earth. Although rain forests cover only 6% of Earth's total land area, ecologists estimate that they are home to between 5 million and 8 million species, representing half to two-thirds of the world's total. For example, a recent survey of a 2.5-acre site in the upper Amazon basin revealed 283 species of trees, most of which were represented by a single individual. In a 5-square-kilometer (about 3-square-mile) tract of rain forest in Peru, scientists counted more than 1300 species of butterfly and 600 species of bird. In comparison, the entire United States is home to only 400 butterfly species and 700 bird species.

FIGURE 42-9 The tropical rain forest biome
(a) Towering trees draped with vines reach for the light in the dense tropical rain forest. Amid their branches dwells the most diverse assortment of animals on Earth, including **(b)** a fruit-eating toucan, **(c)** a golden-eyed leaf frog, **(d)** a tree-dwelling orchid, and **(e)** a howler monkey. **QUESTION** How does a biome with such poor soil support the highest plant productivity (see Fig. 41-3) and the greatest animal diversity on Earth?

Tropical rain forests typically have several layers of vegetation. The tallest trees reach 150 feet (50 meters) and tower above the rest of the forest. Below is a fairly continuous canopy of treetops at about 90 to 120 feet (30 to 40 meters). Another layer of shorter trees typically stands below the canopy. Huge woody vines, commonly 328 feet (100 meters) or more in length, grow up the trees, reaching the sunlight far above. These layers of vegetation block out most of the sunlight. Many of the plants that live in the dim green light that filters through to the forest floor have enormous leaves to trap the little available energy.

Because edible plant material close to the ground in tropical rain forests is relatively scarce, much of the animal life—including numerous birds, monkeys, and insects—is *arboreal*, or tree-dwelling. Competition for the nutrients that do reach the ground is intense among both animals and plants. Even such unlikely sources of food as the droppings of monkeys are in great demand. For example, dung beetles feed and lay their eggs on monkey droppings. When ecologists attempted to collect droppings of the South American howler monkeys to find out what the monkeys had been eating, they found themselves in a race with the beetles, hundreds of which would arrive within minutes after a dropping hit the ground!

Almost as soon as bacteria or fungi release any nutrients from dead plants or animals into the soil, rainforest trees and vines absorb the nutrients. This is one of the reasons why, despite the teeming vegetation, agriculture is very risky and destructive in rain forests. Virtually all the nutrients in a rain forest are already tied up in the vegetation, so the soil is infertile. If the trees are carried away for lumber, few nutrients remain to support crops. Further, even if the nutrients are released by burning the vegetation, the heavy year-round rainfall quickly dissolves and erodes them away, leaving the soil infertile after only a few seasons of cultivation. The exposed soil, which is rich in iron and aluminum, then takes on an impenetrable, bricklike quality as it bakes in the tropical sun. As a result, secondary succession on cleared rainforest land is slowed significantly. Even small forest cuttings take about 70 years to regenerate.

Human Impact Despite their unsuitability for agriculture, rain forests are being felled for lumber or burned down for ranching or farming at an alarming rate (Fig. 42-10). Estimates of rainforest destruction range from about 22,000 square miles to about 52,000 square miles annually (for comparison, the state of Connecticut occupies about 5000 square miles). Based on satellite images, the Brazilian government reported a loss of 10,000 square miles of rain forest in their country alone between 2001 and 2002, 40% more than previous estimates. Rain forest left standing is often in fragments too small to allow reproduction of trees and to provide adequate foraging and breeding habitat for larger animals. Fires in clear-cut areas also spread along the ground into adjacent forest, killing young trees and disrupting community interactions. In West Africa, forest burning is causing acid rain, which damages the remaining trees. Further, much of the rainfall in a rain forest comes from water transpired through the forest's leaves. As large tracts of rain forests disappear, the region becomes drier, more stressed, and more susceptible to fires. Researchers estimate that one-quarter of the carbon dioxide released into the atmosphere during the past decade came from cutting and burning tropical rain forests, exacerbating global warming.

At least 40% of the world's rain forests are now gone. Harvard University biologist Edward O. Wilson estimates that the destruction of tropical rain forests may drive 27,000 species to extinction annually. The impact of these losses of irreplaceable biodiversity is incalculable, but humanity certainly loses access to a wealth of potential drugs and raw materials. For example, in Malaysia, the sap of a rare species of tree yielded a compound called *calanolide A*, which offers promise as an anti-AIDS drug; a synthetic version is now being tested in human clinical trials. A compound isolated from an African rain forest leaf fungus may eventually help control diabetes, a health problem that is becoming more common in the U.S.

Although disastrous losses continue, some areas have been set aside as protected preserves, and some reforestation efforts are under way. Local residents are becoming more involved in conservation efforts, as illustrated in our Case Study and Links to Life. In Brazil, a union of people who harvest natural rubber from tree sap is fighting to preserve large tracts of land for rubber production and harvesting fruits and nuts. These efforts are steps toward the ultimate solution, which is tragically slow in coming: sustainable use. *Sustainable use* means deriving benefits from an ecosystem, whether from tourism or harvesting products, in a way that can be sustained indefinitely without damage to the ecosystem.

FIGURE 42-10 Amazonian rain forest cleared by burning
The burned area will be converted to ranching or agriculture, but both are doomed to failure due to poor soil quality. Spreading fires and the smoke they produce also threaten adjacent uncut forests and their myriad inhabitants.

FIGURE 42-11 The African savanna
(a) Elephants roam beneath a rainbow. **(b)** A red-billed oxpecker looks up at a sleeping white rhino. Oxpeckers feed on parasites that live on rhino skin. **(c)** Large herds of grazing animals, such as zebras, can still be seen on African preserves. The herds of herbivores provide food for the greatest assortment of large carnivores on Earth. **(d)** A cheetah feasts on its prey (both rhinos and cheetahs are endangered).

Tropical Deciduous Forests

Slightly farther from the equator, the rainfall is not nearly as constant, and there are pronounced wet and dry seasons. In these areas, which include much of India as well as parts of southeast Asia, South America and Central America, **tropical deciduous forests** grow. During the dry season, the trees cannot get enough water from the soil to compensate for evaporation from their leaves. As a result, the plants have adapted to the dry season by shedding their leaves, thereby minimizing water loss. If the rains fail to return on schedule, the trees delay the formation of new leaves until the drought passes.

Savanna

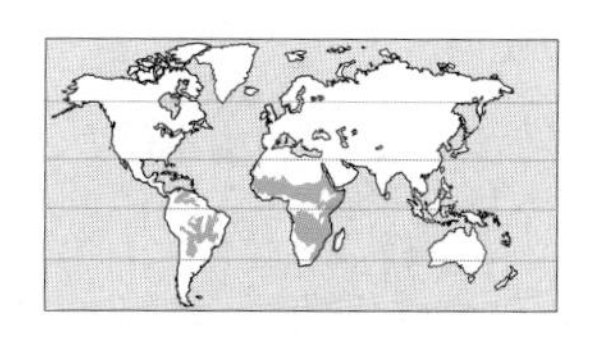

Along the edges of the tropical deciduous forest, the trees gradually become more widely spaced, with grasses growing between them. Eventually, grasses become the dominant vegetation, with only scattered trees and thorny scrub forests here and there; this biome is the **savanna** (Fig. 42-11). Savanna grasslands typically have a rainy season during which virtually all of the year's precipitation falls—12 inches (30 centimeters) or less. When the dry season arrives, it comes with a vengeance. Rain might not fall for months, and the soil becomes hard, dry, and dusty. Grasses are well adapted to this type of climate, growing very rapidly during the rainy season and dying back to drought-resistant roots during the arid times. Only a few specialized trees, such as the thorny acacia or the water-storing baobab, can survive the devastating savanna dry seasons. In areas in which the dry season becomes even more pronounced, virtually no trees can grow, and the savanna imperceptibly grades into tropical grassland.

The African savanna has probably the most diverse and impressive array of large mammals on Earth. These mammals include numerous herbivores such as antelope, wildebeest, water buffalo, elephants, and giraffes and carnivores such as the lion, leopard, hyena, and wild dog.

FIGURE 42-12 Poaching threatens African wildlife
Rhinoceros horns, believed by some to have aphrodisiac properties, fetch staggering prices and encourage poaching. The black rhino is now nearly extinct.

HUMAN IMPACT Africa's rapidly expanding human population threatens the wildlife of the savanna. Poaching has driven the black rhinoceros to the brink of extinction (Fig. 42-12) and endangers the African elephant, a keystone species in this ecosystem. The abundant grasses that make the savanna a suitable habitat for so much wildlife also make it suitable for grazing domestic cattle. As the human population of East Africa increases, so does the pressure of cattle grazing on the savanna. Fences increasingly disrupt the migration of the great herds of herbivores in search of food and water. Ecologists have discovered that the native herbivores are much more efficient at converting grass into meat than are cattle. Perhaps the future African savanna may support herds of domesticated antelope and other large native grazers in place of cattle.

Deserts

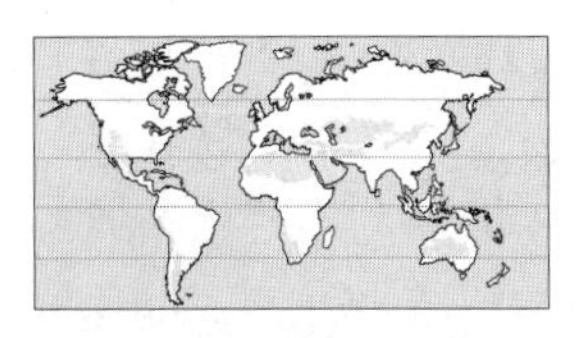

Even drought-resistant grasses need at least 10 to 20 inches (25 to 50 centimeters) of rain a year, depending on its seasonal distribution and the average temperature. When less than 10 inches of rain fall, **desert** biomes result. Although we tend to think of them as hot, deserts are defined by their lack of rain rather than by their temperatures. In the Gobi desert of Asia, for example, temperatures average below freezing for half the year, while summer temperatures average 105 °F to 110 °F (42 °C to 43°C). Desert biomes are found on every continent, typically around 20° to 30° N and S latitude and also in the rain shadows of major mountain ranges.

As with all biomes, deserts include a variety of environments. At one extreme are certain areas of the Sahara Desert or Chile, where it virtually never rains and no vegetation grows (Fig. 42-13a). More commonly, deserts are characterized by widely spaced vegetation and large areas of bare ground (Fig. 42-13b). In many cases, the perennial plants are bushes or cacti with large, shallow root systems. The shallow roots quickly soak up soil moisture after the infrequent desert storms. The rest of the plant is typically covered with a waterproof, waxy coating to prevent evaporation of precious water. The thick stems of cacti and other succulents store water. The spines of cacti are leaves modified for protection and water conservation, presenting almost no surface area for evaporation. In many deserts, all the rain falls in just a few storms, and specialized annual wildflowers take advantage of the brief period of moisture to race through germination, growth, flowering, and seed production in a month or less (Fig. 42-14).

The animals of the deserts, like the plants, are specially adapted to survive on little water. Most deserts appear to be almost completely devoid of animal life during summer days because the animals seek relief

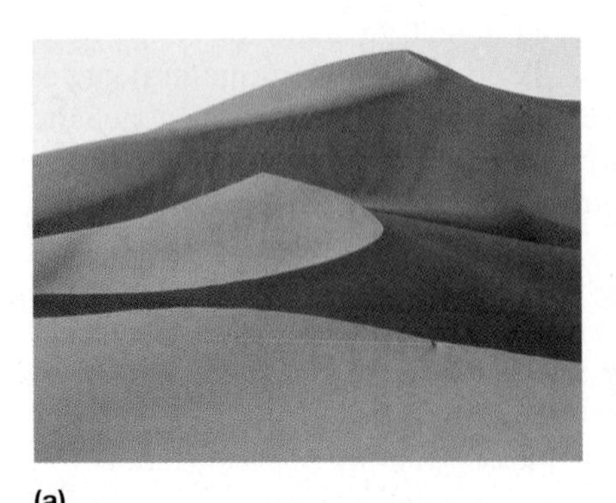

(a)

(b)

(c)

FIGURE 42-13 The desert biome
(a) Under the most extreme conditions of heat and drought, deserts can be almost devoid of life, such as these sand dunes of the Sahara Desert in Africa. **(b)** Throughout much of Utah and Nevada, the Great Basin Desert presents a monotonous landscape of widely spaced shrubs, such as sagebrush and greasewood. These shrubs often secrete a growth inhibitor from their roots, preventing germination of nearby plants and thus reducing competition for water. **(c)** The kangaroo rat is an elusive inhabitant of the deserts of North America.

FIGURE 42-14 The Sonoran Desert
After a relatively wet spring, this Arizona desert is carpeted with wildflowers. Through much of the year—sometimes for several years—annual wildflower seeds lie dormant, waiting for adequate spring rains to fall. **QUESTION** How might desert plant seeds "determine" if rainfall is adequate for germination? (Hint: Refer back to Chapter 25.)

FIGURE 42-15 The chaparral biome
Limited to coastal regions and maintained by fires set by lightning, this hardy biome is characterized by drought-resistant shrubs and small trees. Some of the shrubs in this chaparral near San Francisco, California, turn brilliant red in the fall.

from the sun and heat in cool underground burrows. After dark, when deserts cool down considerably, horned lizards, snakes, and other reptiles emerge to feed, as do mammals such as the kangaroo rat (Fig. 42-13c) and birds such as the burrowing owl. Most of the smaller animals survive without ever drinking, getting all the water they need from their food and from that produced during cellular respiration in their tissues. Larger animals, such as desert bighorn sheep, are dependent on permanent water holes during the driest times of the year.

Human Impact Desert ecosystems are fragile. Ecologists studying the soil of the Mojave Desert in southern California recently found treadmarks left by tanks in 1940 when General Patton trained tank crews in preparation for entry into World War II. The desert soil is stabilized and enriched by microscopic cyanobacteria whose filaments intertwine among sand grains. The tanks, and now numerous off-road vehicles that careen about the desert for recreation, destroy this crucial network. This allows the soil to erode and reduces nutrients available to the desert's slow-growing plants. Ecologists estimate that the desert soil may require hundreds of years to fully recover from heavy vehicle use.

Scientists once believed that human activities were causing a progressive southward spread of the Sahara Desert in Africa, a process called *desertification*. However, data from satellite photographs taken at intervals since 1980 indicate that the Sahara Desert has spread southward and then has retreated back repeatedly over this 20-year period in accordance with changes in the amount of rainfall. Although the dry savanna south of the Sahara Desert has been significantly overgrazed and degraded by a human population that is above its sustainable carrying capacity, the desert biome is not spreading as a result, as was formerly believed.

Chaparral

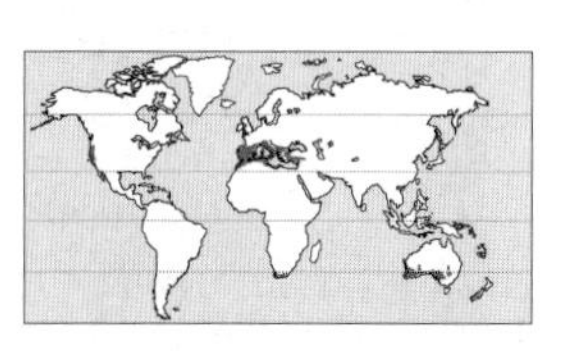

In many coastal regions that border on deserts, such as southern California and much of the Mediterranean, we find a unique type of vegetation called **chaparral**. The annual rainfall in these regions is up to 30 inches, nearly all of which falls during cool, wet winters that alternate with hot, dry summers. The proximity of the sea provides a slightly longer rainy season in the winter and frequent fogs during the spring and fall, reducing evaporation. Chaparral consists of small trees or large bushes with thick waxy or fuzzy leaves that conserve water, such as sages and evergreen oak. These hardy shrubs are also able to withstand the frequent summer fires started by lightning (Fig. 42-15).

FIGURE 42-16 **Tallgrass prairie in Missouri**
In the central U.S., moisture-bearing winds out of the Gulf of Mexico produce summer rains, allowing a lush growth of tall grasses and wildflowers such as these coneflowers. Periodic fires, now carefully managed, prevent the encroachment of forest. **QUESTION** Why is tallgrass prairie one of the most endangered biomes in the world?

Grasslands

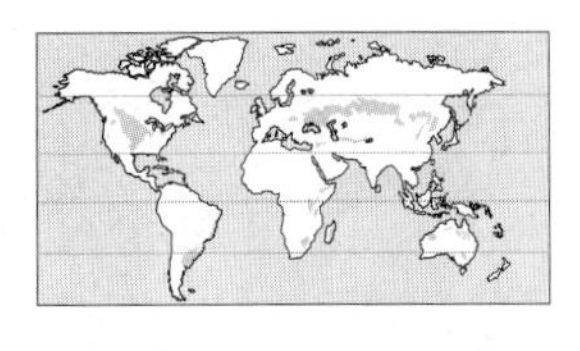

In the temperate regions of North America, deserts occur in the rain shadows east of the mountain ranges, such as the Sierra Nevada and Rocky Mountains. Eastward, as the rainfall gradually increases, the land supports more and more grasses, giving rise to the prairies of the Midwest. Most **grassland**, or **prairie**, biomes are located in the centers of continents, such as North America and Eurasia, and receive 10 to 30 inches (25 to 75 centimeters) of rain annually. In general, they have a continuous cover of grass and virtually no trees except along the rivers. From the tallgrass prairies of Iowa, Missouri, and Illinois where the rainfall is relatively high (Fig. 42-16), to the drier shortgrass prairies of eastern Colorado, Wyoming, and Montana (Fig. 42-17), the North American grassland once stretched across almost half the continent.

FIGURE 42-17 **Shortgrass prairie**
The lands east of the Rocky Mountains receive relatively little rainfall, and **(a)** shortgrass prairie results, characterized by low-growing bunch grasses such as buffalo grass and grama grass. **(b)** Pronghorn antelope, **(c)** prairie dogs, and **(d)** protected bison herds occupy this biome, in which **(e)** wildflowers such as this coneflower abound.

Water and fire are the crucial factors in the competition between grasses and trees. The hot, dry summers and frequent droughts of the shortgrass prairies can be tolerated by grass but are fatal to trees. In the more eastern tallgrass prairies, forests are the climax ecosystems; historically, however, the trees were destroyed by frequent fires, often set by Native Americans to maintain grazing land for the bison. Although the tops of the grasses are destroyed by fire, their root systems usually survive; trees, however, are killed outright. The grasslands of North America once supported huge herds of bison—as many as 60 million in the early nineteenth century. Pronghorn antelope can still be seen in some prairies of the western U.S., where bobcats and coyotes are the major large predators (see the prairie food web illustrated in Fig. 41-5). Grasses growing and decomposing for thousands of years produced what may be the most fertile soil in the world. An acre of natural tallgrass prairie in the U.S. supports 200–400 different native plants.

Human Impact The development of plows that could break through the dense grass turf set the stage for converting the midwestern U.S. prairies into the "breadbasket" of North America, so named because enormous quantities of grain are cultivated in its fertile soil. The tallgrass prairie, now one of the most endangered ecosystems worldwide, has nearly all been converted to agricultural land. Only about 1% remains, in tiny protected remnants maintained by periodic controlled burning.

On the dry western shortgrass prairie, cattle have replaced the bison and pronghorn antelope. As a result of their overgrazing the grasses, the boundary between the cool deserts and the grassland has often been altered in favor of desert plants. Much of the sagebrush desert of the American West is actually overgrazed shortgrass prairie (Fig. 42-18). Cattle prefer grass to sagebrush, so heavy grazing destroys the grass. Consequently, moisture that the grass would have absorbed is left in the soil, encouraging the growth of the woody sagebrush. Thus, the prairie grasses are replaced by plants characteristic of the cool desert.

Temperate Deciduous Forests

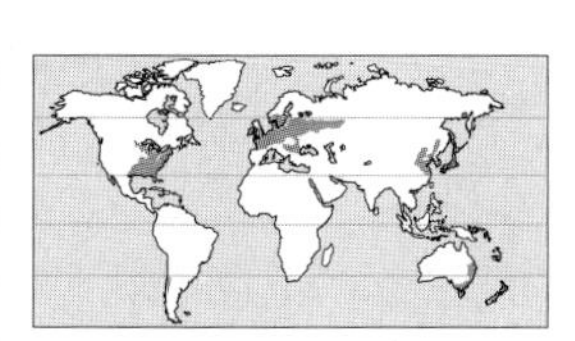

At their eastern edge, the North American grasslands merge into the **temperate deciduous forest** biome, also found in Western Europe and East Asia (Fig. 42-19). More precipitation occurs there than in the grasslands (30 to 60 inches, or 75 to 150 centimeters), particularly during the summer. The soil retains enough moisture for trees to grow, and the resulting forest shades out grasses. In contrast to tropical forests, the temperate deciduous forest biome has cold winters, usually with at least several hard frosts and long periods of below-freezing weather. Winter in this biome has an effect on the trees similar to that of the dry season in the tropical deciduous forests: During periods of subfreezing temperatures, liquid water is not available. To reduce evaporation when water is in short supply, the trees drop their leaves in the fall. They produce leaves again in the spring, when liquid water becomes available. During the brief time in spring when the ground has thawed but the trees have not yet blocked off all the sunlight, abundant wildflowers grace the forest floor.

Insects and other arthropods are numerous and conspicuous in deciduous forests. The decaying leaf litter on the forest floor also provides food and habitat for bacteria, earthworms, fungi, and small plants. Many arthropods feed on these or on each other. A variety of vertebrates, including mice, shrews, squirrels, raccoons, deer, bears, and many species of birds, dwell in the deciduous forests.

Human Impact Large predatory mammals such as black bears, wolves, bobcats, and mountain lions were formerly abundant, but hunting and habitat loss has severely reduced their numbers and effectively eliminated wolves from deciduous forests. In many deciduous forests, deer are plentiful because of lack of natural predators. Clearing for lumber, agriculture, and housing has dramatically reduced deciduous forests in the U.S. from their original extent, and virgin deciduous forests are now almost nonexistent. Over the past 50 years, however, Forest Service data show that forest cover in the U.S. (both evergreen and deciduous) has increased as a result of regrowth of forests on abandoned farms, paper recycling that decreases demand for wood pulp, more efficient lumber milling and tree farming techniques, and the use of alternative building materials.

FIGURE 42-18 Sagebrush desert or shortgrass prairie? Biomes are influenced by human activities as well as by temperature, rainfall, and soil. The shortgrass prairie field on the right has been overgrazed by cattle, causing the grasses to be replaced by sagebrush.

FIGURE 42-19 The temperate deciduous forest biome
(a) In temperate deciduous forests of the eastern U.S., **(b)** the white-tailed deer is the largest herbivore, and **(c)** birds such as the blue jay are abundant. **(d)** In spring, a profusion of woodland wildflowers (such as these hepaticas) blooms briefly before the trees produce leaves.

Temperate Rain Forests

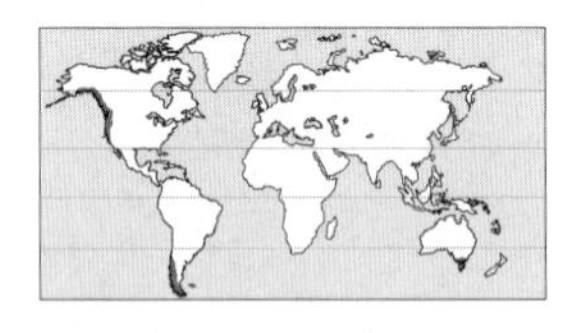

On the U.S. Pacific coast, from the lowlands of the Olympic Peninsula in Washington State to southeast Alaska, lies a **temperate rain forest** biome (Fig. 42-20). Temperate rain forests, which are relatively rare, are also located along the southeastern coast of Australia and the southwestern coasts of New Zealand and Chile. As in the tropical rain forest, there is no shortage of liquid water year-round. This abundance of water is due to two factors. First, there is a tremendous amount of rain. The Hoh River rain forest in Olympic National Park receives more than 160 inches (400 centimeters) of rain annually, more than 24 inches (60 centimeters) in the month of December alone. Second, the moderating influence of the Pacific Ocean prevents severe frost from occurring along the coast, so the ground seldom freezes and liquid water remains available.

The abundance of water means that the trees have no need to shed their leaves in the fall, and almost all the trees are evergreens. In contrast to the broadleaf evergreen trees of the Tropics, temperate rain forests are dominated by conifers. The ground and typically the trunks of the trees are covered with mosses and ferns. As in tropical rain forests, so little light reaches the forest floor that tree seedlings usually cannot become established. Whenever one of the forest giants falls, however, it opens up a patch of light, and new seedlings quickly sprout, commonly right atop the fallen log. This event produces a "nurse log" (see Fig. 42-20b).

Taiga

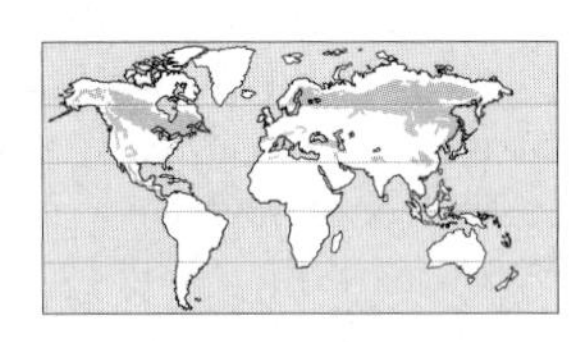

North of the grasslands and temperate forests, the **taiga**, also called the **northern coniferous forest** (Fig. 42-21, p. 880), stretches horizontally across all of North America and

FIGURE 42-20 The temperate rainforest biome
(a) The Hoh River temperate rain forest in Olympic National Park. Coniferous trees do not block the light as effectively as do broadleaf trees, so ferns, mosses, and wildflowers grow in the pale green light of the forest floor. **(b)** The dead feed the living, as new trees grow on this "nurse log," and **(c)** flowering foxglove and **(d)** fungi find ideal conditions amid the moist, decaying vegetation.

Eurasia, including parts of the northern U.S. and much of southern Canada. Conditions in the taiga are harsher than those in the temperate deciduous forest. In the taiga, the winters are longer and colder, and the growing season is shorter. The few months of warm weather are too short to allow trees the luxury of regrowing leaves in the spring. As a result, the taiga is populated almost entirely by evergreen coniferous trees with narrow, waxy needles. The waxy coating and small surface area of the needles reduce water loss by evaporation during the cold months, and the leaves remain on the trees year-round. Thus, the trees are instantly ready to take advantage of good growing conditions when spring arrives, and they can continue slow growth late into the fall.

Because of the harsh climate in the taiga, the diversity of life there is much lower than in many other biomes. Vast stretches of central Alaska, for example, are covered by a somber forest that consists almost exclusively of black spruce and an occasional birch. Large mammals such as the wood bison, grizzly bear, moose, and wolf, which have mostly been eradicated in the southern regions of their original range, still roam the taiga, as do smaller animals such as the wolverine, fox, snowshoe hare, and deer. Outside of Alaska, where these large mammals remain reasonably abundant, small populations of wolves roam the northern U.S., including Idaho, Montana (where they have been reintroduced into Yellowstone National Park), Michigan, Wisconsin, and Minnesota (which hosts the largest wolf population in the lower 48 states).

Human Impact The taiga is a major source of lumber. Clear-cutting, the removal of all the trees in a given area for use in paper-making and construction, has destroyed huge expanses of forest, both in Canada and the U.S. Pacific Northwest (Fig. 42-22). Owing to the remoteness of the northernmost taiga and the severity of its climate, a greater percentage of the taiga remains in undisturbed condition than any other North American biome except the tundra.

FIGURE 42-21 The taiga (or northern coniferous forest) biome
(a) The small needles and pyramidal shape of conifers allow them to shed heavy snows. **(b)** Winter is a challenge not only for the trees but also for animals such as this snowshoe hare and its predators, the bobcat and **(c)** the great horned owl. Taiga animals face diminished food supply but increased energy requirements during subfreezing weather.

FIGURE 42-22 Clear-cutting
Clear-cutting, as seen in this Oregon forest, is relatively simple and cheap, but its environmental costs are high. Erosion diminishes the fertility of the soil, slowing new growth. Further, the dense stands of same-age trees that typically regrow are more vulnerable to attack by parasites than a natural stand of trees of various ages would be.

Tundra

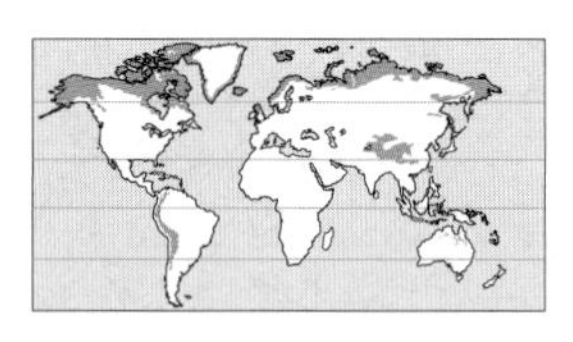

The last biome encountered before we reach the polar ice caps is the arctic **tundra**, a vast treeless region bordering the Arctic Ocean (Fig. 42-23). Conditions in the tundra are severe. Winter temperatures in the arctic tundra often reach −40 °F (−55 °C) or below, winds howl at 30 to 60 miles (50 to 100 kilometers) an hour, and precipitation averages 10 inches (25 centimeters) or less each year, making this a "freezing desert." Even during the summer, the temperatures can drop to freezing, and the growing season may last only a few weeks before a hard frost occurs. Somewhat less cold but similar conditions produce alpine tundra on mountaintops above the altitude where trees can grow.

The cold climate of the arctic tundra results in **permafrost**, a permanently frozen layer of soil typically no more than about 1.5 feet below the surface. As a result, when summer thaws come, the water from melted

FIGURE 42-23 The tundra biome
(a) Life on the tundra is adapted to cold. **(b)** Plants such as dwarf willows and perennial wildflowers (such as this dwarf clover) grow low to the ground, escaping the chilling tundra wind. Tundra animals, such as **(c)** caribou and **(d)** arctic foxes, can regulate blood flow in their legs, keeping them just warm enough to prevent frostbite, while preserving precious body heat for the brain and vital organs.

snow and ice cannot soak into the ground, and the tundra becomes a huge marsh. Trees cannot survive in the tundra; the permafrost limits root growth to the topmost meter or so of soil.

Nevertheless, the tundra supports a surprising abundance and variety of life. The ground is carpeted with small perennial flowers and dwarf willows no more than a few centimeters tall and often covered with a large lichen called "reindeer moss," a favorite food of caribou. The standing water provides a superb habitat for mosquitoes. The mosquitoes and other insects provide food for numerous birds, most of which migrate long distances to nest and raise their young during the brief summer feast. The tundra vegetation supports lemmings, which are eaten by wolves, snowy owls, arctic foxes, and even grizzly bears.

Human Impact The tundra is among the most fragile of all the biomes because of its short growing season. A willow 4 inches (10 centimeters) high may have a trunk 3 inches (7 centimeters) in diameter and be 50 years old. Human activities in the tundra leave scars that persist for centuries. Fortunately for the tundra inhabitants, the impact of civilization is localized around oil drilling sites, pipelines, mines, and military bases.

Rainfall and Temperature Determine the Vegetation a Biome Can Support

Terrestrial biomes are greatly influenced by both temperature and rainfall, whose effects interact. Temperature strongly influences the effectiveness of rainfall in providing soil moisture for plants and standing water for animals to drink. The hotter it is, the more rapidly water evaporates, both from the ground and from plants. As a result of this interaction of temperature with rainfall (and to a lesser extent, the distribution of rain throughout the year), areas that receive almost exactly the same rainfall can have startlingly different vegetation, all the way from desert to taiga. Now that you are familiar with biomes, let's travel from southern Arizona to central Alaska, visiting ecosystems that each receive about 12 inches (28 centimeters) of rain annually.

The Sonoran Desert near Tucson, Arizona (see Fig. 42-14) has an average annual temperature of 68 °F (20 °C) and receives about 12 inches (28 centimeters) of rain each year. The landscape is dominated by giant saguaro cactus and low-growing, drought-resistant bushes. Going 931 miles (1500 kilometers) north will bring you to eastern Montana, where rainfall is about the same; however, you will be amid shortgrass prairie (see Fig. 42-17) largely because the average temperature is much lower, about 45 °F (7 °C). Much farther north, central Alaska receives about the same annual rainfall, yet it is covered with taiga forest (see Fig. 42-21). As a result of the low average annual temperature (about 25 °F, or –4 °C), permafrost underlies much of the ground here. During the summer thaw, the taiga earns its Russian name "swamp forest," although its rainfall is about the same as that of the Sonoran Desert.

42.4 How Is Life in Water Distributed?

Although this chapter has thus far emphasized terrestrial biomes, the saltwater oceans and seas are the largest ecosystems on Earth, covering about 71% of its surface. Freshwater ecosystems, in contrast, cover less than 1%.

The unique properties of water lend three common features to aquatic ecosystems. First, because water is slower to heat and cool than air, temperatures in aquatic ecosystems are more moderate than are those in terrestrial ecosystems. Second, although water may appear quite transparent, it absorbs a considerable amount of the light energy that sustains life. Even in the clearest water, the intensity of light decreases rapidly with depth. At depths of 600 feet (200 meters) or more, little light is left to power photosynthesis. If the water is at all cloudy—for example, because of suspended sediment or microorganisms—the depth to which light can penetrate is greatly reduced. Third, nutrients in aquatic ecosystems tend to be concentrated near the bottom sediments, where light levels are too low to support photosynthesis. This separation of energy and nutrients limits aquatic life. Of the four requirements for life, aquatic ecosystems provide abundant water and appropriate temperatures. Thus, the major factors that determine the quantity and type of life in aquatic ecosystems are the remaining two factors: energy and nutrients.

Although they share some common features, aquatic ecosystems are extremely diverse. Freshwater ecosystems encompass rivers, streams, ponds, lakes, and marshes; marine (saltwater) ecosystems include estuaries, tide pools, coral reefs, the open ocean, and vent communities. In the following sections, we look more closely at some of these important aquatic ecosystems.

Freshwater Lakes Have Distinct Regions of Life

Freshwater lakes vary tremendously in size, depth, and nutrient content. Although each lake is unique, moderate to large lakes in temperate climates share some common features, including distinct zones of life.

Life Zones Are Determined by Access to Light and Nutrients

The distribution of life in lakes depends largely on access to light, nutrients, and in some cases a place for attachment (the bottom). The life zones of lakes, then, correspond to specific locations within the lake: the *littoral zone*, the *limnetic zone*, and the *profundal zone* (Fig. 42-24).

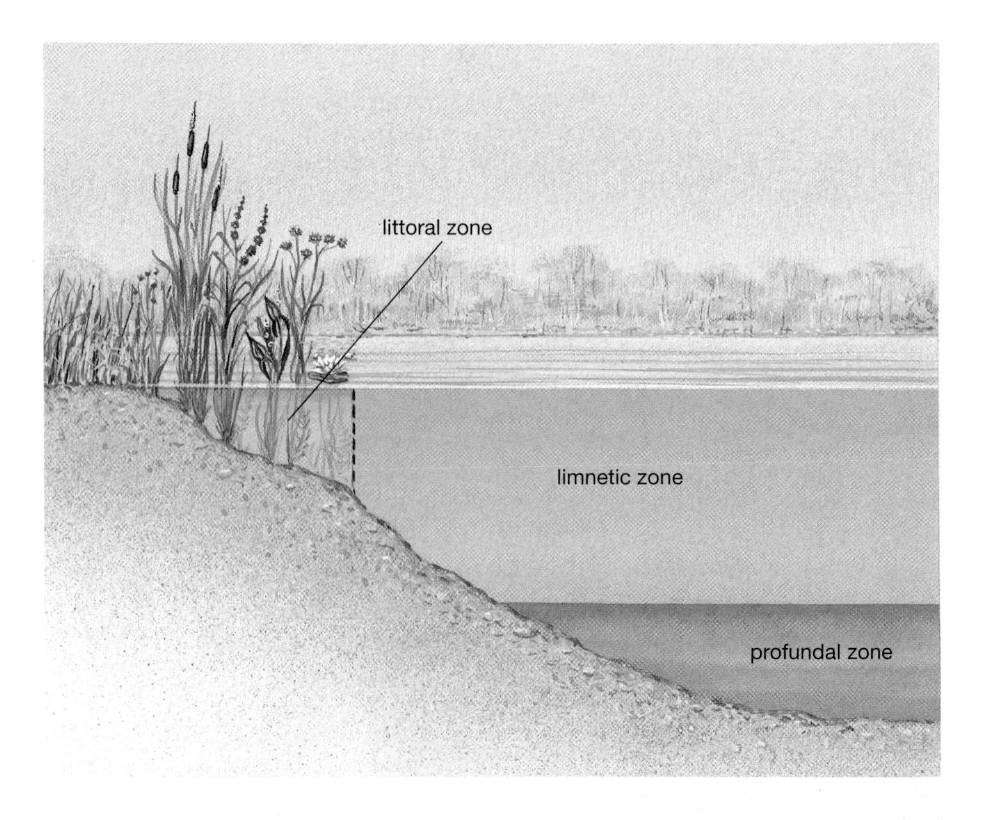

FIGURE 42-24 Lake life zones
There are three life zones in a "typical" large lake: a nearshore littoral zone with rooted plants, an open-water limnetic zone, and a deep, dark profundal zone.

Near the shore is the **littoral zone**. In this zone, the water is shallow, and plants find abundant light, anchorage, and adequate nutrients from the bottom sediments. Not surprisingly, littoral-zone communities are the most diverse. Cattails and bulrushes abound near the shore, while water lilies and entirely submerged vascular plants and algae may flourish at the deepest reaches of the littoral zone. The plants of the littoral zone trap sediments carried in by streams and by runoff from the surrounding land, increasing the nutrient content in this region. Living among the anchored plants are microscopic organisms called **plankton**. There are two forms of plankton: **phytoplankton** (Greek, "drifting plants"), which includes photosynthetic protists, bacteria, and algae; and **zooplankton** (Greek, "drifting animals"), such as protozoa and tiny crustaceans. The greatest diversity of animals in the lake is also found in this zone. Littoral invertebrate animals include small crustaceans, insect larvae, snails, flatworms, and hydra; littoral vertebrates include frogs, minnows, and aquatic snakes and turtles.

As the water increases in depth farther from shore, plants are unable to anchor to the bottom and still collect enough light for photosynthesis. This open-water area is divided into two regions: the upper limnetic zone and the lower profundal zone (see Fig. 42-24). In the **limnetic zone**, enough light penetrates to support photosynthesis. Here phytoplankton including cyanobacteria (also called *blue-green algae*) serve as producers. These are eaten by protozoa and small crustaceans, which in turn are consumed by fish. Below the limnetic zone lies the **profundal zone**, where light is insufficient to support photosynthesis. This area is nourished mainly by detritus that falls from the littoral and limnetic zones and by incoming sediment. It is inhabited primarily by decomposers and detritus feeders, such as bacteria, snails and insect larvae, and fish that swim freely among the different zones.

Freshwater Lakes Are Classified According to Their Nutrient Content

Although each lake is unique, freshwater lakes can be classified on the basis of their nutrient content as either *oligotrophic* or *eutrophic*.

Oligotrophic (Greek, "poorly fed") **lakes** are very low in nutrients. Many are formed by glaciers that scrape depressions in bare rock, and they are fed by mountain streams carrying little sediment. Because there is little sediment or microscopic life to cloud the water, oligotrophic lakes are clear, and light penetrates deeply. Therefore, photosynthesis (and oxygenation, as oxygen is produced as a by-product of photosynthesis) in deeper water is possible, and the limnetic zone may extend all the way to the bottom. Oxygen-loving fish, such as trout, thrive in oligotrophic lakes.

Eutrophic (Greek, "well fed") **lakes** receive larger inputs of sediments, organic material, and inorganic nutrients (such as phosphorus) from their surroundings, allowing them to support dense communities. They are murkier from suspended sediment and dense phytoplankton populations, so the lighted limnetic zone is shallower. Dense "blooms" of algae occur seasonally in the limnetic zone. Their dead bodies fall into the profundal zone, where they are used as food by decomposer organisms. The metabolic activities of these decomposers use oxygen, depleting the oxygen content of the profundal zone in eutrophic lakes.

Lakes are transient ecosystems, although very large lakes may persist for millions of years. Over time, they gradually fill with sediment, undergoing succession to dry land (see Chapter 40). As nutrient-rich sediment accumulates, oligotrophic lakes tend to become eutrophic, a process called *eutrophication*.

HUMAN IMPACT Human activities can greatly accelerate the process of eutrophication, because nutrients are carried into lakes from farms, feedlots, sewage, and even fertilized suburban lawns. Over-enriched lakes become clogged with microorganisms whose dead bodies are attacked by bacteria that deplete the water of oxygen. Normal community interactions are disrupted as organisms in higher trophic levels are smothered. In the early 1960s in Seattle, Washington, local sewage plants released 20 million gallons of waste into Lake Washington daily. Nutrients from human waste and high-phosphate detergents eutrophied the lake, causing foul odors, murky water, and dead fish. Citizen concern led Seattle to divert sewage out of the lake starting in 1963. By 1975 the lake had fully recovered.

The Great Lakes, whose shores host paper mills and a variety of other polluting industries and whose waters carry vessels from distant countries, have been heavily impacted by human activities. Exotic species such as the zebra mussel are altering their community structures. Imported parasitic sea lampreys and overfishing devastated lake trout populations in the 1940s. Although fishing is now regulated and the lampreys are controlled, the trout still are unable to reproduce normally because of persistent endocrine-disrupting chemicals (discussed in Chapter 33) that have accumulated in the water and sediments. Recently, officials were puzzled to find that the beleaguered trout carried high levels of toxaphene, which was used heavily as a substitute for the pesticide DDT (see Chapter 41). The EPA banned toxaphene in 1982 after discovering that it could cause birth defects and cancer. Some scientists hypothesize that the toxaphene is still being carried in the atmosphere and deposited in the Great Lakes from fields in the south, where it was heavily sprayed on cotton 20 years ago.

Acid rain (see Chapter 41), caused primarily by burning fossil fuels, poses a very different threat, particularly to small freshwater lakes and ponds. In the Adirondack Mountains of New York State, roughly 25% of the lakes have been rendered nearly lifeless by acid rain, and the problem is growing and spreading to more southern states.

Marine Ecosystems Cover Much of Earth

In the oceans, the upper layer of water, where the light is strong enough to support photosynthesis (to a depth of about 650 feet, or 200 meters), is called the **photic zone**. Below the photic zone lies the **aphotic zone**, where the only energy comes from the excrement and bodies of organisms that sink or swim in this deep region (Fig. 42-25).

As in lakes, most of the nutrients in the oceans are at or near the bottom, where there is not enough light for photosynthesis. Nutrients dissolved in the water of the photic zone are constantly being incorporated into the bodies of living organisms. When these organisms die, some sink into the aphotic zone, providing its organisms with nutrients. If no additional nutrients entered the photic zone, life there would eventually cease.

Fortunately, there are two sources of nutrients that supply the photic zone: the land, from which rivers constantly remove nutrients and carry them to the oceans, and **upwelling**, an upward flow that brings cold, nutrient-laden water from the ocean depths to the surface. Upwelling occurs along western coastlines—as in California, Peru, and West Africa—where prevailing winds displace surface water, causing it to be replaced by water from below. Upwelling also occurs around Antarctica. Not surprisingly, the major concentrations of life in the oceans are found where abundant light is combined with a source of nutrients, which occurs most commonly in regions of upwelling and in shallow coastal waters.

Coastal Waters Support the Most Abundant Marine Life

The most abundant life in the oceans is found in a narrow strip surrounding Earth's landmasses, where the water is shallow and a steady flow of nutrients washes off the land. Coastal waters consist of two regions: the **intertidal zone**, which is alternately covered and uncovered by water with the rising and falling of the tides; and the **nearshore zone**, relatively shallow but constantly submerged areas including bays and coastal wetland (Fig. 42-26). These coastal wetlands include *salt marshes*, gradually sloping coastal areas that are protected from waves and so contain accumulations of rich silt, and **estuaries**, which are wetlands formed where rivers meet the oceans. The nearshore zone is the only part of the ocean where large salt-tolerant plants or seaweeds can grow, anchored to the bottom. In addition, the abundance of nutrients and sunlight in this zone promotes the growth of a veritable soup of photosynthetic phytoplankton. Associated with these plants and protists are animals from nearly every phylum: annelid worms, sea anemones, jellyfish, sea urchins, sea stars,

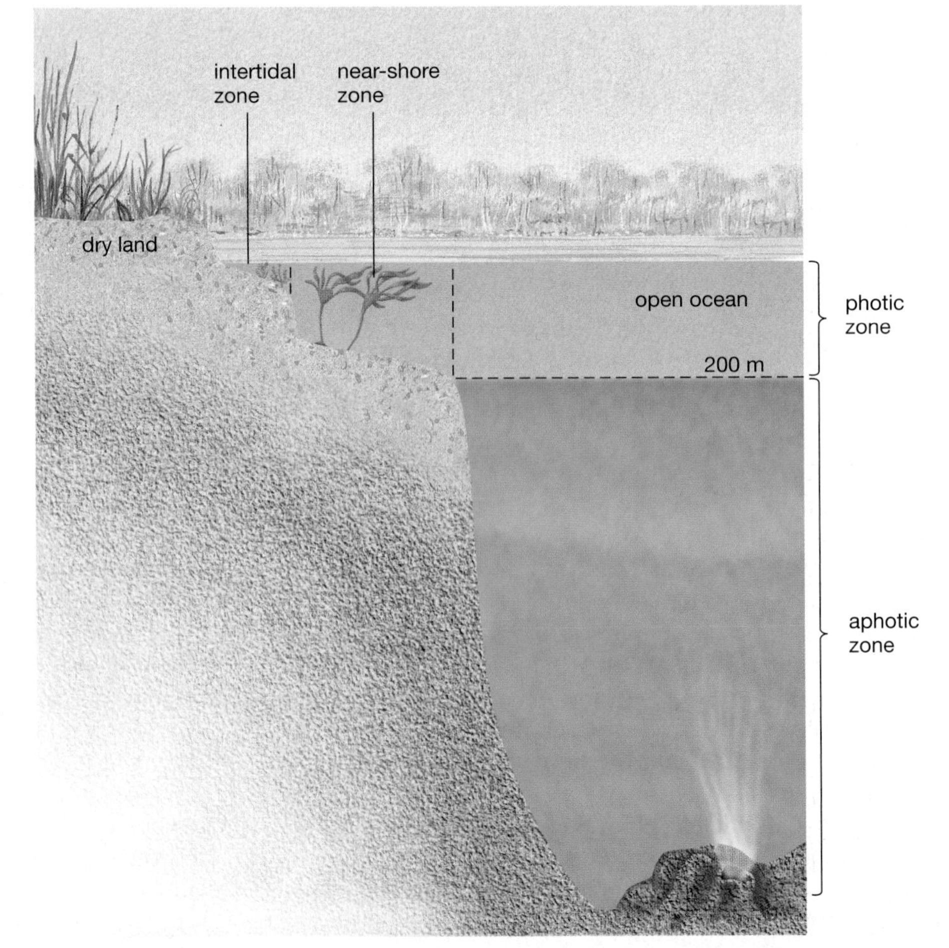

FIGURE 42-25 Ocean life zones
Photosynthesis can occur only in the upper photic zone, which includes the intertidal and near-shore zones and the upper waters of the open ocean. Life in the aphotic zone relies on energy-rich material that drifts down from the photic zone or, in the unique case of hydrothermal vent communities, on energy stored in hydrogen sulfide and trapped by chemosynthesis.

FIGURE 42-26 Nearshore ecosystems
(a) A salt marsh in the eastern United States. Expanses of shallow water fringed by marsh grass (*Spartina*) provide excellent habitat and breeding grounds for many marine organisms and shorebirds. **(b)** Although the shifting sands present a challenge to life, grasses stabilize them, and animals such as (inset) this *Emerita* crab burrow in the sandy intertidal zone. **(c)** A rocky intertidal shore in Oregon, where animals and algae grip the rock against the pounding waves and resist drying during low tide. (Inset) Colorful sea stars cling to the rocks surrounded by a seaweed, fucus. **(d)** Towering kelp sway through the clear water off southern California, providing the basis for a diverse community of invertebrates, fishes, and (inset) an occasional sea otter. **QUESTION** Nearshore ecosystems have the highest productivity in the ocean. What factors explain this? Which of the ecosystems pictured here would you expect to have the highest productivity, and why?

mussels, snails, fish, and sea otters, to name just a few. A large number and variety of organisms live permanently in coastal waters, but many that spend most of their lives in the open ocean come into the coastal waters to reproduce. Bays, salt marshes, and estuaries in particular are the breeding grounds for a wide variety of organisms, such as crabs, shrimp, and an array of fish, including most of our commercially important species, such as salmon. Off the coast of California, great undersea forests of kelp provide food and shelter for a rich assemblage of fish and invertebrates that in turn provide food for otters and seals (see Fig. 42-26d). The productivity of freshwater and saltwater wetlands is second only to rain forests. Coral reefs, another unique form of coastal ecosystem, are discussed below.

Human Impact Coastal regions are of great importance not only to the organisms that live or breed there but also to humans who use them for food sources, recreation, mineral and oil extraction, or living places. Like rain forests, wetlands of all types once covered 6% of Earth's surface, but nearly half of them have been dredged or filled in. In the U.S., only about 100 million acres of wetlands remain of an original 215 million acres. As populations increase in coastal states and resources such as oil become increasingly scarce, the conflict between preservation of coastal wetlands as wildlife and animal habitat and development of these areas for housing, harbors and marinas, and energy extraction will become increasingly intense. Although conservation efforts have slowed the loss of wetlands, the U.S. still loses well over 100,000 acres yearly. Estuaries are also threatened by runoff from farming operations. Pig farms often collect concentrated animal wastes in large holding ponds. When leaks or floods transport this material into local rivers and estuaries, it can be fatal to the estuarian community of phytoplankton, invertebrates, and fish. Because so much ocean life depends on the well-being of the coastal waters, it is essential to protect these fragile, vital areas.

Coral Reefs

In warm tropical waters with just the right combination of bottom depth, wave action, and nutrients, specialized algae and corals build reefs from their own calcium carbonate skeletons. **Coral reefs** are most abundant in tropical waters of the Pacific and Indian Oceans, the Caribbean, and the Gulf of Mexico as far north as southern Florida, where the maximum water temperatures range between 72 °F and 82 °F (22 °C and 28 °C).

Reef-building corals (phylum Cnidaria, described in Chapter 22) harbor photosynthetic unicellular algae called *dinoflagellates* within their tissues in a mutualistic relationship. The algae represent up to half the weight of the coral polyp and give the corals their diverse, brilliant colors (Fig. 42-27). These corals thrive within the photic zone at depths of less than 130 feet (40 meters), where light penetrates the clear water and provides energy for photosynthesis. The algae benefit from the high nitrogen, phosphorus, and carbon dioxide levels in the coral tissues. In return, algae provide food for the coral and help produce calcium carbonate, which forms the coral skeleton. The skeletons of corals accumulate over thousands of years, providing an anchoring place for diverse forms of algae and shelter and food for the most diverse collection of invertebrates and fish in the oceans (see Fig. 42-27). In some ways, coral reefs might be considered the ocean equivalent of rain forests, since they are home to more than 93,000 known species with probably ten times that number yet to be identified. The

FIGURE 42-27 Coral reefs
(a) Coral reefs, composed of the bodies of corals and algae, provide habitat for an extremely diverse community of extravagantly colored animals. **(b)** Many fish, including this blue tang, feed on coral (note the bright yellow corals in the background). A vast array of invertebrates such as **(c)** this sponge and **(d)** blue-ringed octopus live among the corals of Australia's Great Barrier Reef. This tiny octopus (6 inches, fully extended) is one of the world's most venomous creatures. **QUESTION** Why does "bleaching" threaten the life of a coral reef? What causes bleaching?

Great Barrier Reef in Australia supports more than 200 species of coral alone, and a single reef may harbor 3000 identified species of fish, invertebrates, and algae.

Human Impact Coral reefs are extremely sensitive to many types of disturbance. Anything that diminishes the clarity of the water harms the coral's photosynthetic partners and hinders coral growth. When people farm, log, and develop coastal land, erosion carries silt into the water. As sewage and agricultural runoff pollute coastal waters, eutrophication reduces both sunlight and oxygen. Silt has ruined several reefs near Honolulu, Hawaii. In the Philippines, rainforest logging has dramatically increased erosion; here, destruction of the rain forests is also destroying the local coral reefs.

Overfishing threatens reef communities. In many tropical countries, mollusks, turtles, fish, crustaceans, and the corals themselves are being harvested from reefs faster than they can reproduce. Many are sold to shell enthusiasts and aquarium owners in developed countries. Some collectors use dynamite or poisons to stun the fish before collecting them, leaving most dead. Removing predatory fish and invertebrates from reefs may disrupt the ecological balance of the community, allowing an explosion in populations of coral-eating sea urchins or sea stars.

Throughout the Tropics, scientists and concerned citizens have observed a disturbing trend—coral reefs and their communities are becoming "sick." Infections, tumors, and lesions with unknown causes are causing massive mortality of corals and their communities throughout the world. Teams of researchers at Harvard and the University of New Hampshire studying these trends suggest that the declining health of coral reefs results from complex interactions involving both human disturbances and global warming. Higher water temperature combined with runoff from land encourage explosions of microscopic algal populations in nearshore waters. The algae release toxins that damage the health of the entire reef community. Bleaching is another symptom of poor health observed in corals worldwide. When waters become too warm, the corals expel their colorful symbiotic algae, leaving the corals a deathly white. In 2002, Australia's Great Barrier Reef suffered bleaching in almost 60% of its coral. Without their algal partners, the corals will eventually starve.

Florida's famous coral reefs are in trouble. Warm water from a nuclear power plant, nutrients from sewage effluent, and silt carried in from rivers cloud the water and promote the growth of harmful algae. More than a million divers and snorkelers visit the reefs annually and accidentally (or sometimes deliberately) damage the coral. Boat anchors claw the reefs and boat engines release oil into formerly pristine waters. Infections are increasingly common, and the two warmest decades in recorded history have caused extensive coral bleaching. But there is also some good news—the world's first marine park was established off Florida in 1960, and several other marine sanctuaries have been established in Florida in more recent years. A fishing ban over small coral reef reserves in the Florida Keys in 1997 has already produced larger populations of several important marine species such as lobsters and grouper fish.

As with any ecosystem, both protection and sustainable use are crucial to the survival of these fragile and diverse underwater treasures. Carefully regulated harvesting and tourism produce far greater and longer-lasting economic benefits than do activities that destroy the reefs. Establishing preserves is crucial, but preserves will be effective only if we can prevent human activities on the land from damaging the quality of the water on which reef communities rely.

The Open Ocean

Beyond the coastal regions lie vast areas of the ocean in which the bottom is too deep to allow plants to anchor and still receive enough light to grow. Most life in the open ocean (Fig. 42-28a) is limited to the upper photic zone, where life-forms are **pelagic**—that is, free-swimming or floating—for their entire lives (Fig. 42-28b–d). The food web of the open ocean is dependent on phytoplankton consisting of microscopic photosynthetic protists, mainly diatoms and dinoflagellates (Fig. 42-28e). These organisms are consumed by zooplankton, such as tiny crustaceans that are relatives of crabs and lobsters (Fig. 42-28f). Zooplankton in turn serve as food for larger invertebrates, small fish, and even marine mammals such as the humpback whale (see Fig. 42-28d).

One challenge faced by inhabitants of the open ocean is to remain afloat in the photic zone, where sunlight and food are abundant. Many members of the planktonic community have elaborate flotation devices, such as oil droplets in their cells or long projections, which slow their rate of sinking (see Fig. 42-28e). Most fish have swim bladders that can be filled with gas to regulate their buoyancy. Some animals, and even some of the phytoplankton, actively swim to stay in the photic zone. Many small crustaceans migrate to the surface at night to feed, then sink into the dark depths during daylight, thus avoiding visual predators such as fish. The amount of pelagic life varies tremendously from place to place. The blue clarity of tropical waters is a result of a lack of nutrients, which limits the concentration of plankton in the water. Nutrient-rich waters that support a large plankton community are greenish and relatively murky.

Below the photic zone, the only available energy in most regions comes from the excrement and dead bodies that drift down from above. Nevertheless, a surprising quantity and variety of life exist in the aphotic zone, including fishes of bizarre shapes, worms, sea cucumbers, sea stars, and mollusks.

Human Impact Two major threats to the open ocean are pollution and overfishing. Open-ocean pollution takes several forms. Oceangoing vessels such as cruise ships dump millions of plastic containers overboard daily, and plastic six-pack holders, foam cups, and packing material

FIGURE 42-28 The open ocean
(a) The open ocean supports abundant life in the photic zone, where light is available. **(b)** Porpoises skim the surface, **(c)** fish such as the blue jack swim, and **(d)** rare humpback whales leap clear of the water. **(e)** The photosynthetic phytoplankton are the producers on which most other marine life ultimately depends. Phytoplankton are eaten by **(f)** zooplankton, represented by this microscopic crustacean, a copepod. The spiny projections on these planktonic creatures help keep them from sinking below the photic zone.

wash and blow off the land. The plastic looks like food to unsuspecting sea turtles, gulls, porpoises, seals, and whales, many of which die after trying to consume it. Until 1992 New York City placed its refuse and sewage sludge on barges and towed them out to sea, creating a heavily contaminated area covering 40 square miles of open-ocean floor. The open ocean has also served as a dumping ground for radioactive wastes. Oil contaminates the open ocean from many sources, including oil tanker spills, runoff from improper disposal on land, leakage from offshore oil wells, and natural seepage. In the Gulf of Mexico, the Mississippi River deposits nutrient-laden sediment from agricultural runoff onto the seafloor, creating a "dead zone" where oxygen is depleted in the deeper waters and the marine community is eliminated. The dead zone covers approximately 7000 square miles during the warmer months of each year and threatens both the ecological community and the fishing industry that relies on a healthy ecosystem.

The increasing demand for fish to feed a growing human population, coupled with increasingly efficient fishing technologies including satellites, radar, and sonar, has caused the depletion of many commercially exploited fish populations. Most of the world's fisheries, despite these technologies, now harvest fewer fish—evidence that fish are being taken in unsustainable numbers. Many fisheries are heavily subsidized by their governments, a practice that promotes the continuing harvest of a diminishing resource. The cod fishery of the northeastern U.S. and eastern Canada has collapsed from overfishing. Natural populations of lobsters, salmon, haddock, king crab, swordfish, and many other types of seafood have also declined dramatically because of overfishing. While dredging for scallops, fishermen often scrape entire marine communities from the seafloor. "Dolphin safe" tuna netting techniques have dramatically reduced accidental dolphin catches but have trapped more sea turtles and sharks. An estimated 100 million sharks are caught annually; many have their fins cut off (shark-fin soup is an Asian delicacy) and then are dumped back into the ocean to die. These slow-growing predators are a keystone species in ocean food webs, and because many shark species do not breed until they are a decade or more old and produce small numbers of offspring, their populations are slow to recover.

In 1994 a marine reserve was established off the coast of Massachusetts. Within 5 years, scallop and fish populations rebounded. Studies of reserves throughout the world suggest that there are substantial improvements in the diversity, number, and size of marine animals within these areas, and that benefits are considerable after just 3 years. It is politically difficult to establish no-fishing zones due to intense opposition from the fishing industry, but this approach holds the most promise for saving endangered fish populations and, ultimately, fisheries throughout the world. Fish farming, or *aquaculture,* can also help meet the demand for some types of seafood, including shrimp and salmon, but fish and shrimp farms must be carefully designed and managed to avoid damaging local ecosystems.

Hydrothermal Vent Communities

In 1977 a new and unusual source of nutrients, the basis of a spectacular undersea community, was discovered in the deep ocean. Geologists exploring the Galapagos Rift (an area of the Pacific floor where the plates that form Earth's crust are separating) found vents emitting superheated water, black with sulfur and minerals. Surrounding these vents was a rich community of pink fish, blind white crabs, enormous mussels, giant white clams, sea anemones, and giant tube worms (Fig. 42-29). Twenty-two previously undescribed families and 284 new species of organisms have been found in these **hydrothermal vent communities**. Scientists have now identified vent communities in many deep-sea areas where tectonic plates are spreading apart and material from Earth's interior is spewing forth to form new crust.

In this unique ecosystem, sulfur bacteria serve as the primary producers. They harvest energy from an unlikely source that is deadly to most other forms of life—hydrogen sulfide discharged from cracks in Earth's crust. This process, called *chemosynthesis*, replaces photosynthesis in the vent communities, which flourish more than a mile below the ocean surface. Both bacteria and archaea proliferate in the hot water surrounding the vents, covering nearby rocks with thick, matlike colonies. These colonies provide the food on which the animals of the vent community thrive. Many vent animals consume the microorganisms directly. Others, such as the giant tube worm (which lacks a digestive tract), harbor the bacteria within special organs in their bodies. In this mutualistic association, the bacteria provide high-energy carbon compounds, and the tube worm provides hydrogen sulfide. The worm, which can reach a length of 9 feet, derives its red color from a unique form of hemoglobin that transports hydrogen sulfide, and some oxygen, to the symbiotic bacteria. These tube worms hold the record for invertebrate longevity; based on growth rates in the field, even the average-sized worms are 170–250 years old.

The bacteria and archaea that inhabit the vent communities hold the record for survival at high temperatures. One species cannot reproduce at water temperatures below 194 °F (90 °C); another can survive water temperatures of 248 °F (106 °C; water at this depth can reach temperatures much higher than boiling because of the tremendous pressure). Scientists are investigating how the enzymes and other proteins of these heat-loving microbes can continue to function at temperatures that would destroy the proteins in our bodies.

The world still holds wonders and mysteries for those who seek them. We have only begun to explore the versatility and diversity of life on Earth.

FIGURE 42-29 Hydrothermal vent communities
Located in the ocean depths, vent communities include giant tube worms nearly 4 meters (12 feet) long. Parts of these worms are colored red by a hydrogen sulfide-trapping form of hemoglobin.

EARTH WATCH Humans and Ecosystems

"We must consider our planet to be on loan from our children, rather than being a gift from our ancestors."

Gro Harlem Brundtland, former Prime Minister of Norway

The expanding human population has left relatively few ecosystems undisturbed. Human impacts on natural ecosystems are so diverse and wide ranging that they far exceed the scope of this book. However, we can compare some general characteristics of ecosystems dominated by humans to those of undisturbed ecosystems. Listed below are six differences between these two types of ecosystems and a few ideas for minimizing them.

First, ecosystems dominated by humans tend to be simpler—that is, to have fewer species and fewer community interactions—than do undisturbed ecosystems. Although a city street bustles with life and apparent complexity, count the number of species you encounter in an average block and compare it with the number you encounter while hiking in wilderness for a similar distance. As humans enter an ecosystem, animals in the highest trophic levels are the first to go. Carnivores are always relatively rare, and their specialized needs are most easily disrupted. Many big carnivores, such as the wolf and mountain lion, require large, undisturbed hunting territories. Humans often selectively destroy large predators, believing them to be a threat to people or livestock. As a result, even in relatively undisturbed areas, many large carnivores have been eliminated. For example, gray wolves and grizzly bears once roamed throughout most of the lower 48 states. Wolves were nearly driven to extinction in the lower 48 states by the beginning of the twentieth century (reintroductions and protection have now increased their populations), while grizzly bears only occupy about 1% of their original range.

On the western shortgrass prairie, prairie dog towns fall as human towns and ranches rise (Fig. E42-2). The black-footed ferret, a predator of prairie dogs, faces possible extinction. Farm fields have been deliberately simplified from the original prairie biome to eliminate competition and predation and allow the maximum productivity of a single crop species. Nowhere is the contrast between human and natural ecosystems greater than in the tropical rain forests, whose unparalleled diversity is replaced by failed attempts at farming.

There is probably no practical way to restore to human ecosystems the great diversity found in undisturbed areas, nor is doing so always desirable from a human standpoint. Agriculture, for example, demands a simplified ecosystem. Cities concentrate human activities and culture and may lessen the human impact on the surrounding countryside. However, as we recognize the benefits of artificially simplified ecosystems, we must be aware of the need to preserve intact as many natural communities as possible. The undisturbed forest traps and purifies water and can reduce air pollution. Swamps and estuaries contain a wealth of detritus feeders and decomposers that purify water. Coastal wetlands are breeding places for millions of birds and spawning sites for the majority of our commercially important fish and crustacean species. In undisturbed, diverse ecosystems we find aesthetic pleasure as well as a storehouse of species whose commercial or medicinal values are not yet recognized. For example, nearly half the medicines in use today were originally discovered in plants, and we have examined only a small fraction of existing plants for possible medical uses.

FIGURE E42-2 Habitat destruction
Loss of habitat due to human activities is a major threat to most of Earth's wild animals and plants.

Second, whereas natural ecosystems run on sunlight, human ecosystems have become dependent on nonrenewable energy from fossil fuels. From the suburbanite pouring gas into a lawn mower to the farmer driving a tractor, managing a simplified ecosystem is an energy-intensive proposition. Energy must be expended to oppose the tendency of the natural system to restore complexity. Fertilizers also require large amounts of energy to produce, and farms must be heavily fertilized because nutrient cycles have been disrupted.

To counteract this trend, some farmers are returning to organic farming. Plant and animal waste and natural nutrient cycles are used to maintain soil fertility. Alternating legume crops, such as soybeans and alfalfa, with other crops helps maintain nitrogen in the soil. Mulching can help retain fertility, reduce water loss, and control weeds. The use of natural insect predators and limited, well-planned pesticide spraying can dramatically reduce the need for poisons. Organic farms can be as productive as more conventional farms while using 15% to 50% less energy from fossil fuels per quantity of food produced.

In our homes and commercial buildings, better insulation and increased use of solar heat can result in dramatic energy savings. By increasing our use of renewable energy sources, such as wind and especially sunlight, we can conserve fossil fuels, dramatically reduce pollution, and move human ecosystems a step closer to those that occur naturally.

Third, natural ecosystems recycle nutrients, whereas human ecosystems tend to lose nutrients. Walk around some suburban neighborhoods on trash-pickup day. Grass clippings and leaves are packed in plastic bags to be hauled away. To compensate for this loss of natural nutrients, many suburbanites heavily fertilize their lawns and gardens. A similar trend has occurred in modern farming. The exposed soil is eroded away by wind and water, removing crucial nutrients and requiring large inputs of fertilizer to replace them. Runoff from the field, carrying fertile topsoil and artificial fertilizers, may pollute nearby rivers, streams, and lakes. Pesticides may kill detritus feeders and decomposers, further disrupting natural nutrient cycles. Thus, whereas fertile soil accumulates in many natural ecosystems, it tends to be lost in those dominated by humans. Some 3 billion tons of topsoil are eroded annually from farms in the U.S. The Mississippi River alone carries off 40 tons each hour.

Organic farming can help reverse this trend, too. We can apply the same principles to our lawns and gardens by composting organic wastes either in our own gardens or in community compost facilities. Farmers often reduce erosion by using contour planting, in which row crops are oriented so that they slow the flow of water instead of funneling it downslope. Row crops can also be alternated in strips with dense, soil-catching crops such as wheat. Planting rows of trees as windbreaks helps prevent soil loss from blowing wind, and it creates a more diverse ecosystem and a nesting place for insect-eating birds. Farm fields need not be plowed in the fall and left unplanted to erode during the winter; in fact, an increasing number of farmers are planting crops in the stubble from the previous season, reducing erosion.

Fourth, natural ecosystems tend to store water and purify it through biological processes, whereas human ecosystems tend to pollute water and shed it rapidly. A thundershower strikes a forest, an adjacent city, and a farm. The rich soil of the forest sponges up the water, which gradually drains into the ground, filtered by the soil and purified by decomposers that break down organic contaminants. In the nearby city, water pours from sidewalks, rooftops, and streets, picking up soot, silt, oil, heavy metals, and garbage. It races down gutters into storm sewers, and a weakly toxic soup gushes into the nearest stream or river. Farm runoff carries priceless topsoil, expensive fertilizer, and animal manure into rivers and lakes, where these potential resources become pollutants.

Preventing erosion will simultaneously conserve water and reduce water pollution from farm runoff. Manure from cattle feedlots, which is a significant source of both groundwater and surface-water pollution, could be placed on fields, where it would restore needed nutrients.

Although water will continue to run off our cities, the pollutants it carries can be reduced by minimizing our reliance on fossil fuels. We must eliminate leaded gasoline worldwide; U.S. gas is now lead-free, but leaded gas remains available throughout most of the world. We must also tighten standards for emissions from diesel and gasoline engines and smokestacks. Efficient public transportation systems will reduce the pollution (and massive frustration) caused by congested traffic. Insulation will reduce power consumption in our homes and offices, as will increased reliance on solar heat.

Fifth, simple human ecosystems such as farms tend to be unstable, whereas natural ecosystems have many species and tend to remain stable over time. In natural ecosystems, herbivorous insects are controlled by predators, including other insects, birds, and shrews. Insect populations are also limited because their preferred plants are scattered among many other plants rather than growing in a pure stand, as on a farm. On farms, both pest insects and their natural predators are exposed to pesticides. Unfortunately, the pests may develop resistance to the poison, but their predators are killed. In these simplified communities, unfavorable weather conditions or the introduction of an exotic species can be disastrous.

Farmers can counteract this trend by planting smaller fields with a wider variety of crops. Alternating crops helps maintain soil fertility and also helps prevent the proliferation of disease and insect pests that are specialized for a particular crop. Populations of corn borers, for example, will die off during years when the field is planted with alfalfa. The use of biological controls such as natural insect predators and insect diseases can reduce reliance on pesticides.

Finally, human ecosystems are characterized by continuously growing populations, whereas nonhuman populations in natural ecosystems are relatively stable. As our population expands, the spread of human-dominated ecosystems presents a growing threat to the diversity of species and to the delicate balance that has evolved over the 3-billion-year history of life on Earth.

In summary, natural ecosystems tend to be complex, stable, and self-sustaining, powered by solar energy and nourished by recycled nutrients. They provide diverse habitats for wildlife, purify contaminants through the action of decomposers, and build up nutrient-rich soil. Modern human ecosystems are relatively simple and are sustained by large inputs of energy from fossil fuels. They tend to minimize wildlife habitat, contaminate soil and water, and lose nutrients and fertile soil. These problems are compounded by continued population growth, which is causing the expansion of human-dominated ecosystems at the expense of undisturbed ones.

Human ecosystems do not have to be as disruptive and alien to the operation of natural ecosystems as we have allowed them to become. Through understanding, education, commitment, appropriate use of technology, and stabilization of our population, we can reverse many of these destructive trends.

CASE STUDY REVISITED Wings of Hope

The Arabuko-Sokoke forest remains under siege from squatters who want to clear land and establish homes within its confines. But where the farmers gather their butterflies, the forest suffers far less poaching; the farmers now report poachers rather than join them. Over several years of monitoring, project manager Ian Gordon sees no evidence that butterfly populations are being reduced. With full stomachs and money to buy a few minor luxuries, the people can now afford to support the philosophy of one village elder who states, "The forest is here, we found the forest here, and we have to leave it here for our children's generation."

Consider This: Most conservationists agree that "fencing and fining" is not the way to preserve habitat; local residents must actively support and participate in its preservation. Design or research other projects that fit the model of sustainable use of rain forests or other endangered ecosystems, such as coral reefs.

Links to Life: Crave Chocolate and Save Rain Forests?

Cacao trees—the beans of which are used to produce chocolate—are native to the rain forests of Central and South America, but due to tremendous worldwide demand for chocolate (people in the U.S. alone consume about 3 billion pounds each year) cacao trees are now cultivated in equatorial countries throughout the world. In an attempt to increase production, vast stretches of rain forest have been cleared and cacao trees planted in large plantations. The cleared rain forest quickly loses its fertility and eventually can be used only for pasture. Lacking the diverse environment in which cacao trees evolved, pollination rates are low, and about one-third of the cocoa crop is lost each year to pests. In Asia, growers douse their plantations with herbicides, fungicides, and pesticides, indiscriminate killers that reduce biodiversity and create toxic runoff. In 2000 a group of chocolate companies, recognizing the need to protect both cacao trees and their rainforest habitat, established the World Cocoa Foundation (WCF), whose mission includes producing "quality cocoa in a sustainable, environmentally friendly manner." In a WCF-funded project in Ghana, Africa, seeds from the neem tree from India (which possess natural insecticidal properties) are ground up and sprayed on the cacao trees to kill a serious beetle pest without harming beneficial insects, including the pinhead-sized fly that pollinates the cacao flowers. Fortunately, several South American countries have recognized the value of cultivating cacao under the conditions in which it evolved—beneath a dense rainforest canopy. This environment provides outstanding habitat for a variety of rainforest species, including a new species of ovenbird discovered within a natural cacao farm in Brazil in 1994. The golden lion tamarin, a recently discovered and critically endangered primate, needs more habitat than has been set aside for it; officials are now hopeful that the tamarin will be saved by extending its habitat into Brazilian rain forest preserved by cacao farmers. The Atlantic rain forest of Brazil has been reduced to less than 8% of its original size, but some officials hope that, thanks to careful cultivation of cacao, portions of the rain forest might be restored. In Chapter 2, you learned that cocoa powder has several properties that are actually good for you. Thanks to the World Cocoa Foundation, you can feel even better about satisfying your chocolate cravings.

Summary of Key Concepts

42.1 What Factors Influence Earth's Climate?

The availability of sunlight, water, and appropriate temperatures determines the climate of a given region. Sunlight maintains Earth's temperature. Equal amounts of solar energy are spread over a smaller surface at the equator than farther north and south, making the equator relatively warm, while higher latitudes have lower overall temperatures. Earth's tilt on its axis causes dramatic seasonal variation at northern and southern latitudes.

The rising of warm air and sinking of cool air in regular patterns from north to south produce areas of low and high moisture. These patterns are modified by the topography of continents and by ocean currents.

42.2 What Conditions Does Life Require?

The requirements for life on Earth include nutrients, energy, liquid water, and a reasonable temperature. The differences in the form and abundance of living things in various locations on Earth are largely attributable to differences in the interplay of these four factors.

42.3 How Is Life on Land Distributed?

On land, the crucial limiting factors are temperature and liquid water. On continents, large regions with similar climates will have similar vegetation, determined by the interaction of temperature and rainfall or the availability of water. These regions are called biomes.

Tropical forest biomes, located near the equator, are warm and wet, dominated by huge broadleaf evergreen trees. Most nutrients are tied up in vegetation, and most animal life is arboreal. Rain forests, home to at least 50% of all species, are rapidly being cut for agriculture, although the soil is extremely poor.

The African savanna is an extensive grassland with pronounced wet and dry seasons. It is home to the world's most diverse and extensive herds of large mammals.

Most deserts, hot and dry, are located between 20° and 30° N and S latitude and in the rain shadows of mountain ranges. In deserts, plants are widely spaced and have adaptations to conserve water. Animals tend to be small, nocturnal, and also adapted to drought.

Chaparral exists in desertlike conditions that are moderated by their proximity to a coastline, allowing small trees and bushes to thrive. Grasslands, concentrated in the centers of continents, have a continuous grass cover and few trees. They produce the world's richest soils and have largely been converted to agriculture.

Temperate deciduous forests, whose broadleaf trees drop their leaves in winter to conserve moisture, dominate the eastern half of the U.S. and are also found in Western Europe and East Asia. Higher precipitation occurs there than in the grasslands. Wet temperate rain forests, dominated by evergreens, are found on the northern Pacific coast of the U.S. The taiga, or northern coniferous forest, covers much of the northern U.S., southern Canada, and northern Eurasia. It is dominated by conifers whose small, waxy needles are adapted for water conservation and year-round photosynthesis.

The tundra is a frozen desert where permafrost prevents the growth of trees, and bushes remain stunted. Nonetheless, diverse arrays of animal life and perennial plants flourish in this fragile biome, which is found on mountain peaks and the Arctic.

42.4 How Is Life in Water Distributed?

Energy and nutrients are the major limiting factors in the distribution and abundance of life in aquatic ecosystems. Nutrients are found in bottom sediments and are washed in from surrounding land, concentrating them near shore and in deep water.

Freshwater lakes have three life zones. The littoral zone, near shore, is rich in energy and nutrients and supports the most diverse community. The limnetic zone is the lighted region of open water where photosynthesis can occur. The profundal zone is the deep water, where light is inadequate for photosynthesis and the community is dominated by heterotrophic organisms. Oligotrophic lakes are clear and low in nutrients, and they support sparse communities. Eutrophic lakes are rich in nutrients and support dense communities. During succession to dry land, lakes tend to go from an oligotrophic to a eutrophic condition.

Most life in the oceans is found in shallow water, where sunlight can penetrate, and is concentrated near the continents and in areas of upwelling, where nutrients are most plentiful. Coastal waters, consisting of the intertidal zone and the nearshore zone, contain the most abundant life. Producers include aquatic plants anchored to the bottom and photosynthetic protists called phytoplankton. Coral reefs are confined to warm, shallow seas. These calcium carbonate reefs form a complex habitat supporting the most diverse undersea ecosystem, threatened by silt, overfishing, and global warming.

In the open ocean, most life is found in the photic zone, where light supports phytoplankton. In the lower aphotic zone, life is supported by nutrients that drift down from the photic zone. Many ocean fisheries have been overexploited. Specialized vent communities, supported by chemosynthetic bacteria, thrive at great depths in the superheated waters where Earth's crustal plates are separating.

Key Terms

aphotic zone *p. 884*
biodiversity *p. 870*
biome *p. 869*
chaparral *p. 875*
climate *p. 864*
coral reef *p. 886*
desert *p. 874*
El Niño *p. 868*
estuary *p. 884*
eutrophic lake *p. 883*
grassland *p. 876*
gyre *p. 866*
hydrothermal vent community *p. 889*
intertidal zone *p. 884*
La Niña *p. 868*
limnetic zone *p. 883*
littoral zone *p. 883*
nearshore zone *p. 884*
northern coniferous forest *p. 878*
oligotrophic lake *p. 883*
ozone layer *p. 865*
pelagic *p. 887*
permafrost *p. 880*
photic zone *p. 884*
phytoplankton *p. 883*
plankton *p. 883*
prairie *p. 876*
profundal zone *p. 883*
rain shadow *p. 867*
savanna *p. 873*
taiga *p. 878*
temperate deciduous forest *p. 877*
temperate rain forest *p. 878*
tropical deciduous forest *p. 873*
tropical rain forest *p. 870*
tundra *p. 880*
upwelling *p. 884*
weather *p. 864*
zooplankton *p. 883*

Thinking Through the Concepts

To take a multiple-choice quiz with feedback on the contents of this chapter, visit http://www.prenhall.com/audesirk7. *Log in to the Web site selected by your instructor and navigate to the Self Test section for this chapter.*

? Review Questions

1. Explain how air currents contribute to the formation of the Tropics and the large deserts.
2. What are large, roughly circular ocean currents called? What effect do they have on climate, and where is that effect strongest?
3. What are the four major requirements for life? Which two are most often limiting in terrestrial ecosystems? In ocean ecosystems?
4. Explain why traveling up a mountain takes you through biomes similar to those you would encounter traveling north for a long distance.
5. Where are the nutrients of the tropical forest biome concentrated? Why is life in the tropical rain forest concentrated high above the ground?
6. Explain two undesirable effects of agriculture in the tropical rainforest biome.
7. List some adaptations of (a) desert plants and (b) desert animals to heat and drought.
8. What human activities damage deserts?
9. How are trees of the taiga adapted to a lack of water and a short growing season?
10. How do deciduous and coniferous biomes differ?
11. What single environmental factor best explains why there is shortgrass prairie in Colorado, tallgrass prairie in Illinois, and deciduous forest in Ohio?
12. Where are the world's largest populations of large herbivores and carnivores located?
13. Where is life in the oceans most abundant, and why?
14. Why is the diversity of life so high in coral reefs? What human impacts threaten them?
15. Distinguish among the limnetic, littoral, and profundal zones of lakes in terms of their location and the communities they support.
16. Distinguish between oligotrophic and eutrophic lakes. Describe (a) a natural scenario and (b) a human-created scenario under which an oligotrophic lake might be converted to a eutrophic lake.
17. Distinguish between the photic and aphotic zones. How do organisms in the photic zone obtain nutrients? How are nutrients obtained in the aphotic zone?
18. What unusual primary producer forms the basis for hydrothermal vent communities?
19. On the basis of the location of the worst atmospheric ozone depletion, which biomes are likely to be most affected by increased UV penetration?

Applying the Concepts

1. List at least six differences between human-dominated and undisturbed ecosystems, and discuss in some detail how these differences can be minimized.
2. In which terrestrial biome is your college or university located? Discuss similarities and differences between your location and the general description of that biome in the text. If you are living in a city, how has the urban environment modified community interactions within your biome?
3. During the 1960s and 1970s, many parts of the U.S. and Canada banned the use of detergents containing phosphates. Until that time, almost all laundry detergents and many soaps and shampoos had high concentrations of phosphates. What environmental concern do you think prompted these bans, and what ecosystem has benefited most from the bans?
4. Because ozone depletion is expected to get worse, not better, for decades to come, biologists have tried to assess which types of species will be most susceptible to increased UV penetration. Two of the groups that may be most vulnerable—ocean plankton and long-lived birds and mammals—are quite different from each other. Try to think of what makes each of these groups so susceptible to increased UV radiation.
5. Understanding the ways in which the four basic requirements for life determine where different biomes occur can help us predict the consequences of global warming. Global warming is expected to make most areas warmer, but it is also expected to change rainfall in ways that are hard to predict—some areas will get wetter, and others drier. Our ignorance of how rainfall will change is not very important in understanding shifts in more northerly biomes, but it is very important for our understanding of changes in tropical areas. Look at Figure 42-8 and explain why this is true.
6. More northerly forests are far better able to regenerate after logging than are tropical rain forests. Try to explain why this is true. HINT: The cold soils of northern climates greatly slow down decomposition rates.

For More Information

Burroughs, D. "On the Wings of Hope." *International Wildlife*, July–August 2000. Kenyan butterflies are saving a unique forest and its people, the basis for our Case Study.

Chadwick, D. H. "Blue Refuges." *National Geographic*, March 1998. Beautiful photographs highlight 12 national marine sanctuaries set aside to preserve a variety of marine ecosystems along the coastal U.S.

Falkowski, P. G., "The Ocean's Invisible Forest." *Scientific American*, August 2002. Describes the productivity of phytoplankton and their importance in capturing carbon dioxide, and speculates about their impact on global warming.

Ferber, D. "Keeping the Stygian Waters at Bay." *Science*, February 9, 2001. A good overview of the dead zone in the Gulf of Mexico and how it might be reduced.

Helmuth, L. "Can This Swamp Be Saved?" *Science News*, April 17, 1999. Describes the ambitious protection plan for the Florida Everglades.

Hinrichsen, D. "Requiem for Reefs?" *International Wildlife*, March–April 1997. Stunning photographs and a compelling text describe the beauty of Earth's imperiled coral reefs.

Holloway, M. "Sustaining the Amazon." *Scientific American*, July 1993. Describes the threats to the Amazon rain forest and innovative ways to preserve tropical forests and their biodiversity.

Kusler, J. A., Mitsch, W. J., and Larson, J. S. "Wetlands." *Scientific American*, January 1994. Discusses the characteristics and values of these threatened but highly important ecosystems.

Mallin, M. A, "Impacts of Industrial Animal Production on Rivers and Estuaries." *American Scientist*, January–February 2000. Lagoons designed to store waste from large numbers of farm animals pose a threat to nearby freshwater ecosystems.

Mitchell, J. G. "Our National Forests." *National Geographic*, March 1997. Stunning photographs and engaging text describe the threats to U.S. forests.

Pauly, D., and Watson, R. "Counting the Last Fish." *Scientific American*, July 2003. Large predatory fish stocks have been decimated worldwide by overfishing, and fisheries are increasingly exploiting smaller fish that feed lower on the food chain, endangering these populations as well.

Raloff, J. "Underwater Refuge." *Science News,* April 28, 2001. Describes the benefits of coastal no-fishing zones and attempts to expand them.

Raloff, J. "Clipping the Fin Trade." *Science News,* October 12, 2002. Documents the overfishing of sharks, usually for their fins alone, and how difficult it is for shark populations to recover.

Tunnicliffe, V. "Hydrothermal-Vent Communities of the Deep Sea." *American Scientist*, July–August 1992. A comprehensive look at the structure and function of undersea vent communities, with excellent illustrations.

Williams, T. "What Good Is a Wetland?" *Audubon*, November–December 1996. In addition to supporting a diversity of wildlife, purifying drinking water, and protecting people from floods, wetlands are beautiful.

Zimmer, C. "The El Niño Factor." *Discover*, January 1999. Explains the mechanisms and results of El Niño and La Niña weather patterns.

Media Activities

To access a Media Activity visit http://www.prenhall.com/audesirk7. *Log in to the Web site selected by your instructor, navigate to this chapter, and select the appropriate Media Activity number.*

42.1 Tropical Atmospheric Circulation and World Biomes

Estimated time: 5 minutes

In this activity you will learn the effect of solar radiation on air circulation in the earth's atmosphere. Examine how the circulation of air effects the distribution of Earth's biomes.

42.2 Web Investigation: Wings of Hope

Estimated time: 30 minutes

The Kipepeo Butterfly Project in the coastal dry forests of Kenya is an example of one strategy being employed to save forest ecosystems throughout the world. In this activity explore some of the advantages and disadvantages of this project and some related projects.

APPENDIX I
Metric System Conversions

To Convert Metric Units:	Multiply by:	To Get English Equivalent:
	Length	
Centimeters (cm)	0.3937	Inches (in.)
Meters (m)	3.2808	Feet (ft)
Meters (m)	1.0936	Yards (yd)
Kilometers (km)	0.6214	Miles (mi)
	Area	
Square centimeters (cm^2)	0.155	Square inches ($in.^2$)
Square meters (m^2)	10.7639	Square feet (ft^2)
Square meters (m^2)	1.1960	Square yards (yd^2)
Square kilometers (km^2)	0.3831	Square miles (mi^2)
Hectare (ha) (10,000 m^2)	2.4710	Acres (a)
	Volume	
Cubic centimeters (cm^3)	0.06	Cubic inches ($in.^3$)
Cubic meters (m^3)	35.30	Cubic feet (ft^3)
Cubic meters (m^3)	1.3079	Cubic yards (yd^3)
Cubic kilometers (km^3)	0.24	Cubic miles (mi^3)
Liters (L)	1.0567	Quarts (qt), U.S.
Liters (L)	0.26	Gallons (gal), U.S.
	Mass	
Grams (g)	0.03527	Ounces (oz)
Kilograms (kg)	2.2046	Pounds (lb)
Metric ton (tonne) (t)	1.10	Ton (tn), U.S.
	Speed	
Meters/second (mps)	2.24	Miles/hour (mph)
Kilometers/hour (kmph)	0.62	Miles/hour (mph)

To Convert English Units:	Multiply by:	To Get Metric Equivalent:
	Length	
Inches (in.)	2.54	Centimeters (cm)
Feet (ft)	0.3048	Meters (m)
Yards (yd)	0.9144	Meters (m)
Miles (mi)	1.6094	Kilometers (km)
	Area	
Square inches ($in.^2$)	6.45	Square centimeters (cm^2)
Square feet (ft^2)	0.0929	Square meters (m^2)
Square yards (yd^2)	0.8361	Square meters (m^2)
Square miles (mi^2)	2.5900	Square kilometers (km^2)
Acres (a)	0.4047	Hectare (ha) (10,000 m^2)
	Volume	
Cubic inches ($in.^3$)	16.39	Cubic centimeters (cm^3)
Cubic feet (ft^3)	0.028	Cubic meters (m^3)
Cubic yards (yd^3)	0.765	Cubic meters (m^3)
Cubic miles (mi^3)	4.17	Cubic kilometers (km^3)
Quarts (qt), U.S.	0.9463	Liters (L)
Gallons (gal), U.S.	3.8	Liters (L)
	Mass	
Ounces (oz)	28.3495	Grams (g)
Pounds (lb)	0.4536	Kilograms (kg)
Ton (tn), U.S.	0.91	Metric ton (tonne) (t)
	Speed	
Miles/hour (mph)	0.448	Meters/second (mps)
Miles/hour (mph)	1.6094	Kilometers/hour (kmph)

Metric Prefixes

Prefix		Meaning	
giga-	G	10^9 =	1,000,000,000
mega-	M	10^6 =	1,000,000
kilo-	k	10^3 =	1000
hecto-	h	10^2 =	100
deka-	da	10^1 =	10
		10^0 =	1
deci-	d	10^{-1} =	0.1
centi-	c	10^{-2} =	0.01
milli-	m	10^{-3} =	0.001
micro-	μ	10^{-6} =	0.000001

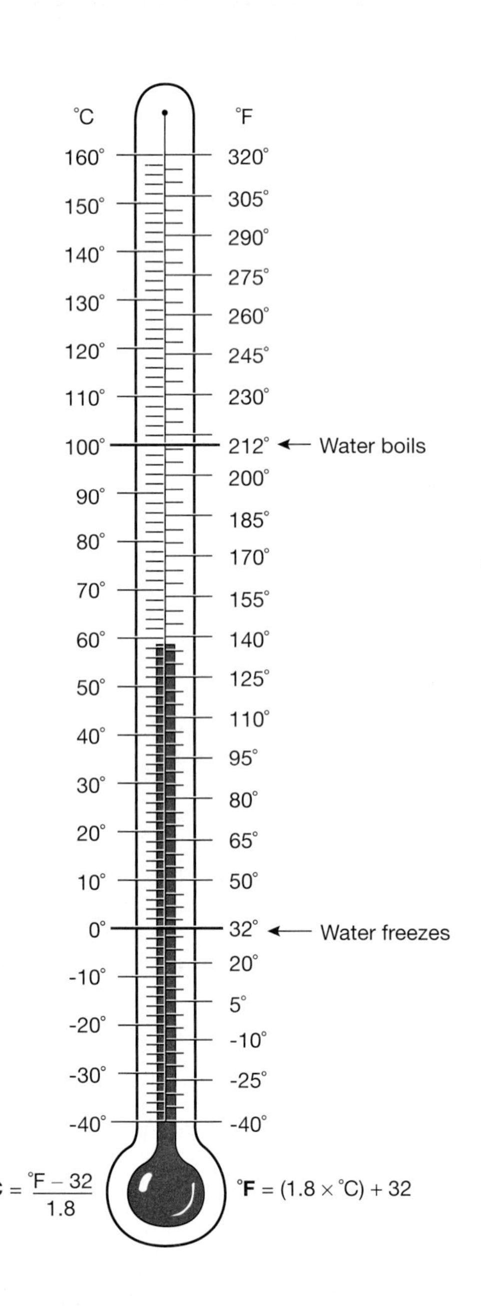

APPENDIX II
Classification of Major Groups of Organisms*

Domain	Kingdom	Phylum	Common Name
Bacteria (prokaryotic, peptidoglycan in cell wall)			bacteria
Archaea (prokaryotic, no peptidoglycan in cell wall)			archaeans
Eukarya (eukaryotic)			
		Rhodophyta	red algae
		Euglenophyta	euglenoids
		Myxomycota	plasmodial slime molds
		Acrasiomycota	cellular slime molds
		Sarcomastigophora	zooflagellates, amoebae
		Chlorophyta	green algae
	Alveolata		alveolates
		Apicomplexa	sporozoans
		Pyrrophyta	dinoflagellates
		Ciliophora	ciliates
	Chromista		chromists
		Oomycota	egg fungi
		Phaeophyta	brown algae
		Bacillariophyta	diatoms
	Fungi (multicellular, heterotrophic, absorb nutrients)		fungi
		Chytridiomycota	chytrids
		Zygomycota	zygote fungi
		Ascomycota	sac fungi
		Basidiomycota	club fungi
	Plantae (multicellular, photosynthetic)		plants
		Bryophyta	liverworts, mosses
		Pteridophyta	ferns
		Coniferophyta	evergreens
		Anthophyta	flowering plants
	Animalia (multicellular, heterotrophic, ingest nutrients)		animals
		Porifera	sponges
		Cnidaria	hydras, sea anemones, jellyfish, corals
		Ctenophora	comb jellies
		Platyhelminthes	flatworms
		Nematoda	roundworms
		Annelida	segmented worms
		Oligochaeta	earthworms
		Polychaeta	tube worms
		Hirudinea	leeches
		Arthropoda	arthropods ("jointed legs")
		Insecta	insects
		Arachnida	spiders, ticks
		Crustacea	crabs, lobsters
		Mollusca	mollusks ("soft-bodied")
		Gastropoda	snails
		Pelecypoda	mussels, clams
		Cephalopoda	squid, octopuses
		Echinodermata	sea stars, sea urchins, sea cucumbers
		Chordata	chordates
		Urochordata	tunicates
		Cephalochordata	lancelets
		Myxini	hagfishes
		Vertebrata	vertebrates
		Pertromyzontiformes	lampreys
		Chondrichthyes	sharks, rays
		Osteichthyes	bony fishes
		Actinopterygii	ray-finned fishes
		Sarcopterygii	lobe-finned fishes
		Amphibia	frogs, salamanders
		Anapsida	turtles
		Diapsida	
		Archosauria	birds, crocodiles
		Squamata	lizards, snakes
		Mammalia	mammals

*This table lists only those taxonomic categories described in the textbook.

Glossary

abdomen: the body segment at the posterior end of an animal with segmentation; contains most of the digestive structures.
abiotic (ā-bī-ah′-tik): nonliving; the abiotic portion of an ecosystem includes soil, rock, water, and the atmosphere.
abortion: the procedure for terminating pregnancy; the cervix is dilated, and the embryo and placenta are removed.
abscisic acid (ab-sis′-ik): a plant hormone that generally inhibits the action of other hormones, enforcing dormancy in seeds and buds and causing the closing of stomata.
abscission layer: a layer of thin-walled cells, located at the base of the petiole of a leaf, that produces an enzyme that digests the cell wall holding leaf to stem, allowing the leaf to fall off.
absorption: the process by which nutrients are taken into cells.
accessory pigments: colored molecules other than chlorophyll that absorb light energy and pass it to chlorophyll.
acellular slime mold: a type of funguslike protist that forms a multinucleate structure that crawls in amoeboid fashion and ingests decaying organic matter; also called *plasmodial slime mold*.
acetylcholine (ah-sēt′-il-kō′-lēn): a neurotransmitter in the brain and in synapses of motor neurons that innervate skeletal muscles.
acid: a substance that releases hydrogen ions (H^+) into solution; a solution with a pH of less than 7.
acid deposition: the deposition of nitric or sulfuric acid, either dissolved in rain (acid rain) or in the form of dry particles, as a result of the production of nitrogen oxides or sulfur dioxide through burning, primarily of fossil fuels.
acidic: with an H^+ concentration exceeding that of OH^-; releasing H^+.
acquired immune deficiency syndrome (AIDS): an infectious disease caused by the human immunodeficiency virus (HIV); attacks and destroys T cells, thus weakening the immune system.
acrosome (ak′-rō-sōm): a vesicle, located at the tip of an animal sperm, that contains enzymes needed to dissolve protective layers around the egg.
actin (ak′-tin): a major muscle protein whose interactions with myosin produce contraction; found in the thin filaments of the muscle fiber; see also *myosin*.
action potential: a rapid change from a negative to a positive electrical potential in a nerve cell. This signal travels along an axon without a change in intensity.
activation energy: in a chemical reaction, the energy needed to force the electron shells of reactants together, prior to the formation of products.
active site: the region of an enzyme molecule that binds substrates and performs the catalytic function of the enzyme.
active transport: the movement of materials across a membrane through the use of cellular energy, normally against a concentration gradient.
adaptation: a trait that increases the ability of an individual to survive and reproduce compared to individuals without the trait.
adaptive radiation: the rise of many new species in a relatively short time as a result of a single species that invades different habitats and evolves under different environmental pressures in those habitats.
adenine: a nitrogenous base found in both DNA and RNA; abbreviated as *A*.
adenosine diphosphate (a-den′-ō-sēn dī-fos′-fāt; ADP): a molecule composed of the sugar ribose, the base adenine, and two phosphate groups; a component of ATP.
adenosine triphosphate (a-den′-ō-sēn trī-fos′-fāt; ATP): a molecule composed of the sugar ribose, the base adenine, and three phosphate groups; the major energy carrier in cells. The last two phosphate groups are attached by "high-energy" bonds.
adipose tissue (a′-di-pōs): tissue composed of fat cells.
adrenal cortex: the outer part of the adrenal gland, which secretes steroid hormones that regulate metabolism and salt balance.
adrenal gland: a mammalian endocrine gland, adjacent to the kidney; secretes hormones that function in water regulation and in the stress response.
adrenal medulla: the inner part of the adrenal gland, which secretes epinephrine (adrenaline) and norepinephrine (noradrenaline).
adrenocorticotropic hormone (a-drēn-ō-kor-tik-ō-trō′-pik; ACTH): a hormone, secreted by the anterior pituitary, that stimulates the release of hormones by the adrenal glands, especially in response to stress.
aerobic: using oxygen.
age structure: the distribution of males and females in a population according to age groups.
agglutination (a-gloo-tin-ā′-shun): the clumping of foreign substances or microbes, caused by binding with antibodies.
aggression: antagonistic behavior, normally among members of the same species, often resulting from competition for resources.
aggressive mimicry (mim′ik-rē): the evolution of a predatory organism to resemble a harmless animal or part of the environment, thus gaining access to prey.
aging: a gradual accumulation of random damage to essential biological molecules, particularly DNA, that begins at an early age. Eventually, the body's ability to repair the damage is exceeded, impairing function at all levels, from cells to tissues to organs.
aldosterone: a hormone, secreted by the adrenal cortex, that helps regulate ion concentration in the blood by stimulating the reabsorption of sodium by the kidneys and sweat glands.
alga (al′-ga; pl., algae, al′-jē): any photosynthetic member of the eukaryotic kingdom Protista.
allantois (al-an-tō′-is): one of the embryonic membranes of reptiles, birds, and mammals; in reptiles and birds, serves as a waste-storage organ; in mammals, forms most of the umbilical cord.
allele (al-ēl′): one of several alternative forms of a particular gene.
allele frequency: for any given gene, the relative proportion of each allele of that gene in a population.
allergy: an inflammatory response produced by the body in response to invasion by foreign materials, such as pollen, that are themselves harmless.
allopatric speciation (al-ō-pat′-rik): speciation that occurs when two populations are separated by a physical barrier that prevents gene flow between them (geographical isolation).
allosteric regulation: the process by which enzyme action is enhanced or inhibited by small organic molecules that act as regulators by binding to the enzyme and altering its active site.
alternation of generations: a life cycle, typical of plants, in which a diploid sporophyte (spore-producing) generation alternates with a haploid gametophyte (gamete-producing) generation.
altruism: a type of behavior that may decrease the reproductive success of the individual performing it but benefits that of other individuals.
alveolate (al-vē′-ō-lāt): a member of the Alveolata, a large assemblage of protists that is assigned kingdom status by many systematists. The alveolates, which are characterized by a system of sacs beneath the cell membrane, include ciliates, foraminiferans, dinoflagellates, and apicomplexans.
alveolus (al-vē′-ō-lus; pl., alveoli): a tiny air sac within the lungs, surrounded by capillaries, where gas exchange with the blood occurs.
amino acid: the individual subunit of which proteins are made, composed of a central carbon atom bonded to an amino group ($-NH_2$), a carboxyl group ($-COOH$), a hydrogen atom, and a variable group of atoms denoted by the letter *R*.
amino acid based hormone: a class of hormone that is synthesized by the body from single amino acids. Examples include epinephrine and thyroxine.
ammonia: NH_3; a highly toxic nitrogen-containing waste product of amino acid breakdown. In the mammalian liver, it is converted to urea.
amniocentesis (am-nē-ō-sen-tē′-sis): a procedure for sampling the amniotic fluid surrounding a fetus: A sterile needle is inserted through the abdominal wall, uterus, and amniotic sac of a pregnant woman; 10 to 20 milliliters of amniotic fluid are withdrawn. Various tests may be performed on the fluid and the fetal cells suspended in it to provide information on the developmental and genetic state of the fetus.
amnion (am′-nē-on): one of the embryonic membranes of reptiles, birds, and mammals; encloses a fluid-filled cavity that envelops the embryo.
amniote egg (am-nē-ōt′): the egg of reptiles and birds; contains an amnion that encloses the embryo in a watery environment, allowing the egg to be laid on dry land.

amoeba: a type of animal-like protist that uses a characteristic streaming mode of locomotion by extending a cellular projection called a *pseudopod*.

amoeboid cell: a protist or animal cell that moves by extending a cellular projection called a pseudopod.

amphibian: a member of the chordate class Amphibia, which includes the frogs, toads, and salamanders, as well as the limbless caecelians.

amplexus (am-plek′-sus): in amphibians, a form of external fertilization in which the male holds the female during spawning and releases his sperm directly onto her eggs.

ampulla: a muscular bulb that is part of the water-vascular system of echinoderms; controls the movement of tube feet, which are used for locomotion.

amygdala (am-ig′-da-la): part of the forebrain of vertebrates that is involved in the production of appropriate behavioral responses to environmental stimuli.

amylase (am′-i-lās): an enzyme, found in saliva and pancreatic secretions, that catalyzes the breakdown of starch.

anaerobe: an organism whose respiration does not require oxygen.

anaerobic: not using oxygen.

analogous structures: structures that have similar functions and superficially similar appearance but very different anatomies, such as the wings of insects and birds. The similarities are due to similar environmental pressures rather than to common ancestry.

anaphase (an′-a-fāz): in mitosis, the stage in which the sister chromatids of each chromosome separate from one another and are moved to opposite poles of the cell; in meiosis I, the stage in which homologous chromosomes, consisting of two sister chromatids, are separated; in meiosis II, the stage in which the sister chromatids of each chromosome separate from one another and are moved to opposite poles of the cell.

androgen: a male sex hormone.

androgen insensitivity: a rare condition in which an individual with XY chromosomes is female in appearance because the body's cells don't respond to the male hormones that are present.

angina (an-jī′-nuh): chest pain associated with reduced blood flow to the heart muscle, caused by the obstruction of coronary arteries.

angiosperm (an′-jē-ō-sperm): a flowering vascular plant.

angiotensin (an-jē-ō-ten′-sun): a hormone that functions in water regulation in mammals by stimulating physiological changes that increase blood volume and blood pressure.

annual ring: a pattern of alternating light (early) and dark (late) xylem of woody stems and roots, formed as a result of the unequal availability of water in different seasons of the year, normally spring and summer.

antagonistic muscles: a pair of muscles, one of which contracts and in so doing extends the other; an arrangement that makes possible movement of the skeleton at joints.

anterior: the front, forward, or head end of an animal.

anterior pituitary: a lobe of the pituitary gland that produces prolactin and growth hormone as well as hormones that regulate hormone production in other glands.

anther (an′-ther): the uppermost part of the stamen, in which pollen develops.

antheridium (an-ther-id′-ē-um): a structure in which male sex cells are produced, found in the bryophytes and certain seedless vascular plants.

antibiotic resistance: the ability of a mutated pathogen to resist the effects of an antibiotic that normally kills it.

antibody: a protein, produced by cells of the immune system, that combines with a specific antigen and normally facilitates the destruction of the antigen.

anticodon: a sequence of three bases in transfer RNA that is complementary to the three bases of a codon of messenger RNA.

antidiuretic hormone (an-tē-dī-ūr-et′-ik; ADH): a hormone produced by the hypothalamus and released into the bloodstream by the posterior pituitary when blood volume is low; increases the permeability of the distal tubule and the collecting duct to water, allowing more water to be reabsorbed into the bloodstream.

antigen: a complex molecule, normally a protein or polysaccharide, that stimulates the production of a specific antibody.

antioxidant: any molecule that reacts with free radicals, neutralizing their ability to damage biological molecules. Vitamins C and E are examples of dietary antioxidants.

aphotic zone: the region of the ocean below 200 m, where sunlight does not penetrate.

apical dominance: the phenomenon whereby a growing shoot tip inhibits the sprouting of lateral buds.

apical meristem (āp′-i-kul mer′-i-stem): the cluster of meristematic cells at the tip of a shoot or root (or one of their branches).

apicomplexan (ā-pē-kom-pleks′-an): a member of the protist phylum Apicomplexa, which includes mostly parasitic, single-celled eukaryotes such as *Plasmodium*, which causes malaria in humans. The apicomplexans are part of a larger group known as the alveolates.

appendicular skeleton (ap-pen-dik′-ū-lur): the portion of the skeleton consisting of the bones of the extremities and their attachments to the axial skeleton; the pectoral and pelvic girdles, the arms, legs, hands, and feet.

aqueous humor (ā′-kwē-us): the clear, watery fluid between the cornea and lens of the eye.

Archaea: one of life's three domains; consists of prokaryotes that are only distantly related to members of the domain Bacteria.

archegonium (ar-ke-gō′-nē-um): a structure in which female sex cells are produced; found in the bryophytes and certain seedless vascular plants.

arteriole (ar-tēr′-ē-ōl): a small artery that empties into capillaries. Contraction of the arteriole regulates blood flow to various parts of the body.

artery (ar′-tuh-rē): a vessel with muscular, elastic walls that conducts blood away from the heart.

arthropod: a member of the animal phylum Arthropoda, which includes the insects, spiders, ticks, mites, scorpions, crustaceans, millipedes, and centipedes.

artificial selection: a selective breeding procedure in which only those individuals with particular traits are chosen as breeders; used mainly to enhance desirable traits in domestic plants and animals; may also be used in evolutionary biology experiments.

ascus (as′-kus): a saclike case in which sexual spores are formed by members of the fungal division Ascomycota.

asexual reproduction: reproduction that does not involve the fusion of haploid sex cells. The parent body may divide and new parts regenerate, or a new, smaller individual may form as an attachment to the parent, to drop off when complete.

association neuron: in a neural network, a nerve cell that is postsynaptic to a sensory neuron and presynaptic to a motor neuron. In actual circuits, there may be many association neurons between individual sensory and motor neurons.

atherosclerosis (ath′-er-ō-skler-ō′-sis): a disease characterized by the obstruction of arteries by cholesterol deposits and thickening of the arterial walls.

atom: the smallest particle of an element that retains the properties of the element.

atomic nucleus: the membrane-bound organelle of eukaryotic cells that contains the cell's genetic material.

atomic number: the number of protons in the nuclei of all atoms of a particular element.

atrial natriuretic peptide (ā′-trē-ul nā-trē-ū-ret′-ik; ANP): a hormone, secreted by cells in the mammalian heart, that reduces blood volume by inhibiting the release of ADH and aldosterone.

atrioventricular (AV) node (ā′-trē-ō-ven-trik′-ū-lar nōd): a specialized mass of muscle at the base of the right atrium through which the electrical activity initiated in the sinoatrial node is transmitted to the ventricles.

atrioventricular valve: a heart valve that separates each atrium from each ventricle, preventing the backflow of blood into the atria during ventricular contraction.

atrium (ā′-trē-um): a chamber of the heart that receives venous blood and passes it to a ventricle.

auditory canal (aw′-di-tor-ē): a canal within the outer ear that conducts sound from the external ear to the tympanic membrane.

auditory nerve: the nerve leading from the mammalian cochlea to the brain, carrying information about sound.

auditory tube: a tube connecting the middle ear with the pharynx that allows pressure to equilibrate between the middle ear and the outside air (also called the Eustachian tube).

autoimmune disease: a disorder in which the immune system produces antibodies against the body's own cells.

autonomic nervous system: the part of the peripheral nervous system of vertebrates that synapses on glands, internal organs, and smooth muscle and produces largely involuntary responses.

autosome (aw′-tō-sōm): a chromosome that occurs in homologous pairs in both males and females and that does not bear the genes determining sex.

autotroph (aw′-tō-trōf): "self-feeder"; normally, a photosynthetic organism; a producer.
auxin (awk′-sin): a plant hormone that influences many plant functions, including phototropism, apical dominance, and root branching; generally stimulates cell elongation and, in some cases, cell division and differentiation.
axial skeleton: the skeleton forming the body axis, including the skull, vertebral column, and rib cage.
axon: a long extension of a nerve cell, extending from the cell body to synaptic endings on other nerve cells or on muscles.
bacillus (buh-sil′-us; pl., bacilli): a rod-shaped bacterium.
Bacteria: one of life's three domains; consists of prokaryotes that are only distantly related to members of the domain Archaea.
bacterial conjugation: the exchange of genetic material between two bacteria.
bacteriophage (bak-tir′-ē-ō-fāj): a virus specialized to attack bacteria.
bacterium (bak-tir′-ē-um; pl., bacteria): an organism consisting of a single prokaryotic cell surrounded by a complex polysaccharide coat.
balanced polymorphism: the prolonged maintenance of two or more alleles in a population, normally because each allele is favored by a separate environmental pressure.
ball-and-socket joint: a joint in which the rounded end of one bone fits into a hollow depression in another, as in the hip; allows movement in several directions.
bark: the outer layer of a woody stem, consisting of phloem, cork cambium, and cork cells.
Barr body: an inactivated X chromosome in cells of female mammals, which have two X chromosomes; normally appears as a dark spot in the nucleus.
basal body: a structure resembling a centriole that produces a cilium or flagellum and anchors this structure within the plasma membrane.
base: (1) a substance capable of combining with and neutralizing H^+ ions in a solution; a solution with a pH of more than 7; (2) in molecular genetics, one of the nitrogen-containing, single- or double-ringed structures that distinguish one nucleotide from another. In DNA, the bases are adenine, guanine, cytosine, and thymine.
basic: with an H^+ concentration less than that of OH^-; combining with H^+.
basidiospore (ba-sid′-ē-ō-spor): a sexual spore formed by members of the fungal division Basidiomycota.
basidium (bas-id′-ē-um): a diploid cell, typically club-shaped, formed by members of the fungal division Basidiomycota; produces basidiospores by meiosis.
basilar membrane (bas′-eh-lar): a membrane in the cochlea that bears hair cells that respond to the vibrations produced by sound.
basophil (bas′-ō-fil): a type of white blood cell that releases both substances that inhibit blood clotting and chemicals that participate in allergic reactions and in responses to tissue damage and microbial invasion.
B cell: a type of lymphocyte that participates in humoral immunity; gives rise to plasma cells, which secrete antibodies into the circulatory system, and to memory cells.
behavior: any observable activity of a living animal.
behavioral isolation: the lack of mating between species of animals that differ substantially in courtship and mating rituals.
bilateral symmetry: a body plan in which only a single plane through the central axis will divide the body into mirror-image halves.
bile (bīl): a liquid secretion, produced by the liver, that is stored in the gallbladder and released into the small intestine during digestion; a complex mixture of bile salts, water, other salts, and cholesterol.
bile salt: a substance that is synthesized in the liver from cholesterol and amino acids and that assists in the breakdown of lipids by dispersing them into small particles on which enzymes can act.
binary fission: the process by which a single bacterium divides in half, producing two identical offspring.
binocular vision: the ability to see objects simultaneously through both eyes, providing greater depth perception and more-accurate judgment of the size and distance of an object from the eyes.
biodegradable: able to be broken down into harmless substances by decomposers.
biodiversity: the total number of species within an ecosystem and the resulting complexity of interactions among them.
biogeochemical cycle: also called a *nutrient cycle*, the process by which a specific nutrient in an ecosystem is transferred between living organisms and the nutrient's reservoir in the nonliving environment.
biological clock: a metabolic timekeeping mechanism found in most organisms, whereby the organism measures the approximate length of a day (24 hours) even without external environmental cues such as light and darkness.
biological magnification: the increasing accumulation of a toxic substance in progressively higher trophic levels.
biomass: the dry weight of organic material in an ecosystem.
biome (bī′-ōm): a terrestrial ecosystem that occupies an extensive geographical area and is characterized by a specific type of plant community: for example, deserts.
biosphere (bī′-ō-sfēr): that part of Earth inhabited by living organisms; includes both living and nonliving components.
biotechnology: any industrial or commercial use or alteration of organisms, cells, or biological molecules to achieve specific practical goals.
biotic (bī-ah′-tik): living.
biotic potential: the maximum rate at which a population could increase, assuming ideal conditions that allow a maximum birth rate and minimum death rate.
birth control pill: a temporary contraceptive method that prevents ovulation by providing a continuing supply of estrogen and progesterone, which in turn suppresses LH release; must be taken daily, normally for 21 days of each menstrual cycle.
bladder: a hollow muscular storage organ for storing urine.
blade: the flat part of a leaf.
blastocyst (blas′-tō-sist): an early stage of human embryonic development, consisting of a hollow ball of cells, enclosing a mass of cells attached to its inner surface, which becomes the embryo.
blastopore: the site at which a blastula indents to form a gastrula.
blastula (blas′-tū-luh): in animals, the embryonic stage attained at the end of cleavage, in which the embryo normally consists of a hollow ball with a wall one or several cell layers thick.
blind spot: the area of the retina at which the axons of the ganglion cell merge to form the optic nerve; the blind spot of the retina.
blood: a fluid consisting of plasma in which blood cells are suspended; carried within the circulatory system.
blood–brain barrier: relatively impermeable capillaries of the brain that protect the cells of the brain from potentially damaging chemicals that reach the bloodstream.
blood clotting: a complex process by which platelets, the protein fibrin, and red blood cells block an irregular surface in or on the body, such as a damaged blood vessel, sealing the wound.
blood vessel: a channel that conducts blood throughout the body.
body mass index (BMI): a number derived from an individual's weight and height used to estimate body fat. The formula is: weight (in kg)/height^2 (in meters^2).
bone: a hard, mineralized connective tissue that is a major component of the vertebrate endoskeleton; provides support and sites for muscle attachment.
book lung: a structure composed of thin layers of tissue, resembling pages in a book, that are enclosed in a chamber and used as a respiratory organ by certain types of arachnids.
boom-and-bust cycle: a population cycle characterized by rapid exponential growth followed by a sudden massive die-off, seen in seasonal species and in some populations of small rodents, such as lemmings.
Bowman's capsule: the cup-shaped portion of the nephron in which blood filtrate is collected from the glomerulus.
bradykinin (brā′-dē-kī′-nin): a chemical, formed during tissue damage, that binds to receptor molecules on pain nerve endings, giving rise to the sensation of pain.
brain: the part of the central nervous system of vertebrates that is enclosed within the skull.
branch root: a root that arises as a branch of a preexisting root, through divisions of pericycle cells and subsequent differentiation of the daughter cells.
bronchiole (bron′-kē-ōl): a narrow tube, formed by repeated branching of the bronchi, that conducts air into the alveoli.
bronchus (bron′-kus): a tube that conducts air from the trachea to each lung.
bryophyte (brī′-ō-fīt): a simple nonvascular plant of the division Bryophyta, including mosses and liverworts.
bud: in animals, a small copy of an adult that develops on the body of the parent and eventually breaks off and becomes independent; in plants, an embryonic shoot, normally very short and consisting of an apical meristem with several leaf primordia.

budding: asexual reproduction by the growth of a miniature copy, or bud, of the adult animal on the body of the parent. The bud breaks off to begin independent existence.
buffer: a compound that minimizes changes in pH by reversibly taking up or releasing H^+ ions.
bulbourethral gland (bul-bō-ū-rē′-thrul): in male mammals, a gland that secretes a basic, mucus-containing fluid that forms part of the semen.
bulk flow: the movement of many molecules of a gas or fluid in unison from an area of higher pressure to an area of lower pressure.
bundle-sheath cell: one of a group of cells that surround the veins of plants; in C_4 (but not in C_3) plants, bundle-sheath cells contain chloroplasts.
C_3 cycle: the cyclic series of reactions whereby carbon dioxide is fixed into carbohydrates during the light-independent reactions of photosynthesis; also called *Calvin-Benson cycle*.
C_4 pathway: the series of reactions in certain plants that fixes carbon dioxide into oxaloacetic acid, which is later broken down for use in the C_3 cycle of photosynthesis.
calcitonin (kal-si-tōn′-in): a hormone, secreted by the thyroid gland, that inhibits the release of calcium from bone.
calorie (kal′-ō-rē): the amount of energy required to raise the temperature of 1 gram of water by 1 degree Celsius.
Calorie: a unit of energy, in which the energy content of foods is measured; the amount of energy required to raise the temperature of 1 liter of water 1 degree Celsius; also called a *kilocalorie*, equal to 1000 calories.
Calvin-Benson cycle: see C_3 *cycle*.
cambium (kam′-bē-um; pl., **cambia):** a lateral meristem, parallel to the long axis of roots and stems, that causes secondary growth of woody plant stems and roots. See *cork cambium; vascular cambium*.
camouflage (cam′-a-flaj): coloration and/or shape that renders an organism inconspicuous in its environment.
cancer: a disease in which some of the body's cells escape from normal regulatory processes and divide without control.
capillary: the smallest type of blood vessel, connecting arterioles with venules. Capillary walls, through which the exchange of nutrients and wastes occurs, are only one cell thick.
capsule: a polysaccharide or protein coating that some disease-causing bacteria secrete outside their cell wall.
carbohydrate: a compound composed of carbon, hydrogen, and oxygen, with the approximate chemical formula $(CH_2O)_n$; includes sugars and starches.
carbon fixation: the initial steps in the C_3 cycle, in which carbon dioxide reacts with ribulose bisphosphate to form a stable organic molecule.
cardiac cycle (kar′-dē-ak): the alternation of contraction and relaxation of the heart chambers.
cardiac muscle (kar′-dē-ak): the specialized muscle of the heart, able to initiate its own contraction, independent of the nervous system.
carnivore (kar′-neh-vor): literally, "meat eater"; a predatory organism that feeds on herbivores or on other carnivores; a secondary (or higher) consumer.
carotenoid (ka-rot′-en-oid): a red, orange, or yellow pigment, found in chloroplasts, that serves as an accessory light-gathering molecule in thylakoid photosystems.
carpel (kar′pel): the female reproductive structure of a flower, composed of stigma, style, and ovary.
carrier: an individual who is heterozygous for a recessive condition; displays the dominant phenotype but can pass on the recessive allele to offspring.
carrier protein: a membrane protein that facilitates the diffusion of specific substances across the membrane. The molecule to be transported binds to the outer surface of the carrier protein; the protein then changes shape, allowing the molecule to move across the membrane through the protein.
carrying capacity: the maximum population size that an ecosystem can support indefinitely; determined primarily by the availability of space, nutrients, water, and light.
cartilage (kar′-teh-lij): a form of connective tissue that forms portions of the skeleton; consists of chondrocytes and their extracellular secretion of collagen; resembles flexible bone.
Casparian strip (kas-par′-ē-un): a waxy, waterproof band, located in the cell walls between endodermal cells in a root, that prevents the movement of water and minerals into and out of the vascular cylinder through the extracellular space.
catalyst (kat′-uh-list): a substance that speeds up a chemical reaction without itself being permanently changed in the process; lowers the activation energy of a reaction.
catastrophism: the hypothesis that Earth has experienced a series of geological catastrophes, probably imposed by a supernatural being, that accounts for the multitude of species, both extinct and modern, and preserves creationism.
cell: the smallest unit of life, consisting, at a minimum, of an outer membrane that encloses a watery medium containing organic molecules, including genetic material composed of DNA.
cell body: the part of a nerve cell in which most of the common cellular organelles are located; typically a site of integration of inputs to the nerve cell.
cell cycle: the sequence of events in the life of a cell, from one division to the next.
cell division: splitting of one cell into two; the process of cellular reproduction.
cell-mediated immunity: an immune response in which foreign cells or substances are destroyed by contact with T cells.
cell plate: in plant cell division, a series of vesicles that fuse to form the new plasma membranes and cell wall separating the daughter cells.
cellular respiration: the oxygen-requiring reactions, occurring in mitochondria, that break down the end products of glycolysis into carbon dioxide and water while capturing large amounts of energy as ATP.
cellular slime mold: a funguslike protist consisting of individual amoeboid cells that can aggregate to form a sluglike mass, which in turn forms a fruiting body.
cellulase: an enzyme that catalyzes the breakdown of the carbohydrate cellulose into its component glucose molecules; almost entirely restricted to microorganisms.
cellulose: an insoluble carbohydrate composed of glucose subunits; forms the cell wall of plants.
cell wall: a layer of material, normally made up of cellulose or cellulose-like materials, that is outside the plasma membrane of plants, fungi, bacteria, and some protists.
central nervous system: in vertebrates, the brain and spinal cord.
central vacuole: a large, fluid-filled vacuole occupying most of the volume of many plant cells; performs several functions, including maintaining turgor pressure.
centriole (sen′-trē-ōl): in animal cells, a short, barrel-shaped ring consisting of nine microtubule triplets; a microtubule-containing structure at the base of each cilium and flagellum; gives rise to the microtubules of cilia and flagella and is involved in spindle formation during cell division.
centromere (sen′-trō-mēr): the region of a replicated chromosome at which the sister chromatids are held together until they separate during cell division.
cephalization (sef-ul-ī-zā′-shun): the tendency of sensory organs and nervous tissue to become concentrated in the head region over evolutionary time.
cerebellum (ser-uh-bel′-um): the part of the hindbrain of vertebrates that is concerned with coordinating movements of the body.
cerebral cortex (ser-ē′-brul kor′-tex): a thin layer of neurons on the surface of the vertebrate cerebrum, in which most neural processing and coordination of activity occurs.
cerebral hemisphere: one of two nearly symmetrical halves of the cerebrum, connected by a broad band of axons, the corpus callosum.
cerebrospinal fluid: a clear fluid, produced within the ventricles of the brain, that fills the ventricles and cushions the brain and spinal cord.
cerebrum (ser-ē′-brum): the part of the forebrain of vertebrates that is concerned with sensory processing, the direction of motor output, and the coordination of most bodily activities; consists of two nearly symmetrical halves (the hemispheres) connected by a broad band of axons, the corpus callosum.
cervical cap: a birth control device consisting of a rubber cap that fits over the cervix, preventing sperm form entering the uterus.
cervix (ser′-viks): a ring of connective tissue at the outer end of the uterus, leading into the vagina.
channel protein: a membrane protein that forms a channel or pore completely through the membrane and that is usually permeable to one or to a few water-soluble molecules, especially ions.
chaparral: a biome that is located in coastal regions but has very low annual rainfall.
chemical bond: the force of attraction between neighboring atoms that holds them together in a molecule.
chemical equilibrium: the condition in which the "forward" reaction of reactants to products proceeds at the same rate as the

"backward" reaction from products to reactants, so that no net change in chemical composition occurs.
chemical reaction: the process that forms and breaks chemical bonds that hold atoms together.
chemiosmosis (ke-mē-oz-mō′-sis): a process of ATP generation in chloroplasts and mitochondria. The movement of electrons down an electron transport system is used to pump hydrogen ions across a membrane, thereby building up a concentration gradient of hydrogen ions across the membrane; the hydrogen ions diffuse back across the membrane through the pores of ATP-synthesizing enzymes; the energy of their movement down their concentration gradient drives ATP synthesis.
chemoreceptor: a sensory receptor that responds to chemicals from the environment; used in the chemical senses of taste and smell.
chemosynthetic (kēm′-ō-sin-the-tik): capable of oxidizing inorganic molecules to obtain energy.
chemotactic (kēm-ō-tak′-tik): moving toward chemicals given off by food or away from toxic chemicals.
chiasma (kī-as′-muh; pl., chiasmata): a point at which a chromatid of one chromosome crosses with a chromatid of the homologous chromosome during prophase I of meiosis; the site of exchange of chromosomal material between chromosomes.
chitin (kī′-tin): a compound found in the cell walls of fungi and the exoskeletons of insects and some other arthropods; composed of chains of nitrogen-containing, modified glucose molecules.
chlamydia (kla-mid′-ē-uh): a sexually transmitted disease, caused by a bacterium, that causes inflammation of the urethra in males and of the urethra and cervix in females.
chlorophyll (klor′-ō-fil): a pigment found in chloroplasts that captures light energy during photosynthesis; absorbs violet, blue, and red light but reflects green light.
chloroplast (klor′-ō-plast): the organelle in plants and plantlike protists that is the site of photosynthesis; surrounded by a double membrane and containing an extensive internal membrane system that bears chlorophyll.
cholecystokinin (kō′-lē-sis-tō-kī′-nin): a digestive hormone, produced by the small intestine, that stimulates the release of pancreatic enzymes.
chondrocyte (kon′-drō-sīt): a living cell of cartilage. With their extracellular secretions of collagen, chondrocytes form cartilage.
chorion (kor′-ē-on): the outermost embryonic membrane in reptiles, birds, and mammals; in birds and reptiles, functions mostly in gas exchange; in mammals, forms most of the embryonic part of the placenta.
chorionic gonadotropin (CG): a hormone, secreted by the chorion (one of the fetal membranes), that maintains the integrity of the corpus luteum during early pregnancy.
chorionic villus (kor-ē-on-ik; pl., chorionic villi): in mammalian embryos, a fingerlike projection of the chorion that penetrates the uterine lining and forms the embryonic portion of the placenta.
chorionic villus sampling (CVS): a procedure for sampling cells from the chorionic villi produced by a fetus: A tube is inserted into the uterus of a pregnant woman, and a small sample of villi are suctioned off for genetic and biochemical analyses.
choroid (kor′-oid): a darkly pigmented layer of tissue, behind the retina, that contains blood vessels and pigment that absorbs stray light.
chromatid (krō′-ma-tid): one of the two identical strands of DNA and protein that forms a replicated chromosome. The two sister chromatids are joined at the centromere.
chromatin (krō′-ma-tin): the complex of DNA and proteins that makes up eukaryotic chromosomes.
chromist: a member of the Chromista, a large assemblage of protists that is assigned kingdom status by many systematists. The chromists include the diatoms, brown algae, and water molds.
chromosome (krō′-mō-sōm): a single DNA double helix together with proteins that help to organize the DNA.
chronic bronchitis: a persistent lung infection characterized by coughing, swelling of the lining of the respiratory tract, an increase in mucus production, and a decrease in the number and activity of cilia.
chyme (kīm): an acidic, souplike mixture of partially digested food, water, and digestive secretions that is released from the stomach into the small intestine.
ciliate (sil′-ē-et): a protozoan characterized by cilia and by a complex unicellular structure, including harpoonlike organelles called trichocysts. Members of the genus *Paramecium* are well-known ciliates.
cilium (sil′-ē-um; pl., cilia): a short, hairlike projection from the surface of certain eukaryotic cells that contains microtubules in a 9 + 2 arrangement. The movement of cilia may propel cells through a fluid medium or move fluids over a stationary surface layer of cells.
circadian rhythm (sir-kā′-dē-un): an event that recurs with a period of about 24 hours, even in the absence of environmental cues.
citric acid cycle: see *Krebs cycle.*
class: the taxonomic category composed of related genera. Closely related classes form a division or phylum.
cleavage: the early cell divisions of embryos, in which little or no growth occurs between divisions; reduces the cell size and distributes gene-regulating substances to the newly formed cell.
climate: patterns of weather that prevail from year to year and even from century to century in a given region.
climax community: a diverse and relatively stable community that forms the endpoint of succession.
clitoris: an external structure of the female reproductive system; composed of erectile tissue; a sensitive point of stimulation during sexual response.
clonal selection: the mechanism by which the immune response gains specificity; an invading antigen elicits a response from only a few lymphocytes, which proliferate to form a clone of cells that attack only the specific antigen that stimulated their production.
clone: offspring that are produced by mitosis and are therefore genetically identical to each other.
cloning: the process of producing many identical copies of a gene; also the production of many genetically identical copies of an organism.
closed circulatory system: the type of circulatory system, found in certain worms and vertebrates, in which the blood is always confined within the heart and vessels.
club fungus: a fungus of the division Basidiomycota, whose members (which include mushrooms, puffballs, and shelf fungi) reproduce by means of basidiospores.
clumped distribution: the distribution characteristic of populations in which individuals are clustered into groups; may be social or based on the need for a localized resource.
cnidocyte (nīd′-ō-sīt): in members of the phylum Cnidaria, a specialized cell that houses a stinging apparatus.
cochlea (kahk′-lē-uh): a coiled, bony, fluid-filled tube found in the mammalian inner ear; contains receptors (hair cells) that respond to the vibration of sound.
codominance: the relation between two alleles of a gene, such that both alleles are phenotypically expressed in heterozygous individuals.
codon: a sequence of three bases of messenger RNA that specifies a particular amino acid to be incorporated into a protein; certain codons also signal the beginning or end of protein synthesis.
coelom (sē′-lōm): a space or cavity that separates the body wall from the inner organs.
coenzyme: an organic molecule that is bound to certain enzymes and is required for the enzymes' proper functioning; typically, a nucleotide bound to a water-soluble vitamin.
coevolution: the evolution of adaptations in two species due to their extensive interactions with one another, such that each species acts as a major force of natural selection on the other.
cohesion: the tendency of the molecules of a substance to stick together.
cohesion–tension theory: a model for the transport of water in xylem, by which water is pulled up the xylem tubes, powered by the force of evaporation of water from the leaves (producing tension) and held together by hydrogen bonds between nearby water molecules (cohesion).
coleoptile (kō-lē-op′-tīl): a protective sheath surrounding the shoot in monocot seeds, allowing the shoot to push aside soil particles as it grows.
collagen (kol′-uh-jen): a fibrous protein in connective tissue such as bone and cartilage.
collar cell: a specialized cell lining the inside channels of sponges. Flagella extend from a sievelike collar, creating a water current that draws microscopic organisms through the collar and into the body, where they become trapped.
collecting duct: a conducting tube, within the kidney, that collects urine from many nephrons and conducts it through the renal medulla into the renal pelvis. Urine may become concentrated in the collecting ducts if ADH is present.

collenchyma (kōl-en′-ki-muh): an elongated, polygonal plant cell type with irregularly thickened primary cell walls that is alive at maturity and that supports the plant body.
colon: the longest part of the large intestine, exclusive of the rectum.
colostrum (kō-los′-trum): a yellowish fluid, high in protein and containing antibodies, that is produced by the mammary glands before milk secretion begins.
commensalism (kum-en′-sal-iz-um): a symbiotic relationship in which one species benefits while another species is neither harmed nor benefited.
communication: the act of producing a signal that causes another animal, normally of the same species, to change its behavior in a way that is beneficial to one or both participants.
community: all the interacting populations within an ecosystem.
compact bone: the hard and strong outer bone; composed of osteons.
companion cell: a cell adjacent to a sieve-tube element in phloem, involved in the control and nutrition of the sieve-tube element.
competition: interaction among individuals who attempt to utilize a resource (for example, food or space) that is limited relative to the demand for it.
competitive exclusion principle: the concept that no two species can simultaneously and continuously occupy the same ecological niche.
competitive inhibition: the process by which two or more molecules that are somewhat similar in structure compete for the active site of an enzyme.
complement: a group of blood-borne proteins that participate in the destruction of foreign cells to which antibodies have bound.
complementary base pair: in nucleic acids, bases that pair by hydrogen bonding. In DNA, adenine is complementary to thymine and guanine is complementary to cytosine; in RNA, adenine is complementary to uracil, and guanine to cytosine.
complement reaction: an interaction among foreign cells, antibodies, and complement proteins that results in the destruction of the foreign cells.
complement system: a series of reactions in which complement proteins bind to antibody stems, attracting to the site phagocytic white blood cells that destroy the invading cell that triggers the reactions.
complete flower: a flower that has all four floral parts (sepals, petals, stamens, and carpels).
compound: a substance whose molecules are formed by different types of atoms; can be broken into its constituent elements by chemical means.
compound eye: a type of eye, found in arthropods, that is composed of numerous independent subunits called *ommatidia*. Each ommatidium apparently contributes a piece of a mosaiclike image perceived by the animal.
concentration: the number of particles of a dissolved substance in a given unit of volume.
concentration gradient: the difference in concentration of a substance between two parts of a fluid or across a barrier such as a membrane.
conclusion: the final operation in the scientific method; a decision made about the validity of a hypothesis on the basis of experimental evidence.
condensation: compaction of eukaryotic chromosomes into discrete units in preparation for mitosis or meiosis.
condom: a contraceptive sheath worn over the penis during intercourse to prevent sperm from being deposited in the vagina.
conducting portion: the portion of the respiratory system in lung-breathing vertebrates that carries air to the lungs.
cone: a cone-shaped photoreceptor cell in the vertebrate retina; not as sensitive to light as are the rods. The three types of cones are most sensitive to different colors of light and provide color vision; see also *rod*.
conifer (kon′-eh-fer): a member of a class of tracheophytes (Coniferophyta) that reproduces by means of seeds formed inside cones and that retains its leaves throughout the year.
conjugation: in prokaryotes, the transfer of DNA from one cell to another via a temporary connection; in single-celled eukaryotes, the mutual exchange of genetic material between two temporarily joined cells.
connective tissue: a tissue type consisting of diverse tissues, including bone, fat, and blood, that generally contain large amounts of extracellular material.
constant region: the part of an antibody molecule that is similar in all antibodies.
consumer: an organism that eats other organisms; a heterotroph.
contest competition: a mechanism for resolving intraspecific competition by using social or chemical interactions.
contraception: the prevention of pregnancy.
contractile vacuole: a fluid-filled vacuole in certain protists that takes up water from the cytoplasm, contracts, and expels the water outside the cell through a pore in the plasma membrane.
control: that portion of an experiment in which all possible variables are held constant; in contrast to the "experimental" portion, in which a particular variable is altered.
convergence: a condition in which a large number of nerve cells provide input to a smaller number of cells.
convergent evolution: the independent evolution of similar structures among unrelated organisms as a result of similar environmental pressures; see *analogous structures*.
convolution: a folding of the cerebral cortex of the vertebrate brain.
copulation: reproductive behavior in which the penis of the male is inserted into the body of the female, where it releases sperm.
coral reef: a biome created by animals (reef-building corals) and plants in warm tropical waters.
cork cambium: a lateral meristem in woody roots and stems that gives rise to cork cells.
cork cell: a protective cell of the bark of woody stems and roots; at maturity, cork cells are dead, with thick, waterproofed cell walls.
cornea (kor′-nē-uh): the clear outer covering of the eye, in front of the pupil and iris.
corona radiata (kuh-rō′-nuh rā-dē-a′-tuh): the layer of cells surrounding an egg after ovulation.
corpus callosum (kor′pus kal-ō′-sum): the band of axons that connect the two cerebral hemispheres of vertebrates.
corpus luteum (kor′-pus loo′-tē-um): in the mammalian ovary, a structure that is derived from the follicle after ovulation and that secretes the hormones estrogen and progesterone.
cortex: the part of a primary root or stem located between the epidermis and the vascular cylinder.
cortisol (kor′-ti-sol): a steroid hormone released into the bloodstream by the adrenal cortex in response to stress. Cortisol helps the body cope with short-term stressors by raising blood glucose levels, and also inhibits the immune response.
cotyledon (kot-ul-ē′don): a leaflike structure within a seed that absorbs food molecules from the endosperm and transfers them to the growing embryo; also called *seed leaf*.
coupled reaction: a pair of reactions, one exergonic and one endergonic, that are linked together such that the energy produced by the exergonic reaction provides the energy needed to drive the endergonic reaction.
covalent bond (kō-vā′-lent): a chemical bond between atoms in which electrons are shared.
crab lice: an arthropod parasite that can infest humans; can be transmitted by sexual contact.
creationism: the hypothesis that all species on Earth were created in essentially their present form by a supernatural being and that significant modification of those species—specifically, their transformation into new species—cannot occur by natural processes.
crista (kris′-tuh; pl., cristae): a fold in the inner membrane of a mitochondrion.
crop: an organ, found in both earthworms and birds, in which ingested food is temporarily stored before being passed to the gizzard, where it is pulverized.
cross-bridge: in muscles, an extension of myosin that binds to and pulls on actin to produce muscle contraction.
cross-fertilization: the union of sperm and egg from two individuals of the same species.
crossing over: the exchange of corresponding segments of the chromatids of two homologous chromosomes during meiosis.
cultural evolution: changes in the behavior of a population of animals, especially humans, by learning behaviors acquired by members of previous generations.
cuticle (kū′-ti-kul): a waxy or fatty coating on the exposed surfaces of epidermal cells of many land plants, which aids in the retention of water.
cyanobacterium: a photosynthetic prokaryotic cell that utilizes chlorophyll and releases oxygen as a photosynthetic by-product; sometimes called *blue-green algae*.
cyclic AMP: a cyclic nucleotide, formed within many target cells as a result of the reception of amino acid derivatives or peptide hormones, that causes metabolic changes in the cell; often called a *second messenger*.
cyclic nucleotide (sik′-lik noo′-klē-ō-tīd): a nucleotide in which the phosphate group is

bonded to the sugar at two points, forming a ring; serves as an intracellular messenger.

cyst (sist): an encapsulated resting stage in the life cycle of certain invertebrates, such as parasitic flatworms and roundworms.

cystic fibrosis: an inherited disorder characterized by the buildup of salt in the lungs and the production of thick, sticky mucus that clogs the airways, restricts air exchange, and promotes infection.

cytokine (sī′-tō-kīn): any of several chemical messenger molecules released by cells that facilitate communication with other cells and transfer signals within and between the various systems of the body. Cytokines are important in cellular differentiation and the immune system.

cytokinesis (sī-tō-ki-nē′-sis): the division of the cytoplasm and organelles into two daughter cells during cell division; normally occurs during telophase of mitosis.

cytokinin (sī-tō-kī′-nin): a plant hormone that promotes cell division, fruit growth, and the sprouting of lateral buds and prevents the aging of plant parts, especially leaves.

cytoplasm (sī′-tō-plaz-um): the material contained within the plasma membrane of a cell, exclusive of the nucleus.

cytosine: a nitrogenous base found in both DNA and RNA; abbreviated as C.

cytoskeleton: a network of protein fibers in the cytoplasm that gives shape to a cell, holds and moves organelles, and is typically involved in cell movement.

cytotoxic T cell: a type of T cell that, upon contacting foreign cells, directly destroys them.

day-neutral plant: a plant in which flowering occurs as soon as the plant has grown and developed, regardless of daylength.

decomposer: an organism, normally a fungus or bacterium, that digests organic material by secreting digestive enzymes into the environment, in the process liberating nutrients into the environment.

deductive reasoning: the process of generating hypotheses about how a specific experiment or observation will turn out.

deforestation: the excessive cutting of forests, primarily rain forests in the Tropics, to clear space for agriculture.

dehydration synthesis: a chemical reaction in which two molecules are joined by a covalent bond with the simultaneous removal of a hydrogen from one molecule and a hydroxyl group from the other, forming water; the reverse of hydrolysis.

deletion mutation: a mutation in which one or more pairs of nucleotides are removed from a gene.

demographic transition: a change in population dynamic in which a stable population experiences rapid growth and then returns to a stable (although much larger) size.

denature: to disrupt the secondary and/or tertiary structure of a protein while leaving its amino acid sequence intact. Denatured proteins can no longer perform their biological functions.

dendrite (den′-drīt): a branched tendril that extends outward from the cell body of a neuron; specialized to respond to signals from the external environment or from other neurons.

denitrifying bacterium (dē-nī′-treh-fī-ing): a bacterium that breaks down nitrates, releasing nitrogen gas to the atmosphere.

density-dependent: referring to any factor, such as predation, that limits population size more effectively as the population density increases.

density-independent: referring to any factor that limits a population's size and growth regardless of its density.

deoxyribonucleic acid (dē-ox-ē-rī-bō-noo-klā′-ik; DNA): a molecule composed of deoxyribose nucleotides; contains the genetic information of all living cells.

dermal tissue system: a plant tissue system that makes up the outer covering of the plant body.

dermis (dur′-mis): the layer of skin beneath the epidermis; composed of connective tissue and containing blood vessels, muscles, nerve endings, and glands.

desert: a biome in which less than 25 to 50 centimeters (10 to 20 inches) of rain falls each year.

desertification: the spread of deserts by human activities.

desmosome (dez′-mō-sōm): a strong cell-to-cell junction that attaches adjacent cells to one another.

detritus feeder (de-trī′-tus): one of a diverse group of organisms, ranging from worms to vultures, that live off the wastes and dead remains of other organisms.

deuterostome (doo′-ter-ō-stōm): an animal with a mode of embryonic development in which the coelom is derived from outpocketings of the gut; characteristic of echinoderms and chordates.

development: the process by which an organism proceeds from fertilized egg through adulthood to eventual death.

diabetes mellitus (dī-uh-bē′-tēs mel-ī′-tus): a disease characterized by defects in the production, release, or reception of insulin; characterized by high blood glucose levels that fluctuate with sugar intake.

dialysis (dī-al′-i-sis): the passive diffusion of substances across an artificial semipermeable membrane.

diaphragm (dī′-uh-fram): in the respiratory system, a dome-shaped muscle forming the floor of the chest cavity that, when it contracts, pulls itself downward, enlarging the chest cavity and causing air to be drawn into the lungs; in a reproductive sense, a contraceptive rubber cap that fits snugly over the cervix, preventing the sperm from entering the uterus and thereby preventing pregnancy.

diatom (dī′-uh-tom): a protist that includes photosynthetic forms with two-part glassy outer coverings; important photosynthetic organisms in fresh water and salt water.

dicot (dī′-kaht): short for dicotyledon; a type of flowering plant characterized by embryos with two cotyledons, or seed leaves, modified for food storage.

differentially permeable: referring to the ability of some substances to pass through a membrane more readily than can other substances.

differential reproduction: differences in reproductive output among individuals of a population, normally as a result of genetic differences.

differentiated cell: a mature cell specialized for a specific function; in plants, differentiated cells normally do not divide.

differentiation: the process whereby relatively unspecialized cells, especially of embryos, become specialized into particular tissue types.

diffusion: the net movement of particles from a region of high concentration of that particle to a region of low concentration, driven by the concentration gradient; may occur entirely within a fluid or across a barrier such as a membrane.

digestion: the process by which food is physically and chemically broken down into molecules that can be absorbed by cells.

digestive system: a group of organs responsible for ingesting and then digesting food substances into simple molecules that can be absorbed and then expelling undigested wastes from the body.

dinoflagellate (dī-nō-fla′-jel-et): a protist that includes photosynthetic forms in which two flagella project through armorlike plates; abundant in oceans; can reproduce rapidly, causing "red tides."

dioecious (dī-ē′-shus): pertaining to organisms in which male and female gametes are produced by separate individuals rather than in the same individual.

diploid (dip′-loid): referring to a cell with pairs of homologous chromosomes.

direct development: a developmental pathway in which the offspring is born as a miniature version of the adult and does not radically change in body form as it grows and matures.

directional selection: a type of natural selection in which one extreme phenotype is favored over all others.

disaccharide (dī-sak′-uh-rīd): a carbohydrate formed by the covalent bonding of two monosaccharides.

disruptive selection: a type of natural selection in which both extreme phenotypes are favored over the average phenotype.

distal tubule: in the nephrons of the mammalian kidney, the last segment of the renal tubule through which the filtrate passes just before it empties into the collecting duct; a site of selective secretion and reabsorption as water and ions pass between the blood and the filtrate across the tubule membrane.

disturbance: any event that disrupts the ecosystem by altering its community, its abiotic structure, or both; disturbance precedes succession.

disulfide bridge: the covalent bond formed between the sulfur atoms of two cysteines in a protein; typically causes the protein to fold by bringing otherwise distant parts of the protein close together.

divergence: a condition in which a small number of nerve cells provide input to a larger number of cells.

divergent evolution: evolutionary change in which the differences between two lineages become more pronounced with the passage of time.

division: the taxonomic category contained within a kingdom and consisting of related classes of plants, fungi, bacteria, or plantlike protists.

DNA–DNA hybridization: a technique by which DNA from two species is separated into single strands and then allowed to re-form; hybrid double-stranded DNA from the two species can occur where the sequence of nucleotides is complementary. The greater the degree of hybridization, the closer the evolutionary relatedness of the two species.
DNA fingerprinting: the use of restriction enzymes to cut DNA segments into a unique set of restriction fragments from one individual that can be distinguished from the restriction fragments of other individuals by gel electrophoresis.
DNA helicase: an enzyme that helps unwind the DNA double helix during DNA replication.
DNA library: a readily accessible, easily duplicable complete set of all the DNA of a particular organism, normally cloned into bacterial plasmids.
DNA ligase: an enzyme that joins the sugars and phosphates in a DNA strand to create a continuous sugar-phosphate backbone.
DNA polymerase: an enzyme that bonds DNA nucleotides together into a continuous strand, using a preexisting DNA strand as a template.
DNA probe: a sequence of nucleotides that is complementary to the nucleotide sequence in a gene under study; used to locate a given gene within a DNA library.
DNA replication: the copying of the double-stranded DNA molecule, producing two identical DNA double helices.
DNA sequencing: the process of determining the chemical composition of a DNA molecule (in particular, the order in which the molecule's constituent nucleic acids are arranged).
domain: the broadest category for classifying organisms; organisms are classified into three domains: Bacteria, Archaea, and Eukarya.
dominance hierarchy: a social arrangement in which a group of animals, usually through aggressive interactions, establishes a rank for some or all of the group members that determines access to resources.
dominant: an allele that can determine the phenotype of heterozygotes completely, such that they are indistinguishable from individuals homozygous for the allele; in the heterozygotes, the expression of the other (recessive) allele is completely masked.
dopamine (dōp′-uh-mēn): a transmitter in the brain whose actions are largely inhibitory. The loss of dopamine-containing neurons causes Parkinson's disease.
dormancy: a state in which an organism does not grow or develop; usually marked by lowered metabolic activity and resistance to adverse environmental conditions.
dorsal (dor′-sul): the top, back, or uppermost surface of an animal oriented with its head forward.
dorsal root ganglion: a ganglion, located on the dorsal (sensory) branch of each spinal nerve, that contains the cell bodies of sensory neurons.
double covalent bond: a covalent bond in which two atoms share two pairs of electrons.
double fertilization: in flowering plants, the fusion of two sperm nuclei with the nuclei of two cells of the female gametophyte. One sperm nucleus fuses with the egg to form the zygote; the second sperm nucleus fuses with the two haploid nuclei of the primary endosperm cell, forming a triploid endosperm cell.
double helix (hē′-liks): the shape of the two-stranded DNA molecule; like a ladder twisted lengthwise into a corkscrew shape.
doubling time: the time it would take a population to double in size at its current growth rate.
douching: washing the vagina; after intercourse, an attempt to wash sperm out of the vagina before they enter the uterus; an ineffective contraceptive method.
Down syndrome: a genetic disorder caused by the presence of three copies of chromosome 21; common characteristics include mental retardation, distinctively shaped eyelids, a small mouth with protruding tongue, heart defects, and low resistance to infectious diseases; also called *trisomy 21*.
duct: a tube or opening through which exocrine secretions are released.
duplicated chromosome: a eukaryotic chromosome following DNA replication; consists of two sister chromatids joined at the centromeres.
echolocation: the use of ultrasonic sounds, which bounce back from nearby objects, to produce an auditory "image" of nearby surroundings; used by bats and porpoises.
ecological footprint: the area of biologically productive space needed to supply the demands of an average human at current levels of technology.
ecological isolation: the lack of mating between organisms belonging to different populations that occupy distinct habitats within the same general area.
ecological niche (nitch): the role of a particular species within an ecosystem, including all aspects of its interaction with the living and nonliving environments.
ecology (ē-kol′-uh-jē): the study of the interrelationships of organisms with each other and with their nonliving environment.
ecosystem (ē′kō-sis-tem): all the organisms and their nonliving environment within a defined area.
ectoderm (ek′-tō-derm): the outermost embryonic tissue layer, which gives rise to structures such as hair, the epidermis of the skin, and the nervous system.
effector (ē-fek′-tor): a part of the body (normally a muscle or gland) that carries out responses as directed by the nervous system.
egg: the haploid female gamete, normally large and nonmotile, containing food reserves for the developing embryo.
electrocardiogram (ECG): the read-out of an instrument that records the electrical activity generated by cardiac muscle action potentials. These electrical events are measured by electrodes placed at specific sites on the surface of the body.
electrolocation: the production of high-frequency electrical signals from an electric organ in front of the tail of weak electrical fish; used to detect and locate nearly objects.
electron: a subatomic particle, found in an electron shell outside the nucleus of an atom, that bears a unit of negative charge and very little mass.
electron carrier: a molecule that can reversibly gain or lose electrons. Electron carriers generally accept high-energy electrons produced during an exergonic reaction and donate the electrons to acceptor molecules that use the energy to drive endergonic reactions.
electron shell: a region within which electrons orbit that corresponds to a fixed energy level at a given distance from the atomic nucleus of an atom.
electron transport system: a series of electron carrier molecules, found in the thylakoid membranes of chloroplasts and the inner membrane of mitochondria, that extract energy from electrons and generate ATP or other energetic molecules.
element: a substance that cannot be broken down, or converted, to a simpler substance by ordinary chemical means.
El Niño (el nēn′-yō): literally "boy child"; a reduction in intensity of Northeast Tradewinds that causes widespread disruption of weather patterns.
embryo: in animals, the stages of development that begin with the fertilization of the egg cell and end with hatching or birth; in mammals in particular, the early stages in which the developing animal does not yet resemble the adult of the species.
embryonic disc: in human embryonic development, the flat, two-layered group of cells that separates the amniotic cavity from the yolk sac.
embryonic stem cell: a cell derived from an early embryo that is capable of differentiating into any of the adult cell types.
embryo sac: the haploid female gametophyte of flowering plants.
emergent property: an intangible attribute that arises as the result of complex ordered interactions among individual parts.
emigration (em-uh-grā′shun): migration of individuals out of an area.
emphysema (em-fuh-sē′-muh): a condition in which the alveoli of the lungs become brittle and rupture, causing decreased area for gas exchange.
endergonic (en-der-gon′-ik): pertaining to a chemical reaction that requires an input of energy to proceed; an "uphill" reaction.
endocrine disruptors: environmental pollutants that interfere with endocrine function, often by disrupting the action of sex hormones.
endocrine gland: a ductless, hormone-producing gland consisting of cells that release their secretions into the extracellular fluid from which the secretions diffuse into nearby capillaries.
endocrine hormones: chemical messages produced by specialized cells and released into the circulatory system. They cause a prolonged or temporary change in target cells bearing specific receptors for these hormones.
endocrine system: an animal's organ system for cell-to-cell communication, composed of hormones and the cells that secrete them and receive them.
endocytosis (en-dō-si-tō′-sis): the process in which the plasma membrane engulfs extracellular material, forming membrane-bound sacs that enter the cytoplasm and thereby move material into the cell.

endoderm (en′-dō-derm): the innermost embryonic tissue layer, which gives rise to structures such as the lining of the digestive and respiratory tracts.
endodermis (en-dō-der′-mis): the innermost layer of small, close-fitting cells of the cortex of a root that form a ring around the vascular cylinder.
endogenous pyrogen: a chemical, produced by the body, that stimulates the production of a fever.
endometrium (en-dō-mē′-trē-um): the nutritive inner lining of the uterus.
endoplasmic reticulum (ER) (en-dō-plaz′-mik re-tik′-ū-lum): a system of membranous tubes and channels within eukaryotic cells; the site of most protein and lipid syntheses.
endorphin (en-dor′-fin): one of a group of peptide neuromodulators in the vertebrate brain that, by reducing the sensation of pain, mimics some of the actions of opiates.
endoskeleton (en′-dō-skel′-uh-tun): a rigid internal skeleton with flexible joints to allow for movement.
endosperm: a triploid food storage tissue in the seeds of flowering plants that nourishes the developing plant embryo.
endospore: a protective resting structure of some rod-shaped bacteria that withstands unfavorable environmental conditions.
endosymbiont hypothesis: the hypothesis that certain organelles, especially chloroplasts and mitochondria, arose as mutually beneficial associations between the ancestors of eukaryotic cells and captured bacteria that lived within the cytoplasm of the pre-eukaryotic cell.
energy: the capacity to do work.
energy-carrier molecule: a molecule that stores energy in "high-energy" chemical bonds and releases the energy to drive coupled endothermic reactions. In cells, ATP is the most common energy-carrier molecule.
energy level: the specific amount of energy characteristic of a given electron shell in an atom.
energy pyramid: a graphical representation of the energy contained in succeeding trophic levels, with maximum energy at the base (primary producers) and steadily diminishing amounts at higher levels.
entropy (en′-trō-pē): a measure of the amount of randomness and disorder in a system.
environmental estrogens: chemicals in the environment that mimic some of the effects of estrogen in animals.
environmental resistance: any factor that tends to counteract biotic potential, limiting population size.
enzyme (en′zīm): a protein catalyst that speeds up the rate of specific biological reactions.
eosinophil (ē-ō-sin′-ō-fil): a type of white blood cell that converges on parasitic invaders and releases substances to kill them.
epicotyl (ep′-ē-kot-ul): the part of the embryonic shoot located above the cotyledons but below the tip of the shoot.
epidermal tissue: dermal tissue in plants that forms the epidermis, the outermost cell layer that covers young plants.
epidermis (ep-uh-der′-mis): in animals, specialized epithelial tissue that forms the outer layer of skin; in plants, the outermost layer of cells of a leaf, young root, or young stem.
epididymis (e-pi-di′-di-mus): a series of tubes that connect with and receive sperm from the seminiferous tubules of the testis.
epiglottis (ep-eh-glah′-tis): a flap of cartilage in the lower pharynx that covers the opening to the larynx during swallowing; directs food down the esophagus.
epinephrine (ep-i-nef′-rin): a hormone, secreted by the adrenal medulla, that is released in response to stress and that stimulates a variety of responses, including the release of glucose from skeletal muscle and an increase in heart rate.
epithelial cell (eh-puh-thē′-lē-ul): a flattened cell that covers the outer body surfaces of a sponge.
epithelial tissue (eh-puh-thē′-lē-ul): a tissue type that forms membranes that cover the body surface and line body cavities, and that also gives rise to glands.
equilibrium population: a population in which allele frequencies and the distribution of genotypes do not change from generation to generation.
erythroblastosis fetalis (eh-rith′-rō-blas-tō′-sis fē-tal′-is): a condition in which the red blood cells of a newborn Rh-positive baby are attacked by antibodies produced by its Rh-negative mother, causing jaundice and anemia. Retardation and death are possible consequences if treatment is inadequate.
erythrocyte (eh-rith′-rō-sīt): a red blood cell, active in oxygen transport, that contains the red pigment hemoglobin.
erythropoietin (eh-rith′-rō-pō-ē′-tin): a hormone produced by the kidneys in response to oxygen deficiency that stimulates the production of red blood cells by the bone marrow.
esophagus (eh-sof′-eh-gus): a muscular passageway that conducts food from the pharynx to the stomach in humans and other mammals.
essential amino acid: an amino acid that is a required nutrient; the body is unable to manufacture essential amino acids, so they must be supplied in the diet.
essential fatty acid: a fatty acid that is a required nutrient; the body is unable to manufacture essential fatty acids, so they must be supplied in the diet.
estrogen: in vertebrates, a female sex hormone, produced by follicle cells of the ovary, that stimulates follicle development, oogenesis, the development of secondary sex characteristics, and growth of the uterine lining.
estuary: a wetland formed where a river meets the ocean; the salinity there is quite variable but lower than in sea water and higher than in fresh water.
ethology (ē-thol′-ō-jē): the study of animal behavior in natural or near-natural conditions.
ethylene: a plant hormone that promotes the ripening of fruits and the dropping of leaves and fruit.
euglenoid (ū′-gle-noid): a protist characterized by one or more whiplike flagella that are used for locomotion and by a photoreceptor that detects light. Euglenoids are photosynthetic, but if deprived of chlorophyll, some are capable of heterotrophic nutrition.
Eukarya (ū-kar′-ē-a): one of life's three domains; consists of all eukaryotes (plants, animals, fungi, and protists).
eukaryote (ū-kar′-ē-ōt): an organism whose cells are eukaryotic; plants, animals, fungi, and protists are eukaryotes.
eukaryotic (ū-kar-ē-ot′-ik): referring to cells of organisms of the domain Eukarya (plants, animals, fungi, and protists). Eukaryotic cells have genetic material enclosed within a membrane-bound nucleus and contain other membrane-bound organelles.
Eustachian tube (ū-stā′-shin): a tube connecting the middle ear with the pharynx; allows pressure between the middle ear and the atmosphere to equilibrate.
eutrophic lake: a lake that receives sufficiently large inputs of sediments, organic material, and inorganic nutrients from its surroundings to support dense communities; murky with poor light penetration.
evergreen: a plant that retains green leaves throughout the year.
evolution: the descent of modern organisms with modification from preexisting life-forms; strictly speaking, any change in the proportions of different genotypes in a population from one generation to the next.
excretion: the elimination of waste substances from the body; can occur from the digestive system, skin glands, urinary system, or lungs.
excretory pore: an opening in the body wall of certain invertebrates, such as the earthworm, through which urine is excreted.
exergonic (ex-er-gon′-ik): pertaining to a chemical reaction that liberates energy (either as heat or in the form of increased entropy); a "downhill" reaction.
exhalation: the act of releasing air from the lungs, which results from a relaxation of the respiratory muscles.
exocrine gland: a gland that releases its secretions into ducts that lead to the outside of the body or into the digestive tract.
exocytosis (ex-ō-si-tō′-sis): the process in which intracellular material is enclosed within a membrane-bound sac that moves to the plasma membrane and fuses with it, releasing the material outside the cell.
exon: a segment of DNA in a eukaryotic gene that codes for amino acids in a protein (see also *intron*).
exoskeleton (ex′-ō-skel′-uh-tun): a rigid external skeleton that supports the body, protects the internal organs, and has flexible joints that allow for movement.
exotic/exotic species: a foreign species introduced into an ecosystem where it did not evolve; such species may flourish and outcompete native species.
experiment: the third operation in the scientific method; the testing of a hypothesis by further observations, leading to a conclusion.
exponential growth: a continuously accelerating increase in population size.
extensor: a muscle that straightens a joint.
external ear: the fleshy portion of the ear that extends outside the skull.
external fertilization: the union of sperm and egg outside the body of either parent.
extinction: the death of all members of a species.

extracellular digestion: the physical and chemical breakdown of food that occurs outside a cell, normally in a digestive cavity.
extraembryonic membrane: in the embryonic development of reptiles, birds, and mammals, either the chorion, amnion, allantois, or yolk sac; functions in gas exchange, provision of the watery environment needed for development, waste storage, and storage of the yolk, respectively.
eyespot: a simple, lensless eye found in various invertebrates, including flatworms and jellyfish. Eyespots can distinguish light from dark and sometimes the direction of light, but they cannot form an image.
facilitated diffusion: the diffusion of molecules across a membrane, assisted by protein pores or carriers embedded in the membrane.
fairy ring: a circular pattern of mushrooms formed when reproductive structures erupt from the underground hyphae of a club fungus that has been growing outward in all directions from its original location.
family: the taxonomic category contained within an order and consisting of related genera.
farsighted: the inability to focus on nearby objects, caused by the eyeball being slightly too short.
fat (molecular): a lipid composed of three saturated fatty acids covalently bonded to glycerol; solid at room temperature.
fat (tissue): adipose tissue; connective tissue that stores the lipid fat; composed of cells packed with triglycerides.
fatty acid: an organic molecule composed of a long chain of carbon atoms, with a carboxylic acid (COOH) group at one end; may be saturated (all single bonds between the carbon atoms) or unsaturated (one or more double bonds between the carbon atoms).
feces: semisolid waste material that remains in the intestine after absorption is complete and is voided through the anus. Feces consist of indigestible wastes and the dead bodies of bacteria.
feedback inhibition: in enzyme-mediated chemical reactions, the condition in which the product of a reaction inhibits one or more of the enzymes involved in synthesizing the product.
fermentation: anaerobic reactions that convert the pyruvic acid produced by glycolysis into lactic acid or alcohol and CO_2.
fertilization: the fusion of male and female haploid gametes, forming a zygote.
fetal alcohol syndrome (FAS): a cluster of symptoms, including retardation and physical abnormalities, that occur in infants born to mothers who consumed large amounts of alcoholic beverages during pregnancy.
fetus: the later stages of mammalian embryonic development (after the second month for humans), when the developing animal has come to resemble the adult of the species.
fever: an elevation in body temperature caused by chemicals (pyrogens) that are released by white blood cells in response to infection.
fibrillation: rapid, uncoordinated, and ineffective contractions of heart muscle cells.
fibrin (fī′-brin): a clotting protein formed in the blood in response to a wound; binds with other fibrin molecules and provides a matrix around which a blood clot forms.
fibrinogen (fī-brin′-ō-jen): the inactive form of the clotting protein fibrin. Fibrinogen is converted into fibrin by the enzyme thrombin, which is produced in response to injury.
fibrous root system: a root system, commonly found in monocots, characterized by many roots of approximately the same size arising from the base of the stem.
filament: in flowers, the stalk of a stamen, which bears an anther at its tip.
filtrate: the fluid produced by filtration; in the kidneys, the fluid produced by the filtration of blood through the glomerular capillaries.
filtration: within Bowman's capsule in each nephron of a kidney, the process by which blood is pumped under pressure through permeable capillaries of the glomerulus, forcing out water, dissolved wastes, and nutrients.
fimbria (fim′-brē-uh; pl., **fimbriae):** in female mammals, the ciliated, fingerlike projections of the oviduct that sweep the ovulated egg from the ovary into the oviduct.
first law of thermodynamics: the principle of physics that states that within any isolated system, energy can be neither created nor destroyed but can be converted from one form to another.
fission: asexual reproduction by dividing the body into two smaller, complete organisms.
fitness: the reproductive success of an organism, usually expressed in relation to the average reproductive success of all individuals in the same population.
flagellum (fla-jel′-um; pl., **flagella):** a long, hairlike extension of the plasma membrane; in eukaryotic cells, it contains microtubules arranged in a 9 + 2 pattern. The movement of flagella propel some cells through fluids.
flame cell: in flatworms, a specialized cell, containing beating cilia, that conducts water and wastes through the branching tubes that serve as an excretory system.
flexor: a muscle that flexes (decreases the angle of) a joint.
florigen: one of a group of plant hormones that can both trigger and inhibit flowering; daylength is a stimulus.
flower: the reproductive structure of an angiosperm plant.
fluid: a liquid or gas.
fluid mosaic model: a model of membrane structure; according to this model, membranes are composed of a double layer of phospholipids in which various proteins are embedded. The phospholipid bilayer is a somewhat fluid matrix that allows the movement of proteins within it.
follicle: in the ovary of female mammals, the oocyte and its surrounding accessory cells.
follicle-stimulating hormone (FSH): a hormone, produced by the anterior pituitary, that stimulates spermatogenesis in males and the development of the follicle in females.
food chain: a linear feeding relationship in a community, using a single representative from each of the trophic levels.
food vacuole: a membranous sac, within a single cell, in which food is enclosed. Digestive enzymes are released into the vacuole, where intracellular digestion occurs.
food web: a representation of the complex feeding relationships (in terms of interacting food chains) within a community, including many organisms at various trophic levels, with many of the consumers occupying more than one level simultaneously.
foraminiferan (for-am-i-nif′-er-un): an aquatic (largely marine) protist characterized by a typically elaborate calcium carbonate shell.
forebrain: during development, the anterior portion of the brain. In mammals, the forebrain differentiates into the thalamus, the limbic system, and the cerebrum. In humans, the cerebrum contains about half of all the neurons in the brain.
fossil: the remains of a dead organism, normally preserved in rock; may be petrified bones or wood; shells; impressions of body forms, such as feathers, skin, or leaves; or markings made by organisms, such as footprints.
fossil fuel: a fuel such as coal, oil, and natural gas, derived from the remains of ancient organisms.
founder effect: a type of genetic drift in which an isolated population founded by a small number of individuals may develop allele frequencies that are very different from those of the parent population as a result of chance inclusion of disproportionate numbers of certain alleles in the founders.
fovea (fō′-vē-uh): in the vertebrate retina, the central region on which images are focused; contains closely packed cones.
free-living: not parasitic.
free nerve ending: on some receptor neurons, a finely branched ending that responds to touch and pressure, to heat and cold, or to pain; produces the sensations of itching and tickling.
free nucleotides: nucleotides that have not been joined together to form a DNA or RNA strand.
free radical: a molecule with an unpaired electron, which makes it highly unstable and reactive with nearby molecules. By stealing an electron from the molecule it attacks, it creates a new free radical and begins a chain reaction that can lead to the destruction of biological molecules crucial to life.
fruit: in flowering plants, the ripened ovary (plus, in some cases, other parts of the flower), which contains the seeds.
fruiting body: a spore-forming reproductive structure of certain protists, bacteria, and fungi.
functional group: one of several groups of atoms commonly found in an organic molecule, including hydrogen, hydroxyl, amino, carboxyl, and phosphate groups, that determine the characteristics and chemical reactivity of the molecule.
gallbladder: a small sac, next to the liver, in which the bile secreted by the liver is stored and concentrated. Bile is released from the gallbladder to the small intestine through the bile duct.
gamete (gam′-ēt): a haploid sex cell formed in sexually reproducing organisms.
gametic incompatibility: the inability of sperm from one species to fertilize eggs of another species.
gametophyte (ga-mēt′-ō-fīt): the multicellular haploid stage in the life cycle of plants.
ganglion (gang′-lē-un): a cluster of neurons.

ganglion cell: a type of cell, of which the innermost layer of the vertebrate retina is composed, whose axons form the optic nerve.
gap junction: a type of cell-to-cell junction in animals in which channels connect the cytoplasm of adjacent cells.
gas-exchange portion: the portion of the respiratory system in lung-breathing vertebrates where gas is exchanged in the alveoli of the lungs.
gastric inhibitory peptide: a hormone, produced by the small intestine, that inhibits the activity of the stomach.
gastrin: a hormone, produced by the stomach, that stimulates acid secretion in response to the presence of food.
gastrovascular cavity: a saclike chamber with digestive functions, found in simple invertebrates; a single opening serves as both mouth and anus, and the chamber provides direct access of nutrients to the cells.
gastrula (gas′-troo-luh): in animal development, a three-layered embryo with ectoderm, mesoderm, and endoderm cell layers. The endoderm layer normally encloses the primitive gut.
gastrulation (gas-troo-la′-shun): the process whereby a blastula develops into a gastrula, including the formation of endoderm, ectoderm, and mesoderm.
gel electrophoresis: a technique in which molecules (such as DNA fragments) are placed on restricted tracks in a thin sheet of gelatinous material and exposed to an electric field; the molecules then migrate at a rate determined by certain characteristics, such as length.
gene: a unit of heredity that encodes the information needed to specify the amino acid sequence of proteins and hence particular traits; a functional segment of DNA located at a particular place on a chromosome.
gene flow: the movement of alleles from one population to another owing to the migration of individual organisms.
gene pool: the total of all alleles of all genes in a population; for a single gene, the total of all the alleles of that gene that occur in a population.
generative cell: in flowering plants, one of the haploid cells of a pollen grain; undergoes mitosis to form two sperm cells.
genetically modified organism (GMO): an organism that has been produced through the techniques of genetic engineering.
genetic code: the collection of codons of mRNA, each of which directs the incorporation of a particular amino acid into a protein during protein synthesis.
genetic drift: a change in the allele frequencies of a small population purely by chance.
genetic engineering: the modification of genetic material to achieve specific goals.
genetic equilibrium: a state in which the allele frequencies and the distribution of genotypes of a population do not change from generation to generation.
genetic recombination: the generation of new combinations of alleles on homologous chromosomes due to the exchange of DNA during crossing over.
genital herpes: a sexually transmitted disease, caused by a virus, that can cause painful blisters on the genitals and surrounding skin.
genital warts: a sexually transmitted disease, caused by a virus, that forms growths or bumps on the external genitalia, in or around the vagina or anus, or on the cervix in females or penis, scrotum, groin, or thigh in males.
genome (jē′-nōm): the entire set of genes carried by a member of any given species.
genotype (jēn′-ō-tīp): the genetic composition of an organism; the actual alleles of each gene carried by the organism.
genus (jē-nus): the taxonomic category contained within a family and consisting of very closely related species.
geographical isolation: the separation of two populations by a physical barrier.
germ layer: a tissue layer formed during early embryonic development.
germination: the growth and development of a seed, spore, or pollen grain.
gibberellin (jib-er-el′-in): a plant hormone that stimulates seed germination, fruit development, and cell division and elongation.
gill: in aquatic animals, a branched tissue richly supplied with capillaries around which water is circulated for gas exchange.
gizzard: a muscular organ, found in earthworms and birds, in which food is mechanically broken down prior to chemical digestion.
gland: a cluster of cells that are specialized to secrete substances such as sweat or hormones.
glial cell: a cell of the nervous system that provides support and insulation for neurons.
global warming: a gradual rise in global atmospheric temperature as a result of an amplification of the natural greenhouse effect due to human activities.
glomerulus (glō-mer′-ū-lus): a dense network of thin-walled capillaries, located within the Bowman's capsule of each nephron of the kidney, where blood pressure forces water and dissolved nutrients through capillary walls for filtration by the nephron.
glucagon (gloo′-ka-gon): a hormone, secreted by the pancreas, that increases blood sugar by stimulating the breakdown of glycogen (to glucose) in the liver.
glucocorticoid (gloo-kō-kor′-tik-oid): a class of hormones, released by the adrenal cortex in response to the presence of ACTH, that make additional energy available to the body by stimulating the synthesis of glucose.
glucose: the most common monosaccharide, with the molecular formula $C_6H_{12}O_6$; most polysaccharides, including cellulose, starch, and glycogen, are made of glucose subunits covalently bonded together.
glycerol (glis′-er-ol): a three-carbon alcohol to which fatty acids are covalently bonded to make fats and oils.
glycogen (glī′-kō-jen): a long, branched polymer of glucose that is stored by animals in the muscles and liver and metabolized as a source of energy.
glycolysis (glī-kol′-i-sis): reactions, carried out in the cytoplasm, that break down glucose into two molecules of pyruvic acid, producing two ATP molecules; does not require oxygen but can proceed when oxygen is present.
glycoprotein: a protein to which a carbohydrate is attached.
goiter: a swelling of the neck caused by iodine deficiency, which affects the functioning of the thyroid gland and its hormones.
Golgi complex (gōl′-jē): a stack of membranous sacs, found in most eukaryotic cells, that is the site of processing and separation of membrane components and secretory materials.
gonad: an organ where reproductive cells are formed; in males, the testes, and in females, the ovaries.
gonadotropin-releasing hormone (GnRH): a hormone produced by the neurosecretory cells of the hypothalamus, which stimulates cells in the anterior pituitary to release FSH and LH. GnRH is involved in the menstrual cycle and in spermatogenesis.
gonorrhea (gon-uh-rē′-uh): a sexually transmitted bacterial infection of the reproductive organs; if untreated, can result in sterility.
gradient: a difference in concentration, pressure, or electrical charge between two regions.
Gram stain: a stain that is selectively taken up by the cell walls of certain types of bacteria (gram-positive bacteria) and rejected by the cell walls of others (gram-negative bacteria); used to distinguish bacteria on the basis of their cell wall construction.
granum (gra′-num; pl., grana): a stack of thylakoids in chloroplasts.
grassland: a biome, located in the centers of continents, that supports grasses; also called *prairie*.
gravitropism: growth with respect to the direction of gravity.
gray crescent: in frog embryonic development, an area of intermediate pigmentation in the fertilized egg; contains gene-regulating substances required for the normal development of the tadpole.
gray matter: the outer portion of the brain and inner region of the spinal cord; composed largely of neuron cell bodies, which give this area a gray color.
greenhouse effect: the process in which certain gases such as carbon dioxide and methane trap sunlight energy in a planet's atmosphere as heat; the glass in a greenhouse does the same. The result, global warming, is being enhanced by the production of these gases by humans.
greenhouse gas: a gas, such as carbon dioxide or methane, that traps sunlight energy in a planet's atmosphere as heat; a gas that participates in the greenhouse effect.
ground tissue system: a plant tissue system consisting of parenchyma, collenchyma, and sclerenchyma cells that makes up the bulk of a leaf or young stem, excluding vascular or dermal tissues. Most ground tissue cells function in photosynthesis, support, or carbohydrate storage.
growth hormone: a hormone, released by the anterior pituitary, that stimulates growth, especially of the skeleton.
growth rate: a measure of the change in population size per individual per unit of time.
guanine: a nitrogenous base found in both DNA and RNA; abbreviated as *G*.
guard cell: one of a pair of specialized epidermal cells surrounding the central opening of a stoma of a leaf, which regulates the size of the opening.
gymnosperm (jim′-nō-sperm): a nonflowering seed plant, such as a conifer, cycad, or gingko.

gyre (jīr): a roughly circular pattern of ocean currents, formed because continents interrupt the currents' flow; rotates clockwise in the Northern Hemisphere and counterclockwise in the Southern Hemisphere.

habituation (heh-bich-oo-ā'-shun): simple learning characterized by a decline in response to a harmless, repeated stimulus.

hair cell: a type of mechanoreceptor cell found in the inner ear that produces an electrical signal when stiff "hairlike" cilia projecting from the surface of the cell are bent. Hair cells in the cochlea respond to sound vibrations; those in the vestibular system respond to motion and gravity.

hair follicle: a gland in the dermis of mammalian skin, formed from epithelial tissue, that produces a hair.

halophile (hā'-lō-fīl): literally, "salt-loving"; a type of archaen that thrives in concentrated salt solutions.

haploid (hap'-loid): referring to a cell that has only one member of each pair of homologous chromosomes.

Hardy-Weinberg principle: a mathematical model proposing that, under certain conditions, the allele frequencies and genotype frequencies in a sexually reproducing population will remain constant over generations.

Haversian system (ha-ver'-sē-un): see *osteon*.

head: the anteriormost segment of an animal with segmentation.

heart: a muscular organ responsible for pumping blood within the circulatory system throughout the body.

heart attack: a severe reduction or blockage of blood flow through a coronary artery, depriving some of the heart muscle of its blood supply.

heartwood: older xylem that contributes to the strength of a tree trunk.

heat of fusion: the energy that must be removed from a compound to transform it from a liquid into a solid at its freezing temperature.

heat of vaporization: the energy that must be supplied to a compound to transform it from a liquid into a gas at its boiling temperature.

heliozoan (hē-lē-ō-zō'-un): an aquatic (largely freshwater) animal-like protist; some have elaborate silica-based shells.

helix (hē'-liks): a coiled, springlike secondary structure of a protein.

helper T cell: a type of T cell that helps other immune cells recognize and act against antigens.

hemocoel (hē'-mō-sēl): a blood cavity within the bodies of certain invertebrates in which blood bathes tissues directly; part of an open circulatory system.

hemodialysis (hē-mō-dī-al'-luh-sis): a procedure that simulates kidney function in individuals with damaged or ineffective kidneys; blood is diverted from the body, artificially filtered, and returned to the body.

hemoglobin (hē'mō-glō-bin): the iron-containing protein that gives red blood cells their color; binds to oxygen in the lungs and releases it to the tissues.

hemophilia: a recessive, sex-linked disease in which the blood fails to clot normally.

herbivore (erb'-i-vor): literally, "plant-eater"; an organism that feeds directly and exclusively on producers; a primary consumer.

hermaphrodite (her-maf'-ruh-dīt'): an organism that possesses both male and female sexual organs.

hermaphroditic (her-maf'-ruh-dit'-ik): possessing both male and female sexual organs. Some hermaphroditic animals can fertilize themselves; others must exchange sex cells with a mate.

heterotroph (het'-er-ō-trōf'): literally, "other-feeder"; an organism that eats other organisms; a consumer.

heterozygous (het-er-ō-zī'-gus): carrying two different alleles of a given gene; also called *hybrid*.

hindbrain: the posterior portion of the brain, containing the medulla, pons, and cerebellum.

hinge joint: a joint at which one bone is moved by muscle and the other bone remains fixed, such as in the knee, elbow, or fingers; allows movement in only two dimensions.

hippocampus (hip-ō-kam'-pus): the part of the forebrain of vertebrates that is important in emotion and especially learning.

histamine: a substance released by certain cells in response to tissue damage and invasion of the body by foreign substances; promotes the dilation of arterioles and the leakiness of capillaries and triggers some of the events of the inflammatory response.

homeobox (hō'-mē-ō-boks): a sequence of DNA coding for special, 60-amino-acid proteins, which activate or inactivate genes that control development; these sequences specify embryonic cell differentiation.

homeostasis (hōm-ē-ō-stā'sis): the maintenance of a relatively constant environment required for the optimal functioning of cells, maintained by the coordinated activity of numerous regulatory mechanisms, including the respiratory, endocrine, circulatory, and excretory systems.

hominid: a human or a prehistoric relative of humans, beginning with the Australopithecines, whose fossils date back at least 4.4 million years.

homologous structures: structures that may differ in function but that have similar anatomy, presumably because the organisms that possess them have descended from common ancestors.

homologue (hō-'mō-log): a chromosome that is similar in appearance and genetic information to another chromosome with which it pairs during meiosis; also called *homologous chromosome*.

homozygous (hō-mō-zī'-gus): carrying two copies of the same allele of a given gene; also called *true-breeding*.

hormone: a chemical that is synthesized by one group of cells, secreted, and then carried in the bloodstream to other cells, whose activity is influenced by reception of the hormone.

host: the prey organism on or in which a parasite lives; is harmed by the relationship.

human immunodeficiency virus (HIV): a pathogenic retrovirus that causes acquired immune deficiency syndrome (AIDS) by attacking and destroying the immune system's T cells.

humoral immunity: an immune response in which foreign substances are inactivated or destroyed by antibodies that circulate in the blood.

Huntington disease: an incurable genetic disorder, caused by a dominant allele, that produces progressive brain deterioration, resulting in the loss of motor coordination, flailing movements, personality disturbances, and eventual death.

hybrid: an organism that is the offspring of parents differing in at least one genetically determined characteristic; also used to refer to the offspring of parents of different species.

hybrid infertility: reduced fertility (typically, complete sterility) in the hybrid offspring of two species.

hybrid inviability: the failure of a hybrid offspring of two species to survive to maturity.

hybridoma: a cell produced by fusing an antibody-producing cell with a myeloma cell; used to produce monoclonal antibodies.

hydrogen bond: the weak attraction between a hydrogen atom that bears a partial positive charge (due to polar covalent bonding with another atom) and another atom, normally oxygen or nitrogen, that bears a partial negative charge; hydrogen bonds may form between atoms of a single molecule or of different molecules.

hydrologic cycle: the water cycle, driven by solar energy; a nutrient cycle in which the main reservoir of water is the ocean and most of the water remains in the form of water throughout the cycle (rather than being used in the synthesis of new molecules).

hydrolysis (hī-drol'-i-sis): the chemical reaction that breaks a covalent bond by means of the addition of hydrogen to the atom on one side of the original bond and a hydroxyl group to the atom on the other side; the reverse of dehydration synthesis.

hydrophilic (hī-drō-fil'-ik): pertaining to a substance that dissolves readily in water or to parts of a large molecule that form hydrogen bonds with water.

hydrophobic (hī-drō-fō'-bik): pertaining to a substance that does not dissolve in water.

hydrophobic interaction: the tendency for hydrophobic molecules to cluster together when immersed in water.

hydrostatic skeleton (hī-drō-stat'-ik): a body type that uses fluid contained in body compartments to provide support and mass against which muscles can contract.

hydrothermal vent community: a community of unusual organisms, living in the deep ocean near hydrothermal vents, that depends on the chemosynthetic activities of sulfur bacteria.

hypertension: arterial blood pressure that is chronically elevated above the normal level.

hypertonic (hī-per-ton'-ik): referring to a solution that has a higher concentration of dissolved particles (and therefore a lower concentration of free water) than has the cytoplasm of a cell.

hypha (hī'-fuh; pl., hyphae): a threadlike structure that consists of elongated cells, typically with many haploid nuclei; many hyphae make up the fungal body.

hypocotyl (hī'-pō-kot-ul): the part of the embryonic shoot located below the cotyledons but above the root.

hypothalamus (hī-pō-thal'-a-mus): a region of the brain that controls the secretory activity of the pituitary gland; synthesizes, stores, and

releases certain peptide hormones; directs autonomic nervous system responses.
hypothesis (hī-poth′-eh-sis): the second operation in the scientific method; a supposition based on previous observations that is offered as an explanation for the observed phenomenon and is used as the basis for further observations, or experiments.
hypotonic (hī-pō-ton′-ik): referring to a solution that has a lower concentration of dissolved particles (and therefore a higher concentration of free water) than has the cytoplasm of a cell.
immigration (im-uh-grā′-shun): migration of individuals into an area.
immune response: a specific response by the immune system to the invasion of the body by a particular foreign substance or microorganism, characterized by the recognition of the foreign substance by immune cells and its subsequent destruction by antibodies or by cellular attack.
immune system: cells such as macrophages, B cells, and T cells and molecules such as antibodies that work together to combat microbial invasion of the body.
imperfect fungus: a fungus of the division Deuteromycota; no species in this division has been observed to form sexual reproductive structures.
implantation: the process whereby the early embryo embeds itself within the lining of the uterus.
imprinting: the process by which an animal forms an association with another animal or object in the environment during a sensitive period of development.
inclusive fitness: the reproductive success of all organisms that bear a given allele, normally expressed in relation to the average reproductive success of all individuals in the same population; compare with *fitness*.
incomplete dominance: a pattern of inheritance in which the heterozygous phenotype is intermediate between the two homozygous phenotypes.
incomplete flower: a flower that is missing one of the four floral parts (sepals, petals, stamens, or carpels).
independent assortment: see *law of independent assortment*.
indirect development: a developmental pathway in which an offspring goes through radical changes in body form as it matures.
induction: the process by which a group of cells causes other cells to differentiate into a specific tissue type.
inductive reasoning: the process of creating a generalization based on many specific observations that support the generalization, coupled with an absence of observations that contradict it.
inflammatory response: a nonspecific, local response to injury to the body, characterized by the phagocytosis of foreign substances and tissue debris by white blood cells and by the walling off of the injury site by the clotting of fluids that escape from nearby blood vessels.
inhalation: the act of drawing air into the lungs by enlarging the chest cavity.
inheritance: the genetic transmission of characteristics from parent to offspring.
inheritance of acquired characteristics: the hypothesis that organisms' bodies change during their lifetimes by use and disuse and that these changes are inherited by their offspring.
inhibiting hormone: a hormone, secreted by the neurosecretory cells of the hypothalamus, that inhibits the release of specific hormones from the anterior pituitary.
innate (in-āt′): inborn; instinctive; determined by the genetic makeup of the individual.
inner cell mass: in human embryonic development, the cluster of cells, on one side of the blastocyst, that will develop into the embryo.
inner ear: the innermost part of the mammalian ear; composed of the bony, fluid-filled tubes of the cochlea and the vestibular apparatus.
inorganic: describing any molecule that does not contain both carbon and hydrogen.
insertion: the site of attachment of a muscle to the relatively movable bone on one side of a joint.
insertion mutation: a mutation in which one or more pairs of nucleotides are inserted into a gene.
insight learning: a complex form of learning that requires the manipulation of mental concepts to arrive at adaptive behavior.
instinctive: innate; inborn; determined by the genetic makeup of the individual.
insulin: a hormone, secreted by the pancreas, that lowers blood sugar by stimulating the conversion of glucose to glycogen in the liver.
integration: in nerve cells, the process of adding up electrical signals from sensory inputs or other nerve cells to determine the appropriate outputs.
integument (in-teg′-ū-ment): in plants, the outer layers of cells of the ovule that surrounds the embryo sac; develops into the seed coat.
intensity: the strength of stimulation or response.
interferon: a protein released by certain virus-infected cells that increases the resistance of other, uninfected, cells to viral attack.
intermediate filament: part of the cytoskeleton of eukaryotic cells that probably functions mainly for support and is composed of several types of proteins.
intermembrane compartment: the fluid-filled space between the inner and outer membranes of a mitochondrion.
internal fertilization: the union of sperm and egg inside the body of the female.
internode: the part of a stem between two nodes.
interphase: the stage of the cell cycle between cell divisions; the stage in which chromosomes are replicated and other cell functions occur, such as growth, movement, and acquisition of nutrients.
interspecific competition: competition among individuals of different species.
interstitial cell (in-ter-sti′-shul): in the vertebrate testis, a testosterone-producing cell located between the seminiferous tubules.
interstitial fluid (in-ter-sti′-shul): fluid, similar in composition to plasma (except lacking large proteins), that leaks from capillaries and acts as a medium of exchange between the body cells and the capillaries.
intertidal zone: an area of the ocean shore that is alternately covered and exposed by the tides.
intervertebral disc (in-ter-ver-tē′-brul): a pad of cartilage between two vertebrae that acts as a shock absorber.
intracellular digestion: the chemical breakdown of food within single cells.
intraspecific competition: competition among individuals of the same species.
intrauterine device (IUD): a small copper or plastic loop, squiggle, or shield that is inserted in the uterus; a contraceptive method that works by irritating the uterine lining so that it cannot receive the embryo.
intron: a segment of DNA in a eukaryotic gene that does not code for amino acids in a protein.
invertebrate (in-vert′-uh-bret): an animal that never possesses a vertebral column.
ion (ī′-on): a charged atom or molecule; an atom or molecule that has either an excess of electrons (and hence is negatively charged) or has lost electrons (and is positively charged).
ionic bond: a chemical bond formed by the electrical attraction between positively and negatively charged ions.
iris: the pigmented muscular tissue of the vertebrate eye that surrounds and controls the size of the pupil, through which light enters.
islet cell: a cluster of cells in the endocrine portion of the pancreas that produce insulin and glucagon.
isolating mechanism: a morphological, physiological, behavioral, or ecological difference that prevents members of two species from interbreeding.
isotonic (ī-sō-ton′-ik): referring to a solution that has the same concentration of dissolved particles (and therefore the same concentration of free water) as has the cytoplasm of a cell.
isotope: one of several forms of a single element, the nuclei of which contain the same number of protons but different numbers of neutrons.
J-curve: the J-shaped growth curve of an exponentially growing population in which increasing numbers of individuals join the population during each succeeding time period.
joint: a flexible region between two rigid units of an exoskeleton or endoskeleton, allowing for movement between the units.
karyotype: a preparation showing the number, sizes, and shapes of all chromosomes within a cell and, therefore, within the individual or species from which the cell was obtained.
keratin (ker′-uh-tin): a fibrous protein in hair, nails, and the epidermis of skin.
keystone species: a species whose influence on community structure is greater than its abundance would suggest.
kidney: one of a pair of organs of the excretory system that is located on either side of the spinal column and filters blood, removing wastes and regulating the composition and water content of the blood.
kinetic energy: the energy of movement; includes light, heat, mechanical movement, and electricity.
kinetochore (ki-net′-ō-kor): a protein structure that forms at the centromere regions of

chromosomes; attaches the chromosomes to the spindle.
kingdom: the second broadest taxonomic category, contained within a domain and consisting of related phyla or divisions.
kin selection: a type of natural selection that favors a certain allele because it increases the survival or reproductive success of relatives that bear the same allele.
Klinefelter syndrome: a set of characteristics typically found in individuals who have two X chromosomes and one Y chromosome; these individuals are phenotypically males but are sterile and have several femalelike traits, including broad hips and partial breast development.
Krebs cycle: a cyclic series of reactions, occurring in the matrix of mitochondria, in which the acetyl groups from the pyruvic acids produced by glycolysis are broken down to CO_2, accompanied by the formation of ATP and electron carriers; also called *citric acid cycle*.
kuru: a degenerative brain disease, first discovered in the cannibalistic Fore tribe of New Guinea, that is caused by a prion.
labium (pl., **labia**)**:** one of a pair of folds of skin of the external structures of the mammalian female reproductive system.
labor: a series of contractions of the uterus that result in birth.
lactation: the secretion of milk from the mammary glands.
lacteal (**lak-tēl′**)**:** a single lymph capillary that penetrates each villus of the small intestine.
lactose (**lak′-tōs**)**:** a disaccharide composed of glucose and galactose; found in mammalian milk.
La Niña (**la nēn′-ya**)**:** literally "girl child"; a reversal of the El Niño weather pattern.
large intestine: the final section of the digestive tract; consists of the colon and the rectum, where feces are formed and stored.
larva (**lar′-vuh**)**:** an immature form of an organism with indirect development prior to metamorphosis into its adult form; includes the caterpillars of moths and butterflies and the maggots of flies.
larynx (**lar′-inks**)**:** that portion of the air passage between the pharynx and the trachea; contains the vocal cords.
lateral bud: a cluster of meristematic cells at the node of a stem; under appropriate conditions, it grows into a branch.
lateral meristem: a meristematic tissue that forms cylinders parallel to the long axis of roots and stems; normally located between the primary xylem and primary phloem (vascular cambium) and just outside the phloem (cork cambium); also called *cambium*.
law of independent assortment: the independent inheritance of two or more distinct traits; states that the alleles for one trait may be distributed to the gametes independently of the alleles for other traits.
law of segregation: Gregor Mendel's conclusion that each gamete receives only one of each parent's pair of genes for each trait.
laws of thermodynamics: the physical laws that define the basic properties and behavior of energy.
leaf: an outgrowth of a stem, normally flattened and photosynthetic.
leaf primordium (**pri-mor′-dē-um;** pl., **primordia**)**:** a cluster of meristem cells, located at the node of a stem, that develops into a leaf.
learning: an adaptive change in behavior as a result of experience.
legume (**leg′-ūm**)**:** a member of a family of plants characterized by root swellings in which nitrogen-fixing bacteria are housed; includes soybeans, lupines, alfalfa, and clover.
lens: a clear object that bends light rays; in eyes, a flexible or movable structure used to focus light on a layer of photoreceptor cells.
leptin: a peptide hormone. One of the functions of leptin, which is released by fat cells, is to help the body monitor its fat stores and regulate weight.
leukocyte (**loo′-kō-sīt**)**:** any of the white blood cells circulating in the blood.
lichen (**lī′-ken**)**:** a symbiotic association between an alga or cyanobacterium and a fungus, resulting in a composite organism.
life cycle: the events in the life of an organism from one generation to the next.
ligament: a tough connective tissue band connecting two bones.
light-dependent reactions: the first stage of photosynthesis, in which the energy of light is captured as ATP and NADPH; occurs in thylakoids of chloroplasts.
light-harvesting complex: in photosystems, the assembly of pigment molecules (chlorophyll and accessory pigments) that absorb light energy and transfer that energy to electrons.
light-independent reactions: the second stage of photosynthesis, in which the energy obtained by the light-dependent reactions is used to fix carbon dioxide into carbohydrates; occurs in the stroma of chloroplasts.
lignin: a hard material that is embedded in the cell walls of vascular plants and provides support in terrestrial species; an early and important adaptation to terrestrial life.
limbic system: a diverse group of brain structures, mostly in the lower forebrain, that includes the thalamus, hypothalamus, amygdala, hippocampus, and parts of the cerebrum and is involved in basic emotions, drives, behaviors, and learning.
limnetic zone: a lake zone in which enough light penetrates to support photosynthesis.
linkage: the inheritance of certain genes as a group because they are parts of the same chromosome. Linked genes do not show independent assortment.
lipase (**lī′-pās**)**:** an enzyme that catalyzes the breakdown of lipids such as fats.
lipid (**li′-pid**)**:** one of a number of organic molecules containing large nonpolar regions composed solely of carbon and hydrogen, which make lipids hydrophobic and insoluble in water; includes oils, fats, waxes, phospholipids, and steroids.
littoral zone: a lake zone, near the shore, in which water is shallow and plants find abundant light, anchorage, and adequate nutrients.
liver: an organ with varied functions, including bile production, glycogen storage, and the detoxification of poisons.
lobefin: a member of the fish order Sarcopterygii, which includes coelacanths and lungfishes. Ancestors of today's lobefins gave rise to the first amphibians, and thus ultimately to all tetrapod vertebrates.
local hormones: a general term for messenger molecules produced by most cells and released into the cells' immediate vicinity in the bloodstream. Local hormones, which include prostaglandins and cytokines, influence nearby cells bearing appropriate receptors.
locus: the physical location of a gene on a chromosome.
long-day plant: a plant that will flower only if the length of daylight is greater than some species-specific duration.
long-night plant: a plant that will flower only if the duration of uninterrupted darkness is longer than some species-specific duration (sometimes called a *short-day plant*).
long-term memory: the second phase of learning; a more-or-less permanent memory formed by a structural change in the brain, brought on by repetition.
loop of Henle (**hen′-lē**)**:** a specialized portion of the tubule of the nephron in birds and mammals that creates an osmotic concentration gradient in the fluid immediately surrounding it. This gradient in turn makes possible the production of urine more osmotically concentrated than blood plasma.
lung: a paired respiratory organ consisting of inflatable chambers within the chest cavity in which gas exchange occurs.
luteinizing hormone (LH): a hormone, produced by the anterior pituitary, that stimulates testosterone production in males and the development of the follicle, ovulation, and the production of the corpus luteum in females.
lymph (limf): a pale fluid, within the lymphatic system, that is composed primarily of interstitial fluid and lymphocytes.
lymphatic system: a system consisting of lymph vessels, lymph capillaries, lymph nodes, and the thymus and spleen; helps protect the body against infection, absorbs fats, and returns excess fluid and small proteins to the blood circulatory system.
lymph node: a small structure that filters lymph; contains lymphocytes and macrophages, which inactivate foreign particles such as bacteria.
lymphocyte (**lim′-fō-sīt**)**:** a type of white blood cell important in the immune response.
lysosome (**lī′-sō-sōm**)**:** a membrane-bound organelle containing intracellular digestive enzymes.
macronutrient: a nutrient needed in relatively large quantities (often defined as making up more than 0.1% of an organism's body).
macrophage (**mak′-rō-fāj**)**:** a type of white blood cell that engulfs microbes and destroys them by phagocytosis; also presents microbial antigens to T cells, helping stimulate the immune response.
magnetotactic: able to detect and respond to Earth's magnetic field.
major histocompatibility complex (MHC): proteins, normally located on the surfaces of body cells, that identify the cell as "self"; also important in stimulating and regulating the immune response.
maltose (**mal′-tōs**)**:** a disaccharide composed of two glucose molecules.

mammal: a member of the chordate class Mammalia, which includes vertebrates with hair and mammary glands.
mammary gland (mam′-uh-rē): a milk-producing gland used by female mammals to nourish their young.
mantle (man′-tul): an extension of the body wall in certain invertebrates, such as mollusks; may secrete a shell, protect the gills, and, as in cephalopods, aid in locomotion.
marsupial (mar-soo′-pē-ul): a mammal whose young are born at an extremely immature stage and undergo further development in a pouch while they remain attached to a mammary gland; includes kangaroos, opossums, and koalas.
mass extinction: the extinction of an extraordinarily large number of species in a short period of geologic time. Mass extinctions have recurred periodically throughout the history of life.
mast cell: a cell of the immune system that synthesizes histamine and other molecules used in the body's response to trauma and that are a factor in allergic reactions.
matrix: the fluid contained within the inner membrane of a mitochondrion.
mechanical incompatibility: the inability of male and female organisms to exchange gametes, normally because their reproductive structures are incompatible.
mechanoreceptor: a receptor that responds to mechanical deformation, such as that caused by pressure, touch, or vibration.
medulla (med-ū′-luh): the part of the hindbrain of vertebrates that controls automatic activities such as breathing, swallowing, heart rate, and blood pressure.
medusa (meh-doo′-suh): a bell-shaped, typically free-swimming stage in the life cycle of many cnidarians; includes jellyfish.
megakaryocyte (meg-a-kar′-ē-ō-sīt): a large cell type that remains in the bone marrow, pinching off pieces of itself that then enter the circulation as platelets.
megaspore: a haploid cell formed by meiosis from a diploid megaspore mother cell; through mitosis and differentiation, develops into the female gametophyte.
megaspore mother cell: a diploid cell, within the ovule of a flowering plant, that undergoes meiosis to produce four haploid megaspores.
meiosis (mī-ō′-sis): a type of cell division, used by eukaryotic organisms, in which a diploid cell divides twice to produce four haploid cells.
meiotic cell division: meiosis followed by cytokinesis.
melanocyte-stimulating hormone (me-lan′-ō-sīt): a hormone, released by the anterior pituitary, that regulates the activity of skin pigments in some vertebrates.
melatonin (mel-uh-tōn′-in): a hormone, secreted by the pineal gland, that is involved in the regulation of circadian cycles.
membrane: in multicellular organism, a continuous sheet of epithelial cells that covers the body and lines body cavities; in a cell, a thin sheet of lipids and proteins that surrounds the cell or its organelles, separating them from their surroundings.
memory B cell: a type of white blood cell that is produced as a result of the binding of an antibody on a B cell to an antigen on an invading microorganism. Memory B cells persist in the bloodstream and provide future immunity to invaders bearing that antigen.
memory T cell: a type of white blood cell that is produced as a result of the binding of a receptor on a T cell to an antigen on an invading microorganism. Memory T cells persist in the bloodstream and provide future immunity to invaders bearing that antigen.
meninges (men-in′-jēz): three layers of connective tissue that surround the brain and spinal cord.
menstrual cycle: in human females, a complex 28-day cycle during which hormonal interactions among the hypothalamus, pituitary gland, and ovary coordinate ovulation and the preparation of the uterus to receive and nourish the fertilized egg. If pregnancy does not occur, the uterine lining is shed during menstruation.
menstruation: in human females, the monthly discharge of uterine tissue and blood from the uterus.
meristem cell (mer′-i-stem): an undifferentiated cell that remains capable of cell division throughout the life of a plant.
mesoderm (mēz′-ō-derm): the middle embryonic tissue layer, lying between the endoderm and ectoderm, and normally the last to develop; gives rise to structures such as muscle and skeleton.
mesoglea (mez-ō-glē′-uh): a middle, jelly-like layer within the body wall of cnidarians.
mesophyll (mez′-ō-fil): loosely packed parenchyma cells beneath the epidermis of a leaf.
messenger RNA (mRNA): a strand of RNA, complementary to the DNA of a gene, that conveys the genetic information in DNA to the ribosomes to be used during protein synthesis; sequences of three bases (codons) in mRNA specify particular amino acids to be incorporated into a protein.
metabolic pathway: a sequence of chemical reactions within a cell, in which the products of one reaction are the reactants for the next reaction.
metabolism: the sum of all chemical reactions that occur within a single cell or within all the cells of a multicellular organism.
metamorphosis (met-a-mor′-fō-sis): in animals with indirect development, a radical change in body form from larva to sexually mature adult, as seen in amphibians (tadpole to frog) and insects (caterpillar to butterfly).
metaphase (met′-a-fāz): the stage of mitosis in which the chromosomes, attached to spindle fibers at kinetochores, are lined up along the equator of the cell.
methanogen (me-than′-ō-jen): a type of anaerobic archaean capable of converting carbon dioxide to methane.
microbe: a microorganism.
microevolution: change over successive generations in the composition of a population's gene pool.
microfilament: part of the cytoskeleton of eukaryotic cells that is composed of the proteins actin and (in some cases) myosin; functions in the movement of cell organelles and in locomotion by extension of the plasma membrane.
micronutrient: a nutrient needed only in small quantities (often defined as making up less than 0.01% of an organism's body).
microsphere: a small, hollow sphere formed from proteins or proteins complexed with other compounds.
microspore: a haploid cell formed by meiosis from a microspore mother cell; through mitosis and differentiation, develops into the male gametophyte.
microspore mother cell: a diploid cell contained within an anther of a flowering plant, which undergoes meiosis to produce four haploid microspores.
microtubule: a hollow, cylindrical strand, found in eukaryotic cells, that is composed of the protein tubulin; part of the cytoskeleton used in the movement of organelles, cell growth, and the construction of cilia and flagella.
microvillus (mī-krō-vi′-lus; pl., microvilli): a microscopic projection of the plasma membrane of each villus; increases the surface area of the villus.
midbrain: during development, the central portion of the brain; contains an important relay center, the reticular formation.
middle ear: the part of the mammalian ear composed of the tympanic membrane, the Eustachian tube, and three bones (hammer, anvil, and stirrup) that transmit vibrations from the auditory canal to the oval window.
middle lamella: a thin layer of sticky polysaccharides, such as pectin, and other carbohydrates that separates and holds together the primary cell walls of adjacent plant cells.
mimicry (mim′-ik-rē): the situation in which a species has evolved to resemble something else—typically another type of organism.
mineral: an inorganic substance, especially one in rocks or soil.
mitochondrion (mī-tō-kon′-drē-un): an organelle, bounded by two membranes, that is the site of the reactions of aerobic metabolism.
mitosis (mī-tō′-sis): a type of nuclear division, used by eukaryotic cells, in which one copy of each chromosome (already duplicated during interphase before mitosis) moves into each of two daughter nuclei; the daughter nuclei are therefore genetically identical to each other.
mitotic cell division: mitosis followed by cytokinesis.
molecule (mol′-e-kūl): a particle composed of one or more atoms held together by chemical bonds; the smallest particle of a compound that displays all the properties of that compound.
molt: to shed an external body covering, such as an exoskeleton, skin, feathers, or fur.
monoclonal antibody: an antibody produced in the lab by the cloning of hybridoma cells; each clone of cells produces a single antibody.
monocot: short for monocotyledon; a type of flowering plant characterized by embryos with one seed leaf, or cotyledon.
monoecious (mon-ē′-shus): pertaining to organisms in which male and female gametes are produced in the same individual.
monomer (mo′-nō-mer): a small organic molecule, several of which may be bonded together to form a chain called a *polymer*.
monophyletic: referring to a group of species that contains all the known descendents of an ancestral species.

monosaccharide (mo-nō-sak′-uh-rīd): the basic molecular unit of all carbohydrates, normally composed of a chain of carbon atoms bonded to hydrogen and hydroxyl groups.
monotreme: a mammal that lays eggs; for example, the platypus.
morula (mor′-ū-luh): in animals, an embryonic stage during cleavage, when the embryo consists of a solid ball of cells.
motor neuron: a neuron that receives instructions from the association neurons and activates effector organs, such as muscles or glands.
motor unit: a single motor neuron and all the muscle fibers on which it forms synapses.
mouth: the opening of a tubular digestive system into which food is first introduced.
mucous membrane: the lining of the inside of the respiratory and digestive tracts.
multicellular: many-celled; most members of the kingdoms Fungi, Plantae, and Animalia are multicellular, with intimate cooperation among cells.
multiple alleles: as many as dozens of alleles produced for every gene as a result of different mutations.
muscle fiber: an individual muscle cell.
mutation: a change in the base sequence of DNA in a gene; normally refers to a genetic change significant enough to alter the appearance or function of the organism.
mutualism (mū′-choo-ul-iz-um): a symbiotic relationship in which both participating species benefit.
mycelium (mī-sēl′-ē-um): the body of a fungus, consisting of a mass of hyphae.
mycorrhiza (mī-kō-rī′zuh; pl., **mycorrhizae):** a symbiotic relationship between a fungus and the roots of a land plant that facilitates mineral extraction and absorption.
myelin (mī′-uh-lin): a wrapping of insulating membranes of specialized nonneural cells around the axon of a vertebrate nerve cell; increases the speed of conduction of action potentials.
myofibril (mī-ō-fī′-bril): a cylindrical subunit of a muscle cell, consisting of a series of sarcomeres; surrounded by sarcoplasmic reticulum.
myometrium (mī-ō-mē′-trē-um): the muscular outer layer of the uterus.
myosin (mī′-ō-sin): one of the major proteins of muscle, the interaction of which with the protein actin produces muscle contraction; found in the thick filaments of the muscle fiber; see also *actin*.
natural causality: the scientific principle that natural events occur as a result of preceding natural causes.
natural killer cell: a type of white blood cell that destroys some virus-infected cells and cancerous cells on contact; part of the immune system's nonspecific internal defense against disease.
natural selection: the unequal survival and reproduction of organisms due to environmental forces, resulting in the preservation of favorable adaptations. Usually, natural selection refers specifically to differential survival and reproduction on the basis of genetic differences among individuals.
near-shore zone: the region of coastal water that is relatively shallow but constantly submerged; includes bays and coastal wetlands and can support large plants or seaweeds.
nearsighted: the inability to focus on distant objects caused by an eyeball that is slightly too long.
negative feedback: a situation in which a change initiates a series of events that tend to counteract the change and restore the original state. Negative feedback in physiological systems maintains homeostasis.
nephridium (nef-rid′-ē-um): an excretory organ found in earthworms, mollusks, and certain other invertebrates; somewhat resembles a single vertebrate nephron.
nephron (nef′-ron): the functional unit of the kidney; where blood is filtered and urine formed.
nephrostome (nef′-rō-stōm): the funnel-shaped opening of the nephridium of some invertebrates such as earthworms; coelomic fluid is drawn into the nephrostome for filtration.
nerve: a bundle of axons of nerve cells, bound together in a sheath.
nerve cord: a paired neural structure in most animals that conducts nervous signals to and from the ganglia; in chordates, a nervous structure lying along the dorsal side of the body; also called spinal cord.
nerve net: a simple form of nervous system, consisting of a network of neurons that extend throughout the tissues of an organism such as a cnidarian.
nerve tissue: the tissue that make up the brain, spinal cord, and nerves; consists of neurons and glial cells.
net primary productivity: the energy stored in the autotrophs of an ecosystem over a given time period.
neural tube: a structure, derived from ectoderm during early embryonic development, that later becomes the brain and spinal cord.
neuromuscular junction: the synapse formed between a motor neuron and a muscle fiber.
neuron (noor′-on): a single nerve cell.
neuropeptide: a small protein molecule with neurotransmitter-like actions.
neurosecretory cell: a specialized nerve cell that synthesizes and releases hormones.
neurotransmitter: a chemical that is released by a nerve cell close to a second nerve cell, a muscle, or a gland cell and that influences the activity of the second cell.
neutral mutation: a mutation that has little or no effect on the function of the encoded protein.
neutralization: the process of covering up or inactivating a toxic substance with antibody.
neutron: a subatomic particle that is found in the nuclei of atoms, bears no charge, and has a mass approximately equal to that of a proton.
neutrophil (nū′-trō-fil): a type of white blood cell that engulfs invading microbes and contributes to the nonspecific defenses of the body against disease.
nitrogen fixation: the process that combines atmospheric nitrogen with hydrogen to form ammonium (NH_4^+).
nitrogen-fixing bacterium: a bacterium that possess the ability to remove nitrogen (N_2) from the atmosphere and combine it with hydrogen to produce ammonium (NH_4^+).
node: in plants, a region of a stem at which leaves and lateral buds are located; in vertebrates, an interruption of the myelin on a myelinated axon, exposing naked membrane at which action potentials are generated.
nodule: a swelling on the root of a legume or other plant that consists of cortex cells inhabited by nitrogen-fixing bacteria.
nondisjunction: an error in meiosis in which chromosomes fail to segregate properly into the daughter cells.
nonpolar covalent bond: a covalent bond with equal sharing of electrons.
norepinephrine (nor-ep-i-nef-rin′): a neurotransmitter, released by neurons of the parasympathetic nervous system, that prepares the body to respond to stressful situations; also called *noradrenaline*.
northern coniferous forest: a biome with long, cold winters and only a few months of warm weather; populated almost entirely by evergreen coniferous trees; also called *taiga*.
notochord (nōt′-ō-kord): a stiff but somewhat flexible, supportive rod found in all members of the phylum Chordata at some stage of development.
nuclear envelope: the double-membrane system surrounding the nucleus of eukaryotic cells; the outer membrane is typically continuous with the endoplasmic reticulum.
nucleic acid (noo-klā′-ik): an organic molecule composed of nucleotide subunits; the two common types of nucleic acids are ribonucleic acid (RNA) and deoxyribonucleic acid (DNA).
nucleoid (noo-klē-oid): the location of the genetic material in prokaryotic cells; not membrane-enclosed.
nucleolus (noo-klē′-ō-lus): the region of the eukaryotic nucleus that is engaged in ribosome synthesis; consists of the genes encoding ribosomal RNA, newly synthesized ribosomal RNA, and ribosomal proteins.
nucleotide: a subunit of which nucleic acids are composed; a phosphate group bonded to a sugar (deoxyribose in DNA), which is in turn bonded to a nitrogen-containing base (adenine, guanine, cytosine, or thymine in DNA). Nucleotides are linked together, forming a strand of nucleic acid, as follows: Bonds between the phosphate of one nucleotide link to the sugar of the next nucleotide.
nucleotide substitution: a mutation that replaces one nucleotide in a DNA molecule with another; for example, a change from an adenine to a guanine.
nucleus (atomic): the central region of an atom, consisting of protons and neutrons.
nucleus (cellular): the membrane-bound organelle of eukaryotic cells that contains the cell's genetic material.
nutrient: a substance acquired from the environment and needed for the survival, growth, and development of an organism.
nutrient cycle: a description of the pathways of a specific nutrient (such as carbon, nitrogen, phosphorus, or water) through the living and nonliving portions of an ecosystem.
nutrition: the process of acquiring nutrients from the environment and, if necessary, processing them into a form that can be used by the body.

observation: the first operation in the scientific method; the noting of a specific phenomenon, leading to the formulation of a hypothesis.
oil: a lipid composed of three fatty acids, some of which are unsaturated, covalently bonded to a molecule of glycerol; liquid at room temperature.
olfaction (ōl-fak′-shun): a chemical sense, the sense of smell; in terrestrial vertebrates, the result of the detection of airborne molecules.
oligotrophic lake: a lake that is very low in nutrients and hence clear with extensive light penetration.
ommatidium (ōm-ma-tid′-ē-um): an individual light-sensitive subunit of a compound eye; consists of a lens and several receptor cells.
omnivore: an organism that consumes both plants and other animals.
one gene, one protein rule: the premise that each gene encodes the information for the synthesis of a single protein.
oogenesis: the process by which egg cells are formed.
oogonium (ō-ō-gō′-nē-um; pl., oogonia): in female animals, a diploid cell that gives rise to a primary oocyte.
open circulatory system: a type of circulatory system found in some invertebrates, such as arthropods and mollusks, that includes an open space (the hemocoel) in which blood directly bathes body tissues.
operant conditioning: a laboratory training procedure in which an animal learns to perform a response (such as pressing a lever) through reward or punishment.
operculum: an external flap, supported by bone, that covers and protects the gills of most fish.
opioid (ōp′-ē-oid): one of a group of peptide neuromodulators in the vertebrate brain that mimic some of the actions of opiates (such as opium) and also seem to influence many other processes, including emotion and appetite.
optic nerve: the nerve leading from the eye to the brain, carrying visual information.
order: the taxonomic category contained within a class and consisting of related families.
organ: a structure (such as the liver, kidney, or skin) composed of two or more distinct tissue types that function together.
organelle (or-guh-nel′): a structure, found in the cytoplasm of eukaryotic cells, that performs a specific function; sometimes refers specifically to membrane-bound structures, such as the nucleus or endoplasmic reticulum.
organic/organic molecule: describing a molecule that contains both carbon and hydrogen.
organism (or′-guh-niz-um): an individual living thing.
organogenesis (or-gan-ō-jen′-uh-sis): the process by which the layers of the gastrula (endoderm, ectoderm, mesoderm) rearrange into organs.
organ system: two or more organs that work together to perform a specific function; for example, the digestive system.
origin: the site of attachment of a muscle to the relatively stationary bone on one side of a joint.
osmosis (oz-mō′-sis): the diffusion of water across a differentially permeable membrane, normally down a concentration gradient of free water molecules. Water moves into the solution that has a lower concentration of free water from a solution with the higher concentration of free water.
osmotic pressure: the pressure required to counterbalance the tendency of water to move from a solution with a higher concentration of free water molecules into a solution with a lower concentration of free water molecules.
osteoblast (os′-tē-ō-blast): a cell type that produces bone.
osteoclast (os′-tē-ō-klast): a cell type that dissolves bone.
osteocyte (os′-tē-ō-sīt): a mature bone cell.
osteon: a unit of hard bone consisting of concentric layers of bone matrix, with embedded osteocytes, surrounding a small central canal that contains a capillary.
osteoporosis (os′-tē-ō-por-ō′-sis): a condition in which bones become porous, weak, and easily fractured; most common in elderly women.
outer ear: the outermost part of the mammalian ear, including the external ear and auditory canal leading to the tympanic membrane.
oval window: the membrane-covered entrance to the inner ear.
ovary: in animals, the gonad of females; in flowering plants, a structure at the base of the carpel that contains one or more ovules and develops into the fruit.
oviduct: in mammals, the tube leading from the ovary to the uterus.
ovulation: the release of a secondary oocyte, ready to be fertilized, from the ovary.
ovule: a structure within the ovary of a flower, inside which the female gametophyte develops; after fertilization, develops into the seed.
oxytocin (oks-ē-tō′-sin): a hormone, released by the posterior pituitary, that stimulates the contraction of uterine and mammary gland muscles.
ozone layer: the ozone-enriched layer of the upper atmosphere that filters out some of the sun's ultraviolet radiation.
pacemaker: a cluster of specialized muscle cells in the upper right atrium of the heart that produce spontaneous electrical signals at a regular rate; the sinoatrial node.
pain receptor: a receptor that has extensive areas of membranes studded with special receptor proteins that respond to light or to a chemical.
palisade cell: a columnar mesophyll cell, containing chloroplasts, just beneath the upper epidermis of a leaf.
pancreas (pan′-krē-us): a combined exocrine and endocrine gland located in the abdominal cavity next to the stomach. The endocrine portion secretes the hormones insulin and glucagon, which regulate glucose concentrations in the blood. The exocrine portion secretes enzymes for fat, carbohydrate, and protein digestion into the small intestine and neutralizes the acidic chyme.
pancreatic juice: a mixture of water, sodium bicarbonate, and enzymes released by the pancreas into the small intestine.
parasite (par′-uh-sīt): an organism that lives in or on a larger prey organism, called a *host*, weakening it.
parasitism: a symbiotic relationship in which one organism (commonly smaller and more numerous than its host) benefits by feeding on the other, which is normally harmed but not immediately killed.
parasympathetic division: the division of the autonomic nervous system that produces largely involuntary responses related to the maintenance of normal body functions, such as digestion.
parathormone: a hormone, secreted by the parathyroid gland, that stimulates the release of calcium from bones.
parathyroid gland: one of four small endocrine glands, embedded in the surface of the thyroid gland, that produces parathormone, which (with calcitonin from the thyroid gland) regulates calcium ion concentration in the blood.
parathyroid hormone: a hormone released by the parathyroid gland that works in conjunction with calcitonin to control calcium levels in the blood and other body fluids.
parenchyma (par-en′-ki-muh): a plant cell type that is alive at maturity, normally with thin primary cell walls, that carries out most of the metabolism of a plant. Most dividing meristem cells in a plant are parenchyma.
parthenogenesis (par-the-nō-jen′uh-sis): a specialization of sexual reproduction, in which a haploid egg undergoes development without fertilization.
passive transport: the movement of materials across a membrane down a gradient of concentration, pressure, or electrical charge without using cellular energy.
pathogenic (path′-ō-jen-ik): capable of producing disease; refers to an organism with such a capability (a pathogen).
pedigree: a diagram showing genetic relationships among a set of individuals, normally with respect to a specific genetic trait.
pelagic (puh-la′-jik): free-swimming or floating.
penis: an external structure of the male reproductive and urinary systems; serves to deposit sperm into the female reproductive system and delivers urine to the exterior.
peptide (pep′-tīd): a chain composed of two or more amino acids linked together by peptide bonds.
peptide bond: the covalent bond between the amino group's nitrogen of one amino acid and the carboxyl group's carbon of a second amino acid, joining the two amino acids together in a peptide or protein.
peptide hormone: a hormone consisting of a chain of amino acids; includes small proteins that function as hormones.
peptidoglycan (pep-tid-ō-glī′-kan): a component of prokaryotic cell walls that consists of chains of sugars cross-linked by short chains of amino acids called *peptides*.
pericycle (per′-i-sī-kul): the outermost layer of cells of the vascular cylinder of a root.
periderm: the outer cell layers of roots and a stem that have undergone secondary growth, consisting primarily of cork cambium and cork cells.
peripheral nerve: a nerve that links the brain and spinal cord to the rest of the body.
peripheral nervous system: in vertebrates, the part of the nervous system that connects

the central nervous system to the rest of the body.

peristalsis: rhythmic coordinated contractions of the smooth muscles of the digestive tract that move substances through the digestive tract.

permafrost: a permanently frozen layer of soil in the arctic tundra that cannot support the growth of trees.

petal: part of a flower, typically brightly colored and fragrant, that attracts potential animal pollinators.

petiole (pet′-ē-ōl): the stalk that connects the blade of a leaf to the stem.

phagocytic cell (fa-gō-sit′-ik): a type of immune system cell that destroys invading microbes by using phagocytosis to engulf and digest the microbes.

phagocytosis (fa-gō-sī-tō′-sis): a type of endocytosis in which extensions of a plasma membrane engulf extracellular particles and transport them into the interior of the cell.

pharyngeal gill slit (far-in′-jē-ul): an opening, located just posterior to the mouth, that connects the digestive tube to the outside environment; present (as some stage of life) in all chordates.

pharynx (far′-inks): in vertebrates, a chamber that is located at the back of the mouth and is shared by the digestive and respiratory systems; in some invertebrates, the portion of the digestive tube just posterior to the mouth.

phenotype (fēn′-ō-tīp): the physical characteristics of an organism; can be defined as outward appearance (such as flower color), as behavior, or in molecular terms (such as glycoproteins on red blood cells).

pheromone (fer′-uh-mōn): a chemical produced by an organism that alters the behavior or physiological state of another member of the same species.

phloem (flō′-um): a conducting tissue of vascular plants that transports a concentrated sugar solution up and down the plant.

phospholipid (fos-fō-li′-pid): a lipid consisting of glycerol bonded to two fatty acids and one phosphate group, which bears another group of atoms, typically charged and containing nitrogen. A double layer of phospholipids is a component of all cellular membranes.

phospholipid bilayer: a double layer of phospholipids that forms the basis of all cellular membranes. The phospholipid heads, which are hydrophilic, face the water of extracellular fluid or the cytoplasm; the tails, which are hydrophobic, are buried in the middle of the bilayer.

photic zone: the region of the ocean where light is strong enough to support photosynthesis.

photon (fō′-ton): the smallest unit of light energy.

photopigment (fō′-tō-pig-ment): a chemical substance in photoreceptor cells that, when struck by light, changes in molecular conformation.

photoreceptor: a receptor cell that responds to light; in vertebrates, rods and cones.

photorespiration: a series of reactions in plants in which O_2 replaces CO_2 during the C_3 cycle, preventing carbon fixation; this wasteful process dominates when C_3 plants are forced to close their stomata to prevent water loss.

photosynthesis: the complete series of chemical reactions in which the energy of light is used to synthesize high-energy organic molecules, normally carbohydrates, from low-energy inorganic molecules, normally carbon dioxide and water.

photosystem: in thylakoid membranes, a light-harvesting complex and its associated electron transport system.

phototactic: capable of detecting and responding to light.

phototropism: growth with respect to the direction of light.

pH scale: a scale, with values from 0 to 14, used for measuring the relative acidity of a solution; at pH 7 a solution is neutral, pH 0 to 7 is acidic, and pH 7 to 14 is basic; each unit on the scale represents a tenfold change in H^+ concentration.

phycocyanin (fī-kō-sī′-uh-nin): a blue or purple pigment that is located in the membranes of chloroplasts and is used as an accessory light-gathering molecule in thylakoid photosystems.

phylogeny (fī-lah′-jen-ē): the evolutionary history of a group of species.

phylum (fī-lum): the taxonomic category of animals and animal-like protists that is contained within a kingdom and consists of related classes.

phytochrome (fī′-tō-krōm): a light-sensitive plant pigment that mediates many plant responses to light, including flowering, stem elongation, and seed germination.

phytoplankton (fī′-tō-plank-ten): photosynthetic protists that are abundant in marine and freshwater environments.

pilus (pil′-us; pl., **pili):** a hairlike projection that is made of protein, located on the surface of certain bacteria, and is typically used to attach a bacterium to another cell.

pineal gland (pi-nē′-al): a small gland within the brain that secretes melatonin; controls the seasonal reproductive cycles of some mammals.

pinocytosis (pi-nō-sī-tō′-sis): the nonselective movement of extracellular fluid, enclosed within a vesicle formed from the plasma membrane, into a cell.

pioneer: an organism that is among the first to colonize an unoccupied habitat in the first stages of succession.

pit: an area in the cell walls between two plant cells in which secondary walls did not form, such that the two cells are separated only by a relatively thin and porous primary cell wall.

pith: cells forming the center of a root or stem.

pituitary gland: an endocrine gland, located at the base of the brain, that produces several hormones, many of which influence the activity of other glands.

placenta (pluh-sen′-tuh): in mammals, a structure formed by a complex interweaving of the uterine lining and the embryonic membranes, especially the chorion; functions in gas, nutrient, and waste exchange between embryonic and maternal circulatory systems and secretes hormones.

placental (pluh-sen′-tul): referring to a mammal, possessing a placenta (that is, species that are not marsupials or monotremes).

plankton: microscopic organisms that live in marine or freshwater environments; includes phytoplankton and zooplankton.

plant hormone: the plant-regulating chemicals auxin, gibberellins, cytokinins, ethylene, and abscisic acid; somewhat resemble animal hormones in that they are chemicals produced by cells in one location that influence the growth or metabolic activity of other cells, typically some distance away in the plant body.

plaque (plak): a deposit of cholesterol and other fatty substances within the wall of an artery.

plasma: the fluid, noncellular portion of the blood.

plasma cell: an antibody-secreting descendant of a B cell.

plasma membrane: the outer membrane of a cell, composed of a bilayer of phospholipids in which proteins are embedded.

plasmid (plaz′-mid): a small, circular piece of DNA located in the cytoplasm of many bacteria; normally does not carry genes required for the normal functioning of the bacterium but may carry genes that assist bacterial survival in certain environments, such as a gene for antibiotic resistance.

plasmodesma (plaz-mō-dez′-muh; pl., **plasmodesmata):** a cell-to-cell junction in plants that connects the cytoplasm of adjacent cells.

plasmodial slime mold: see *acellular slime mold.*

plasmodium (plaz-mō′-dē-um): a sluglike mass of cytoplasm containing thousands of nuclei that are not confined within individual cells.

plastid (plas′-tid): in plant cells, an organelle bounded by two membranes that may be involved in photosynthesis (chloroplasts), pigment storage, or food storage.

platelet (plāt′-let): a cell fragment that is formed from megakaryocytes in bone marrow and lacks a nucleus; circulates in the blood and plays a role in blood clotting.

plate tectonics: the theory that Earth's crust is divided into irregular plates that are converging, diverging, or slipping by one another; these motions cause continental drift, the movement of continents over Earth's surface.

pleated sheet: a form of secondary structure exhibited by certain proteins, such as silk, in which many protein chains lie side-by-side, with hydrogen bonds holding adjacent chains together.

pleiotropy (plē′-ō-trō-pē): a situation in which a single gene influences more than one phenotypic characteristic.

pleural membrane: a membrane that lines the chest cavity and surrounds the lungs.

point mutation: a mutation in which a single base pair in DNA has been changed.

polar body: in oogenesis, a small cell, containing a nucleus but virtually no cytoplasm, produced by the first meiotic division of the primary oocyte.

polar covalent bond: a covalent bond with unequal sharing of electrons, such that one atom is relatively negative and the other is relatively positive.

polar nucleus: in flowering plants, one of two nuclei in the primary endosperm cell of the female gametophyte; formed by the mitotic division of a megaspore.

pollen/pollen grain: the male gametophyte of a seed plant.
pollination: in flowering plants, when pollen grains land on the stigma of a flower of the same species; in conifers, when pollen grains land within the pollen chamber of a female cone of the same species.
polygenic inheritance: a pattern of inheritance in which the interactions of two or more functionally similar genes determine phenotype.
polymer (pah′-li-mer): a molecule composed of three or more (perhaps thousands) smaller subunits called *monomers*, which may be identical (for example, the glucose monomers of starch) or different (for example, the amino acids of a protein).
polymerase chain reaction (PCR): a method of producing virtually unlimited numbers of copies of a specific piece of DNA, starting with as little as one copy of the desired DNA.
polyp (pah′-lip): the sedentary, vase-shaped stage in the life cycle of many cnidarians; includes hydra and sea anemones.
polypeptide: a short polymer of amino acids; often used as a synonym for protein.
polyploidy (pahl′-ē-ploid-ē): having more than two homologous chromosomes of each type.
polysaccharide (pahl-ē-sak′-uh-rīd): a large carbohydrate molecule composed of branched or unbranched chains of repeating monosaccharide subunits, normally glucose or modified glucose molecules; includes starches, cellulose, and glycogen.
pons: a portion of the hindbrain, just above the medulla, that contains neurons that influence sleep and the rate and pattern of breathing.
population: all the members of a particular species within an ecosystem, found in the same time and place and actually or potentially interbreeding.
population bottleneck: a form of genetic drift in which a population becomes extremely small; may lead to differences in allele frequencies as compared with other populations of the species and to a loss in genetic variability.
population cycle: out-of-phase cyclical patterns of predator and prey populations.
population genetics: the study of the frequency, distribution, and inheritance of alleles in a population.
positive feedback: a situation in which a change initiates events that tend to amplify the original change.
post-anal tail: a tail that extends beyond the anus; exhibited by all chordates at some stage of development.
posterior: the tail, hindmost, or rear end of an animal.
posterior pituitary: a lobe of the pituitary gland that is an outgrowth of the hypothalamus and that releases antidiuretic hormone and oxytocin.
postmating isolating mechanism: any structure, physiological function, or developmental abnormality that prevents organisms of two different populations, once mating has occurred, from producing vigorous, fertile offspring.
postsynaptic neuron: at a synapse, the nerve cell that changes its electrical potential in response to a chemical (the neurotransmitter) released by another (presynaptic) cell.
postsynaptic potential (PSP): an electrical signal produced in a postsynaptic cell by transmission across the synapse; it may be excitatory (EPSP), making the cell more likely to produce an action potential, or inhibitory (IPSP), tending to inhibit an action potential.
potential energy: "stored" energy, normally chemical energy or energy of position within a gravitational field.
prairie: a biome, located in the centers of continents, that supports grasses; also called *grassland*.
preadaptation: a feature evolved under one set of environmental conditions that, purely by chance, helps an organism adapt to new environmental conditions.
prebiotic evolution: evolution before life existed; especially, the abiotic synthesis of organic molecules.
precapillary sphincter (sfink′-ter): a ring of smooth muscle between an arteriole and a capillary that regulates the flow of blood into the capillary bed.
predation (pre-dā′-shun): the act of killing and eating another living organism.
predator: an organism that kills and eats other organisms.
premating isolating mechanism: any structure, physiological function, or behavior that prevents organisms of two different populations from exchanging gametes.
pressure-flow theory: a model for the transport of sugars in phloem, by which the movement of sugars into a phloem sieve tube causes water to enter the tube by osmosis, while the movement of sugars out of another part of the same sieve tube causes water to leave by osmosis; the resulting pressure gradient causes the bulk movement of water and dissolved sugars from the end of the tube into which sugar is transported toward the end of the tube from which sugar is removed.
presynaptic neuron: a nerve cell that releases a chemical (the neurotransmitter) at a synapse, causing changes in the electrical activity of another (postsynaptic) cell.
prey: organisms that are killed and eaten by another organism.
primary cell wall: cellulose and other carbohydrates secreted by a young plant cell between the middle lamella and the plasma membrane.
primary consumer: an organism that feeds on producers; an herbivore.
primary endosperm cell: the central cell of the female gametophyte of a flowering plant, containing the polar nuclei (normally two); after fertilization, undergoes repeated mitotic divisions to produce the endosperm of the seed.
primary growth: growth in length and development of the initial structures of plant roots and shoots, due to the cell division of apical meristems and differentiation of the daughter cells.
primary oocyte (ō′-ō-sīt): a diploid cell, derived from the oogonium by growth and differentiation, that undergoes meiosis, producing the egg.
primary phloem: phloem in young stems produced from an apical meristem.
primary root: the first root that develops from a seed.
primary spermatocyte (sper-ma′-tō-sīt): a diploid cell, derived from the spermatogonium by growth and differentiation, that undergoes meiosis, producing four sperm.
primary structure: the amino acid sequence of a protein.
primary succession: succession that occurs in an environment, such as bare rock, in which no trace of a previous community was present.
primary xylem: xylem in young stems produced from an apical meristem.
primate: a mammal characterized by the presence of an opposable thumb, forward-facing eyes, and a well-developed cerebral cortex; includes lemurs, monkeys, apes, and humans.
primitive streak: in reptiles, birds, and mammals, the region of the ectoderm of the two-layered embryonic disc through which cells migrate, forming mesoderm.
prion (prē′-on): a protein that, in mutated form, acts as an infectious agent that causes certain neurodegenerative diseases, including kuru and scrapie.
producer: a photosynthetic organism; an autotroph.
product: an atom or molecule that is formed from reactants in a chemical reaction.
profundal zone: a lake zone in which light is insufficient to support photosynthesis.
progesterone (prō-ge′-ster-ōn): a hormone, produced by the corpus luteum, that promotes the development of the uterine lining in females.
prokaryote (prō-kar′-ē-ōt): an organism whose cells are prokaryotic; bacteria and archaea are prokaryotes.
prokaryotic (prō-kar-ē-ot′-ik): referring to cells of the domains Bacteria or Archaea. Prokaryotic cells have genetic material that is not enclosed in a membrane-bound nucleus; they lack other membrane-bound organelles.
prolactin: a hormone, released by the anterior pituitary, that stimulates milk production in human females.
promoter: a specific sequence of DNA to which RNA polymerase binds, initiating gene transcription.
prophase (prō′-fāz): the first stage of mitosis, in which the chromosomes first become visible in the light microscope as thickened, condensed threads and the spindle begins to form; as the spindle is completed, the nuclear envelope breaks apart, and the spindle fibers invade the nuclear region and attach to the kinetochores of the chromosomes. Also, the first stage of meiosis: In meiosis I, the homologous chromosomes pair up and exchange parts at chiasmata; in meiosis II, the spindle re-forms and chromosomes attach to the microtubules.
prostaglandin (pro-stuh-glan′-din): a family of modified fatty acid hormones manufactured by many cells of the body.
prostate gland (pros′-tāt): a gland that produces part of the fluid component of semen; prostatic fluid is basic and contains a chemical that activates sperm movement.
protease (prō′-tē-ās): an enzyme that digests proteins.

protein: polymer of amino acids joined by peptide bonds.
protist: a eukaryotic organism that is not a plant, animal, or fungus. The term encompasses a diverse array of organisms and does not represent a monophyletic group.
protocell: the hypothetical evolutionary precursor of living cells, consisting of a mixture of organic molecules within a membrane.
proton: a subatomic particle that is found in the nuclei of atoms, bears a unit of positive charge, and has a relatively large mass, roughly equal to the mass of the neutron.
protonephridium (prō-tō-nef-rid′-ē-um; pl., **protonephridia):** an excretory system consisting of tubules that have external opening but lack internal openings; for example, the flame-cell system of flatworms.
protostome (prō′-tō-stōm): an animal with a mode of embryonic development in which the coelom is derived from splits in the mesoderm; characteristic of arthropods, annelids, and mollusks.
protozoan (prō-tuh-zō′-an; pl., **protozoa):** a nonphotosynthetic or animal-like protist.
proximal tubule: in nephrons of the mammalian kidney, the portion of the renal tubule just after the Bowman's capsule; receives filtrate from the capsule and is the site where selective secretion and reabsorption between the filtrate and the blood begins.
pseudocoelom (soo′-dō-sēl′-ōm): "false coelom"; a body cavity that has a different embryological origin than a coelom but serves a similar function; found in roundworms.
pseudoplasmodium (soo′-dō-plaz-mō′-dē-um): an aggregation of individual amoeboid cells that form a sluglike mass.
pseudopod (sood′-ō-pod): an extension of the plasma membrane by which certain cells, such as amoebae, locomote and engulf prey.
puberty: a stage of development (in humans, usually beginning in the early teenage years) characterized by rapid growth and the appearance of secondary sexual characteristics in response to increased secretion of testosterone in males and estrogen in females.
Punnett square method: an intuitive way to predict the genotypes and phenotypes of offspring in specific crosses.
pupa: a developmental stage in some insect species in which the organism stops moving and feeding and may be encased in a cocoon; occurs between the larval and the adult phases.
pupil: the adjustable opening in the center of the iris, through which light enters the eye.
pyloric sphincter (pī-lor′-ik sfink′-ter): a circular muscle, located at the base of the stomach, that regulates the passage of chyme into the small intestine.
pyruvate: a three-carbon molecule that is formed by glycolysis and then used in fermentation or cellular respiration.
quaternary structure (kwat′-er-nuh-rē): the complex three-dimensional structure of a protein composed of more than one peptide chain.
queen substance: a chemical, produced by a queen bee, that can act as both a primer and a pheromone.
radial symmetry: a body plan in which any plane along a central axis will divide the body into approximately mirror-image halves. Cnidarians and many adult echinoderms have radial symmetry.
radioactive: pertaining to an atom with an unstable nucleus that spontaneously disintegrates, with the emission of radiation.
radiolarian (rā-dē-ō-lar′-ē-un): an aquatic protist (largely marine) characterized by typically elaborate silica shells.
radula (ra′-dū-luh): a ribbon of tissue in the mouth of gastropod mollusks; bears numerous teeth on its outer surface and is used to scrape and drag food into the mouth.
rain shadow: a local dry area created by the modification of rainfall patterns by a mountain range.
random distribution: distribution characteristic of populations in which the probability of finding an individual is equal in all parts of an area.
reactant: an atom or molecule that is used up in a chemical reaction to form a product.
reaction center: in the light-harvesting complex of a photosystem, the chlorophyll molecule to which light energy is transferred by the antenna molecules (light-absorbing pigments); the captured energy ejects an electron from the reaction center chlorophyll, and the electron is transferred to the electron transport system.
receptor: a cell that responds to an environmental stimulus (chemicals, sound, light, pH, and so on) by changing its electrical potential; also, a protein molecule in a plasma membrane that binds to another molecule (hormone, neurotransmitter), triggering metabolic or electrical changes in a cell.
receptor-mediated endocytosis: the selective uptake of molecules from the extracellular fluid by binding to a receptor located at a coated pit on the plasma membrane and pinching off the coated pit into a vesicle that moves into the cytoplasm.
receptor potential: an electrical potential change in a receptor cell, produced in response to the reception of an environmental stimulus (chemicals, sound, light, heat, and so on). The size of the receptor potential is proportional to the intensity of the stimulus.
receptor protein: a protein, located on a membrane (or in the cytoplasm), that recognizes and binds to specific molecules. Binding by receptor proteins typically triggers a response by a cell, such as endocytosis, increased metabolic rate, or cell division.
recessive: an allele that is expressed only in homozygotes and is completely masked in heterozygotes.
recognition protein: a protein or glycoprotein protruding from the outside surface of a plasma membrane that identifies a cell as belonging to a particular species, to a specific individual of that species, and in many cases to one specific organ within the individual.
recombinant DNA: DNA that has been altered by the recombination of genes from a different organism, typically from a different species.
recombination: the formation of new combinations of the different alleles of each gene on a chromosome; the result of crossing over.
rectum: the terminal portion of the vertebrate digestive tube, where feces are stored until they can be eliminated.
reflex: a simple, stereotyped movement of part of the body that occurs automatically in response to a stimulus.
regeneration: the regrowth of a body part after loss or damage; also, asexual reproduction by means of the regrowth of an entire body from a fragment.
releasing hormone: a hormone, secreted by the hypothalamus, that causes the release of specific hormones by the anterior pituitary.
renal artery: the artery carrying blood to each kidney.
renal cortex: the outer layer of the kidney; where nephrons are located.
renal medulla: the layer of the kidney just inside the renal cortex; where loops of Henle produce a highly concentrated interstitial fluid, important in the production of concentrated urine.
renal pelvis: the inner chamber of the kidney; where urine from the collecting ducts accumulates before it enters the ureters.
renal vein: the vein carrying cleansed blood away from each kidney.
renin: an enzyme that is released (in mammals) when blood pressure and/or sodium concentration in the blood drops below a set point; initiates a cascade of events that restores blood pressure and sodium concentration.
replacement-level fertility (RLF): the average birthrate at which a reproducing population exactly replaces itself during its lifetime.
replication bubble: the unwound portion of the two parental DNA strands, separated by DNA helicase, in DNA replication.
reproductive isolation: the failure of organisms of one population to breed successfully with members of another; may be due to premating or postmating isolating mechanisms.
reptile: a member of the chordate group that includes the snakes, lizards, turtles, alligators, and crocodiles; not a monophyletic group.
reservoir: the major source and storage site of a nutrient in an ecosystem, normally in the abiotic portion.
resource partitioning: the coexistence of two species with similar requirements, each occupying a smaller niche than either would if it were by itself; a means of minimizing their competitive interactions.
respiratory center: a cluster of neurons, located in the medulla of the brain, that sends rhythmic bursts of nerve impulses to the respiratory muscles, resulting in breathing.
resting potential: a negative electrical potential in unstimulated nerve cells.
restriction enzyme: an enzyme, normally isolated from bacteria, that cuts double-stranded DNA at a specific nucleotide sequence; the nucleotide sequence that is cut differs for different restriction enzymes.
restriction fragment: a piece of DNA that has been isolated by cleaving a larger piece of DNA with restriction enzymes.
restriction fragment length polymorphism (RFLP): a difference in the length of restriction fragments, produced by cutting samples of DNA from different individuals of the same species with the same set of restriction enzymes; the result of differences in nucleotide sequences among individuals of the same species.

reticular formation (reh-tik′-ū-lar): a diffuse network of neurons extending from the hindbrain, through the midbrain, and into the lower reaches of the forebrain; involved in filtering sensory input and regulating what information is relayed to conscious brain centers for further attention.
retina (ret′-in-uh): a multilayered sheet of nerve tissue at the rear of camera-type eyes, composed of photoreceptor cells plus associated nerve cells that refine the photoreceptor information and transmit it to the optic nerve.
retrovirus: a virus that uses RNA as its genetic material. When it invades a eukaryotic cell, a retrovirus "reverse transcribes" its RNA into DNA, which then directs the synthesis of more viruses, using the transcription and translation machinery of the cell.
reverse transcriptase: an enzyme found in retroviruses that catalyzes the synthesis of DNA from an RNA template.
Rh factor: a protein on the red blood cells of some people (Rh-positive) but not others (Rh-negative); the exposure of Rh-negative individuals to Rh-positive blood triggers the production of antibodies to Rh-positive blood cells.
rhizoid (rī′-zoid): a rootlike structure found in bryophytes that anchors the plant and absorbs water and nutrients from the soil.
rhizome (rī′-zōm): an undergound stem, usually horizontal, that stores food.
rhythm method: a contraceptive method involving abstinence from intercourse during ovulation.
ribonucleic acid (rī-bō-noo-klā′-ik; RNA): a molecule composed of ribose nucleotides, each of which consists of a phosphate group, the sugar ribose, and one of the bases adenine, cytosine, guanine, or uracil; transfers hereditary instructions from the nucleus to the cytoplasm; also the genetic material of some viruses.
ribosomal RNA (rRNA): a type of RNA that combines with proteins to form ribosomes.
ribosome: an organelle consisting of two subunits, each composed of ribosomal RNA and protein; the site of protein synthesis, during which the sequence of bases of messenger RNA is translated into the sequence of amino acids in a protein.
ribozyme: an RNA molecule that can catalyze certain chemical reactions, especially those involved in the synthesis and processing of RNA itself.
RNA polymerase: in RNA synthesis, an enzyme that catalyzes the bonding of free RNA nucleotides into a continuous strand, using RNA nucleotides that are complementary to those of a strand of DNA.
rod: a rod-shaped photoreceptor cell in the vertebrate retina, sensitive to dim light but not involved in color vision; see also *cone*.
root: the part of the plant body, normally underground, that provides anchorage, absorbs water and dissolved nutrients and transports them to the stem, produces some hormones, and in some plants serves as a storage site for carbohydrates.
root cap: a cluster of cells at the tip of a growing root, derived from the apical meristem; protects the growing tip from damage as it burrows through the soil.
root hair: a fine projection from an epidermal cell of a young root that increases the absorptive surface area of the root.
root system: the part of a plant, normally below ground, that anchors the plant in the soil, absorbs water and minerals, stores food, transports water, minerals, sugars, and hormones, and produces certain hormones.
rough endoplasmic reticulum: endoplasmic reticulum lined on the outside with ribosomes.
runner: a horizontally growing stem that may develop new plants at nodes that touch the soil.
sac fungus: a fungus of the division Ascomycota, whose members form spores in a saclike case called an *ascus*.
sapwood: young xylem that transports water and minerals in a tree trunk.
saprobe (sap′-rōb): an organism that derives its nutrients from the bodies of dead organisms.
sarcodine (sar-kō′-dīn): a nonphotosynthetic protist (protozoan) characterized by the ability to form pseudopodia; some sarcodines, such as amoebae, are naked, whereas others have elaborate shells.
sarcomere (sark′-ō-mēr): the unit of contraction of a muscle fiber; a subunit of the myofibril, consisting of actin and myosin filaments and bounded by Z lines.
sarcoplasmic reticulum (sark′-ō-plas′-mik re-tik′-ū-lum): the specialized endoplasmic reticulum in muscle cells; forms interconnected hollow tubes. The sarcoplasmic reticulum stores calcium ions and releases them into the interior of the muscle cell, initiating contraction.
saturated: referring to a fatty acid with as many hydrogen atoms as possible bonded to the carbon backbone; a fatty acid with no double bonds in its carbon backbone.
savanna: a biome that is dominated by grasses and supports scattered trees and thorny scrub forests; typically has a rainy season in which all the year's precipitation falls.
scientific method: a rigorous procedure for making observations of specific phenomena and searching for the order underlying those phenomena; consists of four operations: observation, hypothesis, experiment, and conclusion.
scientific name: the name of an organism formed from the two smallest major taxonomic categories—the genus and the species.
scientific theory: a general explanation of natural phenomena developed through extensive and reproducible observations; more general and reliable than a hypothesis.
sclera: a tough, white connective tissue layer that covers the outside of the eyeball and forms the white of the eye.
sclerenchyma (skler-en′-ki-muh): a plant cell type with thick, hardened secondary cell walls that normally dies as the last stage of differentiation and both supports and protects the plant body.
scramble competition: a free-for-all scramble for limited resources among individuals of the same species.
scrotum (skrō′-tum): the pouch of skin containing the testes of male mammals.
S-curve: the S-shaped growth curve that describes a population of long-lived organisms introduced into a new area; consists of an initial period of exponential growth, followed by decreasing growth rate, and, finally, relative stability around a growth rate of zero.
sebaceous gland (se-bā′-shus): a gland in the dermis of skin, formed from epithelial tissue, that produces the oily substance sebum, which lubricates the epidermis.
secondary cell wall: a thick layer of cellulose and other polysaccharides secreted by certain plant cells between the primary cell wall and the plasma membrane.
secondary consumer: an organism that feeds on primary consumers; a carnivore.
secondary growth: growth in the diameter of a stem or root due to cell division in lateral meristems and differentiation of their daughter cells.
secondary oocyte (ō′-ō-sīt): a large haploid cell derived from the first meiotic division of the diploid primary oocyte.
secondary phloem: phloem produced from the cells that arise toward the outside of the vascular cambium.
secondary spermatocyte (sper-ma′-tō-sīt): a large haploid cell derived by meiosis I from the diploid primary spermatocyte.
secondary structure: a repeated, regular structure assumed by protein chains held together by hydrogen bonds; for example, a helix.
secondary succession: succession that occurs after an existing community is disturbed—for example, after a forest fire; much more rapid than primary succession.
secondary xylem: xylem produced from cells that arise at the inside of the vascular cambium.
second law of thermodynamics: the principle of physics that states that any change in an isolated system causes the quantity of concentrated, useful energy to decrease and the amount of randomness and disorder (entropy) to increase.
second messenger: an intracellular chemical, such as cyclic AMP, that is synthesized or released within a cell in response to the binding of a hormone or neurotransmitter (the first messenger) to receptors on the cell surface; brings about specific changes in the metabolism of the cell.
secretin: a hormone, produced by the small intestine, that stimulates the production and release of digestive secretions by the pancreas and liver.
seed: the reproductive structure of a seed plant; protected by a seed coat; contains an embryonic plant and a supply of food for it.
seed coat: the thin, tough, and waterproof outermost covering of a seed, formed from the integuments of the ovule.
segmentation (seg-men-tā′-shun): an animal body plan in which the body is divided into repeated, typically similar units.
segmentation movement: a contraction of the small intestine that results in the mixing of partially digested food and digestive enzymes. Segmentation movements also bring nutrients into contact with the absorptive intestinal wall.
segregation: see *law of segregation*.
selectively permeable: the quality of a membrane that allows certain molecules or ions to move through it more readily than others.
self-fertilization: the union of sperm and egg from the same individual.

selfish gene: the concept that genes promote their own survival in individuals through innate self-sacrificing behavior that enhances the survival of others that carry the same genes; helps explain the evolution of altruism.
semen: the sperm-containing fluid produced by the male reproductive tract.
semiconservative replication: the process of replication of the DNA double helix; the two DNA strands separate, and each is used as a template for the synthesis of a complementary DNA strand. Consequently, each daughter double helix consists of one parental strand and one new strand.
semilunar valve: a paired valve between the ventricles of the heart and the pulmonary artery and aorta; prevents the backflow of blood into the ventricles when they relax.
seminal vesicle: in male mammals, a gland that produces a basic, fructose-containing fluid that forms part of the semen.
seminiferous tubule (sem-i-ni′-fer-us): in the vertebrate testis, a series of tubes in which sperm are produced.
senescence: in plants, a specific aging process, typically including deterioration and the dropping of leaves and flowers.
sensitive period: the particular stage in an animal's life during which it imprints.
sensory neuron: a nerve cell that responds to a stimulus from the internal or external environment.
sensory receptor: a cell (typically, a neuron) specialized to respond to particular internal or external environmental stimuli by producing an electrical potential.
sepal (sē′-pul): the set of modified leaves that surround and protect a flower bud, typically opening into green, leaflike structures when the flower blooms.
septum (pl., **septa**)**:** a partition that separates the fungal hypha into individual cells; pores in septa allow the transfer of materials between cells.
serotonin (ser-uh-tō′-nin): in the central nervous system, a neurotransmitter that is involved in mood, sleep, and the inhibition of pain.
Sertoli cell: in the seminiferous tubule, a large cell that regulates spermatogenesis and nourishes the developing sperm.
sessile (ses′-ul): not free to move about, usually permanently attached to a surface.
severe combined immune deficiency (SCID): a disorder in which no immune cells, or very few, are formed; the immune system is incapable of responding properly to invading disease organisms, and the individual is very vulnerable to common infections.
sex chromosomes: the pair of chromosomes that usually determines the sex of an organism; for example, the X and Y chromosomes in mammals.
sex-linked: referring to a pattern of inheritance characteristic of genes located on one type of sex chromosome (for example, X) and not found on the other type (for example, Y); also called X-linked. In sex-linked inheritance, traits are controlled by genes carried on the X chromosome; females show the dominant trait unless they are homozygous recessive, whereas males express whichever allele is on their single X chromosome.
sexually transmitted disease (STD): a disease that is passed from person to person by sexual contact.
sexual recombination: during sexual reproduction, the formation of new combinations of alleles in offspring as a result of the inheritance of one homologous chromosome from each of two genetically distinct parents.
sexual reproduction: a form of reproduction in which genetic material from two parent organisms is combined in the offspring; normally, two haploid gametes fuse to form a diploid zygote.
sexual selection: a type of natural selection in which the choice of mates by one sex is the selective agent.
shoot system: all the parts of a vascular plant exclusive of the root; normally aboveground, consisting of stem, leaves, buds, and (in season) flowers and fruits; functions include photosynthesis, transport of materials, reproduction, and hormone synthesis.
short-day plant: a plant that will flower only if the length of daylight is shorter than some species-specific duration.
short-night plant: a plant that will flower only if the duration of darkness is shorter than some species-specific duration (sometimes called a *long-day plant*).
sickle-cell anemia: a recessive disease caused by a single amino acid substitution in the hemoglobin molecule. Sickle-cell hemoglobin molecules tend to cluster together, distorting the shape of red blood cell shape and causing them to break and clog capillaries.
sieve plate: in plants, a structure between two adjacent sieve-tube elements in phloem, where holes formed in the primary cell walls interconnect the cytoplasm of the elements; in echinoderms, the opening through which water enters the water-vascular system.
sieve tube: in phloem, a single strand of sieve-tube elements that transports sugar solutions.
sieve-tube element: one of the cells of a sieve tube, which form the phloem.
simple diffusion: the diffusion of water, dissolved gases, or lipid-soluble molecules through the phospholipid bilayer of a cellular membrane.
single covalent bond: a covalent bond in which two atoms share one pair of electrons.
sink: in plants, any structure that uses up sugars or converts sugars to starch and toward which phloem fluids will flow.
sinoatrial (SA) node (sī′-nō-āt′-rē-ul): a small mass of specialized muscle in the wall of the right atrium; generates electrical signals rhythmically and spontaneously and serves as the heart's pacemaker.
skeletal muscle: the type of muscle that is attached to and moves the skeleton and is under the direct, normally voluntary, control of the nervous system; also called *striated muscle*.
skeleton: a supporting structure for the body, on which muscles act to change the body configuration; may be external or internal.
skin: the tissue that makes up the outer surface of an animal body.
slime layer: a sticky polysaccharide or protein coating that some disease-causing bacteria secrete outside their cell wall; helps the cells aggregate and stick to smooth surfaces.
small intestine: the portion of the digestive tract, located between the stomach and large intestine, in which most digestion and absorption of nutrients occur.
smell: the olfactory sense that allows animals to respond to odorous, airborne chemicals in their external environment.
smooth endoplasmic reticulum: endoplasmic reticulum without ribosomes.
smooth muscle: the type of muscle that surrounds hollow organs, such as the digestive tract, bladder, and blood vessels; normally not under voluntary control.
sodium–potassium pump: in nerve cell plasma membranes, a set of active-transport molecules that use the energy of ATP to pump sodium ions out of the cell and potassium ions in, maintaining the concentration gradients of these ions across the membrane.
solvent: a liquid capable of dissolving (uniformly dispersing) other substances in itself.
somatic nervous system: that portion of the peripheral nervous system that controls voluntary movement by activating skeletal muscles.
source: in plants, any structure that actively synthesizes sugar and away from which phloem fluid will be transported.
spawning: a method of external fertilization in which male and female parents shed gametes into the water, and sperm must swim through the water to reach the eggs.
speciation: the process of species formation, in which a single species splits into two or more species.
species (spē′-sēs): the basic unit of taxonomic classification, consisting of a population or series of populations of closely related and similar organisms. In sexually reproducing organisms, a species can be defined as a population or series of populations of organisms that interbreed freely with one another under natural conditions but that do not interbreed with members of other species.
specific heat: the amount of energy required to raise the temperature of 1 gram of a substance by 1 °C.
sperm: the haploid male gamete, normally small, motile, and containing little cytoplasm.
spermatid: a haploid cell derived from the secondary spermatocyte by meiosis II; differentiates into the mature sperm.
spermatogenesis: the process by which sperm cells form.
spermatogonium (pl., **spermatogonia**)**:** a diploid cell, lining the walls of the seminiferous tubules, that gives rise to a primary spermatocyte.
spermatophore: in a variation on internal fertilization in some animals, the males package their sperm in a container that can be inserted into the female reproductive tract.
spermicide: a sperm-killing chemical; used for contraceptive purposes.
spicule (spik′-ūl): a subunit of the endoskeleton of sponges that is made of protein, silica, or calcium carbonate.
spinal cord: the part of the central nervous system of vertebrates that extends from the base of the brain to the hips and is protected by the bones of the vertebral column; contains the cell bodies of motor neurons that form synapses with skeletal muscles, the circuitry

for some simple reflex behaviors, and axons that communicate with the brain.

spindle microtubules: microtubules organized in a spindle shape that separate chromosomes during mitosis or meiosis.

spiracle (spi′-ruh-kul): an opening in the abdominal segment of insects through which air enters the tracheae.

spirillum (spi′-ril-um; pl., **spirilla):** a spiral-shaped bacterium.

spleen: an organ of the lymphatic system in which lymphocytes are produced and blood is filtered past lymphocytes and macrophages, which remove foreign particles and aged red blood cells.

spongy bone: porous, lightweight bone tissue in the interior of bones; the location of bone marrow.

spongy cell: an irregularly shaped mesophyll cell, containing chloroplasts, located just above the lower epidermis of a leaf.

spontaneous generation: the proposal that living organisms can arise from nonliving matter.

sporangium (spor-an′-jē-um; pl., **sporangia):** a structure in which spores are produced.

spore: a haploid reproductive cell capable of developing into an adult without fusing with another cell; in the alternation-of-generation life cycle of plants, a haploid cell that is produced by meiosis and then undergoes repeated mitotic divisions and differentiation of daughter cells to produce the gametophyte, a multicellular, haploid organism.

sporophyte (spor′-ō-fīt): the diploid form of a plant that produces haploid, asexual spores through meiosis.

sporozoan (spor-ō-zō′-un): a parasitic protist with a complex life cycle, typically involving more than one host; named for their ability to form infectious spores. A well-known sporozoan (genus *Plasmodium*) causes malaria.

stabilizing selection: a type of natural selection in which those organisms that display extreme phenotypes are selected against.

stamen (stā′-men): the male reproductive structure of a flower, consisting of a filament and an anther, in which pollen grains develop.

starch: a polysaccharide that is composed of branched or unbranched chains or glucose molecules; used by plants as a carbohydrate-storage molecule.

start codon: the first AUG codon in a messenger RNA molecule.

startle coloration: a form of mimicry in which a color pattern (in many cases resembling large eyes) can be displayed suddenly by a prey organism when approached by a predator.

stem: the portion of the plant body, normally located above ground, that bears leaves and reproductive structures such as flowers and fruit.

stem cell: an undifferentiated cell that is capable of dividing and giving rise to one or more distinct types of differentiated cell(s).

sterilization: a generally permanent method of contraception in which the pathways through which the sperm (vas deferens) or egg (oviducts) must travel are interrupted; the most common form of contraception.

steroid: see *steroid hormone*.

steroid hormone: a class of hormone whose chemical structure (four fused carbon rings with various functional groups) resembles cholesterol; steroids, which are lipids, are secreted by the ovaries and placenta, the testes, and the adrenal cortex.

stigma (stig′-muh): the pollen-capturing tip of a carpel.

stoma (stō′-muh; pl., **stomata):** an adjustable opening in the epidermis of a leaf, surrounded by a pair of guard cells, that regulates the diffusion of carbon dioxide and water into and out of the leaf.

stomach: the muscular sac between the esophagus and small intestine where food is stored and mechanically broken down and in which protein digestion begins.

stop codon: a codon in messenger RNA that stops protein synthesis and causes the completed protein chain to be released from the ribosome.

strand: a single polymer of nucleotides; DNA is composed of two strands.

striated muscle: see *skeletal muscle*.

stroke: an interruption of blood flow to part of the brain caused by the rupture of an artery or the blocking of an artery by a blood clot. Loss of blood supply leads to rapid death of the area of the brain affected.

stroma (strō′-muh): the semi-fluid material inside chloroplasts in which the grana are embedded.

style: a stalk connecting the stigma of a carpel with the ovary at its base.

subatomic particle: the particles of which atoms are made: electrons, protons, and neutrons.

subclimax: a community in which succession is stopped before the climax community is reached and is maintained by regular disturbances—for example, tallgrass prairie maintained by periodic fires.

substrate: the atoms or molecules that are the reactants for an enzyme-catalyzed chemical reaction.

subunit: a small organic molecule, several of which may be bonded together to form a larger molecule. See also *monomer*.

succession (suk-seh′-shun): a structural change in a community and its nonliving environment over time. Community changes alter the ecosystem in ways that favor competitors, and species replace one another in a somewhat predictable manner until a stable, self-sustaining climax community is reached.

sucrose: a disaccharide composed of glucose and fructose.

sugar: a simple carbohydrate molecule, either a monosaccharide or a disaccharide.

sugar-phosphate backbone: a major feature of DNA structure, formed by attaching the sugar of one nucleotide to the phosphate from the adjacent nucleotide in a DNA strand.

suppressor T cell: a type of T cell that depresses the response of other immune cells to foreign antigens.

surface tension: the property of a liquid to resist penetration by objects at its interface with the air, due to cohesion between molecules of the liquid.

survivorship curve: a curve resulting when the number of individuals of each age in a population is graphed against their age, usually expressed as a percentage of their maximum life span.

symbiosis (sim′-bī-ō′sis): a close interaction between organisms of different species over an extended period. Either or both species may benefit from the association, or (in the case of parasitism) one of the participants is harmed. Symbiosis includes parasitism, mutualism, and commensalism.

symbiotic: referring to an ecological relationship based on symbiosis.

sympathetic division: the division of the autonomic nervous system that produces largely involuntary responses that prepare the body for stressful or highly energetic situations.

sympatric speciation (sim-pat′-rik): speciation that occurs in populations that are not physically divided; normally due to ecological isolation or chromosomal aberrations (such as polyploidy).

synapse (sin′-aps): the site of communication between nerve cells. At a synapse, one cell (presynaptic) normally releases a chemical (the neurotransmitter) that changes the electrical potential of the second (postsynaptic) cell.

synaptic terminal: a swelling at the branched ending of an axon; where the axon forms a synapse.

syphilis (si′-ful-is): a sexually transmitted bacterial infection of the reproductive organs; if untreated, can damage the nervous and circulatory systems.

systematics: the branch of biology concerned with reconstructing phylogenies and with naming and classifying species.

taiga (tī′-guh): a biome with long, cold winters and only a few months of warm weather; dominated by evergreen coniferous trees; also called *northern coniferous forest*.

taproot system: a root system, commonly found in dicots, that consists of a long, thick main root and many smaller lateral roots, all of which grow from the primary root.

target cell: a cell on which a particular hormone exerts its effect.

taste: a chemical sense for substances dissolved in water or saliva; in mammals, perceptions of sweet, sour, bitter, or salt produced by the stimulation of receptors on the tongue.

taste bud: a cluster of taste receptor cells and supporting cells that is located in a small pit beneath the surface of the tongue and that communicates with the mouth through a small pore. The human tongue has about 10,000 taste buds.

taxis (taks′-is; pl., **taxes):** an innate behavior that is a directed movement of an organism toward or away from a stimulus such as heat, light, or gravity.

taxonomy (tax-on′-uh-mē): the science by which organisms are classified into hierarchically arranged categories that reflect their evolutionary relationships.

Tay-Sachs disease: a recessive disease caused by a deficiency in enzymes that regulate lipid breakdown in the brain.

T cell: a type of lymphocyte that recognizes and destroys specific foreign cells or substances or that regulates other cells of the immune system.

T-cell receptor: a protein receptor, located on the surface of a T cell, that binds a specific antigen and triggers the immune response of the T cell.

tectorial membrane (tek-tor′-ē-ul): one of the membranes of the cochlea in which the hairs of the hair cells are embedded. In sound reception, movement of the basilar membrane relative to the tectorial membrane bends the cilia.
telomere (tē′-le-mēr): the nucleotides at the end of a chromosome that protect the chromosome from damage, especially during shortening or joining with the end of another chromosome during replication.
telophase (tēl′-ō-fāz): in mitosis, the final stage, in which a nuclear envelope re-forms around each new daughter nucleus, the spindle fibers disappear, and the chromosomes relax from their condensed form; in meiosis I, the stage during which the spindle fibers disappear and the chromosomes normally relax from their condensed form; in meiosis II, the stage during which chromosomes relax into their extended state, nuclear envelopes re-form, and cytokinesis occurs.
temperate deciduous forest: a biome in which winters are cold and summer rainfall is sufficient to allow enough moisture for trees to grow and shade out grasses.
temperate rain forest: a biome in which there is no shortage of liquid water year-round and that is dominated by conifers.
template strand: the strand of the DNA double helix from which RNA is transcribed.
temporal isolation: the inability of organisms to mate if they have significantly different breeding seasons.
tendon: a tough connective tissue band connecting a muscle to a bone.
tendril: a slender outgrowth of a stem that coils about external objects and supports the stem; normally a modified leaf or branch.
tentacle (ten′-te-kul): an elongate, extensible projection of the body of cnidarians and cephalopod mollusks that may be used for grasping, stinging, and immobilizing prey, and for locomotion.
terminal bud: meristem tissue and surrounding leaf primordia that are located at the tip of the plant shoot.
territoriality: the defense of an area in which important resources are located.
tertiary consumer (ter′-shē-er-ē): a carnivore that feeds on other carnivores (secondary consumers).
tertiary structure (ter′-shē-er-ē): the complex three-dimensional structure of a single peptide chain; held in place by disulfide bonds between cysteines.
test cross: a breeding experiment in which an individual showing the dominant phenotype is mated with an individual that is homozygous recessive for the same gene. The ratio of offspring with dominant versus recessive phenotypes can be used to determine the genotype of the phenotypically dominant individual.
testis (pl., **testes**)**:** the gonad of male mammals.
testosterone: in vertebrates, a hormone produced by the interstitial cells of the testis; stimulates spermatogenesis and the development of male secondary sex characteristics.
thalamus: the part of the forebrain that relays sensory information to many parts of the brain.
theory: in science, an explanation for natural events that is based on a large number of observations and is in accord with scientific principles, especially causality.
thermoacidophile (ther-mō-a-sid′-eh-fīl): an archaean that thrives in hot, acidic environments.
thermoreceptor: a sensory receptor that responds to changes in temperature.
thick filament: in the sarcomere, a bundle of myosin that interacts with thin filaments, producing muscle contraction.
thin filament: in the sarcomere, a protein strand that interacts with thick filaments, producing muscle contraction; composed primarily of actin, with accessory proteins.
thorax: the segment between the head and abdomen in animals with segmentation; the segment to which structures used in locomotion are attached.
thorn: a hard, pointed outgrowth of a stem; normally a modified branch.
threshold: the electrical potential (less negative than the resting potential) at which an action potential is triggered.
thrombin: an enzyme produced in the blood as a result of injury to a blood vessel; catalyzes the production of fibrin, a protein that assists in blood clot formation.
thylakoid (thī′-luh-koid): a disk-shaped, membranous sac found in chloroplasts, the membranes of which contain the photosystems and ATP-synthesizing enzymes used in the light-dependent reactions of photosynthesis.
thymine: a nitrogenous base found only in DNA; abbreviated as *T*.
thymosin: a hormone, secreted by the thymus, that stimulates the maturation of cells of the immune system.
thymus (thī′-mus): an organ of the lymphatic system that is located in the upper chest in front of the heart and that secretes thymosin, which stimulates lymphocyte maturation; begins to degenerate at puberty and has little function in the adult.
thyroid gland: an endocrine gland, located in front of the larynx in the neck, that secretes the hormones thyroxine (affecting metabolic rate) and calcitonin (regulating calcium ion concentration in the blood).
thyroid-stimulating hormone (TSH): a hormone, released by the anterior pituitary, that stimulates the thyroid gland to release hormones.
thyroxine (thī-rox′-in): a hormone, secreted by the thyroid gland, that stimulates and regulates metabolism.
tight junction: a type of cell-to-cell junction in animals that prevents the movement of materials through the spaces between cells.
tissue: a group of (normally similar) cells that together carry out a specific function; for example, muscle; may include extracellular material produced by its cells.
tonsil: a patch of lymphatic tissue consisting of connective tissue that contains many lymphocytes; located in the pharynx and throat.
trachea (trā′-kē-uh): in birds and mammals, a rigid but flexible tube, supported by rings of cartilage, that conducts air between the larynx and the bronchi; in insects, an elaborately branching tube that carries air from openings called *spiracles* near each body cell.
tracheid (trā′-kē-id): an elongated xylem cell with tapering ends that contains pits in the cell wall; forms tubes that transport water.
tracheophyte (trā′-kē-ō-fīt): a plant that has conducting vessels; a vascular plant.
transcription: the synthesis of an RNA molecule from a DNA template.
transducer: a device that converts signals from one form to another. Sensory receptors are transducers that convert environmental stimuli, such as heat, light, or vibration, into electrical signals (such as action potentials) recognized by the nervous system.
transfer RNA (tRNA): a type of RNA that binds to a specific amino acid by means of a set of three bases (the anticodon) on the tRNA that are complementary to the mRNA codon for that amino acid; carries its amino acid to a ribosome during protein synthesis, recognizes a codon of mRNA, and positions its amino acid for incorporation into the growing protein chain.
transformation: a method of acquiring new genes, whereby DNA from one bacterium (normally released after the death of the bacterium) becomes incorporated into the DNA of another, living, bacterium.
transgenic: referring to an animal or a plant that expresses DNA derived from another species.
translation: the process whereby the sequence of bases of messenger RNA is converted into the sequence of amino acids of a protein.
transpiration (trans′-per-ā-shun): the evaporation of water through the stomata of a leaf.
transport protein: a protein that regulates the movement of water-soluble molecules through the plasma membrane.
trial-and-error learning: a process by which adaptive responses are learned through rewards or punishments provided by the environment.
trichomoniasis (trik-ō-mō-nī′-uh-sis): a sexually transmitted disease, caused by the protist *Trichomonas*, that causes inflammation of the mucous membranes than line the urinary tract and genitals.
tricuspid valve: the valve between the right ventricle and the right atrium of the heart.
triglyceride (trī-glis′-er-īd): a lipid composed of three fatty-acid molecules bonded to a single glycerol molecule.
triple covalent bond: a covalent bond that occurs when two atoms share three pairs of electrons.
trisomy 21: see *Down syndrome*.
trisomy X: a condition of females who have three X chromosomes instead of the normal two; most such women are phenotypically normal and are fertile.
trophic level: literally, "feeding level"; the categories of organisms in a community, and the position of an organism in a food chain, defined by the organism's source of energy; includes producers, primary consumers, secondary consumers, and so on.
tropical deciduous forest: a biome with pronounced wet and dry seasons and plants that must shed their leaves during the dry season to minimize water loss.

tropical rain forest: a biome with evenly warm, evenly moist conditions; dominated by broadleaf evergreen trees; the most diverse biome.
true-breeding: pertaining to an individual all of whose offspring produced through self-fertilization are identical to the parental type. True-breeding individuals are homozygous for a given trait.
T tubule: a deep infolding of the muscle plasma membrane; conducts the action potential inside a cell.
tubal ligation: a surgical procedure in which a woman's oviducts are cut so that the egg cannot reach the uterus, making her infertile.
tube cell: the outermost cell of a pollen grain; digests a tube through the tissues of the carpel, ultimately penetrating into the female gametophyte.
tube foot: a cylindrical extension of the water-vascular system of echinoderms; used for locomotion, grasping food, and respiration.
tubular reabsorption: the process by which cells of the tubule of the nephron remove water and nutrients from the filtrate within the tubule and return those substances to the blood.
tubular secretion: the process by which cells of the tubule of the nephron remove additional wastes from the blood, actively secreting those wastes into the tubule.
tubule (toob′-ūl): the tubular portion of the nephron; includes a proximal portion, the loop of Henle, and a distal portion. Urine is formed from the blood filtrate as it passes through the tubule.
tumor: a mass that forms in otherwise normal tissue; caused by the uncontrolled growth of cells.
tundra: a biome with severe weather conditions (extreme cold and wind and little rainfall) that cannot support trees.
turgor pressure: pressure developed within a cell (especially the central vacuole of plant cells) as a result of osmotic water entry.
Turner syndrome: a set of characteristics typical of a woman with only one X chromosome: sterile, with a tendency to be very short and to lack normal female secondary sexual characteristics.
tympanic membrane (tim-pan′-ik): the eardrum; a membrane, stretched across the opening of the ear, that transmits vibration of sound waves to bones of the middle ear.
unicellular: single-celled; most members of the domains Bacteria and Archaea and the kingdom Protista are unicellular.
uniform distribution: the distribution characteristic of a population with a relatively regular spacing of individuals, commonly as a result of territorial behavior.
uniformitarianism: the hypothesis that Earth developed gradually through natural processes, similar to those at work today, that occur over long periods of time.
unsaturated: referring to a fatty acid with fewer than the maximum number of hydrogen atoms bonded to its carbon backbone; a fatty acid with one or more double bonds in its carbon backbone.
upwelling: an upward flow that brings cold, nutrient-laden water from the ocean depths to the surface; occurs along western coastlines.
uracil: a nitrogenous base found in RNA; abbreviated as *U*.
urea (ū-rē′-uh): a water-soluble, nitrogen-containing waste product of amino acid breakdown; one of the principal components of mammalian urine.
ureter (ū′-re-ter): a tube that conducts urine from each kidney to the bladder.
urethra (ū-rē′-thruh): the tube leading from the urinary bladder to the outside of the body; in males, the urethra also receives sperm from the vas deferens and conducts both sperm and urine (at different times) to the tip of the penis.
uric acid (ūr′-ik): a nitrogen-containing waste product of amino acid breakdown; a relatively insoluble white crystal excreted by birds, reptiles, and insects.
urinary system: the organ system that produces, stores, and eliminates urine, which contains cellular wastes, excess water and nutrients, and toxic or foreign substances. The urinary system is critical for maintaining homeostatic conditions within the bloodstream. It includes the kidneys, ureters, bladder, and urethra.
urine: the fluid produced and excreted by the urinary system of vertebrates; contains water and dissolved wastes, such as urea.
uterine tube: also called the oviduct, the tube leading out of the ovary to the uterus, into which the secondary oocyte (egg cell) is released.
uterus: in female mammals, the part of the reproductive tract that houses the embryo during pregnancy.
vaccination: an injection into the body that contains antigens characteristic of a particular disease organism and that stimulates an immune response.
vacuole (vak′-ū-ōl): a vesicle that is typically large and consists of a single membrane enclosing a fluid-filled space.
vagina: the passageway leading from the outside of a female mammal's body to the cervix of the uterus.
variable: a condition, particularly in a scientific experiment, that is subject to change.
variable region: the part of an antibody molecule that differs among antibodies; the ends of the variable regions of the light and heavy chains form the specific binding site for antigens.
vascular (vas′-kū-lar): describing tissues that contain vessels for transporting liquids.
vascular bundle a strand of xylem and phloem in leaves and stems; in leaves, commonly called a *vein*.
vascular cambium: a lateral meristem that is located between the xylem and phloem of a woody root or stem and that gives rise to secondary xylem and phloem.
vascular cylinder: the centrally located conducting tissue of a young root, consisting of primary xylem and phloem.
vascular tissue system: a plant tissue system consisting of xylem (which transports water and minerals from root to shoot) and phloem (which transports water and sugars throughout the plant).
vas deferens (vaz de′-fer-enz): the tube connecting the epididymis of the testis with the urethra.
vasectomy: a surgical procedure in which a man's vas deferens are cut, preventing sperm from reaching the penis during ejaculation, thereby making him infertile.
vector: a carrier that introduces foreign genes into cells.
vein: in vertebrates, a large-diameter, thin-walled vessel that carries blood from venules back to the heart; in vascular plants, a vascular bundle, or a strand of xylem and phloem in leaves.
ventral (ven′-trul): the lower side or underside of an animal whose head is oriented forward.
ventricle (ven′-tre-kul): the lower muscular chamber on each side of the heart, which pumps blood out through the arteries. The right ventricle sends blood to the lungs; the left ventricle pumps blood to the rest of the body.
venule (ven′-ūl): a narrow vessel with thin walls that carries blood from capillaries to veins.
vertebral column (ver-tē′-brul): a column of serially arranged skeletal units (the vertebrae) that enclose the nerve cord in vertebrates; the backbone.
vertebrate: an animal that possesses a vertebral column.
vesicle (ves′-i-kul): a small, membrane-bound sac within the cytoplasm.
vessel: a tube of xylem composed of vertically stacked vessel elements with heavily perforated or missing end walls, leaving a continuous, uninterrupted hollow cylinder.
vessel element: one of the cells of a xylem vessel; elongated, dead at maturity, with thick, lignified lateral cell walls for support but with end walls that are either heavily perforated or missing.
vestigial structure (ves-tij′-ē-ul): a structure that serves no apparent purpose but is homologous to functional structures in related organisms and provides evidence of evolution.
villus (vi′-lus; pl., villi): a fingerlike projection of the wall of the small intestine that increases the absorptive surface area.
viroid (vī′-roid): a particle of RNA that is capable of infecting a cell and of directing the production of more viroids; responsible for certain plant diseases.
virus (vī′-rus): a noncellular parasitic particle that consists of a protein coat surrounding a strand of genetic material; multiplies only within a cell of a living organism (the host).
vitamin: one of a group of diverse chemicals that must be present in trace amounts in the diet to maintain health; used by the body in conjunction with enzymes in a variety of metabolic reactions.
vitreous humor (vit′-rē-us): a clear, jellylike substance that fills the large chamber of the eye between the lens and the retina.
vocal cord: one of a pair of bands of elastic tissue that extend across the opening of the larynx and produce sound when air is forced between them. Muscles alter the tension on the vocal cords and control the size and shape of the opening, which in turn determines whether sound is produced and what its pitch will be.
waggle dance: a symbolic form of communication used by honeybee foragers to commu-

nicate the location of a food source to their hivemates.

warning coloration: bright coloration that warns predators that the potential prey is distasteful or even poisonous.

water mold: a funguslike protist that includes some pathogens, such as the downy mildew, which attacks grapes.

water-vascular system: a system in echinoderms that consists of a series of canals through which seawater is conducted and is used to inflate tube feet for locomotion, grasping food, and respiration.

wax: a lipid composed of fatty acids covalently bonded to long-chain alcohols.

weather: short-term fluctuations in temperature, humidity, cloud cover, wind, and precipitation in a region over periods of hours to days.

Werner syndrome: a rare condition in which a defective gene causes premature aging; caused by a mutation in the gene that codes for DNA replication/repair enzymes.

white matter: the portion of the brain and spinal cord that consists largely of myelin-covered axons and that give these areas a white appearance.

withdrawal: the removal of the penis from the vagina just before ejaculation in an attempt to avoid pregnancy; an ineffective contraceptive method.

working memory: the first phase of learning; short-term memory that is electrical or biochemical in nature.

xylem (**zī-lum**): a conducting tissue of vascular plants that transports water and minerals from root to shoot.

yolk: protein-rich or lipid-rich substances contained in eggs that provide food for the developing embryo.

yolk sac: one of the embryonic membranes of reptilian, bird, and mammalian embryos; in birds and reptiles, a membrane surrounding the yolk in the egg; in mammals, forms part of the umbilical cord and the digestive tract but is empty.

Z line: a fibrous protein structure to which the thin filaments of skeletal muscle are attached; forms the boundary of a sarcomere.

zona pellucida (**pel-oo′-si-duh**): a clear, noncellular layer between the corona radiata and the egg.

zooflagellate (**zō-ō-fla′-jel-et**): a nonphotosynthetic protist that moves by using flagella.

zooplankton: nonphotosynthetic protists that are abundant in marine and freshwater environments.

zoospore (**zō′-ō-spor**): a nonsexual reproductive cell that swims by using flagella; formed by members of the protistan division Oomycota.

zygospore (**zī′-gō-spor**): a fungal spore, produced by the division Zygomycota, that is surrounded by a thick, resistant wall and forms from a diploid zygote.

zygote (**zī′-gōt**): in sexual reproduction, a diploid cell (the fertilized egg) formed by the fusion of two haploid gametes.

zygote fungus: a fungus of the division Zygomycota, which includes the species that cause fruit rot and bread mold.

Photo Credits

Chapter 1 opener: NASA Headquarters; 1-1a: Andrew Syred/Science Photo Library/Photo Researchers, Inc.; 1-1b: Craig Tuttle/Corbis/Stock Market; 1-1c: Kim Taylor/Bruce Coleman Inc.; 1-3: Dr. Jeremy Burgess/Science Photo Library/Photo Researchers, Inc.; 1-4: William R. Sallaz/Duomo Photography Incorporated; 1-5: Kim Taylor/Bruce Coleman Inc.; 1-6: Johnny Johnson/DRK Photo; 1-7: Lawrence Livermore National Laboratory/Photo Researchers, Inc.; 1-8a: CNRI/Science Photo Library/Photo Researchers, Inc.; 1-8b: Dr. M. Rohde, Gesellschaft fur Biotechnologische Forschung/Science Photo Library/Photo Researchers, Inc.; 1-8c: Eric V. Grave/Photo Researchers, Inc.; 1-8d: Richard L. Carlton/Photo Researchers, Inc.; 1-8e: Patti Murray/Animals Animals/Earth Scenes; 1-8f: Jeff Rotman/Getty Images Inc. - Stone Allstock; 1-9: John Durham/Science Photo Library/Photo Researchers, Inc.; 1-10: Doug Perrine/DRK Photo; 1-11: Francois Gohier/Photo Researchers, Inc.; 1-13: Teresa and Gerald Audesirk; E1-2: Luiz C. Marigo/Peter Arnold, Inc.; E1-2iR: Gunter Ziesler/Peter Arnold, Inc.; E1-2iL: Joe McDonald/Visuals Unlimited

Unit 1 opener: Manfred Kage/Peter Arnold, Inc.; **Chapter 2 opener:** Stephen Dalton/Photo Researchers, Inc.; Chapter 2 opener inset: Brad Wrobleski/Masterfile Corporation; 2-8a: Robert B. Suter, Vassar College; 2-8b: Teresa and Gerald Audesirk; E2-1c: National Institutes of Health/Science Source/Photo Researchers, Inc.; E2-2: Chocolate Manufacturers Association; E2-2i: J.-C. Carton/Bruce Coleman Inc.

Chapter 3 opener: Spencer Grant/PhotoEdit; 3-3a: Dr. Jeremy Burgess/Science Photo Library/Photo Researchers, Inc.; 3-4L: Larry Ulrich Stock Photography, Inc./DRK Photo; 3-4M: Dr. Jeremy Burgess/Science Photo Library/Photo Researchers, Inc.; 3-4R: Biophoto Associates/Photo Researchers, Inc.; 3-5L: Richard Kolar/Animals Animals/Earth Scenes; 3-7a: Jean-Michel Labat/Auscape International Pty. Ltd.; 3-7b: Donald Specker/Animals Animals/Earth Scenes; 3-10a: Robert Pearcy/Animals Animals/Earth Scenes; 3-10b: Jeff Foott/DRK Photo; 3-10c: Nuridsany et Perennou/Photo Researchers, Inc.

Chapter 4 opener: Tom McHugh/Photo Researchers, Inc.; 4-5a,b,c: Joseph Kurantsin-Mills, The George Washington University Medical Center; 4-8a-d: M.M. Perry and A.B. Gilbert, *Journal of Cell Science*, 39:257-272 (1979). © 1979 The Company of Biologists Limited.; 4-9R: L.A. Hufnagel, Ultrastructural Aspects of Chemoreception in Ciliated Protists (Ciliophora), *Journal of Electron Microscopy Technique*, 1991. Photomicrograph by Jurgen Bohmer and Linda Hufnagel, University of Rhode Island.; 4-13: Stephen J. Krasemann/DRK Photo

Chapter 5 opener: © BBC Photo Library; 5-5b: E. Guth, T. Hashimoto, and S.F. Conti; 5-6: Carolina Biological Supply Company/Phototake NYC; 5-7: Omikron/Science Source/Photo Researchers, Inc.; 5-8UR,WR: Barry F. King/Biological Photo Service; 5-9M: Don W. Fawcett/Photo Researchers, Inc.; 5-11a,b: Thomas Eisner, Cornell University.; 5-12WL,WR: Nigel Cattlin/Holt Studios International/Photo Researchers, Inc.; 5-13R: The Keith R. Porter Endowment for Cell Biology; 5-14R: W.P. Wergin/Biological Photo Service; 5-15UR: Biophoto Associates/Photo Researchers, Inc.; 5-16b: Molecular Probes, Inc.; 5-17UL: William L. Dentler/Biological Photo Service; 5-17WR: E. de Harven/Photo Researchers, Inc.; 5-18aR: Ellen R. Dirksen/Visuals Unlimited; 5-18bR: Yorgos Nikas/Getty Images Inc. - Stone Allstock; E5-1aUL: Biophoto Associates/Photo Researchers, Inc.; E5-1aWR: Cecil Fox/Science Source/Photo Researchers, Inc.; E5-1bUL,WR: Brian J. Ford; E5-1c: Jean-Claude Revy/Phototake NYC; E5-2a: M.I. Walker/Photo Researchers, Inc.; E5-2b: David M. Phillips/Visuals Unlimited; E5-2c: Manfred Kage/Peter Arnold, Inc.; E5-2d: National Library of Medicine

Chapter 6 opener: Sylvain Grandadam/Getty Images Inc. - Stone Allstock; 6-1: Photograph by Dr. Harold E. Edgerton. © Harold & Esther Edgerton Foundation, courtesy of Palm Press, Inc.

Chapter 7 opener: Joe Tucciarone; 7-2a: Ken W. Davis/Tom Stack & Associates, Inc.; 7-5: Colin Milkins/Oxford Scientific Films/Animals Animals/Earth Scenes

Chapter 8 opener: Wayne Lankinen/DRK Photo; 8-3: AP/Wide World Photos; 8-3bL,R: Teresa Audesirk

Unit 2 opener: Ron Kimball Photography; **Chapter 9 opener:** Bill Bachmann/PhotoEdit; 9-2a: Rosalind Franklin/Photo Researchers, Inc.; 9-2b: Cold Spring Harbor Laboratory Archives/Peter Arnold, Inc.; 9-3c: Michael Freeman/Phototake NYC; E9-1: A. Barrington Brown/Photo Researchers, Inc.

Chapter 10 opener: Ernest Braun/Getty Images Inc. - Stone Allstock; 10-9: Estate of Murray L. Barr, M.D., Robert M. Barr, M.D., Executor; 10-10: Frederic Jacana/Photo Researchers, Inc.; E10-3: Reproduced from C.J. Epstein et al., Werner's syndrome: A review of its symptomatology, natural history, pathologic features, genetics, and relationship to the natural aging process. *Medicine* 45(3):177-221 (1966). Copyright © 1966 by Lippincott Willi; E10-2: Howard W. Jones, Jr., M.D., Eastern Virginia Medical School

Chapter 11 opener: AP/Wide World Photos; Chapter 11 opener inset: Stuart Conway Photography; 11-1a: © Biophoto Associates/Photo Researchers, Inc.; 11-1b: John Durham/Science Photo Library/Photo Researchers, Inc.; 11-1c: Carolina Biological Supply Company/Phototake NYC; 11-1d: Teresa and Gerald Audesirk; 11-6: Biophoto Associates/Photo Researchers, Inc.; 11-7: CNRI/Science Photo Library/Photo Researchers, Inc.; 11-8a-hU: M. Abbey/Photo Researchers, Inc.; 11-9b: T.E. Schroeder/Biological Photo Service; E11-1R: Photograph courtesy of The Roslin Institute.

Chapter 12 opener: Laurel Frankel, Sports Imageing/Time Inc. Magazines/Sports Illustrated; 12-1: Archiv/Photo Researchers, Inc.; 12-8: Biophoto Associates/Photo Researchers, Inc.; 12-13: Jane Burton/Bruce Coleman Inc.; 12-15a: P. Marazzi/Science Photo Library/Photo Researchers, Inc.; 12-15b: K.H. Switak/Photo Researchers, Inc.; 12-15c: Gary Retherford/Photo Researchers, Inc.; 12-16a: Dennis Kunkel/Phototake NYC; 12-16b: Walter Reinhart/Phototake NYC; 12-17a: Hart-Davis/Science Photo Library/Photo Researchers, Inc.; 12-18UR: Gunn and Stewart/Mary Evans/Photo Researchers, Inc.; 12-19a: CNRI/Science Photo Library/Photo Researchers, Inc.; 12-19b: Lawrence Migdale/Pix; E12-1: Abraham Menashe Inc. ; E12-2a: ©Gallo Images/CORBIS; E12-2b: Courtesy Dr. Doel Sarjarto.

Chapter 13 opener, left and right: AP/Wide World Photos; 13-1aR: Stanley N. Cohen/Science Photo Library/Photo Researchers, Inc.; 13-8: Pharmacia Corporation; E13-1: Janet Chapple, author of *Yellowstone Treasures: The Traveler's Companion to the National Park*; E13-2: Garth Fletcher

Unit 3 opener: Francois Gohier/Photo Researchers, Inc.; **Chapter 14 opener:** O. Louis Mazzatenta/NGS Image Collection; 14-1: Jim Steinberg/Photo Researchers, Inc.; 14-3a: John Cancalosi/DRK Photo; 14-3b: David M. Dennis/Tom Stack & Associates, Inc.; 14-3c: Chip Clark; 14-9a,b: Stephen Dalton/Photo Researchers, Inc.; 14-9c: Douglas T. Cheeseman, Jr./Peter Arnold, Inc.; 14-9d: Gerard Lacz/Peter Arnold, Inc.; 14-10a-c: Photo Lennart Nilsson/Albert Bonniers Forlag AB; 14-11a: Stephen J. Krasemann/DRK Photo; 14-11b: Timothy O'Keefe/Tom Stack & Associates, Inc.; E14-1: Corbis/Bettmann; E14-2: Frans Lanting/Photo Researchers, Inc.

Chapter 15 opener: John Cole/Science Photo Library/Photo Researchers, Inc.; 15-3b: Luiz C. Marigo/Peter Arnold, Inc.; 15-3c: Y.R. Tymstra/Valan Photos; 15-4: Gregory Dimijian/Photo Researchers, Inc.; 15-5: Ed Degginger/Color-Pic, Inc.; 15-6: W. Perry Conway/Tom Stack & Associates, Inc.; 15-7: D. Cavagnaro/DRK Photo; 15-8a: Patti Murray/Animals Animals/Earth Scenes; 15-8b: Tim Davis/Photo Researchers, Inc.; 15-11: Glen E. Woolfenden/Archbold Biological Station; 15-12: Thomas A. Wiewandt/Wild Horizons Inc.; E15-1: Francois Gohier/Photo Researchers, Inc.

Chapter 16 opener: Alan Rabinowitz; 16-1a: Wayne Lankinen/Valan Photos; 16-1b: Edgar T. Jones/Bruce Coleman Inc.; 16-3a,b: Pat & Tom Leeson/Photo Researchers, Inc.; 16-4: Guy L. Bush, Michigan State University; 16-7U: Mark Smith/Photo Researchers, Inc.; 16-7M: Thomas Kitchin/Tom Stack & Associates, Inc.; 16-7W: Mark Smith/Photo Researchers, Inc.; 16-8: Tim Laman/NGS Image Collection; 16-9: Joy Spurr/Bruce Coleman Inc.; 16-10: Tom McHugh/Steinhart Aquarium/Photo Researchers, Inc.; 16-11: The Kern Company; E16-1: Martin Harvey/Peter Arnold, Inc.

Chapter 17 opener: TSADO/NASA/Tom Stack & Associates, Inc.; 17-3: Sidney Fox/Science VU/Visuals Unlimited; 17-5: Michael Abbey/Visuals Unlimited; 17-6a: Milwaukee Public Museum, Photograph Collection; 17-6b: James L. Amos/Photo Researchers, Inc.; 17-6c: Carolina Biological Supply Company/Phototake NYC; 17-6d: Douglas Faulkner/Photo Researchers, Inc.; 17-7: Illustration by Ludek Pesek/Science Photo Library/Photo Researchers, Inc.; 17-8: Terry Whittaker/Photo Researchers, Inc.; 17-9: Illustration by Chris Butler/Science Photo Library/Photo Researchers, Inc.; 17-12a: Tom McHugh/Chicago Zoological Park/Photo Researchers, Inc.; 17-12b: Frans Lanting/Minden Pictures; 17-12c: Nancy Adams/Tom Stack & Associates, Inc.; 17-13: © Michel Brunet/M.P.F.T.; 17-16: David Frayer, Dept. of Anthropology, University of Kansas; 17-17: Jerome Chatin/Getty Images, Inc - Liaison

Chapter 18 opener: Tom Brakefield/DRK Photo; 18-1a: Wayne Lankinen/Bruce Coleman Inc.; 18-1b: M.C. Chamberlain/DRK Photo; 18-1c: Maslowski/Photo Researchers, Inc.; 18-2a: C. Steven Murphree/Biological Photo Service; 18-2b: Dr. Greg Rouse, Department of Invertebrate Zoology, National Museum of Natural History, Smithsonian Institution; 18-2c: Dr. Jeremy Burgess/Science Photo Library/Photo Researchers, Inc.; 18-4a: Hans Gelderblom/Getty Images Inc. - Stone Allstock; 18-4b: Reprinted by permission of Springer-Verlag from W.J. Jones, J.A. Leigh, F. Mayer, C.R. Woese, and R.S. Wolfe, Methanococcus jannaschii sp. nov., an extremely thermophilic methanogen from a submarine hydrothermal vent. *Archives of Microbiology* 136:254-2; 18-7: Zig Koch/Kino Fotoarquivo/©1992. Reprinted with permission of *Discover Magazine*.

Chapter 19 opener: Photo Researchers, Inc./AP/Wide World Photos; 19-2b: Centers for Disease Control and Prevention/Photo Researchers, Inc.; 19-4: Oliver Meckes/Ottawa/Photo Researchers, Inc.; 19-5: Reproduced by permission from K. Pan et al., Conversion of alpha-helices beta-sheets features in the formation of the scrapie prion proteins. *Proceedings of the National Academy of Sciences* 90:1962-1966 (1993), fig. 4c. Copyright © 1993 National Ac; 19-6a: David M. Phillips/Visuals Unlimited; 19-6b: Karl O. Stetter, University of Regensburg, Germany; 19-6c: CNRI/Science Photo Library/Photo Researchers, Inc.; 19-7a: Karl O. Stetter, University of Regensburg, Germany; 19-8: Photo Lennart Nilsson/Albert Bonniers Forlag; 19-9: A.B. Dowsett/Science Photo Library/Photo Researchers, Inc.; 19-10: CNRI/Science Photo Library/Photo Researchers, Inc.; 19-11: Dennis Kunkel/

Phototake NYC; 19-12: Alan L. Detrick/Photo Researchers, Inc.; 19-13: Biophoto Associates/Photo Researchers, Inc.; 19-14a,b: C.P. Vance/Visuals Unlimited; 19-15a: Carolina Biological Supply Company/Phototake NYC; 19-15b: Eric Grave/Science Source/Photo Researchers, Inc.; 19-16: William Merrill, Penn State University; 19-17: Manfred Kage/Peter Arnold, Inc.; 19-18a: D.P. Wilson/Eric and David Hosking/Photo Researchers, Inc.; 19-18b: Lawrence E. Naylor/Photo Researchers, Inc.; 19-19: David M. Phillips/Visuals Unlimited; 19-20: Kevin Schafer/Peter Arnold, Inc.; 19-23: Oliver Meckes & Nicole Ottawa/ Eye of Science/Photo Researchers, Inc.; 19-24a: P.W. Grace/Science Source/Photo Researchers, Inc.; 19-24b: Cabisco/Visuals Unlimited; 19-25W: Cabisco/ Visuals Unlimited; 19-27: Jeffrey L. Rotman Photography; 19-28: Oliver Meckes/Photo Researchers, Inc.; 19-29: P.M. Motta and F.M. Magliocca/Science Photo Library/Photo Researchers, Inc.; 19-30: M.I. Walker/Photo Researchers, Inc.; 19-31: M.I. Walker/ Science Source/Photo Researchers, Inc.; 19-32a: Ed Degginger/Color-Pic, Inc.; 19-32b: Manfred Kage/Peter Arnold, Inc.; 19-33: Harry Rogers

Chapter 20 opener: Michael W. Beug; 20-1a: Robert & Linda Mitchell Photography; 20-1b: Elmer Koneman/ Visuals Unlimited; 20-2: Jeff Lepore/Photo Researchers, Inc.; 20-3: Thomas J. Volk, University of Wisconsin-La Crosse; 20-4U: Carolina Biological Supply Company/Phototake NYC; 20-4W: Breck P. Kent; 20-5a: W.K. Fletcher/Photo Researchers, Inc.; 20-5b: David Dvorak, Jr.; 20-7a: Scott Camazine/Photo Researchers, Inc.; 20-7b: David M. Dennis/Tom Stack & Associates, Inc.; 20-7c: Hans Reinhard/Bruce Coleman Inc.; 20-8: Robert & Linda Mitchell Photography; 20-10a: Jeff Foott Productions; 20-10b: Robert & Linda Mitchell Photography; 20-11: Stanley L. Flegler/Visuals Unlimited; 20-12: David M. Dennis/Tom Stack & Associates, Inc.; 20-13: Michael Fogden/DRK Photo; 20-14: David M. Phillips/Visuals Unlimited; 20-15: Teresa and Gerald Audesirk; 20-16: M. Viard/ Jacana/Photo Researchers, Inc.; 20-17: G.L. Barron/ Biological Photo Service; 20-18: Cabisco/Visuals Unlimited

Chapter 21 opener: Frans Lanting/Minden Pictures; 21-3a: John Gerlach/Tom Stack & Associates, Inc.; 21-3b: John Shaw/Tom Stack & Associates, Inc.; 21-4UL: Carolina Biological Supply Company/ Phototake NYC; 21-5a: Dwight R. Kuhn Photography; 21-5b: Milton Rand/Tom Stack & Associates, Inc.; 21-5c: Larry Ulrich/DRK Photo; 21-6UL: Milton Rand/ Tom Stack & Associates, Inc.; 21-8UL: Dr. William M. Harlow/Photo Researchers, Inc.; 21-8UR: Gilbert S. Grant/Photo Researchers, Inc.; 21-7c: Andy Roberts/ Getty Images Inc. - Stone Allstock; 21-7d: John Kaprielian/Photo Researchers, Inc.; 21-9a: Maurice Nimmo/A-Z Botanical Collection, Ltd.; 21-9b: Teresa and Gerald Audesirk; 21-10a: Dwight R. Kuhn Photography; 21-10b: David Dare Parker/Auscape International Pty. Ltd.; 21-10bi: Matt Jones/Auscape International Pty. Ltd.; 21-10c: Dwight R. Kuhn Photography; 21-10d: Teresa and Gerald Audesirk; 21-10e: Larry West/Photo Researchers, Inc.; 21-12: Thomas Kitchin/Tom Stack & Associates, Inc.

Chapter 22 opener: Martin A. Collins, University of Aberdeen, Aberdeen, Scotland, U.K.; 22-5a: Larry Lipsky/DRK Photo; 22-5b: Brian Parker/Tom Stack & Associates, Inc.; 22-5c: Charles Seaborn/Odyssey Productions, Inc.; 22-6a: Gregory Ochocki/Photo Researchers, Inc.; 22-6b: Charles Seaborn/Odyssey Productions, Inc.; 22-6c: Teresa and Gerald Audesirk; 22-6d: David B. Fleetham/Innerspace Visions; 22-9a: Carolina Biological Supply Company/Phototake NYC; 22-10UR1: Martin Rotker/Phototake NYC; 22-10UR2: Stanley Flegler/Visuals Unlimited; 22-12a: Kjell B. Sandved/Butterfly Alphabet, Inc.; 22-12b: David Bull Photography; 22-12c: J.H. Robinson/ Photo Researchers, Inc.; 22-14a: Ray Coleman/Photo Researchers, Inc.; 22-14b: Estate of Alex Kerstitch; 22-15a: Fred Bavendam/Peter Arnold, Inc.; 22-15b: Ed Reschke/Peter Arnold, Inc.; 22-16a: Fred Bavendam/ Peter Arnold, Inc., 22-16b: Kjell B. Sandved/Photo Researchers, Inc.; 22-16c: Estate of Alex Kerstitch; 22-17: Teresa and Gerald Audesirk; 22-18: Dwight R. Kuhn Photography; 22-20: David Scharf/Peter Arnold, Inc.; 22-21a: Carolina Biological Supply Company/ Phototake NYC; 22-21b: Peter J. Bryant/Biological Photo Service; 22-21c: Stephen Dalton/Photo Researchers, Inc.; 22-21d: Werner H. Muller/Peter Arnold, Inc.; 22-21e: Stanley Breeden/DRK Photo; 22-22a: Dwight R. Kuhn Photography; 22-22b: Tim Flach/Getty Images Inc. - Stone Allstock; 22-22c: Teresa and Gerald Audesirk; 22-23a: Marty Cordano/ DRK Photo; 22-23b: ©Tom Brakefield/CORBIS; 22-24a: Tom Branch/Photo Researchers, Inc.; 22-24b: Peter J. Bryant/Biological Photo Service; 22-24c: Carolina Biological Supply Company/Phototake NYC; 22-24d: Estate of Alex Kerstitch; 22-25: Tom E. Adams/Peter Arnold, Inc.; 22-26a: Carolina Biological Supply Company/Phototake NYC; 22-26b: Reproduced by permission from Howard Shiang, D.V.M., *Journal of the American Veterinary Medical Association* 163:981, Oct. 1973.; 22-27a: Teresa and Gerald Audesirk; 22-27b: Jeff Foott Productions; 22-27c: Chris Newbert/ Bruce Coleman Inc.; 22-28b: Michael Male/Photo Researchers, Inc.

Chapter 23 opener: Peter Scoones - TCL/Masterfile Corporation; 23-1: John Giannicchi/Science Source/ Photo Researchers, Inc.; 23-2R: Tom McHugh/Photo Researchers, Inc.; 23-3a: Tom McHugh/Steinhart Aquarium/Photo Researchers, Inc.; 23-3b: Tom Stack & Associates, Inc.; 23-3bI: Breck P. Kent; 23-4a: Jeffrey L. Rotman Photography; 23-4b: David Hall/Photo Researchers, Inc.; 23-5a: Peter David/Getty Images Inc. - Hulton Archive Photos; 23-5b: Mike Neumann/Photo Researchers, Inc.; 23-5c: Stephen Frink/Getty Images Inc. - Stone Allstock; 23-6a: Breck P. Kent/Animals Animals/Earth Scenes; 23-6b: Joe McDonald/Tom Stack & Associates, Inc.; 23-6c: Cosmos Blank/National Audubon Society/Photo Researchers, Inc.; 23-7a: David G. Barker/Tom Stack & Associates, Inc.; 23-7b: Roger K. Burnard/Biological Photo Service; 23-7c: Frans Lanting/Minden Pictures; 23-8: Carolina Biological Supply Company/Phototake NYC; 23-9a: Walter E. Harvey/Photo Researchers, Inc.; 23-9b: Carolina Biological Supply Company/Phototake NYC; 23-9c: Ray Ellis/Photo Researchers, Inc.; 23-10: Tom McHugh/Photo Researchers, Inc.; 23-11a: Flip Nicklin/Minden Pictures; 23-11b: Jonathan Watts/Science Photo Library/Photo Researchers, Inc.; 23-11c: C. and M. Denis-Huot/Peter Arnold, Inc.; 23-11d: S.R. Maglione/Photo Researchers, Inc.; 23-12a: Yves Kerban/Jacana/Photo Researchers, Inc.; 23-12b: Mark Newman/Auscape International Pty. Ltd.; 23-12bi: D. Parer and E. Parer-Cook/Auscape International Pty. Ltd.; E23-1: Stanley Breeden/ DRK Photo

Unit 4 opener: John Shaw/Tom Stack & Associates, Inc.; **Chapter 24 opener:** Yva Momatiuk/John Eastcott/ Minden Pictures; 24-4: Dr. Jeremy Burgess/Science Photo Library/Photo Researchers, Inc.; 24-5a,b,c: George Wilder/Visuals Unlimited; 24-9UR: Ed Reschke/Peter Arnold, Inc.; 24-12a: Runk/Schoenberger/Grant Heilman Photography, Inc.; 24-12b: John Cunningham/ Visuals Unlimited; 24-13: John D. Cunningham/Visuals Unlimited; 24-14a: Lynwood M. Chace/Photo Researchers, Inc.; 24-14b: Dwight R. Kuhn Photography; 24-15UR: Ed Reschke/Peter Arnold, Inc.; 24-15WR: E.R. Degginger/Animals Animals/Earth Scenes; 24-16: J. Robert Waaland, University of Washington/Biological Photo Service; 24-17: Teresa and Gerald Audesirk; 24-20a: Biophoto Associates/Science Source/Photo Researchers, Inc.; 24-20b: R. Roncadori/ Visuals Unlimited; 24-21R: Hugh Spencer/ National Audubon Society/Photo Researchers, Inc.; 24-23aW: Ray Simon/Photo Researchers, Inc.; 24-23b: Dr. Jeremy Burgess/Science Photo Library/ Photo Researchers, Inc.; 24-24a,b: Martin H. Zimmermann, Harvard Forest, Harvard University; 24-26a: Teresa Audesirk; 24-26b: Francois Gohier/Photo Researchers, Inc.; 24-27a: Kenneth W. Fink/Photo Researchers, Inc.; 24-27b: Belinda Wright/DRK Photo; 24-28a: Teresa and Gerald Audesirk; 24-28b: Carr Clifton/Minden Pictures; 24-28c: Teresa and Gerald Audesirk; 24-28d: Kim Taylor/Bruce Coleman Inc.; E24-1: Michael Fogden/DRK Photo

Chapter 25 opener: Marty Cordano/DRK Photo; Chapter 25 opener inset: Dan Cheatham/DRK Photo; 25-1L: Jeff Foott/Bruce Coleman Inc.; 25-4b: Heather Audesirk; 25-5: Teresa and Gerald Audesirk; 25-6-WL,WR: Carolina Biological Supply Company/ Phototake NYC; 25-7: Dennis Kunkel Microscopy, Inc.; 25-10L: Carolina Biological Supply Company/ Phototake NYC; 25-14: Teresa and Gerald Audesirk; 25-15a,bU: Thomas Eisner, Cornell University.; 25-16a,b: Nuridsany et Perennou/The National Audubon Society Collection/Photo Researchers, Inc.; 25-17: This item is reproduced by permission of The Huntington Library, San Marino, California.; 25-18a: Merlin D. Tuttle/Bat Conservation International/Photo Researchers, Inc.; 25-18b: Oxford Scientific Films/Animals Animals/Earth Scenes; 25-19: Dr. J.A.L. Cooke/Oxford Scientific Films/ Animals Animals/Earth Scenes; 25-20a: Tom Bean/ DRK Photo; 25-20b: Michael Fogden/DRK Photo; 25-21a: Teresa and Gerald Audesirk; 25-21b: Carolina Biological Supply Company/Phototake NYC; 25-22: Teresa and Gerald Audesirk; 25-23: Scott Camazine/Photo Researchers, Inc.; 25-24: Fogden Natural History Photographs; E25-1a: George Holton/ Photo Researchers, Inc.; E25-1b: Merlin D. Tuttle/Bat Conservation International/Photo Researchers, Inc.

Chapter 26 opener and inset: Barry Runk/Stan Schoenberger/Grant Heilman Photography, Inc.; 26-1c: Runk/Schoenberger/Grant Heilman Photography, Inc.; 26-6: John Cancalosi/DRK Photo; 26-7: Biophoto Associates/Photo Researchers, Inc.; 26-9a,b: David Sieren/Visuals Unlimited; E26-1: Carolina Biological/ Visuals Unlimited; E26-1i: Wayne P. Armstrong, Palomar College

Unit 5 opener: Randy Wells/Getty Images Inc. - Stone Allstock; **Chapter 27 opener:** AP/Wide World Photos; 27-4: Robert Brons/Biological Photo Service; 27-5: Manfred Kage/Peter Arnold, Inc.; 27-6U: © Susumu Nishinaga/SPL/Photo Researchers, Inc.; 27-6W: Fred Bruemmer/OKAPIA/Photo Researchers, Inc.; 27-7: Yorgos Nikas/Science Photo Library/Photo Researchers, Inc.; 27-8: Manfred Kage/Peter Arnold, Inc.

Chapter 28 opener: AP/Wide World Photos; Chapter 28 opener inset: Dr. Shaker Mousa; 28-8: Copyright Dennis Kunkel Microscopy, Inc.; 28-11: Photo Lennart Nilsson/Albert Bonniers Forlag; 28-12: Kenneth Kaushansky, M.D., University of Washington Medical Center; 28-13b: CNRI/Science Photo Library/Photo Researchers, Inc.; 28-16: Photo Lennart Nilsson/Albert Bonniers Forlag; 28-20: John D. Cunningham/Visuals Unlimited; 28-21: World Health Organization/Image Library, Special Programme for Research and Training in Tropical Diseases/Photo by Chandran; E28-1: Gwen Rosenberg/Alliance Pharmaceutical Corp.; E28-2: ©AFP/CORBIS; E28-3a: Lou Lainey/ Getty Images/Time Life Pictures; E28-4a: Spencer Grant/PhotoEdit

Chapter 29 opener: Spencer Grant/Grantpix/Stock Boston; 29-1a: Jeffrey L. Rotman Photography; 29-1b: Brian Parker/Tom Stack & Associates, Inc.; 29-1c: Phil Degginger/Color-Pic, Inc.; 29-5a: E.R. Degginger/Bruce Coleman Inc.; 29-5b: Breck P. Kent; 29-5c: David M. Dennis/Tom Stack & Associates, Inc.; 29-6b: Hans-Rainer Duncker, University of Giessen, Germany; E29-1: Bill Travis, M.D./National Cancer Institute; E29-2: O. Auerbach/Visuals Unlimited

Chapter 30 opener: Barbara Penoya/Getty Images, Inc.- Photodisc.; 30-1: Charles Krebs/Getty Images Inc. - Stone Allstock; 30-2a: Howard Brinton/General Board of Global Ministries; 30-2b: © 205 A.D.A.M., Inc.; 30-2c: Biophoto Associates/Photo Researchers, Inc.; 30-4: Tobi Zausner; 30-6a: E.R. Degginger/Photo Researchers, Inc.; E30-1: De Keerle-Parker/Getty Images, Inc - Liaison; E30-2: Dr. E. Walker/Science Photo Library/Photo Researchers, Inc.

Chapter 31 opener: Argus Fotoarchiv/Peter Arnold, Inc.; 31-9: Getty Images Inc. - Stone Allstock; E31-2: Dan McCoy/Rainbow

Chapter 32 opener: Mednet/Phototake NYC; 32-1U: David Scharf/Peter Arnold, Inc.; 32-2: Photo Lennart Nilsson/Albert Bonniers Forlag; 32-3a: Photo Lennart Nilsson/Albert Bonniers Forlag; 32-3b: S.H.E. Kaufmann & J.R. Golecki/Science Photo Library/Photo Researchers, Inc.; 32-9a: Secchi-Lecaque/Roussel/CNRI/Science Source/Photo Researchers, Inc.; 32-9b: CNRI/Science Photo Library/Photo Researchers, Inc.; 32-10: Oliver Meckes & Nicole Ottawa/Photo Researchers, Inc.; 32-14b,c: National Institute for Biological Standards and Control (U.K.)/Science Photo Library/Photo Researchers, Inc.; E32-1b: CNRI/Science Photo Library/Photo Researchers, Inc.

Chapter 33 opener: Pat Benic/Reuters/Corbis/Bettmann; 33-4: Corbis/Bettmann; 33-6b: Biophoto Associates/Photo Researchers, Inc.; 33-10: Photo courtesy of The Jackson Laboratory, Bar Harbor, Maine.

Chapter 34 opener: Ron Edmonds/AP/Wide World Photos; 34-14: From H. Damasio, T. Grabowski, R. Frank, A.M. Galaburda, A.R. Damasio, The Return of Phineas Gage: Clues about the brain from the skull of a famous patient. *Science* 264:112-115, 1994; 34-16a: Scanning electron micrographs by Dr. Robert S. Preston and Prof. Joseph E. Hawkins, Kresge Hearing Research Institute, University of Michigan Medical School.; 34-17a: Dennis Kunkel Microscopy, Inc.; 34-20: Susan Eichorst; 34-21a: Frans Lanting/Minden Pictures; 34-21b: Larry West/Photo Researchers, Inc.; 34-25a: Charles E. Mohr/Photo Researchers, Inc.; 34-25b: Brian Parker/Tom Stack & Associates, Inc.; E34-2: Ken Graham/Bruce Coleman Inc.; E34-3: Marcus E. Raichle, M.D., Washington University Medical Center; E34-4a: William J. Powers, M.D., Washington University School of Medicine; E34-4b: Monte S. Buchsbaum, M.D., Mount Sinai School of Medicine, New York, N.Y.

Chapter 35 opener: NASA/Johnson Space Center; 35-1T: Eric Grave/Photo Researchers, Inc.; 35-4b: Don W. Fawcett/Science Source/Photo Researchers, Inc.; 35-5aB: Jeremy Burgess/Science Photo Library/Photo Researchers, Inc.; 35-5bB: Kent Wood/Photo Researchers, Inc.; 35-5bT: Lawrence E. Naylor/Photo Researchers, Inc.; 35-7: Photo Researchers, Inc.; 35-9b: Manfred Kage/Peter Arnold, Inc.; E35-2a: Michael Klein/Peter Arnold, Inc.; E35-2b: Yoav Levy/Phototake NYC

Chapter 36 opener: Audubon Nature Institute, New Orleans; Chapter 36 opener inset: AP/Wide World Photos; 36-1: Biophoto Associates/Photo Researchers, Inc.; 36-3: Peter J. Bryant/Biological Photo Service; 36-4: Teresa and Gerald Audesirk; 36-4i: Peter Harrison; 36-5a: Anne et Jacques Six; 36-5b: Rudie Kuiter/Innerspace Visions; 36-6: Michael Fogden/Oxford Scientific Films/Animals Animals/Earth Scenes; 36-7a: Stephen J. Krasemann/DRK Photo; 36-7b: Tom McHugh/Photo Researchers, Inc.; 36-7c: Frans Lanting/Minden Pictures; 36-11b: CNRI/Science Photo Library /Photo Researchers, Inc.; 36-16: C. Edelmann/La Villette/Photo Researchers, Inc.; 36-18a: Photo Lennart Nilsson/Albert Bonniers Forlag; 36-18b: Francis Leroy, Biocosmos/SPL/Custom Medical Stock Photo, Inc.; E36-2: Hank Morgan/Photo Researchers, Inc.; E36-3: Brook Kraft/Corbis/Sygma

Chapter 37 opener: Bob Daemmrich Photography, Inc.; 37-1aL: Image Quest 3-D/NHPA Limited; 37-1aR: Jim Bain/NHPA Limited; 37-1bL: Hans Pfletschinger/Peter Arnold, Inc.; 37-1bR: Larry West/Photo Researchers, Inc.; 37-2a: Paul A. Zahl/Photo Researchers, Inc.; 37-2b: K.H. Switak/Photo Researchers, Inc.; 37-2c: J.A.L. Cooke/Animals Animals/Earth Scenes; 37-2d: Norbert Rosing/Oxford Scientific Films/Animals Animals/Earth Scenes; 37-8b: Photo Lennart Nilsson/Albert Bonniers Forlag; 37-11a: Photo Lennart Nilsson/Albert Bonniers Forlag; 37-11b: Petit Format/Nestle/Science Source/Photo Researchers, Inc.; 37-11c: Photo Lennart Nilsson/Albert Bonniers Forlag; 37-15: Jerry Gay/Getty Images Inc. - Stone Allstock; E37-2: Leonard McCombe/Getty Images/Time Life Pictures; E37-3: Sterling K. Clarren, M.D., Children's Hospital & Regional Medical Center, University of Washington School of Medicine

Unit 6 opener: Paradise, Part One of *The Trilogy of the Earth* by Suzanne Duranceau/Illustratrice, Inc.; **Chapter 38 opener:** Reuters NewMedia Inc./Corbis/Bettmann; Chapter 38 opener inset: Joe McDonald/DRK Photo; 38-1a,b: Eric and David Hosking/Frank Lane Picture Agency Limited; 38-3a-e: Boltin Picture Library; 38-4: Frans Lanting/Minden Pictures; 38-6: Thomas McAvoy/Getty Images/Time Life Pictures; 38-8: Renee Lynn/Photo Researchers, Inc.; 38-9: Ken Cole/Animals Animals/Earth Scenes; 38-10: Richard K. LaVal/Animals Animals/Earth Scenes; 38-11: Nuridsany et Perennou/Photo Researchers, Inc.; 38-12: Robert & Linda Mitchell Photography; 38-13a: Stephen J. Krasemann/Photo Researchers, Inc.; 38-13b: Hans Pfletschinger/Peter Arnold, Inc.; 38-14a: M.P. Kahl/DRK Photo; 38-14b: Marc C. Chamberlain; 38-15: Ray Dove/Visuals Unlimited; 38-16: William Ervin/Natural Imagery; 38-17: Michael K. Nichols/National Geographic Image Collection; 38-18: John D. Cunningham/Visuals Unlimited; 38-20: Joe McDonald/Visuals Unlimited; 38-21a: © Konrad Wothe/Minden Pictures; 38-21b: Frans Lanting/Minden Pictures; 38-22: Anne et Jacques Six; 38-24a: Dwight R. Kuhn Photography; 38-24b: Teresa and Gerald Audesirk; 38-25: Fred Bruemmer/Peter Arnold, Inc.; 38-28: Raymond A. Mendez/Animals Animals/Earth Scenes; 38-29: Photo Lennart Nilsson/Albert Bonniers Forlag; 38-30: William P. Fifer, New York State Psychiatric Institute, Columbia University; 38-31a: Frans Lanting/Minden Pictures; 38-31b: Daniel J. Cox/Getty Images, Inc - Liaison; 38-31c: Michio Hoshino/Minden Pictures

Chapter 39 opener: Michael P. Gadomski/Photo Researchers, Inc.; Chapter 39 opener inset UR: Ken Wilson; Papilio/Corbis/Bettmann; Chapter 39 opener inset WL: W. Perry Conway/Corbis/Bettmann; 39-1a: Prof. S. Cinti/University of Ancona, Italy/CNRI/Phototake NYC; 39-1b: © David G. Atfield, Sr.; 39-6R: Bryan and Cherry Alexander; 39-7: Tom McHugh/Photo Researchers, Inc.; 39-8: Hellio-Van Ingen/Auscape International Pty. Ltd.; 39-10i: NASA/Johnson Space Center; E39-3: Frans Lanting/Minden Pictures; E39-3i: Andy Holbrooke/Corbis/Stock Market

Chapter 40 opener: Ron Peplowski/DTE Energy; Chapter 40 opener inset: Courtesy of GLSGN Exotic Species Library/Ontario Ministry of Natural Resources.; 40-3a: Henry Ausloos/Animals Animals/Earth Scenes; 40-3b: Stephen Dalton/Photo Researchers, Inc.; 40-4a: Art Wolfe, Inc.; 40-4b: Frans Lanting/Minden Pictures; 40-5a: Marty Cordano/DRK Photo; 40-5b: Paul A. Zahl/Photo Researchers, Inc.; 40-5c: Ray Coleman/Photo Researchers, Inc.; 40-5d: Robert & Linda Mitchell Photography; 40-6a: Jean-Paul Ferrero and Jean-Michel Labat/Auscape International Pty. Ltd.; 40-6b: Charles V. Angelo/Photo Researchers, Inc.; 40-7: Dr. James L. Castner; 40-8a: Carmela Leszczynski/Animals Animals/Earth Scenes; 40-8b: Breck P. Kent; 40-8c,d: Ed Degginger/Color-Pic, Inc.; 40-8d: Ed Degginger/Color-Pic, Inc.; 40-9a: Robert P. Carr/Bruce Coleman Inc.; 40-9b: Monica Mather and Bernard Roitberg, Simon Fraser University; 40-10a: Zig Leszczynski/Animals Animals/Earth Scenes; 40-10b: Dr. James L. Castner; 40-10c: Jeff Lepore/Photo Researchers, Inc.; 40-11a: Thomas Eisner and Daniel Aneshansley, Cornell University.; 40-11b: Frans Lanting/Minden Pictures; 40-12a: Teresa Audesirk; 40-12b: W. Gregory Brown/Animals Animals/Earth Scenes; 40-13a: Jim Zipp/Photo Researchers, Inc.; 40-13b: Stephen J. Krasemann/Photo Researchers, Inc.; 40-14aL: Dennis Oda/Getty Images Inc. - Stone Allstock; 40-14aR: Krafft-Explorer/Photo Researchers, Inc.; 40-14bL,R: © 2000 Gary Braasch; 40-14cL: Raymond German/Corbis/Bettmann; 40-14cR: Douglas Faulkner/Corbis/Bettmann; 40-17a,b: Robert & Linda Mitchell Photography; 40-18L,R: Cristina Sandoval, Ph.D.; E40-1: Gilbert Grant/Photo Researchers, Inc.; E40-1i: Carol Hughes/Bruce Coleman Inc.; E40-2a: Chuck Pratt/Bruce Coleman Inc.; E40-2b: Robert & Linda Mitchell Photography; E40-2c: Jack Jeffrey/Photo Resource Hawaii Stock Photography; E40-3: AL SATTERWHITE/Getty Images, Inc. - Taxi

Chapter 41 opener: JOHN GIUSTINA/Getty Images, Inc. - Taxi; 41-12: Tom Walker/Stock Boston; 41-13: William E. Ferguson; 41-14: Will McIntyre/Photo Researchers, Inc.; E41-1: SchneiderStock Photography

Chapter 42 opener and both insets: Don L. Boroughs; 42-2b: NASA/Johnson Space Center; 42-6a: Teresa and Gerald Audesirk; 42-6b: Tom McHugh/Photo Researchers, Inc.; 42-9a: Loren McIntyre; 42-9b: Schafer and Hill/Peter Arnold, Inc.; 42-9c: Michael Fogden/Oxford Scientific Films Ltd.; 42-9d: Shiell, Richard/Animals Animals/Earth Scenes; 42-9e: Frans Lanting/Minden Pictures; 42-10: Jacques Jangoux/Peter Arnold, Inc.; 42-11a: W. Perry Conway/Aerie Nature Series, Inc.; 42-11b: Theo Allofs/Allofs Photography; 42-11c: Stephen J. Krasemann/DRK Photo; 42-11d: Aubry Lang/Valan Photos; 42-12: James Hancock/Photo Researchers, Inc.; 42-13a: Don & Pat Valenti/DRK Photo; 42-13b: Brian Parker/Tom Stack & Associates, Inc.; 42-13c: Tom McHugh/Photo Researchers, Inc.; 42-14: William H. Mullins/Photo Researchers, Inc.; 42-15: Ken Lucas/Visuals Unlimited; 42-16: Harvey Payne; 42-17a: Tom and Susan Bean, Inc./DRK Photo; 42-17b: W. Perry Conway/Tom Stack & Associates, Inc.; 42-17c: Nicholas DeVore III/Network Aspen; 42-17d: Bob Gurr/Valan Photos; 42-17e: Jim Brandenburg/Minden Pictures; 42-18: Teresa and Gerald Audesirk; 42-19a: © 1993 Gary Braasch; 42-19b: Thomas Kitchin/Tom Stack & Associates, Inc.; 42-19c: Stephen J. Krasemann/DRK Photo; 42-19d: Joe McDonald/Animals Animals/Earth Scenes; 42-20a: Susan G. Drinker/Corbis/Stock Market; 42-20b: Teresa and Gerald Audesirk; 42-20c: Jeff Foott Productions; 42-20d: Richard Thom/Tom Stack & Associates, Inc.; 42-21a: Teresa and Gerald Audesirk; 42-21b: Marty Stouffer Productions/Animals Animals/Earth Scenes; 42-21c: Wm. Bacon III/National Audubon Society/Photo Researchers, Inc.; 42-22: Gary Braasch Photography; 42-23a: Stephen J. Krasemann/DRK Photo; 42-23b: Jeff Foott/Tom Stack & Associates, Inc.; 42-23c: Michael Giannechini/Photo Researchers, Inc.; 42-23d: James Simon/Photo Researchers, Inc.; 42-26a: T.P. Dickinson/Comstock Images; 42-26b: Teresa and Gerald Audesirk; 42-26bi: Walter Dawn/National Audubon Society/Photo Researchers, Inc.; 42-26c: Teresa and Gerald Audesirk; 42-26ci: Thomas Kitchin/Tom Stack & Associates, Inc.; 42-26d: Robert R. Given, Marymount College; 42-26di: Stan Wayman/Photo Researchers, Inc.; 42-27a: Franklin J. Viola/Viola's Photo Visions, Inc.; 42-27b: Estate of Alex Kerstitch; 42-27c: Charles Seaborn/Odyssey Productions, Inc.; 42-27d: Estate of Alex Kerstitch; 42-28a: Stephen J. Krasemann/DRK Photo; 42-28b: Paul Humann/Jeffrey L. Rotman Photography; 42-28c: Fred McConnaughey/Photo Researchers, Inc.; 42-28d: James D. Watt/Getty Images Inc. - Hulton Archive Photos; 42-28e: Biophoto Associates/Science Source/Photo Researchers, Inc.; 42-28f: Robert Arnold/Getty Images Inc. - Hulton Archive Photos; 42-29: J. Frederick Grassle, Institute of Marine and Coastal Sciences, Rutgers University; E42-1: NASA Headquarters; E42-2: Roy Morsch/Corbis/Stock Market

Index

***MA* refers to Media Activities.**

Audesirk/Audesirk/Byers

Biology: Life on Earth, 7e Accelerator CD-ROM

CD License Agreement

Pearson Prentice Hall
Pearson Education, Inc.
Upper Saddle River, NJ 07458

YOU SHOULD CAREFULLY READ THE TERMS AND CONDITIONS BEFORE USING THE CD-ROM PACKAGE. USING THIS CD-ROM PACKAGE INDICATES YOUR ACCEPTANCE OF THESE TERMS AND CONDITIONS.

Pearson Education, Inc. provides this program and licenses its use. You assume responsibility for the selection of the program to achieve your intended results, and for the installation, use, and results obtained from the program. This license extends only to use of the program in the United States or countries in which the program is marketed by authorized distributors.

LICENSE GRANT
You hereby accept a nonexclusive, nontransferable, permanent license to install and use the program ON A SINGLE COMPUTER at any given time. You may copy the program solely for backup or archival purposes in support of your use of the program on the single computer. You may not modify, translate, disassemble, decompile, or reverse engineer the program, in whole or in part.

TERM
The License is effective until terminated. Pearson Education, Inc. reserves the right to terminate this License automatically if any provision of the License is violated. You may terminate the License at any time. To terminate this License, you must return the program, including documentation, along with a written warranty stating that all copies in your possession have been returned or destroyed.

LIMITED WARRANTY
THE PROGRAM IS PROVIDED "AS IS" WITHOUT WARRANTY OF ANY KIND, EITHER EXPRESSED OR IMPLIED, INCLUDING, BUT NOT LIMITED TO, THE IMPLIED WARRANTIES OR MERCHANTABILITY AND FITNESS FOR A PARTICULAR PURPOSE. THE ENTIRE RISK AS TO THE QUALITY AND PERFORMANCE OF THE PROGRAM IS WITH YOU. SHOULD THE PROGRAM PROVE DEFECTIVE, YOU (AND NOT PEARSON EDUCATION, INC. OR ANY AUTHORIZED DEALER) ASSUME THE ENTIRE COST OF ALL NECESSARY SERVICING, REPAIR, OR CORRECTION. NO ORAL OR WRITTEN INFORMATION OR ADVICE GIVEN BY PEARSON EDUCATION, INC., ITS DEALERS, DISTRIBUTORS, OR AGENTS SHALL CREATE A WARRANTY OR INCREASE THE SCOPE OF THIS WARRANTY.

SOME STATES DO NOT ALLOW THE EXCLUSION OF IMPLIED WARRANTIES, SO THE ABOVE EXCLUSION MAY NOT APPLY TO YOU. THIS WARRANTY GIVES YOU SPECIFIC LEGAL RIGHTS AND YOU MAY ALSO HAVE OTHER LEGAL RIGHTS THAT VARY FROM STATE TO STATE.

Pearson Education, Inc. does not warrant that the functions contained in the program will meet your requirements or that the operation of the program will be uninterrupted or error-free.

However, Pearson Education, Inc. warrants the CD-ROM(s) on which the program is furnished to be free from defects in material and workmanship under normal use for a period of ninety (90) days from the date of delivery to you as evidenced by a copy of your receipt.

The program should not be relied on as the sole basis to solve a problem whose incorrect solution could result in injury to person or property. If the program is employed in such a manner, it is at the user's own risk and Pearson Education, Inc. explicitly disclaims all liability for such misuse.

LIMITATION OF REMEDIES
Pearson Education, Inc.'s entire liability and your exclusive remedy shall be: 1. the replacement of any CD-ROM not meeting Pearson Education, Inc.'s "LIMITED WARRANTY" and that is returned to Pearson Education, or 2. if Pearson Education is unable to deliver a replacement CD-ROM that is free of defects in materials or workmanship, you may terminate this agreement by returning the program.

IN NO EVENT WILL PEARSON EDUCATION, INC. BE LIABLE TO YOU FOR ANY DAMAGES, INCLUDING ANY LOST PROFITS, LOST SAVINGS, OR OTHER INCIDENTAL OR CONSEQUENTIAL DAMAGES ARISING OUT OF THE USE OR INABILITY TO USE SUCH PROGRAM EVEN IF PEARSON EDUCATION, INC. OR AN AUTHORIZED DISTRIBUTOR HAS BEEN ADVISED OF THE POSSIBILITY OF SUCH DAMAGES, OR FOR ANY CLAIM BY ANY OTHER PARTY.

SOME STATES DO NOT ALLOW FOR THE LIMITATION OR EXCLUSION OF LIABILITY FOR INCIDENTAL OR CONSEQUENTIAL DAMAGES, SO THE ABOVE LIMITATION OR EXCLUSION MAY NOT APPLY TO YOU.

GENERAL
You may not sublicense, assign, or transfer the license of the program. Any attempt to sublicense, assign or transfer any of the rights, duties, or obligations hereunder is void.

This Agreement will be governed by the laws of the State of New York.

Should you have any questions concerning this Agreement, you may contact Pearson Education, Inc. by writing to:

ESM Media Development
Higher Education Division
Pearson Education, Inc.
1 Lake Street
Upper Saddle River, NJ 07458

Should you have any questions concerning technical support, you may write to:

New Media Production
Higher Education Division
Pearson Education, Inc.
1 Lake Street
Upper Saddle River, NJ 07458

YOU ACKNOWLEDGE THAT YOU HAVE READ THIS AGREEMENT, UNDERSTAND IT, AND AGREE TO BE BOUND BY ITS TERMS AND CONDITIONS. YOU FURTHER AGREE THAT IT IS THE COMPLETE AND EXCLUSIVE STATEMENT OF THE AGREEMENT BETWEEN US THAT SUPERSEDES ANY PROPOSAL OR PRIOR AGREEMENT, ORAL OR WRITTEN, AND ANY OTHER COMMUNICATIONS BETWEEN US RELATING TO THE SUBJECT MATTER OF THIS AGREEMENT.